HANDBOOK
of tables for
MATHEMATICS

HANDBOOK
of tables for
MATHEMATICS

REVISED FOURTH EDITION

1975

EDITOR-IN-CHIEF

ROBERT C. WEAST, Ph.D.

Vice President, Research, Consolidated Natural Gas Service Company, Inc.
Formerly Professor of Chemistry at Case Institute of Technology

EDITOR-IN-CHIEF OF MATHEMATICS

SAMUEL M. SELBY, Ph.D. Sc.D.

Distinguished Professor Emeritus of Mathematics and formerly Chairman,
Mathematics Department, University of Akron.
Emeritus Professor of Mathematics, Hiram College, Hiram, Ohio

Published by

CRC PRESS, Inc.

18901 Cranwood Parkway, Cleveland, Ohio 44128

Library of Congress Cataloging in Publication Data

Main entry under title:

Handbook of tables for mathematics.

First-2d ed. (1962–64) published under title:
Handbook of mathematical tables.
At head of title: CRC.
Bibliography:
Includes index.
1. Mathematics – Tables, etc. – Collected works.
I. Weast, Robert C., 1916– ed. II. Selby,
Samuel, M., ed. III. Chemical Rubber Company, Cleveland.
QA47.H32 1975 510'.21'2 75-2190
ISBN 0-87819-655-2

This book represents information obtained from authentic and highly regarded sources. Reprinted material is quoted with permission, and sources are indicated. A wide variety of references are listed. Every reasonable effort has been made to give reliable data and information, but the author and the publisher cannot assume responsibility for the validity of all materials or for the consequences of their use.

Revised Fourth Edition

Second Printing, 1976

International Standard Book Numbers 0-8493-0654-X

Former International Standard Book Number 0-87819-655-2

Library of Congress Card Number 62-15661

Printed in the United States

Direct all inquiries and correspondence to
CRC PRESS, INC., 18901 Cranwood Parkway, Cleveland, Ohio 44128

Preface *

"That's one small step for a man . . . one giant leap for mankind."

These were the immortal words uttered on July 20th, 1969 by Neil Armstrong who first placed man's footstep on the surface of the moon. This was the beginning of a new scientific era as the astronomer, the mathematician, the physicist, the biologist, the scientist, and the engineer take on new challenges to explore the universe of the future.

In the light of this historic accomplishment it is appropriate that the *Fourth Edition of the HANDBOOK of tables for MATHEMATICS* contains a new section under the headings of *Astrodynamics; Basic Orbital Equations* and *Astrodynamical Terminology, Notation and Usage.* This detailed work on the nomenclature and symbolism as used in celestial mechanics was prepared under the guidance and supervision of Dr. Samuel Herrick, Professor of Astronomy and Engineering at the University of California.

Many other features are to be observed in this new edition such as:

- a completely re-organized set of integral tables which include an additional 150 formulas,
- a table on Z-transforms,
- an extended section on hyperbolic functions with tables on inverse hyperbolic functions and the Gudermanian.
- an extension to complement the present financial tables for $(1 + i)^n$ and $(1 + i)^{-n}$ in ¼% intervals from 8% to 20%. This feature is in recognition for the need of interest rates that better serve our present economy.

Other sections have been updated and extended wherever it seemed to be advantageous to users of mathematical tables. The acknowledgement page in the forefront of the tables records the several contributors who authored many of these changes. It is hoped that the additions and revisions will meet with the same popular reception which has been accorded the previous three editions of this Handbook.

The editor and the editorial staff wish to thank the Advisory Board and the many users of the Tables for their constructive suggestions and contributions toward the preparation and production of this new *Fourth Edition Handbook of tables for Mathematics.* The editor is particularly grateful to our Administrative Editor, Mrs. Florence Thomas, for her meticulous handling of details that were so essential in the final preparation of this Fourth Edition.

As in the past, The Chemical Rubber Co. invites and welcomes constructive comments from its users, as this is a most effective means for keeping the editions of these tables updated and abreast of the times.

SAMUEL M. SELBY, PH.D., SC.D.
EDITOR-IN-CHIEF OF MATHEMATICS

June 1970

*In this reprint the financial sections for $(1 + i)^n$ and $(1 + i)^{-n}$ have been extended in 1/4% intervals from 0% to 20% for n from 1 to 100. A table scaled accordingly has been added for $\frac{1}{a_{\overline{n}|i}}$. It is hoped that this additional material will meet the several requests made by past users of these tables.

January 1975

Mathematics Advisory Board

Acknowledgments

Acknowledgment is made to the following authors, editors, and publishers whose material has been used in the Handbook of tables for Mathematics, and for which permission has been received.

American Society for Testing Materials STP-15C;
ASTM Manual on Quality Control of Materials (1951)
Factors for Computing Control Limits

American Statistical Association, Journal of
Vol. 41 (1946) 557–566, W. J. Dixon and A. M. Mood
Critical Values for the Sign Test

*Bargmann, Rolf E., Ph.D.
Matrices and Determinants

Biometrika Trustees, E. S. Pearson
Cambridge University Press
Biometrika Tables for Statisticians and Biometrika, Vol. 32,
Percentage Points, Chi-Square Distribution
Percentage Points, *F*-Distribution

*Cherrington, E. H., Jr., Ph.D.
Table for Conversion of Calendar Date to Julian Date

E. Richard Cohen, A.B., M.S., Ph.D.
Fundamental Physical Constants

*Deskins, W. E., Ph.D.
Basic Concepts in Algebra

*Eves, Howard, Ph.D.
Analytic Geometry
Curves and Surfaces
Mensuration Formula
Trigonometry

*Gaskell, R. E., Ph.D.
The Laplace Transform

*Girling, Brian, M.Sc., F.I.M.A.
Differentiation Formulas
Finite Differences
Sine, Cosine and Exponential Integrals
The Z Transforms

Gupta, Madhu-Sudan, M.Sc. Tech., M.S., A.M.
Inverse Hyperbolic Function Tables

Harper & Rowe, New York, N.Y.
Concepts of Calculus by A. H. Lighston
Bessel Function J_0, J_1

Herrick, Samuel, B.A., Sc.D., Ph.D.
Astrodynamics: Basic Orbital Equations
Astrodynamical Terminology, Notation, and Usage
Condensed from Astrodynamics; N.Y. Van Nostrand Reinhold, 1970

International Business Machines Corporation
IBM Brochure "360 Principles of Operation," Form A 22-6821-3
Hexadecimal-Direct Conversion Table

Masket, A. V. H., Ph.D.
Relativity Contraction-Dilation Factor

Oliver and Boyd, Ltd., Edinburgh, Scotland
Statistical Tables for Biological, Agricultural and Medical Research of Fisher & Yates
Percentage Points, Student's t-Distribution

*Pratt, Richard, A. M.
Explanations to Tables
Use of Logarithms (Law of Exponents)
Integral Tables

The Royal Society, London, England
Vol. 3 (1954) 2 Royal Society Mathematical Tables
Number of Combinations

Springer-Verlag New York, Inc.
Fünfstellige Funktionentafeln (1930), Hayashi, K.
Number of Permutations

CONTENTS

Units for a System of Measures as used Internationally

Quantity Measured	Unit	Abbreviation
Length	meter	m
Mass	kilogram	kg
Time	second	s
Electric current	ampere	A
Temperature	degree Kelvin	°K
Luminous intensity	candela	cd

Supplementary and Derived Units From Base Units as used Internationally

	Supplementary Units		
Plane angle	radian	rad	
Solid angle	steradian	sr	
	Derived Units		
Area	square meter	m^2	
Volume	cubic meter	m^3	
Frequency	hertz	Hz	(s^{-1})
Density	kilogram per cubic meter	kg/m^3	
Velocity	meter per second	m/s	
Angular velocity	radian per second	rad/s	
Acceleration	meter per second squared	m/s^2	
Angular acceleration	radian per second squared	rad/s^2	
Force	newton	N	$(kg \cdot m/s^2)$
Pressure	newton per sq meter	N/m^2	
Kinematic viscosity	sq meter per second	m^2/s	
Dynamic viscosity	newton-second per sq meter	$N \cdot s/m^2$	
Work, energy, quantity of heat	joule	J	$(N \cdot m)$
Power	watt	W	(J/s)
Electric charge	coulomb	C	$(A \cdot s)$
Voltage, potential difference, electromotive force	volt	V	(W/A)
Electric field strength	volt per meter	V/m	
Electric resistance	ohm	Ω	(V/A)
Electric capacitance	farad	F	$(A \cdot s/V)$
Magnetic flux	weber	Wb	$(V \cdot s)$
Inductance	henry	H	$(V \cdot s/A)$
Magnetic flux density	tesla	T	(Wb/m^2)
Magnetic field strength	ampere per meter	A/m	
Magnetomotive force	ampere	A	
Luminous flux	lumen	lm	$(cd \cdot sr)$
Luminance	candela per sq meter	cd/m^2	
Illumination	lux	lx	(lm/m^2)

METRIC CONVERSION TABLE

Linear, Weight

Inches	Centimeters	Centimeters	Inches
1	2.54	1	0.393701
2	5.08	2	0.787402
3	7.62	3	1.181102
4	10.16	4	1.574803
5	12.70	5	1.968504
6	15.24	6	2.362205
7	17.78	7	2.755906
8	20.32	8	3.149606
9	22.86	9	3.543307
Feet	**Meters**	**Meters**	**Feet**
1	0.3048	1	3.280840
2	0.6096	2	6.561680
3	0.9144	3	9.842520
4	1.2192	4	13.123360
5	1.5240	5	16.404199
6	1.8288	6	19.685039
7	2.1336	7	22.965879
8	2.4384	8	26.246719
9	2.7432	9	29.527559
Yards	**Meters**	**Meters**	**Yards**
1	0.9144	1	1.0936133
2	1.8288	2	2.1872266
3	2.7432	3	3.2808399
4	3.6576	4	4.3744532
5	4.5720	5	5.4680665
6	5.4864	6	6.5616798
7	6.4008	7	7.6552931
8	7.3152	8	8.7489064
9	8.2296	9	9.8425197
Miles (Statute)	**Kilometers**	**Kilometers**	**Miles**
1	1.609344	1	0.6213712
2	3.218688	2	1.2427424
3	4.828032	3	1.8641136
4	6.437376	4	2.4854848
5	8.046720	5	3.1068560
6	9.656064	6	3.7282272
7	11.265408	7	4.3495983
8	12.874752	8	4.9709695
9	14.484096	9	5.5923407
Pounds (av)	**Kilograms**	**Kilograms**	**Pounds (av)**
1	0.45359237	1	2.2046226
2	0.90718474	2	4.4092452
3	1.36077711	3	6.6138679
4	1.81436948	4	8.8184905
5	2.26796185	5	11.0231131
6	2.72155422	6	13.2277357
7	3.17514659	7	15.4323584
8	3.62873896	8	17.6369810
9	4.08233133	9	19.8416036

CONVERSION FACTORS

U. S. AND METRIC UNITS

Each unit in bold face type is followed by its equivalent in other units of the same quantity.

Acre—0.0015625 square mile (statute); 4.3560×10^4 square feet; 0.40468564 hectare.

Bushel—(U.S.)—1.244456 cubic feet; 2150.42 cubic inches; 0.035239 cubic meter; 35.23808 liters.

Centimeter—0.0328084 foot; 0.393701 inch.

Circular Mil—7.853982×10^{-7} square inches; 5.067075×10^{-6} square centimeters.

Cubic Centimeter—0.061024 cubic inch; 0.270512 dram (U. S. fluid); 16.230664 minims (U. S.); 0.999972 milliliter.

Cubic Foot—0.803564 bushel (U. S.); 7.480520 gallons (U. S. liquid); 0.028317 cubic meter; 28.31605 liters.

Cubic Inch—16.387064 cubic centimeters.

Cubic Meter—35.314667 cubic feet; 264.17205 gallons (U. S. liquid).

Foot—0.3048 meter.

Gallon (U. S. liquid)—0.1336816 cubic foot; 0.832675 gallon (British); 231 cubic inches; 0.0037854 cubic meter; 3.785306 liters.

Grain—0.06479891 gram.

Gram—0.00220462 pound (avoirdupois); 0.035274 ounce (avoirdupois); 15.432358 grains.

Hectare—2.471054 acres; 1.07639×10^5 square feet.

Inch—2.54 centimeters.

Kilogram—2.204623 pounds (avoirdupois).

Kilometer—0.621371 mile (statute).

Liter—0.264179 gallon (U. S. liquid); 0.0353157 cubic foot; 1.056718 quarts (U. S. liquid).

Meter—1.093613 yards; 3.280840 feet; 39.37008 inches.

Mile (statute)—1.609344 kilometers.

Ounce (U. S. fluid)—1.804688 cubic inches; 29.573730 cubic centimeters.

Ounce (avoirdupois)—28.349523 grams.

Ounce (apothecary or troy)—31.103486 grams.

Pint (U. S. liquid)—0.473163 liter; 473.17647 cubic centimeters.

Pound (avoirdupois)—0.453592 kilogram; 453.59237 grams.

Pound (apothecary or troy)—0.3732417 kilogram, 373.24172 grams.

Quart (U. S. dry)—1.10119 liters.

Quart (liquid)—0.946326 liter.

Radian—57.295779 degrees.

Rod—5.0292 meters.

Square Centimeter—0.155000 square inch.

Square Foot—0.09290304 square meter.

Square Inch—645.16 square millimeters.

Square Meter—10.763910 square feet.

Square Yard—0.836127 square meter.

Ton (short)—907.18474 kilograms.

Yard—0.9144 meter.

See Index for extensive Conversion Factors.

RECOMMENDED UNIT PREFIXES

Multiples and submultiples	Prefixes	Symbols
10^{12}	tera	T
10^{9}	giga	G
10^{6}	mega	M
10^{3}	kilo	k
10^{2}	hecto	h
10	deka	da
10^{-1}	deci	d
10^{-2}	centi	c
10^{-3}	milli	m
10^{-6}	micro	μ (greek mu)
10^{-9}	nano	n
10^{-12}	pico	p
10^{-15}	femto	f
10^{-18}	atto	a

DEFINED VALUES AND EQUIVALENTS

Meter .... **(m)**	1 650 763.73 wave lengths in vacuo of the unperturbed transition $2p_{10} - 5d_5$ in ^{86}Kr
Kilogram .... **(kg)**	mass of the international kilogram at Sèvres, France
Second .... **(s)**	1/31 556 925.974 7 of the tropical year at 12^h ET, 0 January 1900
Degree Kelvin .... **(°K)**	defined in the thermodynamic scale by assigning 273.16°K to the triple point of water (freezing point, 273.15°K = 0°C)
Unified atomic mass unit .... **(u)**	1/12 the mass of an atom of the ^{12}C nuclide
Mole .... **(mol)**	amount of substance containing the same number of atoms as 12 g of pure ^{12}C
Standard acceleration of free fall .... **(g_n)**	9.806 65 M s^{-2}, 980.665 cm s^{-2}
Normal atmospheric pressure .... **(atm)**	101 325 N m^{-2}, 1 013 250 dyn cm^{-2}
Thermochemical calorie .... **(cal_{th})**	4.1840 J, 4.1840 × 10^7 erg
International Steam Table calorie .... **(cal_{IT})**	4.1868 J, 4.1868 × 10^7 erg
Liter .... **(l)**	0.001 000 028 m^3, 1 000.028 cm^3 (recommended by CIPM, 1950)
	0.001 m^3, 1000 cm^3 (recommended by GCWM 1964)
Inch .... **(in)**	0.0254 m, 2.54 cm
Pound (avdp) .... **(lb)**	0.453 592 37 kg, 453.592 37 g

DECIMAL EQUIVALENTS OF COMMON FRACTIONS

		1/64	= 0.015 625		11/32	22/64	= 0.343 75			43/64	= 0.671 875
	1/32	2/64	= .031 25			23/64	= .359 375	11/16	22/32	44/64	= .687 5
		3/64	= .046 875	3/8	12/32	24/64	= .375			45/64	= .703 125
1/16	2/32	4/64	= .062 5			25/64	= .390 625		23/32	46/64	= .718 75
		5/64	= .078 125		13/32	26/64	= .406 25			47/64	= .734 375
	3/32	6/64	= .093 75			27/64	= .421 875	3/4	24/32	48/64	= .75
		7/64	= .109 375	7/16	14/32	28/64	= .437 5			49/64	= .765 625
1/8	4/32	8/64	= .125			29/64	= .453 125		25/32	50/64	= .781 25
		9/64	= .140 625		15/32	30/64	= .468 75			51/64	= .796 875
	5/32	10/64	= .156 25			31/64	= .484 375	13/16	26/32	52/64	= .812 5
		11/64	= .171 875	1/2	16/32	32/64	= .50			53/64	= .828 125
3/16	6/32	12/64	= .187 5			33/64	= .515 625		27/32	54/64	= .843 75
		13/64	= .203 125		17/32	34/64	= .531 25			55/64	= .859 375
	7/32	14/64	= .218 75			35/64	= .546 875	7/8	28/32	56/64	= .875
		15/64	= .234 375	9/16	18/32	36/64	= .562 5			57/64	= .890 625
1/4	8/32	16/64	= .25			37/64	= .578 125		29/32	58/64	= .906 25
		17/64	= .265 625		19/32	38/64	= .593 75			59/64	= .921 875
	9/32	18/64	= .281 25			39/64	= .609 375	15/16	30/32	60/64	= .937 5
		19/64	= .296 875	5/8	20/32	40/64	= .625			61/64	= .953 125
5/16	10/32	20/64	= .312 5			41/64	= .640 625		31/32	62/64	= .968 75
		21/64	= .328 125		21/32	42/64	= .656 25			63/64	= .984 375

FUNDAMENTAL PHYSICAL CONSTANTS

DR. E. RICHARD COHEN

The following table contains data which are a tentative revision of the 1963 values of the fundamental physical constants.

It has become increasingly clear in the last several years that the 1963 analysis of the fundamental physical constants by Cohen and DuMond must be revised and that the values recommended at the time are in error by as much as 100 ppm. The strongest evidence for this revision came from the measurement in 1967 of macroscopic phase coherence in superconductors by Langenberg, Parker and Taylor at the University of Pennsylvania. Their measured value of the quantum of magnetic flux ($h/2e$), measured with an accuracy of 4 ppm, was inconsistent with the 1963 recommendation by 10 times that amount. This verified the growing evidence from spectroscopic and microwave data that the value of the fine structure constant needed a revision of 20 ppm.

It is therefore clear that a complete revision of the 1963 recommendation is necessary. Such a revision will of course include experimental data in addition to that on the fine structure constant, including careful attention to the electrical standards maintained by each national standards laboratory as recalibrated with respect to BIPM in 1968, effective January 1, 1969.

The following table of numerical values of the physical constants is intended as a general indication of the extent of the revision required in the 1963 values. Because of the tentative nature, and since the full effect of experimental correlations between data have not been calculated, no errors are quoted for these values. The numerical values, although tentative, and not representing a full reassessment of the available data are believed to be more reliable than the 1963 values. A more recent discussion of the status of the physical constants as of approximately January, 1969, is given by B. N. Taylor, W. H. Parker, and D. N. Langenberg in Reviews of Modern Physics, July 1969.

FUNDAMENTAL PHYSICAL CONSTANTS

Constant	Symbol	Old value	New value		*Correction ppm
Speed of light in vacuum	c	2.997925_1	2.997925	$\times 10^8$ ms^{-1}	0
Gravitational constant	G	6.670_5	6.670	10^{-11} Nm^{2}kg^{-2}	0
Elementary charge	e	1.60210_2	1.6022	10^{-19}C	+60
		4.80298_7	4.8032	10^{-10} esu	+60
Avogadro constant	N_A	6.02252_9	6.0222	10^{26} kmole^{-1}	−60
Mass unit	u	1.66043_2	1.66053	10^{-27} kg	+60
Electron rest mass	m_e	9.10908_{13}	9.1096	10^{-31} kg	+60
		5.48597_3	5.48593	10^{-4} u	0
Proton rest mass	m_p	1.67252_3	1.67262	10^{-27} kg	+60
		1.00727663_8	1.00727661	u	0
Neutron rest mass	m_n	1.67482_3	1.67492	10^{-27} kg	+60
		1.0086654_4	1.0086652	u	0
Faraday constant	F	9.64870_5	9.6487	10^4C mole^{-1}	0
		2.89261_{25}	2.8926	10^{14} esu	0
Planck constant	h	6.62559_{16}	6.6262	10^{-34} Js	+100
	$h/2\pi$	1.054494_{25}	1.05459	10^{-34} Js	+100
Fine-structure constant	α	7.29720_3	7.29735	10^{-3}	+20
$2\pi e^2/hc$	$1/\alpha$	137.0388_6	137.0360		−20
Charge-to-mass ratio	e/m_e	1.758796_6	1.75880	10^{11}C kg^{-1}	0
for electron		5.27274_2	5.27276	10^{17} esu	0
Quantum of magnetic flux	hc/e	4.13556_4	4.13571	10^{-11} Wb	+40
		1.379474_{13}	1.37952	10^{-17} esu	+40
Rydberg constant	R_∞	1.0973731_1	1.0973731	10^7 m^{-1}	0
Bohr radius	a_0	5.29167_2	5.29177	10^{-11} m	+20
Compton wavelength of	$\lambda_c = h/m_e c$	2.42621_2	2.42631	10^{-12} m	+40
electron	$\lambda_c/2\pi$	3.86144_3	3.86159	10^{-13} m	+40
Gyromagnetic ratio of	γ	2.675192_7	2.67519	10^8 rad s^{-1}T^{-1}	0
proton	$\gamma/2\pi$	4.25770_1	4.2577	10^7 Hz T^{-1}	0
(Uncorrected for	γ'	2.675123_7	†2.67512	10^8 s^{-1}T^{-1}	0
diamagnetism H_2O)	$\gamma'/2\pi$	4.25759_1	†4.257586	10^7 Hz T^{-1}	0
Bohr magneton	μ_B	9.2732_2	9.2741	10^{-24} J T^{-1}	+100
Nuclear magneton	μ_N	5.05050_{13}	5.0510	10^{-27} J T^{-1}	+100
Proton Moment	μ_p	1.41049_4	1.4106	10^{-26} J T^{-1}	+80
	μ_p/μ_N	2.79276_2	2.79278		0
(Uncorrected for diamagnetism in H_2O sample)		2.79268_2	2.79271		0
Gas constant	R_0	8.31434_{35}	8.3143	J deg^{-1} mole^{-1}	0
Boltzmann constant	k	1.38054_6	1.3806	10^{-23} J deg^{-1}	+60
First radiation constant ($2\pi hc^2$)	c_1	3.74150_9	3.7418	10^{-16} W m^2	+80
Second radiation constant (hc/k)	c_2	1.43879_6	1.4388	10^{-2} m deg	0
Stephan-Boltzmann constant	σ	5.6697_{10}	5.6696	10^{-8} W m^{-2} deg^{-4}	−20

* This column gives the correction resulting only from the increase of 20 ppm in the value of the fine structure constant, not the total change from 1963 to the tentative new value.

† The value for the gyromagnetic ratio of the proton has been recommended by the Comité International des Poids et Mesures in their meeting of 14–17 October 1968 for international metrological usage. This value is based on the 1969 BIPM scales of resistance and electromotive force which are in agreement, as exactly as is possible, with the (absolute) definitions of electrical units adopted by the Conférence Generale des Poids et Mesures.

NUMBERS CONTAINING π

	Number	Logarithm		Number	Logarithm
π	3.1415 927	0.4971 499	$2\pi^2$	19.7392 088	1.2953 297
2π	6.2831 853	0.7981 799	$\pi/180$	0.0174 533	8.2418 774 − 10
3π	9.4247 780	0.9742 711	$180/\pi$	57.2957 795	1.7581 226
4π	12.5663 706	1.0992 099	$4\pi^2$	39.4784 176	1.5963 597
8π	25.1327 412	1.4002 399	$1/\pi^2$	0.1013 212	9.0057 003 − 10
$\pi/2$	1.5707 963	0.1961 199	$1/(2\pi^2)$	0.0506 606	8.7046 703 − 10
$\pi/3$	1.0471 976	0.0200 286	$1/(4\pi^2)$	0.0253 303	8.4036 403 − 10
$\pi/4$	0.7853 982	9.8950 899 − 10	$\sqrt{\pi}$	1.7724 539	0.2485 749
$\pi/6$	0.5235 988	9.7189 986 − 10	$\sqrt{\pi/4}$ or		
$\pi/8$	0.3926 991	9.5940 599 − 10	$\sqrt{\pi}/2$	0.8862 269	9.9475 449 − 10
$2\pi/3$	2.0943 951	0.3210 586	$\sqrt{\pi}/4$	0.4431 135	9.6465 149 − 10
$4\pi/3$	4.1887 902	0.6220 886	$\sqrt{\pi/2}$	1.2533 141	0.0980 599
$1/\pi$	0.3183 099	9.5028 501 − 10	$\sqrt{2/\pi}$	0.7978 846	9.9019 401 − 10
$2/\pi$	0.6366 198	9.8038 801 − 10	π^3	31.0062 767	1.4914 496
$4/\pi$	1.2732 395	0.1049 101	$\sqrt[3]{\pi}$	1.4645 919	0.1657 166
$1/(2\pi)$	0.1591 549	9.2018 201 − 10	$1/\sqrt[3]{\pi}$	0.6827 841	9.8342 834 − 10
$1/(4\pi)$	0.0795 775	8.9007 901 − 10	$\sqrt[3]{\pi^2}$	2.1450 294	0.3314 332
$1/(6\pi)$	0.0530 516	8.7246 989 − 10	$1/\sqrt{\pi}$	0.5641 896	9.7514 251 − 10
$1/(8\pi)$	0.0397 887	8.5997 601 − 10	$1/\sqrt{2\pi}$	0.3989 423	9.6009 101 − 10
π^2	9.8696 044	0.9942 997	$2/\sqrt{\pi}$	1.1283 792	0.0524 551

MULTIPLES OF $\frac{\pi}{2}$

n	$n\frac{\pi}{2}$	n	$n\frac{\pi}{2}$	n	$n\frac{\pi}{2}$	n	$n\frac{\pi}{2}$
1	1.57079 63268	26	40.84070 44967	51	80.11061 26665	76	119.38052 08364
2	3.14159 26536	27	42.41150 08235	52	81.68140 89933	77	120.95131 71632
3	4.71238 89804	28	43.98229 71503	53	83.25220 53201	78	122.52211 34900
4	6.28318 53072	29	45.55309 34771	54	84.82300 16469	79	124.09290 98168
5	7.85398 16340	30	47.12388 98038	55	86.39379 79737	80	125.66370 61436
6	9.42477 79608	31	48.69468 61306	56	87.96459 43005	81	127.23450 24704
7	10.99557 42876	32	50.26548 24574	57	89.53539 06273	82	128.80529 87972
8	12.56637 06144	33	51.83627 87842	58	91.10618 69541	83	130.37609 51240
9	14.13716 69412	34	53.40707 51110	59	92.67698 32809	84	131.94689 14508
10	15.70796 32679	35	54.97787 14378	60	94.24777 96077	85	133.51768 77776
11	17.27875 95947	36	56.54866 77646	61	95.81857 59345	86	135.08848 41044
12	18.84955 59215	37	58.11946 40914	62	97.38937 22613	87	136.65928 04312
13	20.42035 22483	38	59.69026 04182	63	98.96016 85881	88	138.23007 67580
14	21.99114 85751	39	61.26105 67450	64	100.53096 49149	89	139.80087 30847
15	23.56194 49019	40	62.83185 30718	65	102.10176 12417	90	141.37166 94115
16	25.13274 12287	41	64.40264 93986	66	103.67255 75685	91	142.94246 57383
17	26.70353 75555	42	65.97344 57254	67	105.24335 38953	92	144.51326 20651
18	28.27433 38823	43	67.54424 20522	68	106.81415 02221	93	146.08405 83919
19	29.84513 02091	44	69.11503 83790	69	108.38494 65488	94	147.65485 47187
20	31.41592 65359	45	70.68583 47058	70	109.95574 28756	95	149.22565 10455
21	32.98672 28627	46	72.25663 10326	71	111.52653 92024	96	150.79644 73723
22	34.55751 91895	47	73.82742 73594	72	113.09733 55292	97	152.36724 36991
23	36.12831 55163	48	75.39822 36862	73	114.66813 18560	98	153.93804 00259
24	37.69911 18431	49	76.96902 00129	74	116.23892 81828	99	155.50883 63527
25	39.26990 81699	50	78.53981 63397	75	117.80972 45096	100	157.07963 26795

MISCELLANEOUS CONSTANTS

π CONSTANTS

π = 3.14159 26535 89793 23846 26433 83279 50288 41971 69399 37511
$1/\pi$ = 0.31830 98861 83790 67153 77675 26745 02872 40689 19291 48091
π^2 = 9.86960 44010 89358 61883 44909 99876 15113 53136 99407 24079
$\log_e \pi$ = 1.14472 98858 49400 17414 34273 51353 05871 16472 94812 91531
$\log_{10} \pi$ = 0.49714 98726 94133 85435 12682 88290 89887 36516 78324 38044
$\log_{10} \sqrt{2\pi}$ = 0.39908 99341 79057 52478 25035 91507 69595 02099 34102 92128

CONSTANTS INVOLVING e

e = 2.71828 18284 59045 23536 02874 71352 66249 77572 47093 69996
$1/e$ = 0.36787 94411 71442 32159 55237 70161 46086 74458 11131 03177
e^2 = 7.38905 60989 30650 22723 04274 60575 00781 31803 15570 55185
$M = \log_{10} e$ = 0.43429 44819 03251 82765 11289 18916 60508 22943 97005 80367
$1/M = \log_e 10$ = 2.30258 50929 94045 68401 79914 54684 36420 76011 01488 62877
$\log_{10} M$ = 9.63778 43113 00536 78912 29674 98645 − 10

π^e AND e^π CONSTANTS

π^e = 22.45915 77183 61045 47342 71522
e^π = 23.14069 26327 79269 00572 90864
$e^{-\pi}$ = 0.04321 39182 63772 24977 44177
$e^{\frac{1}{2}\pi}$ = 4.81047 73809 65351 65547 30357
$i^i = e^{-\frac{1}{2}\pi}$ = 0.20787 95763 50761 90854 69556

NUMERICAL CONSTANTS

$\sqrt{2}$ = 1.41421 35623 73095 04880 16887 24209 69807 85696 71875 37695
$\sqrt[3]{2}$ = 1.25992 10498 94873 16476 72106 07278 22835 05702 51464 70151
$\log_e 2$ = 0.69314 71805 59945 30941 72321 21458 17656 80755 00134 36026
$\log_{10} 2$ = 0.30102 99956 63981 19521 37388 94724 49302 67681 89881 46211
$\sqrt{3}$ = 1.73205 08075 68877 29352 74463 41505 87236 69428 05253 81039
$\sqrt[3]{3}$ = 1.44224 95703 07408 38232 16383 10780 10958 83918 69253 49935
$\log_e 3$ = 1.09861 22886 68109 69139 52452 36922 52570 46474 90557 82275
$\log_{10} 3$ = 0.47712 12547 19662 43729 50279 03255 11530 92001 28864 19070

$\log_e 10 - \log_e 2 = \log_e 5$ = 1.60943 79124 34100 37560 07593 33226 18763 95256 01354 26851

OTHER CONSTANTS

Euler's Constant γ = 0.57721 56649 01532 86061
$\log_e \gamma$ = −0.54953 93129 81644 82234
Golden Ratio ϕ = 1.61803 39887 49894 84820 45868 34365 63811 77203 09180

$= \lim_{n\to\infty} (f_{n+1}/f_n)$, $f_1, f_2, \ldots$ = Fib. sequence

$= \frac{\sqrt{5}+1}{2}$, $\phi - 1 = \frac{1}{\phi}$

MENSURATION FORMULAE

DR. HOWARD EVES

TRIANGLES

In the following: K = area, r = radius of the inscribed circle, R = radius of the circumscribed circle.

Right Triangle

$A + B = C = 90°$

$c^2 = a^2 + b^2$ (*Pythagorean relation*)

$a = \sqrt{(c+b)(c-b)}$

$K = \frac{1}{2}ab$

$r = \dfrac{ab}{a+b+c}, \quad R = \frac{1}{2}c$

$h = \dfrac{ab}{c}, \quad m = \dfrac{b^2}{c}, \quad n = \dfrac{a^2}{c}$

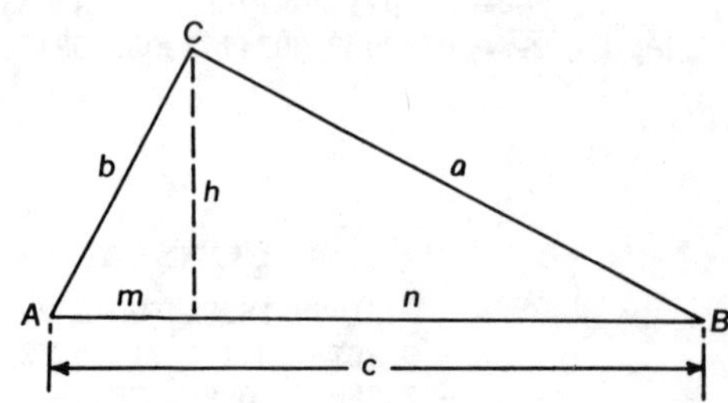

Equilateral Triangle

$A = B = C = 60°$

$K = \frac{1}{4}a^2\sqrt{3}$

$r = \frac{1}{6}a\sqrt{3}, \quad R = \frac{1}{3}a\sqrt{3}$

$h = \frac{1}{2}a\sqrt{3}$

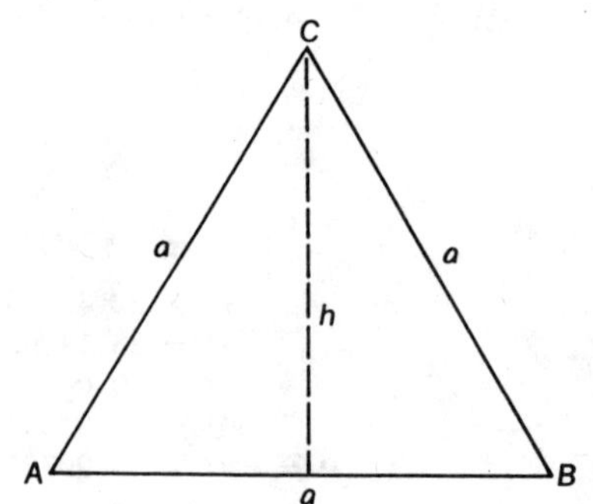

General Triangle

Let $s = \frac{1}{2}(a+b+c)$, h_c = length of altitude on side c, t_c = length of bisector of angle C, m_c = length of median to side c.

$A + B + C = 180°$

$c^2 = a^2 + b^2 - 2ab\cos C$

(*law of cosines*)

$K = \frac{1}{2}h_c c = \frac{1}{2}ab\sin C$

$= \dfrac{c^2 \sin A \sin B}{2\sin C}$

$= rs = \dfrac{abc}{4R}$

$= \sqrt{s(s-a)(s-b)(s-c)}$ (*Heron's formula*)

$G = \frac{1}{4}\sqrt{(a+b+c)(-a+b+c)(a-b+c)(a+b-c)}$

$$r = c \sin \frac{A}{2} \sin \frac{B}{2} \sec \frac{C}{2} = \frac{ab \sin C}{2s} = (s - c) \tan \frac{C}{2}$$

$$= \sqrt{\frac{(s-a)(s-b)(s-c)}{s}} = \frac{K}{s} = 4R \sin \frac{A}{2} \sin \frac{B}{2} \sin \frac{C}{2}$$

$$R = \frac{c}{2 \sin C} = \frac{abc}{4\sqrt{s(s-a)(s-b)(s-c)}} = \frac{abc}{4K}$$

$$h_c = a \sin B = b \sin A = \frac{2K}{c}$$

$$t_c = \frac{2ab}{a+b} \cos \frac{C}{2} = \sqrt{ab\left\{1 - \frac{c^2}{(a+b)^2}\right\}}$$

$$m_c = \sqrt{\frac{a^2}{2} + \frac{b^2}{2} - \frac{c^2}{4}}$$

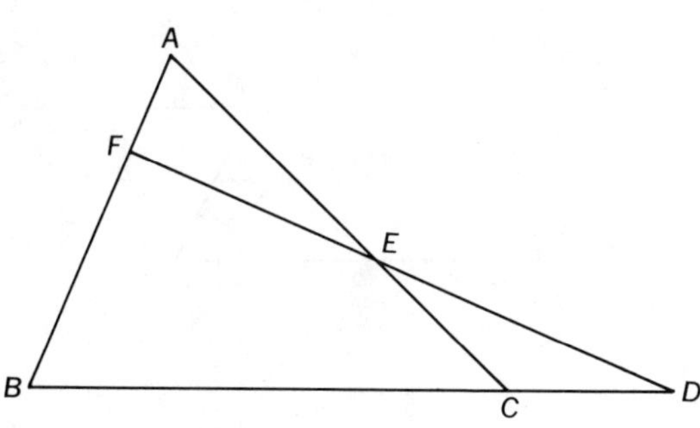

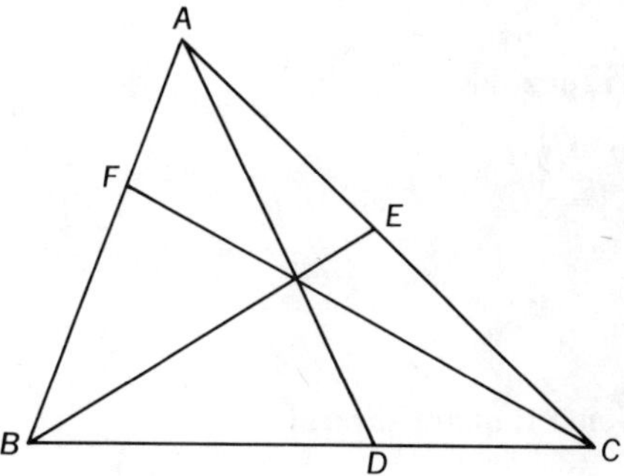

Menelaus' Theorem. A necessary and sufficient condition for points D, E, F on the respective side lines BC, CA, AB of a triangle ABC to be collinear is that

$$BD \cdot CE \cdot AF = - DC \cdot EA \cdot FB,$$

where all segments in the formula are directed segments.

Ceva's Theorem. A necessary and sufficient condition for AD, BE, CF, where D, E, F are points on the respective side lines BC, CA, AB of a triangle ABC, to be concurrent is that

$$BD \cdot CE \cdot AF = + DC \cdot EA \cdot FB,$$

where all segments in the formula are directed segments.

QUADRILATERALS

In the following: K = area, p and q are diagonals.

Rectangle

$A = B = C = D = 90°$

$K = ab, \quad p = \sqrt{a^2 + b^2}$

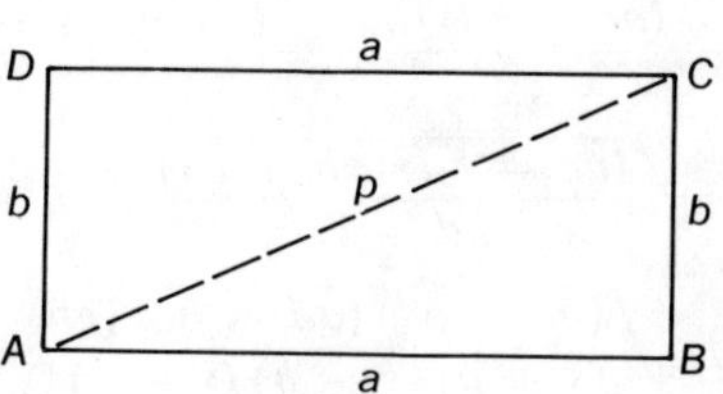

Parallelogram

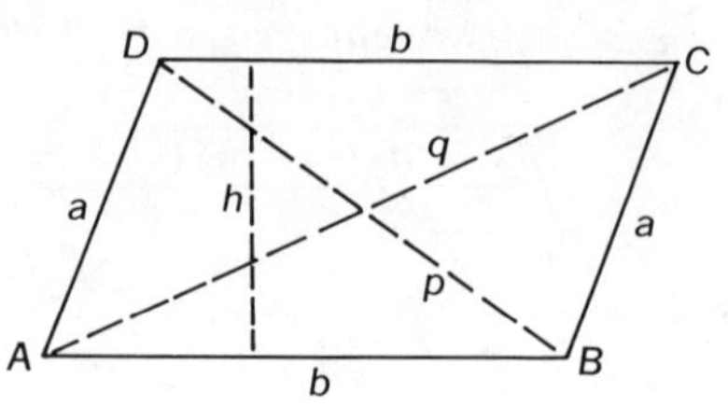

$$A = C, \quad B = D, \quad A + B = 180°$$
$$K = bh = ab \sin A = ab \sin B$$
$$h = a \sin A = a \sin B$$
$$p = \sqrt{a^2 + b^2 - 2ab \cos A}$$
$$q = \sqrt{a^2 + b^2 - 2ab \cos B} = \sqrt{a^2 + b^2 + 2ab \cos A}$$

Rhombus

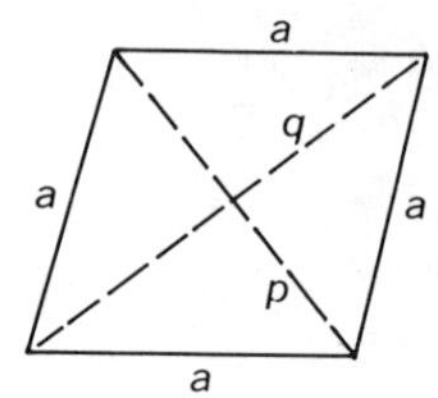

$$p^2 + q^2 = 4a^2$$
$$K = \tfrac{1}{2}pq$$

Trapezoid

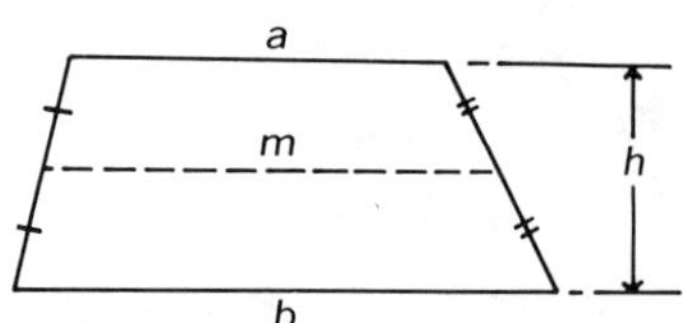

$$m = \tfrac{1}{2}(a + b)$$
$$K = \tfrac{1}{2}(a + b)h = mh$$

General quadrilateral

Let $s = \frac{1}{2}(a + b + c + d)$.

$$K = \tfrac{1}{2}pq \sin\theta$$
$$= \tfrac{1}{4}(b^2 + d^2 - a^2 - c^2) \tan\theta$$
$$= \tfrac{1}{4}\sqrt{4p^2q^2 - (b^2 + d^2 - a^2 - c^2)^2}$$

(*Bretschneider's formula*)

$$= \sqrt{(s - a)(s - b)(s - c)(s - d) - abcd \cos^2\left(\frac{A + B}{2}\right)}$$

Theorem. The diagonals of a quadrilateral with consecutive sides a, b, c, d are perpendicular if and only if $a^2 + c^2 = b^2 + d^2$.

Cyclic Quadrilateral

Let R = radius of the circumscribed circle.

$$A + C = B + D = 180°$$
$$K = \sqrt{(s - a)(s - b)(s - c)(s - d)}$$

(*Brahmagupta's formula*)

$$= \frac{\sqrt{(ac + bd)(ad + bc)(ab + cd)}}{4R}$$

$$p = \sqrt{\frac{(ac + bd)(ab + cd)}{ad + bc}}, \quad q = \sqrt{\frac{(ac + bd)(ad + bc)}{ab + cd}}$$

$$R = \frac{1}{2}\sqrt{\frac{(ac + bd)(ad + bc)(ab + cd)}{(s - a)(s - b)(s - c)(s - d)}}, \quad \sin\theta = \frac{2K}{ac + bd}$$

Ptolemy's Theorem. A convex quadrilateral with consecutive sides a, b, c, d and diagonals p and q is cyclic if and only if $ac + bd = pq$.

Cyclic-inscriptable Quadrilateral

Let r = radius of the inscribed circle, R = radius of the circumscribed circle, m = distance between the centers of the inscribed and the circumscribed circles.

$$A + C = B + D = 180°$$

$$a + c = b + d$$

$$K = \sqrt{abcd}$$

$$\frac{1}{(R - m)^2} + \frac{1}{(R + m)^2} = \frac{1}{r^2}$$

$$r = \frac{\sqrt{abcd}}{s}$$

$$R = \tfrac{1}{2}\sqrt{\frac{(ac + bd)(ad + bc)(ab + cd)}{abcd}}$$

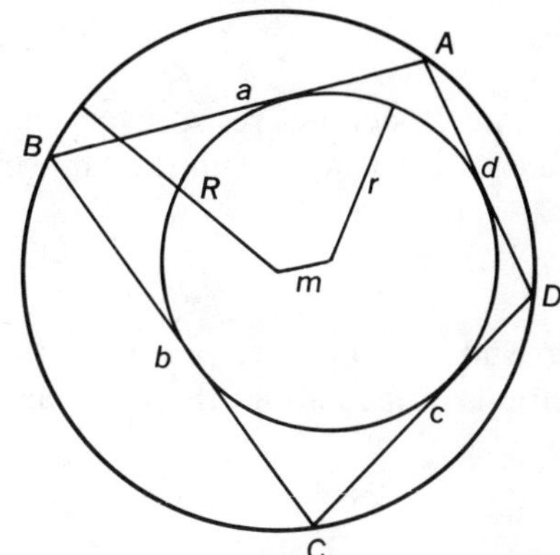

REGULAR POLYGONS

In the following: n = number of sides, s = length of each side, p = perimeter, θ = one of the vertex angles, r = radius of the inscribed circle, R = radius of the circumscribed circle, K = area.

$$\theta = \left(\frac{n - 2}{n}\right) 180°$$

$$s = 2r \tan \frac{180°}{n} = 2R \sin \frac{180°}{n}$$

$$p = ns$$

$$K = \tfrac{1}{4} ns^2 \cot \frac{180°}{n}$$

$$= nr^2 \tan \frac{180°}{n}$$

$$= \tfrac{1}{2} nR^2 \sin \frac{360°}{n}$$

$$r = \tfrac{1}{2} s \cot \frac{180°}{n}, \quad R = \tfrac{1}{2} s \csc \frac{180°}{n}$$

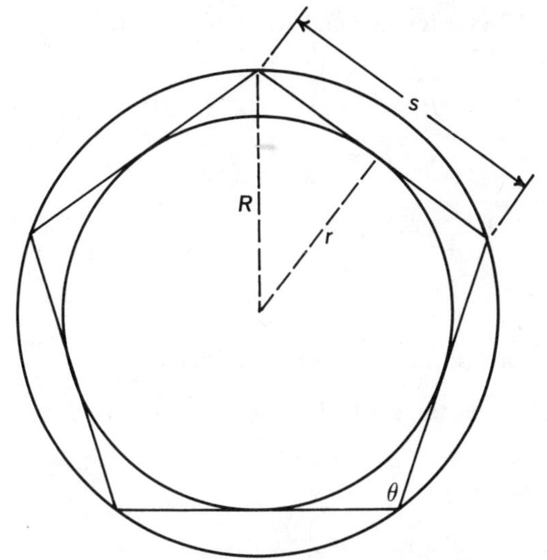

Polygon	n	K	r	R
Triangle (equilateral)	3	$0.43301s^2$	$0.28868s$	$0.57735s$
Square	4	$1.00000s^2$	$0.50000s$	$0.70711s$
Pentagon	5	$1.72048s^2$	$0.68819s$	$0.85065s$
Hexagon	6	$2.59808s^2$	$0.86603s$	$1.00000s$
Heptagon	7	$3.63391s^2$	$1.0383s$	$1.1524s$
Octagon	8	$4.82843s^2$	$1.2071s$	$1.3066s$
Nonagon	9	$6.18182s^2$	$1.3737s$	$1.4619s$
Decagon	10	$7.69421s^2$	$1.5388s$	$1.6180s$
Undecagon	11	$9.36564s^2$	$1.7028s$	$1.7747s$
Dodecagon	12	$11.19615s^2$	$1.8660s$	$1.9319s$

If s_k denotes the side of a regular polygon of k sides inscribed in a circle of radius R, then

$$s_{2n} = \sqrt{2R^2 - R\sqrt{4R^2 - s_n^2}}.$$

If S_k denotes the side of a regular polygon of k sides circumscribed about a circle of radius r, then

$$S_{2n} = \frac{2rS_n}{2r + \sqrt{4r^2 + S_n^2}}$$

If p_k and P_k denote, respectively, the perimeters of regular polygons of k sides inscribed in and circumscribed about the same circle, then

$$P_{2n} = \frac{2p_nP_n}{p_n + P_n} \quad \text{and} \quad p_{2n} = \sqrt{p_nP_{2n}}.$$

If a_k and A_k denote, respectively, the areas of regular polygons of k sides inscribed in and circumscribed about the same circle, then

$$a_{2n} = \sqrt{a_nA_n} \quad \text{and} \quad A_{2n} = \frac{2a_{2n}A_n}{a_{2n} + A_n}.$$

CIRCLES

In the following: R = radius, D = diameter, C = circumference, K = area.

Circumference and Area of a Circle

$$C = 2\pi R = \pi D \qquad (\pi = 3.14159\cdots)$$
$$K = \pi R^2 = \tfrac{1}{4}\pi D^2 = 0.7854D^2$$
$$C = 2\sqrt{\pi K} = \frac{2K}{R}$$
$$K = \frac{C^2}{4\pi} = \tfrac{1}{2}CR$$

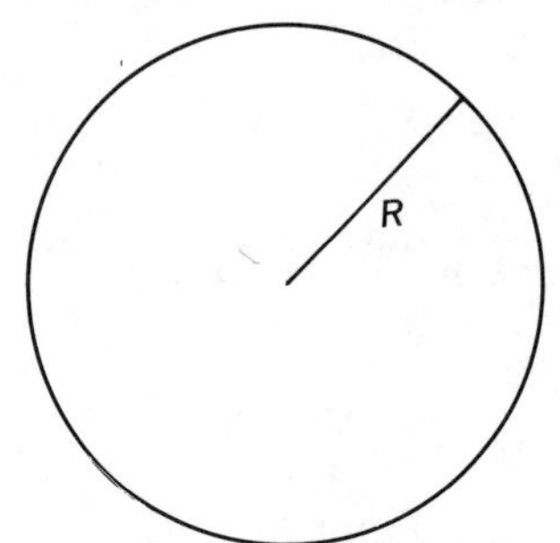

Sector and Segment of a Circle

Let the central angle θ be measured in radians ($\theta < \pi$).

$$h = R - d, \quad d = R - h$$
$$s = R\theta$$
$$d = R\cos\frac{\theta}{2} = \tfrac{1}{2}c\cot\frac{\theta}{2}$$
$$= \tfrac{1}{2}\sqrt{4R^2 - c^2}$$
$$c = 2R\sin\frac{\theta}{2} = 2d\tan\frac{\theta}{2}$$
$$= 2\sqrt{R^2 - d^2} = \sqrt{4h(2R - h)}$$
$$\theta = \frac{s}{R} = 2\,\mathrm{Cos}^{-1}\frac{d}{R} = 2\,\mathrm{Tan}^{-1}\frac{c}{2d} = 2\,\mathrm{Sin}^{-1}\frac{c}{2R}$$
$$K\,(\text{sector}) = \tfrac{1}{2}Rs = \tfrac{1}{2}R^2\theta$$
$$K\,(\text{segment}) = \tfrac{1}{2}R^2(\theta - \sin\theta) = \tfrac{1}{2}(Rs - cd) = R^2\,\mathrm{Cos}^{-1}\frac{d}{R} - d\sqrt{R^2 - d^2}$$
$$= R^2\,\mathrm{Cos}^{-1}\frac{R - h}{R} - (R - h)\sqrt{2Rh - h^2}$$

Sector of an Annulus

$$h = R_1 - R_2$$
$$K = \tfrac{1}{2}\theta(R_1 + R_2)(R_1 - R_2)$$
$$= \tfrac{1}{2}\theta h(R_1 + R_2)$$
$$= \tfrac{1}{2}h(s_1 + s_2)$$

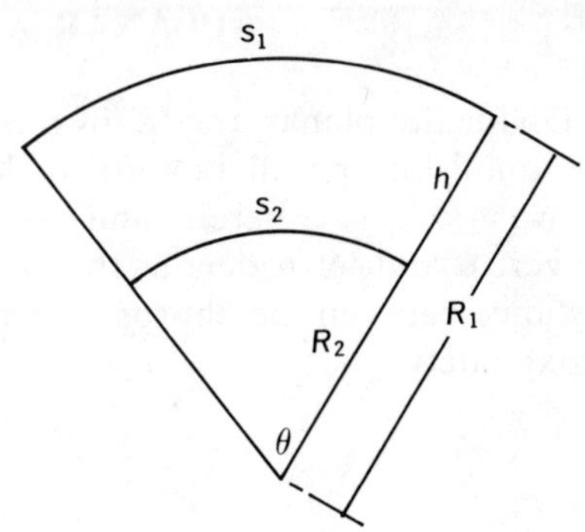

CONIC SECTIONS

Ellipse

Let p = circumference, K = area

$$p = 2\pi\sqrt{\frac{a^2 + b^2}{2}} \quad \text{(approximately)}$$

$$= 4aE \quad \text{(exactly)}$$ See table of elliptic integral for E, using $k = \sqrt{a^2 - b^2}/a$.

$$K = \pi ab$$

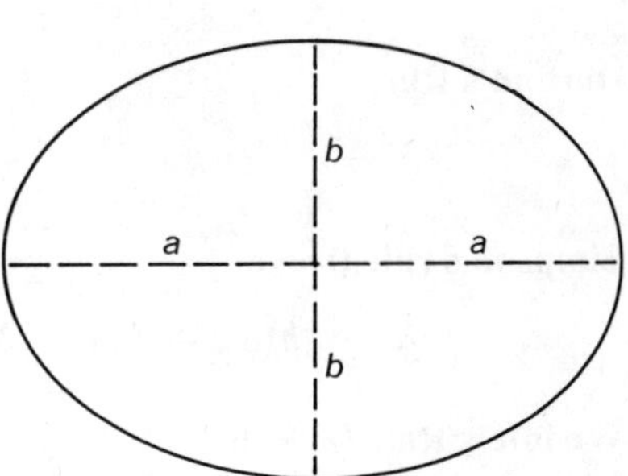

Parabolic Segment

$$s = \sqrt{4x^2 + y^2} + \frac{y^2}{2x}\log_e\left[\frac{2x + \sqrt{4x^2 + y^2}}{y}\right]$$

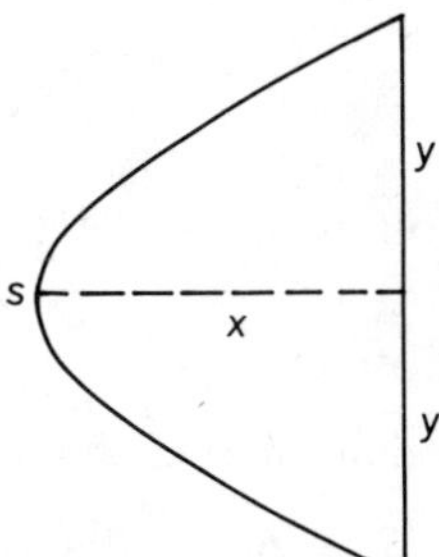

K (right segment) $= \frac{4}{3}xy$

K (oblique segment) $= \frac{4}{3}T$, where T is the area of the triangle with base along the chord of the segment and with opposite vertex at the point on the parabola at which the tangent to the parabola is parallel to the chord of the segment.

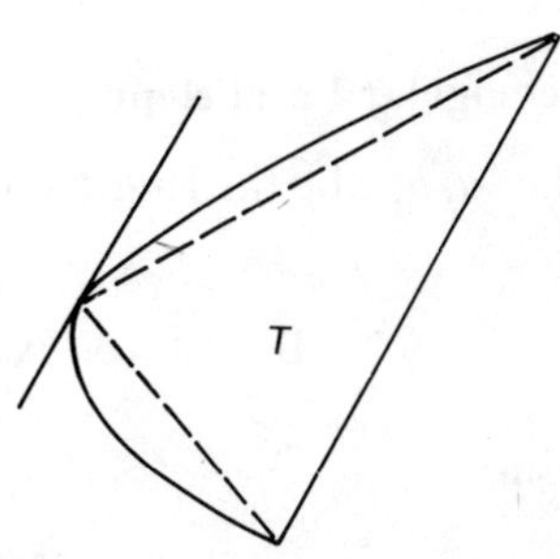

CAVALIERI'S THEOREM FOR THE PLANE

If two planar areas are included between a pair of parallel lines, and if the two segments cut off by the areas on any line parallel to the including lines are equal in length, then the two planar areas are equal.

PLANAR AREAS BY APPROXIMATION

Divide the planar area K into n strips by equidistant parallel chords of lengths $y_0, y_1, y_2, \ldots, y_n$ (where y_0 and/or y_n may be zero), and let h denote the common distance between the chords. Then, approximately:

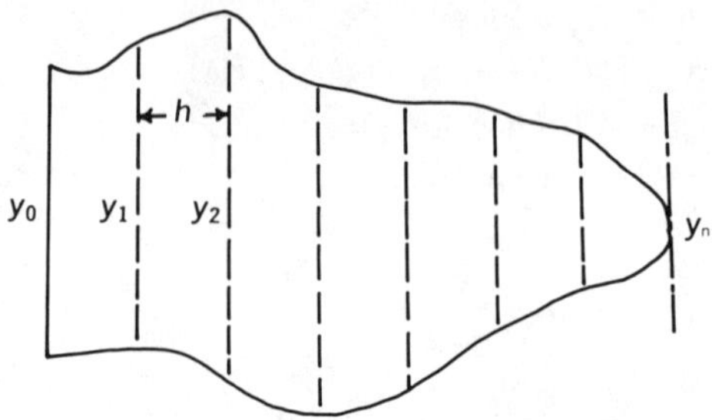

Trapezoidal Rule

$$K = h(\tfrac{1}{2}y_0 + y_1 + y_2 + \cdots + y_{n-1} + \tfrac{1}{2}y_n)$$

Durand's Rule

$$K = h(\tfrac{4}{10}y_0 + \tfrac{11}{10}y_1 + y_2 + y_3 + \cdots + y_{n-2} + \tfrac{11}{10}y_{n-1} + \tfrac{4}{10}y_n)$$

Simpson's rule (n even)

$$K = \tfrac{1}{3}h(y_0 + 4y_1 + 2y_2 + 4y_3 + 2y_4 + \cdots + 2y_{n-2} + 4y_{n-1} + y_n)$$

Weddle's Rule ($n = 6$)

$$K = \tfrac{3}{10}h(y_0 + 5y_1 + y_2 + 6y_3 + y_4 + 5y_5 + y_6)$$

SOLIDS BOUNDED BY PLANES

In the following: S = lateral surface, T = total surface, V = volume.

Cube

Let a = length of each edge.

$$T = 6a^2, \quad \text{diagonal of face} = a\sqrt{2}$$
$$V = a^3, \quad \text{diagonal of cube} = a\sqrt{3}$$

Rectangular Parallelepiped (or box)

Let a, b, c be the lengths of its edges.

$$T = 2(ab + bc + ca), \quad V = abc$$
$$\text{diagonal} = \sqrt{a^2 + b^2 + c^2}$$

Prism

S = (perimeter of right section) × (lateral edge)
V = (area of right section) × (lateral edge)
= (area of base) × (altitude)

Truncated Triangular Prism

V = (area of right section) × $\tfrac{1}{3}$(sum of the three lateral edges)

Pyramid

S of regular pyramid $= \frac{1}{2}$(perimeter of base) $\times$ (slant height)

$V = \frac{1}{3}$(area of base) $\times$ (altitude)

Frustum of Pyramid

Let B_1 = area of lower base, B_2 = area of upper base, h = altitude.

S of regular figure $= \frac{1}{2}$(sum of perimeters of bases) $\times$ (slant height)

$V = \frac{1}{3}h(B_1 + B_2 + \sqrt{B_1 B_2})$

Prismatoid

A *prismatoid* is a polyhedron having for bases two polygons in parallel planes, and for lateral faces triangles or trapezoids with one side lying in one base, and the opposite vertex or side lying in the other base, of the polyhedron. Let B_1 = area of lower base, M = area of midsection, B_2 = area of upper base, h = altitude.

$$V = \tfrac{1}{6}h(B_1 + 4M + B_2) \quad \text{(the } \textit{prismoidal formula}\text{)}$$

Note: Since cubes, rectangular parallelepipeds, prisms, pyramids, and frustums of pyramids are all examples of prismatoids, the formula for the volume of a prismatoid subsumes most of the above volume formulae.

Regular Polyhedra

Let v = number of vertices, e = number of edges, f = number of faces, α = each dihedral angle, a = length of each edge, r = radius of the inscribed sphere, R = radius of the circumscribed sphere, A = area of each face, T = total area, V = volume.

$v - e + f = 2$ (the *Euler-Descartes formula*—actually holds for *any* convex polyhedron)

$$T = fA$$

$$V = \tfrac{1}{3}rfA = \tfrac{1}{3}rT$$

Name	Nature of Surface	T	V
Tetrahedron	4 equilateral triangles	$1.73205a^2$	$0.11785a^3$
Hexahedron (cube)	6 squares	$6.00000a^2$	$1.00000a^3$
Octahedron	8 equilateral triangles	$3.46410a^2$	$0.47140a^3$
Dodecahedron	12 regular pentagons	$20.64573a^2$	$7.66312a^3$
Icosahedron	20 equilateral triangles	$8.66025a^2$	$2.18169a^3$

Name	v	e	f	α	a	r
Tetrahedron	4	6	4	70° 32′	$1.633R$	$0.333R$
Hexahedron	8	12	6	90°	$1.155R$	$0.577R$
Octahedron	6	12	8	109° 28′	$1.414R$	$0.577R$
Dodecahedron	20	30	12	116° 34′	$0.714R$	$0.795R$
Icosahedron	12	30	20	138° 11′	$1.051R$	$0.795R$

Name	A	r	R	V
Tetrahedron	$\frac{1}{4}a^2\sqrt{3}$	$\frac{1}{12}a\sqrt{6}$	$\frac{1}{4}a\sqrt{6}$	$\frac{1}{12}a^3\sqrt{2}$
Hexahedron	a^2	$\frac{1}{2}a$	$\frac{1}{2}a\sqrt{3}$	a^3
Octahedron	$\frac{1}{4}a^2\sqrt{3}$	$\frac{1}{6}a\sqrt{6}$	$\frac{1}{2}a\sqrt{2}$	$\frac{1}{3}a^3\sqrt{2}$
Dodecahedron	$\frac{1}{4}a^2\sqrt{25 + 10\sqrt{5}}$	$\frac{1}{20}a\sqrt{250 + 110\sqrt{5}}$	$\frac{1}{4}a(\sqrt{15} + \sqrt{3})$	$\frac{1}{4}a^3(15 + 7\sqrt{5})$
Icosahedron	$\frac{1}{4}a^2\sqrt{3}$	$\frac{1}{12}a\sqrt{42 + 18\sqrt{5}}$	$\frac{1}{4}a\sqrt{10 + 2\sqrt{5}}$	$\frac{5}{12}a^3(3 + \sqrt{5})$

CYLINDERS AND CONES

In the following: B_1 = area of lower base, B_2 = area of upper base, h = altitude, S = lateral surface, T = total surface, V = volume.

Cylinder

S = (perimeter of right section) × (lateral edge)
V = (area of right section) × (lateral edge)

Right Circular Cylinder

Let R = radius of base.

$$S = 2\pi Rh, \quad T = 2\pi R(R + h), \quad V = \pi R^2 h$$

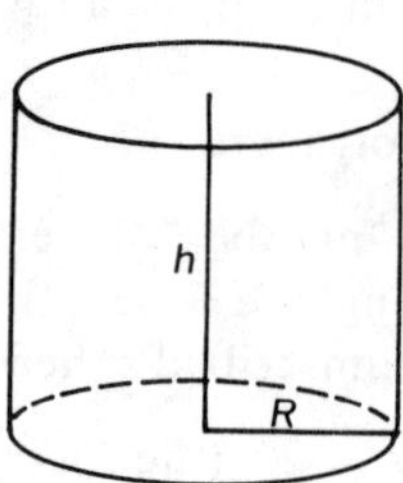

Cone

$$V = \tfrac{1}{3}B_1 h$$

Right Circular Cone

Let R = radius of base, s = slant height.

$$s = \sqrt{R^2 + h^2}$$
$$S = \pi Rs = \pi R\sqrt{R^2 + h^2}$$
$$T = \pi R(R + s) = \pi R(R + \sqrt{R^2 + h^2})$$
$$V = \tfrac{1}{3}\pi R^2 h$$

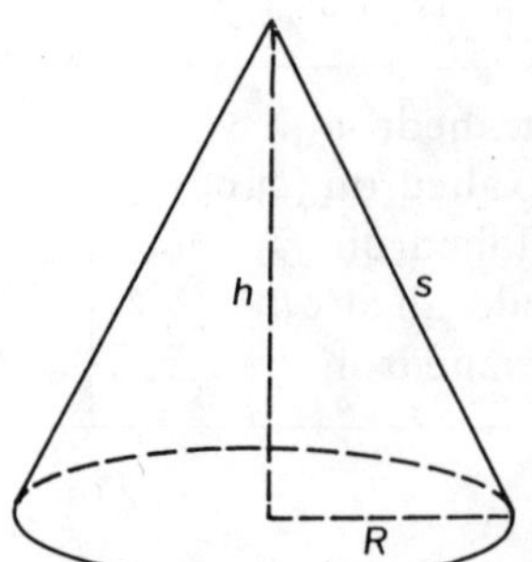

Frustum of Cone

$$V = \tfrac{1}{3}h(B_1 + B_2 + \sqrt{B_1 B_2})$$

Frustum of Right Circular Cone

Let R_1 = radius of lower base, R_2 = radius of upper base, s = slant height.

$$s = \sqrt{(R_1 - R_2)^2 + h^2}$$

$$S = \pi(R_1 + R_2)s$$

$$= \pi(R_1 + R_2)\sqrt{(R_1 - R_2)^2 + h^2}$$

$$T = \pi[R_1^2 + R_2^2 + (R_1 + R_2)s]$$

$$= \pi[R_1^2 + R_2^2 + (R_1 + R_2)\sqrt{(R_1 - R_2)^2 + h^2}]$$

$$V = \tfrac{1}{3}\pi h(R_1^2 + R_2^2 + R_1R_2)$$

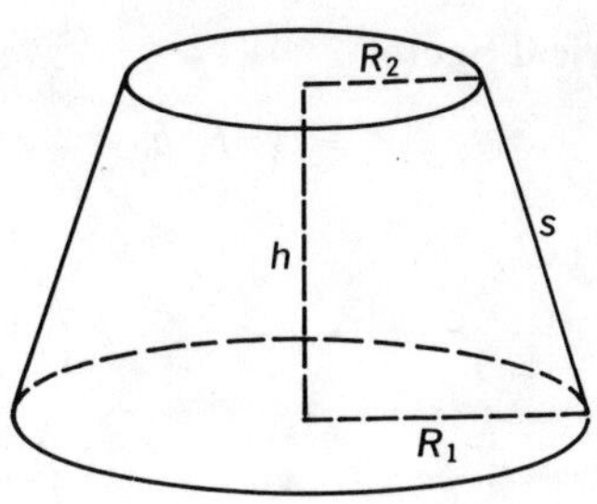

SPHERICAL FIGURES

In the following: R = radius of sphere, D = diameter of sphere, S = surface area, V = volume.

Sphere

$$D = 2R$$

$$S = 4\pi R^2 = \pi D^2 = 12.57R^2$$

$$V = \tfrac{4}{3}\pi R^3 = \tfrac{1}{6}\pi D^3 = 4.189R^3$$

Zone and Segment of One Base

$$S = 2\pi Rh = \pi Dh = \pi p^2$$

$$V = \tfrac{1}{3}\pi h^2(3R - h) = \tfrac{1}{6}\pi h(3a^2 + h^2)$$

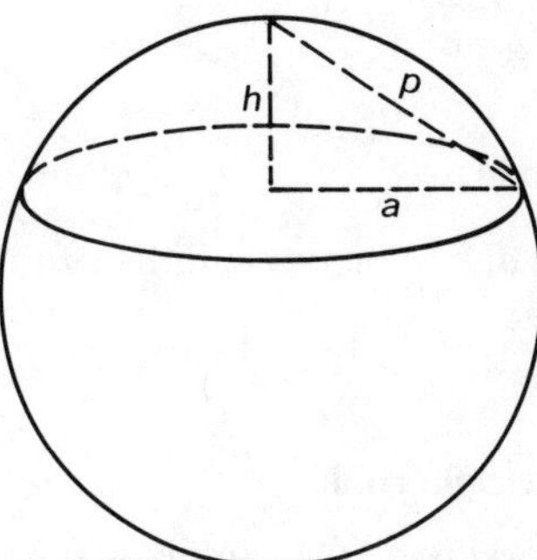

Zone and Segment of Two Bases

$$S = 2\pi Rh = \pi Dh$$

$$V = \tfrac{1}{6}\pi h(3a^2 + 3b^2 + h^2)$$

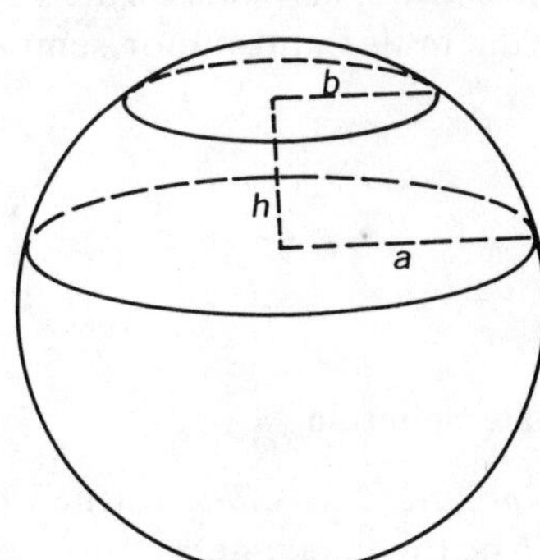

Lune

$$S = 2R^2\theta, \quad \theta \text{ in radians}$$

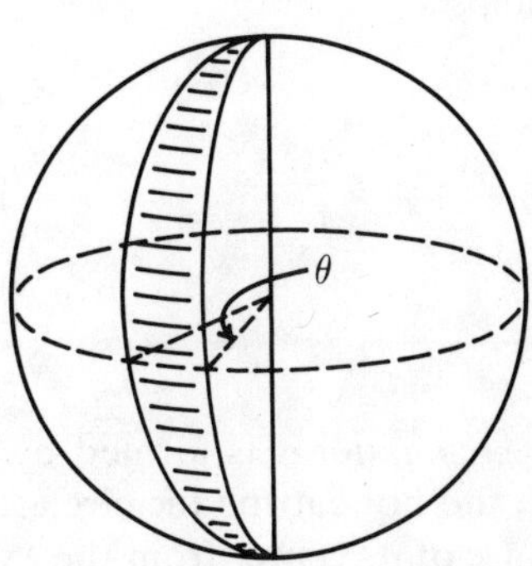

Spherical Sector

$$V = \tfrac{2}{3}\pi R^2 h = \tfrac{1}{6}\pi D^2 h$$

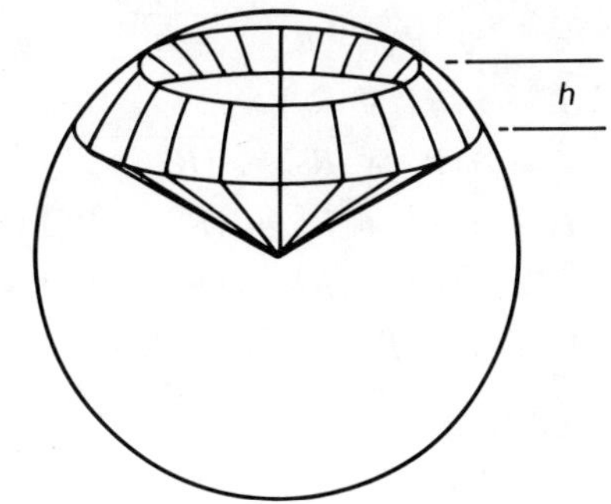

Spherical Triangle and Polygon

Let A, B, C be the angles, in radians, of the triangle; let θ = sum of angles, in radians, of a spherical polygon of n sides.

$$S = (A + B + C - \pi) R^2 = \Omega R^2 \Rightarrow A+B+C = \pi + \Omega$$
$$S = [\theta - (n - 2)\pi] R^2 = \Omega R^2 \Rightarrow \theta = (n-2)\pi + \Omega$$

Ω = solid angle

SPHEROIDS

Ellipsoid

Let a, b, c be the lengths of the semiaxes.

$$V = \tfrac{4}{3}\pi abc$$

Oblate Spheroid

An *oblate spheroid* is formed by the rotation of an ellipse about its minor axis. Let a and b be the major and minor semiaxes, respectively, and ϵ the eccentricity, of the revolving ellipse.

$$S = 2\pi a^2 + \pi \frac{b^2}{\epsilon} \log_e \frac{1 + \epsilon}{1 - \epsilon}$$

$$V = \tfrac{4}{3}\pi a^2 b$$

Prolate Spheroid

A *prolate spheroid* is formed by the rotation of an ellipse about its major axis. Let a and b be the major and minor semiaxes, respectively, and ϵ the eccentricity, of the revolving ellipse.

$$S = 2\pi b^2 + 2\pi \frac{ab}{\epsilon} \sin^{-1} \epsilon$$

$$V = \tfrac{4}{3}\pi a b^2$$

CIRCULAR TORUS

A *circular torus* is formed by the rotation of a circle about an axis in the plane of the circle and not cutting the circle. Let r be the radius of the revolving circle and let R be the distance of its center from the axis of rotation.

$$S = 4\pi^2 Rr$$
$$V = 2\pi^2 Rr^2$$

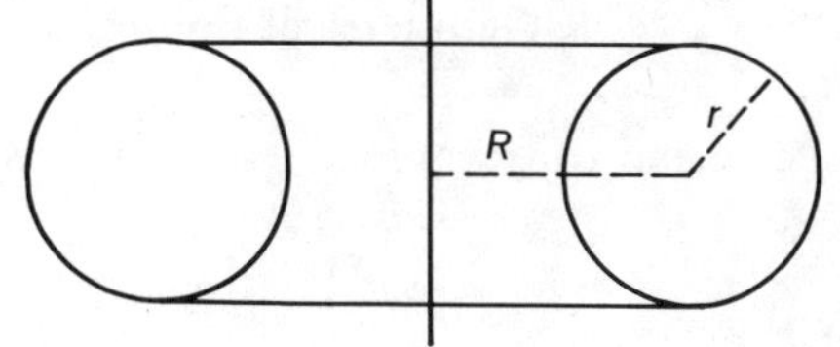

PAPPUS-GULDINUS THEOREMS

1. If a planar arc be revolved about an axis in its plane, but not cutting the arc, the area of the surface of revolution so formed is equal to the product of the length of the arc and the length of the path traced by the centroid of the arc.
2. If a planar area be revolved about an axis in its plane, but not intersecting the area, the volume of the solid of revolution so formed is equal to the product of the area and the length of the path traced by the centroid of the area.

CAVALIERI'S THEOREM FOR SPACE

If two solids are included between a pair of parallel planes, and if the two sections cut by them on any plane parallel to the including planes are equal in area, then the volumes of the solids are equal.

GENERAL PRISMATOID

A *general prismatoid* is a solid such that the area A_y of any section parallel to and distant y from a fixed plane can be expressed as a polynomial in y of degreee not higher than the third. That is,

$$A_y = ay^3 + by^2 + cy + d,$$

where a, b, c, d are constants which may be positive, zero, or negative. Let B_1 = area of lower base, M = area of midsection, B_2 = area of upper base, h = altitude.

$$V = \tfrac{1}{6}h(B_1 + 4M + B_2)$$

Note. All prismatoids, cylinders, cones, spheres, spheroids, and many other solids are general prismatoids.

CENTROIDS

If a geometrical figure possesses a center of symmetry, that point is the centroid of the figure.

If a geometrical figure possesses an axis of symmetry, the centroid of the figure lies on that axis.

Geometrical Figure	Location of Centroid
Perimeter of triangle	Center of the inscribed circle of the triangle whose vertices are the midpoints of the sides of the given triangle.
Arc of semicircle of radius R	Distance from diameter $= \dfrac{2R}{\pi}$
Arc of 2α radians of a circle of radius R	Distance from center of circle $= \dfrac{R \sin \alpha}{\alpha}$
Area of triangle	Intersection of the medians
Area of quadrilateral	Intersection of the diagonals of the parallelogram whose sides pass through adjacent trisection points of pairs of consecutive sides of the quadrilateral
Area of semicircle of radius R	Distance from diameter $= \dfrac{4R}{3\pi}$
Area of circular sector of radius R and central angle 2α radians	Distance from center of circle $= \dfrac{2R \sin \alpha}{3\alpha}$
Area of semiellipse of altitude h	Distance from base $= \dfrac{4h}{3\pi}$
Area of a quadrant of an ellipse of major and minor semiaxes a and b	Distance from minor axis $= \dfrac{4a}{3\pi}$, distance from major axis $= \dfrac{4b}{3\pi}$
Area of right parabolic segment of altitude h	Distance from base $= \frac{2}{5}h$
Lateral area of regular pyramid or right circular cone	Distance from base $= \frac{1}{3}h$
Area of hemisphere of radius R	Distance from base $= \frac{1}{2}R$
Volume of pyramid or cone	One fourth the way from the centroid of the base to the vertex of the pyramid or cone
Volume of frustum of pyramid or cone with d as the distance between the centroids of its bases and k as the ratio of similarity of the upper base to the lower base	On the line joining the centroids of the two bases at a distance from the centroid of the lower base $= \frac{1}{4}d\left(\dfrac{1 + 2k + 3k^2}{1 + k + k^2}\right)$
Volume of hemisphere of radius R	Distance from base $= \frac{3}{8}R$
Volume of revolution of altitude h obtained by revolving a semiellipse about its axis of symmetry	Distance from base $= \frac{3}{8}h$
Volume of paraboloid of revolution of altitude h	Distance from base $= \frac{1}{3}h$

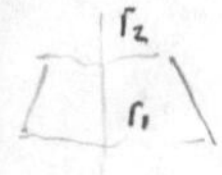

RECIPROCALS, CIRCUMFERENCE AND AREA OF CIRCLES

As a matter of convenience, the values of 1000 × (1/*n*) are given in the table. To obtain the actual value of the reciprocal, shift the decimal point three places to the left.

Circumferences and areas of circles are given for the values of *n* as the diameter.

n = dia	$1000\frac{1}{n}$	Circumference πn	Area $\frac{\pi n^2}{4}$	n = dia	$1000\frac{1}{n}$	Circumference πn	Area $\frac{\pi n^2}{4}$
0	∞	0.000000	.0000000	**50**	20.00000	157.0796	1963.495
1	1000.000	3.141593	.7853982	51	19.60784	160.2212	2042.821
2	500.0000	6.283185	3.141593	52	19.23077	163.3628	2123.717
3	333.3333	9.424778	7.068583	53	18.86792	166.5044	2206.183
4	250.0000	12.56637	12.56637	54	18.51852	169.6460	2290.221
5	200.0000	15.70796	19.63495	55	18.18182	172.7876	2375.829
6	166.6667	18.84956	28.27433	56	17.85714	175.9292	2463.009
7	142.8571	21.99115	38.48451	57	17.54386	179.0708	2551.759
8	125.0000	25.13274	50.26548	58	17.24138	182.2124	2642.079
9	111.1111	28.27433	63.61725	59	16.94915	185.3540	2733.971
10	100.0000	31.41593	78.53982	**60**	16.66667	188.4956	2827.433
11	90.90909	34.55752	95.03318	61	16.39344	191.6372	2922.467
12	83.33333	37.69911	113.0973	62	16.12903	194.7787	3019.071
13	76.92308	40.84070	132.7323	63	15.87302	197.9203	3117.245
14	71.42857	43.98230	153.9380	64	15.62500	201.0619	3216.991
15	66.66667	47.12389	176.7146	65	15.38462	204.2035	3318.307
16	62.50000	50.26548	201.0619	66	15.15152	207.3451	3421.194
17	58.82353	53.40708	226.9801	67	14.92537	210.4867	3525.652
18	55.55556	56.54867	254.4690	68	14.70588	213.6283	3631.681
19	52.63158	59.69026	283.5287	69	14.49275	216.7699	3739.281
20	50.00000	62.83185	314.1593	**70**	14.28571	219.9115	3848.451
21	47.61905	65.97345	346.3606	71	14.08451	223.0531	3959.192
22	45.45455	69.11504	380.1327	72	13.88889	226.1947	4071.504
23	43.47826	72.25663	415.4756	73	13.69863	229.3363	4185.387
24	41.66667	75.39822	452.3893	74	13.51351	232.4779	4300.840
25	40.00000	78.53982	490.8739	75	13.33333	235.6194	4417.865
26	38.46154	81.68141	530.9292	76	13.15789	238.7610	4536.460
27	37.03704	84.82300	572.5553	77	12.98701	241.9026	4656.626
28	35.71429	87.96459	615.7522	78	12.82051	245.0442	4778.362
29	34.48276	91.10619	660.5199	79	12.65823	248.1858	4901.670
30	33.33333	94.24778	706.8583	**80**	12.50000	251.3274	5026.548
31	32.25806	97.38937	754.7676	81	12.34568	254.4690	5152.997
32	31.25000	100.5310	804.2477	82	12.19512	257.6106	5281.017
33	30.30303	103.6726	855.2986	83	12.04819	260.7522	5410.608
34	29.41176	106.8142	907.9203	84	11.90476	263.8938	5541.769
35	28.57143	109.9557	962.1128	85	11.76471	267.0354	5674.502
36	27.77778	113.0973	1017.876	86	11.62791	270.1770	5808.805
37	27.02703	116.2389	1075.210	87	11.49425	273.3186	5944.679
38	26.31579	119.3805	1134.115	88	11.36364	276.4602	6082.123
39	25.64103	122.5221	1194.591	89	11.23596	279.6017	6221.139
40	25.00000	125.6637	1256.637	**90**	11.11111	282.7433	6361.725
41	24.39024	128.8053	1320.254	91	10.98901	285.8849	6503.882
42	23.80952	131.9469	1385.442	92	10.86957	289.0265	6647.610
43	23.25581	135.0885	1452.201	93	10.75269	292.1681	6792.909
44	22.72727	138.2301	1520.531	94	10.63830	295.3097	6939.778
45	22.22222	141.3717	1590.431	95	10.52632	298.4513	7088.218
46	21.73913	144.5133	1661.903	96	10.41667	301.5929	7238.229
47	21.27660	147.6549	1734.945	97	10.30928	304.7345	7389.811
48	20.83333	150.7964	1809.557	98	10.20408	307.8761	7542.964
49	20.40816	153.9380	1885.741	99	10.10101	311.0177	7697.687
50	20.00000	157.0796	1963.495	**100**	10.00000	314.1593	7853.982

RECIPROCALS, CIRCUMFERENCE AND AREA OF CIRCLES (Continued)

n = dia	$1000\frac{1}{n}$	Circum-ference πn	Area $\frac{\pi n^2}{4}$	n = dia	$1000\frac{1}{n}$	Circum-ference πn	Area $\frac{\pi n^2}{4}$
100	10.00000	314.1593	7853.982	**150**	6.666 667	471.2389	17671.46
101	9.900 990	317.3009	8011.847	151	6.622 517	474.3805	17907.86
102	9.803 922	320.4425	8171.282	152	6.578 947	477.5221	18145.84
103	9.708 738	323.5840	8332.289	153	6.535 948	480.6637	18385.39
104	9.615 385	326.7256	8494.867	154	6.493 506	483.8053	18626.50
105	9.523 810	329.8672	8659.015	155	6.451 613	486.9469	18869.19
106	9.433 962	333.0088	8824.734	156	6.410 256	490.0885	19113.45
107	9.345 794	336.1504	8992.024	157	6.369 427	493.2300	19359.28
108	9.259 259	339.2920	9160.884	158	6.329 114	496.3716	19606.68
109	9.174 312	342.4336	9331.316	159	6.289 308	499.5132	19855.65
110	9.090 909	345.5752	9503.318	**160**	6.250 000	502.6548	20106.19
111	9.009 009	348.7168	9676.891	161	6.211 180	505.7964	20358.31
112	8.928 571	351.8584	9852.035	162	6.172 840	508.9380	20611.99
113	8.849 558	355.0000	10028.75	163	6.134 969	512.0796	20867.24
114	8.771 930	358.1416	10207.03	164	6.097 561	515.2212	21124.07
115	8.695 652	361.2832	10386.89	165	6.060 606	518.3628	21382.46
116	8.620 690	364.4247	10568.32	166	6.024 096	521.5044	21642.43
117	8.547 009	367.5663	10751.32	167	5.988 024	524.6460	21903.97
118	8.474 576	370.7079	10935.88	168	5.952 381	527.7876	22167.08
119	8.403 361	373.8495	11122.02	169	5.917 160	530.9292	22431.76
120	8.333 333	376.9911	11309.73	**170**	5.882 353	534.0708	22698.01
121	8.264 463	380.1327	11499.01	171	5.847 953	537.2123	22965.83
122	8.196 721	383.2743	11689.87	172	5.813 953	540.3539	23235.22
123	8.130 081	386.4159	11882.29	173	5.780 347	543.4955	23506.18
124	8.064 516	389.5575	12076.28	174	5.747 126	546.6371	23778.71
125	8.000 000	392.6991	12271.85	175	5.714 286	549.7787	24052.82
126	7.936 508	395.8407	12468.98	176	5.681 818	552.9203	24328.49
127	7.874 016	398.9823	12667.69	177	5.649 718	556.0619	24605.74
128	7.812 500	402.1239	12867.96	178	5.617 978	559.2035	24884.56
129	7.751 938	405.2655	13069.81	179	5.586 592	562.3451	25164.94
130	7.692 308	408.4070	13273.23	**180**	5.555 556	565.4867	25446.90
131	7.633 588	411.5486	13478.22	181	5.524 862	568.6283	25730.43
132	7.575 758	414.6902	13684.78	182	5.494 505	571.7699	26015.53
133	7.518 797	417.8318	13892.91	183	5.464 481	574.9115	26302.20
134	7.462 687	420.9734	14102.61	184	5.434 783	578.0530	26590.44
135	7.407 407	424.1150	14313.88	185	5.405 405	581.1946	26880.25
136	7.352 941	427.2566	14526.72	186	5.376 344	584.3362	27171.63
137	7.299 270	430.3982	14741.14	187	5.347 594	587.4778	27464.59
138	7.246 377	433.5398	14957.12	188	5.319 149	590.6194	27759.11
139	7.194 245	436.6814	15174.68	189	5.291 005	593.7610	28055.21
140	7.142 857	439.8230	15393.80	**190**	5.263 158	596.9026	28352.87
141	7.092 199	442.9646	15614.50	191	5.235 602	600.0442	28652.11
142	7.042 254	446.1062	15836.77	192	5.208 333	603.1858	28952.92
143	6.993 007	449.2477	16060.61	193	5.181 347	606.3274	29255.30
144	6.944 444	452.3893	16286.02	194	5.154 639	609.4690	29559.25
145	6.896 552	455.5309	16513.00	195	5.128 205	612.6106	29864.77
146	6.849 315	458.6725	16741.55	196	5.102 041	615.7522	30171.86
147	6.802 721	461.8141	16971.67	197	5.076 142	618.8938	30480.52
148	6.756 757	464.9557	17203.36	198	5.050 505	622.0353	30790.75
149	6.711 409	468.0973	17436.62	199	5.025 126	625.1769	31102.55
150	6.666 667	471.2389	17671.46	**200**	5.000 000	628.3185	31415.93

RECIPROCALS, CIRCUMFERENCE AND AREA OF CIRCLES (Continued)

n = dia	$1000\frac{1}{n}$	Circumference πn	Area $\frac{\pi n^2}{4}$	n = dia	$1000\frac{1}{n}$	Circumference πn	Area $\frac{\pi n^2}{4}$
200	5.000 000	628.3185	31415.93	**250**	4.000 000	785.3982	49087.39
201	4.975 124	631.4601	31730.87	251	3.984 066	788.5398	49480.87
202	4.950 495	634.6017	32047.39	252	3.968 254	791.6813	49875.92
203	4.926 108	637.7433	32365.47	253	3.952 569	794.8229	50272.55
204	4.901 961	640.8849	32685.13	254	3.937 008	797.9645	50670.75
205	4.878 049	644.0265	33006.36	255	3.921 569	801.1061	51070.52
206	4.854 369	647.1681	33329.16	256	3.906 250	804.2477	51471.85
207	4.830 918	650.3097	33653.53	257	3.891 051	807.3893	51874.76
208	4.807 692	653.4513	33979.47	258	3.875 969	810.5309	52279.24
209	4.784 689	656.5929	34306.98	259	3.861 004	813.6725	52685.29
210	4.761 905	659.7345	34636.06	**260**	3.846 154	816.8141	53092.92
211	4.739 336	662.8760	34966.71	261	3.831 418	819.9557	53502.11
212	4.716 981	666.0176	35298.94	262	3.816 794	823.0973	53912.87
213	4.694 836	669.1592	35632.73	263	3.802 281	826.2389	54325.21
214	4.672 897	672.3008	35968.09	264	3.787 879	829.3805	54739.11
215	4.651 163	675.4424	36305.03	265	3.773 585	832.5221	55154.59
216	4.629 630	678.5840	36643.54	266	3.759 398	835.6636	55571.63
217	4.608 295	681.7256	36983.61	267	3.745 318	838.8052	55990.25
218	4.587 156	684.8672	37325.26	268	3.731 343	841.9468	56410.44
219	4.566 210	688.0088	37668.48	269	3.717 472	845.0884	56832.20
220	4.545 455	691.1504	38013.27	**270**	3.703 704	848.2300	57255.53
221	4.524 887	694.2920	38359.63	271	3.690 037	851.3716	57680.43
222	4.504 505	697.4336	38707.56	272	3.676 471	854.5132	58106.90
223	4.484 305	700.5752	39057.07	273	3.663 004	857.6548	58534.94
224	4.464 286	703.7168	39408.14	274	3.649 635	860.7964	58964.55
225	4.444 444	706.8583	39760.78	275	3.636 364	863.9380	59395.74
226	4.424 779	709.9999	40115.00	276	3.623 188	867.0796	59828.49
227	4.405 286	713.1415	40470.78	277	3.610 108	870.2212	60262.82
228	4.385 965	716.2831	40828.14	278	3.597 122	873.3628	60698.71
229	4.366 812	719.4247	41187.07	279	3.584 229	876.5044	61136.18
230	4.347 826	722.5663	41547.56	**280**	3.571 429	879.6459	61575.22
231	4.329 004	725.7079	41909.63	281	3.558 719	882.7875	62015.82
232	4.310 345	728.8495	42273.27	282	3.546 099	885.9291	62458.00
233	4.291 845	731.9911	42638.48	283	3.533 569	889.0707	62901.75
234	4.273 504	735.1327	43005.26	284	3.521 127	892.2123	63347.07
235	4.255 319	738.2743	43373.61	285	3.508 772	895.3539	63793.97
236	4.237 288	741.4159	43743.54	286	3.496 503	898.4955	64242.43
237	4.219 409	744.5575	44115.03	287	3.484 321	901.6371	64692.46
238	4.201 681	747.6991	44488.09	288	3.472 222	904.7787	65144.07
239	4.184 100	750.8406	44862.73	289	3.460 208	907.9203	65597.24
240	4.166 667	753.9822	45238.93	**290**	3.448 276	911.0619	66051.99
241	4.149 378	757.1238	45616.71	291	3.436 426	914.2035	66508.30
242	4.132 231	760.2654	45996.06	292	3.424 658	917.3451	66966.19
243	4.115 226	763.4070	46376.98	293	3.412 969	920.4866	67425.65
244	4.098 361	766.5486	46759.47	294	3.401 361	923.6282	67886.68
245	4.081 633	769.6902	47143.52	295	3.389 831	926.7698	68349.28
246	4.065 041	772.8318	47529.16	296	3.378 378	929.9114	68813.45
247	4.048 583	775.9734	47916.36	297	3.367 003	933.0530	69279.19
248	4.032 258	779.1150	48305.13	298	3.355 705	936.1946	69746.50
249	4.016 064	782.2566	48695.47	299	3.344 482	939.3362	70215.38
250	4.000 000	785.3982	49087.39	**300**	3.333 333	942.4778	70685.83

RECIPROCALS, CIRCUMFERENCE AND AREA OF CIRCLES (Continued)

n=dia	$1000\frac{1}{n}$	Circumference πn	Area $\frac{\pi n^2}{4}$	n=dia	$1000\frac{1}{n}$	Circumference πn	Area $\frac{\pi n^2}{4}$
300	3.333 333	942.4778	70685.83	**350**	2.857 143	1099.557	96211.28
301	3.322 259	945.6194	71157.86	351	2.849 003	1102.699	96761.84
302	3.311 258	948.7610	71631.45	352	2.840 909	1105.841	97313.97
303	3.300 330	951.9026	72106.62	353	2.832 861	1108.982	97867.68
304	3.289 474	955.0442	72583.36	354	2.824 859	1112.124	98422.96
305	3.278 689	958.1858	73061.66	355	2.816 901	1115.265	98979.80
306	3.267 974	961.3274	73541.54	356	2.808 989	1118.407	99538.22
307	3.257 329	964.4689	74022.99	357	2.801 120	1121.549	100 098.2
308	3.246 753	967.6105	74506.01	358	2.793 296	1124.690	100 659.8
309	3.236 246	970.7521	74990.60	359	2.785 515	1127.832	101 222.9
310	3.225 806	973.8937	75476.76	**360**	2.777 778	1130.973	101 787.6
311	3.215 434	977.0353	75964.50	361	2.770 083	1134.115	102 353.9
312	3.205 128	980.1769	76453.80	362	2.762 431	1137.257	102 921.7
313	3.194 888	983.3185	76944.67	363	2.754 821	1140.398	103 491.1
314	3.184 713	986.4601	77437.12	364	2.747 253	1143.540	104 062.1
315	3.174 603	989.6017	77931.13	365	2.739 726	1146.681	104 634.7
316	3.164 557	992.7433	78426.72	366	2.732 240	1149.823	105 208.8
317	3.154 574	995.8849	78923.88	367	2.724 796	1152.965	105 784.5
318	3.144 654	999.0265	79422.60	368	2.717 391	1156.106	106 361.8
319	3.134 796	1002.168	79922.90	369	2.710 027	1159.248	106 940.6
320	3.125 000	1005.310	80424.77	**370**	2.702 703	1162.389	107 521.0
321	3.115 265	1008.451	80928.21	371	2.695 418	1165.531	108 103.0
322	3.105 590	1011.593	81433.22	372	2.688 172	1168.672	108 686.5
323	3.095 975	1014.734	81939.80	373	2.680 965	1171.814	109 271.7
324	3.086 420	1017.876	82447.96	374	2.673 797	1174.956	109 858.4
325	3.076 923	1021.018	82957.68	375	2.666 667	1178.097	110 446.6
326	3.067 485	1024.159	83468.98	376	2.659 574	1181.239	111 036.5
327	3.058 104	1027.301	83981.84	377	2.652 520	1184.380	111 627.9
328	3.048 780	1030.442	84496.28	378	2.645 503	1187.522	112 220.8
329	3.039 514	1033.584	85012.28	379	2.638 522	1190.664	112 815.4
330	3.030 303	1036.726	85529.86	**380**	2.631 579	1193.805	113 411.5
331	3.021 148	1039.867	86049.01	381	2.624 672	1196.947	114 009.2
332	3.012 048	1043.009	86569.73	382	2.617 801	1200.088	114 608.4
333	3.003 003	1046.150	87092.02	383	2.610 966	1203.230	115 209.3
334	2.994 012	1049.292	87615.88	384	2.604 167	1206.372	115 811.7
335	2.985 075	1052.434	88141.31	385	2.597 403	1209.513	116 415.6
336	2.976 190	1055.575	88668.31	386	2.590 674	1212.655	117 021.2
337	2.967 359	1058.717	89196.88	387	2.583 979	1215.796	117 628.3
338	2.958 580	1061.858	89727.03	388	2.577.320	1218.938	118 237.0
339	2.949 853	1065.000	90258.74	389	2.570 694	1222.080	118 847.2
340	2.941 176	1068.142	90792.03	**390**	2.564 103	1225.221	119 459.1
341	2.932 551	1071.283	91326.88	391	2.557 545	1228.363	120 072.5
342	2.923 977	1074.425	91863.31	392	2.551 020	1231.504	120 687.4
343	2.915 452	1077.566	92401.31	393	2.544 529	1234.646	121 304.0
344	2.906 977	1080.708	92940.88	394	2.538.071	1237.788	121 922.1
345	2.898 551	1083.849	93482.02	395	2.531.646	1240.929	122 541.7
346	2.890 173	1086.991	94024.73	396	2.525.253	1244.071	123 163.0
347	2.881 844	1090.133	94569.01	397	2.518.892	1247.212	123 785.8
348	2.873 563	1093.274	95114.86	398	2.512 563	1250.354	124 410.2
349	2.865 330	1096.416	95662.28	399	2.506 266	1253.495	125 036.2
350	2.857 143	1099.557	96211.28	**400**	2.500 000	1256.637	125 663.7

RECIPROCALS, CIRCUMFERENCE AND AREA OF CIRCLES (Continued)

n=dia	$1000\frac{1}{n}$	Circumference πn	Area $\frac{\pi n^2}{4}$	n=dia	$1000\frac{1}{n}$	Circumference πn	Area $\frac{\pi n^2}{4}$
400	2.500 000	1256.637	125 663.7	**450**	2.222 222	1413.717	159 043.1
401	2.493 766	1259.779	126 292.8	451	2.217 295	1416.858	159 750.8
402	2.487 562	1262.920	126 923.5	452	2.212 389	1420.000	160 460.0
403	2.481 390	1266.062	127 555.7	453	2.207 506	1423.141	161 170.8
404	2.475 248	1269.203	128 189.5	454	2.202 643	1426.283	161 883.1
405	2.469 136	1272.345	128 824.9	455	2.197 802	1429.425	162 597.1
406	2.463 054	1275.487	129 461.9	456	2.192 982	1432.566	163 312.6
407	2.457 002	1278.628	130 100.4	457	2.188 184	1435.708	164 029.6
408	2.450 980	1281.770	130 740.5	458	2.183 406	1438.849	164 748.3
409	2.444 988	1284.911	131 382.2	459	2.178 649	1441.991	165 468.5
410	2.439 024	1288.053	132 025.4	**460**	2.173 913	1445.133	166 190.3
411	2.433 090	1291.195	132 670.2	461	2.169 197	1448.274	166 913.6
412	2.427 184	1294.336	133 316.6	462	2.164 502	1451.416	167 638.5
413	2.421 308	1297.478	133 964.6	463	2.159 827	1454.557	168 365.0
414	2.415 459	1300.619	134 614.1	464	2.155 172	1457.699	169 093.1
415	2.409 639	1303.761	135 265.2	465	2.150 538	1460.841	169 822.7
416	2.403 846	1306.903	135 917.9	466	2.145 923	1463.982	170 553.9
417	2.398 082	1310.044	136 572.1	467	2.141 328	1467.124	171 286.7
418	2.392 344	1313.186	137 227.9	468	2.136 752	1470.265	172 021.0
419	2.386 635	1316.327	137 885.3	469	2.132 196	1473.407	172 757.0
420	2.380 952	1319.469	138 544.2	**470**	2.127 660	1476.549	173 494.5
421	2.375 297	1322.611	139 204.8	471	2.123 142	1479.690	174 233.5
422	2.369 668	1325.752	139 866.8	472	2.118 644	1482.832	174 974.1
423	2.364 066	1328.894	140 530.5	473	2.114 165	1485.973	175 716.3
424	2.358 491	1332.035	141 195.7	474	2.109 705	1489.115	176 460.1
425	2.352 941	1335.177	141 862.5	475	2.105 263	1492.257	177 205.5
426	2.347 418	1338.318	142 530.9	476	2.100 840	1495.398	177 952.4
427	2.341 920	1341.460	143 200.9	477	2.096 436	1498.540	178 700.9
428	2.336 449	1344.602	143 872.4	478	2.092 050	1501.681	179 450.9
429	2.331 002	1347.743	144 545.5	479	2.087 683	1504.823	180 202.5
430	2.325 581	1350.885	145 220.1	**480**	2.083 333	1507.964	180 955.7
431	2.320 186	1354.026	145 896.3	481	2.079 002	1511.106	181 710.5
432	2.314 815	1357.168	146 574.1	482	2.074 689	1514.248	182 466.8
433	2.309 469	1360.310	147 253.5	483	2.070 393	1517.389	183 224.8
434	2.304 147	1363.451	147 934.5	484	2.066 116	1520.531	183 984.2
435	2.298 851	1366.593	148 617.0	485	2.061 856	1523.672	184 745.3
436	2.293 578	1369.734	149 301.0	486	2.057 613	1526.814	185 507.9
437	2.288 330	1372.876	149 986.7	487	2.053 388	1529.956	186 272.1
438	2.283 105	1376.018	150 673.9	488	2.049 180	1533.097	187 037.9
439	2.277 904	1379.159	151 362.7	489	2.044 990	1536.239	187 805.2
440	2.272 727	1382.301	152 053.1	**490**	2.040 816	1539.380	188 574.1
441	2.267 574	1385.442	152 745.0	491	2.036 660	1542.522	189 344.6
442	2.262 443	1388.584	153 438.5	492	2.032 520	1545.664	190 116.6
443	2.257 336	1391.726	154 133.6	493	2.028 398	1548.805	190 890.2
444	2.252 252	1394.867	154 830.3	494	2.024 291	1551.947	191 665.4
445	2.247 191	1398.009	155 528.5	495	2.020 202	1555.088	192 442.2
446	2.242 152	1401.150	156 228.3	496	2.016 129	1558.230	193 220.5
447	2.237 136	1404.292	156 929.6	497	2.012 072	1561.372	194 000.4
448	2.232 143	1407.434	157 632.6	498	2.008 032	1564.513	194 781.9
449	2.227 171	1410.575	158 337.1	499	2.004 008	1567.655	195 564.9
450	2.222 222	1413.717	159 043.1	**500**	2.000 000	1570.796	196 349.5

RECIPROCALS, CIRCUMFERENCE AND AREA OF CIRCLES (Continued)

n=dia	$1000\frac{1}{n}$	Circumference πn	Area $\frac{\pi n^2}{4}$	n=dia	$1000\frac{1}{n}$	Circumference πn	Area $\frac{\pi n^2}{4}$
500	2.000 000	1570.796	196 349.5	**550**	1.818 182	1727.876	237 582.9
501	1.996 008	1573.938	197 135.7	551	1.814 882	1731.018	238 447.7
502	1.992 032	1577.080	197 923.5	552	1.811 594	1734.159	239 314.0
503	1.988 072	1580.221	198 712.8	553	1.808 318	1737.301	240 181.8
504	1.984 127	1583.363	199 503.7	554	1.805 054	1740.442	241 051.3
505	1.980 198	1586.504	200 296.2	555	1.801 802	1743.584	241 922.3
506	1.976 285	1589.646	201 090.2	556	1.798 561	1746.726	242 794.8
507	1.972 387	1592.787	201 885.8	557	1.795 332	1749.867	243 669.0
508	1.968 504	1595.929	202 683.0	558	1.792 115	1753.009	244 544.7
509	1.964 637	1599.071	203 481.7	559	1.788 909	1756.150	245 422.0
510	1.960 784	1602.212	204 282.1	**560**	1.785 714	1759.292	246 300.9
511	1.956 947	1605.354	205 084.0	561	1.782 531	1762.433	247 181.3
512	1.953 125	1608.495	205 887.4	562	1.779 359	1765.575	248 063.3
513	1.949 318	1611.637	206 692.4	563	1.776 199	1768.717	248 946.9
514	1.945 525	1614.779	207 499.1	564	1.773 050	1771.858	249 832.0
515	1.941 748	1617.920	208 307.2	565	1.769 912	1775.000	250 718.7
516	1.937 984	1621.062	209 117.0	566	1.766 784	1778.141	251 607.0
517	1.934 236	1624.203	209 928.3	567	1.763 668	1781.283	252 496.9
518	1.930 502	1627.345	210 741.2	568	1.760 563	1784.425	253 388.3
519	1.926 782	1630.487	211 555.6	569	1.757 469	1787.566	254 281.3
520	1.923 077	1633.628	212 371.7	**570**	1.754 386	1790.708	255 175.9
521	1.919 386	1636.770	213 189.3	571	1.751 313	1793.849	256 072.0
522	1.915 709	1639.911	214 008.4	572	1.748 252	1796.991	256 969.7
523	1.912 046	1643.053	214 829.2	573	1.745 201	1800.133	257 869.0
524	1.908 397	1646.195	215 651.5	574	1.742 160	1803.274	258 769.8
525	1.904 762	1649.336	216 475.4	575	1.739 130	1806.416	259 672.3
526	1.901 141	1652.478	217 300.8	576	1.736 111	1809.557	260 576.3
527	1.897 533	1655.619	218 127.8	577	1.733 102	1812.699	261 481.8
528	1.893 939	1658.761	218 956.4	578	1.730 104	1815.841	262 389.0
529	1.890 359	1661.903	219 786.6	579	1.727 116	1818.982	263 297.7
530	1.886 792	1665.044	220 618.3	**580**	1.724 138	1822.124	264 207.9
531	1.883 239	1668.186	221 451.7	581	1.721 170	1825.265	265 119.8
532	1.879 699	1671.327	222 286.5	582	1.718 213	1828.407	266 033.2
533	1.876 173	1674.469	223 123.0	583	1.715 266	1831.549	266 948.2
534	1.872 659	1677.610	223 961.0	584	1.712 329	1834.690	267 864.8
535	1.869 159	1680.752	224 800.6	585	1.709 402	1837.832	268 782.9
536	1.865 672	1683.894	225 641.8	586	1.706 485	1840.973	269 702.6
537	1.862 197	1687.035	226 484.5	587	1.703 578	1844.115	270 623.9
538	1.858 736	1690.177	227 328.8	588	1.700 680	1847.256	271 546.7
539	1.855 288	1693.318	228 174.7	589	1.697 793	1850.398	272 471.1
540	1.851 852	1696.460	229 022.1	**590**	1.694 915	1853.540	273 397.1
541	1.848 429	1699.602	229 871.1	591	1.692 047	1856.681	274 324.7
542	1.845 018	1702.743	230 721.7	592	1.689 189	1859.823	275 253.8
543	1.841 621	1705.885	231 573.9	593	1.686 341	1862.964	276 184.5
544	1.838 235	1709.026	232 427.6	594	1.683 502	1866.106	277 116.7
545	1.834 862	1712.168	233 282.9	595	1.680 672	1869.248	278 050.6
546	1.831 502	1715.310	234 139.8	596	1.677 852	1872.389	278 986.0
547	1.828 154	1718.451	234 998.2	597	1.675 042	1875.531	279 923.0
548	1.824 818	1721.593	235 858.2	598	1.672 241	1878.672	280 861.5
549	1.821 494	1724.734	236 719.8	599	1.669 449	1881.814	281 801.6
550	1.818 182	1727.876	237 582.9	**600**	1.666 667	1884.956	282 743.3

RECIPROCALS, CIRCUMFERENCE AND AREA OF CIRCLES (Continued)

n=dia	$1000\frac{1}{n}$	Circum-ference πn	Area $\frac{\pi n^2}{4}$	n=dia	$1000\frac{1}{n}$	Circum-ference πn	Area $\frac{\pi n^2}{4}$
600	1.666 667	1884.956	282 743.3	**650**	1.538 462	2042.035	331 830.7
601	1.663 894	1888.097	283 686.6	651	1.536 098	2045.177	332 852.5
602	1.661 130	1891.239	284 631.4	652	1.533 742	2048.318	333 875.9
603	1.658 375	1894.380	285 577.8	653	1.531 394	2051.460	334 900.8
604	1.655 629	1897.522	286 525.8	654	1.529 052	2054.602	335 927.4
605	1.652 893	1900.664	287 475.4	655	1.526 718	2057.743	336 955.4
606	1.650 165	1903.805	288 426.5	656	1.524 390	2060.885	337 985.1
607	1.647 446	1906.947	289 379.2	657	1.522 070	2064.026	339 016.3
608	1.644 737	1910.088	290 333.4	658	1.519 757	2067.168	340 049.1
609	1.642 036	1913.230	291 289.3	659	1.517 451	2070.310	341 083.5
610	1.639 344	1916.372	292 246.7	**660**	1.515 152	2073.451	342 119.4
611	1.636 661	1919.513	293 205.6	661	1.512 859	2076.593	343 157.0
612	1.633 987	1922.655	294 166.2	662	1.510 574	2079.734	344 196.0
613	1.631 321	1925.796	295 128.3	663	1.508 296	2082.876	345 236.7
614	1.628 664	1928.938	296 092.0	664	1.506 024	2086.018	346 278.9
615	1.626 016	1932.079	297 057.2	665	1.503 759	2089.159	347 322.7
616	1.623 377	1935.221	298 024.0	666	1.501 502	2092.301	348 368.1
617	1.620 746	1938.363	298 992.4	667	1.499 250	2095.442	349 415.0
618	1.618 123	1941.504	299 962.4	668	1.497 006	2098.584	350 463.5
619	1.615 509	1944.646	300 933.9	669	1.494 768	2101.725	351 513.6
620	1.612 903	1947.787	301 907.1	**670**	1.492 537	2104.867	352 565.2
621	1.610 306	1950.929	302 881.7	671	1.490 313	2108.009	353 618.5
622	1.607 717	1954.071	303 858.0	672	1.488 095	2111.150	354 673.2
623	1.605 136	1957.212	304 835.8	673	1.485 884	2114.292	355 729.6
624	1.602 564	1960.354	305 815.2	674	1.483 680	2117.433	356 787.5
625	1.600 000	1963.495	306 796.2	675	1.481 481	2120.575	357 847.0
626	1.597 444	1966.637	307 778.7	676	1.479 290	2123.717	358 908.1
627	1.594 896	1969.779	308 762.8	677	1.477 105	2126.858	359 970.8
628	1.592 357	1972.920	309 748.5	678	1.474 926	2130.000	361 035.0
629	1.589 825	1976.062	310 735.7	679	1.472 754	2133.141	362 100.8
630	1.587 302	1979.203	311 724.5	**680**	1.470 588	2136.283	363 168.1
631	1.584 786	1982.345	312 714.9	681	1.468 429	2139.425	364 237.0
632	1.582 278	1985.487	313 706.9	682	1.466 276	2142.566	365 307.5
633	1.579 779	1988.628	314 700.4	683	1.464 129	2145.708	366 379.6
634	1.577 287	1991.770	315 695.5	684	1.461 988	2148.849	367 453.2
635	1.574 803	1994.911	316 692.2	685	1.459 854	2151.991	368 528.5
636	1.572 327	1998.053	317 690.4	686	1.457 726	2155.133	369 605.2
637	1.569 859	2001.195	318 690.2	687	1.455 604	2158.274	370 683.6
638	1.567 398	2004.336	319 691.6	688	1.453 488	2161.416	371 763.5
639	1.564 945	2007.478	320 694.6	689	1.451 379	2164.557	372 845.0
640	1.562 500	2010.619	321 699.1	**690**	1.449 275	2167.699	373 928.1
641	1.560 062	2013.761	322 705.2	691	1.447 178	2170.841	375 012.7
642	1.557 632	2016.902	323 712.8	692	1.445 087	2173.982	376 098.9
643	1.555 210	2020.044	324 722.1	693	1.443 001	2177.124	377 186.7
644	1.552 795	2023.186	325 732.9	694	1.440 922	2180.265	378 276.0
645	1.550 388	2026.327	326 745.3	695	1.438 849	2183.407	379 366.9
646	1.547 988	2029.469	327 759.2	696	1.436 782	2186.548	380 459.4
647	1.545 595	2032.610	328 774.7	697	1.434 720	2189.690	381 553.5
648	1.543 210	2035.752	329 791.8	698	1.432 665	2192.832	382 649.1
649	1.540 832	2038.894	330 810.5	699	1.430 615	2195.973	383 746.3
650	1.538 462	2042.035	331 830.7	**700**	1.428 571	2199.115	384 845.1

RECIPROCALS, CIRCUMFERENCE AND AREA OF CIRCLES (Continued)

n=dia	$1000\frac{1}{n}$	Circumference πn	Area $\frac{\pi n^2}{4}$	n=dia	$1000\frac{1}{n}$	Circumference πn	Area $\frac{\pi n^2}{4}$
700	1.428 571	2199.115	384 845.1	**750**	1.333 333	2356.194	441 786.5
701	1.426 534	2202.256	385 945.4	751	1.331 558	2359.336	442 965.3
702	1.424 501	2205.398	387 047.4	752	1.329 787	2362.478	444 145.8
703	1.422 475	2208 540	388 150.8	753	1.328 021	2365.619	445 327.8
704	1.420 455	2211.681	389 255.9	754	1.326 260	2368.761	446 511.4
705	1.418 440	2214.823	390 362.5	755	1.324 503	2371.902	447 696.6
706	1.416 431	2217.964	391 470.7	756	1.322 751	2375.044	448 883.3
707	1.414 427	2221.106	392 580.5	757	1.321 004	2378.186	450 071.6
708	1.412 429	2224.248	393 691.8	758	1.319 261	2381.327	451 261.5
709	1.410 437	2227.389	394 804.7	759	1.317 523	2384.469	452 453.0
710	1.408 451	2230.531	395 919.2	**760**	1.315 789	2387.610	453 646.0
711	1.406 470	2233.672	397 035.3	761	1.314 060	2390.752	454 840.6
712	1.404 494	2236.814	398 152.9	762	1.312 336	2393.894	456 036.7
713	1.402 525	2239.956	399 272.1	763	1.310 616	2397.035	457 234.5
714	1.400 560	2243.097	400 392.8	764	1.308 901	2400.177	458 433.8
715	1.398 601	2246.239	401 515.2	765	1.307 190	2403.318	459 634.6
716	1.396 648	2249.380	402 639.1	766	1.305 483	2406.460	460 837.1
717	1.394 700	2252.522	403 764.6	767	1.303 781	2409.602	462 041.1
718	1.392 758	2255.664	404 891.6	768	1.302 083	2412.743	463 246.7
719	1.390 821	2258.805	406 020.2	769	1.300 390	2415.885	464 453.8
720	1.388 889	2261.947	407 150.4	**770**	1.298 701	2419.026	465 662.6
721	1.386 963	2265.088	408 282.2	771	1.297 017	2422.168	466 872.9
722	1.385 042	2268.230	409 415.5	772	1.295 337	2425.310	468 084.7
723	1.383 126	2271.371	410 550.4	773	1.293 661	2428.451	469 298.2
724	1.381 215	2274.513	411 686.9	774	1.291 990	2431.593	470 513.2
725	1.379 310	2277.655	412 824.9	775	1.290 323	2434.734	471 729.8
726	1.377 410	2280.796	413 964.5	776	1.288 660	2437.876	472 947.9
727	1.375 516	2283.938	415 105.7	777	1.287 001	2441.017	474 167.6
728	1.373 626	2287.079	416 248.5	778	1.285 347	2444.159	475 388.9
729	1.371 742	2290.221	417 392.8	779	1.283 697	2447.301	476 611.8
730	1.369 863	2293.363	418 538.7	**780**	1.282 051	2450.442	477 836.2
731	1.367 989	2296.504	419 686.1	781	1.280 410	2453.584	479 062.2
732	1.366 120	2299.646	420 835.2	782	1.278 772	2456.725	480 289.8
733	1.364 256	2302.787	421 985.8	783	1.277 139	2459.867	481 519.0
734	1.362 398	2305.929	423 138.0	784	1.275 510	2463.009	482 749.7
735	1.360 544	2309.071	424 291.7	785	1.273 885	2466.150	483 982.0
736	1.358 696	2312.212	425 447.0	786	1.272 265	2469.292	485 215.8
737	1.356 852	2315.354	426 603.9	787	1.270 648	2472.433	486 451.3
738	1.355 014	2318.495	427 762.4	788	1.269 036	2475.575	487 688.3
739	1.353 180	2321.637	428 922.4	789	1.267 427	2478.717	488 926.9
740	1.351 351	2324.779	430 084.0	**790**	1.265 823	2481.858	490 167.0
741	1.349 528	2327.920	431 247.2	791	1.264 223	2485.000	491 408.7
742	1.347 709	2331.062	432 412.0	792	1.262 626	2488.141	492 652.0
743	1.345 895	2334.203	433 578.3	793	1.261 034	2491.283	493 896.8
744	1.344 086	2337.345	434 746.2	794	1.259 446	2494.425	495 143.3
745	1.342 282	2340.487	435 915.6	795	1.257 862	2497.566	496 391.3
746	1.340 483	2343.628	437 086.6	796	1.256 281	2500.708	497 640.8
747	1.338 688	2346.770	438 259.2	797	1.254 705	2503.849	498 892.0
748	1.336 898	2349.911	439 433.4	798	1.253 133	2506.991	500 144.7
749	1.335 113	2353.053	440 609.2	799	1.251 564	2510.133	501 399.0
750	1.333 333	2356.194	441 786.5	**800**	1.250 000	2513.274	502 654.8

RECIPROCALS, CIRCUMFERENCE AND AREA OF CIRCLES (Continued)

n=dia	$1000\frac{1}{n}$	Circumference πn	Area $\frac{\pi n^2}{4}$	n=dia	$1000\frac{1}{n}$	Circumference πn	Area $\frac{\pi n^2}{4}$
800	1.250 000	2513.274	502 654.8	**850**	1.176 471	2670.354	567 450.2
801	1.248 439	2516.416	503 912.2	851	1.175 088	2673.495	568 786.1
802	1.246 883	2519.557	505 171.2	852	1.173 709	2676.637	570 123.7
803	1.245 330	2522.699	506 431.8	853	1.172 333	2679.779	571 462.8
804	1.243 781	2525.840	507 693.9	854	1.170 960	2682.920	572 803.4
805	1.242 236	2528.982	508 957.6	855	1.169 591	2686.062	574 145.7
806	1.240 695	2532.124	510 222.9	856	1.168 224	2689.203	575 489.5
807	1.239 157	2535.265	511 489.8	857	1.166 861	2692.345	576 834.9
808	1.237 624	2538.407	512 758.2	858	1.165 501	2695.486	578 181.9
809	1.236 094	2541.548	514 028.2	859	1.164 144	2698.628	579 530.4
810	1.234 568	2544.690	515 299.7	**860**	1.162 791	2701.770	580 880.5
811	1.233 046	2547.832	516 572.9	861	1.161 440	2704.911	582 232.2
812	1.231 527	2550.973	517 847.6	862	1.160 093	2708.053	583 585.4
813	1.230 012	2554.115	519 123.8	863	1.158 749	2711.194	584 940.2
814	1.228 501	2557.256	520 401.7	864	1.157 407	2714.336	586 296.6
815	1.226 994	2560.398	521 681.1	865	1.156 069	2717.478	587 654.5
816	1.225 490	2563.540	522 962.1	866	1.154 734	2720.619	589 014.1
817	1.223 990	2566.681	524 244.6	867	1.153 403	2723.761	590 375.2
818	1.222 494	2569.823	525 528.8	868	1.152 074	2726.902	591 737.8
819	1.221 001	2572.964	526 814.5	869	1.150 748	2730.044	593 102.1
820	1.219 512	2576.106	528 101.7	**870**	1.149 425	2733.186	594 467.9
821	1.218 027	2579.248	529 390.6	871	1.148 106	2736.327	595 835.2
822	1.216 545	2582.389	530 681.0	872	1.146 789	2739.469	597 204.2
823	1.215 067	2585.531	531 973.0	873	1.145 475	2742.610	598 574.7
824	1.213 592	2588.672	533 266.5	874	1.144 165	2745.752	599 946.8
825	1.212 121	2591.814	534 561.6	875	1.142 857	2748.894	601 320.5
826	1.210 654	2594.956	535 858.3	876	1.141 553	2752.035	602 695.7
827	1.209 190	2598.097	537 156.6	877	1.140 251	2755.177	604 072.5
828	1.207 729	2601.239	538 456.4	878	1.138 952	2758.318	605 450.9
829	1.206 273	2604.380	539 757.8	879	1.137 656	2761.460	606 830.8
830	1.204 819	2607.522	541 060.8	**880**	1.136 364	2764.602	608 212.3
831	1.203 369	2610.663	542 365.3	881	1.135 074	2767.743	609 595.4
832	1.201 923	2613.805	543 671.5	882	1.133 787	2770.885	610 980.1
833	1.200 480	2616.947	544 979.1	883	1.132 503	2774.026	612 366.3
834	1.199 041	2620.088	546 288.4	884	1.131 222	2777.168	613 754.1
835	1.197 605	2623.230	547 599.2	885	1.129 944	2780.309	615 143.5
836	1.196 172	2626.371	548 911.6	886	1.128 668	2783.451	616 534.4
837	1.194 743	2629.513	550 225.6	887	1.127 396	2786.593	617 926.9
838	1.193 317	2632.655	551 541.1	888	1.126 126	2789.734	619 321.0
839	1.191 895	2635.796	552 858.3	889	1.124 859	2792.876	620 716.7
840	1.190 476	2638.938	554 176.9	**890**	1.123 596	2796.017	622 113.9
841	1.189 061	2642.079	555 497.2	891	1.122 334	2799.159	623 512.7
842	1.187 648	2645.221	556 819.0	892	1.121 076	2802.301	624 913.0
843	1.186 240	2648.363	558 142.4	893	1.119 821	2805.442	626 315.0
844	1.184 834	2651.504	559 467.4	894	1.118 568	2808.584	627 718.5
845	1.183 432	2654.646	560 793.9	895	1.117 318	2811.725	629 123.6
846	1.182 033	2657.787	562 122.0	896	1.116 071	2814.867	630 530.2
847	1.180 638	2660.929	563 451.7	897	1.114 827	2818.009	631 938.4
848	1.179 245	2664.071	564 783.0	898	1.113 586	2821.150	633 348.2
849	1.177 856	2667.212	566 115.8	899	1.112 347	2824.292	634 759.6
850	1.176 471	2670.354	567 450.2	**900**	1.111 111	2827.433	636 172.5

RECIPROCALS, CIRCUMFERENCE AND AREA OF CIRCLES (Continued)

n=dia	$1000\frac{1}{n}$	Circumference πn	Area $\frac{\pi n^2}{4}$	n=dia	$1000\frac{1}{n}$	Circumference πn	Area $\frac{\pi n^2}{4}$
900	1.111 111	2827.433	636 172.5	**950**	1.052 632	2984.513	708 821.8
901	1.109 878	2830.575	637 587.0	951	1.051 525	2987.655	710 314.9
902	1.108 647	2833.717	639 003.1	952	1.050 420	2990.796	711 809.5
903	1.107 420	2836.858	640 420.7	953	1.049 318	2993.938	713 305.7
904	1.106 195	2840.000	641 839.9	954	1.048 218	2997.079	714 803.4
905	1.104 972	2843.141	643 260.7	955	1.047 120	3000.221	716 302.8
906	1.103 753	2846.283	644 683.1	956	1.046 025	3003.363	717 803.7
907	1.102 536	2849.425	646 107.0	957	1.044 932	3006.504	719 306.1
908	1.101 322	2852.566	647 532.5	958	1.043 841	3009.646	720 810.2
909	1.100 110	2855.708	648 959.6	959	1.042 753	3012.787	722 315.8
910	1.098 901	2858.849	650 388.2	**960**	1.041 667	3015.929	723 822.9
911	1.097 695	2861.991	651 818.4	961	1.040 583	3019.071	725 331.7
912	1.096 491	2865.133	653 250.2	962	1.039 501	3022.212	726 842.0
913	1.095 290	2868.274	654 683.6	963	1.038 422	3025.354	728 353.9
914	1.094 092	2871.416	656 118.5	964	1.037 344	3028.495	729 867.4
915	1.092 896	2874.557	657 555.0	965	1.036 269	3031.637	731 382.4
916	1.091 703	2877.699	658 993.0	966	1.035 197	3034.779	732 899.0
917	1.090 513	2880.840	660 432.7	967	1.034 126	3037.920	734 417.2
918	1.089 325	2883.982	661 873.9	968	1.033 058	3041.062	735 936.9
919	1.088 139	2887.124	663 316.7	969	1.031 992	3044.203	737 458.2
920	1.086 957	2890.265	664 761.0	**970**	1.030 928	3047.345	738 981.1
921	1.085 776	2893.407	666 206.9	971	1.029 866	3050.486	740 505.6
922	1.084 599	2896.548	667 654.4	972	1.028 807	3053.628	742 031.6
923	1.083 424	2899.690	669 103.5	973	1.027 749	3056.770	743 559.2
924	1.082 251	2902.832	670 554.1	974	1.026 694	3059.911	745 088.4
925	1.081 081	2905.973	672 006.3	975	1.025 641	3063.053	746 619.1
926	1.079 914	2909.115	673 460.1	976	1.024 590	3066.194	748 151.4
927	1.078 749	2912.256	674 915.4	977	1.023 541	3069.336	749 685.3
928	1.077 586	2915.398	676 372.3	978	1.022 495	3072.478	751 220.8
929	1.076 426	2918.540	677 830.8	979	1.021 450	3075.619	752 757.8
930	1.075 269	2921.681	679 290.9	**980**	1.020 408	3078.761	754 296.4
931	1.074 114	2924.823	680 752.5	981	1.019 368	3081.902	755 836.6
932	1.072 961	2927.964	682 215.7	982	1.018 330	3085.044	757 378.3
933	1.071 811	2931.106	683 680.5	983	1.017 294	3088.186	758 921.6
934	1.070 664	2934.248	685 146.8	984	1.016 260	3091.327	760 466.5
935	1.069 519	2937.389	686 614.7	985	1.015 228	3094.469	762 012.9
936	1.068 376	2940.531	688 084.2	986	1.014 199	3097.610	763 561.0
937	1.067 236	2943.672	689 555.2	987	1.013 171	3100.752	765 110.5
938	1.066 098	2946.814	691 027.9	988	1.012 146	3103.894	766 661.7
939	1.064 963	2949.956	692 502.1	989	1.011 122	3107.035	768 214.4
940	1.063 830	2953.097	693 977.8	**990**	1.010 101	3110.177	769 768.7
941	1.062 699	2956.239	695 455.2	991	1.009 082	3113.318	771 324.6
942	1.061 571	2959.380	696 934.1	992	1.008 065	3116.460	772 882.1
943	1.060 445	2962.522	698 414.5	993	1.007 049	3119.602	774 441.1
944	1.059 322	2965.663	699 896.6	994	1.006 036	3122.743	776 001.7
945	1.058 201	2968.805	701 380.2	995	1.005 025	3125.885	777 563.8
946	1.057 082	2971.947	702 865.4	996	1.004 016	3129.026	779 127.5
947	1.055 966	2975.088	704 352.1	997	1.003 009	3132.168	780 692.8
948	1.054 852	2978.230	705 840.5	998	1.002 004	3135.309	782 259.7
949	1.053 741	2981.371	707 330.4	999	1.001 001	3138.451	783 828.2
950	1.052 632	2984.513	708 821.8	**1000**	1.000 000	3141.593	785 398.2

POWERS OF NUMBERS

This table contains the fourth through ninth power of the numbers from 1 to 100. Second and third powers will be found in a later table headed "Squares, Square Root, Cube and Cube Root."

The larger numbers are expressed exponentially to at least seven significant figures.

The approximate value written as a whole number may be obtained by shifting the decimal point to the right by the number of places indicated in the exponent of 10 shown at the head of each group of values. For example: the approximate value of 33^8 is found in the table as 14.064086×10^{11}. Written as a whole number it is 1 406 408 600 000.

POWERS OF NUMBERS

n	n^4	n^5	n^6	n^7	n^8	n^9
1	1	1	1	1	1	1
2	16	32	64	128	256	512
3	81	243	729	2187	6561	19683
4	256	1024	4096	16384	65536	262144
5	625	3125	15625	78125	390625	1953125
6	1296	7776	46656	279936	1679616	10077696
7	2401	16807	117649	823543	5764801	40353607
8	4096	32768	262144	2097152	16777216	134217728
9	6561	59049	531441	4782969	43046721	387420489
					$\times 10^8$	$\times 10^9$
10	10000	100000	1000000	10000000	1.000000	1.000000
11	14641	161051	1771561	19487171	2.143589	2.357948
12	20736	248832	2985984	35831808	4.299817	5.159780
13	28561	371293	4826809	62748517	8.157307	10.604499
14	38416	537824	7529536	105413504	14.757891	20.661047
15	50625	759375	11390625	170859375	25.628906	38.443359
16	65536	1048576	16777216	268435456	42.949673	68.719477
17	83521	1419857	24137569	410338673	69.757574	118.587876
18	104976	1889568	34012224	612220032	110.199606	198.359290
19	130321	2476099	47045881	893871739	169.835630	322.687698
				$\times 10^9$	$\times 10^{10}$	$\times 10^{11}$
20	160000	3200000	64000000	1.280000	2.560000	5.120000
21	194481	4084101	85766121	1.801089	3.782286	7.942800
22	234256	5153632	113379904	2.494358	5.487587	12.072692
23	279841	6436343	148035889	3.404825	7.831099	18.011527
24	331776	7962624	191102976	4.586471	11.007531	26.418075
25	390625	9765625	244140625	6.103516	15.258789	38.146973
26	456976	11881376	308915776	8.031810	20.882706	54.295037
27	531441	14348907	387420489	10.460353	28.242954	76.255975
28	614656	17210368	481890304	13.492929	37.780200	105.784560
29	707281	20511149	594823321	17.249876	50.024641	145.071460
			$\times 10^8$	$\times 10^{10}$	$\times 10^{11}$	$\times 10^{13}$
30	810000	24300000	7.290000	2.187000	6.561000	1.968300
31	923521	28629151	8.875037	2.751261	8.528910	2.643962
32	1048576	33554432	10.737418	3.435974	10.995116	3.518437
33	1185921	39135393	12.914680	4.261844	14.064086	4.641148
34	1336336	45435424	15.448044	5.252335	17.857939	6.071699
35	1500625	52521875	18.382656	6.433930	22.518754	7.881564
36	1679616	60466176	21.767823	7.836416	28.211099	10.155996
37	1874161	69343957	25.657264	9.493188	35.124795	12.996174
38	2085136	79235168	30.109364	11.441558	43.477921	16.521610
39	2313441	90224199	35.187438	13.723101	53.520093	20.872836
			$\times 10^9$	$\times 10^{10}$	$\times 10^{12}$	$\times 10^{14}$
40	2560000	102400000	4.096000	16.384000	6.553600	2.621440
41	2825761	115856201	4.750104	19.475427	7.984925	3.273819
42	3111696	130691232	5.489032	23.053933	9.682652	4.066714
43	3418801	147008443	6.321363	27.181861	11.688200	5.025926
44	3748096	164916224	7.256314	31.927781	14.048224	6.181218
45	4100625	184528125	8.303766	37.366945	16.815125	7.566806
46	4477456	205962976	9.474297	43.581766	20.047612	9.221902
47	4879681	229345007	10.779215	50.662312	23.811287	11.191305
48	5308416	254803968	12.230590	58.706834	28.179280	13.526055
49	5764801	282475249	13.841287	67.822307	33.232931	16.284136
50	6250000	312500000	15.625000	78.125000	39.062500	19.531250

POWERS OF NUMBERS (Continued)

n	n^4	n^5	n^6	n^7	n^8	n^9
			$\times 10^9$	$\times 10^{11}$	$\times 10^{13}$	$\times 10^{14}$
50	6250000	312500000	15.625000	7.812500	3.906250	19.531250
51	6765201	345025251	17.596288	8.974107	4.576794	23.341652
52	7311616	380204032	19.770610	10.280717	5.345973	27.799059
53	7890481	418195493	22.164361	11.747111	6.225969	32.997636
54	8503056	459165024	24.794911	13.389252	7.230196	39.043059
55	9150625	503284375	27.680641	15.224352	8.373394	46.053666
56	9834496	550731776	30.840979	17.270948	9.671731	54.161694
57	10556001	601692057	34.296447	19.548975	11.142916	63.514620
58	11316496	656356768	38.068693	22.079842	12.806308	74.276587
59	12117361	714924299	42.180534	24.886515	14.683044	86.629958
		$\times 10^8$	$\times 10^{10}$	$\times 10^{11}$	$\times 10^{13}$	$\times 10^{16}$
60	12960000	7.776000	4.665600	27.993600	16.796160	1.007770
61	13845841	8.445963	5.152037	31.427428	19.170731	1.169415
62	14776336	9.161328	5.680024	35.216146	21.834011	1.353709
63	15752961	9.924365	6.252350	39.389806	24.815578	1.563381
64	16777216	10.737418	6.871948	43.980465	28.147498	1.801440
65	17850625	11.602906	7.541889	49.022279	31.864481	2.071191
66	18974736	12.523326	8.265395	54.551607	36.004061	2.376268
67	20151121	13.501251	9.045838	60.607116	40.606768	2.720653
68	21381376	14.539336	9.886748	67.229888	45.716324	3.108710
69	22667121	15.640313	10.791816	74.463533	51.379837	3.545209
		$\times 10^8$	$\times 10^{10}$	$\times 10^{12}$	$\times 10^{14}$	$\times 10^{16}$
70	24010000	16.807000	11.764900	8.235430	5.764801	4.035361
71	25411681	18.042294	12.810028	9.095120	6.457535	4.584850
72	26873856	19.349176	13.931407	10.030613	7.222041	5.199870
73	28398241	20.730716	15.133423	11.047399	8.064601	5.887159
74	29986576	22.190066	16.420649	12.151280	8.991947	6.654041
75	31640625	23.730469	17.797852	13.348389	10.011292	7.508469
76	33362176	25.355254	19.269993	14.645195	11.130348	8.459064
77	35153041	27.067842	20.842238	16.048523	12.357363	9.515169
78	37015056	28.871744	22.519960	17.565569	13.701144	10.686892
79	38950081	30.770564	24.308746	19.203909	15.171088	11.985160
		$\times 10^8$	$\times 10^{10}$	$\times 10^{12}$	$\times 10^{14}$	$\times 10^{16}$
80	40960000	32.768000	26.214400	20.971520	16.777216	13.421773
81	43046721	34.867844	28.242954	22.876792	18.530202	15.009464
82	45212176	37.073984	30.400667	24.928547	20.441409	16.761955
83	47458321	39.390406	32.694037	27.136051	22.522922	18.694026
84	49787136	41.821194	35.129803	29.509035	24.787589	20.821575
85	52200625	44.370531	37.714952	32.057709	27.249053	23.161695
86	54700816	47.042702	40.456724	34.792782	29.921793	25.732742
87	57289761	49.842092	43.362620	37.725479	32.821167	28.554415
88	59969536	52.773192	46.440409	40.867560	35.963452	31.647838
89	62742241	55.840594	49.698129	44.231335	39.365888	35.035640
		$\times 10^9$	$\times 10^{11}$	$\times 10^{13}$	$\times 10^{15}$	$\times 10^{17}$
90	65610000	5.904900	5.314410	4.782969	4.304672	3.874205
91	68574961	6.240321	5.678693	5.167610	4.702525	4.279298
92	71639296	6.590815	6.063550	5.578466	5.132189	4.721614
93	74805201	6.956884	6.469902	6.017009	5.595818	5.204111
94	78074896	7.339040	6.898698	6.484776	6.095689	5.729948
95	81450625	7.737809	7.350919	6.983373	6.634204	6.302494
96	84934656	8.153727	7.827578	7.514475	7.213896	6.925340
97	88529281	8.587340	8.329720	8.079828	7.837434	7.602311
98	92236816	9.039208	8.858424	8.681255	8.507630	8.337478
99	96059601	9.509900	9.414801	9.320653	9.227447	9.135172
100	100000000	10.000000	10.000000	10.000000	10.000000	10.000000

POSITIVE POWERS OF TWO

n	2^n	n	2^n
1	2	51	22517 99813 68524 8
2	4	52	45035 99627 37049 6
3	8	53	90071 99254 74099 2
4	16	54	18014 39850 94819 84
5	32	55	36028 79701 89639 68
6	64	56	72057 59403 79279 36
7	128	57	14411 51880 75855 872
8	256	58	28823 03761 51711 744
9	512	59	57646 07523 03423 488
10	1024	60	11529 21504 60684 6976
11	2048	61	23058 43009 21369 3952
12	4096	62	46116 86018 42738 7904
13	8192	63	92233 72036 85477 5808
14	16384	64	18446 74407 37095 51616
15	32768	65	36893 48814 74191 03232
16	65536	66	73786 97629 48382 06464
17	13107 2	67	14757 39525 89676 41292 8
18	26214 4	68	29514 79051 79352 82585 6
19	52428 8	69	59029 58103 58705 65171 2
20	10485 76	70	11805 91620 71741 13034 24
21	20971 52	71	23611 83241 43482 26068 48
22	41943 04	72	47223 66482 86964 52136 96
23	83886 08	73	94447 32965 73929 04273 92
24	16777 216	74	18889 46593 14785 80854 784
25	33554 432	75	37778 93186 29571 61709 568
26	67108 864	76	75557 86372 59143 23419 136
27	13421 7728	77	15111 57274 51828 64683 8272
28	26843 5456	78	30223 14549 03657 29367 6544
29	53687 0912	79	60446 29098 07314 58735 3088
30	10737 41824	80	12089 25819 61462 91747 06176
31	21474 83648	81	24178 51639 22925 83494 12352
32	42949 67296	82	48357 03278 45851 66988 24704
33	85899 34592	83	96714 06556 91703 33976 49408
34	17179 86918 4	84	19342 81311 38340 66795 29881 6
35	34359 73836 8	85	38685 62622 76681 33590 59763 2
36	68719 47673 6	86	77371 25245 53362 67181 19526 4
37	13743 89534 72	87	15474 25049 10672 53436 23905 28
38	27487 79069 44	88	30948 50098 21345 06872 47810 56
39	54975 58138 88	89	61897 00196 42690 13744 95621 12
40	10995 11627 776	90	12379 40039 28538 02748 99124 224
41	21990 23255 552	91	24758 80078 57076 05497 98248 448
42	43980 46511 104	92	49517 60157 14152 10995 96496 896
43	87960 93022 208	93	99035 20314 28304 21991 92993 792
44	17592 18604 4416	94	19807 04062 85660 84398 38598 7584
45	35184 37208 8832	95	39614 08125 71321 68796 77197 5168
46	70368 74417 7664	96	79228 16251 42643 37593 54395 0336
47	14073 74883 55328	97	15845 63250 28528 67518 70879 00672
48	28147 49767 10656	98	31691 26500 57057 35037 41758 01344
49	56294 99534 21312	99	63382 53001 14114 70074 83516 02688
50	11258 99906 84262 4	100	12676 50600 22822 94014 96703 20537 6
		101	25353 01200 45645 88029 93406 41075 2

NEGATIVE POWERS OF TWO

n	2^{-n}
0	1.0
1	0.5
2	0.25
3	0.125
4	0.0625
5	0.03125
6	0.01562 5
7	0.00781 25
8	0.00390 625
9	0.00195 3125
10	0.00097 65625
11	0.00048 82812 5
12	0.00024 41406 25
13	0.00012 20703 125
14	0.00006 10351 5625
15	0.00003 05175 78125
16	0.00001 52587 89062 5
17	0.00000 76293 94531 25
18	0.00000 38146 97265 625
19	0.00000 19073 48632 8125
20	0.00000 09536 74316 40625
21	0.00000 04768 37158 20312 5
22	0.00000 02384 18579 10156 25
23	0.00000 01192 09289 55078 125
24	0.00000 00596 04644 77539 0625
25	0.00000 00298 02322 38769 53125
26	0.00000 00149 01161 19384 76562 5
27	0.00000 00074 50580 59692 38281 25
28	0.00000 00037 25290 29846 19140 625
29	0.00000 00018 62645 14923 09570 3125
30	0.00000 00009 31322 57461 54785 15625
31	0.00000 00004 65661 28730 77392 57812 5
32	0.00000 00002 32830 64365 38696 28906 25
33	0.00000 00001 16415 32182 69348 14453 125
34	0.00000 00000 58207 66091 34674 07226 5625
35	0.00000 00000 29103 83045 67337 03613 28125
36	0.00000 00000 14551 91522 83668 51806 64062 5
37	0.00000 00000 07275 95761 41834 25903 32031 25
38	0.00000 00000 03637 97880 70917 12951 66015 625
39	0.00000 00000 01818 98940 35458 56475 83007 8125
40	0.00000 00000 00909 49470 17729 28237 91503 90625
41	0.00000 00000 00454 74735 08864 64118 95751 95312 5
42	0.00000 00000 00227 37367 54432 32059 47875 97656 25
43	0.00000 00000 00113 68683 77216 16029 73937 98828 125
44	0.00000 00000 00056 84341 88608 08014 86968 99414 0625
45	0.00000 00000 00028 42170 94304 04007 43484 49707 03125
46	0.00000 00000 00014 21085 47152 02003 71742 24853 51562 5
47	0.00000 00000 00007 10542 73576 01001 85871 12426 75781 25
48	0.00000 00000 00003 55271 36788 00500 92935 56213 37890 625
49	0.00000 00000 00001 77635 68394 00250 46467 78106 68945 3125
50	0.00000 00000 00000 88817 84197 00125 23233 89053 34472 65625

SUMS OF POWERS OF INTEGERS, $\sum_{k=1}^{n} k^m$

$(m = 1, 2, 3, 4);\ 1 \leq n \leq 40$

n	Σk	Σk^2	Σk^3	Σk^4
1	1	1	1	1
2	3	5	9	17
3	6	14	36	98
4	10	30	100	354
5	15	55	225	979
6	21	91	441	2275
7	28	140	784	4676
8	36	204	1296	8772
9	45	285	2025	15333
10	55	385	3025	25333
11	66	506	4356	39974
12	78	650	6084	60710
13	91	819	8281	89271
14	105	1015	11025	127687
15	120	1240	14400	178312
16	136	1496	18496	243848
17	153	1785	23409	327369
18	171	2109	29241	432345
19	190	2470	36100	562666
20	210	2870	44100	722666
21	231	3311	53361	917147
22	253	3795	64009	1151403
23	276	4324	76176	1431244
24	300	4900	90000	1763020
25	325	5525	105625	2153645
26	351	6201	123201	2610621
27	378	6930	142884	3142062
28	406	7714	164836	3756718
29	435	8555	189225	4463999
30	465	9455	216225	5273999
31	496	10416	246016	6197520
32	528	11440	278784	7246096
33	561	12529	314721	8432017
34	595	13685	354025	9768353
35	630	14910	396900	11268978
36	666	16206	443556	12948594
37	703	17575	494209	14822755
38	741	19019	549081	16907891
39	780	20540	608400	19221332
40	820	22140	672400	21781332

* SUMS OF POWERS OF THE FIRST *n* INTEGERS

$$\sum_{k=1}^{n} k = 1 + 2 + 3 + \cdots + n = \frac{n(n+1)}{2}$$

$$\sum_{k=1}^{n} k^2 = 1^2 + 2^2 + 3^2 + \cdots + n^2 = \frac{n(n+1)(2n+1)}{6}$$

$$\sum_{k=1}^{n} k^3 = \frac{n^2(n+1)^2}{4}$$

$$\sum_{k=1}^{n} k^4 = \frac{n}{30}(n+1)(2n+1)(3n^2+3n-1).$$

$$\sum_{k=1}^{n} k^5 = \frac{n^2}{12}(n+1)^2(2n^2+2n-1).$$

$$\sum_{k=1}^{n} k^6 = \frac{n}{42}(n+1)(2n+1)(3n^4+6n^3-3n+1).$$

$$\sum_{k=1}^{n} k^7 = \frac{n^2}{24}(n+1)^2(3n^4+6n^3-n^2-4n+2).$$

$$\sum_{k=1}^{n} k^8 = \frac{n}{90}(n+1)(2n+1)(5n^6+15n^5+5n^4-15n^3-n^2+9n-3).$$

$$\sum_{k=1}^{n} k^9 = \frac{n^2}{20}(n+1)^2(2n^6+6n^5+n^4-8n^3+n^2+6n-3).$$

$$\sum_{k=1}^{n} k^{10} = \frac{n}{66}(n+1)(2n+1)(3n^8+12n^7+8n^6-18n^5 - 10n^4+24n^3+2n^2-15n+5).$$

*See page 70 for generating formula.

FACTORS AND PRIMES

The table presents the prime factors of *all* factorable numbers and the mantissas of the common logarithms of all prime numbers from 1 to 2000.

It should be noted that the third digit of the number is given at the top of the page and that the table runs across two facing pages. Thus, the factors of 258 are found, on the right hand page, on a line with 25 and under vertical column 8 to be 2·3·43.

FACTORS AND PRIMES

If n is prime the mantissa of its common logarithm is given. If n is not prime its prime factors are given.

n	0	1	2	3	4
0	....		**3010300**	**4771213**	2^2
1	$2 \cdot 5$	**0413927**	$2^2 \cdot 3$	**1139434**	$2 \cdot 7$
2	$2^2 \cdot 5$	$3 \cdot 7$	$2 \cdot 11$	**3617278**	$2^3 \cdot 3$
3	$2 \cdot 3 \cdot 5$	**4913617**	2^5	$3 \cdot 11$	$2 \cdot 17$
4	$2^3 \cdot 5$	**6127839**	$2 \cdot 3 \cdot 7$	**6334685**	$2^2 \cdot 11$
5	$2 \cdot 5^2$	$3 \cdot 17$	$2^2 \cdot 13$	**7242759**	$2 \cdot 3^3$
6	$2^2 \cdot 3 \cdot 5$	**7853298**	$2 \cdot 31$	$3^2 \cdot 7$	2^6
7	$2 \cdot 5 \cdot 7$	**8512583**	$2^3 \cdot 3^2$	**8633229**	$2 \cdot 37$
8	$2^4 \cdot 5$	3^4	$2 \cdot 41$	**9190781**	$2^2 \cdot 3 \cdot 7$
9	$2 \cdot 3^2 \cdot 5$	$7 \cdot 13$	$2^2 \cdot 23$	$3 \cdot 31$	$2 \cdot 47$
10	$2^2 \cdot 5^2$	**0043214**	$2 \cdot 3 \cdot 17$	**0128372**	$2^3 \cdot 13$
11	$2 \cdot 5 \cdot 11$	$3 \cdot 37$	$2^4 \cdot 7$	**0530784**	$2 \cdot 3 \cdot 19$
12	$2^3 \cdot 3 \cdot 5$	11^2	$2 \cdot 61$	$3 \cdot 41$	$2^2 \cdot 31$
13	$2 \cdot 5 \cdot 13$	**1172713**	$2^2 \cdot 3 \cdot 11$	$7 \cdot 19$	$2 \cdot 67$
14	$2^2 \cdot 5 \cdot 7$	$3 \cdot 47$	$2 \cdot 71$	$11 \cdot 13$	$2^4 \cdot 3^2$
15	$2 \cdot 3 \cdot 5^2$	**1789769**	$2^3 \cdot 19$	$3^2 \cdot 17$	$2 \cdot 7 \cdot 11$
16	$2^5 \cdot 5$	$7 \cdot 23$	$2 \cdot 3^3$	**2121876**	$2^2 \cdot 41$
17	$2 \cdot 5 \cdot 17$	$3^2 \cdot 19$	$2^2 \cdot 43$	**2380461**	$2 \cdot 3 \cdot 29$
18	$2^2 \cdot 3^2 \cdot 5$	**2576786**	$2 \cdot 7 \cdot 13$	$3 \cdot 61$	$2^3 \cdot 23$
19	$2 \cdot 5 \cdot 19$	**2810334**	$2^6 \cdot 3$	**2855573**	$2 \cdot 97$
20	$2^3 \cdot 5^2$	$3 \cdot 67$	$2 \cdot 101$	$7 \cdot 29$	$2^2 \cdot 3 \cdot 17$
21	$2 \cdot 3 \cdot 5 \cdot 7$	**3242825**	$2^2 \cdot 53$	$3 \cdot 71$	$2 \cdot 107$
22	$2^2 \cdot 5 \cdot 11$	$13 \cdot 17$	$2 \cdot 3 \cdot 37$	**3483049**	$2^5 \cdot 7$
23	$2 \cdot 5 \cdot 23$	$3 \cdot 7 \cdot 11$	$2^3 \cdot 29$	**3673559**	$2 \cdot 3^2 \cdot 13$
24	$2^4 \cdot 3 \cdot 5$	**3820170**	$2 \cdot 11^2$	3^5	$2^2 \cdot 61$
25	$2 \cdot 5^3$	**3996737**	$2^2 \cdot 3^2 \cdot 7$	$11 \cdot 23$	$2 \cdot 127$
26	$2^2 \cdot 5 \cdot 13$	$3^2 \cdot 29$	$2 \cdot 131$	**4199557**	$2^3 \cdot 3 \cdot 11$
27	$2 \cdot 3^3 \cdot 5$	**4329693**	$2^4 \cdot 17$	$3 \cdot 7 \cdot 13$	$2 \cdot 137$
28	$2^3 \cdot 5 \cdot 7$	**4487063**	$2 \cdot 3 \cdot 47$	**4517864**	$2^2 \cdot 71$
29	$2 \cdot 5 \cdot 29$	$3 \cdot 97$	$2^2 \cdot 73$	**4668676**	$2 \cdot 3 \cdot 7^2$
30	$2^2 \cdot 3 \cdot 5^2$	$7 \cdot 43$	$2 \cdot 151$	$3 \cdot 101$	$2^4 \cdot 19$
31	$2 \cdot 5 \cdot 31$	**4927604**	$2^3 \cdot 3 \cdot 13$	**4955443**	$2 \cdot 157$
32	$2^6 \cdot 5$	$3 \cdot 107$	$2 \cdot 7 \cdot 23$	$17 \cdot 19$	$2^2 \cdot 3^4$
33	$2 \cdot 3 \cdot 5 \cdot 11$	**5198280**	$2^2 \cdot 83$	$3^2 \cdot 37$	$2 \cdot 167$
34	$2^2 \cdot 5 \cdot 17$	$11 \cdot 31$	$2 \cdot 3^2 \cdot 19$	7^3	$2^3 \cdot 43$
35	$2 \cdot 5^2 \cdot 7$	$3^3 \cdot 13$	$2^5 \cdot 11$	**5477747**	$2 \cdot 3 \cdot 59$
36	$2^3 \cdot 3^2 \cdot 5$	19^2	$2 \cdot 181$	$3 \cdot 11^2$	$2^2 \cdot 7 \cdot 13$
37	$2 \cdot 5 \cdot 37$	$7 \cdot 53$	$2^2 \cdot 3 \cdot 31$	**5717088**	$2 \cdot 11 \cdot 17$
38	$2^2 \cdot 5 \cdot 19$	$3 \cdot 127$	$2 \cdot 191$	**5831988**	$2^7 \cdot 3$
39	$2 \cdot 3 \cdot 5 \cdot 13$	$17 \cdot 23$	$2^3 \cdot 7^2$	$3 \cdot 131$	$2 \cdot 197$
40	$2^4 \cdot 5^2$	**6031444**	$2 \cdot 3 \cdot 67$	$13 \cdot 31$	$2^2 \cdot 101$
41	$2 \cdot 5 \cdot 41$	$3 \cdot 137$	$2^2 \cdot 103$	$7 \cdot 59$	$2 \cdot 3^2 \cdot 23$
42	$2^2 \cdot 3 \cdot 5 \cdot 7$	**6242821**	$2 \cdot 211$	$3^2 \cdot 47$	$2^3 \cdot 53$
43	$2 \cdot 5 \cdot 43$	**6344773**	$2^4 \cdot 3^3$	**6364879**	$2 \cdot 7 \cdot 31$
44	$2^3 \cdot 5 \cdot 11$	$3^2 \cdot 7^2$	$2 \cdot 13 \cdot 17$	**6464037**	$2^2 \cdot 3 \cdot 37$
45	$2 \cdot 3^2 \cdot 5^2$	$11 \cdot 41$	$2^2 \cdot 113$	$3 \cdot 151$	$2 \cdot 227$
46	$2^2 \cdot 5 \cdot 23$	**6637009**	$2 \cdot 3 \cdot 7 \cdot 11$	**6655810**	$2^4 \cdot 29$
47	$2 \cdot 5 \cdot 47$	$3 \cdot 157$	$2^3 \cdot 59$	$11 \cdot 43$	$2 \cdot 3 \cdot 79$
48	$2^5 \cdot 3 \cdot 5$	$13 \cdot 37$	$2 \cdot 241$	$3 \cdot 7 \cdot 23$	$2^2 \cdot 11^2$
49	$2 \cdot 5 \cdot 7^2$	**6910815**	$2^2 \cdot 3 \cdot 41$	$17 \cdot 29$	$2 \cdot 13 \cdot 19$
50	$2^2 \cdot 5^3$	$3 \cdot 167$	$2 \cdot 251$	**7015680**	$2^3 \cdot 3^2 \cdot 7$

FACTORS AND PRIMES (Continued)

n	5	6	7	8	9
0	**6989700**	$2 \cdot 3$	**8450980**	2^3	3^2
1	$3 \cdot 5$	2^4	**2304489**	$2 \cdot 3^2$	**2787536**
2	5^2	$2 \cdot 13$	3^3	$2^2 \cdot 7$	**4623980**
3	$5 \cdot 7$	$2^2 \cdot 3^2$	**5682017**	$2 \cdot 19$	$3 \cdot 13$
4	$3^2 \cdot 5$	$2 \cdot 23$	**6720979**	$2^4 \cdot 3$	7^2
5	$5 \cdot 11$	$2^3 \cdot 7$	$3 \cdot 19$	$2 \cdot 29$	**7708520**
6	$5 \cdot 13$	$2 \cdot 3 \cdot 11$	**8260748**	$2^2 \cdot 17$	$3 \cdot 23$
7	$3 \cdot 5^2$	$2^2 \cdot 19$	$7 \cdot 11$	$2 \cdot 3 \cdot 13$	**8976271**
8	$5 \cdot 17$	$2 \cdot 43$	$3 \cdot 29$	$2^3 \cdot 11$	**9493900**
9	$5 \cdot 19$	$2^5 \cdot 3$	**9867717**	$2 \cdot 7^2$	$3^2 \cdot 11$
10	$3 \cdot 5 \cdot 7$	$2 \cdot 53$	**0293838**	$2^2 \cdot 3^3$	**0374265**
11	$5 \cdot 23$	$2^2 \cdot 29$	$3^2 \cdot 13$	$2 \cdot 59$	$7 \cdot 17$
12	5^3	$2 \cdot 3^2 \cdot 7$	**1038037**	2^7	$3 \cdot 43$
13	$3^3 \cdot 5$	$2^3 \cdot 17$	**1367206**	$2 \cdot 3 \cdot 23$	**1430148**
14	$5 \cdot 29$	$2 \cdot 73$	$3 \cdot 7^2$	$2^2 \cdot 37$	**1731863**
15	$5 \cdot 31$	$2^2 \cdot 3 \cdot 13$	**1958997**	$2 \cdot 79$	$3 \cdot 53$
16	$3 \cdot 5 \cdot 11$	$2 \cdot 83$	**2227165**	$2^3 \cdot 3 \cdot 7$	13^2
17	$5^2 \cdot 7$	$2^4 \cdot 11$	$3 \cdot 59$	$2 \cdot 89$	**2528530**
18	$5 \cdot 37$	$2 \cdot 3 \cdot 31$	$11 \cdot 17$	$2^2 \cdot 47$	$3^3 \cdot 7$
19	$3 \cdot 5 \cdot 13$	$2^2 \cdot 7^2$	**2944662**	$2 \cdot 3^2 \cdot 11$	**2988531**
20	$5 \cdot 41$	$2 \cdot 103$	$3^2 \cdot 23$	$2^4 \cdot 13$	$11 \cdot 19$
21	$5 \cdot 43$	$2^3 \cdot 3^3$	$7 \cdot 31$	$2 \cdot 109$	$3 \cdot 73$
22	$3^2 \cdot 5^2$	$2 \cdot 113$	**3560259**	$2^2 \cdot 3 \cdot 19$	**3598355**
23	$5 \cdot 47$	$2^2 \cdot 59$	$3 \cdot 79$	$2 \cdot 7 \cdot 17$	**3783979**
24	$5 \cdot 7^2$	$2 \cdot 3 \cdot 41$	$13 \cdot 19$	$2^3 \cdot 31$	$3 \cdot 83$
25	$3 \cdot 5 \cdot 17$	2^8	**4099331**	$2 \cdot 3 \cdot 43$	$7 \cdot 37$
26	$5 \cdot 53$	$2 \cdot 7 \cdot 19$	$3 \cdot 89$	$2^2 \cdot 67$	**4297523**
27	$5^2 \cdot 11$	$2^2 \cdot 3 \cdot 23$	**4424798**	$2 \cdot 139$	$3^2 \cdot 31$
28	$3 \cdot 5 \cdot 19$	$2 \cdot 11 \cdot 13$	$7 \cdot 41$	$2^5 \cdot 3^2$	17^2
29	$5 \cdot 59$	$2^3 \cdot 37$	$3^3 \cdot 11$	$2 \cdot 149$	$13 \cdot 23$
30	$5 \cdot 61$	$2 \cdot 3^2 \cdot 17$	**4871384**	$2^2 \cdot 7 \cdot 11$	$3 \cdot 103$
31	$3^2 \cdot 5 \cdot 7$	$2^2 \cdot 79$	**5010593**	$2 \cdot 3 \cdot 53$	$11 \cdot 29$
32	$5^2 \cdot 13$	$2 \cdot 163$	$3 \cdot 109$	$2^3 \cdot 41$	$7 \cdot 47$
33	$5 \cdot 67$	$2^4 \cdot 3 \cdot 7$	**5276299**	$2 \cdot 13^2$	$3 \cdot 113$
34	$3 \cdot 5 \cdot 23$	$2 \cdot 173$	**5403295**	$2^2 \cdot 3 \cdot 29$	**5428254**
35	$5 \cdot 71$	$2^2 \cdot 89$	$3 \cdot 7 \cdot 17$	$2 \cdot 179$	**5550944**
36	$5 \cdot 73$	$2 \cdot 3 \cdot 61$	**5646661**	$2^4 \cdot 23$	$3^2 \cdot 41$
37	$3 \cdot 5^3$	$2^3 \cdot 47$	$13 \cdot 29$	$2 \cdot 3^3 \cdot 7$	**5786392**
38	$5 \cdot 7 \cdot 11$	$2 \cdot 193$	$3^2 \cdot 43$	$2^2 \cdot 97$	**5899496**
39	$5 \cdot 79$	$2^2 \cdot 3^2 \cdot 11$	**5987905**	$2 \cdot 199$	$3 \cdot 7 \cdot 19$
40	$3^4 \cdot 5$	$2 \cdot 7 \cdot 29$	$11 \cdot 37$	$2^3 \cdot 3 \cdot 17$	**6117233**
41	$5 \cdot 83$	$2^5 \cdot 13$	$3 \cdot 139$	$2 \cdot 11 \cdot 19$	**6222140**
42	$5^2 \cdot 17$	$2 \cdot 3 \cdot 71$	$7 \cdot 61$	$2^2 \cdot 107$	$3 \cdot 11 \cdot 13$
43	$3 \cdot 5 \cdot 29$	$2^2 \cdot 109$	$19 \cdot 23$	$2 \cdot 3 \cdot 73$	**6424645**
44	$5 \cdot 89$	$2 \cdot 223$	$3 \cdot 149$	$2^6 \cdot 7$	**6522463**
45	$5 \cdot 7 \cdot 13$	$2^3 \cdot 3 \cdot 19$	**6599162**	$2 \cdot 229$	$3^3 \cdot 17$
46	$3 \cdot 5 \cdot 31$	$2 \cdot 233$	**6693169**	$2^2 \cdot 3^2 \cdot 13$	$7 \cdot 67$
47	$5^2 \cdot 19$	$2^2 \cdot 7 \cdot 17$	$3^2 \cdot 53$	$2 \cdot 239$	**6803355**
48	$5 \cdot 97$	$2 \cdot 3^5$	**6875290**	$2^3 \cdot 61$	$3 \cdot 163$
49	$3^2 \cdot 5 \cdot 11$	$2^4 \cdot 31$	$7 \cdot 71$	$2 \cdot 3 \cdot 83$	**6981005**
50	$5 \cdot 101$	$2 \cdot 11 \cdot 23$	$3 \cdot 13^2$	$2^2 \cdot 127$	**7067178**

FACTORS AND PRIMES (Continued)

n	0	1	2	3	4
50	$2^2 \cdot 5^3$	$3 \cdot 167$	$2 \cdot 251$	**7015680**	$2^3 \cdot 3^2 \cdot 7$
51	$2 \cdot 3 \cdot 5 \cdot 17$	$7 \cdot 73$	2^9	$3^3 \cdot 19$	$2 \cdot 257$
52	$2^3 \cdot 5 \cdot 13$	**7168377**	$2 \cdot 3^2 \cdot 29$	**7185017**	$2^2 \cdot 131$
53	$2 \cdot 5 \cdot 53$	$3^2 \cdot 59$	$2^2 \cdot 7 \cdot 19$	$13 \cdot 41$	$2 \cdot 3 \cdot 89$
54	$2^2 \cdot 3^3 \cdot 5$	**7331973**	$2 \cdot 271$	$3 \cdot 181$	$2^5 \cdot 17$
55	$2 \cdot 5^2 \cdot 11$	$19 \cdot 29$	$2^3 \cdot 3 \cdot 23$	$7 \cdot 79$	$2 \cdot 277$
56	$2^4 \cdot 5 \cdot 7$	$3 \cdot 11 \cdot 17$	$2 \cdot 281$	**7505084**	$2^2 \cdot 3 \cdot 47$
57	$2 \cdot 3 \cdot 5 \cdot 19$	**7566361**	$2^2 \cdot 11 \cdot 13$	$3 \cdot 191$	$2 \cdot 7 \cdot 41$
58	$2^2 \cdot 5 \cdot 29$	$7 \cdot 83$	$2 \cdot 3 \cdot 97$	$11 \cdot 53$	$2^3 \cdot 73$
59	$2 \cdot 5 \cdot 59$	$3 \cdot 197$	$2^4 \cdot 37$	**7730547**	$2 \cdot 3^3 \cdot 11$
60	$2^3 \cdot 3 \cdot 5^2$	**7788745**	$2 \cdot 7 \cdot 43$	$3^2 \cdot 67$	$2^2 \cdot 151$
61	$2 \cdot 5 \cdot 61$	$13 \cdot 47$	$2^2 \cdot 3^2 \cdot 17$	**7874605**	$2 \cdot 307$
62	$2^2 \cdot 5 \cdot 31$	$3^3 \cdot 23$	$2 \cdot 311$	$7 \cdot 89$	$2^4 \cdot 3 \cdot 13$
63	$2 \cdot 3^2 \cdot 5 \cdot 7$	**8000294**	$2^3 \cdot 79$	$3 \cdot 211$	$2 \cdot 317$
64	$2^7 \cdot 5$	**8068580**	$2 \cdot 3 \cdot 107$	**8082110**	$2^2 \cdot 7 \cdot 23$
65	$2 \cdot 5^2 \cdot 13$	$3 \cdot 7 \cdot 31$	$2^2 \cdot 163$	**8149132**	$2 \cdot 3 \cdot 109$
66	$2^2 \cdot 3 \cdot 5 \cdot 11$	**8202015**	$2 \cdot 331$	$3 \cdot 13 \cdot 17$	$2^3 \cdot 83$
67	$2 \cdot 5 \cdot 67$	$11 \cdot 61$	$2^5 \cdot 3 \cdot 7$	**8280151**	$2 \cdot 337$
68	$2^3 \cdot 5 \cdot 17$	$3 \cdot 227$	$2 \cdot 11 \cdot 31$	**8344207**	$2^2 \cdot 3^2 \cdot 19$
69	$2 \cdot 3 \cdot 5 \cdot 23$	**8394780**	$2^2 \cdot 173$	$3^2 \cdot 7 \cdot 11$	$2 \cdot 347$
70	$2^2 \cdot 5^2 \cdot 7$	**8457180**	$2 \cdot 3^3 \cdot 13$	$19 \cdot 37$	$2^6 \cdot 11$
71	$2 \cdot 5 \cdot 71$	$3^2 \cdot 79$	$2^3 \cdot 89$	$23 \cdot 31$	$2 \cdot 3 \cdot 7 \cdot 17$
72	$2^4 \cdot 3^2 \cdot 5$	$7 \cdot 103$	$2 \cdot 19^2$	$3 \cdot 241$	$2^2 \cdot 181$
73	$2 \cdot 5 \cdot 73$	$17 \cdot 43$	$2^2 \cdot 3 \cdot 61$	**8651040**	$2 \cdot 367$
74	$2^2 \cdot 5 \cdot 37$	$3 \cdot 13 \cdot 19$	$2 \cdot 7 \cdot 53$	**8709888**	$2^3 \cdot 3 \cdot 31$
75	$2 \cdot 3 \cdot 5^3$	**8756399**	$2^4 \cdot 47$	$3 \cdot 251$	$2 \cdot 13 \cdot 29$
76	$2^3 \cdot 5 \cdot 19$	**8813847**	$2 \cdot 3 \cdot 127$	$7 \cdot 109$	$2^2 \cdot 191$
77	$2 \cdot 5 \cdot 7 \cdot 11$	$3 \cdot 257$	$2^2 \cdot 193$	**8881795**	$2 \cdot 3^2 \cdot 43$
78	$2^2 \cdot 3 \cdot 5 \cdot 13$	$11 \cdot 71$	$2 \cdot 17 \cdot 23$	$3^3 \cdot 29$	$2^4 \cdot 7^2$
79	$2 \cdot 5 \cdot 79$	$7 \cdot 113$	$2^3 \cdot 3^2 \cdot 11$	$13 \cdot 61$	$2 \cdot 397$
80	$2^5 \cdot 5^2$	$3^2 \cdot 89$	$2 \cdot 401$	$11 \cdot 73$	$2^2 \cdot 3 \cdot 67$
81	$2 \cdot 3^4 \cdot 5$	**9090209**	$2^2 \cdot 7 \cdot 29$	$3 \cdot 271$	$2 \cdot 11 \cdot 37$
82	$2^2 \cdot 5 \cdot 41$	**9143432**	$2 \cdot 3 \cdot 137$	**9153998**	$2^3 \cdot 103$
83	$2 \cdot 5 \cdot 83$	$3 \cdot 277$	$2^6 \cdot 13$	$7^2 \cdot 17$	$2 \cdot 3 \cdot 139$
84	$2^3 \cdot 3 \cdot 5 \cdot 7$	29^2	$2 \cdot 421$	$3 \cdot 281$	$2^2 \cdot 211$
85	$2 \cdot 5^2 \cdot 17$	$23 \cdot 37$	$2^2 \cdot 3 \cdot 71$	**9309490**	$2 \cdot 7 \cdot 61$
86	$2^2 \cdot 5 \cdot 43$	$3 \cdot 7 \cdot 41$	$2 \cdot 431$	**9360108**	$2^5 \cdot 3^3$
87	$2 \cdot 3 \cdot 5 \cdot 29$	$13 \cdot 67$	$2^3 \cdot 109$	$3^2 \cdot 97$	$2 \cdot 19 \cdot 23$
88	$2^4 \cdot 5 \cdot 11$	**9449759**	$2 \cdot 3^2 \cdot 7^2$	**9459607**	$2^2 \cdot 13 \cdot 17$
89	$2 \cdot 5 \cdot 89$	$3^4 \cdot 11$	$2^2 \cdot 223$	$19 \cdot 47$	$2 \cdot 3 \cdot 149$
90	$2^2 \cdot 3^2 \cdot 5^2$	$17 \cdot 53$	$2 \cdot 11 \cdot 41$	$3 \cdot 7 \cdot 43$	$2^3 \cdot 113$
91	$2 \cdot 5 \cdot 7 \cdot 13$	**9595184**	$2^4 \cdot 3 \cdot 19$	$11 \cdot 83$	$2 \cdot 457$
92	$2^3 \cdot 5 \cdot 23$	$3 \cdot 307$	$2 \cdot 461$	$13 \cdot 71$	$2^2 \cdot 3 \cdot 7 \cdot 11$
93	$2 \cdot 3 \cdot 5 \cdot 31$	$7^2 \cdot 19$	$2^2 \cdot 233$	$3 \cdot 311$	$2 \cdot 467$
94	$2^2 \cdot 5 \cdot 47$	**9735896**	$2 \cdot 3 \cdot 157$	$23 \cdot 41$	$2^4 \cdot 59$
95	$2 \cdot 5^2 \cdot 19$	$3 \cdot 317$	$2^3 \cdot 7 \cdot 17$	**9790929**	$2 \cdot 3^2 \cdot 53$
96	$2^6 \cdot 3 \cdot 5$	31^2	$2 \cdot 13 \cdot 37$	$3^2 \cdot 107$	$2^2 \cdot 241$
97	$2 \cdot 5 \cdot 97$	**9872192**	$2^2 \cdot 3^5$	$7 \cdot 139$	$2 \cdot 487$
98	$2^2 \cdot 5 \cdot 7^2$	$3^2 \cdot 109$	$2 \cdot 491$	**9925535**	$2^3 \cdot 3 \cdot 41$
99	$2 \cdot 3^2 \cdot 5 \cdot 11$	**9960737**	$2^5 \cdot 31$	$3 \cdot 331$	$2 \cdot 7 \cdot 71$
100	$2^3 \cdot 5^3$	$7 \cdot 11 \cdot 13$	$2 \cdot 3 \cdot 167$	$17 \cdot 59$	$2^2 \cdot 251$

FACTORS AND PRIMES (Continued)

n	5	6	7	8	9
50	$5 \cdot 101$	$2 \cdot 11 \cdot 23$	$3 \cdot 13^2$	$2^2 \cdot 127$	**7067178**
51	$5 \cdot 103$	$2^2 \cdot 3 \cdot 43$	$11 \cdot 47$	$2 \cdot 7 \cdot 37$	$3 \cdot 173$
52	$3 \cdot 5^2 \cdot 7$	$2 \cdot 263$	$17 \cdot 31$	$2^4 \cdot 3 \cdot 11$	23^2
53	$5 \cdot 107$	$2^3 \cdot 67$	$3 \cdot 179$	$2 \cdot 269$	$7^2 \cdot 11$
54	$5 \cdot 109$	$2 \cdot 3 \cdot 7 \cdot 13$	**7379873**	$2^2 \cdot 137$	$3^2 \cdot 61$
55	$3 \cdot 5 \cdot 37$	$2^2 \cdot 139$	**7458552**	$2 \cdot 3^2 \cdot 31$	$13 \cdot 43$
56	$5 \cdot 113$	$2 \cdot 283$	$3^4 \cdot 7$	$2^3 \cdot 71$	**7551123**
57	$5^2 \cdot 23$	$2^6 \cdot 3^2$	**7611758**	$2 \cdot 17^2$	$3 \cdot 193$
58	$3^2 \cdot 5 \cdot 13$	$2 \cdot 293$	**7686381**	$2^2 \cdot 3 \cdot 7^2$	$19 \cdot 31$
59	$5 \cdot 7 \cdot 17$	$2^2 \cdot 149$	$3 \cdot 199$	$2 \cdot 13 \cdot 23$	**7774268**
60	$5 \cdot 11^2$	$2 \cdot 3 \cdot 101$	**7831887**	$2^5 \cdot 19$	$3 \cdot 7 \cdot 29$
61	$3 \cdot 5 \cdot 41$	$2^3 \cdot 7 \cdot 11$	**7902852**	$2 \cdot 3 \cdot 103$	**7916906**
62	5^4	$2 \cdot 313$	$3 \cdot 11 \cdot 19$	$2^2 \cdot 157$	$17 \cdot 37$
63	$5 \cdot 127$	$2^2 \cdot 3 \cdot 53$	$7^2 \cdot 13$	$2 \cdot 11 \cdot 29$	$3^2 \cdot 71$
64	$3 \cdot 5 \cdot 43$	$2 \cdot 17 \cdot 19$	**8109043**	$2^3 \cdot 3^4$	$11 \cdot 59$
65	$5 \cdot 131$	$2^4 \cdot 41$	$3^2 \cdot 73$	$2 \cdot 7 \cdot 47$	**8188854**
66	$5 \cdot 7 \cdot 19$	$2 \cdot 3^2 \cdot 37$	$23 \cdot 29$	$2^2 \cdot 167$	$3 \cdot 223$
67	$3^3 \cdot 5^2$	$2^2 \cdot 13^2$	**8305887**	$2 \cdot 3 \cdot 113$	$7 \cdot 97$
68	$5 \cdot 137$	$2 \cdot 7^3$	$3 \cdot 229$	$2^4 \cdot 43$	$13 \cdot 53$
69	$5 \cdot 139$	$2^3 \cdot 3 \cdot 29$	$17 \cdot 41$	$2 \cdot 349$	$3 \cdot 233$
70	$3 \cdot 5 \cdot 47$	$2 \cdot 353$	$7 \cdot 101$	$2^2 \cdot 3 \cdot 59$	**8506462**
71	$5 \cdot 11 \cdot 13$	$2^2 \cdot 179$	$3 \cdot 239$	$2 \cdot 359$	**8567289**
72	$5^2 \cdot 29$	$2 \cdot 3 \cdot 11^2$	**8615344**	$2^3 \cdot 7 \cdot 13$	3^6
73	$3 \cdot 5 \cdot 7^2$	$2^5 \cdot 23$	$11 \cdot 67$	$2 \cdot 3^2 \cdot 41$	**8686444**
74	$5 \cdot 149$	$2 \cdot 373$	$3^2 \cdot 83$	$2^2 \cdot 11 \cdot 17$	$7 \cdot 107$
75	$5 \cdot 151$	$2^2 \cdot 3^3 \cdot 7$	**8790959**	$2 \cdot 379$	$3 \cdot 11 \cdot 23$
76	$3^2 \cdot 5 \cdot 17$	$2 \cdot 383$	$13 \cdot 59$	$2^8 \cdot 3$	**8859263**
77	$5^2 \cdot 31$	$2^3 \cdot 97$	$3 \cdot 7 \cdot 37$	$2 \cdot 389$	$19 \cdot 41$
78	$5 \cdot 157$	$2 \cdot 3 \cdot 131$	**8959747**	$2^2 \cdot 197$	$3 \cdot 263$
79	$3 \cdot 5 \cdot 53$	$2^2 \cdot 199$	**9014583**	$2 \cdot 3 \cdot 7 \cdot 19$	$17 \cdot 47$
80	$5 \cdot 7 \cdot 23$	$2 \cdot 13 \cdot 31$	$3 \cdot 269$	$2^3 \cdot 101$	**9079485**
81	$5 \cdot 163$	$2^4 \cdot 3 \cdot 17$	$19 \cdot 43$	$2 \cdot 409$	$3^2 \cdot 7 \cdot 13$
82	$3 \cdot 5^2 \cdot 11$	$2 \cdot 7 \cdot 59$	**9175055**	$2^2 \cdot 3^2 \cdot 23$	**9185545**
83	$5 \cdot 167$	$2^2 \cdot 11 \cdot 19$	$3^3 \cdot 31$	$2 \cdot 419$	**9237620**
84	$5 \cdot 13^2$	$2 \cdot 3^2 \cdot 47$	$7 \cdot 11^2$	$2^4 \cdot 53$	$3 \cdot 283$
85	$3^2 \cdot 5 \cdot 19$	$2^3 \cdot 107$	**9329808**	$2 \cdot 3 \cdot 11 \cdot 13$	**9339932**
86	$5 \cdot 173$	$2 \cdot 433$	$3 \cdot 17^2$	$2^2 \cdot 7 \cdot 31$	$11 \cdot 79$
87	$5^3 \cdot 7$	$2^2 \cdot 3 \cdot 73$	**9429996**	$2 \cdot 439$	$3 \cdot 293$
88	$3 \cdot 5 \cdot 59$	$2 \cdot 443$	**9479236**	$2^3 \cdot 3 \cdot 37$	$7 \cdot 127$
89	$5 \cdot 179$	$2^7 \cdot 7$	$3 \cdot 13 \cdot 23$	$2 \cdot 449$	$29 \cdot 31$
90	$5 \cdot 181$	$2 \cdot 3 \cdot 151$	**9576073**	$2^2 \cdot 227$	$3^2 \cdot 101$
91	$3 \cdot 5 \cdot 61$	$2^2 \cdot 229$	$7 \cdot 131$	$2 \cdot 3^3 \cdot 17$	**9633155**
92	$5^2 \cdot 37$	$2 \cdot 463$	$3^2 \cdot 103$	$2^5 \cdot 29$	**9680157**
93	$5 \cdot 11 \cdot 17$	$2^3 \cdot 3^2 \cdot 13$	**9717396**	$2 \cdot 7 \cdot 67$	$3 \cdot 313$
94	$3^3 \cdot 5 \cdot 7$	$2 \cdot 11 \cdot 43$	**9763500**	$2^2 \cdot 3 \cdot 79$	$13 \cdot 73$
95	$5 \cdot 191$	$2^2 \cdot 239$	$3 \cdot 11 \cdot 29$	$2 \cdot 479$	$7 \cdot 137$
96	$5 \cdot 193$	$2 \cdot 3 \cdot 7 \cdot 23$	**9854265**	$2^3 \cdot 11^2$	$3 \cdot 17 \cdot 19$
97	$3 \cdot 5^2 \cdot 13$	$2^4 \cdot 61$	**9898946**	$2 \cdot 3 \cdot 163$	$11 \cdot 89$
98	$5 \cdot 197$	$2 \cdot 17 \cdot 29$	$3 \cdot 7 \cdot 47$	$2^2 \cdot 13 \cdot 19$	$23 \cdot 43$
99	$5 \cdot 199$	$2^2 \cdot 3 \cdot 83$	**9986952**	$2 \cdot 499$	$3^3 \cdot 37$
100	$3 \cdot 5 \cdot 67$	$2 \cdot 503$	$19 \cdot 53$	$2^4 \cdot 3^2 \cdot 7$	**0038912**

FACTORS AND PRIMES (Continued)

n	0	1	2	3	4
100	$2^3 \cdot 5^3$	$7 \cdot 11 \cdot 13$	$2 \cdot 3 \cdot 167$	$17 \cdot 59$	$2^2 \cdot 251$
101	$2 \cdot 5 \cdot 101$	$3 \cdot 337$	$2^2 \cdot 11 \cdot 23$	**0056094**	$2 \cdot 3 \cdot 13^2$
102	$2^2 \cdot 3 \cdot 5 \cdot 17$	**0090257**	$2 \cdot 7 \cdot 73$	$3 \cdot 11 \cdot 31$	2^{10}
103	$2 \cdot 5 \cdot 103$	**0132587**	$2^3 \cdot 3 \cdot 43$	**0141003**	$2 \cdot 11 \cdot 47$
104	$2^4 \cdot 5 \cdot 13$	$3 \cdot 347$	$2 \cdot 521$	$7 \cdot 149$	$2^2 \cdot 3^2 \cdot 29$
105	$2 \cdot 3 \cdot 5^2 \cdot 7$	**0216027**	$2^2 \cdot 263$	$3^4 \cdot 13$	$2 \cdot 17 \cdot 31$
106	$2^2 \cdot 5 \cdot 53$	**0257154**	$2 \cdot 3^2 \cdot 59$	**0265333**	$2^3 \cdot 7 \cdot 19$
107	$2 \cdot 5 \cdot 107$	$3^2 \cdot 7 \cdot 17$	$2^4 \cdot 67$	$29 \cdot 37$	$2 \cdot 3 \cdot 179$
108	$2^3 \cdot 3^3 \cdot 5$	$23 \cdot 47$	$2 \cdot 541$	$3 \cdot 19^2$	$2^2 \cdot 271$
109	$2 \cdot 5 \cdot 109$	**0378248**	$2^2 \cdot 3 \cdot 7 \cdot 13$	**0386202**	$2 \cdot 547$
110	$2^2 \cdot 5^2 \cdot 11$	$3 \cdot 367$	$2 \cdot 19 \cdot 29$	**0425755**	$2^4 \cdot 3 \cdot 23$
111	$2 \cdot 3 \cdot 5 \cdot 37$	$11 \cdot 101$	$2^3 \cdot 139$	$3 \cdot 7 \cdot 53$	$2 \cdot 557$
112	$2^5 \cdot 5 \cdot 7$	$19 \cdot 59$	$2 \cdot 3 \cdot 11 \cdot 17$	**0503798**	$2^2 \cdot 281$
113	$2 \cdot 5 \cdot 113$	$3 \cdot 13 \cdot 29$	$2^2 \cdot 283$	$11 \cdot 103$	$2 \cdot 3^4 \cdot 7$
114	$2^2 \cdot 3 \cdot 5 \cdot 19$	$7 \cdot 163$	$2 \cdot 571$	$3^2 \cdot 127$	$2^3 \cdot 11 \cdot 13$
115	$2 \cdot 5^2 \cdot 23$	**0610753**	$2^7 \cdot 3^2$	**0618293**	$2 \cdot 577$
116	$2^3 \cdot 5 \cdot 29$	$3^3 \cdot 43$	$2 \cdot 7 \cdot 83$	**0655797**	$2^2 \cdot 3 \cdot 97$
117	$2 \cdot 3^2 \cdot 5 \cdot 13$	**0685569**	$2^2 \cdot 293$	$3 \cdot 17 \cdot 23$	$2 \cdot 587$
118	$2^2 \cdot 5 \cdot 59$	**0722499**	$2 \cdot 3 \cdot 197$	$7 \cdot 13^2$	$2^5 \cdot 37$
119	$2 \cdot 5 \cdot 7 \cdot 17$	$3 \cdot 397$	$2^3 \cdot 149$	**0766404**	$2 \cdot 3 \cdot 199$
120	$2^4 \cdot 3 \cdot 5^2$	**0795430**	$2 \cdot 601$	$3 \cdot 401$	$2^2 \cdot 7 \cdot 43$
121	$2 \cdot 5 \cdot 11^2$	$7 \cdot 173$	$2^2 \cdot 3 \cdot 101$	**0838608**	$2 \cdot 607$
122	$2^2 \cdot 5 \cdot 61$	$3 \cdot 11 \cdot 37$	$2 \cdot 13 \cdot 47$	**0874265**	$2^3 \cdot 3^2 \cdot 17$
123	$2 \cdot 3 \cdot 5 \cdot 41$	**0902581**	$2^4 \cdot 7 \cdot 11$	$3^2 \cdot 137$	$2 \cdot 617$
124	$2^3 \cdot 5 \cdot 31$	$17 \cdot 73$	$2 \cdot 3^3 \cdot 23$	$11 \cdot 113$	$2^2 \cdot 311$
125	$2 \cdot 5^4$	$3^2 \cdot 139$	$2^2 \cdot 313$	$7 \cdot 179$	$2 \cdot 3 \cdot 11 \cdot 19$
126	$2^2 \cdot 3^2 \cdot 5 \cdot 7$	$13 \cdot 97$	$2 \cdot 631$	$3 \cdot 421$	$2^4 \cdot 79$
127	$2 \cdot 5 \cdot 127$	$31 \cdot 41$	$2^3 \cdot 3 \cdot 53$	$19 \cdot 67$	$2 \cdot 7^2 \cdot 13$
128	$2^8 \cdot 5$	$3 \cdot 7 \cdot 61$	$2 \cdot 641$	**1082267**	$2^2 \cdot 3 \cdot 107$
129	$2 \cdot 3 \cdot 5 \cdot 43$	**1109262**	$2^2 \cdot 17 \cdot 19$	$3 \cdot 431$	$2 \cdot 647$
130	$2^2 \cdot 5^2 \cdot 13$	**1142773**	$2 \cdot 3 \cdot 7 \cdot 31$	**1149444**	$2^3 \cdot 163$
131	$2 \cdot 5 \cdot 131$	$3 \cdot 19 \cdot 23$	$2^5 \cdot 41$	$13 \cdot 101$	$2 \cdot 3^2 \cdot 73$
132	$2^3 \cdot 3 \cdot 5 \cdot 11$	**1209028**	$2 \cdot 661$	$3^3 \cdot 7^2$	$2^2 \cdot 331$
133	$2 \cdot 5 \cdot 7 \cdot 19$	11^3	$2^2 \cdot 3^2 \cdot 37$	$31 \cdot 43$	$2 \cdot 23 \cdot 29$
134	$2^2 \cdot 5 \cdot 67$	$3^2 \cdot 149$	$2 \cdot 11 \cdot 61$	$17 \cdot 79$	$2^6 \cdot 3 \cdot 7$
135	$2 \cdot 3^3 \cdot 5^2$	$7 \cdot 193$	$2^3 \cdot 13^2$	$3 \cdot 11 \cdot 41$	$2 \cdot 677$
136	$2^4 \cdot 5 \cdot 17$	**1338581**	$2 \cdot 3 \cdot 227$	$29 \cdot 47$	$2^2 \cdot 11 \cdot 31$
137	$2 \cdot 5 \cdot 137$	$3 \cdot 457$	$2^2 \cdot 7^3$	**1376705**	$2 \cdot 3 \cdot 229$
138	$2^2 \cdot 3 \cdot 5 \cdot 23$	**1401937**	$2 \cdot 691$	$3 \cdot 461$	$2^3 \cdot 173$
139	$2 \cdot 5 \cdot 139$	$13 \cdot 107$	$2^4 \cdot 3 \cdot 29$	$7 \cdot 199$	$2 \cdot 17 \cdot 41$
140	$2^3 \cdot 5^2 \cdot 7$	$3 \cdot 467$	$2 \cdot 701$	$23 \cdot 61$	$2^2 \cdot 3^3 \cdot 13$
141	$2 \cdot 3 \cdot 5 \cdot 47$	$17 \cdot 83$	$2^2 \cdot 353$	$3^2 \cdot 157$	$2 \cdot 7 \cdot 101$
142	$2^2 \cdot 5 \cdot 71$	$7^2 \cdot 29$	$2 \cdot 3^2 \cdot 79$	**1532049**	$2^4 \cdot 89$
143	$2 \cdot 5 \cdot 11 \cdot 13$	$3^3 \cdot 53$	$2^3 \cdot 179$	**1562462**	$2 \cdot 3 \cdot 239$
144	$2^5 \cdot 3^2 \cdot 5$	$11 \cdot 131$	$2 \cdot 7 \cdot 103$	$3 \cdot 13 \cdot 37$	$2^2 \cdot 19^2$
145	$2 \cdot 5^2 \cdot 29$	**1616674**	$2^2 \cdot 3 \cdot 11^2$	**1622656**	$2 \cdot 727$
146	$2^2 \cdot 5 \cdot 73$	$3 \cdot 487$	$2 \cdot 17 \cdot 43$	$7 \cdot 11 \cdot 19$	$2^3 \cdot 3 \cdot 61$
147	$2 \cdot 3 \cdot 5 \cdot 7^2$	**1676127**	$2^6 \cdot 23$	$3 \cdot 491$	$2 \cdot 11 \cdot 67$
148	$2^3 \cdot 5 \cdot 37$	**1705551**	$2 \cdot 3 \cdot 13 \cdot 19$	**1711412**	$2^2 \cdot 7 \cdot 53$
149	$2 \cdot 5 \cdot 149$	$3 \cdot 7 \cdot 71$	$2^2 \cdot 373$	**1740598**	$2 \cdot 3^2 \cdot 83$
150	$2^2 \cdot 3 \cdot 5^3$	$19 \cdot 79$	$2 \cdot 751$	$3^2 \cdot 167$	$2^5 \cdot 47$

FACTORS AND PRIMES (Continued)

n	5	6	7	8	9
100	3 · 5 · 67	2 · 503	19 · 53	$2^4 \cdot 3^2 \cdot 7$	**0038912**
101	5 · 7 · 29	$2^3 \cdot 127$	$3^2 \cdot 113$	2 · 509	**0081742**
102	$5^2 \cdot 41$	$2 \cdot 3^3 \cdot 19$	13 · 79	$2^2 \cdot 257$	$3 \cdot 7^3$
103	$3^2 \cdot 5 \cdot 23$	$2^2 \cdot 7 \cdot 37$	17 · 61	2 · 3 · 173	**0166155**
104	5 · 11 · 19	2 · 523	3 · 349	$2^3 \cdot 131$	**0207755**
105	5 · 211	$2^5 \cdot 3 \cdot 11$	7 · 151	$2 \cdot 23^2$	3 · 353
106	3 · 5 · 71	2 · 13 · 41	11 · 97	$2^2 \cdot 3 \cdot 89$	**0289777**
107	$5^2 \cdot 43$	$2^2 \cdot 269$	3 · 359	$2 \cdot 7^2 \cdot 11$	13 · 83
108	5 · 7 · 31	2 · 3 · 181	**0362295**	$2^6 \cdot 17$	$3^2 \cdot 11^2$
109	3 · 5 · 73	$2^3 \cdot 137$	**0402066**	$2 \cdot 3^2 \cdot 61$	7 · 157
110	5 · 13 · 17	2 · 7 · 79	$3^3 \cdot 41$	$2^2 \cdot 277$	**0449315**
111	5 · 223	$2^2 \cdot 3^2 \cdot 31$	**0480532**	2 · 13 · 43	3 · 373
112	$3^2 \cdot 5^3$	2 · 563	$7^2 \cdot 23$	$2^3 \cdot 3 \cdot 47$	**0526939**
113	5 · 227	$2^4 \cdot 71$	3 · 379	2 · 569	17 · 67
114	5 · 229	2 · 3 · 191	31 · 37	$2^2 \cdot 7 \cdot 41$	3 · 383
115	3 · 5 · 7 · 11	$2^2 \cdot 17^2$	13 · 89	2 · 3 · 193	19 · 61
116	5 · 233	2 · 11 · 53	3 · 389	$2^4 \cdot 73$	7 · 167
117	$5^2 \cdot 47$	$2^3 \cdot 3 \cdot 7^2$	11 · 107	2 · 19 · 31	$3^2 \cdot 131$
118	3 · 5 · 79	2 · 593	**0744507**	$2^2 \cdot 3^3 \cdot 11$	29 · 41
119	5 · 239	$2^3 \cdot 13 \cdot 23$	$3^2 \cdot 7 \cdot 19$	2 · 599	11 · 109
120	5 · 241	$2 \cdot 3^2 \cdot 67$	17 · 71	$2^3 \cdot 51$	3 · 13 · 31
121	$3^5 \cdot 5$	$2^6 \cdot 19$	**0852906**	2 · 3 · 7 · 29	23 · 53
122	$5^2 \cdot 7^2$	2 · 613	3 · 409	$2^2 \cdot 307$	**0895519**
123	5 · 13 · 19	$2^2 \cdot 3 \cdot 103$	**0923697**	2 · 619	3 · 7 · 59
124	3 · 5 · 83	2 · 7 · 89	29 · 43	$2^5 \cdot 3 \cdot 13$	**0965624**
125	5 · 251	$2^3 \cdot 157$	3 · 419	2 · 17 · 37	**1000257**
126	5 · 11 · 23	2 · 3 · 211	7 · 181	$2^2 \cdot 317$	$3^3 \cdot 47$
127	$3 \cdot 5^2 \cdot 17$	$2^2 \cdot 11 \cdot 29$	**1061909**	$2 \cdot 3^2 \cdot 71$	**1068705**
128	5 · 257	2 · 643	$3^2 \cdot 11 \cdot 13$	$2^3 \cdot 7 \cdot 23$	**1102529**
129	5 · 7 · 37	$2^4 \cdot 3^4$	**1129400**	2 · 11 · 59	3 · 433
130	$3^2 \cdot 5 \cdot 29$	2 · 653	**1162756**	$2^2 \cdot 3 \cdot 109$	7 · 11 · 17
131	5 · 263	$2^2 \cdot 7 \cdot 47$	3 · 439	2 · 659	**1202448**
132	$5^2 \cdot 53$	2 · 3 · 13 · 17	**1228709**	$2^4 \cdot 83$	3 · 443
133	3 · 5 · 89	$2^3 \cdot 167$	7 · 191	2 · 3 · 223	13 · 103
134	5 · 269	2 · 673	3 · 449	$2^2 \cdot 337$	19 · 71
135	5 · 271	$2^2 \cdot 3 \cdot 113$	23 · 59	2 · 7 · 97	$3^2 \cdot 151$
136	3 · 5 · 7 · 13	2 · 683	**1357685**	$2^3 \cdot 3^2 \cdot 19$	37^2
137	$5^3 \cdot 11$	$2^5 \cdot 43$	$3^4 \cdot 17$	2 · 13 · 53	7 · 197
138	5 · 277	$2 \cdot 3^2 \cdot 7 \cdot 11$	19 · 73	$2^2 \cdot 347$	3 · 463
139	$3^2 \cdot 5 \cdot 31$	$2^2 \cdot 349$	11 · 127	2 · 3 · 233	**1458177**
140	5 · 281	2 · 19 · 37	3 · 7 · 67	$2^7 \cdot 11$	**1489110**
141	5 · 283	$2^3 \cdot 3 \cdot 59$	13 · 109	2 · 709	3 · 11 · 43
142	$3 \cdot 5^2 \cdot 19$	2 · 23 · 31	**1544240**	$2^2 \cdot 3 \cdot 7 \cdot 17$	**1550322**
143	5 · 7 · 41	$2^2 \cdot 359$	3 · 479	2 · 719	**1580608**
144	$5 \cdot 17^2$	2 · 3 · 241	**1604685**	$2^3 \cdot 181$	$3^2 \cdot 7 \cdot 23$
145	3 · 5 · 97	$2^4 \cdot 7 \cdot 13$	31 · 47	$2 \cdot 3^6$	**1640553**
146	5 · 293	2 · 733	$3^2 \cdot 163$	$2^2 \cdot 367$	13 · 113
147	$5^2 \cdot 59$	$2^2 \cdot 3^2 \cdot 41$	7 · 211	2 · 739	3 · 17 · 29
148	$3^3 \cdot 5 \cdot 11$	2 · 743	**1723110**	$2^4 \cdot 3 \cdot 31$	**1728947**
149	5 · 13 · 23	$2^3 \cdot 11 \cdot 17$	3 · 499	2 · 7 · 107	**1758016**
150	5 · 7 · 43	2 · 3 · 251	11 · 137	$2^2 \cdot 13 \cdot 29$	3 · 503

FACTORS AND PRIMES (Continued)

n	0	1	2	3	4
150	$2^2 \cdot 3 \cdot 5^3$	$19 \cdot 79$	$2 \cdot 751$	$3^2 \cdot 167$	$2^5 \cdot 47$
151	$2 \cdot 5 \cdot 151$	**1792645**	$2^3 \cdot 3^3 \cdot 7$	$17 \cdot 89$	$2 \cdot 757$
152	$2^4 \cdot 5 \cdot 19$	$3^2 \cdot 13^2$	$2 \cdot 761$	**1826999**	$2^2 \cdot 3 \cdot 127$
153	$2 \cdot 3^2 \cdot 5 \cdot 17$	**1849752**	$2^2 \cdot 383$	$3 \cdot 7 \cdot 73$	$2 \cdot 13 \cdot 59$
154	$2^2 \cdot 5 \cdot 7 \cdot 11$	$23 \cdot 67$	$2 \cdot 3 \cdot 257$	**1883659**	$2^3 \cdot 193$
155	$2 \cdot 5^2 \cdot 31$	$3 \cdot 11 \cdot 47$	$2^4 \cdot 97$	**1911715**	$2 \cdot 3 \cdot 7 \cdot 37$
156	$2^3 \cdot 3 \cdot 5 \cdot 13$	$7 \cdot 223$	$2 \cdot 11 \cdot 71$	$3 \cdot 521$	$2^2 \cdot 17 \cdot 23$
157	$2 \cdot 5 \cdot 157$	**1961762**	$2^2 \cdot 3 \cdot 131$	$11^2 \cdot 13$	$2 \cdot 787$
158	$2^2 \cdot 5 \cdot 79$	$3 \cdot 17 \cdot 31$	$2 \cdot 7 \cdot 113$	**1994809**	$2^4 \cdot 3^2 \cdot 11$
159	$2 \cdot 3 \cdot 5 \cdot 53$	$37 \cdot 43$	$2^3 \cdot 199$	$3^3 \cdot 59$	$2 \cdot 797$
160	$2^6 \cdot 5^2$	**2043913**	$2 \cdot 3^2 \cdot 89$	$7 \cdot 229$	$2^2 \cdot 401$
161	$2 \cdot 5 \cdot 7 \cdot 23$	$3^2 \cdot 179$	$2^2 \cdot 13 \cdot 31$	**2076344**	$2 \cdot 3 \cdot 269$
162	$2^2 \cdot 3^4 \cdot 5$	**2097830**	$2 \cdot 811$	$3 \cdot 541$	$2^3 \cdot 7 \cdot 29$
163	$2 \cdot 5 \cdot 163$	$7 \cdot 233$	$2^5 \cdot 3 \cdot 17$	$23 \cdot 71$	$2 \cdot 19 \cdot 43$
164	$2^3 \cdot 5 \cdot 41$	$3 \cdot 547$	$2 \cdot 821$	$31 \cdot 53$	$2^2 \cdot 3 \cdot 137$
165	$2 \cdot 3 \cdot 5^2 \cdot 11$	$13 \cdot 127$	$2^2 \cdot 7 \cdot 59$	$3 \cdot 19 \cdot 29$	$2 \cdot 827$
166	$2^2 \cdot 5 \cdot 83$	$11 \cdot 151$	$2 \cdot 3 \cdot 277$	**2208922**	$2^7 \cdot 13$
167	$2 \cdot 5 \cdot 167$	$3 \cdot 557$	$2^3 \cdot 11 \cdot 19$	$7 \cdot 239$	$2 \cdot 3^3 \cdot 31$
168	$2^4 \cdot 3 \cdot 5 \cdot 7$	41^2	$2 \cdot 29^2$	$3^2 \cdot 11 \cdot 17$	$2^2 \cdot 421$
169	$2 \cdot 5 \cdot 13^2$	$19 \cdot 89$	$2^2 \cdot 3^2 \cdot 47$	**2286570**	$2 \cdot 7 \cdot 11^2$
170	$2^2 \cdot 5^2 \cdot 17$	$3^5 \cdot 7$	$2 \cdot 23 \cdot 37$	$13 \cdot 131$	$2^3 \cdot 3 \cdot 71$
171	$2 \cdot 3^2 \cdot 5 \cdot 19$	$29 \cdot 59$	$2^4 \cdot 107$	$3 \cdot 571$	$2 \cdot 857$
172	$2^3 \cdot 5 \cdot 43$	**2357809**	$2 \cdot 3 \cdot 7 \cdot 41$	**2362853**	$2^2 \cdot 431$
173	$2 \cdot 5 \cdot 173$	$3 \cdot 577$	$2^2 \cdot 433$	**2387986**	$2 \cdot 3 \cdot 17^2$
174	$2^2 \cdot 3 \cdot 5 \cdot 29$	**2407988**	$2 \cdot 13 \cdot 67$	$3 \cdot 7 \cdot 83$	$2^4 \cdot 109$
175	$2 \cdot 5^3 \cdot 7$	$17 \cdot 103$	$2^3 \cdot 3 \cdot 73$	**2437819**	$2 \cdot 877$
176	$2^5 \cdot 5 \cdot 11$	$3 \cdot 587$	$2 \cdot 881$	$41 \cdot 43$	$2^2 \cdot 3^2 \cdot 7^2$
177	$2 \cdot 3 \cdot 5 \cdot 59$	$7 \cdot 11 \cdot 23$	$2^2 \cdot 443$	$3^2 \cdot 197$	$2 \cdot 887$
178	$2^2 \cdot 5 \cdot 89$	$13 \cdot 137$	$2 \cdot 3^4 \cdot 11$	**2511513**	$2^3 \cdot 223$
179	$2 \cdot 5 \cdot 179$	$3^2 \cdot 199$	$2^8 \cdot 7$	$11 \cdot 163$	$2 \cdot 3 \cdot 13 \cdot 23$
180	$2^3 \cdot 3^2 \cdot 5^2$	**2555137**	$2 \cdot 17 \cdot 53$	$3 \cdot 601$	$2^2 \cdot 11 \cdot 41$
181	$2 \cdot 5 \cdot 181$	**2579185**	$2^2 \cdot 3 \cdot 151$	$7^2 \cdot 37$	$2 \cdot 907$
182	$2^2 \cdot 5 \cdot 7 \cdot 13$	$3 \cdot 607$	$2 \cdot 911$	**2607867**	$2^5 \cdot 3 \cdot 19$
183	$2 \cdot 3 \cdot 5 \cdot 61$	**2626883**	$2^3 \cdot 229$	$3 \cdot 13 \cdot 47$	$2 \cdot 7 \cdot 131$
184	$2^4 \cdot 5 \cdot 23$	$7 \cdot 263$	$2 \cdot 3 \cdot 307$	$19 \cdot 97$	$2^2 \cdot 461$
185	$2 \cdot 5^2 \cdot 37$	$3 \cdot 617$	$2^2 \cdot 463$	$17 \cdot 109$	$2 \cdot 3^2 \cdot 103$
186	$2^2 \cdot 3 \cdot 5 \cdot 31$	**2697464**	$2 \cdot 7^2 \cdot 19$	$3^4 \cdot 23$	$2^3 \cdot 233$
187	$2 \cdot 5 \cdot 11 \cdot 17$	**2720738**	$2^4 \cdot 3^2 \cdot 13$	**2725378**	$2 \cdot 937$
188	$2^3 \cdot 5 \cdot 47$	$3^2 \cdot 11 \cdot 19$	$2 \cdot 941$	$7 \cdot 269$	$2^2 \cdot 3 \cdot 157$
189	$2 \cdot 3^3 \cdot 5 \cdot 7$	$31 \cdot 61$	$2^2 \cdot 11 \cdot 43$	$3 \cdot 631$	$2 \cdot 947$
190	$2^2 \cdot 5^2 \cdot 19$	**2789821**	$2 \cdot 3 \cdot 317$	$11 \cdot 173$	$2^4 \cdot 7 \cdot 17$
191	$2 \cdot 5 \cdot 191$	$3 \cdot 7^2 \cdot 13$	$2^3 \cdot 239$	**2817150**	$2 \cdot 3 \cdot 11 \cdot 29$
192	$2^7 \cdot 3 \cdot 5$	$17 \cdot 113$	$2 \cdot 31^2$	$3 \cdot 641$	$2^2 \cdot 13 \cdot 37$
193	$2 \cdot 5 \cdot 193$	**2857823**	$2^2 \cdot 3 \cdot 7 \cdot 23$	**2862319**	$2 \cdot 967$
194	$2^2 \cdot 5 \cdot 97$	$3 \cdot 647$	$2 \cdot 971$	$29 \cdot 67$	$2^3 \cdot 3^5$
195	$2 \cdot 3 \cdot 5^2 \cdot 13$	**2902573**	$2^5 \cdot 61$	$3^2 \cdot 7 \cdot 31$	$2 \cdot 977$
196	$2^3 \cdot 5 \cdot 7^2$	$37 \cdot 53$	$2 \cdot 3^2 \cdot 109$	$13 \cdot 151$	$2^2 \cdot 491$
197	$2 \cdot 5 \cdot 197$	$3^3 \cdot 73$	$2^2 \cdot 17 \cdot 29$	**2951271**	$2 \cdot 3 \cdot 7 \cdot 47$
198	$2^2 \cdot 3^2 \cdot 5 \cdot 11$	$7 \cdot 283$	$2 \cdot 991$	$3 \cdot 661$	$2^6 \cdot 31$
199	$2 \cdot 5 \cdot 199$	$11 \cdot 181$	$2^3 \cdot 3 \cdot 83$	**2995073**	$2 \cdot 997$
200	$2^4 \cdot 5^3$	$3 \cdot 23 \cdot 29$	$2 \cdot 7 \cdot 11 \cdot 13$	**3016809**	$2^2 \cdot 3 \cdot 167$

FACTORS AND PRIMES (Continued)

n	5	6	7	8	9
150	$5 \cdot 7 \cdot 43$	$2 \cdot 3 \cdot 251$	$11 \cdot 137$	$2^2 \cdot 13 \cdot 29$	$3 \cdot 503$
151	$3 \cdot 5 \cdot 101$	$2^2 \cdot 379$	$37 \cdot 41$	$2 \cdot 3 \cdot 11 \cdot 23$	$7^2 \cdot 31$
152	$5^2 \cdot 61$	$2 \cdot 7 \cdot 109$	$3 \cdot 509$	$2^3 \cdot 191$	$11 \cdot 139$
153	$5 \cdot 307$	$2^9 \cdot 3$	$29 \cdot 53$	$2 \cdot 769$	$3^4 \cdot 19$
154	$3 \cdot 5 \cdot 103$	$2 \cdot 773$	$7 \cdot 13 \cdot 17$	$2^2 \cdot 3^2 \cdot 43$	**1900514**
155	$5 \cdot 311$	$2^2 \cdot 389$	$3^2 \cdot 173$	$2 \cdot 19 \cdot 41$	**1928461**
156	$5 \cdot 313$	$2 \cdot 3^3 \cdot 29$	**1950690**	$2^5 \cdot 7^2$	$3 \cdot 523$
157	$3^2 \cdot 5^2 \cdot 7$	$2^3 \cdot 197$	$19 \cdot 83$	$2 \cdot 3 \cdot 263$	**1983821**
158	$5 \cdot 317$	$2 \cdot 13 \cdot 61$	$3 \cdot 23^2$	$2^2 \cdot 397$	$7 \cdot 227$
159	$5 \cdot 11 \cdot 29$	$2^2 \cdot 3 \cdot 7 \cdot 19$	**2033049**	$2 \cdot 17 \cdot 47$	$3 \cdot 13 \cdot 41$
160	$3 \cdot 5 \cdot 107$	$2 \cdot 11 \cdot 73$	**2060159**	$2^3 \cdot 3 \cdot 67$	**2065560**
161	$5 \cdot 17 \cdot 19$	$2^4 \cdot 101$	$3 \cdot 7^2 \cdot 11$	$2 \cdot 809$	**2092468**
162	$5^3 \cdot 13$	$2 \cdot 3 \cdot 271$	**2113876**	$2^2 \cdot 11 \cdot 37$	$3^2 \cdot 181$
163	$3 \cdot 5 \cdot 109$	$2^2 \cdot 409$	**2140487**	$2 \cdot 3^2 \cdot 7 \cdot 13$	$11 \cdot 149$
164	$5 \cdot 7 \cdot 47$	$2 \cdot 823$	$3^3 \cdot 61$	$2^4 \cdot 103$	$17 \cdot 97$
165	$5 \cdot 331$	$2^3 \cdot 3^2 \cdot 23$	**2193225**	$2 \cdot 829$	$3 \cdot 7 \cdot 79$
166	$3^2 \cdot 5 \cdot 37$	$2 \cdot 7^2 \cdot 17$	**2219356**	$2^2 \cdot 3 \cdot 139$	**2224563**
167	$5^2 \cdot 67$	$2^2 \cdot 419$	$3 \cdot 13 \cdot 43$	$2 \cdot 839$	$23 \cdot 73$
168	$5 \cdot 337$	$2 \cdot 3 \cdot 281$	$7 \cdot 241$	$2^3 \cdot 211$	$3 \cdot 563$
169	$3 \cdot 5 \cdot 113$	$2^5 \cdot 53$	**2296818**	$2 \cdot 3 \cdot 283$	**2301934**
170	$5 \cdot 11 \cdot 31$	$2 \cdot 853$	$3 \cdot 569$	$2^2 \cdot 7 \cdot 61$	**2327421**
171	$5 \cdot 7^3$	$2^2 \cdot 3 \cdot 11 \cdot 13$	$17 \cdot 101$	$2 \cdot 859$	$3^2 \cdot 191$
172	$3 \cdot 5^2 \cdot 23$	$2 \cdot 863$	$11 \cdot 157$	$2^6 \cdot 3^3$	$7 \cdot 13 \cdot 19$
173	$5 \cdot 347$	$2^3 \cdot 7 \cdot 31$	$3^2 \cdot 193$	$2 \cdot 11 \cdot 79$	$37 \cdot 47$
174	$5 \cdot 349$	$2 \cdot 3^2 \cdot 97$	**2422929**	$2^2 \cdot 19 \cdot 23$	$3 \cdot 11 \cdot 53$
175	$3^3 \cdot 5 \cdot 13$	$2^2 \cdot 439$	$7 \cdot 251$	$2 \cdot 3 \cdot 293$	**2452658**
176	$5 \cdot 353$	$2 \cdot 883$	$3 \cdot 19 \cdot 31$	$2^3 \cdot 13 \cdot 17$	$29 \cdot 61$
177	$5^2 \cdot 71$	$2^4 \cdot 3 \cdot 37$	**2496874**	$2 \cdot 7 \cdot 127$	$3 \cdot 593$
178	$3 \cdot 5 \cdot 7 \cdot 17$	$2 \cdot 19 \cdot 47$	**2521246**	$2^2 \cdot 3 \cdot 149$	**2526103**
179	$5 \cdot 359$	$2^2 \cdot 449$	$3 \cdot 599$	$2 \cdot 29 \cdot 31$	$7 \cdot 257$
180	$5 \cdot 19^2$	$2 \cdot 3 \cdot 7 \cdot 43$	$13 \cdot 139$	$2^4 \cdot 113$	$3^3 \cdot 67$
181	$3 \cdot 5 \cdot 11^2$	$2^3 \cdot 227$	$23 \cdot 79$	$2 \cdot 3^2 \cdot 101$	$17 \cdot 107$
182	$5^2 \cdot 73$	$2 \cdot 11 \cdot 83$	$3^2 \cdot 7 \cdot 29$	$2^2 \cdot 457$	$31 \cdot 59$
183	$5 \cdot 367$	$2^2 \cdot 3^3 \cdot 17$	$11 \cdot 167$	$2 \cdot 919$	$3 \cdot 613$
184	$3^2 \cdot 5 \cdot 41$	$2 \cdot 13 \cdot 71$	**2664669**	$2^3 \cdot 3 \cdot 7 \cdot 11$	43^2
185	$5 \cdot 7 \cdot 53$	$2^6 \cdot 29$	$3 \cdot 619$	$2 \cdot 929$	$11 \cdot 13^2$
186	$5 \cdot 373$	$2 \cdot 3 \cdot 311$	**2711443**	$2^2 \cdot 467$	$3 \cdot 7 \cdot 89$
187	$3 \cdot 5^4$	$2^2 \cdot 7 \cdot 67$	**2734643**	$2 \cdot 3 \cdot 313$	**2739268**
188	$5 \cdot 13 \cdot 29$	$2 \cdot 23 \cdot 41$	$3 \cdot 17 \cdot 37$	$2^5 \cdot 59$	**2762320**
189	$5 \cdot 379$	$2^3 \cdot 3 \cdot 79$	$7 \cdot 271$	$2 \cdot 13 \cdot 73$	$3^2 \cdot 211$
190	$3 \cdot 5 \cdot 127$	$2 \cdot 953$	**2803507**	$2^2 \cdot 3^2 \cdot 53$	$23 \cdot 83$
191	$5 \cdot 383$	$2^2 \cdot 479$	$3^3 \cdot 71$	$2 \cdot 7 \cdot 137$	$19 \cdot 101$
192	$5^2 \cdot 7 \cdot 11$	$2 \cdot 3^2 \cdot 107$	$41 \cdot 47$	$2^3 \cdot 241$	$3 \cdot 643$
193	$3^2 \cdot 5 \cdot 43$	$2^4 \cdot 11^2$	$13 \cdot 149$	$2 \cdot 3 \cdot 17 \cdot 19$	$7 \cdot 277$
194	$5 \cdot 389$	$2 \cdot 7 \cdot 139$	$3 \cdot 11 \cdot 59$	$2^2 \cdot 487$	**2898118**
195	$5 \cdot 17 \cdot 23$	$2^2 \cdot 3 \cdot 163$	$19 \cdot 103$	$2 \cdot 11 \cdot 89$	$3 \cdot 653$
196	$3 \cdot 5 \cdot 131$	$2 \cdot 983$	$7 \cdot 281$	$2^4 \cdot 3 \cdot 41$	$11 \cdot 179$
197	$5^2 \cdot 79$	$2^3 \cdot 13 \cdot 19$	$3 \cdot 659$	$2 \cdot 23 \cdot 43$	**2964458**
198	$5 \cdot 397$	$2 \cdot 3 \cdot 331$	**2981979**	$2^2 \cdot 7 \cdot 71$	$3^2 \cdot 13 \cdot 17$
199	$3 \cdot 5 \cdot 7 \cdot 19$	$2^2 \cdot 499$	**3003781**	$2 \cdot 3^3 \cdot 37$	**3008128**
200	$5 \cdot 401$	$2 \cdot 17 \cdot 59$	$3^2 \cdot 223$	$2^3 \cdot 251$	$7^2 \cdot 41$

EXTENDED TABLES OF FACTORS AND PRIMES

The following procedure makes possible the determination of a number between 2009 and 19,949 as being either prime, or if not, what its factors will be. It is herewith included with the kind permission of its author, Professor Leonard Caners.

The following symbols will be used:

N is the number whose factors, if any, are to be determined.
A is N with the last digit dropped. Thus if $N = 17{,}873$, $A = 1787$.
R is the range and is the integral part of the square root of N with its last two digits dropped. Thus R of 17,873 = 13.
K is the key number and is found in the table below.
P_1 is a possible factor corresponding to K.
The twin series are given in the table for the sake of completeness but are not written down in actual practice.
There are two steps as outlined below.

Step I

N	K	P_1	Series	Procedure
Any number ending in 1, 3, 7, or 9, within limits stated above	$A - R$	$10R$ + last digit	$K + n$; $P_1 - 10n$ $n = 0, 1, \ldots, 2R + 1$	Beginning with K read table of Factors and Primes from left to right and look for corresponding possible factors

Step II

N ending in	K	P_1	Series	Procedure
1	$(A - 2) - 3R$	$10R + 7$	$K + 3n$; $P_1 - 10n$ $n = 0, 1, \ldots, 2R + 1$	Same as in Step I except that only every *third* entry in table of Factors and Primes is examined for corresponding possible factors
3	$A - 3R$	$10R + 1$	$n = 0, 1, \ldots, 2R + 1$	
7	$(A - 2) - 3R$	$10R + 9$	$n = 0, 1, \ldots, 2R + 1$	
9	$A - 3R$	$10R + 3$	$n = 0, 1, \ldots, 2R + 1$	

As an example consider 7519. Using tables on preceding pages proceed as follows:

Step I

From Table $N = 7519$; $A = 751$; $R = 8$, $K = 743$; $P_1 = 89$ and the series obtained:

743	89
744	79
745	69
746	59
—	—
751	9
752	1
753	11
—	—
760	81

EXTENDED TABLES OF FACTORS AND PRIMES (Continued)

Examining the Table of Factors and Primes beginning with 743 note whether 743 has 89 as a factor; whether 744 has 79 as a factor, etc. Then continue to read from left to right keeping in mind the possible factors 69, 59, . . . 81.

Since none are found one concludes that 7519 has no factors within the given range ending in 9 or 1.

Step II

From Table $K = 727$ and $P_1 = 83$; the series obtained is:

727	83
730	73
733	63
—	—
751	3
754	7
—	—
778	87

Proceed exactly as in Step I. Begin with 727 and note whether it has 83 as a factor. Thereafter examine every *third* entry for the remaining corresponding possible factors 73, 63 . . . 3, 7, . . . 87. Since 730 yields the factor 73 one concludes that 73 is the factor of 7519. By division $7519 = 73 \times 103$. Had no factor been found in either Step I or Step II the conclusion would be that the number under consideration was prime.

Proof for Step I

Let N end in 1. Then, if N has a factor ending in 1, write

$$(10a + 1)(10b + 1) = N = 10A + 1$$
$$10ab + a + b = A$$

or

$$b(10a + 1) = (A - a)$$

This proves that any factor $10a + 1$ of N is also a factor of $A - a$. Let $a = (R - n)$, then $(10a + 1)$ becomes $(10R + 1) - 10n$ and $(A - a)$ becomes $(A - R) + n$, i.e., $P_1 - 10n$ and $K + n$ respectively as given in the table. Since $P_1 - 10n$ becomes a number ending in 9 when $P_1 < 10n$, all possible factors ending in 9 are also provided for.

The proof for Step II is similar. For possible factors ending in 3 or 7 write

$$(10a + 7)(10b + 3) = N = 10A + 1.$$

By identical reasoning this results in $P_1 - 10n$ and $K + 3n$ respectively given in the table. This completes the proof for numbers ending in 1.

Identical reasoning establishes the key numbers and series for ending in 3, 7, or 9.

FACTORIALS, EXACT VALUES

n	n!	n	n!
1	1	11	399 16800
2	2	12	4790 01600
3	6	13	62270 20800
4	24	14	8 71782 91200
5	120	15	130 76743 68000
6	720	16	2092 27898 88000
7	5040	17	35568 74280 96000
8	40320	18	6 40237 37057 28000
9	3 62880	19	121 64510 04088 32000
10	36 28800	20	2432 90200 81766 40000

FACTORIALS AND THEIR COMMON LOGARITHMS

The product $n \times (n - 1) \times (n - 2) \times \cdots \times 1$ is called factorial n, expressed as $n!$ For example: *factorial* $5 = 5! = 5 \times 4 \times 3 \times 2 \times 1 = 120$. Factorials are very often met with in series. For purposes of computation in such cases the table giving the values of the factorials and of their logarithms for numbers from 1 to 100 is provided. The values of the factorials are expressed exponentially to 5 significant figures.

For large values of n, *Stirling's formula* is

$$e^{-n}n^n \sqrt{2\pi n} < n! < e^{-n}n^n \sqrt{2\pi n}\left(1 + \frac{1}{12n - 1}\right).$$

This gives the following approximate formula for calculating factorials of large integers:

$$n! \approx e^{-n}n^n \sqrt{2\pi n}$$

and

$$\log n! \approx \left(n + \frac{1}{2}\right) \log n - n \log e + \log \sqrt{2\pi}.$$

The following table gives the reciprocals of the factorials and their logarithms for numbers from 1 to 100.

Above is a short table giving the exact values of the factorials of the numbers from 1 to 20.

FACTORIALS AND THEIR COMMON LOGARITHMS

n	$n!$	$\log n!$	n	$n!$	$\log n!$
			50	3.0414×10^{64}	64.48307
1	1.0000	0.00000	51	1.5511×10^{66}	66.19065
2	2.0000	0.30103	52	8.0658×10^{67}	67.90665
3	6.0000	0.77815	53	4.2749×10^{69}	69.63092
4	2.4000×10	1.38021	54	2.3084×10^{71}	71.36332
5	1.2000×10^{2}	2.07918	**55**	1.2696×10^{73}	73.10368
6	7.2000×10^{2}	2.85733	56	7.1100×10^{74}	74.85187
7	5.0400×10^{3}	3.70243	57	4.0527×10^{76}	76.60774
8	4.0320×10^{4}	4.60552	58	2.3506×10^{78}	78.37117
9	3.6288×10^{5}	5.55976	59	1.3868×10^{80}	80.14202
10	3.6288×10^{6}	6.55976	**60**	8.3210×10^{81}	81.92017
11	3.9917×10^{7}	7.60116	61	5.0758×10^{83}	83.70550
12	4.7900×10^{8}	8.68034	62	3.1470×10^{85}	85.49790
13	6.2270×10^{9}	9.79428	63	1.9826×10^{87}	87.29724
14	8.7178×10^{10}	10.94041	64	1.2689×10^{89}	89.10342
15	1.3077×10^{12}	12.11650	**65**	8.2477×10^{90}	90.91633
16	2.0923×10^{13}	13.32062	66	5.4434×10^{92}	92.73587
17	3.5569×10^{14}	14.55107	67	3.6471×10^{94}	94.56195
18	6.4024×10^{15}	15.80634	68	2.4800×10^{96}	96.39446
19	1.2165×10^{17}	17.08509	69	1.7112×10^{98}	98.23331
20	2.4329×10^{18}	18.38612	**70**	1.1979×10^{100}	100.07841
21	5.1091×10^{19}	19.70834	71	8.5048×10^{101}	101.92966
22	1.1240×10^{21}	21.05077	72	6.1234×10^{103}	103.78700
23	2.5852×10^{22}	22.41249	73	4.4701×10^{105}	105.65032
24	6.2045×10^{23}	23.79271	74	3.3079×10^{107}	107.51955
25	1.5511×10^{25}	25.19065	**75**	2.4809×10^{109}	109.39461
26	4.0329×10^{26}	26.60562	76	1.8855×10^{111}	111.27543
27	1.0889×10^{28}	28.03698	77	1.4518×10^{113}	113.16192
28	3.0489×10^{29}	29.48414	78	1.1324×10^{115}	115.05401
29	8.8418×10^{30}	30.94654	79	8.9462×10^{116}	116.95164
30	2.6525×10^{32}	32.42366	**80**	7.1569×10^{118}	118.85473
31	8.2228×10^{33}	33.91502	81	5.7971×10^{120}	120.76321
32	2.6313×10^{35}	35.42017	82	4.7536×10^{122}	122.67703
33	8.6833×10^{36}	36.93869	83	3.9455×10^{124}	124.59610
34	2.9523×10^{38}	38.47016	84	3.3142×10^{126}	126.52038
35	1.0333×10^{40}	40.01423	**85**	2.8171×10^{128}	128.44980
36	3.7199×10^{41}	41.57054	86	2.4227×10^{130}	130.38430
37	1.3764×10^{43}	43.13874	87	2.1078×10^{132}	132.32382
38	5.2302×10^{44}	44.71852	88	1.8548×10^{134}	134.26830
39	2.0398×10^{46}	46.30959	89	1.6508×10^{136}	136.21769
40	8.1592×10^{47}	47.91165	**90**	1.4857×10^{138}	138.17194
41	3.3453×10^{49}	49.52443	91	1.3520×10^{140}	140.13098
42	1.4050×10^{51}	51.14768	92	1.2438×10^{142}	142.09476
43	6.0415×10^{52}	52.78115	93	1.1568×10^{144}	144.06325
44	2.6583×10^{54}	54.42460	94	1.0874×10^{146}	146.03638
45	1.1962×10^{56}	56.07781	**95**	1.0330×10^{148}	148.01410
46	5.5026×10^{57}	57.74057	96	9.9168×10^{149}	149.99637
47	2.5862×10^{59}	59.41267	97	9.6193×10^{151}	151.98314
48	1.2414×10^{61}	61.09391	98	9.4269×10^{153}	153.97437
49	6.0828×10^{62}	62.78410	99	9.3326×10^{155}	155.97000
50	3.0414×10^{64}	64.48307	**100**	9.3326×10^{157}	157.97000

$$n! = \left(\frac{n}{e}\right)^n \sqrt{2\pi n} + h; \; n = 1, 2, 3, \ldots \quad \left[0 < \frac{h}{n!} < \frac{1}{12n}\right]$$

$$\lim_{n\to\infty} \frac{n!e^n}{n^{n+\frac{1}{2}}} = \sqrt{2\pi} \qquad \lim_{n\to\infty} \frac{(n!)^{\frac{1}{n}}}{n} = \frac{1}{e}$$

RECIPROCALS OF FACTORIALS AND THEIR COMMON LOGARITHMS

n	$1/n!$	$\log(1/n!)$	n	$1/n!$	$\log(1/n!)$
1	1.	.00000	51	$.64470 \times 10^{-66}$	$\overline{67}.80935$
2	0.5	$\overline{1}.69897$	52	$.12398 \times 10^{-67}$	$\overline{68}.09335$
3	.16667	$\overline{1}.22185$	53	$.23392 \times 10^{-69}$	$\overline{70}.36908$
*4	$.41667 \times 10^{-1}$	$\overline{2}.61979$	54	$.45282 \times 10^{-71}$	$\overline{72}.63668$
5	$.83333 \times 10^{-2}$	$\overline{3}.92082$	55	$.78765 \times 10^{-73}$	$\overline{74}.89632$
6	$.13889 \times 10^{-2}$	$\overline{3}.14267$	56	$.14065 \times 10^{-74}$	$\overline{75}.14813$
7	$.19841 \times 10^{-3}$	$\overline{4}.29757$	57	$.24675 \times 10^{-76}$	$\overline{77}.39226$
8	$.24802 \times 10^{-4}$	$\overline{5}.39448$	58	$.42542 \times 10^{-78}$	$\overline{79}.62883$
9	$.27557 \times 10^{-5}$	$\overline{6}.44024$	59	$.72108 \times 10^{-80}$	$\overline{81}.85798$
10	$.27557 \times 10^{-6}$	$\overline{7}.44024$	60	$.12018 \times 10^{-81}$	$\overline{82}.07983$
11	$.25052 \times 10^{-7}$	$\overline{8}.39884$	61	$.19701 \times 10^{-83}$	$\overline{84}.29450$
12	$.20877 \times 10^{-8}$	$\overline{9}.31966$	62	$.31776 \times 10^{-85}$	$\overline{86}.50210$
13	$.16059 \times 10^{-9}$	$\overline{10}.20572$	63	$.50439 \times 10^{-87}$	$\overline{88}.70276$
14	$.11471 \times 10^{-10}$	$\overline{11}.05959$	64	$.78808 \times 10^{-89}$	$\overline{90}.89658$
15	$.76472 \times 10^{-12}$	$\overline{13}.88350$	65	$.12125 \times 10^{-90}$	$\overline{91}.08367$
16	$.47795 \times 10^{-13}$	$\overline{14}.67938$	66	$.18371 \times 10^{-92}$	$\overline{93}.26413$
17	$.28115 \times 10^{-14}$	$\overline{15}.44893$	67	$.27419 \times 10^{-94}$	$\overline{95}.43805$
18	$.15619 \times 10^{-15}$	$\overline{16}.19365$	68	$.40323 \times 10^{-96}$	$\overline{97}.60554$
19	$.82206 \times 10^{-17}$	$\overline{18}.91491$	69	$.58439 \times 10^{-98}$	$\overline{99}.76668$
20	$.41103 \times 10^{-18}$	$\overline{19}.61388$	70	$.83479 \times 10^{-100}$	$\overline{101}.92159$
21	$.19573 \times 10^{-19}$	$\overline{20}.29165$	71	$.11758 \times 10^{-101}$	$\overline{102}.07034$
22	$.88968 \times 10^{-21}$	$\overline{22}.94923$	72	$.16331 \times 10^{-103}$	$\overline{104}.21300$
23	$.38682 \times 10^{-22}$	$\overline{23}.58751$	73	$.22371 \times 10^{-105}$	$\overline{106}.34968$
24	$.16117 \times 10^{-23}$	$\overline{24}.20728$	74	$.30231 \times 10^{-107}$	$\overline{108}.48045$
25	$.64470 \times 10^{-25}$	$\overline{26}.80935$	75	$.40308 \times 10^{-109}$	$\overline{110}.60538$
26	$.24796 \times 10^{-26}$	$\overline{27}.39438$	76	$.53036 \times 10^{-111}$	$\overline{112}.72457$
27	$.91836 \times 10^{-28}$	$\overline{29}.96302$	77	$.68880 \times 10^{-113}$	$\overline{114}.83808$
28	$.32799 \times 10^{-29}$	$\overline{30}.51586$	78	$.88308 \times 10^{-115}$	$\overline{116}.94599$
29	$.11310 \times 10^{-30}$	$\overline{31}.05346$	79	$.11178 \times 10^{-116}$	$\overline{117}.04836$
30	$.37644 \times 10^{-32}$	$\overline{33}.57634$	80	$.13973 \times 10^{-118}$	$\overline{119}.14527$
31	$.12161 \times 10^{-33}$	$\overline{34}.08498$	81	$.17250 \times 10^{-120}$	$\overline{121}.23679$
32	$.38004 \times 10^{-35}$	$\overline{36}.57983$	82	$.21037 \times 10^{-122}$	$\overline{123}.32297$
33	$.11516 \times 10^{-36}$	$\overline{37}.06131$	83	$.25345 \times 10^{-124}$	$\overline{125}.40390$
34	$.33872 \times 10^{-38}$	$\overline{39}.52984$	84	$.30173 \times 10^{-126}$	$\overline{127}.47962$
35	$.96775 \times 10^{-40}$	$\overline{41}.98577$	85	$.35497 \times 10^{-128}$	$\overline{129}.55020$
36	$.26882 \times 10^{-41}$	$\overline{42}.42946$	86	$.41276 \times 10^{-130}$	$\overline{131}.61570$
37	$.72653 \times 10^{-43}$	$\overline{44}.86126$	87	$.47443 \times 10^{-132}$	$\overline{133}.67618$
38	$.19120 \times 10^{-44}$	$\overline{45}.28148$	88	$.53914 \times 10^{-134}$	$\overline{135}.73170$
39	$.49024 \times 10^{-46}$	$\overline{47}.69041$	89	$.60577 \times 10^{-136}$	$\overline{137}.78231$
40	$.12256 \times 10^{-47}$	$\overline{48}.08835$	90	$.67308 \times 10^{-138}$	$\overline{139}.82806$
41	$.29893 \times 10^{-49}$	$\overline{50}.47557$	91	$.73964 \times 10^{-140}$	$\overline{141}.86902$
42	$.71174 \times 10^{-51}$	$\overline{52}.85232$	92	$.80399 \times 10^{-142}$	$\overline{143}.90524$
43	$.16552 \times 10^{-52}$	$\overline{53}.21885$	93	$.86445 \times 10^{-144}$	$\overline{145}.93675$
44	$.37618 \times 10^{-54}$	$\overline{55}.57540$	94	$.91962 \times 10^{-146}$	$\overline{147}.96362$
45	$.83598 \times 10^{-56}$	$\overline{57}.92219$	95	$.96805 \times 10^{-148}$	$\overline{149}.98590$
46	18173×10^{-57}	$\overline{58}.25943$	96	$.10084 \times 10^{-149}$	$\overline{150}.00363$
47	$.38667 \times 10^{-59}$	$\overline{60}.58733$	97	$.10396 \times 10^{-151}$	$\overline{152}.01686$
48	$.80554 \times 10^{-61}$	$\overline{62}.90609$	98	$.10608 \times 10^{-153}$	$\overline{154}.02563$
49	$.16440 \times 10^{-62}$	$\overline{63}.21590$	99	$.10715 \times 10^{-155}$	$\overline{156}.03000$
50	$.32880 \times 10^{-64}$	$\overline{65}.51693$	100	$.10715 \times 10^{-157}$	$\overline{158}.03000$

* For example $\log \frac{1}{4!} = \overline{2}.61979 = .61979 - 2 = 8.61979 - 10.$

POSITIONAL NOTATION

In our ordinary system of writing numbers, the value of any digit depends on its position in the number. The value of a digit in any position is ten times the value of the same digit one position to the right, or one-tenth the value of the same digit one position to the left. Thus, for example,

$$173.246 = 1 \times 10^2 + 7 \times 10^1 + 3 + 2 \times \frac{1}{10} + 4 \times \frac{1}{10^2} + 6 \times \frac{1}{10^3}.$$

There is no reason that a number other than 10 cannot be used as the *base*, or *radix*, of the number system. In fact, bases of 2, 8, and 16 are commonly used in working with digital computers. When the base used is not clear from the context, it is usually indicated as a parenthesized subscript or merely as a subscript. Thus

$$743_{(8)} = 7 \times 8^2 + 4 \times 8 + 3 = 7 \times 64 + 4 \times 8 + 3 = 448 + 32 + 3 = 483_{(10)}$$

$$1011.101_{(2)} = 1 \times 2^3 + 0 \times 2^2 + 1 \times 2 + 1 + 1 \times \tfrac{1}{2} + 0 \times \tfrac{1}{4} + 1 \times \tfrac{1}{8} = 11.625_{(10)}$$

CHANGE OF BASE

In this section, it is assumed that all calculations will be performed in base 10, since this is the only base in which most people can easily compute. However, there is no logical reason that some other base could not be used for the computations.

To convert a number from another base into base 10:

Simply write down the digits of the number, with each one multiplied by its appropriate positional value. Then perform the indicated computations in base 10, and write down the answer.

For examples, see the two examples in the previous section.

To convert a number from base 10 into another base:

The part of the number to the left of the point and the part to the right must be operated on separately. For the integer part (the part to the left of the point):

a. Divide the number by the new base, getting an integer quotient and remainder.
b. Write down the remainder as the last digit of the number in the new base.
c. Using the quotient from the last division in place of the original number, repeat the above two steps until the quotient becomes zero.

For the fractional part (the part to the right of the point):

a. Multiply the number by the new base.
b. Write down the integral part of the product as the first digit of the fractional part in the new base.
c. Using the fractional part of the last product in place of the original number, repeat the above two steps until the product becomes an integer, or until the desired number of places have been computed.

Examples:

These examples show a convenient method of arranging the computations.

1. Convert $103.118_{(10)}$ to base 8.

8	103	7		.118
8	12	4		8
	1		147.074324...	.944
				8
				7.552
				8
				4.416
				8
				3.328
				8
				2.624
				8
				4.992

The calculation of the fractional part could be carried out as far as desired. It is a non-terminating fraction which will eventually repeat itself.

$$103.118_{(10)} = 147.074324\ldots_{(8)}$$

The calculations may be further shortened by not writing down the multiplier and divisor at each step of the algorithm, as shown in the next example.

2. Convert $275.824_{(10)}$ to base 5.

5	275	0	.824
	55	0	4.120
	11	1	0.600
	2		3.000

$$275.824_{(10)} = 2100.403_{(5)}$$

To convert from one base to another (neither of which is 10):

The easiest procedure is usually to convert first to base 10, and then to the desired base. However, there are two exceptions to this:

1. If computational facility is possessed in either of the bases, it may be used instead of base 10, and the appropriate one of the above methods applied.
2. If the two bases are different powers of the same number, the conversion may be done digit-by-digit to the base which is the common root of both bases, and then digit-by-digit back to the other base.

Example: Convert $127.653_{(8)}$ to base 16. (For base 16, the letters A–F are used for the digits $10_{(10)}$–$15_{(10)}$.)

The first step is to convert the number to base 2, simply by converting each digit to its binary equivalent:

$$127.653_{(8)} = 001\ 010\ 111 \cdot 110\ 101\ 011_{(2)}$$

Now by simply regrouping the binary number into groups of four binary digits, starting at the point, we convert to base 16:

$$127.653_{(8)} = 101\ 0111 \cdot 1101\ 0101\ 1_{(2)} = 57.\mathrm{D}58_{(16)}$$

$10^{\pm n}$ IN OCTAL SCALE

10^n	n	10^{-n}	10^n	n	10^{-n}
1	0	1.000 000 000 000 000	112 402 762 000	10	0.000 000 000 006 676
12	1	0.063 146 314 631 463	1 351 035 564 000	11	0.000 000 000 000 538
144	2	0.005 075 341 217 270	16 432 451 210 000	12	0.000 000 000 000 043
1 750	3	0.000 406 111 564 571	221 441 634 520 000	13	0.000 000 000 000 003
23 420	4	0.000 032 155 613 531	2 657 142 036 440 000	14	0.000 000 000 000 000
303 240	5	0.000 002 476 132 611	34 327 724 461 500 000	15	0.000 000 000 000 000
3 641 100	6	0.000 000 206 157 364	434 157 115 760 200 000	16	0.000 000 000 000 000
46 113 200	7	0.000 000 015 327 745	5 432 127 413 542 400 000	17	0.000 000 000 000 000
575 360 400	8	0.000 000 001 257 144	67 405 553 164 731 000 000	18	0.000 000 000 000 000
7 346 545 000	9	0.000 000 000 104 560			

2^n IN DECIMAL SCALE

n	2^n	n	2^n	n	2^n
0.001	1.00069 33874 62581	0.01	1.00695 55500 56719	0.1	1.07177 34625 36293
0.002	1.00138 72557 11335	0.02	1.01395 94797 90029	0.2	1.14869 83549 97035
0.003	1.00208 16050 79633	0.03	1.02101 21257 07193	0.3	1.23114 44133 44916
0.004	1.00277 64359 01078	0.04	1.02811 38266 56067	0.4	1.31950 79107 72894
0.005	1.00347 17485 09503	0.05	1.03526 49238 41377	0.5	1.41421 35623 73095
0.006	1.00416 75432 38973	0.06	1.04246 57608 41121	0.6	1.51571 65665 10398
0.007	1.00486 38204 23785	0.07	1.04971 66836 23067	0.7	1.62450 47927 12471
0.008	1.00556 05803 98468	0.08	1.05701 80405 61380	0.8	1.74110 11265 92248
0.009	1.00625 78234 97782	0.09	1.06437 01824 53360	0.9	1.86606 59830 73615

$n \log_{10} 2$, $n \log_2 10$ IN DECIMAL SCALE

n	$n \log_{10} 2$	$n \log_2 10$	n	$n \log_{10} 2$	$n \log_2 10$
1	0.30102 99957	3.32192 80949	6	1.80617 99740	19.93156 85693
2	0.60205 99913	6.64385 61898	7	2.10720 99696	23.25349 66642
3	0.90308 99870	9.96578 42847	8	2.40823 99653	26.57542 47591
4	1.20411 99827	13.28771 23795	9	2.70926 99610	29.89735 28540
5	1.50514 99783	16.60964 04744	10	3.01029 99566	33.21928 09489

ADDITION AND MULTIPLICATION TABLES

Binary Scale

Addition

$$0 + 0 = 0$$
$$0 + 1 = 1 + 0 = 1$$
$$1 + 1 = 10$$

Multiplication

$$0 \times 0 = 0$$
$$0 \times 1 = 1 \times 0 = 0$$
$$1 \times 1 = 1$$

Octal Scale

Addition

0	01	02	03	04	05	06	07
1	02	03	04	05	06	07	10
2	03	04	05	06	07	10	11
3	04	05	06	07	10	11	12
4	05	06	07	10	11	12	13
5	06	07	10	11	12	13	14
6	07	10	11	12	13	14	15
7	10	11	12	13	14	15	16

Multiplication

1	02	03	04	05	06	07
2	04	06	10	12	14	16
3	06	11	14	17	22	25
4	10	14	20	24	30	34
5	12	17	24	31	36	43
6	14	22	30	36	44	52
7	16	25	34	43	52	61

MATHEMATICAL CONSTANTS IN OCTAL SCALE

$\pi = (3.11037\ 552421)_{(8)}$	$e = (2.55760\ 521305)_{(8)}$	$\gamma = (0.44742\ 147707)_{(8)}$
$\pi^{-1} = (0.24276\ 301556)_{(8)}$	$e^{-1} = (0.27426\ 530661)_{(8)}$	$\log_e \gamma = -(0.43127\ 233602)_{(8)}$
$\sqrt{\pi} = (1.61337\ 611067)_{(8)}$	$\sqrt{e} = (1.51411\ 230704)_{(8)}$	$\log_2 \gamma = -(0.62573\ 030645)_{(8)}$
$\log_e \pi = (1.11206\ 404435)_{(8)}$	$\log_{10} e = (0.33626\ 754251)_{(8)}$	$\sqrt{2} = (1.32404\ 746320)_{(8)}$
$\log_2 \pi = (1.51544\ 163223)_{(8)}$	$\log_2 e = (1.34252\ 166245)_{(8)}$	$\log_e 2 = (0.54271\ 027760)_{(8)}$
$\sqrt{10} = (3.12305\ 407267)_{(8)}$	$\log_2 10 = (3.24464\ 741136)_{(8)}$	$\log_e 10 = (2.23273\ 067355)_{(8)}$

OCTAL-DECIMAL INTEGER CONVERSION TABLE

	0	1	2	3	4	5	6	7
0000	0000	0001	0002	0003	0004	0005	0006	0007
0010	0008	0009	0010	0011	0012	0013	0014	0015
0020	0016	0017	0018	0019	0020	0021	0022	0023
0030	0024	0025	0026	0027	0028	0029	0030	0031
0040	0032	0033	0034	0035	0036	0037	0038	0039
0050	0040	0041	0042	0043	0044	0045	0046	0047
0060	0048	0049	0050	0051	0052	0053	0054	0055
0070	0056	0057	0058	0059	0060	0061	0062	0063
0100	0064	0065	0066	0067	0068	0069	0070	0071
0110	0072	0073	0074	0075	0076	0077	0078	0079
0120	0080	0081	0082	0083	0084	0085	0086	0087
0130	0088	0089	0090	0091	0092	0093	0094	0095
0140	0096	0097	0098	0099	0100	0101	0102	0103
0150	0104	0105	0106	0107	0108	0109	0110	0111
0160	0112	0113	0114	0115	0116	0117	0118	0119
0170	0120	0121	0122	0123	0124	0125	0126	0127
0200	0128	0129	0130	0131	0132	0133	0134	0135
0210	0136	0137	0138	0139	0140	0141	0142	0143
0220	0144	0145	0146	0147	0148	0149	0150	0151
0230	0152	0153	0154	0155	0156	0157	0158	0159
0240	0160	0161	0162	0163	0164	0165	0166	0167
0250	0168	0169	0170	0171	0172	0173	0174	0175
0260	0176	0177	0178	0179	0180	0181	0182	0183
0270	0184	0185	0186	0187	0188	0189	0190	0191
0300	0192	0193	0194	0195	0196	0197	0198	0199
0310	0200	0201	0202	0203	0204	0205	0206	0207
0320	0208	0209	0210	0211	0212	0213	0214	0215
0330	0216	0217	0218	0219	0220	0221	0222	0223
0340	0224	0225	0226	0227	0228	0229	0230	0231
0350	0232	0233	0234	0235	0236	0237	0238	0239
0360	0240	0241	0242	0243	0244	0245	0246	0247
0370	0248	0249	0250	0251	0252	0253	0254	0255

	0	1	2	3	4	5	6	7
0400	0256	0257	0258	0259	0260	0261	0262	0263
0410	0264	0265	0266	0267	0268	0269	0270	0271
0420	0272	0273	0274	0275	0276	0277	0278	0279
0430	0280	0281	0282	0283	0284	0285	0286	0287
0440	0288	0289	0290	0291	0292	0293	0294	0295
0450	0296	0297	0298	0299	0300	0301	0302	0303
0460	0304	0305	0306	0307	0308	0309	0310	0311
0470	0312	0313	0314	0315	0316	0317	0318	0319
0500	0320	0321	0322	0323	0324	0325	0326	0327
0510	0328	0329	0330	0331	0332	0333	0334	0335
0520	0336	0337	0338	0339	0340	0341	0342	0343
0530	0344	0345	0346	0347	0348	0349	0350	0351
0540	0352	0353	0354	0355	0356	0357	0358	0359
0550	0360	0361	0362	0363	0364	0365	0366	0367
0560	0368	0369	0370	0371	0372	0373	0374	0375
0570	0376	0377	0378	0379	0380	0381	0382	0383
0600	0384	0385	0386	0387	0388	0389	0390	0391
0610	0392	0393	0394	0395	0396	0397	0398	0399
0620	0400	0401	0402	0403	0404	0405	0406	0407
0630	0408	0409	0410	0411	0412	0413	0414	0415
0640	0416	0417	0418	0419	0420	0421	0422	0423
0650	0424	0425	0426	0427	0428	0429	0430	0431
0660	0432	0433	0434	0435	0436	0437	0438	0439
0670	0440	0441	0442	0443	0444	0445	0446	0447
0700	0448	0449	0450	0451	0452	0453	0454	0455
0710	0456	0457	0458	0459	0460	0461	0462	0463
0720	0464	0465	0466	0467	0468	0469	0470	0471
0730	0472	0473	0474	0475	0476	0477	0478	0479
0740	0480	0481	0482	0483	0484	0485	0486	0487
0750	0488	0489	0490	0491	0492	0493	0494	0495
0760	0496	0497	0498	0499	0500	0501	0502	0503
0770	0504	0505	0506	0507	0508	0509	0510	0511

0000 to 0777 (Octal)	0000 to 0511 (Decimal)

Octal	Decimal
10000-	**4096**
20000-	**8192**
30000-	**12288**
40000-	**16384**
50000-	**20480**
60000-	**24576**
70000-	**28672**

	0	1	2	3	4	5	6	7
1000	0512	0513	0514	0515	0516	0517	0518	0519
1010	0520	0521	0522	0523	0524	0525	0526	0527
1020	0528	0529	0530	0531	0532	0533	0534	0535
1030	0536	0537	0538	0539	0540	0541	0542	0543
1040	0544	0545	0546	0547	0548	0549	0550	0551
1050	0552	0553	0554	0555	0556	0557	0558	0559
1060	0560	0561	0562	0563	0564	0565	0566	0567
1070	0568	0569	0570	0571	0572	0573	0574	0575
1100	0576	0577	0578	0579	0580	0581	0582	0583
1110	0584	0585	0586	0587	0588	0589	0590	0591
1120	0592	0593	0594	0595	0596	0597	0598	0599
1130	0600	0601	0602	0603	0604	0605	0606	0607
1140	0608	0609	0610	0611	0612	0613	0614	0615
1150	0616	0617	0618	0619	0620	0621	0622	0623
1160	0624	0625	0626	0627	0628	0629	0630	0631
1170	0632	0633	0634	0635	0636	0637	0638	0639
1200	0640	0641	0642	0643	0644	0645	0646	0647
1210	0648	0649	0650	0651	0652	0653	0654	0655
1220	0656	0657	0658	0659	0660	0661	0662	0663
1230	0664	0665	0666	0667	0668	0669	0670	0671
1240	0672	0673	0674	0675	0676	0677	0678	0679
1250	0680	0681	0682	0683	0684	0685	0686	0687
1260	0688	0689	0690	0691	0692	0693	0694	0695
1270	0696	0697	0698	0699	0700	0701	0702	0703
1300	0704	0705	0706	0707	0708	0709	0710	0711
1310	0712	0713	0714	0715	0716	0717	0718	0719
1320	0720	0721	0722	0723	0724	0725	0726	0727
1330	0728	0729	0730	0731	0732	0733	0734	0735
1340	0736	0737	0738	0739	0740	0741	0742	0743
1350	0744	0745	0746	0747	0748	0749	0750	0751
1360	0752	0753	0754	0755	0756	0757	0758	0759
1370	0760	0761	0762	0763	0764	0765	0766	0767

	0	1	2	3	4	5	6	7
1400	0768	0769	0770	0771	0772	0773	0774	0775
1410	0776	0777	0778	0779	0780	0781	0782	0783
1420	0784	0785	0786	0787	0788	0789	0790	0791
1430	0792	0793	0794	0795	0796	0797	0798	0799
1440	0800	0801	0802	0803	0804	0805	0806	0807
1450	0808	0809	0810	0811	0812	0813	0814	0815
1460	0816	0817	0818	0819	0820	0821	0822	0823
1470	0824	0825	0826	0827	0828	0829	0830	0831
1500	0832	0833	0834	0835	0836	0837	0838	0839
1510	0840	0841	0842	0843	0844	0845	0846	0847
1520	0848	0849	0850	0851	0852	0853	0854	0855
1530	0856	0857	0858	0859	0860	0861	0862	0863
1540	0864	0865	0866	0867	0868	0869	0870	0871
1550	0872	0873	0874	0875	0876	0877	0878	0879
1560	0880	0881	0882	0883	0884	0885	0886	0887
1570	0888	0889	0890	0891	0892	0893	0894	0895
1600	0896	0897	0898	0899	0900	0901	0902	0903
1610	0904	0905	0906	0907	0908	0909	0910	0911
1620	0912	0913	0914	0915	0916	0917	0918	0919
1630	0920	0921	0922	0923	0924	0925	0926	0927
1640	0928	0929	0930	0931	0932	0933	0934	0935
1650	0936	0937	0938	0939	0940	0941	0942	0943
1660	0944	0945	0946	0947	0948	0949	0950	0951
1670	0952	0953	0954	0955	0956	0957	0958	0959
1700	0960	0961	0962	0963	0964	0965	0966	0967
1710	0968	0969	0970	0971	0972	0973	0974	0975
1720	0976	0977	0978	0979	0980	0981	0982	0983
1730	0984	0985	0986	0987	0988	0989	0990	0991
1740	0992	0993	0994	0995	0996	0997	0998	0999
1750	1000	1001	1002	1003	1004	1005	1006	1007
1760	1008	1009	1010	1011	1012	1013	1014	1015
1770	1016	1017	1018	1019	1020	1021	1022	1023

1000 to 1777 (Octal)	0512 to 1023 (Decimal)

OCTAL-DECIMAL INTEGER CONVERSION TABLE (Continued)

	0	1	2	3	4	5	6	7
2000	1024	1025	1026	1027	1028	1029	1030	1031
2010	1032	1033	1034	1035	1036	1037	1038	1039
2020	1040	1041	1042	1043	1044	1045	1046	1047
2030	1048	1049	1050	1051	1052	1053	1054	1055
2040	1056	1057	1058	1059	1060	1061	1062	1063
2050	1064	1065	1066	1067	1068	1069	1070	1071
2060	1072	1073	1074	1075	1076	1077	1078	1079
2070	1080	1081	1082	1083	1084	1085	1086	1087
2100	1088	1089	1090	1091	1092	1093	1094	1095
2110	1096	1097	1098	1099	1100	1101	1102	1103
2120	1104	1105	1106	1107	1108	1109	1110	1111
2130	1112	1113	1114	1115	1116	1117	1118	1119
2140	1120	1121	1122	1123	1124	1125	1126	1127
2150	1128	1129	1130	1131	1132	1133	1134	1135
2160	1136	1137	1138	1139	1140	1141	1142	1143
2170	1144	1145	1146	1147	1148	1149	1150	1151
2200	1152	1153	1154	1155	1156	1157	1158	1159
2210	1160	1161	1162	1163	1164	1165	1166	1167
2220	1168	1169	1170	1171	1172	1173	1174	1175
2230	1176	1177	1178	1179	1180	1181	1182	1183
2240	1184	1185	1186	1187	1188	1189	1190	1191
2250	1192	1193	1194	1195	1196	1197	1198	1199
2260	1200	1201	1202	1203	1204	1205	1206	1207
2270	1208	1209	1210	1211	1212	1213	1214	1215
2300	1216	1217	1218	1219	1220	1221	1222	1223
2310	1224	1225	1226	1227	1228	1229	1230	1231
2320	1232	1233	1234	1235	1236	1237	1238	1239
2330	1240	1241	1242	1243	1244	1245	1246	1247
2340	1248	1249	1250	1251	1252	1253	1254	1255
2350	1256	1257	1258	1259	1260	1261	1262	1263
2360	1264	1265	1266	1267	1268	1269	1270	1271
2370	1272	1273	1274	1275	1276	1277	1278	1279

	0	1	2	3	4	5	6	7
2400	1280	1281	1282	1283	1284	1285	1286	1287
2410	1288	1289	1290	1291	1292	1293	1294	1295
2420	1296	1297	1298	1299	1300	1301	1302	1303
2430	1304	1305	1306	1307	1308	1309	1310	1311
2440	1312	1313	1314	1315	1316	1317	1318	1319
2450	1320	1321	1322	1323	1324	1325	1326	1327
2460	1328	1329	1330	1331	1332	1333	1334	1335
2470	1336	1337	1338	1339	1340	1341	1342	1343
2500	1344	1345	1346	1347	1348	1349	1350	1351
2510	1352	1353	1354	1355	1356	1357	1358	1359
2520	1360	1361	1362	1363	1364	1365	1366	1367
2530	1368	1369	1370	1371	1372	1373	1374	1375
2540	1376	1377	1378	1379	1380	1381	1382	1383
2550	1384	1385	1386	1387	1388	1389	1390	1391
2560	1392	1393	1394	1395	1396	1397	1398	1399
2570	1400	1401	1402	1403	1404	1405	1406	1407
2600	1408	1409	1410	1411	1412	1413	1414	1415
2610	1416	1417	1418	1419	1420	1421	1422	1423
2620	1424	1425	1426	1427	1428	1429	1430	1431
2630	1432	1433	1434	1435	1436	1437	1438	1439
2640	1440	1441	1442	1443	1444	1445	1446	1447
2650	1448	1449	1450	1451	1452	1453	1454	1455
2660	1456	1457	1458	1459	1460	1461	1462	1463
2670	1464	1465	1466	1467	1468	1469	1470	1471
2700	1472	1473	1474	1475	1476	1477	1478	1479
2710	1480	1481	1482	1483	1484	1485	1486	1487
2720	1488	1489	1490	1491	1492	1493	1494	1495
2730	1496	1497	1498	1499	1500	1501	1502	1503
2740	1504	1505	1506	1507	1508	1509	1510	1511
2750	1512	1513	1514	1515	1516	1517	1518	1519
2760	1520	1521	1522	1523	1524	1525	1526	1527
2770	1528	1529	1530	1531	1532	1533	1534	1535

2000 to 2777 (Octal)	1024 to 1535 (Decimal)

Octal	Decimal
10000-	**4096**
20000-	**8192**
30000-	**12288**
40000-	**16384**
50000-	**20480**
60000-	**24576**
70000-	**28672**

3000 to 3777 (Octal)	1536 to 2047 (Decimal)

	0	1	2	3	4	5	6	7
3000	1536	1537	1538	1539	1540	1541	1542	1543
3010	1544	1545	1546	1547	1548	1549	1550	1551
3020	1552	1553	1554	1555	1556	1557	1558	1559
3030	1560	1561	1562	1563	1564	1565	1566	1567
3040	1568	1569	1570	1571	1572	1573	1574	1575
3050	1576	1577	1578	1579	1580	1581	1582	1583
3060	1584	1585	1586	1587	1588	1589	1590	1591
3070	1592	1593	1594	1595	1596	1597	1598	1599
3100	1600	1601	1602	1603	1604	1605	1606	1607
3110	1608	1609	1610	1611	1612	1613	1614	1615
3120	1616	1617	1618	1619	1620	1621	1622	1623
3130	1624	1625	1626	1627	1628	1629	1630	1631
3140	1632	1633	1634	1635	1636	1637	1638	1639
3150	1640	1641	1642	1643	1644	1645	1646	1647
3160	1648	1649	1650	1651	1652	1653	1654	1655
3170	1656	1657	1658	1659	1660	1661	1662	1663
3200	1664	1665	1666	1667	1668	1669	1670	1671
3210	1672	1673	1674	1675	1676	1677	1678	1679
3220	1680	1681	1682	1683	1684	1685	1686	1687
3230	1688	1689	1690	1691	1692	1693	1694	1695
3240	1696	1697	1698	1699	1700	1701	1702	1703
3250	1704	1705	1706	1707	1708	1709	1710	1711
3260	1712	1713	1714	1715	1716	1717	1718	1719
3270	1720	1721	1722	1723	1724	1725	1726	1727
3300	1728	1729	1730	1731	1732	1733	1734	1735
3310	1736	1737	1738	1739	1740	1741	1742	1743
3320	1744	1745	1746	1747	1748	1749	1750	1751
3330	1752	1753	1754	1755	1756	1757	1758	1759
3340	1760	1761	1762	1763	1764	1765	1766	1767
3350	1768	1769	1770	1771	1772	1773	1774	1775
3360	1776	1777	1778	1779	1780	1781	1782	1783
3370	1784	1785	1786	1787	1788	1789	1790	1791

	0	1	2	3	4	5	6	7
3400	1792	1793	1794	1795	1796	1797	1798	1799
3410	1800	1801	1802	1803	1804	1805	1806	1807
3420	1808	1809	1810	1811	1812	1813	1814	1815
3430	1816	1817	1818	1819	1820	1821	1822	1823
3440	1824	1825	1826	1827	1828	1829	1830	1831
3450	1832	1833	1834	1835	1836	1837	1838	1839
3460	1840	1841	1842	1843	1844	1845	1846	1847
3470	1848	1849	1850	1851	1852	1853	1854	1855
3500	1856	1857	1858	1859	1860	1861	1862	1863
3510	1864	1865	1866	1867	1868	1869	1870	1871
3520	1872	1873	1874	1875	1876	1877	1878	1879
3530	1880	1881	1882	1883	1884	1885	1886	1887
3540	1888	1889	1890	1891	1892	1893	1894	1895
3550	1896	1897	1898	1899	1900	1901	1902	1903
3560	1904	1905	1906	1907	1908	1909	1910	1911
3570	1912	1913	1914	1915	1916	1917	1918	1919
3600	1920	1921	1922	1923	1924	1925	1926	1927
3610	1928	1929	1930	1931	1932	1933	1934	1935
3620	1936	1937	1938	1939	1940	1941	1942	1943
3630	1944	1945	1946	1947	1948	1949	1950	1951
3640	1952	1953	1954	1955	1956	1957	1958	1959
3650	1960	1961	1962	1963	1964	1965	1966	1967
3660	1968	1969	1970	1971	1972	1973	1974	1975
3670	1976	1977	1978	1979	1980	1981	1982	1983
3700	1984	1985	1986	1987	1988	1989	1990	1991
3710	1992	1993	1994	1995	1996	1997	1998	1999
3720	2000	2001	2002	2003	2004	2005	2006	2007
3730	2008	2009	2010	2011	2012	2013	2014	2015
3740	2016	2017	2018	2019	2020	2021	2022	2023
3750	2024	2025	2026	2027	2028	2029	2030	2031
3760	2032	2033	2034	2035	2036	2037	2038	2039
3770	2040	2041	2042	2043	2044	2045	2046	2047

OCTAL-DECIMAL INTEGER CONVERSION TABLE (Continued)

	0	1	2	3	4	5	6	7
4000	2048	2049	2050	2051	2052	2053	2054	2055
4010	2056	2057	2058	2059	2060	2061	2062	2063
4020	2064	2065	2066	2067	2068	2069	2070	2071
4030	2072	2073	2074	2075	2076	2077	2078	2079
4040	2080	2081	2082	2083	2084	2085	2086	2087
4050	2088	2089	2090	2091	2092	2093	2094	2095
4060	2096	2097	2098	2099	2100	2101	2102	2103
4070	2104	2105	2106	2107	2108	2109	2110	2111
4100	2112	2113	2114	2115	2116	2117	2118	2119
4110	2120	2121	2122	2123	2124	2125	2126	2127
4120	2128	2129	2130	2131	2132	2133	2134	2135
4130	2136	2137	2138	2139	2140	2141	2142	2143
4140	2144	2145	2146	2147	2148	2149	2150	2151
4150	2152	2153	2154	2155	2156	2157	2158	2159
4160	2160	2161	2162	2163	2164	2165	2166	2167
4170	2168	2169	2170	2171	2172	2173	2174	2175
4200	2176	2177	2178	2179	2180	2181	2182	2183
4210	2184	2185	2186	2187	2188	2189	2190	2191
4220	2192	2193	2194	2195	2196	2197	2198	2199
4230	2200	2201	2202	2203	2204	2205	2206	2207
4240	2208	2209	2210	2211	2212	2213	2214	2215
4250	2216	2217	2218	2219	2220	2221	2222	2223
4260	2224	2225	2226	2227	2228	2229	2230	2231
4270	2232	2233	2234	2235	2236	2237	2238	2239
4300	2240	2241	2242	2243	2244	2245	2246	2247
4310	2248	2249	2250	2251	2252	2253	2254	2255
4320	2256	2257	2258	2259	2260	2261	2262	2263
4330	2264	2265	2266	2267	2268	2269	2270	2271
4340	2272	2273	2274	2275	2276	2277	2278	2279
4350	2280	2281	2282	2283	2284	2285	2286	2287
4360	2288	2289	2290	2291	2292	2293	2294	2295
4370	2296	2297	2298	2299	2300	2301	2302	2303

	0	1	2	3	4	5	6	7
4400	2304	2305	2306	2307	2308	2309	2310	2311
4410	2312	2313	2314	2315	2316	2317	2318	2319
4420	2320	2321	2322	2323	2324	2325	2326	2327
4430	2328	2329	2330	2331	2332	2333	2334	2335
4440	2336	2337	2338	2339	2340	2341	2342	2343
4450	2344	2345	2346	2347	2348	2349	2350	2351
4460	2352	2353	2354	2355	2356	2357	2358	2359
4470	2360	2361	2362	2363	2364	2365	2366	2367
4500	2368	2369	2370	2371	2372	2373	2374	2375
4510	2376	2377	2378	2379	2380	2381	2382	2383
4520	2384	2385	2386	2387	2388	2389	2390	2391
4530	2392	2393	2394	2395	2396	2397	2398	2399
4540	2400	2401	2402	2403	2404	2405	2406	2407
4550	2408	2409	2410	2411	2412	2413	2114	2415
4560	2416	2417	2418	2419	2420	2421	2422	2423
4570	2424	2425	2426	2427	2428	2429	2430	2431
4600	2432	2433	2434	2435	2436	2437	2438	2439
4610	2440	2441	2442	2443	2444	2445	2446	2447
4620	2448	2449	2450	2451	2452	2453	2454	2455
4630	2456	2457	2458	2459	2460	2461	2462	2463
4640	2464	2465	2466	2467	2468	2469	2470	2471
4650	2472	2473	2474	2475	2476	2477	2478	2479
4660	2480	2481	2482	2483	2484	2485	2486	2487
4670	2488	2489	2490	2491	2492	2493	2494	2495
4700	2496	2497	2498	2499	2500	2501	2502	2503
4710	2504	2505	2506	2507	2508	2509	2510	2511
4720	2512	2513	2514	2515	2516	2517	2518	2519
4730	2520	2521	2522	2523	2524	2525	2526	2527
4740	2528	2529	2530	2531	2532	2533	2534	2535
4750	2536	2537	2538	25[illegible]	2540	2541	2542	2543
4760	2544	2545	2546	2547	2548	2549	2550	2551
4770	2552	2553	2554	2555	2556	2557	2558	2559

4000 to 4777 (Octal)	2048 to 2559 (Decimal)

Octal	Decimal
10000-	**4096**
20000-	**8192**
30000-	**12288**
40000-	**16384**
50000-	**20480**
60000-	**24576**
70000-	**28672**

5000 to 5777 (Octal)	2560 to 3071 (Decimal)

	0	1	2	3	4	5	6	7
5000	2560	2561	2562	2563	2564	2565	2566	2567
5010	2568	2569	2570	2571	2572	2573	2574	2575
5020	2576	2577	2578	2579	2580	2581	2582	2583
5030	2584	2585	2586	2587	2588	2589	2590	2591
5040	2592	2593	2594	2595	2596	2597	2598	2599
5050	2600	2601	2602	2603	2604	2605	2606	2607
5060	2608	2609	2610	2611	2612	2613	2614	2615
5070	2616	2617	2618	2619	2620	2621	2622	2623
5100	2624	2625	2626	2627	2628	2629	2630	2631
5110	2632	2633	2634	2635	2636	2637	2638	2639
5120	2640	2641	2642	2643	2644	2645	2646	2647
5130	2648	2649	2650	2651	2652	2653	2654	2655
5140	2656	2657	2658	2659	2660	2661	2662	2663
5150	2664	2665	2666	2667	2668	2669	2670	2671
5160	2672	2673	2674	2675	2676	2677	2678	2679
5170	2680	2681	2682	2683	2684	2685	2686	2687
5200	2688	2689	2690	2691	2692	2693	2694	2695
5210	2696	2697	2698	2699	2700	2701	2702	2703
5220	2704	2705	2706	2707	2708	2709	2710	2711
5230	2712	2713	2714	2715	2716	2717	2718	2719
5240	2720	2721	2722	2723	2724	2725	2726	2727
5250	2728	2729	2730	2731	2732	2733	2734	2735
5260	2736	2737	2738	2739	2740	2741	2742	2743
5270	2744	2745	2746	2747	2748	2749	2750	2751
5300	2752	2753	2754	2755	2756	2757	2758	2759
5310	2760	2761	2762	2763	2764	2765	2766	2767
5320	2768	2769	2770	2771	2772	2773	2774	2775
5330	2776	2777	2778	2779	2780	2781	2782	2783
5340	2784	2785	2786	2787	2788	2789	2790	2791
5350	2792	2793	2794	2795	2796	2797	2798	2799
5360	2800	2801	2802	2803	2804	2805	2806	2807
5370	2808	2809	2810	2811	2812	2813	2814	2815

	0	1	2	3	4	5	6	7
5400	2816	2817	2818	2819	2820	2821	2822	2823
5410	2824	2825	2826	2827	2828	2829	2830	2831
5420	2832	2833	2834	2835	2836	2837	2838	2839
5430	2840	2841	2842	2843	2844	2845	2846	2847
5440	2848	2849	2850	2851	2852	2853	2854	2855
5450	2856	2857	2858	2859	2860	2861	2862	2863
5460	2864	2865	2866	2867	2868	2869	2870	2871
5470	2872	2873	2874	2875	2876	2877	2878	2879
5500	2880	2881	2882	2883	2884	2885	2886	2887
5510	2888	2889	2890	2891	2892	2893	2894	2895
5520	2896	2897	2898	2899	2900	2901	2902	2903
5530	2904	2905	2906	2907	2908	2909	2910	2911
5540	2912	2913	2914	2915	2916	2917	2918	2919
5550	2920	2921	2922	2923	2924	2925	2926	2927
5560	2928	2929	2930	2931	2932	2933	2934	2935
5570	2936	2937	2938	2939	2940	2941	2942	2943
5600	2944	2945	2946	2947	2948	2949	2950	2951
5610	2952	2953	2954	2955	2956	2957	2958	2959
5620	2960	2961	2962	2963	2964	2965	2966	2967
5630	2968	2969	2970	2971	2972	2973	2974	2975
5640	2976	2977	2978	2979	2980	2981	2982	2983
5650	2984	2985	2986	2987	2988	2989	2990	2991
5660	2992	2993	2994	2995	2996	2997	2998	2999
5670	3000	3001	3002	3003	3004	3005	3006	3007
5700	3008	3009	3010	3011	3012	3013	3014	3015
5710	3016	3017	3018	3019	3020	3021	3022	3023
5720	3024	3025	3026	3027	3028	3029	3030	3031
5730	3032	3033	3034	3035	3036	3037	3038	3039
5740	3040	3041	3042	3043	3044	3045	3046	3047
5750	3048	3049	3050	3051	3052	3053	3054	3055
5760	3056	3057	3058	3059	3060	3061	3062	3063
5770	3064	3065	3066	3067	3068	3069	3070	3071

OCTAL-DECIMAL INTEGER CONVERSION TABLE (Continued)

6000 to 6777 (Octal) | 3072 to 3583 (Decimal)

Octal	Decimal
10000-	**4096**
20000-	**8192**
30000-	**12288**
40000-	**16384**
50000-	**20480**
60000-	**24576**
70000-	**28672**

	0	1	2	3	4	5	6	7
6000	3072	3073	3074	3075	3076	3077	3078	3079
6010	3080	3081	3082	3083	3084	3085	3086	3087
6020	3088	3089	3090	3091	3092	3093	3094	3095
6030	3096	3097	3098	3099	3100	3101	3102	3103
6040	3104	3105	3106	3107	3108	3109	3110	3111
6050	3112	3113	3114	3115	3116	3117	3118	3119
6060	3120	3121	3122	3123	3124	3125	3126	3127
6070	3128	3129	3130	3131	3132	3133	3134	3135
6100	3136	3137	3138	3139	3140	3141	3142	3143
6110	3144	3145	3146	3147	3148	3149	3150	3151
6120	3152	3153	3154	3155	3156	3157	3158	3159
6130	3160	3161	3162	3163	3164	3165	3166	3167
6140	3168	3169	3170	3171	3172	3173	3174	3175
6150	3176	3177	3178	3179	3180	3181	3182	3183
6160	3184	3185	3186	3187	3188	3189	3190	3191
6170	3192	3193	3194	3195	3196	3197	3198	3199
6200	3200	3201	3202	3203	3204	3205	3206	3207
6210	3208	3209	3210	3211	3212	3213	3214	3215
6220	3216	3217	3218	3219	3220	3221	3222	3223
6230	3224	3225	3226	3227	3228	3229	3230	3231
6240	3232	3233	3234	3235	3236	3237	3238	3239
6250	3240	3241	3442	3243	3244	3245	3246	3247
6260	3248	3249	3250	3251	3252	3253	3254	3255
6270	3256	3257	3258	3259	3260	3261	3262	3263
6300	3264	3265	3266	3267	3268	3269	3270	3871
6310	3272	3273	3274	3275	3276	3277	3278	3279
6320	3280	3281	3282	3283	3284	3285	3286	3287
6330	3288	3289	3290	3291	3292	3293	3294	3295
6340	3296	3297	3298	3299	3300	3301	3302	3003
6350	3304	3305	3306	3307	3308	3309	3310	3311
6360	3312	3313	3314	3315	3316	3317	3318	3319
6370	3320	3321	3322	3323	3324	3325	3326	3327

	0	1	2	3	4	5	6	7
6400	3328	3329	3330	3331	3332	3333	3334	3335
6410	3336	3337	3338	3339	3340	3341	3342	3343
6420	3344	3345	3346	3347	3348	3349	3350	3351
6430	3352	3353	3354	3355	3356	3357	3358	3359
6440	3360	3361	3362	3363	3364	3365	3366	3367
6450	3368	3369	3370	3371	3372	3373	3374	3375
6460	3376	3377	3378	3379	3380	3381	3382	3383
6470	3384	3385	3386	3387	3388	3389	3390	3391
6500	3392	3393	3394	3395	3396	3397	3398	3399
6510	3400	3401	3402	3403	3404	3405	3406	3407
6520	3408	3409	3410	3411	3412	3413	3414	3415
6530	3416	3417	3418	3419	3420	3421	3422	3423
6540	3424	3425	3426	3427	3428	3429	3430	3431
6550	3432	3433	3434	3435	3436	3437	3438	3439
6560	3440	3441	3442	3443	3444	3445	3446	3447
6570	3448	3449	3450	3451	3452	3453	3454	3455
6600	3456	3457	3458	3459	3460	3461	3462	3463
6610	3464	3465	3466	3467	3468	3469	3470	3471
6620	3472	3473	3474	3475	3476	3477	3478	3479
6630	3480	3481	3482	3483	3484	3485	3486	3487
6640	3488	3489	3490	3491	3492	3493	3494	3495
6650	3496	3497	3498	3499	3500	3501	3502	3503
6660	3504	3505	3506	3507	3508	3509	3510	3511
6670	3512	3513	3514	3515	3516	3517	3518	3519
6700	3520	3521	3522	3523	3524	3525	3526	3527
6710	3528	3529	3530	3531	3532	3533	3534	3535
6720	3536	3537	3538	3539	3540	3541	3542	3543
6730	3544	3545	3546	3547	3548	3549	3550	3551
6740	3552	3553	3554	3555	3556	3557	3558	3559
6750	3560	3561	3562	3563	3564	3655	3566	3567
6760	3568	3569	3570	3571	3572	3573	3574	3575
6770	3576	3577	3578	3579	3580	3581	3582	3583

7000 to 7777 (Octal) | 3584 to 4095 (Decimal)

	0	1	2	3	4	5	6	7
7000	3584	3585	3586	3587	3588	3589	3590	3591
7010	3592	3593	3594	3595	3596	3597	3598	3599
7020	3600	3601	3602	3603	3604	3605	3606	3607
7030	3608	3609	3610	3611	3612	3613	3614	3615
7040	3616	3617	3618	3619	3620	3621	3622	3623
7050	3624	3625	3626	3627	3628	3629	3630	3631
7060	3632	3633	3634	3635	3636	3637	3638	3639
7070	3640	3641	3642	3643	3644	3645	3646	3647
7100	3648	3649	3650	3651	3652	3653	3654	3655
7110	3656	3657	3658	3659	3660	3661	3662	3663
7120	3664	3665	3666	3667	3668	3669	3670	3671
7130	3672	3673	3674	3675	3676	3677	3678	3679
7140	3680	3681	3682	3683	3684	3685	3686	3687
7150	3688	3689	3690	3691	3692	3693	3694	3695
7160	3696	3697	3698	3699	3700	3701	3702	3703
7170	3704	3705	3706	3707	3708	3709	3710	3711
7200	3712	3713	3714	3715	3716	3717	3718	3719
7210	3720	3721	3722	3723	3724	3725	3726	3727
7220	3728	3729	3730	3731	3732	3733	3734	3735
7230	3736	3737	3738	3739	3740	3741	3742	3743
7240	3744	3745	3746	3747	3748	3749	3750	3751
7250	3752	3753	3754	3755	3756	3757	3758	3759
7260	3760	3761	3762	3763	3764	3765	3766	3767
7270	3768	3769	3770	3771	3772	3773	3774	3775
7300	3776	3777	3778	3779	3780	3781	3782	3783
7310	3784	3785	3786	3787	3788	3789	3790	3791
7320	3792	3893	3794	3795	3796	3797	3798	3799
7330	3800	3801	3802	3803	3804	3805	3806	3807
7340	3808	3809	3810	3811	3812	3813	3814	3815
7350	3816	3817	3818	3819	3820	3821	3822	3823
7360	3824	3825	3826	3827	3828	3829	3830	3831
7370	3832	3833	3834	3835	3836	3837	3838	3839

	0	1	2	3	4	5	6	7
7400	3840	3841	3482	3843	3844	3845	3846	3847
7410	3848	3849	3850	3851	3852	3853	3854	3855
7420	3856	3857	3858	3859	3860	3861	3862	3863
7430	3864	3865	3866	3867	3868	3869	3870	3871
7440	3872	3873	3874	3875	3876	3877	3878	3879
7450	3880	3881	3882	3883	3884	3885	3886	3887
7460	3888	3889	3890	3891	3892	3893	3894	3895
7470	3896	3897	3898	3899	3900	3901	3902	3903
7500	3904	3905	3906	3907	3908	3909	3910	3911
7510	3912	3913	3914	3915	3916	3917	3918	3919
7520	3920	3921	3922	3923	3924	3925	3926	3927
7530	3928	3929	3930	3931	3932	3933	3934	3935
7540	3936	3937	3938	3939	3940	3941	3942	3943
7550	3944	3945	3946	3947	3948	3949	3950	3951
7560	3952	3953	3954	3955	3956	3957	3958	3959
7570	3960	3961	3962	3963	3964	3965	3966	3967
7600	3968	3969	4970	3971	3972	3973	3974	3975
7610	3976	3977	3978	3979	3980	3981	3982	3983
7620	3984	3985	3986	3987	3988	3989	3990	3991
7630	3992	3993	3994	3995	3996	3997	3998	3999
7640	4000	4001	4002	4003	4004	4005	4006	4007
7650	4008	4009	4010	4011	4012	4013	4014	4015
7660	4016	4017	4018	4019	4020	4021	4022	4023
7670	4024	4025	4026	4027	4028	4029	4030	4031
7700	4032	4033	4034	4035	4036	4037	4038	4039
7710	4040	4041	4042	4043	4044	4045	4046	4047
7720	4048	4049	4050	4051	4052	4053	4054	4055
7730	4056	4057	4058	4059	4060	4061	4062	4063
7740	4064	4065	4066	4067	4068	4069	4070	4071
7750	4072	4073	4074	4075	4076	4077	4078	4079
7760	4080	4081	4082	4083	4084	4085	4086	4087
7770	4088	4089	4090	4091	4092	4093	4094	4095

OCTAL-DECIMAL FRACTION CONVERSION TABLE

This table covers the entries from $(.000)_8$ to $(.377)_8$. For entries from $(.400)_8$ to $(.777)_8$, cognizance should be made of the fact that $(.400)_8$ is $(.500)_{10}$. Hence if $(.637)_8$ is desired, find $(.237)_8$ in table, namely, $(.310456)_{10}$ and add $(.50000)_{10}$ for $(.400)_8$. Thus

$$(.637)_8 = (.237)_8 + (.400)_8$$
$$= (.310456)_{10} + (.50000)_{10}$$
$$= (.810456)_{10}.$$

OCTAL	DEC.	OCTAL	DEC.	OCTAL	DEC.	OCTAL	DEC.
.000	.000000	.100	.125000	.200	.250000	.300	.375000
.001	.001953	.101	.126953	.201	.251953	.301	.376953
.002	.003906	.102	.128906	.202	.253906	.302	.378906
.003	.005859	.103	.130859	.203	.255859	.303	.380859
.004	.007812	.104	.132812	.204	.257812	.304	.382812
.005	.009765	.105	.134765	.205	.259765	.305	.384765
.006	.011718	.106	.136718	.206	.261718	.306	.386718
.007	.013671	.107	.138671	.207	.263671	.307	.388671
.010	.015625	.110	.140625	.210	.265625	.310	.390625
.011	.017578	.111	.142578	.211	.267578	.311	.392578
.012	.019531	.112	.144531	.212	.269531	.312	.394531
.013	.021484	.113	.146484	.213	.271484	.313	.396484
.014	.023437	.114	.148437	.214	.273437	.314	.398437
.015	.025390	.115	.150390	.215	.275390	.315	.400490
.016	.027343	.116	.152343	.216	.277343	.316	.402343
.017	.029296	.117	.154296	.217	.279296	.317	.404296
.020	.031250	.120	.156250	.220	.281250	.320	.406250
.021	.033203	.121	.158203	.221	.283203	.321	.408203
.022	.035156	.122	.160156	.222	.285156	.322	.410156
.023	.037109	.123	.162109	.223	.287109	.323	.412109
.024	.039062	.124	.164062	.224	.289062	.324	.414062
.025	.041015	.125	.166015	.225	.291015	.325	.416015
.026	.042968	.126	.167968	.226	.292968	.326	.417968
.027	.044921	.127	.169921	.227	.294921	.327	.419921
.030	.046875	.130	.171875	.230	.294875	.330	.421875
.031	.048828	.131	.173828	.231	.298828	.331	.423828
.032	.050781	.132	.175781	.232	.300781	.332	.425781
.033	.052734	.133	.177734	.233	.302734	.333	.427734
.034	.054687	.134	.179687	.234	.304687	.334	.429687
.035	.056640	.135	.181640	.235	.306640	.335	.431640
.036	.058593	.136	.183593	.236	.308593	.336	.433593
.037	.060546	.137	.185546	.237	.310546	.337	.435546
.040	.062500	.140	.187500	.240	.312500	.340	.437500
.041	.064453	.141	.189453	.241	.314453	.341	.439453
.042	.066406	.142	.191406	.242	.316406	.342	.441406
.043	.068359	.143	.193359	.243	.318359	.343	.443359
.044	.070312	.144	.195312	.244	.320312	.344	.445312
.045	.072265	.145	.197265	.245	.322265	.345	.447265
.046	.074218	.146	.199218	.246	.324218	.346	.449218
.047	.076171	.147	.201171	.247	.326171	.347	.451171
.050	.078125	.150	.203125	.250	.328125	.350	.453125
.051	.080078	.151	.205078	.251	.330078	.351	.455078
.052	.082031	.152	.207031	.252	.332031	.352	.457031
.053	.083984	.153	.208984	.253	.333984	.353	.458984
.054	.085937	.154	.210937	.254	.335937	.354	.460937
.055	.087890	.155	.212890	.255	.337890	.355	.462890
.056	.089843	.156	.214843	.256	.339843	.356	.464843
.057	.091796	.157	.216796	.257	.341796	.357	.466796
.060	.093750	.160	.218750	.260	.343750	.360	.468750
.061	.095703	.161	.220703	.261	.345703	.361	.470703
.062	.097656	.162	.222656	.262	.347656	.362	.472656
.063	.099609	.163	.224609	.263	.349609	.363	.474609
.064	.101562	.164	.226562	.264	.351562	.364	.476562
.065	.103515	.165	.228515	.265	.353515	.365	.478515
.066	.105468	.166	.230468	.266	.355468	.366	.480468
.067	.107421	.167	.232421	.267	.357421	.367	.482421
.070	.109375	.170	.234375	.270	.359375	.370	.484375
.071	.111328	.171	.236328	.271	.361328	.371	.486328
.072	.113281	.172	.238281	.272	.363281	.372	.488281
.073	.115234	.173	.240234	.273	.365234	.373	.490234
.074	.117187	.174	.242187	.274	.367187	.374	.492187
.075	.119140	.175	.244140	.275	.369140	.375	.494140
.076	.121093	.176	.246093	.276	.371093	.376	.496093
.077	.123046	.177	.248046	.277	.373046	.377	.498046

OCTAL-DECIMAL FRACTION CONVERSION TABLE (Continued)

OCTAL	DEC.	OCTAL	DEC.	OCTAL	DEC.	OCTAL	DEC.
.000000	.000000	.000100	.000244	.000200	.000488	.000300	.000732
.000001	.000004	.000101	.000247	.000201	.000492	.000301	.000736
.000002	.000007	.000102	.000251	.000202	.000495	.000302	.000740
.000003	.000011	.000103	.000255	.000203	.000499	.000303	.000743
.000004	.000015	.000104	.000259	.000204	.000503	.000304	.000747
.000005	.000019	.000105	.000263	.000205	.000507	.000305	.000751
.000006	.000022	.000106	.000267	.000206	.000511	.000306	.000755
.000007	.000026	.000107	.000270	.000207	.000514	.000307	.000759
.000010	.000030	.000110	.000274	.000210	.000518	.000310	.000762
.000011	.000034	.000111	.000278	.000211	.000522	.000311	.000766
.000012	.000038	.000112	.000282	.000212	.000526	.000312	.000770
.000013	.000041	.000113	.000286	.000213	.000530	.000313	.000774
.000014	.000045	.000114	.000289	.000214	.000534	.000314	.000778
.000015	.000049	.000115	.000293	.000215	.000537	.000315	.000782
.000016	.000053	.000116	.000297	.000216	.000541	.000316	.000785
.000017	.000057	.000117	.000301	.000217	.000545	.000317	.000789
.000020	.000061	.000120	.000305	.000220	.000549	.000320	.000793
.000021	.000064	.000121	.000308	.000221	.000553	.000321	.000797
.000022	.000068	.000122	.000312	.000222	.000556	.000322	.000801
.000023	.000072	.000123	.000316	.000223	.000560	.000323	.000805
.000024	.000076	.000124	.000320	.000224	.000564	.000324	.000808
.000025	.000080	.000125	.000324	.000225	.000568	.000325	.000812
.000026	.000083	.000126	.000328	.000226	.000572	.000326	.000816
.000027	.000087	.000127	.000331	.000227	.000576	.000327	.000820
.000030	.000091	.000130	.000335	.000230	.000579	.000330	.000823
.000031	.000095	.000131	.000339	.000231	.000583	.000331	.000827
.000032	.000099	.000132	.000343	.000232	.000587	.000332	.000831
.000033	.000102	.000133	.000347	.000233	.000591	.000333	.000835
.000034	.000106	.000134	.000350	.000234	.000595	.000334	.000839
.000035	.000110	.000135	.000354	.C00235	.000598	.000335	.000843
.000036	.000114	.000136	.000358	.000236	.000602	.000336	.000846
.000037	.000118	.000137	.000362	.000237	.000606	.000337	.000850
.000040	.000122	.000140	.000366	.000240	.000610	.000340	.000854
.000041	.000125	.000141	.000370	.000241	.000614	.000341	.000858
.000042	.000129	.000142	.000373	.000242	.000617	.000342	.000862
.000043	.000133	.000143	.000377	.000243	.000621	.000343	.000865
.000044	.000137	.000144	.000381	.000244	.000625	.000344	.000869
.000045	.000141	.000145	.000385	.000245	.000629	.000345	.000873
.000046	.000144	.000146	.000389	.000246	.000633	.000346	.000877
.000047	.000148	.000147	.000392	.000247	.000637	.000347	.000881
.000050	.000152	.000150	.000396	.000250	.000640	.000350	.000885
.000051	.000156	.000151	.000400	.000251	.000644	.000351	.000888
.000052	.000160	.000152	.000404	.000252	.000648	.000352	.000892
.000053	.000164	.000153	.000408	.000253	.000652	.000353	.000896
.000054	.000167	.000154	.000411	.000254	.000656	.000354	.000900
.000055	.000171	.000155	.000415	.000255	.000659	.000355	.000904
.000056	.000175	.000156	.000419	.000256	.000663	.000356	.000907
.000057	.000179	.000157	.000423	.000257	.000667	.000357	.000911
.000060	.000183	.000160	.000427	.000260	.000671	.000360	.000915
.000061	.000186	.000161	.000431	.000261	.000675	.000361	.000919
.000062	.000190	.000162	.000434	.000262	.000679	.000362	.000923
.000063	.000194	.000163	.000438	.000263	.000682	.000363	.000926
.000064	.000198	.000164	.000442	.000264	.000686	.000364	.000930
.000065	.000202	.000165	.000446	.000265	.000690	.000365	.000934
.000066	.000205	.000166	.000450	.000266	.000694	.000366	.000938
.000067	.000209	.000167	.000453	.000267	.000698	.000367	.000942
.000070	.000213	.000170	.000457	.000270	.000701	.000370	.000946
.000071	.000217	.000171	.000461	.000271	.000705	.000371	.000949
.000072	.000221	.000172	.000465	.000272	.000709	.000372	.000953
.000073	.000225	.000173	.000469	.000273	.000713	.000373	.000957
.000074	.000228	.000174	.000473	.000274	.000717	.000374	.000961
.000075	.000232	.000175	.000476	.000275	.000720	.000375	.000965
.000076	.000326	.000176	.000480	.000276	.000724	.000376	.000968
.000077	.000240	.000177	.000484	.000277	.000728	.000377	.000972

OCTAL-DECIMAL FRACTION CONVERSION TABLE (Continued)

OCTAL	DEC.	OCTAL	DEC.	OCTAL	DEC.	OCTAL	DEC.
.000400	.000976	.000500	.001220	.000600	.001464	.000700	.001708
.000401	.000980	.000501	.001224	.000601	.001468	.000701	.001712
.000402	.000984	.000502	.001228	.000602	.001472	.000702	.001716
.000403	.000988	.000503	.001232	.000603	.001476	.000703	.001720
.000404	.000991	.000504	.001235	.000604	.001480	.000704	.001724
.000405	.000995	.000505	.001239	.000605	.001483	.000705	.001728
.000406	.000999	.000506	.001243	.000606	.001487	.000706	.001731
.000407	.001003	.000507	.001247	.000607	.001491	.000707	.001735
.000410	.001007	.000510	.001251	.000610	.001495	.000710	.001739
.000411	.001010	.000511	.001255	.000611	.001499	.000711	.001743
.000412	.001014	.000512	.001258	.000612	.001502	.000712	.001747
.000413	.001018	.000513	.001262	.000613	.001506	.000713	.001750
.000414	.001022	.000514	.001266	.000614	.001510	.000714	.001754
.000415	.001026	.000515	.001270	.000615	.001514	.000715	.001758
.000416	.001029	.000516	.001274	.000616	.001518	.000716	.001762
.000417	.001033	.000517	.001277	.000617	.001522	.000717	.001766
.000420	.001037	.000520	.001281	.000620	.001525	.000720	.001770
.000421	.001041	.000521	.001285	.000621	.001529	.000721	.001733
.000422	.001045	.000522	.001289	.000622	.001533	.000722	.001777
.000423	.001049	.000523	.001293	.000623	.001537	.000723	.001781
.000424	.001052	.000524	.001296	.000624	.001541	.000724	.001785
.000425	.001056	.000525	.001300	.000625	.001544	.000725	.001789
.000426	.001060	.000526	.001304	.000626	.001548	.000726	.001792
.000427	.001064	.000527	.001308	.000627	.001552	.000727	.001796
.000430	.001068	.000530	.001312	.000630	.001556	.000730	.001800
.000431	.001071	.000531	.001316	.000631	.001560	.000731	.001804
.000432	.001075	.000532	.001319	.000632	.001564	.000732	.001808
.000433	.001079	.000533	.001323	.000633	.001567	.000733	.001811
.000434	.001083	.000534	.001327	.000634	.001571	.000734	.001815
.000435	.001087	.000535	.001331	.000635	.001575	.000735	.001819
.000436	.001091	.000536	.001335	.000636	.001579	.000736	.001823
.000437	.001094	.000537	.001338	.000637	.001583	.000737	.001827
.000440	.001098	.000540	.001342	.000640	.001586	.000740	.001831
.000441	.001102	.000541	.001346	.000641	.001590	.000741	.001834
.000442	.001106	.000542	.001350	.000642	.001594	.000742	.001838
.000443	.001110	.000543	.001354	.000643	.001598	.000743	.001842
.000444	.001113	.000544	.001358	.000644	.001602	.000744	.001846
.000445	.001117	.000545	.001361	.000645	.001605	.000745	.001850
.000446	.001121	.000546	.001365	.000646	.001609	.000746	.001853
.000447	.001125	.000547	.001369	.000647	.001613	.000747	.001857
.000450	.001129	.000550	.001373	.000650	.001617	.000750	.001861
.000451	.001132	.000551	.001377	.000651	.001621	.000751	.001865
.000452	.001136	.000552	.001380	.000652	.001625	.000752	.001869
.000453	.001140	.000553	.001384	.000653	.001628	.000753	.001873
.000454	.001144	.000554	.001388	.000654	.001632	.000754	.001876
.000455	.001148	.000555	.001392	.000655	.001636	.000755	.001880
.000456	.001152	.000556	.001396	.000656	.001640	.000756	.001884
.000457	.001155	.000557	.001399	.000657	.001644	.000757	.001888
.000460	.001159	.000560	.001403	.000660	.001647	.000760	.001892
.000461	.001163	.000561	.001407	.000661	.001651	.000761	.001895
.000462	.001167	.000562	.001411	.000662	.001655	.000762	.001899
.000463	.001171	.000563	.001415	.000663	.001659	.000763	.001903
.000464	.001174	.000564	.001419	.000664	.001663	.000764	.001907
.000465	.001178	.000565	.001422	.000665	.001667	.000765	.001911
.000466	.001182	.000566	.001426	.000666	.001670	.000766	.001914
.000467	.001186	.000567	.001430	.000667	.001674	.000767	.001918
.000470	.001190	.000570	.001434	.000670	.001678	.000770	.001922
.000471	.001194	.000571	.001438	.000671	.001682	.000771	.001926
.000472	.001197	.000572	.001441	.000672	.001686	.000772	.001930
.000473	.001201	.000573	.001445	.000673	.001689	.000773	.001934
.000474	.001205	.000574	.001449	.000674	.001693	.000774	.001937
.000475	.001209	.000575	.001453	.000675	.001697	.000775	.001941
.000476	.001213	.000576	.001457	.000676	.001701	.000776	.001945
.000477	.001216	.000577	.001461	.000677	.001705	.000777	.001949

I. HEXADECIMAL AND DECIMAL DIRECT CONVERSION TABLE

The following tables aid in converting hexadecimal (base 16) numbers to decimal, and the reverse. Note that the base 16 digits for the decimal values 10–15 are represented by the letters A–F, respectively.

This table provides direct conversion of decimal and hexadecimal numbers in these ranges:

HEXADECIMAL	DECIMAL
000 to FFF	0000 to 4095

For numbers outside the range of the table, add the following values to the table figures:

HEXADECIMAL	DECIMAL
1000	4096
2000	8192
3000	12288
4000	16384
5000	20480
6000	24576
7000	28672
8000	32768
9000	36864
A000	40960
B000	45056
C000	49152
D000	53248
E000	57344
F000	61440

	0	1	2	3	4	5	6	7	8	9	A	B	C	D	E	F
00_	0000	0001	0002	0003	0004	0005	0006	0007	0008	0009	0010	0011	0012	0013	0014	0015
01_	0016	0017	0018	0019	0020	0021	0022	0023	0024	0025	0026	0027	0028	0029	0030	0031
02_	0032	0033	0034	0035	0036	0037	0038	0039	0040	0041	0042	0043	0044	0045	0046	0047
03_	0048	0049	0050	0051	0052	0053	0054	0055	0056	0057	0058	0059	0060	0061	0062	0063
04_	0064	0065	0066	0067	0068	0069	0070	0071	0072	0073	0074	0075	0076	0077	0078	0079
05_	0080	0081	0082	0083	0084	0085	0086	0087	0088	0089	0090	0091	0092	0093	0094	0095
06_	0096	0097	0098	0099	0100	0101	0102	0103	0104	0105	0106	0107	0108	0109	0110	0111
07_	0112	0113	0114	0115	0116	0117	0118	0119	0120	0121	0122	0123	0124	0125	0126	0127
08_	0128	0129	0130	0131	0132	0133	0134	0135	0136	0137	0138	0139	0140	0141	0142	0143
09_	0144	0145	0146	0147	0148	0149	0150	0151	0152	0153	0154	0155	0156	0157	0158	0159
0A_	0160	0161	0162	0163	0164	0165	0166	0167	0168	0169	0170	0171	0172	0173	0174	0175
0B_	0176	0177	0178	0179	0180	0181	0182	0183	0184	0185	0186	0187	0188	0189	0190	0191
0C_	0192	0193	0194	0195	0196	0197	0198	0199	0200	0201	0202	0203	0204	0205	0206	0207
0D_	0208	0209	0210	0211	0212	0213	0214	0215	0216	0217	0218	0219	0220	0221	0222	0223
0E_	0224	0225	0226	0227	0228	0229	0230	0231	0232	0233	0234	0235	0236	0237	0238	0239
0F_	0240	0241	0242	0243	0244	0245	0246	0247	0248	0249	0250	0251	0252	0253	0254	0255
10_	0256	0257	0258	0259	0260	0261	0262	0263	0264	0265	0266	0267	0268	0269	0270	0271
11_	0272	0273	0274	0275	0276	0277	0278	0279	0280	0281	0282	0283	0284	0285	0286	0287
12_	0288	0289	0290	0291	0292	0293	0294	0295	0296	0297	0298	0299	0300	0301	0302	0303
13_	0304	0305	0306	0307	0308	0309	0310	0311	0312	0313	0314	0315	0316	0317	0318	0319
14_	0320	0321	0322	0323	0324	0325	0326	0327	0328	0329	0330	0331	0332	0333	0334	0335
15_	0336	0337	0338	0339	0340	0341	0342	0343	0344	0345	0346	0347	0348	0349	0350	0351
16_	0352	0353	0354	0355	0356	0357	0358	0359	0360	0361	0362	0363	0364	0365	0366	0367
17_	0368	0369	0370	0371	0372	0373	0374	0375	0376	0377	0378	0379	0380	0381	0382	0383
18_	0384	0385	0386	0387	0388	0389	0390	0391	0392	0393	0394	0395	0396	0397	0398	0399
19_	0400	0401	0402	0403	0404	0405	0406	0407	0408	0409	0410	0411	0412	0413	0414	0415
1A_	0416	0417	0418	0419	0420	0421	0422	0423	0424	0425	0426	0427	0428	0429	0430	0431
1B_	0432	0433	0434	0435	0436	0437	0438	0439	0440	0441	0442	0443	0444	0445	0446	0447
1C_	0448	0449	0450	0451	0452	0453	0454	0455	0456	0457	0458	0459	0460	0461	0462	0463
1D_	0464	0465	0466	0467	0468	0469	0470	0471	0472	0473	0474	0475	0476	0477	0478	0479
1E_	0480	0481	0482	0483	0484	0485	0486	0487	0488	0489	0490	0491	0492	0493	0494	0495
1F_	0496	0497	0498	0499	0500	0501	0502	0503	0504	0505	0506	0507	0508	0509	0510	0511

DIRECT CONVERSION TABLE (Continued)

	0	1	2	3	4	5	6	7	8	9	A	B	C	D	E	F
20_	0512	0513	0514	0515	0516	0517	0518	0519	0520	0521	0522	0523	0524	0525	0526	0527
21_	0528	0529	0530	0531	0532	0533	0534	0535	0536	0537	0538	0539	0540	0541	0542	0543
22_	0544	0545	0546	0547	0548	0549	0550	0551	0552	0553	0554	0555	0556	0557	0558	0559
23_	0560	0561	0562	0563	0564	0565	0566	0567	0568	0569	0570	0571	0572	0573	0574	0575
24_	0576	0577	0578	0579	0580	0581	0582	0583	0584	0585	0586	0587	0588	0589	0590	0591
25_	0592	0593	0594	0595	0596	0597	0598	0599	0600	0601	0602	0603	0604	0605	0606	0607
26_	0608	0609	0610	0611	0612	0613	0614	0615	0616	0617	0618	0619	0620	0621	0622	0623
27_	0624	0625	0626	0627	0628	0629	0630	0631	0632	0633	0634	0635	0636	0637	0638	0639
28_	0640	0641	0642	0643	0644	0645	0646	0647	0648	0649	0650	0651	0652	0653	0654	0655
29_	0656	0657	0658	0659	0660	0661	0662	0663	0664	0665	0666	0667	0668	0669	0670	0671
2A_	0672	0673	0674	0675	0676	0677	0678	0679	0680	0681	0682	0683	0684	0685	0686	0687
2B_	0688	0689	0690	0691	0692	0693	0694	0695	0696	0697	0698	0699	0700	0701	0702	0703
2C_	0704	0705	0706	0707	0708	0709	0710	0711	0712	0713	0714	0715	0716	0717	0718	0719
2D_	0720	0721	0722	0723	0724	0725	0726	0727	0728	0729	0730	0731	0732	0733	0734	0735
2E_	0736	0737	0738	0739	0740	0741	0742	0743	0744	0745	0746	0747	0748	0749	0750	0751
2F_	0752	0753	0754	0755	0756	0757	0758	0759	0760	0761	0762	0763	0764	0765	0766	0767
30_	0768	0769	0770	0771	0772	0773	0774	0775	0776	0777	0778	0779	0780	0781	0782	0783
31_	0784	0785	0786	0787	0788	0789	0790	0791	0792	0793	0794	0795	0796	0797	0798	0799
32_	0800	0801	0802	0803	0804	0805	0806	0807	0808	0809	0810	0811	0812	0813	0814	0815
33_	0816	0817	0818	0819	0820	0821	0822	0823	0824	0825	0826	0827	0828	0829	0830	0831
34_	0832	0833	0834	0835	0836	0837	0838	0839	0840	0841	0842	0843	0844	0845	0846	0847
35_	0848	0849	0850	0851	0852	0853	0854	0855	0856	0857	0858	0859	0860	0861	0862	0863
36_	0864	0865	0866	0867	0868	0869	0870	0871	0872	0873	0874	0875	0876	0877	0878	0879
37_	0880	0881	0882	0883	0884	0885	0886	0887	0888	0889	0890	0891	0892	0893	0894	0895
38_	0896	0897	0898	0899	0900	0901	0902	0903	0904	0905	0906	0907	0908	0909	0910	0911
39_	0912	0913	0914	0915	0916	0917	0918	0919	0920	0921	0922	0923	0924	0925	0926	0927
3A_	0928	0929	0930	0931	0932	0933	0934	0935	0936	0937	0938	0939	0940	0941	0942	0943
3B_	0944	0945	0946	0947	0948	0949	0950	0951	0952	0953	0954	0955	0956	0957	0958	0959
3C_	0960	0961	0962	0963	0964	0965	0966	0967	0968	0969	0970	0971	0972	0973	0974	0975
3D_	0976	0977	0978	0979	0980	0981	0982	0983	0984	0985	0986	0987	0988	0989	0990	0991
3E_	0992	0993	0994	0995	0996	0997	0998	0999	1000	1001	1002	1003	1004	1005	1006	1007
3F_	1008	1009	1010	1011	1012	1013	1014	1015	1016	1017	1018	1019	1020	1021	1022	1023
40_	1024	1025	1026	1027	1028	1029	1030	1031	1032	1033	1034	1035	1036	1037	1038	1039
41_	1040	1041	1042	1043	1044	1045	1046	1047	1048	1049	1050	1051	1052	1053	1054	1055
42_	1056	1057	1058	1059	1060	1061	1062	1063	1064	1065	1066	1067	1068	1069	1070	1071
43_	1072	1073	1074	1075	1076	1077	1078	1079	1080	1081	1082	1083	1084	1085	1086	1087
44_	1088	1089	1090	1091	1092	1093	1094	1095	1096	1097	1098	1099	1100	1101	1102	1103
45_	1104	1105	1106	1107	1108	1109	1110	1111	1112	1113	1114	1115	1116	1117	1118	1119
46_	1120	1121	1122	1123	1124	1125	1126	1127	1128	1129	1130	1131	1132	1133	1134	1135
47_	1136	1137	1138	1139	1140	1141	1142	1143	1144	1145	1146	1147	1148	1149	1150	1151
48_	1152	1153	1154	1155	1156	1157	1158	1159	1160	1161	1162	1163	1164	1165	1166	1167
49_	1168	1169	1170	1171	1172	1173	1174	1175	1176	1177	1178	1179	1180	1181	1182	1183
4A_	1184	1185	1186	1187	1188	1189	1190	1191	1192	1193	1194	1195	1196	1197	1198	1199
4B_	1200	1201	1202	1203	1204	1205	1206	1207	1208	1209	1210	1211	1212	1213	1214	1215
4C_	1216	1217	1218	1219	1220	1221	1222	1223	1224	1225	1226	1227	1228	1229	1230	1231
4D_	1232	1233	1234	1235	1236	1237	1238	1239	1240	1241	1242	1243	1244	1245	1246	1247
4E_	1248	1249	1250	1251	1252	1253	1254	1255	1256	1257	1258	1259	1260	1261	1262	1263
4F_	1264	1265	1266	1267	1268	1269	1270	1271	1272	1273	1274	1275	1276	1277	1278	1279
50_	1280	1281	1282	1283	1284	1285	1286	1287	1288	1289	1290	1291	1292	1293	1294	1295
51_	1296	1297	1298	1299	1300	1301	1302	1303	1304	1305	1306	1307	1308	1309	1310	1311
52_	1312	1313	1314	1315	1316	1317	1318	1319	1320	1321	1322	1323	1324	1325	1326	1327
53_	1328	1329	1330	1331	1332	1333	1334	1335	1336	1337	1338	1339	1340	1341	1342	1343
54_	1344	1345	1346	1347	1348	1349	1350	1351	1352	1353	1354	1355	1356	1357	1358	1359
55_	1360	1361	1362	1363	1364	1365	1366	1367	1368	1369	1370	1371	1372	1373	1374	1375
56_	1376	1377	1378	1379	1380	1381	1382	1383	1384	1385	1386	1387	1388	1389	1390	1391
57_	1392	1393	1394	1395	1396	1397	1398	1399	1400	1401	1402	1403	1404	1405	1406	1407
58_	1408	1409	1410	1411	1412	1413	1414	1415	1416	1417	1418	1419	1420	1421	1422	1423
59_	1424	1425	1426	1427	1428	1429	1430	1431	1432	1433	1434	1435	1436	1437	1438	1439
5A_	1440	1441	1442	1443	1444	1445	1446	1447	1448	1449	1450	1451	1452	1453	1454	1455
5B_	1456	1457	1458	1459	1460	1461	1462	1463	1464	1465	1466	1467	1468	1469	1470	1471
5C_	1472	1473	1474	1475	1476	1477	1478	1479	1480	1481	1482	1483	1484	1485	1486	1487
5D_	1488	1489	1490	1491	1492	1493	1494	1495	1496	1497	1498	1499	1500	1501	1502	1503
5E_	1504	1505	1506	1507	1508	1509	1510	1511	1512	1513	1514	1515	1516	1517	1518	1519
5F_	1520	1521	1522	1523	1524	1525	1526	1527	1528	1529	1530	1531	1532	1533	1534	1535

DIRECT CONVERSION TABLE (Continued)

	0	1	2	3	4	5	6	7	8	9	A	B	C	D	E	F
60_	1536	1537	1538	1539	1540	1541	1542	1543	1544	1545	1546	1547	1548	1549	1550	1551
61_	1552	1553	1554	1555	1556	1557	1558	1559	1560	1561	1562	1563	1564	1565	1566	1567
62_	1568	1569	1570	1571	1572	1573	1574	1575	1576	1577	1578	1579	1580	1581	1582	1583
63_	1584	1585	1586	1587	1588	1589	1590	1591	1592	1593	1594	1595	1596	1597	1598	1599
64_	1600	1601	1602	1603	1604	1605	1606	1607	1608	1609	1610	1611	1612	1613	1614	1615
65_	1616	1617	1618	1619	1620	1621	1622	1623	1624	1625	1626	1627	1628	1629	1630	1631
66_	1632	1633	1634	1635	1636	1637	1638	1639	1640	1641	1642	1643	1644	1645	1646	1647
67_	1648	1649	1650	1651	1652	1653	1654	1655	1656	1657	1658	1659	1660	1661	1662	1663
68_	1664	1665	1666	1667	1668	1669	1670	1671	1672	1673	1674	1675	1676	1677	1678	1679
69_	1680	1681	1682	1683	1684	1685	1686	1687	1688	1689	1690	1691	1692	1693	1694	1695
6A_	1696	1697	1698	1699	1700	1701	1702	1703	1704	1705	1706	1707	1708	1709	1710	1711
6B_	1712	1713	1714	1715	1716	1717	1718	1719	1720	1721	1722	1723	1724	1725	1726	1727
6C_	1728	1729	1730	1731	1732	1733	1734	1735	1736	1737	1738	1739	1740	1741	1742	1743
6D_	1744	1745	1746	1747	1748	1749	1750	1751	1752	1753	1754	1755	1756	1757	1758	1759
6E_	1760	1761	1762	1763	1764	1765	1766	1767	1768	1769	1770	1771	1772	1773	1774	1775
6F_	1776	1777	1778	1779	1780	1781	1782	1783	1784	1785	1786	1787	1788	1789	1790	1791
70_	1792	1793	1794	1795	1796	1797	1798	1799	1800	1801	1802	1803	1804	1805	1806	1807
71_	1808	1809	1810	1811	1812	1813	1814	1815	1816	1817	1818	1819	1820	1821	1822	1823
72_	1824	1825	1826	1827	1828	1829	1830	1831	1832	1833	1834	1835	1836	1837	1838	1839
73_	1840	1841	1842	1843	1844	1845	1846	1847	1848	1849	1850	1851	1852	1853	1854	1855
74_	1856	1857	1858	1859	1860	1861	1862	1863	1864	1865	1866	1867	1868	1869	1870	1871
75_	1872	1873	1874	1875	1876	1877	1878	1879	1880	1881	1882	1883	1884	1885	1886	1887
76_	1888	1889	1890	1891	1892	1893	1894	1895	1896	1897	1898	1899	1900	1901	1902	1903
77_	1904	1905	1906	1907	1908	1909	1910	1911	1912	1913	1914	1915	1916	1917	1918	1919
78_	1920	1921	1922	1923	1924	1925	1926	1927	1928	1929	1930	1931	1932	1933	1934	1935
79_	1936	1937	1938	1939	1940	1941	1942	1943	1944	1945	1946	1947	1948	1949	1950	1951
7A_	1952	1953	1954	1955	1956	1957	1958	1959	1960	1961	1962	1963	1964	1965	1966	1967
7B_	1968	1969	1970	1971	1972	1973	1974	1975	1976	1977	1978	1979	1980	1981	1982	1983
7C_	1984	1985	1986	1987	1988	1989	1990	1991	1992	1993	1994	1995	1996	1997	1998	1999
7D_	2000	2001	2002	2003	2004	2005	2006	2007	2008	2009	2010	2011	2012	2013	2014	2015
7E_	2016	2017	2018	2019	2020	2021	2022	2023	2024	2025	2026	2027	2028	2029	2030	2031
7F_	2032	2033	2034	2035	2036	2037	2038	2039	2040	2041	2042	2043	2044	2045	2046	2047
80_	2048	2049	2050	2051	2052	2053	2054	2055	2056	2057	2058	2059	2060	2061	2062	2063
81_	2064	2065	2066	2067	2068	2069	2070	2071	2072	2073	2074	2075	2076	2077	2078	2079
82_	2080	2081	2082	2083	2084	2085	2086	2087	2088	2089	2090	2091	2092	2093	2094	2095
83_	2096	2097	2098	2099	2100	2101	2102	2103	2104	2105	2106	2107	2108	2109	2110	2111
84_	2112	2113	2114	2115	2116	2117	2118	2119	2120	2121	2122	2123	2124	2125	2126	2127
85_	2128	2129	2130	2131	2132	2133	2134	2135	2136	2137	2138	2139	2140	2141	2142	2143
86_	2144	2145	2146	2147	2148	2149	2150	2151	2152	2153	2154	2155	2156	2157	2158	2159
87_	2160	2161	2162	2163	2164	2165	2166	2167	2168	2169	2170	2171	2172	2173	2174	2175
88_	2176	2177	2178	2179	2180	2181	2182	2183	2184	2185	2186	2187	2188	2189	2190	2191
89_	2192	2193	2194	2195	2196	2197	2198	2199	2200	2201	2202	2203	2204	2205	2206	2207
8A_	2208	2209	2210	2211	2212	2213	2214	2215	2216	2217	2218	2219	2220	2221	2222	2223
8B_	2224	2225	2226	2227	2228	2229	2230	2231	2232	2233	2234	2235	2236	2237	2238	2239
8C_	2240	2241	2242	2243	2244	2245	2246	2247	2248	2249	2250	2251	2252	2253	2254	2255
8D_	2256	2257	2258	2259	2260	2261	2262	2263	2264	2265	2266	2267	2268	2269	2270	2271
8E_	2272	2273	2274	2275	2276	2277	2278	2279	2280	2281	2282	2283	2284	2285	2286	2287
8F_	2288	2289	2290	2291	2292	2293	2294	2295	2296	2297	2298	2299	2300	2301	2302	2303
90_	2304	2305	2306	2307	2308	2309	2310	2311	2312	2313	2314	2315	2316	2317	2318	2319
91_	2320	2321	2322	2323	2324	2325	2326	2327	2328	2329	2330	2331	2332	2333	2334	2335
92_	2336	2337	2338	2339	2340	2341	2342	2343	2344	2345	2346	2347	2348	2349	2350	2351
93_	2352	2353	2354	2355	2356	2357	2358	2359	2360	2361	2362	2363	2364	2365	2366	2367
94_	2368	2369	2370	2371	2372	2373	2374	2375	2376	2377	2378	2379	2380	2381	2382	2383
95_	2384	2385	2386	2387	2388	2389	2390	2391	2392	2393	2394	2395	2396	2397	2398	2399
96_	2400	2401	2402	2403	2404	2405	2406	2407	2408	2409	2410	2411	2412	2413	2414	2415
97_	2416	2417	2418	2419	2420	2421	2422	2423	2424	2425	2426	2427	2428	2429	2430	2431
98_	2432	2433	2434	2435	2436	2437	2438	2439	2440	2441	2442	2443	2444	2445	2446	2447
99_	2448	2449	2450	2451	2452	2453	2454	2455	2456	2457	2458	2459	2460	2461	2462	2463
9A_	2464	2465	2466	2467	2468	2469	2470	2471	2472	2473	2474	2475	2476	2477	2478	2479
9B_	2480	2481	2482	2483	2484	2485	2486	2487	2488	2489	2490	2491	2492	2493	2494	2495
9C_	2496	2497	2498	2499	2500	2501	2502	2503	2504	2505	2506	2507	2508	2509	2510	2511
9D_	2512	2513	2514	2515	2516	2517	2518	2519	2520	2521	2522	2523	2524	2525	2526	2527
9E_	2528	2529	2530	2531	2532	2533	2534	2535	2536	2537	2538	2539	2540	2541	2542	2543
9F_	2544	2545	2546	2547	2548	2549	2550	2551	2552	2553	2554	2555	2556	2557	2558	2559

DIRECT CONVERSION TABLE (Continued)

	0	1	2	3	4	5	6	7	8	9	A	B	C	D	E	F
A0__	2560	2561	2562	2563	2564	2565	2566	2567	2568	2569	2570	2571	2572	2573	2574	2575
A1__	2576	2577	2578	2579	2580	2581	2582	2583	2584	2585	2586	2587	2588	2589	2590	2591
A2__	2592	2593	2594	2595	2596	2597	2598	2599	2600	2601	2602	2603	2604	2605	2606	2607
A3__	2608	2609	2610	2611	2612	2613	2614	2615	2616	2617	2618	2619	2620	2621	2622	2623
A4__	2624	2625	2626	2627	2628	2629	2630	2631	2632	2633	2634	2635	2636	2637	2638	2639
A5__	2640	2641	2642	2643	2644	2645	2646	2647	2648	2649	2650	2651	2652	2653	2654	2655
A6__	2656	2657	2658	2659	2660	2661	2662	2663	2664	2665	2666	2667	2668	2669	2670	2671
A7__	2672	2673	2674	2675	2676	2677	2678	2679	2680	2681	2682	2683	2684	2685	2686	2687
A8__	2688	2689	2690	2691	2692	2693	2694	2695	2696	2697	2698	2699	2700	2701	2702	2703
A9__	2704	2705	2706	2707	2708	2709	2710	2711	2712	2713	2714	2715	2716	2717	2718	2719
AA__	2720	2721	2722	2723	2724	2725	2726	2727	2728	2729	2730	2731	2732	3733	2734	2735
AB__	2736	2737	2738	2739	2740	2741	2742	2743	2744	2745	2746	2747	2748	2749	2750	2751
AC__	2752	2753	2754	2755	2756	2757	2758	2759	2760	2761	2762	2763	2764	2765	2766	2767
AD__	2768	2769	2770	2771	2772	2773	2774	2775	2776	2777	2778	2779	2780	2781	2782	2783
AE__	2784	2785	2786	2787	2788	2789	2790	2791	2792	2793	2794	2795	2796	2797	2798	2799
AF__	2800	2801	2802	2803	2804	2805	2806	2807	2808	2809	2810	2811	2812	2813	2814	2815
B0__	2816	2817	2818	2819	2820	2821	2822	2823	2824	2825	2826	2827	2828	2829	2830	2831
B1__	2832	2833	2834	2835	2836	2837	2838	2839	2840	2841	2842	2843	2844	2845	2846	2847
B2__	2848	2849	2850	2851	2852	2853	2854	2855	2856	2857	2858	2859	2860	2861	2862	2863
B3__	2864	2865	2866	2867	2868	2869	2870	2871	2872	2873	2874	2875	2876	2877	2878	2879
B4__	2880	2881	2882	2883	2884	2885	2886	2887	2888	2889	2890	2891	2892	2893	2894	2895
B5__	2896	2897	2898	2899	2900	2901	2902	2903	2904	2905	2906	2907	2908	2909	2910	2911
B6__	2912	2913	2914	2915	2916	2917	2918	2919	2920	2921	2922	2923	2924	2925	2926	2927
B7__	2928	2929	2930	2931	2932	2933	2934	2935	2936	2937	2938	2939	2940	2941	2942	2943
B8__	2944	2945	2946	2947	2948	2949	2950	2951	2952	2953	2954	2955	2956	2957	2958	2959
B9__	2960	2961	2962	2963	2964	2965	2966	2967	2968	2969	2970	2971	2972	2973	2974	2975
BA__	2976	2977	2978	2979	2980	2981	2982	2983	2984	2985	2986	2987	2988	2989	2990	2991
BB__	2992	2993	2994	2995	2996	2997	2998	2999	3000	3001	3002	3003	3004	3005	3006	3007
BC__	3008	3009	3010	3011	3012	3013	3014	3015	3016	3017	3018	3019	3020	3021	3022	3023
BD__	3024	3025	3026	3027	3028	3029	3030	3031	3032	3033	3034	3035	3036	3037	3038	3039
BE__	3040	3041	3042	3043	3044	3045	3046	3047	3048	3049	3050	3051	3052	3053	3054	3055
BF__	3056	3057	3058	3059	3060	3061	3062	3063	3064	3065	3066	3067	3068	3069	3070	3071
C0__	3072	3073	3074	3075	3076	3077	3078	3079	3080	3081	3082	3083	3084	3085	3086	3087
C1__	3088	3089	3090	3091	3092	3093	3094	3095	3096	3097	3098	3099	3100	3101	3102	3103
C2__	3104	3105	3106	3107	3108	3109	3110	3111	3112	3113	3114	3115	3116	3117	3118	3119
C3__	3120	3121	3122	3123	3124	3125	3126	3127	3128	3129	3130	3131	3132	3133	3134	3135
C4__	3136	3137	3138	3139	3140	3141	3142	3143	3144	3145	3146	3147	3148	3149	3150	3151
C5__	3152	3153	3154	3155	3156	3157	3158	3159	3160	3161	3162	3613	3164	3165	3166	3167
C6__	3168	3169	3170	3171	3172	3173	3174	3175	3176	3177	3178	3179	3180	3181	3182	3183
C7__	3184	3185	3186	3187	3188	3189	3190	3191	3192	3193	3194	3195	3196	3197	3198	3199
C8__	3200	3201	3202	3203	3204	3205	3206	3207	3208	3209	3210	3211	3212	3213	3214	3215
C9__	3216	3217	3218	3219	3220	3221	3222	3223	3224	3225	3226	3227	3228	3229	3230	3231
CA__	3232	3233	3234	3235	3236	3237	3238	3239	3240	3241	3242	3243	3244	3245	3246	3247
CB__	3248	3249	3250	3251	3252	3253	3254	3255	3256	3257	3258	3259	3260	3261	3262	3263
CC__	3264	3265	3266	3267	3268	3269	3270	3271	3272	3273	3274	3275	3276	3277	3278	3279
CD__	3280	3281	3282	3283	3284	3285	3286	3287	3288	3289	3290	3291	3292	3293	3294	3295
CE__	3296	3297	3298	3299	3300	3301	3302	3303	3304	3305	3306	3307	3308	3309	3310	3311
CF__	3312	3313	3314	3315	3316	3317	3318	3319	3320	3321	3322	3323	3324	3325	3326	3327
D0__	3328	3329	3330	3331	3332	3333	3334	3335	3336	3337	3338	3339	3340	3341	3342	3343
D1__	3344	3345	3346	3347	3348	3349	3350	3351	3352	3353	3354	3355	3356	3357	3358	3359
D2__	3360	3361	3362	3363	3364	3365	3366	3367	3368	3369	3370	3371	3372	3373	3374	3375
D3__	3376	3377	3378	3379	3380	3381	3382	3383	3384	3385	3386	3387	3388	3389	3390	3391
D4__	3392	3393	3394	3395	3396	3397	3398	3399	3400	3401	3402	3403	3404	3405	3406	3407
D5__	3408	3409	3410	3411	3412	3413	3414	3415	3416	3417	3418	3419	3420	3421	3422	3423
D6__	3424	3425	3426	3427	3428	3429	3430	3431	3432	3433	3434	3435	3436	3437	3438	3439
D7__	3440	3441	3442	3443	3444	3445	3446	3447	3448	3449	3450	3451	3452	3453	3454	3455
D8__	3456	3457	3458	3459	3460	3461	3462	3463	3464	3465	3466	3467	3468	3469	3470	3471
D9__	3472	3473	3474	3475	3476	3477	3478	3479	3480	3481	3482	3483	3484	3485	3486	3487
DA__	3488	3489	3490	3491	3492	3493	3494	3495	3496	3497	3498	3499	3500	3501	3502	3503
DB__	3504	3505	3506	3507	3508	3509	3510	3511	3512	3513	3514	3515	3516	3517	3518	3519
DC__	3520	3521	3522	3523	3524	3525	3526	3527	3528	3529	3530	3531	3532	3533	3534	3535
DD__	3536	3537	3538	3539	3540	3541	3542	3543	3544	3545	3546	3547	3548	3549	3550	3551
DE__	3552	3553	3554	3555	3556	3557	3558	3559	3560	3561	3562	3563	3564	3565	3566	3567
DF__	3568	3569	3570	3571	3572	3573	3574	3575	3576	3577	3578	3579	3580	3581	3582	3583

DIRECT CONVERSION TABLE (Continued)

	0	1	2	3	4	5	6	7	8	9	A	B	C	D	E	F
E0_	3584	3585	3586	3587	3588	3589	3590	3591	3592	3593	3594	3595	3596	3597	3598	3599
E1_	3600	3601	3602	3603	3604	3605	3606	3607	3608	3609	3610	3611	3612	3613	3614	3615
E2_	3616	3617	3618	3619	3620	3621	3622	3623	3624	3625	3626	3627	3628	3629	3630	3631
E3_	3632	3633	3634	3635	3636	3637	3638	3639	3640	3641	3642	3643	3644	3645	3646	3647
E4_	3648	3649	3650	3651	3652	3653	3654	3655	3656	3657	3658	3659	3660	3661	3662	3663
E5_	3664	3665	3666	3667	3668	3669	3670	3671	3672	3673	3674	3675	3676	3677	3678	3679
E6_	3680	3681	3682	3683	3684	3685	3686	3687	3688	3689	3690	3691	3692	3693	3694	3695
E7_	3696	3697	3698	3699	3700	3701	3702	3703	3704	3705	3706	3707	3708	3709	3710	3711
E8_	3712	3713	3714	3715	3716	3717	3718	3719	3720	3721	3722	3723	3724	3725	3726	3727
E9_	3728	3729	3730	3731	3732	3733	3734	3735	3736	3737	3738	3739	3740	3741	3742	3743
EA_	3744	3745	3746	3747	3748	3749	3750	3751	3752	3753	3754	3755	3756	3757	3758	3759
EB_	3760	3761	3762	3763	3764	3765	3766	3767	3768	3769	3770	3771	3772	3773	3774	3775
EC_	3776	3777	3778	3779	3780	3781	3782	3783	3784	3785	3786	3787	3788	3789	3790	3791
ED_	3792	3793	3794	3795	3796	3797	3798	3799	3800	3801	3802	3803	3804	3805	3806	3807
EE_	3808	3809	3810	3811	3812	3813	3814	3815	3816	3817	3818	3819	3820	3821	3822	3823
EF_	3824	3825	3826	3827	3828	3829	3830	3831	3832	3833	3834	3835	3836	3837	3838	3839
F0_	3840	3841	3842	3843	3844	3845	3846	3847	3848	3849	3850	3851	3852	3853	3854	3855
F1_	3856	3857	3858	3859	3860	3861	3862	3863	3864	3865	3866	3867	3868	3869	3870	3871
F2_	3872	3873	3874	3875	3876	3877	3878	3879	3880	3881	3882	3883	3884	3885	3886	3887
F3_	3888	3889	3890	3891	3892	3893	3894	3895	3896	3897	3898	3899	3900	3901	3902	3903
F4_	3904	3905	3906	3907	3908	3909	3910	3911	3912	3913	3914	3915	3916	3917	3918	3919
F5_	3920	3921	3922	3923	3924	3925	3926	3927	3928	3929	3930	3931	3932	3933	3934	3935
F6_	3936	3937	3938	3939	3940	3941	3942	3943	3944	3945	3946	3947	3948	3949	3950	3951
F7_	3952	3953	3954	3955	3956	3957	3958	3959	3960	3961	3962	3963	3964	3965	3966	3967
F8_	3968	3969	3970	3971	3972	3973	3974	3975	3976	3977	3978	3979	3980	3981	3982	3983
F9_	3984	3985	3986	3987	3988	3989	3990	3991	3992	3993	3994	3995	3996	3997	3998	3999
FA_	4000	4001	4002	4003	4004	4005	4006	4007	4008	4009	4010	4011	4012	4013	4014	4015
FB_	4016	4017	4018	4019	4020	4021	4022	4023	4024	4025	4026	4027	4028	4029	4030	4031
FC_	4032	4033	4034	4035	4036	4037	4038	4039	4040	4041	4042	4043	4044	4045	4046	4047
FD_	4048	4049	4050	4051	4052	4053	4054	4055	4056	4057	4058	4059	4060	4061	4062	4063
FE_	4064	4065	4066	4067	4068	4069	4070	4071	4072	4073	4074	4075	4076	4077	4078	4079
FF_	4080	4081	4082	4083	4084	4085	4086	4087	4088	4089	4090	4091	4092	4093	4094	4095

II. HEXADECIMAL AND DECIMAL INTEGER CONVERSION TABLE

8		7		6		5		4		3		2		1	
Hex	Decimal	Hex	Decimal	Hex	Decimal	Hex	Decimal	Hex	Decimal	Hex	Decimal	Hex	Decimal	Hex	Decimal
0	0	0	0	0	0	0	0	0	0	0	0	0	0	0	0
1	268,435,456	1	16,777,216	1	1,048,576	1	65,536	1	4,096	1	256	1	16	1	1
2	536,870,912	2	33,554,432	2	2,097,152	2	131,072	2	8,192	2	512	2	32	2	2
3	805,306,368	3	50,331,648	3	3,145,728	3	196,608	3	12,288	3	768	3	48	3	3
4	1,073,741,824	4	67,108,864	4	4,194,304	4	262,144	4	16,384	4	1,024	4	64	4	4
5	1,342,177,280	5	83,886,080	5	5,242,880	5	327,680	5	20,480	5	1,280	5	80	5	5
6	1,610,612,736	6	100,663,296	6	6,291,456	6	393,216	6	24,576	6	1,536	6	96	6	6
7	1,879,048,192	7	117,440,512	7	7,340,032	7	458,752	7	28,672	7	1,792	7	112	7	7
8	2,147,483,648	8	134,217,728	8	8,388,608	8	524,288	8	32,768	8	2,048	8	128	8	8
9	2,415,919,104	9	150,994,944	9	9,437,184	9	589,824	9	36,864	9	2,304	9	144	9	9
A	2,684,354,560	A	167,772,160	A	10,485,760	A	655,360	A	40,960	A	2,560	A	160	A	10
B	2,952,790,016	B	184,549,376	B	11,534,336	B	720,896	B	45,056	B	2,816	B	176	B	11
C	3,221,225,472	C	201,326,592	C	12,582,912	C	786,432	C	49,152	C	3,072	C	192	C	12
D	3,489,660,928	D	218,103,808	D	13,631,488	D	851,968	D	53,248	D	3,328	D	208	D	13
E	3,758,096,384	E	234,881,024	E	14,680,064	E	917,504	E	57,344	E	3,584	E	224	E	14
F	4,026,531,840	F	251,658,240	F	15,728,640	F	983,040	F	61,440	F	3,840	F	240	F	15
8		7		6		5		4		3		2		1	

INTEGER CONVERSION TABLE (Continued)

TO CONVERT HEXADECIMAL TO DECIMAL

1. Locate the column of decimal numbers corresponding to the left-most digit or letter of the hexadecimal; select from this column and record the number that corresponds to the position of the hexadecimal digit or letter.
2. Repeat step 1 for the next (second from the left) position.
3. Repeat step 1 for the units (third from the left) position.
4. Add the numbers selected from the table to form the decimal number.

EXAMPLE

Conversion of Hexadecimal Value	D34
1. D	3328
2. 3	48
3. 4	4
4. Decimal	3380

To convert integer numbers greater than the capacity of table, use the techniques below:

HEXADECIMAL TO DECIMAL

Successive cumulative multiplication from left to right, adding units position.

Example: $D34_{16} = 3380_{10}$

$$
\begin{array}{rr}
D = & 13 \\
 & \times 16 \\
\hline
 & 208 \\
3 = & +3 \\
\hline
 & 211 \\
 & \times 16 \\
\hline
 & 3376 \\
4 = & +4 \\
\hline
 & 3380
\end{array}
$$

TO CONVERT DECIMAL TO HEXADECIMAL

1. (a) Select from the table the highest decimal number that is equal to or less than the number to be converted.
   (b) Record the hexadecimal of the column containing the selected number.
   (c) Subtract the selected decimal from the number to be converted.
2. Using the remainder from step 1(c) repeat all of step 1 to develop the second position of the hexadecimal (and a remainder).
3. Using the remainder from step 2 repeat all of step 1 to develop the units position of the hexadecimal.
4. Combine terms to form the hexadecimal number.

EXAMPLE

Conversion of Decimal Value	3380
1. D	−3328
	52
2. 3	−48
	4
3. 4	−4
4. Hexadecimal	D34

DECIMAL TO HEXADECIMAL

Divide and collect the remainder in reverse order.

Example: $3380_{10} = X_{16}$

16 | 3380 — remainder
16 | 211 — 4
16 | 13 — 3
D

$3380_{10} = D34_{16}$

INTEGER CONVERSION TABLE (Continued)

POWERS OF 16 TABLE

Example: $268{,}435{,}456_{10} = (2.68435456 \times 10^8)_{10} = 1000\ 0000_{16} = (10^7)_{16}$

16^n	n
1	0
16	1
256	2
4 096	3
65 536	4
1 048 576	5
16 777 216	6
268 435 456	7
4 294 967 296	8
68 719 476 736	9
1 099 511 627 776	10 = A
17 592 186 044 416	11 = B
281 474 976 710 656	12 = C
4 503 599 627 370 496	13 = D
72 057 594 037 927 936	14 = E
1 152 921 504 606 846 976	15 = F

Decimal Values

III. HEXADECIMAL AND DECIMAL FRACTION CONVERSION TABLE

1		2		3		4	
Hex	Decimal	Hex	Decimal	Hex	Decimal	Hex	Decimal Equivalent
.0	.0000	.00	.0000 0000	.000	.0000 0000 0000	.0000	.0000 0000 0000 0000
.1	.0625	.01	.0039 0625	.001	.0002 4414 0625	.0001	.0000 1525 8789 0625
.2	.1250	.02	.0078 1250	.002	.0004 8828 1250	.0002	.0000 3051 7578 1250
.3	.1875	.03	.0117 1875	.003	.0007 3242 1875	.0003	.0000 4577 6367 1875
.4	.2500	.04	.0156 2500	.004	.0009 7656 2500	.0004	.0000 6103 5156 2500
.5	.3125	.05	.0195 3125	.005	.0012 2070 3125	.0005	.0000 7629 3945 3125
.6	.3750	.06	.0234 3750	.006	.0014 6484 3750	.0006	.0000 9155 2734 3750
.7	.4375	.07	.0273 4375	.007	.0017 0898 4375	.0007	.0001 0681 1523 4375
.8	.5000	.08	.0312 5000	.008	.0019 5312 5000	.0008	.0001 2207 0312 5000
.9	.5625	.09	.0351 5625	.009	.0021 9726 5625	.0009	.0001 3732 9101 5625
.A	.6250	.0A	.0390 6250	.00A	.0024 4140 6250	.000A	.0001 5258 7890 6250
.B	.6875	.0B	.0429 6875	.00B	.0026 8554 6875	.000B	.0001 6784 6679 6875
.C	.7500	.0C	.0468 7500	.00C	.0029 2968 7500	.000C	.0001 8310 5468 7500
.D	.8125	.0D	.0507 8125	.00D	.0031 7382 8125	.000D	.0001 9836 4257 8125
.E	.8750	.0E	.0546 8750	.00E	.0034 1796 8750	.000E	.0002 1362 3046 8750
.F	.9375	.0F	.0585 9375	.00F	.0036 6210 9375	.000F	.0002 2888 1835 9375
1		2		3		4	

TO CONVERT .ABC HEXADECIMAL TO DECIMAL

Find .A in position 1	.6250
Find .0B in position 2	.0429 6875
Find .00C in position 3	.0029 2968 7500
.ABC Hex is equal to	.6708 9843 7500

FRACTION CONVERSION TABLE (Continued)

TO CONVERT .13 DECIMAL TO HEXADECIMAL

1. Find .1250 next lowest to	.1300	
subtract	−.1250	= .2 Hex
2. Find .0039 0625 next lowest to	.0050 0000	
	−.0039 0625	= .01
3. Find .0009 7656 2500	.0010 9375 0000	
	−.0009 7656 2500	= .004
4. Find .0001 0681 1523 4375	.0001 1718 7500 0000	
	−.0001 0681 1523 4375	= .0007
	.0000 1037 5976 5625	= .2147 Hex

5. 13 Decimal is approximately equal to ⟶ .2147 Hex

To convert fractions beyond the capacity of table, use techniques below:

HEXADECIMAL FRACTION TO DECIMAL

Convert the hexadecimal fraction to its decimal equivalent using the same technique as for integer numbers. Divide the results by 16^n (n is the number of fraction positions).

Example: $.8A7_{16} = .540771_{10}$

$8A7_{16} = 2215_{10}$

$16^3 = 4096$

$$4096 \overline{)2215.000000} = .540771$$

DECIMAL FRACTION TO HEXADECIMAL

Collect integer parts of product in the order of calculation.

Example: $.5408_{10} = .8A7_{16}$

```
           .5408
            ×16
8  ←  [8]  .6528
            ×16
A  ←  [10] .4448
            ×16
7  ←  [7]  .1168
```

HEXADECIMAL ADDITION AND SUBTRACTION TABLE

Example: 6 + 2 = 8, 8 − 2 = 6, and 8 − 6 = 2

	1	2	3	4	5	6	7	8	9	A	B	C	D	E	F
1	02	03	04	05	06	07	08	09	0A	0B	0C	0D	0E	0F	10
2	03	04	05	06	07	08	09	0A	0B	0C	0D	0E	0F	10	11
3	04	05	06	07	08	09	0A	0B	0C	0D	0E	0F	10	11	12
4	05	06	07	08	09	0A	0B	0C	0D	0E	0F	10	11	12	13
5	06	07	08	09	0A	0B	0C	0D	0E	0F	10	11	12	13	14
6	07	08	09	0A	0B	0C	0D	0E	0F	10	11	12	13	14	15
7	08	09	0A	0B	0C	0D	0E	0F	10	11	12	13	14	15	16
8	09	0A	0B	0C	0D	0E	0F	10	11	12	13	14	15	16	17
9	0A	0B	0C	0D	0E	0F	10	11	12	13	14	15	16	17	18
A	0B	0C	0D	0E	0F	10	11	12	13	14	15	16	17	18	19
B	0C	0D	0E	0F	10	11	12	13	14	15	16	17	18	19	1A
C	0D	0E	0F	10	11	12	13	14	15	16	17	18	19	1A	1B
D	0E	0F	10	11	12	13	14	15	16	17	18	19	1A	1B	1C
E	0F	10	11	12	13	14	15	16	17	18	19	1A	1B	1C	1D
F	10	11	12	13	14	15	16	17	18	19	1A	1B	1C	1D	1E

HEXADECIMAL MULTIPLICATION TABLE

Example: 2 × 4 = 08, F × 2 = 1E

	1	2	3	4	5	6	7	8	9	A	B	C	D	E	F
1	01	02	03	04	05	06	07	08	09	0A	0B	0C	0D	0E	0F
2	02	04	06	08	0A	0C	0E	10	12	14	16	18	1A	1C	1E
3	03	06	09	0C	0F	12	15	18	1B	1E	21	24	27	2A	2D
4	04	08	0C	10	14	18	1C	20	24	28	2C	30	34	38	3C
5	05	0A	0F	14	19	1E	23	28	2D	32	37	3C	41	46	4B
6	06	0C	12	18	1E	24	2A	30	36	3C	42	48	4E	54	5A
7	07	0E	15	1C	23	2A	31	38	3F	46	4D	54	5B	62	69
08	08	10	18	20	28	30	38	40	48	50	58	60	68	70	78
9	09	12	1B	24	2D	36	3F	48	51	5A	63	6C	75	7E	87
A	0A	14	1E	28	32	3C	46	50	5A	64	6E	78	82	8C	96
B	0B	16	21	2C	37	42	4D	58	63	6E	79	84	8F	9A	A5
C	0C	18	24	30	3C	48	54	60	6C	78	84	90	9C	A8	B4
D	0D	1A	27	34	41	4E	5B	68	75	82	8F	9C	A9	B6	C3
E	0E	1C	2A	38	46	54	62	70	7E	8C	9A	A8	B6	C4	D2
F	0F	1E	2D	3C	4B	5A	69	78	87	96	A5	B4	C3	D2	E1

SQUARES, SQUARE ROOT, CUBES AND CUBE ROOT

The squares and cubes from 1 to 1000 are given exactly. The roots are given to seven significant figures. Since the square roots of $10n$ are given, values of the square roots from 1 to 10 000 may be found directly. For the square roots of numbers below and above this range, use may be made of the following relations: $\sqrt{100n} = 10\sqrt{n}$; $\sqrt{1000n} = 10\sqrt{10n}$; $\sqrt{\frac{1}{10}n} = \frac{1}{10}\sqrt{10n}$; $\sqrt{\frac{1}{100}n} = \frac{1}{10}\sqrt{n}$; $\sqrt{\frac{1}{1000}n} = \frac{1}{100}\sqrt{10n}$. For example, the square root of 0.268 may be found by using the form $\sqrt{0.268} = \frac{1}{100}\sqrt{10 \times 268}$. The tabular value for the square root of $10n$ for 268 is 51.76872. Hence, the desired root is 0.5176872.

Values of cube roots for all numbers from 1 to 100 000 will be found directly in the table. Cube roots for numbers above or below this range will be found from the following relations: $\sqrt[3]{1000n} = 10\sqrt[3]{n}$; $\sqrt[3]{10\,000n} = 10\sqrt[3]{10n}$; $\sqrt[3]{100\,000n} = 10\sqrt[3]{100n}$; $\sqrt[3]{\frac{1}{10}n} = \frac{1}{10}\sqrt[3]{100n}$; $\sqrt[3]{\frac{1}{100}n} = \frac{1}{10}\sqrt[3]{10n}$; $\sqrt[3]{\frac{1}{1000}n} = \frac{1}{10}\sqrt[3]{n}$. For example, the cube root of 731 000 may be found by using the form, $\sqrt[3]{731\,000} = 10\sqrt[3]{731}$. The tabular value of the root for 731 is 9.008 223. The desired root is, therefore, 90.08223.

For fourth and fifth roots, see the table following this one. For higher integral powers of numbers, see table "Powers of Numbers."

SQUARES, SQUARE ROOT, CUBES AND CUBE ROOT

n	n^2	$\sqrt{n}$	$\sqrt{10n}$	n^3	$\sqrt[3]{n}$	$\sqrt[3]{10n}$	$\sqrt[3]{100n}$
1	1	1.000 000	3.162 278	1	1.000 000	2.154 435	4.641 589
2	4	1.414 214	4.472 136	8	1.259 921	2.714 418	5.848 035
3	9	1.732 051	5.477 226	27	1.442 250	3.107 233	6.694 330
4	16	2.000 000	6.324 555	64	1.587 401	3.419 952	7.368 063
5	25	2.236 068	7.071 068	125	1.709 976	3.684 031	7.937 005
6	36	2.449 490	7.745 967	216	1.817 121	3.914 868	8.434 327
7	49	2.645 751	8.366 600	343	1.912 931	4.121 285	8.879 040
8	64	2.828 427	8.944 272	512	2.000 000	4.308 869	9.283 178
9	81	3.000 000	9.486 833	729	2.080 084	4.481 405	9.654 894
10	100	3.162 278	10.00000	1 000	2.154 435	4.641 589	10.00000
11	121	3.316 625	10.48809	1 331	2.223 980	4.791 420	10.32280
12	144	3.464 102	10.95445	1 728	2.289 428	4.932 424	10.62659
13	169	3.605 551	11.40175	2 197	2.351 335	5.065 797	10.91393
14	196	3.741 657	11.83216	2 744	2.410 142	5.192 494	11.18689
15	225	3.872 983	12.24745	3 375	2.466 212	5.313 293	11.44714
16	256	4.000 000	12.64911	4 096	2.519 842	5.428 835	11.69607
17	289	4.123 106	13.03840	4 913	2.571 282	5.539 658	11.93483
18	324	4.242 641	13.41641	5 832	2.620 741	5.646 216	12.16440
19	361	4.358 899	13.78405	6 859	2.668 402	5.748 897	12.38562
20	400	4.472 136	14.14214	8 000	2.714 418	5.848 035	12.59921
21	441	4.582 576	14.49138	9 261	2.758 924	5.943 922	12.80579
22	484	4.690 416	14.83240	10 648	2.802 039	6.036 811	13.00591
23	529	4.795 832	15.16575	12 167	2.843 867	6.126 926	13.20006
24	576	4.898 979	15.49193	13 824	2.884 499	6.214 465	13.38866
25	625	5.000 000	15.81139	15 625	2.924 018	6.299 605	13.57209
26	676	5.099 020	16.12452	17 576	2.962 496	6.382 504	13.75069
27	729	5.196 152	16.43168	19 683	3.000 000	6.463 304	13.92477
28	784	5.291 503	16.73320	21 952	3.036 589	6.542 133	14.09460
29	841	5.385 165	17.02939	24 389	3.072 317	6.619 106	14.26043
30	900	5.477 226	17.32051	27 000	3.107 233	6.694 330	14.42250
31	961	5.567 764	17.60682	29 791	3.141 381	6.767 899	14.58100
32	1 024	5.656 854	17.88854	32 768	3.174 802	6.839 904	14.73613
33	1 089	5.744 563	18.16590	35 937	3.207 534	6.910 423	14.88806
34	1 156	5.830 952	18.43909	39 304	3.239 612	6.979 532	15.03695
35	1 225	5.916 080	18.70829	42 875	3.271 066	7.047 299	15.18294
36	1 296	6.000 000	18.97367	46 656	3.301 927	7.113 787	15.32619
37	1 369	6.082 763	19.23538	50 653	3.332 222	7.179 054	15.46680
38	1 444	6.164 414	19.49359	54 872	3.361 975	7.243 156	15.60491
39	1 521	6.244 998	19.74842	59 319	3.391 211	7.306 144	15.74061
40	1 600	6.324 555	20.00000	64 000	3.419 952	7.368 063	15.87401
41	1 681	6.403 124	20.24846	68 921	3.448 217	7.428 959	16.00521
42	1 764	6.480 741	20.49390	74 088	3.476 027	7.488 872	16.13429
43	1 849	6.557 439	20.73644	79 507	3.503 398	7.547 842	16.26133
44	1 936	6.633 250	20.97618	85 184	3.530 348	7.605 905	16.38643
45	2 025	6.708 204	21.21320	91 125	3.556 893	7.663 094	16.50964
46	2 116	6.782 330	21.44761	97 336	3.583 048	7.719 443	16.63103
47	2 209	6.855 655	21.67948	103 823	3.608 826	7.774 980	16.75069
48	2 304	6.928 203	21.90890	110 592	3.634 241	7.829 735	16.86865
49	2 401	7.000 000	22.13594	117 649	3.659 306	7.883 735	16.98499
50	2 500	7.071 068	22.36068	125 000	3.684 031	7.937 005	17.09976

SQUARES, SQUARE ROOT, CUBES AND CUBE ROOT (Continued)

n	n^2	$\sqrt{n}$	$\sqrt{10n}$	n^3	$\sqrt[3]{n}$	$\sqrt[3]{10n}$	$\sqrt[3]{100n}$
50	2 500	7.071 068	22.36068	125 000	3.684 031	7.937 005	17.09976
51	2 601	7.141 428	22.58318	132 651	3.708 430	7.989 570	17.21301
52	2 704	7.211 103	22.80351	140 608	3.732 511	8.041 452	17.32478
53	2 809	7.280 110	23.02173	148 877	3.756 286	8.092 672	17.43513
54	2 916	7.348 469	23.23790	157 464	3.779 763	8.143 253	17.54411
55	3 025	7.416 198	23.45208	166 375	3.802 952	8.193 213	17.65174
56	3 136	7.483 315	23.66432	175 616	3.825 862	8.242 571	17.75808
57	3 249	7.549 834	23.87467	185 193	3.848 501	8.291 344	17.86316
58	3 364	7.615 773	24.08319	195 112	3.870 877	8.339 551	17.96702
59	3 481	7.681 146	24.28992	205 379	3.892 996	8.387 207	18.06969
60	3 600	7.745 967	24.49490	216 000	3.914 868	8.434 327	18.17121
61	3 721	7.810 250	24.69818	226 981	3.936 497	8.480 926	18.27160
62	3 844	7.874 008	24.89980	238 328	3.957 892	8.527 019	18.37091
63	3 969	7.937 254	25.09980	250 047	3.979 057	8.572 619	18.46915
64	4 096	8.000 000	25.29822	262 144	4.000 000	8.617 739	18.56636
65	4 225	8.062 258	25.49510	274 625	4.020 726	8.662 391	18.66256
66	4 356	8.124 038	25.69047	287 496	4.041 240	8.706 588	18.75777
67	4 489	8.185 353	25.88436	300 763	4.061 548	8.750 340	18.85204
68	4 624	8.246 211	26.07681	314 432	4.081 655	8.793 659	18.94536
69	4 761	8.306 624	26.26785	328 509	4.101 566	8.836 556	19.03778
70	4 900	8.366 600	26.45751	343 000	4.121 285	8.879 040	19.12931
71	5 041	8.426 150	26.64583	357 911	4.140 818	8.921 121	19.21997
72	5 184	8.485 281	26.83282	373 248	4.160 168	8.962 809	19.30979
73	5 329	8.544 004	27.01851	389 017	4.179 339	9.004 113	19.39877
74	5 476	8.602 325	27.20294	405 224	4.198 336	9.045 042	19.48695
75	5 625	8.660 254	27.38613	421 875	4.217 163	9.085 603	19.57434
76	5 776	8.717 798	27.56810	438 976	4.235 824	9.125 805	19.66095
77	5 929	8.774 964	27.74887	456 533	4.254 321	9.165 656	19.74681
78	6 084	8.831 761	27.92848	474 552	4.272 659	9.205 164	19.83192
79	6 241	8.888 194	28.10694	493 039	4.290 840	9.244 335	19.91632
80	6 400	8.944 272	28.28427	512 000	4.308 869	9.283 178	20.00000
81	6 561	9.000 000	28.46050	531 441	4.326 749	9.321 698	20.08299
82	6 724	9.055 385	28.63564	551 368	4.344 481	9.359 902	20.16530
83	6 889	9.110 434	28.80972	571 787	4.362 071	9.397 796	20.24694
84	7 056	9.165 151	28.98275	592 704	4.379 519	9.435 388	20.32793
85	7 225	9.219 544	29.15476	614 125	4.396 830	9.472 682	20.40828
86	7 396	9.273 618	29.32576	636 056	4.414 005	9.509 685	20.48800
87	7 569	9.327 379	29.49576	658 503	4.431 048	9.546 403	20.56710
88	7 744	9.380 832	29.66479	681 472	4.447 960	9.582 840	20.64560
89	7 921	9.433 981	29.83287	704 969	4.464 745	9.619 002	20.72351
90	8 100	9.486 833	30.00000	729 000	4.481 405	9.654 894	20.80084
91	8 281	9.539 392	30.16621	753 571	4.497 941	9.690 521	20.87759
92	8 464	9.591 663	30.33150	778 688	4.514 357	9.725 888	20.95379
93	8 649	9.643 651	30.49590	804 357	4.530 655	9.761 000	21.02944
94	8 836	9.695 360	30.65942	830 584	4.546 836	9.795 861	21.10454
95	9 025	9.746 794	30.82207	857 375	4.562 903	9.830 476	21.17912
96	9 216	9.797 959	30.98387	884 736	4.578 857	9.864 848	21.25317
97	9 409	9.848 858	31.14482	912 673	4.594 701	9.898 983	21.32671
98	9 604	9.899 495	31.30495	941 192	4.610 436	9.932 884	21.39975
99	9 801	9.949 874	31.46427	970 299	4.626 065	9.966 555	21.47229
100	10 000	10.00000	31.62278	1 000 000	4.641 589	10.00000	21.54435

SQUARES, SQUARE ROOT, CUBES AND CUBE ROOT (Continued)

n	n^2	$\sqrt{n}$	$\sqrt{10n}$	n^3	$\sqrt[3]{n}$	$\sqrt[3]{10n}$	$\sqrt[3]{100n}$
100	10 000	10.00000	31.62278	1 000 000	4.641 589	10.00000	21.54435
101	10 201	10.04988	31.78050	1 030 301	4.657 010	10.03322	21.61592
102	10 404	10.09950	31.93744	1 061 208	4.672 329	10.06623	21.68703
103	10 609	10.14889	32.09361	1 092 727	4.687 548	10.09902	21.75767
104	10 816	10.19804	32.24903	1 124 864	4.702 669	10.13159	21.82786
105	11 025	10.24695	32.40370	1 157 625	4.717 694	10.16396	21.89760
106	11 236	10.29563	32.55764	1 191 016	4.732 623	10.19613	21.96689
107	11 449	10.34408	32.71085	1 225 043	4.747 459	10.22809	22.03575
108	11 664	10.39230	32.86335	1 259 712	4.762 203	10.25986	22.10419
109	11 881	10.44031	33.01515	1 295 029	4.776 856	10.29142	22.17220
110	12 100	10.48809	33.16625	1 331 000	4.791 420	10.32280	22.23980
111	12 321	10.53565	33.31666	1 367 631	4.805 896	10.35399	22.30699
112	12 544	10.58301	33.46640	1 404 928	4.820 285	10.38499	22.37378
113	12 769	10.63015	33.61547	1 442 897	4.834 588	10.41580	22.44017
114	12 996	10.67708	33.76389	1 481 544	4.848 808	10.44644	22.50617
115	13 225	10.72381	33.91165	1 520 875	4.862 944	10.47690	22.57179
116	13 456	10.77033	34.05877	1 560 896	4.876 999	10.50718	22.63702
117	13 689	10.81665	34.20526	1 601 613	4.890 973	10.53728	22.70189
118	13 924	10.86278	34.35113	1 643 032	4.904 868	10.56722	22.76638
119	14 161	10.90871	34.49638	1 685 159	4.918 685	10.59699	22.83051
120	14 400	10.95445	34.64102	1 728 000	4.932 424	10.62659	22.89428
121	14 641	11.00000	34.78505	1 771 561	4.946 087	10.65602	22.95770
122	14 884	11.04536	34.92850	1 815 848	4.959 676	10.68530	23.02078
123	15 129	11.09054	35.07136	1 860 867	4.973 190	10.71441	23.08350
124	15 376	11.13553	35.21363	1 906 624	4.986 631	10.74337	23.14589
125	15 625	11.18034	35.35534	1 953 125	5.000 000	10.77217	23.20794
126	15 876	11.22497	35.49648	2 000 376	5.013 298	10.80082	23.26967
127	16 129	11.26943	35.63706	2 048 383	5.026 526	10.82932	23.33107
128	16 384	11.31371	35.77709	2 097 152	5.039 684	10.85767	23.39214
129	16 641	11.35782	35.91657	2 146 689	5.052 774	10.88587	23.45290
130	16 900	11.40175	36.05551	2 197 000	5.065 797	10.91393	23.51335
131	17 161	11.44552	36.19392	2 248 091	5.078 753	10.94184	23.57348
132	17 424	11.48913	36.33180	2 299 968	5.091 643	10.96961	23.63332
133	17 689	11.53256	36.46917	2 352 637	5.104 469	10.99724	23.69285
134	17 956	11.57584	36.60601	2 406 104	5.117 230	11.02474	23.75208
135	18 225	11.61895	36.74235	2 460 375	5.129 928	11.05209	23.81102
136	18 496	11.66190	36.87818	2 515 456	5.142 563	11.07932	23.86966
137	18 769	11.70470	37.01351	2 571 353	5.155 137	11.10641	23.92803
138	19 044	11.74734	37.14835	2 628 072	5.167 649	11.13336	23.98610
139	19 321	11.78983	37.28270	2 685 619	5.180 101	11.16019	24.04390
140	19 600	11.83216	37.41657	2 744 000	5.192 494	11.18689	24.10142
141	19 881	11.87434	37.54997	2 803 221	5.204 828	11.21346	24.15867
142	20 164	11.91638	37.68289	2 863 288	5.217 103	11.23991	24.21565
143	20 449	11.95826	37.81534	2 924 207	5.229 322	11.26623	24.27236
144	20 736	12.00000	37.94733	2 985 984	5.241 483	11.29243	24.32881
145	21 025	12.04159	38.07887	3 048 625	5.253 588	11.31851	24.38499
146	21 316	12.08305	38.20995	3 112 136	5.265 637	11.34447	24.44092
147	21 609	12.12436	38.34058	3 176 523	5.277 632	11.37031	24.49660
148	21 904	12.16553	38.47077	3 241 792	5.289 572	11.39604	24.55202
149	22 201	12.20656	38.60052	3 307 949	5.301 459	11.42165	24.60719
150	22.500	12.24745	38.72983	3 375 000	5.313 293	11.44714	24.66212

SQUARES, SQUARE ROOT, CUBES AND CUBE ROOT (Continued)

n	n^2	$\sqrt{n}$	$\sqrt{10n}$	n^3	$\sqrt[3]{n}$	$\sqrt[3]{10n}$	$\sqrt[3]{100n}$
150	22 500	12.24745	38.72983	3 375 000	5.313 293	11.44714	24.66212
151	22 801	12.28821	38.85872	3 442 951	5.325 074	11.47252	24.71680
152	23 104	12.32883	38.98718	3 511 808	5.336 803	11.49779	24.77125
153	23 409	12.36932	39.11521	3 581 577	5.348 481	11.52295	24.82545
154	23 716	12.40967	39.24283	3 652 264	5.360 108	11.54800	24.87942
155	24 025	12.44990	39.37004	3 723 875	5.371 685	11.57295	24.93315
156	24 336	12.49000	39.49684	3 796 416	5.383 213	11.59778	24.98666
157	24 649	12.52996	39.62323	3 869 893	5.394 691	11.62251	25.03994
158	24 964	12.56981	39.74921	3 944 312	5.406 120	11.64713	25.09299
159	25 281	12.60952	39.87480	4 019 679	5.417 502	11.67165	25.14581
160	25 600	12.64911	40.00000	4 096 000	5.428 835	11.69607	25.19842
161	25 921	12.68858	40.12481	4 173 281	5.440 122	11.72039	25.25081
162	26 244	12.72792	40.24922	4 251 528	5.451 362	11.74460	25.30298
163	26 569	12.76715	40.37326	4 330 747	5.462 556	11.76872	25.35494
164	26 896	12.80625	40.49691	4 410 944	5.473 704	11.79274	25.40668
165	27 225	12.84523	40.62019	4 492 125	5.484 807	11.81666	25.45822
166	27 556	12.88410	40.74310	4 574 296	5.495 865	11.84048	25.50954
167	27 889	12.92285	40.86563	4 657 463	5.506 878	11.86421	25.56067
168	28 224	12.96148	40.98780	4 741 632	5.517 848	11.88784	25.61158
169	28 561	13.00000	41.10961	4 826 809	5.528 775	11.91138	25.66230
170	28 900	13.03840	41.23106	4 913 000	5.539 658	11.93483	25.71282
171	29 241	13.07670	41.35215	5 000 211	5.550 499	11.95819	25.76313
172	29 584	13.11488	41.47288	5 088 448	5.561 298	11.98145	25.81326
173	29 929	13.15295	41.59327	5 177 717	5.572 055	12.00463	25.86319
174	30 276	13.19091	41.71331	5 268 024	5.582 770	12.02771	25.91292
175	30 625	13.22876	41.83300	5 359 375	5.593 445	12.05071	25.96247
176	30 976	13.26650	41.95235	5 451 776	5.604 079	12.07362	26.01183
177	31 329	13.30413	42.07137	5 545 233	5.614 672	12.09645	26.06100
178	31 684	13.34166	42.19005	5 639 752	5.625 226	12.11918	26.10999
179	32 041	13.37909	42.30839	5 735 339	5.635 741	12.14184	26.15879
180	32 400	13.41641	42.42641	5 832 000	5.646 216	12.16440	26.20741
181	32 761	13.45362	42.54409	5 929 741	5.656 653	12.18689	26.25586
182	33 124	13.49074	42.66146	6 028 568	5.667 051	12.20929	26.30412
183	33 489	13.52775	42.77850	6 128 487	5.677 411	12.23161	26.35221
184	33 856	13.56466	42.89522	6 229 504	5.687 734	12.25385	26.40012
185	34 225	13.60147	43.01163	6 331 625	5.698 019	12.27601	26.44786
186	34 596	13.63818	43.12772	6 434 856	5.708 267	12.29809	26.49543
187	34 969	13.67479	43.24350	6 539 203	5.718 479	12.32009	26.54283
188	35 344	13.71131	43.35897	6 644 672	5.728 654	12.34201	26.59006
189	35 721	13.74773	43.47413	6 751 269	5.738 794	12.36386	26.63712
190	36 100	13.78405	43.58899	6 859 000	5.748 897	12.38562	26.68402
191	36 481	13.82027	43.70355	6 967 871	5.758 965	12.40731	26.73075
192	36 864	13.85641	43.81780	7 077 888	5.768 998	12.42893	26.77732
193	37 249	13.89244	43.93177	7 189 057	5.778 997	12.45047	26.82373
194	37 636	13.92839	44.04543	7 301 384	5.788 960	12.47194	26.86997
195	38 025	13.96424	44.15880	7 414 875	5.798 890	12.49333	26.91606
196	38 416	14.00000	44.27189	7 529 536	5.808 786	12.51465	26.96199
197	38 809	14.03567	44.38468	7 645 373	5.818 648	12.53590	27.00777
198	39 204	14.07125	44.49719	7 762 392	5.828 477	12.55707	27.05339
199	39 601	14.10674	44.60942	7 880 599	5.838 272	12.57818	27.09886
200	40 000	14.14214	44.72136	8 000 000	5.848 035	12.59921	27.14418

SQUARES, SQUARE ROOT, CUBES AND CUBE ROOT (Continued)

n	n^2	$\sqrt{n}$	$\sqrt{10n}$	n^3	$\sqrt[3]{n}$	$\sqrt[3]{10n}$	$\sqrt[3]{100n}$
200	40 000	14.14214	44.72136	8 000 000	5.848 035	12.59921	27.14418
201	40 401	14.17745	44.83302	8 120 601	5.857 766	12.62017	27.18934
202	40 804	14.21267	44.94441	8 242 408	5.867 464	12.64107	27.23436
203	41 209	14.24781	45.05552	8 365 427	5.877 131	12.66189	27.27922
204	41 616	14.28286	45.16636	8 489 664	5.886 765	12.68265	27.32394
205	42 025	14.31782	45.27693	8 615 125	5.896 369	12.70334	27.36852
206	42 436	14.35270	45.38722	8 741 816	5.905 941	12.72396	27.41295
207	42 849	14.38749	45.49725	8 869 743	5.915 482	12.74452	27.45723
208	43 264	14.42221	45.60702	8 998 912	5.924 992	12.76501	27.50138
209	43 681	14.45683	45.71652	9 129 329	5.934 472	12.78543	27.54538
210	44 100	14.49138	45.82576	9 261 000	5.943 922	12.80579	27.58924
211	44 521	14.52584	45.93474	9 393 931	5.953 342	12.82609	27.63296
212	44 944	14.56022	46.04346	9 528 128	5.962 732	12.84632	27.67655
213	45 369	14.59452	46.15192	9 663 597	5.972 093	12.86648	27.72000
214	45 796	14.62874	46.26013	9 800 344	5.981 424	12.88659	27.76331
215	46 225	14.66288	46.36809	9 938 375	5.990 726	12.90663	27.80649
216	46 656	14.69694	46.47580	10 077 696	6.000 000	12.92661	27.84953
217	47 089	14.73092	46.58326	10 218 313	6.009 245	12.94653	27.89244
218	47 524	14.76482	46.69047	10 360 232	6.018 462	12.96638	27.93522
219	47 961	14.79865	46.79744	10 503 459	6.027 650	12.98618	27.97787
220	48 400	14.83240	46.90416	10 648 000	6.036 811	13.00591	28.02039
221	48 841	14.86607	47.01064	10 793 861	6.045 944	13.02559	28.06278
222	49 284	14.89966	47.11688	10 941 048	6.055 049	13.04521	28.10505
223	49 729	14.93318	47.22288	11 089 567	6.064 127	13.06477	28.14718
224	50 176	14.96663	47.32864	11 239 424	6.073 178	13.08427	28.18919
225	50 625	15.00000	47.43416	11 390 625	6.082 202	13.10371	28.23108
226	51 076	15.03330	47.53946	11 543 176	6.091 199	13.12309	28.27284
227	51 529	15.06652	47.64452	11 697 083	6.100 170	13.14242	28.31448
228	51 984	15.09967	47.74935	11 852 352	6.109 115	13.16169	28.35600
229	52 441	15.13275	47.85394	12 008 989	6.118 033	13.18090	28.39739
230	52 900	15.16575	47.95832	12 167 000	6.126 926	13.20006	28.43867
231	53 361	15.19868	48.06246	12 326 391	6.135 792	13.21916	28.47983
232	53 824	15.23155	48.16638	12 487 168	6.144 634	13.23821	28.52086
233	54 289	15.26434	48.27007	12 649 337	6.153 449	13.25721	28.56178
234	54 756	15.29706	48.37355	12 812 904	6.162 240	13.27614	28.60259
235	55 225	15.32971	48.47680	12 977 875	6.171 006	13.29503	28.64327
236	55 696	15.36229	48.57983	13 144 256	6.179 747	13.31386	28.68384
237	56 169	15.39480	48.68265	13 312 053	6.188 463	13.33264	28.72430
238	56 644	15.42725	48.78524	13 481 272	6.197 154	13.35136	28.76464
239	57 121	15.45962	48.88763	13 651 919	6.205 822	13.37004	28.80487
240	57 600	15.49193	48.98979	13 824 000	6.214 465	13.38866	28.84499
241	58 081	15.52417	49.09175	13 997 521	6.223 084	13.40723	28.88500
242	58 564	15.55635	49.19350	14 172 488	6.231 680	13.42575	28.92489
243	59 049	15.58846	49.29503	14 348 907	6.240 251	13.44421	28.96468
244	59 536	15.62050	49.39636	14 526 784	6.248 800	13.46263	29.00436
245	60 025	15.65248	49.49747	14 706 125	6.257 325	13.48100	29.04393
246	60 516	15.68439	49.59839	14 886 936	6.265 827	13.49931	29.08339
247	61 009	15.71623	49.69909	15 069 223	6.274 305	13.51758	29.12275
248	61 504	15.74802	49.79960	15 252 992	6.282 761	13.53580	29.16199
249	62 001	15.77973	49.89990	15 438 249	6.291 195	13.55397	29.20114
250	62 500	15.81139	50.00000	15 625 000	6.299 605	13.57209	29.24018

SQUARES, SQUARE ROOT, CUBES AND CUBE ROOT (Continued)

n	n^2	$\sqrt{n}$	$\sqrt{10n}$	n^3	$\sqrt[3]{n}$	$\sqrt[3]{10n}$	$\sqrt[3]{100n}$
250	62 500	15.81139	50.00000	15 625 000	6.299 605	13.57209	29.24018
251	63 001	15.84298	50.09990	15 813 251	6.307 994	13.59016	29.27911
252	63 504	15.87451	50.19960	16 003 008	6.316 360	13.60818	29.31794
253	64 009	15.90597	50.29911	16 194 277	6.324 704	13.62616	29.35667
254	64 516	15.93738	50.39841	16 387 064	6.333 026	13.64409	29.39530
255	65 025	15.96872	50.49752	16 581 375	6.341 326	13.66197	29.43383
256	65 536	16.00000	50.59644	16 777 216	6.349 604	13.67981	29.47225
257	66 049	16.03122	50.69517	16 974 593	6.357 861	13.69760	29.51058
258	66 564	16.06238	50.79370	17 173 512	6.366 097	13.71534	29.54880
259	67 081	16.09348	50.89204	17 373 979	6.374 311	13.73304	29.58693
260	67 600	16.12452	50.99020	17 576 000	6.382 504	13.75069	29.62496
261	68 121	16.15549	51.08816	17 779 581	6.390 677	13.76830	29.66289
262	68 644	16.18641	51.18594	17 984 728	6.398 828	13.78586	29.70073
263	69 169	16.21727	51.28353	18 191 447	6.406 959	13.80337	29.73847
264	69 696	16.24808	51.38093	18 399 744	6.415 069	13.82085	29.77611
265	70 225	16.27882	51.47815	18 609 625	6.423 158	13.83828	29.81366
266	70 756	16.30951	51.57519	18 821 096	6.431 228	13.85566	29.85111
267	71 289	16.34013	51.67204	19 034 163	6.439 277	13.87300	29.88847
268	71 824	16.37071	51.76872	19 248 832	6.447 306	13.89030	29.92574
269	72 361	16.40122	51.86521	19 465 109	6.455 315	13.90755	29.96292
270	72 900	16.43168	51.96152	19 683 000	6.463 304	13.92477	30.00000
271	73 441	16.46208	52.05766	19 902 511	6.471 274	13.94194	30.03699
272	73 984	16.49242	52.15362	20 123 648	6.479 224	13.95906	30.07389
273	74 529	16.52271	52.24940	20 346 417	6.487 154	13.97615	30.11070
274	75 076	16.55295	52.34501	20 570 824	6.495 065	13.99319	30.14742
275	75 625	16.58312	52.44044	20 796 875	6.502 957	14.01020	30.18405
276	76 176	16.61325	52.53570	21 024 576	6.510 830	14.02716	30.22060
277	76 729	16.64332	52.63079	21 253 933	6.518 684	14.04408	30.25705
278	77 284	16.67333	52.72571	21 484 952	6.526 519	14.06096	30.29342
279	77 841	16.70329	52.82045	21 717 639	6.534 335	14.07780	30.32970
280	78 400	16.73320	52.91503	21 952 000	6.542 133	14.09460	30.36589
281	78 961	16.76305	53.00943	22 188 041	6.549 912	14.11136	30.40200
282	79 524	16.79286	53.10367	22 425 768	6.557 672	14.12808	30.43802
283	80 089	16.82260	53.19774	22 665 187	6.565 414	14.14476	30.47395
284	80 656	16.85230	53.29165	22 906 304	6.573 138	14.16140	30.50981
285	81 225	16.88194	53.38539	23 149 125	6.580 844	14.17800	30.54557
286	81 796	16.91153	53.47897	23 393 656	6.588 532	14.19456	30.58126
287	82 369	16.94107	53.57238	23 639 903	6.596 202	14.21109	30.61686
288	82 944	16.97056	53.66563	23 887 872	6.603 854	14.22757	30.65238
289	83 521	17.00000	53.75872	24 137 569	6.611 489	14.24402	30.68781
290	84 100	17.02939	53.85165	24 389 000	6.619 106	14.26043	30.72317
291	84 681	17.05872	53.94442	24 642 171	6.626 705	14.27680	30.75844
292	85 264	17.08801	54.03702	24 897 088	6.634 287	14.29314	30.79363
293	85 849	17.11724	54.12947	25 153 757	6.641 852	14.30944	30.82875
294	86 436	17.14643	54.22177	25 412 184	6.649 400	14.32570	30.86378
295	87 025	17.17556	54.31390	25 672 375	6.656 930	14.34192	30.89873
296	87 616	17.20465	54.40588	25 934 336	6.664 444	14.35811	30.93361
297	88 209	17.23369	54.49771	26 198 073	6.671 940	14.37426	30.96840
298	88 804	17.26268	54.58938	26 463 592	6.679 420	14.39037	31.00312
299	89 401	17.29162	54.68089	26 730 899	6.686 883	14.40645	31.03776
300	90 000	17.32051	54.77226	27 000 000	6.694 330	14.42250	31.07233

SQUARES, SQUARE ROOT, CUBES AND CUBE ROOT (Continued)

n	n^2	$\sqrt{n}$	$\sqrt{10n}$	n^3	$\sqrt[3]{n}$	$\sqrt[3]{10n}$	$\sqrt[3]{100n}$
300	90 000	17 32051	54.77226	27 000 000	6.694 330	14.42250	31.07233
301	90 601	17.34935	54.86347	27 270 901	6.701 759	14.43850	31.10681
302	91 204	17.37815	54.95453	27 543 608	6.709 173	14.45447	31.14122
303	91 809	17.40690	55.04544	27 818 127	6.716 570	14.47041	31.17556
304	92 416	17.43560	55.13620	28 094 464	6.723 951	14.48631	31.20982
305	93 025	17.46425	55.22681	28 372 625	6.731 315	14.50218	31.24400
306	93 636	17.49286	55.31727	28 652 616	6.738 664	14.51801	31.27811
307	94 249	17.52142	55.40758	28 934 443	6.745 997	14.53381	31.31214
308	94 864	17.54993	55.49775	29 218 112	6.753 313	14.54957	31.34610
309	95 481	17.57840	55.58777	29 503 629	6.760 614	14.56530	31.37999
310	96 100	17.60682	55.67764	29 791 000	6.767 899	14.58100	31.41381
311	96 721	17.63519	55.76737	30 080 231	6.775 169	14.59666	31.44755
312	97 344	17.66352	55.85696	30 371 328	6.782 423	14.61229	31.48122
313	97 969	17.69181	55.94640	30 664 297	6.789 661	14.62788	31.51482
314	98 596	17.72005	56.03570	30 959 144	6.796 884	14.64344	31.54834
315	99 225	17.74824	56.12486	31 255 875	6.804 092	14.65897	31.58180
316	99 856	17.77639	56.21388	31 554 496	6.811 285	14.67447	31.61518
317	100 489	17.80449	56.30275	31 855 013	6.818 462	14.68993	31.64850
318	101 124	17.83255	56.39149	32 157 432	6.825 624	14.70536	31.68174
319	101 761	17.86057	56.48008	32 461 759	6.832 771	14.72076	31.71492
320	102 400	17.88854	56.56854	32 768 000	6.839 904	14.73613	31.74802
321	103 041	17.91647	56.65686	33 076 161	6.847 021	14.75146	31.78106
322	103 684	17.94436	56.74504	33 386 248	6.854 124	14.76676	31.81403
323	104 329	17.97220	56.83309	33 698 267	6.861 212	14.78203	31.84693
324	104 976	18.00000	56.92100	34 012 224	6.868 285	14.79727	31.87976
325	105 625	18.02776	57.00877	34 328 125	6.875 344	14.81248	31.91252
326	106 276	18.05547	57.09641	34 645 976	6.882 389	14.82766	31.94522
327	106 929	18.08314	57.18391	34 965 783	6.889 419	14.84280	31.97785
328	107 584	18.11077	57.27128	35 287 552	6.896 434	14.85792	32.01041
329	108 241	18.13836	57.35852	35 611 289	6.903 436	14.87300	32.04291
330	108 900	18.16590	57.44563	35 937 000	6.910 423	14.88806	32.07534
331	109 561	18.19341	57.53260	36 264 691	6.917 396	14.90308	32.10771
332	110 224	18.22087	57.61944	36 594 368	6.924 356	14.91807	32.14001
333	110 889	18.24829	57.70615	36 926 037	6.931 301	14.93303	32.17225
334	111 556	18.27567	57.79273	37 259 704	6.938 232	14.94797	32.20442
335	112 225	18.30301	57.87918	37 595 375	6.945 150	14.96287	32.23653
336	112 896	18.33030	57.96551	37 933 056	6.952 053	14.97774	32.26857
337	113 569	18.35756	58.05170	38 272 753	6.958 943	14.99259	32.30055
338	114 244	18.38478	58.13777	38 614 472	6.965 820	15.00740	32.33247
339	114 921	18.41195	58.22371	38 958 219	6.972 683	15.02219	32.36433
340	115 600	18.43909	58.30952	39 304 000	6.979 532	15.03695	32.39612
341	116 281	18.46619	58.39521	39 651 821	6.986 368	15.05167	32.42785
342	116 964	18.49324	58.48077	40 001 688	6.993 191	15.06637	32.45952
343	117 649	18.52026	58.56620	40 353 607	7.000 000	15.08104	32.49112
344	118 336	18.54724	58.65151	40 707 584	7.006 796	15.09568	32.52267
345	119 025	18.57418	58.73670	41 063 625	7.013 579	15.11030	32.55415
346	119 716	18.60108	58.82176	41 421 736	7.020 349	15.12488	32.58557
347	120 409	18.62794	58.90671	41 781 923	7.027 106	15.13944	32.61694
348	121 104	18.65476	58.99152	42 144 192	7.033 850	15.15397	32.64824
349	121 801	18.68154	59.07622	42 508 549	7.040 581	15.16847	32.67948
350	122 500	18.70829	59.16080	42 875 000	7.047 299	15.18294	32.71066

SQUARES, SQUARE ROOT, CUBES AND CUBE ROOT (Continued)

n	n^2	$\sqrt{n}$	$\sqrt{10n}$	n^3	$\sqrt[3]{n}$	$\sqrt[3]{10n}$	$\sqrt[3]{100n}$
350	122 500	18.70829	59.16080	42 875 000	7.047 299	15.18294	32.71066
351	123 201	18.73499	59.24525	43 243 551	7.054 004	15.19739	32.74179
352	123 904	18.76166	59.32959	43 614 208	7.060 697	15.21181	32.77285
353	124 609	18.78829	59.41380	43 986 977	7.067 377	15.22620	32.80386
354	125 316	18.81489	59.49790	44 361 864	7.074 044	15.24057	32.83480
355	126 025	18.84144	59.58188	44 738 875	7.080 699	15.25490	32.86569
356	126 736	18.86796	59.66574	45 118 016	7.087 341	15.26921	32.89652
357	127 449	18.89444	59.74948	45 499 293	7.093 971	15.28350	32.92730
358	128 164	18.92089	59.83310	45 882 712	7.100 588	15.29775	32.95801
359	128 881	18.94730	59.91661	46 268 279	7.107 194	15.31198	32.98867
360	129 600	18.97367	60.00000	46 656 000	7.113 787	15.32619	33.01927
361	130 321	19.00000	60.08328	47 045 881	7.120 367	15.34037	33.04982
362	131 044	19.02630	60.16644	47 437 928	7.126 936	15.35452	33.08031
363	131 769	19.05256	60.24948	47 832 147	7.133 492	15.36864	33.11074
364	132 496	19.07878	60.33241	48 228 544	7.140 037	15.38274	33.14112
365	133 225	19.10497	60.41523	48 627 125	7.146 569	15.39682	33.17144
366	133 956	19.13113	60.49793	49 027 896	7.153 090	15.41087	33.20170
367	134 689	19.15724	60.58052	49 430 863	7.159 599	15.42489	33.23191
368	135 424	19.18333	60.66300	49 836 032	7.166 096	15.43889	33.26207
369	136 161	19.20937	60.74537	50 243 409	7.172 581	15.45286	33.29217
370	136 900	19.23538	60.82763	50 653 000	7.179 054	15.46680	33.32222
371	137 641	19.26136	60.90977	51 064 811	7.185 516	15.48073	33.35221
372	138 384	19.28730	60.99180	51 478 848	7.191 966	15.49462	33.38215
373	139 129	19.31321	61.07373	51 895 117	7.198 405	15.50849	33.41204
374	139 876	19.33908	61.15554	52 313 624	7.204 832	15.52234	33.44187
375	140 625	19.36492	61.23724	52 734 375	7.211 248	15.53616	33.47165
376	141 376	19.39072	61.31884	53 157 376	7.217 652	15.54996	33.50137
377	142 129	19.41649	61.40033	53 582 633	7.224 045	15.56373	33.53105
378	142 884	19.44222	61.48170	54 010 152	7.230 427	15.57748	33.56067
379	143 641	19.46792	61.56298	54 439 939	7.236 797	15.59121	33.59024
380	144 400	19.49359	61.64414	54 872 000	7.243 156	15.60491	33.61975
381	145 161	19.51922	61.72520	55 306 341	7.249 505	15.61858	33.64922
382	145 924	19.54482	61.80615	55 742 968	7.255 842	15.63224	33.67863
383	146 689	19.57039	61.88699	56 181 887	7.262 167	15.64587	33.70800
384	147 456	19.59592	61.96773	56 623 104	7.268 482	15.65947	33.73731
385	148 225	19.62142	62.04837	57 066 625	7.274 786	15.67305	33.76657
386	148 996	19.64688	62.12890	57 512 456	7.281 079	15.68661	33.79578
387	149 769	19.67232	62.20932	57 960 603	7.287 362	15.70014	33.82494
388	150 544	19.69772	62.28965	58 411 072	7.293 633	15.71366	33.85405
389	151 321	19.72308	62.36986	58 863 869	7.299 894	15.72714	33.88310
390	152 100	19.74842	62.44998	59 319 000	7.306 144	15.74061	33.91211
391	152 881	19.77372	62.52999	59 776 471	7.312 383	15.75405	33.94107
392	153 664	19.79899	62.60990	60 236 288	7.318 611	15.76747	33.96999
393	154 449	19.82423	62.68971	60 698 457	7.324 829	15.78087	33.99885
394	155 236	19.84943	62.76942	61 162 984	7.331 037	15.79424	34.02766
395	156 025	19.87461	62.84903	61 629 875	7.337 234	15.80759	34.05642
396	156 816	19.89975	62.92853	62 099 136	7.343 420	15.82092	34.08514
397	157 609	19.92486	63.00794	62 570 773	7.349 597	15.83423	34.11381
398	158 404	19.94994	63.08724	63 044 792	7.355 762	15.84751	34.14242
399	159 201	19.97498	63.16645	63 521 199	7.361 918	15.86077	34.17100
400	160 000	20.00000	63.24555	64 000 000	7.368 063	15.87401	34.19952

SQUARES, SQUARE ROOT, CUBES AND CUBE ROOT (Continued)

n	n^2	$\sqrt{n}$	$\sqrt{10n}$	n^3	$\sqrt[3]{n}$	$\sqrt[3]{10n}$	$\sqrt[3]{100n}$
400	160 000	20.00000	63.24555	64 000 000	7.368 063	15.87401	34.19952
401	160 801	20.02498	63.32456	64 481 201	7.374 198	15.88723	34.22799
402	161 604	20.04994	63.40347	64 964 808	7.380 323	15.90042	34.25642
403	162 409	20.07486	63.48228	65 450 827	7.386 437	15.91360	34.28480
404	163 216	20.09975	63.56099	65 939 264	7.392 542	15.92675	34.31314
405	164 025	20.12461	63.63961	66 430 125	7.398 636	15.93988	34.34143
406	164 836	20.14944	63.71813	66 923 416	7.404 721	15.95299	34.36967
407	165 649	20.17424	63.79655	67 419 143	7.410 795	15.96607	34.39786
408	166 464	20.19901	63.87488	67 917 312	7.416 860	15.97914	34.42601
409	167 281	20.22375	63.95311	68 417 929	7.422 914	15.99218	34.45412
410	168 100	20.24846	64.03124	68 921 000	7.428 959	16.00521	34.48217
411	168 921	20.27313	64.10928	69 426 531	7.434 994	16.01821	34.51018
412	169 744	20.29778	64.18723	69 934 528	7.441 019	16.03119	34.53815
413	170 569	20.32240	64.26508	70 444 997	7.447 034	16.04415	34.56607
414	171 396	20.34699	64.34283	70 957 944	7.453 040	16.05709	34.59395
415	172 225	20.37155	64.42049	71 473 375	7.459 036	16.07001	34.62178
416	173 056	20.39608	64.49806	71 991 296	7.465 022	16.08290	34.64956
417	173 889	20.42058	64.57554	72 511 713	7.470 999	16.09578	34.67731
418	174 724	20.44505	64.65292	73 034 632	7.476 966	16.10864	34.70500
419	175 561	20.46949	64.73021	73 560 059	7.482 924	16.12147	34.73266
420	176 400	20.49390	64.80741	74 088 000	7.488 872	16.13429	34.76027
421	177 241	20.51828	64.88451	74 618 461	7.494 811	16.14708	34.78783
422	178 084	20.54264	64.96153	75 151 448	7.500 741	16.15986	34.81535
423	178 929	20.56696	65.03845	75 686 967	7.506 661	16.17261	34.84283
424	179 776	20.59126	65.11528	76 225 024	7.512 572	16.18534	34.87027
425	180 625	20.61553	65.19202	76 765 625	7.518 473	16.19806	34.89766
426	181 476	20.63977	65.26868	77 308 776	7.524 365	16.21075	34.92501
427	182 329	20.66398	65.34524	77 854 483	7.530 248	16.22343	34.95232
428	183 184	20.68816	65.42171	78 402 752	7.536 122	16.23608	34.97958
429	184 041	20.71232	65.49809	78 953 589	7.541 987	16.24872	35.00680
430	184 900	20.73644	65.57439	79 507 000	7.547 842	16.26133	35.03398
431	185 761	20.76054	65.65059	80 062 991	7.553 689	16.27393	35.06112
432	186 624	20.78461	65.72671	80 621 568	7.559 526	16.28651	35.08821
433	187 489	20.80865	65.80274	81 182 737	7.565 355	16.29906	35.11527
434	188 356	20.83267	65.87868	81 746 504	7.571 174	16.31160	35.14228
435	189 225	20.85665	65.95453	82 312 875	7.576 985	16.32412	35.16925
436	190 096	20.88061	66.03030	82 881 856	7.582 787	16.33662	35.19618
437	190 969	20.90454	66.10598	83 453 453	7.588 579	16.34910	35.22307
438	191 844	20.92845	66.18157	84 027 672	7.594 363	16.36156	35.24991
439	192 721	20.95233	66.25708	84 604 519	7.600 139	16.37400	35.27672
440	193 600	20.97618	66.33250	85 184 000	7.605 905	16.38643	35.30348
441	194 481	21.00000	66.40783	85 766 121	7.611 663	16.39883	35.33021
442	195 364	21.02380	66.48308	86 350 888	7.617 412	16.41122	35.35689
443	196 249	21.04757	66.55825	86 938 307	7.623 152	16.42358	35.38354
444	197 136	21.07131	66.63332	87 528 384	7.628 884	16.43593	35.41014
445	198 025	21.09502	66.70832	88 121 125	7.634 607	16.44826	35.43671
446	198 916	21.11871	66.78323	88 716 536	7.640 321	16.46057	35.46323
447	199 809	21.14237	66.85806	89 314 623	7.646 027	16.47287	35.48971
448	200 704	21.16601	66.93280	89 915 392	7.651 725	16.48514	35.51616
449	201 601	21.18962	67.00746	90 518 849	7.657 414	16.49740	35.54257
450	202 500	21.21320	67.08204	91 125 000	7.663 094	16.50964	35.56893

SQUARES, SQUARE ROOT, CUBES AND CUBE ROOT (Continued)

n	n^2	$\sqrt{n}$	$\sqrt{10n}$	n^3	$\sqrt[3]{n}$	$\sqrt[3]{10n}$	$\sqrt[3]{100n}$
450	202 500	21.21320	67.08204	91 125 000	7.663 094	16.50964	35.56893
451	203 401	21.23676	67.15653	91 733 851	7.668 766	16.52186	35.59526
452	204 304	21.26029	67.23095	92 345 408	7.674 430	16.53406	35.62155
453	205 209	21.28380	67.30527	92 959 677	7.680 086	16.54624	35.64780
454	206 116	21.30728	67.37952	93 576 664	7.685 733	16.55841	35.67401
455	207 025	21.33073	67.45369	94 196 375	7.691 372	16.57056	35.70018
456	207 936	21.35416	67.52777	94 818 816	7.697 002	16.58269	35.72632
457	208 849	21.37756	67.60178	95 443 993	7.702 625	16.59480	35.75242
458	209 764	21.40093	67.67570	96 071 912	7.708 239	16.60690	35.77848
459	210 681	21.42429	67.74954	96 702 579	7.713 845	16.61897	35.80450
460	211 600	21.44761	67.82330	97 336 000	7.719 443	16.63103	35.83048
461	212 521	21.47091	67.89698	97 972 181	7.725 032	16.64308	35.85642
462	213 444	21.49419	67.97058	98 611 128	7.730 614	16.65510	35.88233
463	214 369	21.51743	68.04410	99 252 847	7.736 188	16.66711	35.90820
464	215 296	21.54066	68.11755	99 897 344	7.741 753	16.67910	35.93404
465	216 225	21.56386	68.19091	100 544 625	7.747 311	16.69108	35.95983
466	217 156	21.58703	68.26419	101 194 696	7.752 861	16.70303	35.98559
467	218 089	21.61018	68.33740	101 847 563	7.758 402	16.71497	36.01131
468	219 024	21.63331	68.41053	102 503 232	7.763 936	16.72689	36.03700
469	219 961	21.65641	68.48357	103 161 709	7.769 462	16.73880	36.06265
470	220 900	21.67948	68.55655	103 823 000	7.774 980	16.75069	36.08826
471	221 841	21.70253	68.62944	104 487 111	7.780 490	16.76256	36.11384
472	222 784	21.72556	68.70226	105 154 048	7.785 993	16.77441	36.13938
473	223 729	21.74856	68.77500	105 823 817	7.791 488	16.78625	36.16488
474	224 676	21.77154	68.84766	106 496 424	7.796 975	16.79807	36.19035
475	225 625	21.79449	68.92024	107 171 875	7.802 454	16.80988	36.21578
476	226 576	21.81742	68.99275	107 850 176	7.807 925	16.82167	36.24118
477	227 529	21.84033	69.06519	108 531 333	7.813 389	16.83344	36.26654
478	228 484	21.86321	69 13754	109 215 352	7.818 846	16.84519	36.29187
479	229 441	21.88607	69.20983	109 902 239	7.824 294	16.85693	36.31716
480	230 400	21.90890	69 28203	110 592 000	7.829 735	16.86865	36.34241
481	231 361	21.93171	69.35416	111 284 641	7.835 169	16.88036	36.36763
482	232 324	21.95450	69.42622	111 980 168	7.840 595	16.89205	36.39282
483	233 289	21.97726	69 49820	112 678 587	7.846 013	16.90372	36.41797
484	234 256	22.00000	69.57011	113 379 904	7.851 424	16.91538	36.44308
485	235 225	22.02272	69.64194	114 084 125	7.856 828	16.92702	36.46817
486	236 196	22.04541	69.71370	114 791 256	7.862 224	16.93865	36.49321
487	237 169	22.06808	69.78539	115 501 303	7.867 613	16.95026	36.51822
488	238 144	22.09072	69.85700	116 214 272	7.872 994	16.96185	36.54320
489	239 121	22.11334	69.92853	116 930 169	7.878 368	16.97343	36.56815
490	240 100	22.13594	70.00000	117 649 000	7.883 735	16.98499	36.59306
491	241 081	22.15852	70.07139	118 370 771	7.889 095	16.99654	36.61793
492	242 064	22.18107	70.14271	119 095 488	7.894 447	17.00807	36.64278
493	243 049	22.20360	70.21396	119 823 157	7.899 792	17.01959	36.66758
494	244 036	22.22611	70.28513	120 553 784	7.905 129	17.03108	36.69236
495	245 025	22.24860	70.35624	121 287 375	7.910 460	17.04257	36.71710
496	246 016	22.27106	70.42727	122 023 936	7.915 783	17.05404	36.74181
497	247 009	22.29350	70.49823	122 763 473	7.921 099	17.06549	36.76649
498	248 004	22.31591	70.56912	123 505 992	7.926 408	17.07693	36.79113
499	249 001	22.33831	70.63993	124 251 499	7.931 710	17.08835	36.81574
500	250 000	22.36068	70.71068	125 000 000	7.937 005	17.09976	36.84031

SQUARES, SQUARE ROOT, CUBES AND CUBE ROOT (Continued)

n	n^2	$\sqrt{n}$	$\sqrt{10n}$	n^3	$\sqrt[3]{n}$	$\sqrt[3]{10n}$	$\sqrt[3]{100n}$
500	250 000	22.36068	70.71068	125 000 000	7.937 005	17.09976	36.84031
501	251 001	22.38303	70.78135	125 751 501	7.942 293	17.11115	36.86486
502	252 004	22.40536	70.85196	126 506 008	7.947 574	17.12253	36.88937
503	253 009	22.42766	70.92249	127 263 527	7.952 848	17.13389	36.91385
504	254 016	22.44994	70.99296	128 024 064	7.958 114	17.14524	36.93830
505	255 025	22.47221	71.06335	128 787 625	7.963 374	17.15657	36.96271
506	256 036	22.49444	71.13368	129 554 216	7.968 627	17.16789	36.98709
507	257 049	22.51666	71.20393	130 323 843	7.973 873	17.17919	37.01144
508	258 064	22.53886	71.27412	131 096 512	7.979 112	17.19048	37.03576
509	259 081	22.56103	71.34424	131 872 229	7.984 344	17.20175	37.06004
510	260 100	22.58318	71.41428	132 651 000	7.989 570	17.21301	37.08430
511	261 121	22.60531	71.48426	133 432 831	7.994 788	17.22425	37.10852
512	262 144	22.62742	71.55418	134 217 728	8.000 000	17.23548	37.13271
513	263 169	22.64950	71.62402	135 005 697	8.005 205	17.24669	37.15687
514	264 196	22.67157	71.69379	135 796 744	8.010 403	17.25789	37.18100
515	265 225	22.69361	71.76350	136 590 875	8.015 595	17.26908	37.20509
516	266 256	22.71563	71.83314	137 388 096	8.020 779	17.28025	37.22916
517	267 289	22.73763	71.90271	138 188 413	8.025 957	17.29140	37.25319
518	268 324	22.75961	71.97222	138 991 832	8.031 129	17.30254	37.27720
519	269 361	22.78157	72.04165	139 798 359	8.036 293	17.31367	37.30117
520	270 400	22.80351	72.11103	140 608 000	8.041 452	17.32478	37.32511
521	271 441	22.82542	72.18033	141 420 761	8.046 603	17.33588	37.34902
522	272 484	22.84732	72.24957	142 236 648	8.051 748	17.34696	37.37290
523	273 529	22.86919	72.31874	143 055 667	8.056 886	17.35804	37.39675
524	274 576	22.89105	72.38784	143 877 824	8.062 018	17.36909	37.42057
525	275 625	22.91288	72.45688	144 703 125	8.067 143	17.38013	37.44436
526	276 676	22.93469	72.52586	145 531 576	8.072 262	17.39116	37.46812
527	277 729	22.95648	72.59477	146 363 183	8.077 374	17.40218	37.49185
528	278 784	22.97825	72.66361	147 197 952	8.082 480	17.41318	37.51555
529	279 841	23.00000	72.73239	148 035 889	8.087 579	17.42416	37.53922
530	280 900	23.02173	72.80110	148 877 000	8.092 672	17.43513	37.56286
531	281 961	23.04344	72.86975	149 721 291	8.097 759	17.44609	37.58647
532	283 024	23.06513	72.93833	150 568 768	8.102 839	17.45704	37.61005
533	284 089	23.08679	73.00685	151 419 437	8.107 913	17.46797	37.63360
534	285 156	23.10844	73.07530	152 273 304	8.112 980	17.47889	37.65712
535	286 225	23.13007	73.14369	153 130 375	8.118 041	17.48979	37.68061
536	287 296	23.15167	73.21202	153 990 656	8.123 096	17.50068	37.70407
537	288 369	23.17326	73.28028	154 854 153	8.128 145	17.51156	37.72751
538	289 444	23.19483	73.34848	155 720 872	8.133 187	17.52242	37.75091
539	290 521	23.21637	73.41662	156 590 819	8.138 223	17.53327	37.77429
540	291 600	23.23790	73.48469	157 464 000	8.143 253	17.54411	37.79763
541	292 681	23.25941	73.55270	158 340 421	8.148 276	17.55493	37.82095
542	293 764	23.28089	73.62065	159 220 088	8.153 294	17.56574	37.84424
543	294 849	23.30236	73.68853	160 103 007	8.158 305	17.57654	37.86750
544	295 936	23.32381	73.75636	160 989 184	8.163 310	17.58732	37.89073
545	297 025	23.34524	73.82412	161 878 625	8.168 309	17.59809	37.91393
546	298 116	23.36664	73.89181	162 771 336	8.173 302	17.60885	37.93711
547	299 209	23.38803	73.95945	163 667 323	8.178 289	17.61959	37.96025
548	300 304	23.40940	74.02702	164 566 592	8.183 269	17.63032	37.98337
549	301 401	23.43075	74.09453	165 469 149	8.188 244	17.64104	38.00646
550	302 500	23.45208	74.16198	166 375 000	8.193 213	17.65174	38.02952

SQUARES, SQUARE ROOT, CUBES AND CUBE ROOT (Continued)

n	n^2	$\sqrt{n}$	$\sqrt{10n}$	n^3	$\sqrt[3]{n}$	$\sqrt[3]{10n}$	$\sqrt[3]{100n}$
550	302 500	23.45208	74.16198	166 375 000	8.193 213	17.65174	38.02952
551	303 601	23.47339	74.22937	167 284 151	8.198 175	17.66243	38.05256
552	304 704	23.49468	74.29670	168 196 608	8.203 132	17.67311	38.07557
553	305 809	23.51595	74.36397	169 112 377	8.208 082	17.68378	38.09854
554	306 916	23.53720	74.43118	170 031 464	8.213 027	17.69443	38.12149
555	308 025	23.55844	74.49832	170 953 875	8.217 966	17.70507	38.14442
556	309 136	23.57965	74.56541	171 879 616	8.222 899	17.71570	38.16731
557	310 249	23.60085	74.63243	172 808 693	8.227 825	17.72631	38.19018
558	311 364	23.62202	74.69940	173 741 112	8.232 746	17.73691	38.21302
559	312 481	23.64318	74.76630	174 676 879	8.237 661	17.74750	38.23584
560	313 600	23.66432	74.83315	175 616 000	8.242 571	17.75808	38.25862
561	314 721	23.68544	74.89993	176 558 481	8.247 474	17.76864	38.28138
562	315 844	23.70654	74.96666	177 504 328	8.252 372	17.77920	38.30412
563	316 969	23.72762	75.03333	178 453 547	8.257 263	17.78973	38.32682
564	318 096	23.74868	75.09993	179 406 144	8.262 149	17.80026	38.34950
565	319 225	23.76973	75.16648	180 362 125	8.267 029	17.81077	38.37215
566	320 356	23.79075	75.23297	181 321 496	8.271 904	17.82128	38.39478
567	321 489	23.81176	75.29940	182 284 263	8.276 773	17.83177	38.41737
568	322 624	23.83275	75.36577	183 250 432	8.281 635	17.84224	38.43995
569	323 761	23.85372	75.43209	184 220 009	8.286 493	17.85271	38.46249
570	324 900	23.87467	75.49834	185 193 000	8.291 344	17.86316	38.48501
571	326 041	23.89561	75.56454	186 169 411	8.296 190	17.87360	38.50750
572	327 184	23.91652	75.63068	187 149 248	8.301 031	17.88403	38.52997
573	328 329	23.93742	75.69676	188 132 517	8.305 865	17.89444	38.55241
574	329 476	23.95830	75.76279	189 119 224	8.310 694	17.90485	38.57482
575	330 625	23.97916	75.82875	190 109 375	8.315 517	17.91524	38.59721
576	331 776	24.00000	75.89466	191 102 976	8.320 335	17.92562	38.61958
577	332 929	24.02082	75.96052	192 100 033	8.325 148	17.93599	38.64191
578	334 084	24.04163	76.02631	193 100 552	8.329 954	17.94634	38.66422
579	335 241	24.06242	76.09205	194 104 539	8.334 755	17.95669	38.68651
580	336 400	24.08319	76.15773	195 112 000	8.339 551	17.96702	38.70877
581	337 561	24.10394	76.22336	196 122 941	8.344 341	17.97734	38.73100
582	338 724	24.12468	76.28892	197 137 368	8.349 126	17.98765	38.75321
583	339 889	24.14539	76.35444	198 155 287	8.353 905	17.99794	38.77539
584	341 056	24.16609	76.41989	199 176 704	8.358 678	18.00823	38.79755
585	342 225	24.18677	76.48529	200 201 625	8.363 447	18.01850	38.81968
586	343 396	24.20744	76.55064	201 230 056	8.368 209	18.02876	38.84179
587	344 569	24.22808	76.61593	202 262 003	8.372 967	18.03901	38.86387
588	345 744	24.24871	76.68116	203 297 472	8.377 719	18.04925	38.88593
589	346 921	24.26932	76.74634	204 336 469	8.382 465	18.05947	38.90796
590	348 100	24.28992	76.81146	205 379 000	8.387 207	18.06969	38.92996
591	349 281	24.31049	76.87652	206 425 071	8.391 942	18.07989	38.95195
592	350 464	24.33105	76.94154	207 474 688	8.396 673	18.09008	38.97390
593	351 649	24.35159	77.00649	208 527 857	8.401 398	18.10026	38.99584
594	352 836	24.37212	77.07140	209 584 584	8.406 118	18.11043	39.01774
595	354 025	24.39262	77.13624	210 644 875	8.410 833	18.12059	39.03963
596	355 216	24.41311	77.20104	211 708 736	8.415 542	18.13074	39.06149
597	356 409	24.43358	77.26578	212 776 173	8.420 246	18.14087	39.08332
598	357 604	24.45404	77.33046	213 847 192	8.424 945	18.15099	39.10513
599	358 801	24.47448	77.39509	214 921 799	8.429 638	18.16111	39.12692
600	360 000	24.49490	77.45967	216 000 000	8.434 327	18.17121	39.14868

SQUARES, SQUARE ROOT, CUBES AND CUBE ROOT (Continued)

n	n^2	$\sqrt{n}$	$\sqrt{10n}$	n^3	$\sqrt[3]{n}$	$\sqrt[3]{10n}$	$\sqrt[3]{100n}$
600	360 000	24.49490	77.45967	216 000 000	8.434 327	18.17121	39.14868
601	361 201	24.51530	77.52419	217 081 801	8.439 010	18.18130	39.17041
602	362 404	24.53569	77.58866	218 167 208	8.443 688	18.19137	39.19213
603	363 609	24.55606	77.65307	219 256 227	8.448 361	18.20144	39.21382
604	364 816	24.57641	77.71744	220 348 864	8.453 028	18.21150	39.23548
605	366 025	24.59675	77.78175	221 445 125	8.457 691	18.22154	39.25712
606	367 236	24.61707	77.84600	222 545 016	8.462 348	18.23158	39.27874
607	368 449	24.63737	77.91020	223 648 543	8.467 000	18.24160	39.30033
608	369 664	24.65766	77.97435	224 755 712	8.471 647	18.25161	39.32190
609	370 881	24.67793	78.03845	225 866 529	8.476 289	18.26161	39.34345
610	372 100	24.69818	78.10250	226 981 000	8.480 926	18.27160	39.36497
611	373 321	24.71841	78.16649	228 099 131	8.485 558	18.28158	39.38647
612	374 544	24.73863	78.23043	229 220 928	8.490 185	18.29155	39.40795
613	375 769	24.75884	78.29432	230 346 397	8.494 807	18.30151	39.42940
614	376 996	24.77902	78.35815	231 475 544	8.499 423	18.31145	39.45083
615	378 225	24.79919	78.42194	232 608 375	8.504 035	18.32139	39.47223
616	379 456	24.81935	78.48567	233 744 896	8.508 642	18.33131	39.49362
617	380 689	24.83948	78.54935	234 885 113	8.513 243	18.34123	39.51498
618	381 924	24.85961	78.61298	236 029 032	8.517 840	18.35113	39.53631
619	383 161	24.87971	78.67655	237 176 659	8.522 432	18.36102	39.55763
620	384 400	24.89980	78.74008	238 328 000	8.527 019	18.37091	39.57892
621	385 641	24.91987	78.80355	239 483 061	8.531 601	18.38078	39.60018
622	386 884	24.93993	78.86698	240 641 848	8.536 178	18.39064	39.62143
623	388 129	24.95997	78.93035	241 804 367	8.540 750	18.40049	39.64265
624	389 376	24.97999	78.99367	242 970 624	8.545 317	18.41033	39.66385
625	390 625	25.00000	79.05694	244 140 625	8.549 880	18.42016	39.68503
626	391 876	25.01999	79.12016	245 314 376	8.554 437	18.42998	39.70618
627	393 129	25.03997	79.18333	246 491 883	8.558 990	18.43978	39.72731
628	394 384	25.05993	79.24645	247 673 152	8.563 538	18.44958	39.74842
629	395 641	25.07987	79.30952	248 858 189	8.568 081	18.45937	39.76951
630	396 900	25.09980	79.37254	250 047 000	8.572 619	18.46915	39.79057
631	398 161	25.11971	79.43551	251 239 591	8.577 152	18.47891	39.81161
632	399 424	25.13961	79.49843	252 435 968	8.581 681	18.48867	39.83263
633	400 689	25.15949	79.56130	253 636 137	8.586 205	18.49842	39.85363
634	401 956	25.17936	79.62412	254 840 104	8.590 724	18.50815	39.87461
635	403 225	25.19921	79.68689	256 047 875	8.595 238	18.51788	39.89556
636	404 496	25.21904	79.74961	257 259 456	8.599 748	18.52759	39.91649
637	405 769	25.23886	79.81228	258 474 853	8.604 252	18.53730	39.93740
638	407 044	25.25866	79.87490	259 694 072	8.608 753	18.54700	39.95829
639	408 321	25.27845	79.93748	260 917 119	8.613 248	18.55668	39.97916
640	409 600	25.29822	80.00000	262 144 000	8.617 739	18.56636	40.00000
641	410 881	25.31798	80.06248	263 374 721	8.622 225	18.57602	40.02082
642	412 164	25.33772	80.12490	264 609 288	8.626 706	18.58568	40.04162
643	413 449	25.35744	80.18728	265 847 707	8.631 183	18.59532	40.06240
644	414 736	25 37716	80.24961	267 089 984	8.635 655	18.60495	40.08316
645	416 025	25.39685	80.31189	268 336 125	8.640 123	18.61458	40.10390
646	417 316	25.41653	80.37413	269 586 136	8.644 585	18.62419	40.12461
647	418 609	25.43619	80.43631	270 840 023	8.649 044	18.63380	40.14530
648	419 904	25.45584	80.49845	272 097 792	8.653 497	18.64340	40.16598
649	421 201	25.47548	80.56054	273 359 449	8.657 947	18.65298	40.18663
650	422 500	25.49510	80.62258	274 625 000	8.662 391	18.66256	40.20726

SQUARES, SQUARE ROOT, CUBES AND CUBE ROOT (Continued)

n	n^2	$\sqrt{n}$	$\sqrt{10n}$	n^3	$\sqrt[3]{n}$	$\sqrt[3]{10n}$	$\sqrt[3]{100n}$
650	422 500	25 49510	80.62258	274 625 000	8.662 391	18.66256	40.20726
651	423 801	25.51470	80.68457	275 894 451	8.666 831	18.67212	40.22787
652	425 104	25.53429	80.74652	277 167 808	8.671 266	18.68168	40.24845
653	426 409	25.55386	80.80842	278 445 077	8.675 697	18.69122	40.26902
654	427 716	25.57342	80.87027	279 726 264	8.680 124	18.70076	40.28957
655	429 025	25 59297	80.93207	281 011 375	8.684 546	18.71029	40.31009
656	430 336	25.61250	80.99383	282 300 416	8.688 963	18.71980	40.33059
657	431 649	25.63201	81.05554	283 593 393	8.693 376	18.72931	40.35108
658	432 964	25.65151	81.11720	284 890 312	8.697 784	18.73881	40.37154
659	434 281	25.67100	81.17881	286 191 179	8.702 188	18.74830	40.39198
660	435 600	25.69047	81.24038	287 496 000	8.706 588	18.75777	40.41240
661	436 921	25.70992	81.30191	288 804 781	8.710 983	18.76724	40.43280
662	438 244	25.72936	81.36338	290 117 528	8.715 373	18.77670	40.45318
663	439 569	25.74879	81.42481	291 434 247	8.719 760	18.78615	40.47354
664	440 896	25.76820	81.48620	292 754 944	8.724 141	18.79559	40.49388
665	442 225	25.78759	81.54753	294 079 625	8.728 519	18.80502	40.51420
666	443 556	25.80698	81.60882	295 408 296	8.732 892	18.81444	40.53449
667	444 889	25.82634	81.67007	296 740 963	8.737 260	18.82386	40.55477
668	446 224	25.84570	81.73127	298 077 632	8.741 625	18.83326	40.57503
669	447 561	25.86503	81.79242	299 418 309	8.745 985	18.84265	40.59526
670	448 900	25.88436	81.85353	300 763 000	8.750 340	18.85204	40.61548
671	450 241	25.90367	81.91459	302 111 711	8.754 691	18.86141	40.63568
672	451 584	25.92296	81.97561	303 464 448	8.759 038	18.87078	40.65585
673	452 929	25.94224	82.03658	304 821 217	8.763 381	18.88013	40.67601
674	454 276	25.96151	82.09750	306 182 024	8.767 719	18.88948	40.69615
675	455 625	25.98076	82.15838	307 546 875	8.772 053	18.89882	40.71626
676	456 976	26.00000	82.21922	308 915 776	8.776 383	18.90814	40.73636
677	458 329	26.01922	82.28001	310 288 733	8.780 708	18.91746	40.75644
678	459 684	26.03843	82.34076	311 665 752	8.785 030	18.92677	40.77650
679	461 041	26.05763	82.40146	313 046 839	8.789 347	18.93607	40.79653
680	462 400	26.07681	82.46211	314 432 000	8.793 659	18.94536	40.81655
681	463 761	26.09598	82.52272	315 821 241	8.797 968	18.95465	40.83655
682	465 124	26.11513	82.58329	317 214 568	8.802 272	18.96392	40.85653
683	466 489	26.13427	82.64381	318 611 987	8.806 572	18.97318	40.87649
684	467 856	26.15339	82.70429	320 013 504	8.810 868	18.98244	40.89643
685	469 225	26.17250	82.76473	321 419 125	8.815 160	18.99169	40.91635
686	470 596	26.19160	82.82512	322 828 856	8.819 447	19.00092	40.93625
687	471 969	26.21068	82.88546	324 242 703	8.823 731	19.01015	40.95613
688	473 344	26.22975	82.94577	325 660 672	8.828 010	19.01937	40.97599
689	474 721	26.24881	83.00602	327 082 769	8.832 285	19.02858	40.99584
690	476 100	26.26785	83.06624	328 509 000	8.836 556	19.03778	41.01566
691	477 481	26.28688	83.12641	329 939 371	8.840 823	19.04698	41.03546
692	478 864	26.30589	83.18654	331 373 888	8.845 085	19.05616	41.05525
693	480 249	26.32489	83.24662	332 812 557	8.849 344	19.06533	41.07502
694	481 636	26.34388	83.30666	334 255 384	8.853 599	19.07450	41.09476
695	483 025	26.36285	83.36666	335 702 375	8.857 849	19.08366	41.11449
696	484 416	26.38181	83.42661	337 153 536	8.862 095	19.09281	41.13420
697	485 809	26.40076	83.48653	338 608 873	8.866 338	19.10195	41.15389
698	487 204	26.41969	83.54639	340 068 392	8.870 576	19.11108	41.17357
699	488 601	26.43861	83.60622	341 532 099	8.874 810	19.12020	41.19322
700	490 000	26.45751	83.66600	343 000 000	8.879 040	19.12931	41.21285

SQUARES, SQUARE ROOT, CUBES AND CUBE ROOT (Continued)

n	n^2	$\sqrt{n}$	$\sqrt{10n}$	n^3	$\sqrt[3]{n}$	$\sqrt[3]{10n}$	$\sqrt[3]{100n}$
700	490 000	26.45751	83.66600	343 000 000	8.879 040	19.12931	41.21285
701	491 401	26.47640	83.72574	344 472 101	8.883 266	19.13842	41.23247
702	492 804	26.49528	83.78544	345 948 408	8.887 488	19.14751	41.25207
703	494 209	26.51415	83.84510	347 428 927	8.891 706	19.15660	41.27164
704	495 616	26.53300	83.90471	348 913 664	8.895 920	19.16568	41.29120
705	497 025	26.55184	83.96428	350 402 625	8.900 130	19.17475	41.31075
706	498 436	26.57066	84.02381	351 895 816	8.904 337	19.18381	41.33027
707	499 849	26.58947	84.08329	353 393 243	8.908 539	19.19286	41.34977
708	501 264	26.60827	84.14274	354 894 912	8.912 737	19.20191	41.36926
709	502 681	26.62705	84.20214	356 400 829	8.916 931	19.21095	41.38873
710	504 100	26.64583	84.26150	357 911 000	8.921 121	19.21997	41.40818
711	505 521	26.66458	84.32082	359 425 431	8.925 308	19.22899	41.42761
712	506 944	26.68333	84.38009	360 944 128	8.929 490	19.23800	41.44702
713	508 369	26.70206	84.43933	362 467 097	8.933 669	19.24701	41.46642
714	509 796	26.72078	84.49852	363 994 344	8.937 843	19.25600	41.48579
715	511 225	26.73948	84.55767	365 525 875	8.942 014	19.26499	41.50515
716	512 656	26.75818	84.61678	367 061 696	8.946 181	19.27396	41.52449
717	514 089	26.77686	84.67585	368 601 813	8.950 344	19.28293	41.54382
718	515 524	26.79552	84.73488	370 146 232	8.954 503	19.29189	41.56312
719	516 961	26.81418	84.79387	371 694 959	8.958 658	19.30084	41.58241
720	518 400	26.83282	84.85281	373 248 000	8.962 809	19.30979	41.60168
721	519 841	26.85144	84.91172	374 805 361	8.966 957	19.31872	41.62093
722	521 284	26.87006	84.97058	376 367 048	8.971 101	19.32765	41.64016
723	522 729	26.88866	85.02941	377 933 067	8.975 241	19.33657	41.65938
724	524 176	26.90725	85.08819	379 503 424	8.979 377	19.34548	41.67857
725	525 625	26.92582	85.14693	381 078 125	8.983 509	19.35438	41.69775
726	527 076	26.94439	85.20563	382 657 176	8.987 637	19.36328	41.71692
727	528 529	26.96294	85.26429	384 240 583	8.991 762	19.37216	41.73606
728	529 984	26.98148	85.32292	385 828 352	8.995 883	19.38104	41.75519
729	531 441	27.00000	85.38150	387 420 489	9.000 000	19.38991	41.77430
730	532 900	27.01851	85.44004	389 017 000	9.004 113	19.39877	41.79339
731	534 361	27.03701	85.49854	390 617 891	9.008 223	19.40763	41.81247
732	535 824	27.05550	85.55700	392 223 168	9.012 329	19.41647	41.83152
733	537 289	27.07397	85.61542	393 832 837	9.016 431	19.42531	41.85056
734	538 756	27.09243	85.67380	395 446 904	9.020 529	19.43414	41.86959
735	540 225	27.11088	85.73214	397 065 375	9.024 624	19.44296	41.88859
736	541 696	27.12932	85.79044	398 688 256	9.028 715	19.45178	41.90758
737	543 169	27.14774	85.84870	400 315 553	9.032 802	19.46058	41.92655
738	544 644	27.16616	85.90693	401 947 272	9.036 886	19.46938	41.94551
739	546 121	27.18455	85.96511	403 583 419	9.040 966	19.47817	41.96444
740	547 600	27.20294	86.02325	405 224 000	9.045 042	19.48695	41.98336
741	549 081	27.22132	86.08136	406 869 021	9.049 114	19.49573	42.00227
742	550 564	27.23968	86.13942	408 518 488	9.053 183	19.50449	42.02115
743	552 049	27.25803	86.19745	410 172 407	9.057 248	19.51325	42.04002
744	553 536	27.27636	86.25543	411 830 784	9.061 310	19.52200	42.05887
745	555 025	27.29469	86.31338	413 493 625	9.065 368	19.53074	42.07771
746	556 516	27.31300	86.37129	415 160 936	9.069 422	19.53948	42.09653
747	558 009	27.33130	86.42916	416 832 723	9.073 473	19.54820	42.11533
748	559 504	27.34959	86.48699	418 508 992	9.077 520	19.55692	42.13411
749	561 001	27.36786	86.54479	420 189 749	9.081 563	19.56563	42.15288
750	562 500	27.38613	86.60254	421 875 000	9.085 603	19.57434	42.17163

SQUARES, SQUARE ROOT, CUBES AND CUBE ROOT (Continued)

n	n^2	$\sqrt{n}$	$\sqrt{10n}$	n^3	$\sqrt[3]{n}$	$\sqrt[3]{10n}$	$\sqrt[3]{100n}$
750	562 500	27.38613	86.60254	421 875 000	9.085 603	19.57434	42.17163
751	564 001	27.40438	86.66026	423 564 751	9.089 639	19.58303	42.19037
752	565 504	27.42262	86.71793	425 259 008	9.093 672	19.59172	42.20909
753	567 009	27.44085	86.77557	426 957 777	9.097 701	19.60040	42.22779
754	568 516	27.45906	86.83317	428 661 064	9.101 727	19.60908	42.24647
755	570 025	27.47726	86.89074	430 368 875	9.105 748	19.61774	42.26514
756	571 536	27.49545	86.94826	432 081 216	9.109 767	19.62640	42.28379
757	573 049	27.51363	87.00575	433 798 093	9.113 782	19.63505	42.30243
758	574 564	27.53180	87.06320	435 519 512	9.117 793	19.64369	42.32105
759	576 081	27.54995	87.12061	437 245 479	9.121 801	19.65232	42.33965
760	577 600	27.56810	87.17798	438 976 000	9.125 805	19.66095	42.35824
761	579 121	27.58623	87.23531	440 711 081	9.129 806	19.66957	42.37681
762	580 644	27.60435	87.29261	442 450 728	9.133 803	19.67818	42.39536
763	582 169	27.62245	87.34987	444 194 947	9.137 797	19.68679	42.41390
764	583 696	27.64055	87.40709	445 943 744	9.141 787	19.69538	42.43242
765	585 225	27.65863	87.46428	447 697 125	9.145 774	19.70397	42.45092
766	586 756	27.67671	87.52143	449 455 096	9.149 758	19.71256	42.46941
767	588 289	27.69476	87.57854	451 217 663	9.153 738	19.72113	42.48789
768	589 824	27.71281	87.63561	452 984 832	9.157 714	19.72970	42.50634
769	591 361	27.73085	87.69265	454 756 609	9.161 687	19.73826	42.52478
770	592 900	27.74887	87.74964	456 533 000	9.165 656	19.74681	42.54321
771	594 441	27.76689	87.80661	458 314 011	9.169 623	19.75535	42.56162
772	595 984	27.78489	87.86353	460 099 648	9.173 585	19.76389	42.58001
773	597 529	27.80288	87.92042	461 889 917	9.177 544	19.77242	42.59839
774	599 076	27.82086	87.97727	463 684 824	9.181 500	19.78094	42.61675
775	600 625	27.83882	88.03408	465 484 375	9.185 453	19.78946	42.63509
776	602 176	27.85678	88.09086	467 288 576	9.189 402	19.79797	42.65342
777	603 729	27.87472	88.14760	469 097 433	9.193 347	19.80647	42.67174
778	605 284	27.89265	88.20431	470 910 952	9.197 290	19.81496	42.69004
779	606 841	27.91057	88.26098	472 729 139	9.201 229	19.82345	42.70832
780	608 400	27.92848	88.31761	474 552 000	9.205 164	19.83192	42.72659
781	609 961	27.94638	88.37420	476 379 541	9.209 096	19.84040	42.74484
782	611 524	27.96426	88.43076	478 211 768	9.213 025	19.84886	42.76307
783	613 089	27.98214	88.48729	480 048 687	9.216 950	19.85732	42.78129
784	614 656	28.00000	88.54377	481 890 304	9.220 873	19.86577	42.79950
785	616 225	28.01785	88.60023	483 736 625	9.224 791	19.87421	42.81769
786	617 796	28.03569	88.65664	485 587 656	9.228 707	19.88265	42.83586
787	619 369	28.05352	88.71302	487 443 403	9.232 619	19.89107	42.85402
788	620 944	28.07134	88.76936	489 303 872	9.236 528	19.89950	42.87216
789	622 521	28.08914	88.82567	491 169 069	9.240 433	19.90791	42.89029
790	624 100	28.10694	88.88194	493 039 000	9.244 335	19.91632	42.90840
791	625 681	28.12472	88.93818	494 913 671	9.248 234	19.92472	42.92650
792	627 264	28.14249	88.99438	496 793 088	9.252 130	19.93311	42.94458
793	628 849	28.16026	89.05055	498 677 257	9.256 022	19.94150	42.96265
794	630 436	28.17801	89.10668	500 566 184	9.259 911	19.94987	42.98070
795	632 025	28.19574	89.16277	502 459 875	9.263 797	19.95825	42.99874
796	633 616	28.21347	89.21883	504 358 336	9.267 680	19.96661	43.01676
797	635 209	28.23119	89.27486	506 261 573	9.271 559	19.97497	43.03477
798	636 804	28.24889	89.33085	508 169 592	9.275 435	19.98332	43.05276
799	638 401	28.26659	89.38680	510 082 399	9.279 308	19.99166	43.07073
800	640 000	28.28427	89.44272	512 000 000	9.283 178	20.00000	43.08869

SQUARES, SQUARE ROOT, CUBES AND CUBE ROOT (Continued)

n	n^2	$\sqrt{n}$	$\sqrt{10n}$	n^3	$\sqrt[3]{n}$	$\sqrt[3]{10n}$	$\sqrt[3]{100n}$
800	640 000	28.28427	89.44272	512 000 000	9.283 178	20.00000	43.08869
801	641 601	28.30194	89.49860	513 922 401	9.287 044	20.00833	43.10664
802	643 204	28.31960	89.55445	515 849 608	9.290 907	20.01665	43.12457
803	644 809	28.33725	89.61027	517 781 627	9.294 767	20.02497	43.14249
804	646 416	28.35489	89.66605	519 718 464	9.298 624	20.03328	43.16039
805	648 025	28.37252	89.72179	521 660 125	9.302 477	20.04158	43.17828
806	649 636	28.39014	89.77750	523 606 616	9.306 328	20.04988	43.19615
807	651 249	28.40775	89.83318	525 557 943	9.310 175	20.05816	43.21400
808	652 864	28.42534	89.88882	527 514 112	9.314 019	20.06645	43.23185
809	654 481	28.44293	89.94443	529 475 129	9.317 860	20.07472	43.24967
810	656 100	28.46050	90.00000	531 441 000	9.321 698	20.08299	43.26749
811	657 721	28.47806	90.05554	533 411 731	9.325 532	20.09125	43.28529
812	659 344	28.49561	90.11104	535 387 328	9.329 363	20.09950	43.30307
813	660 969	28.51315	90.16651	537 367 797	9.333 192	20.10775	43.32084
814	662 596	28.53069	90.22195	539 353 144	9.337 017	20.11599	43.33859
815	664 225	28.54820	90.27735	541 343 375	9.340 839	20.12423	43.35633
816	665 856	28.56571	90.33272	543 338 496	9.344 657	20.13245	43.37406
817	667 489	28.58321	90.38805	545 338 513	9.348 473	20.14067	43.39177
818	669 124	28.60070	90.44335	547 343 432	9.352 286	20.14889	43.40947
819	670 761	28.61818	90.49862	549 353 259	9.356 095	20.15710	43.42715
820	672 400	28.63564	90.55385	551 368 000	9.359 902	20.16530	43.44481
821	674 041	28.65310	90.60905	553 387 661	9.363 705	20.17349	43.46247
822	675 684	28.67054	90.66422	555 412 248	9.367 505	20.18168	43.48011
823	677 329	28.68798	90.71935	557 441 767	9.371 302	20.18986	43.49773
824	678 976	28.70540	90.77445	559 476 224	9.375 096	20.19803	43.51534
825	680 625	28.72281	90.82951	561 515 625	9.378 887	20.20620	43.53294
826	682 276	28.74022	90.88454	563 559 976	9.382 675	20.21436	43.55052
827	683 929	28.75761	90.93954	565 609 283	9.386 460	20.22252	43.56809
828	685 584	28.77499	90.99451	567 663 552	9.390 242	20.23066	43.58564
829	687 241	28.79236	91.04944	569 722 789	9.394 021	20.23880	43.60318
830	688 900	28.80972	91.10434	571 787 000	9.397 796	20.24694	43.62071
831	690 561	28.82707	91.15920	573 856 191	9.401 569	20.25507	43.63822
832	692 224	28.84441	91.21403	575 930 368	9.405 339	20.26319	43.65572
833	693 889	28.86174	91.26883	578 009 537	9.409 105	20.27130	43.67320
834	695 556	28.87906	91.32360	580 093 704	9.412 869	20.27941	43.69067
835	697 225	28.89637	91.37833	582 182 875	9.416 630	20.28751	43.70812
836	698 896	28.91366	91.43304	584 277 056	9.420 387	20.29561	43.72556
837	700 569	28.93095	91.48770	586 376 253	9.424 142	20.30370	43.74299
838	702 244	28.94823	91.54234	588 480 472	9.427 894	20.31178	43.76041
839	703 921	28 96550	91.59694	590 589 719	9.431 642	20.31986	43.77781
840	705 600	28.98275	91.65151	592 704 000	9.435 388	20.32793	43.79519
841	707 281	29.00000	91.70605	594 823 321	9.439 131	20.33599	43.81256
842	708 964	29.01724	91.76056	596 947 688	9.442 870	20.34405	43.82992
843	710 649	29.03446	91.81503	599 077 107	9.446 607	20.35210	43.84727
844	712 336	29.05168	91.86947	601 211 584	9.450 341	20.36014	43.86460
845	714 025	29.06888	91.92388	603 351 125	9.454 072	20.36818	43.88191
846	715 716	29.08608	91.97826	605 495 736	9.457 800	20.37621	43.89922
847	717 409	29.10326	92.03260	607 645 423	9.461 525	20.38424	43.91651
848	719 104	29.12044	92.08692	609 800 192	9.465 247	20.39226	43.93378
849	720 801	29.13760	92.14120	611 960 049	9.468 966	20.40027	43.95105
850	722 500	29.15476	92.19544	614 125 000	9.472 682	20.40828	43.96830

SQUARES, SQUARE ROOT, CUBES AND CUBE ROOT (Continued)

n	n^2	$\sqrt{n}$	$\sqrt{10n}$	n^3	$\sqrt[3]{n}$	$\sqrt[3]{10n}$	$\sqrt[3]{100n}$
850	722 500	29.15476	92.19544	614 125 000	9.472 682	20.40828	43.96830
851	724 201	29.17190	92.24966	616 295 051	9.476 396	20.41628	43.98553
852	725 904	29.18904	92.30385	618 470 208	9.480 106	20.42427	44.00275
853	727 609	29.20616	92.35800	620 650 477	9.483 814	20.43226	44.01996
854	729 316	29.22328	92.41212	622 835 864	9.487 518	20.44024	44.03716
855	731 025	29.24038	92.46621	625 026 375	9.491 220	20.44821	44.05434
856	732 736	29.25748	92.52027	627 222 016	9.494 919	20.45618	44.07151
857	734 449	29.27456	92.57429	629 422 793	9.498 615	20.46415	44.08866
858	736 164	29.29164	92.62829	631 628 712	9.502 308	20.47210	44.10581
859	737 881	29.30870	92.68225	633 839 779	9.505 998	20.48005	44.12293
860	739 600	29.32576	92.73618	636 056 000	9.509 685	20.48800	44.14005
861	741 321	29.34280	92.79009	638 277 381	9.513 370	20.49593	44.15715
862	743 044	29.35984	92.84396	640 503 928	9.517 052	20.50387	44.17424
863	744 769	29.37686	92.89779	642 735 647	9.520 730	20.51179	44.19132
864	746 496	29.39388	92.95160	644 972 544	9.524 406	20.51971	44.20838
865	748 225	29.41088	93.00538	647 214 625	9.528 079	20.52762	44.22543
866	749 956	29.42788	93.05912	649 461 896	9.531 750	20.53553	44.24246
867	751 689	29.44486	93.11283	651 714 363	9.535 417	20.54343	44.25949
868	753 424	29.46184	93.16652	653 972 032	9.539 082	20.55133	44.27650
869	755 161	29.47881	93.22017	656 234 909	9.542 744	20.55922	44.29349
870	756 900	29.49576	93.27379	658 503 000	9.546 403	20.56710	44.31048
871	758 641	29.51271	93.32738	660 776 311	9.550 059	20.57498	44.32745
872	760 384	29.52965	93.38094	663 054 848	9.553 712	20.58285	44.34440
873	762 129	29.54657	93.43447	665 338 617	9.557 363	20.59071	44.36135
874	763 876	29.56349	93.48797	667 627 624	9.561 011	20.59857	44.37828
875	765 625	29.58040	93.54143	669 921 875	9.564 656	20.60643	44.39520
876	767 376	29.59730	93.59487	672 221 376	9.568 298	20.61427	44.41211
877	769 129	29.61419	93.64828	674 526 133	9.571 938	20.62211	44.42900
878	770 884	29.63106	93.70165	676 836 152	9.575 574	20.62995	44.44588
879	772 641	29.64793	93.75500	679 151 439	9.579 208	20.63778	44.46275
880	774 400	29.66479	93.80832	681 472 000	9.582 840	20.64560	44.47960
881	776 161	29.68164	93.86160	683 797 841	9.586 468	20.65342	44.49644
882	777 924	29.69848	93.91486	686 128 968	9.590 094	20.66123	44.51327
883	779 689	29.71532	93.96808	688 465 387	9.593 717	20.66904	44.53009
884	781 456	29.73214	94.02127	690 807 104	9.597 337	20.67684	44.54689
885	783 225	29.74895	94.07444	693 154 125	9.600 955	20.68463	44.56368
886	784 996	29.76575	94.12757	695 506 456	9.604 570	20.69242	44.58046
887	786 769	29.78255	94.18068	697 864 103	9.608 182	20.70020	44.59723
888	788 544	29.79933	94.23375	700 227 072	9.611 791	20.70798	44.61398
889	790 321	29.81610	94.28680	702 595 369	9.615 398	20.71575	44.63072
890	792 100	29.83287	94.33981	704 969 000	9.619 002	20.72351	44.64745
891	793 881	29.84962	94.39280	707 347 971	9.622 603	20.73127	44.66417
892	795 664	29.86637	94.44575	709 732 288	9.626 202	20.73902	44.68087
893	797 449	29.88311	94.49868	712 121 957	9.629 797	20.74677	44.69756
894	799 236	29.89983	94.55157	714 516 984	9.633 391	20.75451	44.71424
895	801 025	29.91655	94.60444	716 917 375	9.636 981	20.76225	44.73090
896	802 816	29.93326	94.65728	719 323 136	9.640 569	20.76998	44.74756
897	804 609	29.94996	94.71008	721 734 273	9.644 154	20.77770	44.76420
898	806 404	29.96665	94.76286	724 150 792	9.647 737	20.78542	44.78083
899	808 201	29.98333	94.81561	726 572 699	9.651 317	20.79313	44.79744
900	810 000	30.00000	94.86833	729 000 000	9.654 894	20.80084	44.81405

SQUARES, SQUARE ROOT, CUBES AND CUBE ROOT (Continued)

n	n^2	$\sqrt{n}$	$\sqrt{10n}$	n^3	$\sqrt[3]{n}$	$\sqrt[3]{10n}$	$\sqrt[3]{100n}$
900	810 000	30.00000	94.86833	729 000 000	9.654 894	20.80084	44.81405
901	811 801	30.01666	94.92102	731 432 701	9.658 468	20.80854	44.83064
902	813 604	30.03331	94.97368	733 870 808	9.662 040	20.81623	44.84722
903	815 409	30.04996	95.02631	736 314 327	9.665 610	20.82392	44.86379
904	817 216	30.06659	95.07891	738 763 264	9.669 176	20.83161	44.88034
905	819 025	30.08322	95.13149	741 217 625	9.672 740	20.83929	44.89688
906	820 836	30.09983	95.18403	743 677 416	9.676 302	20.84696	44.91341
907	822 649	30.11644	95.23655	746 142 643	9.679 860	20.85463	44.92993
908	824 464	30.13304	95.28903	748 613 312	9.683 417	20.86229	44.94644
909	826 281	30.14963	95.34149	751 089 429	9.686 970	20.86994	44.96293
910	828 100	30.16621	95.39392	753 571 000	9.690 521	20.87759	44.97941
911	829 921	30.18278	95.44632	756 058 031	9.694 069	20.88524	44.99588
912	831 744	30.19934	95.49869	758 550 528	9.697 615	20.89288	45.01234
913	833 569	30.21589	95.55103	761 048 497	9.701 158	20.90051	45.02879
914	835 396	30.23243	95.60335	763 551 944	9.704 699	20.90814	45.04522
915	837 225	30.24897	95.65563	766 060 875	9.708 237	20.91576	45.06164
916	839 056	30.26549	95.70789	768 575 296	9.711 772	20.92338	45.07805
917	840 889	30.28201	95.76012	771 095 213	9.715 305	20.93099	45.09445
918	842 724	30.29851	95.81232	773 620 632	9.718 835	20.93860	45.11084
919	844 561	30.31501	95.86449	776 151 559	9.722 363	20.94620	45.12721
920	846 400	30.33150	95.91663	778 688 000	9.725 888	20.95379	45.14357
921	848 241	30.34798	95.96874	781 229 961	9.729 411	20.96138	45.15992
922	850 084	30.36445	96.02083	783 777 448	9.732 931	20.96896	45.17626
923	851 929	30.38092	96.07289	786 330 467	9.736 448	20.97654	45.19259
924	853 776	30.39737	96.12492	788 889 024	9.739 963	20.98411	45.20891
925	855 625	30.41381	96.17692	791 453 125	9.743 476	20.99168	45.22521
926	857 476	30.43025	96.22889	794 022 776	9.746 986	20.99924	45.24150
927	859 329	30.44667	96.28084	796 597 983	9.750 493	21.00680	45.25778
928	861 184	30.46309	96.33276	799 178 752	9.753 998	21.01435	45.27405
929	863 041	30.47950	96.38465	801 765 089	9.757 500	21.02190	45.29030
930	864 900	30.49590	96.43651	804 357 000	9.761 000	21.02944	45.30655
931	866 761	30.51229	96.48834	806 954 491	9.764 497	21.03697	45.32278
932	868 624	30.52868	96.54015	809 557 568	9.767 992	21.04450	45.33900
933	870 489	30.54505	96.59193	812 166 237	9.771 485	21.05203	45.35521
934	872 356	30.56141	96.64368	814 780 504	9.774 974	21.05954	45.37141
935	874 225	30.57777	96.69540	817 400 375	9.778 462	21.06706	45.38760
936	876 096	30.59412	96.74709	820 025 856	9.781 946	21.07456	45.40377
937	877 969	30.61046	96.79876	822 656 953	9.785 429	21.08207	45.41994
938	879 844	30.62679	96.85040	825 293 672	9.788 909	21.08956	45.43609
939	881 721	30.64311	96.90201	827 936 019	9.792 386	21.09706	45.45223
940	883 600	30.65942	96.95360	830 584 000	9.795 861	21.10454	45.46836
941	885 481	30.67572	97.00515	833 237 621	9.799 334	21.11202	45.48448
942	887 364	30.69202	97.05668	835 896 888	9.802 804	21.11950	45.50058
943	889 249	30.70831	97.10819	838 561 807	9.806 271	21.12697	45.51668
944	891 136	30.72458	97.15966	841 232 384	9.809 736	21.13444	45.53276
945	893 025	30.74085	97.21111	843 908 625	9.813 199	21.14190	45.54883
946	894 916	30.75711	97.26253	846 590 536	9.816 659	21.14935	45.56490
947	896 809	30.77337	97.31393	849 278 123	9.820 117	21.15680	45.58095
948	898 704	30.78961	97.36529	851 971 392	9.823 572	21.16424	45.59698
949	900 601	30.80584	97.41663	854 670 349	9.827 025	21.17168	45.61301
950	902 500	30.82207	97.46794	857 375 000	9.830 476	21.17912	45.62903

SQUARES, SQUARE ROOT, CUBES AND CUBE ROOT (Continued)

n	n^2	$\sqrt{n}$	$\sqrt{10n}$	n^3	$\sqrt[3]{n}$	$\sqrt[3]{10n}$	$\sqrt[3]{100n}$
950	902 500	30.82207	97.46794	857 375 000	9.830 476	21.17912	45.62903
951	904 401	30.83829	97.51923	860 085 351	9.833 924	21.18655	45.64503
952	906 304	30.85450	97.57049	862 801 408	9.837 369	21.19397	45.66102
953	908 209	30.87070	97.62172	865 523 177	9.840 813	21.20139	45.67701
954	910 116	30.88689	97.67292	868 250 664	9.844 254	21.20880	45.69298
955	912 025	30.90307	97.72410	870 983 875	9.847 692	21.21621	45.70894
956	913 936	30.91925	97.77525	873 722 816	9.851 128	21.22361	45.72489
957	915 849	30.93542	97.82638	876 467 493	9.854 562	21.23101	45.74082
958	917 764	30.95158	97.87747	879 217 912	9.857 993	21.23840	45.75675
959	919 681	30.96773	97.92855	881 974 079	9.861 422	21.24579	45.77267
960	921 600	30.98387	97.97959	884 736 000	9.864 848	21.25317	45.78857
961	923 521	31.00000	98.03061	887 503 681	9.868 272	21.26055	45.80446
962	925 444	31.01612	98.08160	890 277 128	9.871 694	21.26792	45.82035
963	927 369	31.03224	98.13256	893 056 347	9.875 113	21.27529	45.83622
964	929 296	31.04835	98.18350	895 841 344	9.878 530	21.28265	45.85208
965	931 225	31.06445	98.23441	898 632 125	9.881 945	21.29001	45.86793
966	933 156	31.08054	98.28530	901 428 696	9.885 357	21.29736	45.88376
967	935 089	31.09662	98.33616	904 231 063	9.888 767	21.30470	45.89959
968	937 024	31.11270	98.38699	907 039 232	9.892 175	21.31204	45.91541
969	938 961	31.12876	98.43780	909 853 209	9.895 580	21.31938	45.93121
970	940 900	31.14482	98.48858	912 673 000	9.898 983	21.32671	45.94701
971	942 841	31.16087	98.53933	915 498 611	9.902 384	21.33404	45.96279
972	944 784	31.17691	98.59006	918 330 048	9.905 782	21.34136	45.97857
973	946 729	31.19295	98.64076	921 167 317	9.909 178	21.34868	45.99433
974	948 676	31.20897	98.69144	924 010 424	9.912 571	21.35599	46.01008
975	950 625	31.22499	98.74209	926 859 375	9.915 962	21.36329	46.02582
976	952 576	31.24100	98.79271	929 714 176	9.919 351	21.37059	46.04155
977	954 529	31.25700	98.84331	932 574 833	9.922 738	21.37789	46.05727
978	956 484	31.27299	98.89388	935 441 352	9.926 122	21.38518	46.07298
979	958 441	31.28898	98.94443	938 313 739	9.929 504	21.39247	46.08868
980	960 400	31.30495	98.99495	941 192 000	9.932 884	21.39975	46.10436
981	962 361	31.32092	99.04544	944 076 141	9.936 261	21.40703	46.12004
982	964 324	31.33688	99.09591	946 966 168	9.939 636	21.41430	46.13571
983	966 289	31.35283	99.14636	949 862 087	9.943 009	21.42156	46.15136
984	968 256	31.36877	99.19677	952 763 904	9.946 380	21.42883	46.16700
985	970 225	31.38471	99.24717	955 671 625	9.949 748	21.43608	46.18264
986	972 196	31.40064	99.29753	958 585 256	9.953 114	21.44333	46.19826
987	974 169	31.41656	99.34787	961 504 803	9.956 478	21.45058	46.21387
988	976 144	31.43247	99.39819	964 430 272	9.959 839	21.45782	46.22948
989	978 121	31.44837	99.44848	967 361 669	9.963 198	21.46506	46.24507
990	980 100	31.46427	99.49874	970 299 000	9.966 555	21.47229	46.26065
991	982 081	31.48015	99.54898	973 242 271	9.969 910	21.47952	46.27622
992	984 064	31.49603	99.59920	976 191 488	9.973 262	21.48674	46.29178
993	986 049	31.51190	99.64939	979 146 657	9.976 612	21.49396	46.30733
994	988 036	31.52777	99.69955	982 107 784	9.979 960	21.50117	46.32287
995	990 025	31.54362	99.74969	985 074 875	9.983 305	21.50838	46.33840
996	992 016	31.55947	99.79980	988 047 936	9.986 649	21.51558	46.35392
997	994 009	31.57531	99.84989	991 026 973	9.989 990	21.52278	46.36943
998	996 004	31.59114	99.89995	994 011 992	9.993 329	21.52997	46.38492
999	998 001	31.60696	99.94999	997 002 999	9.996 666	21.53716	46.40041
1000	1 000 000	31.62278	100.00000	1 000 000 000	10.000 000	21.54435	46.41589

FOURTH AND FIFTH ROOTS

This table is given as a supplement to the previous table. It contains the fourth and fifth roots of the numbers from 1 to 1000, accurate to seven significant digits.

FOURTH AND FIFTH ROOTS

n	$\sqrt[4]{n}$	$\sqrt[5]{n}$	n	$\sqrt[4]{n}$	$\sqrt[5]{n}$
1	1.0000 00	1.0000 00	51	2.6723 45	2.1954 02
2	1.1892 07	1.1486 98	52	2.6853 50	2.2039 45
3	1.3160 74	1.2457 31	53	2.6981 68	2.2123 57
4	1.4142 14	1.3195 08	54	2.7108 06	2.2206 43
5	1.4953 49	1.3797 30	55	2.7232 70	2.2288 07
6	1.5650 85	1.4309 69	56	2.7355 65	2.2368 54
7	1.6265 77	1.4757 73	57	2.7476 96	2.2447 86
8	1.6817 93	1.5157 17	58	2.7596 69	2.2526 08
9	1.7320 51	1.5518 46	59	2.7714 88	2.2603 22
10	1.7782 79	1.5848 93	60	2.7831 58	2.2679 33
11	1.8211 60	1.6153 94	61	2.7946 82	2.2754 43
12	1.8612 10	1.6437 52	62	2.8060 66	2.2828 55
13	1.8988 29	1.6702 78	63	2.8173 13	2.2901 72
14	1.9343 36	1.6952 18	64	2.8284 27	2.2973 97
15	1.9679 90	1.7187 72	65	2.8394 12	2.3045 32
16	2.0000 00	1.7411 01	66	2.8502 70	2.3115 79
17	2.0305 43	1.7623 40	67	2.8610 06	2.3185 42
18	2.0597 67	1.7826 02	68	2.8716 22	2.3254 22
19	2.0877 98	1.8019 83	69	2.8821 21	2.3322 22
20	2.1147 43	1.8205 64	70	2.8925 08	2.3389 43
21	2.1406 95	1.8384 16	71	2.9027 83	2.3455 88
22	2.1657 37	1.8556 01	72	2.9129 51	2.3521 58
23	2.1899 39	1.8721 71	73	2.9230 13	2.3586 56
24	2.2133 64	1.8881 75	74	2.9329 72	2.3650 83
25	2.2360 68	1.9036 54	75	2.9428 31	2.3714 41
26	2.2581 01	1.9186 45	76	2.9525 92	2.3777 31
27	2.2795 07	1.9331 82	77	2.9622 57	2.3839 56
28	2.3003 27	1.9472 94	78	2.9718 28	2.3901 16
29	2.3205 96	1.9610 09	79	2.9813 08	2.3962 13
30	2.3403 47	1.9743 50	80	2.9906 98	2.4022 49
31	2.3596 11	1.9873 41	81	3.0000 00	2.4082 25
32	2.3784 14	2.0000 00	82	3.0092 17	2.4141 42
33	2.3967 82	2.0123 47	83	3.0183 49	2.4200 01
34	2.4147 36	2.0243 97	84	3.0274 00	2.4258 05
35	2.4322 99	2.0361 68	85	3.0363 70	2.4315 53
36	2.4494 90	2.0476 73	86	3.0452 62	2.4372 48
37	2.4663 26	2.0589 24	87	3.0540 76	2.4428 90
38	2.4828 24	2.0699 35	88	3.0628 14	2.4484 80
39	2.4989 99	2.0807 17	89	3.0714 79	2.4540 19
40	2.5148 67	2.0912 79	90	3.0800 70	2.4595 09
41	2.5304 40	2.1016 32	91	3.0885 91	2.4649 51
42	2.5457 30	2.1117 86	92	3.0970 41	2.4703 45
43	2.5607 50	2.1217 47	93	3.1054 23	2.4756 92
44	2.5755 10	2.1315 26	94	3.1137 37	2.4809 93
45	2.5900 20	2.1411 27	95	3.1219 86	2.4862 50
46	2.6042 91	2.1505 60	96	3.1301 69	2.4914 62
47	2.6183 30	2.1598 30	97	3.1382 89	2.4966 31
48	2.6321 48	2.1689 44	98	3.1463 46	2.5017 58
49	2.6457 51	2.1779 06	99	3.1543 42	2.5068 42
50	2.6591 48	2.1867 24	100	3.1622 78	2.5118 86

FOURTH AND FIFTH ROOTS (Continued)

n	$\sqrt[4]{n}$	$\sqrt[5]{n}$	n	$\sqrt[4]{n}$	$\sqrt[5]{n}$
101	3.1701 54	2.5168 90	151	3.5054 54	2.7276 92
102	3.1779 72	2.5218 55	152	3.5112 43	2.7312 96
103	3.1857 33	2.5267 80	153	3.5170 04	2.7348 80
104	3.1934 37	2.5316 68	154	3.5227 37	2.7384 46
105	3.2010 86	2.5365 17	155	3.5284 42	2.7419 93
106	3.2086 80	2.5413 31	156	3.5341 19	2.7455 22
107	3.2162 21	2.5461 08	157	3.5397 69	2.7490 33
108	3.2237 10	2.5508 49	158	3.5453 92	2.7525 26
109	3.2311 46	2.5555 55	159	3.5509 89	2.7560 01
110	3.2385 32	2.5602 27	160	3.5565 59	2.7594 59
111	3.2458 67	2.5648 65	161	3.5621 03	2.7629 00
112	3.2531 53	2.5694 70	162	3.5676 21	2.7663 24
113	3.2603 90	2.5740 42	163	3.5731 14	2.7697 31
114	3.2675 80	2.5785 82	164	3.5785 82	2.7731 21
115	3.2747 22	2.5830 90	165	3.5840 25	2.7764 94
116	3.2818 18	2.5875 67	166	3.5894 43	2.7798 52
117	3.2888 68	2.5920 13	167	3.5948 36	2.7831 93
118	3.2958 73	2.5964 29	168	3.6002 06	2.7865 18
119	3.3028 34	2.6008 15	169	3.6055 51	2.7898 27
120	3.3097 51	2.6051 71	170	3.6108 73	2.7931 21
121	3.3166 25	2.6094 99	171	3.6161 72	2.7964 00
122	3.3234 56	2.6137 98	172	3.6214 47	2.7996 63
123	3.3302 46	2.6180 69	173	3.6266 99	2.8029 10
124	3.3369 94	2.6223 12	174	3.6319 29	2.8061 43
125	3.3437 02	2.6265 28	175	3.6371 36	2.8093 61
126	3.3503 69	2.6307 17	176	3.6423 21	2.8125 65
127	3.3569 97	2.6348 79	177	3.6474 83	2.8157 54
128	3.3635 86	2.6390 16	178	3.6526 24	2.8189 28
129	3.3701 36	2.6431 26	179	3.6577 44	2.8220 88
130	3.3766 48	2.6472 12	180	3.6628 42	2.8252 35
131	3.3831 23	2.6512 72	181	3.6679 18	2.8283 67
132	3.3895 61	2.6553 07	182	3.6729 74	2.8314 85
133	3.3959 63	2.6593 18	183	3.6780 09	2.8345 90
134	3.4023 28	2.6633 05	184	3.6830 23	2.8376 81
135	3.4086 58	2.6672 69	185	3.6880 17	2.8407 59
136	3.4149 53	2.6712 08	186	3.6929 91	2.8438 23
137	3.4212 13	2.6751 25	187	3.6979 45	2.8468 74
138	3.4274 39	2.6790 19	188	3.7028 79	2.8499 13
139	3.4336 32	2.6828 91	189	3.7077 93	2.8529 38
140	3.4397 91	2.6867 40	190	3.7126 88	2.8559 51
141	3.4459 17	2.6905 67	191	3.7175 63	2.8589 51
142	3.4520 10	2.6943 73	192	3.7224 19	2.8619 38
143	3.4580 72	2.6981 57	193	3.7272 57	2.8649 13
144	3.4641 02	2.7019 20	194	3.7320 76	2.8678 76
145	3.4701 00	2.7056 62	195	3.7368 76	2.8708 26
146	3.4760 68	2.7093 84	196	3.7416 57	2.8737 65
147	3.4820 05	2.7130 85	197	3.7464 21	2.8766 91
148	3.4879 11	2.7167 67	198	3.7511 66	2.8796 06
149	3.4937 88	2.7204 28	199	3.7558 93	2.8825 09
150	3.4996 36	2.7240 70	200	3.7606 03	2.8854 00

FOURTH AND FIFTH ROOTS (Continued)

n	$\sqrt[4]{n}$	$\sqrt[5]{n}$	n	$\sqrt[4]{n}$	$\sqrt[5]{n}$
201	3.7652 95	2.8882 79	251	3.9803 24	3.0194 98
202	3.7699 70	2.8911 48	252	3.9842 83	3.0219 00
203	3.7746 27	2.8940 05	253	3.9882 29	3.0242 95
204	3.7792 67	2.8968 50	254	3.9921 65	3.0266 82
205	3.7838 90	2.8996 85	255	3.9960 88	3.0290 61
206	3.7884 96	2.9025 08	256	4.0000 00	3.0314 33
207	3.7930 85	2.9053 21	257	4.0039 01	3.0337 98
208	3.7976 58	2.9081 22	258	4.0077 90	3.0361 55
209	3.8022 14	2.9109 13	259	4.0116 68	3.0385 05
210	3.8067 54	2.9136 93	260	4.0155 34	3.0408 48
211	3.8112 78	2.9164 63	261	4.0193 90	3.0431 83
212	3.8157 86	2.9192 22	262	4.0232 34	3.0455 12
213	3.8202 77	2.9219 71	263	4.0270 68	3.0478 33
214	3.8247 53	2.9247 10	264	4.0308 90	3.0501 47
215	3.8292 14	2.9274 38	265	4.0347 02	3.0524 54
216	3.8336 59	2.9301 56	266	4.0385 03	3.0547 55
217	3.8380 88	2.9328 64	267	4.0422 93	3.0570 48
218	3.8425 02	2.9355 62	268	4.0460 73	3.0593 34
219	3.8469 01	2.9382 51	269	4.0498 42	3.0616 14
220	3.8512 85	2.9409 29	270	4.0536 00	3.0638 87
221	3.8556 54	2.9435 98	271	4.0573 49	3.0661 53
222	3.8600 08	2.9462 57	272	4.0610 86	3.0684 13
223	3.8643 48	2.9489 06	273	4.0648 14	3.0706 66
224	3.8686 73	2.9515 46	274	4.0685 31	3.0729 12
225	3.8729 83	2.9541 77	275	4.0722 38	3.0751 52
226	3.8772 80	2.9567 98	276	4.0759 35	3.0773 85
227	3.8815 61	2.9594 10	277	4.0796 22	3.0796 12
228	3.8858 29	2.9620 13	278	4.0832 99	3.0818 32
229	3.8900 83	2.9646 07	279	4.0869 66	3.0840 46
230	3.8943 23	2.9671 91	280	4.0906 23	3.0862 54
231	3.8985 49	2.9697 67	281	4.0942 71	3.0884 55
232	3.9027 61	2.9723 34	282	4.0979 09	3.0906 50
233	3.9069 60	2.9748 92	283	4.1015 37	3.0928 39
234	3.9111 45	2.9774 41	284	4.1051 55	3.0950 21
235	3.9153 17	2.9799 82	285	4.1087 64	3.0971 98
236	3.9194 76	2.9825 13	286	4.1123 64	3.0993 68
237	3.9236 21	2.9850 37	287	4.1159 54	3.1015 33
238	3.9277 54	2.9875 51	288	4.1195 34	3.1036 91
239	3.9318 73	2.9900 58	289	4.1231 06	3.1058 44
240	3.9359 79	2.9925 56	290	4.1266 68	3.1079 90
241	3.9400 73	2.9950 45	291	4.1302 21	3.1101 30
242	3.9441 54	2.9975 27	292	4.1337 64	3.1122 65
243	3.9482 22	3.0000 00	293	4.1372 99	3.1143 94
244	3.9522 78	3.0024 65	294	4.1408 25	3.1165 17
245	3.9563 21	3.0049 22	295	4.1443 41	3.1186 34
246	3.9603 52	3.0073 71	296	4.1478 49	3.1207 45
247	3.9643 71	3.0098 12	297	4.1513 48	3.1228 51
248	3.9683 77	3.0122 45	298	4.1548 38	3.1249 51
249	3.9723 71	3.0146 71	299	4.1583 19	3.1270 46
250	3.9763 54	3.0170 88	300	4.1617 91	3.1291 35

FOURTH AND FIFTH ROOTS (Continued)

n	$\sqrt[4]{n}$	$\sqrt[5]{n}$	n	$\sqrt[4]{n}$	$\sqrt[5]{n}$
301	4.1652 55	3.1312 18	351	4.3283 94	3.2289 51
302	4.1687 10	3.1332 96	352	4.3314 74	3.2307 89
303	4.1721 57	3.1353 68	353	4.3345 47	3.2326 22
304	4.1755 95	3.1374 35	354	4.3376 13	3.2344 52
305	4.1790 25	3.1394 96	355	4.3406 73	3.2362 77
306	4.1824 46	3.1415 52	356	4.3437 27	3.2380 98
307	4.1858 59	3.1436 03	357	4.3467 74	3.2399 15
308	4.1892 64	3.1456 48	358	4.3498 15	3.2417 28
309	4.1926 60	3.1476 88	359	4.3528 49	3.2435 37
310	4.1960 48	3.1497 23	360	4.3558 77	3.2453 42
311	4.1994 28	3.1517 52	361	4.3588 99	3.2471 43
312	4.2027 99	3.1537 77	362	4.3619 14	3.2489 40
313	4.2061 63	3.1557 96	363	4.3649 24	3.2507 33
314	4.2095 18	3.1578 10	364	4.3679 27	3.2525 22
315	4.2128 66	3.1598 18	365	4.3709 24	3.2543 07
316	4.2162 06	3.1618 22	366	4.3739 14	3.2560 89
317	4.2195 37	3.1638 21	367	4.3768 99	3.2578 66
318	4.2228 61	3.1658 14	368	4.3798 77	3.2596 39
319	4.2261 77	3.1678 03	369	4.3828 50	3.2614 09
320	4.2294 85	3.1697 86	370	4.3858 16	3.2631 75
321	4.2327 85	3.1717 65	371	4.3887 77	3.2649 37
322	4.2360 78	3.1737 39	372	4.3917 31	3.2666 95
323	4.2393 63	3.1757 08	373	4.3946 80	3.2684 49
324	4.2426 41	3.1776 72	374	4.3976 22	3.2702 00
325	4.2459 11	3.1796 31	375	4.4005 59	3.2719 47
326	4.2491 73	3.1815 85	376	4.4034 89	3.2736 90
327	4.2524 28	3.1835 34	377	4.4064 14	3.2754 30
328	4.2556 75	3.1854 79	378	4.4093 34	3.2771 65
329	4.2589 15	3.1874 19	379	4.4122 47	3.2788 98
330	4.2621 48	3.1893 54	380	4.4151 54	3.2806 26
331	4.2653 73	3.1912 85	381	4.4180 56	3.2823 51
332	4.2685 91	3.1932 11	382	4.4209 52	3.2840 72
333	4.2718 01	3.1951 32	383	4.4238 43	3.2857 90
334	4.2750 05	3.1970 49	384	4.4267 28	3.2875 04
335	4.2782 01	3.1989 61	385	4.4296 07	3.2892 14
336	4.2813 90	3.2008 69	386	4.4324 80	3.2909 21
337	4.2845 72	3.2027 72	387	4.4353 48	3.2926 24
338	4.2877 47	3.2046 70	388	4.4382 11	3.2943 24
339	4.2909 15	3.2065 64	389	4.4410 68	3.2960 21
340	4.2940 76	3.2084 54	390	4.4439 19	3.2977 13
341	4.2972 30	3.2103 39	391	4.4467 65	3.2994 03
342	4.3003 77	3.2122 20	392	4.4496 06	3.3010 89
343	4.3035 17	3.2140 96	393	4.4524 41	3.3027 71
344	4.3066 50	3.2159 68	394	4.4552 70	3.3044 50
345	4.3097 77	3.2178 35	395	4.4580 95	3.3061 26
346	4.3128 96	3.2196 99	396	4.4609 13	3.3077 98
347	4.3160 09	3.2215 58	397	4.4637 27	3.3094 67
348	4.3191 15	3.2234 12	398	4.4665 35	3.3111 33
349	4.3222 15	3.2252 63	399	4.4693 38	3.3127 95
350	4.3253 08	3.2271 09	400	4.4721 36	3.3144 54

FOURTH AND FIFTH ROOTS (Continued)

n	$\sqrt[4]{n}$	$\sqrt[5]{n}$	n	$\sqrt[4]{n}$	$\sqrt[5]{n}$
401	4.4777 16	3.3161 10	451	4.6083 36	3.3949 65
402	4.4749 28	3.3177 62	452	4.6108 88	3.3964 69
403	4.4804 98	3.3194 11	453	4.6134 37	3.3979 71
404	4.4832 75	3.3210 57	454	4.6159 80	3.3994 70
405	4.4860 46	3.3226 99	455	4.6185 20	3.4009 66
406	4.4888 13	3.3243 38	456	4.6210 56	3.4024 60
407	4.4915 74	3.3259 74	457	4.6235 87	3.4039 51
408	4.4943 31	3.3276 07	458	4.6261 14	3.4054 39
409	4.4970 82	3.3292 37	459	4.6286 38	3.4069 25
410	4.4998 29	3.3308 63	460	4.6311 57	3.4084 08
411	4.5025 70	3.3324 86	461	4.6336 71	3.4098 89
412	4.5053 06	3.3341 06	462	4.6361 82	3.4113 67
413	4.5080 37	3.3357 23	463	4.6386 89	3.4128 42
414	4.5107 64	3.3373 37	464	4.6411 92	3.4143 15
415	4.5134 85	3.3389 48	465	4.6436 90	3.4157 85
416	4.5162 02	3.3405 55	466	4.6461 85	3.4172 53
417	4.5189 13	3.3421 60	467	4.6486 75	3.4187 19
418	4.5216 20	3.3437 61	468	4.6511 62	3.4201 82
419	4.5243 22	3.3453 60	469	4.6536 45	3.4216 42
420	4.5270 19	3.3469 55	470	4.6561 23	3.4231 00
421	4.5297 11	3.3485 47	471	4.6585 98	3.4245 55
422	4.5323 99	3.3501 36	472	4.6610 69	3.4260 08
423	4.5350 81	3.3517 23	473	4.6635 35	3.4274 59
424	4.5377 59	3.3533 06	474	4.6659 98	3.4289 07
425	4.5404 33	3.3548 86	475	4.6684 57	3.4303 52
426	4.5431 01	3.3564 63	476	4.6709 13	3.4317 95
427	4.5457 65	3.3580 38	477	4.6733 64	3.4332 36
428	4.5484 24	3.3596 09	478	4.6758 11	3.4346 74
429	4.5510 78	3.3611 78	479	4.6782 55	3.4361 10
430	4.5537 28	3.3627 43	480	4.6806 95	3.4375 44
431	4.5563 74	3.3643 06	481	4.6831 31	3.4389 75
432	4.5590 14	3.3658 65	482	4.6855 63	3.4404 04
433	4.5616 50	3.3674 22	483	4.6879 91	3.4418 30
434	4.5642 82	3.3689 76	484	4.6904 16	3.4432 54
435	4.5669 09	3.3705 27	485	4.6928 37	3.4446 76
436	4.5695 31	3.3720 76	486	4.6952 54	3.4460 95
437	4.5721 49	3.3736 21	487	4.6976 67	3.4475 12
438	4.5747 62	3.3751 64	488	4.7000 77	3.4489 27
439	4.5773 71	3.3767 03	489	4.7024 83	3.4503 39
440	4.5799 76	3.3782 40	490	4.7048 85	3.4517 49
441	4.5825 76	3.3797 74	491	4.7072 84	3.4531 57
442	4.5851 71	3.3813 06	492	4.7096 79	3.4545 62
443	4.5877 63	3.3828 34	493	4.7120 70	3.4559 65
444	4.5903 49	3.3843 60	494	4.7144 58	3.4573 66
445	4.5929 32	3.3858 83	495	4.7168 42	3.4587 65
446	4.5955 10	3.3874 04	496	4.7192 22	3.4601 61
447	4.5980 84	3.3889 21	497	4.7215 99	3.4615 55
448	4.6006 53	3.3904 36	498	4.7239 72	3.4629 47
449	4.6032 18	3.3919 49	499	4.7263 42	3.4643 37
450	4.6057 79	3.3934 58	500	4.7287 08	3.4657 24

FOURTH AND FIFTH ROOTS (Continued)

n	$\sqrt[4]{n}$	$\sqrt[5]{n}$	n	$\sqrt[4]{n}$	$\sqrt[5]{n}$
501	4.7310 71	3.4671 09	551	4.8449 34	3.5337 05
502	4.7334 30	3.4684 92	552	4.8471 31	3.5349 87
503	4.7357 85	3.4698 73	553	4.8493 25	3.5362 67
504	4.7381 37	3.4712 52	554	4.8515 16	3.5375 45
505	4.7404 86	3.4726 28	555	4.8537 04	3.5388 21
506	4.7428 31	3.4740 02	556	4.8558 88	3.5400 95
507	4.7451 72	3.4753 74	557	4.8580 70	3.5413 68
508	4.7475 10	3.4767 44	558	4.8602 49	3.5426 39
509	4.7498 45	3.4781 12	559	4.8624 25	3.5439 07
510	4.7521 76	3.4794 78	560	4.8645 99	3.5451 74
511	4.7545 04	3.4808 41	561	4.8667 69	3.5464 40
512	4.7568 28	3.4822 02	562	4.8689 36	3.5477 03
513	4.7591 49	3.4835 61	563	4.8711 01	3.5489 65
514	4.7614 67	3.4849 18	564	4.8732 62	3.5502 25
515	4.7637 81	3.4862 73	565	4.8754 21	3.5514 83
516	4.7660 92	3.4876 26	566	4.8775 77	3.5527 39
517	4.7684 00	3.4889 77	567	4.8797 30	3.5539 93
518	4.7707 04	3.4903 26	568	4.8818 80	3.5552 46
519	4.7730 04	3.4916 72	569	4.8840 27	3.5564 97
520	4.7753 02	3.4930 17	570	4.8861 72	3.5577 46
521	4.7775 96	3.4943 59	571	4.8883 13	3.5589 94
522	4.7798 87	3.4957 00	572	4.8904 52	3.5602 39
523	4.7844 59	3.4970 38	573	4.8925 88	3.5614 83
524	4.7821 75	3.4983 74	574	4.8947 21	3.5627 26
525	4.7867 40	3.4997 08	575	4.8968 52	3.5639 66
526	4.7890 18	3.5010 41	576	4.8989 79	3.5664 42
527	4.7912 92	3.5023 71	577	4.9011 04	3.5652 05
528	4.7935 63	3.5036 99	578	4.9032 27	3.5676 77
529	4.7958 32	3.5050 25	579	4.9053 46	3.5689 11
530	4.7980 96	3.5063 49	580	4.9074 63	3.5701 43
531	4.8003 58	3.5076 71	581	4.9095 77	3.5713 73
532	4.8026 16	3.5089 92	582	4.9116 88	3.5726 02
533	4.8048 72	3.5103 10	583	4.9137 96	3.5738 29
534	4.8071 24	3.5116 26	584	4.9159 02	3.5750 54
535	4.8093 73	3.5129 40	585	4.9180 05	3.5762 77
536	4.8116 19	3.5142 52	586	4.9201 05	3.5774 99
537	4.8138 61	3.5155 63	587	4.9222 03	3.5787 19
538	4.8161 01	3.5168 71	588	4.9242 98	3.5799 38
539	4.8183 37	3.5181 78	589	4.9263 90	3.5811 55
540	4.8205 71	3.5194 82	590	4.9284 80	3.5823 70
541	4.8228 01	3.5207 85	591	4.9305 67	3.5835 83
542	4.8250 28	3.5220 85	592	4.9326 51	3.5847 95
543	4.8272 52	3.5233 84	593	4.9347 33	3.5860 05
544	4.8294 73	3.5246 81	594	4.9368 12	3.5872 14
545	4.8316 91	3.5259 76	595	4.9388 89	3.5884 21
546	4.8339 06	3.5272 69	596	4.9409 63	3.5896 26
547	4.8361 17	3.5285 60	597	4.9430 34	3.5908 30
548	4.8383 26	3.5298 49	598	4.9451 02	3.5920 32
549	4.8405 32	3.5311 36	599	4.9471 69	3.5932 33
550	4.8427 35	3.5324 22	600	4.9492 32	3.5944 32

FOURTH AND FIFTH ROOTS (Continued)

n	$\sqrt[4]{n}$	$\sqrt[5]{n}$	n	$\sqrt[4]{n}$	$\sqrt[5]{n}$
601	4.9512 93	3.5956 29	651	5.0512 08	3.6535 60
602	4.9533 51	3.5968 25	652	5.0531 47	3.6546 81
603	4.9554 07	3.5980 19	653	5.0550 83	3.6558 02
604	4.9574 60	3.5992 12	654	5.0570 17	3.6569 21
605	4.9595 11	3.6004 03	655	5.0589 49	3.6580 38
606	4.9615 59	3.6015 92	656	5.0608 79	3.6591 55
607	4.9636 05	3.6027 80	657	5.0628 07	3.6602 70
608	4.9656 48	3.6039 66	658	5.0647 32	3.6613 83
609	4.9676 88	3.6051 51	659	5.0666 55	3.6624 95
610	4.9697 26	3.6063 34	660	5.0685 76	3.6636 06
611	4.9717 62	3.6075 16	661	5.0704 95	3.6647 16
612	4.9737 95	3.6086 96	662	5.0724 12	3.6658 24
613	4.9758 25	3.6098 74	663	5.0743 26	3.6669 31
614	4.9778 53	3.6110 51	664	5.0762 39	3.6680 36
615	4.9798 79	3.6122 27	665	5.0781 49	3.6691 40
616	4.9819 02	3.6134 01	666	5.0800 57	3.6702 43
617	4.9839 23	3.6145 73	667	5.0819 63	3.6713 45
618	4.9859 41	3.6157 44	668	5.0838 66	3.6724 45
619	4.9879 57	3.6169 14	669	5.0857 68	3.6735 44
620	4.9899 70	3.6180 81	670	5.0876 67	3.6746 41
621	4.9919 81	3.6192 48	671	5.0895 65	3.6757 38
622	4.9939 89	3.6204 13	672	5.0914 60	3.6768 33
623	4.9959 95	3.6215 76	673	5.0933 53	3.6779 26
624	4.9979 99	3.6227 38	674	5.0952 44	3.6790 19
625	5.0000 00	3.6238 98	675	5.0971 33	3.6801 10
626	5.0019 99	3.6250 57	676	5.0990 20	3.6811 99
627	5.0039 95	3.6262 15	677	5.1009 04	3.6822 88
628	5.0059 89	3.6273 71	678	5.1027 87	3.6833 75
629	5.0079 81	3.6285 25	679	5.1046 67	3.6844 61
630	5.0099 70	3.6296 78	680	5.1065 46	3.6855 46
631	5.0119 57	3.6308 30	681	5.1084 22	3.6866 29
632	5.0139 42	3.6319 80	682	5.1102 96	3.6877 11
633	5.0159 24	3.6331 28	683	5.1121 69	3.6887 92
634	5.0179 04	3.6342 76	684	5.1140 39	3.6898 71
635	5.0198 81	3.6354 21	685	5.1159 07	3.6909 50
636	5.0218 56	3.6365 66	686	5.1177 73	3.6920 27
637	5.0238 29	3.6377 08	687	5.1196 37	3.6931 02
638	5.0258 00	3.6388 50	688	5.1214 99	3.6941 77
639	5.0277 68	3.6399 90	689	5.1233 59	3.6952 50
640	5.0297 34	3.6411 28	690	5.1252 17	3.6963 22
641	5.0316 97	3.6422 66	691	5.1270 73	3.6973 93
642	5.0336 59	3.6434 01	692	5.1289 27	3.6984 62
643	5.0356 18	3.6445 36	693	5.1307 79	3.6995 31
644	5.0375 74	3.6456 68	694	5.1326 29	3.7005 98
645	5.0395 29	3.6468 00	695	5.1344 77	3.7016 64
646	5.0414 81	3.6479 30	696	5.1363 23	3.7027 28
647	5.0434 31	3.6490 59	697	5.1381 67	3.7037 92
648	5.0453 78	3.6501 86	698	5.1400 09	3.7048 54
649	5.0473 24	3.6513 12	699	5.1418 49	3.7059 15
650	5.0492 67	3.6524 36	700	5.1436 87	3.7069 75

FOURTH AND FIFTH ROOTS (Continued)

n	$\sqrt[4]{n}$	$\sqrt[5]{n}$	n	$\sqrt[4]{n}$	$\sqrt[5]{n}$
701	5.1455 23	3.7080 33	751	5.2349 19	3.7594 82
702	5.1473 57	3.7090 90	752	5.2366 61	3.7604 82
703	5.1491 89	3.7101 47	753	5.2384 01	3.7614 82
704	5.1510 19	3.7112 01	754	5.2401 39	3.7624 81
705	5.1528 47	3.7122 55	755	5.2418 76	3.7634 78
706	5.1546 74	3.7133 08	756	5.2436 11	3.7644 74
707	5.1564 98	3.7143 59	757	5.2453 44	3.7654 70
708	5.1583 20	3.7154 09	758	5.2470 75	3.7664 64
709	5.1601 41	3.7164 58	759	5.2488 05	3.7674 57
710	5.1619 59	3.7175 06	760	5.2505 33	3.7684 50
711	5.1637 76	3.7185 53	761	5.2522 59	3.7694 41
712	5.1655 91	3.7195 98	762	5.2539 84	3.7704 31
713	5.1674 04	3.7206 42	763	5.2557 07	3.7714 20
714	5.1692 14	3.7216 85	764	5.2574 28	3.7724 08
715	5.1710 23	3.7227 27	765	5.2591 48	3.7733 95
716	5.1728 31	3.7237 68	766	5.2608 65	3.7743 81
717	5.1746 36	3.7248 07	767	5.2625 82	3.7753 66
718	5.1764 39	3.7258 46	768	5.2642 96	3.7763 50
719	5.1782 41	3.7268 83	769	5.2660 09	3.7773 33
720	5.1800 40	3.7279 19	770	5.2677 20	3.7783 15
721	5.1818 38	3.7289 54	771	5.2694 29	3.7792 96
722	5.1836 34	3.7299 88	772	5.2711 37	3.7802 76
723	5.1854 28	3.7310 21	773	5.2728 43	3.7812 54
724	5.1872 20	3.7320 52	774	5.2745 48	3.7822 32
725	5.1890 10	3.7330 83	775	5.2762 51	3.7832 09
726	5.1907 98	3.7341 12	776	5.2779 52	3.7841 85
727	5.1925 85	3.7351 40	777	5.2796 51	3.7851 60
728	5.1943 70	3.7361 67	778	5.2813 49	3.7861 33
729	5.1961 52	3.7371 93	779	5.2830 46	3.7871 06
730	5.1979 33	3.7382 18	780	5.2847 40	3.7880 78
731	5.1997 13	3.7392 41	781	5.2864 33	3.7890 49
732	5.2014 90	3.7402 64	782	5.2881 25	3.7900 19
733	5.2032 66	3.7412 85	783	5.2898 14	3.7909 87
734	5.2050 39	3.7423 05	784	5.2915 03	3.7919 55
735	5.2068 11	3.7433 24	785	5.2931 89	3.7929 22
736	5.2085 81	3.7443 42	786	5.2948 74	3.7938 88
737	5.2103 50	3.7453 59	787	5.2965 57	3.7948 53
738	5.2121 16	3.7463 75	788	5.2982 39	3.7958 17
739	5.2138 81	3.7473 90	789	5.2999 19	3.7967 80
740	5.2156 44	3.7484 04	790	5.3015 98	3.7977 42
741	5.2174 05	3.7494 16	791	5.3032 75	3.7987 03
742	5.2191 64	3.7504 28	792	5.3049 50	3.7996 63
743	5.2209 22	3.7514 38	793	5.3066 24	3.8006 22
744	5.2226 78	3.7524 47	794	5.3082 96	3.8015 80
745	5.2244 32	3.7534 55	795	5.3099 67	3.8025 37
746	5.2261 84	3.7544 62	796	5.3116 36	3.8034 93
747	5.2279 35	3.7554 68	797	5.3133 03	3.8044 48
748	5.2296 83	3.7564 73	798	5.3149 69	3.8054 02
749	5.2314 30	3.7574 77	799	5.3166 33	3.8063 56
750	5.2331 76	3.7584 80	800	5.3182 96	3.8073 08

FOURTH AND FIFTH ROOTS (Continued)

n	$\sqrt[4]{n}$	$\sqrt[5]{n}$	n	$\sqrt[4]{n}$	$\sqrt[5]{n}$
801	5.3199 57	3.8082 59	851	5.4011 02	3.8546 59
802	5.3216 17	3.8092 10	852	5.4026 88	3.8555 64
803	5.3232 75	3.8101 59	853	5.4042 73	3.8564 69
804	5.3249 31	3.8111 08	854	5.4058 56	3.8573 72
805	5.3265 86	3.8120 55	855	5.4074 38	3.8582 75
806	5.3282 40	3.8130 02	856	5.4090 18	3.8591 77
807	5.3298 92	3.8139 47	857	5.4105 97	3.8600 79
808	5.3315 42	3.8148 92	858	5.4121 75	3.8609 79
809	5.3331 97	3.8158 36	859	5.4137 51	3.8618 79
810	5.3348 38	3.8167 79	860	5.4153 26	3.8627 77
811	5.3364 84	3.8177 21	861	5.4169 00	3.8636 75
812	5.3381 28	3.8186 62	862	5.4184 72	3.8645 72
813	5.3397 71	3.8196 02	863	5.4200 43	3.8654 69
814	5.3414 12	3.8205 41	864	5.4216 12	3.8663 64
815	5.3430 52	3.8214 79	865	5.4231 80	3.8672 59
816	5.3446 90	3.8224 17	866	5.4247 47	3.8681 52
817	5.3463 27	3.8233 53	867	5.4263 12	3.8690 45
818	5.3479 62	3.8242 89	868	5.4278 76	3.8699 37
819	5.3495 96	3.8252 23	869	5.4294 39	3.8708 29
820	5.3512 28	3.8261 57	870	5.4310 00	3.8717 19
821	5.3528 59	3.8270 90	871	5.4325 60	3.8726 09
822	5.3544 88	3.8280 21	872	5.4341 19	3.8734 98
823	5.3561 16	3.8289 52	873	5.4356 76	3.8743 86
824	5.3577 42	3.8298 82	874	5.4372 32	3.8752 73
825	5.3593 67	3.8308 12	875	5.4387 87	3.8761 59
826	5.3609 90	3.8317 40	876	5.4403 40	3.8770 45
827	5.3626 12	3.8326 67	877	5.4418 92	3.8779 30
828	5.3642 32	3.8335 94	878	5.4434 42	3.8788 14
829	5.3658 51	3.8345 19	879	5.4449 92	3.8796 97
830	5.3674 69	3.8354 44	880	5.4465 40	3.8805 79
831	5.3690 85	3.8363 68	881	5.4480 86	3.8814 61
832	5.3706 99	3.8372 90	882	5.4496 32	3.8823 41
833	5.3723 12	3.8382 12	883	5.4511 76	3.8832 21
834	5.3739 24	3.8391 33	884	5.4527 18	3.8841 00
835	5.3755 34	3.8400 54	885	5.4542 60	3.8849 79
836	5.3771 43	3.8409 73	886	5.4558 00	3.8858 56
837	5.3787 50	3.8418 91	887	5.4573 39	3.8867 33
838	5.3803 56	3.8428 09	888	5.4588 76	3.8876 09
839	5.3819 60	3.8437 26	889	5.4604 12	3.8884 84
840	5.3835 63	3.8446 42	890	5.4619 47	3.8893 59
841	5.3851 65	3.8455 57	891	5.4634 81	3.8902 32
842	5.3867 65	3.8464 71	892	5.4650 13	3.8911 05
843	5.3883 64	3.8473 84	893	5.4665 44	3.8919 77
844	5.3899 61	3.8482 96	894	5.4680 74	3.8928 49
845	5.3915 57	3.8492 08	895	5.4696 02	3.8937 19
846	5.3931 51	3.8501 18	896	5.4711 30	3.8945 89
847	5.3947 44	3.8510 28	897	5.4726 56	3.8954 58
848	5.3963 36	3.8519 37	898	5.4741 80	3.8963 26
849	5.3979 26	3.8528 45	899	5.4757 03	3.8971 93
850	5.3995 15	3.8537 52	900	5.4772 26	3.8980 60

FOURTH AND FIFTH ROOTS (Continued)

n	$\sqrt[4]{n}$	$\sqrt[5]{n}$	n	$\sqrt[4]{n}$	$\sqrt[5]{n}$
901	5.4787 46	3.8989 26	951	5.5532 23	3.9412 69
902	5.4802 66	3.8997 91	952	5.5546 82	3.9420 98
903	5.4817 84	3.9006 55	953	5.5561 41	3.9429 26
904	5.4833 01	3.9015 19	954	5.5575 98	3.9437 53
905	5.4848 17	3.9023 81	955	5.5590 53	3.9445 79
906	5.4863 32	3.9032 43	956	5.5605 08	3.9454 05
907	5.4878 45	3.9041 05	957	5.5619 62	3.9462 30
908	5.4893 57	3.9049 65	958	5.5634 14	3.9470 54
909	5.4908 68	3.9058 25	959	5.5648 65	3.9478 78
910	5.4923 77	3.9066 84	960	5.5663 15	3.9487 01
911	5.4938 85	3.9075 42	961	5.5677 64	3.9495 23
912	5.4953 92	3.9084 00	962	5.5692 12	3.9503 45
913	5.4968 98	3.9092 56	963	5.5706 59	3.9511 66
914	5.4984 03	3.9101 12	964	5.5721 05	3.9519 86
915	5.4999 06	3.9109 68	965	5.5735 49	3.9528 06
916	5.5014 08	3.9118 22	966	5.5749 92	3.9536 25
917	5.5029 09	3.9126 76	967	5.5764 35	3.9544 43
918	5.5044 09	3.9135 29	968	5.5778 76	3.9552 60
919	5.5059 07	3.9143 81	969	5.5793 16	3.9560 77
920	5.5074 04	3.9152 33	970	5.5807 55	3.9568 93
921	5.5089 00	3.9160 83	971	5.5821 92	3.9577 09
922	5.5103 95	3.9169 33	972	5.5836 29	3.9585 24
923	5.5118 89	3.9177 83	973	5.5850 65	3.9593 38
924	5.5133 81	3.9186 31	974	5.5864 99	3.9601 51
925	5.5148 72	3.9194 79	975	5.5879 33	3.9609 64
926	5.5163 62	3.9203 26	976	5.5893 65	3.9617 76
927	5.5178 51	3.9211 72	977	5.5907 96	3.9625 88
928	5.5193 38	3.9220 18	978	5.5922 26	3.9633 99
929	5.5208 24	3.9228 63	979	5.5936 55	3.9642 09
930	5.5223 09	3.9237 07	980	5.5950 83	3.9650 18
931	5.5237 93	3.9245 51	981	5.5965 10	3.9658 27
932	5.5252 76	3.9253 93	982	5.5979 35	3.9666 36
933	5.5267 58	3.9262 35	983	5.5993 60	3.9674 43
934	5.5282 38	3.9270 77	984	5.6007 83	3.9682 50
935	5.5297 17	3.9279 17	985	5.6022 06	3.9690 56
936	5.5311 95	3.9287 57	986	5.6036 27	3.9698 62
937	5.5326 72	3.9295 96	987	5.6050 47	3.9706 67
938	5.5341 47	3.9304 35	988	5.6064 67	3.9714 71
939	5.5356 22	3.9312 72	989	5.6078 85	3.9722 75
940	5.5370 95	3.9321 09	990	5.6093 02	3.9730 78
941	5.5385 67	3.9329 45	991	5.6107 18	3.9738 80
942	5.5400 38	3.9337 81	992	5.6121 33	3.9746 82
943	5.5415 07	3.9346 16	993	5.6135 46	3.9754 83
944	5.5429 76	3.9354 50	994	5.6149 59	3.9762 83
945	5.5444 43	3.9362 83	995	5.6163 71	3.9770 83
946	5.5459 10	3.9371 16	996	5.6177 81	3.9778 82
947	5.5473 75	3.9379 48	997	5.6191 91	3.9786 80
948	5.5488 38	3.9387 79	998	5.6205 99	3.9794 78
949	5.5503 01	3.9396 10	999	5.6220 07	3.9802 75
950	5.5517 63	3.9404 40	1000	5.6234 13	3.9810 72

RELATIVITY CONTRACTION DILATION FACTOR

DR. A. V. H. MASKET

$$\beta = \frac{v}{c} = \frac{\text{speed of object}}{\text{speed of light in vacuo}}, \text{ in Special Relativity.}$$

Rest mass energy in ergs is $E = m_0c^2$, where m_0 = rest mass in grams.

Let r = (kinetic energy of object)/(rest mass energy of object), then the necessary value of β for a given value of r is exactly calculated as

$$\beta = \frac{\sqrt{r(r+2)}}{r+1}$$

$$\text{Kinetic Energy} = m_0c^2\left(\frac{1}{\sqrt{1-\beta^2}} - 1\right)$$

RELATIVITY CONTRACTION DILATION FACTOR

β	$\sqrt{1 - \beta^2}$	$\frac{1}{\sqrt{1 - \beta^2}}$	β	$\sqrt{1 - \beta^2}$	$\frac{1}{\sqrt{1 - \beta^2}}$
.0010	1.000000	1.00000	.0510	.998699	1.00130
.0020	.999998	1.00000	.0520	.998647	1.00135
.0030	.999996	1.00000	.0530	.998595	1.00141
.0040	.999992	1.00001	.0540	.998541	1.00146
.0050	.999987	1.00001	.0550	.998486	1.00152
.0060	.999982	1.00002	.0560	.998431	1.00157
.0070	.999975	1.00002	.0570	.998374	1.00163
.0080	.999968	1.00003	.0580	.998317	1.00169
.0090	.999959	1.00004	.0590	.998258	1.00175
.0100	.999950	1.00005	.0600	.998198	1.00180
.0110	.999939	1.00006	.0610	.998138	1.00187
.0120	.999928	1.00007	.0620	.998076	1.00193
.0130	.999915	1.00008	.0630	.998014	1.00199
.0140	.999902	1.00010	.0640	.997950	1.00205
.0150	.999887	1.00011	.0650	.997885	1.00212
.0160	.999872	1.00013	.0660	.997820	1.00219
.0170	.999855	1.00014	.0670	.997753	1.00225
.0180	.999838	1.00016	.0680	.997685	1.00232
.0190	.999819	1.00018	.0690	.997617	1.00239
.0200	.999800	1.00020	.0700	.997547	1.00246
.0210	.999779	1.00022	.0710	.997476	1.00253
.0220	.999758	1.00024	.0720	.997405	1.00260
.0230	.999735	1.00026	.0730	.997332	1.00268
.0240	.999712	1.00029	.0740	.997258	1.00275
.0250	.999687	1.00031	.0750	.997184	1.00282
.0260	.999662	1.00034	.0760	.997108	1.00290
.0270	.999635	1.00036	.0770	.997031	1.00298
.0280	.999608	1.00039	.0780	.996953	1.00306
.0290	.999579	1.00042	.0790	.996875	1.00314
.0300	.999550	1.00045	.0800	.996795	1.00322
.0310	.999519	1.00048	.0810	.996714	1.00330
.0320	.999488	1.00051	.0820	.996632	1.00338
.0330	.999455	1.00054	.0830	.996550	1.00346
.0340	.999422	1.00058	.0840	.996466	1.00355
.0350	.999387	1.00061	.0850	.996381	1.00363
.0360	.999352	1.00065	.0860	.996295	1.00372
.0370	.999315	1.00069	.0870	.996208	1.00381
.0380	.999278	1.00072	.0880	.996120	1.00389
.0390	.999239	1.00076	.0890	.996032	1.00398
.0400	.999200	1.00080	.0900	.995942	1.00407
.0410	.999159	1.00084	.0910	.995851	1.00417
.0420	.999118	1.00088	.0920	.995759	1.00426
.0430	.999075	1.00093	.0930	.995666	1.00435
.0440	.999032	1.00097	.0940	.995572	1.00445
.0450	.998987	1.00101	.0950	.995477	1.00454
.0460	.998941	1.00106	.0960	.995381	1.00464
.0470	.998895	1.00111	.0970	.995284	1.00474
.0480	.998847	1.00115	.0980	.995186	1.00484
.0490	.998799	1.00120	.0990	.995087	1.00494
.0500	.998749	1.00125	.1000	.994987	1.00504

RELATIVITY CONTRACTION DILATION FACTOR (Continued)

β	$\sqrt{1-\beta^2}$	$\frac{1}{\sqrt{1-\beta^2}}$	β	$\sqrt{1-\beta^2}$	$\frac{1}{\sqrt{1-\beta^2}}$
.1005	.994937	1.00509	.1255	.992094	1.00797
.1010	.994886	1.00514	.1260	.992030	1.00803
.1015	.994836	1.00519	.1265	.991967	1.00810
.1020	.994784	1.00524	.1270	.991903	1.00816
.1025	.994733	1.00529	.1275	.991839	1.00823
.1030	.994681	1.00535	.1280	.991774	1.00829
.1035	.994629	1.00540	.1285	.991710	1.00836
.1040	.994577	1.00545	.1290	.991645	1.00843
.1045	.994525	1.00551	.1295	.991579	1.00849
.1050	.994472	1.00556	.1300	.991514	1.00856
.1055	.994419	1.00561	.1305	.991448	1.00863
.1060	.994366	1.00567	.1310	.991382	1.00869
.1065	.994313	1.00572	.1315	.991316	1.00876
.1070	.994259	1.00577	.1320	.991250	1.00883
.1075	.994205	1.00583	.1325	.991183	1.00890
.1080	.994151	1.00588	.1330	.991116	1.00896
.1085	.994096	1.00594	.1335	.991049	1.00903
.1090	.994042	1.00599	.1340	.990981	1.00910
.1095	.993987	1.00605	.1345	.990914	1.00917
.1100	.993932	1.00611	.1350	.990846	1.00924
.1105	.993876	1.00616	.1355	.990777	1.00931
.1110	.993820	1.00622	.1360	.990709	1.00938
.1115	.993764	1.00627	.1365	.990640	1.00945
.1120	.993708	1.00633	.1370	.990571	1.00952
.1125	.993652	1.00639	.1375	.990502	1.00959
.1130	.993595	1.00645	.1380	.990432	1.00966
.1135	.993538	1.00650	.1385	.990362	1.00973
.1140	.993481	1.00656	.1390	.990292	1.00980
.1145	.993423	1.00662	.1395	.990222	1.00987
.1150	.993365	1.00668	.1400	.990152	1.00995
.1155	.993307	1.00674	.1405	.990081	1.01002
.1160	.993249	1.00680	.1410	.990010	1.01009
.1165	.993191	1.00686	.1415	.989938	1.01016
.1170	.993132	1.00692	.1420	.989867	1.01024
.1175	.993073	1.00698	.1425	.989795	1.01031
.1180	.993014	1.00704	.1430	.989723	1.01038
.1185	.992954	1.00710	.1435	.989650	1.01046
.1190	.992894	1.00716	.1440	.989578	1.01053
.1195	.992834	1.00722	.1445	.989505	1.01061
.1200	.992774	1.00728	.1450	.989432	1.01068
.1205	.992713	1.00734	.1455	.989358	1.01076
.1210	.992653	1.00740	.1460	.989285	1.01083
.1215	.992591	1.00746	.1465	.989211	1.01091
.1220	.992530	1.00753	.1470	.989136	1.01098
.1225	.992469	1.00759	.1475	.989062	1.01106
.1230	.992407	1.00765	.1480	.988987	1.01114
.1235	.992345	1.00771	.1485	.988912	1.01121
.1240	.992282	1.00778	.1490	.988837	1.01129
.1245	.992220	1.00784	.1495	.988762	1.01137
.1250	.992157	1.00791	.1500	.988686	1.01144

RELATIVITY CONTRACTION DILATION FACTOR (Continued)

β	$\sqrt{1-\beta^2}$	$\frac{1}{\sqrt{1-\beta^2}}$	β	$\sqrt{1-\beta^2}$	$\frac{1}{\sqrt{1-\beta^2}}$
.1505	.988610	1.01152	.1755	.984479	1.01577
.1510	.988534	1.01160	.1760	.984390	1.01586
.1515	.988457	1.01168	.1765	.984301	1.01595
.1520	.988380	1.01176	.1770	.984211	1.01604
.1525	.988303	1.01183	.1775	.984121	1.01614
.1530	.988226	1.01191	.1780	.984030	1.01623
.1535	.988149	1.01199	.1785	.983940	1.01632
.1540	.988071	1.01207	.1790	.983849	1.01642
.1545	.987993	1.01215	.1795	.983758	1.01651
.1550	.987914	1.01223	.1800	.983667	1.01660
.1555	.987836	1.01231	.1805	.983575	1.01670
.1560	.987757	1.01239	.1810	.983483	1.01679
.1565	.987678	1.01248	.1815	.983391	1.01689
.1570	.987599	1.01256	.1820	.983299	1.01699
.1575	.987519	1.01264	.1825	.983206	1.01708
.1580	.987439	1.01272	.1830	.983113	1.01718
.1585	.987359	1.01280	.1835	.983020	1.01727
.1590	.987279	1.01289	.1840	.982926	1.01737
.1595	.987198	1.01297	.1845	.982833	1.01747
.1600	.987117	1.01305	.1850	.982739	1.01756
.1605	.987036	1.01313	.1855	.982644	1.01766
.1610	.986954	1.01322	.1860	.982550	1.01776
.1615	.986873	1.01330	.1865	.982455	1.01786
.1620	.986791	1.01339	.1870	.982360	1.01796
.1625	.986709	1.01347	.1875	.982265	1.01806
.1630	.986626	1.01356	.1880	.982169	1.01815
.1635	.986543	1.01364	.1885	.982073	1.01825
.1640	.986460	1.01373	.1890	.981977	1.01835
.1645	.986377	1.01381	.1895	.981881	1.01845
.1650	.986294	1.01390	.1900	.981784	1.01855
.1655	.986210	1.01398	.1905	.981687	1.01865
.1660	.986126	1.01407	.1910	.981590	1.01876
.1665	.986041	1.01416	.1915	.981493	1.01886
.1670	.985957	1.01424	.1920	.981395	1.01896
.1675	.985872	1.01433	.1925	.981297	1.01906
.1680	.985787	1.01442	.1930	.981199	1.01916
.1685	.985702	1.01451	.1935	.981100	1.01926
.1690	.985616	1.01459	.1940	.981002	1.01937
.1695	.985530	1.01468	.1945	.980903	1.01947
.1700	.985444	1.01477	.1950	.980803	1.01957
.1705	.985358	1.01486	.1955	.980704	1.01968
.1710	.985271	1.01495	.1960	.980604	1.01978
.1715	.985184	1.01504	.1965	.980504	1.01988
.1720	.985097	1.01513	.1970	.980403	1.01999
.1725	.985010	1.01522	.1975	.980303	1.02009
.1730	.984922	1.01531	.1980	.980202	1.02020
.1735	.984834	1.01540	.1985	.980101	1.02030
.1740	.984746	1.01549	.1990	.979999	1.02041
.1745	.984657	1.01558	.1995	.979898	1.02051
.1750	.984568	1.01567	.2000	.979796	1.02062

RELATIVITY CONTRACTION DILATION FACTOR (Continued)

β	$\sqrt{1-\beta^2}$	$\frac{1}{\sqrt{1-\beta^2}}$	β	$\sqrt{1-\beta^2}$	$\frac{1}{\sqrt{1-\beta^2}}$
.2005	.979694	1.02073	.2255	.974243	1.02644
.2010	.979591	1.02083	.2260	.974127	1.02656
.2015	.979489	1.02094	.2265	.974011	1.02668
.2020	.979386	1.02105	.2270	.973895	1.02680
.2025	.979282	1.02116	.2275	.973778	1.02693
.2030	.979179	1.02126	.2280	.973661	1.02705
.2035	.979075	1.02137	.2285	.973544	1.02718
.2040	.978971	1.02148	.2290	.973426	1.02730
.2045	.978867	1.02159	.2295	.973309	1.02742
.2050	.978762	1.02170	.2300	.973191	1.02755
.2055	.978657	1.02181	.2305	.973072	1.02767
.2060	.978552	1.02192	.2310	.972954	1.02780
.2065	.978447	1.02203	.2315	.972835	1.02792
.2070	.978341	1.02214	.2320	.972716	1.02805
.2075	.978235	1.02225	.2325	.972592	1.02818
.2080	.978129	1.02236	.2330	.972477	1.02830
.2085	.978022	1.02247	.2335	.972357	1.02843
.2090	.977916	1.02258	.2340	.972237	1.02856
.2095	.977809	1.02269	.2345	.972116	1.02868
.2100	.977701	1.02281	.2350	.971995	1.02881
.2105	.977594	1.02292	.2355	.971874	1.02894
.2110	.977486	1.02303	.2360	.971753	1.02907
.2115	.977378	1.02315	.2365	.971631	1.02920
.2120	.977270	1.02326	.2370	.971510	1.02933
.2125	.977161	1.02337	.2375	.971388	1.02946
.2130	.977052	1.02349	.2380	.971265	1.02958
.2135	.976943	1.02360	.2385	.971142	1.02972
.2140	.976834	1.02372	.2390	.971020	1.02985
.2145	.976724	1.02383	.2395	.970896	1.02998
.2150	.976614	1.02395	.2400	.970773	1.03011
.2155	.976504	1.02406	.2405	.970649	1.03024
.2160	.976393	1.02418	.2410	.970525	1.03037
.2165	.976283	1.02429	.2415	.970401	1.03050
.2170	.976172	1.02441	.2420	.970276	1.03063
.2175	.976060	1.02453	.2425	.970151	1.03077
.2180	.975949	1.02464	.2430	.970026	1.03090
.2185	.975837	1.02476	.2435	.969901	1.03103
.2190	.975725	1.02488	.2440	.969775	1.03117
.2195	.975612	1.02500	.2445	.969649	1.03130
.2200	.975500	1.02512	.2450	.969523	1.03143
.2205	.975387	1.02523	.2455	.969397	1.03157
.2210	.975274	1.02535	.2460	.969270	1.03170
.2215	.975160	1.02547	.2465	.969143	1.03184
.2220	.975047	1.02559	.2470	.969015	1.03198
.2225	.974933	1.02571	.2475	.968888	1.03211
.2230	.974818	1.02583	.2480	.968760	1.03225
.2235	.974704	1.02595	.2485	.968632	1.03238
.2240	.974589	1.02607	.2490	.968503	1.03252
.2245	.974474	1.02619	.2495	.968375	1.03266
.2250	.974359	1.02632	.2500	.968246	1.03280

RELATIVITY CONTRACTION DILATION FACTOR (Continued)

β	$\sqrt{1-\beta^2}$	$\frac{1}{\sqrt{1-\beta^2}}$	β	$\sqrt{1-\beta^2}$	$\frac{1}{\sqrt{1-\beta^2}}$
.2505	.968117	1.03293	.2755	.961301	1.04026
.2510	.967987	1.03307	.2760	.961158	1.04041
.2515	.967857	1.03321	.2765	.961014	1.04057
.2520	.967727	1.03335	.2770	.960870	1.04072
.2525	.967597	1.03349	.2775	.960726	1.04088
.2530	.967466	1.03363	.2780	.960581	1.04104
.2535	.967335	1.03377	.2785	.960436	1.04119
.2540	.967204	1.03391	.2790	.960291	1.04135
.2545	.967073	1.03405	.2795	.960146	1.04151
.2550	.966941	1.03419	.2800	.960000	1.04167
.2555	.966809	1.03433	.2805	.959854	1.04183
.2560	.966677	1.03447	.2810	.959708	1.04198
.2565	.966544	1.03461	.2815	.959561	1.04214
.2570	.966411	1.03476	.2820	.959414	1.04230
.2575	.966278	1.03490	.2825	.959267	1.04246
.2580	.966145	1.03504	.2830	.959120	1.04262
.2585	.966011	1.03518	.2835	.958972	1.04278
.2590	.965877	1.03533	.2840	.958824	1.04294
.2595	.965743	1.03547	.2845	.958676	1.04311
.2600	.965609	1.03562	.2850	.958528	1.04327
.2605	.965474	1.03576	.2855	.958379	1.04343
.2610	.965339	1.03591	.2860	.958230	1.04359
.2615	.965203	1.03605	.2865	.958080	1.04375
.2620	.965068	1.03620	.2870	.957931	1.04392
.2625	.964932	1.03634	.2875	.957781	1.04408
.2630	.964796	1.03649	.2880	.957630	1.04424
.2635	.964659	1.03664	.2885	.957480	1.04441
.2640	.964523	1.03678	.2890	.957329	1.04457
.2645	.964386	1.03693	.2895	.957178	1.04474
.2650	.964248	1.03708	.2900	.957027	1.04490
.2655	.964111	1.03723	.2905	.956875	1.04507
.2660	.963973	1.03737	.2910	.956723	1.04523
.2665	.963835	1.03752	.2915	.956571	1.04540
.2670	.963697	1.03767	.2920	.956418	1.04557
.2675	.963558	1.03782	.2925	.956266	1.04573
.2680	.963419	1.03797	.2930	.956112	1.04590
.2685	.963280	1.03812	.2935	.955959	1.04607
.2690	.963140	1.03827	.2940	.955805	1.04624
.2695	.963000	1.03842	.2945	.955651	1.04641
.2700	.962860	1.03857	.2950	.955497	1.04658
.2705	.962720	1.03872	.2955	.955343	1.04674
.2710	.962579	1.03888	.2960	.955188	1.04691
.2715	.962438	1.03903	.2965	.955033	1.04708
.2720	.962297	1.03918	.2970	.954877	1.04725
.2725	.962156	1.03933	.2975	.954722	1.04743
.2730	.962014	1.03949	.2980	.954566	1.04760
.2735	.961872	1.03964	.2985	.954410	1.04777
.2740	.961730	1.03979	.2990	.954253	1.04794
.2745	.961587	1.03995	.2995	.954096	1.04811
.2750	.961444	1.04010	.3000	.953939	1.04828

RELATIVITY CONTRACTION DILATION FACTOR (Continued)

β	$\sqrt{1-\beta^2}$	$\frac{1}{\sqrt{1-\beta^2}}$	β	$\sqrt{1-\beta^2}$	$\frac{1}{\sqrt{1-\beta^2}}$
.3005	.953782	1.04846	.3255	.945542	1.05759
.3010	.953624	1.04863	.3260	.945370	1.05779
.3015	.953466	1.04880	.3265	.945197	1.05798
.3020	.953308	1.04898	.3270	.945024	1.05817
.3025	.953149	1.04915	.3275	.944851	1.05837
.3030	.952991	1.04933	.3280	.944678	1.05856
.3035	.952831	1.04950	.3285	.944504	1.05876
.3040	.952672	1.04968	.3290	.944330	1.05895
.3045	.952512	1.04986	.3295	.944156	1.05915
.3050	.952352	1.05003	.3300	.943981	1.05934
.3055	.952192	1.05021	.3305	.943806	1.05954
.3060	.952032	1.05039	.3310	.943631	1.05974
.3065	.951871	1.05056	.3315	.943455	1.05993
.3070	.951710	1.05074	.3320	.943279	1.06013
.3075	.951548	1.05092	.3325	.943103	1.06033
.3080	.951386	1.05110	.3330	.942927	1.06053
.3085	.951224	1.05128	.3335	.942750	1.06073
.3090	.951062	1.05146	.3340	.942573	1.06093
.3095	.950899	1.05164	.3345	.942396	1.06113
.3100	.950737	1.05182	.3350	.942218	1.06133
.3105	.950573	1.05200	.3355	.942040	1.06153
.3110	.950410	1.05218	.3360	.941862	1.06173
.3115	.950246	1.05236	.3365	.941683	1.06193
.3120	.950082	1.05254	.3370	.941505	1.06213
.3125	.949918	1.05272	.3375	.941326	1.06233
.3130	.949753	1.05291	.3380	.941146	1.06253
.3135	.949588	1.05309	.3385	.940966	1.06274
.3140	.949423	1.05327	.3390	.940786	1.06294
.3145	.949257	1.05345	.3395	.940606	1.06314
.3150	.949092	1.05364	.3400	.940425	1.06335
.3155	.948926	1.05382	.3405	.940245	1.06355
.3160	.948759	1.05401	.3410	.940063	1.06376
.3165	.948593	1.05419	.3415	.939882	1.06396
.3170	.948426	1.05438	.3420	.939700	1.06417
.3175	.948258	1.05457	.3425	.939518	1.06438
.3180	.948091	1.05475	.3430	.939335	1.06458
.3185	.947923	1.05494	.3435	.939153	1.06479
.3190	.947755	1.05513	.3440	.938970	1.06500
.3195	.947586	1.05531	.3445	.938786	1.06521
.3200	.947418	1.05550	.3450	.938603	1.06541
.3205	.947249	1.05569	.3455	.938419	1.06562
.3210	.947079	1.05588	.3460	.938235	1.06583
.3215	.946910	1.05607	.3465	.938050	1.06604
.3220	.946740	1.05626	.3470	.937865	1.06625
.3225	.946569	1.05645	.3475	.937680	1.06646
.3230	.946399	1.05664	.3480	.937495	1.06667
.3235	.946228	1.05683	.3485	.937309	1.06688
.3240	.946057	1.05702	.3490	.937123	1.06710
.3245	.945886	1.05721	.3495	.936936	1.06731
.3250	.945714	1.05740	.3500	.936750	1.06752

RELATIVITY CONTRACTION DILATION FACTOR (Continued)

β	$\sqrt{1-\beta^2}$	$\frac{1}{\sqrt{1-\beta^2}}$	β	$\sqrt{1-\beta^2}$	$\frac{1}{\sqrt{1-\beta^2}}$
.3505	.936563	1.06773	.3755	.926822	1.07896
.3510	.936375	1.06795	.3760	.926620	1.07919
.3515	.936188	1.06816	.3765	.926417	1.07943
.3520	.936000	1.06838	.3770	.926213	1.07966
.3525	.935812	1.06859	.3775	.926010	1.07990
.3530	.935623	1.06881	.3780	.925806	1.08014
.3535	.935435	1.06902	.3785	.925601	1.08038
.3540	.935245	1.06924	.3790	.925397	1.08062
.3545	.935056	1.06945	.3795	.925192	1.08086
.3550	.934866	1.06967	.3800	.924986	1.08110
.3555	.934676	1.06989	.3805	.924781	1.08134
.3560	.934486	1.07011	.3810	.924575	1.08158
.3565	.934295	1.07033	.3815	.924369	1.08182
.3570	.934104	1.07054	.3820	.924162	1.08206
.3575	.933913	1.07076	.3825	.923955	1.08230
.3580	.933722	1.07098	.3830	.923748	1.08255
.3585	.933530	1.07120	.3835	.923541	1.08279
.3590	.933338	1.07142	.3840	.923333	1.08303
.3595	.933145	1.07164	.3845	.923125	1.08328
.3600	.932952	1.07187	.3850	.922917	1.08352
.3605	.932759	1.07209	.3855	.922708	1.08377
.3610	.932566	1.07231	.3860	.922499	1.08401
.3615	.932372	1.07253	.3865	.922289	1.08426
.3620	.932178	1.07276	.3870	.922080	1.08450
.3625	.931984	1.07298	.3875	.921870	1.08475
.3630	.931789	1.07320	.3880	.921659	1.08500
.3635	.931594	1.07343	.3885	.921449	1.08525
.3640	.931399	1.07365	.3890	.921238	1.08550
.3645	.931203	1.07388	.3895	.921026	1.08575
.3650	.931008	1.07411	.3900	.920815	1.08599
.3655	.930811	1.07433	.3905	.920603	1.08624
.3660	.930615	1.07456	.3910	.920391	1.08650
.3665	.930418	1.07479	.3915	.920178	1.08675
.3670	.930221	1.07501	.3920	.919965	1.08700
.3675	.930024	1.07524	.3925	.919752	1.08725
.3680	.929826	1.07547	.3930	.919538	1.08750
.3685	.929628	1.07570	.3935	.919325	1.08776
.3690	.929429	1.07593	.3940	.919110	1.08801
.3695	.929231	1.07616	.3945	.918896	1.08826
.3700	.929032	1.07639	.3950	.918681	1.08852
.3705	.928832	1.07662	.3955	.918466	1.08877
.3710	.928633	1.07685	.3960	.918251	1.08903
.3715	.928433	1.07708	.3965	.918035	1.08928
.3720	.928233	1.07732	.3970	.917819	1.08954
.3725	.928032	1.07755	.3975	917602	1.08980
.3730	.927831	1.07778	.3980	.917385	1.09005
.3735	.927630	1.07802	.3985	.917168	1.09031
.3740	.927429	1.07825	.3990	.916951	1.09057
.3745	.927227	1.07848	.3995	.916733	1.09083
.3750	.927025	1.07872	.4000	.916515	1.09109

RELATIVITY CONTRACTION DILATION FACTOR (Continued)

β	$\sqrt{1-\beta^2}$	$\dfrac{1}{\sqrt{1-\beta^2}}$	β	$\sqrt{1-\beta^2}$	$\dfrac{1}{\sqrt{1-\beta^2}}$
.4005	.916297	1.09135	.4255	.904958	1.10502
.4010	.916078	1.09161	.4260	.904723	1.10531
.4015	.915859	1.09187	.4265	.904488	1.10560
.4020	.915640	1.09213	.4270	.904252	1.10589
.4025	.915420	1.09239	.4275	.904015	1.10618
.4030	.915200	1.09266	.4280	.903779	1.10647
.4035	.914980	1.09292	.4285	.903542	1.10676
.4040	.914759	1.09318	.4290	.903304	1.10705
.4045	.914538	1.09345	.4295	.903067	1.10734
.4050	.914317	1.09371	.4300	.902829	1.10763
.4055	.914095	1.09398	.4305	.902591	1.10792
.4060	.913873	1.09424	.4310	.902352	1.10822
.4065	.913651	1.09451	.4315	.902113	1.10851
.4070	.913428	1.09478	.4320	.901874	1.10880
.4075	.913205	1.09504	.4325	.901634	1.10910
.4080	.912982	1.09531	.4330	.901394	1.10939
.4085	.912758	1.09558	.4335	.901154	1.10969
.4090	.912534	1.09585	.4340	.900913	1.10999
.4095	.912310	1.09612	.4345	.900672	1.11028
.4100	.912086	1.09639	.4350	.900430	1.11058
.4105	.911861	1.09666	.4355	.900189	1.11088
.4110	.911635	1.09693	.4360	.899947	1.11118
.4115	.911410	1.09720	.4365	.899704	1.11148
.4120	.911184	1.09747	.4370	.899462	1.11178
.4125	.910958	1.09775	.4375	.899218	1.11208
.4130	.910731	1.09802	.4380	898975	1.11238
.4135	.910504	1.09829	.4385	.898731	1.11268
.4140	.910277	1.09857	.4390	.898487	1.11298
.4145	.910049	1.09884	.4395	.898243	1.11328
.4150	.909821	1.09912	.4400	.897998	1.11359
.4155	.909593	1.09939	.4405	.897753	1.11389
.4160	.909365	1.09967	.4410	.897507	1.11420
.4165	.909136	1.09995	.4415	.897261	1.11450
.4170	.908906	1.10022	.4420	.897015	1.11481
.4175	.908677	1.10050	.4425	.896769	1.11511
.4180	.908447	1.10078	.4430	.896522	1.11542
.4185	.908217	1.10106	.4435	.896274	1.11573
.4190	.907986	1.10134	.4440	.896027	1.11604
.4195	.907755	1.10162	.4445	.895779	1.11635
.4200	.907524	1.10190	.4450	.895531	1.11666
.4205	.907293	1.10218	.4455	.895282	1.11697
.4210	.907061	1.10246	.4460	.895033	1.11728
.4215	.906828	1.10274	.4465	.894784	1.11759
.4220	.906596	1.10303	.4470	.894534	1.11790
.4225	.906363	1.10331	.4475	.894284	1.11821
.4230	.906130	1.10359	.4480	.894034	1.11853
.4235	.905896	1.10388	.4485	.893783	1.11884
.4240	.905662	1.10416	.4490	.893532	1.11915
.4245	.905428	1.10445	.4495	.893280	1.11947
.4250	.905193	1.10474	.4500	.893029	1.11979

RELATIVITY CONTRACTION DILATION FACTOR (Continued)

β	$\sqrt{1-\beta^2}$	$\frac{1}{\sqrt{1-\beta^2}}$	β	$\sqrt{1-\beta^2}$	$\frac{1}{\sqrt{1-\beta^2}}$
.4505	.892776	1.12010	.4755	.879716	1.13673
.4510	.892524	1.12042	.4760	.879445	1.13708
.4515	.892271	1.12074	.4765	.879174	1.13743
.4520	.892018	1.12105	.4770	.878903	1.13778
.4525	.891764	1.12137	.4775	.878632	1.13813
.4530	.891511	1.12169	.4780	.878360	1.13849
.4535	.891256	1.12201	.4785	.878088	1.13884
.4540	.891002	1.12233	.4790	.877815	1.13919
.4545	.890747	1.12265	.4795	.877542	1.13955
.4550	.890491	1.12298	.4800	.877268	1.13990
.4555	.890236	1.12330	.4805	.876995	1.14026
.4560	.889980	1.12362	.4810	.876721	1.14061
.4565	.889723	1.12394	.4815	.876446	1.14097
.4570	.889467	1.12427	.4820	.876171	1.14133
.4575	.889210	1.12459	.4825	.875896	1.14169
.4580	.888952	1.12492	.4830	.875620	1.14205
.4585	.888694	1.12525	.4835	.875344	1.14241
.4590	.888436	1.12557	.4840	.875068	1.14277
.4595	.888178	1.12490	.4845	.874791	1.14313
.4600	.887919	1.12623	.4850	.874514	1.14349
.4605	.887660	1.12656	.4855	.874237	1.14386
.4610	.887400	1.12689	.4860	.873959	1.14422
.4615	.887140	1.12722	.4865	.873681	1.14458
.4620	.886880	1.12755	.4870	.873402	1.14495
.4625	.886619	1.12788	.4875	.873123	1.14531
.4630	.886358	1.12821	.4880	.872844	1.14568
.4635	.886097	1.12854	.4885	.872564	1.14605
.4640	.885835	1.12888	.4890	.872284	1.14642
.4645	.885573	1.12921	.4895	.872003	1.14678
.4650	.885311	1.12955	.4900	.871722	1.14715
.4655	.885048	1.12988	.4905	.871441	1.14752
.4660	.884785	1.13022	.4910	.871160	1.14790
.4665	.884521	1.13056	.4915	.870878	1.14827
.4670	.884257	1.13089	.4920	.870595	1.14864
.4675	.883993	1.13123	.4925	.870312	1.14901
.4680	.883728	1.13157	.4930	.870029	1.14939
.4685	.883463	1.13191	.4935	.869746	1.14976
.4690	.883198	1.13225	.4940	.869462	1.15014
.4695	.882932	1.13259	.4945	.869178	1.15051
.4700	.882666	1.13293	.4950	.868893	1.15089
.4705	.882400	1.13327	.4955	.868608	1.15127
.4710	.882133	1.13362	.4960	.868323	1.15165
.4715	.881866	1.13396	.4965	.868037	1.15202
.4720	.881599	1.13430	.4970	.867751	1.15240
.4725	.881331	1.13465	.4975	.867464	1.15279
.4730	.881062	1.13499	.4980	.867177	1.15317
.4735	.880794	1.13534	.4985	.866890	1.15355
.4740	.880525	1.13569	.4990	.866602	1.15393
.4745	.880256	1.13603	.4995	.866314	1.15432
.4750	.879986	1.13638	.5000	.866025	1.15470

RELATIVITY CONTRACTION DILATION FACTOR (Continued)

β	$\sqrt{1-\beta^2}$	$\frac{1}{\sqrt{1-\beta^2}}$	β	$\sqrt{1-\beta^2}$	$\frac{1}{\sqrt{1-\beta^2}}$
.5005	.865737	1.15509	.5255	.850794	1.17537
.5010	.865447	1.15547	.5260	.850485	1.17580
.5015	.865158	1.15586	.5265	.850175	1.17623
.5020	.864868	1.15625	.5270	.849865	1.17666
.5025	.864577	1.15663	.5275	.849555	1.17709
.5030	.864286	1.15702	.5280	.849244	1.17752
.5035	.863995	1.15741	.5285	.848933	1.17795
.5040	.863704	1.15780	.5290	.848622	1.17838
.5045	.863412	1.15820	.5295	.848310	1.17881
.5050	.863119	1.15859	.5300	.847998	1.17925
.5055	.862827	1.15898	.5305	.847685	1.17968
.5060	.862533	1.15938	.5310	.847372	1.18012
.5065	.862240	1.15977	.5315	.847058	1.18056
.5070	.861946	1.16017	.5320	.846744	1.18099
.5075	.861652	1.16056	.5325	.846430	1.18143
.5080	.861357	1.16096	.5330	.846115	1.18187
.5085	.861062	1.16136	.5335	.845800	1.18231
.5090	.860767	1.16176	.5340	.845484	1.18275
.5095	.860471	1.16215	.5345	.845168	1.18320
.5100	.860174	1.16255	.5350	.844852	1.18364
.5105	.859878	1.16296	.5355	.844535	1.18408
.5110	.859581	1.16336	.5360	.844218	1.18453
.5115	.859283	1.16376	.5365	.843900	1.18497
.5120	.858985	1.16416	.5370	.843582	1.18542
.5125	.858687	1.16457	.5375	.843264	1.18587
.5130	.858389	1.16497	.5380	.842945	1.18632
.5135	.858090	1.16538	.5385	.842626	1.18677
.5140	.857790	1.16579	.5390	.842306	1.18722
.5145	.857490	1.16619	.5395	.841986	1.18767
.5150	.857190	1.16660	.5400	.841665	1.18812
.5155	.856890	1.16701	.5405	.841344	1.18857
.5160	.856589	1.16742	.5410	.841023	1.18903
.5165	.856287	1.16783	.5415	.840701	1.18948
.5170	.855985	1.16824	.5420	.840378	1.18994
.5175	.855683	1.16866	.5425	.840056	1.19040
.5180	.855381	1.16907	.5430	.839733	1.19086
.5185	.855078	1.16948	.5435	.839409	1.19131
.5190	.854774	1.16990	.5440	.839085	1.19177
.5195	.854470	1.17032	.5445	.838761	1.19223
.5200	.854166	1.17073	.5450	.838436	1.19270
.5205	.853862	1.17115	.5455	.838111	1.19316
.5210	.853557	1.17157	.5460	.837785	1.19362
.5215	.853251	1.17199	.5465	.837459	1.19409
.5220	.852945	1.17241	.5470	.837133	1.19455
.5225	.852639	1.17283	.5475	.836806	1.19502
.5230	.852333	1.17325	.5480	.836478	1.19549
.5235	.852026	1.17367	.5485	.836151	1.19596
.5240	.851718	1.17410	.5490	.835822	1.19643
.5245	.851410	1.17452	.5495	.835494	1.19690
.5250	.851102	1.17495	.5500	.835165	1.19737

RELATIVITY CONTRACTION DILATION FACTOR (Continued)

β	$\sqrt{1-\beta^2}$	$\frac{1}{\sqrt{1-\beta^2}}$	β	$\sqrt{1-\beta^2}$	$\frac{1}{\sqrt{1-\beta^2}}$
.5505	.834835	1.19784	.5755	.817802	1.22279
.5510	.834505	1.19831	.5760	.817450	1.22332
.5515	.834175	1.19879	.5765	.817097	1.22384
.5520	.833844	1.19926	.5770	.816744	1.22437
.5525	.833513	1.19974	.5775	.816391	1.22490
.5530	.833181	1.20022	.5780	.816037	1.22543
.5535	.832849	1.20070	.5785	.815682	1.22597
.5540	.832517	1.20118	.5790	.815328	1.22650
.5545	.832184	1.20166	.5795	.814972	1.22704
.5550	.831850	1.20214	.5800	.814616	1.22757
.5555	.831517	1.20262	.5805	.814260	1.22811
.5560	.831182	1.20311	.5810	.813904	1.22865
.5565	.830848	1.20359	.5815	.813546	1.22919
.5570	.830512	1.20408	.5820	.813189	1.22973
.5575	.830177	1.20456	.5825	.812831	1.23027
.5580	.829841	1.20505	.5830	.812472	1.23081
.5585	.829505	1.20554	.5835	.812113	1.23136
.5590	.829168	1.20603	.5840	.811754	1.23190
.5595	.828830	1.20652	.5845	.811394	1.23245
.5600	.828493	1.20701	.5850	.811033	1.23300
.5605	.828154	1.20750	.5855	.810672	1.23354
.5610	.827816	1.20800	.5860	.810311	1.23409
.5615	.827477	1.20849	.5865	.809949	1.23465
.5620	.827137	1.20899	.5870	.809587	1.23520
.5625	.826797	1.20949	.5875	.809224	1.23575
.5630	.826457	1.20998	.5880	.808861	1.23631
.5635	.826116	1.21048	.5885	.808497	1.23686
.5640	.825775	1.21098	.5890	.808133	1.23742
.5645	.825433	1.21149	.5895	.807768	1.23798
.5650	.825091	1.21199	.5900	.807403	1.23854
.5655	.824748	1.21249	.5905	.807038	1.23910
.5660	.824405	1.21300	.5910	.806672	1.23966
.5665	.824062	1.21350	.5915	.806305	1.24023
.5670	.823718	1.21401	.5920	.805938	1.24079
.5675	.823373	1.21452	.5925	.805570	1.24136
.5680	.823029	1.21502	.5930	.805202	1.24192
.5685	.822683	1.21553	.5935	.804834	1.24249
.5690	.822338	1.21605	.5940	.804465	1.24306
.5695	.821991	1.21656	.5945	.804096	1.24363
.5700	.821645	1.21707	.5950	.803726	1.24421
.5705	.821298	1.21759	.5955	.803355	1.24478
.5710	.820950	1.21810	.5960	.802984	1.24535
.5715	.820602	1.21862	.5965	.802613	1.24593
.5720	.820254	1.21914	.5970	.802241	1.24651
.5725	.819905	1.21965	.5975	.801869	1.24709
.5730	.819555	1.22017	.5980	.801496	1.24767
.5735	.819206	1.22069	.5985	.801123	1.24825
.5740	.818855	1.22122	.5990	.800749	1.24883
.5745	.818505	1.22174	.5995	.800375	1.24941
.5750	.818153	1.22226	.6000	.800000	1.25000

RELATIVITY CONTRACTION DILATION FACTOR (Continued)

β	$\sqrt{1-\beta^2}$	$\frac{1}{\sqrt{1-\beta^2}}$	β	$\sqrt{1-\beta^2}$	$\frac{1}{\sqrt{1-\beta^2}}$
.6005	.799625	1.25059	.6255	.780224	1.28168
.6010	.799249	1.25117	.6260	.779823	1.28234
.6015	.798873	1.25176	.6265	.779421	1.28300
.6020	.798496	1.25235	.6270	.779019	1.28367
.6025	.798119	1.25295	.6275	.778617	1.28433
.6030	.797741	1.25354	.6280	.778213	1.28499
.6035	.797363	1.25413	.6285	.777810	1.28566
.6040	.796984	1.25473	.6290	.777405	1.28633
.6045	.796605	1.25533	.6295	.777000	1.28700
.6050	.796225	1.25593	.6300	.776595	1.28767
.6055	.795845	1.25653	.6305	.776189	1.28835
.6060	.795465	1.25713	.6310	.775783	1.28902
.6065	.795083	1.25773	.6315	.775376	1.28970
.6070	.794702	1.25833	.6320	.774968	1.29038
.6075	.794320	1.25894	.6325	.774560	1.29105
.6080	.793937	1.25955	.6330	.774152	1.29174
.6085	.793554	1.26015	.6335	.773743	1.29242
.6090	.793170	1.26076	.6340	.773333	1.29310
.6095	.792786	1.26137	.6345	.772923	1.29379
.6100	.792401	1.26199	.6350	.772512	1.29448
.6105	.792016	1.26260	.6355	.772101	1.29517
.6110	.791631	1.26322	.6360	.771689	1.29586
.6115	.791244	1.26383	.6365	.771277	1.29655
.6120	.790858	1.26445	.6370	.770864	1.29725
.6125	.790471	1.26507	.6375	.770450	1.29794
.6130	.790083	1.26569	.6380	.770036	1.29864
.6135	.789695	1.26631	.6385	.769622	1.29934
.6140	.789306	1.26694	.6390	.769207	1.30004
.6145	.788917	1.26756	.6395	.768791	1.30074
.6150	.788527	1.26819	.6400	.768375	1.30145
.6155	.788137	1.26882	.6405	.767958	1.30215
.6160	.787746	1.26944	.6410	.767541	1.30286
.6165	.787355	1.27008	.6415	767123	1.30357
.6170	.786963	1.27071	.6420	.766705	1.30428
.6175	.786571	1.27134	.6425	.766286	1.30500
.6180	.786178	1.27198	.6430	.765866	1.30571
.6185	.785785	1.27261	.6435	.765446	1.30643
.6190	.785391	1.27325	.6440	.765025	1.30715
.6195	.784997	1.27389	.6445	.764604	1.30787
.6200	.784602	1.27453	.6450	.764183	1.30859
.6205	.784206	1.27517	.6455	.763760	1.30931
.6210	.783811	1.27582	.6460	.763337	1.31004
.6215	.783414	1.27646	.6465	.762914	1.31076
.6220	.783017	1.27711	.6470	.762490	1.31149
.6225	.782620	1.27776	.6475	.762065	1.31222
.6230	.782222	1.27841	.6480	.761640	1.31296
.6235	.781823	1.27906	.6485	.761215	1.31369
.6240	.781424	1.27971	.6490	.760788	1.31443
.6245	.781025	1.28037	.6495	.760362	1.31516
.6250	.780625	1.28103	.6500	.759934	1.31590

RELATIVITY CONTRACTION DILATION FACTOR (Continued)

β	$\sqrt{1-\beta^2}$	$\frac{1}{\sqrt{1-\beta^2}}$	β	$\sqrt{1-\beta^2}$	$\frac{1}{\sqrt{1-\beta^2}}$
.6505	.759506	1.31664	.6755	.737360	1.35619
.6510	.759078	1.31739	.6760	.736902	1.35703
.6515	.758649	1.31813	.6765	.736443	1.35788
.6520	.758219	1.31888	.6770	.735983	1.35873
.6525	.757789	1.31963	.6775	.735523	1.35958
.6530	.757358	1.32038	.6780	.735062	1.36043
.6535	.756927	1.32113	.6785	.734600	1.36128
.6540	.756495	1.32189	.6790	.734138	1.36214
.6545	.756062	1.32264	.6795	.733676	1.36300
.6550	.755629	1.32340	.6800	.733212	1.36386
.6555	.755195	1.32416	.6805	.732748	1.36473
.6560	.754761	1.32492	.6810	.732283	1.36559
.6565	.754326	1.32569	.6815	.731818	1.36646
.6570	.753891	1.32645	.6820	.731352	1.36733
.6575	.753455	1.32722	.6825	.730886	1.36820
.6580	.753018	1.32799	.6830	.730418	1.36908
.6585	.752581	1.32876	.6835	.729951	1.36996
.6590	.752143	1.32953	.6840	.729482	1.37084
.6595	.751705	1.33031	.6845	.729013	1.37172
.6600	.751266	1.33109	.6850	.728543	1.37260
.6605	.750826	1.33187	.6855	.728073	1.37349
.6610	.750386	1.33265	.6860	.727602	1.37438
.6615	.749945	1.33343	.6865	.727130	1.37527
.6620	.749504	1.33422	.6870	.726657	1.37616
.6625	.749062	1.33500	.6875	.726184	1.37706
.6630	.748619	1.33579	.6880	.725711	1.37796
.6635	.748176	1.33658	.6885	.725236	1.37886
.6640	.747733	1.33738	.6890	.724761	1.37976
.6645	.747288	1.33817	.6895	.724286	1.38067
.6650	.746843	1.33897	.6900	.723809	1.38158
.6655	.746398	1.33977	.6905	.723332	1.38249
.6660	.745952	1.34057	.6910	.722855	1.38340
.6665	.745505	1.34137	.6915	.722376	1.38432
.6670	.745058	1.34218	.6920	.721897	1.38524
.6675	.744610	1.34299	.6925	.721418	1.38616
.6680	.744161	1.34379	.6930	.720938	1.38708
.6685	.743712	1.34461	.6935	.720457	1.38801
.6690	.743262	1.34542	.6940	.719975	1.38894
.6695	.742812	1.34624	.6945	.719493	1.38987
.6700	.742361	1.34705	.6950	.719010	1.39080
.6705	.741910	1.34787	.6955	.718526	1.39174
.6710	.741457	1.34870	.6960	.718042	1.39268
.6715	.741005	1.34952	.6965	.717557	1.39362
.6720	.740551	1.35035	.6970	.717071	1.39456
.6725	.740097	1.35117	.6975	.716585	1.39551
.6730	.739642	1.35200	.6980	.716098	1.39646
.6735	.739187	1.35284	.6985	.715610	1.39741
.6740	.738731	1.35367	.6990	.715122	1.39836
.6745	.738275	1.35451	.6995	.714633	1.39932
.6750	.737818	1.35535	.7000	.714143	1.40028

RELATIVITY CONTRACTION DILATION FACTOR (Continued)

β	$\sqrt{1-\beta^2}$	$\frac{1}{\sqrt{1-\beta^2}}$	β	$\sqrt{1-\beta^2}$	$\frac{1}{\sqrt{1-\beta^2}}$
.7005	.713652	1.40124	.7255	.688222	1.45302
.7010	.713161	1.40221	.7260	.687695	1.45413
.7015	.712669	1.40318	.7265	.687166	1.45525
.7020	.712177	1.40415	.7270	.686637	1.45637
.7025	.711684	1.40512	.7275	.686108	1.45750
.7030	.711190	1.40609	.7280	.685577	1.45863
.7035	.710695	1.40707	.7285	.685046	1.45976
.7040	.710200	1.40805	.7290	.684514	1.46089
.7045	.709704	1.40904	.7295	.683981	1.46203
.7050	.709207	1.41002	.7300	.683447	1.46317
.7055	.708710	1.41101	.7305	.682913	1.46432
.7060	.708212	1.41201	.7310	.682377	1.46546
.7065	.707713	1.41300	.7315	.681841	1.46662
.7070	.707214	1.41400	.7320	.681305	1.46777
.7075	.706713	1.41500	.7325	.680767	1.46893
.7080	.706212	1.41600	.7330	.680229	1.47009
.7085	.705711	1.41701	.7335	.679689	1.47126
.7090	.705208	1.41802	.7340	.679149	1.47243
.7095	.704705	1.41903	.7345	.678609	1.47360
.7100	.704202	1.42005	.7350	.678067	1.47478
.7105	.703697	1.42107	.7355	.677525	1.47596
.7110	.703192	1.42209	.7360	.676982	1.47715
.7115	.702686	1.42311	.7365	.676438	1.47833
.7120	.702179	1.42414	.7370	.675893	1.47952
.7125	.701672	1.42517	.7375	.675347	1.48072
.7130	.701164	1.42620	.7380	.674801	1.48192
.7135	.700655	1.42724	.7385	.674253	1.48312
.7140	.700146	1.42827	.7390	.673705	1.48433
.7145	.699635	1.42932	.7395	.673157	1.48554
.7150	.699124	1.43036	.7400	.672607	1.48675
.7155	.698613	1.43141	.7405	.672056	1.48797
.7160	.698100	1.43246	.7410	.671505	1.48919
.7165	.697587	1.43351	.7415	.670953	1.49042
.7170	.697073	1.43457	.7420	.670400	1.49165
.7175	.696559	1.43563	.7425	.669846	1.49288
.7180	.696043	1.43669	.7430	.669291	1.49412
.7185	.695527	1.43776	.7435	.668736	1.49536
.7190	.695010	1.43883	.7440	.668180	1.49660
.7195	.694492	1.43990	.7445	.667622	1.49785
.7200	.693974	1.44098	.7450	.667064	1.49911
.7205	.693455	1.44205	.7455	.666506	1.50036
.7210	.692935	1.44314	.7460	.665946	1.50162
.7215	.692414	1.44422	.7465	.665385	1.50289
.7220	.691893	1.44531	.7470	.664824	1.50416
.7225	.691371	1.44640	.7475	.664262	1.50543
.7230	.690848	1.44750	.7480	.663699	1.50671
.7235	.690324	1.44859	.7485	.663135	1.50799
.7240	.689800	1.44970	.7490	.662570	1.50927
.7245	.689275	1.45080	.7495	.662004	1.51056
.7250	.688749	1.45191	.7500	.661438	1.51186

RELATIVITY CONTRACTION DILATION FACTOR (Continued)

β	$\sqrt{1-\beta^2}$	$\frac{1}{\sqrt{1-\beta^2}}$	β	$\sqrt{1-\beta^2}$	$\frac{1}{\sqrt{1-\beta^2}}$
.7505	.660870	1.51316	.7755	.631348	1.58391
.7510	.660302	1.51446	.7760	.630733	1.58546
.7515	.659733	1.51576	.7765	.630117	1.58701
.7520	.659163	1.51708	.7770	.629501	1.58856
.7525	.658592	1.51839	.7775	.628883	1.59012
.7530	.658021	1.51971	.7780	.628264	1.59169
.7535	.657448	1.52103	.7785	.627645	1.59326
.7540	.656874	1.52236	.7790	.627024	1.59484
.7545	.656300	1.52369	.7795	.626402	1.59642
.7550	.655725	1.52503	.7800	.625780	1.59801
.7555	.655149	1.52637	.7805	.625156	1.59960
.7560	.654572	1.52772	.7810	.624531	1.60120
.7565	.653994	1.52907	.7815	.623905	1.60281
.7570	.653415	1.53042	.7820	.623278	1.60442
.7575	.652835	1.53178	.7825	.622651	1.60604
.7580	.652255	1.53314	.7830	.622022	1.60766
.7585	.651673	1.53451	.7835	.621392	1.60929
.7590	.651091	1.53588	.7840	.620761	1.61093
.7595	.650507	1.53726	.7845	.620129	1.61257
.7600	.649923	1.53864	.7850	.619496	1.61422
.7605	.649338	1.54003	.7855	.618862	1.61587
.7610	.648752	1.54142	.7860	.618226	1.61753
.7615	.648165	1.54282	.7865	.617590	1.61920
.7620	.647577	1.54422	.7870	.616953	1.62087
.7625	.646988	1.54562	.7875	.616315	1.62255
.7630	.646398	1.54703	.7880	.615675	1.62423
.7635	.645808	1.54845	.7885	.615035	1.62592
.7640	.645216	1.54987	.7890	.614393	1.62762
.7645	.644624	1.55129	.7895	.613751	1.62933
.7650	.644030	1.55272	.7900	.613107	1.63104
.7655	.643436	1.55416	.7905	.612462	1.63275
.7660	.642841	1.55560	.7910	.611816	1.63448
.7665	.642244	1.55704	.7915	.611169	1.63621
.7670	.641647	1.55849	.7920	.610521	1.63795
.7675	.641049	1.55994	.7925	.609872	1.63969
.7680	.640450	1.56140	.7930	.609222	1.64144
.7685	.639850	1.56287	.7935	.608570	1.64320
.7690	.639249	1.56434	.7940	.607918	1.64496
.7695	.638647	1.56581	.7945	.607264	1.64673
.7700	.638044	1.56729	.7950	.606609	1.64851
.7705	.637440	1.56878	.7955	.605954	1.65029
.7710	.636835	1.57027	.7960	.605297	1.65208
.7715	.636229	1.57176	.7965	.604639	1.65388
.7720	.635623	1.57326	.7970	.603979	1.65569
.7725	.635015	1.57477	.7975	.603319	1.65750
.7730	.634406	1.57628	.7980	.602657	1.65932
.7735	.633796	1.57779	.7985	.601995	1.66114
.7740	.633186	1.57932	.7990	.601331	1.66298
.7745	.632574	1.58084	.7995	.600666	1.66482
.7750	.631961	1.58238	.8000	.600000	1.66667

RELATIVITY CONTRACTION DILATION FACTOR (Continued)

β	$\sqrt{1-\beta^2}$	$\frac{1}{\sqrt{1-\beta^2}}$	β	$\sqrt{1-\beta^2}$	$\frac{1}{\sqrt{1-\beta^2}}$
.8005	.599333	1.66852	.8255	.564402	1.77179
.8010	.598664	1.67039	.8260	.563670	1.77409
.8015	.597995	1.67226	.8265	.562937	1.77640
.8020	.597324	1.67413	.8270	.562202	1.77872
.8025	.596652	1.67602	.8275	.561466	1.78105
.8030	.595979	1.67791	.8280	.560728	1.78340
.8035	.595305	1.67981	.8285	.559989	1.78575
.8040	.594629	1.68172	.8290	.559249	1.78811
.8045	.593953	1.68364	.8295	.558507	1.79049
.8050	.593275	1.68556	.8300	.557763	1.79287
.8055	.592596	1.68749	.8305	.557019	1.79527
.8060	.591916	1.68943	.8310	.556272	1.79768
.8065	.591234	1.69138	.8315	.555525	1.80010
.8070	.590551	1.69333	.8320	.554776	1.80253
.8075	.589868	1.69530	.8325	.554025	1.80497
.8080	.589182	1.69727	.8330	.553273	1.80743
.8085	.588496	1.69925	.8335	.552519	1.80989
.8090	.587809	1.70123	.8340	.551764	1.81237
.8095	.587120	1.70323	.8345	.551008	1.81486
.8100	.586430	1.70523	.8350	.550250	1.81736
.8105	.585739	1.70725	.8355	.549490	1.81987
.8110	.585046	1.70927	.8360	.548729	1.82239
.8115	.584352	1.71130	.8365	.547967	1.82493
.8120	.583657	1.71333	.8370	.547203	1.82748
.8125	.582961	1.71538	.8375	.546437	1.83004
.8130	.582264	1.71743	.8380	.545670	1.83261
.8135	.581565	1.71950	.8385	.544902	1.83519
.8140	.580865	1.72157	.8390	.544131	1.83779
.8145	.580164	1.72365	.8395	.543360	1.84040
.8150	.579461	1.72574	.8400	.542586	1.84302
.8155	.578757	1.72784	.8405	.541812	1.84566
.8160	.578052	1.72995	.8410	.541035	1.84831
.8165	.577345	1.73207	.8415	.540257	1.85097
.8170	.576638	1.73419	.8420	.539478	1.85365
.8175	.575929	1.73633	.8425	.538696	1.85633
.8180	.575218	1.73847	.8430	.537914	1.85903
.8185	.574507	1.74062	.8435	.537129	1.86175
.8190	.573794	1.74279	.8440	.536343	1.86448
.8195	.573079	1.74496	.8445	.535556	1.86722
.8200	.572364	1.74714	.8450	.534766	1.86998
.8205	.571647	1.74933	.8455	.533975	1.87275
.8210	.570928	1.75153	.8460	.533183	1.87553
.8215	.570209	1.75374	.8465	.532389	1.87833
.8220	.569487	1.75596	.8470	.531593	1.88114
.8225	.568765	1.75820	.8475	.530795	1.88397
.8230	.568041	1.76044	.8480	.529996	1.88681
.8235	.567316	1.76269	.8485	.529195	1.88966
.8240	.566590	1.76495	.8490	.528393	1.89253
.8245	.565862	1.76722	.8495	.527589	1.89542
.8250	.565133	1.76950	.8500	.526783	1.89832

RELATIVITY CONTRACTION DILATION FACTOR (Continued)

β	$\sqrt{1-\beta^2}$	$\frac{1}{\sqrt{1-\beta^2}}$	β	$\sqrt{1-\beta^2}$	$\frac{1}{\sqrt{1-\beta^2}}$
.8505	.525975	1.90123	.8755	.483218	2.06946
.8510	.525166	1.90416	.8760	.482311	2.07335
.8515	.524355	1.90711	.8765	.481402	2.07727
.8520	.523542	1.91007	.8770	.480490	2.08121
.8525	.522727	1.91304	.8775	.479577	2.08517
.8530	.521911	1.91604	.8780	.478661	2.08916
.8535	.521093	1.91904	.8785	.477742	2.09318
.8540	.520273	1.92207	.8790	.476822	2.09722
.8545	.519451	1.92511	.8795	.475899	2.10129
.8550	.518628	1.92816	.8800	.474974	2.10538
.8555	.517803	1.93124	.8805	.474046	2.10950
.8560	.516976	1.93433	.8810	.473116	2.11365
.8565	.516147	1.93743	.8815	.472184	2.11782
.8570	.515316	1.94056	.8820	.471249	2.12202
.8575	.514484	1.94370	.8825	.470312	2.12625
.8580	.513650	1.94685	.8830	.469373	2.13050
.8585	.512814	1.95003	.8835	.468431	2.13479
.8590	.511976	1.95322	.8840	.467487	2.13910
.8595	.511136	1.95643	.8845	.466540	2.14344
.8600	.510294	1.95965	.8850	.465591	2.14781
.8605	.509450	1.96290	.8855	.464639	2.15221
.8610	.508605	1.96616	.8860	.463685	2.15664
.8615	.507758	1.96944	.8865	.462729	2.16109
.8620	.506908	1.97274	.8870	.461769	2.16558
.8625	.506057	1.97606	.8875	.460808	2.17010
.8630	.505204	1.97940	.8880	.459843	2.17465
.8635	.504349	1.98275	.8885	.458877	2.17924
.8640	.503492	1.98613	.8890	.457907	2.18385
.8645	.502633	1.98952	.8895	.456935	2.18849
.8650	.501772	1.99294	.8900	.455961	2.19317
.8655	.500909	1.99637	.8905	.454983	2.19788
.8660	.500044	1.99982	.8910	.454003	2.20263
.8665	.499177	2.00330	.8915	.453021	2.20740
.8670	.498308	2.00679	.8920	.452035	2.21222
.8675	.497437	2.01030	.8925	.451047	2.21706
.8680	.496564	2.01384	.8930	.450057	2.22194
.8685	.495689	2.01739	.8935	.449063	2.22686
.8690	.494812	2.02097	.8940	.448067	2.23181
.8695	.493933	2.02457	.8945	.447068	2.23680
.8700	.493052	2.02818	.8950	.446066	2.24182
.8705	.492168	2.03182	.8955	.445062	2.24688
.8710	.491283	2.03549	.8960	.444054	2.25198
.8715	.490396	2.03917	.8965	.443044	2.25711
.8720	.489506	2.04288	.8970	.442031	2.26229
.8725	.488614	2.04660	.8975	.441014	2.26750
.8730	.487720	2.05036	.8980	.439995	2.27275
.8735	.486824	2.05413	.8985	.438974	2.27804
.8740	.485926	2.05793	.8990	.437949	2.28337
.8745	.485026	2.06175	.8995	.436921	2.28874
.8750	.484123	2.06559	.9000	.435890	2.29416

RELATIVITY CONTRACTION DILATION FACTOR (Continued)

β	$\sqrt{1-\beta^2}$	$\frac{1}{\sqrt{1-\beta^2}}$	β	$\sqrt{1-\beta^2}$	$\frac{1}{\sqrt{1-\beta^2}}$
.9002	.435477	2.29633	.9102	.414169	2.41447
.9004	.435063	2.29852	.9104	.413729	2.41704
.9006	.434649	2.30071	.9106	.413289	2.41962
.9008	.434234	2.30290	.9108	.412848	2.42220
.9010	.433819	2.30511	.9110	.412406	2.42479
.9012	.433403	2.30732	.9112	.411964	2.42739
.9014	.432987	2.30954	.9114	.411522	2.43001
.9016	.432571	2.31176	.9116	.411078	2.43263
.9018	.432154	2.31399	.9118	.410635	2.43526
.9020	.431736	2.31623	.9120	.410190	2.43789
.9022	.431318	2.31848	.9122	.409745	2.44054
.9024	.430899	2.32073	.9124	.409300	2.44320
.9026	.430480	2.32299	.9126	.408854	2.44586
.9028	.430061	2.32525	.9128	.408407	2.44854
.9030	.429641	2.32753	.9130	.407960	2.45122
.9032	.429220	2.32981	.9132	.407512	2.45392
.9034	.428799	2.33210	.9134	.407063	2.45662
.9036	.428377	2.33439	.9136	.406614	2.45933
.9038	.427955	2.33669	.9138	.406164	2.46206
.9040	.427532	2.33900	.9140	.405714	2.46479
.9042	.427109	2.34132	.9142	.405263	2.46753
.9044	.426686	2.34365	.9144	.404812	2.47028
.9046	.426261	2.34598	.9146	.404360	2.47305
.9048	.425837	2.34832	.9148	.403907	2.47582
.9050	.425412	2.35066	.9150	.403454	2.47860
.9052	.424986	2.35302	.9152	.403000	2.48139
.9054	.424560	2.35538	.9154	.402545	2.48419
.9056	.424133	2.35775	.9156	.402090	2.48700
.9058	.423706	2.36013	.9158	.401635	2.48983
.9060	.423278	2.36252	.9160	.401178	2.49266
.9062	.422849	2.36491	.9162	.400721	2.49550
.9064	.422420	2.36731	.9164	.400264	2.49835
.9066	.421991	2.36972	.9166	.399806	2.50122
.9068	.421561	2.37214	.9168	.399347	2.50409
.9070	.421131	2.37456	.9170	.398887	2.50697
.9072	.420700	2.37699	.9172	.398427	2.50987
.9074	.420268	2.37943	.9174	.397966	2.51278
.9076	.419836	2.38188	.9176	.397505	2.51569
.9078	.419403	2.38434	.9178	.397043	2.51862
.9080	.418970	2.38680	.9180	.396580	2.52156
.9082	.418536	2.38928	.9182	.396117	2.52451
.9084	.418102	2.39176	.9184	.395653	2.52747
.9086	.417667	2.39425	.9186	.395189	2.53044
.9088	.417232	2.39675	.9188	.394723	2.53342
.9090	.416796	2.39925	.9190	.394258	2.53641
.9092	.416360	2.40177	.9192	.393791	2.53942
.9094	.415923	2.40429	.9194	.393324	2.54243
.9096	.415485	2.40683	.9196	.392856	2.54546
.9098	.415047	2.40937	.9198	.392388	2.54850
.9100	.414608	2.41192	.9200	.391918	2.55155

RELATIVITY CONTRACTION DILATION FACTOR (Continued)

β	$\sqrt{1-\beta^2}$	$\frac{1}{\sqrt{1-\beta^2}}$	β	$\sqrt{1-\beta^2}$	$\frac{1}{\sqrt{1-\beta^2}}$
.9202	.391449	2.55461	.9302	.367053	2.72440
.9204	.390978	2.55769	.9304	.366546	2.72817
.9206	.390507	2.56077	.9306	.366038	2.73196
.9208	.390035	2.56387	.9308	.365529	2.73576
.9210	.389563	2.56698	.9310	.365019	2.73958
.9212	.389089	2.57010	.9312	.364509	2.74342
.9214	.388616	2.57324	.9314	.363997	2.74727
.9216	.388141	2.57638	.9316	.363485	2.75114
.9218	.387666	2.57954	.9318	.362972	2.75503
.9220	.387190	2.58271	.9320	.362458	2.75894
.9222	.386713	2.58590	.9322	.361944	2.76286
.9224	.386236	2.58909	.9324	.361428	2.76680
.9226	.385758	2.59230	.9326	.360912	2.77076
.9228	.385279	2.59552	.9328	.360394	2.77474
.9230	.384800	2.59875	.9330	.359876	2.77873
.9232	.384320	2.60200	.9332	.359357	2.78274
.9234	.383839	2.60526	.9334	.358838	2.78678
.9236	.383358	2.60853	.9336	.358317	2.79083
.9238	.382875	2.61182	.9338	.357795	2.79489
.9240	.382392	2.61511	.9340	.357273	2.79898
.9242	.381909	2.61843	.9342	.356750	2.80309
.9244	.381424	2.62175	.9344	.356226	2.80721
.9246	.380939	2.62509	.9346	.355700	2.81135
.9248	.380454	2.62844	.9348	.355175	2.81552
.9250	.379967	2.63181	.9350	.354648	2.81970
.9252	.379480	2.63519	.9352	.354120	2.82390
.9254	.378992	2.63858	.9354	.353591	2.82812
.9256	.378503	2.64199	.9356	.353062	2.83237
.9258	.378014	2.64541	.9358	.352531	2.83663
.9260	.377524	2.64884	.9360	.352000	2.84091
.9262	.377033	2.65229	.9362	.351468	2.84521
.9264	.376541	2.65575	.9364	.350935	2.84953
.9266	.376048	2.65923	.9366	.350400	2.85388
.9268	.375555	2.66272	.9368	.349865	2.85824
.9270	.375061	2.66623	.9370	.349329	2.86263
.9272	.374567	2.66975	.9372	.348792	2.86703
.9274	.374071	2.67329	.9374	.348255	2.87146
.9276	.373575	2.67684	.9376	.347716	2.87591
.9278	.373078	2.68040	.9378	.347176	2.88038
.9280	.372580	2.68399	.9380	.346635	2.88488
.9282	.372082	2.68758	.9382	.346094	2.88939
.9284	.371582	2.69119	.9384	.345551	2.89393
.9286	.371082	2.69482	.9386	.345007	2.89849
.9288	.370581	2.69846	.9388	.344463	2.90307
.9290	.370080	2.70212	.9390	.343917	2.90768
.9292	.369577	2.70579	.9392	.343371	2.91231
.9294	.369074	2.70948	.9394	.342823	2.91696
.9296	.368570	2.71319	.9396	.342275	2.92163
.9298	.368065	2.71691	.9398	.341725	2.92633
.9300	.367560	2.72065	.9400	.341174	2.93105

RELATIVITY CONTRACTION DILATION FACTOR (Continued)

β	$\sqrt{1-\beta^2}$	$\frac{1}{\sqrt{1-\beta^2}}$	β	$\sqrt{1-\beta^2}$	$\frac{1}{\sqrt{1-\beta^2}}$
.9402	.340623	2.93580	.9502	.311641	3.20882
.9404	.340070	2.94057	.9504	.311030	3.21512
.9406	.339517	2.94536	.9506	.310418	3.22146
.9408	.338962	2.95018	.9508	.309805	3.22783
.9410	.338407	2.95503	.9510	.309191	3.23425
.9412	.337850	2.95989	.9512	.308575	3.24070
.9414	.337292	2.96479	.9514	.307958	3.24720
.9416	.336733	2.96971	.9516	.307339	3.25373
.9418	.336174	2.97465	.9518	.306719	3.26031
.9420	.335613	2.97962	.9520	.306098	3.26693
.9422	.335051	2.98462	.9522	.305475	3.27359
.9424	.334488	2.98964	.9524	.304851	3.28029
.9426	.333924	2.99469	.9526	.304226	3.28703
.9428	.333359	2.99977	.9528	.303599	3.29382
.9430	.332793	3.00487	.9530	.302970	3.30065
.9432	.332225	3.01000	.9532	.302340	3.30753
.9434	.331657	3.01516	.9534	.301709	3.31445
.9436	.331088	3.02035	.9536	.301076	3.32142
.9438	.330517	3.02556	.9538	.300442	3.32843
.9440	.329945	3.03080	.9540	.299807	3.33548
.9442	.329373	3.03607	.9542	.299169	3.34259
.9444	.328799	3.04137	.9544	.298531	3.34974
.9446	.328224	3.04670	.9546	.297891	3.35694
.9448	.327648	3.05206	.9548	.297249	3.36418
.9450	.327070	3.05745	.9550	.296606	3.37148
.9452	.326492	3.06286	.9552	.295961	3.37882
.9454	.325912	3.06831	.9554	.295315	3.38622
.9456	.325332	3.07379	.9556	.294667	3.39366
.9458	.324750	3.07929	.9558	.294018	3.40116
.9460	.324167	3.08483	.9560	.293367	3.40870
.9462	.323582	3.09040	.9562	.292714	3.41630
.9464	.322997	3.09600	.9564	.292060	3.42395
.9466	.322410	3.10164	.9566	.291404	3.43166
.9468	.321823	3.10730	.9568	.290747	3.43942
.9470	.321234	3.11300	.9570	.290088	3.44723
.9472	.320643	3.11873	.9572	.289427	3.45510
.9474	.320052	3.12449	.9574	.288765	3.46302
.9476	.319459	3.13029	.9576	.288101	3.47100
.9478	.318865	3.13612	.9578	.287435	3.47904
.9480	.318270	3.14198	.9580	.286768	3.48714
.9482	.317674	3.14788	.9582	.286099	3.49529
.9484	.317076	3.15381	.9584	.285429	3.50350
.9486	.316478	3.15978	.9586	.284756	3.51178
.9488	.315877	3.16578	.9588	.284082	3.52011
.9490	.315276	3.17182	.9590	.283406	3.52851
.9492	.314673	3.17790	.9592	.282728	3.53696
.9494	.314069	3.18401	.9594	.282049	3.54548
.9496	.313464	3.19016	.9596	.281368	3.55407
.9498	.312858	3.19634	.9598	.280685	3.56272
.9500	.312250	3.20256	.9600	.280000	3.57143

RELATIVITY CONTRACTION DILATION FACTOR (Continued)

β	$\sqrt{1-\beta^2}$	$\frac{1}{\sqrt{1-\beta^2}}$	β	$\sqrt{1-\beta^2}$	$\frac{1}{\sqrt{1-\beta^2}}$
.9602	.279313	3.58021	.9702	.242306	4.12702
.9604	.278625	3.58905	.9704	.241503	4.14073
.9606	.277935	3.59797	.9706	.240698	4.15458
.9608	.277242	3.60695	.9708	.239890	4.16857
.9610	.276548	3.61600	.9710	.239079	4.18271
.9612	.275852	3.62513	.9712	.238266	4.19699
.9614	.275155	3.63432	.9714	.237449	4.21143
.9616	.274455	3.64359	.9716	.236629	4.22602
.9618	.273753	3.65293	.9718	.235807	4.24076
.9620	.273049	3.66234	.9720	.234981	4.25567
.9622	.272344	3.67183	.9722	.234152	4.27073
.9624	.271636	3.68139	.9724	.233320	4.28596
.9626	.270927	3.69104	.9726	.232485	4.30135
.9628	.270215	3.70076	.9728	.231647	4.31692
.9630	.269501	3.71056	.9730	.230805	4.33266
.9632	.268786	3.72044	.9732	.229960	4.34858
.9634	.268068	3.73040	.9734	.229112	4.36467
.9636	.267348	3.74044	.9736	.228261	4.38095
.9638	.266626	3.75057	.9738	.227406	4.39742
.9640	.265902	3.76078	.9740	.226548	4.41408
.9642	.265176	3.77108	.9742	.225686	4.43093
.9644	.264448	3.78146	.9744	.224821	4.44798
.9646	.263717	3.79194	.9746	.223953	4.46523
.9648	.262985	3.80250	.9748	.223081	4.48268
.9650	.262250	3.81316	.9750	.222205	4.50035
.9652	.261513	3.82390	.9752	.221325	4.51823
.9654	.260774	3.83474	.9754	.220442	4.53633
.9656	.260032	3.84568	.9756	.219556	4.55466
.9658	.259288	3.85671	.9758	.218665	4.57321
.9660	.258542	3.86784	.9760	.217771	4.59199
.9662	.257794	3.87907	.9762	.216872	4.61101
.9664	.257043	3.89040	.9764	.215970	4.63027
.9666	.256290	3.90183	.9766	.215064	4.64978
.9668	.255534	3.91337	.9768	.214154	4.66955
.9670	.254776	3.92501	.9770	.213239	4.68957
.9672	.254016	3.93676	.9772	.212321	4.70985
.9674	.253253	3.94862	.9774	.211398	4.73041
.9676	.252488	3.96058	.9776	.210471	4.75124
.9678	.251720	3.97266	.9778	.209540	4.77235
.9680	.250950	3.98485	.9780	.208605	4.79375
.9682	.250177	3.99716	.9782	.207665	4.81545
.9684	.249402	4.00959	.9784	.206721	4.83745
.9686	.248624	4.02213	.9786	.205772	4.85975
.9688	.247844	4.03480	.9788	.204818	4.88237
.9690	.247061	4.04759	.9790	.203860	4.90532
.9692	.246275	4.06050	.9792	.202897	4.92860
.9694	.245487	4.07354	.9794	.201930	4.95222
.9696	.244695	4.08671	.9796	.200957	4.97618
.9698	.243902	4.10002	.9798	.199980	5.00050
.9700	.243105	4.11345	.9800	.198997	5.02519

RELATIVITY CONTRACTION DILATION FACTOR (Continued)

β	$\sqrt{1-\beta^2}$	$\frac{1}{\sqrt{1-\beta^2}}$	β	$\sqrt{1-\beta^2}$	$\frac{1}{\sqrt{1-\beta^2}}$
.9802	.198010	5.05025	.9901	.140364	7.12435
.9804	.197017	5.07569	.9902	.139657	7.16042
.9806	.196019	5.10153	.9903	.138946	7.19706
.9808	.195016	5.12778	.9904	.138231	7.23426
.9810	.194008	5.15443	.9905	.137513	7.27205
.9812	.192994	5.18152	.9906	.136790	7.31045
.9814	.191974	5.20904	.9907	.136064	7.34946
.9816	.190949	5.23701	.9908	.135334	7.38911
.9818	.189918	5.26544	.9909	.134600	7.42941
.9820	.188881	5.29434	.9910	.133862	7.47039
.9822	.187838	5.32373	.9911	.133119	7.51205
.9824	.186789	5.35363	.9912	.132373	7.55442
.9826	.185734	5.38403	.9913	.131622	7.59752
.9828	.184673	5.41497	.9914	.130866	7.64138
.9830	.183606	5.44646	.9915	.130107	7.68600
.9832	.182532	5.47851	.9916	.129342	7.73142
.9834	181451	5.51113	.9917	.128573	7.77766
.9836	.180364	5.54435	.9918	.127800	7.82475
.9838	.179270	5.57819	.9919	.127021	7.87270
.9840	.178168	5.61267	.9920	.126238	7.92155
.9842	.177060	5.64779	.9921	.125450	7.97133
.9844	.175945	5.68359	.9922	.124656	8.02207
.9846	.174822	5.72009	.9923	.123858	8.07379
.9848	.173692	5.75731	.9924	.123054	8.12653
.9850	.172554	5.79528	.9925	.122245	8.18032
9852	.171409	5.83401	.9926	.121430	8.23520
.9854	.170255	5.87354	.9927	.120610	8.29120
.9856	.169094	5.91388	.9928	.119784	8.34837
.9858	.167924	5.95509	.9929	.118952	8.40675
.9860	.166745	5.99717	.9930	.118114	8.46637
.9862	.165558	6.04017	.9931	.117271	8.52729
.9864	.164363	6.08411	.9932	.116421	8.58954
.9866	.163158	6.12904	.9933	.115564	8.65319
.9868	.161944	6.17499	.9934	.114702	8.71828
.9870	.160720	6.22199	.9935	.113832	8.78487
.9872	.159487	6.27010	.9936	.112956	8.85301
.9874	.158244	6.31935	.9937	.112073	8.92277
.9876	.156991	6.36978	.9938	.111183	8.99422
.9878	.155728	6.42146	.9939	.110285	9.06741
.9880	.154454	6.47442	.9940	.109380	9.14243
.9882	.153169	6.52873	.9941	.108467	9.21935
.9884	.151873	6.58444	.9942	.107547	9.29826
.9886	.150566	6.64162	.9943	.106619	9.37923
.9888	.149247	6.70032	.9944	.105682	9.46237
.9890	.147916	6.76062	.9945	.104737	9.54776
.9892	.146572	6.82258	.9946	.103783	9.63552
.9894	.145216	6.88630	.9947	.102820	9.72575
.9896	.143847	6.95185	.9948	.101848	9.81858
.9898	.142464	7.01932	.9949	.100866	9.91412
.9900	.141067	7.08881	.9950	.0998749	10.0125

RELATIVITY CONTRACTION DILATION FACTOR (Continued)

β	$\sqrt{1-\beta^2}$	$\frac{1}{\sqrt{1-\beta^2}}$	β	$\sqrt{1-\beta^2}$	$\frac{1}{\sqrt{1-\beta^2}}$
.995025	.0996255	10.0376	.996275	.0862330	11.5965
.995050	.0993755	10.0628	.996300	.0859436	11.6355
.995075	.0991249	10.0883	.996325	.0856533	11.6750
.995100	.0988736	10.1139	.996350	.0853620	11.7148
.995125	.0986217	10.1398	.996375	.0850697	11.7551
.995150	.0983691	10.1658	.996400	.0847764	11.7957
.995175	.0981158	10.1920	.996425	.0844821	11.8368
.995200	.0978619	10.2185	.996450	.0841867	11.8784
.995225	.0976074	10.2451	.996475	.0838903	11.9203
.995250	.0973521	10.2720	.996500	.0835928	11.9628
.995275	.0970962	10.2991	.996525	.0832942	12.0056
.995300	.0968396	10.3264	.996550	.0829946	12.0490
.995325	.0965823	10.3539	.996575	.0826938	12.0928
.995350	.0963243	10.3816	.996600	.0823920	12.1371
.995375	.0960657	10.4095	.996625	.0820890	12.1819
.995400	.0958063	10.4377	996650	.0817849	12.2272
.995425	.0955462	10.4661	.996675	.0814797	12.2730
.995450	.0952853	10.4948	.996700	.0811733	12.3193
.995475	.0950238	10.5237	.996725	.0808658	12.3662
.995500	.0947615	10.5528	.996750	.0805570	12.4136
.995525	.0944985	10.5822	.996775	.0802471	12.4615
.995550	.0942348	10.6118	.996800	.0799360	12.5100
.995575	.0939703	10.6417	.996825	.0796236	12.5591
.995600	.0937051	10.6718	.996850	.0793100	12.6087
.995625	.0934391	10.7022	.996875	.0789952	12.6590
.995650	.0931723	10.7328	.996900	.0786790	12.7099
.995675	.0929048	10.7637	.996925	.0783616	12.7613
.995700	.0926364	10.7949	.996950	.0780429	12.8135
.995725	.0923673	10.8263	.996975	.0777229	12.8662
.995750	.0920974	10.8581	.997000	.0774016	12.9196
.995775	.0918267	10.8901	.997025	.0770789	12.9737
.995800	.0915552	10.9224	.997050	.0767548	13.0285
.995825	.0912829	10.9550	.997075	.0764293	13.0840
.995850	.0910098	10.9878	.997100	.0761025	13.1402
.995875	.0907358	11.0210	.997125	.0757742	13.1971
.995900	.0904610	11.0545	.997150	.0754445	13.2548
.995925	.0901853	11.0883	.997175	.0751134	13.3132
.995950	.0899088	11.1224	.997200	.0747807	13.3724
.995975	.0896315	11.1568	.997225	.0744466	13.4324
.996000	.0893532	11.1915	.997250	.0741110	13.4933
.996025	.0890741	11.2266	.997275	.0737738	13.5549
.996050	.0887941	11.2620	.997300	.0734351	13.6175
.996075	.0885132	11.2977	.997325	.0730948	13.6809
.996100	.0882315	11.3338	.997350	.0727529	13.7452
.996125	.0879488	11.3703	.997375	.0724093	13.8104
.996150	.0876651	11.4070	.997400	.0720641	13.8765
.996175	.0873806	11.4442	.997425	.0717173	13.9436
.996200	.0870951	11.4817	.997450	.0713687	14.0117
.996225	.0868087	11.5196	.997475	.0710185	14.0808
.996250	.0865213	11.5578	.997500	.0706665	14.1510

RELATIVITY CONTRACTION DILATION FACTOR (Continued)

β	$\sqrt{1-\beta^2}$	$\frac{1}{\sqrt{1-\beta^2}}$	β	$\sqrt{1-\beta^2}$	$\frac{1}{\sqrt{1-\beta^2}}$
.997525	.0703127	14.2222	.998775	.0494823	20.2092
.997550	.0699571	14.2945	.998800	.0489751	20.4185
.997575	.0695997	14.3679	.998825	.0484626	20.6345
.997600	.0692405	14.4424	.998850	.0479445	20.8574
.997625	.0688793	14.5181	.998875	.0474208	21.0878
.997650	.0685163	14.5951	.998900	.0468913	21.3259
.997675	.0681513	14.6732	.998925	.0463556	21.5724
.997700	.0677843	14.7527	.998950	.0458137	21.8275
.997725	.0674153	14.8334	.998975	.0452653	22.0920
.997750	.0670443	14.9155	.999000	.0447102	22.3663
.997775	.0666712	14.9990	.999025	.0441480	22.6511
.997800	.0662960	15.0839	.999050	.0435786	22.9470
.997825	.0659187	15.1702	.999075	.0430017	23.2549
.997850	.0655391	15.2581	.999100	.0424169	23.5755
.997875	.0651574	15.3475	.999125	.0418238	23.9098
.997900	.0647734	15.4384	.999150	.0412223	24.2587
.997925	.0643871	15.5311	.999175	.0406118	24.6234
.997950	.0639984	15.6254	.999200	.0399920	25.0050
.997975	.0636074	15.7214	.999225	.0393624	25.4049
.998000	.0632139	15.8193	.999250	.0387226	25.8247
.998025	.0628180	15.9190	.999275	.0380720	26.2660
.998050	.0624195	16.0206	.999300	.0374100	26.7308
.998075	.0620185	16.1242	.999325	.0367361	27.2211
.998100	.0616149	16.2299	.999350	.0360497	27.7395
.998125	.0612085	16.3376	.999375	.0353498	28.2887
.998150	.0607995	16.4475	.999400	.0346358	28.8718
.998175	.0603877	16.5597	.999425	.0339068	29.4926
.998200	.0599730	16.6742	.999450	.0331617	30.1553
.998225	.0595554	16.7911	.999475	.0323995	30.8647
.998250	.0591349	16.9105	.999500	.0316188	31.6267
.998275	.0587114	17.0325	.999525	.0308184	32.4481
.998300	.0582847	17.1572	.999550	.0299966	33.3371
.998325	.0578549	17.2846	.999575	.0291517	34.3034
.998350	.0574219	17.4150	.999600	.0282814	35.3589
.998375	.0569856	17.5483	.999625	.0273836	36.5183
.998400	.0565459	17.6847	.999650	.0264552	37.7998
.998425	.0561028	17.8244	.999675	.0254930	39.2264
.998450	.0556561	17.9675	.999700	.0244931	40.8279
.998475	.0552057	18.1141	.999725	.0234505	42.6431
.998500	.0547517	18.2643	.999750	.0223593	44.7242
.998525	.0542939	18.4183	.999775	.0212120	47.1431
.998550	.0538321	18.5763	.999800	.0199990	50.0025
.998575	.0533664	18.7384	.999825	.0187075	53.4546
.998600	.0528965	18.9048	.999850	.0173199	57.7372
.998625	.0524224	19.0758	.999875	.0158109	63.2475
.998650	.0519440	19.2515	.999900	.0141418	70.7124
.998675	.0514611	19.4322	.999925	.0122472	81.6512
.998700	.0509736	19.6180	.999950	.0099999	100.001
.998725	.0504814	19.8093	.999975	.0070710	141.422
.998750	.0499844	20.0063	1.000000	.0000000	∞

FACTS AND EXPANSIONS

$$(a \pm b)^2 = a^2 \pm 2ab + b^2.$$
$$(a \pm b)^3 = a^3 \pm 3a^2b + 3ab^2 \pm b^3.$$
$$(a \pm b)^4 = a^4 \pm 4a^3b + 6a^2b^2 \pm 4ab^3 + b^4.$$
$$a^2 - b^2 = (a - b)(a + b).$$
$$a^2 + b^2 = (a + b\sqrt{-1})(a - b\sqrt{-1}).$$
$$a^3 - b^3 = (a - b)(a^2 + ab + b^2).$$
$$a^3 + b^3 = (a + b)(a^2 - ab + b^2).$$
$$a^4 + b^4 = (a^2 + ab\sqrt{2} + b^2)(a^2 - ab\sqrt{2} + b^2).$$
$$a^n - b^n = (a - b)(a^{n-1} + a^{n-2}b + \ldots + b^{n-1}).$$
$$a^n - b^n = (a + b)(a^{n-1} - a^{n-2}b + \ldots - b^{n-1}),$$
for even values of n.
$$a^n + b^n = (a + b)(a^{n-1} - a^{n-2}b + \ldots + b^{n-1}),$$
for odd values of n.
$$a^4 + a^2b^2 + b^4 = (a^2 + ab + b^2)(a^2 - ab + b^2).$$
$$(a + b + c)^2 = a^2 + b^2 + c^2 + 2ab + 2ac + 2bc.$$
$$(a + b + c)^3 = a^3 + b^3 + c^3 + 3a^2(b + c) + 3b^2(a + c) + 3c^2(a + b) + 6abc.$$
$$(a + b + c + d + \ldots)^2 = a^2 + b^2 + c^2 + d^2 + \ldots + 2a(b + c + d + \ldots) + 2b(c + d + \ldots) + 2c(d + \ldots) + \ldots$$

See also under Series.

POWERS AND ROOTS

$$a^x \times a^y = a^{(x+y)}. \qquad a^0 = 1\ [\text{if } a \neq 0] \qquad (ab)^x = a^x b^x.$$

$$\frac{a^x}{a^y} = a^{(x-y)}. \qquad a^{-x} = \frac{1}{a^x}. \qquad \left(\frac{a}{b}\right)^x = \frac{a^x}{b^x}.$$

$$(a^x)^y = a^{xy}. \qquad a^{\frac{1}{x}} = \sqrt[x]{a}. \qquad \sqrt[x]{ab} = \sqrt[x]{a}\,\sqrt[x]{b}.$$

$$\sqrt[x]{\sqrt[y]{a}} = \sqrt[xy]{a}. \qquad a^{\frac{x}{y}} = \sqrt[y]{a^x}. \qquad \sqrt[x]{\frac{a}{b}} = \frac{\sqrt[x]{a}}{\sqrt[x]{b}}.$$

PROPORTION

If $\frac{a}{b} = \frac{c}{d}$, then $\frac{a+b}{b} = \frac{c+d}{d}$,

$$\frac{a-b}{b} = \frac{c-d}{d}, \qquad \frac{a-b}{a+b} = \frac{c-d}{c+d}.$$

*ARITHMETIC PROGRESSION

An arithmetic progression is a sequence of numbers such that each number differs from the previous number by a constant amount, called the *common difference*.

If a_1 is the first term; a_n the nth term; d the common difference; n the number of terms; and s_n the sum of n terms—

$$a_n = a_1 + (n - 1)d, \quad s_n = \frac{n}{2}[a_1 + a_n].$$

$$s_n = \frac{n}{2}[2a_1 + (n - 1)d].$$

The arithmetic mean between a and b is given by $\dfrac{a + b}{2}$.

*GEOMETRIC PROGRESSION

A geometric progression is a sequence of numbers such that each number bears a constant ratio, called the *common ratio*, to the previous number.

If a_1 is the first term; a_n the nth term; r the common ratio; n the number of terms; and s_n the sum of n terms

$$a_n = a_1 r^{n-1}; \; s_n = a_1 \frac{1 - r^n}{1 - r}$$

$$= a_1 \frac{r^n - 1}{r - 1}, \quad r \neq 1.$$

$$= \frac{a_1 - ra_n}{1 - r}$$

$$= \frac{ra_n - a_1}{r - 1}$$

If $|r| < 1$, then the sum of an infinite geometrical progression converges to the limiting value

$$\frac{a_1}{1 - r}, \quad \left[s_\infty = \lim_{n \to \infty} \frac{a_1(1 - r^n)}{1 - r} = \frac{a_1}{1 - r}\right]$$

The geometric mean between a and b is given by $\sqrt{ab}$.

*It is customary to represent a_n by l in a finite progression and refer to it as the last term.

HARMONIC PROGRESSION

A sequence of numbers whose reciprocals form an arithmetic progression is called an harmonic progression. Thus

$$\frac{1}{a_1}, \frac{1}{a_1 + d}, \frac{1}{a_1 + 2d}, \ldots, \frac{1}{a_1 + (n - 1)d}, \ldots,$$

where

$$\frac{1}{a_n} = \frac{1}{a_1 + (n - 1)d}$$

forms an harmonic progression. The harmonic mean between a and b is given by $\dfrac{2ab}{a + b}$.

If A, G, H respectively represent the arithmetic mean, geometric mean, and harmonic mean between a and b, then $G^2 = AH$.

FACTORIALS

$\underline{/n} = n! = e^{-n}n^n\sqrt{2\pi n}$, approximately, known as Stirling's formula
$\log_e n! = n\log_e n - n$, approximately

PERMUTATIONS

If $M = {}_nP_r = P_{n:r}$ denotes the number of permutations of n distinct things taken r at a time,—

$$M = n(n-1)(n-2)\cdots(n-r+1) = \frac{n!}{(n-r)!}$$

COMBINATIONS

If $M = {}_nC_r = C_{n:r} = \binom{n}{r}$ denotes the number of combinations of n distinct things taken r at a time,—

$$M = \frac{n(n-1)(n-2)\cdots(n-r+1)}{r!} = \frac{n!}{r!(n-r)!}$$

By definition $\binom{n}{0} = 1$

QUADRATIC EQUATIONS

Any quadratic equation may be reduced to the form,—

$$ax^2 + bx + c = 0.$$

Then

$$x = \frac{-b \pm \sqrt{b^2 - 4ac}}{2a}.$$

If a, b, and c are real then:
If $b^2 - 4ac$ is positive, the roots are real and unequal;
If $b^2 - 4ac$ is zero, the roots are real and equal;
If $b^2 - 4ac$ is negative, the roots are imaginary and unequal.

CUBIC EQUATIONS

A cubic equation, $y^3 + py^2 + qy + r = 0$ may be reduced to the form,—

$$x^3 + ax + b = 0$$

by substituting for y the value, $x - \frac{p}{3}$. Here

$$a = \tfrac{1}{3}(3q - p^2) \text{ and } b = \tfrac{1}{27}(2p^3 - 9pq + 27r).$$

For solution let,—

$$A = \sqrt[3]{-\frac{b}{2} + \sqrt{\frac{b^2}{4} + \frac{a^3}{27}}}, \qquad B = \sqrt[3]{-\frac{b}{2} - \sqrt{\frac{b^2}{4} + \frac{a^3}{27}}},$$

then the values of x will be given by,

$$x = A + B, \quad -\frac{A+B}{2} + \frac{A-B}{2}\sqrt{-3}, \quad -\frac{A+B}{2} - \frac{A-B}{2}\sqrt{-3}.$$

If p, q, r are real, then:

If $\frac{b^2}{4} + \frac{a^3}{27} > 0$, there will be one real root and two conjugate imaginary roots;

If $\frac{b^2}{4} + \frac{a^3}{27} = 0$, there will be three real roots of which at least two are equal;

If $\frac{b^2}{4} + \frac{a^3}{27} < 0$, there will be three real and unequal roots.

Trignometric Solution of the Cubic Equation

The form $x^3 + ax + b = 0$ with $ab \neq 0$ can always be solved by transforming it to the trignometric identity

$$4\cos^3\theta - 3\cos\theta - \cos(3\theta) \equiv 0.$$

Let $x = m\cos\theta$, then

$$x^3 + ax + b \equiv m^3\cos^3\theta + am\cos\theta + b \equiv 4\cos^3\theta - 3\cos\theta - \cos(3\theta) \equiv 0.$$

Hence

$$\frac{4}{m^3} = -\frac{3}{am} = \frac{-\cos(3\theta)}{b},$$

from which follows that

$$m = 2\sqrt{-\frac{a}{3}}, \quad \cos(3\theta) = \frac{3b}{am}.$$

Any solution θ_1 which satisfies $\cos(3\theta) = \frac{3b}{am}$, will also have the solutions

$$\theta_1 + \frac{2\pi}{3} \quad \text{and} \quad \theta_1 + \frac{4\pi}{3}.$$

The roots of the cubic $x^3 + ax + b = 0$ are

$$2\sqrt{-\frac{a}{3}}\cos\theta_1, \quad 2\sqrt{-\frac{a}{3}}\cos\left(\theta_1 + \frac{2\pi}{3}\right), \quad 2\sqrt{-\frac{a}{3}}\cos\left(\theta_1 + \frac{4\pi}{3}\right).$$

Example where hyperbolic functions are necessary for solution with latter procedure

The roots of the equation $x^3 - x + 2 = 0$ may be found as follows:

Here

$$a = -1, \quad b = 2, \quad m = 2\sqrt{\tfrac{1}{3}} = 1.155$$

$$\cos(3\theta) = \frac{6}{-1.155} = -5.196$$

$$\cos(3\theta) = -\cos(3\theta - \pi) = -\cosh[i(3\theta - \pi)] = -5.196.$$

Using hyperbolic function tables for $\cosh [i(3\theta - \pi)] = 5.196$, it is found that

$$i(3\theta - \pi) = 2.332.$$

Thus

$$3\theta - \pi = -i(2.332).$$
$$3\theta = \pi - i(2.332)$$
$$\theta_1 = \frac{\pi}{3} - i(0.777)$$
$$\theta_1 + \frac{2\pi}{3} = \pi - i(0.777)$$
$$\theta_1 + \frac{4\pi}{3} = \frac{5\pi}{3} - i(0.777)$$

$$\begin{aligned}\cos\theta_1 &= \cos\left[\frac{\pi}{3} - i(0.777)\right]\\ &= \left(\cos\frac{\pi}{3}\right)[\cos i(0.777)] + \left(\sin\frac{\pi}{3}\right)[\sin i(0.777)]\\ &= \left(\cos\frac{\pi}{3}\right)(\cosh 0.777) + i\left(\sin\frac{\pi}{3}\right)(\sinh 0.777)\\ &= (0.5)(1.317) + i(0.866)(0.858) = 0.659 + i(0.743).\end{aligned}$$

Note that

$$\cos\mu = \cosh(i\mu) \quad \text{and} \quad \sin\mu = -i\sinh(i\mu).$$

Similarly

$$\begin{aligned}\cos\left(\theta_1 + \frac{2\pi}{3}\right) &= \cos[\pi - i(0.777)]\\ &= (\cos\pi)(\cosh 0.777) + i(\sin\pi)(\sinh 0.777)\\ &= -1.317,\end{aligned}$$

and

$$\begin{aligned}\cos\left(\theta_1 + \frac{4\pi}{3}\right) &= \cos\left[\frac{5\pi}{3} - i(0.777)\right]\\ &= \left(\cos\frac{5\pi}{3}\right)(\cosh 0.777) + i\left(\sin\frac{5\pi}{3}\right)(\sinh 0.777)\\ &= (0.5)(1.317) - i(0.866)(0.858) = 0.659 - i(0.743).\end{aligned}$$

The required roots are

$$\begin{aligned}1.155[0.659 + i(0.743)] &= 0.760 + i(0.858)\\ (1.155)(-1.317) &= -1.520\\ (1.155)[0.659 - i(0.743)] &= 0.760 - i(0.858).\end{aligned}$$

QUARTIC OR BIQUARDATIC EQUATION

A quartic equation,

$$x^4 + ax^3 + bx^2 + cx + d = 0,$$

has the *resolvent cubic equation*

$$y^3 - by^2 + (ac - 4d)y - a^2d + 4bd - c^2 = 0.$$

Let y be any root of this equation, and

$$R = \sqrt{\frac{a^2}{4} - b + y}.$$

If $R \neq 0$, then let

$$D = \sqrt{\frac{3a^2}{4} - R^2 - 2b + \frac{4ab - 8c - a^3}{4R}}$$

and

$$E = \sqrt{\frac{3a^2}{4} - R^2 - 2b - \frac{4ab - 8c - a^3}{4R}}.$$

If $R = 0$, then let

$$D = \sqrt{\frac{3a^2}{4} - 2b + 2\sqrt{y^2 - 4d}}$$

and

$$E = \sqrt{\frac{3a^2}{4} - 2b - 2\sqrt{y^2 - 4d}}.$$

Then the four roots of the original equation are given by

$$x = -\frac{a}{4} + \frac{R}{2} \pm \frac{D}{2}$$

and

$$x = -\frac{a}{4} - \frac{R}{2} \pm \frac{E}{2}.$$

EQUATION $x^n = c$

Using DeMoivre's theorem:

$$(\cos\theta + i\sin\theta)^n = \cos n\theta + i\sin n\theta;\ i = \sqrt{-1},$$

the equation $x^n = c$ has n roots given by

$$x = \sqrt[n]{c}\left(\cos\frac{2m\pi}{n} + i\sin\frac{2m\pi}{n}\right) \text{ if } c > 0,$$

or

$$x = \sqrt[n]{-c}\left(\cos\frac{(2m+1)\pi}{n} + i\sin\frac{(2m+1)\pi}{n}\right) \text{ if } c < 0,$$

where m takes the n values 0,1,2, . . . $(n - 1)$ giving n roots.

PARTIAL FRACTIONS

This section applies only to rational algebraic fractions with numerator of lower degree than the denominator. Improper fractions can be reduced to proper fractions by long division.

Every fraction may be expressed as the sum of component fractions whose denominators are factors of the denominator of the original fraction.

Let $N(x)$ = numerator, a polynomial of the form

$$n_0 + n_1x + n_2x^2 + \cdots + n_ix^i$$

I. *Non-repeated Linear Factors*

$$\frac{N(x)}{(x-a)G(x)} = \frac{A}{x-a} + \frac{F(x)}{G(x)}$$

$$A = \left[\frac{N(x)}{G(x)}\right]_{x=a}$$

$F(x)$ determined by methods discussed in the following sections.

Example:

$$\frac{x^2+3}{x(x-2)(x^2+2x+4)} = \frac{A}{x} + \frac{B}{x-2} + \frac{F(x)}{x^2+2x+4}$$

$$A = \left[\frac{x^2+3}{(x-2)(x^2+2x+4)}\right]_{x=0} = -\frac{3}{8}$$

$$B = \left[\frac{x^2+3}{x(x^2+2x+4)}\right]_{x=2} = \frac{4+3}{2(4+4+4)} = \frac{7}{24}$$

II. *Repeated Linear Factors*

$$\frac{N(x)}{x^mG(x)} = \frac{A_0}{x^m} + \frac{A_1}{x^{m-1}} + \cdots + \frac{A_{m-1}}{x} + \frac{F(x)}{G(x)}$$

$$F(x) = f_0 + f_1x + f_2x^2 + \cdots, \quad G(x) = g_0 + g_1x + g_2x^2 + \cdots$$

$$A_0 = \frac{n_0}{g_0}, \quad A_1 = \frac{n_1 - A_0g_1}{g_0}, \quad A_2 = \frac{n_2 - A_0g_2 - A_1g_1}{g_0}$$

General term:

$$A_k = \frac{1}{g_0}\left[n_k - \sum_{i=0}^{k-1} A_ig_{k-i}\right]$$

$$*m = 1\begin{cases} f_0 = n_1 - A_0g_1 \\ f_1 = n_2 - A_0g_2 \\ f_j = n_{j+1} - A_0g_{j+1} \end{cases}$$

$$m = 2\begin{cases} f_0 = n_2 - A_0g_2 - A_1g_1 \\ f_1 = n_3 - A_0g_3 - A_1g_2 \\ f_j = n_{j+2} - [A_0g_{j+2} + A_1g_{j+1}] \end{cases}$$

$$m = 3 \begin{cases} f_0 = n_3 - A_0 g_3 - A_1 g_2 - A_2 g_1 \\ f_1 = n_3 - A_0 g_4 - A_1 g_3 - A_2 g_2 \\ f_j = n_{j+3} - [A_0 g_{j+3} + A_1 g_{j+2} + A_2 g_{j+1}] \end{cases}$$

$$\text{any } m\text{: } f_j = n_{m+j} - \sum_{i=0}^{m-1} A_i g_{m+j-i}$$

Example:

$$\frac{x^2 + 1}{x^3(x^2 - 3x + 6)} = \frac{A_0}{x^3} + \frac{A_1}{x^2} + \frac{A_2}{x} + \frac{f_1 x + f_0}{x^2 - 3x + 6}$$

$$A_0 = \frac{1}{6}, \quad A_1 = \frac{0 - (\frac{1}{6})(-3)}{6} = \frac{1}{12},$$

$$A_2 = \frac{1 - (\frac{1}{6})(1) - (\frac{1}{12})(-3)}{6} = \frac{13}{72},$$

$$m = 3 \begin{cases} f_0 = 0 - \frac{1}{6}(0) + \frac{1}{12}(1) - \frac{13}{72}(-3) = \frac{11}{24} \\ f_1 = 0 - \frac{1}{6}(0) - \frac{1}{12}(0) - \frac{13}{72}(1) = -\frac{13}{72} \end{cases}$$

*Note: If $G(x)$ contains linear factors, $F(x)$ may be determined by previous section I.

III. *Repeated Linear Factors*

$$\frac{N(x)}{(x-a)^m G(x)} = \frac{A_0}{(x-a)^m} + \frac{A_1}{(x-a)^{m-1}} + \cdots + \frac{A_{m-1}}{(x-a)} + \frac{F(x)}{G(x)}$$

Change to form $\dfrac{N'(y)}{y^m G'(y)}$ by substitution of $x = y + a$. Resolve into partial fractions in terms of y as described in Section II. Then express in terms of x by substitution $y = x - a$.

Example:

$$\frac{x - 3}{(x-2)^2(x^2 + x + 1)}$$

Let $x - 2 = y$, $x = y + 2$

$$\frac{(y+2) - 3}{y^2[(y+2)^2 + (y+2) + 1]} = \frac{y - 1}{y^2(y^2 + 5y + 7)} = \frac{A_0}{y^2} + \frac{A_1}{y} + \frac{f_1 y + f_0}{y^2 + 5y + 7}$$

$$A_0 = -\frac{1}{7}, \quad A_1 = \frac{1 - (-\frac{1}{7})(5)}{7} = \frac{12}{49},$$

$$m = 2 \begin{cases} f_0 = 0 - (-\frac{1}{7})(1) - (\frac{12}{49})(5) = -\frac{53}{49} \\ f_1 = 0 - (-\frac{1}{7})(0) - (\frac{12}{49})(1) = -\frac{12}{49} \end{cases}$$

$$\therefore \quad \frac{y - 1}{y^2(y^2 + 5y + 7)} = \frac{-\frac{1}{7}}{y^2} + \frac{\frac{12}{49}}{y} + \frac{-\frac{12}{49}y - \frac{53}{49}}{y^2 + 5y + 7}$$

Let $y = x - 2$, then

$$\frac{x - 3}{(x-2)^2(x^2 + x + 1)} = \frac{-\frac{1}{7}}{(x-2)^2} + \frac{\frac{12}{35}}{(x-2)} + \frac{-\frac{12}{49}(x-2) - \frac{53}{49}}{x^2 + x + 1}$$

$$= -\frac{1}{7(x-2)^2} + \frac{12}{35(x-2)} + \frac{-12x - 29}{49(x^2 + x + 1)}$$

IV. *Repeated Linear Factors*

Alternative method of determining coefficients:

$$\frac{N(x)}{(x-a)^m G(x)} = \frac{A_0}{(x-a)^m} + \cdots + \frac{A_k}{(x-a)^{m-k}} + \cdots + \frac{A_{m-1}}{x-a} + \frac{F(x)}{G(x)}$$

$$A_k = \frac{1}{k!}\left\{D_x^k\left[\frac{N(x)}{G(x)}\right]\right\}_{x=a}$$

where D_x^k is the differentiating operator, and the derivative of zero order is defined as:

$$D_x^0 u = u.$$

V. *Factors of Higher Degree*

Factors of higher degree have the corresponding numerators indicated.

$$\frac{N(x)}{(x^2 + h_1 x + h_0) G(x)} = \frac{a_1 x + a_0}{x^2 + h_1 x + h_0} + \frac{F(x)}{G(x)}$$

$$\frac{N(x)}{(x^2 + h_1 x + h_0)^2 G(x)} = \frac{a_1 x + a_0}{(x^2 + h_1 x + h_0)^2} + \frac{b_1 x + b_0}{(x^2 + h_1 x + h_0)} + \frac{F(x)}{G(x)}$$

$$\frac{N(x)}{(x^3 + h_2 x^2 + h_1 x + h_0) G(x)} = \frac{a_2 x^2 + a_1 x + a_0}{x^3 + h_2 x^2 + h_1 x + h_0} + \frac{F(x)}{G(x)}$$

etc.

Problems of this type are determined first by solving for the coefficients due to linear factors as shown above, and then determining the remaining coefficients by the general methods given below.

VI. *General Methods for Evaluating Coefficients*

1. $$\frac{N(x)}{D(x)} = \frac{N(x)}{G(x)H(x)L(x)} = \frac{A(x)}{G(x)} + \frac{B(x)}{H(x)} + \frac{C(x)}{L(x)} + \cdots$$

Multiply both sides of equation by $D(x)$ to clear fractions. Then collect terms, equate like powers of x, and solve the resulting simultaneous equations for the unknown coefficients.

2. Clear fractions as above. Then let x assume certain convenient values ($x = 1, 0, -1, \ldots$). Solve the resulting equations for the unknown coefficients.

3. $$\frac{N(x)}{G(x)H(x)} = \frac{A(x)}{G(x)} + \frac{B(x)}{H(x)}$$

Then

$$\frac{N(x)}{G(x)H(x)} - \frac{A(x)}{G(x)} = \frac{B(x)}{H(x)}$$

If $A(x)$ can be determined, such as by Method I, then $B(x)$ can be found as above.

BASIC CONCEPTS IN ALGEBRA

DR. W. E. DESKINS

I. ALGEBRA OF SETS

1. Intuitively a set is a collection of objects called the elements of the set. Set and set membership are generally accepted as basic, undefined terms used to define and construct mathematical systems.

   The notation $a \in A$ indicates that a is an element of the set A. The notation $a \notin A$ means that a is not a member of A.

   A set is sometimes specified by listing its elements within a set of braces: $\{a\}$ is the set containing only the element a.
2. Set A is a subset of set B provided $a \in A$ implies $a \in B$. This is denoted by $A \subseteq B$. Every set has as a subset the empty or null set, denoted by ϕ, which has no elements.
3. Set A equals set B, written $A = B$, if and only if $A \subseteq B$ and $B \subseteq A$. A is a proper subset of B, sometimes indicated by $A \subset B$, if and only if $A \subseteq B$ and $A \neq B$; then B has at least one element which does not belong to A.
4. The Cartesian product of sets A and B, denoted by $A \times B$, is the set of all ordered pairs (a, b) where $a \in A$ and $b \in B$. A subset R of $A \times A$ is a binary relation on A, and this is an equivalence relation on A provided (i) $(a, a) \in R$ for every $a \in A$, (ii) $(a, b) \in R$ implies $(b, a) \in R$, and (iii) $(a, b) \in R$ and $(b, c) \in R$ imply $(a, c) \in R$. Ordinary equality of numbers, equality of sets, and congruence of plane figures are examples of equivalence relations.
5. A subset F of $A \times B$ is a *function* from A to B provided each element of A appears exactly once as the first element of a pair in F. A function F from A to B is *onto* provided each element of B appears at least once as the second element of a pair in F. It is *one-to-one* provided each element of B appears at most once as the second element of a pair in F. A function from $A \times A$ to A is a *binary operation* on A. Addition and multiplication of ordinary numbers are examples of binary operations.
6. If consideration is restricted to elements and subsets of a particular set I, then I is the universal set.
7. Common binary operations on subsets of I are: $A \cup B$, the union or join of sets A and B, is the set of all elements of I which belong to either A or B or both A and B.

   $A \cap B$, the intersection or meet of sets A and B, is the set of all elements of I which belong to both A and B.

   $A \backslash B$, the difference of sets A and B, is the set of elements of I which belong to A but not B.

   The difference $I \backslash A$ is denoted by A' and called the complement of A (relative to I). Except in dealing with the concept of complementation the use of a universal set is not essential to the above ideas.
8. Some theorems basic to the Algebra of Sets:

   Let A, B, and C be arbitrary subsets of a universal set I.

   (a) (Commutativity) $A \cup B = B \cup A$ and $A \cap B = B \cap A$.

   (b) (Associativity) $(A \cup B) \cup C = A \cup (B \cup C)$ and
   $(A \cap B) \cap C = A \cap (B \cap C)$.

(c) (Distributivity) $A \cap (B \cup C) = (A \cap B) \cup (A \cap C)$ and
$A \cup (B \cap C) = (A \cup B) \cap (A \cup C)$.
(d) (Idempotency) $A \cup A = A \cap A = A$.
(e) Properties of I and ϕ: $A \cap I = A \cup \phi = A$,
$A \cup I = I$, and
$A \cap \phi = \phi$.
(f) $(A \cap B) \cup (A \backslash B) = A$.
(g) $(A \backslash B) \cup B = A \cup B$.
(h) $A \subseteq A \cup B$.
(i) $A \cap B \subseteq A$.
(j) $A \cup B = A$ if and only if $B \subseteq A$.
(k) $A \cap B = A$ if and only if $A \subseteq B$.
(m) $A \backslash B = A \backslash (A \cap B)$.
(n) (DeMorgan's Theorem) $(A \backslash B) \cap (A \backslash C) = A \backslash (B \cup C)$ and
$(A \backslash B) \cup (A \backslash C) = A \backslash (B \cap C)$.
(o) $(A \cup B)' = A' \cap B'$ and $(A \cap B)' = A' \cup B'$.
(p) $A \cup A' = I$ and $A \cap A' = \phi$.

9. A mathematical system S is a set $S = \{E, O, A\}$ where E is a nonempty set of elements, O is a set of relations and operations on E, and A is a set of axioms, postulates, or assumptions concerning the elements of E and O.
10. The Algebra of Sets provides an example of a mathematical system called a Boolean Algebra (or Boolean Ring) which is defined as:
Set E of elements $a, b, c, \ldots$;
Set O of 2 binary operations $\oplus$ and $\otimes$; (Here $a \oplus b$ denotes the image of (a, b) under the binary operation.)
Set A of axioms for all a, b, c of E:

A_1. The binary operations are commutative; i.e.,

$$a \oplus b = b \oplus a \quad \text{and} \quad a \otimes b = b \otimes a.$$

A_2. Each binary operation is distributive over the other; i.e.,

$$a \oplus (b \otimes c) = (a \oplus b) \otimes (a \oplus c) \quad \text{and} \quad a \otimes (b \oplus c) = (a \otimes b) \oplus (a \otimes c).$$

A_3. There exist elements e and z in E such that for each $a \in E$, $a \oplus z = a$ and $a \otimes e = a$.

A_4. For each $a \in E$ there exists an element $a' \in E$ such that $a \otimes a' = e$ and $a \oplus a' = z$.

In the algebra of subsets of a (universal) set I, ϕ plays the role of z, I that of e, $\cup$ that of $\oplus$, and $\cap$ that of $\otimes$.

11. A Boolean Algebra has the Principle of Duality: If the interchanges of $\left\{\begin{matrix}\oplus \text{ and } \otimes \\ e \text{ and } z\end{matrix}\right\}$ are made in a correct statement, then the result is also a correct statement.
12. In addition to the Algrebra of Sets which is a Boolean Algebra, other representations of Boolean Algebra that are interesting of themselves and valuable for their applications are:
(a) The Algebra of Symbolic Logic
(b) The Algebra of Switching Currents

Algebra of Sets	*Binary Operator*	*Symbolic Logic*	*Binary Operator*	*Switching Circuits*	*Binary Operator*
Union of 2 sets	$\cup$	Disjunction of 2 propositions	$\vee$	2 switches in ‖	$+$
Intersection of 2 sets	$\cap$	Conjunction of 2 propositions	$\wedge$	2 switches in series	$\times$
Complement of a set A		Negation of a proposition T, F	$\sim$	on-off, or 1, 0	1 or $^-$

Both in symbolic logic and in switching circuits, we can consider "0" and "1" as the elements, in the former representing "False" and "True"; in the latter, "Off" and "On", satisfying the following "rules":

$$\left.\begin{array}{l}0+0=0\\1+1=1\end{array}\right\}\text{ i.e. } a+a=a$$

$$\left.\begin{array}{l}0\times 0=0\\1\times 1=1\end{array}\right\}\text{ i.e. } a\times a=a$$

$$0'=1$$

$$1'=0$$

$$\left.\begin{array}{l}0+0'=0+1=1\\1+1'=1+0=1\end{array}\right\}\text{ i.e. } a+a'=1$$

$$\left.\begin{array}{l}0\times 0'=0\times 1=0\\1\times 1'=1\times 0=0\end{array}\right\} a\times a'=0$$

In switching circuits:

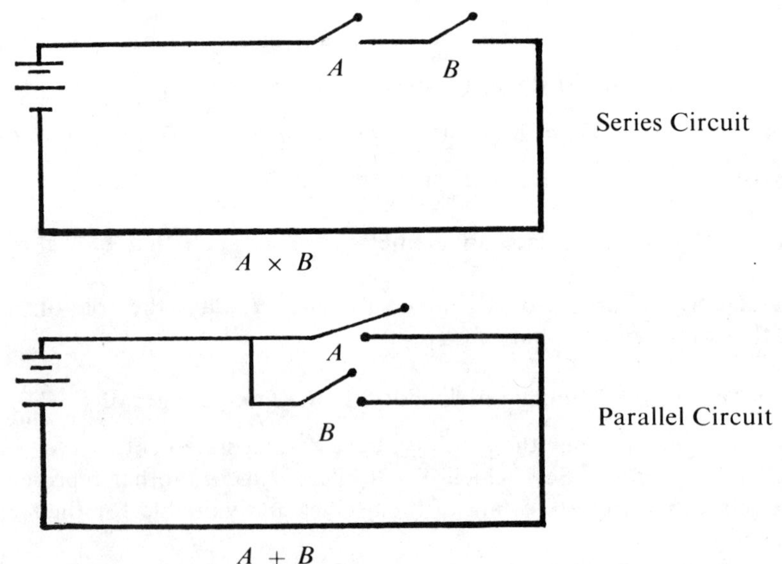

here, "0" represents an open circuit: [open switch] and "1"

$A = 0$

a closed circuit: [closed switch]

$A = 1$

In the Algebra of Symbolic Logic, we use *Truth Tables* to define the operations $\wedge$, $\vee$, $\sim$ as follows:

					Other Operators Used	
p	q	$p \wedge q$	$p \vee q$	$\sim p$	$p \rightarrow q$	$p \leftrightarrow q$
T	T	T	T	F	T	T
T	F	F	T	F	F	F
F	T	F	T	T	T	F
F	F	F	F	T	T	T

13. In order to re-emphasize the use of switching circuits and their relation to truth tables the following is included. Conventionally a "1" represents "True" and a "0" represents "False." The switching circuit symbols are $-$, $\cdot$, $+$, $\rightarrow$, $\equiv$ representing "Not," "And," "Or," "Implies," "Equivalent" respectively and their Truth Table Definitions are

p	q	$p \cdot q$	$p + q$	p	$p \rightarrow q$	$p \equiv q$
0	0	0	0	1	1	1
0	1	0	1	1	1	0
1	0	0	1	0	0	0
1	1	1	1	0	1	1

The comparison with the Algebra of Symbolic Logic being obvious. The "rules" for these circuits are as follows:

$$0 + 0 = 0$$
$$1 + 1 = 1$$
$$1 + 0 = 0 + 1 = 1$$
$$0 \cdot 0 = 0$$
$$1 \cdot 1 = 1$$
$$0 \cdot 1 = 1 \cdot 0 = 0$$
$$\bar{0} = 1$$
$$\bar{1} = 0$$

Mechanical switches or relays are represented by

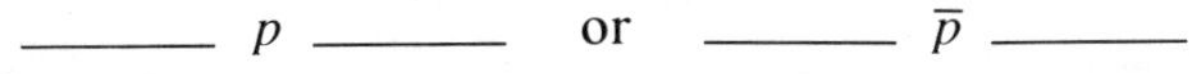

the former indicating that the circuit is closed, i.e. the switch is made, when $p = 1$ and the latter indicating the converse namely that the circuit is closed when $\bar{p} = 1$ or, what amounts to the same thing, when $p = 0$.

Electronic switches or gates are represented by more complex symbols—four in all, three of which are independent and can stand alone

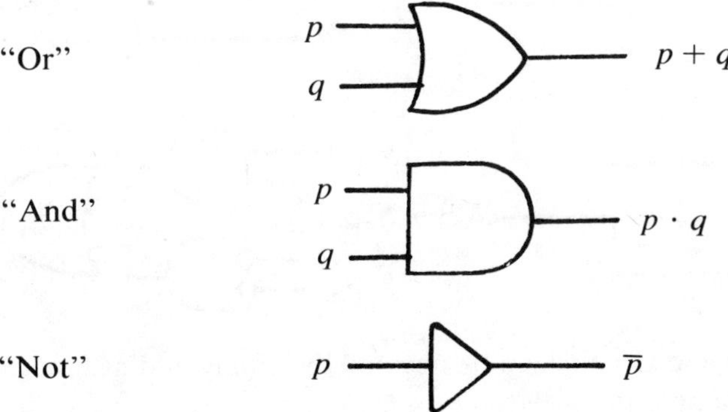

and one which represents the negation of an input or an output and is used with one of the above

"Not" ○

An example of its use on an input line is

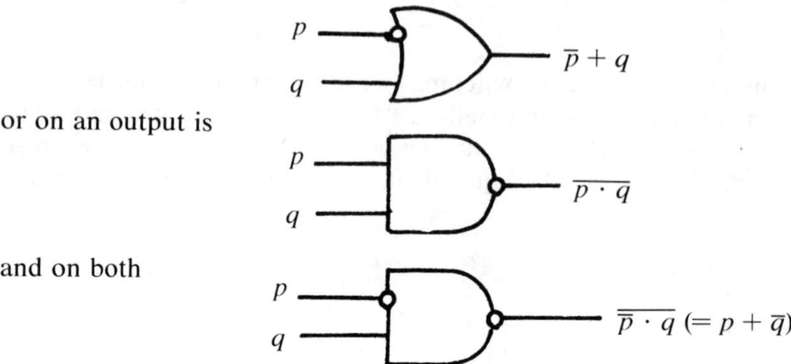

or on an output is

and on both

The basic functions obtained from the two types of switching circuits are

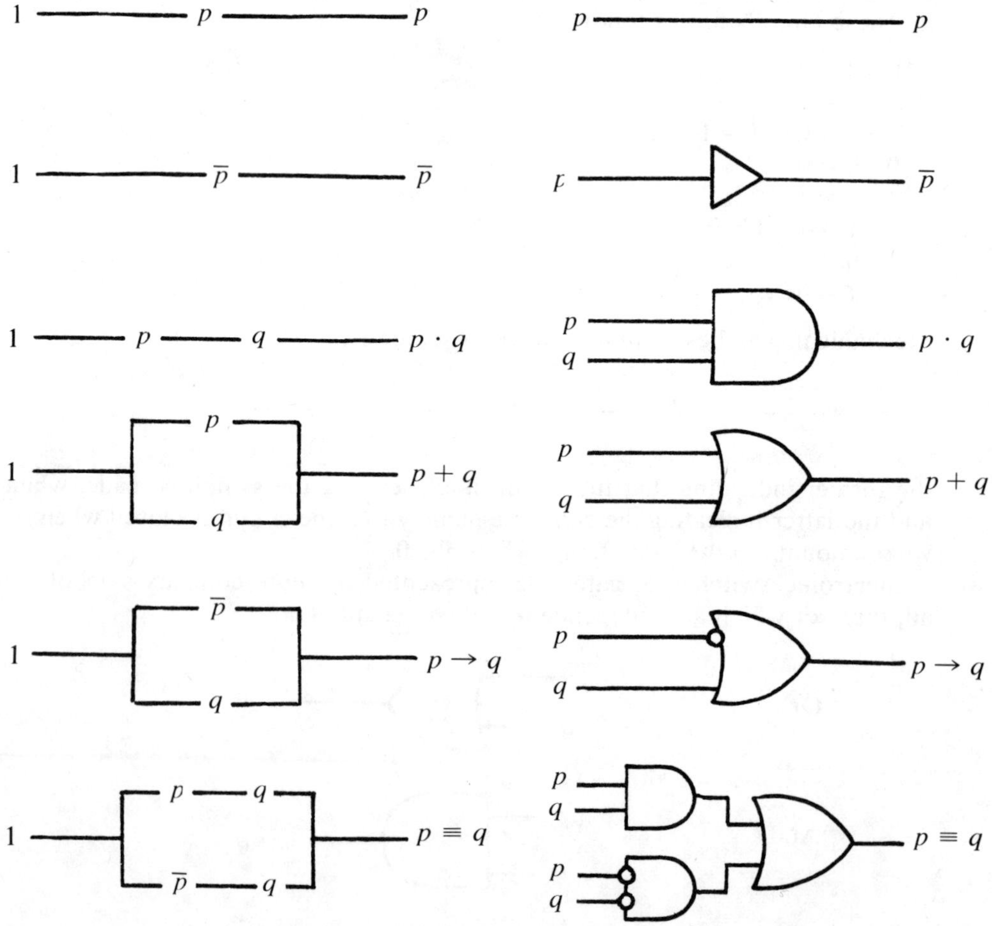

1 —— p —— p p —— p

1 —— $\overline{p}$ —— $\overline{p}$ p —▷— $\overline{p}$

1 —— p —— q —— $p \cdot q$ p, q —D— $p \cdot q$

1 —[p / q]— $p + q$ p, q —OR— $p + q$

1 —[$\overline{p}$ / q]— $p \to q$ p, q —OR— $p \to q$

1 —[p — q / $\overline{p}$ — q]— $p \equiv q$ p, q, p, q — $p \equiv q$

All the above electronic circuits can be negated by simply adding a negating circle to the output as for example in

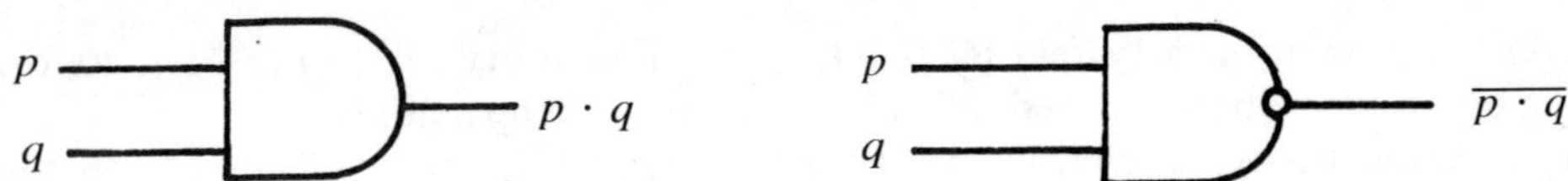

Alternative circuits however, which are direct analogues of their relay switching counterparts, are

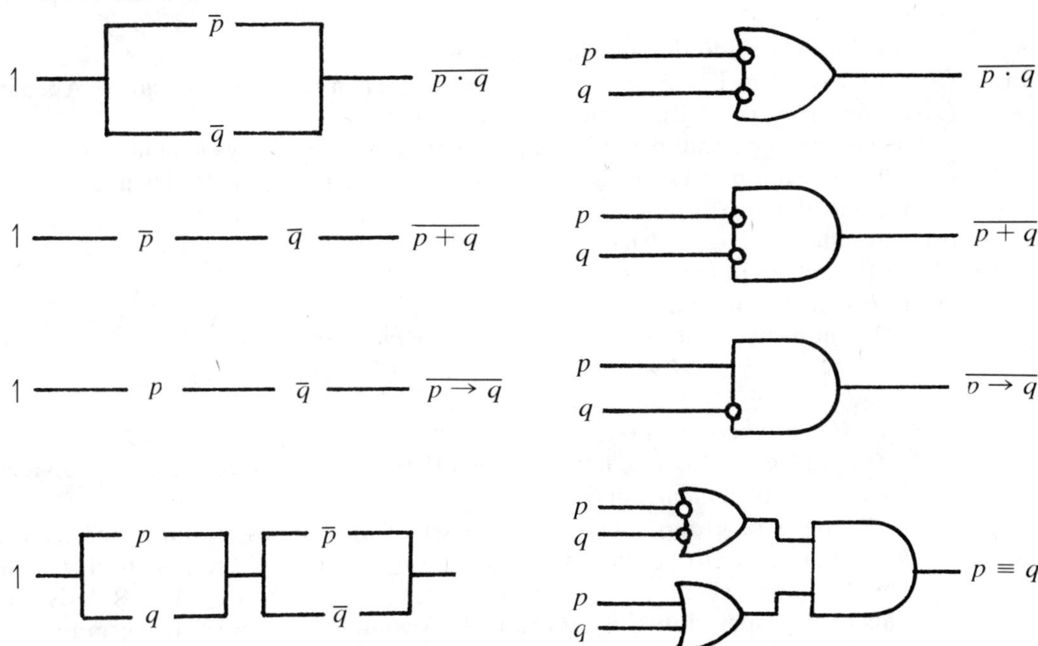

The operation + is sometimes referred to as the "Inclusive Or" and $\not\equiv$ as the "Exclusive Or", the former having the value "True" when both the inputs are "True"—see the truth table. Note that $p \not\equiv q$ is a shorthand for $\overline{p \equiv q}$.

14. Two sets A and B are equivalent (have same cardinal) if and only if there exists a one-to-one correspondence between the elements of the two sets. This is an equivalence relation on the collection of subsets of set I.

 A set is infinite if and only if it is equivalent with a proper subset of itself.

 A set is called countably (denumerably) infinite if it is equivalent with the set of all positive integers. The set of all rational numbers is countably infinite but the set of all real numbers is noncountably infinite. The cardinal of the set of all rational numbers is denoted by (aleph null); the cardinal of the set of reals is denoted by (aleph).

II. ABSTRACT ALGEBRAIC SYSTEMS

1. *Semigroup*. A semigroup is a system $\{S, 0, A\}$; S is a nonempty set $\{a, b, c, \ldots\}$, θ consists of one binary operation on S, denoted by $*$, and A consists of the axiom

 A_1. Associativity: $a * (b * c) = (a * b) * c$ for all $a, b \circ c \in S$.

 Basic Theorem. (*Generalized Associativity*). If $a_1, a_2, \ldots, a_n$ are elements of S then all associations of the n elements yield the same "product". (For example,

$$a * ((b * c) * d) = a * (b * (c * d)) = (a * b) * (c * d), \text{ etc.})$$

2. *Group*. A group is a system $\{G, 0, A\}$; G is a nonempty set $\{a, b, c, \ldots\}$, θ consists of one binary operation denoted by $\circ$, and A consists of the axioms:

A_1. Associativity: $a \circ (b \circ c) = (a \circ b) \circ c$ for all $a, b, c \in G$.

A_2. Identity Element: G contains an element e having the property, $a \circ e = e \circ a = a$ for every $a \in G$.

A_3. Inverse Element: For each $a \in G$ there is an element $a' \in G$ with the property, $a \circ a' = a' \circ a = e$.

If the following additional axiom belongs to A,

A_4. Commutativity: $a \circ b = b \circ a$ for all $a, b \in G$. Then the group is called Abelian (after Niels Henrik Abel). Some basic theorems:

(a) The element e (Axiom A_2) is unique. Then e is *the* identity element of G.

(b) The element a' (Axiom A_3) is unique for each $a \in G$. Then a' is *the* inverse of a in G.

(c) The equation $a \circ x = b$ has a unique solution in G, viz., $x = a' \circ b$.

(d) $(a')' = a$ and $(a \circ b)' = b' \circ a'$.

(e) $a \circ b = a \circ c$ if and only if $b = c$.

If a nonempty subset H of G satisfies the two conditions:

H_1. $a \circ b \in H$ whenever $a, b \in H$. (Closure)

H_2. $a \in H$ if and only if $a' \in H$.

then H is a *subgroup* of G.

(Lagrange). If G is a finite set then the number of elements in H divides the number of elements in G.

Example of group. Let G be the set of all one-to-one functions from a nonempty S onto itself. For any $f, g \in G$, define the function $f \circ g$ as the function which maps s onto $f(g(s))$, for each $s \in S$. Relative to this binary operation G is a group, the *symmetric group* of all permutations on S.

Each group is essentially a subgroup of the symmetric group of some set S.

3. *Ring*. A ring is a system $\{R, \theta, A\}$; R is a nonempty set $\{a, b, c, \ldots\}$, θ consists of two binary operations denoted by $+$ and $\times$, and A consists of the axioms:

A_0. Relative to addition (i.e., $+$) R is an Abelian group in which the identity element is denoted by z and the inverse of a is denoted by $-a$.

M_0. Relative to multiplication (i.e., $\times$) R is a semigroup.

D_1. Left distributive: $a \times (b + c) = (a \times b) + (a \times c)$, all $a, b, c \in R$.

D_2. Right distributive: $(b + c)a = (b \times a) + (c \times a)$, all $a, b, c \in R$.

EXAMPLE 1. The set of all integers (whole numbers) and ordinary addition and multiplication.

EXAMPLE 2. The set of all real functions continuous on the interval $0 \leq y \leq 1$, with addition and multiplication defined by $(f + g)(y) = f(y) + g(y)$, sum of real numbers, and $(f \times g)(y) = f(y) \times g(y)$, product of real numbers.

Special types of rings have been studied extensively.

3.1 *Integral Domain*. An integral domain is a ring R in which multiplication $(\times)$ satisfies the additional assumptions:

M_1. Commutativity: $a \times b = b \times a$ for all a and b in R.

M_2. Multiplicative identity: R contains an element $e \neq z$ with the property $a \times e = e \times a = a$ for all a in R.

M_3. Cancellation: $a \times b = a \times c$ if and only if $b = c$.

An element u of integral domain R is a *unit* provided R contains v such that $u \times v = e$.

An element p of integral domain R is a *prime* (irreducible element) provided $p = a \times b$ implies that exactly one of the elements a or b is a unit.

The elements of integral domain R which differ from z and are neither units nor primes are *composites*.

In some integral domains (such as the ring of integers) each composite can be factored uniquely (up to unit factors) as the product of a finite set of primes. However in the integral domain of all entire functions this is not true.

3.2 *Field*. A field is an integral domain in which every element except z is a unit. In other words, the non-z elements form an Abelian group relative to multiplication ($\times$).

EXAMPLE 1. The rational field consisting of ordinary fractions, addition, and multiplication.

EXAMPLE 2. The set of all real numbers $a + b\sqrt{2}$, a and b rational. Then

$$(a + b\sqrt{2}) + (c + d\sqrt{2}) = (a + c) + (b + d)\sqrt{2} \text{ and}$$
$$(a + b\sqrt{2}) \times (c + d\sqrt{2}) = (ac + 2bd) + (ad + bc)\sqrt{2}.$$

Besides these well-known examples there exist finite fields (sometimes called Galois fields).

EXAMPLE 3. Let p be a prime integer. Denote by $GF(p)$ the p integers $0, 1, \ldots, p - 1$. Define addition($\oplus$) of two of these elements a and b as the remainder of $a + b$ (ordinary addition) after division by p. (Thus $1 \oplus (p - 1) = 0$.)

Define $a \otimes b$, the product, to be the remainder of ab (ordinary multiplication) after division by p. (Thus, when $p = 3$, $2 \otimes 2 = 1$.) The resulting system $\{GF(p), \oplus, \otimes\}$ is the (modular) field of integers modulo p.

3.3 *Skew Field or Division Ring*. A skew field is a ring in which the non-z elements form a group relative to multiplication ($\times$).

The classical example of a skew field is the ring of real *quaternions*, first described by W. R. Hamilton. A quaternion is expressible in the form $ae + bi + cj + dk$ where a, b, c, and d are real numbers and e, i, j, and k are elements which commute with all real numbers and multiply as follows:

$$e \times e = e,\quad e \times i = i \times e = i,\quad e \times j = j \times e = j,\quad e \times k = k \times e = k;$$
$$i \times i = -e,\quad i \times j = k,\quad j \times i = -k,\quad i \times k = -j,\quad k \times i = j,$$
$$j \times j = -e,\quad j \times k = i,\quad k \times j = -i,\quad k \times k = -e.$$

These elements distribute over addition. e is generally identified with and written as the real number 1.

3.4 *Matric Ring*. The matric ring $M_n(R)$ over the ring R, where n is a positive integer, consists of all doubly-ordered sets of n^2 elements of R, written as an array

$$\begin{pmatrix} a_{1,1} & a_{1,2} \cdots a_{1,n} \\ a_{2,1} & a_{2,2} \cdots a_{2,n} \\ \vdots & \\ a_{n,1} & \cdots \quad a_{n,n} \end{pmatrix} = (a_{i,j})$$

with addition and multiplication defined as follows:

$$(a_{i,j}) + (b_{i,j}) = (a_{i,j} + b_{i,j})$$
$$(a_{i,j}) \times (b_{i,j}) = (c_{i,j})$$
where
$$c_{i,j} = \sum_{k=1}^{n} a_{i,k}\, b_{k,j}, \quad i = 1,\ldots,n \quad \text{and} \quad j = 1,\ldots,n.$$

If $n > 1$, then multiplication is noncommutative in general; i.e., $(a_{i,j}) \times (b_{i,j})$ can differ from $(b_{i,j}) \times (a_{i,j})$. Moreover, the product of two nonzero matrices can be the zero matrix (which consists of only the element z in all n^2 positions).

A similar useful method for forming a new ring from a known ring utilizes sequences.

3.5 *Power Series and Polynomial Ring.* Let R be a ring in which multiplication $(\times)$ is commutative. The set $PS(R)$ of all sequences $(a_0, a_1, \ldots)$ with $a_i \in R$ is the power series ring of R, with addition and multiplication defined as

$$(a_0, a_1, \ldots) \oplus (b_0, b_1, \ldots) = (a_0 + b_0, \quad a_1 + b_1, \ldots) \text{ and}$$
$$(a_0, a_1, \ldots) \oplus (b_0, b_1, \ldots) = (c_0, c_1, \ldots)$$

where

$$c_0 = a_0 \times b_0, \quad c_1 = a_0 \times b_1 + a_1 \times b_0, \ldots, \text{and,}$$

generally,

$$c_n = a_0 \times b_n + a_1 \times b_{n-1} + \cdots + a_n \times b_0.$$

The subset $P(R)$ of $PS(R)$ consisting of those sequences $(a_0, a_1, \ldots)$ in which at most only finitely many of the a_i differ from z, form a ring relative to the addition and multiplication just defined. This ring $\{P(R), \oplus, \otimes\}$, is the polynomial ring of R.

Some theorems for rings, fields, etc.

(a) In a ring R, if $a = b$ and $c = d$, then $a + c = b + d$ and $a \times c = b \times d$.
(b) In a ring R, $-(-a) = a$; $(-a) \times b = a \times (-b) = -(a \times b)$; and $(-a) \times (-b) = a \times b$, for all $a,b \in R$.
(c) In a ring R, $a \times z = z \times a = z$, for all $a \in R$.
(d) In a ring R the equation $a + x = b$ has a unique solution, viz., $x = -a + b$.
(f) In a field, skew field, or integral domain, $a \times b = z$ if and only if a and/or b equals z.
(g) A finite integral domain is a field.
(h) The polynomial ring of an integral domain is also an integral domain.
(i) The power series ring of an integral domain is also an integral domain.
(j) A ring is a field provided it is both an integral domain and a skew field.
(k) If R is a (skew) field, then the equation $a \times y = b$, $a \neq z$, has a unique solution $y = a' \times b$.
(l) The polynomial ring and the power series ring of a field are unique factorization domains.

4. *Vector Space.* A vector space $V(F)$ over a field F consists of a nonempty set V (the vectors), a binary operation $(\oplus)$ on V, a function (called *scalar multiplication*) from the product set $F \times V$ onto V with the image of (a,v) denoted by $a \circ v$, and the following axioms:

A_0. Relative to addition $(\oplus)$ V is an Abelian group in which the identity element (vector) is denoted by z and the inverse of v is denoted by $-v$.

M_1. $a \circ (b \circ v) = (ab) \circ v$ for all $a,b \in F$ and $v \in F$. (Here ab denotes the product of a and b in F.)

M_2. $1 \circ v = v$ for all $v \in V$. (Here 1 denotes the multiplicative indentity element of F.)

D_1. $a \circ (\mu \oplus \nu) = (a \circ \mu) \oplus (a \circ \nu)$ for all $a \in F$, $\mu, \nu \in V$.

D_2. $(a + b) \circ \nu = (a \circ \nu) \oplus (b \circ \nu)$ for all $a,b \in F$, $\nu \in V$. (Here + denotes addition in the field F.)

The elements of F are referred to as *scalars*.

EXAMPLE 1. The polynomial ring $P(F)$ of a field F is a vector space over F. In this example scalar multiplication is a special case of the multiplication defined for $P(F)$.

EXAMPLE 2. Denote by $C_n(F)$ the set of all n-tuples, $(a_1, a_2, \ldots, a_n)$, n a positive integer, with all $a_i \in F$. Define

$$(a_1, \ldots, a_n) \oplus (b_1, \ldots, b_n) = (a_1 + b_1, \ldots, a_n + b_n) \quad \text{and}$$
$$c \circ (a_1, \ldots, a_n) = (c \times a_1, \ldots, c \times a_n),$$

where + and $\times$ denote the addition and multiplication, respectively, of the field F. Relative to $\oplus$ and $\circ$, $C_n(F)$ is a vector space, called the *n-dimensional coordinate space* over F.

A vector space $V(F)$ is *n-dimensional* over F provided V contains n elements ν_1, ν_2, $\ldots$, ν_n such that each element $\nu \in V$ is uniquely expressible in the form

$$\nu = a_1 \circ \nu_1 \oplus a_2 \circ \nu_2 \oplus \cdots \oplus a_n \circ \nu_n$$

for some $a_1, a_2, \ldots, a_n \in F$.

Two vector spaces $V(F)$ and $W(F)$ over the field of scalars F are *isomorphic* provided there is a one-to-one correspondence between the elements of V and the elements of W which is preserved under the arithmetic of the two spaces.

Basic Theorem. An n-dimensional vector space $V(F)$ is isomorphic with the coordinate space $C_n(F)$ (of Example 2, above).

MATRICES AND DETERMINANTS

Dr. R. E. Bargmann

1. GENERAL DEFINITIONS

1.1. A matrix is an array of numbers, consisting of m rows and n columns. It is usually denoted by a bold-face capital letter, e.g.,

$$\mathbf{A} \qquad \boldsymbol{\Sigma} \qquad \mathbf{M}$$

1.2. The (i, j) element of a matrix is the element occurring in row i and column j. It is usually denoted by a lower-case letter with subscripts, e.g.,

$$a_{ij} \qquad \sigma_{ij} \qquad m_{ij}$$

Exceptions to this convention will be stated where required.

1.3. A matrix is called rectangular if m (number of rows) $\neq n$ (number of columns).

1.4. A matrix is called square if $m = n$.

1.5a. In the transpose of a matrix $\mathbf{A}$, denoted by $\mathbf{A}'$, the element in the j'th row and i'th column of $\mathbf{A}$ is equal to the element in the i'th row and j'th column of $\mathbf{A}'$. Formally $(\mathbf{A}')_{ij} = (\mathbf{A})_{ji}$ where the symbol $(\mathbf{A}')_{ij}$ denotes the (i, j) element of $\mathbf{A}'$.

1.5b. The Hermitian conjugate of a matrix $\mathbf{A}$, denoted by $\mathbf{A}^{\mathrm{H}}$ or $\mathbf{A}^{\dagger}$ is obtained by transposing $\mathbf{A}$ and replacing each element by its conjugate complex. Hence if

$$a_{kl} = u_{kl} + iv_{kl}$$

then

$$(\mathbf{A}^{\mathrm{H}})_{kl} = u_{lk} - iv_{lk}$$

where typical elements have been denoted by (k, l) to avoid confusion with $i = \sqrt{-1}$.

1.6a. A square matrix is called symmetric if $\mathbf{A} = \mathbf{A}'$.

1.6b. A square matrix is called Hermitian if $\mathbf{A} = \mathbf{A}^{\mathrm{H}}$.

1.7. A matrix with m rows and 1 column is called a column vector and is usually denoted by bold faced, lower-case letters, e.g.,

$$\boldsymbol{\beta} \qquad \mathbf{x} \qquad \mathbf{a}$$

1.8. A matrix with one row and n columns is called a row vector and is usually denoted by a primed, bold faced, lower-case letter, e.g.,

$$\mathbf{a}' \qquad \mathbf{c}' \qquad \boldsymbol{\mu}'$$

1.9. A matrix with one row and one column is called a scalar, and is usually denoted by a lower-case letter, occasionally italicized.

1.10. The diagonal extending from upper left (NW) to lower right (SE) is called the principal diagonal of a square matrix.

1.11a. A matrix with all elements above the principal diagonal equal to zero is called a lower triangular matrix.

Example

$$\mathbf{T} = \begin{bmatrix} t_{11} & 0 & 0 \\ t_{21} & t_{22} & 0 \\ t_{31} & t_{32} & t_{33} \end{bmatrix} \text{ is lower triangular}$$

1.11b. The transpose of a lower triangular matrix is called an upper triangular matrix.

1.12. A square matrix with all off-diagonal elements equal to zero is called a diagonal matrix, denoted by the letter **D** with subscript indicating the typical element in the principal diagonal.

Example

$$\mathbf{D}_a = \begin{bmatrix} a_1 & 0 & 0 \\ 0 & a_2 & 0 \\ 0 & 0 & a_3 \end{bmatrix} \text{ is diagonal}$$

2. ADDITION, SUBTRACTION, AND MULTIPLICATION

2.1. Two matrices **A** and **B** can be added (subtracted) if the number of rows (columns) in **A** equals the number of rows (columns) in **B**.

$$\mathbf{A} \pm \mathbf{B} = \mathbf{C}$$

implies

$$a_{ij} \pm b_{ij} = c_{ij} \qquad \begin{aligned} i &= 1, 2, \ldots m \\ j &= 1, 2, \ldots n \end{aligned}$$

2.2. Multiplication of a matrix or vector by a scalar implies multiplication of each element by the scalar. If

$$\mathbf{B} = \gamma \mathbf{A}$$

then

$$b_{ij} = \gamma a_{ij}$$

for all elements.

2.3a. Two matrices, **A** and **B**, can be multiplied if the number of columns in **A** equals the number of rows in **B**.

2.3b. Let **A** be of order $(m \times n)$ (have m rows and n columns) and **B** of order $(n \times p)$. Then the product of two matrices **C** = **AB**, is a matrix of order $(m \times p)$ with elements

$$c_{ij} = \sum_{k=1}^{n} a_{ik} b_{kj}$$

This states that c_{ij} is the scalar product of the i'th row vector of **A** and the j'th column vector of **B**.

Example

$$\begin{bmatrix} 3 & 4 & 2 \\ 2 & 3 & -1 \end{bmatrix} \begin{bmatrix} 1 & -2 & -4 \\ 0 & -1 & 2 \\ 6 & -3 & 9 \end{bmatrix} = \begin{bmatrix} 15 & -16 & 14 \\ -4 & -4 & -11 \end{bmatrix}$$

e.g.,

$$c_{23} = [2 \quad 3 \quad -1] \begin{bmatrix} -4 \\ 2 \\ 9 \end{bmatrix}$$

$$= 2 \times (-4) + 3 \times 2 + (-1) \times 9 = -11$$

2.3c. In general, matrix multiplication is not commutative

$$\mathbf{AB} \neq \mathbf{BA}$$

2.3d. Matrix multiplication is associative

$$\mathbf{A}(\mathbf{BC}) = (\mathbf{AB})\mathbf{C}$$

2.3e. The distributive law for multiplication and addition holds as in the case of scalars,

$$(\mathbf{A} + \mathbf{B})\mathbf{C} = \mathbf{AC} + \mathbf{BC}$$
$$\mathbf{C}(\mathbf{A} + \mathbf{B}) = \mathbf{CA} + \mathbf{CB}$$

2.4. In some applications, the term-by-term product of two matrices **A** and **B** of identical order is defined as

$$\mathbf{C} = \mathbf{A} * \mathbf{B}$$

where

$$c_{ij} = a_{ij}b_{ij}$$

2.5. $(\mathbf{ABC})' = \mathbf{C}'\mathbf{B}'\mathbf{A}'$
2.6. $(\mathbf{ABC})^{\mathrm{H}} = \mathbf{C}^{\mathrm{H}}\mathbf{B}^{\mathrm{H}}\mathbf{A}^{\mathrm{H}}$
2.7. If both **A** and **B** are symmetric, then $(\mathbf{AB})' = \mathbf{BA}$. Note that the product of two symmetric matrices is generally not symmetric.

3. RECOGNITION RULES AND SPECIAL FORMS

3.1. A column (row) vector with all elements equal to zero is called a null vector, and usually denoted by the symbol **0**.
3.2. A null matrix has all elements equal to zero.
3.3a. A diagonal matrix with all elements equal to one in the principal diagonal is called the identity matrix **I**.
3.3b. $\gamma\mathbf{I}$, i.e., a diagonal matrix with all diagonal elements equal to a constant γ, is called a scalar matrix.
3.4. A matrix which has only one element equal to one and all others equal to zero is called an elementary matrix $(\mathbf{EL})_{ij}$.

Example

$$(\mathbf{EL})_{23} = \begin{bmatrix} 0 & 0 & 0 & 0 & 0 \\ 0 & 0 & 1 & 0 & 0 \\ 0 & 0 & 0 & 0 & 0 \\ 0 & 0 & 0 & 0 & 0 \end{bmatrix}$$

The order of the matrix is usually implicit.
3.5a. The symbol **j** is reserved for a column vector with all elements equal to 1.
3.5b. The symbol **j**′ is reserved for a row vector with all elements equal to 1.
3.6. An expression ending with a column vector is a column vector.

Example

$$\mathbf{ABx} = \mathbf{y}$$

(It is assumed that rule 2.3a is satisfied, else matrix multiplication would not be defined.)
3.7. An expression beginning with a row vector is a row vector.

Example

$$\mathbf{y}'(\mathbf{A} + \mathbf{BC}) = \mathbf{d}'$$

3.8. An expression beginning with a row vector and ending with a column vector, is a scalar.

Example

$$\mathbf{a}'\mathbf{B}\mathbf{c} = \gamma$$

3.9a. If $\mathbf{Q}$ is a square matrix, the scalar $\mathbf{x}'\mathbf{Q}\mathbf{x}$ is called a quadratic form. If $\mathbf{Q}$ is non-symmetric, one can always find a symmetric matrix $\mathbf{Q}^*$ such that

$$\mathbf{x}'\mathbf{Q}\mathbf{x} = \mathbf{x}'\mathbf{Q}^*\mathbf{x}$$

where

$$(\mathbf{Q}^*)_{ij} = \tfrac{1}{2}(q_{ij} + q_{ji})$$

3.9b. If $\mathbf{Q}$ is a square matrix the scalar $\mathbf{x}^H\mathbf{Q}\mathbf{x}$ is called a Hermitian form.

3.10. A scalar $\mathbf{x}'\mathbf{Q}\mathbf{y}$ is called a bilinear form.

3.11. The scalar $\mathbf{x}'\mathbf{x} = \Sigma x_i^2$, i.e., the sum of squares of all elements of $\mathbf{x}$.

3.12. The scalar $\mathbf{x}'\mathbf{y} = \Sigma x_i y_i$, i.e., the sum of products of elements in $\mathbf{x}$ by those in $\mathbf{y}$. $\mathbf{x}$ and $\mathbf{y}$ have the same number of elements.

3.13. The scalar $\mathbf{x}'\mathbf{D}_w\mathbf{x} = \Sigma w_i x_i^2$ is called a weighted sum of squares.

3.14. The scalar $\mathbf{x}'\mathbf{D}_w\mathbf{y} = \Sigma w_i x_i y_i$ is called a weighted sum of products.

3.15a. The vector $\mathbf{A}\mathbf{j}$ is a column vector whose elements are the row sums of $\mathbf{A}$.

3.15b. The vector $\mathbf{j}'\mathbf{A}$ is a row vector whose elements are the column sums of $\mathbf{A}$.

3.15c. The scalar $\mathbf{j}'\mathbf{A}\mathbf{j}$ is the sum of all elements in $\mathbf{A}$. Schematically

A	Aj
j'A	j'Aj

3.16a. If $\mathbf{B} = \mathbf{D}_w\mathbf{A}$; then $b_{ij} = w_i a_{ij}$.

3.16b. If $\mathbf{B} = \mathbf{A}\mathbf{D}_w$; then $b_{ij} = a_{ij} w_j$.

3.17. Interchanging summation and matrix notation:

If

$$\mathbf{ABCD} = \mathbf{E}$$

then

$$e_{ij} = \sum_k \sum_l \sum_m a_{ik} b_{kl} c_{lm} d_{mj}$$

The second subscript of an element must coincide with the first of the next one. Reordering and transposing may be required.

Example

If

$$e_{ij} = \sum_k \sum_l \sum_m a_{kl} b_{ki} c_{jm} d_{ml}$$

$$= \sum_k \sum_l \sum_m b_{ki} a_{kl} d_{ml} c_{jm}$$

Then

$$\mathbf{E} = \mathbf{B}'\mathbf{A}\mathbf{D}'\mathbf{C}'$$

3.18a. $\mathbf{A}'\mathbf{A}$ is a symmetric matrix whose (i, j) element is the scalar product of the i'th column vector and the j'th column vector of $\mathbf{A}$.

3.18b. $\mathbf{A}\mathbf{A}'$ is a symmetric matrix whose (i, j) element is the scalar product of the i'th row vector and the j'th row vector of $\mathbf{A}$.

4. DETERMINANTS

4.1a. A determinant $|\mathbf{A}|$ or $\det(\mathbf{A})$ is a scalar function of a square matrix defined in such a way that

$$|\mathbf{A}|\,|\mathbf{B}| = |\mathbf{AB}|$$

and

$$\begin{vmatrix} a_{11} & a_{12} \\ a_{21} & a_{22} \end{vmatrix} = a_{11}a_{22} - a_{12}a_{21}$$

4.1b. $|\mathbf{A}| = |\mathbf{A}'|$

4.2.
$$\begin{vmatrix} a_{11} & a_{12} & a_{13} \\ a_{21} & a_{22} & a_{23} \\ a_{31} & a_{32} & a_{33} \end{vmatrix} = a_{11}a_{22}a_{33} + a_{12}a_{23}a_{31} + a_{13}a_{21}a_{32} - a_{13}a_{22}a_{31} - a_{11}a_{23}a_{32} - a_{12}a_{21}a_{33}$$

4.3.
$$\begin{vmatrix} a_{11} & a_{12} & \cdots & a_{1n} \\ a_{21} & a_{22} & \cdots & a_{2n} \\ \cdot & \cdot & \cdots & \cdot \\ a_{n1} & a_{n2} & \cdots & a_{nm} \end{vmatrix} = \sum (-1)^{\delta} a_{1i_1} a_{2i_2} \cdots a_{ni_n}$$

where the sum is over all permutations

$$i_1 \neq i_2 \neq \cdots i_n$$

and δ denotes the number of exchanges necessary to bring the sequence $(i_1, i_2, \ldots i_n)$ back into the natural order $(1, 2, \ldots n)$.

4.4. If two rows (columns) in a matrix are exchanged, the determinant will change its sign.

4.5. A determinant does not change its value if a linear combination of other rows (columns) is added to any given row (column).

Example

$$\begin{vmatrix} a_{11} & a_{12} & a_{13} & a_{14} \\ b_{21} & b_{22} & b_{23} & b_{24} \\ a_{31} & a_{32} & a_{33} & a_{34} \\ a_{41} & a_{42} & a_{43} & a_{44} \end{vmatrix} = \begin{vmatrix} a_{11} & a_{12} & a_{13} & a_{14} \\ a_{21} & a_{22} & a_{23} & a_{24} \\ a_{31} & a_{32} & a_{33} & a_{34} \\ a_{41} & a_{42} & a_{43} & a_{44} \end{vmatrix}$$

where

$$b_{2i} = a_{2i} + \gamma_1 a_{1i} + \gamma_3 a_{3i} + \gamma_4 a_{4i}$$
$$i = 1, 2, 3, 4$$

$\gamma_1, \gamma_3, \gamma_4$ arbitrary.

4.6. If the i'th row (column) equals (a constant times) the j'th row (column) of a matrix, its determinant is equal to zero, $(i \neq j)$.

4.7. If, in a matrix $\mathbf{A}$, each element of a row (column) is multiplied by a constant γ, the determinant is multiplied by γ.

4.8. $|\gamma\mathbf{A}| = \gamma^n |\mathbf{A}|$ assuming that $\mathbf{A}$ is of order $(n \times n)$.

4.9. The cofactor of a square matrix $\mathbf{A}$, $\text{cof}_{ij}(\mathbf{A})$ is the determinant of a matrix obtained by striking the i'th row and j'th column of $\mathbf{A}$ and choosing positive (negative) sign if $i + j$ is even (odd).

Example

$$\operatorname{cof}_{23}\begin{bmatrix} 2 & 4 & 3 \\ 6 & 1 & 5 \\ -2 & 1 & 3 \end{bmatrix} = -\begin{vmatrix} 2 & 4 \\ -2 & 1 \end{vmatrix}$$

$$= -(2 + 8) = -10$$

4.10. (Laplace Development)

$$|\mathbf{A}| = a_{i1}\operatorname{cof}_{i1}(\mathbf{A}) + a_{i2}\operatorname{cof}_{i2}(\mathbf{A}) + \cdots + a_{in}\operatorname{cof}_{in}(\mathbf{A})$$

$$= a_{1j}\operatorname{cof}_{1j}(\mathbf{A}) + a_{2j}\operatorname{cof}_{2j}(\mathbf{A}) + \cdots + a_{nj}\operatorname{cof}_{nj}(\mathbf{A})$$

for any row i or any column j.

4.11. Numerical Evaluation of the determinant of a symmetric matrix.

Note: If $\mathbf{A}$ is non-symmetric, form $\mathbf{A}'\mathbf{A}$ or $\mathbf{A}\mathbf{A}'$ by rule 3.18, obtain its determinant, and take the square root.

("Forward Doolittle Scheme", "left side")

Let

$$p_{11} = a_{11}, \quad p_{12} = a_{12} = a_{21}, \ldots p_{1n} = a_{1n}$$

$$\begin{array}{ccccc}
p_{11} & p_{12} & p_{13} & \cdots & p_{1n} \\
1 & u_{12} & u_{13} & \cdots & u_{1n} \\
\hline
 & a_{22} & a_{23} & \cdots & a_{2n} \\
 & p_{22} & p_{23} & \cdots & p_{2n} \\
 & 1 & u_{23} & \cdots & u_{2n} \\
\hline
 & & a_{33} & \cdots & a_{3n} \\
 & & p_{33} & \cdots & p_{3n} \\
 & & 1 & \cdots & u_{3n} \\
\hline
 & & \cdot & \cdots & \cdot \\
 & & & & a_{nn} \\
 & & & & p_{nn} \\
 & & & & 1
\end{array}$$

$$u_{1i} = p_{1i}/p_{11} \qquad i = 1, 2, \ldots n$$

$$p_{2i} = a_{2i} - u_{12}p_{1i} \qquad i = 2, 3, \ldots n$$

$$u_{2i} = p_{2i}/p_{22}$$

$$p_{3i} = a_{3i} - u_{13}p_{1i} - u_{23}p_{2i} \qquad i = 3, 4, \ldots n$$

$$u_{3i} = p_{3i}/p_{33}$$

$$p_{ki} = a_{ki} - u_{1k}p_{1i} - u_{2k}p_{2i} - \cdots - u_{k-1,k}p_{k-1,i} \qquad i = k, k+1, \ldots n \quad k = 2, 3, \ldots n$$

$$u_{ki} = p_{ki}/p_{kk}$$

If, at some stage, $p_{kk} = 0$, reordering of rows and columns may be required. If the matrix is positive-definite (see 8.16) (always true for $\mathbf{AA}'$ or $\mathbf{A}'\mathbf{A}$, see rule 10.24), none of the

p_{kk} will be zero. The p_{ii} are called pivots. Then

$$|\mathbf{A}| = \prod_{i=1}^{n} p_{ii}$$

Further, if **A** is partitioned

$$\mathbf{A} = \begin{bmatrix} \mathbf{A}_{11} & \mathbf{A}_{12} \\ \mathbf{A}'_{12} & \mathbf{A}_{22} \end{bmatrix}$$

where $\mathbf{A}_{11}$ is of order $(k \times k)$, then

$$|\mathbf{A}_{11}| = \prod_{i=1}^{k} p_{ii}$$

(Numerical Examples: see 6.14.)

5. SINGULARITY AND RANK

5.1. A matrix **A** is called singular if there exists a vector $\mathbf{x} \neq \mathbf{0}$ such that $\mathbf{Ax} = \mathbf{0}$ or $\mathbf{A}'\mathbf{x} = \mathbf{0}$. Note $\mathbf{x} \neq \mathbf{0}$ if a single element of $\mathbf{x}$ is unequal 0. If a matrix is not singular, it is called non-singular.

5.2. If a matrix $\mathbf{A}_1$ can be formed by selection of r rows and columns of **A** such that $\mathbf{A}_1\mathbf{x} \neq \mathbf{0}$ or $\mathbf{A}'_1\mathbf{x} \neq \mathbf{0}$ for every $\mathbf{x} \neq \mathbf{0}$, and if addition of an $(r+1)st$ row and column would produce a singular matrix, r is called the rank of **A**.

Example

$$\mathbf{A} = \begin{bmatrix} 2 & 4 & 6 \\ 1 & 3 & 7 \\ 3 & 7 & 13 \\ 1 & 1 & -1 \end{bmatrix}$$

Note that

$$[1, \quad 1, \quad -1] \begin{bmatrix} 2 & 4 & 6 \\ 1 & 3 & 7 \\ 3 & 7 & 13 \end{bmatrix} = [0 \quad 0 \quad 0]$$

and

$$[1, \quad -1, \quad -1] \begin{bmatrix} 2 & 4 & 6 \\ 1 & 3 & 7 \\ 1 & 1 & -1 \end{bmatrix} = [0 \quad 0 \quad 0]$$

but

$$\begin{bmatrix} 2 & 4 \\ 1 & 3 \end{bmatrix} \begin{bmatrix} x_1 \\ x_2 \end{bmatrix} \neq \begin{bmatrix} 0 \\ 0 \end{bmatrix}$$

or

$$[x_1 \quad x_2] \begin{bmatrix} 2 & 4 \\ 1 & 3 \end{bmatrix} \neq [0, \quad 0]$$

for any arbitrary

$$[x_1, \quad x_2] \neq [0, \quad 0].$$

Hence the matrix has rank 2.

5.3. If $\mathbf{A}$ has rank r and if $\mathbf{A}_1$ is a non-singular submatrix consisting of r rows and columns of $\mathbf{A}$, then $\mathbf{A}_1$ is called a basis of $\mathbf{A}$.

5.4a. The determinant of a square singular matrix is 0.

5.4b. The determinant of a non-singular matrix is $\neq 0$.

5.5. $\text{rank}(\mathbf{AB}) \leq \min[\text{rank}(\mathbf{A}), \text{rank}(\mathbf{B})]$.

5.6. $\text{rank}(\mathbf{AA}') = \text{rank}(\mathbf{A}'\mathbf{A}) = \text{rank}(\mathbf{A})$.

5.7. $|\mathbf{A}'\mathbf{A}| = |\mathbf{AA}'| = |\mathbf{A}|^2$ if $\mathbf{A}$ is square.

5.8. $|\mathbf{A}'\mathbf{A}| = |\mathbf{AA}'| \geq 0$ for every $\mathbf{A}$ with real elements.

6. INVERSION

(Regular Case, non-singular matrices)

6.1. If $\mathbf{A}$ is square and non-singular ($|\mathbf{A}| \neq 0$) there exists a unique matrix $\mathbf{A}^{-1}$ such that $\mathbf{AA}^{-1} = \mathbf{A}^{-1}\mathbf{A} = \mathbf{I}$.

6.2. $(\mathbf{ABC})^{-1} = \mathbf{C}^{-1}\mathbf{B}^{-1}\mathbf{A}^{-1}$ (provided that all inverses exist).

6.3. $(\mathbf{A}^{-1})' = (\mathbf{A}')^{-1}$

6.4. $\mathbf{Ax} = \mathbf{b}$ is a system of linear equations. If $\mathbf{A}$ is square and non-singular, there exists a unique solution

$$\mathbf{x} = \mathbf{A}^{-1}\mathbf{b}$$

6.5. $(\gamma\mathbf{A})^{-1} = (1/\gamma)\mathbf{A}^{-1}$

6.6. $|\mathbf{A}^{-1}| = 1/|\mathbf{A}|$

6.7. $\mathbf{D}_w^{-1} = \mathbf{D}_{1/w}$ where $\mathbf{D}$ is a diagonal matrix.

6.8. If

$$\mathbf{A} = \mathbf{B} + \mathbf{uv}'$$

then

$$\mathbf{A}^{-1} = \mathbf{B}^{-1} - \lambda\mathbf{yz}'$$

where

$$\mathbf{y} = \mathbf{B}^{-1}\mathbf{u}, \quad \mathbf{z}' = \mathbf{v}'\mathbf{B}^{-1},$$

and

$$\lambda = 1/(1 + \mathbf{z}'\mathbf{u})$$

Example 6.8.1

$$\mathbf{A} = \begin{bmatrix} 4 & 2 & 4 & 5 \\ 3 & 9 & 12 & 15 \\ 2 & 4 & 11 & 10 \\ 1 & 2 & 4 & 10 \end{bmatrix}$$

This matrix can be written as

$$\begin{bmatrix} 3 & 0 & 0 & 0 \\ 0 & 3 & 0 & 0 \\ 0 & 0 & 3 & 0 \\ 0 & 0 & 0 & 5 \end{bmatrix} + \begin{bmatrix} 1 \\ 3 \\ 2 \\ 1 \end{bmatrix} [1 \quad 2 \quad 4 \quad 5] = \mathbf{B} + \mathbf{uv}'$$

$$\mathbf{B}^{-1} = \begin{bmatrix} 1/3 & 0 & 0 & 0 \\ 0 & 1/3 & 0 & 0 \\ 0 & 0 & 1/3 & 0 \\ 0 & 0 & 0 & 1/5 \end{bmatrix}$$

$$\mathbf{y} = \mathbf{B}^{-1}\mathbf{u} = \begin{bmatrix} 1/3 \\ 1 \\ 2/3 \\ 1/5 \end{bmatrix}$$

$$\mathbf{z}' = \mathbf{v}'\mathbf{B}^{-1} = [1/3 \quad 2/3 \quad 4/3 \quad 1]$$

$$\mathbf{z}'\mathbf{u} = 1/3 \times 1 + 2/3 \times 3 + 4/3 \times 2 + 1 \times 1 = 6$$

$$\lambda = 1/7$$

$$\mathbf{A}^{-1} = \begin{bmatrix} 1/3 & 0 & 0 & 0 \\ 0 & 1/3 & 0 & 0 \\ 0 & 0 & 1/3 & 0 \\ 0 & 0 & 0 & 1/5 \end{bmatrix} - (1/7) \begin{bmatrix} 1/3 \\ 1 \\ 2/3 \\ 1/5 \end{bmatrix} [1/3 \quad 2/3 \quad 4/3 \quad 1]$$

$$= (1/315) \begin{bmatrix} 100 & -10 & -20 & -15 \\ -15 & 75 & -60 & -45 \\ -10 & -20 & 65 & -30 \\ -3 & -6 & -12 & 54 \end{bmatrix}$$

(This rule is especially useful if all off-diagonal elements are equal, then $\mathbf{u} = k\mathbf{j}$ and $\mathbf{v}' = \mathbf{j}'$ and $\mathbf{B}$ is diagonal.)

6.9. Let $\mathbf{B}$ (elements b_{ij}) have a known inverse, $\mathbf{B}^{-1}$ (elements b^{ij}). Let $\mathbf{A} = \mathbf{B}$ except for one element $a_{rs} = b_{rs} + k$. Then the elements of $\mathbf{A}^{-1}$ are

$$a^{ij} = b^{ij} - \frac{kb^{ir}b^{sj}}{1 + kb^{sr}}.$$

6.10. (Partitioning)

Let

$$\mathbf{A} = \begin{matrix} & (p) & (q) \\ (p) & \mathbf{B} & \mathbf{C} \\ (q) & \mathbf{D} & \mathbf{E} \end{matrix}$$

(letters in parentheses denote order of the submatrices)

Let $\mathbf{B}^{-1}$ and $\mathbf{E}^{-1}$ exist. Then

$$\mathbf{A}^{-1} = \begin{bmatrix} \mathbf{X} & \mathbf{Y} \\ \mathbf{Z} & \mathbf{U} \end{bmatrix}$$

where

$$\mathbf{X} = (\mathbf{B} - \mathbf{C}\mathbf{E}^{-1}\mathbf{D})^{-1}$$
$$\mathbf{U} = (\mathbf{E} - \mathbf{D}\mathbf{B}^{-1}\mathbf{C})^{-1}$$
$$\mathbf{Y} = -\mathbf{B}^{-1}\mathbf{C}\mathbf{U}$$
$$\mathbf{Z} = -\mathbf{E}^{-1}\mathbf{D}\mathbf{X}$$

6.11. (Partitioning of Determinants)

Let

$$|\mathbf{A}| = \begin{vmatrix} \mathbf{B} & \mathbf{C} \\ \mathbf{D} & \mathbf{E} \end{vmatrix} \qquad \text{(same structure as in 6.10)}$$

Then

$$|\mathbf{A}| = |\mathbf{E}|\ |(\mathbf{B} - \mathbf{C}\mathbf{E}^{-1}\mathbf{D})| = |\mathbf{B}|\ |(\mathbf{E} - \mathbf{D}\mathbf{B}^{-1}\mathbf{C})|$$

6.12. Let

$$\mathbf{A} = \mathbf{B} + \mathbf{U}\mathbf{V}$$

where $\mathbf{B}(n \times n)$ has an inverse

$\mathbf{U}$ is of order $(n \times k)$, with k usually very small
$\mathbf{V}$ is of order $(k \times n)$

(the special case for $k = 1$ is treated in 6.8).

Then $$\mathbf{A}^{-1} = \mathbf{B}^{-1} - \mathbf{Y}\mathbf{\Lambda}\mathbf{Z}$$

where

$$\mathbf{Y} = \mathbf{B}^{-1}\mathbf{U}(n \times k)$$
$$\mathbf{Z} = \mathbf{V}\mathbf{B}^{-1}(k \times n)$$

and

$$\mathbf{\Lambda}(k \times k) = [\mathbf{I} + \mathbf{Z}\mathbf{U}]^{-1}$$

6.13. Let a_{ij} denote the elements of $\mathbf{A}$ and a^{ij} those of $\mathbf{A}^{-1}$. Then

$$a^{ij} = \mathrm{cof}_{ji}(\mathbf{A})/\,|\mathbf{A}|$$

where cof is the determinant defined in 4.9.

6.14. "Doolittle" Method of inverting symmetric matrices (see also 4.11). Let

$$p_{11} = a_{11}, \quad p_{12} = a_{12} = a_{21}, \ldots p_{1n} = a_{1n} = a_{n1}$$

Forward Solution

p_{11}	p_{12}	p_{13}	$\cdots$	p_{1n}	1				
1	u_{12}	u_{13}	$\cdots$	u_{1n}	$u_{1\mathrm{I}}$				
	a_{22}	a_{23}	$\cdots$	a_{2n}	0	1			
	p_{22}	p_{23}	$\cdots$	p_{2n}	$p_{2\mathrm{I}}$	$p_{2\mathrm{II}}$			
	1	u_{23}	$\cdots$	u_{2n}	$u_{2\mathrm{I}}$	$u_{2\mathrm{II}}$			
		a_{33}	$\cdots$	a_{3n}	0	0	1		
		p_{33}	$\cdots$	p_{3n}	$p_{3\mathrm{I}}$	$p_{3\mathrm{II}}$	$p_{3\mathrm{III}}$		
		1	$\cdots$	u_{3n}	$u_{3\mathrm{I}}$	$u_{3\mathrm{II}}$	$u_{3\mathrm{III}}$		
		$\cdot$	$\cdots$	$\cdot$	$\cdot$	$\cdots$	$\cdot$		
				a_{nn}	0	0	0	$\cdots$	1
				p_{nn}	$p_{n\mathrm{I}}$	$p_{n\mathrm{II}}$	$p_{n\mathrm{III}}$	$\cdots$	p_{nN}
				1	$u_{n\mathrm{I}}$	$u_{n\mathrm{II}}$	$u_{n\mathrm{III}}$	$\cdots$	u_{nN}

$$u_{1i} = p_{1i}/p_{11} \qquad i = 1, 2, \ldots n, \mathrm{I}$$

$$p_{2i} = a_{2i} - u_{12}p_{1i} \qquad i = 2, 3, \ldots n, \mathrm{I}, \mathrm{II}$$

$$u_{2i} = p_{2i}/p_{22}$$

$$p_{3i} = a_{3i} - u_{13}p_{1i} - u_{23}p_{2i} \qquad i = 3, 4, \ldots n, \mathrm{I}, \mathrm{II}, \mathrm{III}$$

$$u_{3i} = p_{3i}/p_{33}$$

$$p_{ki} = a_{ki} - u_{1k}p_{1i} - u_{2k}p_{2i} - \cdots - u_{k-1,k}p_{k-1,i} \qquad i = k, k+1, \ldots n, \mathrm{I}, \mathrm{II}, \ldots K; \quad k = 2, 3, \ldots n$$

$$u_{ki} = p_{ki}/p_{kk}$$

Backward Solution

(j refers to Arabic, J refers to Roman numerals)

The elements of $\mathbf{A}^{-1}$ are a^{ij}

$$a^{nj} = u_{nJ} \qquad j = 1, 2, \ldots n; \quad J = \mathrm{I}, \mathrm{II}, \ldots N$$

$$a^{n-1j} = u_{n-1,J} - u_{n-1,n}a^{nj} \qquad j = 1, 2, \ldots (n-1); \quad J = \mathrm{I}, \mathrm{II}, \ldots (N-1)$$

$$a^{n-2,j} = u_{n-2,J} - u_{n-2,n}a^{nj} - u_{n-2,n-1}a^{n-1,j} \qquad j = 1, 2, \ldots (n-2); \quad J = \mathrm{I}, \mathrm{II}, \ldots (N-2)$$

$$a^{n-k,j} = u_{n-k,J} - u_{n-k,n}a^{nj} - u_{n-k,n-1}a^{n-1,j} - \cdots - u_{n-k,n-k+1}a^{n-k+1,j} \qquad j = 1, 2, \ldots (n-k); \quad J = \mathrm{I}, \mathrm{II}, \ldots (N-k); \quad k = 1, 2, \ldots (n-1),$$

and $a^{ji} = a^{ij}$.

Numerical Example 6.14.1.

Invert the Matrix

$$\begin{bmatrix} 25 & 30 & -10 \\ 30 & 40 & -6 \\ -10 & -6 & 17 \end{bmatrix}$$

a_1	25	30	−10	1		
u_1	1	1.2	−0.4	0.04		
	a_2	40	−6	0	1	
	p_2	4	6	−1.2	1	
	u_2	1	1.5	−0.3	0.25	
		a_3	17	0	0	1
		p_3	4	2.2	−1.5	1
		u_3	1	0.55	−0.375	0.25
	1.61	−1.125	0.55			
	−1.125	0.8125	−0.375			
	0.55	−0.375	0.25			

Enter row a_1.
Elements in u_1 = Elements in a_1 divided by $a_{11}(=25)$.
Enter row a_2.

$$p_{22} = 40 - 1.2 \times 30 = 4$$
$$p_{23} = -6 - 1.2 \times (-10) = 6$$
$$p_{21} = 0 - 1.2 \times 1 = -1.2$$
$$p_{2II} = 1$$

Elements in u_2 = Elements in p_2 divided by $p_{22}(=4)$.
Enter row a_3.

$$p_{33} = 17 - (-0.4) \times (-10) - 1.5 \times 6 = 4$$
$$p_{3I} = 0 - (-0.4) \times 1 - 1.5 \times (-1.2) = 2.2$$
$$p_{3II} = 0 - 1.5 \times 1 = -1.5$$
$$p_{3III} = 1$$

Elements in u_3 = Elements in p_3 divided by $p_{33}(=4)$.

Copy the right-hand side of the last (third) u – row as the last column below the double line.

$$a^{21} = -0.3 - 1.5 \times 0.55 = -1.125$$
$$a^{22} = 0.25 - 1.5 \times (-0.375) = 0.8125$$
$$a^{23} = 0 - 1.5 \times 0.25 = -0.375 \quad \text{(check against } a^{32}\text{)}.$$

These are entered in the next to last (second) column below.

$$a^{11} = 0.04 - (-0.4) \times 0.55 - 1.2 \times (-1.125) = 1.61$$
$$a^{12} = 0 - (-0.4) \times (-0.375) - 1.2 \times 0.8125 = -1.125 \quad \text{(check against } a^{21}\text{)}$$
$$a^{13} = 0 - (-0.4) \times (0.25) - 1.2 \times (-0.375) = 0.55 \quad \text{(check against } a^{31}\text{)}.$$

6.15. A matrix is called orthogonal if $\mathbf{A}' = \mathbf{A}^{-1}$ (or $\mathbf{AA}' = \mathbf{I}$).

7. TRACES

7.1. If $\mathbf{A}$ is a square matrix then the trace of $\mathbf{A}$ is $tr\ \mathbf{A} = \sum_i a_{ii}$, i.e., the sum of the diagonal elements.

7.2. If $\mathbf{A}$ is of order $(m \times k)$ and $\mathbf{B}$ of order $(k \times m)$ then $tr(\mathbf{AB}) = tr(\mathbf{BA})$.

7.3. If $\mathbf{A}$ is of order $(m \times k)$, $\mathbf{B}$ of order $(k \times r)$ and $\mathbf{C}$ of order $(r \times m)$, then

$$tr(\mathbf{ABC}) = tr(\mathbf{BCA}) = tr(\mathbf{CAB}).$$

7.3a. If $\mathbf{b}$ is a column vector and $\mathbf{c}'$ a row vector, then

$$tr(\mathbf{Abc}') = tr(\mathbf{bc}'\mathbf{A}) = \mathbf{c}'\mathbf{Ab}$$

since the trace of a scalar is the scalar.

7.4. $tr(\mathbf{A} + \gamma\mathbf{B}) = tr\mathbf{A} + \gamma tr\mathbf{B}$; where γ is a scalar.

7.5. $tr(\mathbf{EL})_{ij}\mathbf{A} = tr\mathbf{A}(\mathbf{EL})_{ij} = a_{ji}$; where $(\mathbf{EL})_{ij}$ is an elementary matrix as defined in 3.4.

7.6. $tr(\mathbf{EL})_{ij}\mathbf{A}(\mathbf{EL})_{rs}\mathbf{B} = a_{jr}b_{si}$

(These rules are useful in matrix differentiation)

7.7. The trace of the second order of a square matrix $\mathbf{A}$ is the sum of the determinants of all $\binom{n}{2}$ matrices of order (2×2) which can be formed by intersecting rows i and j with columns i and j.

$$tr_2\mathbf{A} = \begin{vmatrix} a_{11} & a_{12} \\ a_{21} & a_{22} \end{vmatrix} + \begin{vmatrix} a_{11} & a_{13} \\ a_{31} & a_{33} \end{vmatrix}$$

$$+ \cdots + \begin{vmatrix} a_{11} & a_{1n} \\ a_{n1} & a_{nn} \end{vmatrix} + \begin{vmatrix} a_{22} & a_{23} \\ a_{32} & a_{33} \end{vmatrix}$$

$$+ \cdots + \begin{vmatrix} a_{22} & a_{2n} \\ a_{n2} & a_{nn} \end{vmatrix} + \cdots + \begin{vmatrix} a_{n-1,n-1} & a_{n-1,n} \\ a_{n,n-1} & a_{nn} \end{vmatrix}.$$

7.8. The trace of the k'th order of a square matrix is the sum of the determinants of all $\binom{n}{k}$ matrices of order $(k \times k)$ which can be formed by intersecting any k rows of $\mathbf{A}$ with the same k columns.

$$tr_k\mathbf{A} = \sum \begin{vmatrix} a_{i_1 i_1} & a_{i_1 i_2} & \cdots & a_{i_1 i_k} \\ a_{i_2 i_1} & a_{i_2 i_2} & \cdots & a_{i_2 i_k} \\ \cdot & \cdot & \cdots & \cdot \\ a_{i_k i_1} & a_{i_k i_2} & \cdots & a_{i_k i_k} \end{vmatrix}$$

where the sum extends over all combinations of n elements taken k at a time in order

$$i_1 < i_2 < \cdots < i_k.$$

7.9. Rules 7.2 and 7.3 (cyclic exchange) are valid for trace of k'th order.
7.10. $tr_n\mathbf{A} = |\mathbf{A}|$ if $\mathbf{A}$ is of order $(n \times n)$.

8. CHARACTERISTIC ROOTS AND VECTORS

8.1. If $\mathbf{A}$ is a square matrix of order $(n \times n)$, then $|\mathbf{A} - \lambda\mathbf{I}| = 0$ is called the characteristic equation of the matrix $\mathbf{A}$. It is a polynomial of the n'th degree in λ.

8.2. The n roots of the characteristic equation (not necessarily distinct) are called the characteristic roots of $\mathbf{A}$

$$ch(\mathbf{A}) = \lambda_1, \lambda_2, \ldots \lambda_n$$

8.3. The characteristic equation of $\mathbf{A}$ can be obtained by the relation

$$\lambda^n - (tr\mathbf{A})\lambda^{n-1} + (tr_2\mathbf{A})\lambda^{n-2} - (tr_3\mathbf{A})\lambda^{n-3} \cdots - (-1)^n(tr_{n-1}\mathbf{A})\lambda + (-1)^n|\mathbf{A}| = 0$$

where tr_k is defined in 7.8.

Example 8.3.1

$$\mathbf{A} = \begin{bmatrix} 25 & 30 & -10 \\ 30 & 40 & -6 \\ -10 & -6 & 17 \end{bmatrix}$$

$tr\mathbf{A} = 25 + 40 + 17 = 82$

$tr_2\mathbf{A} = (25 \times 40 - 30 \times 30) + (25 \times 17 - 10 \times 10) + (40 \times 17 - 6 \times 6) = 1069$

$tr_3\mathbf{A} = |\mathbf{A}| = 25 \times 4 \times 4 = 400$

(cf. 6.14 and procedure stated in 4.11)

Hence

$$\lambda^3 - 82\lambda^2 + 1069\lambda - 400 = 0$$

The solutions (by Newton iteration) are

$$\lambda_1 = 65.86108$$
$$\lambda_2 = 15.75339$$
$$\lambda_3 = 0.38553$$

These are the characteristic roots of **A**.

8.4. $ch(\mathbf{A} + \gamma\mathbf{I}) = \gamma + ch(\mathbf{A})$

8.5. $ch(\mathbf{AB}) = ch(\mathbf{BA})$

(except that **AB** or **BA** may have additional roots equal to zero).

8.6. $ch(\mathbf{A}^{-1}) = 1/ch(\mathbf{A})$

8.7. If $\lambda_1, \lambda_2, \ldots \lambda_n$ are the roots of **A** then

$$\sum_i \lambda_i = tr\mathbf{A}$$

$$\sum_{i<j} \lambda_i\lambda_j = tr_2\mathbf{A}$$

$$\sum_{i<j<k} \lambda_i\lambda_j\lambda_k = tr_3\mathbf{A}$$

$$\prod_i \lambda_i = |\mathbf{A}|$$

8.8. If $\mathbf{x}'$ denotes the radius vector (running coordinates $[x, y, z]$) and if a matrix **Q** is positive-definite, then

$$(\mathbf{x}' - \mathbf{x}_0')\mathbf{Q}^{-1}(\mathbf{x} - \mathbf{x}_0) = 1$$

is the equation of an ellipsoid with center at $[x_0, y_0, z_0] = \mathbf{x}_0'$ and semi-axes equal to the square roots of the characteristic roots of **Q**.

8.9. The characteristic roots of a triangular (or diagonal) matrix are the diagonal elements of the matrix.

8.10. If **A** is a real matrix with positive roots, then

$$ch_{\min}(\mathbf{AA}') \le [ch_{\min}(\mathbf{A})]^2 \le [ch_{\max}(\mathbf{A})]^2 \le ch_{\max}(\mathbf{AA}')$$

where $ch_{\min}$ denotes the smallest and $ch_{\max}$ the largest root.

8.11. The ratio of two quadratic forms (**B** non-singular)

$$u = \frac{\mathbf{x}'\mathbf{A}\mathbf{x}}{\mathbf{x}'\mathbf{B}\mathbf{x}}$$

attains stationary values at the roots of $\mathbf{B}^{-1}\mathbf{A}$. In particular

$$u_{\max} = ch_{\max}(\mathbf{B}^{-1}\mathbf{A}) \qquad \text{and} \qquad u_{\min} = ch_{\min}(\mathbf{B}^{-1}\mathbf{A})$$

8.12. The equation system

$$\mathbf{Ax} = \lambda\mathbf{x}$$

permits non-zero solutions only if λ is one of the characteristic roots of **A**. Such a solution **x** is called a characteristic vector.

8.13. If **x** is a solution to 8.12, so is $\gamma\mathbf{x}$ for an arbitrary scalar γ.

8.14. A solution **x** which has unit length ($\mathbf{x}'\mathbf{x} = 1$) is called the eigenvector associated with the characteristic root λ of **A**. The vector is frequently denoted by **e**.

8.15. A real symmetric matrix has real roots.

8.16. A matrix **A** is called positive-definite (abbreviated p.d.) if the quadratic form $\mathbf{x}'\mathbf{A}\mathbf{x} > 0$ for every $\mathbf{x} \neq \mathbf{0}$.

8.17. A matrix **A** is called positive-semidefinite (abbreviated p.s.d.) if the quadratic form $\mathbf{x}'\mathbf{A}\mathbf{x} > 0$ and/or $\mathbf{x}'\mathbf{A}\mathbf{x} = 0$ for some $\mathbf{x} \neq \mathbf{0}$.

8.18. A positive-definite real symmetric matrix has only positive characteristic roots.

8.19. If a real symmetric matrix is positive-semidefinite, it has no negative roots. The number of non-zero roots equals the rank of the matrix.

8.20. If all roots of a real symmetric matrix are distinct, the associated eigenvectors are distinct.

8.21. The matrix of eigenvectors

$$\mathbf{E} = [\mathbf{e}_1, \mathbf{e}_2, \ldots \mathbf{e}_n]$$

of a real symmetric matrix is (or can be chosen to be) orthogonal.

8.22. $\mathbf{AE} = \mathbf{ED}_\lambda$.

8.23. For a real symmetric matrix, $\mathbf{A} = \mathbf{ED}_\lambda\mathbf{E}'$ (decomposition into matrices of unit rank)

$$\mathbf{E}'\mathbf{AE} = \mathbf{D}_\lambda$$

where $\mathbf{D}_\lambda$ denotes the diagonal matrix of characteristic roots ordered in the same way as the eigenvector columns in **E**.

8.24. If $f(\lambda)$ is a polynomial in λ, then

$$f(\mathbf{A}) = \mathbf{ED}_{f(\lambda)}\mathbf{E}^{-1}$$

where λ are the characteristic roots of **A** and **E** is the matrix of associated eigenvectors. If **A** is symmetric, $\mathbf{E}^{-1} = \mathbf{E}'$.

Example 8.24.1

Consider the matrix in 8.3 (and 6.14).

$$\mathbf{A} = \begin{bmatrix} 25 & 30 & -10 \\ 30 & 40 & -6 \\ -10 & -6 & 17 \end{bmatrix}$$

The characteristic roots were found in Example 8.3.1,

$$\lambda_1 = 65.86108 \qquad \lambda_2 = 15.75339 \qquad \lambda_3 = 0.38553.$$

To find some **x** such that $\mathbf{Ax} = \lambda_1\mathbf{x}$, we arbitrarily set the first element of **x** equal to 1. Using only the first two rows of **A** we solve the equation system

$$25 + 30x_2 - 10x_3 = 65.86108$$
$$30 + 40x_2 - 6x_3 = 65.86108x_2$$

which yields $x_2 = 1.24294$ and $x_3 = -0.35729$. Substitution of these values into the third equation

$$-10 - 6x_2 + 17x_3 = 65.86108x_3$$

yields zero to five decimal places, indicating the accuracy of the first characteristic root. To reduce to unit length the characteristic vector

$$[1 \qquad 1.24294 \qquad -0.35729]$$

we divide each element by

$$\sqrt{1 + 1.24294^2 + 0.35729^2}$$

and thus obtain the first eigenvector

$$[0.61170 \qquad 0.76030 \qquad -0.21855]$$

This, written as a column vector, is $\mathbf{e}_1$. Repeating the same process for the second and third eigenvectors we obtain

$$\mathbf{e}_2 = \begin{bmatrix} -0.08659 \\ 0.33896 \\ 0.93681 \end{bmatrix} \qquad \mathbf{e}_3 = \begin{bmatrix} 0.78634 \\ -0.55412 \\ 0.27318 \end{bmatrix}$$

The three vectors can be placed into the eigenvector matrix **E**, which is easily seen to be orthogonal.

9. CONDITIONAL INVERSES

9.1. Any matrix **A** (singular or non-singular, rectangular or square) has some conditional or generalized inverse $\mathbf{A}^{(-1)}$ defined by the relation

$$\mathbf{A}\mathbf{A}^{(-1)}\mathbf{A} = \mathbf{A}.$$

9.2. If (and only if) **A** is square and non-singular, $\mathbf{A}^{(-1)}$ is unique and equals $\mathbf{A}^{-1}$. Otherwise there will be infinitely many matrices $\mathbf{A}^{(-1)}$ which satisfy the defining relation 9.1.

9.3a. If **A** is rectangular $(n \times m)$ of rank m, with $m < n$, then $\mathbf{A}^{(-1)}$ is of order $(m \times n)$ and $\mathbf{A}^{(-1)}\mathbf{A} = \mathbf{I}(m \times m)$. Then $\mathbf{A}^{(-1)}$ is called an inverse from the left. $\mathbf{A}\mathbf{A}^{(-1)} \neq \mathbf{I}$ in this case.

9.3b. If **A** is rectangular $(n \times m)$ of rank n, with $m > n$, then $\mathbf{A}^{(-1)}$ is of order $(m \times n)$ and $\mathbf{A}\mathbf{A}^{(-1)} = \mathbf{I}(n \times n)$. Then $\mathbf{A}^{(-1)}$ is called an inverse to the right. In this case,

$$\mathbf{A}^{(-1)}\mathbf{A} \neq \mathbf{I}.$$

9.3c. For a square, singular matrix, $\mathbf{A}\mathbf{A}^{(-1)} \neq \mathbf{I}$ and $\mathbf{A}^{(-1)}\mathbf{A} \neq \mathbf{I}$.

Example 9.3.1

$$\mathbf{A} = \begin{bmatrix} 3 \\ 2 \\ 1 \end{bmatrix}$$

The row vector $[1/3 \quad 0 \quad 0]$ is an inverse from the left. The row vector

$$[x \quad y \quad (1 - 3x - 2y)]$$

is a conditional inverse of the above matrix **A** for any values of x and y. It is called the generalized inverse of **A**.

Example 9.3.2

$$\mathbf{A} = \begin{bmatrix} 1 & 2 & 3 \\ 2 & 5 & 6 \\ 3 & 7 & 9 \end{bmatrix}$$

A conditional inverse is

$$\mathbf{A}^{(-1)} = \begin{bmatrix} 5 & -2 & 0 \\ -2 & 1 & 0 \\ 0 & 0 & 0 \end{bmatrix}$$

Here it was obtained by inversion of the basis (the 2×2 matrix in the upper left-hand corner) and replacement of the other elements by zeros.

9.4. A square matrix $\mathbf{A}$ is called idempotent if $\mathbf{AA} = \mathbf{A}^2 = \mathbf{A}$.

9.5. $\mathbf{AA}^{(-1)}$ and $\mathbf{A}^{(-1)}\mathbf{A}$ are idempotent.

9.6. All characteristic roots of idempotent matrices are either zero or one.

9.7. A system of linear equations (m equations in n unknowns)

$$\mathbf{Ax} = \mathbf{b}$$

is called consistent if there exists some solution $\mathbf{x}$ which satisfies the equation system.

Example 9.7.1

The system

$$\begin{aligned} x + y &= 2 \\ 2x + 2y &= 4 \end{aligned} \qquad \text{is consistent.}$$

Example 9.7.2

The system

$$\begin{aligned} x + y &= 2 \\ 2x + 2y &= 5 \end{aligned} \qquad \text{is inconsistent,}$$

for no pair of values (x, y) will satisfy this system.

9.8. If, in a system of equations (rectangular or square)

$$\mathbf{Ax} = \mathbf{b}$$

$\mathbf{AA}^{(-1)}\mathbf{b} = \mathbf{b}$ for some conditional inverse $\mathbf{A}^{(-1)}$, then $\mathbf{AA}^{(-1)}\mathbf{b} = \mathbf{b}$ for every conditional inverse of $\mathbf{A}$, and $\mathbf{Ax} = \mathbf{b}$ is consistent. Conversely, if $\mathbf{AA}^{(-1)}\mathbf{b} \neq \mathbf{b}$ for some conditional inverse $\mathbf{A}^{(-1)}$ then $\mathbf{AA}^{(-1)}\mathbf{b} \neq \mathbf{b}$ for every conditional inverse of $\mathbf{A}$, and $\mathbf{Ax} = \mathbf{b}$ is inconsistent.

9.9. If $\mathbf{Ax} = \mathbf{b}$ is consistent, then $\mathbf{x} = \mathbf{A}^{(-1)}\mathbf{b}$ is a solution (generally a different one for each $\mathbf{A}^{(-1)}$).

9.10. Let $\mathbf{y}$ ($p \times 1$) be a set of linear functions of the solutions $\mathbf{x}$ ($n \times 1$) of a consistent system of equations $\mathbf{Ax} = \mathbf{b}$, given by the relation $\mathbf{y} = \mathbf{Cx}$. Then $\mathbf{y} = \mathbf{Cx}$ is called unique if the same values of $\mathbf{y}$ will result regardless which solution $\mathbf{x}$ is used.

Example 9.10.1

$$\begin{aligned} 3x + 4y + 5z &= 22 \\ x + y + z &= 6 \end{aligned}$$

is a consistent system. One solution would be

$$x = 3 \quad y = 2 \quad z = 1$$

Another solution is

$$x = 2 \quad y = 4 \quad z = 0$$

The linear function

$$[7 \quad 9 \quad 11] \begin{bmatrix} x \\ y \\ z \end{bmatrix} = u$$

$(7x + 9y + 11z = u)$ will have the same value (50) regardless which of the two (or any other) solutions is substituted. Thus u is unique.

9.11. Let $\mathbf{Ax} = \mathbf{b}$ be a consistent system of equations. For $\mathbf{Cx} = \mathbf{y}$ to be a unique linear combination of the solution $\mathbf{x}$, it is necessary and sufficient that $\mathbf{CA}^{(-1)}\mathbf{A} = \mathbf{C}$. If this relation holds for some $\mathbf{A}^{(-1)}$ it will hold for every conditional inverse of $\mathbf{A}$. If it is violated for some $\mathbf{A}^{(-1)}$ it will be violated for every $\mathbf{A}^{(-1)}$, and $\mathbf{y}$ will be non-unique.

9.12. Let $\mathbf{A}$ be of rank r and select r rows and r columns which form a basis of $\mathbf{A}$. Then a conditional inverse of $\mathbf{A}$ can be obtained as follows: Invert the $(r \times r)$ matrix, place the inverse (without transposing) into the r rows corresponding to the column numbers and the r columns corresponding to the row numbers of the basis, and place zero into all remaining elements. Thus, if $\mathbf{A}$ is of order (5×4) and rank 3, and if rows 1, 2, 4 and columns 2, 3, 4 are selected as a basis, $\mathbf{A}^{(-1)}$, of order (4×5) will contain the inverse elements of the basis in rows 2, 3, 4 and columns 1, 2, 4, and zeros elsewhere. (See example 9.3.2)

9.13. If $\mathbf{A}$ is a square, singular matrix of order $(n \times n)$ and rank r, let $\mathbf{M}$ be a matrix of order $[n \times (n - r)]$ and $\mathbf{K}$ another matrix of order $[(n - r) \times n]$ chosen in such a way that $\mathbf{A} + \mathbf{MK}$ is non-singular. Then $(\mathbf{A} + \mathbf{MK})^{-1}$ is a conditional inverse of $\mathbf{A}$.

Example 9.13.1

$$\mathbf{A} = \begin{bmatrix} 3 & -1 & -1 & -1 \\ -1 & 3 & -1 & -1 \\ -1 & -1 & 3 & -1 \\ -1 & -1 & -1 & 3 \end{bmatrix}$$

is of order (4×4) and rank 3. Take $\mathbf{M} = \mathbf{j}$ (column vector of ones) and $\mathbf{K} = \mathbf{j}'$ (row vector of ones). Then $\mathbf{A} + \mathbf{MK} = \mathbf{A} + \mathbf{jj}' = 4\mathbf{I}$. Hence $(1/4)\ \mathbf{I}$ is a conditional inverse of $\mathbf{A}$.

9.14. The "Doolittle" method (see 6.14) can be employed to obtain a conditional inverse of a symmetric matrix. If, at any stage, the leading element of the p-row is zero, that cycle is disregarded.

Example 9.14.1

Invert, conditionally, the matrix

$$\mathbf{A} = \begin{bmatrix} 4 & 2 & -2 & 4 \\ 2 & 17 & 11 & 6 \\ -2 & 11 & 10 & 1 \\ 4 & 6 & 1 & 30 \end{bmatrix}$$

4	2	−2	4	1			
1	.5	−.5	1	.25			
	17	11	6	0	1		
	16	12	4	−.5	1		
	1	.75	.25	−.03125	.0625		
		10	1	0	0	1	
		0					
			30	0	0	0	1
			25	−.875	−.25	0	1
			1	−.035	−.01	0	.04

$$\begin{bmatrix} .29625 & -.0225 & 0 & -.035 \\ -.0225 & .065 & 0 & -.01 \\ 0 & 0 & 0 & 0 \\ -.035 & -.01 & 0 & .04 \end{bmatrix} = \mathbf{A}^{(-1)}$$

10. MATRIX DIFFERENTIATION

10.1a. If the elements of a matrix $\mathbf{Y}$ $(m \times n)$ are functions of a scalar, x, the expression

$$\partial \mathbf{Y}/\partial x$$

denotes a matrix of order $(m \times n)$ with elements $\partial y_{ij}/\partial x$.

10.1b. If the elements of a column (row) vector $\mathbf{y}$ $(\mathbf{y}')$ are functions of a scalar, x, the expression

$$\partial \mathbf{y}/\partial x \quad (\partial \mathbf{y}'/\partial x)$$

denotes a column (row) vector with elements $\partial y_i/\partial x$.

10.2a. If y is a scalar function of $m \times n$ variables, x_{ij}, arranged into a matrix $\mathbf{X}$, the expression

$$\partial y/\partial \mathbf{X}$$

denotes a matrix with elements $\partial y/\partial x_{ij}$.

(Note: Partial differentiation is performed with respect to the element in row i and column j of $\mathbf{X}$. If the same x-variable occurs in another place as, e.g., in a symmetric matrix, differentiation with respect to the distinct (repeated) variable is performed in two stages.)

Example 10.2.1

If $y = \mathbf{j}'\mathbf{X}\mathbf{j}$ (sum of all elements of a square matrix), $\partial y/\partial \mathbf{X}$ is a matrix of ones. If $\mathbf{X}$ is symmetric, one can introduce a new notation $x_{ij} = x_{ji} = z_{ij}$. Then

$$\begin{aligned} \partial y/\partial z_{ij} &= (\partial y/\partial x_{ij})(\partial x_{ij}/\partial z_{ij}) \\ &\quad + (\partial y/\partial x_{ji})(\partial x_{ji}/\partial z_{ij}) \\ &= 1 + 1 = 2 \quad (\text{if } i \neq j) \\ &= 1 \qquad\qquad (\text{if } i = j). \end{aligned}$$

10.2b. If y is a scalar function of n variables, x_i, arranged into a column (row) vector $\mathbf{x}$ $(\mathbf{x}')$, the expression

$$\partial y/\partial \mathbf{x} \qquad (\partial y/\partial \mathbf{x}')$$

denotes a column (row) vector with elements $\partial y/\partial x_i$.

10.3. If $\mathbf{y}$ is a column vector with m elements, each a function of n variables, x_i, arranged into a row vector $\mathbf{x}'$, the expression $\partial \mathbf{y}/\partial \mathbf{x}'$ denotes a matrix with m rows and n columns, with elements $\partial y_i/\partial x_j$.

10.4. $\partial \mathbf{Y}/\partial y_{ij} = (\mathbf{EL})_{ij}$ (see definition of $(\mathbf{EL})$ in 3.4).

10.5. $\partial \mathbf{UV}/\partial x = (\partial \mathbf{U}/\partial x)\mathbf{V} + \mathbf{U}(\partial \mathbf{V}/\partial x)$.

10.6. $\partial \mathbf{AY}/\partial x = \mathbf{A}(\partial \mathbf{Y}/\partial x)$ (if elements of $\mathbf{A}$ are not functions of x).

10.7. $\partial \mathbf{Y}'/\partial y_{ij} = (\mathbf{EL})_{ji}$

10.8. $\partial \mathbf{A}'\mathbf{YA}/\partial x = \mathbf{A}'(\partial \mathbf{Y}/\partial x)\mathbf{A}$

10.9. $\partial \mathbf{Y}'\mathbf{AY}/\partial x = (\partial \mathbf{Y}'/\partial x)\mathbf{AY} + \mathbf{Y}'\mathbf{A}(\partial \mathbf{Y}/\partial x)$

10.10. $\partial \mathbf{a}'\mathbf{x}/\partial \mathbf{x} = \mathbf{a}$

10.11. $\partial \mathbf{x}'\mathbf{x}/\partial \mathbf{x} = 2\mathbf{x}$

10.12. $\partial \mathbf{x}'\mathbf{A}\mathbf{x}/\partial \mathbf{x} = \mathbf{A}\mathbf{x} + \mathbf{A}'\mathbf{x}$
10.13. (Chain Rule No. 1) $\quad \partial \mathbf{y}/\partial \mathbf{x}' = (\partial \mathbf{y}/\partial \mathbf{z}')(\partial \mathbf{z}/\partial \mathbf{x}')$
10.14. $\partial \mathbf{A}\mathbf{x}/\partial \mathbf{x}' = \mathbf{A}$
10.15. $\partial tr\mathbf{X}/\partial \mathbf{X} = \mathbf{I}$
10.16. $\partial tr\mathbf{A}\mathbf{X}/\partial \mathbf{X} = \partial tr\mathbf{X}\mathbf{A}/\partial \mathbf{X} = \mathbf{A}'$
10.17. $\partial tr\mathbf{A}\mathbf{X}\mathbf{B}/\partial \mathbf{X} = \mathbf{A}'\mathbf{B}'$
10.18. $\partial tr\mathbf{X}'\mathbf{A}\mathbf{X}/\partial \mathbf{X} = \mathbf{A}\mathbf{X} + \mathbf{A}'\mathbf{X}$
10.19. $\partial \log |\mathbf{X}| / \partial \mathbf{X} = (\mathbf{X}')^{-1}$ (log to base e).
10.20. $\partial \mathbf{Y}^{-1}/\partial x = -\mathbf{Y}^{-1}(\partial \mathbf{Y}/\partial x)\mathbf{Y}^{-1}$.
10.21. (Chain Rule No. 2)

$$\partial y/\partial x = tr(\partial y/\partial \mathbf{Z})(\partial \mathbf{Z}'/\partial x)$$

where y and x are scalars. The scalar y is a function of $m \times n$ variables z_{ij}, and each of the z_{ij} is a function of x.

Example 10.21.1
Obtain $\log |\mathbf{R} - \mathbf{F}\mathbf{F}'| / \partial \mathbf{F}$, where $\mathbf{R}$ is symmetric.
By Chain Rule No. 2:

$$\begin{aligned}
&\partial \log |\mathbf{R} - \mathbf{F}\mathbf{F}'| / \partial f_{ij} \\
&= tr[\partial \log |\mathbf{R} - \mathbf{F}\mathbf{F}'| / \partial(\mathbf{R} - \mathbf{F}\mathbf{F}')][\partial(\mathbf{R} - \mathbf{F}\mathbf{F}')/\partial f_{ij}] \quad \text{(since } \mathbf{R} \text{ and } \mathbf{F}\mathbf{F}' \text{ are symmetric)} \\
&= tr(\mathbf{R} - \mathbf{F}\mathbf{F}')^{-1}[\partial(\mathbf{R} - \mathbf{F}\mathbf{F}')/\partial f_{ij}] \quad \text{(by 10.19)} \\
&= tr(\mathbf{R} - \mathbf{F}\mathbf{F}')^{-1}[-(\partial \mathbf{F}/\partial f_{ij})\mathbf{F}' - \mathbf{F}(\partial \mathbf{F}'/\partial f_{ij})] \quad \text{(by 10.5)} \\
&= tr(\mathbf{R} - \mathbf{F}\mathbf{F}')^{-1}[-(\mathbf{EL})_{ij}\mathbf{F}' - \mathbf{F}(\mathbf{EL})_{ji}] \quad \text{(by 10.4 and 10.7)} \\
&= -tr(\mathbf{R} - \mathbf{F}\mathbf{F}')^{-1}(\mathbf{EL})_{ij}\mathbf{F}' - tr(\mathbf{R} - \mathbf{F}\mathbf{F}')^{-1}\mathbf{F}(\mathbf{EL})_{ji} \\
&= -tr(\mathbf{EL})_{ij}\mathbf{F}'(\mathbf{R} - \mathbf{F}\mathbf{F}')^{-1} - tr(\mathbf{EL})_{ji}(\mathbf{R} - \mathbf{F}\mathbf{F}')^{-1}\mathbf{F} \quad \text{(by 7.3)} \\
&= -[\mathbf{F}'(\mathbf{R} - \mathbf{F}\mathbf{F}')^{-1}]_{ji} - [(\mathbf{R} - \mathbf{F}\mathbf{F}')^{-1}\mathbf{F}]_{ij},
\end{aligned}$$

where $[\,]_{ij}$ denotes the (i,j) element of the matrix in brackets (by 7.5),

$$\begin{aligned}
&= -[(\mathbf{R} - \mathbf{F}\mathbf{F}')^{-1}\mathbf{F}]_{ij} - [(\mathbf{R} - \mathbf{F}\mathbf{F}')^{-1}\mathbf{F}]_{ij} \quad \text{(since } \mathbf{R} - \mathbf{F}\mathbf{F}' \text{ is symmetric)} \\
&= -2[(\mathbf{R} - \mathbf{F}\mathbf{F}')^{-1}\mathbf{F}]_{ij}.
\end{aligned}$$

Hence, by definition 10.2a

$$\partial \log |\mathbf{R} - \mathbf{F}\mathbf{F}'| / \partial \mathbf{F} = -2(\mathbf{R} - \mathbf{F}\mathbf{F}')^{-1}\mathbf{F}.$$

10.22. $|\partial \mathbf{y}/\partial \mathbf{x}'| = J(\mathbf{y}; \mathbf{x})$ is called the Jacobian or Functional Determinant used in variable transformation of multiple integrals. Formally, if $\mathbf{y}$ is a column vector with m elements, each a function of m variables x_i arranged into a row vector $\mathbf{x}'$,

$$dx_1\, dx_2 \ldots dx_m = |\partial \mathbf{y}/\partial \mathbf{x}'|^{-1} dy_1\, dy_2 \ldots dy_m.$$

10.23. For a scalar y (a function of m variables x_i) to attain a stationary value, it is necessary that

$$\partial y/\partial \mathbf{x} = \mathbf{0}.$$

10.24. For a stationary value to be a minimum (maximum) it is necessary that

$$\partial(\partial y/\partial \mathbf{x})/\partial \mathbf{x}' \qquad (-\partial(\partial y/\partial \mathbf{x})/\partial \mathbf{x}')$$

be a positive-definite matrix for the value of $\mathbf{x}$ satisfying 10.23.

Example 10.24.1
Find the values of β which minimize $u = \mathbf{x}'\mathbf{x}$ (the sum of squares of x_i) where $\mathbf{x} = \mathbf{y} - \mathbf{A}\beta$ (with $\mathbf{y}$ and $\mathbf{A}$ known and fixed).

$$\begin{aligned}\partial u/\partial\boldsymbol{\beta}' &= (\partial u/\partial\mathbf{x}')(\partial\mathbf{x}/\partial\boldsymbol{\beta}') \quad \text{(by Chain Rule No. 1)}\\ &= -2\mathbf{x}'\mathbf{A} \quad \text{(by 10.11 and 10.14)}\end{aligned}$$

Hence

$$\begin{aligned}\partial u/\partial\boldsymbol{\beta} &= -2\mathbf{A}'\mathbf{x}\\ &= -2\mathbf{A}'(\mathbf{y} - \mathbf{A}\boldsymbol{\beta}).\end{aligned}$$

Hence, for a stationary value, by 10.23, it is necessary that

$$\mathbf{A}'\mathbf{A}\hat{\boldsymbol{\beta}} = \mathbf{A}'\mathbf{y}$$

where $\hat{\boldsymbol{\beta}}$ denotes the values which make u stationary. Now,

$$\partial(\partial u/\partial\boldsymbol{\beta})/\partial\boldsymbol{\beta}' = 2\partial(\mathbf{A}'\mathbf{A}\boldsymbol{\beta})/\partial\boldsymbol{\beta}' = 2\mathbf{A}'\mathbf{A}.$$

If $\mathbf{A}$ has real elements, and if $\mathbf{A}'\mathbf{A}$ is non-singular, then it is positive-definite (since, given an arbitrary real $\mathbf{x} \neq \mathbf{0}$, $\mathbf{x}'\mathbf{A}'\mathbf{A}\mathbf{x} = \mathbf{z}'\mathbf{z}$, with $\mathbf{z} = \mathbf{A}\mathbf{x}$; thus this is a sum of squares). Hence $\hat{\boldsymbol{\beta}}$ minimizes u.

10.25. (Generalized Newton Iteration)

Let $\mathbf{x}_0'$ be an initial estimate (m elements) of the roots of the m equations

$$\mathbf{f}(\mathbf{x}') = \mathbf{0}$$

where the m elements of the column vector $\mathbf{f}$ are each functions of $x_1, x_2, \ldots x_m$. Then an improved root is

$$\mathbf{x}_1 = \mathbf{x}_0 - \mathbf{Q}_0^{-1}\mathbf{f}(\mathbf{x}_0'),$$

where $\mathbf{Q}_0$ is the matrix of derivatives $\partial\mathbf{f}/\partial\mathbf{x}'$ evaluated at $\mathbf{x} = \mathbf{x}_0$. The usual procedure consists of evaluating $\mathbf{f}(\mathbf{x}_0')$, then solving $\mathbf{Q}_0\mathbf{u} = \mathbf{f}(\mathbf{x}_0')$ for $\mathbf{u}$. Then $\mathbf{x}_1 = \mathbf{x}_0 - \mathbf{u}$.

Example 10.25.1

Solve

$$\begin{aligned}f_1(x,y) &= x^3 - x^2y + y^2 - 3.526 = 0\\ f_2(x,y) &= x^3 + y^3 - 14.911 = 0\end{aligned}$$

$$\mathbf{Q} = \begin{bmatrix} 3x^2 - 2xy & 2y - x^2 \\ 3x^2 & 3y^2 \end{bmatrix}$$

Take $x_0 = 1, \quad y_0 = 2$

$$\begin{aligned}f_1(x_0, y_0) &= -0.526\\ f_2(x_0, y_0) &= -5.911\end{aligned}$$

$$\mathbf{Q}_0 = \begin{bmatrix} -1 & 3 \\ 3 & 12 \end{bmatrix}$$

$$\begin{aligned}-u + 3v &= -0.526\\ 3u + 12v &= -5.911\end{aligned}$$

yields $u = -0.55, \quad v = -0.36.$

Then,

$$\begin{aligned}x_1 &= x_0 - u = 1.55\\ y_1 &= y_0 - v = 2.36\\ f_1(x_1, y_1) &= 0.0976\\ f_2(x_1, y_1) &= 1.9572\end{aligned}$$

$$\mathbf{Q}_1 = \begin{bmatrix} -0.1085 & 2.3175 \\ 7.2075 & 16.7088 \end{bmatrix}$$

$$-0.1085u + 2.3175v = 0.0976$$
$$7.2075u + 16.7088v = 1.9572$$
$$\text{yields } u = 0.157, \quad v = 0.049.$$

Then

$$x_2 = x_1 - u = 1.393$$
$$y_2 = y_1 - v = 2.311$$
$$f_1(x_2, y_2) = 0.03337$$
$$f_2(x_2, y_2) = 0.13443$$

$$\mathbf{Q}_2 = \begin{bmatrix} -0.61710 & 2.68155 \\ 5.82135 & 16.02216 \end{bmatrix}$$

$$-0.61710u + 2.68155v = 0.03337$$
$$5.82135u + 16.02216v = 0.13443$$
$$\text{yields } u = -0.0068, \quad v = 0.0109.$$

Then,

$$x_3 = x_2 - u = 1.3998$$
$$y_3 = y_2 - v = 2.3001$$

(The exact roots are $x = 1.4$ and $y = 2.3$).

11. STATISTICAL MATRIX FORMS

11.1. Let E denote the expectation operator, and let $\mathbf{y}$ be a set of p random variables. Then

$$E(\mathbf{y}) = \boldsymbol{\mu}$$

states that $E(y_i) = \mu_i \quad (i = 1, 2, \ldots p)$.

11.2. Let var denote variance. Then

$$\text{var}(\mathbf{y}) = \boldsymbol{\Sigma}$$

denotes a $p \times p$ symmetric matrix whose elements are $\text{cov}(y_i, y_j)$, and whose diagonal elements are $\text{var}(y_i)$, where cov denotes covariance.

11.3. $E(\mathbf{AY} + \mathbf{b}) = \mathbf{A}E(\mathbf{y}) + \mathbf{b} = \mathbf{A}\boldsymbol{\mu} + \mathbf{b}$

11.4. $\text{var}(\mathbf{Ay} + \mathbf{b}) = \mathbf{A}\,\text{var}(\mathbf{y})\,\mathbf{A}' = \mathbf{A}\boldsymbol{\Sigma}\mathbf{A}'$

11.5. $\text{cov}(\mathbf{y}, \mathbf{z}')$ denotes a matrix with elements $\text{cov}(y_i, z_j)$.
$\text{cov}(\mathbf{z}, \mathbf{y}') = [\text{cov}(\mathbf{y}, \mathbf{z}')]'$.

11.6. $\text{cov}(\mathbf{Ay} + \mathbf{b}, \mathbf{z}'\mathbf{C} + \mathbf{d}') = \mathbf{A}\,\text{cov}(\mathbf{y}, \mathbf{z}')\,\mathbf{C}$

11.7. $\text{var}(\mathbf{y}) = E(\mathbf{yy}') - E(\mathbf{y})E(\mathbf{y}')$

11.8. $\text{cov}(\mathbf{y}, \mathbf{z}') = E(\mathbf{yz}') - E(\mathbf{y})E(\mathbf{z}')$

11.9. (Expected "sum of squares")

$$E(\mathbf{y}'\mathbf{Q}\mathbf{y}) = tr[\mathbf{Q}\,\text{var}\,(\mathbf{y})] + E(\mathbf{y}')\mathbf{Q}E(\mathbf{y}).$$

11.10. If a matrix $\mathbf{Q}$ is symmetric and positive-definite, one can find a lower triangular matrix $\mathbf{T}$ (with positive diagonal terms, for uniqueness) such that $\mathbf{TT}' = \mathbf{Q}$. The matrices $\mathbf{T}$ and $\mathbf{T}^{-1}$ can be obtained from the Doolittle pattern (6.14) (Gauss elimination or square-root method) as follows: In each cycle, divide the p-row (left and right hand side) by $\sqrt{p_{ii}}$ (instead of p_{ii} for the u-row). Thus obtain rows designated as t-rows. The left-hand side (Arabic subscripts) is $\mathbf{T}'$, and the right-hand side (Roman subscripts) is $\mathbf{T}^{-1}$

11.11. If a coordinate system $\mathbf{x}$ is oblique, and if the cosines between reference vectors (scalar products of basis vectors of unit length) are stated in a symmetric matrix $\mathbf{Q}$, then $\mathbf{T}^{-1}\mathbf{x} = \mathbf{y}$ is an orthogonal system, where $\mathbf{T}$ is obtained from $\mathbf{Q}$ by 11.10.

11.12. The likelihood function of a sample of size n from a multivariate normal distribution (p responses), with common variance-covariance matrix $\Sigma(p \times p)$, and with means or main effects replaced by maximum-likelihood or least-squares estimates, can be written as

$$\log L = -\frac{np}{2}\log 2\pi - \frac{n}{2}\log|\Sigma| - \frac{n}{2}\, tr\, \Sigma^{-1}\mathbf{S}$$

where $\Sigma(p \times p)$ is the common variance-covariance matrix, and $\mathbf{S}$ is its maximum-likelihood estimate (matrix of sums of squares and products due to error, divided by sample size n).

11.13. If Σ has a structure under a model or null hypothesis, and if elements of Σ are to be estimated, by maximum-likelihood, two cases can be distinguished: (11.14) Σ^{-1} has the same structure (intraclass correlation, mixed model, compound symmetry, factor analysis). (11.15) Σ^{-1} has a different structure (autocorrelation, Simplex structure).

11.14. If the structure of Σ and Σ^{-1} is identical, and if u and v are elements (or functions of elements) of Σ^{-1} then estimates of Σ can be obtained from the relations (usually requiring Newton iteration, see 10.25):

$$\partial \log L/\partial u = \frac{n}{2}\, tr\, \mathbf{A}(\Sigma - \mathbf{S})$$

where $\mathbf{A} = \partial\Sigma^{-1}/\partial u$, is frequently an elementary matrix (see 3.4 and, especially, the rules 7.5 and 7.6).

$$\partial^2 \log L/\partial u \partial v = \frac{n}{2}\, tr(\partial\mathbf{A}/\partial v)(\Sigma - \mathbf{S}) + \frac{n}{2}\, tr\, \mathbf{A}\Sigma^{-1}\mathbf{B}\Sigma^{-1}$$

where $\mathbf{B} = \partial\Sigma^{-1}/\partial v$. These rules are useful to obtain Newton iterations and asymptotic variance-covariance matrices of the estimates.

11.15. If the structures of Σ and Σ^{-1} are different, then an estimate of Σ can be obtained from the relations

$$\partial \log L/\partial x = -\frac{n}{2}\, tr\, \mathbf{A}(\Sigma^{-1} - \mathbf{Q}),$$

where

$$\mathbf{Q} = \Sigma^{-1}\mathbf{S}\Sigma^{-1}$$

and

$$\mathbf{A} = \partial\Sigma/\partial x \qquad \text{(see comments in 11.14).}$$

$$\partial^2 \log L/\partial x \partial y = -\frac{n}{2}\, tr(\partial\mathbf{A}/\partial y)(\Sigma^{-1} - \mathbf{Q}) + \frac{n}{2}\, tr\, \mathbf{A}\Sigma^{-1}\mathbf{B}(\Sigma^{-1} - \mathbf{Q}) - \frac{n}{2}\, tr\, \mathbf{AQB}\Sigma^{-1},$$

where

$$\mathbf{B} = \partial\Sigma/\partial y$$

x and y are elements (or functions of elements) of Σ. The comments of 11.14 apply, but the iterative procedure is considerably more complex.

Suggestions for further reading:

A. S. Householder, *The Theory of Matrices in Numerical Analysis*, Blaisdell, 1964.
S. R. Searle, *Matrix Algebra for the Biological Sciences*, Wiley, 1966.
P. H. Schonemann, Matrix differentiation of traces and determinants, *Psychometrika*, 1966.

USE OF LOGARITHMS

LAWS OF EXPONENTS

For a any real number and m a positive integer, the exponential a^m is defined as

$$\underbrace{a \cdot a \cdot a \cdot \cdots \cdot a}_{m \text{ terms}}$$

Using this definition, it is easy to show that the following three *laws of exponents* hold:

I. $a^m \cdot a^n = a^{m+n}$

II. $$\frac{a^m}{a^n} = \begin{cases} a^{m-n} & \text{if } m > n \\ 1 & \text{if } m = n \\ \dfrac{1}{a^{n-m}} & \text{if } m < n \end{cases}$$

III. $(a^m)^n = a^{mn}$

The n-th root function is defined as the inverse of the n-th power function; that is, if

$$b^n = a, \quad \text{then} \quad b = \sqrt[n]{a}.$$

If n is odd, there will be a unique real number satisfying the above definition for $\sqrt[n]{a}$, for any real value of a. If n is even, for positive values of a there will be two real values for $\sqrt[n]{a}$, one positive and one negative. By convention, the symbol $\sqrt[n]{a}$ is understood to mean the positive value in this case. If n is even and a is negative, there are no real values for $\sqrt[n]{a}$.

If we now attempt to extend the definition of the exponential a^t to all rational values of the exponent t, in such a way that the three laws of exponents continue to hold, it is easily shown that the required definitions are:

$$a^0 = 1$$

$$a^{p/q} = \sqrt[q]{a^p}$$

$$a^{-t} = \frac{1}{a^t}$$

In order to avoid difficulties with imaginary numbers and division by zero, a must now be restricted to be positive.

With this extended definition, it is possible to restate the second law of exponents in a simpler form:

II′. $\dfrac{a^m}{a^n} = a^{m-n}$

It is shown in advanced calculus that this definition may be further extended so that the exponent may be any real number, and the laws of exponents continue to hold. When the quantity a^x thus defined is viewed as a function of the exponent x, with the base a held constant, it is a continuous function. Also, if $a > 1$, the exponential function is monotone increasing, and if $0 < a < 1$, it is monotone decreasing.

LOGARITHMS

Any monotone function has a single-valued inverse function, which is also monotone. Furthermore, if the original function is continuous, so is the inverse. Therefore, the inverse function to the exponential function a^x exists for all positive values of a, except $a = 1$. This function is given the name *logarithm to the base a*, abbreviated $\log_a$. That is, if

$$x = a^y, \quad \text{then} \quad y = \log_a x.$$

This function is defined and continuous for all positive values of x. It is monotone increasing if $a > 1$, and monotone decreasing if $a < 1$.

If the laws of exponents are rewritten in terms of logarithms, they become the *laws of logarithms*:

I. $\log_a(xy) = \log_a x + \log_a y$

II. $\log_a\left(\dfrac{x}{y}\right) = \log_a x - \log_a y$

III. $\log_a(x^n) = n \log_a x$

Logarithms derive their main usefulness in computation from the above laws, since they allow multiplication, division, and exponentiation to be replaced by the simpler operations of addition, subtraction, and multiplication, respectively. See the examples which follow.

Further recourse to the definition of logarithm leads to the following formula for change of base

$$\log_a x = \log_b x/\log_b a = (\log_b x)\cdot(\log_a b)$$

Two numbers are commonly used as bases for logarithms. Logarithms to the base 10 are most convenient for use in computation. These logarithm are called common or Briggsian logarithms.

The other usual base for logarithms is an irrational number denoted by e, whose value is approximately 2.71828 .... These logarithms are called natural, Naperian, or hyperbolic logarithms, and occur in many formulas of higher mathematics. The abbreviation ln is frequently used for the natural logarithm function.

Other bases for logarithms, such as 2 and 3, occur in certain applications. These applications are rather specialized and separate tables for these bases are not given. Instead, the formulas for change of base are applied to common or natural logarithms.

If the formulas for change of base are applied to the two usual bases, the following formulas result:

$$\begin{aligned}\log_{10} x = \log_e x/\log_e 10 = (\log_{10} e)(\log_e x) &= M \log_e x \\ &= 0.43429\ 44819 \log_e x \\ \log_e x = \log_{10} x/\log_{10} e = (\log_e 10)(\log_{10} x) &= \frac{1}{M} \log_{10} x \\ &= 2.30258\ 50930 \log_{10} x\end{aligned}$$

The following remarks apply to common logarithms.

Since most numbers are irrational powers of ten, a common logarithm, in general, consists of an integer, which is called the characteristic, and an endless decimal, the mantissa.

It is to be observed that the common logarithms of all numbers expressed by the same figures in the same order with the decimal point in different positions have different characteristics but the same mantissa. To illustrate:—if the decimal point stand after the first figure of a number, counting from the left, the characteristic is 0; if after two figures, it is 1; if after three figures, it is 2; and so forth. If the decimal point stand before the first significant figure the characteristic is -1, usually written $\bar{1}$; if there is one zero between the decimal point and the first significant figure it is $\bar{2}$, and so on. For example: log 256 = 2.408240, log 2.56 = 0.408240, log 0.256 = $\bar{1}$.408240, log 0.00256 = $\bar{3}$.408240. The two latter are often written log 0.256 = 9.408240 − 10, log 0.00256 = 7.408240 − 10.

Notice that, although the common logarithm of a number less than one is a negative number, it is customarily written as a negative characteristic and a *positive* mantissa, since the mantissas are usually given in tables as positive numbers. This is the reason that the negative sign is written above the characteristic, since it does not apply to the mantissa. Thus log 0.00256 = $\bar{3}$.408240 = 7.408240 − 10 = −2.591760.

A method of determining characteristics of logarithms is to write the number with one figure to the left of the decimal point multiplied by the appropriate power of 10. The characteristic is then the exponent used. For example:

$$256\ 000\ 000 = 2.56 \times 10^{8} \qquad \log = 8.408240$$

$$0.000\ 000\ 256 = 2.56 \times 10^{-7} \qquad \log = \bar{7}.408240 \quad \text{or} \quad 3.408240 - 10$$

Inasmuch as the characteristic may be determined by inspection, the mantissas only are given in tables of common logarithms.

USE OF LOGARITHM TABLES

To find the common logarithm of a number:

(*Note:* This description and examples refer specifically to the table entitled "Six-Place Mantissas for Common Logarithms." For the other tables, there will be minor differences from this description. Most of these differences are obvious. Notes with the individual tables explain any differences which are not immediately obvious.)

For a number of four figures, take out the tabular mantissa on a line with the first three figures of the number and under its fourth figure. The characteristic is determined as previously explained.

For a number of less than four figures, supply zeros to make a four figure number and take the value of the mantissa from the tables as before. For example: log 2 = log 2.000 = 0.301030.

(Notice that in some of the tables not all of the digits of the logarithm are given for every value. For example, in the table of six-place common logarithms, the first two digits of each mantissa are given only once for each line. The remaining four digits of each mantissa are given in the correct place in the table. When the leading two digits are not given on a line, they should be taken from the last line above it on which they do appear. When a mantissa is marked with an asterisk, it indicates that the value for the leading digits is to be taken from the *next* line instead of the present line. Similar remarks apply to the other tables in which this method of presenting the values are used.)

For a number of more than four figures, interpolation must be used. There are several precautions that must be observed when interpolating:

1. Linear interpolation, as described below, may only be used to add one extra digit to the argument (i.e., in the present case, for a five-digit argument).
2. Even though the mantissas given in the table are accurate to six decimal places, interpolated values are accurate only to the same number of places as in the argument, i.e. five places.
3. Because of the rapidly changing values in this region of the table, interpolation is not accurate if the first two digits of the argument are 11 or 12. For this reason, the table has been extended at the end so that such values may be read directly from

the table with five-digit arguments, without interpolation. (The four-place table is not so extended. If interpolation is required in this section of the four-place table, the value should be read instead from the six-place table without interpolation.)

If the above precautions cannot be observed. then higher order interpolation should be used.

Where applicable, linear interpolation is carried out as follows:

Take the tabular value of the mantissa for the first four figures; find the difference between the mantissa and the next greater tabular mantissa and multiply the difference so found by the remaining figures of the number as a decimal and add the product to the mantissa of the first four figures. For example, to find log 46.762:

$$\log 46.76 = 1.669875$$

Tabular difference between this mantissa and that for 4677 is .000092

$$\begin{aligned}\therefore \log 46.762 &= 1.669875 + .2 \times .000092 \\ &= 1.669875 + .000018 \\ &= 1.669893\end{aligned}$$

The last digit is not accurate, so must be rounded out. Thus

$$\log 46.762 = 1.66989$$

The accuracy will not ordinarily be affected by more than 1 in the last place if the mantissas are rounded to five decimal places before interpolation, and this makes the computation somewhat easier.

In the four-place logarithm table, a column of proportional parts is given at the end of each line. The number in the column under the fourth digit of the argument is the amount that must be added to any mantissa in that line to interpolate for the fourth digit. This number is to be added to the last place of the mantissa. These numbers are averages for the entire line, so may be off by 1 in the last place.

For example, to find log 33.74

$$\begin{aligned}\log 33.7 &= 1.5276 \\ \text{proportional part for } 4 &= \underline{5} \\ \therefore \log 33.74 &= 1.5281\end{aligned}$$

To find the number corresponding to a given logarithm:

(*Note*: This number is called the antilogarithm, and is denoted by $\log^{-1}$. Since the logarithm function is the inverse of the exponential function, $\log_a^{-1} x = a^x$. Therefore, any procedure or table which calculates antilogarithms may also be used to calculate exponentials, and vice-versa. In particular, tables of e^x may be used to compute antilogarithms to the base e.)

The procedure given below refers to the six-place logarithm table. As before, any significant deviation for other tables will be noted.

If the mantissa is found exactly in the table, join the figure at the top which is directly above the given mantissa to the three figures on the line at the left and place the decimal point according to the characteristic of the logarithm. For example,

$$\log^{-1} 3.399674 = \text{antilog } 3.399674 = 2510.$$

If the mantissa is not found exactly in the table it is necessary to interpolate. For example, to find antilog 3.400280, we find in the table

$$\begin{aligned}\text{antilog } 3.400192 &= 2513. \\ \text{antilog } \underline{3.400365} &= 2514. \\ \text{tabular difference} \quad 173 &\end{aligned}$$

The required difference is 88, so we must add $\frac{88}{173} = .5$.

$$\therefore \text{ antilog } 3.400280 = 2513.5$$

The same precautions must be observed for interpolation in finding antilogarithms as in finding logarithms.

A four-place antilogarithm table is also provided. When using this table, the mantissa of the logarithm is looked up on the margins of the table, and the significant digits of the antilogarithm are read from the body of the table. Lookup and interpolation are done in the same manner as when looking up logarithms in a table of logarithms.

Tables of natural logarithms are used in the same way as tables of common logarithms, except that they contain both the characteristics and the mantissas of the logarithms.

Examples of the use of logarithms in computation follow. Almost all computation with logarithms is done with common logarithms, since the computation of the characteristic is simpler, and since only the significant digits of the argument need be given in the table, without regard for the decimal point location. These examples all use the table of six-place common logarithms.

1. $52600 \times 0.00381 \times 2.74 = 549.11$

$$\begin{array}{lll} \log 52600 & = & 4.720986 \\ \log \quad 0.00381 & = & \bar{3}.580925 \\ \log \quad 2.74 & = & \underline{0.437751} \\ \text{adding} & = & 2.739662 \\ \text{antilog} & = & 549.11 \end{array}$$

The sum is the logarithm of the product, the mantissa of which is 739662. On looking up this mantissa in the logarithm tables we see that it corresponds to the digits 54911. The characteristic is 2, hence there are three figures before the decimal point. The number corresponding to the logarithm, called the antilogarithm, is 549.11.

2. $0.00123 \div 52.7 = 0.000\ 023\ 34$

$$\begin{array}{lll} \log \ 0.00123 & = & \bar{3}.089905 \\ \log 52.7 & = & \underline{1.721811} \\ \text{subtracting} & = & \bar{5}.368094 \\ \text{antilog} & = & 0.000\ 023\ 34 \end{array}$$

An alternative method:

$$\begin{array}{lll} \log \ 0.00123 & = & 7.089905 - 10 \\ \log 52.7 & = & \underline{1.721811 } \\ & & 5.368094 - 10 \end{array}$$

The characteristic 5(5. − 10) shows four zeros after the decimal point before the first significant figure.

3. $\dfrac{273 \times 780}{292 \times 760} \times 15 \times 0.09 = 1.2954$

$$\begin{array}{lll} \log 273 & = & 2.436163 \\ \log 780 & = & 2.892095 \\ \log \ 15 & = & 1.176091 \\ \log \ 0.09 & = & \underline{\bar{2}.954243} \\ \log \text{numerator} & = & 5.458592 \end{array}$$

$$\begin{array}{lll} \log 292 & = & 2.465383 \\ \log 760 & = & \underline{2.880814} \\ \log \text{denominator} & = & 5.346197 \end{array}$$

$$\begin{array}{lll} \log \text{numerator} & = & 5.458592 \\ \log \text{denominator} & = & \underline{5.346197} \\ \text{subtracting} & = & 0.112395 \\ \text{antilog} & = & 1.2954 \end{array}$$

As division may be accomplished by multiplying by the reciprocal of a number, the above may be considerably simplified. The logarithm of the reciprocal of a number, called the cologarithm, is readily obtained from the table by subtracting the logarithm of the number from zero. This may be readily read off from the table of mantissas. Change the sign of the characteristic algebraically adding to it -1, then mentally subtract each figure of the mantissa from 9 proceeding from left to right, the last figure being subtracted from 10. The example then is:

$$
\begin{aligned}
\log 273 &= 2.436163\\
\log 780 &= 2.892095\\
\log 15 &= 1.176091\\
\log 0.09 &= \bar{2}.954243\\
\text{colog } 292 &= \bar{3}.534617\\
\text{colog } 760 &= \underline{\bar{3}.119186}\\
& 0.112395
\end{aligned}
$$

4. $(0.00098)^4 = 9.2237 \times 10^{-13}$

$$
\begin{aligned}
\log 0.00098 &= \bar{4}.991226\\
& \underline{\qquad\quad 4}\\
& 3.964904 \quad (a)\\
\bar{4} \times 4 &= \bar{16}. \qquad\quad (b)\\
\log (0.00098)^4 &= \bar{13}.964904 \quad (c)\\
\text{antilog} &= 9.2237 \times 10^{-13}
\end{aligned}
$$

An alternative method:

$$
\begin{aligned}
\log 0.00098 &= 6.991226 - 10\\
& \underline{\qquad\qquad\quad 4}\\
& 27.964904 - 40\\
\text{or } & 7.964904 - 20\\
\text{or } & \bar{13}.964904\\
\text{antilog} &= 9.2237 \times 10^{-13}
\end{aligned}
$$

In the above it will be noted that the mantissa is always positive, hence the multiplication of the mantissa shown at (a), while (b) shows the multiplication of the characteristic. (c) is the algebraic sum.

5. $\sqrt[5]{492} = 3.4546$
$\log 492 = 2.691965$

Dividing the logarithm by 5 gives as the logarithm of the root 0.538393 the antilogarithm of which is 3.4546 both characteristic and mantissa being positive. When the characteristic is negative and not evenly divisible by the root to be taken, a modification of the logarithm is necessary, as the following example shows:

6. $\sqrt[3]{0.000372} = 0.07192$

$$
\begin{aligned}
\log 3.72 \times 10^{-4} &= \bar{4}.570543 \qquad\quad (a)\\
&= 26.570543 - 30 \quad (b)
\end{aligned}
$$

Dividing (b) by 3 gives $8.856848 - 10$ which may be written $\bar{2}.856848$ and is the logarithm of the root sought, the antilogarithm of which is 0.07192.

7. $(0.000\ 372)^{1.2} = 0.000\ 076\ 675$

$$
\begin{aligned}
\log 0.000\ 372 &= \bar{4}.570543\\
\text{or } & 6.570543 - 10\\
& \underline{\qquad\qquad\quad 1.2}\\
& 7.884652 - 12\\
\text{antilog} &= 0.000\ 076\ 675
\end{aligned}
$$

8. $(0.000372)^{-1.32} = 33642$

$$
\begin{aligned}
\text{colog } 0.000372 &= 3.429457\\
& \underline{\qquad\quad 1.32}\\
& 4.526883\\
\text{antilog} &= 33642
\end{aligned}
$$

*FOUR-PLACE MANTISSAS FOR COMMON LOGARITHMS OF DECIMAL FRACTIONS

N	0	1	2	3	4	5	6	7	8	9
.10	−1.000	−.9957	−.9914	−.9872	−.9830	−.9788	−.9747	−.9706	−.9666	−.9626
.11	−.9586	−.9547	−.9508	−.9469	−.9431	−.9393	−.9355	−.9318	−.9281	−.9245
.12	−.9208	−.9172	−.9136	−.9101	−.9066	−.9031	−.8996	−.8962	−.8928	−.8894
.13	−.8861	−.8827	−.8794	−.8761	−.8729	−.8697	−.8665	−.8633	−.8601	−.8570
.14	−.8539	−.8508	−.8477	−.8447	−.8416	−.8386	−.8356	−.8327	−.8297	−.8268
.15	−.8239	−.8210	−.8182	−.8153	−.8125	−.8097	−.8069	−.8041	−.8013	−.7986
.16	−.7959	−.7932	−.7905	−.7878	−.7852	−.7825	−.7799	−.7773	−.7747	−.7721
.17	−.7696	−.7670	−.7645	−.7620	−.7595	−.7570	−.7545	−.7520	−.7496	−.7471
.18	−.7447	−.7423	−.7399	−.7375	−.7352	−.7328	−.7305	−.7282	−.7258	−.7235
.19	−.7212	−.7190	−.7167	−.7144	−.7122	−.7100	−.7077	−.7055	−.7033	−.7011
.20	−.6990	−.6968	−.6946	−.6925	−.6904	−.6882	−.6861	−.6840	−.6819	−.6799
.21	−.6778	−.6757	−.6737	−.6716	−.6696	−.6676	−.6655	−.6635	−.6615	−.6596
.22	−.6576	−.6556	−.6536	−.6517	−.6498	−.6478	−.6459	−.6440	−.6421	−.6402
.23	−.6383	−.6364	−.6345	−.6326	−.6308	−.6289	−.6271	−.6253	−.6234	−.6216
.24	−.6198	−.6180	−.6162	−.6144	−.6126	−.6108	−.6091	−.6073	−.6055	−.6038
.25	−.6021	−.6003	−.5986	−.5969	−.5952	−.5935	−.5918	−.5901	−.5884	−.5867
.26	−.5850	−.5834	−.5817	−.5800	−.5784	−.5768	−.5751	−.5735	−.5719	−.5702
.27	−.5686	−.5670	−.5654	−.5638	−.5622	−.5607	−.5591	−.5575	−.5560	−.5544
.28	−.5528	−.5513	−.5498	−.5482	−.5467	−.5452	−.5436	−.5421	−.5406	−.5391
.29	−.5376	−.5361	−.5346	−.5331	−.5317	−.5302	−.5287	−.5272	−.5258	−.5243
.30	−.5229	−.5214	−.5200	−.5186	−.5171	−.5157	−.5143	−.5129	−.5114	−.5100
.31	−.5086	−.5072	−.5058	−.5045	−.5031	−.5017	−.5003	−.4989	−.4976	−.4962
.32	−.4949	−.4935	−.4921	−.4908	−.4895	−.4881	−.4868	−.4855	−.4841	−.4828
.33	−.4815	−.4802	−.4789	−.4776	−.4763	−.4750	−.4737	−.4724	−.4711	−.4698
.34	−.4685	−.4672	−.4660	−.4647	−.4634	−.4622	−.4609	−.4597	−.4584	−.4572
.35	−.4559	−.4547	−.4535	−.4522	−.4510	−.4498	−.4486	−.4473	−.4461	−.4449
.36	−.4437	−.4425	−.4413	−.4401	−.4389	−.4377	−.4365	−.4353	−.4342	−.4330
.37	−.4318	−.4306	−.4295	−.4283	−.4271	−.4260	−.4248	−.4237	−.4225	−.4214
.38	−.4202	−.4191	−.4179	−.4168	−.4157	−.4145	−.4134	−.4123	−.4112	−.4101
.39	−.4089	−.4078	−.4067	−.4056	−.4045	−.4034	−.4023	−.4012	−.4001	−.3990
.40	−.3979	−.3969	−.3958	−.3947	−.3936	−.3925	−.3915	−.3904	−.3893	−.3883
.41	−.3872	−.3862	−.3851	−.3840	−.3830	−.3820	−.3809	−.3799	−.3788	−.3778
.42	−.3768	−.3757	−.3747	−.3737	−.3726	−.3716	−.3706	−.3696	−.3686	−.3675
.43	−.3665	−.3655	−.3645	−.3635	−.3625	−.3615	−.3605	−.3595	−.3585	−.3575
.44	−.3565	−.3556	−.3546	−.3536	−.3526	−.3516	−.3507	−.3497	−.3487	−.3478
.45	−.3468	−.3458	−.3449	−.3439	−.3429	−.3420	−.3410	−.3401	−.3391	−.3382
.46	−.3372	−.3363	−.3354	−.3344	−.3335	−.3325	−.3316	−.3307	−.3298	−.3288
.47	−.3279	−.3270	−.3261	−.3251	−.3242	−.3233	−.3224	−.3215	−.3206	−.3197
.48	−.3188	−.3179	−.3170	−.3161	−.3152	−.3143	−.3134	−.3125	−.3116	−.3107
.49	−.3098	−.3089	−.3080	−.3072	−.3063	−.3054	−.3045	−.3036	−.3028	−.3019
.50	−.3010	−.3002	−.2993	−.2984	−.2976	−.2967	−.2958	−.2950	−.2941	−.2933
.51	−.2924	−.2916	−.2907	−.2899	−.2890	−.2882	−.2874	−.2865	−.2857	−.2848
.52	−.2840	−.2832	−.2823	−.2815	−.2807	−.2798	−.2790	−.2782	−.2774	−.2765
.53	−.2757	−.2749	−.2741	−.2733	−.2725	−.2716	−.2708	−.2700	−.2692	−.2684
.54	−.2676	−.2668	−.2660	−.2652	−.2644	−.2636	−.2628	−.2620	−.2612	−.2604

* This table gives the logarithms of the decimal fractions which are negative numbers.

For example log 0.61 = −0.2147 = 9.7853 − 10. It should be noted that the entries as given can be used conveniently to find cologarithms of positive numbers. Every positive number $N = P \cdot (10)^k$, where $0 < P \leqq 1$. Since colog N = − log N, it follows that colog N = − log P − k.

For example colog 0.61 = − log 0.61 = 0.2147; colog 61 = 0.2147 − 2; and colog 0.00061 = 3.2147.

*FOUR-PLACE MANTISSAS FOR COMMON LOGARITHMS OF DECIMAL FRACTIONS (Continued)

N	0	1	2	3	4	5	6	7	8	9
.55	−.2596	−.2588	−.2581	−.2573	−.2565	−.2557	−.2549	−.2541	−.2534	−.2526
.56	−.2518	−.2510	−.2503	−.2495	−.2487	−.2480	−.2472	−.2464	−.2457	−.2449
.57	−.2441	−.2434	−.2426	−.2418	−.2411	−.2403	−.2396	−.2388	−.2381	−.2373
.58	−.2366	−.2358	−.2351	−.2343	−.2336	−.2328	−.2321	−.2314	−.2306	−.2299
.59	−.2291	−.2284	−.2277	−.2269	−.2262	−.2255	−.2248	−.2240	−.2233	−.2226
.60	−.2218	−.2211	−.2204	−.2197	−.2190	−.2182	−.2175	−.2168	−.2161	−.2154
.61	−.2147	−.2140	−.2132	−.2125	−.2118	−.2111	−.2104	−.2097	−.2090	−.2083
.62	−.2076	−.2069	−.2062	−.2055	−.2048	−.2041	−.2034	−.2027	−.2020	−.2013
.63	−.2007	−.2000	−.1993	−.1986	−.1979	−.1972	−.1965	−.1959	−.1952	−.1945
.64	−.1938	−.1931	−.1925	−.1918	−.1911	−.1904	−.1898	−.1891	−.1884	−.1878
.65	−.1871	−.1864	−.1858	−.1851	−.1844	−.1838	−.1831	−.1824	−.1818	−.1811
.66	−.1805	−.1798	−.1791	−.1785	−.1778	−.1772	−.1765	−.1759	−.1752	−.1746
.67	−.1739	−.1733	−.1726	−.1720	−.1713	−.1707	−.1701	−.1694	−.1688	−.1681
.68	−.1675	−.1669	−.1662	−.1656	−.1649	−.1643	−.1637	−.1630	−.1624	−.1618
.69	−.1612	−.1605	−.1599	−.1593	−.1586	−.1580	−.1574	−.1568	−.1561	−.1555
.70	−.1549	−.1543	−.1537	−.1530	−.1524	−.1518	−.1512	−.1506	−.1500	−.1494
.71	−.1487	−.1481	−.1475	−.1469	−.1463	−.1457	−.1451	−.1445	−.1439	−.1433
.72	−.1427	−.1421	−.1415	−.1409	−.1403	−.1397	−.1391	−.1385	−.1379	−.1373
.73	−.1367	−.1361	−.1355	−.1349	−.1343	−.1337	−.1331	−.1325	−.1319	−.1314
.74	−.1308	−.1302	−.1296	−.1290	−.1284	−.1278	−.1273	−.1267	−.1261	−.1255
.75	−.1249	−.1244	−.1238	−.1232	−.1226	−.1221	−.1215	−.1209	−.1203	−.1198
.76	−.1192	−.1186	−.1180	−.1175	−.1169	−.1163	−.1158	−.1152	−.1146	−.1141
.77	−.1135	−.1129	−.1124	−.1118	−.1113	−.1107	−.1101	−.1096	−.1090	−.1085
.78	−.1079	−.1073	−.1068	−.1062	−.1057	−.1051	−.1046	−.1040	−.1035	−.1029
.79	−.1024	−.1018	−.1013	−.1007	−.1002	−.0996	−.0991	−.0985	−.0980	−.0975
.80	−.0969	−.0964	−.0958	−.0953	−.0947	−.0942	−.0937	−.0931	−.0926	−.0921
.81	−.0915	−.0910	−.0904	−.0899	−.0894	−.0888	−.0883	−.0878	−.0872	−.0867
.82	−.0862	−.0857	−.0851	−.0846	−.0841	−.0835	−.0830	−.0825	−.0820	−.0814
.83	−.0809	−.0804	−.0799	−.0794	−.0788	−.0783	−.0778	−.0773	−.0768	−.0762
.84	−.0757	−.0752	−.0747	−.0742	−.0737	−.0731	−.0726	−.0721	−.0716	−.0711
.85	−.0706	−.0701	−.0696	−.0691	−.0685	−.0680	−.0675	−.0670	−.0665	−.0660
.86	−.0655	−.0650	−.0645	−.0640	−.0635	−.0630	−.0625	−.0620	−.0615	−.0610
.87	−.0605	−.0600	−.0595	−.0590	−.0585	−.0580	−.0575	−.0570	−.0565	−.0560
.88	−.0555	−.0550	−.0545	−.0540	−.0535	−.0531	−.0526	−.0521	−.0516	−.0511
.89	−.0506	−.0501	−.0496	−.0491	−.0487	−.0482	−.0477	−.0472	−.0467	−.0462
.90	−.0458	−.0453	−.0448	−.0443	−.0438	−.0434	−.0429	−.0424	−.0419	−.0414
.91	−.0410	−.0405	−.0400	−.0395	−.0391	−.0386	−.0381	−.0376	−.0372	−.0367
.92	−.0362	−.0357	−.0353	−.0348	−.0343	−.0339	−.0334	−.0329	−.0325	−.0320
.93	−.0315	−.0311	−.0306	−.0301	−.0297	−.0292	−.0287	−.0283	−.0278	−.0273
.94	−.0269	−.0264	−.0259	−.0255	−.0250	−.0246	−.0241	−.0237	−.0232	−.0227
.95	−.0223	−.0218	−.0214	−.0209	−.0205	−.0200	−.0195	−.0191	−.0186	−.0182
.96	−.0177	−.0173	−.0168	−.0164	−.0159	−.0155	−.0150	−.0146	−.0141	−.0137
.97	−.0132	−.0128	−.0123	−.0119	−.0114	−.0110	−.0106	−.0101	−.0097	−.0092
.98	−.0088	−.0083	−.0079	−.0074	−.0070	−.0066	−.0061	−.0057	−.0052	−.0048
.99	−.0044	−.0039	−.0035	−.0031	−.0026	−.0022	−.0017	−.0013	−.0009	−.0004

* See footnote page 175.

FOUR-PLACE MANTISSAS FOR COMMON LOGARITHMS

N	0	1	2	3	4	5	6	7	8	9	Proportional Parts 1	2	3	4	5	6	7	8	9
10	0000	0043	0086	0128	0170	0212	0253	0294	0334	0374	*4	8	12	17	21	25	29	33	37
11	0414	0453	0492	0531	0569	0607	0645	0682	0719	0755	4	8	11	15	19	23	26	30	34
12	0792	0828	0864	0899	0934	0969	1004	1038	1072	1106	3	7	10	14	17	21	24	28	31
13	1139	1173	1206	1239	1271	1303	1335	1367	1399	1430	3	6	10	13	16	19	23	26	29
14	1461	1492	1523	1553	1584	1614	1644	1673	1703	1732	3	6	9	12	15	18	21	24	27
15	1761	1790	1818	1847	1875	1903	1931	1959	1987	2014	*3	6	8	11	14	17	20	22	25
16	2041	2068	2095	2122	2148	2175	2201	2227	2253	2279	3	5	8	11	13	16	18	21	24
17	2304	2330	2355	2380	2405	2430	2455	2480	2504	2529	2	5	7	10	12	15	17	20	22
18	2553	2577	2601	2625	2648	2672	2695	2718	2742	2765	2	5	7	9	12	14	16	19	21
19	2788	2810	2833	2856	2878	2900	2923	2945	2967	2989	2	4	7	9	11	13	16	18	20
20	3010	3032	3054	3075	3096	3118	3139	3160	3181	3201	2	4	6	8	11	13	15	17	19
21	3222	3243	3263	3284	3304	3324	3345	3365	3385	3404	2	4	6	8	10	12	14	16	18
22	3424	3444	3464	3483	3502	3522	3541	3560	3579	3598	2	4	6	8	10	12	14	15	17
23	3617	3636	3655	3674	3692	3711	3729	3747	3766	3784	2	4	6	7	9	11	13	15	17
24	3802	3820	3838	3856	3874	3892	3909	3927	3945	3962	2	4	5	7	9	11	12	14	16
25	3979	3997	4014	4031	4048	4065	4082	4099	4116	4133	2	3	5	7	9	10	12	14	15
26	4150	4166	4183	4200	4216	4232	4249	4265	4281	4298	2	3	5	7	8	10	11	13	15
27	4314	4330	4346	4362	4378	4393	4409	4425	4440	4456	2	3	5	6	8	9	11	13	14
28	4472	4487	4502	4518	4533	4548	4564	4579	4594	4609	2	3	5	6	8	9	11	12	14
29	4624	4639	4654	4669	4683	4698	4713	4728	4742	4757	1	3	4	6	7	9	10	12	13
30	4771	4786	4800	4814	4829	4843	4857	4871	4886	4900	1	3	4	6	7	9	10	11	13
31	4914	4928	4942	4955	4969	4983	4997	5011	5024	5038	1	3	4	6	7	8	10	11	12
32	5051	5065	5079	5092	5105	5119	5132	5145	5159	5172	1	3	4	5	7	8	9	11	12
33	5185	5198	5211	5224	5237	5250	5263	5276	5289	5302	1	3	4	5	6	8	9	10	12
34	5315	5328	5340	5353	5366	5378	5391	5403	5416	5428	1	3	4	5	6	8	9	10	11
35	5441	5453	5465	5478	5490	5502	5514	5527	5539	5551	1	2	4	5	6	7	9	10	11
36	5563	5575	5587	5599	5611	5623	5635	5647	5658	5670	1	2	4	5	6	7	8	10	11
37	5682	5694	5705	5717	5729	5740	5752	5763	5775	5786	1	2	3	5	6	7	8	9	10
38	5798	5809	5821	5832	5843	5855	5866	5877	5888	5899	1	2	3	5	6	7	8	9	10
39	5911	5922	5933	5944	5955	5966	5977	5988	5999	6010	1	2	3	4	5	7	8	9	10
40	6021	6031	6042	6053	6064	6075	6085	6096	6107	6117	1	2	3	4	5	6	8	9	10
41	6128	6138	6149	6160	6170	6180	6191	6201	6212	6222	1	2	3	4	5	6	7	8	9
42	6232	6243	6253	6263	6274	6284	6294	6304	6314	6325	1	2	3	4	5	6	7	8	9
43	6335	6345	6355	6365	6375	6385	6395	6405	6415	6425	1	2	3	4	5	6	7	8	9
44	6435	6444	6454	6464	6474	6484	6493	6503	6513	6522	1	2	3	4	5	6	7	8	9
45	6532	6542	6551	6561	6571	6580	6590	6599	6609	6618	1	2	3	4	5	6	7	8	9
46	6628	6637	6646	6656	6665	6675	6684	6693	6702	6712	1	2	3	4	5	6	7	7	8
47	6721	6730	6739	6749	6758	6767	6776	6785	6794	6803	1	2	3	4	5	5	6	7	8
48	6812	6821	6830	6839	6848	6857	6866	6875	6884	6893	1	2	3	4	4	5	6	7	8
49	6902	6911	6920	6928	6937	6946	6955	6964	6972	6981	1	2	3	4	4	5	6	7	8
50	6990	6998	7007	7016	7024	7033	7042	7050	7059	7067	1	2	3	3	4	5	6	7	8
51	7076	7084	7093	7101	7110	7118	7126	7135	7143	7152	1	2	3	3	4	5	6	7	8
52	7160	7168	7177	7185	7193	7202	7210	7218	7226	7235	1	2	2	3	4	5	6	7	7
53	7243	7251	7259	7267	7275	7284	7292	7300	7308	7316	1	2	2	3	4	5	6	6	7
54	7324	7332	7340	7348	7356	7364	7372	7380	7388	7396	1	2	2	3	4	5	6	6	7
N	0	1	2	3	4	5	6	7	8	9	1	2	3	4	5	6	7	8	9

*The use of proportional parts for log 1.01 to log 1.59 is less accurate than usual.

FOUR-PLACE MANTISSAS FOR COMMON LOGARITHMS (Continued)

N	0	1	2	3	4	5	6	7	8	9	Proportional Parts								
											1	2	3	4	5	6	7	8	9
55	7404	7412	7419	7427	7435	7443	7451	7459	7466	7474	1	2	2	3	4	5	5	6	7
56	7482	7490	7497	7505	7513	7520	7528	7536	7543	7551	1	2	2	3	4	5	5	6	7
57	7559	7566	7574	7582	7589	7597	7604	7612	7619	7627	1	2	2	3	4	5	5	6	7
58	7634	7642	7649	7657	7664	7672	7679	7686	7694	7701	1	1	2	3	4	4	5	6	7
59	7709	7716	7723	7731	7738	7745	7752	7760	7767	7774	1	1	2	3	4	4	5	6	7
60	7782	7789	7796	7803	7810	7818	7825	7832	7839	7846	1	1	2	3	4	4	5	6	6
61	7853	7860	7868	7875	7882	7889	7896	7903	7910	7917	1	1	2	3	4	4	5	6	6
62	7924	7931	7938	7945	7952	7959	7966	7973	7980	7987	1	1	2	3	3	4	5	6	6
63	7993	8000	8007	8014	8021	8028	8035	8041	8048	8055	1	1	2	3	3	4	5	5	6
64	8062	8069	8075	8082	8089	8096	8102	8109	8116	8122	1	1	2	3	3	4	5	5	6
65	8129	8136	8142	8149	8156	8162	8169	8176	8182	8189	1	1	2	3	3	4	5	5	6
66	8195	8202	8209	8215	8222	8228	8235	8241	8248	8254	1	1	2	3	3	4	5	5	6
67	8261	8267	8274	8280	8287	8293	8299	8306	8312	8319	1	1	2	3	3	4	5	5	6
68	8325	8331	8338	8344	8351	8357	8363	8370	8376	8382	1	1	2	3	3	4	4	5	6
69	8388	8395	8401	8407	8414	8420	8426	8432	8439	8445	1	1	2	2	3	4	4	5	6
70	8451	8457	8463	8470	8476	8482	8488	8494	8500	8506	1	1	2	2	3	4	4	5	6
71	8513	8519	8525	8531	8537	8543	8549	8555	8561	8567	1	1	2	2	3	4	4	5	5
72	8573	8579	8585	8591	8597	8603	8609	8615	8621	8627	1	1	2	2	3	4	4	5	5
73	8633	8639	8645	8651	8657	8663	8669	8675	8681	8686	1	1	2	2	3	4	4	5	5
74	8692	8698	8704	8710	8716	8722	8727	8733	8739	8745	1	1	2	2	3	4	4	5	5
75	8751	8756	8762	8768	8774	8779	8785	8791	8797	8802	1	1	2	2	3	3	4	5	5
76	8808	8814	8820	8825	8831	8837	8842	8848	8854	8859	1	1	2	2	3	3	4	5	5
77	8865	8871	8876	8882	8887	8893	8899	8904	8910	8915	1	1	2	2	3	3	4	4	5
78	8921	8927	8932	8938	8943	8949	8954	8960	8965	8971	1	1	2	2	3	3	4	4	5
79	8976	8982	8987	8993	8998	9004	9009	9015	9020	9025	1	1	2	2	3	3	4	4	5
80	9031	9036	9042	9047	9053	9058	9063	9069	9074	9079	1	1	2	2	3	3	4	4	5
81	9085	9090	9096	9101	9106	9112	9117	9122	9128	9133	1	1	2	2	3	3	4	4	5
82	9138	9143	9149	9154	9159	9165	9170	9175	9180	9186	1	1	2	2	3	3	4	4	5
83	9191	9196	9201	9206	9212	9217	9222	9227	9232	9238	1	1	2	2	3	3	4	4	5
84	9243	9248	9253	9258	9263	9269	9274	9279	9284	9289	1	1	2	2	3	3	4	4	5
85	9294	9299	9304	9309	9315	9320	9325	9330	9335	9340	1	1	2	2	3	3	4	4	5
86	9345	9350	9355	9360	9365	9370	9375	9380	9385	9390	1	1	2	2	3	3	4	4	5
87	9395	9400	9405	9410	9415	9420	9425	9430	9435	9440	0	1	1	2	2	3	3	4	4
88	9445	9450	9455	9460	9465	9469	9474	9479	9484	9489	0	1	1	2	2	3	3	4	4
89	9494	9499	9504	9509	9513	9518	9523	9528	9533	9538	0	1	1	2	2	3	3	4	4
90	9542	9547	9552	9557	9562	9566	9571	9576	9581	9586	0	1	1	2	2	3	3	4	4
91	9590	9595	9600	9605	9609	9614	9619	9624	9628	9633	0	1	1	2	2	3	3	4	4
92	9638	9643	9647	9652	9657	9661	9666	9671	9675	9680	0	1	1	2	2	3	3	4	4
93	9685	9689	9694	9699	9703	9708	9713	9717	9722	9727	0	1	1	2	2	3	3	4	4
94	9731	9736	9741	9745	9750	9754	9759	9763	9768	9773	0	1	1	2	2	3	3	4	4
95	9777	9782	9786	9791	9795	9800	9805	9809	9814	9818	0	1	1	2	2	3	3	4	4
96	9823	9827	9832	9836	9841	9845	9850	9854	9859	9863	0	1	1	2	2	3	3	4	4
97	9868	9872	9877	9881	9886	9890	9894	9899	9903	9908	0	1	1	2	2	3	3	4	4
98	9912	9917	9921	9926	9930	9934	9939	9943	9948	9952	0	1	1	2	2	3	3	4	4
99	9956	9961	9965	9969	9974	9978	9983	9987	9991	9996	0	1	1	2	2	3	3	3	4
N	0	1	2	3	4	5	6	7	8	9	1	2	3	4	5	6	7	8	9

ANTILOGARITHMS

	0	1	2	3	4	5	6	7	8	9	Proportional Parts 1	2	3	4	5	6	7	8	9
.00	1000	1002	1005	1007	1009	1012	1014	1016	1019	1021	0	0	1	1	1	1	2	2	2
.01	1023	1026	1028	1030	1033	1035	1038	1040	1042	1045	0	0	1	1	1	1	2	2	2
.02	1047	1050	1052	1054	1057	1059	1062	1064	1067	1069	0	0	1	1	1	1	2	2	2
.03	1072	1074	1076	1079	1081	1084	1086	1089	1091	1094	0	0	1	1	1	1	2	2	2
.04	1096	1099	1102	1104	1107	1109	1112	1114	1117	1119	0	1	1	1	1	2	2	2	2
.05	1122	1125	1127	1130	1132	1135	1138	1140	1143	1146	0	1	1	1	1	2	2	2	2
.06	1148	1151	1153	1156	1159	1161	1164	1167	1169	1172	0	1	1	1	1	2	2	2	2
.07	1175	1178	1180	1183	1186	1189	1191	1194	1197	1199	0	1	1	1	1	2	2	2	2
.08	1202	1205	1208	1211	1213	1216	1219	1222	1225	1227	0	1	1	1	1	2	2	2	3
.09	1230	1233	1236	1239	1242	1245	1247	1250	1253	1256	0	1	1	1	1	2	2	2	3
.10	1259	1262	1265	1268	1271	1274	1276	1279	1282	1285	0	1	1	1	1	2	2	2	3
.11	1288	1291	1294	1297	1300	1303	1306	1309	1312	1315	0	1	1	1	2	2	2	3	3
.12	1318	1321	1324	1327	1330	1334	1337	1340	1343	1346	0	1	1	1	2	2	2	2	3
.13	1349	1352	1355	1358	1361	1365	1368	1371	1374	1377	0	1	1	1	2	2	2	3	3
.14	1380	1384	1387	1390	1393	1396	1400	1403	1406	1409	0	1	1	1	2	2	2	3	3
.15	1413	1416	1419	1422	1426	1429	1432	1435	1439	1442	0	1	1	1	2	2	2	3	3
.16	1445	1449	1452	1455	1459	1462	1466	1469	1472	1476	0	1	1	1	2	2	2	3	3
.17	1479	1483	1486	1489	1493	1496	1500	1503	1507	1510	0	1	1	1	2	2	2	3	3
.18	1514	1517	1521	1524	1528	1531	1535	1538	1542	1545	0	1	1	1	2	2	2	3	3
.19	1549	1552	1556	1560	1563	1567	1570	1574	1578	1581	0	1	1	1	2	2	3	3	3
.20	1585	1589	1592	1596	1600	1603	1607	1611	1614	1618	0	1	1	1	2	2	3	3	3
.21	1622	1626	1629	1633	1637	1641	1644	1648	1652	1656	0	1	1	2	2	2	3	3	3
.22	1660	1663	1667	1671	1675	1679	1683	1687	1690	1694	0	1	1	2	2	2	3	3	3
.23	1698	1702	1706	1710	1714	1718	1722	1726	1730	1734	0	1	1	2	2	2	3	3	4
.24	1738	1742	1746	1750	1754	1758	1762	1766	1770	1774	0	1	1	2	2	2	3	3	4
.25	1778	1782	1786	1791	1795	1799	1803	1807	1811	1816	0	1	1	2	2	2	3	3	4
.26	1820	1824	1828	1832	1837	1841	1845	1849	1854	1858	0	1	1	2	2	3	3	3	4
.27	1862	1866	1871	1875	1879	1884	1888	1892	1897	1901	0	1	1	2	2	3	3	3	4
.28	1905	1910	1914	1919	1923	1928	1932	1936	1941	1945	0	1	1	2	2	3	3	4	4
.29	1950	1954	1959	1963	1968	1972	1977	1982	1986	1991	0	1	1	2	2	3	3	4	4
.30	1995	2000	2004	2009	2014	2018	2023	2028	2032	2037	0	1	1	2	2	3	3	4	4
.31	2042	2046	2051	2056	2061	2065	2070	2075	2080	2084	0	1	1	2	2	3	3	4	4
.32	2089	2094	2099	2104	2109	2113	2118	2123	2128	2133	0	1	1	2	2	3	3	4	4
.33	2138	2143	2148	2153	2158	2163	2168	2173	2178	2183	0	1	1	2	2	3	3	4	4
.34	2188	2193	2198	2203	2208	2213	2218	2223	2228	2234	1	1	2	2	3	3	4	4	5
.35	2239	2244	2249	2254	2259	2265	2270	2275	2280	2286	1	1	2	2	3	3	4	4	5
.36	2291	2296	2301	2307	2312	2317	2323	2328	2333	2339	1	1	2	2	3	3	4	4	5
.37	2344	2350	2355	2360	2366	2371	2377	2382	2388	2393	1	1	2	2	3	3	4	4	5
.38	2399	2404	2410	2415	2421	2427	2432	2438	2443	2449	1	1	2	2	3	3	4	4	5
.39	2455	2460	2466	2472	2477	2483	2489	2495	2500	2506	1	1	2	2	3	3	4	5	5
.40	2512	2518	2523	2529	2535	2541	2547	2553	2559	2564	1	1	2	2	3	4	4	5	5
.41	2570	2576	2582	2588	2594	2600	2606	2612	2618	2624	1	1	2	2	3	4	4	5	5
.42	2630	2636	2642	2649	2655	2661	2667	2673	2679	2685	1	1	2	2	3	4	4	5	6
.43	2692	2698	2704	2710	2716	2723	2729	2735	2742	2748	1	1	2	3	3	4	4	5	6
.44	2754	2761	2767	2773	2780	2786	2793	2799	2805	2812	1	1	2	3	3	4	4	5	6
.45	2818	2825	2831	2838	2844	2851	2858	2864	2871	2877	1	1	2	3	3	4	5	5	6
.46	2884	2891	2897	2904	2911	2917	2924	2931	2938	2944	1	1	2	3	3	4	5	5	6
.47	2951	2958	2965	2972	2979	2985	2992	2999	3006	3013	1	1	2	3	3	4	5	5	6
.48	3020	3027	3034	3041	3048	3055	3062	3069	3076	3083	1	1	2	3	4	4	5	6	6
.49	3090	3097	3105	3112	3119	3126	3133	3141	3148	3155	1	1	2	3	4	4	5	6	6
	0	1	2	3	4	5	6	7	8	9	1	2	3	4	5	6	7	8	9

ANTILOGARITHMS (Continued)

	0	1	2	3	4	5	6	7	8	9	Proportional Parts 1	2	3	4	5	6	7	8	9
.50	3162	3170	3177	3184	3192	3199	3206	3214	3221	3228	1	1	2	3	4	4	5	6	7
.51	3236	3243	3251	3258	3266	3273	3281	3289	3296	3304	1	2	2	3	4	5	5	6	7
.52	3311	3319	3327	3334	3342	3350	3357	3365	3373	3381	1	2	2	3	4	5	5	6	7
.53	3388	3396	3404	3412	3420	3428	3436	3443	3451	3459	1	2	2	3	4	5	6	6	7
.54	3467	3475	3483	3491	3499	3508	3516	3524	3532	3540	1	2	2	3	4	5	6	6	7
.55	3548	3556	3565	3573	3581	3589	3597	3606	3614	3622	1	2	2	3	4	5	6	7	7
.56	3631	3639	3648	3656	3664	3673	3681	3690	3698	3707	1	2	3	3	4	5	6	7	8
.57	3715	3724	3733	3741	3750	3758	3767	3776	3784	3793	1	2	3	3	4	5	6	7	8
.58	3802	3811	3819	3828	3837	3846	3855	3864	3873	3882	1	2	3	4	4	5	6	7	8
.59	3890	3899	3908	3917	3926	3936	3945	3954	3963	3972	1	2	3	4	5	5	6	7	8
.60	3981	3990	3999	4009	4018	4027	4036	4046	4055	4064	1	2	3	4	5	6	6	7	8
.61	4074	4083	4093	4102	4111	4121	4130	4140	4150	4159	1	2	3	4	5	6	7	8	9
.62	4169	4178	4188	4198	4207	4217	4227	4236	4246	4256	1	2	3	4	5	6	7	8	9
.63	4266	4276	4285	4295	4305	4315	4325	4335	4345	4355	1	2	3	4	5	6	7	8	9
.64	4365	4375	4385	4395	4406	4416	4426	4436	4446	4457	1	2	3	4	5	6	7	8	9
.65	4467	4477	4487	4498	4508	4519	4529	4539	4550	4560	1	2	3	4	5	6	7	8	9
.66	4571	4581	4592	4603	4613	4624	4634	4645	4656	4667	1	2	3	4	5	6	7	9	10
.67	4677	4688	4699	4710	4721	4732	4742	4753	4764	4775	1	2	3	4	5	7	8	9	10
.68	4786	4797	4808	4819	4831	4842	4853	4864	4875	4887	1	2	3	4	6	7	8	9	10
.69	4898	4909	4920	4932	4943	4955	4966	4977	4989	5000	1	2	3	5	6	7	8	9	10
.70	5012	5023	5035	5047	5058	5070	5082	5093	5105	5117	1	2	4	5	6	7	8	9	11
.71	5129	5140	5152	5164	5176	5188	5200	5212	5224	5236	1	2	4	5	6	7	8	10	11
.72	5248	5260	5272	5284	5297	5309	5321	5333	5346	5358	1	2	4	5	6	7	9	10	11
.73	5370	5383	5395	5408	5420	5433	5445	5458	5470	5483	1	3	4	5	6	8	9	10	11
.74	5495	5508	5521	5534	5546	5559	5572	5585	5598	5610	1	3	4	5	6	8	9	10	12
.75	5623	5636	5649	5662	5675	5689	5702	5715	5728	5741	1	3	4	5	7	8	9	10	12
.76	5754	5768	5781	5794	5808	5821	5834	5848	5861	5875	1	3	4	5	7	8	9	11	12
.77	5888	5902	5916	5929	5943	5957	5970	5984	5998	6012	1	3	4	5	7	8	10	11	12
.78	6026	6039	6053	6067	6081	6095	6109	6124	6138	6152	1	3	4	6	7	8	10	11	13
.79	6166	6180	6194	6209	6223	6237	6252	6266	6281	6295	1	3	4	6	7	9	10	11	13
.80	6310	6324	6339	6353	6368	6383	6397	6412	6427	6442	1	3	4	6	7	9	10	12	13
.81	6457	6471	6486	6501	6516	6531	6546	6561	6577	6592	2	3	5	6	8	9	11	12	14
.82	6607	6622	6637	6653	6668	6683	6699	6714	6730	6745	2	3	5	6	8	9	11	12	14
.83	6761	6776	6792	6808	6823	6839	6855	6871	6887	6902	2	3	5	6	8	9	11	13	14
.84	6918	6934	6950	6966	6982	6998	7015	7031	7047	7063	2	3	5	6	8	10	11	13	15
.85	7079	7096	7112	7129	7145	7161	7178	7194	7211	7228	2	3	5	7	8	10	12	13	15
.86	7244	7261	7278	7295	7311	7328	7345	7362	7379	7396	2	3	5	7	8	10	12	13	15
.87	7413	7430	7447	7464	7482	7499	7516	7534	7551	7568	2	3	5	7	9	10	12	14	16
.88	7586	7603	7621	7638	7656	7674	7691	7709	7727	7745	2	4	5	7	9	11	12	14	16
.89	7762	7780	7798	7816	7834	7852	7870	7889	7907	7925	2	4	5	7	9	11	13	14	16
.90	7943	7962	7980	7998	8017	8035	8054	8072	8091	8110	2	4	6	7	9	11	13	15	17
.91	8128	8147	8166	8185	8204	8222	8241	8260	8279	8299	2	4	6	8	9	11	13	15	17
.92	8318	8337	8356	8375	8395	8414	8433	8453	8472	8492	2	4	6	8	10	12	14	15	17
.93	8511	8531	8551	8570	8590	8610	8630	8650	8670	8690	2	4	6	8	10	12	14	16	18
.94	8710	8730	8750	8770	8790	8810	8831	8851	8872	8892	2	4	6	8	10	12	14	16	18
.95	8913	8933	8954	8974	8995	9016	9036	9057	9078	9099	2	4	6	8	10	12	15	17	19
.96	9120	9141	9162	9183	9204	9226	9247	9268	9290	9311	2	4	6	8	11	13	15	17	19
.97	9333	9354	9376	9397	9419	9441	9462	9484	9506	9528	2	4	7	9	11	13	15	17	20
.98	9550	9572	9594	9616	9638	9661	9683	9705	9727	9750	2	4	7	9	11	13	16	18	20
.99	9772	9795	9817	9840	9863	9886	9908	9931	9954	9977	2	5	7	9	11	14	16	18	20
	0	1	2	3	4	5	6	7	8	9	1	2	3	4	5	6	7	8	9

SIX-PLACE MANTISSAS FOR COMMON LOGARITHMS

N	0	1	2	3	4	5	6	7	8	9
100	00 0000	0434	0868	1301	1734	2166	2598	3029	3461	3891
01	4321	4751	5181	5609	6038	6466	6894	7321	7748	8174
02	00 8600	9026	9451	9876	*0300	*0724	*1147	*1570	*1993	*2415
03	01 2837	3259	3680	4100	4521	4940	5360	5779	6197	6616
04	01 7033	7451	7868	8284	8700	9116	9532	9947	*0361	*0775
05	02 1189	1603	2016	2428	2841	3252	3664	4075	4486	4896
06	5306	5715	6125	6533	6942	7350	7757	8164	8571	8978
07	02 9384	9789	*0195	*0600	*1004	*1408	*1812	*2216	*2619	*3021
08	03 3424	3826	4227	4628	5029	5430	5830	6230	6629	7028
09	03 7426	7825	8223	8620	9017	9414	9811	*0207	*0602	*0998
110	04 1393	1787	2182	2576	2969	3362	3755	4148	4540	4932
11	5323	5714	6105	6495	6885	7275	7664	8053	8442	8830
12	04 9218	9606	9993	*0380	*0766	*1153	*1538	*1924	*2309	*2694
13	05 3078	3463	3846	4230	4613	4996	5378	5760	6142	6524
14	05 6905	7286	7666	8046	8426	8805	9185	9563	9942	*0320
15	06 0698	1075	1452	1829	2206	2582	2958	3333	3709	4083
16	4458	4832	5206	5580	5953	6326	6699	7071	7443	7815
17	06 8186	8557	8928	9298	9668	*0038	*0407	*0776	*1145	*1514
18	07 1882	2250	2617	2985	3352	3718	4085	4451	4816	5182
19	5547	5912	6276	6640	7004	7368	7731	8094	8457	8819
120	07 9181	9543	9904	*0266	*0626	*0987	*1347	*1707	*2067	*2426
21	08 2785	3144	3503	3861	4219	4576	4934	5291	5647	6004
22	6360	6716	7071	7426	7781	8136	8490	8845	9198	9552
23	08 9905	*0258	*0611	*0963	*1315	*1667	*2018	*2370	*2721	*3071
24	09 3422	3772	4122	4471	4820	5169	5518	5866	6215	6562
25	09 6910	7257	7604	7951	8298	8644	8990	9335	9681	*0026
26	10 0371	0715	1059	1403	1747	2091	2434	2777	3119	3462
27	3804	4146	4487	4828	5169	5510	5851	6191	6531	6871
28	10 7210	7549	7888	8227	8565	8903	9241	9579	9916	*0253
29	11 0590	0926	1263	1599	1934	2270	2605	2940	3275	3609
130	3943	4277	4611	4944	5278	5611	5943	6276	6608	6940
31	11 7271	7603	7934	8265	8595	8926	9256	9586	9915	*0245
32	12 0574	0903	1231	1560	1888	2216	2544	2871	3198	3525
33	3852	4178	4504	4830	5156	5481	5806	6131	6456	6781
34	12 7105	7429	7753	8076	8399	8722	9045	9368	9690	*0012
35	13 0334	0655	0977	1298	1619	1939	2260	2580	2900	3219
36	3539	3858	4177	4496	4814	5133	5451	5769	6086	6403
37	6721	7037	7354	7671	7987	8303	8618	8934	9249	9564
38	13 9879	*0194	*0508	*0822	*1136	*1450	*1763	*2076	*2389	*2702
39	14 3015	3327	3639	3951	4263	4574	4885	5196	5507	5818
140	6128	6438	6748	7058	7367	7676	7985	8294	8603	8911
41	14 9219	9527	9835	*0142	*0449	*0756	*1063	*1370	*1676	*1982
42	15 2288	2594	2900	3205	3510	3815	4120	4424	4728	5032
43	5336	5640	5943	6246	6549	6852	7154	7457	7759	8061
44	15 8362	8664	8965	9266	9567	9868	*0168	*0469	*0769	*1068
45	16 1368	1667	1967	2266	2564	2863	3161	3460	3758	4055
46	4353	4650	4947	5244	5541	5838	6134	6430	6726	7022
47	16 7317	7613	7908	8203	8497	8792	9086	9380	9674	9968
48	17 0262	0555	0848	1141	1434	1726	2019	2311	2603	2895
49	3186	3478	3769	4060	4351	4641	4932	5222	5512	5802
150	17 6091	6381	6670	6959	7248	7536	7825	8113	8401	8689
N	**0**	**1**	**2**	**3**	**4**	**5**	**6**	**7**	**8**	**9**

SIX-PLACE MANTISSAS FOR COMMON LOGARITHMS (Continued)

N	0	1	2	3	4	5	6	7	8	9
150	17 6091	6381	6670	6959	7248	7536	7825	8113	8401	8689
51	17 8977	9264	9552	9839	*0126	*0413	*0699	*0986	*1272	*1558
52	18 1844	2129	2415	2700	2985	3270	3555	3839	4123	4407
53	4691	4975	5259	5542	5825	6108	6391	6674	6956	7239
54	18 7521	7803	8084	8366	8647	8928	9209	9490	9771	*0051
55	19 0332	0612	0892	1171	1451	1730	2010	2289	2567	2846
56	3125	3403	3681	3959	4237	4514	4792	5069	5346	5623
57	5900	6176	6453	6729	7005	7281	7556	7832	8107	8382
58	19 8657	8932	9206	9481	9755	*0029	*0303	*0577	*0850	*1124
59	20 1397	1670	1943	2216	2488	2761	3033	3305	3577	3848
160	4120	4391	4663	4934	5204	5475	5746	6016	6286	6556
61	6826	7096	7365	7634	7904	8173	8441	8710	8979	9247
62	20 9515	9783	*0051	*0319	*0586	*0853	*1121	*1388	*1654	*1921
63	21 2188	2454	2720	2986	3252	3518	3783	4049	4314	4579
64	4844	5109	5373	5638	5902	6166	6430	6694	6957	7221
65	21 7484	7747	8010	8273	8536	8798	9060	9323	9585	9846
66	22 0108	0370	0631	0892	1153	1414	1675	1936	2196	2456
67	2716	2976	3236	3496	3755	4015	4274	4533	4792	5051
68	5309	5568	5826	6084	6342	6600	6858	7115	7372	7630
69	22 7887	8144	8400	8657	8913	9170	9426	9682	9938	*0193
170	23 0449	0704	0960	1215	1470	1724	1979	2234	2488	2742
71	2996	3250	3504	3757	4011	4264	4517	4770	5023	5276
72	5528	5781	6033	6285	6537	6789	7041	7292	7544	7795
73	23 8046	8297	8548	8799	9049	9299	9550	9800	*0050	*0300
74	24 0549	0799	1048	1297	1546	1795	2044	2293	2541	2790
75	3038	3286	3534	3782	4030	4277	4525	4772	5019	5266
76	5513	5759	6006	6252	6499	6745	6991	7237	7482	7728
77	24 7973	8219	8464	8709	8954	9198	9443	9687	9932	*0176
78	25 0420	0664	0908	1151	1395	1638	1881	2125	2368	2610
79	2853	3096	3338	3580	3822	4064	4306	4548	4790	5031
180	5273	5514	5755	5996	6237	6477	6718	6958	7198	7439
81	25 7679	7918	8158	8398	8637	8877	9116	9355	9594	9833
82	26 0071	0310	0548	0787	1025	1263	1501	1739	1976	2214
83	2451	2688	2925	3162	3399	3636	3873	4109	4346	4582
84	4818	5054	5290	5525	5761	5996	6232	6467	6702	6937
85	7172	7406	7641	7875	8110	8344	8578	8812	9046	9279
86	26 9513	9746	9980	*0213	*0446	*0679	*0912	*1144	*1377	*1609
87	27 1842	2074	2306	2538	2770	3001	3233	3464	3696	3927
88	4158	4389	4620	4850	5081	5311	5542	5772	6002	6232
89	6462	6692	6921	7151	7380	7609	7838	8067	8296	8525
190	27 8754	8982	9211	9439	9667	9895	*0123	*0351	*0578	*0806
91	28 1033	1261	1488	1715	1942	2169	2396	2622	2849	3075
92	3301	3527	3753	3979	4205	4431	4656	4882	5107	5332
93	5557	5782	6007	6232	6456	6681	6905	7130	7354	7578
94	28 7802	8026	8249	8473	8696	8920	9143	9366	9589	9812
95	29 0035	0257	0480	0702	0925	1147	1369	1591	1813	2034
96	2256	2478	2699	2920	3141	3363	3584	3804	4025	4246
97	4466	4687	4907	5127	5347	5567	5787	6007	6226	6446
98	6665	6884	7104	7323	7542	7761	7979	8198	8416	8635
99	29 8853	9071	9289	9507	9725	9943	*0161	*0378	*0595	*0813
200	30 1030	1247	1464	1681	1898	2114	2331	2547	2764	2980
N	**0**	**1**	**2**	**3**	**4**	**5**	**6**	**7**	**8**	**9**

SIX-PLACE MANTISSAS FOR COMMON LOGARITHMS (Continued)

N	0	1	2	3	4	5	6	7	8	9
200	30 1030	1247	1464	1681	1898	2114	2331	2547	2764	2980
01	3196	3412	3628	3844	4059	4275	4491	4706	4921	5136
02	5351	5566	5781	5996	6211	6425	6639	6854	7068	7282
03	7496	7710	7924	8137	8351	8564	8778	8991	9204	9417
04	30 9630	9843	*0056	*0268	*0481	*0693	*0906	*1118	*1330	*1542
05	31 1754	1966	2177	2389	2600	2812	3023	3234	3445	3656
06	3867	4078	4289	4499	4710	4920	5130	5340	5551	5760
07	5970	6180	6390	6599	6809	7018	7227	7436	7646	7854
08	31 8063	8272	8481	8689	8898	9106	9314	9522	9730	9938
09	32 0146	0354	0562	0769	0977	1184	1391	1598	1805	2012
210	2219	2426	2633	2839	3046	3252	3458	3665	3871	4077
11	4282	4488	4694	4899	5105	5310	5516	5721	5926	6131
12	6336	6541	6745	6950	7155	7359	7563	7767	7972	8176
13	32 8380	8583	8787	8991	9194	9398	9601	9805	*0008	*0211
14	33 0414	0617	0819	1022	1225	1427	1630	1832	2034	2236
15	2438	2640	2842	3044	3246	3447	3649	3850	4051	4253
16	4454	4655	4856	5057	5257	5458	5658	5859	6059	6260
17	6460	6660	6860	7060	7260	7459	7659	7858	8058	8257
18	33 8456	8656	8855	9054	9253	9451	9650	9849	*0047	*0246
19	34 0444	0642	0841	1039	1237	1435	1632	1830	2028	2225
220	2423	2620	2817	3014	3212	3409	3606	3802	3999	4196
21	4392	4589	4785	4981	5178	5374	5570	5766	5962	6157
22	6353	6549	6744	6939	7135	7330	7525	7720	7915	8110
23	34 8305	8500	8694	8889	9083	9278	9472	9666	9860	*0054
24	35 0248	0442	0636	0829	1023	1216	1410	1603	1796	1989
25	2183	2375	2568	2761	2954	3147	3339	3532	3724	3916
26	4108	4301	4493	4685	4876	5068	5260	5452	5643	5834
27	6026	6217	6408	6599	6790	6981	7172	7363	7554	7744
28	7935	8125	8316	8506	8696	8886	9076	9266	9456	9646
29	35 9835	*0025	*0215	*0404	*0593	*0783	*0972	*1161	*1350	*1539
230	36 1728	1917	2105	2294	2482	2671	2859	3048	3236	3424
31	3612	3800	3988	4176	4363	4551	4739	4926	5113	5301
32	5488	5675	5862	6049	6236	6423	6610	6796	6983	7169
33	7356	7542	7729	7915	8101	8287	8473	8659	8845	9030
34	36 9216	9401	9587	9772	9958	*0143	*0328	*0513	*0698	*0883
35	37 1068	1253	1437	1622	1806	1991	2175	2360	2544	2728
36	2912	3096	3280	3464	3647	3831	4015	4198	4382	4565
37	4748	4932	5115	5298	5481	5664	5846	6029	6212	6394
38	6577	6759	6942	7124	7306	7488	7670	7852	8034	8216
39	37 8398	8580	8761	8943	9124	9306	9487	9668	9849	*0030
240	38 0211	0392	0573	0754	0934	1115	1296	1476	1656	1837
41	2017	2197	2377	2557	2737	2917	3097	3277	3456	3636
42	3815	3995	4174	4353	4533	4712	4891	5070	5249	5428
43	5606	5785	5964	6142	6321	6499	6677	6856	7034	7212
44	7390	7568	7746	7924	8101	8279	8456	8634	8811	8989
45	38 9166	9343	9520	9698	9875	*0051	*0228	*0405	*0582	*0759
46	39 0935	1112	1288	1464	1641	1817	1993	2169	2345	2521
47	2697	2873	3048	3224	3400	3575	3751	3926	4101	4277
48	4452	4627	4802	4977	5152	5326	5501	5676	5850	6025
49	6199	6374	6548	6722	6896	7071	7245	7419	7592	7766
250	39 7940	8114	8287	8461	8634	8808	8981	9154	9328	9501
N	**0**	**1**	**2**	**3**	**4**	**5**	**6**	**7**	**8**	**9**

SIX-PLACE MANTISSAS FOR COMMON LOGARITHMS (Continued)

N	0	1	2	3	4	5	6	7	8	9
250	39 7940	8114	8287	8461	8634	8808	8981	9154	9328	9501
51	39 9674	9847	*0020	*0192	*0365	*0538	*0711	*0883	*1056	*1228
52	40 1401	1573	1745	1917	2089	2261	2433	2605	2777	2949
53	3121	3292	3464	3635	3807	3978	4149	4320	4492	4663
54	4834	5005	5176	5346	5517	5688	5858	6029	6199	6370
55	6540	6710	6881	7051	7221	7391	7561	7731	7901	8070
56	8240	8410	8579	8749	8918	9087	9257	9426	9595	9764
57	40 9933	*0102	*0271	*0440	*0609	*0777	*0946	*1114	*1283	*1451
58	41 1620	1788	1956	2124	2293	2461	2629	2796	2964	3132
59	3300	3467	3635	3803	3970	4137	4305	4472	4639	4806
260	4973	5140	5307	5474	5641	5808	5974	6141	6308	6474
61	6641	6807	6973	7139	7306	7472	7638	7804	7970	8135
62	8301	8467	8633	8798	8964	9129	9295	9460	9625	9791
63	41 9956	*0121	*0286	*0451	*0616	*0781	*0945	*1110	*1275	*1439
64	42 1604	1768	1933	2097	2261	2426	2590	2754	2918	3082
65	3246	3410	3574	3737	3901	4065	4228	4392	4555	4718
66	4882	5045	5208	5371	5534	5697	5860	6023	6186	6349
67	6511	6674	6836	6999	7161	7324	7486	7648	7811	7973
68	8135	8297	8459	8621	8783	8944	9106	9268	9429	9591
69	42 9752	9914	*0075	*0236	*0398	*0559	*0720	*0881	*1042	*1203
270	43 1364	1525	1685	1846	2007	2167	2328	2488	2649	2809
71	2969	3130	3290	3450	3610	3770	3930	4090	4249	4409
72	4569	4729	4888	5048	5207	5367	5526	5685	5844	6004
73	6163	6322	6481	6640	6799	6957	7116	7275	7433	7592
74	7751	7909	8067	8226	8384	8542	8701	8859	9017	9175
75	43 9333	9491	9648	9806	9964	*0122	*0279	*0437	*0594	*0752
76	44 0909	1066	1224	1381	1538	1695	1852	2009	2166	2323
77	2480	2637	2793	2950	3106	3263	3419	3576	3732	3889
78	4045	4201	4357	4513	4669	4825	4981	5137	5293	5449
79	5604	5760	5915	6071	6226	6382	6537	6692	6848	7003
280	7158	7313	7468	7623	7778	7933	8088	8242	8397	8552
81	44 8706	8861	9015	9170	9324	9478	9633	9787	9941	*0095
82	45 0249	0403	0557	0711	0865	1018	1172	1326	1479	1633
83	1786	1940	2093	2247	2400	2553	2706	2859	3012	3165
84	3318	3471	3624	3777	3930	4082	4235	4387	4540	4692
85	4845	4997	5150	5302	5454	5606	5758	5910	6062	6214
86	6366	6518	6670	6821	6973	7125	7276	7428	7579	7731
87	7882	8033	8184	8336	8487	8638	8789	8940	9091	9242
88	45 9392	9543	9694	9845	9995	*0146	*0296	*0447	*0597	*0748
89	46 0898	1048	1198	1348	1499	1649	1799	1948	2098	2248
290	2398	2548	2697	2847	2997	3146	3296	3445	3594	3744
91	3893	4042	4191	4340	4490	4639	4788	4936	5085	5234
92	5383	5532	5680	5829	5977	6126	6274	6423	6571	6719
93	6868	7016	7164	7312	7460	7608	7756	7904	8052	8200
94	8347	8495	8643	8790	8938	9085	9233	9380	9527	9675
95	46 9822	9969	*0116	*0263	*0410	*0557	*0704	*0851	*0998	*1145
96	47 1292	1438	1585	1732	1878	2025	2171	2318	2464	2610
97	2756	2903	3049	3195	3341	3487	3633	3779	3925	4071
98	4216	4362	4508	4653	4799	4944	5090	5235	5381	5526
99	5671	5816	5962	6107	6252	6397	6542	6687	6832	6976
300	47 7121	7266	7411	7555	7700	7844	7989	8133	8278	8422
N	0	1	2	3	4	5	6	7	8	9

SIX-PLACE MANTISSAS FOR COMMON LOGARITHMS (Continued)

N	0	1	2	3	4	5	6	7	8	9
300	47 7121	7266	7411	7555	7700	7844	7989	8133	8278	8422
01	47 8566	8711	8855	8999	9143	9287	9431	9575	9719	9863
02	48 0007	0151	0294	0438	0582	0725	0869	1012	1156	1299
03	1443	1586	1729	1872	2016	2159	2302	2445	2588	2731
04	2874	3016	3159	3302	3445	3587	3730	3872	4015	4157
05	4300	4442	4585	4727	4869	5011	5153	5295	5437	5579
06	5721	5863	6005	6147	6289	6430	6572	6714	6855	6997
07	7138	7280	7421	7563	7704	7845	7986	8127	8269	8410
08	8551	8692	8833	8974	9114	9255	9396	9537	9677	9818
09	48 9958	*0099	*0239	*0380	*0520	*0661	*0801	*0941	*1081	*1222
310	49 1362	1502	1642	1782	1922	2062	2201	2341	2481	2621
11	2760	2900	3040	3179	3319	3458	3597	3737	3876	4015
12	4155	4294	4433	4572	4711	4850	4989	5128	5267	5406
13	5544	5683	5822	5960	6099	6238	6376	6515	6653	6791
14	6930	7068	7206	7344	7483	7621	7759	7897	8035	8173
15	8311	8448	8586	8724	8862	8999	9137	9275	9412	9550
16	49 9687	9824	9962	*0099	*0236	*0374	*0511	*0648	*0785	*0922
17	50 1059	1196	1333	1470	1607	1744	1880	2017	2154	2291
18	2427	2564	2700	2837	2973	3109	3246	3382	3518	3655
19	3791	3927	4063	4199	4335	4471	4607	4743	4878	5014
320	5150	5286	5421	5557	5693	5828	5964	6099	6234	6370
21	6505	6640	6776	6911	7046	7181	7316	7451	7586	7721
22	7856	7991	8126	8260	8395	8530	8664	8799	8934	9068
23	50 9203	9337	9471	9606	9740	9874	*0009	*0143	*0277	*0411
24	51 0545	0679	0813	0947	1081	1215	1349	1482	1616	1750
25	1883	2017	2151	2284	2418	2551	2684	2818	2951	3084
26	3218	3351	3484	3617	3750	3883	4016	4149	4282	4415
27	4548	4681	4813	4946	5079	5211	5344	5476	5609	5741
28	5874	6006	6139	6271	6403	6535	6668	6800	6932	7064
29	7196	7328	7460	7592	7724	7855	7987	8119	8251	8382
330	8514	8646	8777	8909	9040	9171	9303	9434	9566	9697
31	51 9828	9959	*0090	*0221	*0353	*0484	*0615	*0745	*0876	*1007
32	52 1138	1269	1400	1530	1661	1792	1922	2053	2183	2314
33	2444	2575	2705	2835	2966	3096	3226	3356	3486	3616
34	3746	3876	4006	4136	4266	4396	4526	4656	4785	4915
35	5045	5174	5304	5434	5563	5693	5822	5951	6081	6210
36	6339	6469	6598	6727	6856	6985	7114	7243	7372	7501
37	7630	7759	7888	8016	8145	8274	8402	8531	8660	8788
38	52 8917	9045	9174	9302	9430	9559	9687	9815	9943	*0072
39	53 0200	0328	0456	0584	0712	0840	0968	1096	1223	1351
340	1479	1607	1734	1862	1990	2117	2245	2372	2500	2627
41	2754	2882	3009	3136	3264	3391	3518	3645	3772	3899
42	4026	4153	4280	4407	4534	4661	4787	4914	5041	5167
43	5294	5421	5547	5674	5800	5927	6053	6180	6306	6432
44	6558	6685	6811	6937	7063	7189	7315	7441	7567	7693
45	7819	7945	8071	8197	8322	8448	8574	8699	8825	8951
46	53 9076	9202	9327	9452	9578	9703	9829	9954	*0079	*0204
47	54 0329	0455	0580	0705	0830	0955	1080	1205	1330	1454
48	1579	1704	1829	1953	2078	2203	2327	2452	2576	2701
49	2825	2950	3074	3199	3323	3447	3571	3696	3820	3944
350	54 4068	4192	4316	4440	4564	4688	4812	4936	5060	5183
N	**0**	**1**	**2**	**3**	**4**	**5**	**6**	**7**	**8**	**9**

SIX-PLACE MANTISSAS FOR COMMON LOGARITHMS (Continued)

N	0	1	2	3	4	5	6	7	8	9
350	54 4068	4192	4316	4440	4564	4688	4812	4936	5060	5183
51	5307	5431	5555	5678	5802	5925	6049	6172	6296	6419
52	6543	6666	6789	6913	7036	7159	7282	7405	7529	7652
53	7775	7898	8021	8144	8267	8389	8512	8635	8758	8881
54	54 9003	9126	9249	9371	9494	9616	9739	9861	9984	*0106
55	55 0228	0351	0473	0595	0717	0840	0962	1084	1206	1328
56	1450	1572	1694	1816	1938	2060	2181	2303	2425	2547
57	2668	2790	2911	3033	3155	3276	3398	3519	3640	3762
58	3883	4004	4126	4247	4368	4489	4610	4731	4852	4973
59	5094	5215	5336	5457	5578	5699	5820	5940	6061	6182
360	6303	6423	6544	6664	6785	6905	7026	7146	7267	7387
61	7507	7627	7748	7868	7988	8108	8228	8349	8469	8589
62	8709	8829	8948	9068	9188	9308	9428	9548	9667	9787
63	55 9907	*0026	*0146	*0265	*0385	*0504	*0624	*0743	*0863	*0982
64	56 1101	1221	1340	1459	1578	1698	1817	1936	2055	2174
65	2293	2412	2531	2650	2769	2887	3006	3125	3244	3362
66	3481	3600	3718	3837	3955	4074	4192	4311	4429	4548
67	4666	4784	4903	5021	5139	5257	5376	5494	5612	5730
68	5848	5966	6084	6202	6320	6437	6555	6673	6791	6909
69	7026	7144	7262	7379	7497	7614	7732	7849	7967	8084
370	8202	8319	8436	8554	8671	8788	8905	9023	9140	9257
71	56 9374	9491	9608	9725	9842	9959	*0076	*0193	*0309	*0426
72	57 0543	0660	0776	0893	1010	1126	1243	1359	1476	1592
73	1709	1825	1942	2058	2174	2291	2407	2523	2639	2755
74	2872	2988	3104	3220	3336	3452	3568	3684	3800	3915
75	4031	4147	4263	4379	4494	4610	4726	4841	4957	5072
76	5188	5303	5419	5534	5650	5765	5880	5996	6111	6226
77	6341	6457	6572	6687	6802	6917	7032	7147	7262	7377
78	7492	7607	7722	7836	7951	8066	8181	8295	8410	8525
79	8639	8754	8868	8983	9097	9212	9326	9441	9555	9669
380	57 9784	9898	*0012	*0126	*0241	*0355	*0469	*0583	*0697	*0811
81	58 0925	1039	1153	1267	1381	1495	1608	1722	1836	1950
82	2063	2177	2291	2404	2518	2631	2745	2858	2972	3085
83	3199	3312	3426	3539	3652	3765	3879	3992	4105	4218
84	4331	4444	4557	4670	4783	4896	5009	5122	5235	5348
85	5461	5574	5686	5799	5912	6024	6137	6250	6362	6475
86	6587	6700	6812	6925	7037	7149	7262	7374	7486	7599
87	7711	7823	7935	8047	8160	8272	8384	8496	8608	8720
88	8832	8944	9056	9167	9279	9391	9503	9615	9726	9838
89	58 9950	*0061	*0173	*0284	*0396	*0507	*0619	*0730	*0842	*0953
390	59 1065	1176	1287	1399	1510	1621	1732	1843	1955	2066
91	2177	2288	2399	2510	2621	2732	2843	2954	3064	3175
92	3286	3397	3508	3618	3729	3840	3950	4061	4171	4282
93	4393	4503	4614	4724	4834	4945	5055	5165	5276	5386
94	5496	5606	5717	5827	5937	6047	6157	6267	6377	6487
95	6597	6707	6817	6927	7037	7146	7256	7366	7476	7586
96	7695	7805	7914	8024	8134	8243	8353	8462	8572	8681
97	8791	8900	9009	9119	9228	9337	9446	9556	9665	9774
98	9883	9992	*0101	*0210	*0319	*0428	*0537	*0646	*0755	*0864
99	60 0973	1082	1191	1299	1408	1517	1625	1734	1843	1951
400	60 2060	2169	2277	2386	2494	2603	2711	2819	2928	3036
N	**0**	**1**	**2**	**3**	**4**	**5**	**6**	**7**	**8**	**9**

SIX-PLACE MANTISSAS FOR COMMON LOGARITHMS (Continued)

N	0	1	2	3	4	5	6	7	8	9
400	60 2060	2169	2277	2386	2494	2603	2711	2819	2928	3036
01	3144	3253	3361	3469	3577	3686	3794	3902	4010	4118
02	4226	4334	4442	4550	4658	4766	4874	4982	5089	5197
03	5305	5413	5521	5628	5736	5844	5951	6059	6166	6274
04	6381	6489	6596	6704	6811	6919	7026	7133	7241	7348
05	7455	7562	7669	7777	7884	7991	8098	8205	8312	8419
06	8526	8633	8740	8847	8954	9061	9167	9274	9381	9488
07	60 9594	9701	9808	9914	*0021	*0128	*0234	*0341	*0447	*0554
08	61 0660	0767	0873	0979	1086	1192	1298	1405	1511	1617
09	1723	1829	1936	2042	2148	2254	2360	2466	2572	2678
410	2784	2890	2996	3102	3207	3313	3419	3525	3630	3736
11	3842	3947	4053	4159	4264	4370	4475	4581	4686	4792
12	4897	5003	5108	5213	5319	5424	5529	5634	5740	5845
13	5950	6055	6160	6265	6370	6476	6581	6686	6790	6895
14	7000	7105	7210	7315	7420	7525	7629	7734	7839	7943
15	8048	8153	8257	8362	8466	8571	8676	8780	8884	8989
16	61 9093	9198	9302	9406	9511	9615	9719	9824	9928	*0032
17	62 0136	0240	0344	0448	0552	0656	0760	0864	0968	1072
18	1176	1280	1384	1488	1592	1695	1799	1903	2007	2110
19	2214	2318	2421	2525	2628	2732	2835	2939	3042	3146
420	3249	3353	3456	3559	3663	3766	3869	3973	4076	4179
21	4282	4385	4488	4591	4695	4798	4901	5004	5107	5210
22	5312	5415	5518	5621	5724	5827	5929	6032	6135	6238
23	6340	6443	6546	6648	6751	6853	6956	7058	7161	7263
24	7366	7468	7571	7673	7775	7878	7980	8082	8185	8287
25	8389	8491	8593	8695	8797	8900	9002	9104	9206	9308
26	62 9410	9512	9613	9715	9817	9919	*0021	*0123	*0224	*0326
27	63 0428	0530	0631	0733	0835	0936	1038	1139	1241	1342
28	1444	1545	1647	1748	1849	1951	2052	2153	2255	2356
29	2457	2559	2660	2761	2862	2963	3064	3165	3266	3367
430	3468	3569	3670	3771	3872	3973	4074	4175	4276	4376
31	4477	4578	4679	4779	4880	4981	5081	5182	5283	5383
32	5484	5584	5685	5785	5886	5986	6087	6187	6287	6388
33	6488	6588	6688	6789	6889	6989	7089	7189	7290	7390
34	7490	7590	7690	7790	7890	7990	8090	8190	8290	8389
35	8489	8589	8689	8789	8888	8988	9088	9188	9287	9387
36	63 9486	9586	9686	9785	9885	9984	*0084	*0183	*0283	*0382
37	64 0481	0581	0680	0779	0879	0978	1077	1177	1276	1375
38	1474	1573	1672	1771	1871	1970	2069	2168	2267	2366
39	2465	2563	2662	2761	2860	2959	3058	3156	3255	3354
440	3453	3551	3650	3749	3847	3946	4044	4143	4242	4340
41	4439	4537	4636	4734	4832	4931	5029	5127	5226	5324
42	5422	5521	5619	5717	5815	5913	6011	6110	6208	6306
43	6404	6502	6600	6698	6796	6894	6992	7089	7187	7285
44	7383	7481	7579	7676	7774	7872	7969	8067	8165	8262
45	8360	8458	8555	8653	8750	8848	8945	9043	9140	9237
46	64 9335	9432	9530	9627	9724	9821	9919	*0016	*0113	*0210
47	65 0308	0405	0502	0599	0696	0793	0890	0987	1084	1181
48	1278	1375	1472	1569	1666	1762	1859	1956	2053	2150
49	2246	2343	2440	2536	2633	2730	2826	2923	3019	3116
450	65 3213	3309	3405	3502	3598	3695	3791	3888	3984	4080
N	**0**	**1**	**2**	**3**	**4**	**5**	**6**	**7**	**8**	**9**

SIX-PLACE MANTISSAS FOR COMMON LOGARITHMS (Continued)

N	0	1	2	3	4	5	6	7	8	9
450	65 3213	3309	3405	3502	3598	3695	3791	3888	3984	4080
51	4177	4273	4369	4465	4562	4658	4754	4850	4946	5042
52	5138	5235	5331	5427	5523	5619	5715	5810	5906	6002
53	6098	6194	6290	6386	6482	6577	6673	6769	6864	6960
54	7056	7152	7247	7343	7438	7534	7629	7725	7820	7916
55	8011	8107	8202	8298	8393	8488	8584	8679	8774	8870
56	8965	9060	9155	9250	9346	9441	9536	9631	9726	9821
57	65 9916	*0011	*0106	*0201	*0296	*0391	*0486	*0581	*0676	*0771
58	66 0865	0960	1055	1150	1245	1339	1434	1529	1623	1718
59	1813	1907	2002	2096	2191	2286	2380	2475	2569	2663
460	2758	2852	2947	3041	3135	3230	3324	3418	3512	3607
61	3701	3795	3889	3983	4078	4172	4266	4360	4454	4548
62	4642	4736	4830	4924	5018	5112	5206	5299	5393	5487
63	5581	5675	5769	5862	5956	6050	6143	6237	6331	6424
64	6518	6612	6705	6799	6892	6986	7079	7173	7266	7360
65	7453	7546	7640	7733	7826	7920	8013	8106	8199	8293
66	8386	8479	8572	8665	8759	8852	8945	9038	9131	9224
67	66 9317	9410	9503	9596	9689	9782	9875	9967	*0060	*0153
68	67 0246	0339	0431	0524	0617	0710	0802	0895	0988	1080
69	1173	1265	1358	1451	1543	1636	1728	1821	1913	2005
470	2098	2190	2283	2375	2467	2560	2652	2744	2836	2929
71	3021	3113	3205	3297	3390	3482	3574	3666	3758	3850
72	3942	4034	4126	4218	4310	4402	4494	4586	4677	4769
73	4861	4953	5045	5137	5228	5320	5412	5503	5595	5687
74	5778	5870	5962	6053	6145	6236	6328	6419	6511	6602
75	6694	6785	6876	6968	7059	7151	7242	7333	7424	7516
76	7607	7698	7789	7881	7972	8063	8154	8245	8336	8427
77	8518	8609	8700	8791	8882	8973	9064	9155	9246	9337
78	67 9428	9519	9610	9700	9791	9882	9973	*0063	*0154	*0245
79	68 0336	0426	0517	0607	0698	0789	0879	0970	1060	1151
480	1241	1332	1422	1513	1603	1693	1784	1874	1964	2055
81	2145	2235	2326	2416	2506	2596	2686	2777	2867	2957
82	3047	3137	3227	3317	3407	3497	3587	3677	3767	3857
83	3947	4037	4127	4217	4307	4396	4486	4576	4666	4756
84	4845	4935	5025	5114	5204	5294	5383	5473	5563	5652
85	5742	5831	5921	6010	6100	6189	6279	6368	6458	6547
86	6636	6726	6815	6904	6994	7083	7172	7261	7351	7440
87	7529	7618	7707	7796	7886	7975	8064	8153	8242	8331
88	8420	8509	8598	8687	8776	8865	8953	9042	9131	9220
89	68 9309	9398	9486	9575	9664	9753	9841	9930	*0019	*0107
490	69 0196	0285	0373	0462	0550	0639	0728	0816	0905	0993
91	1081	1170	1258	1347	1435	1524	1612	1700	1789	1877
92	1965	2053	2142	2230	2318	2406	2494	2583	2671	2759
93	2847	2935	3023	3111	3199	3287	3375	3463	3551	3639
94	3727	3815	3903	3991	4078	4166	4254	4342	4430	4517
95	4605	4693	4781	4868	4956	5044	5131	5219	5307	5394
96	5482	5569	5657	5744	5832	5919	6007	6094	6182	6269
97	6356	6444	6531	6618	6706	6793	6880	6968	7055	7142
98	7229	7317	7404	7491	7578	7665	7752	7839	7926	8014
99	8101	8188	8275	8362	8449	8535	8622	8709	8796	8883
500	69 8970	9057	9144	9231	9317	9404	9491	9578	9664	9751
N	**0**	**1**	**2**	**3**	**4**	**5**	**6**	**7**	**8**	**9**

SIX-PLACE MANTISSAS FOR COMMON LOGARITHMS (Continued)

N	0	1	2	3	4	5	6	7	8	9
500	69 8970	9057	9144	9231	9317	9404	9491	9578	9664	9751
01	69 9838	9924	*0011	*0098	*0184	*0271	*0358	*0444	*0531	*0617
02	70 0704	0790	0877	0963	1050	1136	1222	1309	1395	1482
03	1568	1654	1741	1827	1913	1999	2086	2172	2258	2344
04	2431	2517	2603	2689	2775	2861	2947	3033	3119	3205
05	3291	3377	3463	3549	3635	3721	3807	3893	3979	4065
06	4151	4236	4322	4408	4494	4579	4665	4751	4837	4922
07	5008	5094	5179	5265	5350	5463	5522	5607	5693	5778
08	5864	5949	6035	6120	6206	6291	6376	6462	6547	6632
09	6718	6803	6888	6974	7059	7144	7229	7315	7400	7485
510	7570	7655	7740	7826	7911	7996	8081	8166	8251	8336
11	8421	8506	8591	8676	8761	8846	8931	9015	9100	9185
12	70 9270	9355	9440	9524	9609	9694	9779	9863	9948	*0033
13	71 0117	0202	0287	0371	0456	0540	0625	0710	0794	0879
14	0963	1048	1132	1217	1301	1385	1470	1554	1639	1723
15	1807	1892	1976	2060	2144	2229	2313	2397	2481	2566
16	2650	2734	2818	2902	2986	3070	3154	3238	3323	3407
17	3491	3575	3659	3742	3826	3910	3994	4078	4162	4246
18	4330	4414	4497	4581	4665	4749	4833	4916	5000	5084
19	5167	5251	5335	5418	5502	5586	5669	5753	5836	5920
520	6003	6087	6170	6254	6337	6421	6504	6588	6671	6754
21	6838	6921	7004	7088	7171	7254	7338	7421	7504	7587
22	7671	7754	7837	7920	8003	8086	8169	8253	8336	8419
23	8502	8585	8668	8751	8834	8917	9000	9083	9165	9248
24	71 9331	9414	9497	9580	9663	9745	9828	9911	9994	*0077
25	72 0159	0242	0325	0407	0490	0573	0655	0738	0821	0903
26	0986	1068	1151	1233	1316	1398	1481	1563	1646	1728
27	1811	1893	1975	2058	2140	2222	2305	2387	2469	2552
28	2634	2716	2798	2881	2963	3045	3127	3209	3291	3374
29	3456	3538	3620	3702	3784	3866	3948	4030	4112	4194
530	4276	4358	4440	4522	4604	4685	4767	4849	4931	5013
31	5095	5176	5258	5340	5422	5503	5585	5667	5748	5830
32	5912	5993	6075	6156	6238	6320	6401	6483	6564	6646
33	6727	6809	6890	6972	7053	7134	7216	7297	7379	7460
34	7541	7623	7704	7785	7866	7948	8029	8110	8191	8273
35	8354	8435	8516	8597	8678	8759	8841	8922	9003	9084
36	9165	9246	9327	9408	9489	9570	9651	9732	9813	9893
37	72 9974	*0055	*0136	*0217	*0298	*0378	*0459	*0540	*0621	*0702
38	73 0782	0863	0944	1024	1105	1186	1266	1347	1428	1508
39	1589	1669	1750	1830	1911	1991	2072	2152	2233	2313
540	2394	2474	2555	2635	2715	2796	2876	2956	3037	3117
41	3197	3278	3358	3438	3518	3598	3679	3759	3839	3919
42	3999	4079	4160	4240	4320	4400	4480	4560	4640	4720
43	4800	4880	4960	5040	5120	5200	5279	5359	5439	5519
44	5599	5679	5759	5838	5918	5998	6078	6157	6237	6317
45	6397	6476	6556	6635	6715	6795	6874	6954	7034	7113
46	7193	7272	7352	7431	7511	7590	7670	7749	7829	7908
47	7987	8067	8146	8225	8305	8384	8463	8543	8622	8701
48	8781	8860	8939	9018	9097	9177	9256	9335	9414	9493
49	73 9572	9651	9731	9810	9889	9968	*0047	*0126	*0205	*0284
550	74 0363	0442	0521	0600	0678	0757	0836	0915	0994	1073
N	**0**	**1**	**2**	**3**	**4**	**5**	**6**	**7**	**8**	**9**

SIX-PLACE MANTISSAS FOR COMMON LOGARITHMS (Continued)

N	0	1	2	3	4	5	6	7	8	9
550	74 0363	0442	0521	0600	0678	0757	0836	0915	0994	1073
51	1152	1230	1309	1388	1467	1546	1624	1703	1782	1860
52	1939	2018	2096	2175	2254	2332	2411	2489	2568	2647
53	2725	2804	2882	2961	3039	3118	3196	3275	3353	3431
54	3510	3588	3667	3745	3823	3902	3980	4058	4136	4215
55	4293	4371	4449	4528	4606	4684	4762	4840	4919	4997
56	5075	5153	5231	5309	5387	5465	5543	5621	5699	5777
57	5855	5933	6011	6089	6167	6245	6323	6401	6479	6556
58	6634	6712	6790	6868	6945	7023	7101	7179	7256	7334
59	7412	7489	7567	7645	7722	7800	7878	7955	8033	8110
560	8188	8266	8343	8421	8498	8576	8653	8731	8808	8885
61	8963	9040	9118	9195	9272	9350	9427	9504	9582	9659
62	74 9736	9814	9891	9968	*0045	*0123	*0200	*0277	*0354	*0431
63	75 0508	0586	0663	0740	0817	0894	0971	1048	1125	1202
64	1279	1356	1433	1510	1587	1664	1741	1818	1895	1972
65	2048	2125	2202	2279	2356	2433	2509	2586	2663	2740
66	2816	2893	2970	3047	3123	3200	3277	3353	3430	3506
67	3583	3660	3736	3813	3889	3966	4042	4119	4195	4272
68	4348	4425	4501	4578	4654	4730	4807	4883	4960	5036
69	5112	5189	5265	5341	5417	5494	5570	5646	5722	5799
570	5875	5951	6027	6103	6180	6256	6332	6408	6484	6560
71	6636	6712	6788	6864	6940	7016	7092	7168	7244	7320
72	7396	7472	7548	7624	7700	7775	7851	7927	8003	8079
73	8155	8230	8306	8382	8458	8533	8609	8685	8761	8836
74	8912	8988	9063	9139	9214	9290	9366	9441	9517	9592
75	75 9668	9743	9819	9894	9970	*0045	*0121	*0196	*0272	*0347
76	76 0422	0498	0573	0649	0724	0799	0875	0950	1025	1101
77	1176	1251	1326	1402	1477	1552	1627	1702	1778	1853
78	1928	2003	2078	2153	2228	2303	2378	2453	2529	2604
79	2679	2754	2829	2904	2978	3053	3128	3203	3278	3353
580	3428	3503	3578	3653	3727	3802	3877	3952	4027	4101
81	4176	4251	4326	4400	4475	4550	4624	4699	4774	4848
82	4923	4998	5072	5147	5221	5296	5370	5445	5520	5594
83	5669	5743	5818	5892	5966	6041	6115	6190	6264	6338
84	6413	6487	6562	6636	6710	6785	6859	6933	7007	7082
85	7156	7230	7304	7379	7453	7527	7601	7675	7749	7823
86	7898	7972	8046	8120	8194	8268	8342	8416	8490	8564
87	8638	8712	8786	8860	8934	9008	9082	9156	9230	9303
88	76 9377	9451	9525	9599	9673	9746	9820	9894	9968	*0042
89	77 0115	0189	0263	0336	0410	0484	0557	0631	0705	0778
590	0852	0926	0999	1073	1146	1220	1293	1367	1440	1514
91	1587	1661	1734	1808	1881	1955	2028	2102	2175	2248
92	2322	2395	2468	2542	2615	2688	2762	2835	2908	2981
93	3055	3128	3201	3274	3348	3421	3494	3567	3640	3713
94	3786	3860	3933	4006	4079	4152	4225	4298	4371	4444
95	4517	4590	4663	4736	4809	4882	4955	5028	5100	5173
96	5246	5319	5392	5465	5538	5610	5683	5756	5829	5902
97	5974	6047	6120	6193	6265	6338	6411	6483	6556	6629
98	6701	6774	6846	6919	6992	7064	7137	7209	7282	7354
99	7427	7499	7572	7644	7717	7789	7862	7934	8006	8079
600	77 8151	8224	8296	8368	8441	8513	8585	8658	8730	8802
N	**0**	**1**	**2**	**3**	**4**	**5**	**6**	**7**	**8**	**9**

SIX-PLACE MANTISSAS FOR COMMON LOGARITHMS (Continued)

N	0	1	2	3	4	5	6	7	8	9
600	77 8151	8224	8296	8368	8441	8513	8585	8658	8730	8802
01	8874	8947	9019	9091	9163	9236	9308	9380	9452	9524
02	77 9596	9669	9741	9813	9885	9957	*0029	*0101	*0173	*0245
03	78 0317	0389	0461	0533	0605	0677	0749	0821	0893	0965
04	1037	1109	1181	1253	1324	1396	1468	1540	1612	1684
05	1755	1827	1899	1971	2042	2114	2186	2258	2329	2401
06	2473	2544	2616	2688	2759	2831	2902	2974	3046	3117
07	3189	3260	3332	3403	3475	3546	3618	3689	3761	3832
08	3904	3975	4046	4118	4189	4261	4332	4403	4475	4546
09	4617	4689	4760	4831	4902	4974	5045	5116	5187	5259
610	5330	5401	5472	5543	5615	5686	5757	5828	5899	5970
11	6041	6112	6183	6254	6325	6396	6467	6538	6609	6680
12	6751	6822	6893	6964	7035	7106	7177	7248	7319	7390
13	7460	7531	7602	7673	7744	7815	7885	7956	8027	8098
14	8168	8239	8310	8381	8451	8522	8593	8663	8734	8804
15	8875	8946	9016	9087	9157	9228	9299	9369	9440	9510
16	78 9581	9651	9722	9762	9863	9933	*0004	*0074	*0144	*0215
17	79 0285	0356	0426	0496	0567	0637	0707	0778	0848	0918
18	0988	1059	1129	1199	1269	1340	1410	1480	1550	1620
19	1691	1761	1831	1901	1971	2041	2111	2181	2252	2322
620	2392	2462	2532	2602	2672	2742	2812	2882	2952	3022
21	3092	3162	3231	3301	3371	3441	3511	3581	3651	3721
22	3790	3860	3930	4000	4070	4139	4209	4279	4349	4418
23	4488	4558	4627	4697	4767	4836	4906	4976	5045	5115
24	5185	5254	5324	5393	5463	5532	5602	5672	5741	5811
25	5880	5949	6019	6088	6158	6227	6297	6366	6436	6505
26	6574	6644	6713	6782	6852	6921	6990	7060	7129	7198
27	7268	7337	7406	7475	7545	7614	7683	7752	7821	7890
28	7960	8029	8098	8167	8236	8305	8374	8443	8513	8582
29	8651	8720	8789	8858	8927	8996	9065	9134	9203	9272
630	79 9341	9409	9478	9547	9616	9685	9754	9823	9892	9961
31	80 0029	0098	0167	0236	0305	0373	0442	0511	0580	0648
32	0717	0786	0854	0923	0992	1061	1129	1198	1266	1335
33	1404	1472	1541	1609	1678	1747	1815	1884	1952	2021
34	2089	2158	2226	2295	2363	2432	2500	2568	2637	2705
35	2774	2842	2910	2979	3047	3116	3184	3252	3321	3389
36	3457	3525	3594	3662	3730	3798	3867	3935	4003	4071
37	4139	4208	4276	4344	4412	4480	4548	4616	4685	4753
38	4821	4889	4957	5025	5093	5161	5229	5297	5365	5433
39	5501	5569	5637	5705	5773	5841	5908	5976	6044	6112
640	6180	6248	6316	6384	6451	6519	6587	6655	6723	6790
41	6858	6926	6994	7061	7129	7197	7264	7332	7400	7467
42	7535	7603	7670	7738	7806	7873	7941	8008	8076	8143
43	8211	8279	8346	8414	8481	8549	8616	8684	8751	8818
44	8886	8953	9021	9088	9156	9223	9290	9358	9425	9492
45	80 9560	9627	9694	9762	9829	9896	9964	*0031	*0098	*0165
46	81 0233	0300	0367	0434	0501	0569	0636	0703	0770	0837
47	0904	0971	1039	1106	1173	1240	1307	1374	1441	1508
48	1575	1642	1709	1776	1843	1910	1977	2044	2111	2178
49	2245	2312	2379	2445	2512	2579	2646	2713	2780	2847
650	81 2913	2980	3047	3114	3181	3247	3314	3381	3448	3514
N	**0**	**1**	**2**	**3**	**4**	**5**	**6**	**7**	**8**	**9**

SIX-PLACE MANTISSAS FOR COMMON LOGARITHMS (Continued)

N	0	1	2	3	4	5	6	7	8	9
650	81 2913	2980	3047	3114	3181	3247	3314	3381	3448	3514
51	3581	3648	3714	3781	3848	3914	3981	4048	4114	4181
52	4248	4314	4381	4447	4514	4581	4647	4714	4780	4847
53	4913	4980	5046	5113	5179	5246	5312	5378	5445	5511
54	5578	5644	5711	5777	5843	5910	5976	6042	6109	6175
55	6241	6308	6374	6440	6506	6573	6639	6705	6771	6838
56	6904	6970	7036	7102	7169	7235	7301	7367	7433	7499
57	7565	7631	7698	7764	7830	7896	7962	8028	8094	8160
58	8226	8292	8358	8424	8490	8556	8622	8688	8754	8820
59	8885	8951	9017	9083	9149	9215	9281	9346	9412	9478
660	81 9544	9610	9676	9741	9807	9873	9939	*0004	*0070	*0136
61	82 0201	0267	0333	0399	0464	0530	0595	0661	0727	0792
62	0858	0924	0989	1055	1120	1186	1251	1317	1382	1448
63	1514	1579	1645	1710	1775	1841	1906	1972	2037	2103
64	2168	2233	2299	2364	2430	2495	2560	2626	2691	2756
65	2822	2887	2952	3018	3083	3148	3213	3279	3344	3409
66	3474	3539	3605	3670	3735	3800	3865	3930	3996	4061
67	4126	4191	4256	4321	4386	4451	4516	4581	4646	4711
68	4776	4841	4906	4971	5036	5101	5166	5231	5296	5361
69	5426	5491	5556	5621	5686	5751	5815	5880	5945	6010
670	6075	6140	6204	6269	6334	6399	6464	6528	6593	6658
71	6723	6787	6852	6917	6981	7046	7111	7175	7240	7305
72	7369	7434	7499	7563	7628	7692	7757	7821	7886	7951
73	8015	8080	8144	8209	8273	8338	8402	8467	8531	8595
74	8660	8724	8789	8853	8918	8982	9046	9111	9175	9239
75	9304	9368	9432	9497	9561	9625	9690	9754	9818	9882
76	82 9947	*0011	*0075	*0139	*0204	*0268	*0332	*0396	*0460	*0525
77	83 0589	0653	0717	0781	0845	0909	0973	1037	1102	1166
78	1230	1294	1358	1422	1486	1550	1614	1678	1742	1806
79	1870	1934	1998	2062	2126	2189	2253	2317	2381	2445
680	2509	2573	2637	2700	2764	2828	2892	2956	3020	3083
81	3147	3211	3275	3338	3402	3466	3530	3593	3657	3721
82	3784	3848	3912	3975	4039	4103	4166	4230	4294	4357
83	4421	4484	4548	4611	4675	4739	4802	4866	4929	4993
84	5056	5120	5183	5247	5310	5373	5437	5500	5564	5627
85	5691	5754	5817	5881	5944	6007	6071	6134	6197	6261
86	6324	6387	6451	6514	6577	6641	6704	6767	6830	6894
87	6957	7020	7083	7146	7210	7273	7336	7399	7462	7525
88	7588	7652	7715	7778	7841	7904	7967	8030	8093	8156
89	8219	8282	8345	8408	8471	8534	8597	8660	8723	8786
690	8849	8912	8975	9038	9101	9164	9227	9289	9352	9415
91	83 9478	9541	9604	9667	9729	9792	9855	9918	9981	*0043
92	84 0106	0169	0232	0294	0357	0420	0482	0545	0608	0671
93	0733	0796	0859	0921	0984	1046	1109	1172	1234	1297
94	1359	1422	1485	1547	1610	1672	1735	1797	1860	1922
95	1985	2047	2110	2172	2235	2297	2360	2422	2484	2547
96	2609	2672	2734	2796	2859	2921	2983	3046	3108	3170
97	3233	3295	3357	3420	3482	3544	3606	3669	3731	3793
98	3855	3918	3980	4042	4104	4166	4229	4291	4353	4415
99	4477	4539	4601	4664	4726	4788	4850	4912	4974	5036
700	84 5098	5160	5222	5284	5346	5408	5470	5532	5594	5656
N	**0**	**1**	**2**	**3**	**4**	**5**	**6**	**7**	**8**	**9**

SIX-PLACE MANTISSAS FOR COMMON LOGARITHMS (Continued)

N	0	1	2	3	4	5	6	7	8	9
700	84 5098	5160	5222	5284	5346	5408	5470	5532	5594	5656
01	5718	5780	5842	5904	5966	6028	6090	6151	6213	6275
02	6337	6399	6461	6523	6585	6646	6708	6770	6832	6894
03	6955	7017	7079	7141	7202	7264	7326	7388	7449	7511
04	7573	7634	7696	7758	7819	7881	7943	8004	8066	8128
05	8189	8251	8312	8374	8435	8497	8559	8620	8682	8743
06	8805	8866	8928	8989	9051	9112	9174	9235	9297	9358
07	84 9419	9481	9542	9604	9665	9726	9788	9849	9911	9972
08	85 0033	0095	0156	0217	0279	0340	0401	0462	0524	0585
09	0646	0707	0769	0830	0891	0952	1014	1075	1136	1197
710	1258	1320	1381	1442	1503	1564	1625	1686	1747	1809
11	1870	1931	1992	2053	2114	2175	2236	2297	2358	2419
12	2480	2541	2602	2663	2724	2785	2846	2907	2968	3029
13	3090	3150	3211	3272	3333	3394	3455	3516	3577	3637
14	3698	3759	3820	3881	3941	4002	4063	4124	4185	4245
15	4306	4367	4428	4488	4549	4610	4670	4731	4792	4852
16	4913	4974	5034	5095	5156	5216	5277	5337	5398	5459
17	5519	5580	5640	5701	5761	5822	5882	5943	6003	6064
18	6124	6185	6245	6306	6366	6427	6487	6548	6608	6668
19	6729	6789	6850	6910	6970	7031	7091	7152	7212	7272
720	7332	7393	7453	7513	7574	7634	7694	7755	7815	7875
21	7935	7995	8056	8116	8176	8236	8297	8357	8417	8477
22	8537	8597	8657	8718	8778	8838	8898	8958	9018	9078
23	9138	9198	9258	9318	9379	9439	9499	9559	9619	9679
24	85 9739	9799	9859	9918	9978	*0038	*0098	*0158	*0218	*0278
25	86 0338	0398	0458	0518	0578	0637	0697	0757	0817	0877
26	0937	0996	1056	1116	1176	1236	1295	1355	1415	1475
27	1534	1594	1654	1714	1773	1833	1893	1952	2012	2072
28	2131	2191	2251	2310	2370	2430	2489	2549	2608	2668
29	2728	2787	2847	2906	2966	3025	3085	3144	3204	3263
730	3323	3382	3442	3501	3561	3620	3680	3739	3799	3858
31	3917	3977	4036	4096	4155	4214	4274	4333	4392	4452
32	4511	4570	4630	4689	4748	4808	4867	4926	4985	5045
33	5104	5163	5222	5282	5341	5400	5459	5519	5578	5637
34	5696	5755	5814	5874	5933	5992	6051	6110	6169	6228
35	6287	6346	6405	6465	6524	6583	6642	6701	6760	6819
36	6878	6937	6996	7055	7114	7173	7232	7291	7350	7409
37	7467	7526	7585	7644	7703	7762	7821	7880	7939	7998
38	8056	8115	8174	8233	8292	8350	8409	8468	8527	8586
39	8644	8703	8762	8821	8879	8938	8997	9056	9114	9173
740	9232	9290	9349	9408	9466	9525	9584	9642	9701	9760
41	86 9818	9877	9935	9994	*0053	*0111	*0170	*0228	*0287	*0345
42	87 0404	0462	0521	0579	0638	0696	0755	0813	0872	0930
43	0989	1047	1106	1164	1223	1281	1339	1398	1456	1515
44	1573	1631	1690	1748	1806	1865	1923	1981	2040	2098
45	2156	2215	2273	2331	2389	2448	2506	2564	2622	2681
46	2739	2797	2855	2913	2972	3030	3088	3146	3204	3262
47	3321	3379	3437	3495	3553	3611	3669	3727	3785	3844
48	3902	3960	4018	4076	4134	4192	4250	4308	4366	4424
49	4482	4540	4598	4656	4714	4772	4830	4888	4945	5003
750	87 5061	5119	5177	5235	5293	5351	5409	5466	5524	5582
N	**0**	**1**	**2**	**3**	**4**	**5**	**6**	**7**	**8**	**9**

SIX-PLACE MANTISSAS FOR COMMON LOGARITHMS (Continued)

N	0	1	2	3	4	5	6	7	8	9
750	87 5061	5119	5177	5235	5293	5351	5409	5466	5524	5582
51	5640	5698	5756	5813	5871	5929	5987	6045	6102	6160
52	6218	6276	6333	6391	6449	6507	6564	6622	6680	6737
53	6795	6853	6910	6968	7026	7083	7141	7199	7256	7314
54	7371	7429	7487	7544	7602	7659	7717	7774	7832	7889
55	7947	8004	8062	8119	8177	8234	8292	8349	8407	8464
56	8522	8579	8637	8694	8752	8809	8866	8924	8981	9039
57	9096	9153	9211	9268	9325	9383	9440	9497	9555	9612
58	87 9669	9726	9784	9841	9898	9956	*0013	*0070	*0127	*0185
59	88 0242	0299	0356	0413	0417	0528	0585	0642	0699	0756
760	0814	0871	0928	0985	1042	1099	1156	1213	1271	1328
61	1385	1442	1499	1556	1613	1670	1727	1784	1841	1898
62	1955	2012	2069	2126	2183	2240	2297	2354	2411	2468
63	2525	2581	2638	2695	2752	2809	2866	2923	2980	3037
64	3093	3150	3207	3264	3321	3377	3434	3491	3548	3605
65	3661	3718	3775	3832	3888	3945	4002	4059	4115	4172
66	4229	4285	4342	4399	4455	4512	4569	4625	4682	4739
67	4795	4852	4909	4965	5022	5078	5135	5192	5248	5305
68	5361	5418	5474	5531	5587	5644	5700	5757	5813	5870
69	5926	5983	6039	6096	6152	6209	6265	6321	6378	6434
770	6491	6547	6604	6660	6716	6773	6829	6885	6942	6998
71	7054	7111	7167	7223	7280	7336	7392	7449	7505	7561
72	7617	7674	7730	7786	7842	7898	7955	8011	8067	8123
73	8179	8236	8292	8348	8404	8460	8516	8573	8629	8685
74	8741	8797	8853	8909	8965	9021	9077	9134	9190	9246
75	9302	9358	9414	9470	9526	9582	9638	9694	9750	9806
76	88 9862	9918	9974	*0030	*0086	*0141	*0197	*0253	*0309	*0365
77	89 0421	0477	0533	0589	0645	0700	0756	0812	0868	0924
78	0980	1035	1091	1147	1203	1259	1314	1370	1426	1482
79	1537	1593	1649	1705	1760	1816	1872	1928	1983	2039
780	2095	2150	2206	2262	2317	2373	2429	2484	2540	2595
81	2651	2707	2762	2818	2873	2929	2985	3040	3096	3151
82	3207	3262	3318	3373	3429	3484	3540	3595	3651	3706
83	3762	3817	3873	3928	3984	4039	4094	4150	4205	4261
84	4316	4371	4427	4482	4538	4593	4648	4704	4759	4814
85	4870	4925	4980	5036	5091	5146	5201	5257	5312	5367
86	5423	5478	5533	5588	5644	5699	5754	5809	5864	5920
87	5975	6030	6085	6140	6195	6251	6306	6361	6416	6471
88	6526	6581	6636	6692	6747	6802	6857	6912	6967	7022
89	7077	7132	7187	7242	7297	7352	7407	7462	7517	7572
790	7627	7682	7737	7792	7847	7902	7957	8012	8067	8122
91	8176	8231	8286	8341	8396	8451	8506	8561	8615	8670
92	8725	8780	8835	8890	8944	8999	9054	9109	9164	9218
93	9273	9328	9383	9437	9492	9547	9602	9656	9711	9766
94	89 9821	9875	9930	9985	*0039	*0094	*0149	*0203	*0258	*0312
95	90 0367	0422	0476	0531	0586	0640	0695	0749	0804	0859
96	0913	0968	1022	1077	1131	1186	1240	1295	1349	1404
97	1458	1513	1567	1622	1676	1731	1785	1840	1894	1948
98	2003	2057	2112	2166	2221	2275	2329	2384	2438	2492
99	2547	2601	2655	2710	2764	2818	2873	2927	2981	3036
800	90 3090	3144	3199	3253	3307	3361	3416	3470	3524	3578
N	**0**	**1**	**2**	**3**	**4**	**5**	**6**	**7**	**8**	**9**

SIX-PLACE MANTISSAS FOR COMMON LOGARITHMS (Continued)

N	0	1	2	3	4	5	6	7	8	9
800	90 3090	3144	3199	3253	3307	3361	3416	3470	3524	3578
01	3633	3687	3741	3795	3849	3904	3958	4012	4066	4120
02	4174	4229	4283	4337	4391	4445	4499	4553	4607	4661
03	4716	4770	4824	4878	4932	4986	5040	5094	5148	5202
04	5256	5310	5364	5418	5472	5526	5580	5634	5688	5742
05	5796	5850	5904	5958	6012	6066	6119	6173	6227	6281
06	6335	6389	6443	6497	6551	6604	6658	6712	6766	6820
07	6874	6927	6981	7035	7089	7143	7196	7250	7304	7358
08	7411	7465	7519	7573	7626	7680	7734	7787	7841	7895
09	7949	8002	8056	8110	8163	8217	8270	8324	8378	8431
810	8485	8539	8592	8646	8699	8753	8807	8860	8914	8967
11	9021	9074	9128	9181	9235	9289	9342	9396	9449	9503
12	90 9556	9610	9663	9716	9770	9823	9877	9930	9984	*0037
13	91 0091	0144	0197	0251	0304	0358	0411	0464	0518	0571
14	0624	0678	0731	0784	0838	0891	0944	0998	1051	1104
15	1158	1211	1264	1317	1371	1424	1477	1530	1584	1637
16	1690	1743	1797	1850	1903	1956	2009	2063	2116	2169
17	2222	2275	2328	2381	2435	2488	2541	2594	2647	2700
18	2753	2806	2859	2913	2966	3019	3072	3125	3178	3231
19	3284	3337	3390	3443	3496	3549	3602	3655	3708	3761
820	3814	3867	3920	3973	4026	4079	4132	4184	4237	4290
21	4343	4396	4449	4502	4555	4608	4660	4713	4766	4819
22	4872	4925	4977	5030	5083	5136	5189	5241	5294	5347
23	5400	5453	5505	5558	5611	5664	5716	5769	5822	5875
24	5927	5980	6033	6085	6138	6191	6243	6296	6349	6401
25	6454	6507	6559	6612	6664	6717	6770	6822	6875	6927
26	6980	7033	7085	7138	7190	7243	7295	7348	7400	7453
27	7506	7558	7611	7663	7716	7768	7820	7873	7925	7978
28	8030	8083	8135	8188	8240	8293	8345	8397	8450	8502
29	8555	8607	8659	8712	8764	8816	8869	8921	8973	9026
830	9078	9130	9183	9235	9287	9340	9392	9444	9496	9549
31	91 9601	9653	9706	9758	9810	9862	9914	9967	*0019	*0071
32	92 0123	0176	0228	0280	0332	0384	0436	0489	0541	0593
33	0645	0697	0749	0801	0853	0906	0958	1010	1062	1114
34	1166	1218	1270	1322	1374	1426	1478	1530	1582	1634
35	1686	1738	1790	1842	1894	1946	1998	2050	2102	2154
36	2206	2258	2310	2362	2414	2466	2518	2570	2622	2674
37	2725	2777	2829	2881	2933	2985	3037	3089	3140	3192
38	3244	3296	3348	3399	3451	3503	3555	3607	3658	3710
39	3762	3814	3865	3917	3969	4021	4072	4124	4176	4228
840	4279	4331	4383	4434	4486	4538	4589	4641	4693	4744
41	4796	4848	4899	4951	5003	5054	5106	5157	5209	5261
42	5312	5364	5415	5467	5518	5570	5621	5673	5725	5776
43	5828	5879	5931	5982	6034	6085	6137	6188	6240	6291
44	6342	6394	6445	6497	6548	6600	6651	6702	6754	6805
45	6857	6908	6959	7011	7062	7114	7165	7216	7268	7319
46	7370	7422	7473	7524	7576	7627	7678	7730	7781	7832
47	7883	7935	7986	8037	8088	8140	8191	8242	8293	8345
48	8396	8447	8498	8549	8601	8652	8703	8754	8805	8857
49	8908	8959	9010	9061	9112	9163	9215	9266	9317	9368
850	92 9419	9470	9521	9572	9623	9674	9725	9776	9827	9879
N	**0**	**1**	**2**	**3**	**4**	**5**	**6**	**7**	**8**	**9**

SIX-PLACE MANTISSAS FOR COMMON LOGARITHMS (Continued)

N	0	1	2	3	4	5	6	7	8	9
850	92 9419	9470	9521	9572	9623	9674	9725	9776	9827	9879
51	92 9930	9981	*0032	*0083	*0134	*0185	*0236	*0287	*0338	*0389
52	93 0440	0491	0542	0592	0643	0694	0745	0796	0847	0898
53	0949	1000	1051	1102	1153	1204	1254	1305	1356	1407
54	1458	1509	1560	1610	1661	1712	1763	1814	1865	1915
55	1966	2017	2068	2118	2169	2220	2271	2322	2372	2423
56	2474	2524	2575	2626	2677	2727	2778	2829	2879	2930
57	2981	3031	3082	3133	3183	3234	3285	3335	3386	3437
58	3487	3538	3589	3639	3690	3740	3791	3841	3892	3943
59	3993	4044	4094	4145	4195	4246	4296	4347	4397	4448
860	4498	4549	4599	4650	4700	4751	4801	4852	4902	4953
61	5003	5054	5104	5154	5205	5255	5306	5356	5406	5457
62	5507	5558	5608	5658	5709	5759	5809	5860	5910	5960
63	6011	6061	6111	6162	6212	6262	6313	6363	6413	6463
64	6514	6564	6614	6665	6715	6765	6815	6865	6916	6966
65	7016	7066	7166	7167	7217	7267	7317	7367	7418	7468
66	7518	7568	7618	7668	7718	7769	7819	7869	7919	7969
67	8019	8069	8119	8169	8219	8269	8320	8370	8420	8470
68	8520	8570	8620	8670	8720	8770	8820	8870	8920	8970
69	9020	9070	9120	9170	9220	9270	9320	9369	9419	9469
870	93 9519	9569	9619	9669	9719	9769	9819	9869	9918	9968
71	94 0018	0068	0118	0168	0218	0267	0317	0367	0417	0467
72	0516	0566	0616	0666	0716	0765	0815	0865	0915	0964
73	1014	1064	1114	1163	1213	1263	1313	1362	1412	1462
74	1511	1561	1611	1660	1710	1760	1809	1859	1909	1958
75	2008	2058	2107	2157	2207	2256	2306	2355	2405	2455
76	2504	2554	2603	2653	2702	2752	2801	2851	2901	2950
77	3000	3049	3099	3148	3198	3247	3297	3346	3396	3445
78	3495	3544	3593	3643	3692	3742	3791	3841	3890	3939
79	3989	4038	4088	4137	4186	4236	4285	4335	4384	4433
880	4483	4532	4581	4631	4680	4729	4779	4828	4877	4927
81	4976	5025	5074	5124	5173	5222	5272	5321	5370	5419
82	5469	5518	5567	5616	5665	5715	5764	5813	5862	5912
83	5961	6010	6059	6108	6157	6207	6256	6305	6354	6403
84	6452	6501	6551	6600	6649	6698	6747	6796	6845	6894
85	6943	6992	7041	7090	7139	7189	7238	7287	7336	7385
86	7434	7483	7532	7581	7630	7679	7728	7777	7826	7875
87	7924	7973	8022	8070	8119	8168	8217	8266	8315	8364
88	8413	8462	8511	8560	8608	8657	8706	8755	8804	8853
89	8902	8951	8999	9048	9097	9146	9195	9244	9292	9341
890	9390	9439	9488	9536	9585	9634	9683	9731	9780	9829
91	94 9878	9926	9975	*0024	*0073	*0121	*0170	*0219	*0267	*0316
92	95 0365	0414	0462	0511	0560	0608	0657	0706	0754	0803
93	0851	0900	0949	0997	1046	1095	1143	1192	1240	1289
94	1338	1386	1435	1483	1532	1580	1629	1677	1726	1775
95	1823	1872	1920	1969	2017	2066	2114	2163	2211	2260
96	2308	2356	2405	2453	2502	2550	2599	2647	2696	2744
97	2792	2841	2889	2938	2986	3034	3083	3131	3180	3228
98	3276	3325	3373	3421	3470	3518	3566	3615	3663	3711
99	3760	3808	3856	3905	3953	4001	4049	4098	4146	4194
900	95 4243	4291	4339	4387	4435	4484	4532	4580	4628	4677
N	**0**	**1**	**2**	**3**	**4**	**5**	**6**	**7**	**8**	**9**

SIX-PLACE MANTISSAS FOR COMMON LOGARITHMS (Continued)

N	0	1	2	3	4	5	6	7	8	9
900	95 4243	4291	4339	4387	4435	4484	4532	4580	4628	4677
01	4725	4773	4821	4869	4918	4966	5014	5062	5110	5158
02	5207	5255	5303	5351	5399	5447	5495	5543	5592	5640
03	5688	5736	5784	5832	5880	5928	5976	6024	6072	6120
04	6168	6216	6265	6313	6361	6409	6457	6505	6553	6601
05	6649	6697	6745	6793	6840	6888	6936	6984	7032	7080
06	7128	7176	7224	7272	7320	7368	7416	7464	7512	7559
07	7607	7655	7703	7751	7799	7847	7894	7942	7990	8038
08	8086	8134	8181	8229	8277	8325	8373	8421	8468	8516
09	8564	8612	8659	8707	8755	8803	8850	8898	8946	8994
910	9041	9089	9137	9185	9232	9280	9328	9375	9423	9471
11	9518	9566	9614	9661	9709	9757	9804	9852	9900	9947
12	95 9995	*0042	*0090	*0138	*0185	*0233	*0280	*0328	*0376	*0423
13	96 0471	0518	0566	0613	0661	0709	0756	0804	0851	0899
14	0946	0994	1041	1089	1136	1184	1231	1279	1326	1374
15	1421	1469	1516	1563	1611	1658	1706	1753	1801	1848
16	1895	1943	1990	2038	2085	2132	2180	2227	2275	2322
17	2369	2417	2464	2511	2559	2606	2653	2701	2748	2795
18	2843	2890	2937	2985	3032	3079	3126	3174	3221	3268
19	3316	3363	3410	3457	3504	3552	3599	3646	3693	3741
920	3788	3835	3882	3929	3977	4024	4071	4118	4165	4212
21	4260	4307	4354	4401	4448	4495	4542	4590	4637	4684
22	4731	4778	4825	4872	4919	4966	5013	5061	5108	5155
23	5202	5249	5296	5343	5390	5437	5484	5531	5578	5625
24	5672	5719	5766	5813	5860	5907	5954	6001	6048	6095
25	6142	6189	6236	6283	6329	6376	6423	6470	6517	6564
26	6611	6658	6705	6752	6799	6845	6892	6939	6986	7033
27	7080	7127	7173	7220	7267	7314	7361	7408	7454	7501
28	7548	7595	7642	7688	7735	7782	7829	7875	7922	7969
29	8016	8062	8109	8156	8203	8249	8296	8343	8390	8436
930	8483	8530	8576	8623	8670	8716	8763	8810	8856	8903
31	8950	8996	9043	9090	9136	9183	9229	9276	9323	9369
32	9416	9463	9509	9556	9602	9649	9695	9742	9789	9835
33	96 9882	9928	9975	*0021	*0068	*0114	*0161	*0207	*0254	*0300
34	97 0347	0393	0440	0486	0533	0579	0626	0672	0719	0765
35	0812	0858	0904	0951	0997	1044	1090	1137	1183	1229
36	1276	1322	1369	1415	1461	1508	1554	1601	1647	1693
37	1740	1786	1832	1879	1925	1971	2018	2064	2110	2157
38	2203	2249	2295	2342	2388	2434	2481	2527	2573	2619
39	2666	2712	2758	2804	2851	2897	2943	2989	3035	3082
940	3128	3174	3220	3266	3313	3359	3405	3451	3497	3543
41	3590	3636	3682	3728	3774	3820	3866	3913	3959	4005
42	4051	4097	4143	4189	4235	4281	4327	4374	4420	4466
43	4512	4558	4604	4650	4696	4742	4788	4834	4880	4926
44	4972	5018	5064	5110	5156	5202	5248	5294	5340	5386
45	5432	5478	5524	5570	5616	5662	5707	5753	5799	5845
46	5891	5937	5983	6029	6075	6121	6167	6212	6258	6304
47	6350	6396	6442	6488	6533	6579	6625	6671	6717	6763
48	6808	6854	6900	6946	6992	7037	7083	7129	7175	7220
49	7266	7312	7358	7403	7449	7495	7541	7586	7632	7678
950	97 7724	7769	7815	7861	7906	7952	7998	8043	8089	8135
N	**0**	**1**	**2**	**3**	**4**	**5**	**6**	**7**	**8**	**9**

SIX-PLACE MANTISSAS FOR COMMON LOGARITHMS (Continued)

N	0	1	2	3	4	5	6	7	8	9
950	97 7724	7769	7815	7861	7906	7952	7998	8043	8089	8135
51	8181	8226	8272	8317	8363	8409	8454	8500	8546	8591
52	8637	8683	8728	8774	8819	8865	8911	8956	9002	9047
53	9093	9138	9184	9230	9275	9321	9366	9412	9457	9503
54	97 9548	9594	9639	9685	9730	9776	9821	9867	9912	9958
55	98 0003	0049	0094	0140	0185	0231	0276	0322	0367	0412
56	0458	0503	0549	0594	0640	0685	0730	0776	0821	0867
57	0912	0957	1003	1048	1093	1139	1184	1229	1275	1320
58	1366	1411	1456	1501	1547	1592	1637	1683	1728	1773
59	1819	1864	1909	1954	2000	2045	2090	2135	2181	2226
960	2271	2316	2362	2407	2452	2497	2543	2588	2633	2678
61	2723	2769	2814	2859	2904	2949	2994	3040	3085	3130
62	3175	3220	3265	3310	3356	3401	3446	3491	3536	3581
63	3626	3671	3716	3762	3807	3852	3897	3942	3987	4032
64	4077	4122	4167	4212	4257	4302	4347	4392	4437	4482
65	4527	4572	4617	4662	4707	4752	4797	4842	4887	4932
66	4977	5022	5067	5112	5157	5202	5247	5292	5337	5382
67	5426	5471	5516	5561	5606	5651	5696	5741	5786	5830
68	5875	5920	5965	6010	6055	6100	6144	6189	6234	6279
69	6324	6369	6413	6458	6503	6548	6593	6637	6682	6727
970	6772	6817	6861	6906	6951	6996	7040	7085	7130	7175
71	7219	7264	7309	7353	7398	7443	7488	7532	7577	7622
72	7666	7711	7756	7800	7845	7890	7934	7979	8024	8068
73	8113	8157	8202	8247	8291	8336	8381	8425	8470	8514
74	8559	8604	8648	8693	8737	8782	8826	8871	8916	8960
75	9005	9049	9094	9138	9183	9227	9272	9316	9361	9405
76	9450	9494	9539	9583	9628	9672	9717	9761	9806	9850
77	98 9895	9939	9983	*0028	*0072	*0117	*0161	*0206	*0250	*0294
78	99 0339	0383	0428	0472	0516	0561	0605	0650	0694	0738
79	0783	0827	0871	0916	0960	1004	1049	1093	1137	1182
980	1226	1270	1315	1359	1403	1448	1492	1536	1580	1625
81	1669	1713	1758	1802	1846	1890	1935	1979	2023	2067
82	2111	2156	2200	2244	2288	2333	2377	2421	2465	2509
83	2554	2598	2642	2686	2730	2774	2819	2863	2907	2951
84	2995	3039	3083	3127	3172	3216	3260	3304	3348	3392
85	3436	3480	3524	3568	3613	3657	3701	3745	3789	3833
86	3877	3921	3965	4009	4053	4097	4141	4185	4229	4273
87	4317	4361	4405	4449	4493	4537	4581	4625	4669	4713
88	4757	4801	4845	4889	4933	4977	5021	5065	5108	5152
89	5196	5240	5284	5328	5372	5416	5460	5504	5547	5591
990	5635	5679	5723	5767	5811	5854	5898	5942	5986	6030
91	6074	6117	6161	6205	6249	6293	6337	6380	6424	6468
92	6512	6555	6599	6643	6687	6731	6774	6818	6862	6906
93	6949	6993	7037	7080	7124	7168	7212	7255	7299	7343
94	7386	7430	7474	7517	7561	7605	7648	7692	7736	7779
95	7823	7867	7910	7954	7998	8041	8085	8129	8172	8216
96	8259	8303	8347	8390	8434	8477	8521	8564	8608	8652
97	8695	8739	8782	8826	8869	8913	8956	9000	9043	9087
98	9131	9174	9218	9261	9305	9348	9392	9435	9479	9522
99	99 9565	9609	9652	9696	9739	9783	9826	9870	9913	9957
1000	00 0000	0043	0087	0130	0174	0217	0260	0304	0347	0391
N	**0**	**1**	**2**	**3**	**4**	**5**	**6**	**7**	**8**	**9**

SEVEN-PLACE MANTISSAS FOR COMMON LOGARITHMS:

N	0	1	2	3	4	5	6	7	8	9
1000	000 0000	0434	0869	1303	1737	2171	2605	3039	3473	3907
1001	4341	4775	5208	5642	6076	6510	6943	7377	7810	8244
1002	8677	9111	9544	9977	*0411	*0844	*1277	*1710	*2143	*2576
1003	001 3009	3442	3875	4308	4741	5174	5607	6039	6472	6905
1004	7337	7770	8202	8635	9067	9499	9932	*0364	*0796	*1228
1005	002 1661	2093	2525	2957	3389	3821	4253	4685	5116	5548
1006	5980	6411	6843	7275	7706	8138	8569	9001	9432	9863
1007	003 0295	0726	1157	1588	2019	2451	2882	3313	3744	4174
1008	4605	5036	5467	5898	6328	6759	7190	7620	8051	8481
1009	8912	9342	9772	*0203	*0633	*1063	*1493	*1924	*2354	*2784
1010	004 3214	3644	4074	4504	4933	5363	5793	6223	6652	7082
1011	7512	7941	8371	8800	9229	9659	*0088	*0517	*0947	*1376
1012	005 1805	2234	2663	3092	3521	3950	4379	4808	5237	5666
1013	6094	6523	6952	7380	7809	8238	8666	9094	9523	9951
1014	006 0380	0808	1236	1664	2092	2521	2949	3377	3805	4233
1015	4660	5088	5516	5944	6372	6799	7227	7655	8082	8510
1016	8937	9365	9792	*0219	*0647	*1074	*1501	*1928	*2355	*2782
1017	007 3210	3637	4064	4490	4917	5344	5771	6198	6624	7051
1018	7478	7904	8331	8757	9184	9610	*0037	*0463	*0889	*1316
1019	008 1742	2168	2594	3020	3446	3872	4298	4724	5150	5576
1020	6002	6427	6853	7279	7704	8130	8556	8981	9407	9832
1021	009 0257	0683	1108	1533	1959	2384	2809	3234	3659	4084
1022	4509	4934	5359	5784	6208	6633	7058	7483	7907	8332
1023	8756	9181	9605	*0030	*0454	*0878	*1303	*1727	*2151	*2575
1024	010 3000	3424	3848	4272	4696	5120	5544	5967	6391	6815
1025	7239	7662	8086	8510	8933	9357	9780	*0204	*0627	*1050
1026	011 1474	1897	2320	2743	3166	3590	4013	4436	4859	5282
1027	5704	6127	6550	6973	7396	7818	8241	8664	9086	9509
1028	9931	*0354	*0776	*1198	*1621	*2043	*2465	*2887	*3310	*3732
1029	012 4154	4576	4998	5420	5842	6264	6685	7107	7529	7951
1030	8372	8794	9215	9637	*0059	*0480	*0901	*1323	*1744	*2165
1031	013 2587	3008	3429	3850	4271	4692	5113	5534	5955	6376
1032	6797	7218	7639	8059	8480	8901	9321	9742	*0162	*0583
1033	014 1003	1424	1844	2264	2685	3105	3525	3945	4365	4785
1034	5205	5625	6045	6465	6885	7305	7725	8144	8564	8984
1035	9403	9823	*0243	*0662	*1082	*1501	*1920	*2340	*2759	*3178
1036	015 3598	4017	4436	4855	5274	5693	6112	6531	6950	7369
1037	7788	8206	8625	9044	9462	9881	*0300	*0718	*1137	*1555
1038	016 1974	2392	2810	3229	3647	4065	4483	4901	5319	5737
1039	6155	6573	6991	7409	7827	8245	8663	9080	9498	9916
1040	017 0333	0751	1168	1586	2003	2421	2838	3256	3673	4090
1041	4507	4924	5342	5759	6176	6593	7010	7427	7844	8260
1042	8677	9094	9511	9927	*0344	*0761	*1177	*1594	*2010	*2427
1043	018 2843	3259	3676	4092	4508	4925	5341	5757	6173	6589
1044	7005	7421	7837	8253	8669	9084	9500	9916	*0332	*0747
1045	019 1163	1578	1994	2410	2825	3240	3656	4071	4486	4902
1046	5317	5732	6147	6562	6977	7392	7807	8222	8637	9052
1047	9467	9882	*0296	*0711	*1126	*1540	*1955	*2369	*2784	*3198
1048	020 3613	4027	4442	4856	5270	5684	6099	6513	6927	7341
1049	7755	8169	8583	8997	9411	9824	*0238	*0652	*1066	*1479
1050	021 1893	2307	2720	3134	3547	3961	4374	4787	5201	5614
N	**0**	**1**	**2**	**3**	**4**	**5**	**6**	**7**	**8**	**9**

SEVEN-PLACE MANTISSAS FOR COMMON LOGARITHMS: (Continued)

N	0	1	2	3	4	5	6	7	8	9
1050	021 1893	2307	2720	3134	3547	3961	4374	4787	5201	5614
1051	6027	6440	6854	7267	7680	8093	8506	8919	9332	9745
1052	022 0157	0570	0983	1396	1808	2221	2634	3046	3459	3871
1053	4284	4696	5109	5521	5933	6345	6758	7170	7582	7994
1054	8406	8818	9230	9642	*0054	*0466	*0878	*1289	*1701	*2113
1055	023 2525	2936	3348	3759	4171	4582	4994	5405	5817	6228
1056	6639	7050	7462	7873	8284	8695	9106	9517	9928	*0339
1057	024 0750	1161	1572	1982	2393	2804	3214	3625	4036	4446
1058	4857	5267	5678	6088	6498	6909	7319	7729	8139	8549
1059	8960	9370	9780	*0190	*0600	*1010	*1419	*1829	*2239	*2649
1060	025 3059	3468	3878	4288	4697	5107	5516	5926	6335	6744
1061	7154	7563	7972	8382	8791	9200	9609	*0018	*0427	*0836
1062	026 1245	1654	2063	2472	2881	3289	3698	4107	4515	4924
1063	5333	5741	6150	6558	6967	7375	7783	8192	8600	9008
1064	9416	9824	*0233	*0641	*1049	*1457	*1865	*2273	*2680	*3088
1065	027 3496	3904	4312	4719	5127	5535	5942	6350	6757	7165
1066	7572	7979	8387	8794	9201	9609	*0016	*0423	*0830	*1237
1067	028 1644	2051	2458	2865	3272	3679	4086	4492	4899	5306
1068	5713	6119	6526	6932	7339	7745	8152	8558	8964	9371
1069	9777	*0183	*0590	*0996	*1402	*1808	*2214	*2620	*3026	*3432
1070	029 3838	4244	4649	5055	5461	5867	6272	6678	7084	7489
1071	7895	8300	8706	9111	9516	9922	*0327	*0732	*1138	*1543
1072	030 1948	2353	2758	3163	3568	3973	4378	4783	5188	5592
1073	5997	6402	6807	7211	7616	8020	8425	8830	9234	9638
1074	031 0043	0447	0851	1256	1660	2064	2468	2872	3277	3681
1075	4085	4489	4893	5296	5700	6104	6508	6912	7315	7719
1076	8123	8526	8930	9333	9737	*0140	*0544	*0947	*1350	*1754
1077	032 2157	2560	2963	3367	3770	4173	4576	4979	5382	5785
1078	6188	6590	6993	7396	7799	8201	8604	9007	9409	9812
1079	033 0214	0617	1019	1422	1824	2226	2629	3031	3433	3835
1080	4238	4640	5042	5444	5846	6248	6650	7052	7453	7855
1081	8257	8659	9060	9462	9864	*0265	*0667	*1068	*1470	*1871
1082	034 2273	2674	3075	3477	3878	4279	4680	5081	5482	5884
1083	6285	6686	7087	7487	7888	8289	8690	9091	9491	9892
1084	035 0293	0693	1094	1495	1895	2296	2696	3096	3497	3897
1085	4297	4698	5098	5498	5898	6298	6698	7098	7498	7898
1086	8298	8698	9098	9498	9898	*0297	*0697	*1097	*1496	*1896
1087	036 2295	2695	3094	3494	3893	4293	4692	5091	5491	5890
1088	6289	6688	7087	7486	7885	8284	8683	9082	9481	9880
1089	037 0279	0678	1076	1475	1874	2272	2671	3070	3468	3867
1090	4265	4663	5062	5460	5858	6257	6655	7053	7451	7849
1091	8248	8646	9044	9442	9839	*0237	*0635	*1033	*1431	*1829
1092	038 2226	2624	3022	3419	3817	4214	4612	5009	5407	5804
1093	6202	6599	6996	7393	7791	8188	8585	8982	9379	9776
1094	039 0173	0570	0967	1364	1761	2158	2554	2951	3348	3745
1095	4141	4538	4934	5331	5727	6124	6520	6917	7313	7709
1096	8106	8502	8898	9294	9690	*0086	*0482	*0878	*1274	*1670
1097	040 2066	2462	2858	3254	3650	4045	4441	4837	5232	5628
1098	6023	6419	6814	7210	7605	8001	8396	8791	9187	9582
1099	9977	*0372	*0767	*1162	*1557	*1952	*2347	*2742	*3137	*3532
1100	041 3927	4322	4716	5111	5506	5900	6295	6690	7084	7479
N	**0**	**1**	**2**	**3**	**4**	**5**	**6**	**7**	**8**	**9**

SEVEN-PLACE MANTISSAS FOR COMMON LOGARITHMS: (Continued)

N	0	1	2	3	4	5	6	7	8	9
1100	041 3927	4322	4716	5111	5506	5900	6295	6690	7084	7479
1101	• 7873	8268	8662	9056	9451	9845	*0239	*0633	*1028	*1422
1102	042 1816	2210	2604	2998	3392	3786	4180	4574	4968	5361
1103	5755	6149	6543	6936	7330	7723	8117	8510	8904	9297
1104	9691	*0084	*0477	*0871	*1264	*1657	*2050	*2444	*2837	*3230
1105	043 3623	4016	4409	4802	5195	5587	5980	6373	6766	7159
1106	7551	7944	8337	8729	9122	9514	9907	*0299	*0692	*1084
1107	044 1476	1869	2261	2653	3045	3437	3829	4222	4614	5006
1108	5398	5790	6181	6573	6965	7357	7749	8140	8532	8924
1109	9315	9707	*0099	*0490	*0882	*1273	*1664	*2056	*2447	*2839
1110	045 3230	3621	4012	4403	4795	5186	5577	5968	6359	6750
1111	7141	7531	7922	8313	8704	9095	9485	9876	*0267	*0657
1112	046 1048	1438	1829	2219	2610	3000	3391	3781	4171	4561
1113	4952	5342	5732	6122	6512	6902	7292	7682	8072	8462
1114	8852	9242	9632	*0021	*0411	*0801	*1190	*1580	*1970	*2359
1115	047 2749	3138	3528	3917	4306	4696	5085	5474	5864	6253
1116	6642	7031	7420	7809	8198	8587	8976	9365	9754	*0143
1117	048 0532	0921	1309	1698	2087	2475	2864	3253	3641	4030
1118	4418	4806	5195	5583	5972	6360	6748	7136	7525	7913
1119	8301	8689	9077	9465	9853	*0241	*0629	*1017	*1405	*1792
1120	049 2180	2568	2956	3343	3731	4119	4506	4894	5281	5669
1121	6056	6444	6831	7218	7606	7993	8380	8767	9154	9541
1122	9929	*0316	*0703	*1090	*1477	*1863	*2250	*2637	*3024	*3411
1123	050 3798	4184	4571	4958	5344	5731	6117	6504	6890	7277
1124	7663	8049	8436	8822	9208	9595	9981	*0367	*0753	*1139
1125	051 1525	1911	2297	2683	3069	3455	3841	4227	4612	4998
1126	5384	5770	6155	6541	6926	7312	7697	8083	8468	8854
1127	9239	9624	*0010	*0395	*0780	*1166	*1551	*1936	*2321	*2706
1128	052 3091	3476	3861	4246	4631	5016	5400	5785	6170	6555
1129	6939	7324	7709	8093	8478	8862	9247	9631	*0016	*0400
1130	053 0784	1169	1553	1937	2321	2706	3090	3474	3858	4242
1131	4626	5010	5394	5778	6162	6546	6929	7313	7697	8081
1132	8464	8848	9232	9615	9999	*0382	*0766	*1149	*1532	*1916
1133	054 2299	2682	3066	3449	3832	4215	4598	4981	5365	5748
1134	6131	6514	6896	7279	7662	8045	8428	8811	9193	9576
1135	9959	*0341	*0724	*1106	*1489	*1871	*2254	*2636	*3019	*3401
1136	055 3783	4166	4548	4930	5312	5694	6077	6459	6841	7223
1137	7605	7987	8369	8750	9132	9514	9896	*0278	*0659	*1041
1138	056 1423	1804	2186	2567	2949	3330	3712	4093	4475	4856
1139	5237	5619	6000	6381	6762	7143	7524	7905	8287	8668
1140	9049	9429	9810	*0191	*0572	*0953	*1334	*1714	*2095	*2476
1141	057 2856	3237	3618	3998	4379	4759	5140	5520	5900	6281
1142	6661	7041	7422	7802	8182	8562	8942	9322	9702	*0082
1143	058 0462	0842	1222	1602	1982	2362	2741	3121	3501	3881
1144	4260	4640	5019	5399	5778	6158	6537	6917	7296	7676
1145	8055	8434	8813	9193	9572	9951	*0330	*0709	*1088	*1467
1146	059 1846	2225	2604	2983	3362	3741	4119	4498	4877	5256
1147	5634	6013	6391	6770	7148	7527	7905	8284	8662	9041
1148	9419	9797	*0175	*0554	*0932	*1310	*1688	*2066	*2444	*2822
1149	060 3200	3578	3956	4334	4712	5090	5468	5845	6223	6601
1150	6978	7356	7734	8111	8489	8866	9244	9621	9999	*0376
N	**0**	**1**	**2**	**3**	**4**	**5**	**6**	**7**	**8**	**9**

SEVEN-PLACE MANTISSAS FOR COMMON LOGARITHMS: (Continued)

N	0	1	2	3	4	5	6	7	8	9
1150	060 6978	7356	7734	8111	8489	8866	9244	9621	9999	*0376
1151	061 0753	1131	1508	1885	2262	2639	3017	3394	3771	4148
1152	4525	4902	5279	5656	6032	6409	6786	7163	7540	7916
1153	8293	8670	9046	9423	9799	*0176	*0552	*0929	*1305	*1682
1154	062 2058	2434	2811	3187	3563	3939	4316	4692	5068	5444
1155	5820	6196	6572	6948	7324	7699	8075	8451	8827	9203
1156	9578	9954	*0330	*0705	*1081	*1456	*1832	*2207	*2583	*2958
1157	063 3334	3709	4084	4460	4835	5210	5585	5960	6335	6711
1158	7086	7461	7836	8211	8585	8960	9335	9710	*0085	*0460
1159	064 0834	1209	1584	1958	2333	2708	3082	3457	3831	4205
1160	4580	4954	5329	5703	6077	6451	6826	7200	7574	7948
1161	8322	8696	9070	9444	9818	*0192	*0566	*0940	*1314	*1688
1162	065 2061	2435	2809	3182	3556	3930	4303	4677	5050	5424
1163	5797	6171	6544	6917	7291	7664	8037	8410	8784	9157
1164	9530	9903	*0276	*0649	*1022	*1395	*1768	*2141	*2514	*2886
1165	066 3259	3632	4005	4377	4750	5123	5495	5868	6241	6613
1166	6986	7358	7730	8103	8475	8847	9220	9592	9964	*0336
1167	067 0709	1081	1453	1825	2197	2569	2941	3313	3685	4057
1168	4428	4800	5172	5544	5915	6287	6659	7030	7402	7774
1169	8145	8517	8888	9259	9631	*0002	*0374	*0745	*1116	*1487
1170	068 1859	2230	2601	2972	3343	3714	4085	4456	4827	5198
1171	5569	5940	6311	6681	7052	7423	7794	8164	8535	8906
1172	9276	9647	*0017	*0388	*0758	*1129	*1499	*1869	*2240	*2610
1173	069 2980	3350	3721	4091	4461	4831	5201	5571	5941	6311
1174	6681	7051	7421	7791	8160	8530	8900	9270	9639	*0009
1175	070 0379	0748	1118	1487	1857	2226	2596	2965	3335	3704
1176	4073	4442	4812	5181	5550	5919	6288	6658	7027	7396
1177	7765	8134	8503	8871	9240	9609	9978	*0347	*0715	*1084
1178	071 1453	1822	2190	2559	2927	3296	3664	4033	4401	4770
1179	5138	5506	5875	6243	6611	6979	7348	7716	8084	8452
1180	8820	9188	9556	9924	*0292	*0660	*1028	*1396	*1763	*2131
1181	072 2499	2867	3234	3602	3970	4337	4705	5072	5440	5807
1182	6175	6542	6910	7277	7644	8011	8379	8746	9113	9480
1183	9847	*0215	*0582	*0949	*1316	*1683	*2050	*2416	*2783	*3150
1184	073 3517	3884	4251	4617	4984	5351	5717	6084	6450	6817
1185	7184	7550	7916	8283	8649	9016	9382	9748	*0114	*0481
1186	074 0847	1213	1579	1945	2311	2677	3043	3409	3775	4141
1187	4507	4873	5239	5605	5970	6336	6702	7068	7433	7799
1188	8164	8530	8895	9261	9626	9992	*0357	*0723	*1088	*1453
1189	075 1819	2184	2549	2914	3279	3644	4010	4375	4740	5105
1190	5470	5835	6199	6564	6929	7294	7659	8024	8388	8753
1191	9118	9482	9847	*0211	*0576	*0940	*1305	*1669	*2034	*2398
1192	076 2763	3127	3491	3855	4220	4584	4948	5312	5676	6040
1193	6404	6768	7132	7496	7860	8224	8588	8952	9316	9680
1194	077 0043	0407	0771	1134	1498	1862	2225	2589	2952	3316
1195	3679	4042	4406	4769	5133	5496	5859	6222	6585	6949
1196	7312	7675	8038	8401	8764	9127	9490	9853	*0216	*0579
1197	078 0942	1304	1667	2030	2393	2755	3118	3480	3843	4206
1198	4568	4931	5293	5656	6018	6380	6743	7105	7467	7830
1199	8192	8554	8916	9278	9640	*0003	*0365	*0727	*1089	*1451
1200	079 1812	2174	2536	2898	3260	3622	3983	4345	4707	5068
N	**0**	**1**	**2**	**3**	**4**	**5**	**6**	**7**	**8**	**9**

NATURAL OR NAPERIAN LOGARITHMS

0.000–0.499

N	0	1	2	3	4	5	6	7	8	9
0.00	$-\infty$	−6‡ .90776	−6 .21461	−5 .80914	−5 .52146	−5 .29832	−5 .11600	−4 .96185	−4 .82831	−4 .71053
.01	−4.60517	.50986	.42285	.34281	.26870	.19971	.13517	.07454	.01738	*.96332
.02	−3.91202	.86323	.81671	.77226	.72970	.68888	.64966	.61192	.57555	.54046
.03	.50656	.47377	.44202	.41125	.38139	.35241	.32424	.29684	.27017	.24419
.04	.21888	.19418	.17009	.14656	.12357	.10109	.07911	.05761	.03655	.01593
.05	−2.99573	.97593	.95651	.93746	.91877	.90042	.88240	.86470	.84731	.83022
.06	.81341	.79688	.78062	.76462	.74887	.73337	.71810	.70306	.68825	.67365
.07	.65926	.64508	.63109	.61730	.60369	.59027	.57702	.56395	.55105	.53831
.08	.52573	.51331	.50104	.48891	.47694	.46510	.45341	.44185	.43042	.41912
.09	.40795	.39690	.38597	.37516	.36446	.35388	.34341	.33304	.32279	.31264
0.10	−2.30259	.29263	.28278	.27303	.26336	.25379	.24432	.23493	.22562	.21641
.11	.20727	.19823	.18926	.18037	.17156	.16282	.15417	.14558	.13707	.12863
.12	.12026	.11196	.10373	.09557	.08747	.07944	.07147	.06357	.05573	.04794
.13	.04022	.03256	.02495	.01741	.00992	.00248	*.99510	*.98777	*.98050	*.97328
.14	−1.96611	.95900	.95193	.94491	.93794	.93102	.92415	.91732	.91054	.90381
.15	.89712	.89048	.88387	.87732	.87080	.86433	.85790	.85151	.84516	.83885
.16	.83258	.82635	.82016	.81401	.80789	.80181	.79577	.78976	.78379	.77786
.17	.77196	.76609	.76026	.75446	.74870	.74297	.73727	.73161	.72597	.72037
.18	.71480	.70926	.70375	.69827	.69282	.68740	.68201	.67665	.67131	.66601
.19	.66073	.65548	.65026	.64507	.63990	.63476	.62964	.62455	.61949	.61445
0.20	−1.60944	.60445	.59949	.59455	.58964	.58475	.57988	.57504	.57022	.56542
.21	.56065	.55590	.55117	.54646	.54178	.53712	.53248	.52786	.52326	.51868
.22	.51413	.50959	.50508	.50058	.49611	.49165	.48722	.48281	.47841	.47403
.23	.46968	.46534	.46102	.45672	.45243	.44817	.44392	.43970	.43548	.43129
.24	.42712	.42296	.41882	.41469	.41059	.40650	.40242	.39837	.39433	.39030
.25	.38629	.38230	.37833	.37437	.37042	.36649	.36258	.35868	.35480	.35093
.26	.34707	.34323	.33941	.33560	.33181	.32803	.32426	.32051	.31677	.31304
.27	.30933	.30564	.30195	.29828	.29463	.29098	.28735	.28374	.28013	.27654
.28	.27297	.26940	.26585	.26231	.25878	.25527	.25176	.24827	.24479	.24133
.29	.23787	.23443	.23100	.22758	.22418	.22078	.21740	.21402	.21066	.20731
0.30	−1.20397	.20065	.19733	.19402	.19073	.18744	.18417	.18091	.17766	.17441
.31	.17118	.16796	.16475	.16155	.15836	.15518	.15201	.14885	.14570	.14256
.32	.13943	.13631	.13320	.13010	.12701	.12393	.12086	.11780	.11474	.11170
.33	.10866	.10564	.10262	.09961	.09661	.09362	.09064	.08767	.08471	.08176
.34	.07881	.07587	.07294	.07002	.06711	.06421	.06132	.05843	.05555	.05268
.35	−1.04982	.04697	.04412	.04129	.03846	.03564	.03282	.03002	.02722	.02443
.36	.02165	.01888	.01611	.01335	.01060	.00786	.00512	.00239	*.99967	*.99696
.37	−0.99425	.99155	.98886	.98618	.98350	.98083	.97817	.97551	.97286	.97022
.38	.96758	.96496	.96233	.95972	.95711	.95451	.95192	.94933	.94675	.94418
.39	.94161	.93905	.93649	.93395	.93140	.92887	.92634	.92382	.92130	.91879
0.40	−0.91629	.91379	.91130	.90882	.90634	.90387	.90140	.89894	.89649	.89404
.41	.89160	.88916	.88673	.88431	.88189	.87948	.87707	.87467	.87227	.86988
.42	.86750	.86512	.86275	.86038	.85802	.85567	.85332	.85097	.84863	.84630
.43	.84397	.84165	.83933	.83702	.83471	.83241	.83011	.82782	.82554	.82326
.44	.82098	.81871	.81645	.81419	.81193	.80968	.80744	.80520	.80296	.80073
.45	.79851	.79629	.79407	.79186	.78966	.78746	.78526	.78307	.78089	.77871
.46	.77653	.77436	.77219	.77003	.76787	.76572	.76357	.76143	.75929	.75715
.47	.75502	.75290	.75078	.74866	.74655	.74444	.74234	.74024	.73814	.73605
.48	.73397	.73189	.72981	.72774	.72567	.72361	.72155	.71949	.71744	.71539
.49	.71335	.71131	.70928	.70725	.70522	.70320	.70118	.69917	.69716	.69515

‡ Note that the whole number values are given above the decimal values for the first line. In the second and following lines they are given at the left. All decimal values are negative on this page.

NATURAL OR NAPERIAN LOGARITHMS (Continued)

0.500–0.999

N	0	1	2	3	4	5	6	7	8	9
0.50	−0.69315	.69115	.68916	.68717	.68518	.68320	.68122	.67924	.67727	.67531
.51	.67334	.67139	.66943	.66748	.66553	.66359	.66165	.65971	.65778	.65585
.52	.65393	.65201	.65009	.64817	.64626	.64436	.64245	.64055	.63866	.63677
.53	.63488	.63299	.63111	.62923	.62736	.62549	.62362	.62176	.61990	.61804
.54	.61619	.61434	.61249	.61065	.60881	.60697	.60514	.60331	.60148	.59966
.55	.59784	.59602	.59421	.59240	.59059	.58879	.58699	.58519	.58340	.58161
.56	.57982	.57803	.57625	.57448	.57270	.57093	.56916	.56740	.56563	.56387
.57	.56212	.56037	.55862	.55687	.55513	.55339	.55165	.54991	.54818	.54645
.58	.54473	.54300	.54128	.53957	.53785	.53614	.53444	.53273	.53103	.52933
.59	.52763	.52594	.52425	.52256	.52088	.51919	.51751	.51584	.51416	.51249
0.60	−0.51083	.50916	.50750	.50584	.50418	.50253	.50088	.49923	.49758	.49594
.61	.49430	.49266	.49102	.48939	.48776	.48613	.48451	.48289	.48127	.47965
.62	.47804	.47642	.47482	.47321	.47160	.47000	.46840	.46681	.46522	.46362
.63	.46204	.46045	.45887	.45728	.45571	.45413	.45256	.45099	.44942	.44785
.64	.44629	.44473	.44317	.44161	.44006	.43850	.43696	.43541	.43386	.43232
.65	.43078	.42925	.42771	.42618	.42465	.42312	.42159	.42007	.41855	.41703
.66	.41552	.41400	.41249	.41098	.40947	.40797	.40647	.40497	.40347	.40197
.67	.40048	.39899	.39750	.39601	.39453	.39304	.39156	.39008	.38861	.38713
.68	.38566	.38419	.38273	.38126	.37980	.37834	.37688	.37542	.37397	.37251
.69	.37106	.36962	.36817	.36673	.36528	.36384	.36241	.36097	.35954	.35810
0.70	−0.35667	.35525	.35382	.35240	.35098	.34956	.34814	.34672	.34531	.34390
.71	.34249	.34108	.33968	.33827	.33687	.33547	.33408	.33268	.33129	.32989
.72	.32850	.32712	.32573	.32435	.32296	.32158	.32021	.31883	.31745	.31608
.73	.31471	.31334	.31197	.31061	.30925	.30788	.30653	.30517	.30381	.30246
.74	.30111	.29975	.29841	.29706	.29571	.29437	.29303	.29169	.29035	.28902
.75	.28768	.28635	.28502	.28369	.28236	.28104	.27971	.27839	.27707	.27575
.76	.27444	.27312	.27181	.27050	.26919	.26788	.26657	.26527	.26397	.26266
.77	.26136	.26007	.25877	.25748	.25618	.25489	.25360	.25231	.25103	.24974
.78	.24846	.24718	.24590	.24462	.24335	.24207	.24080	.23953	.23826	.23699
.79	.23572	.23446	.23319	.23193	.23067	.22941	.22816	.22690	.22565	.22439
0.80	−0.22314	.22189	.22065	.21940	.21816	.21691	.21567	.21433	.21319	.21196
.81	.21072	.20949	.20825	.20702	.20579	.20457	.20334	.20212	.20089	.19967
.82	.19845	.19723	.19601	.19480	.19358	.19237	.19116	.18995	.18874	.18754
.83	.18633	.18513	.18392	.18272	.18152	.18032	.17913	.17793	.17674	.17554
.84	.17435	.17316	.17198	.17079	.16960	.16842	.16724	.16605	.16487	.16370
.85	−0.16252	.16134	.16017	.15900	.15782	.15665	.15548	.15432	.15315	.15199
.86	.15082	.14966	.14850	.14734	.14618	.14503	.14387	.14272	.14156	.14041
.87	.13926	.13811	.13697	.13582	.13467	.13353	.13239	.13125	.13011	.12897
.88	.12783	.12670	.12556	.12443	.12330	.12217	.12104	.11991	.11878	.11766
.89	.11653	.11541	.11429	.11317	.11205	.11093	.10981	.10870	.10759	.10647
0.90	−0.10536	.10425	.10314	.10203	.10093	.09982	.09872	.09761	.09651	.09541
.91	.09431	.09321	.09212	.09102	.08992	.08883	.08774	.08665	.08556	.08447
.92	.08338	.08230	.08121	.08013	.07904	.07796	.07688	.07580	.07472	.07365
.93	.07257	.07150	.07042	.06935	.06828	.06721	.06614	.06507	.06401	.06294
.94	.06188	.06081	.05975	.05869	.05763	.05657	.05551	.05446	.05340	.05235
.95	.05129	.05024	.04919	.04814	.04709	.04604	.04500	.04395	.04291	.04186
.96	.04082	.03978	.03874	.03770	.03666	.03563	.03459	.03356	.03252	.03149
.97	.03046	.02943	.02840	.02737	.02634	.02532	.02429	.02327	.02225	.02122
.98	.02020	.01918	.01816	.01715	.01613	.01511	.01410	.01309	.01207	.01106
.99	.01005	.00904	.00803	.00702	.00602	.00501	.00401	.00300	.00200	.00100

NATURAL OR NAPERIAN LOGARITHMS (Continued)

To find the natural logarithm of a number which is $\frac{1}{10}$, $\frac{1}{100}$, $\frac{1}{1000}$, etc. of a number whose logarithm is given, subtract from the given logarithm $\log_e 10$, $2 \log_e 10$, $3 \log_e 10$, etc.

To find the natural logarithm of a number which is 10, 100, 1000, etc. times a number whose logarithm is given, add to the given logarithm $\log_e 10$, $2 \log_e 10$, $3 \log_e 10$, etc.

$\log_e 10 = 2.30258\ 50930$	$6 \log_e 10 = 13.81551\ 05580$
$2 \log_e 10 = 4.60517\ 01860$	$7 \log_e 10 = 16.11809\ 56510$
$3 \log_e 10 = 6.90775\ 52790$	$8 \log_e 10 = 18.42068\ 07440$
$4 \log_e 10 = 9.21034\ 03720$	$9 \log_e 10 = 20.72326\ 58369$
$5 \log_e 10 = 11.51292\ 54650$	$10 \log_e 10 = 23.02585\ 09299$

See preceding table for logarithms for numbers between 0.000 and 0.999.

1.00–4.99

N	0	1	2	3	4	5	6	7	8	9
1.0	0.00000	.00995	.01980	.02956	.03922	.04879	.05827	.06766	.07696	.08618
.1	.09531	.10436	.11333	.12222	.13103	.13976	.14842	.15700	.16551	.17395
.2	.18232	.19062	.19885	.20701	.21511	.22314	.23111	.23902	.24686	.25464
.3	.26236	.27003	.27763	.28518	.29267	.30010	.30748	.31481	.32208	.32930
.4	.33647	.34359	.35066	.35767	.36464	.37156	.37844	.38526	.39204	.39878
.5	.40547	.41211	.41871	.42527	.43178	.43825	.44469	.45108	.45742	.46373
.6	.47000	.47623	.48243	.48858	.49470	.50078	.50682	.51282	.51879	.52473
.7	.53063	.53649	.54232	.54812	.55389	.55962	.56531	.57098	.57661	.58222
.8	.58779	.59333	.59884	.60432	.60977	.61519	.62058	.62594	.63127	.63658
.9	.64185	.64710	.65233	.65752	.66269	.66783	.67294	.67803	.68310	.68813
2.0	0.69315	.69813	.70310	.70804	.71295	.71784	.72271	.72755	.73237	.73716
.1	.74194	.74669	.75142	.75612	.76081	.76547	.77011	.77473	.77932	.78390
.2	.78846	.79299	.79751	.80200	.80648	.81093	.81536	.81978	.82418	.82855
.3	.83291	.83725	.84157	.84587	.85015	.85442	.85866	.86289	.86710	.87129
.4	.87547	.87963	.88377	.88789	.89200	.89609	.90016	.90422	.90826	.91228
.5	.91629	.92028	.92426	.92822	.93216	.93609	.94001	.94391	.94779	.95166
.6	.95551	.95935	.96317	.96698	.97078	.97456	.97833	.98208	.98582	.98954
.7	.99325	.99695	*.00063	*.00430	*.00796	*.01160	*.01523	*.01885	*.02245	*.02604
.8	1.02962	.03318	.03674	.04028	.04380	.04732	.05082	.05431	.05779	.06126
.9	.06471	.06815	.07158	.07500	.07841	.08181	.08519	.08856	.09192	.09527
3.0	1.09861	.10194	.10526	.10856	.11186	.11514	.11841	.12168	.12493	.12817
.1	.13140	.13462	.13783	.14103	.14422	.14740	.15057	.15373	.15688	.16002
.2	.16315	.16627	.16938	.17248	.17557	.17865	.18173	.18479	.18784	.19089
.3	.19392	.19695	.19996	.20297	.20597	.20896	.21194	.21491	.21788	.22083
.4	.22378	.22671	.22964	.23256	.23547	.23837	.24127	.24415	.24703	.24990
.5	.25276	.25562	.25846	.26130	.26413	.26695	.26976	.27257	.27536	.27815
.6	.28093	.28371	.28647	.28923	.29198	.29473	.29746	.30019	.30291	.30563
.7	.30833	.31103	.31372	.31641	.31909	.32176	.32442	.32708	.32972	.33237
.8	.33500	.33763	.34025	.34286	.34547	.34807	.35067	.35325	.35584	.35841
.9	.36098	.36354	.36609	.36864	.37118	.37372	.37624	.37877	.38128	.38379
4.0	1.38629	.38879	.39128	.39377	.39624	.39872	.40118	.40364	.40610	.40854
.1	.41099	.41342	.41585	.41828	.42070	.42311	.42552	.42792	.43031	.43270
.2	.43508	.43746	.43984	.44220	.44456	.44692	.44927	.45161	.45395	.45629
.3	.45862	.46094	.46326	.46557	.46787	.47018	.47247	.47476	.47705	.47933
.4	.48160	.48387	.48614	.48840	.49065	.49290	.49515	.49739	.49962	.50185
.5	.50408	.50630	.50851	.51072	.51293	.51513	.51732	.51951	.52170	.52388
.6	.52606	.52823	.53039	.53256	.53471	.53687	.53902	.54116	.54330	.54543
.7	.54756	.54969	.55181	.55393	.55604	.55814	.56025	.56235	.56444	.56653
.8	.56862	.57070	.57277	.57485	.57691	.57898	.58104	.58309	.58515	.58719
.9	.58924	.59127	.59331	.59534	.59737	.59939	.60141	.60342	.60543	.60744

NATURAL OR NAPERIAN LOGARITHMS (Continued)

5.00–9.99

N	0	1	2	3	4	5	6	7	8	9
5.0	1.60944	.61144	.61343	.61542	.61741	.61939	.62137	.62334	.62531	.62728
.1	.62924	.63120	.63315	.63511	.63705	.63900	.64094	.64287	.64481	.64673
.2	.64866	.65058	.65250	.65441	.65632	.65823	.66013	.66203	.66393	.66582
.3	.66771	.66959	.67147	.67335	.67523	.67710	.67896	.68083	.68269	.68455
.4	.68640	.68825	.69010	.69194	.69378	.69562	.69745	.69928	.70111	.70293
.5	.70475	.70656	.70838	.71019	.71199	.71380	.71560	.71740	.71919	.72098
.6	.72277	.72455	.72633	.72811	.72988	.73166	.73342	.73519	.73695	.73871
.7	.74047	.74222	.74397	.74572	.74746	.74920	.75094	.75267	.75440	.75613
.8	.75786	.75958	.76130	.76302	.76473	.76644	.76815	.76985	.77156	.77326
.9	.77495	.77665	.77834	.78002	.78171	.78339	.78507	.78675	.78842	.79009
6.0	1.79176	.79342	.79509	.79675	.79840	.80006	.80171	.80336	.80500	.80665
.1	.80829	.80993	.81156	.81319	.81482	.81645	.81808	.81970	.82132	.82294
.2	.82455	.82616	.82777	.82938	.83098	.83258	.83418	.83578	.83737	.83896
.3	.84055	.84214	.84372	.84530	.84688	.84845	.85003	.85160	.85317	.85473
.4	.85630	.85786	.85942	.86097	.86253	.86408	.86563	.86718	.86872	.87026
.5	.87180	.87334	.87487	.87641	.87794	.87947	.88099	.88251	.88403	.88555
.6	.88707	.88858	.89010	.89160	.89311	.89462	.89612	.89762	.89912	.90061
.7	.90211	.90360	.90509	.90658	.90806	.90954	.91102	.91250	.91398	.91545
.8	.91692	.91839	.91986	.92132	.92279	.92425	.92571	.92716	.92862	.93007
.9	.93152	.93297	.93442	.93586	.93730	.93874	.94018	.94162	.94305	.94448
7.0	1.94591	.94734	.94876	.95019	.95161	.95303	.95445	.95586	.95727	.95869
.1	.96009	.96150	.96291	.96431	.96571	.96711	.96851	.96991	.97130	.97269
.2	.97408	.97547	.97685	.97824	.97962	.98100	.98238	.98376	.98513	.98650
.3	.98787	.98924	.99061	.99198	.99334	.99470	.99606	.99742	.99877	*.00013
.4	2.00148	.00283	.00418	.00553	.00687	.00821	.00956	.01089	.01223	.01357
.5	.01490	.01624	.01757	.01890	.02022	.02155	.02287	.02419	.02551	.02683
.6	.02815	.02946	.03078	.03209	.03340	.03471	.03601	.03732	.03862	.03992
.7	.04122	.04252	.04381	.04511	.04640	.04769	.04898	.05027	.05156	.05284
.8	.05412	.05540	.05668	.05796	.05924	.06051	.06179	.06306	.06433	.06560
.9	.06686	.06813	.06939	.07065	.07191	.07317	.07443	.07568	.07694	.07819
8.0	2.07944	.08069	.08194	.08318	.08443	.08567	.08691	.08815	.08939	.09063
.1	.09186	.09310	.09433	.09556	.09679	.09802	.09924	.10047	.10169	.10291
.2	.10413	.10535	.10657	.10779	.10900	.11021	.11142	.11263	.11384	.11505
.3	.11626	.11746	.11866	.11986	.12106	.12226	.12346	.12465	.12585	.12704
.4	.12823	.12942	.13061	.13180	.13298	.13417	.13535	.13653	.13771	.13889
.5	.14007	.14124	.14242	.14359	.14476	.14593	.14710	.14827	.14943	.15060
.6	.15176	.15292	.15409	.15524	.15640	.15756	.15871	.15987	.16102	.16217
.7	.16332	.16447	.16562	.16677	.16791	.16905	.17020	.17134	.17248	.17361
.8	.17475	.17589	.17702	.17816	.17929	.18042	.18155	.18267	.18380	.18493
.9	.18605	.18717	.18830	.18942	.19054	.19165	.19277	.19389	.19500	.19611
9.0	2.19722	.19834	.19944	.20055	.20166	.20276	.20387	.20497	.20607	.20717
.1	.20827	.20937	.21047	.21157	.21266	.21375	.21485	.21594	.21703	.21812
.2	.21920	.22029	.22138	.22246	.22354	.22462	.22570	.22678	.22786	.22894
.3	.23001	.23109	.23216	.23324	.23431	.23538	.23645	.23751	.23858	.23965
.4	.24071	.24177	.24284	.24390	.24496	.24601	.24707	.24813	.24918	.25024
.5	.25129	.25234	.25339	.25444	.25549	.25654	.25759	.25863	.25968	.26072
.6	.26176	.26280	.26384	.26488	.26592	.26696	.26799	.26903	.27006	.27109
.7	.27213	.27316	.27419	.27521	.27624	.27727	.27829	.27932	.28034	.28136
.8	.28238	.28340	.28442	.28544	.28646	.28747	.28849	.28950	.29051	.29152
.9	.29253	.29354	.29455	.29556	.29657	.29757	.29858	.29958	.30058	.30158

NATURAL OR NAPERIAN LOGARITHMS (Continued)

Constants

$\log_e 10 = 2.30258\ 50930$	$6 \log_e 10 = 13.81551\ 05580$
$2 \log_e 10 = 4.60517\ 01860$	$7 \log_e 10 = 16.11809\ 56510$
$3 \log_e 10 = 6.90775\ 52790$	$8 \log_e 10 = 18.42068\ 07440$
$4 \log_e 10 = 9.21034\ 03720$	$9 \log_e 10 = 20.72326\ 58369$
$5 \log_e 10 = 11.51292\ 54650$	$10 \log_e 10 = 23.02585\ 09299$

10.0–49.9

N	0	1	2	3	4	5	6	7	8	9
10.	2.30259	.31254	.32239	.33214	.34181	.35138	.36085	.37024	.37955	.38876
11.	.39790	.40695	.41591	.42480	.43361	.44235	.45101	.45959	.46810	.47654
12.	.48491	.49321	.50144	.50960	.51770	.52573	.53370	.54160	.54945	.55723
13.	.56495	.57261	.58022	.58776	.59525	.60269	.61007	.61740	.62467	.63189
14.	.63906	.64617	.65324	.66026	.66723	.67415	.68102	.68785	.69463	.70136
15.	.70805	.71469	.72130	.72785	.73437	.74084	.74727	.75366	.76001	.76632
16.	.77259	.77882	.78501	.79117	.79728	.80336	.80940	.81541	.82138	.82731
17.	.83321	.83908	.84491	.85071	.85647	.86220	.86790	.87356	.87920	.88480
18.	.89037	.89591	.90142	.90690	.91235	.91777	.92316	.92852	.93386	.93916
19.	.94444	.94969	.95491	.96011	.96527	.97041	.97553	.98062	.98568	.99072
20.	2.99573	*.00072	*.00568	*.01062	*.01553	*.02042	*.02529	*.03013	*.03495	*.03975
21.	3.04452	.04927	.05400	.05871	.06339	.06805	.07269	.07731	.08191	.08649
22.	.09104	.09558	.10009	.10459	.10906	.11352	.11795	.12236	.12676	.13114
23.	.13549	.13983	.14415	.14845	.15274	.15700	.16125	.16548	.16969	.17388
24.	.17805	.18221	.18635	.19048	.19458	.19867	.20275	.20680	.21084	.21487
25.	.21888	.22287	.22684	.23080	.23475	.23868	.24259	.24649	.25037	.25424
26.	.25810	.26194	.26576	.26957	.27336	.27714	.28091	.28466	.28840	.29213
27.	.29584	.29953	.30322	.30689	.31054	.31419	.31782	.32143	.32504	.32863
28.	.33220	.33577	.33932	.34286	.34639	.34990	.35341	.35690	.36038	.36384
29.	.36730	.37074	.37417	.37759	.38099	.38439	.38777	.39115	.39451	.39786
30.	3.40120	.40453	.40784	.41115	.41444	.41773	.42100	.42426	.42751	.43076
31.	.43399	.43721	.44042	.44362	.44681	.44999	.45316	.45632	.45947	.46261
32.	.46574	.46886	.47197	.47507	.47816	.48124	.48431	.48738	.49043	.49347
33.	.49651	.49953	.50255	.50556	.50856	.51155	.51453	.51750	.52046	.52342
34.	.52636	.52930	.53223	.53515	.53806	.54096	.54385	.54674	.54962	.55249
35.	.55535	.55820	.56105	.56388	.56671	.56953	.57235	.57515	.57795	.58074
36.	.58352	.58629	.58906	.59182	.59457	.59731	.60005	.60278	.60550	.60821
37.	.61092	.61362	.61631	.61899	.62167	.62434	.62700	.62966	.63231	.63495
38.	.63759	.64021	.64284	.64545	.64806	.65066	.65325	.65584	.65842	.66099
39.	.66356	.66612	.66868	.67122	.67377	.67630	.67883	.68135	.68387	.68638
40.	3.68888	.69138	.69387	.69635	.69883	.70130	.70377	.70623	.70868	.71113
41.	.71357	.71601	.71844	.72086	.72328	.72569	.72810	.73050	.73290	.73529
42.	.73767	.74005	.74242	.74479	.74715	.74950	.75185	.75420	.75654	.75887
43.	.76120	.76352	.76584	.76815	.77046	.77276	.77506	.77735	.77963	.78191
44.	.78419	.78646	.78872	.79098	.79324	.79549	.79773	.79997	.80221	.80444
45.	.80666	.80888	.81110	.81331	.81551	.81771	.81991	.82210	.82428	.82647
46.	.82864	.83081	.83298	.83514	.83730	.83945	.84160	.84374	.84588	.84802
47.	.85015	.85227	.85439	.85651	.85862	.86073	.86283	.86493	.86703	.86912
48.	.87120	.87328	.87536	.87743	.87950	.88156	.88362	.88568	.88773	.88978
49.	.89182	.89386	.89589	.89792	.89995	.90197	.90399	.90600	.90801	.91002

NATURAL OR NAPERIAN LOGARITHMS (Continued)

50.0–99.9

N	0	1	2	3	4	5	6	7	8	9
50.	3.91202	.91402	.91602	.91801	.91999	.92197	.92395	.92593	.92790	.92986
51.	.93183	.93378	.93574	.93769	.93964	.94158	.94352	.94546	.94739	.94932
52.	.95124	.95316	.95508	.95700	.95891	.96081	.96272	.96462	.96651	.96840
53.	.97029	.97218	.97406	.97594	.97781	.97968	.98155	.98341	.98527	.98713
54.	.98898	.99083	.99268	.99452	.99636	.99820	*.00003	*.00186	*.00369	*.00551
55.	4.00733	.00915	.01096	.01277	.01458	.01638	.01818	.01998	.02177	.02356
56.	.02535	.02714	.02892	.03069	.03247	.03424	.03601	.03777	.03954	.04130
57.	.04305	.04480	.04655	.04830	.05004	.05178	.05352	.05526	.05699	.05872
58.	.06044	.06217	.06389	.06560	.06732	.06903	.07073	.07244	.07414	.07584
59.	.07754	.07923	.08092	.08261	.08429	.08598	.08766	.08933	.09101	.09268
60.	4.09434	.09601	.09767	.09933	.10099	.10264	.10429	.10594	.10759	.10923
61.	.11087	.11251	.11415	.11578	.11741	.11904	.12066	.12228	.12390	.12552
62.	.12713	.12875	.13036	.13196	.13357	.13517	.13677	.13836	.13996	.14155
63.	.14313	.14472	.14630	.14789	.14946	.15104	.15261	.15418	.15575	.15732
64.	.15888	.16044	.16200	.16356	.16511	.16667	.16821	.16976	.17131	.17285
65.	.17439	.17592	.17746	.17899	.18052	.18205	.18358	.18510	.18662	.18814
66.	.18965	.19117	.19268	.19419	.19570	.19720	.19870	.20020	.20170	.20320
67.	.20469	.20618	.20767	.20916	.21065	.21213	.21361	.21509	.21656	.21804
68.	.21951	.22098	.22244	.22391	.22537	.22683	.22829	.22975	.23120	.23266
69.	.23411	.23555	.23700	.23844	.23989	.24133	.24276	.24420	.24563	.24707
70.	4.24850	.24992	.25135	.25277	.25419	.25561	.25703	.25845	.25986	.26127
71.	.26268	.26409	.26549	.26690	.26830	.26970	.27110	.27249	.27388	.27528
72.	.27667	.27805	.27944	.28082	.28221	.28359	.28496	.28634	.28772	.28909
73.	.29046	.29183	.29320	.29456	.29592	.29729	.29865	.30000	.30136	.30271
74.	.30407	.30542	.30676	.30811	.30946	.31080	.31214	.31348	.31482	.31615
75.	.31749	.31882	.32015	.32149	.32281	.32413	.32546	.32678	.32810	.32942
76.	.33073	.33205	.33336	.33467	.33598	.33729	.33860	.33990	.34120	.34251
77.	.34381	.34510	.34640	.34769	.34899	.35028	.35157	.35286	.35414	.35543
78.	.35671	.35800	.35927	.36055	.36182	.36310	.36437	.36564	.36691	.36818
79.	.36945	.37071	.37198	.37324	.37450	.37576	.37701	.37827	.37952	.38078
80.	4.38203	.38328	.38452	.38577	.38701	.38826	.38950	.39074	.39198	.39321
81.	.39445	.39568	.39692	.39815	.39938	.40060	.40183	.40305	.40428	.40550
82.	.40672	.40794	.40916	.41037	.41159	.41280	.41401	.41522	.41643	.41764
83.	.41884	.42004	.42125	.42245	.42365	.42485	.42604	.42724	.42843	.42963
84.	.43082	.43201	.43319	.43438	.43557	.43675	.43793	.43912	.44030	.44147
85.	.44265	.44383	.44500	.44617	.44735	.44852	.44969	.45085	.45202	.45318
86.	.45435	.45551	.45667	.45783	.45899	.46014	.46130	.46245	.46361	.46476
87.	.46591	.46706	.46820	.46935	.47050	.47164	.47278	.47392	.47506	.47620
88.	.47734	.47847	.47961	.48074	.48187	.48300	.48413	.48526	.48639	.48751
89.	.48864	.48976	.49088	.49200	.49312	.49424	.49536	.49647	.49758	.49870
90.	4.49981	.50092	.50203	.50314	.50424	.50535	.50645	.50756	.50866	.50976
91.	.51086	.51196	.51305	.51415	.51525	.51634	.51743	.51852	.51961	.52070
92.	.52179	.52287	.52396	.52504	.52613	.52721	.52829	.52937	.53045	.53152
93.	.53260	.53367	.53475	.53582	.53689	.53796	.53903	.54010	.54116	.54223
94.	.54329	.54436	.54542	.54648	.54754	.54860	.54966	.55071	.55177	.55282
95.	.55388	.55493	.55598	.55703	.55808	.55913	.56017	.56122	.56226	.56331
96.	.56435	.56539	.56643	.56747	.56851	.56954	.57058	.57161	.57265	.57368
97.	.57471	.57574	.57677	.57780	.57883	.57985	.58088	.58190	.58292	.58395
98.	.58497	.58599	.58701	.58802	.58904	.59006	.59107	.59208	.59310	.59411
99.	.59512	.59613	.59714	.59815	.59915	.60016	.60116	.60217	.60317	.60417

NATURAL OR NAPERIAN LOGARITHMS (Continued)

0–499

N	0	1	2	3	4	5	6	7	8	9
0	− ∞	0.00000	0.69315	1.09861	.38629	.60944	.79176	.94591	*.07944	*.19722
1	2.30259	.39790	.48491	.56495	.63906	.70805	.77259	.83321	.89037	.94444
2	.99573	*.04452	*.09104	*.13549	*.17805	*.21888	*.25810	*.29584	*.33220	*.36730
3	3.40120	.43399	.46574	.49651	.52636	.55535	.58352	.61092	.63759	.66356
4	.68888	.71357	.73767	.76120	.78419	.80666	.82864	.85015	.87120	.89182
5	.91202	.93183	.95124	.97029	.98898	*.00733	*02535	*.04305	*.06044	*.07754
6	4.09434	.11087	.12713	.14313	.15888	.17439	.18965	.20469	.21951	.23411
7	.24850	.26268	.27667	.29046	.30407	.31749	.33073	.34381	.35671	.36945
8	.38203	.39445	.40672	.41884	.43082	.44265	.45435	.46591	.47734	.48864
9	.49981	.51086	.52179	.53260	.54329	.55388	.56435	.57471	.58497	.59512
10	4.60517	.61512	.62497	.63473	.64439	.65396	.66344	.67283	.68213	.69135
11	.70048	.70953	.71850	.72739	.73620	.74493	.75359	.76217	.77068	.77912
12	.78749	.79579	.80402	.81218	.82028	.82831	.83628	.84419	.85203	.85981
13	.86753	.87520	.88280	.89035	.89784	.90527	.91265	.91998	.92725	.93447
14	.94164	.94876	.95583	.96284	.96981	.97673	.98361	.99043	.99721	* 00395
15	5.01064	.01728	.02388	.03044	.03695	.04343	.04986	.05625	.06260	.06890
16	.07517	.08140	.08760	.09375	.09987	.10595	.11199	.11799	.12396	.12990
17	.13580	.14166	.14749	.15329	.15906	.16479	.17048	.17615	.18178	.18739
18	.19296	.19850	.20401	.20949	.21494	.22036	.22575	.23111	.23644	.24175
19	.24702	.25227	.25750	.26269	.26786	.27300	.27811	.28320	.28827	.29330
20	5.29832	.30330	.30827	.31321	.31812	.32301	.32788	.33272	.33754	.34233
21	.34711	.35186	.35659	.36129	.36598	.37064	.37528	.37990	.38450	.38907
22	.39363	.39816	.40268	.40717	.41165	.41610	.42053	.42495	.42935	.43372
23	.43808	.44242	.44674	.45104	.45532	.45959	.46383	.46806	.47227	.47646
24	.48064	.48480	.48894	.49306	.49717	.50126	.50533	.50939	.51343	.51745
25	.52146	.52545	.52943	.53339	.53733	.54126	.54518	.54908	.55296	.55683
26	.56068	.56452	.56834	.57215	.57595	.57973	.58350	.58725	.59099	.59471
27	.59842	.60212	.60580	.60947	.61313	.61677	.62040	.62402	.62762	.63121
28	.63479	.63835	.64191	.64545	.64897	.65249	.65599	.65948	.66296	.66643
29	.66988	.67332	.67675	.68017	.68358	.68698	.69036	.69373	.69709	.70044
30	5.70378	.70711	.71043	.71373	.71703	.72031	.72359	.72685	.73010	.73334
31	.73657	.73979	.74300	.74620	.74939	.75257	.75574	.75890	.76205	.76519
32	.76832	.77144	.77455	.77765	.78074	.78383	.78690	.78996	.79301	.79606
33	.79909	.80212	.80513	.80814	.81114	.81413	.81711	.82008	.82305	.82600
34	.82895	.83188	.83481	.83773	.84064	.84354	.84644	.84932	.85220	.85507
35	.85793	.86079	.86363	.86647	.86930	.87212	.87493	.87774	.88053	.88332
36	.88610	.88888	.89164	.89440	.89715	.89990	.90263	.90536	.90808	.91080
37	.91350	.91620	.91889	.92158	.92426	.92693	.92959	.93225	.93489	.93754
38	.94017	.94280	.94542	.94803	.95064	.95324	.95584	.95842	.96101	.96358
39	.96615	.96871	.97126	.97381	.97635	.97889	.98141	.98394	.98645	.98896
40	5.99146	.99396	.99645	.99894	*.00141	*.00389	*.00635	*.00881	*.01127	*.01372
41	6.01616	.01859	.02102	.02345	.02587	.02828	.03069	.03309	.03548	.03787
42	.04025	.04263	.04501	.04737	.04973	.05209	.05444	.05678	.05912	.06146
43	.06379	.06611	.06843	.07074	.07304	.07535	.07764	.07993	.08222	.08450
44	.08677	.08904	.09131	.09357	.09582	.09807	.10032	.10256	.10479	.10702
45	.10925	.11147	.11368	.11589	.11810	.12030	.12249	.12468	.12687	.12905
46	.13123	.13340	.13556	.13773	.13988	.14204	.14419	.14633	.14847	.15060
47	.15273	.15486	.15698	.15910	.16121	.16331	.16542	.16752	.16961	.17170
48	.17379	.17587	.17794	.18002	.18208	.18415	.18621	.18826	.19032	.19236
49	.19441	.19644	.19848	.20051	.20254	.20456	.20658	.20859	.21060	.21261

NATURAL OR NAPERIAN LOGARITHMS (Continued)

500–999

N	0	1	2	3	4	5	6	7	8	9
50	6.21461	.21661	.21860	.22059	.22258	.22456	.22654	.22851	.23048	.23245
51	.23441	.23637	.23832	.24028	.24222	.24417	.24611	.24804	.24998	.25190
52	.25383	.25575	.25767	.25958	.26149	.26340	.26530	.26720	.26910	.27099
53	.27288	.27476	.27664	.27852	.28040	.28227	.28413	.28600	.28786	.28972
54	.29157	.29342	.29527	.29711	.29895	.30079	.30262	.30445	.30628	.30810
55	.30992	.31173	.31355	.31536	.31716	.31897	.32077	.32257	.32436	.32615
56	.32794	.32972	.33150	.33328	.33505	.33683	.33859	.34036	.34212	.34388
57	.34564	.34739	.34914	.35089	.35263	.35437	.35611	.35784	.35957	.36130
58	.36303	.36475	.36647	.36819	.36990	.37161	.37332	.37502	.37673	.37843
59	.38012	.38182	.38351	.38519	.38688	.38856	.39024	.39192	.39359	.39526
60	6.39693	.39859	.40026	.40192	.40357	.40523	.40688	.40853	.41017	.41182
61	.41346	.41510	.41673	.41836	.41999	.42162	.42325	.42487	.42649	.42811
62	.42972	.43133	.43294	.43455	.43615	.43775	.43935	.44095	.44254	.44413
63	.44572	.44731	.44889	.45047	.45205	.45362	.45520	.45677	.45834	.45990
64	.46147	.46303	.46459	.46614	.46770	.46925	.47080	.47235	.47389	.47543
65	.47697	.47851	.48004	.48158	.48311	.48464	.48616	.48768	.48920	.49072
66	.49224	.49375	.49527	.49677	.49828	.49979	.50129	.50279	.50429	.50578
67	.50728	.50877	.51026	.51175	.51323	.51471	.51619	.51767	.51915	.52062
68	.52209	.52356	.52503	.52649	.52796	.52942	.53088	.53233	.53379	.53524
69	.53669	.53814	.53959	.54103	.54247	.54391	.54535	.54679	.54822	.54965
70	6.55108	.55251	.55393	.55536	.55678	.55820	.55962	.56103	.56244	.56386
71	.56526	.56667	.56808	.56948	.57088	.57228	.57368	.57508	.57647	.57786
72	.57925	.58064	.58203	.58341	.58479	.58617	.58755	.58893	.59030	.59167
73	.59304	.59441	.59578	.59715	.59851	.59987	.60123	.60259	.60394	.60530
74	.60665	.60800	.60935	.61070	.61204	.61338	.61473	.61607	.61740	.61874
75	.62007	.62141	.62274	.62407	.62539	.62672	.62804	.62936	.63068	.63200
76	.63332	.63463	.63595	.63726	.63857	.63988	.64118	.64249	.64379	.64509
77	.64639	.64769	.64898	.65028	.65157	.65286	.65415	.65544	.65673	.65801
78	.65929	.66058	.66185	.66313	.66441	.66568	.66696	.66823	.66950	.67077
79	.67203	.67330	.67456	.67582	.67708	.67834	.67960	.68085	.68211	.68336
80	6.68461	.68586	.68711	.68835	.68960	.69084	.69208	.69332	.69456	.69580
81	.69703	.69827	.69950	.70073	.70196	.70319	.70441	.70564	.70686	.70808
82	.70930	.71052	.71174	.71296	.71417	.71538	.71659	.71780	.71901	.72022
83	.72143	.72263	.72383	.72503	.72623	.72743	.72863	.72982	.73102	.73221
84	.73340	.73459	.73578	.73697	.73815	.73934	.74052	.74170	.74288	.74406
85	.74524	.74641	.74759	.74876	.74993	.75110	.75227	.75344	.75460	.75577
86	.75693	.75809	.75926	.76041	.76157	.76273	.76388	.76504	.76619	.76734
87	.76849	.76964	.77079	.77194	.77308	.77422	.77537	.77651	.77765	.77878
88	.77992	.78106	.78219	.78333	.78446	.78559	.78672	.78784	.78897	.79010
89	.79122	.79234	.79347	.79459	.79571	.79682	.79794	.79906	.80017	.80128
90	6.80239	.80351	.80461	.80572	.80683	.80793	.80904	.81014	.81124	.81235
91	.81344	.81454	.81564	.81674	.81783	.81892	.82002	.82111	.82220	.82329
92	.82437	.82546	.82655	.82763	.82871	.82979	.83087	.83195	.83303	.83411
93	.83518	.83626	.83733	.83841	.83948	.84055	.84162	.84268	.84375	.84482
94	.84588	.84694	.84801	.84907	.85013	.85118	.85224	.85330	.85435	.85541
95	.85646	.85751	.85857	.85961	.86066	.86171	.86276	.86380	.86485	.86589
96	.86693	.86797	.86901	.87005	.87109	.87213	.87316	.87420	.87523	.87626
97	.87730	.87833	.87936	.88038	.88141	.88244	.88346	.88449	.88551	.88653
98	.88755	.88857	.88959	.89061	.89163	.89264	.89366	.89467	.89568	.89669
99	.89770	.89871	.89972	.90073	.90174	.90274	.90375	.90475	.90575	.90675

TEN PLACE MANTISSAS FOR NATURAL LOGARITHMS

x	$\log_e x$	x	$\log_e x$	x	$\log_e x$	x	$\log_e x$
1.000	0.0000000 000	1.050	0.0487901 642	1.100	0.0953101 798	1.150	0.1397619 424
1.001	0.0009995 003	1.051	0.0497420 919	1.101	0.0962188 577	1.151	0.1406311 297
1.002	0.0019980 027	1.052	0.0506931 143	1.102	0.0971267 107	1.152	0.1414995 623
1.003	0.0029955 090	1.053	0.0516432 332	1.103	0.0980337 403	1.153	0.1423672 413
1.004	0.0039920 213	1.054	0.0525924 501	1.104	0.0989399 479	1.154	0.1432341 681
1.005	0.0049875 415	1.055	0.0535407 669	1.105	0.0998453 350	1.155	0.1441003 440
1.006	0.0059820 717	1.056	0.0544881 853	1.106	0.1007499 031	1.156	0.1449657 703
1.007	0.0069756 137	1.057	0.0554347 069	1.107	0.1016536 537	1.157	0.1458304 482
1.008	0.0079681 696	1.058	0.0563803 334	1.108	0.1025565 883	1.158	0.1466943 792
1.009	0.0089597 414	1.059	0.0573250 666	1.109	0.1034587 084	1.159	0.1475575 644
1.010	0.0099503 309	1.060	0.0582689 081	1.110	0.1043600 153	1.160	0.1484200 051
1.011	0.0109399 400	1.061	0.0592118 596	1.111	0.1052605 107	1.161	0.1492817 027
1.012	0.0119285 709	1.062	0.0601539 228	1.112	0.1061601 958	1.162	0.1501426 584
1.013	0.0129162 253	1.063	0.0610950 994	1.113	0.1070590 723	1.163	0.1510028 735
1.014	0.0139029 052	1.064	0.0620353 909	1.114	0.1079571 415	1.164	0.1518623 493
1.015	0.0148886 125	1.065	0.0629747 992	1.115	0.1088544 049	1.165	0.1527210 870
1.016	0.0158733 492	1.066	0.0639133 257	1.116	0.1097508 640	1.166	0.1535790 879
1.017	0.0168571 171	1.067	0.0648509 723	1.117	0.1106465 201	1.167	0.1544363 533
1.018	0.0178399 181	1.068	0.0657877 405	1.118	0.1115413 747	1.168	0.1552928 844
1.019	0.0188217 542	1.069	0.0667236 320	1.119	0.1124354 293	1.169	0.1561486 825
1.020	0.0198026 273	1.070	0.0676586 485	1.120	0.1133286 853	1.170	0.1570037 488
1.021	0.0207825 392	1.071	0.0685927 915	1.121	0.1142211 441	1.171	0.1578580 846
1.022	0.0217614 918	1.072	0.0695260 626	1.122	0.1151128 071	1.172	0.1587116 912
1.023	0.0227394 870	1.073	0.0704584 636	1.123	0.1160036 758	1.173	0.1595645 697
1.024	0.0237165 266	1.074	0.0713899 961	1.124	0.1168937 515	1.174	0.1604167 214
1.025	0.0246926 126	1.075	0.0723206 616	1.125	0.1177830 357	1.175	0.1612681 476
1.026	0.0256677 467	1.076	0.0732504 617	1.126	0.1186715 297	1.176	0.1621188 495
1.027	0.0266419 309	1.077	0.0741793 982	1.127	0.1195592 351	1.177	0.1629688 283
1.028	0.0276151 670	1.078	0.0751074 725	1.128	0.1204461 531	1.178	0.1638180 852
1.029	0.0285874 569	1.079	0.0760346 863	1.129	0.1213322 852	1.179	0.1646666 216
1.030	0.0295588 022	1.080	0.0769610 411	1.130	0.1222176 327	1.180	0.1655144 385
1.031	0.0305292 050	1.081	0.0778865 387	1.131	0.1231021 971	1.181	0.1663615 372
1.032	0.0314986 671	1.082	0.0788111 804	1.132	0.1239859 798	1.182	0.1672079 190
1.033	0.0324671 901	1.083	0.0797349 680	1.133	0.1248689 820	1.183	0.1680535 850
1.034	0.0334347 761	1.084	0.0806579 030	1.134	0.1257512 053	1.184	0.1688985 365
1.035	0.0344014 267	1.085	0.0815799 870	1.135	0.1266326 509	1.185	0.1697427 746
1.036	0.0353671 438	1.086	0.0825012 215	1.136	0.1275133 203	1.186	0.1705863 006
1.037	0.0363319 292	1.087	0.0834216 081	1.137	0.1283932 148	1.187	0.1714291 156
1.038	0.0372957 847	1.088	0.0843411 484	1.138	0.1292723 357	1.188	0.1722712 209
1.039	0.0382587 121	1.089	0.0852598 440	1.139	0.1301506 845	1.189	0.1731126 177
1.040	0.0392207 132	1.090	0.0861776 962	1.140	0.1310282 624	1.190	0.1739533 071
1.041	0.0401817 896	1.091	0.0870947 069	1.141	0.1319050 709	1.191	0.1747932 904
1.042	0.0411419 433	1.092	0.0880108 773	1.142	0.1327811 112	1.192	0.1756325 686
1.043	0.0421011 760	1.093	0.0889262 092	1.143	0.1336563 848	1.193	0.1764711 431
1.044	0.0430594 895	1.094	0.0898407 040	1.144	0.1345308 930	1.194	0.1773090 150
1.045	0.0440168 854	1.095	0.0907543 633	1.145	0.1354046 370	1.195	0.1781461 854
1.046	0.0449733 656	1.096	0.0916671 885	1.146	0.1362776 183	1.196	0.1789826 555
1.047	0.0459289 319	1.097	0.0925791 813	1.147	0.1371498 381	1.197	0.1798184 266
1.048	0.0468835 859	1.098	0.0934903 431	1.148	0.1380212 979	1.198	0.1806534 997
1.049	0.0478373 294	1.099	0.0944006 754	1.149	0.1388919 989	1.199	0.1814878 760

TEN PLACE MANTISSAS FOR NATURAL LOGARITHMS (Continued)

x	$\log_e x$	x	$\log_e x$	x	$\log_e x$	x	$\log_e x$
1.200	0.1823215 568	1.250	0.2231435 513	1.300	0.2623642 645	1.350	0.3001045 925
1.201	0.1831545 431	1.251	0.2239432 315	1.301	0.2631331 995	1.351	0.3008450 590
1.202	0.1839868 361	1.252	0.2247422 727	1.302	0.2639015 438	1.352	0.3015849 776
1.203	0.1848184 370	1.253	0.2255406 759	1.303	0.2646692 981	1.353	0.3023243 492
1.204	0.1856493 469	1.254	0.2263384 422	1.304	0.2654364 635	1.354	0.3030631 745
1.205	0.1864795 669	1.255	0.2271355 726	1.305	0.2662030 408	1.355	0.3038014 543
1.206	0.1873090 983	1.256	0.2279320 680	1.306	0.2669690 309	1.356	0.3045391 895
1.207	0.1881379 421	1.257	0.2287279 296	1.307	0.2677344 346	1.357	0.3052763 809
1.208	0.1889660 995	1.258	0.2295231 583	1.308	0.2684992 530	1.358	0.3060130 291
1.209	0.1897935 716	1.259	0.2303177 551	1.309	0.2692634 869	1.359	0.3067491 352
1.210	0.1906203 596	1.260	0.2311117 210	1.310	0.2700271 372	1.360	0.3074846 997
1.211	0.1914464 646	1.261	0.2319050 570	1.311	0.2707902 048	1.361	0.3082197 237
1.212	0.1922718 876	1.262	0.2326977 641	1.312	0.2715526 905	1.362	0.3089542 077
1.213	0.1930966 300	1.263	0.2334898 434	1.313	0.2723145 953	1.363	0.3096881 527
1.214	0.1939206 926	1.264	0.2342812 957	1.314	0.2730759 201	1.364	0.3104215 594
1.215	0.1947440 768	1.265	0.2350721 222	1.315	0.2738366 656	1.365	0.3111544 286
1.216	0.1955667 835	1.266	0.2358623 237	1.316	0.2745968 329	1.366	0.3118867 611
1.217	0.1963888 140	1.267	0.2366519 013	1.317	0.2753564 228	1.367	0.3126185 577
1.218	0.1972101 693	1.268	0.2374408 560	1.318	0.2761154 361	1.368	0.3133498 192
1.219	0.1980308 505	1.269	0.2382291 887	1.319	0.2768738 737	1.369	0.3140805 463
1.220	0.1988508 587	1.270	0.2390169 005	1.320	0.2776317 366	1.370	0.3148107 398
1.221	0.1996701 951	1.271	0.2398039 922	1.321	0.2783890 255	1.371	0.3155404 006
1.222	0.2004888 607	1.272	0.2405904 649	1.322	0.2791457 414	1.372	0.3162695 293
1.223	0.2013068 567	1.273	0.2413763 196	1.323	0.2799018 851	1.373	0.3169981 268
1.224	0.2021241 841	1.274	0.2421615 571	1.324	0.2806574 575	1.374	0.3177261 938
1.225	0.2029408 440	1.275	0.2429461 786	1.325	0.2814124 594	1.375	0.3184537 311
1.226	0.2037568 375	1.276	0.2437301 849	1.326	0.2821668 918	1.376	0.3191807 395
1.227	0.2045721 657	1.277	0.2445135 771	1.327	0.2829207 554	1.377	0.3199072 197
1.228	0.2053868 297	1.278	0.2452963 560	1.328	0.2836740 511	1.378	0.3206331 726
1.229	0.2062008 306	1.279	0.2460785 226	1.329	0.2844267 797	1.379	0.3213585 988
1.230	0.2070141 694	1.280	0.2468600 779	1.330	0.2851789 422	1.380	0.3220834 992
1.231	0.2078268 472	1.281	0.2476410 229	1.331	0.2859305 394	1.381	0.3228078 744
1.232	0.2086388 651	1.282	0.2484213 585	1.332	0.2866815 721	1.382	0.3235317 253
1.233	0.2094502 242	1.283	0.2492010 856	1.333	0.2874320 412	1.383	0.3242550 527
1.234	0.2102609 255	1.284	0.2499802 053	1.334	0.2881819 475	1.384	0.3249778 572
1.235	0.2110709 701	1.285	0.2507587 183	1.335	0.2889312 919	1.385	0.3257001 396
1.236	0.2118803 590	1.286	0.2515366 258	1.336	0.2896800 751	1.386	0.3264219 008
1.237	0.2126890 934	1.287	0.2523139 286	1.337	0.2904282 981	1.387	0.3271431 413
1.238	0.2134971 743	1.288	0.2530906 277	1.338	0.2911759 617	1.388	0.3278638 621
1.239	0.2143046 026	1.289	0.2538667 240	1.339	0.2919230 667	1.389	0.3285840 638
1.240	0.2151113 796	1.290	0.2546422 184	1.340	0.2926696 140	1.390	0.3293037 471
1.241	0.2159175 062	1.291	0.2554171 119	1.341	0.2934156 043	1.391	0.3300229 129
1.242	0.2167229 835	1.292	0.2561914 054	1.342	0.2941610 385	1.392	0.3307415 619
1.243	0.2175278 125	1.293	0.2569650 998	1.343	0.2949059 175	1.393	0.3314596 948
1.244	0.2183319 943	1.294	0.2577381 961	1.344	0.2956502 421	1.394	0.3321773 123
1.245	0.2191355 299	1.295	0.2585106 952	1.345	0.2963940 131	1.395	0.3328944 153
1.246	0.2199384 204	1.296	0.2592825 979	1.346	0.2971372 312	1.396	0.3336110 043
1.247	0.2207406 667	1.297	0.2600539 053	1.347	0.2978798 974	1.397	0.3343270 803
1.248	0.2215422 699	1.298	0.2608246 183	1.348	0.2986220 125	1.398	0.3350426 438
1.249	0.2223432 311	1.299	0.2615947 377	1.349	0.2993635 772	1.399	0.3357576 957

TEN PLACE MANTISSAS FOR NATURAL LOGARITHMS (Continued)

x	$\log_e x$	x	$\log_e x$	x	$\log_e x$	x	$\log_e x$
1.400	0.3364722 366	1.450	0.3715635 564	1.500	0.4054651 081	1.550	0.4382549 309
1.401	0.3371862 674	1.451	0.3722529 739	1.501	0.4061315 527	1.551	0.4388998 842
1.402	0.3378997 886	1.452	0.3729419 164	1.502	0.4067975 533	1.551	0.4395444 218
1.403	0.3386128 011	1.453	0.3736303 846	1.503	0.4074631 108	1.553	0.4401885 442
1.404	0.3393253 056	1.454	0.3743183 791	1.504	0.4081282 255	1.554	0.4408322 519
1.405	0.3400373 028	1.455	0.3750059 006	1.505	0.4087928 982	1.555	0.4414755 456
1.406	0.3407487 934	1.456	0.3756929 498	1.506	0.4094571 294	1.556	0.4421184 258
1.407	0.3414597 781	1.457	0.3763795 272	1.507	0.4101209 196	1.557	0.4427608 929
1.408	0.3421702 577	1.458	0.3770656 336	1.508	0.4107842 696	1.558	0.4434029 474
1.409	0.3428802 329	1.459	0.3777512 695	1.509	0.4114471 798	1.559	0.4440445 901
1.410	0.3435897 044	1.460	0.3784364 357	1.510	0.4121096 508	1.560	0.4446858 213
1.411	0.3442986 729	1.461	0.3791211 328	1.511	0.4127716 833	1.561	0.4453266 415
1.412	0.3450071 391	1.462	0.3798053 613	1.512	0.4134332 778	1.562	0.4459670 514
1.413	0.3457151 037	1.463	0.3804891 220	1.513	0.4140944 348	1.563	0.4466070 514
1.414	0.3464225 675	1.464	0.3811724 155	1.514	0.4147551 550	1.564	0.4472466 421
1.415	0.3471295 311	1.465	0.3818552 425	1.515	0.4154154 390	1.565	0.4478858 240
1.416	0.3478359 953	1.466	0.3825376 035	1.516	0.4160752 872	1.566	0.4485245 976
1.417	0.3485419 607	1.467	0.3832194 992	1.517	0.4167347 004	1.567	0.4491629 634
1.418	0.3492474 281	1.468	0.3839009 302	1.518	0.4173936 790	1.568	0.4498009 219
1.419	0.3499523 982	1.469	0.3845818 972	1.519	0.4180522 236	1.569	0.4504384 738
1.420	0.3506568 716	1.470	0.3852624 008	1.520	0.4187103 349	1.570	0.4510756 194
1.421	0.3513608 491	1.471	0.3859424 416	1.521	0.4193680 133	1.571	0.4517123 593
1.422	0.3520643 314	1.472	0.3866220 203	1.522	0.4200252 594	1.572	0.4523486 940
1.423	0.3527673 191	1.473	0.3873011 375	1.523	0.4206820 739	1.573	0.4529846 241
1.424	0.3534698 130	1.474	0.3879797 938	1.524	0.4213384 573	1.574	0.4536201 500
1.425	0.3541718 137	1.475	0.3886579 898	1.525	0.4219944 101	1.575	0.4542552 723
1.426	0.3548733 220	1.476	0.3893357 262	1.526	0.4226499 329	1.576	0.4548899 914
1.427	0.3555743 385	1.477	0.3900130 035	1.527	0.4233050 262	1.577	0.4555243 080
1.428	0.3562748 639	1.478	0.3906898 225	1.528	0.4239596 907	1.578	0.4561582 224
1.429	0.3569748 989	1.479	0.3913661 837	1.529	0.4246139 269	1.579	0.4567917 353
1.430	0.3576744 443	1.480	0.3920420 878	1.530	0.4252677 354	1.580	0.4574248 470
1.431	0.3583735 006	1.481	0.3927175 353	1.531	0.4259211 167	1.581	0.4580575 582
1.432	0.3590720 685	1.482	0.3933925 269	1.532	0.4265740 713	1.582	0.4586898 693
1.433	0.3597701 488	1.483	0.3940670 632	1.533	0.4272265 999	1.583	0.4593217 809
1.434	0.3604677 422	1.484	0.3947411 447	1.534	0.4278787 029	1.584	0.4599532 934
1.435	0.3611648 492	1.485	0.3954147 723	1.535	0.4285303 810	1.585	0.4605844 073
1.436	0.3618614 706	1.486	0.3960879 463	1.536	0.4291816 347	1.586	0.4612151 232
1.437	0.3625576 071	1.487	0.3967606 675	1.537	0.4298324 646	1.587	0.4618454 415
1.438	0.3632532 593	1.488	0.3974329 364	1.538	0.4304828 711	1.588	0.4624753 628
1.439	0.3639484 279	1.489	0.3981047 537	1.539	0.4311328 549	1.589	0.4631048 876
1.440	0.3646431 136	1.490	0.3987761 200	1.540	0.4317824 164	1.590	0.4637340 162
1.441	0.3653373 170	1.491	0.3994470 358	1.541	0.4324315 563	1.591	0.4643627 494
1.442	0.3660310 389	1.492	0.4001175 018	1.542	0.4330802 751	1.592	0.4649910 874
1.443	0.3667242 798	1.493	0.4007875 186	1.543	0.4337285 734	1.593	0.4656190 309
1.444	0.3674170 405	1.494	0.4014570 867	1.544	0.4343764 516	1.594	0.4662465 804
1.445	0.3681093 216	1.495	0.4021262 068	1.545	0.4350239 103	1.595	0.4668737 362
1.446	0.3688011 237	1.496	0.4027948 796	1.546	0.4356709 502	1.596	0.4675004 990
1.447	0.3694924 476	1.497	0.4034631 054	1.547	0.4363175 716	1.597	0.4681268 692
1.448	0.3701832 940	1.498	0.4041308 851	1.548	0.4369637 752	1.598	0.4687528 473
1.449	0.3708736 633	1.499	0.4047982 191	1.549	0.4376095 614	1.599	0.4693784 339

TEN PLACE MANTISSAS FOR NATURAL LOGARITHMS (Continued)

x	$\log_e x$	x	$\log_e x$	x	$\log_e x$	x	$\log_e x$
1.600	0.4700036 292	1.650	0.5007752 879	1.700	0.5306282 511	1.750	0.5596157 879
1.601	0.4706284 340	1.651	0.5013811 649	1.701	0.5312163 134	1.751	0.5601870 533
1.602	0.4712528 486	1.652	0.5019866 751	1.702	0.5318040 302	1.752	0.5607579 925
1.603	0.4718768 736	1.653	0.5025918 188	1.703	0.5323914 017	1.753	0.5613286 059
1.604	0.4725005 094	1.654	0.5031965 966	1.704	0.5329784 284	1.754	0.5618988 939
1.605	0.4731237 566	1.655	0.5038010 088	1.705	0.5335651 107	1.755	0.5624688 569
1.606	0.4737466 155	1.656	0.5044050 560	1.706	0.5341514 491	1.756	0.5630384 952
1.607	0.4743690 868	1.657	0.5050087 384	1.707	0.5347374 438	1.757	0.5636078 092
1.608	0.4749911 708	1.658	0.5056120 567	1.708	0.5353230 954	1.758	0.5641767 993
1.609	6.4756128 680	1.659	0.5062150 112	1.709	0.5359084 041	1.759	0.5647454 658
1.610	0.4762341 790	1.660	0.5068176 024	1.710	0.5364933 705	1.760	0.5653138 091
1.611	0.4768551 042	1.661	0.5074198 306	1.711	0.5370779 949	1.761	0.5658818 295
1.612	0.4774756 441	1.662	0.5080216 964	1.712	0.5376622 777	1.762	0.5664495 275
1.613	0.4780957 991	1.663	0.5086232 002	1.713	0.5382462 193	1.763	0.5670169 034
1.614	0.4787155 698	1.664	0.5092243 424	1.714	0.5388298 202	1.764	0.5675839 576
1.615	0.4793349 567	1.665	0.5098251 234	1.715	0.5394130 806	1.765	0.5681506 904
1.616	0.4799539 601	1,666	0.5104255 437	1.716	0.5399960 011	1.766	0.5687171 022
1.617	0.4805725 806	1.667	0.5110256 038	1.717	0.5405785 819	1.767	0.5692831 933
1.618	0.4811908 186	1.668	0.5116253 039	1.718	0.5411608 236	1.768	0.5698489 642
1.619	0.4818086 747	1.669	0.5122246 447	1.719	0.5417427 264	1.769	0.5704144 152
1.620	0.4824261 492	1.670	0.5128236 264	1.720	0.5423242 908	1.770	0.5709795 466
1.621	0.4830432 428	1.671	0.5134222 496	1.721	0.5429055 172	1.771	0.5715443 588
1.622	0.4836599 557	1.672	0.5140205 147	1.722	0.5434864 061	1.772	0.5721088 522
1.623	0.4842762 885	1.673	0.5146184 220	1.723	0.5440669 575	1.773	0.5726730 271
1.624	0.4848922 417	1.674	0.5152159 721	1.724	0.5446471 722	1.774	0.5732368 839
1.625	0.4855078 158	1.675	0.5158131 653	1.725	0.5452270 505	1.775	0.5738004 229
1.626	0.4861230 111	1.676	0.5164100 021	1.726	0.5458065 927	1.776	0.5743636 446
1.627	0.4867378 282	1.677	0.5170064 828	1.727	0.5463857 992	1.777	0.5749265 492
1.628	0.4873522 676	1.678	0.5176026 080	1.728	0.5469646 704	1.778	0.5754891 371
1.629	0.4879663 296	1.679	0.5181983 781	1.729	0.5475432 067	1.779	0.5760514 087
1.630	0.4885800 148	1.680	0.5187937 934	1.730	0.5481214 085	1.780	0.5766133 643
1.631	0.4891933 236	1.681	0.5193888 544	1.731	0.5486992 762	1.781	0.5771750 043
1.632	0.4898062 565	1.682	0.5199835 616	1.732	0.5492768 101	1.782	0.5777363 290
1.633	0.4904188 140	1.683	0.5205779 152	1.733	0.5498540 107	1.783	0.5782973 389
1.634	0.4910309 964	1.684	0.5211719 158	1.734	0.5504308 784	1.784	0.5788580 342
1.635	0.4916428 043	1.685	0.5217655 638	1.735	0.5510074 134	1.785	0.5794184 152
1.636	0.4922542 382	1.686	0.5223588 596	1.736	0.5515836 162	1.786	0.5799784 825
1.637	0.4928652 984	1.687	0.5229518 036	1.737	0.5521594 873	1.787	0.5805382 362
1.638	0.4934759 854	1.688	0.5235443 962	1.738	0.5527350 268	1.788	0.5810976 768
1.639	0.4940862 998	1.689	0.5241366 378	1.739	0.5533102 354	1.789	0.5816568 045
1.640	0.4946962 418	1.690	0.5247285 289	1.740	0.5538851 132	1.790	0.5822156 199
1.641	0.4953058 121	1.691	0.5253200 699	1.741	0.5544596 608	1.791	0.5827741 231
1.642	0.4959150 110	1.692	0.5259112 612	1.742	0.5550338 784	1.792	0.5833323 146
1.643	0.4965238 391	1.693	0.5265021 032	1.743	0.5556077 665	1.793	0.5838901 946
1.644	0.4971322 966	1.694	0.5270925 962	1.744	0.5561813 255	1.794	0.5844477 636
1.645	0.4977403 842	1.695	0.5276827 408	1.745	0.5567545 557	1.795	0.5850050 219
1.646	0.4983481 023	1.696	0.5282725 374	1.746	0.5573274 574	1.796	0.5855619 699
1.647	0.4989554 512	1.697	0.5288619 863	1.747	0.5579000 312	1.797	0.5861186 078
1.648	0.4995624 315	1.698	0.5294510 879	1.748	0.5584722 772	1.798	0.5866749 361
1.649	0.5001690 436	1.699	0.5300398 427	1.749	0.5590441 960	1.799	0.5872309 550

TEN PLACE MANTISSAS FOR NATURAL LOGARITHMS (Continued)

x	$\log_e x$	x	$\log_e x$	x	$\log_e x$	x	$\log_e x$
1.800	0.5877866 649	1.850	0.6151856 391	1.900	0.6418538 862	1.950	0.6678293 726
1.801	0.5883420 662	1.851	0.6157260 336	1.901	0.6423800 635	1.951	0.6683420 616
1.802	0.5888971 592	1.852	0.6162661 362	1.902	0.6429059 641	1.952	0.6688544 880
1.803	0.5894519 442	1.853	0.6168059 473	1.903	0.6434315 883	1.953	0.6693666 519
1.804	0.5900064 216	1.854	0.6173454 671	1.904	0.6439569 364	1.954	0.6698785 536
1.805	0.5905605 918	1.855	0.6178846 961	1.905	0.6444820 086	1.955	0.6703901 934
1.806	0.5911144 550	1.856	0.6184236 344	1.906	0.6450068 052	1.956	0.6709015 716
1.807	0.5916680 116	1.857	0.6189622 824	1.907	0.6455313 266	1.957	0.6714126 884
1.808	0.5922212 620	1.858	0.6195006 404	1.908	0.6460555 730	1.958	0.6719235 441
1.809	0.5927742 064	1.859	0.6200387 087	1.909	0.6465795 447	1.959	0.6724341 390
1.810	0.5933268 453	1.860	0.6205764 877	1.910	0.6471032 421	1.960	0.6729444 732
1.811	0.5938791 789	1.861	0.6211139 777	1.911	0.6476266 653	1.961	0.6734545 472
1.812	0.5944312 076	1.862	0.6216511 789	1.912	0.6481498 146	1.962	0.6739643 611
1.813	0.5949829 318	1.863	0.6221880 916	1.913	0.6486726 905	1.963	0.6744739 153
1.814	0.5955343 517	1.864	0.6227247 163	1.914	0.6491952 930	1.964	0.6749832 099
1.815	0.5960854 677	1.865	0.6232610 531	1.915	0.6497176 226	1.965	0.6754922 453
1.816	0.5966362 802	1.866	0.6237971 024	1.916	0.6502396 795	1.966	0.6760010 217
1.817	0.5971867 894	1.867	0.6243328 646	1.917	0.6507614 641	1.967	0.6765095 394
1.818	0.5977369 958	1.868	0.6248683 398	1.918	0.6512829 765	1.968	0.6770177 986
1.819	0.5982868 995	1.869	0.6254035 285	1.919	0.6518042 170	1.969	0.6775257 997
1.820	0.5988365 011	1.870	0.6259384 309	1.920	0.6523251 860	1.970	0.6780335 427
1.821	0.5993858 007	1.871	0.6264730 473	1.921	0.6528458 838	1.971	0.6785410 282
1.822	0.5999347 988	1.872	0.6270073 781	1.922	0.6533663 105	1.972	0.6790482 562
1.823	0.6004834 957	1.873	0.6275414 235	1.923	0.6538864 666	1.973	0.6795552 270
1.824	0.6010318 917	1.874	0.6280751 838	1.924	0.6544063 522	1.974	0.6800619 410
1.825	0.6015799 870	1.875	0.6286086 594	1.925	0.6549259 677	1.975	0.6805683 984
1.826	0.6021277 822	1.876	0.6291418 506	1.926	0.6554453 134	1.976	0.6810745 993
1.827	0.6026752 774	1.877	0.6296747 576	1.927	0.6559643 894	1.977	0.6815805 442
1.828	0.6032224 730	1.878	0.6302073 808	1.928	0.6564831 962	1.978	0.6820862 332
1.829	0.6037693 694	1.879	0.6307397 204	1.929	0.6570017 339	1.979	0.6825916 666
1.830	0.6043159 669	1.880	0.6312717 768	1.930	0.6575200 029	1.980	0.6830968 447
1.831	0.6048622 657	1.881	0.6318035 503	1.931	0.6580380 034	1.981	0.6836017 677
1.832	0.6054082 663	1.882	0.6323350 412	1.932	0.6585557 358	1.982	0.6841064 359
1.833	0.6059539 689	1.883	0.6328662 497	1.933	0.6590732 002	1.983	0.6846108 495
1.834	0.6064993 738	1.884	0.6333971 762	1.934	0.6595903 970	1.984	0.6851150 089
1.835	0.6070444 815	1.885	0.6339278 209	1.935	0.6601073 265	1.985	0.6856189 141
1.836	0.6075892 922	1.886	0.6344581 842	1.936	0.6606239 889	1.986	0.6861225 656
1.837	0.6081338 062	1.887	0.6349882 664	1.937	0.6611403 844	1.987	0.6866259 636
1.838	0.6086780 239	1.888	0.6355180 677	1.938	0.6616565 135	1.988	0.6871291 082
1.839	0.6092219 456	1.889	0.6360475 885	1.939	0.6621723 763	1.989	0.6876319 999
1.840	0.6097655 716	1.890	0.6365768 291	1.940	0.6626879 731	1.990	0.6881346 387
1.841	0.6103089 023	1.891	0.6371057 897	1.941	0.6632033 042	1.991	0.6886370 251
1.842	0.6108519 378	1.892	0.6376344 706	1.942	0.6637183 699	1.992	0.6891391 592
1.843	0.6113946 787	1.893	0.6381628 722	1.943	0.6642331 704	1.993	0.6896410 412
1.844	0.6119371 251	1.894	0.6386909 948	1.944	0.6647477 061	1.994	0.6901426 715
1.845	0.6124792 775	1.895	0.6392188 385	1.945	0.6652619 771	1.995	0.6906440 503
1.846	0.6130211 361	1.896	0.6397464 038	1.946	0.6657759 838	1.996	0.6911451 779
1.847	0.6135627 012	1.897	0.6402736 910	1.947	0.6662897 264	1.997	0.6916460 544
1.848	0.6141039 732	1.898	0.6408007 002	1.948	0.6668032 052	1.998	0.6921466 802
1.849	0.6146449 524	1.899	0.6413274 318	1.949	0.6673164 205	1.999	0.6926470 555

TEN PLACE MANTISSAS FOR NATURAL LOGARITHMS (Continued)

x	$\log_e x$	x	$\log_e x$	x	$\log_e x$	x	$\log_e x$
2.000	0.6931471 806	2.050	0.7178397 932	2.100	0.7419373 447	2.150	0.7654678 421
2.001	0.6936470 556	2.051	0.7183274 791	2.101	0.7424134 219	2.151	0.7659328 503
2.002	0.6941466 809	2.052	0.7188149 273	2.102	0.7428892 725	2.152	0.7663976 423
2.003	0.6946460 567	2.053	0.7193021 380	2.103	0.7433648 967	2.153	0.7668622 184
2.004	0.6951451 832	2.054	0.7197891 115	2.104	0.7438402 949	2.154	0.7673265 787
2.005	0.6956440 608	2.055	0.7202758 479	2.105	0.7443154 671	2.155	0.7677907 236
2.006	0.6961426 895	2.056	0.7207623 476	2.106	0.7447904 137	2.156	0.7682546 530
2.007	0.6966410 698	2.057	0.7212486 107	2.107	0.7452651 348	2.157	0.7687183 674
2.008	0.6971392 018	2.058	0.7217346 374	2.108	0.7457396 307	2.158	0.7691818 668
2.009	0.6976370 858	2.059	0.7222204 280	2.109	0.7462139 015	2.159	0.7696451 515
2.010	0.6981347 221	2.060	0.7227059 828	2.110	0.7466879 475	2.160	0.7701082 217
2.011	0.6986321 108	2.061	0.7231913 019	2.111	0.7471617 689	2.161	0.7705710 775
2.012	0.6991292 522	2.062	0.7236763 856	2.112	0.7476353 658	2.162	0.7710337 192
2.013	0.6996261 467	2.063	0.7241612 341	2.113	0.7481087 386	2.163	0.7714961 470
2.014	0.7001227 943	2.064	0.7246458 476	2.114	0.7485818 874	2.164	0.7719583 610
2.015	0.7006191 954	2.065	0.7251302 264	2.115	0.7490548 125	2.165	0.7724203 615
2.016	0.7011153 502	2.066	0.7256143 707	2.116	0.7495275 140	2.166	0.7728821 486
2.017	0.7016112 590	2.067	0.7260982 807	2.117	0.7499999 922	2.167	0.7733437 226
2.018	0.7021069 219	2.068	0.7265819 566	2.118	0.7504722 472	2.168	0.7738050 836
2.019	0.7026023 393	2.069	0.7270653 988	2.119	0.7509442 793	2.169	0.7742662 318
2.020	0.7030975 114	2.070	0.7275486 073	2.120	0.7514160 887	2.170	0.7747271 676
2.021	0.7035924 384	2.071	0.7280315 824	2.121	0.7518876 756	2.171	0.7751878 909
2.022	0.7040871 206	2.072	0.7285143 244	2.122	0.7523590 402	2.172	0.7756484 002
2.023	0.7045815 582	2.073	0.7289968 335	2.123	0.7528301 827	2.173	0.7761087 013
2.024	0.7050757 514	2.074	0.7294791 098	2.124	0.7533011 034	2.174	0.7765687 887
2.025	0.7055697 006	2.075	0.7299611 537	2.125	0.7537718 024	2.175	0.7770286 645
2.026	0.7060634 058	2.076	0.7304429 653	2.126	0.7542422 799	2.176	0.7774883 290
2.027	0.7065568 675	2.077	0.7309245 449	2.127	0.7547125 362	2.177	0.7779477 823
2.028	0.7070500 857	2.078	0.7314058 927	2.128	0.7551825 715	2.178	0.7784070 245
2.029	0.7075430 608	2.079	0.7318870 089	2.129	0.7556523 859	2.179	0.7788660 560
2.030	0.7080357 931	2.080	0.7323678 937	2.130	0.7561219 797	2.180	0.7793248 768
2.031	0.7085282 826	2.081	0.7328485 474	2.131	0.7565913 531	2.181	0.7797834 872
2.032	0.7090205 297	2.082	0.7333289 702	2.132	0.7570605 063	2.182	0.7802418 874
2.033	0.7095125 346	2.083	0.7338091 623	2.133	0.7575294 395	2.183	0.7807000 776
2.034	0.7100042 976	2.084	0.7342891 239	2.134	0.7579981 529	2.184	0.7811580 579
2.035	0.7104958 189	2.085	0.7347688 553	2.135	0.7584666 467	2.185	0.7816158 285
2.036	0.7109870 987	2.086	0.7352483 566	2.136	0.7589349 211	2.186	0.7820733 898
2.037	0.7114781 372	2.087	0.7357276 281	2.137	0.7594029 763	2.187	0.7825307 417
2.038	0.7119689 348	2.088	0.7362066 700	2.138	0.7598708 126	2.188	0.7829878 846
2.039	0.7124594 916	2.089	0.7366854 826	2.139	0.7603384 301	2.189	0.7834448 185
2.040	0.7129498 079	2.090	0.7371640 660	2.140	0.7608058 290	2.190	0.7839015 438
2.041	0.7134398 838	2.091	0.7376424 204	2.141	0.7612730 096	2.191	0.7843580 606
2.042	0.7139297 197	2.092	0.7381205 462	2.142	0.7617399 720	2.192	0.7848143 691
2.043	0.7144193 158	2.093	0.7385984 435	2.143	0.7622067 165	2.193	0.7852704 694
2.044	0.7149086 723	2.094	0.7390761 124	2.144	0.7626732 432	2.194	0.7857263 619
2.045	0.7153977 895	2.095	0.7395535 534	2.145	0.7631395 524	2.195	0.7861820 465
2.046	0.7158866 675	2.096	0.7400307 665	2.146	0.7636056 442	2.196	0.7866375 236
2.047	0.7163753 067	2.097	0.7405077 519	2.147	0.7640715 189	2.197	0.7870927 934
2.048	0.7168637 072	2.098	0.7409845 100	2.148	0.7645371 766	2.198	0.7875478 560
2.049	0.7173518 693	2.099	0.7414610 408	2.149	0.7650026 177	2.199	0.7880027 116

TEN PLACE MANTISSAS FOR NATURAL LOGARITHMS (Continued)

x	$\log_e x$	x	$\log_e x$	x	$\log_e x$	x	$\log_e x$
2.200	0.7884573 604	2.250	0.8109302 162	2.300	0.8329091 229	2.350	0.8544153 282
2.201	0.7889118 025	2.251	0.8113745 619	2.301	0.8333438 111	2.351	0.8548407 696
2.202	0.7893660 383	2.252	0.8118187 103	2.302	0.8337783 103	2.352	0.8552660 300
2.203	0.7898200 678	2.253	0.8122626 615	2.303	0.8342126 208	2.353	0.8556911 097
2.204	0.7902738 913	2.254	0.8127064 156	2.304	0.8346467 428	2.354	0.8561160 088
2.205	0.7907275 089	2.255	0.8131499 730	2.305	0.8350806 764	2.355	0.8565407 275
2.206	0.7911809 208	2.256	0.8135933 336	2.306	0.8355144 218	2.356	0.8569652 658
2.207	0.7916341 273	2.257	0.8140364 978	2.307	0.8359479 792	2.357	0.8573896 240
2.208	0.7920871 284	2.258	0.8144794 657	2.308	0.8363813 486	2.358	0.8578138 021
2.209	0.7925399 244	2.259	0.8149222 375	2.309	0.8368145 304	2.359	0.8582378 004
2.210	0.7929925 155	2.260	0.8153648 133	2.310	0.8372475 245	2.360	0.8586616 190
2.211	0.7934449 019	2.261	0.8158071 933	2.311	0.8376803 313	2.361	0.8590852 581
2.212	0.7938970 837	2.262	0.8162493 777	2.312	0.8381129 508	2.362	0.8595087 178
2.213	0.7943490 611	2.263	0.8166913 667	2.313	0.8385453 832	2.363	0.8599319 982
2.214	0.7948008 343	2.264	0.8171331 603	2.314	0.8389776 288	2.364	0.8603550 995
2.215	0.7952524 035	2.265	0.8175747 589	2.315	0.8394096 875	2.365	0.8607780 219
2.216	0.7957037 689	2.266	0.8180161 626	2.316	0.8398415 597	2.366	0.8612007 656
2.217	0.7961549 306	2.267	0.8184573 715	2.317	0.8402732 455	2.367	0.8616233 305
2.218	0.7966058 889	2.268	0.8188983 859	2.318	0.8407047 449	2.368	0.8620457 170
2.219	0.7970566 440	2.269	0.8193392 058	2.319	0.8411360 583	2.369	0.8624679 252
2.220	0.7975071 959	2.270	0.8197798 315	2.320	0.8415671 857	2.370	0.8628899 551
2.221	0.7979575 449	2.271	0.8202202 631	2.321	0.8419981 273	2.371	0.8633118 071
2.222	0.7984076 912	2.272	0.8206605 009	2.322	0.8424288 833	2.372	0.8637334 811
2.223	0.7988576 350	2.273	0.8211005 449	2.323	0.8428594 538	2.373	0.8641549 775
2.224	0.7993073 764	2.274	0.8215403 953	2.324	0.8432898 390	2.374	0.8645762 962
2.225	0.7997569 156	2.275	0.8219800 524	2.325	0.8437200 390	2.375	0.8649974 375
2.226	0.8002062 529	2.276	0.8224195 163	2.326	0.8441500 541	2.376	0.8654184 015
2.227	0.8006553 883	2.277	0.8228587 871	2.327	0.8445798 843	2.377	0.8658391 884
2.228	0.8011043 221	2.278	0.8232978 650	2.328	0.8450095 299	2.378	0.8662597 983
2.229	0.8015530 544	2.279	0.8237367 503	2.329	0.8454389 909	2.379	0.8666802 313
2.230	0.8020015 855	2.280	0.8241754 430	2.330	0.8458682 676	2.380	0.8671004 877
2.231	0.8024499 155	2.281	0.8246139 433	2.331	0.8462973 601	2.381	0.8675205 675
2.232	0.8028980 445	2.282	0.8250522 514	2.332	0.8467262 685	2.382	0.8679404 709
2.233	0.8033459 729	2.283	0.8254903 675	2.333	0.8471549 930	2.383	0.8683601 981
2.234	0.8037937 006	2.284	0.8259282 918	2.334	0.8475835 339	2.384	0.8687797 492
2.235	0.8042412 281	2.285	0.8263660 243	2.335	0.8480118 911	2.385	0.8691991 243
2.236	0.8046885 553	2.286	0.8268035 654	2.336	0.8484400 650	2.386	0.8696183 237
2.237	0.8051356 825	2.287	0.8272409 150	2.337	0.8488680 556	2.387	0.8700373 474
2.238	0.8055826 099	2.288	0.8276780 735	2.338	0.8492958 630	2.388	0.8704561 955
2.239	0.8060293 376	2.289	0.8281150 410	2.339	0.8497234 876	2.389	0.8708748 683
2.240	0.8064758 659	2.290	0.8285518 176	2.340	0.8501509 294	2.390	0.8712933 659
2.241	0.8069221 948	2.291	0.8289884 035	2.341	0.8505781 885	2.391	0.8717116 885
2.242	0.8073683 246	2.292	0.8294247 989	2.342	0.8510052 652	2.392	0.8721298 361
2.243	0.8078142 555	2.293	0.8298610 039	2.343	0.8514321 595	2.393	0.8725478 089
2.244	0.8082599 877	2.294	0.8302970 187	2.344	0.8518588 717	2.394	0.8729656 071
2.245	0.8087055 212	2.295	0.8307328 435	2.345	0.8522854 019	2.395	0.8733832 309
2.246	0.8091508 563	2.296	0.8311684 785	2.346	0.8527117 502	2.396	0.8738006 803
2.247	0.8095959 932	2.297	0.8316039 237	2.347	0.8531379 169	2.397	0.8742179 555
2.248	0.8100409 320	2.298	0.8320391 794	2.348	0.8535639 020	2.398	0.8746350 566
2.249	0.8104856 730	2.299	0.8324742 458	2.349	0.8539897 057	2.399	0.8750519 839

TEN PLACE MANTISSAS FOR NATURAL LOGARITHMS (Continued)

x	$\log_e x$	x	$\log_e x$	x	$\log_e x$	x	$\log_e x$
2.400	0.8754687 374	2.450	0.8960880 246	2.500	0.9162907 319	2.550	0.9360933 592
2.401	0.8758853 172	2.451	0.8964961 045	2.501	0.9166906 519	2.551	0.9364854 392
2.402	0.8763017 237	2.452	0.8969040 181	2.502	0.9170904 120	2.552	0.9368773 655
2.403	0.8767179 568	2.453	0.8973117 653	2.503	0.9174900 124	2.553	0.9372691 383
2.404	0.8771340 167	2.454	0.8977193 463	2.504	0.9178894 532	2.554	0.9376607 576
2.405	0.8775499 036	2.455	0.8981267 612	2.505	0.9182887 345	2.555	0.9380522 237
2.406	0.8779656 176	2.456	0.8985340 103	2.506	0.9186878 565	2.556	0.9384435 365
2.407	0.8783811 588	2.457	0.8989410 935	2.507	0.9190868 192	2.557	0.9388346 963
2.408	0.8787965 274	2.458	0.8993480 111	2.508	0.9194856 228	2.558	0.9392257 032
2.409	0.8792117 236	2.459	0.8997547 632	2.509	0.9198842 674	2.559	0.9396165 572
2.410	0.8796267 475	2.460	0.9001613 499	2.510	0.9202827 531	2.560	0.9400072 585
2.411	0.8800415 992	2.461	0.9005677 714	2.511	0.9206810 802	2.561	0.9403978 072
2.412	0.8804562 789	2.462	0.9009740 278	2.512	0.9210792 486	2.562	0.9407882 035
2.413	0.8808707 866	2.463	0.9013801 191	2.513	0.9214772 586	2.563	0.9411784 474
2.414	0.8812851 227	2.464	0.9017860 457	2.514	0.9218751 102	2.564	0.9415685 391
2.415	0.8816992 871	2.465	0.9021918 075	2.515	0.9222728 036	2.565	0.9419584 786
2.416	0.8821132 801	2.466	0.9025974 047	2.516	0.9226703 388	2.566	0.9423482 662
2.417	0.8825271 017	2.467	0.9030028 375	2.517	0.9230677 162	2.567	0.9427379 019
2.418	0.8829407 522	2.468	0.9034081 060	2.518	0.9234649 356	2.568	0.9431273 858
2.419	0.8833542 316	2.469	0.9038132 103	2.519	0.9238619 974	2.569	0.9435167 181
2.420	0.8837675 402	2.470	0.9042181 506	2.520	0.9242589 015	2.570	0.9439058 989
2.421	0.8841806 780	2.471	0.9046229 270	2.521	0.9246556 482	2.571	0.9442949 283
2.422	0.8845936 451	2.472	0.9050275 396	2.522	0.9250522 375	2.572	0.9446838 064
2.423	0.8850064 418	2.473	0.9054319 885	2.523	0.9254486 697	2.573	0.9450725 333
2.424	0.8854190 682	2.474	0.9058362 740	2.524	0.9258449 447	2.574	0.9454611 092
2.425	0.8858315 244	2.475	0.9062403 960	2.525	0.9262410 627	2.575	0.9458495 341
2.426	0.8862438 105	2.476	0.9066443 548	2.526	0.9266370 239	2.576	0.9462378 082
2.427	0.8866559 267	2.477	0.9070481 505	2.527	0.9270328 284	2.577	0.9466259 317
2.428	0.8870678 732	2.478	0.9074517 832	2.528	0.9274284 763	2.578	0.9470139 045
2.429	0.8874796 500	2.479	0.9078552 531	2.529	0.9278239 676	2.579	0.9474017 269
2.430	0.8878912 574	2.480	0.9082585 602	2.530	0.9282193 027	2.580	0.9477893 989
2.431	0.8883026 953	2.481	0.9086617 047	2.531	0.9286144 816	2.581	0.9481769 207
2.432	0.8887139 641	2.482	0.9090646 868	2.532	0.9290095 043	2.582	0.9485642 924
2.433	0.8891250 638	2.483	0.9094675 065	2.533	0.9294043 710	2.583	0.9489515 141
2.434	0.8895359 946	2.484	0.9098701 641	2.534	0.9297990 819	2.584	0.9493385 859
2.435	0.8899467 565	2.485	0.9102726 595	2.535	0.9301936 370	2.585	0.9497255 080
2.436	0.8903573 498	2.486	0.9106749 931	2.536	0.9305880 366	2.586	0.9501122 803
2.437	0.8907677 746	2.487	0.9110771 648	2.537	0.9309822 806	2.587	0.9504989 032
2.438	0.8911780 311	2.488	0.9114791 749	2.538	0.9313763 693	2.588	0.9508853 766
2.439	0.8915881 192	2.489	0.9118810 234	2.539	0.9317703 027	2.589	0.9512717 008
2.440	0.8919980 393	2.490	0.9122827 105	2.540	0.9321640 810	2.590	0.9516578 757
2.441	0.8924077 914	2.491	0.9126842 363	2.541	0.9325577 043	2.591	0.9520439 016
2.442	0.8928173 757	2.492	0.9130856 009	2.542	0.9329511 728	2.592	0.9524297 785
2.443	0.8932267 923	2.493	0.9134868 045	2.543	0.9333444 864	2.593	0.9528155 066
2.444	0.8936360 413	2.494	0.9138878 473	2.544	0.9337376 455	2.594	0.9532010 859
2.445	0.8940451 229	2.495	0.9142887 292	2.545	0.9341306 500	2.595	0.9535865 166
2.446	0.8944540 373	2.496	0.9146894 505	2.546	0.9345235 001	2.596	0.9539717 988
2.447	0.8948627 845	2.497	0.9150900 113	2.547	0.9349161 960	2.597	0.9543569 327
2.448	0.8952713 647	2.498	0.9154904 117	2.548	0.9353087 377	2.598	0.9547419 182
2.449	0.8956797 780	2.499	0.9158906 519	2.549	0.9357011 254	2.599	0.9551267 557

TEN PLACE MANTISSAS FOR NATURAL LOGARITHMS (Continued)

x	$\log_e x$	x	$\log_e x$	x	$\log_e x$	x	$\log_e x$
2.600	0.9555114 450	2.650	0.9745596 400	2.700	0.9932517 730	2.750	1.0116009 117
2.601	0.9558959 865	2.651	0.9749369 273	2.701	0.9936220 748	2.751	1.0119644 819
2.602	0.9562803 801	2.652	0.9753140 723	2.702	0.9939922 395	2.752	1.0123279 201
2.603	0.9566646 260	2.653	0.9756910 752	2.703	0.9943622 673	2.753	1.0126912 262
2.604	0.9570487 243	2.654	0.9760679 359	2.704	0.9947321 582	2.754	1.0130544 003
2.605	0.9574326 752	2.655	0.9764446 547	2.705	0.9951019 123	2.755	1.0134174 426
2.606	0.9578164 787	2.656	0.9768212 316	2.706	0.9954715 297	2.756	1.0137803 532
2.607	0.9582001 350	2.657	0.9771976 668	2.707	0.9958410 106	2.757	1.0141431 320
2.608	0.9585836 441	2.658	0.9775739 603	2.708	0.9962103 551	2.758	1.0145057 794
2.609	0.9589670 062	2.659	0.9779501 123	2.709	0.9965795 631	2.759	1.0148682 952
2.610	0.9593502 213	2.660	0.9783261 228	2.710	0.9969486 349	2.760	1.0152306 797
2.611	0.9597332 897	2.661	0.9787019 920	2.711	0.9973175 705	2.761	1.0155929 329
2.612	0.9601162 114	2.662	0.9790777 200	2.712	0.9976863 701	2.762	1.0159550 550
2.613	0.9604989 865	2.663	0.9794533 068	2.713	0.9980550 337	2.763	1.0163170 459
2.614	0.9608816 152	2.664	0.9798287 527	2.714	0.9984235 614	2.764	1.0166789 059
2.615	0.9612640 975	2.665	0.9802040 576	2.715	0.9987919 534	2.765	1.0170406 350
2.616	0.9616464 336	2.666	0.9805792 218	2.716	0.9991602 097	2.766	1.0174022 332
2.617	0.9620286 235	2.667	0.9809542 452	2.717	0.9995283 304	2.767	1.0177637 008
2.618	0.9624106 675	2.668	0.9813291 281	2.718	0.9998963 157	2.768	1.0181250 378
2.619	0.9627925 655	2.669	0.9817038 704	2.719	1.0002641 657	2.769	1.0184862 442
2.620	0.9631743 178	2.670	0.9820784 724	2.720	1.0006318 803	2.770	1.0188473 202
2.621	0.9635559 243	2.671	0.9824529 341	2.721	1.0009994 598	2.771	1.0192082 659
2.622	0.9639373 853	2.672	0.9828272 557	2.722	1.0013669 042	2.772	1.0195690 813
2.623	0.9643187 009	2.673	0.9832014 372	2.723	1.0017342 137	2.773	1.0199297 666
2.624	0.9646998 711	2.674	0.9835754 787	2.724	1.0021013 882	2.774	1.0202903 219
2.625	0.9650808 960	2.675	0.9839493 803	2.725	1.0024684 281	2.775	1.0206507 472
2.626	0.9654617 759	2.676	0.9843231 423	2.726	1.0028353 333	2.776	1.0210110 426
2.627	0.9658425 107	2.677	0.9846967 645	2.727	1.0032021 039	2.777	1.0213712 083
2.628	0.9662231 006	2.678	0.9850702 473	2.728	1.0035687 400	2.778	1.0217312 443
2.629	0.9666035 457	2.679	0.9854435 906	2.729	1.0039352 417	2.779	1.0220911 508
2.630	0.9669838 462	2.680	0.9858167 945	2.730	1.0043016 092	2.780	1.0224509 277
2.631	0.9673640 021	2.681	0.9861898 593	2.731	1.0046678 425	2.781	1.0228105 753
2.632	0.9677440 135	2.682	0.9865627 849	2.732	1.0050339 417	2.782	1.0231700 935
2.633	0.9681238 805	2.683	0.9869355 714	2.733	1.0053999 069	2.783	1.0235294 825
2.634	0.9685036 033	2.684	0.9873082 191	2.734	1.0057657 383	2.784	1.0238887 425
2.635	0.9688831 820	2.685	0.9876807 280	2.735	1.0061314 359	2.785	1.0242478 734
2.636	0.9692626 166	2.686	0.9880530 981	2.736	1.0064969 998	2.786	1.0246068 754
2.637	0.9696419 074	2.687	0.9884253 296	2.737	1.0068624 301	2.787	1.0249657 485
2.638	0.9700210 543	2.688	0.9887974 227	2.738	1.0072277 269	2.788	1.0253244 929
2.639	0.9704000 575	2.689	0.9891693 773	2.739	1.0075928 903	2.789	1.0256831 086
2.640	0.9707789 172	2.690	0.9895411 936	2.740	1.0079579 204	2.790	1.0260415 958
2.641	0.9711576 333	2.691	0.9899128 717	2.741	1.0083228 173	2.791	1.0263999 546
2.642	0.9715362 061	2.692	0.9902844 118	2.742	1.0086875 811	2.792	1.0267581 849
2.643	0.9719146 356	2.693	0.9906558 138	2.743	1.0090522 120	2.793	1.0271162 870
2.644	0.9722929 220	2.694	0.9910270 780	2.744	1.0094167 099	2.794	1.0274742 608
2.645	0.9726710 653	2.695	0.9913982 044	2.745	1.0097810 750	2.795	1.0278321 066
2.646	0.9730490 657	2.696	0.9917691 931	2.746	1.0101453 073	2.796	1.0281898 244
2.647	0.9734269 232	2.697	0.9921400 442	2.747	1.0105094 071	1.797	1.0285474 142
2.648	0.9738046 381	2.698	0.9925107 578	2.748	1.0108733 744	2.798	1.0289048 762
2.649	0.9741822 103	2.699	0.9928813 340	2.749	1.0112372 092	2.799	1.0292622 105

TEN PLACE MANTISSAS FOR NATURAL LOGARITHMS (Continued)

x	$\log_e x$	x	$\log_e x$	x	$\log_e x$	x	$\log_e x$
2.800	1.0296194 172	2.850	1.0473189 943	2.900	1.0647107 370	2.950	1.0818051 704
2.801	1.0299764 963	2.851	1.0476698 099	2.901	1.0650555 051	2.951	1.0821440 960
2.802	1.0303334 479	2.852	1.0480205 026	2.902	1.0654001 545	2.952	1.0824829 067
2.803	1.0306902 722	2.853	1.0483710 722	2.903	1.0657446 850	2.953	1.0828216 028
2.804	1.0310469 692	2.854	1.0487215 191	2.904	1.0660890 970	2.954	1.0831601 841
2.805	1.0314035 390	2.855	1.0490718 431	2.905	1.0664333 903	2.955	1.0834986 509
2.806	1.0317599 817	2.856	1.0494220 445	2.906	1.0667775 651	2.956	1.0838370 031
2.807	1.0321162 974	2.857	1.0497721 232	2.907	1.0671216 216	2.957	1.0841752 409
2.808	1.0324724 862	2.858	1.0501220 795	2.908	1.0674655 597	2.958	1.0845133 643
2.809	1.0328285 481	2.859	1.0504719 133	2.909	1.0678093 795	2.959	1.0848513 734
2.810	1.0331844 833	2.860	1.0508216 248	2.910	1.0681530 812	2.960	1.0851892 683
2.811	1.0335402 919	2.861	1.0511712 141	2.911	1.0684966 648	2.961	1.0855270 491
2.812	1.0338959 739	2.862	1.0515206 811	2.912	1.0688401 303	2.962	1.0858647 158
2.813	1.0342515 295	2.863	1.0518700 261	2.913	1.0691834 780	2.963	1.0862022 686
2.814	1.0346069 587	2.864	1.0522192 491	2.914	1.0695267 078	2.964	1.0865397 074
2.815	1.0349622 616	2.865	1.0525683 502	2.915	1.0698698 198	2.965	1.0868770 324
2.816	1.0353174 383	2.866	1.0529173 294	2.916	1.0702128 141	2.966	1.0872142 437
2.817	1.0356724 889	2.867	1.0532661 869	2.917	1.0705556 909	2.967	1.0875513 413
2.818	1.0360274 135	2.868	1.0536149 227	2.918	1.0708984 501	2.968	1.0878883 253
2.819	1.0363822 121	2.869	1.0539635 370	2.919	1.0712410 919	2.969	1.0882251 958
2.820	1.0367368 850	2.870	1.0543120 298	2.920	1.0715836 163	2.970	1.0885619 528
2.821	1.0370914 320	2.871	1.0546604 011	2.921	1.0719260 234	2.971	1.0888985 965
2.822	1.0374458 534	2.872	1.0550086 512	2.922	1.0722683 133	2.972	1.0892351 269
2.823	1.0378001 493	2.873	1.0553567 800	2.923	1.0726104 861	2.973	1.0895715 440
2.824	1.0381543 196	2.874	1.0557047 877	2.924	1.0729525 419	2.974	1.0899078 480
2.825	1.0385083 646	2.875	1.0560526 742	2.925	1.0732944 807	2.975	1.0902440 390
2.826	1.0388622 843	2.876	1.0564004 399	2.926	1.0736363 026	2.976	1.0905801 170
2.827	1.0392160 787	2.877	1.0567480 846	2.927	1.0739780 077	2.977	1.0909160 820
2.828	1.0395697 480	2.878	1.0570956 085	2.928	1.0743195 961	2.978	1.0912519 343
2.829	1.0399232 923	2.879	1.0574430 116	2.929	1.0746610 678	2.979	1.0915876 737
2.830	1.0402767 117	2.880	1.0577902 941	2.930	1.0750024 230	2.980	1.0919233 005
2.831	1.0406300 061	2.881	1.0581374 561	2.931	1.0753436 617	2.981	1.0922588 147
2.832	1.0409831 758	2.882	1.0584844 976	2.932	1.0756847 840	2.982	1.0925942 163
2.833	1.0413362 208	2.883	1.0588314 187	2.933	1.0760257 900	2.983	1.0929295 055
2.834	1.0416891 413	2.884	1.0591782 194	2.934	1.0763666 797	2.984	1.0932646 823
2.835	1.0420419 372	2.885	1.0595249 000	2.935	1.0767074 533	2.985	1.0935997 468
2.836	1.0423946 087	2.886	1.0598714 604	2.936	1.0770481 108	2.986	1.0939346 991
2.837	1.0427471 558	2.887	1.0602179 007	2.937	1.0773886 522	2.987	1.0942695 392
2.838	1.0430995 787	2.888	1.0605642 210	2.938	1.0777290 778	2.988	1.0946042 673
2.839	1.0434518 775	2.889	1.0609104 215	2.939	1.0780693 874	2.989	1.0949388 833
2.840	1.0438040 522	2.890	1.0612565 021	2.940	1.0784095 814	2.990	1.0952733 874
2.841	1.0441561 029	2.891	1.0616024 630	2.941	1.0787496 596	2.991	1.0956077 796
2.842	1.0445080 297	2.892	1.0619483 043	2.942	1.0790896 222	2.992	1.0959420 601
2.843	1.0448598 327	2.893	1.0622940 260	2.943	1.0794294 693	2.993	1.0962762 289
2.844	1.0452115 119	2.894	1.0626396 282	2.944	1.0797692 009	2.994	1.0966102 860
2.845	1.0455630 676	2.895	1.0629851 110	2.945	1.0801088 171	2.995	1.0969442 316
2.846	1.0459144 997	2.896	1.0633304 745	2.946	1.0804483 180	2.996	1.0972780 657
2.847	1.0462658 083	2.897	1.0636757 188	2.947	1.0807877 038	2.997	1.0976117 883
2.848	1.0466169 935	2.898	1.0640208 439	2.948	1.0811269 743	2.998	1.0979453 997
2.849	1.0469680 555	2.899	1.0643658 499	2.949	1.0814661 298	2.999	1.0982788 998

TEN PLACE MANTISSAS FOR NATURAL LOGARITHMS (Continued)

x	$\log_e x$	x	$\log_e x$	x	$\log_e x$	x	$\log_e x$
3.000	1.0986122 887	3.050	1.1151415 906	3.100	1.1314021 115	3.150	1.1474024 528
3.001	1.0989455 665	3.051	1.1154694 057	3.101	1.1317246 401	3.151	1.1477198 628
3.002	1.0992787 332	3.052	1.1157971 134	3.102	1.1320470 648	3.152	1.1480371 720
3.003	1.0996117 890	3.053	1.1161247 138	3.103	1.1323693 855	3.153	1.1483543 806
3.004	1.0999447 339	3.054	1.1164522 068	3.104	1.1326916 023	3.154	1.1486714 885
3.005	1.1002775 680	3.055	1.1167795 926	3.105	1.1330137 154	3.155	1.1489884 960
3.006	1.1006102 913	3.056	1.1171068 713	3.106	1.1333357 247	3.156	1.1493054 030
3.007	1.1009429 040	3.057	1.1174340 429	3.107	1.1336576 304	3.157	1.1496222 096
3.008	1.1012754 061	3.058	1.1177611 075	3.108	1.1339794 325	3.158	1.1499389 158
3.009	1.1016077 976	3.059	1.1180880 652	3.109	1.1343011 311	3.159	1.1502555 218
3.010	1.1019400 788	3.060	1.1184149 160	3.110	1.1346227 262	3.160	1.1505720 276
3.011	1.1022722 495	3.061	1.1187416 600	3.111	1.1349442 179	3.161	1.1508884 332
3.012	1.1026043 099	3.062	1.1190682 972	3.112	1.1352656 063	3.162	1.1512047 388
3.013	1.1029362 601	3.063	1.1193948 279	3.113	1.1355868 915	3.163	1.1515209 443
3.014	1.1032681 002	3.064	1.1197212 519	3.114	1.1359080 734	3.164	1.1518370 499
3.015	1.1035998 302	3.065	1.1200475 694	3.115	1.1362291 522	3.165	1.1521530 556
3.016	1.1039314 501	3.066	1.1203737 804	3.116	1.1365501 280	3.166	1.1524689 615
3.017	1.1042629 602	3.067	1.1206998 851	3.117	1.1368710 008	3.167	1.1527847 676
3.018	1.1045943 603	3.068	1.1210258 835	3.118	1.1371917 706	3.168	1.1531004 740
3.019	1.1049256 507	3.069	1.1213517 756	3.119	1.1375124 376	3.169	1.1534160 807
3.020	1.1052568 314	3.070	1.1216775 616	3.120	1.1378330 018	3.170	1.1537315 879
3.021	1.1055879 024	3.071	1.1220032 415	3.121	1.1381534 633	3.171	1.1540469 956
3.022	1.1059188 639	3.072	1.1223288 153	3.122	1.1384738 221	3.172	1.1543623 038
3.023	1.1062497 158	3.073	1.1226542 831	3.123	1.1387940 783	3.173	1.1546775 126
3.024	1.1065804 583	3.074	1.1229796 451	3.124	1.1391142 320	3.174	1.1549926 221
3.025	1.1069110 915	3.075	1.1233049 013	3.125	1.1394342 832	3.175	1.1553076 323
3.026	1.1072416 154	3.076	1.1236300 516	3.126	1.1397542 320	3.176	1.1556225 434
3.027	1.1075720 300	3.077	1.1239550 963	3.127	1.1400740 785	3.177	1.1559373 553
3.028	1.1079023 356	3.078	1.1242800 354	3.128	1.1403938 227	3.178	1.1562520 681
3.029	1.1082325 320	3.079	1.1246048 689	3.129	1.1407134 647	3.179	1.1565666 819
3.030	1.1085626 195	3.080	1.1249295 970	3.130	1.1410330 046	3.180	1.1568811 968
3.031	1.1088925 981	3.081	1.1252542 196	3.131	1.1413524 423	3.181	1.1571956 128
3.032	1.1092224 678	3.082	1.1255787 369	3.132	1.1416717 781	3.182	1.1575099 299
3.033	1.1095522 287	3.083	1.1259031 489	3.133	1.1419910 120	3.183	1.1578241 483
3.034	1.1098818 809	3.084	1.1262274 557	3.134	1.1423101 439	3.184	1.1581382 680
3.035	1.1102114 245	3.085	1.1265516 574	3.135	1.1426291 741	3.185	1.1584522 890
3.036	1.1105408 595	3.086	1.1268757 539	3.136	1.1429481 025	3.186	1.1587662 115
3.037	1.1108701 861	3.087	1.1271997 455	3.137	1.1432669 292	3.187	1.1590800 354
3.038	1.1111994 042	3.088	1.1275236 322	3.138	1.1435856 543	3.188	1.1593937 609
3.039	1.1115285 139	3.089	1.1278474 139	3.139	1.1439042 779	3.189	1.1597073 880
3.040	1.1118575 154	3.090	1.1281710 909	3.140	1.1442227 999	3.190	1.1600209 168
3.041	1.1121864 087	3.091	1.1284946 631	3.141	1.1445412 206	3.191	1.1603343 473
3.042	1.1125151 938	3.092	1.1288181 307	3.142	1.1448595 398	3.192	1.1606476 796
3.043	1.1128438 709	3.093	1.1291414 937	3.143	1.1451777 578	3.193	1.1609609 137
3.044	1.1131724 400	3.094	1.1294647 522	3.144	1.1454958 746	3.194	1.1612740 498
3.045	1.1135009 012	3.095	1.1297879 061	3.145	1.1458138 902	3.195	1.1615870 878
3.046	1.1138292 545	3.096	1.1301109 557	3.146	1.1461318 046	3.196	1.1619000 279
3.047	1.1141575 000	3.097	1.1304339 010	3.147	1.1464496 181	3.197	1.1622128 701
3.048	1.1144856 378	3.098	1.1307567 420	3.148	1.1467673 306	3.198	1.1625256 144
3.049	1.1148136 680	3.099	1.1310794 788	3.149	1.1470849 421	3.199	1.1628382 610

TEN PLACE MANTISSAS FOR NATURAL LOGARITHMS (Continued)

x	$\log_e x$	x	$\log_e x$	x	$\log_e x$	x	$\log_e x$
3.200	1.1631508 098	3.250	1.1786549 963	3.300	1.1939224 685	3.350	1.2089603 458
3.201	1.1634632 610	3.351	1.1789626 413	3.301	1.1942254 529	3.351	1.2092588 088
3.202	1.1637756 146	3.252	1.1792701 917	3.302	1.1945283 455	3.352	1.2095571 826
3.203	1.1640878 706	3.253	1.1795776 475	3.303	1.1948311 464	3.353	1.2098554 675
3.204	1.1644000 292	3.254	1.1798850 088	3.304	1.1951338 557	3.354	1.2101536 634
3.205	1.1647120 904	3.255	1.1801922 757	3.305	1.1954364 733	3.355	1.2104517 704
3.206	1.1650240 542	3.256	1.1804994 481	3.306	1.1957389 994	3.356	1.2107497 886
3.207	1.1653359 207	3.257	1.1808065 263	3.307	1.1960414 340	3.357	1.2110477 180
3.208	1.1656476 900	3.258	1.1811135 102	3.308	1.1963437 772	3.358	1.2113455 587
3.209	1.1659593 621	3.259	1.1814203 999	3.309	1.1966460 289	3.359	1.2116433 106
3.210	1.1662709 371	3.260	1.1817271 954	3.310	1.1969481 894	3.360	1.2119409 740
3.211	1.1665824 151	3.261	1.1820338 968	3.311	1.1972502 586	3.361	1.2122385 487
3.212	1.1668937 961	3.262	1.1823405 042	3.312	1.1975522 365	3.362	1.2125360 350
3.213	1.1672050 801	3.263	1.1826470 176	3.313	1.1978541 233	3.363	1.2128334 328
3.214	1.1675162 673	3.264	1.1829534 371	3.314	1.1981559 190	3.364	1.2131307 421
3.215	1.1678273 577	3.265	1.1832597 627	3.315	1.1984576 236	3.365	1.2134279 631
3.216	1.1681383 513	3.266	1.1835659 945	3.316	1.1987592 373	3.366	1.2137250 958
3.217	1.1684492 483	3.267	1.1838721 326	3.317	1.1990607 600	3.367	1.2140221 402
3.218	1.1687600 486	3.268	1.1841781 770	3.318	1.1993621 918	3.368	1.2143190 964
3.219	1.1690707 523	3.269	1.1844841 277	3.319	1.1996635 327	3.369	1.2146159 644
3.220	1.1693813 596	3.270	1.1847899 849	3.320	1.1999647 829	3.370	1.2149127 444
3.221	1.1696918 703	3.271	1.1850957 486	3.321	1.2002659 424	3.371	1.2152094 363
3.222	1.1700022 848	3.272	1.1854014 187	3.322	1.2005670 112	3.372	1.2155060 401
3.223	1.1703126 028	3.273	1.1857069 955	3.323	1.2008679 894	3.373	1.2158025 561
3.224	1.1706228 246	3.274	1.1860124 789	3.324	1.2011688 770	3.374	1.2160989 841
3.225	1.1709329 502	3.275	1.1863178 691	3.325	1.2014696 741	3.375	1.2163953 243
3.226	1.1712429 797	3.276	1.1866231 660	3.326	1.2017703 808	3.376	1.2166915 767
3.227	1.1715529 131	3.277	1.1869283 697	3.327	1.2020709 970	3.377	1.2169877 414
3.228	1.1718627 504	3.278	1.1872334 803	3.328	1.2023715 230	3.378	1.2172838 184
3.229	1.1721724 918	3.279	1.1875384 979	3.329	1.2026719 586	3.379	1.2175798 077
3.230	1.1724821 372	3.280	1.1878434 224	3.330	1.2029723 040	3.380	1.2178757 095
3.231	1.1727916 868	3.281	1.1881482 540	3.331	1.2032725 592	3.381	1.2181715 237
3.232	1.1731011 407	3.282	1.1884529 927	3.332	1.2035727 243	3.382	1.2184672 505
3.233	1.1734104 987	3.283	1.1887576 385	3.333	1.2038727 993	3.383	1.2187628 898
3.234	1.1737197 612	3.284	1.1890621 916	3.334	1.2041727 843	3.384	1.2190584 417
3.235	1.1740289 280	3.285	1.1893666 519	3.335	1.2044726 794	3.385	1.2193539 064
3.236	1.1743379 992	3.286	1.1896710 196	3.336	1.2047724 845	3.386	1.2196492 837
3.237	1.1746469 749	3.287	1.1899752 947	3.337	1.2050721 998	3.387	1.2199445 738
3.238	1.1749558 553	3.288	1.1902794 772	3.338	1.2053718 252	3.388	1.2202397 768
3.239	1.1752646 402	3.289	1.1905835 672	3.339	1.2056713 610	3.389	1.2205348 926
3.240	1.1755733 298	3.290	1.1908875 648	3.340	1.2059708 070	3.390	1.2208299 214
3.241	1.1758819 242	3.291	1.1911914 700	3.341	1.2062701 634	3.391	1.2211248 631
3.242	1.1761904 233	3.292	1.1914952 828	3.342	1.2065694 302	3.392	1.2214197 179
3.243	1.1764988 273	3.293	1.1917990 034	3.343	1.2068686 074	3.393	1.2217144 858
3.244	1.1768071 363	3.294	1.1921026 318	3.344	1.2071676 952	3.394	1.2220091 668
3.245	1.1771153 502	3.295	1.1924061 680	3.345	1.2074666 936	3.395	1.2223037 610
3.246	1.1774234 691	3.296	1.1927096 120	3.346	1.2077656 026	3.396	1.2225982 684
3.247	1.1777314 931	3.297	1.1930129 641	3.347	1.2080644 222	3.397	1.2228926 892
3.248	1.1780394 223	3.298	1.1933162 241	3.348	1.2083631 526	3.398	1.2231870 232
3.249	1.1783472 567	3.299	1.1936193 922	3.349	1.2086617 938	3.399	1.2234812 707

TEN PLACE MANTISSAS FOR NATURAL LOGARITHMS (Continued)

x	$\log_e x$	x	$\log_e x$	x	$\log_e x$	x	$\log_e x$
3.400	1.2237754 316	3.450	1.2383742 310	3.500	1.2527629 685	3.550	1.2669476 035
3.401	1.2240695 060	3.451	1.2386640 441	3.501	1.2530486 420	3.551	1.2672292 540
3.402	1.2243634 940	3.452	1.2389537 732	3.502	1.2533342 339	3.552	1.2675108 251
3.403	1.2246573 955	3.453	1.2392434 184	3.503	1.2536197 442	3.553	1.2677923 170
3.404	1.2249512 107	3.454	1.2395329 797	3.504	1.2539051 731	3.554	1.2680737 297
3.405	1.2252449 396	3.455	1.2398224 572	3.505	1.2541905 205	3.555	1.2683550 633
3.406	1.2255385 822	3.456	1.2401118 509	3.506	1.2544757 865	3.556	1.2686363 177
3.407	1.2258321 387	3.457	1.2404011 609	3.507	1.2547609 712	3.557	1.2689174 930
3.408	1.2261256 090	3.458	1.2406903 873	3.508	1.2550460 745	3.558	1.2691985 892
3.409	1.2264189 932	3.459	1.2409795 300	3.509	1.2553310 966	3.559	1.2694796 065
3.410	1.2267122 913	3.460	1.2412685 891	3.510	1.2556160 375	3.560	1.2697605 449
3.411	1.2270055 034	3.461	1.2415575 647	3.511	1.2559008 972	3.561	1.2700414 043
3.412	1.2272986 296	3.462	1.2418464 568	3.512	1.2561856 758	3.562	1.2703221 849
3.413	1.2275916 699	3.463	1.2421352 654	3.513	1.2564703 733	3.563	1.2706028 866
3.414	1.2278846 244	3.464	1.2424239 907	3.514	1.2567549 898	3.564	1.2708835 096
3.415	1.2281774 930	3.465	1.2427126 326	3.515	1.2570395 253	3.565	1.2711640 539
3.416	1.2284702 759	3.466	1.2430011 913	3.516	1.2573239 798	3.566	1.2714445 194
3.417	1.2287629 731	3.467	1.2432896 667	3.517	1.2576083 535	3.567	1.2717249 064
3.418	1.2290555 847	3.468	1.2435780 589	3.518	1.2578926 463	3.568	1.2720052 147
3.419	1.2293481 107	3.469	1.2438663 680	3.519	1.2581768 583	3.569	1.2722854 445
3.420	1.2296405 511	3.470	1.2441545 940	3.520	1.2584609 896	3.570	1.2725655 958
3.421	1.2299329 060	3.471	1.2444427 369	3.521	1.2587450 402	3.571	1.2728456 686
3.422	1.2302251 755	3.472	1.2447307 968	3.522	1.2590290 101	3.572	1.2731256 630
3.423	1.2305173 595	3.473	1.2450187 738	3.523	1.2593128 994	3.573	1.2734055 790
3.424	1.2308094 583	3.474	1.2453066 678	3.524	1.2595967 081	3.574	1.2736854 167
3.425	1.2311014 717	3.475	1.2455944 790	3.525	1.2598804 363	3.575	1.2739651 761
3.426	1.2313933 999	3.476	1.2458822 074	3.526	1.2601640 840	3.576	1.2742448 573
3.427	1.2316852 429	3.477	1.2461698 530	3.527	1.2604476 513	3.577	1.2745244 603
3.428	1.2319770 007	3.478	1.2464574 159	3.528	1.2607311 381	3.578	1.2748039 851
3.429	1.2322686 735	3.479	1.2467448 962	3.529	1.2610145 447	3.579	1.2750834 318
3.430	1.2325602 612	3.480	1.2470322 938	3.530	1.2612978 709	3.580	1.2753628 004
3.431	1.2328517 639	3.481	1.2473196 088	3.531	1.2615811 169	3.581	1.2756420 910
3.432	1.2331431 816	3.482	1.2476068 413	3.532	1.2618642 827	3.582	1.2759213 036
3.433	1.2334345 145	3.483	1.2478939 914	3.533	1.2621473 684	3.583	1.2762004 383
3.434	1.2337257 625	3.484	1.2481810 590	3.534	1.2624303 739	3.584	1.2764794 951
3.435	1.2340169 257	3.485	1.2484680 442	3.535	1.2627132 993	3.585	1.2767584 741
3.436	1.2343080 041	3.486	1.2487549 471	3.536	1.2629961 448	3.586	1.2770373 752
3.437	1.2345989 979	3.487	1.2490417 677	3.537	1.2632789 102	3.587	1.2773161 986
3.438	1.2348899 070	3.488	1.2493285 060	3.538	1.2635615 957	3.588	1.2775949 442
3.439	1.2351807 315	3.489	1.2496151 622	3.539	1.2638442 014	3.589	1.2778736 122
3.440	1.2354714 714	3.490	1.2499017 362	3.540	1.2641267 271	3.590	1.2781522 025
3.441	1.2357621 268	3.491	1.2501882 281	3.541	1.2644091 731	3.591	1.2784307 152
3.442	1.2360526 978	3.492	1.2504746 380	3.542	1.2646915 394	3.592	1.2787091 504
3.443	1.2363431 844	3.493	1.2507609 658	3.543	1.2649738 259	3.593	1.2789875 081
3.444	1.2366335 866	3.494	1.2510472 117	3.544	1.2652560 327	3.594	1.2792657 884
3.445	1.2369239 045	3.495	1.2513333 757	3.545	1.2655381 600	3.595	1.2795439 912
3.446	1.2372141 381	3.496	1.2516194 578	3.546	1.2658202 077	3.596	1.2798221 166
3.447	1.2375042 875	3.497	1.2519054 581	3.547	1.2661021 758	3.597	1.2801001 647
3.448	1.2377943 528	3.498	1.2521913 766	3.548	1.2663840 644	3.598	1.2803781 355
3.449	1.2380843 340	3.499	1.2524772 134	3.549	1.2666658 737	3.599	1.2806560 291

TEN PLACE MANTISSAS FOR NATURAL LOGARITHMS (Continued)

x	$\log_e x$	x	$\log_e x$	x	$\log_e x$	x	$\log_e x$
3.600	1.2809338 455	3.650	1.2947271 676	3.700	1.3083328 197	3.750	1.3217558 400
3.601	1.2812115 847	3.651	1.2950011 027	3.701	1.3086030 534	3.751	1.3220224 711
3.602	1.2814892 468	3.652	1.2952749 627	3.702	1.3088732 142	3.752	1.3222890 311
3.603	1.2817668 318	3.653	1.2955487 478	3.703	1.3091433 019	3.753	1.3225555 202
3.604	1.2820443 397	3.654	1.2958224 580	3.704	1.3094133 168	3.754	1.3228219 382
3.605	1.2823217 707	3.655	1.2960960 932	3.705	1.3096832 587	3.755	1.3230882 852
3.606	1.2825991 248	3.656	1.2963696 536	3.706	1.3099531 279	3.756	1.3233545 613
3.607	1.2828764 019	3.657	1.2966431 392	3.707	1.3102229 242	3.757	1.3236207 666
3.608	1.2831536 022	3.658	1.2969165 500	3.708	1.3104926 477	3.758	1.3238869 010
3.609	1.2834307 257	3.659	1.2971898 860	3.709	1.3107622 985	3.759	1.3241529 646
3.610	1.2837077 723	3.660	1.2974631 474	3.710	1.3110318 766	3.760	1.3244189 574
3.611	1.2839847 423	3.661	1.2977363 341	3.711	1.3113013 821	3.761	1.3246848 795
3.612	1.2842616 356	3.662	1.2980094 463	3.712	1.3115708 149	3.762	1.3249507 309
3.613	1.2845384 522	3.663	1.2982824 838	3.713	1.3118401 752	3.763	1.3252165 116
3.614	1.2848151 922	3.664	1.2985554 468	3.714	1.3121094 629	3.764	1.3254822 217
3.615	1.2850918 556	3.665	1.2988283 353	3.715	1.3123786 782	3.765	1.3257478 613
3.616	1.2853684 425	3.666	1.2991011 494	3.716	1.3126478 210	3.766	1.3260134 302
3.617	1.2856449 530	3.667	1.2993738 891	3.717	1.3129168 913	3.767	1.3262789 287
3.618	1.2859213 870	3.668	1.2996465 544	3.718	1.3131858 893	3.768	1.3265443 567
3.619	1.2861977 446	3.669	1.2999191 454	3.719	1.3134548 149	3.769	1.3268097 143
3.620	1.2864740 258	3.670	1.3001916 621	3.720	1.3137236 683	3.770	1.3270750 015
3.621	1.2867502 308	3.671	1.3004641 045	3.721	1.3139924 494	3.771	1.3273402 183
3.622	1.2870263 595	3.672	1.3007364 728	3.722	1.3142611 582	3.772	1.3276053 648
3.623	1.2873024 119	3.673	1.3010087 668	3.723	1.3145297 949	3.773	1.3278704 410
3.624	1.2875783 882	3.674	1.3012809 868	3.724	1.3147983 594	3.774	1.3281354 469
3.625	1.2878542 883	3.675	1.3015531 327	3.725	1.3150668 518	3.775	1.3284003 827
3.626	1.2881301 123	3.676	1.3018252 045	3.726	1.3153352 722	3.776	1.3286652 483
3.627	1.2884058 603	3.677	1.3020972 023	3.727	1.3156036 205	3.777	1.3289300 437
3.628	1.2886815 323	3.678	1.3023691 262	3.728	1.3158718 968	3.778	1.3291947 691
3.629	1.2889571 282	3.679	1.3026409 761	3.729	1.3161401 012	3.779	1.3294594 244
3.630	1.2892326 483	3.680	1.3029127 522	3.730	1.3164082 337	3.780	1.3297240 096
3.631	1.2895080 924	3.681	1.3031844 544	3.731	1.3166762 942	3.781	1.3299885 249
3.632	1.2897834 607	3.682	1.3034560 828	3.732	1.3169442 830	3.782	1.3302529 702
3.633	1.2900587 532	3.683	1.3037276 375	3.733	1.3172121 999	3.783	1.3305173 457
3.634	1.2903339 700	3.684	1.3039991 184	3.734	1.3174800 451	3.784	1.3307816 512
3.635	1.2906091 110	3.685	1.3042705 256	3.735	1.3177478 186	3.785	1.3310458 869
3.636	1.2908841 763	3.686	1.3045418 592	3.736	1.3180155 204	3.786	1.3313100 528
3.637	1.2911591 660	3.687	1.3048131 193	3.737	1.3182831 505	3.787	1.3315741 489
3.638	1.2914340 801	3.688	1.3050843 057	3.738	1.3185507 090	3.788	1.3318381 753
3.639	1.2917089 186	3.689	1.3053554 186	3.739	1.3188181 960	3.789	1.3321021 320
3.640	1.2919836 816	3.690	1.3056264 581	3.740	1.3190856 114	3.790	1.3323660 191
3.641	1.2922583 692	3.691	1.3058974 240	3.741	1.3193529 554	3.791	1.3326298 365
3.642	1.2925329 813	3.692	1.3061683 166	3.742	1.3196202 279	3.792	1.3328935 844
3.643	1.2928075 180	3.693	1.3064391 359	3.743	1.3198874 289	3.793	1.3331572 627
3.644	1.2930819 794	3.694	1.3067098 818	3.744	1.3201545 586	3.794	1.3334208 715
3.645	1.2933563 655	3.695	1.3069805 544	3.745	1.3204216 170	3.795	1.3336844 108
3.646	1.2936306 763	3.696	1.3072511 538	3.746	1.3206886 040	3.796	1.3339478 807
3.647	1.2939049 118	3.697	1.3075216 800	3.747	1.3209555 198	3.797	1.3342112 812
3.648	1.2941790 722	3.698	1.3077921 330	3.748	1.3212223 644	3.798	1.3344746 124
3.649	1.2944531 575	3.699	1.3080625 129	3.749	1.3214891 378	3.799	1.3347378 742

TEN PLACE MANTISSAS FOR NATURAL LOGARITHMS (Continued)

x	$\log_e x$	x	$\log_e x$	x	$\log_e x$	x	$\log_e x$
3.800	1.3350010 667	3.850	1.3480731 483	3.900	1.3609765 531	3.950	1.3737155 789
3.801	1.3352641 900	3.851	1.3483328 548	3.901	1.3612329 305	3.951	1.3739687 114
3.802	1.3355272 441	3.852	1.3485924 939	3.902	1.3614892 422	3.952	1.3742217 799
3.803	1.3357902 289	3.853	1.3488520 656	3.903	1.3617454 882	3.953	1.3744747 843
3.804	1.3360531 447	3.854	1.3491115 700	3.904	1.3620016 686	3.954	1.3747277 247
3.805	1.3363159 913	3.855	1.3493710 070	3.905	1.3622577 833	3.955	1.3749806 012
3.806	1.3365787 689	3.856	1.3496303 767	3.906	1.3625138 325	3.956	1.3752334 138
3.807	1.3368414 774	4.857	1.3498896 792	3.907	1.3627698 161	3.957	1.3754861 624
3.808	1.3371041 169	3.858	1.3501489 145	3.908	1.3630257 342	3.958	1.3757388 472
3.809	1.3373666 875	3.859	1.3504080 826	3.909	1.3632815 868	3.959	1.3759914 681
3.810	1.3376291 891	3.860	1.3506671 835	3.910	1.3635373 740	3.960	1.3762440 253
3.811	1.3378916 219	3.861	1.3509262 173	3.911	1.3637930 958	3.961	1.3764965 186
3.812	1.3381539 858	3.862	1.3511851 840	3.912	1.3640487 522	3.962	1.3767489 483
3.813	1.3384162 809	3.863	1.3514440 837	3.913	1.3643043 432	3.963	1.3770013 142
3.814	1.3386785 072	3.864	1.3517029 164	3.914	1.3645598 690	3.964	1.3772536 165
3.815	1.3389406 647	3.865	1.3519616 820	3.915	1.3648153 294	3.965	1.3775058 551
3.816	1.3392027 536	3.866	1.3522203 808	3.916	1.3650707 247	3.966	1.3777580 301
3.817	1.3394647 738	3.867	1.3524790 126	3.917	1.3653260 547	3.967	1.3780101 415
3.818	1.3397267 253	3.868	1.3527375 776	3.918	1.3655813 195	3.968	1.3782621 894
3.819	1.3399886 082	3.869	1.3529960 757	3.919	1.3658365 192	3.969	1.3785141 738
3.820	1.3402504 226	3.870	1.3532545 070	3.920	1.3660916 538	3.970	1.3787660 947
3.821	1.3405121 685	3.871	1.3535128 716	3.921	1.3663467 233	3.971	1.3790179 521
3.822	1.3407738 458	3.872	1.3537711 694	3.922	1.3666017 278	3.972	1.3792697 462
3.823	1.3410354 547	3.873	1.3540294 005	3.923	1.3668566 672	3.973	1.3795214 768
3.824	1.3412969 952	3.874	1.3542875 650	3.924	1.3671115 417	3.974	1.3797731 441
3.825	1.3415584 673	3.875	1.3545456 628	3.925	1.3673663 512	3.975	1.3800247 481
3.826	1.3418198 710	3.876	1.3548036 940	3.926	1.3676210 959	3.976	1.3802762 888
3.827	1.3420812 064	3.877	1.3550616 587	3.927	1.3678757 756	3.977	1.3805277 662
3.828	1.3423424 736	3.878	1.3553195 568	3.928	1.3681303 905	3.978	1.3807791 804
3.829	1.3426036 725	3.879	1.3555773 885	3.929	1.3683849 406	3.979	1.3810305 314
3.830	1.3428648 032	3.880	1.3558351 536	3.930	1.3686394 259	3.980	1.3812818 193
3.831	1.3431258 657	3.881	1.3560928 524	3.931	1.3688938 464	3.981	1.3815330 440
3.832	1.3433868 601	3.882	1.3563504 847	3.932	1.3691482 023	3.982	1.3817842 056
3.833	1.3436477 864	3.883	1.3566080 507	3.933	1.3694024 934	3.983	1.3820353 042
3.834	1.3439086 446	3.884	1.3568655 504	3.934	1.3696567 200	3.984	1.3822863 397
3.835	1.3441694 348	3.885	1.3571229 838	3.935	1.3699108 819	3.985	1.3825373 122
3.836	1.3444301 570	3.886	1.3573803 510	3.936	1.3701649 792	3.986	1.3827882 218
3.837	1.3446908 113	3.887	1.3576376 519	3.937	1.3704190 120	3.987	1.3830390 684
3.838	1.3449513 976	3.888	1.3578948 866	3.938	1.3706729 802	3.988	1.3832898 521
3.839	1.3452119 160	3.889	1.3581520 552	3.939	1.3709268 840	3.989	1.3835405 729
3.840	1.3454723 666	3.890	1.3584091 576	3.940	1.3711807 233	3.990	1.3837912 309
3.841	1.3457327 494	3.891	1.3586661 940	3.941	1.3714344 982	3.991	1.3840418 261
3.842	1.3459930 643	3.892	1.3589231 643	3.942	1.3716882 087	3.992	1.3842923 584
3.843	1.3462533 116	3.893	1.3591800 686	3.943	1.3719418 549	3.993	1.3845428 281
3.844	1.3465134 911	3.894	1.3594369 070	3.944	1.3721954 367	3.994	1.3847932 350
3.845	1.3467736 030	3.895	1.3596936 793	3.945	1.3724489 543	3.995	1.3850435 792
3.846	1.3470336 472	3.896	1.3599503 858	3.946	1.3727024 076	3.996	1.3852938 608
3.847	1.3472936 238	3.897	1.3602070 264	3.947	1.3729557 967	3.997	1.3855440 797
3.848	1.3475535 328	3.898	1.3604636 011	3.948	1.3732091 216	3.998	1.3857942 361
3.849	1.3478133 743	3.899	1.3607201 100	3.949	1.3734623 823	3.999	1.3860443 299

TEN PLACE MANTISSAS FOR NATURAL LOGARITHMS (Continued)

x	$\log_e x$	x	$\log_e x$	x	$\log_e x$	x	$\log_e x$
4.000	1.3862943 611	4.050	1.3987168 811	4.100	1.4109869 737	4.150	1.4231083 342
4.001	1.3865443 299	4.051	1.3989637 642	4.101	1.4112308 464	4.151	1.4233492 691
4.002	1.3867942 362	4.052	1.3992105 864	4.102	1.4114746 597	4.152	1.4235901 459
4.003	1.3870440 800	4.053	1.3994573 476	4.103	1.4117184 135	4.153	1.4238309 646
4.004	1.3872938 615	4.054	1.3997040 480	4.104	1.4119621 079	4.154	1.4240717 255
4.005	1.3875435 805	4.055	1.3999506 876	4.105	1.4122057 429	4.155	1.4243124 283
4.006	1.3877932 372	4.056	1.4001972 663	4.106	1.4124493 186	4.156	1.4245530 732
4.007	1.3880428 317	4.057	1.4004437 842	4.107	1.4126928 350	4.157	1.4247936 603
4.008	1.3882923 638	4.058	1.4006902 414	4.108	1.4129362 921	4.158	1.4250341 894
4.009	1.3885418 337	4.059	1.4009366 379	4.109	1.4131796 899	4.159	1.4252746 608
4.010	1.3887912 413	4.060	1.4011829 736	4.110	1.4134230 285	4.160	1.4255150 743
4.011	1.3890405 868	4.061	1.4014292 487	4.111	1.4136663 079	4.161	1.4257554 300
4.012	1.3892898 701	4.062	1.4016754 632	4.112	1.4139095 282	4.162	1.4259957 280
4.013	1.3895390 913	4.063	1.4019216 170	4.113	1.4141526 892	4.163	1.4262359 682
4.014	1.3897882 504	4.064	1.4021677 103	4.114	1.4143957 912	4.164	1.4264761 508
4.015	1.3900373 474	4.065	1.4024137 430	4.115	1.4146388 341	4.165	1.4267162 756
4.016	1.3902863 824	4.066	1.4026597 152	4.116	1.4148818 180	4.166	1.4269563 428
4.017	1.3905353 554	4.067	1.4029056 269	4.117	1.4151247 428	4.167	1.4271963 524
4.018	1.3907842 664	4.068	1.4031514 782	4.118	1.4153676 086	4.168	1.4274363 045
4.019	1.3910331 155	4.069	1.4033972 690	4.119	1.4156104 155	4.169	1.4276761 989
4.020	1.3912819 026	4.070	1.4036429 995	4.120	1.4158531 634	4.170	1.4279160 358
4.021	1.3915306 279	4.071	1.4038886 695	4.121	1.4160958 524	4.171	1.4281558 152
4.022	1.3917792 914	4.072	1.4041342 792	4.122	1.4163384 825	4.172	1.4283955 371
4.023	1.3920278 930	4.073	1.4043798 287	4.123	1.4165810 537	4.173	1.4286352 016
4.024	1.3922764 328	4.074	1.4046253 178	4.124	1.4168235 662	4.174	1.4288748 087
4.025	1.3925249 109	4.075	1.4048707 467	4.125	1.4170660 198	4.175	1.4291143 583
4.026	1.3927733 272	4.076	1.4051161 154	4.126	1.4173084 146	4.176	1.4293538 506
4.027	1.3930216 819	4.077	1.4053614 238	4.127	1.4175507 508	4.177	1.4295932 855
4.028	1.3932699 749	4.078	1.4056066 722	4.128	1.4177930 282	4.178	1.4298326 631
4.029	1.3935182 062	4.079	1.4058518 603	4.129	1.4180352 469	4.179	1.4300719 835
4.030	1.3937663 760	4.080	1.4060969 884	4.130	1.4182774 070	4.180	1.4303112 465
4.031	1.3940144 841	4.081	1.4063420 564	4.131	1.4185195 084	4.181	1.4305504 524
4.032	1.3942625 308	4.082	1.4065870 644	4.132	1.4187615 513	4.182	1.4307896 010
4.033	1.3945105 159	4.083	1.4068320 123	4.133	1.4190035 355	4.183	1.4310286 925
4.034	1.3947584 395	4.084	1.4070769 003	4.134	1.4192454 613	4.184	1.4312677 268
4.035	1.3950063 017	4.085	1.4073217 283	4.135	1.4194873 285	4.185	1.4315067 039
4.036	1.3952541 025	4.086	1.4075664 964	4.136	1.4197291 372	4.186	1.4317456 240
4.037	1.3955018 419	4.087	1.4078112 046	4.137	1.4199708 875	4.187	1.4319844 870
4.038	1.3957495 199	4.088	1.4080558 529	4.138	1.4202125 793	4.188	1.4322232 930
4.039	1.3959971 366	4.089	1.4083004 414	4.139	1.4204542 128	4.189	1.4324620 420
4.040	1.3962446 920	4.090	1.4085449 701	4.140	1.4206957 878	4.190	1.4327007 339
4.041	1.3964921 861	4.091	1.4087894 389	4.141	1.4209373 046	4.191	1.4329393 689
4.042	1.3967396 190	4.092	1.4090338 481	4.142	1.4211787 630	4.192	1.4331779 470
4.043	1.3969869 907	4.093	1.4092781 975	4.143	1.4214201 631	4.193	1.4334164 682
4.044	1.3972343 012	4.094	1.4095224 872	4.144	1.4216615 050	4.194	1.4336549 325
4.045	1.3974815 505	4.095	1.4097667 173	4.145	1.4219027 886	4.195	1.4338933 399
4.046	1.3977287 387	4.096	1.4100108 877	4.146	1.4221440 140	4.196	1.4341316 905
4.047	1.3979758 659	4.097	1.4102549 986	4.147	1.4223851 813	4.197	1.4343699 844
4.048	1.3982229 320	4.098	1.4104990 498	4.148	1.4226262 904	4.198	1.4346082 214
4.049	1.3984699 371	4.099	1.4107430 415	4.149	1.4228673 414	4.199	1.4348464 017

TEN PLACE MANTISSAS FOR NATURAL LOGARITHMS (Continued)

x	$\log_e x$	x	$\log_e x$	x	$\log_e x$	x	$\log_e x$
4.200	1.4350845 253	4.250	1.4469189 829	4.300	1.4586150 227	4.350	1.4701758 451
4.201	1.4353225 922	4.251	1.4471542 494	4.301	1.4588475 538	4.351	1.4704057 037
4.202	1.4355606 024	4.252	1.4473894 605	4.302	1.4590800 308	4.352	1.4706355 096
4.203	1.4357985 560	4.253	1.4476246 163	4.303	1.4593124 539	4.353	1.4708652 626
4.204	1.4360364 530	4.254	1.4478597 168	4.304	1.4595448 229	4.354	1.4710949 628
4.205	1.4362742 934	4.255	1.4480947 620	4.305	1.4597771 379	4.355	1.4713246 103
4.206	1.4365120 773	4.256	1.4483297 520	4.306	1.4600093 989	4.356	1.4715542 051
4.207	1.4367498 046	4.257	1.4485646 868	4.307	1.4602416 061	4.357	1.4717837 471
4.208	1.4369874 754	4.258	1.4487995 665	4.308	1.4604737 593	4.358	1.4720132 365
4.209	1.4372250 898	4.259	1.4490343 909	4.309	1.4607058 586	4.359	1.4722426 733
4.210	1.4374626 477	4.260	1.4492691 603	4.310	1.4609379 041	4.360	1.4724720 574
4.211	1.4377001 492	4.261	1.4495038 745	4.311	1.4611698 958	4.361	1.4727013 889
4.212	1.4379375 943	4.262	1.4497385 337	4.312	1.4614018 336	4.362	1.4729306 678
4.213	1.4381749 830	4.263	1.4499731 378	4.313	1.4616337 177	4.363	1.4731598 941
4.214	1.4384123 154	4.264	1.4502076 869	4.314	1.4618655 480	4.364	1.4733890 680
4.215	1.4386495 915	4.265	1.4504421 809	4.315	1.4620973 245	4.365	1.4736181 893
4.216	1.4388868 112	4.266	1.4506766 200	4.316	1.4623290 474	4.366	1.4738472 581
4.217	1.4391239 748	4.267	1.4509110 042	4.317	1.4625607 166	4.367	1.4740762 745
4.218	1.4393610 821	4.268	1.4511453 334	4.318	1.4627923 321	4.368	1.4743052 384
4.219	1.4395981 331	4.269	1.4513796 078	4.319	1.4630238 940	4.369	1.4745341 500
4.220	1.4398351 280	4.270	1.4516138 272	4.320	1.4632554 023	4.370	1.4747630 091
4.221	1.4400720 668	4.271	1.4518479 919	4.321	1.4634868 569	4.371	1.4749918 159
4.222	1.4403089 494	4.272	1.4520821 017	4.322	1.4637182 581	4.372	1.4752205 703
4.223	1.4405457 760	4.273	1.4523161 567	4.323	1.4639496 057	4.373	1.4754492 724
4.224	1.4407825 464	4.274	1.4525501 569	4.324	1.4641808 998	4.374	1.4756779 223
4.225	1.4410192 608	4.275	1.4527841 024	4.325	1.4644121 404	4.375	1.4759065 198
4.226	1.4412559 192	4.276	1.4530179 932	4.326	1.4646433 275	4.376	1.4761350 651
4.227	1.4414925 216	4.277	1.4532518 292	4.327	1.4648744 612	4.377	1.4763635 582
4.228	1.4417290 680	4.278	1.4534856 107	4.328	1.4651055 415	4.378	1.4765919 991
4.229	1.4419655 585	4.279	1.4537193 374	4.329	1.4653365 685	4.379	1.4768203 878
4.230	1.4422019 931	4.280	1.4539530 096	4.330	1.4655675 420	4.380	1.4770487 244
4.231	1.4424383 717	4.281	1.4541866 272	4.331	1.4657984 622	4.381	1.4772770 088
4.232	1.4426746 946	4.282	1.4544201 902	4.332	1.4660293 291	4.382	1.4775052 412
4.233	1.4429109 615	4.283	1.4546536 986	4.333	1.4662601 428	4.383	1.4777334 214
4.234	1.4431471 727	4.284	1.4548871 526	4.334	1.4664909 031	4.384	1.4779615 496
4.235	1.4433833 281	4.285	1.4551205 520	4.335	1.4667216 102	4.385	1.4781896 258
4.236	1.4436194 277	4.286	1.4553538 971	4.336	1.4669522 641	4.386	1.4784176 500
4.237	1.4438554 716	4.287	1.4555871 876	4.337	1.4671828 649	4.387	1.4786456 222
4.238	1.4440914 598	4.288	1.4558204 238	4.338	1.4674134 124	4.388	1.4788735 424
4.239	1.4443273 924	4.289	1.4560536 055	4.339	1.4676439 068	4.389	1.4791014 107
4.240	1.4445632 692	4.290	1.4562867 329	4.340	1.4678743 481	4.390	1.4793292 271
4.241	1.4447990 905	4.291	1.4565198 060	4.341	1.4681047 363	4.391	1.4795569 916
4.242	1.4450348 561	4.292	1.4567528 248	4.342	1.4683350 715	4.392	1.4797847 042
4.243	1.4452705 662	4.293	1.4569857 892	4.343	1.4685653 536	4.393	1.4800123 650
4.244	1.4455062 208	4.294	1.4572186 995	4.344	1.4687955 826	4.394	1.4802399 740
4.245	1.4457418 198	4.295	1.4574515 554	4.345	1.4690257 587	4.395	1.4804675 311
4.246	1.4459773 633	4.296	1.4576843 572	4.346	1.4692558 818	4.396	1.4806950 365
4.247	1.4462128 513	4.297	1.4579171 048	4.347	1.4694859 520	4.397	1.4809224 902
4.248	1.4464482 839	4.298	1.4581497 982	4.348	1.4697159 693	4.398	1.4811498 921
4.249	1.4466836 611	4.299	1.4583824 375	4.349	1.4699459 336	4.399	1.4813772 424

TEN PLACE MANTISSAS FOR NATURAL LOGARITHMS (Continued)

x	$\log_e x$	x	$\log_e x$	x	$\log_e x$	x	$\log_e x$
4.400	1.4816045 409	4.450	1.4929040 962	4.500	1.5040773 968	4.550	1.5151272 330
4.401	1.4818317 878	4.451	1.4931287 900	4.501	1.5042995 943	4.551	1.5153469 890
4.402	1.4820589 831	4.452	1.4933534 334	4.502	1.5045217 425	4.552	1.5155666 968
4.403	1.4822861 268	4.453	1.4935780 263	4.503	1.5047438 413	4.553	1.5157863 564
4.404	1.4825132 189	4.454	1.4938025 688	4.504	1.5049658 908	4.554	1.5160059 676
4.405	1.4827402 594	4.455	1.4940270 609	4.505	1.5051878 911	4.555	1.5162255 307
4.406	1.4829672 484	4.456	1.4942515 026	4.506	1.5054098 420	4.556	1.5164450 456
4.407	1.4831941 859	4.457	1.4944758 940	4.507	1.5056317 437	4.557	1.5166645 123
4.408	1.4834210 719	4.458	1.4947002 350	4.508	1.5058535 962	4.558	1.5168839 308
4.409	1.4836479 064	4.459	1.4949245 256	4.509	1.5060753 994	4.559	1.5171033 012
4.410	1.4838746 895	4.460	1.4951487 660	4.510	1.5062971 535	4.560	1.5173226 235
4.411	1.4841014 211	4.461	1.4953729 561	4.511	1.5065188 584	4.561	1.5175418 977
4.412	1.4843281 014	4.462	1.4955970 960	4.512	1.5067405 142	4.562	1.5177611 239
4.413	1.4845547 303	4.463	1.4958211 856	4.513	1.5069621 208	4.563	1.5179803 019
4.414	1.4847813 078	4.464	1.4960452 251	4.514	1.5071836 784	4.564	1.5181994 320
4.415	1.4850078 341	4.465	1.4962692 143	4.515	1.5074051 869	4.565	1.5184185 140
4.416	1.4852343 090	4.466	1.4964931 534	4.516	1.5076266 463	4.566	1.5186375 481
4.417	1.4854607 326	4.467	1.4967170 424	4.517	1.5078480 567	4.567	1.5188565 342
4.418	1.4856871 050	4.468	1.4969408 812	4.518	1.5080694 180	4.568	1.5190754 724
4.419	1.4859134 261	4.469	1.4971646 700	4.519	1.5082907 304	4.569	1.5192943 626
4.420	1.4861396 961	4.470	1.4973884 086	4.520	1.5085119 938	4.570	1.5195132 049
4.421	1.4863659 148	4.471	1.4976120 973	4.521	1.5087332 083	4.571	1.5197319 993
4.422	1.4865920 824	4.472	1.4978357 359	4.522	1.5089543 739	4.572	1.5199507 459
4.423	1.4868181 989	4.473	1.4980593 245	4.523	1.5091754 905	4.573	1.5201694 447
4.424	1.4870442 642	4.474	1.4982828 631	4.524	1.5093965 583	4.574	1.5203880 956
4.425	1.4872702 785	4.475	1.4985063 517	4.525	1.5096175 772	4.575	1.5206066 987
4.426	1.4874962 416	4.476	1.4987297 904	4.526	1.5098385 472	4.576	1.5208252 541
4.427	1.4877221 538	4.477	1.4989531 793	4.527	1.5100594 685	4.577	1.5210437 617
4.428	1.4879480 148	4.478	1.4991765 182	4.528	1.5102803 409	4.578	1.5212622 215
4.429	1.4881738 249	4.479	1.4993998 072	4.529	1.5105011 646	4.579	1.5214806 337
4.430	1.4883995 841	4.480	1.4996230 464	4.530	1.5107219 395	4.580	1.5216989 981
4.431	1.4886252 922	4.481	1.4998462 358	4.531	1.5109426 657	4.581	1.5219173 149
4.432	1.4888509 494	4.482	1.5000693 754	4.532	1.5111633 432	4.582	1.5221355 840
4.433	1.4890765 558	4.483	1.5002924 652	4.533	1.5113839 720	4.583	1.5223538 055
4.434	1.4893021 112	4.484	1.5005155 052	4.534	1.5116045 521	4.584	1.5225719 794
4.435	1.4895276 158	4.485	1.5007384 955	4.535	1.5118250 836	4.585	1.5227901 057
4.436	1.4897530 695	4.486	1.5009614 361	4.536	1.5120455 664	4.586	1.5230081 844
4.437	1.4899784 724	4.487	1.5011843 270	4.537	1.5122660 007	4.587	1.5232262 156
4.438	1.4902038 245	4.488	1.5014071 682	4.538	1.5124863 864	4.588	1.5234441 993
4.439	1.4904291 259	4.489	1.5016299 598	4.539	1.5127067 235	4.589	1.5236621 354
4.440	1.4906543 764	4.490	1.5018527 018	4.540	1.5129270 121	4.590	1.5238800 241
4.441	1.4908795 763	4.491	1.5020753 941	4.541	1.5131472 521	4.591	1.5240978 653
4.442	1.4911047 255	4.492	1.5022980 369	4.542	1.5133674 437	4.592	1.5243156 590
4.443	1.4913298 240	4.493	1.5025206 301	4.543	1.5135875 868	4.593	1.5245334 053
4.444	1.4915548 718	4.494	1.5027431 738	4.544	1.5138076 814	4.594	1.5247511 043
4.445	1.4917798 690	4.495	1.5029656 679	4.545	1.5140277 276	4.595	1.5249687 558
4.446	1.4920048 155	4.496	1.5031881 126	4.546	1.5142477 254	4.596	1.5251863 600
4.447	1.4922297 115	4.497	1.5034105 078	4.547	1.5144676 748	4.597	1.5254039 168
4.448	1.4924545 569	4.498	1.5036328 535	4.548	1.5146875 759	4.598	1.5256214 263
4.449	1.4926793 518	4.499	1.5038551 499	4.549	1.5149074 286	4.599	1.5258388 886

TEN PLACE MANTISSAS FOR NATURAL LOGARITHMS (Continued)

x	$\log_e x$	x	$\log_e x$	x	$\log_e x$	x	$\log_e x$
4.600	1.5260563 035	4.650	1.5368672 196	4.700	1.5475625 087	4.750	1.5581446 180
4.601	1.5262736 712	4.651	1.5370822 502	4.701	1.5477752 520	4.751	1.5583551 222
4.602	1.5264909 916	4.652	1.5372972 347	4.702	1.5479879 501	4.752	1.5585655 821
4.603	1.5267082 648	4.653	1.5375121 729	4.703	1.5482006 030	4.753	1.5587759 976
4.604	1.5269254 909	4.654	1.5377270 649	4.704	1.5484132 106	4.754	1.5589863 689
4.605	1.5271426 697	4.655	1.5379419 107	4.705	1.5486257 730	4.755	1.5591966 960
4.606	1.5273598 014	4.656	1.5381567 104	4.706	1.5488382 903	4.756	1.5594069 788
4.607	1.5275768 860	4.657	1.5383714 640	4.707	1.5490507 624	4.757	1.5596172 175
4.608	1.5277939 234	4.658	1.5385861 715	4.708	1.5492631 894	4.758	1.5598274 119
4.609	1.5280109 137	4.659	1.5388008 328	4.709	1.5494755 713	4.759	1.5600375 621
4.610	1.5282278 570	4.660	1.5390154 481	4.710	1.5496879 080	4.760	1.5602476 682
4.611	1.5284447 532	4.661	1.5392300 174	4.711	1.5499001 997	4.761	1.5604577 302
4.612	1.5286616 024	4.662	1.5394445 406	4.712	1.5501124 463	4.762	1.5606677 481
4.613	1.5288784 046	4.663	1.5396590 178	4.713	1.5503246 479	4.763	1.5608777 218
4.614	1.5290951 598	4.664	1.5398734 490	4.714	1.5505368 045	4.764	1.5610876 515
4.615	1.5293118 680	4.665	1.5400878 343	4.715	1.5507489 161	4.765	1.5612975 371
4.616	1.5295285 292	4.666	1.5403021 736	4.716	1.5509609 827	4.766	1.5615073 787
4.617	1.5297451 435	4.667	1.5405164 670	4.717	1.5511730 043	4.767	1.5617171 762
4.618	1.5299617 109	4.668	1.5407307 144	4.718	1.5513849 810	4.768	1.5619269 298
4.619	1.5301782 314	4.669	1.5409449 160	4.719	1.5515969 127	4.769	1.5621366 393
4.620	1.5303947 051	4.670	1.5411590 717	4.720	1.5518087 996	4.770	1.5623463 049
4.621	1.5306111 319	4.671	1.5413731 815	4.721	1.5520206 416	4.771	1.5625559 265
4.622	1.5308275 119	4.672	1.5415872 455	4.722	1.5522324 387	4.772	1.5627655 042
4.623	1.5310438 450	4.673	1.5418012 637	4.723	1.5524441 909	4.773	1.5629750 380
4.624	1.5312601 314	4.674	1.5420152 361	4.724	1.5526558 983	4.774	1.5631845 279
4.625	1.5314763 710	4.675	1.5422291 627	4.725	1.5528675 609	4.775	1.5633939 739
4.626	1.5316925 638	4.676	1.5424430 436	4.726	1.5530791 788	4.776	1.5636033 761
4.627	1.5319087 099	4.677	1.5426568 787	4.727	1.5532907 518	4.777	1.5638127 344
4.628	1.5321248 093	4.678	1.5428706 682	4.728	1.5535022 801	4.778	1.5640220 489
4.629	1.5323408 620	4.679	1.5430844 119	4.729	1.5537137 637	4.779	1.5642313 196
4.630	1.5325568 681	4.680	1.5432981 099	4.730	1.5539252 025	4.780	1.5644405 465
4.631	1.5327728 275	4.681	1.5435117 623	4.731	1.5541365 966	4.781	1.5646497 296
4.632	1.5329887 403	4.682	1.5437253 691	4.732	1.5543479 461	4.782	1.5648588 690
4.633	1.5332046 064	4.683	1.5439389 302	4.733	1.5545592 509	4.783	1.5650679 647
4.634	1.5334204 260	4.684	1.5441524 457	4.734	1.5547705 111	4.784	1.5652770 166
4.635	1.5336361 990	4.685	1.5443659 157	4.735	1.5549817 266	4.785	1.5654860 249
4.636	1.5338519 255	4.686	1.5445793 401	4.736	1.5551928 976	4.786	1.5656949 895
4.637	1.5340676 054	4.687	1.5447927 189	4.737	1.5554040 239	4.787	1.5659039 104
4.638	1.5342832 388	4.688	1.5450060 523	4.738	1.5556151 057	4.788	1.5661127 877
4.639	1.5344988 258	4.689	1.5452193 401	4.739	1.5558261 430	4.789	1.5663216 214
4.640	1.5347143 662	4.690	1.5454325 825	4.740	1.5560371 357	4.790	1.5665304 114
4.641	1.5349298 603	4.691	1.5456457 793	4.741	1.5562480 839	4.791	1.5667391 579
4.642	1.5351453 079	4.692	1.5458589 308	4.742	1.5564589 876	4.792	1.5669478 608
4.643	1.5353607 090	4.693	1.5460720 368	4.743	1.5566698 469	4.793	1.5671565 202
4.644	1.5355760 638	4.694	1.5462850 974	4.744	1.5568806 617	4.794	1.5673651 360
4.645	1.5357913 723	4.695	1.5464981 127	4.745	1.5570914 321	4.795	1.5675737 083
4.646	1.5360066 343	4.696	1.5467110 825	4.746	1.5573021 580	4.796	1.5677822 372
4.647	1.5362218 501	4.697	1.5469240 070	4.747	1.5575128 396	4.797	1.5679907 225
4.648	1.5364370 195	4.698	1.5471368 862	4.748	1.5577234 767	4.798	1.5681991 644
4.649	1.5366521 427	4.699	1.5473497 201	4.749	1.5579340 696	4.799	1.5684075 629

TEN PLACE MANTISSAS FOR NATURAL LOGARITHMS (Continued)

x	$\log_e x$	x	$\log_e$	x	$\log_e x$	x	$\log_e x$
4.800	1.5686159 179	4.850	1.5789787 049	4.900	1.5892352 051	4.950	1.5993875 766
4.801	1.5688242 295	4.851	1.5791848 693	4.901	1.5894392 659	4.951	1.5995895 764
4.802	1.5690324 978	4.852	1.5793909 911	4.902	1.5896432 851	4.952	1.5997915 354
4.803	1.5692407 227	4.853	1.5795970 704	4.903	1.5898472 627	4.953	1.5999934 536
4.804	1.5694489 042	4.854	1.5798031 073	4.904	1.5900511 986	4.954	1.6001953 311
4.805	1.5696570 424	4.855	1.5800091 017	4.905	1.5902550 930	4.955	1.6003971 678
4.806	1.5698651 373	4.856	1.5802150 538	4.906	1.5904589 458	4.956	1.6005989 638
4.807	1.5700731 889	4.857	1.5804209 634	4.907	1.5906627 571	4.957	1.6008007 190
4.808	1.5702811 972	4.858	1.5806268 306	4.908	1.5908665 268	4.958	1.6010024 336
4.809	1.5704891 623	4.859	1.5808326 554	4.909	1.5910702 551	4.959	1.6012041 075
4.810	1.5706970 841	4.860	1.5810384 379	4.910	1.5912739 418	4.960	1.6014057 407
4.811	1.5709049 627	4.861	1.5812441 781	4.911	1.5914775 871	4.961	1.6016073 333
4.812	1.5711127 981	4.862	1.5814498 759	4.912	1.5916811 908	4.962	1.6018088 853
4.813	1.5713205 903	4.863	1.5816555 314	4.913	1.5918847 532	4.963	1.6020103 966
4.814	1.5715283 394	4.864	1.5818611 447	4.914	1.5920882 741	4.964	1.6022118 673
4.815	1.5717360 453	4.865	1.5820667 156	4.915	1.5922917 536	4.965	1.6024132 975
4.816	1.5719437 080	4.866	1.5822722 444	4.916	1.5924951 917	4.966	1.6026146 871
4.817	1.5721513 276	4.867	1.5824777 309	4.917	1.5926985 884	4.967	1.6028160 361
4.818	1.5723589 042	4.868	1.5826831 751	4.918	1.5929019 438	4.968	1.6030173 446
4.819	1.5725664 377	4.869	1.5828885 772	4.919	1.5931052 578	4.969	1.6032186 126
4.820	1.5727739 281	4.870	1.5830939 371	4.920	1.5933085 305	4.970	1.6034198 401
4.821	1.5729813 754	4.871	1.5832992 548	4.921	1.5935117 619	4.971	1.6036210 271
4.822	1.5731887 798	4.872	1.5835045 304	4.922	1.5937149 520	4.972	1.6038221 736
4.823	1.5733961 411	4.873	1.5837097 639	4.923	1.5939181 008	4.973	1.6040232 797
4.824	1.5736034 594	4.874	1.5839149 552	4.924	1.5941212 083	4.974	1.6042243 454
4.825	1.5738107 348	4.875	1.5841201 044	4.925	1.5943242 746	4.975	1.6044253 706
4.826	1.5740179 672	4.876	1.5843252 116	4.926	1.5945272 997	4.976	1.6046263 554
4.827	1.5742251 567	4.877	1.5845302 767	4.927	1.5947302 836	4.977	1.6048272 999
4.828	1.5744323 032	4.878	1.5847352 998	4.928	1.5949332 262	4.978	1.6050282 039
4.829	1.5746394 069	4.879	1.5849402 808	4.929	1.5951361 277	4.979	1.6052290 677
4.830	1.5748464 677	4.880	1.5851452 199	4.930	1.5953389 881	4.980	1.6054298 910
4.831	1.5750534 856	4.881	1.5853501 169	4.931	1.5955418 072	4.981	1.6056306 741
4.832	1.5752604 606	4.882	1.5855549 720	4.932	1.5957445 853	4.982	1.6058314 168
4.833	1.5754673 929	4.883	1.5857597 851	4.933	1.5959473 223	4.983	1.6060321 193
4.834	1.5756742 823	4.884	1.5859645 562	4.934	1.5961500 181	4.984	1.6062327 815
4.835	1.5758811 289	4.885	1.5861692 855	4.935	1.5963526 729	4.985	1.6064334 034
4.836	1.5760879 328	4.886	1.5863739 728	4.936	1.5965552 866	4.986	1.6066339 851
4.837	1.5762946 938	4.887	1.5865786 183	4.937	1.5967578 593	4.987	1.6068345 266
4.838	1.5765014 122	4.888	1.5867832 219	4.938	1.5969603 909	4.988	1.6070350 278
4.839	1.5767080 878	4.889	1.5869877 836	4.939	1.5971628 816	4.989	1.6072354 889
4.840	1.5769147 207	4.890	1.5871923 035	4.940	1.5973653 312	4.990	1.6074359 098
4.841	1.5771213 110	4.891	1.5873967 816	4.941	1.5975677 399	4.991	1.6076362 905
4.842	1.5773278 585	4.892	1.5876012 178	4.942	1.5977701 076	4.992	1.6078366 311
4.843	1.5775343 634	4.893	1.5878056 123	4.943	1.5979724 343	4.993	1.6080369 315
4.844	1.5777408 257	4.894	1.5880099 650	4.944	1.5981747 202	4.994	1.6082371 919
4.845	1.5779472 453	4.895	1.5882142 760	4.945	1.5983769 651	4.995	1.6084374 121
4.846	1.5781536 224	4.896	1.5884185 452	4.946	1.5985791 691	4.996	1.6086375 923
4.847	1.5783599 569	4.897	1.5886227 727	4.947	1.5987813 322	4.997	1.6088377 324
4.848	1.5785662 488	4.898	1.5888269 585	4.948	1.5989834 545	4.998	1.6090378 324
4.849	1.5787724 981	4.899	1.5890311 027	4.949	1.5991855 360	4.999	1.6092378 924

TEN PLACE MANTISSAS FOR NATURAL LOGARITHMS (Continued)

x	$\log_e x$	x	$\log_e x$	x	$\log_e x$	x	$\log_e x$
5.000	1.6094379 124	5.050	1.6193882 433	5.100	1.6292405 397	5.150	1.6389967 147
5.001	1.6096378 924	5.051	1.6195862 435	5.101	1.6294365 989	5.151	1.6391908 706
5.002	1.6098378 325	5.052	1.6197842 045	5.102	1.6296326 197	5.152	1.6393849 888
5.003	1.6100377 325	5.053	1.6199821 263	5.103	1.6298286 021	5.153	1.6395790 693
5.004	1.6102375 926	5.054	1.6201800 090	5.104	1.6300245 460	5.154	1.6397731 122
5.005	1.6104374 128	5.055	1.6203778 525	5.105	1.6302204 516	5.155	1.6399671 175
5.006	1.6106371 930	5.056	1.6205756 568	5.106	1.6304163 188	5.156	1.6401610 851
5.007	1.6108369 333	5.057	1.6207734 221	5.107	1.6306121 477	5.157	1.6403550 151
5.008	1.6110366 338	5.058	1.6209711 482	5.108	1.6308079 382	5.158	1.6405489 075
5.009	1.6112362 944	5.059	1.6211688 353	5.109	1.6310036 903	5.159	1.6407427 623
5.010	1.6114359 151	5.060	1.6213664 833	5.110	1.6311994 042	5.160	1.6409365 795
5.011	1.6116354 960	5.061	1.6215640 922	5.111	1.6313950 798	5.161	1.6411303 592
5.012	1.6118350 370	5.062	1.6217616 621	5.112	1.6315907 171	5.162	1.6413241 013
5.013	1.6120345 383	5.063	1.6219591 930	5.113	1.6317863 161	5.163	1.6415178 059
5.014	1.6122339 997	5.064	1.6221566 848	5.114	1.6319818 769	5.164	1.6417114 730
5.015	1.6124334 214	5.065	1.6223541 377	5.115	1.6321773 994	5.165	1.6419051 026
5.016	1.6126328 033	5.066	1.6225515 516	5.116	1.6323728 837	5.166	1.6420986 947
5.017	1.6128321 455	5.067	1.6227489 265	5.117	1.6325683 298	5.167	1.6422922 493
5.018	1.6130314 479	5.068	1.6229462 625	5.118	1.6327637 377	5.168	1.6424857 665
5.019	1.6132307 107	5.069	1.6231435 595	5.119	1.6329591 075	5.169	1.6426792 462
5.020	1.6134299 337	5.070	1.6233408 176	5.120	1.6331544 391	5.170	1.6428726 885
5.021	1.6136291 171	5.071	1.6235380 368	5.121	1.6333497 325	5.171	1.6430660 934
5.022	1.6138282 607	5.072	1.6237352 171	5.122	1.6335449 878	5.172	1.6432594 609
5.023	1.6140273 648	5.073	1.6239323 586	5.123	1.6337402 050	5.173	1.6434527 910
5.024	1.6142264 292	5.074	1.6241294 612	5.124	1.6339353 840	5.174	1.6436460 838
5.025	1.6144254 539	5.075	1.6243265 249	5.125	1.6341305 250	5.175	1.6438393 392
5.026	1.6146244 391	5.076	1.6245235 499	5.126	1.6343256 279	5.176	1.6440325 572
5.027	1.6148233 847	5.077	1.6247205 360	5.127	1.6345206 928	5.177	1.6442257 379
5.028	1.6150222 907	5.078	1.6249174 833	5.128	1.6347157 196	5.178	1.6444188 813
5.029	1.6152211 572	5.079	1.6251143 918	5.129	1.6349107 084	5.179	1.6446119 874
5.030	1.6154199 841	5.080	1.6253112 616	5.130	1.6351056 592	5.180	1.6448050 563
5.031	1.6156187 715	5.081	1.6255080 926	5.131	1.6353005 720	5.181	1.6449980 878
5.032	1.6158175 194	5.082	1.6257048 849	5.132	1.6354954 468	5.182	1.6451910 821
5.033	1.6160162 278	5.083	1.6259016 385	5.133	1.6356902 836	5.183	1.6453840 392
5.034	1.6162148 967	5.084	1.6260983 533	5.134	1.6358850 824	5.184	1.6455769 590
5.035	1.6164135 262	5.085	1.6262950 295	5.135	1.6360798 434	5.185	1.6457698 417
5.036	1.6166121 162	5.086	1.6264916 670	5.136	1.6362745 664	5.186	1.6459626 871
5.037	1.6168106 668	5.087	1.6266882 658	5.137	1.6364692 515	5.187	1.6461554 954
5.038	1.6170091 779	5.088	1.6268848 260	5.138	1.6366638 987	5.188	1.6463482 665
5.039	1.6172076 497	5.089	1.6270813 476	5.139	1.6368585 080	5.189	1.6465410 004
5.040	1.6174060 821	5.090	1.6272778 306	5.140	1.6370530 795	5.190	1.6467336 972
5.041	1.6176044 751	5.091	1.6274742 749	5.141	1.6372476 131	5.191	1.6469263 568
5.042	1.6178028 288	5.092	1.6276706 807	5.142	1.6374421 088	5.192	1.6471189 794
5.043	1.6180011 431	5.093	1.6278670 479	5.143	1.6376365 668	5.193	1.6473115 649
5.044	1.6181994 181	5.094	1.6280633 766	5.144	1.6378309 869	5.194	1.6475041 132
5.045	1.6183976 538	5.095	1.6282596 667	5.145	1.6380253 693	5.195	1.6476966 246
5.046	1.6185958 502	5.096	1.6284559 183	5.146	1.6382197 139	5.196	1.6478890 988
5.047	1.6187940 074	5.097	1.6286521 314	5.147	1.6384140 207	5.197	1.6480815 360
5.048	1.6189921 252	5.098	1.6288483 060	5.148	1.6386082 897	5.198	1.6482739 362
5.049	1.6191902 039	5.099	1.6290444 421	5.149	1.6388025 211	5.199	1.6484662 994

TEN PLACE MANTISSAS FOR NATURAL LOGARITHMS (Continued)

x	$\log_e x$	x	$\log_e x$	x	$\log_e x$	x	$\log_e x$
5.200	1.6486586 256	5.250	1.6582280 766	5.300	1.6677068 206	5.350	1.6770965 609
5.201	1.6488509 148	5.251	1.6584185 347	5.301	1.6678954 820	5.351	1.6772834 593
5.202	1.6490431 670	5.252	1.6586089 564	5.302	1.6680841 079	5.352	1.6774703 228
5.203	1.6492353 823	5.253	1.6587993 420	5.303	1.6682726 982	5.353	1.6776571 514
5.204	1.6494275 607	5.254	1.6589896 913	5.304	1.6684612 529	5.354	1.6778439 451
5.205	1.6496197 021	5.255	1.6591800 043	5.305	1.6686497 721	5.355	1.6780307 039
5.206	1.6498118 066	5.256	1.6593702 812	5.306	1.6688382 557	5.356	1.6782174 278
5.207	1.6500038 742	5.257	1.6595605 218	5.307	1.6690267 038	5.357	1.6784041 169
5.208	1.6501959 049	5.258	1.6597507 263	5.308	1.6692151 165	5.358	1.6785907 711
5.209	1.6503878 988	5.259	1.6599408 946	5.309	1.6694034 936	5.359	1.6787773 905
5.210	1.6505798 558	5.260	1.6601310 267	5.310	1.6695918 353	5.360	1.6789639 751
5.211	1.6507717 759	5.261	1.6603211 227	5.311	1.6697801 414	5.361	1.6791505 248
5.212	1.6509636 593	5.262	1.6605111 826	5.312	1.6699684 122	5.362	1.6793370 398
5.213	1.6511555 058	5.263	1.6607012 064	5.313	1.6701566 475	5.363	1.6795235 500
5.214	1.6513473 155	5.264	1.6608911 940	5.314	1.6703448 473	5.364	1.6797099 654
5.215	1.6515390 885	5.265	1.6610811 456	5.315	1.6705330 118	5.365	1.6798963 761
5.216	1.6517308 246	5.266	1.6612710 611	5.316	1.6707211 409	5.366	1.6800827 520
5.217	1.6519225 240	5.267	1.6614609 405	5.317	1.6709092 345	5.367	1.6802690 932
5.218	1.6521141 867	5.268	1.6616507 839	5.318	1.6710972 928	5.368	1.6804553 997
5.219	1.6523058 126	5.269	1.6618405 912	5.319	1.6712853 158	5.369	1.6806416 714
5.220	1.6524974 019	5.270	1.6620303 626	5.320	1.6714733 034	5.370	1.6808279 085
5.221	1.6526889 544	5.271	1.6622200 979	5.321	1.6716612 556	5.371	1.6810141 109
5.222	1.6528804 703	5.272	1.6624097 972	5.322	1.6718491 726	5.372	1.6812002 787
5.223	1.6530719 495	5.273	1.6625994 605	5.323	1.6720370 542	5.373	1.6813864 117
5.224	1.6532633 920	5.274	1.6627890 879	5.324	1.6722249 005	5.374	1.6815725 102
5.225	1.6534547 979	5.275	1.6629786 794	5.325	1.6724127 116	5.375	1.6817585 740
5.226	1.6536461 671	5.276	1.6631682 349	5.326	1.6726004 874	5.376	1.6819446 032
5.227	1.6538374 997	5.277	1.6633577 544	5.327	1.6727882 279	5.377	1.6821305 978
5.228	1.6540287 958	5.278	1.6635472 381	5.328	1.6729759 332	5.378	1.6823165 578
5.229	1.6542200 552	5.279	1.6637366 858	5.329	1.6731636 033	5.379	1.6825024 833
5.230	1.6544112 781	5.280	1.6639260 977	5.330	1.6733512 382	5.380	1.6826883 742
5.231	1.6546024 644	5.281	1.6641154 737	5.331	1.6735388 378	5.381	1.6828742 305
5.232	1.6547936 142	5.282	1.6643048 139	5.332	1.6737264 023	5.382	1.6830600 523
5.233	1.6549847 274	5.283	1.6644941 182	5.333	1.6739139 316	5.383	1.6832458 396
5.234	1.6551758 041	5.284	1.6646833 867	5.334	1.6741014 258	5.384	1.6834315 923
5.235	1.6553668 443	5.285	1.6648726 193	5.335	1.6742888 848	5.385	1.6836173 106
5.236	1.6555578 480	5.286	1.6650618 162	5.336	1.6744763 086	5.386	1.6838029 944
5.237	1.6557488 153	5.287	1.6652509 773	5.337	1.6746636 974	5.387	1.6839886 437
5.238	1.6559397 461	5.288	1.6654401 025	5.338	1.6748510 510	5.388	1.6841742 585
5.239	1.6561306 404	5.289	1.6656291 921	5.339	1.6750383 695	5.389	1.6843598 390
5.240	1.6563214 983	5.290	1.6658182 459	5.340	1.6752256 530	5.390	1.6845453 849
5.241	1.6565123 198	5.291	1.6660072 639	5.341	1.6754129 014	5.391	1.6847308 965
5.242	1.6567031 049	5.292	1.6661962 463	5.342	1.6756001 147	5.392	1.6849163 736
5.243	1.6568938 536	5.293	1.6663851 929	5.343	1.6757872 930	5.393	1.6851018 164
5.244	1.6570845 659	5.294	1.6665741 038	5.344	1.6759744 362	5.394	1.6852872 247
5.245	1.6572752 418	5.295	1.6667629 791	5.345	1.6761615 445	5.395	1.6854725 987
5.246	1.6574658 814	5.296	1.6669518 186	5.346	1.6763486 177	5.396	1.6856579 383
5.247	1.6576564 847	5.297	1.6671406 226	5.347	1.6765356 560	5.397	1.6858432 436
5.248	1.6578470 516	5.298	1.6673293 908	5.348	1.6767226 592	5.398	1.6860285 146
5.249	1.6580375 823	5.299	1.6675181 235	5.349	1.6769096 275	5.399	1.6862137 512

TEN PLACE MANTISSAS FOR NATURAL LOGARITHMS (Continued)

x	$\log_e x$	x	$\log_e x$	x	$\log_e x$	x	$\log_e x$
5.400	1.6863989 536	5.450	1.6956156 087	5.500	1.7047480 922	5.550	1.7137979 278
5.401	1.6865841 216	5.451	1.6957990 781	5.501	1.7049298 939	5.551	1.7139780 917
5.402	1.6867692 554	5.452	1.6959825 138	5.502	1.7051116 625	5.552	1.7141582 232
5.403	1.6869543 549	5.453	1.6961659 159	5.503	1.7052933 981	5.553	1.7143383 223
5.404	1.6871394 201	5.454	1.6963492 844	5.504	1.7054751 006	5.554	1.7145183 889
5.405	1.6873244 511	5.455	1.6965326 193	5.505	1.7056567 702	5.555	1.7146984 231
5.406	1.6875094 479	5.456	1.6967159 205	5.506	1.7058384 067	5.556	1.7148784 249
5.407	1.6876944 104	5.457	1.6968991 882	5.507	1.7060200 103	5.557	1.7150583 943
5.408	1.6878793 387	5.458	1.6970824 223	5.508	1.7062015 809	5.558	1.7152383 313
5.409	1.6880642 329	5.459	1.6972656 228	5.509	1.7063831 185	5.559	1.7154182 360
5.410	1.6882490 929	5.460	1.6974487 898	5.510	1.7065646 232	5.560	1.7155981 083
5.411	1.6884339 187	5.461	1.6976319 232	5.511	1.7067460 949	5.561	1.7157779 482
5.412	1.6886187 103	5.462	1.6978150 231	5.512	1.7069275 337	5.562	1.7159577 558
5.413	1.6888034 678	5.463	1.6979980 894	5.513	1.7071089 396	5.563	1.7161375 311
5.414	1.6889881 912	5.464	1.6981811 223	5.514	1.7072903 126	5.564	1.7163172 741
5.415	1.6891728 805	5.465	1.6983641 216	5.515	1.7074716 527	5.565	1.7164969 847
5.416	1.6893575 356	5.466	1.6985470 875	5.516	1.7076529 599	5.566	1.7166766 631
5.417	1.6895421 567	5.467	1.6987300 199	5.517	1.7078342 343	5.567	1.7168563 092
5.418	1.6897267 437	5.468	1.6989129 189	5.518	1.7080154 758	5.568	1.7170359 230
5.419	1.6899112 966	5.469	1.6990957 844	5.519	1.7081966 845	5.569	1.7172155 046
5.420	1.6900958 155	5.470	1.6992786 164	5.520	1.7083778 603	5.570	1.7173950 539
5.421	1.6902803 003	5.471	1.6994614 151	5.521	1.7085590 033	5.571	1.7175745 710
5.422	1.6904647 511	5.472	1.6996441 803	5.522	1.7087401 135	5.572	1.7177540 559
5.423	1.6906491 679	5.473	1.6998269 122	5.523	1.7089211 909	5.573	1.7179335 086
5.424	1.6908335 506	5.474	1.7000096 106	5.524	1.7091022 355	5.574	1.7181129 291
5.425	1.6910178 994	5.475	1.7001922 757	5.525	1.7092832 474	5.575	1.7182923 173
5.426	1.6912022 142	5.476	1.7003749 074	5.526	1.7094642 265	5.576	1.7184716 735
5.427	1.6913864 951	5.477	1.7005575 058	5.527	1.7096451 729	5.577	1.7186509 974
5.428	1.6915707 420	5.478	1.7007400 708	5.528	1.7098260 865	5.578	1.7188302 892
5.429	1.6917549 549	5.479	1.7009226 026	5.529	1.7100069 674	5.579	1.7190095 489
5.430	1.6919391 339	5.480	1.7011051 010	5.530	1.7101878 155	5.580	1.7191887 764
5.431	1.6921232 791	5.481	1.7012875 661	5.531	1.7103686 310	5.581	1.7193679 718
5.432	1.6923073 903	5.482	1.7014699 979	5.532	1.7105494 138	5.582	1.7195471 351
5.433	1.6924914 676	5.483	1.7016523 964	5.533	1.7107301 639	5.583	1.7197262 663
5.434	1.6926755 110	5.484	1.7018347 617	5.534	1.7109108 814	5.584	1.7199053 655
5.435	1.6928595 206	5.485	1.7020170 937	5.535	1.7110915 662	5.585	1.7200844 325
5.436	1.6930434 963	5.486	1.7021993 925	5.536	1.7112722 183	5.586	1.7202634 675
5.437	1.6932274 382	5.487	1.7023816 581	5.537	1.7114528 378	5.587	1.7204424 705
5.438	1.6934113 462	5.488	1.7025638 904	5.538	1.7116334 247	5.588	1.7206214 414
5.439	1.6935952 204	5.489	1.7027460 896	5.539	1.7118139 791	5.589	1.7208003 803
5.440	1.6937790 609	5.490	1.7029282 555	5.540	1.7119945 008	5.590	1.7209792 872
5.441	1.6939628 675	5.491	1.7031103 883	5.541	1.7121749 899	5.591	1.7211581 620
5.442	1.6941466 404	5.492	1.7032924 879	5.542	1.7123554 464	5.592	1.7213370 049
5.443	1.6943303 795	5.493	1.7034745 544	5.543	1.7125358 704	5.593	1.7215158 158
5.444	1.6945140 848	5.494	1.7036565 877	5.544	1.7127162 619	5.594	1.7216945 948
5.445	1.6946977 564	5.495	1.7038385 879	5.545	1.7128966 208	5.595	1.7218733 418
5.446	1.6948813 943	5.496	1.7040205 549	5.546	1.7130769 472	5.596	1.7220520 568
5.447	1.6950649 984	5.497	1.7042024 889	5.547	1.7132572 411	5.597	1.7222307 399
5.448	1.6952485 688	5.498	1.7043843 897	5.548	1.7134375 025	5.598	1.7224093 911
5.449	1.6954321 056	5.499	1.7045662 575	5.549	1.7136177 313	5.599	1.7225880 104

TEN PLACE MANTISSAS FOR NATURAL LOGARITHMS (Continued)

x	$\log_e x$	x	$\log_e x$	x	$\log_e x$	x	$\log_e x$
5.600	1.7227665 977	5.650	1.7316555 452	5.700	1.7404661 748	5.750	1.7491998 548
5.601	1.7229451 532	5.651	1.7318325 206	5.701	1.7406415 980	5.751	1.7493737 527
5.602	1.7231236 768	5.652	1.7320094 648	5.702	1.7408169 905	5.752	1.7495476 204
5.603	1.7233021 686	5.653	1.7321863 777	5.703	1.7409923 522	5.753	1.7497214 579
5.604	1.7234806 285	5.654	1.7323632 593	5.704	1.7411676 831	5.754	1.7498952 651
5.605	1.7236590 565	5.655	1.7325401 096	5.705	1.7413429 833	5.755	1.7500690 422
5.606	1.7238374 527	5.656	1.7327169 286	5.706	1.7415182 528	5.756	1.7502427 890
5.607	1.7240158 171	5.657	1.7328937 164	5.707	1.7416934 916	5.757	1.7504165 057
5.608	1.7241941 497	5.658	1.7330704 729	5.708	1.7418686 996	5.758	1.7505901 922
5.609	1.7243724 505	5.659	1.7332471 982	5.709	1.7420438 770	5.759	1.7507638 485
5.610	1.7245507 195	5.660	1.7334238 922	5.710	1.7422190 237	5.760	1.7509374 747
5.611	1.7247289 568	5.661	1.7336005 551	5.711	1.7423941 397	5.761	1.7511110 708
5.612	1.7249071 622	5.662	1.7337771 867	5.712	1.7425692 251	5.762	1.7512846 367
5.613	1.7250853 360	5.663	1.7339537 871	5.713	1.7427442 797	5.763	1.7514581 725
5.614	1.7252634 779	5.664	1.7341303 564	5.714	1.7429193 038	5.764	1.7516316 781
5.615	1.7254415 882	5.665	1.7343068 945	5.715	1.7430942 972	5.765	1.7518051 537
5.616	1.7256196 667	5.666	1.7344834 014	5.716	1.7432692 601	5.766	1.7519785 992
5.617	1.7257977 136	5.667	1.7346598 772	5.717	1.7434441 923	5.767	1.7521520 146
5.618	1.7259757 287	5.668	1.7348363 218	5.718	1.7436190 939	5.768	1.7523254 000
5.619	1.7261537 121	5.669	1.7350127 353	5.719	1.7437939 649	5.769	1.7524987 553
5.620	1.7263316 639	5.670	1.7351891 177	5.720	1.7439688 054	5.770	1.7526720 805
5.621	1.7265095 840	5.671	1.7353654 690	5.721	1.7441436 153	5.771	1.7528453 757
5.622	1.7266874 725	5.672	1.7355417 892	5.722	1.7443183 946	5.772	1.7530186 409
5.623	1.7268653 293	5.673	1.7357180 783	5.723	1.7444931 434	5.773	1.7531918 761
5.624	1.7270431 545	5.674	1.7358943 364	5.724	1.7446678 617	5.774	1.7533650 812
5.625	1.7272209 481	5.675	1.7360705 634	5.725	1.7448425 494	5.775	1.7535382 564
5.626	1.7273987 101	5.676	1.7362467 593	5.726	1.7450172 067	5.776	1.7537114 016
5.627	1.7275764 405	5.677	1.7364229 242	5.727	1.7451918 334	5.777	1.7538845 168
5.628	1.7277541 393	5.678	1.7365990 581	5.728	1.7453664 297	5.778	1.7540576 020
5.629	1.7279318 065	5.679	1.7367751 609	5.729	1.7455409 954	5.779	1.7542306 573
5.630	1.7281094 422	5.680	1.7369512 327	5.730	1.7457155 307	5.780	1.7544036 827
5.631	1.7282870 463	5.681	1.7371272 736	5.731	1.7458900 356	5.781	1.7545766 781
5.632	1.7284646 189	5.682	1.7373032 834	5.732	1.7460645 100	5.782	1.7547496 436
5.633	1.7286421 599	5.683	1.7374792 623	5.733	1.7462389 539	5.783	1.7549225 792
5.634	1.7288196 695	5.684	1.7376552 102	5.734	1.7464133 675	5.784	1.7550954 849
5.635	1.7289971 475	5.685	1.7378311 272	5.735	1.7465877 506	5.785	1.7552683 606
5.636	1.7291745 940	5.686	1.7380070 132	5.736	1.7467621 033	5.786	1.7554412 066
5.637	1.7293520 091	5.687	1.7381828 683	5.737	1.7469364 256	5.787	1.7556140 226
5.638	1.7295293 927	5.688	1.7383586 925	5.738	1.7471107 176	5.788	1.7557868 088
5.639	1.7297067 448	5.689	1.7385344 858	5.739	1.7472849 791	5.789	1.7559595 651
5.640	1.7298840 655	5.690	1.7387102 481	5.740	1.7474592 103	5.790	1.7561322 916
5.641	1.7300613 548	5.691	1.7388859 796	5.741	1.7476334 112	5.791	1.7563049 882
5.642	1.7302386 126	5.692	1.7390616 802	5.742	1.7478075 817	5.792	1.7564776 551
5.643	1.7304158 390	5.693	1.7392373 500	5.743	1.7479817 219	5.793	1.7566502 921
5.644	1.7305930 340	5.694	1.7394129 889	5.744	1.7481558 317	5.794	1.7568228 993
5.645	1.7307701 976	5.695	1.7395885 969	5.745	1.7483299 113	5.795	1.7569954 768
5.646	1.7309473 298	5.696	1.7397641 741	5.746	1.7485039 606	5.796	1.7571680 245
5.647	1.7311244 307	5.697	1.7399397 205	5.747	1.7486779 795	5.797	1.7573405 424
5.648	1.7313015 022	5.698	1.7401152 361	5.748	1.7488519 682	5.798	1.7575130 305
5.649	1.7314785 383	5.699	1.7402907 209	5.749	1.7490259 266	5.799	1.7576854 889

TEN PLACE MANTISSAS FOR NATURAL LOGARITHMS (Continued)

x	$\log_e x$	x	$\log_e x$	x	$\log_e x$	x	$\log_e x$
5.800	1.7578579 176	5.850	1.7664416 612	5.900	1.7749523 509	5.950	1.7833912 196
5.801	1.7580303 165	5.851	1.7666125 868	5.901	1.7751218 281	5.951	1.7835592 727
5.802	1.7582026 857	5.852	1.7667834 832	5.902	1.7752912 765	5.952	1.7837272 975
5.803	1.7583750 252	5.853	1.7669543 503	5.903	1.7754606 963	5.953	1.7838952 942
5.804	1.7585473 350	5.854	1.7671251 883	5.904	1.7756300 873	5.954	1.7840632 626
5.805	1.7587196 151	5.855	1.7672959 970	5.905	1.7757994 496	5.955	1.7842312 028
5.806	1.7588918 656	5.856	1.7674667 767	5.906	1.7759687 833	5.956	1.7843991 148
5.807	1.7590640 864	5.857	1.7676375 271	5.907	1.7761380 883	5.957	1.7845669 986
5.808	1.7592362 775	5.858	1.7678082 484	5.908	1.7763073 647	5.958	1.7847348 543
5.809	1.7594084 390	5.859	1.7679789 406	5.909	1.7764766 124	5.959	1.7849026 818
5.810	1.7595805 709	5.860	1.7681496 036	5.910	1.7766458 314	5.960	1.7850704 811
5.811	1.7597526 731	5.861	1.7683202 375	5.911	1.7768150 218	5.961	1.7852382 522
5.812	1.7599247 457	5.862	1.7684908 423	5.912	1.7769841 836	5.962	1.7854059 953
5.813	1.7600967 887	5.863	1.7686614 180	5.913	1.7771533 168	5.963	1.7855737 101
5.814	1.7602688 021	5.864	1.7688319 646	5.914	1.7773224 214	5.964	1.7857413 969
5.815	1.7604407 860	5.865	1.7690024 821	5.915	1.7774914 974	5.965	1.7859090 555
5.816	1.7606127 402	5.866	1.7691729 706	5.916	1.7776605 448	5.966	1.7860766 861
5.817	1.7607846 649	5.867	1.7693434 299	5.917	1.7778295 637	5.967	1.7862442 885
5.818	1.7609565 601	5.868	1.7695138 603	5.918	1.7779985 540	5.968	1.7864118 629
5.819	1.7611284 257	5.869	1.7696842 616	5.919	1.7781675 157	5.969	1.7865794 092
5.820	1.7613002 617	5.870	1.7698546 338	5.920	1.7783364 489	5.970	1.7867469 274
5.821	1.7614720 683	5.871	1.7700249 771	5.921	1.7785053 535	5.971	1.7869144 176
5.822	1.7616438 453	5.872	1.7701952 913	5.922	1.7786742 297	5.972	1.7870818 797
5.823	1.7618155 929	5.873	1.7703655 765	5.923	1.7788430 773	5.973	1.7872493 138
5.824	1.7619873 109	5.874	1.7705358 328	5.924	1.7790118 964	5.974	1.7874167 198
5.825	1.7621589 995	5.875	1.7707060 600	5.925	1.7791806 870	5.975	1.7875840 978
5.826	1.7623306 585	5.876	1.7708762 583	5.926	1.7793494 492	5.976	1.7877514 478
5.827	1.7625022 882	5.877	1.7710464 276	5.927	1.7795181 828	5.977	1.7879187 698
5.828	1.7626738 883	5.878	1.7712165 680	9.928	1.7796868 880	5.978	1.7880860 639
5.829	1.7628454 591	5.879	1.7713866 794	5.929	1.7798555 647	5.979	1.7882533 299
5.830	1.7630170 004	5.880	1.7715567 619	5.930	1.7800242 130	5.980	1.7884205 680
5.831	1.7631885 122	5.881	1.7717268 155	5.931	1.7801928 329	5.981	1.7885877 781
5.832	1.7633599 947	5.882	1.7718968 401	5.932	1.7803614 243	5.982	1.7887549 602
5.833	1.7635314 478	5.883	1.7720668 359	5.933	1.7805299 873	5.983	1.7889221 144
5.834	1.7637028 714	5.884	1.7722368 027	5.934	1.7806985 219	5.984	1.7890892 407
5.835	1.7638742 657	5.885	1.7724067 407	5.935	1.7808670 281	5.985	1.7892563 390
5.836	1.7640456 307	5.886	1.7725766 498	5.936	1.7810355 059	5.986	1.7894234 094
5.837	1.7642169 662	5.887	1.7727465 300	5.937	1.7812039 553	5.987	1.7895904 519
5.838	1.7643882 724	5.888	1.7729163 814	5.938	1.7813723 763	5.988	1.7897574 666
5.839	1.7645595 493	5.889	1.7730862 040	5.939	1.7815407 690	5.989	1.7899244 533
5.840	1.7647307 968	5.890	1.7732559 977	5.940	1.7817091 334	5.990	1.7900914 121
5.841	1.7649020 151	5.891	1.7734257 625	5.941	1.7818774 694	5.991	1.7902583 431
5.842	1.7650732 040	5.892	1.7735954 986	5.942	1.7820457 770	5.992	1.7904252 462
5.843	1.7652443 636	5.893	1.7737652 059	5.943	1.7822140 564	5.993	1.7905921 215
5.844	1.7654154 939	5.894	1.7739348 843	5.944	1.7823823 074	5.994	1.7907589 689
5.845	1.7655865 949	5.895	1.7741045 340	5.945	1.7825505 301	5.995	1.7909257 885
5.846	1.7657576 667	5.896	1.7742741 549	5.946	1.7827187 246	5.996	1.7910925 802
5.847	1.7659287 092	5.897	1.7744437 470	5.947	1.7828868 907	5.997	1.7912593 442
5.848	1.7660997 224	5.898	1.7746133 104	5.948	1.7830550 286	5.998	1.7914260 803
5.849	1.7662707 065	5.899	1.7747828 450	5.949	1.7832231 382	5.999	1.7915927 887

TEN PLACE MANTISSAS FOR NATURAL LOGARITHMS (Continued)

x	$\log_e x$	x	$\log_e x$	x	$\log_e x$	x	$\log_e x$
6.000	1.7917594 692	6.050	1.8000582 720	6.100	1.8082887 712	6.150	1.8164520 818
6.001	1.7919261 220	5.051	1.8002235 476	6.101	1.8084526 922	6.151	1.8166146 702
6.002	1.7920927 470	6.052	1.8003887 959	6.102	1.8086165 863	6.152	1.8167772 322
6.003	1.7922593 443	6.053	1.8005540 169	6.103	1.8087804 536	6.153	1.8169397 678
6.004	1.7924259 138	6.054	1.8007192 106	6.104	1.8089442 940	6.154	1.8171022 769
6.005	1.7925924 555	6.055	1.8008843 770	6.105	1.8091081 076	6.155	1.8172647 596
6.006	1.7927589 696	6.056	1.8010495 161	6.106	1.8092718 943	6.156	1.8174272 160
6.007	1.7929254 559	6.057	1.8012146 280	6.107	1.8094356 542	6.157	1.8175896 459
6.008	1.7930919 145	6.058	1.8013797 126	6.108	1.8095993 874	6.158	1.8177520 495
6.009	1.7932583 454	6.059	1.8015447 700	6.109	1.8097630 937	6.159	1.8179144 267
6.010	1.7934247 485	6.060	1.8017098 001	6.110	1.8099267 732	6.160	1.8180767 775
6.011	1.7935911 241	6.061	1.8018748 030	6.111	1.8100904 259	6.161	1.8182391 020
6.012	1.7937574 719	6.062	1.8020397 786	6.112	1.8102540 519	6.162	1.8184014 002
6.013	1.7939237 921	6.063	1.8022047 271	6.113	1.8104176 510	6.163	1.8185636 720
6.014	1.7940900 846	6.064	1.8023696 483	6.114	1.8105812 235	6.164	1.8187259 175
6.015	1.7942563 494	6.065	1.8025345 424	6.115	1.8107447 691	6.165	1.8188881 366
6.016	1.7944225 866	6.066	1.8026994 093	6.116	1.8109082 881	6.166	1.8190503 295
6.017	1.7945887 962	6.067	1.8028642 490	6.117	1.8110717 803	6.167	1.8192124 960
6.018	1.7947549 782	6.068	1.8030290 615	6.118	1.8112352 457	6.168	1.8193746 363
6.019	1.7949211 326	6.069	1.8031938 469	6.119	1.8113986 845	6.169	1.8195367 502
6.020	1.7950872 593	6.070	1.8033586 051	6.120	1.8115620 965	6.170	1.8196988 379
6.021	1.7952533 585	6.071	1.8035233 361	6.121	1.8117254 819	6.171	1.8198608 993
6.022	1.7954194 301	6.072	1.8036880 401	6.122	1.8118888 405	6.172	1.8200229 345
6.023	1.7955854 741	6.073	1.8038527 169	6.123	1.8120521 725	6.173	1.8201849 434
6.024	1.7957514 905	6.074	1.8040173 666	6.124	1.8122154 778	6.174	1.8203469 261
6.025	1.7959174 794	6.075	1.8041819 892	6.125	1.8123787 564	6.175	1.8205088 825
6.026	1.7960834 407	6.076	1.8043465 847	6.126	1.8125420 084	6.176	1.8206708 127
6.027	1.7962493 745	6.077	1.8045111 532	6.127	1.8127052 337	6.177	1.8208327 167
6.028	1.7964152 808	6.078	1.8046756 945	6.128	1.8128684 324	6.178	1.8209945 945
6.029	1.7965811 595	6.079	1.8048402 088	6.129	1.8130316 045	6.179	1.8211564 461
6.030	1.7967470 107	6.080	1.8050046 960	6.130	1.8131947 499	6.180	1.8213182 715
6.031	1.7969128 345	6.081	1.8051691 561	6.131	1.8133578 688	6.181	1.8214800 707
6.032	1.7970786 307	6.082	1.8053335 893	6.132	1.8135209 610	6.182	1.8216418 437
6.033	1.7972443 995	6.083	1.8054979 953	6.133	1.8136840 266	6.183	1.8218035 906
6.034	1.7974101 407	6.084	1.8056623 744	6.134	1.8138470 657	6.184	1.8219653 113
6.035	1.7975758 545	6.085	1.8058267 264	6.135	1.8140100 782	6.185	1.8221270 058
6.036	1.7977415 409	6.086	1.8059910 515	6.136	1.8141730 641	6.186	1.8222886 743
6.037	1.7979071 998	6.087	1.8061553 495	6.137	1.8143360 234	6.187	1.8224503 165
6.038	1.7980728 313	6.088	1.8063196 206	6.138	1.8144989 562	6.188	1.8226119 327
6.039	1.7982384 353	6.089	1.8064838 646	6.139	1.8146618 624	6.189	1.8227735 228
6.040	1.7984040 119	6.090	1.8066480 817	6.140	1.8148247 422	6.190	1.8229350 867
6.041	1.7985695 612	6.091	1.8068122 719	6.141	1.8149875 953	6.191	1.8230966 245
6.042	1.7987350 830	6.092	1.8069764 350	6.142	1.8151504 220	6.192	1.8232581 363
6.043	1.7989005 774	6.093	1.8071405 713	6.143	1.8153132 222	6.193	1.8234196 220
6.044	1.7990660 444	6.094	1.8073046 806	6.144	1.8154759 958	6.194	1.8235810 816
6.045	1.7992314 841	6.095	1.8074687 629	6.145	1.8156387 430	6.195	1.8237425 151
6.046	1.7993968 964	6.096	1.8076328 184	6.146	1.8158014 637	6.196	1.8239039 226
6.047	1.7995622 813	6.097	1.8077968 469	6.147	1.8159641 579	6.197	1.8240653 040
6.048	1.7997276 389	6.098	1.8079608 486	6.148	1.8161268 257	6.198	1.8242266 594
6.049	1.7998929 691	6.099	1.8081248 233	6.149	1.8162894 670	6.199	1.8243879 887

TEN PLACE MANTISSAS FOR NATURAL LOGARITHMS (Continued)

x	$\log_e x$	x	$\log_e x$	x	$\log_e x$	x	$\log_e x$
6.200	1.8245492 921	6.250	1.8325814 637	6.300	1.8405496 334	6.350	1.8484548 129
6.201	1.8247105 694	6.251	1.8327414 509	6.301	1.8407083 510	6.351	1.8486122 808
6.202	1.8248718 207	6.252	1.8329014 126	6.302	1.8408670 433	6.352	1.8487697 239
6.203	1.8250330 460	6.253	1.8330613 486	6.303	1.8410257 105	6.353	1.8489271 423
6.204	1.8251942 453	6.254	1.8332212 590	6.304	1.8411843 526	6.354	1.8490845 358
6.205	1.8253554 187	6.255	1.8333811 439	6.305	1.8413429 694	6.355	1.8492419 046
6.206	1.8255165 660	6.256	1.8335410 032	6.306	1.8415015 612	6.356	1.8493992 487
6.207	1.8256776 874	6.257	1.8337008 370	6.307	1.8416601 277	6.357	1.8495565 680
6.208	1.8258387 829	6.258	1.8338606 452	6.308	1.8418186 691	6.358	1.8497138 625
6.209	1.8259998 524	6.259	1.8340204 279	6.309	1.8419771 854	6.359	1.8498711 323
6.210	1.8261608 959	6.260	1.8341801 851	6.310	1.8421356 766	6.360	1.8500283 774
6.211	1.8263219 136	6.261	1.8343399 168	6.311	1.8422941 426	6.361	1.8501855 977
6.212	1.8264829 053	6.262	1.8344996 229	6.312	1.8424525 835	6.362	1.8503427 933
6.213	1.8266438 711	6.263	1.8346593 035	6.313	1.8426109 994	6.363	1.8504999 643
6.214	1.8268048 110	6.264	1.8348189 587	6.314	1.8427693 901	6.364	1.8506571 105
6.215	1.8269657 250	6.265	1.8349785 883	6.315	1.8429277 558	6.365	1.8508142 320
6.216	1.8271266 131	6.266	1.8351381 925	6.316	1.8430860 964	6.366	1.8509713 289
6.217	1.8272874 753	6.267	1.8352977 712	6.317	1.8432444 119	6.367	1.8511284 010
6.218	1.8274483 116	6.268	1.8354573 245	6.318	1.8434027 024	6.368	1.8512854 485
6.219	1.8276091 221	6.269	1.8356168 523	6.319	1.8435609 678	6.369	1.8514424 714
6.220	1.8277699 068	6.270	1.8357763 546	6.320	1.8437192 082	6.370	1.8515994 696
6.221	1.8279306 655	6.271	1.8359358 316	6.321	1.8438774 235	6.371	1.8517564 431
6.222	1.8280913 985	6.272	1.8360952 830	6.322	1.8440356 138	6.372	1.8519133 920
6.223	1.8282521 056	6.273	1.8362547 091	6.323	1.8441937 791	6.373	1.8520703 163
6.224	1.8284127 869	6.274	1.8364141 098	6.324	1.8443519 193	6.374	1.8522272 160
6.225	1.8285734 424	6.275	1.8365734 850	6.325	1.8445100 346	6.375	1.8523840 910
6.226	1.8287340 720	6.276	1.8367328 349	6.326	1.8446681 249	6.376	1.8525409 415
6.227	1.8288946 759	6.277	1.8368921 593	6.327	1.8448261 902	6.377	1.8526977 673
6.228	1.8290552 540	6.278	1.8370514 584	6.328	1.8449842 305	6.378	1.8528545 686
6.229	1.8292158 063	6.279	1.8372107 321	6.329	1.8451422 458	6.379	1.8530113 453
6.230	1.8293763 328	6.280	1.8373699 805	6.330	1.8453002 362	6.380	1.8531680 974
6.231	1.8295368 336	6.281	1.8375292 035	6.331	1.8454582 016	6.381	1.8533248 249
6.232	1.8296973 086	6.282	1.8376884 011	6.332	1.8456161 420	6.382	1.8534815 279
6.233	1.8298577 578	6.283	1.8378475 734	6.333	1.8457740 575	6.383	1.8536382 063
6.234	1.8300181 813	6.284	1.8380067 204	6.334	1.8459319 481	6.384	1.8537948 601
6.235	1.8301785 791	6.285	1.8381658 420	6.335	1.8460898 138	6.385	1.8539514 895
6.236	1.8303389 512	6.286	1.8383249 384	6.336	1.8462476 545	6.386	1.8541080 943
6.237	1.8304992 975	6.287	1.8384840 094	6.337	1.8464054 703	6.387	1.8542646 746
6.238	1.8306596 182	6.288	1.8386430 551	6.338	1.8465632 613	6.388	1.8544212 304
6.239	1.8308199 131	6.289	1.8388020 756	6.339	1.8467210 273	6.389	1.8545777 616
6.240	1.8309801 824	6.290	1.8389610 707	6.340	1.8468787 684	6.390	1.8547342 684
6.241	1.8311404 260	6.291	1.8391200 406	6.341	1.8470364 847	6.391	1.8548907 507
6.242	1.8313006 438	6.292	1.8392789 852	6.342	1.8471941 761	6.392	1.8550472 085
6.243	1.8314608 361	6.293	1.8394379 045	6.343	1.8473518 427	6.393	1.8552036 418
6.244	1.8316210 027	6.294	1.8395967 986	6.344	1.8475094 843	6.394	1.8553600 506
6.245	1.8317811 436	6.295	1.8397556 675	6.345	1.8476671 012	6.395	1.8555164 350
6.246	1.8319412 589	6.296	1.8399145 111	6.346	1.8478246 932	6.396	1.8556727 950
6.247	1.8321013 485	6.297	1.8400733 295	6.347	1.8479822 603	6.397	1.8558291 305
6.248	1.8322614 125	6.298	1.8402321 227	6.348	1.8481398 027	6.398	1.8559854 415
6.249	1.8324214 509	6.299	1.8403908 906	6.349	1.8482973 202	6.399	1.8561417 282

TEN PLACE MANTISSAS FOR NATURAL LOGARITHMS (Continued)

x	$\log_e x$	x	$\log_e x$	x	$\log_e x$	x	$\log_e x$
6.400	1.8562979 904	6.450	1.8640801 308	6.500	1.8718021 769	6.550	1.8794650 496
6.401	1.8564542 282	6.451	1.8642351 576	6.501	1.8719560 112	6.551	1.8796177 097
6.402	1.8566104 415	6.452	1.8643901 603	6.502	1.8721098 219	6.552	1.8797703 466
6.403	1.8567666 305	6.453	1.8645451 390	6.503	1.8722636 089	6.553	1.8799229 601
6.404	1.8569227 951	6.454	1.8647000 936	6.504	1.8724173 722	6.554	1.8800755 503
6.405	1.8570789 353	6.455	1.8648550 243	6.505	1.8725711 120	6.555	1.8802281 172
6.406	1.8572350 512	6.456	1.8650099 310	6.506	1.8727248 281	6.556	1.8803806 609
6.407	1.8573911 427	6.457	1.8651648 136	6.507	1.8728785 205	6.557	1.8805331 813
6.408	1.8575472 098	6.458	1.8653196 723	6.508	1.8730321 894	6.558	1.8806856 784
6.409	1.8577032 525	6.459	1.8654745 071	6.509	1.8731858 346	6.559	1.8808381 523
6.410	1.8578592 709	6.460	1.8656293 178	6.510	1.8733394 562	6.560	1.8809906 030
6.411	1.8580152 650	6.461	1.8657841 046	6.511	1.8734930 543	6.561	1.8811430 304
6.412	1.8581712 347	6.462	1.8659388 674	6.512	1.8736466 287	6.562	1.8812954 345
6.413	1.8583271 802	6.463	1.8660936 063	6.513	1.8738001 796	6.563	1.8814478 155
6.414	1.8584831 013	6.464	1.8662483 212	6.514	1.8739537 069	6.564	1.8816001 732
6.415	1.8586389 981	6.465	1.8664030 122	6.515	1.8741072 106	6.565	1.8817525 078
6.416	1.8587948 706	6.466	1.8665576 793	6.516	1.8742606 907	6.566	1.8819048 191
6.417	1.8589507 188	6.467	1.8667123 225	6.517	1.8744141 474	6.567	1.8820571 072
6.418	1.8591065 427	6.468	1.8668669 417	6.518	1.8745675 804	6.568	1.8822093 722
6.419	1.8592623 423	6.469	1.8670215 371	6.519	1.8747209 899	6.569	1.8823616 139
6.420	1.8594181 177	6.470	1.8671761 085	6.520	1.8748743 759	6.570	1.8825138 325
6.421	1.8595738 688	6.471	1.8673306 561	6.521	1.8750277 384	6.571	1.8826660 279
6.422	1.8597295 957	6.472	1.8674851 798	6.522	1.8751810 774	6.572	1.8828182 002
6.423	1.8598852 983	6.473	1.8676396 796	6.523	1.8753343 928	6.573	1.8829703 493
6.424	1.8600409 766	6.474	1.8677941 555	6.524	1.8754876 848	6.574	1.8831224 752
6.425	1.8601966 308	6.475	1.8679486 076	6.525	1.8756409 532	6.575	1.8832745 781
6.426	1.8603522 607	6.476	1.8681030 358	6.526	1.8757941 982	6.576	1.8834266 578
6.427	1.8605078 664	6.477	1.8682574 402	6.527	1.8759474 197	6.577	1.8835787 143
6.428	1.8606634 479	6.478	1.8684118 207	6.528	1.8761006 177	6.578	1.8837307 478
6.429	1.8608190 052	6.479	1.8685661 775	6.529	1.8762537 922	6.579	1.8838827 581
6.430	1.8609745 382	6.480	1.8687205 104	6.530	1.8764069 433	6.580	1.8840347 453
6.431	1.8611300 472	6.481	1.8688748 194	6.531	1.8765600 709	6.581	1.8841867 095
6.432	1.8612855 319	6.482	1.8690291 047	6.532	1.8767131 751	6.582	1.8843386 505
6.433	1.8614409 924	6.483	1.8691833 662	6.533	1.8768662 559	6.583	1.8844905 685
6.434	1.8615964 288	6.484	1.8693376 039	6.534	1.8770193 132	6.584	1.8846424 634
6.435	1.8617518 410	6.485	1.8694918 178	6.535	1.8771723 471	6.585	1.8847943 352
6.436	1.8619072 291	6.486	1.8696460 079	6.536	1.8773253 576	6.586	1.8849461 840
6.437	1.8620625 931	6.487	1.8698001 742	6.537	1.8774783 446	6.587	1.8850980 097
6.438	1.8622179 329	6.488	1.8699543 168	6.538	1.8776313 083	6.588	1.8852498 123
6.439	1.8623732 486	6.489	1.8701084 356	6.539	1.8777842 486	6.589	1.8854015 919
6.440	1.8625285 401	6.490	1.8702625 307	6.540	1.8779371 655	6.590	1.8855533 485
6.441	1.8626838 076	6.491	1.8704166 021	6.541	1.8780900 590	6.591	1.8857050 821
6.442	1.8628390 509	6.492	1.8705706 497	6.542	1.8782429 291	6.592	1.8858567 926
6.443	1.8629942 702	6.493	1.8707246 735	6.543	1.8783957 759	6.593	1.8860084 801
6.444	1.8631494 653	6.494	1.8708786 737	6.544	1.8785485 993	6.594	1.8861601 446
6.445	1.8633046 364	6.495	1.8710326 501	6.545	1.8787013 994	6.595	1.8863117 862
6.446	1.8634597 834	6.496	1.8711866 029	6.546	1.8788541 761	6.596	1.8864634 047
6.447	1.8636149 063	6.497	1.8713405 319	6.547	1.8790069 295	6.597	1.8866150 002
6.448	1.8637700 052	6.498	1.8714944 372	6.548	1.8791596 595	6.598	1.8867665 728
6.449	1.8639250 800	6.499	1.8716483 189	6.549	1.8793123 662	6.599	1.8869181 224

TEN PLACE MANTISSAS FOR NATURAL LOGARITHMS (Continued)

x	log_e x	x	log_e x	x	log_e x	x	log_e x
6.600	1.8870696 490	6.650	1.8946168 547	6.700	1.9021075 264	6.750	1.9095425 049
6.601	1.8872211 527	6.651	1.8947672 193	6.701	1.9022567 690	6.751	1.9096906 421
6.602	1.8873726 334	6.652	1.8949175 613	6.702	1.9024059 893	6.752	1.9098387 573
6.603	1.8875240 912	6.653	1.8950678 808	6.703	1.9025551 874	6.753	1.9099868 506
6.604	1.8876755 261	6.654	1.8952181 776	6.704	1.9027043 632	6.754	1.9101349 220
6.605	1.8878269 380	6.655	1.8953684 518	6.705	1.9028535 167	6.755	1.9102829 714
6.606	1.8879783 270	6.656	1.8955187 035	6.706	1.9030026 480	6.756	1.9104309 989
6.607	1.8881296 930	6.657	1.8956689 326	6.707	1.9031517 571	6.757	1.9105790 046
6.608	1.8882810 362	6.658	1.8958191 392	6.708	1.9033008 440	6.758	1.9107269 883
6.609	1.8884323 565	6.659	1.8959693 231	6.709	1.9034499 086	6.759	1.9108749 501
6.610	1.8885836 539	6.660	1.8961194 846	6.710	1.9035989 510	6.760	1.9110228 901
6.611	1.8887349 284	6.661	1.8962696 234	6.711	1.9037479 712	6.761	1.9111708 081
6.612	1.8888861 800	6.662	1.8964197 398	6.712	1.9038969 692	6.762	1.9113187 043
6.613	1.8890374 087	6.663	1.8965698 336	6.713	1.9040459 450	6.763	1.9114665 786
6.614	1.8891886 146	6.664	1.8967199 049	6.714	1.9041948 986	6.764	1.9116144 310
6.615	1.8893397 976	6.665	1.8968699 536	6.715	1.9043438 300	6.765	1.9117622 616
6.616	1.8894909 577	6.666	1.8970199 799	6.716	1.9044927 392	6.766	1.9119100 704
6.617	1.8896420 950	6.667	1.8971699 836	6.717	1.9046416 263	6.767	1.9120578 573
6.618	1.8897932 095	6.668	1.8973199 649	6.718	1.9047904 912	6.768	1.9122056 223
6.619	1.8899443 011	6.669	1.8974699 237	6.719	1.9049393 339	6.769	1.9123533 655
6.620	1.8900953 699	6.670	1.8976198 599	6.720	1.9050881 545	6.770	1.9125010 869
6.621	1.8902464 159	6.671	1.8977697 737	6.721	1.9052369 530	6.771	1.9126487 865
6.622	1.8903974 391	6.672	1.8979196 651	6.722	1.9053857 293	6.772	1.9127964 643
6.623	1.8905484 395	6.673	1.8980695 339	6.723	1.9055344 835	6.773	1.9129441 202
6.624	1.8906994 171	6.674	1.8982193 803	6.724	1.9056832 155	6.774	1.9130917 544
6.625	1.8908503 719	6.675	1.8983692 043	6.725	1.9058319 255	6.775	1.9132393 668
6.626	1.8910013 039	6.676	1.8985190 058	6.726	1.9059806 133	6.776	1.9133869 573
6.627	1.8911522 131	6.677	1.8986687 849	6.727	1.9061292 790	6.777	1.9135345 262
6.628	1.8913030 996	6.678	1.8988185 415	6.728	1.9062779 227	6.778	1.9136820 732
6.629	1.8914539 633	6.679	1.8989682 757	6.729	1.9064265 442	6.779	1.9138295 984
6.630	1.8916048 042	6.680	1.8991179 875	6.730	1.9065751 437	6.780	1.9139771 020
6.631	1.8917556 224	6.681	1.8992676 769	6.731	1.9067237 210	6.781	1.9141245 837
6.632	1.8919064 178	6.682	1.8994173 439	6.732	1.9068722 763	6.782	1.9142720 437
6.633	1.8920571 905	6.683	1.8995669 885	6.733	1.9070208 095	6.783	1.9144194 820
6.634	1.8922079 405	6.684	1.8997166 107	6.734	1.9071693 207	6.784	1.9145668 985
6.635	1.8923586 678	6.685	1.8998662 106	6.735	1.9073178 099	6.785	1.9147142 933
6.636	1.8925093 723	6.686	1.9000157 880	6.736	1.9074662 769	6.786	1.9148616 664
6.637	1.8926600 542	6.687	1.9001653 431	6.737	1.9076147 220	6.787	1.9150090 177
6.638	1.8928107 133	6.688	1.9003148 758	6.738	1.9077631 450	6.788	1.9151563 474
6.639	1.8929613 497	6.689	1.9004643 861	6.739	1.9079115 460	6.789	1.9153036 553
6.640	1.8931119 635	6.690	1.9006138 741	6.740	1.9080599 249	6.790	1.9154509 416
6.641	1.8932625 546	6.691	1.9007633 398	6.741	1.9082082 819	6.791	1.9155982 061
6.642	1.8934131 230	6.692	1.9009127 831	6.742	1.9083566 168	6.792	1.9157454 490
6.643	1.8935636 687	6.693	1.9010622 041	6.743	1.9085049 298	6.793	1.9158926 702
6.644	1.8937141 918	6.694	1.9012116 028	6.744	1.9086532 207	6.794	1.9160398 697
6.645	1.8938646 922	6.695	1.9013609 791	6.745	1.9088014 897	6.795	1.9161870 476
6.646	1.8940151 699	6.696	1.9015103 332	6.746	1.9089497 366	6.796	1.9163342 038
6.647	1.8941656 251	6.697	1.9016596 649	6.747	1.9090979 616	6.797	1.9164813 384
6.648	1.8943160 576	6.698	1.9018089 744	6.748	1.9092461 647	6.798	1.9166284 513
6.649	1.8944664 674	6.699	1.9019582 615	6.749	1.9093943 458	6.799	1.9167755 425

TEN PLACE MANTISSAS FOR NATURAL LOGARITHMS (Continued)

x	$\log_e x$	x	$\log_e x$	x	$\log_e x$	x	$\log_e x$
6.800	1.9169226 122	6.850	1.9242486 523	6.900	1.9315214 116	6.950	1.9387416 596
6.801	1.9170696 602	6.851	1.9243946 270	6.901	1.9316663 286	6.951	1.9388855 341
6.802	1.9172166 866	6.852	1.9245405 805	6.902	1.9318112 247	6.952	1.9390293 880
6.803	1.9173636 914	6.853	1.9246865 126	6.903	1.9319560 997	6.953	1.9391732 211
6.804	1.9175106 745	6.854	1.9248324 235	6.904	1.9321009 538	6.954	1.9393170 336
6.805	1.9176576 361	6.855	1.9249783 130	6.905	1.9322457 869	6.955	1.9394608 254
6.806	1.9178045 761	6.856	1.9251241 813	6.906	1.9323905 990	6.956	1.9396045 965
6.807	1.9179514 945	6.857	1.9252700 283	6.907	1.9325353 901	6.657	1.9397483 469
6.808	1.9180983 913	6.858	1.9254158 540	6.908	1.9326801 603	6.958	1.9398920 767
6.809	1.9182452 665	6.859	1.9255616 585	6.909	1.9328249 095	6.959	1.9400357 859
6.810	1.9183921 202	6.860	1.9257074 417	6.910	1.9329696 378	6.960	1.9401794 743
6.811	1.9185389 523	6.861	1.9258532 037	6.911	1.9331143 451	6.961	1.9403231 422
6.812	1.9186857 628	6.862	1.9259989 444	6.912	1.9332590 315	6.962	1.9404667 894
6.813	1.9188325 518	6.863	1.9261446 639	6.913	1.9334036 970	6.963	1.9406104 160
6.814	1.9189793 192	6.864	1.9262903 622	6.914	1.9335483 415	6.964	1.9407540 219
6.815	1.9191260 651	6.865	1.9264360 392	6.915	1.9336929 651	6.965	1.9408976 072
6.816	1.9192727 895	6.866	1.9265816 950	6.916	1.9338375 678	6.966	1.9410411 719
6.817	1.9194194 924	6.867	1.9267273 296	6.917	1.9339821 496	6.967	1.9411847 160
6.818	1.9195661 737	6.868	1.9268729 430	6.918	1.9341267 105	6.968	1.9413282 396
6.819	1.9197128 335	6.869	1.9270185 352	6.919	1.9342712 505	6.969	1.9414717 425
6.820	1.9198594 719	6.870	1.9271641 062	6.920	1.9344157 696	6.970	1.9416152 248
6.821	1.9200060 887	6.871	1.9273096 560	6.921	1.9345602 679	6.971	1.9417586 865
6.822	1.9201526 840	6.872	1.9274551 847	6.922	1.9347047 452	6.972	1.9419021 277
6.823	1.9202992 578	6.873	1.9276006 921	6.923	1.9348492 017	6.973	1.9420455 482
6.824	1.9204458 102	6.874	1.9277461 784	6.924	1.9349936 373	6.974	1.9421889 483
6.825	1.9205923 411	6.875	1.9278916 436	6.925	1.9351380 521	6.975	1.9423323 277
6.826	1.9207388 505	6.876	1.9280370 875	6.926	1.9352824 460	6.976	1.9424756 866
6.827	1.9208853 384	6.877	1.9281825 103	6.927	1.9354268 190	6.977	1.9426190 250
6.828	1.9210318 049	6.878	1.9283279 120	6.928	1.9355711 713	6.978	1.9427623 428
6.829	1.9211782 500	6.879	1.9284732 925	6.929	1.9357155 026	6.979	1.9429056 400
6.830	1.9213246 736	6.880	1.9286186 519	6.930	1.9358598 132	6.980	1.9430489 168
6.831	1.9214710 757	6.881	1.9287639 902	6.931	1.9360041 029	6.981	1.9431921 730
6.832	1.9216174 565	6.882	1.9289093 074	6.932	1.9361483 719	6.982	1.9433354 087
6.833	1.9217638 158	6.883	1.9290546 034	6.933	1.9362926 200	6.983	1.9434786 239
6.834	1.9219101 537	6.884	1.9291998 783	6.934	1.9364368 473	6.984	1.9436218 185
6.835	1.9220564 702	6.885	1.9293451 322	6.935	1.9365810 538	6.985	1.9437649 927
6.836	1.9222027 653	6.886	1.9294903 649	6.936	1.9367252 395	6.986	1.9439081 464
6.837	1.9223490 389	6.887	1.9296355 766	6.937	1.9368694 044	6.987	1.9440512 796
6.838	1.9224952 912	6.888	1.9297807 671	6.938	1.9370135 485	6.988	1.9441943 923
6.839	1.9226415 221	6.889	1.9299259 366	6.939	1.9371576 719	6.989	1.9443374 845
6.840	1.9227877 316	6.890	1.9300710 850	6.940	1.9373017 745	6.990	1.9444805 562
6.841	1.9229339 198	6.891	1.9302162 124	6.941	1.9374458 564	6.991	1.9446236 075
6.842	1.9230800 866	6.892	1.9303613 187	6.942	1.9375899 174	6.992	1.9447666 384
6.843	1.9232262 320	6.893	1.9305064 039	6.943	1.9377339 578	6.993	1.9449096 487
6.844	1.9233723 560	6.894	1.9306514 681	6.944	1.9378779 774	6.994	1.9450526 386
6.845	1.9235184 587	6.895	1.9307965 112	6.945	1.9380219 762	6.995	1.9451956 081
6.846	1.9236645 401	6.896	1.9309415 334	6.946	1.9381659 543	6.996	1.9453385 572
6.847	1.9238106 001	6.897	1.9310865 344	6.947	1.9383099 117	6.997	1.9454814 858
6.848	1.9239566 388	6.898	1.9312315 145	6.948	1.9384538 484	6.998	1.9456243 939
6.849	1.9241026 562	6.899	1.9313764 736	6.949	1.9385977 643	6.999	1.9457672 817

TEN PLACE MANTISSAS FOR NATURAL LOGARITHMS (Continued)

x	$\log_e x$	x	$\log_e x$	x	$\log_e x$	x	$\log_e x$
7.000	1.9459101 491	7.050	1.9530276 168	7.100	1.9600947 840	7.150	1.9671123 567
7.001	1.9460529 960	7.051	1.9531694 507	7.101	1.9602356 192	7.151	1.9672522 071
7.002	1.9461958 225	7.052	1.9533112 645	7.102	1.9603764 345	7.152	1.9673920 379
7.003	1.9463386 287	7.053	1.9534530 582	7.103	1.9605172 300	7.153	1.9675318 491
7.004	1.9464814 144	7.054	1.9535948 318	7.104	1.9606580 057	7.154	1.9676716 408
7.005	1.9466241 798	7.055	1.9537365 853	7.105	1.9607987 615	7.155	1.9678114 130
7.006	1.9467669 248	7.056	1.9538783 187	7.106	1.9609394 976	7.156	1.9679511 656
7.007	1.9469096 494	7.057	1.9540200 320	7.107	1.9610802 138	7.157	1.9680908 988
7.008	1.9470523 536	7.058	1.9541617 252	7.108	1.9612209 103	7.158	1.9682306 123
7.009	1.9471950 375	7.059	1.9543033 984	7.109	1.9613615 869	7.159	1.9683703 064
7.010	1.9473377 010	7.060	1.9544450 515	7.110	1.9615022 438	7.160	1.9685099 810
7.011	1.9474803 442	7.061	1.9545866 845	7.111	1.9616428 809	7.161	1.9686496 360
7.012	1.9476229 671	7.062	1.9547282 975	7.112	1.9617834 982	7.162	1.9687892 716
7.013	1.9477655 696	7.063	1.9548698 904	7.113	1.9619240 958	7.163	1.9689288 876
7.014	1.9479081 517	7.064	1.9550114 633	7.114	1.9620646 735	7.164	1.9690684 842
7.015	1.9480507 136	7.065	1.9551530 161	7.115	1.9622052 315	7.165	1.9692080 613
7.016	1.9481932 551	7.066	1.9552945 489	7.116	1.9623457 698	7.166	1.9693476 189
7.017	1.9483357 763	7.067	1.9554360 617	7.117	1.9624862 883	7.167	1.9694871 570
7.018	1.9484782 772	7.068	1.9555775 545	7.118	1.9626267 871	7.168	1.9696266 757
7.019	1.9486207 577	7.069	1.9557190 272	7.119	1.9627672 661	7.169	1.9697661 749
7.020	1.9487632 180	7.070	1.9558604 799	7.120	1.9629077 254	7.170	1.9699056 546
7.021	1.9489056 580	7.071	1.9560019 126	7.121	1.9630481 650	7.171	1.9700451 149
7.022	1.9490480 777	7.072	1.9561433 253	7.122	1.9631885 849	7.172	1.9701845 557
7.023	1.9491904 772	7.073	1.9562847 181	7.123	1.9633289 850	7.173	1.9703239 771
7.024	1.9493328 563	7.074	1.9564260 908	7.124	1.9634693 655	7.174	1.9704633 791
7.025	1.9494752 152	7.075	1.9565674 435	7.125	1.9636097 262	7.175	1.9706027 616
7.026	1.9496175 538	7.076	1.9567087 763	7.126	1.9637500 672	7.176	1.9707421 248
7.027	1.9497598 722	7.077	1.9568500 891	7.127	1.9638903 885	7.177	1.9708814 684
7.028	1.9499021 703	7.078	1.9569913 819	7.128	1.9640306 901	7.178	1.9710207 927
7.029	1.9500444 482	7.079	1.9571326 548	7.129	1.9641709 721	7.179	1.9711600 976
7.030	1.9501867 058	7.080	1.9572739 077	7.130	1.9643112 344	7.180	1.9712993 831
7.031	1.9503289 432	7.081	1.9574151 407	7.131	1.9644514 770	7.181	1.9714386 491
7.032	1.9504711 604	7.082	1.9575563 537	7.132	1.9645917 000	7.182	1.9715778 958
7.033	1.9506133 573	7.083	1.9576975 468	7.133	1.9647319 033	7.183	1.9717171 231
7.034	1.9507555 341	7.084	1.9578387 199	7.134	1.9648720 869	7.184	1.9718563 310
7.035	1.9508976 906	7.085	1.9579798 731	7.135	1.9650122 509	7.185	1.9719955 195
7.036	1.9510398 269	7.086	1.9581210 064	7.136	1.9651523 953	7.186	1.9721346 887
7.037	1.9511819 430	7.087	1.9582621 198	7.137	1.9652925 200	7.187	1.9722738 385
7.038	1.9513240 389	7.088	1.9584032 133	7.138	1.9654326 251	7.188	1.9724129 689
7.039	1.9514661 146	7.089	1.9585442 869	7.139	1.9655727 105	7.189	1.9725520 800
7.040	1.9516081 702	7.090	1.9586853 405	7.140	1.9657127 764	7.190	1.9726911 717
7.041	1.9517502 055	7.091	1.9588263 743	7.141	1.9658528 226	7.191	1.9728302 441
7.042	1.9518922 207	7.092	1.9589673 882	7.142	1.9659928 492	7.192	1.9729692 972
7.043	1.9520342 158	7.093	1.9591083 822	7.143	1.9661328 562	7.193	1.9731083 309
7.044	1.9521761 906	7.094	1.9592493 564	7.144	1.9662728 436	7.194	1.9732473 453
7.045	1.9523181 454	7.095	1.9593903 106	7.145	1.9664128 114	7.195	1.9733863 403
7.046	1.9524600 799	7.096	1.9595312 450	7.146	1.9665527 596	7.196	1.9735253 161
7.047	1.9526019 943	7.097	1.9596721 595	7.147	1.9666926 882	7.197	1.9736642 725
7.048	1.9527438 886	7.098	1.9598130 542	7.148	1.9668325 973	7.198	1.9738032 097
7.049	1.9528857 628	7.099	1.9599539 291	7.149	1.9669724 868	7.199	1.9739421 275

TEN PLACE MANTISSAS FOR NATURAL LOGARITHMS (Continued)

x	$\log_e x$	x	$\log_e x$	x	$\log_e x$	x	$\log_e x$
7.200	1.9740810 260	7.250	1.9810014 689	7.300	1.9878743 482	7.350	1.9947003 132
7.201	1.9742199 053	7.251	1.9811393 904	7.301	1.9880113 251	7.351	1.9948363 584
7.202	1.9743587 652	7.252	1.9812772 929	7.302	1.9881482 832	7.352	1.9949723 851
7.203	1.9744976 059	7.253	1.9814151 764	7.303	1.9882852 226	7.353	1.9951083 932
7.204	1.9746364 273	7.254	1.9815530 409	7.304	1.9884221 433	7.354	1.9952443 829
7.205	1.9747752 295	7.255	1.9816908 863	7.305	1.9885590 452	7.355	1.9953803 541
7.206	1.9749140 123	7.256	1.9818287 128	7.306	1.9886959 284	7.356	1.9955163 067
7.207	1.9750527 759	7.257	1.9819665 203	7.307	1.9888327 928	7.357	1.9956522 410
7.208	1.9751915 203	7.258	1.9821043 088	7.308	1.9889696 385	7.358	1.9957881 567
7.209	1.9753302 454	7.259	1.9822420 783	7.309	1.9891064 655	7.359	1.9959240 539
7.210	1.9754689 513	7.260	1.9823798 288	7.310	1.9892432 738	7.360	1.9960599 327
7.211	1.9756076 379	7.261	1.9825175 604	7.311	1.9893800 633	7.361	1.9961957 931
7.212	1.9757463 053	7.262	1.9826552 730	7.312	1.9895168 342	7.362	1.9963316 350
7.213	1.9758849 535	7.263	1.9827929 666	7.313	1.9896535 863	7.363	1.9964674 584
7.214	1.9760235 825	7.264	1.9829306 413	7.314	1.9897903 197	7.364	1.9966032 634
7.215	1.9761621 922	7.265	1.9830682 970	7.315	1.9899270 345	7.365	1.9967390 499
7.216	1.9763007 828	7.266	1.9832059 338	7.316	1.9900637 305	7.366	1.9968748 180
7.217	1.9764393 541	7.267	1.9833435 516	7.317	1.9902004 079	7.367	1.9970105 677
7.218	1.9765779 062	7.268	1.9834811 505	7.318	1.9903370 666	7.368	1.9971462 990
7.219	1.9767164 392	7.269	1.9836187 305	7.319	1.9904737 066	7.369	1.9972820 118
7.220	1.9768549 529	7.270	1.9837562 915	7.320	1.9906103 280	7.370	1.9974177 062
7.221	1.9769934 475	7.271	1.9838938 337	7.321	1.9907469 307	7.371	1.9975533 822
7.222	1.9771319 229	7.272	1.9840313 569	7.322	1.9908835 147	7.372	1.9976890 398
7.223	1.9772703 791	7.273	1.9841688 612	7.323	1.9910200 801	7.373	1.9978246 790
7.224	1.9774088 161	7.274	1.9843063 466	7.324	1.9911566 268	7.374	1.9979602 998
7.225	1.9775472 340	7.275	1.9844438 131	7.325	1.9912931 549	7.375	1.9980959 022
7.226	1.9776856 327	7.276	1.9845812 607	7.326	1.9914296 644	7.376	1.9982314 863
7.227	1.9778240 123	7.277	1.9847186 894	7.327	1.9915661 552	7.377	1.9983670 519
7.228	1.9779623 727	7.278	1.9848560 992	7.328	1.9917026 274	7.378	1.9985025 992
7.229	1.9781007 140	7.279	1.9849934 901	7.329	1.9918390 809	7.379	1.9986381 281
7.230	1.9782390 362	7.280	1.9851308 622	7.330	1.9919755 159	7.380	1.9987736 386
7.231	1.9783773 392	7.281	1.9852682 154	7.331	1.9921119 322	7.381	1.9989091 308
7.232	1.9785156 231	7.282	1.9854055 498	7.332	1.9922485 300	7.382	1.9990446 046
7.233	1.9786538 879	7.283	1.9855428 652	7.333	1.9923847 091	7.383	1.9991800 601
7.234	1.9787921 335	7.284	1.9856801 619	7.334	1.9925210 696	7.384	1.9993154 972
7.235	1.9789303 601	7.285	1.9858174 396	7.335	1.9926574 116	7.385	1.9994509 160
7.236	1.9790685 675	7.286	1.9859546 986	7.336	1.9927937 350	7.386	1.9995863 164
7.237	1.9792067 559	7.287	1.9860919 387	7.337	1.9929300 397	7.387	1.9997216 985
7.238	1.9793449 251	7.288	1.9862291 600	7.338	1.9930663 259	7.388	1.9998570 623
7.239	1.9794830 753	7.289	1.9863663 624	7.339	1.9932025 936	7.389	1.9999924 078
7.240	1.9796212 064	7.290	1.9865035 460	7.340	1.9933388 426	7.390	2.0001277 350
7.241	1.9797593 184	7.291	1.9866407 108	7.341	1.9934750 731	7.391	2.0002630 438
7.242	1.9798974 113	7.292	1.9867778 568	7.342	1.9936112 851	7.392	2.0003983 343
7.243	1.9800354 852	7.293	1.9869149 840	7.343	1.9937474 785	7.393	2.0005336 066
7.244	1.9801735 400	7.294	1.9870520 924	7.344	1.9938836 533	7.394	2.0006688 605
7.245	1.9803115 758	7.295	1.9871891 820	7.345	1.9940198 096	7.395	2.0008040 962
7.246	1.9804495 925	7.296	1.9873262 528	7.346	1.9941559 474	7.396	2.0009393 135
7.247	1.9805875 901	7.297	1.9874633 048	7.347	1.9942920 666	7.397	2.0010745 126
7.248	1.9807255 687	7.298	1.9876003 380	7.348	1.9944281 674	7.398	2.0012096 934
7.249	1.9808635 283	7.299	1.9877373 525	7.349	1.9945642 495	7.399	2.0013448 559

TEN PLACE MANTISSAS FOR NATURAL LOGARITHMS (Continued)

x	$\log_e x$	x	$\log_e x$	x	$\log_e x$	x	$\log_e x$
7.400	2.0014800 002	7.450	2.0082140 324	7.500	2.0149030 205	7.550	2.0215475 633
7.401	2.0016151 262	7.451	2.0083482 516	7.501	2.0150363 450	7.551	2.0216800 048
7.402	2.0017502 340	7.452	2.0084824 527	7.502	2.0151696 517	7.552	2.0218124 288
7.403	2.0018853 235	7.453	2.0086166 359	7.503	2.0153029 406	7.553	2.0219448 353
7.404	2.0020203 947	7.454	2.0087508 011	7.504	2.0154362 117	7.554	2.0220772 243
7.405	2.0021554 477	7.455	2.0088849 482	7.505	2.0155694 651	7.555	2.0222095 957
7.406	2.0022904 825	7.456	2.0090190 774	7.506	2.0157027 007	7.556	2.0223419 496
7.407	2.0024254 990	7.457	2.0091531 886	7.507	2.0158359 186	7.557	2.0224742 860
7.408	2.0025604 973	7.458	2.0092872 818	7.508	2.0159691 187	7.558	2.0226066 049
7.409	2.0026954 774	7.459	2.0094213 570	7.509	2.0161023 011	7.559	2.0227389 063
7.410	2.0028304 393	7.460	2.0095554 142	7.510	2.0162354 658	7.560	2.0228711 902
7.411	2.0029653 830	7.461	2.0096894 535	7.511	2.0163686 127	7.561	2.0230034 566
7.412	2.0031003 084	7.462	2.0098234 748	7.512	2.0165017 419	7.562	2.0231357 055
7.413	2.0032352 157	7.463	2.0099574 781	7.513	2.0166348 534	7.563	2.0232679 369
7.414	2.0033701 047	7.464	2.0100914 635	7.514	2.0167679 472	7.564	2.0234001 508
7.415	2.0035049 756	7.465	2.0102254 310	7.515	2.0169010 232	7.565	2.0235323 472
7.416	2.0036398 283	7.466	2.0103593 805	7.516	2.0170340 816	7.566	2.0236645 262
7.417	2.0037746 628	7.467	2.0104933 121	7.517	2.0171671 222	7.567	2.0237966 877
7.418	2.0039094 791	7.468	2.0106272 257	7.518	2.0173001 451	7.568	2.0239288 317
7.419	2.0040442 772	7.469	2.0107611 214	7.519	2.0174331 504	7.569	2.0240609 583
7.420	2.0041790 572	7.470	2.0108949 991	7.520	2.0175661 380	7.570	2.0241930 674
7.421	2.0043138 190	7.471	2.0110288 590	7.521	2.0176991 078	7.571	2.0243251 591
7.422	2.0044485 626	7.472	2.0111627 009	7.522	2.0178320 600	7.572	2.0244572 333
7.423	2.0045832 881	7.473	2.0112965 249	7.523	2.0179649 946	7.573	2.0245892 901
7.424	2.0047179 955	7.474	2.0114303 311	7.524	2.0180979 114	7.574	2.0247213 295
7.425	2.0048526 847	7.475	2.0115641 193	7.525	2.0182308 106	7.575	2.0248533 514
7.426	2.0049873 558	7.476	2.0116978 896	7.526	2.0183636 922	7.576	2.0249853 559
7.427	2.0051220 087	7.477	2.0118316 420	7.527	2.0184965 561	7.577	2.0251173 429
7.428	2.0052566 435	7.478	2.0119653 766	7.528	2.0186294 023	7.578	2.0252493 126
7.429	2.0053912 602	7.479	2.0120990 932	7.529	2.0187622 309	7.579	2.0253812 648
7.430	2.0055258 587	7.480	2.0122327 920	7.530	2.0188950 418	7.580	2.0255131 997
7.431	2.0056604 392	7.481	2.0123664 729	7.531	2.0190278 351	7.581	2.0256451 171
7.432	2.0057950 015	7.482	2.0125001 359	7.532	2.0191606 108	7.582	2.0257770 171
7.433	2.0059295 457	7.483	2.0126337 811	7.533	2.0192933 688	7.583	2.0259088 997
7.434	2.0060640 719	7.484	2.0127674 084	7.534	2.0194261 093	7.584	2.0260407 650
7.435	2.0061985 799	7.485	2.0129010 179	7.535	2.0195588 321	7.585	2.0261726 128
7.436	2.0063330 699	7.486	2.0130346 095	7.536	2.0196915 373	7.586	2.0263044 433
7.437	2.0064675 417	7.487	2.0131681 832	7.537	2.0198242 249	7.587	2.0264362 564
7.438	2.0066019 955	7.488	2.0133017 392	7.538	2.0199568 948	7.588	2.0265680 521
7.439	2.0067364 312	7.489	2.0134352 773	7.539	2.0200895 472	7.589	2.0266998 304
7.440	2.0068708 488	7.490	2.0135687 975	7.540	2.0202221 820	7.590	2.0268315 914
7.441	2.0070052 484	7.491	2.0137023 000	7.541	2.0203547 992	7.591	2.0269633 350
7.442	2.0071396 299	7.492	2.0138357 846	7.542	2.0204873 988	7.592	2.0270950 613
7.443	2.0072739 934	7.493	2.0139692 514	7.543	2.0206199 809	7.593	3.0272267 702
7.444	2.0074083 388	7.494	2.0141027 004	7.544	2.0207525 453	7.594	2.0273584 618
7.445	2.0075426 661	7.495	2.0142361 316	7.545	2.0208850 922	7.595	2.0274901 360
7.446	2.0076769 755	7.496	2.0143695 449	7.546	2.0210176 215	7.596	2.0276217 930
7.447	2.0078112 667	7.497	2.0145029 405	7.547	2.0211501 333	7.597	2.0277534 325
7.448	2.0079455 400	7.498	2.0146363 183	7.548	2.0212826 275	7.598	2.0278850 548
7.449	2.0080797 952	7.499	2.0147696 783	7.549	2.0214151 042	7.599	2.0280166 597

TEN PLACE MANTISSAS FOR NATURAL LOGARITHMS (Continued)

x	$\log_e x$	x	$\log_e x$	x	$\log_e x$	x	$\log_e x$
7.600	2.0281482 473	7.650	2.0347056 478	7.700	2.0412203 289	7.750	2.0476928 434
7.601	2.0282798 176	7.651	2.0348363 582	7.701	2.0413501 906	7.751	2.0478218 673
7.602	2.0284113 706	7.652	2.0349670 516	7.702	2.0414800 354	7.752	2.0479508 746
7.603	2.0285429 062	7.653	2.0350977 278	7.703	2.0416098 634	7.753	2.0480798 652
7.604	2.0286744 246	7.654	2.0352283 870	7.704	2.0417396 745	7.754	2.0482088 392
7.605	2.0288059 257	7.655	2.0353590 291	7.705	2.0418694 688	7.755	2.0483377 966
7.606	2.0289374 095	7.656	2.0354896 542	7.706	2.0419992 462	7.756	2.0484667 374
7.607	2.0290688 760	7.657	2.0356202 621	7.707	2.0421290 068	7.757	2.0485956 615
7.608	2.0292003 252	7.658	2.0357508 531	7.708	2.0422587 506	7.758	2.0487245 690
7.609	2.0293317 572	7.659	2.0358814 269	7.709	2.0423884 775	7.759	2.0488534 599
7.610	2.0294631 719	7.660	2.0360119 838	7.710	2.0425181 876	7.760	2.0489823 342
7.611	2.0295945 693	7.661	2.0361425 235	7.711	2.0426478 809	7.761	2.0491111 919
7.612	2.0297259 494	7.662	2.0362730 463	7.712	2.0427775 573	7.762	2.0492400 329
7.613	2.0298573 123	7.663	2.0364035 520	7.713	2.0429072 170	7.763	2.0493688 574
7.614	2.0299886 580	7.664	2.0365340 407	7.714	2.0430368 598	7.764	2.0494976 653
7.615	2.0301199 863	7.665	2.0366645 123	7.715	2.0431664 858	7.765	2.0496264 566
7.616	2.0302512 975	7.666	2.0367949 670	7.716	2.0432960 950	7.766	2.0497552 313
7.617	2.0303825 914	7.667	2.0369254 046	7.717	2.0434256 875	7.767	2.0498839 894
7.618	2.0305138 681	7.668	2.0370558 252	7.718	2.0435552 631	7.768	2.0500127 310
7.619	2.0306451 275	7.669	2.0371862 288	7.719	2.0436848 220	7.769	2.0501414 560
7.620	2.0307763 697	7.670	2.0373166 154	7.720	2.0438143 640	7.770	2.0502701 644
7.621	2.0309075 947	7.671	2.0374469 850	7.721	2.0439438 893	7.771	2.0503988 562
7.622	2.0310388 025	7.672	2.0375773 376	7.722	2.0440733 978	7.772	2.0505275 315
7.623	2.0311699 930	7.673	2.0377076 732	7.723	2.0442028 896	7.773	2.0506561 902
7.624	2.0313011 664	7.674	2.0378379 918	7.724	2.0443323 646	7.774	2.0507848 324
7.625	2.0314323 225	7.675	2.0379682 935	7.725	2.0444618 228	7.775	2.0509134 581
7.626	2.0315634 614	7.676	2.0380985 781	7.726	2.0445912 642	7.776	2.0510420 672
7.627	2.0316945 832	7.677	2.0382288 458	7.727	2.0447206 890	7.777	2.0511706 597
7.628	2.0318256 877	7.678	2.0383590 966	7.728	2.0448500 969	7.778	2.0512992 357
7.629	2.0319567 751	7.679	2.0384893 303	7.729	2.0449794 881	7.779	2.0514277 952
7.630	2.0320878 453	7.680	2.0386195 472	7.730	2.0451088 626	7.780	2.0515563 382
7.631	2.0322188 983	7.681	2.0387497 470	7.731	2.0452382 203	7.781	2.0516848 646
7.632	2.0323499 341	7.682	2.0388799 299	7.732	2.0453675 613	7.782	2.0518133 746
7.633	2.0324809 528	7.683	2.0390100 959	7.733	2.0454968 856	7.783	2.0519418 680
7.634	2.0326119 543	7.684	2.0391402 449	7.734	2.0456261 932	7.784	2.0520703 449
7.635	2.0327429 387	7.685	2.0392703 770	7.735	2.0457554 840	7.785	2.0521988 053
7.636	2.0328739 059	7.686	2.0394004 921	7.736	2.0458847 582	7.786	2.0523272 492
7.637	2.0330048 559	7.687	2.0395305 904	7.737	2.0460140 156	7.787	2.0524556 766
7.638	2.0331357 888	7.688	2.0396606 717	7.738	2.0461432 563	7.788	2.0525840 875
7.639	2.0332667 046	7.689	2.0397907 361	7.739	2.0462724 803	7.789	2.0527124 819
7.640	2.0333976 032	7.690	2.0399207 835	7.740	2.0464016 876	7.790	2.0528408 599
7.641	2.0335284 847	7.691	2.0400508 141	7.741	2.0465308 782	7.791	2.0529692 213
7.642	2.0336593 490	7.692	2.0401808 277	7.742	2.0466600 522	7.792	2.0530975 663
7.643	2.0337901 963	7.693	2.0403108 245	7.743	2.0467892 094	7.793	2.0532258 949
7.644	2.0339210 264	7.694	2.0404408 043	7.744	2.0469183 500	7.794	2.0533542 069
7.645	2.0340518 394	7.695	2.0405707 673	7.745	2.0470474 739	7.795	2.0534825 025
7.646	2.0341826 353	7.696	2.0407007 134	7.746	2.0471765 811	7.796	2.0536107 816
7.647	2.0343134 141	7.697	2.0408306 426	7.747	2.0473056 716	7.797	2.0537390 443
7.648	2.0344441 757	7.698	2.0409605 549	7.748	2.0474347 455	7.798	2.0538672 906
7.649	2.0345749 203	7.699	2.0410904 503	7.749	2.0475638 028	7.799	2.0539955 203

TEN PLACE MANTISSAS FOR NATURAL LOGARITHMS (Continued)

x	$\log_e x$	x	$\log_e x$	x	$\log_e x$	x	$\log_e x$
7.800	2.0541237 337	7.850	2.0605135 318	7.900	2.0668627 595	7.950	2.0731719 287
7.801	2.0542519 306	7.851	2.0606409 122	7.901	2.0669893 337	7.951	2.0732977 069
7.802	2.0543801 111	7.852	2.0607682 764	7.902	2.0671158 920	7.952	2.0734234 693
7.803	2.0545082 751	7.853	2.0608956 244	7.903	2.0672424 342	7.953	2.0735492 160
7.804	2.0546364 228	7.854	2.0610229 562	7.904	2.0673689 604	7.954	2.0736749 468
7.805	2.0547645 540	7.855	2.0611502 717	7.905	2.0674954 707	7.955	2.0738006 618
7.806	2.0548926 688	7.856	2.0612775 711	7.906	2.0676219 649	7.956	2.0739263 610
7.807	2.0550207 671	7.857	2.0614048 542	7.907	2.0677484 431	7.957	2.0740520 444
7.808	2.0551488 491	7.858	2.0615321 211	7.908	2.0678749 053	7.958	2.0741777 120
7.809	2.0552769 147	7.859	2.0616593 719	7.909	2.0680013 515	7.959	2.0743033 638
7.810	2.0554049 639	7.860	2.0617866 064	7.910	2.0681277 818	7.960	2.0744289 999
7.811	2.0555329 966	7.861	2.0619138 248	7.911	2.0682541 960	7.961	2.0745546 201
7.812	2.0556610 130	7.862	2.0620410 270	7.912	2.0683805 943	7.962	2.0746802 246
7.813	2.0557890 130	7.863	2.0621682 130	7.913	2.0685069 766	7.963	2.0748058 133
7.814	2.0559169 966	7.864	2.0622953 828	7.914	2.0686333 430	7.964	2.0749313 862
7.815	2.0560449 639	7.865	2.0624225 365	7.915	2.0687596 933	7.965	2.0750569 433
7.816	2.0561729 147	7.866	2.0625496 740	7.916	2.0688860 277	7.966	2.0751824 848
7.817	2.0563008 492	7.867	2.0626767 953	7.917	2.0690123 462	7.967	2.0753080 104
7.818	2.0564287 674	7.868	2.0628039 005	7.918	2.0691386 487	7.968	2.0754335 203
7.819	2.0565566 691	7.869	2.0629309 896	7.919	2.0692649 352	7.969	2.0755590 144
7.820	2.0566845 546	7.870	2.0630580 624	7.920	2.0693912 058	7.970	2.0756844 928
7.821	2.0568124 236	7.871	2.0631851 192	7.921	2.0695174 605	7.971	2.0758099 554
7.822	2.0569402 763	7.872	2.0633121 597	7.922	2.0696436 992	7.972	2.0759354 024
7.823	2.0570681 127	7.873	2.0634391 842	7.923	2.0697699 220	7.973	2.0760608 335
7.824	2.0571959 327	7.874	2.0635661 925	7.924	2.0698961 288	7.974	2.0761862 490
7.825	2.0573237 364	7.875	2.0636931 847	7.925	2.0700223 198	7.975	2.0763116 487
7.826	2.0574515 238	7.876	2.0638201 608	7.926	2.0701484 948	7.976	2.0764370 327
7.827	2.0575792 948	7.877	2.0639471 207	7.927	2.0702746 539	7.977	2.0765624 009
7.828	2.0577070 495	7.878	2.0640740 645	7.928	2.0704007 970	7.978	2.0766877 535
7.829	2.0578347 879	7.879	2.0642009 923	7.929	2.0705269 243	7.979	2.0768130 903
7.830	2.0579625 100	7.880	2.0643279 039	7.930	2.0706530 356	7.980	2.0769384 115
7.831	2.0580902 158	7.881	2.0644547 994	7.931	2.0707791 311	7.981	2.0770637 169
7.832	2.0582179 052	7.882	2.0645816 788	7.932	2.0709052 107	7.982	2.0771890 067
7.833	2.0583455 784	7.883	2.0647085 421	7.933	2.0710312 743	7.983	2.0773142 807
7.834	2.0584732 352	7.884	2.0648353 893	7.934	2.0711573 221	7.984	2.0774395 390
7.835	2.0586008 758	7.885	2.0649622 204	7.935	2.0712833 540	7.985	2.0775647 817
7.836	2.0587285 001	7.886	2.0650890 355	7.936	2.0714093 700	7.986	2.0776900 086
7.837	2.0588561 081	7.887	2.0652158 344	7.937	2.0715353 701	7.987	2.0778152 199
7.838	2.0589836 998	7.888	2.0653426 173	7.938	2.0716613 544	7.988	2.0779404 156
7.839	2.0591112 752	7.889	2.0654693 841	7.939	2.0717873 227	7.989	2.0780655 955
7.840	2.0592388 344	7.890	2.0655961 349	7.940	2.0719132 753	7.990	2.0781907 598
7.841	2.0593663 772	7.891	2.0657228 695	7.941	2.0720392 119	7.991	2.0783159 084
7.842	2.0594939 039	7.892	2.0658495 882	7.942	2.0721651 327	7.992	2.0784410 413
7.843	2.0596214 142	7.893	2.0659762 907	7.943	2.0722910 377	7.993	2.0785661 586
7.844	2.0597489 083	7.894	2.0661029 772	7.944	2.0724169 267	7.994	2.0786912 603
7.845	2.0598763 862	7.895	2.0662296 477	7.945	2.0725428 000	7.995	2.0788163 463
7.846	2.0600038 478	7.896	2.0663563 021	7.946	2.0726686 574	7.996	2.0789414 166
7.847	2.0601312 931	7.897	2.0664829 405	7.947	2.0727944 990	7.997	2.0790664 713
7.848	2.0602587 223	7.898	2.0666095 629	7.948	2.0729203 247	7.998	2.0791915 104
7.849	2.0603861 351	7.899	2.0667361 692	7.949	2.0730461 346	7.999	2.0793165 339

TEN PLACE MANTISSAS FOR NATURAL LOGARITHMS (Continued)

x	$\log_e x$	x	$\log_e x$	x	$\log_e x$	x	$\log_e x$
8.000	2.0794415 417	8.050	2.0856720 914	8.100	2.0918640 617	8.150	2.0980179 273
8.001	2.0795665 339	8.051	2.0857963 073	8.101	2.0919875 108	8.151	2.0981406 191
8.002	2.0796915 104	8.052	2.0859205 078	8.102	2.0921109 448	8.152	2.0982632 959
8.003	2.0798164 714	8.053	2.0860446 928	8.103	2.0922343 635	8.153	2.0983859 577
8.004	2.0799414 167	8.054	2.0861688 624	8.104	2.0923577 669	8.154	2.0985086 044
8.005	2.0800663 464	8.055	2.0862930 166	8.105	2.0924811 552	8.155	2.0086312 361
8.006	2.0801912 606	8.056	2.0864171 554	8.106	2.0926045 282	8.156	2.0987538 527
8.007	2.0803161 591	8.057	2.0865412 788	8.107	2.0927278 860	8.157	2.0988764 543
8.008	2.0804410 420	8.058	2.0866653 868	8.108	2.0928512 286	8.158	2.0989990 409
8.009	2.0805659 093	8.059	2.0867894 793	8.109	2.0929745 560	8.159	2.0991216 124
8.010	2.0806907 611	8.060	2.0869135 565	8.110	2.0930978 681	8.160	2.0992441 690
8.011	2.0808155 972	8.061	2.0870376 183	8.111	2.0932211 651	8.161	2.0993667 105
8.012	2.0809404 178	8.062	2.0871616 647	8.112	2.0933444 468	8.162	2.0994892 370
8.013	2.0810652 228	8.063	2.0872856 957	8.113	2.0934677 134	8.163	2.0996117 485
8.014	2.0811900 122	8.064	2.0874097 113	8.114	2.0935909 648	8.164	2.0997342 449
8.015	2.0813147 861	8.065	2.0875337 116	8.115	2.0937142 010	8.165	2.0998567 264
8.016	2.0814395 443	8.066	2.0876576 965	8.116	2.0938374 220	8.166	2.0999791 929
8.017	2.0815642 871	8.067	2.0877816 660	8.117	2.0939606 278	8.167	2.1001016 444
8.018	2.0816890 142	8.068	2.0879056 201	8.118	2.0940838 184	8.168	2.1002240 809
8.019	2.0818137 258	8.069	2.0880295 589	8.119	2.0942069 939	8.169	2.1003465 024
8.020	2.0819384 219	8.070	2.0881534 823	8.120	2.0943301 542	8.170	2.1004689 089
8.021	2.0820631 024	8.071	2.0882773 903	8.121	2.0944532 993	8.171	2.1005913 004
8.022	2.0821877 673	8.072	2.0884012 831	8.122	2.0945764 293	8.172	2.1007136 770
8.023	2.0823124 168	8.073	2.0885251 604	8.123	2.0946995 441	8.173	2.1008360 385
8.024	2.0824370 507	8.074	2.0886490 224	8.124	2.0948226 437	8.174	2.1009583 851
8.025	2.0825616 690	8.075	2.0887728 691	8.125	2.0949457 282	8.175	2.1010807 168
8.026	2.0826862 718	8.076	2.0888967 005	8.126	2.0950687 976	8.176	2.1012030 335
8.027	2.0828108 591	8.077	2.0890205 165	8.127	2.0951918 518	8.177	2.1013253 352
8.028	2.0829354 309	8.078	2.0891443 171	8.128	2.0953148 908	8.178	2.1014476 219
8.029	2.0830599 872	8.079	2.0892681 025	8.129	2.0954379 148	8.179	2.1015698 938
8.030	2.0831845 280	8.080	2.0893918 725	8.130	2.0955609 236	8.180	2.1016921 506
8.031	2.0833090 532	8.081	2.0895156 273	8.131	2.0956839 172	8.181	2.1018143 925
8.032	2.0834335 629	8.082	2.0896393 667	8.132	2.0958068 958	8.182	2.1019366 195
8.033	2.0835580 572	8.083	2.0897630 908	8.133	2.0959298 592	8.183	2.1020588 315
8.034	2.0836825 359	8.084	2.0898867 995	8.134	2.0960528 075	8.184	2.1021810 286
8.035	2.0838069 992	8.085	2.0900104 930	8.135	2.0961757 407	8.185	2.1023032 108
8.036	2.0839314 470	8.086	2.0901341 712	8.136	2.0962986 587	8.186	2.1024253 781
8.037	2.0840558 792	8.087	2.0902578 341	8.137	2.0964215 617	8.187	2.1025475 304
8.038	2.0841802 960	8.088	2.0903814 817	8.138	2.0965444 496	8.188	2.1026696 678
8.039	2.0843046 973	8.089	2.0905051 140	8.139	2.0966673 223	8.189	2.1027917 903
8.040	2.0844290 832	8.090	2.0906287 311	8.140	2.0967901 800	8.190	2.1029138 979
8.041	2.0845534 536	8.091	2.0907523 328	8.141	2.0969130 226	8.191	2.1030359 905
8.042	2.0846778 085	8.092	2.0908759 193	8.142	2.0970358 501	8.192	2.1031580 683
8.043	2.0848021 479	8.093	2.0909994 905	8.143	2.0971586 625	8.193	2.1032801 312
8.044	2.0849264 719	8.094	2.0911230 465	8.144	2.0972814 598	8.194	2.1034021 791
8.045	2.0850507 804	8.095	2.0912465 871	8.145	2.0974042 421	8.195	2.1035242 122
8.046	2.0851750 735	8.096	2.0913701 125	8.146	2.0975270 092	8.196	2.1036462 304
8.047	2.0852993 512	8.097	2.0914936 227	8.147	2.0976497 613	8.197	2.1037682 337
8.048	2.0854236 134	8.098	2.0916171 176	8.148	2.0977724 984	8.198	2.1038902 221
8.049	2.0855478 601	8.099	2.0917405 973	8.149	2.0978952 203	8.199	2.1040121 956

TEN PLACE MANTISSAS FOR NATURAL LOGARITHMS (Continued)

x	$\log_e x$	x	$\log_e x$	x	$\log_e x$	x	$\log_e x$
8.200	2.1041341 543	8.250	2.1102132 003	8.300	2.1162555 148	8.350	2.1222615 389
8.201	2.1042560 981	8.251	2.1103344 051	8.301	2.1163759 895	8.351	2.1223812 922
8.202	2.1043780 270	8.252	2.1104555 952	8.302	2.1164964 496	8.352	2.1225010 311
8.203	2.1044999 410	8.253	2.1105767 706	8.303	2.1166168 953	8.353	2.1226207 558
8.204	2.1046218 402	8.254	2.1106979 313	8.304	2.1167373 264	8.354	2.1227404 661
8.205	2.1047437 245	8.255	2.1108190 774	8.305	2.1168577 431	8.355	2.1228601 620
8.206	2.1048655 940	8.256	2.1109402 087	8.306	2.1169781 452	8.356	2.1229798 437
8.207	2.1049874 486	8.257	2.1110613 254	8.307	2.1170985 329	8.357	2.1230995 110
8.208	2.1051092 884	8.258	2.1111824 275	8.308	2.1172189 060	8.358	2.1232191 640
8.209	2.1052311 134	8.259	2.1113035 148	8.309	2.1173392 647	8.359	2.1233388 027
8.210	2.1053529 235	8.260	2.1114245 875	8.310	2.1174596 089	8.360	2.1234584 271
8.211	2.1054747 187	8.261	2.1115456 456	8.311	2.1175799 386	8.361	2.1235780 372
8.212	2.1055964 992	8.262	2.1116666 890	8.312	2.1177002 538	8.362	2.1236976 329
8.213	2.1057182 648	8.263	2.1117877 177	8.313	2.1178205 545	8.363	2.1238172 144
8.214	2.1058400 155	8.264	2.1119087 318	8.314	2.1179408 408	8.364	2.1239367 816
8.215	2.1059617 515	8.265	2.1120297 313	8.315	2.1180611 126	8.365	2.1240563 344
8.216	2.1060834 726	8.266	2.1121507 161	8.316	2.1181813 700	8.366	2.1241758 730
8.217	2.1062051 789	8.267	2.1122716 863	8.317	2.1183016 129	8.367	2.1242953 973
8.218	2.1063268 705	8.268	2.1123926 418	8.318	2.1184218 413	8.368	2.1244149 073
8.219	2.1064485 472	8.269	2.1125135 827	8.319	2.1185420 553	8.369	2.1245344 031
8.220	2.1065702 091	8.270	2.1126345 090	8.320	2.1186622 548	8.370	2.1246538 845
8.221	2.1066918 562	8.271	2.1127554 207	8.321	2.1187824 399	8.371	2.1247733 517
8.222	2.1068134 885	8.272	2.1128763 178	8.322	2.1189026 106	8.372	2.1248928 046
8.223	2.1069351 060	8.273	2.1129972 002	8.323	2.1190227 668	8.373	2.1250122 432
8.224	2.1070567 087	8.274	2.1131180 680	8.324	2.1191429 085	8.374	2.1251316 676
8.225	2.1071782 967	8.275	2.1132389 213	8.325	2.1192630 359	8.375	2.1252510 777
8.226	2.1072998 698	8.276	2.1133597 599	8.326	2.1193831 488	8.376	2.1253704 736
8.227	2.1074214 282	8.277	2.1134805 839	8.327	2.1195032 473	8.377	2.1254898 552
8.228	2.1075429 718	8.278	2.1136013 933	8.328	2.1196233 313	8.378	2.1256092 225
8.229	2.1076645 006	8.279	2.1137221 882	8.329	2.1197434 010	8.379	2.1257285 756
8.230	2.1077860 147	8.280	2.1138429 684	8.330	2.1198634 562	8.380	2.1258479 145
8.231	2.1079075 140	8.281	2.1139637 341	8.331	2.1199834 970	8.381	2.1259672 391
8.232	2.1080289 985	8.282	2.1140844 851	8.332	2.1201035 234	8.382	2.1260865 495
8.233	2.1081504 683	8.283	2.1142052 216	8.333	2.1202235 354	8.383	2.1262058 457
8.234	2.1082719 233	8.284	2.1143259 435	8.334	2.1203435 330	8.384	2.1263251 276
8.235	2.1083933 636	8.285	2.1144466 509	8.335	2.1204635 162	8.385	2.1264443 953
8.236	2.1085147 892	8.286	2.1145673 437	8.336	2.1205834 850	8.386	2.1265636 487
8.237	2.1086362 000	8.287	2.1146880 219	8.337	2.1207034 394	8.387	2.1266828 880
8.238	2.1087575 960	8.288	2.1148086 855	8.338	2.1208233 795	8.388	2.1268021 130
8.239	2.1088789 773	8.289	2.1149293 346	8.339	2.1209433 051	8.389	2.1269213 239
8.240	2.1090003 439	8.290	2.1150499 691	8.340	2.1210632 164	8.390	2.1270405 205
8.241	2.1091216 958	8.291	2.1151705 891	8.341	2.1211831 133	8.391	2.1271597 029
8.242	2.1092430 329	8.292	2.1152911 946	8.342	2.1213029 958	8.392	2.1272788 711
8.243	2.1093643 553	8.293	2.1154117 855	8.343	2.1214228 639	8.393	2.1273980 251
8.244	2.1094856 630	8.294	2.1155323 618	8.344	2.1215427 177	8.394	2.1275171 649
8.245	2.1096069 560	8.295	2.1156529 236	8.345	2.1216625 571	8.395	2.1276362 905
8.246	2.1097282 343	8.296	2.1157734 709	8.346	2.1217823 822	8.396	2.1277554 020
8.247	2.1098494 979	8.297	2.1158940 037	8.347	2.1219021 929	8.397	2.1278744 992
8.248	2.1099707 467	8.298	2.1160145 219	8.348	2.1220219 892	8.398	2.1279935 823
8.249	2.1100919 809	8.299	2.1161350 256	8.349	2.1221417 712	8.399	2.1281126 511

TEN PLACE MANTISSAS FOR NATURAL LOGARITHMS (Continued)

x	$\log_e x$	x	$\log_e x$	x	$\log_e x$	x	$\log_e x$
8.400	2.1282317 058	8.450	2.1341664 414	8.500	2.1400661 635	8.550	2.1459312 829
8.401	2.1283507 464	8.451	2.1342847 776	8.501	2.1401838 036	8.551	2.1460482 352
8.402	2.1284697 727	8.452	2.1344030 998	8.502	2.1403014 299	8.552	2.1461651 737
8.403	2.1285887 849	8.453	2.1345214 079	8.503	2.1404190 424	8.553	2.1462820 986
8.404	2.1287077 830	8.454	2.1346397 021	8.504	2.1405366 410	8.554	2.1463990 098
8.405	2.1288267 669	8.455	2.1347579 824	8.505	2.1406542 258	8.555	2.1465159 073
8.406	2.1289457 366	8.456	2.1348762 486	8.506	2.1407717 968	8.556	2.1466327 912
8.407	2.1290646 922	8.457	2.1349945 008	8.507	2.1408893 540	8.557	2.1467496 614
8.408	2.1291836 336	8.458	2.1351127 391	8.508	2.1410068 973	8.558	2.1468665 180
8.409	2.1293025 609	8.459	2.1352309 633	8.509	2.1411244 269	8.559	2.1469833 609
8.410	2.1294214 740	8.460	2.1353491 736	8.510	2.1412419 426	8.560	2.1471001 902
8.411	2.1295403 730	8.461	2.1354673 699	8.511	2.1413594 445	8.561	2.1472170 058
8.412	2.1296592 578	8.462	2.1355855 523	8.512	2.1414769 326	8.562	2.1473338 077
8.413	2.1297781 286	8.463	2.1357037 207	8.513	2.1415944 069	8.563	2.1474505 960
8.414	2.1298969 852	8.464	2.1358218 751	8.514	2.1417118 674	8.564	2.1475673 707
8.415	2.1300158 276	8.465	2.1359400 156	8.515	2.1418293 141	8.565	2.1476841 318
8.416	2.1301346 560	8.466	2.1360581 420	8.516	2.1419467 470	8.566	2.1478008 792
8.417	2.1302534 702	8.467	2.1361762 547	8.517	2.1420641 662	8.567	2.1479176 130
8.418	2.1303722 703	8.468	2.1362943 533	8.518	2.1421815 715	8.568	2.1480343 331
8.419	2.1304910 564	8.469	2.1364124 379	8.519	2.1422989 631	8.569	2.1481510 397
8.420	2.1306098 283	8.470	2.1365305 087	8.520	2.1424163 408	8.570	2.1482677 326
8.421	2.1307285 860	8.471	2.1366485 654	8.521	2.1425337 048	8.571	2.1483844 119
8.422	2.1308473 297	8.472	2.1367666 083	8.522	2.1426510 551	8.572	2.1485010 776
8.423	2.1309660 593	8.473	2.1368846 372	8.523	2.1427683 915	8.573	2.1486177 297
8.424	2.1310847 748	8.474	2.1370026 522	8.524	2.1428857 142	8.574	2.1487343 682
8.425	2.1312034 762	8.475	2.1371206 533	8.525	2.1430030 232	8.575	2.1488509 931
8.426	2.1313221 636	8.476	2.1372386 404	8.526	2.1431203 183	8.576	2.1489676 043
8.427	2.1314408 368	8.477	2.1373566 136	8.527	2.1432375 998	8.577	2.1490842 020
8.428	2.1315594 959	8.478	2.1374745 729	8.528	2.1433548 674	8.578	2.1492007 861
8.429	2.1316781 410	8.479	2.1375925 183	8.529	2.1434721 213	8.579	2.1493173 566
8.430	2.1317967 720	8.480	2.1377104 498	8.530	2.1435893 615	8.580	2.1494339 135
8.431	2.1319153 889	8.481	2.1378283 674	8.531	2.1437065 879	8.581	2.1495504 568
8.432	2.1320339 918	8.482	2.1379462 711	8.532	2.1438238 006	8.582	2.1496669 866
8.433	2.1321525 806	8.483	2.1380641 608	8.533	2.1439409 996	8.583	2.1497835 027
8.434	2.1322711 553	8.484	2.1381820 367	8.534	2.1440581 848	8.584	2.1499000 053
8.435	2.1323897 160	8.485	2.1382998 987	8.535	2.1441753 562	8.585	2.1500164 943
8.436	2.1325082 626	8.486	2.1384177 468	8.536	2.1442925 140	8.586	2.1501329 698
8.437	2.1326267 952	8.487	2.1385355 810	8.537	2.1444096 580	8.587	2.1502494 317
8.438	2.1327453 137	8.488	2.1386534 013	8.538	2.1445267 883	8.588	2.1503658 800
8.439	2.1328638 182	8.489	2.1387712 078	8.539	2.1446439 049	8.589	2.1504823 148
8.440	2.1329823 086	8.490	2.1388890 003	8.540	2.1447610 078	8.590	2.1505987 360
8.441	2.1331007 850	8.491	2.1390067 790	8.541	2.1448780 969	8.591	2.1507151 437
8.442	2.1332192 474	8.492	2.1391245 438	8.542	2.1449951 724	8.592	2.1508315 378
8.443	2.1333376 957	8.493	2.1392422 948	8.543	2.1451122 342	8.593	2.1509479 183
8.444	2.1334561 300	8.494	2.1393600 319	8.544	2.1452292 822	8.594	2.1510642 854
8.445	2.1335745 503	8.495	2.1394777 551	8.545	2.1453463 166	8.595	2.1511806 388
8.446	2.1336929 565	8.496	2.1395954 645	8.546	2.1454633 372	8.596	2.1512969 788
8.447	2.1338113 487	8.497	2.1397131 600	8.547	2.1455803 442	8.597	2.1514133 052
8.448	2.1339297 270	8.498	2.1398308 417	8.548	2.1456973 375	8.598	2.1515296 181
8.449	2.1340480 912	8.499	2.1399485 095	8.549	2.1458143 170	8.599	2.1516459 174

TEN PLACE MANTISSAS FOR NATURAL LOGARITHMS (Continued)

x	$\log_e x$	x	$\log_e x$	x	$\log_e x$	x	$\log_e x$
8.600	2.1517622 033	8.650	2.1575593 209	8.700	2.1633230 257	8.750	2.1690537 004
8.601	2.1518784 756	8.651	2.1576749 212	8.701	2.1634379 616	8.751	2.1691679 796
8.602	2.1519947 344	8.652	2.1577905 081	8.702	2.1635528 843	8.752	2.1692822 457
8.603	2.1521109 796	8.653	2.1579060 816	8.703	2.1636677 938	8.753	2.1693964 988
8.604	2.1522272 114	8.654	2.1580216 418	8.704	2.1637826 901	8.754	2.1695107 388
8.605	2.1523434 297	8.655	2.1581371 886	8.705	2.1638975 732	8.755	2.1696249 657
8.606	2.1524596 344	8.656	2.1582527 221	8.706	2.1640124 431	8.756	2.1697391 797
8.607	2.1525758 257	8.657	2.1583682 422	8.707	2.1641272 998	8.757	2.1698533 805
8.608	2.1526920 034	8.658	2.1584837 490	8.708	2.1642421 434	8.758	2.1699675 684
8.609	2.1528081 677	8.659	2.1585992 425	8.709	2.1643569 737	8.759	2.1700817 432
8.610	2.1529243 184	8.660	2.1587147 226	8.710	2.1644717 909	8.760	2.1701959 049
8.611	2.1530404 557	8.661	2.1588301 893	8.711	2.1645865 948	8.761	2.1703100 537
8.612	2.1531565 795	8.662	2.1589456 428	8.712	2.1647013 856	8.762	2.1704241 894
8.613	2.1532726 898	8.663	2.1590610 829	8.713	2.1648161 632	8.763	2.1705383 121
8.614	2.1533887 866	8.664	2.1591765 097	8.714	2.1649309 277	8.764	2.1706524 217
8.615	2.1535048 700	8.665	2.1592919 232	8.715	2.1650456 790	8.765	2.1707665 184
8.616	2.1536209 399	8.666	2.1594073 233	8.716	2.1651604 171	8.766	2.1708806 020
8.617	2.1537369 963	8.667	2.1595227 102	8.717	2.1652751 420	8.767	2.1709946 726
8.618	2.1538530 392	8.668	2.1596380 837	8.718	2.1653898 538	8.768	2.1711087 302
8.619	2.1539690 687	8.669	2.1597534 439	8.719	2.1655045 524	8.769	2.1712227 748
8.620	2.1540850 847	8.670	2.1598687 908	8.720	2.1656192 379	8.770	2.1713368 064
8.621	2.1542010 872	8.671	2.1599841 244	8.721	2.1657339 102	8.771	2.1714508 250
8.622	2.1543170 763	8.672	2.1600994 447	8.722	2.1658485 694	8.772	2.1715648 306
8.623	2.1544330 520	8.673	2.1602147 517	8.723	2.1659632 155	8.773	2.1716788 231
8.624	2.1545490 142	8.674	2.1603300 454	8.724	2.1660778 483	8.774	2.1717928 027
8.625	2.1546649 629	8.675	2.1604453 258	8.725	2.1661924 681	8.775	2.1719067 694
8.626	2.1547808 982	8.676	2.1605605 930	8.726	2.1663070 747	8.776	2.1720207 230
8.627	2.1548968 201	8.677	2.1606758 468	8.727	2.1664216 682	8.777	2.1721346 636
8.628	2.1550127 285	8.678	2.1607910 874	8.728	2.1665362 485	8.778	2.1722485 913
8.629	2.1551286 235	8.679	2.1609063 147	8.729	2.1666508 158	8.779	2.1723625 059
8.630	2.1552445 051	8.680	2.1610215 287	8.730	2.1667653 699	8.780	2.1724764 076
8.631	2.1553603 732	8.681	2.1611367 294	8.731	2.1668799 108	8.781	2.1725902 964
8.632	2.1554762 280	8.682	2.1612519 169	8.732	2.1669944 387	8.782	2.1727041 721
8.633	2.1555920 693	8.683	2.1613670 911	8.733	2.1671089 534	8.783	2.1728180 349
8.634	2.1557078 971	8.684	2.1614822 520	8.734	2.1672234 551	8.784	2.1729318 848
8.635	2.1558237 116	8.685	2.1615973 997	8.735	2.1673379 436	8.785	2.1730457 216
8.636	2.1559395 127	8.686	2.1617125 341	8.736	2.1674524 190	8.786	2.1731595 456
8.637	2.1560553 003	8.687	2.1618276 553	8.737	2.1675668 813	8.787	2.1732733 565
8.638	2.1561710 745	8.688	2.1619427 632	8.738	2.1676813 305	8.788	2.1733871 545
8.639	2.1562868 354	8.689	2.1620578 579	8.739	2.1677957 666	8.789	2.1735009 396
8.640	2.1564025 828	8.690	2.1621729 393	8.740	2.1679101 897	8.790	2.1736147 117
8.641	2.1565183 169	8.691	2.1622880 075	8.741	2.1680245 996	8.791	2.1737284 709
8.642	2.1566340 375	8.692	2.1624030 624	8.742	2.1681389 964	8.792	2.1738422 171
8.643	2.1567497 448	8.693	2.1625181 041	8.743	2.1682533 802	8.793	2.1739559 504
8.644	2.1568654 386	8.694	2.1626331 326	8.744	2.1683677 509	8.794	2.1740696 708
8.645	2.1569811 191	8.695	2.1627481 478	8.745	2.1684821 085	8.795	2.1741833 782
8.646	2.1570967 862	8.696	2.1628631 498	8.746	2.1685964 530	8.796	2.1742970 727
8.647	2.1572124 400	8.697	2.1629781 386	8.747	2.1687107 844	8.797	2.1744107 543
8.648	2.1573280 803	8.698	2.1630931 142	8.748	2.1688251 028	8.798	2.1745244 229
8.649	2.1574437 073	8.699	2.1632080 765	8.749	2.1689394 081	8.799	2.1746380 786

TEN PLACE MANTISSAS FOR NATURAL LOGARITHMS (Continued)

x	$\log_e x$	x	$\log_e x$	x	$\log_e x$	x	$\log_e x$
8.800	2.1747517 215	8.850	2.1804174 590	8.900	2.1860512 767	8.950	2.1916535 323
8.801	2.1748653 514	8.851	2.1805304 470	8.901	2.1861636 300	8.951	2.1917652 579
8.802	2.1749789 684	8.852	2.1806434 222	8.902	2.1862759 706	8.952	2.1918769 710
8.803	2.1750925 725	8.853	2.1807563 846	8.903	2.1863882 986	8.953	2.1919886 717
8.804	2.1752061 637	8.854	2.1808693 343	8.904	2.1865006 140	8.954	2.1921003 598
8.805	2.1753197 419	8.855	2.1809822 712	8.905	2.1866129 167	8.955	2.1922120 355
8.806	2.1754333 073	8.856	2.1810951 954	8.906	2.1867252 069	8.956	2.1923236 987
8.807	2.1755468 598	8.857	2.1812081 068	8.907	2.1868374 844	8.957	2.1924353 495
8.808	2.1756603 994	8.858	2.1813210 055	8.908	2.1869497 494	8.958	2.1925469 878
8.809	2.1757739 261	8.859	2.1814338 914	8.909	2.1870620 017	8.959	2.1926586 136
8.810	2.1758874 399	8.860	2.1815467 646	8.910	2.1871742 415	8.960	2.1927702 270
8.811	2.1760009 409	8.861	2.1816596 251	8.911	2.1872864 686	8.961	2.1928818 279
8.812	2.1761144 289	8.862	2.1817724 728	8.912	2.1873986 832	8.962	2.1929934 164
8.813	2.1762279 041	8.863	2.1818853 078	8.913	2.1875108 851	8.963	2.1931049 924
8.814	2.1763413 664	8.864	2.1819981 300	8.914	2.1876230 745	8.964	2.1932165 559
8.815	2.1764548 158	8.865	2.1821109 395	8.915	2.1877352 513	8.965	2.1933281 071
8.816	2.1765682 524	8.866	2.1822237 363	8.916	2.1878474 155	8.966	2.1934396 457
8.817	2.1766816 761	8.867	2.1823365 204	8.917	2.1879595 672	8.967	2.1935511 720
8.818	2.1767950 869	8.868	2.1824492 918	8.918	2.1880717 062	8.968	2.1936626 858
8.819	2.1769084 849	8.869	2.1825620 504	8.919	2.1881838 327	8.969	2.1937741 871
8.820	2.1770218 700	8.870	2.1826747 963	8.920	2.1882959 466	8.970	2.1938856 761
8.821	2.1771352 423	8.871	2.1827875 295	8.921	2.1884080 479	8.971	2.1939971 526
8.822	2.1772486 017	8.872	2.1829002 500	8.922	2.1885201 367	8.972	2.1941086 167
8.823	2.1773619 482	8.873	2.1830129 579	8.923	2.1886322 129	8.973	2.1942200 683
8.824	2.1774752 820	8.874	2.1831256 530	8.924	2.1887442 766	8.974	2.1943315 076
8.825	2.1775886 028	8.875	2.1832383 354	8.925	2.1888563 277	8.975	2.1944429 344
8.826	2.1777019 108	8.876	2.1833510 051	8.926	2.1889683 662	8.976	2.1945543 488
8.827	2.1778152 060	8.877	2.1834636 621	8.927	2.1890803 922	8.977	2.1946657 508
8.828	2.1779284 884	8.878	2.1835763 064	8.928	2.1891924 056	8.978	2.1947771 404
8.829	2.1780417 579	8.879	2.1836889 381	8.929	2.1893044 065	8.979	2.1948885 175
8.830	2.1781550 146	8.880	2.1838015 570	8.930	2.1894163 949	8.980	2.1949998 823
8.831	2.1782682 585	8.881	2.1839141 633	8.931	2.1895283 707	8.981	2.1951112 347
8.832	2.1783814 895	8.882	2.1840267 569	8.932	2.1896403 340	8.982	2.1952225 747
8.833	2.1784947 078	8.883	2.1841393 378	8.933	2.1897522 847	8.983	2.1953339 022
8.834	2.1786079 132	8.884	2.1842519 060	8.934	2.1898642 229	8.984	2.1954452 174
8.835	2.1787211 058	8.885	2.1843644 616	8.935	2.1899761 486	8.985	2.1955565 202
8.836	2.1788342 856	8.886	2.1844770 045	8.936	2.1900880 618	8.986	2.1956678 106
8.837	2.1789474 525	8.887	2.1845895 348	8.937	2.1901999 624	8.987	2.1957790 887
8.838	2.1790606 067	8.888	2.1847020 523	8.938	2.1903118 505	8.988	2.1958903 543
8.839	2.1791737 481	8.889	2.1848145 573	8.939	2.1904237 261	8.989	2.1960016 076
8.840	2.1792868 766	8.890	2.1849270 495	8.940	2.1905355 892	8.990	2.1961128 485
8.841	2.1793999 924	8.891	2.1850395 291	8.941	2.1906474 398	8.991	2.1962240 770
8.842	2.1795130 954	8.892	2.1851519 961	8.942	2.1907592 778	8.992	2.1963352 932
8.843	2.1796261 856	8.893	2.1852644 504	8.943	2.1908711 034	8.993	2.1964464 969
8.844	2.1797392 630	8.894	2.1853768 921	8.944	2.1909829 164	8.994	2.1965576 883
8.845	2.1798523 276	8.895	2.1854893 211	8.945	2.1910947 170	8.995	2.1966688 674
8.846	2.1799653 794	8.896	2.1856017 375	8.946	2.1912065 050	8.996	2.1967800 341
8.847	2.1800784 185	8.897	2.1857141 413	8.947	2.1913182 806	8.997	2.1968911 884
8.848	2.1801914 448	8.898	2.1858265 324	8.948	2.1914300 436	8.998	2.1970023 304
8.849	2.1803044 583	8.899	2.1859389 109	8.949	2.1915417 942	8.999	2.1971134 601

TEN PLACE MANTISSAS FOR NATURAL LOGARITHMS (Continued)

x	$\log_e x$	x	$\log_e x$	x	$\log_e x$	x	$\log_e x$
9.000	2.1972245 773	9.050	2.2027647 577	9.100	2.2082744 135	9.150	2.2137538 793
9.001	2.1973356 823	9.051	2.2028752 488	9.101	2.2083842 976	9.151	2.2138631 629
9.002	2.1974467 749	9.052	2.2029857 278	9.102	2.2084941 696	9.152	2.2139724 346
9.003	2.1975578 551	9.053	2.2030961 945	9.103	2.2086040 295	9.153	2.2140816 944
9.004	2.1976689 230	9.054	2.2032066 490	9.104	2.2087138 774	9.154	2.2141909 422
9.005	2.1977799 786	9.055	2.2033170 913	9.105	2.2088237 132	9.155	2.2143001 781
9.006	2.1978910 219	9.056	2.2034275 215	9.106	2.2089335 369	9.156	2.2144094 021
9.007	2.1980020 528	9.057	2.2035379 394	9.107	2.2090433 486	9.157	2.2145186 141
9.008	2.1981130 714	9.058	2.2036483 451	9.108	2.2091531 482	9.158	2.2146278 142
9.009	2.1982240 777	9.059	2.2037587 387	9.109	2.2092629 358	9.159	2.2147370 024
9.010	2.1983350 716	9.060	2.2038691 201	9.110	2.2093727 113	9.160	2.2148461 787
9.011	2.1984460 533	9.061	2.2039794 892	9.111	2.2094824 747	9.161	2.2149553 430
9.012	2.1985570 226	9.062	2.2040898 462	9.112	2.2095922 261	9.162	2.2150644 955
9.013	2.1986679 796	9.063	2.2042001 911	9.113	2.2097019 655	9.163	2.2151736 360
9.014	2.1987789 243	9.064	2.2043105 237	9.114	2.2098116 928	9.164	2.2152827 646
9.015	2.1988898 567	9.065	2.2044208 442	9.115	2.2099214 081	9.165	2.2153918 813
9.016	2.1990007 767	9.066	2.2045311 525	9.116	2.2100311 114	9.166	2.2155009 861
9.017	2.1991116 845	9.067	2.2046414 487	9.117	2.2101408 026	9.167	2.2156100 790
9.018	2.1992225 800	9.068	2.2047517 326	9.118	2.2102504 818	9.168	2.2157191 600
9.019	2.1993334 632	9.069	2.2048620 045	9.119	2.2103601 490	9.169	2.2158282 291
9.020	2.1994443 341	9.070	2.2049722 641	9.120	2.2104698 041	9.170	2.2159372 863
9.021	2.1995551 927	9.071	2.2050825 116	9.121	2.2105794 472	9.171	2.2160463 316
9.022	2.1996660 390	9.072	2.2051927 470	9.122	2.2106890 783	9.172	2.2161553 650
9.023	2.1997768 730	9.073	2.2053029 702	9.123	2.2107986 974	9.173	2.2162643 865
9.024	2.1998876 948	9.074	2.2054131 812	9.124	2.2109083 044	9.174	2.2163733 962
9.025	2.1999985 042	9.075	2.2055233 802	9.125	2.2110178 995	9.175	2.2164823 939
9.026	2.2001093 014	9.076	2.2056355 669	9.126	2.2111274 825	9.176	2.2165913 798
9.027	2.2002200 863	9.077	2.2057437 415	9.127	2.2112370 535	9.177	2.2167003 538
9.028	2.2003308 590	9.078	2.2058539 040	9.128	2.2113466 126	9.178	2.2168093 160
9.029	2.2004416 193	9.079	2.2059640 544	9.129	2.2114561 596	9.179	2.2169182 662
9.030	2.2005523 674	9.080	2.2060741 926	9.130	2.2115656 946	9.180	2.2170272 046
9.031	2.2006631 033	9.081	2.2061843 187	9.131	2.2116752 176	9.181	2.2171361 312
9.032	2.2007738 268	9.082	2.2062944 327	9.132	2.2117847 287	9.182	2.2172450 458
9.033	2.2008845 382	9.083	2.2064045 345	9.133	2.2118942 277	9.183	2.2173539 486
9.034	2.2009952 372	9.084	2.2065146 242	9.134	2.2120037 148	9.184	2.2174628 396
9.035	2.2011059 240	9.085	2.2066247 018	9.135	2.2121131 898	9.185	2.2175717 187
9.036	2.2012165 986	9.086	2.2067347 673	9.136	2.2122226 529	9.186	2.2176805 859
9.037	2.2013272 609	9.087	2.2068448 207	9.137	2.2123321 040	9.187	2.2177894 413
9.038	2.2014379 110	9.088	2.2069548 620	9.138	2.2124415 431	9.188	2.2178982 848
9.039	2.2015485 488	9.089	2.2070648 911	9.139	2.2125509 703	9.189	2.2180071 165
9.040	2.2016591 744	9.090	2.2071749 082	9.140	2.2126603 855	9.190	2.2181159 364
9.041	2.2017697 878	9.091	2.2072849 131	9.141	2.2127697 887	9.191	2.2182247 444
9.042	2.2018803 889	9.092	2.2073949 060	9.142	2.2128791 799	9.192	2.2183335 405
9.043	2.2019909 778	9.093	2.2075018 867	9.143	2.2129885 592	9.193	2.2184423 249
9.044	2.2021015 544	9.094	2.2076148 554	9.144	2.2130979 265	9.194	2.2185510 974
9.045	2.2022121 188	9.095	2.2077248 120	9.145	2.2132072 818	9.195	2.2186598 581
9.046	2.2023226 711	9.096	2.2078347 564	9.146	2.2133166 252	9.196	2.2187686 069
9.047	2.2024332 110	9.097	2.2079446 888	9.147	2.2134259 567	9.197	2.2188773 439
9.048	2.2025437 388	9.098	2.2080546 091	9.148	2.2135352 762	9.198	2.2189860 691
9.049	2.2026542 544	9.099	2.2081645 174	9.149	2.2136445 837	9.199	2.2190947 825

TEN PLACE MANTISSAS FOR NATURAL LOGARITHMS (Continued)

x	$\log_e x$	x	$\log_e x$	x	$\log_e x$	x	$\log_e x$
9.200	2.2192034 841	9.250	2.2246235 515	9.300	2.2300144 002	9.350	2.2353763 433
9.201	2.2193121 738	9.251	2.2247316 538	9.301	2.2301219 213	9.351	2.2354832 895
9.202	2.2194208 517	9.252	2.2248397 444	9.302	2.2302294 308	9.352	2.2355902 242
9.203	2.2195295 179	9.253	2.2249478 233	9.303	2.2303369 288	9.353	2.2356971 475
9.204	2.2196381 722	9.254	2.2250558 905	9.304	2.2304444 152	9.354	2.2358040 593
9.205	2.2197468 147	9.255	2.2251639 460	9.305	2.2305518 901	9.355	2.2359109 597
9.206	2.2198554 454	9.256	2.2252719 899	9.306	2.2306593 534	9.356	2.2360178 487
9.207	2.2199640 643	9.257	2.2253800 221	9.307	2.2307668 052	9.357	2.2361247 263
9.208	2.2200726 714	9.258	2.2254880 426	9.308	2.2308742 454	9.358	2.2362315 924
9.209	2.2201812 667	9.259	2.2255960 515	9.309	2.2309816 741	9.359	2.2363384 472
9.210	2.2202898 503	9.260	2.2257040 487	9.310	2.2310890 913	9.360	2.2364452 905
9.211	2.2203984 220	9.261	2.2258120 342	9.311	2.2311964 969	9.361	2.2365521 224
9.212	2.2205069 820	9.262	2.2259200 081	9.312	2.2313038 910	9.362	2.2366589 429
9.213	2.2206155 301	9.263	2.2260279 703	9.313	2.2314112 735	9.363	2.2367657 520
9.214	2.2207240 665	9.264	2.2261359 208	9.314	2.2315186 446	9.364	2.2368725 496
9.215	2.2208325 911	9.265	2.2262438 597	9.315	2.2316260 041	9.365	2.2369793 359
9.216	2.2209411 040	9.266	2.2263517 870	9.316	2.2317333 520	9.366	2.2370861 108
9.217	2.2210496 050	9.267	2.2264597 026	9.317	2.2318406 885	9.367	2.2371928 742
9.218	2.2211580 943	9.268	2.2265676 066	9.318	2.2319480 134	9.368	2.2372996 263
9.219	2.2212665 718	9.269	2.2266754 989	9.319	2.2320553 268	9.369	2.2374063 670
9.220	2.2213750 376	9.270	2.2267833 796	9.320	2.2321626 287	9.370	2.2375130 963
9.221	2.2214834 916	9.271	2.2268912 486	9.321	2.2322699 191	9.371	2.2376198 141
9.222	2.2215919 338	9.272	2.2269991 060	9.322	2.2323771 980	9.372	2.2377265 206
9.223	2.2217003 643	9.273	2.2271069 518	9.323	2.2324844 653	9.373	2.2378332 158
9.224	2.2218087 830	9.274	2.2272147 860	9.324	2.2325917 212	9.374	2.2379398 995
9.225	2.2219171 899	9.275	2.2273226 085	9.325	2.2326989 655	9.375	2.2380465 719
9.226	2.2220255 851	9.276	2.2274304 194	9.326	2.2328061 984	9.376	2.2381532 328
9.227	2.2221339 686	9.277	2.2275382 187	9.327	2.2329134 197	9.377	2.2382598 824
9.228	2.2222423 403	9.278	2.2276460 063	9.328	2.2330206 296	9.378	2.2383665 207
9.229	2.2223507 003	9.279	2.2277537 824	9.329	2.2331278 280	9.379	2.2384731 475
9.230	2.2224590 485	9.280	2.2278615 468	9.330	2.2332350 149	9.380	2.2385797 630
9.231	2.2225673 850	9.281	2.2279692 996	9.331	2.2333421 903	9.381	2.2386863 671
9.232	2.2226757 098	9.282	2.2280770 408	9.332	2.2334493 542	9.382	2.2387929 599
9.233	2.2227840 228	9.283	2.2281847 704	9.333	2.2335565 066	9.383	2.2388995 413
9.234	2.2228923 241	9.284	2.2282924 884	9.334	2.2336636 475	9.384	2.2390061 114
9.235	2.2230006 137	9.285	2.2284001 948	9.335	2.2337707 770	9.385	2.2391126 700
9.236	2.2231088 915	9.286	2.2285078 896	9.336	2.2338778 950	9.386	2.2392192 174
9.237	2.2232171 576	9.287	2.2286155 728	9.337	2.2339850 015	9.387	2.2393257 534
9.238	2.2233254 120	9.288	2.2287232 444	9.338	2.2340920 965	9.388	2.2394322 780
9.239	2.2234336 547	9.289	2.2288309 044	9.339	2.2341991 801	9.389	2.2395387 913
9.240	2.2235418 857	9.290	2.2289385 528	9.340	2.2343062 522	9.390	2.2396452 932
9.241	2.2236501 049	9.291	2.2290461 897	9.341	2.2344133 129	9.391	2.2397517 838
9.242	2.2237583 124	9.292	2.2291538 149	9.342	2.2345203 621	9.392	2.2398582 631
9.243	2.2238665 083	9.293	2.2292614 286	9.343	2.2346273 998	9.393	2.2399647 310
9.244	2.2239746 924	9.294	2.2293690 307	9.344	2.2347344 261	9.394	2.2400711 876
9.245	2.2240828 648	9.295	2.2294766 212	9.345	2.2348414 409	9.395	2.2401776 329
9.246	2.2241910 256	9.296	2.2295842 001	9.346	2.2349484 443	9.396	2.2402840 668
9.247	2.2242991 746	9.297	2.2296917 675	9.347	2.2350554 362	9.397	2.2403904 894
9.248	2.2244073 119	9.298	2.2297993 233	9.348	2.2351624 167	9.398	2.2404969 007
9.249	2.2245154 376	9.299	2.2299068 675	9.349	2.2352693 857	9.399	2.2406033 006

TEN PLACE MANTISSAS FOR NATURAL LOGARITHMS (Continued)

x	$\log_e x$	x	$\log_e x$	x	$\log_e x$	x	$\log_e x$
9.400	2.2407096 893	9.450	2.2460147 415	9.500	2.2512917 986	9.550	2.2565411 545
9.401	2.2408160 666	9.451	2.2461205 560	9.501	2.2513970 562	9.551	2.2566458 611
9.402	2.2409224 326	9.452	2.2462263 593	9.502	2.2515023 028	9.552	2.2567505 567
9.403	2.2410287 873	9.453	2.2463321 514	9.503	2.2516075 382	9.553	2.2568552 413
9.404	2.2411351 307	9.454	2.2464379 324	9.504	2.2517127 626	9.554	2.2569599 150
9.405	2.2412414 628	9.455	2.2465437 021	9.505	2.2518179 759	9.555	2.2570645 777
9.406	2.2413477 835	9.456	2.2466494 607	9.506	2.2519231 782	9.556	2.2571692 295
9.407	2.2414540 930	9.457	2.2467552 080	9.507	2.2520283 694	9.557	2.2572738 703
9.408	2.2415603 912	9.458	2.2468609 442	9.508	2.2521335 495	9.558	2.2573785 002
9.409	2.2416666 780	9.459	2.2469666 692	9.509	2.2522387 186	9.559	2.2574831 191
9.410	2.2417729 536	9.460	2.2470723 831	9.510	2.2523438 766	9.560	2.2575877 271
9.411	2.2418792 179	9.461	2.2471780 857	9.511	2.2524490 235	9.561	2.2576923 241
9.412	2.2419854 709	9.462	2.2472837 772	9.512	2.2525541 594	9.562	2.2577969 102
9.413	2.2420917 126	9.463	2.2473894 575	9.513	2.2526592 842	9.563	2.2579014 854
9.414	2.2421979 430	9.464	2.2474951 267	0.514	2.2527643 980	9.564	2.2580060 496
9.415	2.2423041 621	9.465	2.2476007 847	9.515	2.2528695 007	9.565	2.2581106 029
9.416	2.2424103 700	9.466	2.2477064 315	9.516	2.2529745 924	9.566	2.2582151 453
9.417	2.2425165 665	9.467	2.2478120 671	9.517	2.2530796 731	9.567	2.2583196 767
9.418	2.2426227 518	9.468	2.2479176 917	9.518	2.2531847 427	9.568	2.2584241 972
9.419	2.2427289 258	9.469	2.2480233 050	9.519	2.2532898 013	9.569	2.2585287 068
9.420	2.2428350 886	9.470	2.2481289 072	9.520	2.2533948 488	9.570	2.2586332 055
9.421	2.2429412 401	9.471	2.2482344 982	9.521	2.2534998 853	9.571	2.2587376 932
9.422	2.2430473 803	9.472	2.2483400 781	9.522	2.2536049 108	9.572	2.2588421 700
9.423	2.2431535 092	9.473	2.2484456 469	9.523	2.2537099 252	9.573	2.2589466 360
9.424	2.2432596 269	9.474	2.2485512 045	9.524	2.2538149 286	9.574	2.2590510 910
9.425	2.2433657 333	9.475	2.2486567 510	9.525	2.2539199 210	9.575	2.2591555 351
9.426	2.2434718 285	9.476	2.2487622 863	9.526	2.2540249 024	9.576	2.2592599 683
9.427	2.2435779 124	9.477	2.2488678 105	0.527	2.2541298 727	9.577	2.2593643 905
9.428	2.2436839 851	9.478	2.2489733 235	9.528	2.2542348 321	9.578	2.2594688 019
9.429	2.2437900 465	9.479	2.2490788 255	9.529	2.2543397 804	9.579	2.2595732 024
9.430	2.2438960 966	9.480	2.2491843 163	9.530	2.2544447 177	9.580	2.2596775 920
9.431	2.2440021 356	9.481	2.2492897 959	9.531	2.2545496 440	9.581	2.2597819 707
9.432	2.2441081 632	9.482	2.2493952 645	9.532	2.2546545 592	9.582	2.2598863 385
9.433	2.2442141 797	9.483	2.2495007 219	9.533	2.2547594 635	9.583	2.2599906 954
9.434	2.2443201 849	9.484	2.2496061 682	9.534	2.2548643 568	9.584	2.2600950 414
9.435	2.2444261 788	9.485	2.2497116 034	9.535	2.2549692 391	9.585	2.2601993 765
9.436	2.2445321 615	9.486	2.2498170 275	9.536	2.2550741 103	9.586	2.2603037 007
9.437	2.2446381 330	9.487	2.2499224 404	9.537	2.2551789 706	9.587	2.2604080 141
9.438	2.2447440 933	9.488	2.2500278 423	9.538	2.2552838 199	9.588	2.2605123 166
9.439	2.2448500 423	9.489	2.2501332 330	9.539	2.2553886 582	9.589	2.2606166 082
9.440	2.2449559 802	9.490	2.2502386 126	9.540	2.2554934 855	9.590	2.2607208 889
9.441	2.2450619 068	9.491	2.2503439 811	9.541	2.2555983 018	9.591	2.2608251 587
9.442	2.2451678 221	9.492	2.2504493 386	9.542	2.2557031 071	9.592	2.2609294 177
9.443	2.2452737 263	9.493	2.2505546 849	9.543	2.2558079 014	9.593	2.2610336 658
9.444	2.2453796 192	9.494	2.2506600 201	9.544	2.2559126 848	9.594	2.2611379 031
9.445	2.2454855 010	9.495	2.2507653 443	9.545	2.2560174 572	9.595	2.2612421 295
9.446	2.2455913 715	9.496	2.2508706 573	9.546	2.2561222 186	9.596	2.2613463 450
9.447	2.2456972 308	9.497	2.2509759 593	9.547	2.2562269 690	9.597	2.2614505 496
9.448	2.2458030 789	9.498	2.2510812 501	9.548	2.2563317 085	9.598	2.2615547 434
9.449	2.2459089 158	9.499	2.2511865 299	9.549	2.2564364 370	9.599	2.2616589 264

TEN PLACE MANTISSAS FOR NATURAL LOGARITHMS (Continued)

x	$\log_e x$	x	$\log_e x$	x	$\log_e x$	x	$\log_e x$
9.600	2.2617630 985	9.650	2.2669579 154	9.700	2.2721258 855	9.750	2.2772672 850
9.601	2.2618672 597	9.651	2.2670615 369	9.701	2.2722289 730	9.751	2.2773698 439
9.602	2.2619714 101	9.652	2.2671651 478	9.702	2.2723320 498	9.752	2.2774723 922
9.603	2.2620755 497	9.653	2.2672687 479	9.703	2.2724351 160	9.753	2.2775749 300
9.604	2.2621796 784	9.654	2.2673723 372	9.704	2.2725381 716	9.754	2.2776774 573
9.605	2.2622837 962	9.655	2.2674759 159	9.705	2.2726412 166	9.755	2.2777799 741
9.606	2.2623879 032	9.656	2.2675794 838	9.706	2.2727442 510	9.756	2.2778824 804
9.607	2.2624919 994	9.657	2.2676830 410	9.707	2.2728472 747	9.757	2.2779849 761
9.608	2.2625960 848	9.658	2.2677865 875	9.708	2.2729502 879	9.758	2.2780874 614
9.609	2.2627001 593	9.659	2.2678901 232	9.709	2.2730532 904	9.759	2.2781899 362
9.610	2.2628042 230	9.660	2.2679936 482	9.710	2.2731562 823	9.760	2.2782924 004
9.611	2.2629082 758	9.661	2.2680971 625	9.711	2.2732592 636	9.761	2.2783948 542
9.612	2.2630123 179	9.662	2.2682006 661	9.712	2.2733622 343	9.762	2.2784972 975
9.613	2.2631163 491	9.663	2.2683041 590	9.713	2.2734651 944	9.763	2.2785997 302
9.614	2.2632203 695	9.664	2.2684076 412	9.714	2.2735681 439	9.764	2.2787021 525
9.615	2.2633243 790	9.665	2.2685111 127	9.715	2.2736710 828	9.765	2.2788045 643
9.616	2.2634283 778	9.666	2.2686145 734	9.716	2.2737740 111	9.766	2.2789069 656
9.617	2.2635323 657	9.667	2.2687180 235	9.717	2.2738769 289	9.767	2.2790093 565
9.618	2.2636363 429	9.668	2.2688214 628	9.718	2.2739798 360	9.768	2.2791117 368
9.619	2.2637403 092	9.669	2.2689248 915	9.719	2.2740827 325	9.769	2.2792141 067
9.620	2.2638442 647	9.670	2.2690283 095	9.720	2.2741856 185	9.770	2.2793164 661
9.621	2.2639482 094	9.671	2.2691317 167	9.721	2.2742884 938	9.771	2.2794188 150
9.622	2.2640521 433	9.672	2.2692351 133	9.722	2.2743913 586	9.772	2.2795211 534
9.623	2.2641560 664	9.673	2.2693384 992	9.723	2.2744942 128	9.773	2.2796234 814
9.624	2.2642599 787	9.674	2.2694418 744	9.724	2.2745970 565	9.774	2.2797257 988
9.625	2.2643638 802	9.675	2.2695452 389	9.725	2.2746998 895	9.775	2.2798281 059
9.626	2.2644677 709	9.676	2.2696485 927	9.726	2.2748027 120	9.776	2.2799304 024
9.627	2.2645716 508	9.677	2.2697519 359	9.727	2.2749055 239	9.777	2.2800326 885
9.628	2.2646755 199	9.678	2.2698552 684	9.728	2.2750083 252	9.778	2.2801349 642
9.629	2.2647793 783	9.679	2.2699585 902	9.729	2.2751111 160	9.779	2.2802372 293
9.630	2.2648832 258	9.680	2.2700619 013	9.730	2.2752138 962	9.780	2.2803394 840
9.631	2.2649870 626	9.681	2.2701652 017	9.731	2.2753166 658	9.781	2.2804417 283
9.632	2.2650908 886	9.682	2.2702684 915	9.732	2.2754194 249	9.782	2.2805439 621
9.633	2.2651947 038	9.683	2.2703717 706	9.733	2.2755221 734	9.783	2.2806461 855
9.634	2.2652985 082	9.684	2.2704750 391	9.734	2.2756249 114	9.784	2.2807483 984
9.635	2.2654023 019	9.685	2.2705782 969	9.735	2.2757276 388	9.785	2.2808506 008
9.636	2.2655060 848	9.686	2.2706815 440	9.736	2.2758303 557	9.786	2.2809527 929
9.637	2.2656098 569	9.687	2.2707847 804	9.737	2.2759330 620	9.787	2.2810549 744
9.638	2.2657136 182	9.688	2.2708880 063	9.738	2.2760357 578	9.788	2.2811571 456
9.639	2.2658173 688	9.689	2.2709912 214	9.739	2.2761384 430	9.789	2.2812593 063
9.640	2.2659211 086	9.690	2.2710944 259	9.740	2.2762411 177	9.790	2.2813614 565
9.641	2.2660248 377	9.691	2.2711976 198	9.741	2.2763437 818	9.791	2.2814635 964
9.642	2.2661285 560	9.692	2.2713008 030	9.742	2.2764464 354	9.792	2.2815657 258
9.643	2.2662322 635	9.693	2.2714039 755	9.743	2.2765490 784	9.793	2.2816678 447
9.644	2.2663359 603	9.694	2.2715071 374	9.744	2.2766517 110	9.794	2.2817699 533
9.645	2.2664396 464	9.695	2.2716102 887	9.745	2.2767543 330	9.795	2.2818720 514
9.646	2.2665433 216	9.696	2.2717134 293	9.746	2.2768569 444	9.796	2.2819741 391
9.647	2.2666469 862	9.697	2.2718165 593	9.747	2.2769595 454	9.797	2.2820762 164
9.648	2.2667506 400	9.698	2.2719196 787	9.748	2.2770621 358	9.798	2.2821782 832
9.649	2.2668542 830	9.699	2.2720227 874	9.749	2.2771647 156	9.799	2.2822803 397

TEN PLACE MANTISSAS FOR NATURAL LOGARITHMS (Continued)

x	$\log_e x$	x	$\log_e x$	x	$\log_e x$	x	$\log_e x$
9.800	2.2823823 857	9.850	2.2874714 552	9.900	2.2925347 571	9.950	2.2975725 512
9.801	2.2824844 213	9.851	2.2875729 729	9.901	2.2926357 621	9.951	2.2976730 486
9.802	2.2825864 465	9.852	2.2876744 803	9.902	2.2927367 569	9.952	2.2977735 360
9.803	2.2826884 613	9.853	2.2877759 773	9.903	2.2928377 415	9.953	2.2978740 133
9.804	2.2827904 657	9.854	2.2878774 641	9.904	2.2929387 159	9.954	2.2979744 804
9.805	2.2828924 596	9.855	2.2879789 406	9.905	2.2930396 802	9.955	2.2980749 375
9.806	2.2829944 432	9.856	2.2880804 068	9.906	2.2931406 342	9.956	2.2981753 845
9.807	2.2830964 164	9.857	2.2881818 627	9.907	2.2932415 780	9.957	2.2982758 214
9.808	2.2831983 792	9.858	2.2882833 083	9.908	2.2933425 116	9.958	2.2983762 482
9.809	2.2833003 316	9.859	2.2883847 436	9.909	2.2934434 351	9.959	2.2984766 649
9.810	2.2834022 736	9.860	2.2884861 686	9.910	2.2935443 483	9.960	2.2985770 716
9.811	2.2835042 052	9.861	2.2885875 834	9.911	2.2936452 514	9.961	2.2986774 682
9.812	2.2836061 264	9.862	2.2886889 878	9.912	2.2937461 443	9.962	2.2987778 547
9.813	2.2837080 372	9.863	2.2887903 820	9.913	2.2938470 271	9.963	2.2988782 311
9.814	2.2838099 377	9.864	2.2888917 659	9.914	2.2939478 996	9.964	2.2989785 974
9.815	2.2839118 277	9.865	2.2889931 395	9.915	2.2940487 620	9.965	2.2990789 537
9.816	2.2840137 074	9.866	2.2890945 028	9.916	2.2941496 142	9.966	2.2991792 999
9.817	2.2841155 767	9.867	2.2891958 559	9.917	2.2942504 562	9.967	2.2992796 360
9.818	2.2842174 356	9.868	2.2892971 987	9.918	2.2943512 881	9.968	2.2993799 620
9.819	2.2843192 842	9.869	2.2893985 312	9.919	2.2944521 098	9.969	2.2994802 780
9.820	2.2844211 224	9.870	2.2894998 534	9.920	2.2945529 213	9.970	2.2995805 840
9.821	2.2845229 502	9.871	2.2896011 654	9.921	2.2946537 227	9.971	2.2996808 798
9.822	2.2846247 676	9.872	2.2897024 672	9.922	2.2947545 139	9.972	2.2997811 657
9.823	2.2847265 747	9.873	2.2898037 587	9.923	2.2948552 949	9.973	2.2998814 414
9.824	2.2848283 714	9.874	2.2899050 398	9.924	2.2949560 658	9.974	2.2999817 071
9.825	2.2849301 578	9.875	2.2900063 108	9.925	2.2950568 266	9.975	2.3000819 628
9.826	2.2850319 337	9.876	2.2901075 715	9.926	2.2951575 772	9.976	2.3001822 084
9.827	2.2851336 994	9.877	2.2902088 219	9.927	2.2952583 176	9.977	2.3002824 439
9.828	2.2852354 547	9.878	2.2903100 621	9.928	2.2953590 479	9.978	2.3003826 694
9.829	2.2853371 996	9.879	2.2904112 921	9.929	2.2954597 681	9.979	2.3004828 849
9.830	2.2854389 342	9.880	2.2905125 118	9.930	2.2955604 781	9.980	2.3005830 903
9.831	2.2855406 584	9.881	2.2906137 212	9.931	2.2956611 779	9.981	2.3006832 857
9.832	2.2856423 723	9.882	2.2907149 204	9.932	2.2957618 676	9.982	2.3007834 710
9.833	2.2857440 758	9.883	2.2908161 094	9.933	2.2958625 472	9.983	2.3008836 464
9.834	2.2858457 690	9.884	2.2909172 881	9.934	2.2959632 167	9.984	2.3009838 116
9.835	2.2859474 518	9.885	2.2910184 566	9.935	2.2960638 760	9.985	2.3010839 669
9.836	2.2860491 244	9.886	2.2911196 149	9.936	2.2961645 252	9.986	2.3011841 121
9.837	2.2861507 865	9.887	2.2912207 629	9.937	2.2962651 642	9.987	2.3012842 473
9.838	2.2862524 384	9.888	2.2913219 007	9.938	2.2963657 932	9.988	2.3013843 724
9.839	2.2863540 799	9.889	2.2914230 283	9.939	2.2964664 120	9.989	2.3014844 876
9.840	2.2864557 111	9.890	2.2915241 456	9.940	2.2965670 207	9.990	2.3015845 927
9.841	2.2865573 319	9.891	2.2916252 528	9.941	2.2966676 192	9.991	2.3016846 878
9.842	2.2866589 424	9.892	2.2917263 497	9.942	2.2967682 077	9.992	2.3017847 728
9.843	2.2867605 426	9.893	2.2918274 363	9.943	2.2968687 860	9.993	2.3018848 479
9.844	2.2868621 325	9.894	2.2919285 128	9.944	2.2969693 542	9.994	2.3019849 129
9.845	2.2869637 121	9.895	2.2920295 791	9.945	2.2970699 123	9.995	2.3020849 680
9.846	2.2870652 813	9.896	2.2921306 351	9.946	2.2971704 603	9.996	2.3021850 130
9.847	2.2871668 403	9.897	2.2922316 809	9.947	2.2972709 982	9.997	2.3022850 480
9.848	2.2872683 889	9.898	2.2923327 165	9.948	2.2973715 259	9.998	2.3023850 730
9.849	2.2873699 272	9.899	2.2924337 419	9.949	2.2974720 436	9.999	2.3024850 880

RADIX TABLE OF NATURAL LOGARITHMS

x	n	$\log[1 + x(10^{-n})]$	$-\log[1 - x(10^{-n})]$
1	10	0.00000 00000 99999 99999 50000	0.00000 00001 00000 00000 50000
2	10	0.00000 00001 99999 99998 00000	0.00000 00002 00000 00002 00000
3	10	0.00000 00002 99999 99995 50000	0.00000 00003 00000 00004 50000
4	10	0.00000 00003 99999 99992 00000	0.00000 00004 00000 00008 00000
5	10	0.00000 00004 99999 99987 50000	0.00000 00005 00000 00012 50000
6	10	0.00000 00005 99999 99982 00000	0.00000 00006 00000 00018 00000
7	10	0.00000 00006 99999 99975 50000	0.00000 00007 00000 00024 50000
8	10	0.00000 00007 99999 99968 00000	0.00000 00008 00000 00032 00000
9	10	0.00000 00008 99999 99959 50000	0.00000 00009 00000 00040 50000
1	9	0.00000 00009 99999 99950 00000	0.00000 00010 00000 00050 00000
2	9	0.00000 00019 99999 99800 00000	0.00000 00020 00000 00200 00000
3	9	0.00000 00029 99999 99550 00000	0.00000 00030 00000 00450 00000
4	9	0.00000 00039 99999 99200 00000	0.00000 00040 00000 00800 00000
5	9	0.00000 00049 99999 98750 00000	0.00000 00050 00000 01250 00000
6	9	0.00000 00059 99999 98200 00001	0.00000 00060 00000 01800 00001
7	9	0.00000 00069 99999 97550 00001	0.00000 00070 00000 02450 00001
8	9	0.00000 00079 99999 96800 00002	0.00000 00080 00000 03200 00002
9	9	0.00000 00089 99999 95950 00002	0.00000 00090 00000 04050 00002
1	8	0.00000 00099 99999 95000 00003	0.00000 00100 00000 05000 00003
2	8	0.00000 00199 99999 80000 00027	0.00000 00200 00000 20000 00027
3	8	0.00000 00299 99999 55000 00090	0.00000 00300 00000 45000 00090
4	8	0.00000 00399 99999 20000 00213	0.00000 00400 00000 80000 00213
5	8	0.00000 00499 99998 75000 00417	0.00000 00500 00001 25000 00417
6	8	0.00000 00599 99998 20000 00720	0.00000 00600 00001 80000 00720
7	8	0.00000 00699 99997 55000 01143	0.00000 00700 00002 45000 01143
8	8	0.00000 00799 99996 80000 01707	0.00000 00800 00003 20000 01707
9	8	0.00000 00899 99995 95000 02430	0.00000 00900 00004 05000 02430
1	7	0.00000 00999 99995 00000 03333	0.00000 01000 00005 00000 03333
2	7	0.00000 01999 99980 00000 26667	0.00000 02000 00020 00000 26667
3	7	0.00000 02999 99955 00000 90000	0.00000 03000 00045 00000 90000
4	7	0.00000 03999 99920 00002 13333	0.00000 04000 00080 00002 13333
5	7	0.00000 04999 99875 00004 16667	0.00000 05000 00125 00004 16667
6	7	0.00000 05999 99820 00007 20000	0.00000 06000 00180 00007 20000
7	7	0.00000 06999 99755 00011 43333	0.00000 07000 00245 00011 43334
8	7	0.00000 07999 99680 00017 06666	0.00000 08000 00320 00017 06668
9	7	0.00000 08999 99595 00024 29998	0.00000 09000 00405 00024 30002
1	6	0.00000 09999 99500 00033 33331	0.00000 10000 00500 00033 33336
2	6	0.00000 19999 98000 00266 66627	0.00000 20000 02000 00266 66707
3	6	0.00000 29999 95500 00899 99798	0.00000 30000 04500 00900 00203
4	6	0.00000 39999 92000 02133 32693	0.00000 40000 08000 02133 33973
5	6	0.00000 49999 87500 04166 65104	0.00000 50000 12500 04166 68229
6	6	0.00000 59999 82000 07199 96760	0.00000 60000 18000 07200 03240
7	6	0.00000 69999 75500 11433 27331	0.00000 70000 24500 11433 39336
8	6	0.00000 79999 68000 17066 56427	0.00000 80000 32000 17066 76907
9	6	0.00000 89999 59500 24299 83598	0.00000 90000 40500 24300 16403

For $n > 10$, $\log[1 \pm x(10^{-n})] = \pm x(10^{-n}) - \frac{1}{2}x^2(10^{-2n})$ to 25 decimals.

RADIX TABLE OF NATURAL LOGARITHMS (Continued)

x	n	$\log[1 + x(10^{-n})]$	$-\log[1 - x)10^{-n})]$
1	5	0.00000 99999 50000 33333 08334	0.00001 00000 50000 33333 58334
2	5	0.00001 99998 00002 66662 66673	0.00002 00002 00002 66670 66673
3	5	0.00002 99995 50008 99979 75049	0.00003 00004 50009 00020 25049
4	5	0.00003 99992 00021 33269 33538	0.00004 00008 00021 33397 33538
5	5	0.00004 99987 50041 66510 42292	0.00005 00012 50041 66822 92292
6	5	0.00005 99982 00071 99676 01555	0.00006 00018 00072 00324 01555
7	5	0.00006 99975 50114 32733 11695	0.00007 00024 50114 33933 61695
8	5	0.00007 99968 00170 65642 73220	0.00008 00032 00170 67690 73221
9	5	0.00008 99959 50242 98359 86809	0.00009 00040 50243 01640 36811
1	4	0.00009 99950 00333 30833 53332	0.00010 00050 00333 35833 53335
2	4	0.00019 99800 02666 26673 06560	0.00020 00200 02667 06673 06773
3	4	0.00029 99550 08997 97548 58785	0.00030 00450 09002 02548 61215
4	4	0.00039 99200 21326 93538 06509	0.00040 00800 21339 73538 20162
5	4	0.00049 98750 41651 04791 40636	0.00050 01250 41682 29791 92719
6	4	0.00059 98200 71967 61554 42280	0.00060 01800 72032 41555 97800
7	4	0.00069 97551 14273 34192 77369	0.00070 02451 14393 39196 69533
8	4	0.00079 96801 70564 33215 90059	0.00080 03201 70769 13224 63873
9	4	0.00089 95952 42836 09300 94948	0.00090 04052 43164 14318 66419
1	3	0.00099 95003 33083 53316 68094	0.00100 05003 33583 53350 01430
2	3	0.00199 80026 62673 05601 82538	0.00200 20026 70673 07735 16511
3	3	0.00299 55089 79798 47881 16106	0.00300 45090 20298 72181 32509
4	3	0.00399 20212 69537 45299 90751	0.00400 80213 97538 81834 87927
5	3	0.00498 75415 11039 07361 21022	0.00501 25418 23544 28204 30937
6	3	0.00598 20716 77547 46378 20189	0.00601 80723 25563 01620 19350
7	3	0.00697 56137 36425 24209 95222	0.00702 46149 36964 45987 41123
8	3	0.00796 81696 49176 87351 07973	0.00803 21716 97264 25903 86494
9	3	0.00895 97413 71471 90444 31465	0.00904 07446 52149 06220 55241
1	2	0.00995 03308 53168 08284 82154	0.01005 03358 53501 44118 35489
2	2	0.01980 26272 96179 71302 60291	0.02020 27073 17519 44840 80453
3	2	0.02955 88022 41544 40273 26194	0.03045 92074 84708 54591 92613
4	2	0.03922 07131 53281 29626 92009	0.04082 19945 20255 12955 45771
5	2	0.04879 01641 69432 00306 53744	0.05129 32943 87550 53342 61961
6	2	0.05826 89081 23975 77552 57184	0.06187 54037 18087 47179 78001
7	2	0.06765 86484 73814 80526 84159	0.07257 06928 34835 43071 15733
8	2	0.07696 10411 36128 32498 42170	0.08338 16089 39051 05839 47658
9	2	0.08617 76962 41052 33234 13335	0.09431 06794 71241 32687 71427
1	1	0.09531 01798 04324 86004 39521	0.10536 05156 57826 30122 75010
2	1	0.18232 15567 93954 62621 17180	0.22314 35513 14209 75576 62951
3	1	0.26236 42644 67491 05203 54960	0.35667 49439 38732 37891 26387
4	1	0.33647 22366 21212 93050 45934	0.51082 56237 65990 68320 55141
5	1	0.40546 51081 08164 38197 80131	0.69314 71805 59945 30941 72321
6	1	0.47000 36292 45735 55365 09370	0.91629 07318 74155 06518 35272
7	1	0.53062 82510 62170 39623 15432	1.20397 28043 25935 99262 27462
8	1	0.58778 66649 02119 00818 97311	1.60943 79124 34100 37460 07593
9	1	0.64185 38861 72394 77599 10360	2.30258 50929 94045 68401 79915
1	0	0.69314 71805 59945 30941 72321	∞

For $n > 10$, $\log[1 \pm x(10^{-n})] = \pm x(10^{-n}) - \frac{1}{2}x^2(10^{-2n})$ to 25 decimals.

SQUARE ROOT OF $[1 + x^2]$ AND ITS NATURAL LOGARITHM

x	$\sqrt{1+x^2}$	$\log_e \sqrt{1+x^2}$	x	$\sqrt{1+x^2}$	$\log_e \sqrt{1+x^2}$
0.000	1.00000 00000	0.00000 00000	0.050	1.00124 92197	0.00124 84402
0.001	1.00000 05000	0.00000 05000	0.051	1.00129 96554	0.00129 88117
0.002	1.00000 20000	0.00000 20000	0.052	1.00135 10873	0.00135 01755
0.003	1.00000 45000	0.00000 45000	0.053	1.00140 35151	0.00140 25311
0.004	1.00000 80000	0.00000 79999	0.054	1.00145 69387	0.00145 58784
0.005	1.00001 24999	0.00001 24998	0.055	1.00151 13579	0.00151 02170
0.006	1.00001 79998	0.00001 79997	0.056	1.00156 67726	0.00156 55465
0.007	1.00002 44997	0.00002 44994	0.057	1.00162 31826	0.00162 18667
0.008	1.00003 19995	0.00003 19990	0.058	1.00168 05878	0.00167 91773
0.009	1.00004 04992	0.00004 04984	0.059	1.00173 89880	0.00173 74777
0.010	1.00004 99988	0.00004 99975	0.060	1.00179 83829	0.00179 67678
0.011	1.00006 04982	0.00006 04964	0.061	1.00185 87725	0.00185 70472
0.012	1.00007 19974	0.00007 19948	0.062	1.00192 01565	0.00191 83154
0.013	1.00008 44964	0.00008 44929	0.063	1.00198 25348	0.00198 05722
0.014	1.00009 79952	0.00009 79904	0.064	1.00204 59071	0.00204 38171
0.015	1.00011 24937	0.00011 24874	0.065	1.00211 02734	0.00210 80499
0.016	1.00012 79918	0.00012 79837	0.066	1.00217 56333	0.00217 32701
0.017	1.00014 44896	0.00014 44792	0.067	1.00224 19868	0.00223 94773
0.018	1.00016 19869	0.00016 19738	0.068	1.00230 93335	0.00230 66711
0.019	1.00018 04837	0.00018 04675	0.069	1.00237 76733	0.00237 48512
0.020	1.00019 99800	0.00019 99601	0.070	1.00244 70061	0.00244 40171
0.021	1.00022 04757	0.00022 04514	0.071	1.00251 73315	0.00251 41684
0.022	1.00024 19707	0.00024 19415	0.072	1.00258 86494	0.00258 53047
0.023	1.00026 44650	0.00026 44301	0.073	1.00266 09597	0.00265 74256
0.024	1.00028 79585	0.00028 79171	0.074	1.00273 42619	0.00273 05307
0.025	1.00031 24512	0.00031 24024	0.075	1.00280 85560	0.00280 46194
0.026	1.00033 79429	0.00033 78859	0.076	1.00288 38417	0.00287 96915
0.027	1.00036 44336	0.00036 43672	0.077	1.00296 01188	0.00295 57463
0.028	1.00039 19232	0.00039 18465	0.078	1.00303 73871	0.00303 27836
0.029	1.00042 04116	0.00042 03233	0.079	1.00311 56464	0.00311 08028
0.030	1.00044 98988	0.00044 97976	0.080	1.00319 48963	0.00318 98036
0.031	1.00048 03846	0.00048 02693	0.081	1.00327 51367	0.00326 97852
0.032	1.00051 18690	0.00051 17380	0.082	1.00335 63674	0.00335 07474
0.033	1.00054 43518	0.00054 42037	0.083	1.00343 85881	0.00343 26896
0.034	1.00057 78331	0.00057 76662	0.084	1.00352 17985	0.00351 56114
0.035	1.00061 23125	0.00061 21252	0.085	1.00360 59984	0.00359 95124
0.036	1.00064 77902	0.00064 75805	0.086	1.00369 11876	0.00368 43919
0.037	1.00068 42659	0.00068 40319	0.087	1.00377 73658	0.00377 02495
0.038	1.00072 17396	0.00072 14793	0.088	1.00386 45327	0.00385 70846
0.039	1.00076 02110	0.00075 99223	0.089	1.00395 26881	0.00394 48968
0.040	1.00079 96803	0.00079 93608	0.090	1.00404 18318	0.00403 36855
0.041	1.00084 01471	0.00083 97944	0.091	1.00413 19634	0.00412 34503
0.042	1.00088 16114	0.00088 12231	0.092	1.00422 30828	0.00421 41906
0.043	1.00092 40730	0.00092 36464	0.093	1.00431 51896	0.00430 59059
0.044	1.00096 75320	0.00096 70642	0.094	1.00440 82835	0.00439 85955
0.045	1.00101 19879	0.00101 14762	0.095	1.00450 23644	0.00449 22591
0.046	1.00105 74409	0.00105 68823	0.096	1.00459 74318	0.00458 68960
0.047	1.00110 38907	0.00110 32819	0.097	1.00469 34856	0.00468 25056
0.048	1.00115 13372	0.00115 06750	0.098	1.00479 05254	0.00477 90875
0.049	1.00119 97803	0.00119 90612	0.099	1.00488 85510	0.00487 66409

SQUARE ROOT OF $[1 + x^2]$ AND ITS NATURAL LOGARITHM (Continued)

x	$\sqrt{1+x^2}$	$\log_e \sqrt{1+x^2}$	x	$\sqrt{1+x^2}$	$\log_e \sqrt{1+x^2}$
0.100	1.00498 75621	0.00497 51654	0.150	1.01118 74208	0.01112 53046
0.101	1.00508 75584	0.00507 46605	0.151	1.01133 62448	0.01127 24711
0.102	1.00518 85395	0.00517 51255	0.152	1.01148 60355	0.01142 05718
0.103	1.00529 05053	0.00527 65598	0.153	1.01163 67925	0.01156 96058
0.104	1.00539 34553	0.00537 89628	0.154	1.01178 85155	0.01171 95723
0.105	1.00549 73894	0.00548 23338	0.155	1.01194 12038	0.01187 04702
0.106	1.00560 23071	0.00558 66725	0.156	1.01209 48572	0.01202 22989
0.107	1.00570 82082	0.00569 19781	0.157	1.01224 94752	0.01217 50575
0.108	1.00581 50924	0.00579 82500	0.158	1.01240 50573	0.01232 87450
0.109	1.00592 29593	0.00590 54875	0.159	1.01256 16031	0.01248 33607
0.110	1.00603 18086	0.00601 36901	0.160	1.01271 91121	0.01263 89037
0.111	1.00614 16401	0.00612 28572	0.161	1.01287 75839	0.01279 53729
0.112	1.00625 24534	0.00623 29880	0.162	1.01303 70181	0.01295 27676
0.113	1.00636 42482	0.00634 40819	0.163	1.01319 74141	0.01311 10870
0.114	1.00647 70241	0.00645 61383	0.164	1.01335 87716	0.01327 03300
0.115	1.00659 07808	0.00656 91566	0.165	1.01352 10901	0.01343 04958
0.116	1.00670 55180	0.00668 31361	0.166	1.01368 43690	0.01359 15836
0.117	1.00682 12354	0.00679 80761	0.167	1.01384 86080	0.01375 35923
0.118	1.00693 79326	0.00691 39759	0.168	1.01401 38066	0.01391 65211
0.119	1.00705 56092	0.00703 08349	0.169	1.01417 99643	0.01408 03691
0.120	1.00717 42650	0.00714 86524	0.170	1.01434 70806	0.01424 51352
0.121	1.00729 38995	0.00726 74277	0.171	1.01451 51551	0.01441 08188
0.122	1.00741 45125	0.00738 71602	0.172	1.01468 41873	0.01457 74187
0.123	1.00753 61036	0.00750 78490	0.173	1.01485 41767	0.01474 49339
0.124	1.00765 86724	0.00762 94937	0.174	1.01502 51228	0.01491 33638
0.125	1.00778 22185	0.00775 20933	0.175	1.01519 70252	0.01508 27072
0.126	1.00790 67417	0.00787 56473	0.176	1.01536 98833	0.01525 29632
0.127	1.00803 22415	0.00800 01548	0.177	1.01554 36967	0.01542 41309
0.128	1.00815 87177	0.00812 56153	0.178	1.01571 84649	0.01559 62094
0.129	1.00828 61697	0.00825 20279	0.179	1.01589 41874	0.01576 91975
0.130	1.00841 45973	0.00837 93919	0.180	1.01607 08637	0.01594 30945
0.131	1.00854 40000	0.00850 77067	0.181	1.01624 84932	0.01611 78992
0.132	1.00867 43776	0.00863 69714	0.182	1.01642 70756	0.01629 36108
0.133	1.00880 57296	0.00876 71853	0.183	1.01660 66102	0.01647 02284
0.134	1.00893 80556	0.00889 83476	0.184	1.01678 70967	0.01664 77507
0.135	1.00907 13552	0.00903 04577	0.185	1.01696 85344	0.01682 61770
0.136	1.00920 56282	0.00916 35147	0.186	1.01715 09229	0.01700 55063
0.137	1.00934 08740	0.00929 75179	0.187	1.01733 42617	0.01718 57373
0.138	1.00947 70924	0.00943 24664	0.188	1.01751 85502	0.01736 68694
0.139	1.00961 42828	0.00956 83597	0.189	1.01770 37880	0.01754 89013
0.140	1.00975 24449	0.00970 51968	0.190	1.01788 99744	0.01773 18321
0.141	1.00989 15783	0.00984 29769	0.191	1.01807 71091	0.01791 56609
0.142	1.01003 16827	0.00998 16994	0.192	1.01826 51914	0.01810 03866
0.143	1.01017 27575	0.01012 13633	0.193	1.01845 42209	0.01828 60081
0.144	1.01031 48024	0.01026 19679	0.194	1.01864 41970	0.01847 25245
0.145	1.01045 78170	0.01040 35124	0.195	1.01883 51191	0.01865 99346
0.146	1.01060 18009	0.01054 59960	0.196	1.01902 69869	0.01884 82376
0.147	1.01074 67537	0.01068 94178	0.197	1.01921 97997	0.01903 74323
0.148	1.01089 26748	0.01083 37770	0.198	1.01941 35569	0.01922 75178
0.149	1.01103 95640	0.01097 90729	0.199	1.01960 82581	0.01941 84929

SQUARE ROOT OF $[1 + x^2]$ AND ITS NATURAL LOGARITHM (Continued)

x	$\sqrt{1+x^2}$	$\log_e \sqrt{1+x^2}$	x	$\sqrt{1+x^2}$	$\log_e \sqrt{1+x^2}$
0.200	1.01980 39027	0.01961 03566	0.250	1.03077 64064	0.03031 23110
0.201	1.02000 04902	0.01980 31079	0.251	1.03101 93985	0.03054 80201
0.202	1.02019 80200	0.01999 67457	0.252	1.03126 33029	0.03078 45585
0.203	1.02039 64916	0.02019 12690	0.253	1.03150 81192	0.03102 19248
0.204	1.02059 59044	0.02038 66767	0.254	1.03175 38466	0.03126 01179
0.205	1.02079 62578	0.02058 29677	0.255	2.03200 04845	0.03149 91366
0.206	1.02099 75514	0.02078 01410	0.256	1.03224 80322	0.03173 89795
0.207	1.02119 97846	0.02097 81954	0.257	1.03249 64891	0.03197 96454
0.208	1.02140 29567	0.02117 71300	0.258	1.03274 58545	0.03222 11333
0.209	1.02160 70673	0.02137 69437	0.259	1.03299 61278	0.03246 34416
0.210	1.02181 21158	0.02157 76352	0.260	1.03324 73082	0.03270 65693
0.211	1.02201 81016	0.02177 92036	0.261	1.03349 93953	0.03295 05150
0.212	1.02222 50242	0.02198 16477	0.262	1.03375 23881	0.03319 52776
0.213	1.02243 28829	0.02218 49666	0.263	1.03400 62862	0.03344 08556
0.214	1.02264 16772	0.02238 91590	0.264	1.03426 10889	0.03368 72480
0.215	1.02285 14066	0.02259 42238	0.265	1.03451 67954	0.03393 44534
0.216	1.02306 20705	0.02280 01601	0.266	1.03477 34051	0.03418 24706
0.217	1.02327 36682	0.02300 69665	0.267	1.03503 09174	0.03443 12982
0.218	1.02348 61992	0.02321 46422	0.268	1.03528 93315	0.03468 09350
0.219	1.02369 96630	0.02342 31857	0.269	1.03554 86469	0.03493 13799
0.220	1.02391 40589	0.02363 25963	0.270	1.03580 88627	0.03518 26313
0.221	1.02412 93864	0.02384 28725	0.271	1.03606 99783	0.03543 46882
0.222	1.02434 56448	0.02405 40134	0.272	1.03633 19931	0.03568 75492
0.223	1.02456 28336	0.02426 60179	0.273	1.03659 49064	0.03594 12131
0.224	1.02478 09522	0.02447 88846	0.274	1.03685 87175	0.03619 56784
0.225	1.02500 00000	0.02469 26127	0.275	1.03712 34256	0.03645 09441
0.226	1.02521 99764	0.02490 72007	0.276	1.03738 90302	0.03670 70086
0.227	1.02544 08808	0.02512 26478	0.277	1.03765 55305	0.03696 38709
0.228	1.02566 27126	0.02533 89525	0.278	1.03792 29259	0.03722 15295
0.229	1.02588 54712	0.02555 61140	0.279	1.03819 12155	0.03747 99831
0.230	1.02610 91560	0.02577 41310	0.280	1.03846 03989	0.03773 92307
0.231	1.02633 37664	0.02599 30023	0.281	1.03873 04751	0.03799 92706
0.232	1.02655 93017	0.02621 27267	0.282	1.03900 14437	0.03826 01017
0.233	1.02678 57615	0.02643 33031	0.283	1.03927 33038	0.03852 17226
0.234	1.02701 31450	0.02665 47303	0.284	1.03954 60548	0.03878 41321
0.235	1.02724 14517	0.02687 70072	0.285	1.03981 96959	0.03904 73289
0.236	1.02747 06808	0.02710 01326	0.286	1.04009 42265	0.03931 13115
0.237	1.02770 08320	0.02732 41052	0.287	1.04036 96458	0.03957 60787
0.238	1.02793 19044	0.02754 89240	0.288	1.04064 59533	0.03984 16292
0.239	1.02816 38974	0.02777 45877	0.289	1.04092 31480	0.04010 79618
0.240	1.02839 68106	0.02800 10951	0.290	1.04120 12294	0.04037 50749
0.241	1.02863 06431	0.02822 84450	0.291	1.04148 01966	0.04064 29673
0.242	1.02886 53945	0.02845 66364	0.292	1.04176 00491	0.04091 16377
0.243	1.02910 10640	0.02868 56678	0.293	1.04204 07862	0.04118 10848
0.244	1.02933 76511	0.02891 55383	0.294	1.04232 24069	0.04145 13072
0.245	1.02957 51551	0.02914 62464	0.295	1.04260 49108	0.04172 23035
0.246	1.02981 35754	0.02937 77911	0.296	1.04288 82970	0.04199 40725
0.247	1.03005 29113	0.02961 01711	0.297	1.04317 25648	0.04226 66128
0.248	1.03029 31622	0.02984 33852	0.298	1.04345 77136	0.04253 99231
0.249	1.03053 43274	0.03007 74322	0.299	1.04374 37425	0.04281 40020

SQUARE ROOT OF $[1 + x^2]$ AND ITS NATURAL LOGARITHM (Continued)

x	$\sqrt{1+x^2}$	$\log_e \sqrt{1+x^2}$	x	$\sqrt{1+x^2}$	$\log_e \sqrt{1+x^2}$
0.300	1.04403 06509	0.04308 88481	0.350	1.05948 10050	0.05777 91704
0.301	1.04431 84380	0.04336 44603	0.351	1.05981 17758	0.05809 13223
0.302	1.04460 71032	0.04364 08369	0.352	1.06014 33865	0.05840 41693
0.303	1.04489 66456	0.04391 79768	0.353	1.06047 58366	0.05871 77099
0.304	1.04518 70646	0.04419 58786	0.354	1.06080 91251	0.05903 19426
0.305	1.04547 83594	0.04447 45409	0.355	1.06114 32514	0.05934 68660
0.306	1.04577 05293	0.04475 39624	0.356	1.06147 82146	0.05966 24787
0.307	1.04606 35736	0.04503 41416	0.357	1.06181 40138	0.05997 87793
0.308	1.04635 74915	0.04531 50774	0.358	1.06215 06484	0.06029 57663
0.309	1.04665 22823	0.04559 67682	0.359	1.06248 81176	0.06061 34384
0.310	1.04694 79452	0.04587 92126	0.360	1.06282 64204	0.06093 17939
0.311	1.04724 44796	0.04616 24095	0.361	1.06316 55563	0.06125 08316
0.312	1.04754 18846	0.04644 63574	0.362	1.06350 55242	0.06157 05500
0.313	1.04784 01596	0.04673 10548	0.363	1.06384 63235	0.06189 09477
0.314	1.04813 93037	0.04701 65005	0.364	1.06418 79533	0.06221 20232
0.315	1.04843 93163	0.04730 26930	0.365	1.06453 04129	0.06253 37751
0.316	1.04874 01966	0.04758 96311	0.366	1.06487 37014	0.06285 62019
0.317	1.04904 19439	0.04787 73133	0.367	1.06521 78181	0.06317 93023
0.318	1.04934 45573	0.04816 57381	0.368	1.06556 27621	0.06350 30747
0.319	1.04964 80362	0.04845 49044	0.369	1.06590 85327	0.06382 75178
0.320	1.04995 23799	0.04874 48107	0.370	1.06625 51289	0.06415 26301
0.321	1.05025 75875	0.04903 54555	0.371	1.06660 25502	0.06447 84102
0.322	1.05056 36582	0.04932 68376	0.372	1.06695 07955	0.06480 48565
0.323	1.05087 05915	0.04961 89554	0.373	1.06729 98641	0.06513 19677
0.324	1.05117 83864	0.04991 18076	0.374	1.06764 97553	0.06545 97423
0.325	1.05148 70422	0.05020 53930	0.375	1.06800 04682	0.06578 81790
0.326	1.05179 65583	0.05049 97100	0.376	1.06835 20019	0.06611 72761
0.327	1.05210 69337	0.05079 47572	0.377	1.06870 43558	0.06644 70323
0.328	1.05241 81678	0.05109 05333	0.378	1.06905 75289	0.06677 74462
0.329	1.05273 02598	0.05138 70369	0.379	1.06941 15204	0.06710 85163
0.330	1.05304 32090	0.05168 42665	0.380	1.06976 63296	0.06744 02412
0.331	1.05335 70145	0.05198 22209	0.381	1.07012 19557	0.06777 26193
0.332	1.05367 16756	0.05228 08984	0.382	1.07047 83977	0.06810 56493
0.333	1.05398 71916	0.05258 02979	0.383	1.07083 56550	0.06843 93297
0.334	1.05430 35616	0.05288 04178	0.384	1.07119 37266	0.06877 36590
0.335	1.05462 07849	0.05318 12568	0.385	1.07155 26119	0.06910 86359
0.336	1.05493 88608	0.05348 28134	0.386	1.07191 23098	0.06944 42588
0.337	1.05525 77884	0.05378 50863	0.387	1.07227 28198	0.06978 05264
0.338	1.05557 75670	0.05408 80740	0.388	1.07263 41408	0.07011 74370
0.339	1.05589 81959	0.05439 17752	0.389	1.07299 62721	0.07045 49894
0.340	1.05621 96741	0.05469 61885	0.390	1.07335 92129	0.07079 31821
0.341	1.05654 20011	0.05500 13122	0.391	1.07372 29624	0.07113 20135
0.342	1.05686 51759	0.05530 71452	0.392	1.07408 75197	0.07147 14823
0.343	1.05718 91978	0.05561 36860	0.393	1.07445 28840	0.07181 15870
0.344	1.05751 40661	0.05592 09331	0.394	1.07481 90545	0.07215 23261
0.345	1.05783 97799	0.05622 88852	0.395	1.07518 60304	0.07249 36982
0.346	1.05816 63385	0.05653 75409	0.396	1.07555 38108	0.07283 57018
0.347	1.05849 37411	0.05684 68987	0.397	1.07592 23950	0.07317 83355
0.348	1.05882 19869	0.05715 69571	0.398	1.07629 17820	0.07352 15979
0.349	1.05915 10752	0.05746 77148	0.399	1.07666 19711	0.07386 54874

SQUARE ROOT OF $[1 + x^2]$ AND ITS NATURAL LOGARITHM (Continued)

x	$\sqrt{1 + x^2}$	$\log_e \sqrt{1 + x^2}$	x	$\sqrt{1 + x^2}$	$\log_e \sqrt{1 + x^2}$
0.400	1.07703 29614	0.07421 00025	0.450	1.09658 56100	0.09220 13615
0.401	1.07740 47522	0.07455 51420	0.451	1.09699 63537	0.09257 58574
0.402	1.07777 73425	0.07490 09043	0.452	1.09740 78549	0.09295 09034
0.403	1.07815 07316	0.07524 72879	0.453	1.09782 01128	0.09332 64979
0.404	1.07852 49186	0.07559 42914	0.454	1.09823 31264	0.09370 26397
0.405	1.07889 99027	0.07594 19134	0.455	1.09864 68951	0.09407 93272
0.406	1.07927 56830	0.07629 01523	0.456	1.09906 14178	0.09445 65590
0.407	1.07965 22588	0.07663 90067	0.457	1.09947 66937	0.09483 43336
0.408	1.08002 96292	0.07698 84753	0.458	1.09989 27221	0.09521 26496
0.409	1.08040 77934	0.07733 85564	0.459	1.10030 95019	0.09559 15056
0.410	1.08078 67505	0.07768 92486	0.460	1.10072 70325	0.09597 09001
0.411	1.08116 64997	0.07804 05506	0.461	1.10114 53128	0.09635 08317
0.412	1.08154 70401	0.07839 24608	0.462	1.10156 43422	0.09673 12989
0.413	1.08192 83710	0.07874 49777	0.463	1.10198 41197	0.09711 23002
0.414	1.08231 04915	0.07909 81000	0.464	1.10240 46444	0.09749 38343
0.415	1.08269 34007	0.07945 18262	0.465	1.10282 59156	0.09787 58997
0.416	1.08307 70979	0.07980 61547	0.466	1.10324 79322	0.09825 84949
0.417	1.08346 15822	0.08016 10842	0.467	1.10367 06936	0.09864 16186
0.418	1.08384 68527	0.08051 66132	0.468	1.10409 41989	0.09902 52693
0.419	1.08423 29086	0.08087 27403	0.469	1.10451 84471	0.09940 94455
0.420	1.08461 97490	0.08122 94639	0.470	1.10494 34375	0.09979 41459
0.421	1.08500 73732	0.08158 67826	0.471	1.10536 91691	0.10017 93688
0.422	1.08539 57803	0.08194 46950	0.472	1.10579 56412	0.10056 51131
0.423	1.08578 49695	0.08230 31996	0.473	1.10622 28528	0.10095 13772
0.424	1.08617 49399	0.08266 22950	0.474	1.10665 08031	0.10133 81596
0.425	1.08656 56906	0.08302 19797	0.475	1.10707 94913	0.10172 54590
0.426	1.08695 72209	0.08338 22522	0.476	1.10750 89165	0.10211 32739
0.427	1.08734 95298	0.08374 31111	0.477	1.10793 90777	0.10250 16028
0.428	1.08774 26166	0.08410 45549	0.478	1.10836 99743	0.10289 04445
0.429	1.08813 64804	0.08446 65821	0.479	1.10880 16053	0.10327 97973
0.430	1.08853 11204	0.08482 91915	0.480	1.10923 39699	0.10366 96599
0.431	1.08892 65356	0.08519 23813	0.481	1.10966 70672	0.10406 00309
0.432	1.08932 27254	0.08555 61503	0.482	1.11010 08963	0.10445 09088
0.433	1.08971 96887	0.08592 04969	0.483	1.11053 54564	0.10484 22922
0.434	1.09011 74249	0.08628 54197	0.484	1.11097 07467	0.10523 41797
0.435	1.09051 59330	0.08665 09172	0.485	1.11140 67662	0.10562 65699
0.436	1.09091 52121	0.08701 69881	0.486	1.11184 35142	0.10601 94613
0.437	1.09131 52615	0.08738 36307	0.487	1.11228 09897	0.10641 28525
0.438	1.09171 60803	0.08775 08438	0.488	1.11271 91919	0.10680 67421
0.439	1.09211 76677	0.08811 86258	0.489	1.11315 81199	0.10720 11286
0.440	1.09252 00227	0.08848 69753	0.490	1.11359 77730	0.10759 60108
0.441	1.09292 31446	0.08885 58908	0.491	1.11403 81502	0.10799 13870
0.442	1.09332 70325	0.08922 53708	0.492	1.11447 92506	0.10838 72560
0.433	1.09373 16856	0.08959 54139	0.493	1.11492 10734	0.10878 36162
0.444	1.09413 71029	0.08996 60188	0.494	1.11536 36178	0.10918 04664
0.445	1.09454 32837	0.09033 71838	0.495	1.11580 68829	0.10957 78050
0.446	1.09495 02272	0.09070 89076	0.496	1.11625 08679	0.10997 56307
0.447	1.09535 79324	0.09108 11887	0.497	1.11669 55718	0.11037 39420
0.448	1.09576 63985	0.09145 40258	0.498	1.11714 09938	0.11077 27375
0.449	1.09617 56246	0.09182 74172	0.499	1.11758 71331	0.11117 20159

SQUARE ROOT OF $[1 + x^2]$ AND ITS NATURAL LOGARITHM (Continued)

x	$\sqrt{1+x^2}$	$\log_e \sqrt{1+x^2}$	x	$\sqrt{1+x^2}$	$\log_e \sqrt{1+x^2}$
0.500	1.11803 39888	0.11157 17756	0.550	1.14127 12211	0.13214 27474
0.501	1.11848 15600	0.11197 20154	0.551	1.14175 34760	0.13256 52176
0.502	1.11892 98459	0.11237 27338	0.552	1.14223 64029	0.13298 80976
0.503	1.11937 88456	0.11277 39293	0.553	1.14272 00007	0.13341 13861
0.504	1.11982 85583	0.11317 56006	0.554	1.14320 42687	0.13383 50816
0.505	1.12027 89831	0.11357 77464	0.555	1.14368 92060	0.13425 91830
0.506	1.12073 01192	0.11398 03650	0.556	1.14417 48118	0.13468 36888
0.507	1.12118 19656	0.11438 34553	0.557	1.14466 10852	0.13510 85978
0.508	1.12163 45216	0.11478 70157	0.558	1.14514 80254	0.13553 39085
0.509	1.12208 77862	0.11519 10449	0.559	1.14563 56314	0.13595 96196
0.510	1.12254 17587	0.11559 55415	0.560	1.14612 39026	0.13638 57299
0.511	1.12299 64381	0.11600 05040	0.561	1.14661 28379	0.13681 22379
0.512	1.12345 18236	0.11640 59311	0.562	1.14710 24366	0.13723 91424
0.513	1.12390 79144	0.11681 18214	0.563	1.14759 26978	0.13766 64421
0.514	1.12436 47095	0.11721 81735	0.564	1.14808 36206	0.13809 41355
0.515	1.12482 22082	0.11762 49860	0.565	1.14857 52043	0.13852 22215
0.516	1.12528 04095	0.11803 22575	0.566	1.14906 74480	0.13895 06986
0.517	1.12573 93126	0.11843 99866	0.567	1.14956 03507	0.13937 95656
0.518	1.12619 89167	0.11884 81719	0.568	1.15005 39118	0.13980 88211
0.519	1.12665 92209	0.11925 68121	0.569	1.15054 81302	0.14023 84638
0.520	1.12712 02243	0.11966 59058	0.570	1.15104 30053	0.14066 84924
0.521	1.12758 19261	0.12007 54515	0.571	1.15153 85360	0.14109 89057
0.522	1.12804 43254	0.12048 54479	0.572	1.15203 47217	0.14152 97022
0.523	1.12850 74213	0.12089 58936	0.573	1.15253 15614	0.14196 08807
0.524	1.12897 12131	0.12130 67872	0.574	1.15302 90543	0.14239 24398
0.525	1.12943 56998	0.12171 81273	0.575	1.15352 71995	0.14282 43784
0.526	1.12990 08806	0.12212 99126	0.576	1.15402 59962	0.14325 66949
0.527	1.13036 67546	0.12254 21417	0.577	1.15452 54436	0.14368 93882
0.528	1.13083 33211	0.12295 48132	0.578	1.15502 55408	0.14412 24570
0.529	1.13130 05790	0.12336 79257	0.579	1.15552 62870	0.14455 59000
0.530	1.13176 85276	0.12378 14779	0.580	1.15602 76813	0.14498 97158
0.531	1.13223 71660	0.12419 54684	0.581	1.15652 97229	0.14542 39031
0.532	1.13270 64933	0.12460 98958	0.582	1.15703 24110	0.14585 84608
0.533	1.13317 65088	0.12502 47588	0.583	1.15753 57446	0.14629 33874
0.534	1.13364 72114	0.12544 00559	0.584	1.15803 97230	0.14672 86817
0.535	1.13411 86005	0.12585 57858	0.585	1.15854 43453	0.14716 43423
0.536	1.13459 06751	0.12627 19472	0.586	1.15904 96107	0.14760 03682
0.537	1.13506 34344	0.12668 85387	0.587	1.15955 55183	0.14803 67578
0.538	1.13553 68774	0.12710 55589	0.588	1.16006 20673	0.14847 35100
0.539	1.13601 10035	0.12752 30064	0.589	1.16056 92569	0.14891 06234
0.540	1.13648 58116	0.12794 08800	0.590	1.16107 70862	0.14934 80969
0.541	1.13696 13010	0.12835 91782	0.591	1.16158 55543	0.14978 59290
0.542	1.13743 74708	0.12877 78997	0.592	1.16209 46605	0.15022 41186
0.543	1.13791 43201	0.12919 70431	0.593	1.16260 44039	0.15066 26643
0.544	1.13839 18482	0.12961 66070	0.594	1.16311 47837	0.15110 15649
0.545	1.13887 00540	0.13003 65902	0.595	1.16362 57990	0.05154 08191
0.546	1.13934 89369	0.13045 69913	0.596	1.16413 74489	0.15198 04256
0.547	1.13982 84959	0.13087 78089	0.597	1.16464 97328	0.15242 03833
0.548	1.14030 87301	0.13129 90417	0.598	1.16516 26496	0.15286 06907
0.549	1.14078 96388	0.13172 06883	0.599	1.16567 61986	0.15330 13466

SQUARE ROOT OF [1 + x^2] AND ITS NATURAL LOGARITHM (Continued)

x	$\sqrt{1 + x^2}$	$\log_e \sqrt{1 + x^2}$	x	$\sqrt{1 + x^2}$	$\log_e \sqrt{1 + x^2}$
0.600	1.16619 03790	0.15374 23498	0.650	1.19268 60442	0.17620 79436
0.601	1.16670 51898	0.15418 36992	0.651	1.19323 13271	0.17666 50280
0.602	1.16722 06304	0.15462 53932	0.652	1.19377 71986	0.17712 23968
0.603	1.16773 66998	0.15506 74307	0.653	1.19432 36580	0.17758 00486
0.604	1.16825 33972	0.15550 98105	0.654	1.19487 07043	0.17803 79823
0.605	1.16877 07217	0.15595 25313	0.655	1.19541 83368	0.17849 61968
0.606	1.16928 86727	0.15639 55918	0.656	1.19596 65547	0.17895 46909
0.607	1.16980 72491	0.15683 89909	0.657	1.19651 53572	0.17941 34634
0.608	1.17032 64502	0.15728 27271	0.658	1.19706 47434	0.17987 25132
0.609	1.17084 62751	0.15772 67994	0.659	1.19761 47127	0.18033 18392
0.610	1.17136 67231	0.15817 12065	0.660	1.19816 52641	0.18079 14402
0.611	1.17188 77933	0.15861 59471	0.661	1.19871 63968	0.18125 13150
0.612	1.17240 94848	0.15906 10199	0.662	1.19926 81101	0.18171 14625
0.613	1.17293 17968	0.15950 64238	0.663	1.19982 04032	0.18217 18816
0.614	1.17345 47286	0.15995 21574	0.664	1.20037 32753	0.18263 25712
0.615	1.17397 82792	0.16039 82197	0.665	1.20092 67255	0.18309 35300
0.616	1.17450 24478	0.16084 46093	0.666	1.20148 07531	0.18355 47570
0.617	1.17502 72337	0.16129 13250	0.667	1.20203 53572	0.18401 62510
0.618	1.17555 26360	0.16173 83655	0.668	1.20259 05371	0.18447 80109
0.619	1.17607 86538	0.16218 57298	0.669	1.20314 62920	0.18494 00356
0.620	1.17660 52864	0.16263 34164	0.670	1.20370 26211	0.18540 23239
0.621	1.17713 25329	0.16308 14243	0.671	1.20425 95235	0.18586 48748
0.622	1.17766 03925	0.16352 97520	0.672	1.20481 69986	0.18632 76870
0.623	1.17818 88643	0.16397 83986	0.673	1.20537 50454	0.18679 07595
0.624	1.17871 79476	0.16442 73628	0.674	1.20593 36632	0.18725 40911
0.625	1.17924 76415	0.16487 66432	0.675	1.20649 28512	0.18771 76809
0.626	1.17977 79452	0.16532 62387	0.676	1.20705 26086	0.18818 15275
0.627	1.18030 88579	0.16577 61482	0.677	1.20761 29347	0.18864 56299
0.628	1.18084 03787	0.16622 63704	0.678	1.20817 38286	0.18910 99870
0.629	1.18137 25069	0.16667 69040	0.679	1.20873 52895	0.18957 45977
0.630	1.18190 52415	0.16712 77479	0.680	1.20929 73166	0.19003 94609
0.631	1.18243 85819	0.16757 89009	0.681	1.20985 99093	0.19050 45754
0.632	1.18297 25272	0.16803 03617	0.682	1.21042 30665	0.19096 99403
0.633	1.18350 70764	0.16848 21292	0.683	1.21098 67877	0.19143 55543
0.634	1.18404 22290	0.16893 42022	0.684	1.21155 10720	0.19190 14164
0.635	1.18457 79839	0.16938 65795	0.685	1.21211 59185	0.19236 75255
0.636	1.18511 43405	0.16983 92598	0.686	1.21268 13266	0.19283 38804
0.637	1.18565 12978	0.17029 22420	0.687	1.21324 72955	0.19330 04802
0.638	1.18618 88551	0.17074 55249	0.688	1.21381 38243	0.19376 73236
0.639	1.18672 70116	0.17119 91073	0.689	1.21438 09122	0.19423 44097
0.640	1.18726 57664	0.17165 29881	0.690	1.21494 85586	0.19470 17373
0.641	1.18780 51187	0.17210 71660	0.691	1.21551 67625	0.19516 93053
0.642	1.18834 50677	0.17256 16398	0.692	1.21608 55233	0.19563 71127
0.643	1.18888 56127	0.17301 64085	0.693	1.21665 48401	0.19610 51585
0.644	1.18942 67527	0.17347 14706	0.694	1.21722 47122	0.19657 34414
0.645	1.18996 84870	0.17392 68253	0.695	1.21779 51388	0.19704 19604
0.646	1.19051 08147	0.17438 24711	0.696	1.21836 61190	0.19751 07145
0.647	1.19105 37351	0.17483 84070	0.697	1.21893 76522	0.19797 97025
0.648	1.19159 72474	0.17529 46318	0.698	1.21950 97376	0.19844 89235
0.649	1.19214 13507	0.17575 11444	0.699	1.22008 23743	0.19891 83764

SQUARE ROOT OF [1 + x^2] AND ITS NATURAL LOGARITHM (Continued)

x	$\sqrt{1+x^2}$	$\log_e \sqrt{1+x^2}$	x	$\sqrt{1+x^2}$	$\log_e \sqrt{1+x^2}$
0.700	1.22065 55616	0.19938 80600	0.750	1.25000 00000	0.22314 35513
0.701	1.22122 92987	0.19985 79733	0.751	1.25060 02559	0.22362 36407
0.702	1.22180 35849	0.20032 81153	0.752	1.25120 10230	0.22410 39085
0.703	1.22237 84193	0.20079 84849	0.753	1.25180 23007	0.22458 43534
0.704	1.22295 38013	0.20126 90811	0.754	1.25240 40881	0.22506 49747
0.705	1.22352 97299	0.20173 99027	0.755	1.25300 63847	0.22554 57714
0.706	1.22410 62045	0.20221 09488	0.756	1.25360 91895	0.22602 67425
0.707	1.22468 32243	0.20268 22182	0.757	1.25421 25019	0.22650 78871
0.708	1.22526 07886	0.20315 37100	0.758	1.25481 63212	0.22698 92043
0.709	1.22583 88964	0.20360 54231	0.759	1.25542 06466	0.22747 06930
0.710	1.22641 75472	0.20409 73563	0.760	1.25602 54774	0.22795 23524
0.711	1.22699 67400	0.20456 95089	0.761	1.25663 08129	0.22843 41816
0.712	1.22757 64742	0.20504 18795	0.762	1.25723 66523	0.22891 61795
0.713	1.22815 67490	0.20551 44673	0.763	1.25784 29950	0.22939 83452
0.714	1.22873 75635	0.20598 72712	0.764	1.25844 98401	0.22988 06779
0.715	1.22931 89171	0.20646 02901	0.765	1.25905 71869	0.23036 31765
0.716	1.22990 08090	0.20693 35231	0.766	1.25966 50348	0.23084 58403
0.717	1.23048 32384	0.20740 69690	0.767	1.26027 33830	0.23132 86681
0.718	1.23106 62046	0.20788 06269	0.768	1.26088 22308	0.23181 16592
0.719	1.23164 97067	0.20835 44957	0.769	1.26149 15775	0.23229 48125
0.720	1.23223 37441	0.20882 85745	0.770	1.26210 14222	0.23277 81271
0.721	1.23281 83159	0.20930 28621	0.771	1.26271 17644	0.23326 16023
0.722	1.23340 34214	0.20977 73576	0.772	1.26332 26033	0.23374 52369
0.723	1.23398 90599	0.21025 20599	0.773	1.26393 39382	0.23422 90302
0.724	1.23457 52306	0.21072 69681	0.774	1.26454 57683	0.23471 29812
0.725	1.23516 19327	0.21120 20811	0.775	1.26515 80929	0.23519 70890
0.726	1.23574 91655	0.21167 73979	0.776	1.26577 09114	0.23568 13526
0.727	1.23633 69282	0.21215 29175	0.777	1.26638 42229	0.23616 57713
0.728	1.23692 52201	0.21262 86389	0.778	1.26699 80268	0.23665 03440
0.729	1.23751 40403	0.21310 45611	0.779	1.26761 23224	0.23713 50699
0.730	1.23810 33883	0.21358 06831	0.780	1.26822 71090	0.23761 99480
0.731	1.23869 32631	0.21405 70039	0.781	1.26884 23858	0.23810 49775
0.732	1.23928 36641	0.21453 35224	0.782	1.26945 81521	0.23859 01576
0.733	1.23987 45904	0.21501 02378	0.783	1.27007 44073	0.23907 54871
0.734	1.24046 60415	0.21548 71489	0.784	1.27069 11505	0.23956 09655
0.735	1.24105 80164	0.21596 42548	0.785	1.27130 83812	0.24004 65916
0.736	1.24165 05144	0.21644 15546	0.786	1.27192 60985	0.24053 23645
0.737	1.24224 35349	0.21691 90472	0.787	1.27254 43018	0.24101 82836
0.738	1.24283 70770	0.21739 67316	0.788	1.27316 29904	0.24150 43478
0.739	1.24343 11400	0.21787 46068	0.789	1.27378 21635	0.24199 05562
0.740	1.24402 57232	0.21835 26720	0.790	1.27440 18205	0.24247 69081
0.741	1.24462 08258	0.21883 09259	0.791	1.27502 19606	0.24296 34025
0.742	1.24521 64470	0.21930 93679	0.792	1.27564 25832	0.24345 00384
0.743	1.24581 25862	0.21978 79967	0.793	1.27626 36875	0.24393 68152
0.744	1.24640 92426	0.22026 68115	0.794	1.27688 52728	0.24442 37318
0.745	1.24700 64154	0.22074 58113	0.795	1.27750 73385	0.24491 07875
0.746	1.24760 41039	0.22122 49952	0.796	1.27812 98839	0.24539 79814
0.747	1.24820 23073	0.22170 43620	0.797	1.27875 29081	0.24588 53124
0.748	1.24880 10250	0.22218 39110	0.798	1.27937 64106	0.24637 27800
0.749	1.24940 02561	0.22266 36411	0.799	1.28000 03906	0.24686 03831

SQUARE ROOT OF $[1 + x^2]$ AND ITS NATURAL LOGARITHM (Continued)

x	$\sqrt{1 + x^2}$	$\log_e \sqrt{1 + x^2}$	x	$\sqrt{1 + x^2}$	$\log_e \sqrt{1 + x^2}$
0.800	1.28062 48475	0.24734 81209	0.850	1.31244 04748	0.27188 83620
0.801	1.28124 97805	0.24783 59926	0.851	1.31308 83443	0.27238 18744
0.802	1.28187 51889	0.24832 39973	0.852	1.31373 66555	0.27287 54856
0.803	1.28250 10721	0.24881 21340	0.853	1.31438 54077	0.27336 91859
0.804	1.28312 74294	0.24930 04021	0.854	1.31503 46003	0.27386 29773
0.805	1.28375 42600	0.24978 88006	0.855	1.31568 42326	0.27435 68593
0.806	1.28438 15632	0.25027 73288	0.856	1.31633 43040	0.27485 08311
0.807	1.28500 93385	0.25076 59856	0.857	1.31698 48139	0.27534 48919
0.808	1.28563 75850	0.25125 47704	0.858	1.31763 57615	0.27583 90409
0.809	1.28626 63021	0.25174 36823	0.859	1.31828 71463	0.27633 32775
0.810	1.28689 54892	0.25223 27203	0.860	1.31893 89675	0.27682 76009
0.811	1.28752 51454	0.25272 18838	0.861	1.31959 12246	0.27732 20103
0.812	1.28815 52701	0.25321 11718	0.862	1.32024 39169	0.27781 65050
0.813	1.28878 58628	0.25370 05835	0.863	1.32089 70437	0.27831 10844
0.814	1.28941 69225	0.25419 01181	0.864	1.32155 06044	0.27880 57475
0.815	1.29004 84487	0.25467 97748	0.865	1.32220 45984	0.27930 04937
0.816	1.29068 04407	0.25516 95527	0.866	1.32285 90250	0.27979 53224
0.817	1.29131 28978	0.25565 94511	0.867	1.32351 38836	0.28029 02327
0.818	1.29194 58193	0.25614 94690	0.868	1.32416 91735	0.28078 52239
0.819	1.29257 92045	0.25663 96056	0.869	1.32482 48941	0.28128 02953
0.820	1.29321 30528	0.25712 98603	0.870	1.32548 10448	0.28177 54462
0.821	1.29384 73635	0.25762 02320	0.871	1.32613 76248	0.28227 06759
0.822	1.29448 21358	0.25811 07201	0.872	1.32679 46337	0.28276 59836
0.823	1.29511 73692	0.25860 13236	0.873	1.32745 20707	0.28326 13685
0.824	1.29575 30629	0.25909 20419	0.874	1.32810 99352	0.28375 68302
0.825	1.29638 92162	0.25958 28740	0.875	1.32876 82266	0.28425 23677
0.826	1.29702 58286	0.26007 38193	0.876	1.32942 69442	0.28474 79804
0.827	1.29766 28992	0.26056 48767	0.877	1.33008 60874	0.28524 36676
0.828	1.29830 04275	0.26105 60457	0.878	1.33074 56556	0.28573 94285
0.829	1.29893 84127	0.26154 73253	0.879	1.33140 56482	0.28623 52625
0.830	1.29957 68542	0.26203 87147	0.880	1.33206 60644	0.28673 11688
0.831	1.30021 57513	0.26253 02132	0.881	1.33272 69038	0.28722 71468
0.832	1.30085 51034	0.26302 18201	0.882	1.33338 81655	0.28772 31958
0.833	1.30149 49097	0.26351 35344	0.883	1.33404 98491	0.28821 93150
0.834	1.30213 51696	0.26400 53553	0.884	1.33471 19540	0.28871 55037
0.835	1.30277 58825	0.26449 72822	0.885	1.33537 44793	0.28921 17614
0.836	1.30341 70476	0.26498 93142	0.886	1.33603 74246	0.28970 80871
0.837	1.30405 86643	0.26548 14505	0.887	1.33670 07893	0.29020 44805
0.838	1.30470 07320	0.26597 36903	0.888	1.33736 45726	0.29070 09405
0.839	1.30534 32499	0.26646 60329	0.889	1.33802 87740	0.29119 74667
0.840	1.30598 62174	0.26695 84775	0.890	1.33869 33928	0.29169 40583
0.841	1.30662 96338	0.26745 10233	0.891	1.33935 84285	0.29219 07146
0.842	1.30727 34985	0.26794 36695	0.892	1.34002 38804	0.29268 74350
0.843	1.30791 78109	0.26843 64154	0.893	1.34068 97479	0.29318 42188
0.844	1.30856 25702	0.26892 92601	0.894	1.34135 60303	0.29368 10652
0.845	1.30920 77757	0.26942 22029	0.895	1.34202 27271	0.29417 79730
0.846	1.30985 34269	0.26991 52431	0.896	1.34268 98376	0.29467 49434
0.847	1.31049 95231	0.27040 83799	0.897	1.34335 73612	0.29517 19739
0.848	1.31114 60636	0.27090 16124	0.898	1.34402 52974	0.29566 90644
0.849	1.31179 30477	0.27139 49400	0.899	1.34469 36454	0.29616 62142

SQUARE ROOT OF $[1 + x^2]$ AND ITS NATURAL LOGARITHM (Continued)

x	$\sqrt{1+x^2}$	$\log_e \sqrt{1+x^2}$	x	$\sqrt{1+x^2}$	$\log_e \sqrt{1+x^2}$
0.900	1.34536 24047	0.29666 34226	0.950	1.37931 14224	0.32158 44054
0.901	1.34603 15747	0.29716 06891	0.951	1.38000 03623	0.32208 37617
0.902	1.34670 11547	0.29765 80129	0.952	1.38068 96827	0.32258 31444
0.903	1.34737 11441	0.29815 53933	0.953	1.38137 93831	0.32308 25529
0.904	1.34804 15424	0.29865 28298	0.954	1.38206 94628	0.32358 19866
0.905	1.34871 23489	0.29915 03216	0.955	1.38275 99213	0.32408 14449
0.906	1.34938 35630	0.29964 78681	0.956	1.38345 07581	0.32458 09273
0.907	1.35005 51841	0.30014 54687	0.957	1.38414 19725	0.32508 04332
0.908	1.35072 72115	0.30064 31227	0.958	1.38483 35640	0.32557 99620
0.909	1.35139 96448	0.30114 08293	0.959	1.38552 55321	0.32607 95132
0.910	1.35207 24833	0.30163 85881	0.960	1.38621 78761	0.32657 90862
0.911	1.35274 57263	0.30213 63983	0.961	1.38691 05955	0.32707 86804
0.912	1.35341 93733	0.30263 42593	0.962	1.38760 36898	0.32757 82952
0.913	1.35409 34237	0.30313 21704	0.963	1.38829 71584	0.32807 79302
0.914	1.35476 78768	0.30363 01310	0.964	1.38899 10007	0.32857 75848
0.915	1.35544 27321	0.30412 81405	0.965	1.38968 52162	0.32907 72583
0.916	1.35611 79890	0.30462 61982	0.966	1.39037 98042	0.32957 69503
0.917	1.35679 36468	0.30512 43036	0.967	1.39107 47643	0.33007 66601
0.918	1.35746 97050	0.30562 24558	0.968	1.39177 00960	0.33057 63873
0.919	1.35814 61630	0.30612 06544	0.969	1.39246 57985	0.33107 61313
0.920	1.35882 30201	0.30661 88987	0.970	1.39316 18714	0.33157 58915
0.921	1.35950 02758	0.30711 71880	0.971	1.39385 83142	0.33207 56674
0.922	1.36017 79295	0.30761 55217	0.972	1.39455 51262	0.33257 54585
0.923	1.36085 59806	0.30811 38993	0.973	1.39525 23009	0.33307 52641
0.924	1.36153 44285	0.30861 23200	0.974	1.39594 98558	0.33357 50838
0.925	1.36221 32726	0.30911 07833	0.975	1.39664 77723	0.33407 49170
0.926	1.36289 25123	0.30960 92885	0.976	1.39734 60559	0.33457 47632
0.927	1.36357 21470	0.31010 78350	0.977	1.39804 47060	0.33507 46218
0.928	1.36425 21761	0.31060 64222	0.978	1.39874 37221	0.33557 44924
0.929	1.36493 25991	0.31110 50494	0.979	1.39944 31035	0.33607 43743
0.930	1.36561 34153	0.31160 37162	0.980	1.40014 28499	0.33657 42670
0.931	1.36629 46242	0.31210 24218	0.981	1.40084 29605	0.33707 41701
0.932	1.36697 62251	0.31260 11656	0.982	1.40154 34349	0.33757 40829
0.933	1.36765 82175	0.31309 99470	0.983	1.40224 42726	0.33807 40050
0.934	1.36834 06009	0.31359 87655	0.984	1.40294 54729	0.33857 39357
0.935	1.36902 33745	0.31409 76203	0.985	1.40364 70354	0.33907 38747
0.936	1.36970 65379	0.31459 65110	0.986	1.40434 89595	0.33957 38214
0.937	1.37039 00904	0.31509 54369	0.987	1.40505 12446	0.34007 37752
0.938	1.37107 40316	0.31559 43973	0.988	1.40575 38903	0.34057 37356
0.939	1.37175 83606	0.31609 33918	0.989	1.40645 68959	0.34107 37021
0.940	1.37244 30771	0.31659 24197	0.990	1.40716 02610	0.34157 36742
0.941	1.37312 81805	0.31709 14804	0.991	1.40786 39849	0.34207 36514
0.942	1.37381 36700	0.31759 05733	0.992	1.40856 80672	0.34257 36332
0.943	1.37449 95453	0.31808 96978	0.993	1.40927 25074	0.34307 36190
0.944	1.37518 58056	0.31858 88534	0.994	1.40997 73048	0.34357 36083
0.945	1.37587 24505	0.31908 80393	0.995	1.41068 24590	0.34407 36007
0.946	1.37655 94793	0.31958 72552	0.996	1.41138 79694	0.34457 35956
0.947	1.37724 68914	0.32008 65003	0.997	1.41209 38354	0.34507 35925
0.948	1.37793 46864	0.32058 57741	0.998	1.41280 00566	0.34557 35910
0.949	1.37862 28636	0.32108 50760	0.999	1.41350 66325	0.34607 35904
				1.41421 35624	0.34657 35903

EXPONENTIAL FUNCTIONS

Values of e^x, log e^x and e^{-x} where e is the base of the natural system of logarithms 2.71828... and x has values from 0 to 10. Facilitating the solution of exponential equations, these tables also serve as a table of natural or Naperian antilogarithms. For instance, if the logarithm or exponent x = 3.26, the corresponding number or value of e^x is 26.050. Its reciprocal e^{-x} is .038388.

x	e^x	$\text{Log}_{10}(e^x)$	e^{-x}	x	e^x	$\text{Log}_{10}(e^x)$	e^{-x}
0.00	1.0000	0.00000	1.000000	**0.50**	1.6487	0.21715	0.606531
0.01	1.0101	.00434	0.990050	0.51	1.6653	.22149	.600496
0.02	1.0202	.00869	.980199	0.52	1.6820	.22583	.594521
0.03	1.0305	.01303	.970446	0.53	1.6989	.23018	.588605
0.04	1.0408	.01737	.960789	0.54	1.7160	.23452	.582748
0.05	1.0513	0.02171	0.951229	**0.55**	1.7333	0.23886	0.576950
0.06	1.0618	.02606	.941765	0.56	1.7507	.24320	.571209
0.07	1.0725	.03040	.932394	0.57	1.7683	.24755	.565525
0.08	1.0833	.03474	.923116	0.58	1.7860	.25189	.559898
0.09	1.0942	.03909	.913931	0.59	1.8040	.25623	.554327
0.10	1.1052	0.04343	0.904837	**0.60**	1.8221	0.26058	0.548812
0.11	1.1163	.04777	.895834	0.61	1.8404	.26492	.543351
0.12	1.1275	.05212	.886920	0.62	1.8589	.26926	.537944
0.13	1.1388	.05646	.878095	0.63	1.8776	.27361	.532592
0.14	1.1503	.06080	.869358	0.64	1.8965	.27795	.527292
0.15	1.1618	0.06514	0.860708	**0.65**	1.9155	0.28229	0.522046
0.16	1.1735	.06949	.852144	0.66	1.9348	.28663	.516851
0.17	1.1853	.07383	.843665	0.67	1.9542	.29098	.511709
0.18	1.1972	.07817	.835270	0.68	1.9739	.29532	.506617
0.19	1.2092	.08252	.826959	0.69	1.9937	.29966	.501576
0.20	1.2214	0.08686	0.818731	**0.70**	2.0138	0.30401	0.496585
0.21	1.2337	.09120	.810584	0.71	2.0340	.30835	.491644
0.22	1.2461	.09554	.802519	0.72	2.0544	.31269	.486752
0.23	1.2586	.09989	.794534	0.73	2.0751	.31703	.481909
0.24	1.2712	.10423	.786628	0.74	2.0959	.32138	.477114
0.25	1.2840	0.10857	0.778801	**0.75**	2.1170	0.32572	0.472367
0.26	1.2969	.11292	.771052	0.76	2.1383	.33006	.467666
0.27	1.3100	.11726	.763379	0.77	2.1598	.33441	.463013
0.28	1.3231	.12160	.755784	0.78	2.1815	.33875	.458406
0.29	1.3364	.12595	.748264	0.79	2.2034	.34309	.453845
0.30	1.3499	0.13029	0.740818	**0.80**	2.2255	0.34744	0.449329
0.31	1.3634	.13463	.733447	0.81	2.2479	.35178	.444858
0.32	1.3771	.13897	.726149	0.82	2.2705	.35612	.440432
0.33	1.3910	.14332	.718924	0.83	2.2933	.36046	.436049
0.34	1.4049	.14766	.711770	0.84	2.3164	.36481	.431711
0.35	1.4191	0.15200	0.704688	**0.85**	2.3396	0.36915	0.427415
0.36	1.4333	.15635	.697676	0.86	2.3632	.37349	.423162
0.37	1.4477	.16069	.690734	0.87	2.3869	.37784	.418952
0.38	1.4623	.16503	.683861	0.88	2.4109	.38218	.414783
0.39	1.4770	.16937	.677057	0.89	2.4351	.38652	.410656
0.40	1.4918	0.17372	0.670320	**0.90**	2.4596	0.39087	0.406570
0.41	1.5068	.17806	.663650	0.91	2.4843	.39521	.402524
0.42	1.5220	.18240	.657047	0.92	2.5093	.39955	.398519
0.43	1.5373	.18675	.650509	0.93	2.5345	.40389	.394554
0.44	1.5527	.19109	.644036	0.94	2.5600	.40824	.390628
0.45	1.5683	0.19543	0.637628	**0.95**	2.5857	0.41258	0.386741
0.46	1.5841	.19978	.631284	0.96	2.6117	.41692	.382893
0.47	1.6000	.20412	.625002	0.97	2.6379	.42127	.379083
0.48	1.6161	.20846	.618783	0.98	2.6645	.42561	.375311
0.49	1.6323	.21280	.612626	0.99	2.6912	.42995	.371577
0.50	1.6487	0.21715	0.606531	**1.00**	2.7183	0.43429	0.367879

EXPONENTIAL FUNCTIONS (Continued)

x	e^x	$\text{Log}_{10}(e^x)$	e^{-x}
1.00	2.7183	0.43429	0.367879
1.01	2.7456	.43864	.364219
1.02	2.7732	.44298	.360595
1.03	2.8011	.44732	.357007
1.04	2.8292	.45167	.353455
1.05	2.8577	0.45601	0.349938
1.06	2.8864	.46035	.346456
1.07	2.9154	.46470	.343009
1.08	2.9447	.46904	.339596
1.09	2.9743	.47338	.336216
1.10	3.0042	0.47772	0.332871
1.11	3.0344	.48207	.329559
1.12	3.0649	.48641	.326280
1.13	3.0957	.49075	.323033
1.14	3.1268	.49510	.319819
1.15	3.1582	0.49944	0.316637
1.16	3.1899	.50378	.313486
1.17	3.2220	.50812	.310367
1.18	3.2544	.51247	.307279
1.19	3.2871	.51681	.304221
1.20	3.3201	0.52115	0.301194
1.21	3.3535	.52550	.298197
1.22	3.3872	.52984	.295230
1.23	3.4212	.53418	.292293
1.24	3.4556	.53853	.289384
1.25	3.4903	0.54287	0.286505
1.26	3.5254	.54721	.283654
1.27	3.5609	.55155	.280832
1.28	3.5966	.55590	.278037
1.29	3.6328	.56024	.275271
1.30	3.6693	0.56458	0.272532
1.31	3.7062	.56893	.269820
1.32	3.7434	.57327	.267135
1.33	3.7810	.57761	.264477
1.34	3.8190	.58195	.261846
1.35	3.8574	0.58630	0.259240
1.36	3.8962	.59064	.256661
1.37	3.9354	.59498	.254107
1.38	3.9749	.59933	.251579
1.39	4.0149	.60367	.249075
1.40	4.0552	0.60801	0.246597
1.41	4.0960	.61236	.244143
1.42	4.1371	.61670	.241714
1.43	4.1787	.62104	.239309
1.44	4.2207	.62538	.236928
1.45	4.2631	0.62973	0.234570
1.46	4.3060	.63407	.232236
1.47	4.3492	.63841	.229925
1.48	4.3929	.64276	.227638
1.49	4.4371	.64710	.225373
1.50	4.4817	0.65144	0.223130

x	e^x	$\text{Log}_{10}(e^x)$	e^{-x}
1.50	4.4817	0.65144	0.223130
1.51	4.5267	.65578	.220910
1.52	4.5722	.66013	.218712
1.53	4.6182	.66447	.216536
1.54	4.6646	.66881	.214381
1.55	4.7115	0.67316	0.212248
1.56	4.7588	.67750	.210136
1.57	4.8066	.68184	.208045
1.58	4.8550	.68619	.205975
1.59	4.9037	.69053	.203926
1.60	4.9530	0.69487	0.201897
1.61	5.0028	.69921	.199888
1.62	5.0531	.70356	.197899
1.63	5.1039	.70790	.195930
1.64	5.1552	.71224	.193980
1.65	5.2070	0.71659	0.192050
1.66	5.2593	.72093	.190139
1.67	5.3122	.72527	.188247
1.68	5.3656	.72961	.186374
1.69	5.4195	.73396	.184520
1.70	5.4739	0.73830	0.182684
1.71	5.5290	.74264	.180866
1.72	5.5845	.74699	.179066
1.73	5.6407	.75133	.177284
1.74	5.6973	.75567	.175520
1.75	5.7546	0.76002	0.173774
1.76	5.8124	.76436	.172045
1.77	5.8709	.76870	.170333
1.78	5.9299	.77304	.168638
1.79	5.9895	.77739	.166960
1.80	6.0496	0.78173	0.165299
1.81	6.1104	.78607	.163654
1.82	6.1719	.79042	.162026
1.83	6.2339	.79476	.160414
1.84	6.2965	.79910	.158817
1.85	6.3598	0.80344	0.157237
1.86	6.4237	.80779	.155673
1.87	6.4883	.81213	.154124
1.88	6.5535	.81647	.152590
1.89	6.6194	.82082	.151072
1.90	6.6859	0.82516	0.149569
1.91	6.7531	.82950	.148080
1.92	6.8210	.83385	.146607
1.93	6.8895	.83819	.145148
1.94	6.9588	.84253	.143704
1.95	7.0287	0.84687	0.142274
1.96	7.0993	.85122	.140858
1.97	7.1707	.85556	.139457
1.98	7.2427	.85990	.138069
1.99	7.3155	.86425	.136695
2.00	7.3891	0.86859	0.135335

EXPONENTIAL FUNCTIONS (Continued)

x	e^x	$\text{Log}_{10}(e^x)$	e^{-x}	x	e^x	$\text{Log}_{10}(e^x)$	e^{-x}
2.00	7.3891	0.86859	0.135335	**2.50**	12.182	1.08574	0.082085
2.01	7.4633	.87293	.133989	2.51	12.305	1.09008	.081268
2.02	7.5383	.87727	.132655	2.52	12.429	1.09442	.080460
2.03	7.6141	.88162	.131336	2.53	12.554	1.09877	.079659
2.04	7.6906	.88596	.130029	2.54	12.680	1.10311	.078866
2.05	7.7679	0.89030	0.128735	**2.55**	12.807	1.10745	0.078082
2.06	7.8460	.89465	.127454	2.56	12.936	1.11179	.077305
2.07	7.9248	.89899	.126186	2.57	13.066	1.11614	.076536
2.08	8.0045	.90333	.124930	2.58	13.197	1.12048	.075774
2.09	8.0849	.90768	.123687	2.59	13.330	1.12482	.075020
2.10	8.1662	0.91202	0.122456	**2.60**	13.464	1.12917	0.074274
2.11	8.2482	.91636	.121238	2.61	13.599	1.13351	.073535
2.12	8.3311	.92070	.120032	2.62	13.736	1.13785	.072803
2.13	8.4149	.92505	.118837	2.63	13.874	1.14219	.072078
2.14	8.4994	.92939	.117655	2.64	14.013	1.14654	.071361
2.15	8.5849	0.93373	0.116484	**2.65**	14.154	1.15088	0.070651
2.16	8.6711	.93808	.115325	2.66	14.296	1.15522	.069948
2.17	8.7583	.94242	.114178	2.67	14.440	1.15957	.069252
2.18	8.8463	.94676	.113042	2.68	14.585	1.16391	.068563
2.19	8.9352	.95110	.111917	2.69	14.732	1.16825	.067881
2.20	9.0250	0.95545	0.110803	**2.70**	14.880	1.17260	0.067206
2.21	9.1157	.95979	.109701	2.71	15.029	1.17694	.066537
2.22	9.2073	.96413	.108609	2.72	15.180	1.18128	.065875
2.23	9.2999	.96848	.107528	2.73	15.333	1.18562	.065219
2.24	9.3933	.97282	.106459	2.74	15.487	1.18997	.064570
2.25	9.4877	0.97716	0.105399	**2.75**	15.643	1.19431	0.063928
2.26	9.5831	.98151	.104350	2.76	15.800	1.19865	.063292
2.27	9.6794	.98585	.103312	2.77	15.959	1.20300	.062662
2.28	9.7767	.99019	.102284	2.78	16.119	1.20734	.062039
2.29	9.8749	.99453	.101266	2.79	16.281	1.21168	.061421
2.30	9.9742	0.99888	0.100259	**2.80**	16.445	1.21602	0.060810
2.31	10.074	1.00322	.099261	2.81	16.610	1.22037	.060205
2.32	10.176	1.00756	.098274	2.82	16.777	1.22471	.059606
2.33	10.278	1.01191	.097296	2.83	16.945	1.22905	.059013
2.34	10.381	1.01625	.096328	2.84	17.116	1.23340	.058426
2.35	10.486	1.02059	0.095369	**2.85**	17.288	1.23774	0.057844
2.36	10.591	1.02493	.094420	2.86	17.462	1.24208	.057269
2.37	10.697	1.02928	.093481	2.87	17.637	1.24643	.056699
2.38	10.805	1.03362	.092551	2.88	17.814	1.25077	.056135
2.39	10.913	1.03796	.091630	2.89	17.993	1.25511	.055576
2.40	11.023	1.04231	0.090718	**2.90**	18.174	1.25945	0.055023
2.41	11.134	1.04665	.089815	2.91	18.357	1.26380	.054476
2.42	11.246	1.05099	.088922	2.92	18.541	1.26814	.053934
2.43	11.359	1.05534	.088037	2.93	18.728	1.27248	.053397
2.44	11.473	1.05968	.087161	2.94	18.916	1.27683	.052866
2.45	11.588	1.06402	0.086294	**2.95**	19.106	1.28117	0.052340
2.46	11.705	1.06836	.085435	2.96	19.298	1.28551	.051819
2.47	11.822	1.07271	.084585	2.97	19.492	1.28985	.051303
2.48	11.941	1.07705	.083743	2.98	19.688	1.29420	.050793
2.49	12.061	1.08139	.082910	2.99	19.886	1.29854	.050287
2.50	12.182	1.08574	0.082085	**3.00**	20.086	1.30288	0.049787

EXPONENTIAL FUNCTIONS (Continued)

x	e^x	$\text{Log}_{10}(e^x)$	e^{-x}
3.00	20.086	1.30288	0.049787
3.01	20.287	1.30723	.049292
3.02	20.491	1.31157	.048801
3.03	20.697	1.31591	.048316
3.04	20.905	1.32026	.047835
3.05	21.115	1.32460	0.047359
3.06	21.328	1.32894	.046888
3.07	21.542	1.33328	.046421
3.08	21.758	1.33763	.045959
3.09	21.977	1.34197	.045502
3.10	22.198	1.34631	0.045049
3.11	22.421	1.35066	.044601
3.12	22.646	1.35500	.044157
3.13	22.874	1.35934	.043718
3.14	23.104	1.36368	.043283
3.15	23.336	1.36803	0.042852
3.16	23.571	1.37237	.042426
3.17	23.807	1.37671	.042004
3.18	24.047	1.38106	.041586
3.19	24.288	1.38540	.041172
3.20	24.533	1.38974	0.040762
3.21	24.779	1.39409	.040357
3.22	25.028	1.39843	.039955
3.23	25.280	1.40277	.039557
3.24	25.534	1.40711	.039164
3.25	25.790	1.41146	0.038774
3.26	26.050	1.41580	.038388
3.27	26.311	1.42014	.038006
3.28	26.576	1.42449	.037628
3.29	26.843	1.42883	.037254
3.30	27.113	1.43317	0.036883
3.31	27.385	1.43751	.036516
3.32	27.660	1.44186	.036153
3.33	27.938	1.44620	.035793
3.34	28.219	1.45054	.035437
3.35	28.503	1.45489	0.035084
3.36	28.789	1.45923	.034735
3.37	29.079	1.46357	.034390
3.38	29.371	1.46792	.034047
3.39	29.666	1.47226	.033709
3.40	29.964	1.47660	0.033373
3.41	30.265	1.48094	.033041
3.42	30.569	1.48529	.032712
3.43	30.877	1.48963	.032387
3.44	31.187	1.49397	.032065
3.45	31.500	1.49832	0.031746
3.46	31.817	1.50266	.031430
3.47	32.137	1.50700	.031117
3.48	32.460	1.51134	.030807
3.49	32.786	1.51569	.030501
3.50	33.115	1.52003	0.030197

x	e^x	$\text{Log}_{10}(e^x)$	e^{-x}
3.50	33.115	1.52003	0.030197
3.51	33.448	1.52437	.029897
3.52	33.784	1.52872	.029599
3.53	34.124	1.53306	.029305
3.54	34.467	1.53740	.029013
3.55	34.813	1.54175	0.028725
3.56	35.163	1.54609	.028439
3.57	35.517	1.55043	.028156
3.58	35.874	1.55477	.027876
3.59	36.234	1.55912	.027598
3.60	36.598	1.56346	0.027324
3.61	36.966	1.56780	.027052
3.62	37.338	1.57215	.026783
3.63	37.713	1.57649	.026516
3.64	38.092	1.58083	.026252
3.65	38.475	1.58517	0.025991
3.66	38.861	1.58952	.025733
3.67	39.252	1.59386	.025476
3.68	39.646	1.59820	.025223
3.69	40.045	1.60255	.024972
3.70	40.447	1.60689	0.024724
3.71	40.854	1.61123	.024478
3.72	41.264	1.61558	.024234
3.73	41.679	1.61992	.023993
3.74	42.098	1.62426	.023754
3.75	42.521	1.62860	0.023518
3.76	42.948	1.63295	.023284
3.77	43.380	1.63729	.023052
3.78	43.816	1.64163	.022823
3.79	44.256	1.64598	.022596
3.80	44.701	1.65032	0.022371
3.81	45.150	1.65466	.022148
3.82	45.604	1.65900	.021928
3.83	46.063	1.66335	.021710
3.84	46.525	1.66769	.021494
3.85	46.993	1.67203	0.021280
3.86	47.465	1.67638	.021068
3.87	47.942	1.68072	.020858
3.88	48.424	1.68506	.020651
3.89	48.911	1.68941	.020445
3.90	49.402	1.69375	0.020242
3.91	49.899	1.69809	.020041
3.92	50.400	1.70243	.019841
3.93	50.907	1.70678	.019644
3.94	51.419	1.71112	.019448
3.95	51.935	1.71546	0.019255
3.96	52.457	1.71981	.019063
3.97	52.985	1.72415	.018873
3.98	53.517	1.72849	.018686
3.99	54.055	1.73283	.018500
4.00	54.598	1.73718	0.018316

EXPONENTIAL FUNCTIONS (Continued)

x	e^x	$\text{Log}_{10}(e^x)$	e^{-x}
4.00	54.598	1.73718	0.018316
4.01	55.147	1.74152	.018133
4.02	55.701	1.74586	.017953
4.03	56.261	1.75021	.017774
4.04	56.826	1.75455	.017597
4.05	57.397	1.75889	0.017422
4.06	57.974	1.76324	.017249
4.07	58.557	1.76758	.017077
4.08	59.145	1.77192	.016907
4.09	59.740	1.77626	.016739
4.10	60.340	1.78061	0.016573
4.11	60.947	1.78495	.016408
4.12	61.559	1.78929	.016245
4.13	62.178	1.79364	.016083
4.14	62.803	1.79798	.015923
4.15	63.434	1.80232	0.015764
4.16	64.072	1.80667	.015608
4.17	64.715	1.81101	.015452
4.18	65.366	1.81535	.015299
4.19	66.023	1.81969	.015146
4.20	66.686	1.82404	0.014996
4.21	67.357	1.82838	.014846
4.22	68.033	1.83272	.014699
4.23	68.717	1.83707	.014552
4.24	69.408	1.84141	.014408
4.25	70.105	1.84575	0.014264
4.26	70.810	1.85009	.014122
4.27	71.522	1.85444	.013982
4.28	72.240	1.85878	.013843
4.29	72.966	1.86312	.013705
4.30	73.700	1.86747	0.013569
4.31	74.440	1.87181	.013434
4.32	75.189	1.87615	.013300
4.33	75.944	1.88050	.013168
4.34	76.708	1.88484	.013037
4.35	77.478	1.88918	0.012907
4.36	78.257	1.89352	.012778
4.37	79.044	1.89787	.012651
4.38	79.838	1.90221	.012525
4.39	80.640	1.90655	.012401
4.40	81.451	1.91090	0.012277
4.41	82.269	1.91524	.012155
4.42	83.096	1.91958	.012034
4.43	83.931	1.92392	.011914
4.44	84.775	1.92827	.011796
4.45	85.627	1.93261	0.011679
4.46	86.488	1.93695	.011562
4.47	87.357	1.94130	.011447
4.48	88.235	1.94564	.011333
4.49	89.121	1.94998	.011221
4.50	90.017	1.95433	0.011109

x	e^x	$\text{Log}_{10}(e^x)$	e^{-x}
4.50	90.017	1.95433	0.011109
4.51	90.922	1.95867	.010998
4.52	91.836	1.96301	.010889
4.53	92.759	1.96735	.010781
4.54	93.691	1.97170	.010673
4.55	94.632	1.97604	0.010567
4.56	95.583	1.98038	.010462
4.57	96.544	1.98473	.010358
4.58	97.514	1.98907	.010255
4.59	98.494	1.99341	.010153
4.60	99.484	1.99775	0.010052
4.61	100.48	2.00210	.009952
4.62	101.49	2.00644	.009853
4.63	102.51	2.01078	.009755
4.64	103.54	2.01513	.009658
4.65	104.58	2.01947	0.009562
4.66	105.64	2.02381	.009466
4.67	106.70	2.02816	.009372
4.68	107.77	2.03250	.009279
4.69	108.85	2.03684	.009187
4.70	109.95	2.04118	0.009095
4.71	111.05	2.04553	.009005
4.72	112.17	2.04987	.008915
4.73	113.30	2.05421	.008826
4.74	114.43	2.05856	.008739
4.75	115.58	2.06290	0.008652
4.76	116.75	2.06724	.008566
4.77	117.92	2.07158	.008480
4.78	119.10	2.07593	.008396
4.79	120.30	2.08027	.008312
4.80	121.51	2.08461	0.008230
4.81	122.73	2.08896	.008148
4.82	123.97	2.09330	.008067
4.83	125.21	2.09764	.007987
4.84	126.47	2.10199	.007907
4.85	127.74	2.10633	0.007828
4.86	129.02	2.11067	.007750
4.87	130.32	2.11501	.007673
4.88	131.63	2.11936	.007597
4.89	132.95	2.12370	.007521
4.90	134.29	2.12804	0.007447
4.91	135.64	2.13239	.007372
4.92	137.00	2.13673	.007299
4.93	138.38	2.14107	.007227
4.94	139.77	2.14541	.007155
4.95	141.17	2.14976	0.007083
4.96	142.59	2.15410	.007013
4.97	144.03	2.15844	.006943
4.98	145.47	2.16279	.006874
4.99	146.94	2.16713	.006806
5.00	148.41	2.17147	0.006738

EXPONENTIAL FUNCTIONS (Continued)

x	e^x	$\text{Log}_{10}(e^x)$	e^{-x}
5.00	148.41	2.17147	0.006738
5.01	149.90	2.17582	.006671
5.02	151.41	2.18016	.006605
5.03	152.93	2.18450	.006539
5.04	154.47	2.18884	.006474
5.05	156.02	2.19319	0.006409
5.06	157.59	2.19753	.006346
5.07	159.17	2.20187	.006282
5.08	160.77	2.20622	.006220
5.09	162.39	2.21056	.006158
5.10	164.02	2.21490	0.006097
5.11	165.67	2.21924	.006036
5.12	167.34	2.22359	.005976
5.13	169.02	2.22793	.005917
5.14	170.72	2.23227	.005858
5.15	172.43	2.23662	0.005799
5.16	174.16	2.24096	.005742
5.17	175.91	2.24530	.005685
5.18	177.68	2.24965	.005628
5.19	179.47	2.25399	.005572
5.20	181.27	2.25833	0.005517
5.21	183.09	2.26267	.005462
5.22	184.93	2.26702	.005407
5.23	186.79	2.27136	.005354
5.24	188.67	2.27570	.005300
5.25	190.57	2.28005	0.005248
5.26	192.48	2.28439	.005195
5.27	194.42	2.28873	.005144
5.28	196.37	2.29307	.005092
5.29	198.34	2.29742	.005042
5.30	200.34	2.30176	0.004992
5.31	202.35	2.30610	.004942
5.32	204.38	2.31045	.004893
5.33	206.44	2.31479	.004844
5.34	208.51	2.31913	.004796
5.35	210.61	2.32348	0.004748
5.36	212.72	2.32782	.004701
5.37	214.86	2.33216	.004654
5.38	217.02	2.33650	.004608
5.39	219.20	2.34085	.004562
5.40	221.41	2.34519	0.004517
5.41	223.63	2.34953	.004472
5.42	225.88	2.35388	.004427
5.43	228.15	2.35822	.004383
5.44	230.44	2.36256	.004339
5.45	232.76	2.36690	0.004296
5.46	235.10	2.37125	.004254
5.47	237.46	2.37559	.004211
5.48	239.85	2.37993	.004169
5.49	242.26	2.38428	.004128
5.50	244.69	2.38862	0.004087

x	e^x	$\text{Log}_{10}(e^x)$	e^{-x}
5.50	244.69	2.38862	0.0040868
5.55	257.24	2.41033	.0038875
5.60	270.43	2.43205	.0036979
5.65	284.29	2.45376	.0035175
5.70	298.87	2.47548	.0033460
5.75	314.19	2.49719	0.0031828
5.80	330.30	2.51891	.0030276
5.85	347.23	2.54062	.0028799
5.90	365.04	2.56234	.0027394
5.95	383.75	2.58405	.0026058
6.00	403.43	2.60577	0.0024788
6.05	424.11	2.62748	.0023579
6.10	445.86	2.64920	.0022429
6.15	468.72	2.67091	.0021335
6.20	492.75	2.69263	.0020294
6.25	518.01	2.71434	0.0019305
6.30	544.57	2.73606	.0018363
6.35	572.49	2.75777	.0017467
6.40	601.85	2.77948	.0016616
6.45	632.70	2.80120	.0015805
6.50	665.14	2.82291	0.0015034
6.55	699.24	2.84463	.0014301
6.60	735.10	2.86634	.0013604
6.65	772.78	2.88806	.0012940
6.70	812.41	2.90977	.0012309
6.75	854.06	2.93149	0.0011709
6.80	897.85	2.95320	.0011138
6.85	943.88	2.97492	.0010595
6.90	992.27	2.99663	.0010078
6.95	1043.1	3.01835	.0009586
7.00	1096.6	3.04006	0.0009119
7.05	1152.9	3.06178	.0008674
7.10	1212.0	3.08349	.0008251
7.15	1274.1	3.10521	.0007849
7.20	1339.4	3.12692	.0007466
7.25	1408.1	3.14863	0.0007102
7.30	1480.3	3.17035	.0006755
7.35	1556.2	3.19206	.0006426
7.40	1636.0	3.21378	.0006113
7.45	1719.9	3.23549	.0005814
7.50	1808.0	3.25721	0.0005531
7.55	1900.7	3.27892	.0005261
7.60	1998.2	3.30064	.0005005
7.65	2100.6	3.32235	.0004760
7.70	2208.3	3.34407	.0004528
7.75	2321.6	3.36578	0.0004307
7.80	2440.6	3.38750	.0004097
7.85	2565.7	3.40921	.0003898
7.90	2697.3	3.43093	.0003707
7.95	2835.6	3.45264	.0003527
8.00	2981.0	3.47436	0.0003355

EXPONENTIAL FUNCTIONS (Continued)

x	e^x	$\text{Log}_{10}(e^x)$	e^{-x}
8.00	2981.0	3.47436	0.0003355
8.05	3133.8	3.49607	.0003191
8.10	3294.5	3.51779	.0003035
8.15	3463.4	3.53950	.0002887
8.20	3641.0	3.56121	.0002747
8.25	3827.6	3.58293	0.0002613
8.30	4023.9	3.60464	.0002485
8.35	4230.2	3.62636	.0002364
8.40	4447.1	3.64807	.0002249
8.45	4675.1	3.66979	.0002139
8.50	4914.8	3.69150	0.0002035
8.55	5166.8	3.71322	.0001935
8.60	5431.7	3.73493	.0001841
8.65	5710.1	3.75665	.0001751
8.70	6002.9	3.77836	.0001666
8.75	6310.7	3.80008	0.0001585
8.80	6634.2	3.82179	.0001507
8.85	6974.4	3.84351	.0001434
8.90	7332.0	3.86522	.0001364
8.95	7707.9	3.88694	.0001297
9.00	8103.1	3.90865	0.0001234
9.05	8518.5	3.93037	.0001174
9.10	8955.3	3.95208	.0001117
9.15	9414.4	3.97379	.0001062
9.20	9897.1	3.99551	.0001010
9.25	10405	4.01722	0.0000961
9.30	10938	4.03894	.0000914
9.35	11499	4.06065	.0000870
9.40	12088	4.08237	.0000827
9.45	12708	4.10408	.0000787
9.50	13360	4.12580	0.0000749
9.55	14045	4.14751	.0000712
9.60	14765	4.16923	.0000677
9.65	15522	4.19094	.0000644
9.70	16318	4.21266	.0000613
9.75	17154	4.23437	0.0000583
9.80	18034	4.25609	.0000555
9.85	18958	4.27780	.0000527
9.90	19930	4.29952	.0000502
9.95	20952	4.32123	0.0000477
10.00	22026	4.34294	0.0000454

TEN-PLACE EXPONENTIAL FUNCTIONS e^x AND e^{-x}

x	e^x	e^{-x}	x	e^x	e^{-x}
0.000	1.00000 00000	1.00000 00000	0.050	1.05127 10964	0.95122 94245
0.001	1.00100 05002	0.99900 04998	0.051	1.05232 28933	0.95027 86705
0.002	1.00200 20013	0.99800 19987	0.052	1.05337 57425	0.94932 88668
0.003	1.00300 45045	0.99700 44955	0.053	1.05442 96451	0.94838 00125
0.004	1.00400 80107	0.99600 79893	0.054	1.05548 46022	0.94743 21065
0.005	1.00501 25209	0.99501 24792	0.055	1.05654 06147	0.94648 51480
0.006	1.00601 80361	0.99401 79641	0.056	1.05759 76837	0.94553 91359
0.007	1.00702 45573	0.99302 44429	0.057	1.05865 58104	0.94459 40694
0.008	1.00803 20855	0.99203 19148	0.058	1.05971 49957	0.94364 99474
0.009	1.00904 06218	0.99104 03788	0.059	1.06077 52407	0.94270 67692
0.010	1.01005 01671	0.99004 98337	0.060	1.06183 65465	0.94176 45336
0.011	1.01106 07224	0.98906 02788	0.061	1.06289 89142	0.94082 32398
0.012	1.01207 22889	0.98807 17129	0.062	1.06396 23447	0.93988 28868
0.013	1.01308 48674	0.98708 41350	0.063	1.06502 68392	0.93894 34737
0.014	1.01409 84589	0.98609 75443	0.064	1.06609 23988	0.93800 49995
0.015	1.01511 30646	0.98511 19396	0.065	1.06715 90244	0.93706 74634
0.016	1.01612 86854	0.98412 73201	0.066	1.06822 67172	0.93613 08643
0.017	1.01714 53223	0.98314 36846	0.067	1.06929 54782	0.93519 52013
0.018	1.01816 29764	0.98216 10324	0.068	1.07036 53085	0.93426 04736
0.019	1.01918 16486	0.98117 93622	0.069	1.07143 62091	0.93332 66801
0.020	1.02020 13400	0.98019 86733	0.070	1.07250 81813	0.93239 38199
0.021	1.02122 20516	0.97921 89646	0.071	1.07358 12259	0.93146 18921
0.022	1.02224 37845	0.97824 02351	0.072	1.07465 53441	0.93053 08958
0.023	1.02326 65395	0.97726 24838	0.073	1.07573 05369	0.92960 08300
0.024	1.02429 03179	0.97628 57098	0.074	1.07680 68055	0.92867 16938
0.025	1.02531 51205	0.97530 99120	0.075	1.07788 41509	0.92774 34863
0.026	1.02634 09485	0.97433 50896	0.076	1.07896 25742	0.92681 62066
0.027	1.02736 78028	0.97336 12415	0.077	1.08004 20764	0.92588 98536
0.028	1.02839 56844	0.97238 83668	0.078	1.08112 26587	0.92496 44265
0.029	1.02942 45945	0.97141 64645	0.079	1.08220 43221	0.92403 99244
0.030	1.03045 45340	0.97044 55335	0.080	1.08328 70677	0.92311 63464
0.031	1.03148 55039	0.96947 55731	0.081	1.08437 08966	0.92219 36914
0.032	1.03251 75053	0.96850 65821	0.082	1.08545 58098	0.92127 19587
0.033	1.03355 05392	0.96753 85596	0.083	1.08654 18085	0.92035 11472
0.034	1.03458 46067	0.96657 15046	0.084	1.08762 88938	0.91943 12561
0.035	1.03561 97088	0.96560 54163	0.085	1.08871 70667	0.91851 22844
0.036	1.03665 58465	0.96464 02935	0.086	1.08980 63283	0.91759 42312
0.037	1.03769 30208	0.96367 61353	0.087	1.09089 66797	0.91667 70956
0.038	1.03873 12329	0.96271 29409	0.088	1.09198 81220	0.91576 08767
0.039	1.03977 04837	0.96175 07091	0.089	1.09308 06563	0.91484 55736
0.040	1.04081 07742	0.96078 94392	0.090	1.09417 42837	0.91393 11853
0.041	1.04185 21055	0.95982 91299	0.091	1.09526 90053	0.91301 77109
0.042	1.04289 44788	0.95886 97806	0.092	1.09636 48221	0.91210 51495
0.043	1.04393 78949	0.95791 13901	0.093	1.09746 17353	0.91119 35003
0.044	1.04498 23549	0.95695 39575	0.094	1.09855 97459	0.91028 27622
0.045	1.04602 78599	0.95599 74818	0.095	1.09965 88551	0.90937 29345
0.046	1.04707 44110	0.95504 19622	0.096	1.10075 90640	0.90846 40161
0.047	1.04812 20091	0.95408 73976	0.097	1.10186 03736	0.90755 60061
0.048	1.04917 06553	0.95313 37871	0.098	1.10296 27851	0.90664 89038
0.049	1.05022 03507	0.95218 11297	0.099	1.10406 62996	0.90574 27080
0.050	1.05127 10964	0.95122 94245	0.100	1.10517 09181	0.90483 74180

TEN-PLACE EXPONENTIAL FUNCTIONS e^x AND e^{-x} (Continued)

x	e^x	e^{-x}	x	e^x	e^{-x}
0.100	1.10517 09181	0.90483 74180	0.150	1.16183 42427	0.86070 79764
0.101	1.10627 66418	0.90393 30329	0.151	1.16299 66581	0.85984 76987
0.102	1.10738 34717	0.90302 95517	0.152	1.16416 02364	0.85898 82807
0.103	1.10849 14091	0.90212 69735	0.153	1.16532 49789	0.85812 97218
0.104	1.10960 04549	0.90122 52974	0.154	1.16649 08868	0.85727 20210
0.105	1.11071 06104	0.90032 45226	0.155	1.16765 79611	0.85641 51775
0.106	1.11182 18765	0.89942 46481	0.156	1.16882 62031	0.85555 91904
0.107	1.11293 42545	0.89852 56730	0.157	1.16999 56139	0.85470 40588
0.108	1.11404 77454	0.89762 75964	0.158	1.17116 61947	0.85384 97820
0.109	1.11516 23503	0.89673 04175	0.159	1.17233 79467	0.85299 63590
0.110	1.11627 80705	0.89583 41353	0.160	1.17351 08710	0.85214 37890
0.111	1.11739 49069	0.89493 87489	0.161	1.17468 49688	0.85129 20711
0.112	1.11851 28606	0.89404 42575	0.162	1.17586 02413	0.85044 12045
0.113	1.11963 19329	0.89315 06601	0.163	1.17703 66897	0.84959 11884
0.114	1.12075 21249	0.89225 79559	0.164	1.17821 43151	0.84874 20219
0.115	1.12187 34376	0.89136 61439	0.165	1.17939 31187	0.84789 37041
0.116	1.12299 58721	0.89047 52233	0.166	1.18057 31017	0.84704 62342
0.117	1.12411 94297	0.88958 51932	0.167	1.18175 42653	0.84619 96113
0.118	1.12524 41114	0.88869 60526	0.168	1.18293 66106	0.84535 38347
0.119	1.12636 99183	0.88780 78008	0.169	1.18412 01389	0.84450 89034
0.120	1.12749 68516	0.88692 04367	0.170	1.18530 48513	0.84366 48166
0.121	1.12862 49124	0.88603 39596	0.171	1.18649 07490	0.84282 15735
0.122	1.12975 41018	0.88514 83685	0.172	1.18767 78332	0.84197 91732
0.123	1.13088 44209	0.88426 36626	0.173	1.18886 61051	0.84113 76148
0.124	1.13201 58710	0.88337 98409	0.174	1.19005 55658	0.84029 68977
0.125	1.13314 84531	0.88249 69026	0.175	1.19124 62166	0.83945 70208
0.126	1.13428 21683	0.88161 48468	0.176	1.19243 80587	0.83861 79833
0.127	1.13541 70178	0.88073 36726	0.177	1.19363 10931	0.83777 97845
0.128	1.13655 30027	0.87985 33791	0.178	1.19482 53212	0.83694 24235
0.129	1.13769 01242	0.87897 39655	0.179	1.19602 07442	0.83610 58994
0.130	1.13882 83833	0.87809 54309	0.180	1.19721 73631	0.83527 02114
0.131	1.13996 77813	0.87721 77744	0.181	1.19841 51793	0.83443 53587
0.132	1.14110 83193	0.87634 09951	0.182	1.19961 41939	0.83360 13404
0.133	1.14224 99983	0.87546 50921	0.183	1.20081 44081	0.83276 81557
0.134	1.14339 28196	0.87459 00646	0.184	1.20201 58231	0.83193 58038
0.135	1.14453 67844	0.87371 59117	0.185	1.20321 84401	0.83110 42839
0.136	1.14568 18936	0.87284 26325	0.186	1.20442 22604	0.83027 35950
0.137	1.14682 81485	0.87197 02261	0.187	1.20562 72850	0.82944 37364
0.138	1.14797 55503	0.87109 86917	0.188	1.20683 35153	0.82861 47072
0.139	1.14912 41000	0.87022 80285	0.189	1.20804 09525	0.82778 65067
0.140	1.15027 37989	0.86935 82354	0.190	1.20924 95977	0.82695 91339
0.141	1.15142 46480	0.86848 93117	0.191	1.21045 94521	0.82613 25882
0.142	1.15257 66485	0.86762 12565	0.192	1.21167 05170	0.82530 68685
0.143	1.15372 98017	0.86675 40689	0.193	1.21288 27935	0.82448 19741
0.144	1.15488 41085	0.86588 77481	0.194	1.21409 62830	0.82365 79043
0.145	1.15603 95703	0.86502 22931	0.195	1.21531 09865	0.82283 46581
0.146	1.15719 61881	0.86415 77032	0.196	1.21652 69053	0.82201 22347
0.147	1.15835 39630	0.86329 39774	0.197	1.21774 40407	0.82119 06333
0.148	1.15951 28964	0.86243 11149	0.198	1.21896 23938	0.82036 98531
0.149	1.16067 29892	0.86156 91149	0.199	1.22018 19659	0.81954 98933
0.150	1.16183 42427	0.86070 79764	0.200	1.22140 27582	0.81873 07531

TEN-PLACE EXPONENTIAL FUNCTIONS e^x AND e^{-x} (Continued)

x	e^x	e^{-x}	x	e^x	e^{-x}
0.200	1.22140 27582	0.81873 07531	0.250	1.28402 54167	0.77880 07831
0.201	1.22262 47718	0.81791 24316	0.251	1.28531 00843	0.77802 23716
0.202	1.22384 80081	0.81709 49279	0.252	1.28659 60373	0.77724 47381
0.203	1.22507 24682	0.81627 82414	0.253	1.28788 32768	0.77646 78818
0.204	1.22629 81535	0.81546 23712	0.254	1.28917 18043	0.77569 18020
0.205	1.22752 50650	0.81464 73164	0.255	1.29046 16209	0.77491 64980
0.206	1.22875 32040	0.81383 30763	0.256	1.29175 27279	0.77414 19688
0.207	1.22998 25718	0.81301 96500	0.257	1.29304 51268	0.77336 82138
0.208	1.23121 31695	0.81220 70367	0.258	1.29433 88186	0.77259 52321
0.209	1.23244 49985	0.81139 52356	0.259	1.29563 38048	0.77182 30230
0.210	1.23367 80600	0.81058 42460	0.260	1.29693 00867	0.77105 15858
0.211	1.23491 23551	0.80977 40669	0.261	1.29822 76654	0.77028 09196
0.212	1.23614 78851	0.80896 46976	0.262	1.29952 65424	0.76951 10237
0.213	1.23738 46512	0.80815 61372	0.263	1.30082 67190	0.76874 18973
0.214	1.23862 26548	0.80734 83850	0.264	1.30212 81963	0.76797 35397
0.215	1.23986 18970	0.80654 14402	0.265	1.30343 09758	0.76720 59500
0.216	1.24110 23790	0.80573 53019	0.266	1.30473 50587	0.76643 91275
0.217	1.24234 41021	0.80492 99693	0.267	1.30604 04463	0.76567 30715
0.218	1.24358 70676	0.80412 54417	0.268	1.30734 71400	0.76490 77811
0.219	1.24483 12767	0.80332 17182	0.269	1.30865 51410	0.76414 32556
0.220	1.24607 67306	0.80251 87980	0.270	1.30996 44507	0.76337 94943
0.221	1.24732 34306	0.80171 66803	0.271	1.31127 50704	0.76261 64964
0.222	1.24857 13779	0.80091 53643	0.272	1.31258 70013	0.76185 42611
0.223	1.24982 05737	0.80011 48493	0.273	1.31390 02448	0.76109 27876
0.224	1.25107 10194	0.79931 51344	0.274	1.31521 48022	0.76033 20753
0.225	1.25232 27162	0.79851 62188	0.275	1.31653 06749	0.75957 21232
0.226	1.25357 56653	0.79771 81017	0.276	1.31784 78640	0.75881 29308
0.227	1.25482 98679	0.79692 07823	0.277	1.31916 63710	0.75805 44971
0.228	1.25608 53254	0.79612 42598	0.278	1.32048 61972	0.75729 68215
0.229	1.25734 20390	0.79532 85335	0.279	1.32180 73439	0.75653 99032
0.230	1.25860 00099	0.79453 36025	0.280	1.32312 98123	0.75578 37415
0.231	1.25985 92394	0.79373 94660	0.281	1.32445 36039	0.75502 83355
0.232	1.26111 97288	0.79294 61233	0.282	1.32577 87200	0.75427 36845
0.233	1.26238 14793	0.79215 35735	0.283	1.32710 51618	0.75351 97879
0.234	1.26364 44922	0.79136 18159	0.284	1.32843 29308	0.75276 66447
0.235	1.26490 87687	0.79057 08496	0.285	1.32976 20281	0.75201 42543
0.236	1.26617 43102	0.78978 06739	0.286	1.33109 24553	0.75126 26159
0.237	1.26744 11178	0.78899 12880	0.287	1.33242 42135	0.75051 17288
0.238	1.26870 91928	0.78820 26911	0.288	1.33375 73041	0.74976 15922
0.239	1.26997 85366	0.78741 48824	0.289	1.33509 17285	0.74901 22054
0.240	1.27124 91503	0.78662 78611	0.290	1.33642 74880	0.74826 35676
0.241	1.27252 10353	0.78584 16264	0.291	1.33776 45840	0.74751 56780
0.242	1.27379 41928	0.78505 61776	0.292	1.33910 30176	0.74676 85360
0.243	1.27506 86241	0.78427 15138	0.293	1.34044 27904	0.74602 21407
0.244	1.27634 43305	0.78348 76343	0.294	1.34178 39037	0.74527 64914
0.245	1.27762 13132	0.78270 45382	0.295	1.34312 63587	0.74453 15875
0.246	1.27889 95735	0.78192 22249	0.296	1.34447 01568	0.74378 74280
0.247	1.28017 91128	0.78114 06935	0.297	1.34581 52994	0.74304 40124
0.248	1.28145 99322	0.78035 99433	0.298	1.34716 17879	0.74230 13397
0.249	1.28274 20331	0.77957 99734	0.299	1.34850 96235	0.74155 94094
0.250	1.28402 54167	0.77880 07381	0.300	1.34985 88076	0.74081 82207

TEN-PLACE EXPONENTIAL FUNCTIONS e^x AND e^{-x} (Continued)

x	e^x	e^{-x}	x	e^x	e^{-x}
0.300	1.34985 88076	0.74081 82207	0.350	1.41906 75486	0.70468 80898
0.301	1.35120 93415	0.74007 77727	0.351	1.42048 73259	0.70398 37539
0.302	1.35256 12267	0.73933 80649	0.352	1.42190 85237	0.70328 01220
0.303	1.35391 44644	0.73859 90964	0.353	1.42333 11434	0.70257 71934
0.304	1.35526 90561	0.73786 08665	0.354	1.42475 51865	0.70187 49674
0.305	1.35662 50030	0.73712 33744	0.355	1.42618 06543	0.70117 34432
0.306	1.35798 23065	0.73638 66195	0.356	1.42760 75483	0.70047 26202
0.307	1.35934 09681	0.73565 06009	0.357	1.42903 58699	0.69977 24977
0.308	1.36070 09889	0.73491 53180	0.358	1.43046 56205	0.69907 30750
0.309	1.36206 23705	0.73418 07700	0.359	1.43189 68016	0.69837 43514
0.310	1.36342 51141	0.73344 69562	0.360	1.43332 94146	0.69767 63261
0.311	1.36478 92212	0.73271 38759	0.361	1.43476 34609	0.69697 89985
0.312	1.36615 46930	0.73198 15282	0.362	1.43619 89420	0.69628 23678
0.313	1.36752 15310	0.73124 99126	0.363	1.43763 58592	0.69558 64335
0.314	1.36888 97365	0.73051 90282	0.364	1.43907 42142	0.69489 11947
0.315	1.37025 93110	0.72978 88743	0.365	1.44051 40081	0.69419 66509
0.316	1.37163 02556	0.72905 94502	0.366	1.44195 52427	0.69350 28012
0.317	1.37300 25719	0.72833 07551	0.367	1.44339 79191	0.69280 96450
0.318	1.37437 62612	0.72760 27884	0.368	1.44484 20390	0.69211 71817
0.319	1.37575 13249	0.72687 55493	0.369	1.44628 76037	0.69142 54105
0.320	1.37712 77643	0.72614 90371	0.370	1.44773 46147	0.69073 43306
0.321	1.37850 55809	0.72542 32510	0.371	1.44918 30734	0.69004 39416
0.322	1.37988 47760	0.72469 81903	0.372	1.45063 29813	0.68935 42425
0.323	1.38126 53509	0.72397 38544	0.373	1.45208 43398	0.68866 52328
0.324	1.38264 73071	0.72325 02424	0.374	1.45353 71505	0.68797 69118
0.325	1.38403 06460	0.72252 73536	0.375	1.45499 14146	0.68728 92788
0.326	1.38541 53689	0.72180 51874	0.376	1.45644 71338	0.68660 23330
0.327	1.38680 14772	0.72108 37430	0.377	1.45790 43094	0.68591 60739
0.328	1.38818 89723	0.72036 30197	0.378	1.45936 29429	0.68523 05007
0.329	1.38957 78556	0.71964 30167	0.379	1.46082 30357	0.68454 56127
0.330	1.39096 81285	0.71892 37334	0.380	1.46228 45894	0.68386 14092
0.331	1.39235 97923	0.71820 51690	0.381	1.46374 76054	0.68317 78896
0.332	1.39375 28485	0.71748 73229	0.382	1.46521 20851	0.68249 50532
0.333	1.39514 72985	0.71677 01942	0.383	1.46667 80301	0.68181 28993
0.334	1.39654 31436	0.71605 37822	0.384	1.46814 54417	0.68113 14272
0.335	1.39794 03852	0.71533 80864	0.385	1.46961 43214	0.68045 06362
0.336	1.39933 90248	0.71462 31058	0.386	1.47108 46708	0.67977 05257
0.337	1.40073 90637	0.71390 88399	0.387	1.47255 64913	0.67909 10949
0.338	1.40214 05034	0.71319 52879	0.388	1.47402 97843	0.67841 23433
0.339	1.40354 33452	0.71248 24491	0.389	1.47550 45513	0.67773 42700
0.340	1.40494 75906	0.71177 03228	0.390	1.47698 07939	0.67705 68745
0.341	1.40635 32409	0.71105 89082	0.391	1.47845 85134	0.67638 01560
0.342	1.40776 02975	0.71034 82047	0.392	1.47993 77114	0.67570 41140
0.343	1.40916 87619	0.70963 82116	0.393	1.48141 83893	0.67502 87476
0.344	1.41057 86355	0.70892 89280	0.394	1.48290 05487	0.67435 40562
0.345	1.41198 99197	0.70822 03535	0.395	1.48438 41909	0.67368 00392
0.346	1.41340 26158	0.70751 24871	0.396	1.48586 93176	0.67300 66959
0.347	1.41481 67254	0.70680 53283	0.397	1.48735 59301	0.67233 40256
0.348	1.41623 22497	0.70609 88762	0.398	1.48884 40299	0.67166 20277
0.349	1.41764 91903	0.70539 31303	0.399	1.49033 36186	0.67099 07014
0.350	1.41906 75486	0.70468 80897	0.400	1.49182 46976	0.67032 00460

TEN-PLACE EXPONENTIAL FUNCTIONS e^x AND e^{-x} (Continued)

x	e^x	e^{-x}	x	e^x	e^{-x}
0.400	1.49182 46976	0.67032 00460	0.450	1.56831 21855	0.63762 81516
0.401	1.49331 72685	0.66965 00610	0.451	1.56988 12821	0.63699 08422
0.402	1.49481 13327	0.66898 07457	0.452	1.57145 19486	0.63635 41697
0.403	1.49630 68917	0.66831 20993	0.453	1.57302 41865	0.63571 81336
0.404	1.49780 39470	0.66764 41213	0.454	1.57459 79975	0.63508 27332
0.405	1.49930 25001	0.66697 68109	0.455	1.57617 33830	0.63444 79679
0.406	1.50080 25525	0.66631 01674	0.456	1.57775 03448	0.63381 38371
0.407	1.50230 41057	0.66564 41903	0.457	1.57932 88842	0.63318 03401
0.408	1.50380 71612	0.66497 88788	0.458	1.58090 90031	0.63254 74762
0.409	1.50531 17205	0.66431 42323	0.459	1.58249 07028	0.63191 52449
0.410	1.50681 77851	0.66365 02501	0.460	1.58407 39850	0.63128 36455
0.411	1.50832 53566	0.66298 69316	0.461	1.58565 88513	0.63065 26774
0.412	1.50983 44363	0.66232 42761	0.462	1.58724 53032	0.63002 23399
0.413	1.51134 50259	0.66166 22828	0.463	1.58883 33424	0.62939 26325
0.414	1.51285 71269	0.66100 09513	0.464	1.59042 29704	0.62876 35545
0.415	1.51437 07407	0.66034 02807	0.465	1.59201 41889	0.62813 51052
0.416	1.51588 58689	0.65968 02705	0.466	1.59360 69993	0.62750 72840
0.417	1.51740 25129	0.65902 09199	0.467	1.59520 14034	0.62688 00904
0.418	1.51892 06744	0.65836 22284	0.468	1.59679 74027	0.62625 35237
0.419	1.52044 03548	0.65770 41953	0.469	1.59839 49988	0.62562 75832
0.420	1.52196 15556	0.65704 68198	0.470	1.59999 41932	0.62500 22683
0.421	1.52348 42784	0.65639 01014	0.471	1.60159 49877	0.62437 75784
0.422	1.52500 85247	0.65573 40394	0.472	1.60319 73837	0.62375 35129
0.423	1.52653 42960	0.65507 86331	0.473	1.60480 13830	0.62313 00712
0.424	1.52806 15938	0.65442 38819	0.474	1.60640 69870	0.62250 72526
0.425	1.52959 04197	0.65376 97851	0.475	1.60801 41975	0.62188 50565
0.426	1.53112 07751	0.65311 63421	0.476	1.60962 30160	0.62126 34822
0.427	1.53265 26617	0.65246 35522	0.477	1.61123 34441	0.62064 25293
0.428	1.53418 60810	0.65181 14148	0.478	1.61284 54834	0.62002 21970
0.429	1.53572 10344	0.65115 99292	0.479	1.61445 91356	0.61940 24847
0.430	1.53725 75235	0.65050 90947	0.480	1.61607 44022	0.61878 33918
0.431	1.53879 55500	0.64985 89108	0.481	1.61769 12849	0.61816 49177
0.432	1.54033 51152	0.64920 93767	0.482	1.61930 97853	0.61754 70618
0.433	1.54187 62207	0.64856 04918	0.483	1.62092 99050	0.61692 98234
0.434	1.54341 88681	0.64791 22555	0.484	1.62255 16457	0.61631 32019
0.435	1.54496 30590	0.64726 46671	0.485	1.62417 50088	0.61569 71968
0.436	1.54650 87947	0.64661 77259	0.486	1.62579 99962	0.61508 18073
0.437	1.54805 60771	0.64597 14314	0.487	1.62742 66094	0.61446 70329
0.438	1.54960 49074	0.64532 57829	0.488	1.62905 48500	0.61385 28730
0.439	1.55115 52874	0.64468 07796	0.489	1.63068 47196	0.61323 93270
0.440	1.55270 72185	0.64403 64211	0.490	1.63231 62200	0.61262 63942
0.441	1.55426 07023	0.64339 27066	0.491	1.63394 93526	0.61201 40740
0.442	1.55581 57404	0.64274 96355	0.492	1.63558 41192	0.61140 23658
0.443	1.55737 23343	0.64210 72071	0.493	1.63722 05214	0.61079 12691
0.444	1.55893 04856	0.64146 54208	0.494	1.63885 85608	0.61018 07831
0.445	1.56049 01958	0.64082 42760	0.495	1.64049 82391	0.60957 09073
0.446	1.56205 14665	0.64018 37721	0.496	1.64213 95578	0.60896 16411
0.447	1.56361 42993	0.63954 39083	0.497	1.64378 25187	0.60835 29838
0.448	1.56517 86957	0.63890 46840	0.498	1.64542 71234	0.60774 49349
0.449	1.56674 46572	0.63826 60987	0.499	1.64707 33735	0.60713 74937
0.450	1.56831 21855	0.63762 81516	0.500	1.64872 12707	0.60653 06597

TEN-PLACE EXPONENTIAL FUNCTIONS e^x AND e^{-x} (Continued)

x	e^x	e^{-x}	x	e^x	e^{-x}
0.500	1.64872 12707	0.60653 06597	0.550	1.73325 30179	0.57694 98104
0.501	1.65037 08166	0.60592 44322	0.551	1.73498 71378	0.57637 31489
0.502	1.65202 20129	0.60531 88106	0.552	1.73672 29927	0.57579 70639
0.503	1.65367 48612	0.60471 37944	0.553	1.73846 05844	0.57522 15546
0.504	1.65532 93632	0.60410 93829	0.554	1.74019 99145	0.57464 66206
0.505	1.65698 55205	0.60350 55754	0.555	1.74194 09848	0.57407 22612
0.506	1.65864 33348	0.60290 23715	0.556	1.74368 37970	0.57349 84759
0.507	1.66030 28077	0.60229 97705	0.557	1.74542 83529	0.57292 52641
0.508	1.66196 39409	0.60169 77718	0.558	1.74717 46543	0.57235 26252
0.509	1.66362 67361	0.60109 63747	0.559	1.74892 27028	0.57178 05586
0.510	1.66529 11949	0.60049 55788	0.560	1.75067 25003	0.57120 90638
0.511	1.66695 73191	0.59989 53834	0.561	1.75242 40484	0.57063 81403
0.512	1.66862 51101	0.59929 57878	0.562	1.75417 73490	0.57006 77874
0.513	1.67029 45698	0.59869 67916	0.563	1.75593 24037	0.56949 80045
0.514	1.67196 56998	0.59809 83941	0.564	1.75768 92144	0.56892 87912
0.515	1.67363 85018	0.59750 05946	0.565	1.75944 77827	0.56836 01468
0.516	1.67531 29774	0.59690 33927	0.566	1.76120 81105	0.56779 20707
0.517	1.67698 91283	0.59630 67876	0.567	1.76297 01995	0.56722 45624
0.518	1.67866 69562	0.59571 07789	0.568	1.76473 40515	0.56665 76214
0.519	1.68034 64628	0.59511 53659	0.569	1.76649 96682	0.56609 12470
0.520	1.68202 76497	0.59452 05480	0.570	1.76826 70514	0.56552 54387
0.521	1.68371 05186	0.59392 63246	0.571	1.77003 62029	0.56496 01959
0.522	1.68539 50713	0.59333 26951	0.572	1.77180 71244	0.56439 55181
0.523	1.68708 13093	0.59273 96590	0.573	1.77357 98178	0.56383 14047
0.524	1.68876 92345	0.59214 72156	0.574	1.77535 42847	0.56326 78551
0.525	1.69045 88484	0.59155 53644	0.575	1.77713 05269	0.56270 48688
0.526	1.69215 01527	0.59096 41047	0.576	1.77890 85463	0.56214 24452
0.527	1.69384 31492	0.59037 34360	0.577	1.78068 83446	0.56158 05837
0.528	1.69553 78396	0.58978 33576	0.578	1.78246 99236	0.56101 92838
0.529	1.69723 42255	0.58919 38690	0.579	1.78425 32850	0.56045 85450
0.530	1.69893 23086	0.58860 49697	0.580	1.78603 84308	0.55989 83666
0.531	1.70063 20907	0.58801 66589	0.581	1.78782 53625	0.55933 87481
0.532	1.70233 35734	0.58742 89362	0.582	1.78961 40821	0.55877 96889
0.533	1.70403 67584	0.58684 18008	0.583	1.79140 45913	0.55822 11885
0.534	1.70574 16475	0.58625 52524	0.584	1.79319 68919	0.55766 32463
0.535	1.70744 82423	0.58566 92901	0.585	1.79499 09856	0.55710 58618
0.536	1.70915 65445	0.58508 39136	0.586	1.79678 68744	0.55654 90344
0.537	1.71086 65559	0.58449 91221	0.587	1.79858 45600	0.55599 27636
0.538	1.71257 82782	0.58391 49152	0.588	1.80038 40441	0.55543 70487
0.539	1.71429 17130	0.58333 12921	0.589	1.80218 53287	0.55488 18893
0.540	1.71600 68622	0.58274 82524	0.590	1.80398 84154	0.55432 72847
0.541	1.71772 37273	0.58216 57954	0.591	1.80579 33061	0.55377 32345
0.542	1.71944 23102	0.58158 39206	0.592	1.80760 00026	0.55321 97381
0.543	1.72116 26125	0.58100 26274	0.593	1.80940 85067	0.55266 67949
0.544	1.72288 46360	0.58042 19151	0.594	1.81121 88202	0.55211 44043
0.545	1.72460 83824	0.57984 17833	0.595	1.81303 09450	0.55156 25659
0.546	1.72633 38534	0.57926 22314	0.596	1.81484 48827	0.55101 12790
0.547	1.72806 10507	0.57868 32587	0.597	1.81666 06353	0.55046 05431
0.548	1.72978 99760	0.57810 48647	0.598	1.81847 82046	0.54991 03577
0.549	1.73152 06312	0.57752 70488	0.599	1.82029 75923	0.54936 07222
0.550	1.73325 30179	0.57694 98104	0.600	1.82211 88004	0.54881 16361

TEN-PLACE EXPONENTIAL FUNCTIONS e^x AND e^{-x} (Continued)

x	e^x	e^{-x}	x	e^x	e^{-x}
0.600	1.82211 88004	0.54881 16361	0.650	1.91554 08290	0.52204 57768
0.601	1.82394 18306	0.54826 30988	0.651	1.91745 73279	0.52152 39919
0.602	1.82576 66847	0.54771 51097	0.652	1.91937 57443	0.52100 27286
0.603	1.82759 33645	0.54716 76684	0.653	1.92129 60801	0.52048 19863
0.604	1.82942 18720	0.54662 07742	0.654	1.92321 83371	0.51996 17645
0.605	1.83125 22089	0.54607 44266	0.655	1.92514 25174	0.51944 20626
0.606	1.83308 43770	0.54552 86252	0.656	1.92706 86228	0.51892 28802
0.607	1.83491 83783	0.54498 33692	0.657	1.92899 66553	0.51840 42167
0.608	1.83675 42144	0.54443 86582	0.658	1.93092 66167	0.51788 60716
0.609	1.83859 18873	0.54389 44917	0.659	1.93285 85091	0.51736 84443
0.610	1.84043 13988	0.54335 08691	0.660	1.93479 23344	0.51685 13345
0.611	1.84227 27507	0.54280 77898	0.661	1.93672 80945	0.51633 47415
0.612	1.84411 59449	0.54226 52533	0.662	1.93866 57912	0.51581 86648
0.613	1.84596 09832	0.54172 32591	0.663	1.94060 54267	0.51530 31040
0.614	1.84780 78675	0.54118 18066	0.664	1.94254 70027	0.51478 80585
0.615	1.84965 65996	0.54064 08953	0.665	1.94449 05213	0.51427 35277
0.616	1.85150 71813	0.54010 05246	0.666	1.94643 59844	0.51375 95112
0.617	1.85335 96145	0.53956 06941	0.667	1.94838 33940	0.51324 60085
0.618	1.85521 39011	0.53902 14031	0.668	1.95033 27519	0.51273 30190
0.619	1.85707 00430	0.53848 26511	0.669	1.95228 40601	0.51222 05423
0.620	1.85892 80418	0.53794 44376	0.670	1.95423 73206	0.51170 85778
0.621	1.86078 78997	0.53740 67620	0.671	1.95619 25354	0.51119 71250
0.622	1.86264 96183	0.53686 96239	0.672	1.95814 97064	0.51068 61834
0.623	1.86451 31995	0.53633 30226	0.673	1.96010 88355	0.51017 57524
0.624	1.86637 86453	0.53579 69577	0.674	1.96206 99247	0.50966 58317
0.625	1.86824 59574	0.53526 14285	0.675	1.96403 29760	0.50915 64206
0.626	1.87011 51378	0.53472 64346	0.676	1.96599 79913	0.50864 75187
0.627	1.87198 61883	0.53419 19755	0.677	1.96796 49726	0.50813 91254
0.628	1.87385 91108	0.53365 80505	0.678	1.96993 39219	0.50763 12403
0.629	1.87573 39072	0.53312 46592	0.679	1.97190 48411	0.50712 38628
0.630	1.87761 05793	0.53259 18010	0.680	1.97387 77322	0.50661 69924
0.631	1.87948 91290	0.53205 94754	0.681	1.97585 25972	0.50611 06286
0.632	1.88136 95581	0.53152 76819	0.682	1.97782 94381	0.50560 47709
0.633	1.88325 18687	0.53099 64199	0.683	1.97980 82568	0.50509 94189
0.634	1.88513 60625	0.53046 56889	0.684	1.98178 90553	0.50459 45719
0.635	1.88702 21415	0.52993 54883	0.685	1.98377 18355	0.50409 02296
0.636	1.88891 01074	0.52940 58177	0.686	1.98575 65996	0.50358 63913
0.637	1.89079 99623	0.52887 66765	0.687	1.98774 33494	0.50308 30566
0.638	1.89269 17080	0.52834 80642	0.688	1.98973 20870	0.50258 02250
0.639	1.89458 53464	0.52781 99802	0.689	1.99172 28142	0.50207 78960
0.640	1.89648 08793	0.52729 24240	0.690	1.99371 55332	0.50157 60691
0.641	1.89837 83087	0.52676 53952	0.691	1.99571 02460	0.50107 47437
0.642	1.90027 76366	0.52623 88931	0.692	1.99770 69544	0.50057 39194
0.643	1.90217 88646	0.52571 29172	0.693	1.99970 56605	0.50007 35957
0.644	1.90408 19949	0.52518 74671	0.694	2.00170 63664	0.49957 37721
0.645	1.90598 70293	0.52466 25421	0.695	2.00370 90739	0.49907 44480
0.646	1.90789 39696	0.52413 81418	0.696	2.00571 37852	0.49857 56230
0.647	1.90980 28178	0.52361 42656	0.697	2.00772 05022	0.49807 72966
0.648	1.91171 35759	0.52309 09131	0.698	2.00972 92269	0.49757 94682
0.649	1.91362 62456	0.52256 80836	0.699	2.01173 99613	0.49708 21375
0.650	1.91554 08290	0.52204 57768	0.700	2.01375 27075	0.49658 53038

TEN-PLACE EXPONENTIAL FUNCTIONS e^x AND e^{-x} (Continued)

x	e^x	e^{-x}	x	e^x	e^{-x}
0.700	2.01375 27075	0.49658 53038	0.750	2.11700 00166	0.47236 65527
0.701	2.01576 74674	0.49608 89667	0.751	2.11911 80755	0.47189 44223
0.702	2.01778 42431	0.49559 31257	0.752	2.12123 82535	0.47142 27637
0.703	2.01980 30365	0.49509 77803	0.753	2.12336 05527	0.47095 15766
0.704	2.02182 38498	0.49460 29300	0.754	2.12548 49753	0.47048 08604
0.705	2.02384 66849	0.49410 85743	0.755	2.12761 15234	0.47001 06147
0.706	2.02587 15439	0.49361 47127	0.756	2.12974 01990	0.46954 08390
0.707	2.02789 84287	0.49312 13447	0.757	2.13187 10045	0.46907 15329
0.708	2.02992 73414	0.49262 84698	0.758	2.13400 39418	0.46860 26958
0.709	2.03195 82840	0.49213 60876	0.759	2.13613 90131	0.46813 43273
0.710	2.03399 12586	0.49164 41975	0.760	2.13827 62205	0.46766 64270
0.711	2.03602 62672	0.49115 27990	0.761	2.14041 55662	0.46719 89943
0.712	2.03806 33119	0.49066 18917	0.762	2.14255 70523	0.46673 20289
0.713	2.04010 23945	0.49017 14751	0.763	2.14470 06810	0.46626 55301
0.714	2.04214 35173	0.48968 15486	0.764	2.14684 64544	0.46579 94976
0.715	2.04418 66823	0.48919 21118	0.765	2.14899 43747	0.46533 39310
0.716	2.04623 18914	0.48870 31642	0.766	2.15114 44439	0.46486 88296
0.717	2.04827 91467	0.48821 47053	0.767	2.15329 66643	0.46440 41932
0.718	2.05032 84504	0.48772 67346	0.768	2.15545 10379	0.46394 00211
0.719	2.05237 98043	0.48723 92517	0.769	2.15760 75671	0.46347 63130
0.720	2.05443 32106	0.48675 22560	0.770	2.15976 62538	0.46301 30683
0.721	2.05648 86714	0.48626 57470	0.771	2.16192 71003	0.46255 02867
0.722	2.05854 61887	0.48577 97243	0.772	2.16409 01087	0.46208 79676
0.723	2.06060 57645	0.48529 41874	0.773	2.16625 52812	0.46162 61106
0.724	2.06266 74009	0.48480 91358	0.774	2.16842 26200	0.46116 47152
0.725	2.06473 11000	0.48432 45690	0.775	2.17059 21272	0.46070 37810
0.726	2.06679 68638	0.48384 04865	0.776	2.17276 38050	0.46024 33075
0.727	2.06886 46944	0.48335 68878	0.777	2.17493 76555	0.45978 32942
0.728	2.07093 45939	0.48287 37725	0.778	2.17711 36810	0.45932 37408
0.729	2.07300 65643	0.48239 11401	0.779	2.17929 18836	0.45886 46466
0.730	2.07508 06077	0.48190 89901	0.780	2.18147 22655	0.45840 60113
0.731	2.07715 67262	0.48142 73220	0.781	2.18365 48289	0.45794 78344
0.732	2.07923 49218	0.48094 61353	0.782	2.18583 95759	0.45749 01155
0.733	2.08131 51967	0.48046 54295	0.783	2.18802 65087	0.45703 28540
0.734	2.08339 75529	0.47998 52043	0.784	2.19021 56296	0.45657 60496
0.735	2.08548 19925	0.47950 54590	0.785	2.19240 69407	0.45611 97018
0.736	2.08756 85176	0.47902 61932	0.786	2.19460 04442	0.45566 38101
0.737	2.08965 71302	0.47854 74064	0.787	2.19679 61424	0.45520 83740
0.738	2.09174 78325	0.47806 90982	0.788	2.19899 40373	0.45475 33932
0.739	2.09384 06266	0.47759 12681	0.789	2.20119 41312	0.45429 88671
0.740	2.09593 55145	0.47711 39155	0.790	2.20339 64263	0.45384 47953
0.741	2.09803 24983	0.47663 70401	0.791	2.20560 09247	0.45339 11773
0.742	2.10013 15802	0.47616 06413	0.792	2.20780 76288	0.45293 80128
0.743	2.10223 27622	0.47568 47186	0.793	2.21001 65407	0.45248 53012
0.744	2.10433 60464	0.47520 92717	0.794	2.21222 76627	0.45203 30420
0.745	2.10644 14350	0.47473 42999	0.795	2.21444 09968	0.45158 12349
0.746	2.10854 89300	0.47425 98029	0.796	2.21665 65454	0.45112 98794
0.747	2.11065 85335	0.47378 57802	0.797	2.21887 43106	0.45067 89750
0.748	2.11277 02478	0.47331 22312	0.798	2.22109 42948	0.45022 85213
0.749	2.11488 40747	0.47283 91556	0.799	2.22331 65000	0.44977 85178
0.750	2.11700 00166	0.47236 65527	0.800	2.22554 09285	0.44932 89641

TEN-PLACE EXPONENTIAL FUNCTIONS e^x AND e^{-x} (Continued)

x	e^x	e^{-x}	x	e^x	e^{-x}
0.800	2.22554 09285	0.44932 89641	0.850	2.33964 68519	0.42741 49319
0.801	2.22776 75826	0.44887 98597	0.851	2.34198 76690	0.42698 77307
0.802	2.22999 64644	0.44843 12042	0.852	2.34433 08280	0.42656 09563
0.803	2.23222 75762	0.44798 29972	0.853	2.34667 63314	0.42613 46086
0.804	2.23446 09203	0.44753 52381	0.854	2.34902 41815	0.42570 86870
0.805	2.23669 64988	0.44708 79266	0.855	2.35137 43806	0.42528 31911
0.806	2.23893 43140	0.44664 10621	0.856	2.35372 69310	0.42485 81205
0.807	2.24117 43682	0.44619 46443	0.857	2.35608 18352	0.42443 34747
0.808	2.24341 66635	0.44574 86727	0.858	2.35843 90955	0.42400 92534
0.809	2.24566 12023	0.44530 31468	0.859	2.36079 87142	0.42358 54561
0.810	2.24790 79867	0.44485 80662	0.860	2.36316 06937	0.42316 20823
0.811	2.25015 70190	0.44441 34305	0.861	2.36552 50364	0.42273 91317
0.812	2.25240 83015	0.44396 92392	0.862	2.36789 17446	0.42231 66039
0.813	2.25466 18363	0.44352 54919	0.863	2.37026 08207	0.42189 44984
0.814	2.25691 76259	0.44308 21881	0.864	2.37263 22670	0.42147 28148
0.815	2.25917 56723	0.44263 93274	0.865	2.37500 60860	0.42105 15526
0.816	2.26143 59780	0.44219 69093	0.866	2.37738 22800	0.42063 07115
0.817	2.26369 85451	0.44175 49334	0.867	2.37976 08513	0.42021 02911
0.818	2.26596 33758	0.44131 33993	0.868	2.38214 18025	0.41979 02908
0.819	2.26823 04726	0.44087 23064	0.869	2.38452 51357	0.41937 07103
0.820	2.27049 98375	0.44043 16545	0.870	2.38691 08535	0.41895 15492
0.821	2.27277 14730	0.43999 14430	0.871	2.38929 89582	0.41853 28071
0.822	2.27504 53812	0.43955 16715	0.872	2.39168 94522	0.41811 44835
0.823	2.27732 15645	0.43911 23395	0.873	2.39408 23379	0.41769 65780
0.824	2.27960 00251	0.43867 34466	0.874	2.39647 76177	0.41727 90902
0.825	2.28188 07653	0.43823 49925	0.875	2.39887 52940	0.41686 20197
0.826	2.28416 37874	0.43779 69765	0.876	2.40127 53691	0.41644 53660
0.827	2.28644 90937	0.43735 93984	0.877	2.40367 78455	0.41602 91288
0.828	2.28873 66864	0.43692 22576	0.878	2.40608 27256	0.41561 33076
0.829	2.29102 65678	0.43648 55537	0.879	2.40849 00118	0.41519 79021
0.830	2.29331 87403	0.43604 92863	0.880	2.41089 97064	0.41478 29117
0.831	2.29561 32060	0.43561 34550	0.881	2.41331 18120	0.41436 83361
0.832	2.29790 99674	0.43517 80593	0.882	2.41572 63308	0.41395 41749
0.833	2.30020 90267	0.43474 30987	0.883	2.41814 32654	0.41354 04276
0.834	2.30251 03863	0.43430 85729	0.884	2.42056 26182	0.41312 70939
0.835	2.30481 40483	0.43387 44814	0.885	2.42298 43915	0.41271 41733
0.836	2.30712 00151	0.43344 08238	0.886	2.42540 85878	0.41230 16654
0.837	2.30942 82891	0.43300 75996	0.887	2.42783 52095	0.41188 95698
0.838	2.31173 88725	0.43257 48085	0.888	2.43026 42590	0.41147 78861
0.839	2.31405 17676	0.43214 24499	0.889	2.43269 57388	0.41106 66139
0.840	2.31636 69768	0.43171 05234	0.890	2.43512 96513	0.41065 57528
0.841	2.31868 45023	0.43127 90287	0.891	2.43756 59989	0.41024 53023
0.842	2.32100 43466	0.43084 79652	0.892	2.44000 47841	0.40983 52620
0.843	2.32332 65118	0.43041 73326	0.893	2.44244 60093	0.40942 56316
0.844	2.32565 10004	0.42998 71304	0.894	2.44488 96769	0.40901 64106
0.845	2.32797 78146	0.42955 73582	0.895	2.44733 57895	0.40860 75986
0.846	2.33030 69568	0.42912 80156	0.896	2.44978 43493	0.40819 91953
0.847	2.33263 84293	0.42869 91020	0.897	2.45223 53590	0.40779 12001
0.848	2.33497 22344	0.42827 06172	0.898	2.45468 88209	0.40738 36127
0.849	2.33730 83745	0.42784 25607	0.899	2.45714 47374	0.40697 64328
0.850	2.33964 68519	0.42741 49319	0.900	2.45960 31112	0.40656 96597

TEN-PLACE EXPONENTIAL FUNCTIONS e^x AND e^{-x} (Continued)

x	e^x	e^{-x}	x	e^x	e^{-x}
0.900	2.45960 31112	0.40656 96597	0.950	2.58570 96593	0.38674 10235
0.901	2.46206 39445	0.40616 32933	0.951	2.58829 66623	0.38635 44757
0.902	2.46452 72399	0.40575 73330	0.952	2.59088 62535	0.38596 83144
0.903	2.46699 29998	0.40535 17785	0.953	2.59347 84356	0.38558 25390
0.904	2.46946 12267	0.40494 66293	0.954	2.59607 32112	0.38519 71492
0.905	2.47193 19231	0.40454 18851	0.955	2.59867 05829	0.38481 21446
0.906	2.47440 50914	0.40413 75454	0.956	2.60127 05533	0.38442 75248
0.907	2.47688 07341	0.40373 36099	0.957	2.60387 31249	0.38404 32894
0.908	2.47935 88537	0.40333 00781	0.958	2.60647 83004	0.38365 94380
0.909	2.48183 94526	0.40292 69496	0.959	2.60908 60824	0.38327 59704
0.910	2.48432 25334	0.40252 42240	0.960	2.61169 64734	0.38289 28860
0.911	2.48680 80985	0.40212 19010	0.961	2.61430 94762	0.38251 01845
0.912	2.48929 61504	0.40171 99801	0.962	2.61692 50932	0.38212 78655
0.913	2.49178 66916	0.40131 84609	0.963	2.61954 33272	0.38174 59286
0.914	2.49427 97246	0.40091 73430	0.964	2.62216 41808	0.38136 43735
0.915	2.49677 52519	0.40051 66261	0.965	2.62478 76565	0.38098 31997
0.916	2.49927 32760	0.40011 63097	0.966	2.62741 37570	0.38060 24070
0.917	2.50177 37993	0.39971 63933	0.967	2.63004 24849	0.38022 19948
0.918	2.50427 68244	0.39931 68767	0.968	2.63267 38428	0.37984 19629
0.919	2.50678 23538	0.39891 77595	0.969	2.63530 78334	0.37946 23107
0.920	2.50929 03899	0.39851 90411	0.970	2.63794 44594	0.37908 30381
0.921	2.51180 09354	0.39812 07212	0.971	2.64058 37232	0.37870 41445
0.922	2.51431 39926	0.39772 27995	0.972	2.64322 56277	0.37832 56297
0.923	2.51682 95642	0.39732 52755	0.973	2.64587 01754	0.37794 74932
0.924	2.51934 76526	0.39692 81488	0.974	2.64851 73689	0.37756 97346
0.925	2.52186 82604	0.39653 14191	0.975	2.65116 72110	0.37719 23536
0.926	2.52439 13900	0.39613 50859	0.976	2.65381 97042	0.37681 53497
0.927	2.52691 70440	0.39573 91488	0.977	2.65647 48513	0.37643 87227
0.928	2.52944 52249	0.39534 36074	0.978	2.65913 26548	0.37606 24722
0.929	2.53197 59353	0.39494 84614	0.979	2.66179 31175	0.37568 65977
0.930	2.53450 91776	0.39455 37104	0.980	2.66445 62419	0.37531 10989
0.931	2.53704 49545	0.39415 93539	0.981	2.66712 20308	0.37493 59753
0.932	2.53958 32684	0.39376 53915	0.982	2.66979 04869	0.37456 12268
0.933	2.54212 41219	0.39337 18230	0.983	2.67246 16127	0.37418 68528
0.934	2.54466 75175	0.39297 86478	0.984	2.67513 54110	0.37381 28529
0.935	2.54721 34577	0.39258 58655	0.985	2.67781 18844	0.37343 92269
0.936	2.54976 19452	0.39219 34759	0.986	2.68049 10357	0.37306 59744
0.937	2.55231 29825	0.39180 14784	0.987	2.68317 28674	0.37269 30949
0.938	2.55486 65720	0.39140 98728	0.988	2.68585 73823	0.37232 05881
0.939	2.55742 27165	0.39101 86586	0.989	2.68854 45830	0.37194 84536
0.940	2.55998 14183	0.39062 78354	0.990	2.69123 44723	0.37157 66910
0.941	2.56254 26802	0.39023 74028	0.991	2.69392 70529	0.37120 53001
0.942	2.56510 65045	0.38984 73604	0.992	2.69662 23274	0.37083 42803
0.943	2.56767 28940	0.38945 77079	0.993	2.69932 02984	0.37046 36314
0.944	2.57024 18512	0.38906 84449	0.994	2.70202 09688	0.37009 33529
0.945	2.57281 33786	0.38867 95709	0.995	2.70472 43413	0.36972 34445
0.946	2.57538 74788	0.38829 10856	0.996	2.70743 04184	0.36935 39059
0.947	2.57796 41544	0.38790 29886	0.997	2.71013 92030	0.36898 47366
0.948	2.58054 34080	0.38751 52795	0.998	2.71285 06977	0.36861 59363
0.949	2.58312 52421	0.38712 79579	0.999	2.71556 49053	0.36824 75046
0.950	2.58570 96593	0.38674 10235	1.000	2.71828 18285	0.36787 94412

TEN-PLACE EXPONENTIAL FUNCTIONS
e^x AND e^{-x} (Continued)

x	e^x	e^{-x}
0.0	1.00000 00000	1.00000 00000
0.1	1.10517 09181	0.90483 74180
0.2	1.22140 27582	0.81873 07531
0.3	1.34985 88076	0.74081 82207
0.4	1.49182 46976	0.67032 00460
0.5	1.64872 12707	0.60653 06597
0.6	1.82211 88004	0.54881 16361
0.7	2.01375 27075	0.49658 53038
0.8	2.22554 09285	0.44932 89641
0.9	2.45960 31112	0.40656 96597
1.0	2.71828 18285	0.36787 94412
1.1	2.00416 60239	0.33287 10837
1.2	3.32011 69227	0.30119 42119
1.3	3.66929 66676	0.27253 17930
1.4	4.05519 99668	0.24659 69639
1.5	4.48168 90703	0.22313 01601
1.6	4.95303 24244	0.20189 65180
1.7	5.47394 73917	0.18268 35241
1.8	6.04964 74644	0.16529 88882
1.9	6.68589 44423	0.14956 86192
2.0	7.38905 60989	0.13533 52832
2.1	8.16616 99126	0.12245 64283
2.2	9.02501 34994	0.11080 31584
2.3	9.97418 24548	0.10025 88437
2.4	11.02317 63806	0.09071 79533
2.5	12.18249 39607	0.08208 49986
2.6	13.46373 80350	0.07427 35782
2.7	14.87973 17249	0.06720 55127
2.8	16.44464 67711	0.06081 00626
2.9	18.17414 53694	0.05502 32201
3.0	20.08553 69232	0.04978 70684
3.1	22.19795 12814	0.04504 92024
3.2	24.53253 01971	0.04076 22040
3.3	27.11263 89207	0.03688 31674
3.4	29.96410 00474	0.03337 32700
3.5	33.11545 19587	0.03019 73834
3.6	36.59823 44437	0.02732 37224
3.7	40.44730 43601	0.02472 35265
3.8	44.70118 44933	0.02237 07719
3.9	49.40244 91055	0.02024 19114
4.0	54.59815 00331	0.01831 56389
4.1	60.34028 75974	0.01657 26754
4.2	66.68633 10409	0.01499 55768
4.3	73.69979 36996	0.01356 85590
4.4	81.45086 86650	0.01227 73399
4.5	90.01713 13005	0.01110 89965
4.6	99.48431 56419	0.01005 18357
4.7	109.94717 24521	0.00909 52771
4.8	121.51041 75187	0.00822 97470
4.9	134.28977 96849	0.00744 65831
5.0	148.41315 91026	0.00673 79470

RADIX TABLE OF THE EXPONENTIAL FUNCTION

x	n	$e^{[x(10^{-n})]}$	$e^{-[x(10^{-n})]}$
1	10	1.00000 00001 00000 00000 50000	0.99999 99999 00000 00000 50000
2	10	1.00000 00002 00000 00002 00000	0.99999 99998 00000 00002 00000
3	10	1.00000 00003 00000 00004 50000	0.99999 99997 00000 00004 50000
4	10	1.00000 00004 00000 00008 00000	0.99999 99996 00000 00008 00000
5	10	1.00000 00005 00000 00012 50000	0.99999 99995 00000 00012 50000
6	10	1.00000 00006 00000 00018 00000	0.99999 99994 00000 00018 00000
7	10	1.00000 00007 00000 00024 50000	0.99999 99993 00000 00024 50000
8	10	1.00000 00008 00000 00032 00000	0.99999 99992 00000 00032 00000
9	10	1.00000 00009 00000 00040 50000	0.99999 99991 00000 00040 50000
1	9	1.00000 00010 00000 00050 00000	0.99999 99990 00000 00050 00000
2	9	1.00000 00020 00000 00200 00000	0.99999 99980 00000 00200 00000
3	9	1.00000 00030 00000 00450 00000	0.99999 99970 00000 00450 00000
4	9	1.00000 00040 00000 00800 00000	0.99999 99960 00000 00800 00000
5	9	1.00000 00050 00000 01250 00000	0.99999 99950 00000 01250 00000
6	9	1.00000 00060 00000 01800 00000	0.99999 99940 00000 01800 00000
7	9	1.00000 00070 00000 02450 00001	0.99999 99930 00000 02449 99999
8	9	1.00000 00080 00000 03200 00001	0.99999 99920 00000 03199 99999
9	9	1.00000 00090 00000 04050 00001	0.99999 99910 00000 04049 99999
1	8	1.00000 00100 00000 05000 00002	0.99999 99900 00000 04999 99998
2	8	1.00000 00200 00000 20000 00013	0.99999 99800 00000 19999 99987
3	8	1.00000 00300 00000 45000 00045	0.99999 99700 00000 44999 99955
4	8	1.00000 00400 00000 80000 00107	0.99999 99600 00000 79999 99893
5	8	1.00000 00500 00001 25000 00208	0.99999 99500 00001 24999 99792
6	8	1.00000 00600 00001 80000 00360	0.99999 99400 00001 79999 99640
7	8	1.00000 00700 00002 45000 00572	0.99999 99300 00002 44999 99428
8	8	1.00000 00800 00003 20000 00853	0.99999 99200 00003 19999 99147
9	8	1.00000 00900 00004 05000 01215	0.99999 99100 00004 04999 98785
1	7	1.00000 01000 00005 00000 01667	0.99999 99000 00004 99999 98333
2	7	1.00000 02000 00020 00000 13333	0.99999 98000 00019 99999 86667
3	7	1.00000 03000 00045 00000 45000	0.99999 97000 00044 99999 55000
4	7	1.00000 04000 00080 00001 06667	0.99999 96000 00079 99998 93333
5	7	1.00000 05000 00125 00002 08333	0.99999 95000 00124 99997 91667
6	7	1.00000 06000 00180 00003 60000	0.99999 94000 00179 99996 40000
7	7	1.00000 07000 00245 00005 71667	0.99999 93000 00244 99994 28333
8	7	1.00000 08000 00320 00008 53334	0.99999 92000 00319 99991 46667
9	7	1.00000 09000 00405 00012 15000	0.99999 91000 00404 99987 85000
1	6	1.00000 10000 00500 00016 66667	0.99999 90000 00499 99983 33334
2	6	1.00000 20000 02000 00133 33340	0.99999 80000 01999 99866 66673
3	6	1.00000 30000 04500 00450 00034	0.99999 70000 04499 99550 00034
4	6	1.00000 40000 08000 01066 66773	0.99999 60000 07999 98933 33440
5	6	1.00000 50000 12500 02083 33594	0.99999 50000 12499 97916 66927
6	6	1.00000 60000 18000 03600 00540	0.99999 40000 17999 96400 00540
7	6	1.00000 70000 24500 05716 67667	0.99999 30000 24499 94283 34334
8	6	1.00000 80000 32000 08533 35040	0.99999 20000 31999 91466 68373
9	6	1.00000 90000 40500 12150 02734	0.99999 10000 40499 87850 02734

For $n > 10$, $e^{\pm[x(10^{-n})]} = 1 \pm x(10^{-n}) + \frac{1}{2}x^2(10^{-2n})$ to 25 decimals.

RADIX TABLE OF THE EXPONENTIAL FUNCTION

x	n	$e^{[x(10^{-n})]}$	$e^{-[x(10^{-n})]}$
1	5	1.00001 00000 50000 16666 70833	0.99999 00000 49999 83333 37500
2	5	1.00002 00002 00001 33334 00000	0.99998 00001 99998 66667 33333
3	5	1.00003 00004 50004 50003 37502	0.99997 00004 49995 50003 37498
4	5	1.00004 00008 00010 66677 33342	0.99996 00007 99989 33343 99991
5	5	1.00005 00012 50020 83359 37526	0.99995 00012 49979 16692 70807
6	5	1.00006 00018 00036 00054 00065	0.99994 00017 99964 00053 99935
7	5	1.00007 00024 50057 16766 70973	0.99993 00024 49942 83433 37360
8	5	1.00008 00032 00085 33504 00273	0.99992 00031 99914 66837 33060
9	5	1.00009 00040 50121 50273 37992	0.99991 00040 49878 50273 37008
1	4	1.00010 00050 00166 67083 34167	0.99990 00049 99833 33749 99167
2	4	1.00020 00200 01333 40000 26668	0.99980 00199 98666 73333 06668
3	4	1.00030 00450 04500 33752 02510	0.99970 00449 95500 33747 97510
4	4	1.00040 00800 10667 73341 86724	0.99960 00799 89334 39991 46724
5	4	1.00050 01250 20835 93776 04384	0.99950 01249 79169 27057 29384
6	4	1.00060 01800 36005 40064 80648	0.99940 01799 64005 39935 20648
7	4	1.00070 02450 57176 67223 40801	0.99930 02449 42843 33609 95801
8	4	1.00080 03200 85350 40273 10308	0.99920 03199 14683 73060 30307
9	4	1.00090 04051 21527 34242 14882	0.99910 04048 78527 33257 99880
1	3	1.00100 05001 66708 34166 80558	0.99900 04998 33374 99166 80554
2	3	1.00200 20013 34000 26675 55810	0.99800 19986 67333 06675 55302
3	3	1.00300 45045 03377 02601 29341	0.99700 44955 03372 97601 20662
4	3	1.00400 80106 77341 87235 88080	0.99600 79893 43991 47235 23064
5	3	1.00501 25208 59401 06338 35662	0.99501 24791 92682 31335 25642
6	3	1.00601 80360 54064 86485 55845	0.99401 79640 53935 26474 44988
7	3	1.00702 45572 66848 55523 16000	0.99302 44429 33235 10490 47970
8	3	1.00803 20855 04273 43117 20736	0.99203 19148 37060 63033 98697
9	3	1.00904 06217 73867 81406 25705	0.99104 03787 72883 66216 45648
1	2	1.01005 01670 84168 05754 21655	0.99004 98337 49168 05357 39060
2	2	1.02020 13400 26755 81016 01439	0.98019 86733 06755 30222 08141
3	2	1.03045 45339 53516 85561 24400	0.97044 55335 48508 17693 25284
4	2	1.04081 07741 92388 22675 70448	0.96078 94391 52323 20943 92107
5	2	1.05127 10963 76024 03969 75176	0.95122 94245 00714 00909 14253
6	2	1.06183 65465 45359 62222 46849	0.94176 45335 84248 70953 71528
7	2	1.07250 81812 54216 47905 31039	0.93239 38199 05948 22885 79726
8	2	1.08328 70676 74958 55443 59878	0.92311 63463 86635 78291 07598
9	2	1.09417 42837 05210 35787 28976	0.91393 11852 71228 18674 73535
1	1	1.10517 09180 75647 62481 17078	0.90483 74180 35959 57316 42491
2	1	1.22140 27581 60169 83392 10720	0.81873 07530 77981 85866 99355
3	1	1.34985 88075 76003 10398 37443	0.74081 82206 81717 86606 68738
4	1	1.49182 46976 41270 31782 48530	0.67032 00460 35639 30074 44329
5	1	1.64872 12707 00128 14684 86508	0.60653 06597 12633 42360 37995
6	1	1.82211 88003 90508 97487 53677	0.54881 16360 94026 43262 84589
7	1	2.01375 27074 70476 52162 45494	0.49658 53037 91409 51470 48001
8	1	2.22554 09284 92467 60457 95375	0.44932 89641 17221 59143 01024
9	1	2.45960 31111 56949 66380 01266	0.40656 96597 40599 11188 34542
1	0	2.71828 18284 59045 23536 02875	0.36787 94411 71442 32159 55238

For $n > 10$, $e^{\pm[x(10^{-n})]} = 1 \pm x(10^{-n}) + \frac{1}{2}x^2(10^{-2n})$ to 25 decimals.

TRIGONOMETRY

DR. HOWARD EVES

PLANE TRIGONOMETRY

Angle

That part of a straight line lying entirely to one side of a point O on the line is called a *ray* (or a *half-line*); the point O is called the *origin* of the ray. A ray of origin O is identified by the notation OA, where A is any point of the ray.

If a ray OA is rotated, in a plane, about its origin O onto ray OB, an *angle AOB* is said to be generated. Ray OA is called the *initial side*, ray OB the *terminal side*, and point O the *vertex* of the angle. The angle is said to be *positive* or *negative* according as the generating rotation is counterclockwise or clockwise.

An angle is said to be in *standard position* if its vertex is at the origin O and its initial side is on the positive x-axis of a rectangular Cartesian coordinate system (see p. 496). If the terminal side of the angle falls on a coordinate axis, the angle is called a *quadrantal angle*; otherwise the angle is called a *first, second, third,* or *fourth quadrant angle* according as the terminal side falls in the first, second, third, or fourth quadrant of the coordinate system.

An angle of one *degree* is an angle in which the rotation is 1/360 of one complete rotation.

A *straight angle* is an angle of 180° (180 degrees).

A *right angle* is an angle of 90°.

An *acute angle* is an angle between 0° and 90°.

An *obtuse angle* is an angle between 90° and 180°.

A *radian* is an angle subtended at the center of a circle by an arc whose length is equal to that of the radius.

$$180^\circ = \pi \text{ radians}; \quad 1^\circ = \frac{\pi}{180} \text{ radians}; \quad 1 \text{ radian} = \frac{180}{\pi} \text{ degrees.}$$

The Trigonometric Functions of an Acute Angle

In the right triangle ABC,

sine A = $\sin A$ = a/c

cosine A = $\cos a$ = b/c

tangent A = $\tan A$ = a/b

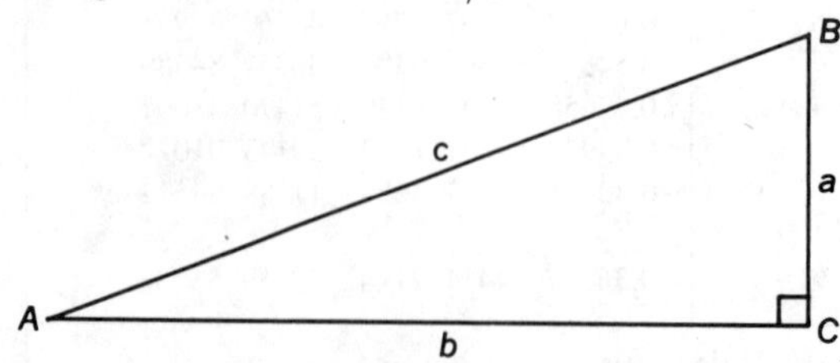

cosecant A = $\csc A$ = c/a

secant A = $\sec A$ = c/b

cotangent A = $\cot A$ = $\operatorname{ctn} A$ = b/a

exsecant A = $\operatorname{exsec} A$ = $\sec A - 1$

versine A = $\operatorname{vers} A$ = $1 - \cos A$

coversine A = $\operatorname{covers} A$ = $1 - \sin A$

haversine A = $\operatorname{hav} A$ = $\frac{1}{2} \operatorname{vers} A$

The Trigonometric Functions of an Arbitrary Angle

Let α be any angle in standard position and let $P(x,y)$ be any point on the terminal side of the angle. Denote the positive distance OP by r. Then

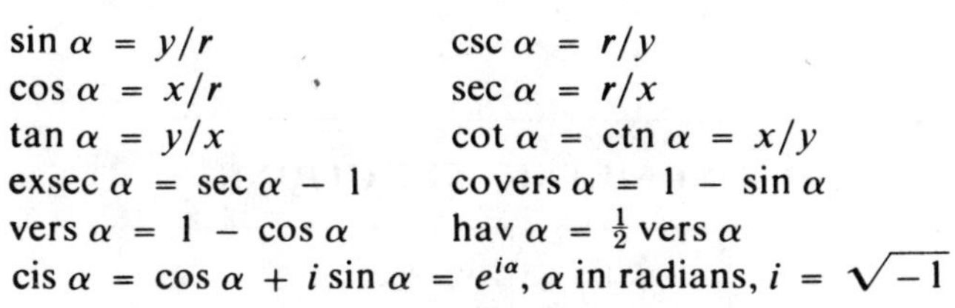

$\sin \alpha = y/r$ $\qquad \csc \alpha = r/y$
$\cos \alpha = x/r$ $\qquad \sec \alpha = r/x$
$\tan \alpha = y/x$ $\qquad \cot \alpha = \operatorname{ctn} \alpha = x/y$
$\operatorname{exsec} \alpha = \sec \alpha - 1$ $\qquad \operatorname{covers} \alpha = 1 - \sin \alpha$
$\operatorname{vers} \alpha = 1 - \cos \alpha$ $\qquad \operatorname{hav} \alpha = \frac{1}{2} \operatorname{vers} \alpha$
$\operatorname{cis} \alpha = \cos \alpha + i \sin \alpha = e^{i\alpha}$, α in radians, $i = \sqrt{-1}$

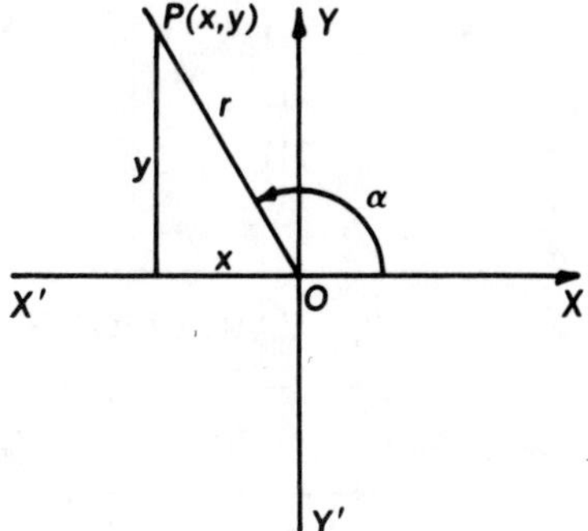

RELATIONS BETWEEN CIRCULAR (OR INVERSE CIRCULAR) FUNCTIONS

$$\left(0 \le x \le \frac{\pi}{2}\right)$$

	$\sin x = a$	$\cos x = a$	$\tan x = a$
$\sin x$	a	$(1 - a^2)^{1/2}$	$a(1 + a^2)^{-1/2}$
$\cos x$	$(1 - a^2)^{1/2}$	a	$(1 + a^2)^{-1/2}$
$\tan x$	$a(1 - a^2)^{-1/2}$	$a^{-1}(1 - a^2)^{1/2}$	a
$\csc x$	a^{-1}	$(1 - a^2)^{-1/2}$	$a^{-1}(1 + a^2)^{1/2}$
$\sec x$	$(1 - a^2)^{-1/2}$	a^{-1}	$(1 + a^2)^{1/2}$
$\cot x$	$a^{-1}(1 - a^2)^{1/2}$	$a(1 - a^2)^{-1/2}$	a^{-1}

	$\csc x = a$	$\sec x = a$	$\cot x = a$
$\sin x$	a^{-1}	$a^{-1}(a^2 - 1)^{1/2}$	$(1 + a^2)^{-1/2}$
$\cos x$	$a^{-1}(a^2 - 1)^{1/2}$	a^{-1}	$a(1 + a^2)^{-1/2}$
$\tan x$	$(a^2 - 1)^{-1/2}$	$(a^2 - 1)^{1/2}$	a^{-1}
$\csc x$	a	$a(a^2 - 1)^{-1/2}$	$(1 + a^2)^{1/2}$
$\sec x$	$a(a^2 - 1)^{-1/2}$	a	$a^{-1}(1 + a^2)^{1/2}$
$\cot x$	$(a^2 - 1)^{1/2}$	$(a^2 - 1)^{-1/2}$	a

Examples:

If $\sec x = a$, then $\tan x = (a^2 - 1)^{1/2}$
$\arctan a = \arccos (1 + a^2)^{-1/2}$

SIGNS OF THE TRIGONOMETRIC FUNCTIONS

Quadrant	sin	cos	tan	cot	sec	csc
I	+	+	+	+	+	+
II	+	−	−	−	−	+
III	−	−	+	+	−	−
IV	−	+	−	−	+	−

VARIATIONS OF THE TRIGONOMETRIC FUNCTIONS

Quadrant	sin	cos	tan	cot	sec	csc
I	$0 \to +1$	$+1 \to 0$	$0 \to +\infty$	$+\infty \to 0$	$+1 \to +\infty$	$+\infty \to +1$
II	$+1 \to 0$	$0 \to -1$	$-\infty \to 0$	$0 \to -\infty$	$-\infty \to -1$	$+1 \to +\infty$
III	$0 \to -1$	$-1 \to 0$	$0 \to +\infty$	$+\infty \to 0$	$-1 \to -\infty$	$-\infty \to -1$
IV	$-1 \to 0$	$0 \to +1$	$-\infty \to 0$	$0 \to -\infty$	$+\infty \to +1$	$-1 \to -\infty$

TRIGONOMETRIC FUNCTIONS OF SOME SPECIAL ANGLES

Angle	sin	cos	tan	cot	sec	csc
$0° = 0$	0	1	0	$\cdots$	1	$\cdots$
$15° = \frac{\pi}{12}$	$\frac{\sqrt{2}}{4}(\sqrt{3}-1)$	$\frac{\sqrt{2}}{4}(\sqrt{3}+1)$	$2-\sqrt{3}$	$2+\sqrt{3}$	$\sqrt{2}(\sqrt{3}-1)$	$\sqrt{2}(\sqrt{3}+1)$
$30° = \frac{\pi}{6}$	$1/2$	$\sqrt{3}/2$	$\sqrt{3}/3$	$\sqrt{3}$	$2\sqrt{3}/3$	2
$45° = \frac{\pi}{4}$	$\sqrt{2}/2$	$\sqrt{2}/2$	1	1	$\sqrt{2}$	$\sqrt{2}$
$60° = \frac{\pi}{3}$	$\sqrt{3}/2$	$1/2$	$\sqrt{3}$	$\sqrt{3}/3$	2	$2\sqrt{3}/3$
$75° = \frac{5\pi}{12}$	$\frac{\sqrt{2}}{4}(\sqrt{3}+1)$	$\frac{\sqrt{2}}{4}(\sqrt{3}-1)$	$2+\sqrt{3}$	$2-\sqrt{3}$	$\sqrt{2}(\sqrt{3}+1)$	$\sqrt{2}(\sqrt{3}-1)$
$90° = \frac{\pi}{2}$	1	0	$\cdots$	0	$\cdots$	1
$105° = \frac{7\pi}{12}$	$\frac{\sqrt{2}}{4}(\sqrt{3}+1)$	$-\frac{\sqrt{2}}{4}(\sqrt{3}-1)$	$-(2+\sqrt{3})$	$-(2-\sqrt{3})$	$-2(\sqrt{3}+1)$	$\sqrt{2}(\sqrt{3}-1)$
$120° = \frac{2\pi}{3}$	$\sqrt{3}/2$	$-1/2$	$-\sqrt{3}$	$-\sqrt{3}/3$	-2	$2\sqrt{3}/3$
$135° = \frac{3\pi}{4}$	$\sqrt{2}/2$	$-\sqrt{2}/2$	-1	-1	$-\sqrt{2}$	$\sqrt{2}$
$150° = \frac{5\pi}{6}$	$1/2$	$-\sqrt{3}/2$	$-\sqrt{3}/3$	$-\sqrt{3}$	$-2\sqrt{3}/3$	2
$165° = \frac{11\pi}{12}$	$\frac{\sqrt{2}}{4}(\sqrt{3}-1)$	$-\frac{\sqrt{2}}{4}(\sqrt{3}+1)$	$-(2-\sqrt{3})$	$-(2+\sqrt{3})$	$-\sqrt{2}(\sqrt{3}-1)$	$\sqrt{2}(\sqrt{3}+1)$
$180° = \pi$	0	-1	0	$\cdots$	-1	$\cdots$
$270° = \frac{3\pi}{2}$	-1	0	$\cdots$	0	$\cdots$	-1

Relations of the Functions

$$\sin x = \frac{1}{\csc x} \qquad \csc x = \frac{1}{\sin x}$$

$$\cos x = \frac{1}{\sec x} \qquad \sec x = \frac{1}{\cos x}$$

$$\tan x = \frac{1}{\cot x} = \frac{\sin x}{\cos x} \qquad \sin^2 x + \cos^2 x = 1$$

$$1 + \tan^2 x = \sec^2 x$$

$$\cot x = \frac{1}{\tan x} = \frac{\cos x}{\sin x} \qquad 1 + \cot^2 x = \csc^2 x$$

$$^*\sin x = \pm\sqrt{1 - \cos^2 x} \qquad ^*\cos x = \pm\sqrt{1 - \sin^2 x}$$

$$^*\tan x = \pm\sqrt{\sec^2 x - 1} \qquad ^*\sec x = \pm\sqrt{\tan^2 x + 1}$$

$$^*\cot x = \pm\sqrt{\csc^2 x - 1} \qquad ^*\csc x = \pm\sqrt{\cot^2 x + 1}$$

$$\sin x = \cos(90° - x) = \sin(180° - x)$$

$$\cos x = \sin(90° - x) = -\cos(180° - x)$$

$$\tan x = \cot(90° - x) = -\tan(180° - x)$$

$$\cot x = \tan(90° - x) = -\cot(180° - x)$$

$$\csc x = \cot\frac{x}{2} - \cot x$$

*The sign in front of radical depends on quadrant in which x falls.

Reduction Formulae

$$\sin\alpha = +\cos(\alpha - 90°) = -\sin(\alpha - 180°) = -\cos(\alpha - 270°)$$

$$\cos\alpha = -\sin(\alpha - 90°) = -\cos(\alpha - 180°) = +\sin(\alpha - 270°)$$

$$\tan\alpha = -\cot(\alpha - 90°) = +\tan(\alpha - 180°) = -\cot(\alpha - 270°)$$

$$\cot\alpha = -\tan(\alpha - 90°) = +\cot(\alpha - 180°) = -\tan(\alpha - 270°)$$

$$\sec\alpha = -\csc(\alpha - 90°) = -\sec(\alpha - 180°) = +\csc(\alpha - 270°)$$

$$\csc\alpha = +\sec(\alpha - 90°) = -\csc(\alpha - 180°) = -\sec(\alpha - 270°)$$

FURTHER REDUCTION FORMULAE

	sin	cos	tan	cot	sec	csc
$-\alpha$	$-\sin\alpha$	$+\cos\alpha$	$-\tan\alpha$	$-\cot\alpha$	$+\sec\alpha$	$-\csc\alpha$
$90° + \alpha$	$+\cos\alpha$	$-\sin\alpha$	$-\cot\alpha$	$-\tan\alpha$	$-\csc\alpha$	$+\sec\alpha$
$90° - \alpha$	$+\cos\alpha$	$+\sin\alpha$	$+\cot\alpha$	$+\tan\alpha$	$+\csc\alpha$	$+\sec\alpha$
$180° + \alpha$	$-\sin\alpha$	$-\cos\alpha$	$+\tan\alpha$	$+\cot\alpha$	$-\sec\alpha$	$-\csc\alpha$
$180° - \alpha$	$+\sin\alpha$	$-\cos\alpha$	$-\tan\alpha$	$-\cot\alpha$	$-\sec\alpha$	$+\csc\alpha$
$270° + \alpha$	$-\cos\alpha$	$+\sin\alpha$	$-\cot\alpha$	$-\tan\alpha$	$+\csc\alpha$	$-\sec\alpha$
$270° - \alpha$	$-\cos\alpha$	$-\sin\alpha$	$+\cot\alpha$	$+\tan\alpha$	$-\csc\alpha$	$-\sec\alpha$
$360° + \alpha$	$+\sin\alpha$	$+\cos\alpha$	$+\tan\alpha$	$+\cot\alpha$	$+\sec\alpha$	$+\csc\alpha$
$360° - \alpha$	$-\sin\alpha$	$+\cos\alpha$	$-\tan\alpha$	$-\cot\alpha$	$+\sec\alpha$	$-\csc\alpha$

The above table may be summarized and extended by the following easily remembered rule:

$$f(\pm\alpha + n90°) = \pm g(\alpha)$$

where n may be any integer, positive, negative, or zero
f is any one of the six trigonometric functions: sin, cos, tan, cot, sec, or csc
α may be any real angle measure

If n is even, then g is the same function as f. If n is odd, then g is the *cofunction* of f.

(Sine and cosine, tangent and cotangent, secant and cosecant, are cofunctions of each other.)

The second $\pm$ sign is not necessarily the same as the first one, but is determined as follows: For a given function f, a given value of n, and a given choice of the first $\pm$ sign, the second $\pm$ sign will be the same for all values of α. Thus it is only necessary to check the sign for any one value of α, and the formula will be complete.

EXAMPLES:

$\tan(\alpha + 270°) = \pm \cot \alpha$. Since $n(=3)$ is odd, we use the cofunction. To determine the sign, assume a value of α in the first quadrant. Then $\alpha + 270°$ is in the fourth quadrant, where the tangent is negative, so a minus sign is required. Thus the formula becomes

$$\tan(\alpha + 270°) = -\cot \alpha, \quad \text{valid for all values of } \alpha.$$

$\cos(\alpha - 450°) = \pm \sin \alpha$. Again assuming a value of α in the first quadrant, we find that $\alpha - 450°$ is in the fourth quadrant. Thus $\cos(\alpha - 450°)$ would be positive, and no minus sign is needed. Hence the formula becomes

$$\cos(\alpha - 450°) = \sin \alpha.$$

$\sec(180° - \alpha) = \pm \sec \alpha$. Here $n(=2)$ is even, so we use the same function. To determine the sign, again assume a value of α in the first quadrant. Then $180° - \alpha$ is in the second quadrant, where the secant is negative. Thus the formula, valid for all values of α, is

$$\sec(180° - \alpha) = -\sec \alpha.$$

Fundamental Identities

Where a double sign appears in the following, the choice of sign depends upon the quadrant in which the angle terminates.

Reciprocal relations

$$\sin \alpha = \frac{1}{\csc \alpha}, \quad \cos \alpha = \frac{1}{\sec \alpha}, \quad \tan \alpha = \frac{1}{\cot \alpha}$$

$$\csc \alpha = \frac{1}{\sin \alpha}, \quad \sec \alpha = \frac{1}{\cos \alpha}, \quad \cot \alpha = \frac{1}{\tan \alpha}$$

Product relations

$$\sin \alpha = \tan \alpha \cos \alpha, \quad \cos \alpha = \cot \alpha \sin \alpha$$

$$\tan \alpha = \sin \alpha \sec \alpha, \quad \cot \alpha = \cos \alpha \csc \alpha$$

$$\sec \alpha = \csc \alpha \tan \alpha, \quad \csc \alpha = \sec \alpha \cot \alpha$$

Quotient relations

$$\sin \alpha = \frac{\tan \alpha}{\sec \alpha}, \quad \cos \alpha = \frac{\cot \alpha}{\csc \alpha}, \quad \tan \alpha = \frac{\sin \alpha}{\cos \alpha}$$

$$\csc \alpha = \frac{\sec \alpha}{\tan \alpha}, \quad \sec \alpha = \frac{\csc \alpha}{\cot \alpha}, \quad \cot \alpha = \frac{\cos \alpha}{\sin \alpha}$$

Pythagorean relations

$$\sin^2\alpha + \cos^2\alpha = 1, \qquad 1 + \tan^2\alpha = \sec^2\alpha, \qquad 1 + \cot^2\alpha = \csc^2\alpha$$

Angle-sum and angle-difference relations

$$\sin(\alpha + \beta) = \sin\alpha\cos\beta + \cos\alpha\sin\beta$$
$$\sin(\alpha - \beta) = \sin\alpha\cos\beta - \cos\alpha\sin\beta$$
$$\cos(\alpha + \beta) = \cos\alpha\cos\beta - \sin\alpha\sin\beta$$
$$\cos(\alpha - \beta) = \cos\alpha\cos\beta + \sin\alpha\sin\beta$$
$$\tan(\alpha + \beta) = \frac{\tan\alpha + \tan\beta}{1 - \tan\alpha\tan\beta}$$
$$\tan(\alpha - \beta) = \frac{\tan\alpha - \tan\beta}{1 + \tan\alpha\tan\beta}$$
$$\cot(\alpha + \beta) = \frac{\cot\beta\cot\alpha - 1}{\cot\beta + \cot\alpha}$$
$$\cot(\alpha - \beta) = \frac{\cot\beta\cot\alpha + 1}{\cot\beta - \cot\alpha}$$
$$\sin(\alpha + \beta)\sin(\alpha - \beta) = \sin^2\alpha - \sin^2\beta = \cos^2\beta - \cos^2\alpha$$
$$\cos(\alpha + \beta)\cos(\alpha - \beta) = \cos^2\alpha - \sin^2\beta = \cos^2\beta - \sin^2\alpha$$

Double-angle relations

$$\sin 2\alpha = 2\sin\alpha\cos\alpha = \frac{2\tan\alpha}{1 + \tan^2\alpha}$$
$$\cos 2\alpha = \cos^2\alpha - \sin^2\alpha = 2\cos^2\alpha - 1 = 1 - 2\sin^2\alpha = \frac{1 - \tan^2\alpha}{1 + \tan^2\alpha}$$
$$\tan 2\alpha = \frac{2\tan\alpha}{1 - \tan^2\alpha}, \qquad \cot 2\alpha = \frac{\cot^2\alpha - 1}{2\cot\alpha}$$

Multiple-angle relations

$$\sin 3\alpha = 3\sin\alpha - 4\sin^3\alpha$$
$$\cos 3\alpha = 4\cos^3\alpha - 3\cos\alpha$$
$$\sin 4\alpha = 4\sin\alpha\cos\alpha - 8\sin^3\alpha\cos\alpha$$
$$\cos 4\alpha = 8\cos^4\alpha - 8\cos^2\alpha + 1$$
$$\sin 5\alpha = 5\sin\alpha - 20\sin^3\alpha + 16\sin^5\alpha$$
$$\cos 5\alpha = 16\cos^5\alpha - 20\cos^3\alpha + 5\cos\alpha$$
$$\sin 6\alpha = 32\cos^5\alpha\sin\alpha - 32\cos^3\alpha\sin\alpha + 6\cos\alpha\sin\alpha$$
$$\cos 6\alpha = 32\cos^6\alpha - 48\cos^4\alpha + 18\cos^2\alpha - 1$$
$$\sin n\alpha = 2\sin(n - 1)\alpha\cos\alpha - \sin(n - 2)\alpha$$
$$\cos n\alpha = 2\cos(n - 1)\alpha\cos\alpha - \cos(n - 2)\alpha$$
$$\tan 3\alpha = \frac{3\tan\alpha - \tan^3\alpha}{1 - 3\tan^2\alpha}$$
$$\tan 4\alpha = \frac{4\tan\alpha - 4\tan^3\alpha}{1 - 6\tan^2\alpha + \tan^4\alpha}$$
$$\tan n\alpha = \frac{\tan(n - 1)\alpha + \tan\alpha}{1 - \tan(n - 1)\alpha\tan\alpha}$$

Function-product relations

$$\sin\alpha\sin\beta = \tfrac{1}{2}\cos(\alpha-\beta) - \tfrac{1}{2}\cos(\alpha+\beta)$$
$$\cos\alpha\cos\beta = \tfrac{1}{2}\cos(\alpha-\beta) + \tfrac{1}{2}\cos(\alpha+\beta)$$
$$\sin\alpha\cos\beta = \tfrac{1}{2}\sin(\alpha+\beta) + \tfrac{1}{2}\sin(\alpha-\beta)$$
$$\cos\alpha\sin\beta = \tfrac{1}{2}\sin(\alpha+\beta) - \tfrac{1}{2}\sin(\alpha-\beta)$$

Function-sum and function-difference relations

$$\sin\alpha + \sin\beta = 2\sin\tfrac{1}{2}(\alpha+\beta)\cos\tfrac{1}{2}(\alpha-\beta)$$
$$\sin\alpha - \sin\beta = 2\cos\tfrac{1}{2}(\alpha+\beta)\sin\tfrac{1}{2}(\alpha-\beta)$$
$$\cos\alpha + \cos\beta = 2\cos\tfrac{1}{2}(\alpha+\beta)\cos\tfrac{1}{2}(\alpha-\beta)$$
$$\cos\alpha - \cos\beta = -2\sin\tfrac{1}{2}(\alpha+\beta)\sin\tfrac{1}{2}(\alpha-\beta)$$
$$\tan\alpha + \tan\beta = \frac{\sin(\alpha+\beta)}{\cos\alpha\cos\beta}, \qquad \tan\alpha - \tan\beta = \frac{\sin(\alpha-\beta)}{\cos\alpha\cos\beta}$$
$$\cot\alpha + \cot\beta = \frac{\sin(\alpha+\beta)}{\sin\alpha\sin\beta}, \qquad \cot\alpha - \cot\beta = \frac{\sin(\beta-\alpha)}{\sin\alpha\sin\beta}$$
$$\frac{\sin\alpha+\sin\beta}{\sin\alpha-\sin\beta} = \frac{\tan\frac{1}{2}(\alpha+\beta)}{\tan\frac{1}{2}(\alpha-\beta)}, \qquad \frac{\sin\alpha+\sin\beta}{\cos\alpha-\cos\beta} = \cot\tfrac{1}{2}(\beta-\alpha)$$
$$\frac{\sin\alpha+\sin\beta}{\cos\alpha+\cos\beta} = \tan\tfrac{1}{2}(\alpha+\beta), \qquad \frac{\sin\alpha-\sin\beta}{\cos\alpha+\cos\beta} = \tan\tfrac{1}{2}(\alpha-\beta)$$

Half-angle relations

$$\sin\frac{\alpha}{2} = \pm\sqrt{\frac{1-\cos\alpha}{2}}, \qquad \cos\frac{\alpha}{2} = \pm\sqrt{\frac{1+\cos\alpha}{2}}$$
$$\tan\frac{\alpha}{2} = \pm\sqrt{\frac{1-\cos\alpha}{1+\cos\alpha}} = \frac{1-\cos\alpha}{\sin\alpha} = \frac{\sin\alpha}{1+\cos\alpha}$$
$$\cot\frac{\alpha}{2} = \pm\sqrt{\frac{1+\cos\alpha}{1-\cos\alpha}} = \frac{1+\cos\alpha}{\sin\alpha} = \frac{\sin\alpha}{1-\cos\alpha}$$

Power relations

$$\sin^2\alpha = \tfrac{1}{2}(1-\cos 2\alpha), \qquad \sin^3\alpha = \tfrac{1}{4}(3\sin\alpha - \sin 3\alpha)$$
$$\sin^4\alpha = \tfrac{1}{8}(3 - 4\cos 2\alpha + \cos 4\alpha)$$
$$\cos^2\alpha = \tfrac{1}{2}(1+\cos 2\alpha), \qquad \cos^3\alpha = \tfrac{1}{4}(3\cos\alpha + \cos 3\alpha)$$
$$\cos^4\alpha = \tfrac{1}{8}(3 + 4\cos 2\alpha + \cos 4\alpha)$$
$$\tan^2\alpha = \frac{1-\cos 2\alpha}{1+\cos 2\alpha}, \qquad \cot^2\alpha = \frac{1+\cos 2\alpha}{1-\cos 2\alpha}$$

Exponential relations (α in radians), Euler's equation

$$e^{i\alpha} = \cos\alpha + i\sin\alpha, \qquad i = \sqrt{-1}$$
$$\sin\alpha = \frac{e^{i\alpha} - e^{-i\alpha}}{2i}, \qquad \cos\alpha = \frac{e^{i\alpha} + e^{-i\alpha}}{2}$$
$$\tan\alpha = -i\left(\frac{e^{i\alpha} - e^{-i\alpha}}{e^{i\alpha} + e^{-i\alpha}}\right) = -i\left(\frac{e^{2i\alpha} - 1}{e^{2i\alpha} + 1}\right)$$

THE TRIGONOMETRIC FUNCTIONS IN TERMS OF ONE ANOTHER

Function	$\sin\alpha$	$\cos\alpha$	$\tan\alpha$	$\cot\alpha$	$\sec\alpha$	$\csc\alpha$
$\sin\alpha$	$\sin\alpha$	$\pm\sqrt{1-\cos^2\alpha}$	$\dfrac{\tan\alpha}{\pm\sqrt{1+\tan^2\alpha}}$	$\dfrac{1}{\pm\sqrt{1+\cot^2\alpha}}$	$\dfrac{\pm\sqrt{\sec^2\alpha-1}}{\sec\alpha}$	$\dfrac{1}{\csc\alpha}$
$\cos\alpha$	$\pm\sqrt{1-\sin^2\alpha}$	$\cos\alpha$	$\dfrac{1}{\pm\sqrt{1+\tan^2\alpha}}$	$\dfrac{\cot\alpha}{\pm\sqrt{1+\cot^2\alpha}}$	$\dfrac{1}{\sec\alpha}$	$\dfrac{\pm\sqrt{\csc^2\alpha-1}}{\csc\alpha}$
$\tan\alpha$	$\dfrac{\sin\alpha}{\pm\sqrt{1-\sin^2\alpha}}$	$\dfrac{\pm\sqrt{1-\cos^2\alpha}}{\cos\alpha}$	$\tan\alpha$	$\dfrac{1}{\cot\alpha}$	$\pm\sqrt{\sec^2\alpha-1}$	$\dfrac{1}{\pm\sqrt{\csc^2\alpha-1}}$
$\cot\alpha$	$\dfrac{\pm\sqrt{1-\sin^2\alpha}}{\sin\alpha}$	$\dfrac{\cos\alpha}{\pm\sqrt{1-\cos^2\alpha}}$	$\dfrac{1}{\tan\alpha}$	$\cot\alpha$	$\dfrac{1}{\pm\sqrt{\sec^2\alpha-1}}$	$\pm\sqrt{\csc^2\alpha-1}$
$\sec\alpha$	$\dfrac{1}{\pm\sqrt{1-\sin^2\alpha}}$	$\dfrac{1}{\cos\alpha}$	$\pm\sqrt{1+\tan^2\alpha}$	$\dfrac{\pm\sqrt{1+\cot^2\alpha}}{\cot\alpha}$	$\sec\alpha$	$\dfrac{\csc\alpha}{\pm\sqrt{\csc^2\alpha-1}}$
$\csc\alpha$	$\dfrac{1}{\sin\alpha}$	$\dfrac{1}{\pm\sqrt{1-\cos^2\alpha}}$	$\dfrac{\pm\sqrt{1+\tan^2\alpha}}{\tan\alpha}$	$\pm\sqrt{1+\cot^2\alpha}$	$\dfrac{\sec\alpha}{\pm\sqrt{\sec^2\alpha-1}}$	$\csc\alpha$

Note. The choice of sign depends upon the quadrant in which the angle terminates.

Principal Values of the Inverse Trigonometric Functions

The notation arcsin x (or $\sin^{-1}x$) is used to denote any angle whose sine is x; Arcsin x (or $\mathrm{Sin}^{-1}x$) is usually used to denote the *principal value*. Similar notation is used for the other inverse trigonometric functions. The principal values of the inverse trigonometric functions are defined as follows:

$-\pi/2 \leqq \mathrm{Arcsin}\, x \leqq \pi/2,$	$-1 \leqq x \leqq 1$
$0 \leqq \mathrm{Arccos}\, x \leqq \pi,$	$-1 \leqq x \leqq 1$
$-\pi/2 < \mathrm{Arctan}\, x < \pi/2,$	$-\infty < x < \infty$
$0 < \mathrm{Arccsc}\, x \leqq \pi/2,$ $-\pi < \mathrm{Arccsc}\, x \leqq -\pi/2,$	$x \geqq 1$ $x \leqq -1$
$0 \leqq \mathrm{Arcsec}\, x < \pi/2,$ $-\pi \leqq \mathrm{Arcsec}\, x < -\pi/2,$	$x \geqq 1$ $x \leqq -1$
$0 < \mathrm{Arccot}\, x < \pi,$	$-\infty < x < \infty$

Note. There is no uniform agreement on the definitions of Arccsc x, Arcsec x, Arccot x for negative values of x.

Fundamental Identities Involving Principal Values

$$\mathrm{Arcsin}\, x + \mathrm{Arccos}\, x = \pi/2$$
$$\mathrm{Arctan}\, x + \mathrm{Arccot}\, x = \pi/2$$

If $\alpha = \mathrm{Arcsin}\, x$, then

$$\sin\alpha = x, \qquad \cos\alpha = \sqrt{1-x^2}, \qquad \tan\alpha = \frac{x}{\sqrt{1-x^2}}$$

$$\csc\alpha = \frac{1}{x}, \qquad \sec\alpha = \frac{1}{\sqrt{1-x^2}}, \qquad \cot\alpha = \frac{\sqrt{1-x^2}}{x}$$

If $\alpha = \text{Arccos}\, x$, then

$$\sin\alpha = \sqrt{1-x^2}, \qquad \cos\alpha = x, \qquad \tan\alpha = \frac{\sqrt{1-x^2}}{x}$$

$$\csc\alpha = \frac{1}{\sqrt{1-x^2}}, \qquad \sec\alpha = \frac{1}{x}, \qquad \cot\alpha = \frac{x}{\sqrt{1-x^2}}$$

If $\alpha = \text{Arctan}\, x$, then

$$\sin\alpha = \frac{x}{\sqrt{1+x^2}}, \qquad \cos\alpha = \frac{1}{\sqrt{1+x^2}}, \qquad \tan\alpha = x$$

$$\csc\alpha = \frac{\sqrt{1+x^2}}{x}, \qquad \sec\alpha = \sqrt{1+x^2}, \qquad \cot\alpha = \frac{1}{x}$$

Relations Between Principal Values of Inverse Trigonometric Functions

The following additional relations between principal values of the inverse trigonometric functions are useful:

$$\begin{aligned}
\text{Arcsin}\, x &= \text{Arccos}(-x) - \frac{\pi}{2} = \frac{\pi}{2} - \text{Arccos}\, x \\
&= -\,\text{Arcsin}(-x) = \text{Arctan}\, \frac{x}{\sqrt{1-x^2}} \\
&= \frac{\pi}{2} - \text{Arccot}\, \frac{x}{\sqrt{1-x^2}}
\end{aligned}$$

$$\begin{aligned}
\text{Arctan}\, x &= \text{Arccot}(-x) - \frac{\pi}{2} = \frac{\pi}{2} - \text{Arccot}\, x \\
&= -\,\text{Arctan}(-x) = \text{Arcsin}\, \frac{x}{\sqrt{x^2+1}} \\
&= \frac{\pi}{2} - \text{Arccos}\, \frac{x}{\sqrt{x^2+1}}
\end{aligned}$$

$$\begin{aligned}
\text{Arccos}\, x &= \frac{\pi}{2} + \text{Arcsin}(-x) = \frac{\pi}{2} - \text{Arcsin}\, x \\
&= \pi - \text{Arccos}(-x) = \text{Arccot}\, \frac{x}{\sqrt{1-x^2}} \\
&= \frac{\pi}{2} - \text{Arctan}\, \frac{x}{\sqrt{1-x^2}}
\end{aligned}$$

$$\begin{aligned}
\text{Arccot}\, x &= \frac{\pi}{2} + \text{Arctan}(-x) = \frac{\pi}{2} - \text{Arctan}\, x \\
&= \pi - \text{Arccot}(-x) = \text{Arccos}\, \frac{x}{\sqrt{x^2+1}} \\
&= \frac{\pi}{2} - \text{Arcsin}\, \frac{x}{\sqrt{x^2+1}}
\end{aligned}$$

$$\text{Arccsc}\, x = \text{Arcsec}\, \frac{x}{\sqrt{x^2 - 1}}$$

$$\text{Arcsec}\, x = \text{Arccsc}\, \frac{x}{\sqrt{x^2 - 1}}$$

$$\cos(\text{Arcsin}\, x) = \sin(\text{Arccos}\, x) = \sqrt{1 - x^2}$$

$$\sec(\text{Arctan}\, x) = \csc(\text{Arccot}\, x) = \sqrt{x^2 + 1}$$

$$\tan(\text{Arcsec}\, x) = \cot(\text{Arccsc}\, x) = \sqrt{x^2 - 1}$$

The above are valid for x positive or negative. The following are valid only for $x \geq 0$.

$$\text{Arcsin}\, x = \text{Arccos}\, \sqrt{1 - x^2} = \text{Arccot}\, \frac{\sqrt{1 - x^2}}{x}$$

$$= \text{Arcsec}\, \frac{1}{\sqrt{1 - x^2}} = \text{Arccsc}\, \frac{1}{x}$$

$$= \frac{\pi}{2} - \text{Arcsin}\, \sqrt{1 - x^2}$$

$$\text{Arccos}\, x = \text{Arcsin}\, \sqrt{1 - x^2} = \text{Arctan}\, \frac{\sqrt{1 - x^2}}{x}$$

$$= \text{Arcsec}\, \frac{1}{x} = \text{Arccsc}\, \frac{1}{\sqrt{1 - x^2}}$$

$$= \frac{\pi}{2} - \text{Arccos}\, \sqrt{1 - x^2}$$

$$\text{Arctan}\, x = \text{Arccot}\, \frac{1}{x} = \text{Arccos}\, \frac{1}{\sqrt{x^2 + 1}}$$

$$= \text{Arcsec}\, \sqrt{x^2 + 1} = \text{Arccsc}\, \frac{\sqrt{x^2 + 1}}{x}$$

$$= \frac{\pi}{2} - \text{Arctan}\, \frac{1}{x}$$

$$\text{Arccot}\, x = \text{Arctan}\, \frac{1}{x} = \text{Arcsin}\, \frac{1}{\sqrt{x^2 + 1}}$$

$$= \text{Arcsec}\, \frac{\sqrt{x^2 + 1}}{x} = \text{Arccsc}\, \sqrt{x^2 + 1}$$

$$= \frac{\pi}{2} - \text{Arccot}\, \frac{1}{x}$$

$$\text{Arcsec}\, x = \text{Arctan}\, \sqrt{x^2 - 1} = \text{Arccot}\, \frac{1}{\sqrt{x^2 - 1}}$$

$$= \text{Arcsin}\, \frac{\sqrt{x^2 - 1}}{x} = \text{Arccos}\, \frac{1}{x} = \pi + \text{Arcsec}(-x)$$

$$= \frac{\pi}{2} - \text{Arccsc}\, x = -\frac{\pi}{2} - \text{Arccsc}(-x)$$

$$\operatorname{Arccsc} x = \operatorname{Arctan} \frac{1}{\sqrt{x^2-1}} = \operatorname{Arccot} \sqrt{x^2-1}$$

$$= \operatorname{Arcsin} \frac{1}{x} = \operatorname{Arccos} \frac{\sqrt{x^2-1}}{x} = \pi + \operatorname{Arccsc}(-x)$$

$$= \frac{\pi}{2} - \operatorname{Arcsec} x = -\frac{\pi}{2} - \operatorname{Arcsec}(-x)$$

The following are valid if $x < 0$.

$$\operatorname{Arcsin} x = -\operatorname{Arccos} \sqrt{1-x^2} = \operatorname{Arccot} \frac{\sqrt{1-x^2}}{x} - \pi$$

$$= -\pi - \operatorname{Arccsc} \frac{1}{x} = -\operatorname{Arcsec} \frac{1}{\sqrt{1-x^2}}$$

$$= \operatorname{Arcsin} \sqrt{1-x^2} - \frac{\pi}{2}$$

$$\operatorname{Arccos} x = \pi - \operatorname{Arcsin} \sqrt{1-x^2} = \pi + \operatorname{Arctan} \frac{\sqrt{1-x^2}}{x}$$

$$= \pi - \operatorname{Arccsc} \frac{1}{\sqrt{1-x^2}} = -\operatorname{Arcsec} \frac{1}{x}$$

$$= \operatorname{Arccos} \sqrt{1-x^2} + \frac{\pi}{2}$$

$$\operatorname{Arctan} x = \operatorname{Arccot} \frac{1}{x} - \pi = -\operatorname{Arccos} \frac{1}{\sqrt{x^2+1}}$$

$$= -\operatorname{Arcsec} \sqrt{x^2+1} = -\pi - \operatorname{Arccsc} \frac{\sqrt{x^2+1}}{x}$$

$$= -\frac{\pi}{2} - \operatorname{Arctan} \frac{1}{x}$$

$$\operatorname{Arccot} x = \pi + \operatorname{Arctan} \frac{1}{x} = \pi - \operatorname{Arcsin} \frac{1}{\sqrt{x^2+1}}$$

$$= -\operatorname{Arcsec} \frac{\sqrt{x^2+1}}{x} = \pi - \operatorname{Arccsc} \sqrt{x^2+1}$$

$$= \frac{3\pi}{2} - \operatorname{Arccot} \frac{1}{x}$$

$$\operatorname{Arcsec} x = \operatorname{Arctan} \sqrt{x^2-1} - \pi = \operatorname{Arccot} \frac{1}{\sqrt{x^2-1}} - \pi$$

$$= -\pi - \operatorname{Arcsin} \frac{\sqrt{x^2-1}}{x} = -\operatorname{Arccos} \frac{1}{x} = \operatorname{Arcsec}(-x) - \pi$$

$$= -\frac{3\pi}{2} - \operatorname{Arccsc} x = -\frac{\pi}{2} - \operatorname{Arccsc}(-x)$$

$$\operatorname{Arccsc} x = \operatorname{Arctan} \frac{1}{\sqrt{x^2-1}} - \pi = \operatorname{Arccot} \sqrt{x^2-1} - \pi$$

$$= -\pi - \operatorname{Arcsin} \frac{1}{x} = -\operatorname{Arccos} \frac{\sqrt{x^2-1}}{x} = \operatorname{Arccsc}(-x) - \pi$$

$$= -\frac{3\pi}{2} - \operatorname{Arcsec} x = -\frac{\pi}{2} - \operatorname{Arcsec}(-x)$$

Plane Triangle Formulae

In the following, A, B, C denote the angles of any plane triangle, a, b, c the corresponding opposite sides, and $s = \frac{1}{2}(a + b + c)$.

Radius of inscribed circle:

$$r = \sqrt{\frac{(s-a)(s-b)(s-c)}{s}}$$

Radius of circumscribed circle:

$$R = \frac{a}{2\sin A} = \frac{b}{2\sin B} = \frac{c}{2\sin C}$$

Law of sines:

$$\frac{a}{\sin A} = \frac{b}{\sin B} = \frac{c}{\sin C}$$

Law of cosines:

$$a^2 = b^2 + c^2 - 2bc\cos A, \qquad \cos A = \frac{b^2 + c^2 - a^2}{2bc}$$

$$b^2 = c^2 + a^2 - 2ca\cos B, \qquad \cos B = \frac{c^2 + a^2 - b^2}{2ca}$$

$$c^2 = a^2 + b^2 - 2ab\cos C, \qquad \cos C = \frac{a^2 + b^2 - c^2}{2ab}$$

Law of tangents:

$$\frac{b-c}{b+c} = \frac{\tan\frac{1}{2}(B-C)}{\tan\frac{1}{2}(B+C)}, \qquad \frac{c-a}{c+a} = \frac{\tan\frac{1}{2}(C-A)}{\tan\frac{1}{2}(C+A)}$$

$$\frac{a-b}{a+b} = \frac{\tan\frac{1}{2}(A-B)}{\tan\frac{1}{2}(A+B)}$$

Half-angle formulae:

$$\tan\tfrac{1}{2}A = \frac{r}{s-a}, \qquad \tan\tfrac{1}{2}B = \frac{r}{s-b}, \qquad \tan\tfrac{1}{2}C = \frac{r}{s-c}$$

$$\sin\tfrac{1}{2}A = \sqrt{\frac{(s-b)(s-c)}{bc}}, \qquad \cos\tfrac{1}{2}A = \sqrt{\frac{s(s-a)}{bc}}$$

$$\sin\tfrac{1}{2}B = \sqrt{\frac{(s-c)(s-a)}{ca}}, \qquad \cos\tfrac{1}{2}B = \sqrt{\frac{s(s-b)}{ca}}$$

$$\sin\tfrac{1}{2}C = \sqrt{\frac{(s-a)(s-b)}{ab}}, \qquad \cos\tfrac{1}{2}C = \sqrt{\frac{s(s-c)}{ab}}$$

Area:

$$K = \tfrac{1}{2}bc \sin A = \tfrac{1}{2}ca \sin B = \tfrac{1}{2}ab \sin C$$

$$K = \frac{a^2 \sin B \sin C}{2 \sin A} = \frac{b^2 \sin C \sin A}{2 \sin B} = \frac{c^2 \sin A \sin B}{2 \sin C}$$

$$K = \sqrt{s(s-a)(s-b)(s-c)} = rs = \frac{abc}{4R}$$

Mollweide's formulae:

$$\frac{b-c}{a} = \frac{\sin \frac{1}{2}(B-C)}{\cos \frac{1}{2}A}, \qquad \frac{c-a}{b} = \frac{\sin \frac{1}{2}(C-A)}{\cos \frac{1}{2}B}$$

$$\frac{a-b}{c} = \frac{\sin \frac{1}{2}(A-B)}{\cos \frac{1}{2}C}$$

Newton's formulae:

$$\frac{b+c}{a} = \frac{\cos \frac{1}{2}(B-C)}{\sin \frac{1}{2}A}, \qquad \frac{c+a}{b} = \frac{\cos \frac{1}{2}(C-A)}{\sin \frac{1}{2}B}$$

$$\frac{a+b}{c} = \frac{\cos \frac{1}{2}(A-B)}{\sin \frac{1}{2}C}$$

Solution of Right Triangles

(**a**) Given acute angle A and opposite leg a.

$$B = 90° - A, \quad b = a/\tan A = a \cot A, \quad c = a/\sin A = a \csc A$$

(**b**) Given acute angle A and adjacent leg b.

$$B = 90° - A, \quad a = b \tan A, \quad c = b/\cos A = b \sec A$$

(**c**) Given acute angle A and hypotenuse c.

$$B = 90° - A, \quad a = c \sin A, \quad b = c \cos A$$

(**d**) Given legs a and b.

$$c = \sqrt{a^2 + b^2}, \quad \tan A = a/b, \quad B = 90° - A$$

(**e**) Given hypotenuse c and leg a.

$$b = \sqrt{(c+a)(c-a)}, \quad \sin A = a/c, \quad B = 90° - A$$

Solution of Oblique Triangles

(**a**) Given sides b and c and included angle A.

Nonlogarithmic solution

$$a^2 = b^2 + c^2 - 2bc \cos A, \quad \cos B = (c^2 + a^2 - b^2)/2ca,$$
$$\cos C = (a^2 + b^2 - c^2)/2ab$$

Logarithmic solution

$$\tfrac{1}{2}(B + C) = 90° - \tfrac{1}{2}A, \quad \tan \tfrac{1}{2}(B - C) = \frac{b-c}{b+c} \tan \tfrac{1}{2}(B + C),$$
$$B = \tfrac{1}{2}(B + C) + \tfrac{1}{2}(B - C), \quad C = \tfrac{1}{2}(B + C) - \tfrac{1}{2}(B - C),$$
$$a = (b \sin A)/\sin B, \quad K = \tfrac{1}{2}bc \sin A$$

Check. $A + B + C = 180°$, or use Newton's formula or law of sines.

(b) Given angles B and C and included side a.

$$A = 180° - (B + C), \quad b = (a \sin B)/\sin A,$$
$$c = (a \sin C)/\sin A, \quad K = \frac{a^2 \sin B \sin C}{2 \sin A}$$

Check. $a = b \cos C + c \cos B$, or use Newton's formula or law of tangents.

(c) Given sides a and c and opposite angle A.

$$\sin C = (c \sin A)/a, \quad B = 180° - (A + C),$$
$$b = (a \sin B)/\sin A, \quad K = \tfrac{1}{2}ac \sin B$$

Check. $a = b \cos C + c \cos B$, or use Newton's formula or law of tangents.
Note. In this case there may be two solutions, for C may have two values:

$C_1 < 90°$ and $C_2 = 180° - C_1 > 90°$. If $A + C_2 > 180°$, use only C_1.

(d) Given the three sides a, b, c.

Nonlogarithmic solution

$$\cos A = (b^2 + c^2 - a^2)/2bc, \quad \cos B = (c^2 + a^2 - b^2)/2ca,$$
$$\cos C = (a^2 + b^2 - c^2)/2ab$$

Logarithmic solution

$$s = \tfrac{1}{2}(a + b + c), \quad r = \sqrt{\frac{(s - a)(s - b)(s - c)}{s}},$$
$$\tan \tfrac{1}{2}A = \frac{r}{s - a}, \quad \tan \tfrac{1}{2}B = \frac{r}{s - b}, \quad \tan \tfrac{1}{2}C = \frac{r}{s - c},$$
$$K = \sqrt{s(s - a)(s - b)(s - c)}$$

Check. $A + B + C = 180°$.

Relations Between Accuracy of Computed Lengths and Angles

When solving a triangle for any of its parts, the following should be observed:

Significant figures for sides	Angles to the nearest
2	degree
3	ten minutes
4	minute
5	tenth of a minute

SPHERICAL TRIGONOMETRY

Right Spherical Triangles

Let a, b, c be the sides of a right spherical triangle with opposite angles $A, B, C = 90°$, respectively, where each side is measured by the angle subtended at the center of the sphere.

$$\begin{aligned}
\sin a &= \tan b \cot B, & \sin a &= \sin A \sin c \\
\sin b &= \tan a \cot A, & \sin b &= \sin B \sin c \\
\cos A &= \tan b \cot c, & \cos A &= \cos a \sin B \\
\cos B &= \tan a \cot c, & \cos B &= \cos b \sin A \\
\cos c &= \cot A \cot B, & \cos c &= \cos a \cos b
\end{aligned}$$

Napier's Rules of Circular Parts

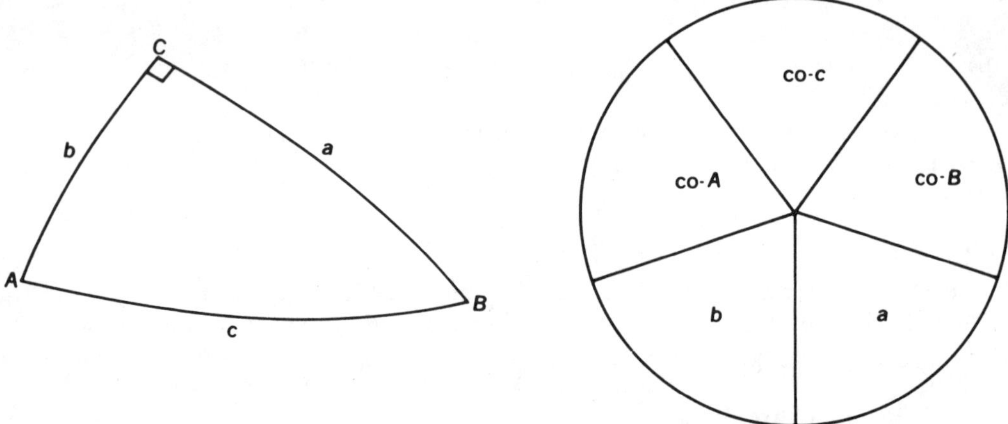

Arrange the five quantities a, b, co-A (complement of A), co-c, co-B of a right spherical triangle right-angled at C in cyclic order as pictured. If any one of these quantities is designated a *middle* part, then two of the other parts are *adjacent* to it, and the remaining two parts are *opposite* to it. The above formulae for a right spherical triangle may be recalled by the following two rules:

(a) The sine of any middle part is equal to the product of the *tangents* of the two *adjacent* parts.

(b) The sine of any middle part is equal to the product of the *cosines* of the two *opposite* parts.

Rules for Determining the Quadrant of a Calculated Part of a Right Spherical Triangle

(a) A leg and the angle opposite it are always of the same quadrant.
(b) If the hypotenuse if less than 90°, the legs are of the same quadrant.
(c) If the hypotenuse is greater than 90°, the legs are of unlike quadrants.

Oblique Spherical Triangles

In the following, a, b, c represent the sides of any spherical triangle, A, B, C the corresponding opposite angles, $s = \frac{1}{2}(a + b + c)$, $S = \frac{1}{2}(A + B + C)$, Δ = area of triangle, E = spherical excess of triangle, R = radius of the sphere upon which the triangle lies, and a', b', c', A', B', C' are the corresponding parts of the polar triangle.

$$0° < a + b + c < 360°, \qquad 180° < A + B + C < 540°$$
$$E = A + B + C - 180°, \qquad \Delta = \pi R^2 E/180$$
$$\tan \tfrac{1}{4}E = \sqrt{\tan \tfrac{s}{2} \tan \tfrac{1}{2}(s - a) \tan \tfrac{1}{2}(s - b) \tan \tfrac{1}{2}(s - c)}$$
$$A = 180° - a', \qquad B = 180° - b', \qquad C = 180° - c'$$
$$a = 180° - A', \qquad b = 180° - B', \qquad c = 180° - C'$$

Law of sines:

$$\frac{\sin a}{\sin A} = \frac{\sin b}{\sin B} = \frac{\sin c}{\sin C}$$

Law of cosines for sides:

$$\cos a = \cos b \cos c + \sin b \sin c \cos A$$
$$\cos b = \cos c \cos a + \sin c \sin a \cos B$$
$$\cos c = \cos a \cos b + \sin a \sin b \cos C$$

Law of cosines for angles:

$$\cos A = -\cos B \cos C + \sin B \sin C \cos a$$
$$\cos B = -\cos C \cos A + \sin C \sin A \cos b$$
$$\cos C = -\cos A \cos B + \sin A \sin B \cos c$$

Law of tangents:

$$\frac{\tan \frac{1}{2}(B - C)}{\tan \frac{1}{2}(B + C)} = \frac{\tan \frac{1}{2}(b - c)}{\tan \frac{1}{2}(b + c)}, \quad \frac{\tan \frac{1}{2}(C - A)}{\tan \frac{1}{2}(C + A)} = \frac{\tan \frac{1}{2}(c - a)}{\tan \frac{1}{2}(c + a)}$$
$$\frac{\tan \frac{1}{2}(A - B)}{\tan \frac{1}{2}(A + B)} = \frac{\tan \frac{1}{2}(a - b)}{\tan \frac{1}{2}(a + b)}$$

Half-angle formulae:

$$\tan \tfrac{1}{2}A = \frac{k}{\sin(s - a)}, \quad \tan \tfrac{1}{2}B = \frac{k}{\sin(s - b)}, \quad \tan \tfrac{1}{2}C = \frac{k}{\sin(s - c)},$$

where

$$k^2 = \frac{\sin(s - a)\sin(s - b)\sin(s - c)}{\sin s}.$$

Half-side formulae:

$$\tan \tfrac{1}{2}\alpha = K \cos(S - A), \quad \tan \tfrac{1}{2}b = K \cos(S - B),$$
$$\tan \tfrac{1}{2}c = K \cos(S - C),$$

where

$$K^2 = \frac{-\cos S}{\cos(S - A)\cos(S - B)\cos(S - C)}.$$

Gauss's formulae:

$$\frac{\sin \frac{1}{2}(a - b)}{\sin \frac{1}{2}c} = \frac{\sin \frac{1}{2}(A - B)}{\cos \frac{1}{2}C}, \quad \frac{\cos \frac{1}{2}(a - b)}{\cos \frac{1}{2}c} = \frac{\sin \frac{1}{2}(A + B)}{\cos \frac{1}{2}C}$$
$$\frac{\sin \frac{1}{2}(a + b)}{\sin \frac{1}{2}c} = \frac{\cos \frac{1}{2}(A - B)}{\sin \frac{1}{2}C}, \quad \frac{\cos \frac{1}{2}(a + b)}{\cos \frac{1}{2}c} = \frac{\cos \frac{1}{2}(A + B)}{\sin \frac{1}{2}C}$$

Napier's analogies:

$$\frac{\sin \frac{1}{2}(A - B)}{\sin \frac{1}{2}(A + B)} = \frac{\tan \frac{1}{2}(a - b)}{\tan \frac{1}{2}c}, \quad \frac{\sin \frac{1}{2}(a - b)}{\sin \frac{1}{2}(a + b)} = \frac{\tan \frac{1}{2}(A - B)}{\cot \frac{1}{2}C}$$
$$\frac{\cos \frac{1}{2}(A - B)}{\cos \frac{1}{2}(A + B)} = \frac{\tan \frac{1}{2}(a + b)}{\tan \frac{1}{2}c}, \quad \frac{\cos \frac{1}{2}(a - b)}{\cos \frac{1}{2}(a + b)} = \frac{\tan \frac{1}{2}(A + B)}{\cot \frac{1}{2}C}$$

Haversine formulae:

$$\text{hav}\, a = \text{hav}(b - c) + \sin b \sin c \,\text{hav}\, A$$
$$\text{hav}\, A = \frac{\sin(s - b)\sin(s - c)}{\sin b \sin c}$$
$$= \frac{\text{hav}\, a - \text{hav}(b - c)}{\sin b \sin c}$$
$$= \text{hav}[180° - (B + C)] + \sin B \sin C \,\text{hav}\, a$$

Rules for Determining the Quadrant of a Calculated Part of an Oblique Spherical Triangle

(a) If $A > B > C$, then $a > b > c$.

(b) A side (angle) which differs more from 90° than does another side (angle) is in the same quadrant as its opposite angle (side).

(c) Half the sum of any two sides and half the sum of the opposite angles are in the same quadrant.

SUMMARY OF SOLUTION OF OBLIQUE SPHERICAL TRIANGLES

Given	Solution	Check
Three sides	Half-angle formulae.	Law of sines
Three angles	Half-side formulae.	Law of sines
Two sides and included angle	Napier's analogies (to find sum and difference of unknown angles), then law of sines (to find remaining side).	Gauss's formula
Two angles and included side	Napier's analogies (to find sum and difference of unknown sides), then law of sines (to find remaining angle).	Gauss's formula
Two sides and an opposite angle	Law of sines (to find an angle), then Napier's analogies (to find remaining angle and side). Note number of solutions.	Gauss's formula
Two angles and an opposite side	Law of sines (to find a side), then Napier's analogies (to find remaining side and angle). Note number of solutions.	Gauss's formula

HYPERBOLIC FUNCTIONS

Geometrical Definition

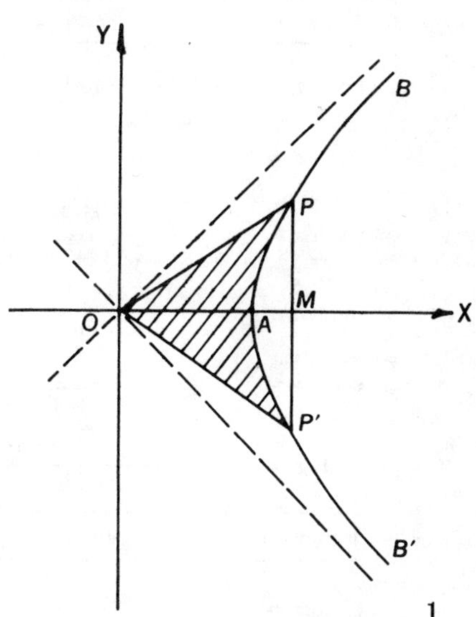

Let O be the center, A the vertex, and P any point of the branch $B'AB$ of a rectangular hyperbola. Set $OM = x$, $MP = y$, $OA = a$, and

$$u = \frac{\text{area } OPAP'}{a^2}.$$

Then

hyperbolic sine of $u = \sinh u = y/a$,
hyperbolic cosine of $u = \cosh u = x/a$.

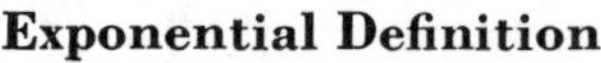

Exponential Definition

hyperbolic sine of $u = \sinh u = \frac{1}{2}(e^u - e^{-u})$ $\qquad \operatorname{csch} u = \dfrac{1}{\sinh u}$,

hyperbolic cosine of $u = \cosh u = \frac{1}{2}(e^u + e^{-u})$ = cos(iu) $\qquad \operatorname{sech} u = \dfrac{1}{\cosh u}$,

hyperbolic tangent of $u = \tanh u = \dfrac{\sinh u}{\cosh u} = \dfrac{e^u - e^{-u}}{e^u + e^{-u}}$ $\qquad \coth u = \dfrac{1}{\tanh u}$

Domain and Range for Real Argument

Function	Domain (interval of u)	Range (interval of function)	Remarks
$\sinh u$	$(-\infty, +\infty)$	$(-\infty, +\infty)$	
$\cosh u$	$(-\infty, +\infty)$	$[1, +\infty)$	
$\tanh u$	$(-\infty, +\infty)$	$(-1, +1)$	
cosech u	$(-\infty, 0)$	$(0, -\infty)$	Two branches, pole
	$(0, +\infty)$	$(+\infty, 0)$	at $u = 0$.
sech u	$(-\infty, +\infty)$	$(0, 1]$	
$\coth u$	$(-\infty, 0)$	$(-1, -\infty)$	Two branches, pole
	$(0, +\infty)$	$(+\infty, 1)$	at $u = 0$

Hyperbolic Functions In Terms of One Another

Function	sinh x	cosh x	tanh x
sinh x =	sinh x	$\pm\sqrt{\cosh^2 x - 1}$	$\dfrac{\tanh x}{\sqrt{1-\tanh^2 x}}$
cosh x =	$\sqrt{1+\sinh^2 x}$	cosh x	$\dfrac{1}{\sqrt{1-\tanh^2 x}}$
tanh x =	$\dfrac{\sinh x}{\sqrt{1+\sinh^2 x}}$	$\pm\dfrac{\sqrt{\cosh^2 x - 1}}{\cosh x}$	tanh x
cosech x =	$\dfrac{1}{\sinh x}$	$\pm\dfrac{1}{\sqrt{\cosh^2 x - 1}}$	$\dfrac{\sqrt{1-\tanh^2 x}}{\tanh x}$
sech x =	$\dfrac{1}{\sqrt{1+\sinh^2 x}}$	$\dfrac{1}{\cosh x}$	$\sqrt{1-\tanh^2 x}$
coth x =	$\dfrac{\sqrt{1+\sinh^2 x}}{\sinh x}$	$\dfrac{\pm\cosh x}{\sqrt{\cosh^2 x - 1}}$	$\dfrac{1}{\tanh x}$

Function	cosech x	sech x	coth x
sinh x =	$\dfrac{1}{\operatorname{cosech} x}$	$\pm\dfrac{\sqrt{1-\operatorname{sech}^2 x}}{\operatorname{sech} x}$	$\dfrac{\pm 1}{\sqrt{\coth^2 x - 1}}$
cosh x =	$\pm\dfrac{\sqrt{\operatorname{cosech}^2 x + 1}}{\operatorname{cosech} x}$	$\dfrac{1}{\operatorname{sech} x}$	$\pm\dfrac{\pm\coth x}{\sqrt{\coth^2 x - 1}}$
tanh x =	$\dfrac{1}{\sqrt{\operatorname{cosech}^2 x + 1}}$	$\pm\sqrt{1-\operatorname{sech}^2 x}$	$\dfrac{1}{\coth x}$
cosech x =	cosech x	$\pm\dfrac{\operatorname{sech} x}{\sqrt{1-\operatorname{sech}^2 x}}$	$\pm\dfrac{\sqrt{\coth^2 x - 1}}{1}$
sech x =	$\pm\dfrac{\operatorname{cosec} x}{\sqrt{\operatorname{cosech}^2 x + 1}}$	sech x	$\pm\dfrac{\sqrt{\coth^2 x - 1}}{\coth x}$
coth x =	$\sqrt{\operatorname{cosech}^2 x + 1}$	$\pm\dfrac{1}{\sqrt{1-\operatorname{sech}^2 x}}$	coth x

Whenever two signs are shown, choose + sign if x is positive, − sign if x is negative.

Special Values of Hyperbolic Functions

x	0	$\frac{\pi}{2}i$	πi	$\frac{3\pi}{2}i$	∞
sinh x	0	i	0	$-i$	∞
cosh x	1	0	-1	0	∞
tanh x	0	∞i	0	$-\infty i$	1
csch x	∞	$-i$	∞	i	0
sech x	1	∞	-1	∞	0
coth x	∞	0	∞	0	1

Symmetry and Periodicity

$$\sinh(-u) = -\sinh u, \qquad \operatorname{csch}(-u) = -\operatorname{csch} u$$
$$\cosh(-u) = \cosh u, \qquad \operatorname{sech}(-u) = \operatorname{sech} u$$
$$\tanh(-u) = -\tanh u, \qquad \coth(-u) = -\coth u$$

sinh u, cosh u, cosech u and sech u are periodic with a period $2\pi i$; tanh u and coth u are periodic with a period πi.

Fundamental Identities

Reciprocal Relations

$$\operatorname{csch} u = \frac{1}{\sinh u}, \qquad \operatorname{sech} u = \frac{1}{\cosh u}, \qquad \coth u = \frac{1}{\tanh u}$$

Product Relations

$$\begin{aligned} \sinh u &= \tanh u \cosh u & \cosh u &= \coth u \sinh u \\ \tanh u &= \sinh u \operatorname{sech} u & \coth u &= \cosh u \operatorname{cosech} u \\ \operatorname{sech} u &= \operatorname{cosech} u \tanh u & \operatorname{cosech} u &= \operatorname{sech} u \coth u \end{aligned}$$

Quotient Relations

$$\sinh u = \frac{\tanh u}{\operatorname{sech} u} \qquad \cosh u = \frac{\coth u}{\operatorname{cosech} u} \qquad \tanh u = \frac{\sinh u}{\cosh u}$$

$$\operatorname{cosech} u = \frac{\operatorname{sech} u}{\tanh u} \qquad \operatorname{sech} u = \frac{\operatorname{cosech} u}{\coth u} \qquad \coth u = \frac{\cosh u}{\sinh u}$$

Relations Between Squares of Functions

$$\begin{aligned} \cosh^2 u - \sinh^2 u &= 1, & \tanh^2 u + \operatorname{sech}^2 u &= 1 \\ \coth^2 u - \operatorname{csch}^2 u &= 1, & \operatorname{csch}^2 u - \operatorname{sech}^2 u &= \operatorname{csch}^2 u \operatorname{sech}^2 u \end{aligned}$$

Angle-Sum and Angle-Difference Relations

$$\begin{aligned} \sinh (u + v) &= \sinh u \cosh v + \cosh u \sinh v \\ \sinh (u - v) &= \sinh u \cosh v - \cosh u \sinh v \\ \cosh (u + v) &= \cosh u \cosh v + \sinh u \sinh v \\ \cosh (u - v) &= \cosh u \cosh v - \sinh u \sinh v \end{aligned}$$

$$\tanh (u + v) = \frac{\tanh u + \tanh v}{1 + \tanh u \tanh v} = \frac{\sinh 2u + \sinh 2v}{\cosh 2u + \cosh 2v}$$

$$\tanh (u - v) = \frac{\tanh u - \tanh v}{1 - \tanh u \tanh v} = \frac{\sinh 2u - \sinh 2v}{\cosh 2u - \cosh 2v}$$

$$\coth (u + v) = \frac{1 + \coth u \coth v}{\coth u + \coth v} = \frac{\sinh 2u - \sinh 2v}{\cosh 2u - \cosh 2v}$$

$$\coth (u - v) = \frac{1 - \coth u \coth v}{\coth u - \coth v} = \frac{\sinh 2u + \sinh 2v}{\cosh 2u - \cosh 2v}$$

Multiple Angle Relations

$$\sinh 2u = 2 \sinh u \cosh u = \frac{2 \tanh u}{1 - \tanh^2 u}$$

$$\cosh 2u = \cosh^2 u + \sinh^2 u = 2 \cosh^2 u - 1 = 1 + 2 \sinh^2 u = \frac{1 + \tanh^2 u}{1 - \tanh^2 u}$$

$$\tanh 2u = \frac{2 \tanh u}{1 + \tanh^2 u}$$

$$\coth 2u = \frac{\coth^2 u + 1}{2 \coth u}$$

$$\begin{aligned} \sinh 3u &= 3 \sinh u + 4 \sinh^3 u = \sinh u \,(4 \cosh^2 u - 1) \\ \cosh 3u &= 4 \cosh^3 u - 3 \cosh u = \cosh u \,(1 + 4 \sinh^2 u) \end{aligned}$$

$$\tanh 3u = \frac{3 \tanh u + \tanh^3 u}{1 + 3 \tanh^2 u}$$

$$\coth 3u = \frac{3 \coth u + \coth^3 u}{1 + 3 \coth^2 u}$$

$$\begin{aligned}
\sinh 4u &= 4 \sinh u \cosh u(2 \cosh^2 u - 1) \\
&= 4 \sinh u \cosh u(1 + 2 \sinh^2 u) \\
&= 4 \sinh u \cosh u(\cosh^2 u + \sinh^2 u) \\
\cosh 4u &= 1 + 8 \cosh^2 u(\cosh^2 u - 1) \\
&= 1 + 8 \sinh^2 u(\sinh^2 u + 1) \\
&= \cosh^4 u + 6 \sinh^2 u \cosh^2 u + \sinh^4 u
\end{aligned}$$

$$\tanh 4u = \frac{4 \tanh u(1 + \tanh^2 u)}{1 + 6 \tanh^2 u + \tanh^4 u}$$

$$\coth 4u = \frac{\coth^4 u + 6 \coth^2 u + 1}{4 \coth u(\coth^2 u + 1)}$$

$$\begin{aligned}
\sinh 5u &= \sinh u(16 \sinh^4 u + 20 \sinh^2 u + 5) \\
&= \sinh u(16 \cosh^4 u - 12 \cosh^2 u + 1) \\
\cosh 5u &= \cosh u(16 \cosh^4 u - 20 \cosh^2 u + 5) \\
&= \cosh u(16 \sinh^4 u + 12 \sinh^2 u + 1) \\
\sinh 6u &= 2 \sinh u \cosh u(16 \cosh^4 u - 16 \cosh^2 u + 3) \\
&= 2 \sinh u \cosh u(16 \sinh^4 u + 16 \sinh^2 u + 3) \\
\cosh 6u &= 32 \cosh^6 u - 48 \cosh^4 u + 18 \cosh^2 u - 1 \\
&= 32 \sinh^6 u + 48 \sinh^4 u + 18 \sinh^2 u + 1
\end{aligned}$$

$$\begin{aligned}
\sinh nu = \sinh u \Big[&(2 \cosh u)^{n-1} - \frac{(n-2)}{1!} \cdot (2 \cosh u)^{n-3} + \frac{(n-3)(n-4)}{2!} \\
&\cdot (2 \cosh u)^{n-5} - \frac{(n-4)(n-5)(n-6)}{3!} (2 \cosh u)^{n-7} + \cdots \Big].
\end{aligned}$$

$$\begin{aligned}
\cosh nu = \frac{1}{2} \Big[&(2 \cosh u)^n - \frac{n}{1!} (2 \cosh u)^{n-2} + \frac{n(n-3)}{2!} (2 \cosh u)^{n-4} \\
&- \frac{n(n-4)(n-5)}{3!} (2 \cosh u)^{n-6} + \cdots \Big].
\end{aligned}$$

Half Angle Relations

$$\begin{aligned}
\sinh \tfrac{1}{2}u &= \pm \sqrt{\tfrac{1}{2}(\cosh u - 1)} \\
\cosh \tfrac{1}{2}u &= \sqrt{\tfrac{1}{2}(\cosh u + 1)}
\end{aligned}$$

$$\tanh \frac{u}{2} = \frac{\sinh u}{1 + \cosh u} = \frac{\cosh u - 1}{\sinh u} = \pm \sqrt{\frac{\cosh u - 1}{\cosh u + 1}}$$

$$\coth \frac{u}{2} = \frac{1 + \cosh u}{\sinh u} = \frac{\sinh u}{\cosh u - 1} = \pm \sqrt{\frac{\cosh u + 1}{\cosh u - 1}}$$

Choose + sign if u is positive, otherwise choose the − sign.

Function Sum and Function Difference Relations

$$\begin{aligned}
\sinh u + \sinh v &= 2 \sinh \tfrac{1}{2}(u + v) \cosh \tfrac{1}{2}(u - v) \\
\sinh u - \sinh v &= 2 \cosh \tfrac{1}{2}(u + v) \sinh \tfrac{1}{2}(u - v) \\
\cosh u + \cosh v &= 2 \cosh \tfrac{1}{2}(u + v) \cosh \tfrac{1}{2}(u - v) \\
\cosh u - \cosh v &= 2 \sinh \tfrac{1}{2}(u + v) \sinh \tfrac{1}{2}(u - v)
\end{aligned}$$

$$\tanh u + \tanh v = (1 + \tanh u \tanh v) \tanh (u + v) = \frac{\sinh (u + v)}{\cosh u \cosh v}$$

$$\tanh u - \tanh v = (1 - \tanh u \tanh v) \tanh (u - v) = \frac{\sinh (u - v)}{\cosh u \cosh v}$$

$$\coth u + \coth v = \frac{1 + \coth u \coth v}{\coth(u+v)} = \frac{\sinh(u+v)}{\sinh u \sinh v}$$

$$\coth u - \coth v = \frac{1 - \coth u \coth v}{\coth(u-v)} = \frac{\sinh(u-v)}{\sinh u \sinh v}$$

$$\sinh u + \cosh u = \frac{1 + \tanh \frac{1}{2}u}{1 - \tanh \frac{1}{2}u} = e^u \qquad \cosh u - \sinh u = \frac{1 - \tanh \frac{1}{2}u}{1 + \tanh \frac{1}{2}u} = e^{-u}$$

Function Product Relations

$$\sinh u \cosh v = \tfrac{1}{2}\sinh(u+v) + \tfrac{1}{2}\sinh(u-v)$$
$$\cosh u \sinh v = \tfrac{1}{2}\sinh(u+v) - \tfrac{1}{2}\sinh(u-v)$$
$$\cosh u \cosh v = \tfrac{1}{2}\cosh(u+v) + \tfrac{1}{2}\cosh(u-v)$$
$$\sinh u \sinh v = \tfrac{1}{2}\cosh(u+v) - \tfrac{1}{2}\cosh(u-v)$$
$$\sinh(u+v)\sinh(u-v) = \sinh^2 u - \sinh^2 v = \cosh^2 u - \cosh^2 v$$
$$\cosh(u+v)\cosh(u-v) = \sinh^2 u + \cosh^2 v = \cosh^2 u + \sinh^2 v$$

Power Relations

$$\sinh^2 u = \tfrac{1}{2}(\cosh 2u - 1)$$
$$\cosh^2 u = \tfrac{1}{2}(\cosh 2u + 1)$$
$$\sinh^3 u = \tfrac{1}{4}(-3\sinh u + \sinh 3u)$$
$$\cosh^3 u = \tfrac{1}{4}(3\cosh u + \cosh 3u)$$
$$\sinh^4 u = \tfrac{1}{8}(3 - 4\cosh 2u + \cosh 4u)$$
$$\cosh^4 u = \tfrac{1}{8}(3 + 4\cosh 2u + \cosh 4u)$$
$$\sinh^5 u = \tfrac{1}{16}(10\sinh u - 5\sinh 3u + \sinh 5u)$$
$$\cosh^5 u = \tfrac{1}{16}(10\cosh u + 5\cosh 3u + \cosh 5u)$$
$$\sinh^6 u = \tfrac{1}{32}(-10 + 15\cosh 2u - 6\cosh 4u + \cosh 6u)$$
$$\cosh^6 u = \tfrac{1}{32}(10 + 15\cosh 2u + 6\cosh 4u + \cosh 6u)$$
$$(\cosh u \pm \sinh u)^n = \cosh nu \pm \sinh nu$$

Relations with Circular Functions

$$\sinh iu = i\sin u, \qquad \sinh u = -i\sin iu$$
$$\cosh iu = \cos u, \qquad \cosh u = \cos iu$$
$$\tanh iu = i\tan u, \qquad \tanh u = -i\tan iu$$
$$\operatorname{cosech} iu = -i\operatorname{cosec} u \qquad \operatorname{cosech} u = i\operatorname{cosec} iu$$
$$\operatorname{sech} iu = \sec u \qquad \operatorname{sech} u = \sec iu$$
$$\coth iu = -i\cot u \qquad \coth u = i\coth iu$$

Hyperbolic Functions of Complex Argument

$$\sinh(u+iv) = \sinh u \cos v + i\cosh u \sin v$$
$$\sinh(u-iv) = \sinh u \cos v - i\cosh u \sin v$$
$$\cosh(u+iv) = \cosh u \cos v + i\sinh u \sin v$$
$$\cosh(u-iv) = \cosh u \cos v - i\sinh u \sin v$$

$$\tanh(u+iv) = \frac{\sinh 2u + i\sin 2v}{\cosh 2u + \cos 2v}$$

$$\tanh(u-iv) = \frac{\sinh 2u - i\sin 2v}{\cosh 2u + \cos 2v}$$

$$\coth(u+iv) = \frac{\sinh 2u - i\sin 2v}{\cosh 2u - \cos 2v}$$

$$\coth(u-iv) = \frac{\sinh 2u + i\sin 2v}{\cosh 2u - \cos 2v}$$

$$\sinh(u + \tfrac{1}{2}\pi i) = i\cosh u, \qquad \cosh(u + \tfrac{1}{2}\pi i) = i\sinh u$$
$$\sinh(u + \pi i) = -\sinh u, \qquad \cosh(u + \pi i) = -\cosh u$$
$$\sinh(u + 2\pi i) = \sinh u, \qquad \cosh(u + 2\pi i) = \cosh u$$

Series for Hyperbolic Functions

(see series expansions for sinh nu and cosh nu under multiple angle relations).

$$\sinh x = x + \frac{x^3}{3!} + \frac{x^5}{5!} + \frac{x^7}{7!} + \cdots + \frac{x^{2n+1}}{(2n+1)!} + \cdots \qquad |x| < \infty$$

$$\sinh ax = \frac{2}{\pi}\sinh \pi a\left[\frac{\sin x}{a^2+1^2} - \frac{2\sin 2x}{a^2+2^2} + \frac{3\sin 3x}{a^2+3^2} - + \cdots\right]$$

$$= \frac{2}{\pi}\sinh \pi a \sum_{n=1}^{\infty} (-1)^{n+1}\frac{n\sin nx}{n^2+a^2}, \qquad |x| < \pi$$

$$\cosh x = 1 + \frac{x^2}{2!} + \frac{x^4}{4!} + \frac{x^6}{6!} + \cdots + \frac{x^{2n}}{(2n)!} + \cdots \qquad |x| < \infty$$

$$\cosh ax = \frac{2a}{\pi}\sinh \pi a\left[\frac{1}{2a^2} - \frac{\cos x}{a^2+1^2} + \frac{\cos 2x}{a^2+2^2} - \frac{\cos 3x}{a^2+3^2} + - \cdots\right]$$

$$= \frac{\sinh \pi a}{a\pi} + \frac{2a}{\pi}\sinh \pi a \sum_{n=1}^{\infty} (-1)^n \frac{\cos nx}{n^2+a^2}, \qquad |x| < \pi$$

$$\tanh x = x - \frac{1}{3}x^3 + \frac{2}{15}x^5 - \frac{17}{315}x^7 + \frac{62}{2835}x^9 - \cdots + \frac{(-1)^{n+1}2^{2n}(2^{2n}-1)}{(2n)!}B_{2n}x^{2n-1} \pm \cdots^{(1)} \qquad |x| < \frac{\pi}{2}$$

$$\tanh x = 1 - 2e^{-2x} + 2e^{-4x} - 2e^{-6x} + - \cdots$$

$$= 1 + 2\sum_{n=1}^{\infty} (-1)^n e^{-2nx}, \qquad \text{Re}(x) > 0$$

$$\tanh x = 2x\left[\frac{1}{\left(\frac{\pi}{2}\right)^2 + x^2} + \frac{1}{\left(\frac{3\pi}{2}\right)^2 + x^2} + \frac{1}{\left(\frac{5\pi}{2}\right)^2 + x^2} + \cdots\right]$$

$$= 2x\sum_{n=0}^{\infty} \frac{1}{\left(n+\frac{1}{2}\right)^2\pi^2 + x^2}$$

$$\coth x = \frac{1}{x} + \frac{x}{3} - \frac{x^3}{45} + \frac{2x^5}{945} - \frac{x^7}{4725} + \cdots + \frac{(-1)^{n+1}2^{2n}}{(2n)!}B_{2n}x^{2n-1} \pm \cdots^{(1)} \qquad 0 < |x| < \pi$$

$$\coth x = 1 + 2e^{-2x} + 2e^{-4x} + 2e^{-6x} + \cdots$$

$$= 1 + 2\sum_{n=1}^{\infty} e^{-2nx} \qquad \text{Re}(x) > 0$$

$$\coth x = \frac{1}{x} + 2x\left[\frac{1}{\pi^2+x^2} + \frac{1}{(2\pi)^2+x^2} + \frac{1}{(3\pi)^2+x^2} + \cdots\right]$$

$$= \frac{1}{x} + 2x\sum_{n=1}^{\infty}\frac{1}{(n\pi)^2+x^2}$$

(1) B_{2n} denotes Bernoulli's numbers.

$$\operatorname{sech} x = 1 - \frac{1}{2!}x^2 + \frac{5}{4!}x^4 - \frac{61}{6!}x^6 + \frac{1385}{8!}x^8 - \cdots + \frac{E_{2n}x^{2n}}{(2n)!} \pm \cdots^{(2)} \qquad |x| < \frac{\pi}{2}$$

$$\operatorname{sech} x = 2e^{-x} - 2e^{-3x} + 2e^{-5x} - 2e^{-7x} + - \cdots$$

$$= 2\sum_{n=0}^{\infty}(-1)^n e^{-(2n+1)x}, \ \operatorname{Re}(x) > 0$$

$$\operatorname{sech} x = 4\pi\left[\frac{1}{\pi^2 + 4x^2} - \frac{3}{(3\pi)^2 + 4x^2} + \frac{5}{(5\pi)^2 + 4x^2} - + \cdots\right]$$

$$= 4\pi\sum_{n=0}^{\infty}\frac{(-1)^n(2n+1)}{(2n+1)^2\pi^2 + 4x^2}$$

$$\operatorname{cosech} x = \frac{1}{x} - \frac{x}{6} + \frac{7x^3}{360} - \frac{31x^5}{15{,}120} + \cdots + \frac{2(2^{2n-1} - 1)}{(2n)!} B_{2n}x^{2n-1} + \cdots^{(1)} \qquad 0 < |x| < \pi$$

$$\operatorname{cosech} x = 2e^{-x} + 2e^{-3x} + 2e^{-5x} + \cdots$$

$$= 2\sum_{n=0}^{\infty}e^{-(2n+1)x}, \ \operatorname{Re}(x) > 0$$

$$\operatorname{cosech} x = \frac{1}{x} - \frac{2x}{\pi^2 + x^2} + \frac{2x}{(2\pi)^2 + x^2} - \frac{2x}{(3\pi)^2 + x^2} + - \cdots$$

$$= \frac{1}{x} + 2x\sum_{n=1}^{\infty}\frac{(-1)^n}{(n\pi)^2 + x^2}$$

INVERSE HYPERBOLIC FUNCTIONS

Definitions

If $x = \sinh y$, then $y = \sinh^{-1} x$. Other inverse functions are defined similarly.

Domain and Range

Function	Domain	Range	Remarks
$\sinh^{-1} x$	$(-\infty, +\infty)$	$(-\infty, +\infty)$	odd function
$\cosh^{-1} x$	$[1, +\infty)$	$(-\infty, +\infty)$	even function double valued
$\tanh^{-1} x$	$(-1, 1)$	$(-\infty, +\infty)$	odd function
$\operatorname{cosech}^{-1} x$	$(-\infty, 0)$, $(0, \infty)$	$(0, -\infty)$, $(\infty, 0)$	odd function two branches, pole at $x = 0$
$\operatorname{sech}^{-1} x$	$(0, 1]$	$(-\infty, +\infty)$	double valued
$\coth^{-1} x$	$(-\infty, -1)$, $(1, \infty)$	$(0, -\infty)$, $(\infty, 0)$	odd function two branches

(2) E_{2n} denotes Euler's numbers.

Graph of the Inverse Functions

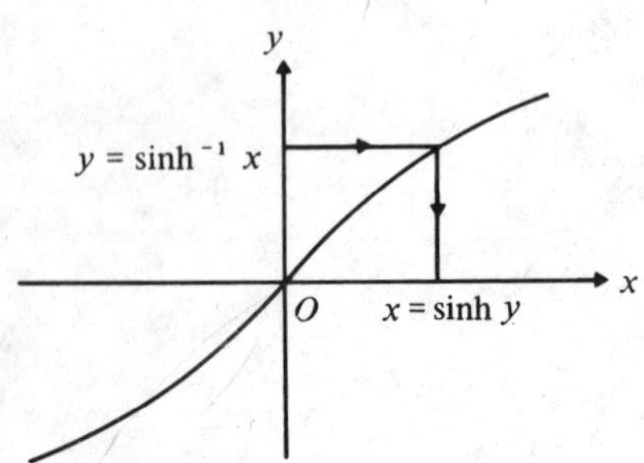

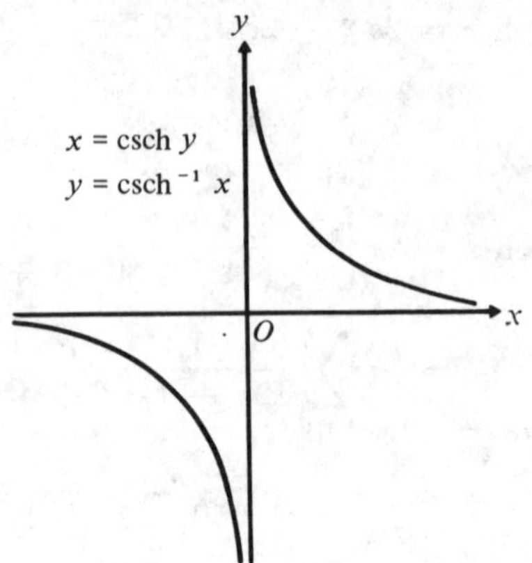

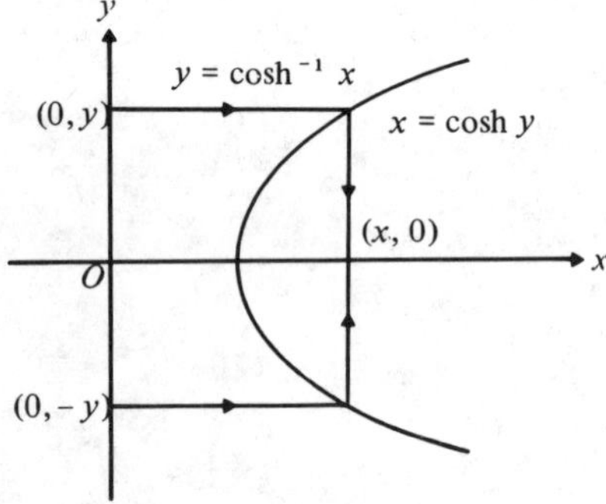

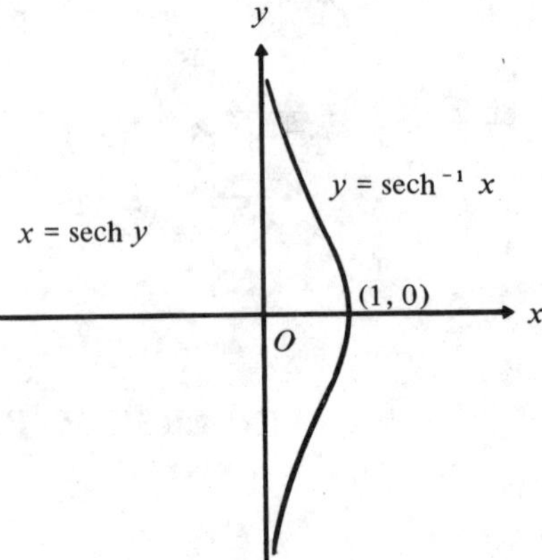

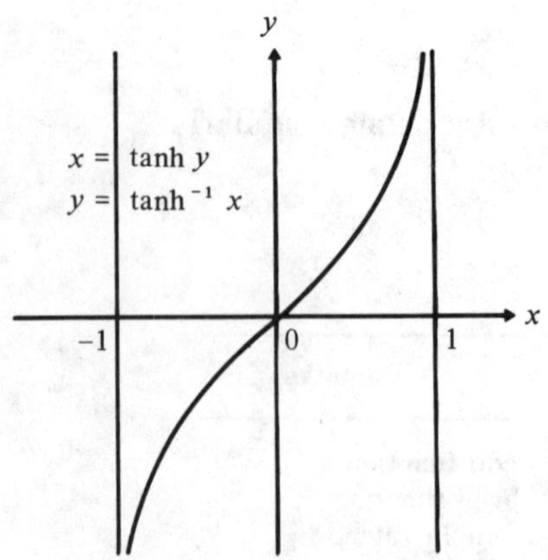

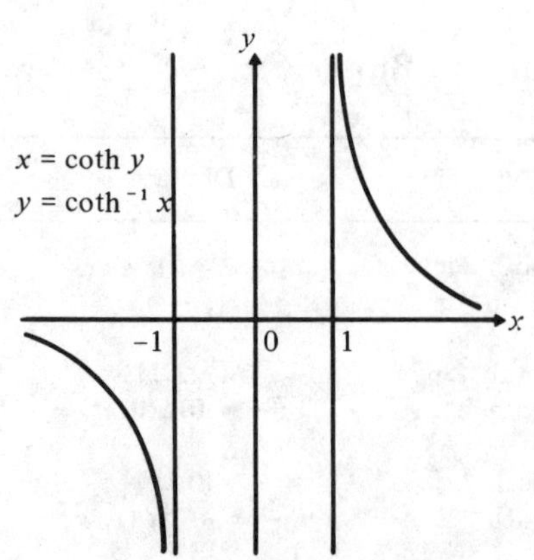

Inverse Hyperbolic Functions In Terms of One Another

Function	$\sinh^{-1} x$	$\cosh^{-1} x$ + if $x > 0$ − if $x < 0$	$\tanh^{-1} x$
$\sinh^{-1} x =$	$\sinh^{-1} x$	$\pm \cosh^{-1} \sqrt{x^2 + 1}$	$\tanh^{-1} \dfrac{x}{\sqrt{1 + x^2}}$
$\cosh^{-1} x =$	$\pm \sinh^{-1} \sqrt{x^2 - 1}$	$\cosh^{-1} x$	$\pm \tanh^{-1} \dfrac{\sqrt{x^2 - 1}}{x}$
$\tanh^{-1} x =$	$\sinh^{-1} \dfrac{x}{\sqrt{1 - x^2}}$	$\pm \cosh^{-1} \dfrac{1}{\sqrt{1 - x^2}}$	$\tanh^{-1} x$
$\operatorname{cosech}^{-1} x =$	$\sinh^{-1} \dfrac{1}{x}$	$\pm \cosh^{-1} \dfrac{\sqrt{x^2 + 1}}{x}$	$\tanh^{-1} \dfrac{1}{\sqrt{1 + x^2}}$
$\operatorname{sech}^{-1} x =$	$\pm \sinh^{-1} \dfrac{\sqrt{1 - x^2}}{x}$	$\cosh^{-1} \dfrac{1}{x}$	$\pm \tanh^{-1} \sqrt{1 - x^2}$
$\coth^{-1} x =$	$\sinh^{-1} \dfrac{1}{\sqrt{x^2 - 1}}$	$\pm \cosh^{-1} \dfrac{x}{\sqrt{x^2 - 1}}$	$\tanh^{-1} \dfrac{1}{x}$

Function	$\operatorname{cosech}^{-1} x$	$\operatorname{sech}^{-1} x$ + if $x > 0$ − if $x < 0$	$\coth^{-1} x$
$\sinh^{-1} x =$	$\operatorname{cosech}^{-1} \dfrac{1}{x}$	$\pm \operatorname{sech}^{-1} \dfrac{1}{\sqrt{1 + x^2}}$	$\coth^{-1} \dfrac{\sqrt{1 + x^2}}{x}$
$\cosh^{-1} x =$	$\pm \operatorname{cosech}^{-1} \dfrac{1}{\sqrt{x^2 - 1}}$	$\operatorname{sech}^{-1} \dfrac{1}{x}$	$\pm \coth^{-1} \dfrac{x}{\sqrt{x^2 - 1}}$
$\tanh^{-1} x =$	$\operatorname{cosech}^{-1} \dfrac{\sqrt{1 - x^2}}{x}$	$\pm \operatorname{sech}^{-1} \sqrt{1 - x^2}$	$\coth^{-1} \dfrac{1}{x}$
$\operatorname{cosech}^{-1} x =$	$\operatorname{cosech}^{-1} x$	$\pm \operatorname{sech}^{-1} \dfrac{x}{\sqrt{x^2 + 1}}$	$\coth^{-1} \sqrt{1 + x^2}$
$\operatorname{sech}^{-1} x =$	$\pm \operatorname{cosech}^{-1} \dfrac{x}{\sqrt{1 - x^2}}$	$\operatorname{sech}^{-1} x$	$\pm \coth^{-1} \dfrac{1}{\sqrt{1 - x^2}}$
$\coth^{-1} x =$	$\operatorname{cosech}^{-1} \sqrt{x^2 - 1}$	$\operatorname{sech}^{-1} \dfrac{\sqrt{x^2 - 1}}{x}$	$\coth^{-1} x$

Fundamental Identities

Relations with Logarithmic Functions

$$\sinh^{-1} x = \log_e (x + \sqrt{x^2+1})$$

$$\cosh^{-1} x = \log_e (x \pm \sqrt{x^2-1}), \quad x \geqq 1.$$ The plus sign is used for the principal value.

$$\tanh^{-1} x = \tfrac{1}{2} \log_e \left(\frac{1+x}{1-x}\right), \quad x^2 < 1$$

$$\operatorname{csch}^{-1} x = \log_e \left(\frac{1 \pm \sqrt{1+x^2}}{x}\right).$$ The plus sign is used if $x > 0$, the minus sign if $x < 0$.

$$\operatorname{sech}^{-1} x = \log_e \left(\frac{1 \pm \sqrt{1-x^2}}{x}\right), \quad 0 < x \leq 1.$$ The plus sign is used for the principal values.

$$\coth^{-1} x = \tfrac{1}{2} \log_e \left(\frac{x+1}{x-1}\right), \quad x^2 > 1$$

Relations with Circular Functions

$$\sinh^{-1} x = -i \sin^{-1} (ix) \qquad \sinh^{-1} (ix) = i \sin^{-1} x$$

$$\cosh^{-1} x = \pm i \cos^{-1} x \qquad \cosh^{-1} (ix) = \pm i \cos^{-1} (ix)$$

$$\tanh^{-1} x = -i \tan^{-1} (ix) \qquad \tanh^{-1} (ix) = i \tan^{-1} x$$

$$\operatorname{cosech}^{-1} x = i \operatorname{cosec}^{-1} (ix) \qquad \operatorname{cosech}^{-1} (ix) = -i \operatorname{cosec}^{-1} x$$

$$\operatorname{sech}^{-1} x = \pm i \sec^{-1} x \qquad \operatorname{sech}^{-1} (ix) = \pm i \sec^{-1} (ix)$$

$$\coth^{-1} x = i \coth^{-1} (ix) \qquad \coth^{-1} (ix) = -i \cot^{-1} (x)$$

Function Sum and Function Difference Relations

$$\sinh^{-1} x + \sinh^{-1} y = \sinh^{-1} (x\sqrt{1+y^2} + y\sqrt{1+x^2})$$

$$\sinh^{-1} x - \sinh^{-1} y = \sinh^{-1} (x\sqrt{1+y^2} - y\sqrt{1+x^2})$$

$$\cosh^{-1} x + \cosh^{-1} y = \cosh^{-1} (xy + \sqrt{(x^2-1)(y^2-1)})$$

$$\cosh^{-1} x - \cosh^{-1} y = \cosh^{-1} (xy - \sqrt{(x^2-1)(y^2-1)})$$

$$\tanh^{-1} x + \tanh^{-1} y = \tanh^{-1} \left(\frac{x+y}{1+xy}\right)$$

$$\tanh^{-1} x - \tanh^{-1} y = \tanh^{-1} \left(\frac{x-y}{1-xy}\right)$$

$$\sinh^{-1} x + \cosh^{-1} y = \sinh^{-1} (xy + \sqrt{(1+x^2)(y^2-1)})$$

$$= \cosh^{-1} (y\sqrt{1+x^2} + x\sqrt{y^2-1})$$

$$\sinh^{-1} x - \cosh^{-1} y = \sinh^{-1} (xy - \sqrt{(1+x^2)(y^2-1)})$$

$$= \cosh^{-1} (y\sqrt{1+x^2} - x\sqrt{y^2-1})$$

$$\tanh^{-1} x + \coth^{-1} y = \tanh^{-1} \left(\frac{xy+1}{y+x}\right)$$

$$= \coth^{-1} \left(\frac{y+x}{xy+1}\right)$$

$$\tanh^{-1} x - \coth^{-1} y = \tanh^{-1} \left(\frac{xy-1}{y-x}\right)$$

$$= \coth^{-1} \left(\frac{y-x}{xy-1}\right)$$

Series Expansions

$$\sinh^{-1} x = x - \frac{1}{2\cdot 3}x^3 + \frac{1\cdot 3}{2\cdot 4\cdot 5}x^5 - \frac{1\cdot 3\cdot 5}{2\cdot 4\cdot 6\cdot 7}x^7 + \cdots + (-1)^n \cdot \frac{1\cdot 3\cdot 5\ldots(2n-1)}{2\cdot 4\cdot 6\ldots 2n(2n+1)}x^{2n+1} \pm \cdots \qquad |x| < 1$$

$$\sinh^{-1} x = \ln(2x) + \frac{1}{2}\cdot\frac{1}{2x^2} - \frac{1\cdot 3}{2\cdot 4}\cdot\frac{1}{4x^4} + \frac{1\cdot 3\cdot 5}{2\cdot 4\cdot 6}\cdot\frac{1}{6x^6} - \cdots$$

$$= \ln(2x) + \sum_{n=1}^{\infty} (-1)^{n+1}\frac{(2n)!x^{-2n}}{2^{2n}(n!)^2 2n}, \qquad |x| > 1$$

$$\cosh^{-1} x = \pm\left[\ln(2x) - \frac{1}{2\cdot 2x^2} - \frac{1\cdot 3}{2\cdot 4\cdot 4x^4} - \frac{1\cdot 3\cdot 5}{2\cdot 4\cdot 6\cdot 6x^6} - \cdots\right] \qquad x > 1$$

$$\operatorname{cosech}^{-1} x = \frac{1}{x} - \frac{1}{2}\cdot\frac{1}{3x^3} + \frac{1\cdot 3}{2\cdot 4}\cdot\frac{1}{5x^5} - \frac{1\cdot 3\cdot 5}{2\cdot 4\cdot 6}\cdot\frac{1}{7x^7} + - \cdots$$

$$= \sum_{n=0}^{\infty} (-1)^n \frac{(2n)!x^{-2n-1}}{2^{2n}(n!)^2(2n+1)}, \qquad |x| > 1$$

$$\operatorname{cosech}^{-1} x = \ln\frac{2}{x} + \frac{1}{2}\cdot\frac{x^2}{2} - \frac{1\cdot 3}{2\cdot 4}\cdot\frac{x^4}{4} + \frac{1\cdot 3\cdot 5}{2\cdot 4\cdot 6}\cdot\frac{x^6}{6} - + \cdots$$

$$= \ln\frac{2}{x} + \sum_{n=1}^{\infty}\frac{(-1)^{n+1}(2n)!x^{2n}}{2^{2n}(n!)^2 2n}, \qquad 0 < x < 1$$

$$\operatorname{sech}^{-1} x = \ln\frac{2}{x} - \frac{1}{2}\frac{x^2}{2} - \frac{1\cdot 3}{2\cdot 4}\cdot\frac{x^4}{4} - \frac{1\cdot 3\cdot 5}{2\cdot 4\cdot 6}\cdot\frac{x^6}{6} - - \cdots$$

$$= \ln\frac{2}{x} - \sum_{n=1}^{\infty}\frac{(2n)!x^{2n}}{2^{2n}(n!)^2 2n}, \qquad 0 < x < 1$$

$$\tanh^{-1} x = x + \frac{x^3}{3} + \frac{x^5}{5} + \frac{x^7}{7} + \cdots + \frac{x^{2n+1}}{2n+1} + \cdots \qquad |x| < 1$$

$${}^*\coth^{-1} x = \frac{1}{x} + \frac{1}{3x^3} + \frac{1}{5x^5} + \frac{1}{7x^7} + \cdots + \frac{1}{(2n+1)x^{2n+1}} + \cdots \qquad |x| > 1$$

GUDERMANNIAN FUNCTION

This function is useful because it relates circular and hyperbolic functions without the use of functions of imaginary argument.

Definition

$$\gamma = \text{the gudermannian of } x, \text{ written as gd } x$$

$$= 2\tan^{-1} e^x - \frac{\pi}{2}$$

$$x = \text{gd}^{-1}\gamma$$

$$= \ln\tan\left(\frac{\pi}{4} + \frac{\gamma}{2}\right)$$

$$= \ln(\sec\gamma + \tan\gamma)$$

*** Can also be written as arg tanh x and arg coth x respectively.**

Special Values of Gudermannian Function

x	0	∞	$-\infty$
gd x	0	$\frac{1}{2}\pi$	$-\frac{1}{2}\pi$

Relations with Hyperbolic and Circular Functions

$$\sinh x = \tan(\text{gd } x) \qquad \text{cosech } x = \cot(\text{gd } x)$$
$$\cosh x = \sec(\text{gd } x) \qquad \text{sech } (x) = \cos(\text{gd } x)$$
$$\tanh x = \sin(\text{gd } x) \qquad \coth(x) = \text{cosec } (\text{gd } x)$$

Derivatives

$$\frac{d}{dx}(\text{gd } x) = \text{sech } x$$

$$\frac{d}{dx}(\text{gd}^{-1}\gamma) = \sec\gamma$$

Fundamental Identities

$$\tanh(\tfrac{1}{2}x) = \tan(\tfrac{1}{2}\text{gd } x)$$

$$e^x = \cosh x + \sinh x = \sec(\text{gd } x) + \tan(\text{gd } x)$$

$$= \tan\left(\frac{\pi}{4} + \frac{1}{2}\text{ gd } x\right) = \frac{1 + \sin(\text{gd } x)}{\cos(\text{gd } x)} = \frac{1 + \tan(\frac{1}{2}\text{gd } x)}{1 - \tan(\frac{1}{2}\text{gd } x)}$$

$$\text{gd } x = 2\tan^{-1}\left(\tanh\frac{1}{2}x\right) = \int_0^x \frac{dt}{\cosh t}$$

$$\text{gd}^{-1}\gamma = \int_0^\gamma \frac{dt}{\cos t}$$

$$i\,\text{gd}^{-1}\gamma = \text{gd } i\gamma, \text{ where } i = \sqrt{-1}.$$

If

$$\alpha + i\beta = \text{gd}(x + iy)$$

then

$$\tan\alpha = \frac{\sinh x}{\cos y} \qquad \tanh\beta = \frac{\sin y}{\cosh x}$$

$$\tanh x = \frac{\sin\alpha}{\cosh\beta} \qquad \tan y = \frac{\sinh\beta}{\cosh\alpha}$$

Series Expansions

$$\text{gd } x = x - \frac{1}{6}x^3 + \frac{1}{24}x^5 - \frac{61}{5040}x^7 + - \cdots$$

$$= x - \sum_{n=1}^{\infty} \frac{E_{2n}}{(2n+1)!} x^{2n+1} \qquad |x| < 1 \qquad (1)$$

$$\text{gd } x = \frac{\pi}{2} - \text{sech } x - \frac{1}{2}\frac{\text{sech}^3 x}{3} - \frac{1 \cdot 3}{2 \cdot 4}\frac{\text{sech}^5 x}{5} + - \cdots$$

$$= \frac{\pi}{2} - \sum_{n=0}^{\infty} \frac{(2n)!}{2^{2n}(n!)^2}\frac{\text{sech}^{2n+1} x}{(2n+1)} \qquad x \text{ large}$$

$$\operatorname{gd} x = \frac{2}{1}\tanh\frac{x}{2} - \frac{2}{3}\tanh^3\frac{x}{2} + \frac{2}{5}\tanh^5\frac{x}{2} - \frac{2}{7}\tanh^7\frac{x}{2} + - \cdots$$

$$= 2\sum_{n=0}^{\infty}\frac{(-1)^n}{2n+1}\tanh^{2n+1}\frac{x}{2}$$

$$\operatorname{gd}^{-1}\gamma = \gamma + \frac{1}{6}\gamma^3 + \frac{1}{24}\gamma^5 + \frac{61}{5040}\gamma^7 + \cdots$$

$$= \gamma + \sum_{n=1}^{\infty}\frac{(-1)^n E_{2n}\,\gamma^{2n+1}}{(2n+1)!}, \qquad |\gamma| < \frac{\pi}{2} \quad (1)$$

$$\operatorname{gd}^{-1}\gamma = \frac{2}{1}\tan\frac{\gamma}{2} + \frac{2}{3}\tan^3\frac{\gamma}{2} + \frac{2}{5}\tan^5\frac{\gamma}{2} + \cdots$$

$$= 2\sum_{n=0}^{\infty}\frac{1}{2n+1}\tan^{2n+1}\frac{\gamma}{2}$$

[1] E_n are Euler's numbers.

References

1. For extensive tables of inverse hyperbolic functions: Tables of Inverse Hyperbolic Functions, Harvard University Computation Laboratory (Cambridge, Harvard University Press, 1949).
2. For tables of hyperbolic functions with complex argument: Tables of Complex Hyperbolic and Circular Functions, A. E. Kennelly (Cambridge, Harvard University Press, 1914).

HYPERBOLIC FUNCTIONS AND THEIR COMMON LOGARITHMS

The logarithms given below show the mantissa only. The proper characteristic must be added.

x	Sinh x		Cosh x		Tanh x		Coth x	
	Value	$\log_{10}$	Value	$\log_{10}$	Value	$\log_{10}$	Value	$\log_{10}$
0.00	0.00000	$-\infty$	1.00000	.00000	0.00000	$-\infty$	∞	∞
0.01	.01000	.00001	1.00005	.00002	.01000	.99999	100.003	.00001
0.02	.02000	.30106	1.00020	.00009	.02000	.30097	50.007	.69903
0.03	.03000	.47719	1.00045	.00020	.02999	.47699	33.343	.52301
0.04	.04001	.60218	1.00080	.00035	.03998	.60183	25.013	.39817
0.05	0.05002	.69915	1.00125	.00054	0.04996	.69861	20.017	.30139
0.06	.06004	.77841	1.00180	.00078	.05993	.77763	16.687	.22237
0.07	.07006	.84545	1.00245	.00106	.06989	.84439	14.309	.15561
0.08	.08009	.90355	1.00320	.00139	.07983	.90216	12.527	.09784
0.09	.09012	.95483	1.00405	.00176	.08976	.95307	11.141	.04693
0.10	0.10017	.00072	1.00500	.00217	0.09967	.99856	10.0333	.00144
0.11	.11022	.04227	1.00606	.00262	.10956	.03965	9.1275	.96035
0.12	.12029	.08022	1.00721	.00312	.11943	.07710	8.3733	.92290
0.13	.13037	.11517	1.00846	.00366	.12927	.11151	7.7356	.88849
0.14	.14046	.14755	1.00982	.00424	.13909	.14330	7.1895	.85670
0.15	0.15056	.17772	1.01127	.00487	0.14889	.17285	6.7166	.82715
0.16	.16068	.20597	1.01283	.00554	.15865	.20044	6.3032	.79956
0.17	.17082	.23254	1.01448	.00625	.16838	.22629	5.9389	.77371
0.18	.18097	.25762	1.01624	.00700	.17808	.25062	5.6154	.74938
0.19	.19115	.28136	1.01810	.00779	.18775	.27357	5.3263	.72643
0.20	0.20134	.30392	1.02007	.00863	0.19738	.29529	5.0665	.70471
0.21	.21155	.32541	1.02213	.00951	.20697	.31590	4.8317	.68410
0.22	.22178	.34592	1.02430	.01043	.21652	.33549	4.6186	.66451
0.23	.23203	.36555	1.02657	.01139	.22603	.35416	4.4242	.64584
0.24	.24231	.38437	1.02894	.01239	.23550	.37198	4.2464	.62802
0.25	0.25261	.40245	1.03141	.01343	0.24492	.38902	4.0830	.61098
0.26	.26294	.41986	1.03399	.01452	.25430	.40534	3.9324	.59466
0.27	.27329	.43663	1.03667	.01564	.26362	.42099	3.7933	.57901
0.28	.28367	.45282	1.03946	.01681	.27291	.43601	3.6643	.56399
0.29	.29408	.46847	1.04235	.01801	.28213	.45046	3.5444	.54954
0.30	0.30452	.48362	1.04534	.01926	0.29131	.46436	3.4327	.53564
0.31	.31499	.49830	1.04844	.02054	.30044	.47775	3.3285	.52225
0.32	.32549	.51254	1.05164	.02187	.30951	.49067	3.2309	.50933
0.33	.33602	.52637	1.05495	.02323	.31852	.50314	3.1395	.49686
0.34	.34659	.53981	1.05836	.02463	.32748	.51518	3.0536	.48482
0.35	0.35719	.55290	1.06188	.02607	0.33638	.52682	2.9729	.47318
0.36	.36783	.56564	1.06550	.02755	.34521	.53809	2.8968	.46191
0.37	.37850	.57807	1.06923	.02907	.35399	.54899	2.8249	.45101
0.38	.38921	.59019	1.07307	.03063	.36271	.55956	2.7570	.44044
0.39	.39996	.60202	1.07702	.03222	.37136	.56980	2.6928	.43020
0.40	0.41075	.61358	1.08107	.03385	0.37995	.57973	2.6319	.42027
0.41	.42158	.62488	1.08523	.03552	.38847	.58936	2.5742	.41064
0.42	.43246	.63594	1.08950	.03723	.39693	.59871	2.5193	.40129
0.43	.44337	.64677	1.09388	.03897	.40532	.60780	2.4672	.39220
0.44	.45434	.65738	1.09837	.04075	.41364	.61663	2.4175	.38337
0.45	0.46534	.66777	1.10297	.04256	0.42190	.62521	2.3702	.37479
0.46	.47640	.67797	1.10768	.04441	.43008	.63355	2.3251	.36645
0.47	.48750	.68797	1.11250	.04630	.43820	.64167	2.2821	.35833
0.48	.49865	.69779	1.11743	.04822	.44624	.64957	2.2409	.35043
0.49	.50984	.70744	1.12247	.05018	.45422	.65726	2.2016	.34274

HYPERBOLIC FUNCTIONS AND THEIR COMMON LOGARITHMS (Continued)

x	Sinh x		Cosh x		Tanh x		Coth x	
	Value	$\log_{10}$	Value	$\log_{10}$	Value	$\log_{10}$	Value	$\log_{10}$
0.50	0.52110	.71692	1.12763	.05217	0.46212	.66475	2.1640	.33525
0.51	.53240	.72624	1.13289	.05419	.46995	.67205	2.1279	.32795
0.52	.54375	.73540	1.13827	.05625	.47770	.67916	2.0934	.32084
0.53	.55516	.74442	1.14377	.05834	.48538	.68608	2.0602	.31392
0.54	.56663	.75330	1.14938	.06046	.49299	.69284	2.0284	.30716
0.55	0.57815	.76204	1.15510	.06262	0.50052	.69942	1.9979	.30058
0.56	.58973	.77065	1.16094	.06481	.50798	.70584	1.9686	.29416
0.57	.60137	.77914	1.16690	.06703	.51536	.71211	1.9404	.28789
0.58	.61307	.78751	1.17297	.06929	.52267	.71822	1.9133	.28178
0.59	.62483	.79576	1.17916	.07157	.52990	.72419	1.8872	.27581
0.60	0.63665	.80390	1.18547	.07389	0.53705	.73001	1.8620	.26999
0.61	.64854	.81194	1.19189	.07624	.54413	.73570	1.8378	.26430
0.62	.66049	.81987	1.19844	.07861	.55113	.74125	1.8145	.25875
0.63	.67251	.82770	1.20510	.08102	.55805	.74667	1.7919	.25333
0.64	.68459	.83543	1.21189	.08346	.56490	.75197	1.7702	.24803
0.65	0.69675	.84308	1.21879	.08593	0.57167	.75715	1.7493	.24285
0.66	.70897	.85063	1.22582	.08843	.57836	.76220	1.7290	.23780
0.67	.72126	.85809	1.23297	.09095	.58498	.76714	1.7095	.23286
0.68	.73363	.86548	1.24025	.09351	.59152	.77197	1.6906	.22803
0.69	.74607	.87278	1.24765	.09609	.59798	.77669	1.6723	.22331
0.70	0.75858	.88000	1.25517	.09870	0.60437	.78130	1.6546	.21870
0.71	.77117	.88715	1.26282	.10134	.61068	.78581	1.6375	.21419
0.72	.78384	.89423	1.27059	.10401	.61691	.79022	1.6210	.20978
0.73	.79659	.90123	1.27849	.10670	.62307	.79453	1.6050	.20547
0.74	.80941	.90817	1.28652	.10942	.62915	.79875	1.5895	.20125
0.75	0.82232	.91504	1.29468	.11216	0.63515	.80288	1.5744	.19712
0.76	.83530	.92185	1.30297	.11493	.64108	.80691	1.5599	.19309
0.77	.84838	.92859	1.31139	.11773	.64693	.81086	1.5458	.18914
0.78	.86153	.93527	1.31994	.12055	.65271	.81472	1.5321	.18528
0.79	.87478	.94190	1.32862	.12340	.65841	.81850	1.5188	.18150
0.80	0.88811	.94846	1.33743	.12627	0.66404	.82219	1.5059	.17781
0.81	.90152	.95498	1.34638	.12917	.66959	.82581	1.4935	.17419
0.82	.91503	.96144	1.35547	.13209	.67507	.82935	1.4813	.17065
0.83	.92863	.96784	1.36468	.13503	.68048	.83281	1.4696	.16719
0.84	.94233	.97420	1.37404	.13800	.68581	.83620	1.4581	.16380
0.85	0.95612	.98051	1.38353	.14099	0.69107	.83952	1.4470	.16048
0.86	.97000	.98677	1.39316	.14400	.69626	.84277	1.4362	.15723
0.87	.98398	.99299	1.40293	.14704	.70137	.84595	1.4258	.15405
0.88	.99806	.99916	1.41284	.15009	.70642	.84906	1.4156	.15094
0.89	1.01224	.00528	1.42289	.15317	.71139	.85211	1.4057	.14789
0.90	1.02652	.01137	1.43309	.15627	0.71630	.85509	1.3961	.14491
0.91	1.04090	.01741	1.44342	.15939	.72113	.85801	1.3867	.14199
0.92	1.05539	.02341	1.45390	.16254	.72590	.86088	1.3776	.13912
0.93	1.06998	.02937	1.46453	.16570	.73059	.86368	1.3687	.13632
0.94	1.08468	.03530	1.47530	.16888	.73522	.86642	1.3601	.13358
0.95	1.09948	.04119	1.48623	.17208	0.73978	.86910	1.3517	.13090
0.96	1.11440	.04704	1.49729	.17531	.74428	.87173	1.3436	.12827
0.97	1.12943	.05286	1.50851	.17855	.74870	.87431	1.3356	.12569
0.98	1.14457	.05864	1.51988	.18181	.75307	.87683	1.3279	.12317
0.99	1.15983	.06439	1.53141	.18509	.75736	.87930	1.3204	.12070

HYPERBOLIC FUNCTIONS AND THEIR COMMON LOGARITHMS (Continued)

x	Sinh x		Cosh x		Tanh x		Coth x	
	Value	$\log_{10}$	Value	$\log_{10}$	Value	$\log_{10}$	Value	$\log_{10}$
1.00	1.17520	.07011	1.54308	.18839	0.76159	.88172	1.3130	.11828
1.01	1.19069	.07580	1.55491	.19171	.76576	.88409	1.3059	.11591
1.02	1.20630	.08146	1.56689	.19504	.76987	.88642	1.2989	.11358
1.03	1.22203	.08708	1.57904	.19839	.77391	.88869	1.2921	.11131
1.04	1.23788	.09268	1.59134	.20176	.77789	.89092	1.2855	.10908
1.05	1.25386	.09825	1.60379	.20515	0.78181	.89310	1.2791	.10690
1.06	1.26996	.10379	1.61641	.20855	.78566	.89524	1.2728	.10476
1.07	1.28619	.10930	1.62919	.21197	.78946	.89733	1.2667	.10267
1.08	1.30254	.11479	1.64214	.21541	.79320	.89938	1.2607	.10062
1.09	1.31903	.12025	1.65525	.21886	.79688	.90139	1.2549	.09861
1.10	1.33565	.12569	1.66852	.22233	0.80050	.90336	1.2492	.09664
1.11	1.35240	.13111	1.68196	.22582	.80406	.90529	1.2437	.09471
1.12	1.36929	.13649	1.69557	.22931	.80757	.90718	1.2383	.09282
1.13	1.38631	.14186	1.70934	.23283	.81102	.90903	1.2330	.09097
1.14	1.40347	.14720	1.72329	.23636	.81441	.91085	1.2279	.08915
1.15	1.42078	.15253	1.73741	.23990	0.81775	.91262	1.2229	.08738
1.16	1.43822	.15783	1.75171	.24346	.82104	.91436	1.2180	.08564
1.17	1.45581	.16311	1.76618	.24703	.82427	.91607	1.2132	.08393
1.18	1.47355	.16836	1.78083	.25062	.82745	.91774	1.2085	.08226
1.19	1.49143	.17360	1.79565	.25422	.83058	.91938	1.2040	.08062
1.20	1.50946	.17882	1.81066	.25784	0.83365	.92099	1.1995	.07901
1.21	1.52764	.18402	1.82584	.26146	.83668	.92256	1.1952	.07744
1.22	1.54598	.18920	1.84121	.26510	.83965	.92410	1.1910	.07590
1.23	1.56447	.19437	1.85676	.26876	.84258	.92561	1.1868	.07439
1.24	1.58311	.19951	1.87250	.27242	.84546	.92709	1.1828	.07291
1.25	1.60192	.20464	1.88842	.27610	0.84828	.92854	1.1789	.07146
1.26	1.62088	.20975	1.90454	.27979	.85106	.92996	1.1750	.07004
1.27	1.64001	.21485	1.92084	.28349	.85380	.93135	1.1712	.06865
1.28	1.65930	.21993	1.93734	.28721	.85648	.93272	1.1676	.06728
1.29	1.67876	.22499	1.95403	.29093	.85913	.93406	1.1640	.06594
1.30	1.69838	.23004	1.97091	.29467	0.86172	.93537	1.1605	.06463
1.31	1.71818	.23507	1.98800	.29842	.86428	.93665	1.1570	.06335
1.32	1.73814	.24009	2.00528	.30217	.86678	.93791	1.1537	.06209
1.33	1.75828	.24509	2.02276	.30594	.86925	.93914	1.1504	.06086
1.34	1.77860	.25008	2.04044	.30972	.87167	.94035	1.1472	.05965
1.35	1.79909	.25505	2.05833	.31352	0.87405	.94154	1.1441	.05846
1.36	1.81977	.26002	2.07643	.31732	.87639	.94270	1.1410	.05730
1.37	1.84062	.26496	2.09473	.32113	.87869	.94384	1.1381	.05616
1.38	1.86166	.26990	2.11324	.32495	.88095	.94495	1.1351	.05505
1.39	1.88289	.27482	2.13196	.32878	.88317	.94604	1.1323	.05396
1.40	1.90430	.27974	2.15090	.33262	0.88535	.94712	1.1295	.05288
1.41	1.92591	.28464	2.17005	.33647	.88749	.94817	1.1268	.05183
1.42	1.94770	.28952	2.18942	.34033	.88960	.94919	1.1241	.05081
1.43	1.96970	.29440	2.20900	.34420	.89167	.95020	1.1215	.04980
1.44	1.99188	.29926	2.22881	.34807	.89370	.95119	1.1189	.04881
1.45	2.01427	.30412	2.24884	.35196	0.89569	.95216	1.1165	.04784
1.46	2.03686	.30896	2.26910	.35585	.89765	.95311	1.1140	.04689
1.47	2.05965	.31379	2.28958	.35976	.89958	.95404	1.1116	.04596
1.48	2.08265	.31862	2.31029	.36367	.90147	.95495	1.1093	.04505
1.49	2.10586	.32343	2.33123	.36759	.90332	.95584	1.1070	.04416

HYPERBOLIC FUNCTIONS AND THEIR COMMON LOGARITHMS (Continued)

x	Sinh x		Cosh x		Tanh x		Coth x	
	Value	$\log_{10}$	Value	$\log_{10}$	Value	$\log_{10}$	Value	$\log_{10}$
1.50	2.12928	.32823	2.35241	.37151	0.90515	.95672	1.1048	.04328
1.51	2.15291	.33303	2.37382	.37545	.90694	.95758	1.1026	.04242
1.52	2.17676	.33781	2.39547	.37939	.90870	.95842	1.1005	.04158
1.53	2.20082	.34258	2.41736	.38334	.91042	.95924	1.0984	.04076
1.54	2.22510	.34735	2.43949	.38730	.91212	.96005	1.0963	.03995
1.55	2.24961	.35211	2.46186	.39126	0.91379	.96084	1.0943	.03916
1.56	2.27434	.35686	2.48448	.39524	.91542	.96162	1.0924	.03838
1.57	2.29930	.36160	2.50735	.39921	.91703	.96238	1.0905	.03762
1.58	2.32449	.36633	2.53047	.40320	.91860	.96313	1.0886	.03687
1.59	2.34991	.37105	2.55384	.40719	.92015	.96386	1.0868	.03614
1.60	2.37557	.37577	2.57746	.41119	0.92167	.96457	1.0850	.03543
1.61	2.40146	.38048	2.60135	.41520	.92316	.96528	1.0832	.03472
1.62	2.42760	.38518	2.62549	.41921	.92462	.96597	1.0815	.03403
1.63	2.45397	.38987	2.64990	.42323	.92606	.96664	1.0798	.03336
1.64	2.48059	.39456	2.67457	.42725	.92747	.96730	1.0782	.03270
1.65	2.50746	.39923	2.69951	.43129	0.92886	.96795	1.0766	.03205
1.66	2.53459	.40391	2.72472	.43532	.93022	.96858	1.0750	.03142
1.67	2.56196	.40857	2.75021	.43937	.93155	.96921	1.0735	.03079
1.68	2.58959	.41323	2.77596	.44341	.93286	.96982	1.0720	.03018
1.69	2.61748	.41788	2.80200	.44747	.93415	.97042	1.0705	.02958
1.70	2.64563	.42253	2.82832	.45153	.93541	.97100	1.0691	.02900
1.71	2.67405	.42717	2.85491	.45559	.93665	.97158	1.0676	.02842
1.72	2.70273	.43180	2.88180	.45966	.93786	.97214	1.0663	.02786
1.73	2.73168	.43643	2.90897	.46374	.93906	.97269	1.0649	.02731
1.74	2.76091	.44105	2.93643	.46782	.94023	.97323	1.0636	.02677
1.75	2.79041	.44567	2.96419	.47191	0.94138	.97376	1.0623	.02624
1.76	2.82020	.45028	2.99224	.47600	.94250	.97428	1.0610	.02572
1.77	2.85026	.45488	3.02059	.48009	.94361	.97479	1.0598	.02521
1.78	2.88061	.45948	3.04925	.48419	.94470	.97529	1.0585	.02471
1.79	2.91125	.46408	3.07821	.48830	.94576	.97578	1.0574	.02422
1.80	2.94217	.46867	3.10747	.49241	0.94681	.97626	1.0562	.02374
1.81	2.97340	.47325	3.13705	.49652	.94783	.97673	1.0550	.02327
1.82	3.00492	.47783	3.16694	.50064	.94884	.97719	1.0539	.02281
1.83	3.03674	.48241	3.19715	.50476	.94983	.97764	1.0528	.02236
1.84	3.06886	.48698	3.22768	.50889	.95080	.97809	1.0518	.02191
1.85	3.10129	.49154	3.25853	.51302	0.95175	.97852	1.0507	.02148
1.86	3.13403	.49610	3.28970	.51716	.95268	.97895	1.0497	.02105
1.87	3.16709	.50066	3.32121	.52130	.95359	.97936	1.0487	.02064
1.88	3.20046	.50521	3.35305	.52544	.95449	.97977	1.0477	.02023
1.89	3.23415	.50976	3.38522	.52959	.95537	.98017	1.0467	.01983
1.90	3.26816	.51430	3.41773	.53374	0.95624	.98057	1.0458	.01943
1.91	3.30250	.51884	3.45058	.53789	.95709	.98095	1.0448	.01905
1.92	3.33718	.52338	3.48378	.54205	.95792	.98133	1.0439	.01867
1.93	3.37218	.52791	3.51733	.54621	.95873	.98170	1.0430	.01830
1.94	3.40752	.53244	3.55123	.55038	.95953	.98206	1.0422	.01794
1.95	3.44321	.53696	3.58548	.55455	0.96032	.98242	1.0413	.01758
1.96	3.47923	.54148	3.62009	.55872	.96109	.98276	1.0405	.01724
1.97	3.51561	.54600	3.65507	.56290	.96185	.98311	1.0397	.01689
1.98	3.55234	.55051	3.69041	.56707	.96259	.98344	1.0389	.01656
1.99	3.58942	.55502	3.72611	.57126	.96331	.98377	1.0381	.01623

HYPERBOLIC FUNCTIONS AND THEIR COMMON LOGARITHMS (Continued)

x	Sinh x		Cosh x		Tanh x		Coth x	
	Value	$\log_{10}$	Value	$\log_{10}$	Value	$\log_{10}$	Value	$\log_{10}$
2.00	3.62686	.55953	3.76220	.57544	0.96403	.98409	1.0373	.01591
2.01	3.66466	.56403	3.79865	.57963	.96473	.98440	1.0366	.01560
2.02	3.70283	.56853	3.83549	.58382	.96541	.98471	1.0358	.01529
2.03	3.74138	.57303	3.87271	.58802	.96609	.98502	1.0351	.01498
2.04	3.78029	.57753	3.91032	.59221	.96675	.98531	1.0344	.01469
2.05	3.81958	.58202	3.94832	.59641	0.96740	.98560	1.0337	.01440
2.06	3.85926	.58650	3.98671	.60061	.96803	.98589	1.0330	.01411
2.07	3.89932	.59099	4.02550	.60482	.96865	.98617	1.0324	.01383
2.08	3.93977	.59547	4.06470	.60903	.96926	.98644	1.0317	.01356
2.09	3.98061	.59995	4.10430	.61324	.96986	.98671	1.0311	.01329
2.10	4.02186	.60443	4.14431	.61745	0.97045	.98697	1.0304	.01303
2.11	4.06350	.60890	4.18474	.62167	.97103	.98723	1.0298	.01277
2.12	4.10555	.61337	4.22558	.62589	.97159	.98748	1.0292	.01252
2.13	4.14801	.61784	4.26685	.63011	.97215	.98773	1.0286	.01227
2.14	4.19089	.62231	4.30855	.63433	.97269	.98798	1.0281	.01202
2.15	4.23419	.62677	4.35067	.63856	0.97323	.98821	1.0275	.01179
2.16	4.27791	.63123	4.39323	.64278	.97375	.98845	1.0270	.01155
2.17	4.32205	.63569	4.43623	.64701	.97426	.98868	1.0264	.01132
2.18	4.36663	.64015	4.47967	.65125	.97477	.98890	1.0259	.01110
2.19	4.41165	.64460	4.52356	.65548	.97526	.98912	1.0254	.01088
2.20	4.45711	.64905	4.56791	.65972	0.97574	.98934	1.0249	.01066
2.21	4.50301	.65350	4.61271	.66396	.97622	.98955	1.0244	.01045
2.22	4.54936	.65795	4.65797	.66820	.97668	.98975	1.0239	.01025
2.23	4.59617	.66240	4.70370	.67244	.97714	.98996	1.0234	.01004
2.24	4.64344	.66684	4.74989	.67668	.97759	.99016	1.0229	.00984
2.25	4.69117	.67128	4.79657	.68093	0.97803	.99035	1.0225	.00965
2.26	4.73937	.67572	4.84372	.68518	.97846	.99054	1.0220	.00946
2.27	4.78804	.68016	4.89136	.68943	.97888	.99073	1.0216	.00927
2.28	4.83720	.68459	4.93948	.69368	.97929	.99091	1.0211	.00909
2.29	4.88684	.68903	4.98810	.69794	.97970	.99109	1.0207	.00891
2.30	4.93696	.69346	5.03722	.70219	0.98010	.99127	1.0203	.00873
2.31	4.98758	.69789	5.08684	.70645	.98049	.99144	1.0199	.00856
2.32	5.03870	.70232	5.13697	.71071	.98087	.99161	1.0195	00839
2.33	5.09032	.70675	5.18762	.71497	.98124	.99178	1.0191	.00822
2.34	5.14245	.71117	5.23878	.71923	.98161	.99194	1.0187	.00806
2.35	5.19510	.71559	5.29047	.72349	0.98197	.99210	1.0184	.00790
2.36	5.24827	.72002	5.34269	.72776	.98233	.99226	1.0180	.00774
2.37	5.30196	.72444	5.39544	.73203	.98267	.99241	1.0176	.00759
2.38	5.35618	.72885	5.44873	.73630	.98301	.99256	1.0173	.00744
2.39	5.41093	.73327	5.50256	.74056	.98335	.99271	1.0169	.00729
2.40	5.46623	.73769	5.55695	.74484	0.98367	.99285	1.0166	.00715
2.41	5.52207	.74210	5.61189	.74911	.98400	.99299	1.0163	.00701
2.42	5.57847	.74652	5.66739	.75338	.98431	.99313	1.0159	.00687
2.43	5.63542	.75093	5.72346	.75766	.98462	.99327	1.0156	.00673
2.44	5.69294	.75534	5.78010	.76194	.98492	.99340	1.0153	.00660
2.45	5.75103	.75975	5.83732	.76621	0.98522	.99353	1.0150	.00647
2.46	5.80969	.75415	5.89512	.77049	.98551	.99366	1.0147	.00634
2.47	5.86893	.76856	5.95352	.77477	.98579	.99379	1.0144	.00621
2.48	5.92876	.77296	6.01250	.77906	.98607	.99391	1.0141	.00609
2.49	5.98918	.77737	6.07209	.78334	.98635	.99403	1.0138	.00597

HYPERBOLIC FUNCTIONS AND THEIR COMMON LOGARITHMS (Continued)

x	Sinh x Value	Sinh x $\log_{10}$	Cosh x Value	Cosh x $\log_{10}$	Tanh x Value	Tanh x $\log_{10}$	Coth x Value	Coth x $\log_{10}$
2.50	6.05020	.78177	6.13229	.78762	0.98661	.99415	1.0136	.00585
2.51	6.11183	.78617	6.19310	.79191	.98688	.99426	1.0133	.00574
2.52	6.17407	.79057	6.25453	.79619	.98714	.99438	1.0130	.00562
2.53	6.23692	.79497	6.31658	.80048	.98739	.99449	1.0128	.00551
2.54	6.30040	.79937	6.37927	.80477	.98764	.99460	1.0125	.00540
2.55	6.36451	.80377	6.44259	.80906	0.98788	.99470	1.0123	.00530
2.56	6.42926	.80816	6.50656	.81335	.98812	.99481	1.0120	.00519
2.57	6.49464	.81256	6.57118	.81764	.98835	.99491	1.0118	.00509
2.58	6.56068	.81695	6.63646	.82194	.98858	.99501	1.0115	.00499
2.59	6.62738	.82134	6.70240	.82623	.98881	.99511	1.0113	.00489
2.60	6.69473	.82573	6.76901	.83052	0.98903	.99521	1.0111	.00479
2.61	6.76276	.83012	6.83629	.83482	.98924	.99530	1.0109	.00470
2.62	6.83146	.83451	6.90426	.83912	.98946	.99540	1.0107	.00460
2.63	6.90085	.83890	6.97292	.84341	.98966	.99549	1.0104	.00451
2.64	6.97092	.84329	7.04228	.84771	.98987	.99558	1.0102	.00442
2.65	7.04169	.84768	7.11234	.85201	0.99007	.99566	1.0100	.00434
2.66	7.11317	.85206	7.18312	.85631	.99026	.99575	1.0098	.00425
2.67	7.18536	.85645	7.25461	.86061	.99045	.99583	1.0096	.00417
2.68	7.25827	.86083	7.32683	.86492	.99064	.99592	1.0094	.00408
2.69	7.33190	.86522	7.39978	.86922	.99083	.99600	1.0093	.00400
2.70	7.40626	.86960	7.47347	.87352	0.99101	.99608	1.0091	.00392
2.71	7.48137	.87398	7.54791	.87783	.99118	.99615	1.0089	.00385
2.72	7.55722	.87836	7.62310	.88213	.99136	.99623	1.0087	.00377
2.73	7.63383	.88274	7.69905	.88644	.99153	.99631	1.0085	.00369
2.74	7.71121	.88712	7.77578	.89074	.99170	.99638	1.0084	.00362
2.75	7.78935	.89150	7.85328	.89505	0.99186	.99645	1.0082	.00355
2.76	7.86828	.89588	7.93157	.89936	.99202	.99652	1.0080	.00348
2.77	7.94799	.90026	8.01065	.90367	.99218	.99659	1.0079	.00341
2.78	8.02849	.90463	8.09053	.90798	.99233	.99666	1.0077	.00334
2.79	8.10980	.90901	8.17122	.91229	.99248	.99672	1.0076	.00328
2.80	8.19192	.91339	8.25273	.91660	0.99263	.99679	1.0074	.00321
2.81	8.27486	.91776	8.33506	.92091	.99278	.99685	1.0073	.00315
2.82	8.35862	.92213	8.41823	.92522	.99292	.99691	1.0071	.00309
2.83	8.44322	.92651	8.50224	.92953	.99306	.99698	1.0070	.00302
2.84	8.52867	.93088	8.58710	.93385	.99320	.99704	1.0069	.00296
2.85	8.61497	.93525	8.67281	.93816	0.99333	.99709	1.0067	.00291
2.86	8.70213	.93963	8.75940	.94247	.99346	.99715	1.0066	.00285
2.87	8.79016	.94400	8.84686	.94679	.99359	.99721	1.0065	.00279
2.88	8.87907	.94837	8.93520	.95110	.99372	.99726	1.0063	.00274
2.89	8.96887	.95274	9.02444	.95542	.99384	.99732	1.0062	.00268
2.90	9.05956	.95711	9.11458	.95974	0.99396	.99737	1.0061	.00263
2.91	9.15116	.96148	9.20564	.96405	.99408	.99742	1.0060	.00258
2.92	9.24368	.96584	9.29761	.96837	.99420	.99747	1.0058	.00253
2.93	9.33712	.97021	9.39051	.97269	.99431	.99752	1.0057	.00248
2.94	9.43149	.97458	9.48436	.97701	.99443	.99757	1.0056	.00243
2.95	9.52681	.97895	9.57915	.98133	0.99454	.99762	1.0055	.00238
2.96	9.62308	.98331	9.67490	.98565	.99464	.99767	1.0054	.00233
2.97	9.72031	.98768	9.77161	.98997	.99475	.99771	1.0053	.00229
2.98	9.81851	.99205	9.86930	.99429	.99485	.99776	1.0052	.00224
2.99	9.91770	.99641	9.96798	.99861	.99496	.99780	1.0051	.00220

HYPERBOLIC FUNCTIONS AND THEIR COMMON LOGARITHMS (Continued)

x	Sinh x		Cosh x		Tanh x		Coth x	
	Value	$\log_{10}$	Value	$\log_{10}$	Value	$\log_{10}$	Value	$\log_{10}$
3.0	10.0179	.00078	10.0677	.00293	0.99505	.99785	1.0050	.00215
3.1	11.0765	.04440	11.1215	.04616	.99595	.99824	1.0041	.00176
3.2	12.2459	.08799	12.2866	.08943	.99668	.99856	1.0033	.00144
3.3	13.5379	.13155	13.5748	.13273	.99728	.99882	1.0027	.00118
3.4	14.9654	.17509	14.9987	.17605	.99777	.99903	1.0022	.00097
3.5	16.5426	.21860	16.5728	.21940	0.99818	.99921	1.0018	.00079
3.6	18.2855	.26211	18.3128	.26275	.99851	.99935	1.0015	.00065
3.7	20.2113	.30559	20.2360	.30612	.99878	.99947	1.0012	.00053
3.8	22.3394	.34907	22.3618	.34951	.99900	.99957	1.0010	.00043
3.9	24.6911	.39254	24.7113	.39290	.99918	.99964	1.0008	.00036
4.0	27.2899	.43600	27.3082	.43629	0.99933	.99971	1.0007	.00029
4.1	30.1619	.47946	30.1784	.47970	.99945	.99976	1.0005	.00024
4.2	33.3357	.52291	33.3507	.52310	.99955	.99980	1.0004	.00020
4.3	36.8431	.56636	36.8567	.56652	.99963	.99984	1.0004	.00016
4.4	40.7193	.60980	40.7316	.60993	.99970	.99987	1.0003	.00013
4.5	45.0030	.65324	45.0141	.65335	0.99975	.99989	1.0002	.00011
4.6	49.7371	.69668	49.7472	.69677	.99980	.99991	1.0002	.00009
4.7	54.9690	.74012	54.9781	.74019	.99983	.99993	1.0002	.00007
4.8	60.7511	.78355	60.7593	.78361	.99986	.99994	1.0001	.00006
4.9	67.1412	.82699	67.1486	.82704	.99989	.99995	1.0001	.00005
5.0	74.2032	.87042	74.2099	.87046	0.99991	.99996	1.0001	.00004
5.1	82.008	.91386	82.014	.91389	.99993	.99997	1.0001	.00003
5.2	90.633	.95729	90.639	.95731	.99994	.99997	1.0001	.00003
5.3	100.17	.00072	100.17	.00074	.99995	.99998	1.0000	.00002
5.4	110.70	.04415	110.71	.04417	.99996	.99998	1.0000	.00002
5.5	122.34	.08758	122.35	.08760	0.99997	.99999	1.0000	.00001
5.6	135.21	.13101	135.22	.13103	.99997	.99999	1.0000	.00001
5.7	149.43	.17444	149.44	.17445	.99998	.99999	1.0000	.00001
5.8	165.15	.21787	165.15	.21788	.99998	.99999	1.0000	.00001
5.9	182.52	.26130	182.52	.26131	.99998	.99999	1.0000	.00001
6.0	201.71	.30473	201.72	.30474	0.99999	.00000	1.0000	.00000
6.1	222.93	.34817	222.93	.34817	.99999	.00000	1.0000	.00000
6.2	246.37	.39159	246.38	.39161	.99999	.00000	1.0000	.00000
6.3	272.29	.43503	272.29	.43503	.99999	.00000	1.0000	.00000
6.4	300.92	.47845	300.92	.47845	.99999	.00000	1.0000	.00000
6.5	332.57	.52188	332.57	.52188	1.0000	.00000	1.0000	.00000
6.6	367.55	.56532	367.55	.56532	1.0000	.00000	1.0000	.00000
6.7	406.20	.60874	406.20	.60874	1.0000	.00000	1.0000	.00000
6.8	448.92	.65217	448.92	.65217	1.0000	.00000	1.0000	.00000
6.9	496.14	.69560	496.14	.69560	1.0000	.00000	1.0000	.00000
7.0	548.32	.73903	548.32	.73903	1.0000	.00000	1.0000	.00000
7.1	605.98	.78246	605.98	.78246	1.0000	.00000	1.0000	.00000
7.2	669.72	.82589	669.72	.82589	1.0000	.00000	1.0000	.00000
7.3	740.15	.86932	740.15	.86932	1.0000	.00000	1.0000	.00000
7.4	817.99	.91275	817.99	.91275	1.0000	.00000	1.0000	.00000
7.5	904.02	.95618	904.02	.95618	1.0000	.00000	1.0000	.00000
7.6	999.10	.99961	999.10	.99961	1.0000	.00000	1.0000	.00000
7.7	1104.2	.04305	1104.2	.04305	1.0000	.00000	1.0000	.00000
7.8	1220.3	.08647	1220.3	.08647	1.0000	.00000	1.0000	.00000
7.9	1348.6	.12988	1348.6	.12988	1.0000	.00000	1.0000	.00000

HYPERBOLIC FUNCTIONS AND THEIR COMMON LOGARITHMS (Continued)

x	Sinh x		Cosh x		Tanh x		Coth x	
	Value	$\log_{10}$	Value	$\log_{10}$	Value	$\log_{10}$	Value	$\log_{10}$
8.0	1490.5	.17333	1490.5	.17333	1.0000	.00000	1.0000	.00000
8.1	1647.2	.21675	1647.2	.21675	1.0000	.00000	1.0000	.00000
8.2	1820.5	.26019	1820.5	.26019	1.0000	.00000	1.0000	.00000
8.3	2011.9	.30360	2011.9	.30360	1.0000	.00000	1.0000	.00000
8.4	2223.5	.34704	2223.5	.34704	1.0000	.00000	1.0000	.00000
8.5	2457.4	.39048	2457.4	.39048	1.0000	.00000	1.0000	.00000
8.6	2715.8	.43390	2715.8	.43390	1.0000	.00000	1.0000	.00000
8.7	3001.5	.47734	3001.5	.47734	1.0000	.00000	1.0000	.00000
8.8	3317.1	.52076	3317.1	.52076	1.0000	.00000	1.0000	.00000
8.9	3666.0	.56419	3666.0	.56419	1.0000	.00000	1.0000	.00000
9.0	4051.5	.60762	4051.5	.60762	1.0000	.00000	1.0000	.00000
9.1	4477.6	.65105	4477.6	.65105	1.0000	.00000	1.0000	.00000
9.2	4948.6	.69448	4948.6	.69448	1.0000	.00000	1.0000	.00000
9.3	5469.0	.73791	5469.0	.73791	1.0000	.00000	1.0000	.00000
9.4	6044.2	.78134	6044.2	.78134	1.0000	.00000	1.0000	.00000
9.5	6679.9	.82477	6679.9	.82477	1.0000	.00000	1.0000	.00000
9.6	7382.4	.86820	7382.4	.86820	1.0000	.00000	1.0000	.00000
9.7	8158.8	.91163	8158.8	.91163	1.0000	.00000	1.0000	.00000
9.8	9016.9	.95506	9016.9	.95506	1.0000	.00000	1.0000	.00000
9.9	9965.2	.99849	9965.2	.99849	1.0000	.00000	1.0000	.00000
10.0	11013.2	.04191	11013.2	.04191	1.0000	.00000	1.0000	.00000

EXPONENTIAL AND HYPERBOLIC FUNCTIONS FOR πx

x	$e^{\pi x}$	$e^{-\pi x}$	$\sinh \pi x$	$\cosh \pi x$	$\tanh \pi x$
0.00	1.00000	1.00000	0.00000	1.00000	0.00000
0.01	1.03191	0.96907	0.03142	1.00049	0.03141
0.02	1.06485	0.93910	0.06287	1.00197	0.06275
0.03	1.09883	0.91006	0.09439	1.00444	0.09397
0.04	1.13390	0.88191	0.12599	1.00791	0.12501
0.05	1.17009	0.85464	0.15773	1.01236	0.15580
0.06	1.20743	0.82820	0.18961	1.01782	0.18629
0.07	1.24597	0.80259	0.22169	1.02428	0.21643
0.08	1.28573	0.77777	0.25398	1.03175	0.24617
0.09	1.32676	0.75371	0.28653	1.04024	0.27544
0.10	1.36911	0.73040	0.31935	1.04976	0.30422
0.11	1.41280	0.70781	0.35249	1.06031	0.33245
0.12	1.45789	0.68592	0.38598	1.07191	0.36009
0.13	1.50442	0.66471	0.41986	1.08456	0.38712
0.14	1.55243	0.64415	0.45414	1.09829	0.41350
0.15	1.60198	0.62423	0.48887	1.11310	0.43920
0.16	1.65310	0.60492	0.52409	1.12901	0.46420
0.17	1.70586	0.58621	0.55982	1.14604	0.48849
0.18	1.76030	0.56808	0.59611	1.16419	0.51204
0.19	1.81648	0.55051	0.63298	1.18350	0.53484
0.20	1.87446	0.53349	0.67048	1.20397	0.55689
0.21	1.93428	0.51699	0.70865	1.22563	0.57819
0.22	1.99601	0.50100	0.74751	1.24850	0.59872
0.23	2.05971	0.48550	0.78710	1.27261	0.61850
0.24	2.12545	0.47049	0.82748	1.29797	0.63752
0.25	2.19328	0.45594	0.86867	1.32461	0.65579
0.26	2.26328	0.44184	0.91072	1.35256	0.67333
0.27	2.33551	0.42817	0.95367	1.38184	0.69014
0.28	2.41005	0.41493	0.99756	1.41249	0.70624
0.29	2.48696	0.40210	1.04243	1.44453	0.72164
0.30	2.56633	0.38966	1.08834	1.47800	0.73636
0.31	2.64824	0.37761	1.13531	1.51292	0.75041
0.32	2.73275	0.36593	1.18341	1.54934	0.76382
0.33	2.81997	0.35461	1.23268	1.58729	0.77659
0.34	2.90997	0.34365	1.28316	1.62681	0.78876
0.35	3.00284	0.33302	1.33491	1.66793	0.80034
0.36	3.09867	0.32272	1.38798	1.71070	0.81135
0.37	3.19756	0.31274	1.44241	1.75515	0.82182
0.38	3.29961	0.30307	1.49827	1.80134	0.83176
0.39	3.40492	0.29369	1.55561	1.84931	0.84119
0.40	3.51359	0.28461	1.61449	1.89910	0.85013
0.41	3.62572	0.27581	1.67496	1.95076	0.85862
0.42	3.74143	0.26728	1.73708	2.00436	0.86665
0.43	3.86084	0.25901	1.80091	2.05993	0.87426
0.44	3.98406	0.25100	1.86653	2.11753	0.88147
0.45	4.11121	0.24324	1.93398	2.17722	0.88828
0.46	4.24241	0.23571	2.00335	2.23906	0.89473
0.47	4.37781	0.22842	2.07469	2.30312	0.90082
0.48	4.51753	0.22136	2.14808	2.36944	0.90658
0.49	4.66170	0.21451	2.22359	2.43811	0.91202
0.50	4.81048	0.20788	2.30130	2.50918	0.91715
0.51	4.96400	0.20145	2.38128	2.58273	0.92200
0.52	5.12243	0.19522	2.46360	2.65882	0.92658
0.53	5.28591	0.18918	2.54836	2.73754	0.93089
0.54	5.45460	0.18333	2.63564	2.81897	0.93497

EXPONENTIAL, HYPERBOLIC FUNCTIONS FOR πx (Continued)

x	$e^{\pi x}$	$e^{-\pi x}$	πx	$\cosh \pi x$	$\tanh \pi x$
0.55	5.62869	0.17766	2.72551	2.90317	0.93880
0.56	5.80832	0.17217	2.81808	2.99024	0.94242
0.57	5.99369	0.16684	2.91343	3.08027	0.94584
0.58	6.18498	0.16168	3.01165	3.17333	0.94905
0.59	6.38237	0.15668	3.11284	3.26953	0.95208
0.60	6.58606	0.15184	3.21711	3.36895	0.95493
0.61	6.79625	0.14714	3.32456	3.47170	0.95762
0.62	7.01315	0.14259	3.43528	3.57787	0.96015
0.63	7.23698	0.13818	3.54940	3.68758	0.96253
0.64	7.46794	0.13391	3.66702	3.80092	0.96477
0.65	7.70628	0.12976	3.78826	3.91802	0.96688
0.66	7.95222	0.12575	3.91323	4.03899	0.96887
0.67	8.20601	0.12186	4.04208	4.16394	0.97073
0.68	8.46790	0.11809	4.17491	4.29300	0.97249
0.69	8.73815	0.11444	4.31186	4.42630	0.97415
0.70	9.01703	0.11090	4.45306	4.56396	0.97570
0.71	9.30480	0.10747	4.59867	4.70614	0.97716
0.72	9.60176	0.10415	4.74881	4.85296	0.97854
0.73	9.90820	0.10093	4.90364	5.00456	0.97983
0.74	10.22442	0.09781	5.06331	5.16111	0.98105
0.75	10.55072	0.09478	5.22797	5.32275	0.98219
0.76	10.88745	0.09185	5.39780	5.48965	0.98327
0.77	11.23492	0.08901	5.57295	5.66196	0.98428
0.78	11.59347	0.08626	5.75361	5.83986	0.98523
0.79	11.96347	0.08359	5.93994	6.02353	0.98612
0.80	12.34528	0.08100	6.13214	6.21314	0.98696
0.81	12.73928	0.07850	6.33039	6.40889	0.98775
0.82	13.14585	0.07607	6.53489	6.61096	0.98849
0.83	13.56539	0.07372	6.74584	6.81955	0.98919
0.84	13.99833	0.07144	6.96344	7.03488	0.98985
0.85	14.44508	0.06923	7.18793	7.25715	0.99046
0.86	14.90609	0.06709	7.41950	7.48659	0.99104
0.87	15.38181	0.06501	7.65840	7.72341	0.99158
0.88	15.87271	0.06300	7.90486	7.96786	0.99209
0.89	16.37929	0.06105	8.15912	8.22017	0.99257
0.90	16.90202	0.05916	8.41243	8.48059	0.99302
0.91	17.44145	0.05733	8.69206	8.74939	0.99345
0.92	17.99808	0.05556	8.97126	9.02682	0.99384
0.93	18.57248	0.05384	9.25932	9.31316	0.99422
0.94	19.16522	0.05218	9.55652	9.60870	0.99457
0.95	19.77687	0.05056	9.86315	9.91372	0.99490
0.96	20.40804	0.04900	10.17952	10.22852	0.99521
0.97	21.05935	0.04748	10.50594	10.55342	0.99550
0.98	21.73146	0.04602	10.84272	10.88874	0.99577
0.99	22.42501	0.04459	11.19021	11.23480	0.99603
1.00	23.14069	0.04321	11.54874	11.59195	0.99627

INVERSE HYPERBOLIC FUNCTIONS

x	$\sinh^{-1} x$	$\tanh^{-1} x$	$\text{cosech}^{-1} x$	$\text{sech}^{-1} x$
0.0	0.00000	0.00000	∞	∞
0.01	0.01000	0.01000	5.29834	5.29829
0.02	0.02000	0.02000	4.60527	4.60507
0.03	0.03000	0.03001	4.19993	4.19948
0.04	0.03999	0.04002	3.91242	3.91162
0.05	0.04998	0.05004	3.68950	3.68825
0.06	0.05996	0.06007	3.50746	3.50566
0.07	0.06994	0.07011	3.35363	3.35118
0.08	0.07991	0.08017	3.22047	3.21727
0.09	0.08988	0.09024	3.10311	3.09906
0.10	0.09983	0.10034	2.99822	2.99322
0.11	0.10978	0.11045	2.90343	2.89738
0.12	0.11971	0.12058	2.81699	2.80979
0.13	0.12964	0.13074	2.73757	2.72912
0.14	0.13955	0.14093	2.66412	2.65432
0.15	0.14944	0.15114	2.59585	2.58459
0.16	0.15933	0.16139	2.53207	2.51927
0.17	0.16919	0.17167	2.47225	2.45780
0.18	0.17904	0.18198	2.41595	2.39975
0.19	0.18888	0.19234	2.36278	2.34473
0.20	0.19869	0.20273	2.31244	2.29243
0.21	0.20849	0.21317	2.26464	2.24258
0.22	0.21826	0.22366	2.21916	2.19495
0.23	0.22802	0.23419	2.17579	2.14933
0.24	0.23775	0.24477	2.13436	2.10554
0.25	0.24747	0.25541	2.09471	2.06344
0.26	0.25716	0.26611	2.05671	2.02288
0.27	0.26682	0.27686	2.02023	1.98374
0.28	0.27646	0.28768	1.98516	1.94591
0.29	0.28608	0.29857	1.95141	1.90930
0.30	0.29567	0.30952	1.91890	1.87382
0.31	0.30524	0.32055	1.88753	1.83939
0.32	0.31478	0.33165	1.85725	1.80594
0.33	0.32429	0.34283	1.82799	1.77340
0.34	0.33377	0.35409	1.79968	1.74172
0.35	0.34322	0.36544	1.77228	1.71083
0.36	0.35265	0.37689	1.74573	1.68070
0.37	0.36204	0.38842	1.71999	1.65127
0.38	0.37140	0.40006	1.69502	1.62250
0.39	0.38073	0.41180	1.67078	1.59436
0.40	0.39004	0.42365	1.64723	1.56680
0.41	0.39930	0.43561	1.62434	1.53979
0.42	0.40854	0.44769	1.60209	1.51331
0.43	0.41774	0.45990	1.58043	1.48731
0.44	0.42691	0.47223	1.55935	1.46178
0.45	0.43605	0.48470	1.53882	1.43669
0.46	0.44515	0.49731	1.51881	1.41200
0.47	0.45422	0.51007	1.49931	1.38771
0.48	0.46325	0.52298	1.48029	1.36379
0.49	0.47225	0.53606	1.46174	1.34021
0.50	0.48121	0.54931	1.44364	1.31696

INVERSE HYPERBOLIC FUNCTIONS (Continued)

x	$\sinh^{-1} x$	$\tanh^{-1} x$	$\text{cosech}^{-1} x$	$\text{sech}^{-1} x$
0.50	0.48121	0.54931	1.44364	1.31696
0.51	0.49014	0.56273	1.42596	1.29401
0.52	0.49903	0.57634	1.40870	1.27136
0.53	0.50788	0.59015	1.39183	1.24898
0.54	0.51670	0.60416	1.37535	1.22686
0.55	0.52548	0.61838	1.35924	1.20497
0.56	0.53422	0.63283	1.34348	1.18331
0.57	0.54293	0.64752	1.32807	1.16186
0.58	0.55160	0.66246	1.31299	1.14060
0.59	0.56023	0.67767	1.29824	1.11952
0.60	0.56882	0.69315	1.28380	1.09861
0.61	0.57738	0.70892	1.26965	1.07785
0.62	0.58590	0.72501	1.25580	1.05723
0.63	0.59438	0.74142	1.24223	1.03673
0.64	0.60282	0.75817	1.22894	1.01635
0.65	0.61122	0.77530	1.21591	0.99606
0.66	0.61959	0.79281	1.20314	0.97585
0.67	0.62792	0.81074	1.19062	0.95572
0.68	0.63620	0.82911	1.17834	0.93564
0.69	0.64446	0.84796	1.16629	0.91560
0.70	0.65267	0.86730	1.15448	0.89559
0.71	0.66084	0.88718	1.14288	0.87559
0.72	0.66897	0.90764	1.13151	0.85558
0.73	0.67707	0.92873	1.12034	0.83555
0.74	0.68513	0.95048	1.10938	0.81549
0.75	0.69315	0.97296	1.09861	0.79537
0.76	0.70113	0.99622	1.08804	0.77517
0.77	0.70907	1.02033	1.07766	0.75487
0.78	0.71697	1.04537	1.06746	0.73445
0.79	0.72484	1.07143	1.05744	0.71388
0.80	0.73267	1.09861	1.04759	0.69315
0.81	0.74046	1.12703	1.03792	0.67221
0.82	0.74821	1.15682	1.02840	0.65103
0.83	0.75592	1.18814	1.01905	0.62958
0.84	0.76360	1.22117	1.00986	0.60781
0.85	0.77124	1.25615	1.00082	0.58568
0.86	0.77884	1.29334	0.99193	0.56313
0.87	0.78640	1.33308	0.98319	0.54008
0.88	0.79393	1.37577	0.97459	0.51647
0.89	0.80142	1.42193	0.96613	0.49220
0.90	0.80887	1.47222	0.95780	0.46715
0.91	0.81628	1.52752	0.94961	0.44116
0.92	0.82366	1.58903	0.94154	0.41406
0.93	0.83100	1.65839	0.93361	0.38560
0.94	0.83830	1.73805	0.92580	0.35542
0.95	0.84557	1.83178	0.91810	0.32304
0.96	0.85280	1.94591	0.91053	0.28768
0.97	0.86000	2.09230	0.90307	0.24807
0.98	0.86716	2.29756	0.89573	0.20169
0.99	0.87428	2.64665	0.88850	0.14201
1.00	0.88137	∞	0.88137	0.00000

INVERSE HYPERBOLIC FUNCTIONS (Continued)

x	$\sinh^{-1} x$	$\cosh^{-1} x$	$\text{cosech}^{-1} x$	$\coth^{-1} x$
1.00	0.88137	0.00000	0.88137	∞
1.01	0.88843	0.14130	0.87436	2.65165
1.02	0.89545	0.19967	0.86744	2.30756
1.03	0.90243	0.24434	0.86063	2.10730
1.04	0.90938	0.28191	0.85391	1.96591
1.05	0.91629	0.31492	0.84730	1.85679
1.06	0.92317	0.34470	0.84078	1.76806
1.07	0.93002	0.37202	0.83435	1.69340
1.08	0.93683	0.39738	0.82801	1.62905
1.09	0.94360	0.42114	0.82177	1.57255
1.10	0.95035	0.44357	0.81561	1.52226
1.11	0.95706	0.46485	0.80954	1.47698
1.12	0.96373	0.48513	0.80355	1.43584
1.13	0.97038	0.50453	0.79764	1.39817
1.14	0.97699	0.52316	0.79182	1.36346
1.15	0.98357	0.54110	0.78607	1.33129
1.16	0.99011	0.55840	0.78041	1.30134
1.17	0.99663	0.57514	0.77482	1.27334
1.18	1.00311	0.59135	0.76930	1.24706
1.19	1.00956	0.60708	0.76386	1.22232
1.20	1.01597	0.62236	0.75849	1.19895
1.21	1.02236	0.63724	0.75319	1.17682
1.22	1.02871	0.65173	0.74796	1.15582
1.23	1.03504	0.66586	0.74279	1.13584
1.24	1.04133	0.67966	0.73770	1.11680
1.25	1.04759	0.69315	0.73267	1.09861
1.26	1.05382	0.70634	0.72770	1.08122
1.27	1.06003	0.71924	0.72280	1.06456
1.28	1.06620	0.73189	0.71796	1.04857
1.29	1.07234	0.74428	0.71318	1.03321
1.30	1.07845	0.75643	0.70846	1.01844
1.31	1.08453	0.76836	0.70380	1.00422
1.32	1.09059	0.78007	0.69920	0.99050
1.33	1.09661	0.79157	0.69465	0.97727
1.34	1.10261	0.80288	0.69016	0.96448
1.35	1.10857	0.81400	0.68572	0.95212
1.36	1.11451	0.82494	0.68134	0.94016
1.37	1.12042	0.83570	0.67701	0.92857
1.38	1.12630	0.84630	0.67273	0.91734
1.39	1.13216	0.85673	0.66851	0.90645
1.40	1.13798	0.86701	0.66433	0.89588
1.41	1.14378	0.87715	0.66020	0.88561
1.42	1.14955	0.88714	0.65612	0.87563
1.43	1.15530	0.89699	0.65209	0.86593
1.44	1.16101	0.90670	0.64811	0.85649
1.45	1.16670	0.91629	0.64417	0.84730
1.46	1.17237	0.92575	0.64028	0.83835
1.47	1.17801	0.93509	0.63643	0.82962
1.48	1.18362	0.94432	0.63263	0.82111
1.49	1.18920	0.95343	0.62886	0.81282
1.50	1.19476	0.96242	0.62515	0.80472

INVERSE HYPERBOLIC FUNCTIONS (Continued)

x	$\sinh^{-1} x$	$\cosh^{-1} x$	$\operatorname{cosech}^{-1} x$	$\coth^{-1} x$
1.50	1.19476	0.96242	0.62515	0.80472
1.51	1.20030	0.97131	0.62147	0.79681
1.52	1.20581	0.98010	0.61783	0.78909
1.53	1.21129	0.98879	0.61424	0.78155
1.54	1.21675	0.99737	0.61068	0.77418
1.55	1.22218	1.00587	0.60716	0.76697
1.56	1.22759	1.01426	0.60368	0.75991
1.57	1.23298	1.02257	0.60024	0.75301
1.58	1.23834	1.03079	0.59684	0.74626
1.59	1.24367	1.03892	0.59347	0.73965
1.60	1.24898	1.04697	0.59014	0.73317
1.61	1.25427	1.05493	0.58685	0.72682
1.62	1.25954	1.06282	0.58359	0.72061
1.63	1.26478	1.07063	0.58036	0.71451
1.64	1.26999	1.07836	0.57717	0.70853
1.65	1.27519	1.08601	0.57401	0.70267
1.66	1.28036	1.09360	0.57089	0.69692
1.67	1.28551	1.10111	0.56780	0.69128
1.68	1.29064	1.10855	0.56474	0.68574
1.69	1.29574	1.11592	0.56171	0.68030
1.70	1.30082	1.12323	0.55871	0.67496
1.71	1.30588	1.13047	0.55574	0.66972
1.72	1.31092	1.13765	0.55281	0.66457
1.73	1.31593	1.14476	0.54990	0.65951
1.74	1.32093	1.15182	0.54702	0.65453
1.75	1.32590	1.15881	0.54417	0.64964
1.76	1.33085	1.16574	0.54135	0.64483
1.77	1.33578	1.17262	0.53856	0.64011
1.78	1.34069	1.17944	0.53579	0.63546
1.79	1.34557	1.18620	0.53305	0.63088
1.80	1.35044	1.19291	0.53034	0.62638
1.81	1.35529	1.19957	0.52766	0.62195
1.82	1.36011	1.20617	0.52500	0.61759
1.83	1.36492	1.21272	0.52237	0.61330
1.84	1.36970	1.21922	0.51976	0.60908
1.85	1.37447	1.22567	0.51718	0.60492
1.86	1.37921	1.23207	0.51462	0.60082
1.87	1.38394	1.23842	0.51208	0.59679
1.88	1.38864	1.24473	0.50957	0.59281
1.89	1.39333	1.25098	0.50709	0.58890
1.90	1.39800	1.25720	0.50462	0.58504
1.91	1.40265	1.26336	0.50218	0.58123
1.92	1.40728	1.26949	0.49977	0.57748
1.93	1.41188	1.27557	0.49737	0.57379
1.94	1.41648	1.28160	0.49500	0.57014
1.95	1.42105	1.28760	0.49265	0.56655
1.96	1.42560	1.29355	0.49032	0.56301
1.97	1.43014	1.29946	0.48801	0.55951
1.98	1.43466	1.30533	0.48572	0.55606
1.99	1.43915	1.31117	0.48346	0.55266
2.00	1.44364	1.31696	0.48121	0.54931

INVERSE HYPERBOLIC FUNCTIONS (Continued)

x	$\sinh^{-1} x$	$\cosh^{-1} x$	$\operatorname{cosech}^{-1} x$	$\coth^{-1} x$
2.00	1.44364	1.31696	0.48121	0.54931
2.10	1.48748	1.37286	0.45982	0.51805
2.20	1.52966	1.42542	0.44019	0.49041
2.30	1.57028	1.47504	0.42213	0.46578
2.40	1.60944	1.52208	0.40547	0.44365
2.50	1.64723	1.56680	0.39004	0.42365
2.60	1.68374	1.60944	0.37571	0.40547
2.70	1.71905	1.65019	0.36239	0.38885
2.80	1.75323	1.68924	0.34996	0.37361
2.90	1.78634	1.72671	0.33834	0.35956
3.00	1.81845	1.76275	0.32745	0.34657
3.10	1.84960	1.79746	0.31723	0.33452
3.20	1.87986	1.83094	0.30763	0.32331
3.30	1.90927	1.86328	0.29857	0.31285
3.40	1.93788	1.89456	0.29003	0.30307
3.50	1.96572	1.92485	0.28196	0.29389
3.60	1.99284	1.95421	0.27432	0.28527
3.70	2.01926	1.98270	0.26708	0.27716
3.80	2.04503	2.01037	0.26021	0.26950
3.90	2.07017	2.03727	0.25368	0.26226
4.00	2.09471	2.06344	0.24747	0.25541
4.10	2.11869	2.08892	0.24155	0.24892
4.20	2.14211	2.11375	0.23590	0.24275
4.30	2.16502	2.13796	0.23051	0.23689
4.40	2.18742	2.16158	0.22536	0.23131
4.50	2.20935	2.18464	0.22043	0.22599
4.60	2.23081	2.20717	0.21571	0.22092
4.70	2.25184	2.22920	0.21119	0.21607
4.80	2.27244	2.25073	0.20685	0.21143
4.90	2.29264	2.27180	0.20269	0.20699
5.00	2.31244	2.29243	0.19869	0.20273
5.10	2.33186	2.31263	0.19484	0.19865
5.20	2.35093	2.33243	0.19114	0.19473
5.30	2.36964	2.35183	0.18758	0.19097
5.40	2.38801	2.37086	0.18414	0.18735
5.50	2.40606	2.38953	0.18083	0.18386
5.60	2.42379	2.40784	0.17764	0.18051
5.70	2.44122	2.42583	0.17455	0.17727
5.80	2.45836	2.44349	0.17157	0.17415
5.90	2.47521	2.46084	0.16869	0.17114
6.00	2.49178	2.47789	0.16590	0.16824
6.10	2.50809	2.49465	0.16321	0.16543
6.20	2.52414	2.51113	0.16060	0.16271
6.30	2.53994	2.52734	0.15807	0.16008
6.40	2.55549	2.54329	0.15562	0.15754
6.50	2.57081	2.55898	0.15325	0.15508
6.60	2.58591	2.57443	0.15094	0.15269
6.70	2.60078	2.58964	0.14871	0.15038
6.80	2.61543	2.60462	0.14653	0.14813
6.90	2.62988	2.61938	0.14442	0.14596
7.00	2.64412	2.63392	0.14238	0.14384

GUDERMANNIAN FUNCTION

x	0	1	2	3	4	5	6	7	8	9
0.0	0.00000	0.01000	0.02000	0.03000	0.03999	0.04998	0.05996	0.06994	0.07991	0.08988
0.1	0.09983	0.10978	0.11971	0.12964	0.13954	0.14944	0.15932	0.16919	0.17904	0.18887
0.2	0.19868	0.20847	0.21825	0.22800	0.23773	0.24744	0.25712	0.26678	0.27641	0.28602
0.3	0.29560	0.30515	0.31467	0.32417	0.33363	0.34307	0.35247	0.36184	0.37117	0.38047
0.4	0.38974	0.39897	0.40817	0.41733	0.42645	0.43554	0.44459	0.45359	0.46256	0.47149
0.5	0.48038	0.48923	0.49803	0.50680	0.51552	0.52420	0.53284	0.54143	0.54997	0.55848
0.6	0.56694	0.57535	0.58372	0.59204	0.60031	0.60854	0.61672	0.62486	0.63294	0.64098
0.7	0.64897	0.65692	0.66481	0.67266	0.68045	0.68820	0.69590	0.70355	0.71115	0.71870
0.8	0.72620	0.73366	0.74106	0.74841	0.75571	0.76297	0.77017	0.77732	0.78443	0.79148
0.9	0.79848	0.80544	0.81234	0.81919	0.82599	0.83275	0.83945	0.84611	0.85271	0.85926
1.0	0.86577	0.87223	0.87863	0.88499	0.89130	0.89756	0.90377	0.90993	0.91604	0.92211
1.1	0.92813	0.93410	0.94002	0.94589	0.95172	0.95750	0.96323	0.96892	0.97455	0.98015
1.2	0.98569	0.99119	0.99665	1.00205	1.00742	1.01274	1.01801	1.02324	1.02842	1.03356
1.3	1.03866	1.04371	1.04872	1.05368	1.05860	1.06348	1.06832	1.07312	1.07787	1.08258
1.4	1.08725	1.09188	1.09647	1.10101	1.10552	1.10999	1.11441	1.11880	1.12315	1.12746
1.5	1.13173	1.13596	1.14015	1.14431	1.14843	1.15251	1.15655	1.16056	1.16453	1.16846
1.6	1.17236	1.17622	1.18005	1.18384	1.18760	1.19132	1.19500	1.19866	1.20228	1.20586
1.7	1.20941	1.21293	1.21642	1.21987	1.22330	1.22668	1.23004	1.23337	1.23666	1.23993
1.8	1.24316	1.24636	1.24954	1.25268	1.25579	1.25888	1.26193	1.26496	1.26795	1.27092
1.9	1.27386	1.27677	1.27966	1.28251	1.28534	1.28815	1.29092	1.29367	1.29639	1.29909
2.0	1.30176	1.30441	1.30703	1.30962	1.31219	1.31473	1.31726	1.31975	1.32222	1.32467
2.1	1.32710	1.32950	1.33188	1.33423	1.33656	1.33887	1.34116	1.34343	1.34567	1.34789
2.2	1.35009	1.35227	1.35443	1.35656	1.35868	1.36077	1.36285	1.36490	1.36694	1.36895
2.3	1.37095	1.37292	1.37488	1.37682	1.37873	1.38063	1.38251	1.38438	1.38622	1.38805
2.4	1.38986	1.39165	1.39342	1.39518	1.39691	1.39864	1.40034	1.40203	1.40370	1.40535
2.5	1.40699	1.40862	1.41022	1.41181	1.41339	1.41495	1.41649	1.41802	1.41954	1.42104
2.6	1.42252	1.42399	1.42545	1.42689	1.42832	1.42973	1.43113	1.43251	1.43388	1.43524
2.7	1.43659	1.43792	1.43924	1.44054	1.44183	1.44311	1.44438	1.44564	1.44688	1.44811
2.8	1.44933	1.45053	1.45173	1.45291	1.45408	1.45524	1.45638	1.45752	1.45864	1.45976
2.9	1.46086	1.46195	1.46303	1.46410	1.46516	1.46621	1.46725	1.46828	1.46930	1.47031
3.0	1.47130	1.47229	1.47327	1.47424	1.47520	1.47615	1.47709	1.47802	1.47894	1.47986
3.1	1.48076	1.48165	1.48254	1.48342	1.48428	1.48514	1.48600	1.48684	1.48767	1.48850
3.2	1.48932	1.49013	1.49093	1.49172	1.49251	1.49329	1.49406	1.49482	1.49558	1.49632
3.3	1.49706	1.49780	1.49852	1.49924	1.49995	1.50066	1.50135	1.50204	1.50273	1.50340
3.4	1.50407	1.50474	1.50539	1.50605	1.50669	1.50733	1.50796	1.50858	1.50920	1.50981
3.5	1.51042	1.51102	1.51161	1.51220	1.51279	1.51336	1.51393	1.51450	1.51506	1.51561
3.6	1.51616	1.51671	1.51724	1.51778	1.51830	1.51883	1.51934	1.51985	1.52036	1.52086
3.7	1.52136	1.52185	1.52234	1.52282	1.52330	1.52377	1.52424	1.52470	1.52516	1.52561
3.8	1.52606	1.52651	1.52695	1.52738	1.52782	1.52824	1.52867	1.52909	1.52950	1.52991
3.9	1.53032	1.53072	1.53112	1.53151	1.53190	1.53229	1.53267	1.53305	1.53343	1.53380
4.0	1.53417	1.53453	1.53489	1.53525	1.53561	1.53596	1.53630	1.53664	1.53698	1.53732
4.1	1.53765	1.53798	1.53831	1.53863	1.53895	1.53927	1.53958	1.53989	1.54020	1.54051
4.2	1.54081	1.54111	1.54140	1.54169	1.54198	1.54227	1.54255	1.54283	1.54311	1.54339
4.3	1.54366	1.54393	1.54420	1.54446	1.54472	1.54498	1.54524	1.54550	1.54575	1.54600
4.4	1.54624	1.54649	1.54673	1.54697	1.54721	1.54744	1.54767	1.54790	1.54813	1.54836
4.5	1.54858	1.54880	1.54902	1.54924	1.54945	1.54966	1.54987	1.55008	1.55029	1.55049
4.6	1.55069	1.55089	1.55109	1.55129	1.55148	1.55167	1.55186	1.55205	1.55224	1.55242
4.7	1.55261	1.55279	1.55297	1.55314	1.55332	1.55349	1.55367	1.55384	1.55400	1.55417
4.8	1.55434	1.55450	1.55466	1.55482	1.55498	1.55514	1.55530	1.55545	1.55560	1.55575
4.9	1.55590	1.55605	1.55620	1.55634	1.55649	1.55663	1.55677	1.55691	1.55705	1.55719

GUDERMANNIAN FUNCTION (Continued)

x	0	1	2	3	4	5	6	7	8	9
5.0	1.55732	1.55745	1.55759	1.55772	1.55785	1.55798	1.55811	1.55823	1.55836	1.55848
5.1	1.55860	1.55872	1.55884	1.55896	1.55908	1.55920	1.55931	1.55943	1.55954	1.55965
5.2	1.55976	1.55987	1.55998	1.56009	1.56020	1.56030	1.56041	1.56051	1.56061	1.56071
5.3	1.56081	1.56091	1.56101	1.56111	1.56120	1.56130	1.56139	1.56149	1.56158	1.56167
5.4	1.56176	1.56185	1.56194	1.56203	1.56212	1.56220	1.56229	1.56237	1.56246	1.56254
5.5	1.56262	1.56270	1.56278	1.56286	1.56294	1.56302	1.56310	1.56318	1.56325	1.56333
5.6	1.56340	1.56347	1.56355	1.56362	1.56369	1.56376	1.56383	1.56390	1.56397	1.56404
5.7	1.56410	1.56417	1.56424	1.56430	1.56437	1.56443	1.56449	1.56456	1.56462	1.56468
5.8	1.56474	1.56480	1.56486	1.56492	1.56498	1.56504	1.56509	1.56515	1.56521	1.56526
5.9	1.56532	1.56537	1.56543	1.56548	1.56553	1.56558	1.56564	1.56569	1.56574	1.56579
6.0	1.56584	1.56589	1.56594	1.56599	1.56603	1.56608	1.56613	1.56617	1.56622	1.56627
6.1	1.56631	1.56636	1.56640	1.56644	1.56649	1.56653	1.56657	1.56661	1.56666	1.56670
6.2	1.56674	1.56678	1.56682	1.56686	1.56690	1.56694	1.56697	1.56701	1.56705	1.56709
6.3	1.56712	1.56716	1.56720	1.56723	1.56727	1.56730	1.56734	1.56737	1.56741	1.56744
6.4	1.56747	1.56751	1.56754	1.56757	1.56760	1.56764	1.56767	1.56770	1.56773	1.56776
6.5	1.56779	1.56782	1.56785	1.56788	1.56791	1.56794	1.56796	1.56799	1.56802	1.56805
6.6	1.56808	1.56810	1.56813	1.56816	1.56818	1.56821	1.56823	1.56826	1.56828	1.56831
6.7	1.56833	1.56836	1.56838	1.56841	1.56843	1.56845	1.56848	1.56850	1.56852	1.56855
6.8	1.56857	1.56859	1.56861	1.56863	1.56866	1.56868	1.56870	1.56872	1.56874	1.56876
6.9	1.56878	1.56880	1.56882	1.56884	1.56886	1.56888	1.56890	1.56892	1.56894	1.56895
7.0	1.56897	1.56899	1.56901	1.56903	1.56904	1.56906	1.56908	1.56910	1.56911	1.56913
7.1	1.56915	1.56916	1.56918	1.56919	1.56921	1.56923	1.56924	1.56926	1.56927	1.56929
7.2	1.56930	1.56932	1.56933	1.56935	1.56936	1.56938	1.56939	1.56940	1.56942	1.56943
7.3	1.56945	1.56946	1.56947	1.56949	1.56950	1.56951	1.56952	1.56954	1.56955	1.56956
7.4	1.56957	1.56959	1.56960	1.56961	1.56962	1.56963	1.56965	1.56966	1.56967	1.56968
7.5	1.56969	1.56970	1.56971	1.56972	1.56973	1.56974	1.56975	1.56976	1.56978	1.56979
7.6	1.56980	1.56981	1.56982	1.56983	1.56983	1.56984	1.56985	1.56986	1.56987	1.56988
7.7	1.56989	1.56990	1.56991	1.56992	1.56993	1.56993	1.56994	1.56995	1.56996	1.56997
7.8	1.56998	1.56999	1.56999	1.57000	1.57001	1.57002	1.57002	1.57003	1.57004	1.57005
7.9	1.57005	1.57006	1.57007	1.57008	1.57008	1.57008	1.57010	1.57010	1.57011	1.57012
8.0	1.57013	1.57013	1.57014	1.57015	1.57015	1.57016	1.57016	1.57017	1.57018	1.57018
8.1	1.57019	1.57020	1.57020	1.57021	1.57021	1.57022	1.57022	1.57023	1.57024	1.57024
8.2	1.57025	1.57025	1.57026	1.57026	1.57027	1.57027	1.57028	1.57028	1.57029	1.57029
8.3	1.57030	1.57030	1.57031	1.57031	1.57032	1.57032	1.57033	1.57033	1.57034	1.57034
8.4	1.57035	1.57035	1.57036	1.57036	1.57036	1.57037	1.57037	1.57038	1.57038	1.57039
8.5	1.57039	1.57039	1.57040	1.57040	1.57041	1.57041	1.57041	1.57042	1.57042	1.57042
8.6	1.57043	1.57043	1.57044	1.57044	1.57044	1.57045	1.57045	1.57045	1.57046	1.57046
8.7	1.57046	1.57047	1.57047	1.57047	1.57048	1.57048	1.57048	1.57049	1.57049	1.57049
8.8	1.57049	1.57050	1.57050	1.57050	1.57051	1.57051	1.57051	1.57052	1.57052	1.57052
8.9	1.57052	1.57053	1.57053	1.57053	1.57053	1.57054	1.57054	1.57054	1.57054	1.57055
9.0	1.57055	1.57055	1.57055	1.57056	1.57056	1.57056	1.57056	1.57057	1.57057	1.57057
9.1	1.57057	1.57058	1.57058	1.57058	1.57058	1.57058	1.57059	1.57059	1.57059	1.57059
9.2	1.57059	1.57060	1.57060	1.57060	1.57060	1.57060	1.57061	1.57061	1.57061	1.57061
9.3	1.57061	1.57062	1.57062	1.57062	1.57062	1.57062	1.57062	1.57063	1.57063	1.57063
9.4	1.57063	1.57063	1.57063	1.57064	1.57064	1.57064	1.57064	1.57064	1.57064	1.57065
9.5	1.57065	1.57065	1.57065	1.57065	1.57065	1.57065	1.57066	1.57066	1.57066	1.57066
9.6	1.57066	1.57066	1.57066	1.57066	1.57067	1.57067	1.57067	1.57067	1.57067	1.57067
9.7	1.57067	1.57067	1.57068	1.57068	1.57068	1.57068	1.57068	1.57068	1.57068	1.57068
9.8	1.57069	1.57069	1.57069	1.57069	1.57069	1.57069	1.57069	1.57069	1.57069	1.57069
9.9	1.57070	1.57070	1.57070	1.57070	1.57070	1.57070	1.57070	1.57070	1.57070	1.57070

INVERSE GUDERMANNIAN FUNCTION

x	0	1	2	3	4	5	6	7	8	9
0.0	0.00000	0.01000	0.02000	0.03000	0.04001	0.05002	0.06004	0.07006	0.08009	0.09012
0.1	0.10017	0.11022	0.12029	0.13037	0.14046	0.15057	0.16069	0.17082	0.18098	0.19115
0.2	0.20135	0.21156	0.22180	0.23206	0.24234	0.25265	0.26298	0.27334	0.28373	0.29415
0.3	0.30460	0.31509	0.32561	0.33616	0.34675	0.35737	0.36804	0.37874	0.38949	0.40028
0.4	0.41111	0.42199	0.43292	0.44390	0.45493	0.46600	0.47714	0.48833	0.49957	0.51087
0.5	0.52224	0.53366	0.54515	0.55671	0.56834	0.58003	0.59180	0.60364	0.61555	0.62755
0.6	0.63962	0.65178	0.66402	0.67636	0.68878	0.70129	0.71390	0.72661	0.73942	0.75233
0.7	0.76535	0.77848	0.79172	0.80508	0.81856	0.83217	0.84590	0.85976	0.87376	0.88790
0.8	0.90218	0.91660	0.93118	0.94592	0.96082	0.97589	0.99113	1.00654	1.02215	1.03794
0.9	1.05392	1.07011	1.08651	1.10313	1.11997	1.13704	1.15435	1.17192	1.18974	1.20783
1.0	1.22619	1.24485	1.26380	1.28306	1.30265	1.32258	1.34285	1.36349	1.38451	1.40593
1.1	1.42776	1.45003	1.47275	1.49594	1.51963	1.54384	1.56860	1.59394	1.61987	1.64645
1.2	1.67370	1.70166	1.73037	1.75987	1.79022	1.82147	1.85367	1.88689	1.92120	1.95667
1.3	1.99340	2.03147	2.07100	2.11210	2.15491	2.19959	2.24630	2.29524	2.34666	2.40080
1.4	2.45800	2.51861	2.58307	2.65193	2.72583	2.80558	2.89219	2.98695	3.09160	3.20843
1.5	3.34068	3.49307	3.67286	3.89217	4.17343	4.56609	5.22169	7.82865		

DEGREES, MINUTES, AND SECONDS TO RADIANS

Units in degrees, minutes *or* seconds	Degrees to Radians	Minutes to Radians	Seconds to Radians
10	0.174 5329	0.002 9089	0.000 0485
20	0.349 0659	0.005 8178	0.000 0970
30	0.523 5988	0.008 7266	0.000 1454
40	0.698 1317	0.011 6355	0.000 1939
50	0.872 6646	0.014 5444	0.000 2424
60	1.047 1976	0.017 4533	0.000 2909
70	1.221 7305	(0.020 3622)	(0.000 3394)
80	1.396 2634	(0.023 2711)	(0.000 3879)
90	1.570 7963	(0.026 1799)	(0.000 4363)
100	1.745 3293	....	....
200	3.490 6585	....	....
300	5.235 9878	....	....

where n = 1, 2, 3, 4, etc. n (100°) = n (1.745 3293)

*RADIANS TO DEGREES, MINUTES, AND SECONDS

Radians	1.0	0.1	0.01	0.001	0.0001
1	57° 17′ 44.8″	5° 43′ 46.5″	0° 34′ 22.6″	0° 03′ 26.3″	0° 00′ 20.6″
2	114° 35′ 29.6″	11° 27′ 33.0″	1° 08′ 45.3″	0° 06′ 52.5″	0° 00′ 41.3″
3	171° 53′ 14.4″	17° 11′ 19.4″	1° 43′ 07.9′	0° 10′ 18.8″	0° 01′ 01.9″
4	229° 10′ 59.2″	22° 55′ 05.9″	2° 17′ 30.6″	0° 13′ 45.1′	0° 01′ 22.5″
5	286° 28′ 44.0″	28° 38′ 52.4″	2° 51′ 53.2″	0° 17′ 11.3″	0° 01′ 43.1″
6	343° 46′ 28.8″	34° 22′ 38.9″	3° 26′ 15.9″	0° 20′ 37.6″	0° 02′ 03.8″
7	401° 04′ 13.6″	40° 06′ 25.4″	4° 00′ 38.5″	0° 24′ 03.9″	0° 02′ 24.4″
8	458° 21′ 58.4″	45° 50′ 11.8″	4° 35′ 01.2″	0° 27′ 30.1″	0° 02′ 45.0″
9	515° 39′ 43.3″	51° 33′ 58.3″	5° 09′ 23.8″	0° 30′ 56.4″	0° 03′ 05.6″

*If 3.214 is desired in degrees, minutes and seconds it is obtained as follows:

3 = 171° 53′ 14.4″
.2 = 11° 27′ 33.0″
.01 = 0° 34′ 22.6″
.004 = 0° 13′ 45.1″
3.214 = 184° 8′ 55.1″

MILS—RADIANS—DEGREES

1 mil = 0.00098175 radians = 0.05625° = 3.375′ = 202.5″
1000 mils = 0.98175 radians = 56.25°
6400 mils = 360° = 2π radians
1 radian = 1018.6 mils
1° = 17.777778 mils
1′ = 0.296296 mils
1″ = 0.0049383 mils

DEGREES—RADIANS

1 radian = 57° 17′ 44″ .80625

		log
1 radian	= 57.29577 95131 degrees	1.75812 26324
1 radian	= 3437.74677 07849 minutes	3.53627 38828
1 radian	= 206264.80625 seconds	5.31442 51332
1 degree	= 0.01745 32925 19943 radians	8.24187 73676–10
1 minute	= 0.00029 08882 08666 radians	6.46372 61172–10
1 second	= 0.00000 48481 36811 radians	4.68557 48668–10

DEGREES—RADIANS

The table gives in radians the angle which is expressed in degrees and minutes at the side and top. Angles expressed to the nearest minute and second can readily be converted to radians by adding to the equivalent of the whole number of degrees the equivalents of the minutes and seconds found on the third page of this table.

°	00′	10	20	30	40	50
0	0.00000	0.00291	0.00582	0.00873	0.01164	0.01454
1	0.01745	0.02036	0.02327	0.02618	0.02909	0.03200
2	0.03491	0.03782	0.04072	0.04363	0.04654	0.04945
3	0.05236	0.05527	0.05818	0.06109	0.06400	0.06690
4	0.06981	0.07272	0.07563	0.07854	0.08145	0.08436
5	0.08727	0.09018	0.09308	0.09599	0.09890	0.10181
6	0.10472	0.10763	0.11054	0.11345	0.11636	0.11926
7	0.12217	0.12508	0.12799	0.13090	0.13381	0.13672
8	0.13963	0.14254	0.14544	0.14835	0.15126	0.15417
9	0.15708	0.15999	0.16290	0.16581	0.16872	0.17162
10	0.17453	0.17744	0.18035	0.18326	0.18617	0.18908
11	0.19199	0.19490	0.19780	0.20071	0.20362	0.20653
12	0.20944	0.21235	0.21526	0.21817	0.22108	0.22398
13	0.22689	0.22980	0.23271	0.23562	0.23853	0.24144
14	0.24435	0.24725	0.25016	0.25307	0.25598	0.25889
15	0.26180	0.26471	0.26762	0.27053	0.27343	0.27634
16	0.27925	0.28216	0.28507	0.28798	0.29089	0.29380
17	0.29671	0.29961	0.30252	0.30543	0.30834	0.31125
18	0.31416	0.31707	0.31998	0.32289	0.32579	0.32870
19	0.33161	0.33452	0.33743	0.34034	0.34325	0.34616
20	0.34907	0.35197	0.35488	0.35779	0.36070	0.36361
21	0.36652	0.36943	0.37234	0.37525	0.37815	0.38106
22	0.38397	0.38688	0.38979	0.39270	0.39561	0.39852
23	0.40143	0.40433	0.40724	0.41015	0.41306	0.41597
24	0.41888	0.42179	0.42470	0.42761	0.43051	0.43342
25	0.43633	0.43924	0.44215	0.44506	0.44797	0.45088
26	0.45379	0.45669	0.45960	0.46251	0.46542	0.46833
27	0.47124	0.47415	0.47706	0.47997	0.48287	0.48578
28	0.48869	0.49160	0.49451	0.49742	0.50033	0.50324
29	0.50615	0.50905	0.51196	0.51487	0.51778	0.52069
30	0.52360	0.52651	0.52942	0.53233	0.53523	0.53814
31	0.54105	0.54396	0.54687	0.54978	0.55269	0.55560
32	0.55851	0.56141	0.56432	0.56723	0.57014	0.57305
33	0.57596	0.57887	0.58178	0.58469	0.58759	0.59050
34	0.59341	0.59632	0.59923	0.60214	0.60505	0.60796
35	0.61087	0.61377	0.61668	0.61959	0.62250	0.62541
36	0.62832	0.63123	0.63414	0.63705	0.63995	0.64286
37	0.64577	0.64868	0.65159	0.65450	0.65741	0.66032
38	0.66323	0.66613	0.66904	0.67195	0.67486	0.67777
39	0.68068	0.68359	0.68650	0.68941	0.69231	0.69522
40	0.69813	0.70104	0.70395	0.70686	0.70977	0.71268
41	0.71558	0.71849	0.72140	0.72431	0.72722	0.73013
42	0.73304	0.73595	0.73886	0.74176	0.74467	0.74758
43	0.75049	0.75340	0.75631	0.75922	0.76213	0.76504
44	0.76794	0.77085	0.77376	0.77667	0.77958	0.78249
45	0.78540	0.78831	0.79122	0.79412	0.79703	0.79994
46	0.80285	0.80576	0.80867	0.81158	0.81449	0.81740
47	0.82030	0.82321	0.82612	0.82903	0.83194	0.83485
48	0.83776	0.84067	0.84358	0.84648	0.84939	0.85230
49	0.85521	0.85812	0.86103	0.86394	0.86685	0.86976

DEGREES—RADIANS (Continued)

°	00′	10	20	30	40	50
50	0.87266	0.87557	0.87848	0.88139	0.88430	0.88721
51	0.89012	0.89303	0.89594	0.89884	0.90175	0.90466
52	0.90757	0.91048	0.91339	0.91630	0.91921	0.92212
53	0.92502	0.92793	0.93084	0.93375	0.93666	0.93957
54	0.94248	0.94539	0.94830	0.95120	0.95411	0.95702
55	0.95993	0.96284	0.96575	0.96866	0.97157	0.97448
56	0.97738	0.98029	0.98320	0.98611	0.98902	0.99193
57	0.99484	0.99775	1.00066	1.00356	1.00647	1.00938
58	1.01229	1.01520	1.01811	1.02102	1.02393	1.02684
59	1.02974	1.03265	1.03556	1.03847	1.04138	1.04429
60	1.04720	1.05011	1.05302	1.05592	1.05883	1.06174
61	1.06465	1.06756	1.07047	1.07338	1.07629	1.07920
62	1.08210	1.08501	1.08792	1.09083	1.09374	1.09665
63	1.09956	1.10247	1.10538	1.10828	1.11119	1.11410
64	1.11701	1.11992	1.12283	1.12574	1.12865	1.13156
65	1.13446	1.13737	1.14028	1.14319	1.14610	1.14901
66	1.15192	1.15483	1.15774	1.16064	1.16355	1.16646
67	1.16937	1.17228	1.17519	1.17810	1.18101	1.18392
68	1.18682	1.18973	1.19264	1.19555	1.19846	1.20137
69	1.20428	1.20719	1.21009	1.21300	1.21591	1.21882
70	1.22173	1.22464	1.22755	1.23046	1.23337	1.23627
71	1.23918	1.24209	1.24500	1.24791	1.25082	1.25373
72	1.25664	1.25955	1.26245	1.26536	1.26827	1.27118
73	1.27409	1.27700	1.27991	1.28282	1.28573	1.28863
74	1.29154	1.29445	1.29736	1.30027	1.30318	1.30609
75	1.30900	1.31191	1.31481	1.31772	1.32063	1.32354
76	1.32645	1.32936	1.33227	1.33518	1.33809	1.34099
77	1.34390	1.34681	1.34972	1.35263	1.35554	1.35845
78	1.36136	1.36427	1.36717	1.37008	1.37299	1.37590
79	1.37881	1.38172	1.38463	1.38754	1.39045	1.39335
80	1.39626	1.39917	1.40208	1.40499	1.40790	1.41081
81	1.41372	1.41663	1.41953	1.42244	1.42535	1.42826
82	1.43117	1.43408	1.43699	1.43990	1.44281	1.44571
83	1.44862	1.45153	1.45444	1.45735	1.46026	1.46317
84	1.46608	1.46899	1.47189	1.47480	1.47771	1.48062
85	1.48353	1.48644	1.48935	1.49226	1.49517	1.49807
86	1.50098	1.50389	1.50680	1.50971	1.51262	1.51553
87	1.51844	1.52135	1.52425	1.52716	1.53007	1.53298
88	1.53589	1.53880	1.54171	1.54462	1.54753	1.55043
89	1.55334	1.55625	1.55916	1.56207	1.56498	1.56789
90	1.57080	1.57371	1.57661	1.57952	1.58243	1.58534
91	1.58825	1.59116	1.59407	1.59698	1.59989	1.60279
92	1.60570	1.60861	1.61152	1.61443	1.61734	1.62025
93	1.62316	1.62607	1.62897	1.63188	1.63479	1.63770
94	1.64061	1.64352	1.64643	1.64934	1.65225	1.65515
95	1.65806	1.66097	1.66388	1.66679	1.66970	1.67261
96	1.67552	1.67842	1.68133	1.68424	1.68715	1.69006
97	1.69297	1.69588	1.69879	1.70170	1.70460	1.70751
98	1.71042	1.71333	1.71624	1.71915	1.72206	1.72497
99	1.72788	1.73078	1.73369	1.73660	1.73951	1.74242
100	1.74533	1.74824	1.75115	1.75406	1.75696	1.75987
101	1.76278	1.76569	1.76860	1.77151	1.77442	1.77733
102	1.78024	1.78314	1.78605	1.78896	1.79187	1.79478
103	1.79769	1.80060	1.80351	1.80642	1.80932	1.81223
104	1.81514	1.81805	1.82096	1.82387	1.82678	1.82969
105	1.83260	1.83550	1.83841	1.84132	1.84423	1.84714
106	1.85005	1.85296	1.85587	1.85878	1.86168	1.86459
107	1.86750	1.87041	1.87332	1.87623	1.87914	1.88205
108	1.88496	1.88786	1.89077	1.89368	1.89659	1.89950
109	1.90241	1.90532	1.90823	1.91114	1.91404	1.91695
110	1.91986	1.92277	1.92568	1.92859	1.93150	1.93441

DEGREES—RADIANS (Continued)

Deg.	Radians	Deg.	Radians	Min.	Radians	Sec.	Radians
90	1.57080	**150**	2.61799	**0**	0.00000	**0**	0.00000
91	1.58825	151	2.63545	1	0.00029	1	0.00000
92	1.60570	152	2.65290	2	0.00058	2	0.00001
93	1.62316	153	2.67035	3	0.00087	3	0.00001
94	1.64061	154	2.68781	4	0.00116	4	0.00002
95	1.65806	**155**	2.70526	**5**	0.00145	**5**	0.00002
96	1.67552	156	2.72271	6	0.00175	6	0.00003
97	1.69297	157	2.74017	7	0.00204	7	0.00003
98	1.71042	158	2.75762	8	0.00233	8	0.00004
99	1.72788	159	2.77507	9	0.00262	9	0.00004
100	1.74533	**160**	2.79253	**10**	0.00291	**10**	0.00005
101	1.76278	161	2.80998	11	0.00320	11	0.00005
102	1.78024	162	2.82743	12	0.00349	12	0.00006
103	1.79769	163	2.84489	13	0.00378	13	0.00006
104	1.81514	164	2.86234	14	0.00407	14	0.00007
105	1.83260	**165**	2.87979	**15**	0.00436	**15**	0.00007
106	1.85005	166	2.89725	16	0.00465	16	0.00008
107	1.86750	167	2.91470	17	0.00495	17	0.00008
108	1.88496	168	2.93215	18	0.00524	18	0.00009
109	1.90241	169	2.94961	19	0.00553	19	0.00009
110	1.91986	**170**	2.96706	**20**	0.00582	**20**	0.00010
111	1.93732	171	2.98451	21	0.00611	21	0.00010
112	1.95477	172	3.00197	22	0.00640	22	0.00011
113	1.97222	173	3.01942	23	0.00669	23	0.00011
114	1.98968	174	3.03687	24	0.00698	24	0.00012
115	2.00713	**175**	3.05433	**25**	0.00727	**25**	0.00012
116	2.02458	176	3.07178	26	0.00756	26	0.00013
117	2.04204	177	3.08923	27	0.00785	27	0.00013
118	2.05949	178	3.10669	28	0.00814	28	0.00014
119	2.07694	179	3.12414	29	0.00844	29	0.00014
120	2.09440	**180**	3.14159	**30**	0.00873	**30**	0.00015
121	2.11185	190	3.31613	31	0.00902	31	0.00015
122	2.12930	200	3.49066	32	0.00931	32	0.00016
123	2.14675	210	3.66519	33	0.00960	33	0.00016
124	2.16421	220	3.83972	34	0.00989	34	0.00016
125	2.18166	**230**	4.01426	**35**	0.01018	**35**	0.00017
126	2.19911	240	4.18879	36	0.01047	36	0.00017
127	2.21657	250	4.36332	37	0.01076	37	0.00018
128	2.23402	260	4.53786	38	0.01105	38	0.00018
129	2.25147	270	4.71239	39	0.01134	39	0.00019
130	2.26893	**280**	4.88692	**40**	0.01164	**40**	0.00019
131	2.28638	290	5.06145	41	0.01193	41	0.00020
132	2.30383	300	5.23599	42	0.01222	42	0.00020
133	2.32129	310	5.41052	43	0.01251	43	0.00021
134	2.33874	320	5.58505	44	0.01280	44	0.00021
135	2.35619	**330**	5.75959	**45**	0.01309	**45**	0.00022
136	2.37365	340	5.93412	46	0.01338	46	0.00022
137	2.39110	350	6.10865	47	0.01367	47	0.00023
138	2.40855	360	6.28319	48	0.01396	48	0.00023
139	2.42601	370	6.45772	49	0.01425	49	0.00024
140	2.44346	**380**	6.63225	**50**	0.01454	**50**	0.00024
141	2.46091	390	6.80678	51	0.01484	51	0.00025
142	2.47837	400	6.98132	52	0.01513	52	0.00025
143	2.49582	410	7.15585	53	0.01542	53	0.00026
144	2.51327	420	7.33038	54	0.01571	54	0.00026
145	2.53073	**430**	7.50492	**55**	0.01600	**55**	0.00027
146	2.54818	440	7.67945	56	0.01629	56	0.00027
147	2.56563	450	7.85398	57	0.01658	57	0.00028
148	2.58309	460	8.02851	58	0.01687	58	0.00028
149	2.60054	470	8.20305	59	0.01716	59	0.00029
150	2.61799	**480**	8.37758	**60**	0.01745	**60**	0.00029

DEGREES AND DECIMAL FRACTIONS TO RADIANS

The table below facilitates conversion of an angle expressed in degrees and decimal fractions into radians. To convert 25.78 into radians, find the equivalents, successively, of 20°, 5°, 0°.7, 0°.08 and add.

Deg.	Radians	Deg.	Radians	Deg.	Radians	Deg.	Radians	Deg.	Radians
10	0.174533	1	0.017453	0.1	0.001745	0.01	0.000175	0.001	0.000017
20	0.349066	2	.034907	.2	.003491	.02	.000349	.002	.000035
30	0.523599	3	.052360	.3	.005236	.03	.000524	.003	.000052
40	0.698132	4	.069813	.4	.006981	.04	.000698	.004	.000070
50	0.872665	5	.087266	.5	.008727	.05	.000873	.005	.000087
60	1.047198	6	.104720	.6	.010472	.06	.001047	.006	.000105
70	1.221730	7	.122173	.7	.012217	.07	.001222	.007	.000122
80	1.396263	8	.139626	.8	.013963	.08	.001396	.008	.000140
90	1.570796	9	.157080	.9	.015708	.09	.001571	.009	.000157

RADIANS TO DEGREES AND DECIMALS

Radians	Degrees	Radians	Degrees	Radians	Degrees	Radians	Degrees
1	57.2958	0.1	5.7296	0.01	0.5730	0.001	0.0573
2	114.5916	.2	11.4592	.02	1.1459	.002	.1146
3	171.8873	.3	17.1887	.03	1.7189	.003	.1719
4	229.1831	.4	22.9183	.04	2.2918	.004	.2292
5	286.4789	.5	28.6479	.05	2.8648	.005	.2865
6	343.7747	.6	34.3775	.06	3.4377	.006	.3438
7	401.0705	.7	40.1070	.07	4.0107	.007	.4011
8	458.3662	.8	45.8366	.08	4.5837	.008	.4584
9	515.6620	.9	51.5662	.09	5.1566	.009	.5157
10	572.9578	1.0	57.2958	.10	5.7296	.010	.5730

RADIANS—DEGREES

Multiples and Fractions of π Radians in Degrees

Radians	Radians	Deg.	Radians	Radians	Deg.	Radians	Radians	Deg.
π	3.1416	180	$\pi/2$	1.5708	90	$2\pi/3$	2.0944	120
2π	6.2832	360	$\pi/3$	1.0472	60	$3\pi/4$	2.3562	135
3π	9.4248	540	$\pi/4$	0.7854	45	$5\pi/6$	2.6180	150
4π	12.5664	720	$\pi/5$	0.6283	36	$7\pi/6$	3.6652	210
5π	15.7080	900	$\pi/6$	0.5236	30	$5\pi/4$	3.9270	225
6π	18.8496	1080	$\pi/7$	0.4488	25.714	$4\pi/3$	4.1888	240
7π	21.9911	1260	$\pi/8$	0.3927	22.5	$3\pi/2$	4.7124	270
8π	25.1327	1440	$\pi/9$	0.3491	20	$5\pi/3$	5.2360	300
9π	28.2743	1620	$\pi/10$	0.3142	18	$7\pi/4$	5.4978	315
10π	31.4159	1800	$\pi/12$	0.2618	15	$11\pi/6$	5.7596	330

CONVERSION OF ANGLES FROM ARC TO TIME

Arc	Time	Arc	Time	Arc	Time	Arc	Time
°	h m	°	h m	″	s	″	s
′	m s	′	m s				
0	0 00	20	1 20	0	0.00	8	0.53
1	0 04	30	2 00	1	0.07	9	0.60
2	0 08	40	2 40	2	0.13	10	0.67
3	0 12	50	3 20	3	0.20	20	1.33
4	0 16	60	4 00	4	0.27	30	2.00
5	0 20	70	4 40	5	0.33	40	2.67
6	0 24	80	5 20	6	0.40	50	3.33
7	0 28	90	6 00	7	0.47	60	4.00
8	0 32	100	6 40				
9	0 36	200	13 20				
10	0 40	300	20 00				

MINUTES AND SECONDS TO DECIMAL PARTS OF A DEGREE

MINUTES AND SECONDS TO DECIMAL PARTS OF A DEG.				DECIMAL PARTS OF A DEGREE TO MINUTES AND SECONDS					
Min.	Degrees	Sec.	Degrees	Deg.	′	″	Deg.	′	″
0	0.00000	**0**	0.00000	**0.00**	0	00	**0.60**	36	
1	.01667	1	.00028	.01	0	36	.61	36	36
2	.03333	2	.00056	.02	1	12	.62	37	12
3	.05	3	.00083	.03	1	48	.63	37	48
4	.06667	4	.00111	.04	2	24	.64	38	24
5	.08333	5	.00139	.05	3		.65	39	
6	.10	6	.00167	.06	3	36	.66	39	36
7	.11667	7	.00194	.07	4	12	.67	40	12
8	.13333	8	.00222	.08	4	48	.68	40	48
9	.15	9	.0025	.09	5	24	.69	41	24
10	0.16667	**10**	0.00278	**0.10**	6		**0.70**	42	
11	.18333	11	.00306	.11	6	36	.71	42	36
12	.20	12	.00333	.12	7	12	.72	43	12
13	.21667	13	.00361	.13	7	48	.73	43	48
14	.23333	14	.00389	.14	8	24	.74	44	24
15	.25	15	.00417	.15	9		.75	45	
16	.26667	16	.00444	.16	9	36	.76	45	36
17	.28333	17	.00472	.17	10	12	.77	46	12
18	.30	18	.005	.18	10	48	.78	46	48
19	.31667	19	.00528	.19	11	24	.79	47	24
20	0.33333	**20**	0.00556	**0.20**	12		**0.80**	48	
21	.35	21	.00583	.21	12	36	.81	48	36
22	.36667	22	.00611	.22	13	12	.82	49	12
23	.38333	23	.00639	.23	13	48	.83	49	48
24	.40	24	.00667	.24	14	24	.84	50	24
25	.41667	25	.00694	.25	15		.85	51	
26	.43333	26	.00722	.26	15	36	.86	51	36
27	.45	27	.0075	.27	16	12	.87	52	12
28	.46667	28	.00778	.28	16	48	.88	52	48
29	.48333	29	.00806	.29	17	24	.89	53	24
30	0.50	**30**	0.00833	**0.30**	18		**0.90**	54	
31	.51667	31	.00861	.31	18	36	.91	54	36
32	.53333	32	.00889	.32	19	12	.92	55	12
33	.55	33	.00917	.33	19	48	.93	55	48
34	.56667	34	.00944	.34	20	24	.94	56	24
35	.58333	35	.00972	.35	21		.95	57	
36	.60	36	.01	.36	21	36	.96	57	36
37	.61667	37	.01028	.37	22	12	.97	58	12
38	.63333	38	.01056	.38	22	48	.98	58	48
39	.65	39	.01083	.39	23	24	.99	59	24
40	0.66667	**40**	0.01111	**0.40**	24		**1.00**	60	
41	.68333	41	.01139	.41	24	36			
42	.70	42	.01167	.42	25	12			
43	.71667	43	.01194	.43	25	48			
44	.73333	44	.01222	.44	26	24			
45	.75	45	.0125	.45	27				
46	.76667	46	.01278	.46	27	36			
47	.78333	47	.01306	.47	28	12			
48	.80	48	.01333	.48	28	48			
49	.81667	49	.01361	.49	29	24			
50	0.83333	**50**	0.01389	**0.50**	30				
51	.85	51	.01417	.51	30	36			
52	.86667	52	.01444	.52	31	12			
53	.88333	53	.01472	.53	31	48			
54	.90	54	.015	.54	32	24			
55	.91667	55	.01528	.55	33				
56	.93333	56	.01556	.56	33	36			
57	.95	57	.01583	.57	34	12			
58	.96667	58	.01611	.58	34	48			
59	.98333	59	.01639	.59	35	24			
60	1.00	**60**	0.01667	**0.60**	36				

Deg.	Sec.
0.000	0.0
.001	3.6
.002	7.2
.003	10.8
.004	14.4
.005	18.
.006	21.6
.007	25.2
.008	28.8
.009	32.4
0.010	36.

NATURAL TRIGONOMETRIC FUNCTIONS TO FIVE PLACES

Values of the trigonometric functions of angles for each minute from 0–360°.

For degrees indicated at the top of the page use the column headings at the top. For degrees indicated at the bottom use the column indications at the bottom.

With degrees at the left of each block (top or bottom), use the minute column at the left and with degrees at the right of each block use the minute column at the right.

If natural trigonometric function tables are used for angle measures greater than 90° and less than 360°, appropriate signs for the functions must be supplied in accordance with the quadrant in which the angle measure belongs.

Linear interpolation may be used to obtain values of the functions for tenths of degrees, except in regions where the functions are rapidly changing. See section on "Use of Logarithm Tables" for further discussion in interpolation practice.

NATURAL TRIGONOMETRIC FUNCTIONS TO FIVE PLACES

0° (180°) (359°) **179°**

′	Sin	Tan	Cot	Cos	Sec	Csc	′
0	.00000	.00000	———	1.0000	1.0000	———	**60**
1	.00029	.00029	3437.7	1.0000	1.0000	3437.7	59
2	.00058	.00058	1718.9	1.0000	1.0000	1718.9	58
3	.00087	.00087	1145.9	1.0000	1.0000	1145.9	57
4	.00116	.00116	859.44	1.0000	1.0000	859.44	56
5	.00145	.00145	687.55	1.0000	1.0000	687.55	**55**
6	.00175	.00175	572.96	1.0000	1.0000	572.96	54
7	.00204	.00204	491.11	1.0000	1.0000	491.11	53
8	.00233	.00233	429.72	1.0000	1.0000	429.72	52
9	.00262	.00262	381.97	1.0000	1.0000	381.97	51
10	.00291	.00291	343.77	1.0000	1.0000	343.78	**50**
11	.00320	.00320	312.52	.99999	1.0000	312.52	49
12	.00349	.00349	286.48	.99999	1.0000	286.48	48
13	.00378	.00378	264.44	.99999	1.0000	264.44	47
14	.00407	.00407	245.55	.99999	1.0000	245.55	46
15	.00436	.00436	229.18	.99999	1.0000	229.18	**45**
16	.00465	.00465	214.86	.99999	1.0000	214.86	44
17	.00495	.00495	202.22	.99999	1.0000	202.22	43
18	.00524	.00524	190.98	.99999	1.0000	190.99	42
19	.00553	.00553	180.93	.99998	1.0000	180.93	41
20	.00582	.00582	171.89	.99998	1.0000	171.89	**40**
21	.00611	.00611	163.70	.99998	1.0000	163.70	39
22	.00640	.00640	156.26	.99998	1.0000	156.26	38
23	.00669	.00669	149.47	.99998	1.0000	149.47	37
24	.00698	.00698	143.24	.99998	1.0000	143.24	36
25	.00727	.00727	137.51	.99997	1.0000	137.51	**35**
26	.00756	.00756	132.22	.99997	1.0000	132.22	34
27	.00785	.00785	127.32	.99997	1.0000	127.33	33
28	.00814	.00815	122.77	.99997	1.0000	122.78	32
29	.00844	.00844	118.54	.99996	1.0000	118.54	31
30	.00873	.00873	114.59	.99996	1.0000	114.59	**30**
31	.00902	.00902	110.89	.99996	1.0000	110.90	29
32	.00931	.00931	107.43	.99996	1.0000	107.43	28
33	.00960	.00960	104.17	.99995	1.0000	104.18	27
34	.00989	.00989	101.11	.99995	1.0000	101.11	26
35	.01018	.01018	98.218	.99995	1.0001	98.223	**25**
36	.01047	.01047	95.489	.99995	1.0001	95.495	24
37	.01076	.01076	92.908	.99994	1.0001	92.914	23
38	.01105	.01105	90.463	.99994	1.0001	90.469	22
39	.01134	.01135	88.144	.99994	1.0001	88.149	21
40	.01164	.01164	85.940	.99993	1.0001	85.946	**20**
41	.01193	.01193	83.844	.99993	1.0001	83.849	19
42	.01222	.01222	81.847	.99993	1.0001	81.853	18
43	.01251	.01251	79.943	.99992	1.0001	79.950	17
44	.01280	.01280	78.126	.99992	1.0001	78.133	16
45	.01309	.01309	76.390	.99991	1.0001	76.397	**15**
46	.01338	.01338	74.729	.99991	1.0001	74.736	14
47	.01367	.01367	73.139	.99991	1.0001	73.146	13
48	.01396	.01396	71.615	.99990	1.0001	71.622	12
49	.01425	.01425	70.153	.99990	1.0001	70.160	11
50	.01454	.01455	68.750	.99989	1.0001	68.757	**10**
51	.01483	.01484	67.402	.99989	1.0001	67.409	9
52	.01513	.01513	66.105	.99989	1.0001	66.113	8
53	.01542	.01542	64.858	.99988	1.0001	64.866	7
54	.01571	.01571	63.657	.99988	1.0001	63.665	6
55	.01600	.01600	62.499	.99987	1.0001	62.507	**5**
56	.01629	.01629	61.383	.99987	1.0001	61.391	4
57	.01658	.01658	60.306	.99986	1.0001	60.314	3
58	.01687	.01687	59.266	.99986	1.0001	59.274	2
59	.01716	.01716	58.261	.99985	1.0001	58.270	1
60	.01745	.01746	57.290	.99985	1.0002	57.299	**0**
′	Cos	Cot	Tan	Sin	Csc	Sec	′

90° (270°) (269°) **89°**

1° (181°) (358°) **178°**

′	Sin	Tan	Cot	Cos	Sec	Csc	′
0	.01745	.01746	57.290	.99985	1.0002	57.299	**60**
1	.01774	.01775	56.351	.99984	1.0002	56.359	59
2	.01803	.01804	55.442	.99984	1.0002	55.451	58
3	.01832	.01833	54.561	.99983	1.0002	54.570	57
4	.01862	.01862	53.709	.99983	1.0002	53.718	56
5	.01891	.01891	52.882	.99982	1.0002	52.892	**55**
6	.01920	.01920	52.081	.99982	1.0002	52.090	54
7	.01949	.01949	51.303	.99981	1.0002	51.313	53
8	.01978	.01978	50.549	.99980	1.0002	50.558	52
9	.02007	.02007	49.816	.99980	1.0002	49.826	51
10	.02036	.02036	49.104	.99979	1.0002	49.114	**50**
11	.02065	.02066	48.412	.99979	1.0002	48.422	49
12	.02094	.02095	47.740	.99978	1.0002	47.750	48
13	.02123	.02124	47.085	.99977	1.0002	47.096	47
14	.02152	.02153	46.449	.99977	1.0002	46.460	46
15	.02181	.02182	45.829	.99976	1.0002	45.840	**45**
16	.02211	.02211	45.226	.99976	1.0002	45.237	44
17	.02240	.02240	44.639	.99975	1.0003	44.650	43
18	.02269	.02269	44.066	.99974	1.0003	44.077	42
19	.02298	.02298	43.508	.99974	1.0003	43.520	41
20	.02327	.02328	42.964	.99973	1.0003	42.976	**40**
21	.02356	.02357	42.433	.99972	1.0003	42.445	39
22	.02385	.02386	41.916	.99972	1.0003	41.928	38
23	.02414	.02415	41.411	.99971	1.0003	41.423	37
24	.02443	.02444	40.917	.99970	1.0003	40.930	36
25	.02472	.02473	40.436	.99969	1.0003	40.448	**35**
26	.02501	.02502	39.965	.99969	1.0003	39.978	34
27	.02530	.02531	39.506	.99968	1.0003	39.519	33
28	.02560	.02560	39.057	.99967	1.0003	39.070	32
29	.02589	.02589	38.618	.99966	1.0003	38.631	31
30	.02618	.02619	38.188	.99966	1.0003	38.202	**30**
31	.02647	.02648	37.769	.99965	1.0004	37.782	29
32	.02676	.02677	37.358	.99964	1.0004	37.371	28
33	.02705	.02706	36.956	.99963	1.0004	36.970	27
34	.02734	.02735	36.563	.99963	1.0004	36.576	26
35	.02763	.02764	36.178	.99962	1.0004	36.191	**25**
36	.02792	.02793	35.801	.99961	1.0004	35.815	24
37	.02821	.02822	35.431	.99960	1.0004	35.445	23
38	.02850	.02851	35.070	.99959	1.0004	35.084	22
39	.02879	.02881	34.715	.99959	1.0004	34.730	21
40	.02908	.02910	34.368	.99958	1.0004	34.382	**20**
41	.02938	.02939	34.027	.99957	1.0004	34.042	19
42	.02967	.02968	33.694	.99956	1.0004	33.708	18
43	.02996	.02997	33.366	.99955	1.0004	33.381	17
44	.03025	.03026	33.045	.99954	1.0005	33.060	16
45	.03054	.03055	32.730	.99953	1.0005	32.746	**15**
46	.03083	.03084	32.421	.99952	1.0005	32.437	14
47	.03112	.03114	32.118	.99952	1.0005	32.134	13
48	.03141	.03143	31.821	.99951	1.0005	31.836	12
49	.03170	.03172	31.528	.99950	1.0005	31.544	11
50	.03199	.03201	31.242	.99949	1.0005	31.258	**10**
51	.03228	.03230	30.960	.99948	1.0005	30.976	9
52	.03257	.03259	30.683	.99947	1.0005	30.700	8
53	.03286	.03288	30.412	.99946	1.0005	30.428	7
54	.03316	.03317	30.145	.99945	1.0006	30.161	6
55	.03345	.03346	29.882	.99944	1.0006	29.899	**5**
56	.03374	.03376	29.624	.99943	1.0006	29.641	4
57	.03403	.03405	29.371	.99942	1.0006	29.388	3
58	.03432	.03434	29.122	.99941	1.0006	29.139	2
59	.03461	.03463	28.877	.99940	1.0006	28.894	1
60	.03490	.03492	28.636	.99939	1.0006	28.654	**0**
′	Cos	Cot	Tan	Sin	Csc	Sec	′

91° (271°) (268°) **88°**

NATURAL TRIGONOMETRIC FUNCTIONS TO FIVE PLACES

2° (182°) (357°) **177°**

′	Sin	Tan	Cot	Cos	Sec	Csc	′
0	.03490	.03492	28.636	.99939	1.0006	28.654	**60**
1	.03519	.03521	28.399	.99938	1.0006	28.417	59
2	.03548	.03550	28.166	.99937	1.0006	28.184	58
3	.03577	.03579	27.937	.99936	1.0006	27.955	57
4	.03606	.03609	27.712	.99935	1.0007	27.730	56
5	.03635	.03638	27.490	.99934	1.0007	27.508	**55**
6	.03664	.03667	27.271	.99933	1.0007	27.290	54
7	.03693	.03696	27.057	.99932	1.0007	27.075	53
8	.03723	.03725	26.845	.99931	1.0007	26.864	52
9	.03752	.03754	26.637	.99930	1.0007	26.655	51
10	.03781	.03783	26.432	.99929	1.0007	26.451	**50**
11	.03810	.03812	26.230	.99927	1.0007	26.249	49
12	.03839	.03842	26.031	.99926	1.0007	26.050	48
13	.03868	.03871	25.835	.99925	1.0007	25.854	47
14	.03897	.03900	25.642	.99924	1.0008	25.661	46
15	.03926	.03929	25.452	.99923	1.0008	25.471	**45**
16	.03955	.03958	25.264	.99922	1.0008	25.284	44
17	.03984	.03987	25.080	.99921	1.0008	25.100	43
18	.04013	.04016	24.898	.99919	1.0008	24.918	42
19	.04042	.04046	24.719	.99918	1.0008	24.739	41
20	.04071	.04075	24.542	.99917	1.0008	24.562	**40**
21	.04100	.04104	24.368	.99916	1.0008	24.388	39
22	.04129	.04133	24.196	.99915	1.0009	24.216	38
23	.04159	.04162	24.026	.99913	1.0009	24.047	37
24	.04188	.04191	23.859	.99912	1.0009	23.880	36
25	.04217	.04220	23.695	.99911	1.0009	23.716	**35**
26	.04246	.04250	23.532	.99910	1.0009	23.553	34
27	.04275	.04279	23.372	.99909	1.0009	23.393	33
28	.04304	.04308	23.214	.99907	1.0009	23.235	32
29	.04333	.04337	23.058	.99906	1.0009	23.079	31
30	.04362	.04366	22.904	.99905	1.0010	22.926	**30**
31	.04391	.04395	22.752	.99904	1.0010	22.774	29
32	.04420	.04424	22.602	.99902	1.0010	22.624	28
33	.04449	.04454	22.454	.99901	1.0010	22.476	27
34	.04478	.04483	22.308	.99900	1.0010	22.330	26
35	.04507	.04512	22.164	.99898	1.0010	22.187	**25**
36	.04536	.04541	22.022	.99897	1.0010	22.044	24
37	.04565	.04570	21.881	.99896	1.0010	21.904	23
38	.04594	.04599	21.743	.99894	1.0011	21.766	22
39	.04623	.04628	21.606	.99893	1.0011	21.629	21
40	.04653	.04658	21.470	.99892	1.0011	21.494	**20**
41	.04682	.04687	21.337	.99890	1.0011	21.360	19
42	.04711	.04716	21.205	.99889	1.0011	21.229	18
43	.04740	.04745	21.075	.99888	1.0011	21.098	17
44	.04769	.04774	20.946	.99886	1.0011	20.970	16
45	.04798	.04803	20.819	.99885	1.0012	20.843	**15**
46	.04827	.04833	20.693	.99883	1.0012	20.717	14
47	.04856	.04862	20.569	.99882	1.0012	20.593	13
48	.04885	.04891	20.446	.99881	1.0012	20.471	12
49	.04914	.04920	20.325	.99879	1.0012	20.350	11
50	.04943	.04949	20.206	.99878	1.0012	20.230	**10**
51	.04972	.04978	20.087	.99876	1.0012	20.112	9
52	.05001	.05007	19.970	.99875	1.0013	19.995	8
53	.05030	.05037	19.855	.99873	1.0013	19.880	7
54	.05059	.05066	19.740	.99872	1.0013	19.766	6
55	.05088	.05095	19.627	.99870	1.0013	19.653	**5**
56	.05117	.05124	19.516	.99869	1.0013	19.541	4
57	.05146	.05153	19.405	.99867	1.0013	19.431	3
58	.05175	.05182	19.296	.99866	1.0013	19.322	2
59	.05205	.05212	19.188	.99864	1.0014	19.214	1
60	.05234	.05241	19.081	.99863	1.0014	19.107	**0**
′	Cos	Cot	Tan	Sin	Csc	Sec	′

92° (272°) (267°) **87°**

3° (183°) (356°) **176°**

′	Sin	Tan	Cot	Cos	Sec	Csc	′
0	.05234	.05241	19.081	.99863	1.0014	19.107	**60**
1	.05263	.05270	18.976	.99861	1.0014	19.002	59
2	.05292	.05299	18.871	.99860	1.0014	18.898	58
3	.05321	.05328	18.768	.99858	1.0014	18.794	57
4	.05350	.05357	18.666	.99857	1.0014	18.692	56
5	.05379	.05387	18.564	.99855	1.0014	18.591	**55**
6	.05408	.05416	18.464	.99854	1.0015	18.492	54
7	.05437	.05445	18.366	.99852	1.0015	18.393	53
8	.05466	.05474	18.268	.99851	1.0015	18.295	52
9	.05495	.05503	18.171	.99849	1.0015	18.198	51
10	.05524	.05533	18.075	.99847	1.0015	18.103	**50**
11	.05553	.05562	17.980	.99846	1.0015	18.008	49
12	.05582	.05591	17.886	.99844	1.0016	17.914	48
13	.05611	.05620	17.793	.99842	1.0016	17.822	47
14	.05640	.05649	17.702	.99841	1.0016	17.730	46
15	.05669	.05678	17.611	.99839	1.0016	17.639	**45**
16	.05698	.05708	17.521	.99838	1.0016	17.549	44
17	.05727	.05737	17.431	.99836	1.0016	17.460	43
18	.05756	.05766	17.343	.99834	1.0017	17.372	42
19	.05785	.05795	17.256	.99833	1.0017	17.285	41
20	.05814	.05824	17.169	.99831	1.0017	17.198	**40**
21	.05844	.05854	17.084	.99829	1.0017	17.113	39
22	.05873	.05883	16.999	.99827	1.0017	17.028	38
23	.05902	.05912	16.915	.99826	1.0017	16.945	37
24	.05931	.05941	16.832	.99824	1.0018	16.862	36
25	.05960	.05970	16.750	.99822	1.0018	16.779	**35**
26	.05989	.05999	16.668	.99821	1.0018	16.698	34
27	.06018	.06029	16.587	.99819	1.0018	16.618	33
28	.06047	.06058	16.507	.99817	1.0018	16.538	32
29	.06076	.06087	16.428	.99815	1.0019	16.459	31
30	.06105	.06116	16.350	.99813	1.0019	16.380	**30**
31	.06134	.06145	16.272	.99812	1.0019	16.303	29
32	.06163	.06175	16.195	.99810	1.0019	16.226	28
33	.06192	.06204	16.119	.99808	1.0019	16.150	27
34	.06221	.06233	16.043	.99806	1.0019	16.075	26
35	.06250	.06262	15.969	.99804	1.0020	16.000	**25**
36	.06279	.06291	15.895	.99803	1.0020	15.926	24
37	.06308	.06321	15.821	.99801	1.0020	15.853	23
38	.06337	.06350	15.748	.99799	1.0020	15.780	22
39	.06366	.06379	15.676	.99797	1.0020	15.708	21
40	.06395	.06408	15.605	.99795	1.0021	15.637	**20**
41	.06424	.06438	15.534	.99793	1.0021	15.566	19
42	.06453	.06467	15.464	.99792	1.0021	15.496	18
43	.06482	.06496	15.394	.99790	1.0021	15.427	17
44	.06511	.06525	15.325	.99788	1.0021	15.358	16
45	.06540	.06554	15.257	.99786	1.0021	15.290	**15**
46	.06569	.06584	15.189	.99784	1.0022	15.222	14
47	.06598	.06613	15.122	.99782	1.0022	15.155	13
48	.06627	.06642	15.056	.99780	1.0022	15.089	12
49	.06656	.06671	14.990	.99778	1.0022	15.023	11
50	.06685	.06700	14.924	.99776	1.0022	14.958	**10**
51	.06714	.06730	14.860	.99774	1.0023	14.893	9
52	.06743	.06759	14.795	.99772	1.0023	14.829	8
53	.06773	.06788	14.732	.99770	1.0023	14.766	7
54	.06802	.06817	14.669	.99768	1.0023	14.703	6
55	.06831	.06847	14.606	.99766	1.0023	14.640	**5**
56	.06860	.06876	14.544	.99764	1.0024	14.578	4
57	.06889	.06905	14.482	.99762	1.0024	14.517	3
58	.06918	.06934	14.421	.99760	1.0024	14.456	2
59	.06947	.06963	14.361	.99758	1.0024	14.395	1
60	.06976	.06993	14.301	.99756	1.0024	14.336	**0**
′	Cos	Cot	Tan	Sin	Csc	Sec	′

93° (273°) (266°) **86°**

NATURAL TRIGONOMETRIC FUNCTIONS TO FIVE PLACES

4° (184°) (355°) **175°**

′	Sin	Tan	Cot	Cos	Sec	Csc	′
0	.06976	.06993	14.301	.99756	1.0024	14.336	**60**
1	.07005	.07022	14.241	.99754	1.0025	14.276	59
2	.07034	.07051	14.182	.99752	1.0025	14.217	58
3	.07063	.07080	14.124	.99750	1.0025	14.159	57
4	.07092	.07110	14.065	.99748	1.0025	14.101	56
5	.07121	.07139	14.008	.99746	1.0025	14.044	**55**
6	.07150	.07168	13.951	.99744	1.0026	13.987	54
7	.07179	.07197	13.894	.99742	1.0026	13.930	53
8	.07208	.07227	13.838	.99740	1.0026	13.874	52
9	.07237	.07256	13.782	.99738	1.0026	13.818	51
10	.07266	.07285	13.727	.99736	1.0027	13.763	**50**
11	.07295	.07314	13.672	.99734	1.0027	13.708	49
12	.07324	.07344	13.617	.99731	1.0027	13.654	48
13	.07353	.07373	13.563	.99729	1.0027	13.600	47
14	.07382	.07402	13.510	.99727	1.0027	13.547	46
15	.07411	.07431	13.457	.99725	1.0028	13.494	**45**
16	.07440	.07461	13.404	.99723	1.0028	13.441	44
17	.07469	.07490	13.352	.99721	1.0028	13.389	43
18	.07498	.07519	13.300	.99719	1.0028	13.337	42
19	.07527	.07548	13.248	.99716	1.0028	13.286	41
20	.07556	.07578	13.197	.99714	1.0029	13.235	**40**
21	.07585	.07607	13.146	.99712	1.0029	13.184	39
22	.07614	.07636	13.096	.99710	1.0029	13.134	38
23	.07643	.07665	13.046	.99708	1.0029	13.084	37
24	.07672	.07695	12.996	.99705	1.0030	13.035	36
25	.07701	.07724	12.947	.99703	1.0030	12.985	**35**
26	.07730	.07753	12.898	.99701	1.0030	12.937	34
27	.07759	.07782	12.850	.99699	1.0030	12.888	33
28	.07788	.07812	12.801	.99696	1.0030	12.840	32
29	.07817	.07841	12.754	.99694	1.0031	12.793	31
30	.07846	.07870	12.706	.99692	1.0031	12.745	**30**
31	.07875	.07899	12.659	.99689	1.0031	12.699	29
32	.07904	.07929	12.612	.99687	1.0031	12.652	28
33	.07933	.07958	12.566	.99685	1.0032	12.606	27
34	.07962	.07987	12.520	.99683	1.0032	12.560	26
35	.07991	.08017	12.474	.99680	1.0032	12.514	**25**
36	.08020	.08046	12.429	.99678	1.0032	12.469	24
37	.08049	.08075	12.384	.99676	1.0033	12.424	23
38	.08078	.08104	12.339	.99673	1.0033	12.379	22
39	.08107	.08134	12.295	.99671	1.0033	12.335	21
40	.08136	.08163	12.251	.99668	1.0033	12.291	**20**
41	.08165	.08192	12.207	.99666	1.0034	12.248	19
42	.08194	.08221	12.163	.99664	1.0034	12.204	18
43	.08223	.08251	12.120	.99661	1.0034	12.161	17
44	.08252	.08280	12.077	.99659	1.0034	12.119	16
45	.08281	.08309	12.035	.99657	1.0034	12.076	**15**
46	.08310	.08339	11.992	.99654	1.0035	12.034	14
47	.08339	.08368	11.950	.99652	1.0035	11.992	13
48	.08368	.08397	11.909	.99649	1.0035	11.951	12
49	.08397	.08427	11.867	.99647	1.0035	11.909	11
50	.08426	.08456	11.826	.99644	1.0036	11.868	**10**
51	.08455	.08485	11.785	.99642	1.0036	11.828	9
52	.08484	.08514	11.745	.99639	1.0036	11.787	8
53	.08513	.08544	11.705	.99637	1.0036	11.747	7
54	.08542	.08573	11.664	.99635	1.0037	11.707	6
55	.08571	.08602	11.625	.99632	1.0037	11.668	**5**
56	.08600	.08632	11.585	.99630	1.0037	11.628	4
57	.08629	.08661	11.546	.99627	1.0037	11.589	3
58	.08658	.08690	11.507	.99625	1.0038	11.551	2
59	.08687	.08720	11.468	.99622	1.0038	11:512	1
60	.08716	.08749	11.430	.99619	1.0038	11.474	**0**
′	Cos	Cot	Tan	Sin	Csc	Sec	′

94° (274°) (265°) **85°**

5° (185°) (354°) **174°**

′	Sin	Tan	Cot	Cos	Sec	Csc	′
0	.08716	.08749	11.430	.99619	1.0038	11.474	**60**
1	.08745	.08778	11.392	.99617	1.0038	11.436	59
2	.08774	.08807	11.354	.99614	1.0039	11.398	58
3	.08803	.08837	11.316	.99612	1.0039	11.360	57
4	.08831	.08866	11.279	.99609	1.0039	11.323	56
5	.08860	.08895	11.242	.99607	1.0039	11.286	**55**
6	.08889	.08925	11.205	.99604	1.0040	11.249	54
7	.08918	.08954	11.168	.99602	1.0040	11.213	53
8	.08947	.08983	11.132	.99599	1.0040	11.176	52
9	.08976	.09013	11.095	.99596	1.0041	11.140	51
10	.09005	.09042	11.059	.99594	1.0041	11.105	**50**
11	.09034	.09071	11.024	.99591	1.0041	11.069	49
12	.09063	.09101	10.988	.99588	1.0041	11.034	48
13	.09092	.09130	10.953	.99586	1.0042	10.998	47
14	.09121	.09159	10.918	.99583	1.0042	10.963	46
15	.09150	.09189	10.883	.99580	1.0042	10.929	**45**
16	.09179	.09218	10.848	.99578	1.0042	10.894	44
17	.09208	.09247	10.814	.99575	1.0043	10.860	43
18	.09237	.09277	10.780	.99572	1.0043	10.826	42
19	.09266	.09306	10.746	.99570	1.0043	10.792	41
20	.09295	.09335	10.712	.99567	1.0043	10.758	**40**
21	.09324	.09365	10.678	.99564	1.0044	10.725	39
22	.09353	.09394	10.645	.99562	1.0044	10.692	38
23	.09382	.09423	10.612	.99559	1.0044	10.659	37
24	.09411	.09453	10.579	.99556	1.0045	10.626	36
25	.09440	.09482	10.546	.99553	1.0045	10.593	**35**
26	.09469	.09511	10.514	.99551	1.0045	10.561	34
27	.09498	.09541	10.481	.99548	1.0045	10.529	33
28	.09527	.09570	10.449	.99545	1.0046	10.497	32
29	.09556	.09600	10.417	.99542	1.0046	10.465	31
30	.09585	.09629	10.385	.99540	1.0046	10.433	**30**
31	.09614	.09658	10.354	.99537	1.0047	10.402	29
32	.09642	.09688	10.322	.99534	1.0047	10.371	28
33	.09671	.09717	10.291	.99531	1.0047	10.340	27
34	.09700	.09746	10.260	.99528	1.0047	10.309	26
35	.09729	.09776	10.229	.99526	1.0048	10.278	**25**
36	.09758	.09805	10.199	.99523	1.0048	10.248	24
37	.09787	.09834	10.168	.99520	1.0048	10.217	23
38	.09816	.09864	10.138	.99517	1.0049	10.187	22
39	.09845	.09893	10.108	.99514	1.0049	10.157	21
40	.09874	.09923	10.078	.99511	1.0049	10.128	**20**
41	.09903	.09952	10.048	.99508	1.0049	10.098	19
42	.09932	.09981	10.019	.99506	1.0050	10.068	18
43	.09961	.10011	9.9893	.99503	1.0050	10.039	17
44	.09990	.10040	9.9601	.99500	1.0050	10.010	16
45	.10019	.10069	9.9310	.99497	1.0051	9.9812	**15**
46	.10048	.10099	9.9021	.99494	1.0051	9.9525	14
47	.10077	.10128	9.8734	.99491	1.0051	9.9239	13
48	.10106	.10158	9.8448	.99488	1.0051	9.8955	12
49	.10135	.10187	9.8164	.99485	1.0052	9.8672	11
50	.10164	.10216	9.7882	.99482	1.0052	9.8391	**10**
51	.10192	.10246	9.7601	.99479	1.0052	9.8112	9
52	.10221	.10275	9.7322	.99476	1.0053	9.7834	8
53	.10250	.10305	9.7044	.99473	1.0053	9.7558	7
54	.10279	.10334	9.6768	.99470	1.0053	9.7283	6
55	.10308	.10363	9.6493	.99467	1.0054	9.7010	**5**
56	.10337	.10393	9.6220	.99464	1.0054	9.6739	4
57	.10366	.10422	9.5949	.99461	1.0054	9.6469	3
58	.10395	.10452	9.5679	.99458	1.0054	9.6200	2
59	.10424	.10481	9.5411	.99455	1.0055	9.5933	1
60	.10453	.10510	9.5144	.99452	1.0055	9.5668	**0**
′	Cos	Cot	Tan	Sin	Csc	Sec	′

95° (275°) (264°) **84°**

NATURAL TRIGONOMETRIC FUNCTIONS TO FIVE PLACES

6° (186°) (353°) **173°**

′	Sin	Tan	Cot	Cos	Sec	Csc	′
0	.10453	.10510	9.5144	.99452	1.0055	9.5668	**60**
1	.10482	.10540	9.4878	.99449	1.0055	9.5404	59
2	.10511	.10569	9.4614	.99446	1.0056	9.5141	58
3	.10540	.10599	9.4352	.99443	1.0056	9.4880	57
4	.10569	.10628	9.4090	.99440	1.0056	9.4620	56
5	.10597	.10657	9.3831	.99437	1.0057	9.4362	**55**
6	.10626	.10687	9.3572	.99434	1.0057	9.4105	54
7	.10655	.10716	9.3315	.99431	1.0057	9.3850	53
8	.10684	.10746	9.3060	.99428	1.0058	9.3596	52
9	.10713	.10775	9.2806	.99424	1.0058	9.3343	51
10	.10742	.10805	9.2553	.99421	1.0058	9.3092	**50**
11	.10771	.10834	9.2302	.99418	1.0059	9.2842	49
12	.10800	.10863	9.2052	.99415	1.0059	9.2593	48
13	.10829	.10893	9.1803	.99412	1.0059	9.2346	47
14	.10858	.10922	9.1555	.99409	1.0059	9.2100	46
15	.10887	.10952	9.1309	.99406	1.0060	9.1855	**45**
16	.10916	.10981	9.1065	.99402	1.0060	9.1612	44
17	.10945	.11011	9.0821	.99399	1.0060	9.1370	43
18	.10973	.11040	9.0579	.99396	1.0061	9.1129	42
19	.11002	.11070	9.0338	.99393	1.0061	9.0890	41
20	.11031	.11099	9.0098	.99390	1.0061	9.0652	**40**
21	.11060	.11128	8.9860	.99386	1.0062	9.0415	39
22	.11089	.11158	8.9623	.99383	1.0062	9.0179	38
23	.11118	.11187	8.9387	.99380	1.0062	8.9944	37
24	.11147	.11217	8.9152	.99377	1.0063	8.9711	36
25	.11176	.11246	8.8919	.99374	1.0063	8.9479	**35**
26	.11205	.11276	8.8686	.99370	1.0063	8.9248	34
27	.11234	.11305	8.8455	.99367	1.0064	8.9019	33
28	.11263	.11335	8.8225	.99364	1.0064	8.8790	32
29	.11291	.11364	8.7996	.99360	1.0064	8.8563	31
30	.11320	.11394	8.7769	.99357	1.0065	8.8337	**30**
31	.11349	.11423	8.7542	.99354	1.0065	8.8112	29
32	.11378	.11452	8.7317	.99351	1.0065	8.7888	28
33	.11407	.11482	8.7093	.99347	1.0066	8.7665	27
34	.11436	.11511	8.6870	.99344	1.0066	8.7444	26
35	.11465	.11541	8.6648	.99341	1.0066	8.7223	**25**
36	.11494	.11570	8.6427	.99337	1.0067	8.7004	24
37	.11523	.11600	8.6208	.99334	1.0067	8.6786	23
38	.11552	.11629	8.5989	.99331	1.0067	8.6569	22
39	.11580	.11659	8.5772	.99327	1.0068	8.6353	21
40	.11609	.11688	8.5555	.99324	1.0068	8.6138	**20**
41	.11638	.11718	8.5340	.99320	1.0068	8.5924	19
42	.11667	.11747	8.5126	.99317	1.0069	8.5711	18
43	.11696	.11777	8.4913	.99314	1.0069	8.5500	17
44	.11725	.11806	8.4701	.99310	1.0069	8.5289	16
45	.11754	.11836	8.4490	.99307	1.0070	8.5079	**15**
46	.11783	.11865	8.4280	.99303	1.0070	8.4871	14
47	.11812	.11895	8.4071	.99300	1.0070	8.4663	13
48	.11840	.11924	8.3863	.99297	1.0071	8.4457	12
49	.11869	.11954	8.3656	.99293	1.0071	8.4251	11
50	.11898	.11983	8.3450	.99290	1.0072	8.4047	**10**
51	.11927	.12013	8.3245	.99286	1.0072	8.3843	9
52	.11956	.12042	8.3041	.99283	1.0072	8.3641	8
53	.11985	.12072	8.2838	.99279	1.0073	8.3439	7
54	.12014	.12101	8.2636	.99276	1.0073	8.3238	6
55	.12043	.12131	8.2434	.99272	1.0073	8.3039	**5**
56	.12071	.12160	8.2234	.99269	1.0074	8.2840	4
57	.12100	.12190	8.2035	.99265	1.0074	8.2642	3
58	.12129	.12219	8.1837	.99262	1.0074	8.2446	2
59	.12158	.12249	8.1640	.99258	1.0075	8.2250	1
60	.12187	.12278	8.1443	.99255	1.0075	8.2055	**0**
′	Cos	Cot	Tan	Sin	Csc	Sec	′

96° (276°) (263°) **83°**

7° (187°) (352°) **172°**

′	Sin	Tan	Cot	Cos	Sec	Csc	′
0	.12187	.12278	8.1443	.99255	1.0075	8.2055	**60**
1	.12216	.12308	8.1248	.99251	1.0075	8.1861	59
2	.12245	.12338	8.1054	.99248	1.0076	8.1668	58
3	.12274	.12367	8.0860	.99244	1.0076	8.1476	57
4	.12302	.12397	8.0667	.99240	1.0077	8.1285	56
5	.12331	.12426	8.0476	.99237	1.0077	8.1095	**55**
6	.12360	.12456	8.0285	.99233	1.0077	8.0905	54
7	.12389	.12485	8.0095	.99230	1.0078	8.0717	53
8	.12418	.12515	7.9906	.99226	1.0078	8.0529	52
9	.12447	.12544	7.9718	.99222	1.0078	8.0342	51
10	.12476	.12574	7.9530	.99219	1.0079	8.0156	**50**
11	.12504	.12603	7.9344	.99215	1.0079	7.9971	49
12	.12533	.12633	7.9158	.99211	1.0079	7.9787	48
13	.12562	.12662	7.8973	.99208	1.0080	7.9604	47
14	.12591	.12692	7.8789	.99204	1.0080	7.9422	46
15	.12620	.12722	7.8606	.99200	1.0081	7.9240	**45**
16	.12649	.12751	7.8424	.99197	1.0081	7.9059	44
17	.12678	.12781	7.8243	.99193	1.0081	7.8879	43
18	.12706	.12810	7.8062	.99189	1.0082	7.8700	42
19	.12735	.12840	7.7882	.99186	1.0082	7.8522	41
20	.12764	.12869	7.7704	.99182	1.0082	7.8344	**40**
21	.12793	.12899	7.7525	.99178	1.0083	7.8168	39
22	.12822	.12929	7.7348	.99175	1.0083	7.7992	38
23	.12851	.12958	7.7171	.99171	1.0084	7.7817	37
24	.12880	.12988	7.6996	.99167	1.0084	7.7642	36
25	.12908	.13017	7.6821	.99163	1.0084	7.7469	**35**
26	.12937	.13047	7.6647	.99160	1.0085	7.7296	34
27	.12966	.13076	7.6473	.99156	1.0085	7.7124	33
28	.12995	.13106	7.6301	.99152	1.0086	7.6953	32
29	.13024	.13136	7.6129	.99148	1.0086	7.6783	31
30	.13053	.13165	7.5958	.99144	1.0086	7.6613	**30**
31	.13081	.13195	7.5787	.99141	1.0087	7.6444	29
32	.13110	.13224	7.5618	.99137	1.0087	7.6276	28
33	.13139	.13254	7.5449	.99133	1.0087	7.6109	27
34	.13168	.13284	7.5281	.99129	1.0088	7.5942	26
35	.13197	.13313	7.5113	.99125	1.0088	7.5776	**25**
36	.13226	.13343	7.4947	.99122	1.0089	7.5611	24
37	.13254	.13372	7.4781	.99118	1.0089	7.5446	23
38	.13283	.13402	7.4615	.99114	1.0089	7.5282	22
39	.13312	.13432	7.4451	.99110	1.0090	7.5119	21
40	.13341	.13461	7.4287	.99106	1.0090	7.4957	**20**
41	.13370	.13491	7.4124	.99102	1.0091	7.4795	19
42	.13399	.13521	7.3962	.99098	1.0091	7.4635	18
43	.13427	.13550	7.3800	.99094	1.0091	7.4474	17
44	.13456	.13580	7.3639	.99091	1.0092	7.4315	16
45	.13485	.13609	7.3479	.99087	1.0092	7.4156	**15**
46	.13514	.13639	7.3319	.99083	1.0093	7.3998	14
47	.13543	.13669	7.3160	.99079	1.0093	7.3840	13
48	.13572	.13698	7.3002	.99075	1.0093	7.3684	12
49	.13600	.13728	7.2844	.99071	1.0094	7.3527	11
50	.13629	.13758	7.2687	.99067	1.0094	7.3372	**10**
51	.13658	.13787	7.2531	.99063	1.0095	7.3217	9
52	.13687	.13817	7.2375	.99059	1.0095	7.3063	8
53	.13716	.13846	7.2220	.99055	1.0095	7.2909	7
54	.13744	.13876	7.2066	.99051	1.0096	7.2757	6
55	.13773	.13906	7.1912	.99047	1.0096	7.2604	**5**
56	.13802	.13935	7.1759	.99043	1.0097	7.2453	4
57	.13831	.13965	7.1607	.99039	1.0097	7.2302	3
58	.13860	.13995	7.1455	.99035	1.0097	7.2152	2
59	.13889	.14024	7.1304	.99031	1.0098	7.2002	1
60	.13917	.14054	7.1154	.99027	1.0098	7.1853	**0**
′	Cos	Cot	Tan	Sin	Csc	Sec	′

97° (277°) (262°) **82°**

NATURAL TRIGONOMETRIC FUNCTIONS TO FIVE PLACES

8° (188°) (351°) **171°**

′	Sin	Tan	Cot	Cos	Sec	Csc	′
0	.13917	.14054	7.1154	.99027	1.0098	7.1853	**60**
1	.13946	.14084	7.1004	.99023	1.0099	7.1705	59
2	.13975	.14113	7.0855	.99019	1.0099	7.1557	58
3	.14004	.14143	7.0706	.99015	1.0100	7.1410	57
4	.14033	.14173	7.0558	.99011	1.0100	7.1263	56
5	.14061	.14202	7.0410	.99006	1.0100	7.1117	**55**
6	.14090	.14232	7.0264	.99002	1.0101	7.0972	54
7	.14119	.14262	7.0117	.98998	1.0101	7.0827	53
8	.14148	.14291	6.9972	.98994	1.0102	7.0683	52
9	.14177	.14321	6.9827	.98990	1.0102	7.0539	51
10	.14205	.14351	6.9682	.98986	1.0102	7.0396	**50**
11	.14234	.14381	6.9538	.98982	1.0103	7.0254	49
12	.14263	.14410	6.9395	.98978	1.0103	7.0112	48
13	.14292	.14440	6.9252	.98973	1.0104	6.9971	47
14	.14320	.14470	6.9110	.98969	1.0104	6.9830	46
15	.14349	.14499	6.8969	.98965	1.0105	6.9690	**45**
16	.14378	.14529	6.8828	.98961	1.0105	6.9550	44
17	.14407	.14559	6.8687	.98957	1.0105	6.9411	43
18	.14436	.14588	6.8548	.98953	1.0106	6.9273	42
19	.14464	.14618	6.8408	.98948	1.0106	6.9135	41
20	.14493	.14648	6.8269	.98944	1.0107	6.8998	**40**
21	.14522	.14678	6.8131	.98940	1.0107	6.8861	39
22	.14551	.14707	6.7994	.98936	1.0108	6.8725	38
23	.14580	.14737	6.7856	.98931	1.0108	6.8589	37
24	.14608	.14767	6.7720	.98927	1.0108	6.8454	36
25	.14637	.14796	6.7584	.98923	1.0109	6.8320	**35**
26	.14666	.14826	6.7448	.98919	1.0109	6.8186	34
27	.14695	.14856	6.7313	.98914	1.0110	6.8052	33
28	.14723	.14886	6.7179	.98910	1.0110	6.7919	32
29	.14752	.14915	6.7045	.98906	1.0111	6.7787	31
30	.14781	.14945	6.6912	.98902	1.0111	6.7655	**30**
31	.14810	.14975	6.6779	.98897	1.0112	6.7523	29
32	.14838	.15005	6.6646	.98893	1.0112	6.7392	28
33	.14867	.15034	6.6514	.98889	1.0112	6.7262	27
34	.14896	.15064	6.6383	.98884	1.0113	6.7132	26
35	.14925	.15094	6.6252	.98880	1.0113	6.7003	**25**
36	.14954	.15124	6.6122	.98876	1.0114	6.6874	24
37	.14982	.15153	6.5992	.98871	1.0114	6.6745	23
38	.15011	.15183	6.5863	.98867	1.0115	6.6618	22
39	.15040	.15213	6.5734	.98863	1.0115	6.6490	21
40	.15069	.15243	6.5606	.98858	1.0116	6.6363	**20**
41	.15097	.15272	6.5478	.98854	1.0116	6.6237	19
42	.15126	.15302	6.5350	.98849	1.0116	6.6111	18
43	.15155	.15332	6.5223	.98845	1.0117	6.5986	17
44	.15184	.15362	6.5097	.98841	1.0117	6.5861	16
45	.15212	.15391	6.4971	.98836	1.0118	6.5736	**15**
46	.15241	.15421	6.4846	.98832	1.0118	6.5612	14
47	.15270	.15451	6.4721	.98827	1.0119	6.5489	13
48	.15299	.15481	6.4596	.98823	1.0119	6.5366	12
49	.15327	.15511	6.4472	.98818	1.0120	6.5243	11
50	.15356	.15540	6.4348	.98814	1.0120	6.5121	**10**
51	.15385	.15570	6.4225	.98809	1.0120	6.4999	9
52	.15414	.15600	6.4103	.98805	1.0121	6.4878	8
53	.15442	.15630	6.3980	.98800	1.0121	6.4757	7
54	.15471	.15660	6.3859	.98796	1.0122	6.4637	6
55	.15500	.15689	6.3737	.98791	1.0122	6.4517	**5**
56	.15529	.15719	6.3617	.98787	1.0123	6.4398	4
57	.15557	.15749	6.3496	.98782	1.0123	6.4279	3
58	.15586	.15779	6.3376	.98778	1.0124	6.4160	2
59	.15615	.15809	6.3257	.98773	1.0124	6.4042	1
60	.15643	.15838	6.3138	.98769	1.0125	6.3925	**0**
′	Cos	Cot	Tan	Sin	Csc	Sec	′

98° (278°) (261°) **81°**

9° (189°) (350°) **170°**

′	Sin	Tan	Cot	Cos	Sec	Csc	′
0	.15643	.15838	6.3138	.98769	1.0125	6.3925	**60**
1	.15672	.15868	6.3019	.98764	1.0125	6.3807	59
2	.15701	.15898	6.2901	.98760	1.0126	6.3691	58
3	.15730	.15928	6.2783	.98755	1.0126	6.3574	57
4	.15758	.15958	6.2666	.98751	1.0127	6.3458	56
5	.15787	.15988	6.2549	.98746	1.0127	6.3343	**55**
6	.15816	.16017	6.2432	.98741	1.0127	6.3228	54
7	.15845	.16047	6.2316	.98737	1.0128	6.3113	53
8	.15873	.16077	6.2200	.98732	1.0128	6.2999	52
9	.15902	.16107	6.2085	.98728	1.0129	6.2885	51
10	.15931	.16137	6.1970	.98723	1.0129	6.2772	**50**
11	.15959	.16167	6.1856	.98718	1.0130	6.2659	49
12	.15988	.16196	6.1742	.98714	1.0130	6.2546	48
13	.16017	.16226	6.1628	.98709	1.0131	6.2434	47
14	.16046	.16256	6.1515	.98704	1.0131	6.2323	46
15	.16074	.16286	6.1402	.98700	1.0132	6.2211	**45**
16	.16103	.16316	6.1290	.98695	1.0132	6.2100	44
17	.16132	.16346	6.1178	.98690	1.0133	6.1990	43
18	.16160	.16376	6.1066	.98686	1.0133	6.1880	42
19	.16189	.16405	6.0955	.98681	1.0134	6.1770	41
20	.16218	.16435	6.0844	.98676	1.0134	6.1661	**40**
21	.16246	.16465	6.0734	.98671	1.0135	6.1552	39
22	.16275	.16495	6.0624	.98667	1.0135	6.1443	38
23	.16304	.16525	6.0514	.98662	1.0136	6.1335	37
24	.16333	.16555	6.0405	.98657	1.0136	6.1227	36
25	.16361	.16585	6.0296	.98652	1.0137	6.1120	**35**
26	.16390	.16615	6.0188	.98648	1.0137	6.1013	34
27	.16419	.16645	6.0080	.98643	1.0138	6.0906	33
28	.16447	.16674	5.9972	.98638	1.0138	6.0800	32
29	.16476	.16704	5.9865	.98633	1.0139	6.0694	31
30	.16505	.16734	5.9758	.98629	1.0139	6.0589	**30**
31	.16533	.16764	5.9651	.98624	1.0140	6.0483	29
32	.16562	.16794	5.9545	.98619	1.0140	6.0379	28
33	.16591	.16824	5.9439	.98614	1.0141	6.0274	27
34	.16620	.16854	5.9333	.98609	1.0141	6.0170	26
35	.16648	.16884	5.9228	.98604	1.0142	6.0067	**25**
36	.16677	.16914	5.9124	.98600	1.0142	5.9963	24
37	.16706	.16944	5.9019	.98595	1.0143	5.9860	23
38	.16734	.16974	5.8915	.98590	1.0143	5.9758	22
39	.16763	.17004	5.8811	.98585	1.0144	5.9656	21
40	.16792	.17033	5.8708	.98580	1.0144	5.9554	**20**
41	.16820	.17063	5.8605	.98575	1.0145	5.9452	19
42	.16849	.17093	5.8502	.98570	1.0145	5.9351	18
43	.16878	.17123	5.8400	.98565	1.0146	5.9250	17
44	.16906	.17153	5.8298	.98561	1.0146	5.9150	16
45	.16935	.17183	5.8197	.98556	1.0147	5.9049	**15**
46	.16964	.17213	5.8095	.98551	1.0147	5.8950	14
47	.16992	.17243	5.7994	.98546	1.0148	5.8850	13
48	.17021	.17273	5.7894	.98541	1.0148	5.8751	12
49	.17050	.17303	5.7794	.98536	1.0149	5.8652	11
50	.17078	.17333	5.7694	.98531	1.0149	5.8554	**10**
51	.17107	.17363	5.7594	.98526	1.0150	5.8456	9
52	.17136	.17393	5.7495	.98521	1.0150	5.8358	8
53	.17164	.17423	5.7396	.98516	1.0151	5.8261	7
54	.17193	.17453	5.7297	.98511	1.0151	5.8164	6
55	.17222	.17483	5.7199	.98506	1.0152	5.8067	**5**
56	.17250	.17513	5.7101	.98501	1.0152	5.7970	4
57	.17279	.17543	5.7004	.98496	1.0153	5.7874	3
58	.17308	.17573	5.6906	.98491	1.0153	5.7778	2
59	.17336	.17603	5.6809	.98486	1.0154	5.7683	1
60	.17365	.17633	5.6713	.98481	1.0154	5.7588	**0**
′	Cos	Cot	Tan	Sin	Csc	Sec	′

99° (279°) (260°) **80°**

NATURAL TRIGONOMETRIC FUNCTIONS TO FIVE PLACES

10° (190°) (349°) **169°**

′	Sin	Tan	Cot	Cos	Sec	Csc	′
0	.17365	.17633	5.6713	.98481	1.0154	5.7588	**60**
1	.17393	.17663	5.6617	.98476	1.0155	5.7493	59
2	.17422	.17693	5.6521	.98471	1.0155	5.7398	58
3	.17451	.17723	5.6425	.98466	1.0156	5.7304	57
4	.17479	.17753	5.6329	.98461	1.0156	5.7210	56
5	.17508	.17783	5.6234	.98455	1.0157	5.7117	**55**
6	.17537	.17813	5.6140	.98450	1.0157	5.7023	54
7	.17565	.17843	5.6045	.98445	1.0158	5.6930	53
8	.17594	.17873	5.5951	.98440	1.0158	5.6838	52
9	.17623	.17903	5.5857	.98435	1.0159	5.6745	51
10	.17651	.17933	5.5764	.98430	1.0160	5.6653	**50**
11	.17680	.17963	5.5671	.98425	1.0160	5.6562	49
12	.17708	.17993	5.5578	.98420	1.0161	5.6470	48
13	.17737	.18023	5.5485	.98414	1.0161	5.6379	47
14	.17766	.18053	5.5393	.98409	1.0162	5.6288	46
15	.17794	.18083	5.5301	.98404	1.0162	5.6198	**45**
16	.17823	.18113	5.5209	.98399	1.0163	5.6107	44
17	.17852	.18143	5.5118	.98394	1.0163	5.6017	43
18	.17880	.18173	5.5026	.98389	1.0164	5.5928	42
19	.17909	.18203	5.4936	.98383	1.0164	5.5838	41
20	.17937	.18233	5.4845	.98378	1.0165	5.5749	**40**
21	.17966	.18263	5.4755	.98373	1.0165	5.5660	39
22	.17995	.18293	5.4665	.98368	1.0166	5.5572	38
23	.18023	.18323	5.4575	.98362	1.0166	5.5484	37
24	.18052	.18353	5.4486	.98357	1.0167	5.5396	36
25	.18081	.18384	5.4397	.98352	1.0168	5.5308	**35**
26	.18109	.18414	5.4308	.98347	1.0168	5.5221	34
27	.18138	.18444	5.4219	.98341	1.0169	5.5134	33
28	.18166	.18474	5.4131	.98336	1.0169	5.5047	32
29	.18195	.18504	5.4043	.98331	1.0170	5.4960	31
30	.18224	.18534	5.3955	.98325	1.0170	5.4874	**30**
31	.18252	.18564	5.3868	.98320	1.0171	5.4788	29
32	.18281	.18594	5.3781	.98315	1.0171	5.4702	28
33	.18309	.18624	5.3694	.98310	1.0172	5.4617	27
34	.18338	.18654	5.3607	.98304	1.0173	5.4532	26
35	.18367	.18684	5.3521	.98299	1.0173	5.4447	**25**
36	.18395	.18714	5.3435	.98294	1.0174	5.4362	24
37	.18424	.18745	5.3349	.98288	1.0174	5.4278	23
38	.18452	.18775	5.3263	.98283	1.0175	5.4194	22
39	.18481	.18805	5.3178	.98277	1.0175	5.4110	21
40	.18509	.18835	5.3093	.98272	1.0176	5.4026	**20**
41	.18538	.18865	5.3008	.98267	1.0176	5.3943	19
42	.18567	.18895	5.2924	.98261	1.0177	5.3860	18
43	.18595	.18925	5.2839	.98256	1.0178	5.3777	17
44	.18624	.18955	5.2755	.98250	1.0178	5.3695	16
45	.18652	.18986	5.2672	.98245	1.0179	5.3612	**15**
46	.18681	.19016	5.2588	.98240	1.0179	5.3530	14
47	.18710	.19046	5.2505	.98234	1.0180	5.3449	13
48	.18738	.19076	5.2422	.98229	1.0180	5.3367	12
49	.18767	.19106	5.2339	.98223	1.0181	5.3286	11
50	.18795	.19136	5.2257	.98218	1.0181	5.3205	**10**
51	.18824	.19166	5.2174	.98212	1.0182	5.3124	9
52	.18852	.19197	5.2092	.98207	1.0183	5.3044	8
53	.18881	.19227	5.2011	.98201	1.0183	5.2963	7
54	.18910	.19257	5.1929	.98196	1.0184	5.2883	6
55	.18938	.19287	5.1848	.98190	1.0184	5.2804	**5**
56	.18967	.19317	5.1767	.98185	1.0185	5.2724	4
57	.18995	.19347	5.1686	.98179	1.0185	5.2645	3
58	.19024	.19378	5.1606	.98174	1.0186	5.2566	2
59	.19052	.19408	5.1526	.98168	1.0187	5.2487	1
60	.19081	.19438	5.1446	.98163	1.0187	5.2408	**0**
′	Cos	Cot	Tan	Sin	Csc	Sec	′

100° (280°) (259°) **79°**

11° (191°) (348°) **168°**

′	Sin	Tan	Cot	Cos	Sec	Csc	′
0	.19081	.19438	5.1446	.98163	1.0187	5.2408	**60**
1	.19109	.19468	5.1366	.98157	1.0188	5.2330	59
2	.19138	.19498	5.1286	.98152	1.0188	5.2252	58
3	.19167	.19529	5.1207	.98146	1.0189	5.2174	57
4	.19195	.19559	5.1128	.98140	1.0189	5.2097	56
5	.19224	.19589	5.1049	.98135	1.0190	5.2019	**55**
6	.19252	.19619	5.0970	.98129	1.0191	5.1942	54
7	.19281	.19649	5.0892	.98124	1.0191	5.1865	53
8	.19309	.19680	5.0814	.98118	1.0192	5.1789	52
9	.19338	.19710	5.0736	.98112	1.0192	5.1712	51
10	.19366	.19740	5.0658	.98107	1.0193	5.1636	**50**
11	.19395	.19770	5.0581	.98101	1.0194	5.1560	49
12	.19423	.19801	5.0504	.98096	1.0194	5.1484	48
13	.19452	.19831	5.0427	.98090	1.0195	5.1409	47
14	.19481	.19861	5.0350	.98084	1.0195	5.1333	46
15	.19509	.19891	5.0273	.98079	1.0196	5.1258	**45**
16	.19538	.19921	5.0197	.98073	1.0197	5.1183	44
17	.19566	.19952	5.0121	.98067	1.0197	5.1109	43
18	.19595	.19982	5.0045	.98061	1.0198	5.1034	42
19	.19623	.20012	4.9969	.98056	1.0198	5.0960	41
20	.19652	.20042	4.9894	.98050	1.0199	5.0886	**40**
21	.19680	.20073	4.9819	.98044	1.0199	5.0813	39
22	.19709	.20103	4.9744	.98039	1.0200	5.0739	38
23	.19737	.20133	4.9669	.98033	1.0201	5.0666	37
24	.19766	.20164	4.9594	.98027	1.0201	5.0593	36
25	.19794	.20194	4.9520	.98021	1.0202	5.0520	**35**
26	.19823	.20224	4.9446	.98016	1.0202	5.0447	34
27	.19851	.20254	4.9372	.98010	1.0203	5.0375	33
28	.19880	.20285	4.9298	.98004	1.0204	5.0302	32
29	.19908	.20315	4.9225	.97998	1.0204	5.0230	31
30	.19937	.20345	4.9152	.97992	1.0205	5.0159	**30**
31	.19965	.20376	4.9078	.97987	1.0205	5.0087	29
32	.19994	.20406	4.9006	.97981	1.0206	5.0016	28
33	.20022	.20436	4.8933	.97975	1.0207	4.9944	27
34	.20051	.20466	4.8860	.97969	1.0207	4.9873	26
35	.20079	.20497	4.8788	.97963	1.0208	4.9803	**25**
36	.20108	.20527	4.8716	.97958	1.0209	4.9732	24
37	.20136	.20557	4.8644	.97952	1.0209	4.9662	23
38	.20165	.20588	4.8573	.97946	1.0210	4.9591	22
39	.20193	.20618	4.8501	.97940	1.0210	4.9521	21
40	.20222	.20648	4.8430	.97934	1.0211	4.9452	**20**
41	.20250	.20679	4.8359	.97928	1.0212	4.9382	19
42	.20279	.20709	4.8288	.97922	1.0212	4.9313	18
43	.20307	.20739	4.8218	.97916	1.0213	4.9244	17
44	.20336	.20770	4.8147	.97910	1.0213	4.9175	16
45	.20364	.20800	4.8077	.97905	1.0214	4.9106	**15**
46	.20393	.20830	4.8007	.97899	1.0215	4.9037	14
47	.20421	.20861	4.7937	.97893	1.0215	4.8969	13
48	.20450	.20891	4.7867	.97887	1.0216	4.8901	12
49	.20478	.20921	4.7798	.97881	1.0217	4.8833	11
50	.20507	.20952	4.7729	.97875	1.0217	4.8765	**10**
51	.20535	.20982	4.7659	.97869	1.0218	4.8697	9
52	.20563	.21013	4.7591	.97863	1.0218	4.8630	8
53	.20592	.21043	4.7522	.97857	1.0219	4.8563	7
54	.20620	.21073	4.7453	.97851	1.0220	4.8496	6
55	.20649	.21104	4.7385	.97845	1.0220	4.8429	**5**
56	.20677	.21134	4.7317	.97839	1.0221	4.8362	4
57	.20706	.21164	4.7249	.97833	1.0222	4.8296	3
58	.20734	.21195	4.7181	.97827	1.0222	4.8229	2
59	.20763	.21225	4.7114	.97821	1.0223	4.8163	1
60	.20791	.21256	4.7046	.97815	1.0223	4.8097	**0**
′	Cos	Cot	Tan	Sin	Csc	Sec	′

101° (281°) (258°) **78°**

NATURAL TRIGONOMETRIC FUNCTIONS TO FIVE PLACES

12° (192°) (347°) **167°**

′	Sin	Tan	Cot	Cos	Sec	Csc	′
0	.20791	.21256	4.7046	.97815	1.0223	4.8097	**60**
1	.20820	.21286	4.6979	.97809	1.0224	4.8032	59
2	.20848	.21316	4.6912	.97803	1.0225	4.7966	58
3	.20877	.21347	4.6845	.97797	1.0225	4.7901	57
4	.20905	.21377	4.6779	.97791	1.0226	4.7836	56
5	.20933	.21408	4.6712	.97784	1.0227	4.7771	**55**
6	.20962	.21438	4.6646	.97778	1.0227	4.7706	54
7	.20990	.21469	4.6580	.97772	1.0228	4.7641	53
8	.21019	.21499	4.6514	.97766	1.0228	4.7577	52
9	.21047	.21529	4.6448	.97760	1.0229	4.7512	51
10	.21076	.21560	4.6382	.97754	1.0230	4.7448	**50**
11	.21104	.21590	4.6317	.97748	1.0230	4.7384	49
12	.21132	.21621	4.6252	.97742	1.0231	4.7321	48
13	.21161	.21651	4.6187	.97735	1.0232	4.7257	47
14	.21189	.21682	4.6122	.97729	1.0232	4.7194	46
15	.21218	.21712	4.6057	.97723	1.0233	4.7130	**45**
16	.21246	.21743	4.5993	.97717	1.0234	4.7067	44
17	.21275	.21773	4.5928	.97711	1.0234	4.7004	43
18	.21303	.21804	4.5864	.97705	1.0235	4.6942	42
19	.21331	.21834	4.5800	.97698	1.0236	4.6879	41
20	.21360	.21864	4.5736	.97692	1.0236	4.6817	**40**
21	.21388	.21895	4.5673	.97686	1.0237	4.6755	39
22	.21417	.21925	4.5609	.97680	1.0238	4.6693	38
23	.21445	.21956	4.5546	.97673	1.0238	4.6631	37
24	.21474	.21986	4.5483	.97667	1.0239	4.6569	36
25	.21502	.22017	4.5420	.97661	1.0240	4.6507	**35**
26	.21530	.22047	4.5357	.97655	1.0240	4.6446	34
27	.21559	.22078	4.5294	.97648	1.0241	4.6385	33
28	.21587	.22108	4.5232	.97642	1.0241	4.6324	32
29	.21616	.22139	4.5169	.97636	1.0242	4.6263	31
30	.21644	.22169	4.5107	.97630	1.0243	4.6202	**30**
31	.21672	.22200	4.5045	.97623	1.0243	4.6142	29
32	.21701	.22231	4.4983	.97617	1.0244	4.6081	28
33	.21729	.22261	4.4922	.97611	1.0245	4.6021	27
34	.21758	.22292	4.4860	.97604	1.0245	4.5961	26
35	.21786	.22322	4.4799	.97598	1.0246	4.5901	**25**
36	.21814	.22353	4.4737	.97592	1.0247	4.5841	24
37	.21843	.22383	4.4676	.97585	1.0247	4.5782	23
38	.21871	.22414	4.4615	.97579	1.0248	4.5722	22
39	.21899	.22444	4.4555	.97573	1.0249	4.5663	21
40	.21928	.22475	4.4494	.97566	1.0249	4.5604	**20**
41	.21956	.22505	4.4434	.97560	1.0250	4.5545	19
42	.21985	.22536	4.4373	.97553	1.0251	4.5486	18
43	.22013	.22567	4.4313	.95747	1.0251	4.5428	17
44	.22041	.22597	4.4253	.97541	1.0252	4.5369	16
45	.22070	.22628	4.4194	.97534	1.0253	4.5311	**15**
46	.22098	.22658	4.4134	.97528	1.0253	4.5253	14
47	.22126	.22689	4.4075	.97521	1.0254	4.5195	13
48	.22155	.22719	4.4015	.97515	1.0255	4.5137	12
49	.22183	.22750	4.3956	.97508	1.0256	4.5079	11
50	.22212	.22781	4.3897	.97502	1.0256	4.5022	**10**
51	.22240	.22811	4.3838	.97496	1.0257	4.4964	9
52	.22268	.22842	4.3779	.97489	1.0258	4.4907	8
53	.22297	.22872	4.3721	.97483	1.0258	4.4850	7
54	.22325	.22903	4.3662	.97476	1.0259	4.4793	6
55	.22353	.22934	4.3604	.97470	1.0260	4.4736	**5**
56	.22382	.22964	4.3546	.97463	1.0260	4.4679	4
57	.22410	.22995	4.3488	.97457	1.0261	4.4623	3
58	.22438	.23026	4.3430	.97450	1.0262	4.4566	2
59	.22467	.23056	4.3372	.97444	1.0262	4.4510	1
60	.22495	.23087	4.3315	.97437	1.0263	4.4454	**0**
′	Cos	Cot	Tan	Sin	Csc	Sec	′

102° (282°) (257°) **77°**

13° (193°) (346°) **166°**

′	Sin	Tan	Cot	Cos	Sec	Csc	′
0	.22495	.23087	4.3315	.97437	1.0263	4.4454	**60**
1	.22523	.23117	4.3257	.97430	1.0264	4.4398	59
2	.22552	.23148	4.3200	.97424	1.0264	4.4342	58
3	.22580	.23179	4.3143	.97417	1.0265	4.4287	57
4	.22608	.23209	4.3086	.97411	1.0266	4.4231	56
5	.22637	.23240	4.3029	.97404	1.0266	4.4176	**55**
6	.22665	.23271	4.2972	.97398	1.0267	4.4121	54
7	.22693	.23301	4.2916	.97391	1.0268	4.4066	53
8	.22722	.23332	4.2859	.97384	1.0269	4.4011	52
9	.22750	.23363	4.2803	.97378	1.0269	4.3956	51
10	.22778	.23393	4.2747	.97371	1.0270	4.3901	**50**
11	.22807	.23424	4.2691	.97365	1.0271	4.3847	49
12	.22835	.23455	4.2635	.97358	1.0271	4.3792	48
13	.22863	.23485	4.2580	.97351	1.0272	4.3738	47
14	.22892	.23516	4.2524	.97345	1.0273	4.3684	46
15	.22920	.23547	4.2468	.97338	1.0273	4.3630	**45**
16	.22948	.23578	4.2413	.97331	1.0274	4.3576	44
17	.22977	.23608	4.2358	.97325	1.0275	4.3522	43
18	.23005	.23639	4.2303	.97318	1.0276	4.3469	42
19	.23033	.23670	4.2248	.97311	1.0276	4.3415	41
20	.23062	.23700	4.2193	.97304	1.0277	4.3362	**40**
21	.23090	.23731	4.2139	.97298	1.0278	4.3309	39
22	.23118	.23762	4.2084	.97291	1.0278	4.3256	38
23	.23146	.23793	4.2030	.97284	1.0279	4.3203	37
24	.23175	.23823	4.1976	.97278	1.0280	4.3150	36
25	.23203	.23854	4.1922	.97271	1.0281	4.3098	**35**
26	.23231	.23885	4.1868	.97264	1.0281	4.3045	34
27	.23260	.23916	4.1814	.97257	1.0282	4.2993	33
28	.23288	.23946	4.1760	.97251	1.0283	4.2941	32
29	.23316	.23977	4.1706	.97244	1.0283	4.2889	31
30	.23345	.24008	4.1653	.97237	1.0284	4.2837	**30**
31	.23373	.24039	4.1600	.97230	1.0285	4.2785	29
32	.23401	.24069	4.1547	.97223	1.0286	4.2733	28
33	.23429	.24100	4.1493	.97217	1.0286	4.2681	27
34	.23458	.24131	4.1441	.97210	1.0287	4.2630	26
35	.23486	.24162	4.1388	.97203	1.0288	4.2579	**25**
36	.23514	.24193	4.1335	.97196	1.0288	4.2527	24
37	.23542	.24223	4.1282	.97189	1.0289	4.2476	23
38	.23571	.24254	4.1230	.97182	1.0290	4.2425	22
39	.23599	.24285	4.1178	.97176	1.0291	4.2375	21
40	.23627	.24316	4.1126	.97169	1.0291	4.2324	**20**
41	.23656	.24347	4.1074	.97162	1.0292	4.2273	19
42	.23684	.24377	4.1022	.97155	1.0293	4.2223	18
43	.23712	.24408	4.0970	.97148	1.0294	4.2173	17
44	.23740	.24439	4.0918	.97141	1.0294	4.2122	16
45	.23769	.24470	4.0867	.97134	1.0295	4.2072	**15**
46	.23797	.24501	4.0815	.97127	1.0296	4.2022	14
47	.23825	.24532	4.0764	.97120	1.0297	4.1973	13
48	.23853	.24562	4.0713	.97113	1.0297	4.1923	12
49	.23882	.24593	4.0662	.97106	1.0298	4.1873	11
50	.23910	.24624	4.0611	.97100	1.0299	4.1824	**10**
51	.23938	.24655	4.0560	.97093	1.0299	4.1774	9
52	.23966	.24686	4.0509	.97086	1.0300	4.1725	8
53	.23995	.24717	4.0459	.97079	1.0301	4.1676	7
54	.24023	.24747	4.0408	.97072	1.0302	4.1627	6
55	.24051	.24778	4.0358	.97065	1.0302	4.1578	**5**
56	.24079	.24809	4.0308	.97058	1.0303	4.1529	4
57	.24108	.24840	4.0257	.97051	1.0304	4.1481	3
58	.24136	.24871	4.0207	.97044	1.0305	4.1432	2
59	.24164	.24902	4.0158	.97037	1.0305	4.1384	1
60	.24192	.24933	4.0108	.97030	1.0306	4.1336	**0**
′	Cos	Cot	Tan	Sin	Csc	Sec	′

103° (283°) (256°) **76°**

NATURAL TRIGONOMETRIC FUNCTIONS TO FIVE PLACES

14° (194°) — (345°) **165°**

′	Sin	Tan	Cot	Cos	Sec	Csc	′
0	.24192	.24933	4.0108	.97030	1.0306	4.1336	**60**
1	.24220	.24964	4.0058	.97023	1.0307	4.1287	59
2	.24249	.24995	4.0009	.97015	1.0308	4.1239	58
3	.24277	.25026	3.9959	.97008	1.0308	4.1191	57
4	.24305	.25056	3.9910	.97001	1.0309	4.1144	56
5	.24333	.25087	3.9861	.96994	1.0310	4.1096	**55**
6	.24362	.25118	3.9812	.96987	1.0311	4.1048	54
7	.24390	.25149	3.9763	.96980	1.0311	4.1001	53
8	.24418	.25180	3.9714	.96973	1.0312	4.0954	52
9	.24446	.25211	3.9665	.96966	1.0313	4.0906	51
10	.24474	.25242	3.9617	.96959	1.0314	4.0859	**50**
11	.24503	.25273	3.9568	.96952	1.0314	4.0812	49
12	.24531	.25304	3.9520	.96945	1.0315	4.0765	48
13	.24559	.25335	3.9471	.96937	1.0316	4.0718	47
14	.24587	.25366	3.9423	.96930	1.0317	4.0672	46
15	.24615	.25397	3.9375	.96923	1.0317	4.0625	**45**
16	.24644	.25428	3.9327	.96916	1.0318	4.0579	44
17	.24672	.25459	3.9279	.96909	1.0319	4.0532	43
18	.24700	.25490	3.9232	.96902	1.0320	4.0486	42
19	.24728	.25521	3.9184	.96894	1.0321	4.0440	41
20	.24756	.25552	3.9136	.96887	1.0321	4.0394	**40**
21	.24784	.25583	3.9089	.96880	1.0322	4.0348	39
22	.24813	.25614	3.9042	.96873	1.0323	4.0302	38
23	.24841	.25645	3.8995	.96866	1.0324	4.0256	37
24	.24869	.25676	3.8947	.96858	1.0324	4.0211	36
25	.24897	.25707	3.8900	.96851	1.0325	4.0165	**35**
26	.24925	.25738	3.8854	.96844	1.0326	4.0120	34
27	.24954	.25769	3.8807	.96837	1.0327	4.0075	33
28	.24982	.25800	3.8760	.96829	1.0327	4.0029	32
29	.25010	.25831	3.8714	.96822	1.0328	3.9984	31
30	.25038	.25862	3.8667	.96815	1.0329	3.9939	**30**
31	.25066	.25893	3.8621	.96807	1.0330	3.9894	29
32	.25094	.25924	3.8575	.96800	1.0331	3.9850	28
33	.25122	.25955	3.8528	.96793	1.0331	3.9805	27
34	.25151	.25986	3.8482	.96786	1.0332	3.9760	26
35	.25179	.26017	3.8436	.96778	1.0333	3.9716	**25**
36	.25207	.26048	3.8391	.96771	1.0334	3.9672	24
37	.25235	.26079	3.8345	.96764	1.0334	3.9627	23
38	.25263	.26110	3.8299	.96756	1.0335	3.9583	22
39	.25291	.26141	3.8254	.96749	1.0336	3.9539	21
40	.25320	.26172	3.8208	.96742	1.0337	3.9495	**20**
41	.25348	.26203	3.8163	.96734	1.0338	3.9451	19
42	.25376	.26235	3.8118	.96727	1.0338	3.9408	18
43	.25404	.26266	3.8073	.96719	1.0339	3.9364	17
44	.25432	.26297	3.8028	.96712	1.0340	3.9320	16
45	.25460	.26328	3.7983	.96705	1.0341	3.9277	**15**
46	.25488	.26359	3.7938	.96697	1.0342	3.9234	14
47	.25516	.26390	3.7893	.96690	1.0342	3.9190	13
48	.25545	.26421	3.7848	.96682	1.0343	3.9147	12
49	.25573	.26452	3.7804	.96675	1.0344	3.9104	11
50	.25601	.26483	3.7760	.96667	1.0345	3.9061	**10**
51	.25629	.26515	3.7715	.96660	1.0346	3.9018	9
52	.25657	.26546	3.7671	.96653	1.0346	3.8976	8
53	.25685	.26577	3.7627	.96645	1.0347	3.8933	7
54	.25713	.26608	3.7583	.96638	1.0348	3.8890	6
55	.25741	.26639	3.7539	.96630	1.0349	3.8848	**5**
56	.25769	.26670	3.7495	.96623	1.0350	3.8806	4
57	.25798	.26701	3.7451	.96615	1.0350	3.8763	3
58	.25826	.26733	3.7408	.96608	1.0351	3.8721	2
59	.25854	.26764	3.7364	.96600	1.0352	3.8679	1
60	.25882	.26795	3.7321	.96593	1.0353	3.8637	**0**
′	Cos	Cot	Tan	Sin	Csc	Sec	′

104° (284°) — (255°) **75°**

15° (195°) — (344°) **164**

′	Sin	Tan	Cot	Cos	Sec	Csc	′
0	.25882	.26795	3.7321	.96593	1.0353	3.8637	**60**
1	.25910	.26826	3.7277	.96585	1.0354	3.8595	59
2	.25938	.26857	3.7234	.96578	1.0354	3.8553	58
3	.25966	.26888	3.7191	.96570	1.0355	3.8512	57
4	.25994	.26920	3.7148	.96562	1.0356	3.8470	56
5	.26022	.26951	3.7105	.96555	1.0357	3.8428	**55**
6	.26050	.26982	3.7062	.96547	1.0358	3.8387	54
7	.26079	.27013	3.7019	.96540	1.0358	3.8346	53
8	.26107	.27044	3.6976	.96532	1.0359	3.8304	52
9	.26135	.27076	3.6933	.96524	1.0360	3.8263	51
10	.26163	.27107	3.6891	.96517	1.0361	3.8222	**50**
11	.26191	.27138	3.6848	.96509	1.0362	3.8181	49
12	.26219	.27169	3.6806	.96502	1.0363	3.8140	48
13	.26247	.27201	3.6764	.96494	1.0363	3.8100	47
14	.26275	.27232	3.6722	.96486	1.0364	3.8059	46
15	.26303	.27263	3.6680	.96479	1.0365	3.8018	**45**
16	.26331	.27294	3.6638	.96471	1.0366	3.7978	44
17	.26359	.27326	3.6596	.96463	1.0367	3.7937	43
18	.26387	.27357	3.6554	.96456	1.0367	3.7897	42
19	.26415	.27388	3.6512	.96448	1.0368	3.7857	41
20	.26443	.27419	3.6470	.96440	1.0369	3.7817	**40**
21	.26471	.27451	3.6429	.96433	1.0370	3.7777	39
22	.26500	.27482	3.6387	.96425	1.0371	3.7737	38
23	.26528	.27513	3.6346	.96417	1.0372	3.7697	37
24	.26556	.27545	3.6305	.96410	1.0372	3.7657	36
25	.26584	.27576	3.6264	.94602	1.0373	3.7617	**35**
26	.26612	.27607	3.6222	.96394	1.0374	3.7577	34
27	.26640	.27638	3.6181	.96386	1.0375	3.7538	33
28	.26668	.27670	3.6140	.96379	1.0376	3.7498	32
29	.26696	.27701	3.6100	.96371	1.0377	3.7459	31
30	.26724	.27732	3.6059	.96363	1.0377	3.7420	**30**
31	.26752	.27764	3.6018	.96355	1.0378	3.7381	29
32	.26780	.27795	3.5978	.96347	1.0379	3.7341	28
33	.26808	.27826	3.5937	.96340	1.0380	3.7302	27
34	.26836	.27858	3.5897	.96332	1.0381	3.7263	26
35	.26864	.27889	3.5856	.96324	1.0382	3.7225	**25**
36	.26892	.27921	3.5816	.96316	1.0382	3.7186	24
37	.26920	.27952	3.5776	.96308	1.0383	3.7147	23
38	.26948	.27983	3.5736	.96301	1.0384	3.7108	22
39	.26976	.28015	3.5696	.96293	1.0385	3.7070	21
40	.27004	.28046	3.5656	.96285	1.0386	3.7032	**20**
41	.27032	.28077	3.5616	.96277	1.0387	3.6993	19
42	.27060	.28109	3.5576	.96269	1.0388	3.6955	18
43	.27088	.28140	3.5536	.96261	1.0388	3.6917	17
44	.27116	.28172	3.5497	.96253	1.0389	3.6879	16
45	.27144	.28203	3.5457	.96246	1.0390	3.6840	**15**
46	.27172	.28234	3.5418	.96238	1.0391	3.6803	14
47	.27200	.28266	3.5379	.96230	1.0392	3.6765	13
48	.27228	.28297	3.5339	.96222	1.0393	3.6727	12
49	.27256	.28329	3.5300	.96214	1.0394	3.6689	11
50	.27284	.28360	3.5261	.96206	1.0394	3.6652	**10**
51	.27312	.28391	3.5222	.96198	1.0395	3.6614	9
52	.27340	.28423	3.5183	.96190	1.0396	3.6575	8
53	.27368	.28454	3.5144	.96182	1.0397	3.6539	7
54	.27396	.28486	3.5105	.96174	1.0398	3.6502	6
55	.27424	.28517	3.5067	.96166	1.0399	3.6465	**5**
56	.27452	.28549	3.5028	.96158	1.0400	3.6427	4
57	.27480	.28580	3.4989	.96150	1.0400	3.6390	3
58	.27508	.28612	3.4951	.96142	1.0401	3.6353	2
59	.27536	.28643	3.4912	.96134	1.0402	3.6316	1
60	.27564	.28675	3.4874	.96126	1.0403	3.6280	**0**
′	Cos	Cot	Tan	Sin	Csc	Sec	′

105° (285°) — (254°) **74**

NATURAL TRIGONOMETRIC FUNCTIONS TO FIVE PLACES

16° (196°) (343°) **163°**

′	Sin	Tan	Cot	Cos	Sec	Csc	′
0	.27564	.28675	3.4874	.96126	1.0403	3.6280	**60**
1	.27592	.28706	3.4836	.96118	1.0404	3.6243	59
2	.27620	.28738	3.4798	.96110	1.0405	3.6206	58
3	.27648	.28769	3.4760	.96102	1.0406	3.6169	57
4	.27676	.28801	3.4722	.96094	1.0406	3.6133	56
5	.27704	.28832	3.4684	.96086	1.0407	3.6097	**55**
6	.27731	.28864	3.4646	.96078	1.0408	3.6060	54
7	.27759	.28895	3.4608	.96070	1.0409	3.6024	53
8	.27787	.28927	3.4570	.96062	1.0410	3.5988	52
9	.27815	.28958	3.4533	.96054	1.0411	3.5951	51
10	.27843	.28990	3.4495	.96046	1.0412	3.5915	**50**
11	.27871	.29021	3.4458	.96037	1.0413	3.5879	49
12	.27899	.29053	3.4420	.96029	1.0413	3.5843	48
13	.27927	.29084	3.4383	.96021	1.0414	3.5808	47
14	.27955	.29116	3.4346	.96013	1.0415	3.5772	46
15	.27983	.29147	3.4308	.96005	1.0416	3.5736	**45**
16	.28011	.29179	3.4271	.95997	1.0417	3.5700	44
17	.28039	.29210	3.4234	.95989	1.0418	3.5665	43
18	.28067	.29242	3.4197	.95981	1.0419	3.5629	42
19	.28095	.29274	3.4160	.95972	1.0420	3.5594	41
20	.28123	.29305	3.4124	.95964	1.0421	3.5559	**40**
21	.28150	.29337	3.4087	.95956	1.0421	3.5523	39
22	.28178	.29368	3.4050	.95948	1.0422	3.5488	38
23	.28206	.29400	3.4014	.95940	1.0423	3.5453	37
24	.28234	.29432	3.3977	.95931	1.0424	3.5418	36
25	.28262	.29463	3.3941	.95923	1.0425	3.5383	**35**
26	.28290	.29495	3.3904	.95915	1.0426	3.5348	34
27	.28318	.29526	3.3868	.95907	1.0427	3.5313	33
28	.28346	.29558	3.3832	.95898	1.0428	3.5279	32
29	.28374	.29590	3.3796	.95890	1.0429	3.5244	31
30	.28402	.29621	3.3759	.95882	1.0429	3.5209	**30**
31	.28429	.29653	3.3723	.95874	1.0430	3.5175	29
32	.28457	.29685	3.3687	.95865	1.0431	3.5140	28
33	.28485	.29716	3.3652	.95857	1.0432	3.5106	27
34	.28513	.29748	3.3616	.95849	1.0433	3.5072	26
35	.28541	.29780	3.3580	.95841	1.0434	3.5037	**25**
36	.28569	.29811	3.3544	.95832	1.0435	3.5003	24
37	.28597	.29843	3.3509	.95824	1.0436	3.4969	23
38	.28625	.29875	3.3473	.95816	1.0437	3.4935	22
39	.28652	.29906	3.3438	.95807	1.0438	3.4901	21
40	.28680	.29938	3.3402	.95799	1.0439	3.4867	**20**
41	.28708	.29970	3.3367	.95791	1.0439	3.4833	19
42	.28736	.30001	3.3332	.95782	1.0440	3.4799	18
43	.28764	.30033	3.3297	.95774	1.0441	3.4766	17
44	.28792	.30065	3.3261	.95766	1.0442	3.4732	16
45	.28820	.30097	3.3226	.95757	1.0443	3.4699	**15**
46	.28847	.30128	3.3191	.95749	1.0444	3.4665	14
47	.28875	.30160	3.3156	.95740	1.0445	3.4632	13
48	.28903	.30192	3.3122	.95732	1.0446	3.4598	12
49	.28931	.30224	3.3087	.95724	1.0447	3.4565	11
50	.28959	.30255	3.3052	.95715	1.0448	3.4532	**10**
51	.28987	.30287	3.3017	.95707	1.0449	3.4499	9
52	.29015	.30319	3.2983	.95698	1.0450	3.4465	8
53	.29042	.30351	3.2948	.95690	1.0450	3.4432	7
54	.29070	.30382	3.2914	.95681	1.0451	3.4399	6
55	.29098	.30414	3.2879	.95673	1.0452	3.4367	**5**
56	.29126	.30446	3.2845	.95664	1.0453	3.4334	4
57	.29154	.30478	3.2811	.95656	1.0454	3.4301	3
58	.29182	.30509	3.2777	.95647	1.0455	3.4268	2
59	.29209	.30541	3.2743	.95639	1.0456	3.4236	1
60	.29237	.30573	3.2709	.95630	1.0457	3.4203	**0**
′	Cos	Cot	Tan	Sin	Csc	Sec	′

106° (286°) (253°) **73°**

17° (197°) (342°) **162°**

′	Sin	Tan	Cot	Cos	Sec	Csc	′
0	.29237	.30573	3.2709	.95630	1.0457	3.4203	**60**
1	.29265	.30605	3.2675	.95622	1.0458	3.4171	59
2	.29293	.30637	3.2641	.95613	1.0459	3.4138	58
3	.29321	.30669	3.2607	.95605	1.0460	3.4106	57
4	.29348	.30700	3.2573	.95596	1.0461	3.4073	56
5	.29376	.30732	3.2539	.95588	1.0462	3.4041	**55**
6	.29404	.30764	3.2506	.95579	1.0463	3.4009	54
7	.29432	.30796	3.2472	.95571	1.0463	3.3977	53
8	.29460	.30828	3.2438	.95562	1.0464	3.3945	52
9	.29487	.30860	3.2405	.95554	1.0465	3.3913	51
10	.29515	.30891	3.2371	.95545	1.0466	3.3881	**50**
11	.29543	.30923	3.2338	.95536	1.0467	3.3849	49
12	.29571	.30955	3.2305	.95528	1.0468	3.3817	48
13	.29599	.30987	3.2272	.95519	1.0469	3.3785	47
14	.29626	.31019	3.2238	.95511	1.0470	3.3754	46
15	.29654	.31051	3.2205	.95502	1.0471	3.3722	**45**
16	.29682	.31083	3.2172	.95493	1.0472	3.3691	44
17	.29710	.31115	3.2139	.95485	1.0473	3.3659	43
18	.29737	.31147	3.2106	.95476	1.0474	3.3628	42
19	.29765	.31178	3.2073	.95467	1.0475	3.3596	41
20	.29793	.31210	3.2041	.95459	1.0476	3.3565	**40**
21	.29821	.31242	3.2008	.95450	1.0477	3.3534	39
22	.29849	.31274	3.1975	.95441	1.0478	3.3502	38
23	.29876	.31306	3.1943	.95433	1.0479	3.3471	37
24	.29904	.31338	3.1910	.95424	1.0480	3.3440	36
25	.29932	.31370	3.1878	.95415	1.0480	3.3409	**35**
26	.29960	.31402	3.1845	.95407	1.0481	3.3378	34
27	.29987	.31434	3.1813	.95398	1.0482	3.3347	33
28	.30015	.31466	3.1780	.95389	1.0483	3.3317	32
29	.30043	.31498	3.1748	.95380	1.0484	3.3286	31
30	.30071	.31530	3.1716	.95372	1.0485	3.3255	**30**
31	.30098	.31562	3.1684	.95363	1.0486	3.3224	29
32	.30126	.31594	3.1652	.95354	1.0487	3.3194	28
33	.30154	.31626	3.1620	.95345	1.0488	3.3163	27
34	.30182	.31658	3.1588	.95337	1.0489	3.3133	26
35	.30209	.31690	3.1556	.95328	1.0490	3.3102	**25**
36	.30237	.31722	3.1524	.95319	1.0491	3.3072	24
37	.30265	.31754	3.1492	.95310	1.0492	3.3042	23
38	.30292	.31786	3.1460	.95301	1.0493	3.3012	22
39	.30320	.31818	3.1429	.95293	1.0494	3.2981	21
40	.30348	.31850	3.1397	.95284	1.0495	3.2951	**20**
41	.30376	.31882	3.1366	.95275	1.0496	3.2921	19
42	.30403	.31914	3.1334	.95266	1.0497	3.2891	18
43	.30431	.31946	3.1303	.95257	1.0498	3.2861	17
44	.30459	.31978	3.1271	.95248	1.0499	3.2831	16
45	.30486	.32010	3.1240	.95240	1.0500	3.2801	**15**
46	.30514	.32042	3.1209	.95231	1.0501	3.2772	14
47	.30542	.32074	3.1178	.95222	1.0502	3.2742	13
48	.30570	.32106	3.1146	.95213	1.0503	3.2712	12
49	.30597	.32139	3.1115	.95204	1.0504	3.2683	11
50	.30625	.32171	3.1084	.95195	1.0505	3.2653	**10**
51	.30653	.32203	3.1053	.95186	1.0506	3.2624	9
52	.30680	.32235	3.1022	.95177	1.0507	3.2594	8
53	.30708	.32267	3.0991	.95168	1.0508	3.2565	7
54	.30736	.32299	3.0961	.95159	1.0509	3.2535	6
55	.30763	.32331	3.0930	.95150	1.0510	3.2506	**5**
56	.30791	.32363	3.0899	.95142	1.0511	3.2477	4
57	.30819	.32396	3.0868	.95133	1.0512	3.2448	3
58	.30846	.32428	3.0838	.95124	1.0513	3.2419	2
59	.30874	.32460	3.0807	.95115	1.0514	3.2390	1
60	.30902	.32492	3.0777	.95106	1.0515	3.2361	**0**
′	Cos	Cot	Tan	Sin	Csc	Sec	′

107° (287°) (252°) **72°**

NATURAL TRIGONOMETRIC FUNCTIONS TO FIVE PLACES

18° (198°) (341°) **161°**

′	Sin	Tan	Cot	Cos	Sec	Csc	′
0	.30902	.32492	3.0777	.95106	1.0515	3.2361	**60**
1	.30929	.32524	3.0746	.95097	1.0516	3.2332	59
2	.30957	.32556	3.0716	.95088	1.0517	3.2303	58
3	.30985	.32588	3.0686	.95079	1.0518	3.2274	57
4	.31012	.32621	3.0655	.95070	1.0519	3.2245	56
5	.31040	.32653	3.0625	.95061	1.0520	3.2217	**55**
6	.31068	.32685	3.0595	.95052	1.0521	3.2188	54
7	.31095	.32717	3.0565	.95043	1.0522	3.2159	53
8	.31123	.32749	3.0535	.95033	1.0523	3.2131	52
9	.31151	.32782	3.0505	.95024	1.0524	3.2102	51
10	.31178	.32814	3.0475	.95015	1.0525	3.2074	**50**
11	.31206	.32846	3.0445	.95006	1.0526	3.2045	49
12	.31233	.32878	3.0415	.94997	1.0527	3.2017	48
13	.31261	.32911	3.0385	.94988	1.0528	3.1989	47
14	.31289	.32943	3.0356	.94979	1.0529	3.1960	46
15	.31316	.32975	3.0326	.94970	1.0530	3.1932	**45**
16	.31344	.33007	3.0296	.94961	1.0531	3.1904	44
17	.31372	.33040	3.0267	.94952	1.0532	3.1876	43
18	.31399	.33072	3.0237	.94943	1.0533	3.1848	42
19	.31427	.33104	3.0208	.94933	1.0534	3.1820	41
20	.31454	.33136	3.0178	.94924	1.0535	3.1792	**40**
21	.31482	.33169	3.0149	.94915	1.0536	3.1764	39
22	.31510	.33201	3.0120	.94906	1.0537	3.1736	38
23	.31537	.33233	3.0090	.94897	1.0538	3.1708	37
24	.31565	.33266	3.0061	.94888	1.0539	3.1681	36
25	.31593	.33298	3.0032	.94878	1.0540	3.1653	**35**
26	.31620	.33330	3.0003	.94869	1.0541	3.1625	34
27	.31648	.33363	2.9974	.94860	1.0542	3.1598	33
28	.31675	.33395	2.9945	.94851	1.0543	3.1570	32
29	.31703	.33427	2.9916	.94842	1.0544	3.1543	31
30	.31730	.33460	2.9887	.94832	1.0545	3.1515	**30**
31	.31758	.33492	2.9858	.94823	1.0546	3.1488	29
32	.31786	.33524	2.9829	.94814	1.0547	3.1461	28
33	.31813	.33557	2.9800	.94805	1.0548	3.1433	27
34	.31841	.33589	2.9772	.94795	1.0549	3.1406	26
35	.31868	.33621	2.9743	.94786	1.0550	3.1379	**25**
36	.31896	.33654	2.9714	.94777	1.0551	3.1352	24
37	.31923	.33686	2.9686	.94768	1.0552	3.1325	23
38	.31951	.33718	2.9657	.94758	1.0553	3.1298	22
39	.31979	.33751	2.9629	.94749	1.0554	3.1271	21
40	.32006	.33783	2.9600	.94740	1.0555	3.1244	**20**
41	.32034	.33816	2.9572	.94730	1.0556	3.1217	19
42	.32061	.33848	2.9544	.94721	1.0557	3.1190	18
43	.32089	.33881	2.9515	.94712	1.0558	3.1163	17
44	.32116	.33913	2.9487	.94702	1.0559	3.1137	16
45	.32144	.33945	2.9459	.94693	1.0560	3.1110	**15**
46	.32171	.33978	2.9431	.94684	1.0561	3.1083	14
47	.32199	.34010	2.9403	.94674	1.0563	3.1057	13
48	.32227	.34043	2.9375	.94665	1.0564	3.1030	12
49	.32254	.34075	2.9347	.94656	1.0565	3.1004	11
50	.32282	.34108	2.9319	.94646	1.0566	3.0977	**10**
51	.32309	.34140	2.9291	.94637	1.0567	3.0951	9
52	.32337	.34173	2.9263	.94627	1.0568	3.0925	8
53	.32364	.34205	2.9235	.94618	1.0569	3.0898	7
54	.32392	.34238	2.9208	.94609	1.0570	3.0872	6
55	.32419	.34270	2.9180	.94599	1.0571	3.0846	**5**
56	.32447	.34303	2.9152	.94590	1.0572	3.0820	4
57	.32474	.34335	2.9125	.94580	1.0573	3.0794	3
58	.32502	.34368	2.9097	.94571	1.0574	3.0768	2
59	.32529	.34400	2.9070	.94561	1.0575	3.0742	1
60	.32557	.34433	2.9042	.94552	1.0576	3.0716	**0**
′	Cos	Cot	Tan	Sin	Csc	Sec	′

108° (288°) (251°) **71°**

19° (199°) (340°) **160°**

′	Sin	Tan	Cot	Cos	Sec	Csc	′
0	.32557	.34433	2.9042	.94552	1.0576	3.0716	**60**
1	.32584	.34465	2.9015	.94542	1.0577	3.0690	59
2	.32612	.34498	2.8987	.94533	1.0578	3.0664	58
3	.32639	.34530	2.8960	.94523	1.0579	3.0638	57
4	.32667	.34563	2.8933	.94514	1.0580	3.0612	56
5	.32694	.34596	2.8905	.94504	1.0582	3.0586	**55**
6	.32722	.34628	2.8878	.94495	1.0583	3.0561	54
7	.32749	.34661	2.8851	.94485	1.0584	3.0535	53
8	.32777	.34693	2.8824	.94476	1.0585	3.0509	52
9	.32804	.34726	2.8797	.94466	1.0586	3.0484	51
10	.32832	.34758	2.8770	.94457	1.0587	3.0458	**50**
11	.32859	.34791	2.8743	.94447	1.0588	3.0433	49
12	.32887	.34824	2.8716	.94438	1.0589	3.0407	48
13	.32914	.34856	2.8689	.94428	1.0590	3.0382	47
14	.32942	.34889	2.8662	.94418	1.0591	3.0357	46
15	.32969	.34922	2.8636	.94409	1.0592	3.0331	**45**
16	.32997	.34954	2.8609	.94399	1.0593	3.0306	44
17	.33024	.34987	2.8582	.94390	1.0594	3.0281	43
18	.33051	.35020	2.8556	.94380	1.0595	3.0256	42
19	.33079	.35052	2.8529	.94370	1.0597	3.0231	41
20	.33106	.35085	2.8502	.94361	1.0598	3.0206	**40**
21	.33134	.35118	2.8476	.94351	1.0599	3.0181	39
22	.33161	.35150	2.8449	.94342	1.0600	3.0156	38
23	.33189	.35183	2.8423	.94332	1.0601	3.0131	37
24	.33216	.35216	2.8397	.94322	1.0602	3.0106	36
25	.33244	.35248	2.8370	.94313	1.0603	3.0081	**35**
26	.33271	.35281	2.8344	.94303	1.0604	3.0056	34
27	.33298	.35314	2.8318	.94293	1.0605	3.0031	33
28	.33326	.35346	2.8291	.94284	1.0606	3.0007	32
29	.33353	.35379	2.8265	.94274	1.0607	2.9982	31
30	.33381	.35412	2.8239	.94264	1.0608	2.9957	**30**
31	.33408	.35445	2.8213	.94254	1.0610	2.9933	29
32	.33436	.35477	2.8187	.94245	1.0611	2.9908	28
33	.33463	.35510	2.8161	.94235	1.0612	2.9884	27
34	.33490	.35543	2.8135	.94225	1.0613	2.9859	26
35	.33518	.35576	2.8109	.94215	1.0614	2.9835	**25**
36	.33545	.35608	2.8083	.94206	1.0615	2.9811	24
37	.33573	.35641	2.8057	.94196	1.0616	2.9786	23
38	.33600	.35674	2.8032	.94186	1.0617	2.9762	22
39	.33627	.35707	2.8006	.94176	1.0618	2.9738	21
40	.33655	.35740	2.7980	.94167	1.0619	2.9713	**20**
41	.33682	.35772	2.7955	.94157	1.0621	2.9689	19
42	.33710	.35805	2.7929	.94147	1.0622	2.9665	18
43	.33737	.35838	2.7903	.94137	1.0623	2.9641	17
44	.33764	.35871	2.7878	.94127	1.0624	2.9617	16
45	.33792	.35904	2.7852	.94118	1.0625	2.9593	**15**
46	.33819	.35937	2.7827	.94108	1.0626	2.9569	14
47	.33846	.35969	2.7801	.94098	1.0627	2.9545	13
48	.33874	.36002	2.7776	.94088	1.0628	2.9521	12
49	.33901	.36035	2.7751	.94078	1.0629	2.9498	11
50	.33929	.36068	2.7725	.94068	1.0631	2.9474	**10**
51	.33956	.36101	2.7700	.94058	1.0632	2.9450	9
52	.33983	.36134	2.7675	.94049	1.0633	2.9426	8
53	.34011	.36167	2.7650	.94039	1.0634	2.9403	7
54	.34038	.36199	2.7625	.94029	1.0635	2.9379	6
55	.34065	.36232	2.7600	.94019	1.0636	2.9355	**5**
56	.34093	.36265	2.7575	.94009	1.0637	2.9332	4
57	.34120	.36298	2.7550	.93999	1.0638	2.9308	3
58	.34147	.36331	2.7525	.93989	1.0640	2.9285	2
59	.34175	.36364	2.7500	.93979	1.0641	2.9261	1
60	.34202	.36397	2.7475	.93969	1.0642	2.9238	**0**
′	Cos	Cot	Tan	Sin	Csc	Sec	′

109° (289°) (250°) **70°**

NATURAL TRIGONOMETRIC FUNCTIONS TO FIVE PLACES

20° (200°) (339°) **159°**

′	Sin	Tan	Cot	Cos	Sec	Csc	′
0	.34202	.36397	2.7475	.93969	1.0642	2.9238	**60**
1	.34229	.36430	2.7450	.93959	1.0643	2.9215	59
2	.34257	.36463	2.7425	.93949	1.0644	2.9191	58
3	.34284	.36496	2.7400	.93939	1.0645	2.9168	57
4	.34311	.36529	2.7376	.93929	1.0646	2.9145	56
5	.34339	.36562	2.7351	.93919	1.0647	2.9122	**55**
6	.34366	.36595	2.7326	.93909	1.0649	2.9099	54
7	.34393	.36628	2.7302	.93899	1.0650	2.9075	53
8	.34421	.36661	2.7277	.93889	1.0651	2.9052	52
9	.34448	.36694	2.7253	.93879	1.0652	2.9029	51
10	.34475	.36727	2.7228	.93869	1.0653	2.9006	**50**
11	.34503	.36760	2.7204	.93859	1.0654	2.8983	49
12	.34530	.36793	2.7179	.93849	1.0655	2.8960	48
13	.34557	.36826	2.7155	.93839	1.0657	2.8938	47
14	.34584	.36859	2.7130	.93829	1.0658	2.8915	46
15	.34612	.36892	2.7106	.93819	1.0659	2.8892	**45**
16	.34639	.36925	2.7082	.93809	1.0660	2.8869	44
17	.34666	.36958	2.7058	.93799	1.0661	2.8846	43
18	.34694	.36991	2.7034	.93789	1.0662	2.8824	42
19	.34721	.37024	2.7009	.93779	1.0663	2.8801	41
20	.34748	.37057	2.6985	.93769	1.0665	2.8779	**40**
21	.34775	.37090	2.6961	.93759	1.0666	2.8756	39
22	.34803	.37123	2.6937	.93748	1.0667	2.8733	38
23	.34830	.37157	2.6913	.93738	1.0668	2.8711	37
24	.34857	.37190	2.6889	.93728	1.0669	2.8688	36
25	.34884	.37223	2.6865	.93718	1.0670	2.8666	**35**
26	.34912	.37256	2.6841	.93708	1.0671	2.8644	34
27	.34939	.37289	2.6818	.93698	1.0673	2.8621	33
28	.34966	.37322	2.6794	.93688	1.0674	2.8599	32
29	.34993	.37355	2.6770	.93677	1.0675	2.8577	31
30	.35021	.37388	2.6746	.93667	1.0676	2.8555	**30**
31	.35048	.37422	2.6723	.93657	1.0677	2.8532	29
32	.35075	.37455	2.6699	.93647	1.0678	2.8510	28
33	.35102	.37488	2.6675	.93637	1.0680	2.8488	27
34	.35130	.37521	2.6652	.93626	1.0681	2.8466	26
35	.35157	.37554	2.6628	.93616	1.0682	2.8444	**25**
36	.35184	.37588	2.6605	.93606	1.0683	2.8422	24
37	.35211	.37621	2.6581	.93596	1.0684	2.8400	23
38	.35239	.37654	2.6558	.93585	1.0685	2.8378	22
39	.35266	.37687	2.6534	.93575	1.0687	2.8356	21
40	.35293	.37720	2.6511	.93565	1.0688	2.8334	**20**
41	.35320	.37754	2.6488	.93555	1.0689	2.8312	19
42	.35347	.37787	2.6464	.93544	1.0690	2.8291	18
43	.35375	.37820	2.6441	.93534	1.0691	2.8269	17
44	.35402	.37853	2.6418	.93524	1.0692	2.8247	16
45	.35429	.37887	2.6395	.93514	1.0694	2.8225	**15**
46	.35456	.37920	2.6371	.93503	1.0695	2.8204	14
47	.35484	.37953	2.6348	.93493	1.0696	2.8182	13
48	.35511	.37986	2.6325	.93483	1.0697	2.8161	12
49	.35538	.38020	2.6302	.93472	1.0698	2.8139	11
50	.35565	.38053	2.6279	.93462	1.0700	2.8117	**10**
51	.35592	.38086	2.6256	.93452	1.0701	2.8096	9
52	.35619	.38120	2.6233	.93441	1.0702	2.8075	8
53	.35647	.38153	2.6210	.93431	1.0703	2.8053	7
54	.35674	.38186	2.6187	.93420	1.0704	2.8032	6
55	.35701	.38220	2.6165	.93410	1.0705	2.8010	**5**
56	.35728	.38253	2.6142	.93400	1.0707	2.7989	4
57	.35755	.38286	2.6119	.93389	1.0708	2.7968	3
58	.35782	.38320	2.6096	.93379	1.0709	2.7947	2
59	.35810	.38353	2.6074	.93368	1.0710	2.7925	1
60	.35837	.38386	2.6051	.93358	1.0711	2.7904	**0**
′	Cos	Cot	Tan	Sin	Csc	Sec	′

110° (290°) (249°) **69°**

21° (201°) (338°) **158°**

′	Sin	Tan	Cot	Cos	Sec	Csc	′
0	.35837	.38386	2.6051	.93358	1.0711	2.7904	**60**
1	.35864	.38420	2.6028	.93348	1.0713	2.7883	59
2	.35891	.38453	2.6006	.93337	1.0714	2.7862	58
3	.35918	.38487	2.5983	.93327	1.0715	2.7841	57
4	.35945	.38520	2.5961	.93316	1.0716	2.7820	56
5	.35973	.38553	2.5938	.93306	1.0717	2.7799	**55**
6	.36000	.38587	2.5916	.93295	1.0719	2.7778	54
7	.36027	.38620	2.5893	.93285	1.0720	2.7757	53
8	.36054	.38654	2.5871	.93274	1.0721	2.7736	52
9	.36081	.38687	2.5848	.93264	1.0722	2.7715	51
10	.36108	.38721	2.5826	.93253	1.0723	2.7695	**50**
11	.36135	.38754	2.5804	.93243	1.0725	2.7674	49
12	.36162	.38787	2.5782	.93232	1.0726	2.7653	48
13	.36190	.38821	2.5759	.93222	1.0727	2.7632	47
14	.36217	.38854	2.5737	.93211	1.0728	2.7612	46
15	.36244	.38888	2.5715	.93201	1.0730	2.7591	**45**
16	.36271	.38921	2.5693	.93190	1.0731	2.7570	44
17	.36298	.38955	2.5671	.93180	1.0732	2.7550	43
18	.36325	.38988	2.5649	.93169	1.0733	2.7529	42
19	.36352	.39022	2.5627	.93159	1.0734	2.7509	41
20	.36379	.39055	2.5605	.93148	1.0736	2.7488	**40**
21	.36406	.39089	2.5583	.93137	1.0737	2.7468	39
22	.36434	.39122	2.5561	.93127	1.0738	2.7447	38
23	.36461	.39156	2.5539	.93116	1.0739	2.7427	37
24	.36488	.39190	2.5517	.93106	1.0740	2.7407	36
25	.36515	.39223	2.5495	.93095	1.0742	2.7386	**35**
26	.36542	.39257	2.5473	.93084	1.0743	2.7366	34
27	.36569	.39290	2.5452	.93074	1.0744	2.7346	33
28	.36596	.39324	2.5430	.93063	1.0745	2.7325	32
29	.36623	.39357	2.5408	.93052	1.0747	2.7305	31
30	.36650	.39391	2.5386	.93042	1.0748	2.7285	**30**
31	.36677	.39425	2.5365	.93031	1.0749	2.7265	29
32	.36704	.39458	2.5343	.93020	1.0750	2.7245	28
33	.36731	.39492	2.5322	.93010	1.0752	2.7225	27
34	.36758	.39526	2.5300	.92999	1.0753	2.7205	26
35	.36785	.39559	2.5279	.92988	1.0754	2.7185	**25**
36	.36812	.39593	2.5257	.92978	1.0755	2.7165	24
37	.36839	.39626	2.5236	.92967	1.0757	2.7145	23
38	.36867	.39660	2.5214	.92956	1.0758	2.7125	22
39	.36894	.39694	2.5193	.92945	1.0759	2.7105	21
40	.36921	.39727	2.5172	.92935	1.0760	2.7085	**20**
41	.36948	.39761	2.5150	.92924	1.0761	2.7065	19
42	.36975	.39795	2.5129	.92913	1.0763	2.7046	18
43	.37002	.39829	2.5108	.92902	1.0764	2.7026	17
44	.37029	.39862	2.5086	.92892	1.0765	2.7006	16
45	.37056	.39896	2.5065	.92881	1.0766	2.6986	**15**
46	.37083	.39930	2.5044	.92870	1.0768	2.6967	14
47	.37110	.39963	2.5023	.92859	1.0769	2.6947	13
48	.37137	.39997	2.5002	.92849	1.0770	2.6927	12
49	.37164	.40031	2.4981	.92838	1.0771	2.6908	11
50	.37191	.40065	2.4960	.92827	1.0773	2.6888	**10**
51	.37218	.40098	2.4939	.92816	1.0774	2.6869	9
52	.37245	.40132	2.4918	.92805	1.0775	2.6849	8
53	.37272	.40166	2.4897	.92794	1.0777	2.6830	7
54	.37299	.40200	2.4876	.92784	1.0778	2.6811	6
55	.37326	.40234	2.4855	.92773	1.0779	2.6791	**5**
56	.37353	.40267	2.4834	.92762	1.0780	2.6772	4
57	.37380	.40301	2.4813	.92751	1.0782	2.6752	3
58	.37407	.40335	2.4792	.92740	1.0783	2.6733	2
59	.37434	.40369	2.4772	.92729	1.0784	2.6714	1
60	.37461	.40403	2.4751	.92718	1.0785	2.6695	**0**
′	Cos	Cot	Tan	Sin	Csc	Sec	′

111° (291°) (248°) **68°**

NATURAL TRIGONOMETRIC FUNCTIONS TO FIVE PLACES

22° (202°) — (337°) **157°**

′	Sin	Tan	Cot	Cos	Sec	Csc	′
0	.37461	.40403	2.4751	.92718	1.0785	2.6695	**60**
1	.37488	.40436	2.4730	.92707	1.0787	2.6675	59
2	.37515	.40470	2.4709	.92697	1.0788	2.6656	58
3	.37542	.40504	2.4689	.92686	1.0789	2.6637	57
4	.37569	.40538	2.4668	.92675	1.0790	2.6618	56
5	.37595	.40572	2.4648	.92664	1.0792	2.6599	**55**
6	.37622	.40606	2.4627	.92653	1.0793	2.6580	54
7	.37649	.40640	2.4606	.92642	1.0794	2.6561	53
8	.37676	.40674	2.4586	.92631	1.0796	2.6542	52
9	.37703	.40707	2.4566	.92620	1.0797	2.6523	51
10	.37730	.40741	2.4545	.92609	1.0798	2.6504	**50**
11	.37757	.40775	2.4525	.92598	1.0799	2.6485	49
12	.37784	.40809	2.4504	.92587	1.0801	2.6466	48
13	.37811	.40843	2.4484	.92576	1.0802	2.6447	47
14	.37838	.40877	2.4464	.92565	1.0803	2.6429	46
15	.37865	.40911	2.4443	.92554	1.0804	2.6410	**45**
16	.37892	.40945	2.4423	.92543	1.0806	2.6391	44
17	.37919	.40979	2.4403	.92532	1.0807	2.6372	43
18	.37946	.41013	2.4383	.92521	1.0808	2.6354	42
19	.37973	.41047	2.4362	.92510	1.0810	2.6335	41
20	.37999	.41081	2.4342	.92499	1.0811	2.6316	**40**
21	.38026	.41115	2.4322	.92488	1.0812	2.6298	39
22	.38053	.41149	2.4302	.92477	1.0814	2.6279	38
23	.38080	.41183	2.4282	.92466	1.0815	2.6260	37
24	.38107	.41217	2.4262	.92455	1.0816	2.6242	36
25	.38134	.41251	2.4242	.92444	1.0817	2.6223	**35**
26	.38161	.41285	2.4222	.92432	1.0819	2.6205	34
27	.38188	.41319	2.4202	.92421	1.0820	2.6186	33
28	.38215	.41353	2.4182	.92410	1.0821	2.6168	32
29	.38241	.41387	2.4162	.92399	1.0823	2.6150	31
30	.38268	.41421	2.4142	.92388	1.0824	2.6131	**30**
31	.38295	.41455	2.4122	.92377	1.0825	2.6113	29
32	.38322	.41490	2.4102	.92366	1.0827	2.6095	28
33	.38349	.41524	2.4083	.92355	1.0828	2.6076	27
34	.38376	.41558	2.4063	.92343	1.0829	2.6058	26
35	.38403	.41592	2.4043	.92332	1.0830	2.6040	**25**
36	.38430	.41626	2.4023	.92321	1.0832	2.6022	24
37	.38456	.41660	2.4004	.92310	1.0833	2.6003	23
38	.38483	.41694	2.3984	.92299	1.0834	2.5985	22
39	.38510	.41728	2.3964	.92287	1.0836	2.5967	21
40	.38537	.41763	2.3945	.92276	1.0837	2.5949	**20**
41	.38564	.41797	2.3925	.92265	1.0838	2.5931	19
42	.38591	.41831	2.3906	.92254	1.0840	2.5913	18
43	.38617	.41865	2.3886	.92243	1.0841	2.5895	17
44	.38644	.41899	2.3867	.92231	1.0842	2.5877	16
45	.38671	.41933	2.3847	.92220	1.0844	2.5859	**15**
46	.38698	.41968	2.3828	.92209	1.0845	2.5841	14
47	.38725	.42002	2.3808	.92198	1.0846	2.5823	13
48	.38752	.42036	2.3789	.92186	1.0848	2.5805	12
49	.38778	.42070	2.3770	.92175	1.0849	2.5788	11
50	.38805	.42105	2.3750	.92164	1.0850	2.5770	**10**
51	.38832	.42139	2.3731	.92152	1.0852	2.5752	9
52	.38859	.42173	2.3712	.92141	1.0853	2.5734	8
53	.38886	.42207	2.3693	.92130	1.0854	2.5716	7
54	.38912	.42242	2.3673	.92119	1.0856	2.5699	6
55	.38939	.42276	2.3654	92107	1.0857	2.5681	**5**
56	.38966	.42310	2.3635	92096	1.0858	2.5663	4
57	.38993	.42345	2.3616	.92085	1.0860	2.5646	3
58	.39020	.42379	2.3597	.92073	1.0861	2.5628	2
59	.39046	.42413	2.3578	.92062	1.0862	2.5611	1
60	.39073	.42447	2.3559	.92050	1.0864	2.5593	**0**
′	Cos	Cot	Tan	Sin	Csc	Sec	′

112° (292°) — (247°) **67°**

23° (203°) — (336°) **156°**

′	Sin	Tan	Cot	Cos	Sec	Csc	′
0	.39073	.42447	2.3559	.92050	1.0864	2.5593	**60**
1	.39100	.42482	2.3539	.92039	1.0865	2.5576	59
2	.39127	.42516	2.3520	.92028	1.0866	2.5558	58
3	.39153	.42551	2.3501	.92016	1.0868	2.5541	57
4	.39180	.42585	2 3483	92005	1.0869	2.5523	56
5	.39207	.42619	2.3464	.91994	1.0870	2.5506	**55**
6	.39234	.42654	2.3445	.91982	1.0872	2.5488	54
7	.39260	.42688	2.3426	.91971	1.0873	2.5471	53
8	.39287	.42722	2.3407	.91959	1.0874	2.5454	52
9	.39314	.42757	2.3388	.91948	1.0876	2.5436	51
10	.39341	.42791	2.3369	.91936	1.0877	2.5419	**50**
11	.39367	.42826	2.3351	.91925	1.0878	2.5402	49
12	.39394	.42860	2.3332	.91914	1.0880	2.5384	48
13	.39421	.42894	2.3313	.91902	1.0881	2.5367	47
14	.39448	.42929	2.3294	.91891	1.0883	2.5350	46
15	.39474	.42963	2.3276	.91879	1.0884	2.5333	**45**
16	.39501	.42998	2.3257	.91868	1.0885	2.5316	44
17	.39528	.43032	2.3238	.91856	1.0887	2.5299	43
18	.39555	.43067	2.3220	.91845	1.0888	2.5282	42
19	.39581	.43101	2.3201	.91833	1.0889	2.5264	41
20	.39608	.43136	2.3183	.91822	1.0891	2.5247	**40**
21	.39635	.43170	2.3164	.91810	1.0892	2.5230	39
22	.39661	.43205	2.3146	.91799	1.0893	2.5213	38
23	.39688	.43239	2.3127	.91787	1.0895	2.5196	37
24	.39715	.43274	2.3109	.91775	1.0896	2.5180	36
25	.39741	.43308	2.3090	.91764	1.0898	2.5163	**35**
26	.39768	.43343	2.3072	.91752	1.0899	2.5146	34
27	.39795	.43378	2.3053	.91741	1.0900	2.5129	33
28	.39822	.43412	2.3035	.91729	1.0902	2.5112	32
29	.39848	.43447	2.3017	.91718	1.0903	2.5095	31
30	.39875	.43481	2.2998	.91706	1.0904	2.5078	**30**
31	.39902	.43516	2.2980	.91694	1.0906	2.5062	29
32	.39928	.43550	2.2962	.91683	1.0907	2.5045	28
33	.39955	.43585	2.2944	.91671	1.0909	2.5028	27
34	.39982	.43620	2.2925	.91660	1.0910	2.5012	26
35	.40008	.43654	2.2907	.91648	1.0911	2.4995	**25**
36	.40035	.43689	2.2889	.91636	1.0913	2.4978	24
37	.40062	.43724	2.2871	.91625	1.0914	2.4962	23
38	.40088	.43758	2.2853	.91613	1.0915	2.4945	22
39	.40115	.43793	2.2835	.91601	1.0917	2.4928	21
40	.40141	.43828	2.2817	.91590	1.0918	2.4912	**20**
41	.40168	.43862	2.2799	.91578	1.0920	2.4895	19
42	.40195	.43897	2.2781	.91566	1.0921	2.4879	18
43	.40221	.43932	2.2763	.91555	1.0922	2.4862	17
44	.40248	.43966	2.2745	.91543	1.0924	2.4846	16
45	.40275	.44001	2.2727	.91531	1.0925	2.4830	**15**
46	.40301	.44036	2.2709	.91519	1.0927	2.4813	14
47	.40328	.44071	2.2691	.91508	1.0928	2.4797	13
48	.40355	.44105	2.2673	.91496	1.0929	2.4780	12
49	.40381	.44140	2.2655	.91484	1.0931	2.4764	11
50	.40408	.44175	2.2637	.91472	1.0932	2.4748	**10**
51	.40434	.44210	2.2620	.91461	1.0934	2.4731	9
52	.40461	.44244	2.2602	.91449	1.0935	2.4715	8
53	.40488	.44279	2.2584	.91437	1.0936	2.4699	7
54	.40514	.44314	2.2566	.91425	1.0938	2.4683	6
55	.40541	.44349	2.2549	.91414	1.0939	2.4667	**5**
56	.40567	.44384	2.2531	.91402	1.0941	2.4650	4
57	.40594	.44418	2.2513	.91390	1.0942	2.4634	3
58	.40621	.44453	2.2496	.91378	1.0944	2.4618	2
59	.40647	.44488	2.2478	.91366	1.0945	2.4602	1
60	.40674	.44523	2.2460	.91355	1.0946	2.4586	**0**
′	Cos	Cot	Tan	Sin	Csc	Sec	′

113° (293°) — (246°) **66°**

NATURAL TRIGONOMETRIC FUNCTIONS TO FIVE PLACES

24° (204°) (335°) **155°**

′	Sin	Tan	Cot	Cos	Sec	Csc	′
0	.40674	.44523	2.2460	.91355	1.0946	2.4586	**60**
1	.40700	.44558	2.2443	.91343	1.0948	2.4570	59
2	.40727	.44593	2.2425	.91331	1.0949	2.4554	58
3	.40753	.44627	2.2408	.91319	1.0951	2.4538	57
4	.40780	.44662	2.2390	.91307	1.0952	2.4522	56
5	.40806	.44697	2.2373	.91295	1.0953	2.4506	**55**
6	.40833	.44732	2.2355	.91283	1.0955	2.4490	54
7	.40860	.44767	2.2338	.91272	1.0956	2.4474	53
8	.40886	.44802	2.2320	.91260	1.0958	2.4458	52
9	.40913	.44837	2.2303	.91248	1.0959	2.4442	51
10	.40939	.44872	2.2286	.91236	1.0961	2.4426	**50**
11	.40966	.44907	2.2268	.91224	1.0962	2.4411	49
12	.40992	.44942	2.2251	.91212	1.0963	2.4395	48
13	.41019	.44977	2.2234	.91200	1.0965	2.4379	47
14	.41045	.45012	2.2216	.91188	1.0966	2.4363	46
15	.41072	.45047	2.2199	.91176	1.0968	2.4348	**45**
16	.41098	.45082	2.2182	.91164	1.0969	2.4332	44
17	.41125	.45117	2.2165	.91152	1.0971	2.4316	43
18	.41151	.45152	2.2148	.91140	1.0972	2.4300	42
19	.41178	.45187	2.2130	.91128	1.0974	2.4285	41
20	.41204	.45222	2.2113	.91116	1.0975	2.4269	**40**
21	.41231	.45257	2.2096	.91104	1.0976	2.4254	39
22	.41257	.45292	2.2079	.91092	1.0978	2.4238	38
23	.41284	.45327	2.2062	.91080	1.0979	2.4222	37
24	.41310	.45362	2.2045	.91068	1.0981	2.4207	36
25	.41337	.45397	2.2028	.91056	1.0982	2.4191	**35**
26	.41363	.45432	2.2011	.91044	1.0984	2.4176	34
27	.41390	.45467	2.1994	.91032	1.0985	2.4160	33
28	.41416	.45502	2.1977	.91020	1.0987	2.4145	32
29	.41443	.45538	2.1960	.91008	1.0988	2.4130	31
30	.41469	.45573	2.1943	.90996	1.0989	2.4114	**30**
31	.41496	.45608	2.1926	.90984	1.0991	2.4099	29
32	.41522	.45643	2.1909	.90972	1.0992	2.4083	28
33	.41549	.45678	2.1892	.90960	1.0994	2.4068	27
34	.41575	.45713	2.1876	.90948	1.0995	2.4053	26
35	.41602	.45748	2.1859	.90936	1.0997	2.4038	**25**
36	.41628	.45784	2.1842	.90924	1.0998	2.4022	24
37	.41655	.45819	2.1825	.90911	1.1000	2.4007	23
38	.41681	.45854	2.1808	.90899	1.1001	2.3992	22
39	.41707	.45889	2.1792	.90887	1.1003	2.3977	21
40	.41734	.45924	2.1775	.90875	1.1004	2.3961	**20**
41	.41760	.45960	2.1758	.90863	1.1006	2.3946	19
42	.41787	.45995	2.1742	.90851	1.1007	2.3931	18
43	.41813	.46030	2.1725	.90839	1.1009	2.3916	17
44	.41840	.46065	2.1708	.90826	1.1010	2.3901	16
45	.41866	.46101	2.1692	.90814	1.1011	2.3886	**15**
46	.41892	.46136	2.1675	.90802	1.1013	2.3871	14
47	.41919	.46171	2.1659	.90790	1.1014	2.3856	13
48	.41945	.46206	2.1642	.90778	1.1016	2.3841	12
49	.41972	.46242	2.1625	.90766	1.1017	2.3826	11
50	.41998	.46277	2.1609	.90753	1.1019	2.3811	**10**
51	.42024	.46312	2.1592	.90741	1.1020	2.3796	9
52	.42051	.46348	2.1576	.90729	1.1022	2.3781	8
53	.42077	.46383	2.1560	.90717	1.1023	2.3766	7
54	.42104	.46418	2.1543	.90704	1.1025	2.3751	6
55	.42130	.46454	2.1527	.90692	1.1026	2.3736	**5**
56	.42156	.46489	2.1510	.90680	1.1028	2.3721	4
57	.42183	.46525	2.1494	.90668	1.1029	2.3706	3
58	.42209	.46560	2.1478	.90655	1.1031	2.3692	2
59	.42235	.46595	2.1461	.90643	1.1032	2.3677	1
60	.42262	.46631	2.1445	.90631	1.1034	2.3662	**0**
′	Cos	Cot	Tan	Sin	Csc	Sec	′

114° (294°) (245°) **65°**

25° (205°) (334°) **154°**

′	Sin	Tan	Cot	Cos	Sec	Csc	′
0	.42262	.46631	2.1445	.90631	1.1034	2.3662	**60**
1	.42288	.46666	2.1429	.90618	1.1035	2.3647	59
2	.42315	.46702	2.1413	.90606	1.1037	2.3633	58
3	.42341	.46737	2.1396	.90594	1.1038	2.3618	57
4	.42367	.46772	2.1380	.90582	1.1040	2.3603	56
5	.42394	.46808	2.1364	.90569	1.1041	2.3588	**55**
6	.42420	.46843	2.1348	.90557	1.1043	2.3574	54
7	.42446	.46879	2.1332	.90545	1.1044	2.3559	53
8	.42473	.46914	2.1315	.90532	1.1046	2.3545	52
9	.42499	.46950	2.1299	.90520	1.1047	2.3530	51
10	.42525	.46985	2.1283	.90507	1.1049	2.3515	**50**
11	.42552	.47021	2.1267	.90495	1.1050	2.3501	49
12	.42578	.47056	2.1251	.90483	1.1052	2.3486	48
13	.42604	.47092	2.1235	.90470	1.1053	2.3472	47
14	.42631	.47128	2.1219	.90458	1.1055	2.3457	46
15	.42657	.47163	2.1203	.90446	1.1056	2.3443	**45**
16	.42683	.47199	2.1187	.90433	1.1058	2.3428	44
17	.42709	.47234	2.1171	.90421	1.1059	2.3414	43
18	.42736	.47270	2.1155	.90408	1.1061	2.3400	42
19	.42762	.47305	2.1139	.90396	1.1062	2.3385	41
20	.42788	.47341	2.1123	.90383	1.1064	2.3371	**40**
21	.42815	.47377	2.1107	.90371	1.1066	2.3356	39
22	.42841	.47412	2.1092	.90358	1.1067	2.3342	38
23	.42867	.47448	2.1076	.90346	1.1069	2.3328	37
24	.42894	.47483	2.1060	.90334	1.1070	2.3314	36
25	.42920	.47519	2.1044	.90321	1.1072	2.3299	**35**
26	.42946	.47555	2.1028	.90309	1.1073	2.3285	34
27	.42972	.47590	2.1013	.90296	1.1075	2.3271	33
28	.42999	.47626	2.0997	.90284	1.1076	2.3257	32
29	.43025	.47662	2.0981	.90271	1.1078	2.3242	31
30	.43051	.47698	2.0965	.90259	1.1079	2.3228	**30**
31	.43077	.47733	2.0950	.90246	1.1081	2.3214	29
32	.43104	.47769	2.0934	.90233	1.1082	2.3200	28
33	.43130	.47805	2.0918	.90221	1.1084	2.3186	27
34	.43156	.47840	2.0903	.90208	1.1085	2.3172	26
35	.43182	.47876	2.0887	.90196	1.1087	2.3158	**25**
36	.43209	.47912	2.0872	.90183	1.1089	2.3144	24
37	.43235	.47948	2.0856	.90171	1.1090	2.3130	23
38	.43261	.47984	2.0840	.90158	1.1092	2.3115	22
39	.43287	.48019	2.0825	.90146	1.1093	2.3101	21
40	.43313	.48055	2.0809	.90133	1.1095	2.3088	**20**
41	.43340	.48091	2.0794	.90120	1.1096	2.3074	19
42	.43366	.48127	2.0778	.90108	1.1098	2.3060	18
43	.43392	.48163	2.0763	.90095	1.1099	2.3046	17
44	.43418	.48198	2.0748	.90082	1.1101	2.3032	16
45	.43445	.48234	2.0732	.90070	1.1102	2.3018	**15**
46	.43471	.48270	2.0717	.90057	1.1104	2.3004	14
47	.43497	.48306	2.0701	.90045	1.1106	2.2990	13
48	.43523	.48342	2.0686	.90032	1.1107	2.2976	12
49	.43549	.48378	2.0671	.90019	1.1109	2.2962	11
50	.43575	.48414	2.0655	.90007	1.1110	2.2949	**10**
51	.43602	.48450	2.0640	.89994	1.1112	2.2935	9
52	.43628	.48486	2.0625	.89981	1.1113	2.2921	8
53	.43654	.48521	2.0609	.89968	1.1115	2.2907	7
54	.43680	.48557	2.0594	.89956	1.1117	2.2894	6
55	.43706	.48593	2.0579	.89943	1.1118	2.2880	**5**
56	.43733	.48629	2.0564	.89930	1.1120	2.2866	4
57	.43759	.48665	2.0549	.89918	1.1121	2.2853	3
58	.43785	.48701	2.0533	.89905	1.1123	2.2839	2
59	.43811	.48737	2.0518	.89892	1.1124	2.2825	1
60	.43837	.48773	2.0503	.89879	1.1126	2.2812	**0**
′	Cos	Cot	Tan	Sin	Csc	Sce	′

115° (295°) (244°) **64°**

NATURAL TRIGONOMETRIC FUNCTIONS TO FIVE PLACES

26° (206°) (333°) **153°**

′	Sin	Tan	Cot	Cos	Sec	Csc	′
0	.43837	.48773	2.0503	.89879	1.1126	2.2812	**60**
1	.43863	.48809	2.0488	.89867	1.1128	2.2798	59
2	.43889	.48845	2.0473	.89854	1.1129	2.2785	58
3	.43916	.48881	2.0458	.89841	1.1131	2.2771	57
4	.43942	.48917	2.0443	.89828	1.1132	2.2757	56
5	.43968	.48953	2.0428	.89816	1.1134	2.2744	**55**
6	.43994	.48989	2.0413	.89803	1.1136	2.2730	54
7	.44020	.49026	2.0398	.89790	1.1137	2.2717	53
8	.44046	.49062	2.0383	.89777	1.1139	2.2703	52
9	.44072	.49098	2.0368	.89764	1.1140	2.2690	51
10	.44098	.49134	2.0353	.89752	1.1142	2.2677	**50**
11	.44124	.49170	2.0338	.89739	1.1143	2.2663	49
12	.44151	.49206	2.0323	.89726	1.1145	2.2650	48
13	.44177	.49242	2.0308	.89713	1.1147	2.2636	47
14	.44203	.49278	2.0293	.89700	1.1148	2.2623	46
15	.44229	.49315	2.0278	.89687	1.1150	2.2610	**45**
16	.44255	.49351	2.0263	.89674	1.1151	2.2596	44
17	.44281	.49387	2.0248	.89662	1.1153	2.2583	43
18	.44307	.49423	2.0233	.89649	1.1155	2.2570	42
19	.44333	.49459	2.0219	.89636	1.1156	2.2556	41
20	.44359	.49495	2.0204	.89623	1.1158	2.2543	**40**
21	.44385	.49532	2.0189	.89610	1.1159	2.2530	39
22	.44411	.49568	2.0174	.89597	1.1161	2.2517	38
23	.44437	.49604	2.0160	.89584	1.1163	2.2504	37
24	.44464	.49640	2.0145	.89571	1.1164	2.2490	36
25	.44490	.49677	2.0130	.89558	1.1166	2.2477	**35**
26	.44516	.49713	2.0115	.89545	1.1168	2.2464	34
27	.44542	.49749	2.0101	.89532	1.1169	2.2451	33
28	.44568	.49786	2.0086	.89519	1.1171	2.2438	32
29	.44594	.49822	2.0072	.89506	1.1172	2.2425	31
30	.44620	.49858	2.0057	.89493	1.1174	2.2412	**30**
31	.44646	.49894	2.0042	.89480	1.1176	2.2399	29
32	.44672	.49931	2.0028	.89467	1.1177	2.2385	28
33	.44698	.49967	2.0013	.89454	1.1179	2.2372	27
34	.44724	.50004	1.9999	.89441	1.1180	2.2359	26
35	.44750	.50040	1.9984	.89428	1.1182	2.2346	**25**
36	.44776	.50076	1.9970	.89415	1.1184	2.2333	24
37	.44802	.50113	1.9955	.89402	1.1185	2.2320	23
38	.44828	.50149	1.9941	.89389	1.1187	2.2308	22
39	.44854	.50185	1.9926	.89376	1.1189	2.2295	21
40	.44880	.50222	1.9912	.89363	1.1190	2.2282	**20**
41	.44906	.50258	1.9897	.89350	1.1192	2.2269	19
42	.44932	.50295	1.9883	.89337	1.1194	2.2256	18
43	.44958	.50331	1.9868	.89324	1.1195	2.2243	17
44	.44984	.50368	1.9854	.89311	1.1197	2.2230	16
45	.45010	.50404	1.9840	.89298	1.1198	2.2217	**15**
46	.45036	.50441	1.9825	.89285	1.1200	2.2205	14
47	.45062	.50477	1.9811	.89272	1.1202	2.2192	13
48	.45088	.50514	1.9797	.89259	1.1203	2.2179	12
49	.45114	.50550	1.9782	.89245	1.1205	2.2166	11
50	.45140	.50587	1.9768	.89232	1.1207	2.2153	**10**
51	.45166	.50623	1.9754	.89219	1.1208	2.2141	9
52	.45192	.50660	1.9740	.89206	1.1210	2.2128	8
53	.45218	.50696	1.9725	.89193	1.1212	2.2115	7
54	.45243	.50733	1.9711	.89180	1.1213	2.2103	6
55	.45269	.50769	1.9697	.89167	1.1215	2.2090	**5**
56	.45295	.50806	1.9683	.89153	1.1217	2.2077	4
57	.45321	.50843	1.9669	.89140	1.1218	2.2065	3
58	.45347	.50879	1.9654	.89127	1.1220	2.2052	2
59	.45373	.50916	1.9640	.89114	1.1222	2.2039	1
60	.45399	.50953	1.9626	.89101	1.1223	2.2027	**0**
′	Cos	Cot	Tan	Sin	Csc	Sec	′

116° (296°) (243°) **63°**

27° (207°) (332°) **152°**

′	Sin	Tan	Cot	Cos	Sec	Csc	′
0	.45399	.50953	1.9626	.89101	1.1223	2.2027	**60**
1	.45425	.50989	1.9612	.89087	1.1225	2.2014	59
2	.45451	.51026	1.9598	.89074	1.1227	2.2002	58
3	.45477	.51063	1.9584	.89061	1.1228	2.1989	57
4	.45503	.51099	1.9570	.89048	1.1230	2.1977	56
5	.45529	.51136	1.9556	.89035	1.1232	2.1964	**55**
6	.45554	.51173	1.9542	.89021	1.1233	2.1952	54
7	.45580	.51209	1.9528	.89008	1.1235	2.1939	53
8	.45606	.51246	1.9514	.88995	1.1237	2.1927	52
9	.45632	.51283	1.9500	.88981	1.1238	2.1914	51
10	.45658	.51319	1.9486	.88968	1.1240	2.1902	**50**
11	.45684	.51356	1.9472	.88955	1.1242	2.1890	49
12	.45710	.51393	1.9458	.88942	1.1243	2.1877	48
13	.45736	.51430	1.9444	.88928	1.1245	2.1865	47
14	.45762	.51467	1.9430	.88915	1.1247	2.1852	46
15	.45787	.51503	1.9416	.88902	1.1248	2.1840	**45**
16	.45813	.51540	1.9402	.88888	1.1250	2.1828	44
17	.45839	.51577	1.9388	.88875	1.1252	2.1815	43
18	.45865	.51614	1.9375	.88862	1.1253	2.1803	42
19	.45891	.51651	1.9361	.88848	1.1255	2.1791	41
20	.45917	.51688	1.9347	.88835	1.1257	2.1779	**40**
21	.45942	.51724	1.9333	.88822	1.1259	2.1766	39
22	.45968	.51761	1.9319	.88808	1.1260	2.1754	38
23	.45994	.51798	1.9306	.88795	1.1262	2.1742	37
24	.46020	.51835	1.9292	.88782	1.1264	2.1730	36
25	.46046	.51872	1.9278	.88768	1.1265	2.1718	**35**
26	.46072	.51909	1.9265	.88755	1.1267	2.1705	34
27	.46097	.51946	1.9251	.88741	1.1269	2.1693	33
28	.46123	.51983	1.9237	.88728	1.1270	2.1681	32
29	.46149	.52020	1.9223	.88715	1.1272	2.1669	31
30	.46175	.52057	1.9210	.88701	1.1274	2.1657	**30**
31	.46201	.52094	1.9196	.88688	1.1276	2.1645	29
32	.46226	.52131	1.9183	.88674	1.1277	2.1633	28
33	.46252	.52168	1.9169	.88661	1.1279	2.1621	27
34	.46278	.52205	1.9155	.88647	1.1281	2.1609	26
35	.46304	.52242	1.9142	.88634	1.1282	2.1596	**25**
36	.46330	.52279	1.9128	.88620	1.1284	2.1584	24
37	.46355	.52316	1.9115	.88607	1.1286	2.1572	23
38	.46381	.52353	1.9101	.88593	1.1288	2.1560	22
39	.46407	.52390	1.9088	.88580	1.1289	2.1549	21
40	.46433	.52427	1.9074	.88566	1.1291	2.1537	**20**
41	.46458	.52464	1.9061	.88553	1.1293	2.1525	19
42	.46484	.52501	1.9047	.88539	1.1294	2.1513	18
43	.46510	.52538	1.9034	.88526	1.1296	2.1501	17
44	.46536	.52575	1.9020	.88512	1.1298	2.1489	16
45	.46561	.52613	1.9007	.88499	1.1300	2.1477	**15**
46	.46587	.52650	1.8993	.88485	1.1301	2.1465	14
47	.46613	.52687	1.8980	.88472	1.1303	2.1453	13
48	.46639	.52724	1.8967	.88458	1.1305	2.1441	12
49	.46664	.52761	1.8953	.88445	1.1307	2.1430	11
50	.46690	.52798	1.8940	.88431	1.1308	2.1418	**10**
51	.46716	.52836	1.8927	.88417	1.1310	2.1406	9
52	.46742	.52873	1.8913	.88404	1.1312	2.1394	8
53	.46767	.52910	1.8900	.88390	1.1313	2.1382	7
54	.46793	.52947	1.8887	.88377	1.1315	2.1371	6
55	.46819	.52985	1.8873	.88363	1.1317	2.1359	**5**
56	.46844	.53022	1.8860	.88349	1.1319	2.1347	4
57	.46870	.53059	1.8847	.88336	1.1320	2.1336	3
58	.46896	.53096	1.8834	.88322	1.1322	2.1324	2
59	.46921	.53134	1.8820	.88308	1.1324	2.1312	1
60	.46947	.53171	1.8807	.88295	1.1326	2.1301	**0**
′	Cos	Cot	Tan	Sin	Csc	Sec	′

117° (297°) (242°) **62°**

NATURAL TRIGONOMETRIC FUNCTIONS TO FIVE PLACES

28° (208°) (331°) **151°**

′	Sin	Tan	Cot	Cos	Sec	Csc	′
0	.46947	.53171	1.8807	.88295	1.1326	2.1301	**60**
1	.46973	.53208	1.8794	.88281	1.1327	2.1289	59
2	.46999	.53246	1.8781	.88267	1.1329	2.1277	58
3	.47024	.53283	1.8768	.88254	1.1331	2.1266	57
4	.47050	.53320	1.8755	.88240	1.1333	2.1254	56
5	.47076	.53358	1.8741	.88226	1.1334	2.1242	**55**
6	.47101	.53395	1.8728	.88213	1.1336	2.1231	54
7	.47127	.53432	1.8715	.88199	1.1338	2.1219	53
8	.47153	.53470	1.8702	.88185	1.1340	2.1208	52
9	.47178	.53507	1.8689	.88172	1.1342	2.1196	51
10	.47204	.53545	1.8676	.88158	1.1343	2.1185	**50**
11	.47229	.53582	1.8663	.88144	1.1345	2.1173	49
12	.47255	.53620	1.8650	.88130	1.1347	2.1162	48
13	.47281	.53657	1.8637	.88117	1.1349	2.1150	47
14	.47306	.53694	1.8624	.88103	1.1350	2.1139	46
15	.47332	.53732	1.8611	.88089	1.1352	2.1127	**45**
16	.47358	.53769	1.8598	.88075	1.1354	2.1116	44
17	.47383	.53807	1.8585	.88062	1.1356	2.1105	43
18	.47409	.53844	1.8572	.88048	1.1357	2.1093	42
19	.47434	.53882	1.8559	.88034	1.1359	2.1082	41
20	.47460	.53920	1.8546	.88020	1.1361	2.1070	**40**
21	.47486	.53957	1.8533	.88006	1.1363	2.1059	39
22	.47511	.53995	1.8520	.87993	1.1365	2.1048	38
23	.47537	.54032	1.8507	.87979	1.1366	2.1036	37
24	.47562	.54070	1.8495	.87965	1.1368	2.1025	36
25	.47588	.54107	1.8482	.87951	1.1370	2.1014	**35**
26	.47614	.54145	1.8469	.87937	1.1372	2.1002	34
27	.47639	.54183	1.8456	.87923	1.1374	2.0991	33
28	.47665	.54220	1.8443	.87909	1.1375	2.0980	32
29	.47690	.54258	1.8430	.87896	1.1377	2.0969	31
30	.47716	.54296	1.8418	.87882	1.1379	2.0957	**30**
31	.47741	.54333	1.8405	.87868	1.1381	2.0946	29
32	.47767	.54371	1.8392	.87854	1.1383	2.0935	28
33	.47793	.54409	1.8379	.87840	1.1384	2.0924	27
34	.47818	.54446	1.8367	.87826	1.1386	2.0913	26
35	.47844	.54484	1.8354	.87812	1.1388	2.0901	**25**
36	.47869	.54522	1.8341	.87798	1.1390	2.0890	24
37	.47895	.54560	1.8329	.87784	1.1392	2.0879	23
38	.47920	.54597	1.8316	.87770	1.1393	2.0868	22
39	.47946	.54635	1.8303	.87756	1.1395	2.0857	21
40	.47971	.54673	1.8291	.87743	1.1397	2.0846	**20**
41	.47997	.54711	1.8278	.87729	1.1399	2.0835	19
42	.48022	.54748	1.8265	.87715	1.1401	2.0824	18
43	.48048	.54786	1.8253	.87701	1.1402	2.0813	17
44	.48073	.54824	1.8240	.87687	1.1404	2.0802	16
45	.48099	.54862	1.8228	.87673	1.1406	2.0791	**15**
46	.48124	.54900	1.8215	.87659	1.1408	2.0779	14
47	.48150	.54938	1.8202	.87645	1.1410	2.0768	13
48	.48175	.54975	1.8190	.87631	1.1412	2.0757	12
49	.48201	.55013	1.8177	.87617	1.1413	2.0747	11
50	.48226	.55051	1.8165	.87603	1.1415	2.0736	**10**
51	.48252	.55089	1.8152	.87589	1.1417	2.0725	9
52	.48277	.55127	1.8140	.87575	1.1419	2.0714	8
53	.48303	.55165	1.8127	.87561	1.1421	2.0703	7
54	.48328	.55203	1.8115	.87546	1.1423	2.0692	6
55	.48354	.55241	1.8103	.87532	1.1424	2.0681	**5**
56	.48379	.55279	1.8090	.87518	1.1426	2.0670	4
57	.48405	.55317	1.8078	.87504	1.1428	2.0659	3
58	.48430	.55355	1.8065	.87490	1.1430	2.0648	2
59	.48456	.55393	1.8053	.87476	1.1432	2.0637	1
60	.48481	.55431	1.8040	.87462	1.1434	2.0627	**0**
′	Cos	Cot	Tan	Sin	Csc	Sec	′

118° (298°) (241°) **61°**

29° (209°) (330)° **150°**

′	Sin	Tan	Cot	Cos	Sec	Csc	′
0	.48481	.55431	1.8040	.87462	1.1434	2.0627	**60**
1	.48506	.55469	1.8028	.87448	1.1435	2.0616	59
2	.48532	.55507	1.8016	.87434	1.1437	2.0605	58
3	.48557	.55545	1.8003	.87420	1.1439	2.0594	57
4	.48583	.55583	1.7991	.87406	1.1441	2.0583	56
5	.48608	.55621	1.7979	.87391	1.1443	2.0573	**55**
6	.48634	.55659	1.7966	.87377	1.1445	2.0562	54
7	.48659	.55697	1.7954	.87363	1.1446	2.0551	53
8	.48684	.55736	1.7942	.87349	1.1448	2.0540	52
9	.48710	.55774	1.7930	.87335	1.1450	2.0530	51
10	.48735	.55812	1.7917	.87321	1.1452	2.0519	**50**
11	.48761	.55850	1.7905	.87306	1.1454	2.0508	49
12	.48786	.55888	1.7893	.87292	1.1456	2.0498	48
13	.48811	.55926	1.7881	.87278	1.1458	2.0487	47
14	.48837	.55964	1.7868	.87264	1.1460	2.0476	46
15	.48862	.56003	1.7856	.87250	1.1461	2.0466	**45**
16	.48888	.56041	1.7844	.87235	1.1463	2.0455	44
17	.48913	.56079	1.7832	.87221	1.1465	2.0445	43
18	.48938	.56117	1.7820	.87207	1.1467	2.0434	42
19	.48964	.56156	1.7808	.87193	1.1469	2.0423	41
20	.48989	.56194	1.7796	.87178	1.1471	2.0413	**40**
21	.49014	.56232	1.7783	.87164	1.1473	2.0402	39
22	.49040	.56270	1.7771	.87150	1.1474	2.0392	38
23	.49065	.56309	1.7759	.87136	1.1476	2.0381	37
24	.49090	.56347	1.7747	.87121	1.1478	2.0371	36
25	.49116	.56385	1.7735	.87107	1.1480	2.0360	**35**
26	.49141	.56424	1.7723	.87093	1.1482	2.0350	34
27	.49166	.56462	1.7711	.87079	1.1484	2.0339	33
28	.49192	.56501	1.7699	.87064	1.1486	2.0329	32
29	.49217	.56539	1.7687	.87050	1.1488	2.0318	31
30	.49242	.56577	1.7675	.87036	1.1490	2.0308	**30**
31	.49268	.56616	1.7663	.87021	1.1491	2.0297	29
32	.49293	.56654	1.7651	.87007	1.1493	2.0287	28
33	.49318	.56693	1.7639	.86993	1.1495	2.0276	27
34	.49344	.56731	1.7627	.86978	1.1497	2.0266	26
35	.49369	.56769	1.7615	.86964	1.1499	2.0256	**25**
36	.49394	.56808	1.7603	.86949	1.1501	2.0245	24
37	.49419	.56846	1.7591	.86935	1.1503	2.0235	23
38	.49445	.56885	1.7579	.86921	1.1505	2.0225	22
39	.49470	.56923	1.7567	.86906	1.1507	2.0214	21
40	.49495	.56962	1.7556	.86892	1.1509	2.0204	**20**
41	.49521	.57000	1.7544	.86878	1.1510	2.0194	19
42	.49546	.57039	1.7532	.86863	1.1512	2.0183	18
43	.49571	.57078	1.7520	.86849	1.1514	2.0173	17
44	.49596	.57116	1.7508	.86834	1.1516	2.0163	16
45	.49622	.57155	1.7496	.86820	1.1518	2.0152	**15**
46	.49647	.57193	1.7485	.86805	1.1520	2.0142	14
47	.49672	.57232	1.7473	.86791	1.1522	2.0132	13
48	.49697	.57271	1.7461	.86777	1.1524	2.0122	12
49	.49723	.57309	1.7449	.86762	1.1526	2.0112	11
50	.49748	.57348	1.7437	.86748	1.1528	2.0101	**10**
51	.49773	.57386	1.7426	.86733	1.1530	2.0091	9
52	.49798	.57425	1.7414	.86719	1.1532	2.0081	8
53	.49824	.57464	1.7402	.86704	1.1533	2.0071	7
54	.49849	.57503	1.7391	.86690	1.1535	2.0061	6
55	.49874	.57541	1.7379	.86675	1.1537	2.0051	**5**
56	.49899	.57580	1.7367	.86661	1.1539	2.0040	4
57	.49924	.57619	1.7355	.86646	1.1541	2.0030	3
58	.49950	.57657	1.7344	.86632	1.1543	2.0020	2
59	.49975	.57696	1.7332	.86617	1.1545	2.0010	1
60	.50000	.57735	1.7321	.86603	1.1547	2.0000	**0**
′	Cos	Cot	Tan	Sin	Csc	Sec	′

119° (299°) (240°) **60°**

NATURAL TRIGONOMETRIC FUNCTIONS TO FIVE PLACES

30° (210°) (329°) **149°**

′	Sin	Tan	Cot	Cos	Sec	Csc	′
0	.50000	.57735	1.7321	.86603	1.1547	2.0000	**60**
1	.50025	.57774	1.7309	.86588	1.1549	1.9990	59
2	.50050	.57813	1.7297	.86573	1.1551	1.9980	58
3	.50076	.57851	1.7286	.86559	1.1553	1.9970	57
4	50101	.57890	1.7274	.86544	1.1555	1.9960	56
5	.50126	.57929	1.7262	.86530	1.1557	1.9950	**55**
6	.50151	.57968	1.7251	.86515	1.1559	1.9940	54
7	.50176	.58007	1.7239	.86501	1.1561	1.9930	53
8	.50201	.58046	1.7228	.86486	1.1563	1.9920	52
9	.50227	.58085	1.7216	.86471	1.1565	1.9910	51
10	.50252	.58124	1.7205	.86457	1.1566	1.9900	**50**
11	.50277	.58162	1.7193	.86442	1.1568	1.9890	49
12	.50302	.58201	1.7182	.86427	1.1570	1.9880	48
13	.50327	.58240	1.7170	.86413	1.1572	1.9870	47
14	.50352	.58279	1.7159	.86398	1.1574	1.9860	46
15	.50377	.58318	1.7147	.86384	1.1576	1.9850	**45**
16	.50403	.58357	1.7136	.86369	1.1578	1.9840	44
17	.50428	.58396	1.7124	.86354	1.1580	1.9830	43
18	.50453	.58435	1.7113	.86340	1.1582	1.9821	42
19	.50478	.58474	1.7102	.86325	1.1584	1.9811	41
20	.50503	.58513	1.7090	.86310	1.1586	1.9801	**40**
21	.50528	.58552	1.7079	.86295	1.1588	1.9791	39
22	.50553	.58591	1.7067	.86281	1.1590	1.9781	38
23	.50578	.58631	1.7056	.86266	1.1592	1.9771	37
24	.50603	.58670	1.7045	.86251	1.1594	1.9762	36
25	.50628	.58709	1.7033	.86237	1.1596	1.9752	**35**
26	.50654	.58748	1.7022	.86222	1.1598	1.9742	34
27	.50679	.58787	1.7011	.86207	1.1600	1.9732	33
28	.50704	.58826	1.6999	.86192	1.1602	1.9722	32
29	.50729	.58865	1.6988	.86178	1.1604	1.9713	31
30	.50754	.58905	1.6977	.86163	1.1606	1.9703	**30**
31	.50779	.58944	1.6965	.86148	1.1608	1.9693	29
32	.50804	.58983	1.6954	.86133	1.1610	1.9684	28
33	.50829	.59022	1.6943	.86119	1.1612	1.9674	27
34	.50854	.59061	1.6932	.86104	1.1614	1.9664	26
35	.50879	.59101	1.6920	.86089	1.1616	1.9654	**25**
36	.50904	.59140	1.6909	.86074	1.1618	1.9645	24
37	.50929	.59179	1.6898	.86059	1.1620	1.9635	23
38	.50954	.59218	1.6887	.86045	1.1622	1.9625	22
39	.50979	.59258	1.6875	.86030	1.1624	1.9616	21
40	.51004	.59297	1.6864	.86015	1.1626	1.9606	**20**
41	.51029	.59336	1.6853	.86000	1.1628	1.9597	19
42	.51054	.59376	1.6842	.85985	1.1630	1.9587	18
43	.51079	.59415	1.6831	.85970	1.1632	1.9577	17
44	.51104	.59454	1.6820	.85956	1.1634	1.9568	16
45	.51129	.59494	1.6808	.85941	1.1636	1.9558	**15**
46	.51154	.59533	1.6797	.85926	1.1638	1.9549	14
47	.51179	.59573	1.6786	.85911	1.1640	1.9539	13
48	.51204	.59612	1.6775	.85896	1.1642	1.9530	12
49	.51229	.59651	1.6764	.85881	1.1644	1.9520	11
50	.51254	.59691	1.6753	.85866	1.1646	1.9511	**10**
51	.51279	.59730	1.6742	.85851	1.1648	1.9501	9
52	.51304	.59770	1.6731	.85836	1.1650	1.9492	8
53	.51329	.59809	1.6720	.85821	1.1652	1.9482	7
54	.51354	.59849	1.6709	.85806	1.1654	1.9473	6
55	.51379	.59888	1.6698	.85792	1.1656	1.9463	**5**
56	.51404	.59928	1.6687	.85777	1.1658	1.9454	4
57	.51429	.59967	1.6676	.85762	1.1660	1.9444	3
58	.51454	.60007	1.6665	.85747	1.1662	1.9435	2
59	.51479	.60046	1.6654	.85732	1.1664	1.9425	1
60	.51504	.60086	1.6643	.85717	1.1666	1.9416	**0**
′	Cos	Cot	Tan	Sin	Csc	Sec	′

120° (300°) (239°) **59°**

31° (211°) (328°) **148°**

′	Sin	Tan	Cot	Cos	Sec	Csc	′
0	.51504	.60086	1.6643	.85717	1.1666	1.9416	**60**
1	.51529	.60126	1.6632	.85702	1.1668	1.9407	59
2	.51554	.60165	1.6621	.85687	1.1670	1.9397	58
3	.51579	.60205	1.6610	.85672	1.1672	1.9388	57
4	.51604	.60245	1.6599	.85657	1.1675	1.9379	56
5	.51628	.60284	1.6588	.85642	1.1677	1.9369	**55**
6	.51653	.60324	1.6577	.85627	1.1679	1.9360	54
7	.51678	.60364	1.6566	.85612	1.1681	1.9351	53
8	.51703	.60403	1.6555	.85597	1.1683	1.9341	52
9	.51728	.60443	1.6545	.85582	1.1685	1.9332	51
10	.51753	.60483	1.6534	.85567	1.1687	1.9323	**50**
11	.51778	.60522	1.6523	.85551	1.1689	1.9313	49
12	.51803	.60562	1.6512	.85536	1.1691	1.9304	48
13	.51828	.60602	1.6501	.85521	1.1693	1.9295	47
14	.51852	.60642	1.6490	.85506	1.1695	1.9285	46
15	.51877	.60681	1.6479	.85491	1.1697	1.9276	**45**
16	.51902	.60721	1.6469	.85476	1.1699	1.9267	44
17	.51927	.60761	1.6458	.85461	1.1701	1.9258	43
18	.51952	.60801	1.6447	.85446	1.1703	1.9249	42
19	.51977	.60841	1.6436	.85431	1.1705	1.9239	41
20	.52002	.60881	1.6426	.85416	1.1707	1.9230	**40**
21	.52026	.60921	1.6415	.85401	1.1710	1.9221	39
22	.52051	.60960	1.6404	.85385	1.1712	1.9212	38
23	.52076	.61000	1.6393	.85370	1.1714	1.9203	37
24	.52101	.61040	1.6383	.85355	1.1716	1.9194	36
25	.52126	.61080	1.6372	.85340	1.1718	1.9184	**35**
26	.52151	.61120	1.6361	.85325	1.1720	1.9175	34
27	.52175	.61160	1.6351	.85310	1.1722	1.9166	33
28	.52200	.61200	1.6340	.85294	1.1724	1.9157	32
29	.52225	.61240	1.6329	.85279	1.1726	1.9148	31
30	.52250	.61280	1.6319	.85264	1.1728	1.9139	**30**
31	.52275	.61320	1.6308	.85249	1.1730	1.9130	29
32	.52299	.61360	1.6297	.85234	1.1732	1.9121	28
33	.52324	.61400	1.6287	.85218	1.1735	1.9112	27
34	.52349	.61440	1.6276	.85203	1.1737	1.9103	26
35	.52374	.61480	1.6265	.85188	1.1739	1.9094	**25**
36	.52399	.61520	1.6255	.85173	1.1741	1.9084	24
37	.52423	.61561	1.6244	.85157	1.1743	1.9075	23
38	.52448	.61601	1.6234	.85142	1.1745	1.9066	22
39	.52473	.61641	1.6223	.85127	1.1747	1.9057	21
40	.52498	.61681	1.6212	.85112	1.1749	1.9048	**20**
41	.52522	.61721	1.6202	.85096	1.1751	1.9039	19
42	.52547	.61761	1.6191	.85081	1.1753	1.9031	18
43	.52572	.61801	1.6181	.85066	1.1756	1.9022	17
44	.52597	.61842	1.6170	.85051	1.1758	1.9013	16
45	.52621	.61882	1.6160	.85035	1.1760	1.9004	**15**
46	.52646	.61922	1.6149	.85020	1.1762	1.8995	14
47	.52671	.61962	1.6139	.85005	1.1764	1.8986	13
48	.52696	.62003	1.6128	.84989	1.1766	1.8977	12
49	.52720	.62043	1.6118	.84974	1.1768	1.8968	11
50	.52745	.62083	1.6107	.84959	1.1770	1.8959	**10**
51	.52770	.62124	1.6097	.84943	1.1773	1.8950	9
52	.52794	.62164	1.6087	.84928	1.1775	1.8941	8
53	.52819	.62204	1.6076	.84913	1.1777	1.8933	7
54	.52844	.62245	1.6066	.84897	1.1779	1.8924	6
55	.52869	.62285	1.6055	.84882	1.1781	1.8915	**5**
56	.52893	.62325	1.6045	.84866	1.1783	1.8906	4
57	.52918	.62366	1.6034	.84851	1.1785	1.8897	3
58	.52943	.62406	1.6024	.84836	1.1788	1.8888	2
59	.52967	.62446	1.6014	.84820	1.1790	1.8880	1
60	.52992	.62487	1.6003	.84805	1.1792	1.8871	**0**
′	Cos	Cot	Tan	Sin	Csc	Sec	′

121° (301°) (238°) **58°**

NATURAL TRIGONOMETRIC FUNCTIONS TO FIVE PLACES

32° (212°) — (327°) **147°**

′	Sin	Tan	Cot	Cos	Sec	Csc	′
0	.52992	.62487	1.6003	.84805	1.1792	1.8871	**60**
1	.53017	.62527	1.5993	.84789	1.1794	1.8862	59
2	.53041	.62568	1.5983	.84774	1.1796	1.8853	58
3	.53066	.62608	1.5972	.84759	1.1798	1.8844	57
4	.53091	.62649	1.5962	.84743	1.1800	1.8836	56
5	.53115	.62689	1.5952	.84728	1.1803	1.8827	**55**
6	.53140	.62730	1.5941	.84712	1.1805	1.8818	54
7	.53164	.62770	1.5931	.84697	1.1807	1.8810	53
8	.53189	.62811	1.5921	.84681	1.1809	1.8801	52
9	.53214	.62852	1.5911	.84666	1.1811	1.8792	51
10	.53238	.62892	1.5900	.84650	1.1813	1.8783	**50**
11	.53263	.62933	1.5890	.84635	1.1815	1.8775	49
12	.53288	.62973	1.5880	.84619	1.1818	1.8766	48
13	.53312	.63014	1.5869	.84604	1.1820	1.8757	47
14	.53337	.63055	1.5859	.84588	1.1822	1.8749	46
15	.53361	.63095	1.5849	.84573	1.1824	1.8740	**45**
16	.53386	.63136	1.5839	.84557	1.1826	1.8731	44
17	.53411	.63177	1.5829	.84542	1.1828	1.8723	43
18	.53435	.63217	1.5818	.84526	1.1831	1.8714	42
19	.53460	.63258	1.5808	.84511	1.1833	1.8706	41
20	.53484	.63299	1.5798	.84495	1.1835	1.8697	**40**
21	.53509	.63340	1.5788	.84480	1.1837	1.8688	39
22	.53534	.63380	1.5778	.84464	1.1830	1.8680	38
23	.53558	.63421	1.5768	.84448	1.1842	1.8671	37
24	.53583	.63462	1.5757	.84433	1.1844	1.8663	36
25	.53607	.63503	1.5747	.84417	1.1846	1.8654	**35**
26	.53632	.63544	1.5737	.84402	1.1848	1.8646	34
27	.53656	.63584	1.5727	.84386	1.1850	1.8637	33
28	.53681	.63625	1.5717	.84370	1.1852	1.8629	32
29	.53705	.63666	1.5707	.84355	1.1855	1.8620	31
30	.53730	.63707	1.5697	.84339	1.1857	1.8612	**30**
31	.53754	.63748	1.5687	.84324	1.1859	1.8603	29
32	.53779	.63789	1.5677	.84308	1.1861	1.8595	28
33	.53804	.63830	1.5667	.84292	1.1863	1.8586	27
34	.53828	.63871	1.5657	.84277	1.1866	1.8578	26
35	.53853	.63912	1.5647	.84261	1.1868	1.8569	**25**
36	.53877	.63953	1.5637	.84245	1.1870	1.8561	24
37	.53902	.63994	1.5627	.84230	1.1872	1.8552	23
38	.53926	.64035	1.5617	.84214	1.1875	1.8544	22
39	.53951	.64076	1.5607	.84198	1.1877	1.8535	21
40	.53975	.64117	1.5597	.84182	1.1879	1.8527	**20**
41	.54000	.64158	1.5587	.84167	1.1881	1.8519	19
42	.54024	.64199	1.5577	.84151	1.1883	1.8510	18
43	.54049	.64240	1.5567	.84135	1.1886	1.8502	17
44	.54073	.64281	1.5557	.84120	1.1888	1.8494	16
45	.54097	.64322	1.5547	.84104	1.1890	1.8485	**15**
46	.54122	.64363	1.5537	.84088	1.1892	1.8477	14
47	.54146	.64404	1.5527	.84072	1.1895	1.8468	13
48	.54171	.64446	1.5517	.84057	1.1897	1.8460	12
49	.54195	.64487	1.5507	.84041	1.1899	1.8452	11
50	.54220	.64528	1.5497	.84025	1.1901	1.8443	**10**
51	.54244	.64569	1.5487	.84009	1.1903	1.8435	9
52	.54269	.64610	1.5477	.83994	1.1906	1.8427	8
53	.54293	.64652	1.5468	.83978	1.1908	1.8419	7
54	.54317	.64693	1.5458	.83962	1.1910	1.8410	6
55	.54342	.64734	1.5448	.83946	1.1912	1.8402	**5**
56	.54366	.64775	1.5438	.83930	1.1915	1.8394	4
57	.54391	.64817	1.5428	.83915	1.1917	1.8385	3
58	.54415	.64858	1.5418	.83899	1.1919	1.8377	2
59	.54440	.64899	1.5408	.83883	1.1921	1.8369	1
60	.54464	.64941	1.5399	.83867	1.1924	1.8361	**0**
′	Cos	Cot	Tan	Sin	Csc	Sec	′

122° (302°) — (237°) **57°**

33° (213°) — (326°) **146°**

′	Sin	Tan	Cot	Cos	Sec	Csc	′
0	.54464	.64941	1.5399	.83867	1.1924	1.8361	**60**
1	.54488	.64982	1.5389	.83851	1.1926	1.8353	59
2	.54513	.65024	1.5379	.83835	1.1928	1.8344	58
3	.54537	.65065	1.5369	.83819	1.1930	1.8336	57
4	.54561	.65106	1.5359	.83804	1.1933	1.8328	56
5	.54586	.65148	1.5350	.83788	1.1935	1.8320	**55**
6	.54610	.65189	1.5340	.83772	1.9137	1.8312	54
7	.54635	.65231	1.5330	.83756	1.1939	1.8303	53
8	.54659	.65272	1.5320	.83740	1.1942	1.8295	52
9	.54683	.65314	1.5311	.83724	1.1944	1.8287	51
10	.54708	.65355	1.5301	.83708	1.1946	1.8279	**50**
11	.54732	.65397	1.5291	.83692	1.1949	1.8271	49
12	.54756	.65438	1.5282	.83676	1.1951	1.8263	48
13	.54781	.65480	1.5272	.83660	1.1953	1.8255	47
14	.54805	.65521	1.5262	.83645	1.1955	1.8247	46
15	.54829	.65563	1.5253	.83629	1.1958	1.8238	**45**
16	.54854	.65604	1.5243	.83613	1.1960	1.8230	44
17	.54878	.65646	1.5233	.83597	1.1962	1.8222	43
18	.54902	.65688	1.5224	.83581	1.1964	1.8214	42
19	.54927	.65729	1.5214	.83565	1.1967	1.8206	41
20	.54951	.65771	1.5204	.83549	1.1969	1.8198	**40**
21	.54975	.65813	1.5195	.83533	1.1971	1.8190	39
22	.54999	.65854	1.5185	.83517	1.1974	1.8182	38
23	.55024	.65896	1.5175	.83501	1.1976	1.8174	37
24	.55048	.65938	1.5166	.83485	1.1978	1.8166	36
25	.55072	.65980	1.5156	.83469	1.1981	1.8158	**35**
26	.55097	.66021	1.5147	.83453	1.1983	1.8150	34
27	.55121	.66063	1.5137	.83437	1.1985	1.8142	33
28	.55145	.66105	1.5127	.83421	1.1987	1.8134	32
29	.55169	.66147	1.5118	.83405	1.1990	1.8126	31
30	.55194	.66189	1.5108	.83389	1.1992	1.8118	**30**
31	.55218	.66230	1.5099	.83373	1.1994	1.8110	29
32	.55242	.66272	1.5089	.83356	1.1997	1.8102	28
33	.55266	.66314	1.5080	.83340	1.1999	1.8094	27
34	.55291	.66356	1.5070	.83324	1.2001	1.8086	26
35	.55315	.66398	1.5061	.83308	1.2004	1.8078	**25**
36	.55339	.66440	1.5051	.83292	1.2006	1.8070	24
37	.55363	.66482	1.5042	.83276	1.2008	1.8062	23
38	.55388	.66524	1.5032	.83260	1.2011	1.8055	22
39	.55412	.66566	1.5023	.83244	1.2013	1.8047	21
40	.55436	.66608	1.5013	.83228	1.2015	1.8039	**20**
41	.55460	.66650	1.5004	.83212	1.2018	1.8031	19
42	.55484	.66692	1.4994	.83195	1.2020	1.8023	18
43	.55509	.66734	1.4985	.83179	1.2022	1.8015	17
44	.55533	.66776	1.4975	.83163	1.2025	1.8007	16
45	.55557	.66818	1.4966	.83147	1.2027	1.8000	**15**
46	.55581	.66860	1.4957	.83131	1.2029	1.7992	14
47	.55605	.66902	1.4947	.83115	1.2032	1.7984	13
48	.55630	.66944	1.4938	.83098	1.2034	1.7976	12
49	.55654	.66986	1.4928	.83082	1.2036	1.7968	11
50	.55678	.67028	1.4919	.83066	1.2039	1.7960	**10**
51	.55702	.67071	1.4910	.83050	1.2041	1.7953	9
52	.55726	.67113	1.4900	.83034	1.2043	1.7945	8
53	.55750	.67155	1.4891	.83017	1.2046	1.7937	7
54	.55775	.67197	1.4882	.83001	1.2048	1.7929	6
55	.55799	.67239	1.4872	.82985	1.2050	1.7922	**5**
56	.55823	.67282	1.4863	.82969	1.2053	1.7914	4
57	.55847	.67324	1.4854	.82953	1.2055	1.7906	3
58	.55871	.67366	1.4844	.82936	1.2057	1.7898	2
59	.55895	.67409	1.4835	.82920	1.2060	1.7891	1
60	.55919	.67451	1.4826	.82904	1.2062	1.7883	**0**
′	Cos	Cot	Tan	Sin	Csc	Sec	′

123° (303°) — (236°) **56°**

NATURAL TRIGONOMETRIC FUNCTIONS TO FIVE PLACES

34° (214°) — (325°) **145°**

′	Sin	Tan	Cot	Cos	Sec	Csc	′
0	.55919	.67451	1.4826	.82904	1.2062	1.7883	**60**
1	.55943	.67493	1.4816	.82887	1.2065	1.7875	59
2	.55968	.67536	1.4807	.82871	1.2067	1.7868	58
3	.55992	.67578	1.4798	.82855	1.2069	1.7860	57
4	.56016	.67620	1.4788	.82839	1.2072	1.7852	56
5	.56040	.67663	1.4779	.82822	1.2074	1.7844	**55**
6	.56064	.67705	1.4770	.82806	1.2076	1.7837	54
7	.56088	.67748	1.4761	.82790	1.2079	1.7829	53
8	.56112	.67790	1.4751	.82773	1.2081	1.7821	52
9	.56136	.67832	1.4742	.82757	1.2084	1.7814	51
10	.56160	.67875	1.4733	.82741	1.2086	1.7806	**50**
11	.56184	.67917	1.4724	.82724	1.2088	1.7799	49
12	.56208	.67960	1.4715	.82708	1.2091	1.7791	48
13	.56232	.68002	1.4705	.82692	1.2093	1.7783	47
14	.56256	.68045	1.4696	.82675	1.2096	1.7776	46
15	.56280	.68088	1.4687	.82659	1.2098	1.7768	**45**
16	.56305	.68130	1.4678	.82643	1.2100	1.7761	44
17	.56329	.68173	1.4669	.82626	1.2103	1.7753	43
18	.56353	.68215	1.4659	.82610	1.2105	1.7745	42
19	.56377	.68258	1.4650	.82593	1.2108	1.7738	41
20	.56401	.68301	1.4641	.82577	1.2110	1.7730	**40**
21	.56425	.68343	1.4632	.82561	1.2112	1.7723	39
22	.56449	.68386	1.4623	.82544	1.2115	1.7715	38
23	.56473	.68429	1.4614	.82528	1.2117	1.7708	37
24	.56497	.68471	1.4605	.82511	1.2120	1.7700	36
25	.56521	.68514	1.4596	.82495	1.2122	1.7693	**35**
26	.56545	.68557	1.4586	.82478	1.2124	1.7685	34
27	.56569	.68600	1.4577	.82462	1.2127	1.7678	33
28	.56593	.68642	1.4568	.82446	1.2129	1.7670	32
29	.56617	.68685	1.4559	.82429	1.2132	1.7663	31
30	.56641	.68728	1.4550	.82413	1.2134	1.7655	**30**
31	.56665	.68771	1.4541	.82396	1.2136	1.7648	29
32	.56689	.68814	1.4532	.82380	1.2139	1.7640	28
33	.56713	.68857	1.4523	.82363	1.2141	1.7633	27
34	.56736	.68900	1.4514	.82347	1.2144	1.7625	26
35	.56760	.68942	1.4505	.82330	1.2146	1.7618	**25**
36	.56784	.68985	1.4496	.82314	1.2149	1.7610	24
37	.56808	.69028	1.4487	.82297	1.2151	1.7603	23
38	.56832	.69071	1.4478	.82281	1.2154	1.7596	22
39	.56856	.69114	1.4469	.82264	1.2156	1.7588	21
40	.56880	.69157	1.4460	.82248	1.2158	1.7581	**20**
41	.56904	.69200	1.4451	.82231	1.2161	1.7573	19
42	.56928	.69243	1.4442	.82214	1.2163	1.7566	18
43	.56952	.69286	1.4433	.82198	1.2166	1.7559	17
44	.56976	.69329	1.4424	.82181	1.2168	1.7551	16
45	.57000	.69372	1.4415	.82165	1.2171	1.7544	**15**
46	.57024	.69416	1.4406	.82148	1.2173	1.7537	14
47	.57047	.69459	1.4397	.82132	1.2176	1.7529	13
48	.57071	.69502	1.4388	.82115	1.2178	1.7522	12
49	.57095	.69545	1.4379	.82098	1.2181	1.7515	11
50	.57119	.69588	1.4370	.82082	1.2183	1.7507	**10**
51	.57143	.69631	1.4361	.82065	1.2185	1.7500	9
52	.57167	.69675	1.4352	.82048	1.2188	1.7493	8
53	.57191	.69718	1.4344	.82032	1.2190	1.7485	7
54	.57215	.69761	1.4335	.82015	1.2193	1.7478	6
55	.57238	.69804	1.4326	.81999	1.2195	1.7471	**5**
56	.57262	.69847	1.4317	.81982	1.2198	1.7463	4
57	.57286	.69891	1.4308	.81965	1.2200	1.7456	3
58	.57310	.69934	1.4299	.81949	1.2203	1.7449	2
59	.57334	.69977	1.4290	.81932	1.2205	1.7442	1
60	.57358	.70021	1.4281	.81915	1.2208	1.7434	**0**
′	Cos	Cot	Tan	Sin	Csc	Sec	′

124° (304°) — (235°) **55°**

35° (215°) — (324°) **144°**

′	Sin	Tan	Cot	Cos	Sec	Csc	′
0	.57358	.70021	1.4281	.81915	1.2208	1.7434	**60**
1	.57381	.70064	1.4273	.81899	1.2210	1.7427	59
2	.57405	.70107	1.4264	.81882	1.2213	1.7420	58
3	.57429	.70151	1.4255	.81865	1.2215	1.7413	57
4	.57453	.70194	1.4246	.81848	1.2218	1.7406	56
5	.57477	.70238	1.4237	.81832	1.2220	1.7398	**55**
6	.57501	.70281	1.4229	.81815	1.2223	1.7391	54
7	.57524	.70325	1.4220	.81798	1.2225	1.7384	53
8	.57548	.70368	1.4211	.81782	1.2228	1.7377	52
9	.57572	.70412	1.4202	.81765	1.2230	1.7370	51
10	.57596	.70455	1.4193	.81748	1.2233	1.7362	**50**
11	.57619	.70499	1.4185	.81731	1.2235	1.7355	49
12	.57643	.70542	1.4176	.81714	1.2238	1.7348	48
13	.57667	.70586	1.4167	.81698	1.2240	1.7341	47
14	.57691	.70629	1.4158	.81681	1.2243	1.7334	46
15	.57715	.70673	1.4150	.81664	1.2245	1.7327	**45**
16	.57738	.70717	1.4141	.81647	1.2248	1.7320	44
17	.57762	.70760	1.4132	.81631	1.2250	1.7312	43
18	.57786	.70804	1.4124	.81614	1.2253	1.7305	42
19	.57810	.70848	1.4115	.81597	1.2255	1.7298	41
20	.57833	.70891	1.4106	.81580	1.2258	1.7291	**40**
21	.57857	.70935	1.4097	.81563	1.2260	1.7284	39
22	.57881	.70979	1.4089	.81546	1.2263	1.7277	38
23	.57904	.71023	1.4080	.81530	1.2265	1.7270	37
24	.57928	.71066	1.4071	.81513	1.2268	1.7263	36
25	.57952	.71110	1.4063	.81496	1.2271	1.7256	**35**
26	.57976	.71154	1.4054	.81479	1.2273	1.7249	34
27	.57999	.71198	1.4045	.81462	1.2276	1.7242	33
28	.58023	.71242	1.4037	.81445	1.2278	1.7235	32
29	.58047	.71285	1.4028	.81428	1.2281	1.7228	31
30	.58070	.71329	1.4019	.81412	1.2283	1.7221	**30**
31	.58094	.71373	1.4011	.81395	1.2286	1.7213	29
32	.58118	.71417	1.4002	.81378	1.2288	1.7206	28
33	.58141	.71461	1.3994	.81361	1.2291	1.7199	27
34	.58165	.71505	1.3985	.81344	1.2293	1.7192	26
35	.58189	.71549	1.3976	.81327	1.2296	1.7185	**25**
36	.58212	.71593	1.3968	.81310	1.2299	1.7179	24
37	.58236	.71637	1.3959	.81293	1.2301	1.7172	23
38	.58260	.71681	1.3951	.81276	1.2304	1.7165	22
39	.58283	.71725	1.3942	.81259	1.2306	1.7158	21
40	.58307	.71769	1.3934	.81242	1.2309	1.7151	**20**
41	.58330	.71813	1.3925	.81225	1.2311	1.7144	19
42	.58354	.71857	1.3916	.81208	1.2314	1.7137	18
43	.58378	.71901	1.3908	.81191	1.2317	1.7130	17
44	.58401	.71946	1.3899	.81174	1.2319	1.7123	16
45	.58425	.71990	1.3891	.81157	1.2322	1.7116	**15**
46	.58449	.72034	1.3882	.81140	1.2324	1.7109	14
47	.58472	.72078	1.3874	.81123	1.2327	1.7102	13
48	.58496	.72122	1.3865	.81106	1.2329	1.7095	12
49	.58519	.72167	1.3857	.81089	1.2332	1.7088	11
50	.58543	.72211	1.3848	.81072	1.2335	1.7081	**10**
51	.58567	.72255	1.3840	.81055	1.2337	1.7075	9
52	.58590	.72299	1.3831	.81038	1.2340	1.7068	8
53	.58614	.72344	1.3823	.81021	1.2342	1.7061	7
54	.58637	.72388	1.3814	.81004	1.2345	1.7054	6
55	.58661	.72432	1.3806	.80987	1.2348	1.7047	**5**
56	.58684	.72477	1.3798	.80970	1.2350	1.7040	4
57	.58708	.72521	1.3789	.80953	1.2353	1.7033	3
58	.58731	.72565	1.3781	.80936	1.2355	1.7027	2
59	.58755	.72610	1.3772	.80919	1.2358	1.7020	1
60	.58779	.72654	1.3764	.80902	1.2361	1.7013	**0**
′	Cos	Cot	Tan	Sin	Csc	Sec	′

125° (305°) — (234°) **54°**

NATURAL TRIGONOMETRIC FUNCTIONS TO FIVE PLACES

36° (216°) (323°) **143°**

′	Sin	Tan	Cot	Cos	Sec	Csc	′
0	.58779	.72654	1.3764	.80902	1.2361	1.7013	**60**
1	.58802	.72699	1.3755	.80885	1.2363	1.7006	59
2	.58826	.72743	1.3747	.80867	1.2366	1.6999	58
3	.58849	.72788	1.3739	.80850	1.2369	1.6993	57
4	.58873	.72832	1.3730	.80833	1.2371	1.6986	56
5	.58896	.72877	1.3722	.80816	1.2374	1.6979	**55**
6	.58920	.72921	1.3713	.80799	1.2376	1.6972	54
7	.58943	.72966	1.3705	.80782	1.2379	1.6966	53
8	.58967	.73010	1.3697	.80765	1.2382	1.6959	52
9	.58990	.73055	1.3688	.80748	1.2384	1.6952	51
10	.59014	.73100	1.3680	.80730	1.2387	1.6945	**50**
11	.59037	.73144	1.3672	.80713	1.2390	1.6939	49
12	.59061	.73189	1.3663	.80696	1.2392	1.6932	48
13	.59084	.73234	1.3655	.80679	1.2395	1.6925	47
14	.59108	.73278	1.3647	.80662	1.2397	1.6918	46
15	.59131	.73323	1.3638	.80644	1.2400	1.6912	**45**
16	.59154	.73368	1.3630	.80627	1.2403	1.6905	44
17	.59178	.73413	1.3622	.80610	1.2405	1.6898	43
18	.59201	.73457	1.3613	.80593	1.2408	1.6892	42
19	.59225	.73502	1.3605	.80576	1.2411	1.6885	41
20	.59248	.73547	1.3597	.80558	1.2413	1.6878	**40**
21	.59272	.73592	1.3588	.80541	1.2416	1.6871	39
22	.59295	.73637	1.3580	.80524	1.2419	1.6865	38
23	.59318	.73681	1.3572	.80507	1.2421	1.6858	37
24	.59342	.73726	1.3564	.80489	1.2424	1.6852	36
25	.59365	.73771	1.3555	.80472	1.2427	1.6845	**35**
26	.59389	.73816	1.3547	.80455	1.2429	1.6838	34
27	.59412	.73861	1.3539	.80438	1.2432	1.6832	33
28	.59436	.73906	1.3531	.80420	1.2435	1.6825	32
29	.59459	.73951	1.3522	.80403	1.2437	1.6818	31
30	.59482	.73996	1.3514	.80386	1.2440	1.6812	**30**
31	.59506	.74041	1.3506	.80368	1.2443	1.6805	29
32	.59529	.74086	1.3498	.80351	1.2445	1.6799	28
33	.59552	.74131	1.3490	.80334	1.2448	1.6792	27
34	.59576	.74176	1.3481	.80316	1.2451	1.6785	26
35	.59599	.74221	1.3473	.80299	1.2453	1.6779	**25**
36	.59622	.74267	1.3465	.80282	1.2456	1.6772	24
37	.59646	.74312	1.3457	.80264	1.2459	1.6766	23
38	.59669	.74357	1.3449	.80247	1.2462	1.6759	22
39	.59693	.74402	1.3440	.80230	1.2464	1.6753	21
40	.59716	.74447	1.3432	.80212	1.2467	1.6746	**20**
41	.59739	.74492	1.3424	.80195	1.2470	1.6739	19
42	.59763	.74538	1.3416	.80178	1.2472	1.6733	18
43	.59786	.74583	1.3408	.80160	1.2475	1.6726	17
44	.59809	.74628	1.3400	.80143	1.2478	1.6720	16
45	.59832	.74674	1.3392	.80125	1.2480	1.6713	**15**
46	.59856	.74719	1.3384	.80108	1.2483	1.6707	14
47	.59879	.74764	1.3375	.80091	1.2486	1.6700	13
48	.59902	.74810	1.3367	.80073	1.2489	1.6694	12
49	.59926	.74855	1.3359	.80056	1.2491	1.6687	11
50	.59949	.74900	1.3351	.80038	1.2494	1.6681	**10**
51	.59972	.74946	1.3343	.80021	1.2497	1.6674	9
52	.59995	.74991	1.3335	.80003	1.2499	1.6668	8
53	.60019	.75037	1.3327	.79986	1.2502	1.6661	7
54	.60042	.75082	1.3319	.79968	1.2505	1.6655	6
55	.60065	.75128	1.3311	.79951	1.2508	1.6649	**5**
56	.60089	.75173	1.3303	.79934	1.2510	1.6642	4
57	.60112	.75219	1.3295	.79916	1.2513	1.6636	3
58	.60135	.75264	1.3287	.79899	1.2516	1.6629	2
59	.60158	.75310	1.3278	.79881	1.2519	1.6623	1
60	.60182	.75355	1.3270	.79864	1.2521	1.6616	**0**
′	Cos	Cot	Tan	Sin	Csc	Sec	′

126° (306°) (233°) **53°**

37° (217°) (322°) **142°**

′	Sin	Tan	Cot	Cos	Sec	Csc	′
0	.60182	.75355	1.3270	.79864	1.2521	1.6616	**60**
1	.60205	.75401	1.3262	.79846	1.2524	1.6610	59
2	.60228	.75447	1.3254	.79829	1.2527	1.6604	58
3	.60251	.75492	1.3246	.79811	1.2530	1.6597	57
4	.60274	.75538	1.3238	.79793	1.2532	1.6591	56
5	.60298	.75584	1.3230	.79776	1.2535	1.6584	**55**
6	.60321	.75629	1.3222	.79758	1.2538	1.6578	54
7	.60344	.75675	1.3214	.79741	1.2541	1.6572	53
8	.60367	.75721	1.3206	.79723	1.2543	1.6565	52
9	.60390	.75767	1.3198	.79706	1.2546	1.6559	51
10	.60414	.75812	1.3190	.79688	1.2549	1.6553	**50**
11	.60437	.75858	1.3182	.79671	1.2552	1.6546	49
12	.60460	.75904	1.3175	.79653	1.2554	1.6540	48
13	.60483	.75950	1.3167	.79635	1.2557	1.6534	47
14	.60506	.75996	1.3159	.79618	1.2560	1,6527	46
15	.60529	.76042	1.3151	.79600	1.2563	1.6521	**45**
16	.60553	.76088	1.3143	.79583	1.2566	1.6515	44
17	.60576	.76134	1.3135	.79565	1.2568	1.6508	43
18	.60599	.76180	1.3127	.79547	1.2571	1.6502	42
19	.60622	.76226	1.3119	.79530	1.2574	1.6496	41
20	.60645	.76272	1.3111	.79512	1.2577	1.6489	**40**
21	.60668	.76318	1.3103	.79494	1.2579	1.6483	39
22	.60691	.76364	1.3095	.79477	1.2582	1.6477	38
23	.60714	.76410	1.3087	.79459	1.2585	1.6471	37
24	.60738	.76456	1.3079	.79441	1.2588	1.6464	36
25	.60761	.76502	1.3072	.79424	1.2591	1.6458	**35**
26	.60784	.76548	1.3064	.79406	1.2593	1.6452	34
27	.60807	.76594	1.3056	.79388	1.2596	1.6446	33
28	.60830	.76640	1.3048	.79371	1.2599	1.6439	32
29	.60853	.76686	1.3040	.79353	1.2602	1.6433	31
30	.60876	.76733	1.3032	.79335	1.2605	1.6427	**30**
31	.60899	.76779	1.3024	.79318	1.2608	1.6421	29
32	.60922	.76825	1.3017	.79300	1.2610	1.6414	28
33	.60945	.76871	1.3009	.79282	1.2613	1.6408	27
34	.60968	.76918	1.3001	.79264	1.2616	1.6402	26
35	.60991	.76964	1.2993	.79247	1.2619	1.6396	**25**
36	.61015	.77010	1.2985	.79229	1.2622	1.6390	24
37	.61038	.77057	1.2977	.79211	1.2624	1.6383	23
38	.61061	.77103	1.2970	.79193	1.2627	1.6377	22
39	.61084	.77149	1.2962	.79176	1.2630	1.6371	21
40	.61107	.77196	1.2954	.79158	1.2633	1.6365	**20**
41	.61130	.77242	1.2946	.79140	1.2636	1.6359	19
42	.61153	.77289	1.2938	.79122	1.2639	1.6353	18
43	.61176	.77335	1.2931	.79105	1.2641	1.6346	17
44	.61199	.77382	1.2923	.79087	1.2644	1.6340	16
45	.61222	.77428	1.2915	.79069	1.2647	1.6334	**15**
46	.61245	.77475	1.2907	.79051	1.2650	1.6328	14
47	.61268	.77521	1.2900	.79033	1.2653	1.6322	13
48	.61291	.77568	1.2892	.79016	1.2656	1.6316	12
49	.61314	.77615	1.2884	.78998	1.2659	1.6310	11
50	.61337	.77661	1.2876	.78980	1.2661	1.6303	**10**
51	.61360	.77708	1.2869	.78962	1.2664	1.6297	9
52	.61383	.77754	1.2861	.78944	1.2667	1.6291	8
53	.61406	.77801	1.2853	.78926	1.2670	1.6285	7
54	.61429	.77848	1.2846	.78908	1.2673	1.6279	6
55	.61451	.77895	1.2838	.78891	1.2676	1.6273	**5**
56	.61474	.77941	1.2830	.78873	1.2679	1.6267	4
57	.61497	.77988	1.2822	.78855	1.2682	1.6261	3
58	.61520	.78035	1.2815	.78837	1.2684	1.6255	2
59	.61543	.78082	1.2807	.78819	1.2687	1.6249	1
60	.61566	.78129	1.2799	.78801	1.2690	1.6243	**0**
′	Cos	Cot	Tan	Sin	Csc	Sec	′

127° (307°) (232°) **52°**

NATURAL TRIGONOMETRIC FUNCTIONS TO FIVE PLACES

38° (218°) (321°) **141°**

′	Sin	Tan	Cot	Cos	Sec	Csc	′
0	.61566	.78129	1.2799	.78801	1.2690	1.6243	**60**
1	.61589	.78175	1.2792	.78783	1.2693	1.6237	59
2	.61612	.78222	1.2784	.78765	1.2696	1.6231	58
3	.61635	.78269	1.2776	.78747	1.2699	1.6225	57
4	.61658	.78316	1.2769	.78729	1.2702	1.6219	56
5	.61681	.78363	1.2761	.78711	1.2705	1.6213	**55**
6	.61704	.78410	1.2753	.78694	1.2708	1.6207	54
7	.61726	.78457	1.2746	.78676	1.2710	1.6201	53
8	.61749	.78504	1.2738	.78658	1.2713	1.6195	52
9	.61772	.78551	1.2731	.78640	1.2716	1.6189	51
10	.61795	.78598	1.2723	.78622	1.1719	1.6183	**50**
11	.61818	.78645	1.2715	.78604	1.2722	1.6177	49
12	.61841	.78692	1.2708	.78586	1.2725	1.6171	48
13	.61864	.78739	1.2700	.78568	1.2728	1.6165	47
14	.61887	.78786	1.2693	.78550	1.2731	1.6159	46
15	.61909	.78834	1.2685	.78532	1.2734	1.6153	**45**
16	.61932	.78881	1.2677	.78514	1.2737	1.6147	44
17	.61955	.78928	1.2670	.78496	1.2740	1.6141	43
18	.61978	.78975	1.2662	.78478	1.2742	1.6135	42
19	.62001	.79022	1.2655	.78460	1.2745	1.6129	41
20	.62024	.79070	1.2647	.78442	1.2748	1.6123	**40**
21	.62046	.79117	1.2640	.78424	1.2751	1.6117	39
22	.62069	.79164	1.2632	.78405	1.2754	1.6111	38
23	.62092	.79212	1.2624	.78387	1.2757	1.6105	37
24	.62115	.79259	1.2617	.78369	1.2760	1.6099	36
25	.62138	.79306	1.2609	.78351	1.2763	1.6093	**35**
26	.62160	.79354	1.2602	.78333	1.2766	1.6087	34
27	.62183	.79401	1.2594	.78315	1.2769	1.6082	33
28	.62206	.79449	1.2587	.78297	1.2772	1.6076	32
29	.62229	.79496	1.2579	.78279	1.2775	1.6070	31
30	.62251	.79544	1.2572	.78261	1.2778	1.6064	**30**
31	.62274	.79591	1.2564	.78243	1.2781	1.6058	29
32	.62297	.79639	1.2557	.78225	1.2784	1.6052	28
33	.62320	.79686	1.2549	.78206	1.2787	1.6046	27
34	.62342	.79734	1.2542	.78188	1.2790	1.6040	26
35	.62365	.79781	1.2534	.78170	1.2793	1.6035	**25**
36	.62388	.79829	1.2527	.78152	1.2796	1.6029	24
37	.62411	.79877	1.2519	.78134	1.2799	1.6023	23
38	.62433	.79924	1.2512	.78116	1.2802	1.6017	22
39	.62456	.79972	1.2504	.78098	1.2804	1.6011	21
40	.62479	.80020	1.2497	.78079	1.2807	1.6005	**20**
41	.62502	.80067	1.2489	.78061	1.2810	1.6000	19
42	.62524	.80115	1.2482	.78043	1.2813	1.5994	18
43	.62547	.80163	1.2475	.78025	1.2816	1.5988	17
44	.62570	.80211	1.2467	.78007	1.2819	1.5982	16
45	.62592	.80258	1.2460	.77988	1.2822	1.5976	**15**
46	.62615	.80306	1.2452	.77970	1.2825	1.5971	14
47	.62638	.80354	1.2445	.77952	1.2828	1.5965	13
48	.62660	.80402	1.2437	.77934	1.2831	1.5959	12
49	.62683	.80450	1.2430	.77916	1.2834	1.5953	11
50	.62706	.80498	1.2423	.77897	1.2837	1.5948	**10**
51	.62728	.80546	1.2415	.77879	1.2840	1.5942	9
52	.62751	.80594	1.2408	.77861	1.2843	1.5936	8
53	.62774	.80642	1.2401	.77843	1.2846	1.5930	7
54	.62796	.80690	1.2393	.77824	1.2849	1.5925	6
55	.62819	.80738	1.2386	.77806	1.2852	1.5919	**5**
56	.62842	.80786	1.2378	.77788	1.2855	1.5913	4
57	.62864	.80834	1.2371	.77769	1.2859	1.5907	3
58	.62887	.80882	1.2364	.77751	1.2862	1.5902	2
59	.62909	.80930	1.2356	.77733	1.2865	1.5896	1
60	.62932	.80978	1.2349	.77715	1.2868	1.5890	**0**
′	Cos	Cot	Tan	Sin	Csc	Sec	′

128° (308°) (231°) **51°**

39° (219°) (320°) **140°**

′	Sin	Tan	Cot	Cos	Sec	Csc	′
0	.62932	.80978	1.2349	.77715	1.2868	1.5890	**60**
1	.62955	.81027	1.2342	.77696	1.2871	1.5884	59
2	.62977	.81075	1.2334	.77678	1.2874	1.5879	58
3	.63000	.81123	1.2327	.77660	1.2877	1.5873	57
4	.63022	.81171	1.2320	.77641	1.2880	1.5867	56
5	.63045	.81220	1.2312	.77623	1.2883	1.5862	**55**
6	.63068	.81268	1.2305	.77605	1.2886	1.5856	54
7	.63090	.81316	1.2298	.77586	1.2889	1.5850	53
8	.63113	.81364	1.2290	.77568	1.2892	1.5845	52
9	.63135	.81413	1.2283	.77550	1.2895	1.5839	51
10	.63158	.81461	1.2276	.77531	1.2898	1.5833	**50**
11	.63180	.81510	1.2268	.77513	1.2901	1.5828	49
12	.63203	.81558	1.2261	.77494	1.2904	1.5822	48
13	.63225	.81606	1.2254	.77476	1.2907	1.5816	47
14	.63248	.81655	1.2247	.77458	1.2910	1.5811	46
15	.63271	.81703	1.2239	.77439	1.2913	1.5805	**45**
16	.63293	.81752	1.2232	.77421	1.2916	1.5800	44
17	.63316	.81800	1.2225	.77402	1.2919	1.5794	43
18	.63338	.81849	1.2218	.77384	1.2923	1.5788	42
19	.63361	.81898	1.2210	.77366	1.2926	1.5783	41
20	.63383	.81946	1.2203	.77347	1.2929	1.5777	**40**
21	.63406	.81995	1.2196	.77329	1.2932	1.5771	39
22	.63428	.82044	1.2189	.77310	1.2935	1.5766	38
23	.63451	.82092	1.2181	.77292	1.2938	1.5760	37
24	.63473	.82141	1.2174	.77273	1.2941	1.5755	36
25	.63496	.82190	1.2167	.77255	1.2944	1.5749	**35**
26	.63518	.82238	1.2160	.77236	1.2947	1.5744	34
27	.63540	.82287	1.2153	.77218	1.2950	1.5738	33
28	.63563	.82336	1.2145	.77199	1.2953	1.5732	32
29	.63585	.82385	1.2138	.77181	1.2957	1.5727	31
30	.63608	.82434	1.2131	.77162	1.2960	1.5721	**30**
31	.63630	.82483	1.2124	.77144	1.2963	1.5716	29
32	.63653	.82531	1.2117	.77125	1.2966	1.5710	28
33	.63675	.82580	1.2109	.77107	1.2969	1.5705	27
34	.63698	.82629	1.2102	.77088	1.2972	1.5699	26
35	.63720	.82678	1.2095	.77070	1.2975	1.5694	**25**
36	.63742	.82727	1.2088	.77051	1.2978	1.5688	24
37	.63765	.82776	1.2081	.77033	1.2981	1.5683	23
38	.63787	.82825	1.2074	.77014	1.2985	1.5677	22
39	.63810	.82874	1.2066	.76996	1.2988	1.5672	21
40	.63832	.82923	1.2059	.76977	1.2991	1.5666	**20**
41	.63854	.82972	1.2052	.76959	1.2994	1.5661	19
42	.63877	.83022	1.2045	.76940	1.2997	1.5655	18
43	.63899	.83071	1.2038	.76921	1.3000	1.5650	17
44	.63922	.83120	1.2031	.76903	1.3003	1.5644	16
45	.63944	.83169	1.2024	.76884	1.3007	1.5639	**15**
46	.63966	.83218	1.2017	.76866	1.3010	1.5633	14
47	.63989	.83268	1.2009	.76847	1.3013	1.5628	13
48	.64011	.83317	1.2002	.76828	1.3016	1.5622	12
49	.64033	.83366	1.1995	.76810	1.3019	1.5617	11
50	.64056	.83415	1.1988	.76791	1.3022	1.5611	**10**
51	.64078	.83465	1.1981	.76772	1.3026	1.5606	9
52	.64100	.83514	1.1974	.76754	1.3029	1.5601	8
53	.64123	.83564	1.1967	.76735	1.3032	1.5595	7
54	.64145	.83613	1.1960	.76717	1.3035	1.5590	6
55	.64167	.83662	1.1953	.76698	1.3038	1.5584	**5**
56	.64190	.83712	1.1946	.76679	1.3041	1.5579	4
57	.64212	.83761	1.1939	.76661	1.3045	1.5573	3
58	.64234	.83811	1.1932	.76642	1.3048	1.5568	2
59	.64256	.83860	1.1925	.76623	1.3051	1.5563	1
60	.64279	.83910	1.1918	.76604	1.3054	1.5557	**0**
′	Cos	Cot	Tan	Sin	Csc	Sec	′

129° (309°) (230°) **50°**

NATURAL TRIGONOMETRIC FUNCTIONS TO FIVE PLACES

40° (220°) — (319°) **139°**

′	Sin	Tan	Cot	Cos	Sec	Csc	′
0	.64279	.83910	1.1918	.76604	1.3054	1.5557	**60**
1	.64301	.83960	1.1910	.76586	1.3057	1.5552	59
2	.64323	.84009	1.1903	.76567	1.3060	1.5546	58
3	.64346	.84059	1.1896	.76548	1.3064	1.5541	57
4	.64368	.84108	1.1889	.76530	1.3067	1.5536	56
5	.64390	.84158	1.1882	.76511	1.3070	1.5530	**55**
6	.64412	.84208	1.1875	.76492	1.3073	1.5525	54
7	.64435	.84258	1.1868	.76473	1.3076	1.5520	53
8	.64457	.84307	1.1861	.76455	1.3080	1.5514	52
9	.64479	.84357	1.1854	.76436	1.3083	1.5509	51
10	.64501	.84407	1.1847	.76417	1.3086	1.5504	**50**
11	.64524	.84457	1.1840	.76398	1.3089	1.5498	49
12	.64546	.84507	1.1833	.76380	1.3093	1.5493	48
13	.64568	.84556	1.1826	.76361	1.3096	1.5488	47
14	.64590	.84606	1.1819	.76342	1.3099	1.5482	46
15	.64612	.84656	1.1812	.76323	1.3102	1.5477	**45**
16	.64635	.84706	1.1806	.76304	1.3105	1.5472	44
17	.64657	.84756	1.1799	.76286	1.3109	1.5466	43
18	.64679	.84806	1.1792	.76267	1.3112	1.5461	42
19	.64701	.84856	1.1785	.76248	1.3115	1.5456	41
20	.64723	.84906	1.1778	.76229	1.3118	1.5450	**40**
21	.64746	.84956	1.1771	.76210	1.3122	1.5445	39
22	.64768	.85006	1.1764	.76192	1.3125	1.5440	38
23	.64790	.85057	1.1757	.76173	1.3128	1.5435	37
24	.64812	.85107	1.1750	.76154	1.3131	1.5429	36
25	.64834	.85157	1.1743	.76135	1.3135	1.5424	**35**
26	.64856	.85207	1.1736	.76116	1.3138	1.5419	34
27	.64878	.85257	1.1729	.76097	1.3141	1.5413	33
28	.64901	.85308	1.1722	.76078	1.3144	1.5408	32
29	.64923	.85358	1.1715	.76059	1.3148	1.5403	31
30	.64945	.85408	1.1708	.76041	1.3151	1.5398	**30**
31	.64967	.85458	1.1702	.76022	1.3154	1.5392	29
32	.64989	.85509	1.1695	.76003	1.3157	1.5387	28
33	.65011	.85559	1.1688	.75984	1.3161	1.5382	27
34	.65033	.85609	1.1681	.75965	1.3164	1.5377	26
35	.65055	.85660	1.1674	.75946	1.3167	1.5372	**25**
36	.65077	.85710	1.1667	.75927	1.3171	1.5366	24
37	.65100	.85761	1.1660	.75908	1.3174	1.5361	23
38	.65122	.85811	1.1653	.75889	1.3177	1.5356	22
39	.65144	.85862	1.1647	.75870	1.3180	1.5351	21
40	.65166	.85912	1.1640	.75851	1.3184	1.5345	**20**
41	.65188	.85963	1.1633	.75832	1.3187	1.5340	19
42	.65210	.86014	1.1626	.75813	1.3190	1.5335	18
43	.65232	.86064	1.1619	.75794	1.3194	1.5330	17
44	.65254	.86115	1.1612	.75775	1.3197	1.5325	16
45	.65276	.86166	1.1606	.75756	1.3200	1.5320	**15**
46	.65298	.86216	1.1599	.75738	1.3203	1.5314	14
47	.65320	.86267	1.1592	.75719	1.3207	1.5309	13
48	.65342	.86318	1.1585	.75700	1.3210	1.5304	12
49	.65364	.86368	1.1578	.75680	1.3213	1.5299	11
50	.65386	.86419	1.1571	.75661	1.3217	1.5294	**10**
51	.65408	.86470	1.1565	.75642	1.3220	1.5289	9
52	.65430	.86521	1.1558	.75623	1.3223	1.5283	8
53	.65452	.86572	1.1551	.75604	1.3227	1.5278	7
54	.65474	.86623	1.1544	.75585	1.3230	1.5273	6
55	.65496	.86674	1.1538	.75566	1.3233	1.5268	**5**
56	.65518	.86725	1.1531	.75547	1.3237	1.5263	4
57	.65540	.86776	1.1524	.75528	1.3240	1.5258	3
58	.65562	.86827	1.1517	.75509	1.3243	1.5253	2
59	.65584	.86878	1.1510	.75490	1.3247	1.5248	1
60	.65606	.86929	1.1504	.75471	1.3250	1.5243	**0**
′	Cos	Cot	Tan	Sin	Csc	Sec	′

130° (310°) — (229°) **49°**

41° (221°) — (318°) **138°**

′	Sin	Tan	Cot	Cos	Sec	Csc	′
0	.65606	.86929	1.1504	.75471	1.3250	1.5243	**60**
1	.65628	.86980	1.1497	.75452	1.3253	1.5237	59
2	.65650	.87031	1.1490	.75433	1.3257	1.5232	58
3	.65672	.87082	1.1483	.75414	1.3260	1.5227	57
4	.65694	.87133	1.1477	.75395	1.3264	1.5222	56
5	.65716	.87184	1.1470	.75375	1.3267	1.5217	**55**
6	.65738	.87236	1.1463	.75356	1.3270	1.5212	54
7	.65759	.87287	1.1456	.75337	1.3274	1.5207	53
8	.65781	.87338	1.1450	.75318	1.3277	1.5202	52
9	.65803	.87389	1.1443	.75299	1.3280	1.5197	51
10	.65825	.87441	1.1436	.75280	1.3284	1.5192	**50**
11	.65847	.87492	1.1430	.75261	1.3287	1.5187	49
12	.65869	.87543	1.1423	.75241	1.3291	1.5182	48
13	.65891	.87595	1.1416	.75222	1.3294	1.5177	47
14	.65913	.87646	1.1410	.75203	1.3297	1.5172	46
15	.65935	.87698	1.1403	.75184	1.3301	1.5167	**45**
16	.65956	.87749	1.1396	.75165	1.3304	1.5162	44
17	.65978	.87801	1.1389	.75146	1.3307	1.5156	43
18	.66000	.87852	1.1383	.75126	1.3311	1.5151	42
19	.66022	.87904	1.1376	.75107	1.3314	1.5146	41
20	.66044	.87955	1.1369	.75088	1.3318	1.5141	**40**
21	.66066	.88007	1.1363	.75069	1.3321	1.5136	39
22	.66088	.88059	1.1356	.75050	1.3325	1.5131	38
23	.66109	.88110	1.1349	.75030	1.3328	1.5126	37
24	.66131	.88162	1.1343	.75011	1.3331	1.5121	36
25	.66153	.88214	1.1336	.74992	1.3335	1.5116	**35**
26	.66175	.88265	1.1329	.74973	1.3338	1.5111	34
27	.66197	.88317	1.1323	.74953	1.3342	1.5107	33
28	.66218	.88369	1.1316	.74934	1.3345	1.5102	32
29	.66240	.88421	1.1310	.74915	1.3348	1.5097	31
30	.66262	.88473	1.1303	.74896	1.3352	1.5092	**30**
31	.66284	.88524	1.1296	.74876	1.3355	1.5087	29
32	.66306	.88576	1.1290	.74857	1.3359	1.5082	28
33	.66327	.88628	1.1283	.74838	1.3362	1.5077	27
34	.66349	.88680	1.1276	.74818	1.3366	1.5072	26
35	.66371	.88732	1.1270	.74799	1.3369	1.5067	**25**
36	.66393	.88784	1.1263	.74780	1.3373	1.5062	24
37	.66414	.88836	1.1257	.74760	1.3376	1.5057	23
38	.66436	.88888	1.1250	.74741	1.3380	1.5052	22
39	.66458	.88940	1.1243	.74722	1.3383	1.5047	21
40	.66480	.88992	1.1237	.74703	1.3386	1.5042	**20**
41	.66501	.89045	1.1230	.74683	1.3390	1.5037	19
42	.66523	.89097	1.1224	.74664	1.3393	1.5032	18
43	.66545	.89149	1.1217	.74644	1.3397	1.5027	17
44	.66566	.89201	1.1211	.74625	1.3400	1.5023	16
45	.66588	.89253	1.1204	.74606	1.3404	1.5018	**15**
46	.66610	.89306	1.1197	.74586	1.3407	1.5013	14
47	.66632	.89358	1.1191	.74567	1.3411	1.5008	13
48	.66653	.89410	1.1184	.74548	1.3414	1.5003	12
49	.66675	.89463	1.1178	.74528	1.3418	1.4998	11
50	.66697	.89515	1.1171	.74509	1.3421	1.4993	**10**
51	.66718	.89567	1.1165	.74489	1.3425	1.4988	9
52	.66740	.89620	1.1158	.74470	1.3428	1.4984	8
53	.66762	.89672	1.1152	.74451	1.3432	1.4979	7
54	.66783	.89725	1.1145	.74431	1.3435	1.4974	6
55	.66805	.89777	1.1139	.74412	1.3439	1.4969	**5**
56	.66827	.89830	1.1132	.74392	1.3442	1.4964	4
57	.66848	.89883	1.1126	.74373	1.3446	1.4959	3
58	.66870	.89935	1.1119	.74353	1.3449	1.4954	2
59	.66891	.89988	1.1113	.74334	1.3453	1.4950	1
60	.66913	.90040	1.1106	.74314	1.3456	1.4945	**0**
′	Cos	Cot	Tan	Sin	Csc	Sec	′

131° (311°) — (228°) **48°**

NATURAL TRIGONOMETRIC FUNCTIONS TO FIVE PLACES

42° (222°) — (317°) **137°**

′	Sin	Tan	Cot	Cos	Sec	Csc	′
0	.66913	.90040	1.1106	.74314	1.3456	1.4945	**60**
1	.66935	.90093	1.1100	.74295	1.3460	1.4940	59
2	.66956	.90146	1.1093	.74276	1.3463	1.4935	58
3	.66978	.90199	1.1087	.74256	1.3467	1.4930	57
4	.66999	.90251	1.1080	.74237	1.3470	1.4925	56
5	.67021	.90304	1.1074	.74217	1.3474	1.4921	**55**
6	.67043	.90357	1.1067	.74198	1.3478	1.4916	54
7	.67064	.90410	1.1061	.74178	1.3481	1.4911	53
8	.67086	.90463	1.1054	.74159	1.3485	1.4906	52
9	.67107	.90516	1.1048	.74139	1.3488	1.4901	51
10	.67129	.90569	1.1041	.74120	1.3492	1.4897	**50**
11	.67151	.90621	1.1035	.74100	1.3495	1.4892	49
12	.67172	.90674	1.1028	.74080	1.3499	1.4887	48
13	.67194	.90727	1.1022	.74061	1.3502	1.4882	47
14	.67215	.90781	1.1016	.74041	1.3506	1.4878	46
15	.67237	.90834	1.1009	.74022	1.3510	1.4873	**45**
16	.67258	.90887	1.1003	.74002	1.3513	1.4868	44
17	.67280	.90940	1.0996	.73983	1.3517	1.4863	43
18	.67301	.90993	1.0990	.73963	1.3520	1.4859	42
19	.67323	.91046	1.0983	.73944	1.3524	1.4854	41
20	.67344	.91099	1.0977	.73924	1.3527	1.4849	**40**
21	.67366	.91153	1.0971	.73904	1.3531	1.4844	39
22	.67387	.91206	1.0964	.73885	1.3535	1.4840	38
23	.67409	.91259	1.0958	.73865	1.3538	1.4835	37
24	.67430	.91313	1.0951	.73846	1.3542	1.4830	36
25	.67452	.91366	1.0945	.73826	1.3545	1.4825	**35**
26	.67473	.91419	1.0939	.73806	1.3549	1.4821	34
27	.67495	.91473	1.0932	.73787	1.3553	1.4816	33
28	.67516	.91526	1.0926	.73767	1.3556	1.4811	32
29	.67538	.91580	1.0919	.73747	1.3560	1.4807	31
30	.67559	.91633	1.0913	.73728	1.3563	1.4802	**30**
31	.67580	.91687	1.0907	.73708	1.3567	1.4797	29
32	.67602	.91740	1.0900	.73688	1.3571	1.4792	28
33	.67623	.91794	1.0894	.73669	1.3574	1.4788	27
34	.67645	.91847	1.0888	.73649	1.3578	1.4783	26
35	.67666	.91901	1.0881	.73629	1.3582	1.4778	**25**
36	.67688	.91955	1.0875	.73610	1.3585	1.4774	24
37	.67709	.92008	1.0869	.73590	1.3589	1.4769	23
38	.67730	.92062	1.0862	.73570	1.3592	1.4764	22
39	.67752	.92116	1.0856	.73551	1.3596	1.4760	21
40	.67773	.92170	1.0850	.73531	1.3600	1.4755	**20**
41	.67795	.92224	1.0843	.73511	1.3603	1.4750	19
42	.67816	.92277	1.0837	.73491	1.3607	1.4746	18
43	.67837	.92331	1.0831	.73472	1.3611	1.4741	17
44	.67859	.92385	1.0824	.73452	1.3614	1.4737	16
45	.67880	.92439	1.0818	.73432	1.3618	1.4732	**15**
46	.67901	.92493	1.0812	.73413	1.3622	1.4727	14
47	.67923	.92547	1.0805	.73393	1.3625	1.4723	13
48	.67944	.92601	1.0799	.73373	1.3629	1.4718	12
49	.67965	.92655	1.0793	.73353	1.3633	1.4713	11
50	.67987	.92709	1.0786	.73333	1.3636	1.4709	**10**
51	.68008	.92763	1.0780	.73314	1.3640	1.4704	9
52	.68029	.92817	1.0774	.73294	1.3644	1.4700	8
53	.68051	.92872	1.0768	.73274	1.3647	1.4695	7
54	.68072	.92926	1.0761	.73254	1.3651	1.4690	6
55	.68093	.92980	1.0755	.73234	1.3655	1.4686	**5**
56	.68115	.93034	1.0749	.73215	1.3658	1.4681	4
57	.68136	.93088	1.0742	.73195	1.3662	1.4677	3
58	.68157	.93143	1.0736	.73175	1.3666	1.4672	2
59	.68179	.93197	1.0730	.73155	1.3670	1.4667	1
60	.68200	.93252	1.0724	.73135	1.3673	1.4663	**0**
′	Cos	Cot	Tan	Sin	Csc	Sec	′

132° (312°) — (227°) **47°**

43° (223°) — (316°) **136°**

′	Sin	Tan	Cot	Cos	Sec	Csc	′
0	.68200	.93252	1.0724	.73135	1.3673	1.4663	**60**
1	.68221	.93306	1.0717	.73116	1.3677	1.4658	59
2	.68242	.93360	1.0711	.73096	1.3681	1.4654	58
3	.68264	.93415	1.0705	.73076	1.3684	1.4649	57
4	.68285	.93469	1.0699	.73056	1.3688	1.4645	56
5	.68306	.93524	1.0692	.73036	1.3692	1.4640	**55**
6	.68327	.93578	1.0686	.73016	1.3696	1.4635	54
7	.68349	.93633	1.0680	.72996	1.3699	1.4631	53
8	.68370	.93688	1.0674	.72976	1.3703	1.4626	52
9	.68391	.93742	1.0668	.72957	1.3707	1.4622	51
10	.68412	.93797	1.0661	.72937	1.3711	1.4617	**50**
11	.68434	.93852	1.0655	.72917	1.3714	1.4613	49
12	.68455	.93906	1.0649	.72897	1.3718	1.4608	48
13	.68476	.93961	1.0643	.72877	1.3722	1.4604	47
14	.68497	.94016	1.0637	.72857	1.3726	1.4599	46
15	.68518	.94071	1.0630	.72837	1.3729	1.4595	**45**
16	.68539	.94125	1.0624	.72817	1.3733	1.4590	44
17	.68561	.94180	1.0618	.72797	1.3737	1.4586	43
18	.68582	.94235	1.0612	.72777	1.3741	1.4581	42
19	.68603	.94290	1.0606	.72757	1.3744	1.4577	41
20	.68624	.94345	1.0599	.72737	1.3748	1.4572	**40**
21	.68645	.94400	1.0593	.72717	1.3752	1.4568	39
22	.68666	.94455	1.0587	.72697	1.3756	1.4563	38
23	.68688	.94510	1.0581	.72677	1.3759	1.4559	37
24	.68709	.94565	1.0575	.72657	1.3763	1.4554	36
25	.68730	.94620	1.0569	.72637	1.3767	1.4550	**35**
26	.68751	.94676	1.0562	.72617	1.3771	1.4545	34
27	.68772	.94731	1.0556	.72597	1.3775	1.4541	33
28	.68793	.94786	1.0550	.72577	1.3778	1.4536	32
29	.68814	.94841	1.0544	.72557	1.3782	1.4532	31
30	.68835	.94896	1.0538	.72537	1.3786	1.4527	**30**
31	.68857	.94952	1.0532	.72517	1.3790	1.4523	29
32	.68878	.95007	1.0526	.72497	1.3794	1.4518	28
33	.68899	.95062	1.0519	.72477	1.3797	1.4514	27
34	.68920	.95118	1.0513	.72457	1.3801	1.4510	26
35	.68941	.95173	1.0507	.72437	1.3805	1.4505	**25**
36	.68962	.95229	1.0501	.72417	1.3809	1.4501	24
37	.68983	.95284	1.0495	.72397	1.3813	1.4496	23
38	.69004	.95340	1.0489	.72377	1.3817	1.4492	22
39	.69025	.95395	1.0483	.72357	1.3820	1.4487	21
40	.69046	.95451	1.0477	.72337	1.3824	1.4483	**20**
41	.69067	.95506	1.0470	.72317	1.3828	1.4479	19
42	.69088	.95562	1.0464	.72297	1.3832	1.4474	18
43	.69109	.95618	1.0458	.72277	1.3836	1.4470	17
44	.69130	.95673	1.0452	.72257	1.3840	1.4465	16
45	.69151	.95729	1.0446	.72236	1.3843	1.4461	**15**
46	.69172	.95785	1.0440	.72216	1.3847	1.4457	14
47	.69193	.95841	1.0434	.72196	1.3851	1.4452	13
48	.69214	.95897	1.0428	.72176	1.3855	1.4448	12
49	.69235	.95952	1.0422	.72156	1.3859	1.4443	11
50	.69256	.96008	1.0416	.72136	1.3863	1.4439	**10**
51	.69277	.96064	1.0410	.72116	1.3867	1.4435	9
52	.69298	.96120	1.0404	.72095	1.3871	1.4430	8
53	.69319	.96176	1.0398	.72075	1.3874	1.4426	7
54	.69340	.96232	1.0392	.72055	1.3878	1.4422	6
55	.69361	.96288	1.0385	.72035	1.3882	1.4417	**5**
56	.69382	.96344	1.0379	.72015	1.3886	1.4413	4
57	.69403	.96400	1.0373	.71995	1.3890	1.4409	3
58	.69424	.96457	1.0367	.71974	1.3894	1.4404	2
59	.69445	.96513	1.0361	.71954	1.3898	1.4400	1
60	.69466	.96569	1.0355	.71934	1.3902	1.4396	**0**
′	Cos	Cot	Tan	Sin	Csc	Sec	′

133° (313°) — (226°) **46°**

NATURAL TRIGONOMETRIC FUNCTIONS TO FIVE PLACES

44° (224°) (315°) **135°**

′	Sin	Tan	Cot	Cos	Sec	Csc	′
0	.69466	.96569	1.0355	.71934	1.3902	1.4396	**60**
1	.69487	.96625	1.0349	.71914	1.3906	1.4391	59
2	.69508	.96681	1.0343	.71894	1.3909	1.4387	58
3	.69529	.96738	1.0337	.71873	1.3913	1.4383	57
4	.69549	.96794	1.0331	.71853	1.3917	1.4378	56
5	.69570	.96850	1.0325	.71833	1.3921	1.4374	**55**
6	.69591	.96907	1.0319	.71813	1.3925	1.4370	54
7	.69612	.96963	1.0313	.71792	1.3929	1.4365	53
8	.69633	.97020	1.0307	.71772	1.3933	1.4361	52
9	.69654	.97076	1.0301	.71752	1.3937	1.4357	51
10	.69675	.97133	1.0295	.71732	1.3941	1.4352	**50**
11	.69696	.97189	1.0289	.71711	1.3945	1.4348	49
12	.69717	.97246	1.0283	.71691	1.3949	1.4344	48
13	.69737	.97302	1.0277	.71671	1.3953	1.4340	47
14	.69758	.97359	1.0271	.71650	1.3957	1.4335	46
15	.69779	.97416	1.0265	.71630	1.3961	1.4331	**45**
16	.69800	.97472	1.0259	.71610	1.3965	1.4327	44
17	.69821	.97529	1.0253	.71590	1.3969	1.4322	43
18	.69842	.97586	1.0247	.71569	1.3972	1.4318	42
19	.69862	.97643	1.0241	.71549	1.3976	1.4314	41
20	.69883	.97700	1.0235	.71529	1.3980	1.4310	**40**
21	.69904	.97756	1.0230	.71508	1.3984	1.4305	39
22	.69925	.97813	1.0224	.71488	1.3988	1.4301	38
23	.69946	.97870	1.0218	.71468	1.3992	1.4297	37
24	.69966	.97927	1.0212	.71447	1.3996	1.4293	36
25	.69987	.97984	1.0206	.71427	1.4000	1.4288	**35**
26	.70008	.98041	1.0200	.71407	1.4004	1.4284	34
27	.70029	.98098	1.0194	.71386	1.4008	1.4280	33
28	.70049	.98155	1.0188	.71366	1.4012	1.4276	32
29	.70070	.98213	1.0182	.71345	1.4016	1.4271	31
30	.70091	.98270	1.0176	.71325	1.4020	1.4267	**30**
31	.70112	.98327	1.0170	.71305	1.4024	1.4263	29
32	.70132	.98384	1.0164	.71284	1.4028	1.4259	28
33	.70153	.98441	1.0158	.71264	1.4032	1.4255	27
34	.70174	.98499	1.0152	.71243	1.4036	1.4250	26
35	.70195	.98556	1.0147	.71223	1.4040	1.4246	**25**
36	.70215	.98613	1.0141	.71203	1.4044	1.4242	24
37	.70236	.98671	1.0135	.71182	1.4048	1.4238	23
38	.70257	.98728	1.0129	.71162	1.4052	1.4234	22
39	.70277	.98786	1.0123	.71141	1.4057	1.4229	21
40	.70298	.98843	1.0117	.71121	1.4061	1.4225	**20**
41	.70319	.98901	1.0111	.71100	1.4065	1.4221	19
42	.70339	.98958	1.0105	.71080	1.4069	1.4217	18
43	.70360	.99016	1.0099	.71059	1.4073	1.4213	17
44	.70381	.99073	1.0094	.71039	1.4077	1.4208	16
45	.70401	.99131	1.0088	.71019	1.4081	1.4204	**15**
46	.70422	.99189	1.0082	.70998	1.4085	1.4200	14
47	.70443	.99247	1.0076	.70978	1.4089	1.4196	13
48	.70463	.99304	1.0070	.70957	1.4093	1.4192	12
49	.70484	.99362	1.0064	.70937	1.4097	1.4188	11
50	.70505	.99420	1.0058	.70916	1.4101	1.4183	**10**
51	.70525	.99478	1.0052	.70896	1.4105	1.4179	9
52	.70546	.99536	1.0047	.70875	1.4109	1.4175	8
53	.70567	.99594	1.0041	.70855	1.4113	1.4171	7
54	.70587	.99652	1.0035	.70834	1.4118	1.4167	6
55	.70608	.99710	1.0029	.70813	1.4122	1.4163	**5**
56	.70628	.99768	1.0023	.70793	1.4126	1.4159	4
57	.70649	.99826	1.0017	.70772	1.4130	1.4154	3
58	.70670	.99884	1.0012	.70752	1.4134	1.4150	2
59	.70690	.99942	1.0006	.70731	1.4138	1.4146	1
60	.70711	1.0000	1.0000	.70711	1.4142	1.4142	**0**
′	Cos	Cot	Tan	Sin	Csc	Sec	′

134° (314°) (225°) **45°**

NATURAL TRIGONOMETRIC FUNCTIONS SINE, COSINE, TANGENT, COTANGENT, FOR ANGLES IN DEGREES AND DECIMALS

The table that follows, NATURAL TRIGONOMETRIC FUNCTIONS, SINE, COSINE, TANGENT, COTANGENT, FOR ANGLES IN DEGREES AND DECIMALS, gives the values of the sine, cosine, tangent, and cotangent for each tenth of a degree from 0° to 90° where the entries are correct to five significant digits.

For angle values in the left column, use the column headings at the top of each column. For angle values in the right column, use the column headings at the bottom of each column.

Linear interpolation may be used to obtain functional values for angles expressed to hundredths of a degree, except where the functions are rapidly varying.

This is followed by the table NATURAL TRIGONOMETRIC FUNCTIONS TO FIVE TENTHS OF A DEGREE (SECANTS AND COSECANTS), which gives the values of the secant and cosecant for equally spaced intervals of one-half degree from 0° to 90°, and the entries are correct to six significant digits.

NATURAL TRIGONOMETRIC FUNCTIONS SINE, COSINE, TANGENT, COTANGENT, FOR ANGLES IN DEGREES AND DECIMALS

Deg.	Sin	Tan	* Cot	Cos	
0.0	0 00000	0.00000	∞	1.0000	**90.0**
.1	.00175	00175	573.0	1.0000	89.9
.2	.00349	00349	286.5	1.0000	.8
.3	.00524	00524	191.0	1.0000	.7
.4	.00698	.00698	143.24	1 0000	.6
.5	00873	.00873	114.59	1.0000	.5
.6	.01047	.01047	95.49	0.9999	.4
.7	.01222	.01222	81 85	.9999	.3
.8	.01396	.01396	71.62	.9999	.2
.9	.01571	.01571	63.66	.9999	89.1
1.0	0.01745	0.01746	57.29	0.9998	**89.0**
.1	.01920	.01920	52.08	9998	88.9
.2	02094	02095	47.74	.9998	.8
.3	.02269	.02269	44.07	.9997	.7
.4	.02443	.02444	40.92	.9997	.6
.5	.02618	.02619	38.19	.9997	.5
.6	.02792	.02793	35.80	.9996	.4
.7	.02967	.02968	33.69	.9996	.3
.8	.03141	.03143	31.82	.9995	.2
.9	.03316	.03317	30.14	.9995	88.1
2.0	0.03490	0.03492	28.64	0.9994	**88.0**
.1	.03664	.03667	27.27	9993	87.9
.2	.03839	03842	26.03	.9993	.8
.3	.04013	.04016	24.90	.9992	.7
.4	.04188	04191	23.86	.9991	.6
.5	.04362	.04366	22.90	.9990	.5
.6	.04536	.04541	22 02	.9990	.4
.7	.04711	.04716	21.20	.9989	.3
.8	.04885	.04891	20.45	.9988	.2
.9	.05059	.05066	19.74	.9987	87.1
3.0	0.05234	0.05241	19.081	0.9986	**87.0**
.1	.05408	.05416	18.464	.9985	86.9
.2	.05582	.05591	17.886	.9984	.8
.3	.05756	.05766	17.343	.9983	.7
.4	.05931	.05941	16.832	9982	.6
.5	.06105	.06116	16.350	.9981	.5
.6	.06279	.06291	15.895	.9980	.4
.7	.06453	.06467	15.464	.9979	.3
.8	.06627	.06642	15.056	.9978	.2
.9	.06802	.06817	14.669	.9977	86.1
4.0	0.06976	0.06993	14.301	0.9976	**86.0**
.1	.07150	.07168	13.951	.9974	85.9
.2	.07324	.07344	13.617	.9973	.8
.3	.07498	.07519	13.300	.9972	.7
.4	.07672	.07695	12.996	.9971	.6
.5	.07846	.07870	12.706	.9969	.5
.6	.08020	.08046	12.429	.9968	.4
.7	.08194	.08221	12.163	.9966	.3
.8	.08368	.08397	11.909	.9965	.2
.9	.08542	.08573	11.664	.9963	85.1
5.0	0.08716	0.08749	11.430	0.9962	**85.0**
.1	.08889	.08925	11.205	.9960	84.9
.2	.09063	.09101	10.988	.9959	.8
.3	.09237	.09277	10.780	.9957	.7
.4	.09411	.09453	10.579	.9956	.6
.5	.09585	.09629	10.385	.9954	.5
.6	.09758	.09805	10.199	.9952	4
.7	.09932	.09981	10.019	.9951	.3
.8	.10106	.10158	9.845	.9949	.2
.9	.10279	.10334	9.677	.9947	84.1
6.0	0.10453	0.10510	9.514	0.9945	**84.0**
	Cos	Cot	* Tan	Sin	Deg.

Deg.	Sin	Tan	Cot	Cos	
6.0	0.10453	0.10510	9 514	0.9945	**84.0**
.1	.10626	.10687	9.357	.9943	83.9
.2	.10800	10863	9.205	.9942	.8
.3	.10973	.11040	9.058	9940	.7
.4	.11147	.11217	8.915	9938	.6
.5	.11320	.11394	8.777	.9936	.5
.6	.11494	.11570	8.643	9934	.4
.7	.11667	11747	8.513	9932	.3
.8	.11840	.11924	8.386	.9930	.2
.9	.12014	.12101	8.264	.9928	83.1
7.0	0.12187	0.12278	8.144	0.9925	**83.0**
.1	.12360	.12456	8.028	.9923	82.9
.2	.12533	.12633	7.916	.9921	.8
.3	.12706	.12810	7.806	.9919	.7
.4	.12880	.12988	7.700	.9917	.6
.5	.13053	.13165	7 596	.9914	.5
.6	.13226	13343	7.495	.9912	.4
.7	.13399	.13521	7.396	.9910	.3
.8	.13572	.13698	7.300	.9907	.2
.9	.13744	.13876	7.207	.9905	82.1
8.0	0.13917	0.14054	7.115	0.9903	**82.0**
.1	.14090	.14232	7.026	.9900	81.9
.2	.14263	.14410	6.940	9898	.8
.3	.14436	.14588	6.855	.9895	.7
.4	.14608	.14767	6.772	.9893	.6
.5	.14781	.14945	6.691	.9890	.5
.6	.14954	.15124	6.612	.9888	.4
.7	.15126	.15302	6.535	.9885	.3
.8	.15299	.15481	6.460	.9882	.2
.9	.15471	.15660	6.386	.9880	81.1
9.0	0.15643	0.15838	6.314	0.9877	**81.0**
.1	.15816	.16017	6.243	.9874	80.9
.2	.15988	.16196	6.174	.9871	.8
.3	.16160	.16376	6.107	.9869	.7
.4	.16333	.16555	6.041	.9866	.6
.5	.16505	.16734	5.976	.9863	.5
.6	.16677	.16914	5.912	.9860	.4
.7	.16849	.17093	5.850	.9857	.3
.8	.17021	.17273	5.789	.9854	.2
.9	.17193	.17453	5.730	.9851	80.1
10.0	0.1736	0.1763	5.671	0.9848	**80.0**
.1	.1754	.1781	5.614	.9845	79.9
.2	.1771	.1799	5.558	.9842	.8
.3	.1788	.1817	5.503	.9839	.7
.4	.1805	.1835	5.449	.9836	.6
.5	.1822	.1853	5.396	.9833	.5
.6	.1840	.1871	5.343	.9829	.4
.7	.1857	.1890	5.292	.9826	.3
.8	.1874	.1908	5.242	.9823	.2
.9	.1891	.1926	5.193	.9820	79.1
11.0	0.1908	0.1944	5.145	0.9816	**79.0**
.1	.1925	.1962	5.097	.9813	78.9
.2	.1942	.1980	5.050	.9810	.8
.3	.1959	.1998	5.005	.9806	.7
.4	.1977	.2016	4.959	.9803	.6
.5	.1994	.2035	4.915	.9799	.5
.6	.2011	.2053	4.872	.9796	.4
.7	.2028	.2071	4.829	.9792	.3
.8	.2045	.2089	4.787	.9789	.2
.9	.2062	.2107	4.745	.9785	78.1
12.0	0.2079	0.2126	4.705	0.9781	**78.0**
	Cos	Cot	Tan	Sin	Deg.

*** Interpolation in this section of the table is inaccurate.**

NATURAL TRIGONOMETRIC FUNCTIONS SINE, COSINE, TANGENT, COTANGENT, FOR ANGLES IN DEGREES AND DECIMALS (Continued)

Deg.	Sin	Tan	Cot	Cos	
12.0	0.2079	0.2126	4.705	0.9781	**78.0**
.1	.2096	.2144	4.665	.9778	77.9
.2	.2113	.2162	4.625	.9774	.8
.3	.2130	.2180	4.586	.9770	.7
.4	.2147	.2199	4.548	.9767	.6
.5	.2164	.2217	4.511	.9763	.5
.6	.2181	.2235	4.474	.9759	.4
.7	.2198	.2254	4.437	.9755	.3
.8	.2215	.2272	4.402	.9751	.2
.9	.2233	.2290	4.366	.9748	77.1
13.0	0.2250	0.2309	4.331	0.9744	**77.0**
.1	.2267	.2327	4.297	.9740	76.9
.2	.2284	.2345	4.264	.9736	.8
.3	.2300	.2364	4.230	.9732	.7
.4	.2317	.2382	4.198	.9728	.6
.5	.2334	.2401	4.165	.9724	.5
.6	.2351	.2419	4.134	.9720	.4
.7	.2368	.2438	4.102	.9715	.3
.8	.2385	.2456	4.071	.9711	.2
.9	.2402	.2475	4.041	.9707	76.1
14.0	0.2419	0.2493	4.011	0.9703	**76.0**
.1	.2436	.2512	3.981	.9699	75.9
.2	.2453	.2530	3.952	.9694	.8
.3	.2470	.2549	3.923	.9690	.7
.4	.2487	.2568	3.895	.9686	.6
.5	.2504	.2586	3.867	.9681	.5
.6	.2521	.2605	3.839	.9677	.4
.7	.2538	.2623	3.812	.9673	.3
.8	.2554	.2642	3.785	.9668	.2
.9	.2571	.2661	3.758	.9664	75.1
15.0	0.2588	0.2679	3.732	0.9659	**75.0**
.1	.2605	.2698	3.706	.9655	74.9
.2	.2622	.2717	3.681	.9650	.8
.3	.2639	.2736	3.655	.9646	.7
.4	.2656	.2754	3.630	.9641	.6
.5	.2672	.2773	3.606	.9636	.5
.6	.2689	.2792	3.582	.9632	.4
.7	.2706	.2811	3.558	.9627	.3
.8	.2723	.2830	3.534	.9622	.2
.9	.2740	.2849	3.511	.9617	74.1
16.0	0.2756	0.2867	3.487	0.9613	**74.0**
.1	.2773	.2886	3.465	.9608	73.9
.2	.2790	.2905	3.442	.9603	.8
.3	.2807	.2924	3.420	.9598	.7
.4	.2823	.2943	3.398	.9593	.6
.5	.2840	.2962	3.376	.9588	.5
.6	.2857	.2981	3.354	.9583	.4
.7	.2874	.3000	3.333	.9578	.3
.8	.2890	.3019	3.312	.9573	.2
.9	.2907	.3038	3.291	.9568	73.1
17.0	0.2924	0.3057	3.271	0.9563	**73.0**
.1	.2940	.3076	3.251	.9558	72.9
.2	.2957	.3096	3.230	.9553	.8
.3	.2974	.3115	3.211	.9548	.7
.4	.2990	.3134	3.191	.9542	.6
.5	.3007	.3153	3.172	.9537	.5
.6	.3024	.3172	3.152	.9532	.4
.7	.3040	.3191	3.133	.9527	.3
.8	.3057	.3211	3.115	.9521	.2
.9	.3074	.3230	3.096	.9516	72.1
18.0	0.3090	0.3249	3.078	0.9511	**72.0**
	Cos	Cot	Tan	Sin	Deg.

Deg.	Sin	Tan	Cot	Cos	
18.0	0.3090	0.3249	3.078	0.9511	**72.0**
.1	.3107	.3269	3.060	.9505	71.9
.2	.3123	.3288	3.042	.9500	.8
.3	.3140	.3307	3.024	.9494	.7
.4	.3156	.3327	3.006	.9489	.6
.5	.3173	.3346	2.989	.9483	.5
.6	.3190	.3365	2.971	.9478	.4
.7	.3206	.3385	2.954	.9472	.3
.8	.3223	.3404	2.937	.9466	.2
.9	.3239	.3424	2.921	.9461	71.1
19.0	0.3256	0.3443	2.904	0.9455	**71.0**
.1	.3272	.3463	2.888	.9449	70.9
.2	.3289	.3482	2.872	.9444	.8
.3	.3305	.3502	2.856	.9438	.7
.4	.3322	.3522	2.840	.9432	.6
.5	.3338	.3541	2.824	.9426	.5
.6	.3355	.3561	2.808	.9421	.4
.7	.3371	.3581	2.793	.9415	.3
.8	.3387	.3600	2.778	.9409	.2
.9	.3404	.3620	2.762	.9403	70.1
20.0	0.3420	0.3640	2.747	0.9397	**70.0**
.1	.3437	.3659	2.733	.9391	69.9
.2	.3453	.3679	2.718	.9385	.8
.3	.3469	.3699	2.703	.9379	.7
.4	.3486	.3719	2.689	.9373	.6
.5	.3502	.3739	2.675	.9367	.5
.6	.3518	.3759	2.660	.9361	.4
.7	.3535	.3779	2.646	.9354	.3
.8	.3551	.3799	2.633	.9348	.2
.9	.3567	.3819	2.619	.9342	69.1
21.0	0.3584	0.3839	2.605	0.9336	**69.0**
.1	.3600	.3859	2.592	.9330	68.9
.2	.3616	.3879	2.578	.9323	.8
.3	.3633	.3899	2.565	.9317	.7
.4	.3649	.3919	2.552	.9311	.6
.5	.3665	.3939	2.539	.9304	.5
.6	.3681	.3959	2.526	.9298	.4
.7	.3697	.3979	2.513	.9291	.3
.8	.3714	.4000	2.500	.9285	.2
.9	.3730	.4020	2.488	.9278	68.1
22.0	0.3746	0.4040	2.475	0.9272	**68.0**
.1	.3762	.4061	2.463	.9265	67.9
.2	.3778	.4081	2.450	.9259	.8
.3	.3795	.4101	2.438	.9252	.7
.4	.3811	.4122	2.426	.9245	.6
.5	.3827	.4142	2.414	.9239	.5
.6	.3843	.4163	2.402	.9232	.4
.7	.3859	.4183	2.391	.9225	.3
.8	.3875	.4204	2.379	.9219	.2
.9	.3891	.4224	2.367	.9212	67.1
23.0	0.3907	0.4245	2.356	0.9205	**67.0**
.1	.3923	.4265	2.344	.9198	66.9
.2	.3939	.4286	2.333	.9191	.8
.3	.3955	.4307	2.322	.9184	.7
.4	.3971	.4327	2.311	.9178	.6
.5	.3987	.4348	2.300	.9171	.5
.6	.4003	.4369	2.289	.9164	.4
.7	.4019	.4390	2.278	.9157	.3
.8	.4035	.4411	2.267	.9150	.2
.9	.4051	.4431	2.257	.9143	66.1
24.0	0.4067	0.4452	2.246	0.9135	**66.0**
	Cos	Cot	Tan	Sin	Deg.

NATURAL TRIGONOMETRIC FUNCTIONS SINE, COSINE, TANGENT, COTANGENT, FOR ANGLES IN DEGREES AND DECIMALS (Continued)

Deg.	Sin	Tan	Cot	Cos	
24.0	0.4067	0.4452	2.246	0.9135	**66.0**
.1	.4083	.4473	2.236	.9128	65.9
.2	.4099	.4494	2.225	.9121	.8
.3	.4115	.4515	2.215	.9114	.7
.4	.4131	.4536	2.204	.9107	.6
.5	.4147	.4557	2.194	.9100	.5
.6	.4163	.4578	2.184	.9092	.4
.7	.4179	.4599	2.174	.9085	.3
.8	.4195	.4621	2.164	.9078	.2
.9	.4210	.4642	2.154	.9070	65.1
25.0	0.4226	0.4663	2.145	0.9063	**65.0**
.1	.4242	.4684	2.135	.9056	64.9
.2	.4258	.4706	2.125	.9048	.8
.3	.4274	.4727	2.116	.9041	.7
.4	.4289	.4748	2.106	.9033	.6
.5	.4305	.4770	2.097	.9026	.5
.6	.4321	.4791	2.087	.9018	.4
.7	.4337	.4813	2.078	.9011	.3
.8	.4352	.4834	2.069	.9003	.2
.9	.4368	.4856	2.059	.8996	64.1
26.0	0.4384	0.4877	2.050	0.8988	**64.0**
.1	.4399	.4899	2.041	.8980	63.9
.2	.4415	.4921	2.032	.8973	.8
.3	.4431	.4942	2.023	.8965	.7
.4	.4446	.4964	2.014	.8957	.6
.5	.4462	.4986	2.006	.8949	.5
.6	.4478	.5008	1.997	.8942	.4
.7	.4493	.5029	1.988	.8934	.3
.8	.4509	.5051	1.980	.8926	.2
.9	.4524	.5073	1.971	.8918	63.1
27.0	0.4540	0.5095	1.963	0.8910	**63.0**
.1	.4555	.5117	1.954	.8902	62.9
.2	.4571	.5139	1.946	.8894	.8
.3	.4586	.5161	1.937	.8886	.7
.4	.4602	.5184	1.929	.8878	.6
.5	.4617	.5206	1.921	.8870	.5
.6	.4633	.5228	1.913	.8862	.4
.7	.4648	.5250	1.905	.8854	.3
.8	.4664	.5272	1.897	.8846	.2
.9	.4679	.5295	1.889	.8838	62.1
28.0	0.4695	0.5317	1.881	0.8829	**62.0**
.1	.4710	.5340	1.873	.8821	61.9
.2	.4726	.5362	1.865	.8813	.8
.3	.4741	.5384	1.857	.8805	.7
.4	.4756	.5407	1.849	.8796	.6
.5	.4772	.5430	1.842	.8788	.5
.6	.4787	.5452	1.834	.8780	.4
.7	.4802	.5475	1.827	.8771	.3
.8	.4818	.5498	1.819	.8763	.2
.9	.4833	.5520	1.811	.8755	61.1
29.0	0.4848	0.5543	1.804	0.8746	**61.0**
.1	.4863	.5566	1.797	.8738	60.9
.2	.4879	.5589	1.789	.8729	.8
.3	.4894	.5612	1.782	.8721	.7
.4	.4909	.5635	1.775	.8712	.6
.5	.4924	.5658	1.767	.8704	.5
.6	.4939	.5681	1.760	.8695	.4
.7	.4955	.5704	1.753	.8686	.3
.8	.4970	.5727	1.746	.8678	.2
.9	.4985	.5750	1.739	.8669	60.1
30.0	0.5000	0.5774	1.732	0.8660	**60.0**
	Cos	Cot	Tan	Sin	Deg.

Deg.	Sin	Tan	Cot	Cos	
30.0	0.5000	0.5774	1.7321	0.8660	**60.0**
.1	.5015	.5797	1.7251	.8652	59.9
.2	.5030	.5820	1.7182	.8643	.8
.3	.5045	.5844	1.7113	.8634	.7
.4	.5060	.5867	1.7045	.8625	.6
.5	.5075	.5890	1.6977	.8616	.5
.6	.5090	.5914	1.6909	.8607	.4
.7	.5105	.5938	1.6842	.8599	.3
.8	.5120	.5961	1.6775	.8590	.2
.9	.5135	.5985	1.6709	.8581	59.1
31.0	0.5150	0.6009	1.6643	0.8572	**59.0**
.1	.5165	.6032	1.6577	.8563	58.9
.2	.5180	.6056	1.6512	.8554	.8
.3	.5195	.6080	1.6447	.8545	.7
.4	.5210	.6104	1.6383	.8536	.6
.5	.5225	.6128	1.6319	.8526	.5
.6	.5240	.6152	1.6255	.8517	.4
.7	.5255	.6176	1.6191	.8508	.3
.8	.5270	.6200	1.6128	.8499	.2
.9	.5284	.6224	1.6066	.8490	58.1
32.0	0.5299	0.6249	1.6003	0.8480	**58.0**
.1	.5314	.6273	1.5941	.8471	57.9
.2	.5329	.6297	1.5880	.8462	.8
.3	.5344	.6322	1.5818	.8453	.7
.4	.5358	.6346	1.5757	.8443	.6
.5	.5373	.6371	1.5697	.8434	.5
.6	.5388	.6395	1.5637	.8425	.4
.7	.5402	.6420	1.5577	.8415	.3
.8	.5417	.6445	1.5517	.8406	.2
.9	.5432	.6469	1.5458	.8396	57.1
33.0	0.5446	0.6494	1.5399	0.8387	**57.0**
.1	.5461	.6519	1.5340	.8377	56.9
.2	.5476	.6544	1.5282	.8368	.8
.3	.5490	.6569	1.5224	.8358	.7
.4	.5505	.6594	1.5166	.8348	.6
.5	.5519	.6619	1.5108	.8339	.5
.6	.5534	.6644	1.5051	.8329	.4
.7	.5548	.6669	1.4994	.8320	.3
.8	.5563	.6694	1.4938	.8310	.2
.9	.5577	.6720	1.4882	.8300	56.1
34.0	0.5592	0.6745	1.4826	0.8290	**56.0**
.1	.5606	.6771	1.4770	.8281	55.9
.2	.5621	.6796	1.4715	.8271	.8
.3	.5635	.6822	1.4659	.8261	.7
.4	.5650	.6847	1.4605	.8251	.6
.5	.5664	.6873	1.4550	.8241	.5
.6	.5678	.6899	1.4496	.8231	.4
.7	.5693	.6924	1.4442	.8221	.3
.8	.5707	.6950	1.4388	.8211	.2
.9	.5721	.6976	1.4335	.8202	55.1
35.0	0.5736	0.7002	1.4281	0.8192	**55.0**
.1	.5750	.7028	1.4229	.8181	54.9
.2	.5764	.7054	1.4176	.8171	.8
.3	.5779	.7080	1.4124	.8161	.7
.4	.5793	.7107	1.4071	.8151	.6
.5	.5807	.7133	1.4019	.8141	.5
.6	.5821	.7159	1.3968	.8131	.4
.7	.5835	.7186	1.3916	.8121	.3
.8	.5850	.7212	1.3865	.8111	.2
.9	.5864	.7239	1.3814	.8100	54.1
36.0	0.5878	0.7265	1.3764	0.8090	**54.0**
	Cos	Cot	Tan	Sin	Deg.

NATURAL TRIGONOMETRIC FUNCTIONS SINE, COSINE, TANGENT, COTANGENT, FOR ANGLES IN DEGREES AND DECIMALS (Continued)

Deg.	Sin	Tan	Cot	Cos	
36.0	0.5878	0.7265	1.3764	0.8090	**54.0**
.1	.5892	.7292	1.3713	.8080	53.9
.2	.5906	.7319	1.3663	.8070	.8
.3	.5920	.7346	1.3613	.8059	.7
.4	.5934	.7373	1.3564	.8049	.6
.5	.5948	.7400	1.3514	.8039	.5
.6	.5962	.7427	1.3465	.8028	.4
.7	.5976	.7454	1.3416	.8018	.3
.8	.5990	.7481	1.3367	.8007	.2
.9	.6004	.7508	1.3319	.7997	53.1
37.0	0.6018	0.7536	1.3270	0.7986	**53.0**
.1	.6032	.7563	1.3222	7976	52.9
.2	.6046	.7590	1.3175	.7965	.8
.3	.6060	.7618	1.3127	.7955	.7
.4	.6074	.7646	1.3079	.7944	.6
.5	.6088	.7673	1.3032	.7934	.5
.6	.6101	.7701	1.2985	.7923	.4
.7	.6115	.7729	1.2938	.7912	.3
.8	.6129	.7757	1.2892	.7902	.2
.9	.6143	.7785	1.2846	.7891	52.1
38.0	0.6157	0.7813	1.2799	0.7880	**52.0**
.1	.6170	.7841	1.2753	.7869	51.9
.2	.6184	.7869	1.2708	.7859	.8
.3	.6198	.7898	1.2662	.7848	.7
.4	.6211	.7926	1.2617	.7837	.6
.5	.6225	.7954	1.2572	.7826	.5
.6	.6239	.7983	1.2527	.7815	.4
.7	.6252	.8012	1.2482	.7804	.3
.8	.6266	.8040	1.2437	.7793	.2
.9	.6280	.8069	1.2393	.7782	51.1
39.0	0.6293	0.8098	1.2349	0.7771	**51.0**
.1	.6307	.8127	1.2305	.7760	50.9
.2	.6320	.8156	1.2261	.7749	.8
.3	.6334	.8185	1.2218	.7738	.7
.4	.6347	.8214	1.2174	.7727	.6
.5	.6361	.8243	1.2131	.7716	.5
.6	.6374	.8273	1.2088	.7705	.4
.7	.6388	.8302	1.2045	.7694	.3
.8	.6401	.8332	1.2002	.7683	.2
.9	.6414	.8361	1.1960	.7672	50.1
40.0	0.6428	0.8391	1.1918	0.7660	**50.0**
.1	.6441	.8421	1.1875	.7649	49.9
.2	.6455	.8451	1.1833	.7638	.8
.3	.6468	.8481	1.1792	.7627	.7
.4	.6481	.8511	1.1750	.7615	.6
40.5	0.6494	0.8541	1.1708	0.7604	**49.5**
	Cos	Cot	Tan	Sin	Deg.

Deg.	Sin	Tan	Cot	Cos	
40.5	0.6494	0.8541	1.1708	0.7604	**49.5**
.6	.6508	.8571	1.1667	.7593	.4
.7	.6521	.8601	1.1626	.7581	.3
.8	.6534	.8632	1.1585	.7570	.2
.9	.6547	.8662	1.1544	.7559	49.1
41.0	0.6561	0.8693	1.1504	0.7547	**49.0**
.1	.6574	.8724	1.1463	.7536	48.9
.2	.6587	.8754	1.1423	.7524	.8
.3	.6600	.8785	1.1383	.7513	.7
.4	.6613	.8816	1.1343	.7501	.6
.5	.6626	.8847	1.1303	.7490	.5
.6	.6639	.8878	1.1263	.7478	.4
.7	.6652	.8910	1.1224	.7466	.3
.8	.6665	.8941	1.1184	.7455	.2
.9	.6678	.8972	1.1145	.7443	48.1
42.0	0.6691	0.9004	1.1106	0.7431	**48.0**
.1	.6704	.9036	1.1067	.7420	47.9
.2	.6717	.9067	1.1028	.7408	.8
.3	.6730	.9099	1.0990	.7396	.7
.4	.6743	.9131	1.0951	.7385	.6
.5	.6756	.9163	1.0913	.7373	.5
.6	.6769	.9195	1.0875	.7361	.4
.7	.6782	.9228	1.0837	.7349	.3
.8	.6794	.9260	1.0799	.7337	.2
.9	.6807	.9293	1.0761	.7325	47.1
43.0	0.6820	0.9325	1.0724	0.7314	**47.0**
.1	.6833	.9358	1.0686	.7302	46.9
.2	.6845	.9391	1.0649	.7290	.8
.3	.6858	.9424	1.0612	.7278	.7
.4	.6871	9457	1.0575	.7266	.6
.5	.6884	.9490	1.0538	.7254	.5
.6	.6896	.9523	1.0501	.7242	.4
.7	.6909	.9556	1.0464	.7230	.3
.8	.6921	.9590	1.0428	.7218	.2
.9	.6934	.9623	1.0392	.7206	46.1
44.0	0.6947	0.9657	1.0355	0.7193	**46.0**
.1	.6959	.9691	1.0319	.7181	45.9
.2	.6972	.9725	1.0283	.7169	.8
.3	.6984	.9759	1.0247	.7157	.7
.4	.6997	.9793	1.0212	.7145	.6
.5	.7009	.9827	1.0176	.7133	.5
.6	.7022	.9861	1.0141	.7120	.4
.7	.7034	.9896	1.0105	.7108	.3
.8	.7046	.9930	1.0070	.7096	.2
.9	.7059	.9965	1.0035	.7083	45.1
45.0	0.7071	1.0000	1.0000	0.7071	**45.0**
	Cos	Cot	Tan	Sin	Deg.

NATURAL TRIGONOMETRIC FUNCTIONS TO FIVE TENTHS OF A DEGREE (SECANTS AND COSECANTS)

Deg.	Sec	Csc	Deg.	Deg.	Sec	Csc	Deg.
0.0	1.00000	∞	90.0	22.5	1.08239	2.61313	67.5
0.5	1.00004	114.59301	89.5	23.0	1.08636	2.55930	67.0
1.0	1.00015	57.29869	89.0	23.5	1.09044	2.50784	66.5
1.5	1.00034	38.20155	88.5	24.0	1.09464	2.45859	66.0
2.0	1.00061	28.65371	88.0	24.5	1.09895	2.41142	65.5
2.5	1.00095	22.92559	87.5	25.0	1.10338	2.36620	65.0
3.0	1.00137	19.10732	87.0	25.5	1.10793	2.32282	64.5
3.5	1.00187	16.38041	86.5	26.0	1.11260	2.28117	64.0
4.0	1.00244	14.33559	86.0	26.5	1.11740	2.24116	63.5
4.5	1.00309	12.74549	85.5	27.0	1.12233	2.20269	63.0
5.0	1.00382	11.47371	85.0	27.5	1.12738	2.16568	62.5
5.5	1.00463	10.43343	84.5	28.0	1.13257	2.13005	62.0
6.0	1.00551	9.56677	84.0	28.5	1.13789	2.09574	61.5
6.5	1.00647	8.83367	83.5	29.0	1.14335	2.06267	61.0
7.0	1.00751	8.20551	83.0	29.5	1.14896	2.03077	60.5
7.5	1.00863	7.66130	82.5	30.0	1.15470	2.00000	60.0
8.0	1.00983	7.18530	82.0	30.5	1.16059	1.97029	59.5
8.5	1.01111	6.76547	81.5	31.0	1.16663	1.94160	59.0
9.0	1.01247	6.39245	81.0	31.5	1.17283	1.91388	58.5
9.5	1.01391	6.05886	80.5	32.0	1.17918	1.88708	58.0
10.0	1.01543	5.75877	80.0	32.5	1.18569	1.86116	57.5
10.5	1.01703	5.48740	79.5	33.0	1.19236	1.83608	57.0
11.0	1.01872	5.24084	79.0	33.5	1.19920	1.81180	56.5
11.5	1.02049	5.01585	78.5	34.0	1.20622	1.78829	56.0
12.0	1.02234	4.80973	78.0	34.5	1.21341	1.76552	55.5
12.5	1.02428	4.62023	77.5	35.0	1.22077	1.74345	55.0
13.0	1.02630	4.44541	77.0	35.5	1.22833	1.72205	54.5
13.5	1.02842	4.28366	76.5	36.0	1.23607	1.70130	54.0
14.0	1.03061	4.13357	76.0	36.5	1.24400	1.68117	53.5
14.5	1.03290	3.99393	75.5	37.0	1.25214	1.66164	53.0
15.0	1.03528	3.86370	75.0	37.5	1.26047	1.64268	52.5
15.5	1.03774	3.74198	74.5	38.0	1.26902	1.62427	52.0
16.0	1.04030	3.62796	74.0	38.5	1.27778	1.60639	51.5
16.5	1.04295	3.52094	73.5	39.0	1.28676	1.58902	51.0
17.0	1.04569	3.42030	73.0	39.5	1.29597	1.57213	50.5
17.5	1.04853	3.32551	72.5	40.0	1.30541	1.55572	50.0
18.0	1.05146	3.23607	72.0	40.5	1.31509	1.53977	49.5
18.5	1.05449	3.15155	71.5	41.0	1.32501	1.52425	49.0
19.0	1.05762	3.07155	71.0	41.5	1.33519	1.50916	48.5
19.5	1.06085	2.99574	70.5	42.0	1.34563	1.49448	48.0
20.0	1.06418	2.92380	70.0	42.5	1.35634	1.48019	47.5
20.5	1.06761	2.85545	69.5	43.0	1.36733	1.46628	47.0
21.0	1.07114	2.79043	69.0	43.5	1.37860	1.45274	46.5
21.5	1.07479	2.72850	68.5	44.0	1.39016	1.43956	46.0
22.0	1.07853	2.66947	68.0	44.5	1.40203	1.42672	45.5
22.5	1.08239	2.61313	67.5	45.0	1.41421	1.41421	45.0
	Csc	Sec			Csc	Sec	

NATURAL TRIGONOMETRIC FUNCTIONS

FOR ANGLES IN *x* RADIANS

x	Sin	Tan	Cot	Cos	*x*	Sin	Tan	Cot	Cos
.00	.00000	.00000	∞	1.00000	**.50**	.47943	.54630	1.8305	.87758
01	.01000	.01000	99.997	0.99995	.51	.48818	.55936	1.7878	.87274
.02	.02000	.02000	49.993	.99980	.52	.49688	.57256	1.7465	.86782
.03	.03000	.03001	33.323	.99955	.53	.50553	.58592	1.7067	.86281
.04	.03999	.04002	24.987	.99920	.54	.51414	.59943	1.6683	.85771
.05	.04998	.05004	19.983	.99875	.55	.52269	.61311	1.6310	.85252
.06	.05996	.06007	16.647	.99820	.56	.53119	.62695	1.5950	.84726
.07	.06994	.07011	14.262	.99755	.57	.53963	.64097	1.5601	.84190
.08	.07991	.08017	12.473	.99680	.58	.54802	.65517	1.5263	.83646
.09	.08988	.09024	11.081	.99595	.59	.55636	.66956	1.4935	.83094
.10	.09983	.10033	9.9666	.99500	**.60**	.56464	.68414	1.4617	.82534
.11	.10978	.11045	9.0542	.99396	.61	.57287	.69892	1.4308	.81965
.12	.11971	.12058	8.2933	.99281	.62	.58104	.71391	1.4007	.81388
.13	.12963	.13074	7.6489	.99156	.63	.58914	.72911	1.3715	.80803
.14	.13954	.14092	7.0961	.99022	.64	.59720	.74454	1.3431	.80210
.15	.14944	.15114	6.6166	.98877	.65	.60519	.76020	1.3154	.79608
.16	.15932	.16138	6.1966	.98723	.66	.61312	.77610	1.2885	.78999
.17	.16918	.17166	5.8256	.98558	.67	.62099	.79225	1.2622	.78382
.18	.17903	.18197	5.4954	.98384	.68	.62879	.80866	1.2366	.77757
.19	.18886	.19232	5.1997	.98200	.69	.63654	.82534	1.2116	.77125
.20	.19867	.20271	4.9332	.98007	**.70**	.64422	.84229	1.1872	.76484
.21	.20846	.21314	4.6917	.97803	.71	.65183	.85953	1.1634	.75836
.22	.21823	.22362	4.4719	.97590	.72	.65938	.87707	1.1402	.75181
.23	.22798	.23414	4.2709	.97367	.73	.66687	.89492	1.1174	.74517
.24	.23770	.24472	4.0864	.97134	.74	.67429	.91309	1.0952	.73847
.25	.24740	.25534	3.9163	.96891	.75	.68164	.93160	1.0734	.73169
.26	.25708	.26602	3.7591	.96639	.76	.68892	.95045	1.0521	.72484
.27	.26673	.27676	3.6133	.96377	.77	.69614	.96967	1.0313	.71791
.28	.27636	.28755	3.4776	.96106	.78	.70328	.98926	1.0109	.71091
.29	.28595	.29841	3.3511	.95824	.79	.71035	1.0092	.99084	.70385
.30	.29552	.30934	3.2327	.95534	**.80**	.71736	1.0296	.97121	.69671
.31	.30506	.32033	3.1218	.95233	.81	.72429	1.0505	.95197	.68950
.32	.31457	.33139	3.0176	.94924	.82	.73115	1.0717	.93309	.68222
.33	.32404	.34252	2.9195	.94604	.83	.73793	1.0934	.91455	.67488
.34	.33349	.35374	2.8270	.94275	.84	.74464	1.1156	.89635	.66746
.35	.34290	.36503	2.7395	.93937	.85	.75128	1.1383	.87848	.65998
.36	.35227	.37640	2.6567	.93590	.86	.75784	1.1616	.86091	.65244
.37	.36162	.38786	2.5782	.93233	.87	.76433	1.1853	.84365	.64483
.38	.37092	.39941	2.5037	.92866	.88	.77074	1.2097	.82668	.63715
.39	.38019	.41105	2.4328	.92491	.89	.77707	1.2346	.80998	.62941
.40	.38942	.42279	2.3652	.92106	**.90**	.78333	1.2602	.79355	.62161
.41	.39861	.43463	2.3008	.91712	.91	.78950	1.2864	.77738	.61375
.42	.40776	.44657	2.2393	.91309	.92	.79560	1.3133	.76146	.60582
.43	.41687	.45862	2.1804	.90897	.93	.80162	1.3409	.74578	.59783
.44	.42594	.47078	2.1241	.90475	.94	.80756	1.3692	.73034	.58979
.45	.43497	.48306	2.0702	.90045	.95	.81342	1.3984	.71511	.58168
.46	.44395	.49545	2.0184	.89605	.96	.81919	1.4284	.70010	.57352
.47	.45289	.50797	1.9686	.89157	.97	.82489	1.4592	.68531	.56530
.48	.46178	.52061	1.9208	.88699	.98	.83050	1.4910	.67071	.55702
.49	.47063	.53339	1.8748	.88233	.99	.83603	1.5237	.65631	.54869
.50	.47943	.54630	1.8305	.87758	**1.00**	.84147	1.5574	.64209	.54030
x	Sin	Tan	Cot	Cos	*x*	Sin	Tan	Cot	Cos

NATURAL TRIGONOMETRIC FUNCTIONS
FOR ANGLES IN x RADIANS (Continued)

x	Sin	Tan	Cot	Cos	x	Sin	Tan	Cot	Cos
1.00	.84147	1.5574	.64209	.54030	**1.50**	.99749	14.101	.07091	.07074
1.01	.84683	1.5922	.62806	.53186	1.51	.99815	16.428	.06087	.06076
1.02	.85211	1.6281	.61420	.52337	1.52	.99871	19.670	.05084	.05077
1.03	.85730	1.6652	.60051	.51482	1.53	.99917	24.498	.04082	.04079
1.04	.86240	1.7036	.58699	.50622	1.54	.99953	32.461	.03081	.03079
1.05	.86742	1.7433	.57362	.49757	1.55	.99978	48.078	.02080	.02079
1.06	.87236	1.7844	.56040	.48887	1.56	.99994	92.620	.01080	.01080
1.07	.87720	1.8270	.54734	.48012	1.57	1.00000	1255.8	.00080	.00080
1.08	.88196	1.8712	.53441	.47133	1.58	.99996	−108.65	−.00920	−.00920
1.09	.88663	1.9171	.52162	.46249	1.59	.99982	−52.067	−.01921	−.01920
1.10	.89121	1.9648	.50897	.45360	**1.60**	.99957	−34.233	−.02921	−.02920
1.11	.89570	2.0143	.49644	.44466	1.61	.99923	−25.495	−.03922	−.03919
1.12	.90010	2.0660	.48404	.43568	1.62	.99879	−20.307	−.04924	−.04918
1.13	.90441	2.1198	.47175	.42666	1.63	.99825	−16.871	−.05927	−.05917
1.14	.90863	2.1759	.45959	.41759	1.64	.99761	−14.427	−.06931	−.06915
1.15	.91276	2.2345	.44753	.40849	1.65	.99687	−12.599	−.07937	−.07912
1.16	.91680	2.2958	.43558	.39934	1.66	.99602	−11.181	−.08944	−.08909
1.17	.92075	2.3600	.42373	.39015	1.67	.99508	−10.047	−.09953	−.09904
1.18	.92461	2.4273	.41199	.38092	1.68	.99404	− 9.1208	−.10964	−.10899
1.19	.92837	2.4979	.40034	.37166	1.69	.99290	− 8.3492	−.11977	−.11892
1.20	.93204	2.5722	.38878	.36236	**1.70**	99166	− 7.6966	−.12993	−.12884
1.21	.93562	2.6503	.37731	.35302	1.71	.99033	− 7.1373	−.14011	−.13875
1.22	.93910	2.7328	.36593	.34365	1.72	.98889	− 6.6524	−.15032	−.14865
1.23	.94249	2.8198	.35463	.33424	1.73	.98735	− 6.2281	−.16056	−.15853
1.24	.94578	2.9119	.34341	.32480	1.74	.98572	− 5.8535	−.17084	−.16840
1.25	.94898	3.0096	.33227	.31532	1.75	.98399	− 5.5204	−.18115	−.17825
1.26	.95209	3.1133	.32121	.30582	1.76	.98215	− 5.2221	−.19149	−.18808
1.27	.95510	3.2236	.31021	.29628	1.77	.98022	− 4.9534	−.20188	−.19789
1.28	.95802	3.3413	.29928	.28672	1.78	.97820	− 4.7101	−.21231	−.20768
1.29	.96084	3.4672	.28842	.27712	1.79	.97607	− 4.4887	−.22278	−.21745
1.30	.96356	3.6021	.27762	.26750	**1.80**	.97385	− 4.2863	−.23330	−.22720
1.31	.96618	3.7471	.26687	.25785	1.81	.97153	− 4.1005	−.24387	−.23693
1.32	.96872	3.9033	.25619	.24818	1.82	.96911	− 3.9294	−.25449	−.24663
1.33	.97115	4.0723	.24556	.23848	1.83	.96659	− 3.7712	−.26517	−.25631
1.34	.97348	4.2556	.23498	.22875	1.84	.96398	− 3.6245	−.27590	−.26596
1.35	.97572	4.4552	.22446	.21901	1.85	.96128	− 3.4881	−.28669	−.27559
1.36	.97786	4.6734	.21398	.20924	1.86	.95847	− 3.3608	−.29755	−.28519
1.37	.97991	4.9131	.20354	.19945	1.87	.95557	− 2.2419	−.30846	−.29476
1.38	.98185	5.1774	.19315	.18964	1.88	.95258	− 3.1304	−.31945	−.30430
1.39	.98370	5.4707	.18279	.17981	1.89	.94949	− 3.0257	−.33051	−.31381
1.40	.98545	5.7979	.17248	.16997	**1.90**	.94630	− 2.9271	−.34164	−.32329
1.41	.98710	6.1654	.16220	.16010	1.91	.94302	− 2.8341	−.35284	−.33274
1.42	.98865	6.5811	.15195	.15023	1.92	.93965	− 2.7463	−.36413	−.34215
1.43	.99010	7.0555	.14173	.14033	1.93	.93618	− 2.6632	−.37549	−.35153
1.44	.99146	7.6018	.13155	.13042	1.94	.93262	− 2.5843	−.38695	−.36087
1.45	.99271	8.2381	.12139	.12050	1.95	.92896	− 2.5095	−.39849	−.37018
1.46	.99387	8.9886	.11125	.11057	1.96	.92521	− 2.4383	−.41012	−.37945
1.47	.99492	9.8874	.10114	.10063	1.97	.92137	− 2.3705	−.42185	−.38868
1.48	.99588	10.983	.09105	.09067	1.98	.91744	− 2.3058	−.43368	−.39788
1.49	.99674	12.350	.08097	.08071	1.99	.91341	− 2.2441	−.44562	−.40703
1.50	.99749	14.101	.07091	.07074	**2.00**	.90930	− 2.1850	−.45766	−.41615
x	Sin	Tan	Cot	Cos	x	Sin	Tan	Cot	Cos

NATURAL TRIGONOMETRIC FUNCTIONS SECANTS AND COSECANTS FOR ANGLES IN RADIANS

x	sec x	csc x	x	sec x	csc x	x	sec x	csc x
0.00	1.00000	∞	0.55	1.17299	1.91319	1.10	2.20460	1.12207
0.01	1.00005	100.00167	0.56	1.18028	1.88258	1.11	2.24890	1.11644
0.02	1.00020	50.00333	0.57	1.18779	1.85311	1.12	2.29525	1.11099
0.03	1.00045	33.33833	0.58	1.19551	1.82474	1.13	2.34379	1.10569
0.04	1.00080	25.00667	0.59	1.20345	1.79739	1.14	2.39467	1.10055
0.05	1.00125	20.00834	0.60	1.21163	1.77103	1.15	2.44806	1.09557
0.06	1.00180	16.67667	0.61	1.22003	1.74560	1.16	2.50413	1.09075
0.07	1.00246	14.29739	0.62	1.22868	1.72107	1.17	2.56311	1.08607
0.08	1.00321	12.51334	0.63	1.23758	1.69738	1.18	2.62519	1.08154
0.09	1.00406	11.12613	0.64	1.24673	1.67449	1.19	2.69063	1.07716
0.10	1.00502	10.01669	0.65	1.25615	1.65238	1.20	2.75970	1.07292
0.11	1.00608	9.10927	0.66	1.26583	1.63101	1.21	2.83271	1.06881
0.12	1.00724	8.35337	0.67	1.27580	1.61034	1.22	2.90997	1.06485
0.13	1.00851	7.71402	0.68	1.28605	1.59035	1.23	2.99188	1.06102
0.14	1.00988	7.16624	0.69	1.29660	1.57100	1.24	3.07885	1.05732
0.15	1.01136	6.69173	0.70	1.30746	1.55227	1.24	3.17136	1.05376
0.16	1.01294	6.27675	0.71	1.31863	1.53413	1.26	3.26993	1.05032
0.17	1.01463	5.91078	0.72	1.33013	1.51657	1.27	3.37518	1.04701
0.18	1.01642	5.58567	0.73	1.34197	1.49954	1.28	3.48778	1.04382
0.19	1.01833	5.29496	0.74	1.35415	1.48305	1.29	3.60853	1.04076
0.20	1.02034	5.03349	0.75	1.36670	1.46705	1.30	3.73833	1.03782
0.21	1.02246	4.79709	0.76	1.37962	1.45154	1.31	3.87822	1.03500
0.22	1.02470	4.58233	0.77	1.39293	1.43650	1.32	4.02941	1.03230
0.23	1.02705	4.38640	0.78	1.40664	1.42191	1.33	4.19329	1.02971
0.24	1.02951	4.20692	0.79	1.42077	1.40775	1.34	4.37153	1.02724
0.25	1.03208	4.04197	0.80	1.43532	1.39401	1.35	4.56607	1.02428
0.26	1.03478	3.88983	0.81	1.45033	1.38067	1.36	4.77923	1.02264
0.27	1.03759	3.74909	0.82	1.46580	1.36772	1.37	5.01379	1.02050
0.28	1.04052	3.61853	0.83	1.48175	1.35514	1.38	5.27313	1.01848
0.29	1.04358	3.49709	0.84	1.49821	1.34292	1.39	5.56133	1.01657
0.30	1.04675	3.38386	0.85	1.51519	1.33106	1.40	5.88349	1.01477
0.31	1.05005	3.27806	0.86	1.53271	1.31954	1.41	6.24593	1.01307
0.32	1.05348	3.17898	0.87	1.55080	1.30834	1.42	6.65666	1.01148
0.33	1.05704	3.08601	0.88	1.56949	1.29746	1.43	7.12598	1.00999
0.34	1.06072	2.99862	0.89	1.58878	1.28688	1.44	7.66732	1.00862
0.35	1.06454	2.91632	0.90	1.60873	1.27661	1.45	8.29856	1.00734
0.36	1.06849	2.83870	0.91	1.62934	1.26662	1.46	9.04406	1.00617
0.37	1.07258	2.76537	0.92	1.65065	1.25691	1.47	9.93781	1.00510
0.38	1.07682	2.69600	0.93	1.67270	1.24747	1.48	11.02881	1.00414
0.39	1.08119	2.63027	0.94	1.69552	1.23830	1.49	12.39028	1.00327
0.40	1.08570	2.56793	0.95	1.71915	1.22938	1.50	14.13683	1.00251
0.41	1.09037	2.50872	0.96	1.74362	1.22072	1.51	16.45850	1.00185
0.42	1.09518	2.45242	0.97	1.76897	1.21229	1.52	19.69493	1.00129
0.43	1.10015	2.39882	0.98	1.79526	1.20410	1.53	24.51881	1.00083
0.44	1.10528	2.34775	0.99	1.82252	1.19614	1.54	32.47654	1.00047
0.45	1.11056	2.29903	1.00	1.85082	1.18840	1.55	48.08888	1.00022
0.46	1.11601	2.25252	1.01	1.88019	1.18087	1.56	92.62589	1.00006
0.47	1.12162	2.20806	1.02	1.91071	1.17356	1.57	+1255.76599	1.00000
0.48	1.12740	2.16553	1.03	1.94243	1.16645	1.58	− 108.65381	1.00004
0.49	1.13336	2.12483	1.04	1.97542	1.15955	1.59	− 52.07657	1.00018
0.50	1.13949	2.08583	1.05	2.00976	1.15284	1.60	− 34.24714	1.00043
0.51	1.14581	2.04844	1.06	2.04552	1.14632			
0.52	1.15231	2.01256	1.07	2.08279	1.13999			
0.53	1.15901	1.97811	1.08	2.12166	1.13384			
0.54	1.16580	1.94501	1.09	2.16223	1.12787			
0.55	1.17299	1.91319	1.10	2.20460	1.12207			

TEN-PLACE SINES AND COSINES FOR RADIAN MEASURE

x	$\sin x$	$\cos x$	x	$\sin x$	$\cos x$
0.000	0.00000 00000	1.00000 00000	0.050	0.04997 91693	0.99875 02604
0.001	0.00099 99998	0.99999 95000	0.051	0.05097 78944	0.99869 97819
0.002	0.00199 99987	0.99999 80000	0.052	0.05197 65685	0.99864 83046
0.003	0.00299 99955	0.99999 55000	0.053	0.05297 51907	0.99859 58287
0.004	0.00399 99893	0.99999 20000	0.054	0.05397 37598	0.99854 23543
0.005	0.00499 99792	0.99998 75000	0.055	0.05497 22750	0.99848 78812
0.006	0.00599 99640	0.99998 20001	0.056	0.05597 07353	0.99843 24097
0.007	0.00699 99428	0.99997 55001	0.057	0.05696 91395	0.99837 59398
0.008	0.00799 99147	0.99996 80002	0.058	0.05796 74868	0.99831 84715
0.009	0.00899 98785	0.99995 95003	0.059	0.05896 57761	0.99826 00048
0.010	0.00999 98333	0.99995 00004	0.060	0.05996 40065	0.99820 05399
0.011	0.01099 97782	0.99993 95006	0.061	0.06096 21769	0.99814 00768
0.012	0.01199 97120	0.99992 80009	0.062	0.06196 02863	0.99807 86156
0.013	0.01299 96338	0.99991 55012	0.063	0.06295 83338	0.99801 61563
0.014	0.01399 95427	0.99990 20016	0.064	0.06395 63183	0.99795 26990
0.015	0.01499 94375	0.99988 75021	0.065	0.06495 42388	0.99788 82437
0.016	0.01599 93173	0.99987 20027	0.066	0.06595 20944	0.99782 27905
0.017	0.01699 91812	0.99985 55035	0.067	0.06694 98841	0.99775 63395
0.018	0.01799 90280	0.99983 80044	0.068	0.06794 76068	0.99768 88908
0.019	0.01899 88569	0.99981 95054	0.069	0.06894 52615	0.99672 04443
0.020	0.01999 86667	0.99980 00067	0.070	0.06994 28473	0.99755 10003
0.021	0.02099 84565	0.99977 95081	0.071	0.07094 03632	0.99748 05586
0.022	0.02199 82254	0.99975 80098	0.072	0.07193 78081	0.99740 91196
0.023	0.02299 79722	0.99973 55117	0.073	0.07293 51811	0.99733 66830
0.024	0.02399 76961	0.99971 20138	0.074	0.07393 24812	0.99726 32492
0.025	0.02499 73959	0.99968 75163	0.075	0.07492 97073	0.99718 88181
0.026	0.02599 70708	0.99966 20190	0.076	0.07592 68585	0.99711 33898
0.027	0.02699 67196	0.99963 55221	0.077	0.07692 39337	0.99703 69644
0.028	0.02799 63415	0.99960 80256	0.078	0.07792 09321	0.99695 95420
0.029	0.02899 59353	0.99957 95295	0.079	0.07891 78525	0.99688 11226
0.030	0.02999 55002	0.99955 00337	0.080	0.07991 46940	0.99680 17063
0.031	0.03099 50351	0.99951 95385	0.081	0.08091 14556	0.99672 12932
0.032	0.03199 45389	0.99948 80437	0.082	0.08190 81362	0.99663 98834
0.033	0.03299 40108	0.99945 55494	0.083	0.08290 47350	0.99655 74770
0.034	0.03399 34497	0.99942 20557	0.084	0.08390 12508	0.99647 40740
0.035	0.03499 28546	0.99938 75625	0.085	0.08489 76828	0.99638 96745
0.036	0.03599 22245	0.99935 20700	0.086	0.08589 40299	0.99630 42786
0.037	0.03699 15584	0.99931 55781	0.087	0.08689 02910	0.99621 78865
0.038	0.03799 08553	0.99927 80869	0.088	0.08788 64653	0.99613 04981
0.039	0.03899 01143	0.99923 95964	0.089	0.08888 25517	0.99604 21136
0.040	0.03998 93342	0.99920 01067	0.090	0.08987 85492	0.99595 27330
0.041	0.04098 85141	0.99915 96177	0.091	0.09087 44568	0.99586 23565
0.042	0.04198 76531	0.99911 81296	0.092	0.09187 02736	0.99577 09841
0.043	0.04298 67501	0.99907 56424	0.093	0.09286 59985	0.99567 86160
0.044	0.04398 58040	0.99903 21562	0.094	0.09386 16305	0.99558 52522
0.045	0.04498 48140	0.99898 76708	0.095	0.09485 71686	0.99549 08928
0.046	0.04598 37790	0.99894 21865	0.096	0.09585 26119	0.99539 55379
0.047	0.04698 26981	0.99889 57033	0.097	0.09684 79594	0.99529 91876
0.048	0.04798 15701	0.99884 82212	0.098	0.09784 32100	0.99520 18420
0.049	0.04898 03942	0.99879 97402	0.099	0.09883 83627	0.99510 35012
0.050	0.04997 91693	0.99875 02604	0.100	0.09983 34166	0.99500 41653

TEN-PLACE SINES AND COSINES FOR RADIAN MEASURE (Continued)

x	sin x	cos x	x	sin x	cos x
0.100	0.09983 34166	0.99500 41653	0.150	0.14943 81325	0.98877 10779
0.101	0.10082 83707	0.99490 38344	0.151	0.15042 68287	0.98862 11454
0.102	0.10182 32240	0.99480 25086	0.152	0.15141 53744	0.98847 02243
0.103	0.10281 79754	0.99470 01880	0.153	0.15240 37688	0.98831 83147
0.104	0.10381 26240	0.99459 68727	0.154	0.15339 20107	0.98816 54168
0.105	0.10480 71688	0.99449 25627	0.155	0.15438 00993	0.98801 15308
0.106	0.10580 16088	0.99438 72584	0.156	0.15536 80335	0.98785 66567
0.107	0.10679 59430	0.99428 09596	0.157	0.15635 58123	0.98770 07948
0.108	0.10779 01704	0.99417 36665	0.158	0.15734 34347	0.98754 39451
0.109	0.10878 42900	0.99406 53793	0.159	0.15833 08998	0.98738 61079
0.110	0.10977 83008	0.99395 60980	0.160	0.15930 82066	0.98722 72834
0.111	0.11077 22019	0.99384 58227	0.161	0.16030 53541	0.98706 74716
0.112	0.11176 59922	0.99373 45536	0.162	0.16129 23412	0.98690 66727
0.113	0.11275 96707	0.99362 22907	0.163	0.16227 91671	0.98674 48870
0.114	0.11375 32364	0.99350 90343	0.164	0.16326 58307	0.98658 21144
0.115	0.11474 66884	0.99339 47843	0.165	0.16425 23310	0.98641 83553
0.116	0.11574 00256	0.99327 95409	0.166	0.16523 86671	0.98625 36098
0.117	0.11673 32471	0.99316 33043	0.167	0.16622 48379	0.98608 78781
0.118	0.11772 63519	0.99304 60745	0.168	0.16721 08425	0.98592 11602
0.119	0.11871 93390	0.99292 78516	0.169	0.16819 66799	0.98575 34564
0.120	0.11971 22073	0.99280 86359	0.170	0.16918 23491	0.98558 47669
0.121	0.12070 49559	0.99268 84273	0.171	0.17016 78491	0.98541 50918
0.122	0.12169 75838	0.99256 72260	0.172	0.17115 31789	0.98524 44313
0.123	0.12269 00900	0.99244 50321	0.173	0.17213 83376	0.98507 27855
0.124	0.12368 24735	0.99232 18458	0.174	0.17312 33242	0.98490 01547
0.125	0.12467 47334	0.99219 76672	1.175	0.17410 81376	0.98472 65389
0.126	0.12566 68685	0.99207 24964	0.176	0.17509 27769	0.98455 19384
0.127	0.12665 88780	0.99194 63335	0.177	0.17607 72411	0.98437 63534
0.128	0.12765 07609	0.99181 91787	0.178	0.17706 15293	0.98419 97840
0.129	0.12864 25161	0.99169 10321	0.179	0.17804 56404	0.98402 22304
0.130	0.12963 41426	0.99156 18937	0.180	0.17902 95734	0.98384 36928
0.131	0.13062 56395	0.99143 17638	0.181	0.18001 33274	0.98366 41713
0.132	0.13161 70058	0.99130 06425	0.182	0.18099 69014	0.98348 36662
0.133	0.13260 82405	0.99116 85298	0.183	0.18198 02944	0.98330 21776
0.134	0.13359 93425	0.99103 54260	0.184	0.18296 35055	0.98311 97057
0.135	0.13459 03110	0.99090 13312	0.185	0.18394 65335	0.98293 62506
0.136	0.13558 11449	0.99076 62455	0.186	0.18492 93776	0.98275 18127
0.137	0.13657 18432	0.99063 01690	0.187	0.18591 20368	0.98256 63919
0.138	0.13756 24049	0.99049 31018	0.188	0.18689 45101	0.98237 99886
0.139	0.13855 28290	0.99035 50442	0.189	0.18787 67965	0.98219 26030
0.140	0.13954 31146	0.99021 59962	0.190	0.18885 88950	0.98200 42351
0.141	0.14053 32607	0.99007 59580	0.191	0.18984 08046	0.98181 48853
0.142	0.14152 32662	0.98993 49297	0.192	0.19082 25244	0.98162 45536
0.143	0.14251 31302	0.98979 29115	0.193	0.19180 40534	0.98143 32403
0.144	0.14350 28517	0.98964 99035	0.194	0.19278 53906	0.98124 09455
0.145	0.14449 24297	0.98950 59059	0.195	0.19376 65350	0.98104 76695
0.146	0.14548 18632	0.98936 09187	0.196	0.19474 74856	0.98085 34125
0.147	0.14647 11512	0.98921 49422	0.197	0.19572 82415	0.98065 81746
0.148	0.14746 02928	0.98906 79765	0.198	0.19670 88016	0.98046 19561
0.149	0.14844 92868	0.98892 00217	0.199	0.19768 91650	0.98026 47571
0.150	0.14943 81325	0.98877 10779	0.200	0.19866 93308	0.98006 65778

TEN-PLACE SINES AND COSINES FOR RADIAN MEASURE (Continued)

x	sin x	cos x	x	sin x	cos x
0.200	0.19866 93308	0.98006 65778	0.250	0.24740 39593	0.96891 24217
0.201	0.19964 92979	0.97986 74185	0.251	0.24837 27478	0.96866 45333
0.202	0.20062 90653	0.97966 72793	0.252	0.24934 12880	0.96841 56763
0.203	0.20160 86321	0.97946 61604	0.253	0.25030 95788	0.96816 58508
0.204	0.20258 79973	0.97926 40621	0.254	0.25127 76194	0.96791 50572
0.205	0.20356 71599	0.97906 09845	0.255	0.25224 54086	0.96766 32957
0.206	0.20454 61189	0.97885 69279	0.256	0.25321 29456	0.96741 05665
0.207	0.20552 48734	0.97865 18923	0.257	0.25418 02294	0.96715 68699
0.208	0.20650 34224	0.97844 58782	0.258	0.25514 72591	0.96690 22061
0.209	0.20748 17649	0.97823 88856	0.259	0.25611 40335	0.96664 65754
0.210	0.20845 98998	0.97803 09147	0.260	0.25708 05519	0.96638 99781
0.211	0.20943 78264	0.97782 19658	0.261	0.25804 68132	0.96613 24144
0.212	0.21041 55435	0.97761 20391	0.262	0.25901 28164	0.96587 38846
0.213	0.21139 30501	0.97740 11348	0.263	0.25997 85606	0.96561 43889
0.214	0.21237 03454	0.97718 92531	0.264	0.26094 40449	0.96535 39276
0.215	0.21334 74283	0.97697 63942	0.265	0.26190 92681	0.96509 25009
0.216	0.21432 42979	0.97676 25583	0.266	0.26287 42295	0.96483 01091
0.217	0.21530 09531	0.97654 77457	0.267	0.26383 89280	0.96456 67525
0.218	0.21627 73930	0.97633 19565	0.268	0.26480 33627	0.96430 24313
0.219	0.21725 36167	0.97611 51910	0.269	0.26576 75326	0.96403 71459
0.220	0.21822 96231	0.97589 74493	0.270	0.26673 14367	0.96377 08964
0.221	0.21920 54113	0.97567 87318	0.271	0.26769 50741	0.96350 36831
0.222	0.22018 09802	0.97545 90386	0.272	0.26865 84437	0.96323 55063
0.223	0.22115 63290	0.97523 83699	0.273	0.26962 15448	0.96296 63663
0.224	0.22213 14566	0.97501 67260	0.274	0.27058 43761	0.96269 62633
0.225	0.22310 63621	0.97479 41071	0.275	0.27154 69370	0.96242 51976
0.226	0.22408 10445	0.97457 05133	0.276	0.27250 92262	0.96215 31695
0.227	0.22505 55028	0.97434 59451	0.277	0.27347 12430	0.96188 01793
0.228	0.22602 97361	0.97412 04024	0.278	0.27443 29863	0.96160 62271
0.229	0.22700 37433	0.97389 38857	0.279	0.27539 44551	0.96133 13134
0.230	0.22797 75235	0.97366 63950	0.280	0.27635 56486	0.96105 54383
0.231	0.22895 10758	0.97343 79307	0.281	0.27731 65657	0.96077 86022
0.232	0.22992 43991	0.97320 84929	0.282	0.27827 72054	0.96050 08053
0.233	0.23089 74924	0.97297 80820	0.283	0.27923 75670	0.96022 20479
0.234	0.23187 03549	0.97274 66980	0.284	0.28019 76492	0.95994 23302
0.235	0.23284 29855	0.97251 43413	0.285	0.28115 74513	0.95966 16527
0.236	0.23381 53833	0.97228 10121	0.286	0.28211 69722	0.95938 00154
0.237	0.23478 75472	0.97204 67106	0.287	0.28307 62110	0.95909 74188
0.238	0.23575 94764	0.97181 14371	0.288	0.28403 51667	0.95881 38631
0.239	0.23673 11698	0.97157 51918	0.289	0.28499 38384	0.95852 93486
0.240	0.23770 26264	0.97133 79749	0.290	0.28595 22251	0.95824 38755
0.241	0.23867 38454	0.97109 97866	0.291	0.28691 03258	0.95795 74442
0.242	0.23964 48257	0.97086 06272	0.292	0.28786 81397	0.95767 00550
0.243	0.24061 55663	0.97062 04970	0.293	0.28882 56656	0.95738 17080
0.244	0.24158 60663	0.97037 93962	0.294	0.28978 29028	0.95709 24037
0.245	0.24255 63248	0.97013 73250	0.295	0.29073 98501	0.95680 21423
0.246	0.24352 63407	0.96989 42836	0.296	0.29169 65067	0.95651 09241
0.247	0.24449 61130	0.96965 02724	0.297	0.29265 28717	0.95621 87494
0.248	0.24546 56409	0.96940 52915	0.298	0.29360 89439	0.95592 56185
0.249	0.24643 49233	0.96915 93412	0.299	0.29456 47226	0.95563 15316
0.250	0.24740 39593	0.96891 24217	0.300	0.29552 02067	0.95533 64891

TEN-PLACE SINES AND COSINES FOR RADIAN MEASURE (Continued)

x	$\sin x$	$\cos x$	x	$\sin x$	$\cos x$
0.300	0.29552 02067	0.95533 64891	0.350	0.34289 78075	0.93937 27128
0.301	0.29647 53952	0.95504 04913	0.351	0.34383 70086	0.93902 93454
0.302	0.29743 02873	0.95474 35384	0.352	0.34477 58658	0.93868 50389
0.303	0.29838 48820	0.95444 56308	0.353	0.34571 43783	0.93833 97938
0.304	0.29933 91783	0.95414 67688	0.354	0.34665 25451	0.93799 36103
0.305	0.30029 31752	0.95384 69526	0.355	0.34759 03652	0.93764 64888
0.306	0.30124 68719	0.95354 61825	0.356	0.34852 78378	0.93729 84297
0.307	0.30220 02673	0.95324 44589	0.357	0.34946 49618	0.93694 94333
0.308	0.30315 33605	0.95294 17821	0.358	0.35040 17363	0.93659 94999
0.309	0.30410 61505	0.95263 81523	0.359	0.35133 81605	0.93624 86299
0.310	0.30505 86364	0.95233 35699	0.360	0.35227 42333	0.93589 68237
0.311	0.30601 08173	0.95202 80351	0.361	0.35320 99538	0.93554 40816
0.312	0.30696 26922	0.95172 15484	0.362	0.35414 53211	0.93519 04039
0.313	0.30791 42601	0.95141 41099	0.363	0.35508 03343	0.93483 57910
0.314	0.30886 55201	0.95110 57199	0.364	0.35601 49924	0.93448 02433
0.315	0.30981 64712	0.95079 63789	0.365	0.35694 92945	0.93412 37612
0.316	0.31076 71125	0.95048 60871	0.366	0.35788 32396	0.93376 63449
0.317	0.31171 74431	0.95017 48448	0.367	0.35881 68269	0.93340 79948
0.318	0.31266 74619	0.94986 26523	0.368	0.35975 00553	0.93304 87113
0.319	0.31361 71681	0.94954 95100	0.369	0.36068 29240	0.93268 84948
0.320	0.31456 65606	0.94923 54181	0.370	0.36161 54320	0.93232 73456
0.321	0.31551 56386	0.94892 03770	0.371	0.36254 75783	0.93196 52641
0.322	0.31646 44011	0.94860 43869	0.372	0.36347 93622	0.93160 22506
0.323	0.31741 28471	0.94828 74483	0.373	0.36441 07825	0.93123 83055
0.324	0.31836 09756	0.94796 95613	0.374	0.36534 18385	0.93087 34291
0.325	0.31930 87859	0.94765 07264	0.375	0.36627 25291	0.93050 76219
0.326	0.32025 62768	0.94733 09439	0.376	0.36720 28534	0.93014 08842
0.327	0.32120 34474	0.94701 02140	0.377	0.36813 28105	0.92977 32163
0.328	0.32215 02969	0.94668 85371	0.378	0.36906 23995	0.92940 46187
0.329	0.32309 68242	0.94636 59135	0.379	0.36999 16195	0.92903 50917
0.330	0.32404 30284	0.94604 23435	0.380	0.37092 04694	0.92866 46356
0.331	0.32498 89086	0.94571 78275	0.381	0.37184 89484	0.92829 32508
0.332	0.32593 44637	0.94539 23658	0.382	0.37277 70556	0.92792 09378
0.333	0.32687 96930	0.94506 59587	0.383	0.37370 47900	0.92754 76968
0.334	0.32782 45953	0.94473 86065	0.384	0.37463 21507	0.92717 35283
0.335	0.32876 91699	0.94441 03096	0.385	0.37555 91367	0.92679 84327
0.336	0.32971 34156	0.94408 10683	0.386	0.37648 57472	0.92642 24102
0.337	0.33065 73317	0.94375 08829	0.387	0.37741 19813	0.92604 54613
0.338	0.33160 09171	0.94341 97538	0.388	0.37833 78379	0.92566 75864
0.339	0.33254 41709	0.94308 76812	0.389	0.37926 33161	0.92528 87858
0.340	0.33348 70921	0.94275 46655	0.390	0.38018 84151	0.92490 90599
0.341	0.33442 96799	0.94242 07071	0.391	0.38111 31339	0.92452 84091
0.342	0.33537 19332	0.94208 58063	0.392	0.38203 74716	0.92414 68337
0.343	0.33631 38512	0.94174 99634	0.393	0.38296 14273	0.92376 43342
0.344	0.33725 54329	0.94141 31787	0.394	0.38388 50000	0.92338 09110
0.345	0.33819 66772	0.94107 54526	0.395	0.38480 81888	0.92299 65644
0.346	0.33913 75834	0.94073 67854	0.396	0.38573 09928	0.92261 12947
0.347	0.34007 81505	0.94039 71776	0.397	0.38665 34111	0.92222 51025
0.348	0.34101 83775	0.94005 66293	0.398	0.38757 54427	0.92183 79880
0.349	0.34195 82635	0.93971 51409	0.399	0.38849 70868	0.92144 99517
0.350	0.34289 78075	0.93937 27128	0.400	0.38941 83423	0.92106 09940

TEN-PLACE SINES AND COSINES FOR RADIAN MEASURE (Continued)

x	$\sin x$	$\cos x$	x	$\sin x$	$\cos x$
0.400	0.38941 83423	0.92106 09940	0.450	0.43496 55341	0.90044 71024
0.401	0.39033 92084	0.92067 11152	0.451	0.43586 57636	0.90001 16867
0.402	0.39125 96842	0.92028 03157	0.452	0.43676 55572	0.89957 53710
0.403	0.39217 97688	0.91988 85960	0.453	0.43766 49140	0.89913 81557
0.404	0.39309 94611	0.91949 59563	0.454	0.43856 38332	0.89870 00413
0.405	0.39401 87604	0.91910 23972	0.455	0.43946 23138	0.89826 10282
0.406	0.39493 76656	0.91870 79189	0.456	0.44036 03550	0.89782 11168
0.407	0.39585 61759	0.91831 25220	0.457	0.44125 79557	0.89738 03076
0.408	0.39677 42903	0.91791 62067	0.458	0.44215 51153	0.89693 86010
0.409	0.39769 20080	0.91751 89735	0.459	0.44305 18326	0.89649 59975
0.410	0.39860 93280	0.91712 08228	0.460	0.44394 81070	0.89605 24975
0.411	0.39952 62493	0.91672 17550	0.461	0.44484 39373	0.89560 81015
0.412	0.40044 27712	0.91632 17705	0.462	0.44573 93229	0.89516 28098
0.413	0.40135 88926	0.91592 08696	0.463	0.44663 42627	0.89471 66230
0.414	0.40227 46126	0.91551 90528	0.464	0.44752 87558	0.89426 95414
0.415	0.40318 99304	0.91511 63205	0.465	0.44842 28014	0.89382 15656
0.416	0.40410 48450	0.91471 26731	0.466	0.44931 63987	0.89337 26960
0.417	0.40501 93554	0.91430 81109	0.467	0.45020 95465	0.89292 29330
0.418	0.40593 34609	0.91390 26345	0.468	0.45110 22442	0.89247 22770
0.419	0.40684 71604	0.91349 62442	0.469	0.45199 44908	0.89202 07286
0.420	0.40776 04531	0.91308 89403	0.470	0.45288 62854	0.89156 82882
0.421	0.40867 33380	0.91268 07234	0.471	0.45377 76271	0.89111 49562
0.422	0.40958 58142	0.91227 15938	0.472	0.45466 85150	0.89066 07331
0.423	0.41049 78809	0.91186 15519	0.473	0.45555 89482	0.89020 56193
0.424	0.41140 95370	0.91145 05981	0.474	0.45644 89259	0.88974 96153
0.425	0.41232 07817	0.91103 87330	0.475	0.45733 84472	0.88929 27216
0.426	0.41323 16142	0.91062 59567	0.476	0.45822 75111	0.88883 49386
0.427	0.41414 20334	0.91021 22699	0.477	0.45911 61168	0.88837 62668
0.428	0.41505 20384	0.90979 76728	0.478	0.46000 42633	0.88791 67065
0.429	0.41596 16284	0.90938 21659	0.479	0.46089 19499	0.88745 62584
0.430	0.41687 08024	0.90896 57497	0.480	0.46177 91755	0.88699 49228
0.431	0.41777 95596	0.90854 84245	0.481	0.46266 59394	0.88653 27002
0.432	0.41868 78990	0.90813 01907	0.482	0.46355 22406	0.88606 95911
0.433	0.41959 58197	0.90771 10488	0.483	0.46443 80783	0.88560 55959
0.434	0.42050 33208	0.90729 09992	0.484	0.46532 34515	0.88514 07151
0.435	0.42141 04014	0.90687 00423	0.485	0.46620 83594	0.88467 49491
0.436	0.42231 70606	0.90644 81785	0.486	0.46709 28011	0.88420 82985
0.437	0.42322 32974	0.90602 54083	0.487	0.46797 67758	0.88374 07637
0.438	0.42412 91111	0.90560 17321	0.488	0.46886 02824	0.88327 23451
0.439	0.42503 45006	0.90517 71502	0.489	0.46974 33201	0.88280 30433
0.440	0.42593 94651	0.90475 16632	0.490	0.47062 58882	0.88233 28586
0.441	0.42684 40036	0.90432 52715	0.491	0.47150 79856	0.88186 17916
0.442	0.42774 81153	0.90389 79754	0.492	0.47238 96115	0.88138 98428
0.443	0.42865 17993	0.90346 97754	0.493	0.47327 07650	0.88091 70126
0.444	0.42955 50546	0.90304 06719	0.494	0.47415 14452	0.88044 33014
0.445	0.43045 78803	0.90261 06654	0.495	0.47503 16513	0.87996 87098
0.446	0.43136 02756	0.90217 97563	0.496	0.47591 13823	0.87949 32383
0.447	0.43226 22395	0.90174 79450	0.497	0.47679 06375	0.87901 68872
0.448	0.43316 37712	0.90131 52319	0.498	0.47766 94158	0.87853 96572
0.449	0.43406 48697	0.90088 16176	0.499	0.47854 77165	0.87806 15486
0.450	0.43496 55341	0.90044 71024	0.500	0.47942 55386	0.87758 25619

TEN-PLACE SINES AND COSINES FOR RADIAN MEASURE (Continued)

x	sin x	cos x	x	sin x	cos x
0.500	0.47942 55386	0.87758 25619	0.550	0.52268 72289	0.85252 45221
0.501	0.48030 28813	0.87710 26976	0.551	0.52353 94920	0.85200 14087
0.502	0.48117 97437	0.87662 19563	0.552	0.52439 12315	0.85147 74433
0.503	0.48205 61249	0.87614 03383	0.553	0.52524 24466	0.85095 26264
0.504	0.48293 20241	0.87565 78442	0.554	0.52609 31364	0.85042 69585
0.505	0.48380 74403	0.87517 44744	0.555	0.52694 33002	0.84990 04403
0.506	0.48468 23727	0.87469 02295	0.556	0.52779 29370	0.84937 30721
0.507	0.48555 68205	0.87420 51098	0.557	0.52864 20461	0.84884 48546
0.508	0.48643 07827	0.87371 91160	0.558	0.52949 06265	0.84831 57882
0.509	0.48730 42584	0.87323 22484	0.559	0.53033 86774	0.84778 58735
0.510	0.48817 72469	0.87274 45076	0.560	0.53118 61979	0.84725 51110
0.511	0.48904 97472	0.87225 58941	0.561	0.53203 31873	0.84672 35013
0.512	0.48992 17584	0.87176 64083	0.562	0.53287 96446	0.84619 10448
0.513	0.49079 32797	0.87127 60508	0.563	0.53372 55691	0.84565 77422
0.514	0.49166 43102	0.87078 48219	0.564	0.53457 09598	0.84512 35939
0.515	0.49253 48490	0.87029 27223	0.565	0.53541 58160	0.84458 86004
0.516	0.49340 48953	0.86979 97524	0.566	0.53626 01368	0.84405 27624
0.517	0.49427 44483	0.86930 59127	0.567	0.53710 39213	0.84351 60803
0.518	0.49514 35069	0.86881 12037	0.568	0.53794 71686	0.84297 85547
0.519	0.49601 20704	0.86831 56258	0.569	0.53878 98781	0.84244 01862
0.520	0.49688 01378	0.86781 91797	0.570	0.53963 20487	0.84190 09752
0.521	0.49774 77084	0.86732 18657	0.571	0.54047 36798	0.84136 09223
0.522	0.49861 47813	0.86682 36844	0.572	0.54131 47703	0.84082 00280
0.523	0.49948 13555	0.86632 46363	0.573	0.54215 53195	0.84027 82929
0.524	0.50034 74303	0.86582 47219	0.574	0.54299 53266	0.83973 57175
0.525	0.50121 30047	0.86532 39416	0.575	0.54383 47907	0.83919 23024
0.526	0.50207 80779	0.86482 22960	0.576	0.54467 37109	0.83864 80481
0.527	0.50294 26490	0.86431 97856	0.577	0.54551 20865	0.83810 29552
0.528	0.50380 67171	0.86381 64109	0.578	0.54634 99166	0.83755 70241
0.529	0.50467 02815	0.86331 21724	0.579	0.54718 72003	0.83701 02555
0.530	0.50553 33412	0.86280 70705	0.580	0.54802 39368	0.83646 26499
0.531	0.50639 58954	0.86230 11059	0.581	0.54886 01253	0.83591 42078
0.532	0.50725 79431	0.86179 42789	0.582	0.54969 57649	0.83536 49298
0.533	0.50811 94836	0.86128 65901	0.583	0.55053 08549	0.83481 48165
0.534	0.50898 05160	0.86077 80401	0.584	0.55136 53943	0.83426 38683
0.535	0.50984 10394	0.86026 86293	0.585	0.55219 93823	0.83371 20859
0.536	0.51070 10530	0.85975 83582	0.586	0.55303 28182	0.83315 94697
0.537	0.51156 05559	0.85924 72273	0.587	0.55386 57010	0.83260 60204
0.538	0.51241 95472	0.85873 52372	0.588	0.55469 80299	0.83205 17385
0.539	0.51327 80260	0.85822 23884	0.589	0.55552 98042	0.83149 66246
0.540	0.51413 59917	0.85770 86814	0.590	0.55636 10229	0.83094 06791
0.541	0.51499 34431	0.85719 41166	0.591	0.55719 16853	0.83038 39027
0.542	0.51585 03796	0.85667 86946	0.592	0.55802 17904	0.82982 62959
0.543	0.51670 68002	0.85616 24160	0.593	0.55885 13376	0.82926 78593
0.544	0.51756 27041	0.85564 52812	0.594	0.55968 03259	0.82870 85934
0.545	0.51841 80905	0.85512 72908	0.595	0.56050 87545	0.82814 84988
0.546	0.51927 29584	0.85460 84452	0.596	0.56133 66226	0.82758 75761
0.547	0.52012 73071	0.85408 87450	0.597	0.56216 39294	0.82702 58258
0.548	0.52098 11356	0.85356 81908	0.598	0.56299 06740	0.82646 32484
0.549	0.52183 44432	0.85304 67829	0.599	0.56381 68556	0.82589 98446
0.550	0.52268 72289	0.85252 45221	0.600	0.56464 24734	0.82533 56149

TEN-PLACE SINES AND COSINES FOR RADIAN MEASURE (Continued)

x	sin x	cos x	x	sin x	cos x
0.600	0.56464 24734	0.82533 56149	0.650	0.60518 64057	0.79608 37985
0.601	0.56546 75266	0.82477 05599	0.651	0.60598 21868	0.79547 82142
0.602	0.56629 20142	0.82420 46800	0.652	0.60677 73619	0.79487 18344
0.603	0.56711 59356	0.82363 79760	0.653	0.60757 19302	0.79426 46597
0.604	0.56793 92899	0.82307 04484	0.654	0.60836 58910	0.79365 66907
0.605	0.56876 20763	0.82250 20976	0.655	0.60915 92433	0.79304 79281
0.606	0.56958 42938	0.82193 29244	0.656	0.60995 19865	0.79243 83724
0.607	0.57040 59418	0.82136 29292	0.657	0.61074 41198	0.79182 80243
0.608	0.57122 70194	0.82079 21127	0.658	0.61153 56423	0.79121 68844
0.609	0.57204 75258	0.82022 04754	0.659	0.61232 65533	0.79060 49533
0.610	0.57286 74601	0.81964 80178	0.660	0.61311 68520	0.78999 22315
0.611	0.57368 68215	0.81907 47407	0.661	0.61390 65375	0.78937 87198
0.612	0.57450 56093	0.81850 06444	0.662	0.61469 56091	0.78876 44186
0.613	0.57532 38226	0.81792 57296	0.663	0.61548 40661	0.78814 93287
0.614	0.57614 14605	0.81734 99969	0.664	0.61627 19075	0.78753 34507
0.615	0.57695 85223	0.81677 34469	0.665	0.61705 91327	0.78691 67851
0.616	0.57777 50071	0.81619 60801	0.666	0.61784 57409	0.78629 93326
0.617	0.57859 09142	0.81561 78971	0.667	0.61863 17311	0.78568 10939
0.618	0.57940 62426	0.81503 88985	0.668	0.61941 71028	0.78506 20694
0.619	0.58022 09917	0.81445 90848	0.669	0.62020 18550	0.78444 22599
0.620	0.58103 51605	0.81387 84567	0.670	0.62098 59870	0.78382 16659
0.621	0.58184 87483	0.81329 70147	0.671	0.62176 94981	0.78320 02881
0.622	0.58266 17543	0.81271 47594	0.672	0.62255 23874	0.78257 81271
0.623	0.58347 41776	0.81213 16913	0.673	0.62333 46541	0.78195 51835
0.624	0.58428 60174	0.81154 78112	0.674	0.62411 62975	0.78133 14580
0.625	0.58509 72729	0.81096 31195	0.675	0.62489 73167	0.78070 69511
0.626	0.58590 79434	0.81037 76168	0.676	0.62567 77111	0.78008 16636
0.627	0.58671 80279	0.80979 13038	0.677	0.62645 74798	0.77945 55959
0.628	0.58752 75257	0.80920 41810	0.678	0.62723 66220	0.77882 87488
0.629	0.58833 64360	0.80861 62490	0.679	0.62801 51370	0.77820 11229
0.630	0.58914 47579	0.80802 75083	0.680	0.62879 30240	0.77757 27188
0.631	0.58995 24907	0.80743 79596	0.681	0.62957 02822	0.77694 35370
0.632	0.59075 96336	0.80684 76035	0.682	0.63034 69108	0.77631 35784
0.633	0.59156 61857	0.80625 64406	0.683	0.63112 29091	0.77568 28434
0.634	0.59237 21462	0.80566 44714	0.684	0.63189 82763	0.77505 13328
0.635	0.59317 75144	0.80507 16965	0.685	0.63267 30115	0.77441 90471
0.636	0.59398 22893	0.80447 81165	0.686	0.63344 71141	0.77378 59870
0.637	0.59478 64703	0.80388 37321	0.687	0.63422 05832	0.77315 21531
0.638	0.59559 00565	0.80328 85438	0.688	0.63499 34181	0.77251 75460
0.639	0.59639 30471	0.80269 25522	0.689	0.63576 56181	0.77188 21665
0.640	0.59719 54414	0.80209 57579	0.690	0.63653 71822	0.77124 60150
0.641	0.59799 72384	0.80149 81615	0.691	0.63730 81098	0.77060 90923
0.642	0.59879 84374	0.80089 97636	0.692	0.63807 84001	0.76997 13990
0.643	0.59959 90376	0.80030 05648	0.693	0.63884 80524	0.76933 29357
0.644	0.60039 90383	0.79970 05657	0.694	0.63961 70658	0.76869 37031
0.645	0.60119 84385	0.79909 97669	0.695	0.64038 54395	0.76805 37018
0.646	0.60199 72375	0.79849 81691	0.696	0.64115 31729	0.76741 29324
0.647	0.60279 54346	0.79789 57727	0.697	0.64192 02651	0.76677 13957
0.648	0.60359 30288	0.79729 25784	0.698	0.64268 67154	0.76612 90921
0.649	0.60439 00195	0.79668 85868	0.699	0.64345 25231	0.76548 60224
0.650	0.60518 64057	0.79608 37985	0.700	0.64421 76872	0.76484 21873

TEN-PLACE SINES AND COSINES FOR RADIAN MEASURE (Continued)

x	sin x	cos x	x	sin x	cos x
0.700	0.64421 76872	0.76484 21873	0.750	0.68163 87600	0.73168 88689
0.701	0.64498 22072	0.76419 75873	0.751	0.68237 01080	0.73100 68644
0.702	0.64574 60822	0.76355 22231	0.752	0.68310 07735	0.73032 41289
0.703	0.64650 93114	0.76290 60953	0.753	0.68383 07560	0.72964 06631
0.704	0.64727 18941	0.76225 92047	0.754	0.68456 00546	0.72895 64676
0.705	0.64803 38295	0.76161 15518	0.755	0.68528 86687	0.72827 15432
0.706	0.64879 51169	0.76096 31372	0.756	0.68601 65974	0.72758 58905
0.707	0.64955 57556	0.76031 39617	0.757	0.68674 38402	0.72689 95102
0.708	0.65031 57446	0.75966 40259	0.758	0.68747 03962	0.72621 24030
0.709	0.65107 50834	0.75901 33305	0.759	0.68819 62648	0.72552 45697
0.710	0.65183 37710	0.75836 18760	0.760	0.68892 14451	0.72483 60107
0.711	0.65259 18069	0.75770 96631	0.761	0.68964 59365	0.72414 67270
0.712	0.65334 91901	0.75705 66926	0.762	0.69036 97383	0.72345 67191
0.713	0.65410 59200	0.75640 29650	0.763	0.69109 28497	0.72276 59877
0.714	0.65486 19958	0.75574 84810	0.764	0.69181 52701	0.72207 45336
0.715	0.65561 74167	0.75509 32412	0.765	0.69253 69986	0.72138 23574
0.716	0.65637 21820	0.75443 72464	0.766	0.69325 80345	0.72068 94599
0.717	0.65712 62909	0.75378 04971	0.767	0.69397 83772	0.71999 58416
0.718	0.65787 97427	0.75312 29940	0.768	0.69469 80260	0.71930 15033
0.719	0.65863 25367	0.75246 47378	0.769	0.69541 69800	0.71860 64458
0.720	0.65938 46720	0.75180 57291	0.770	0.69613 52386	0.71791 06696
0.721	0.66013 61479	0.75114 59687	0.771	0.69685 28011	0.71721 41755
0.722	0.66088 69637	0.75048 54571	0.772	0.69756 96667	0.71651 69642
0.723	0.66163 71185	0.74982 41950	0.773	0.69828 58348	0.71581 90364
0.724	0.66238 66118	0.74916 21830	0.774	0.69900 13046	0.71512 03928
0.725	0.66313 54427	0.74849 94220	0.775	0.69971 60753	0.71442 10341
0.726	0.66388 36104	0.74783 59124	0.776	0.70043 01464	0.71372 09609
0.727	0.66463 11142	0.74717 16550	0.777	0.70114 35170	0.71302 01740
0.728	0.66537 79535	0.74650 66504	0.778	0.70185 61865	0.71231 86741
0.729	0.66612 41273	0.74584 08993	0.779	0.70256 81541	0.71161 64619
0.730	0.66686 96350	0.74517 44023	0.780	0.70327 94192	0.71091 35380
0.731	0.66761 44758	0.74450 71602	0.781	0.70398 99810	0.71020 99033
0.732	0.66835 86491	0.74383 91736	0.782	0.70469 98388	0.70950 55583
0.733	0.66910 21539	0.74317 04432	0.783	0.70540 89919	0.70880 05038
0.734	0.66984 49897	0.74250 09695	0.784	0.70611 74395	0.70809 47405
0.735	0.67058 71556	0.74183 07534	0.785	0.70682 51811	0.70738 82692
0.736	0.67132 86510	0.74115 97954	0.786	0.70753 22158	0.70668 10904
0.737	0.67206 94750	0.74048 80963	0.787	0.70823 85431	0.70597 32050
0.738	0.67280 96269	0.73981 56567	0.788	0.70894 41620	0.70526 46136
0.739	0.67354 91060	0.73914 24773	0.789	0.70964 90720	0.70455 53169
0.740	0.67428 79116	0.73846 85587	0.790	0.71035 32724	0.70384 53157
0.741	0.67502 60429	0.73779 39017	0.791	0.71105 67624	0.70313 46106
0.742	0.67576 34992	0.73711 85069	0.792	0.71175 95414	0.70242 32024
0.743	0.67650 02797	0.73644 23749	0.793	0.71246 16086	0.70171 10917
0.744	0.67723 63837	0.73576 55065	0.794	0.71316 29634	0.70099 82794
0.745	0.67797 18105	0.73508 79024	0.795	0.71386 36049	0.70028 47660
0.746	0.67870 65592	0.73440 95631	0.796	0.71456 35327	0.69957 05524
0.747	0.67944 06293	0.73373 04895	0.797	0.71526 27458	0.69885 56392
0.748	0.68017 40200	0.73305 06821	0.798	0.71596 12437	0.69814 00272
0.749	0.68090 67305	0.73237 01417	0.799	0.71665 90256	0.69742 37170
0.750	0.68163 87600	0.73168 88689	0.800	0.71735 60909	0.69670 67093

TEN-PLACE SINES AND COSINES FOR RADIAN MEASURE (Continued)

x	sin x	cos x	x	sin x	cos x
0.800	0.71735 60909	0.69670 67093	0.850	0.75128 04051	0.65998 31459
0.801	0.71805 24388	0.69598 90050	0.851	0.75194 00125	0.65923 15356
0.802	0.71874 80687	0.69527 06047	0.852	0.75259 88680	0.65847 92661
0.803	0.71944 29798	0.69455 15091	0.853	0.75325 69708	0.65772 63381
0.804	0.72013 71715	0.69383 17190	0.854	0.75391 43204	0.65697 27524
0.805	0.72083 06430	0.69311 12350	0.855	0.75457 09161	0.65621 85097
0.806	0.72152 33937	0.69239 00579	0.856	0.75522 67572	0.65546 36108
0.807	0.72221 54229	0.69166 81885	0.857	0.75588 18431	0.65470 80565
0.808	0.72290 67298	0.69094 56273	0.858	0.75653 61731	0.65395 18474
0.809	0.72359 73139	0.69022 23753	0.859	0.75718 97466	0.65319 49844
0.810	0.72428 71744	0.68949 84330	0.860	0.75784 25629	0.65243 74682
0.811	0.72497 63105	0.68877 38011	0.861	0.75849 46213	0.65167 92995
0.812	0.72566 47217	0.68804 84806	0.862	0.75914 59213	0.65092 04792
0.813	0.72635 24073	0.68732 24719	0.863	0.75979 64621	0.65016 10079
0.814	0.72703 93665	0.68659 57760	0.864	0.76044 62431	0.64940 08865
0.815	0.72772 55986	0.68586 83935	0.865	0.76109 52636	0.64864 01157
0.816	0.72841 11030	0.68514 03250	0.866	0.76174 35231	0.64787 86962
0.817	0.72909 58790	0.68441 15715	0.867	0.76239 10208	0.64711 66289
0.818	0.72977 99259	0.68368 21335	0.868	0.76303 77561	0.64635 39144
0.819	0.73046 32431	0.68295 20119	0.869	0.76368 37284	0.64559 05536
0.820	0.73114 58297	0.68222 12073	0.870	0.76432 89370	0.64482 65472
0.821	0.73182 76852	0.68148 97205	0.871	0.76497 33813	0.64406 18960
0.822	0.73250 88089	0.68075 75522	0.872	0.76561 70606	0.64329 66007
0.823	0.73318 92001	0.68002 47031	0.873	0.76625 99743	0.64253 06622
0.824	0.73386 88581	0.67929 11740	0.874	0.76690 21217	0.64176 40810
0.825	0.73454 77822	0.67855 69656	0.875	0.76754 35022	0.64099 68582
0.826	0.73522 59718	0.67782 20787	0.876	0.76818 41152	0.64022 89943
0.827	0.73590 34262	0.67708 65139	0.877	0.76882 39600	0.63946 04902
0.828	0.73658 01446	0.67635 02721	0.878	0.76946 30360	0.63869 13466
0.829	0.73725 61265	0.67561 33539	0.879	0.77010 13425	0.63792 15644
0.830	0.73793 13711	0.67487 57601	0.880	0.77073 88789	0.63715 11442
0.831	0.73860 58778	0.67413 74914	0.881	0.77137 56446	0.63638 00869
0.832	0.73927 96459	0.67339 85486	0.882	0.77201 16389	0.63560 83932
0.833	0.73995 26747	0.67265 89323	0.883	0.77264 68611	0.63483 60639
0.834	0.74062 49635	0.67191 86435	0.884	0.77328 13108	0.63406 30997
0.835	0.74129 65117	0.67117 76827	0.885	0.77391 49871	0.63328 95015
0.836	0.74196 73187	0.67043 60507	0.886	0.77454 78896	0.63251 52700
0.837	0.74263 73836	0.66969 37483	0.887	0.77518 00175	0.63174 04060
0.838	0.74330 67059	0.66895 07762	0.888	0.77581 13702	0.63096 49102
0.839	0.74397 52849	0.66820 71351	0.889	0.77644 19471	0.63018 87835
0.840	0.74464 31200	0.66746 28258	0.890	0.77707 17475	0.62941 20266
0.841	0.74531 02104	0.66671 78491	0.891	0.77770 07709	0.62863 46402
0.842	0.74597 65554	0.66597 22057	0.892	0.77832 90166	0.62785 66253
0.843	0.74664 21546	0.66522 58963	0.893	0.77895 64840	0.62707 79825
0.844	0.74730 70070	0.66447 89216	0.894	0.77958 31724	0.62629 87126
0.845	0.74797 11122	0.66373 12825	0.895	0.78020 90812	0.62551 88164
0.846	0.74863 44694	0.66298 29796	0.896	0.78083 42098	0.62473 82947
0.847	0.74929 70779	0.66223 40138	0.897	0.78145 85576	0.62395 71482
0.848	0.74995 89372	0.66148 43857	0.898	0.78208 21239	0.62317 53778
0.849	0.75062 00465	0.66073 40962	0.899	0.78270 49081	0.62239 29842
0.850	0.75128 04051	0.65998 31459	0.900	0.78332 69096	0.62160 99683

TEN-PLACE SINES AND COSINES FOR RADIAN MEASURE (Continued)

x	sin x	cos x	x	sin x	cos x
0.900	0.78332 69096	0.62160 99683	0.950	0.81341 55048	0.58168 30895
0.901	0.78394 81278	0.62082 63307	0.951	0.81399 67811	0.58086 93833
0.902	0.78456 85621	0.62004 20723	0.952	0.81457 72434	0.58005 50962
0.903	0.78518 82118	0.61925 71938	0.953	0.81515 68911	0.57924 02290
0.904	0.78580 70763	0.61847 16961	0.954	0.81573 57236	0.57842 47827
0.905	0.78642 51550	0.61768 55799	0.955	0.81631 37404	0.57760 87579
0.906	0.78704 24472	0.61689 88461	0.956	0.81689 09410	0.57679 21555
0.907	0.78765 89524	0.61611 14953	0.957	0.81746 73246	0.57597 49763
0.908	0.78827 46700	0.61532 35284	0.958	0.81804 28907	0.57515 72211
0.909	0.78888 95993	0.61453 49462	0.959	0.81861 76388	0.57433 88907
0.910	0.78950 37397	0.61374 57495	0.960	0.81919 15683	0.57351 99861
0.911	0.79011 70906	0.61295 59390	0.961	0.81976 46786	0.57270 05079
0.912	0.79072 96514	0.61216 55156	0.962	0.82033 69691	0.57188 04570
0.913	0.79134 14214	0.61137 44800	0.963	0.82090 84393	0.57105 98342
0.914	0.79195 24001	0.61058 28330	0.964	0.82147 90886	0.57023 86404
0.915	0.79256 25869	0.60979 05754	0.965	0.82204 89164	0.56941 68763
0.916	0.79317 19811	0.60899 77081	0.966	0.82261 79222	0.56859 45428
0.917	0.79378 05821	0.60820 42317	0.967	0.82318 61053	0.56777 16407
0.918	0.79438 83893	0.60741 01472	0.968	0.82375 34653	0.56694 81709
0.919	0.79499 54022	0.60661 54552	0.969	0.82432 00015	0.56612 41341
0.920	0.79560 16200	0.60582 01566	0.970	0.82488 57133	0.56529 95312
0.921	0.79620 70423	0.60502 42522	0.971	0.82545 06003	0.56447 43629
0.922	0.79681 16683	0.60422 77428	0.972	0.82601 46619	0.56364 86302
0.923	0.79741 54976	0.60343 06292	0.973	0.82657 78974	0.56282 23339
0.924	0.79801 85294	0.60263 29121	0.974	0.82714 03064	0.56199 54747
0.925	0.79862 07632	0.60183 45924	0.975	0.82770 18882	0.56116 80535
0.926	0.79922 21984	0.60103 56708	0.976	0.82826 26423	0.56034 00712
0.927	0.79982 28343	0.60023 61483	0.977	0.82882 25681	0.55951 15285
0.928	0.80042 26705	0.59943 60254	0.978	0.82938 16651	0.55868 24264
0.929	0.80102 17062	0.59863 53032	0.979	0.82993 99328	0.55785 27655
0.930	0.80161 99409	0.59783 39823	0.980	0.83049 73705	0.55702 25468
0.931	0.80221 73740	0.59703 20636	0.981	0.83105 39777	0.55619 17710
0.932	0.80281 40048	0.59622 95478	0.982	0.83160 97538	0.55536 04391
0.933	0.80340 98329	0.59542 64358	0.983	0.83216 46984	0.55452 85518
0.934	0.80400 48575	0.59462 27284	0.984	0.83271 88108	0.55369 61100
0.935	0.80459 90781	0.59381 84264	0.985	0.83327 20904	0.55286 31144
0.936	0.80519 24941	0.59301 35305	0.986	0.83382 45368	0.55202 95661
0.937	0.80578 51050	0.59220 80417	0.987	0.83437 61494	0.55119 54657
0.938	0.80637 69100	0.59140 19606	0.988	0.83492 69276	0.55036 08140
0.939	0.80696 79087	0.59059 52881	0.989	0.83547 68708	0.54952 56121
0.940	0.80755 81004	0.58978 80250	0.990	0.83602 59786	0.54868 98606
0.941	0.80814 74846	0.58898 01722	0.991	0.83657 42504	0.54785 35604
0.942	0.80873 60606	0.58817 17303	0.992	0.83712 16855	0.54701 67124
0.943	0.80932 38278	0.58736 27003	0.993	0.83766 82836	0.54617 93173
0.944	0.80991 07858	0.58655 30829	0.994	0.83821 40440	0.54534 13761
0.945	0.81049 69338	0.58574 28790	0.995	0.83875 89662	0.54450 28895
0.946	0.81108 22713	0.58493 20893	0.996	0.83930 30496	0.54366 38584
0.947	0.81166 67978	0.58412 07147	0.997	0.83984 62937	0.54282 42837
0.948	0.81225 05126	0.58330 87560	0.998	0.84038 86980	0.54198 41661
0.949	0.81283 34151	0.58249 62140	0.999	0.84093 02619	0.54114 35066
0.950	0.81341 55048	0.58168 30895	1.000	0.84147 09848	0.54030 23059

TEN-PLACE SINES AND COSINES FOR RADIAN MEASURE (Continued)

x	$\sin x$	$\cos x$	x	$\sin x$	$\cos x$
1.000	0.84147 09848	0.54030 23059	1.050	0.86742 32256	0.49757 10479
1.001	0.84201 08663	0.53946 05649	1.051	0.86792 03628	0.49670 33760
1.002	0.84254 99058	0.53861 82844	1.052	0.86841 66322	0.49583 52075
1.003	0.84308 81027	0.53777 54653	1.053	0.86891 20331	0.49496 65431
1.004	0.84362 54565	0.53693 21085	1.054	0.86940 65651	0.49409 73837
1.005	0.84416 19667	0.53608 82147	1.055	0.86990 02277	0.49322 77302
1.006	0.84469 76328	0.53524 37848	1.056	0.87039 30204	0.49235 75835
1.007	0.84523 24541	0.53439 88197	1.057	0.87088 49427	0.49148 69445
1.008	0.84576 64302	0.53355 33202	1.058	0.87137 59941	0.49061 58139
1.009	0.84629 95606	0.53270 72871	1.059	0.87186 61742	0.48974 41928
1.010	0.84683 18446	0.53186 07214	1.060	0.87235 54823	0.48887 20819
1.011	0.84736 32818	0.53101 36237	1.061	0.87284 39182	0.48799 94821
1.012	0.84789 38717	0.53016 59951	1.062	0.87333 14811	0.48712 63943
1.013	0.84842 36136	0.52931 78363	1.063	0.87381 81708	0.48625 28194
1.014	0.84895 25072	0.52846 91481	1.064	0.87430 39866	0.48537 87583
1.015	0.84948 05518	0.52761 99315	1.065	0.87478 89281	0.48450 42117
1.016	0.85000 77469	0.52677 01873	1.066	0.87527 29949	0.48362 91807
1.017	0.85053 40920	0.52591 99163	1.067	0.87575 61863	0.48275 36660
1.018	0.85105 95865	0.52506 91194	1.068	0.87623 85021	0.48187 76686
1.019	0.85158 42300	0.52421 77974	1.069	0.87671 99415	0.48100 11893
1.020	0.85210 80219	0.52336 59513	1.070	0.87720 05043	0.48012 42290
1.021	0.85263 09618	0.52251 35817	1.071	0.87768 01898	0.47924 67886
1.022	0.85315 30489	0.52166 06896	1.072	0.87815 89977	0.47836 88689
1.023	0.85367 42830	0.52080 72759	1.073	0.87863 69274	0.47749 04709
1.024	0.85419 46633	0.51995 33413	1.074	0.87911 39785	0.47661 15954
1.025	0.85471 41895	0.51909 88868	1.075	0.87959 01504	0.47573 22432
1.026	0.85523 28609	0.51824 39132	1.076	0.88006 54428	0.47485 24154
1.027	0.85575 06771	0.51738 84214	1.077	0.88053 98551	0.47397 21126
1.028	0.85626 76376	0.51653 24122	1.078	0.88101 33869	0.47309 13360
1.029	0.85678 37418	0.51567 58864	1.079	0.88148 60376	0.47221 00862
1.030	0.85729 89892	0.51481 88450	1.080	0.88195 78069	0.47132 83642
1.031	0.85781 33793	0.51396 12887	1.081	0.88242 86942	0.47044 61708
1.032	0.85832 69116	0.51310 32185	1.082	0.88289 86991	0.46956 35071
1.033	0.85883 95856	0.51224 46352	1.083	0.88336 78210	0.46868 03737
1.034	0.85935 14007	0.51138 55396	1.084	0.88383 60597	0.46779 67717
1.035	0.85986 23565	0.51052 59327	1.085	0.88430 34144	0.46691 27019
1.036	0.86037 24524	0.50966 58152	1.086	0.88476 98849	0.46602 81652
1.037	0.86088 16879	0.50880 51880	1.087	0.88523 54706	0.46514 31624
1.038	0.86139 00626	0.50794 40521	1.088	0.88570 01711	0.46425 76946
1.039	0.86189 75759	0.50708 24082	1.089	0.88616 39858	0.46337 17624
1.040	0.86240 42272	0.50622 02572	1.090	0.88662 69144	0.46248 53669
1.041	0.86291 00162	0.50535 76000	1.091	0.88708 89564	0.46159 85089
1.042	0.86341 49423	0.50449 44375	1.092	0.88755 01113	0.46071 11893
1.043	0.86391 90049	0.50363 07704	1.093	0.88801 03787	0.45982 34089
1.044	0.86442 22036	0.50276 65998	1.094	0.88846 97580	0.45893 51688
1.045	0.86492 45380	0.50190 19263	1.095	0.88892 82488	0.45804 64697
1.046	0.86542 60073	0.50103 67510	1.096	0.88938 58508	0.45715 73126
1.047	0.86592 66113	0.50017 10746	1.097	0.88984 25633	0.45626 76983
1.048	0.86642 63493	0.49930 48980	1.098	0.89029 83860	0.45537 76278
1.049	0.86692 52209	0.49843 82222	1.099	0.89075 33184	0.45448 71018
1.050	0.86742 32256	0.49757 10479	1.100	0.89120 73601	0.45359 61214

TEN-PLACE SINES AND COSINES FOR RADIAN MEASURE (Continued)

x	sin x	cos x	x	sin x	cos x
1.100	0.89120 73601	0.45359 61214	1.150	0.91276 39403	0.40848 74409
1.101	0.89166 05105	0.45270 46874	1.151	0.91317 19713	0.40757 44729
1.102	0.89211 27693	0.45181 28007	1.152	0.91357 90891	0.40666 10972
1.103	0.89256 41360	0.45092 04622	1.153	0.91398 52933	0.40574 73150
1.104	0.89301 46101	0.45002 76727	1.154	0.91439 05836	0.40483 31270
1.105	0.89346 41911	0.44913 44333	1.155	0.91479 49594	0.40391 85341
1.106	0.89391 28788	0.44824 07446	1.156	0.91519 84205	0.40300 35374
1.107	0.89436 06725	0.44734 66078	1.157	0.91560 09664	0.40208 81376
1.108	0.89480 75718	0.44645 20236	1.158	0.91600 25966	0.40117 23357
1.109	0.89525 35764	0.44555 69929	1.159	0.91640 33109	0.40025 61327
1.110	0.89569 86857	0.44466 15167	1.160	0.91680 31088	0.39933 95294
1.111	0.89614 28993	0.44376 55959	1.161	0.91720 19898	0.39842 25268
1.112	0.89658 62167	0.44286 92312	1.162	0.91759 99537	0.39750 51257
1.113	0.89702 86376	0.44197 24237	1.163	0.91799 70000	0.39658 73272
1.114	0.89747 01614	0.44107 51743	1.164	0.91839 31282	0.39566 91320
1.115	0.89791 07878	0.44017 74837	1.165	0.91878 83381	0.39475 05412
1.116	0.89835 05162	0.43927 93530	1.166	0.91918 26292	0.39383 15557
1.117	0.89878 93463	0.43838 07830	1.167	0.91957 60011	0.39291 21763
1.118	0.89922 72777	0.43748 17746	1.168	0.91996 84534	0.39199 24040
1.119	0.89966 43098	0.43658 23287	1.169	0.92035 99857	0.39107 22397
1.120	0.90010 04422	0.43568 24463	1.170	0.92075 05977	0.39015 16843
1.121	0.90053 56745	0.43478 21281	1.171	0.92114 02890	0.38923 07388
1.122	0.90097 00063	0.43388 13752	1.172	0.92152 90591	0.38830 94040
1.123	0.90140 34371	0.43298 01884	1.173	0.92191 69077	0.38738 76810
1.124	0.90183 59665	0.43207 85687	1.174	0.92230 38343	0.38646 55705
1.125	0.90226 75941	0.43117 65168	1.175	0.92268 98387	0.38554 30736
1.126	0.90269 83194	0.43027 40338	1.176	0.92307 49203	0.38462 01912
1.127	0.90312 81420	0.42937 11205	1.177	0.92345 90789	0.38369 69241
1.128	0.90355 70615	0.42846 77778	1.178	0.92384 23141	0.38277 32733
1.129	0.90398 50774	0.42756 40066	1.179	0.92422 46253	0.38184 92398
1.130	0.90441 21894	0.42665 98079	1.180	0.92460 60124	0.38092 48244
1.131	0.90483 83969	0.42575 51826	1.181	0.92498 64749	0.38000 00280
1.132	0.90526 36996	0.42485 01314	1.182	0.92536 60123	0.37907 48517
1.133	0.90568 80970	0.42394 46555	1.183	0.92574 46244	0.37814 92963
1.134	0.90611 15888	0.42303 87555	1.184	0.92612 23108	0.37722 33628
1.135	0.90653 41744	0.42213 24326	1.185	0.92649 90710	0.37629 70520
1.136	0.90695 58535	0.42122 56875	1.186	0.92687 49048	0.37537 03650
1.137	0.90737 66256	0.42031 85212	1.187	0.92724 98116	0.37444 33025
1.138	0.90779 64904	0.41941 09346	1.188	0.92762 37913	0.37351 58656
1.139	0.90821 54474	0.41850 29285	1.189	0.92799 68433	0.37258 80552
1.140	0.90863 34961	0.41759 45040	1.190	0.92836 89672	0.37165 98723
1.141	0.90905 06362	0.41668 56618	1.191	0.92874 01629	0.37073 13176
1.142	0.90946 68673	0.41577 64030	1.192	0.92911 04298	0.36980 23922
1.143	0.90988 21889	0.41486 67284	1.193	0.92947 97675	0.36887 30971
1.144	0.91029 66006	0.41395 66389	1.194	0.92984 81758	0.36794 34330
1.145	0.91071 01020	0.41304 61355	1.195	0.93021 56543	0.36701 34010
1.146	0.91112 26928	0.41213 52190	1.196	0.93058 22025	0.36608 30020
1.147	0.91153 43723	0.41122 38904	1.197	0.93094 78202	0.36515 22369
1.148	0.91194 51404	0.41031 21506	1.198	0.93131 25069	0.36422 11067
1.149	0.91235 49965	0.40940 00004	1.199	0.93167 62623	0.36328 96122
1.150	0.91276 39403	0.40848 74409	1.200	0.93203 90860	0.36235 77545

TEN-PLACE SINES AND COSINES FOR RADIAN MEASURE (Continued)

x	sin x	cos x	x	sin x	cos x
1.200	0.93203 90860	0.36235 77545	1.250	0.94898 46194	0.31532 23624
1.201	0.93240 09776	0.36142 55344	1.251	0.94929 94672	0.31437 32203
1.202	0.93276 19369	0.36049 29528	1.252	0.94961 33657	0.31342 37638
1.203	0.93312 19634	0.35956 00108	1.253	0.94992 63146	0.31247 39939
1.204	0.93348 10568	0.35862 67092	1.254	0.95023 83136	0.31152 39115
1.205	0.93383 92167	0.35769 30490	1.255	0.95054 93623	0.31057 35175
1.206	0.93419 64428	0.35675 90311	1.256	0.95085 94605	0.30962 28131
1.207	0.93455 27347	0.35582 46564	1.257	0.95116 86078	0.30867 17989
1.208	0.93490 80920	0.35488 99259	1.258	0.95147 68040	0.30772 04762
1.209	0.93526 25144	0.35395 48406	1.259	0.95178 40487	0.30676 88457
1.210	0.93561 60016	0.35301 94012	1.260	0.95209 03416	0.30581 69084
1.211	0.93596 85531	0.35208 36089	1.261	0.95239 56824	0.30486 46653
1.212	0.93632 01687	0.35114 74644	1.262	0.95270 00708	0.30391 21173
1.213	0.93667 08479	0.35021 09688	1.263	0.95300 35065	0.30295 92655
1.214	0.93702 05905	0.34927 41230	1.264	0.95330 59892	0.30200 61106
1.215	0.93736 93960	0.34833 69280	1.265	0.95360 75187	0.30105 26538
1.216	0.93771 72642	0.34739 93846	1.266	0.95390 80945	0.30009 88959
1.217	0.93806 41947	0.34646 14938	1.267	0.95420 77163	0.29914 48379
1.218	0.93841 01871	0.34552 32565	1.268	0.95450 63840	0.29819 04808
1.219	0.93875 52411	0.34458 46737	1.269	0.95480 40972	0.29723 58255
1.220	0.93909 93563	0.34364 57463	1.270	0.95510 08556	0.29628 08729
1.221	0.93944 25325	0.34270 64753	1.271	0.95539 66589	0.29532 56241
1.222	0.93978 47692	0.34176 68616	1.272	0.95569 15067	0.29437 00799
1.223	0.94012 60661	0.34082 69061	1.273	0.95598 53989	0.29341 42414
1.224	0.94046 64229	0.33988 66097	1.274	0.95627 83351	0.29245 81094
1.225	0.94080 58392	0.33894 59735	1.275	0.95657 03150	0.29150 16850
1.226	0.94114 43147	0.33800 49984	1.276	0.95686 13384	0.29054 49691
1.227	0.94148 18491	0.33706 36852	1.277	0.95715 14049	0.28958 79627
1.228	0.94181 84419	0.33612 20350	1.278	0.95744 05142	0.28863 06666
1.229	0.94215 40930	0.33518 00487	1.279	0.95772 86661	0.28767 30820
1.230	0.94248 88019	0.33423 77271	1.280	0.95801 58603	0.28671 52096
1.231	0.94282 25684	0.33329 50714	1.281	0.95830 20964	0.28575 70506
1.232	0.94315 53920	0.33235 20823	1.282	0.95858 73743	0.28479 86058
1.233	0.94348 72724	0.33140 87609	1.283	0.95887 16936	0.28383 98761
1.234	0.94381 82094	0.33046 51081	1.284	0.95915 50540	0.28288 08627
1.235	0.94414 82025	0.32952 11248	1.285	0.95943 74552	0.28192 15664
1.236	0.94447 72515	0.32857 68120	1.286	0.95971 88970	0.28096 19881
1.237	0.94480 53560	0.32763 21706	1.287	0.95999 93791	0.28000 21289
1.238	0.94513 25157	0.32668 72016	1.288	0.96027 89012	0.27904 19897
1.239	0.94545 87303	0.32574 19059	1.289	0.96055 74630	0.27808 15714
1.240	0.94578 39994	0.32479 62844	1.290	0.96083 50642	0.27712 08751
1.241	0.94610 83228	0.32385 03382	1.291	0.96111 17046	0.27615 99016
1.242	0.94643 17000	0.32290 40681	1.292	0.96138 73839	0.27519 86520
1.243	0.94675 41308	0.32195 74751	1.293	0.96166 21018	0.27423 71271
1.244	0.94707 56149	0.32101 05602	1.294	0.96193 58581	0.27327 53281
1.245	0.94739 61518	0.32006 33242	1.295	0.96220 86524	0.27231 32557
1.246	0.94771 57414	0.31911 57682	1.296	0.96248 04845	0.27135 09111
1.247	0.94803 43833	0.31816 78930	1.297	0.96275 13541	0.27038 82951
1.248	0.94835 20771	0.31721 96997	1.298	0.96302 12610	0.26942 54087
1.249	0.94866 88226	0.31627 11892	1.299	0.96329 02049	0.26846 22529
1.250	0.94898 46194	0.31532 23624	1.300	0.96355 81854	0.26749 88286

TEN-PLACE SINES AND COSINES FOR RADIAN MEASURE (Continued)

x	sin x	cos x	x	sin x	cos x
1.300	0.96355 81854	0.26749 88286	1.350	0.97572 33578	0.21900 66871
1.301	0.96382 52024	0.26653 51369	1.351	0.97594 18766	0.21803 08544
1.302	0.96409 12556	0.26557 11785	1.352	0.97615 94195	0.21705 48037
1.303	0.96435 63447	0.26460 69547	1.353	0.97637 59862	0.21607 85359
1.304	0.96462 04694	0.26364 24662	1.354	0.97659 15765	0.21510 20520
1.305	0.96488 36295	0.26267 77140	1.355	0.97680 61902	0.21412 53531
1.306	0.96514 58248	0.26171 26992	1.356	0.97701 98271	0.21314 84400
1.307	0.96540 70548	0.26074 74227	1.357	0.97723 24870	0.21217 13137
1.308	0.96566 73195	0.25978 18854	1.358	0.97744 41697	0.21119 39753
1.309	0.96592 66185	0.25881 60884	1.359	0.97765 48749	0.21021 64257
1.310	0.96618 49516	0.25785 00325	1.360	0.97786 46024	0.20923 86659
1.311	0.96644 23185	0.25688 37188	1.361	0.97807 33521	0.20826 06968
1.312	0.96669 87190	0.25591 71482	1.362	0.97828 11238	0.20728 25195
1.313	0.96695 41527	0.25495 03217	1.363	0.97848 79171	0.20630 41349
1.314	0.96720 86195	0.25398 32402	1.364	0.97869 37320	0.20532 55440
1.315	0.96746 21191	0.25301 59048	1.365	0.97889 85681	0.20434 67478
1.316	0.96771 46512	0.25204 83163	1.366	0.97910 24254	0.20336 77472
1.317	0.96796 62157	0.25108 04758	1.367	0.97930 53036	0.20238 85432
1.318	0.96821 68121	0.25011 23842	1.368	0.97950 72024	0.20140 91369
1.319	0.96846 64404	0.24914 40425	1.369	0.97970 81218	0.20042 95292
1.320	0.96871 51001	0.24817 54517	1.370	0.97990 80614	0.19944 97210
1.321	0.96896 27912	0.24720 66126	1.371	0.98010 70211	0.19846 97134
1.322	0.96920 95133	0.24623 75264	1.372	0.98030 50008	0.19748 95073
1.323	0.96945 52661	0.24526 81939	1.373	0.98050 20001	0.19650 91037
1.324	0.96970 00496	0.24429 86162	1.374	0.98069 80189	0.19552 85036
1.325	0.96994 38633	0.24332 87941	1.375	0.98089 30570	0.19454 77080
1.326	0.97018 67071	0.24235 87288	1.376	0.98108 71143	0.19356 67178
1.327	0.97042 85807	0.24138 84211	1.377	0.98128 01904	0.19258 55341
1.328	0.97066 94838	0.24041 78719	1.378	0.98147 22853	0.19160 41578
1.329	0.97090 94163	0.23944 70824	1.379	0.98166 33986	0.19062 25898
1.330	0.97114 83779	0.23847 60534	1.380	0.98185 35304	0.18964 08313
1.331	0.97138 63684	0.23750 47860	1.381	0.98204 26802	0.18865 88831
1.332	0.97162 33874	0.23653 32810	1.382	0.98223 08481	0.18767 67463
1.333	0.97185 94348	0.23556 15395	1.383	0.98241 80337	0.18669 44217
1.334	0.97209 45104	0.23458 95625	1.384	0.98260 42369	0.18571 19105
1.335	0.97232 86139	0.23361 73508	1.385	0.98278 94574	0.18472 92136
1.336	0.97256 17450	0.23264 49056	1.386	0.98297 36952	0.18374 63319
1.337	0.97279 39036	0.23167 22277	1.387	0.98315 69500	0.18276 32665
1.338	0.97302 50894	0.23069 93181	1.388	0.98333 92217	0.18178 00184
1.339	0.97325 53022	0.22972 61778	1.389	0.98352 05100	0.18079 65884
1.340	0.97348 45417	0.22875 28078	1.390	0.98370 08148	0.17981 29777
1.341	0.97371 28077	0.22777 92091	1.391	0.98388 01359	0.17882 91871
1.342	0.97394 01000	0.22680 53825	1.392	0.98405 84731	0.17784 52177
1.343	0.97416 64184	0.22583 13292	1.393	0.98423 58263	0.17686 10705
1.344	0.97439 17626	0.22485 70500	1.394	0.98441 21952	0.17587 67464
1.345	0.97461 61324	0.22388 25460	1.395	0.98458 75797	0.17489 22464
1.346	0.97483 95276	0.22290 78181	1.396	0.98476 19796	0.17390 75716
1.347	0.97506 19480	0.22193 28672	1.397	0.98493 53948	0.17292 27228
1.348	0.97528 33933	0.22095 76945	1.398	0.98510 78250	0.17193 77011
1.349	0.97550 38633	0.21998 23008	1.399	0.98527 92701	0.17095 25075
1.350	0.97572 33578	0.21900 66871	1.400	0.98544 97300	0.16996 71429

TEN-PLACE SINES AND COSINES FOR RADIAN MEASURE (Continued)

x	$\sin x$	$\cos x$	x	$\sin x$	$\cos x$
1.400	0.98544 97300	0.16996 71429	1.450	0.99271 29910	0.12050 27694
1.401	0.98561 92044	0.16898 16084	1.451	0.99283 29974	0.11950 99963
1.402	0.98578 76931	0.16799 59048	1.452	0.99295 20110	0.11851 71037
1.403	0.98595 51961	0.16701 00333	1.453	0.99307 00316	0.11752 40926
1.404	0.98612 17132	0.16602 39948	1.454	0.99318 70591	0.11653 09640
1.405	0.98628 72441	0.16503 77902	1.455	0.99330 30935	0.11553 77188
1.406	0.98645 17887	0.16405 14206	1.456	0.99341 81345	0.11454 43581
1.407	0.98661 53469	0.16306 48869	1.457	0.99353 21822	0.11355 08829
1.408	0.98677 79184	0.16207 81902	1.458	0.99364 52363	0.11255 72941
1.409	0.98693 95032	0.16109 13314	1.459	0.99375 72967	0.11156 35927
1.410	0.98710 01010	0.16010 43116	1.460	0.99386 83634	0.11056 97798
1.411	0.98725 97117	0.15911 71316	1.461	0.99397 84362	0.10957 58563
1.412	0.98741 83352	0.15812 97925	1.462	0.99408 75151	0.10858 18233
1.413	0.98757 59713	0.15714 22952	1.463	0.99419 55998	0.10758 76816
1.414	0.98773 26198	0.15615 46408	1.464	0.99430 26904	0.10659 34324
1.415	0.98788 82805	0.15516 68303	1.465	0.99440 87867	0.10559 90766
1.416	0.98804 29534	0.15417 88646	1.466	0.99451 38885	0.10460 46152
1.417	0.98819 66382	0.15319 07447	1.467	0.99461 79959	0.10361 00491
1.418	0.98834 93348	0.15220 24717	1.468	0.99472 11086	0.10261 53795
1.419	0.98850 10431	0.15121 40464	1.469	0.99482 32266	0.10162 06073
1.420	0.98865 17629	0.15022 54699	1.470	0.99492 43498	0.10062 57334
1.421	0.98880 14940	0.14923 67432	1.471	0.99502 44780	0.09963 07589
1.422	0.98895 02363	0.14824 78673	1.472	0.99512 36113	0.09863 56848
1.423	0.98909 79897	0.14725 88431	1.473	0.99522 17494	0.09764 05120
1.424	0.98924 47539	0.14626 96716	1.474	0.99531 88923	0.09664 52416
1.425	0.98939 05290	0.14528 03539	1.475	0.99541 50398	0.09564 98745
1.426	0.98953 53146	0.14429 08909	1.476	0.99551 01920	0.09465 44118
1.427	0.98967 91107	0.14330 12836	1.477	0.99560 43486	0.09365 88545
1.428	0.98982 19171	0.14231 15330	1.478	0.99569 75096	0.09266 32035
1.429	0.98996 37337	0.14132 16401	1.479	0.99578 96750	0.09166 74598
1.430	0.99010 45603	0.14033 16058	1.480	0.99588 08445	0.09067 16245
1.431	0.99024 43969	0.13934 14313	1.481	0.99597 10182	0.08967 56984
1.432	0.99038 32432	0.13835 11174	1.482	0.99606 01959	0.08867 96828
1.433	0.99052 10991	0.13736 06651	1.483	0.99614 83775	0.08768 35784
1.434	0.99065 79644	0.13637 00755	1.484	0.99623 55630	0.08668 73863
1.435	0.99079 38392	0.13537 93495	1.485	0.99632 17523	0.08569 11076
1.436	0.99092 87231	0.13438 84882	1.486	0.99640 69452	0.08469 47432
1.437	0.99106 26161	0.13339 74924	1.487	0.99649 11417	0.08369 82940
1.438	0.99119 55180	0.13240 63633	1.488	0.99657 43418	0.08270 17612
1.439	0.99132 74288	0.13141 51017	1.489	0.99665 65452	0.08170 51457
1.440	0.99145 83482	0.13042 37087	1.490	0.99673 77520	0.08070 84485
1.441	0.99158 82761	0.12943 21853	1.491	0.99681 79621	0.07971 16705
1.442	0.99171 72125	0.12844 05325	1.492	0.99689 71754	0.07871 48129
1.443	0.99184 51572	0.12744 87512	1.493	0.99697 53917	0.07771 78765
1.444	0.99197 21100	0.12645 68425	1.494	0.99705 26111	0.07672 08624
1.445	0.99209 80708	0.12546 48074	1.495	0.99712 88334	0.07572 37716
1.446	0.99222 30396	0.12447 26467	1.496	0.99720 40586	0.07472 66051
1.447	0.99234 70161	0.12348 03616	1.497	0.99727 82866	0.07372 93638
1.448	0.99247 00002	0.12248 79530	1.498	0.99735 15173	0.07273 20488
1.449	0.99259 19919	0.12149 54219	1.499	0.99742 37507	0.07173 46611
1.450	0.99271 29910	0.12050 27694	1.500	0.99749 49866	0.07073 72017

TEN-PLACE SINES AND COSINES FOR RADIAN MEASURE (Continued)

x	$\sin x$	$\cos x$
1.500	0.99749 49866	0.07073 72017
1.501	0.99756 52250	0.06973 96715
1.502	0.99763 44659	0.06874 20716
1.503	0.99770 27092	0.06774 44029
1.504	0.99776 99547	0.06674 66665
1.505	0.99783 62025	0.06574 88633
1.506	0.99790 14524	0.06475 09944
1.507	0.99796 57044	0.06375 30607
1.508	0.99802 89585	0.06275 50633
1.509	0.99809 12146	0.06175 70031
1.510	0.99815 24725	0.06075 88812
1.511	0.99821 27323	0.05976 06985
1.512	0.99827 19939	0.05876 24561
1.513	0.99833 02572	0.05776 41549
1.514	0.99838 75222	0.05676 57959
1.515	0.99844 37888	0.05576 73802
1.516	0.99849 90569	0.05476 89087
1.517	0.99855 33266	0.05377 03824
1.518	0.99860 65977	0.05277 18023
1.519	0.99865 88701	0.05177 31695
1.520	0.99871 01440	0.05077 44849
1.521	0.99876 04191	0.04977 57496
1.522	0.99880 96955	0.04877 69644
1.523	0.99885 79730	0.04777 81305
1.524	0.99890 52517	0.04677 92488
1.525	0.99895 15315	0.04578 03203
1.526	0.99899 68123	0.04478 13461
1.527	0.99904 10942	0.04378 23270
1.528	0.99908 43770	0.04278 32642
1.529	0.99912 66607	0.04178 41586
1.530	0.99916 79453	0.04078 50112
1.531	0.99920 82307	0.03978 58231
1.532	0.99924 75169	0.03878 65951
1.533	0.99928 58039	0.03778 73284
1.534	0.99932 30915	0.03678 80238
1.535	0.99935 93799	0.03578 86825
1.536	0.99939 46689	0.03478 93054
1.537	0.99942 89585	0.03378 98935
1.538	0.99946 22487	0.03279 04478
1.539	0.99949 45394	0.03179 09694
1.540	0.99952 58306	0.03079 14591
1.541	0.99955 61223	0.02979 19180
1.542	0.99958 54144	0.02879 23472
1.543	0.99961 37070	0.02779 27475
1.544	0.99964 09999	0.02679 31201
1.545	0.99966 72932	0.02579 34659
1.546	0.99969 25868	0.02479 37858
1.547	0.99971 68808	0.02379 40810
1.548	0.99974 01750	0.02279 43524
1.549	0.99976 24695	0.02179 46010
1.550	0.99978 37642	0.02079 48278

RADIX TABLE FOR CIRCULAR SINES AND COSINES IN RADIANS

x	n	$\sin [x(10^{-n})]$	$\cos [x(10^{-n})]$
1	10	0.00000 00001 00000 00000 00000	0.99999 99999 99999 99999 50000
2	10	0.00000 00002 00000 00000 00000	0.99999 99999 99999 99998 00000
3	10	0.00000 00003 00000 00000 00000	0.99999 99999 99999 99995 50000
4	10	0.00000 00004 00000 00000 00000	0.99999 99999 99999 99992 00000
5	10	0.00000 00005 00000 00000 00000	0.99999 99999 99999 99987 50000
6	10	0.00000 00006 00000 00000 00000	0.99999 99999 99999 99982 00000
7	10	0.00000 00007 00000 00000 00000	0.99999 99999 99999 99975 50000
8	10	0.00000 00008 00000 00000 00000	0.99999 99999 99999 99968 00000
9	10	0.00000 00009 00000 00000 00000	0.99999 99999 99999 99959 50000
1	9	0.00000 00010 00000 00000 00000	0.99999 99999 99999 99950 00000
2	9	0.00000 00020 00000 00000 00000	0.99999 99999 99999 99800 00000
3	9	0.00000 00030 00000 00000 00000	0.99999 99999 99999 99550 00000
4	9	0.00000 00040 00000 00000 00000	0.99999 99999 99999 99200 00000
5	9	0.00000 00050 00000 00000 00000	0.99999 99999 99999 98750 00000
6	9	0.00000 00060 00000 00000 00000	0.99999 99999 99999 98200 00000
7	9	0.00000 00069 99999 99999 99999	0.99999 99999 99999 97550 00000
8	9	0.00000 00079 99999 99999 99999	0.99999 99999 99999 96800 00000
9	9	0.00000 00089 99999 99999 99999	0.99999 99999 99999 95950 00000
1	8	0.00000 00099 99999 99999 99998	0.99999 99999 99999 95000 00000
2	8	0.00000 00199 99999 99999 99987	0.99999 99999 99999 80000 00000
3	8	0.00000 00299 99999 99999 99955	0.99999 99999 99999 55000 00000
4	8	0.00000 00399 99999 99999 99893	0.99999 99999 99999 20000 00000
5	8	0.00000 00499 99999 99999 99792	0.99999 99999 99998 75000 00000
6	8	0.00000 00599 99999 99999 99640	0.99999 99999 99998 20000 00000
7	8	0.00000 00699 99999 99999 99428	0.99999 99999 99997 55000 00000
8	8	0.00000 00799 99999 99999 99147	0.99999 99999 99996 80000 00000
9	8	0.00000 00899 99999 99999 98785	0.99999 99999 99995 95000 00000
1	7	0.00000 00999 99999 99999 98333	0.99999 99999 99995 00000 00000
2	7	0.00000 01999 99999 99999 86667	0.99999 99999 99980 00000 00000
3	7	0.00000 02999 99999 99999 55000	0.99999 99999 99955 00000 00000
4	7	0.00000 03999 99999 99998 93333	0.99999 99999 99920 00000 00000
5	7	0.00000 04999 99999 99997 91667	0.99999 99999 99875 00000 00000
6	7	0.00000 05999 99999 99996 40000	0.99999 99999 99820 00000 00000
7	7	0.00000 06999 99999 99994 28333	0.99999 99999 99755 00000 00000
8	7	0.00000 07999 99999 99991 46667	0.99999 99999 99680 00000 00000
9	7	0.00000 08999 99999 99987 85000	0.99999 99999 99595 00000 00000
1	6	0.00000 09999 99999 99983 33333	0.99999 99999 99500 00000 00000
2	6	0.00000 19999 99999 99866 66667	0.99999 99999 98000 00000 00007
3	6	0.00000 29999 99999 99550 00000	0.99999 99999 95500 00000 00034
4	6	0.00000 39999 99999 98933 33333	0.99999 99999 92000 00000 00107
5	6	0.00000 49999 99999 97916 66667	0.99999 99999 87500 00000 00260
6	6	0.00000 59999 99999 96400 00000	0.99999 99999 82000 00000 00540
7	6	0.00000 69999 99999 94283 33333	0.99999 99999 75500 00000 01000
8	6	0.00000 79999 99999 91466 66667	0.99999 99999 68000 00000 01707
9	6	0.00000 89999 99999 87850 00000	0.99999 99999 59500 00000 02734

For $n > 10$, $\sin [x(10^{-n})] = x(10^{-n})$; $\cos [x(10^{-n})] = 1 - \frac{1}{2}x^2(10^{-2n})$; to 25 decimals.

RADIX TABLE FOR CIRCULAR SINES AND COSINES IN RADIANS

x	n	$\sin [x(10^{-n})]$	$\cos [x(10^{-n})]$
1	5	0.00000 99999 99999 83333 33333	0.99999 99999 50000 00000 04167
2	5	0.00001 99999 99998 66666 66667	0.99999 99998 00000 00000 66667
3	5	0.00002 99999 99995 50000 00002	0.99999 99995 50000 00003 37500
4	5	0.00003 99999 99989 33333 33342	0.99999 99992 00000 00010 66667
5	5	0.00004 99999 99979 16666 66693	0.99999 99987 50000 00026 04167
6	5	0.00005 99999 99964 00000 00065	0.99999 99982 00000 00054 00000
7	5	0.00006 99999 99942 83333 33473	0.99999 99975 50000 00100 04167
8	5	0.00007 99999 99914 66666 66940	0.99999 99968 00000 00170 66667
9	5	0.00008 99999 99878 50000 00492	0.99999 99959 50000 00273 37500
1	4	0.00009 99999 99833 33333 34167	0.99999 99950 00000 00416 66667
2	4	0.00019 99999 98666 66666 93333	0.99999 99800 00000 06666 66666
3	4	0.00029 99999 95500 00002 02500	0.99999 99550 00000 33749 99990
4	4	0.00039 99999 89333 33341 86667	0.99999 99200 00001 06666 66610
5	4	0.00049 99999 79166 66692 70833	0.99999 98750 00002 60416 66450
6	4	0.00059 99999 64000 00064 80000	0.99999 98200 00005 39999 99352
7	4	0.00069 99999 42833 33473 39167	0.99999 97550 00010 00416 65033
8	4	0.00079 99999 14666 66939 73333	0.99999 96800 00017 06666 63026
9	4	0.00089 99998 78500 00492 07499	0.99999 95950 00027 33749 92619
1	3	0.00099 99998 33333 34166 66665	0.99999 95000 00041 66666 52778

For $n > 10$, $\sin [x(10^{-n})] = x(10^{-n})$; $\cos [x(10^{-n})] = 1 - \frac{1}{2}x^2(10^{-2n})$; to 25 decimals.

TEN-PLACE SINES AND COSINES FOR LARGE RADIAN MEASURES

x	sin x	cos x	x	sin x	cos x
0	0.00000 00000	1.00000 00000	50	−0.26237 48537	+0.96496 60285
1	+0.84147 09848	0.54030 23059	51	+0.67022 91758	+0.74215 41968
2	+0.90929 74268	−0.41614 68365	52	+0.98662 75920	−0.16299 07808
3	+0.14112 00081	−0.98999 24966	53	+0.39592 51502	−0.91828 27862
4	−0.75680 24953	−0.65364 36209	54	−0.55878 90489	−0.82930 98329
5	−0.95892 42747	+0.28366 21855	55	−0.99975 51734	+0.02212 67563
6	−0.27941 54982	+0.96017 02867	56	−0.52155 10021	+0.85322 01077
7	+0.65698 65987	+0.75390 22543	57	+0.43616 47552	+0.89986 68270
8	+0.98935 82466	−0.14550 00338	58	+0.99287 26481	+0.11918 01354
9	+0.41211 84852	−0.91113 02619	59	+0.63673 80071	−0.77108 02230
10	−0.54402 11109	−0.83907 15291	60	−0.30481 06211	−0.95241 29804
11	−0.99999 02066	+0.00442 56980	61	−0.96611 77700	−0.25810 16359
12	−0.53657 29180	+0.84385 39587	62	−0.73918 06966	+0.67350 71623
13	+0.42016 70368	+0.90744 67815	63	+0.16735 57003	+0.98589 65816
14	+0.99060 73557	+0.13673 72182	64	+0.92002 60382	+0.39185 72304
15	+0.65028 78402	−0.75968 79129	65	+0.82682 86795	−0.56245 38512
16	−0.28790 33167	−0.95765 94803	66	−0.02655 11540	−0.99964 74560
17	−0.96139 74919	−0.27516 33381	67	−0.85551 99790	−0.51776 97998
18	−0.75098 72468	+0.66031 67082	68	−0.89792 76807	+0.44014 30225
19	+0.14987 72097	+0.98870 46182	69	−0.11478 48138	+0.99339 03797
20	+0.91294 52507	+0.40808 20618	70	+0.77389 06816	+0.63331 92031
21	+0.83665 56385	−0.54772 92602	71	+0.95105 46533	−0.30902 27282
22	−0.00885 13093	−0.99996 08264	72	+0.25382 33628	−0.96725 05883
23	−0.84622 04042	−0.53283 30203	73	−0.67677 19569	−0.73619 27182
24	−0.90557 83620	+0.42417 90073	74	−0.98514 62605	+0.17171 73418
25	−0.13235 17501	+0.99120 28119	75	−0.38778 16354	+0.92175 12697
26	+0.76255 84505	+0.64691 93223	76	+0.56610 76369	+0.82433 13311
27	+0.95637 59284	−0.29213 88087	77	+0.99952 01586	−0.03097 50317
28	+0.27090 57883	−0.96260 58663	78	+0.51397 84560	−0.85780 30932
29	−0.66363 38842	−0.74805 75297	79	−0.44411 26687	−0.89597 09468
30	−0.98803 16241	+0.15425 14499	80	−0.99388 86539	−0.11038 72438
31	−0.40403 76453	+0.91474 23578	81	−0.62988 79943	+0.77668 59820
32	+0.55142 66812	+0.83422 33605	82	+0.31322 87824	+0.94967 76979
33	+0.99991 18601	−0.01327 67472	83	+0.96836 44611	+0.24954 01180
34	+0.52908 26861	−0.84857 02748	84	+0.73319 03201	−0.68002 34956
35	−0.42818 26695	−0.90369 22051	85	−0.17607 56199	−0.98437 66434
36	−0.99177 88534	−0.12796 36896	86	−0.92345 84470	−0.38369 84449
37	−0.64353 81334	+0.76541 40519	87	−0.82181 78366	+0.56975 03343
38	+0.29636 85787	+0.95507 36440	88	+0.03539 83027	+0.99937 32837
39	+0.96379 53863	+0.26664 29324	89	+0.86006 94058	+0.51017 70449
40	+0.74511 31605	−0.66693 80617	90	+0.89399 66636	−0.44807 36161
41	−0.15862 26688	−0.98733 92775	91	+0.10598 75118	−0.99436 74609
42	−0.91652 15479	−0.39998 53150	92	−0.77946 60696	−0.62644 44479
43	−0.83177 47426	+0.55511 33015	93	−0.94828 21413	+0.31742 87015
44	+0.01770 19251	+0.99984 33086	94	−0.24525 19855	+0.96945 93667
45	+0.85090 35245	+0.52532 19888	95	+0.68326 17147	+0.73017 35610
46	+0.90178 83476	−0.43217 79449	96	+0.98358 77454	−0.18043 04493
47	+0.12357 31227	−0.99233 54692	97	+0.37960 77390	−0.92514 75366
48	−0.76825 46613	−0.64014 43395	98	−0.57338 18720	−0.81928 82453
49	−0.95375 26528	+0.30059 25437	99	−0.99920 68342	+0.03982 08804
50	−0.26237 48537	+0.96496 60285	100	−0.50636 56411	+0.86231 88723

TEN-PLACE SINES AND COSINES FOR LARGE RADIAN MEASURES

x	sin x	cos x	x	sin x	cos x
100	−0.50636 564	+0.86231 887	150	−0.71487 643	+0.69925 081
101	+0.45202 579	+0.89200 487	151	+0.20214 988	+0.97935 460
102	+0.99482 679	+0.10158 570	152	+0.93332 052	+0.35904 429
103	+0.62298 863	−0.78223 089	153	+0.80640 058	−0.59136 968
104	−0.32162 240	−0.94686 801	154	−0.06192 034	−0.99808 109
105	−0.97053 528	−0.24095 905	155	−0.87331 198	−0.48716 135
106	−0.72714 250	+0.68648 655	156	−0.88178 462	+0.47165 229
107	+0.18478 174	+0.98277 958	157	−0.07954 854	+0.99683 099
108	+0.92681 851	+0.37550 960	158	+0.79582 410	+0.60552 787
109	+0.81674 261	−0.57700 218	159	+0.93951 973	−0.34249 478
110	−0.04424 268	−0.99902 081	160	+0.21942 526	−0.97562 931
111	−0.86455 145	−0.50254 432	161	−0.70240 779	−0.71177 476
112	−0.88999 560	+0.45596 910	162	−0.97845 035	+0.20648 223
113	−0.09718 191	+0.99526 664	163	−0.35491 018	+0.93490 040
114	+0.78498 039	+0.61952 061	164	+0.59493 278	+0.80377 546
115	+0.94543 533	−0.32580 981	165	+0.99779 728	−0.06633 694
116	+0.23666 139	−0.97159 219	166	+0.48329 156	−0.87545 946
117	−0.68969 794	−0.72409 720	167	−0.47555 019	−0.87968 859
118	−0.98195 217	+0.18912 942	168	−0.99717 329	−0.07513 609
119	−0.37140 410	+0.92847 132	169	−0.60199 987	+0.79849 619
120	+0.58061 118	+0.81418 097	170	+0.34664 946	+0.93799 475
121	+0.99881 522	−0.04866 361	171	+0.97659 087	+0.21510 527
122	+0.49871 315	−0.86676 709	172	+0.70865 914	−0.70555 101
123	−0.45990 349	−0.88796 891	173	−0.21081 053	−0.97752 694
124	−0.99568 699	−0.09277 620	174	−0.93646 197	−0.35076 911
125	−0.61604 046	+0.78771 451	175	−0.80113 460	+0.59848 422
126	+0.32999 083	+0.94398 414	176	+0.07075 224	+0.99749 392
127	+0.97263 007	+0.23235 910	177	+0.87758 979	+0.47941 231
128	+0.72103 771	−0.69289 582	178	+0.87757 534	−0.47943 877
129	−0.19347 339	−0.98110 552	179	+0.07072 217	−0.99749 605
130	−0.93010 595	−0.36729 133	180	−0.80115 264	−0.59846 007
131	−0.81160 339	+0.58420 882	181	−0.93645 140	+0.35079 734
132	+0.05308 359	+0.99859 007	182	−0.21078 107	+0.97753 329
133	+0.86896 576	+0.49487 222	183	+0.70868 041	+0.70552 964
134	+0.88592 482	−0.46382 887	184	+0.97658 438	−0.21513 471
135	+0.08836 869	−0.99608 784	185	+0.34662 118	−0.93800 520
136	−0.79043 321	−0.61254 824	186	−0.60202 394	−0.79847 804
137	−0.94251 445	+0.33416 538	187	−0.99717 102	+0.07516 615
138	−0.22805 226	+0.97364 889	188	−0.47552 367	+0.87970 293
139	+0.69608 013	+0.71796 410	189	+0.48331 795	+0.87544 489
140	+0.98023 966	−0.19781 357	190	+0.99779 928	+0.06630 686
141	+0.36317 137	−0.93172 236	191	+0.59490 855	−0.80379 339
142	−0.58779 501	−0.80900 991	192	−0.35493 836	−0.93488 971
143	−0.99834 536	+0.05750 253	193	−0.97845 657	−0.20645 273
144	−0.49102 159	+0.87114 740	194	−0.70238 633	+0.71179 593
145	+0.46774 516	+0.88386 337	195	+0.21945 467	+0.97562 270
146	+0.99646 917	+0.08395 944	196	+0.93953 006	+0.34246 646
147	+0.60904 402	−0.79313 642	197	+0.79580 584	−0.60555 186
148	−0.33833 339	−0.94102 631	198	−0.07957 859	−0.99682 859
149	−0.97464 865	−0.22374 095	199	−0.88179 884	−0.47162 571
150	−0.71487 643	+0.69925 081	200	−0.87329 730	+0.48718 768

TEN-PLACE SINES AND COSINES FOR LARGE RADIAN MEASURES

x	$\sin x$	$\cos x$	x	$\sin x$	$\cos x$
200	−0.87329 730	+0.48718 768	250	−0.97052 802	+0.24098 831
201	−0.06189 025	+0.99808 296	251	−0.32159 386	+0.94687 771
202	+0.80641 841	+0.59134 538	252	+0.62301 221	+0.78221 211
203	+0.93330 970	−0.35907 242	253	+0.99482 373	−0.10161 569
204	+0.20212 036	−0.97936 069	254	+0.45199 890	−0.89201 850
205	−0.71489 751	−0.69922 926	255	−0.50639 163	−0.86230 361
206	−0.97464 190	+0.22377 033	256	−0.99920 803	−0.03979 076
207	−0.33830 503	+0.94103 651	257	−0.57335 717	+0.81930 553
208	+0.60906 793	+0.79311 806	258	+0.37963 563	+0.92513 609
209	+0.99646 664	−0.08398 947	259	+0.98359 318	+0.18040 080
210	+0.46771 852	−0.88387 747	260	+0.68323 970	−0.73019 416
211	−0.49104 785	−0.87113 260	261	−0.24528 121	−0.96945 197
212	−0.99834 709	−0.05747 243	262	−0.94829 171	−0.31740 012
213	−0.58777 062	+0.80902 763	263	−0.77944 719	+0.62646 794
214	+0.36319 945	+0.93171 141	264	+0.10601 749	+0.99436 427
215	+0.98024 562	+0.19778 403	265	+0.89401 017	+0.44804 667
216	+0.69605 849	−0.71798 508	266	+0.86005 403	−0.51020 297
217	−0.22808 161	−0.97364 202	267	+0.03536 818	−0.99937 435
218	−0.94252 453	−0.33413 697	268	−0.82183 501	−0.56972 556
219	−0.79041 474	+0.61257 207	269	−0.92344 688	+0.38372 628
220	+0.08839 871	+0.99608 517	270	−0.17604 595	+0.98438 195
221	+0.88593 880	+0.46380 216	271	+0.73321 082	+0.68000 139
222	+0.86895 084	−0.49489 841	272	+0.96835 694	−0.24956 931
223	+0.05305 349	−0.99859 167	273	+0.31320 015	−0.94968 714
224	−0.81162 100	−0.58418 435	274	−0.62991 141	−0.77666 699
225	−0.93009 488	+0.36731 937	275	−0.99388 533	+0.11041 720
226	−0.19344 382	+0.98111 135	276	−0.44408 566	+0.89598 433
227	+0.72105 860	+0.69287 409	277	+0.51400 431	+0.85778 760
228	+0.97262 306	−0.23238 842	278	+0.99952 109	+0.03094 490
229	+0.32996 237	−0.94399 409	279	+0.56608 279	−0.82434 840
230	−0.61606 420	−0.78769 594	280	−0.38780 942	−0.92173 958
231	−0.99568 419	+0.09280 622	281	−0.98515 144	−0.17168 765
232	−0.45987 672	+0.88798 277	282	−0.67674 976	+0.73621 312
233	+0.49873 928	+0.86675 206	283	+0.25385 252	+0.96724 294
234	+0.99881 669	+0.04863 350	284	+0.95106 397	+0.30899 406
235	+0.58058 664	−0.81419 847	285	+0.77387 159	−0.63334 253
236	−0.37143 209	−0.92846 012	286	−0.11481 476	−0.99338 692
237	−0.98195 787	−0.18909 982	287	−0.89794 095	−0.44011 595
238	−0.68967 611	+0.72411 799	288	−0.85550 437	+0.51779 559
239	+0.23669 068	+0.97158 506	289	−0.02652 102	+0.99964 826
240	+0.94544 515	+0.32578 131	290	+0.82684 563	+0.56242 893
241	+0.78496 171	−0.61954 428	291	+0.92001 423	−0.39188 496
242	−0.09721 191	−0.99526 371	292	+0.16732 598	−0.98590 163
243	−0.89000 935	−0.45594 228	293	−0.73920 100	−0.67348 488
244	−0.86453 630	+0.50257 038	294	−0.96610 999	+0.25813 076
245	−0.04421 256	+0.99902 215	295	−0.30478 191	+0.95242 217
246	+0.81676 000	+0.57697 756	296	+0.63676 125	+0.77106 103
247	+0.92680 719	−0.37553 754	297	+0.99286 906	−0.11921 006
248	+0.18475 212	−0.98278 515	298	+0.43613 763	−0.89987 997
249	−0.72716 319	−0.68646 463	299	−0.52157 672	−0.85320 439
250	−0.97052 802	+0.24098 831	300	−0.99975 584	−0.02209 662

TEN-PLACE SINES AND COSINES FOR LARGE RADIAN MEASURES

x	$\sin x$	$\cos x$	x	$\sin x$	$\cos x$
300	−0.99975 584	−0.02209 662	350	−0.95893 283	−0.28363 328
301	−0.55876 405	+0.82932 668	351	−0.75678 279	+0.65366 643
302	+0.39595 283	+0.91827 085	352	+0.14114 985	+0.98998 824
303	+0.98663 250	+0.16296 104	353	+0.90930 997	+0.41611 943
304	+0.67020 680	−0.74217 440	354	+0.84145 470	−0.54032 767
305	−0.26240 394	−0.96495 812	355	−0.00003 014	−1.00000 000
306	−0.95376 171	−0.30056 379	356	−0.84148 727	−0.54027 694
307	−0.76823 536	+0.64016 750	357	−0.90928 488	+0.41617 425
308	+0.12360 304	+0.99233 174	358	−0.14109 017	+0.98999 675
309	+0.90180 137	+0.43215 076	359	+0.75682 220	+0.65362 081
310	+0.85088 769	−0.52534 764	360	+0.95891 572	−0.28369 109
311	+0.01767 179	−0.99984 384	361	+0.27938 655	−0.96017 871
312	−0.83179 148	−0.55508 823	362	−0.65700 932	−0.75388 245
313	−0.91650 949	+0.40001 294	363	−0.98935 386	+0.14552 986
314	−0.15859 291	+0.98734 406	364	−0.41209 102	+0.91114 268
315	+0.74513 326	+0.66691 560	365	+0.54404 640	+0.83905 513
316	+0.96378 735	−0.26667 199	366	+0.99999 007	−0.00445 584
317	+0.29633 979	−0.95508 258	367	+0.53654 748	−0.84387 013
318	−0.64356 121	−0.76539 465	368	−0.42019 439	−0.90743 412
319	−0.99177 500	+0.12799 359	369	−0.99061 148	−0.13670 736
320	−0.42815 543	+0.90370 511	370	−0.65026 494	+0.75970 752
321	+0.52910 827	+0.84855 433	371	+0.28793 218	+0.95765 080
322	+0.99991 226	+0.01324 661	372	+0.96140 579	+0.27513 436
323	+0.55140 153	−0.83423 998	373	+0.75096 734	−0.66033 935
324	−0.40406 522	−0.91473 018	374	−0.14990 701	−0.98870 010
325	−0.98803 627	−0.15422 167	375	−0.91295 755	−0.40805 454
326	−0.66361 133	+0.74807 753	376	−0.83663 913	+0.54775 448
327	+0.27093 481	+0.96259 770	377	+0.00888 145	+0.99996 056
328	+0.95638 473	+0.29210 998	378	+0.84623 647	+0.53280 751
329	+0.76253 895	−0.64694 231	379	+0.90556 557	−0.42420 631
330	−0.13238 163	−0.99119 882	380	+0.13232 187	−0.99120 680
331	−0.90559 115	−0.42415 171	381	−0.76257 795	−0.64689 634
332	−0.84620 434	+0.53285 853	382	−0.95636 712	+0.29216 764
333	−0.00882 117	+0.99996 109	383	−0.27087 677	+0.96261 403
344	+0.83667 215	+0.54770 404	384	+0.66365 643	+0.74803 752
335	+0.91293 295	−0.40810 958	385	+0.98802 697	−0.15428 123
336	+0.14984 741	−0.98870 914	386	+0.40401 007	−0.91475 454
337	−0.75100 715	−0.66029 407	387	−0.55145 183	−0.83420 674
338	−0.96138 920	+0.27519 232	388	−0.99991 146	+0.01330 689
339	−0.28787 445	+0.95766 816	389	−0.52905 711	+0.84858 622
340	+0.65031 074	+0.75966 831	390	+0.42820 991	+0.90367 930
341	+0.99060 323	−0.13676 708	391	+0.99178 271	+0.12793 379
342	+0.42013 968	−0.90745 945	392	+0.64351 506	−0.76543 345
343	−0.53659 836	−0.84383 778	393	−0.29639 737	−0.95506 471
344	−0.99999 034	−0.00439 555	394	−0.96380 342	−0.26661 388
345	−0.54399 582	+0.83908 793	395	−0.74509 306	+0.66696 052
346	+0.41214 595	+0.91111 784	396	+0.15865 243	+0.98733 450
347	+0.98936 263	+0.14547 021	397	+0.91653 361	+0.39995 769
348	+0.65696 387	−0.75392 206	398	+0.83175 801	−0.55513 837
349	−0.27944 444	−0.96016 186	399	−0.17773 206	−0.99984 277
350	−0.95893 283	−0.28363 328	400	−0.85091 936	−0.52529 634

TEN-PLACE SINES AND COSINES FOR LARGE RADIAN MEASURES

x	$\sin x$	$\cos x$	x	$\sin x$	$\cos x$
400	−0.85091 936	−0.52529 634	450	−0.68328 373	−0.73015 296
401	−0.90177 532	+0.43220 513	451	−0.98358 231	+0.18046 010
402	−0.12354 321	+0.99233 919	452	−0.37957 985	+0.92515 898
403	+0.76827 396	+0.64012 118	453	+0.57340 657	+0.81927 096
404	+0.95374 359	−0.30062 129	454	+0.99920 563	−0.03985 100
405	+0.26234 577	−0.96497 394	455	+0.50633 965	−0.86233 414
406	−0.67025 155	−0.74213 399	456	−0.45205 268	−0.89199 124
407	−0.98662 268	+0.16302 052	457	−0.99482 985	−0.10155 572
408	−0.39589 747	+0.91829 472	458	−0.62296 505	+0.78224 967
409	+0.55881 405	+0.82929 299	459	+0.32165 095	+0.94685 832
410	+0.99975 451	−0.02215 689	460	+0.97054 255	+0.24092 979
411	+0.52152 528	−0.85323 583	461	+0.72712 181	−0.68650 847
412	−0.43619 188	−0.89985 368	462	−0.18481 137	−0.98277 401
413	−0.99287 624	−0.11915 021	463	−0.92682 982	−0.37548 166
414	−0.63671 476	+0.77109 942	464	−0.81672 521	+0.57702 680
415	+0.30483 933	+0.95240 379	465	+0.04427 279	+0.99901 948
416	+0.96612 555	+0.25807 251	466	+0.86456 660	+0.50251 826
417	+0.73916 039	−0.67352 944	467	+0.88998 186	−0.45599 593
418	−0.16738 542	−0.98589 154	468	+0.09715 190	−0.99526 957
419	−0.92003 785	−0.39182 950	469	−0.78499 906	−0.61949 695
420	−0.82681 172	+0.56247 878	470	−0.94542 551	+0.32583 830
421	+0.02658 129	+0.99964 666	471	−0.23663 211	+0.97159 932
422	+0.85553 559	+0.51774 401	472	+0.68971 977	+0.72407 641
423	+0.89791 441	−0.44017 009	473	+0.98194 647	−0.18915 902
424	+0.11475 487	−0.99339 384	474	+0.37137 611	−0.92848 252
425	−0.77390 977	−0.63329 587	475	−0.58063 573	−0.81416 347
426	−0.95104 534	+0.30905 140	476	−0.99881 376	+0.04869 372
427	−0.25379 421	+0.96725 824	477	−0.49868 703	+0.86678 212
428	+0.67679 415	+0.73617 232	478	+0.45993 026	+0.88795 504
429	+0.98514 108	−0.17174 704	479	+0.99568 978	+0.09274 619
430	+0.38775 385	−0.92176 296	480	+0.61601 671	−0.78773 308
431	−0.56613 249	−0.82431 427	481	−0.33001 928	−0.94397 419
432	−0.99951 922	+0.03100 516	482	−0.97263 707	−0.23232 978
433	−0.51395 260	+0.85781 859	483	−0.72101 682	+0.69291 756
434	+0.44413 968	+0.89595 756	484	+0.19350 297	+0.98109 969
435	+0.99389 198	+0.11035 728	485	+0.93011 702	+0.36726 329
436	+0.62986 458	−0.77670 497	486	+0.81158 578	−0.58423 328
437	−0.31325 741	−0.94966 826	487	−0.05311 369	−0.99858 847
438	−0.96837 198	−0.24951 093	488	−0.86898 067	−0.49484 603
439	−0.73316 982	+0.68004 560	489	−0.88591 083	+0.46385 557
440	+0.17610 529	+0.98437 134	490	−0.08833 866	+0.99609 050
441	+0.92347 001	+0.38367 061	491	+0.79045 167	+0.61252 441
442	+0.82180 066	−0.56977 511	492	+0.94250 438	−0.33419 379
443	−0.03542 843	−0.99937 222	493	+0.22802 291	−0.97365 577
444	−0.86008 478	−0.51015 112	494	−0.69610 177	−0.71794 312
445	−0.89398 316	+0.44810 056	495	−0.98023 370	+0.19784 312
446	−0.10595 754	+0.99437 066	496	−0.36314 328	+0.93173 331
447	+0.77948 495	+0.62642 095	497	+0.58781 939	+0.80899 219
448	+0.94827 257	−0.31745 729	498	+0.99834 363	−0.05753 262
449	+0.24522 276	−0.96946 676	499	+0.49099 533	−0.87116 220
450	−0.68328 373	−0.73015 296	500	−0.46777 181	−0.88384 927

TEN-PLACE SINES AND COSINES FOR LARGE RADIAN MEASURES

x	$\sin x$	$\cos x$	x	$\sin x$	$\cos x$
500	−0.46777 181	−0.88384 927	550	−0.21948 408	−0.97561 608
501	−0.99647 170	−0.08392 940	551	−0.93954 038	−0.34243 814
502	−0.60902 011	+0.79315 478	552	−0.79578 759	+0.60557 585
503	+0.33836 176	+0.94101 611	553	+0.07960 864	+0.99682 620
504	+0.97465 539	+0.22371 157	554	+0.88181 305	+0.47159 913
505	+0.71485 535	−0.69927 236	555	+0.87328 261	−0.48721 400
506	−0.20217 940	−0.97934 850	556	+0.06186 016	−0.99808 483
507	−0.93333 135	−0.35901 615	557	−0.80643 623	−0.59132 107
508	−0.80638 275	+0.59139 399	558	−0.93329 888	+0.35910 055
509	+0.06195 042	+0.99807 923	559	−0.20209 084	+0.97936 678
510	+0.87332 667	+0.48713 502	560	+0.71491 859	+0.69920 771
511	+0.88177 040	−0.47167 887	561	+0.97463 516	−0.22379 971
512	+0.07951 849	−0.99683 339	562	+0.33827 666	−0.94104 671
513	−0.79584 235	−0.60550 389	563	−0.60909 184	−0.79309 970
514	−0.93950 941	+0.34252 310	564	−0.99646 411	+0.08401 951
515	−0.21939 585	+0.97563 593	565	−0.46769 187	+0.88389 157
516	+0.70242 924	+0.71175 358	566	+0.49107 411	+0.87111 780
517	+0.97844 413	−0.20651 172	567	+0.99834 883	+0.05744 234
518	+0.35488 199	−0.93491 110	568	+0.58774 623	−0.80904 534
519	−0.59495 701	−0.80375 753	569	−0.36322 754	−0.93170 046
520	−0.99779 528	+0.06636 701	570	−0.98025 158	−0.19775 448
521	−0.48326 517	+0.87547 403	571	−0.69603 684	+0.71800 607
522	+0.47557 670	+0.87967 426	572	+0.22811 096	+0.97363 514
523	+0.99717 555	+0.07510 603	573	+0.94253 460	+0.33410 856
524	+0.60197 580	−0.79851 433	574	+0.79039 628	−0.61259 589
525	−0.34667 773	−0.93798 430	575	−0.08842 874	−0.99608 251
526	−0.97659 735	−0.21507 583	576	−0.88595 278	−0.46377 546
527	−0.70863 787	+0.70557 237	577	−0.86893 592	+0.49492 461
528	+0.21084 000	+0.97752 059	578	−0.05302 338	+0.99859 327
529	+0.93647 255	+0.35074 088	579	+0.81163 861	+0.58415 989
530	+0.80111 655	−0.59850 837	580	+0.93008 380	−0.36734 740
531	−0.07078 230	−0.99749 179	581	+0.19341 424	−0.98111 719
532	−0.87760 424	−0.47938 586	582	−0.72107 948	−0.69285 235
533	−0.87756 088	+0.47946 522	583	−0.97261 606	+0.23241 774
534	−0.07069 210	+0.99749 818	584	−0.32993 391	+0.94400 403
535	+0.80177 068	+0.59843 592	585	+0.61608 795	+0.78767 737
536	+0.93644 083	−0.35082 557	586	+0.99568 139	−0.09283 623
537	+0.21075 160	−0.97753 965	587	+0.45984 996	−0.88799 663
538	−0.70870 168	−0.70550 828	588	−0.49876 541	−0.86673 702
539	−0.97657 790	+0.21516 415	589	−0.99881 816	−0.04860 339
540	−0.34659 290	+0.93801 565	590	−0.58056 210	+0.81421 597
541	+0.60204 801	+0.79845 989	591	+0.37146 008	+0.92844 893
542	+0.99716 876	−0.07519 621	592	+0.98196 357	+0.18907 022
543	+0.47549 715	−0.87971 726	593	+0.68965 428	+0.72413 878
544	−0.48334 434	−0.87543 032	594	−0.23671 997	−0.97157 792
545	−0.99780 128	−0.06627 678	595	−0.94545 497	−0.32575 281
546	−0.59488 432	+0.80381 133	596	−0.78494 304	+0.61956 794
547	+0.35496 654	+0.93487 901	597	+0.09724 191	+0.99526 078
548	+0.97846 280	+0.20642 324	598	+0.89002 309	+0.45591 545
549	+0.70236 487	−0.71181 710	599	+0.86452 115	−0.50259 644
550	−0.21948 408	−0.97561 608	600	+0.04418 245	−0.99902 348

TEN-PLACE SINES AND COSINES FOR LARGE RADIAN MEASURES

x	sin x	cos x	x	sin x	cos x
600	+0.04418 245	−0.99902 348	650	+0.30475 320	−0.95243 136
601	−0.81677 739	−0.57695 294	651	−0.63678 449	−0.77104 183
602	−0.92679 586	+0.37556 547	652	−0.99286 546	+0.11923 999
603	−0.18472 249	+0.98279 072	653	−0.43611 050	+0.89989 312
604	+0.72718 389	+0.68644 271	654	+0.52160 244	+0.85318 866
605	+0.97052 075	−0.24101 756	655	+0.99975 651	+0.02206 648
606	+0.32156 532	−0.94688 740	656	+0.55873 905	−0.82934 352
607	−0.62303 579	−0.78219 333	657	−0.39598 051	−0.91825 891
608	−0.99482 067	+0.10164 568	658	−0.98663 742	−0.16293 130
609	−0.45197 201	+0.89203 212	659	−0.67018 443	+0.74219 460
610	+0.50641 763	+0.86228 834	660	+0.26243 303	+0.96495 021
611	+0.99920 923	+0.03976 064	661	+0.95377 077	+0.30053 504
612	+0.57333 248	−0.81932 281	662	+0.76821 607	−0.64019 066
613	−0.37966 351	−0.92512 465	663	−0.12363 295	−0.99232 802
614	−0.98359 862	−0.18037 115	664	−0.90181 440	−0.43212 358
615	−0.68321 769	+0.73021 475	665	−0.85087 185	+0.52537 329
616	+0.24531 043	+0.96944 458	666	−0.01764 165	+0.99984 437
617	+0.94830 128	+0.31737 153	667	+0.83180 821	+0.55506 315
618	+0.77942 830	−0.62649 144	668	+0.91649 743	−0.40004 057
619	−0.10604 746	−0.99436 107	669	+0.15856 314	−0.98734 884
620	−0.89402 368	−0.44801 972	670	−0.74515 337	−0.66689 314
621	−0.86003 865	+0.51022 890	671	−0.96377 931	+0.26670 104
622	−0.03533 805	+0.99937 542	672	−0.29631 100	+0.95509 151
623	+0.82185 218	+0.56970 079	673	+0.64358 428	+0.76537 525
624	+0.92343 531	−0.38375 412	674	+0.99177 114	−0.12802 348
625	+0.17601 627	−0.98438 726	675	+0.42812 819	−0.90371 802
626	−0.73323 132	−0.67997 929	676	−0.52913 384	−0.84853 838
627	−0.96834 941	+0.24959 850	677	−0.99991 266	−0.01321 646
628	−0.31317 153	+0.94969 658	678	−0.55137 639	+0.83425 660
629	+0.62993 482	+0.77664 801	679	+0.40409 279	+0.91471 800
630	+0.99388 200	−0.11044 716	680	+0.98804 092	+0.15419 188
631	+0.44405 865	−0.89599 772	681	+0.66358 878	−0.74809 754
632	−0.51403 017	−0.85777 210	682	−0.27096 382	−0.96258 953
633	−0.99952 202	−0.03091 477	683	−0.95639 354	−0.29208 115
634	−0.56605 794	+0.82436 546	684	−0.76251 945	+0.64696 529
635	+0.38783 721	+0.92172 789	685	+0.13241 151	+0.99119 483
636	+0.98515 661	+0.17165 795	686	+0.90560 393	+0.42412 441
637	+0.67672 757	−0.73623 352	687	+0.84618 828	−0.53288 404
638	−0.25388 168	−0.96723 528	688	+0.00879 102	−0.99996 136
639	−0.95107 328	−0.30896 539	689	−0.83668 866	−0.54767 882
640	−0.77385 250	+0.63336 586	690	−0.91292 065	+0.40813 710
641	+0.11484 470	+0.99338 346	691	−0.14981 760	+0.98871 365
642	+0.89795 421	+0.44008 889	692	+0.75102 706	+0.66027 143
643	+0.85548 876	−0.51782 138	693	+0.96138 090	−0.27522 130
644	+0.02649 089	−0.99964 905	694	+0.28784 558	−0.95767 684
645	−0.82686 259	−0.56240 400	695	−0.65033 364	−0.75964 871
646	−0.92000 241	+0.39191 270	696	−0.99059 911	+0.13679 694
647	−0.16729 626	+0.98590 667	697	−0.42011 233	+0.90747 211
648	+0.73922 130	+0.67346 260	698	+0.53662 379	+0.84382 161
649	+0.96610 221	−0.25815 988	699	+0.99999 047	+0.00436 541
650	+0.30475 320	−0.95243 136	700	+0.54397 052	−0.83910 433

TEN-PLACE SINES AND COSINES FOR LARGE RADIAN MEASURES

x	$\sin x$	$\cos x$	x	$\sin x$	$\cos x$
700	+0.54397 052	−0.83910 433	750	+0.74507 295	−0.66698 298
701	−0.41217 342	−0.91110 541	751	−0.15868 219	−0.98732 971
702	−0.98936 702	−0.14544 039	752	−0.91654 566	−0.39993 006
703	−0.65694 115	+0.75394 186	753	−0.83174 127	+0.55516 345
704	+0.27947 339	+0.96015 344	754	+0.01776 220	+0.99984 224
705	+0.95894 137	+0.28360 437	755	+0.85093 519	+0.52527 069
706	+0.75676 309	−0.65368 925	756	+0.90176 229	−0.43223 231
707	−0.14117 969	−0.98998 399	757	+0.12351 330	−0.99234 292
708	−0.90932 251	−0.41609 202	758	−0.76829 325	−0.64009 802
709	−0.84143 841	+0.54035 304	759	−0.95373 453	+0.30065 004
710	+0.00006 029	+1.00000 000	760	−0.26231 668	+0.96498 184
711	+0.84150 356	+0.54025 157	761	+0.67027 392	+0.74211 379
712	+0.90927 234	−0.41620 166	762	+0.98661 776	−0.16305 026
713	+0.14106 032	−0.99000 100	763	+0.39586 979	−0.91830 665
714	−0.75684 190	−0.65359 799	764	−0.55883 905	−0.82927 614
715	−0.95890 717	+0.28372 000	765	−0.99975 384	+0.02218 703
716	−0.27935 761	+0.96018 713	766	−0.52149 956	+0.85325 155
717	+0.65703 205	+0.75386 264	767	+0.43621 901	+0.89984 053
718	+0.98934 947	−0.14555 968	768	+0.99287 983	+0.11912 028
719	+0.41206 355	−0.91115 511	769	+0.63669 152	−0.77111 861
720	−0.54407 170	−0.83903 873	770	−0.30486 804	−0.95239 460
721	−0.99998 994	+0.00448 599	771	−0.96613 333	−0.25804 339
722	−0.53652 204	+0.84388 631	772	−0.73914 009	+0.67355 173
723	+0.42022 174	+0.90742 145	773	+0.16741 514	+0.98588 649
724	+0.99061 560	+0.13667 750	774	+0.92004 966	+0.39180 176
725	+0.65024 204	−0.75972 712	775	+0.82679 477	−0.56250 370
726	−0.28796 105	−0.95764 212	776	−0.02661 142	−0.99964 585
727	−0.96141 408	−0.27510 538	777	−0.85555 119	−0.51771 822
728	−0.75094 744	+0.66036 198	778	−0.89790 114	+0.44019 716
729	+0.14993 682	+0.98869 558	779	−0.11472 492	+0.99339 730
730	+0.91296 985	+0.40802 702	780	+0.77392 886	+0.63327 255
731	+0.83662 262	−0.54777 970	781	+0.95103 602	−0.30908 007
732	−0.00891 160	−0.99996 029	782	+0.25376 505	−0.96726 589
733	−0.84625 253	−0.53278 200	783	−0.67681 634	−0.73615 192
734	−0.90555 279	+0.42423 360	784	−0.98513 591	+0.17177 673
735	−0.13229 199	+0.99121 079	785	−0.38772 606	+0.92177 465
736	+0.76259 745	+0.64687 335	786	+0.56615 733	+0.82429 720
737	+0.95635 831	−0.29219 647	787	+0.99951 829	−0.03103 529
738	+0.27084 775	−0.96262 220	788	+0.51392 674	−0.85783 408
739	−0.66367 898	−0.74801 752	789	−0.44416 668	−0.89594 417
740	−0.98802 232	+0.15431 102	790	−0.99389 531	−0.11032 732
741	−0.40398 250	+0.91476 672	791	−0.62984 117	+0.77672 396
742	+0.55147 697	+0.83419 011	792	+0.31328 604	+0.94965 881
743	+0.99991 106	−0.01333 703	793	+0.96837 950	+0.24948 174
744	+0.52903 153	−0.84860 217	794	+0.73314 932	−0.68006 770
745	−0.42823 715	−0.90366 639	795	−0.17613 497	−0.98436 603
746	−0.99178 657	−0.12790 390	796	−0.92348 158	−0.38364 277
747	−0.64349 199	+0.76545 285	797	−0.82178 349	+0.56979 988
748	+0.29642 616	+0.95505 577	798	+0.03545 855	+0.99937 115
749	+0.96381 146	+0.26658 483	799	+0.86010 016	+0.51012 519
750	+0.74507 295	−0.66698 298	800	+0.89396 965	−0.44812 751

TEN-PLACE SINES AND COSINES FOR LARGE RADIAN MEASURES

x	sin x	cos x	x	sin x	cos x
800	+0.89396 965	−0.44812 751	850	+0.98022 773	−0.19787 267
801	+0.10592 756	−0.99437 385	851	+0.36311 519	−0.93174 426
802	−0.77950 384	−0.62639 745	852	−0.58784 378	−0.80897 447
803	−0.94826 300	+0.31748 587	853	−0.99834 189	+0.05756 271
804	−0.24519 354	+0.96947 415	854	−0.49096 907	+0.87117 700
805	+0.68330 573	+0.73013 237	855	+0.46779 845	+0.88383 517
806	+0.98357 687	−0.18048 975	856	+0.99647 423	+0.08389 936
807	+0.37955 196	−0.92517 042	857	+0.60899 620	−0.79317 314
808	−0.57343 126	−0.81925 368	858	−0.33839 013	−0.94100 591
809	−0.99920 443	+0.03988 112	859	−0.97466 214	−0.22368 219
810	−0.50631 365	+0.86234 940	860	−0.71483 427	+0.69929 390
811	+0.45207 956	+0.89197 762	861	+0.20220 893	+0.97934 241
812	+0.99483 291	+0.10152 573	862	+0.93334 217	+0.35898 802
813	+0.62294 147	−0.78226 845	863	+0.80636 493	−0.59141 830
814	−0.32167 949	−0.94684 862	864	−0.06198 051	−0.99807 736
815	−0.97054 981	−0.24090 054	865	−0.87334 135	−0.48710 870
816	−0.72710 111	+0.68653 039	866	−0.88175 618	+0.47170 545
817	+0.18484 099	+0.98276 844	867	−0.07948 845	+0.99683 579
818	+0.92684 114	+0.37545 372	868	+0.79586 060	+0.60547 989
819	+0.81670 782	−0.57705 142	869	+0.93949 908	−0.34255 142
820	−0.04430 291	−0.99901 814	870	+0.21936 644	−0.97564 254
821	−0.86458 174	−0.50249 220	871	−0.70245 070	−0.71173 241
822	−0.88996 811	+0.45602 276	872	−0.97843 790	+0.20654 122
823	−0.09712 190	+0.99527 249	873	−0.35485 381	+0.93492 180
824	+0.78501 774	+0.61947 329	874	+0.59498 124	+0.80373 959
825	+0.94541 569	−0.32586 680	875	+0.99779 328	−0.06639 709
826	+0.23660 282	−0.97160 646	876	+0.48323 878	−0.87548 859
827	−0.68974 159	−0.72405 561	877	−0.47560 322	−0.87965 992
828	−0.98194 076	+0.18918 862	878	−0.99717 782	−0.07507 597
829	−0.37134 812	+0.92849 371	879	−0.60195 173	+0.79853 248
830	+0.58066 027	+0.81414 596	880	+0.34670 601	+0.93797 385
831	+0.99881 229	−0.04872 383	881	+0.97660 383	+0.21504 639
832	+0.49866 090	−0.86679 716	882	+0.70861 660	−0.70559 373
833	−0.45995 702	−0.88794 118	883	−0.21086 947	−0.97751 423
834	−0.99569 258	−0.09271 618	884	−0.93648 312	−0.35071 265
835	−0.61599 297	+0.78775 165	885	−0.80109 851	+0.59853 252
836	+0.33004 774	+0.94396 424	886	+0.07081 237	+0.99748 965
837	+0.97264 407	+0.23230 046	887	+0.87761 869	+0.47935 940
838	+0.72099 594	−0.69293 929	888	+0.87754 643	−0.47949 167
839	−0.19353 254	−0.98109 386	889	+0.07066 203	−0.99750 031
840	−0.93012 809	−0.36723 525	890	−0.80118 871	−0.59841 177
841	−0.81156 816	+0.58425 775	891	−0.93643 025	+0.35085 380
842	+0.05314 379	+0.99858 687	892	−0.21072 213	+0.97754 600
843	+0.86899 559	+0.49481 983	893	+0.70872 294	+0.70548 692
844	+0.88589 685	−0.46388 228	894	+0.97657 141	−0.21519 358
845	+0.08830 863	−0.99609 316	895	+0.34656 463	−0.93802 610
846	−0.79047 014	−0.61250 058	896	−0.60207 208	−0.79844 174
847	−0.94249 431	+0.33422 221	897	−0.99716 649	+0.07522 627
848	−0.22799 356	+0.97366 264	898	−0.47547 063	+0.87973 159
849	+0.69612 342	+0.71792 213	899	+0.48337 073	+0.87541 575
850	+0.98022 773	−0.19787 267	900	+0.99780 327	+0.06624 670

TEN-PLACE SINES AND COSINES FOR LARGE RADIAN MEASURES

x	sin x	cos x	x	sin x	cos x
900	+0.99780 327	+0.06624 670	950	+0.94546 479	+0.32572 431
901	+0.59486 009	−0.80382 926	951	+0.78492 436	−0.61959 160
902	−0.35499 472	−0.93486 831	952	−0.09727 191	−0.99525 784
903	−0.97846 902	−0.20639 374	953	−0.89003 684	−0.45588 862
904	−0.70234 341	+0.71183 827	954	−0.86450 600	+0.50262 250
905	+0.21951 349	+0.97560 947	955	−0.04415 233	+0.99902 481
906	+0.93955 070	+0.34240 981	956	+0.81679 478	+0.57692 832
907	+0.79576 933	−0.60559 984	957	+0.92678 454	−0.37559 341
908	−0.07963 869	−0.99682 380	958	+0.18469 287	−0.98279 629
909	−0.88182 727	−0.47157 255	959	−0.72720 458	−0.68642 079
910	−0.87326 792	+0.48724 032	960	−0.97051 349	+0.24104 682
911	−0.06183 008	+0.99808 669	961	−0.32153 677	+0.94689 709
912	+0.80645 406	+0.59129 676	962	+0.62305 937	+0.78217 455
913	+0.93328 805	−0.35912 869	963	+0.99481 760	−0.10167 567
914	+0.20206 131	−0.97937 287	964	+0.45194 512	−0.89204 574
915	−0.71493 966	−0.69918 616	965	−0.50644 362	−0.86227 308
916	−0.97462 841	+0.22382 909	966	−0.99921 043	−0.03973 052
917	−0.33824 829	+0.94105 690	967	−0.57330 778	+0.81934 009
918	+0.60911 575	+0.79308 134	968	+0.37969 140	+0.92511 320
919	+0.99646 158	−0.08404 955	969	+0.98360 406	+0.18034 150
920	+0.46766 523	−0.88390 567	970	+0.68319 568	−0.73023 535
921	−0.49110 037	−0.87110 299	971	−0.24533 966	−0.96943 718
922	−0.99835 056	−0.05741 224	972	−0.94831 084	−0.31734 294
923	−0.58772 184	+0.80906 306	973	−0.77940 942	+0.62651 493
924	+0.36325 562	+0.93168 952	974	+0.10607 744	+0.99435 787
925	+0.98025 754	+0.19772 493	975	+0.89403 718	+0.44799 277
926	+0.69601 520	−0.71802 705	976	+0.86002 327	−0.51025 482
927	−0.22814 031	−0.97362 827	977	+0.03530 793	−0.99937 648
928	−0.94254 467	−0.33408 015	978	−0.82186 936	−0.56967 601
929	−0.79037 781	+0.61261 972	979	−0.92342 374	+0.38378 195
930	+0.08845 877	+0.99607 984	980	−0.17598 660	+0.98439 256
931	+0.88596 676	+0.46364 875	981	+0.73325 181	+0.67995 719
932	+0.86892 100	−0.49495 080	982	+0.96834 189	−0.24962 769
933	+0.05299 328	−0.99859 487	983	+0.31314 290	−0.94970 602
934	−0.81165 622	−0.58413 542	984	−0.62995 823	−0.77662 902
935	−0.93007 273	+0.36736 544	985	−0.99387 867	+0.11047 712
936	−0.19338 467	+0.98112 302	986	−0.44403 164	+0.89601 111
937	+0.72110 037	+0.69283 061	987	+0.51405 603	+0.85775 661
938	+0.97260 905	−0.23244 706	988	+0.99952 296	+0.03088 464
939	+0.32990 546	−0.94401 398	989	+0.56603 309	−0.82438 252
940	−0.61611 169	−0.78765 880	990	−0.38786 499	−0.92171 620
941	−0.99567 859	+0.09286 625	991	−0.98516 179	−0.17162 825
942	−0.45982 319	+0.88801 049	992	−0.67670 538	+0.73625 392
943	+0.49879 154	+0.86672 199	993	+0.25391 083	+0.96722 763
944	+0.99881 962	+0.04857 328	994	+0.95108 260	+0.30893 672
945	+0.58053 755	−0.81423 347	995	+0.77383 341	−0.63338 919
946	−0.37148 806	−0.92843 773	996	−0.11487 465	−0.99338 000
947	−0.98196 927	−0.18904 062	997	−0.89796 748	−0.44006 182
948	−0.68963 246	+0.72415 957	998	−0.85547 315	+0.51784 716
949	+0.23674 926	+0.96157 078	999	−0.02646 075	+0.99964 985
950	+0.94546 479	+0.32572 431	1000	+0.82687 954	+0.56237 908

NATURAL TRIGONOMETRIC FUNCTIONS IN x RADIANS

x	$\sin \frac{\pi x}{2}$	$\tan \frac{\pi x}{2}$	$\cot \frac{\pi x}{2}$	$1 - x$
0.00	0.00000	0.00000	∞	1.00
0.01	0.01571	0.01571	63.657	0.99
0.02	0.03141	0.03143	31.821	0.98
0.03	0.04711	0.04716	21.205	0.97
0.04	0.06279	0.06291	15.895	0.96
0.05	0.07846	0.07870	12.706	0.95
0.06	0.09411	0.09453	10.579	0.94
0.07	0.10973	0.11040	9.0579	0.93
0.08	0.12533	0.12633	7.9158	0.92
0.09	0.14090	0.14232	7.0264	0.91
0.10	0.15643	0.15838	6.3138	0.90
0.11	0.17193	0.17453	5.7297	0.89
0.12	0.18738	0.19076	5.2422	0.88
0.13	0.20279	0.20709	4.8288	0.87
0.14	0.21814	0.22353	4.4737	0.86
0.15	0.23345	0.24008	4.1653	0.85
0.16	0.24869	0.25676	3.8947	0.84
0.17	0.26387	0.27357	3.6554	0.83
0.18	0.27899	0.29053	3.4420	0.82
0.19	0.29404	0.30764	3.2506	0.81
0.20	0.30902	0.32492	3.0777	0.80
0.21	0.32392	0.34238	2.9208	0.79
0.22	0.33874	0.36002	2.7776	0.78
0.23	0.35347	0.37787	2.6464	0.77
0.24	0.36812	0.39593	2.5257	0.76
0.25	0.38268	0.41421	2.4142	0.75
0.26	0.39715	0.43274	2.3109	0.74
0.27	0.41151	0.45152	2.2148	0.73
0.28	0.42578	0.47056	2.1251	0.72
0.29	0.43994	0.48989	2.0413	0.71
0.30	0.45399	0.50953	1.9626	0.70
0.31	0.46793	0.52947	1.8887	0.69
0.32	0.48175	0.54975	1.8190	0.68
0.33	0.49546	0.57039	1.7532	0.67
0.34	0.50904	0.59140	1.6909	0.66
0.35	0.55250	0.61280	1.6319	0.65
0.36	0.53583	0.63462	1.5757	0.64
0.37	0.54902	0.65688	1.5224	0.63
0.38	0.56208	0.67960	1.4715	0.62
0.39	0.57501	0.70281	1.4229	0.61
0.40	0.58779	0.72654	1.3764	0.60
0.41	0.60042	0.75082	1.3319	0.59
0.42	0.61291	0.77568	1.2892	0.58
0.43	0.62524	0.80115	1.2482	0.57
0.44	0.63742	0.82727	1.2088	0.56
0.45	0.64945	0.85408	1.1708	0.55
0.46	0.66131	0.88162	1.1343	0.54
0.47	0.67301	0.90993	1.0990	0.53
0.48	0.68455	0.93906	1.0649	0.52
0.49	0.69591	0.96907	1.0319	0.51
0.50	0.70711	1.0000	1.0000	0.50
$1 - x$	$\cos \frac{\pi x}{2}$	$\cot \frac{\pi x}{2}$	$\tan \frac{\pi x}{2}$	x

NATURAL TRIGONOMETRIC FUNCTIONS IN x RADIANS

x	$\cos \frac{\pi x}{2}$	$\sec \frac{\pi x}{2}$	$\csc \frac{\pi x}{2}$	$1 - x$
0.00	1.0000	1.0000	∞	1.00
0.01	0.99988	1.0001	63.665	0.99
0.02	0.99951	1.0005	31.836	0.98
0.03	0.99889	1.0011	21.229	0.97
0.04	0.99803	1.0020	15.926	0.96
0.05	0.99692	1.0031	12.745	0.95
0.06	0.99556	1.0045	10.626	0.94
0.07	0.99396	1.0061	9.1129	0.93
0.08	0.99211	1.0079	7.9787	0.92
0.09	0.99002	1.0101	7.0972	0.91
0.10	0.98769	1.0125	6.3925	0.90
0.11	0.98511	1.0151	5.8164	0.89
0.12	0.98229	1.0180	5.3367	0.88
0.13	0.97922	1.0212	4.9313	0.87
0.14	0.97592	1.0247	4.5841	0.86
0.15	0.97237	1.0284	4.2837	0.85
0.16	0.96858	1.0324	4.0211	0.84
0.17	0.96456	1.0367	3.7897	0.83
0.18	0.96029	1.0413	3.5843	0.82
0.19	0.95579	1.0463	3.4009	0.81
0.20	0.95106	1.0515	3.2361	0.80
0.21	0.94609	1.0570	3.0872	0.79
0.22	0.94088	1.0628	2.9521	0.78
0.23	0.93544	1.0690	2.8291	0.77
0.24	0.92978	1.0755	2.7165	0.76
0.25	0.92388	1.0824	2.6131	0.75
0.26	0.91775	1.0896	2.5180	0.74
0.27	0.91140	1.0972	2.4300	0.73
0.28	0.90483	1.1052	2.3486	0.72
0.29	0.89803	1.1136	2.2730	0.71
0.30	0.89101	1.1223	2.2027	0.70
0.31	0.88377	1.1315	2.1371	0.69
0.32	0.87631	1.1412	2.0757	0.68
0.33	0.86863	1.1512	2.0183	0.67
0.34	0.86074	1.1618	1.9645	0.66
0.35	0.85264	1.1728	1.9139	0.65
0.36	0.84433	1.1844	1.8663	0.64
0.37	0.83581	1.1964	1.8214	0.63
0.38	0.82708	1.2091	1.7791	0.62
0.39	0.81815	1.2223	1.7391	0.61
0.40	0.80902	1.2361	1.7013	0.60
0.41	0.79968	1.2505	1.6655	0.59
0.42	0.79016	1.2656	1.6316	0.58
0.43	0.78043	1.2813	1.5994	0.57
0.44	0.77051	1.2978	1.5688	0.56
0.45	0.76041	1.3151	1.5398	0.55
0.46	0.75011	1.3331	1.5121	0.54
0.47	0.73963	1.3520	1.4859	0.53
0.48	0.72897	1.3718	1.4608	0.52
0.49	0.71813	1.3925	1.4370	0.51
0.50	0.70711	1.4142	1.4142	0.50
$1 - x$	$\sin \frac{\pi x}{2}$	$\csc \frac{\pi x}{2}$	$\sec \frac{\pi x}{2}$	x

NATURAL TRIGONOMETRIC FUNCTIONS SINE, TANGENT, COTANGENT, COSINE FOR ANGLES IN π RADIANS

x		Sin (πx)	Tan (πx)	Cot (πx)	Cos (πx)
.00 or	1.00	.00000	.00000	inf	1.00000
.01	.99	.03141	.03143	31.821	.99951
.02	.98	.06279	.06291	15.895	.99803
.03	.97	.09411	.09453	10.579	.99556
.04	.96	.12533	.12633	7.9158	.99211
.05	.95	.15643	.15838	6.3138	.98769
.06	.94	.18738	.19076	5.2422	.98229
.07	.93	.21814	.22353	4.4737	.97592
.08	.92	.24869	.25676	3.8947	.96858
.09	.91	.27899	.29053	3.4420	.96029
.10	.90	.30902	.32492	3.0777	.95106
.11	.89	.33874	.36002	2.7776	.94088
.12	.88	.36812	.39593	2.5257	.92978
.13	.87	.39715	.43274	2.3109	.91775
.14	.86	.42578	.47056	2.1251	.90483
.15	.85	.45399	.50953	1.9626	.89101
.16	.84	.48175	.54975	1.8190	.87631
.17	.83	.50904	.59140	1.6909	.86074
.18	.82	.53583	.63462	1.5757	.84433
.19	.81	.56208	.67960	1.4715	.82708
.20	.80	.58779	.72654	1.3764	.80902
.21	.79	.61291	.77568	1.2892	.79016
.22	.78	.63742	.82727	1.2088	.77051
.23	.77	.66131	.88162	1.1343	.75011
.24	.76	.68455	.93906	1.0649	.72897
.25	.75	.70711	1.0000	1.0000	.70711
.26	.74	.72897	1.0649	.93906	.68455
.27	.73	.75011	1.1343	.88162	.66131
.28	.72	.77051	1.2088	.82727	.63742
.29	.71	.79016	1.2892	.77568	.61291
.30	.70	.80902	1.3764	.72654	.58779
.31	.69	.82708	1.4715	.67960	.56208
.32	.68	.84433	1.5757	.63462	.53583
.33	.67	.86074	1.6909	.59140	.50904
.34	.66	.87631	1.8190	.54975	.48175
.35	.65	.89101	1.9626	.50953	.45399
.36	.64	.90483	2.1251	.47056	.42578
.37	.63	.91775	2.3109	.43274	.39715
.38	.62	.92978	2.5257	.39593	.36812
.39	.61	.94088	2.7776	.36002	.33874
.40	.60	.95106	3.0777	.32492	.30902
.41	.59	.96029	3.4420	.29053	.27899
.42	.58	.96858	3.8947	.25676	.24869
.43	.57	.97592	4.4737	.22353	.21814
.44	.56	.98229	5.2422	.19076	.18738
.45	.55	.98769	6.3138	.15838	.15643
.46	.54	.99211	7.9158	.12633	.12533
.47	.53	.99556	10.579	.09453	.09411
.48	.52	.99803	15.895	.06291	.06279
.49	.51	.99951	31.821	.03143	.03141
.50	.50	1.0000	inf	.00000	.00000

These functions are useful in the solution of wave equations such as the displacement equation of a sound wave in the form:

$$y = A \sin 2\pi nx$$

without the necessity of reducing the angular rotation either to radians or to degrees in order to find the value of the function.

The algebraic sign of the function follows the familiar Quadrant Law for the particular function desired. Thus a numerical value of 9.13π radians becomes (by the subtraction of the greatest multiple of 2π radians) 1.13π radians. This is the same as $.13\pi$ radians in the 3rd Quadrant, which would give, from the tables above, a value of the sine function of $-.39715$.

Submitted by J. A. Blythe Jr.

COMMON LOGARITHMS OF THE TRIGONOMETRIC FUNCTIONS

Logarithms of the functions are given for each minute from 0–360°.

The quantity −10 is to be appended to all logarithms of the sine and cosine, to logarithms of the tangent from 0–45° and of the cotangent from 45–90°.

With degrees indicated at either side of top of the page use the column headings at the top. With degrees stated at bottom of the page use the column designations at the bottom.

With degrees at the left (top or bottom) use the minute column at the left, and with degrees on the right side of the page use the minute column at the right.

To illustrate the proper employment of headings for angles in the four quadrants—

log sin 6° 23′ = 9.04603 − 10	log sin 186° 23′ = 9.04603 − 10
log sin 83° 15′ = 9.99698 − 10	log sin 263° 15′ = 9.99698 − 10
log cos 96° 41′ = 9.06589 − 10	log cos 276° 41′ = 9.06589 − 10
log cos 173° 49′ = 9.99747 − 10	log cos 353° 49′ = 9.99747 − 10

For the accurate determination of values where the tabular differences are large, the values of CS and CT are given. The following equations indicate their use.

To find the logarithm of the functions of an angle:

For angles 0–3°	For angles 87–90°
$\log \sin \theta = \log \theta'' - CS$	$\log \cos \theta = \log (90° - \theta)'' - CS$
$\log \tan \theta = \log \theta'' - CT$	$\log \cot \theta = \log (90° - \theta)'' - CT$
$\log \cot \theta = \text{colog} \tan \theta$	$\log \tan \theta = \text{colog} \cot \theta$

To find the angle:

For angles 0–3°	For angles 87–90°
$\log \theta'' = \log \sin \theta + CS$	$\log (90° - \theta)'' = \log \cos \theta + CS$
$\log \theta'' = \log \tan \theta + CT$	$\log (90° - \theta)'' = \log \cot \theta + CT$

In the above expressions, θ'' and $(90° - \theta)''$ are used to indicate the value of the angles expressed in seconds. The values in the body of the table are the cologarithms and should be used as indicated above.

The values of the logarithms S and T are also given in a separate table. For these the following relations hold:

To find the function of an angle:

$\log \sin \theta = \log \theta'' + S$	$\log \cos \theta = \log (90° - \theta)'' + S$
$\log \tan \theta = \log \theta'' + T$	$\log \cot \theta = \log (90° - \theta)'' + T$

To find the angle:

$\log \theta'' = \log \sin \theta - S$	$\log (90° - \theta)'' = \log \cos \theta - S$
$\log \theta'' = \log \tan \theta - T$	$\log (90° - \theta)'' = \log \cot \theta - T$

Where the values of CS and CT are given, the angles expressed in seconds are given in the supplementary column at the left.

The tabular differences are given under the headings "d." and "c.d.", the latter referring to the common difference for the tangent and cotangent. Tables of proportional parts ("p.p.") facilitate interpolation. At the bottom of each column will be found special proportional parts between the tabular differences for the tangent or cotangent and those for the sine or cosine. These are useful when one function is to be obtained directly from the other without determining the angle.

For example, suppose log tan θ is given as 9.67644 and log cos θ is required. The difference between the given logarithm and that given in the table, 9.67622 (opposite 25° 23′), is 22. The tabular differences of the two logarithmic functions at this place are 32 and 6. In the proportional table for $\frac{6}{32}$, 22 corresponds to 4; this, subtracted from the tabular logarithmic cosine 9.95591, gives the required log cos θ = 9.95587.

The symbols $\bar{5}$ and $\dot{5}$ are used to indicate how the terminal 5 has been derived. For example, the logarithm 8.8307$\bar{5}$ is more fully given as 8.8307495 while the value 9.4082$\dot{5}$ is derived from 9.4082539. Thus, in rounding off to four places, a number ending in $\bar{5}$ should be decreased but a number ending in $\dot{5}$ should be increased.

COMMON LOGARITHMS OF THE TRIGONOMETRIC FUNCTIONS

Values of S and T

Min.	Values of S, − 10 to be appended					Values of T, − 10 to be appended					Sec.
	0°	1°	2°	3°	4°	0°	1°	2°	3°	4°	
0′	4.68 557	555	549	538	522	4.68 557	562	575	597	628	**0″**
1	557	555	549	537	522	557	562	575	598	629	60
2	557	555	548	537	522	557	562	576	598	629	120
3	557	555	548	537	521	557	562	576	599	630	180
4	557	555	548	537	521	558	563	576	599	631	240
5	557	555	548	537	521	558	563	577	599	631	300
6	557	555	548	536	520	558	563	577	600	632	360
7	557	555	548	536	520	558	563	577	600	632	420
8	557	555	548	536	520	558	563	578	601	633	480
9	557	555	547	536	520	558	563	578	601	634	540
10	4.68 557	555	547	535	519	4.68 558	564	578	602	634	**600**
11	557	554	547	535	519	558	564	579	602	635	660
12	557	554	547	535	519	558	564	579	603	635	720
13	557	554	547	535	518	558	564	579	603	636	780
14	557	554	547	534	518	558	564	580	604	637	840
15	557	554	546	534	518	558	564	580	604	637	900
16	557	554	546	534	517	558	565	580	605	638	960
17	557	554	546	534	517	558	565	581	605	639	1020
18	557	554	546	534	517	558	565	581	606	639	1080
19	557	554	546	533	516	558	565	581	606	640	1140
20	4.68 557	554	546	533	516	4.68 558	565	582	607	640	**1200**
21	557	554	545	533	516	558	566	582	607	641	1260
22	557	553	545	533	515	558	566	582	608	642	1320
23	557	553	545	532	515	558	566	583	608	642	1380
24	557	553	545	532	515	558	566	583	609	643	1440
25	557	553	545	532	515	558	566	583	609	644	1500
26	557	553	544	532	514	558	567	584	610	644	1560
27	557	553	544	531	514	558	567	584	610	645	1620
28	557	553	544	531	514	558	567	584	611	646	1680
29	557	553	544	531	513	559	567	585	611	646	1740
30	4.68 557	553	544	531	513	4.68 559	567	585	612	647	**1800**
31	557	552	544	530	513	559	568	585	612	648	1860
32	557	552	543	530	512	559	568	586	613	648	1920
33	557	552	543	530	512	559	568	586	613	649	1980
34	557	552	543	529	512	559	568	587	614	650	2040
35	557	552	543	529	511	559	569	587	614	650	2100
36	557	552	543	529	511	559	569	587	615	651	2160
37	557	552	542	529	511	559	569	588	615	652	2220
38	557	552	542	528	510	559	569	588	616	652	2280
39	557	552	542	528	510	559	570	589	616	653	2340
40	4.68 557	551	542	528	510	4.68 559	570	589	617	654	**2400**
41	556	551	542	528	509	560	570	589	617	654	2460
42	556	551	541	527	509	560	570	590	618	655	2520
43	556	551	541	527	508	560	571	590	619	656	2580
44	556	551	541	527	508	560	571	591	619	656	2640
45	556	551	541	527	508	560	571	591	620	657	2700
46	556	551	541	526	507	560	571	591	620	658	2760
47	556	551	540	526	507	560	572	592	621	659	2820
48	556	550	540	526	507	560	572	592	621	659	2880
49	556	550	540	525	506	560	572	593	622	660	2940
50	4.68 556	550	540	525	506	4.68 561	572	593	622	661	**3000**
51	556	550	540	525	506	561	573	593	623	661	3060
52	556	550	539	525	505	561	573	594	624	662	3120
53	556	550	539	524	505	561	573	594	624	663	3180
54	556	550	539	524	505	561	573	595	625	664	3240
55	556	549	539	524	504	561	574	595	625	664	3300
56	556	549	539	523	504	561	574	596	626	665	3360
57	556	549	538	523	503	562	574	596	626	666	3420
58	555	549	538	523	503	562	575	596	627	667	3480
59	555	549	538	523	503	562	575	597	628	667	3540
60	4.68 555	549	538	522	502	4.68 562	575	597	628	668	**3600**

COMMON LOGARITHMS OF THE TRIGONOMETRIC FUNCTIONS (Continued)

0° (180°) (359°) **179°**

″	′	L. Sin.	*d.	C. S.	C. T.	L. Tan	*c.d.	L. Cot.	L. Cos.	′
0	**0**	—		—	—	—		—	0.00 000	**60**
60	1	6.46 373	30103	5.31 443	5.31 443	6.46 373	30103	3.53 627	0.00 000	59
120	2	6.76 476	17609	5.31 443	5.31 443	6.76 476	17609	3.23 524	0.00 000	58
180	3	6.94 085	12494	5.31 443	5.31 443	6.94 085	12494	3.05 915	0.00 000	57
240	4	7.06 579	9691	5.31 443	5.31 442	7.06 579	9691	2.93 421	0.00 000	56
300	5	7.16 270	7918	5.31 443	5.31 442	7.16 270	7918	2.83 730	0.00 000	55
360	6	7.24 188	6694	5.31 443	5.31 442	7.24 188	6694	2.75 812	0.00 000	54
420	7	7.30 882	5800	5.31 443	5.31 442	7.30 882	5800	2.69 118	0.00 000	53
480	8	7.36 682	5115	5.31 443	5.31 442	7.36 682	5115	2.63 318	0.00 000	52
540	9	7.41 797	4576	5.31 443	5.31 442	7.41 797	4576	2.58 203	0.00 000	51
600	**10**	7.46 373	4139	5.31 443	5.31 442	7.46 373	4139	2.53 627	0.00 000	**50**
660	11	7.50 512	3779	5.31 443	5.31 442	7.50 512	3779	2.49 488	0.00 000	49
720	12	7.54 291	3476	5.31 443	5.31 442	7.54 291	3476	2.45 709	0.00 000	48
780	13	7.57 767	3218	5.31 443	5.31 442	7.57 767	3219	2.42 233	0.00 000	47
840	14	7.60 985	2997	5.31 443	5.31 442	7.60 986	2996	2.39 014	0.00 000	46
900	15	7.63 982	2802	5.31 443	5.31 442	7.63 982	2803	2.36 018	0.00 000	45
960	16	7.66 784	2633	5.31 443	5.31 442	7.66 785	2633	2.33 215	0.00 000	44
1020	17	7.69 417	2483	5.31 443	5.31 442	7.69 418	2482	2.30 582	9.99 999	43
1080	18	7.71 900	2348	5.31 443	5.31 442	7.71 900	2348	2.28 100	9.99 999	42
1140	19	7.74 248	2227	5.31 443	5.31 442	7.74 248	2228	2.25 752	9.99 999	41
1200	**20**	7.76 475	2119	5.31 443	5.31 442	7.76 476	2119	2.23 524	9.99 999	**40**
1260	21	7.78 594	2021	5.31 443	5.31 442	7.78 595	2020	2.21 405	9.99 999	39
1320	22	7.80 615	1930	5.31 443	5.31 442	7.80 615	1931	2.19 385	9.99 999	38
1380	23	7.82 545	1848	5.31 443	5.31 442	7.82 546	1848	2.17 454	9.99 999	37
1440	24	7.84 393	1773	5.31 443	5.31 442	7.84 394	1773	2.15 606	9.99 999	36
1500	25	7.86 166	1704	5.31 443	5.31 442	7.86 167	1704	2.13 833	9.99 999	35
1560	26	7.87 870	1639	5.31 443	5.31 442	7.87 871	1639	2.12 129	9.99 999	34
1620	27	7.89 509	1579	5.31 443	5.31 442	7.89 510	1579	2.10 490	9.99 999	33
1680	28	7.91 088	1524	5.31 443	5.31 442	7.91 089	1524	2.08 911	9.99 999	32
1740	29	7.92 612	1472	5.31 443	5.31 441	7.92 613	1473	2.07 387	9.99 998	31
1800	**30**	7.94 084	1424	5.31 443	5.31 441	7.94 086	1424	2.05 914	9.99 998	**30**
1860	31	7.95 508	1379	5.31 443	5.31 441	7.95 510	1379	2.04 490	9.99 998	29
1920	32	7.96 887	1336	5.31 443	5.31 441	7.96 889	1336	2.03 111	9.99 998	28
1980	33	7.98 223	1297	5.31 443	5.31 441	7.98 225	1297	2.01 775	9.99 998	27
2040	34	7.99 520	1259	5.31 443	5.31 441	7.99 522	1259	2.00 478	9.99 998	26
2100	35	8.00 779	1223	5.31 443	5.31 441	8.00 781	1223	1.99 219	9.99 998	25
2160	36	8.02 002	1190	5.31 443	5.31 441	8.02 004	1190	1.97 996	9.99 998	24
2220	37	8.03 192	1158	5.31 443	5.31 441	8.03 194	1159	1.96 806	9.99 997	23
2280	38	8.04 350	1128	5.31 443	5.31 441	8.04 353	1128	1.95 647	9.99 997	22
2340	39	8.05 478	1100	5.31 443	5.31 441	8.05 481	1100	1.94 519	9.99 997	21
2400	**40**	8.06 578	1072	5.31 443	5.31 441	8.06 581	1072	1.93 419	9.99 997	**20**
2460	41	8.07 650	1046	5.31 444	5.31 440	8.07 653	1047	1.92 347	9.99 997	19
2520	42	8.08 696	1022	5.31 444	5.31 440	8.08 700	1022	1.91 300	9.99 997	18
2580	43	8.09 718	999	5.31 444	5.31 440	8.09 722	998	1.90 278	9.99 997	17
2640	44	8.10 717	976	5.31 444	5.31 440	8.10 720	976	1.89 280	9.99 996	16
2700	45	8.11 693	954	5.31 444	5.31 440	8.11 696	955	1.88 304	9.99 996	15
2760	46	8.12 647	934	5.31 444	5.31 440	8.12 651	934	1.87 349	9.99 996	14
2820	47	8.13 581	914	5.31 444	5.31 440	8.13 585	915	1.86 415	9.99 996	13
2880	48	8.14 495	896	5.31 444	5.31 440	8.14 500	895	1.85 500	9.99 996	12
2940	49	8.15 391	877	5.31 444	5.31 440	8.15 395	878	1.84 605	9.99 996	11
3000	**50**	8.16 268	860	5.31 444	5.31 439	8.16 273	860	1.83 727	9.99 995	**10**
3060	51	8.17 128	843	5.31 444	5.31 439	8.17 133	843	1.82 867	9.99 995	9
3120	52	8.17 971	827	5.31 444	5.31 439	8.17 976	828	1.82 024	9.99 995	8
3180	53	8.18 798	812	5.31 444	5.31 439	8.18 804	812	1.81 196	9.99 995	7
3240	54	8.19 610	797	5.31 444	5.31 439	8.19 616	797	1.80 384	9.99 995	6
3300	55	8.20 407	782	5.31 444	5.31 439	8.20 413	782	1.79 587	9.99 994	5
3360	56	8.21 189	769	5.31 444	5.31 439	8.21 195	769	1.78 805	9.99 994	4
3420	57	8.21 958	755	5.31 445	5.31 439	8.21 964	756	1.78 036	9.99 994	3
3480	58	8.22 713	743	5.31 445	5.31 438	8.22 720	742	1.77 280	9.99 994	2
3540	59	8.23 456	730	5.31 445	5.31 438	8.23 462	730	1.76 538	9.99 994	1
3600	**60**	8.24 186		5.31 445	5.31 438	8.24 192		1.75 808	9.99 993	**0**
″	′	L. Cos.	d.	C. S.	C. T.	L. Cot.	c. d.	L. Tan.	L. Sin.	′

90° (270°) (269°) **89°**

* Interpolation in this section of the table is not accurate.

COMMON LOGARITHMS OF THE TRIGONOMETRIC FUNCTIONS (Continued)

1° (181°) (358°) **178°**

″	′	L. Sin.	*d.	C. S.	C. T.	L. Tan.	*c.d.	L. Cot.	L. Cos.	′
3600	**0**	8.24 186	717	5.31 445	5.31 438	8.24 192	718	1.75 808	9.99 993	**60**
3660	1	8.24 903	706	5.31 445	5.31 438	8.24 910	706	1.75 090	9.99 993	59
3720	2	8.25 609	695	5.31 445	5.31 438	8.25 616	696	1.74 384	9.99 993	58
3780	3	8.26 304	684	5.31 445	5.31 438	8.26 312	684	1.73 688	9.99 993	57
3840	4	8.26 988	673	5.31 445	5.31 437	8.26 996	673	1.73 004	9.99 992	56
3900	5	8.27 661	663	5.31 445	5.31 437	8.27 669	663	1.72 331	9.99 992	55
3960	6	8.28 324	653	5.31 445	5.31 437	8.28 332	654	1.71 668	9.99 992	54
4020	7	8.28 977	644	5.31 445	5.31 437	8.28 986	643	1.71 014	9.99 992	53
4080	8	8.29 621	634	5.31 445	5.31 437	8.29 629	634	1.70 371	9.99 992	52
4140	9	8.30 255	624	5.31 445	5.31 437	8.30 263	625	1.69 737	9.99 991	51
4200	**10**	8.30 879	616	5.31 446	5.31 437	8.30 888	617	1.69 112	9.99 991	**50**
4260	11	8.31 495	608	5.31 446	5.31 436	8.31 505	607	1.68 495	9.99 991	49
4320	12	8.32 103	599	5.31 446	5.31 436	8.32 112	599	1.67 888	9.99 990	48
4380	13	8.32 702	590	5.31 446	5.31 436	8.32 711	591	1.67 289	9.99 990	47
4440	14	8.33 292	583	5.31 446	5.31 436	8.33 302	584	1.66 698	9.99 990	46
4500	15	8.33 875	575	5.31 446	5.31 436	8.33 886	575	1.66 114	9.99 990	45
4560	16	8.34 450	568	5.31 446	5.31 435	8.34 461	568	1.65 539	9.99 989	44
4620	17	8.35 018	560	5.31 446	5.31 435	8.35 029	561	1.64 971	9.99 989	43
4680	18	8.35 578	553	5.31 446	5.31 435	8.35 590	553	1.64 410	9.99 989	42
4740	19	8.36 131	547	5.31 446	5.31 435	8.36 143	546	1.63 857	9.99 989	41
4800	**20**	8.36 678	539	5.31 446	5.31 435	8.36 689	540	1.63 311	9.99 988	**40**
4860	21	8.37 217	533	5.31 447	5.31 434	8.37 229	533	1.62 771	9.99 988	39
4920	22	8.37 750	526	5.31 447	5.31 434	8.37 762	527	1.62 238	9.99 988	38
4980	23	8.38 276	520	5.31 447	5.31 434	8.38 289	520	1.61 711	9.99 987	37
5040	24	8.38 796	514	5.31 447	5.31 434	8.38 809	514	1.61 191	9.99 987	36
5100	25	8.39 310	508	5.31 447	5.31 434	8.39 323	509	1.60 677	9.99 987	35
5160	26	8.39 818	502	5.31 447	5.31 433	8.39 832	502	1.60 168	9.99 986	34
5220	27	8.40 320	496	5.31 447	5.31 433	8.40 334	496	1.59 666	9.99 986	33
5280	28	8.40 816	491	5.31 447	5.31 433	8.40 830	491	1.59 170	9.99 986	32
5340	29	8.41 307	485	5.31 447	5.31 433	8.41 321	486	1.58 679	9.99 985	31
5400	**30**	8.41 792	480	5.31 447	5.31 433	8.41 807	480	1.58 193	9.99 985	**30**
5460	31	8.42 272	474	5.31 448	5.31 432	8.42 287	475	1.57 713	9.99 985	29
5520	32	8.42 746	470	5.31 448	5.31 432	8.42 762	470	1.57 238	9.99 984	28
5580	33	8.43 216	464	5.31 448	5.31 432	8.43 232	464	1.56 768	9.99 984	27
5640	34	8.43 680	459	5.31 448	5.31 432	8.43 696	460	1.56 304	9.99 984	26
5700	35	8.44 139	455	5.31 448	5.31 431	8.44 156	455	1.55 844	9.99 983	25
5760	36	8.44 594	450	5.31 448	5.31 431	8.44 611	450	1.55 389	9.99 983	24
5820	37	8.45 044	445	5.31 448	5.31 431	8.45 061	446	1.54 939	9.99 983	23
5880	38	8.45 489	441	5.31 448	5.31 431	8.45 507	441	1.54 493	9.99 982	22
5940	39	8.45 930	436	5.31 449	5.31 431	8.45 948	437	1.54 052	9.99 982	21
6000	**40**	8.46 366	433	5.31 449	5.31 430	8.46 385	432	1.53 615	9.99 982	**20**
6060	41	8.46 799	427	5.31 449	5.31 430	8.46 817	428	1.53 183	9.99 981	19
6120	42	8.47 226	424	5.31 449	5.31 430	8.47 245	424	1.52 755	9.99 981	18
6180	43	8.47 650	419	5.31 449	5.31 430	8.47 669	420	1.52 331	9.99 981	17
6240	44	8.48 069	416	5.31 449	5.31 429	8.48 089	416	1.51 911	9.99 980	16
6300	45	8.48 485	411	5.31 449	5.31 429	8.48 505	412	1.51 495	9.99 980	15
6360	46	8.48 896	408	5.31 449	5.31 429	8.48 917	408	1.51 083	9.99 979	14
6420	47	8.49 304	404	5.31 450	5.31 428	8.49 325	404	1.50 675	9.99 979	13
6480	48	8.49 708	400	5.31 450	5.31 428	8.49 729	401	1.50 271	9.99 979	12
6540	49	8.50 108	396	5.31 450	5.31 428	8.50 130	397	1.49 870	9.99 978	11
6600	**50**	8.50 504	393	5.31 450	5.31 428	8.50 527	393	1.49 473	9.99 978	**10**
6660	51	8.50 897	390	5.31 450	5.31 427	8.50 920	390	1.49 080	9.99 977	9
6720	52	8.51 287	386	5.31 450	5.31 427	8.51 310	386	1.48 690	9.99 977	8
6780	53	8.51 673	382	5.31 450	5.31 427	8.51 696	383	1.48 304	9.99 977	7
6840	54	8.52 055	379	5.31 450	5.31 427	8.52 079	380	1.47 921	9.99 976	6
6900	55	8.52 434	376	5.31 451	5.31 426	8.52 459	376	1.47 541	9.99 976	5
6960	56	8.52 810	373	5.31 451	5.31 426	8.52 835	373	1.47 165	9.99 975	4
7020	57	8.53 183	369	5.31 451	5.31 426	8.53 208	370	1.46 792	9.99 975	3
7080	58	8.53 552	367	5.31 451	5.31 425	8.53 578	367	1.46 422	9.99 974	2
7140	59	8.53 919	363	5.31 451	5.31 425	8.53 945	363	1.46 055	9.99 974	1
7200	**60**	8.54 282		5.31 451	5.31 425	8.54 308		1.45 692	9.99 974	**0**
″	′	L. Cos.	d.	C. S.	C. T.	L. Cot.	c.d.	L. Tan.	L. Sin.	′

91° (271°) (268°) **88°**

* Interpolation in this section of the table is not accurate.

COMMON LOGARITHMS OF THE TRIGONOMETRIC FUNCTIONS (Continued)

2° (182°) (357°) **177°**

"	'	L. Sin.	*d.	C. S.	C. T.	L. Tan.	*c.d.	L. Cot.	L. Cos.	'
7200	**0**	8.54 282	360	5.31 451	5.31 425	8.54 308	361	1.45 692	9.99 974	**60**
7260	1	8.54 642	357	5.31 451	5.31 425	8.54 669	358	1.45 331	9.99 973	59
7320	2	8.54 999	355	5.31 452	5.31 424	8.55 027	355	1.44 973	9.99 973	58
7380	3	8.55 354	351	5.31 452	5.31 424	8.55 382	352	1.44 618	9.99 972	57
7440	4	8.55 705	349	5.31 452	5.31 424	8.55 734	349	1.44 266	9.99 972	56
7500	5	8.56 054	346	5.31 452	5.31 423	8.56 083	346	1.43 917	9.99 971	55
7560	6	8.56 400	343	5.31 452	5.31 423	8.56 429	344	1.43 571	9.99 971	54
7620	7	8.56 743	341	5.31 452	5.31 423	8.56 773	341	1.43 227	9.99 970	53
7680	8	8.57 084	337	5.31 453	5.31 422	8.57 114	338	1.42 886	9.99 970	52
7740	9	8.57 421	336	5.31 453	5.31 422	8.57 452	336	1.42 548	9.99 969	51
7800	**10**	8.57 757	332	5.31 453	5.31 422	8.57 788	333	1.42 212	9.99 969	**50**
7860	11	8.58 089	330	5.31 453	5.31 421	8.58 121	330	1.41 879	9.99 968	49
7920	12	8.58 419	328	5.31 453	5.31 421	8.58 451	328	1.41 549	9.99 968	48
7980	13	8.58 747	325	5.31 453	5.31 421	8.58 779	326	1.41 221	9.99 967	47
8040	14	8.59 072	323	5.31 454	5.31 421	8.59 105	323	1.40 895	9.99 967	46
8100	15	8.59 395	320	5.31 454	5.31 420	8.59 428	321	1.40 572	9.99 967	45
8160	16	8.59 715	318	5.31 454	5.31 420	8.59 749	319	1.40 251	9.99 966	44
8220	17	8.60 033	316	5.31 454	5.31 420	8.60 068	316	1.39 932	9.99 966	43
8280	18	8.60 349	313	5.31 454	5.31 419	8.60 384	314	1.39 616	9.99 965	42
8340	19	8.60 662	311	5.31 454	5.31 419	8.60 698	311	1.39 302	9.99 964	41
8400	**20**	8.60 973	309	5.31 455	5.31 418	8.61 009	310	1.38 991	9.99 964	**40**
8460	21	8.61 282	307	5.31 455	5.31 418	8.61 319	307	1.38 681	9.99 963	39
8520	22	8.61 589	305	5.31 455	5.31 418	8.61 626	305	1.38 374	9.99 963	38
8580	23	8.61 894	302	5.31 455	5.31 417	8.61 931	303	1.38 069	9.99 962	37
8640	24	8.62 196	301	5.31 455	5.31 417	8.62 234	301	1.37 766	9.99 962	36
8700	25	8.62 497	298	5.31 455	5.31 417	8.62 535	299	1.37 465	9.99 961	35
8760	26	8.62 795	296	5.31 456	5.31 416	8.62 834	297	1.37 166	9.99 961	34
8820	27	8.63 091	294	5.31 456	5.31 416	8.63 131	295	1.36 869	9.99 960	33
8880	28	8.63 385	293	5.31 456	5.31 416	8.63 426	292	1.36 574	9.99 960	32
8940	29	8.63 678	290	5.31 456	5.31 415	8.63 718	291	1.36 282	9.99 959	31
9000	**30**	8.63 968	288	5.31 456	5.31 415	8.64 009	289	1.35 991	9.99 959	**30**
9060	31	8.64 256	287	5.31 456	5.31 415	8.64 298	287	1.35 702	9.99 958	29
9120	32	8.64 543	284	5.31 457	5.31 414	8.64 585	285	1.35 415	9.99 958	28
9180	33	8.64 827	283	5.31 457	5.31 414	8.64 870	284	1.35 130	9.99 957	27
9240	34	8.65 110	281	5.31 457	5.31 413	8.65 154	281	1.34 846	9.99 956	26
9300	35	8.65 391	279	5.31 457	5.31 413	8.65 435	280	1.34 565	9.99 956	25
9360	36	8.65 670	277	5.31 457	5.31 413	8.65 715	278	1.34 285	9.99 955	24
9420	37	8.65 947	276	5.31 458	5.31 412	8.65 993	276	1.34 007	9.99 955	23
9480	38	8.66 223	274	5.31 458	5.31 412	8.66 269	274	1.33 731	9.99 954	22
9540	39	8.66 497	272	5.31 458	5.31 412	8.66 543	273	1.33 457	9.99 954	21
9600	**40**	8.66 769	270	5.31 458	5.31 411	8.66 816	271	1.33 184	9.99 953	**20**
9660	41	8.67 039	269	5.31 458	5.31 411	8.67 087	269	1.32 913	9.99 952	19
9720	42	8.67 308	267	5.31 459	5.31 410	8.67 356	268	1.32 644	9.99 952	18
9780	43	8.67 575	266	5.31 459	5.31 410	8.67 624	266	1.32 376	9.99 951	17
9840	44	8.67 841	263	5.31 459	5.31 410	8.67 890	264	1.32 110	9.99 951	16
9900	45	8.68 104	263	5.31 459	5.31 409	8.68 154	263	1.31 846	9.99 950	15
9960	46	8.68 367	260	5.31 459	5.31 409	8.68 417	261	1.31 583	9.99 949	14
10020	47	8.68 627	259	5.31 460	5.31 408	8.68 678	260	1.31 322	9.99 949	13
10080	48	8.68 886	258	5.31 460	5.31 408	8.68 938	258	1.31 062	9.99 948	12
10140	49	8.69 144	256	5.31 460	5.31 408	8.69 196	257	1.30 804	9.99 948	11
10200	**50**	8.69 400	254	5.31 460	5.31 407	8.69 453	255	1.30 547	9.99 947	**10**
10260	51	8.69 654	253	5.31 460	5.31 407	8.69 708	254	1.30 292	9.99 946	9
10320	52	8.69 907	252	5.31 461	5.31 406	8.69 962	252	1.30 038	9.99 946	8
10380	53	8.70 159	250	5.31 461	5.31 406	8.70 214	251	1.29 786	9.99 945	7
10440	54	8.70 409	249	5.31 461	5.31 405	8.70 465	249	1.29 535	9.99 944	6
10500	55	8.70 658	247	5.31 461	5.31 405	8.70 714	248	1.29 286	9.99 944	5
10560	56	8.70 905	246	5.31 461	5.31 405	8.70 962	246	1.29 038	9.99 943	4
10620	57	8.71 151	244	5.31 462	5.31 404	8.71 208	245	1.28 792	9.99 942	3
10680	58	8.71 395	243	5.31 462	5.31 404	8.71 453	244	1.28 547	9.99 942	2
10740	59	8.71 638	242	5.31 462	5.31 403	8.71 697	243	1.28 303	9.99 941	1
10800	**60**	8.71 880		5.31 462	5.31 403	8.71 940		1.28 060	9.99 940	**0**
"	'	L. Cos.	d.	C. S.	C. T.	L. Cot.	c.d.	L. Tan.	L. Sin.	

92° (272°) (267°) **87°**

* Interpolation in this section of the table is not accurate.

COMMON LOGARITHMS OF THE TRIGONOMETRIC FUNCTIONS (Continued)

3° (183°) (356°) **176°**

′	L. Sin.	*d.	L. Tan.	*c.d.	L. Cot.	L. Cos.	′	P. P.					
0	8.71 880	240	8.71 940	241	1.28 060	9.99 940	**60**	″	**241**	**239**	**237**	**235**	**234**
1	8.72 120	239	8.72 181	239	1.27 819	9.99 940	59	1	4.0	4.0	4.0	3.9	3.9
2	8.72 359	238	8.72 420	239	1.27 580	9.99 939	58	2	8.0	8.0	7.9	7.8	7.8
3	8.72 597	237	8.72 659	237	1.27 341	9.99 938	57	3	12.0	12.0	11.8	11.8	11.7
4	8.72 834	235	8.72 896	236	1.27 104	9.99 938	56	4	16.1	15.9	15.8	15.7	15.6
5	8.73 069	234	8.73 132	234	1.26 868	9.99 937	55	5	20.1	19.9	19.8	19.6	19.5
6	8.73 303	232	8.73 366	234	1.26 634	9.99 936	54	6	24.1	23.9	23.7	23.5	23.4
7	8.73 535	232	8.73 600	232	1.26 400	9.99 936	53	7	28.1	27.9	27.6	27.4	27.3
8	8.73 767	230	8.73 832	231	1.26 168	9.99 935	52	8	32.1	31.9	31.6	31.3	31.2
9	8.73 997	229	8.74 063	229	1.25 937	9.99 934	51	9	36.2	35.8	35.6	35.2	35.1
10	8.74 226	228	8.74 292	229	1.25 708	9.99 934	**50**	″	**232**	**229**	**227**	**225**	**223**
11	8.74 454	226	8.74 521	227	1.25 479	9.99 933	49	1	3.9	3.8	3.8	3.8	3.7
12	8.74 680	226	8.74 748	226	1.25 252	9.99 932	48	2	7.7	7.6	7.6	7.5	7.4
13	8.74 906	224	8.74 974	225	1.25 026	9.99 932	47	3	11.6	11.4	11.4	11.2	11.2
14	8.75 130	223	8.75 199	224	1.24 801	9.99 931	46	4	15.5	15.3	15.1	15.0	14.9
15	8.75 353	222	8.75 423	222	1.24 577	9.99 930	45	5	19.3	19.1	18.9	18.8	18.6
16	8.75 575	220	8.75 645	222	1.24 355	9.99 929	44	6	23.2	22.9	22.7	22.5	22.3
17	8.75 795	220	8.75 867	220	1.24 133	9.99 929	43	7	27.1	26.7	26.5	26.2	26.0
18	8.76 015	219	8.76 087	219	1.23 913	9.99 928	42	8	30.9	30.5	30.3	30.0	29.7
19	8.76 234	217	8.76 306	219	1.23 694	9.99 927	41	9	34.8	34.4	34.0	33.8	33.4
20	8.76 451	216	8.76 525	217	1.23 475	9.99 926	**40**	″	**222**	**220**	**217**	**215**	**213**
21	8.76 667	216	8.76 742	216	1.23 258	9.99 926	39	1	3.7	3.7	3.6	3.6	3.6
22	8.76 883	214	8.76 958	215	1.23 042	9.99 925	38	2	7.4	7.3	7.2	7.2	7.1
23	8.77 097	213	8.77 173	214	1.22 827	9.99 924	37	3	11.1	11.0	10.8	10.8	10.6
24	8.77 310	212	8.77 387	213	1.22 613	9.99 923	36	4	14.8	14.7	14.5	14.3	14.2
25	8.77 522	211	8.77 600	211	1.22 400	9.99 923	35	5	18.5	18.3	18.1	17.9	17.8
26	8.77 733	210	8.77 811	211	1.22 189	9.99 922	34	6	22.2	22.0	21.7	21.5	21.3
27	8.77 943	209	8.78 022	210	1.21 978	9.99 921	33	7	25.9	25.7	25.3	25.1	24.8
28	8.78 152	208	8.78 232	209	1.21 768	9.99 920	32	8	29.6	29.3	28.9	28.7	28.4
29	8.78 360	208	8.78 441	208	1.21 559	9.99 920	31	9	33.3	33.0	32.6	32.2	32.0
30	8.78 568	206	8.78 649	206	1.21 351	9.99 919	**30**	″	**211**	**208**	**206**	**203**	**201**
31	8.78 774	205	8.78 855	206	1.21 145	9.99 918	29	1	3.5	3.5	3.4	3.4	3.4
32	8.78 979	204	8.79 061	205	1.20 939	9.99 917	28	2	7.0	6.9	6.9	6.8	6.7
33	8.79 183	203	8.79 266	204	1.20 734	9.99 917	27	3	10.6	10.4	10.3	10.2	10.0
34	8.79 386	202	8.79 470	203	1.20 530	9.99 916	26	4	14.1	13.9	13.7	13.5	13.4
35	8.79 588	201	8.79 673	202	1.20 327	9.99 915	25	5	17.6	17.3	17.2	16.9	16.8
36	8.79 789	201	8.79 875	201	1.20 125	9.99 914	24	6	21.1	20.8	20.6	20.3	20.1
37	8.79 990	199	8.80 076	201	1.19 924	9.99 913	23	7	24.6	24.3	24.0	23.7	23.4
38	8.80 189	199	8.80 277	199	1.19 723	9.99 913	22	8	28.1	27.7	27.3	27.1	26.8
39	8.80 388	197	8.80 476	198	1.19 524	9.99 912	21	9	31.6	31.2	30.9	30.4	30.2
40	8.80 585	197	8.80 674	198	1.19 326	9.99 911	**20**	″	**199**	**197**	**195**	**193**	**192**
41	8.80 782	196	8.80 872	196	1.19 128	9.99 910	19	1	3.3	3.3	3.2	3.2	3.2
42	8.80 978	195	8.81 068	196	1.18 932	9.99 909	18	2	6.6	6.6	6.5	6.4	6.4
43	8.81 173	194	8.81 264	195	1.18 736	9.99 909	17	3	10.0	9.8	9.8	9.6	9.6
44	8.81 367	193	8.81 459	194	1.18 541	9.99 908	16	4	13.3	13.1	13.0	12.9	12.8
45	8.81 560	192	8.81 653	193	1.18 347	9.99 907	15	5	16.6	16.4	16.2	16.1	16.0
46	8.81 752	192	8.81 846	192	1.18 154	9.99 906	14	6	19.9	19.7	19.5	19.3	19.2
47	8.81 944	190	8.82 038	192	1.17 962	9.99 905	13	7	23.2	23.0	22.8	22.5	22.4
48	8.82 134	190	8.82 230	190	1.17 770	9.99 904	12	8	26.5	26.3	26.0	25.7	25.6
49	8.82 324	189	8.82 420	190	1.17 580	9.99 904	11	9	29.8	29.6	29.2	29.0	28.8
50	8.82 513	188	8.82 610	189	1.17 390	9.99 903	**10**	″	**189**	**187**	**185**	**183**	**181**
51	8.82 701	187	8.82 799	188	1.17 201	9.99 902	9	1	3.2	3.1	3.1	3.0	3.0
52	8.82 888	187	8.82 987	188	1.17 013	9.99 901	8	2	6.3	6.2	6.2	6.1	6.0
53	8.83 075	186	8.83 175	186	1.16 825	9.99 900	7	3	9.4	9.4	9.2	9.2	9.0
54	8.83 261	185	8.83 361	186	1.16 639	9.99 899	6	4	12.6	12.5	12.3	12.2	12.1
55	8.83 446	184	8.83 547	185	1.16 453	9.99 898	5	5	15.8	15.6	15.4	15.2	15.1
56	8.83 630	183	8.83 732	184	1.16 268	9.99 898	4	6	18.9	18.7	18.5	18.3	18.1
57	8.83 813	183	8.83 916	184	1.16 084	9.99 897	3	7	22.0	21.8	21.6	21.4	21.1
58	8.83 996	181	8.84 100	182	1.15 900	9.99 896	2	8	25.2	24.9	24.7	24.4	24.1
59	8.84 177	181	8.84 282	182	1.15 718	9.99 895	1	9	28.4	28.0	27.8	27.4	27.2
60	8.84 358		8.84 464		1.15 536	9.99 894	**0**	10	31.5	31.2	30.8	30.5	30.2
′	L. Cos.	d.	L. Cot.	c.d	L. Tan.	L. Sin.	′	P. P.					

93° (273°) (266°) **86°**

* Interpolation in this section of the table is not accurate.

COMMON LOGARITHMS OF THE TRIGONOMETRIC FUNCTIONS (Continued)

4° (184°) (355°) **175°**

′	L. Sin.	*d.	L. Tan.	*c.d.	L. Cot.	L. Cos.	′	P. P.					
0	8.84 358	181	8.84 464	182	1.15 536	9.99 894	**60**	″	**182**	**181**	**179**	**178**	**177**
1	8.84 539	179	8.84 646	180	1.15 354	9.99 893	59	1	3.0	3.0	3.0	3.0	3.0
2	8.84 718	179	8.84 826	180	1.15 174	9.99 892	58	2	6.1	6.0	6.0	5.9	5.9
3	8.84 897	178	8.85 006	179	1.14 994	9.99 891	57	3	9.1	9.0	9.0	8.9	8.8
4	8.85 075	177	8.85 185	178	1.14 815	9.99 891	56	4	12.1	12.1	11.9	11.9	11.8
5	8.85 252	177	8.85 363	177	1.14 637	9.99 890	55	5	15.2	15.1	14.9	14.8	14.8
6	8.85 429	176	8.85 540	177	1.14 460	9.99 889	54	6	18.2	18.1	17.9	17.8	17.7
7	8.85 605	175	8.85 717	176	1.14 283	9.99 888	53	7	21.2	21.1	20.9	20.8	20.6
8	8.85 780	175	8.85 893	176	1.14 107	9.99 887	52	8	24.3	24.1	23.9	23.7	23.6
9	8.85 955	173	8.86 069	174	1.13 931	9.99 886	51	9	27.3	27.2	26.8	26.7	26.6
10	8.86 128	173	8.86 243	174	1.13 757	9.99 885	**50**	″	**176**	**175**	**174**	**173**	**172**
11	8.86 301	173	8.86 417	174	1.13 583	9.99 884	49	1	2.9	2.9	2.9	2.9	2.9
12	8.86 474	171	8.86 591	172	1.13 409	9.99 883	48	2	5.9	5.8	5.8	5.8	5.7
13	8.86 645	171	8.86 763	172	1.13 237	9.99 882	47	3	8.8	8.8	8.7	8.6	8.6
14	8.86 816	171	8.86 935	171	1.13 065	9.99 881	46	4	11.7	11.7	11.6	11.5	11.5
15	8.86 987	169	8.87 106	171	1.12 894	9.99 880	45	5	14.7	14.6	14.5	14.4	14.3
16	8.87 156	169	8.87 277	170	1.12 723	9.99 879	44	6	17.6	17.5	17.4	17.3	17.2
17	8.87 325	169	8.87 447	169	1.12 553	9.99 879	43	7	20.5	20.4	20.3	20.2	20.1
18	8.87 494	167	8.87 616	169	1.12 384	9.99 878	42	8	23.5	23.3	23.2	23.1	22.9
19	8.87 661	168	8.87 785	168	1.12 215	9.99 877	41	9	26.4	26.2	26.1	26.0	25.8
20	8.87 829	166	8.87 953	167	1.12 047	9.99 876	**40**	″	**171**	**170**	**169**	**168**	**167**
21	8.87 995	166	8.88 120	167	1.11 880	9.99 875	39	1	2.8	2.8	2.8	2.8	2.8
22	8.88 161	165	8.88 287	166	1.11 713	9.99 874	38	2	5.7	5.7	5.6	5.6	5.6
23	8.88 326	164	8.88 453	165	1.11 547	9.99 873	37	3	8.6	8.5	8.4	8.4	8.4
24	8.88 490	164	8.88 618	165	1.11 382	9.99 872	36	4	11.4	11.3	11.3	11.2	11.1
25	8.88 654	163	8.88 783	165	1.11 217	9.99 871	35	5	14.2	14.2	14.1	14.0	13.9
26	8.88 817	163	8.88 948	163	1.11 052	9.99 870	34	6	17.1	17.0	16.9	16.8	16.7
27	8.88 980	162	8.89 111	163	1.10 889	9.99 869	33	7	20.0	19.8	19.7	19.6	19.5
28	8.89 142	162	8.89 274	163	1.10 726	9.99 868	32	8	22.8	22.7	22.5	22.4	22.3
29	8.89 304	160	8.89 437	161	1.10 563	9.99 867	31	9	25.6	25.5	25.4	25.2	25.0
30	8.89 464	161	8.89 598	162	1.10 402	9.99 866	**30**	″	**166**	**165**	**164**	**163**	**162**
31	8.89 625	159	8.89 760	160	1.10 240	9.99 865	29	1	2.8	2.8	2.7	2.7	2.7
32	8.89 784	159	8.89 920	160	1.10 080	9.99 864	28	2	5.5	5.5	5.5	5.4	5.4
33	8.89 943	159	8.90 080	160	1.09 920	9.99 863	27	3	8.3	8.2	8.2	8.2	8.1
34	8.90 102	158	8.90 240	159	1.09 760	9.99 862	26	4	11.1	11.0	10.9	10.9	10.8
35	8.90 260	157	8.90 399	158	1.09 601	9.99 861	25	5	13.8	13.8	13.7	13.6	13.5
36	8.90 417	157	8.90 557	158	1.09 443	9.99 860	24	6	16.6	16.5	16.4	16.3	16.2
37	8.90 574	156	8.90 715	157	1.09 285	9.99 859	23	7	19.4	19.2	19.1	19.0	18.9
38	8.90 730	155	8.90 872	157	1.09 128	9.99 858	22	8	22.1	22.0	21.9	21.7	21.6
39	8.90 885	155	8.91 029	156	1.08 971	9.99 857	21	9	24.9	24.8	24.6	24.4	24.3
40	8.91 040	155	8.91 185	155	1.08 815	9.99 856	**20**	″	**161**	**160**	**159**	**158**	**157**
41	8.91 195	154	8.91 340	155	1.08 660	9.99 855	19	1	2.7	2.7	2.6	2.6	2.6
42	8.91 349	153	8.91 495	155	1.08 505	9.99 854	18	2	5.4	5.3	5.3	5.3	5.2
43	8.91 502	153	8.91 650	153	1.08 350	9.99 853	17	3	8.0	8.0	8.0	7.9	7.8
44	8.91 655	152	8.91 803	154	1.08 197	9.99 852	16	4	10.7	10.7	10.6	10.5	10.5
45	8.91 807	152	8.91 957	153	1.08 043	9.99 851	15	5	13.4	13.3	13.2	13.2	13.1
46	8.91 959	151	8.92 110	152	1.07 890	9.99 850	14	6	16.1	16.0	15.9	15.8	15.7
47	8.92 110	151	8.92 262	152	1.07 738	9.99 848	13	7	18.8	18.7	18.6	18.4	18.3
48	8.92 261	150	8.92 414	151	1.07 586	9.99 847	12	8	21.5	21.3	21.2	21.1	20.9
49	8.92 411	150	8.92 565	151	1.07 435	9.99 846	11	9	24.2	24.0	23.8	23.7	23.6
50	8.92 561	149	8.92 716	150	1.07 284	9.99 845	**10**	″	**156**	**155**	**154**	**153**	**152**
51	8.92 710	149	8.92 866	150	1.07 134	9.99 844	9	1	2.6	2.6	2.6	2.6	2.5
52	8.92 859	148	8.93 016	149	1.06 984	9.99 843	8	2	5.2	5.2	5.1	5.1	5.1
53	8.93 007	147	8.93 165	148	1.06 835	9.99 842	7	3	7.8	7.8	7.7	7.6	7.6
54	8.93 154	147	8.93 313	149	1.06 687	9.99 841	6	4	10.4	10.3	10.3	10.2	10.1
55	8.93 301	147	8.93 462	147	1.06 538	9.99 840	5	5	13.0	12.9	12.8	12.8	12.7
56	8.93 448	146	8.93 609	147	1.06 391	9.99 839	4	6	15.6	15.5	15.4	15.3	15.2
57	8.93 594	146	8.93 756	147	1.06 244	9.99 838	3	7	18.2	18.1	18.0	17.8	17.7
58	8.93 740	145	8.93 903	146	1.06 097	9.99 837	2	8	20.8	20.7	20.5	20.4	20.3
59	8.93 885	145	8.94 049	146	1.05 951	9.99 836	1	9	23.4	23.2	23.1	23.0	22.8
60	8.94 030		8.94 195		1.05 805	9.99 834	**0**	10	26.0	25.8	25.7	25.5	25.3
′	L. Cos.	d.	L. Cot.	c.d.	L. Tan.	L. Sin.	′	P. P.					

94° (274°) (265°) **85°**

* **Interpolation in this section of the table is not accurate.**

COMMON LOGARITHMS OF THE TRIGONOMETRIC FUNCTIONS (Continued)

5° (185°) (354°) **174°**

′	L. Sin.	d.	L. Tan.	c.d.	L. Cot.	L. Cos.	′	″	P. P.				
0	8.94 030	144	8.94 195	145	1.05 805	9.99 834	**60**	″	**151**	**149**	**148**	**147**	**146**
1	8.94 174	143	8.94 340	145	1.05 660	9.99 833	59	1	2.5	2.5	2.5	2.4	2.4
2	8.94 317	144	8.94 485	145	1.05 515	9.99 832	58	2	5.0	5.0	4.9	4.9	4.9
3	8.94 461	142	8.94 630	143	1.05 370	9.99 831	57	3	7.6	7.4	7.4	7.4	7.3
4	8.94 603	143	8.94 773	144	1.05 227	9.99 830	56	4	10.1	9.9	9.9	9.8	9.7
5	8.94 746	141	8.94 917	143	1.05 083	9.99 829	55	5	12.6	12.4	12.3	12.2	12.2
6	8.94 887	142	8.95 060	142	1.04 940	9.99 828	54	6	15.1	14.9	14.8	14.7	14.6
7	8.95 029	141	8.95 202	142	1.04 798	9.99 827	53	7	17.6	17.4	17.3	17.2	17.0
8	8.95 170	140	8.95 344	142	1.04 656	9.99 825	52	8	20.1	19.9	19.7	19.6	19.5
9	8.95 310	140	8.95 486	141	1.04 514	9.99 824	51	9	22.6	22.4	22.2	22.0	21.9
10	8.95 450	139	8.95 627	140	1.04 373	9.99 823	**50**	″	**145**	**144**	**143**	**142**	**141**
11	8.95 589	139	8.95 767	141	1.04 233	9.99 822	49	1	2.4	2.4	2.4	2.4	2.4
12	8.95 728	139	8.95 908	139	1.04 092	9.99 821	48	2	4.8	4.8	4.8	4.7	4.7
13	8.95 867	138	8.96 047	140	1.03 953	9.99 820	47	3	7.2	7.2	7.2	7.1	7.0
14	8.96 005	138	8.96 187	138	1.03 813	9.99 819	46	4	9.7	9.6	9.5	9.5	9.4
15	8.96 143	137	8.96 325	139	1.03 675	9.99 817	45	5	12.1	12.0	11.9	11.8	11.8
16	8.96 280	137	8.96 464	138	1.03 536	9.99 816	44	6	14.5	14.4	14.3	14.2	14.1
17	8.96 417	136	8.96 602	137	1.03 398	9.99 815	43	7	16.9	16.8	16.7	16.6	16.4
18	8.96 553	136	8.96 739	138	1.03 261	9.99 814	42	8	19.3	19.2	19.1	18.9	18.8
19	8.96 689	136	8.96 877	136	1.03 123	9.99 813	41	9	21.8	21.6	21.4	21.3	21.2
20	8.96 825	135	8.97 013	137	1.02 987	9.99 812	**40**	″	**140**	**139**	**138**	**137**	**136**
21	8.96 960	135	8.97 150	135	1.02 850	9.99 810	39	1	2.3	2.3	2.3	2.3	2.3
22	8.97 095	134	8.97 285	136	1.02 715	9.99 809	38	2	4.7	4.6	4.6	4.6	4.5
23	8.97 229	134	8.97 421	135	1.02 579	9.99 808	37	3	7.0	7.0	6.9	6.8	6.8
24	8.97 363	133	8.97 556	135	1.02 444	9.99 807	36	4	9.3	9.3	9.2	9.1	9.1
25	8.97 496	133	8.97 691	134	1.02 309	9.99 806	35	5	11.7	11.6	11.5	11.4	11.3
26	8.97 629	133	8.97 825	134	1.02 175	9.99 804	34	6	14.0	13.9	13.8	13.7	13.6
27	8.97 762	132	8.97 959	133	1.02 041	9.99 803	33	7	16.3	16.2	16.1	16.0	15.9
28	8.97 894	132	8.98 092	133	1.01 908	9.99 802	32	8	18.7	18.5	18.4	18.3	18.1
29	8.98 026	131	8.98 225	133	1.01 775	9.99 801	31	9	21.0	20.8	20.7	20.6	20.4
30	8.98 157	131	8.98 358	132	1.01 642	9.99 800	**30**	″	**135**	**134**	**133**	**132**	**131**
31	8.98 288	131	8.98 490	132	1.01 510	9.99 798	29	1	2.2	2.2	2.2	2.2	2.2
32	8.98 419	130	8.98 622	131	1.01 378	9.99 797	28	2	4.5	4.5	4.4	4.4	4.4
33	8.98 549	130	8.98 753	131	1.01 247	9.99 796	27	3	6.8	6.7	6.6	6.6	6.6
34	8.98 679	129	8.98 884	131	1.01 116	9.99 795	26	4	9.0	8.9	8.9	8.8	8.7
35	8.98 808	129	8.99 015	130	1.00 985	9.99 793	25	5	11.2	11.2	11.1	11.0	10.9
36	8.98 937	129	8.99 145	130	1.00 855	9.99 792	24	6	13.5	13.4	13.3	13.2	13.1
37	8.99 066	128	8.99 275	130	1.00 725	9.99 791	23	7	15.8	15.6	15.5	15.4	15.3
38	8.99 194	128	8.99 405	129	1.00 595	9.99 790	22	8	18.0	17.9	17.7	17.6	17.5
39	8.99 322	128	8.99 534	128	1.00 466	9.99 788	21	9	20.2	20.1	20.0	19.8	19.6
40	8.99 450	127	8.99 662	129	1.00 338	9.99 787	**20**	″	**130**	**129**	**128**	**127**	**126**
41	8.99 577	127	8.99 791	128	1.00 209	9.99 786	19	1	2.2	2.2	2.1	2.1	2.1
42	8.99 704	126	8.99 919	127	1.00 081	9.99 785	18	2	4.3	4.3	4.3	4.2	4.2
43	8.99 830	126	9.00 046	128	0.99 954	9.99 783	17	3	6.5	6.4	6.4	6.4	6.3
44	8.99 956	126	9.00 174	127	0.99 826	9.99 782	16	4	8.7	8.6	8.5	8.5	8.4
45	9.00 082	125	9.00 301	126	0.99 699	9.99 781	15	5	10.8	10.8	10.7	10.6	10.5
46	9.00 207	125	9.00 427	126	0.99 573	9.99 780	14	6	13.0	12.9	12.8	12.7	12.6
47	9.00 332	124	9.00 553	126	0.99 447	9.99 778	13	7	15.2	15.0	14.9	14.8	14.7
48	9.00 456	125	9.00 679	126	0.99 321	9.99 777	12	8	17.3	17.2	17.1	16.9	16.8
49	9.00 581	123	9.00 805	125	0.99 195	9.99 776	11	9	19.5	19.4	19.2	19.0	18.9
50	9.00 704	124	9.00 930	125	0.99 070	9.99 775	**10**	″	**125**	**124**	**123**	**122**	**121**
51	9.00 828	123	9.01 055	124	0.98 945	9.99 773	9	1	2.1	2.1	2.0	2.0	2.0
52	9.00 951	123	9.01 179	124	0.98 821	9.99 772	8	2	4.2	4.1	4.1	4.1	4.0
53	9.01 074	122	9.01 303	124	0.98 697	9.99 771	7	3	6.2	6.2	6.2	6.1	6.0
54	9.01 196	122	9.01 427	123	0.98 573	9.99 769	6	4	8.3	8.3	8.2	8.1	8.1
55	9.01 318	122	9.01 550	123	0.98 450	9.99 768	5	5	10.4	10.3	10.2	10.2	10.1
56	9.01 440	121	9.01 673	123	0.98 327	9.99 767	4	6	12.5	12.4	12.3	12.2	12.1
57	9.01 561	121	9.01 796	122	0.98 204	9.99 765	3	7	14.6	14.5	14.4	14.2	14.1
58	9.01 682	121	9.01 918	122	0.98 082	9.99 764	2	8	16.7	16.5	16.4	16.3	16.1
59	9.01 803	120	9.02 040	122	0.97 960	9.99 763	1	9	18.8	18.6	18.4	18.3	18.2
60	9.01 923		9.02 162		0.97 838	9.99 761	**0**	10	20.8	20.7	20.5	20.3	20.2
′	L. Cos.	d.	L. Cot.	c.d.	L. Tan.	L. Sin.	′		P. P.				

95° (275°) (264°) **84°**

COMMON LOGARITHMS OF THE TRIGONOMETRIC FUNCTIONS (Continued)

6° (186°) (353°) **173°**

′	L. Sin.	d.	L. Tan.	c.d.	L. Cot.	L. Cos.	′	″	P. P.			
0	9.01 923	120	9.02 162	121	0.97 838	9.99 761	**60**	″	**121**	**120**	**119**	**118**
1	9.02 043	120	9.02 283	121	0.97 717	9.99 760	59	1	2.0	2.0	2.0	2.0
2	9.02 163	120	9.02 404	121	0.97 596	9.99 759	58	2	4.0	4.0	4.0	3.9
3	9.02 283	119	9.02 525	120	0.97 475	9.99 757	57	3	6.0	6.0	6.0	5.9
4	9.02 402	118	9.02 645	121	0.97 355	9.99 756	56	4	8.1	8.0	7.9	7.9
5	9.02 520	119	9.02 766	119	0.97 234	9.99 755	55	5	10.1	10.0	9.9	9.8
6	9.02 639	118	9.02 885	120	0.97 115	9.99 753	54	6	12.1	12.0	11.9	11.8
7	9.02 757	117	9.03 005	119	0.96 995	9.99 752	53	7	14.1	14.0	13.9	13.8
8	9.02 874	118	9.03 124	118	0.96 876	9.99 751	52	8	16.1	16.0	15.9	15.7
9	9.02 992	117	9.03 242	119	0.96 758	9.99 749	51	9	18.2	18.0	17.8	17.7
10	9.03 109	117	9.03 361	118	0.96 639	9.99 748	**50**	10	20.2	20.0	19.8	19.7
11	9.03 226	116	9.03 479	118	0.96 521	9.99 747	49	20	40.3	40.0	39.7	39.3
12	9.03 342	116	9.03 597	117	0.96 403	9.99 745	48	30	60.5	60.0	59.5	59.0
13	9.03 458	116	9.03 714	118	0.96 286	9.99 744	47	40	80.7	80.0	79.3	78.7
14	9.03 574	116	9.03 832	116	0.96 168	9.99 742	46	50	100.8	100.0	99.2	98.3
15	9.03 690	115	9.03 948	117	0.96 052	9.99 741	45	″	**117**	**116**	**115**	**114**
16	9.03 805	115	9.04 065	116	0.95 935	9.99 740	44	1	2.0	1.9	1.9	1.9
17	9.03 920	114	9.04 181	116	0.95 819	9.99 738	43	2	3.9	3.9	3.8	3.8
18	9.04 034	115	9.04 297	116	0.95 703	9.99 737	42	3	5.8	5.8	5.8	5.7
19	9.04 149	113	9.04 413	115	0.95 587	9.99 736	41	4	7.8	7.7	7.7	7.6
20	9.04 262	114	9.04 528	115	0.95 472	9.99 734	**40**	5	9.8	9.7	9.6	9.5
21	9.04 376	114	9.04 643	115	0.95 357	9.99 733	39	6	11.7	11.6	11.5	11.4
22	9.04 490	113	9.04 758	115	0.95 242	9.99 731	38	7	13.6	13.5	13.4	13.3
23	9.04 603	112	9.04 873	114	0.95 127	9.99 730	37	8	15.6	15.5	15.3	15.2
24	9.04 715	113	9.04 987	114	0.95 013	9.99 728	36	9	17.6	17.4	17.2	17.1
25	9.04 828	112	9.05 101	113	0.94 899	9.99 727	35	10	19.5	19.3	19.2	19.0
26	9.04 940	112	9.05 214	114	0.94 786	9.99 726	34	20	39.0	38.7	38.3	38.0
27	9.05 052	112	9.05 328	113	0.94 672	9.99 724	33	30	58.5	58.0	57.5	57.0
28	9.05 164	111	9.05 441	112	0.94 559	9.99 723	32	40	78.0	77.3	76.7	76.0
29	9.05 275	111	9.05 553	113	0.94 447	9.99 721	31	50	97.5	96.7	95.8	95.0
30	9.05 386	111	9.05 666	112	0.94 334	9.99 720	**30**	″	**113**	**112**	**111**	**110**
31	9.05 497	110	9.05 778	112	0.94 222	9.99 718	29	1	1.9	1.9	1.8	1.8
32	9.05 607	110	9.05 890	112	0.94 110	9.99 717	28	2	3.8	3.7	3.7	3.7
33	9.05 717	110	9.06 002	111	0.93 998	9.99 716	27	3	5.6	5.6	5.6	5.5
34	9.05 827	110	9.06 113	111	0.93 887	9.99 714	26	4	7.5	7.5	7.4	7.3
35	9.05 937	109	9.06 224	111	0.93 776	9.99 713	25	5	9.4	9.3	9.2	9.2
36	9.06 046	109	9.06 335	110	0.93 665	9.99 711	24	6	11.3	11.2	11.1	11.0
37	9.06 155	109	9.06 445	111	0.93 555	9.99 710	23	7	13.2	13.1	13.0	12.8
38	9.06 264	108	9.06 556	110	0.93 444	9.99 708	22	8	15.1	14.9	14.8	14.7
39	9.06 372	109	9.06 666	109	0.93 334	9.99 707	21	9	17.0	16.8	16.6	16.5
40	9.06 481	108	9.06 775	110	0.93 225	9.99 705	**20**	10	18.8	18.7	18.5	18.3
41	9.06 589	107	9.06 885	109	0.93 115	9.99 704	19	20	37.7	37.3	37.0	36.7
42	9.06 696	108	9.06 994	109	0.93 006	9.99 702	18	30	56.5	56.0	55.5	55.0
43	9.06 804	107	9.07 103	108	0.92 897	9.99 701	17	40	75.3	74.7	74.0	73.3
44	9.06 911	107	9.07 211	109	0.92 789	9.99 699	16	50	94.2	93.3	92.5	91.7
45	9.07 018	106	9.07 320	108	0.92 680	9.99 698	15	″	**109**	**108**	**107**	**106**
46	9.07 124	107	9.07 428	108	0.92 572	9.99 696	14	1	1.8	1.8	1.8	1.8
47	9.07 231	106	9.07 536	107	0.92 464	9.99 695	13	2	3.6	3.6	3.6	3.5
48	9.07 337	105	9.07 643	108	0.92 357	9.99 693	12	3	5.4	5.4	5.4	5.3
49	9.07 442	106	9.07 751	107	0.92 249	9.99 692	11	4	7.3	7.2	7.1	7.1
50	9.07 548	105	9.07 858	106	0.92 142	9.99 690	**10**	5	9.1	9.0	8.9	8.8
51	9.07 653	105	9.07 964	107	0.92 036	9.99 689	9	6	10.9	10.8	10.7	10.6
52	9.07 758	105	9.08 071	106	0.91 929	9.99 687	8	7	12.7	12.6	12.5	12.4
53	9.07 863	105	9.08 177	106	0.91 823	9.99 686	7	8	14.5	14.4	14.3	14.1
54	9.07 968	104	9.08 283	106	0.91 717	9.99 684	6	9	16.4	16.2	16.0	15.9
55	9.08 072	104	9.08 389	106	0.91 611	9.99 683	5	10	18.2	18.0	17.8	17.7
56	9.08 176	104	9.08 495	105	0.91 505	9.99 681	4	20	36.3	36.0	35.7	35.3
57	9.08 280	103	9.08 600	105	0.91 400	9.99 680	3	30	54.5	54.0	53.5	53.0
58	9.08 383	103	9.08 705	105	0.91 295	9.99 678	2	40	72.7	72.0	71.3	70.7
59	9.08 486	103	9.08 810	104	0.91 190	9.99 677	1	50	90.8	90.0	89.2	88.3
60	9.08 589		9.08 914		0.91 086	9.99 675	**0**					
′	L. Cos.	d.	L. Cot.	c.d.	L. Tan.	L. Sin.	′		P. P.			

96° (276°) (263°) **83°**

COMMON LOGARITHMS OF THE TRIGONOMETRIC FUNCTIONS (Continued)

7° (187°) (352°) **172°**

′	L. Sin.	d.	L. Tan.	c.d.	L. Cot.	L. Cos.	′	P. P.				
0	9.08 589	103	9.08 914	105	0.91 086	9.99 675	**60**	″	**105**	**104**	**103**	**102**
1	9.08 692	103	9.09 019	104	0.90 981	9.99 674	59	1	1.8	1.7	1.7	1.7
2	9.08 795	102	9.09 123	104	0.90 877	9.99 672	58	2	3.5	3.5	3.4	3.4
3	9.08 897	102	9.09 227	103	0.90 773	9.99 670	57	3	5.2	5.2	5.2	5.1
4	9.08 999	102	9.09 330	104	0.90 670	9.99 669	56	4	7.0	6.9	6.9	6.8
5	9.09 101	101	9.09 434	103	0.90 566	9.99 667	55	5	8.8	8.7	8.6	8.5
6	9.09 202	102	9.09 537	103	0.90 463	9.99 666	54	6	10.5	10.4	10.3	10.2
7	9.09 304	101	9.09 640	102	0.90 360	9.99 664	53	7	12.2	12.1	12.0	11.9
8	9.09 405	101	9.09 742	103	0.90 258	9.99 663	52	8	14.0	13.9	13.7	13.6
9	9.09 506	100	9.09 845	102	0.90 155	9.99 661	51	9	15.8	15.6	15.4	15.3
10	9.09 606	101	9.09 947	102	0.90 053	9.99 659	**50**	10	17.5	17.3	17.2	17.0
11	9.09 707	100	9.10 049	101	0.89 951	9.99 658	49	20	35.0	34.7	34.3	34.0
12	9.09 807	100	9.10 150	102	0.89 850	9.99 656	48	30	52.5	52.0	51.5	51.0
13	9.09 907	99	9.10 252	101	0.89 748	9.99 655	47	40	70.0	69.3	68.7	68.0
14	9.10 006	100	9.10 353	101	0.89 647	9.99 653	46	50	87.5	86.7	85.8	85.0
15	9.10 106	99	9.10 454	101	0.89 546	9.99 651	45	″	**101**	**100**	**99**	**98**
16	9.10 205	99	9.10 555	101	0.89 445	9.99 650	44	1	1.7	1.7	1.6	1.6
17	9.10 304	98	9.10 656	100	0.89 344	9.99 648	43	2	3.4	3.3	3.3	3.3
18	9.10 402	99	9.10 756	100	0.89 244	9.99 647	42	3	5.0	5.0	5.0	4.9
19	9.10 501	98	9.10 856	100	0.89 144	9.99 645	41	4	6.7	6.7	6.6	6.5
20	9.10 599	98	9.10 956	100	0.89 044	9.99 643	**40**	5	8.4	8.3	8.2	8.2
21	9.10 697	98	9.11 056	99	0.88 944	9.99 642	39	6	10.1	10.0	9.9	9.8
22	9.10 795	98	9.11 155	99	0.88 845	9.99 640	38	7	11.8	11.7	11.6	11.4
23	9.10 893	97	9.11 254	99	0.88 746	9.99 638	37	8	13.5	13.3	13.2	13.1
24	9.10 990	97	9.11 353	99	0.88 647	9.99 637	36	9	15.2	15.0	14.8	14.7
25	9.11 087	97	9.11 452	99	0.88 548	9.99 635	35	10	16.8	16.7	16.5	16.3
26	9.11 184	97	9.11 551	98	0.88 449	9.99 633	34	20	33.7	33.3	33.0	32.7
27	9.11 281	96	9.11 649	98	0.88 351	9.99 632	33	30	50.5	50.0	49.5	49.0
28	9.11 377	97	9.11 747	98	0.88 253	9.99 630	32	40	67.3	66.7	66.0	65.3
29	9.11 474	96	9.11 845	98	0.88 155	9.99 629	31	50	84.2	83.3	82.5	81.7
30	9.11 570	96	9.11 943	97	0.88 057	9.99 627	**30**	″	**97**	**96**	**95**	**94**
31	9.11 666	95	9.12 040	98	0.87 960	9.99 625	29	1	1.6	1.6	1.6	1.6
32	9.11 761	96	9.12 138	97	0.87 862	9.99 624	28	2	3.2	3.2	3.2	3.1
33	9.11 857	95	9.12 235	97	0.87 765	9.99 622	27	3	4.8	4.8	4.8	4.7
34	9.11 952	95	9.12 332	96	0.87 668	9.99 620	26	4	6.5	6.4	6.3	6.3
35	9.12 047	95	9.12 428	97	0.87 572	9.99 618	25	5	8.1	8.0	7.9	7.8
36	9.12 142	94	9.12 525	96	0.87 475	9.99 617	24	6	9.7	9.6	9.5	9.4
37	9.12 236	95	9.12 621	96	0.87 379	9.99 615	23	7	11.3	11.2	11.1	11.0
38	9.12 331	94	9.12 717	96	0.87 283	9.99 613	22	8	12.9	12.8	12.7	12.5
39	9.12 425	94	9.12 813	96	0.87 187	9.99 612	21	9	14.6	14.4	14.2	14.1
40	9.12 519	93	9.12 909	95	0.87 091	9.99 610	**20**	10	16.2	16.0	15.8	15.7
41	9.12 612	94	9.13 004	95	0.86 996	9.99 608	19	20	32.3	32.0	31.7	31.3
42	9.12 706	93	9.13 099	95	0.86 901	9.99 607	18	30	48.5	48.0	47.5	47.0
43	9.12 799	93	9.13 194	95	0.86 806	9.99 605	17	40	64.7	64.0	63.3	62.7
44	9.12 892	93	9.13 289	95	0.86 711	9.99 603	16	50	80.8	80.0	79.2	78.3
45	9.12 985	93	9.13 384	94	0.86 616	9.99 601	15	″	**93**	**92**	**91**	**90**
46	9.13 078	93	9.13 478	95	0.86 522	9.99 600	14	1	1.6	1.5	1.5	1.5
47	9.13 171	92	9.13 573	94	0.86 427	9.99 598	13	2	3.1	3.1	3.0	3.0
48	9.13 263	92	9.13 667	94	0.86 333	9.99 596	12	3	4.6	4.6	4.6	4.5
49	9.13 355	92	9.13 761	93	0.86 239	9.99 595	11	4	6.2	6.1	6.1	6.0
50	9.13 447	92	9.13 854	94	0.86 146	9.99 593	**10**	5	7.8	7.7	7.6	7.5
51	9.13 539	91	9.13 948	93	0.86 052	9.99 591	9	6	9.3	9.2	9.1	9.0
52	9.13 630	92	9.14 041	93	0.85 959	9.99 589	8	7	10.8	10.7	10.6	10.5
53	9.13 722	91	9.14 134	93	0.85 866	9.99 588	7	8	12.4	12.3	12.1	12.0
54	9.13 813	91	9.14 227	93	0.85 773	9.99 586	6	9	14.0	13.8	13.6	13.5
55	9.13 904	90	9.14 320	92	0.85 680	9.99 584	5	10	15.5	15.3	15.2	15.0
56	9.13 994	91	9.14 412	92	0.85 588	9.99 582	4	20	31.0	30.7	30.3	30.0
57	9.14 085	90	9.14 504	93	0.85 496	9.99 581	3	30	46.5	46.0	45.5	45.0
58	9.14 175	91	9.14 597	91	0.85 403	9.99 579	2	40	62.0	61.3	60.7	60.0
59	9.14 266	90	9.14 688	92	0.85 312	9.99 577	1	50	77.5	76.7	75.8	75.0
60	9.14 356		9.14 780		0.85 220	9.99 575	**0**					
′	L. Cos.	d.	L. Cot.	c.d.	L. Tan.	L. Sin.	′	P. P.				

97° (277°) (262°) **82°**

COMMON LOGARITHMS OF THE TRIGONOMETRIC FUNCTIONS (Continued)

8° (188°) (351°) **171°**

′	L. Sin.	d.	L. Tan.	c.d.	L. Cot.	L. Cos.	′
0	9.14 356	89	9.14 780	92	0.85 220	9.99 575	**60**
1	9.14 445	90	9.14 872	91	0.85 128	9.99 574	59
2	9.14 535	89	9.14 963	91	0.85 037	9.99 572	58
3	9.14 624	90	9.15 054	91	0.84 946	9.99 570	57
4	9.14 714	89	9.15 145	91	0.84 855	9.99 568	56
5	9.14 803	88	9.15 236	91	0.84 764	9.99 566	55
6	9.14 891	89	9.15 327	90	0.84 673	9.99 565	54
7	9.14 980	89	9.15 417	91	0.84 583	9.99 563	53
8	9.15 069	88	9.15 508	90	0.84 492	9.99 561	52
9	9.15 157	88	9.15 598	90	0.84 402	9.99 559	51
10	9.15 245	88	9.15 688	89	0.84 312	9.99 557	**50**
11	9.15 333	88	9.15 777	90	0.84 223	9.99 556	49
12	9.15 421	87	9.15 867	89	0.84 133	9.99 554	48
13	9.15 508	88	9.15 956	90	0.84 044	9.99 552	47
14	9.15 596	87	9.16 046	89	0.83 954	9.99 550	46
15	9.15 683	87	9.16 135	89	0.83 865	9.99 548	45
16	9.15 770	87	9.16 224	88	0.83 776	9.99 546	44
17	9.15 857	87	9.16 312	89	0.83 688	9.99 545	43
18	9.15 944	86	9.16 401	88	0.83 599	9.99 543	42
19	9.16 030	86	9.16 489	88	0.83 511	9.99 541	41
20	9.16 116	87	9.16 577	88	0.83 423	9.99 539	**40**
21	9.16 203	86	9.16 665	88	0.83 335	9.99 537	39
22	9.16 289	85	9.16 753	88	0.83 247	9.99 535	38
23	9.16 374	86	9.16 841	87	0.83 159	9.99 533	37
24	9.16 460	85	9.16 928	88	0.83 072	9.99 532	36
25	9.16 545	86	9.17 016	87	0.82 984	9.99 530	35
26	9.16 631	85	9.17 103	87	0.82 897	9.99 528	34
27	9.16 716	85	9.17 190	87	0.82 810	9.99 526	33
28	9.16 801	85	9.17 277	86	0.82 723	9.99 524	32
29	9.16 886	84	9.17 363	87	0.82 637	9.99 522	31
30	9.16 970	85	9.17 450	86	0.82 550	9.99 520	**30**
31	9.17 055	84	9.17 536	86	0.82 464	9.99 518	29
32	9.17 139	84	9.17 622	86	0.82 378	9.99 517	28
33	9.17 223	84	9.17 708	86	0.82 292	9.99 515	27
34	9.17 307	84	9.17 794	86	0.82 206	9.99 513	26
35	9.17 391	83	9.17 880	85	0.82 120	9.99 511	25
36	9.17 474	84	9.17 965	86	0.82 035	9.99 509	24
37	9.17 558	83	9.18 051	85	0.81 949	9.99 507	23
38	9.17 641	83	9.18 136	85	0.81 864	9.99 505	22
39	9.17 724	83	9.18 221	85	0.81 779	9.99 503	21
40	9.17 807	83	9.18 306	85	0.81 694	9.99 501	**20**
41	9.17 890	83	9.18 391	84	0.81 609	9.99 499	19
42	9.17 973	82	9.18 475	85	0.81 525	9.99 497	18
43	9.18 055	82	9.18 560	84	0.81 440	9.99 495	17
44	9.18 137	83	9.18 644	84	0.81 356	9.99 494	16
45	9.18 220	82	9.18 728	84	0.81 272	9.99 492	15
46	9.18 302	81	9.18 812	84	0.81 188	9.99 490	14
47	9.18 383	82	9.18 896	83	0.81 104	9.99 488	13
48	9.18 465	82	9.18 979	84	0.81 021	9.99 486	12
49	9.18 547	81	9.19 063	83	0.80 937	9.99 484	11
50	9.18 628	81	9.19 146	83	0.80 854	9.99 482	**10**
51	9.18 709	81	9.19 229	83	0.80 771	9.99 480	9
52	9.18 790	81	9.19 312	83	0.80 688	9.99 478	8
53	9.18 871	81	9.19 395	83	0.80 605	9.99 476	7
54	9.18 952	81	9.19 478	83	0.80 522	9.99 474	6
55	9.19 033	80	9.19 561	82	0.80 439	9.99 472	5
56	9.19 113	80	9.19 643	82	0.80 357	9.99 470	4
57	9.19 193	80	9.19 725	82	0.80 275	9.99 468	3
58	9.19 273	80	9.19 807	82	0.80 193	9.99 466	2
59	9.19 353	80	9.19 889	82	0.80 111	9.99 464	1
60	9.19 433		9.19 971		0.80 029	9.99 462	**0**
′	L. Cos.	d.	L. Cot.	c.d.	L. Tan.	L. Sin.	′

P. P.

″	**92**	**91**	**90**
1	1.5	1.5	1.5
2	3.1	3.0	3.0
3	4.6	4.6	4.5
4	6.1	6.1	6.0
5	7.7	7.6	7.5
6	9.2	9.1	9.0
7	10.7	10.6	10.5
8	12.3	12.1	12.0
9	13.8	13.6	13.5
10	15.3	15.2	15.0
20	30.7	30.3	30.0
30	46.0	45.5	45.0
40	61.3	60.7	60.0
50	76.7	75.8	75.0
″	**89**	**88**	**87**
1	1.5	1.5	1.4
2	3.0	2.9	2.9
3	4.4	4.4	4.4
4	5.9	5.9	5.8
5	7.4	7.3	7.2
6	8.9	8.8	8.7
7	10.4	10.3	10.2
8	11.9	11.7	11.6
9	13.4	13.2	13.0
10	14.8	14.7	14.5
20	29.7	29.3	29.0
30	44.5	44.0	43.5
40	59.3	58.7	58.0
50	74.2	73.3	72.5
″	**86**	**85**	**84**
1	1.4	1.4	1.4
2	2.9	2.8	2.8
3	4.3	4.2	4.2
4	5.7	5.7	5.6
5	7.2	7.1	7.0
6	8.6	8.5	8.4
7	10.0	9.9	9.8
8	11.5	11.3	11.2
9	12.9	12.8	12.6
10	14.3	14.2	14.0
20	28.7	28.3	28.0
30	43.0	42.5	42.0
40	57.3	56.7	56.0
50	71.7	70.8	70.0
″	**83**	**82**	**81**
1	1.4	1.4	1.4
2	2.8	2.7	2.7
3	4.2	4.1	4.0
4	5.5	5.5	5.4
5	6.9	6.8	6.8
6	8.3	8.2	8.1
7	9.7	9.6	9.4
8	11.1	10.9	10.8
9	12.4	12.3	12.2
10	13.8	13.7	13.5
20	27.7	27.3	27.0
30	41.5	41.0	40.5
40	55.3	54.7	54.0
50	69.2	68.3	67.5

98° (278°) (261°) **81°**

COMMON LOGARITHMS OF THE TRIGONOMETRIC FUNCTIONS (Continued)

9° (189°) (350°) **170°**

′	L. Sin.	d.	L. Tan.	c.d.	L. Cot.	L. Cos.	′
0	9.19 433	80	9.19 971	82	0.80 029	9.99 462	**60**
1	9.19 513	79	9.20 053	81	0.79 947	9.99 460	59
2	9.19 592	80	9.20 134	82	0.79 866	9.99 458	58
3	9.19 672	79	9.20 216	81	0.79 784	9.99 456	57
4	9.19 751	79	9.20 297	81	0.79 703	9.99 454	56
5	9.19 830	79	9.20 378	81	0.79 622	9.99 452	55
6	9.19 909	79	9.20 459	81	0.79 541	9.99 450	54
7	9.19 988	79	9.20 540	81	0.79 460	9.99 448	53
8	9.20 067	78	9.20 621	80	0.79 379	9.99 446	52
9	9.20 145	78	9.20 701	81	0.79 299	9.99 444	51
10	9.20 223	79	9.20 782	80	0.79 218	9.99 442	**50**
11	9.20 302	78	9.20 862	80	0.79 138	9.99 440	49
12	9.20 380	78	9.20 942	80	0.79 058	9.99 438	48
13	9.20 458	77	9.21 022	80	0.78 978	9.99 436	47
14	9.20 535	78	9.21 102	80	0.78 898	9.99 434	46
15	9.20 613	78	9.21 182	79	0.78 818	9.99 432	45
16	9.20 691	77	9.21 261	80	0.78 739	9.99 429	44
17	9.20 768	77	9.21 341	79	0.78 659	9.99 427	43
18	9.20 845	77	9.21 420	79	0.78 580	9.99 425	42
19	9.20 922	77	9.21 499	79	0.78 501	9.99 423	41
20	9.20 999	77	9.21 578	79	0.78 422	9.99 421	**40**
21	9.21 076	77	9.21 657	79	0.78 343	9.99 419	39
22	9.21 153	76	9.21 736	78	0.78 264	9.99 417	38
23	9.21 229	77	9.21 814	79	0.78 186	9.99 415	37
24	9.21 306	76	9.21 893	78	0.78 107	9.99 413	36
25	9 21 382	76	9.21 971	78	0.78 029	9.99 411	35
26	9.21 458	76	9.22 049	78	0.77 951	9.99 409	34
27	9.21 534	76	9.22 127	78	0.77 873	9.99 407	33
28	9.21 610	75	9.22 205	78	0.77 795	9.99 404	32
29	9.21 685	76	9.22 283	78	0.77 717	9.99 402	31
30	9.21 761	75	9.22 361	77	0.77 639	9.99 400	**30**
31	9.21 836	76	9.22 438	78	0.77 562	9.99 398	29
32	9.21 912	75	9.22 516	77	0.77 484	9.99 396	28
33	9.21 987	75	9.22 593	77	0.77 407	9.99 394	27
34	9.22 062	75	9.22 670	77	0.77 330	9.99 392	26
35	9.22 137	74	9.22 747	77	0.77 253	9.99 390	25
36	9.22 211	75	9.22 824	77	0.77 176	9.99 388	24
37	9.22 286	75	9.22 901	76	0.77 099	9.99 385	23
38	9.22 361	74	9.22 977	77	0.77 023	9.99 383	22
39	9.22 435	74	9.23 054	76	0.76 946	9.99 381	21
40	9.22 509	74	9.23 130	76	0.76 870	9.99 379	**20**
41	9.22 583	74	9.23 206	77	0.76 794	9.99 377	19
42	9.22 657	74	9.23 283	76	0.76 717	9.99 375	18
43	9.22 731	74	9.23 359	76	0.76 641	9.99 372	17
44	9.22 805	73	9.23 435	75	0.76 565	9.99 370	16
45	9.22 878	74	9.23 510	76	0.76 490	9.99 368	15
46	9.22 952	73	9.23 586	75	0.76 414	9.99 366	14
47	9.23 025	73	9.23 661	76	0.76 339	9.99 364	13
48	9.23 098	73	9.23 737	75	0.76 263	9.99 362	12
49	9.23 171	73	9.23 812	75	0.76 188	9.99 359	11
50	9.23 244	73	9.23 887	75	0.76 113	9.99 357	**10**
51	9.23 317	73	9.23 962	75	0.76 038	9.99 355	9
52	9.23 390	72	9.24 037	75	0.75 963	9.99 353	8
53	9.23 462	73	9.24 112	74	0.75 888	9.99 351	7
54	9.23 535	72	9.24 186	75	0.75 814	9.99 348	6
55	9.23 607	72	9.24 261	74	0.75 739	9.99 346	5
56	9.23 679	73	9.24 335	75	0.75 665	9.99 344	4
57	9.23 752	71	9.24 410	74	0.75 590	9.99 342	3
58	9.23 823	72	9.24 484	74	0.75 516	9.99 340	2
59	9.23 895	72	9.24 558	74	0.75 442	9.99 337	1
60	9.23 967		9.24 632		0.75 368	9.99 335	**0**
′	L. Cos.	d.	L. Cot.	c.d.	L. Tan.	L. Sin.	′

P. P.

″	**80**	**79**	**78**	**77**
1	1.3	1.3	1.3	1.3
2	2.7	2.6	2.6	2.6
3	4.0	4.0	3.9	3.8
4	5.3	5.3	5.2	5.1
5	6.7	6.6	6.5	6.4
6	8.0	7.9	7.8	7.7
7	9.3	9.2	9.1	9.0
8	10.7	10.5	10.4	10.3
9	12.0	11.8	11.7	11.6
10	13.3	13.2	13.0	12.8
20	26.7	26.3	26.0	25.7
30	40.0	39.5	39.0	38.5
40	53.3	52.7	52.0	51.3
50	66.7	65.8	65.0	64.2

″	**76**	**75**	**74**	**73**
1	1.3	1.2	1.2	1.2
2	2.5	2.5	2.5	2.4
3	3.8	3.8	3.7	3.6
4	5.1	5.0	4.9	4.9
5	6.3	6.2	6.2	6.1
6	7.6	7.5	7.4	7.3
7	8.9	8.8	8.6	8.5
8	10.1	10.0	9.9	9.7
9	11.4	11.2	11.1	11.0
10	12.7	12.5	12.3	12.2
20	25.3	25.0	24.7	24.3
30	38.0	37.5	37.0	36.5
40	50.7	50.0	49.3	48.7
50	63.3	62.5	61.7	60.8

″	**72**	**71**	**3**	**2**
1	1.2	1.2	0.0	0.0
2	2.4	2.4	0.1	0.1
3	3.6	3.6	0.2	0.1
4	4.8	4.7	0.2	0.1
5	6.0	5.9	0.2	0.2
6	7.2	7.1	0.3	0.2
7	8.4	8.3	0.4	0.2
8	9.6	9.5	0.4	0.3
9	10.8	10.6	0.4	0.3
10	12.0	11.8	0.5	0.3
20	24.0	23.7	1.0	0.7
30	36.0	35.5	1.5	1.0
40	48.0	47.3	2.0	1.3
50	60.0	59.2	2.5	1.7

	3/79	**3/78**	**3/77**
0			
	13.2	13.0	12.8
1			
	39.5	39.0	38.5
2			
	65.8	65.0	64.2
3			

	3/76	**3/75**	**3/74**
0			
	12.7	12.5	12.3
1			
	38.0	37.5	37.0
2			
	63.3	62.5	61.7
3			

99° (279°) (260°) **80°**

COMMON LOGARITHMS OF THE TRIGONOMETRIC FUNCTIONS (Continued)

10° (190°) (349°) **169°**

′	L. Sin.	d.	L. Tan.	c.d.	L. Cot.	L. Cos.	d.	′
0	9.23 967	72	9.24 632	74	0.75 368	9.99 335	2	**60**
1	9.24 039	71	9.24 706	73	0.75 294	9.99 333	2	59
2	9.24 110	71	9.24 779	74	0.75 221	9.99 331	3	58
3	9.24 181	72	9.24 853	73	0.75 147	9.99 328	2	57
4	9.24 253	71	9.24 926	74	0.75 074	9.99 326	2	56
5	9.24 324	71	9.25 000	73	0.75 000	9.99 324	2	55
6	9.24 395	71	9.25 073	73	0.74 927	9.99 322	3	54
7	9.24 466	70	9.25 146	73	0.74 854	9.99 319	2	53
8	9.24 536	71	9.25 219	73	0.74 781	9.99 317	2	52
9	9.24 607	70	9.25 292	73	0.74 708	9.99 315	2	51
10	9.24 677	71	9.25 365	72	0.74 635	9.99 313	3	**50**
11	9.24 748	70	9.25 437	73	0.74 563	9.99 310	2	49
12	9.24 818	70	9.25 510	72	0.74 490	9.99 308	2	48
13	9.24 888	70	9.25 582	73	0.74 418	9.99 306	2	47
14	9.24 958	70	9.25 655	72	0.74 345	9.99 304	3	46
15	9.25 028	70	9.25 727	72	0.74 273	9.99 301	2	45
16	9.25 098	70	9.25 799	72	0.74 201	9.99 299	2	44
17	9.25 168	69	9.25 871	72	0.74 129	9.99 297	3	43
18	9.25 237	70	9.25 943	72	0.74 057	9.99 294	2	42
19	9.25 307	69	9.26 015	71	0.73 985	9.99 292	2	41
20	9.25 376	69	9.26 086	72	0.73 914	9.99 290	2	**40**
21	9.25 445	69	9.26 158	71	0.73 842	9.99 288	3	39
22	9.25 514	69	9.26 229	72	0.73 771	9.99 285	2	38
23	9.25 583	69	9.26 301	71	0.73 699	9.99 283	2	37
24	9.25 652	69	9.26 372	71	0.73 628	9.99 281	3	36
25	9.25 721	69	9.26 443	71	0.73 557	9.99 278	2	35
26	9.25 790	68	9.26 514	71	0.73 486	9.99 276	2	34
27	9.25 858	69	9.26 585	70	0.73 415	9.99 274	3	33
28	9.25 927	68	9.26 655	71	0.73 345	9.99 271	2	32
29	9.25 995	68	9.26 726	71	0.73 274	9.99 269	2	31
30	9.26 063	68	9.26 797	70	0.73 203	9.99 267	3	**30**
31	9.26 131	68	9.26 867	70	0.73 133	9.99 264	2	29
32	9.26 199	68	9.26 937	71	0.73 063	9.99 262	2	28
33	9.26 267	68	9.27 008	70	0.72 992	9.99 260	3	27
34	9.26 335	68	9.27 078	70	0.72 922	9.99 257	2	26
35	9.26 403	67	9.27 148	70	0.72 852	9.99 255	3	25
36	9.26 470	68	9.27 218	70	0.72 782	9.99 252	2	24
37	9.26 538	67	9.27 288	69	0.72 712	9.99 250	2	23
38	9.26 605	67	9.27 357	70	0.72 643	9.99 248	3	22
39	9.26 672	67	9.27 427	69	0.72 573	9.99 245	2	21
40	9.26 739	67	9.27 496	70	0.72 504	9.99 243	2	**20**
41	9.26 806	67	9.27 566	69	0.72 434	9.99 241	3	19
42	9.26 873	67	9.27 635	69	0.72 365	9.99 238	2	18
43	9.26 940	67	9.27 704	69	0.72 296	9.99 236	3	17
44	9.27 007	66	9.27 773	69	0.72 227	9.99 233	2	16
45	9.27 073	67	9.27 842	69	0.72 158	9.99 231	2	15
46	9.27 140	66	9.27 911	69	0.72 089	9.99 229	3	14
47	9.27 206	67	9.27 980	69	0.72 020	9.99 226	2	13
48	9.27 273	66	9.28 049	68	0.71 951	9.99 224	3	12
49	9.27 339	66	9.28 117	69	0.71 883	9.99 221	2	11
50	9.27 405	66	9.28 186	68	0.71 814	9.99 219	2	**10**
51	9.27 471	66	9.28 254	69	0.71 746	9.99 217	3	9
52	9.27 537	65	9.28 323	68	0.71 677	9.99 214	2	8
53	9.27 602	66	9.28 391	68	0.71 609	9.99 212	3	7
54	9.27 668	66	9.28 459	68	0.71 541	9.99 209	2	6
55	9.27 734	65	9.28 527	68	0.71 473	9.99 207	3	5
56	9.27 799	65	9.28 595	67	0.71 405	9.99 204	2	4
57	9.27 864	66	9.28 662	68	0.71 338	9.99 202	2	3
58	9.27 930	65	9.28 730	68	0.71 270	9.99 200	3	2
59	9.27 995	65	9.28 798	67	0.71 202	9.99 197	2	1
60	9.28 060		9.28 865		0.71 135	9.99 195		**0**
′	L. Cos.	d.	L. Cot.	c.d.	L. Tan.	L. Sin.	d.	′

100° (280°) (259°) **79°**

P. P.

″	**74**	**73**	**72**
1	1.2	1.2	1.2
2	2.5	2.4	2.4
3	3.7	3.6	3.6
4	4.9	4.9	4.8
5	6.2	6.1	6.0
6	7.4	7.3	7.2
7	8.6	8.5	8.4
8	9.9	9.7	9.6
9	11.1	11.0	10.8
10	12.3	12.2	12.0
20	24.7	24.3	24.0
30	37.0	36.5	36.0
40	49.3	48.7	48.0
50	61.7	60.8	60.0

″	**71**	**70**	**69**
1	1.2	1.2	1.2
2	2.4	2.3	2.3
3	3.6	3.5	3.4
4	4.7	4.7	4.6
5	5.9	5.8	5.8
6	7.1	7.0	6.9
7	8.3	8.2	8.0
8	9.5	9.3	9.2
9	10.6	10.5	10.4
10	11.8	11.7	11.5
20	23.7	23.3	23.0
30	35.5	35.0	34.5
40	47.3	46.7	46.0
50	59.2	58.3	57.5

″	**68**	**67**	**66**
1	1.1	1.1	1.1
2	2.3	2.2	2.2
3	3.4	3.4	3.3
4	4.5	4.5	4.4
5	5.7	5.6	5.5
6	6.8	6.7	6.6
7	7.9	7.8	7.7
8	9.1	8.9	8.8
9	10.2	10.0	9.9
10	11.3	11.2	11.0
20	22.7	22.3	22.0
30	34.0	33.5	33.0
40	45.3	44.7	44.0
50	56.7	55.8	55.0

	3/74	**3/73**	**3/72**
0			
1	12.3	12.2	12.0
2	37.0	36.5	36.0
3	61.7	60.8	60.0

	3/71	**3/70**	**3/69**	**3/68**
0				
1	11.8	11.7	11.5	11.3
2	35.5	35.0	34.5	34.0
3	59.2	58.3	57.5	56.7

P. P.

COMMON LOGARITHMS OF THE TRIGONOMETRIC FUNCTIONS (Continued)

11° (191°) (348°) **168°**

′	L. Sin.	d.	L. Tan.	c.d.	L. Cot.	L. Cos.	d.	′
0	9.28 060	65	9.28 865	68	0.71 135	9.99 195	3	**60**
1	9.28 125	65	9.28 933	67	0.71 067	9.99 192	2	59
2	9.28 190	64	9.29 000	67	0.71 000	9.99 190	3	58
3	9.28 254	65	9.29 067	67	0.70 933	9.99 187	2	57
4	9.28 319	65	9.29 134	67	0.70 866	9.99 185	3	56
5	9.28 384	64	9.29 201	67	0.70 799	9.99 182	2	55
6	9.28 448	64	9.29 268	67	0.70 732	9.99 180	3	54
7	9.28 512	65	9.29 335	67	0.70 665	9.99 177	2	53
8	9.28 577	64	9.29 402	66	0.70 598	9.99 175	3	52
9	9.28 641	64	9.29 468	67	0.70 532	9.99 172	2	51
10	9.28 705	64	9.29 535	66	0.70 465	9.99 170	3	**50**
11	9.28 769	64	9.29 601	67	0.70 399	9.99 167	2	49
12	9.28 833	63	9.29 668	66	0.70 332	9.99 165	3	48
13	9.28 896	64	9.29 734	66	0.70 266	9.99 162	2	47
14	9.28 960	64	9.29 800	66	0.70 200	9.99 160	3	46
15	9.29 024	63	9.29 866	66	0.70 134	9.99 157	2	45
16	9.29 087	63	9.29 932	66	0.70 068	9.99 155	3	44
17	9.29 150	64	9.29 998	66	0.70 002	9.99 152	2	43
18	9.29 214	63	9.30 064	66	0.69 936	9.99 150	3	42
19	9.29 277	63	9.30 130	65	0.69 870	9.99 147	2	41
20	9.29 340	63	9.30 195	66	0.69 805	9.99 145	3	**40**
21	9.29 403	63	9.30 261	65	0.69 739	9.99 142	2	39
22	9.29 466	63	9.30 326	65	0.69 674	9.99 140	3	38
23	9.29 529	62	9.30 391	66	0.69 609	9.99 137	2	37
24	9.29 591	63	9.30 457	65	0.69 543	9.99 135	3	36
25	9.29 654	62	9.30 522	65	0.69 478	9.99 132	2	35
26	9.29 716	63	9.30 587	65	0.69 413	9.99 130	3	34
27	9.29 779	62	9.30 652	65	0.69 348	9.99 127	3	33
28	9.29 841	62	9.30 717	65	0.69 283	9.99 124	2	32
29	9.29 903	63	9.30 782	64	0.69 218	9.99 122	3	31
30	9.29 966	62	9.30 846	65	0.69 154	9.99 119	2	**30**
31	9.30 028	62	9.30 911	64	0.69 089	9.99 117	3	29
32	9.30 090	61	9.30 975	65	0.69 025	9.99 114	2	28
33	9.30 151	62	9.31 040	64	0.68 960	9.99 112	3	27
34	9.30 213	62	9.31 104	64	0.68 896	9.99 109	3	26
35	9.30 275	61	9.31 168	65	0.68 832	9.99 106	2	25
36	9.30 336	62	9.31 233	64	0.68 767	9.99 104	3	24
37	9.30 398	61	9.31 297	64	0.68 703	9.99 101	2	23
38	9.30 459	62	9.31 361	64	0.68 639	9.99 099	3	22
39	9.30 521	61	9.31 425	64	0.68 575	9.99 096	3	21
40	9.30 582	61	9.31 489	63	0.68 511	9.99 093	2	**20**
41	9.30 643	61	9.31 552	64	0.68 448	9.99 091	3	19
42	9.30 704	61	9.31 616	63	0.68 384	9.99 088	2	18
43	9.30 765	61	9.31 679	64	0.68 321	9.99 086	3	17
44	9.30 826	61	9.31 743	63	0.68 257	9.99 083	3	16
45	9.30 887	60	9.31 806	64	0.68 194	9.99 080	2	15
46	9.30 947	61	9.31 870	63	0.68 130	9.99 078	3	14
47	9.31 008	60	9.31 933	63	0.68 067	9.99 075	3	13
48	9.31 068	61	9.31 996	63	0.68 004	9.99 072	2	12
49	9.31 129	60	9.32 059	63	0.67 941	9.99 070	3	11
50	9.31 189	61	9.32 122	63	0.67 878	9.99 067	3	**10**
51	9.31 250	60	9.32 185	63	0.67 815	9.99 064	2	9
52	9.31 310	60	9.32 248	63	0.67 752	9.99 062	3	8
53	9.31 370	60	9.32 311	62	0.67 689	9.99 059	3	7
54	9.31 430	60	9.32 373	63	0.67 627	9.99 056	2	6
55	9.31 490	59	9.32 436	62	0.67 564	9.99 054	3	5
56	9.31 549	60	9.32 498	63	0.67 502	9.99 051	3	4
57	9.31 609	60	9.32 561	62	0.67 439	9.99 048	2	3
58	9.31 669	59	9.32 623	62	0.67 377	9.99 046	3	2
59	9.31 728	60	9.32 685	62	0.67 315	9.99 043	3	1
60	9.31 788		9.32 747		0.67 253	9.99 040		**0**
′	L. Cos.	d.	L. Cot.	c.d.	L. Tan.	L. Sin.	d.	′

101° (281°) (258°) **78°**

P. P.

″	**65**	**64**	**63**
1	1.1	1.1	1.0
2	2.2	2.1	2.1
3	3.2	3.2	3.2
4	4.3	4.3	4.2
5	5.4	5.3	5.2
6	6.5	6.4	6.3
7	7.6	7.5	7.4
8	8.7	8.5	8.4
9	9.8	9.6	9.4
10	10.8	10.7	10.5
20	21.7	21.3	21.0
30	32.5	32.0	31.5
40	43.3	42.7	42.0
50	54.2	53.3	52.5

″	**62**	**61**	**60**
1	1.0	1.0	1.0
2	2.1	2.0	2.0
3	3.1	3.0	3.0
4	4.1	4.1	4.0
5	5.2	5.1	5.0
6	6.2	6.1	6.0
7	7.2	7.1	7.0
8	8.3	8.1	8.0
9	9.3	9.2	9.0
10	10.3	10.2	10.0
20	20.7	20.3	20.0
30	31.0	30.5	30.0
40	41.3	40.7	40.0
50	51.7	50.8	50.0

″	**59**	**3**	**2**
1	1.0	0.0	0.0
2	2.0	0.1	0.1
3	3.0	0.2	0.1
4	3.9	0.2	0.1
5	4.9	0.2	0.2
6	5.9	0.3	0.2
7	6.9	0.4	0.2
8	7.9	0.4	0.3
9	8.8	0.4	0.3
10	9.8	0.5	0.3
20	19.7	1.0	0.7
30	29.5	1.5	1.0
40	39.3	2.0	1.3
50	49.2	2.5	1.7

	3/67	**3/66**	**3/65**
0			
1	11.2	11.0	10.8
2	33.5	33.0	32.5
3	55.8	55.0	54.2

	3/64	**3/63**	**3/62**
0			
1	10.7	10.5	10.3
2	32.0	31.5	31.0
3	53.3	52.5	51.7

COMMON LOGARITHMS OF THE TRIGONOMETRIC FUNCTIONS (Continued)

12° (192°) (347°) **167°**

′	L. Sin.	d.	L. Tan.	c.d.	L. Cot.	L. Cos.	d.	′
0	9.31 788	59	9.32 747	63	0.67 253	9.99 040	2	**60**
1	9.31 847	60	9.32 810	62	0.67 190	9.99 038	3	59
2	9.31 907	59	9.32 872	61	0.67 128	9.99 035	3	58
3	9.31 966	59	9.32 933	62	0.67 067	9.99 032	2	57
4	9.32 025	59	9.32 995	62	0.67 005	9.99 030	3	56
5	9.32 084	59	9.33 057	62	0.66 943	9.99 027	3	55
6	9.32 143	59	9.33 119	61	0.66 881	9.99 024	2	54
7	9.32 202	59	9.33 180	62	0.66 820	9.99 022	3	53
8	9.32 261	58	9.33 242	61	0.66 758	9.99 019	3	52
9	9.32 319	59	9.33 303	62	0.66 697	9.99 016	3	51
10	9.32 378	59	9.33 365	61	0.66 635	9.99 013	2	**50**
11	9.32 437	58	9.33 426	61	0.66 574	9.99 011	3	49
12	9.32 495	58	9.33 487	61	0.66 513	9.99 008	3	48
13	9.32 553	59	9.33 548	61	0.66 452	9.99 005	3	47
14	9.32 612	58	9.33 609	61	0.66 391	9.99 002	2	46
15	9.32 670	58	9.33 670	61	0.66 330	9.99 000	3	45
16	9.32 728	58	9.33 731	61	0.66 269	9.98 997	3	44
17	9.32 786	58	9.33 792	61	0.66 208	9.98 994	3	43
18	9.32 844	58	9.33 853	60	0.66 147	9.98 991	2	42
19	9.32 902	58	9.33 913	61	0.66 087	9.98 989	3	41
20	9.32 960	58	9.33 974	60	0.66 026	9.98 986	3	**40**
21	9.33 018	57	9.34 034	61	0.65 966	9.98 983	3	39
22	9.33 075	58	9.34 095	60	0.65 905	9.98 980	2	38
23	9.33 133	57	9.34 155	60	0.65 845	9.98 978	3	37
24	9.33 190	58	9.34 215	61	0.65 785	9.98 975	3	36
25	9.33 248	57	9.34 276	60	0.65 724	9.98 972	3	35
26	9.33 305	57	9.34 336	60	0.65 664	9.98 969	2	34
27	9.33 362	58	9.34 396	60	0.65 604	9.98 967	3	33
28	9.33 420	57	9.34 456	60	0.65 544	9.98 964	3	32
29	9.33 477	57	9.34 516	60	0.65 484	9.98 961	3	31
30	9.33 534	57	9.34 576	59	0.65 424	9.98 958	3	**30**
31	9.33 591	56	9.34 635	60	0.65 365	9.98 955	2	29
32	9.33 647	57	9.34 695	60	0.65 305	9.98 953	3	28
33	9.33 704	57	9.34 755	59	0.65 245	9.98 950	3	27
34	9.33 761	57	9.34 814	60	0.65 186	9.98 947	3	26
35	9.33 818	56	9.34 874	59	0.65 126	9.98 944	3	25
36	9.33 874	57	9.34 933	59	0.65 067	9.98 941	3	24
37	9.33 931	56	9.34 992	59	0.65 008	9.98 938	2	23
38	9.33 987	56	9.35 051	60	0.64 949	9.98 936	3	22
39	9.34 043	57	9.35 111	59	0.64 889	9.98 933	3	21
40	9.34 100	56	9.35 170	59	0.64 830	9.98 930	3	**20**
41	9.34 156	56	9.35 229	59	0.64 771	9.98 927	3	19
42	9.34 212	56	9.35 288	59	0.64 712	9.98 924	3	18
43	9.34 268	56	9.35 347	58	0.64 653	9.98 921	2	17
44	9.34 324	56	9.35 405	59	0.64 595	9.98 919	3	16
45	9.34 380	56	9.35 464	59	0.64 536	9.98 916	3	15
46	9.34 436	55	9.35 523	58	0.64 477	9.98 913	3	14
47	9.34 491	56	9.35 581	59	0.64 419	9.98 910	3	13
48	9.34 547	55	9.35 640	58	0.64 360	9.98 907	3	12
49	9.34 602	56	9.35 698	59	0.64 302	9.98 904	3	11
50	9.34 658	55	9.35 757	58	0.64 243	9.98 901	3	**10**
51	9.34 713	56	9.35 815	58	0.64 185	9.98 898	2	9
52	9.34 769	55	9.35 873	58	0.64 127	9.98 896	3	8
53	9.34 824	55	9.35 931	58	0.64 069	9.98 893	3	7
54	9.34 879	55	9.35 989	58	0.64 011	9.98 890	3	6
55	9.34 934	55	9.36 047	58	0.63 953	9.98 887	3	5
56	9.34 989	55	9.36 105	58	0.63 895	9.98 884	3	4
57	9.35 044	55	9.36 163	58	0.63 837	9.98 881	3	3
58	9.35 099	55	9.36 221	58	0.63 779	9.98 878	3	2
59	9.35 154	55	9.36 279	57	0.63 721	9.98 875	3	1
60	9.35 209		9.36 336		0.63 664	9.98 872		**0**
′	L. Cos.	d.	L. Cot.	c.d.	L. Tan.	L. Sin.	d.	′

102° (282°) (257°) **77°**

P. P.

″	**63**	**62**	**61**
1	1.0	1.0	1.0
2	2.1	2.1	2.0
3	3.2	3.1	3.0
4	4.2	4.1	4.1
5	5.2	5.2	5.1
6	6.3	6.2	6.1
7	7.4	7.2	7.1
8	8.4	8.3	8.1
9	9.4	9.3	9.2
10	10.5	10.3	10.2
20	21.0	20.7	20.3
30	31.5	31.0	30.5
40	42.0	41.3	40.7
50	52.5	51.7	50.8

″	**60**	**59**	**58**
1	1.0	1.0	1.0
2	2.0	2.0	1.9
3	3.0	3.0	2.9
4	4.0	3.9	3.8
5	5 0	4.9	4.8
6	6.0	5.9	5.8
7	7.0	6.9	6.8
8	8.0	7.9	7.7
9	9.0	8.8	8.7
10	10.0	9.8	9.7
20	20.0	19.7	19.3
30	30.0	29.5	29.0
40	40.0	39.3	38.7
50	50.0	49.2	48.3

″	**57**	**56**	**55**
1	1.0	0.9	0.9
2	1.9	1.9	1.8
3	2.8	2.8	2.8
4	3.8	3.7	3.7
5	4.8	4.7	4.6
6	5.7	5.6	5.5
7	6.6	6.5	6.4
8	7.6	7.5	7.3
9	8.6	8.4	8.2
10	9.5	9.3	9.2
20	19.0	18.7	18.3
30	28.5	28.0	27.5
40	38.0	37.3	36.7
50	47.5	46.7	45.8

	3/62	**3/61**	**3/60**
0	10.3	10.2	10.0
1	31.0	30.5	30.0
2	51.7	50.8	50.0
3			

	3/59	**3/58**	**3/57**
0	9.8	9.7	9.5
1	29.5	29.0	28.5
2	49.2	48.3	47.5
3			

P. P.

COMMON LOGARITHMS OF THE TRIGONOMETRIC FUNCTIONS (Continued)

13° (193°) (346°) **166°**

′	L. Sin.	d.	L. Tan.	c.d.	L. Cot.	L. Cos.	d.	′
0	9.35 209	54	9.36 336	58	0.63 664	9.98 872	3	**60**
1	9.35 263	55	9.36 394	58	0.63 606	9.98 869	2	59
2	9.35 318	55	9.36 452	57	0.63 548	9.98 867	3	58
3	9.35 373	54	9.36 509	57	0.63 491	9.98 864	3	57
4	9.35 427	54	9.36 566	58	0.63 434	9.98 861	3	56
5	9.35 481	55	9.36 624	57	0.63 376	9.98 858	3	55
6	9.35 536	54	9.36 681	57	0.63 319	9.98 855	3	54
7	9.35 590	54	9.36 738	57	0.63 262	9.98 852	3	53
8	9.35 644	54	9.36 795	57	0.63 205	9.98 849	3	52
9	9.35 698	54	9.36 852	57	0.63 148	9.98 846	3	51
10	9.35 752	54	9.36 909	57	0.63 091	9.98 843	3	**50**
11	9.35 806	54	9.36 966	57	0.63 034	9.98 840	3	49
12	9.35 860	54	9.37 023	57	0.62 977	9.98 837	3	48
13	9.35 914	54	9.37 080	57	0.62 920	9.98 834	3	47
14	9.35 968	54	9.37 137	56	0.62 863	9.98 831	3	46
15	9.36 022	53	9.37 193	57	0.62 807	9.98 828	3	45
16	9.36 075	54	9.37 250	56	0.62 750	9.98 825	3	44
17	9.36 129	53	9.37 306	57	0.62 694	9.98 822	3	43
18	9.36 182	54	9.37 363	56	0.62 637	9.98 819	3	42
19	9.36 236	53	9.37 419	57	0.62 581	9.98 816	3	41
20	9.36 289	53	9.37 476	56	0.62 524	9.98 813	3	**40**
21	9.36 342	53	9.37 532	56	0.62 468	9.98 810	3	39
22	9.36 395	54	9.37 588	56	0.62 412	9.98 807	3	38
23	9.36 449	53	9.37 644	56	0.62 356	9.98 804	3	37
24	9.36 502	53	9.37 700	56	0.62 300	9.98 801	3	36
25	9.36 555	53	9.37 756	56	0.62 244	9.98 798	3	35
26	9.36 608	52	9.37 812	56	0.62 188	9.98 795	3	34
27	9.36 660	53	9.37 868	56	0.62 132	9.98 792	3	33
28	9.36 713	53	9.37 924	56	0.62 076	9.98 789	3	32
29	9.36 766	53	9.37 980	55	0.62 020	9.98 786	3	31
30	9.36 819	52	9.38 035	56	0.61 965	9.98 783	3	**30**
31	9.36 871	53	9.38 091	56	0.61 909	9.98 780	3	29
32	9.36 924	52	9.38 147	55	0.61 853	9.98 777	3	28
33	9.36 976	52	9.38 202	55	0.61 798	9.98 774	3	27
34	9.37 028	53	9.38 257	56	0.61 743	9.98 771	3	26
35	9.37 081	52	9.38 313	55	0.61 687	9.98 768	3	25
36	9.37 133	52	9.38 368	55	0.61 632	9.98 765	3	24
37	9.37 185	52	9.38 423	56	0.61 577	9.98 762	3	23
38	9.37 237	52	9.38 479	55	0.61 521	9.98 759	3	22
39	9.37 289	52	9.38 534	55	0.61 466	9.98 756	3	21
40	9.37 341	52	9.38 589	55	0.61 411	9.98 753	3	**20**
41	9.37 393	52	9.38 644	55	0.61 356	9.98 750	4	19
42	9.37 445	52	9.38 699	55	0.61 301	9.98 746	3	18
43	9.37 497	52	9.38 754	54	0.61 246	9.98 743	3	17
44	9.37 549	51	9.38 808	55	0.61 192	9.98 740	3	16
45	9.37 600	52	9.38 863	55	0.61 137	9.98 737	3	15
46	9.37 652	51	9.38 918	54	0.61 082	9.98 734	3	14
47	9.37 703	52	9.38 972	55	0.61 028	9.98 731	3	13
48	9.37 755	51	9.39 027	55	0.60 973	9.98 728	3	12
49	9.37 806	52	9.39 082	54	0.60 918	9.98 725	3	11
50	9.37 858	51	9.39 136	54	0.60 864	9.98 722	3	**10**
51	9.37 909	51	9.39 190	55	0.60 810	9.98 719	4	9
52	9.37 960	51	9.39 245	54	0.60 755	9.98 715	3	8
53	9.38 011	51	9.39 299	54	0.60 701	9.98 712	3	7
54	9.38 062	51	9.39 353	54	0.60 647	9.98 709	3	6
55	9.38 113	51	9.39 407	54	0.60 593	9.98 706	3	5
56	9.38 164	51	9.39 461	54	0.60 539	9.98 703	3	4
57	9.38 215	51	9.39 515	54	0.60 485	9.98 700	3	3
58	9.38 266	51	9.39 569	54	0.60 431	9.98 697	3	2
59	9.38 317	51	9.39 623	54	0.60 377	9.98 694	4	1
60	9.38 368		9.39 677		0.60 323	9.98 690		**0**
′	L. Cos.	d.	L. Cot.	c.d.	L. Tan.	L. Sin.	d.	′

103° (283°) (256°) **76°**

P. P.

″	**57**	**56**	**55**
1	1.0	0.9	0.9
2	1.9	1.9	1.8
3	2.8	2.8	2.8
4	3.8	3.7	3.7
5	4.8	4.7	4.6
6	5.7	5.6	5.5
7	6.6	6.5	6.4
8	7.6	7.5	7.3
9	8.6	8.4	8.2
10	9.5	9.3	9.2
20	19.0	18.7	18.3
30	28.5	28.0	27.5
40	38.0	37.3	36.7
50	47.5	46.7	45.8

″	**54**	**53**	**52**
1	0.9	0.9	0.9
2	1.8	1.8	1.7
3	2.7	2.6	2.6
4	3.6	3.5	3.5
5	4.5	4.4	4.3
6	5.4	5.3	5.2
7	6.3	6.2	6.1
8	7.2	7.1	6.9
9	8.1	8.0	7.8
10	9.0	8.8	8.7
20	18.0	17.7	17.3
30	27 0	26.5	26.0
40	36.0	35.3	34.7
50	45.0	44.2	43.3

″	**51**	**4**	**3**	**2**
1	0.8	0.1	0.0	0.0
2	1.7	0.1	0.1	0.1
3	2.6	0.2	0.2	0.1
4	3.4	0.3	0.2	0.1
5	4.2	0.3	0.2	0.2
6	5.1	0.4	0.3	0.2
7	6.0	0.5	0.4	0.2
8	6.8	0.5	0.4	0.3
9	7.6	0.6	0.4	0.3
10	8.5	0.7	0.5	0.3
20	17.0	1.3	1.0	0.7
30	25.5	2.0	1.5	1.0
40	34.0	2.7	2.0	1.3
50	42.5	3.3	2.5	1.7

	4/55	**4/54**	**3/58**	**3/57**
0	6.9	6.8	9.7	9.5
1	20.6	20.2	29.0	28.5
2	34.4	33.8	48.3	47.5
3	48.1	47.2	—	—
4				

	3/56	**3/55**	**3/54**
0	9.3	9.2	9.0
1	28.0	27.5	27.0
2	46.7	45.8	45.0
3			

P P.

COMMON LOGARITHMS OF THE TRIGONOMETRIC FUNCTIONS (Continued)

14° (194°) (345°) **165°**

′	L. Sin.	d.	L. Tan.	c.d.	L. Cot.	L. Cos.	d.	′
0	9.38 368	50	9.39 677	54	0.60 323	9.98 690	3	**60**
1	9.38 418	51	9.39 731	54	0.60 269	9.98 687	3	59
2	9.38 469	50	9.39 785	53	0.60 215	9.98 684	3	58
3	9.38 519	51	9.39 838	54	0.60 162	9.98 681	3	57
4	9.38 570	50	9.39 892	53	0.60 108	9.98 678	3	56
5	9.38 620	50	9.39 945	54	0.60 055	9.98 675	4	55
6	9.38 670	51	9.39 999	53	0.60 001	9.98 671	3	54
7	9.38 721	50	9.40 052	54	0.59 948	9.98 668	3	53
8	9.38 771	50	9.40 106	53	0.59 894	9.98 665	3	52
9	9.38 821	50	9.40 159	53	0.59 841	9.98 662	3	51
10	9.38 871	50	9.40 212	54	0.59 788	9.98 659	3	**50**
11	9.38 921	50	9.40 266	53	0.59 734	9.98 656	4	49
12	9.38 971	50	9.40 319	53	0.59 681	9.98 652	3	48
13	9.39 021	50	9.40 372	53	0.59 628	9.98 649	3	47
14	9.39 071	50	9.40 425	53	0.59 575	9.98 646	3	46
15	9.39 121	49	9.40 478	53	0.59 522	9.98 643	3	45
16	9.39 170	50	9.40 531	53	0.59 469	9.98 640	4	44
17	9.39 220	50	9.40 584	52	0.59 416	9.98 636	3	43
18	9.39 270	49	9.40 636	53	0.59 364	9.98 633	3	42
19	9.39 319	50	9.40 689	53	0.59 311	9.98 630	3	41
20	9.39 369	49	9.40 742	53	0.59 258	9.98 627	4	**40**
21	9.39 418	49	9.40 795	52	0.59 205	9.98 623	3	39
22	9.39 467	50	9.40 847	53	0.59 153	9.98 620	3	38
23	9.39 517	49	9.40 900	52	0.59 100	9.98 617	3	37
24	9.39 566	49	9.40 952	53	0.59 048	9.98 614	4	36
25	9.39 615	49	9.41 005	52	0.58 995	9.98 610	3	35
26	9.39 664	49	9.41 057	52	0.58 943	9.98 607	3	34
27	9.39 713	49	9.41 109	52	0.58 891	9.98 604	3	33
28	9.39 762	49	9.41 161	53	0.58 839	9.98 601	4	32
29	9.39 811	49	9.41 214	52	0.58 786	9.98 597	3	31
30	9.39 860	49	9.41 266	52	0.58 734	9.98 594	3	**30**
31	9.39 909	49	9.41 318	52	0.58 682	9.98 591	3	29
32	9.39 958	48	9.41 370	52	0.58 630	9.98 588	4	28
33	9.40 006	49	9.41 422	52	0.58 578	9.98 584	3	27
34	9.40 055	48	9.41 474	52	0.58 526	9.98 581	3	26
35	9.40 103	49	9.41 526	52	0.58 474	9.98 578	4	25
36	9.40 152	48	9.41 578	51	0.58 422	9.98 574	3	24
37	9.40 200	49	9.41 629	52	0.58 371	9.98 571	3	23
38	9.40 249	48	9.41 681	52	0.58 319	9.98 568	3	22
39	9.40 297	49	9.41 733	51	0.58 267	9.98 565	4	21
40	9.40 346	48	9.41 784	52	0.58 216	9.98 561	3	**20**
41	9.40 394	48	9.41 836	51	0.58 164	9.98 558	3	19
42	9.40 442	48	9.41 887	52	0.58 113	9.98 555	4	18
43	9.40 490	48	9.41 939	51	0.58 061	9.98 551	3	17
44	9.40 538	48	9.41 990	51	0.58 010	9.98 548	3	16
45	9.40 586	48	9.42 041	52	0.57 959	9.98 545	4	15
46	9.40 634	48	9.42 093	51	0.57 907	9.98 541	3	14
47	9.40 682	48	9.42 144	51	0.57 856	9.98 538	3	13
48	9.40 730	48	9.42 195	51	0.57 805	9.98 535	4	12
49	9.40 778	47	9.42 246	51	0.57 754	9.98 531	3	11
50	9.40 825	48	9.42 297	51	0.57 703	9.98 528	3	**10**
51	9.40 873	48	9.42 348	51	0.57 652	9.98 525	4	9
52	9.40 921	47	9.42 399	51	0.57 601	9.98 521	3	8
53	9.40 968	48	9.42 450	51	0.57 550	9.98 518	3	7
54	9.41 016	47	9.42 501	51	0.57 499	9.98 515	4	6
55	9.41 063	48	9.42 552	51	0.57 448	9.98 511	3	5
56	9.41 111	47	9.42 603	50	0.57 397	9.98 508	3	4
57	9.41 158	47	9.42 653	51	0.57 347	9.98 505	4	3
58	9.41 205	47	9.42 704	51	0.57 296	9.98 501	3	2
59	9.41 252	48	9.42 755	50	0.57 245	9.98 498	4	1
60	9.41 300		9.42 805		0.57 195	9.98 494		**0**
′	L. Cos.	d.	L. Cot.	c.d.	L. Tan.	L. Sin.	d.	′

P. P.

″	**54**	**53**	**52**
1	0.9	0.9	0.9
2	1.8	1.8	1.7
3	2.7	2.6	2.6
4	3.6	3.5	3.5
5	4.5	4.4	4.3
6	5.4	5.3	5.2
7	6.3	6.2	6.1
8	7.2	7.1	6.9
9	8.1	8.0	7.8
10	9.0	8.8	8.7
20	18.0	17.7	17.3
30	27.0	26.5	26.0
40	36.0	35.3	34.7
50	45.0	44.2	43.3

″	**51**	**50**	**49**
1	0.8	0.8	0.8
2	1.7	1.7	1.6
3	2.6	2.5	2.4
4	3.4	3.3	3.3
5	4.2	4.2	4.1
6	5.1	5.0	4.9
7	6.0	5.8	5.7
8	6.8	6.7	6.5
9	7.6	7.5	7.4
10	8.5	8.3	8.2
20	17.0	16.7	16.3
30	25.5	25.0	24.5
40	34.0	33.3	32.7
50	42.5	41.7	40.8

″	**48**	**47**	**4**	**3**
1	0.8	0.8	0.1	0.0
2	1.6	1.6	0.1	0.1
3	2.4	2.4	0.2	0.2
4	3.2	3.1	0.3	0.2
5	4.0	3.9	0.3	0.2
6	4.8	4.7	0.4	0.3
7	5.6	5.5	0.5	0.4
8	6.4	6.3	0.5	0.4
9	7.2	7.0	0.6	0.4
10	8.0	7.8	0.7	0.5
20	16.0	15.7	1.3	1.0
30	24.0	23.5	2.0	1.5
40	32.0	31.3	2.7	2.0
50	40.0	39.2	3.3	2.5

	4/54	**4/53**	**4/52**	**4/51**
0				
1	6.8	6.6	6.5	6.4
2	20.2	19.9	19.5	19.1
3	33.8	33.1	32.5	31.9
4	47.2	46.4	45.5	44.6

	3/54	**3/53**	**3/52**	**3/51**
0				
1	9.0	8.8	8.7	8.5
2	27.0	26.5	26.0	25.5
3	45.0	44.2	43.3	42.5

104° (284°) (255°) **75°**

COMMON LOGARITHMS OF THE TRIGONOMETRIC FUNCTIONS (Continued)

15° (195°) (344°) **164°**

′	L. Sin.	d.	L. Tan.	c.d.	L.Cot.	L. Cos.	d.	′
0	9.41 300	47	9.42 805	51	0.57 195	9.98 494	3	**60**
1	9.41 347	47	9.42 856	50	0.57 144	9.98 491	3	59
2	9.41 394	47	9.42 906	51	0.57 094	9.98 488	4	58
3	9.41 441	47	9.42 957	50	0.57 043	9.98 484	3	57
4	9.41 488	47	9.43 007	50	0.56 993	9.98 481	4	56
5	9.41 535	47	9.43 057	51	0.56 943	9.98 477	3	55
6	9.41 582	46	9.43 108	50	0.56 892	9.98 474	3	54
7	9.41 628	47	9.43 158	50	0.56 842	9.98 471	4	53
8	9.41 675	47	9.43 208	50	0.56 792	9.98 467	3	52
9	9.41 722	46	9.43 258	50	0.56 742	9.98 464	4	51
10	9.41 768	47	9.43 308	50	0.56 692	9.98 460	3	**50**
11	9.41 815	46	9.43 358	50	0.56 642	9.98 457	4	49
12	9.41 861	47	9.43 408	50	0.56 592	9.98 453	3	48
13	9.41 908	46	9.43 458	50	0.56 542	9.98 450	3	47
14	9.41 954	47	9.43 508	50	0.56 492	9.98 447	4	46
15	9.42 001	46	9.43 558	49	0.56 442	9.98 443	3	45
16	9.42 047	46	9.43 607	50	0.56 393	9.98 440	4	44
17	9.42 093	47	9.43 657	50	0.56 343	9.98 436	3	43
18	9.42 140	46	9.43 707	49	0.56 293	9.98 433	4	42
19	9.42 186	46	9.43 756	50	0.56 244	9.98 429	3	41
20	9.42 232	46	9.43 806	49	0.56 194	9.98 426	4	**40**
21	9.42 278	46	9.43 855	50	0.56 145	9.98 422	3	39
22	9.42 324	46	9.43 905	49	0.56 095	9.98 419	4	38
23	9.42 370	46	9.43 954	50	0.56 046	9.98 415	3	37
24	9.42 416	45	9.44 004	49	0.55 996	9.98 412	3	36
25	9.42 461	46	9.44 053	49	0.55 947	9.98 409	4	35
26	9.42 507	46	9.44 102	49	0.55 898	9.98 405	3	34
27	9.42 553	46	9.44 151	50	0.55 849	9.98 402	4	33
28	9.42 599	45	9.44 201	49	0.55 799	9.98 398	3	32
29	9.42 644	46	9.44 250	49	0.55 750	9.98 395	4	31
30	9.42 690	45	9.44 299	49	0.55 701	9.98 391	3	**30**
31	9.42 735	46	9.44 348	49	0.55 652	9.98 388	4	29
32	9.42 781	45	9.44 397	49	0.55 603	9.98 384	3	28
33	9.42 826	46	9.44 446	49	0.55 554	9.98 381	4	27
34	9.42 872	45	9.44 495	49	0.55 505	9.98 377	4	26
35	9.42 917	45	9.44 544	48	0.55 456	9.98 373	3	25
36	9.42 962	46	9.44 592	49	0.55 408	9.98 370	4	24
37	9.43 008	45	9.44 641	49	0.55 359	9.98 366	3	23
38	9.43 053	45	9.44 690	48	0.55 310	9.98 363	4	22
39	9.43 098	45	9.44 738	49	0.55 262	9.98 359	3	21
40	9.43 143	45	9.44 787	49	0.55 213	9.98 356	4	**20**
41	9.43 188	45	9.44 836	48	0.55 164	9.98 352	3	19
42	9.43 233	45	9.44 884	49	0.55 116	9.98 349	4	18
43	9.43 278	45	9.44 933	48	0.55 067	9.98 345	3	17
44	9.43 323	44	9.44 981	48	0.55 019	9.98 342	4	16
45	9.43 367	45	9.45 029	49	0.54 971	9.98 338	4	15
46	9.43 412	45	9.45 078	48	0.54 922	9.98 334	3	14
47	9.43 457	45	9.45 126	48	0.54 874	9.98 331	4	13
48	9.43 502	44	9.45 174	48	0.54 826	9.98 327	3	12
49	9.43 546	45	9.45 222	49	0.54 778	9.98 324	4	11
50	9.43 591	44	9.45 271	48	0.54 729	9.98 320	3	**10**
51	9.43 635	45	9.45 319	48	0.54 681	9.98 317	4	9
52	9.43 680	44	9.45 367	48	0.54 633	9.98 313	4	8
53	9.43 724	45	9.45 415	48	0.54 585	9.98 309	3	7
54	9.43 769	44	9.45 463	48	0.54 537	9.98 306	4	6
55	9.43 813	44	9.45 511	48	0.54 489	9.98 302	3	5
56	9.43 857	44	9 45 559	47	0.54 441	9.98 299	4	4
57	9.43 901	45	9.45 606	48	0.54 394	9.98 295	4	3
58	9.43 946	44	9.45 654	48	0.54 346	9.98 291	3	2
59	9.43 990	44	9.45 702	48	0.54 298	9.98 288	4	1
60	9.44 034		9.45 750		0.54 250	9.98 284		**0**
′	L. Cos.	d.	L. Cot.	c.d.	L. Tan.	L. Sin.	d.	′

105° (285°) (254°) **74°**

P. P.

″	**51**	**50**	**49**
1	0.8	0.8	0.8
2	1.7	1.7	1.6
3	2.6	2.5	2.4
4	3.4	3.3	3.3
5	4.2	4.2	4.1
6	5.1	5.0	4.9
7	6.0	5.8	5.7
8	6.8	6.7	6.5
9	7.6	7.5	7.4
10	8.5	8.3	8.2
20	17.0	16.7	16.3
30	25.5	25.0	24.5
40	34.0	33.3	32.7
50	42.5	41.7	40.8

″	**48**	**47**	**46**
1	0.8	0.8	0.8
2	1.6	1.6	1.5
3	2.4	2.4	2.3
4	3.2	3.1	3.1
5	4.0	3.9	3.8
6	4.3	4.7	4.6
7	5.6	5.5	5.4
8	6.4	6.3	6.1
9	7.2	7.0	6.9
10	8.0	7.8	7.7
20	16.0	15.7	15.3
30	24.0	23.5	23.0
40	32.0	31.3	30.7
50	40.0	39.2	38.3

″	**45**	**44**	**4**	**3**
1	0.8	0.7	0.1	0.0
2	1.5	1.5	0.1	0.1
3	2.2	2.2	0.2	0.2
4	3.0	2.9	0.3	0.2
5	3.8	3.7	0.3	0.2
6	4.5	4.4	0.4	0.3
7	5.2	5.1	0.5	0.4
8	6.0	5.9	0.5	0.4
9	6.8	6.6	0.6	0.4
10	7.5	7.3	0.7	0.5
20	15.0	14.7	1.3	1.0
30	22.5	22.0	2.0	1.5
40	30.0	29.3	2.7	2.0
50	37.5	36.7	3.3	2.5

	4/50	**4/49**	**4/48**	**4/47**
0				
1	6.2	6.1	6.0	5.9
2	18.8	18.4	18.0	17.6
3	31.2	30.6	30.0	29.4
4	43.8	42.9	42.0	41.1

	3/51	**3/50**	**3/49**	**3/48**
0				
1	8.5	8.3	8.2	8.0
2	25.5	25.0	24.5	24.0
3	42.5	41.7	40.8	40.0

P. P.

COMMON LOGARITHMS OF THE TRIGONOMETRIC FUNCTIONS (Continued)

16° (196°) (343°) **163°**

′	L. Sin.	d.	L. Tan.	c.d.	L. Cot.	L. Cos.	d.	′
0	9.44 034	44	9.45 750	47	0.54 250	9.98 284	3	**60**
1	9.44 078	44	9.45 797	48	0.54 203	9.98 281	4	59
2	9.44 122	44	9.45 845	47	0.54 155	9.98 277	4	58
3	9.44 166	44	9.45 892	48	0.54 108	9.98 273	3	57
4	9.44 210	43	9.45 940	47	0.54 060	9.98 270	4	56
5	9.44 253	44	9.45 987	48	0.54 013	9.98 266	4	55
6	9.44 297	44	9.46 035	47	0.53 965	9.98 262	3	54
7	9.44 341	44	9.46 082	48	0.53 918	9.98 259	4	53
8	9.44 385	43	9.46 130	47	0.53 870	9.98 255	4	52
9	9.44 428	44	9.46 177	47	0.53 823	9.98 251	3	51
10	9.44 472	44	9.46 224	47	0.53 776	9.98 248	4	**50**
11	9.44 516	43	9.46 271	48	0.53 729	9.98 244	4	49
12	9.44 559	43	9.46 319	47	0.53 681	9.98 240	3	48
13	9.44 602	44	9.46 366	47	0.53 634	9.98 237	4	47
14	9.44 646	43	9.46 413	47	0.53 587	9.98 233	4	46
15	9.44 689	44	9.46 460	47	0.53 540	9.98 229	3	45
16	9.44 733	43	9.46 507	47	0.53 493	9.98 226	4	44
17	9.44 776	43	9.46 554	47	0.53 446	9.98 222	4	43
18	9.44 819	43	9.46 601	47	0.53 399	9.98 218	3	42
19	9.44 862	43	9.46 648	46	0.53 352	9.98 215	4	41
20	9.44 905	43	9.46 694	47	0.53 306	9.98 211	4	**40**
21	9.44 948	44	9.46 741	47	0.53 259	9.98 207	3	39
22	9.44 992	43	9.46 788	47	0.53 212	9.98 204	4	38
23	9.45 035	42	9.46 835	46	0.53 165	9.98 200	4	37
24	9.45 077	43	9.46 881	47	0.53 119	9.98 196	4	36
25	9.45 120	43	9.46 928	47	0.53 072	9.98 192	3	35
26	9.45 163	43	9.46 975	46	0.53 025	9.98 189	4	34
27	9.45 206	43	9.47 021	47	0.52 979	9.98 185	4	33
28	9.45 249	43	9.47 068	46	0.52 932	9.98 181	4	32
29	9.45 292	42	9.47 114	46	0.52 886	9.98 177	3	31
30	9.45 334	43	9.47 160	47	0.52 840	9.98 174	4	**30**
31	9.45 377	42	9.47 207	46	0.52 793	9.98 170	4	29
32	9.45 419	43	9.47 253	46	0.52 747	9.98 166	4	28
33	9.45 462	42	9.47 299	47	0.52 701	9.98 162	3	27
34	9.45 504	43	9.47 346	46	0.52 654	9.98 159	4	26
35	9.45 547	42	9.47 392	46	0.52 608	9.98 155	4	25
36	9.45 589	43	9.47 438	46	0.52 562	9.98 151	4	24
37	9.45 632	42	9.47 484	46	0.52 516	9.98 147	3	23
38	9.45 674	42	9.47 530	46	0.52 470	9.98 144	4	22
39	9.45 716	42	9.47 576	46	0.52 424	9.98 140	4	21
40	9.45 758	43	9.47 622	46	0.52 378	9.98 136	4	**20**
41	9.45 801	42	9.47 668	46	0.52 332	9.98 132	3	19
42	9.45 843	42	9.47 714	46	0.52 286	9.98 129	4	18
43	9.45 885	42	9.47 760	46	0.52 240	9.98 125	4	17
44	9.45 927	42	9.47 806	46	0.52 194	9.98 121	4	16
45	9.45 969	42	9.47 852	45	0.52 148	9.98 117	4	15
46	9.46 011	42	9.47 897	46	0.52 103	9.98 113	3	14
47	9.46 053	42	9.47 943	46	0.52 057	9.98 110	4	13
48	9.46 095	41	9.47 989	46	0.52 011	9.98 106	4	12
49	9.46 136	42	9.48 035	45	0.51 965	9.98 102	4	11
50	9.46 178	42	9.48 080	46	0.51 920	9.98 098	4	**10**
51	9.46 220	42	9.48 126	45	0.51 874	9.98 094	4	9
52	9.46 262	41	9.48 171	46	0.51 829	9.98 090	3	8
53	9.46 303	42	9.48 217	45	0.51 783	9.98 087	4	7
54	9.46 345	41	9.48 262	45	0.51 738	9.98 083	4	6
55	9.46 386	42	9.48 307	46	0.51 693	9.98 079	4	5
56	9.46 428	41	9.48 353	45	0.51 647	9.98 075	4	4
57	9.46 469	42	9.48 398	45	0.51 602	9.98 071	4	3
58	9.46 511	41	9.48 443	46	0.51 557	9.98 067	4	2
59	9.46 552	42	9.48 489	45	0.51 511	9.98 063	3	1
60	9.46 594		9.48 534		0.51 466	9.98 060		**0**
′	L. Cos.	d.	L. Cot.	c.d.	L. Tan.	L. Sin.	d.	′

P. P.

″	**48**	**47**	**46**
1	0.8	0.8	0.8
2	1.6	1.6	1.5
3	2.4	2.4	2.3
4	3.2	3.1	3.1
5	4.0	3.9	3.8
6	4.8	4.7	4.6
7	5.6	5.5	5.4
8	6.4	6.3	6.1
9	7.2	7.0	6.9
10	8.0	7.8	7.7
20	16.0	15.7	15.3
30	24.0	23.5	23.0
40	32.0	31.3	30.7
50	40.0	39.2	38.3

″	**45**	**44**	**43**
1	0.8	0.7	0.7
2	1.5	1.5	1.4
3	2.2	2.2	2.2
4	3.0	2.9	2.9
5	3.8	3.7	3.6
6	4.5	4.4	4.3
7	5.2	5.1	5.0
8	6.0	5.9	5.7
9	6.8	6.6	6.4
10	7.5	7.3	7.2
20	15.0	14.7	14.3
30	22.5	22.0	21.5
40	30.0	29.3	28.7
50	37.5	36.7	35.8

″	**42**	**41**	**4**	**3**
1	0.7	0.7	0.1	0.0
2	1.4	1.4	0.1	0.1
3	2.1	2.0	0.2	0.2
4	2.8	2.7	0.3	0.2
5	3.5	3.4	0.3	0.2
6	4.2	4.1	0.4	0.3
7	4.9	4.8	0.5	0.4
8	5.6	5.5	0.5	0.4
9	6.3	6.2	0.6	0.4
10	7.0	6.8	0.7	0.5
20	14.0	13.7	1.3	1.0
30	21.0	20.5	2.0	1.5
40	28.0	27.3	2.7	2.0
50	35.0	34.2	3.3	2.5

	4/48	**4/47**	**4/46**	**4/45**
0	6.0	5.9	5.8	5.6
1	18.0	17.6	17.2	16.9
2	30.0	29.4	28.8	28.1
3	42.0	41.1	40.2	39.4
4				

	3/48	**3/47**	**3/46**	**3/45**
0	8.0	7.8	7.7	7.5
1	24.0	23.5	23.0	22.5
2	40.0	39.2	38.3	37.5
3				

P. P.

106° (286°) (253°) **73°**

COMMON LOGARITHMS OF THE TRIGONOMETRIC FUNCTIONS (Continued)

17° (197°) (342°) **162°**

′	L. Sin.	d.	L. Tan.	c.d.	L. Cot.	L. Cos.	d.	′
0	9.46 694	41	9.48 534	45	0.51 466	9.98 060	4	**60**
1	9.46 635	41	9.48 579	45	0.51 421	9.98 056	4	59
2	9.46 676	41	9.48 624	45	0.51 376	9.98 052	4	58
3	9.46 717	41	9.48 669	45	0.51 331	9.98 048	4	57
4	9.46 758	42	9.48 714	45	0.51 286	9.98 044	4	56
5	9.46 800	41	9.48 759	45	0.51 241	9.98 040	4	55
6	9.46 841	41	9.48 804	45	0.51 196	9.98 036	4	54
7	9.46 882	41	9.48 849	45	0.51 151	9.98 032	3	53
8	9.46 923	41	9.48 894	45	0.51 106	9.98 029	4	52
9	9.46 964	41	9.48 939	45	0.51 061	9.98 025	4	51
10	9.47 005	40	9.48 984	45	0.51 016	9.98 021	4	**50**
11	9.47 045	41	9.49 029	44	0.50 971	9.98 017	4	49
12	9.47 086	41	9.49 073	45	0.50 927	9.98 013	4	48
13	9.47 127	41	9.49 118	45	0.50 882	9.98 009	4	47
14	9.47 168	41	9.49 163	44	0.50 837	9.98 005	4	46
15	9.47 209	40	9.49 207	45	0.50 793	9.98 001	4	45
16	9.47 249	41	9.49 252	44	0.50 748	9.97 997	4	44
17	9.47 290	40	9.49 296	45	0.50 704	9.97 993	4	43
18	9.47 330	41	9.49 341	44	0.50 659	9.97 989	3	42
19	9.47 371	40	9.49 385	45	0.50 615	9.97 986	4	41
20	9.47 411	41	9.49 430	44	0.50 570	9.97 982	4	**40**
21	9.47 452	40	9.49 474	45	0.50 526	9.97 978	4	39
22	9.47 492	41	9.49 519	44	0.50 481	9.97 974	4	38
23	9.47 533	40	9.49 563	44	0.50 437	9.97 970	4	37
24	9.47 573	40	9.49 607	45	0.50 393	9.97 966	4	36
25	9.47 613	41	9.49 652	44	0.50 348	9.97 962	4	35
26	9.47 654	40	9.49 696	44	0.50 304	9.97 958	4	34
27	9.47 694	40	9.49 740	44	0.50 260	9.97 954	4	33
28	9.47 734	40	9.49 784	44	0.50 216	9.97 950	4	32
29	9.47 774	40	9.49 828	44	0.50 172	9.97 946	4	31
30	9.47 814	40	9.49 872	44	0.50 128	9.97 942	4	**30**
31	9.47 854	40	9.49 916	44	0.50 084	9.97 938	4	29
32	9.47 894	40	9.49 960	44	0.50 040	9.97 934	4	28
33	9.47 934	40	9.50 004	44	0.49 996	9.97 930	4	27
34	9.47 974	40	9.50 048	44	0.49 952	9.97 926	4	26
35	9.48 014	40	9.50 092	44	0.49 908	9.97 922	4	25
36	9.48 054	40	9.50 136	44	0.49 864	9.97 918	4	24
37	9.48 094	39	9.50 180	43	0.49 820	9.97 914	4	23
38	9.48 133	40	9.50 223	44	0.49 777	9.97 910	4	22
39	9.48 173	40	9.50 267	44	0.49 733	9.97 906	4	21
40	9.48 213	39	9.50 311	44	0.49 689	9.97 902	4	**20**
41	9.48 252	40	9.50 355	43	0.49 645	9.97 898	4	19
42	9.48 292	40	9.50 398	44	0.49 602	9.97 894	4	18
43	9.48 332	39	9.50 442	43	0.49 558	9.97 890	4	17
44	9.48 371	40	9.50 485	44	0.49 515	9.97 886	4	16
45	9.48 411	39	9.50 529	43	0.49 471	9.97 882	4	15
46	9.48 450	40	9.50 572	44	0.49 428	9.97 878	4	14
47	9.48 490	39	9.50 616	43	0.49 384	9.97 874	4	13
48	9.48 529	39	9.50 659	44	0.49 341	9.97 870	4	12
49	9.48 568	39	9.50 703	43	0.49 297	9.97 866	5	11
50	9.48 607	40	9.50 746	43	0.49 254	9.97 861	4	**10**
51	9.48 647	39	9.50 789	44	0.49 211	9.97 857	4	9
52	9.48 686	39	9.50 833	43	0.49 167	9.97 853	4	8
53	9.48 725	39	9.50 876	43	0.49 124	9.97 849	4	7
54	9.48 764	39	9.50 919	43	0.49 081	9.97 845	4	6
55	9.48 803	39	9.50 962	43	0.49 038	9.97 841	4	5
56	9.48 842	39	9.51 005	43	0.48 995	9.97 837	4	4
57	9.48 881	39	9.51 048	44	0.48 952	9.97 833	4	3
58	9.48 920	39	9.51 092	43	0.48 908	9.97 829	4	2
59	9.48 959	39	9.51 135	43	0.48 865	9.97 825	4	1
60	9.48 998		9.51 178		0.48 822	9.97 821		**0**
′	L. Cos.	d.	L. Cot.	c.d.	L. Tan.	L. Sin.	d.	′

P. P.

″	**45**	**44**	**43**
1	0.8	0.7	0.7
2	1.5	1.5	1.4
3	2.2	2.2	2.2
4	3.0	2.9	2.9
5	3.8	3.7	3.6
6	4.5	4.4	4.3
7	5.2	5.1	5.0
8	6.0	5.9	5.7
9	6.8	6.6	6.4
10	7.5	7.3	7.2
20	15.0	14.7	14.3
30	22.5	22.0	21.5
40	30.0	29.3	28.7
50	37.5	36.7	35.8

″	**42**	**41**	**40**
1	0.7	0.7	0.7
2	1.4	1.4	1.3
3	2.1	2.0	2.0
4	2.8	2.7	2.7
5	3.5	3.4	3.3
6	4.2	4.1	4.0
7	4.9	4.8	4.7
8	5.6	5.5	5.3
9	6.3	6.2	6.0
10	7.0	6.8	6.7
20	14.0	13.7	13.3
30	21.0	20.5	20.0
40	28.0	27.3	26.7
50	35.0	34.2	33.3

″	**39**	**5**	**4**	**3**
1	0.6	0.1	0.1	0.0
2	1.3	0.2	0.1	0.1
3	2.0	0.2	0.2	0.2
4	2.6	0.3	0.3	0.2
5	3.2	0.4	0.3	0.2
6	3.9	0.5	0.4	0.3
7	4.6	0.6	0.5	0.4
8	5.2	0.7	0.5	0.4
9	5.8	0.8	0.6	0.4
10	6.5	0.8	0.7	0.5
20	13.0	1.7	1.3	1.0
30	19.5	2.5	2.0	1.5
40	26.0	3.3	2.7	2.0
50	32.5	4.2	3.3	2.5

	5/43	**4/45**	**4/44**
0	4.3	5.6	5.5
1	12.9	16.9	16.5
2	21.5	28.1	27.5
3	30.1	39.4	38.5
4	38.7	—	—
5			

	4/43	**3/45**	**3/44**
0	5.4	7.5	7.3
1	16.1	22.5	22.0
2	26.9	37.5	36.7
3	37.6	—	—
4			

107° (287°) (252°) **72°**

COMMON LOGARITHMS OF THE TRIGONOMETRIC FUNCTIONS (Continued)

18° (198°) (341°) **161°**

′	L. Sin.	d.	L. Tan.	c.d.	L. Cot.	L. Cos.	d.	′
0	9.48 998	39	9.51 178	43	0.48 822	9.97 821	4	**60**
1	9.49 037	39	9.51 221	43	0.48 779	9.97 817	5	59
2	9.49 076	39	9.51 264	42	0.48 736	9.97 812	4	58
3	9.49 115	38	9.51 306	43	0.48 694	9.97 808	4	57
4	9.49 153	39	9.51 349	43	0.48 651	9.97 804	4	56
5	9.49 192	39	9.51 392	43	0.48 608	9.97 800	4	55
6	9.49 231	38	9.51 435	43	0.48 565	9.97 796	4	54
7	9.49 269	39	9.51 478	42	0.48 522	9.97 792	4	53
8	9.49 308	39	9.51 520	43	0.48 480	9.97 788	4	52
9	9.49 347	38	9.51 563	43	0.48 437	9.97 784	5	51
10	9.49 385	39	9.51 606	42	0.48 394	9.97 779	4	**50**
11	9.49 424	38	9.51 648	43	0.48 352	9.97 775	4	49
12	9.49 462	38	9.51 691	43	0.48 309	9.97 771	4	48
13	9.49 500	39	9.51 734	42	0.48 266	9.97 767	4	47
14	9.49 539	38	9.51 776	43	0.48 224	9.97 763	4	46
15	9.49 577	38	9.51 819	42	0.48 181	9.97 759	5	45
16	9.49 615	39	9.51 861	42	0.48 139	9.97 754	4	44
17	9.49 654	38	9.51 903	43	0.48 097	9.97 750	4	43
18	9.49 692	38	9.51 946	42	0.48 054	9.97 746	4	42
19	9.49 730	38	9.51 988	43	0.48 012	9.97 742	4	41
20	9.49 768	38	9.52 031	42	0.47 969	9.97 738	4	**40**
21	9.49 806	38	9.52 073	42	0.47 927	9.97 734	5	39
22	9.49 844	38	9.52 115	42	0.47 885	9.97 729	4	38
23	9.49 882	38	9.52 157	43	0.47 843	9.97 725	4	37
24	9.49 920	38	9.52 200	42	0.47 800	9.97 721	4	36
25	9.49 958	38	9.52 242	42	0.47 758	9.97 717	4	35
26	9.49 996	38	9.52 284	42	0.47 716	9.97 713	5	34
27	9.50 034	38	9.52 326	42	0.47 674	9.97 708	4	33
28	9.50 072	38	9.52 368	42	0.47 632	9.97 704	4	32
29	9.50 110	38	9.52 410	42	0.47 590	9.97 700	4	31
30	9.50 148	37	9.52 452	42	0.47 548	9.97 696	5	**30**
31	9.50 185	38	9.52 494	42	0.47 506	9.97 691	4	29
32	9.50 223	38	9.52 536	42	0.47 464	9.97 687	4	28
33	9.50 261	37	9.52 578	42	0.47 422	9.97 683	4	27
34	9.50 298	38	9.52 620	41	0.47 380	9.97 679	5	26
35	9.50 336	38	9.52 661	42	0.47 339	9.97 674	4	25
36	9.50 374	37	9.52 703	42	0.47 297	9.97 670	4	24
37	9.50 411	38	9.52 745	42	0.47 255	9.97 666	4	23
38	9.50 449	37	9.52 787	42	0.47 213	9.97 662	5	22
39	9.50 486	37	9.52 829	41	0.47 171	9.97 657	4	21
40	9.50 523	38	9.52 870	42	0.47 130	9.97 653	4	**20**
41	9.50 561	37	9.52 912	41	0.47 088	9.97 649	4	19
42	9.50 598	37	9.52 953	42	0.47 047	9.97 645	5	18
43	9.50 635	38	9.52 995	42	0.47 005	9.97 640	4	17
44	9.50 673	37	9.53 037	41	0.46 963	9.97 636	4	16
45	9.50 710	37	9.53 078	42	0.46 922	9.97 632	4	15
46	9.50 747	37	9.53 120	41	0.46 880	9.97 628	5	14
47	9.50 784	37	9.53 161	41	0.46 839	9.97 623	4	13
48	9.50 821	37	9.53 202	42	0.46 798	9.97 619	4	12
49	9.50 858	38	9.53 244	41	0.46 756	9.97 615	5	11
50	9.50 896	37	9.53 285	42	0.46 715	9.97 610	4	**10**
51	9.50 933	37	9.53 327	41	0.46 673	9.97 606	4	9
52	9.50 970	37	9.53 368	41	0.46 632	9.97 602	5	8
53	9.51 007	36	9.53 409	41	0.46 591	9.97 597	4	7
54	9.51 043	37	9.53 450	42	0.46 550	9.97 593	4	6
55	9.51 080	37	9.53 492	41	0.46 508	9.97 589	5	5
56	9.51 117	37	9.53 533	41	0.46 467	9.97 584	4	4
57	9.51 154	37	9.53 574	41	0.46 426	9.97 580	4	3
58	9.51 191	36	9.53 615	41	0.46 385	9.97 576	5	2
59	9.51 227	37	9.53 656	41	0.46 344	9.97 571	4	1
60	9.51 264		9.53 697		0.46 303	9.97 567		**0**
′	L. Cos.	d.	L. Cot.	c.d.	L. Tan.	L. Sin.	d.	′

108° (288°) (251°) **71°**

P. P.

″	**43**	**42**	**41**
1	0.7	0.7	0.7
2	1.4	1.4	1.4
3	2.2	2.1	2.0
4	2.9	2.8	2.7
5	3.6	3.5	3.4
6	4.3	4.2	4.1
7	5.0	4.9	4.8
8	5.7	5.6	5.5
9	6.4	6.3	6.2
10	7.2	7.0	6.8
20	14.3	14.0	13.7
30	21.5	21.0	20.5
40	28.7	28.0	27.3
50	35.8	35.0	34.2

″	**39**	**38**	**37**
1	0.6	0.6	0.6
2	1.3	1.3	1.2
3	2.0	1.9	1.8
4	2.6	2.5	2.5
5	3.2	3.2	3.1
6	3.9	3.8	3.7
7	4.6	4.4	4.3
8	5.2	5.1	4.9
9	5.8	5.7	5.6
10	6.5	6.3	6.2
20	13.0	12.7	12.3
30	19.5	19.0	18.5
40	26.0	25.3	24.7
50	32.5	31.7	30.8

″	**36**	**5**	**4**
1	0.6	0.1	0.1
2	1.2	0.2	0.1
3	1.8	0.2	0.2
4	2.4	0.3	0.3
5	3.0	0.4	0.3
6	3.6	0.5	0.4
7	4.2	0.6	0.5
8	4.8	0.7	0.5
9	5.4	0.8	0.6
10	6.0	0.8	0.7
20	12.0	1.7	1.3
30	18.0	2.5	2.0
40	24.0	3.3	2.7
50	30.0	4.2	3.3

	5/43	**5/42**	**5/41**
0	4.3	4.2	4.1
1	12.9	12.6	12.3
2	21.5	21.0	20.5
3	30.1	29.4	28.7
4	38.7	37.8	36.9
5			

	4/43	**4/42**	**4/41**
0	5.4	5.2	5.1
1	16.1	15.8	15.4
2	26.9	26.2	25.6
3	37.6	36.8	35.9
4			

P. P.

COMMON LOGARITHMS OF THE TRIGONOMETRIC FUNCTIONS (Continued)

19° (199°) (340°) **160°**

′	L. Sin.	d.	L. Tan.	c.d.	L. Cot.	L. Cos.	d.	′
0	9.51 264	37	9.53 697	41	0.46 303	9.97 567	4	**60**
1	9.51 301	37	9.53 738	41	0.46 262	9.97 563	5	59
2	9.51 338	36	9.53 779	41	0.46 221	9.97 558	4	58
3	9.51 374	37	9.53 820	41	0.46 180	9.97 554	4	57
4	9.51 411	36	9.53 861	41	0.46 139	9.97 550	5	56
5	9.51 447	37	9.53 902	41	0.46 098	9.97 545	4	55
6	9.51 484	36	9.53 943	41	0.46 057	9.97 541	5	54
7	9.51 520	37	9.53 984	41	0.46 016	9.97 536	4	53
8	9.51 557	36	9.54 025	40	0.45 975	9.97 532	4	52
9	9.51 593	36	9.54 065	41	0.45 935	9.97 528	5	51
10	9.51 629	37	9.54 106	41	0.45 894	9.97 523	4	**50**
11	9.51 666	36	9.54 147	40	0.45 853	9.97 519	4	49
12	9.51 702	36	9.54 187	41	0.45 813	9.97 515	5	48
13	9.51 738	36	9.54 228	41	0.45 772	9.97 510	4	47
14	9.51 774	37	9.54 269	40	0.45 731	9.97 506	5	46
15	9.51 811	36	9.54 309	41	0.45 691	9.97 501	4	45
16	9.51 847	36	9.54 350	40	0.45 650	9.97 497	5	44
17	9.51 883	36	9.54 390	41	0.45 610	9.97 492	4	43
18	9.51 919	36	9.54 431	40	0.45 569	9.97 488	4	42
19	9.51 955	36	9.54 471	41	0.45 529	9.97 484	5	41
20	9.51 991	36	9.54 512	40	0.45 488	9.97 479	4	**40**
21	9.52 027	36	9.54 552	41	0.45 448	9.97 475	5	39
22	9.52 063	36	9.54 593	40	0.45 407	9.97 470	4	38
23	9.52 099	36	9.54 633	40	0.45 367	9.97 466	5	37
24	9.52 135	36	9.54 673	41	0.45 327	9.97 461	4	36
25	9.52 171	36	9.54 714	40	0.45 286	9.97 457	4	35
26	9.52 207	35	9.54 754	40	0.45 246	9.97 453	5	34
27	9.52 242	36	9.54 794	41	0.45 206	9.97 448	4	33
28	9.52 278	36	9.54 835	40	0.45 165	9.97 444	5	32
29	9.52 314	36	9.54 875	40	0.45 125	9.97 439	4	31
30	9.52 350	35	9.54 915	40	0.45 085	9.97 435	5	**30**
31	9.52 385	36	9.54 955	40	0.45 045	9.97 430	4	29
32	9.52 421	35	9.54 995	40	0.45 005	9.97 426	5	28
33	9.52 456	36	9.55 035	40	0.44 965	9.97 421	4	27
34	9.52 492	35	9.55 075	40	0.44 925	9.97 417	5	26
35	9.52 527	36	9.55 115	40	0.44 885	9.97 412	4	25
36	9.52 563	35	9.55 155	40	0.44 845	9.97 408	5	24
37	9.52 598	36	9.55 195	40	0.44 805	9.97 403	4	23
38	9.52 634	35	9.55 235	40	0.44 765	9.97 399	5	22
39	9.52 669	36	9.55 275	40	0.44 725	9.97 394	4	21
40	9.52 705	35	9.55 315	40	0.44 685	9.97 390	5	**20**
41	9.52 740	35	9.55 355	40	0.44 645	9.97 385	4	19
42	9.52 775	36	9.55 395	39	0.44 605	9.97 381	5	18
43	9.52 811	35	9.55 434	40	0.44 566	9.97 376	4	17
44	9.52 846	35	9.55 474	40	0.44 526	9.97 372	5	16
45	9.52 881	35	9.55 514	40	0.44 486	9.97 367	4	15
46	9.52 916	35	9.55 554	39	0.44 446	9.97 363	5	14
47	9.52 951	35	9.55 593	40	0.44 407	9.97 358	5	13
48	9.52 986	35	9.55 633	40	0.44 367	9.97 353	4	12
49	9.53 021	35	9.55 673	39	0.44 327	9.97 349	5	11
50	9.53 056	36	9.55 712	40	0.44 288	9.97 344	4	**10**
51	9.53 092	34	9.55 752	39	0.44 248	9.97 340	5	9
52	9.53 126	35	9.55 791	40	0.44 209	9.97 335	4	8
53	9.53 161	35	9.55 831	39	0.44 169	9.97 331	5	7
54	9.53 196	35	9.55 870	40	0.44 130	9.97 326	4	6
55	9.53 231	35	9.55 910	39	0.44 090	9.97 322	5	5
56	9.53 266	35	9.55 949	40	0.44 051	9.97 317	5	4
57	9.53 301	35	9.55 989	39	0.44 011	9.97 312	4	3
58	9.53 336	34	9.56 028	39	0.43 972	9.97 308	5	2
59	9.53 370	35	9.56 067	40	0.43 933	9.97 303	4	1
60	9.53 405		9.56 107		0.43 893	9.97 299		**0**
′	L. Cos.	d.	L. Cot.	c.d.	L. Tan.	L. Sin.	d.	′

P. P.

″	**41**	**40**	**39**
1	0.7	0.7	0.6
2	1.4	1.3	1.3
3	2.0	2.0	2.0
4	2.7	2.7	2.6
5	3.4	3.3	3.2
6	4.1	4.0	3.9
7	4.8	4.7	4.6
8	5.5	5.3	5.2
9	6.2	6.0	5.8
10	6.8	6.7	6.5
20	13.7	13.3	13.0
30	20.5	20.0	19.5
40	27.3	26.7	26.0
50	34.2	33.3	32.5

″	**37**	**36**	**35**
1	0.6	0.6	0.6
2	1.2	1.2	1.2
3	1.8	1.8	1.8
4	2.5	2.4	2.3
5	3.1	3.0	2.9
6	3.7	3.6	3.5
7	4.3	4.2	4.1
8	4.9	4.8	4.7
9	5.6	5.4	5.2
10	6.2	6.0	5.8
20	12.3	12.0	11.7
30	18.5	18.0	17.5
40	24.7	24.0	23.3
50	30.8	30.0	29.2

″	**34**	**5**	**4**
1	0.6	0.1	0.1
2	1.1	0.2	0.1
3	1.7	0.2	0.2
4	2.3	0.3	0.3
5	2.8	0.4	0.3
6	3.4	0.5	0.4
7	4.0	0.6	0.5
8	4.5	0.7	0.5
9	5.1	0.8	0.6
10	5.7	0.8	0.7
20	11.3	1.7	1.3
30	17.0	2.5	2.0
40	22.7	3.3	2.7
50	28.3	4.2	3.3

	5/41	**5/40**	**5/39**
0–1	4.1	4.0	3.9
1–2	12.3	12.0	11.7
2–3	20.5	20.0	19.5
3–4	28.7	28.0	27.3
4–5	36.9	36.0	35.1

	4/41	**4/40**	**4/39**
0–1	5.1	5.0	4.9
1–2	15.4	15.0	14.6
2–3	25.6	25.0	24.4
3–4	35.9	35.0	34.1

109° (289°) (250°) **70°**

COMMON LOGARITHMS OF THE TRIGONOMETRIC FUNCTIONS (Continued)

20° (200°) (339°) **159°**

′	L. Sin.	d.	L. Tan.	c.d.	L. Cot.	L. Cos.	d.	′
0	9.53 405	35	9.56 107	39	0.43 893	9.97 299	5	**60**
1	9.53 440	35	9.56 146	39	0.43 854	9.97 294	5	59
2	9.53 475	34	9.56 185	39	0.43 815	9.97 289	4	58
3	9.53 509	35	9.56 224	40	0.43 776	9.97 285	5	57
4	9.53 544	34	9.56 264	39	0.43 736	9.97 280	4	56
5	9.53 578	35	9.56 303	39	0.43 697	9.97 276	5	55
6	9.53 613	34	9.56 342	39	0.43 658	9.97 271	5	54
7	9.53 647	35	9.56 381	39	0.43 619	9.97 266	4	53
8	9.53 682	34	9.56 420	39	0.43 580	9.97 262	5	52
9	9.53 716	35	9.56 459	39	0.43 541	9.97 257	5	51
10	9.53 751	34	9.56 498	39	0.43 502	9.97 252	4	**50**
11	9.53 785	34	9.56 537	39	0.43 463	9.97 248	5	49
12	9.53 819	35	9.56 576	39	0.43 424	9.97 243	5	48
13	9.53 854	34	9.56 615	39	0.43 385	9.97 238	4	47
14	9.53 888	34	9.56 654	39	0.43 346	9.97 234	5	46
15	9.53 922	35	9.56 693	39	0.43 307	9.97 229	5	45
16	9.53 957	34	9.56 732	39	0.43 268	9.97 224	4	44
17	9.53 991	34	9.56 771	39	0.43 229	9.97 220	5	43
18	9.54 025	34	9.56 810	39	0.43 190	9.97 215	5	42
19	9.54 059	34	9.56 849	38	0.43 151	9.97 210	4	41
20	9.54 093	34	9.56 887	39	0.43 113	9.97 206	5	**40**
21	9.54 127	34	9.56 926	39	0.43 074	9.97 201	5	39
22	9.54 161	34	9.56 965	39	0.43 035	9.97 196	4	38
23	9.54 195	34	9.57 004	38	0.42 996	9.97 192	5	37
24	9.54 229	34	9.57 042	39	0.42 958	9.97 187	5	36
25	9.54 263	34	9.57 081	39	0.42 919	9.97 182	4	35
26	9.54 297	34	9.57 120	38	0.42 880	9.97 178	5	34
27	9.54 331	34	9.57 158	39	0.42 842	9.97 173	5	33
28	9.54 365	34	9.57 197	38	0.42 803	9.97 168	5	32
29	9.54 399	34	9.57 235	39	0.42 765	9.97 163	4	31
30	9.54 433	33	9.57 274	38	0.42 726	9.97 159	5	**30**
31	9.54 466	34	9.57 312	39	0.42 688	9.97 154	5	29
32	9.54 500	34	9.57 351	38	0.42 649	9.97 149	4	28
33	9.54 534	33	9.57 389	39	0.42 611	9.97 145	5	27
34	9.54 567	34	9.57 428	38	0.42 572	9.97 140	5	26
35	9.54 601	34	9.57 466	38	0.42 534	9.97 135	5	25
36	9.54 635	33	9.57 504	39	0.42 496	9.97 130	4	24
37	9.54 668	34	9.57 543	38	0.42 457	9.97 126	5	23
38	9.54 702	33	9.57 581	38	0.42 419	9.97 121	5	22
39	9.54 735	34	9.57 619	39	0.42 381	9.97 116	5	21
40	9.54 769	33	9.57 658	38	0.42 342	9.97 111	4	**20**
41	9.54 802	34	9.57 696	38	0.42 304	9.97 107	5	19
42	9.54 836	33	9.57 734	38	0.42 266	9.97 102	5	18
43	9.54 869	34	9.57 772	38	0.42 228	9.97 097	5	17
44	9.54 903	33	9.57 810	39	0.42 190	9.97 092	5	16
45	9.54 936	33	9.57 849	38	0.42 151	9.97 087	4	15
46	9.54 969	34	9.57 887	38	0.42 113	9.97 083	5	14
47	9.55 003	33	9.57 925	38	0.42 075	9.97 078	5	13
48	9.55 036	33	9.57 963	38	0.42 037	9.97 073	5	12
49	9.55 069	33	9.58 001	38	0.41 999	9.97 068	5	11
50	9.55 102	34	9.58 039	38	0.41 961	9.97 063	4	**10**
51	9.55 136	33	9.58 077	38	0.41 923	9.97 059	5	9
52	9.55 169	33	9.58 115	38	0.41 885	9.97 054	5	8
53	9.55 202	33	9.58 153	38	0.41 847	9.97 049	5	7
54	9.55 235	33	9.58 191	38	0.41 809	9.97 044	5	6
55	9.55 268	33	9.58 229	38	0.41 771	9.97 039	4	5
56	9.55 301	33	9.58 267	37	0.41 733	9.97 035	5	4
57	9.55 334	33	9.58 304	38	0.41 696	9.97 030	5	3
58	9.55 367	33	9.58 342	38	0.41 658	9.97 025	5	2
59	9.55 400	33	9.58 380	38	0.41 620	9.97 020	5	1
60	9.55 433		9.58 418		0.41 582	9.97 015		**0**
′	L. Cos.	d.	L. Cot.	c.d.	L. Tan.	L. Sin.	d.	′

P. P.

″	**40**	**39**	**38**
1	0.7	0.6	0.6
2	1.3	1.3	1.3
3	2.0	2.0	1.9
4	2.7	2.6	2.5
5	3.3	3.2	3.2
6	4.0	3.9	3.8
7	4.7	4.6	4.4
8	5.3	5.2	5.1
9	6.0	5.8	5.7
10	6.7	6.5	6.3
20	13.3	13.0	12.7
30	20.0	19.5	19.0
40	26.7	26.0	25.3
50	33.3	32.5	31.7
″	**37**	**35**	**34**
1	0.6	0.6	0.6
2	1.2	1.2	1.1
3	1.8	1.8	1.7
4	2.5	2.3	2.3
5	3.1	2.9	2.8
6	3.7	3.5	3.4
7	4.3	4.1	4.0
8	4.9	4.7	4.5
9	5.6	5.2	5.1
10	6.2	5.8	5.7
20	12.3	11.7	11.3
30	18.5	17.5	17.0
40	24.7	23.3	22.7
50	30.8	29.2	28.3
″	**33**	**5**	**4**
1	0.6	0.1	0.1
2	1.1	0.2	0.1
3	1.6	0.2	0.2
4	2.2	0.3	0.3
5	2.8	0.4	0.3
6	3.3	0.5	0.4
7	3.8	0.6	0.5
8	4.4	0.7	0.5
9	5.0	0.8	0.6
10	5.5	0.8	0.7
20	11.0	1.7	1.3
30	16.5	2.5	2.0
40	22.0	3.3	2.7
50	27.5	4.2	3.3
	5/40	**5/39**	**5/38**
0	4.0	3.9	3.8
1	12.0	11.7	11.4
2	20.0	19.5	19.0
3	28.0	27.3	26.6
4	36.0	35.1	34.2
5			
	5/37	**4/39**	**4/38**
0	3.7	4.9	4.8
1	11.1	14.6	14.2
2	18.5	24.4	23.8
3	25.9	34.1	33.2
4	33.3	—	—
5			

110° (290°) (249°) **69°**

COMMON LOGARITHMS OF THE TRIGONOMETRIC FUNCTIONS (Continued)

21° (201°) (338°) **158°**

′	L. Sin.	d.	L. Tan.	c.d.	L. Cot.	L. Cos.	d.	′
0	9.55 433	33	9.58 418	37	0.41 582	9.97 015	5	**60**
1	9.55 466	33	9.58 455	38	0.41 545	9.97 010	5	59
2	9.55 499	33	9.58 493	38	0.41 507	9.97 005	4	58
3	9.55 532	32	9.58 531	38	0.41 469	9.97 001	5	57
4	9.55 564	33	9.58 569	37	0.41 431	9.96 996	5	56
5	9.55 597	33	9.58 606	38	0.41 394	9.96 991	5	55
6	9.55 630	33	9.58 644	37	0.41 356	9.96 986	5	54
7	9.55 663	32	9.58 681	38	0.41 319	9.96 981	5	53
8	9.55 695	33	9.58 719	38	0.41 281	9.96 976	5	52
9	9.55 728	33	9.58 757	37	0.41 243	9.96 971	5	51
10	9.55 761	32	9.58 794	38	0.41 206	9.96 966	4	**50**
11	9.55 793	33	9.58 832	37	0.41 168	9.96 962	5	49
12	9.55 826	32	9.58 869	38	0.41 131	9.96 957	5	48
13	9.55 858	33	9.58 907	37	0.41 093	9.96 952	5	47
14	9.55 891	32	9.58 944	37	0.41 056	9.96 947	5	46
15	9.55 923	33	9.58 981	38	0.41 019	9.96 942	5	45
16	9.55 956	32	9.59 019	37	0.40 981	9.96 937	5	44
17	9.55 988	33	9.59 056	38	0.40 944	9.96 932	5	43
18	9.56 021	32	9.59 094	37	0.40 906	9.96 927	5	42
19	9.56 053	32	9.59 131	37	0.40 869	9.96 922	5	41
20	9.56 085	33	9.59 168	37	0.40 832	9.96 917	5	**40**
21	9.56 118	32	9.59 205	38	0.40 795	9.96 912	5	39
22	9.56 150	32	9.59 243	37	0.40 757	9.96 907	4	38
23	9.56 182	33	9.59 280	37	0.40 720	9.96 903	5	37
24	9.56 215	32	9.59 317	37	0.40 683	9.96 898	5	36
25	9.56 247	32	9.59 354	37	0.40 646	9.96 893	5	35
26	9.56 279	32	9.59 391	38	0.40 609	9.96 888	5	34
27	9.56 311	32	9.59 429	37	0.40 571	9.96 883	5	33
28	9.56 343	32	9.59 466	37	0.40 534	9.96 878	5	32
29	9.56 375	33	9.59 503	37	0.40 497	9.96 873	5	31
30	9.56 408	32	9.59 540	37	0.40 460	9.96 868	5	**30**
31	9.56 440	32	9.59 577	37	0.40 423	9.96 863	5	29
32	9.56 472	32	9.59 614	37	0.40 386	9.96 858	5	28
33	9.56 504	32	9.59 651	37	0.40 349	9.96 853	5	27
34	9.56 536	32	9.59 688	37	0.40 312	9.96 848	5	26
35	9.56 568	31	9.59 725	37	0.40 275	9.96 843	5	25
36	9.56 599	32	9.59 762	37	0.40 238	9.96 838	5	24
37	9.56 631	32	9.59 799	36	0.40 201	9.96 833	5	23
38	9.56 663	32	9.59 835	37	0.40 165	9.96 828	5	22
39	9.56 695	32	9.59 872	37	0.40 128	9.96 823	5	21
40	9.56 727	32	9.59 909	37	0.40 091	9.96 818	5	**20**
41	9.56 759	31	9.59 946	37	0.40 054	9.96 813	5	19
42	9.56 790	32	9.59 983	36	0.40 017	9.96 808	5	18
43	9.56 822	32	9.60 019	37	0.39 981	9.96 803	5	17
44	9.56 854	32	9.60 056	37	0.39 944	9.96 798	5	16
45	9.56 886	31	9.60 093	37	0.39 907	9.96 793	5	15
46	9.56 917	32	9.60 130	36	0.39 870	9.96 788	5	14
47	9.56 949	31	9.60 166	37	0.39 834	9.96 783	5	13
48	9.56 980	32	9.60 203	37	0.39 797	9.96 778	6	12
49	9.57 012	32	9.60 240	36	0.39 760	9.96 772	5	11
50	9.57 044	31	9.60 276	37	0.39 724	9.96 767	5	**10**
51	9.57 075	32	9.60 313	36	0.39 687	9.96 762	5	9
52	9.57 107	31	9.60 349	37	0.39 651	9.96 757	5	8
53	9.57 138	31	9.60 386	36	0.39 614	9.96 752	5	7
54	9.57 169	32	9.60 422	37	0.39 578	9.96 747	5	6
55	9.57 201	31	9.60 459	36	0.39 541	9.96 742	5	5
56	9.57 232	32	9.60 495	37	0.39 505	9.96 737	5	4
57	9.57 264	31	9.60 532	36	0.39 468	9.96 732	5	3
58	9.57 295	31	9.60 568	37	0.39 432	9.96 727	5	2
59	9.57 326	32	9.60 605	36	0.39 395	9.96 722	5	1
60	9.57 358		9.60 641		0.39 359	9.96 717		**0**
′	L. Cos.	d.	L. Cot.	c.d.	L. Tan.	L. Sin.	d.	′

111° (291°) (248°) **68°**

P. P.

″	**38**	**37**	**36**
1	0.6	0.6	0.6
2	1.3	1.2	1.2
3	1.9	1.8	1.8
4	2.5	2.5	2.4
5	3.2	3.1	3.0
6	3.8	3.7	3.6
7	4.4	4.3	4.2
8	5.1	4.9	4.8
9	5.7	5.6	5.4
10	6.3	6.2	6.0
20	12.7	12.3	12.0
30	19.0	18.5	18.0
40	25.3	24.7	24.0
50	31.7	30.8	30.0
″	**33**	**32**	**31**
1	0.6	0.5	0.5
2	1.1	1.1	1.0
3	1.6	1.6	1.6
4	2.2	2.1	2.1
5	2.8	2.7	2.6
6	3.3	3.2	3.1
7	3.8	3.7	3.6
8	4.4	4.3	4.1
9	5.0	4.8	4.6
10	5.5	5.3	5.2
20	11.0	10.7	10.3
30	16.5	16.0	15.5
40	22.0	21.3	20.7
50	27.5	26.7	25.8
″	**6**	**5**	**4**
1	0.1	0.1	0.1
2	0.2	0.2	0.1
3	0.3	0.2	0.2
4	0.4	0.3	0.3
5	0.5	0.4	0.3
6	0.6	0.5	0.4
7	0.7	0.6	0.5
8	0.8	0.7	0.5
9	0.9	0.8	0.6
10	1.0	0.8	0.7
20	2.0	1.7	1.3
30	3.0	2.5	2.0
40	4.0	3.3	2.7
50	5.0	4.2	3.3

	6/37	**5/38**	**5/37**
0	3.1	3.8	3.7
1	9.2	11.4	11.1
2	15.4	19.0	18.5
3	21.6	26.6	25.9
4	27.8	34.2	33.3
5	33.9	—	—
6			

	5/36	**4/38**	**4/37**
0	3.6	4.8	4.6
1	10.8	14.2	13.9
2	18.0	23.8	23.1
3	25.2	33.2	32.4
4	32.4	—	—
5			

P. P.

COMMON LOGARITHMS OF THE TRIGONOMETRIC FUNCTIONS (Continued)

22° (202°) | (337°) **157°**

′	L. Sin.	d.	L. Tan.	c.d.	L. Cot.	L. Cos.	d.	′
0	9.57 358	31	9.60 641	36	0.39 359	9.96 717	6	**60**
1	9.57 389	31	9.60 677	37	0.39 323	9.96 711	5	59
2	9.57 420	31	9.60 714	36	0.39 286	9.96 706	5	58
3	9.57 451	31	9.60 750	36	0.39 250	9.96 701	5	57
4	9.57 482	32	9.60 786	37	0.39 214	9.96 696	5	56
5	9.57 514	31	9.60 823	36	0.39 177	9.96 691	5	55
6	9.57 545	31	9.60 859	36	0.39 141	9.96 686	5	54
7	9.57 576	31	9.60 895	36	0.39 105	9.96 681	5	53
8	9.57 607	31	9.60 931	36	0.39 069	9.96 676	6	52
9	9.57 638	31	9.60 967	37	0.39 033	9.96 670	5	51
10	9.57 669	31	9.61 004	36	0.38 996	9.96 665	5	**50**
11	9.57 700	31	9.61 040	36	0.38 960	9.96 660	5	49
12	9.57 731	31	9.61 076	36	0.38 924	9.96 655	5	48
13	9.57 762	31	9.61 112	36	0.38 888	9.96 650	5	47
14	9.57 793	31	9.61 148	36	0.38 852	9.96 645	5	46
15	9.57 824	31	9.61 184	36	0.38 816	9.96 640	6	45
16	9.57 855	30	9.61 220	36	0.38 780	9.96 634	5	44
17	9.57 885	31	9.61 256	36	0.38 744	9.96 629	5	43
18	9.57 916	31	9.61 292	36	0.38 708	9.96 624	5	42
19	9.57 947	31	9.61 328	36	0.38 672	9.96 619	5	41
20	9.57 978	30	9.61 364	36	0.38 636	9.96 614	6	**40**
21	9.58 008	31	9.61 400	36	0.38 600	9.96 608	5	39
22	9.58 039	31	9.61 436	36	0.38 564	9.96 603	5	38
23	9.58 070	31	9.61 472	36	0.38 528	9.96 598	5	37
24	9.58 101	30	9.61 508	36	0.38 492	9.96 593	5	36
25	9.58 131	31	9.61 544	35	0.38 456	9.96 588	6	35
26	9.58 162	30	9.61 579	36	0.38 421	9.96 582	5	34
27	9.58 192	31	9.61 615	36	0.38 385	9.96 577	5	33
28	9.58 223	30	9.61 651	36	0.38 349	9.96 572	5	32
29	9.58 253	31	9.61 687	35	0.38 313	9.96 567	5	31
30	9.58 284	30	9.61 722	36	0.38 278	9.96 562	6	**30**
31	9.58 314	31	9.61 758	36	0.38 242	9.96 556	5	29
32	9.58 345	30	9.61 794	36	0.38 206	9.96 551	5	28
33	9.58 375	31	9.61 830	35	0.38 170	9.96 546	5	27
34	9.58 406	30	9.61 865	36	0.38 135	9.96 541	6	26
35	9.58 436	31	9.61 901	35	0.38 099	9.96 535	5	25
36	9.58 467	30	9.61 936	36	0.38 064	9.96 530	5	24
37	9.58 497	30	9.61 972	36	0.38 028	9.96 525	5	23
38	9.58 527	30	9.62 008	35	0.37 992	9.96 520	6	22
39	9.58 557	31	9.62 043	36	0.37 957	9.96 514	5	21
40	9.58 588	30	9.62 079	35	0.37 921	9.96 509	5	**20**
41	9.58 618	30	9.62 114	36	0.37 886	9.96 504	6	19
42	9.58 648	30	9.62 150	35	0.37 850	9.96 498	5	18
43	9.58 678	31	9.62 185	36	0.37 815	9.96 493	5	17
44	9.58 709	30	9.62 221	35	0.37 779	9.96 488	5	16
45	9.58 739	30	9.62 256	36	0.37 744	9.96 483	6	15
46	9.58 769	30	9.62 292	35	0.37 708	9.96 477	5	14
47	9.58 799	30	9.62 327	35	0.37 673	9.96 472	5	13
48	9.58 829	30	9.62 362	36	0.37 638	9.96 467	6	12
49	9.58 859	30	9.62 398	35	0.37 602	9.96 461	5	11
50	9.58 889	30	9.62 433	35	0.37 567	9.96 456	5	**10**
51	9.58 919	30	9.62 468	36	0.37 532	9.96 451	6	9
52	9.58 949	30	9.62 504	35	0.37 496	9.96 445	5	8
53	9.58 979	30	9.62 539	35	0.37 461	9.96 440	5	7
54	9.59 009	30	9.62 574	35	0.37 426	9.96 435	6	6
55	9.59 039	30	9.62 609	36	0.37 391	9.96 429	5	5
56	9.59 069	29	9.62 645	35	0.37 355	9.96 424	5	4
57	9.59 098	30	9.62 680	35	0.37 320	9.96 419	6	3
58	9.59 128	30	9.62 715	35	0.37 285	9.96 413	5	2
59	9.59 158	30	9.62 750	35	0.37 250	9.96 408	5	1
60	9.59 188		9.62 785		0.37 215	9.96 403		**0**
′	L. Cos.	d.	L. Cot.	c.d.	L. Tan.	L. Sin.	d.	′

112° (292°) | (247°) **67°**

P. P.

″	**37**	**36**	**35**
1	0.6	0.6	0.6
2	1.2	1.2	1.2
3	1.8	1.8	1.8
4	2.5	2.4	2.3
5	3.1	3.0	2.9
6	3.7	3.6	3.5
7	4.3	4.2	4.1
8	4.9	4.8	4.7
9	5.6	5.4	5.2
10	6.2	6.0	5.8
20	12.3	12.0	11.7
30	18.5	18.0	17.5
40	24.7	24.0	23.3
50	30.8	30.0	29.2

″	**32**	**31**	**30**
1	0.5	0.5	0.5
2	1.1	1.0	1.0
3	1.6	1.6	1.5
4	2.1	2.1	2.0
5	2.7	2.6	2.5
6	3.2	3.1	3.0
7	3.7	3.6	3.5
8	4.3	4.1	4.0
9	4.8	4.6	4.5
10	5.3	5.2	5.0
20	10.7	10.3	10.0
30	16.0	15.5	15.0
40	21.3	20.7	20.0
50	26.7	25.8	25.0

″	**29**	**6**	**5**
1	0.5	0.1	0.1
2	1.0	0.2	0.2
3	1.4	0.3	0.2
4	1.9	0.4	0.3
5	2.4	0.5	0.4
6	2.9	0.6	0.5
7	3.4	0.7	0.6
8	3.9	0.8	0.7
9	4.4	0.9	0.8
10	4.8	1.0	0.8
20	9.7	2.0	1.7
30	14.5	3.0	2.5
40	19.3	4.0	3.3
50	24.2	5.0	4.2

	6/36	**6/35**
0	3.0	2.9
1	9.0	8.8
2	15.0	14.6
3	21.0	20.4
4	27.0	26.2
5	33.0	32.1
6		

	5/37	**5/36**	**5/35**
0	3.7	3.6	3.5
1	11.1	10.8	10.5
2	18.5	18.0	17.5
3	25.9	25.2	24.5
4	33.3	32.4	31.5
5			

COMMON LOGARITHMS OF THE TRIGONOMETRIC FUNCTIONS (Continued)

23° (203°) (336°) **156°**

′	L. Sin.	d.	L. Tan.	c.d.	L. Cot.	L. Cos.	d.	′
0	9.59 188	30	9.62 785	35	0.37 215	9.96 403	6	**60**
1	9.59 218	29	9.62 820	35	0.37 180	9.96 397	5	59
2	9.59 247	30	9.62 855	35	0.37 145	9.96 392	5	58
3	9.59 277	30	9.62 890	36	0.37 110	9.96 387	6	57
4	9.59 307	29	9.62 926	35	0.37 074	9.96 381	5	56
5	9.59 336	30	9.62 961	35	0.37 039	9.96 376	6	55
6	9.59 366	30	9.62 996	35	0.37 004	9.96 370	5	54
7	9.59 396	29	9.63 031	35	0.36 969	9.96 365	5	53
8	9.59 425	30	9.63 066	35	0.36 934	9.96 360	6	52
9	9.59 455	29	9.63 101	34	0.36 899	9.96 354	5	51
10	9.59 484	30	9.63 135	35	0.36 865	9.96 349	6	**50**
11	9.59 514	29	9.63 170	35	0.36 830	9.96 343	5	49
12	9.59 543	30	9.63 205	35	0.36 795	9.96 338	5	48
13	9.59 573	29	9.63 240	35	0.36 760	9.96 333	6	47
14	9.59 602	30	9.63 275	35	0.36 725	9.96 327	5	46
15	9.59 632	29	9.63 310	35	0.36 690	9.96 322	6	45
16	9.59 661	29	9.63 345	34	0.36 655	9.96 316	5	44
17	9.59 690	30	9.63 379	35	0.36 621	9.96 311	6	43
18	9.59 720	29	9.63 414	35	0.36 586	9.96 305	5	42
19	9.59 749	29	9.63 449	35	0.36 551	9.96 300	6	41
20	9.59 778	30	9.63 484	35	0.36 516	9.96 294	5	**40**
21	9.59 808	29	9.63 519	34	0.36 481	9.96 289	5	39
22	9.59 837	29	9.63 553	35	0.36 447	9.96 284	6	38
23	9.59 866	29	9.63 588	35	0.36 412	9.96 278	5	37
24	9.59 895	29	9.63 623	34	0.36 377	9.96 273	6	36
25	9.59 924	30	9.63 657	35	0.36 343	9.96 267	5	35
26	9.59 954	29	9.63 692	34	0.36 308	9.96 262	6	34
27	9.59 983	29	9.63 726	35	0.36 274	9.96 256	5	33
28	9.60 012	29	9.63 761	35	0.36 239	9.96 251	6	32
29	9.60 041	29	9.63 796	34	0.36 204	9.96 245	5	31
30	9.60 070	29	9.63 830	35	0.36 170	9.96 240	6	**30**
31	9.60 099	29	9.63 865	34	0.36 135	9.96 234	5	29
32	9.60 128	29	9.63 899	35	0.36 101	9.96 229	6	28
33	9.60 157	29	9.63 934	34	0.36 066	9.96 223	5	27
34	9.60 186	29	9.63 968	35	0.36 032	9.96 218	6	26
35	9.60 215	29	9.64 003	34	0.35 997	9.96 212	5	25
36	9.60 244	29	9.64 037	35	0.35 963	9.96 207	6	24
37	9.60 273	29	9.64 072	34	0.35 928	9.96 201	5	23
38	9.60 302	29	9.64 106	34	0.35 894	9.96 196	6	22
39	9.60 331	28	9.64 140	35	0.35 860	9.96 190	5	21
40	9.60 359	29	9.64 175	34	0.35 825	9.96 185	6	**20**
41	9.60 388	29	9.64 209	34	0.35 791	9.96 179	5	19
42	9.60 417	29	9.64 243	35	0.35 757	9.96 174	6	18
43	9.60 446	28	9.64 278	34	0.35 722	9.96 168	6	17
44	9.60 474	29	9.64 312	34	0.35 688	9.96 162	5	16
45	9.60 503	29	9.64 346	35	0.35 654	9.96 157	6	15
46	9.60 532	29	9.64 381	34	0.35 619	9.96 151	5	14
47	9.60 561	28	9.64 415	34	0.35 585	9.96 146	6	13
48	9.60 589	29	9.64 449	34	0.35 551	9.96 140	5	12
49	9.60 618	28	9.64 483	34	0.35 517	9.96 135	6	11
50	9.60 646	29	9.64 517	35	0.35 483	9.96 129	6	**10**
51	9.60 675	29	9.64 552	34	0.35 448	9.96 123	5	9
52	9.60 704	28	9.64 586	34	0.35 414	9.96 118	6	8
53	9.60 732	29	9.64 620	34	0.35 380	9.96 112	5	7
54	9.60 761	28	9.64 654	34	0.35 346	9.96 107	6	6
55	9.60 789	29	9.64 688	34	0.35 312	9.96 101	6	5
56	9.60 818	28	9.64 722	34	0.35 278	9.96 095	5	4
57	9.60 846	29	9.64 756	34	0.35 244	9.96 090	6	3
58	9.60 875	28	9.64 790	34	0.35 210	9.96 084	5	2
59	9.60 903	28	9.64 824	34	0.35 176	9.96 079	6	1
60	9.60 931		9.64 858		0.35 142	9.96 073		**0**
′	L. Cos.	d.	L. Cot.	c.d.	L. Tan.	L. Sin.	d.	′

113° (293°) (246°) **66°**

P. P.

″	36	35	34
1	0.6	0.6	0.6
2	1.2	1.2	1.1
3	1.8	1.8	1.7
4	2.4	2.3	2.3
5	3.0	2.9	2.8
6	3.6	3.5	3.4
7	4.2	4.1	4.0
8	4.8	4.7	4.5
9	5.4	5.2	5.1
10	6.0	5.8	5.7
20	12.0	11.7	11.3
30	18.0	17.5	17.0
40	24.0	23.3	22.7
50	30.0	29.2	28.3

″	30	29	28
1	0.5	0.5	0.5
2	1.0	1.0	0.9
3	1.5	1.4	1.4
4	2.0	1.9	1.9
5	2.5	2.4	2.3
6	3.0	2.9	2.8
7	3.5	3.4	3.3
8	4.0	3.9	3.7
9	4.5	4.4	4.2
10	5.0	4.8	4.7
20	10.0	9.7	9.3
30	15.0	14.5	14.0
40	20.0	19.3	18.7
50	25.0	24.2	23.3

″	6	5
1	0.1	0.1
2	0.2	0.2
3	0.3	0.2
4	0.4	0.3
5	0.5	0.4
6	0.6	0.5
7	0.7	0.6
8	0.8	0.7
9	0.9	0.8
10	1.0	0.8
20	2.0	1.7
30	3.0	2.5
40	4.0	3.3
50	5.0	4.2

	6/36	6/35	6/34
0			
1	3.0	2.9	2.8
2	9.0	8.8	8.5
3	15.0	14.6	14.2
4	21.0	20.4	19.8
5	27.0	26.2	25.5
6	33.0	32.1	31.2

	5/35	5/34
0		
1	3.5	3.4
2	10.5	10.2
3	17.5	17.0
4	24.5	23.8
5	31.5	30.6

P. P.

COMMON LOGARITHMS OF THE TRIGONOMETRIC FUNCTIONS (Continued)

24° (204°) (335°) **155°**

′	L. Sin.	d.	L. Tan.	c.d.	L. Cot.	L. Cos.	d.	′
0	9.60 931	29	9.64 858	34	0.35 142	9.96 073	6	**60**
1	9.60 960	28	9.64 892	34	0.35 108	9.96 067	5	59
2	9.60 988	28	9.64 926	34	0.35 074	9.96 062	6	58
3	9.61 016	29	9.64 960	34	0.35 040	9.96 056	6	57
4	9.61 045	28	9.64 994	34	0.35 006	9.96 050	5	56
5	9.61 073	28	9.65 028	34	0.34 972	9.96 045	6	55
6	9.61 101	28	9.65 062	34	0.34 938	9.96 039	5	54
7	9.61 129	29	9.65 096	34	0.34 904	9.96 034	6	53
8	9.61 158	28	9.65 130	34	0.34 870	9.96 028	6	52
9	9.61 186	28	9.65 164	33	0.34 836	9.96 022	5	51
10	9.61 214	28	9.65 197	34	0.34 803	9.96 017	6	**50**
11	9.61 242	28	9.65 231	34	0.34 769	9.96 011	6	49
12	9.61 270	28	9.65 265	34	0.34 735	9.96 005	5	48
13	9.61 298	28	9.65 299	34	0.34 701	9.96 000	6	47
14	9.61 326	28	9.65 333	33	0.34 667	9.95 994	6	46
15	9.61 354	28	9.65 366	34	0.34 634	9.95 988	6	45
16	9.61 382	29	9.65 400	34	0.34 600	9.95 982	5	44
17	9.61 411	27	9.65 434	33	0.34 566	9.95 977	6	43
18	9.61 438	28	9.65 467	34	0.34 533	9.95 971	6	42
19	9.61 466	28	9.65 501	34	0.34 499	9.95 965	5	41
20	9.61 494	28	9.65 535	33	0.34 465	9.95 960	6	**40**
21	9.61 522	28	9.65 568	34	0.34 432	9.95 954	6	39
22	9.61 550	28	9.65 602	34	0.34 398	9.95 948	6	38
23	9.61 578	28	9.65 636	33	0.34 364	9.95 942	5	37
24	9.61 606	28	9.65 669	34	0.34 331	9.95 937	6	36
25	9.61 634	28	9.65 703	33	0.34 297	9.95 931	6	35
26	9.61 662	27	9.65 736	34	0.34 264	9.95 925	5	34
27	9.61 689	28	9.65 770	33	0.34 230	9.95 920	6	33
28	9.61 717	28	9.65 803	34	0.34 197	9.95 914	6	32
29	9.61 745	28	9.65 837	33	0.34 163	9.95 908	6	31
30	9.61 773	27	9.65 870	34	0.34 130	9.95 902	5	**30**
31	9.61 800	28	9.65 904	33	0.34 096	9.95 897	6	29
32	9.61 828	28	9.65 937	34	0.34 063	9.95 891	6	28
33	9.61 856	27	9.65 971	33	0.34 029	9.95 885	6	27
34	9.61 883	28	9.66 004	34	0.33 996	9.95 879	6	26
35	9.61 911	28	9.66 038	33	0.33 962	9.95 873	5	25
36	9.61 939	27	9.66 071	33	0.33 929	9.95 868	6	24
37	9.61 966	28	9.66 104	34	0.33 896	9.95 862	6	23
38	9.61 994	27	9.66 138	33	0.33 862	9.95 856	6	22
39	9.62 021	28	9.66 171	33	0.33 829	9.95 850	6	21
40	9.62 049	27	9.66 204	34	0.33 796	9.95 844	5	**20**
41	9.62 076	28	9.66 238	33	0.33 762	9.95 839	6	19
42	9.62 104	27	9.66 271	33	0.33 729	9.95 833	6	18
43	9.62 131	28	9.66 304	33	0.33 696	9.95 827	6	17
44	9.62 159	27	9.66 337	34	0.33 663	9.95 821	6	16
45	9.62 186	28	9.66 371	33	0.33 629	9.95 815	5	15
46	9.62 214	27	9.66 404	33	0.33 596	9.95 810	6	14
47	9.62 241	27	9.66 437	33	0.33 563	9.95 804	6	13
48	9.62 268	28	9.66 470	33	0.33 530	9.95 798	6	12
49	9.62 296	27	9.66 503	34	0.33 497	9.95 792	6	11
50	9.62 323	27	9.66 537	33	0.33 463	9.95 786	6	**10**
51	9.62 350	27	9.66 570	33	0.33 430	9.95 780	5	9
52	9.62 377	28	9.66 603	33	0.33 397	9.95 775	6	8
53	9.62 405	27	9.66 636	33	0.33 364	9.95 769	6	7
54	9.62 432	27	9.66 669	33	0.33 331	9.95 763	6	6
55	9.62 459	27	9.66 702	33	0.33 298	9.95 757	6	5
56	9.62 486	27	9.66 735	33	0.33 265	9.95 751	6	4
57	9.62 513	28	9.66 768	33	0.33 232	9.95 745	6	3
58	9.62 541	27	9.66 801	33	0.33 199	9.95 739	6	2
59	9.62 568	27	9.66 834	33	0.33 166	9.95 733	5	1
60	9.62 595		9.66 867		0.33 133	9.95 728		**0**
′	L. Cos.	d.	L. Cot.	c.d.	L. Tan.	L. Sin.	d.	′

114° (294°) (245°) **65°**

P. P.

″	**34**	**33**
1	0.6	0.6
2	1.1	1.1
3	1.7	1.6
4	2.3	2.2
5	2.8	2.8
6	3.4	3.3
7	4.0	3.8
8	4.5	4.4
9	5.1	5.0
10	5.7	5.5
20	11.3	11.0
30	17.0	16.5
40	22.7	22.0
50	28.3	27.5

″	**29**	**28**	**27**
1	0.5	0.5	0.4
2	1.0	0.9	0.9
3	1.4	1.4	1.4
4	1.9	1.9	1.8
5	2.4	2.3	2.2
6	2.9	2.8	2.7
7	3.4	3.3	3.2
8	3.9	3.7	3.6
9	4.4	4.2	4.0
10	4.8	4.7	4.5
20	9.7	9.3	9.0
30	14.5	14.0	13.5
40	19.3	18.7	18.0
50	24.2	23.3	22.5

″	**6**	**5**
1	0.1	0.1
2	0.2	0.2
3	0.3	0.2
4	0.4	0.3
5	0.5	0.4
6	0.6	0.5
7	0.7	0.6
8	0.8	0.7
9	0.9	0.8
10	1.0	0.8
20	2.0	1.7
30	3.0	2.5
40	4.0	3.3
50	5.0	4.2

	6/34	**6/33**	**5/34**
0	2.8	2.8	3.4
1	8.5	8.2	10.2
2	14.2	13.8	17.0
3	19.8	19.2	23.8
4	25.5	24.8	30.6
5	31.2	30.2	—
6			

COMMON LOGARITHMS OF THE TRIGONOMETRIC FUNCTIONS (Continued)

25° (205°) (334°) **154°**

′	L. Sin.	d.	L. Tan.	c.d.	L. Cot.	L. Cos.	d	′
0	9.62 595	27	9.66 867	33	0.33 133	9.95 728	6	**60**
1	9.62 622	27	9.66 900	33	0.33 100	9.95 722	6	59
2	9.62 649	27	9.66 933	33	0.33 067	9.95 716	6	58
3	9.62 676	27	9.66 966	33	0.33 034	9.95 710	6	57
4	9.62 703	27	9.66 999	33	0.33 001	9.95 704	6	56
5	9.62 730	27	9.67 032	33	0.32 968	9.95 698	6	55
6	9.62 757	27	9.67 065	33	0.32 935	9.95 692	6	54
7	9.62 784	27	9.67 098	33	0.32 902	9.95 686	6	53
8	9.62 811	27	9.67 131	32	0.32 869	9.95 680	6	52
9	9.62 838	27	9.67 163	33	0.32 837	9.95 674	6	51
10	9.62 865	27	9.67 196	33	0.32 804	9.95 668	5	**50**
11	9.62 892	26	9.67 229	33	0.32 771	9.95 663	6	49
12	9.62 918	27	9.67 262	33	0.32 738	9.95 657	6	48
13	9.62 945	27	9.67 295	32	0.32 705	9.95 651	6	47
14	9.62 972	27	9.67 327	33	0.32 673	9.95 645	6	46
15	9.62 999	27	9.67 360	33	0.32 640	9.95 639	6	45
16	9.63 026	26	9.67 393	33	0.32 607	9.95 633	6	44
17	9.63 052	27	9.67 426	32	0.32 574	9.95 627	6	43
18	9.63 079	27	9.67 458	33	0.32 542	9.95 621	6	42
19	9.63 106	27	9.67 491	33	0.32 509	9.95 615	6	41
20	9.63 133	26	9.67 524	32	0.32 476	9.95 609	6	**40**
21	9.63 159	27	9.67 556	33	0.32 444	9.95 603	6	39
22	9.63 186	27	9.67 589	33	0.32 411	9.95 597	6	38
23	9.63 213	26	9.67 622	32	0.32 378	9.95 591	6	37
24	9.63 239	27	9.67 654	33	0.32 346	9.95 585	6	36
25	9.63 266	26	9.67 687	32	0.32 313	9.95 579	6	35
26	9.63 292	27	9.67 719	33	0.32 281	9.95 573	6	34
27	9.63 319	26	9.67 752	33	0.32 248	9.95 567	6	33
28	9.63 345	27	9.67 785	32	0.32 215	9.95 561	6	32
29	9.63 372	26	9.67 817	33	0.32 183	9.95 555	6	31
30	9.63 398	27	9.67 850	32	0.32 150	9.95 549	6	**30**
31	9.63 425	26	9.67 882	33	0.32 118	9.95 543	6	29
32	9.63 451	27	9.67 915	32	0.32 085	9.95 537	6	28
33	9.63 478	26	9.67 947	33	0.32 053	9.95 531	6	27
34	9.63 504	27	9.67 980	32	0.32 020	9.95 525	6	26
35	9.63 531	26	9.68 012	32	0.31 988	9.95 519	6	25
36	9.63 557	26	9.68 044	33	0.31 956	9.95 513	6	24
37	9.63 583	27	9.68 077	32	0.31 923	9.95 507	7	23
38	9.63 610	26	9.68 109	33	0.31 891	9.95 500	6	22
39	9.63 636	26	9.68 142	32	0.31 858	9.95 494	6	21
40	9.63 662	27	9.68 174	32	0.31 826	9.95 488	6	**20**
41	9.63 689	26	9.68 206	33	0.31 794	9.95 482	6	19
42	9.63 715	26	9.68 239	32	0.31 761	9.95 476	6	18
43	9.63 741	26	9.68 271	32	0.31 729	9.95 470	6	17
44	9.63 767	27	9.68 303	33	0.31 697	9.95 464	6	16
45	9.63 794	26	9.68 336	32	0.31 664	9.95 458	6	15
46	9.63 820	26	9.68 368	32	0.31 632	9.95 452	6	14
47	9.63 846	26	9.68 400	32	0.31 600	9.95 446	6	13
48	9.63 872	26	9.68 432	33	0.31 568	9.95 440	6	12
49	9.63 898	26	9.68 465	32	0.31 535	9.95 434	7	11
50	9.63 924	26	9.68 497	32	0.31 503	9.95 427	6	**10**
51	9.63 950	26	9.68 529	32	0.31 471	9.95 421	6	9
52	9.63 976	26	9.68 561	32	0.31 439	9.95 415	6	8
53	9.64 002	26	9.68 593	33	0.31 407	9.95 409	6	7
54	9.64 028	26	9.68 626	32	0.31 374	9.95 403	6	6
55	9.64 054	26	9.68 658	32	0.31 342	9.95 397	6	5
56	9.64 080	26	9.68 690	32	0.31 310	9.95 391	7	4
57	9.64 106	26	9.68 722	32	0.31 278	9.95 384	6	3
58	9.64 132	26	9.68 754	32	0.31 246	9.95 378	6	2
59	9.64 158	26	9.68 786	32	0.31 214	9.95 372	6	1
60	9.64 184		9.68 818		0.31 182	9.95 366		**0**
′	L. Cos.	d.	L. Cot.	c.d.	L. Tan.	L. Sin.	d.	′

115° (295°) (244°) **64°**

P. P.

″	**33**	**32**
1	0.6	0.5
2	1.1	1.1
3	1.6	1.6
4	2.2	2.1
5	2.8	2.7
6	3.3	3.2
7	3.8	3.7
8	4.4	4.3
9	5.0	4.8
10	5.5	5.3
20	11.0	10.7
30	16.5	16.0
40	22.0	21.3
50	27.5	26.7

″	**27**	**26**
1	0.4	0.4
2	0.9	0.9
3	1.4	1.3
4	1.8	1.7
5	2.2	2.2
6	2.7	2.6
7	3.2	3.0
8	3.6	3.5
9	4.0	3.9
10	4.5	4.3
20	9.0	8.7
30	13.5	13.0
40	18.0	17.3
50	22.5	21.7

″	**7**	**6**	**5**
1	0.1	0.1	0.1
2	0.2	0.2	0.2
3	0.4	0.3	0.2
4	0.5	0.4	0.3
5	0.6	0.5	0.4
6	0.7	0.6	0.5
7	0.8	0.7	0.6
8	0.9	0.8	0.7
9	1.0	0.9	0.8
10	1.2	1.0	0.8
20	2.3	2.0	1.7
30	3.5	3.0	2.5
40	4.7	4.0	3.3
50	5.8	5.0	4.2

	7 / **32**	**6** / **32**	**5** / **33**
0	2.3	2.7	3.3
1	6.9	8.0	9.9
2	11.4	13.3	16.5
3	16.0	18.7	23.1
4	20.6	24.0	29.7
5	25.1	29.3	—
6	29.7	—	—
7			

P. P.

COMMON LOGARITHMS OF THE TRIGONOMETRIC FUNCTIONS (Continued)

26° (206°) (333°) **153°**

′	L. Sin.	d.	L. Tan.	c.d.	L. Cot.	L. Cos.	d.	′
0	9.64 184	26	9.68 818	32	0.31 182	9.95 366	6	**60**
1	9.64 210	26	9.68 850	32	0.31 150	9.95 360	6	59
2	9.64 236	26	9.68 882	32	0.31 118	9.95 354	6	58
3	9.64 262	26	9.68 914	32	0.31 086	9.95 348	7	57
4	9.64 288	25	9.68 946	32	0.31 054	9.95 341	6	56
5	9.64 313	26	9.68 978	32	0.31 022	9.95 335	6	55
6	9.64 339	26	9.69 010	32	0.30 990	9.95 329	6	54
7	9.64 365	26	9.69 042	32	0.30 958	9.95 323	6	53
8	9.64 391	26	9.69 074	32	0.30 926	9.95 317	7	52
9	9.64 417	25	9.69 106	32	0.30 894	9.95 310	6	51
10	9.64 442	26	9.69 138	32	0.30 862	9.95 304	6	**50**
11	9.64 468	26	9.69 170	32	0.30 830	9.95 298	6	49
12	9.64 494	25	9.69 202	32	0.30 798	9.95 292	6	48
13	9.64 519	26	9.69 234	32	0.30 766	9.95 286	7	47
14	9.64 545	26	9.69 266	32	0.30 734	9.95 279	6	46
15	9.64 571	25	9.69 298	31	0.30 702	9.95 273	6	45
16	9.64 596	26	9.69 329	32	0.30 671	9.95 267	6	44
17	9.64 622	25	9.69 361	32	0.30 639	9.95 261	7	43
18	9.64 647	26	9.69 393	32	0.30 607	9.95 254	6	42
19	9.64 673	25	9.69 425	32	0.30 575	9.95 248	6	41
20	9.64 698	26	9.69 457	31	0.30 543	9.95 242	6	**40**
21	9.64 724	25	9.69 488	32	0.30 512	9.95 236	7	39
22	9.64 749	26	9.69 520	32	0.30 480	9.95 229	6	38
23	9.64 775	25	9.69 552	32	0.30 448	9.95 223	6	37
24	9.64 800	26	9.69 584	31	0.30 416	9.95 217	6	36
25	9.64 826	25	9.69 615	32	0.30 385	9.95 211	7	35
26	9.64 851	26	9.69 647	32	0.30 353	9.95 204	6	34
27	9.64 877	25	9.69 679	31	0.30 321	9.95 198	6	33
28	9.64 902	25	9.69 710	32	0.30 290	9.95 192	7	32
29	9.64 927	26	9.69 742	32	0.30 258	9.95 185	6	31
30	9.64 953	25	9.69 774	31	0.30 226	9.95 179	6	**30**
31	9.64 978	25	9.69 805	32	0.30 195	9.95 173	6	29
32	9.65 003	26	9.69 837	31	0.30 163	9.95 167	7	28
33	9.65 029	25	9.69 868	32	0.30 132	9.95 160	6	27
34	9.65 054	25	9.69 900	32	0.30 100	9.95 154	6	26
35	9.65 079	25	9.69 932	31	0.30 068	9.95 148	7	25
36	9.65 104	26	9.69 963	32	0.30 037	9.95 141	6	24
37	9.65 130	25	9.69 995	31	0.30 005	9.95 135	6	23
38	9.65 155	25	9.70 026	32	0.29 974	9.95 129	7	22
39	9.65 180	25	9.70 058	31	0.29 942	9.95 122	6	21
40	9.65 205	25	9.70 089	32	0.29 911	9.95 116	6	**20**
41	9.65 230	25	9.70 121	31	0.29 879	9.95 110	7	19
42	9.65 255	26	9.70 152	32	0.29 848	9.95 103	6	18
43	9.65 281	25	9.70 184	31	0.29 816	9.95 097	7	17
44	9.65 306	25	9.70 215	32	0.29 785	9.95 090	6	16
45	9.65 331	25	9.70 247	31	0.29 753	9.95 084	6	15
46	9.65 356	25	9.70 278	31	0.29 722	9.95 078	7	14
47	9.65 381	25	9.70 309	32	0.29 691	9.95 071	6	13
48	9.65 406	25	9.70 341	31	0.29 659	9.95 065	6	12
49	9.65 431	25	9.70 372	32	0.29 628	9.95 059	7	11
50	9.65 456	25	9.70 404	31	0.29 596	9.95 052	6	**10**
51	9.65 481	25	9.70 435	31	0.29 565	9.95 046	7	9
52	9.65 506	25	9.70 466	32	0.29 534	9.95 039	6	8
53	9.65 531	25	9.70 498	31	0.29 502	9.95 033	6	7
54	9.65 556	24	9.70 529	31	0.29 471	9.95 027	7	6
55	9.65 580	25	9.70 560	32	0.29 440	9.95 020	6	5
56	9.65 605	25	9.70 592	31	0.29 408	9.95 014	7	4
57	9.65 630	25	9.70 623	31	0.29 377	9.95 007	6	3
58	9.65 655	25	9.70 654	31	0.29 346	9.95 001	6	2
59	9.65 680	25	9.70 685	32	0.29 315	9.94 995	7	1
60	9.65 705		9.70 717		0.29 283	9.94 988		**0**
′	L. Cos.	d.	L. Cot.	c.d.	L. Tan.	L. Sin.	d.	′

116° (296°) (243°) **63°**

P. P.

″	**32**	**31**
1	0.5	0.5
2	1.1	1.0
3	1.6	1.6
4	2.1	2.1
5	2.7	2.6
6	3.2	3.1
7	3.7	3.6
8	4.3	4.1
9	4.8	4.6
10	5.3	5.2
20	10.7	10.3
30	16.0	15.5
40	21.3	20.7
50	26.7	25.8

″	**26**	**25**	**24**
1	0.4	0.4	0.4
2	0.9	0.8	0.8
3	1.3	1.2	1.2
4	1.7	1.7	1.6
5	2.2	2.1	2.0
6	2.6	2.5	2.4
7	3.0	2.9	2.8
8	3.5	3.3	3.2
9	3.9	3.8	3.6
10	4.3	4.2	4.0
20	8.7	8.3	8.0
30	13.0	12.5	12.0
40	17.3	16.7	16.0
50	21.7	20.8	20.0

″	**7**	**6**
1	0.1	0.1
2	0.2	0.2
3	0.4	0.3
4	0.5	0.4
5	0.6	0.5
6	0.7	0.6
7	0.8	0.7
8	0.9	0.8
9	1.0	0.9
10	1.2	1.0
20	2.3	2.0
30	3.5	3.0
40	4.7	4.0
50	5.8	5.0

	$\frac{7}{32}$	$\frac{7}{31}$	$\frac{6}{32}$
0	2.3	2.2	2.7
1	6.9	6.6	8.0
2	11.4	11.1	13.3
3	16.0	15.5	18.7
4	20.6	19.9	24.0
5	25.1	24.4	29.3
6	29.7	28.8	—
7			

COMMON LOGARITHMS OF THE TRIGONOMETRIC FUNCTIONS (Continued)

27° (207°) (332°) **152°**

′	L. Sin.	d.	L. Tan.	c.d.	L. Cot.	L. Cos.	d.	′
0	9.65 705̄	24	9.70 717	31	0.29 283	9.94 988	6	**60**
1	9.65 729	25	9.70 748	31	0.29 252	9.94 982	7	59
2	9.65 754	25	9.70 779	31	0.29 221	9.94 975̄	6	58
3	9.65 779	25	9.70 810	31	0.29 190	9.94 969	7	57
4	9.65 804	24	9.70 841	32	0.29 159	9.94 962	6	56
5	9.65 828	25	9.70 873	31	0.29 127	9.94 956	7	55
6	9.65 853	25	9.70 904	31	0.29 096	9.94 949	6	54
7	9.65 878	24	9.70 935̄	31	0.29 065̄	9.94 943	7	53
8	9.65 902	25	9.70 966	31	0.29 034	9.94 936	6	52
9	9.65 927	25	9.70 997	31	0.29 003	9.94 930	7	51
10	9.65 952	24	9.71 028	31	0.28 972	9.94 923	6	**50**
11	9.65 976	25	9.71 059	31	0.28 941	9.94 917	6	49
12	9.66 001	24	9.71 090	31	0.28 910	9.94 911	7	48
13	9.66 025̄	25	9.71 121	32	0.28 879	9.94 904	6	47
14	9.66 050	25	9.71 153	31	0.28 847	9.94 898	7	46
15	9.66 075̄	24	9.71 184	31	0.28 816	9.94 891	6	45
16	9.66 099	25	9.71 215̄	31	0.28 785̄	9.94 885̄	7	44
17	9.66 124	24	9.71 246	31	0.28 754	9.94 878	7	43
18	9.66 148	25	9.71 277	31	0.28 723	9.94 871	6	42
19	9.66 173	24	9.71 308	31	0.28 692	9.94 865̄	7	41
20	9.66 197	24	9.71 339	31	0.28 661	9.94 858	6	**40**
21	9.66 221	25	9.71 370	31	0.28 630	9.94 852	7	39
22	9.66 246	24	9.71 401	30	0.28 599	9.94 845̄	6	38
23	9.66 270	25	9.71 431	31	0.28 569	9.94 839	7	37
24	9.66 295̄	24	9.71 462	31	0.28 538	9.94 832	6	36
25	9.66 319	24	9.71 493	31	0.28 507	9.94 826	7	35
26	9.66 343	25	9.71 524	31	0.28 476	9.94 819	6	34
27	9.66 368	24	9.71 555̄	31	0.28 445̄	9.94 813	7	33
28	9.66 392	24	9.71 586	31	0.28 414	9.94 806	7	32
29	9.66 416	25	9.71 617	31	0.28 383	9.94 799	6	31
30	9.66 441	24	9.71 648	31	0.28 352	9.94 793	7	**30**
31	9.66 465̄	24	9.71 679	30	0.28 321	9.94 786	6	29
32	9.66 489	24	9.71 709	31	0.28 291	9.94 780	7	28
33	9.66 513	24	9.71 740	31	0.28 260	9.94 773	6	27
34	9.66 537	25	9.71 771	31	0.28 229	9.94 767	7	26
35	9.66 562	24	9.71 802	31	0.28 198	9.94 760	7	25
36	9.66 586	24	9.71 833	30	0.28 167	9.94 753	6	24
37	9.66 610	24	9.71 863	31	0.28 137	9.94 747	7	23
38	9.66 634	24	9.71 894	31	0.28 106	9.94 740	6	22
39	9.66 658	24	9.71 925̄	30	0.28 075̄	9.94 734	7	21
40	9.66 682	24	9.71 955̄	31	0.28 045̄	9.94 727	7	**20**
41	9.66 706	25	9.71 986	31	0.28 014	9.94 720	6	19
42	9.66 731	24	9.72 017	31	0.27 983	9.94 714	7	18
43	9.66 755̄	24	9.72 048	30	0.27 952	9.94 707	7	17
44	9.66 779	24	9.72 078	31	0.27 922	9.94 700	6	16
45	9.66 803	24	9.72 109	31	0.27 891	9.94 694	7	15
46	9.66 827	24	9.72 140	30	0.27 860	9.94 687	7	14
47	9.66 851	24	9.72 170	31	0.27 830	9.94 680	6	13
48	9.66 875̄	24	9.72 201	30	0.27 799	9.94 674	7	12
49	9.66 899	23	9.72 231	31	0.27 769	9.94 667	7	11
50	9.66 922	24	9.72 262	31	0.27 738	9.94 660	6	**10**
51	9.66 946	24	9.72 293	30	0.27 707	9.94 654	7	9
52	9.66 970	24	9.72 323	31	0.27 677	9.94 647	7	8
53	9.66 994	24	9.72 354	30	0.27 646	9.94 640	6	7
54	9.67 018	24	9.72 384	31	0.27 616	9.94 634	7	6
55	9.67 042	24	9.72 415̄	30	0.27 585̄	9.94 627	7	5
56	9.67 066	24	9.72 445̄	31	0.27 555	9.94 620	6	4
57	9.67 090	23	9.72 476	30	0.27 524	9.94 614	7	3
58	9.67 113	24	9.72 506	31	0.27 494	9.94 607	7	2
59	9.67 137	24	9.72 537	30	0.27 463	9.94 600	7	1
60	9.67 161		9.72 567		0.27 433	9.94 593		**0**
′	L. Cos.	d.	L. Cot.	c.d.	L. Tan.	L. Sin.	d.	′

117° (297°) (242°) **62°**

P. P.

″	**32**	**31**	**30**
1	0.5	0.5	0.5
2	1.1	1.0	1.0
3	1.6	1.6	1.5
4	2.1	2.1	2.0
5	2.7	2.6	2.5
6	3.2	3.1	3.0
7	3.7	3.6	3.5
8	4.3	4.1	4.0
9	4.8	4.6	4.5
10	5.3	5.2	5.0
20	10.7	10.3	10.0
30	16.0	15.5	15.0
40	21.3	20.7	20.0
50	26.7	25.8	25.0

″	**25**	**24**	**23**
1	0.4	0.4	0.4
2	0.8	0.8	0.8
3	1.2	1.2	1.2
4	1.7	1.6	1.5
5	2.1	2.0	1.9
6	2.5	2.4	2.3
7	2.9	2.8	2.7
8	3.3	3.2	3.1
9	3.8	3.6	3.4
10	4.2	4.0	3.8
20	8.3	8.0	7.7
30	12.5	12.0	11.5
40	16.7	16.0	15.3
50	20.8	20.0	19.2

″	**7**	**6**
1	0.1	0.1
2	0.2	0.2
3	0.4	0.3
4	0.5̄	0.4
5	0.6	0.5
6	0.7	0.6
7	0.8	0.7
8	0.9	0.8
9	1.0	0.9
10	1.2	1.0
20	2.3	2.0
30	3.5	3.0
40	4.7	4.0
50	5.8	5.0

	7/30	**6/31**	**6/30**
0	2.1	2.6	2.5
1	6.4	7.8	7.5
2	10.7	12.9	12.5
3	15.0	18.1	17.5
4	19.3	23.2	22.5
5	23.6	28.4	27.5
6	27.9	—	—
7			

P. P.

COMMON LOGARITHMS OF THE TRIGONOMETRIC FUNCTIONS (Continued)

28° (208°) (331°) **151°**

′	L. Sin.	d.	L. Tan.	c.d.	L. Cot.	L. Cos.	d.	′
0	9.67 161	24	9.72 567	31	0.27 433	9.94 593	6	**60**
1	9.67 185	23	9.72 598	30	0.27 402	9.94 587	7	59
2	9.67 208	24	9.72 628	31	0.27 372	9.94 580	7	58
3	9.67 232	24	9.72 659	30	0.27 341	9.94 573	6	57
4	9.67 256	24	9.72 689	31	0.27 311	9.94 567	7	56
5	9.67 280	23	9.72 720	30	0.27 280	9.94 560	7	55
6	9.67 303	24	9.72 750	30	0.27 250	9.94 553	7	54
7	9.67 327	23	9.72 780	31	0.27 220	9.94 546	6	53
8	9.67 350	24	9.72 811	30	0.27 189	9.94 540	7	52
9	9.67 374	24	9.72 841	31	0.27 159	9.94 533	7	51
10	9.67 398	23	9.72 872	30	0.27 128	9.94 526	7	**50**
11	9.67 421	24	9.72 902	30	0.27 098	9.94 519	6	49
12	9.67 445	23	9.72 932	31	0.27 068	9.94 513	7	48
13	9.67 468	24	9.72 963	30	0.27 037	9.94 506	7	47
14	9.67 492	23	9.72 993	30	0.27 007	9.94 499	7	46
15	9.67 515	24	9.73 023	31	0.26 977	9.94 492	7	45
16	9.67 539	23	9.73 054	30	0.26 946	9.94 485	6	44
17	9.67 562	24	9.73 084	30	0.26 916	9.94 479	7	43
18	9.67 586	23	9.73 114	30	0.26 886	9.94 472	7	42
19	9.67 609	24	9.73 144	31	0.26 856	9.94 465	7	41
20	9.67 633	23	9.73 175	30	0.26 825	9.94 458	7	**40**
21	9.67 656	24	9.73 205	30	0.26 795	9.94 451	6	39
22	9.67 680	23	9.73 235	30	0.26 765	9.94 445	7	38
23	9.67 703	23	9.73 265	30	0.26 735	9.94 438	7	37
24	9.67 726	24	9.73 295	31	0.26 705	9.94 431	7	36
25	9.67 750	23	9.73 326	30	0.26 674	9.94 424	7	35
26	9.67 773	23	9.73 356	30	0.26 644	9.94 417	7	34
27	9.67 796	24	9.73 386	30	0.26 614	9.94 410	6	33
28	9.67 820	23	9.73 416	30	0.26 584	9.94 404	7	32
29	9.67 843	23	9.73 446	30	0.26 554	9.94 397	7	31
30	9.67 866	24	9.73 476	31	0.26 524	9.94 390	7	**30**
31	9.67 890	23	9.73 507	30	0.26 493	9.94 383	7	29
32	9.67 913	23	9.73 537	30	0.26 463	9.94 376	7	28
33	9.67 936	23	9.73 567	30	0.26 433	9.94 369	7	27
34	9.67 959	23	9.73 597	30	0.26 403	9.94 362	7	26
35	9.67 982	24	9.73 627	30	0.26 373	9.94 355	6	25
36	9.68 006	23	9.73 657	30	0.26 343	9.94 349	7	24
37	9.68 029	23	9.73 687	30	0.26 313	9.94 342	7	23
38	9.68 052	23	9.73 717	30	0.26 283	9.94 335	7	22
39	9.68 075	23	9.73 747	30	0.26 253	9.94 328	7	21
40	9.68 098	23	9.73 777	30	0.26 223	9.94 321	7	**20**
41	9.68 121	23	9.73 807	30	0.26 193	9.94 314	7	19
42	9.68 144	23	9.73 837	30	0.26 163	9.94 307	7	18
43	9.68 167	23	9.73 867	30	0.26 133	9.94 300	7	17
44	9.68 190	23	9.73 897	30	0.26 103	9.94 293	7	16
45	9.68 213	24	9.73 927	30	0.26 073	9.94 286	7	15
46	9.68 237	23	9.73 957	30	0.26 043	9.94 279	6	14
47	9.68 260	23	9.73 987	30	0.26 013	9.94 273	7	13
48	9.68 283	22	9.74 017	30	0.25 983	9.94 266	7	12
49	9.68 305	23	9.74 047	30	0.25 953	9.94 259	7	11
50	9.68 328	23	9.74 077	30	0.25 923	9.94 252	7	**10**
51	9.68 351	23	9.74 107	30	0.25 893	9.94 245	7	9
52	9.68 374	23	9.74 137	29	0.25 863	9.94 238	7	8
53	9.68 397	23	9.74 166	30	0.25 834	9.94 231	7	7
54	9.68 420	23	9.74 196	30	0.25 804	9.94 224	7	6
55	9.68 443	23	9.74 226	30	0.25 774	9.94 217	7	5
56	9.68 466	23	9.74 256	30	0.25 744	9.94 210	7	4
57	9.68 489	23	9.74 286	30	0.25 714	9.94 203	7	3
58	9.68 512	22	9.74 316	29	0.25 684	9.94 196	7	2
59	9.68 534	23	9.74 345	30	0.25 655	9.94 189	7	1
60	9.68 557		9.74 375		0.25 625	9.94 182		**0**
′	L. Cos.	d.	L. Cot.	c.d.	L. Tan.	L. Sin.	d.	′

P. P.

″	**31**	**30**	**29**
1	0.5	0.5	0.5
2	1.0	1.0	1.0
3	1.6	1.5	1.4
4	2.1	2.0	1.9
5	2.6	2.5	2.4
6	3.1	3.0	2.9
7	3.6	3.5	3.4
8	4.1	4.0	3.9
9	4.6	4.5	4.4
10	5.2	5.0	4.8
20	10.3	10.0	9.7
30	15.5	15.0	14.5
40	20.7	20.0	19.3
50	25.8	25.0	24.2

″	**24**	**23**	**22**
1	0.4	0.4	0.4
2	0.8	0.8	0.7
3	1.2	1.2	1.1
4	1.6	1.5	1.5
5	2.0	1.9	1.8
6	2.4	2.3	2.2
7	2.8	2.7	2.6
8	3.2	3.1	2.9
9	3.6	3.4	3.3
10	4.0	3.8	3.7
20	8.0	7.7	7.3
30	12.0	11.5	11.0
40	16.0	15.3	14.7
50	20.0	19.2	18.3

″	**7**	**6**
1	0.1	0.1
2	0.2	0.2
3	0.4	0.3
4	0.5	0.4
5	0.6	0.5
6	0.7	0.6
7	0.8	0.7
8	0.9	0.8
9	1.0	0.9
10	1.2	1.0
20	2.3	2.0
30	3.5	3.0
40	4.7	4.0
50	5.8	5.0

	7/31	**6/31**	**6/30**
0			
	2.2	2.6	2.5
1			
	6.6	7.8	7.5
2			
	11.1	12.9	12.5
3			
	15.5	18.1	17.5
4			
	19.9	23.2	22.5
5			
	24.4	28.4	27.5
6			
	28.8	—	—
7			

P. P.

118° (298°) (241°) **61°**

COMMON LOGARITHMS OF THE TRIGONOMETRIC FUNCTIONS (Continued)

29° (209°) (330°) **150°**

′	L. Sin.	d.	L. Tan.	c.d.	L. Cot.	L. Cos.	d.	′
0	9.68 557	23	9.74 375	30	0.25 625	9.94 182	7	**60**
1	9.68 580	23	9.74 405	30	0.25 595	9.94 175	7	59
2	9.68 603	22	9.74 435	30	0.25 565	9.94 168	7	58
3	9.68 625	23	9.74 465	29	0.25 535	9.94 161	7	57
4	9.68 648	23	9.74 494	30	0.25 506	9.94 154	7	56
5	9.68 671	23	9.74 524	30	0.25 476	9.94 147	7	55
6	9.68 694	22	9.74 554	29	0.25 446	9.94 140	7	54
7	9.68 716	23	9.74 583	30	0.25 417	9.94 133	7	53
8	9.68 739	23	9.74 613	30	0.25 387	9.94 126	7	52
9	9.68 762	22	9.74 643	30	0.25 357	9.94 119	7	51
10	9.68 784	23	9.74 673	29	0.25 327	9.94 112	7	**50**
11	9.68 807	22	9.74 702	30	0.25 298	9.94 105	7	49
12	9.68 829	23	9.74 732	30	0.25 268	9.94 098	8	48
13	9.68 852	23	9.74 762	29	0.25 238	9.94 090	7	47
14	9.68 875	22	9.74 791	30	0.25 209	9.94 083	7	46
15	9.68 897	23	9.74 821	30	0.25 179	9.94 076	7	45
16	9.68 920	22	9.74 851	29	0.25 149	9.94 069	7	44
17	9.68 942	23	9.74 880	30	0.25 120	9.94 062	7	43
18	9.68 965	22	9.74 910	29	0.25 090	9.94 055	7	42
19	9.68 987	23	9.74 939	30	0.25 061	9.94 048	7	41
20	9.69 010	22	9.74 969	29	0.25 031	9.94 041	7	**40**
21	9.69 032	23	9.74 998	30	0.25 002	9.94 034	7	39
22	9.69 055	22	9.75 028	30	0.24 972	9.94 027	7	38
23	9.69 077	23	9.75 058	29	0.24 942	9.94 020	8	37
24	9.69 100	22	9.75 087	30	0.24 913	9.94 012	7	36
25	9.69 122	22	9.75 117	29	0.24 883	9.94 005	7	35
26	9.69 144	23	9.75 146	30	0.24 854	9.93 998	7	34
27	9.69 167	22	9.75 176	29	0.24 824	9.93 991	7	33
28	9.69 189	23	9.75 205	30	0.24 795	9.93 984	7	32
29	9.69 212	22	9.75 235	29	0.24 765	9.93 977	7	31
30	9.69 234	22	9.75 264	30	0.24 736	9.93 970	7	**30**
31	9.69 256	23	9.75 294	29	0.24 706	9.93 963	8	29
32	9.69 279	22	9.75 323	30	0.24 677	9.93 955	7	28
33	9.69 301	22	9.75 353	29	0.24 647	9.93 948	7	27
34	9.69 323	22	9.75 382	29	0.24 618	9.93 941	7	26
35	9.69 345	23	9.75 411	30	0.24 589	9.93 934	7	25
36	9.69 368	22	9.75 441	29	0.24 559	9.93 927	7	24
37	9.69 390	22	9.75 470	30	0.24 530	9.93 920	8	23
38	9.69 412	22	9.75 500	29	0.24 500	9.93 912	7	22
39	9.69 434	22	9.75 529	29	0.24 471	9.93 905	7	21
40	9.69 456	23	9.75 558	30	0.24 442	9.93 898	7	**20**
41	9.69 479	22	9.75 588	29	0.24 412	9.93 891	7	19
42	9.69 501	22	9.75 617	30	0.24 383	9.93 884	8	18
43	9.69 523	22	9.75 647	29	0.24 353	9.93 876	7	17
44	9.69 545	22	9.75 676	29	0.24 324	9.93 869	7	16
45	9.69 567	22	9.75 705	30	0.24 295	9.93 862	7	15
46	9.69 589	22	9.75 735	29	0.24 265	9.93 855	8	14
47	9.69 611	22	9.75 764	29	0.24 236	9.93 847	7	13
48	9.69 633	22	9.75 793	29	0.24 207	9.93 840	7	12
49	9.69 655	22	9.75 822	30	0.24 178	9.93 833	7	11
50	9.69 677	22	9.75 852	29	0.24 148	9.93 826	7	**10**
51	9.69 699	22	9.75 881	29	0.24 119	9.93 819	8	9
52	9.69 721	22	9.75 910	29	0.24 090	9.93 811	7	8
53	9.69 743	22	9.75 939	30	0.24 061	9.93 804	7	7
54	9.69 765	22	9.75 969	29	0.24 031	9.93 797	8	6
55	9.69 787	22	9.75 998	29	0.24 002	9.93 789	7	5
56	9.69 809	22	9.76 027	29	0.23 973	9.93 782	7	4
57	9.69 831	22	9.76 056	30	0.23 944	9.93 775	7	3
58	9.69 853	22	9.76 086	29	0.23 914	9.93 768	8	2
59	9.69 875	22	9.76 115	29	0.23 885	9.93 760	7	1
60	9.69 897		9.76 144		0.23 856	9.93 753		**0**
′	L. Cos.	d.	L. Cot.	c.d.	L. Tan.	L. Sin.	d.	′

P. P.

″	**30**	**29**	**23**
1	0.5	0.5	0.4
2	1.0	1.0	0.8
3	1.5	1.4	1.2
4	2.0	1.9	1.5
5	2.5	2.4	1.9
6	3.0	2.9	2.3
7	3.5	3.4	2.7
8	4.0	3.9	3.1
9	4.5	4.4	3.4
10	5.0	4.8	3.8
20	10.0	9.7	7.7
30	15.0	14.5	11.5
40	20.0	19.3	15.3
50	25.0	24.2	19.2

″	**22**	**8**	**7**
1	0.4	0.1	0.1
2	0.7	0.3	0.2
3	1.1	0.4	0.4
4	1.5	0.5	0.5
5	1.8	0.7	0.6
6	2.2	0.8	0.7
7	2.6	0.9	0.8
8	2.9	1.1	0.9
9	3.3	1.2	1.0
10	3.7	1.3	1.2
20	7.3	2.7	2.3
30	11.0	4.0	3.5
40	14.7	5.3	4.7
50	18.3	6.7	5.8

	8/30	**8/29**
0	1.9	1.8
1	5.6	5.4
2	9.4	9.1
3	13.1	12.7
4	16.9	16.3
5	20.6	19.9
6	24.4	23.6
7	28.1	27.2
8		

	7/30	**7/29**
0	2.1	2.1
1	6.4	6.2
2	10.7	10.4
3	15.0	14.5
4	19.3	18.6
5	23.6	22.8
6	27.9	26.9
7		

119° (299°) (240°) **60°**

COMMON LOGARITHMS OF THE TRIGONOMETRIC FUNCTIONS (Continued)

30° (210°) (329°) **149°**

′	L. Sin.	d.	L. Tan.	c.d.	L. Cot.	L. Cos.	d.	′
0	9.69 897	22	9.76 144	29	0.23 856	9.93 753	7	**60**
1	9.69 919	22	9.76 173	29	0.23 827	9.93 746	8	59
2	9.69 941	22	9.76 202	29	0.23 798	9.93 738	7	58
3	9.69 963	21	9.76 231	30	0.23 769	9.93 731	7	57
4	9.69 984	22	9.76 261	29	0.23 739	9.93 724	7	56
5	9.70 006	22	9.76 290	29	0.23 710	9.93 717	8	55
6	9.70 028	22	9.76 319	29	0.23 681	9.93 709	7	54
7	9.70 050	22	9.76 348	29	0.23 652	9.93 702	7	53
8	9.70 072	21	9.76 377	29	0.23 623	9.93 695	8	52
9	9.70 093	22	9.76 406	29	0.23 594	9.93 687	7	51
10	9.70 115	22	9.76 435	29	0.23 565	9.93 680	7	**50**
11	9.70 137	22	9.76 464	29	0.23 536	9.93 673	8	49
12	9.70 159	21	9.76 493	29	0.23 507	9.93 665	7	48
13	9.70 180	22	9.76 522	29	0.23 478	9.93 658	8	47
14	9.70 202	22	9.76 551	29	0.23 449	9.93 650	7	46
15	9.70 224	21	9.76 580	29	0.23 420	9.93 643	7	45
16	9.70 245	22	9.76 609	30	0.23 391	9.93 636	8	44
17	9.70 267	21	9.76 639	29	0.23 361	9.93 628	7	43
18	9.70 288	22	9.76 668	29	0.23 332	9.93 621	7	42
19	9.70 310	22	9.76 697	28	0.23 303	9.93 614	8	41
20	9.70 332	21	9.76 725	29	0.23 275	9.93 606	7	**40**
21	9.70 353	22	9.76 754	29	0.23 246	9.93 599	8	39
22	9.70 375	21	9.76 783	29	0.23 217	9.93 591	7	38
23	9.70 396	22	9.76 812	29	0.23 188	9.93 584	7	37
24	9.70 418	21	9.76 841	29	0.23 159	9.93 577	8	36
25	9.70 439	22	9.76 870	29	0.23 130	9.93 569	7	35
26	9.70 461	21	9.76 899	29	0.23 101	9.93 562	8	34
27	9.70 482	22	9.76 928	29	0.23 072	9.93 554	7	33
28	9.70 504	21	9.76 957	29	0.23 043	9.93 547	8	32
29	9.70 525	22	9.76 986	29	0.23 014	9.93 539	7	31
30	9.70 547	21	9.77 015	29	0.22 985	9.93 532	7	**30**
31	9.70 568	22	9.77 044	29	0.22 956	9.93 525	8	29
32	9.70 590	21	9.77 073	28	0.22 927	9.93 517	7	28
33	9.70 611	22	9.77 101	29	0.22 899	9.93 510	8	27
34	9.70 633	21	9.77 130	29	0.22 870	9.93 502	7	26
35	9.70 654	21	9.77 159	29	0.22 841	9.93 495	8	25
36	9.70 675	22	9.77 188	29	0.22 812	9.93 487	7	24
37	9.70 697	21	9.77 217	29	0.22 783	9.93 480	8	23
38	9.70 718	21	9.77 246	28	0.22 754	9.93 472	7	22
39	9.70 739	22	9.77 274	29	0.22 726	9.93 465	8	21
40	9.70 761	21	9.77 303	29	0.22 697	9.93 457	7	**20**
41	9.70 782	21	9.77 332	29	0.22 668	9.93 450	8	19
42	9.70 803	21	9.77 361	29	0.22 639	9.93 442	7	18
43	9.70 824	22	9.77 390	28	0.22 610	9.93 435	8	17
44	9.70 846	21	9.77 418	29	0.22 582	9.93 427	7	16
45	9.70 867	21	9.77 447	29	0.22 553	9.93 420	8	15
46	9.70 888	21	9.77 476	29	0.22 524	9.93 412	7	14
47	9.70 909	22	9.77 505	28	0.22 495	9.93 405	8	13
48	9.70 931	21	9.77 533	29	0.22 467	9.93 397	7	12
49	9.70 952	21	9.77 562	29	0.22 438	9.93 390	8	11
50	9.70 973	21	9.77 591	28	0.22 409	9.93 382	7	**10**
51	9.70 994	21	9.77 619	29	0.22 381	9.93 375	8	9
52	9.71 015	21	9.77 648	29	0.22 352	9.93 367	7	8
53	9.71 036	22	9.77 677	29	0.22 323	9.93 360	8	7
54	9.71 058	21	9.77 706	28	0.22 294	9.93 352	8	6
55	9.71 079	21	9.77 734	29	0.22 266	9.93 344	7	5
56	9.71 100	21	9.77 763	28	0.22 237	9.93 337	8	4
57	9.71 121	21	9.77 791	29	0.22 209	9.93 329	7	3
58	9.71 142	21	9.77 820	29	0.22 180	9.93 322	8	2
59	9.71 163	21	9.77 849	28	0.22 151	9.93 314	7	1
60	9.71 184		9.77 877		0.22 123	9.93 307		**0**
′	L. Cos.	d.	L. Cot.	c.d.	L. Tan.	L. Sin.	d.	′

120° (300°) (239°) **59°**

P. P.

″	**30**	**29**	**28**
1	0.5	0.5	0.5
2	1.0	1.0	0.9
3	1.5	1.4	1.4
4	2.0	1.9	1.9
5	2.5	2.4	2.3
6	3.0	2.9	2.8
7	3.5	3.4	3.3
8	4.0	3.9	3.7
9	4.5	4.4	4.2
10	5.0	4.8	4.7
20	10.0	9.7	9.3
30	15.0	14.5	14.0
40	20.0	19.3	18.7
50	25.0	24.2	23.3

″	**22**	**21**
1	0.4	0.4
2	0.7	0.7
3	1.1	1.0
4	1.5	1.4
5	1.8	1.8
6	2.2	2.1
7	2.6	2.4
8	2.9	2.8
9	3.3	3.2
10	3.7	3.5
20	7.3	7.0
30	11.0	10.5
40	14.7	14.0
50	18.3	17.5

″	**8**	**7**
1	0.1	0.1
2	0.3	0.2
3	0.4	0.4
4	0.5	0.5
5	0.7	0.6
6	0.8	0.7
7	0.9	0.8
8	1.1	0.9
9	1.2	1.0
10	1.3	1.2
20	2.7	2.3
30	4.0	3.5
40	5.3	4.7
50	6.7	5.8

	7/**30**	**7**/**29**	**7**/**28**
0			
	2.1	2.1	2.0
1			
	6.4	6.2	6.0
2			
	10.7	10.4	10.0
3			
	15.0	14.5	14.0
4			
	19.3	18.6	18.0
5			
	23.6	22.8	22.0
6			
	27.9	26.9	26.0
7			

P. P.

COMMON LOGARITHMS OF THE TRIGONOMETRIC FUNCTIONS (Continued)

31° (211°) (328°) **148°**

′	L. Sin.	d.	L. Tan.	c.d.	L. Cot.	L. Cos.	d.	′
0	9.71 184	21	9.77 877	29	0.22 123	9.93 307	8	**60**
1	9.71 205	21	9.77 906	29	0.22 094	9.93 299	8	59
2	9.71 226	21	9.77 935	28	0.22 065	9.93 291	7	58
3	9.71 247	21	9.77 963	29	0.22 037	9.93 284	8	57
4	9.71 268	21	9.77 992	28	0.22 008	9.93 276	7	56
5	9.71 289	21	9.78 020	29	0.21 980	9.93 269	8	55
6	9.71 310	21	9.78 049	28	0.21 951	9.93 261	8	54
7	9.71 331	21	9.78 077	29	0.21 923	9.93 253	7	53
8	9.71 352	21	9.78 106	29	0.21 894	9.93 246	8	52
9	9.71 373	20	9.78 135	28	0.21 865	9.93 238	8	51
10	9.71 393	21	9.78 163	29	0.21 837	9.93 230	7	**50**
11	9.71 414	21	9.78 192	28	0.21 808	9.93 223	8	49
12	9.71 435	21	9.78 220	29	0.21 780	9.93 215	8	48
13	9.71 456	21	9.78 249	28	0.21 751	9.93 207	7	47
14	9.71 477	21	9.78 277	29	0.21 723	9.93 200	8	46
15	9.71 498	21	9.78 306	28	0.21 694	9.93 192	8	45
16	9.71 519	20	9.78 334	29	0.21 666	9.93 184	7	44
17	9.71 539	21	9.78 363	28	0.21 637	9.93 177	8	43
18	9.71 560	21	9.78 391	28	0.21 609	9.93 169	8	42
19	9.71 581	21	9.78 419	29	0.21 581	9.93 161	7	41
20	9.71 602	20	9.78 448	28	0.21 552	9.93 154	8	**40**
21	9.71 622	21	9.78 476	29	0.21 524	9.93 146	8	39
22	9.71 643	21	9.78 505	28	0.21 495	9.93 138	7	38
23	9.71 664	21	9.78 533	29	0.21 467	9.93 131	8	37
24	9.71 685	20	9.78 562	28	0.21 438	9.93 123	8	36
25	9.71 705	21	9.78 590	28	0.21 410	9.93 115	7	35
26	9.71 726	21	9.78 618	29	0.21 382	9.93 108	8	34
27	9.71 747	20	9.78 647	28	0.21 353	9.93 100	8	33
28	9.71 767	21	9.78 675	29	0.21 325	9.93 092	8	32
29	9.71 788	21	9.78 704	28	0.21 296	9.93 084	7	31
30	9.71 809	20	9.78 732	28	0.21 268	9.93 077	8	**30**
31	9.71 829	21	9.78 760	29	0.21 240	9.93 069	8	29
32	9.71 850	20	9.78 789	28	0.21 211	9.93 061	8	28
33	9.71 870	21	9.78 817	28	0.21 183	9.93 053	7	27
34	9.71 891	20	9.78 845	29	0.21 155	9.93 046	8	26
35	9.71 911	21	9.78 874	28	0.21 126	9.93 038	8	25
36	9.71 932	20	9.78 902	28	0.21 098	9.93 030	8	24
37	9.71 952	21	9.78 930	29	0.21 070	9.93 022	8	23
38	9.71 973	21	9.78 959	28	0.21 041	9.93 014	7	22
39	9.71 994	20	9.78 987	28	0.21 013	9.93 007	8	21
40	9.72 014	20	9.79 015	28	0.20 985	9.92 999	8	**20**
41	9.72 034	21	9.79 043	29	0.20 957	9.92 991	8	19
42	9.72 055	20	9.79 072	28	0.20 928	9.92 983	7	18
43	9.72 075	21	9.79 100	28	0.20 900	9.92 976	8	17
44	9.72 096	20	9.79 128	28	0.20 872	9.92 968	8	16
45	9.72 116	21	9.79 156	29	0.20 844	9.92 960	8	15
46	9.72 137	20	9.79 185	28	0.20 815	9.92 952	8	14
47	9.72 157	20	9.79 213	28	0.20 787	9.92 944	8	13
48	9.72 177	21	9.79 241	28	0.20 759	9.92 936	7	12
49	9.72 198	20	9.79 269	28	0.20 731	9.92 929	8	11
50	9.72 218	20	9.79 297	29	0.20 703	9.92 921	8	**10**
51	9.72 238	21	9.79 326	28	0.20 674	9.92 913	8	9
52	9.72 259	20	9.79 354	28	0.20 646	9.92 905	8	8
53	9.72 279	20	9.79 382	28	0.20 618	9.92 897	8	7
54	9.72 299	21	9.79 410	28	0.20 590	9.92 889	8	6
55	9.72 320	20	9.79 438	28	0.20 562	9.92 881	7	5
56	9.72 340	20	9.79 466	29	0.20 534	9.92 874	8	4
57	9.72 360	21	9.79 495	28	0.20 505	9.92 866	8	3
58	9.72 381	20	9.79 523	28	0.20 477	9.92 858	8	2
59	9.72 401	20	9.79 551	28	0.20 449	9.92 850	8	1
60	9.72 421		9.79 579		0.20 421	9.92 842		**0**
′	L. Cos.	d.	L. Cot.	c.d.	L. Tan.	L.Sin.	d.	′

121° (301°) (238°) **58°**

P. P.

″	**29**	**28**
1	0.5	0.5
2	1.0	0.9
3	1.4	1.4
4	1.9	1.9
5	2.4	2.3
6	2.9	2.8
7	3.4	3.3
8	3.9	3.7
9	4.4	4.2
10	4.8	4.7
20	9.7	9.3
30	14.5	14.0
40	19.3	18.7
50	24.2	23.3

″	**21**	**20**
1	0.4	0.3
2	0.7	0.7
3	1.0	1.0
4	1.4	1.3
5	1.8	1.7
6	2.1	2.0
7	2.4	2.3
8	2.8	2.7
9	3.2	3.0
10	3.5	3.3
20	7.0	6.7
30	10.5	10.0
40	14.0	13.3
50	17.5	16.7

″	**8**	**7**
1	0.1	0.1
2	0.3	0.2
3	0.4	0.4
4	0.5	0.5
5	0.7	0.6
6	0.8	0.7
7	0.9	0.8
8	1.1	0.9
9	1.2	1.0
10	1.3	1.2
20	2.7	2.3
30	4.0	3.5
40	5.3	4.7
50	6.7	5.8

	$\frac{8}{30}$	$\frac{8}{29}$	$\frac{8}{28}$
0	1.9	1.8	1.8
1	5.6	5.4	5.2
2	9.4	9.1	8.8
3	13.1	12.7	12.2
4	16.9	16.3	15.8
5	20.6	19.9	19.2
6	24.4	23.6	22.8
7	28.1	27.2	26.2
8			

COMMON LOGARITHMS OF THE TRIGONOMETRIC FUNCTIONS (Continued)

32° (212°) (327°) **147°**

′	L. Sin.	d.	L. Tan.	c.d.	L. Cot.	L. Cos.	d.	′
0	9.72 421	20	9.79 579	28	0.20 421	9.92 842	8	**60**
1	9.72 441	20	9.79 607	28	0.20 393	9.92 834	8	59
2	9.72 461	21	9.79 635	28	0.20 365	9.92 826	8	58
3	9.72 482	20	9.79 663	28	0.20 337	9.92 818	8	57
4	9.72 502	20	9.79 691	28	0.20 309	9.92 810	7	56
5	9.72 522	20	9.79 719	28	0.20 281	9.92 803	8	55
6	9.72 542	20	9.79 747	29	0.20 253	9.92 795	8	54
7	9.72 562	20	9.79 776	28	0.20 224	9.92 787	8	53
8	9.72 582	20	9.79 804	28	0.20 196	9.92 779	8	52
9	9.72 602	20	9.79 832	28	0.20 168	9.92 771	8	51
10	9.72 622	21	9.79 860	28	0.20 140	9.92 763	8	**50**
11	9.72 643	20	9.79 888	28	0.20 112	9.92 755	8	49
12	9.72 663	20	9.79 916	28	0.20 084	9.92 747	8	48
13	9.72 683	20	9.79 944	28	0.20 056	9.92 739	8	47
14	9.72 703	20	9.79 972	28	0.20 028	9.92 731	8	46
15	9.72 723	20	9.80 000	28	0.20 000	9.92 723	8	45
16	9.72 743	20	9.80 028	28	0.19 972	9.92 715	8	44
17	9.72 763	20	9.80 056	28	0.19 944	9.92 707	8	43
18	9.72 783	20	9.80 084	28	0.19 916	9.92 699	8	42
19	9.72 803	20	9.80 112	28	0.19 888	9.92 691	8	41
20	9.72 823	20	9.80 140	28	0.19 860	9.92 683	8	**40**
21	9.72 843	20	9.80 168	27	0.19 832	9.92 675	8	39
22	9.72 863	20	9.80 195	28	0.19 805	9.92 667	8	38
23	9.72 883	19	9.80 223	28	0.19 777	9.92 659	8	37
24	9.72 902	20	9.80 251	28	0.19 749	9.92 651	8	36
25	9.72 922	20	9.80 279	28	0.19 721	9.92 643	8	35
26	9.72 942	20	9.80 307	28	0.19 693	9.92 635	8	34
27	9.72 962	20	9.80 335	28	0.19 665	9.92 627	8	33
28	9.72 982	20	9.80 363	28	0.19 637	9.92 619	8	32
29	9.73 002	20	9.80 391	28	0.19 609	9.92 611	8	31
30	9.73 022	19	9.80 419	28	0.19 581	9.92 603	8	**30**
31	9.73 041	20	9.80 447	27	0.19 553	9.92 595	8	29
32	9.73 061	20	9.80 474	28	0.19 526	9.92 587	8	28
33	9.73 081	20	9.80 502	28	0.19 498	9.92 579	8	27
34	9.73 101	20	9.80 530	28	0.19 470	9.92 571	8	26
35	9.73 121	19	9.80 558	28	0.19 442	9.92 563	8	25
36	9.73 140	20	9.80 586	28	0.19 414	9.92 555	9	24
37	9.73 160	20	9.80 614	28	0.19 386	9.92 546	8	23
38	9.73 180	20	9.80 642	27	0.19 358	9.92 538	8	22
39	9.73 200	19	9.80 669	28	0.19 331	9.92 530	8	21
40	9.73 219	20	9.80 697	28	0.19 303	9.92 522	8	**20**
41	9.73 239	20	9.80 725	28	0.19 275	9.92 514	8	19
42	9.73 259	19	9.80 753	28	0.19 247	9.92 506	8	18
43	9.73 278	20	9.80 781	27	0.19 219	9.92 498	8	17
44	9.73 298	20	9.80 808	28	0.19 192	9.92 490	8	16
45	9.73 318	19	9.80 836	28	0.19 164	9.92 482	9	15
46	9.73 337	20	9.80 864	28	0.19 136	9.92 473	8	14
47	9.73 357	20	9.80 892	27	0.19 108	9.92 465	8	13
48	9.73 377	19	9.80 919	28	0.19 081	9.92 457	8	12
49	9.73 396	20	9.80 947	28	0.19 053	9.92 449	8	11
50	9.73 416	19	9.80 975	28	0.19 025	9.92 441	8	**10**
51	9.73 435	20	9.81 003	27	0.18 997	9.92 433	8	9
52	9.73 455	19	9.81 030	28	0.18 970	9.92 425	9	8
53	9.73 474	20	9.81 058	28	0.18 942	9.92 416	8	7
54	9.73 494	19	9.81 086	27	0.18 914	9.92 408	8	6
55	9.73 513	20	9.81 113	28	0.18 887	9.92 400	8	5
56	9.73 533	19	9.81 141	28	0.18 859	9.92 392	8	4
57	9.73 552	20	9.81 169	27	0.18 831	9.92 384	8	3
58	9.73 572	19	9.81 196	28	0.18 804	9.92 376	9	2
59	9.73 591	20	9.81 224	28	0.18 776	9.92 367	8	1
60	9.73 611		9.81 252		0.18 748	9.92 359		**0**
′	L. Cos.	d.	L. Cot.	c.d.	L. Tan.	L. Sin.	d.	′

P. P.

″	**29**	**28**	**27**
1	0.5	0.5	0.4
2	1.0	0.9	0.9
3	1.4	1.4	1.4
4	1.9	1.9	1.8
5	2.4	2.3	2.2
6	2.9	2.8	2.7
7	3.4	3.3	3.2
8	3.9	3.7	3.6
9	4.4	4.2	4.0
10	4.8	4.7	4.5
20	9.7	9.3	9.0
30	14.5	14.0	13.5
40	19.3	18.7	18.0
50	24.2	23.3	22.5

″	**21**	**20**	**19**
1	0.4	0.3	0.3
2	0.7	0.7	0.6
3	1.0	1.0	1.0
4	1.4	1.3	1.3
5	1.8	1.7	1.6
6	2.1	2.0	1.9
7	2.4	2.3	2.2
8	2.8	2.7	2.5
9	3.2	3.0	2.8
10	3.5	3.3	3.2
20	7.0	6.7	6.3
30	10.5	10.0	9.5
40	14.0	13.3	12.7
50	17.5	16.7	15.8

″	**9**	**8**	**7**
1	0.2	0.1	0.1
2	0.3	0.3	0.2
3	0.4	0.4	0.4
4	0.6	0.5	0.5
5	0.8	0.7	0.6
6	0.9	0.8	0.7
7	1.0	0.9	0.8
8	1.2	1.1	0.9
9	1.4	1.2	1.0
10	1.5	1.3	1.2
20	3.0	2.7	2.3
30	4.5	4.0	3.5
40	6.0	5.3	4.7
50	7.5	6.7	5.8

	8/29	**8/28**	**7/28**
0	1.8	1.8	2.0
1	5.4	5.2	6.0
2	9.1	8.8	10.0
3	12.7	12.2	14.0
4	16.3	15.8	18.0
5	19.9	19.2	22.0
6	23.6	22.8	26.0
7	27.2	26.2	—
8			

P. P.

122° (302°) (237°) **57°**

COMMON LOGARITHMS OF THE TRIGONOMETRIC FUNCTIONS (Continued)

33° (213°) (326°) **146°**

′	L. Sin.	d.	L. Tan.	c.d.	L. Cot.	L. Cos.	d.	′
0	9.73 611	19	9.81 252	27	0.18 748	9.92 359	8	**60**
1	9.73 630	20	9.81 279	28	0.18 721	9.92 351	8	59
2	9.73 650	19	9.81 307	28	0.18 693	9.92 343	8	58
3	9.73 669	20	9.81 335	27	0.18 665	9.92 335	9	57
4	9.73 689	19	9.81 362	28	0.18 638	9.92 326	8	56
5	9.73 708	19	9.81 390	28	0.18 610	9.92 318	8	55
6	9.73 727	20	9.81 418	27	0.18 582	9.92 310	8	54
7	9.73 747	19	9.81 445	28	0.18 555	9.92 302	9	53
8	9.73 766	19	9.81 473	27	0.18 527	9.92 293	8	52
9	9.73 785	20	9.81 500	28	0.18 500	9.92 285	8	51
10	9.73 805	19	9.81 528	28	0.18 472	9.92 277	8	**50**
11	9.73 824	19	9.81 556	27	0.18 444	9.92 269	9	49
12	9.73 843	20	9.81 583	28	0.18 417	9.92 260	8	48
13	9.73 863	19	9.81 611	27	0.18 389	9.92 252	8	47
14	9.73 882	19	9.81 638	28	0.18 362	9.92 244	9	46
15	9.73 901	20	9.81 666	27	0.18 334	9.92 235	8	45
16	9.73 921	19	9.81 693	28	0.18 307	9.92 227	8	44
17	9.73 940	19	9.81 721	27	0.18 279	9.92 219	8	43
18	9.73 959	19	9.81 748	28	0.18 252	9.92 211	9	42
19	9.73 978	19	9.81 776	27	0.18 224	9.92 202	8	41
20	9.73 997	20	9.81 803	28	0.18 197	9.92 194	8	**40**
21	9.74 017	19	9.81 831	27	0.18 169	9.92 186	9	39
22	9.74 036	19	9.81 858	28	0.18 142	9.92 177	8	38
23	9.74 055	19	9.81 886	27	0.18 114	9.92 169	8	37
24	9.74 074	19	9.81 913	28	0.18 087	9.92 161	9	36
25	9.74 093	20	9.81 941	27	0.18 059	9.92 152	8	35
26	9.74 113	19	9.81 968	28	0.18 032	9.92 144	8	34
27	9.74 132	19	9.81 996	27	0.18 004	9.92 136	9	33
28	9.74 151	19	9.82 023	28	0.17 977	9.92 127	8	32
29	9.74 170	19	9.82 051	27	0.17 949	9.92 119	8	31
30	9.74 189	19	9.82 078	28	0.17 922	9.92 111	9	**30**
31	9.74 208	19	9.82 106	27	0.17 894	9.92 102	8	29
32	9.74 227	19	9.82 133	28	0.17 867	9.92 094	8	28
33	9.74 246	19	9.82 161	27	0.17 839	9.92 086	9	27
34	9.74 265	19	9.82 188	27	0.17 812	9.92 077	8	26
35	9.74 284	19	9.82 215	28	0.17 785	9.92 069	9	25
36	9.74 303	19	9.82 243	27	0.17 757	9.92 060	8	24
37	9.74 322	19	9.82 270	28	0.17 730	9.92 052	8	23
38	9.74 341	19	9.82 298	27	0.17 702	9.92 044	9	22
39	9.74 360	19	9.82 325	27	0.17 675	9.92 035	8	21
40	9.74 379	19	9.82 352	28	0.17 648	9.92 027	9	**20**
41	9.74 398	19	9.82 380	27	0.17 620	9.92 018	8	19
42	9.74 417	19	9.82 407	28	0.17 593	9.92 010	8	18
43	9.74 436	19	9.82 435	27	0.17 565	9.92 002	9	17
44	9.74 455	19	9.82 462	27	0.17 538	9.91 993	8	16
45	9.74 474	19	9.82 489	28	0.17 511	9.91 985	9	15
46	9.74 493	19	9.82 517	27	0.17 483	9.91 976	8	14
47	9.74 512	19	9.82 544	27	0.17 456	9.91 968	9	13
48	9.74 531	18	9.82 571	28	0.17 429	9.91 959	8	12
49	9.74 549	19	9.82 599	27	0.17 401	9.91 951	9	11
50	9.74 568	19	9.82 626	27	0.17 374	9.91 942	8	**10**
51	9.74 587	19	9.82 653	28	0.17 347	9.91 934	9	9
52	9.74 606	19	9.82 681	27	0.17 319	9.91 925	8	8
53	9.74 625	19	9.82 708	27	0.17 292	9.91 917	9	7
54	9.74 644	18	9.82 735	27	0.17 265	9.91 908	8	6
55	9.74 662	19	9.82 762	28	0.17 238	9.91 900	9	5
56	9.74 681	19	9.82 790	27	0.17 210	9.91 891	8	4
57	9.74 700	19	9.82 817	27	0.17 183	9.91 883	9	3
58	9.74 719	18	9.82 844	27	0.17 156	9.91 874	8	2
59	9.74 737	19	9.82 871	28	0.17 129	9.91 866	9	1
60	9.74 756		9.82 899		0.17 101	9.91 857		**0**
′	L. Cos.	d.	L. Cot.	c.d.	L. Tan.	L. Sin.	d.	′

123° (303°) (236°) **56°**

P. P.

″	**28**	**27**
1	0.5	0.4
2	0.9	0.9
3	1.4	1.4
4	1.9	1.8
5	2.3	2.2
6	2.8	2.7
7	3.3	3.2
8	3.7	3.6
9	4.2	4.0
10	4.7	4.5
20	9.3	9.0
30	14.0	13.5
40	18.7	18.0
50	23.3	22.5

″	**20**	**19**	**18**
1	0.3	0.3	0.3
2	0.7	0.6	0.6
3	1.0	1.0	0.9
4	1.3	1.3	1.2
5	1.7	1.6	1.5
6	2.0	1.9	1.8
7	2.3	2.2	2.1
8	2.7	2.5	2.4
9	3.0	2.8	2.7
10	3.3	3.2	3.0
20	6.7	6.3	6.0
30	10.0	9.5	9.0
40	13.3	12.7	12.0
50	16.7	15.8	15.0

″	**9**	**8**
1	0.2	0.1
2	0.3	0.3
3	0.4	0.4
4	0.6	0.5
5	0.8	0.7
6	0.9	0.8
7	1.0	0.9
8	1.2	1.1
9	1.4	1.2
10	1.5	1.3
20	3.0	2.7
30	4.5	4.0
40	6.0	5.3
50	7.5	6.7

	9/**28**	**9**/**27**	**8**/**27**
0	1.6	1.5	1.7
1	4.7	4.5	5.1
2	7.8	7.5	8.4
3	10.9	10.5	11.8
4	14.0	13.5	15.2
5	17.1	16.5	18.6
6	20.2	19.5	21.9
7	23.3	22.5	25.3
8	26.4	25.5	—
9			

COMMON LOGARITHMS OF THE TRIGONOMETRIC FUNCTIONS (Continued)

34° (214°) (325°) **145°**

′	L. Sin.	d.	L. Tan.	c.d.	L. Cot.	L. Cos.	d.	′
0	9.74 756	19	9.82 899	27	0.17 101	9.91 857	8	**60**
1	9.74 775	19	9.82 926	27	0.17 074	9.91 849	9	59
2	9.74 794	18	9.82 953	27	0.17 047	9.91 840	8	58
3	9.74 812	19	9.82 980	28	0.17 020	9.91 832	9	57
4	9.74 831	19	9.83 008	27	0.16 992	9.91 823	8	56
5	9.74 850	18	9.83 035	27	0.16 965	9.91 815	9	55
6	9.74 868	19	9.83 062	27	0.16 938	9.91 806	8	54
7	9.74 887	19	9.83 089	28	0.16 911	9.91 798	9	53
8	9.74 906	18	9.83 117	27	0.16 883	9.91 789	8	52
9	9.74 924	19	9.83 144	27	0.16 856	9.91 781	9	51
10	9.74 943	18	9.83 171	27	0.16 829	9.91 772	9	**50**
11	9.74 961	19	9.83 198	27	0.16 802	9.91 763	8	49
12	9.74 980	19	9.83 225	27	0.16 775	9.91 755	9	48
13	9.74 999	18	9.83 252	28	0.16 748	9.91 746	8	47
14	9.75 017	19	9.83 280	27	0.16 720	9.91 738	9	46
15	9.75 036	18	9.83 307	27	0.16 693	9.91 729	9	45
16	9.75 054	19	9.83 334	27	0.16 666	9.91 720	8	44
17	9.75 073	18	9.83 361	27	0.16 639	9.91 712	9	43
18	9.75 091	19	9.83 388	27	0.16 612	9.91 703	8	42
19	9.75 110	18	9.83 415	27	0.16 585	9.91 695	9	41
20	9.75 128	19	9.83 442	28	0.16 558	9.91 686	9	**40**
21	9.75 147	18	9.83 470	27	0.16 530	9.91 677	8	39
22	9.75 165	19	9.83 497	27	0.16 503	9.91 669	9	38
23	9.75 184	18	9.83 524	27	0.16 476	9.91 660	9	37
24	9.75 202	19	9.83 551	27	0.16 449	9.91 651	8	36
25	9.75 221	18	9.83 578	27	0.16 422	9.91 643	9	35
26	9.75 239	19	9.83 605	27	0.16 395	9.91 634	9	34
27	9.75 258	18	9.83 632	27	0.16 368	9.91 625	8	33
28	9.75 276	18	9.83 659	27	0.16 341	9.91 617	9	32
29	9.75 294	19	9.83 686	27	0.16 314	9.91 608	9	31
30	9.75 313	18	9.83 713	27	0.16 287	9.91 599	8	**30**
31	9.75 331	19	9.83 740	28	0.16 260	9.91 591	9	29
32	9.75 350	18	9.83 768	27	0.16 232	9.91 582	9	28
33	9.75 368	18	9.83 795	27	0.16 205	9.91 573	8	27
34	9.75 386	19	9.83 822	27	0.16 178	9.91 565	9	26
35	9.75 405	18	9.83 849	27	0.16 151	9.91 556	9	25
36	9.75 423	18	9.83 876	27	0.16 124	9.91 547	9	24
37	9.75 441	18	9.83 903	27	0.16 097	9.91 538	8	23
38	9.75 459	19	9.83 930	27	0.16 070	9.91 530	9	22
39	9.75 478	18	9.83 957	27	0.16 043	9.91 521	9	21
40	9.75 496	18	9.83 984	27	0.16 016	9.91 512	8	**20**
41	9.75 514	19	9.84 011	27	0.15 989	9.91 504	9	19
42	9.75 533	18	9.84 038	27	0.15 962	9.91 495	9	18
43	9.75 551	18	9.84 065	27	0.15 935	9.91 486	9	17
44	9.75 569	18	9.84 092	27	0.15 908	9.91 477	8	16
45	9.75 587	18	9.84 119	27	0.15 881	9.91 469	9	15
46	9.75 605	19	9.84 146	27	0.15 854	9.91 460	9	14
47	9.75 624	18	9.84 173	27	0.15 827	9.91 451	9	13
48	9.75 642	18	9.84 200	27	0.15 800	9.91 442	9	12
49	9.75 660	18	9.84 227	27	0.15 773	9.91 433	8	11
50	9.75 678	18	9.84 254	26	0.15 746	9.91 425	9	**10**
51	9.75 696	18	9.84 280	27	0.15 720	9.91 416	9	9
52	9.75 714	19	9.84 307	27	0.15 693	9.91 407	9	8
53	9.75 733	18	9.84 334	27	0.15 666	9.91 398	9	7
54	9.75 751	18	9.84 361	27	0.15 639	9.91 389	8	6
55	9.75 769	18	9.84 388	27	0.15 612	9.91 381	9	5
56	9.75 787	18	9.84 415	27	0.15 585	9.91 372	9	4
57	9.75 805	18	9.84 442	27	0.15 558	9.91 363	9	3
58	9.75 823	18	9.84 469	27	0.15 531	9.91 354	9	2
59	9.75 841	18	9.84 496	27	0.15 504	9.91 345	9	1
60	9.75 859		9.84 523		0.15 477	9.91 336		**0**
′	L. Cos.	d.	L. Cot.	c.d.	L. Tan.	L. Sin.	d.	′

124° (304°) (235°) **55°**

P. P.

″	**28**	**27**	**26**
1	0.5	0.4	0.4
2	0.9	0.9	0.9
3	1.4	1.4	1.3
4	1.9	1.8	1.7
5	2.3	2.2	2.2
6	2.8	2.7	2.6
7	3.3	3.2	3.0
8	3.7	3.6	3.5
9	4.2	4.0	3.9
10	4.7	4.5	4.3
20	9.3	9.0	8.7
30	14.0	13.5	13.0
40	18.7	18.0	17.3
50	23.3	22.5	21.7

″	**19**	**18**
1	0.3	0.3
2	0.6	0.6
3	1.0	0.9
4	1.3	1.2
5	1.6	1.5
6	1.9	1.8
7	2.2	2.1
8	2.5	2.4
9	2.8	2.7
10	3.2	3.0
20	6.3	6.0
30	9.5	9.0
40	12.7	12.0
50	15.8	15.0

″	**9**	**8**
1	0.2	0.1
2	0.3	0.3
3	0.4	0.4
4	0.6	0.5
5	0.8	0.7
6	0.9	0.8
7	1.0	0.9
8	1.2	1.1
9	1.4	1.2
10	1.5	1.3
20	3.0	2.7
30	4.5	4.0
40	6.0	5.3
50	7.5	6.7

	9/28	**8/28**	**8/27**
0	1.6	1.8	1.7
1	4.7	5.2	5.1
2	7.8	8.8	8.4
3	10.9	12.2	11.8
4	14.0	15.8	15.2
5	17.1	19.2	18.6
6	20.2	22.8	21.9
7	23.3	26.2	25.3
8	26.4	—	—
9			

P. P.

COMMON LOGARITHMS OF THE TRIGONOMETRIC FUNCTIONS (Continued)

35° (215°) (324°) **144°**

′	L. Sin.	d.	L. Tan.	c.d.	L. Cot.	L. Cos.	d.	′
0	9.75 859	18	9.84 523	27	0.15 477	9.91 336	8	**60**
1	9.75 877	18	9.84 550	26	0.15 450	9.91 328	9	59
2	9.75 895	18	9.84 576	27	0.15 424	9.91 319	9	58
3	9.75 913	18	9.84 603	27	0.15 397	9.91 310	9	57
4	9.75 931	18	9.84 630	27	0.15 370	9.91 301	9	56
5	9.75 949	18	9.84 657	27	0.15 343	9.91 292	9	55
6	9.75 967	18	9.84 684	27	0.15 316	9.91 283	9	54
7	9.75 985	18	9.84 711	27	0.15 289	9.91 274	8	53
8	9.76 003	18	9.84 738	26	0.15 262	9.91 266	9	52
9	9.76 021	18	9.84 764	27	0.15 236	9.91 257	9	51
10	9.76 039	18	9.84 791	27	0.15 209	9.91 248	9	**50**
11	9.76 057	18	9.84 818	27	0.15 182	9.91 239	9	49
12	9.76 075	18	9.84 845	27	0.15 155	9.91 230	9	48
13	9.76 093	18	9.84 872	27	0.15 128	9.91 221	9	47
14	9.76 111	18	9.84 899	26	0.15 101	9.91 212	9	46
15	9.76 129	17	9.84 925	27	0.15 075	9.91 203	9	45
16	9.76 146	18	9.84 952	27	0.15 048	9.91 194	9	44
17	9.76 164	18	9.84 979	27	0.15 021	9.91 185	9	43
18	9.76 182	18	9.85 006	27	0.14 994	9.91 176	9	42
19	9.76 200	18	9.85 033	26	0.14 967	9.91 167	9	41
20	9.76 218	18	9.85 059	27	0.14 941	9.91 158	9	**40**
21	9.76 236	17	9.85 086	27	0.14 914	9.91 149	8	39
22	9.76 253	18	9.85 113	27	0.14 887	9.91 141	9	38
23	9.76 271	18	9.85 140	26	0.14 860	9.91 132	9	37
24	9.76 289	18	9.85 166	27	0.14 834	9.91 123	9	36
25	9.76 307	17	9.85 193	27	0.14 807	9.91 114	9	35
26	9.76 324	18	9.85 220	27	0.14 780	9.91 105	9	34
27	9.76 342	18	9.85 247	26	0.14 753	9.91 096	9	33
28	9.76 360	18	9.85 273	27	0.14 727	9.91 087	9	32
29	9.76 378	17	9.85 300	27	0.14 700	9.91 078	9	31
30	9.76 395	18	9.85 327	27	0.14 673	9.91 069	9	**30**
31	9.76 413	18	9.85 354	26	0.14 646	9.91 060	9	29
32	9.76 431	17	9.85 380	27	0.14 620	9.91 051	9	28
33	9.76 448	18	9.85 407	27	0.14 593	9.91 042	9	27
34	9.76 466	18	9.85 434	26	0.14 566	9.91 033	10	26
35	9.76 484	17	9.85 460	27	0.14 540	9.91 023	9	25
36	9.76 501	18	9.85 487	27	0.14 513	9.91 014	9	24
37	9.76 519	18	9.85 514	26	0.14 486	9.91 005	9	23
38	9.76 537	17	9.85 540	27	0.14 460	9.90 996	9	22
39	9.76 554	18	9.85 567	27	0.14 433	9.90 987	9	21
40	9.76 572	18	9.85 594	26	0.14 406	9.90 978	9	**20**
41	9.76 590	17	9.85 620	27	0.14 380	9.90 969	9	19
42	9.76 607	18	9.85 647	27	0.14 353	9.90 960	9	18
43	9.76 625	17	9.85 674	26	0.14 326	9.90 951	9	17
44	9.76 642	18	9.85 700	27	0.14 300	9.90 942	9	16
45	9.76 660	17	9.85 727	27	0.14 273	9.90 933	9	15
46	9.76 677	18	9.85 754	26	0.14 246	9.90 924	9	14
47	9.76 695	17	9.85 780	27	0.14 220	9.90 915	9	13
48	9.76 712	18	9.85 807	27	0.14 193	9.90 906	10	12
49	9.76 730	17	9.85 834	26	0.14 166	9.90 896	9	11
50	9.76 747	18	9.85 860	27	0.14 140	9.90 887	9	**10**
51	9.76 765	17	9.85 887	26	0.14 113	9.90 878	9	9
52	9.76 782	18	9.85 913	27	0.14 087	9.90 869	9	8
53	9.76 800	17	9.85 940	27	0.14 060	9.90 860	9	7
54	9.76 817	18	9.85 967	26	0.14 033	9.90 851	9	6
55	9.76 835	17	9.85 993	27	0.14 007	9.90 842	10	5
56	9.76 852	18	9.86 020	26	0.13 980	9.90 832	9	4
57	9.76 870	17	9.86 046	27	0.13 954	9.90 823	9	3
58	9.76 887	17	9.86 073	27	0.13 927	9.90 814	9	2
59	9.76 904	18	9.86 100	26	0.13 900	9.90 805	9	1
60	9.76 922		9.86 126		0.13 874	9.90 796		**0**
′	L. Cos.	d.	L. Cot.	c.d.	L. Tan.	L. Sin.	d.	′

125° (305°) (234°) **54°**

P. P.

″	27	26	18
1	0.4	0.4	0.3
2	0.9	0.9	0.6
3	1.4	1.3	0.9
4	1.8	1.7	1.2
5	2.2	2.2	1.5
6	2.7	2.6	1.8
7	3.2	3.0	2.1
8	3.6	3.5	2.4
9	4.0	3.9	2.7
10	4.5	4.3	3.0
20	9.0	8.7	6.0
30	13.5	13.0	9.0
40	18.0	17.3	12.0
50	22.5	21.7	15.0

″	17	10	9	8
1	0.3	0.2	0.2	0.1
2	0.6	0.3	0.3	0.3
3	0.8	0.5	0.4	0.4
4	1.1	0.7	0.6	0.5
5	1.4	0.8	0.8	0.7
6	1.7	1.0	0.9	0.8
7	2.0	1.2	1.0	0.9
8	2.3	1.3	1.2	1.1
9	2.6	1.5	1.4	1.2
10	2.8	1.7	1.5	1.3
20	5.7	3.3	3.0	2.7
30	8.5	5.0	4.5	4.0
40	11.3	6.7	6.0	5.3
50	14.2	8.3	7.5	6.7

	10/27	10/26
0	1.4	1.3
1	4.1	3.9
2	6.8	6.5
3	9.4	9.1
4	12.2	11.7
5	14.8	14.3
6	17.6	16.9
7	20.2	19.5
8	22.9	22.1
9	25.6	24.7
10		

	9/27	9/26
0	1.5	1.4
1	4.5	4.3
2	7.5	7.2
3	10.5	10.1
4	13.5	13.0
5	16.5	15.9
6	19.5	18.8
7	22.5	21.7
8	25.5	24.6
9		

P. P.

COMMON LOGARITHMS OF THE TRIGONOMETRIC FUNCTIONS (Continued)

36° (216°) (323°) **143°**

′	L. Sin.	d.	L. Tan.	c.d.	L. Cot	L. Cos.	d.	′
0	9.76 922	17	9.86 126	27	0.13 874	9.90 796	9	**60**
1	9.76 939	18	9.86 153	26	0.13 847	9.90 787	10	59
2	9.76 957	17	9.86 179	27	0.13 821	9.90 777	9	58
3	9.76 974	17	9.86 206	26	0.13 794	9.90 768	9	57
4	9.76 991	18	9.86 232	27	0.13 768	9.90 759	9	56
5	9.77 009	17	9.86 259	26	0.13 741	9.90 750	9	55
6	9.77 026	17	9.86 285	27	0.13 715	9.90 741	10	54
7	9.77 043	18	9.86 312	26	0.13 688	9.90 731	9	53
8	9.77 061	17	9.86 338	27	0.13 662	9.90 722	9	52
9	9.77 078	17	9.86 365	27	0.13 635	9.90 713	9	51
10	9.77 095	17	9.86 392	26	0.13 608	9.90 704	10	**50**
11	9.77 112	18	9.86 418	27	0.13 582	9.90 694	9	49
12	9.77 130	17	9.86 445	26	0.13 555	9.90 685	9	48
13	9.77 147	17	9.86 471	27	0.13 529	9.90 676	9	47
14	9.77 164	17	9.86 498	26	0.13 502	9.90 667	10	46
15	9.77 181	18	9.86 524	27	0.13 476	9.90 657	9	45
16	9.77 199	17	9.86 551	26	0.13 449	9.90 648	9	44
17	9.77 216	17	9.86 577	26	0.13 423	9.90 639	9	43
18	9.77 233	17	9.86 603	27	0.13 397	9.90 630	10	42
19	9.77 250	18	9.86 630	26	0.13 370	9.90 620	9	41
20	9.77 268	17	9.86 656	27	0.13 344	9.90 611	9	**40**
21	9.77 285	17	9.86 683	26	0.13 317	9.90 602	10	39
22	9.77 302	17	9.86 709	27	0.13 291	9.90 592	9	38
23	9.77 319	17	9.86 736	26	0.13 264	9.90 583	9	37
24	9.77 336	17	9.86 762	27	0.13 238	9.90 574	9	36
25	9.77 353	17	9.86 789	26	0.13 211	9.90 565	10	35
26	9.77 370	17	9.86 815	27	0.13 185	9.90 555	9	34
27	9.77 387	18	9.86 842	26	0.13 158	9.90 546	9	33
28	9.77 405	17	9.86 868	26	0.13 132	9.90 537	10	32
29	9.77 422	17	9.86 894	27	0.13 106	9.90 527	9	31
30	9.77 439	17	9.86 921	26	0.13 079	9.90 518	9	**30**
31	9.77 456	17	9.86 947	27	0.13 053	9.90 509	10	29
32	9.77 473	17	9.86 974	26	0.13 026	9.90 499	9	28
33	9.77 490	17	9.87 000	27	0.13 000	9.90 490	10	27
34	9.77 507	17	9.87 027	26	0.12 973	9.90 480	9	26
35	9.77 524	17	9.87 053	26	0.12 947	9.90 471	9	25
36	9.77 541	17	9.87 079	27	0.12 921	9.90 462	10	24
37	9.77 558	17	9.87 106	26	0.12 894	9.90 452	9	23
38	9.77 575	17	9.87 132	26	0.12 868	9.90 443	9	22
39	9.77 592	17	9.87 158	27	0.12 842	9.90 434	10	21
40	9.77 609	17	9.87 185	26	0.12 815	9.90 424	9	**20**
41	9.77 626	17	9.87 211	27	0.12 789	9.90 415	10	19
42	9.77 643	17	9.87 238	26	0.12 762	9.90 405	9	18
43	9.77 660	17	9.87 264	26	0.12 736	9.90 396	10	17
44	9.77 677	17	9.87 290	27	0.12 710	9.90 386	9	16
45	9.77 694	17	9.87 317	26	0.12 683	9.90 377	9	15
46	9.77 711	17	9.87 343	26	0.12 657	9.90 368	10	14
47	9.77 728	16	9.87 369	27	0.12 631	9.90 358	9	13
48	9.77 744	17	9.87 396	26	0.12 604	9.90 349	10	12
49	9.77 761	17	9.87 422	26	0.12 578	9.90 339	9	11
50	9.77 778	17	9.87 448	27	0.12 552	9.90 330	10	**10**
51	9.77 795	17	9.87 475	26	0.12 525	9.90 320	9	9
52	9.77 812	17	9.87 501	26	0.12 499	9.90 311	10	8
53	9.77 829	17	9.87 527	27	0.12 473	9.90 301	9	7
54	9.77 846	16	9.87 554	26	0.12 446	9.90 292	10	6
55	9.77 862	17	9.87 580	26	0.12 420	9.90 282	9	5
56	9.77 879	17	9.87 606	27	0.12 394	9.90 273	10	4
57	9.77 896	17	9.87 633	26	0.12 367	9.90 263	9	3
58	9.77 913	17	9.87 659	26	0.12 341	9.90 254	10	2
59	9.77 930	16	9.87 685	26	0.12 315	9.90 244	9	1
60	9.77 946		9.87 711		0.12 289	9.90 235		**0**
′	L. Cos.	d.	L. Cot.	c.d.	L. Tan.	L. Sin.	d.	′

P. P.

″	**27**	**26**
1	0.4	0.4
2	0.9	0.9
3	1.4	1.3
4	1.8	1.7
5	2.2	2.2
6	2.7	2.6
7	3.2	3.0
8	3.6	3.5
9	4.0	3.9
10	4.5	4.3
20	9.0	8.7
30	13.5	13.0
40	18.0	17.3
50	22.5	21.7

″	**18**	**17**	**16**
1	0.3	0.3	0.3
2	0.6	0.6	0.5
3	0.9	0.8	0.8
4	1.2	1.1	1.1
5	1.5	1.4	1.3
6	1.8	1.7	1.6
7	2.1	2.0	1.9
8	2.4	2.3	2.1
9	2.7	2.6	2.4
10	3.0	2.8	2.7
20	6.0	5.7	5.3
30	9.0	8.5	8.0
40	12.0	11.3	10.7
50	15.0	14.2	13.3

″	**10**	**9**
1	0.2	0.2
2	0.3	0.3
3	0.5	0.4
4	0.7	0.6
5	0.8	0.8
6	1.0	0.9
7	1.2	1.0
8	1.3	1.2
9	1.5	1.4
10	1.7	1.5
20	3.3	3.0
30	5.0	4.5
40	6.7	6.0
50	8.3	7.5

	9/27	**9/26**
0	1.5	1.4
1	4.5	4.3
2	7.5	7.2
3	10.5	10.1
4	13.5	13.0
5	16.5	15.9
6	19.5	18.8
7	22.5	21.7
8	25.5	24.6
9		

126° (306°) (233°) **53°**

COMMON LOGARITHMS OF THE TRIGONOMETRIC FUNCTIONS (Continued)

37° (217°) (322°) **142°**

′	L. Sin.	d.	L. Tan.	c.d.	L. Cot.	L. Cos.	d.	′
0	9.77 946	17	9.87 711	27	0.12 289	9.90 235	10	**60**
1	9.77 963	17	9.87 738	26	0.12 262	9.90 225	9	59
2	9.77 980	17	9.87 764	26	0.12 236	9.90 216	10	58
3	9.77 997	16	9.87 790	27	0.12 210	9.90 206	9	57
4	9.78 013	17	9.87 817	26	0.12 183	9.90 197	10	56
5	9.78 030	17	9.87 843	26	0.12 157	9.90 187	9	55
6	9.78 047	16	9.87 869	26	0.12 131	9.90 178	10	54
7	9.78 063	17	9.87 895	27	0.12 105	9.90 168	9	53
8	9.78 080	17	9.87 922	26	0.12 078	9.90 159	10	52
9	9.78 097	16	9.87 948	26	0.12 052	9.90 149	10	51
10	9.78 113	17	9.87 974	26	0.12 026	9.90 139	9	**50**
11	9.78 130	17	9.88 000	27	0.12 000	9.90 130	10	49
12	9.78 147	16	9.88 027	26	0.11 973	9.90 120	9	48
13	9.78 163	17	9.88 053	26	0.11 947	9.90 111	10	47
14	9.78 180	17	9.88 079	26	0.11 921	9.90 101	10	46
15	9.78 197	16	9.88 105	26	0.11 895	9.90 091	9	45
16	9.78 213	17	9.88 131	27	0.11 869	9.90 082	10	44
17	9.78 230	16	9.88 158	26	0.11 842	9.90 072	9	43
18	9.78 246	17	9.88 184	26	0.11 816	9.90 063	10	42
19	9.78 263	17	9.88 210	26	0.11 790	9.90 053	10	41
20	9.78 280	16	9.88 236	26	0.11 764	9.90 043	9	**40**
21	9.78 296	17	9.88 262	27	0.11 738	9.90 034	10	39
22	9.78 313	16	9.88 289	26	0.11 711	9.90 024	10	38
23	9.78 329	17	9.88 315	26	0.11 685	9.90 014	9	37
24	9.78 346	16	9.88 341	26	0.11 659	9.90 005	10	36
25	9.78 362	17	9.88 367	26	0.11 633	9.89 995	10	35
26	9.78 379	16	9.88 393	27	0.11 607	9.89 985	9	34
27	9.78 395	17	9.88 420	26	0.11 580	9.89 976	10	33
28	9.78 412	16	9.88 446	26	0.11 554	9.89 966	10	32
29	9.78 428	17	9.88 472	26	0.11 528	9.89 956	9	31
30	9.78 445	16	9.88 498	26	0.11 502	9.89 947	10	**30**
31	9.78 461	17	9.88 524	26	0.11 476	9.89 937	10	29
32	9.78 478	16	9.88 550	27	0.11 450	9.89 927	9	28
33	9.78 494	16	9.88 577	26	0.11 423	9.89 918	10	27
34	9.78 510	17	9.88 603	26	0.11 397	9.89 908	10	26
35	9.78 527	16	9.88 629	26	0.11 371	9.89 898	10	25
36	9.78 543	17	9.88 655	26	0.11 345	9.89 888	9	24
37	9.78 560	16	9.88 681	26	0.11 319	9.89 879	10	23
38	9.78 576	16	9.88 707	26	0.11 293	9.89 869	10	22
39	9.78 592	17	9.88 733	26	0.11 267	9.89 859	10	21
40	9.78 609	16	9.88 759	27	0.11 241	9.89 849	9	**20**
41	9.78 625	17	9.88 786	26	0.11 214	9.89 840	10	19
42	9.78 642	16	9.88 812	26	0.11 188	9.89 830	10	18
43	9.78 658	16	9.88 838	26	0.11 162	9.89 820	10	17
44	9.78 674	17	9.88 864	26	0.11 136	9.89 810	9	16
45	9.78 691	16	9.88 890	26	0.11 110	9.89 801	10	15
46	9.78 707	16	9.88 916	26	0.11 084	9.89 791	10	14
47	9.78 723	16	9.88 942	26	0.11 058	9.89 781	10	13
48	9.78 739	17	9.88 968	26	0.11 032	9.89 771	10	12
49	9.78 756	16	9.88 994	26	0.11 006	9.89 761	9	11
50	9.78 772	16	9.89 020	26	0.10 980	9.89 752	10	**10**
51	9.78 788	17	9.89 046	27	0.10 954	9.89 742	10	9
52	9.78 805	16	9.89 073	26	0.10 927	9.89 732	10	8
53	9.78 821	16	9.89 099	26	0.10 901	9.89 722	10	7
54	9.78 837	16	9.89 125	26	0.10 875	9.89 712	10	6
55	9.78 853	16	9.89 151	26	0.10 849	9.89 702	9	5
56	9.78 869	17	9.89 177	26	0.10 823	9.89 693	10	4
57	9.78 886	16	9.89 203	26	0.10 797	9.89 683	10	3
58	9.78 902	16	9.89 229	26	0.10 771	9.89 673	10	2
59	9.78 918	16	9.89 255	26	0.10 745	9.89 663	10	1
60	9.78 934		9.89 281		0.10 719	9.89 653		**0**
′	L. Cos.	d.	L. Cot.	c.d.	L. Tan.	L. Sin.	d.	′

127° (307°) (232°) **52°**

P. P.

″	**27**	**26**
1	0.4	0.4
2	0.9	0.9
3	1.4	1.3
4	1.8	1.7
5	2.2	2.2
6	2.7	2.6
7	3.2	3.0
8	3.6	3.5
9	4.0	3.9
10	4.5	4.3
20	9.0	8.7
30	13.5	13.0
40	18.0	17.3
50	22.5	21.7
″	**17**	**16**
1	0.3	0.3
2	0.6	0.5
3	0.8	0.8
4	1.1	1.1
5	1.4	1.3
6	1.7	1.6
7	2.0	1.9
8	2.3	2.1
9	2.6	2.4
10	2.8	2.7
20	5.7	5.3
30	8.5	8.0
40	11.3	10.7
50	14.2	13.3
″	**10**	**9**
1	0.2	0.2
2	0.3	0.3
3	0.5	0.4
4	0.7	0.6
5	0.8	0.8
6	1.0	0.9
7	1.2	1.0
8	1.3	1.2
9	1.5	1.4
10	1.7	1.5
20	3.3	3.0
30	5.0	4.5
40	6.7	6.0
50	8.3	7.5

	$\frac{10}{27}$	$\frac{10}{26}$
0	1.4	1.3
1	4.1	3.9
2	6.8	6.5
3	9.4	9.1
4	12.2	11.7
5	14.8	14.3
6	17.6	16.9
7	20.2	19.5
8	22.9	22.1
9	25.6	24.7
10		

P. P.

COMMON LOGARITHMS OF THE TRIGONOMETRIC FUNCTIONS (Continued)

38° (218°) (321°) **141°**

′	L. Sin.	d.	L. Tan.	c.d.	L. Cot.	L. Cos.	d.	′
0	9.78 934	16	9.89 281	26	0.10 719	9.89 653	10	**60**
1	9.78 950	17	9.89 307	26	0.10 693	9.89 643	10	59
2	9.78 967	16	9.89 333	26	0.10 667	9.89 633	9	58
3	9.78 983	16	9.89 359	26	0.10 641	9.89 624	10	57
4	9.78 999	16	9.89 385	26	0.10 615	9.89 614	10	56
5	9.79 015	16	9.89 411	26	0.10 589	9.89 604	10	55
6	9.79 031	16	9.89 437	26	0.10 563	9.89 594	10	54
7	9.79 047	16	9.89 463	26	0.10 537	9.89 584	10	53
8	9.79 063	16	9.89 489	26	0.10 511	9.89 574	10	52
9	9.79 079	16	9.89 515	26	0.10 485	9.89 564	10	51
10	9.79 095	16	9.89 541	26	0.10 459	9.89 554	10	**50**
11	9.79 111	17	9.89 567	26	0.10 433	9.89 544	10	49
12	9.79 128	16	9.89 593	26	0.10 407	9.89 534	10	48
13	9.79 144	16	9.89 619	26	0.10 381	9.89 524	10	47
14	9.79 160	16	9.89 645	26	0.10 355	9.89 514	10	46
15	9.79 176	16	9.89 671	26	0.10 329	9.89 504	9	45
16	9.79 192	16	9.89 697	26	0.10 303	9.89 495	10	44
17	9.79 208	16	9.89 723	26	0.10 277	9.89 485	10	43
18	9.79 224	16	9.89 749	26	0.10 251	9.89 475	10	42
19	9.79 240	16	9.89 775	26	0.10 225	9.89 465	10	41
20	9.79 256	16	9.89 801	26	0.10 199	9.89 455	10	**40**
21	9.79 272	16	9.89 827	26	0.10 173	9.89 445	10	39
22	9.79 288	16	9.89 853	26	0.10 147	9.89 435	10	38
23	9.79 304	15	9.89 879	26	0.10 121	9.89 425	10	37
24	9.79 319	16	9.89 905	26	0.10 095	9.89 415	10	36
25	9.79 335	16	9.89 931	26	0.10 069	9.89 405	10	35
26	9.79 351	16	9.89 957	26	0.10 043	9.89 395	10	34
27	9.79 367	16	9.89 983	26	0.10 017	9.89 385	10	33
28	9.79 383	16	9.90 009	26	0.09 991	9.89 375	11	32
29	9.79 399	16	9.90 035	26	0.09 965	9.89 364	10	31
30	9.79 415	16	9.90 061	25	0.09 939	9.89 354	10	**30**
31	9.79 431	16	9.90 086	26	0.09 914	9.89 344	10	29
32	9.79 447	16	9.90 112	26	0.09 888	9.89 334	10	28
33	9.79 463	15	9.90 138	26	0.09 862	9.89 324	10	27
34	9.79 478	16	9.90 164	26	0.09 836	9.89 314	10	26
35	9.79 494	16	9.90 190	26	0.09 810	9.89 304	10	25
36	9.79 510	16	9.90 216	26	0.09 784	9.89 294	10	24
37	9.79 526	16	9.90 242	26	0.09 758	9.89 284	10	23
38	9.79 542	16	9.90 268	26	0.09 732	9.89 274	10	22
39	9.79 558	15	9.90 294	26	0.09 706	9.89 264	10	21
40	9.79 573	16	9.90 320	26	0.09 680	9.89 254	10	**20**
41	9.79 589	16	9.90 346	25	0.09 654	9.89 244	11	19
42	9.79 605	16	9.90 371	26	0.09 629	9.89 233	10	18
43	9.79 621	15	9.90 397	26	0.09 603	9.89 223	10	17
44	9.79 636	16	9.90 423	26	0.09 577	9.89 213	10	16
45	9.79 652	16	9.90 449	26	0.09 551	9.89 203	10	15
46	9.79 668	16	9.90 475	26	0.09 525	9.89 193	10	14
47	9.79 684	15	9.90 501	26	0.09 499	9.89 183	10	13
48	9.79 699	16	9.90 527	26	0.09 473	9.89 173	11	12
49	9.79 715	16	9.90 553	25	0.09 447	9.89 162	10	11
50	9.79 731	15	9.90 578	26	0.09 422	9.89 152	10	**10**
51	9.79 746	16	9.90 604	26	0.09 396	9.89 142	10	9
52	9.79 762	16	9.90 630	26	0.09 370	9.89 132	10	8
53	9.79 778	15	9.90 656	26	0.09 344	9.89 122	10	7
54	9.79 793	16	9.90 682	26	0.09 318	9.89 112	11	6
55	9.79 809	16	9.90 708	26	0.09 292	9.89 101	10	5
56	9.79 825	15	9.90 734	25	0.09 266	9.89 091	10	4
57	9.79 840	16	9.90 759	26	0.09 241	9.89 081	10	3
58	9.79 856	16	9.90 785	26	0.09 215	9.89 071	11	2
59	9.79 872	15	9.90 811	26	0.09 189	9.89 060	10	1
60	9.79 887		9.90 837		0.09 163	9.89 050		**0**
′	L. Cos.	d.	L. Cot.	c.d.	L. Tan.	L. Sin.	d.	′

128° (308°) (231°) **51°**

P. P.

″	**26**	**25**
1	0.4	0.4
2	0.9	0.8
3	1.3	1.2
4	1.7	1.7
5	2.2	2.1
6	2.6	2.5
7	3.0	2.9
8	3.5	3.3
9	3.9	3.8
10	4.3	4.2
20	8.7	8.3
30	13.0	12.5
40	17.3	16.7
50	21.7	20.8

″	**17**	**16**	**15**
1	0.3	0.3	0.2
2	0.6	0.5	0.5
3	0.8	0.8	0.8
4	1.1	1.1	1.0
5	1.4	1.3	1.2
6	1.7	1.6	1.5
7	2.0	1.9	1.8
8	2.3	2.1	2.0
9	2.6	2.4	2.2
10	2.8	2.7	2.5
20	5.7	5.3	5.0
30	8.5	8.0	7.5
40	11.3	10.7	10.0
50	14.2	13.3	12.5

″	**11**	**10**	**9**
1	0.2	0.2	0.2
2	0.4	0.3	0.3
3	0.6	0.5	0.4
4	0.7	0.7	0.6
5	0.9	0.8	0.8
6	1.1	1.0	0.9
7	1.3	1.2	1.0
8	1.5	1.3	1.2
9	1.6	1.5	1.4
10	1.8	1.7	1.5
20	3.7	3.3	3.0
30	5.5	5.0	4.5
40	7.3	6.7	6.0
50	9.2	8.3	7.5

	10/26	**10/25**	**9/26**
0	1.3	1.2	1.4
1	3.9	3.8	4.3
2	6.5	6.2	7.2
3	9.1	8.8	10.1
4	11.7	11.2	13.0
5	14.3	13.8	15.9
6	16.9	16.2	18.8
7	19.5	18.8	21.7
8	22.1	21.2	24.6
9	24.7	23.8	—
10			

P. P.

COMMON LOGARITHMS OF THE TRIGONOMETRIC FUNCTIONS (Continued)

39° (219°) (320°) **140°**

′	L. Sin.	d.	L. Tan.	c.d.	L. Cot.	L. Cos.	d.	′
0	9.79 887	16	9.90 837	26	0.09 163	9.89 050	10	**60**
1	9.79 903	15	9.90 863	26	0.09 137	9.89 040	10	59
2	9.79 918	16	9.90 889	25	0.09 111	9.89 030	10	58
3	9.79 934	16	9.90 914	26	0.09 086	9.89 020	11	57
4	9.79 950	15	9.90 940	26	0.09 060	9.89 009	10	56
5	9.79 965	16	9.90 966	26	0.09 034	9.88 999	10	55
6	9.79 981	15	9.90 992	26	0.09 008	9.88 989	11	54
7	9.79 996	16	9.91 018	25	0.08 982	9.88 978	10	53
8	9.80 012	15	9.91 043	26	0.08 957	9.88 968	10	52
9	9.80 027	16	9.91 069	26	0.08 931	9.88 958	10	51
10	9.80 043	15	9.91 095	26	0.08 905	9.88 948	11	**50**
11	9.80 058	16	9.91 121	26	0.08 879	9.88 937	10	49
12	9.80 074	15	9.91 147	25	0.08 853	9.88 927	10	48
13	9.80 089	16	9.91 172	26	0.08 828	9.88 917	11	47
14	9.80 105	15	9.91 198	26	0.08 802	9.88 906	10	46
15	9.80 120	16	9.91 224	26	0.08 776	9.88 896	10	45
16	9.80 136	15	9.91 250	26	0.08 750	9.88 886	11	44
17	9.80 151	15	9.91 276	25	0.08 724	9.88 875	10	43
18	9.80 166	16	9.91 301	26	0.08 699	9.88 865	10	42
19	9.80 182	15	9.91 327	26	0.08 673	9.88 855	11	41
20	9.80 197	16	9.91 353	26	0.08 647	9.88 844	10	**40**
21	9.80 213	15	9.91 379	25	0.08 621	9.88 834	10	39
22	9.80 228	16	9.91 404	26	0.08 596	9.88 824	11	38
23	9.80 244	15	9.91 430	26	0.08 570	9.88 813	10	37
24	9.80 259	15	9.91 456	26	0.08 544	9.88 803	10	36
25	9.80 274	16	9.91 482	25	0.08 518	9.88 793	11	35
26	9.80 290	15	9.91 507	26	0.08 493	9.88 782	10	34
27	9.80 305	15	9.91 533	26	0.08 467	9.88 772	11	33
28	9.80 320	16	9.91 559	26	0.08 441	9.88 761	10	32
29	9.80 336	15	9.91 585	25	0.08 415	9.88 751	10	31
30	9.80 351	15	9.91 610	26	0.08 390	9.88 741	11	**30**
31	9.80 366	16	9.91 636	26	0.08 364	9.88 730	10	29
32	9.80 382	15	9.91 662	26	0.08 338	9.88 720	11	28
33	9.80 397	15	9.91 688	25	0.08 312	9.88 709	10	27
34	9.80 412	16	9.91 713	26	0.08 287	9.88 699	11	26
35	9.80 428	15	9.91 739	26	0.08 261	9.88 688	10	25
36	9.80 443	15	9.91 765	26	0.08 235	9.88 678	10	24
37	9.80 458	15	9.91 791	25	0.08 209	9.88 668	11	23
38	9.80 473	16	9.91 816	26	0.08 184	9.88 657	10	22
39	9.80 489	15	9.91 842	26	0.08 158	9.88 647	11	21
40	9.80 504	15	9.91 868	25	0.08 132	9.88 636	10	**20**
41	9.80 519	15	9.91 893	26	0.08 107	9.88 626	11	19
42	9.80 534	16	9.91 919	26	0.08 081	9.88 615	10	18
43	9.80 550	15	9.91 945	26	0.08 055	9.88 605	11	17
44	9.80 565	15	9.91 971	25	0.08 029	9.88 594	10	16
45	9.80 580	15	9.91 996	26	0.08 004	9.88 584	11	15
46	9.80 595	15	9.92 022	26	0.07 978	9.88 573	10	14
47	9.80 610	15	9.92 048	25	0.07 952	9.88 563	11	13
48	9.80 625	16	9.92 073	26	0.07 927	9.88 552	10	12
49	9.80 641	15	9.92 099	26	0.07 901	9.88 542	11	11
50	9.80 656	15	9.92 125	25	0.07 875	9.88 531	10	**10**
51	9.80 671	15	9.92 150	26	0.07 850	9.88 521	11	9
52	9.80 686	15	9.92 176	26	0.07 824	9.88 510	11	8
53	9.80 701	15	9.92 202	25	0.07 798	9.88 499	10	7
54	9.80 716	15	9.92 227	26	0.07 773	9.88 489	11	6
55	9.80 731	15	9.92 253	26	0.07 747	9.88 478	10	5
56	9.80 746	16	9.92 279	25	0.07 721	9.88 468	11	4
57	9.80 762	15	9.92 304	26	0.07 696	9.88 457	10	3
58	9.80 777	15	9.92 330	26	0.07 670	9.88 447	11	2
59	9.80 792	15	9.92 356	25	0.07 644	9.88 436	11	1
60	9.80 807		9.92 381		0.07 619	9.88 425		**0**
′	L. Cos.	d.	L. Cot.	c.d.	L. Tan.	L. Sin.	d.	′

P. P.

″	**26**	**25**
1	0.4	0.4
2	0.9	0.8
3	1.3	1.2
4	1.7	1.7
5	2.2	2.1
6	2.6	2.5
7	3.0	2.9
8	3.5	3.3
9	3.9	3.8
10	4.3	4.2
20	8.7	8.3
30	13.0	12.5
40	17.3	16.7
50	21.7	20.8

″	**16**	**15**
1	0.3	0.2
2	0.5	0.5
3	0.8	0.8
4	1.1	1.0
5	1.3	1.2
6	1.6	1.5
7	1.9	1.8
8	2.1	2.0
9	2.4	2.2
10	2.7	2.5
20	5.3	5.0
30	8.0	7.5
40	10.7	10.0
50	13.3	12.5

″	**11**	**10**
1	0.2	0.2
2	0.4	0.3
3	0.6	0.5
4	0.7	0.7
5	0.9	0.8
6	1.1	1.0
7	1.3	1.2
8	1.5	1.3
9	1.6	1.5
10	1.8	1.7
20	3.7	3.3
30	5.5	5.0
40	7.3	6.7
50	9.2	8.3

	11/26	**11/25**
0	1.2	1.1
1	3.5	3.4
2	5.9	5.7
3	8.3	7.9
4	10.6	10.2
5	13.0	12.5
6	15.4	14.8
7	17.7	17.1
8	20.1	19.3
9	22.5	21.6
10	24.8	23.9
11		

129° (309°) (230°) **50°**

COMMON LOGARITHMS OF THE TRIGONOMETRIC FUNCTIONS (Continued)

40° (220°) (319°) **139°**

′	L. Sin.	d.	L. Tan.	c.d.	L. Cot.	L. Cos.	d.	′
0	9.80 807	15	9.92 381	26	0.07 619	9.88 425	10	**60**
1	9.80 822	15	9.92 407	26	0.07 593	9.88 415	11	59
2	9.80 837	15	9.92 433	25	0.07 567	9.88 404	10	58
3	9.80 852	15	9.92 458	26	0.07 542	9.88 394	11	57
4	9.80 867	15	9.92 484	26	0.07 516	9.88 383	11	56
5	9.80 882	15	9.92 510	25	0.07 490	9.88 372	10	55
6	9.80 897	15	9.92 535	26	0.07 465	9.88 362	11	54
7	9.80 912	15	9.92 561	26	0.07 439	9.88 351	11	53
8	9.80 927	15	9.92 587	25	0.07 413	9.88 340	10	52
9	9.80 942	15	9.92 612	26	0.07 388	9.88 330	11	51
10	9.80 957	15	9.92 638	25	0.07 362	9.88 319	11	**50**
11	9.80 972	15	9.92 663	26	0.07 337	9.88 308	10	49
12	9.80 987	15	9.92 689	26	0.07 311	9.88 298	11	48
13	9.81 002	15	9.92 715	25	0.07 285	9.88 287	11	47
14	9.81 017	15	9.92 740	26	0.07 260	9.88 276	10	46
15	9.81 032	15	9.92 766	26	0.07 234	9.88 266	11	45
16	9.81 047	14	9.92 792	25	0.07 208	9.88 255	11	44
17	9.81 061	15	9.92 817	26	0.07 183	9.88 244	10	43
18	9.81 076	15	9.92 843	25	0.07 157	9.88 234	11	42
19	9.81 091	15	9.92 868	26	0.07 132	9.88 223	11	41
20	9.81 106	15	9.92 894	26	0.07 106	9.88 212	11	**40**
21	9.81 121	15	9.92 920	25	0.07 080	9.88 201	10	39
22	9.81 136	15	9.92 945	26	0.07 055	9.88 191	11	38
23	9.81 151	15	9.92 971	25	0.07 029	9.88 180	11	37
24	9.81 166	14	9.92 996	26	0.07 004	9.88 169	11	36
25	9.81 180	15	9.93 022	26	0.06 978	9.88 158	10	35
26	9.81 195	15	9.93 048	25	0.06 952	9.88 148	11	34
27	9.81 210	15	9.93 073	26	0.06 927	9.88 137	11	33
28	9.81 225	15	9.93 099	25	0.06 901	9.88 126	11	32
29	9.81 240	14	9.93 124	26	0.06 876	9.88 115	10	31
30	9.81 254	15	9.93 150	25	0.06 850	9.88 105	11	**30**
31	9.81 269	15	9.93 175	26	0.06 825	9.88 094	11	29
32	9.81 284	15	9.93 201	26	0.06 799	9.88 083	11	28
33	9.81 299	15	9.93 227	25	0.06 773	9.88 072	11	27
34	9.81 314	14	9.93 252	26	0.06 748	9.88 061	10	26
35	9.81 328	15	9.93 278	25	0.06 722	9.88 051	11	25
36	9.81 343	15	9.93 303	26	0.06 697	9.88 040	11	24
37	9.81 358	14	9.93 329	25	0.06 671	9.88 029	11	23
38	9.81 372	15	9.93 354	26	0.06 646	9.88 018	11	22
39	9.81 387	15	9.93 380	26	0.06 620	9.88 007	11	21
40	9.81 402	15	9.93 406	25	0.06 594	9.87 996	11	**20**
41	9.81 417	14	9.93 431	26	0.06 569	9.87 985	10	19
42	9.81 431	15	9.93 457	25	0.06 543	9.87 975	11	18
43	9.81 446	15	9.93 482	26	0.06 518	9.87 964	11	17
44	9.81 461	14	9.93 508	25	0.06 492	9.87 953	11	16
45	9.81 475	15	9.93 533	26	0.06 467	9.87 942	11	15
46	9.81 490	15	9.93 559	25	0.06 441	9.87 931	11	14
47	9.81 505	14	9.93 584	26	0.06 416	9.87 920	11	13
48	9.81 519	15	9.93 610	26	0.06 390	9.87 909	11	12
49	9.81 534	15	9.93 636	25	0.06 364	9.87 898	11	11
50	9.81 549	14	9.93 661	26	0.06 339	9.87 887	10	**10**
51	9.81 563	15	9.93 687	25	0.06 313	9.87 877	11	9
52	9.81 578	14	9.93 712	26	0.06 288	9.87 866	11	8
53	9.81 592	15	9.93 738	25	0.06 262	9.87 855	11	7
54	9.81 607	15	9.93 763	26	0.06 237	9.87 844	11	6
55	9.81 622	14	9.93 789	25	0.06 211	9.87 833	11	5
56	9.81 636	15	9.93 814	26	0.06 186	9.87 822	11	4
57	9.81 651	14	9.93 840	25	0.06 160	9.87 811	11	3
58	9.81 665	15	9.93 865	26	0.06 135	9.87 800	11	2
59	9.81 680	14	9.93 891	25	0.06 109	9.87 789	11	1
60	9.81 694		9.93 916		0.06 084	9.87 778		**0**
′	L. Cos.	d.	L. Cot.	c.d.	L. Tan.	L. Sin.	d.	′

130° (310°) (229°) **49°**

P. P.

″	**26**	**25**
1	0.4	0.4
2	0.9	0.8
3	1.3	1.2
4	1.7	1.7
5	2.2	2.1
6	2.6	2.5
7	3.0	2.9
8	3.5	3.3
9	3.9	3.8
10	4.3	4.2
20	8.7	8.3
30	13.0	12.5
40	17.3	16.7
50	21.7	20.8

″	**15**	**14**
1	0.2	0.2
2	0.5	0.5
3	0.8	0.7
4	1.0	0.9
5	1.2	1.2
6	1.5	1.4
7	1.8	1.6
8	2.0	1.9
9	2.2	2.1
10	2.5	2.3
20	5.0	4.7
30	7.5	7.0
40	10.0	9.3
50	12.5	11.7

″	**11**	**10**
1	0.2	0.2
2	0.4	0.3
3	0.6	0.5
4	0.7	0.7
5	0.9	0.8
6	1.1	1.0
7	1.3	1.2
8	1.5	1.3
9	1.6	1.5
10	1.8	1.7
20	3.7	3.3
30	5.5	5.0
40	7.3	6.7
50	9.2	8.3

	11/26	**10/26**	**10/25**
0	1.2	1.3	1.2
1	3.5	3.9	3.8
2	5.9	6.5	6.2
3	8.3	9.1	8.8
4	10.6	11.7	11.2
5	13.0	14.3	13.8
6	15.4	16.9	16.2
7	17.7	19.5	18.8
8	20.1	22.1	21.2
9	22.5	24.7	23.8
10	24.8	—	—
11			

COMMON LOGARITHMS OF THE TRIGONOMETRIC FUNCTIONS (Continued)

41° (221°) (318°) **138°**

′	L. Sin.	d.	L. Tan.	c.d.	L. Cot.	L. Cos.	d.	′
0	9.81 694	15	9.93 916	26	0.06 084	9.87 778	11	**60**
1	9.81 709	14	9.93 942	25	0.06 058	9.87 767	11	59
2	9.81 723	15	9.93 967	26	0.06 033	9.87 756	11	58
3	9.81 738	14	9.93 993	25	0.06 007	9.87 745	11	57
4	9.81 752	15	9.94 018	26	0.05 982	9.87 734	11	56
5	9.81 767	14	9.94 044	25	0.05 956	9.87 723	11	55
6	9.81 781	15	9.94 069	26	0.05 931	9.87 712	11	54
7	9.81 796	14	9.94 095	25	0.05 905	9.87 701	11	53
8	9.81 810	15	9.94 120	26	0.05 880	9.87 690	11	52
9	9.81 825	14	9.94 146	25	0.05 854	9.87 679	11	51
10	9.81 839	15	9.94 171	26	0.05 829	9.87 668	11	**50**
11	9.81 854	14	9.94 197	25	0.05 803	9.87 657	11	49
12	9.81 868	14	9.94 222	26	0.05 778	9.87 646	11	48
13	9.81 882	15	9.94 248	25	0.05 752	9.87 635	11	47
14	9.81 897	14	9.94 273	26	0.05 727	9.87 624	11	46
15	9.81 911	15	9.94 299	25	0.05 701	9.87 613	12	45
16	9.81 926	14	9.94 324	26	0.05 676	9.87 601	11	44
17	9.81 940	15	9.94 350	25	0.05 650	9.87 590	11	43
18	9.81 955	14	9.94 375	26	0.05 625	9.87 579	11	42
19	9.81 969	14	9.94 401	25	0.05 599	9.87 568	11	41
20	9.81 983	15	9.94 426	26	0.05 574	9.87 557	11	**40**
21	9.81 998	14	9.94 452	25	0.05 548	9.87 546	11	39
22	9.82 012	14	9.94 477	26	0.05 523	9.87 535	11	38
23	9.82 026	15	9.94 503	25	0.05 497	9.87 524	11	37
24	9.82 041	14	9.94 528	26	0.05 472	9.87 513	12	36
25	9.82 055	14	9.94 554	25	0.05 446	9.87 501	11	35
26	9.82 069	15	9.94 579	25	0.05 421	9.87 490	11	34
27	9.82 084	14	9.94 604	26	0.05 396	9.87 479	11	33
28	9.82 098	14	9.94 630	25	0.05 370	9.87 468	11	32
29	9.82 112	14	9.94 655	26	0.05 345	9.87 457	11	31
30	9.82 126	15	9.94 681	25	0.05 319	9.87 446	12	**30**
31	9.82 141	14	9.94 706	26	0.05 294	9.87 434	11	29
32	9.82 155	14	9.94 732	25	0.05 268	9.87 423	11	28
33	9.82 169	15	9.94 757	26	0.05 243	9.87 412	11	27
34	9.82 184	14	9.94 783	25	0.05 217	9.87 401	11	26
35	9.82 198	14	9.94 808	26	0.05 192	9.87 390	12	25
36	9.82 212	14	9.94 834	25	0.05 166	9.87 378	11	24
37	9.82 226	14	9.94 859	25	0.05 141	9.87 367	11	23
38	9.82 240	15	9.94 884	26	0.05 116	9.87 356	11	22
39	9.82 255	14	9.94 910	25	0.05 090	9.87 345	11	21
40	9.82 269	14	9.94 935	26	0.05 065	9.87 334	12	**20**
41	9.82 283	14	9.94 961	25	0.05 039	9.87 322	11	19
42	9.82 297	14	9.94 986	26	0.05 014	9.87 311	11	18
43	9.82 311	15	9.95 012	25	0.04 988	9.87 300	12	17
44	9.82 326	14	9.95 037	25	0.04 963	9.87 288	11	16
45	9.82 340	14	9.95 062	26	0.04 938	9.87 277	11	15
46	9.82 354	14	9.95 088	25	0.04 912	9.87 266	11	14
47	9.82 368	14	9.95 113	26	0.04 887	9.87 255	12	13
48	9.82 382	14	9.95 139	25	0.04 861	9.87 243	11	12
49	9.82 396	14	9.95 164	26	0.04 836	9.87 232	11	11
50	9.82 410	14	9.95 190	25	0.04 810	9.87 221	12	**10**
51	9.82 424	15	9.95 215	25	0.04 785	9.87 209	11	9
52	9.82 439	14	9.95 240	26	0.04 760	9.87 198	11	8
53	9.82 453	14	9.95 266	25	0.04 734	9.87 187	12	7
54	9.82 467	14	9.95 291	26	0.04 709	9.87 175	11	6
55	9.82 481	14	9.95 317	25	0.04 683	9.87 164	11	5
56	9.82 495	14	9.95 342	26	0.04 658	9.87 153	12	4
57	9.82 509	14	9.95 368	25	0.04 632	9.87 141	11	3
58	9.82 523	14	9.95 393	25	0.04 607	9.87 130	11	2
59	9.82 537	14	9.95 418	26	0.04 582	9.87 119	12	1
60	9.82 551		9.95 444		0.04 556	9.87 107		**0**
′	L. Cos.	d.	L. Cot.	c.d.	L. Tan.	L. Sin.	d.	′

131° (311°) (228°) **48°**

P. P.

″	**26**	**25**
1	0.4	0.4
2	0.9	0.8
3	1.3	1.2
4	1.7	1.7
5	2.2	2.1
6	2.6	2.5
7	3.0	2.9
8	3.5	3.3
9	3.9	3.8
10	4.3	4.2
20	8.7	8.3
30	13.0	12.5
40	17.3	16.7
50	21.7	20.8

″	**15**	**14**
1	0.2	0.2
2	0.5	0.5
3	0.8	0.7
4	1.0	0.9
5	1.2	1.2
6	1.5	1.4
7	1.8	1.6
8	2.0	1.9
9	2.2	2.1
10	2.5	2.3
20	5.0	4.7
30	7.5	7.0
40	10.0	9.3
50	12.5	11.7

″	**12**	**11**
1	0.2	0.2
2	0.4	0.4
3	0.6	0.6
4	0.8	0.7
5	1.0	0.9
6	1.2	1.1
7	1.4	1.3
8	1.6	1.5
9	1.8	1.6
10	2.0	1.8
20	4.0	3.7
30	6.0	5.5
40	8.0	7.3
50	10.0	9.2

	12/26	**12/25**	**11/25**
0			
1	1.1	1.1	1.1
2	3.2	3.1	3.4
3	5.4	5.2	5.7
4	7.6	7.3	7.9
5	9.8	9.4	10.2
6	11.9	11.5	12.5
7	14.1	13.5	14.8
8	16.2	15.6	17.1
9	18.4	17.7	19.3
10	20.6	19.8	21.6
11	22.8	21.9	23.9
12	24.9	23.9	—

COMMON LOGARITHMS OF THE TRIGONOMETRIC FUNCTIONS (Continued)

42° (222°) (317°) **137°**

′	L. Sin.	d.	L. Tan.	c.d.	L. Cot.	L. Cos.	d.	′
0	9.82 551	14	9.95 444	25	0.04 556	9.87 107	11	**60**
1	9.82 565	14	9.95 469	26	0.04 531	9.87 096	11	59
2	9.82 579	14	9.95 495	25	0.04 505	9.87 085	12	58
3	9.82 593	14	9.95 520	25	0.04 480	9.87 073	11	57
4	9.82 607	14	9.95 545	26	0.04 455	9.87 062	12	56
5	9.82 621	14	9.95 571	25	0.04 429	9.87 050	11	55
6	9.82 635	14	9.95 596	26	0.04 404	9.87 039	11	54
7	9.82 649	14	9.95 622	25	0.04 378	9.87 028	12	53
8	9.82 663	14	9.95 647	25	0.04 353	9.87 016	11	52
9	9.82 677	14	9.95 672	26	0.04 328	9.87 005	12	51
10	9.82 691	14	9.95 698	25	0.04 302	9.86 993	11	**50**
11	9.82 705	14	9.95 723	25	0.04 277	9.86 982	12	49
12	9.82 719	14	9.95 748	26	0.04 252	9.86 970	11	48
13	9.82 733	14	9.95 774	25	0.04 226	9.86 959	12	47
14	9.82 747	14	9.95 799	26	0.04 201	9.86 947	11	46
15	9.82 761	14	9.95 825	25	0.04 175	9.86 936	12	45
16	9.82 775	13	9.95 850	25	0.04 150	9.86 924	11	44
17	9.82 788	14	9.95 875	26	0.04 125	9.86 913	11	43
18	9.82 802	14	9.95 901	25	0.04 099	9.86 902	12	42
19	9.82 816	14	9.95 926	26	0.04 074	9.86 890	11	41
20	9.82 830	14	9.95 952	25	0.04 048	9.86 879	12	**40**
21	9.82 844	14	9.95 977	25	0.04 023	9.86 867	12	39
22	9.82 858	14	9.96 002	26	0.03 998	9.86 855	11	38
23	9.82 872	13	9.96 028	25	0.03 972	9.86 844	12	37
24	9.82 885	14	9.96 053	25	0.03 947	9.86 832	11	36
25	9.82 899	14	9.96 078	26	0.03 922	9.86 821	12	35
26	9.82 913	14	9.96 104	25	0.03 896	9.86 809	11	34
27	9.82 927	14	9.96 129	26	0.03 871	9.86 798	12	33
28	9.82 941	14	9.96 155	25	0.03 845	9.86 786	11	32
29	9.82 955	13	9.96 180	25	0.03 820	9.86 775	12	31
30	9.82 968	14	9.96 205	26	0.03 795	9.86 763	11	**30**
31	9.82 982	14	9.96 231	25	0.03 769	9.86 752	12	29
32	9.82 996	14	9.96 256	25	0.03 744	9.86 740	12	28
33	9.83 010	13	9.96 281	26	0.03 719	9.86 728	11	27
34	9.83 023	14	9.96 307	25	0.03 693	9.86 717	12	26
35	9.83 037	14	9.96 332	25	0.03 668	9.86 705	11	25
36	9.83 051	14	9.96 357	26	0.03 643	9.86 694	12	24
37	9.83 065	13	9.96 383	25	0.03 617	9.86 682	12	23
38	9.83 078	14	9.96 408	25	0.03 592	9.86 670	11	22
39	9.83 092	14	9.96 433	26	0.03 567	9.86 659	12	21
40	9.83 106	14	9.96 459	25	0.03 541	9.86 647	12	**20**
41	9.83 120	13	9.96 484	26	0.03 516	9.86 635	11	19
42	9.83 133	14	9.96 510	25	0.03 490	9.86 624	12	18
43	9.83 147	14	9.96 535	25	0.03 465	9.86 612	12	17
44	9.83 161	13	9.96 560	26	0.03 440	9.86 600	11	16
45	9.83 174	14	9.96 586	25	0.03 414	9.86 589	12	15
46	9.83 188	14	9.96 611	25	0.03 389	9.86 577	12	14
47	9.83 202	13	9.96 636	26	0.03 364	9.86 565	11	13
48	9.83 215	14	9.96 662	25	0.03 338	9.86 554	12	12
49	9.83 229	13	9.96 687	25	0.03 313	9.86 542	12	11
50	9.83 242	14	9.96 712	26	0.03 288	9.86 530	12	**10**
51	9.83 256	14	9.96 738	25	0.03 262	9.86 518	11	9
52	9.83 270	13	9.96 763	25	0.03 237	9.86 507	12	8
53	9.83 283	14	9.96 788	26	0.03 212	9.86 495	12	7
54	9.83 297	13	9.96 814	25	0.03 186	9.86 483	11	6
55	9.83 310	14	9.96 839	25	0.03 161	9.86 472	12	5
56	9.83 324	14	9.96 864	26	0.03 136	9.86 460	12	4
57	9.83 338	13	9.96 890	25	0.03 110	9.86 448	12	3
58	9.83 351	14	9.96 915	25	0.03 085	9.86 436	11	2
59	9.83 365	13	9.96 940	26	0.03 060	9.86 425	12	1
60	9.83 378		9.96 966		0.03 034	9.86 413		**0**
′	L. Cos.	d.	L. Cot.	c.d.	L. Tan.	L. Sin.	d.	′

132° (312°) (227°) **47°**

P. P.

″	26	25
1	0.4	0.4
2	0.9	0.8
3	1.3	1.2
4	1.7	1.7
5	2.2	2.1
6	2.6	2.5
7	3.0	2.9
8	3.5	3.3
9	3.9	3.8
10	4.3	4.2
20	8.7	8.3
30	13.0	12.5
40	17.3	16.7
50	21.7	20.8

″	14	13
1	0.2	0.2
2	0.5	0.4
3	0.7	0.6
4	0.9	0.9
5	1.2	1.1
6	1.4	1.3
7	1.6	1.5
8	1.9	1.7
9	2.1	2.0
10	2.3	2.2
20	4.7	4.3
30	7.0	6.5
40	9.3	8.7
50	11.7	10.8

″	12	11
1	0.2	0.2
2	0.4	0.4
3	0.6	0.6
4	0.8	0.7
5	1.0	0.9
6	1.2	1.1
7	1.4	1.3
8	1.6	1.5
9	1.8	1.6
10	2.0	1.8
20	4.0	3.7
30	6.0	5.5
40	8.0	7.3
50	10.0	9.2

	12/26	11/26	11/25
0	1.1	1.2	1.1
1	3.2	3.5	3.4
2	5.4	5.9	5.7
3	7.6	8.3	7.9
4	9.8	10.6	10.2
5	11.9	13.0	12.5
6	14.1	15.4	14.8
7	16.2	17.7	17.1
8	18.4	20.1	19.3
9	20.6	22.5	21.6
10	22.8	24.8	23.9
11	24.9	—	—
12			

P. P.

COMMON LOGARITHMS OF THE TRIGONOMETRIC FUNCTIONS (Continued)

43° (223°) (316°) **136°**

′	L. Sin.	d.	L. Tan.	c.d.	L. Cot.	L. Cos.	d.	′
0	9.83 378	14	9.96 966	25	0.03 034	9.86 413	12	**60**
1	9.83 392	13	9.96 991	25	0.03 009	9.86 401	12	59
2	9.83 405	14	9.97 016	26	0.02 984	9.86 389	12	58
3	9.83 419	13	9.97 042	25	0.02 958	9.86 377	11	57
4	9.83 432	14	9.97 067	25	0.02 933	9.86 366	12	56
5	9.83 446	13	9.97 092	26	0.02 908	9.86 354	12	55
6	9.83 459	14	9.97 118	25	0.02 882	9.86 342	12	54
7	9.83 473	13	9.97 143	25	0.02 857	9.86 330	12	53
8	9.83 486	14	9.97 168	25	0.02 832	9.86 318	12	52
9	9.83 500	13	9.97 193	26	0.02 807	9.86 306	11	51
10	9.83 513	14	9.97 219	25	0.02 781	9.86 295	12	**50**
11	9.83 527	13	9.97 244	25	0.02 756	9.86 283	12	49
12	9.83 540	14	9.97 269	26	0.02 731	9.86 271	12	48
13	9.83 554	13	9.97 295	25	0.02 705	9.86 259	12	47
14	9.83 567	14	9.97 320	25	0.02 680	9.86 247	12	46
15	9.83 581	13	9.97 345	26	0.02 655	9.86 235	12	45
16	9.83 594	14	9.97 371	25	0.02 629	9.86 223	12	44
17	9.83 608	13	9.97 396	25	0.02 604	9.86 211	11	43
18	9.83 621	13	9.97 421	26	0.02 579	9.86 200	12	42
19	9.83 634	14	9.97 447	25	0.02 553	9.86 188	12	41
20	9.83 648	13	9.97 472	25	0.02 528	9.86 176	12	**40**
21	9.83 661	13	9.97 497	26	0.02 503	9.86 164	12	39
22	9.83 674	14	9.97 523	25	0.02 477	9.86 152	12	38
23	9.83 688	13	9.97 548	25	0.02 452	9.86 140	12	37
24	9.83 701	14	9.97 573	25	0.02 427	9.86 128	12	36
25	9.83 715	13	9.97 598	26	0.02 402	9.86 116	12	35
26	9.83 728	13	9.97 624	25	0.02 376	9.86 104	12	34
27	9.83 741	14	9.97 649	25	0.02 351	9.86 092	12	33
28	9.83 755	13	9.97 674	26	0.02 326	9.86 080	12	32
29	9.83 768	13	9.97 700	25	0.02 300	9.86 068	12	31
30	9.83 781	14	9.97 725	25	0.02 275	9.86 056	12	**30**
31	9.83 795	13	9.97 750	26	0.02 250	9.86 044	12	29
32	9.83 808	13	9.97 776	25	0.02 224	9.86 032	12	28
33	9.83 821	13	9.97 801	25	0.02 199	9.86 020	12	27
34	9.83 834	14	9.97 826	25	0.02 174	9.86 008	12	26
35	9.83 848	13	9.97 851	26	0.02 149	9.85 996	12	25
36	9.83 861	13	9.97 877	25	0.02 123	9.85 984	12	24
37	9.83 874	13	9.97 902	25	0.02 098	9.85 972	12	23
38	9.83 887	14	9.97 927	26	0.02 073	9.85 960	12	22
39	9.83 901	13	9.97 953	25	0.02 047	9.85 948	12	21
40	9.83 914	13	9.97 978	25	0.02 022	9.85 936	12	**20**
41	9.83 927	13	9.98 003	26	0.01 997	9.85 924	12	19
42	9.83 940	14	9.98 029	25	0.01 971	9.85 912	12	18
43	9.83 954	13	9.98 054	25	0.01 946	9.85 900	12	17
44	9.83 967	13	9.98 079	25	0.01 921	9.85 888	12	16
45	9.83 980	13	9.98 104	26	0.01 896	9.85 876	12	15
46	9.83 993	13	9.98 130	25	0.01 870	9.85 864	13	14
47	9.84 006	14	9.98 155	25	0.01 845	9.85 851	12	13
48	9.84 020	13	9.98 180	26	0.01 820	9.85 839	12	12
49	9.84 033	13	9.98 206	25	0.01 794	9.85 827	12	11
50	9.84 046	13	9.98 231	25	0.01 769	9.85 815	12	**10**
51	9.84 059	13	9.98 256	25	0.01 744	9.85 803	12	9
52	9.84 072	13	9.98 281	26	0.01 719	9.85 791	12	8
53	9.84 085	13	9.98 307	25	0.01 693	9.85 779	13	7
54	9.84 098	14	9.98 332	25	0.01 668	9.85 766	12	6
55	9.84 112	13	9.98 357	26	0.01 643	9.85 754	12	5
56	9.84 125	13	9.98 383	25	0.01 617	9.85 742	12	4
57	9.84 138	13	9.98 408	25	0.01 592	9.85 730	12	3
58	9.84 151	13	9.98 433	25	0.01 567	9.85 718	12	2
59	9.84 164	13	9.98 458	26	0.01 542	9.85 706	13	1
60	9.84 177		9.98 484		0.01 516	9.85 693		**0**
′	L. Cos.	d.	L. Cot.	c.d.	L. Tan.	L. Sin.	d.	′

P. P.

″	**26**	**25**
1	0.4	0.4
2	0.9	0.8
3	1.3	1.2
4	1.7	1.7
5	2.2	2.1
6	2.6	2.5
7	3.0	2.9
8	3.5	3.3
9	3.9	3.8
10	4.3	4.2
20	8.7	8.3
30	13.0	12.5
40	17.3	16.7
50	21.7	20.8

″	**14**	**13**
1	0.2	0.2
2	0.5	0.4
3	0.7	0.6
4	0.9	0.9
5	1.2	1.1
6	1.4	1.3
7	1.6	1.5
8	1.9	1.7
9	2.1	2.0
10	2.3	2.2
20	4.7	4.3
30	7.0	6.5
40	9.3	8.7
50	11.7	10.8

″	**12**	**11**
1	0.2	0.2
2	0.4	0.4
3	0.6	0.6
4	0.8	0.7
5	1.0	0.9
6	1.2	1.1
7	1.4	1.3
8	1.6	1.5
9	1.8	1.6
10	2.0	1.8
20	4.0	3.7
30	6.0	5.5
40	8.0	7.3
50	10.0	9.2

	13/26	**13/25**	**12/25**
0			
1	1.0	0.9	1.1
2	3.0	2.9	3.1
3	5.0	4.8	5.2
4	7.0	6.7	7.3
5	9.0	8.7	9.4
6	11.0	10.6	11.5
7	13.0	12.5	13.5
8	15.0	14.4	15.6
9	17.0	16.3	17.7
10	19.0	18.3	19.8
11	21.0	20.2	21.9
12	23.0	22.1	23.9
13	25.0	24.1	—

133° (313°) (226°) **46°**

COMMON LOGARITHMS OF THE TRIGONOMETRIC FUNCTIONS (Continued)

44° (224°) (315°) **135°**

′	L. Sin.	d.	L. Tan.	c.d.	L. Cot.	L. Cos.	d.	′
0	9.84 177	13	9.98 484	25	0.01 516	9.85 693	12	**60**
1	9.84 190	13	9.98 509	25	0.01 491	9.85 681	12	59
2	9.84 203	13	9.98 534	26	0.01 466	9.85 669	12	58
3	9.84 216	13	9.98 560	25	0.01 440	9.85 657	12	57
4	9.84 229	13	9.98 585	25	0.01 415	9.85 645	13	56
5	9.84 242	13	9.98 610	25	0.01 390	9.85 632	12	55
6	9.84 255	14	9.98 635	26	0.01 365	9.85 620	12	54
7	9.84 269	13	9.98 661	25	0.01 339	9.85 608	12	53
8	9.84 282	13	9.98 686	25	0.01 314	9.85 596	13	52
9	9.84 295	13	9.98 711	26	0.01 289	9.85 583	12	51
10	9.84 308	13	9.98 737	25	0.01 263	9.85 571	12	**50**
11	9.84 321	13	9.98 762	25	0.01 238	9.85 559	12	49
12	9.84 334	13	9.98 787	25	0.01 213	9.85 547	13	48
13	9.84 347	13	9.98 812	26	0.01 188	9.85 534	12	47
14	9.84 360	13	9.98 838	25	0.01 162	9.85 522	12	46
15	9.84 373	12	9.98 863	25	0.01 137	9.85 510	13	45
16	9.84 385	13	9.98 888	25	0.01 112	9.85 497	12	44
17	9.84 398	13	9.98 913	26	0.01 087	9.85 485	12	43
18	9.84 411	13	9.98 939	25	0.01 061	9.85 473	13	42
19	9.84 424	13	9.98 964	25	0.01 036	9.85 460	12	41
20	9.84 437	13	9.98 989	26	0.01 011	9.85 448	12	**40**
21	9.84 450	13	9.99 015	25	0.00 985	9.85 436	13	39
22	9.84 463	13	9.99 040	25	0.00 960	9.85 423	12	38
23	9.84 476	13	9.99 065	25	0.00 935	9.85 411	12	37
24	9.84 489	13	9.99 090	26	0.00 910	9.85 399	13	36
25	9.84 502	13	9.99 116	25	0.00 884	9.85 386	12	35
26	9.84 515	13	9.99 141	25	0.00 859	9.85 374	13	34
27	9.84 528	12	9.99 166	25	0.00 834	9.85 361	12	33
28	9.84 540	13	9.99 191	26	0.00 809	9.85 349	12	32
29	9.84 553	13	9.99 217	25	0.00 783	9.85 337	13	31
30	9.84 566	13	9.99 242	25	0.00 758	9.85 324	12	**30**
31	9.84 579	13	9.99 267	26	0.00 733	9.85 312	13	29
32	9.84 592	13	9.99 293	25	0.00 707	9.85 299	12	28
33	9.84 605	13	9.99 318	25	0.00 682	9.85 287	13	27
34	9.84 618	12	9.99 343	25	0.00 657	9.85 274	12	26
35	9.84 630	13	9.99 368	26	0.00 632	9.85 262	12	25
36	9.84 643	13	9.99 394	25	0.00 606	9.85 250	13	24
37	9.84 656	13	9.99 419	25	0.00 581	9.85 237	12	23
38	9.84 669	13	9.99 444	25	0.00 556	9.85 225	13	22
39	9.84 682	12	9.99 469	26	0.00 531	9.85 212	12	21
40	9.84 694	13	9.99 495	25	0.00 505	9.85 200	13	**20**
41	9.84 707	13	9.99 520	25	0.00 480	9.85 187	12	19
42	9 84 720	13	9.99 545	25	0.00 455	9.85 175	13	18
43	9.84 733	12	9.99 570	26	0.00 430	9.85 162	12	17
44	9.84 745	13	9.99 596	25	0.00 404	9.85 150	13	16
45	9.84 758	13	9.99 621	25	0.00 379	9.85 137	12	15
46	9.84 771	13	9.99 646	26	0.00 354	9.85 125	13	14
47	9.84 784	12	9.99 672	25	0.00 328	9.85 112	12	13
48	9.84 796	13	9.99 697	25	0.00 303	9.85 100	13	12
49	9.84 809	13	9.99 722	25	0.00 278	9.85 087	13	11
50	9.84 822	13	9.99 747	26	0.00 253	9.85 074	12	**10**
51	9.84 835	12	9.99 773	25	0.00 227	9.85 062	13	9
52	9.84 847	13	9.99 798	25	0.00 202	9.85 049	12	8
53	9.84 860	13	9.99 823	25	0.00 177	9.85 037	13	7
54	9.84 873	12	9.99 848	26	0.00 152	9.85 024	12	6
55	9.84 885	13	9.99 874	25	0.00 126	9.85 012	13	5
56	9.84 898	13	9.99 899	25	0.00 101	9.84 999	13	4
57	9.84 911	12	9.99 924	25	0.00 076	9.84 986	12	3
58	9.84 923	13	9.99 949	26	0.00 051	9.84 974	13	2
59	9.84 936	13	9.99 975	25	0.00 025	9.84 961	12	1
60	9.84 949		0.00 000		0.00 000	9.84 949		**0**
′	L. Cos.	d.	L. Cot.	c.d.	L. Tan.	L. Sin.	d.	′

134° (314°) (225°) **45°**

P. P.

″	26	25
1	0.4	0.4
2	0.9	0.8
3	1.3	1.2
4	1.7	1.7
5	2.2	2.1
6	2.6	2.5
7	3.0	2.9
8	3.5	3.3
9	3.9	3.8
10	4.3	4.2
20	8.7	8.3
30	13.0	12.5
40	17.3	16.7
50	21.7	20.8

″	14	13	12
1	0.2	0.2	0.2
2	0.5	0.4	0.4
3	0.7	0.6	0.6
4	0.9	0.9	0.8
5	1.2	1.1	1.0
6	1.4	1.3	1.2
7	1.6	1.5	1.4
8	1.9	1.7	1.6
9	2.1	2.0	1.8
10	2.3	2.2	2.0
20	4.7	4.3	4.0
30	7.0	6.5	6.0
40	9.3	8.7	8.0
50	11.7	10.8	10.0

	13/26	13/25
0	1.0	0.9
1	3.0	2.9
2	5.0	4.8
3	7.0	6.7
4	9.0	8.7
5	11.0	10.6
6	13.0	12.5
7	15.0	14.4
8	17.0	16.3
9	19.0	18.3
10	21.0	20.2
11	23.0	22.1
12	25.0	24.1
13		

	12/26	12/25
0	1.1	1.1
1	3.2	3.1
2	5.4	5.2
3	7.6	7.3
4	9.8	9.4
5	11.9	11.5
6	14.1	13.5
7	16.2	15.6
8	18.4	17.7
9	20.6	19.8
10	22.8	21.9
11	24.9	23.9
12		

P. P.

COMMON LOGARITHMS OF THE TRIGONOMETRIC FUNCTIONS FOR ANGLES IN DEGREES AND DECIMALS

The table that follows gives the common logarithms of the functions sine, cosine, tangent and cotangent for each tenth of a degree from 0° to 90°. Linear interpolation may be used to obtain functional values for angles expressed in hundredths of a degree, except where the values of the functions are rapidly changing.

For argument values found in the left column, use the column headings at the top of the columns. For argument values found in the right column, use the column headings at the bottom of the columns.

The quantity -10 is to be appended to the logarithm of all sines and cosines, and to logarithms of tangents from 0° to 45°, and cotangents from 45° to 90°.

COMMON LOGARITHMS OF THE TRIGONOMETRIC FUNCTIONS FOR ANGLES IN DEGREES AND DECIMALS

Deg.	L. Sin	* L. Tan	* L. Cot	L. Cos	
0.0	$-\infty$	$-\infty$	∞	0.00000	**90.0**
.1	7.24188	7.24188	2.75812	0.00000	89.9
.2	7.54291	7.54291	2.45709	0.00000	.8
.3	7.71900	7.71900	2.28100	9.99999	.7
.4	7.84393	7.84394	2.15606	9.99999	.6
.5	7.94084	7.94086	2.05914	9.99998	.5
.6	8.02002	8.02004	1.97996	9.99998	.4
.7	8.08696	8.08700	1.91300	9.99997	.3
.8	8.14495	8.14500	1.85500	9.99996	.2
.9	8.19610	8.19616	1.80384	9.99995	89.1
1.0	8.24186	8.24192	1.75808	9.99993	**89.0**
.1	8.28324	8.28332	1.71668	9.99992	88.9
.2	8.32103	8.32112	1.67888	9.99990	.8
.3	8.35578	8.35590	1.64410	9.99989	.7
.4	8.38796	8.38809	1.61191	9.99987	.6
.5	8.41792	8.41807	1.58193	9.99985	.5
.6	8.44594	8.44611	1.55389	9.99983	.4
.7	8.47226	8.47245	1.52755	9.99981	.3
.8	8.49708	8.49729	1.50271	9.99979	.2
.9	8.52055	8.52079	1.47921	9 99976	88.1
2.0	8.54282	8.54308	1.45692	9.99974	**88.0**
.1	8.56400	8.56429	1.43571	9.99971	87.9
.2	8.58419	8.58451	1.41549	9.99968	.8
.3	8.60349	8.60384	1.39616	9.99965	.7
.4	8.62196	8.62234	1.37766	9.99962	.6
.5	8.63968	8.64009	1.35991	9.99959	.5
.6	8.65670	8.65715	1.34285	9.99955	.4
.7	8.67308	8.67356	1.32644	9.99952	.3
.8	8.68886	8.68938	1.31062	9.99948	.2
.9	8.70409	8.70465	1.29535	9.99944	87.1
3.0	8.71880	8.71940	1.28060	9.99940	**87.0**
.1	8.73303	8.73366	1.26634	9.99936	86.9
.2	8.74680	8.74748	1.25252	9.99932	.8
.3	8.76015	8.76087	1.23913	9.99928	.7
.4	8.77310	8.77387	1.22613	9.99923	.6
.5	8.78568	8.78649	1.21351	9.99919	.5
.6	8.79789	8.79875	1.20125	9.99914	.4
.7	8.80978	8.81068	1.18932	9.99909	.3
.8	8.82134	8.82230	1.17770	9.99904	.2
.9	8.83261	8.83361	1.16639	9.99899	86.1
4.0	8.84358	8.84464	1.15536	9.99894	**86.0**
.1	8.85429	8.85540	1.14460	9.99889	85.9
.2	8.86474	8.86591	1.13409	9.99883	.8
.3	8.87494	8.87616	1.12384	9.99878	.7
.4	8.88490	8.88618	1.11382	9.99872	.6
.5	8.89464	8.89598	1.10402	9.99866	.5
.6	8.90417	8.90557	1.09443	9.99860	.4
.7	8.91349	8.91495	1.08505	9.99854	.3
.8	8.92261	8.92414	1.07586	9.99847	.2
.9	8.93154	8.93313	1.06687	9.99841	85.1
5.0	8.94030	8.94195	1.05805	9.99834	**85.0**
.1	8.94887	8.95060	1.04940	9.99828	84.9
.2	8.95728	8.95908	1.04092	9.99821	.8
.3	8.96553	8.96739	1.03261	9.99814	.7
.4	8.97363	8.97556	1.02444	9.99807	.6
.5	8.98157	8.98358	1.01642	9.99800	.5
.6	8.98937	8.99145	1.00855	9.99792	.4
.7	8.99704	8.99919	1.00081	9.99785	.3
.8	9.00456	9.00679	0.99321	9.99777	.2
.9	9.01196	9.01427	0.98573	9.99769	84.1
6.0	9.01923	9.02162	0.97838	9.99761	**84.0**
	L. Cos	* L. Cot	* L. Tan	L. Sin	Deg.

Deg.	L. Sin	L. Tan	L. Cot	L. Cos	
6.0	9.01923	9.02162	0.97838	9.99761	**84.0**
.1	9.02639	9.02885	0.97115	9.99753	83.9
.2	9.03342	9.03597	0.96403	9.99745	.8
.3	9.04034	9.04297	0.95703	9.99737	.7
.4	9.04715	9.04987	0.95013	9.99728	.6
.5	9.05386	9.05666	0.94334	9.99720	.5
.6	9.06046	9.06335	0.93665	9.99711	.4
.7	9.06696	9.06994	0.93006	9.99702	.3
.8	9.07337	9.07643	0.92357	9.99693	.2
.9	9.07968	9.08283	0.91717	9.99684	83.1
7.0	9.08589	9.08914	0.91086	9.99675	**83.0**
.1	9.09202	9.09537	0.90463	9.99666	82.9
.2	9.09807	9.10150	0.89850	9.99656	.8
.3	9.10402	9.10756	0.89244	9.99647	.7
.4	9.10990	9.11353	0.88647	9.99637	.6
.5	9.11570	9.11943	0.88057	9.99627	.5
.6	9.12142	9.12525	0.87475	9.99617	.4
.7	9.12706	9.13099	0.86901	9.99607	.3
8	9.13263	9.13667	0.86333	9.99596	.2
.9	9.13813	9.14227	0.85773	9.99586	82.1
8.0	9.14356	9.14780	0.85220	9.99575	**82.0**
.1	9.14891	9.15327	0.84673	9.99565	81.9
2	9.15421	9.15867	0.84133	9.99554	.8
.3	9.15944	9.16401	0.83599	9.99543	.7
.4	9.16460	9.16928	0.83072	9.99532	.6
.5	9.16970	9.17450	0.82550	9.99520	.5
.6	9.17474	9.17965	0.82035	9.99509	.4
.7	9.17973	9.18475	0.81525	9.99497	.3
.8	9.18465	9.18979	0.81021	9.99486	.2
.9	9.18952	9.19478	0.80522	9.99474	81.1
9.0	9.19433	9.19971	0.80029	9.99462	**81.0**
.1	9.19909	9.20459	0.79541	9.99450	80.9
.2	9.20380	9.20942	0.79058	9.99438	.8
.3	9.20845	9.21420	0.78580	9.99425	.7
.4	9.21306	9.21893	0.78107	9.99413	.6
.5	9.21761	9.22361	0.77639	9.99400	.5
6	9.22211	9.22824	0.77176	9.99388	.4
.7	9.22657	9.23283	0.76717	9.99375	.3
.8	9.23098	9.23737	0.76263	9.99362	.2
.9	9.23535	9.24186	0.75814	9.99348	80.1
10.0	9.23967	9.24632	0.75368	9.99335	**80.0**
.1	9.24395	9.25073	0.74927	9.99322	79.9
2	9.24818	9.25510	0.74490	9.99308	.8
3	9.25237	9.25943	0.74057	9.99294	.7
.4	9.25652	9.26372	0.73628	9.99281	6
.5	9.26063	9.26797	0.73203	9.99267	.5
.6	9.26470	9.27218	0.72782	9.99252	.4
.7	9.26873	9.27635	0.72365	9.99238	.3
.8	9.27273	9.28049	0.71951	9.99224	.2
.9	9.27668	9.28459	0.71541	9.99209	79.1
11.0	9.28060	9.28865	0.71135	9.99195	**79.0**
.1	9.28448	9.29268	0.70732	9.99180	78.9
.2	9.28833	9.29668	0.70332	9.99165	.8
.3	9.29214	9.30064	0.69936	9.99150	.7
.4	9.29591	9.30457	0.69543	9.99135	.6
.5	9.29966	9.30846	0.69154	9.99119	.5
.6	9.30336	9.31233	0.68767	9.99104	.4
.7	9.30704	9.31616	0.68384	9.99088	.3
.8	9.31068	9.31996	0.68004	9.99072	.2
.9	9.31430	9.32373	0.67627	9.99056	78.1
12.0	9.31788	9.32747	0 67253	9.99040	**78.0**
	L. Cos	L. Cot	L. Tan	L. Sin	Deg.

*Linear interpolation in this section of the table is inaccurate.

COMMON LOGARITHMS OF THE TRIGONOMETRIC FUNCTIONS FOR ANGLES IN DEGREES AND DECIMALS (Continued)

Deg.	L. Sin	L. Tan	L. Cot	L. Cos	
12.0	9.31788	9.32747	0.67253	9.99040	**78.0**
.1	9.32143	9.33119	0.66881	9.99024	77.9
.2	9.32495	9.33487	0.66513	9.99008	.8
.3	9.32844	9.33853	0.66147	9.98991	.7
.4	9.33190	9.34215	0.65785	9.98975	.6
.5	9.33534	9.34576	0.65424	9.98958	.5
.6	9.33874	9.34933	0.65067	9.98941	.4
.7	9.34212	9.35288	0.64712	9.98924	.3
.8	9.34547	9.35640	0.64360	9.98907	.2
.9	9.34879	9.35989	0.64011	9.98890	77.1
13.0	9.35209	9.36336	0.63664	9.98872	**77.0**
.1	9.35536	9.36681	0.63319	9.98855	76.9
.2	9.35860	9.37023	0.62977	9.98837	.8
.3	9.36182	9.37363	0.62637	9.98819	.7
.4	9.36502	9.37700	0.62300	9.98801	.6
.5	9.36819	9.38035	0.61965	6.98783	.5
.6	9.37133	9.38368	0.61632	9.98765	.4
.7	9.37445	9.38699	0.61301	9.98746	.3
.8	9.37755	9.39027	0.60973	9.98728	.2
.9	9.38062	9.39353	0.60647	9.98709	76.1
14.0	9.38368	9.39677	0.60323	9.98690	**76.0**
.1	9.38670	9.39999	0.60001	9.98671	75.9
.2	9.38971	9.40319	0.59681	9.98652	.8
.3	9.39270	9.40636	0.59364	9.98633	.7
.4	9.39566	9.40952	0.59048	9.98614	.6
.5	9.39860	9.41266	0.58734	9.98594	.5
.6	9.40152	9.41578	0.58422	9.98574	.4
.7	9.40442	9.41887	0.58113	9.98555	.3
.8	9.40730	9.42195	0.57805	9.98535	.2
.9	9.41016	9.42501	0.57499	9.98515	75.1
15.0	9.41300	9.42805	0.57195	9.98494	**75.0**
.1	9.41582	9.43108	0.56892	9.98474	74.9
.2	9.41861	9.43408	0.56592	9.98453	.8
.3	9.42140	9.43707	0.56293	9.98433	.7
.4	9.42416	9.44004	0.55996	9.98412	.6
.5	9.42690	9.44299	0.55701	9.98391	.5
.6	9.42962	9.44592	0.55408	9.98370	.4
.7	9.43233	9.44884	0.55116	9.98349	.3
.8	9.43502	9.45174	0.54826	9.98327	.2
.9	9.43769	9.45463	0.54537	9.98306	74.1
16.0	9.44034	9.45750	0.54250	9.98284	**74.0**
.1	9.44297	9.46035	0.53965	9.98262	73.9
.2	9.44559	9.46319	0.53681	9.98240	.8
.3	9.44819	9.46601	0.53399	9.98218	.7
.4	9.45077	9.46881	0.53119	9.98196	.6
.5	9.45334	9.47160	0.52840	9.98174	.5
.6	9.45589	9.47438	0.52562	9.98151	.4
.7	9.45843	9.47714	0.52286	9.98129	.3
.8	9.46095	9.47989	0.52011	9.98106	.2
.9	9.46345	9.48262	0.51738	9.98083	73.1
17.0	9.46594	9.48534	0.51466	9.98060	**73.0**
.1	9.46841	9.48804	0.51196	9.98036	72.9
.2	9.47086	9.49073	0.50927	9.98013	.8
.3	9.47330	9.49341	0.50659	9.97989	.7
.4	9.47573	9.49607	0.50393	9.97966	.6
.5	9.47814	9.49872	0.50128	9.97942	.5
.6	9.48054	9.50136	0.49864	9.97918	.4
.7	9.48292	9.50398	0.49602	9.97894	.3
.8	9.48529	9.50659	0.49341	9.97870	.2
.9	9.48764	9.50919	0.49081	9.97845	72.1
18.0	9.48998	9.51178	0.48822	9.97821	**72.0**
	L. Cos	L. Cot	L. Tan	L. Sin	Deg.

Deg.	L. Sin	L. Tan	L. Cot	L. Cos	
18.0	9.48998	9.51178	0.48822	9.97821	**72.0**
.1	9.49231	9.51435	0.48565	9.97796	71.9
.2	9.49462	9.51691	0.48309	9.97771	.8
.3	9.49692	9.51946	0.48054	9.97746	.7
.4	9.49920	9.52200	0.47800	9.97721	.6
.5	9.50148	9.52452	0.47548	9.97696	.5
.6	9.50374	9.52703	0.47297	9.97670	.4
.7	9.50598	9.52953	0.47047	9.97645	.3
.8	9.50821	9.53202	0.46798	9.97619	.2
.9	9.51043	9.53450	0.46550	9.97593	71.1
19.0	9 51264	9.53697	0.46303	9.97567	**71.0**
.1	9.51484	9.53943	0.46057	9.97541	70.9
.2	9.51702	9.54187	0.45813	9.97515	.8
.3	9.51919	9.54431	0.45569	9.97488	.7
.4	9 52135	9.54673	0.45327	9.97461	.6
.5	9.52350	9.54915	0.45085	9.97435	.5
.6	9.52563	9.55155	0.44845	9.97408	.4
.7	9.52775	9.55395	0.44605	9.97381	.3
.8	9.52986	9.55633	0.44367	9.97353	.2
.9	9.53196	9.55870	0.44130	9.97326	70.1
20.0	9.53405	9.56107	0.43893	9.97299	**70.0**
.1	9.53613	9.56342	0.43658	9.97271	69.9
.2	9.53819	9.56576	0.43424	9.97243	.8
.3	9.54025	9.56810	0.43190	9.97215	.7
.4	9.54229	9.57042	0.42958	9.97187	.6
.5	9.54433	9.57274	0.42726	9.97159	.5
.6	9.54635	9.57504	0.42496	9.97130	.4
.7	9.54836	9.57734	0.42266	9.97102	.3
.8	9.55036	9.57963	0.42037	9.97073	.2
.9	9.55235	9.58191	0.41809	9.97044	69.1
21.0	9.55433	9.58418	0.41582	9.97015	**69.0**
.1	9.55630	9.58644	0.41356	9.96986	68.9
.2	9.55826	9.58869	0.41131	9.96957	.8
.3	9.56021	9.59094	0.40906	9.96927	.7
.4	9.56215	9.59317	0.40683	9.96898	.6
.5	9.56408	9.59540	0.40460	9.96868	.5
.6	9.56599	9.59762	0.40238	9.96838	.4
.7	9.56790	9.59983	0.40017	9.96808	.3
.8	9.56980	9.60203	0.39797	9.96778	.2
.9	9.57169	9.60422	0.39578	9.96747	68.1
22.0	9.57358	9.60641	0.39359	9.96717	**68.0**
.1	9.57545	9.60859	0.39141	9.96686	67.9
.2	9.57731	9.61076	0.38924	9.96655	.8
.3	9.57916	9.61292	0.38708	9.96624	.7
.4	9.58101	9.61508	0.38492	9.96593	.6
.5	9.58284	9.61722	0.38278	9.96562	.5
.6	9.58467	9.61936	0.38064	9.96530	.4
.7	9.58648	9.62150	0.37850	9.96498	.3
.8	9.58829	9.62362	0.37638	9.96467	.2
.9	9.59009	9.62574	0.37426	9.96435	67.1
23.0	9.59188	9.62785	0.37215	9.96403	**67.0**
.1	9.59366	9.62996	0.37004	9.96370	66.9
.2	9.59543	9.63205	0.36795	9.96338	.8
.3	9.59720	9.63414	0.36586	9.96305	.7
.4	9.59895	9.63623	0.36377	9.96273	.6
.5	9.60070	9.63830	0.36170	9.96240	.5
.6	9.60244	9.64037	0.35963	9.96207	.4
.7	9.60417	9.64243	0.35757	9.96174	.3
.8	9.60589	9.64449	0.35551	9.96140	.2
.9	9.60761	9.64654	0.35346	9.96107	66.1
24.0	9.60931	9.64858	0.35142	9.96073	**66.0**
	L. Cos	L. Cot	L. Tan	L. Sin	Deg.

COMMON LOGARITHMS OF THE TRIGONOMETRIC FUNCTIONS FOR ANGLES IN DEGREES AND DECIMALS (Continued)

Deg.	L. Sin	L. Tan	L. Cot	L. Cos	
24.0	9.60931	9.64858	0.35142	9.96073	**66.0**
.1	9.61101	9.65062	0.34938	9.96039	65.9
.2	9.61270	9.65265	0.34735	9.96005	.8
.3	9.61438	9.65467	0.34533	9.95971	.7
.4	9.61606	9.65669	0.34331	9.95937	.6
.5	9.61773	9.65870	0.34130	9.95902	.5
.6	9.61939	9.66071	0.33929	9.95868	.4
.7	9.62104	9.66271	0.33729	9.95833	.3
.8	9.62268	9.66470	0.33530	9.95798	.2
.9	9.62432	9.66669	0.33331	9.95763	65.1
25.0	9.62595	9.66867	0.33133	9.95728	**65.0**
.1	9.62757	9.67065	0.32935	9.95692	64.9
.2	9.62918	9.67262	0.32738	9.95657	.8
.3	9.63079	9.67458	0.32542	9.95621	.7
.4	9.63239	9.67654	0.32346	9.95585	.6
.5	9.63398	9.67850	0.32150	9.95549	.5
.6	9.63557	9.68044	0.31956	9.95513	.4
.7	9.63715	9.68239	0.31761	9.95476	.3
.8	9.63872	9.68432	0.31568	9.95440	.2
.9	9.64028	9.68626	0.31374	9.95403	64.1
26.0	9.64184	9.68818	0.31182	9.95366	**64.0**
.1	9.64339	9.69010	0.30990	9.95329	63.9
.2	9.64494	9.69202	0.30798	9.95292	.8
.3	9.64647	9.69393	0.30607	9.95254	.7
.4	9.64800	9.69584	0.30416	9.95217	.6
.5	9.64953	9.69774	0.30226	9.95179	.5
.6	9.65104	9.69963	0.30037	9.95141	.4
.7	9.65255	9.70152	0.29848	9.95103	.3
.8	9.65406	9.70341	0.29659	9.95065	.2
.9	9.65556	0.70529	0.29471	9.95027	63.1
27.0	9.65705	9.70717	0.29283	9.94988	**63.0**
.1	9.65853	9.70904	0.29096	9.94949	62.9
.2	9.66001	9.71090	0.28910	9.94911	.8
.3	9.66148	9.71277	0.28723	9.94871	.7
.4	9.66295	9.71462	0.28538	9.94832	.6
.5	9.66441	9.71648	0.28352	9.94793	.5
.6	9.66586	9.71833	0.28167	9.94753	.4
.7	9.66731	9.72017	0.27983	9.94714	.3
.8	9.66875	9.72201	0.27799	9.94674	.2
.9	9.67018	9.72384	0.27616	9.94634	62.1
28.0	9.67161	9.72567	0.27433	9.94593	**62.0**
.1	9.67303	9.72750	0.27250	9.94553	61.9
.2	9.67445	9.72932	0.27068	9.94513	.8
.3	9.67586	9.73114	0.26886	9.94472	.7
.4	9.67726	9.73295	0.26705	9.94431	.6
.5	9.67866	9.73476	0.26524	9.94390	.5
.6	9.68006	9.73657	0.26343	9.94349	.4
.7	9.68144	9.73837	0.26163	9.94307	.3
.8	9.68283	9.74017	0.25983	9.94266	.2
.9	9.68420	9.74196	0.25804	9.94224	61.1
29.0	9.68557	9.74375	0.25625	9.94182	**61.0**
.1	9.68694	9.74554	0.25446	9.94140	60.9
.2	9.68829	9.74732	0.25268	9.94098	.8
.3	9.68965	9.74910	0.25090	9.94055	.7
.4	9.69100	9.75087	0.24913	9.94012	.6
.5	9.69234	9.75264	0.24736	9.93970	.5
.6	9.69368	9.75441	0.24559	9.93927	.4
.7	9.69501	9.75617	0.24383	9.93884	.3
.8	9.69633	9.75793	0.24207	9.93840	.2
.9	9.69765	9.75969	0.24031	9.93797	60.1
30.0	9.69897	9.76144	0.23856	9.93753	**60.0**
	L. Cos	L. Cot	L. Tan	L. Sin	Deg.

Deg.	L. Sin	L. Tan	L. Cot	L. Cos	
30.0	9.69897	9.76144	0.23856	9.93753	**60.0**
.1	9.70028	9.76319	0.23681	9.93709	59.9
.2	9.70159	9.76493	0.23507	9.93665	.8
.3	9.70288	9.76668	0.23332	9.93621	.7
.4	9.70418	9.76841	0.23159	9.93577	.6
.5	9.70547	9.77015	0.22985	9.93532	.5
.6	9.70675	9.77188	0.22812	9.93487	.4
.7	9.70803	9.77361	0.22639	9.93442	.3
.8	9.70931	9.77533	0.22467	9.93397	.2
.9	9.71058	9.77706	0.22294	9.93352	59.1
31.0	9.71184	9.77877	0.22123	9.93307	**59.0**
.1	9.71310	9.78049	0.21951	9.93261	58.9
.2	9.71435	9.78220	0.21780	9.93215	.8
.3	9.71560	9.78391	0.21609	9.93169	.7
.4	9.71685	9.78562	0.21438	9.93123	.6
.5	9.71809	9.78732	0.21268	9.93077	.5
.6	9.71932	9.78902	0.21098	9.93030	.4
.7	9.72055	9.79072	0.20928	9.92983	.3
.8	9.72177	9.79241	0.20759	9.92936	.2
.9	9.72299	9.79410	0.20590	9.92889	58.1
32.0	9.72421	9.79579	0.20421	9.92842	**58.0**
.1	9.72542	9.79747	0.20253	9.92795	57.9
.2	9.72663	9.79916	0.20084	9.92747	.8
.3	9.72783	9.80084	0.19916	9.92699	.7
.4	9.72902	9.80251	0.19749	9.92651	.6
.5	9.73022	9.80419	0.19581	9.92603	.5
.6	9.73140	9.80586	0.19414	9.92555	.4
.7	9.73259	9.80753	0.19247	9.92506	.3
.8	9.73377	9.80919	0.19081	9.92457	.2
.9	9.73494	9.81086	0.18914	9.92408	57.1
33.0	9.73611	9.81252	0.18748	9.92359	**57.0**
.1	9.73727	9.81418	0.18582	9.92310	56.9
.2	9.73843	9.81583	0.18417	9.92260	.8
.3	9.73959	9.81748	0.18252	9.92211	.7
.4	9.74074	9.81913	0.18087	9.92161	.6
.5	9.74189	9.82078	0.17922	9.92111	.5
.6	9.74303	9.82243	0.17757	9.92060	.4
.7	9.74417	9.82407	0.17593	9.92010	.3
.8	9.74531	9.82571	0.17429	9.91959	.2
.9	9.74644	9.82735	0.17265	9.91908	56.1
34.0	9.74756	9.82899	0.17101	9.91857	**56.0**
.1	9.74868	9.83062	0.16938	9.91806	55.9
.2	9.74980	9.83225	0.16775	9.91755	.8
.3	9.75091	9.83388	0.16612	9.91703	.7
.4	9.75202	9.83551	0.16449	9.91651	.6
.5	9.75313	9.83713	0.16287	9.91599	.5
.6	9.75423	9.83876	0.16124	9.91547	.4
.7	9.75533	9.84038	0.15962	9.91495	.3
.8	9.75642	9.84200	0.15800	9.91442	.2
.9	9.75751	9.84361	0.15639	9.91389	55.1
35.0	9.75859	9.84523	0.15477	9.91336	**55.0**
.1	9.75967	9.84684	0.15316	9.91283	54.9
.2	9.76075	9.84845	0.15155	9.91230	.8
.3	9.76182	9.85006	0.14994	9.91176	.7
.4	9.76289	9.85166	0.14834	9.91123	.6
.5	9.76395	9.85327	0.14673	9.91069	.5
.6	9.76501	9.85487	0.14513	9.91014	.4
.7	9.76607	9.85647	0.14353	9.90960	.3
.8	9.76712	9.85807	0.14193	9.90906	.2
.9	9.76817	9.85967	0.14033	9.90851	54.1
36.0	9.76922	9.86126	0.13874	9.90796	**54.0**
	L. Cos	L. Cot	L. Tan	L. Sin	Deg.

COMMON LOGARITHMS OF THE TRIGONOMETRIC FUNCTIONS FOR ANGLES IN DEGREES AND DECIMALS (Continued)

Deg.	L. Sin	L. Tan	L. Cot	L. Cos	
36.0	9 76922	9.86126	0.13874	9.90796	**54.0**
.1	9.77026	9.86285	0.13715	9.90741	53.9
.2	9.77130	9.86445	0.13555	9.90685	.8
.3	9.77233	9.86603	0.13397	9.90630	.7
.4	9.77336	9.86762	0.13238	9.90574	.6
.5	9.77439	9.86921	0.13079	9.90518	.5
.6	9.77541	9.87079	0.12921	9.90462	.4
.7	9.77643	9.87238	0.12762	9.90405	.3
.8	9.77744	9.87396	0.12604	9.90349	.2
.9	9.77846	9.87554	0.12446	9.90292	53.1
37.0	9.77946	9 87711	0.12289	9.90235	**53.0**
.1	9.78047	9.87869	0.12131	9.90178	52.9
.2	9 78147	9 88027	0.11973	9 90120	.8
.3	9.78246	9 88184	0.11816	9.90063	.7
.4	9.78346	9.88341	0.11659	9.90005	.6
.5	9.78445	9.88498	0.11502	9.89947	.5
.6	9 78543	9.88655	0.11345	9.89888	.4
.7	9 78642	9.88812	0.11188	9 89830	.3
.8	9.78739	9.88968	0.11032	9.89771	.2
.9	9.78837	9.89125	0.10875	9.89712	52.1
38.0	9 78934	9 89281	0.10719	9.89653	**52.0**
.1	9.79031	9.89437	0.10563	9.89594	51.9
.2	9.79128	9 89593	0.10407	9.89534	.8
.3	9.79224	9.89749	0.10251	9.89475	.7
.4	9.79319	9 89905	0.10095	9.89415	.6
.5	9.79415	9.90061	0.09939	9.89354	.5
.6	9.79510	9.90216	0.09784	9.89294	.4
.7	9.79605	9.90371	0.09629	9.89233	.3
.8	9 79699	9.90527	0.09473	9.89173	.2
.9	9.79793	9 90682	0.09318	9.89112	51.1
39.0	9 79887	9.90837	0.09163	9.89050	**51.0**
.1	9 79981	9.90992	0.09008	9.88989	50.9
.2	9.80074	9 91147	0 08853	9.88927	.8
.3	9.80166	9.91301	0.08699	9.88865	.7
.4	9.80259	9.91456	0 08544	9.88803	.6
.5	9.80351	9.91610	0.08390	9.88741	.5
.6	9 80443	9.91765	0 08235	9.88678	.4
.7	9.80534	9 91919	0 08081	9 88615	.3
.8	9 80625	9.92073	0.07927	9.88552	.2
.9	9.80716	9.92227	0.07773	9.88489	50.1
40.0	9 80807	9.92381	0.07619	9 88425	**50.0**
.1	9.80897	9.92535	0.07465	9.88362	49.9
.2	9.80987	9.92689	0.07311	9.88298	.8
.3	9 81076	9.92843	0.07157	9.88234	.7
.4	9.81166	9.92996	0.07004	9.88169	.6
.5	9.81254	9.93150	0.06850	9.88105	.5
.6	9 81343	9.93303	0.06697	9.88040	.4
.7	9.81431	9.93457	0.06543	9 87975	.3
.8	9.81519	9.93610	0.06390	9.87909	.2
.9	9.81607	9.93763	0.06237	9.87844	49.1
41.0	9.81694	9.93916	0.06084	9.87778	**49.0**
	L. Cos	L. Cot	L. Tan	L. Sin	Deg.

Deg.	D Sin	L. Tan	L. Cot	L. Cos	
41.0	9.81694	9.93916	0.06084	9.87778	**49.0**
.1	9.81781	9.94069	0.05931	9.87712	48.9
.2	9 81868	9.94222	0.05778	9.87646	.8
.3	9.81955	9.94375	0.05625	9.87579	.7
.4	9.82041	9.94528	0.05472	9.87513	.6
.5	9 82126	9.94681	0.05319	9.87446	.5
.6	9.82212	9.94834	0 05166	9.87378	.4
.7	9.82297	9.94986	0.05014	9.87311	.3
.8	9.82382	9.95139	0.04861	9.87243	.2
.9	9.82467	9.95291	0.04709	9.87175	48.1
42.0	9 82551	9 95444	0.04556	9.87107	**48.0**
.1	9.82635	9.95596	0.04404	9.87039	47.9
.2	9 82719	9.95748	0.04252	9 86970	.8
.3	9.82802	9 95901	0.04099	9 86902	.7
.4	9.82885	9.96053	0.03947	9 86832	.6
.5	9.82968	9.96205	0 03795	9.86763	.5
.6	9 83051	9.96357	0 03643	9 86694	.4
.7	9.83133	9.96510	0 03490	9.86624	.3
.8	9 83215	9.96662	0.03338	9.86554	.2
.9	9.83297	9.96814	0.03186	9.86483	47.1
43.0	9.83378	9.96966	0.03034	9.86413	**47.0**
.1	9 83459	9.97118	0.02882	9.86342	46.9
.2	9.83540	9.97269	0.02731	9.86271	.8
.3	9 83621	9 97421	0.02579	9.86200	7
.4	9.83701	9 97573	0.02427	9 86128	.6
.5	9.83781	9.97725	0.02275	9.86056	.5
.6	9.83861	9 97877	0 02123	9.85984	.4
.7	9.83940	9.98029	0.01971	9.85912	.3
.8	9.84020	9 98180	0.01820	9.85839	.2
.9	9.84098	9.98332	0.01668	9.85766	46.1
44.0	9.84177	9 98484	0.01516	9 85693	**46.0**
.1	9.84255	9.98635	0 01365	9.85620	45.9
.2	9 84334	9.98787	0 01213	9 85547	.8
.3	9 84411	9 98939	0 01061	9.85473	.7
.4	9.84489	9.99090	0 00910	9.85399	.6
.5	9.84566	9.99242	0.00758	9.85324	.5
.6	9 84643	9.99394	0.00606	9.85250	.4
.7	9.84720	9.99545	0.00455	9 85175	.3
.8	9.84796	9.99697	0 00303	9.85100	.2
.9	9.84873	9.99848	0.00152	9.85024	45.1
45.0	9.84949	0.00000	0.00000	9.84949	**45.0**
	L. Cos	L. Cot	L. Tan	L. Sin	Deg.

COMMON LOGARITHMS OF THE TRIGONOMETRIC FUNCTIONS FOR ANGLES IN RADIANS

The table that follows gives the common logarithms of the functions sine, cosine, tangent and cotangent for each hundredth of a radian from 0 to 2.

The quantity – 10 is to be appended to all logarithms of the sine and cosine, to logarithms of the tangent from 0 to .78, and of the cotangent from .79 to 2.

Rad.	L. Sin	L. Tan	L. Cot	L. Cos
.00	– ∞	– ∞	∞	0.00000
.01	7.99999	8.00001	1.99999	9.99998
.02	8.30100	8.30109	1.69891	9.99991
.03	8.47706	8.47725	1.52275	9.99980
.04	8.60194	8.60229	1.39771	9.99965
.05	8.69879	8.69933	1.30067	9.99946
.06	8.77789	8.77867	1.22133	9.99922
.07	8.84474	8.84581	1.15419	9.99894
.08	8.90263	8.90402	1.09598	9.99861
.09	8.95366	8.95542	1.04458	9.99824
.10	8.99928	9.00145	0.99855	9.99782
.11	9.04052	9.04315	0.95685	9.99737
.12	9.07814	9.08127	0.91873	9.99687
.13	9.11272	9.11640	0.88360	9.99632
.14	9.14471	9.14898	0.85102	9.99573
.15	9.17446	9.17937	0.82063	9.99510
.16	9.20227	9.20785	0.79215	9.99442
.17	9.22836	9.23466	0.76534	9.99369
.18	9.25292	9.26000	0.74000	9.99293
.19	9.27614	9.28402	0.71598	9.99211
.20	9.29813	9.30688	0.69312	9.99126
.21	9.31902	9.32867	0.67133	9.99035
.22	9.33891	9.34951	0.65049	9.98940
.23	9.35789	9.36948	0.63052	9.98841
.24	9.37603	9.38866	0.61134	9.98737
.25	9.39341	9.40712	0.59288	9.98628
.26	9.41007	9.42492	0.57508	9.98515
.27	9.42607	9.44210	0.55790	9.98397
.28	9.44147	9.45872	0.54128	9.98275
.29	9.45629	9.47482	0.52518	9.98148
.30	9.47059	9.49043	0.50957	9.98016
.31	9.48438	9.50559	0.49441	9.97879
.32	9.49771	9.52034	0.47966	9.97737
.33	9.51060	9.53469	0.46531	9.97591
.34	9.52308	9.54868	0.45132	9.97440
.35	9.53516	9.56233	0.43767	9.97284
.36	9.54688	9.57565	0.42435	9.97123
.37	9.55825	9.58868	0.41132	9.96957
.38	9.56928	9.60142	0.39858	9.96786
.39	9.58000	9.61390	0.38610	9.96610
.40	9.59042	9.62613	0.37387	9.96429
.41	9.60055	9.63812	0.36188	9.96243
.42	9.61041	9.64989	0.35011	9.96051
.43	9.62000	9.66145	0.33855	9.95855
.44	9.62935	9.67282	0.32718	9.95653
.45	9.63845	9.68400	0.31600	9.95446
.46	9.64733	9.69500	0.30500	9.95233
.47	9.65599	9.70583	0.29417	9.95015
.48	9.66443	9.71651	0.28349	9.94792
.49	9.67268	9.72704	0.27296	9.94563
.50	9.68072	9.73743	0.26257	9.94329
Rad.	L. Sin	L. Tan	L. Cot	L. Cos

Rad.	L. Sin	L. Tan	L. Cot	L. Cos
.50	9.68072	9.73743	0.26257	9.94329
.51	9.68858	9.74769	0.25231	9.94089
.52	9.69625	9.75782	0.24218	9.93843
.53	9.70375	9.76784	0.23216	9.93591
.54	9.71108	9.77774	0.22226	9.93334
.55	9.71824	9.78754	0.21246	9.93071
.56	9.72525	9.79723	0.20277	9.92801
.57	9.73210	9.80684	0.19316	9.92526
.58	9.73880	9.81635	0.18365	9.92245
.59	9.74536	9.82579	0.17421	9.91957
.60	9.75177	9.83514	0.16486	9.91663
.61	9.75805	9.84443	0.15557	9.91363
.62	9.76420	9.85364	0.14636	9.91056
.63	9.77022	9.86280	0.13720	9.90743
.64	9.77612	9.87189	0.12811	9.90423
.65	9.78189	9.88093	0.11907	9.90096
.66	9.78754	9.88992	0.11008	9.89762
.67	9.79308	9.89886	0.10114	9.89422
.68	9.79851	9.90777	0.09223	9.89074
.69	9.80382	9.91663	0.08337	9.88719
.70	9.80903	9.92546	0.07454	9.88357
.71	9.81414	9.93426	0.06574	9.87988
.72	9.81914	9.94303	0.05697	9.87611
.73	9.82404	9.95178	0.04822	9.87226
.74	9.82885	9.96051	0.03949	9.86833
.75	9.83355	9.96923	0.03077	9.86433
.76	9.83817	9.97793	0.02207	9.86024
.77	9.84269	9.98662	0.01338	9.85607
.78	9.84713	9.99531	0.00469	9.85182
.79	9.85147	0.00400	9.99600	9.84748
.80	9.85573	0.01268	9.98732	9.84305
.81	9.85991	0.02138	9.97862	9.83853
.82	9.86400	0.03008	9.96992	9.83393
.83	9.86802	0.03879	9.96121	9.82922
.84	9.87195	0.04752	9.95248	9.82443
.85	9.87580	0.05627	9.94373	9.81953
.86	9.87958	0.06504	9.93496	9.81454
.87	9.88328	0.07384	9.92616	9.80944
.88	9.88691	0.08266	9.91734	9.80424
.89	9.89046	0.09153	9.90847	9.79894
.90	9.89394	0.10043	9.89957	9.79352
.91	9.89735	0.10937	9.89063	9.78799
.92	9.90070	0.11835	9.88165	9.78234
.93	9.90397	0.12739	9.87261	9.77658
.94	9.90717	0.13648	9.86352	9.77070
.95	9.91031	0.14563	9.85437	9.76469
.96	9.91339	0.15484	9.84516	9.75855
.97	9.91639	0.16412	9.83588	9.75228
.98	9.91934	0.17347	9.82653	9.74587
.99	9.92222	0.18289	9.81711	9.73933
1.00	9.92504	0.19240	9.80760	9.73264
Rad.	L. Sin	L. Tan	L. Cot	L. Cos

COMMON LOGARITHMS OF THE TRIGONOMETRIC FUNCTIONS FOR ANGLES IN RADIANS (Continued)

Rad.	L. Sin	L. Tan	L. Cot	L. Cos	Rad.	L. Sin	L. Tan	L. Cot	L. Cos
1.00	9.92504	0.19240	9.80760	9.73264	**1.50**	9.99891	1.14926	8.85074	8.84965
1.01	9.92780	0.20200	9.79800	9.72580	1.51	9.99920	1.21559	8.78441	8.78361
1.02	9.93049	0.21169	9.78831	9.71881	1.52	9.99944	1.29379	8.70621	8.70565
1.03	9.93313	0.22148	9.77852	9.71165	1.53	9.99964	1.38914	8.61086	8.61050
1.04	9.93571	0.23137	9.76863	9.70434	1.54	9.99979	1.51136	8.48864	8.48843
1.05	9.93823	0.24138	9.75862	9.69686	1.55	9.99991	1.68195	8.31805	8.31796
1.06	9.94069	0.25150	9.74850	9.68920	1.56	9.99997	1.96671	8.03329	8.03327
1.07	9.94310	0.26175	9.73825	9.68135	1.57	0.00000	3.09891	6.90109	6.90109
1.08	9.94545	0.27212	9.72788	9.67332	1.58	9.99998	2.03603*	7.96397*	7.96396*
1.09	9.94774	0.28264	9.71736	9.66510	1.59	9.99992	1.71656	8.28344	8.28336
1.10	9.94998	0.29331	9.70669	9.65667	**1.60**	9.99981	1.53444	8.46556	8.46538
1.11	9.95216	0.30413	9.69587	9.64803	1.61	9.99967	1.40645	8.59355	8.59322
1.12	9.95429	0.31512	9.68488	9.63917	1.62	9.99947	1.30765	8.69235	8.69182
1.13	9.95637	0.32628	9.67372	9.63008	1.63	9.99924	1.22714	8.77286	8.77209
1.14	9.95839	0.33763	9.66237	9.62075	1.64	9.99896	1.15918	8.84082	8.83978
1.15	9.96036	0.34918	9.65082	9.61118	1.65	9.99864	1.10035	8.89965	8.89829
1.16	9.96228	0.36093	9.63907	9.60134	1.66	9.99827	1.04846	8.95154	8.94981
1.17	9.96414	0.37291	9.62709	9.59123	1.67	9.99786	1.00204	8.99796	8.99582
1.18	9.96596	0.38512	9.61488	9.58084	1.68	9.99741	0.96003	9.03997	9.03737
1.19	9.96772	0.39757	9.60243	9.57015	1.69	9.99691	0.92165	9.07835	9.07526
1.20	9.96943	0.41030	9.58970	9.55914	**1.70**	9.99636	0.88630	9.11370	9.11007
1.21	9.97110	0.42330	9.57670	9.54780	1.71	9.99578	0.85353	9.14647	9.14225
1.22	9.97271	0.43660	9.56340	9.53611	1.72	9.99515	0.82298	9.17702	9.17217
1.23	9.97428	0.45022	9.54978	9.52406	1.73	9.99447	0.79436	9.20564	9.20012
1.24	9.97579	0.46418	9.53582	9.51161	1.74	9.99375	0.76742	9.23258	9.22634
1.25	9.97726	0.47850	9.52150	9.49875	1.75	9.99299	0.74197	9.25803	9.25102
1.26	9.97868	0.49322	9.50678	9.48546	1.76	9.99218	0.71784	9.28216	9.27434
1.27	9.98005	0.50835	9.49165	9.47170	1.77	9.99133	0.69490	9.30510	9.29642
1.28	9.98137	0.52392	9.47608	9.45745	1.78	9.99043	0.67303	9.32697	9.31740
1.29	9.98265	0.53998	9.46002	9.44267	1.79	9.98948	0.65212	9.34788	9.33736
1.30	9.98388	0.55656	9.44344	9.42732	**1.80**	9.98849	0.63208	9.36792	9.35641
1.31	9.98506	0.57369	9.42631	9.41137	1.81	9.98745	0.61284	9.38716	9.37462
1.32	9.98620	0.59144	9.40856	9.39476	1.82	9.98637	0.59432	9.40568	9.39205
1.33	9.98729	0.60984	9.39016	9.37744	1.83	9.98524	0.57648	9.42352	9.40877
1.34	9.98833	0.62896	9.37104	9.35937	1.84	9.98407	0.55925	9.44075	9.42482
1.35	9.98933	0.64887	9.35113	9.34046	1.85	9.98285	0.54258	9.45742	9.44026
1.36	9.99028	0.66964	9.33036	9.32064	1.86	9.98158	0.52645	9.47355	9.45513
1.37	9.99119	0.69135	9.30865	9.29983	1.87	9.98026	0.51080	9.48920	9.46947
1.38	9.99205	0.71411	9.28589	9.27793	1.88	9.97890	0.49560	9.50440	9.48330
1.39	9.99286	0.73804	9.26196	9.25482	1.89	9.97749	0.48082	9.51918	9.49667
1.40	9.99363	0.76327	9.23673	9.23036	**1.90**	9.97603	0.46644	9.53356	9.50959
1.41	9.99436	0.78996	9.21004	9.20440	1.91	9.97452	0.45242	9.54758	9.52210
1.42	9.99504	0.81830	9.18170	9.17674	1.92	9.97296	0.43875	9.56125	9.53422
1.43	9.99568	0.84853	9.15147	9.14716	1.93	9.97136	0.42540	9.57460	9.54596
1.44	9.99627	0.88092	9.11908	9.11536	1.94	9.96970	0.41235	9.58765	9.55735
1.45	9.99682	0.91583	9.08417	9.08100	1.95	9.96800	0.39958	9.60042	9.56841
1.46	9.99733	0.95369	9.04631	9.04364	1.96	9.96624	0.38708	9.61292	9.57916
1.47	9.99779	0.99508	9.00492	9.00271	1.97	9.96443	0.37484	9.62516	9.58960
1.48	9.99821	1.04074	8.95926	8.95747	1.98	9.96258	0.36283	9.63717	9.59975
1.49	9.99858	1.09166	8.90834	8.90692	1.99	9.96067	0.35104	9.64896	9.60963
1.50	9.99891	1.14926	8.85074	8.84965	**2.00**	9.95871	0.33946	9.66054	9.61925
Rad.	L. Sin	L. Tan	L. Cot	L. Cos	Rad.	L. Sin	L. Tan	L. Cot	L. Cos

* Values of the cosine, tangent and cotangent for angles in the table, 1.58 radians and above, are negative. As a consequence the logarithms of the functions involved are for their absolute values. For example log |cos 1.90| = 9.50959 − 10.

SINE AND COSINE FUNCTIONS FOR SPECIAL MULTIPLES OF π RADIANS

k	$\sin\frac{2\pi k}{m}$	$\cos\frac{2\pi k}{m}$	$\sin\frac{2\pi k}{m}$	$\cos\frac{2\pi k}{m}$	$\sin\frac{2\pi k}{m}$	$\cos\frac{2\pi k}{m}$
	$m = 3$		$m = 4$		$m = 5$	
1	0.86603	−0.50000	1.00000	+0.00000	0.95106	+0.30902
2			0.00000	−1.00000	0.58779	−0.80902
	$m = 6$		$m = 7$		$m = 8$	
1	0.86603	+0.50000	0.78183	+0.62349	0.70711	0.70711
2	0.86603	−0.50000	0.97493	−0.22252	1.00000	+0.00000
3	0.00000	−1.00000	0.43388	−0.90097	0.70711	+0.70711
4					0.00000	−1.00000
	$m = 9$		$m = 10$		$m = 11$	
1	0.64279	0.76604	0.58779	0.80902	0.54064	0.84125
2	0.98481	+0.17365	0.95106	+0.30902	0.90963	+0.41542
3	0.86603	−0.50000	0.95106	−0.30902	0.98982	−0.14231
4	0.34202	−0.93969	0.58779	−0.80902	0.75575	−0.65486
5			0.00000	−1.00000	0.28173	−0.95949
	$m = 12$		$m = 13$		$m = 14$	
1	0.50000	0.86603	0.46472	0.88546	0.43388	0.90097
2	0.86603	0.50000	0.82298	0.56806	0.78183	0.62349
3	1.00000	+0.00000	0.99271	+0.12054	0.97493	+0.22252
4	0.86603	−0.50000	0.93502	−0.35460	0.97493	−0.22252
5	0.50000	−0.86603	0.66312	−0.74851	0.78183	−0.62349
6	0.00000	−1.00000	0.23932	−0.97094	0.43388	−0.90097
7					0.00000	−1.00000
	$m = 15$		$m = 16$		$m = 17$	
1	0.40674	0.91355	0.38268	0.92388	0.36124	0.93247
2	0.74314	0.66913	0.70711	0.70711	0.67370	0.73901
3	0.95106	+0.30902	0.92388	0.38268	0.89516	0.44574
4	0.99452	−0.10453	1.00000	+0.00000	0.99573	+0.09227
5	0.86603	−0.50000	0.92388	−0.38268	0.96183	−0.27366
6	0.58779	−0.80902	0.70711	−0.70711	0.79802	−0.60263
7	0.20791	−0.97815	0.38268	−0.92388	0.52643	−0.85022
8			0.00000	−1.00000	0.18375	−0.98297
	$m = 18$		$m = 19$		$m = 20$	
1	0.34202	0.93969	0.32470	0.94582	0.30902	0.95106
2	0.64279	0.76604	0.61421	0.78914	0.58779	0.80902
3	0.86603	0.50000	0.83717	0.54695	0.80902	0.58779
4	0.98481	+0.17365	0.96940	+0.24549	0.95106	0.30902
5	0.98481	−0.17365	0.99658	−0.08258	1.00000	+0.00000
6	0.86603	−0.50000	0.91577	−0.40170	0.95106	−0.30902
7	0.64279	−0.76604	0.73572	−0.67728	0.80902	−0.58779
8	0.34202	−0.93969	0.47595	−0.87947	0.58779	−0.80902
9	0.00000	−1.00000	0.16459	−0.98636	0.30902	−0.95106
10					0.00000	−1.00000
	$m = 21$		$m = 22$		$m = 23$	
1	0.29476	0.95557	0.28173	0.95949	0.26980	0.96292
2	0.56332	0.82624	0.54064	0.84125	0.51958	0.85442
3	0.78183	0.62349	0.75575	0.65486	0.73084	0.68255
4	0.93087	0.36534	0.90963	0.41542	0.88789	0.46007
5	0.99720	+0.07473	0.98982	+0.14231	0.97908	+0.20346
6	0.97493	−0.22252	0.98982	−0.14231	0.99767	−0.06824
7	0.86603	−0.50000	0.90963	−0.41542	0.94226	−0.33488
8	0.68017	−0.73305	0.75575	−0.65486	0.81697	−0.57668
9	0.43388	−0.90097	0.54064	−0.84125	0.63109	−0.77571
10	0.14904	−0.98883	0.28173	−0.95949	0.39840	−0.91721
11			0.00000	−1.00000	0.13617	−0.99069
	$m = 24$		$m = 25$			
1	0.25882	0.96593	0.24869	0.96858		
2	0.50000	0.86603	0.48175	0.87631		
3	0.70711	0.70711	0.68455	0.72897		
4	0.86603	0.50000	0.84433	0.53583		
5	0.96593	0.25882	0.95106	0.30902		
6	1.00000	+0.00000	0.99803	+0.06279		
7	0.96593	−0.25882	0.98229	−0.18738		
8	0.86603	−0.50000	0.90483	−0.42578		
9	0.70711	−0.70711	0.77051	−0.63742		
10	0.50000	−0.86603	0.58779	−0.80902		
11	0.25882	−0.96593	0.36812	−0.92978		
12	0.00000	−1.00000	0.12533	−0.99211		

TEN-PLACE VALUES FOR INVERSE SINES AND TANGENTS

x	arc sin x	arc tan x	x	arc sin x	arc tan x
0.000	0.00000 00000	0.00000 00000	0.050	0.05002 08568	0.04995 83957
0.001	0.00100 00002	0.00099 99997	0.051	0.05102 21344	0.05095 58519
0.002	0.00200 00013	0.00199 99973	0.052	0.05202 34632	0.05195 32066
0.003	0.00300 00045	0.00299 99910	0.053	0.05302 48443	0.05295 04578
0.004	0.00400 00107	0.00399 99787	0.054	0.05402 62785	0.05394 76036
0.005	0.00500 00208	0.00499 99583	0.055	0.05502 77670	0.05494 46421
0.006	0.00600 00360	0.00599 99280	0.056	0.05602 93107	0.05594 15712
0.007	0.00700 00572	0.00699 98857	0.057	0.05703 09107	0.05693 83891
0.008	0.00800 00853	0.00799 98293	0.058	0.05803 25680	0.05793 50936
0.009	0.00900 01215	0.00899 97570	0.059	0.05903 42836	0.05893 16830
0.010	0.01000 01667	0.00999 96667	0.060	0.06003 60584	0.05992 81551
0.011	0.01100 02218	0.01099 95564	0.061	0.06103 78937	0.06092 45081
0.012	0.01200 02880	0.01199 94241	0.062	0.06203 97902	0.06192 07401
0.013	0.01300 03662	0.01299 92677	0.063	0.06304 17491	0.06291 68489
0.014	0.01400 04574	0.01399 90854	0.064	0.06404 37714	0.06391 28328
0.015	0.01500 05626	0.01499 88752	0.065	0.06504 58581	0.06490 86897
0.016	0.01600 06827	0.01599 86349	0.066	0.06604 80102	0.06590 44177
0.017	0.01700 08189	0.01699 83626	0.067	0.06705 02287	0.06690 00148
0.018	0.01800 09721	0.01799 80564	0.068	0.06805 25147	0.06789 54792
0.019	0.01900 11434	0.01899 77142	0.069	0.06905 48691	0.06889 08087
0.020	0.02000 13336	0.01999 73340	0.070	0.07005 72931	0.06988 60016
0.021	0.02100 15438	0.02099 69138	0.071	0.07105 97876	0.07088 10559
0.022	0.02200 17751	0.02199 64517	0.072	0.07206 23536	0.07187 59696
0.023	0.02300 20283	0.02299 59456	0.073	0.07306 49921	0.07287 07407
0.024	0.02400 23046	0.02399 53936	0.074	0.07406 77043	0.07386 53674
0.025	0.02500 26049	0.02499 47936	0.075	0.07507 04911	0.07485 98477
0.026	0.02600 29302	0.02599 41437	0.076	0.07607 33535	0.07585 41797
0.027	0.02700 32816	0.02699 34419	0.077	0.07707 62926	0.07684 83614
0.028	0.02800 36600	0.02799 26861	0.078	0.07807 93093	0.07784 23909
0.029	0.02900 40664	0.02899 18744	0.079	0.07908 24048	0.07883 62663
0.030	0.03000 45018	0.02999 10049	0.080	0.08008 55800	0.07982 99857
0.031	0.03100 49673	0.03099 00754	0.081	0.08108 88360	0.08082 35471
0.032	0.03200 54639	0.03198 90840	0.082	0.08209 21738	0.08181 69486
0.033	0.03300 59924	0.03298 80288	0.083	0.08309 55945	0.08281 01883
0.034	0.03400 65541	0.03398 69077	0.084	0.08409 90990	0.08380 32642
0.035	0.03500 71498	0.03498 57188	0.085	0.08510 26884	0.08479 61745
0.036	0.03600 77805	0.03598 44601	0.086	0.08610 63637	0.08578 89172
0.037	0.03700 84474	0.03698 31295	0.087	0.08711 01260	0.08678 14905
0.038	0.03800 91513	0.03798 17252	0.088	0.08811 39763	0.08777 38923
0.039	0.03900 98933	0.03898 02450	0.089	0.08911 79156	0.08876 61209
0.040	0.04001 06744	0.03997 86871	0.090	0.09012 19450	0.08975 81742
0.041	0.04101 14955	0.04097 70495	0.091	0.09112 60655	0.09075 00504
0.042	0.04201 23578	0.04197 53301	0.092	0.09213 02781	0.09174 17476
0.043	0.04301 32622	0.04297 35270	0.093	0.09313 45840	0.09273 32638
0.044	0.04401 42097	0.04397 16383	0.094	0.09413 89840	0.09372 45973
0.045	0.04501 52014	0.04496 96619	0.095	0.09514 34793	0.09471 57460
0.046	0.04601 62381	0.04596 75958	0.096	0.09614 80709	0.09570 67081
0.047	0.04701 73211	0.04696 54381	0.097	0.09715 27598	0.09669 74817
0.048	0.04801 84511	0.04796 31869	0.098	0.09815 75472	0.09768 80649
0.049	0.04901 96294	0.04896 08401	0.099	0.09916 24339	0.09867 84558
0.050	0.05002 08568	0.04995 83957	0.100	0.10016 74212	0.09966 86525

$$\frac{\pi}{2} = 1.57079\ 63267\ 95$$

TEN-PLACE VALUES FOR INVERSE SINES AND TANGENTS (Continued)

x	arc sin x	arc tan x	x	arc sin x	arc tan x
0.100	0.10016 74212	0.09966 86525	0.150	0.15056 82728	0.14888 99476
0.101	0.10117 25099	0.10065 86532	0.151	0.15157 97940	0.14986 77990
0.102	0.10217 77012	0.10164 84559	0.152	0.15259 14716	0.15084 53616
0.103	0.10318 29962	0.10263 80588	0.153	0.15360 33066	0.15182 26339
0.104	0.10418 83957	0.10362 74600	0.154	0.15461 53002	0.15279 96139
0.105	0.10519 39010	0.10461 66576	0.155	0.15562 74533	0.15377 63001
0.106	0.10619 95131	0.10560 56498	0.156	0.15663 97673	0.15475 26907
0.107	0.10720 52330	0.10659 44347	0.157	0.15765 22431	0.15572 87839
0.108	0.10821 10617	0.10758 30104	0.158	0.15866 48819	0.15670 45780
0.109	0.10921 70004	0.10857 13750	0.159	0.15967 76848	0.15768 00714
0.110	0.11022 30500	0.10955 95268	0.160	0.16069 06530	0.15865 52622
0.111	0.11122 92116	0.11054 74637	0.161	0.16170 37874	0.15963 01488
0.112	0.11223 54864	0.11153 51841	0.162	0.16271 70894	0.16060 47295
0.113	0.11324 18753	0.11252 26859	0.163	0.16373 05599	0.16157 90025
0.114	0.11424 83793	0.11350 99674	0.164	0.16474 42002	0.16255 29662
0.115	0.11525 49997	0.11449 70268	0.165	0.16575 80113	0.16352 66188
0.116	0.11626 17373	0.11548 38621	0.166	0.16677 19944	0.16449 99587
0.117	0.11726 85934	0.11647 04715	0.167	0.16778 61506	0.16547 29842
0.118	0.11827 55688	0.11745 68532	0.168	0.16880 04810	0.16644 56935
0.119	0.11928 26648	0.11844 30053	0.169	0.16981 49868	0.16741 80851
0.120	0.12028 98824	0.11942 89260	0.170	0.17082 96691	0.16839 01571
0.121	0.12129 72226	0.12041 46135	0.171	0.17184 45291	0.16936 19080
0.122	0.12230 46865	0.12140 00659	0.172	0.17285 95678	0.17033 33361
0.123	0.12331 22752	0.12238 52815	0.173	0.17387 47865	0.17130 44396
0.124	0.12431 99897	0.12337 02583	0.174	0.17489 01862	0.17227 52170
0.125	0.12532 78312	0.12435 49945	0.175	0.17590 57682	0.17324 56665
0.126	0.12633 58006	0.12533 94884	0.176	0.17692 15335	0.17421 57864
0.127	0.12734 38991	0.12632 37382	0.177	0.17793 74833	0.17518 55753
0.128	0.12835 21277	0.12730 77419	0.178	0.17895 36187	0.17615 50313
0.129	0.12936 04876	0.12829 14978	0.179	0.17996 99410	0.17712 41528
0.130	0.13036 89797	0.12927 50040	0.180	0.18098 64512	0.17809 29382
0.131	0.13137 76052	0.13025 82589	0.181	0.18200 31506	0.17906 13859
0.132	0.13238 63651	0.13124 12605	0.182	0.18302 00402	0.18002 94942
0.133	0.13339 52606	0.13222 40071	0.183	0.18403 71213	0.18099 72614
0.134	0.13440 42927	0.13320 64968	0.184	0.18505 43949	0.18196 46860
0.135	0.13541 34625	0.13418 87280	0.185	0.18607 18623	0.18293 17662
0.136	0.13642 27710	0.13517 06986	0.186	0.18708 95247	0.18389 85006
0.137	0.13743 22194	0.13615 24071	0.187	0.18810 73831	0.18486 48874
0.138	0.13844 18088	0.13713 38516	0.188	0.18912 54387	0.18583 09251
0.139	0.13945 15402	0.13811 50303	0.189	0.19014 36928	0.18679 66120
0.140	0.14046 14147	0.13909 59415	0.190	0.19116 21465	0.18776 19465
0.141	0.14147 14335	0.14007 65833	0.191	0.19218 08010	0.18872 69271
0.142	0.14248 15975	0.14105 69540	0.192	0.19319 96574	0.18969 15520
0.143	0.14349 19080	0.14203 70518	0.193	0.19421 87170	0.19065 58198
0.144	0.14450 23659	0.14301 68750	0.194	0.19523 79808	0.19161 97288
0.145	0.14551 29725	0.14399 64217	0.195	0.19625 74502	0.19258 32775
0.146	0.14652 37288	0.14497 56903	0.196	0.19727 71262	0.19354 64642
0.147	0.14753 46358	0.14595 46789	0.197	0.19829 70101	0.19450 92873
0.148	0.14854 56948	0.14693 33858	0.198	0.19931 71030	0.19547 17454
0.149	0.14955 69067	0.14791 18093	0.199	0.20033 74062	0.19643 38367
0.150	0.15056 82728	0.14888 99476	0.200	0.20135 79208	0.19739 55599

$\frac{\pi}{2} = 1.57079\ 63267\ 95$

TEN-PLACE VALUES FOR INVERSE SINES AND TANGENTS (Continued)

x	arc sin x	arc tan x	x	arc sin x	arc tan x
0.200	0.20135 79208	0.19739 55599	0.250	0.25268 02551	0.24497 86631
0.201	0.20237 86480	0.19835 69131	0.251	0.25371 31886	0.24591 96179
0.202	0.20339 95891	0.19931 78950	0.252	0.25474 63988	0.24686 01285
0.203	0.20442 07452	0.20027 85040	0.253	0.25577 98871	0.24780 01934
0.204	0.20544 21175	0.20123 87385	0.254	0.25681 36548	0.24873 98114
0.205	0.20646 37073	0.20219 85969	0.255	0.25784 77032	0.24967 89810
0.206	0.20748 55156	0.20315 80777	0.256	0.25888 20337	0.25061 77011
0.207	0.20850 75439	0.20411 71794	0.257	0.25991 66475	0.25155 59702
0.208	0.20952 97932	0.20507 59004	0.258	0.26095 15461	0.25249 37870
0.209	0.21055 22647	0.20603 42392	0.259	0.26198 67308	0.25343 11503
0.210	0.21157 49598	0.20699 21942	0.260	0.26302 22029	0.25436 80586
0.211	0.21259 78795	0.20794 97640	0.261	0.26405 79638	0.25530 45106
0.212	0.21362 10251	0.20890 69470	0.262	0.26509 40148	0.25624 05052
0.213	0.21464 43979	0.20986 37417	0.263	0.26613 03574	0.25717 60408
0.214	0.21566 79991	0.21082 01465	0.264	0.26716 69927	0.25811 11164
0.215	0.21669 18298	0.21177 61600	0.265	0.26820 39223	0.25904 57305
0.216	0.21771 58914	0.21273 17807	0.266	0.26924 11475	0.25997 98819
0.217	0.21874 01850	0.21368 70070	0.267	0.27027 86696	0.26091 35692
0.218	0.21976 47119	0.21464 18375	0.268	0.27131 64901	0.26184 67913
0.219	0.22078 94733	0.21559 62707	0.269	0.27235 46102	0.26277 95468
0.220	0.22181 44705	0.21655 03050	0.270	0.27339 30315	0.26371 18345
0.221	0.22283 97047	0.21750 39390	0.271	0.27443 17552	0.26464 36530
0.222	0.22386 51771	0.21845 71712	0.272	0.27547 07827	0.26557 50012
0.223	0.22489 08890	0.21941 00002	0.273	0.27651 01155	0.26650 58777
0.224	0.22591 68416	0.22036 24244	0.274	0.27754 97549	0.26743 62814
0.225	0.22694 30362	0.22131 44423	0.275	0.27858 97024	0.26836 62109
0.266	0.22796 94740	0.22226 60527	0.276	0.27962 99593	0.26929 56650
0.227	0.22899 61563	0.22321 72538	0.277	0.28067 05270	0.27022 46426
0.228	0.23002 30844	0.22416 80444	0.278	0.28171 14069	0.27115 31422
0.229	0.23105 02595	0.22511 84230	0.279	0.28275 26005	0.27208 11628
0.230	0.23207 76829	0.22606 83880	0.280	0.28379 41092	0.27300 87031
0.231	0.23310 53558	0.22701 79381	0.281	0.28483 59344	0.27393 57618
0.232	0.23413 32795	0.22796 70718	0.282	0.28587 80774	0.27486 23378
0.233	0.23516 14552	0.22891 57877	0.283	0.28692 05398	0.27578 84298
0.234	0.23618 98843	0.22986 40844	0.284	0.28796 33229	0.27671 40367
0.235	0.23721 85681	0.23081 19604	0.285	0.28900 64282	0.27763 91571
0.236	0.23824 75077	0.23175 94143	0.286	0.29004 98571	0.27856 37900
0.237	0.23927 67046	0.23270 64447	0.287	0.29109 36111	0.27948 79341
0.238	0.24030 61599	0.23365 30502	0.288	0.29213 76915	0.28041 15883
0.239	0.24133 58749	0.23459 92293	0.289	0.29318 20999	0.28133 47513
0.240	0.24236 58510	0.23554 49807	0.290	0.29422 68377	0.28225 74220
0.241	0.24339 60895	0.23649 03030	0.291	0.29527 19064	0.28317 95992
0.242	0.24442 65915	0.23743 51947	0.292	0.29631 73074	0.28410 12817
0.243	0.24545 73585	0.23837 96545	0.293	0.29736 30421	0.28502 24683
0.244	0.24648 83918	0.23932 36810	0.294	0.29840 91120	0.28594 31580
0.245	0.24751 96925	0.24026 72728	0.295	0.29945 55187	0.28686 33495
0.246	0.24855 12621	0.24121 04285	0.296	0.30050 22635	0.28778 30417
0.247	0.24958 31019	0.24215 31467	0.297	0.30154 93479	0.28870 22335
0.248	0.25061 52131	0.24309 54262	0.298	0.30259 67735	0.28962 09236
0.249	0.25164 75971	0.24403 72654	0.299	0.30364 45417	0.29053 91110
0.250	0.25268 02551	0.24497 86631	0.300	0.30469 26540	0.29145 67945

$$\frac{\pi}{2} = 1.57079\ 63267\ 95$$

TEN-PLACE VALUES FOR INVERSE SINES AND TANGENTS (Continued)

x	arc sin x	arc tan x	x	arc sin x	arc tan x
0.300	0.30469 26540	0.29145 67945	0.350	0.35757 11036	0.33667 48194
0.301	0.30574 11119	0.29237 39730	0.351	0.35863 88379	0.33756 54101
0.302	0.30678 99169	0.29329 06453	0.352	0.35970 69996	0.33845 54443
0.303	0.30783 90704	0.29420 68105	0.353	0.36077 55906	0.33934 49212
0.304	0.30888 85740	0.29512 24672	0.354	0.36184 46126	0.34023 38399
0.305	0.30993 84293	0.29603 76145	0.355	0.36291 40673	0.34112 21994
0.306	0.31098 86376	0.29695 22512	0.356	0.36398 39565	0.34200 99991
0.307	0.31203 92006	0.29786 63761	0.357	0.36505 42819	0.34289 72379
0.308	0.31309 01197	0.29877 99884	0.358	0.36612 50454	0.34378 39149
0.309	0.31414 13965	0.29969 30867	0.359	0.36719 62486	0.34467 00295
0.310	0.31519 30324	0.30060 56700	0.360	0.36826 78934	0.34555 55806
0.311	0.31624 50291	0.30151 77374	0.361	0.36933 99816	0.34644 05674
0.312	0.31729 73881	0.30242 92875	0.362	0.37041 25148	0.34732 49892
0.313	0.31835 01109	0.30334 03195	0.363	0.37148 54949	0.34820 88450
0.314	0.31940 31990	0.30425 08322	0.364	0.37255 89237	0.34909 21340
0.315	0.32045 66541	0.30516 08246	0.365	0.37363 28031	0.34997 48553
0.316	0.32151 04775	0.30607 02956	0.366	0.37470 71347	0.35085 70083
0.317	0.32256 46710	0.30697 92441	0.367	0.37578 19205	0.35173 85919
0.318	0.32361 92361	0.30788 76692	0.368	0.37685 71622	0.35261 96055
0.319	0.32467 41744	0.30879 55696	0.369	0.37793 28616	0.35350 00482
0.320	0.32572 94873	0.30970 29445	0.370	0.37900 90207	0.35437 99191
0.321	0.32678 51765	0.31060 97928	0.371	0.38008 56412	0.35525 92176
0.322	0.32784 12436	0.31151 61134	0.372	0.38116 27250	0.35613 79427
0.323	0.32889 76902	0.31242 19054	0.373	0.38224 02739	0.35701 60937
0.324	0.32995 45178	0.31332 71676	0.374	0.38331 82898	0.35789 36698
0.325	0.33101 17281	0.31423 18991	0.375	0.38439 67745	0.35877 06703
0.326	0.33206 93226	0.31513 60988	0.376	0.38547 57299	0.35964 70942
0.327	0.33312 73029	0.31603 97659	0.377	0.38655 51580	0.36052 29410
0.328	0.33418 56707	0.31694 28991	0.378	0.38763 50605	0.36139 82097
0.329	0.33524 44276	0.31784 54976	0.379	0.38871 54394	0.36227 28996
0.330	0.33630 35752	0.31874 75604	0.380	0.38979 62965	0.36314 70099
0.331	0.33736 31150	0.31964 90865	0.381	0.39087 76337	0.36402 05400
0.332	0.33842 30488	0.32055 00748	0.382	0.39195 94531	0.36489 34890
0.333	0.33948 33782	0.32145 05244	0.383	0.39304 17564	0.36576 58562
0.334	0.34054 41047	0.32235 04343	0.384	0.39412 45456	0.36663 76408
0.335	0.34160 52301	0.32324 98036	0.385	0.39520 78226	0.36750 88422
0.336	0.34266 67560	0.32414 86313	0.386	0.39629 15893	0.36837 94595
0.337	0.34372 86840	0.32504 69164	0.387	0.39737 58477	0.36924 94920
0.338	0.34479 10158	0.32594 46580	0.388	0.39846 05998	0.37011 89391
0.339	0.34585 37531	0.32684 18551	0.389	0.39954 58475	0.37098 77999
0.340	0.34691 68975	0.32773 85068	0.390	0.40063 15927	0.37185 60738
0.341	0.34798 04507	0.32863 46121	0.391	0.40171 78374	0.37272 37601
0.342	0.34904 44144	0.32953 01700	0.392	0.40280 45836	0.37359 08580
0.343	0.35010 87902	0.33042 51798	0.393	0.40389 18333	0.37445 73669
0.344	0.35117 35799	0.33131 96403	0.394	0.40497 95885	0.37532 32860
0.345	0.35223 87851	0.33221 35507	0.395	0.40606 78511	0.37618 86147
0.346	0.35330 44075	0.33310 69102	0.396	0.40715 66231	0.37705 33522
0.347	0.35437 04489	0.33399 97176	0.397	0.40824 59066	0.37791 74978
0.348	0.35543 69109	0.33489 19723	0.398	0.40933 57036	0.37878 10510
0.349	0.35650 37952	0.33578 36732	0.399	0.41042 60161	0.37964 40110
0.350	0.35757 11036	0.33667 48194	0.400	0.41151 68461	0.38050 63771

$$\frac{\pi}{2} = 1.57079\ 63267\ 95$$

TEN-PLACE VALUES FOR INVERSE SINES AND TANGENTS (Continued)

x	arc sin x	arc tan x	x	arc sin x	arc tan x
0.400	0.41151 68461	0.38050 63771	0.450	0.46676 53390	0.42285 39261
0.401	0.41260 81956	0.38136 81487	0.451	0.46788 54404	0.42368 52157
0.402	0.41370 00668	0.38222 93251	0.452	0.46900 61761	0.42451 58824
0.403	0.41479 24617	0.38308 99056	0.453	0.47012 75486	0.42534 59258
0.404	0.41588 53823	0.38394 98897	0.454	0.47124 95605	0.42617 53455
0.405	0.41697 88306	0.38480 92765	0.455	0.47237 22141	0.42700 41409
0.406	0.41807 28089	0.38566 80656	0.456	0.47349 55122	0.42783 23118
0.407	0.41916 73190	0.38652 62562	0.457	0.47461 94571	0.42865 98577
0.408	0.42026 23632	0.38738 38478	0.458	0.47574 40514	0.42948 67780
0.409	0.42135 79436	0.38824 08396	0.459	0.47686 92977	0.43031 30725
0.410	0.42245 40622	0.38909 72311	0.460	0.47799 51985	0.43113 87407
0.411	0.42355 07211	0.38995 30216	0.461	0.47912 17565	0.43196 37822
0.412	0.42464 79225	0.39080 82105	0.462	0.48024 89741	0.43278 81966
0.413	0.42574 56686	0.39166 27972	0.463	0.48137 68540	0.43361 19834
0.414	0.42684 39613	0.39251 67810	0.464	0.48250 53989	0.43443 51423
0.415	0.42794 28030	0.39337 01615	0.465	0.48363 46112	0.43525 76729
0.416	0.42904 21957	0.39422 29379	0.466	0.48476 44937	0.43607 95747
0.417	0.43014 21415	0.39507 51098	0.467	0.48589 50490	0.43690 08475
0.418	0.43124 26428	0.39592 66763	0.468	0.48702 62797	0.43772 14907
0.419	0.43234 37016	0.39677 76371	0.469	0.48815 81885	0.43854 15041
0.420	0.43344 53201	0.39762 79915	0.470	0.48929 07780	0.43936 08873
0.421	0.43454 75005	0.39847 77389	0.471	0.49042 40510	0.44017 96398
0.422	0.43565 02451	0.39932 68788	0.472	0.49155 80102	0.44099 77614
0.423	0.43675 35559	0.40017 54106	0.473	0.49269 26582	0.44181 52515
0.424	0.43785 74354	0.40102 33336	0.474	0.49382 79978	0.44263 21100
0.425	0.43896 18856	0.40187 06474	0.475	0.49496 40317	0.44344 83364
0.426	0.44006 69088	0.40271 73514	0.476	0.49610 07627	0.44426 39304
0.427	0.44117 25073	0.40356 34451	0.477	0.49723 81935	0.44507 88916
0.428	0.44227 86833	0.40440 89278	0.478	0.49837 63268	0.44589 32197
0.429	0.44338 54391	0.40525 37991	0.479	0.49951 51655	0.44670 69143
0.430	0.44449 27769	0.40609 80583	0.480	0.50065 47124	0.44751 99752
0.431	0.44560 06991	0.40694 17050	0.481	0.50179 49702	0.44833 24019
0.432	0.44670 92078	0.40778 47387	0.482	0.50293 59418	0.44914 41941
0.433	0.44781 83055	0.40862 71587	0.483	0.50407 76301	0.44995 53516
0.434	0.44892 79944	0.40946 89646	0.484	0.50522 00377	0.45076 58739
0.435	0.45003 82768	0.41031 01559	0.485	0.50636 31677	0.45157 57608
0.436	0.45114 91550	0.41115 07320	0.486	0.50750 70228	0.45238 50120
0.437	0.45226 06315	0.41199 06924	0.487	0.50865 16060	0.45319 36272
0.438	0.45337 27084	0.41283 00367	0.488	0.50979 69202	0.45400 16059
0.439	0.45448 53883	0.41366 87642	0.489	0.51094 29682	0.45480 89481
0.440	0.45559 86734	0.41450 68746	0.490	0.51208 97529	0.45561 56532
0.441	0.45671 25661	0.41534 43673	0.491	0.51323 72774	0.45642 17211
0.442	0.45782 70688	0.41618 12418	0.492	0.51438 55446	0.45722 71515
0.443	0.45894 21839	0.41701 74976	0.493	0.51553 45573	0.45803 19440
0.444	0.46005 79138	0.41785 31343	0.494	0.51668 43187	0.45883 60984
0.445	0.46117 42608	0.41868 81514	0.495	0.51783 48316	0.45963 96144
0.446	0.46229 12275	0.41952 25484	0.496	0.51898 60992	0.46044 24918
0.447	0.46340 88162	0.42035 63249	0.497	0.52013 81243	0.46124 47302
0.448	0.46452 70294	0.42118 94803	0.498	0.52129 09100	0.46204 63293
0.449	0.46564 58695	0.42202 20142	0.499	0.52244 44594	0.46284 72890
0.450	0.46676 53390	0.42285 39261	0.500	0.52359 87756	0.46364 76090

$\frac{\pi}{2} = 1.57079\ 63267\ 95$

TEN-PLACE VALUES FOR INVERSE SINES AND TANGENTS (Continued)

x	arc sin x	arc tan x	x	arc sin x	arc tan x
0.500	0.52359 87756	0.46364 76090	0.550	0.58236 42379	0.50284 32109
0.501	0.52475 38616	0.46444 72890	0.551	0.58356 20793	0.50361 06410
0.502	0.52590 97204	0.46524 63287	0.552	0.58476 08688	0.50437 74227
0.503	0.52706 63552	0.46604 47279	0.553	0.58596 06105	0.50514 35558
0.504	0.52822 37692	0.46684 24863	0.554	0.58716 13082	0.50590 90402
0.505	0.52938 19653	0.46763 96038	0.555	0.58836 29661	0.50667 38760
0.506	0.53054 09469	0.46843 60800	0.556	0.58956 55882	0.50743 80630
0.507	0.53170 07170	0.46923 19147	0.557	0.59076 91785	0.50820 16011
0.508	0.53286 12788	0.47002 71078	0.558	0.59197 37412	0.50896 44904
0.509	0.53402 26355	0.47082 16589	0.559	0.59317 92803	0.50972 67306
0.510	0.53518 47903	0.47161 55679	0.560	0.59438 58000	0.51048 83219
0.511	0.53634 77464	0.47240 88344	0.561	0.59559 33044	0.51124 92641
0.512	0.53751 15071	0.47320 14584	0.562	0.59680 17978	0.51200 95572
0.513	0.53867 60757	0.47399 34396	0.563	0.59801 12843	0.51276 92011
0.514	0.53984 14553	0.47478 47778	0.564	0.59922 17681	0.51352 81958
0.515	0.54100 76492	0.47557 54727	0.565	0.60043 32536	0.51428 65413
0.516	0.54217 46609	0.47636 55242	0.566	0.60164 57449	0.51504 42374
0.517	0.54334 24935	0.47715 49321	0.567	0.60285 92464	0.51580 12843
0.518	0.54451 11505	0.47794 36961	0.568	0.60407 37624	0.51655 76817
0.519	0.54568 06351	0.47873 18162	0.569	0.60528 92972	0.51731 34298
0.520	0.54685 09507	0.47951 92920	0.570	0.60650 58552	0.51806 85285
0.521	0.54802 21007	0.48030 61234	0.571	0.60772 34408	0.51882 29777
0.522	0.54919 40886	0.48109 23103	0.572	0.60894 20585	0.51957 67774
0.523	0.55036 69176	0.48187 78524	0.573	0.61016 17126	0.52032 99277
0.524	0.55154 05913	0.48266 27495	0.574	0.61138 24076	0.52108 24285
0.525	0.55271 51131	0.48344 70016	0.575	0.61260 41480	0.52183 42798
0.526	0.55389 04864	0.48423 06084	0.576	0.61382 69384	0.52258 54816
0.527	0.55506 67148	0.48501 35697	0.577	0.61505 07833	0.52333 60339
0.528	0.55624 38018	0.48579 58854	0.578	0.61627 56872	0.52408 59366
0.529	0.55742 17508	0.48657 75554	0.579	0.61750 16548	0.52483 51898
0.530	0.55860 05653	0.48735 85795	0.580	0.61872 86907	0.52558 37936
0.531	0.55978 02491	0.48813 89575	0.581	0.61995 67995	0.52633 17478
0.532	0.56096 08055	0.48891 86893	0.582	0.62118 59858	0.52707 90525
0.533	0.56214 22383	0.48969 77748	0.583	0.62241 62545	0.52782 57077
0.534	0.56332 45509	0.49047 62137	0.584	0.62364 76102	0.52857 17134
0.535	0.56450 77471	0.49125 40060	0.585	0.62488 00578	0.52931 70697
0.536	0.56569 18305	0.49203 11516	0.586	0.62611 36019	0.53006 17766
0.537	0.56687 68047	0.49280 76502	0.587	0.62734 82474	0.53080 58340
0.538	0.56806 26735	0.49358 35018	0.588	0.62858 39991	0.53154 92421
0.539	0.56924 94405	0.49435 87063	0.589	0.62982 08619	0.53229 20008
0.540	0.57043 71094	0.49513 32635	0.590	0.63105 88408	0.53303 41102
0.541	0.57162 56840	0.49590 71733	0.591	0.63229 79406	0.53377 55703
0.542	0.57281 51681	0.49668 04355	0.592	0.63353 81663	0.53451 63811
0.543	0.57400 55653	0.49745 30502	0.593	0.63477 95228	0.53525 65428
0.544	0.57519 68796	0.49822 50172	0.594	0.63602 20153	0.53599 60552
0.545	0.57638 91147	0.49899 63363	0.595	0.63726 56487	0.53673 49186
0.546	0.57758 22745	0.49976 70075	0.596	0.63851 04281	0.53747 31328
0.547	0.57877 63629	0.50053 70306	0.597	0.63975 63587	0.53821 06981
0.548	0.57997 13836	0.50130 64056	0.598	0.64100 34456	0.53894 76144
0.549	0.58116 73406	0.50207 51324	0.599	0.64225 16939	0.53968 38817
0.550	0.58236 42379	0.50284 32109	0.600	0.64350 11088	0.54041 95003

$\frac{\pi}{2} = 1.57079\ 63267\ 95$

TEN-PLACE VALUES FOR INVERSE SINES AND TANGENTS (Continued)

x	arc sin x	arc tan x	x	arc sin x	arc tan x
0.600	0.64350 11088	0.54041 95003	0.650	0.70758 44367	0.57637 52206
0.601	0.64475 16956	0.54115 44700	0.651	0.70890 10819	0.57707 78871
0.602	0.64600 34596	0.54188 87910	0.652	0.71021 92155	0.57777 99113
0.603	0.64725 64060	0.54262 24634	0.653	0.71153 88448	0.57848 12935
0.604	0.64851 05401	0.54335 54871	0.654	0.71285 99773	0.57918 20338
0.605	0.64976 58674	0.54408 78624	0.655	0.71418 26205	0.57988 21324
0.606	0.65102 23933	0.54481 95892	0.656	0.71550 67818	0.58058 15895
0.607	0.65228 01230	0.54555 06677	0.657	0.71683 24688	0.58128 04053
0.608	0.65353 90622	0.54628 10979	0.658	0.71815 96892	0.58197 85800
0.609	0.65479 92164	0.54701 08798	0.659	0.71948 84507	0.58267 61139
0.610	0.65606 05909	0.54774 00137	0.660	0.72081 87609	0.58337 30070
0.611	0.65732 31915	0.54846 84996	0.661	0.72215 06276	0.58406 92596
0.612	0.65858 70237	0.54919 63375	0.662	0.72348 40588	0.58476 48720
0.613	0.65985 20931	0.54992 35276	0.663	0.72481 90622	0.58545 98443
0.614	0.66111 84055	0.55065 00700	0.664	0.72615 56460	0.58615 41768
0.615	0.66238 59665	0.55137 59647	0.665	0.72749 38180	0.58684 78697
0.616	0.66365 47819	0.55210 12119	0.666	0.72883 35864	0.58754 09231
0.617	0.66492 48574	0.55282 58116	0.667	0.73017 49593	0.58823 33373
0.618	0.66619 61989	0.55354 97640	0.668	0.73151 79449	0.58892 51125
0.619	0.66746 88122	0.55427 30692	0.669	0.73286 25515	0.58961 62490
0.620	0.66874 27032	0.55499 57273	0.670	0.73420 87875	0.59030 67469
0.621	0.67001 78779	0.55571 77384	0.671	0.73555 66610	0.59099 66066
0.622	0.67129 43422	0.55643 91027	0.672	0.73690 61808	0.59168 58281
0.623	0.67257 21021	0.55715 98202	0.673	0.73825 73551	0.59237 44119
0.624	0.67385 11636	0.55787 98910	0.674	0.73961 01927	0.59306 23580
0.625	0.67513 15329	0.55859 93153	0.675	0.74096 47022	0.59374 96667
0.626	0.67641 32161	0.55931 80933	0.067	0.74232 08922	0.59443 63383
0.627	0.67769 62194	0.56003 62250	0.677	0.74367 87716	0.59512 23730
0.628	0.67898 05488	0.56075 37106	0.678	0.74503 83492	0.59580 77710
0.629	0.68026 62108	0.56147 05502	0.679	0.74639 96339	0.59649 25326
0.630	0.68155 32116	0.56218 67439	0.680	0.74776 26347	0.59717 66581
0.631	0.68284 15574	0.56290 22919	0.681	0.74912 73606	0.59786 01476
0.623	0.68413 12548	0.56361 71944	0.682	0.75049 38207	0.59854 30015
0.633	0.68542 23100	0.56433 14514	0.683	0.75186 20243	0.59922 52199
0.634	0.68671 47296	0.56504 50631	0.684	0.75323 19805	0.59990 68031
0.635	0.68800 85200	0.56575 80297	0.685	0.75460 36988	0.60058 77514
0.636	0.68930 36879	0.56647 03514	0.686	0.75597 71886	0.60126 80651
0.637	0.69060 02397	0.56718 20282	0.687	0.75735 24593	0.60194 77443
0.638	0.69189 81821	0.56789 30603	0.688	0.75872 95204	0.60262 67894
0.639	0.69319 75219	0.56860 34479	0.689	0.76010 83817	0.60330 52007
0.640	0.69449 82656	0.56931 31911	0.690	0.76148 90527	0.60398 29783
0.641	0.69580 04202	0.57002 22901	0.691	0.76287 15434	0.60466 01225
0.642	0.69710 39923	0.57073 07452	0.692	0.76425 58636	0.60533 66337
0.643	0.69840 89889	0.57143 85563	0.693	0.76564 20232	0.60601 25120
0.644	0.69971 54169	0.57214 57237	0.694	0.76703 00322	0.60668 77578
0.645	0.70102 32832	0.57285 22477	0.695	0.76841 99008	0.60736 23713
0.646	0.70233 25949	0.57355 81282	0.696	0.76981 16391	0.60803 63528
0.647	0.70364 33589	0.57426 33656	0.697	0.77120 52575	0.60870 97026
0.648	0.70495 55825	0.57496 79601	0.698	0.77260 07662	0.60938 24209
0.649	0.70626 92727	0.57567 19116	0.699	0.77399 81757	0.61005 45081
0.650	0.70758 44367	0.57637 52206	0.700	0.77539 74966	0.61072 59644

$\frac{\pi}{2}$ = 1.57079 63267 95

TEN-PLACE VALUES FOR INVERSE SINES AND TANGENTS (Continued)

x	arc sin x	arc tan x	x	arc sin x	arc tan x
0.700	0.77539 74966	0.61072 59644	0.750	0.84806 20790	0.64350 11088
0.701	0.77679 87395	0.61139 67901	0.751	0.84957 52356	0.64414 08017
0.702	0.77820 19150	0.61206 69854	0.752	0.85109 10008	0.64477 98805
0.703	0.77960 70339	0.61273 65508	0.753	0.85260 93917	0.64541 83456
0.704	0.78101 41072	0.61340 54864	0.754	0.85413 04254	0.64605 61975
0.705	0.78242 31458	0.61407 37925	0.755	0.85565 41195	0.64669 34363
0.706	0.78383 41608	0.61474 14695	0.756	0.85718 04914	0.64733 00626
0.707	0.78524 71634	0.61540 85176	0.757	0.85870 95589	0.64796 60767
0.708	0.78666 21647	0.61607 49372	0.758	0.86024 13399	0.64860 14790
0.709	0.78807 91762	0.61674 07285	0.759	0.86177 58525	0.64923 62697
0.710	0.78949 82093	0.61740 58918	0.760	0.86331 31150	0.64987 04494
0.711	0.79091 92756	0.61807 04274	0.761	0.86485 31460	0.65050 40183
0.712	0.79234 23866	0.61873 43356	0.762	0.86639 59640	0.65113 69769
0.713	0.79376 75542	0.61939 76168	0.763	0.86794 15880	0.65176 93255
0.714	0.79519 47902	0.62006 02712	0.764	0.86949 00370	0.65240 10645
0.715	0.79662 41065	0.62072 22992	0.765	0.87104 13304	0.65303 21943
0.716	0.79805 55152	0.62138 37010	0.766	0.87259 54876	0.65366 27152
0.717	0.79948 90285	0.62204 44770	0.767	0.87415 25283	0.65429 26276
0.718	0.80092 46586	0.62270 46274	0.768	0.87571 24725	0.65492 19320
0.719	0.80236 24179	0.62336 41526	0.769	0.87727 53401	0.65555 06287
0.720	0.80380 23189	0.62402 30530	0.770	0.87884 11517	0.65617 87180
0.721	0.80524 43742	0.62468 13287	0.771	0.88040 99276	0.65680 62004
0.722	0.80668 85965	0.62533 89802	0.772	0.88198 16888	0.65743 30762
0.723	0.80813 49987	0.62599 60077	0.773	0.88355 64563	0.65805 93459
0.724	0.80958 35936	0.62665 24117	0.774	0.88513 42512	0.65868 50098
0.725	0.81103 43943	0.62730 81923	0.775	0.88671 50950	0.65931 00683
0.726	0.81248 74140	0.62796 33499	0.776	0.88829 90095	0.65993 45219
0.727	0.81394 26660	0.62861 78849	0.777	0.88988 60167	0.66055 83708
0.728	0.81540 01638	0.62927 17976	0.778	0.89147 61387	0.66118 16155
0.729	0.81685 99207	0.62992 50882	0.779	0.89306 93979	0.66180 42564
0.730	0.81832 19506	0.63057 77572	0.780	0.89466 58172	0.66242 62938
0.731	0.81978 62672	0.63122 98049	0.781	0.89626 54195	0.66304 77283
0.732	0.82125 28845	0.63188 12315	0.782	0.89786 82280	0.66366 85601
0.733	0.82272 18163	0.63253 20375	0.783	0.89947 42662	0.66428 87897
0.734	0.82419 30771	0.63318 22232	0.784	0.90108 35578	0.66490 84174
0.735	0.82566 66810	0.63383 17889	0.785	0.90269 61270	0.66552 74437
0.736	0.82714 26425	0.63448 07349	0.786	0.90431 19981	0.66614 58690
0.737	0.82862 09762	0.63512 90616	0.787	0.90593 11956	0.66676 36937
0.738	0.83010 16968	0.63577 67693	0.788	0.90755 37445	0.66738 09181
0.739	0.83158 48192	0.63642 38585	0.789	0.90917 96699	0.66799 75427
0.740	0.83307 03583	0.63707 03293	0.790	0.91080 89974	0.66861 35679
0.741	0.83455 83294	0.63771 61822	0.791	0.91244 17527	0.66922 89941
0.742	0.83604 87477	0.63836 14175	0.792	0.91407 79620	0.66984 38217
0.743	0.83754 16287	0.63900 60355	0.793	0.91571 76517	0.67045 80511
0.744	0.83903 69879	0.63965 00367	0.794	0.91736 08485	0.67107 16828
0.745	0.84053 48411	0.64029 34213	0.795	0.91900 75795	0.67168 47170
0.746	0.84203 52042	0.64093 61898	0.796	0.92065 78721	0.67229 71543
0.747	0.84353 80932	0.64157 83424	0.797	0.92231 17539	0.67290 89951
0.748	0.84504 35244	0.64221 98795	0.798	0.92396 92532	0.67352 02397
0.749	0.84655 15142	0.64286 08015	0.799	0.92563 03983	0.67413 08886
0.750	0.84806 20790	0.64350 11088	0.800	0.92729 52180	0.67474 09422

$$\frac{\pi}{2} = 1.57079\ 63267\ 95$$

TEN-PLACE VALUES FOR INVERSE SINES AND TANGENTS (Continued)

x	arc sin x	arc tan x	x	arc sin x	arc tan x
0.800	0.92729 52180	0.67474 09422	0.850	1.01598 52938	0.70449 40642
0.801	0.92896 37414	0.67535 04009	0.851	1.01788 65272	0.70507 43294
0.802	0.93063 59981	0.67595 92652	0.852	1.01979 36362	0.70565 40220
0.803	0.93231 20179	0.67656 75354	0.853	1.02170 66824	0.70623 31425
0.804	0.93399 18310	0.67717 52120	0.854	1.02362 57289	0.70681 16915
0.805	0.93567 54682	0.67778 22954	0.855	1.02555 08396	0.70738 96693
0.806	0.93736 29605	0.67838 87860	0.856	1.02748 20794	0.70796 70764
0.807	0.93905 43392	0.67899 46842	0.857	1.02941 95147	0.70854 39134
0.808	0.94074 96363	0.67959 99905	0.858	1.03136 32127	0.70912 01806
0.809	0.94244 88841	0.68020 47052	0.859	1.03331 32421	0.70969 58784
0.810	0.94415 21152	0.68080 88289	0.860	1.03526 96725	0.71027 10075
0.811	0.94585 93626	0.68141 23619	0.861	1.03723 25750	0.71084 55682
0.812	0.94757 06601	0.68201 53047	0.862	1.03920 20218	0.71141 95610
0.813	0.94928 60416	0.68261 76577	0.863	1.04117 80867	0.71199 29863
0.814	0.95100 55416	0.68321 94212	0.864	1.04316 08445	0.71256 58447
0.815	0.95272 91949	0.68382 05959	0.865	1.04515 03717	0.71313 81365
0.816	0.95445 70371	0.68442 11820	0.866	1.04714 67459	0.71370 98623
0.817	0.95618 91039	0.68502 11800	0.867	1.04915 00464	0.71428 10225
0.818	0.95792 54318	0.68562 05903	0.868	1.05116 03539	0.71485 16177
0.819	0.95966 60576	0.68621 94134	0.869	1.05317 77507	0.71542 16481
0.820	0.96141 10188	0.68681 76498	0.870	1.05520 23205	0.71599 11144
0.821	0.96316 03531	0.68741 52997	0.871	1.05723 41490	0.71656 00170
0.822	0.96491 40992	0.68801 23638	0.872	1.05927 33231	0.71712 83563
0.823	0.96667 22959	0.68860 88423	0.873	1.06131 99317	0.71769 61329
0.824	0.96843 49828	0.68920 47358	0.874	1.06337 40654	0.71826 33472
0.825	0.97020 21999	0.68980 00447	0.875	1.06543 58165	0.71882 99996
0.826	0.97197 39881	0.69039 47695	0.876	1.06750 52793	0.71939 60907
0.827	0.97375 03884	0.69098 89104	0.877	1.06958 25500	0.71996 16209
0.828	0.97553 14428	0.69158 24681	0.878	1.07166 77267	0.72052 65907
0.829	0.97731 71938	0.69217 54430	0.879	1.07376 09094	0.72109 10005
0.830	0.97910 76844	0.69276 78354	0.880	1.07586 22005	0.72165 48509
0.831	0.98090 29583	0.69335 96459	0.881	1.07797 17042	0.72221 81422
0.832	0.98270 30600	0.69395 08748	0.882	1.08008 95271	0.72278 08751
0.833	0.98450 80345	0.69454 15226	0.883	1.08221 57780	0.72334 30499
0.834	0.98631 79274	0.69513 15898	0.884	1.08435 05682	0.72390 46671
0.835	0.98813 27853	0.69572 10769	0.885	1.08649 40110	0.72446 57272
0.836	0.98995 26551	0.69630 99841	0.886	1.08864 62227	0.72502 62307
0.837	0.99177 75848	0.69689 83121	0.887	1.09080 73218	0.72558 61781
0.838	0.99360 76229	0.69748 60612	0.888	1.09297 74295	0.72614 55697
0.839	0.99544 28187	0.69807 32320	0.889	1.09515 66699	0.72670 44062
0.840	0.99728 32224	0.69865 98247	0.890	1.09734 51695	0.72726 26880
0.841	0.99912 88847	0.69924 58400	0.891	1.09954 30582	0.72782 04155
0.842	1.00097 98574	0.69983 12782	0.892	1.10175 04685	0.72837 75893
0.843	1.00283 61930	0.70041 61398	0.893	1.10396 75362	0.72893 42098
0.844	1.00469 79448	0.70100 04253	0.894	1.10619 44003	0.72949 02775
0.845	1.00656 51671	0.70158 41350	0.895	1.10843 12028	0.73004 57929
0.846	1.00843 79148	0.70216 72695	0.896	1.11067 80894	0.73060 07564
0.847	1.01031 62439	0.70274 98293	0.897	1.11293 52093	0.73115 51686
0.848	1.01220 02115	0.70333 18146	0.898	1.11520 27152	0.73170 90299
0.849	1.01408 98752	0.70391 32262	0.899	1.11748 07636	0.73226 23408
0.850	1.01598 52938	0.70449 40642	0.900	1.11976 95150	0.73281 51018

$$\frac{\pi}{2} = 1.57079\ 63267\ 95$$

TEN-PLACE VALUES FOR INVERSE SINES AND TANGENTS (Continued)

x	arc sin x	arc tan x	x	arc sin x	arc tan x
0.900	1.11976 95150	0.73281 51018	0.950	1.25323 58975	0.75976 27549
0.901	1.12206 91338	0.73336 73133	0.951	1.25645 42223	0.76028 81167
0.902	1.12437 97886	0.73391 89759	0.952	1.25970 47250	0.76081 29540
0.903	1.12670 16524	0.73447 00901	0.953	1.26298 84259	0.76133 72674
0.904	1.12903 49026	0.73502 06562	0.954	1.26630 64001	0.76186 10574
0.905	1.13137 97213	0.73557 06749	0.955	1.26965 97812	0.76238 43244
0.906	1.13373 62954	0.73612 01465	0.956	1.27304 97667	0.76290 70690
0.907	1.13610 48167	0.73666 90716	0.957	1.27647 76223	0.76342 92916
0.908	1.13848 54823	0.73721 74506	0.958	1.27994 46879	0.76395 09928
0.909	1.14087 84947	0.73776 52841	0.959	1.28345 23838	0.76447 21730
0.910	1.14328 40619	0.73831 25725	0.960	1.28700 22176	0.76499 28327
0.911	1.14570 23977	0.73885 93163	0.961	1.29059 57918	0.76551 29725
0.912	1.14813 37220	0.73940 55160	0.962	1.29423 48124	0.76603 25928
0.913	1.15057 82610	0.73995 11721	0.963	1.29792 10987	0.76655 16941
0.914	1.15303 62474	0.74049 62851	0.964	1.30165 65939	0.76707 02770
0.915	1.15550 79207	0.74104 08554	0.965	1.30544 33772	0.76758 83418
0.916	1.15799 35274	0.74158 48835	0.966	1.30928 36776	0.76810 58892
0.917	1.16049 33216	0.74212 83700	0.967	1.31317 98897	0.76862 29197
0.918	1.16300 75647	0.74267 13153	0.968	1.31713 45907	0.76913 94336
0.919	1.16553 65266	0.74321 37199	0.969	1.32115 05616	0.76965 54315
0.920	1.16808 04852	0.74375 55843	0.970	1.32523 08093	0.77017 09140
0.921	1.17063 97273	0.74429 69090	0.971	1.32937 85941	0.77068 58815
0.922	1.17321 45488	0.74483 76944	0.972	1.33359 74601	0.77120 03345
0.923	1.17580 52551	0.74537 79411	0.973	1.33789 12712	0.77171 42735
0.924	1.17841 21615	0.74591 76496	0.974	1.34226 42528	0.77222 76990
0.925	1.18103 55940	0.74645 68203	0.975	1.34672 10415	0.77274 06116
0.926	1.18367 58892	0.74699 54537	0.976	1.35126 67425	0.77325 30116
0.927	1.18633 33953	0.74753 35504	0.977	1.35590 69997	0.77376 48996
0.928	1.18900 84726	0.74807 11108	0.978	1.36064 80778	0.77427 62762
0.929	1.19170 14936	0.74860 81353	0.979	1.36549 69629	0.77478 71418
0.930	1.19441 28445	0.74914 46246	0.980	1.37046 14845	0.77529 74968
0.931	1.19714 29249	0.74968 05791	0.981	1.37555 04644	0.77580 73419
0.932	1.19989 21493	0.75021 59992	0.982	1.38077 39033	0.77631 66774
0.933	1.20266 09473	0.75075 08855	0.983	1.38614 32130	0.77682 55040
0.934	1.20544 97648	0.75128 52385	0.984	1.39167 15119	0.77733 38221
0.935	1.20825 90645	0.75181 90586	0.985	1.39737 40057	0.77784 16322
0.936	1.21108 93272	0.75235 23464	0.986	1.40326 84833	0.77834 89347
0.937	1.21394 10525	0.75288 51023	0.987	1.40937 59766	0.77885 57303
0.938	1.21681 47598	0.75341 73268	0.988	1.41572 16538	0.77936 20194
0.939	1.21971 09899	0.75394 90205	0.989	1.42233 60558	0.77986 78025
0.940	1.22263 03055	0.75448 01838	0.990	1.42925 68535	0.78037 30801
0.941	1.22557 32933	0.75501 08173	0.991	1.43653 14208	0.78087 78526
0.942	1.22854 05646	0.75554 09213	0.992	1.44422 07408	0.78138 21207
0.943	1.23153 27575	0.75607 04964	0.993	1.45240 56013	0.78188 58848
0.944	1.23455 05382	0.75659 95432	0.994	1.46119 69690	0.78238 91454
0.945	1.23759 46028	0.75712 80620	0.995	1.47075 46132	0.78289 19030
0.946	1.24066 56792	0.75765 60534	0.996	1.48132 37666	0.78339 41580
0.947	1.24376 45292	0.75818 35179	0.997	1.49331 72819	0.78389 59111
0.948	1.24689 19510	0.75871 04560	0.998	1.50754 02279	0.78439 71627
0.949	1.25004 87811	0.75923 68681	0.999	1.52607 12396	0.78489 79133
0.950	1.25323 58975	0.75976 27549	1.000	1.57079 63268	0.78539 81634

$\frac{\pi}{2} = 1.57079\ 63267\ 95$

SQUARE OF THE SINE AND COSINE AND THEIR PRODUCT

The table that follows gives the squares of the sine and cosine and their product for each minute from 0 to 360.

For angles at the top of the page, use the column headings at the top of the page. For angles at the bottom of the page, use the column headings at the bottom of the page.

For degrees on the left, use the minute column on the left. For degrees on the right, use the minute column on the right.

For angles in the second and fourth quadrants, the product of the sine and cosine is negative.

If $\sin^2\theta\cos^2\theta$ is desired, it can be obtained by using the identity

$$\sin^2\theta\cos^2\theta = \tfrac{1}{4}\sin^2 2\theta.$$

Thus

$$\sin^2 10^\circ \cos^2 10^\circ = \tfrac{1}{4}\sin^2 20^\circ = \frac{0.11698}{4} = 0.02924.$$

SQUARE OF THE SINE AND COSINE AND THEIR PRODUCT

Compiled by Niel F. Beardsley.

0° (180°) — (359°) **179°**

′	Sin²	Sin · Cos	Cos²	′
0	.00000	.00000	1.00000	**60**
1	.00000	.00029	1.00000	59
2	.00000	.00058	1.00000	58
3	.00000	.00087	1.00000	57
4	.00000	.00116	1.00000	56
5	.00000	.00145	1.00000	**55**
6	.00000	.00175	1.00000	54
7	.00000	.00204	1.00000	53
8	.00001	.00233	.99999	52
9	.00001	.00262	.99999	51
10	.00001	.00291	.99999	**50**
11	.00001	.00320	.99999	49
12	.00001	.00349	.99999	48
13	.00001	.00378	.99999	47
14	.00002	.00407	.99998	46
15	.00002	.00436	.99998	**45**
16	.00002	.00465	.99998	44
17	.00002	.00495	.99998	43
18	.00003	.00524	.99997	42
19	.00003	.00553	.99997	41
20	.00003	.00582	.99997	**40**
21	.00004	.00611	.99996	39
22	.00004	.00640	.99996	38
23	.00004	.00669	.99996	37
24	.00005	.00698	.99995	36
25	.00005	.00727	.99995	**35**
26	.00006	.00756	.99994	34
27	.00006	.00785	.99994	33
28	.00007	.00814	.99993	32
29	.00007	.00844	.99993	31
30	.00008	.00873	.99992	**30**
31	.00008	.00902	.99992	29
32	.00009	.00931	.99991	28
33	.00009	.00960	.99991	27
34	.00010	.00989	.99990	26
35	.00010	.01018	.99990	**25**
36	.00011	.01047	.99989	24
37	.00012	.01076	.99988	23
38	.00012	.01105	.99988	22
39	.00013	.01134	.99987	21
40	.00014	.01163	.99986	**20**
41	.00014	.01193	.99986	19
42	.00015	.01222	.99985	18
43	.00016	.01251	.99984	17
44	.00016	.01280	.99984	16
45	.00017	.01309	.99983	**15**
46	.00018	.01338	.99982	14
47	.00019	.01367	.99981	13
48	.00019	.01396	.99981	12
49	.00020	.01425	.99980	11
50	.00021	.01454	.99979	**10**
51	.00022	.01483	.99978	9
52	.00023	.01512	.99977	8
53	.00024	.01541	.99976	7
54	.00025	.01571	.99975	6
55	.00026	.01600	.99974	**5**
56	.00027	.01629	.99973	4
57	.00027	.01658	.99973	3
58	.00028	.01687	.99972	2
59	.00029	.01716	.99971	1
60	.00030	.01745	.99970	**0**
′	Cos²	Sin · Cos	Sin²	′

90° (270°) — (269°) **89°**

1° (181°) — (358°) **178°**

′	Sin²	Sin · Cos	Cos²	′
0	.00030	.01745	.99970	**60**
1	.00031	.01774	.99969	59
2	.00033	.01803	.99967	58
3	.00034	.01832	.99966	57
4	.00035	.01861	.99965	56
5	.00036	.01890	.99964	**55**
6	.00037	.01919	.99963	54
7	.00038	.01948	.99962	53
8	.00039	.01978	.99961	52
9	.00040	.02007	.99960	51
10	.00041	.02036	.99959	**50**
11	.00043	.02065	.99957	49
12	.00044	.02094	.99956	48
13	.00045	.02123	.99955	47
14	.00046	.02152	.99954	46
15	.00048	.02181	.99952	**45**
16	.00049	.02210	.99951	44
17	.00050	.02239	.99950	43
18	.00051	.02268	.99949	42
19	.00053	.02297	.99947	41
20	.00054	.02326	.99946	**40**
21	.00056	.02355	.99944	39
22	.00057	.02384	.99943	38
23	.00058	.02413	.99942	37
24	.00060	.02442	.99940	36
25	.00061	.02472	.99939	**35**
26	.00063	.02501	.99937	34
27	.00064	.02530	.99936	33
28	.00066	.02559	.99934	32
29	.00067	.02588	.99933	31
30	.00069	.02617	.99931	**30**
31	.00070	.02646	.99930	29
32	.00072	.02675	.99928	28
33	.00073	.02704	.99927	27
34	.00075	.02733	.99925	26
35	.00076	.02762	.99924	**25**
36	.00078	.02791	.99922	24
37	.00080	.02820	.99920	23
38	.00081	.02849	.99919	22
39	.00083	.02878	.99917	21
40	.00085	.02907	.99915	**20**
41	.00086	.02936	.99914	19
42	.00088	.02965	.99912	18
43	.00090	.02994	.99910	17
44	.00091	.03023	.99909	16
45	.00093	.03052	.99907	**15**
46	.00095	.03081	.99905	14
47	.00097	.03110	.99903	13
48	.00099	.03140	.99901	12
49	.00100	.03169	.99900	11
50	.00102	.03198	.99898	**10**
51	.00104	.03227	.99896	9
52	.00106	.03256	.99894	8
53	.00108	.03285	.99892	7
54	.00110	.03314	.99890	6
55	.00112	.03343	.99888	**5**
56	.00114	.03372	.99886	4
57	.00116	.03401	.99884	3
58	.00118	.03430	.99882	2
59	.00120	.03459	.99880	1
60	.00122	.03488	.99878	**0**
′	Cos²	Sin · Cos	Sin²	′

91° (271°) — (268°) **88°**

SQUARE OF THE SINE AND COSINE AND THEIR PRODUCT (Continued)

2° (182°) — (357°) **177°**

′	Sin²	Sin · Cos	Cos²	′
0	.00122	.03488	.99878	**60**
1	.00124	.03517	.99876	59
2	.00126	.03546	.99874	58
3	.00128	.03575	.99872	57
4	.00130	.03604	.99870	56
5	.00132	.03633	.99868	**55**
6	.00134	.03662	.99866	54
7	.00136	.03691	.99864	53
8	.00139	.03720	.99861	52
9	.00141	.03749	.99859	51
10	.00143	.03778	.99857	**50**
11	.00145	.03807	.99855	49
12	.00147	.03836	.99853	48
13	.00150	.03865	.99850	47
14	.00152	.03894	.99848	46
15	.00154	.03923	.99846	**45**
16	.00156	.03952	.99844	44
17	.00159	.03981	.99841	43
18	.00161	.04010	.99839	42
19	.00163	.04039	.99837	41
20	.00166	.04068	.99834	**40**
21	.00168	.04097	.99832	39
22	.00171	.04126	.99829	38
23	.00173	.04155	.99827	37
24	.00175	.04184	.99825	36
25	.00178	.04213	.99822	**35**
26	.00180	.04242	.99820	34
27	.00183	.04271	.99817	33
28	.00185	.04300	.99815	32
29	.00188	.04329	.99812	31
30	.00190	.04358	.99810	**30**
31	.00193	.04387	.99807	29
32	.00195	.04416	.99805	28
33	.00198	.04445	.99802	27
34	.00201	.04474	.99799	26
35	.00203	.04503	.99797	**25**
36	.00206	.04532	.99794	24
37	.00208	.04561	.99792	23
38	.00211	.04590	.99789	22
39	.00214	.04619	.99786	21
40	.00216	.04647	.99784	**20**
41	.00219	.04676	.99781	19
42	.00222	.04705	.99778	18
43	.00225	.04734	.99775	17
44	.00227	.04763	.99773	16
45	.00230	.04792	.99770	**15**
46	.00233	.04821	.99767	14
47	.00236	.04850	.99764	13
48	.00239	.04879	.99761	12
49	.00241	.04908	.99759	11
50	.00244	.04937	.99756	**10**
51	.00247	.04966	.99753	9
52	.00250	.04995	.99750	8
53	.00253	.05024	.99747	7
54	.00256	.05053	.99744	6
55	.00259	.05082	.99741	**5**
56	.00262	.05111	.99738	4
57	.00265	.05140	.99735	3
58	.00268	.05169	.99732	2
59	.00271	.05197	.99729	1
60	.00274	.05226	.99726	**0**
′	Cos²	Sin · Cos	Sin²	′

92° (272°) — (267°) **87°**

3° (183°) — (356°) **176°**

′	Sin²	Sin · Cos	Cos²	′
0	.00274	.05226	.99726	**60**
1	.00277	.05255	.99723	59
2	.00280	.05284	.99720	58
3	.00283	.05313	.99717	57
4	.00286	.05342	.99714	56
5	.00289	.05371	.99711	**55**
6	.00292	.05400	.99708	54
7	.00296	.05429	.99704	53
8	.00299	.05458	.99701	52
9	.00302	.05487	.99698	51
10	.00305	.05516	.99695	**50**
11	.00308	.05545	.99692	49
12	.00312	.05573	.99688	48
13	.00315	.05602	.99685	47
14	.00318	.05631	.99682	46
15	.00321	.05660	.99679	**45**
16	.00325	.05689	.99675	44
17	.00328	.05718	.99672	43
18	.00331	.05747	.99669	42
19	.00335	.05776	.99665	41
20	.00338	.05805	.99662	**40**
21	.00341	.05834	.99659	39
22	.00345	.05862	.99655	38
23	.00348	.05891	.99652	37
24	.00352	.05920	.99648	36
25	.00355	.05949	.99645	**35**
26	.00359	.05978	.99641	34
27	.00362	.06007	.99638	33
28	.00366	.06036	.99634	32
29	.00369	.06065	.99631	31
30	.00373	.06093	.99627	**30**
31	.00376	.06122	.99624	29
32	.00380	.06151	.99620	28
33	.00383	.06180	.99617	27
34	.00387	.06209	.99613	26
35	.00391	.06238	.99609	**25**
36	.00394	.06267	.99606	24
37	.00398	.06296	.99602	23
38	.00402	.06324	.99598	22
39	.00405	.06353	.99595	21
40	.00409	.06382	.99591	**20**
41	.00413	.06411	.99587	19
42	.00416	.06440	.99584	18
43	.00420	.06469	.99580	17
44	.00424	.06497	.99576	16
45	.00428	.06526	.99572	**15**
46	.00432	.06555	.99568	14
47	.00435	.06584	.99565	13
48	.00439	.06613	.99561	12
49	.00443	.06642	.99557	11
50	.00447	.06670	.99553	**10**
51	.00451	.06699	.99549	9
52	.00455	.06728	.99545	8
53	.00459	.06757	.99541	7
54	.00463	.06786	.99537	6
55	.00467	.06815	.99533	**5**
56	.00471	.06843	.99529	4
57	.00475	.06872	.99525	3
58	.00479	.06901	.99521	2
59	.00483	.06930	.99517	1
60	.00487	.06959	.99513	**0**
′	Cos²	Sin · Cos	Sin²	′

93° (273°) — (266°) 86°

SQUARE OF THE SINE AND COSINE AND THEIR PRODUCT (Continued)

4° (184°) — (355°) **175°**

′	Sin²	Sin · Cos	Cos²	′
0	.00487	.06959	.99513	**60**
1	.00491	.06987	.99509	59
2	.00495	.07016	.99505	58
3	.00499	.07045	.99501	57
4	.00503	.07074	.99497	56
5	.00507	.07103	.99493	**55**
6	.00511	.07131	.99489	54
7	.00515	.07160	.99485	53
8	.00520	.07189	.99480	52
9	.00524	.07218	.99476	51
10	.00528	.07247	.99472	**50**
11	.00532	.07275	.99468	49
12	.00536	.07304	.99464	48
13	.00541	.07333	.99459	47
14	.00545	.07362	.99455	46
15	.00549	.07390	.99451	**45**
16	.00554	.07419	.99446	44
17	.00558	.07448	.99442	43
18	.00562	.07477	.99438	42
19	.00567	.07506	.99433	41
20	.00571	.07534	.99429	**40**
21	.00575	.07563	.99425	39
22	.00580	.07592	.99420	38
23	.00584	.07621	.99416	37
24	.00589	.07649	.99411	36
25	.00593	.07678	.99407	**35**
26	.00598	.07707	.99402	34
27	.00602	.07736	.99398	33
28	.00607	.07764	.99393	32
29	.00611	.07793	.99389	31
30	.00616	.07822	.99384	**30**
31	.00620	.07850	.99380	29
32	.00625	.07879	.99375	28
33	.00629	.07908	.99371	27
34	.00634	.07937	.99366	26
35	.00639	.07965	.99361	**25**
36	.00643	.07994	.99357	24
37	.00648	.08023	.99352	23
38	.00653	.08051	.99347	22
39	.00657	.08080	.99343	21
40	.00662	.08109	.99338	**20**
41	.00667	.08138	.99333	19
42	.00671	.08166	.99329	18
43	.00676	.08195	.99324	17
44	.00681	.08224	.99319	16
45	.00686	.08252	.99314	**15**
46	.00691	.08281	.99309	14
47	.00695	.08310	.99305	13
48	.00700	.08338	.99300	12
49	.00705	.08367	.99295	11
50	.00710	.08396	.99290	**10**
51	.00715	.08424	.99285	9
52	.00720	.08453	.99280	8
53	.00725	.08482	.99275	7
54	.00730	.08510	.99270	6
55	.00735	.08539	.99265	**5**
56	.00740	.08568	.99260	4
57	.00745	.08596	.99255	3
58	.00750	.08625	.99250	2
59	.00755	.08654	.99245	1
60	.00760	.08682	.99240	**0**
′	Cos²	Sin · Cos	Sin²	′

94° (274°) — (265°) **85°**

5° (185°) — (354°) (**174°**)

′	Sin²	Sin · Cos	Cos²	′
0	.00760	.08682	.99240	**60**
1	.00765	.08711	.99235	59
2	.00770	.08740	.99230	58
3	.00775	.08768	.99225	57
4	.00780	.08797	.99220	56
5	.00785	.08826	.99215	**55**
6	.00790	.08854	.99210	54
7	.00795	.08883	.99205	53
8	.00801	.08911	.99199	52
9	.00806	.08940	.99194	51
10	.00811	.08969	.99189	**50**
11	.00816	.08997	.99184	49
12	.00821	.09026	.99179	48
13	.00827	.09055	.99173	47
14	.00832	.09083	.99168	46
15	.00837	.09112	.99163	**45**
16	.00843	.09140	.99157	44
17	.00848	.09169	.99152	43
18	.00853	.09198	.99147	42
19	.00859	.09226	.99141	41
20	.00864	.09255	.99136	**40**
21	.00869	.09283	.99131	39
22	.00875	.09312	.99125	38
23	.00880	.09340	.99120	37
24	.00886	.09369	.99114	36
25	.00891	.09398	.99109	**35**
26	.00897	.09426	.99103	34
27	.00902	.09455	.99098	33
28	.00908	.09483	.99092	32
29	.00913	.09512	.99087	31
30	.00919	.09540	.99081	**30**
31	.00924	.09569	.99076	29
32	.00930	.09598	.99070	28
33	.00935	.09626	.99065	27
34	.00941	.09655	.99059	26
35	.00947	.09683	.99053	**25**
36	.00952	.09712	.99048	24
37	.00958	.09740	.99042	23
38	.00964	.09769	.99036	22
39	.00969	.09797	.99031	21
40	.00975	.09826	.99025	**20**
41	.00981	.09854	.99019	19
42	.00986	.09883	.99014	18
43	.00992	.09911	.99008	17
44	.00998	.09940	.99002	16
45	.01004	.09968	.98996	**15**
46	.01010	.09997	.98990	14
47	.01015	.10025	.98985	13
48	.01021	.10054	.98979	12
49	.01027	.10082	.98973	11
50	.01033	.10111	.98967	**10**
51	.01039	.10139	.98961	9
52	.01045	.10168	.98955	8
53	.01051	.10196	.98949	7
54	.01057	.10225	.98943	6
55	.01063	.10253	.98937	**5**
56	.01069	.10282	.98931	4
57	.01075	.10310	.98925	3
58	.01081	.10339	.98919	2
59	.01087	.10367	.98913	1
60	.01093	.10396	.98907	**0**
′	Cos²	Sin · Cos	Sin²	′

95° (275°) — (264°) **84°**

SQUARE OF THE SINE AND COSINE AND THEIR PRODUCT (Continued)

6° (186°) (353°) **173°**

′	Sin²	Sin · Cos	Cos²	′
0	.01093	.10396	.98907	**60**
1	.01099	.10424	.98901	59
2	.01105	.10452	.98895	58
3	.01111	.10481	.98889	57
4	.01117	.10509	.98883	56
5	.01123	.10538	.98877	**55**
6	.01129	.10566	.98871	54
7	.01135	.10595	.98865	53
8	.01142	.10623	.98858	52
9	.01148	.10652	.98852	51
10	.01154	.10680	.98846	**50**
11	.01160	.10708	.98840	49
12	.01166	.10737	.98834	48
13	.01173	.10765	.98827	47
14	.01179	.10794	.98821	46
15	.01185	.10822	.98815	**45**
16	.01192	.10850	.98808	44
17	.01198	.10879	.98802	43
18	.01204	.10907	.98796	42
19	.01211	.10936	.98789	41
20	.01217	.10964	.98783	**40**
21	.01223	.10992	.98777	39
22	.01230	.11021	.98770	38
23	.01236	.11049	.98764	37
24	.01243	.11077	.98757	36
25	.01249	.11106	.98751	**35**
26	.01255	.11134	.98745	34
27	.01262	.11163	.98738	33
28	.01268	.11191	.98732	32
29	.01275	.11219	.98725	31
30	.01281	.11248	.98719	**30**
31	.01288	.11276	.98712	29
32	.01295	.11304	.98705	28
33	.01301	.11333	.98699	27
34	.01308	.11361	.98692	26
35	.01314	.11389	.98686	**25**
36	.01321	.11418	.98679	24
37	.01328	.11446	.98672	23
38	.01334	.11474	.98666	22
39	.01341	.11502	.98659	21
40	.01348	.11531	.98652	**20**
41	.01354	.11559	.98646	19
42	.01361	.11587	.98639	18
43	.01368	.11616	.98632	17
44	.01375	.11644	.98625	16
45	.01382	.11672	.98618	**15**
46	.01388	.11701	.98612	14
47	.01395	.11729	.98605	13
48	.01402	.11757	.98598	12
49	.01409	.11785	.98591	11
50	.01416	.11814	.98584	**10**
51	.01423	.11842	.98577	9
52	.01429	.11870	.98571	8
53	.01436	.11898	.98564	7
54	.01443	.11927	.98557	6
55	.01450	.11955	98550	**5**
56	.01457	.11983	.98543	4
57	.01464	.12011	.98536	3
58	.01471	.12040	.98529	2
59	.01478	.12068	.98522	1
60	.01485	.12096	.98515	**0**
′	Cos²	Sin · Cos	Sin²	′

96° (276°) (263°) **83°**

7° (187°) (352°) **172°**

′	Sin²	Sin · Cos	Cos²	′
0	.01485	.12096	.98515	**60**
1	.01492	.12124	.98508	59
2	.01499	.12153	.98501	58
3	.01506	.12181	.98494	57
4	.01513	.12209	.98487	56
5	.01521	.12237	.98479	**55**
6	.01528	.12265	.98472	54
7	.01535	.12294	.98465	53
8	.01542	.12322	.98458	52
9	.01549	.12350	.98451	51
10	.01556	.12378	.98444	**50**
11	.01564	.12406	.98436	49
12	.01571	.12434	.98429	48
13	.01578	.12463	.98422	47
14	.01585	.12491	.98415	46
15	.01593	.12519	.98407	**45**
16	.01600	.12547	.98400	44
17	.01607	.12575	.98393	43
18	.01615	.12603	.98385	42
19	.01622	.12632	.98378	41
20	.01629	.12660	.98371	**40**
21	.01637	.12688	.98363	39
22	.01644	.12716	.98356	38
23	.01651	.12744	.98349	37
24	.01659	.12772	.98341	36
25	.01666	.12800	.98334	**35**
26	.01674	.12829	.98326	34
27	.01681	.12857	.98319	33
28	.01689	.12885	.98311	32
29	.01696	.12913	.98304	31
30	.01704	.12941	.98296	**30**
31	.01711	.12969	.98289	29
32	.01719	.12997	.98281	28
33	.01726	.13025	.98274	27
34	.01734	.13053	.98266	26
35	.01742	.13081	.98258	**25**
36	.01749	.13109	.98251	24
37	.01757	.13138	.98243	23
38	.01764	.13166	.98236	22
39	.01772	.13194	.98228	21
40	.01780	.13222	.98220	**20**
41	.01788	.13250	.98212	19
42	.01795	.13278	.98205	18
43	.01803	.13306	.98197	17
44	.01811	.13334	.98189	16
45	.01818	.13362	.98182	**15**
46	.01826	.13390	.98174	14
47	.01834	.13418	.98166	13
48	.01842	.13446	.98158	12
49	.01850	.13474	.98150	11
50	.01858	.13502	.98142	**10**
51	.01865	.13530	.98135	9
52	.01873	.13558	.98127	8
53	.01881	.13586	.98119	7
54	.01889	.13614	.98111	6
55	.01897	.13642	.98103	**5**
56	.01905	.13670	.98095	4
57	.01913	.13698	.98087	3
58	.01921	.13726	.98079	2
59	.01929	.13754	.98071	1
60	.01937	.13782	.98063	**0**
′	Cos²	Sin · Cos	Sin²	′

97° (277°) (262°) **82°**

SQUARE OF THE SINE AND COSINE AND THEIR PRODUCT (Continued)

8° (188°) — (351°) **171°**

′	Sin²	Sin · Cos	Cos²	′
0	.01937	.13782	.98063	**60**
1	.01945	.13810	.98055	59
2	.01953	.13838	.98047	58
3	.01961	.13866	.98039	57
4	.01969	.13894	.98031	56
5	.01977	.13922	.98023	**55**
6	.01985	.13950	.98015	54
7	.01993	.13977	.98007	53
8	.02002	.14005	.97998	52
9	.02010	.14033	.97990	51
10	.02018	.14061	.97982	**50**
11	.02026	.14089	.97974	49
12	.02034	.14117	.97966	48
13	.02043	.14145	.97957	47
14	.02051	.14173	.97949	46
15	.02059	.14201	.97941	**45**
16	.02067	.14229	.97933	44
17	.02076	.14257	.97924	43
18	.02084	.14284	.97916	42
19	.02092	.14312	.97908	41
20	.02101	.14340	.97899	**40**
21	.02109	.14368	.97891	39
22	.02117	.14396	.97883	38
23	.02126	.14424	.97874	37
24	.02134	.14452	.97866	36
25	.02142	.14479	.97858	**35**
26	.02151	.14507	.97849	34
27	.02159	.14535	.97841	33
28	.02168	.14563	.97832	32
29	.02176	.14591	.97824	31
30	.02185	.14619	.97815	**30**
31	.02193	.14646	.97807	29
32	.02202	.14674	.97798	28
33	.02210	.14702	.97790	27
34	.02219	.14730	.97781	26
35	.02227	.14758	.97773	**25**
36	.02236	.14785	.97764	24
37	.02245	.14813	.97755	23
38	.02253	.14841	.97747	22
39	.02262	.14869	.97738	21
40	.02271	.14897	.97729	**20**
41	.02279	.14924	.97721	19
42	.02288	.14952	.97712	18
43	.02297	.14980	.97703	17
44	.02305	.15008	.97695	16
45	.02314	.15035	.97686	**15**
46	.02323	.15063	.97677	14
47	.02332	.15091	.97668	13
48	.02340	.15118	.97660	12
49	.02349	.15146	.97651	11
50	.02358	.15174	.97642	**10**
51	.02367	.15202	.97633	9
52	.02376	.15229	.97624	8
53	.02385	.15257	.97615	7
54	.02394	.15285	.97606	6
55	.02402	.15312	.97598	**5**
56	.02411	.15340	.97589	4
57	.02420	.15368	.97580	3
58	.02429	.15396	.97571	2
59	.02438	.15423	.97562	1
60	.02447	.15451	.97553	**0**
′	Cos²	Sin · Cos	Sin²	′

98° (278°) — (261°) **81°**

9° (189°) — (350°) **170°**

′	Sin²	Sin · Cos	Cos²	′
0	.02447	.15451	.97553	**60**
1	.02456	.15479	.97544	59
2	.02465	.15506	.97535	58
3	.02474	.15534	.97526	57
4	.02483	.15561	.97517	56
5	.02492	.15589	.97508	**55**
6	.02501	.15617	.97499	54
7	.02510	.15644	.97490	53
8	.02520	.15672	.97480	52
9	.02529	.15700	.97471	51
10	.02538	.15727	.97462	**50**
11	.02547	.15755	.97453	49
12	.02556	.15782	.97444	48
13	.02565	.15810	.97435	47
14	.02575	.15838	.97425	46
15	.02584	.15865	.97416	**45**
16	.02593	.15893	.97407	44
17	.02602	.15920	.97398	43
18	.02612	.15948	.97388	42
19	.02621	.15976	.97379	41
20	.02630	.16003	.97370	**40**
21	.02639	.16031	.97361	39
22	.02649	.16058	.97351	38
23	.02658	.16086	.97342	37
24	.02668	.16113	.97332	36
25	.02677	.16141	.97323	**35**
26	.02686	.16168	.97314	34
27	.02696	.16196	.97304	33
28	.02705	.16223	.97295	32
29	.02715	.16251	.97285	31
30	.02724	.16278	.97276	**30**
31	.02734	.16306	.97266	29
32	.02743	.16333	.97257	28
33	.02753	.16361	.97247	27
34	.02762	.16388	.97238	26
35	.02772	.16416	.97228	**25**
36	.02781	.16443	.97219	24
37	.02791	.16471	.97209	23
38	.02800	.16498	.97200	22
39	.02810	.16526	.97190	21
40	.02820	.16553	.97180	**20**
41	.02829	.16581	.97171	19
42	.02839	.16608	.97161	18
43	.02849	.16635	.97151	17
44	.02858	.16663	.97142	16
45	.02868	.16690	.97132	**15**
46	.02878	.16718	.97122	14
47	.02887	.16745	.97113	13
48	.02897	.16773	.97103	12
49	.02907	.16800	.97093	11
50	.02917	.16827	.97083	**10**
51	.02926	.16855	.97074	9
52	.02936	.16882	.97064	8
53	.02946	.16910	.97054	7
54	.02956	.16937	.97044	6
55	.02966	.16964	.97034	**5**
56	.02976	.16992	.97024	4
57	.02986	.17019	.97014	3
58	.02996	.17046	.97004	2
59	.03005	.17074	.96995	1
60	.03015	.17101	.96985	**0**
′	Cos²	Sin · Cos	Sin²	′

99° (279°) — (260°) **80°**

SQUARE OF THE SINE AND COSINE AND THEIR PRODUCT (Continued)

10° (190°) (349°) **169°**

′	Sin²	Sin · Cos	Cos²	′
0	.03015	.17101	.96985	**60**
1	.03025	.17128	.96975	59
2	.03035	.17156	.96965	58
3	.03045	.17183	.96955	57
4	.03055	.17210	.96945	56
5	.03065	.17238	.96935	**55**
6	.03075	.17265	.96925	54
7	.03085	.17292	.96915	53
8	.03095	.17319	.96905	52
9	.03106	.17347	.96894	51
10	.03116	.17374	.96884	**50**
11	.03126	.17401	.96874	49
12	.03136	.17429	.96864	48
13	.03146	.17456	.96854	47
14	.03156	.17483	.96844	46
15	.03166	.17510	.96834	**45**
16	.03177	.17538	.96823	44
17	.03187	.17565	.96813	43
18	.03197	.17592	.96803	42
19	.03207	.17619	.96793	41
20	.03218	.17647	.96782	**40**
21	.03228	.17674	.96772	39
22	.03238	.17701	.96762	38
23	.03248	.17728	.96752	37
24	.03259	.17755	.96741	36
25	.03269	.17783	.96731	**35**
26	.03279	.17810	.96721	34
27	.03290	.17837	.96710	33
28	.03300	.17864	.96700	32
29	.03311	.17891	.96689	31
30	.03321	.17918	.96679	**30**
31	.03331	.17946	.96669	29
32	.03342	.17973	.96658	28
33	.03352	.18000	.96648	27
34	.03363	.18027	.96637	26
35	.03373	.18054	.96627	**25**
36	.03384	.18081	.96616	24
37	.03394	.18108	.96606	23
38	.03405	.18135	.96595	22
39	.03415	.18163	.96585	21
40	.03426	.18190	.96574	**20**
41	.03437	.18217	.96563	19
42	.03447	.18244	.96553	18
43	.03458	.18271	.96542	17
44	.03468	.18298	.96532	16
45	.03479	.18325	.96521	**15**
46	.03490	.18352	.96510	14
47	.03500	.18379	.96500	13
48	.03511	.18406	.96489	12
49	.03522	.18433	.96478	11
50	.03533	.18460	.96467	**10**
51	.03543	.18487	.96457	9
52	.03554	.18514	.96446	8
53	.03565	.18541	.96435	7
54	.03576	.18568	.96424	6
55	.03587	.18595	.96413	**5**
56	.03597	.18622	.96403	4
57	.03608	.18649	.96392	3
58	.03619	.18676	.96381	2
59	.03630	.18703	.96370	1
60	.03641	.18730	.96359	**0**
′	Cos²	Sin · Cos	Sin²	′

100° (280°) (259°) **79°**

11° (191°) (348°) **168°**

′	Sin²	Sin · Cos	Cos²	′
0	.03641	.18730	.96359	**60**
1	.03652	.18757	.96348	59
2	.03663	.18784	.96337	58
3	.03674	.18811	.96326	57
4	.03685	.18838	.96315	56
5	.03695	.18865	.96305	**55**
6	.03706	.18892	.96294	54
7	.03717	.18919	.96283	53
8	.03728	.18946	.96272	52
9	.03740	.18973	.96260	51
10	.03751	.19000	.96249	**50**
11	.03762	.19027	.96238	49
12	.03773	.19054	.96227	48
13	.03784	.19080	.96216	47
14	.03795	.19107	.96205	46
15	.03806	.19134	.96194	**45**
16	.03817	.19161	.96183	44
17	.03828	.19188	.96172	43
18	.03839	.19215	.96161	42
19	.03851	.19242	.96149	41
20	.03862	.19268	.96138	**40**
21	.03873	.19295	.96127	39
22	.03884	.19322	.96116	38
23	.03896	.19349	.96104	37
24	.03907	.19376	.96093	36
25	.03918	.19403	.96082	**35**
26	.03929	.19429	.96071	34
27	.03941	.19456	.96059	33
28	.03952	.19483	.96048	32
29	.03963	.19510	.96037	31
30	.03975	.19537	.96025	**30**
31	.03986	.19563	.96014	29
32	.03998	.19590	.96002	28
33	.04009	.19617	.95991	27
34	.04020	.19644	.95980	26
35	.04032	.19670	.95968	**25**
36	.04043	.19697	.95957	24
37	.04055	.19724	.95945	23
38	.04066	.19751	.95934	22
39	.04078	.19777	.95922	21
40	.04089	.19804	.95911	**20**
41	.04101	.19831	.95899	19
42	.04112	.19857	.95888	18
43	.04124	.19884	.95876	17
44	.04135	.19911	.95865	16
45	.04147	.19937	.95853	**15**
46	.04159	.19964	.95841	14
47	.04170	.19991	.95830	13
48	.04182	.20017	.95818	12
49	.04194	.20044	.95806	11
50	.04205	.20071	.95795	**10**
51	.04217	.20097	.95783	9
52	.04229	.20124	.95771	8
53	.04240	.20151	.95760	7
54	.04252	.20177	.95748	6
55	.04264	.20204	.95736	**5**
56	.04276	.20230	.95724	4
57	.04287	.20257	.95713	3
58	.04299	.20284	.95701	2
59	.04311	.20310	.95689	1
60	.04323	.20337	.95677	**0**
′	Cos²	Sin · Cos	Sin²	′

101° (281°) (258°) **78°**

SQUARE OF THE SINE AND COSINE AND THEIR PRODUCT (Continued)

12° (192°) (347°) **167°**

′	Sin²	Sin · Cos	Cos²	′
0	.04323	.20337	.95677	**60**
1	.04335	.20363	.95665	59
2	.04346	.20390	.95654	58
3	.04358	.20417	.95642	57
4	.04370	.20443	.95630	56
5	.04382	.20470	.95618	**55**
6	.04394	.20496	.95606	54
7	.04406	.20523	.95594	53
8	.04418	.20549	.95582	52
9	.04430	.20576	.95570	51
10	.04442	.20602	.95558	**50**
11	.04454	.20629	.95546	49
12	.04466	.20655	.95534	48
13	.04478	.20682	.95522	47
14	.04490	.20708	.95510	46
15	.04502	.20735	.95498	**45**
16	.04514	.20761	.95486	44
17	.04526	.20788	.95474	43
18	.04538	.20814	.95462	42
19	.04550	.20840	.95450	41
20	.04562	.20867	.95438	**40**
21	.04575	.20893	.95425	39
22	.04587	.20920	.95413	38
23	.04599	.20946	.95401	37
24	.04611	.20973	.95389	36
25	.04623	.20999	.95377	**35**
26	.04636	.21025	.95364	34
27	.04648	.21052	.95352	33
28	.04660	.21078	.95340	32
29	.04672	.21105	.95328	31
30	.04685	.21131	.95315	**30**
31	.04697	.21157	.95303	29
32	.04709	.21184	.95291	28
33	.04722	.21210	.95278	27
34	.04734	.21236	.95266	26
35	.04746	.21263	.95254	**25**
36	.04759	.21289	.95241	24
37	.04771	.21315	.95229	23
38	.04783	.21342	.95217	22
39	.04796	.21368	.95204	21
40	.04808	.21394	.95192	**20**
41	.04821	.21420	.95179	19
42	.04833	.21447	.95167	18
43	.04846	.21473	.95154	17
44	.04858	.21499	.95142	16
45	.04871	.21526	.95129	**15**
46	.04883	.21552	.95117	14
47	.04896	.21578	.95104	13
48	.04908	.21604	.95092	12
49	.04921	.21631	.95079	11
50	.04934	.21657	.95066	**10**
51	.04946	.21683	.95054	9
52	.04959	.21709	.95041	8
53	.04971	.21735	.95029	7
54	.04984	.21762	.95016	6
55	.04997	.21788	.95003	**5**
56	.05009	.21814	.94991	4
57	.05022	.21840	.94978	3
58	.05035	.21866	.94965	2
59	.05048	.21892	.94952	1
60	.05060	.21919	.94940	**0**
′	Cos²	Sin · Cos	Sin²	′

102° (282°) (257°) **77°**

13° (193°) (346°) **166°**

′	Sin²	Sin · Cos	Cos²	′
0	.05060	.21919	.94940	**60**
1	.05073	.21945	.94927	59
2	.05086	.21971	.94914	58
3	.05099	.21997	.94901	57
4	.05111	.22023	.94889	56
5	.05124	.22049	.94876	**55**
6	.05137	.22075	.94863	54
7	.05150	.22101	.94850	53
8	.05163	.22127	.94837	52
9	.05176	.22154	.94824	51
10	.05189	.22180	.94811	**50**
11	.05201	.22206	.94799	49
12	.05214	.22232	.94786	48
13	.05227	.22258	.94773	47
14	.05240	.22284	.94760	46
15	.05253	.22310	.94747	**45**
16	.05266	.22336	.94734	44
17	.05279	.22362	.94721	43
18	.05292	.22388	.94708	42
19	.05305	.22414	.94695	41
20	.05318	.22440	.94682	**40**
21	.05331	.22466	.94669	39
22	.05345	.22492	.94655	38
23	.05358	.22518	.94642	37
24	.05371	.22544	.94629	36
25	.05384	.22570	.94616	**35**
26	.05397	.22596	.94603	34
27	.05410	.22622	.94590	33
28	.05423	.22648	.94577	32
29	.05436	.22674	.94564	31
30	.05450	.22700	.94550	**30**
31	.05463	.22725	.94537	29
32	.05476	.22751	.94524	28
33	.05489	.22777	.94511	27
34	.05503	.22803	.94497	26
35	.05516	.22829	.94484	**25**
36	.05529	.22855	.94471	24
37	.05542	.22881	.94458	23
38	.05556	.22907	.94444	22
39	.05569	.22932	.94431	21
40	.05582	.22958	.94418	**20**
41	.05596	.22984	.94404	19
42	.05609	.23010	.94391	18
43	.05623	.23036	.94377	17
44	.05636	.23062	.94364	16
45	.05649	.23087	.94351	**15**
46	.05663	.23113	.94337	14
47	.05676	.23139	.94324	13
48	.05690	.23165	.94310	12
49	.05703	.23191	.94297	11
50	.05717	.23216	.94283	**10**
51	.05730	.23242	.94270	9
52	.05744	.23268	.94256	8
53	.05757	.23294	.94243	7
54	.05771	.23319	.94229	6
55	.05785	.23345	.94215	**5**
56	.05798	.23371	.94202	4
57	.05812	.23396	.94188	3
58	.05825	.23422	.94175	2
59	.05839	.23448	.94161	1
60	.05853	.23474	.94147	**0**
′	Cos²	Sin · Cos	Sin²	′

103° (283°) (256°) **76°**

SQUARE OF THE SINE AND COSINE AND THEIR PRODUCT (Continued)

14° (194°) (345°) **165°**

′	Sin²	Sin · Cos	Cos²	′
0	.05853	.23474	.94147	**60**
1	.05866	.23499	.94134	59
2	.05880	.23525	.94120	58
3	.05894	.23551	.94106	57
4	.05907	.23576	.94093	56
5	.05921	.23602	.94079	**55**
6	.05935	.23628	.94065	54
7	.05949	.23653	.94051	53
8	.05962	.23679	.94038	52
9	.05976	.23704	.94024	51
10	.05990	.23730	.94010	**50**
11	.06004	.23756	.93996	49
12	.06018	.23781	.93982	48
13	.06031	.23807	.93969	47
14	.06045	.23832	.93955	46
15	.06059	.23858	.93941	**45**
16	.06073	.23883	.93927	44
17	.06087	.23909	.93913	43
18	.06101	.23935	.93899	42
19	.06115	.23960	.93885	41
20	.06129	.23986	.93871	**40**
21	.06143	.24011	.93857	39
22	.06157	.24037	.93843	38
23	.06171	.24062	.93829	37
24	.06185	.24088	.93815	36
25	.06199	.24113	.93801	**35**
26	.06213	.24139	.93787	34
27	.06227	.24164	.93773	33
28	.06241	.24190	.93759	32
29	.06255	.24215	.93745	31
30	.06269	.24240	.93731	**30**
31	.06283	.24266	.93717	29
32	.06297	.24291	.93703	28
33	.06311	.24317	.93689	27
34	.06326	.24342	.93674	26
35	.06340	.24368	.93660	**25**
36	.06354	.24393	.93646	24
37	.06368	.24418	.93632	23
38	.06382	.24444	.93618	22
39	.06397	.24469	.93603	21
40	.06411	.24494	.93589	**20**
41	.06425	.24520	.93575	19
42	.06439	.24545	.93561	18
43	.06454	.24571	.93546	17
44	.06468	.24596	.93532	16
45	.06482	.24621	.93518	**15**
46	.06497	.24646	.93503	14
47	.06511	.24672	.93489	13
48	.06525	.24697	.93475	12
49	.06540	.24722	.93460	11
50	.06554	.24748	.93446	**10**
51	.06568	.24773	.93432	9
52	.06583	.24798	.93417	8
53	.06597	.24823	.93403	7
54	.06612	.24849	.93388	6
55	.06626	.24874	.93374	**5**
56	.06641	.24899	.93359	4
57	.06655	.24924	.93345	3
58	.06670	.24950	.93330	2
59	.06684	.24975	.93316	1
60	.06699	.25000	.93301	**0**
′	Cos²	Sin · Cos	Sin²	′

104° (284°) (255°) **75°**

15° (195°) (344°) **164°**

′	Sin²	Sin · Cos	Cos²	′
0	.06699	.25000	.93301	**60**
1	.06713	.25025	.93287	59
2	.06728	.25050	.93272	58
3	.06742	.25076	.93258	57
4	.06757	.25101	.93243	56
5	.06772	.25126	.93228	**55**
6	.06786	.25151	.93214	54
7	.06801	.25176	.93199	53
8	.06816	.25201	.93184	52
9	.06830	.25226	.93170	51
10	.06845	.25251	.93155	**50**
11	.06860	.25277	.93140	49
12	.06874	.25302	.93126	48
13	.06889	.25327	.93111	47
14	.06904	.25352	.93096	46
15	.06919	.25377	.93081	**45**
16	.06933	.25402	.93067	44
17	.06948	.25427	.93052	43
18	.06963	.25452	.93037	42
19	.06978	.25477	.93022	41
20	.06993	.25502	.93007	**40**
21	.07007	.25527	.92993	39
22	.07022	.25552	.92978	38
23	.07037	.25577	.92963	37
24	.07052	.25602	.92948	36
25	.07067	.25627	.92933	**35**
26	.07082	.25652	.92918	34
27	.07097	.25677	.92903	33
28	.07112	.25702	.92888	32
29	.07127	.25727	.92873	31
30	.07142	.25752	.92858	**30**
31	.07157	.25777	.92843	29
32	.07172	.25802	.92828	28
33	.07187	.25827	.92813	27
34	.07202	.25852	.92798	26
35	.07217	.25876	.92783	**25**
36	.07232	.25901	.92768	24
37	.07247	.25926	.92753	23
38	.07262	.25951	.92738	22
39	.07277	.25976	.92723	21
40	.07292	.26001	.92708	**20**
41	.07307	.26026	.92693	19
42	.07322	.26050	.92678	18
43	.07338	.26075	.92662	17
44	.07353	.26100	.92647	16
45	.07368	.26125	.92632	**15**
46	.07383	.26150	.92617	14
47	.07398	.26175	.92602	13
48	.07414	.26199	.92586	12
49	.07429	.26224	.92571	11
50	.07444	.26249	.92556	**10**
51	.07459	.26274	.92541	9
52	.07475	.26298	.92525	8
53	.07490	.26323	.92510	7
54	.07505	.26348	.92495	6
55	.07521	.26373	.92479	**5**
56	.07536	.26397	.92464	4
57	.07551	.26422	.92449	3
58	.07567	.26447	.92433	2
59	.07582	.26471	.92418	1
60	.07598	.26496	.92402	**0**
′	Cos²	Sin · Cos	Sin²	′

105° (285°) (254°) **74°**

SQUARE OF THE SINE AND COSINE AND THEIR PRODUCT (Continued)

16° (196°) (343°) **163°**

′	Sin²	Sin · Cos	Cos²	′
0	.07598	.26494	.92402	**60**
1	.07613	.26521	.92387	59
2	.07628	.26545	.92372	58
3	.07644	.26570	.92356	57
4	.07659	.26595	.92341	56
5	.07675	.26619	.92325	**55**
6	.07690	.26644	.92310	54
7	.07706	.26668	.92294	53
8	.07721	.26693	.92279	52
9	.07737	.26718	.92263	51
10	.07752	.26742	.92248	**50**
11	.07768	.26767	.92232	49
12	.07784	.26791	.92216	48
13	.07799	.26816	.92201	47
14	.07815	.26840	.92185	46
15	.07830	.26865	.92170	**45**
16	.07846	.26890	.92154	44
17	.07862	.26914	.92138	43
18	.07877	.26939	.92123	42
19	.07893	.26963	.92107	41
20	.07909	.26988	.92091	**40**
21	.07924	.27012	.92076	39
22	.07940	.27036	.92060	38
23	.07956	.27061	.92044	37
24	.07972	.27085	.92028	36
25	.07987	.27110	.92013	**35**
26	.08003	.27134	.91997	34
27	.08019	.27159	.91981	33
28	.08035	.27183	.91965	32
29	.08051	.27208	.91949	31
30	.08066	.27232	.91934	**30**
31	.08082	.27256	.91918	29
32	.08098	.27281	.91902	28
33	.08114	.27305	.91886	27
34	.08130	.27329	.91870	26
35	.08146	.27354	.91854	**25**
36	.08162	.27378	.91838	24
37	.08178	.27402	.91822	23
38	.08194	.27427	.91806	22
39	.08210	.27451	.91790	21
40	.08226	.27475	.91774	**20**
41	.08242	.27500	.91758	19
42	.08258	.27524	.91742	18
43	.08274	.27548	.91726	17
44	.08290	.27573	.91710	16
45	.08306	.27597	.91694	**15**
46	.08322	.27621	.91678	14
47	.08338	.27645	.91662	13
48	.08354	.27670	.91646	12
49	.08370	.27694	.91630	11
50	.08386	.27718	.91614	**10**
51	.08402	.27742	.91598	9
52	.08418	.27766	.91582	8
53	.08435	.27791	.91565	7
54	.08451	.27815	.91549	6
55	.08467	.27839	.91533	**5**
56	.08483	.27863	.91517	4
57	.08499	.27887	.91501	3
58	.08516	.27911	.91484	2
59	.08532	.27936	.91468	1
60	.08548	.27960	.91452	**0**
′	Cos²	Sin · Cos	Sin²	′

106° (286°) (253°) **73°**

17° (197°) (342°) **162°**

′	Sin²	Sin · Cos	Cos²	′
0	.08548	.27960	.91452	**60**
1	.08564	.27984	.91436	59
2	.08581	.28008	.91419	58
3	.08597	.28032	.91403	57
4	.08613	.28056	.91387	56
5	.08630	.28080	.91370	**55**
6	.08646	.28104	.91354	54
7	.08662	.28128	.91338	53
8	.08679	.28152	.91321	52
9	.08695	.28176	.91305	51
10	.08711	.28200	.91289	**50**
11	.08728	.28224	.91272	49
12	.08744	.28248	.91256	48
13	.08761	.28272	.91239	47
14	.08777	.28296	.91223	46
15	.08794	.28320	.91206	**45**
16	.08810	.28344	.91190	44
17	.08827	.28368	.91173	43
18	.08843	.28392	.91157	42
19	.08860	.28416	.91140	41
20	.08876	.28440	.91124	**40**
21	.08893	.28464	.91107	39
22	.08909	.28488	.91091	38
23	.08926	.28512	.91074	37
24	.08943	.28536	.91057	36
25	.08959	.28560	.91041	**35**
26	.08976	.28583	.91024	34
27	.08992	.28607	.91008	33
28	.09009	.28631	.90991	32
29	.09026	.28655	.90974	31
30	.09042	.28679	.90958	**30**
31	.09059	.28703	.90941	29
32	.09076	.28726	.90924	28
33	.09093	.28750	.90907	27
34	.09109	.28774	.90891	26
35	.09126	.28798	.90874	**25**
36	.09143	.28822	.90857	24
37	.09160	.28845	.90840	23
38	.09176	.28869	.90824	22
39	.09193	.28893	.90807	21
40	.09210	.28917	.90790	**20**
41	.09227	.28940	.90773	19
42	.09244	.28964	.90756	18
43	.09260	.28988	.90740	17
44	.09277	.29011	.90723	16
45	.09294	.29035	.90706	**15**
46	.09311	.29059	.90689	14
47	.09328	.29082	.90672	13
48	.09345	.29106	.90655	12
49	.09362	.29130	.90638	11
50	.09379	.29153	.90621	**10**
51	.09396	.29177	.90604	9
52	.09413	.29201	.90587	8
53	.09430	.29224	.90570	7
54	.09447	.29248	.90553	6
55	.09464	.29271	.90536	**5**
56	.09481	.29295	.90519	4
57	.09498	.29319	.90502	3
58	.09515	.29342	.90485	2
59	.09532	.29366	.90468	1
60	.09549	.29389	.90451	**0**
′	Cos²	Sin · Cos	Sin²	′

107° (287°) (252°) **72°**

SQUARE OF THE SINE AND COSINE AND THEIR PRODUCT (Continued)

18° (198°) (341°) **161°**

′	Sin²	Sin · Cos	Cos²	′
0	.09549	.29389	.90451	**60**
1	.09566	.29413	.90434	59
2	.09583	.29436	.90417	58
3	.09601	.29460	.90399	57
4	.09618	.29483	.90382	56
5	.09635	.29507	.90365	**55**
6	.09652	.29530	.90348	54
7	.09669	.29554	.90331	53
8	.09686	.29577	.90314	52
9	.09704	.29601	.90296	51
10	.09721	.29624	.90279	**50**
11	.09738	.29648	.90262	49
12	.09755	.29671	.90245	48
13	.09773	.29694	.90227	47
14	.09790	.29718	.90210	46
15	.09807	.29741	.90193	**45**
16	.09824	.29765	.90176	44
17	.09842	.29788	.90158	43
18	.09859	.29811	.90141	42
19	.09876	.29835	.90124	41
20	.09894	.29858	.90106	**40**
21	.09911	.29881	.90089	39
22	.09929	.29905	.90071	38
23	.09946	.29928	.90054	37
24	.09963	.29951	.90037	36
25	.09981	.29974	.90019	**35**
26	.09998	.29998	.90002	34
27	.10016	.30021	.89984	33
28	.10033	.30044	.89967	32
29	.10051	.30068	.89949	31
30	.10068	.30091	.89932	**30**
31	.10086	.30114	.89914	29
32	.10103	.30137	.89897	28
33	.10121	.30160	.89879	27
34	.10138	.30184	.89862	26
35	.10156	.30207	.89844	**25**
36	.10174	.30230	.89826	24
37	.10191	.30253	.89809	23
38	.10209	.30276	.89791	22
39	.10226	.30299	.89774	21
40	.10244	.30323	.89756	**20**
41	.10262	.30346	.89738	19
42	.10279	.30369	.89721	18
43	.10297	.30392	.89703	17
44	.10315	.30415	.89685	16
45	.10332	.30438	.89668	**15**
46	.10350	.30461	.89650	14
47	.10368	.30484	.89632	13
48	.10386	.30507	.89614	12
49	.10403	.30530	.89597	11
50	.10421	.30553	.89579	**10**
51	.10439	.30576	.89561	9
52	.10457	.30599	.89543	8
53	.10474	.30622	.89526	7
54	.10492	.30645	.89508	6
55	.10510	.30668	.89490	**5**
56	.10528	.30691	.89472	4
57	.10546	.30714	.89454	3
58	.10564	.30737	.89436	2
59	.10582	.30760	.89418	1
60	.10599	.30783	.89401	**0**
′	Cos²	Sin · Cos	Sin²	′

108° (288°) (251°) **71°**

19° (199°) (340°) **160°**

′	Sin²	Sin · Cos	Cos²	′
0	.10599	.30783	.89401	**60**
1	.10617	.30806	.89383	59
2	.10635	.30829	.89365	58
3	.10653	.30852	.89347	57
4	.10671	.30875	.89329	56
5	.10689	.30898	.89311	**55**
6	.10707	.30920	.89293	54
7	.10725	.30943	.89275	53
8	.10743	.30966	.89257	52
9	.10761	.30989	.89239	51
10	.10779	.31012	.89221	**50**
11	.10797	.31035	.89203	49
12	.10815	.31057	.89185	48
13	.10833	.31080	.89167	47
14	.10851	.31103	.89149	46
15	.10870	.31126	.89130	**45**
16	.10888	.31148	.89112	44
17	.10906	.31171	.89094	43
18	.10924	.31194	.89076	42
19	.10942	.31217	.89058	41
20	.10960	.31239	.89040	**40**
21	.10978	.31262	.89022	39
22	.10997	.31285	.89003	38
23	.11015	.31308	.88985	37
24	.11033	.31330	.88967	36
25	.11051	.31353	.88949	**35**
26	.11070	.31376	.88930	34
27	.11088	.31398	.88912	33
28	.11106	.31421	.88894	32
29	.11124	.31443	.88876	31
30	.11143	.31466	.88857	**30**
31	.11161	.31489	.88839	29
32	.11179	.31511	.88821	28
33	.11198	.31534	.88802	27
34	.11216	.31556	.88784	26
35	.11234	.31579	.88766	**25**
36	.11253	.31601	.88747	24
37	.11271	.31624	.88729	23
38	.11290	.31647	.88710	22
39	.11308	.31669	.88692	21
40	.11326	.31692	.88674	**20**
41	.11345	.31714	.88655	19
42	.11363	.31737	.88637	18
43	.11382	.31759	.88618	17
44	.11400	.31781	.88600	16
45	.11419	.31804	.88581	**15**
46	.11437	.31826	.88563	14
47	.11456	.31849	.88544	13
48	.11474	.31871	.88526	12
49	.11493	.31894	.88507	11
50	.11511	.31916	.88489	**10**
51	.11530	.31938	.88470	9
52	.11549	.31961	.88451	8
53	.11567	.31983	.88433	7
54	.11586	.32005	.88414	6
55	.11604	.32028	.88396	**5**
56	.11623	.32050	.88377	4
57	.11642	.32072	.88358	3
58	.11660	.32095	.88340	2
59	.11679	.32117	.88321	1
60	.11698	.32139	.88302	**0**
′	Cos²	Sin · Cos	Sin²	′

109° (289°) (250°) **70°**

SQUARE OF THE SINE AND COSINE AND THEIR PRODUCT (Continued)

20° (200°) — (339°) **159°**

′	Sin²	Sin · Cos	Cos²	′
0	.11698	.32139	.88302	**60**
1	.11716	.32162	.88284	59
2	.11735	.32184	.88265	58
3	.11754	.32206	.88246	57
4	.11773	.32228	.88227	56
5	.11791	.32251	.88209	**55**
6	.11810	.32273	.88190	54
7	.11829	.32295	.88171	53
8	.11848	.32317	.88152	52
9	.11867	.32339	.88133	51
10	.11885	.32362	.88115	**50**
11	.11904	.32384	.88096	49
12	.11923	.32406	.88077	48
13	.11942	.32428	.88058	47
14	.11961	.32450	.88039	46
15	.11980	.32472	.88020	**45**
16	.11999	.32495	.88001	44
17	.12018	.32517	.87982	43
18	.12036	.32539	.87964	42
19	.12055	.32561	.87945	41
20	.12074	.32583	.87926	**40**
21	.12093	.32605	.87907	39
22	.12112	.32627	.87888	38
23	.12131	.32649	.87869	37
24	.12150	.32671	.87850	36
25	.12169	.32693	.87831	**35**
26	.12188	.32715	.87812	34
27	.12207	.32737	.87793	33
28	.12226	.32759	.87774	32
29	.12245	.32781	.87755	31
30	.12265	.32803	.87735	**30**
31	.12284	.32825	.87716	29
32	.12303	.32847	.87697	28
33	.12322	.32869	.87678	27
34	.12341	.32891	.87659	26
35	.12360	.32913	.87640	**25**
36	.12379	.32934	.87621	24
37	.12398	.32956	.87602	23
38	.12418	.32978	.87582	22
39	.12437	.33000	.87563	21
40	.12456	.33022	.87544	**20**
41	.12475	.33044	.87525	19
42	.12494	.33066	.87506	18
43	.12514	.33087	.87486	17
44	.12533	.33109	.87467	16
45	.12552	.33131	.87448	**15**
46	.12571	.33153	.87429	14
47	.12591	.33175	.87409	13
48	.12610	.33196	.87390	12
49	.12629	.33218	.87371	11
50	.12649	.33240	.87351	**10**
51	.12668	.33262	.87332	9
52	.12687	.33283	.87313	8
53	.12707	.33305	.87293	7
54	.12726	.33327	.87274	6
55	.12746	.33348	.87254	**5**
56	.12765	.33370	.87235	4
57	.12784	.33392	.87216	3
58	.12804	.33413	.87196	2
59	.12823	.33435	.87177	1
60	.12843	33457	.87157	**0**
′	Cos²	Sin · Cos	Sin²	′

110° (290°) — (249°) **69°**

21° (201°) — (338°) **158°**

′	Sin²	Sin · Cos	Cos²	′
0	.12843	.33457	.87157	**60**
1	.12862	.33478	.87138	59
2	.12882	.33500	.87118	58
3	.12901	.33521	.87099	57
4	.12921	.33543	.87079	56
5	.12940	.33564	.87060	**55**
6	.12960	.33586	.87040	54
7	.12979	.33608	.87021	53
8	.12999	.33629	.87001	52
9	.13018	.33651	.86982	51
10	.13038	.33672	.86962	**50**
11	.13058	.33694	.86942	49
12	.13077	.33715	.86923	48
13	.13097	.33737	.86903	47
14	.13116	.33758	.86884	46
15	.13136	.33780	.86864	**45**
16	.13156	.33801	.86844	44
17	.13175	.33822	.86825	43
18	.13195	.33844	.86805	42
19	.13215	.33865	.86785	41
20	.13235	.33887	.86765	**40**
21	.13254	.33908	.86746	39
22	.13274	.33929	.86726	38
23	.13294	.33951	.86706	37
24	.13314	.33972	.86686	36
25	.13333	.33993	.86667	**35**
26	.13353	.34015	.86647	34
27	.13373	.34036	.86627	33
28	.13393	.34057	.86607	32
29	.13412	.34079	.86588	31
30	.13432	.34100	.86568	**30**
31	.13452	.34121	.86548	29
32	.13472	.34142	.86528	28
33	.13492	.34164	.86508	27
34	.13512	.34185	.86488	26
35	.13532	.34206	.86468	**25**
36	.13552	.34227	.86448	24
37	.13571	.34249	.86429	23
38	.13591	.34270	.86409	22
39	.13611	.34291	.86389	21
40	.13631	.34312	.86369	**20**
41	.13651	.34333	.86349	19
42	.13671	.34354	.86329	18
43	.13691	.34376	.86309	17
44	.13711	.34397	.86289	16
45	.13731	.34418	.86269	**15**
46	.13751	.34439	.86249	14
47	.13771	.34460	.86229	13
48	.13791	.34481	.86209	12
49	.13811	.34502	.86189	11
50	.13832	.34523	.86168	**10**
51	.13852	.34544	.86148	9
52	.13872	.34565	.86128	8
53	.13892	.34586	.86108	7
54	.13912	.34607	.86088	6
55	.13932	.34628	.86068	**5**
56	.13952	.34649	.86048	4
57	.13972	.34670	.86028	3
58	.13993	.34691	.86007	2
59	.14013	.34712	.85987	1
60	.14033	.34733	.85967	**0**
′	Cos²	Sin · Cos	Sin²	′

111° (291°) — (248°) **68°**

SQUARE OF THE SINE AND COSINE AND THEIR PRODUCT (Continued)

22° (202°) (337°) **157°**

′	Sin²	Sin · Cos	Cos²	′
0	.14033	.34733	.85967	**60**
1	.14053	.34754	.85947	59
2	.14073	.34775	.85927	58
3	.14094	.34796	.85906	57
4	.14114	.34817	.85886	56
5	.14134	.34837	.85866	**55**
6	.14154	.34858	.85846	54
7	.14175	.34879	.85825	53
8	.14195	.34900	.85805	52
9	.14215	.34921	.85785	51
10	.14236	.34942	.85764	**50**
11	.14256	.34962	.85744	49
12	.14276	.34983	.85724	48
13	.14297	.35004	.85703	47
14	.14317	.35025	.85683	46
15	.14337	.35045	.85663	**45**
16	.14358	.35066	.85642	44
17	.14378	.35087	.85622	43
18	.14399	.35108	.85601	42
19	.14419	.35128	.85581	41
20	.14440	.35149	.85560	**40**
21	.14460	.35170	.85540	39
22	.14480	.35190	.85520	38
23	.14501	.35211	.85499	37
24	.14521	.35232	.85479	36
25	.14542	.35252	.85458	**35**
26	.14562	.35273	.85438	34
27	.14583	.35294	.85417	33
28	.14604	.35314	.85396	32
29	.14624	.35335	.85376	31
30	.14645	.35355	.85355	**30**
31	.14665	.35376	.85335	29
32	.14686	.35396	.85314	28
33	.14706	.35417	.85294	27
34	.14727	.35438	.85273	26
35	.14748	.35458	.85252	**25**
36	.14768	.35479	.85232	24
37	.14789	.35499	.85211	23
38	.14810	.35520	.85190	22
39	.14830	.35540	.85170	21
40	.14851	.35560	.85149	**20**
41	.14872	.35581	.85128	19
42	.14892	.35601	.85108	18
43	.14913	.35622	.85087	17
44	.14934	.35642	.85066	16
45	.14955	.35663	.85045	**15**
46	.14975	.35683	.85025	14
47	.14996	.35703	.85004	13
48	.15017	.35724	.84983	12
49	.15038	.35744	.84962	11
50	.15058	.35764	.84942	**10**
51	.15079	.35785	.84921	9
52	.15100	.35805	.84900	8
53	.15121	.35825	.84879	7
54	.15142	.35846	.84858	6
55	.15163	.35866	.84837	**5**
56	.15183	.35886	.84817	4
57	.15204	.35906	.84796	3
58	.15225	.35927	.84775	2
59	.15246	.35947	.84754	1
60	.15267	.35967	.84733	**0**
′	Cos²	Sin · Cos	Sin²	′

112° (292°) (247°) **67°**

23° (203°) (336°) **156°**

′	Sin²	Sin · Cos	Cos²	′
0	.15267	.35967	.84733	**60**
1	.15288	.35987	.84712	59
2	.15309	.36007	.84691	58
3	.15330	.36028	.84670	57
4	.15351	.36048	.84649	56
5	.15372	.36068	.84628	**55**
6	.15393	.36088	.84607	54
7	.15414	.36108	.84586	53
8	.15435	.36128	.84565	52
9	.15456	.36148	.84544	51
10	.15477	.36168	.84523	**50**
11	.15498	.36189	.84502	49
12	.15519	.36209	.84481	48
13	.15540	.36229	.84460	47
14	.15561	.36249	.84439	46
15	.15582	.36269	.84418	**45**
16	.15603	.36289	.84397	44
17	.15624	.36309	.84376	43
18	.15646	.36329	.84354	42
19	.15667	.36349	.84333	41
20	.15688	.36369	.84312	**40**
21	.15709	.36389	.84291	39
22	.15730	.36409	.84270	38
23	.15751	.36429	.84249	37
24	.15773	.36448	.84227	36
25	.15794	.36468	.84206	**35**
26	.15815	.36488	.84185	34
27	.15836	.36508	.84164	33
28	.15858	.36528	.84142	32
29	.15879	.36548	.84121	31
30	.15900	.36568	.84100	**30**
31	.15921	.36588	.84079	29
32	.15943	.36607	.84057	28
33	.15964	.36627	.84036	27
34	.15985	.36647	.84015	26
35	.16007	.36667	.83993	**25**
36	.16028	.36686	.83972	24
37	.16049	.36706	.83951	23
38	.16071	.36726	.83929	22
39	.16092	.36746	.83908	21
40	.16113	.36765	.83887	**20**
41	.16135	.36785	.83865	19
42	.16156	.36805	.83844	18
43	.16178	.36825	.83822	17
44	.16199	.36844	.83801	16
45	.16220	.36864	.83780	**15**
46	.16242	.36884	.83758	14
47	.16263	.36903	.83737	13
48	.16285	.36923	.83715	12
49	.16306	.36942	.83694	11
50	.16328	.36962	.83672	**10**
51	.16349	.36982	.83651	9
52	.16371	.37001	.83629	8
53	.16392	.37021	.83608	7
54	.16414	.37040	.83586	6
55	.16436	.37060	.83564	**5**
56	.16457	.37079	.83543	4
57	.16479	.37099	.83521	3
58	.16500	.37118	.83500	2
59	.16522	.37138	.83478	1
60	.16543	.37157	.83457	**0**
′	Cos²	Sin · Cos	Sin²	′

113° (293°) (246°) **66°**

SQUARE OF THE SINE AND COSINE AND THEIR PRODUCT (Continued)

24° (204°) — (335°) **155°**

′	Sin²	Sin · Cos	Cos²	′
0	.16543	.37157	.83457	**60**
1	.16565	.37177	.83435	59
2	.16587	.37196	.83413	58
3	.16608	.37216	.83392	57
4	.16630	.37235	.83370	56
5	.16652	.37254	.83348	**55**
6	.16673	.37274	.83327	54
7	.16695	.37293	.83305	53
8	.16717	.37313	.83283	52
9	.16738	.37332	.83262	51
10	.16760	.37351	.83240	**50**
11	.16782	.37371	.83218	49
12	.16804	.37390	.83196	48
13	.16825	.37409	.83175	47
14	.16847	.37429	.83153	46
15	.16869	.37448	.83131	**45**
16	.16891	.37467	.83109	44
17	.16913	.37486	.83087	43
18	.16934	.37506	.83066	42
19	.16956	.37525	.83044	41
20	.16978	.37544	.83022	**40**
21	.17000	.37563	.83000	39
22	.17022	.37582	.82978	38
23	.17044	.37602	.82956	37
24	.17066	.37621	.82934	36
25	.17087	.37640	.82913	**35**
26	.17109	.37659	.82891	34
27	.17131	.37678	.82869	33
28	.17153	.37697	.82847	32
29	.17175	.37716	.82825	31
30	.17197	.37735	.82803	**30**
31	.17219	.37755	.82781	29
32	.17241	.37774	.82759	28
33	.17263	.37793	.82737	27
34	.17285	.37812	.82715	26
35	.17307	.37831	.82693	**25**
36	.17329	.37850	.82671	24
37	.17351	.37869	.82649	23
38	.17373	.37888	.82627	22
39	.17395	.37907	.82605	21
40	.17417	.37926	.82583	**20**
41	.17439	.37945	.82561	19
42	.17461	.37964	.82539	18
43	.17483	.37982	.82517	17
44	.17505	.38001	.82495	16
45	.17528	.38020	.82472	**15**
46	.17550	.38039	.82450	14
47	.17572	.38058	.82428	13
48	.17594	.38077	.82406	12
49	.17616	.38096	.82384	11
50	.17638	.38115	.82362	**10**
51	.17661	.38133	.82339	9
52	.17683	.38152	.82317	8
53	.17705	.38171	.82295	7
54	.17727	.38190	.82273	6
55	.17749	.38209	.82251	**5**
56	.17772	.38227	.82228	4
57	.17794	.38246	.82206	3
58	.17816	.38265	.82184	2
59	.17838	.38284	.82162	1
60	.17861	.38302	.82139	**0**
′	Cos²	Sin · Cos	Sin²	′

114° (294°) — (245°) **65°**

25° (205°) — (334°) **154°**

′	Sin²	Sin · Cos	Cos²	′
0	.17861	.38302	.82139	**60**
1	.17883	.38321	.82117	59
2	.17905	.38340	.82095	58
3	.17928	.38358	.82072	57
4	.17950	.38377	.82050	56
5	.17972	.38396	.82028	**55**
6	.17995	.38414	.82005	54
7	.18017	.38433	.81983	53
8	.18039	.38451	.81961	52
9	.18062	.38470	.81938	51
10	.18084	.38489	.81916	**50**
11	.18106	.38507	.81894	49
12	.18129	.38526	.81871	48
13	.18151	.38544	.81849	47
14	.18174	.38563	.81826	46
15	.18196	.38581	.81804	**45**
16	.18219	.38600	.81781	44
17	.18241	.38618	.81759	43
18	.18263	.38637	.81737	42
19	.18286	.38655	.81714	41
20	.18308	.38674	.81692	**40**
21	.18331	.38692	.81669	39
22	.18353	.38710	.81647	38
23	.18376	.38729	.81624	37
24	.18399	.38747	.81601	36
25	.18421	.38766	.81579	**35**
26	.18444	.38784	.81556	34
27	.18466	.38802	.81534	33
28	.18489	.38821	.81511	32
29	.18511	.38839	.81489	31
30	.18534	.38857	.81466	**30**
31	.18557	.38876	.81443	29
32	.18579	.38894	.81421	28
33	.18602	.38912	.81398	27
34	.18624	.38930	.81376	26
35	.18647	.38949	.81353	**25**
36	.18670	.38967	.81330	24
37	.18692	.38985	.81308	23
38	.18715	.39003	.81285	22
39	.18738	.39022	.81262	21
40	.18761	.39040	.81239	**20**
41	.18783	.39058	.81217	19
42	.18806	.39076	.81194	18
43	.18829	.39094	.81171	17
44	.18852	.39112	.81148	16
45	.18874	.39130	.81126	**15**
46	.18897	.39149	.81103	14
47	.18920	.39167	.81080	13
48	.18943	.39185	.81057	12
49	.18965	.39203	.81035	11
50	.18988	.39221	.81012	**10**
51	.19011	.39239	.80989	9
52	.19034	.39257	.80966	8
53	.19057	.39275	.80943	7
54	.19080	.39293	.80920	6
55	.19102	.39311	.80898	**5**
56	.19125	.39329	.80875	4
57	.19148	.39347	.80852	3
58	.19171	.39365	.80829	2
59	.19194	.39383	.80806	1
60	.19217	.39401	.80783	**0**
′	Cos²	Sin · Cos	Sin²	′

115° (295°) — (244°) **64°**

SQUARE OF THE SINE AND COSINE AND THEIR PRODUCT (Continued)

26° (206°) (333°) **153°**

′	Sin²	Sin · Cos	Cos²	′
0	.19217	.39401	.80783	**60**
1	.19240	.39418	.80760	59
2	.19263	.39436	.80737	58
3	.19286	.39454	.80714	57
4	.19309	.39472	.80691	56
5	.19332	.39490	.80668	**55**
6	.19355	.39508	.80645	54
7	.19378	.39526	.80622	53
8	.19401	.39543	.80599	52
9	.19424	.39561	.80576	51
10	.19447	.39579	.80553	**50**
11	.19470	.39597	.80530	49
12	.19493	.39614	.80507	48
13	.19516	.39632	.80484	47
14	.19539	.39650	.80461	46
15	.19562	.39668	.80438	**45**
16	.19585	.39685	.80415	44
17	.19608	.39703	.80392	43
18	.19631	.39721	.80369	42
19	.19654	.39738	.80346	41
20	.19677	.39756	.80323	**40**
21	.19701	.39774	.80299	39
22	.19724	.39791	.80276	38
23	.19747	.39809	.80253	37
24	.19770	.39826	.80230	36
25	.19793	.39844	.80207	**35**
26	.19816	.39862	.80184	34
27	.19840	.39879	.80160	33
28	.19863	.39897	.80137	32
29	.19886	.39914	.80114	31
30	.19909	.39932	.80091	**30**
31	.19932	.39949	.80068	29
32	.19956	.39967	.80044	28
33	.19979	.39984	.80021	27
34	.20002	.40002	.79998	26
35	.20026	.40019	.79974	**25**
36	.20049	.40037	.79951	24
37	.20072	.40054	.79928	23
38	.20095	.40071	.79905	22
39	.20119	.40089	.79881	21
40	.20142	.40106	.79858	**20**
41	.20165	.40124	.79835	19
42	.20189	.40141	.79811	18
43	.20212	.40158	.79788	17
44	.20235	.40176	.79765	16
45	.20259	.40193	.79741	**15**
46	.20282	.40210	.79718	14
47	.20306	.40227	.79694	13
48	.20329	.40245	.79671	12
49	.20352	.40262	.79648	11
50	.20376	.40279	.79624	**10**
51	.20399	.40296	.79601	9
52	.20423	.40314	.79577	8
53	.20446	.40331	.79554	7
54	.20470	.40348	.79530	6
55	.20493	.40365	.79507	**5**
56	.20517	.40382	.79483	4
57	.20540	.40399	.79460	3
58	.20564	.40417	.79436	2
59	.20587	.40434	.79413	1
60	.20611	.40451	.79389	**0**
′	Cos²	Sin · Cos	Sin²	′

116° (296°) (243°) **63°**

27° (207°) (332°) **152°**

′	Sin²	Sin · Cos	Cos²	′
0	.20611	.40451	.79389	**60**
1	.20634	.40468	.79366	59
2	.20658	.40485	.79342	58
3	.20681	.40502	.79319	57
4	.20705	.40519	.79295	56
5	.20729	.40536	.79271	**55**
6	.20752	.40553	.79248	54
7	.20776	.40570	.79224	53
8	.20799	.40587	.79201	52
9	.20823	.40604	.79177	51
10	.20847	.40621	.79153	**50**
11	.20870	.40638	.79130	49
12	.20894	.40655	.79106	48
13	.20918	.40672	.79082	47
14	.20941	.40689	.79059	46
15	.20965	.40706	.79035	**45**
16	.20989	.40723	.79011	44
17	.21012	.40740	.78988	43
18	.21036	.40756	.78964	42
19	.21060	.40773	.78940	41
20	.21083	.40790	.78917	**40**
21	.21107	.40807	.78893	39
22	.21131	.40824	.78869	38
23	.21155	.40840	.78845	37
24	.21178	.40857	.78822	36
25	.21202	.40874	.78798	**35**
26	.21226	.40891	.78774	34
27	.21250	.40907	.78750	33
28	.21274	.40924	.78726	32
29	.21297	.40941	.78703	31
30	.21321	.40958	.78679	**30**
31	.21345	.40974	.78655	29
32	.21369	.40991	.78631	28
33	.21393	.41008	.78607	27
34	.21417	.41024	.78583	26
35	.21440	.41041	.78560	**25**
36	.21464	.41057	.78536	24
37	.21488	.41074	.78512	23
38	.21512	.41091	.78488	22
39	.21536	.41107	.78464	21
40	.21560	.41124	.78440	**20**
41	.21584	.41140	.78416	19
42	.21608	.41157	.78392	18
43	.21632	.41173	.78368	17
44	.21656	.41190	.78344	16
45	.21680	.41206	.78320	**15**
46	.21704	.41223	.78296	14
47	.21728	.41239	.78272	13
48	.21752	.41256	.78248	12
49	.21776	.41272	.78224	11
50	.21800	.41289	.78200	**10**
51	.21824	.41305	.78176	9
52	.21848	.41321	.78152	8
53	.21872	.41338	.78128	7
54	.21896	.41354	.78104	6
55	.21920	.41370	.78080	**5**
56	.21944	.41387	.78056	4
57	.21968	.41403	.78032	3
58	.21992	.41419	.78008	2
59	.22016	.41436	.77984	1
60	.22040	.41452	.77960	**0**
′	Cos²	Sin · Cos	Sin²	′

117° (297°) (242°) **62°**

SQUARE OF THE SINE AND COSINE AND THEIR PRODUCT (Continued)

28° (208°) — (331°) **151°**

′	Sin²	Sin · Cos	Cos²	′
0	.22040	.41452	.77960	**60**
1	.22064	.41468	.77936	59
2	.22089	.41484	.77911	58
3	.22113	.41501	.77887	57
4	.22137	.41517	.77863	56
5	.22161	.41533	.77839	**55**
6	.22185	.41549	.77815	54
7	.22209	.41565	.77791	53
8	.22234	.41582	.77766	52
9	.22258	.41598	.77742	51
10	.22282	.41614	.77718	**50**
11	.22306	.41630	.77694	49
12	.22330	.41646	.77670	48
13	.22355	.41662	.77645	47
14	.22379	.41678	.77621	46
15	.22403	.41694	.77597	**45**
16	.22427	.41710	.77573	44
17	.22452	.41726	.77548	43
18	.22476	.41742	.77524	42
19	.22500	.41758	.77500	41
20	.22525	.41774	.77475	**40**
21	.22549	.41790	.77451	39
22	.22573	.41806	.77427	38
23	.22598	.41822	.77402	37
24	.22622	.41838	.77378	36
25	.22646	.41854	.77354	**35**
26	.22671	.41870	.77329	34
27	.22695	.41886	.77305	33
28	.22719	41902	.77281	32
29	.22744	.41918	.77256	31
30	.22768	.41934	.77232	**30**
31	.22792	.41949	.77208	29
32	.22817	.41965	.77183	28
33	.22841	.41981	.77159	27
34	.22866	.41997	.77134	26
35	.22890	.42013	.77110	**25**
36	.22915	.42028	.77085	24
37	.22939	.42044	.77061	23
38	.22964	.42060	.77036	22
39	.22988	.42076	.77012	21
40	.23012	.42091	.76988	**20**
41	.23037	.42107	.76963	19
42	.23061	.42123	.76939	18
43	.23086	.42138	.76914	17
44	.23110	.42154	.76890	16
45	.23135	.42170	.76865	**15**
46	.23160	.42185	.76840	14
47	.23184	.42201	.76816	13
48	.23209	.42216	.76791	12
49	.23233	.42232	.76767	11
50	.23258	.42248	.76742	**10**
51	.23282	.42263	.76718	9
52	.23307	.42279	.76693	8
53	.23332	.42294	.76668	7
54	.23356	.42310	.76644	6
55	.23381	.42325	.76619	**5**
56	.23405	.42341	.76595	4
57	.23430	.42356	.76570	3
58	.23455	.42372	.76545	2
59	.23479	.42387	.76521	1
60	.23504	.42402	.76496	**0**
′	Cos²	Sin · Cos	Sin²	′

118° (298°) — (241°) **61°**

29° (209°) — (330°) **150°**

′	Sin²	Sin · Cos	Cos²	′
0	.23504	.42402	.76496	**60**
1	.23529	.42418	.76471	59
2	.23553	.42433	.76447	58
3	.23578	.42449	.76422	57
4	.23603	.42464	.76397	56
5	.23627	.42479	.76373	**55**
6	.23652	.42495	.76348	54
7	.23677	.42510	.76323	53
8	.23702	.42525	.76298	52
9	.23726	.42541	.76274	51
10	.23751	.42556	.76249	**50**
11	.23776	.42571	.76224	49
12	.23801	.42586	.76199	48
13	.23825	.42602	.76175	47
14	.23850	.42617	.76150	46
15	.23875	.42632	.76125	**45**
16	.23900	.42647	.76100	44
17	.23925	.42662	.76075	43
18	.23950	.42678	.76050	42
19	.23974	.42693	.76026	41
20	.23999	.42708	.76001	**40**
21	.24024	.42723	.75976	39
22	.24049	.42738	.75951	38
23	.24074	.42753	.75926	37
24	.24099	.42768	.75901	36
25	.24124	.42783	.75876	**35**
26	.24148	.42798	.75852	34
27	.24173	.42813	.75827	33
28	.24198	.42828	.75802	32
29	.24223	.42843	.75777	31
30	.24248	.42858	.75752	**30**
31	.24273	.42873	.75727	29
32	.24298	.42888	.75702	28
33	.24323	.42903	.75677	27
34	.24348	.42918	.75652	26
35	.24373	.42933	.75627	**25**
36	.24398	.42948	.75602	24
37	.24423	.42963	.75577	23
38	.24448	.42978	.75552	22
39	.24473	.42993	.75527	21
40	.24498	.43007	.75502	**20**
41	.24523	.43022	.75477	19
42	.24548	.43037	.75452	18
43	.24573	.43052	.75427	17
44	.24598	.43067	.75402	16
45	.24623	.43081	.75377	**15**
46	.24648	.43096	.75352	14
47	.24673	.43111	.75327	13
48	.24698	.43126	.75302	12
49	.24723	.43140	.75277	11
50	.24749	.43155	.75251	**10**
51	.24774	.43170	.75226	9
52	.24799	.43184	.75201	8
53	.24824	.43199	.75176	7
54	.24849	.43214	.75151	6
55	.24874	.43228	.75126	**5**
56	.24899	.43243	.75101	4
57	.24924	.43258	.75076	3
58	.24950	.43272	.75050	2
59	.24975	.43287	.75025	1
60	.25000	.43301	.75000	**0**
′	Cos²	Sin · Cos	Sin²	′

119° (299°) — (240°) **60°**

SQUARE OF THE SINE AND COSINE AND THEIR PRODUCT (Continued)

30° (210°) (329°) **149°**

′	Sin²	Sin · Cos	Cos²	′
0	.25000	.43301	.75000	**60**
1	.25025	.43316	.74975	59
2	.25050	.43330	.74950	58
3	.25076	.43345	.74924	57
4	.25101	.43359	.74899	56
5	.25126	.43374	.74874	**55**
6	.25151	.43388	.74849	54
7	.25177	.43403	.74823	53
8	.25202	.43417	.74798	52
9	.25227	.43432	.74773	51
10	.25252	.43446	.74748	**50**
11	.25278	.43460	.74722	49
12	.25303	.43475	.74697	48
13	.25328	.43489	.74672	47
14	.25354	.43503	.74646	46
15	.25379	.43518	.74621	**45**
16	.25404	.43532	.74596	44
17	.25429	.43546	.74571	43
18	.25455	.43561	.74545	42
19	.25480	.43575	.74520	41
20	.25506	.43589	.74494	**40**
21	.25531	.43603	.74469	39
22	.25556	.43618	.74444	38
23	.25582	.43632	.74418	37
24	.25607	.43646	.74393	36
25	.25632	.43660	.74368	**35**
26	.25658	.43674	.74342	34
27	.25683	.43689	.74317	33
28	.25709	.43703	.74291	32
29	.25734	.43717	.74266	31
30	.25760	.43731	.74240	**30**
31	.25785	.43745	.74215	29
32	.25810	.43759	.74190	28
33	.25836	.43773	.74164	27
34	.25861	.43787	.74139	26
35	.25887	.43801	.74113	**25**
36	.25912	.43815	.74088	24
37	.25938	.43829	.74062	23
38	.25963	.43843	.74037	22
39	.25989	.43857	.74011	21
40	.26014	.43871	.73986	**20**
41	.26040	.43885	.73960	19
42	.26065	.43899	.73935	18
43	.26091	.43913	.73909	17
44	.26117	.43927	.73883	16
45	.26142	.43941	.73858	**15**
46	.26168	.43955	.73832	14
47	.26193	.43969	.73807	13
48	.26219	.43982	.73781	12
49	.26244	.43996	.73756	11
50	.26270	.44010	.73730	**10**
51	.26296	.44024	.73704	9
52	.26321	.44038	.73679	8
53	.26347	.44051	.73653	7
54	.26372	.44065	.73628	6
55	.26398	.44079	.73602	**5**
56	.26424	.44093	.73576	4
57	.26449	.44106	.73551	3
58	.26475	.44120	.73525	2
59	.26501	.44134	.73499	1
60	.26526	.44147	.73474	**0**
′	Cos²	Sin · Cos	Sin²	′

120° (300°) (239°) **59°**

31° (211°) (328°) **148°**

′	Sin²	Sin · Cos	Cos²	′
0	.26526	.44147	.73474	**60**
1	.26552	.44161	.73448	59
2	.26578	.44175	.73422	58
3	.26604	.44188	.73396	57
4	.26629	.44202	.73371	56
5	.26655	.44215	.73345	**55**
6	.26681	.44229	.73319	54
7	.26706	.44243	.73294	53
8	.26732	.44256	.73268	52
9	.26758	.44270	.73242	51
10	.26784	.44283	.73216	**50**
11	.26809	.44297	.73191	49
12	.26835	.44310	.73165	48
13	.26861	.44324	.73139	47
14	.26887	.44337	.73113	46
15	.26913	.44351	.73087	**45**
16	.26938	.44364	.73062	44
17	.26964	.44377	.73036	43
18	.26990	.44391	.73010	42
19	.27016	.44404	.72984	41
20	.27042	.44418	.72958	**40**
21	.27068	.44431	.72932	39
22	.27093	.44444	.72907	38
23	.27119	.44458	.72881	37
24	.27145	.44471	.72855	36
25	.27171	.44484	.72829	**35**
26	.27197	.44497	.72803	34
27	.27223	.44511	.72777	33
28	.27249	.44524	.72751	32
29	.27275	.44537	.72725	31
30	.27300	.44550	.72700	**30**
31	.27326	.44564	.72674	29
32	.27352	.44577	.72648	28
33	.27378	.44590	.72622	27
34	.27404	.44603	.72596	26
35	.27430	.44616	.72570	**25**
36	.27456	.44629	.72544	24
37	.27482	.44642	.72518	23
38	.27508	.44655	.72492	22
39	.27534	.44669	.72466	21
40	.27560	.44682	.72440	**20**
41	.27586	.44695	.72414	19
42	.27612	.44708	.72388	18
43	.27638	.44721	.72362	17
44	.27664	.44734	.72336	16
45	.27690	.44747	.72310	**15**
46	.27716	.44760	.72284	14
47	.27742	.44773	.72258	13
48	.27768	.44786	.72232	12
49	.27794	.44799	.72206	11
50	.27820	.44811	.72180	**10**
51	.27846	.44824	.72154	9
52	.27873	.44837	.72127	8
53	.27899	.44850	.72101	7
54	.27925	.44863	.72075	6
55	.27951	.44876	.72049	**5**
56	.27977	.44889	.72023	4
57	.28003	.44901	.71997	3
58	.28029	.44914	.71971	2
59	.28055	.44927	.71945	1
60	.28081	.44940	.71919	**0**
′	Cos²	Sin · Cos	Sin²	′

121° (301°) (238°) **58°**

SQUARE OF THE SINE AND COSINE AND THEIR PRODUCT (Continued)

32° (212°) (327°) **147°**

′	Sin²	Sin · Cos	Cos²	′
0	.28081	.44940	.71919	**60**
1	.28108	.44952	.71892	59
2	.28134	.44965	.71866	58
3	.28160	.44978	.71840	57
4	.28186	.44991	.71814	56
5	.28212	.45003	.71788	**55**
6	.28238	.45016	.71762	54
7	.28265	.45029	.71735	53
8	.28291	.45041	.71709	52
9	.28317	.45054	.71683	51
10	.28343	.45066	.71657	**50**
11	.28369	.45079	.71631	49
12	.28396	.45092	.71604	48
13	.28422	.45104	.71578	47
14	.28448	.45117	.71552	46
15	.28474	.45129	.71526	**45**
16	.28501	.45142	.71499	44
17	.28527	.45154	.71473	43
18	.28553	.45167	.71447	42
19	.28580	.45179	.71420	41
20	.28606	.45192	.71394	**40**
21	.28632	.45204	.71368	39
22	.28658	.45217	.71342	38
23	.28685	.45229	.71315	37
24	.28711	.45241	.71289	36
25	.28737	.45254	.71263	**35**
26	.28764	.45266	.71236	34
27	.28790	.45278	.71210	33
28	.28816	.45291	.71184	32
29	.28843	.45303	.71157	31
30	.28869	.45315	.71131	**30**
31	.28895	.45328	.71105	29
32	.28922	.45340	.71078	28
33	.28948	.45352	.71052	27
34	.28975	.45364	.71025	26
35	.29001	.45377	.70999	**25**
36	.29027	.45389	.70973	24
37	.29054	.45401	.70946	23
38	.29080	.45413	.70920	22
39	.29107	.45425	.70893	21
40	.29133	.45438	.70867	**20**
41	.29160	.45450	.70840	19
42	.29186	.45462	.70814	18
43	.29212	.45474	.70788	17
44	.29239	.45486	.70761	16
45	.29265	.45498	.70735	**15**
46	.29292	.45510	.70708	14
47	.29318	.45522	.70682	13
48	.29345	.45534	.70655	12
49	.29371	.45546	.70629	11
50	.29398	.45558	.70602	**10**
51	.29424	.45570	.70576	9
52	.29451	.45582	.70549	8
53	.29477	.45594	.70523	7
54	.29504	.45606	.70496	6
55	.29530	.45618	.70470	**5**
56	.29557	.45630	.70443	4
57	.29583	.45642	.70417	3
58	.29610	.45654	.70390	2
59	.29637	.45665	.70363	1
60	.29663	.45677	.70337	**0**
′	Cos²	Sin · Cos	Sin²	′

122° (302°) (237°) **57°**

33° (213°) (326°) **146°**

′	Sin²	Sin · Cos	Cos²	′
0	.29663	.45677	.70337	**60**
1	.29690	.45689	.70310	59
2	.29716	.45701	.70284	58
3	.29743	.45713	.70257	57
4	.29770	.45724	.70230	56
5	.29796	.45736	.70204	**55**
6	.29823	.45748	.70177	54
7	.29849	.45760	.70151	53
8	.29876	.45771	.70124	52
9	.29903	.45783	.70097	51
10	.29929	.45795	.70071	**50**
11	.29956	.45806	.70044	49
12	.29983	.45818	.70017	48
13	.30009	.45830	.69991	47
14	.30036	.45841	.69964	46
15	.30063	.45853	.69937	**45**
16	.30089	.45865	.69911	44
17	.30116	.45876	.69884	43
18	.30143	.45888	.69857	42
19	.30169	.45899	.69831	41
20	.30196	.45911	.69804	**40**
21	.30223	.45922	.69777	39
22	.30249	.45934	.69751	38
23	.30276	.45945	.69724	37
24	.30303	.45957	.69697	36
25	.30330	.45968	.69670	**35**
26	.30356	.45980	.69644	34
27	.30383	.45991	.69617	33
28	.30410	.46002	.69590	32
29	.30437	.46014	.69563	31
30	.30463	.46025	.69537	**30**
31	.30490	.46037	.69510	29
32	.30517	.46048	.69483	28
33	.30544	.46059	.69456	27
34	.30571	.46071	.69429	26
35	.30597	.46082	.69403	**25**
36	.30624	.46093	.69376	24
37	.30651	.46104	.69349	23
38	.30678	.46116	.69322	22
39	.30705	.46127	.69295	21
40	.30732	.46138	.69268	**20**
41	.30758	.46149	.69242	19
42	.30785	.46161	.69215	18
43	.30812	.46172	.69188	17
44	.30839	.46183	.69161	16
45	.30866	.46194	.69134	**15**
46	.30893	.46205	.69107	14
47	.30920	.46216	.69080	13
48	.30946	.46227	.69054	12
49	.30973	.46238	.69027	11
50	.31000	.46249	.69000	**10**
51	.31027	.46260	.68973	9
52	.31054	.46272	.68946	8
53	.31081	.46283	.68919	7
54	.31108	.46294	.68892	6
55	.31135	.46305	.68865	**5**
56	.31162	.46315	.68838	4
57	.31189	.46326	.68811	3
58	.31216	.46337	.68784	2
59	.31243	.46348	.68757	1
60	.31270	.46359	.68730	**0**
′	Cos²	Sin · Cos	Sin²	′

123° (303°) (236°) **56°**

SQUARE OF THE SINE AND COSINE AND THEIR PRODUCT (Continued)

34° (214°) (325°) **145°**

′	Sin²	Sin · Cos	Cos²	′
0	.31270	.46359	.68730	**60**
1	.31297	.46370	.68703	59
2	.31324	.46381	.68676	58
3	.31351	.46392	.68649	57
4	.31378	.46403	.68622	56
5	.31405	.46413	.68595	**55**
6	.31432	.46424	.68568	54
7	.31459	.46435	.68541	53
8	.31486	.46446	.68514	52
9	.31513	.46457	.68487	51
10	.31540	.46467	.68460	**50**
11	.31567	.46478	.68433	49
12	.31594	.46489	.68406	48
13	.31621	.46500	.68379	47
14	.31648	.46510	.68352	46
15	.31675	.46521	.68325	**45**
16	.31702	.46532	.68298	44
17	.31729	.46542	.68271	43
18	.31756	.46553	.68244	42
19	.31783	.46563	.68217	41
20	.31810	.46574	.68190	**40**
21	.31837	.46585	.68163	39
22	.31865	.46595	.68135	38
23	.31892	.46606	.68108	37
24	.31919	.46616	.68081	36
25	.31946	.46627	.68054	**35**
26	.31973	.46637	.68027	34
27	.32000	.46648	.68000	33
28	.32027	.46658	.67973	32
29	.32054	.46669	.67946	31
30	.32082	.46679	.67918	**30**
31	.32109	.46689	.67891	29
32	.32136	.46700	.67864	28
33	.32163	.46710	.67837	27
34	.32190	.46721	.67810	26
35	.32217	.46731	.67783	**25**
36	.32245	.46741	.67755	24
37	.32272	.46752	.67728	23
38	.32299	.46762	.67701	22
39	.32326	.46772	.67674	21
40	.32353	.46782	.67647	**20**
41	.32381	.46793	.67619	19
42	.32408	.46803	.67592	18
43	.32435	.46813	.67565	17
44	.32462	.46823	.67538	16
45	.32490	.46834	.67510	**15**
46	.32517	.46844	.67483	14
47	.32544	.46854	.67456	13
48	.32571	.46864	.67429	12
49	.32599	.46874	.67401	11
50	.32626	.46884	.67374	**10**
51	.32653	.46894	.67347	9
52	.32681	.46905	.67319	8
53	.32708	.46915	.67292	7
54	.32735	.46925	.67265	6
55	.32762	.46935	.67238	**5**
56	.32790	.46945	.67210	4
57	.32817	.46955	.67183	3
58	.32844	.46965	.67156	2
59	.32872	.46975	.67128	1
60	.32899	.46985	.67101	**0**
′	Cos²	Sin · Cos	Sin²	′

124° (304°) (235°) **55°**

35° (215°) (324°) **144°**

′	Sin²	Sin · Cos	Cos²	′
0	.32899	.46985	.67101	**60**
1	.32926	.46995	.67074	59
2	.32954	.47004	.67046	58
3	.32981	.47014	.67019	57
4	.33008	.47024	.66992	56
5	.33036	.47034	.66964	**55**
6	.33063	.47044	.66937	54
7	.33090	.47054	.66910	53
8	.33118	.47064	.66882	52
9	.33145	.47074	.66855	51
10	.33173	.47083	.66827	**50**
11	.33200	.47093	.66800	49
12	.33227	.47103	.66773	48
13	.33255	.47113	.66745	47
14	.33282	.47122	.66718	46
15	.33310	.47132	.66690	**45**
16	.33337	.47142	.66663	44
17	.33365	.47151	.66635	43
18	.33392	.47161	.66608	42
19	.33419	.47171	.66581	41
20	.33447	.47180	.66553	**40**
21	.33474	.47190	.66526	39
22	.33502	.47200	.66498	38
23	.33529	.47209	.66471	37
24	.33557	.47219	.66443	36
25	.33584	.47228	.66416	**35**
26	.33612	.47238	.66388	34
27	.33639	.47247	.66361	33
28	.33667	.47257	.66333	32
29	.33694	.47266	.66306	31
30	.33722	.47276	.66278	**30**
31	.33749	.47285	.66251	29
32	.33777	.47295	.66223	28
33	.33804	.47304	.66196	27
34	.33832	.47314	.66168	26
35	.33859	.47323	.66141	**25**
36	.33887	.47332	.66113	24
37	.33914	.47342	.66086	23
38	.33942	.47351	.66058	22
39	.33969	.47361	.66031	21
40	.33997	.47370	.66003	**20**
41	.34024	.47379	.65976	19
42	.34052	.47388	.65948	18
43	.34080	.47398	.65920	17
44	.34107	.47407	.65893	16
45	.34135	.47416	.65865	**15**
46	.34162	.47425	.65838	14
47	.34190	.47435	.65810	13
48	.34218	.47444	.65782	12
49	.34245	.47453	.65755	11
50	.34273	.47462	.65727	**10**
51	.34300	.47471	.65700	9
52	.34328	.47480	.65672	8
53	.34356	.47490	.65644	7
54	.34383	.47499	.65617	6
55	.34411	.47508	.65589	**5**
56	.34439	.47517	.65561	4
57	.34466	.47526	.65534	3
58	.34494	.47535	.65506	2
59	.34521	.47544	.65479	1
60	.34549	.47553	.65451	**0**
′	Cos²	Sin · Cos	Sin²	′

125° (305°) (234°) **54°**

SQUARE OF THE SINE AND COSINE AND THEIR PRODUCT (Continued)

36° (216°) (323°) **143°**

′	Sin²	Sin · Cos	Cos²	′
0	.34549	.47553	.65451	**60**
1	.34577	.47562	.65423	59
2	.34604	.47571	.65396	58
3	.34632	.47580	.65368	57
4	.34660	.47589	.65340	56
5	.34688	.47598	.65312	**55**
6	.34715	.47606	.65285	54
7	.34743	.47615	.65257	53
8	.34771	.47624	.65229	52
9	.34798	.47633	.65202	51
10	.34826	.47642	.65174	**50**
11	.34854	.47651	.65146	49
12	.34882	.47660	.65118	48
13	.34909	.47668	.65091	47
14	.34937	.47677	.65063	46
15	.34965	.47686	.65035	**45**
16	.34992	.47695	.65008	44
17	.35020	.47703	.64980	43
18	.35048	.47712	.64952	42
19	.35076	.47721	.64924	41
20	.35103	.47729	.64897	**40**
21	.35131	.47738	.64869	39
22	.35159	.47747	.64841	38
23	.35187	.47755	.64813	37
24	.35215	.47764	.64785	36
25	.35242	.47773	.64758	**35**
26	.35270	.47781	.64730	34
27	.35298	.47790	.64702	33
28	.35326	.47798	.64674	32
29	.35354	.47807	.64646	31
30	.35381	.47815	.64619	**30**
31	.35409	.47824	.64591	29
32	.35437	.47832	.64563	28
33	.35465	.47841	.64535	27
34	.35493	.47849	.64507	26
35	.35521	.47858	.64479	**25**
36	.35548	.47866	.64452	24
37	.35576	.47874	.64424	23
38	.35604	.47883	.64396	22
39	.35632	.47891	.64368	21
40	.35660	.47899	.64340	**20**
41	.35688	.47908	.64312	19
42	.35716	.47916	.64284	18
43	.35743	.47924	.64257	17
44	.35771	.47933	.64229	16
45	.35799	.47941	.64201	**15**
46	.35827	.47949	.64173	14
47	.35855	.47957	.64145	13
48	.35883	.47966	.64117	12
49	.35911	.47974	.64089	11
50	.35939	.47982	.64061	**10**
51	.35967	.47990	.64033	9
52	.35995	.47998	.64005	8
53	.36023	.48007	.63977	7
54	.36050	.48015	.63950	6
55	.36078	.48023	.63922	**5**
56	.36106	.48031	.63894	4
57	.36134	.48039	.63866	3
58	.36162	.48047	.63838	2
59	.36190	.48055	.63810	1
60	.36218	.48063	.63782	**0**
′	Cos²	Sin · Cos	Sin²	′

126° (306°) (233°) **53°**

37° (217°) (322°) **142°**

′	Sin²	Sin · Cos	Cos²	′
0	.36218	.48063	.63782	**60**
1	.36246	.48071	.63754	59
2	.36274	.48079	.63726	58
3	.36302	.48087	.63698	57
4	.36330	.48095	.63670	56
5	.36358	.48103	.63642	**55**
6	.36386	.48111	.63614	54
7	.36414	.48119	.63586	53
8	.36442	.48127	.63558	52
9	.36470	.48135	.63530	51
10	.36498	.48142	.63502	**50**
11	.36526	.48150	.63474	49
12	.36554	.48158	.63446	48
13	.36582	.48166	.63418	47
14	.36610	.48174	.63390	46
15	.36638	.48182	.63362	**45**
16	.36666	.48189	.63334	44
17	.36694	.48197	.63306	43
18	.36722	.48205	.63278	42
19	.36750	.48212	.63250	41
20	.36778	.48220	.63222	**40**
21	.36806	.48228	.63194	39
22	.36834	.48236	.63166	38
23	.36862	.48243	.63138	37
24	.36891	.48251	.63109	36
25	.36919	.48258	.63081	**35**
26	.36947	.48266	.63053	34
27	.36975	.48274	.63025	33
28	.37003	.48281	.62997	32
29	.37031	.48289	.62969	31
30	.37059	.48296	.62941	**30**
31	.37087	.48304	.62913	29
32	.37115	.48311	.62885	28
33	.37143	.48319	.62857	27
34	.37171	.48326	.62829	26
35	.37200	.48334	.62800	**25**
36	.37228	.48341	.62772	24
37	.37256	.48349	.62744	23
38	.37284	.48356	.62716	22
39	.37312	.48363	.62688	21
40	.37340	.48371	.62660	**20**
41	.37368	.48378	.62632	19
42	.37397	.48385	.62603	18
43	.37425	.48393	.62575	17
44	.37453	.48400	.62547	16
45	.37481	.48407	.62519	**15**
46	.37509	.48415	.62491	14
47	.37537	.48422	.62463	13
48	.37566	.48429	.62434	12
49	.37594	.48436	.62406	11
50	.37622	.48444	.62378	**10**
51	.37650	.48451	.62350	9
52	.37678	.48458	.62322	8
53	.37706	.48465	.62294	7
54	.37735	.48472	.62265	6
55	.37763	.48479	.62237	**5**
56	.37791	.48487	.62209	4
57	.37819	.48494	.62181	3
58	.37847	.48501	.62153	2
59	.37876	.48508	.62124	1
60	.37904	.48515	.62096	**0**
′	Cos²	Sin · Cos	Sin²	′

127° (307°) (232°) **52°**

SQUARE OF THE SINE AND COSINE AND THEIR PRODUCT (Continued)

38° (218°) — (321°) **141°**

′	Sin^2	Sin · Cos	Cos^2	′
0	.37904	.48515	.62096	**60**
1	.37932	.48522	.62068	59
2	.37960	.48529	.62040	58
3	.37989	.48536	.62011	57
4	.38017	.48543	.61983	56
5	.38045	.48550	.61955	**55**
6	.38073	.48557	.61927	54
7	.38102	.48564	.61898	53
8	.38130	.48571	.61870	52
9	.38158	.48577	.61842	51
10	.38186	.48584	.61814	**50**
11	.38215	.48591	.61785	49
12	.38243	.48598	.61757	48
13	.38271	.48605	.61729	47
14	.38299	.48612	.61701	46
15	.38328	.48618	.61672	**45**
16	.38356	.48625	.61644	44
17	.38384	.48632	.61616	43
18	.38413	.48639	.61587	42
19	.38441	.48646	.61559	41
20	.38469	.48652	.61531	**40**
21	.38498	.48659	.61502	39
22	.38526	.48666	.61474	38
23	.38554	.48672	.61446	37
24	.38582	.48679	.61418	36
25	.38611	.48686	.61389	**35**
26	.38639	.48692	.61361	34
27	.38667	.48699	.61333	33
28	.38696	.48705	.61304	32
29	.38724	.48712	.61276	31
30	.38752	.48719	.61248	**30**
31	.38781	.48725	.61219	29
32	.38809	.48732	.61191	28
33	.38837	.48738	.61163	27
34	.38866	.48745	.61134	26
35	.38894	.48751	.61106	**25**
36	.38923	.48757	.61077	24
37	.38951	.48764	.61049	23
38	.38979	.48770	.61021	22
39	.39008	.48777	.60992	21
40	.39036	.48783	.60964	**20**
41	.39064	.48789	.60936	19
42	.39093	.48796	.60907	18
43	.39121	.48802	.60879	17
44	.39150	.48808	.60850	16
45	.39178	.48815	.60822	**15**
46	.39206	.48821	.60794	14
47	.39235	.48827	.60765	13
48	.39263	.48834	.60737	12
49	.39292	.48840	.60708	11
50	.39320	.48846	.60680	**10**
51	.39348	.48852	.60652	9
52	.39377	.48858	.60623	8
53	.39405	.48865	.60595	7
54	.39434	.48871	.60566	6
55	.39462	.48877	.60538	**5**
56	.39491	.48883	.60509	4
57	.39519	.48889	.60481	3
58	.39548	.48895	.60452	2
59	.39576	.48901	.60424	1
60	.39604	.48907	.60396	**0**
′	Cos^2	Sin · Cos	Sin^2	′

128° (308°) — (231°) **51°**

39° (219°) — (320°) **140°**

′	Sin^2	Sin · Cos	Cos^2	′
0	.39604	.48907	.60396	**60**
1	.39633	.48913	.60367	59
2	.39661	.48919	.60339	58
3	.39690	.48925	.60310	57
4	.39718	.48931	.60282	56
5	.39747	.48937	.60253	**55**
6	.39775	.48943	.60225	54
7	.39804	.48949	.60196	53
8	.39832	.48955	.60168	52
9	.39861	.48961	.60139	51
10	.39889	.48967	.60111	**50**
11	.39918	.48973	.60082	49
12	.39946	.48979	.60054	48
13	.39975	.48985	.60025	47
14	.40003	.48990	.59997	46
15	.40032	.48996	.59968	**45**
16	.40060	.49002	.59940	44
17	.40089	.49008	.59911	43
18	.40117	.49014	.59883	42
19	.40146	.49019	.59854	41
20	.40174	.49025	.59826	**40**
21	.40203	.49031	.59797	39
22	.40231	.49036	.59769	38
23	.40260	.49042	.59740	37
24	.40288	.49048	.59712	36
25	.40317	.49053	.59683	**35**
26	.40345	.49059	.59655	34
27	.40374	.49065	.59626	33
28	.40402	.49070	.59598	32
29	.40431	.49076	.59569	31
30	.40460	.49081	.59540	**30**
31	.40488	.49087	.59512	29
32	.40517	.49092	.59483	28
33	.40545	.49098	.59455	27
34	.40574	.49103	.59426	26
35	.40602	.49109	.59398	**25**
36	.40631	.49114	.59369	24
37	.40660	.49120	.59340	23
38	.40688	.49125	.59312	22
39	.40717	.49131	.59283	21
40	.40745	.49136	.59255	**20**
41	.40774	.49141	.59226	19
42	.40802	.49147	.59198	18
43	.40831	.49152	.59169	17
44	.40860	.49157	.59140	16
45	.40888	.49163	.59112	**15**
46	.40917	.49168	.59083	14
47	.40945	.49173	.59055	13
48	.40974	.49179	.59026	12
49	.41003	.49184	.58997	11
50	.41031	.49189	.58969	**10**
51	.41060	.49194	.58940	9
52	.41089	.49199	.58911	8
53	.41117	.49205	.58883	7
54	.41146	.49210	.58854	6
55	.41174	.49215	.58826	**5**
56	.41203	.49220	.58797	4
57	.41232	.49225	.58768	3
58	.41260	.49230	.58740	2
59	.41289	.49235	.58711	1
60	.41318	.49240	.58682	**0**
′	Cos^2	Sin · Cos	Sin^2	′

129° (309°) — (230°) **50°**

SQUARE OF THE SINE AND COSINE AND THEIR PRODUCT (Continued)

40° (220°) (319°) **139°**

′	Sin^2	Sin · Cos	Cos^2	′
0	.41318	.49240	.58682	**60**
1	.41346	.49245	.58654	59
2	.41375	.49250	.58625	58
3	.41404	.49255	.58596	57
4	.41432	.49260	.58568	56
5	.41461	.49265	.58539	**55**
6	.41490	.49270	.58510	54
7	.41518	.49275	.58482	53
8	.41547	.49280	.58453	52
9	.41576	.49285	.58424	51
10	.41604	.49290	.58396	**50**
11	.41633	.49295	.58367	49
12	.41662	.49300	.58338	48
13	.41690	.49305	.58310	47
14	.41719	.49309	.58281	46
15	.41748	.49314	.58252	**45**
16	.41776	.49319	.58224	44
17	.41805	.49324	.58195	43
18	.41834	.49329	.58166	42
19	.41862	.49333	.58138	41
20	.41891	.49338	.58109	**40**
21	.41920	.49343	.58080	39
22	.41949	.49347	.58051	38
23	.41977	.49352	.58023	37
24	.42006	.49357	.57994	36
25	.42035	.49361	.57965	**35**
26	.42063	.49366	.57937	34
27	.42092	.49371	.57908	33
28	.42121	.49375	.57879	32
29	.42150	.49380	.57850	31
30	.42178	.49384	.57822	**30**
31	.42207	.49389	.57793	29
32	.42236	.49393	.57764	28
33	.42264	.49398	.57736	27
34	.42293	.49402	.57707	26
35	.42322	.49407	.57678	**25**
36	.42351	.49411	.57649	24
37	.42379	.49416	.57621	23
38	.42408	.49420	.57592	22
39	.42437	.49425	.57563	21
40	.42466	.49429	.57534	**20**
41	.42494	.49433	.57506	19
42	.42523	.49438	.57477	18
43	.42552	.49442	.57448	17
44	.42581	.49446	.57419	16
45	.42610	.49451	.57390	**15**
46	.42638	.49455	.57362	14
47	.42667	.49459	.57333	13
48	.42696	.49464	.57304	12
49	.42725	.49468	.57275	11
50	.42753	.49472	.57247	**10**
51	.42782	.49476	.57218	9
52	.42811	.49480	.57189	8
53	.42840	.49485	.57160	7
54	.42869	.49489	.57131	6
55	.42897	.49493	.57103	**5**
56	.42926	.49497	.57074	4
57	.42955	.49501	.57045	3
58	.42984	.49505	.57016	2
59	.43013	.49509	.56987	1
60	.43041	.49513	.56959	**0**
′	Cos^2	Sin · Cos	Sin^2	′

130° (310°) (229°) **49°**

41° (221°) (318°) **138°**

′	Sin^2	Sin · Cos	Cos^2	′
0	.43041	.49513	.56959	**60**
1	.43070	.49517	.56930	59
2	.43099	.49521	.56901	58
3	.43128	.49525	.56872	57
4	.43157	.49529	.56843	56
5	.43185	.49533	.56815	**55**
6	.43214	.49537	.56786	54
7	.43243	.49541	.56757	53
8	.43272	.49545	.56728	52
9	.43301	.49549	.56699	51
10	.43330	.49553	.56670	**50**
11	.43358	.49557	.56642	49
12	.43387	.49561	.56613	48
13	.43416	.49565	.56584	47
14	.43445	.49568	.56555	46
15	.43474	.49572	.56526	**45**
16	.43503	.49576	.56497	44
17	.43531	.49580	.56469	43
18	.43560	.49584	.56440	42
19	.43589	.49587	.56411	41
20	.43618	.49591	.56382	**40**
21	.43647	.49595	.56353	39
22	.43676	.49598	.56324	38
23	.43704	.49602	.56296	37
24	.43733	.49606	.56267	36
25	.43762	.49609	.56238	**35**
26	.43791	.49613	.56209	34
27	.43820	.49617	.56180	33
28	.43849	.49620	.56151	32
29	.43878	.49624	.56122	31
30	.43907	.49627	.56093	**30**
31	.43935	.49631	.56065	29
32	.43964	.49634	.56036	28
33	.43993	.49638	.56007	27
34	.44022	.49641	.55978	26
35	.44051	.49645	.55949	**25**
36	.44080	.49648	.55920	24
37	.44109	.49652	.55891	23
38	.44138	.49655	.55862	22
39	.44166	.49659	.55834	21
40	.44195	.49662	.55805	**20**
41	.44224	.49665	.55776	19
42	.44253	.49669	.55747	18
43	.44282	.49672	.55718	17
44	.44311	.49675	.55689	16
45	.44340	.49679	.55660	**15**
46	.44369	.49682	.55631	14
47	.44398	.49685	.55602	13
48	.44427	.49688	.55573	12
49	.44455	.49692	.55545	11
50	.44484	.49695	.55516	**10**
51	.44513	.49698	.55487	9
52	.44542	.49701	.55458	8
53	.44571	.49704	.55429	7
54	.44600	.49708	.55400	6
55	.44629	.49711	.55371	**5**
56	.44658	.49714	.55342	4
57	.44687	.49717	.55313	3
58	.44716	.49720	.55284	2
59	.44745	.49723	.55255	1
60	.44774	.49726	.55226	**0**
′	Cos^2	Sin · Cos	Sin^2	′

131° (311°) (228°) **48°**

SQUARE OF THE SINE AND COSINE AND THEIR PRODUCT (Continued)

42° (222°) (317°) **137°**

′	Sin²	Sin · Cos	Cos²	′
0	.44774	.49726	.55226	**60**
1	.44803	.49729	.55197	59
2	.44831	.49732	.55169	58
3	.44860	.49735	.55140	57
4	.44889	.49738	.55111	56
5	.44918	.49741	.55082	**55**
6	.44947	.49744	.55053	54
7	.44976	.49747	.55024	53
8	.45005	.49750	.54995	52
9	.45034	.49753	.54966	51
10	.45063	.49756	.54937	**50**
11	.45092	.49759	.54908	49
12	.45121	.49761	.54879	48
13	.45150	.49764	.54850	47
14	.45179	.49767	.54821	46
15	.45208	.49770	.54792	**45**
16	.45237	.49773	.54763	44
17	.45266	.49775	.54734	43
18	.45295	.49778	.54705	42
19	.45324	.49781	.54676	41
20	.45353	.49784	.54647	**40**
21	.45381	.49786	.54619	39
22	.45410	.49789	.54590	38
23	.45439	.49792	.54561	37
24	.45468	.49794	.54532	36
25	.45497	.49797	.54503	**35**
26	.45526	.49799	.54474	34
27	.45555	.49802	.54445	33
28	.45584	.49805	.54416	32
29	.45613	.49807	.54387	31
30	.45642	.49810	.54358	**30**
31	.45671	.49812	.54329	29
32	.45700	.49815	.54300	28
33	.45729	.49817	.54271	27
34	.45758	.49820	.54242	26
35	.45787	.49822	.54213	**25**
36	.45816	.49825	.54184	24
37	.45845	.49827	.54155	23
38	.45874	.49829	.54126	22
39	.45903	.49832	.54097	21
40	.45932	.49834	.54068	**20**
41	.45961	.49837	.54039	19
42	.45990	.49839	.54010	18
43	.46019	.49841	.53981	17
44	.46048	.49844	.53952	16
45	.46077	.49846	.53923	**15**
46	.46106	.49848	.53894	14
47	.46135	.49850	.53865	13
48	.46164	.49853	.53836	12
49	.46193	.49855	.53807	11
50	.46222	.49857	.53778	**10**
51	.46251	.49859	.53749	9
52	.46280	.49861	.53720	8
53	.46309	.49864	.53691	7
54	.46338	.49866	.53662	6
55	.46367	.49868	.53633	**5**
56	.46396	.49870	.53604	4
57	.46425	.49872	.53575	3
58	.46454	.49874	.53546	2
59	.46483	.49876	.53517	1
60	.46512	.49878	.53488	**0**
′	Cos²	Sin · Cos	Sin²	′

132° (312°) (227°) **47°**

43° (223°) (316°) **136°**

′	Sin²	Sin · Cos	Cos²	′
0	.46512	.49878	.53488	**60**
1	.46541	.49880	.53459	59
2	.46570	.49882	.53430	58
3	.46599	.49884	.53401	57
4	.46628	.49886	.53372	56
5	.46657	.49888	.53343	**55**
6	.46686	.49890	.53314	54
7	.46715	.49892	.53285	53
8	.46744	.49894	.53256	52
9	.46773	.49896	.53227	51
10	.46802	.49898	.53198	**50**
11	.46831	.49900	.53169	49
12	.46860	.49901	.53140	48
13	.46890	.49903	.53110	47
14	.46919	.49905	.53081	46
15	.46948	.49907	.53052	**45**
16	.46977	.49909	.53023	44
17	.47006	.49910	.52994	43
18	.47035	.49912	.52965	42
19	.47064	.49914	.52936	41
20	.47093	.49915	.52907	**40**
21	.47122	.49917	.52878	39
22	.47151	.49919	.52849	38
23	.47180	.49920	.52820	37
24	.47209	.49922	.52791	36
25	.47238	.49924	.52762	**35**
26	.47267	.49925	.52733	34
27	.47296	.49927	.52704	33
28	.47325	.49928	.52675	32
29	.47354	.49930	.52646	31
30	.47383	.49931	.52617	**30**
31	.47412	.49933	.52588	29
32	.47441	.49934	.52559	28
33	.47470	.49936	.52530	27
34	.47499	.49937	.52501	26
35	.47528	.49939	.52472	**25**
36	.47558	.49940	.52442	24
37	.47587	.49942	.52413	23
38	.47616	.49943	.52384	22
39	.47645	.49944	.52355	21
40	.47674	.49946	.52326	**20**
41	.47703	.49947	.52297	19
42	.47732	.49949	.52268	18
43	.47761	.49950	.52239	17
44	.47790	.49951	.52210	16
45	.47819	.49952	.52181	**15**
46	.47848	.49954	.52152	14
47	.47877	.49955	.52123	13
48	.47906	.49956	.52094	12
49	.47935	.49957	.52065	11
50	.47964	.49959	.52036	**10**
51	.47993	.49960	.52007	9
52	.48022	.49961	.51978	8
53	.48052	.49962	.51948	7
54	.48081	.49963	.51919	6
55	.48110	.49964	.51890	**5**
56	.48139	.49965	.51861	4
57	.48168	.49966	.51832	3
58	.48197	.49967	.51803	2
59	.48226	.49969	.51774	1
60	.48255	.49970	.51745	**0**
′	Cos²	Sin · Cos	Sin²	′

133° (313°) (226°) **46°**

SQUARE OF THE SINE AND COSINE AND THEIR PRODUCT (Continued)

44° (224°) (315°) **135°**

′	Sin^2	Sin · Cos	Cos^2	′
0	.48255	.49970	.51745	**60**
1	.48284	.49971	.51716	59
2	.48313	.49972	.51687	58
3	.48342	.49973	.51658	57
4	.48371	.49973	.51629	56
5	.48400	.49974	.51600	**55**
6	.48429	.49975	.51571	54
7	.48459	.49976	.51541	53
8	.48488	.49977	.51512	52
9	.48517	.49978	.51483	51
10	.48546	.49979	.51454	**50**
11	.48575	.49980	.51425	49
12	.48604	.49981	.51396	48
13	.48633	.49981	.51367	47
14	.48662	.49982	.51338	46
15	.48691	.49983	.51309	**45**
16	.48720	.49984	.51280	44
17	.48749	.49984	.51251	43
18	.48778	.49985	.51222	42
19	.48807	.49986	.51193	41
20	.48837	.49986	.51163	**40**
21	.48866	.49987	.51134	39
22	.48895	.49988	.51105	38
23	.48924	.49988	.51076	37
24	.48953	.49989	.51047	36
25	.48982	.49990	.51018	**35**
26	.49011	.49990	.50989	34
27	.49040	.49991	.50960	33
28	.49069	.49991	.50931	32
29	.49098	.49992	.50902	31
30	.49127	.49992	.50873	**30**
31	.49156	.49993	.50844	29
32	.49186	.49993	.50814	28
33	.49215	.49994	.50785	27
34	.49244	.49994	50756	26
35	.49273	.49995	.50727	**25**
36	.49302	.49995	.50698	24
37	.49331	.49996	.50669	23
38	.49360	.49996	.50640	22
39	.49389	.49996	.50611	21
40	.49418	.49997	.50582	**20**
41	.49447	.49997	.50553	19
42	.49476	.49997	.50524	18
43	.49505	.49998	.50495	17
44	.49535	.49998	.50465	16
45	.49564	.49998	.50436	**15**
46	.49593	.49998	.50407	14
47	.49622	.49999	.50378	13
48	.49651	.49999	.50349	12
49	.49680	.49999	.50320	11
50	.49709	.49999	.50291	**10**
51	.49738	.49999	.50262	9
52	.49767	.49999	.50233	8
53	.49796	.50000	.50204	7
54	.49825	.50000	.50175	6
55	.49855	.50000	.50145	**5**
56	.49884	.50000	.50116	4
57	.49913	.50000	.50087	3
58	.49942	.50000	.50058	2
59	.49971	.50000	.50029	1
60	.50000	.50000	.50000	**0**
′	Cos^2	Sin · Cos	Sin^2	′

134° (314°) (225°) **45°**

HAVERSINES

The table that follows gives –values of $\frac{1}{2}(1 - \cos \theta)$ for angles between 0 and 180° are given to five significant figures. The five-place mantissas of the logarithms of the haversines are also given. The correct characteristic must be provided in each case.

The listed values of the haversines were derived from values which were computed to seven significant figures. The logarithms were independently derived from the more exact value of the haversines and are, therefore, in many cases not the exact value of the logarithm of the haversine as listed. This is notably true at the beginning of the table where the logarithm can be given with more exactness than the function.

HAVERSINES

$\text{hav } \theta = \frac{1}{2} \text{ vers } \theta = \frac{1}{2}(1 - \cos \theta) = \sin^2 \frac{1}{2}\theta$
$\text{hav } (-\theta) = \text{hav } \theta$
$\text{hav } (180° - \theta) = \text{hav } (180° + \theta) = 1 - \text{hav } \theta$
Characteristics of the logarithms are omitted.

$\theta°$	0′ Value	0′ Log	10′ Value	10′ Log	20′ Value	20′ Log	30′ Value	30′ Log	40′ Value	40′ Log	50′ Value	50′ Log
0	.00000	—	.00000	$\bar{6}$.32539	.00001	$\bar{6}$.92745	.00002	.27963	.00003	.52951	.00005	.72332
1	.00008	.88168	.00010	.01557	.00014	.13155	.00017	.23385	.00021	.32536	.00026	.40814
2	.00030	.48371	.00036	.55323	.00041	.61759	.00048	.67751	.00054	.73355	.00061	.78620
3	.00069	.83584	.00076	.88279	.00085	.92733	.00093	.96970	.00102	.01009	.00112	.04869
4	.00122	.08564	.00132	.12108	.00143	.15513	.00154	.18790	.00166	.21947	.00178	.24993
5	.00190	.27936	.00203	.30782	.00216	.33538	.00230	.36209	.00244	.38800	.00259	.41315
6	.00274	.43760	.00289	.46138	.00305	.48452	.00321	.50706	.00338	.52902	.00355	.55044
7	.00373	.57135	.00391	.59176	.00409	.61170	.00428	.63120	.00447	.65026	.00467	.66891
8	.00487	.68717	.00507	.70505	.00528	.72257	.00549	.73974	.00571	.75657	.00593	.77308
9	.00616	.78929	.00639	.80519	.00662	.82081	.00686	.83615	.00710	.85122	.00735	.86603
10	.00760	.88059	.00785	.89491	.00811	.90900	.00837	.92286	.00864	.93650	.00891	.94993
11	.00919	.96315	.00947	.97617	.00975	.98899	.01004	.00163	.01033	.01409	.01063	.02636
12	.01093	.03847	.01123	.05041	.01154	.06218	.01185	.07379	.01217	.08525	.01249	.09656
13	.01281	.10772	.01314	.11873	.01348	.12961	.01382	.14035	.01416	.15096	.01450	.16144
14	.01485	.17179	.01521	.18202	.01556	.19212	.01593	.20211	.01629	.21198	.01666	.22175
15	.01704	.23140	.01742	.24094	.01780	.25037	.01818	.25971	.01858	.26894	.01897	.27807
16	.01937	.28711	.01977	.29605	.02018	.30490	.02059	.31366	.02101	.32233	.02142	.33091
17	.02185	.33940	.02227	.34782	.02271	.35614	.02314	.36439	.02358	.37256	.02402	.38065
18	.02447	.38867	.02492	.39660	.02538	.40447	.02584	.41226	.02630	.41998	.02677	.42764
19	.02724	.43522	.02772	.44273	.02820	.45018	.02868	.45757	.02917	.46489	.02966	.47215
20	.03015	.47934	.03065	.48647	.03116	.49355	.03166	.50056	.03218	.50752	.03269	.51442
21	.03321	.52127	.03373	.52805	.03426	.53479	.03479	.54147	.03533	.54810	.03587	.55467
22	.03641	.56120	.03695	.56767	.03751	.57410	.03806	.58047	.03862	.58680	.03918	.59308
23	.03975	.59931	.04032	.60550	.04089	.61164	.04147	.61773	.04205	.62379	.04264	.62979
24	.04323	.63576	.04382	.64168	.04442	.64756	.04502	.65340	.04562	.65920	.04623	.66496
25	.04685	.67067	.04746	.67635	.04808	.68199	.04871	.68759	.04934	.69316	.04997	.69869
26	.05060	.70418	.05124	.70963	.05189	.71505	.05253	.72043	.05318	.72578	.05384	.73109
27	.05450	.73637	.05516	.74162	.05582	.74683	.05649	.75201	.05717	.75715	.05785	.76227
28	.05853	.76735	.05921	.77240	.05990	.77742	.06059	.78241	.06129	.78737	.06199	.79230
29	.06269	.79720	.06340	.80207	.06411	.80691	.06482	.81172	.06554	.81651	.06626	.82126
30	.06699	.82599	.06772	.83069	.06845	.83537	.06919	.84001	.06993	.84464	.07067	.84923
31	.07142	.85380	.07217	.85834	.07292	.86286	.07368	.86735	.07444	.87182	.07521	.87626
32	.07598	.88068	.07675	.88507	.07752	.88944	.07830	.89379	.07909	.89811	.07987	.90241
33	.08066	.90668	.08146	.91094	.08226	.91517	.08306	.91938	.08386	.92356	.08467	.92773
34	.08548	.93187	.08630	.93599	.08711	.94009	.08794	.94417	.08876	.94823	.08959	.95227
35	.09042	.95628	.09126	.96028	.09210	.96426	.09294	.96821	.09379	.97215	.09464	.97607
36	.09549	.97996	.09635	.98384	.09721	.98770	.09807	.99154	.09894	.99536	.09981	.99917
37	.10068	.00295	.10156	.00672	.10244	.01047	.10332	.01420	.10421	.01791	.10510	.02161
38	.10599	.02528	.10689	.02894	.10779	.03259	.10870	.03621	.10960	.03982	.11051	.04341
39	.11143	.04699	.11234	.05055	.11326	.05409	.11419	.05762	.11511	.06113	.11604	.06462
40	.11698	.06810	.11791	.07157	.11885	.07501	.11980	.07845	.12074	.08186	.12169	.08526
41	.12265	.08865	.12360	.09202	.12456	.09538	.12552	.09872	.12649	.10205	.12746	.10536
42	.12843	.10866	.12940	.11194	.13038	.11521	.13136	.11847	.13235	.12171	.13333	.12494
43	.13432	.12815	.13532	.13135	.13631	.13454	.13731	.13771	.13832	.14087	.13932	.14402
44	.14033	.14715	.14134	.15027	.14236	.15338	.14337	.15647	.14440	.15955	.14542	.16262
45	.14645	.16568	.14748	.16872	.14851	.17175	.14955	.17477	.15058	.17778	.15163	.18077
46	.15267	.18376	.15372	.18673	.15477	.18968	.15582	.19263	.15688	.19557	.15794	.19849
47	.15900	.20140	.16007	.20430	.16113	.20719	.16220	.21006	.16328	.21293	.16436	.21578
48	.16543	.21863	.16652	.22146	.16760	.22428	.16869	.22709	.16978	.22989	.17087	.23268
49	.17197	.23545	.17307	.23822	.17417	.24098	.17528	.24372	.17638	.24646	.17749	.24918
50	.17861	.25190	.17972	.25460	.18084	.25729	.18196	.25998	.18308	.26265	.18421	.26532
51	.18534	.26797	.18647	.27061	.18761	.27325	.18874	.27587	.18988	.27848	.19102	.28109
52	.19217	.28368	.19332	.28627	.19447	.28885	.19562	.29141	.19677	.29397	.19793	.29652
53	.19909	.29905	.20026	.30158	.20142	.30410	.20259	.30662	.20376	.30912	.20493	.31161
54	.20611	.31409	.20729	.31657	.20847	.31903	.20965	.32149	.21083	.32394	.21202	.32638
55	.21321	.32881	.21440	.33123	.21560	.33365	.21680	.33605	.21800	.33845	.21920	.34084
56	.22040	.34322	.22161	.34559	.22282	.34795	.22403	.35031	.22525	.35266	.22646	.35499
57	.22768	.35733	.22890	.35965	.23012	.36196	.23135	.36427	.23258	.36657	.23381	.36886
58	.23504	.37114	.23627	.37342	.23751	.37569	.23875	.37794	.23999	.38020	.24124	.38244
59	.24248	.38468	.24373	.38691	.24498	.38913	.24623	.39134	.24749	.39355	.24874	.39575
60	.25000	.39794	.25126	.40012	.25252	.40230	.25379	.40447	.25506	.40663	.25632	.40879

HAVERSINES (Continued)

Characteristics of the logarithms are omitted.

θ°	0′ Value	0′ Log	10′ Value	10′ Log	20′ Value	20′ Log	30′ Value	30′ Log	40′ Value	40′ Log	50′ Value	50′ Log
60	.25000	.39794	.25126	.40012	.25252	.40230	.25379	.40447	.25506	.40663	.25632	.40879
61	.25760	.41094	.25887	.41308	.26014	.41521	.26142	.41734	.26270	.41946	.26398	.42157
62	.26526	.42368	.26655	.42578	.26784	.42787	.26913	.42996	.27042	.43203	.27171	.43411
63	.27300	.43617	.27430	.43823	.27560	.44028	.27690	.44232	.27820	.44436	.27951	.44639
64	.28081	.44842	.28212	.45044	.28343	.45245	.28474	.45446	.28606	.45645	.28737	.45845
65	.28869	.46043	.29001	.46241	.29133	.46439	.29265	.46635	.29398	.46831	.29530	.47027
66	.29663	.47222	.29796	.47416	.29929	.47610	.30063	.47802	.30196	.47995	.30330	.48187
67	.30463	.48378	.30597	.48568	.30732	.48758	.30866	.48948	.31000	.49137	.31135	.49325
68	.31270	.49512	.31405	.49699	.31540	.49886	.31675	.50072	.31810	.50257	.31946	.50442
69	.32082	.50625	.32217	.50809	.32353	.50992	.32490	.51174	.32626	.51356	.32762	.51538
70	.32899	.51718	.33036	.51898	.33173	.52078	.33310	.52257	.33447	.52435	.33584	.52613
71	.33722	.52791	.33859	.52968	.33997	.53144	.34135	.53320	.34273	.53495	.34411	.53670
72	.34549	.53844	.34688	.54017	.34826	.54190	.34965	.54363	.35103	.54535	.35242	.54707
73	.35381	.54878	.35521	.55048	.35660	.55218	.35799	.55387	.35939	.55556	.36078	.55725
74	.36218	.55893	.36358	.56060	.36498	.56227	.36638	.56393	.36778	.56559	.36919	.56725
75	.37059	.56889	.37200	.57054	.37340	.57218	.37481	.57381	.37622	.57544	.37763	.57706
76	.37904	.57868	.38045	.58030	.38186	.58191	.38328	.58351	.38469	.58511	.38611	.58671
77	.38752	.58830	.38894	.58988	.39036	.59147	.39178	.59304	.39320	.59461	.39462	.59618
78	.39604	.59774	.39747	.59929	.39889	.60085	.40032	.60240	.40174	.60395	.40317	.60549
79	.40460	.60702	.40602	.60855	.40745	.61008	.40888	.61160	.41031	.61311	.41174	.61463
80	.41318	.61613	.41461	.61764	.41604	.61914	.41748	.62063	.41891	.62212	.42035	.62361
81	.42178	.62509	.42322	.62657	.42466	.62804	.42610	.62951	.42753	.63097	.42897	.63243
82	.43041	.63389	.43185	.63534	.43330	.63678	.43474	.63823	.43618	.63966	.43762	.64110
83	.43907	.64253	.44051	.64395	.44195	.64538	.44340	.64679	.44484	.64821	.44629	.64962
84	.44774	.65102	.44918	.65242	.45063	.65382	.45208	.65521	.45353	.65660	.45497	.65799
85	.45642	.65937	.45787	.66074	.45932	.66212	.46077	.66348	.46222	.66485	.46367	.66621
86	.46512	.66757	.46657	.66892	.46802	.67027	.46948	.67161	.47093	.67295	.47238	.67429
87	.47383	.67562	.47528	.67695	.47674	.67828	.47819	.67960	.47964	.68092	.48110	.68223
88	.48255	.68354	.48400	.68485	.48546	.68615	.48691	.68745	.48837	.68874	.48982	.69004
89	.49127	.69132	.49273	.69261	.49418	.69389	.49564	.69516	.49709	.69644	.49855	.69770
90	.50000	.69897	.50145	.70023	.50291	.70149	.50436	.70274	.50582	.70399	.50727	.70524
91	.50873	.70648	.51018	.70772	.51163	.70896	.51309	.71019	.51454	.71142	.51600	.71265
92	.51745	.71387	.51890	.71509	.52036	.71630	.52181	.71751	.52326	.71872	.52472	.71992
93	.52617	.72112	.52762	.72232	.52907	.72352	.53052	.72471	.53198	.72589	.53343	.72708
94	.53488	.72825	.53633	.72943	.53778	.73060	.53923	.73177	.54068	.73294	.54213	.73410
95	.54358	.73526	.54503	.73642	.54647	.73757	.54792	.73872	.54937	.73987	.55082	.74101
96	.55226	.74215	.55371	.74328	.55516	.74442	.55660	.74554	.55805	.74667	.55949	.74779
97	.56093	.74891	.56238	.75003	.56382	.75114	.56526	.75225	.56670	.75336	.56815	.75446
98	.56959	.75556	.57103	.75666	.57247	.75775	.57390	.75884	.57534	.75993	.57678	.76101
99	.57822	.76209	.57965	.76317	.58109	.76424	.58252	.76531	.58396	.76638	.58539	.76745
100	.58682	.76851	.58826	.76957	.58969	.77062	.59112	.77167	.59255	.77272	.59398	.77377
101	.59540	.77481	.59683	.77585	.59826	.77689	.59968	.77792	.60111	.77895	.60253	.77998
102	.60396	.78101	.60538	.78203	.60680	.78305	.60822	.78406	.60964	.78507	.61106	.78608
103	.61248	.78709	.61389	.78809	.61531	.78909	.61672	.79009	.61814	.79108	.61955	.79208
104	.62096	.79306	.62237	.79405	.62378	.79503	.62519	.79601	.62660	.79699	.62800	.79796
105	.62941	.79893	.63081	.79990	.63222	.80087	.63362	.80183	.63502	.80279	.63642	.80374
106	.63782	.80470	.63922	.80565	.64061	.80660	.64201	.80754	.64340	.80848	.64479	.80942
107	.64619	.81036	.64758	.81129	.64897	.81222	.65035	.81315	.65174	.81407	.65312	.81500
108	.65451	.81592	.65589	.81683	.65727	.81775	.65865	.81866	.66003	.81956	.66141	.82047
109	.66278	.82137	.66416	.82227	.66553	.82317	.66690	.82406	.66827	.82495	.66964	.82584
110	.67101	.82673	.67238	.82761	.67374	.82849	.67510	.82937	.67647	.83025	.67783	.83112
111	.67918	.83199	.68054	.83285	.68190	.83372	.68325	.83458	.68460	.83544	.68595	.83629
112	.68730	.83715	.68865	.83800	.69000	.83885	.69134	.83969	.69268	.84054	.69403	.84138
113	.69537	.84221	.69670	.84305	.69804	.84388	.69937	.84471	.70071	.84554	.70204	.84636
114	.70337	.84718	.70470	.84800	.70602	.84882	.70735	.84963	.70867	.85044	.70999	.85125
115	.71131	.85206	.71263	.85286	.71394	.85366	.71526	.85446	.71657	.85526	.71788	.85605
116	.71919	.85684	.72049	.85763	.72180	.85841	.72310	.85920	.72440	.85998	.72570	.86076
117	.72700	.86153	.72829	.86230	.72958	.86307	.73087	.86384	.73216	.86461	.73345	.86537
118	.73474	.86613	.73602	.86689	.73730	.86764	.73858	.86840	.73986	.86915	.74113	.86990
119	.74240	.87064	.74368	.87138	.74494	.87212	.74621	.87286	.74748	.87360	.74874	.87433
120	.75000	.87506	.75126	.87579	.75251	.87652	.75377	.87724	.75502	.87796	.75627	.87868

HAVERSINES (Continued)

Characteristics of the logarithms are omitted.

θ°	0′ Value	0′ Log	10′ Value	10′ Log	20′ Value	20′ Log	30′ Value	30′ Log	40′ Value	40′ Log	50′ Value	50′ Log
120	.75000	.87506	.75126	.87579	.75251	.87652	.75377	.87724	.75502	.87796	.75627	.87868
121	.75752	.87939	.75876	.88011	.76001	.88082	.76125	.88153	.76249	.88223	.76373	.88294
122	.76496	.88364	.76619	.88434	.76742	.88503	.76865	.88573	.76988	.88642	.77110	.88711
123	.77232	.88780	.77354	.88848	.77475	.88916	.77597	.88984	.77718	.89052	.77839	.89120
124	.77960	.89187	.78080	.89254	.78200	.89321	.78320	.89387	.78440	.89454	.78560	.89520
125	.78679	.89586	.78798	.89651	.78917	.89717	.79035	.89782	.79153	.89847	.79271	.89912
126	.79389	.89976	.79507	.90040	.79624	.90104	.79741	.90168	.79858	.90232	.79974	.90295
127	.80091	.90358	.80207	.90421	.80323	.90484	.80438	.90546	.80553	.90608	.80668	.90670
128	.80783	.90732	.80898	.90794	.81012	.90855	.81126	.90916	.81239	.90977	.81353	.91037
129	.81466	.91098	.81579	.91158	.81692	.91218	.81804	.91277	.81916	.91337	.82028	.91396
130	.82139	.91455	.82251	.91514	.82362	.91573	.82472	.91631	.82583	.91689	.82693	.91747
131	.82803	.91805	.82913	.91862	.83022	.91919	.83131	.91976	.83240	.92033	.83348	.92090
132	.83457	.92146	.83564	.92202	.83672	.92258	.83780	.92314	.83887	.92369	.83993	.92425
133	.84100	.92480	.84206	.92534	.84312	.92589	.84418	.92643	.84523	.92698	.84628	.92751
134	.84733	.92805	.84837	.92859	.84942	.92912	.85045	.92965	.85149	.93018	.85252	.93071
135	.85355	.93123	.85458	.93175	.85560	.93227	.85663	.93279	.85764	.93331	.85866	.93382
136	.85967	.93433	.86068	.93484	.86168	.93535	.86269	.93585	.86369	.93636	.86468	.93686
137	.86568	.93736	.86667	.93785	.86765	.93835	.86864	.93884	.86962	.93933	.87060	.93982
138	.87157	.94030	.87254	.94079	.87351	.94127	.87448	.94175	.87544	.94223	.87640	.94270
139	.87735	.94318	.87831	.94365	.87926	.94412	.88020	.94458	.88115	.94505	.88209	.94551
140	.88302	.94597	.88396	.94643	.88489	.94689	.88581	.94734	.88674	.94779	.88766	.94824
141	.88857	.94869	.88949	.94914	.89040	.94958	.89130	.95003	.89221	.95047	.89311	.95090
142	.89401	.95134	.89490	.95177	.89579	.95221	.89668	.95264	.89756	.95306	.89844	.95349
143	.89932	.95391	.90019	.95433	.90106	.95475	.90193	.95517	.90279	.95559	.90365	.95600
144	.90451	.95641	.90536	.95682	.90621	.95723	.90706	.95763	.90790	.95804	.90874	.95844
145	.90958	.95884	.91041	.95924	.91124	.95963	.91206	.96002	.91289	.96042	.91370	.96081
146	.91452	.96119	.91533	.96158	.91614	.96196	.91694	.96234	.91774	.96272	.91854	.96310
147	.91934	.96347	.92013	.96385	.92091	.96422	.92170	.96459	.92248	.96495	.92325	.96532
148	.92402	.96568	.92479	.96604	.92556	.96640	.92632	.96676	.92708	.96712	.92783	.96747
149	.92858	.96782	.92933	.96817	.93007	.96852	.93081	.96886	.93155	.96921	.93228	.96955
150	.93301	.96989	.93374	.97023	.93446	.97056	.93518	.97089	.93589	.97123	.93660	.97156
151	.93731	.97188	.93801	.97221	.93871	.97253	.93941	.97285	.94010	.97317	.94079	.97349
152	.94147	.97381	.94215	.97412	.94283	.97443	.94351	.97475	.94418	.97505	.94484	.97536
153	.94550	.97566	.94616	.97597	.94682	.97627	.94747	.97656	.94811	.97686	.94876	.97716
154	.94940	.97745	.95003	.97774	.95066	.97803	.95129	.97831	.95194	.97861	.95254	.97888
155	.95315	.97916	.95377	.97944	.95438	.97972	.95498	.97999	.95558	.98027	.95618	.98054
156	.95677	.98081	.95736	.98108	.95795	.98134	.95853	.98161	.95911	.98187	.95968	.98213
157	.96025	.98239	.96082	.98264	.96138	.98290	.96194	.98315	.96249	.98340	.96305	.98365
158	.96359	.98389	.96413	.98414	.96467	.98438	.96521	.98462	.96574	.98486	.96627	.98510
159	.96679	.98533	.96731	.98557	.96782	.98580	.96834	.98603	.96884	.98625	.96935	.98648
160	.96985	.98670	.97034	.98692	.97083	.98714	.97132	.98736	.97180	.98758	.97228	.98779
161	.97276	.98801	.97323	.98822	.97370	.98842	.97416	.98863	.97462	.98884	.97508	.98904
162	.97553	.98924	.97598	.98944	.97642	.98964	.97686	.98983	.97729	.99003	.97773	.99022
163	.97815	.99041	.97858	.99059	.97899	.99078	.97941	.99096	.97982	.99115	.98023	.99133
164	.98063	.99151	.98103	.99168	.98142	.99186	.98182	.99203	.98220	.99220	.98258	.99237
165	.98296	.99254	.98334	.99270	.98371	.99287	.98407	.99303	.98444	.99319	.98479	.99335
166	.98515	.99350	.98550	.99366	.98584	.99381	.98618	.99396	.98652	.99411	.98686	.99425
167	.98719	.99440	.98751	.99454	.98783	.99468	.98815	.99482	.98846	.99496	.98877	.99509
168	.98907	.99523	.98937	.99536	.98967	.99549	.98996	.99562	.99025	.99574	.99053	.99587
169	.99081	.99599	.99109	.99611	.99136	.99623	.99163	.99635	.99189	.99646	.99215	.99658
170	.99240	.99669	.99265	.99680	.99290	.99691	.99314	.99701	.99338	.99712	.99361	.99722
171	.99384	.99732	.99407	.99742	.99429	.99751	.99451	.99761	.99472	.99770	.99493	.99779
172	.99513	.99788	.99533	.99797	.99553	.99805	.99572	.99814	.99591	.99822	.99609	.99830
173	.99627	.99838	.99645	.99845	.99662	.99853	.99679	.99860	.99695	.99867	.99711	.99874
174	.99726	.99881	.99741	.99887	.99756	.99894	.99770	.99900	.99784	.99906	.99797	.99912
175	.99810	.99917	.99822	.99923	.99834	.99928	.99846	.99933	.99857	.99938	.99868	.99943
176	.99878	.99947	.99888	.99951	.99898	.99956	.99907	.99959	.99915	.99963	.99924	.99967
177	.99931	.99970	.99939	.99973	.99946	.99976	.99952	.99979	.99959	.99982	.99964	.99984
178	.99970	.99987	.99974	.99989	.99979	.99991	.99983	.99993	.99986	.99994	.99990	.99995
179	.99992	.99997	.99995	.99998	.99997	.99999	.99998	.99999	.99999	.00000	1.00000	.00000
180	1.00000	.00000										

ANALYTIC GEOMETRY

DR. HOWARD EVES

RECTANGULAR COORDINATES IN A PLANE

Rectangular (Cartesian) Coordinates

Let $X'X$ (called the *x-axis*) and $Y'Y$ (called the *y-axis*) be two perpendicular lines (here taken horizontally and vertically, respectively) intersecting in point O (called the *origin*). Then any point P in the plane of the axes is located by the distance x (called the *abscissa*) and the distance y (called the *ordinate*) from $Y'Y$ and $X'X$, respectively, to P, where x is taken as positive to the right and negative to the left of $Y'Y$, and y is taken as positive above and negative below $X'X$. The ordered pair of numbers, (x,y), are called *rectangular coordinates* of the point P.

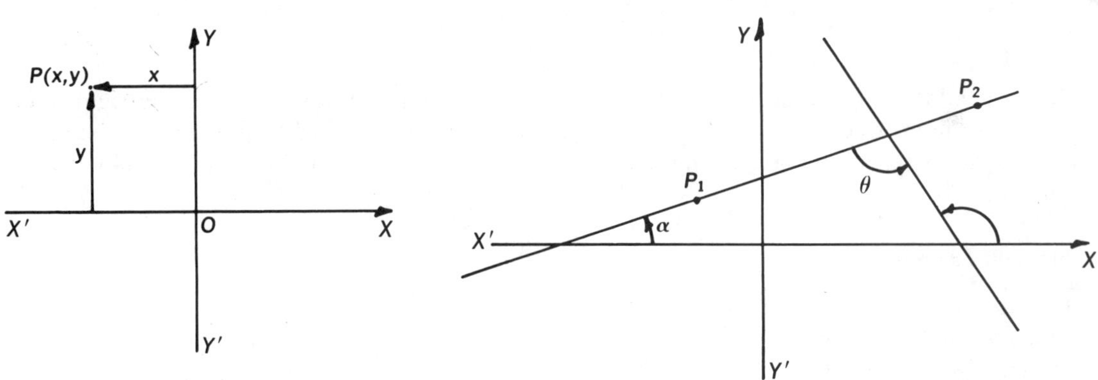

Points, Slopes, Angles

Let $P_1(x_1,y_1)$ and $P_2(x_2,y_2)$ be any two points and let α be the angle measured counterclockwise from $X'X$ to P_1P_2.

Distance between P_1 and P_2: $\sqrt{(x_2 - x_1)^2 + (y_2 - y_1)^2}$

Point dividing P_1P_2 in ratio $\frac{r}{s}$: $\left(\frac{rx_2 + sx_1}{r + s}, \frac{ry_2 + sy_1}{r + s}\right)$

Midpoint of P_1P_2: $\left(\frac{x_1 + x_2}{2}, \frac{y_1 + y_2}{2}\right)$

Slope m of P_1P_2: $m = \tan\alpha = \frac{y_2 - y_1}{x_2 - x_1}$

Angle θ between two lines of slopes m_1 and m_2: $\tan\theta = \frac{m_2 - m_1}{1 + m_1m_2}$

For parallel lines: $m_1 = m_2$

For perpendicular lines: $m_1m_2 = -1$

Points P_1, P_2, P_3 are collinear if and only if $\begin{vmatrix} x_1 & y_1 & 1 \\ x_2 & y_2 & 1 \\ x_3 & y_3 & 1 \end{vmatrix} = 0.$

Polygonal Areas

Area of triangle $P_1P_2P_3$:

$$\frac{1}{2}\begin{vmatrix} x_1 & y_1 & 1 \\ x_2 & y_2 & 1 \\ x_3 & y_3 & 1 \end{vmatrix} = \tfrac{1}{2}(x_1y_2 + x_2y_3 + x_3y_1 - y_1x_2 - y_2x_3 - y_3x_1)$$

Area of polygon $P_1P_2 \cdots P_n$:

$$\tfrac{1}{2}(x_1y_2 + x_2y_3 + \cdots + x_{n-1}y_n + x_ny_1 - y_1x_2 - y_2x_3 - \cdots - y_{n-1}x_n - y_nx_1)$$

Note. The parenthesis in the last formula is remembered by the device

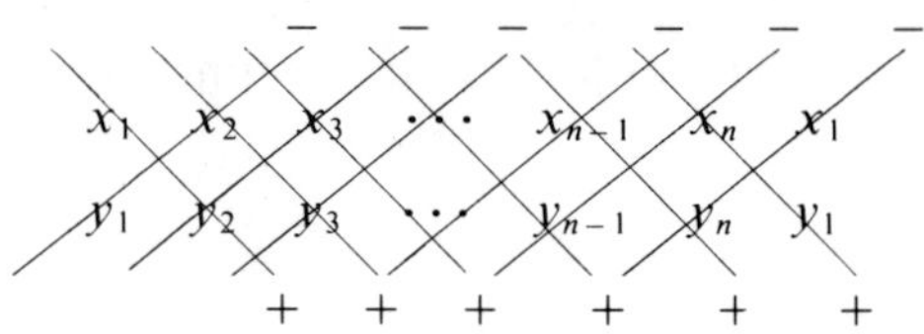

Here one adds the products of coordinates on the lines slanting downward to the right and subtracts the products of coordinates on the lines slanting upward to the right. The area is positive or negative according as $P_1P_2 \cdots P_n$ is a counterclockwise or clockwise polygon.

Straight Lines

Line parallel to y-axis: $x = a$

Line parallel to x-axis: $y = b$

Slope y-intercept form: $y = mx + b$

Intercept form: $\dfrac{x}{a} + \dfrac{y}{b} = 1$

Point-slope form: $y - y_1 = m(x - x_1)$

Two-point form: $\dfrac{y - y_1}{x - x_1} = \dfrac{y_2 - y_1}{x_2 - x_1}$

or

$$\begin{vmatrix} x & y & 1 \\ x_1 & y_1 & 1 \\ x_2 & y_2 & 1 \end{vmatrix} = 0$$

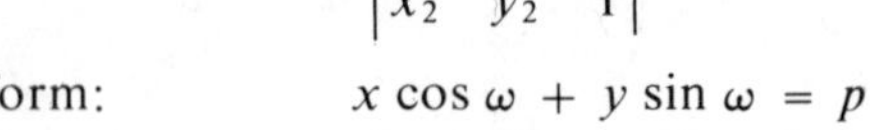

Normal form: $x \cos \omega + y \sin \omega = p$

General form: $Ax + By + C = 0$

Slope: $m = -\dfrac{A}{B}$

Intercepts: $a = -\dfrac{C}{A}, \quad b = -\dfrac{C}{B}$

To reduce $Ax + By + C = 0$ to normal form, divide by $\pm\sqrt{A^2 + B^2}$, where the sign of the radical is chosen opposite to the sign of C when $C \neq 0$ and the same as the sign of B when $C = 0$.

Distance from $Ax + By + C = 0$ to P_1: $\dfrac{Ax_1 + By_1 + C}{\pm\sqrt{A^2 + B^2}}$

Angle θ between lines $A_1x + B_1y + C_1 = 0$ and $A_2x + B_2y + C_2 = 0$
$$\tan\theta = \frac{A_1B_2 - A_2B_1}{A_1A_2 + B_1B_2}$$

Lines parallel: $A_1B_2 = A_2B_1$

Lines perpendicular: $A_1A_2 = -B_1B_2$

Lines $A_1x + B_1y + C_1 = 0, A_2x + B_2y + C_2 = 0, A_3x + B_3y + C_3 = 0$ are concurrent if and only if

$$\begin{vmatrix} A_1 & B_1 & C_1 \\ A_2 & B_2 & C_2 \\ A_3 & B_3 & C_3 \end{vmatrix} = 0.$$

Line of Best Fit

In seeking the straight line which best fits a given set of n points $P_1(x_1, y_1)$, $P_2(x_2, y_2)$, $\cdots$, $P_n(x_n, y_n)$, calculate

$$\overline{x} = \frac{x_1 + x_2 + \cdots + x_n}{n}, \quad \overline{y} = \frac{y_1 + y_2 + \cdots + y_n}{n},$$

$$m = \frac{(x_1y_1 + x_2y_2 + \cdots + x_ny_n) - n\overline{x}\overline{y}}{(x_1^2 + x_2^2 + \cdots + x_n^2) - n\overline{x}^2}.$$

Then the sought line is given by

$$y - \overline{y} = m(x - \overline{x}).$$

Circles

Center at origin, radius r: $x^2 + y^2 = r^2$

Center at (h,k), radius r: $(x - h)^2 + (y - k)^2 = r^2$

General form: $\begin{cases} Ax^2 + Ay^2 + Dx + Ey + F = 0, A \neq 0 \\ x^2 + y^2 + 2dx + 2ey + f = 0 \end{cases}$

Center: $(-d, -e)$

Radius: $r = \sqrt{d^2 + e^2 - f}$

Circle on P_1P_2 as diameter: $(x - x_1)(x - x_2) + (y - y_1)(y - y_2) = 0$

Three-point form:
$$\begin{vmatrix} x^2 + y^2 & x & y & 1 \\ x_1^2 + y_1^2 & x_1 & y_1 & 1 \\ x_2^2 + y_2^2 & x_2 & y_2 & 1 \\ x_3^2 + y_3^2 & x_3 & y_3 & 1 \end{vmatrix} = 0$$

Conic Sections

A *conic section* is the locus of a point P that moves in the plane of a fixed point F (called a *focus*) and a fixed line d (called a *directrix*), F not on d, such that the ratio of the distance of P from F to its distance from d is a constant e (called the *eccentricity*).

If $e = 1$, the conic is a *parabola*; if $e < 1$, an *ellipse*; if $e > 1$, a *hyperbola*.

Focus, (0,0); directrix, $x = -a$: $x^2 + y^2 = e^2(x + a)^2$

Parabolas (e = 1)

Let p = distance from the vertex to the focus, e = eccentricity.

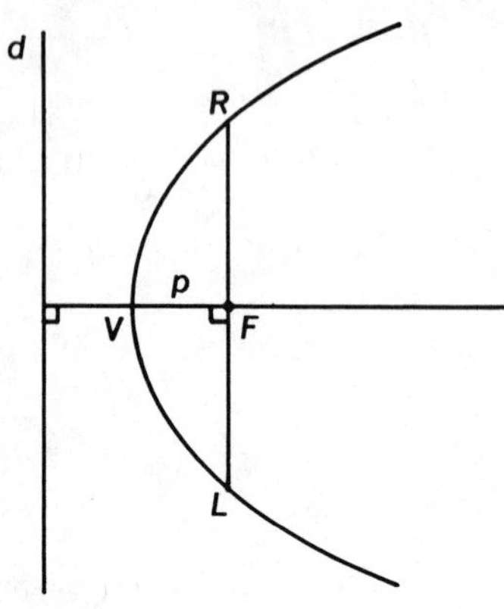

V: vertex
F: focus
d: directrix
LR: latus rectum
line VF: axis

Latus rectum:	$4p$
Distance from vertex to directrix:	p
Vertex at origin, focus at $(p,0)$:	$y^2 = 4px$
Vertex at origin, focus at $(-p,0)$:	$y^2 = -4px$
Vertex at origin, focus at $(0,p)$:	$x^2 = 4py$
Vertex at origin, focus at $(0,-p)$:	$x^2 = -4py$
Vertex, (h,k); focus, $(h + p,k)$:	$(y - k)^2 = 4p(x - h)$
Vertex, (h,k); focus, $(h - p,k)$:	$(y - k)^2 = -4p(x - h)$
Vertex, (h,k); focus, $(h,k + p)$:	$(x - h)^2 = 4p(y - k)$
Vertex, (h,k); focus, $(h,k - p)$:	$(x - h)^2 = -4p(y - k)$
General form, axis parallel to $X'X$:	$Cy^2 + Dx + Ey + F = 0$
General form, axis parallel to $Y'Y$:	$\begin{cases} Ax^2 + Dx + Ey + F = 0 \\ y = ax^2 + bx + c \end{cases}$
General form, axis oblique to coordinate axes:	$Ax^2 + Bxy + Cy^2 + Dx + Ey + F = 0$, $B^2 - 4AC = 0$

Ellipses (e < 1)

Let $2a$ = major axis, $2b$ = minor axis, e = eccentricity.

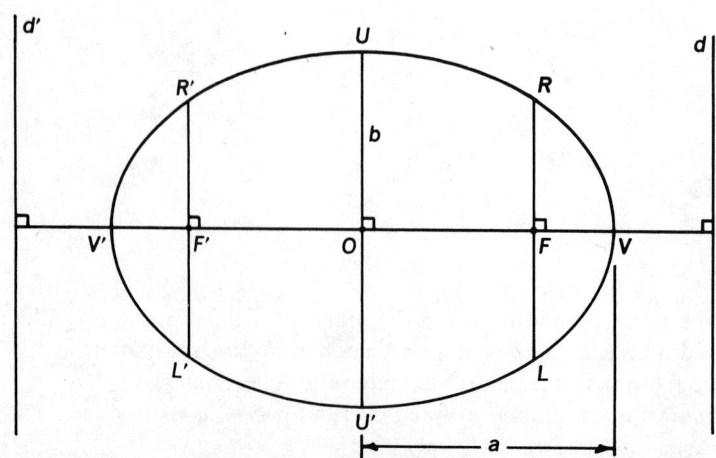

O: center
V, V': vertices
$V'V$: major axis = $2a$
$U'U$: minor axis = $2b$
F, F': foci
d, d': directrices
LR, $L'R'$: latera recta

Eccentricity:	$e = \dfrac{\sqrt{a^2 - b^2}}{a}$
Latus rectum:	$\dfrac{2b^2}{a}$
Distance from center to either focus:	$\sqrt{a^2 - b^2}$
Distance from center to either directrix:	$\dfrac{a}{e}.$
Sum of distances from any point on ellipse to the foci:	$2a$
Center at origin, foci on $X'X$:	$\dfrac{x^2}{a^2} + \dfrac{y^2}{b^2} = 1$
Center at origin, foci on $Y'Y$:	$\dfrac{x^2}{b^2} + \dfrac{y^2}{a^2} = 1$
Center at (h,k), major axis parallel to $X'X$:	$\dfrac{(x-h)^2}{a^2} + \dfrac{(y-k)^2}{b^2} = 1$
Center at (h,k), major axis parallel to $Y'Y$:	$\dfrac{(x-h)^2}{b^2} + \dfrac{(y-k)^2}{a^2} = 1$
General form, axes parallel to coordinate axes:	$Ax^2 + Cy^2 + Dx + Ey + F = 0, AC > 0$
General form, axes oblique to coordinate axes:	$Ax^2 + Bxy + Cy^2 + Dx + Ey + F = 0,$ $B^2 - 4AC < 0$

For a *circle*: $a = b$, $e = 0$, foci coincide at the center of the circle, directrices are at infinity.

Hyperbolas (e > 1)

Let $2a$ = transverse axis, $2b$ = conjugate axis, e = eccentricity.

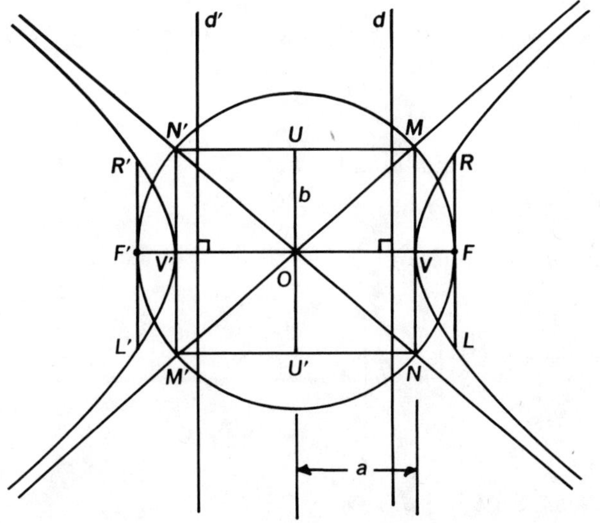

O: center
V, V': vertices
$V'V$: transverse axis $= 2a$
$U'U$: conjugate axis $= 2b$
F, F': foci
d, d': directrices
$LR, L'R'$: latera recta
lines $M'M$ and $N'N$: asymptotes

Eccentricity:	$e = \dfrac{\sqrt{a^2 + b^2}}{a}$
Latus rectum:	$\dfrac{2b^2}{a}$
Distance from center to either focus:	$\sqrt{a^2 + b^2}$
Distance from center to either directrix:	$\dfrac{a}{e}$
Difference of distances of any point on hyperbola from foci:	$2a$
Center at origin, foci on $X'X$:	$\dfrac{x^2}{a^2} - \dfrac{y^2}{b^2} = 1$
Slopes of asymptotes:	$\pm \dfrac{b}{a}$
Center at origin, foci on $Y'Y$:	$\dfrac{y^2}{a^2} - \dfrac{x^2}{b^2} = 1$
Slopes of asymptotes:	$\pm \dfrac{a}{b}$
Center at (h,k), transverse axis parallel to $X'X$:	$\dfrac{(x-h)^2}{a^2} - \dfrac{(y-k)^2}{b^2} = 1$
Slopes of asymptotes:	$\pm \dfrac{b}{a}$
Center at (h, k), transverse axis parallel to $Y'Y$:	$\dfrac{(y-k)^2}{a^2} - \dfrac{(x-h)^2}{b^2} = 1$
Slopes of asymptotes:	$\pm \dfrac{a}{b}$
Center at origin, $X'X$ and $Y'Y$ for asymptotes:	$xy = c$
Center at (h,k), asymptotes parallel to $X'X$ and $Y'Y$:	$(x-h)(y-k) = c$
General form, axes parallel to coordinate axes:	$Ax^2 + Cy^2 + Dx + Ey + F = 0, AC < 0$
General form, axes oblique to coordinate axes:	$Ax^2 + Bxy + Cy^2 + Dx + Ey + F = 0,$ $B^2 - 4AC > 0$

For a *rectangular hyperbola*: $a = b$, $e = \sqrt{2}$, asymptotes are perpendicular.

General Equation of Second Degree

The nature of the graph of the general quadratic equation in x and y,

$$ax^2 + 2hxy + by^2 + 2gx + 2fy + c = 0,$$

is described in the following table in terms of the values of

$$\Delta = \begin{vmatrix} a & h & g \\ h & b & f \\ g & f & c \end{vmatrix}, \quad J = \begin{vmatrix} a & h \\ h & b \end{vmatrix},$$

$$I = a + b, \quad K = \begin{vmatrix} a & g \\ g & c \end{vmatrix} + \begin{vmatrix} b & f \\ f & c \end{vmatrix}.$$

Case	$\triangle$	J	$\triangle/I$	K	Conic
1	$\neq 0$	> 0	< 0		real ellipse
2	$\neq 0$	> 0	> 0		imaginary ellipse
3	$\neq 0$	< 0			hyperbola
4	$\neq 0$	0			parabola
5	0	< 0			real intersecting lines
6	0	> 0			conjugate complex intersecting lines
7	0	0		< 0	real distinct parallel lines
8	0	0		> 0	conjugate complex parallel lines
9	0	0		0	coincident lines

In cases 1, 2, and 3, the center (x_0,y_0) of the conic is given by the simultaneous solution of the equations

$$ax + hy + g = 0, \quad hx + by + f = 0.$$

The equations of the axes of the conic are

$$y - y_0 = m(x - x_0), \quad y - y_0 = -\frac{1}{m}(x - x_0),$$

where m is the positive root of

$$hm^2 + (a - b)m - h = 0.$$

Transformation of Coordinates

To transform an equation of a curve from an old system of rectangular coordinates (x,y) to a new system of rectangular coordinates (x',y'), substitute for each old variable in the equation of the curve its expression in terms of the new variables.

Translation: $\begin{cases} x = x' + h \\ y = y' + k \end{cases}$ The new axes are parallel to the old axes and the coordinates of the new origin in terms of the old system are (h,k).

Rotation: $\begin{cases} x = x' \cos\theta - y' \sin\theta \\ y = x' \sin\theta + y' \cos\theta \end{cases}$ The new origin is coincident with the old origin and the new axes make an angle θ with the old axes.

To remove the xy-term from the equation

$$ax^2 + 2hxy + by^2 + 2gx + 2fy + c = 0,$$

rotate the coordinate axes about the origin through the acute angle $\theta = \arctan m$, where m is the positive root of

$$hm^2 + (a - b)m - h = 0.$$

OBLIQUE COORDINATES IN A PLANE

Oblique (Cartesian) Coordinates

Let $X'X$ (called the *x-axis*, here taken horizontally) and $Y'Y$ (called the *y-axis*) be two lines intersecting in point O (called the *origin*), and denote by ω the counterclockwise angle from $X'X$ to $Y'Y$. Then any point P in the plane of the axes is located by the distance x (called the *abscissa*) measured parallel to the x-axis and the distance y (called the *ordinate*) measured parallel to the y-axis from $Y'Y$ and $X'X$, respectively, to P, where x is taken as positive to the right and negative to the left of $Y'Y$, and y is taken as positive above and negative below $X'X$. The ordered pair of numbers, (x,y), are called *oblique coordinates* of the point P. If $\omega = 90°$, this coordinate system becomes a rectangular (Cartesian) coordinate system.

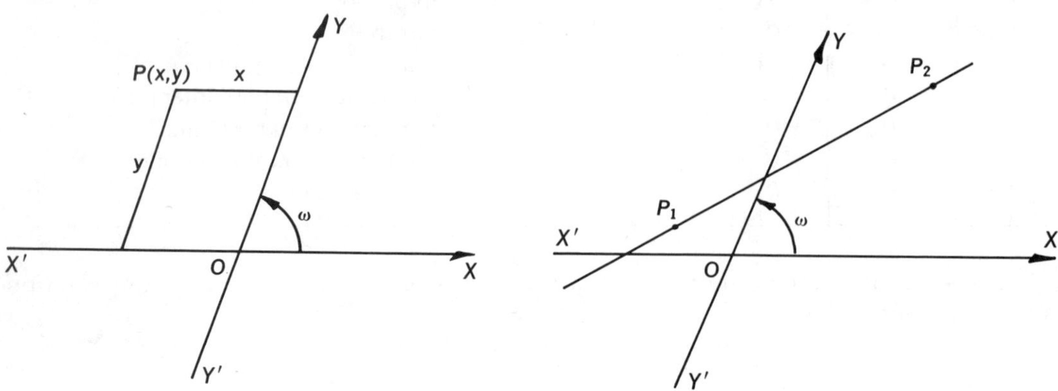

Points

Let $P_1(x_1,y_1)$ and $P_2(x_2,y_2)$ be any two points.
Distance between P_1 and P_2:

$$\sqrt{(x_2 - x_1)^2 + (y_2 - y_1)^2 + 2(x_2 - x_1)(y_2 - y_1)\cos\omega}$$

Point dividing P_1P_2 in ratio $\frac{r}{s}$: $\quad \left(\frac{rx_2 + sx_1}{r+s}, \frac{ry_2 + sy_1}{r+s}\right)$

Midpoint of P_1P_2: $\quad \left(\frac{x_1 + x_2}{2}, \frac{y_1 + y_2}{2}\right)$

Points P_1, P_2, P_3 are collinear if and only if $\begin{vmatrix} x_1 & y_1 & 1 \\ x_2 & y_2 & 1 \\ x_3 & y_3 & 1 \end{vmatrix} = 0.$

Polygonal Areas

Area of triangle $P_1P_2P_3$:

$$\sin\omega \begin{vmatrix} x_1 & y_1 & 1 \\ x_2 & y_2 & 1 \\ x_3 & y_3 & 1 \end{vmatrix} = \tfrac{1}{2}\sin\omega(x_1y_2 + x_2y_3 + x_3y_1 - y_1x_2 - y_2x_3 - y_2x_1)$$

Area of polygon $P_1P_2\cdots P_n$:

$$\tfrac{1}{2}\sin\omega(x_1y_2 + x_2y_3 + \cdots + x_{n-1}y_n + x_ny_1 - y_1x_2 - y_2x_3 - \cdots - y_{n-1}x_n - y_nx_1)$$

The area is positive or negative according as $P_1P_2\cdots P_n$ is a counterclockwise or clockwise polygon.

Straight Lines

Line parallel to y-axis: $\quad x = a$

Line parallel to x-axis: $\quad y = b$

Intercept form: $\dfrac{x}{a} + \dfrac{y}{b} = 1$

Two-point form: $\dfrac{y - y_1}{x - x_1} = \dfrac{y_2 - y_1}{x_2 - x_1}$, or $\begin{vmatrix} x & y & 1 \\ x_1 & y_1 & 1 \\ x_2 & y_2 & 1 \end{vmatrix} = 0$

General form: $Ax + By + C = 0$

Intercepts: $a = -\dfrac{C}{A}, \quad b = -\dfrac{C}{B}$

Distance from $Ax + By + C = 0$ to P_1: $\dfrac{(Ax_1 + By_1 + C)\sin\omega}{\pm\sqrt{A^2 + B^2 - 2AB\cos\omega}}$

Angle θ between lines $A_1x + B_1y + C_1 = 0$ and $A_2x + B_2y + C_2 = 0$:

$$\tan\theta = \frac{(A_1B_2 - A_2B_1)\sin\omega}{A_1A_2 + B_1B_2 - (A_1B_2 + A_2B_1)\cos\omega}$$

Lines parallel: $A_1B_2 = A_2B_1$

Lines perpendicular: $A_1A_2 + B_1B_2 = (A_1B_2 + A_2B_1)\cos\omega$

Lines $A_1x + B_1y + C_1 = 0$, $A_2x + B_2y + C_2 = 0$, $A_3x + B_3y + C_3 = 0$ are concurrent if and only if.

$$\begin{vmatrix} A_1 & B_1 & C_1 \\ A_2 & B_2 & C_2 \\ A_3 & B_3 & C_3 \end{vmatrix} = 0.$$

Circles

Center at (h, k), radius r: $(x - h)^2 + (y - k)^2 + 2(x - h)(y - k)\cos\omega = r^2$

Transformation of Coordinates

Translation: $\begin{cases} x = x' + h \\ y = y' + k \end{cases}$ The new axes are parallel to the old axes and the coordinates of the new origin in terms of the old system are (h, k).

From one oblique system to another, origin fixed:

$$\begin{cases} x = \dfrac{x'\sin(\omega - \theta) + y'\sin(\omega - \omega' - \theta)}{\sin\omega} \\[2ex] y = \dfrac{x'\sin\theta + y'\sin(\omega' + \theta)}{\sin\omega} \end{cases}$$

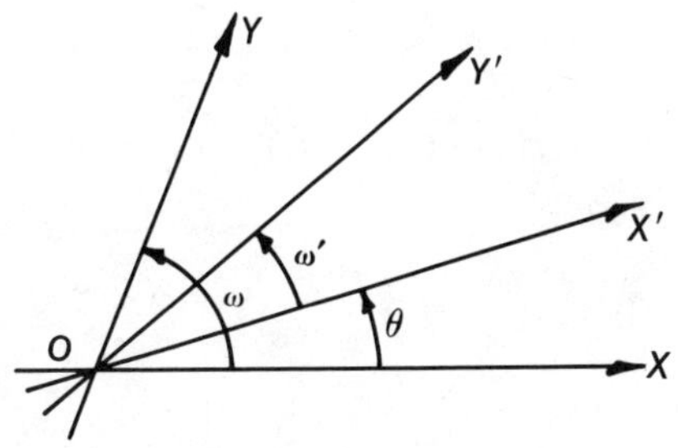

The old and new origins coincide; the old axes intersect at angle ω; the new axes intersect at angle ω'; the counterclockwise angle from x-axis to x'-axis is θ.

POLAR COORDINATES IN A PLANE

Polar Coordinates

In a plane, let OX (called the *initial line*) be a fixed ray radiating from point O (called the *pole* or *origin*). Then any point P, other than O, in the plane is located by angle θ (called the *vectorial angle*) measured from OX to the line determined by O and P and the distance r (called the *radius vector*) from O to P, where θ is taken as positive if measured counterclockwise and negative if measured clockwise, and r is taken as positive if measured along the terminal side of angle θ and negative if measured along the terminal side of θ produced through the pole. Such an ordered pair of numbers, (r, θ), are called *polar coordinates* of the point P. The polar coordinates of the pole O are taken as $(0, \theta)$, where θ is arbitrary. It follows that, for a given initial line and pole, each point of the plane has infinitely many polar coordinates.

Example

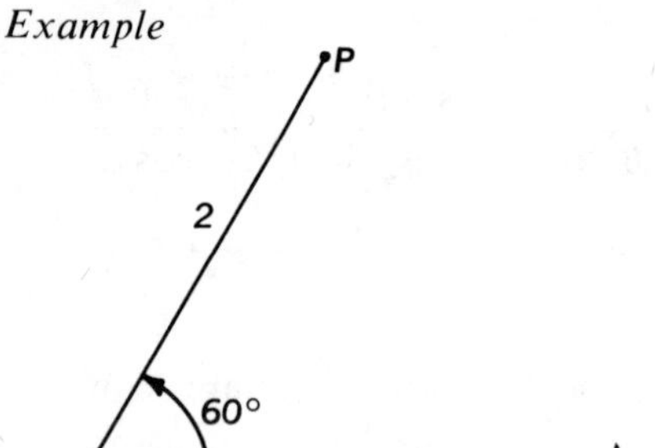

Some polar coordinates of P are: $(2, 60°)$, $(2, 420°)$, $(2, -300°)$, $(-2, 240°)$, $(-2, -120°)$.

Points

Distance between P_1 and P_2: $\sqrt{r_1^2 + r_2^2 - 2r_1r_2\cos(\theta_1 - \theta_2)}$

Points P_1, P_2, P_3 are collinear if and only if

$$r_2r_3\sin(\theta_3 - \theta_2) + r_3r_1\sin(\theta_1 - \theta_3) + r_1r_2\sin(\theta_2 - \theta_1) = 0.$$

Polygonal Areas

Area of triangle $P_1P_2P_3$:

$$\tfrac{1}{2}[r_1r_2\sin(\theta_2 - \theta_1) + r_2r_3\sin(\theta_3 - \theta_2) + r_3r_1\sin(\theta_1 - \theta_3)]$$

Area of polygon $P_1P_2\cdots P_n$:

$$\tfrac{1}{2}[r_1r_2\sin(\theta_2 - \theta_1) + r_2r_3\sin(\theta_3 - \theta_2) + \cdots + r_{n-1}r_n\sin(\theta_n - \theta_{n-1}) + r_nr_1\sin(\theta_1 - \theta_n)]$$

The area is positive or negative according as $P_1P_2\cdots P_n$ is a counterclockwise or clockwise polygon.

Straight Lines

Let p = distance of line from O, ω = counterclockwise angle from OX to the perpendicular through O to the line.

Normal form: $r\cos(\theta - \omega) = p$

Two-point form: $r[r_1\sin(\theta - \theta_1) + r_2\sin(\theta - \theta_2)] = r_1r_2\sin(\theta_2 - \theta_1)$

Circles

Center at pole, radius a: $r = a$

Center at $(a, 0)$ and passing through the pole: $r = 2a\cos\theta$

Center at $\left(a, \frac{\pi}{2}\right)$ and passing through the pole: $r = 2a \sin \theta$

Center (h, α), radius a: $r^2 - 2hr \cos(\theta - \alpha) + h^2 - a^2 = 0$

Conics

Let $2p$ = distance from directrix to focus, e = eccentricity.

Focus at pole, directrix to left of pole: $r = \dfrac{2ep}{1 - e \cos \theta}$

Focus at pole, directrix to right of pole: $r = \dfrac{2ep}{1 + e \cos \theta}$

Focus at pole, directrix below pole: $r = \dfrac{2ep}{1 - e \sin \theta}$

Focus at pole, directrix above pole: $r = \dfrac{2ep}{1 + e \sin \theta}$

Parabola with vertex at pole, directrix to left of pole: $r = \dfrac{4p \cos \theta}{\sin^2 \theta}$

Ellipse with center at pole, semiaxes a and b horizontal and vertical, respectively: $r^2 = \dfrac{a^2 b^2}{a^2 \sin^2 \theta + b^2 \cos^2 \theta}$

Hyperbola with center at pole, semiaxes a and b horizontal and vertical, respectively: $r^2 = \dfrac{a^2 b^2}{a^2 \sin^2 \theta - b^2 \cos^2 \theta}$

Relations Between Rectangular and Polar Coordinates

Let the positive x-axis coincide with the initial line and let r be nonnegative.

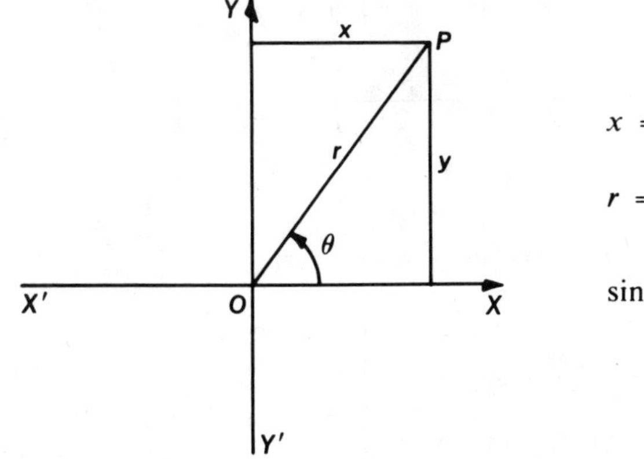

$$x = r \cos \theta, \quad y = r \sin \theta,$$

$$r = \sqrt{x^2 + y^2}, \quad \theta = \arctan \frac{y}{x},$$

$$\sin \theta = \frac{y}{\sqrt{x^2 + y^2}}, \quad \cos \theta = \frac{x}{\sqrt{x^2 + y^2}}$$

RECTANGULAR COORDINATES IN SPACE

Rectangular (Cartesian) Coordinates

Let $X'X$, $Y'Y$, $Z'Z$ (called the *x-axis*, the *y-axis*, and the *z-axis*, respectively) be three mutually perpendicular lines in space intersecting in a point O (called the *origin*), forming in this way three mutually perpendicular planes XOY, XOZ, YOZ (called the *xy*-

plane, the *xz-plane*, and the *yz-plane*, respectively). Then any point P of space is located by its signed distances x, y, z from the yz-plane, the xz-plane, and the xy-plane, respectively, where x and y are the rectangular coordinates with respect to the axes $X'X$ and $Y'Y$ of the orthogonal projection P' of P on the xy-plane (here taken horizontally) and z is taken as positive above and negative below the xy-plane. The ordered triple of numbers, (x, y, z), are called *rectangular coordinates* of the point P.

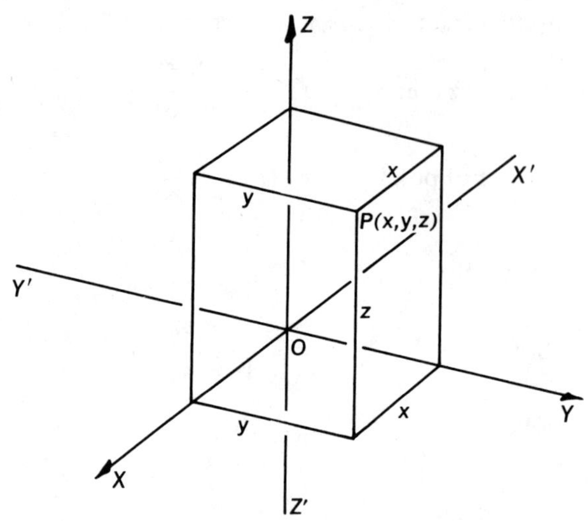

Points

Let $P_1(x_1, y_1, z_1)$ and $P_2(x_2, y_2, z_2)$ be any two points.

Distance between P_1 and P_2: $\sqrt{(x_2 - x_1)^2 + (y_2 - y_1)^2 + (z_2 - z_1)^2}$

Point dividing P_1P_2 in ratio $\frac{r}{s}$: $\left(\frac{rx_2 + sx_1}{r + s}, \frac{ry_2 + sy_1}{r + s}, \frac{rz_2 + sz_1}{r + s}\right)$

Midpoint of P_1P_2: $\left(\frac{x_1 + x_2}{2}, \frac{y_1 + y_2}{2}, \frac{z_1 + z_2}{2}\right)$

Points P_1, P_2, P_3 are collinear if and only if

$$x_2 - x_1 : y_2 - y_1 : z_2 - z_1 = x_3 - x_1 : y_3 - y_1 : z_3 - z_1.$$

Points P_1, P_2, P_3, P_4 are coplanar if and only if $\begin{vmatrix} x_1 & y_1 & z_1 & 1 \\ x_2 & y_2 & z_2 & 1 \\ x_3 & y_3 & z_3 & 1 \\ x_4 & y_4 & z_4 & 1 \end{vmatrix} = 0.$

Area of triangle $P_1P_2P_3$:

$$\frac{1}{2}\sqrt{\begin{vmatrix} y_1 & z_1 & 1 \\ y_2 & z_2 & 1 \\ y_3 & z_3 & 1 \end{vmatrix}^2 + \begin{vmatrix} z_1 & x_1 & 1 \\ z_2 & x_2 & 1 \\ z_3 & x_3 & 1 \end{vmatrix}^2 + \begin{vmatrix} x_1 & y_1 & 1 \\ x_2 & y_2 & 1 \\ x_3 & y_3 & 1 \end{vmatrix}^2}$$

Volume of tetrahedron $P_1P_2P_3P_4$: $\frac{1}{6}\begin{vmatrix} x_1 & y_1 & z_1 & 1 \\ x_2 & y_2 & z_2 & 1 \\ x_3 & y_3 & z_3 & 1 \\ x_4 & y_4 & z_4 & 1 \end{vmatrix}$

Direction Numbers and Direction Cosines

Let α, β, γ (called *direction angles*) be the angles that P_1P_2, or any line parallel to P_1P_2, makes with the x-, y-, and z-axis, respectively. Let d = distance between P_1 and P_2.

Direction cosines of P_1P_2:

$$\cos\alpha = \frac{x_2 - x_1}{d}, \quad \cos\beta = \frac{y_2 - y_1}{d}, \quad \cos\gamma = \frac{z_2 - z_1}{d}$$

$$\cos^2\alpha + \cos^2\beta + \cos^2\gamma = 1$$

If a, b, c are direction numbers of P_1P_2, then:

$$a:b:c = x_2 - x_1 : y_2 - y_1 : z_2 - z_1$$
$$= \cos\alpha : \cos\beta : \cos\gamma$$

$$\cos\alpha = \frac{a}{\pm\sqrt{a^2 + b^2 + c^2}}, \quad \cos\beta = \frac{b}{\pm\sqrt{a^2 + b^2 + c^2}},$$

$$\cos\gamma = \frac{c}{\pm\sqrt{a^2 + b^2 + c^2}}$$

Angle between two lines with direction angles α_1, β_1, γ_1 and α_2, β_2, γ_2:

$$\cos\theta = \cos\alpha_1\cos\alpha_2 + \cos\beta_1\cos\beta_2 + \cos\gamma_1\cos\gamma_2$$

For parallel lines: $\alpha_1 = \alpha_2$, $\beta_1 = \beta_2$, $\gamma_1 = \gamma_2$

For perpendicular lines:

$$\cos\alpha_1\cos\alpha_2 + \cos\beta_1\cos\beta_2 + \cos\gamma_1\cos\gamma_2 = 0$$

Angle between two lines with directions (a_1,b_1,c_1) and (a_2,b_2,c_2):

$$\cos\theta = \frac{a_1a_2 + b_1b_2 + c_1c_2}{\sqrt{a_1^2 + b_1^2 + c_1^2}\sqrt{a_2^2 + b_2^2 + c_2^2}}$$

$$\sin\theta = \frac{\sqrt{(b_1c_2 - c_1b_2)^2 + (c_1a_2 - a_1c_2)^2 + (a_1b_2 - b_1a_2)^2}}{\sqrt{a_1^2 + b_1^2 + c_1^2}\sqrt{a_2^2 + b_2^2 + c_2^2}}$$

For parallel lines: $a_1 : b_1 : c_1 = a_2 : b_2 : c_2$

For perpendicular lines: $a_1a_2 + b_1b_2 + c_1c_2 = 0$

The direction

$$(b_1c_2 - c_1b_2,\ c_1a_2 - a_1c_2,\ a_1b_2 - b_1a_2)$$

is perpendicular to both directions (a_1,b_1,c_1) and (a_2,b_2,c_2).

The directions (a_1,b_1,c_1), (a_2,b_2,c_2), (a_3,b_3,c_3) are parallel to a common plane if and only if

$$\begin{vmatrix} a_1 & b_1 & c_1 \\ a_2 & b_2 & c_2 \\ a_3 & b_3 & c_3 \end{vmatrix} = 0.$$

Straight Lines

Point-direction form: $$\frac{x - x_1}{a} = \frac{y - y_1}{b} = \frac{z - z_1}{c}$$

Two-point form: $$\frac{x - x_1}{x_2 - x_1} = \frac{y - y_1}{y_2 - y_1} = \frac{z - z_1}{z_2 - z_1}$$

Parametric form: $x = x_1 + ta,\ y = y_1 + tb,\ z = z_1 + tc$

General form: $$\begin{cases} A_1x + B_1y + C_1z + D_1 = 0 \\ A_2x + B_2y + C_2z + D_2 = 0 \end{cases}$$

Direction of line: $(B_1C_2 - C_1B_2, C_1A_2 - A_1C_2, A_1B_2 - B_1A_2)$

Projection of segment P_1P_2 on direction (a, b, c):

$$\frac{(x_2 - x_1)a + (y_2 - y_1)b + (z_2 - z_1)c}{\sqrt{a^2 + b^2 + c^2}}$$

Distance from point P_0 to line through P_1 in direction (a, b, c):

$$\sqrt{\frac{\begin{vmatrix} y_0 - y_1 & z_0 - z_1 \\ b & c \end{vmatrix}^2 + \begin{vmatrix} z_0 - z_1 & x_0 - x_1 \\ c & a \end{vmatrix}^2 + \begin{vmatrix} x_0 - x_1 & y_0 - y_1 \\ a & b \end{vmatrix}^2}{a^2 + b^2 + c^2}}$$

Distance between line through P_1 in direction (a_1, b_1, c_1) and line through P_2 in direction (a_2, b_2, c_2):

$$\pm \frac{\begin{vmatrix} x_2 - x_1 & y_2 - y_1 & z_2 - z_1 \\ a_1 & b_1 & c_1 \\ a_2 & b_2 & c_2 \end{vmatrix}}{\sqrt{\begin{vmatrix} b_1 & c_1 \\ b_2 & c_2 \end{vmatrix}^2 + \begin{vmatrix} c_1 & a_1 \\ c_2 & a_2 \end{vmatrix}^2 + \begin{vmatrix} a_1 & b_1 \\ a_2 & b_2 \end{vmatrix}^2}}$$

The line through P_1 in direction (a_1, b_1, c_1) and the line through P_2 in direction (a_2, b_2, c_2) intersect if and only if

$$\begin{vmatrix} x_2 - x_1 & y_2 - y_1 & z_2 - z_1 \\ a_1 & b_1 & c_1 \\ a_2 & b_2 & c_2 \end{vmatrix} = 0.$$

Planes

General form: $Ax + By + Cz + D = 0$

Direction of normal: (A, B, C)

Perpendicular to yz-plane: $By + Cz + D = 0$

Perpendicular to xz-plane: $Ax + Cz + D = 0$

Perpendicular to xy-plane: $Ax + By + D = 0$

Perpendicular to x-axis: $Ax + D = 0$

Perpendicular to y-axis: $By + D = 0$

Perpendicular to z-axis: $Cz + D = 0$

Intercept form: $$\frac{x}{a} + \frac{y}{b} + \frac{z}{c} = 1$$

Plane through point P_1 and perpendicular to direction (a, b, c):

$$a(x - x_1) + b(y - y_1) + c(z - z_1) = 0$$

Plane through point P_1 and parallel to directions (a_1,b_1,c_1) and (a_2,b_2,c_2):

$$\begin{vmatrix} x - x_1 & y - y_1 & z - z_1 \\ a_1 & b_1 & c_1 \\ a_2 & b_2 & c_2 \end{vmatrix} = 0$$

Plane through points P_1 and P_2 parallel to direction (a,b,c):

$$\begin{vmatrix} x - x_1 & y - y_1 & z - z_1 \\ x_2 - x_1 & y_2 - y_1 & z_2 - z_1 \\ a & b & c \end{vmatrix} = 0$$

Three-point form:

$$\begin{vmatrix} x & y & z & 1 \\ x_1 & y_1 & z_1 & 1 \\ x_2 & y_2 & z_2 & 1 \\ x_3 & y_3 & z_3 & 1 \end{vmatrix} = 0 \quad \text{or} \quad \begin{vmatrix} x - x_1 & y - y_1 & z - z_1 \\ x_2 - x_1 & y_2 - y_1 & z_2 - z_1 \\ x_3 - x_1 & y_3 - y_1 & z_3 - z_1 \end{vmatrix} = 0$$

Normal form (p = distance from origin to plane; α, β, γ are direction angles of perpendicular to plane from origin):

$$x \cos \alpha + y \cos \beta + z \cos \gamma = p$$

To reduce $Ax + By + Cz + D = 0$ to normal form, divide by $\pm\sqrt{A^2 + B^2 + C^2}$, where the sign of the radical is chosen opposite to the sign of D when $D \neq 0$, the same as the sign of C when $D = 0$ and $C \neq 0$, the same as the sign of B when $C = D = 0$.

Distance from point P_1 to plane $Ax + By + Cz + D = 0$:

$$\frac{Ax_1 + By_1 + Cz_1 + D}{\pm\sqrt{A^2 + B^2 + C^2}}$$

Angle θ between planes $A_1x + B_1y + C_1z + D_1 = 0$ and $A_2x + B_2y + C_2z + D_2 = 0$:

$$\cos \theta = \frac{A_1A_2 + B_1B_2 + C_1C_2}{\sqrt{A_1^2 + B_1^2 + C_1^2}\sqrt{A_2^2 + B_2^2 + C_2^2}}$$

Planes parallel: $A_1 : B_1 : C_1 = A_2 : B_2 : C_2$

Planes perpendicular: $A_1A_2 + B_1B_2 + C_1C_2 = 0$

Spheres

Center at origin, radius r: $x^2 + y^2 + z^2 = r^2$

Center at (g,h,k), radius r: $(x - g)^2 + (y - h)^2 + (z - k)^2 = r^2$

General form: $\begin{cases} Ax^2 + Ay^2 + Az^2 + Dx + Ey + Fz + M = 0, \quad A \neq 0 \\ x^2 + y^2 + z^2 + 2dx + 2ey + 2fz + m = 0 \end{cases}$

Center: $(-d, -e, -f)$

Radius: $r = \sqrt{d^2 + e^2 + f^2 - m}$

Sphere on P_1P_2 as diameter:

$$(x - x_1)(x - x_2) + (y - y_1)(y - y_2) + (z - z_1)(z - z_2) = 0$$

Four-point form:

$$\begin{vmatrix} x^2 + y^2 + z^2 & x & y & z & 1 \\ x_1^2 + y_1^2 + z_1^2 & x_1 & y_1 & z_1 & 1 \\ x_2^2 + y_2^2 + z_2^2 & x_2 & y_2 & z_2 & 1 \\ x_3^2 + y_3^2 + z_3^2 & x_3 & y_3 & z_3 & 1 \\ x_4^2 + y_4^2 + z_4^2 & x_4 & y_4 & z_4 & 1 \end{vmatrix} = 0$$

The Seventeen Quadric Surfaces in Standard Form

1. Real ellipsoid: $x^2/a^2 + y^2/b^2 + z^2/c^2 = 1$
2. Imaginary ellipsoid: $x^2/a^2 + y^2/b^2 + z^2/c^2 = -1$
3. Hyperboloid of one sheet: $x^2/a^2 + y^2/b^2 - z^2/c^2 = 1$
4. Hyperboloid of two sheets: $x^2/a^2 + y^2/b^2 - z^2/c^2 = -1$
5. Real quadric cone: $x^2/a^2 + y^2/b^2 - z^2/c^2 = 0$
6. Imaginary quadric cone: $x^2/a^2 + y^2/b^2 + z^2/c^2 = 0$
7. Elliptic paraboloid: $x^2/a^2 + y^2/b^2 + 2z = 0$
8. Hyperbolic paraboloid: $x^2/a^2 - y^2/b^2 + 2z = 0$
9. Real elliptic cylinder: $x^2/a^2 + y^2/b^2 = 1$
10. Imaginary elliptic cylinder: $x^2/a^2 + y^2/b^2 = -1$
11. Hyperbolic cylinder: $x^2/a^2 - y^2/b^2 = -1$
12. Real intersecting planes: $x^2/a^2 - y^2/b^2 = 0$
13. Imaginary intersecting planes: $x^2/a^2 + y^2/b^2 = 0$
14. Parabolic cylinder: $x^2 + 2rz = 0$
15. Real parallel planes: $x^2 = a^2$
16. Imaginary parallel planes: $x^2 = -a^2$
17. Coincident planes: $x^2 = 0$

Sphere = special case of (1)

General Equation of Second Degree

The nature of the graph of the general quadratic equation in x, y, z,

$$ax^2 + by^2 + cz^2 + 2fyz + 2gzx + 2hxy + 2px + 2qy + 2rz + d = 0,$$

is described in the following table in terms of $\rho_3, \rho_4, \Delta, k_1, k_2, k_3$, where

$$e = \begin{bmatrix} a & h & g \\ h & b & f \\ g & f & c \end{bmatrix}, \quad E = \begin{bmatrix} a & h & g & p \\ h & b & f & q \\ g & f & c & r \\ p & q & r & d \end{bmatrix},$$

$$\rho_3 = \text{rank } e, \quad \rho_4 = \text{rank } E,$$

$$\Delta = \text{determinant of } E,$$

k_1, k_2, k_3 are the roots of $\begin{vmatrix} a - x & h & g \\ h & b - x & f \\ g & f & c - x \end{vmatrix} = 0.$

Case	ρ_3	ρ_4	Sign of Δ	Nonzero k's same sign?	Quadric Surface
1	3	4	−	yes	Real ellipsoid
2	3	4	+	yes	Imaginary ellipsoid
3	3	4	+	no	Hyperboloid of one sheet
4	3	4	−	no	Hyperboloid of two sheets
5	3	3		no	Real quadric cone
6	3	3		yes	Imaginary quadric cone
7	2	4	−	yes	Elliptic paraboloid
8	2	4	+	no	Hyperbolic paraboloid
9	2	3		yes	Real elliptic cylinder
10	2	3		yes	Imaginary elliptic cylinder
11	2	3		no	Hyperbolic cylinder
12	2	2		no	Real intersecting planes
13	2	2		yes	Imaginary intersecting planes
14	1	3			Parabolic cylinder
15	1	2			Real parallel planes
16	1	2			Imaginary parallel planes
17	1	1			Coincident planes

Cylindrical and Conical Surfaces

Any equation in just two of the variables x, y, z represents a *cylindrical surface* whose elements are parallel to the axis of the missing variable.

Any equation homogeneous in the variables x, y, z represents a *conical surface* whose vertex is at the origin.

Transformation of Coordinates

To transform an equation of a surface from an old system of rectangular coordinates (x, y, z) to a new system of rectangular coordinates (x', y', z'), substitute for each old variable in the equation of the surface its expression in terms of the new variables.

Translation:

$$x = x' + h \quad y = y' + k \quad z = z' + l$$

The new axes are parallel to the old axes and the coordinates of the new origin in terms of the old system are (h, k, l).

Rotation about the origin:

$$x = \lambda_1 x' + \lambda_2 y' + \lambda_3 z'$$
$$y = \mu_1 x' + \mu_2 y' + \mu_3 z'$$
$$z = \nu_1 x' + \nu_2 y' + \nu_3 z'$$

The new origin is coincident with the old origin and the x'-axis, y'-axis, z'-axis have direction cosines $(\lambda_1, \mu_1, \nu_1)$, $(\lambda_2, \mu_2, \nu_2)$, $(\lambda_3, \mu_3, \nu_3)$, respectively, with respect to the old system of axes.

$$x' = \lambda_1 x + \mu_1 y + \nu_1 z$$
$$y' = \lambda_2 x + \mu_2 y + \nu_2 z$$
$$z' = \lambda_3 x + \mu_3 y + \nu_3 z$$

Cylindrical Coordinates

If (r, θ, z) are the cylindrical coordinates and (x, y, z) the rectangular coordinates of a point P, then

$$x = r\cos\theta, \quad r = \sqrt{x^2 + y^2},$$

$$y = r\sin\theta, \quad \theta = \arctan\frac{y}{x},$$

$$z = z, \quad z = z.$$

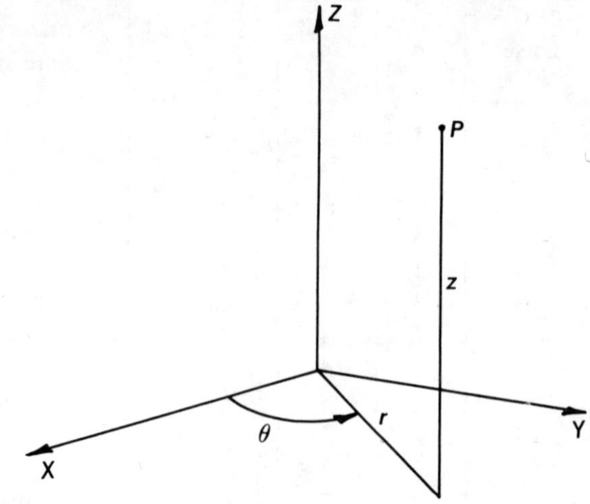

Spherical Coordinates

If (r, θ, ϕ) are the spherical coordinates and (x, y, z) the rectangular coordinates of a point P, then

$$x = r\sin\theta\cos\phi,$$

$$y = r\sin\theta\sin\phi,$$

$$z = r\cos\theta;$$

$$r = \sqrt{x^2 + y^2 + z^2},$$

$$\theta = \arccos\frac{z}{\sqrt{x^2 + y^2 + z^2}},$$

$$\phi = \arctan\frac{y}{x}.$$

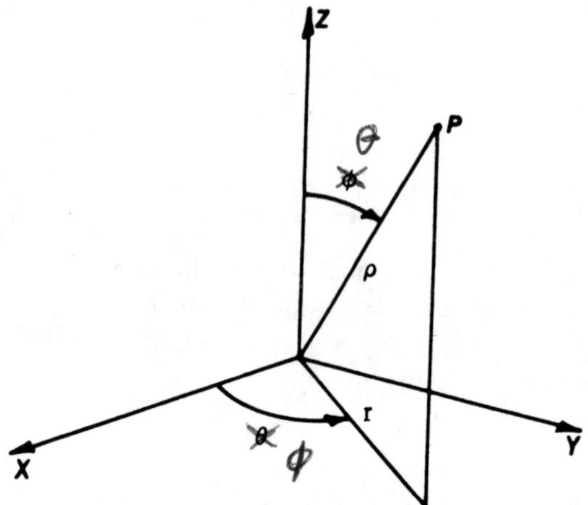

CURVES AND SURFACES

DR. HOWARD EVES

The curves and surfaces collected here for reference appear frequently in mathematical literature. The equations most generally associated with each figure are given. The equation of a plane curve when placed otherwise on the coordinate frame of reference may often be found from the given equation by the following rules.

RECTANGULAR COORDINATES

1. If a given curve is reflected in the x-axis, the new equation is obtained from the old by replacing y by $-y$.
2. If a given curve is reflected in the y-axis, the new equation is obtained from the old by replacing x by $-x$.
3. If a given curve is reflected in the origin, the new equation is obtained from the old by replacing x by $-x$ and y by $-y$.
4. If a given curve is reflected in the line $y = x$, the new equation is obtained from the old by interchanging x and y.
5. If a given curve is rotated about the origin through 90°, the new equation is obtained from the old by replacing x by y and y by $-x$.
6. If a given curve is rotated about the origin through −90°, the new equation is obtained from the old by replacing x by $-y$ and y by x.
7. If a given curve is translated a distance h in the x-direction, the new equation is obtained from the old by replacing x by $x - h$.
8. If a given curve is translated a distance k in the y-direction, the new equation is obtained from the old by replacing y by $y - k$.
9. If a given curve is altered by multiplying all the abscissas by a, the new equation is obtained from the old by replacing x by x/a.
10. If a given curve is altered by multiplying all the ordinates by b, the new equation is obtained from the old by replacing y by y/b.

POLAR COORDINATES

1. If a given curve is reflected in the polar axis, the new equation is obtained from the old by replacing θ by $-\theta$, or by replacing r by $-r$ and θ by $\pi - \theta$.
2. If a given curve is reflected in the 90° axis, the new equation is obtained from the old by replacing θ by $\pi - \theta$, or by replacing r by $-r$ and θ by $-\theta$.
3. If a given curve is reflected in the pole, the new equation is obtained from the old by replacing θ by $\pi + \theta$, or by replacing r by $-r$.
4. If a given curve is rotated about the pole through an angle α, the new equation is obtained from the old by replacing θ by $\theta - \alpha$.

PLANE CURVES

Archimedean spiral
See: Spiral of Archimedes

Astroid
See: Hypocycloid of four cusps

Bifolium

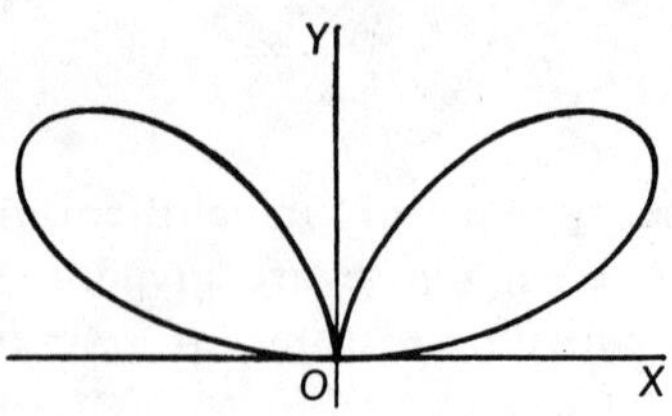

$$(x^2 + y^2)^2 = ax^2y$$
$$r = a \sin\theta \cos^2\theta$$

Cardioid

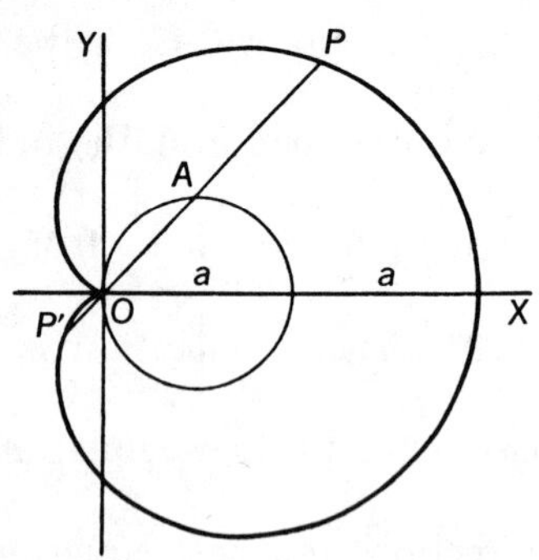

$$(x^2 + y^2 - ax)^2 = a^2(x^2 + y^2)$$
$$r = a(\cos\theta + 1)$$
or
$$r = a(\cos\theta - 1)$$
$$[P'A = AP = a]$$

Cassinian curves
See: Ovals of Cassini

Catenary, Hyperbolic cosine

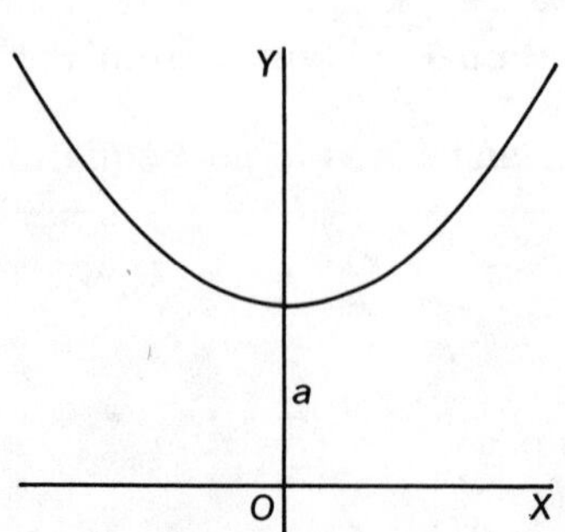

$$y = \frac{x}{a}\,(e^{x/a} + e^{-x/a}) = a \cosh \frac{x}{a}$$

Circle

(a)

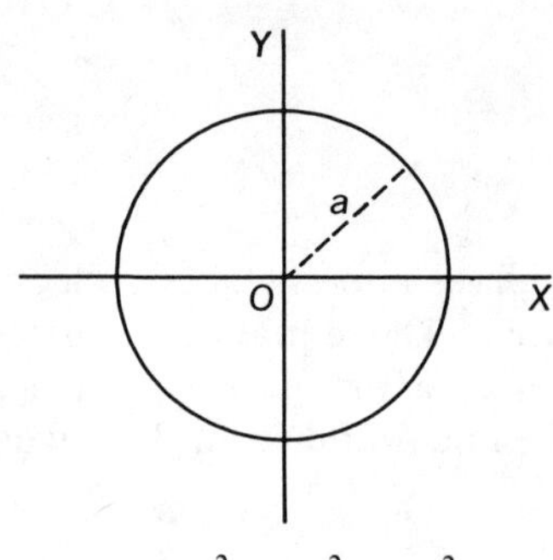

$$x^2 + y^2 = a^2$$
$$r = a$$

(b)

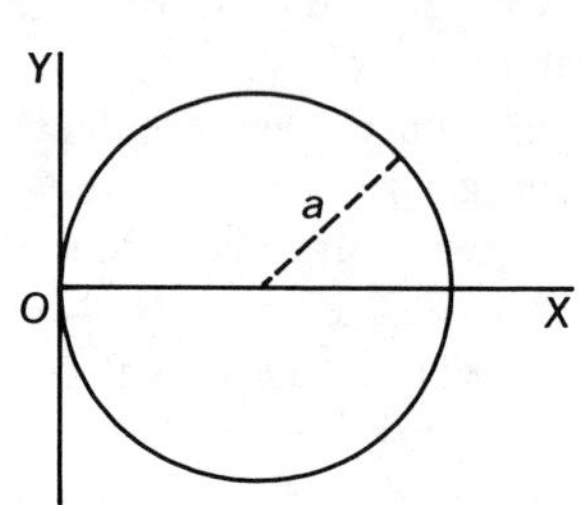

$$x^2 + y^2 = 2ax$$
$$r = 2a\cos\theta$$

(c)

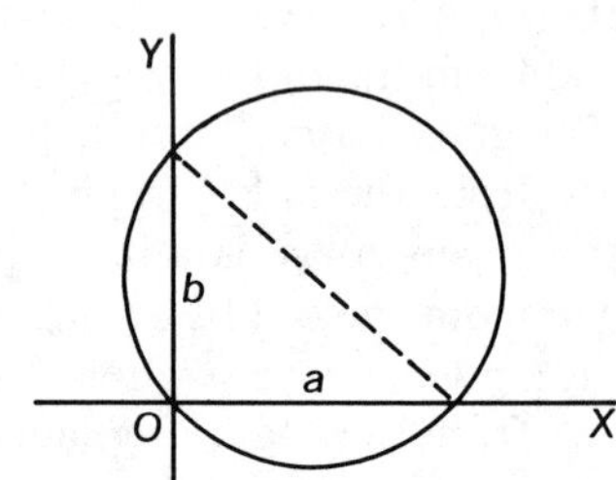

$$x^2 + y^2 = ax + by$$
$$r = a\cos\theta + b\sin\theta$$

Cissoid of Diocles

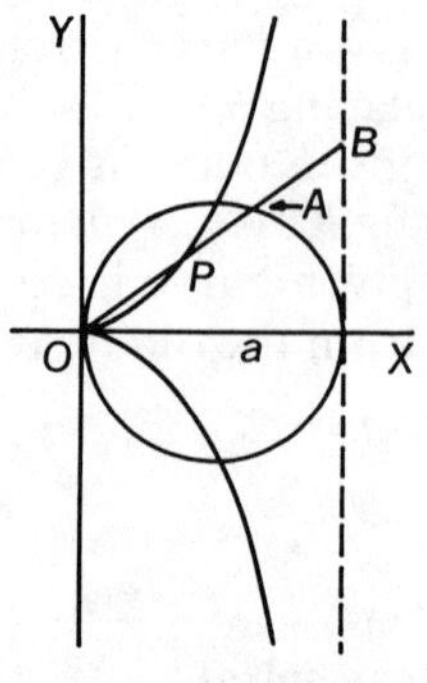

$$y^2(a - x) = x^3$$
$$r = a \sin\theta \tan\theta$$
$$[OP = AB]$$

Cochleoid, Oui-ja board curve

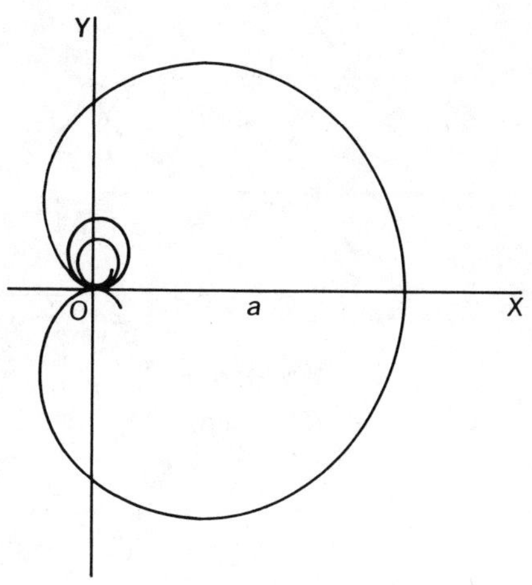

$$(x^2 + y^2)\tan^{-1}(y/x) = ay$$
$$r\theta = a \sin\theta$$

Companion to the cycloid

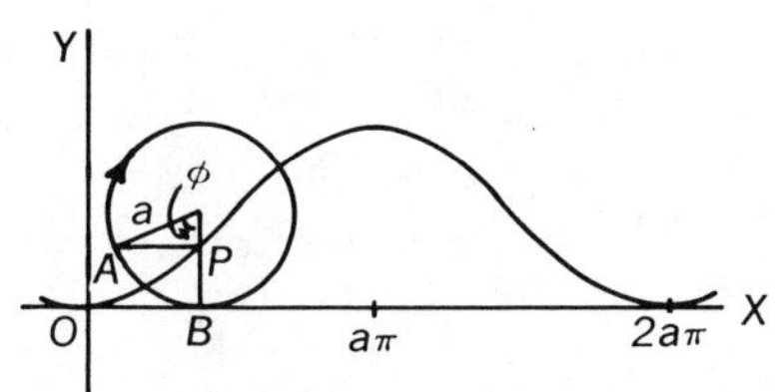

$$\begin{cases} x = a\phi \\ y = a(1 - \cos\phi) \end{cases}$$
$$[OB = \widehat{AB}]$$
(This is a sinusoid)

Conchoid of Nicomedes

(a) $a < b$

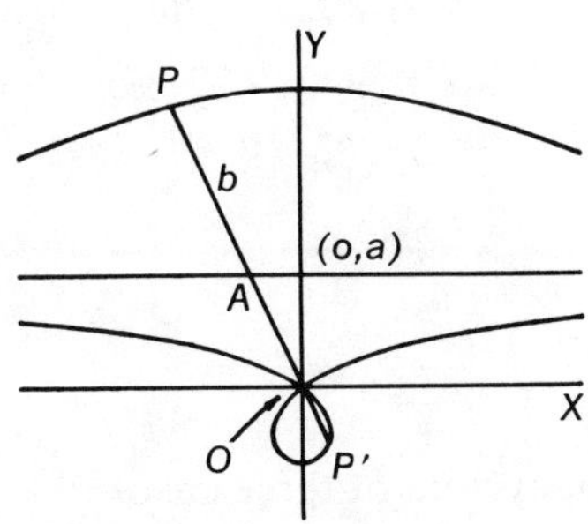

(b) $a > b$

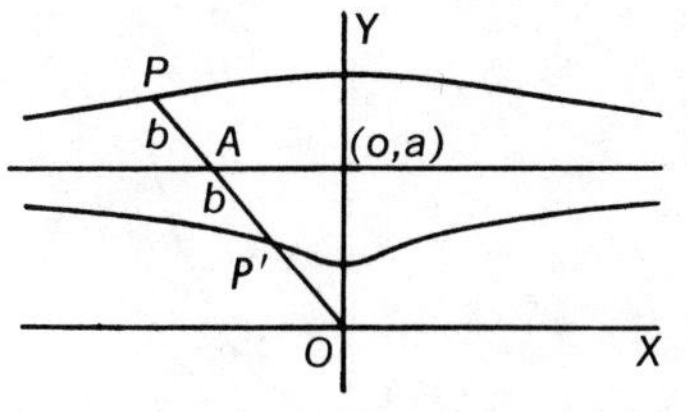

$$(y - a)^2(x^2 + y^2) = b^2y^2$$
$$r = a \csc\theta \pm b$$
$$[P'A = AP = b]$$

Conic sections

See: Circle; Ellipse; Hyperbola; Parabola

Cosecant curve

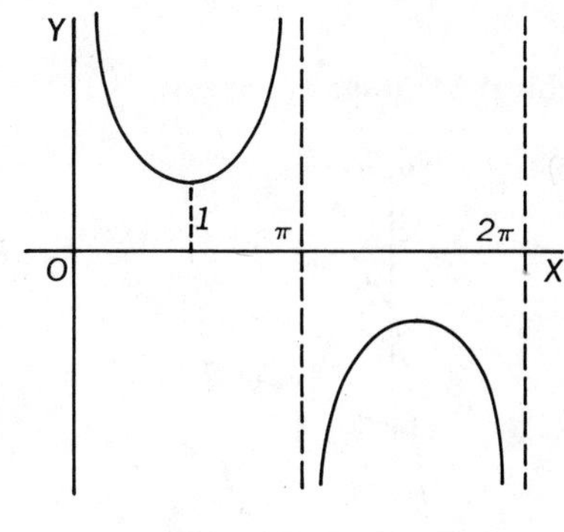

$y = \csc x$

Cosine curve

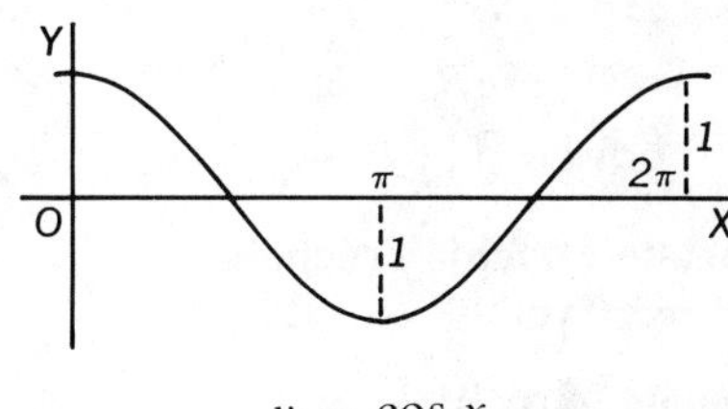

$y = \cos x$

Cotangent curve

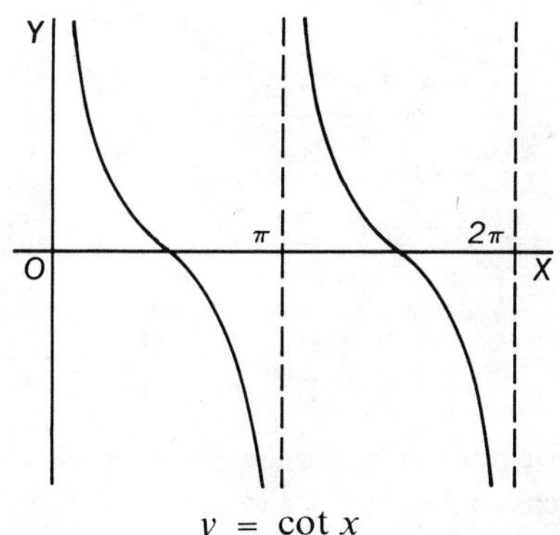

$y = \cot x$

Cubical parabola (special)

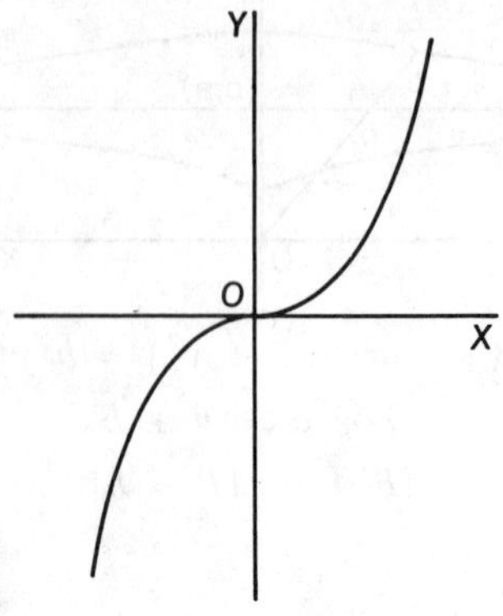

$$y = ax^3, \quad a > 0$$

$$r^2 = \frac{1}{a} \sec^2 \theta \tan \theta, \quad a > 0$$

Cubical parabola (general)

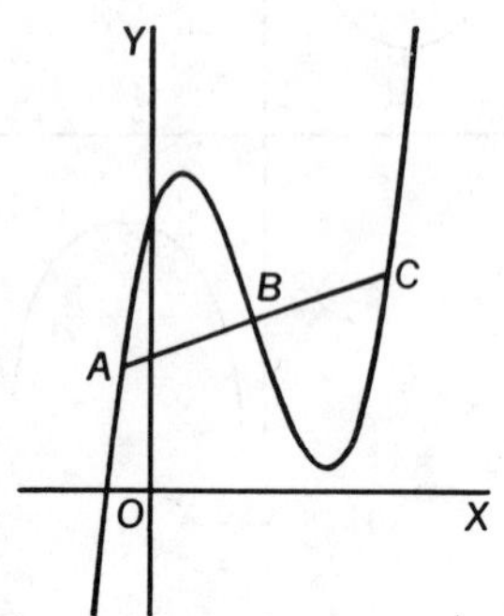

$$y = ax^3 + bx^2 + cx + d, \quad a > 0$$

$$[AB = BC]$$

(abscissa of $B = -b/3a$)

Curtate cycloid, Trochoids
See: Cycloid, curtate

Cycloid (cusp at origin)

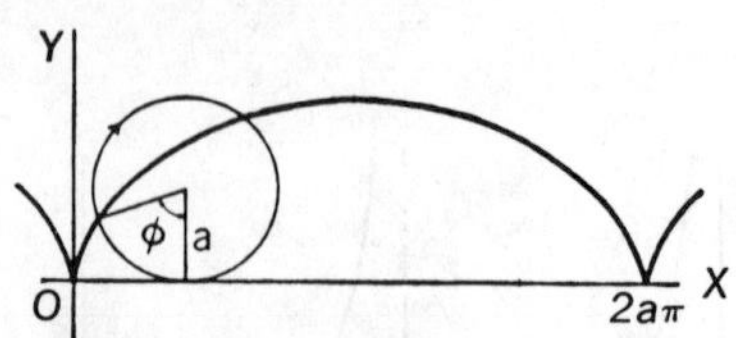

$$x = a \arccos \frac{a - y}{a} \mp \sqrt{2ay - y^2}$$

$$\begin{cases} x = a(\phi - \sin \phi) \\ y = a(1 - \cos \phi) \end{cases}$$

(For one arch: arc length $= 8a$; area $= 3\pi a^2$)

Cycloid (vertex at origin)

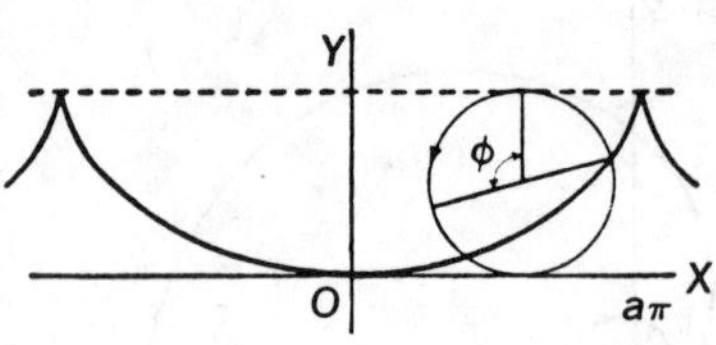

$$x = 2a \arcsin \sqrt{y/2a} + \sqrt{2ay - y^2}$$

$$\begin{cases} x = a(\phi + \sin \phi) \\ y = a(1 - \cos \phi) \end{cases}$$

Cycloid, curtate

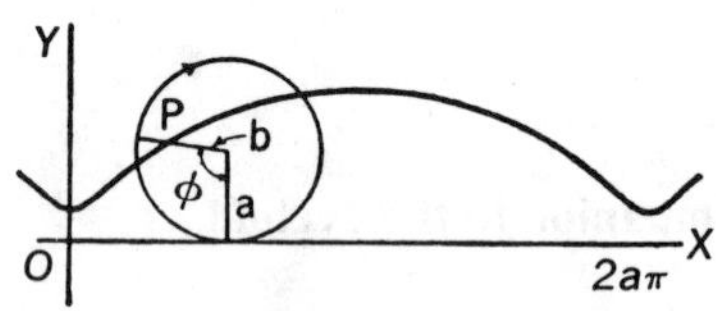

$$\begin{cases} x = a\phi - b \sin \phi \\ y = a - b \cos \phi \end{cases}$$

$$a > b$$

Cycloid, prolate

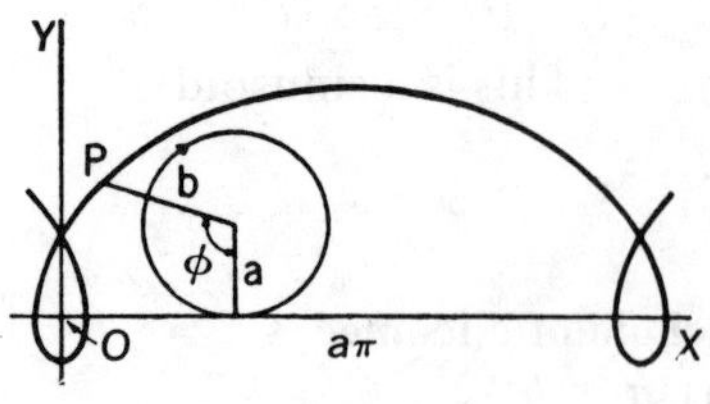

$$\begin{cases} x = a\phi - b \sin \phi \\ y = a - b \cos \phi \end{cases}$$

$$a < b$$

Deltoid
See: Hypocycloid of three cusps

Ellipse

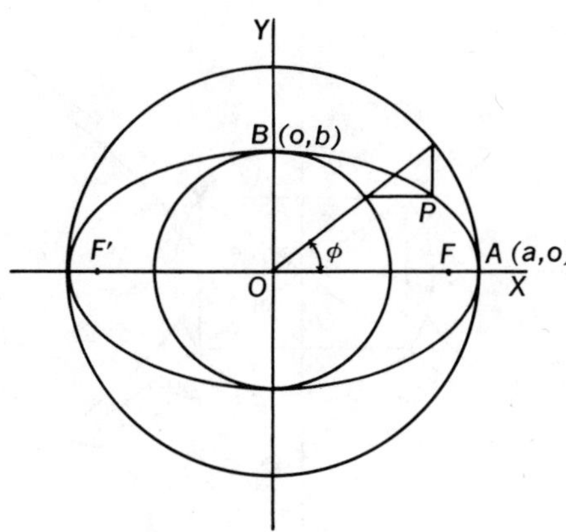

$$x^2/a^2 + y^2/b^2 = 1$$

$$\begin{cases} x = a \cos \phi \\ y = b \sin \phi \end{cases}$$

$$[BF' = BF = a, \quad PF' + PF = 2a]$$

Epicycloid

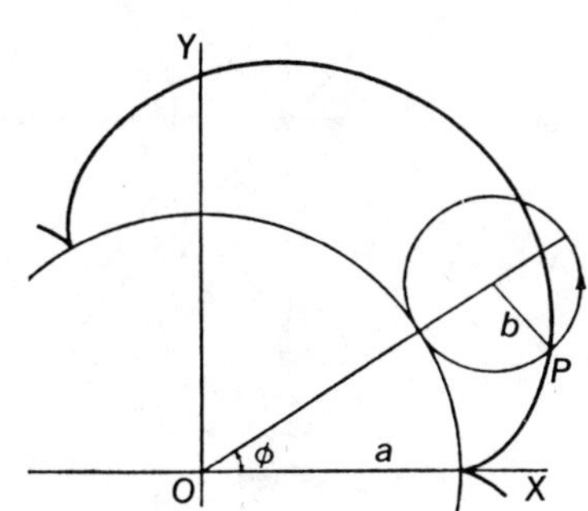

$$\begin{cases} x = (a + b) \cos \phi - b \cos\left(\dfrac{a + b}{b} \phi\right) \\ y = (a + b) \sin \phi - b \sin\left(\dfrac{a + b}{b} \phi\right) \end{cases}$$

Equiangular spiral

See: Spiral, logarithmic or equiangular

Equilateral hyperbola

See: Hyperbola, equilateral or rectangular

Evolute of ellipse

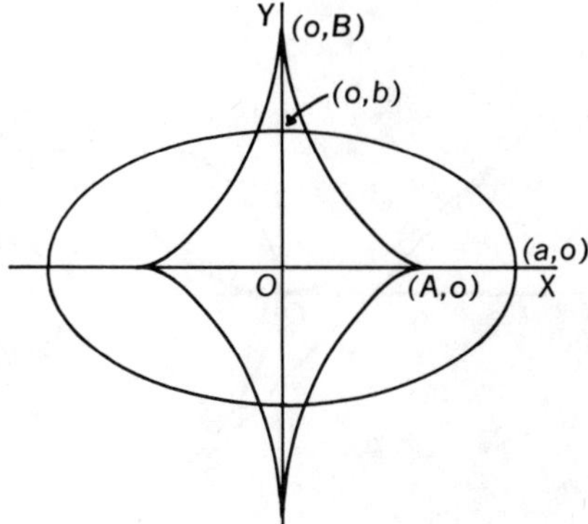

$$(ax)^{2/3} + (by)^{2/3} = (a^2 - b^2)^{2/3}$$

$$\begin{cases} x = A \cos^3 \phi \\ y = B \sin^3 \phi \end{cases}$$

$$[A = (a^2 - b^2)/a, \quad B = (a^2 - b^2)/b]$$

Exponential curve

(1) $a > 0$

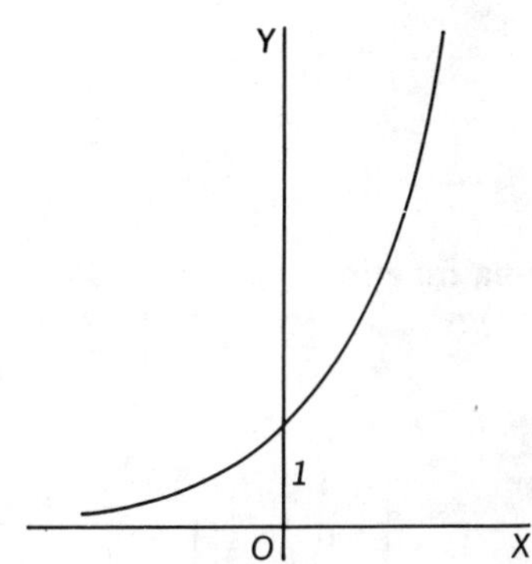

(2) $a < 0$

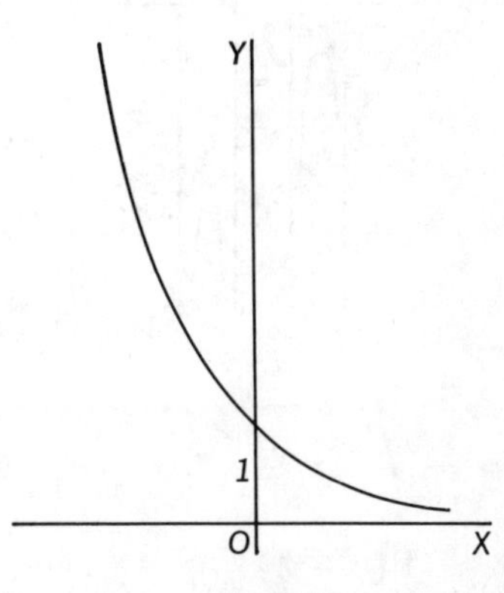

$$y = e^{ax}$$

Folium of Descartes

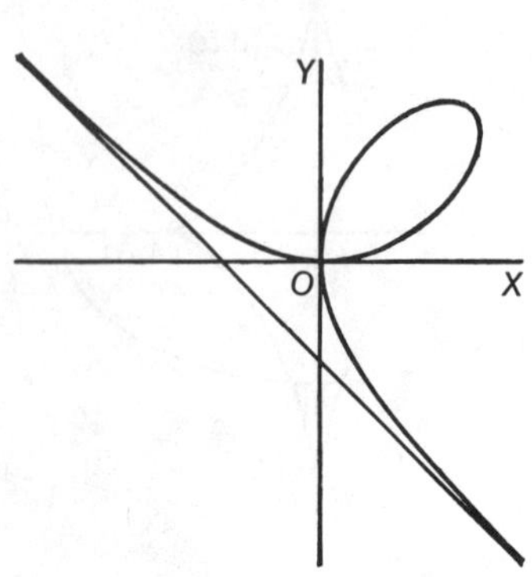

$$x^3 + y^3 - 3axy = 0$$

$$\begin{cases} x = 3a\phi/(1 + \phi^3) \\ y = 3a\phi^2/(1 + \phi^3) \end{cases}$$

$$r = \frac{3a \sin\theta \cos\theta}{\sin^3\theta + \cos^3\theta}$$

[asymptote: $x + y + a = 0$]

Gamma function

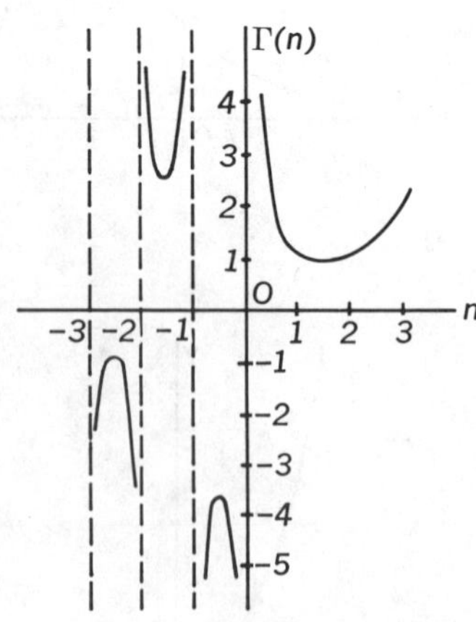

$$\Gamma(n) = \int_0^\infty x^{n-1}e^{-x}dx \quad (n > 0)$$

$$\Gamma(n) = \frac{\Gamma(n+1)}{n} \quad (0 > n \neq -1, -2, -3, \ldots)$$

Hyperbola

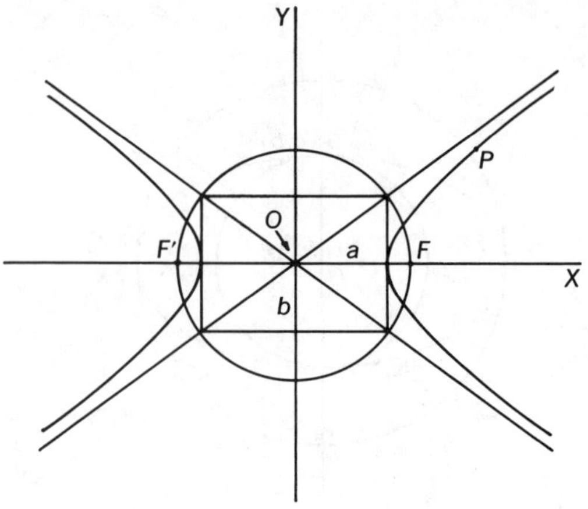

$$x^2/a^2 - y^2/b^2 = 1$$

$$[F'P - FP = 2a]$$

Hyperbola, equilateral or **rectangular**

(1)

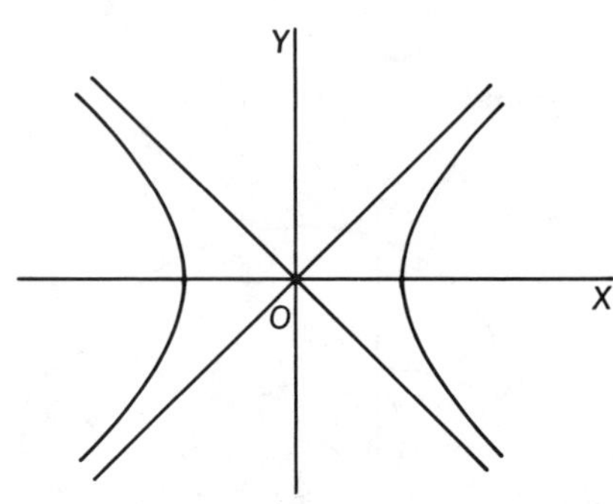

$$x^2 - y^2 = a^2$$

(2)

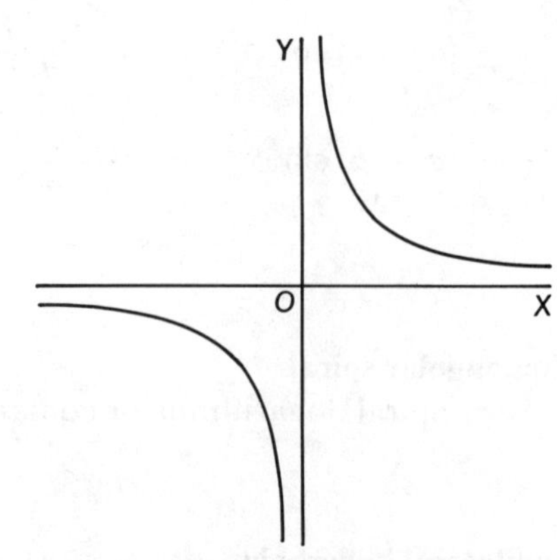

$$xy = k, \quad k > 0$$

(3)

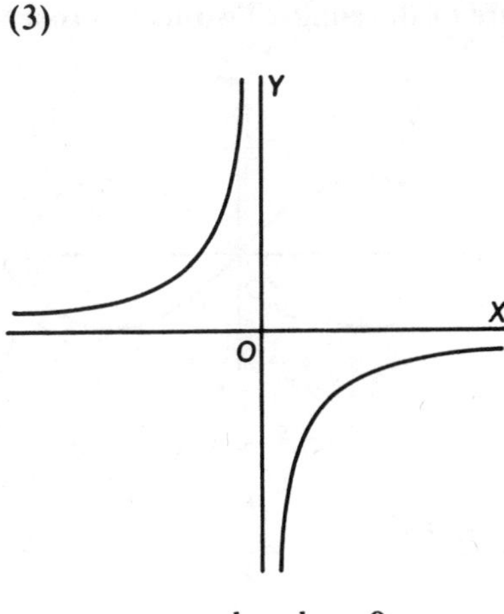

$xy = k, \quad k < 0$

Hyperbolic functions*

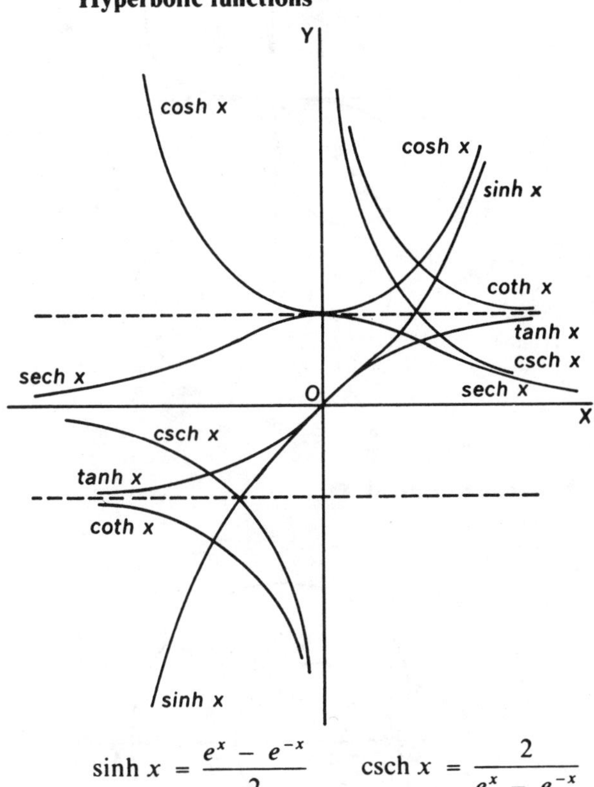

$$\sinh x = \frac{e^x - e^{-x}}{2} \qquad \operatorname{csch} x = \frac{2}{e^x - e^{-x}}$$

$$\cosh x = \frac{e^x + e^{-x}}{2} \qquad \operatorname{sech} x = \frac{2}{e^x + e^{-x}}$$

$$\tanh x = \frac{e^x - e^{-x}}{e^x + e^{-x}} \qquad \coth x = \frac{e^x + e^{-x}}{e^x - e^{-x}}$$

Hyperbolic spiral

See: Spiral, hyperbolic or reciprocal

*See page 312 for inverse hyperbolic functions.

Hypocycloid of three cusps, Deltoid

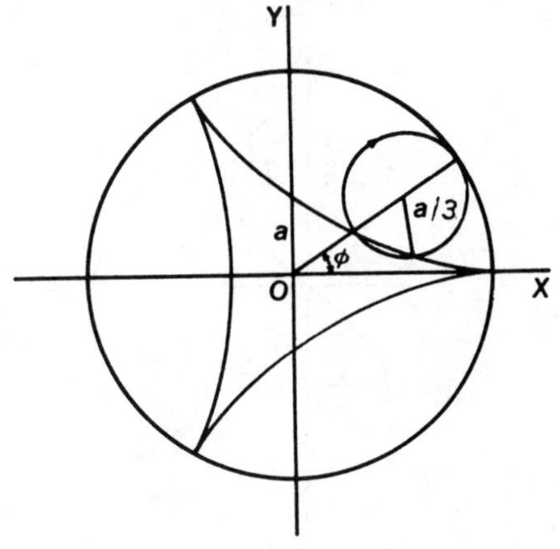

$$\begin{cases} x = 2a\cos\phi + a\cos 2\phi \\ y = 2a\sin\phi - a\sin 2\phi \end{cases}$$

Hypocycloid of four cusps, Astroid

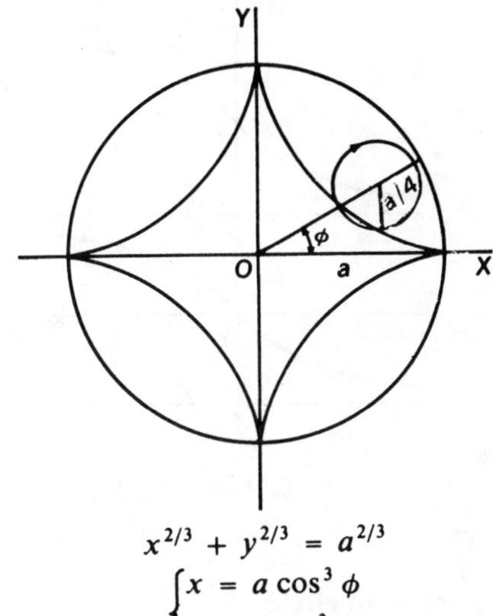

$$x^{2/3} + y^{2/3} = a^{2/3}$$

$$\begin{cases} x = a\cos^3\phi \\ y = a\sin^3\phi \end{cases}$$

Inverse cosine curve

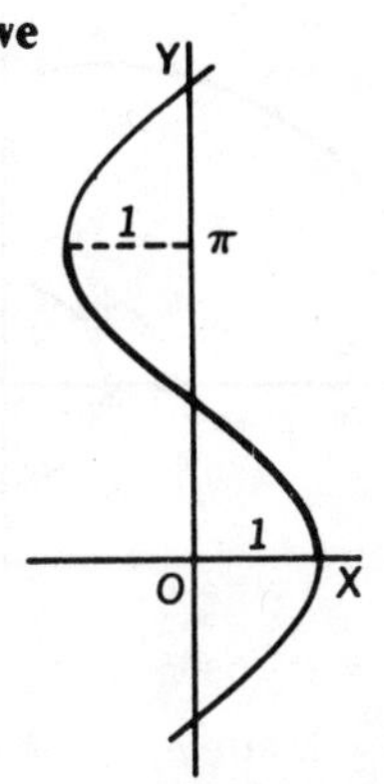

$y = \arccos x$

Inverse sine curve

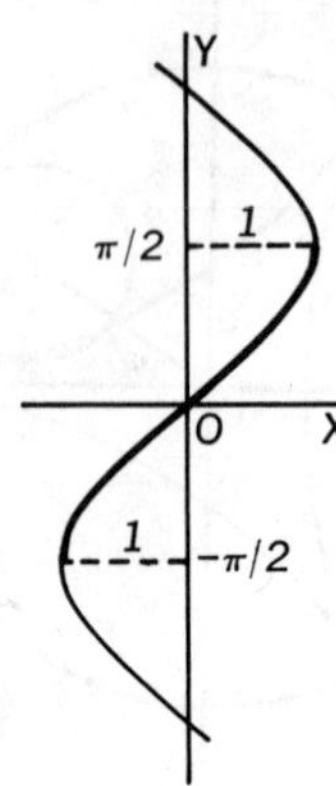

$$y = \arcsin x$$

Inverse tangent curve

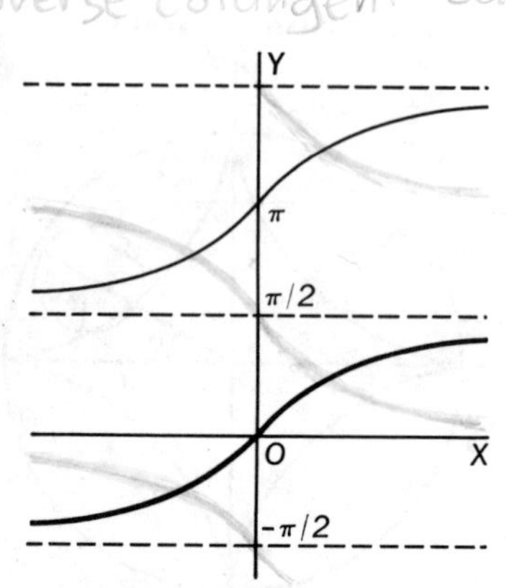

$$y = \arctan x$$

Involute of circle

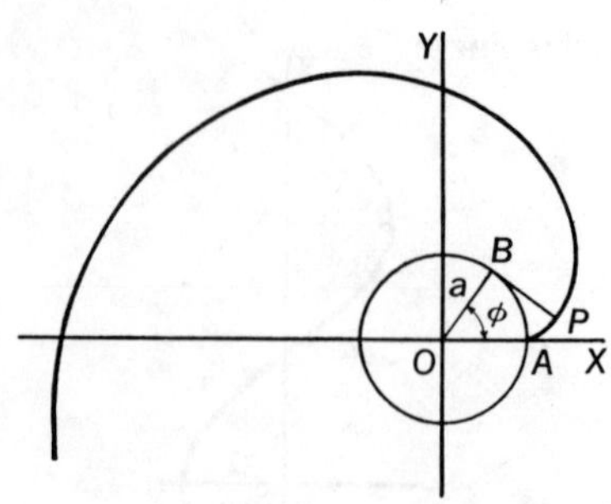

$$\begin{cases} x = a\cos\phi + a\phi\sin\phi \\ y = a\sin\phi - a\phi\cos\phi \end{cases}$$

$$[BP = \widehat{BA}]$$

Lemniscate of Bernoulli, Two-leaved rose

(a)

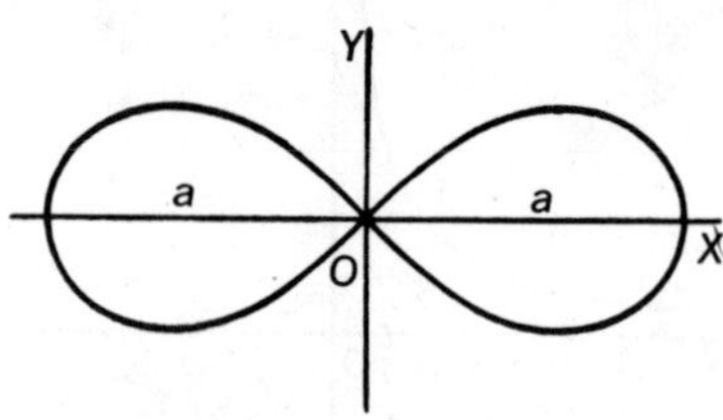

$$(x^2 + y^2)^2 = a^2(x^2 - y^2)$$
$$r^2 = a^2 \cos 2\theta$$

(b)

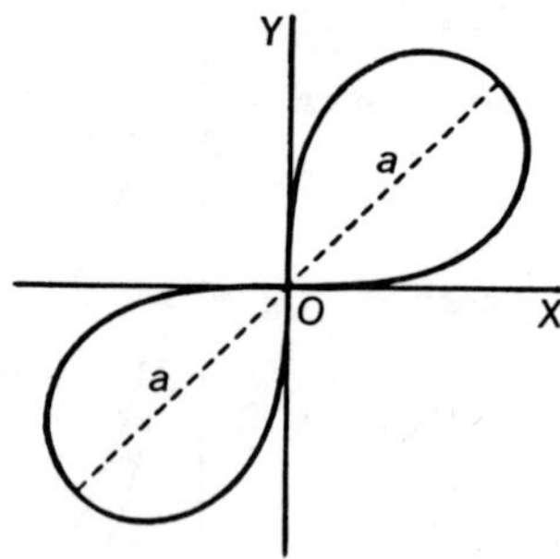

$$(x^2 + y^2)^2 = 2a^2xy$$
$$r^2 = a^2 \sin 2\theta$$

Limaçon of Pascal

(1) $a > b$

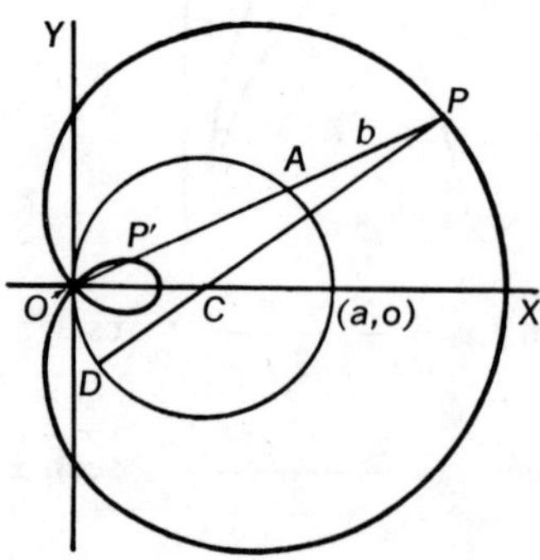

[If $a = 2b$, the curve is called the *trisectrix*, since then $\sphericalangle OPD = \frac{1}{3} \sphericalangle OCD$.]

(2) $a = b$

See: Cardioid

(3) $a < b$

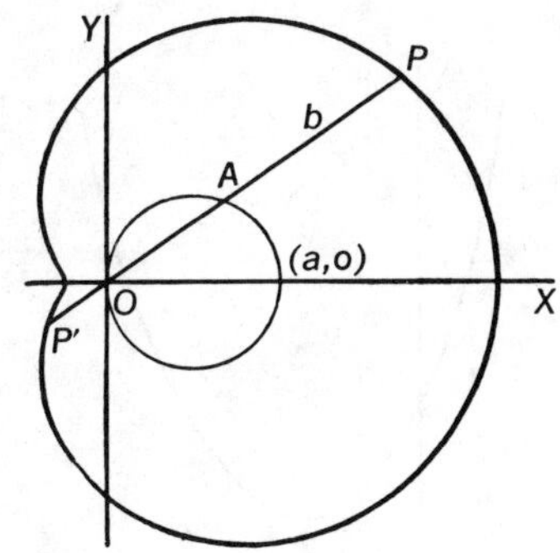

$$(x^2 + y^2 - ax)^2 = b^2(x^2 + y^2)$$
$$r = b + a\cos\theta$$
$$[P'A = AP = b]$$

Lituus

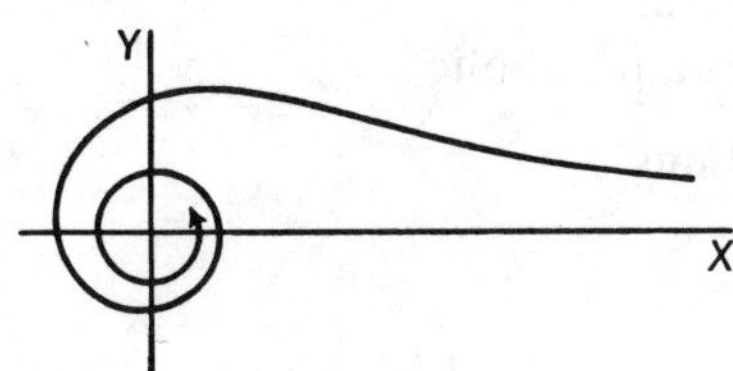

$$r^2\theta = a^2$$

Logarithmic curve

(1) $a > 1$

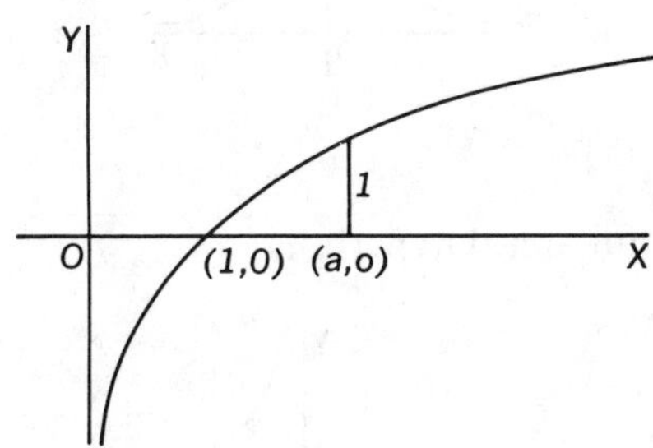

(2) $0 < a < 1$

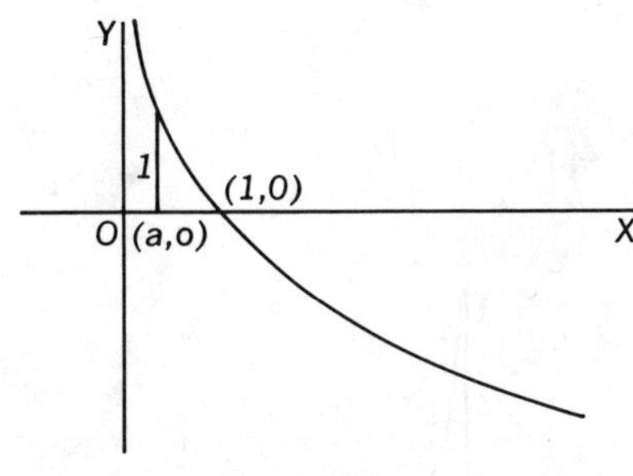

$$y = \log_a x$$

Logarithmic spiral

See: Spiral, logarithmic or equiangular

Nephroid

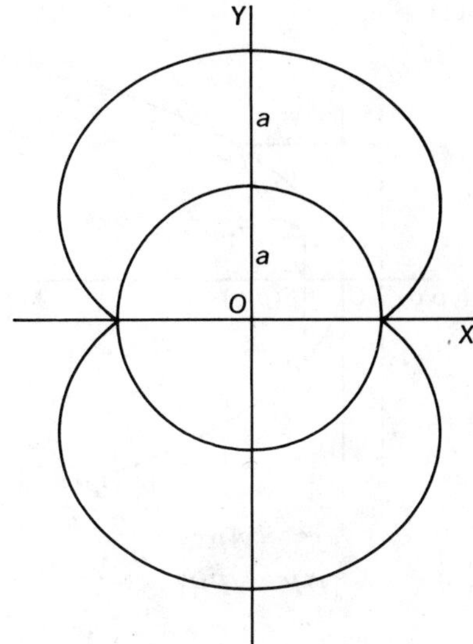

$$\begin{cases} x = \frac{1}{2}a(3\cos\phi - \cos 3\phi) \\ y = \frac{1}{2}a(3\sin\phi - \sin 3\phi) \end{cases}$$

[The nephroid is a 2-cusped epicycloid.]

Oui-ja board curve

See: Cochleoid

Ovals of Cassini

(1) $b > k$

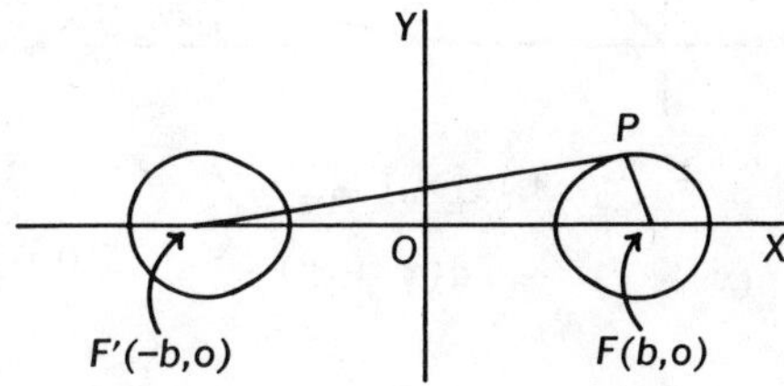

(2) $b = k$

See: Lemniscate of Bernoulli

(3) $b < k$

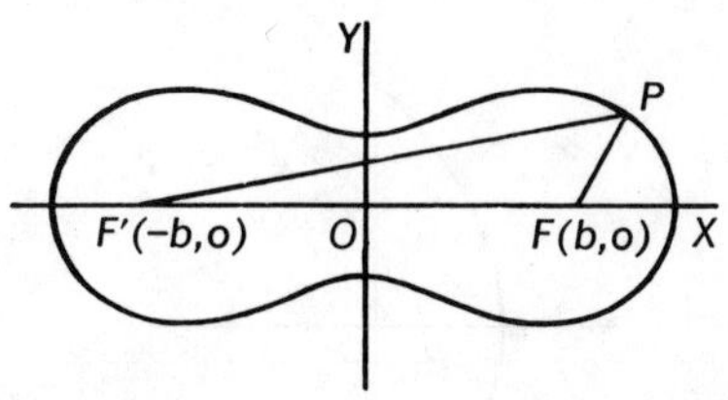

$$(x^2 + y^2 + b^2)^2 - 4b^2x^2 = k^4$$
$$r^4 + b^4 - 2r^2b^2\cos 2\theta = k^4$$
$$[F'P \cdot FP = k^2]$$

[These curves are sections of a torus on planes parallel to the axis of the torus.]

Parabola

(1)

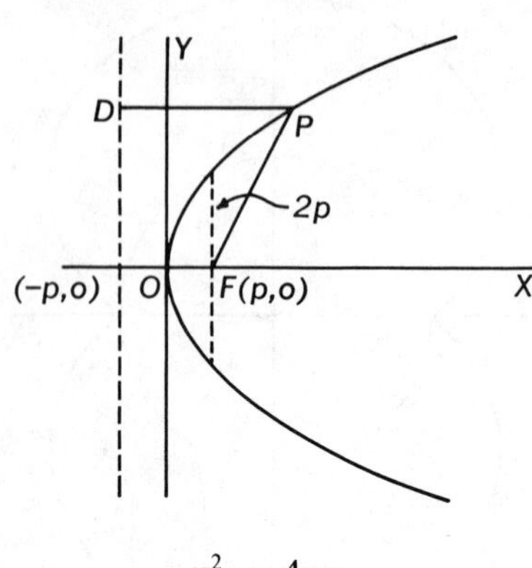

$y^2 = 4px$

$[DP = FP]$

(2)

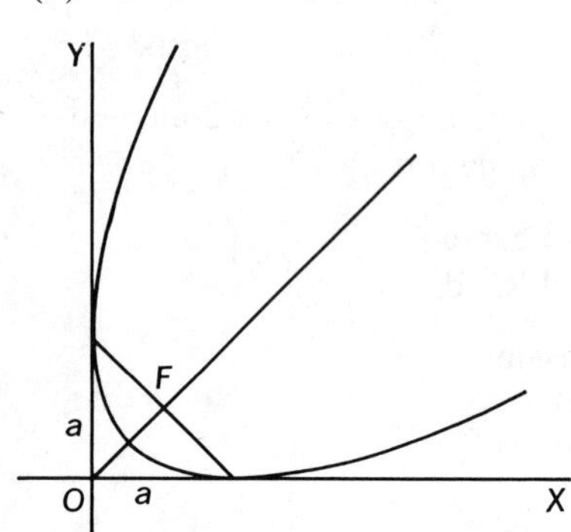

$\pm x^{1/2} \pm y^{1/2} = a^{1/2}$

$(x - y)^2 - 2a(x + y) + a^2 = 0$

(3)

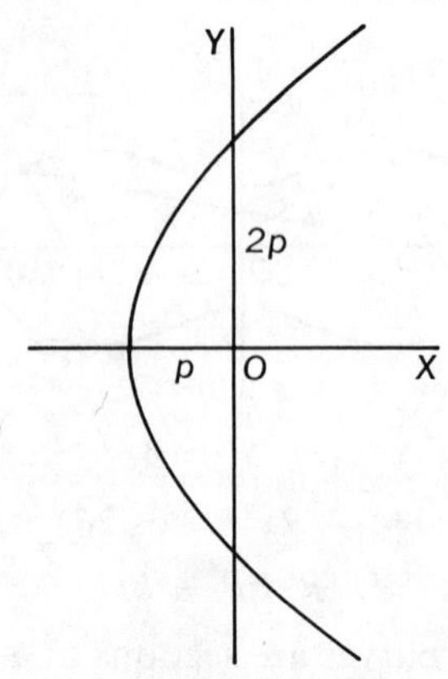

$r = 2p/(1 - \cos\theta)$

(4)

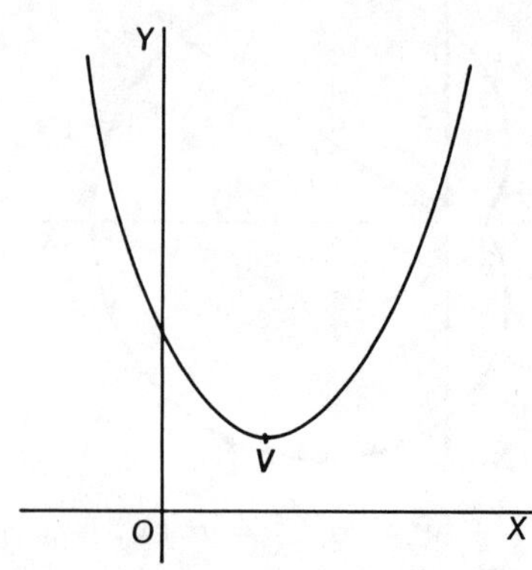

$y = ax^2 + bx + c, \quad a > 0$

[abscissa of vertex $= -b/2a$]

Parabolic spiral

See: Spiral, parabolic

Power functions

(1)

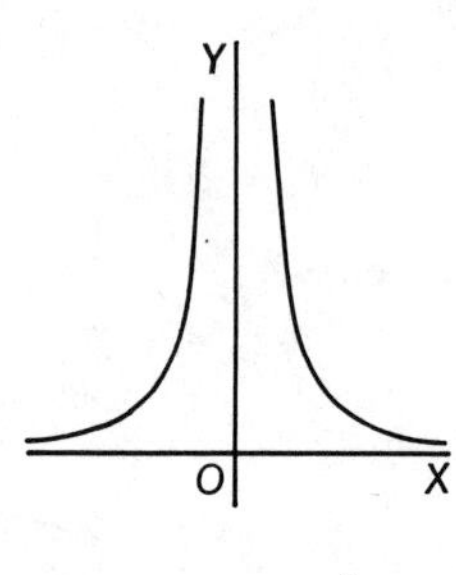

$y = x^{-2}$

(2) Equilateral hyperbola

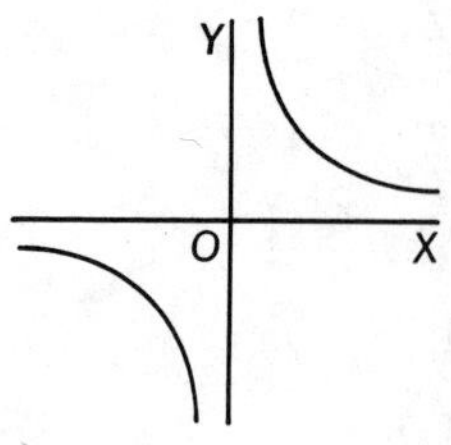

$y = x^{-1}$

(3)

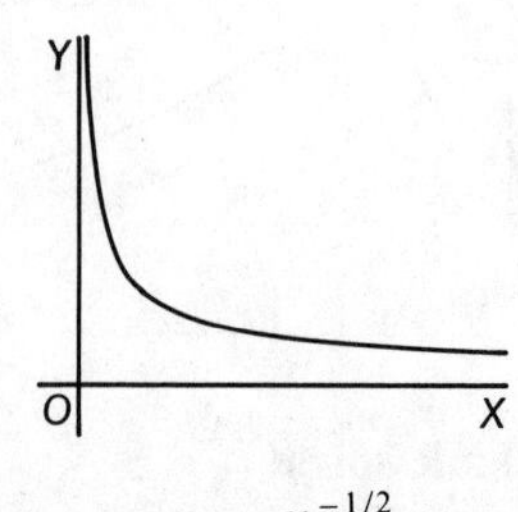

$y = x^{-1/2}$

(4) Cubical parabola

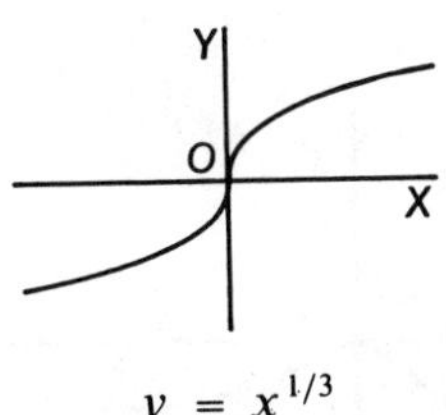

$y = x^{1/3}$

(5) Half of a parabola

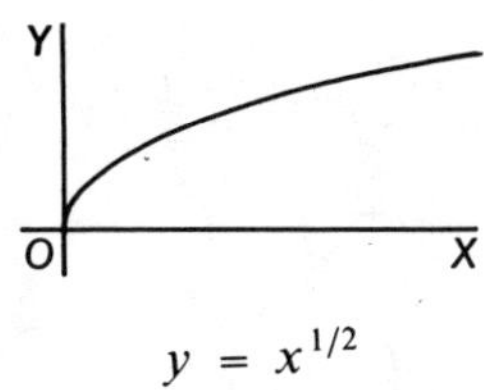

$y = x^{1/2}$

(6) Semicubical parabola

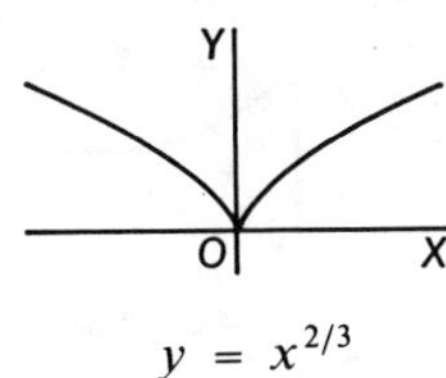

$y = x^{2/3}$

(7) Half of semicubical parabola

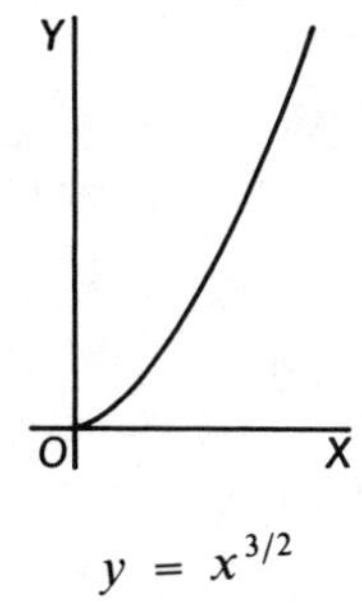

$y = x^{3/2}$

(8) Parabola

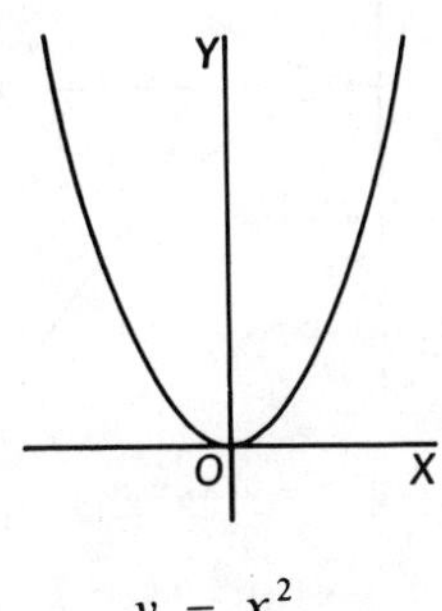

$y = x^2$

(9) Cubical parabola

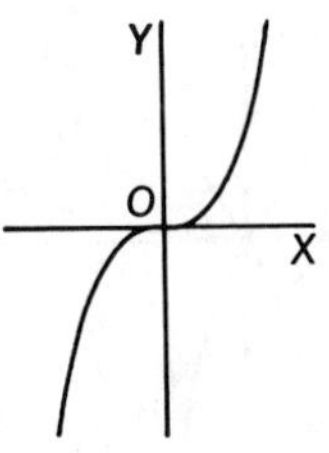

$y = x^3$

Probability curve

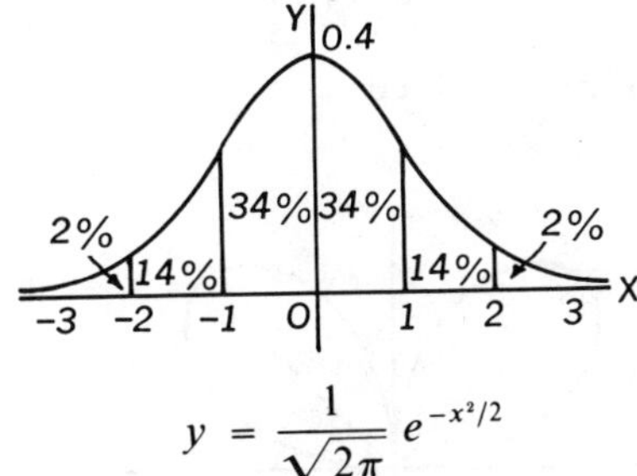

$$y = \frac{1}{\sqrt{2\pi}} e^{-x^2/2}$$

Prolate cycloid, Trochoids
See: Cycloid, prolate

Pursuit curve
See: Tractrix

Quadratrix of Hippias

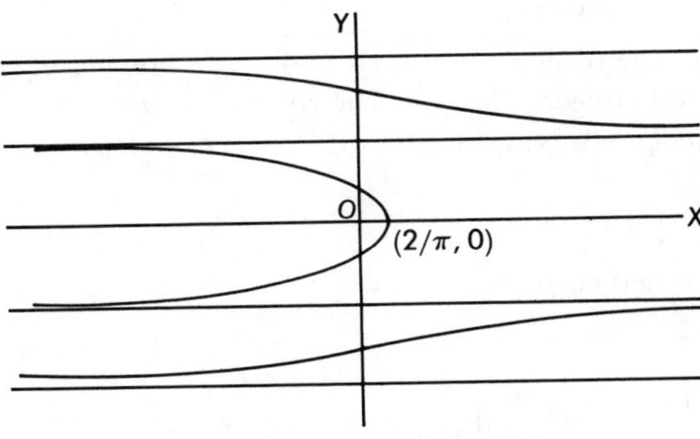

$y = x \tan(\pi y/2)$

Reciprocal spiral
See: Spiral, hyperbolic or reciprocal

Rectangular hyperbola
See: Hyperbola, equilateral or rectangular

Rose curves
(1) Two-leaved
See: Lemniscate of Bernoulli, Two-leaved rose

(2) Three-leaved

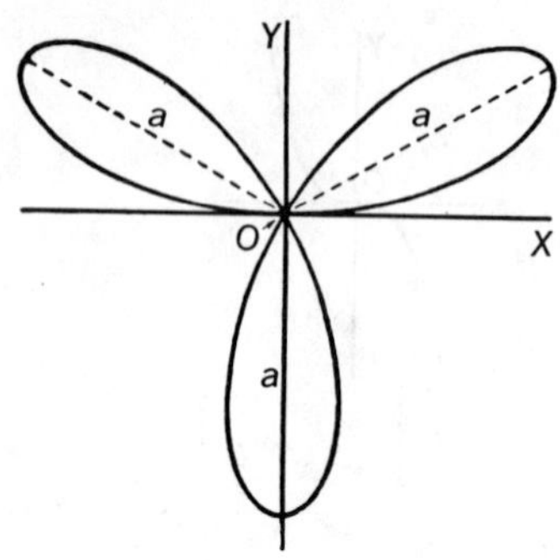

$r = a \sin 3\theta$

(3) Three-leaved

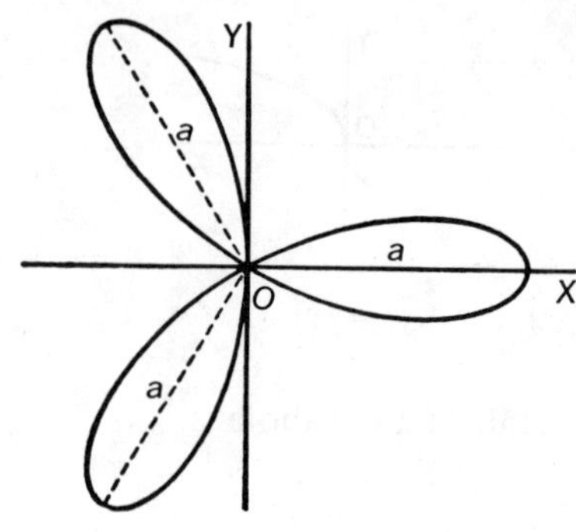

$r = a \cos 3\theta$

(4) Four-leaved

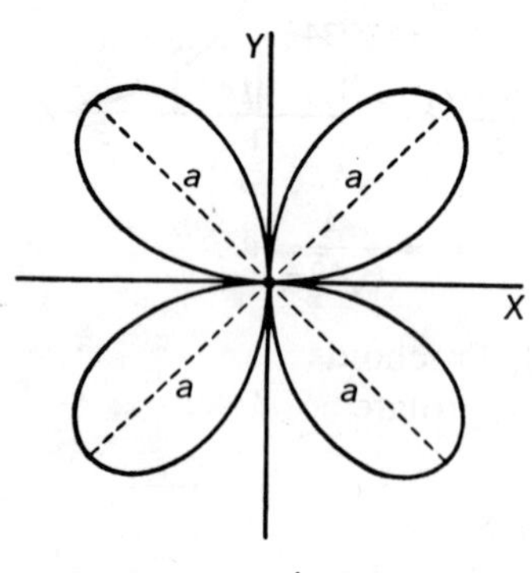

$r = a \sin 2\theta$

(5) Four-leaved

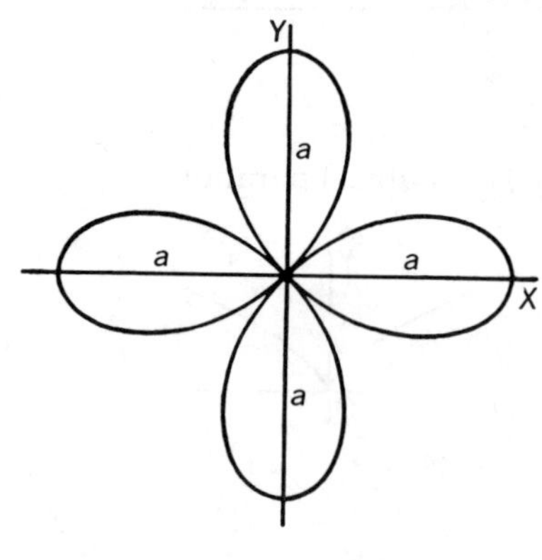

$r = a \cos 2\theta$

(6) n-leaved

The roses $r = a \sin n\theta$ and $r = a \cos n\theta$, have, for n an even integer, $2n$ leaves; for n an odd integer, n leaves. The roses $r^2 = a \sin n\theta$ and $r^2 = a \cos n\theta$, have, for n an even integer, n leaves; for n an odd integer, $2n$ leaves.

Secant curve

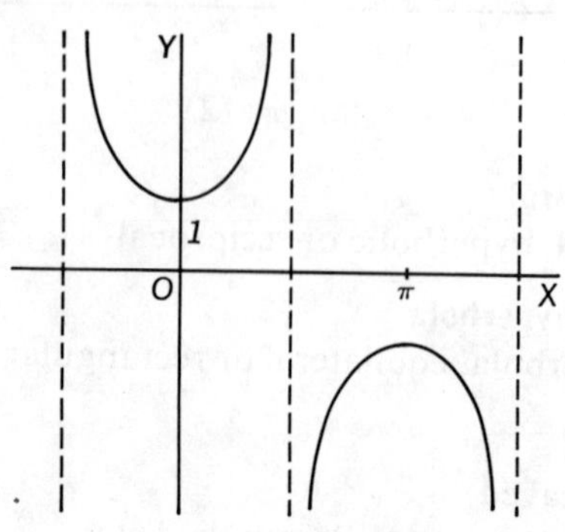

$y = \sec x$

Semicubical parabola

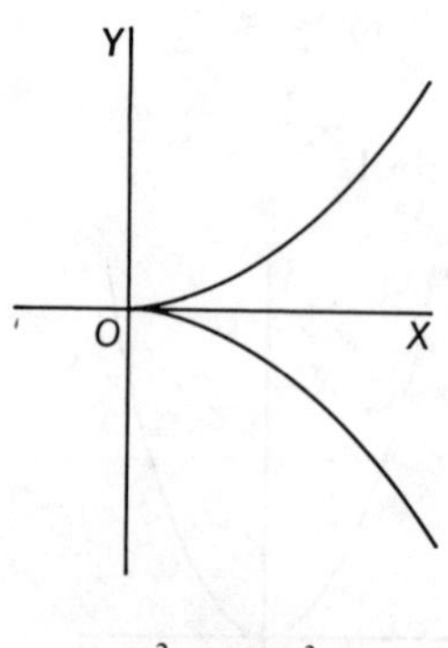

$y^2 = ax^3$

$r = \frac{1}{a} \tan^2 \theta \sec \theta$

Serpentine curve

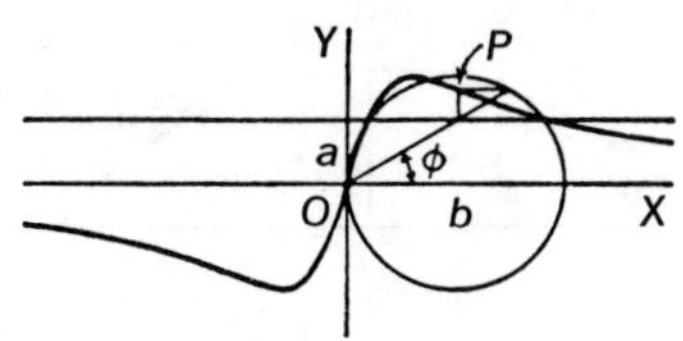

$$(a^2 + x^2)y = abx$$

$$\begin{cases} x = a \cot \phi \\ y = b \sin \phi \cos \phi \end{cases}$$

Sine curve

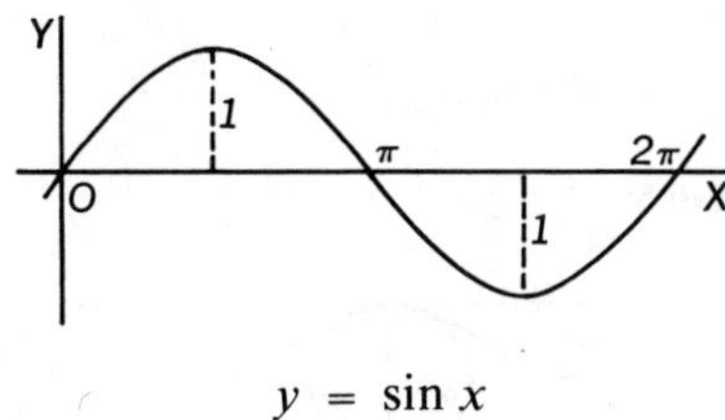

$$y = \sin x$$

Sinusoid

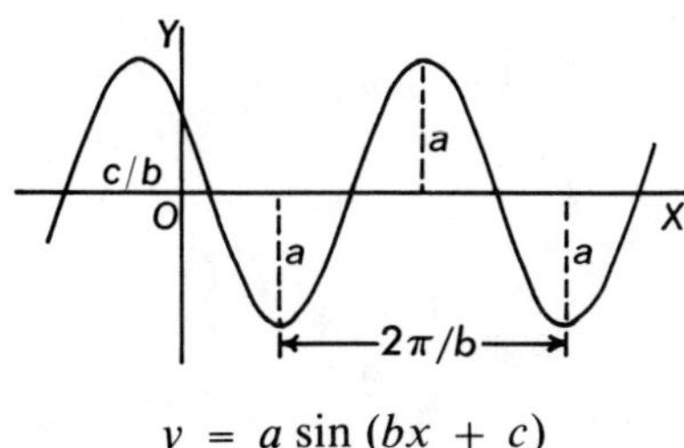

$$y = a \sin (bx + c)$$

Spiral of Archimedes

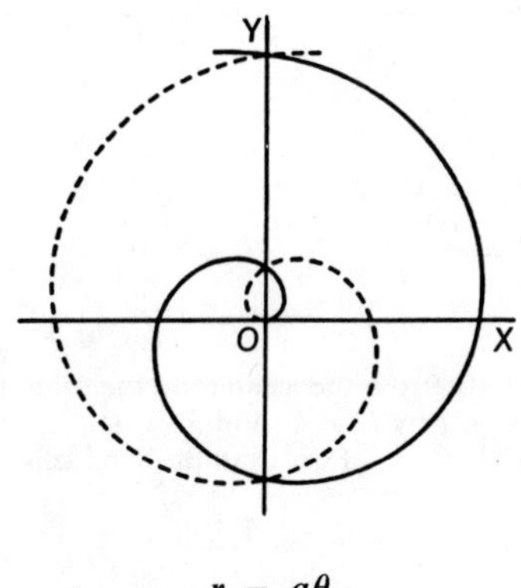

$$r = a\theta$$

Spiral, hyperbolic or reciprocal

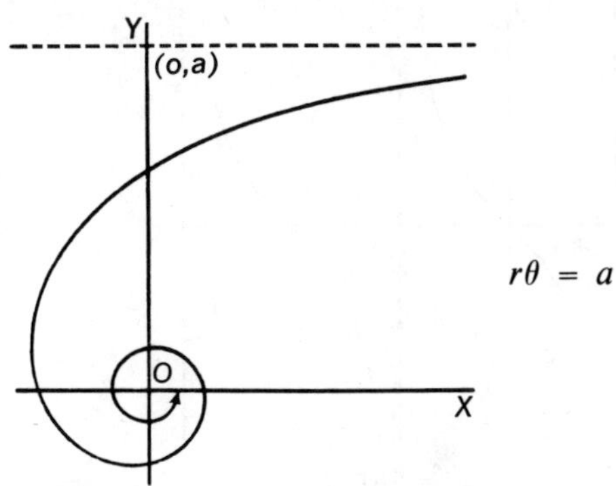

$$r\theta = a$$

Spiral, logarithmic or equiangular

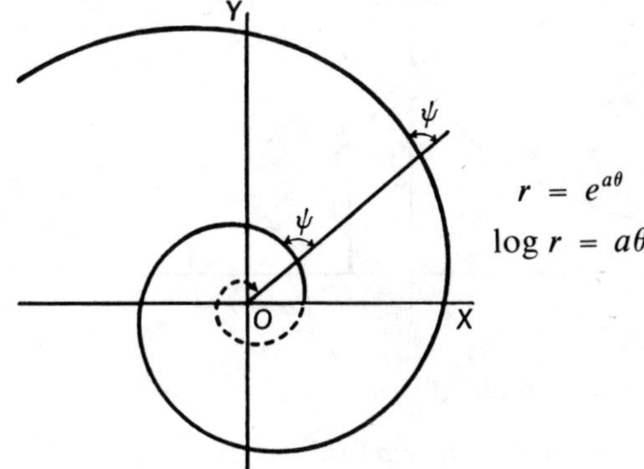

$$r = e^{a\theta}$$

$$\log r = a\theta$$

Spiral, parabolic

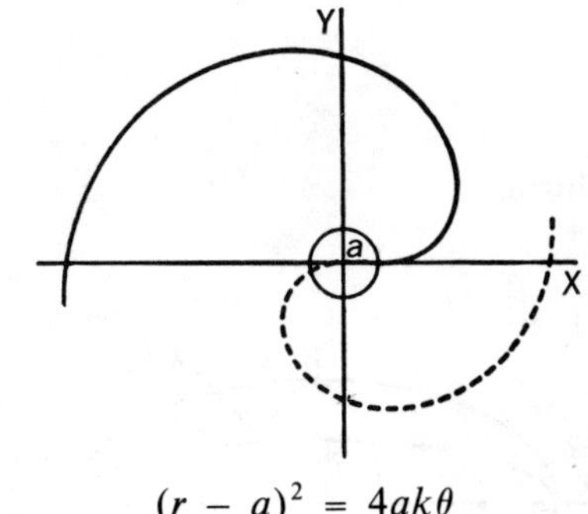

$$(r - a)^2 = 4ak\theta$$

Strophoid

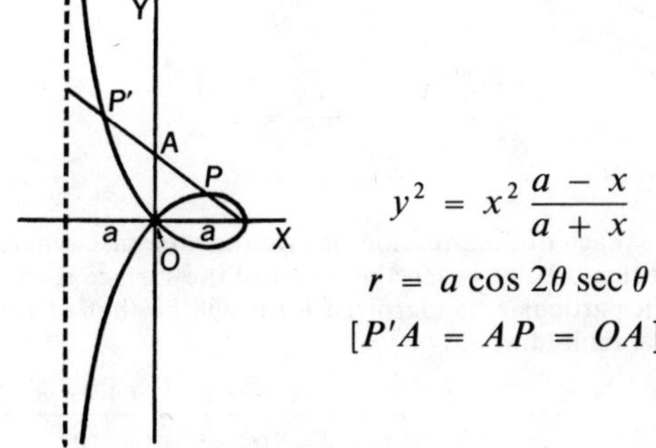

$$y^2 = x^2 \frac{a - x}{a + x}$$

$$r = a \cos 2\theta \sec \theta$$

$$[P'A = AP = OA]$$

Tangent curve

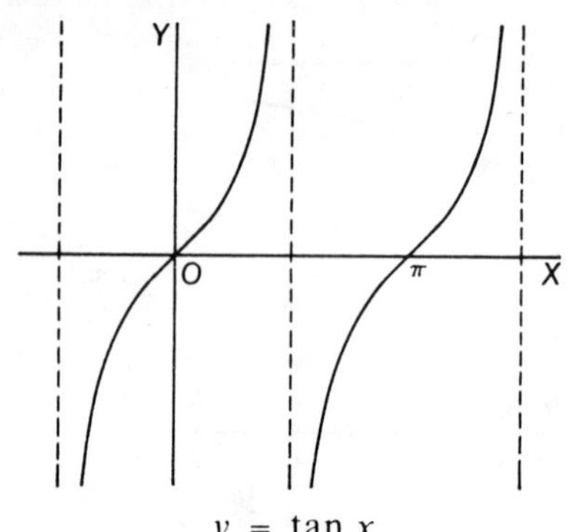

$$y = \tan x$$

Tractrix, Pursuit curve

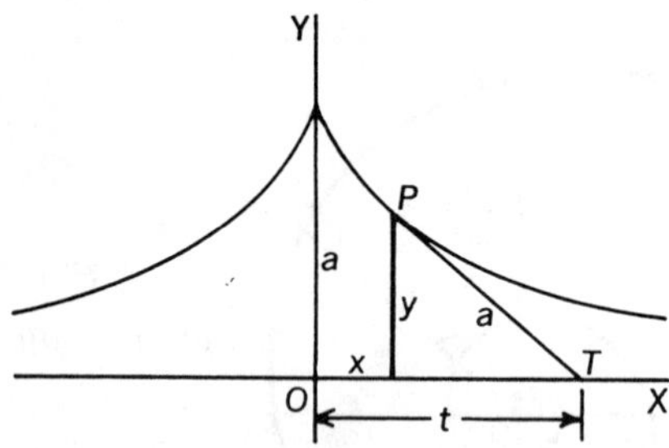

$$x = a\,\text{sech}^{-1}(y/a) - \sqrt{a^2 - y^2}$$

$$\begin{cases} x = t - a\tanh(t/a) \\ y = a\,\text{sech}\,(t/a) \end{cases}$$

$$[PT = a]$$

Trajectory (a parabola)

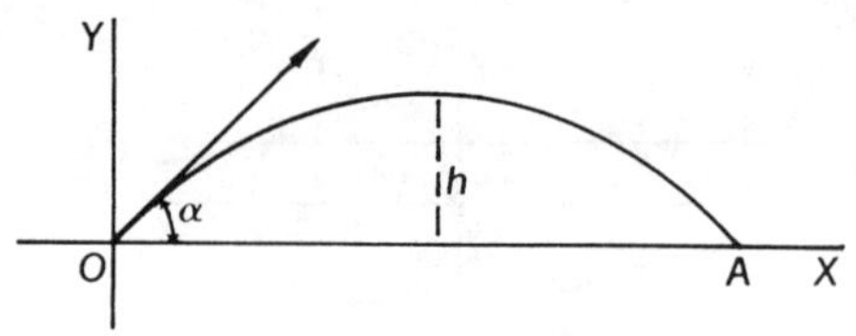

$$y = x\tan\alpha - gx^2/(2v_0^2\cos^2\alpha)$$

$$x = (v_0\cos\alpha)t$$

$$y = (v_0\sin\alpha)t - gt^2/2$$

Trigonometric functions

See: Cosecant curve; Cosine curve; Cotangent curve; Secant curve; Sine curve; Tangent curve

Trisectrix

See: Limaçon of Pascal (1)

Witch of Agnesi

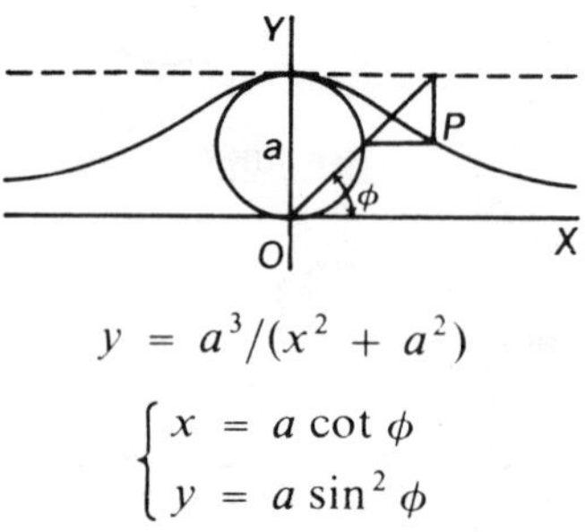

$$y = a^3/(x^2 + a^2)$$

$$\begin{cases} x = a\cot\phi \\ y = a\sin^2\phi \end{cases}$$

*QUADRIC SURFACES

Ellipsoid

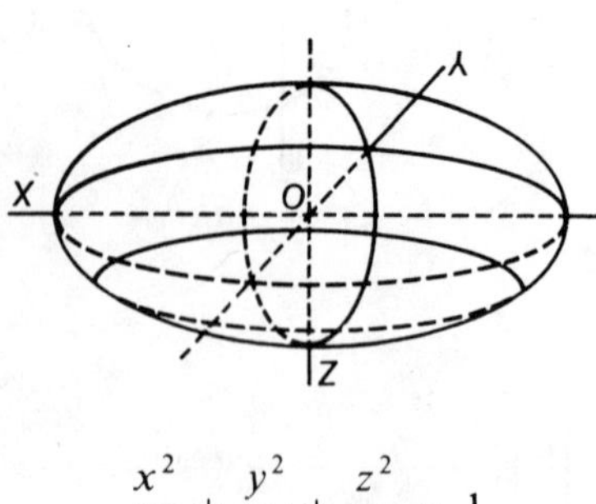

$$\frac{x^2}{a^2} + \frac{y^2}{b^2} + \frac{z^2}{c^2} = 1$$

Elliptic cone

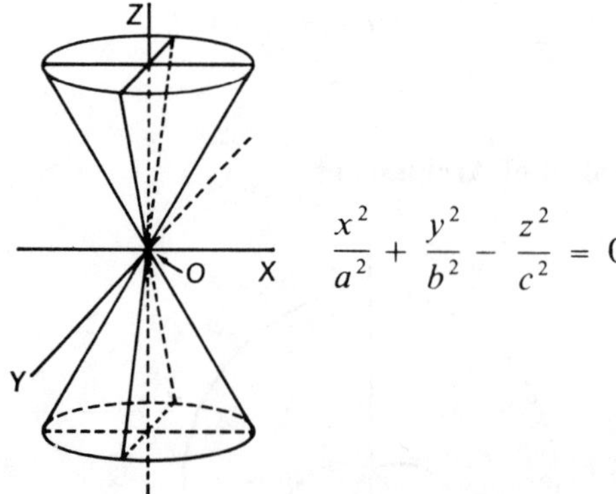

$$\frac{x^2}{a^2} + \frac{y^2}{b^2} - \frac{z^2}{c^2} = 0$$

*Each of the equations is given for the case where the origin is located at (0, 0, 0), the center of the quadric surface. If, however, the center of the surface is at (h, k, l), replace x by $x - h$, y by $y - k$, and z by $z - l$, and the particular standardized form will be that of the surface with center at (h, k, l). For example, the elliptic paraboloid would be

$$\frac{(x - h)^2}{a^2} - \frac{(y - k)^2}{b^2} = c(z - l).$$

Elliptic cylinder

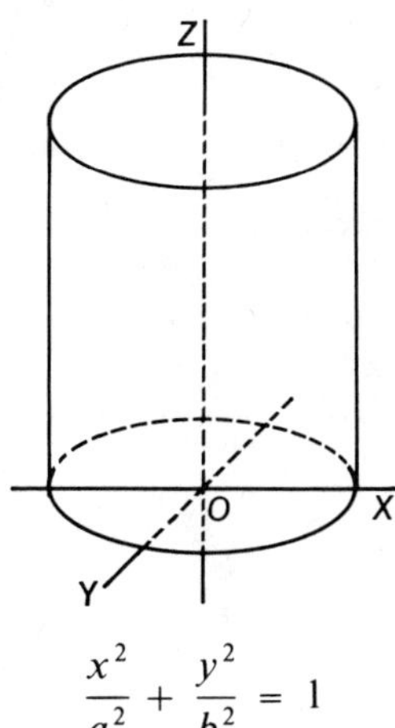

$$\frac{x^2}{a^2} + \frac{y^2}{b^2} = 1$$

Elliptic paraboloid

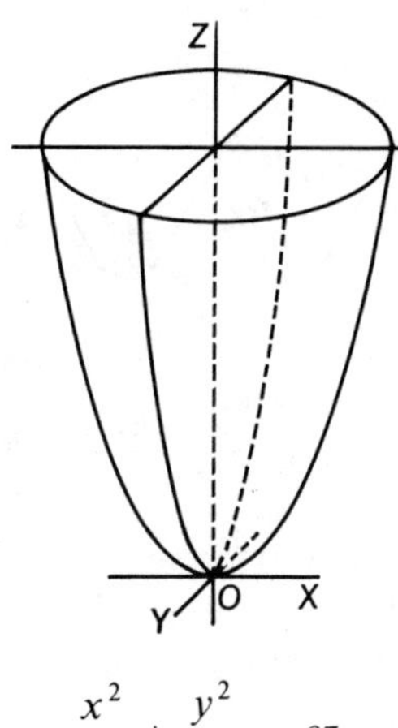

$$\frac{x^2}{a^2} + \frac{y^2}{b^2} = cz$$

Hyperbolic paraboloid

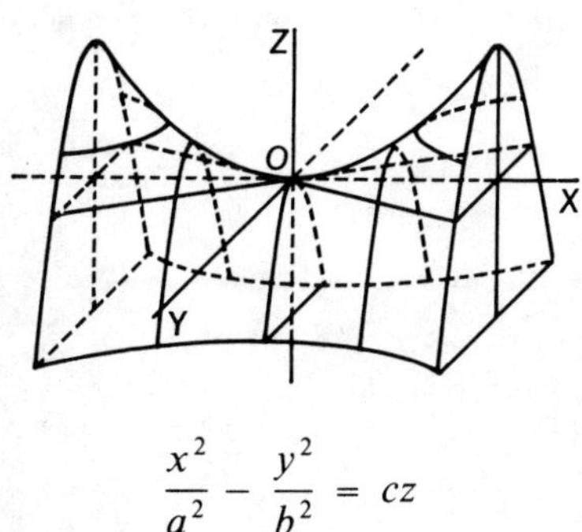

$$\frac{x^2}{a^2} - \frac{y^2}{b^2} = cz$$

Hyperboloid of one sheet

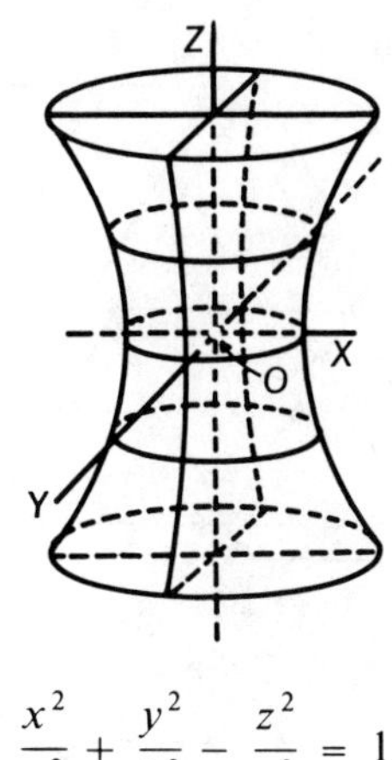

$$\frac{x^2}{a^2} + \frac{y^2}{b^2} - \frac{z^2}{c^2} = 1$$

Hyperboloid of two sheets

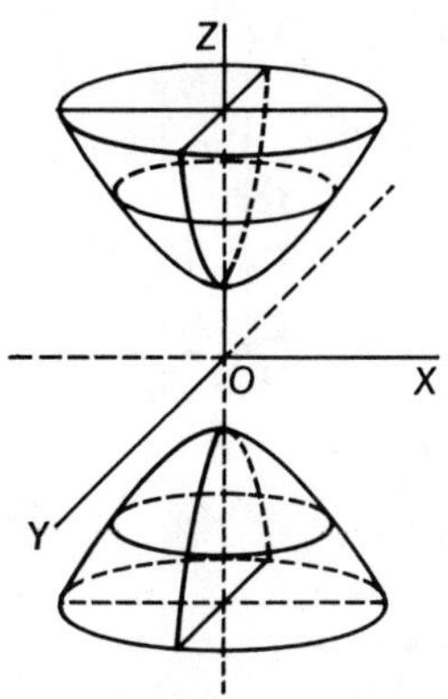

$$\frac{z^2}{c^2} - \frac{x^2}{a^2} - \frac{y^2}{b^2} = 1$$

Sphere

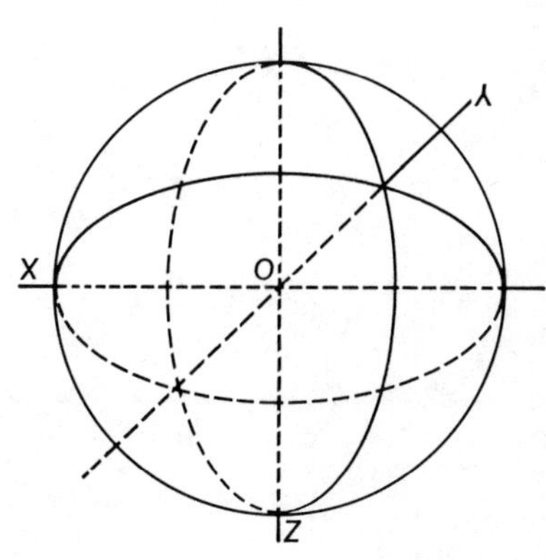

$$x^2 + y^2 + z^2 = a^2$$

PATTERNS OF REGULAR POLYHEDRA

Tetrahedron

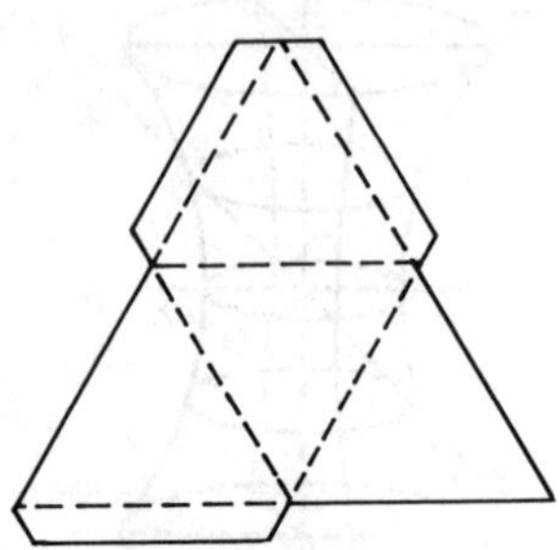

Octahedron

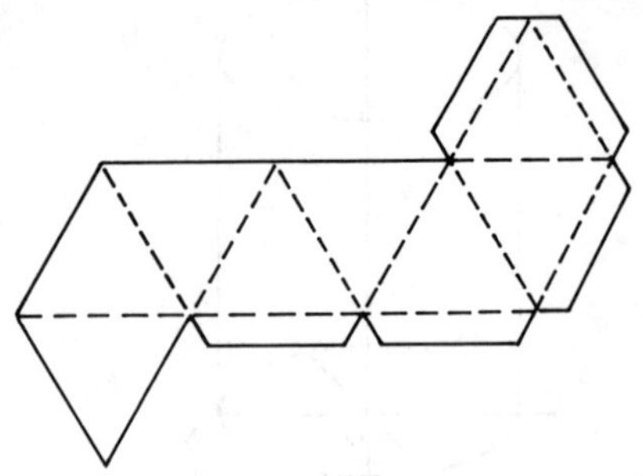

Hexahedron, or **Cube**

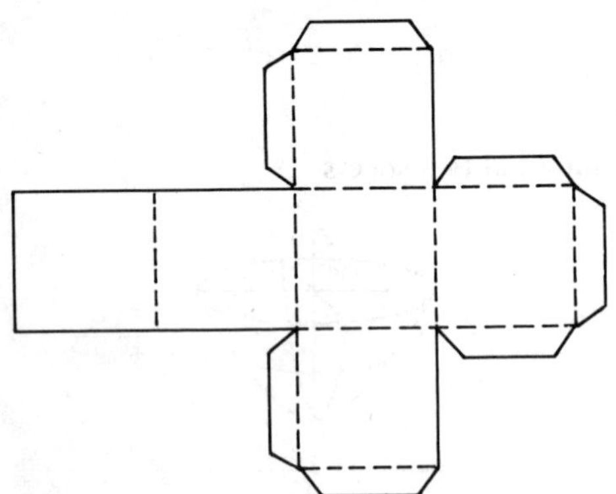

Icosahedron

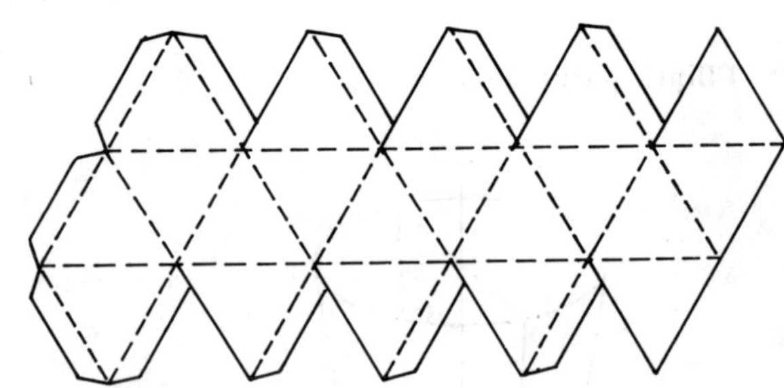

Dodecahedron

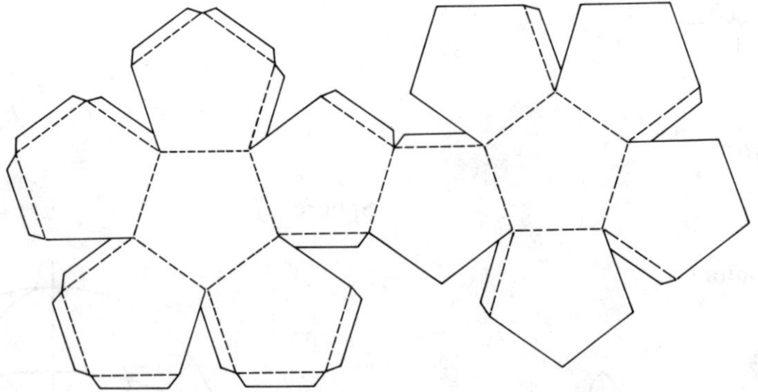

*Derivatives

In the following formulas u, v, w represent functions of x, while a, c, n represent fixed real numbers. All arguments in the trigonometric functions are measured in radians, and all inverse trigonometric and hyperbolic functions represent principal values.

1. $\frac{d}{dx}(a) = 0$

2. $\frac{d}{dx}(x) = 1$

3. $\frac{d}{dx}(au) = a\frac{du}{dx}$

4. $\frac{d}{dx}(u + v - w) = \frac{du}{dx} + \frac{dv}{dx} - \frac{dw}{dx}$

5. $\frac{d}{dx}(uv) = u\frac{dv}{dx} + v\frac{du}{dx}$

6. $\frac{d}{dx}(uvw) = uv\frac{dw}{dx} + vw\frac{du}{dx} + uw\frac{dv}{dx}$

7. $\frac{d}{dx}\left(\frac{u}{v}\right) = \frac{v\frac{du}{dx} - u\frac{dv}{dx}}{v^2} = \frac{1}{v}\frac{du}{dx} - \frac{u}{v^2}\frac{dv}{dx}$

8. $\frac{d}{dx}(u^n) = nu^{n-1}\frac{du}{dx}$

9. $\frac{d}{dx}(\sqrt{u}) = \frac{1}{2\sqrt{u}}\frac{du}{dx}$

10. $\frac{d}{dx}\left(\frac{1}{u}\right) = -\frac{1}{u^2}\frac{du}{dx}$

11. $\frac{d}{dx}\left(\frac{1}{u^n}\right) = -\frac{n}{u^{n+1}}\frac{du}{dx}$

12. $\frac{d}{dx}\left(\frac{u^n}{v^m}\right) = \frac{u^{n-1}}{v^{m+1}}\left(nv\frac{du}{dx} - mu\frac{dv}{dx}\right)$

13. $\frac{d}{dx}(u^n v^m) = u^{n-1}v^{m-1}\left(nv\frac{du}{dx} + mu\frac{dv}{dx}\right)$

14. $\frac{d}{dx}[f(u)] = \frac{d}{du}[f(u)] \cdot \frac{du}{dx}$

* Let $y = f(x)$ and $\frac{dy}{dx} = \frac{d[f(x)]}{dx} = f'(x)$ define respectively a function and its derivative for any value x in their common domain. The differential for the function at such a value x is accordingly defined as

$$dy = d[f(x)] = \frac{dy}{dx}dx = \frac{d[f(x)]}{dx}dx = f'(x)\,dx$$

Each derivative formula has an associated differential formula. For example, formula 6 above has the differential formula

$$d(uvw) = uv\,dw + vw\,du + uw\,dv$$

15. $\dfrac{d^2}{dx^2}[f(u)] = \dfrac{df(u)}{du}\cdot\dfrac{d^2u}{dx^2} + \dfrac{d^2f(u)}{du^2}\cdot\left(\dfrac{du}{dx}\right)^2$

16. $\dfrac{d^n}{dx^n}[uv] = \binom{n}{0}v\dfrac{d^nu}{dx^n} + \binom{n}{1}\dfrac{dv}{dx}\dfrac{d^{n-1}u}{dx^{n-1}} + \binom{n}{2}\dfrac{d^2v}{dx^2}\dfrac{d^{n-2}u}{dx^{n-2}}$

$$+ \cdots + \binom{n}{k}\frac{d^kv}{dx^k}\frac{d^{n-k}u}{dx^{n-k}} + \cdots + \binom{n}{n}u\frac{d^nv}{dx^n}$$

where $\binom{n}{r} = \dfrac{n!}{r!(n-r)!}$ the binomial coefficient, n non-negative integer and $\binom{n}{0} = 1$.

17. $\dfrac{du}{dx} = \dfrac{1}{\dfrac{dx}{du}}$ if $\dfrac{dx}{du} \neq 0$

18. $\dfrac{d}{dx}(\log_a u) = (\log_a e)\dfrac{1}{u}\dfrac{du}{dx}$

19. $\dfrac{d}{dx}(\log_e u) = \dfrac{1}{u}\dfrac{du}{dx}$

20. $\dfrac{d}{dx}(a^u) = a^u(\log_e a)\dfrac{du}{dx}$

21. $\dfrac{d}{dx}(e^u) = e^u\dfrac{du}{dx}$

22. $\dfrac{d}{dx}(u^v) = vu^{v-1}\dfrac{du}{dx} + (\log_e u)u^v\dfrac{dv}{dx}$

23. $\dfrac{d}{dx}(\sin u) = \dfrac{du}{dx}(\cos u)$

24. $\dfrac{d}{dx}(\cos u) = -\dfrac{du}{dx}(\sin u)$

25. $\dfrac{d}{dx}(\tan u) = \dfrac{du}{dx}(\sec^2 u)$

26. $\dfrac{d}{dx}(\cot u) = -\dfrac{du}{dx}(\csc^2 u)$

27. $\dfrac{d}{dx}(\sec u) = \dfrac{du}{dx}\sec u \cdot \tan u$

28. $\dfrac{d}{dx}(\csc u) = -\dfrac{du}{dx}\csc u \cdot \cot u$

29. $\dfrac{d}{dx}(\text{vers } u) = \dfrac{du}{dx}\sin u$

30. $\dfrac{d}{dx}(\text{arc sin } u) = \dfrac{1}{\sqrt{1-u^2}}\dfrac{du}{dx}$, $\left(-\dfrac{\pi}{2} \leq \text{arc sin } u \leq \dfrac{\pi}{2}\right)$

31. $\frac{d}{dx}(\text{arc}\cos u) = -\frac{1}{\sqrt{1-u^2}}\frac{du}{dx}, \qquad (0 \le \text{arc}\cos u \le \pi)$

32. $\frac{d}{dx}(\text{arc}\tan u) = \frac{1}{1+u^2}\frac{du}{dx}, \qquad \left(-\frac{\pi}{2} < \text{arc}\tan u < \frac{\pi}{2}\right)$

33. $\frac{d}{dx}(\text{arc}\cot u) = -\frac{1}{1+u^2}\frac{du}{dx}, \qquad (0 \le \text{arc}\cot u \le \pi)$

34. $\frac{d}{dx}(\text{arc}\sec u) = \frac{1}{u\sqrt{u^2-1}}\frac{du}{dx}, \qquad \left(0 \le \text{arc}\sec u < \frac{\pi}{2}, -\pi \le \text{arc}\sec u < -\frac{\pi}{2}\right)$

35. $\frac{d}{dx}(\text{arc}\csc u) = -\frac{1}{u\sqrt{u^2-1}}\frac{du}{dx}, \qquad \left(0 < \text{arc}\csc u \le \frac{\pi}{2}, -\pi < \text{arc}\csc u \le -\frac{\pi}{2}\right)$

36. $\frac{d}{dx}(\text{arc vers}\, u) = \frac{1}{\sqrt{2u-u^2}}\frac{du}{dx}, \qquad (0 \le \text{arc vers}\, u \le \pi)$

37. $\frac{d}{dx}(\sinh u) = \frac{du}{dx}(\cosh u)$

38. $\frac{d}{dx}(\cosh u) = \frac{du}{dx}(\sinh u)$

39. $\frac{d}{dx}(\tanh u) = \frac{du}{dx}(\text{sech}^2\, u)$

40. $\frac{d}{dx}(\coth u) = -\frac{du}{dx}(\text{csch}^2\, u)$

41. $\frac{d}{dx}(\text{sech}\, u) = -\frac{du}{dx}(\text{sech}\, u \cdot \tanh u)$

42. $\frac{d}{dx}(\text{csch}\, u) = -\frac{du}{dx}(\text{csch}\, u \cdot \coth u)$

43. $\frac{d}{dx}(\sinh^{-1} u) = \frac{d}{dx}[\log(u + \sqrt{u^2+1})] = \frac{1}{\sqrt{u^2+1}}\frac{du}{dx}$

44. $\frac{d}{dx}(\cosh^{-1} u) = \frac{d}{dx}[\log(u + \sqrt{u^2-1})] = \frac{1}{\sqrt{u^2-1}}\frac{du}{dx}, \qquad (u > 1, \cosh^{-1} u > 0)$

45. $\frac{d}{dx}(\tanh^{-1} u) = \frac{d}{dx}\left[\frac{1}{2}\log\frac{1+u}{1-u}\right] = \frac{1}{1-u^2}\frac{du}{dx}, \qquad (u^2 < 1)$

46. $\frac{d}{dx}(\coth^{-1} u) = \frac{d}{dx}\left[\frac{1}{2}\log\frac{u+1}{u-1}\right] = \frac{1}{1-u^2}\frac{du}{dx}, \qquad (u^2 > 1)$

47. $\frac{d}{dx}(\text{sech}^{-1} u) = \frac{d}{dx}\left[\log\frac{1+\sqrt{1-u^2}}{u}\right] = -\frac{1}{u\sqrt{1-u^2}}\frac{du}{dx}, \qquad (0 < u < 1, \text{sech}^{-1} u > 0)$

48. $\frac{d}{dx}(\text{csch}^{-1} u) = \frac{d}{dx}\left[\log\frac{1+\sqrt{1+u^2}}{u}\right] = -\frac{1}{|u|\sqrt{1+u^2}}\frac{du}{dx}$

49. $\dfrac{d}{dq}\displaystyle\int_p^q f(x)\,dx = f(q), \qquad [p \text{ constant}]$

50. $\dfrac{d}{dp}\displaystyle\int_p^q f(x)\,dx = -f(p), \qquad [q \text{ constant}]$

51. $\dfrac{d}{da}\displaystyle\int_p^q f(x,a)\,dx = \int_p^q \frac{\partial}{\partial a}[f(x,a)]\,dx + f(q,a)\frac{dq}{da} - f(p,a)\frac{dp}{da}$

INTEGRATION

The following is a brief discussion of some integration techniques. A more complete discussion can be found in a number of good text books. However, the purpose of this introduction is simply to discuss a few of the important techniques which may be used, in conjunction with the integral table which follows, to integrate particular functions.

No matter how extensive the integral table, it is a fairly uncommon occurrence to find in the table the exact integral desired. Usually some form of transformation will have to be made. The simplest type of transformation, and yet the most general, is substitution. Simple forms of substitution, such as $y = ax$, are employed almost unconsciously by experienced users of integral tables. Other substitutions may require more thought. In some sections of the tables, appropriate substitutions are suggested for integrals which are similar to, but not exactly like, integrals in the table. Finding the right substitution is largely a matter of intuition and experience.

Several precautions must be observed when using substitutions:

1. Be sure to make the substitution in the dx term, as well as everywhere else in the integral.
2. Be sure that the function substituted is one-to-one and continuous. If this is not the case, the integral must be restricted in such a way as to make it true. See the example following.
3. With definite integrals, the limits should also be expressed in terms of the new dependent variable. With indefinite integrals, it is necessary to perform the reverse substitution to obtain the answer in terms of the original independent variable. This may also be done for definite integrals, but it is usually easier to change the limits.

Example:

$$\int \frac{x^4}{\sqrt{a^2 - x^2}} dx$$

Here we make the substitution $x = |a| \sin \theta$. Then $dx = |a| \cos \theta \, d\theta$, and

$$\sqrt{a^2 - x^2} = \sqrt{a^2 - a^2 \sin^2 \theta} = |a|\sqrt{1 - \sin^2 \theta} = |a \cos \theta|$$

Notice the absolute value signs. It is very important to keep in mind that a square root radical always denotes the positive square root, and to assure the sign is always kept positive. Thus $\sqrt{x^2} = |x|$. Failure to observe this is a common cause of errors in integration.

Notice also that the indicated substitution is not a one-to-one function, that is, it does not have a unique inverse. Thus we must restrict the range of θ in such a way as to make the function one-to-one. Fortunately, this is easily done by solving for θ

$$\theta = \sin^{-1} \frac{x}{|a|}$$

and restricting the inverse sine to the principal values, $-\frac{\pi}{2} \le \theta \le \frac{\pi}{2}$.

Thus the integral becomes

$$\int \frac{a^4 \sin^4 \theta |a| \cos \theta \, d\theta}{|a| \, |\cos \theta|}$$

Now, however, in the range of values chosen for θ, $\cos \theta$ is always positive. Thus we may remove the absolute value signs from $\cos \theta$ in the denominator. (This is one of the reasons that the principal values of the inverse trigonometric functions are defined as they are.)

Then the cos θ terms cancel, and the integral becomes

$$a^4 \int \sin^4 \theta \, d\theta$$

By application of integral formulas 299 and 296, we integrate this to

$$-a^4 \frac{\sin^3 \theta \cos \theta}{4} - \frac{3a^4}{8} \cos \theta \sin \theta + \frac{3a^4}{8} \theta + C$$

We now must perform the inverse substitution to get the result in terms of x. We have

$$\theta = \sin^{-1} \frac{x}{|a|}$$

$$\sin \theta = \frac{x}{|a|}$$

Then

$$\cos \theta = \pm\sqrt{1 - \sin^2 \theta} = \pm\sqrt{1 - \frac{x^2}{a^2}} = \pm\frac{\sqrt{a^2 - x^2}}{|a|}.$$

Because of the previously mentioned fact that cos θ is positive, we may omit the $\pm$ sign. The reverse substitution then produces the final answer

$$\int \frac{x^4}{\sqrt{a^2 - x^2}} dx = -\tfrac{1}{4}x^3\sqrt{a^2 - x^2} - \tfrac{3}{8}a^2 x\sqrt{a^2 - x^2} + \frac{3a^4}{8} \sin^{-1} \frac{x}{|a|} + C.$$

Any rational function of x may be integrated, if the denominator is factored into linear and irreducible quadratic factors. The function may then be broken into partial fractions, and the individual partial fractions integrated by use of the appropriate formula from the integral table. See the section on partial fractions for further information.

Many integrals may be reduced to rational functions by proper substitutions. For example,

$$z = \tan \frac{x}{2}$$

will reduce any rational function of the six trigonometric functions of x to a rational function of z. (Frequently there are other substitutions which are simpler to use, but this one will always work. See integral formula number 484.)

Any rational function of x and $\sqrt{ax + b}$ may be reduced to a rational function of z by making the substitution

$$z = \sqrt{ax + b}.$$

Other likely substitutions will be suggested by looking at the form of the integrand.

The other main method of transforming integrals is integration by parts. This involves applying formula number 5 or 6 in the accompanying integral table. The critical factor in this method is the choice of the functions u and v. In order for the method to be successful, $v = \int dv$ and $\int v \, du$ must be easier to integrate than the original integral. Again, this choice is largely a matter of intuition and experience.

Example:

$$\int x \sin x \, dx$$

Two obvious choices are $u = x$, $dv = \sin x \, dx$, or $u = \sin x$, $dv = x \, dx$. Since a preliminary mental calculation indicates that $\int v \, du$ in the second choice would be more, rather than less, complicated than the original integral (it would contain x^2), we use the first choice.

$$u = x \qquad du = dx$$

$$dv = \sin x\,dx \qquad v = -\cos x$$

$$\int x \sin x\,dx = \int u\,dv = uv - \int v\,du = -x\cos x + \int \cos x\,dx$$

$$= \sin x - x\cos x$$

Of course, this result could have been obtained directly from the integral table, but it provides a simple example of the method. In more complicated examples the choice of u and v may not be so obvious, and several different choices may have to be tried. Of course, there is no guarantee that any of them will work.

Integration by parts may be applied more than once, or combined with substitution. A fairly common case is illustrated by the following example.

Example:

$$\int e^x \sin x\,dx$$

Let

$$u = e^x \qquad \text{Then} \quad du = e^x\,dx$$

$$dv = \sin x\,dx \qquad v = -\cos x$$

$$\int e^x \sin x\,dx = \int u\,dv = uv - \int v\,du = -e^x\cos x + \int e^x \cos x\,dx$$

In this latter integral,

$$\text{let} \quad u = e^x \qquad \text{Then} \quad du = e^x\,dx$$

$$dv = \cos x\,dx \qquad v = \sin x$$

$$\int e^x \sin x\,dx = -e^x\cos x + \int e^x\cos x\,dx = -e^x\cos x + \int u\,dv$$

$$= -e^x\cos x + uv - \int v\,du$$

$$= -e^x\cos x + e^x\sin x - \int e^x\sin x\,dx$$

This looks as if a circular transformation has taken place, since we are back at the same integral we started from. However, the above equation can be solved algebraically for the required integral:

$$\int e^x \sin x\,dx = \tfrac{1}{2}(e^x\sin x - e^x\cos x)$$

In the second integration by parts, if the parts had been chosen as $u = \cos x$, $dv = e^x\,dx$, we would indeed have made a circular transformation, and returned to the starting place. In general, when doing repeated integration by parts, one should never choose the function u at any stage to be the same as the function v at the previous stage, or a constant times the previous v.

The following rule is called the extended rule for integration by parts. It is the result of $n + 1$ successive applications of integration by parts.

If

$$g_1(x) = \int g(x)\,dx, \qquad g_2(x) = \int g_1(x)\,dx,$$

$$g_3(x) = \int g_2(x)\,dx, \ldots, g_m(x) = \int g_{m-1}(x)\,dx, \ldots,$$

then

$$\int f(x)\cdot g(x)\,dx = f(x)\cdot g_1(x) - f'(x)\cdot g_2(x) + f''(x)\cdot g_3(x) - + \cdots$$
$$+ (-1)^n f^{(n)}(x) g_{n+1}(x) + (-1)^{n+1} \int f^{(n+1)}(x) g_{n+1}(x)\,dx.$$

A useful special case of the above rule is when $f(x)$ is a polynomial of degree n. Then $f^{(n+1)}(x) = 0$, and

$$\int f(x)\cdot g(x)\,dx = f(x)\cdot g_1(x) - f'(x)\cdot g_2(x) + f''(x)\cdot g_3(x) - + \cdots + (-1)^n f^{(n)}(x) g_{n+1}(x) + C$$

Example:
If $f(x) = x^2$, $g(x) = \sin x$

$$\int x^2 \sin x\,dx = -x^2 \cos x + 2x \sin x + 2\cos x + C$$

Another application of this formula occurs if

$$f''(x) = af(x) \quad \text{and} \quad g''(x) = bg(x),$$

where a and b are unequal constants. In this case, by a process similar to that used in the above example for $\int e^x \sin x\,dx$, we get the formula

$$\int f(x) g(x)\,dx = \frac{f(x)\cdot g'(x) - f'(x)\cdot g(x)}{b - a} + C$$

This formula could have been used in the example mentioned. Here is another example.

Example:
If $f(x) = e^{2x}$, $g(x) = \sin 3x$, then $a = 4$, $b = -9$, and

$$\int e^{2x} \sin 3x\,dx = \frac{3\,e^{2x}\cos 3x - 2\,e^{2x}\sin 3x}{-9 - 4} + C = \frac{e^{2x}}{13}(2\sin 3x - 3\cos 3x) + C$$

The following additional points should be observed when using this table.

1. A constant of integration is to be supplied with the answers for indefinite integrals.
2. Logarithmic expressions are to base $e = 2.71828\ldots$, unless otherwise specified, and are to be evaluated for the absolute value of the arguments involved therein.
3. All angles are measured in radians, and inverse trigonometric and hyperbolic functions represent principal values, unless otherwise indicated.
4. If the application of a formula produces either a zero denominator or the square root of a negative number in the result, there is usually available another form of the answer which avoids this difficulty. In many of the results, the excluded values are specified, but when such are omitted it is presumed that one can tell what these should be, especially when difficulties of the type herein mentioned are obtained.
5. When inverse trigonometric functions occur in the integrals, be sure that any replacements made for them are strictly in accordance with the rules for such functions. This causes

little difficulty when the argument of the inverse trigonometric function is positive, since then all angles involved are in the first quadrant. However, if the argument is negative, special care must be used. Thus if $u > 0$,

$$\sin^{-1} u = \cos^{-1}\sqrt{1-u^2} = \csc^{-1}\frac{1}{u}, \text{ etc.}$$

However, if $u < 0$,

$$\sin^{-1} u = -\cos^{-1}\sqrt{1-u^2} = -\pi - \csc^{-1}\frac{1}{u}, \text{ etc.}$$

See the section on inverse trigonometric functions for a full treatment of the allowable substitutions.

6. In integrals 340–345 and some others, the right side includes expressions of the form

$$A \tan^{-1}[B + C \tan f(x)].$$

In these formulas, the $\tan^{-1}$ does not necessarily represent the principal value. Instead of always employing the principal branch of the inverse tangent function, one must instead use that branch of the inverse tangent function upon which $f(x)$ lies for any particular choice of x.

Example:

$$\int_0^{4\pi} \frac{dx}{2+\sin x} = \frac{2}{\sqrt{3}}\tan^{-1}\frac{2\tan\frac{x}{2}+1}{\sqrt{3}}\Bigg]_0^{4\pi}$$

$$= \frac{2}{\sqrt{3}}\left[\tan^{-1}\frac{2\tan 2\pi + 1}{\sqrt{3}} - \tan^{-1}\frac{2\tan 0 + 1}{\sqrt{3}}\right]$$

$$= \frac{2}{\sqrt{3}}\left[\frac{13\pi}{6} - \frac{\pi}{6}\right] = \frac{4\pi}{\sqrt{3}} = \frac{4\sqrt{3}\pi}{3}$$

Here

$$\tan^{-1}\frac{2\tan 2\pi + 1}{\sqrt{3}} = \tan^{-1}\frac{1}{\sqrt{3}} = \frac{13\pi}{6},$$

since $f(x) = 2\pi$; and

$$\tan^{-1}\frac{2\tan 0 + 1}{\sqrt{3}} = \tan^{-1}\frac{1}{\sqrt{3}} = \frac{\pi}{6},$$

since $f(x) = 0$.

7. B_n and E_n where used in Integrals represents the Bernoulli and Euler numbers as defined in the tables contained from pages 613 to 619.

INTEGRALS

ELEMENTARY FORMS

1. $\int a\,dx = ax$

2. $\int a \cdot f(x)\,dx = a\int f(x)\,dx$

3. $\int \phi(y)\,dx = \int \frac{\phi(y)}{y'}\,dy, \qquad \text{where } y' = \frac{dy}{dx}$

4. $\int (u + v)\,dx = \int u\,dx + \int v\,dx, \qquad$ where u and v are any functions of x

5. $\int u\,dv = u\int dv - \int v\,du = uv - \int v\,du$

6. $\int u\frac{dv}{dx}dx = uv - \int v\frac{du}{dx}dx$

7. $\int x^n\,dx = \frac{x^{n+1}}{n+1}, \qquad \text{except } n = -1$

8. $\int \frac{f'(x)\,dx}{f(x)} = \log f(x), \qquad (df(x) = f'(x)\,dx)$

9. $\int \frac{dx}{x} = \log x$

10. $\int \frac{f'(x)\,dx}{2\sqrt{f(x)}} = \sqrt{f(x)}, \qquad (df(x) = f'(x)\,dx)$

11. $\int e^x\,dx = e^x$

12. $\int e^{ax}\,dx = e^{ax}/a$

13. $\int b^{ax}\,dx = \frac{b^{ax}}{a\log b}, \qquad (b > 0)$

14. $\int \log x\,dx = x\log x - x$

15. $\int a^x \log a\,dx = a^x, \qquad (a > 0)$

16. $\int \frac{dx}{a^2 + x^2} = \frac{1}{a}\tan^{-1}\frac{x}{a}$

17. $$\int \frac{dx}{a^2 - x^2} = \begin{cases} \dfrac{1}{a}\tanh^{-1}\dfrac{x}{a} \\ \text{or} \\ \dfrac{1}{2a}\log\dfrac{a+x}{a-x}, \qquad (a^2 > x^2) \end{cases}$$

18. $$\int \frac{dx}{x^2 - a^2} = \begin{cases} -\dfrac{1}{a}\coth^{-1}\dfrac{x}{a} \\ \text{or} \\ \dfrac{1}{2a}\log\dfrac{x-a}{x+a}, \qquad (x^2 > a^2) \end{cases}$$

19. $$\int \frac{dx}{\sqrt{a^2 - x^2}} = \begin{cases} \sin^{-1}\dfrac{x}{|a|} \\ \text{or} \\ -\cos^{-1}\dfrac{x}{|a|}, \qquad (a^2 > x^2) \end{cases}$$

20. $$\int \frac{dx}{\sqrt{x^2 \pm a^2}} = \log\left(x + \sqrt{x^2 \pm a^2}\right)$$

21. $$\int \frac{dx}{x\sqrt{x^2 - a^2}} = \frac{1}{|a|}\sec^{-1}\frac{x}{a}$$

22. $$\int \frac{dx}{x\sqrt{a^2 \pm x^2}} = -\frac{1}{a}\log\left(\frac{a + \sqrt{a^2 \pm x^2}}{x}\right)$$

FORMS CONTAINING $(a + bx)$

For forms containing $a + bx$, but not listed in the table, the substitution $u = \dfrac{a + bx}{x}$ may prove helpful.

23. $$\int (a + bx)^n\,dx = \frac{(a + bx)^{n+1}}{(n + 1)b}, \qquad (n \neq -1)$$

24. $$\int x(a + bx)^n\,dx$$

$$= \frac{1}{b^2(n + 2)}(a + bx)^{n+2} - \frac{a}{b^2(n + 1)}(a + bx)^{n+1}, \qquad (n \neq -1, -2)$$

25. $$\int x^2(a + bx)^n\,dx = \frac{1}{b^3}\left[\frac{(a + bx)^{n+3}}{n + 3} - 2a\frac{(a + bx)^{n+2}}{n + 2} + a^2\frac{(a + bx)^{n+1}}{n + 1}\right]$$

$$26.\ \int x^m(a+bx)^n\,dx = \begin{cases} \dfrac{x^{m+1}(a+bx)^n}{m+n+1} + \dfrac{an}{m+n+1}\displaystyle\int x^m(a+bx)^{n-1}\,dx \\ \text{or} \\ \dfrac{1}{a(n+1)}\left[-x^{m+1}(a+bx)^{n+1} \right. \\ \qquad \left. + (m+n+2)\displaystyle\int x^m(a+bx)^{n+1}\,dx\right] \\ \text{or} \\ \dfrac{1}{b(m+n+1)}\left[x^m(a+bx)^{n+1} - ma\displaystyle\int x^{m-1}(a+bx)^n\,dx\right] \end{cases}$$

$$27.\ \int \frac{dx}{a+bx} = \frac{1}{b}\log(a+bx)$$

$$28.\ \int \frac{dx}{(a+bx)^2} = -\frac{1}{b(a+bx)}$$

$$29.\ \int \frac{dx}{(a+bx)^3} = -\frac{1}{2b(a+bx)^2}$$

$$30.\ \int \frac{x\,dx}{a+bx} = \begin{cases} \dfrac{1}{b^2}[a+bx-a\log(a+bx)] \\ \text{or} \\ \dfrac{x}{b} - \dfrac{a}{b^2}\log(a+bx) \end{cases}$$

$$31.\ \int \frac{x\,dx}{(a+bx)^2} = \frac{1}{b^2}\left[\log(a+bx) + \frac{a}{a+bx}\right]$$

$$32.\ \int \frac{x\,dx}{(a+bx)^n} = \frac{1}{b^2}\left[\frac{-1}{(n-2)(a+bx)^{n-2}} + \frac{a}{(n-1)(a+bx)^{n-1}}\right], \qquad n \neq 1, 2$$

$$33.\ \int \frac{x^2\,dx}{a+bx} = \frac{1}{b^3}\left[\frac{1}{2}(a+bx)^2 - 2a(a+bx) + a^2\log(a+bx)\right]$$

$$34.\ \int \frac{x^2\,dx}{(a+bx)^2} = \frac{1}{b^3}\left[a+bx-2a\log(a+bx) - \frac{a^2}{a+bx}\right]$$

$$35.\ \int \frac{x^2\,dx}{(a+bx)^3} = \frac{1}{b^3}\left[\log(a+bx) + \frac{2a}{a+bx} - \frac{a^2}{2(a+bx)^2}\right]$$

$$36.\ \int \frac{x^2\,dx}{(a+bx)^n} = \frac{1}{b^3}\left[\frac{-1}{(n-3)(a+bx)^{n-3}} \right.$$
$$\left. + \frac{2a}{(n-2)(a+bx)^{n-2}} - \frac{a^2}{(n-1)(a+bx)^{n-1}}\right], \qquad n \neq 1, 2, 3$$

37. $\int \frac{dx}{x(a+bx)} = -\frac{1}{a}\log\frac{a+bx}{x}$

38. $\int \frac{dx}{x(a+bx)^2} = \frac{1}{a(a+bx)} - \frac{1}{a^2}\log\frac{a+bx}{x}$

39. $\int \frac{dx}{x(a+bx)^3} = \frac{1}{a^3}\left[\frac{1}{2}\left(\frac{2a+bx}{a+bx}\right)^2 + \log\frac{x}{a+bx}\right]$

40. $\int \frac{dx}{x^2(a+bx)} = -\frac{1}{ax} + \frac{b}{a^2}\log\frac{a+bx}{x}$

41. $\int \frac{dx}{x^3(a+bx)} = \frac{2bx-a}{2a^2x^2} + \frac{b^2}{a^3}\log\frac{x}{a+bx}$

42. $\int \frac{dx}{x^2(a+bx)^2} = -\frac{a+2bx}{a^2x(a+bx)} + \frac{2b}{a^3}\log\frac{a+bx}{x}$

FORMS CONTAINING $c^2 \pm x^2$, $x^2 - c^2$

43. $\int \frac{dx}{c^2+x^2} = \frac{1}{c}\tan^{-1}\frac{x}{c}$

44. $\int \frac{dx}{c^2-x^2} = \frac{1}{2c}\log\frac{c+x}{c-x}, \qquad (c^2 > x^2)$

45. $\int \frac{dx}{x^2-c^2} = \frac{1}{2c}\log\frac{x-c}{x+c}, \qquad (x^2 > c^2)$

46. $\int \frac{x\,dx}{c^2 \pm x^2} = \pm\frac{1}{2}\log(c^2 \pm x^2)$

47. $\int \frac{x\,dx}{(c^2 \pm x^2)^{n+1}} = \mp\frac{1}{2n(c^2 \pm x^2)^n}$

48. $\int \frac{dx}{(c^2 \pm x^2)^n} = \frac{1}{2c^2(n-1)}\left[\frac{x}{(c^2 \pm x^2)^{n-1}} + (2n-3)\int\frac{dx}{(c^2 \pm x^2)^{n-1}}\right]$

49. $\int \frac{dx}{(x^2 - c^2)^n} = \frac{1}{2c^2(n-1)}\left[-\frac{x}{(x^2 - c^2)^{n-1}} - (2n-3)\int\frac{dx}{(x^2 - c^2)^{n-1}}\right]$

50. $\int \frac{x\,dx}{x^2-c^2} = \frac{1}{2}\log(x^2-c^2)$

51. $\int \frac{x\,dx}{(x^2-c^2)^{n+1}} = -\frac{1}{2n(x^2-c^2)^n}$

FORMS CONTAINING $a + bx$ and $c + dx$

$$u = a + bx, \qquad v = c + dx, \qquad k = ad - bc$$

If $k = 0$, then $v = \frac{c}{a}u$

52. $$\int \frac{dx}{u \cdot v} = \frac{1}{k} \cdot \log\left(\frac{v}{u}\right)$$

53. $$\int \frac{x\,dx}{u \cdot v} = \frac{1}{k}\left[\frac{a}{b}\log(u) - \frac{c}{d}\log(v)\right]$$

54. $$\int \frac{dx}{u^2 \cdot v} = \frac{1}{k}\left(\frac{1}{u} + \frac{d}{k}\log\frac{v}{u}\right)$$

55. $$\int \frac{x\,dx}{u^2 \cdot v} = \frac{-a}{bku} - \frac{c}{k^2}\log\frac{v}{u}$$

56. $$\int \frac{x^2\,dx}{u^2 \cdot v} = \frac{a^2}{b^2ku} + \frac{1}{k^2}\left[\frac{c^2}{d}\log(v) + \frac{a(k - bc)}{b^2}\log(u)\right]$$

57. $$\int \frac{dx}{u^n \cdot v^m} = \frac{1}{k(m-1)}\left[\frac{-1}{u^{n-1} \cdot v^{m-1}} - (m + n - 2)b\int \frac{dx}{u^n \cdot v^{m-1}}\right]$$

58. $$\int \frac{u}{v}\,dx = \frac{bx}{d} + \frac{k}{d^2}\log(v)$$

59. $$\int \frac{u^m\,dx}{v^n} = \begin{cases} \dfrac{-1}{k(n-1)}\left[\dfrac{u^{m+1}}{v^{n-1}} + b(n - m - 2)\displaystyle\int \dfrac{u^m}{v^{n-1}}\,dx\right] \\ \text{or} \\ \dfrac{-1}{d(n-m-1)}\left[\dfrac{u^m}{v^{n-1}} + mk\displaystyle\int \dfrac{u^{m-1}}{v^n}\,dx\right] \\ \text{or} \\ \dfrac{-1}{d(n-1)}\left[\dfrac{u^m}{v^{n-1}} - mb\displaystyle\int \dfrac{u^{m-1}}{v^{n-1}}\,dx\right] \end{cases}$$

FORMS CONTAINING $(a + bx^n)$

60. $$\int \frac{dx}{a + bx^2} = \frac{1}{\sqrt{ab}}\tan^{-1}\frac{x\sqrt{ab}}{a}, \qquad (ab > 0)$$

61. $$\int \frac{dx}{a + bx^2} = \begin{cases} \dfrac{1}{2\sqrt{-ab}}\log\dfrac{a + x\sqrt{-ab}}{a - x\sqrt{-ab}}, & (ab < 0) \\ \text{or} \\ \dfrac{1}{\sqrt{-ab}}\tanh^{-1}\dfrac{x\sqrt{-ab}}{a}, & (ab < 0) \end{cases}$$

62. $\displaystyle\int \frac{dx}{a^2 + b^2x^2} = \frac{1}{ab}\tan^{-1}\frac{bx}{a}$

63. $\displaystyle\int \frac{x\,dx}{a + bx^2} = \frac{1}{2b}\log(a + bx^2)$

64. $\displaystyle\int \frac{x^2\,dx}{a + bx^2} = \frac{x}{b} - \frac{a}{b}\int \frac{dx}{a + bx^2}$

65. $\displaystyle\int \frac{dx}{(a + bx^2)^2} = \frac{x}{2a(a + bx^2)} + \frac{1}{2a}\int \frac{dx}{a + bx^2}$

66. $\displaystyle\int \frac{dx}{a^2 - b^2x^2} = \frac{1}{2ab}\log\frac{a + bx}{a - bx}$

67. $$\int \frac{dx}{(a + bx^2)^{m+1}} = \begin{cases} \dfrac{1}{2ma}\dfrac{x}{(a + bx^2)^m} + \dfrac{2m - 1}{2ma}\displaystyle\int \frac{dx}{(a + bx^2)^m} \\ \text{or} \\ \dfrac{(2m)!}{(m!)^2}\left[\dfrac{x}{2a}\displaystyle\sum_{r=1}^{m}\frac{r!(r - 1)!}{(4a)^{m-r}(2r)!(a + bx^2)^r} + \frac{1}{(4a)^m}\int \frac{dx}{a + bx^2}\right] \end{cases}$$

68. $\displaystyle\int \frac{x\,dx}{(a + bx^2)^{m+1}} = -\frac{1}{2bm(a + bx^2)^m}$

69. $\displaystyle\int \frac{x^2\,dx}{(a + bx^2)^{m+1}} = \frac{-x}{2mb(a + bx^2)^m} + \frac{1}{2mb}\int \frac{dx}{(a + bx^2)^m}$

70. $\displaystyle\int \frac{dx}{x(a + bx^2)} = \frac{1}{2a}\log\frac{x^2}{a + bx^2}$

71. $\displaystyle\int \frac{dx}{x^2(a + bx^2)} = -\frac{1}{ax} - \frac{b}{a}\int \frac{dx}{a + bx^2}$

72. $$\int \frac{dx}{x(a + bx^2)^{m+1}} = \begin{cases} \dfrac{1}{2am(a + bx^2)^m} + \dfrac{1}{a}\displaystyle\int \frac{dx}{x(a + bx^2)^m} \\ \text{or} \\ \dfrac{1}{2a^{m+1}}\left[\displaystyle\sum_{r=1}^{m}\frac{a^r}{r(a + bx^2)^r} + \log\frac{x^2}{a + bx^2}\right] \end{cases}$$

73. $\displaystyle\int \frac{dx}{x^2(a + bx^2)^{m+1}} = \frac{1}{a}\int \frac{dx}{x^2(a + bx^2)^m} - \frac{b}{a}\int \frac{dx}{(a + bx^2)^{m+1}}$

74. $\displaystyle\int \frac{dx}{a + bx^3} = \frac{k}{3a}\left[\frac{1}{2}\log\frac{(k + x)^3}{a + bx^3} + \sqrt{3}\tan^{-1}\frac{2x - k}{k\sqrt{3}}\right], \qquad \left(k = \sqrt[3]{\frac{a}{b}}\right)$

75. $\displaystyle\int \frac{x\,dx}{a + bx^3} = \frac{1}{3bk}\left[\frac{1}{2}\log\frac{a + bx^3}{(k + x)^3} + \sqrt{3}\tan^{-1}\frac{2x - k}{k\sqrt{3}}\right], \qquad \left(k = \sqrt[3]{\frac{a}{b}}\right)$

76. $$\int \frac{x^2\,dx}{a+bx^3} = \frac{1}{3b}\log(a+bx^3)$$

77. $$\int \frac{dx}{a+bx^4} = \frac{k}{2a}\left[\frac{1}{2}\log\frac{x^2+2kx+2k^2}{x^2-2kx+2k^2} + \tan^{-1}\frac{2kx}{2k^2-x^2}\right], \qquad \left(ab>0,\ k=\sqrt[4]{\frac{a}{4b}}\right)$$

78. $$\int \frac{dx}{a+bx^4} = \frac{k}{2a}\left[\frac{1}{2}\log\frac{x+k}{x-k} + \tan^{-1}\frac{x}{k}\right], \qquad \left(ab<0,\ k=\sqrt[4]{-\frac{a}{b}}\right)$$

79. $$\int \frac{x\,dx}{a+bx^4} = \frac{1}{2bk}\tan^{-1}\frac{x^2}{k}, \qquad \left(ab>0,\ k=\sqrt{\frac{a}{b}}\right)$$

80. $$\int \frac{x\,dx}{a+bx^4} = \frac{1}{4bk}\log\frac{x^2-k}{x^2+k}, \qquad \left(ab<0,\ k=\sqrt{-\frac{a}{b}}\right)$$

81. $$\int \frac{x^2\,dx}{a+bx^4} = \frac{1}{4bk}\left[\frac{1}{2}\log\frac{x^2-2kx+2k^2}{x^2+2kx+2k^2} + \tan^{-1}\frac{2kx}{2k^2-x^2}\right], \qquad \left(ab>0,\ k=\sqrt[4]{\frac{a}{4b}}\right)$$

82. $$\int \frac{x^2\,dx}{a+bx^4} = \frac{1}{4bk}\left[\log\frac{x-k}{x+k} + 2\tan^{-1}\frac{x}{k}\right], \qquad \left(ab<0,\ k=\sqrt[4]{-\frac{a}{b}}\right)$$

83. $$\int \frac{x^3\,dx}{a+bx^4} = \frac{1}{4b}\log(a+bx^4)$$

84. $$\int \frac{dx}{x(a+bx^n)} = \frac{1}{an}\log\frac{x^n}{a+bx^n}$$

85. $$\int \frac{dx}{(a+bx^n)^{m+1}} = \frac{1}{a}\int\frac{dx}{(a+bx^n)^m} - \frac{b}{a}\int\frac{x^n\,dx}{(a+bx^n)^{m+1}}$$

86. $$\int \frac{x^m\,dx}{(a+bx^n)^{p+1}} = \frac{1}{b}\int\frac{x^{m-n}\,dx}{(a+bx^n)^p} - \frac{a}{b}\int\frac{x^{m-n}\,dx}{(a+bx^n)^{p+1}}$$

87. $$\int \frac{dx}{x^m(a+bx^n)^{p+1}} = \frac{1}{a}\int\frac{dx}{x^m(a+bx^n)^p} - \frac{b}{a}\int\frac{dx}{x^{m-n}(a+bx^n)^{p+1}}$$

88. $$\int x^m(a+bx^n)^p\,dx = \begin{cases} \dfrac{1}{b(np+m+1)}\Big[x^{m-n+1}(a+bx^n)^{p+1} \\ \qquad - a(m-n+1)\displaystyle\int x^{m-n}(a+bx^n)^p\,dx\Big] \\ \text{or} \\ \dfrac{1}{np+m+1}\Big[x^{m+1}(a+bx^n)^p \\ \qquad + anp\displaystyle\int x^m(a+bx^n)^{p-1}\,dx\Big] \\ \text{or} \\ \dfrac{1}{a(m+1)}\Big[x^{m+1}(a+bx^n)^{p+1} \\ \qquad - (m+1+np+n)b\displaystyle\int x^{m+n}(a+bx^n)^p\,dx\Big] \\ \text{or} \\ \dfrac{1}{an(p+1)}\Big[-x^{m+1}(a+bx^n)^{p+1} \\ \qquad + (m+1+np+n)\displaystyle\int x^m(a+bx^n)^{p+1}\,dx\Big] \end{cases}$$

FORMS CONTAINING $c^3 \pm x^3$

89. $$\int \frac{dx}{c^3 \pm x^3} = \pm\frac{1}{6c^2}\log\frac{(c\pm x)^3}{c^3\pm x^3} + \frac{1}{c^2\sqrt{3}}\tan^{-1}\frac{2x\mp c}{c\sqrt{3}}$$

90. $$\int \frac{dx}{(c^3\pm x^3)^2} = \frac{x}{3c^3(c^3\pm x^3)} + \frac{2}{3c^3}\int\frac{dx}{c^3\pm x^3}$$

91. $$\int \frac{dx}{(c^3\pm x^3)^{n+1}} = \frac{1}{3nc^3}\left[\frac{x}{(c^3\pm x^3)^n} + (3n-1)\int\frac{dx}{(c^3\pm x^3)^n}\right]$$

92. $$\int \frac{x\,dx}{c^3\pm x^3} = \frac{1}{6c}\log\frac{c^3\pm x^3}{(c\pm x)^3} \pm \frac{1}{c\sqrt{3}}\tan^{-1}\frac{2x\mp c}{c\sqrt{3}}$$

93. $$\int \frac{x\,dx}{(c^3\pm x^3)^2} = \frac{x^2}{3c^3(c^3\pm x^3)} + \frac{1}{3c^3}\int\frac{x\,dx}{c^3\pm x^3}$$

94. $$\int \frac{x\,dx}{(c^3\pm x^3)^{n+1}} = \frac{1}{3nc^3}\left[\frac{x^2}{(c^3\pm x^3)^n} + (3n-2)\int\frac{x\,dx}{(c^3\pm x^3)^n}\right]$$

95. $$\int \frac{x^2\,dx}{c^3\pm x^3} = \pm\frac{1}{3}\log(c^3\pm x^3)$$

96. $$\int \frac{x^2\,dx}{(c^3 \pm x^3)^{n+1}} = \mp \frac{1}{3n(c^3 \pm x^3)^n}$$

97. $$\int \frac{dx}{x(c^3 \pm x^3)} = \frac{1}{3c^3} \log \frac{x^3}{c^3 \pm x^3}$$

98. $$\int \frac{dx}{x(c^3 \pm x^3)^2} = \frac{1}{3c^3(c^3 \pm x^3)} + \frac{1}{3c^6} \log \frac{x^3}{c^3 \pm x^3}$$

99. $$\int \frac{dx}{x(c^3 \pm x^3)^{n+1}} = \frac{1}{3nc^3(c^3 \pm x^3)^n} + \frac{1}{c^3} \int \frac{dx}{x(c^3 \pm x^3)^n}$$

100. $$\int \frac{dx}{x^2(c^3 \pm x^3)} = -\frac{1}{c^3 x} \mp \frac{1}{c^3} \int \frac{x\,dx}{c^3 \pm x^3}$$

101. $$\int \frac{dx}{x^2(c^3 \pm x^3)^{n+1}} = \frac{1}{c^3} \int \frac{dx}{x^2(c^3 \pm x^3)^n} \mp \frac{1}{c^3} \int \frac{x\,dx}{(c^3 \pm x^3)^{n+1}}$$

FORMS CONTAINING $c^4 \pm x^4$

102. $$\int \frac{dx}{c^4 + x^4} = \frac{1}{2c^3\sqrt{2}} \left[\frac{1}{2} \log \frac{x^2 + cx\sqrt{2} + c^2}{x^2 - cx\sqrt{2} + c^2} + \tan^{-1} \frac{cx\sqrt{2}}{c^2 - x^2} \right]$$

103. $$\int \frac{dx}{c^4 - x^4} = \frac{1}{2c^3} \left[\frac{1}{2} \log \frac{c + x}{c - x} + \tan^{-1} \frac{x}{c} \right]$$

104. $$\int \frac{x\,dx}{c^4 + x^4} = \frac{1}{2c^2} \tan^{-1} \frac{x^2}{c^2}$$

105. $$\int \frac{x\,dx}{c^4 - x^4} = \frac{1}{4c^2} \log \frac{c^2 + x^2}{c^2 - x^2}$$

106. $$\int \frac{x^2\,dx}{c^4 + x^4} = \frac{1}{2c\sqrt{2}} \left[\frac{1}{2} \log \frac{x^2 - cx\sqrt{2} + c^2}{x^2 + cx\sqrt{2} + c^2} + \tan^{-1} \frac{cx\sqrt{2}}{c^2 - x^2} \right]$$

107. $$\int \frac{x^2\,dx}{c^4 - x^4} = \frac{1}{2c} \left[\frac{1}{2} \log \frac{c + x}{c - x} - \tan^{-1} \frac{x}{c} \right]$$

108. $$\int \frac{x^3\,dx}{c^4 \pm x^4} = \pm \frac{1}{4} \log (c^4 \pm x^4)$$

FORMS CONTAINING $(a + bx + cx^2)$

$$X = a + bx + cx^2 \text{ and } q = 4ac - b^2$$

If $q = 0$, then $X = c\left(x + \frac{b}{2c}\right)^2$, and formulas starting with 23 should be used in place of these.

109. $$\int \frac{dx}{X} = \frac{2}{\sqrt{q}} \tan^{-1} \frac{2cx + b}{\sqrt{q}}, \qquad (q > 0)$$

110. $$\int \frac{dx}{X} = \begin{cases} \dfrac{-2}{\sqrt{-q}} \tanh^{-1} \dfrac{2cx+b}{\sqrt{-q}} \\ \text{or} \\ \dfrac{1}{\sqrt{-q}} \log \dfrac{2cx+b-\sqrt{-q}}{2cx+b+\sqrt{-q}}, \qquad (q<0) \end{cases}$$

111. $$\int \frac{dx}{X^2} = \frac{2cx+b}{qX} + \frac{2c}{q} \int \frac{dx}{X}$$

112. $$\int \frac{dx}{X^3} = \frac{2cx+b}{q}\left(\frac{1}{2X^2} + \frac{3c}{qX}\right) + \frac{6c^2}{q^2} \int \frac{dx}{X}$$

113. $$\int \frac{dx}{X^{n+1}} = \begin{cases} \dfrac{2cx+b}{nqX^n} + \dfrac{2(2n-1)c}{qn} \displaystyle\int \frac{dx}{X^n} \\ \text{or} \\ \dfrac{(2n)!}{(n!)^2}\left(\dfrac{c}{q}\right)^n \left[\dfrac{2cx+b}{q} \displaystyle\sum_{r=1}^{n} \left(\frac{q}{cX}\right)^r \left(\frac{(r-1)!r!}{(2r)!}\right) + \int \frac{dx}{X}\right] \end{cases}$$

114. $$\int \frac{x\,dx}{X} = \frac{1}{2c} \log X - \frac{b}{2c} \int \frac{dx}{X}$$

115. $$\int \frac{x\,dx}{X^2} = -\frac{bx+2a}{qX} - \frac{b}{q} \int \frac{dx}{X}$$

116. $$\int \frac{x\,dx}{X^{n+1}} = -\frac{2a+bx}{nqX^n} - \frac{b(2n-1)}{nq} \int \frac{dx}{X^n}$$

117. $$\int \frac{x^2}{X}\,dx = \frac{x}{c} - \frac{b}{2c^2} \log X + \frac{b^2-2ac}{2c^2} \int \frac{dx}{X}$$

118. $$\int \frac{x^2}{X^2}\,dx = \frac{(b^2-2ac)x+ab}{cqX} + \frac{2a}{q} \int \frac{dx}{X}$$

119. $$\int \frac{x^m\,dx}{X^{n+1}} = -\frac{x^{m-1}}{(2n-m+1)cX^n} - \frac{n-m+1}{2n-m+1}\cdot\frac{b}{c} \int \frac{x^{m-1}\,dx}{X^{n+1}} + \frac{m-1}{2n-m+1}\cdot\frac{a}{c} \int \frac{x^{m-2}\,dx}{X^{n+1}}$$

120. $$\int \frac{dx}{xX} = \frac{1}{2a} \log \frac{x^2}{X} - \frac{b}{2a} \int \frac{dx}{X}$$

121. $$\int \frac{dx}{x^2X} = \frac{b}{2a^2} \log \frac{X}{x^2} - \frac{1}{ax} + \left(\frac{b^2}{2a^2} - \frac{c}{a}\right) \int \frac{dx}{X}$$

122. $$\int \frac{dx}{xX^n} = \frac{1}{2a(n-1)X^{n-1}} - \frac{b}{2a} \int \frac{dx}{X^n} + \frac{1}{a} \int \frac{dx}{xX^{n-1}}$$

123. $$\int \frac{dx}{x^m X^{n+1}} = -\frac{1}{(m-1)ax^{m-1}X^n} - \frac{n+m-1}{m-1}\cdot\frac{b}{a}\int \frac{dx}{x^{m-1}X^{n+1}} - \frac{2n+m-1}{m-1}\cdot\frac{c}{a}\int \frac{dx}{x^{m-2}X^{n+1}}$$

FORMS CONTAINING $\sqrt{a+bx}$

124. $$\int \sqrt{a+bx}\,dx = \frac{2}{3b}\sqrt{(a+bx)^3}$$

125. $$\int x\sqrt{a+bx}\,dx = -\frac{2(2a-3bx)\sqrt{(a+bx)^3}}{15b^2}$$

126. $$\int x^2\sqrt{a+bx}\,dx = \frac{2(8a^2-12abx+15b^2x^2)\sqrt{(a+bx)^3}}{105b^3}$$

127. $$\int x^m\sqrt{a+bx}\,dx = \begin{cases} \dfrac{2}{b(2m+3)}\left[x^m\sqrt{(a+bx)^3} - ma\displaystyle\int x^{m-1}\sqrt{a+bx}\,dx\right] \\ \text{or} \\ \dfrac{2}{b^{m+1}}\sqrt{a+bx}\displaystyle\sum_{r=0}^{m}\frac{m!(-a)^{m-r}}{r!(m-r)!(2r+3)}(a+bx)^{r+1} \end{cases}$$

128. $$\int \frac{\sqrt{a+bx}}{x}dx = 2\sqrt{a+bx} + a\int \frac{dx}{x\sqrt{a+bx}}$$

129. $$\int \frac{\sqrt{a+bx}}{x^2}dx = -\frac{\sqrt{a+bx}}{x} + \frac{b}{2}\int \frac{dx}{x\sqrt{a+bx}}$$

130. $$\int \frac{\sqrt{a+bx}}{x^m}dx = -\frac{1}{(m-1)a}\left[\frac{\sqrt{(a+bx)^3}}{x^{m-1}} + \frac{(2m-5)b}{2}\int \frac{\sqrt{a+bx}}{x^{m-1}}dx\right]$$

131. $$\int \frac{dx}{\sqrt{a+bx}} = \frac{2\sqrt{a+bx}}{b}$$

132. $$\int \frac{x\,dx}{\sqrt{a+bx}} = -\frac{2(2a-bx)}{3b^2}\sqrt{a+bx}$$

133. $$\int \frac{x^2\,dx}{\sqrt{a+bx}} = \frac{2(8a^2-4abx+3b^2x^2)}{15b^3}\sqrt{a+bx}$$

134. $$\int \frac{x^m\,dx}{\sqrt{a+bx}} = \begin{cases} \dfrac{2}{(2m+1)b}\left[x^m\sqrt{a+bx} - ma\displaystyle\int \frac{x^{m-1}\,dx}{\sqrt{a+bx}}\right] \\ \text{or} \\ \dfrac{2(-a)^m\sqrt{a+bx}}{b^{m+1}} \displaystyle\sum_{r=0}^{m} \frac{(-1)^r m!(a+bx)^r}{(2r+1)r!(m-r)!a^r} \end{cases}$$

135. $$\int \frac{dx}{x\sqrt{a+bx}} = \frac{1}{\sqrt{a}}\log\left(\frac{\sqrt{a+bx}-\sqrt{a}}{\sqrt{a+bx}+\sqrt{a}}\right), \qquad (a>0)$$

136. $$\int \frac{dx}{x\sqrt{a+bx}} = \frac{2}{\sqrt{-a}}\tan^{-1}\sqrt{\frac{a+bx}{-a}}, \qquad (a<0)$$ ERROR

137. $$\int \frac{dx}{x^2\sqrt{a+bx}} = -\frac{\sqrt{a+bx}}{ax} - \frac{b}{2a}\int \frac{dx}{x\sqrt{a+bx}}$$

138. $$\int \frac{dx}{x^n\sqrt{a+bx}} = \begin{cases} -\dfrac{\sqrt{a+bx}}{(n-1)ax^{n-1}} - \dfrac{(2n-3)b}{(2n-2)a}\displaystyle\int \frac{dx}{x^{n-1}\sqrt{a+bx}} \\ \text{or} \\ \dfrac{(2n-2)!}{[(n-1)!]^2}\left[-\dfrac{\sqrt{a+bx}}{a}\displaystyle\sum_{r=1}^{n-1}\frac{r!(r-1)!}{x^r(2r)!}\left(-\frac{b}{4a}\right)^{n-r-1} \right. \\ \qquad \left. +\left(-\dfrac{b}{4a}\right)^{n-1}\displaystyle\int \frac{dx}{x\sqrt{a+bx}}\right] \end{cases}$$

139. $$\int (a+bx)^{\pm\frac{n}{2}}\,dx = \frac{2(a+bx)^{\frac{2\pm n}{2}}}{b(2\pm n)}$$

140. $$\int x(a+bx)^{\pm\frac{n}{2}}\,dx = \frac{2}{b^2}\left[\frac{(a+bx)^{\frac{4\pm n}{2}}}{4\pm n} - \frac{a(a+bx)^{\frac{2\pm n}{2}}}{2\pm n}\right]$$

141. $$\int \frac{dx}{x(a+bx)^{\frac{m}{2}}} = \frac{1}{a}\int \frac{dx}{x(a+bx)^{\frac{m-2}{2}}} - \frac{b}{a}\int \frac{dx}{(a+bx)^{\frac{m}{2}}}.$$

142. $$\int \frac{(a+bx)^{\frac{n}{2}}\,dx}{x} = b\int (a+bx)^{\frac{n-2}{2}}\,dx + a\int \frac{(a+bx)^{\frac{n-2}{2}}}{x}\,dx$$

143. $$\int f(x,\sqrt{a+bx})\,dx = \frac{2}{b}\int f\left(\frac{z^2-a}{b}, z\right)z\,dz, \qquad (z=\sqrt{a+bx})$$

FORMS CONTAINING $\sqrt{a + bx}$ and $\sqrt{c + dx}$

$$u = a + bx \qquad v = c + dx \qquad k = ad - bc$$

If $k = 0$, then $v = \frac{c}{a}u$, and formulas starting with 124 should be used in place of these.

144. $$\int \frac{dx}{\sqrt{uv}} = \begin{cases} \frac{2}{\sqrt{bd}} \tanh^{-1} \frac{\sqrt{bduv}}{bv}, (bd > 0, k < 0) \\ \text{or} \\ \frac{2}{\sqrt{bd}} \tanh^{-1} \frac{\sqrt{bduv}}{du}, (bd > 0, k > 0) \\ \text{or} \\ \frac{1}{\sqrt{bd}} \log \frac{(bv + \sqrt{bduv})^2}{v}, \qquad (bd > 0) \end{cases}$$

145. $$\int \frac{dx}{\sqrt{uv}} = \begin{cases} \frac{2}{\sqrt{-bd}} \tan^{-1} \frac{\sqrt{-bduv}}{bv} \\ \text{or} \\ -\frac{1}{\sqrt{-bd}} \sin^{-1} \left(\frac{2bdx + ad + bc}{|k|} \right), \qquad (bd < 0) \end{cases}$$

146. $$\int \sqrt{uv}\, dx = \frac{k + 2bv}{4bd} \sqrt{uv} - \frac{k^2}{8bd} \int \frac{dx}{\sqrt{uv}}$$

147. $$\int \frac{dx}{v\sqrt{u}} = \begin{cases} \frac{1}{\sqrt{kd}} \log \frac{d\sqrt{u} - \sqrt{kd}}{d\sqrt{u} + \sqrt{kd}} \\ \text{or} \\ \frac{1}{\sqrt{kd}} \log \frac{(d\sqrt{u} - \sqrt{kd})^2}{v}, \qquad (kd > 0) \end{cases}$$

148. $$\int \frac{dx}{v\sqrt{u}} = \frac{2}{\sqrt{-kd}} \tan^{-1} \frac{d\sqrt{u}}{\sqrt{-kd}}, \qquad (kd < 0)$$

149. $$\int \frac{x\, dx}{\sqrt{uv}} = \frac{\sqrt{uv}}{bd} - \frac{ad + bc}{2bd} \int \frac{dx}{\sqrt{uv}}$$

150. $$\int \frac{dx}{v\sqrt{uv}} = \frac{-2\sqrt{uv}}{kv}$$

151. $\displaystyle\int \frac{v\,dx}{\sqrt{uv}} = \frac{\sqrt{uv}}{b} - \frac{k}{2b}\int \frac{dx}{\sqrt{uv}}$

152. $\displaystyle\int \sqrt{\frac{v}{u}}\,dx = \frac{v}{|v|}\int \frac{v\,dx}{\sqrt{uv}}$

153. $\displaystyle\int v^m\sqrt{u}\,dx = \frac{1}{(2m+3)d}\left(2v^{m+1}\sqrt{u} + k\int \frac{v^m\,dx}{\sqrt{u}}\right)$

154. $\displaystyle\int \frac{dx}{v^m\sqrt{u}} = -\frac{1}{(m-1)k}\left(\frac{\sqrt{u}}{v^{m-1}} + \left(m - \frac{3}{2}\right)b\int \frac{dx}{v^{m-1}\sqrt{u}}\right)$

155. $\displaystyle\int \frac{v^m\,dx}{\sqrt{u}} = \begin{cases} \dfrac{2}{b(2m+1)}\left[v^m\sqrt{u} - mk\displaystyle\int \frac{v^{m-1}}{\sqrt{u}}\,dx\right] \\ \text{or} \\ \dfrac{2(m!)^2\sqrt{u}}{b(2m+1)!}\displaystyle\sum_{r=0}^{m}\left(-\frac{4k}{b}\right)^{m-r}\frac{(2r)!}{(r!)^2}v^r \end{cases}$

FORMS CONTAINING $\sqrt{x^2 \pm a^2}$

156. $\displaystyle\int \sqrt{x^2 \pm a^2}\,dx = \tfrac{1}{2}[x\sqrt{x^2 \pm a^2} \pm a^2 \log(x + \sqrt{x^2 \pm a^2})]$

157. $\displaystyle\int \frac{dx}{\sqrt{x^2 \pm a^2}} = \log(x + \sqrt{x^2 \pm a^2})$

158. $\displaystyle\int \frac{dx}{x\sqrt{x^2 - a^2}} = \frac{1}{|a|}\sec^{-1}\frac{x}{a}$

159. $\displaystyle\int \frac{dx}{x\sqrt{x^2 + a^2}} = -\frac{1}{a}\log\left(\frac{a + \sqrt{x^2 + a^2}}{x}\right)$

160. $\displaystyle\int \frac{\sqrt{x^2 + a^2}}{x}\,dx = \sqrt{x^2 + a^2} - a\log\left(\frac{a + \sqrt{x^2 + a^2}}{x}\right)$

161. $\displaystyle\int \frac{\sqrt{x^2 - a^2}}{x}\,dx = \sqrt{x^2 - a^2} - |a|\sec^{-1}\frac{x}{a}$

162. $\displaystyle\int \frac{x\,dx}{\sqrt{x^2 \pm a^2}} = \sqrt{x^2 \pm a^2}$

163. $\displaystyle\int x\sqrt{x^2 \pm a^2}\,dx = \tfrac{1}{3}\sqrt{(x^2 \pm a^2)^3}$

164. $\int \sqrt{(x^2 \pm a^2)^3}\,dx = \frac{1}{4}\left[x\sqrt{(x^2 \pm a^2)^3} \pm \frac{3a^2x}{2}\sqrt{x^2 \pm a^2} + \frac{3a^4}{2}\log\left(x + \sqrt{x^2 \pm a^2}\right)\right]$

165. $\int \frac{dx}{\sqrt{(x^2 \pm a^2)^3}} = \frac{\pm x}{a^2\sqrt{x^2 \pm a^2}}$

166. $\int \frac{x\,dx}{\sqrt{(x^2 \pm a^2)^3}} = \frac{-1}{\sqrt{x^2 \pm a^2}}$

167. $\int x\sqrt{(x^2 \pm a^2)^3}\,dx = \frac{1}{5}\sqrt{(x^2 \pm a^2)^5}$

168. $\int x^2\sqrt{x^2 \pm a^2}\,dx = \frac{x}{4}\sqrt{(x^2 \pm a^2)^3} \mp \frac{a^2}{8}x\sqrt{x^2 \pm a^2} - \frac{a^4}{8}\log\left(x + \sqrt{x^2 \pm a^2}\right)$

169. $\int x^3\sqrt{x^2 + a^2}\,dx = \left(\frac{1}{5}x^2 - \frac{2}{15}a^2\right)\sqrt{(a^2 + x^2)^3}$

170. $\int x^3\sqrt{x^2 - a^2}\,dx = \frac{1}{5}\sqrt{(x^2 - a^2)^5} + \frac{a^2}{3}\sqrt{(x^2 - a^2)^3}$

171. $\int \frac{x^2\,dx}{\sqrt{x^2 \pm a^2}} = \frac{x}{2}\sqrt{x^2 \pm a^2} \mp \frac{a^2}{2}\log\left(x + \sqrt{x^2 \pm a^2}\right)$

172. $\int \frac{x^3\,dx}{\sqrt{x^2 \pm a^2}} = \frac{1}{3}\sqrt{(x^2 \pm a^2)^3} \mp a^2\sqrt{x^2 \pm a^2}$

173. $\int \frac{dx}{x^2\sqrt{x^2 \pm a^2}} = \mp\frac{\sqrt{x^2 \pm a^2}}{a^2x}$

174. $\int \frac{dx}{x^3\sqrt{x^2 + a^2}} = -\frac{\sqrt{x^2 + a^2}}{2a^2x^2} + \frac{1}{2a^3}\log\frac{a + \sqrt{x^2 + a^2}}{x}$

175. $\int \frac{dx}{x^3\sqrt{x^2 - a^2}} = \frac{\sqrt{x^2 - a^2}}{2a^2x^2} + \frac{1}{2|a^3|}\sec^{-1}\frac{x}{a}$

176. $\int x^2\sqrt{(x^2 \pm a^2)^3}\,dx = \frac{x}{6}\sqrt{(x^2 \pm a^2)^5} \mp \frac{a^2x}{24}\sqrt{(x^2 \pm a^2)^3} - \frac{a^4x}{16}\sqrt{x^2 \pm a^2} \mp \frac{a^6}{16}\log\left(x + \sqrt{x^2 \pm a^2}\right)$

177. $\int x^3\sqrt{(x^2 \pm a^2)^3}\,dx = \frac{1}{7}\sqrt{(x^2 \pm a^2)^7} \mp \frac{a^2}{5}\sqrt{(x^2 \pm a^2)^5}$

178. $\displaystyle\int \frac{\sqrt{x^2 \pm a^2}\,dx}{x^2} = -\frac{\sqrt{x^2 \pm a^2}}{x} + \log(x + \sqrt{x^2 \pm a^2})$

179. $\displaystyle\int \frac{\sqrt{x^2 + a^2}}{x^3}\,dx = -\frac{\sqrt{x^2 + a^2}}{2x^2} - \frac{1}{2a}\log\frac{a + \sqrt{x^2 + a^2}}{x}$

180. $\displaystyle\int \frac{\sqrt{x^2 - a^2}}{x^3}\,dx = -\frac{\sqrt{x^2 - a^2}}{2x^2} + \frac{1}{2|a|}\sec^{-1}\frac{x}{a}$

181. $\displaystyle\int \frac{\sqrt{x^2 \pm a^2}}{x^4}\,dx = \mp\frac{\sqrt{(x^2 \pm a^2)^3}}{3a^2x^3}$

182. $\displaystyle\int \frac{x^2\,dx}{\sqrt{(x^2 \pm a^2)^3}} = \frac{-x}{\sqrt{x^2 \pm a^2}} + \log(x + \sqrt{x^2 \pm a^2})$

183. $\displaystyle\int \frac{x^3\,dx}{\sqrt{(x^2 \pm a^2)^3}} = \sqrt{x^2 \pm a^2} \pm \frac{a^2}{\sqrt{x^2 \pm a^2}}$

184. $\displaystyle\int \frac{dx}{x\sqrt{(x^2 + a^2)^3}} = \frac{1}{a^2\sqrt{x^2 + a^2}} - \frac{1}{a^3}\log\frac{a + \sqrt{x^2 + a^2}}{x}$

185. $\displaystyle\int \frac{dx}{x\sqrt{(x^2 - a^2)^3}} = -\frac{1}{a^2\sqrt{x^2 - a^2}} - \frac{1}{|a^3|}\sec^{-1}\frac{x}{a}$

186. $\displaystyle\int \frac{dx}{x^2\sqrt{(x^2 \pm a^2)^3}} = -\frac{1}{a^4}\left[\frac{\sqrt{x^2 \pm a^2}}{x} + \frac{x}{\sqrt{x^2 \pm a^2}}\right]$

187. $\displaystyle\int \frac{dx}{x^3\sqrt{(x^2 + a^2)^3}} = -\frac{1}{2a^2x^2\sqrt{x^2 + a^2}} - \frac{3}{2a^4\sqrt{x^2 + a^2}} + \frac{3}{2a^5}\log\frac{a + \sqrt{x^2 + a^2}}{x}$

188. $\displaystyle\int \frac{dx}{x^3\sqrt{(x^2 - a^2)^3}} = \frac{1}{2a^2x^2\sqrt{x^2 - a^2}} - \frac{3}{2a^4\sqrt{x^2 - a^2}} - \frac{3}{2|a^5|}\sec^{-1}\frac{x}{a}$

189. $\displaystyle\int \frac{x^m}{\sqrt{x^2 \pm a^2}}\,dx = \frac{1}{m}x^{m-1}\sqrt{x^2 \pm a^2} \mp \frac{m-1}{m}a^2\int \frac{x^{m-2}}{\sqrt{x^2 \pm a^2}}\,dx$

190. $\displaystyle\int \frac{x^{2m}}{\sqrt{x^2 \pm a^2}}\,dx = \frac{(2m)!}{2^{2m}(m!)^2}\left[\sqrt{x^2 \pm a^2}\sum_{r=1}^{m}\frac{r!(r-1)!}{(2r)!}(\mp a^2)^{m-r}(2x)^{2r-1} + (\mp a^2)^m \log(x + \sqrt{x^2 \pm a^2})\right]$

191. $\displaystyle\int \frac{x^{2m+1}}{\sqrt{x^2 \pm a^2}}\,dx = \sqrt{x^2 \pm a^2}\sum_{r=0}^{m}\frac{(2r)!(m!)^2}{(2m+1)!(r!)^2}(\mp 4a^2)^{m-r}x^{2r}$

192. $$\int \frac{dx}{x^m\sqrt{x^2 \pm a^2}} = \mp \frac{\sqrt{x^2 \pm a^2}}{(m-1)a^2x^{m-1}} \mp \frac{(m-2)}{(m-1)a^2}\int \frac{dx}{x^{m-2}\sqrt{x^2 \pm a^2}}$$

193. $$\int \frac{dx}{x^{2m}\sqrt{x^2 \pm a^2}} = \sqrt{x^2 \pm a^2}\sum_{r=0}^{m-1} \frac{(m-1)!m!(2r)!2^{2m-2r-1}}{(r!)^2(2m)!(\mp a^2)^{m-r}x^{2r+1}}$$

194. $$\int \frac{dx}{x^{2m+1}\sqrt{x^2 + a^2}} = \frac{(2m)!}{(m!)^2}\left[\frac{\sqrt{x^2 + a^2}}{a^2}\sum_{r=1}^{m} (-1)^{m-r+1} \frac{r!(r-1)!}{2(2r)!(4a^2)^{m-r}x^{2r}} + \frac{(-1)^{m+1}}{2^{2m}a^{2m+1}}\log \frac{\sqrt{x^2 + a^2} + a}{x}\right]$$

195. $$\int \frac{dx}{x^{2m+1}\sqrt{x^2 - a^2}} = \frac{(2m)!}{(m!)^2}\left[\frac{\sqrt{x^2 - a^2}}{a^2}\sum_{r=1}^{m} \frac{r!(r-1)!}{2(2r)!(4a^2)^{m-r}x^{2r}} + \frac{1}{2^{2m}|a|^{2m+1}}\sec^{-1}\frac{x}{a}\right]$$

196. $$\int \frac{dx}{(x-a)\sqrt{x^2 - a^2}} = -\frac{\sqrt{x^2 - a^2}}{a(x-a)}$$

197. $$\int \frac{dx}{(x+a)\sqrt{x^2 - a^2}} = \frac{\sqrt{x^2 - a^2}}{a(x+a)}$$

198. $$\int f(x, \sqrt{x^2 + a^2})\,dx = a\int f(a\tan u, a\sec u)\sec^2 u\,du, \qquad \left(u = \tan^{-1}\frac{x}{a}, a > 0\right)$$

199. $$\int f(x, \sqrt{x^2 - a^2})\,dx = a\int f(a\sec u, a\tan u)\sec u\tan u\,du, \qquad \left(u = \sec^{-1}\frac{x}{a}, a > 0\right)$$

FORMS CONTAINING $\sqrt{a^2 - x^2}$

200. $$\int \sqrt{a^2 - x^2}\,dx = \frac{1}{2}\left[x\sqrt{a^2 - x^2} + a^2\sin^{-1}\frac{x}{|a|}\right]$$

201. $$\int \frac{dx}{\sqrt{a^2 - x^2}} = \begin{cases} \sin^{-1}\dfrac{x}{|a|} \\ \text{or} \\ -\cos^{-1}\dfrac{x}{|a|} \end{cases}$$

202. $$\int \frac{dx}{x\sqrt{a^2 - x^2}} = -\frac{1}{a}\log\left(\frac{a + \sqrt{a^2 - x^2}}{x}\right)$$

203. $\displaystyle\int \frac{\sqrt{a^2 - x^2}}{x}\,dx = \sqrt{a^2 - x^2} - a\log\left(\frac{a + \sqrt{a^2 - x^2}}{x}\right)$

204. $\displaystyle\int \frac{x\,dx}{\sqrt{a^2 - x^2}} = -\sqrt{a^2 - x^2}$

205. $\displaystyle\int x\sqrt{a^2 - x^2}\,dx = -\tfrac{1}{3}\sqrt{(a^2 - x^2)^3}$

206. $\displaystyle\int \sqrt{(a^2 - x^2)^3}\,dx = \frac{1}{4}\left[x\sqrt{(a^2 - x^2)^3} + \frac{3a^2x}{2}\sqrt{a^2 - x^2} + \frac{3a^4}{2}\sin^{-1}\frac{x}{|a|}\right]$

207. $\displaystyle\int \frac{dx}{\sqrt{(a^2 - x^2)^3}} = \frac{x}{a^2\sqrt{a^2 - x^2}}$

208. $\displaystyle\int \frac{x\,dx}{\sqrt{(a^2 - x^2)^3}} = \frac{1}{\sqrt{a^2 - x^2}}$

209. $\displaystyle\int x\sqrt{(a^2 - x^2)^3}\,dx = -\tfrac{1}{5}\sqrt{(a^2 - x^2)^5}$

210. $\displaystyle\int x^2\sqrt{a^2 - x^2}\,dx = -\frac{x}{4}\sqrt{(a^2 - x^2)^3} + \frac{a^2}{8}\left(x\sqrt{a^2 - x^2} + a^2\sin^{-1}\frac{x}{|a|}\right)$

211. $\displaystyle\int x^3\sqrt{a^2 - x^2}\,dx = \left(-\tfrac{1}{5}x^2 - \tfrac{2}{15}a^2\right)\sqrt{(a^2 - x^2)^3}$

212. $\displaystyle\int x^2\sqrt{(a^2 - x^2)^3}\,dx = -\frac{1}{6}x\sqrt{(a^2 - x^2)^5} + \frac{a^2x}{24}\sqrt{(a^2 - x^2)^3}$

$$+ \frac{a^4x}{16}\sqrt{a^2 - x^2} + \frac{a^6}{16}\sin^{-1}\frac{x}{|a|}$$

213. $\displaystyle\int x^3\sqrt{(a^2 - x^2)^3}\,dx = \frac{1}{7}\sqrt{(a^2 - x^2)^7} - \frac{a^2}{5}\sqrt{(a^2 - x^2)^5}$

214. $\displaystyle\int \frac{x^2\,dx}{\sqrt{a^2 - x^2}} = -\frac{x}{2}\sqrt{a^2 - x^2} + \frac{a^2}{2}\sin^{-1}\frac{x}{|a|}$

215. $\displaystyle\int \frac{dx}{x^2\sqrt{a^2 - x^2}} = -\frac{\sqrt{a^2 - x^2}}{a^2x}$

216. $\displaystyle\int \frac{\sqrt{a^2 - x^2}}{x^2}\,dx = -\frac{\sqrt{a^2 - x^2}}{x} - \sin^{-1}\frac{x}{|a|}$

217. $\displaystyle\int \frac{\sqrt{a^2 - x^2}}{x^3}\,dx = -\frac{\sqrt{a^2 - x^2}}{2x^2} + \frac{1}{2a}\log\frac{a + \sqrt{a^2 - x^2}}{x}$

218. $\displaystyle\int \frac{\sqrt{a^2 - x^2}}{x^4}\,dx = -\frac{\sqrt{(a^2 - x^2)^3}}{3a^2x^3}$

219. $$\int \frac{x^2\,dx}{\sqrt{(a^2-x^2)^3}} = \frac{x}{\sqrt{a^2-x^2}} - \sin^{-1}\frac{x}{|a|}$$

220. $$\int \frac{x^3\,dx}{\sqrt{a^2-x^2}} = -\frac{2}{3}(a^2-x^2)^{\frac{3}{2}} - x^2(a^2-x^2)^{\frac{1}{2}} = -\frac{1}{3}\sqrt{a^2-x^2}(x^2+2a^2)$$

221. $$\int \frac{x^3\,dx}{\sqrt{(a^2-x^2)^3}} = 2(a^2-x^2)^{\frac{1}{2}} + \frac{x^2}{(a^2-x^2)^{\frac{1}{2}}} = \frac{a^2}{\sqrt{a^2-x^2}} + \sqrt{a^2-x^2}$$

222. $$\int \frac{dx}{x^3\sqrt{a^2-x^2}} = -\frac{\sqrt{a^2-x^2}}{2a^2x^2} - \frac{1}{2a^3}\log\frac{a+\sqrt{a^2-x^2}}{x}$$

223. $$\int \frac{dx}{x\sqrt{(a^2-x^2)^3}} = \frac{1}{a^2\sqrt{a^2-x^2}} - \frac{1}{a^3}\log\frac{a+\sqrt{a^2-x^2}}{x}$$

224. $$\int \frac{dx}{x^2\sqrt{(a^2-x^2)^3}} = \frac{1}{a^4}\left[-\frac{\sqrt{a^2-x^2}}{x} + \frac{x}{\sqrt{a^2-x^2}}\right]$$

225. $$\int \frac{dx}{x^3\sqrt{(a^2-x^2)^3}} = -\frac{1}{2a^2x^2\sqrt{a^2-x^2}} + \frac{3}{2a^4\sqrt{a^2-x^2}} - \frac{3}{2a^5}\log\frac{a+\sqrt{a^2-x^2}}{x}$$

226. $$\int \frac{x^m}{\sqrt{a^2-x^2}}\,dx = -\frac{x^{m-1}\sqrt{a^2-x^2}}{m} + \frac{(m-1)a^2}{m}\int\frac{x^{m-2}}{\sqrt{a^2-x^2}}\,dx$$

227. $$\int \frac{x^{2m}}{\sqrt{a^2-x^2}}\,dx = \frac{(2m)!}{(m!)^2}\left[-\sqrt{a^2-x^2}\sum_{r=1}^{m}\frac{r!(r-1)!}{2^{2m-2r+1}(2r)!}a^{2m-2r}x^{2r-1} + \frac{a^{2m}}{2^{2m}}\sin^{-1}\frac{x}{|a|}\right]$$

228. $$\int \frac{x^{2m+1}}{\sqrt{a^2-x^2}}\,dx = -\sqrt{a^2-x^2}\sum_{r=0}^{m}\frac{(2r)!(m!)^2}{(2m+1)!(r!)^2}(4a^2)^{m-r}x^{2r}$$

229. $$\int \frac{dx}{x^m\sqrt{a^2-x^2}} = -\frac{\sqrt{a^2-x^2}}{(m-1)a^2x^{m-1}} + \frac{m-2}{(m-1)a^2}\int\frac{dx}{x^{m-2}\sqrt{a^2-x^2}}$$

230. $$\int \frac{dx}{x^{2m}\sqrt{a^2-x^2}} = -\sqrt{a^2-x^2}\sum_{r=0}^{m-1}\frac{(m-1)!m!(2r)!2^{2m-2r-1}}{(r!)^2(2m)!a^{2m-2r}x^{2r+1}}$$

231. $$\int \frac{dx}{x^{2m+1}\sqrt{a^2-x^2}} = \frac{(2m)!}{(m!)^2}\left[-\frac{\sqrt{a^2-x^2}}{a^2}\sum_{r=1}^{m}\frac{r!(r-1)!}{2(2r)!(4a^2)^{m-r}x^{2r}} + \frac{1}{2^{2m}a^{2m+1}}\log\frac{a-\sqrt{a^2-x^2}}{x}\right]$$

232. $$\int \frac{dx}{(b^2 - x^2)\sqrt{a^2 - x^2}} = \frac{1}{2b\sqrt{a^2 - b^2}} \log \frac{(b\sqrt{a^2 - x^2} + x\sqrt{a^2 - b^2})^2}{b^2 - x^2}, \quad (a^2 > b^2)$$

233. $$\int \frac{dx}{(b^2 - x^2)\sqrt{a^2 - x^2}} = \frac{1}{b\sqrt{b^2 - a^2}} \tan^{-1} \frac{x\sqrt{b^2 - a^2}}{b\sqrt{a^2 - x^2}}, \quad (b^2 > a^2)$$

234. $$\int \frac{dx}{(b^2 + x^2)\sqrt{a^2 - x^2}} = \frac{1}{b\sqrt{a^2 + b^2}} \tan^{-1} \frac{x\sqrt{a^2 + b^2}}{b\sqrt{a^2 - x^2}}$$

235. $$\int \frac{\sqrt{a^2 - x^2}}{b^2 + x^2} dx = \frac{\sqrt{a^2 + b^2}}{|b|} \sin^{-1} \frac{x\sqrt{a^2 + b^2}}{|a|\sqrt{x^2 + b^2}} - \sin^{-1} \frac{x}{|a|}$$

236. $$\int f(x, \sqrt{a^2 - x^2})\, dx = a \int f(a \sin u, a \cos u) \cos u\, du, \quad \left(u = \sin^{-1} \frac{x}{a},\ a > 0\right)$$

FORMS CONTAINING $\sqrt{a + bx + cx^2}$

$$X = a + bx + cx^2,\ q = 4ac - b^2, \text{ and } k = \frac{4c}{q}$$

If $q = 0$, then $\sqrt{X} = \sqrt{c}\left|x + \frac{b}{2c}\right|$

237. $$\int \frac{dx}{\sqrt{X}} = \begin{cases} \frac{1}{\sqrt{c}} \log (2\sqrt{cX} + 2cx + b) \\ \text{or} \\ \frac{1}{\sqrt{c}} \sinh^{-1} \frac{2cx + b}{\sqrt{q}}, \quad (c > 0) \end{cases}$$

238. $$\int \frac{dx}{\sqrt{X}} = -\frac{1}{\sqrt{-c}} \sin^{-1} \frac{2cx + b}{\sqrt{-q}}, \quad (c < 0)$$

239. $$\int \frac{dx}{X\sqrt{X}} = \frac{2(2cx + b)}{q\sqrt{X}}$$

240. $$\int \frac{dx}{X^2\sqrt{X}} = \frac{2(2cx + b)}{3q\sqrt{X}}\left(\frac{1}{X} + 2k\right)$$

241. $$\int \frac{dx}{X^n\sqrt{X}} = \begin{cases} \frac{2(2cx + b)\sqrt{X}}{(2n - 1)qX^n} + \frac{2k(n - 1)}{2n - 1} \int \frac{dx}{X^{n-1}\sqrt{X}} \\ \text{or} \\ \frac{(2cx + b)(n!)(n - 1)!4^n k^{n-1}}{q[(2n)!]\sqrt{X}} \sum_{r=0}^{n-1} \frac{(2r)!}{(4kX)^r (r!)^2} \end{cases}$$

242. $$\int \sqrt{X}\,dx = \frac{(2cx+b)\sqrt{X}}{4c} + \frac{1}{2k}\int \frac{dx}{\sqrt{X}}$$

243. $$\int X\sqrt{X}\,dx = \frac{(2cx+b)\sqrt{X}}{8c}\left(X + \frac{3}{2k}\right) + \frac{3}{8k^2}\int \frac{dx}{\sqrt{X}}$$

244. $$\int X^2\sqrt{X}\,dx = \frac{(2cx+b)\sqrt{X}}{12c}\left(X^2 + \frac{5X}{4k} + \frac{15}{8k^2}\right) + \frac{5}{16k^3}\int \frac{dx}{\sqrt{X}}$$

245. $$\int X^n\sqrt{X}\,dx = \begin{cases} \dfrac{(2cx+b)X^n\sqrt{X}}{4(n+1)c} + \dfrac{2n+1}{2(n+1)k}\displaystyle\int X^{n-1}\sqrt{X}\,dx \\ \text{or} \\ \dfrac{(2n+2)!}{[(n+1)!]^2(4k)^{n+1}}\left[\dfrac{k(2cx+b)\sqrt{X}}{c}\displaystyle\sum_{r=0}^{n}\dfrac{r!(r+1)!(4kX)^r}{(2r+2)!} + \displaystyle\int \dfrac{dx}{\sqrt{X}}\right] \end{cases}$$

246. $$\int \frac{x\,dx}{\sqrt{X}} = \frac{\sqrt{X}}{c} - \frac{b}{2c}\int \frac{dx}{\sqrt{X}}$$

247. $$\int \frac{x\,dx}{X\sqrt{X}} = -\frac{2(bx+2a)}{q\sqrt{X}}$$

248. $$\int \frac{x\,dx}{X^n\sqrt{X}} = -\frac{\sqrt{X}}{(2n-1)cX^n} - \frac{b}{2c}\int \frac{dx}{X^n\sqrt{X}}$$

249. $$\int \frac{x^2\,dx}{\sqrt{X}} = \left(\frac{x}{2c} - \frac{3b}{4c^2}\right)\sqrt{X} + \frac{3b^2-4ac}{8c^2}\int \frac{dx}{\sqrt{X}}$$

250. $$\int \frac{x^2\,dx}{X\sqrt{X}} = \frac{(2b^2-4ac)x+2ab}{cq\sqrt{X}} + \frac{1}{c}\int \frac{dx}{\sqrt{X}}$$

251. $$\int \frac{x^2\,dx}{X^n\sqrt{X}} = \frac{(2b^2-4ac)x+2ab}{(2n-1)cqX^{n-1}\sqrt{X}} + \frac{4ac+(2n-3)b^2}{(2n-1)cq}\int \frac{dx}{X^{n-1}\sqrt{X}}$$

252. $$\int \frac{x^3\,dx}{\sqrt{X}} = \left(\frac{x^2}{3c} - \frac{5bx}{12c^2} + \frac{5b^2}{8c^3} - \frac{2a}{3c^2}\right)\sqrt{X} + \left(\frac{3ab}{4c^2} - \frac{5b^3}{16c^3}\right)\int \frac{dx}{\sqrt{X}}$$

253. $$\int \frac{x^n\,dx}{\sqrt{X}} = \frac{1}{nc}x^{n-1}\sqrt{X} - \frac{(2n-1)b}{2nc}\int \frac{x^{n-1}\,dx}{\sqrt{X}} - \frac{(n-1)a}{nc}\int \frac{x^{n-2}\,dx}{\sqrt{X}}$$

254. $\int x\sqrt{X}\,dx = \frac{X\sqrt{X}}{3c} - \frac{b(2cx + b)}{8c^2}\sqrt{X} - \frac{b}{4ck}\int \frac{dx}{\sqrt{X}}$

255. $\int xX\sqrt{X}\,dx = \frac{X^2\sqrt{X}}{5c} - \frac{b}{2c}\int X\sqrt{X}\,dx$

256. $\int xX^n\sqrt{X}\,dx = \frac{X^{n+1}\sqrt{X}}{(2n + 3)c} - \frac{b}{2c}\int X^n\sqrt{X}\,dx$

257. $\int x^2\sqrt{X}\,dx = \left(x - \frac{5b}{6c}\right)\frac{X\sqrt{X}}{4c} + \frac{5b^2 - 4ac}{16c^2}\int \sqrt{X}\,dx$

258. $\int \frac{dx}{x\sqrt{X}} = -\frac{1}{\sqrt{a}}\log\frac{2\sqrt{aX} + bx + 2a}{x}, \qquad (a > 0)$

259. $\int \frac{dx}{x\sqrt{X}} = \frac{1}{\sqrt{-a}}\sin^{-1}\left(\frac{bx + 2a}{|x|\sqrt{-q}}\right), \qquad (a < 0)$

260. $\int \frac{dx}{x\sqrt{X}} = -\frac{2\sqrt{X}}{bx}, \qquad (a = 0)$

261. $\int \frac{dx}{x^2\sqrt{X}} = -\frac{\sqrt{X}}{ax} - \frac{b}{2a}\int \frac{dx}{x\sqrt{X}}$

262. $\int \frac{\sqrt{X}\,dx}{x} = \sqrt{X} + \frac{b}{2}\int \frac{dx}{\sqrt{X}} + a\int \frac{dx}{x\sqrt{X}}$

263. $\int \frac{\sqrt{X}\,dx}{x^2} = -\frac{\sqrt{X}}{x} + \frac{b}{2}\int \frac{dx}{x\sqrt{X}} + c\int \frac{dx}{\sqrt{X}}$

FORMS INVOLVING $\sqrt{2ax - x^2}$

264. $\int \sqrt{2ax - x^2}\,dx = \frac{1}{2}\left[(x - a)\sqrt{2ax - x^2} + a^2\sin^{-1}\frac{x - a}{|a|}\right]$

265. $\int \frac{dx}{\sqrt{2ax - x^2}} = \begin{cases} \cos^{-1}\frac{a - x}{|a|} \\ \text{or} \\ \sin^{-1}\frac{x - a}{|a|} \end{cases}$

266. $$\int x^n\sqrt{2ax-x^2}\,dx = \begin{cases} -\dfrac{x^{n-1}(2ax-x^2)^{\frac{3}{2}}}{n+2} + \dfrac{(2n+1)a}{n+2}\displaystyle\int x^{n-1}\sqrt{2ax-x^2}\,dx \\ \text{or} \\ \sqrt{2ax-x^2}\left[\dfrac{x^{n+1}}{n+2} - \displaystyle\sum_{r=0}^{n}\dfrac{(2n+1)!(r!)^2a^{n-r+1}}{2^{n-r}(2r+1)!(n+2)!n!}x^r\right] \\ \qquad + \dfrac{(2n+1)!a^{n+2}}{2^n n!(n+2)!}\sin^{-1}\dfrac{x-a}{|a|} \end{cases}$$

267. $$\int\frac{\sqrt{2ax-x^2}}{x^n}dx = \frac{(2ax-x^2)^{\frac{3}{2}}}{(3-2n)ax^n} + \frac{n-3}{(2n-3)a}\int\frac{\sqrt{2ax-x^2}}{x^{n-1}}dx$$

268. $$\int\frac{x^n\,dx}{\sqrt{2ax-x^2}} = \begin{cases} \dfrac{-x^{n-1}\sqrt{2ax-x^2}}{n} + \dfrac{a(2n-1)}{n}\displaystyle\int\frac{x^{n-1}}{\sqrt{2ax-x^2}}dx \\ \text{or} \\ -\sqrt{2ax-x^2}\displaystyle\sum_{r=1}^{n}\dfrac{(2n)!r!(r-1)!a^{n-r}}{2^{n-r}(2r)!(n!)^2}x^{r-1} \\ \qquad + \dfrac{(2n)!a^n}{2^n(n!)^2}\sin^{-1}\dfrac{x-a}{|a|} \end{cases}$$

269. $$\int\frac{dx}{x^n\sqrt{2ax-x^2}} = \begin{cases} \dfrac{\sqrt{2ax-x^2}}{a(1-2n)x^n} + \dfrac{n-1}{(2n-1)a}\displaystyle\int\frac{dx}{x^{n-1}\sqrt{2ax-x^2}} \\ \text{or} \\ -\sqrt{2ax-x^2}\displaystyle\sum_{r=0}^{n-1}\dfrac{2^{n-r}(n-1)!n!(2r)!}{(2n)!(r!)^2a^{n-r}x^{r+1}} \end{cases}$$

270. $$\int\frac{dx}{(2ax-x^2)^{\frac{3}{2}}} = \frac{x-a}{a^2\sqrt{2ax-x^2}}$$

271. $$\int\frac{x\,dx}{(2ax-x^2)^{\frac{3}{2}}} = \frac{x}{a\sqrt{2ax-x^2}}$$

MISCELLANEOUS ALGEBRAIC FORMS

272. $$\int\frac{dx}{\sqrt{2ax+x^2}} = \log(x+a+\sqrt{2ax+x^2})$$

273. $$\int\sqrt{ax^2+c}\,dx = \frac{x}{2}\sqrt{ax^2+c} + \frac{c}{2\sqrt{a}}\log(x\sqrt{a}+\sqrt{ax^2+c}), \qquad (a>0)$$

274. $$\int\sqrt{ax^2+c}\,dx = \frac{x}{2}\sqrt{ax^2+c} + \frac{c}{2\sqrt{-a}}\sin^{-1}\left(x\sqrt{-\frac{a}{c}}\right), \qquad (a<0)$$

275. $$\int \sqrt{\frac{1+x}{1-x}}\,dx = \sin^{-1} x - \sqrt{1-x^2}$$

276. $$\int \frac{dx}{x\sqrt{ax^n+c}} = \begin{cases} \dfrac{1}{n\sqrt{c}} \log \dfrac{\sqrt{ax^n+c}-\sqrt{c}}{\sqrt{ax^n+c}+\sqrt{c}} \\ \text{or} \\ \dfrac{2}{n\sqrt{c}} \log \dfrac{\sqrt{ax^n+c}-\sqrt{c}}{\sqrt{x^n}}, \qquad (c > 0) \end{cases}$$

277. $$\int \frac{dx}{x\sqrt{ax^n+c}} = \frac{2}{n\sqrt{-c}} \sec^{-1} \sqrt{-\frac{ax^n}{c}}, \qquad (c < 0)$$

278. $$\int \frac{dx}{\sqrt{ax^2+c}} = \frac{1}{\sqrt{a}} \log (x\sqrt{a} + \sqrt{ax^2+c}), \qquad (a > 0)$$

279. $$\int \frac{dx}{\sqrt{ax^2+c}} = \frac{1}{\sqrt{-a}} \sin^{-1} \left(x\sqrt{-\frac{a}{c}} \right), \qquad (a < 0)$$

280. $$\int (ax^2+c)^{m+\frac{1}{2}}\,dx = \begin{cases} \dfrac{x(ax^2+c)^{m+\frac{1}{2}}}{2(m+1)} + \dfrac{(2m+1)c}{2(m+1)} \displaystyle\int (ax^2+c)^{m-\frac{1}{2}}\,dx \\ \text{or} \\ x\sqrt{ax^2+c} \displaystyle\sum_{r=0}^{m} \dfrac{(2m+1)!(r!)^2 c^{m-r}}{2^{2m-2r+1} m!(m+1)!(2r+1)!} (ax^2+c)^r \\ \qquad + \dfrac{(2m+1)!c^{m+1}}{2^{2m+1} m!(m+1)!} \displaystyle\int \dfrac{dx}{\sqrt{ax^2+c}} \end{cases}$$

281. $$\int x(ax^2+c)^{m+\frac{1}{2}}\,dx = \frac{(ax^2+c)^{m+\frac{3}{2}}}{(2m+3)a}$$

282. $$\int \frac{(ax^2+c)^{m+\frac{1}{2}}}{x}\,dx = \begin{cases} \dfrac{(ax^2+c)^{m+\frac{1}{2}}}{2m+1} + c \displaystyle\int \dfrac{(ax^2+c)^{m-\frac{1}{2}}}{x}\,dx \\ \text{or} \\ \sqrt{ax^2+c} \displaystyle\sum_{r=0}^{m} \dfrac{c^{m-r}(ax^2+c)^r}{2r+1} + c^{m+1} \displaystyle\int \dfrac{dx}{x\sqrt{ax^2+c}} \end{cases}$$

283. $$\int \frac{dx}{(ax^2+c)^{m+\frac{1}{2}}} = \begin{cases} \dfrac{x}{(2m-1)c(ax^2+c)^{m-\frac{1}{2}}} + \dfrac{2m-2}{(2m-1)c} \displaystyle\int \dfrac{dx}{(ax^2+c)^{m-\frac{1}{2}}} \\ \text{or} \\ \dfrac{x}{\sqrt{ax^2+c}} \displaystyle\sum_{r=0}^{m-1} \dfrac{2^{2m-2r-1}(m-1)!m!(2r)!}{(2m)!(r!)^2 c^{m-r}(ax^2+c)^r} \end{cases}$$

284. $\displaystyle\int \frac{dx}{x^m\sqrt{ax^2+c}} = -\frac{\sqrt{ax^2+c}}{(m-1)cx^{m-1}} - \frac{(m-2)a}{(m-1)c}\int \frac{dx}{x^{m-2}\sqrt{ax^2+c}}$

285. $\displaystyle\int \frac{1+x^2}{(1-x^2)\sqrt{1+x^4}}\,dx = \frac{1}{\sqrt{2}}\log\frac{x\sqrt{2}+\sqrt{1+x^4}}{1-x^2}$

286. $\displaystyle\int \frac{1-x^2}{(1+x^2)\sqrt{1+x^4}}\,dx = \frac{1}{\sqrt{2}}\tan^{-1}\frac{x\sqrt{2}}{\sqrt{1+x^4}}$

287. $\displaystyle\int \frac{dx}{x\sqrt{x^n+a^2}} = -\frac{2}{na}\log\frac{a+\sqrt{x^n+a^2}}{\sqrt{x^n}}$

288. $\displaystyle\int \frac{dx}{x\sqrt{x^n-a^2}} = -\frac{2}{na}\sin^{-1}\frac{a}{\sqrt{x^n}}$

289. $\displaystyle\int \sqrt{\frac{x}{a^3-x^3}}\,dx = \frac{2}{3}\sin^{-1}\left(\frac{x}{a}\right)^{\frac{3}{2}}$

FORMS INVOLVING TRIGONOMETRIC FUNCTIONS

290. $\displaystyle\int (\sin ax)\,dx = -\frac{1}{a}\cos ax$

291. $\displaystyle\int (\cos ax)\,dx = \frac{1}{a}\sin ax$

292. $\displaystyle\int (\tan ax)\,dx = -\frac{1}{a}\log\cos ax = \frac{1}{a}\log\sec ax$

293. $\displaystyle\int (\cot ax)\,dx = \frac{1}{a}\log\sin ax = -\frac{1}{a}\log\csc ax$

294. $\displaystyle\int (\sec ax)\,dx = \frac{1}{a}\log(\sec ax + \tan ax) = \frac{1}{a}\log\tan\left(\frac{\pi}{4}+\frac{ax}{2}\right)$

295. $\displaystyle\int (\csc ax)\,dx = \frac{1}{a}\log(\csc ax - \cot ax) = \frac{1}{a}\log\tan\frac{ax}{2}$

296. $\displaystyle\int (\sin^2 ax)\,dx = -\frac{1}{2a}\cos ax\sin ax + \frac{1}{2}x = \frac{1}{2}x - \frac{1}{4a}\sin 2ax$

297. $\displaystyle\int (\sin^3 ax)\,dx = -\frac{1}{3a}(\cos ax)(\sin^2 ax + 2)$

298. $\displaystyle\int (\sin^4 ax)\,dx = \frac{3x}{8} - \frac{\sin 2ax}{4a} + \frac{\sin 4ax}{32a}$

299. $\displaystyle\int (\sin^n ax)\,dx = -\frac{\sin^{n-1} ax\cos ax}{na} + \frac{n-1}{n}\int (\sin^{n-2} ax)\,dx$

300. $$\int (\sin^{2m} ax)\,dx = -\frac{\cos ax}{a}\sum_{r=0}^{m-1}\frac{(2m)!(r!)^2}{2^{2m-2r}(2r+1)!(m!)^2}\sin^{2r+1} ax + \frac{(2m)!}{2^{2m}(m!)^2}x$$

301. $$\int (\sin^{2m+1} ax)\,dx = -\frac{\cos ax}{a}\sum_{r=0}^{m}\frac{2^{2m-2r}(m!)^2(2r)!}{(2m+1)!(r!)^2}\sin^{2r} ax$$

302. $$\int (\cos^2 ax)\,dx = \frac{1}{2a}\sin ax\cos ax + \frac{1}{2}x = \frac{1}{2}x + \frac{1}{4a}\sin 2ax$$

303. $$\int (\cos^3 ax)\,dx = \frac{1}{3a}(\sin ax)(\cos^2 ax + 2)$$

304. $$\int (\cos^4 ax)\,dx = \frac{3x}{8} + \frac{\sin 2ax}{4a} + \frac{\sin 4ax}{32a}$$

305. $$\int (\cos^n ax)\,dx = \frac{1}{na}\cos^{n-1} ax\sin ax + \frac{n-1}{n}\int (\cos^{n-2} ax)\,dx$$

306. $$\int (\cos^{2m} ax)\,dx = \frac{\sin ax}{a}\sum_{r=0}^{m-1}\frac{(2m)!(r!)^2}{2^{2m-2r}(2r+1)!(m!)^2}\cos^{2r+1} ax + \frac{(2m)!}{2^{2m}(m!)^2}x$$

307. $$\int (\cos^{2m+1} ax)\,dx = \frac{\sin ax}{a}\sum_{r=0}^{m}\frac{2^{2m-2r}(m!)^2(2r)!}{(2m+1)!(r!)^2}\cos^{2r} ax$$

308. $$\int \frac{dx}{\sin^2 ax} = \int (\csc^2 ax)\,dx = -\frac{1}{a}\cot ax$$

309. $$\int \frac{dx}{\sin^m ax} = \int (\csc^m ax)\,dx = -\frac{1}{(m-1)a}\cdot\frac{\cos ax}{\sin^{m-1} ax} + \frac{m-2}{m-1}\int \frac{dx}{\sin^{m-2} ax}$$

310. $$\int \frac{dx}{\sin^{2m} ax} = \int (\csc^{2m} ax)\,dx = -\frac{1}{a}\cos ax\sum_{r=0}^{m-1}\frac{2^{2m-2r-1}(m-1)!m!(2r)!}{(2m)!(r!)^2\sin^{2r+1} ax}$$

311. $$\int \frac{dx}{\sin^{2m+1} ax} = \int (\csc^{2m+1} ax)\,dx =$$

$$-\frac{1}{a}\cos ax\sum_{r=0}^{m-1}\frac{(2m)!(r!)^2}{2^{2m-2r}(m!)^2(2r+1)!\sin^{2r+2} ax} + \frac{1}{a}\cdot\frac{(2m)!}{2^{2m}(m!)^2}\log\tan\frac{ax}{2}$$

312. $$\int \frac{dx}{\cos^2 ax} = \int (\sec^2 ax)\,dx = \frac{1}{a}\tan ax$$

313. $$\int \frac{dx}{\cos^n ax} = \int (\sec^n ax)\,dx = \frac{1}{(n-1)a}\cdot\frac{\sin ax}{\cos^{n-1} ax} + \frac{n-2}{n-1}\int \frac{dx}{\cos^{n-2} ax}$$

314. $$\int \frac{dx}{\cos^{2m} ax} = \int (\sec^{2m} ax)\,dx = \frac{1}{a}\sin ax\sum_{r=0}^{m-1}\frac{2^{2m-2r-1}(m-1)!m!(2r)!}{(2m)!(r!)^2\cos^{2r+1} ax}$$

315. $$\int \frac{dx}{\cos^{2m+1} ax} = \int (\sec^{2m+1} ax)\, dx =$$

$$\frac{1}{a} \sin ax \sum_{r=0}^{m-1} \frac{(2m)!(r!)^2}{2^{2m-2r}(m!)^2(2r+1)! \cos^{2r+2} ax}$$

$$+ \frac{1}{a} \cdot \frac{(2m)!}{2^{2m}(m!)^2} \log (\sec ax + \tan ax)$$

316. $$\int (\sin mx)(\sin nx)\, dx = \frac{\sin(m-n)x}{2(m-n)} - \frac{\sin(m+n)x}{2(m+n)}, \qquad (m^2 \neq n^2)$$

317. $$\int (\cos mx)(\cos nx)\, dx = \frac{\sin(m-n)x}{2(m-n)} + \frac{\sin(m+n)x}{2(m+n)}, \qquad (m^2 \neq n^2)$$

318. $$\int (\sin ax)(\cos ax)\, dx = \frac{1}{2a} \sin^2 ax$$

319. $$\int (\sin mx)(\cos nx)\, dx = -\frac{\cos(m-n)x}{2(m-n)} - \frac{\cos(m+n)x}{2(m+n)}, \qquad (m^2 \neq n^2)$$

320. $$\int (\sin^2 ax)(\cos^2 ax)\, dx = -\frac{1}{32a} \sin 4ax + \frac{x}{8}$$

321. $$\int (\sin ax)(\cos^m ax)\, dx = -\frac{\cos^{m+1} ax}{(m+1)a}$$

322. $$\int (\sin^m ax)(\cos ax)\, dx = \frac{\sin^{m+1} ax}{(m+1)a}$$

323. $$\int (\cos^m ax)(\sin^n ax)\, dx = \begin{cases} \dfrac{\cos^{m-1} ax \sin^{n+1} ax}{(m+n)a} \\ \qquad + \dfrac{m-1}{m+n} \displaystyle\int (\cos^{m-2} ax)(\sin^n ax)\, dx \\ \text{or} \\ -\dfrac{\sin^{n-1} ax \cos^{m+1} ax}{(m+n)a} \\ \qquad + \dfrac{n-1}{m+n} \displaystyle\int (\cos^m ax)(\sin^{n-2} ax)\, dx \end{cases}$$

324. $$\int \frac{\cos^m ax}{\sin^n ax}\, dx = \begin{cases} -\dfrac{\cos^{m+1} ax}{(n-1)a \sin^{n-1} ax} - \dfrac{m-n+2}{n-1} \displaystyle\int \frac{\cos^m ax}{\sin^{n-2} ax}\, dx \\ \text{or} \\ \dfrac{\cos^{m-1} ax}{a(m-n) \sin^{n-1} ax} + \dfrac{m-1}{m-n} \displaystyle\int \frac{\cos^{m-2} ax}{\sin^n ax}\, dx \end{cases}$$

325. $$\int \frac{\sin^m ax}{\cos^n ax}\,dx = \begin{cases} \dfrac{\sin^{m+1} ax}{a(n-1)\cos^{n-1} ax} + \dfrac{m-n+2}{n-1}\displaystyle\int \frac{\sin^m ax}{\cos^{n-2} ax}\,dx \\ \text{or} \\ -\dfrac{\sin^{m-1} ax}{a(m-n)\cos^{n-1} ax} - \dfrac{m-1}{m-n}\displaystyle\int \frac{\sin^{m-2} ax}{\cos^n ax}\,dx \end{cases}$$

326. $$\int \frac{\sin ax}{\cos^2 ax}\,dx = \frac{1}{a\cos ax} = \frac{\sec ax}{a}$$

327. $$\int \frac{\sin^2 ax}{\cos ax}\,dx = -\frac{1}{a}\sin ax + \frac{1}{a}\log\tan\left(\frac{\pi}{4} + \frac{ax}{2}\right)$$

328. $$\int \frac{\cos ax}{\sin^2 ax}\,dx = -\frac{1}{a\sin ax} = -\frac{\csc ax}{a}$$

329. $$\int \frac{dx}{(\sin ax)(\cos ax)} = \frac{1}{a}\log\tan ax$$

330. $$\int \frac{dx}{(\sin ax)(\cos^2 ax)} = \frac{1}{a}\left(\sec ax + \log\tan\frac{ax}{2}\right)$$

331. $$\int \frac{dx}{(\sin ax)(\cos^n ax)} = \frac{1}{a(n-1)\cos^{n-1} ax} + \int \frac{dx}{(\sin ax)(\cos^{n-2} ax)}$$

332. $$\int \frac{dx}{(\sin^2 ax)(\cos ax)} = -\frac{1}{a}\csc ax + \frac{1}{a}\log\tan\left(\frac{\pi}{4} + \frac{ax}{2}\right)$$

333. $$\int \frac{dx}{(\sin^2 ax)(\cos^2 ax)} = -\frac{2}{a}\cot 2ax$$

334. $$\int \frac{dx}{\sin^m ax\cos^n ax} = \begin{cases} -\dfrac{1}{a(m-1)(\sin^{m-1} ax)(\cos^{n-1} ax)} \\ \qquad + \dfrac{m+n-2}{m-1}\displaystyle\int \frac{dx}{(\sin^{m-2} ax)(\cos^n ax)} \\ \text{or} \\ \dfrac{1}{a(n-1)\sin^{m-1} ax\cos^{n-1} ax} \\ \qquad - \dfrac{m+n-2}{n-1}\displaystyle\int \frac{dx}{\sin^m ax\cos^{n-2} ax} \end{cases}$$

335. $$\int \sin(a+bx)\,dx = -\frac{1}{b}\cos(a+bx)$$

336. $$\int \cos(a+bx)\,dx = \frac{1}{b}\sin(a+bx)$$

337. $$\int \frac{dx}{1 \pm \sin ax} = \mp\frac{1}{a}\tan\left(\frac{\pi}{4} \mp \frac{ax}{2}\right)$$

338. $$\int \frac{dx}{1+\cos ax} = \frac{1}{a}\tan\frac{ax}{2}$$

339. $$\int \frac{dx}{1-\cos ax} = -\frac{1}{a}\cot\frac{ax}{2}$$

***340.** $$\int \frac{dx}{a+b\sin x} = \begin{cases} \dfrac{2}{\sqrt{a^2-b^2}}\tan^{-1}\dfrac{a\tan\frac{x}{2}+b}{\sqrt{a^2-b^2}} \\ \text{or} \\ \dfrac{1}{\sqrt{b^2-a^2}}\log\dfrac{a\tan\frac{x}{2}+b-\sqrt{b^2-a^2}}{a\tan\frac{x}{2}+b+\sqrt{b^2-a^2}} \end{cases}$$

***341.** $$\int \frac{dx}{a+b\cos x} = \begin{cases} \dfrac{2}{\sqrt{a^2-b^2}}\tan^{-1}\dfrac{\sqrt{a^2-b^2}\tan\frac{x}{2}}{a+b} \\ \text{or} \\ \dfrac{1}{\sqrt{b^2-a^2}}\log\left(\dfrac{\sqrt{b^2-a^2}\tan\frac{x}{2}+a+b}{\sqrt{b^2-a^2}\tan\frac{x}{2}-a-b}\right) \end{cases}$$

***342.** $$\int \frac{dx}{a+b\sin x+c\cos x}$$

$$= \begin{cases} \dfrac{1}{\sqrt{b^2+c^2-a^2}}\log\dfrac{b-\sqrt{b^2+c^2-a^2}+(a-c)\tan\frac{x}{2}}{b+\sqrt{b^2+c^2-a^2}+(a-c)\tan\frac{x}{2}}, & \text{if } a^2<b^2+c^2, a\neq c \\ \text{or} \\ \dfrac{2}{\sqrt{a^2-b^2-c^2}}\tan^{-1}\dfrac{b+(a-c)\tan\frac{x}{2}}{\sqrt{a^2-b^2-c^2}}, & \text{if } a^2>b^2+c^2 \\ \text{or} \\ \dfrac{1}{a}\left[\dfrac{a-(b+c)\cos x-(b-c)\sin x}{a-(b-c)\cos x+(b+c)\sin x}\right], & \text{if } a^2=b^2+c^2, a\neq c. \end{cases}$$

***343.** $$\int \frac{\sin^2 x\,dx}{a+b\cos^2 x} = \frac{1}{b}\sqrt{\frac{a+b}{a}}\tan^{-1}\left(\sqrt{\frac{a}{a+b}}\tan x\right)-\frac{x}{b}, \qquad (ab>0, \text{ or } |a|>|b|)$$

*See note 6 – page 541.

***344.** $$\int \frac{dx}{a^2 \cos^2 x + b^2 \sin^2 x} = \frac{1}{ab} \tan^{-1}\left(\frac{b \tan x}{a}\right)$$

***345.** $$\int \frac{\cos^2 cx}{a^2 + b^2 \sin^2 cx}\, dx = \frac{\sqrt{a^2 + b^2}}{ab^2 c} \tan^{-1} \frac{\sqrt{a^2 + b^2} \tan cx}{a} - \frac{x}{b^2}$$

346. $$\int \frac{\sin cx \cos cx}{a \cos^2 cx + b \sin^2 cx}\, dx = \frac{1}{2c(b - a)} \log (a \cos^2 cx + b \sin^2 cx)$$

347. $$\int \frac{\cos cx}{a \cos cx + b \sin cx}\, dx = \int \frac{dx}{a + b \tan cx} = \frac{1}{c(a^2 + b^2)}[acx + b \log (a \cos cx + b \sin cx)]$$

348. $$\int \frac{\sin cx}{a \sin cx + b \cos cx}\, dx = \int \frac{dx}{a + b \cot cx} = \frac{1}{c(a^2 + b^2)}[acx - b \log (a \sin cx + b \cos cx)]$$

***349.** $$\int \frac{dx}{a \cos^2 x + 2b \cos x \sin x + c \sin^2 x} = \begin{cases} \dfrac{1}{2\sqrt{b^2 - ac}} \log \dfrac{c \tan x + b - \sqrt{b^2 - ac}}{c \tan x + b + \sqrt{b^2 - ac}}, & (b^2 > ac) \\ \text{or} & \\ \dfrac{1}{\sqrt{ac - b^2}} \tan^{-1} \dfrac{c \tan x + b}{\sqrt{ac - b^2}}, & (b^2 < ac) \\ \text{or} & \\ -\dfrac{1}{c \tan x + b}, & (b^2 = ac) \end{cases}$$

350. $$\int \frac{\sin ax}{1 \pm \sin ax}\, dx = \pm x + \frac{1}{a} \tan\left(\frac{\pi}{4} \mp \frac{ax}{2}\right)$$

351. $$\int \frac{dx}{(\sin ax)(1 \pm \sin ax)} = \frac{1}{a} \tan\left(\frac{\pi}{4} \mp \frac{ax}{2}\right) + \frac{1}{a} \log \tan \frac{ax}{2}$$

352. $$\int \frac{dx}{(1 + \sin ax)^2} = -\frac{1}{2a} \tan\left(\frac{\pi}{4} - \frac{ax}{2}\right) - \frac{1}{6a} \tan^3\left(\frac{\pi}{4} - \frac{ax}{2}\right)$$

353. $$\int \frac{dx}{(1 - \sin ax)^2} = \frac{1}{2a} \cot\left(\frac{\pi}{4} - \frac{ax}{2}\right) + \frac{1}{6a} \cot^3\left(\frac{\pi}{4} - \frac{ax}{2}\right)$$

354. $$\int \frac{\sin ax}{(1 + \sin ax)^2}\, dx = -\frac{1}{2a} \tan\left(\frac{\pi}{4} - \frac{ax}{2}\right) + \frac{1}{6a} \tan^3\left(\frac{\pi}{4} - \frac{ax}{2}\right).$$

*See note 6 – page 541.

355. $$\int \frac{\sin ax}{(1-\sin ax)^2}\,dx = -\frac{1}{2a}\cot\left(\frac{\pi}{4}-\frac{ax}{2}\right) + \frac{1}{6a}\cot^3\left(\frac{\pi}{4}-\frac{ax}{2}\right)$$

356. $$\int \frac{\sin x\,dx}{a+b\sin x} = \frac{x}{b} - \frac{a}{b}\int \frac{dx}{a+b\sin x}$$

357. $$\int \frac{dx}{(\sin x)(a+b\sin x)} = \frac{1}{a}\log\tan\frac{x}{2} - \frac{b}{a}\int \frac{dx}{a+b\sin x}$$

358. $$\int \frac{dx}{(a+b\sin x)^2} = \frac{b\cos x}{(a^2-b^2)(a+b\sin x)} + \frac{a}{a^2-b^2}\int \frac{dx}{a+b\sin x}$$

359. $$\int \frac{\sin x\,dx}{(a+b\sin x)^2} = \frac{a\cos x}{(b^2-a^2)(a+b\sin x)} + \frac{b}{b^2-a^2}\int \frac{dx}{a+b\sin x}$$

*360. $$\int \frac{dx}{a^2+b^2\sin^2 cx} = \frac{1}{ac\sqrt{a^2+b^2}}\tan^{-1}\frac{\sqrt{a^2+b^2}\tan cx}{a}$$

*361. $$\int \frac{dx}{a^2-b^2\sin^2 cx} = \begin{cases} \dfrac{1}{ac\sqrt{a^2-b^2}}\tan^{-1}\dfrac{\sqrt{a^2-b^2}\tan cx}{a}, & (a^2 > b^2) \\ \text{or} & \\ \dfrac{1}{2ac\sqrt{b^2-a^2}}\log\dfrac{\sqrt{b^2-a^2}\tan cx + a}{\sqrt{b^2-a^2}\tan cx - a}, & (a^2 < b^2) \end{cases}$$

362. $$\int \frac{\cos ax}{1+\cos ax}\,dx = x - \frac{1}{a}\tan\frac{ax}{2}$$

363. $$\int \frac{\cos ax}{1-\cos ax}\,dx = -x - \frac{1}{a}\cot\frac{ax}{2}$$

364. $$\int \frac{dx}{(\cos ax)(1+\cos ax)} = \frac{1}{a}\log\tan\left(\frac{\pi}{4}+\frac{ax}{2}\right) - \frac{1}{a}\tan\frac{ax}{2}$$

365. $$\int \frac{dx}{(\cos ax)(1-\cos ax)} = \frac{1}{a}\log\tan\left(\frac{\pi}{4}+\frac{ax}{2}\right) - \frac{1}{a}\cot\frac{ax}{2}$$

366. $$\int \frac{dx}{(1+\cos ax)^2} = \frac{1}{2a}\tan\frac{ax}{2} + \frac{1}{6a}\tan^3\frac{ax}{2}$$

367. $$\int \frac{dx}{(1-\cos ax)^2} = -\frac{1}{2a}\cot\frac{ax}{2} - \frac{1}{6a}\cot^3\frac{ax}{2}$$

368. $$\int \frac{\cos ax}{(1+\cos ax)^2}\,dx = \frac{1}{2a}\tan\frac{ax}{2} - \frac{1}{6a}\tan^3\frac{ax}{2}$$

369. $$\int \frac{\cos ax}{(1-\cos ax)^2}\,dx = \frac{1}{2a}\cot\frac{ax}{2} - \frac{1}{6a}\cot^3\frac{ax}{2}$$

*See note 6 – page 541.

370. $$\int \frac{\cos x\, dx}{a + b\cos x} = \frac{x}{b} - \frac{a}{b}\int \frac{dx}{a + b\cos x}$$

371. $$\int \frac{dx}{(\cos x)(a + b\cos x)} = \frac{1}{a}\log\tan\left(\frac{x}{2} + \frac{\pi}{4}\right) - \frac{b}{a}\int \frac{dx}{a + b\cos x}$$

372. $$\int \frac{dx}{(a + b\cos x)^2} = \frac{b\sin x}{(b^2 - a^2)(a + b\cos x)} - \frac{a}{b^2 - a^2}\int \frac{dx}{a + b\cos x}$$

373. $$\int \frac{\cos x}{(a + b\cos x)^2}\, dx = \frac{a\sin x}{(a^2 - b^2)(a + b\cos x)} - \frac{b}{a^2 - b^2}\int \frac{dx}{a + b\cos x}$$

*374. $$\int \frac{dx}{a^2 + b^2 - 2ab\cos cx} = \frac{2}{c(a^2 - b^2)}\tan^{-1}\left(\frac{a + b}{a - b}\tan\frac{cx}{2}\right)$$

*375. $$\int \frac{dx}{a^2 + b^2\cos^2 cx} = \frac{1}{ac\sqrt{a^2 + b^2}}\tan^{-1}\frac{a\tan cx}{\sqrt{a^2 + b^2}}$$

*376. $$\int \frac{dx}{a^2 - b^2\cos^2 cx} = \begin{cases} \dfrac{1}{ac\sqrt{a^2 - b^2}}\tan^{-1}\dfrac{a\tan cx}{\sqrt{a^2 - b^2}}, & (a^2 > b^2) \\ \text{or} & \\ \dfrac{1}{2ac\sqrt{b^2 - a^2}}\log\dfrac{a\tan cx - \sqrt{b^2 - a^2}}{a\tan cx + \sqrt{b^2 - a^2}}, & (b^2 > a^2) \end{cases}$$

377. $$\int \frac{\sin ax}{1 \pm \cos ax}\, dx = \mp \frac{1}{a}\log(1 \pm \cos ax)$$

378. $$\int \frac{\cos ax}{1 \pm \sin ax}\, dx = \pm \frac{1}{a}\log(1 \pm \sin ax)$$

379. $$\int \frac{dx}{(\sin ax)(1 \pm \cos ax)} = \pm \frac{1}{2a(1 \pm \cos ax)} + \frac{1}{2a}\log\tan\frac{ax}{2}$$

380. $$\int \frac{dx}{(\cos ax)(1 \pm \sin ax)} = \mp \frac{1}{2a(1 \pm \sin ax)} + \frac{1}{2a}\log\tan\left(\frac{\pi}{4} + \frac{ax}{2}\right)$$

381. $$\int \frac{\sin ax}{(\cos ax)(1 \pm \cos ax)}\, dx = \frac{1}{a}\log(\sec ax \pm 1)$$

382. $$\int \frac{\cos ax}{(\sin ax)(1 \pm \sin ax)}\, dx = -\frac{1}{a}\log(\csc ax \pm 1)$$

383. $$\int \frac{\sin ax}{(\cos ax)(1 \pm \sin ax)}\, dx = \frac{1}{2a(1 \pm \sin ax)} \pm \frac{1}{2a}\log\tan\left(\frac{\pi}{4} + \frac{ax}{2}\right)$$

384. $$\int \frac{\cos ax}{(\sin ax)(1 \pm \cos ax)}\, dx = -\frac{1}{2a(1 \pm \cos ax)} \pm \frac{1}{2a}\log\tan\frac{ax}{2}$$

*See note 6 – page 541.

385. $$\int \frac{dx}{\sin ax \pm \cos ax} = \frac{1}{a\sqrt{2}} \log \tan\left(\frac{ax}{2} \pm \frac{\pi}{8}\right)$$

386. $$\int \frac{dx}{(\sin ax \pm \cos ax)^2} = \frac{1}{2a} \tan\left(ax \mp \frac{\pi}{4}\right)$$

387. $$\int \frac{dx}{1 + \cos ax \pm \sin ax} = \pm \frac{1}{a} \log\left(1 \pm \tan \frac{ax}{2}\right)$$

388. $$\int \frac{dx}{a^2 \cos^2 cx - b^2 \sin^2 cx} = \frac{1}{2abc} \log \frac{b \tan cx + a}{b \tan cx - a}$$

389. $$\int x(\sin ax)\, dx = \frac{1}{a^2} \sin ax - \frac{x}{a} \cos ax$$

390. $$\int x^2(\sin ax)\, dx = \frac{2x}{a^2} \sin ax - \frac{a^2x^2 - 2}{a^3} \cos ax$$

391. $$\int x^3(\sin ax)\, dx = \frac{3a^2x^2 - 6}{a^4} \sin ax - \frac{a^2x^3 - 6x}{a^3} \cos ax$$

392. $$\int x^m \sin ax\, dx = \begin{cases} -\frac{1}{a} x^m \cos ax + \frac{m}{a} \int x^{m-1} \cos ax\, dx \\ \text{or} \\ \cos ax \sum_{r=0}^{\left[\frac{m}{2}\right]} (-1)^{r+1} \frac{m!}{(m-2r)!} \cdot \frac{x^{m-2r}}{a^{2r+1}} \\ + \sin ax \sum_{r=0}^{\left[\frac{m-1}{2}\right]} (-1)^r \frac{m!}{(m-2r-1)!} \cdot \frac{x^{m-2r-1}}{a^{2r+2}} \end{cases}$$

Note: $[s]$ means greatest integer $\leq s$; $[3\frac{1}{2}] = 3$, $[\frac{1}{2}] = 0$, etc.

393. $$\int x(\cos ax)\, dx = \frac{1}{a^2} \cos ax + \frac{x}{a} \sin ax$$

394. $$\int x^2(\cos ax)\, dx = \frac{2x \cos ax}{a^2} + \frac{a^2x^2 - 2}{a^3} \sin ax$$

395. $$\int x^3(\cos ax)\, dx = \frac{3a^2x^2 - 6}{a^4} \cos ax + \frac{a^2x^3 - 6x}{a^3} \sin ax$$

396. $$\int x^m(\cos ax)\, dx = \begin{cases} \frac{x^m \sin ax}{a} - \frac{m}{a} \int x^{m-1} \sin ax\, dx \\ \text{or} \\ \sin ax \sum_{r=0}^{\left[\frac{m}{2}\right]} (-1)^r \frac{m!}{(m-2r)!} \cdot \frac{x^{m-2r}}{a^{2r+1}} \\ + \cos ax \sum_{r=0}^{\left[\frac{m-1}{2}\right]} (-1)^r \frac{m!}{(m-2r-1)!} \cdot \frac{x^{m-2r-1}}{a^{2r+2}} \end{cases}$$

See note integral 392.

397. $$\int \frac{\sin ax}{x}\,dx = \sum_{n=0}^{\infty} (-1)^n \frac{(ax)^{2n+1}}{(2n+1)(2n+1)!}$$

398. $$\int \frac{\cos ax}{x}\,dx = \log x + \sum_{n=1}^{\infty} (-1)^n \frac{(ax)^{2n}}{2n(2n)!}$$

399. $$\int x(\sin^2 ax)\,dx = \frac{x^2}{4} - \frac{x \sin 2ax}{4a} - \frac{\cos 2ax}{8a^2}$$

400. $$\int x^2(\sin^2 ax)\,dx = \frac{x^3}{6} - \left(\frac{x^2}{4a} - \frac{1}{8a^3}\right) \sin 2ax - \frac{x \cos 2ax}{4a^2}$$

401. $$\int x(\sin^3 ax)\,dx = \frac{x \cos 3ax}{12a} - \frac{\sin 3ax}{36a^2} - \frac{3x \cos ax}{4a} + \frac{3 \sin ax}{4a^2}$$

402. $$\int x(\cos^2 ax)\,dx = \frac{x^2}{4} + \frac{x \sin 2ax}{4a} + \frac{\cos 2ax}{8a^2}$$

403. $$\int x^2(\cos^2 ax)\,dx = \frac{x^3}{6} + \left(\frac{x^2}{4a} - \frac{1}{8a^3}\right) \sin 2ax + \frac{x \cos 2ax}{4a^2}$$

404. $$\int x(\cos^3 ax)\,dx = \frac{x \sin 3ax}{12a} + \frac{\cos 3ax}{36a^2} + \frac{3x \sin ax}{4a} + \frac{3 \cos ax}{4a^2}$$

405. $$\int \frac{\sin ax}{x^m}\,dx = -\frac{\sin ax}{(m-1)x^{m-1}} + \frac{a}{m-1}\int \frac{\cos ax}{x^{m-1}}\,dx$$

406. $$\int \frac{\cos ax}{x^m}\,dx = -\frac{\cos ax}{(m-1)x^{m-1}} - \frac{a}{m-1}\int \frac{\sin ax}{x^{m-1}}\,dx$$

407. $$\int \frac{x}{1 \pm \sin ax}\,dx = \mp \frac{x \cos ax}{a(1 \pm \sin ax)} + \frac{1}{a^2} \log (1 \pm \sin ax)$$

408. $$\int \frac{x}{1 + \cos ax}\,dx = \frac{x}{a} \tan \frac{ax}{2} + \frac{2}{a^2} \log \cos \frac{ax}{2}$$

409. $$\int \frac{x}{1 - \cos ax}\,dx = -\frac{x}{a} \cot \frac{ax}{2} + \frac{2}{a^2} \log \sin \frac{ax}{2}$$

410. $$\int \frac{x + \sin x}{1 + \cos x}\,dx = x \tan \frac{x}{2}$$

411. $$\int \frac{x - \sin x}{1 - \cos x}\,dx = -x \cot \frac{x}{2}$$

412. $$\int \sqrt{1 - \cos ax}\,dx = -\frac{2 \sin ax}{a\sqrt{1 - \cos ax}} = \frac{-2\sqrt{2}}{a} \cos \frac{ax}{2}$$

413. $$\int \sqrt{1 + \cos ax}\,dx = \frac{2 \sin ax}{a\sqrt{1 + \cos ax}} = \frac{2\sqrt{2}}{a} \sin \frac{ax}{2}$$

414. $\int \sqrt{1+\sin x}\,dx = \pm 2\left(\sin\frac{x}{2} - \cos\frac{x}{2}\right),$

[use + if $(8k-1)\frac{\pi}{2} < x \le (8k+3)\frac{\pi}{2}$, otherwise − ; k an integer]

415. $\int \sqrt{1-\sin x}\,dx = \pm 2\left(\sin\frac{x}{2} + \cos\frac{x}{2}\right),$

[use + if $(8k-3)\frac{\pi}{2} < x \le (8k+1)\frac{\pi}{2}$, otherwise − ; k an integer]

416. $\int \frac{dx}{\sqrt{1-\cos x}} = \pm\sqrt{2}\log\tan\frac{x}{4},$

[use + if $4k\pi < x < (4k+2)\pi$, otherwise − ; k an integer]

417. $\int \frac{dx}{\sqrt{1+\cos x}} = \pm\sqrt{2}\log\tan\left(\frac{x+\pi}{4}\right),$

[use + if $(4k-1)\pi < x < (4k+1)\pi$, otherwise − ; k an integer]

418. $\int \frac{dx}{\sqrt{1-\sin x}} = \pm\sqrt{2}\log\tan\left(\frac{x}{4} - \frac{\pi}{8}\right),$

[use + if $(8k+1)\frac{\pi}{2} < x < (8k+5)\frac{\pi}{2}$, otherwise − ; k an integer]

419. $\int \frac{dx}{\sqrt{1+\sin x}} = \pm\sqrt{2}\log\tan\left(\frac{x}{4} + \frac{\pi}{8}\right),$

[use + if $(8k-1)\frac{\pi}{2} < x < (8k+3)\frac{\pi}{2}$, otherwise − ; k an integer]

420. $\int (\tan^2 ax)\,dx = \frac{1}{a}\tan ax - x$

421. $\int (\tan^3 ax)\,dx = \frac{1}{2a}\tan^2 ax + \frac{1}{a}\log\cos ax$

422. $\int (\tan^4 ax)\,dx = \frac{\tan^3 ax}{3a} - \frac{1}{a}\tan x + x$

423. $\int (\tan^n ax)\,dx = \frac{\tan^{n-1} ax}{a(n-1)} - \int (\tan^{n-2} ax)\,dx$

424. $\int (\cot^2 ax)\,dx = -\frac{1}{a}\cot ax - x$

425. $\int (\cot^3 ax)\,dx = -\frac{1}{2a}\cot^2 ax - \frac{1}{a}\log\sin ax$

426. $\int (\cot^4 ax)\,dx = -\frac{1}{3a}\cot^3 ax + \frac{1}{a}\cot ax + x$

427. $$\int (\cot^n ax)\, dx = -\frac{\cot^{n-1} ax}{a(n-1)} - \int (\cot^{n-2} ax)\, dx$$

428. $$\int \frac{x}{\sin^2 ax}\, dx = \int x(\csc^2 ax)\, dx = -\frac{x \cot ax}{a} + \frac{1}{a^2} \log \sin ax$$

429. $$\int \frac{x}{\sin^n ax}\, dx = \int x(\csc^n ax)\, dx = -\frac{x \cos ax}{a(n-1) \sin^{n-1} ax} - \frac{1}{a^2(n-1)(n-2) \sin^{n-2} ax} + \frac{(n-2)}{(n-1)} \int \frac{x}{\sin^{n-2} ax}\, dx$$

430. $$\int \frac{x}{\cos^2 ax}\, dx = \int x(\sec^2 ax)\, dx = \frac{1}{a} x \tan ax + \frac{1}{a^2} \log \cos ax$$

431. $$\int \frac{x}{\cos^n ax}\, dx = \int x(\sec^n ax)\, dx = \frac{x \sin ax}{a(n-1) \cos^{n-1} ax} - \frac{1}{a^2(n-1)(n-2) \cos^{n-2} ax} + \frac{n-2}{n-1} \int \frac{x}{\cos^{n-2} ax}\, dx$$

432. $$\int \frac{\sin ax}{\sqrt{1 + b^2 \sin^2 ax}}\, dx = -\frac{1}{ab} \sin^{-1} \frac{b \cos ax}{\sqrt{1 + b^2}}$$

433. $$\int \frac{\sin ax}{\sqrt{1 - b^2 \sin^2 ax}}\, dx = -\frac{1}{ab} \log (b \cos ax + \sqrt{1 - b^2 \sin^2 ax})$$

434. $$\int (\sin ax)\sqrt{1 + b^2 \sin^2 ax}\, dx = -\frac{\cos ax}{2a} \sqrt{1 + b^2 \sin^2 ax} - \frac{1 + b^2}{2ab} \sin^{-1} \frac{b \cos ax}{\sqrt{1 + b^2}}$$

435. $$\int (\sin ax)\sqrt{1 - b^2 \sin^2 ax}\, dx = -\frac{\cos ax}{2a} \sqrt{1 - b^2 \sin^2 ax} - \frac{1 - b^2}{2ab} \log (b \cos ax + \sqrt{1 - b^2 \sin^2 ax})$$

436. $$\int \frac{\cos ax}{\sqrt{1 + b^2 \sin^2 ax}}\, dx = \frac{1}{ab} \log (b \sin ax + \sqrt{1 + b^2 \sin^2 ax})$$

437. $$\int \frac{\cos ax}{\sqrt{1 - b^2 \sin^2 ax}}\, dx = \frac{1}{ab} \sin^{-1} (b \sin ax)$$

438. $$\int (\cos ax) \sqrt{1 + b^2 \sin^2 ax}\, dx = \frac{\sin ax}{2a} \sqrt{1 + b^2 \sin^2 ax} + \frac{1}{2ab} \log (b \sin ax + \sqrt{1 + b^2 \sin^2 ax})$$

439. $\int (\cos ax)\sqrt{1 - b^2 \sin^2 ax}\, dx = \frac{\sin ax}{2a}\sqrt{1 - b^2 \sin^2 ax} + \frac{1}{2ab}\sin^{-1}(b \sin ax)$

440. $\int \frac{dx}{\sqrt{a + b\tan^2 cx}} = \frac{\pm 1}{c\sqrt{a - b}} \sin^{-1}\left(\sqrt{\frac{a - b}{a}} \sin cx\right), \qquad (a > |b|)$

[use + if $(2k - 1)\frac{\pi}{2} < x \le (2k + 1)\frac{\pi}{2}$, otherwise − ; k an integer]

FORMS INVOLVING INVERSE TRIGONOMETRIC FUNCTIONS

441. $\int (\sin^{-1} ax)\, dx = x \sin^{-1} ax + \frac{\sqrt{1 - a^2x^2}}{a}$

442. $\int (\cos^{-1} ax)\, dx = x \cos^{-1} ax - \frac{\sqrt{1 - a^2x^2}}{a}$

443. $\int (\tan^{-1} ax)\, dx = x \tan^{-1} ax - \frac{1}{2a}\log(1 + a^2x^2)$

444. $\int (\cot^{-1} ax)\, dx = x \cot^{-1} ax + \frac{1}{2a}\log(1 + a^2x^2)$

445. $\int (\sec^{-1} ax)\, dx = x \sec^{-1} ax - \frac{1}{a}\log(ax + \sqrt{a^2x^2 - 1})$

446. $\int (\csc^{-1} ax)\, dx = x \csc^{-1} ax + \frac{1}{a}\log(ax + \sqrt{a^2x^2 - 1})$

447. $\int \left(\sin^{-1}\frac{x}{a}\right) dx = x \sin^{-1}\frac{x}{a} + \sqrt{a^2 - x^2}, \qquad (a > 0)$

448. $\int \left(\cos^{-1}\frac{x}{a}\right) dx = x \cos^{-1}\frac{x}{a} - \sqrt{a^2 - x^2}, \qquad (a > 0)$

449. $\int \left(\tan^{-1}\frac{x}{a}\right) dx = x \tan^{-1}\frac{x}{a} - \frac{a}{2}\log(a^2 + x^2)$

450. $\int \left(\cot^{-1}\frac{x}{a}\right) dx = x \cot^{-1}\frac{x}{a} + \frac{a}{2}\log(a^2 + x^2)$

451. $\int x[\sin^{-1}(ax)]\, dx = \frac{1}{4a^2}[(2a^2x^2 - 1)\sin^{-1}(ax) + ax\sqrt{1 - a^2x^2}]$

452. $\int x[\cos^{-1}(ax)]\, dx = \frac{1}{4a^2}[(2a^2x^2 - 1)\cos^{-1}(ax) - ax\sqrt{1 - a^2x^2}]$

453. $\int x^n[\sin^{-1}(ax)]\,dx = \dfrac{x^{n+1}}{n+1}\sin^{-1}(ax) - \dfrac{a}{n+1}\int \dfrac{x^{n+1}\,dx}{\sqrt{1-a^2x^2}}, \qquad (n \neq -1)$

454. $\int x^n[\cos^{-1}(ax)]\,dx = \dfrac{x^{n+1}}{n+1}\cos^{-1}(ax) + \dfrac{a}{n+1}\int \dfrac{x^{n+1}\,dx}{\sqrt{1-a^2x^2}}, \qquad (n \neq -1)$

455. $\int x(\tan^{-1} ax)\,dx = \dfrac{1+a^2x^2}{2a^2}\tan^{-1} ax - \dfrac{x}{2a}$

456. $\int x^n(\tan^{-1} ax)\,dx = \dfrac{x^{n+1}}{n+1}\tan^{-1} ax - \dfrac{a}{n+1}\int \dfrac{x^{n+1}}{1+a^2x^2}\,dx$

457. $\int x(\cot^{-1} ax)\,dx = \dfrac{1+a^2x^2}{2a^2}\cot^{-1} ax + \dfrac{x}{2a}$

458. $\int x^n(\cot^{-1} ax)\,dx = \dfrac{x^{n+1}}{n+1}\cot^{-1} ax + \dfrac{a}{n+1}\int \dfrac{x^{n+1}}{1+a^2x^2}\,dx$

459. $\int \dfrac{\sin^{-1}(ax)}{x^2}\,dx = a\log\left(\dfrac{1-\sqrt{1-a^2x^2}}{x}\right) - \dfrac{\sin^{-1}(ax)}{x}$

460. $\int \dfrac{\cos^{-1}(ax)\,dx}{x^2} = -\dfrac{1}{x}\cos^{-1}(ax) + a\log\dfrac{1+\sqrt{1-a^2x^2}}{x}$

461. $\int \dfrac{\tan^{-1}(ax)\,dx}{x^2} = -\dfrac{1}{x}\tan^{-1}(ax) - \dfrac{a}{2}\log\dfrac{1+a^2x^2}{x^2}$

462. $\int \dfrac{\cot^{-1} ax}{x^2}\,dx = -\dfrac{1}{x}\cot^{-1} ax - \dfrac{a}{2}\log\dfrac{x^2}{a^2x^2+1}$

463. $\int (\sin^{-1} ax)^2\,dx = x(\sin^{-1} ax)^2 - 2x + \dfrac{2\sqrt{1-a^2x^2}}{a}\sin^{-1} ax$

464. $\int (\cos^{-1} ax)^2\,dx = x(\cos^{-1} ax)^2 - 2x - \dfrac{2\sqrt{1-a^2x^2}}{a}\cos^{-1} ax$

465. $$\int (\sin^{-1} ax)^n\,dx = \begin{cases} x(\sin^{-1} ax)^n + \dfrac{n\sqrt{1-a^2x^2}}{a}(\sin^{-1} ax)^{n-1} \\ \qquad\qquad -n(n-1)\displaystyle\int (\sin^{-1} ax)^{n-2}\,dx \\ \qquad\text{or} \\ \displaystyle\sum_{r=0}^{\left[\frac{n}{2}\right]} (-1)^r \dfrac{n!}{(n-2r)!}\,x(\sin^{-1} ax)^{n-2r} \\ \qquad + \displaystyle\sum_{r=0}^{\left[\frac{n-1}{2}\right]} (-1)^r \dfrac{n!\sqrt{1-a^2x^2}}{(n-2r-1)!\,a}(\sin^{-1} ax)^{n-2r-1} \end{cases}$$

Note: $[s]$ means greatest integer $\leq s$. Thus $[3.5]$ means 3; $[5] = 5$, $[\frac{1}{2}] = 0$.

466. $$\int (\cos^{-1} ax)^n\, dx = \begin{cases} x(\cos^{-1} ax)^n - \dfrac{n\sqrt{1-a^2x^2}}{a}(\cos^{-1} ax)^{n-1} \\ \qquad -n(n-1)\displaystyle\int (\cos^{-1} ax)^{n-2}\, dx \\ \quad \text{or} \\ \displaystyle\sum_{r=0}^{\left[\frac{n}{2}\right]} (-1)^r \frac{n!}{(n-2r)!} x(\cos^{-1} ax)^{n-2r} \\ \qquad - \displaystyle\sum_{r=0}^{\left[\frac{n-1}{2}\right]} (-1)^r \frac{n!\sqrt{1-a^2x^2}}{(n-2r-1)!a}(\cos^{-1} ax)^{n-2r-1} \end{cases}$$

467. $$\int \frac{1}{\sqrt{1-a^2x^2}}(\sin^{-1} ax)\, dx = \frac{1}{2a}(\sin^{-1} ax)^2$$

468. $$\int \frac{x^n}{\sqrt{1-a^2x^2}}(\sin^{-1} ax)\, dx = -\frac{x^{n-1}}{na^2}\sqrt{1-a^2x^2}\sin^{-1} ax + \frac{x^n}{n^2a} + \frac{n-1}{na^2}\int \frac{x^{n-2}}{\sqrt{1-a^2x^2}}\sin^{-1} ax\, dx$$

469. $$\int \frac{1}{\sqrt{1-a^2x^2}}(\cos^{-1} ax)\, dx = -\frac{1}{2a}(\cos^{-1} ax)^2$$

470. $$\int \frac{x^n}{\sqrt{1-a^2x^2}}(\cos^{-1} ax)\, dx = -\frac{x^{n-1}}{na^2}\sqrt{1-a^2x^2}\cos^{-1} ax - \frac{x^n}{n^2a} + \frac{n-1}{na^2}\int \frac{x^{n-2}}{\sqrt{1-a^2x^2}}\cos^{-1} ax\, dx$$

471. $$\int \frac{\tan^{-1} ax}{a^2x^2+1}\, dx = \frac{1}{2a}(\tan^{-1} ax)^2$$

472. $$\int \frac{\cot^{-1} ax}{a^2x^2+1}\, dx = -\frac{1}{2a}(\cot^{-1} ax)^2$$

473. $$\int x\sec^{-1} ax\, dx = \frac{x^2}{2}\sec^{-1} ax - \frac{1}{2a^2}\sqrt{a^2x^2-1}$$

474. $$\int x^n\sec^{-1} ax\, dx = \frac{x^{n+1}}{n+1}\sec^{-1} ax - \frac{1}{n+1}\int \frac{x^n\, dx}{\sqrt{a^2x^2-1}}$$

475. $$\int \frac{\sec^{-1} ax}{x^2}\, dx = -\frac{\sec^{-1} ax}{x} + \frac{\sqrt{a^2x^2-1}}{x}$$

476. $$\int x\csc^{-1} ax\, dx = \frac{x^2}{2}\csc^{-1} ax + \frac{1}{2a^2}\sqrt{a^2x^2-1}$$

477. $$\int x^n\csc^{-1} ax\, dx = \frac{x^{n+1}}{n+1}\csc^{-1} ax + \frac{1}{n+1}\int \frac{x^n\, dx}{\sqrt{a^2x^2-1}}$$

478. $\int \frac{\csc^{-1} ax}{x^2} dx = -\frac{\csc^{-1} ax}{x} - \frac{\sqrt{a^2x^2 - 1}}{x}$

FORMS INVOLVING TRIGONOMETRIC SUBSTITUTIONS

479. $\int f(\sin x)\, dx = 2\int f\left(\frac{2z}{1 + z^2}\right)\frac{dz}{1 + z^2}, \qquad \left(z = \tan\frac{x}{2}\right)$

480. $\int f(\cos x)\, dx = 2\int f\left(\frac{1 - z^2}{1 + z^2}\right)\frac{dz}{1 + z^2}, \qquad \left(z = \tan\frac{x}{2}\right)$

***481.** $\int f(\sin x)\, dx = \int f(u)\frac{du}{\sqrt{1 - u^2}}, \qquad (u = \sin x)$

***482.** $\int f(\cos x)\, dx = -\int f(u)\frac{du}{\sqrt{1 - u^2}}, \qquad (u = \cos x)$

***483.** $\int f(\sin x, \cos x)\, dx = \int f(u, \sqrt{1 - u^2})\frac{du}{\sqrt{1 - u^2}}, \qquad (u = \sin x)$

484. $\int f(\sin x, \cos x)\, dx = 2\int f\left(\frac{2z}{1 + z^2}, \frac{1 - z^2}{1 + z^2}\right)\frac{dz}{1 + z^2}, \qquad \left(z = \tan\frac{x}{2}\right)$

LOGARITHMIC FORMS

485. $\int (\log x)\, dx = x \log x - x$

486. $\int x(\log x)\, dx = \frac{x^2}{2}\log x - \frac{x^2}{4}$

487. $\int x^2(\log x)\, dx = \frac{x^3}{3}\log x - \frac{x^3}{9}$

488. $\int x^n(\log ax)\, dx = \frac{x^{n+1}}{n + 1}\log ax - \frac{x^{n+1}}{(n + 1)^2}$

489. $\int (\log x)^2\, dx = x(\log x)^2 - 2x\log x + 2x$

490. $$\int (\log x)^n\, dx = \begin{cases} x(\log x)^n - n\int (\log x)^{n-1}\, dx, & (n \neq -1) \\ \text{or} & \\ (-1)^n n!\, x \sum_{r=0}^{n} \frac{(-\log x)^r}{r!} & \end{cases}$$

* The square roots appearing in these formulas may be plus or minus, depending on the quadrant of x. Care must be used to give them the proper sign.

491. $\displaystyle\int \frac{(\log x)^n}{x}\,dx = \frac{1}{n+1}(\log x)^{n+1}$

492. $\displaystyle\int \frac{dx}{\log x} = \log(\log x) + \log x + \frac{(\log x)^2}{2\cdot 2!} + \frac{(\log x)^3}{3\cdot 3!} + \cdots$

493. $\displaystyle\int \frac{dx}{x\log x} = \log(\log x)$

494. $\displaystyle\int \frac{dx}{x(\log x)^n} = -\frac{1}{(n-1)(\log x)^{n-1}}$

495. $\displaystyle\int \frac{x^m\,dx}{(\log x)^n} = -\frac{x^{m+1}}{(n-1)(\log x)^{n-1}} + \frac{m+1}{n-1}\int \frac{x^m\,dx}{(\log x)^{n-1}}$

496. $\displaystyle\int x^m(\log x)^n\,dx = \begin{cases} \dfrac{x^{m+1}(\log x)^n}{m+1} - \dfrac{n}{m+1}\displaystyle\int x^m(\log x)^{n-1}\,dx \\ \text{or} \\ (-1)^n \dfrac{n!}{m+1} x^{m+1} \displaystyle\sum_{r=0}^{n} \frac{(-\log x)^r}{r!(m+1)^{n-r}} \end{cases}$

497. $\displaystyle\int x^p \sin(b\,ln\,x)\,dx = \frac{x^{p+1}}{(p+1)^2+b^2}\cdot\left[(p+1)\sin(b\,ln\,x) - b\cos(b\,ln\,x)\right] + c$

498. $\displaystyle\int x^p \cos(b\,ln\,x)\,dx = \frac{x^{p+1}}{(p+1)^2+b^2}\cdot\left[b\sin(b\,ln\,x) + (p+1)\cos(b\,ln\,x)\right] + c$

499. $\displaystyle\int [\log(ax+b)]\,dx = \frac{ax+b}{a}\log(ax+b) - x$

500. $\displaystyle\int \frac{\log(ax+b)}{x^2}\,dx = \frac{a}{b}\log x - \frac{ax+b}{bx}\log(ax+b)$

501. $\displaystyle\int x^m[\log(ax+b)]\,dx = \frac{1}{m+1}\left[x^{m+1} - \left(-\frac{b}{a}\right)^{m+1}\right]\log(ax+b) - \frac{1}{m+1}\left(-\frac{b}{a}\right)^{m+1}\sum_{r=1}^{m+1}\frac{1}{r}\left(-\frac{ax}{b}\right)^r$

502. $\displaystyle\int \frac{\log(ax+b)}{x^m}\,dx = -\frac{1}{m-1}\frac{\log(ax+b)}{x^{m-1}} + \frac{1}{m-1}\left(-\frac{a}{b}\right)^{m-1}\log\frac{ax+b}{x} + \frac{1}{m-1}\left(-\frac{a}{b}\right)^{m-1}\sum_{r=1}^{m-2}\frac{1}{r}\left(-\frac{b}{ax}\right)^r, (m>2)$

503. $\displaystyle\int \left[\log\frac{x+a}{x-a}\right]dx = (x+a)\log(x+a) - (x-a)\log(x-a)$

504. $\displaystyle\int x^m\left[\log\frac{x+a}{x-a}\right]dx = \frac{x^{m+1}-(-a)^{m+1}}{m+1}\log(x+a) - \frac{x^{m+1}-a^{m+1}}{m+1}\log(x-a) + \frac{2a^{m+1}}{m+1}\sum_{r=1}^{\left[\frac{m+1}{2}\right]}\frac{1}{m-2r+2}\left(\frac{x}{a}\right)^{m-2r+2}$

See note integral 392.

505. $\displaystyle\int \frac{1}{x^2}\left[\log\frac{x+a}{x-a}\right]dx = \frac{1}{x}\log\frac{x-a}{x+a} - \frac{1}{a}\log\frac{x^2-a^2}{x^2}$

506. $$\int (\log X)\,dx = \begin{cases} \left(x+\dfrac{b}{2c}\right)\log X - 2x + \dfrac{\sqrt{4ac-b^2}}{c}\tan^{-1}\dfrac{2cx+b}{\sqrt{4ac-b^2}}, \\ \qquad\qquad (b^2-4ac<0) \\ \text{or} \\ \left(x+\dfrac{b}{2c}\right)\log X - 2x + \dfrac{\sqrt{b^2-4ac}}{c}\tanh^{-1}\dfrac{2cx+b}{\sqrt{b^2-4ac}}, \\ \qquad\qquad (b^2-4ac>0) \\ \text{where} \\ X = a+bx+cx^2 \end{cases}$$

507. $\displaystyle\int x^n(\log X)\,dx = \frac{x^{n+1}}{n+1}\log X - \frac{2c}{n+1}\int\frac{x^{n+2}}{X}dx - \frac{b}{n+1}\int\frac{x^{n+1}}{X}dx$

where $X = a + bx + cx^2$

508. $\displaystyle\int [\log(x^2+a^2)]\,dx = x\log(x^2+a^2) - 2x + 2a\tan^{-1}\frac{x}{a}$

509. $\displaystyle\int [\log(x^2-a^2)]\,dx = x\log(x^2-a^2) - 2x + a\log\frac{x+a}{x-a}$

510. $\displaystyle\int x[\log(x^2\pm a^2)]\,dx = \tfrac{1}{2}(x^2\pm a^2)\log(x^2\pm a^2) - \tfrac{1}{2}x^2$

511. $\displaystyle\int [\log(x+\sqrt{x^2\pm a^2})]\,dx = x\log(x+\sqrt{x^2\pm a^2}) - \sqrt{x^2\pm a^2}$

512. $\displaystyle\int x[\log(x+\sqrt{x^2\pm a^2})]\,dx = \left(\frac{x^2}{2}\pm\frac{a^2}{4}\right)\log(x+\sqrt{x^2\pm a^2}) - \frac{x\sqrt{x^2\pm a^2}}{4}$

513. $\displaystyle\int x^m[\log(x+\sqrt{x^2\pm a^2})]\,dx = \frac{x^{m+1}}{m+1}\log(x+\sqrt{x^2\pm a^2})$

$$-\frac{1}{m+1}\int\frac{x^{m+1}}{\sqrt{x^2\pm a^2}}dx$$

514. $\displaystyle\int \frac{\log(x+\sqrt{x^2+a^2})}{x^2}dx = -\frac{\log(x+\sqrt{x^2+a^2})}{x} - \frac{1}{a}\log\frac{a+\sqrt{x^2+a^2}}{x}$

515. $\displaystyle\int \frac{\log(x+\sqrt{x^2-a^2})}{x^2}dx = -\frac{\log(x+\sqrt{x^2-a^2})}{x} + \frac{1}{|a|}\sec^{-1}\frac{x}{a}$

516. $$\int x^n \log(x^2 - a^2)\,dx = \frac{1}{n+1}\Bigg[x^{n+1}\log(x^2 - a^2) - a^{n+1}\log(x - a) - (-a)^{n+1}\log(x + a) - 2\sum_{r=0}^{\left[\frac{n}{2}\right]} \frac{a^{2r}x^{n-2r+1}}{n - 2r + 1}\Bigg]$$

See note integral 392.

EXPONENTIAL FORMS

517. $$\int e^x\,dx = e^x$$

$$\int x^m e^{-ax}\,dx = -e^{-ax}\left[\frac{x^m}{a} + m\frac{x^{m-1}}{a^2} + m(m-1)\frac{x^{m-2}}{a^3} \cdots + \frac{m!x}{1!a^m} + \frac{m!}{a^{m+1}}\right]$$

518. $$\int e^{-x}\,dx = -e^{-x}$$

519. $$\int e^{ax}\,dx = \frac{e^{ax}}{a}$$

520. $$\int x\,e^{ax}\,dx = \frac{e^{ax}}{a^2}(ax - 1)$$

521. $$\int x^m e^{ax}\,dx = \begin{cases} \dfrac{x^m e^{ax}}{a} - \dfrac{m}{a}\displaystyle\int x^{m-1}e^{ax}\,dx \\ \text{or} \\ e^{ax}\displaystyle\sum_{r=0}^{m}(-1)^r \frac{m!x^{m-r}}{(m-r)!a^{r+1}} \end{cases}$$

522. $$\int \frac{e^{ax}\,dx}{x} = \log x + \frac{ax}{1!} + \frac{a^2x^2}{2\cdot 2!} + \frac{a^3x^3}{3\cdot 3!} + \cdots$$

523. $$\int \frac{e^{ax}}{x^m}\,dx = -\frac{1}{m-1}\frac{e^{ax}}{x^{m-1}} + \frac{a}{m-1}\int \frac{e^{ax}}{x^{m-1}}\,dx$$

524. $$\int e^{ax}\log x\,dx = \frac{e^{ax}\log x}{a} - \frac{1}{a}\int \frac{e^{ax}}{x}\,dx$$

525. $$\int \frac{dx}{1 + e^x} = x - \log(1 + e^x) = \log\frac{e^x}{1 + e^x}$$

526. $$\int \frac{dx}{a + be^{px}} = \frac{x}{a} - \frac{1}{ap}\log(a + be^{px})$$

527. $$\int \frac{dx}{ae^{mx} + be^{-mx}} = \frac{1}{m\sqrt{ab}}\tan^{-1}\left(e^{mx}\sqrt{\frac{a}{b}}\right), \qquad (a > 0, b > 0)$$

528. $$\int \frac{dx}{ae^{mx} - be^{-mx}} = \begin{cases} \dfrac{1}{2m\sqrt{ab}}\log\dfrac{\sqrt{a}\,e^{mx} - \sqrt{b}}{\sqrt{a}\,e^{mx} + \sqrt{b}} \\ \text{or} \\ \dfrac{-1}{m\sqrt{ab}}\tanh^{-1}\left(\sqrt{\dfrac{a}{b}}\,e^{mx}\right), \qquad (a > 0, b > 0) \end{cases}$$

529. $\displaystyle\int (a^x - a^{-x})\, dx = \frac{a^x + a^{-x}}{\log a}$

530. $\displaystyle\int \frac{e^{ax}}{b + ce^{ax}}\, dx = \frac{1}{ac}\log(b + ce^{ax})$

531. $\displaystyle\int \frac{x\, e^{ax}}{(1 + ax)^2}\, dx = \frac{e^{ax}}{a^2(1 + ax)}$

532. $\displaystyle\int x\, e^{-x^2}\, dx = -\tfrac{1}{2} e^{-x^2}$

533. $\displaystyle\int e^{ax}[\sin(bx)]\, dx = \frac{e^{ax}[a \sin(bx) - b \cos(bx)]}{a^2 + b^2}$

534. $\displaystyle\int e^{ax}[\sin(bx)][\sin(cx)]\, dx = \frac{e^{ax}[(b - c)\sin(b - c)x + a\cos(b - c)x]}{2[a^2 + (b - c)^2]} - \frac{e^{ax}[(b + c)\sin(b + c)x + a\cos(b + c)x]}{2[a^2 + (b + c)^2]}$

535. $$\int e^{ax}[\sin(bx)][\cos(cx)]\, dx = \begin{cases} \dfrac{e^{ax}[a\sin(b - c)x - (b - c)\cos(b - c)x]}{2[a^2 + (b - c)^2]} + \dfrac{e^{ax}[a\sin(b + c)x - (b + c)\cos(b + c)x]}{2[a^2 + (b + c)^2]} \\ \text{or} \\ \dfrac{e^{ax}}{\rho}[(a\sin bx - b\cos bx)[\cos(cx - \alpha)] - c(\sin bx)\sin(cx - \alpha)] \\ \text{where} \\ \rho = \sqrt{(a^2 + b^2 - c^2)^2 + 4a^2c^2}, \\ \rho\cos\alpha = a^2 + b^2 - c^2, \qquad \rho\sin\alpha = 2ac \end{cases}$$

536. $\displaystyle\int e^{ax}[\sin(bx)][\sin(bx + c)]\, dx$

$$= \frac{e^{ax}\cos c}{2a} - \frac{e^{ax}[a\cos(2bx + c) + 2b\sin(2bx + c)]}{2(a^2 + 4b^2)}$$

537. $\displaystyle\int e^{ax}[\sin(bx)][\cos(bx + c)]\, dx$

$$= \frac{-e^{ax}\sin c}{2a} + \frac{e^{ax}[a\sin(2bx + c) - 2b\cos(2bx + c)]}{2(a^2 + 4b^2)}$$

538. $\displaystyle\int e^{ax}[\cos(bx)]\, dx = \frac{e^{ax}}{a^2 + b^2}[a\cos(bx) + b\sin(bx)]$

539. $$\int e^{ax}[\cos(bx)][\cos(cx)]\,dx = \frac{e^{ax}[(b-c)\sin(b-c)x + a\cos(b-c)x]}{2[a^2+(b-c)^2]} + \frac{e^{ax}[(b+c)\sin(b+c)x + a\cos(b+c)x]}{2[a^2+(b+c)^2]}$$

540. $$\int e^{ax}[\cos(bx)][\cos(bx+c)]\,dx = \frac{e^{ax}\cos c}{2a} + \frac{e^{ax}[a\cos(2bx+c) + 2b\sin(2bx+c)]}{2(a^2+4b^2)}$$

541. $$\int e^{ax}[\cos(bx)][\sin(bx+c)]\,dx = \frac{e^{ax}\sin c}{2a} + \frac{e^{ax}[a\sin(2bx+c) - 2b\cos(2bx+c)]}{2(a^2+4b^2)}$$

542. $$\int e^{ax}[\sin^n bx]\,dx = \frac{1}{a^2+n^2b^2}\left[(a\sin bx - nb\cos bx)\,e^{ax}\sin^{n-1} bx + n(n-1)b^2\int e^{ax}[\sin^{n-2} bx]\,dx\right]$$

543. $$\int e^{ax}[\cos^n bx]\,dx = \frac{1}{a^2+n^2b^2}\left[(a\cos bx + nb\sin bx)\,e^{ax}\cos^{n-1} bx + n(n-1)b^2\int e^{ax}[\cos^{n-2} bx]\,dx\right]$$

544. $$\int x^m e^x \sin x\,dx = \frac{1}{2}x^m e^x(\sin x - \cos x) - \frac{m}{2}\int x^{m-1} e^x \sin x\,dx + \frac{m}{2}\int x^{m-1} e^x \cos x\,dx$$

545. $$\int x^m e^{ax}[\sin bx]\,dx = \begin{cases} x^m e^{ax}\dfrac{a\sin bx - b\cos bx}{a^2+b^2} - \dfrac{m}{a^2+b^2}\displaystyle\int x^{m-1} e^{ax}(a\sin bx - b\cos bx)\,dx \\ \text{or} \\ e^{ax}\displaystyle\sum_{r=0}^{m} \frac{(-1)^r m!\,x^{m-r}}{\rho^{r+1}(m-r)!}\sin[bx - (r+1)\alpha] \\ \text{where} \\ \rho = \sqrt{a^2+b^2}, \qquad \rho\cos\alpha = a, \qquad \rho\sin\alpha = b \end{cases}$$

546. $$\int x^m e^x \cos x\,dx = \frac{1}{2}x^m e^x(\sin x + \cos x) - \frac{m}{2}\int x^{m-1} e^x \sin x\,dx - \frac{m}{2}\int x^{m-1} e^x \cos x\,dx$$

547. $$\int x^m e^{ax}\cos bx\,dx = \begin{cases} x^m e^{ax}\dfrac{a\cos bx + b\sin bx}{a^2+b^2} \\ \qquad -\dfrac{m}{a^2+b^2}\displaystyle\int x^{m-1}e^{ax}(a\cos bx + b\sin bx)\,dx \\ \qquad \text{or} \\ e^{ax}\displaystyle\sum_{r=0}^{m}\dfrac{(-1)^r m!\,x^{m-r}}{\rho^{r+1}(m-r)!}\cos[bx-(r+1)\alpha] \\ \qquad \text{where} \\ \rho = \sqrt{a^2+b^2}, \quad \rho\cos\alpha = a, \quad \rho\sin\alpha = b \end{cases}$$

548. $$\int e^{ax}(\cos^m x)(\sin^n x)\,dx = \begin{cases} \dfrac{e^{ax}\cos^{m-1}x\sin^n x[a\cos x + (m+n)\sin x]}{(m+n)^2+a^2} \\ \qquad -\dfrac{na}{(m+n)^2+a^2}\displaystyle\int e^{ax}(\cos^{m-1}x)(\sin^{n-1}x)\,dx \\ \qquad +\dfrac{(m-1)(m+n)}{(m+n)^2+a^2}\displaystyle\int e^{ax}(\cos^{m-2}x)(\sin^n x)\,dx \\ \qquad \text{or} \\ \dfrac{e^{ax}\cos^m x\sin^{n-1}x[a\sin x - (m+n)\cos x]}{(m+n)^2+a^2} \\ \qquad +\dfrac{ma}{(m+n)^2+a^2}\displaystyle\int e^{ax}(\cos^{m-1}x)(\sin^{n-1}x)\,dx \\ \qquad +\dfrac{(n-1)(m+n)}{(m+n)^2+a^2}\displaystyle\int e^{ax}(\cos^m x)(\sin^{n-2}x)\,dx \\ \qquad \text{or} \\ \dfrac{e^{ax}(\cos^{m-1}x)(\sin^{n-1}x)(a\sin x\cos x + m\sin^2 x - n\cos^2 x)}{(m+n)^2+a^2} \\ \qquad +\dfrac{m(m-1)}{(m+n)^2+a^2}\displaystyle\int e^{ax}(\cos^{m-2}x)(\sin^n x)\,dx \\ \qquad +\dfrac{n(n-1)}{(m+n)^2+a^2}\displaystyle\int e^{ax}(\cos^m x)(\sin^{n-2}x)\,dx \\ \qquad \text{or} \\ \dfrac{e^{ax}(\cos^{m-1}x)(\sin^{n-1}x)(a\cos x\sin x + m\sin^2 x - n\cos^2 x)}{(m+n)^2+a^2} \\ \qquad +\dfrac{m(m-1)}{(m+n)^2+a^2}\displaystyle\int e^{ax}(\cos^{m-2}x)(\sin^{n-2}x)\,dx \\ \qquad +\dfrac{(n-m)(n+m-1)}{(m+n)^2+a^2}\displaystyle\int e^{ax}(\cos^m x)(\sin^{n-2}x)\,dx \end{cases}$$

549. $$\int x\,e^{ax}(\sin bx)\,dx = \frac{x\,e^{ax}}{a^2+b^2}(a\sin bx - b\cos bx) - \frac{e^{ax}}{(a^2+b^2)^2}[(a^2-b^2)\sin bx - 2ab\cos bx]$$

550. $$\int x\,e^{ax}(\cos bx)\,dx = \frac{x\,e^{ax}}{a^2+b^2}(a\cos bx + b\sin bx) - \frac{e^{ax}}{(a^2+b^2)^2}[(a^2-b^2)\cos bx + 2ab\sin bx]$$

551. $$\int \frac{e^{ax}}{\sin^n x}dx = -\frac{e^{ax}[a\sin x + (n-2)\cos x]}{(n-1)(n-2)\sin^{n-1}x} + \frac{a^2+(n-2)^2}{(n-1)(n-2)}\int \frac{e^{ax}}{\sin^{n-2}x}dx$$

552. $$\int \frac{e^{ax}}{\cos^n x}dx = -\frac{e^{ax}[a\cos x - (n-2)\sin x]}{(n-1)(n-2)\cos^{n-1}x} + \frac{a^2+(n-2)^2}{(n-1)(n-2)}\int \frac{e^{ax}}{\cos^{n-2}x}dx$$

553. $$\int e^{ax}\tan^n x\,dx = e^{ax}\frac{\tan^{n-1}x}{n-1} - \frac{a}{n-1}\int e^{ax}\tan^{n-1}x\,dx - \int e^{ax}\tan^{n-2}x\,dx$$

HYPERBOLIC FORMS

554. $$\int (\sinh x)\,dx = \cosh x$$

555. $$\int (\cosh x)\,dx = \sinh x$$

556. $$\int (\tanh x)\,dx = \log\cosh x$$

557. $$\int (\coth x)\,dx = \log\sinh x$$

558. $$\int (\operatorname{sech} x)\,dx = \tan^{-1}(\sinh x)$$

559. $$\int \operatorname{csch} x\,dx = \log\tanh\left(\frac{x}{2}\right)$$

560. $$\int x(\sinh x)\,dx = x\cosh x - \sinh x$$

561. $$\int x^n(\sinh x)\,dx = x^n\cosh x - n\int x^{n-1}(\cosh x)\,dx$$

562. $$\int x(\cosh x)\,dx = x\sinh x - \cosh x$$

563. $$\int x^n(\cosh x)\,dx = x^n\sinh x - n\int x^{n-1}(\sinh x)\,dx$$

564. $\int (\operatorname{sech} x)(\tanh x)\,dx = -\operatorname{sech} x$

565. $\int (\operatorname{csch} x)(\coth x)\,dx = -\operatorname{csch} x$

566. $\int (\sinh^2 x)\,dx = \frac{\sinh 2x}{4} - \frac{x}{2}$

567. $$\int (\sinh^m x)(\cosh^n x)\,dx = \begin{cases} \frac{1}{m+n}(\sinh^{m+1} x)(\cosh^{n-1} x) \\ \qquad + \frac{n-1}{m+n}\int (\sinh^m x)(\cosh^{n-2} x)\,dx \\ \qquad \text{or} \\ \frac{1}{m+n}\sinh^{m-1} x \cosh^{n+1} x \\ \qquad - \frac{m-1}{m+n}\int (\sinh^{m-2} x)(\cosh^n x)\,dx, \qquad (m+n \neq 0) \end{cases}$$

568. $$\int \frac{dx}{(\sinh^m x)(\cosh^n x)} = \begin{cases} -\frac{1}{(m-1)(\sinh^{m-1} x)(\cosh^{n-1} x)} \\ \qquad - \frac{m+n-2}{m-1}\int \frac{dx}{(\sinh^{m-2} x)(\cosh^n x)}, \qquad (m \neq 1) \\ \qquad \text{or} \\ \frac{1}{(n-1)\sinh^{m-1} x \cosh^{n-1} x} \\ \qquad + \frac{m+n-2}{n-1}\int \frac{dx}{(\sinh^m x)(\cosh^{n-2} x)}, \qquad (n \neq 1) \end{cases}$$

569. $\int (\tanh^2 x)\,dx = x - \tanh x$

570. $\int (\tanh^n x)\,dx = -\frac{\tanh^{n-1} x}{n-1} + \int (\tanh^{n-2} x)\,dx, \qquad (n \neq 1)$

571. $\int (\operatorname{sech}^2 x)\,dx = \tanh x$

572. $\int (\cosh^2 x)\,dx = \frac{\sinh 2x}{4} + \frac{x}{2}$

573. $\int (\coth^2 x)\,dx = x - \coth x$

574. $\int (\coth^n x)\,dx = -\frac{\coth^{n-1} x}{n-1} + \int \coth^{n-2} x\,dx, \qquad (n \neq 1)$

575. $\int (\operatorname{csch}^2 x)\, dx = -\operatorname{ctnh} x$

576. $\int (\sinh mx)(\sinh nx)\, dx = \frac{\sinh (m+n)x}{2(m+n)} - \frac{\sinh (m-n)x}{2(m-n)}, \qquad (m^2 \neq n^2)$

577. $\int (\cosh mx)(\cosh nx)\, dx = \frac{\sinh (m+n)x}{2(m+n)} + \frac{\sinh (m-n)x}{2(m-n)}, \qquad (m^2 \neq n^2)$

578. $\int (\sinh mx)(\cosh nx)\, dx = \frac{\cosh (m+n)x}{2(m+n)} + \frac{\cosh (m-n)x}{2(m-n)}, \qquad (m^2 \neq n^2)$

579. $\int \left(\sinh^{-1} \frac{x}{a}\right) dx = x \sinh^{-1} \frac{x}{a} - \sqrt{x^2 + a^2}, \qquad (a > 0)$

580. $\int x\left(\sinh^{-1} \frac{x}{a}\right) dx = \left(\frac{x^2}{2} + \frac{a^2}{4}\right) \sinh^{-1} \frac{x}{a} - \frac{x}{4}\sqrt{x^2 + a^2}, \qquad (a > 0)$

581. $\int x^n(\sinh^{-1} x)\, dx = \frac{x^{n+1}}{n+1} \sinh^{-1} x - \frac{1}{n+1} \int \frac{x^{n+1}}{(1+x^2)^{\frac{1}{2}}} dx, \qquad (n \neq -1)$

582. $$\int \left(\cosh^{-1} \frac{x}{a}\right) dx = \begin{cases} x \cosh^{-1} \frac{x}{a} - \sqrt{x^2 - a^2}, & \left(\cosh^{-1} \frac{x}{a} > 0\right) \\ \text{or} & \\ x \cosh^{-1} \frac{x}{a} + \sqrt{x^2 - a^2}, & \left(\cosh^{-1} \frac{x}{a} < 0\right), \qquad (a > 0) \end{cases}$$

583. $\int x\left(\cosh^{-1} \frac{x}{a}\right) dx = \frac{2x^2 - a^2}{4} \cosh^{-1} \frac{x}{a} - \frac{x}{4}(x^2 - a^2)^{\frac{1}{2}}$

584. $\int x^n(\cosh^{-1} x)\, dx = \frac{x^{n+1}}{n+1} \cosh^{-1} x - \frac{1}{n+1} \int \frac{x^{n+1}}{(x^2-1)^{\frac{1}{2}}} dx, \qquad (n \neq -1)$

585. $\int \left(\tanh^{-1} \frac{x}{a}\right) dx = x \tanh^{-1} \frac{x}{a} + \frac{a}{2} \log (a^2 - x^2), \qquad \left(\left|\frac{x}{a}\right| < 1\right)$

586. $\int \left(\coth^{-1} \frac{x}{a}\right) dx = x \coth^{-1} \frac{x}{a} + \frac{a}{2} \log (x^2 - a^2), \qquad \left(\left|\frac{x}{a}\right| > 1\right)$

587. $\int x\left(\tanh^{-1} \frac{x}{a}\right) dx = \frac{x^2 - a^2}{2} \tanh^{-1} \frac{x}{a} + \frac{ax}{2}, \qquad \left(\left|\frac{x}{a}\right| < 1\right)$

588. $\int x^n\left(\tanh^{-1} x\right) dx = \frac{x^{n+1}}{n+1} \tanh^{-1} x - \frac{1}{n+1} \int \frac{x^{n+1}}{1-x^2} dx, \qquad (n \neq -1)$

589. $\int x\left(\coth^{-1} \frac{x}{a}\right) dx = \frac{x^2 - a^2}{2} \coth^{-1} \frac{x}{a} + \frac{ax}{2}, \qquad \left(\left|\frac{x}{a}\right| > 1\right)$

590. $\int x^n(\coth^{-1} x)\, dx = \frac{x^{n+1}}{n+1} \coth^{-1} x + \frac{1}{n+1} \int \frac{x^{n+1}}{x^2-1} dx, \qquad (n \neq -1)$

591. $\int (\text{sech}^{-1} x)\, dx = x \,\text{sech}^{-1} x + \sin^{-1} x$

592. $\int x \,\text{sech}^{-1} x \, dx = \frac{x^2}{2} \text{sech}^{-1} x - \frac{1}{2}(1 - x^2)$

593. $\int x^n \,\text{sech}^{-1} x \, dx = \frac{x^{n+1}}{n+1} \text{sech}^{-1} x + \frac{1}{n+1} \int \frac{x^n}{(1 - x^2)^{\frac{1}{2}}} dx, \qquad (n \neq -1)$

594. $\int \text{csch}^{-1} x \, dx = x \,\text{csch}^{-1} x + \frac{x}{|x|} \sinh^{-1} x$

595. $\int x \,\text{csch}^{-1} x \, dx = \frac{x^2}{2} \text{csch}^{-1} x + \frac{1}{2} \frac{x}{|x|} \sqrt{1 + x^2}$

596. $\int x^n \,\text{csch}^{-1} x \, dx = \frac{x^{n+1}}{n+1} \text{csch}^{-1} x + \frac{1}{n+1} \frac{x}{|x|} \int \frac{x^n}{(x^2 + 1)^{\frac{1}{2}}} dx, \qquad (n \neq -1)$

DEFINITE INTEGRALS

597. $\int_0^\infty x^{n-1} e^{-x} \, dx = \int_0^1 \left(\log \frac{1}{x}\right)^{n-1} dx = \frac{1}{n} \prod_{m=1}^{\infty} \frac{\left(1 + \frac{1}{m}\right)^n}{1 + \frac{n}{m}}$

$= \Gamma(n),\ n \neq 0, -1, -2, -3, \ldots \qquad$ (Gamma Function)

598. $\int_0^\infty t^n p^{-t} \, dt = \frac{n!}{(\log p)^{n+1}}, \qquad (n = 0, 1, 2, 3, \ldots \text{ and } p > 0)$

599. $\int_0^\infty t^{n-1} e^{-(a+1)t} \, dt = \frac{\Gamma(n)}{(a+1)^n}, \qquad (n > 0, a > -1)$

600. $\int_0^1 x^m \left(\log \frac{1}{x}\right)^n dx = \frac{\Gamma(n+1)}{(m+1)^{n+1}}, \qquad (m > -1, n > -1)$

601. $\Gamma(n)$ is finite if $n > 0$, $\Gamma(n+1) = n\Gamma(n)$

602. $\Gamma(n) \cdot \Gamma(1 - n) = \frac{\pi}{\sin n\pi}$

603. $\Gamma(n) = (n - 1)!$ if $n =$ integer > 0

604. $\Gamma(\frac{1}{2}) = 2 \int_0^\infty e^{-t^2} \, dt = \sqrt{\pi} = 1.7724538509 \cdots = (-\frac{1}{2})!$

605. $\Gamma(n + \frac{1}{2}) = \frac{1 \cdot 3 \cdot 5 \ldots (2n - 1)}{2^n} \sqrt{\pi} \qquad n = 1, 2, 3, \ldots$

606. $\Gamma(-n + \frac{1}{2}) = \frac{(-1)^n 2^n \sqrt{\pi}}{1 \cdot 3 \cdot 5 \ldots (2n - 1)} \qquad n = 1, 2, 3, \ldots$

607. $\displaystyle\int_0^1 x^{m-1}(1-x)^{n-1}\,dx = \int_0^\infty \frac{x^{m-1}}{(1+x)^{m+n}}\,dx = \frac{\Gamma(m)\Gamma(n)}{\Gamma(m+n)} = B(m, n)$

(Beta function)

608. $B(m, n) = B(n, m) = \dfrac{\Gamma(m)\Gamma(n)}{\Gamma(m+n)}$, where m and n are any positive real numbers.

609. $\displaystyle\int_a^b (x-a)^m(b-x)^n\,dx = (b-a)^{m+n+1}\frac{\Gamma(m+1)\cdot\Gamma(n+1)}{\Gamma(m+n+2)},$

$(m > -1, n > -1, b > a)$

610. $\displaystyle\int_1^\infty \frac{dx}{x^m} = \frac{1}{m-1}, \qquad [m > 1]$

611. $\displaystyle\int_0^\infty \frac{dx}{(1+x)x^p} = \pi \csc p\pi, \qquad [p < 1]$

612. $\displaystyle\int_0^\infty \frac{dx}{(1-x)x^p} = -\pi \cot p\pi, \qquad [p < 1]$

613. $\displaystyle\int_0^\infty \frac{x^{p-1}\,dx}{1+x} = \frac{\pi}{\sin p\pi}$

$= B(p, 1-p) = \Gamma(p)\Gamma(1-p), \qquad [0 < p < 1]$

614. $\displaystyle\int_0^\infty \frac{x^{m-1}\,dx}{1+x^n} = \frac{\pi}{n \sin \dfrac{m\pi}{n}}, \qquad [0 < m < n]$

615. $\displaystyle\int_0^\infty \frac{x^a\,dx}{(m+x^b)^c} = \frac{m^{\frac{a+1-bc}{b}}}{b}\left[\frac{\Gamma\left(\frac{a+1}{b}\right)\Gamma\left(c-\frac{a+1}{b}\right)}{\Gamma(c)}\right]$

$\left(a > -1, b > 0, m > 0, c > \dfrac{a+1}{b}\right)$

616. $\displaystyle\int_0^\infty \frac{dx}{(1+x)\sqrt{x}} = \pi$

617. $\displaystyle\int_0^\infty \frac{a\,dx}{a^2+x^2} = \frac{\pi}{2}$, if $a > 0$; 0, if $a = 0$; $-\dfrac{\pi}{2}$, if $a < 0$

618. $\displaystyle\int_0^a (a^2-x^2)^{\frac{n}{2}}\,dx = \frac{1}{2}\int_{-a}^a (a^2-x^2)^{\frac{n}{2}}\,dx = \frac{1\cdot 3\cdot 5\ldots n}{2\cdot 4\cdot 6\ldots(n+1)}\cdot\frac{\pi}{2}\cdot a^{n+1} \qquad (n \text{ odd})$

619. $\displaystyle\int_0^a x^m(a^2-x^2)^{\frac{n}{2}}\,dx = \begin{cases} \dfrac{1}{2}a^{m+n+1}B\left(\dfrac{m+1}{2}, \dfrac{n+2}{2}\right) \\ \text{or} \\ \dfrac{1}{2}a^{m+n+1}\dfrac{\Gamma\left(\frac{m+1}{2}\right)\Gamma\left(\frac{n+2}{2}\right)}{\Gamma\left(\frac{m+n+3}{2}\right)} \end{cases}$

620. $$\int_0^{\pi/2} (\sin^n x)\,dx = \begin{cases} \displaystyle\int_0^{\pi/2} (\cos^n x)\,dx \\ \text{or} \\ \dfrac{1\cdot 3\cdot 5\cdot 7\ldots(n-1)}{2\cdot 4\cdot 6\cdot 8\ldots(n)}\dfrac{\pi}{2}, & (n \text{ an even integer, } n \neq 0) \\ \text{or} \\ \dfrac{2\cdot 4\cdot 6\cdot 8\ldots(n-1)}{1\cdot 3\cdot 5\cdot 7\ldots(n)}, & (n \text{ an odd integer, } n \neq 1) \\ \text{or} \\ \dfrac{\sqrt{\pi}}{2}\dfrac{\Gamma\left(\dfrac{n+1}{2}\right)}{\Gamma\left(\dfrac{n}{2}+1\right)}, & (n > -1) \end{cases}$$

621. $$\int_0^\infty \frac{\sin mx\,dx}{x} = \frac{\pi}{2}, \text{ if } m > 0;\ 0, \text{ if } m = 0;\ -\frac{\pi}{2}, \text{ if } m < 0$$

622. $$\int_0^\infty \frac{\cos x\,dx}{x} = \infty$$

623. $$\int_0^\infty \frac{\tan x\,dx}{x} = \frac{\pi}{2}$$

624. $$\int_0^\pi \sin ax \cdot \sin bx\,dx = \int_0^\pi \cos ax \cdot \cos bx\,dx = 0, \qquad (a \neq b;\, a, b \text{ integers})$$

625. $$\int_0^{\pi/a} [\sin(ax)][\cos(ax)]\,dx = \int_0^\pi [\sin(ax)][\cos(ax)]\,dx = 0$$

626. $$\int_0^\pi [\sin(ax)][\cos(bx)]\,dx = \frac{2a}{a^2 - b^2}, \text{ if } a - b \text{ is odd, or 0 if } a - b \text{ is even}$$

627. $$\int_0^\infty \frac{\sin x \cos mx\,dx}{x}$$

$$= 0, \text{ if } m < -1 \text{ or } m > 1;\ \frac{\pi}{4}, \text{ if } m = \pm 1;\ \frac{\pi}{2}, \text{ if } m^2 < 1$$

628. $$\int_0^\infty \frac{\sin ax \sin bx}{x^2}\,dx = \frac{\pi a}{2}, \qquad (a \leq b)$$

629. $$\int_0^\pi \sin^2 mx\,dx = \int_0^\pi \cos^2 mx\,dx = \frac{\pi}{2}$$

630. $$\int_0^\infty \frac{\sin^2 (px)}{x^2}\,dx = \frac{\pi p}{2}$$

631. $$\int_0^\infty \frac{\sin x}{x^p}\,dx = \frac{\pi}{2\Gamma(p)\sin(p\pi/2)}, \qquad 0 < p < 1$$

632. $$\int_0^\infty \frac{\cos x}{x^p}\,dx = \frac{\pi}{2\Gamma(p)\cos(p\pi/2)}, \qquad 0 < p < 1$$

633. $$\int_0^\infty \frac{1 - \cos px}{x^2}\,dx = \frac{\pi p}{2}$$

634. $$\int_0^\infty \frac{\sin px \cos qx}{x}\,dx = \left\{0, \quad q > p > 0; \quad \frac{\pi}{2}, \quad p > q > 0; \quad \frac{\pi}{4}, \quad p = q > 0\right\}$$

635. $$\int_0^\infty \frac{\cos(mx)}{x^2 + a^2}\,dx = \frac{\pi}{2|a|} e^{-|ma|}$$

636. $$\int_0^\infty \cos(x^2)\,dx = \int_0^\infty \sin(x^2)\,dx = \frac{1}{2}\sqrt{\frac{\pi}{2}}$$ Fresnel integrals

637. $$\int_0^\infty \sin ax^n\,dx = \frac{1}{na^{1/n}}\Gamma(1/n)\sin\frac{\pi}{2n}, \qquad n > 1$$

638. $$\int_0^\infty \cos ax^n\,dx = \frac{1}{na^{1/n}}\Gamma(1/n)\cos\frac{\pi}{2n}, \qquad n > 1$$

639. $$\int_0^\infty \frac{\sin x}{\sqrt{x}}\,dx = \int_0^\infty \frac{\cos x}{\sqrt{x}}\,dx = \sqrt{\frac{\pi}{2}}$$

640. $$\int_0^\infty \frac{\sin^3 x}{x^2}\,dx = \frac{3}{4}\log 3$$

641. $$\int_0^\infty \frac{\sin^3 x}{x^3}\,dx = \frac{3\pi}{8}$$

642. $$\int_0^\infty \frac{\sin^4 x}{x^4}\,dx = \frac{\pi}{3}$$

643. $$\int_0^{\pi/2} \frac{dx}{1 + a\cos x} = \frac{\cos^{-1} a}{\sqrt{1 - a^2}}, \qquad (a < 1)$$

644. $$\int_0^{\pi} \frac{dx}{a + b\cos x} = \frac{\pi}{\sqrt{a^2 - b^2}}, \qquad (a > b \geq 0)$$

645. $$\int_0^{2\pi} \frac{dx}{1 + a\cos x} = \frac{2\pi}{\sqrt{1 - a^2}}, \qquad (a^2 < 1)$$

646. $$\int_0^\infty \frac{\cos ax - \cos bx}{x}\,dx = \log\frac{b}{a}$$

647. $$\int_0^{\pi/2} \frac{dx}{a^2\sin^2 x + b^2\cos^2 x} = \frac{\pi}{2ab}$$

648. $\displaystyle\int_0^{\pi/2} \frac{dx}{(a^2 \sin^2 x + b^2 \cos^2 x)^2} = \frac{\pi(a^2 + b^2)}{4a^3b^3}, \qquad (a, b > 0)$

649. $\displaystyle\int_0^{\pi/2} \sin^{n-1} x \cos^{m-1} x \, dx = \frac{1}{2} \mathrm{B}\left(\frac{n}{2}, \frac{m}{2}\right),$ m and n positive integers

650. $\displaystyle\int_0^{\pi/2} (\sin^{2n+1} \theta) \, d\theta = \frac{2 \cdot 4 \cdot 6 \ldots (2n)}{1 \cdot 3 \cdot 5 \ldots (2n+1)}, \qquad (n = 1, 2, 3 \ldots)$

651. $\displaystyle\int_0^{\pi/2} (\sin^{2n} \theta) \, d\theta = \frac{1 \cdot 3 \cdot 5 \ldots (2n-1)}{2 \cdot 4 \ldots (2n)} \left(\frac{\pi}{2}\right), \qquad (n = 1, 2, 3 \ldots)$

652. $\displaystyle\int_0^{\pi/2} \frac{x}{\sin x} \, dx = 2\left\{\frac{1}{1^2} - \frac{1}{3^2} + \frac{1}{5^2} - \frac{1}{7^2} + \cdots\right\}$

653. $\displaystyle\int_0^{\pi/2} \frac{dx}{1 + \tan^m x} = \frac{\pi}{4}$

654. $\displaystyle\int_0^{\pi/2} \sqrt{\cos \theta} \, d\theta = \frac{(2\pi)^{\frac{3}{2}}}{[\Gamma(\frac{1}{4})]^2}$

655. $\displaystyle\int_0^{\pi/2} (\tan^h \theta) \, d\theta = \frac{\pi}{2 \cos\left(\frac{h\pi}{2}\right)}, \qquad (0 < h < 1)$

656. $\displaystyle\int_0^{\infty} \frac{\tan^{-1}(ax) - \tan^{-1}(bx)}{x} \, dx = \frac{\pi}{2} \log \frac{a}{b}, \qquad (a, b > 0)$

657. The area enclosed by a curve defined through the equation $x^{\frac{b}{c}} + y^{\frac{b}{c}} = a^{\frac{b}{c}}$ where $a > 0$, c a positive odd integer and b a positive even integer is given by

$$\frac{\left[\Gamma\left(\frac{c}{b}\right)\right]^2}{\Gamma\left(\frac{2c}{b}\right)} \left(\frac{2ca^2}{b}\right)$$

658. $I = \displaystyle\iiint\limits_R x^{h-1} y^{m-1} z^{n-1} \, dv$, where R denotes the region of space bounded by the co-ordinate planes and that portion of the surface $\left(\frac{x}{a}\right)^p + \left(\frac{y}{b}\right)^q + \left(\frac{z}{c}\right)^k = 1,$ which lies in the first octant and where $h, m, n, p, q, k, a, b, c,$ denote positive real numbers is given by

$$\int_0^a x^{h-1} \, dx \int_0^{b\left[1 - \left(\frac{x}{a}\right)^p\right]^{\frac{1}{q}}} y^m \, dy \int_0^{c\left[1 - \left(\frac{x}{a}\right)^p - \left(\frac{y}{b}\right)^q\right]^{\frac{1}{k}}} z^{n-1} \, dz$$

$$= \frac{a^h b^m c^n}{pqk} \frac{\Gamma\left(\frac{h}{p}\right) \Gamma\left(\frac{m}{q}\right) \Gamma\left(\frac{n}{k}\right)}{\Gamma\left(\frac{h}{p} + \frac{m}{q} + \frac{n}{k} + 1\right)}$$

659. $\int_0^\infty e^{-ax}\,dx = \frac{1}{a}, \quad (a > 0)$

660. $\int_0^\infty \frac{e^{-ax} - e^{-bx}}{x}\,dx = \log\frac{b}{a}, \quad (a, b > 0)$

661. $\int_0^\infty x^n e^{-ax}\,dx = \begin{cases} \frac{\Gamma(n+1)}{a^{n+1}}, & (n > -1, a > 0) \\ \text{or} & \\ \frac{n!}{a^{n+1}}, & (a > 0, n \text{ positive integer}) \end{cases}$

662. $\int_0^\infty x^n \exp(-ax^p)\,dx = \frac{\Gamma(k)}{pa^k}, \quad \left(n > -1, p > 0, a > 0, k = \frac{n+1}{p}\right)$

663. $\int_0^\infty e^{-a^2x^2}\,dx = \frac{1}{2a}\sqrt{\pi} = \frac{1}{2a}\Gamma\left(\frac{1}{2}\right), \quad (a > 0)$

664. $\int_0^\infty x\,e^{-x^2}\,dx = \frac{1}{2}$

665. $\int_0^\infty x^2\,e^{-x^2}\,dx = \frac{\sqrt{\pi}}{4}$

666. $\int_0^\infty x^{2n}\,e^{-ax^2}\,dx = \frac{1 \cdot 3 \cdot 5 \ldots (2n-1)}{2^{n+1}a^n}\sqrt{\frac{\pi}{a}}$

667. $\int_0^\infty x^{2n+1}\,e^{-ax^2}\,dx = \frac{n!}{2a^{n+1}}, \quad (a > 0)$

668. $\int_0^1 x^m\,e^{-ax}\,dx = \frac{m!}{a^{m+1}}\left[1 - e^{-a}\sum_{r=0}^{m}\frac{a^r}{r!}\right]$

669. $\int_0^\infty e^{\left(-x^2 - \frac{a^2}{x^2}\right)}\,dx = \frac{e^{-2a}\sqrt{\pi}}{2}, \quad (a \geq 0)$

670. $\int_0^\infty e^{-nx}\sqrt{x}\,dx = \frac{1}{2n}\sqrt{\frac{\pi}{n}}$

671. $\int_0^\infty \frac{e^{-nx}}{\sqrt{x}}\,dx = \sqrt{\frac{\pi}{n}}$

672. $\int_0^\infty e^{-ax}(\cos mx)\,dx = \frac{a}{a^2 + m^2}, \quad (a > 0)$

673. $\int_0^\infty e^{-ax}(\sin mx)\,dx = \frac{m}{a^2 + m^2}, \quad (a > 0)$

674. $\displaystyle\int_0^{\infty} x\, e^{-ax}[\sin(bx)]\, dx = \frac{2ab}{(a^2+b^2)^2}, \qquad (a > 0)$

675. $\displaystyle\int_0^{\infty} x\, e^{-ax}[\cos(bx)]\, dx = \frac{a^2-b^2}{(a^2+b^2)^2}, \qquad (a > 0)$

676. $\displaystyle\int_0^{\infty} x^n\, e^{-ax}[\sin(bx)]\, dx = \frac{n![(a-ib)^{n+1}-(a+ib)^{n+1}]}{2i(a^2+b^2)^{n+1}}, \qquad (i^2 = -1, a > 0)$

677. $\displaystyle\int_0^{\infty} x^n\, e^{-ax}[\cos(bx)]\, dx = \frac{n![(a-ib)^{n+1}+(a+ib)^{n+1}]}{2(a^2+b^2)^{n+1}}, \qquad (i^2 = -1, a > 0)$

678. $\displaystyle\int_0^{\infty} \frac{e^{-ax}\sin x}{x}\, dx = \cot^{-1} a, \qquad (a > 0)$

679. $\displaystyle\int_0^{\infty} e^{-a^2x^2}\cos bx\, dx = \frac{\sqrt{\pi}}{2a}\exp\left(-\frac{b^2}{4a^2}\right), \qquad (ab \neq 0)$

680. $\displaystyle\int_0^{\infty} e^{-t\cos\phi}\, t^{b-1}\sin(t\sin\phi)\, dt = [\Gamma(b)]\sin(b\phi), \qquad \left(b > 0, -\frac{\pi}{2} < \phi < \frac{\pi}{2}\right)$

681. $\displaystyle\int_0^{\infty} e^{-t\cos\phi}\, t^{b-1}[\cos(t\sin\phi)]\, dt = [\Gamma(b)]\cos(b\phi), \qquad \left(b > 0, -\frac{\pi}{2} < \phi < \frac{\pi}{2}\right)$

682. $\displaystyle\int_0^{\infty} t^{b-1}\cos t\, dt = [\Gamma(b)]\cos\left(\frac{b\pi}{2}\right), \qquad (0 < b < 1)$

683. $\displaystyle\int_0^{\infty} t^{b-1}(\sin t)\, dt = [\Gamma(b)]\sin\left(\frac{b\pi}{2}\right), \qquad (0 < b < 1)$

684. $\displaystyle\int_0^{1} (\log x)^n\, dx = (-1)^n \cdot n!$

685. $\displaystyle\int_0^{1} \left(\log\frac{1}{x}\right)^{\frac{1}{2}} dx = \frac{\sqrt{\pi}}{2}$

686. $\displaystyle\int_0^{1} \left(\log\frac{1}{x}\right)^{-\frac{1}{2}} dx = \sqrt{\pi}$

687. $\displaystyle\int_0^{1} \left(\log\frac{1}{x}\right)^{n} dx = n!$

688. $\displaystyle\int_0^{1} x\log(1-x)\, dx = -\tfrac{3}{4}$

689. $\displaystyle\int_0^{1} x\log(1+x)\, dx = \tfrac{1}{4}$

690. $\displaystyle\int_0^{1} x^m(\log x)^n\, dx = \frac{(-1)^n n!}{(m+1)^{n+1}}, \qquad m > -1, n = 0, 1, 2, \ldots$

If $n \neq 0, 1, 2, \ldots$ replace $n!$ by $\Gamma(n+1)$.

691. $\int_0^1 \frac{\log x}{1+x}\,dx = -\frac{\pi^2}{12}$

692. $\int_0^1 \frac{\log x}{1-x}\,dx = -\frac{\pi^2}{6}$

693. $\int_0^1 \frac{\log(1+x)}{x}\,dx = \frac{\pi^2}{12}$

694. $\int_0^1 \frac{\log(1-x)}{x}\,dx = -\frac{\pi^2}{6}$

695. $\int_0^1 (\log x)[\log(1+x)]\,dx = 2 - 2\log 2 - \frac{\pi^2}{12}$

696. $\int_0^1 (\log x)[\log(1-x)]\,dx = 2 - \frac{\pi^2}{6}$

697. $\int_0^1 \frac{\log x}{1-x^2}\,dx = -\frac{\pi^2}{8}$

698. $\int_0^1 \log\left(\frac{1+x}{1-x}\right)\cdot\frac{dx}{x} = \frac{\pi^2}{4}$

699. $\int_0^1 \frac{\log x\,dx}{\sqrt{1-x^2}} = -\frac{\pi}{2}\log 2$

700. $\int_0^1 x^m\left[\log\left(\frac{1}{x}\right)\right]^n dx = \frac{\Gamma(n+1)}{(m+1)^{n+1}},$ if $m+1>0,\ n+1>0$

701. $\int_0^1 \frac{(x^p - x^q)\,dx}{\log x} = \log\left(\frac{p+1}{q+1}\right),$ $(p+1>0, q+1>0)$

702. $\int_0^1 \frac{dx}{\sqrt{\log\left(\frac{1}{x}\right)}} = \sqrt{\pi}$, (same as integral 686)

703. $\int_0^\infty \log\left(\frac{e^x+1}{e^x-1}\right)dx = \frac{\pi^2}{4}$

704. $\int_0^{\pi/2} (\log\sin x)\,dx = \int_0^{\pi/2} \log\cos x\,dx = -\frac{\pi}{2}\log 2$

705. $\int_0^{\pi/2} (\log\sec x)\,dx = \int_0^{\pi/2} \log\csc x\,dx = \frac{\pi}{2}\log 2$

706. $\int_0^{\pi} x(\log\sin x)\,dx = -\frac{\pi^2}{2}\log 2$

707. $\int_0^{\pi/2} (\sin x)(\log\sin x)\,dx = \log 2 - 1$

708. $\int_0^{\pi/2} (\log \tan x)\, dx = 0$

709. $\int_0^{\pi} \log (a \pm b \cos x)\, dx = \pi \log \left(\dfrac{a + \sqrt{a^2 - b^2}}{2} \right), \qquad (a \geq b)$

710. $\int_0^{\pi} \log (a^2 - 2ab \cos x + b^2)\, dx = \begin{cases} 2\pi \log a, & a \geq b > 0 \\ 2\pi \log b, & b \geq a > 0 \end{cases}$

711. $\int_0^{\infty} \dfrac{\sin ax}{\sinh bx}\, dx = \dfrac{\pi}{2b} \tanh \dfrac{a\pi}{2b}$

712. $\int_0^{\infty} \dfrac{\cos ax}{\cosh bx}\, dx = \dfrac{\pi}{2b} \operatorname{sech} \dfrac{a\pi}{2b}$

713. $\int_0^{\infty} \dfrac{dx}{\cosh ax} = \dfrac{\pi}{2a}$

714. $\int_0^{\infty} \dfrac{x\, dx}{\sinh ax} = \dfrac{\pi^2}{4a^2}$

715. $\int_0^{\infty} e^{-ax}(\cosh bx)\, dx = \dfrac{a}{a^2 - b^2}, \qquad (0 \leq |b| < a)$

716. $\int_0^{\infty} e^{-ax}(\sinh bx)\, dx = \dfrac{b}{a^2 - b^2}, \qquad (0 \leq |b| < a)$

717. $\int_0^{\infty} \dfrac{\sinh ax}{e^{bx} + 1}\, dx = \dfrac{\pi}{2b} \csc \dfrac{a\pi}{b} - \dfrac{1}{2a}$

718. $\int_0^{\infty} \dfrac{\sinh ax}{e^{bx} - 1}\, dx = \dfrac{1}{2a} - \dfrac{\pi}{2b} \cot \dfrac{a\pi}{b}$

719. $$\int_0^{\pi/2} \frac{dx}{\sqrt{1 - k^2 \sin^2 x}} = \frac{\pi}{2}\left[1 + \left(\frac{1}{2}\right)^2 k^2 + \left(\frac{1 \cdot 3}{2 \cdot 4}\right)^2 k^4 + \left(\frac{1 \cdot 3 \cdot 5}{2 \cdot 4 \cdot 6}\right)^2 k^6 + \cdots \right], \text{ if } k^2 < 1$$

720. $$\int_0^{\pi/2} \sqrt{1 - k^2 \sin^2 x}\, dx = \frac{\pi}{2}\left[1 - \left(\frac{1}{2}\right)^2 k^2 - \left(\frac{1 \cdot 3}{2 \cdot 4}\right)^2 \frac{k^4}{3} - \left(\frac{1 \cdot 3 \cdot 5}{2 \cdot 4 \cdot 6}\right)^2 \frac{k^6}{5} - \cdots \right], \text{ if } k^2 < 1$$

721. $\int_0^{\infty} e^{-x} \log x\, dx = -\gamma = -0.5772157\ldots$

722. $\int_0^{\infty} e^{-x^2} \log x\, dx = -\dfrac{\sqrt{\pi}}{4}(\gamma + 2 \log 2)$

723. $\int_0^\infty \left(\frac{1}{1-e^{-x}} - \frac{1}{x}\right) e^{-x}\,dx = \gamma = 0.5772157\ldots$ [Euler's Constant]

724. $\int_0^\infty \frac{1}{x}\left(\frac{1}{1+x} - e^{-x}\right) dx = \gamma = 0.5772157\ldots$

For n even:

725. $\cos^n x = \frac{1}{2^{n-1}} \sum_{k=0}^{\frac{n}{2}-1} \binom{n}{k} \cos(n-2k)x + \frac{1}{2^n}\binom{n}{n/2}.$

726. $\sin^n x = \frac{1}{2^{n-1}} \sum_{k=0}^{\frac{n}{2}-1} (-1)^k \binom{n}{k} \cos(n-2k)x + (-1)^{n/2} \cdot \binom{n}{n/2} \cdot \frac{1}{2^n}.$

For n odd:

727. $\cos^n x = \frac{1}{2^{n-1}} \sum_{k=0}^{\frac{n-1}{2}} \binom{n}{k} \cos(n-2k)x.$

728. $\sin^n x = \frac{1}{2^{n-1}} \sum_{k=0}^{\frac{n-1}{2}} (-1)^k \binom{n}{k} \sin(n-2k)x.$

For n even:

729. $\int \cos^n x\,dx = \frac{1}{2^{n-1}} \sum_{k=0}^{\frac{n}{2}-1} \binom{n}{k} \frac{\sin(n-2k)x}{(n-2k)} + \frac{1}{2^n}\binom{n}{n/2} x.$

730. $\int \sin^n x\,dx = \frac{1}{2^{n-1}} \sum_{k=0}^{\frac{n}{2}-1} (-1)^k \binom{n}{k} \frac{\sin(n-2k)x}{(n-2k)} + (-1)^{n/2} \cdot \binom{n}{n/2} \cdot \frac{1}{2^n} x.$

For n odd:

731. $\int \cos^n x\,dx = \frac{1}{2^{n-1}} \sum_{k=0}^{\frac{n-1}{2}} \binom{n}{k} \frac{\sin(n-2k)x}{(n-2k)}.$

732. $\int \sin^n x\,dx = \frac{1}{2^{n-1}} \sum_{k=0}^{\frac{n-1}{2}} (-1)^{k+1} \binom{n}{k} \frac{\cos(n-2k)x}{(n-2k)}.$

SERIES

The expression in parentheses following certain of the series indicates the region of convergence. If not otherwise indicated it is to be understood that the series converges for all finite values of x.

BINOMIAL

$$(x+y)^n = x^n + nx^{n-1}y + \frac{n(n-1)}{2!}x^{n-2}y^2 + \frac{n(n-1)(n-2)}{3!}x^{n-3}y^3 + \cdots \quad (y^2 < x^2)$$

$$(1 \pm x)^n = 1 \pm nx + \frac{n(n-1)x^2}{2!} \pm \frac{n(n-1)(n-2)x^3}{3!} + \cdots \text{ etc.} \quad (x^2 < 1)$$

$$(1 \pm x)^{-n} = 1 \mp nx + \frac{n(n+1)x^2}{2!} \mp \frac{n(n+1)(n+2)x^3}{3!} + \cdots \text{ etc.} \quad (x^2 < 1)$$

$$(1 \pm x)^{-1} = 1 \mp x + x^2 \mp x^3 + x^4 \mp x^5 + \cdots \quad (x^2 < 1)$$

$$(1 \pm x)^{-2} = 1 \mp 2x + 3x^2 \mp 4x^3 + 5x^4 \mp 6x^5 + \cdots \quad (x^2 < 1)$$

REVERSION OF SERIES

Let a series be represented by

$$y = a_1x + a_2x^2 + a_3x^3 + a_4x^4 + a_5x^5 + a_6x^6 + \cdots \quad (a_1 \neq 0)$$

to find the coefficients of the series

$$x = A_1y + A_2y^2 + A_3y^3 + A_4y^4 + \cdots$$

$$A_1 = \frac{1}{a_1} \qquad A_2 = -\frac{a_2}{a_1^3} \qquad A_3 = \frac{1}{a_1^5}(2a_2^2 - a_1a_3)$$

$$A_4 = \frac{1}{a_1^7}(5a_1a_2a_3 - a_1^2a_4 - 5a_2^3)$$

$$A_5 = \frac{1}{a_1^9}(6a_1^2a_2a_4 + 3a_1^2a_3^2 + 14a_2^4 - a_1^3a_5 - 21a_1a_2^2a_3)$$

$$A_6 = \frac{1}{a_1^{11}}(7a_1^3a_2a_5 + 7a_1^3a_3a_4 + 84a_1a_2^3a_3 - a_1^4a_6 - 28a_1^2a_2^2a_4 - 28a_1^2a_2a_3^2 - 42a_2^5)$$

$$A_7 = \frac{1}{a_1^{13}}(8a_1^4a_2a_6 + 8a_1^4a_3a_5 + 4a_1^4a_4^2 + 120a_1^2a_2^3a_4 + 180a_1^2a_2^2a_3^2 + 132a_2^6 - a_1^5a_7 - 36a_1^3a_2^2a_5 - 72a_1^3a_2a_3a_4 - 12a_1^3a_3^3 - 330a_1a_2^4a_3)$$

TAYLOR

1. $$f(x) = f(a) + (x-a)f'(a) + \frac{(x-a)^2}{2!}f''(a) + \frac{(x-a)^3}{3!}f'''(a) + \cdots + \frac{(x-a)^n}{n!}f^{(n)}(a) + \cdots \text{ (Taylor's Series)}$$

(Increment form)

2. $f(x+h) = f(x) + hf'(x) + \frac{h^2}{2!} f''(x) + \frac{h^3}{3!} f'''(x) + \cdots$

$$= f(h) + xf'(h) + \frac{x^3}{2!} f''(h) + \frac{x^3}{3!} f'''(h) + \cdots$$

3. If $f(x)$ is a function possessing derivatives of all orders throughout the interval $a \leqq x \leqq b$, then there is a value X, with $a < X < b$, such that

$$f(b) = f(a) + (b-a)f'(a) + \frac{(b-a)^2}{2!} f''(a) + \cdots + \frac{(b-a)^{n-1}}{(n-1)!} f^{(n-1)}(a) + \frac{(b-a)^n}{n!} f^{(n)}(X)$$

$$f(a+h) = f(a) + hf'(a) + \frac{h^2}{2!} f''(a) + \cdots + \frac{h^{n-1}}{(n-1)!} f^{(n-1)}(a) + \frac{h^n}{n!} f^{(n)}(a + \theta h), \quad b = a + h, \ 0 < \theta < 1.$$

or

$$f(x) = f(a) + (x-a)f'(a) + \frac{(x-a)^2}{2!} f''(a) + \cdots + (x-a)^{n-1} \frac{f^{(n-1)}(a)}{(n-1)!} + R_n,$$

where

$$R_n = \frac{f^{(n)}[a + \theta \cdot (x-a)]}{n!} (x-a)^n, \quad 0 < \theta < 1.$$

The above forms are known as Taylor's series with the remainder term.

4. *Taylor's series for a function of two variables*

If $\left(h \frac{\partial}{\partial x} + k \frac{\partial}{\partial y}\right) f(x,y) = h \frac{\partial f(x,y)}{\partial x} + k \frac{\partial f(x,y)}{\partial y}$;

$$\left(h \frac{\partial}{\partial x} + k \frac{\partial}{\partial y}\right)^2 f(x,y) = h^2 \frac{\partial^2 f(x,y)}{\partial x^2} + 2hk \frac{\partial^2 f(x,y)}{\partial x \, \partial y} + k^2 \frac{\partial^2 f(x,y)}{\partial y^2}$$

etc., and if $h\left(\frac{\partial}{\partial x} + k \frac{\partial}{\partial y}\right)^n f(x,y)\Big|_{\substack{x=a \\ y=b}}$ with the bar and subscripts means that after differentiation we are to replace x by a and y by b,

$$f(a+h, b+k) = f(a,b) + \left(h \frac{\partial}{\partial x} + k \frac{\partial}{\partial y}\right) f(x,y)\Big|_{\substack{x=a \\ y=b}} + \cdots + \frac{1}{n!}\left(h \frac{\partial}{\partial x} + k \frac{\partial}{\partial y}\right)^n f(x,y)\Big|_{\substack{x=a \\ y=b}} + \cdots$$

MACLAURIN

$$f(x) = f(0) + xf'(0) + \frac{x^2}{2!} f''(0) + \frac{x^3}{3!} f'''(0) + \cdots + x^{n-1} \frac{f^{(n-1)}(0)}{(n-1)!} + R_n,$$

where

$$R_n = \frac{x^n f^{(n)}(\theta x)}{n!}, \quad 0 < \theta < 1.$$

EXPONENTIAL

$$e = 1 + \frac{1}{1!} + \frac{1}{2!} + \frac{1}{3!} + \frac{1}{4!} + \cdots$$

$$e^x = 1 + x + \frac{x^2}{2!} + \frac{x^3}{3!} + \frac{x^4}{4!} + \cdots \qquad \text{(all real values of } x\text{)}$$

$$a^x = 1 + x \log_e a + \frac{(x \log_e a)^2}{2!} + \frac{(x \log_e a)^3}{3!} + \cdots$$

$$e^x = e^a\left[1 + (x - a) + \frac{(x - a)^2}{2!} + \frac{(x - a)^3}{3!} + \cdots\right]$$

LOGARITHMIC

$$\log_e x = \frac{x - 1}{x} + \frac{1}{2}\left(\frac{x - 1}{x}\right)^2 + \frac{1}{3}\left(\frac{x - 1}{x}\right)^3 + \cdots \qquad (x > \tfrac{1}{2})$$

$$\log_e x = (x - 1) - \tfrac{1}{2}(x - 1)^2 + \tfrac{1}{3}(x - 1)^3 - \cdots \qquad (2 \geq x > 0)$$

$$\log_e x = 2\left[\frac{x - 1}{x + 1} + \frac{1}{3}\left(\frac{x - 1}{x + 1}\right)^3 + \frac{1}{5}\left(\frac{x - 1}{x + 1}\right)^5 + \cdots\right] \qquad (x > 0)$$

$$\log_e(1 + x) = x - \tfrac{1}{2}x^2 + \tfrac{1}{3}x^3 - \tfrac{1}{4}x^4 + \cdots \qquad (-1 < x \leq 1)$$

$$\log_e(n + 1) - \log_e(n - 1) = 2\left[\frac{1}{n} + \frac{1}{3n^3} + \frac{1}{5n^5} + \cdots\right]$$

$$\log_e(a + x) = \log_e a + 2\left[\frac{x}{2a + x} + \frac{1}{3}\left(\frac{x}{2a + x}\right)^3 + \frac{1}{5}\left(\frac{x}{2a + x}\right)^5 + \cdots\right]$$

$$(a > 0, \ -a < x < +\infty)$$

$$\log_e \frac{1 + x}{1 - x} = 2\left[x + \frac{x^3}{3} + \frac{x^5}{5} + \cdots + \frac{x^{2n-1}}{2n - 1} + \cdots\right], \qquad -1 < x < 1$$

$$\log_e x = \log_e a + \frac{(x - a)}{a} - \frac{(x - a)^2}{2a^2} + \frac{(x - a)^3}{3a^3} - + \cdots, \qquad 0 < x \leqq 2a$$

TRIGONOMETRIC

$$\sin x = x - \frac{x^3}{3!} + \frac{x^5}{5!} - \frac{x^7}{7!} + \cdots \qquad \text{(all real values of } x\text{)}$$

$$\cos x = 1 - \frac{x^2}{2!} + \frac{x^4}{4!} - \frac{x^6}{6!} + \cdots \qquad \text{(all real values of } x\text{)}$$

$$\tan x = x + \frac{x^3}{3} + \frac{2x^5}{15} + \frac{17x^7}{315} + \frac{62x^9}{2835} + \cdots + \frac{(-1)^{n+1}(2^{2n} - 1)\, 2^{2n} B_{2n}}{(2n)!} x^{2n-1} + \cdots$$

$\left[x^2 < \frac{\pi^2}{4}\right.$, and B_{2n} represents the n'th Bernoulli number.$\left.\right]$

$$\cot x = \frac{1}{x} - \frac{x}{3} - \frac{x^3}{45} - \frac{2x^5}{945} - \frac{x^7}{4725} - \cdots - \frac{(-1)^{n+1} 2^{2n}}{(2n)!} B_{2n}\, x^{2n-1} - \cdots,$$

[$x^2 < \pi^2$, and B_n represents the n'th Bernoulli number.]

$$\sec x = 1 + \frac{x^2}{2} + \frac{5}{24}x^4 + \frac{61}{720}x^6 + \frac{277}{8064}x^8 + \cdots + \frac{(-1)^n}{(2n)!}E_{2n}x^{2n} + \ldots,$$

$\left[x^2 < \frac{\pi^2}{4}\right.$, and E_n represents the n'th Euler number.$\left.\right]$

$$\csc x = \frac{1}{x} + \frac{x}{6} + \frac{7}{360}x^3 + \frac{31}{15{,}120}x^5 + \frac{127}{604{,}800}x^7 + \cdots$$

$$\frac{(-1)^{n+1}2(2^{2n-1} - 1)}{(2n)!}B_{2n}x^{2n-1} + \cdots,$$

[$x^2 < \pi^2$, and B_n represents n'th Bernoulli number.]

$$\sin x = x\left(1 - \frac{x^2}{\pi^2}\right)\left(1 - \frac{x^2}{2^2\pi^2}\right)\left(1 - \frac{x^2}{3^2\pi^2}\right)\cdots \qquad (x^2 < \infty)$$

$$\cos x = \left(1 - \frac{4x^2}{\pi^2}\right)\left(1 - \frac{4x^2}{3^2\pi^2}\right)\left(1 - \frac{4x^2}{5^2\pi^2}\right)\cdots \qquad (x^2 < \infty)$$

$$\sin^{-1} x = x + \frac{x^3}{2\cdot 3} + \frac{1\cdot 3}{2\cdot 4\cdot 5}x^5 + \frac{1\cdot 3\cdot 5}{2\cdot 4\cdot 6\cdot 7}x^7 + \cdots \qquad \left(x^2 < 1, -\frac{\pi}{2} < \sin^{-1} x < \frac{\pi}{2}\right)$$

$$\cos^{-1} x = \frac{\pi}{2} - \left(x + \frac{x^3}{2\cdot 3} + \frac{1\cdot 3}{2\cdot 4\cdot 5}x^5 + \frac{1\cdot 3\cdot 5x^7}{2\cdot 4\cdot 6\cdot 7} + \cdots\right) \qquad (x^2 < 1, 0 < \cos^{-1} x < \pi)$$

$$\tan^{-1} x = x - \frac{x^3}{3} + \frac{x^5}{5} - \frac{x^7}{7} + \cdots \qquad (x^2 < 1)$$

$$\tan^{-1} x = \frac{\pi}{2} - \frac{1}{x} + \frac{1}{3x^2} - \frac{1}{5x^5} + \frac{1}{7x^7} - \cdots \qquad (x > 1)$$

$$\tan^{-1} x = -\frac{\pi}{2} - \frac{1}{x} + \frac{1}{3x^2} - \frac{1}{5x^5} + \frac{1}{7x^7} - \cdots \qquad (x < -1)$$

$$\cot^{-1} x = \frac{\pi}{2} - x + \frac{x^3}{3} - \frac{x^5}{5} + \frac{x^7}{7} - \cdots \qquad (x^2 < 1)$$

$$\log_e \sin x = \log_e x - \frac{x^2}{6} - \frac{x^4}{180} - \frac{x^6}{2835} - \cdots \qquad (x^2 < \pi^2)$$

$$\log_e \cos x = -\frac{x^2}{2} - \frac{x^4}{12} - \frac{x^6}{45} - \frac{17x^8}{2520} - \cdots \qquad \left(x^2 < \frac{\pi^2}{4}\right)$$

$$\log_e \tan x = \log_e x + \frac{x^2}{3} + \frac{7x^4}{90} + \frac{62x^6}{2835} + \cdots \qquad \left(x^2 < \frac{\pi^2}{4}\right)$$

$$e^{\sin x} = 1 + x + \frac{x^2}{2!} - \frac{3x^4}{4!} - \frac{8x^5}{5!} - \frac{3x^6}{6!} + \frac{56x^7}{7!} + \cdots$$

$$e^{\cos x} = e\left(1 - \frac{x^2}{2!} + \frac{4x^4}{4!} - \frac{31x^6}{6!} + \cdots\right)$$

$$e^{\tan x} = 1 + x + \frac{x^2}{2!} + \frac{3x^3}{3!} + \frac{9x^4}{4!} + \frac{37x^5}{5!} + \cdots \qquad \left(x^2 < \frac{\pi^2}{4}\right)$$

$$\sin x = \sin a + (x - a)\cos a - \frac{(x - a)^2}{2!}\sin a$$

$$- \frac{(x - a)^3}{3!}\cos a + \frac{(x - a)^4}{4!}\sin a + \cdots$$

HYPERBOLIC AND INVERSE HYPERBOLIC

See page 310 for section on Hyperbolic Series

FOURIER

(Also see Index for Cosine and Sine Transforms)

1. If $f(x)$ is a bounded periodic function of period $2L$ (i.e. $f(x + 2L) = f(x)$), and satisfies the *Dirichlet conditions*:

   a) In any period $f(x)$ is continuous, except possibly for a finite number of jump discontinuities.

   b) In any period $f(x)$ has only a finite number of maxima and minima.

Then $f(x)$ may be represented by the *Fourier series*

$$\frac{a_0}{2} + \sum_{n=1}^{\infty}\left(a_n \cos\frac{n\pi x}{L} + b_n \sin\frac{n\pi x}{L}\right),$$

where a_n and b_n are as determined below. This series will converge to $f(x)$ at every point where $f(x)$ is continuous, and to

$$\frac{f(x^+) + f(x^-)}{2}$$

(i.e. the average of the left-hand and right-hand limits) at every point where $f(x)$ has a jump discontinuity.

$$a_n = \frac{1}{L}\int_{-L}^{L} f(x)\cos\frac{n\pi x}{L}\,dx, \quad n = 0, 1, 2, 3, \ldots;$$

$$b_n = \frac{1}{L}\int_{-L}^{L} f(x)\sin\frac{n\pi x}{L}\,dx, \quad n = 1, 2, 3, \ldots$$

We may also write

$$a_n = \frac{1}{L}\int_{\alpha}^{\alpha+2L} f(x)\cos\frac{n\pi x}{L}\,dx \text{ and } b_n = \frac{1}{L}\int_{\alpha}^{\alpha+2L} f(x)\sin\frac{n\pi x}{L}\,dx,$$

where α is any real number. Thus if $\alpha = 0$,

$$a_n = \frac{1}{L}\int_{0}^{2L} f(x)\cos\frac{n\pi x}{L}\,dx, \quad n = 0, 1, 2, 3, \ldots,$$

$$b_n = \frac{1}{L}\int_{0}^{2L} f(x)\sin\frac{n\pi x}{L}\,dx, \quad n = 1, 2, 3, \ldots$$

2. If in addition to the above restrictions, $f(x)$ is even (i.e. $f(-x) = f(x)$), the Fourier series reduces to

$$\frac{a_0}{2} + \sum_{n=1}^{\infty} a_n \cos\frac{n\pi x}{L}.$$

That is, $b_n = 0$. In this case, a simpler formula for a_n is

$$a_n = \frac{2}{L}\int_0^L f(x)\cos\frac{n\pi x}{L}\,dx, \quad n = 0, 1, 2, 3, \ldots$$

3. If in addition to the restrictions in (1), $f(x)$ is an odd function (i.e. $f(-x) = -f(x)$), then the Fourier series reduces to

$$\sum_{n=1}^{\infty} b_n \sin\frac{n\pi x}{L}.$$

That is, $a_n = 0$. In this case, a simpler formula for the b_n is

$$b_n = \frac{2}{L}\int_0^L f(x)\sin\frac{n\pi x}{L}\,dx, \quad n = 1, 2, 3, \ldots$$

4. If in addition to the restrictions in (2) above, $f(x) = -f(L - x)$, then a_n will be 0 for all even values of n, including $n = 0$. Thus in this case, the expansion reduces to

$$\sum_{m=1}^{\infty} a_{2m-1}\cos\frac{(2m-1)\pi x}{L}.$$

5. If in addition to the restrictions in (3) above, $f(x) = f(L - x)$, then b_n will be 0 for all even values of n. Thus in this case, the expansion reduces to

$$\sum_{m=1}^{\infty} b_{2m-1}\sin\frac{(2m-1)\pi x}{L}.$$

(The series in (4) and (5) are known as *odd-harmonic series*, since only the odd harmonics appear. Similar rules may be stated for even-harmonic series, but when a series appears in the even-harmonic form, it means that $2L$ has not been taken as the smallest period of $f(x)$. Since any integral multiple of a period is also a period, series obtained in this way will also work, but in general computation is simplified if $2L$ is taken to be the smallest period.)

6. If we write the Euler definitions for $\cos\theta$ and $\sin\theta$, we obtain the complex form of the Fourier Series known either as the "Complex Fourier Series" or the "Exponential Fourier Series" of $f(x)$. It is represented as

$$f(x) = \tfrac{1}{2}\sum_{n=-\infty}^{n=+\infty} c_n e^{i\omega_n x},$$

where

$$c_n = \frac{1}{L}\int_{-L}^{L} f(x)\,e^{-i\omega_n x}dx, \quad n = 0, \pm 1, \pm 2, \pm 3, \ldots$$

with $\omega_n = \frac{n\pi}{L}$, $n = 0, \pm 1, \pm 2, \ldots$

The set of coefficients $\{c_n\}$ is often referred to as the Fourier spectrum.

7. If both sine and cosine terms are present and if $f(x)$ is of period $2L$ and expandable by a Fourier series, it can be represented as

$$f(x) = \frac{a_0}{2} + \sum_{n=1}^{\infty} c_n \cos\left(\frac{n\pi x}{L} + \phi_n\right),$$

$$\text{where } a_n = c_n \cos \phi_n, b_n = c_n \cos \phi_n, \quad c_n = \sqrt{a_n^2 + b_n^2}, \quad \phi_n = \arctan\left(\frac{a_n}{b_n}\right)$$

It can also be represented as

$$f(x) = \frac{a_0}{2} + \sum_{n=1}^{\infty} c_n \sin\left(\frac{n\pi x}{L} + \phi_n\right),$$

$$\text{where } a_n = c_n \sin \phi_n, b_n = -c_n \sin \phi_n, \quad c_n = \sqrt{a_n^2 + b_n^2}, \quad \phi_n = \arctan\left(-\frac{b_n}{a_n}\right)$$

where the quadrant of ϕ_n is chosen so as to make a_n, b_n, and c_n hold.

8. The following table of trigonometric identities should be helpful for developing Fourier Series.

	n**	n even	n odd	$n/2$ odd	$n/2$ even
$\sin n\pi$	0	0	0	0	0
$\cos n\pi$	$(-1)^n$	$+1$	-1	$+1$	$+1$
*$\sin \frac{n\pi}{2}$		0	$(-1)^{(n-1)/2}$	0	0
*$\cos \frac{n\pi}{2}$		$(-1)^{n/2}$	0	-1	$+1$
$\sin \frac{n\pi}{4}$			$\frac{\sqrt{2}}{2}(-1)^{(n^2+4n+11)/8}$	$(-1)^{(n-2)/4}$	0

*A useful formula for $\sin \frac{n\pi}{2}$ and $\cos \frac{n\pi}{2}$ is given by

$$\sin \frac{n\pi}{2} = \frac{(i)^{n+1}}{2}[(-1)^n - 1] \text{ and } \cos \frac{n\pi}{2} = \frac{(i)^n}{2}[(-1)^n + 1], \text{ where } i^2 = -1.$$

**n any integer.

AUXILIARY FORMULAS FOR FOURIER SERIES

$$1 = \frac{4}{\pi}\left[\sin\frac{\pi x}{k} + \frac{1}{3}\sin\frac{3\pi x}{k} + \frac{1}{5}\sin\frac{5\pi x}{k} + \cdots\right] \qquad [0 < x < k]$$

$$x = \frac{2k}{\pi}\left[\sin\frac{\pi x}{k} - \frac{1}{2}\sin\frac{2\pi x}{k} + \frac{1}{3}\sin\frac{3\pi x}{k} - \cdots\right] \qquad [-k < x < k]$$

$$x = \frac{k}{2} - \frac{4k}{\pi^2}\left[\cos\frac{\pi x}{k} + \frac{1}{3^2}\cos\frac{3\pi x}{k} + \frac{1}{5^2}\cos\frac{5\pi x}{k} + \cdots\right] \qquad [0 < x < k]$$

$$x^2 = \frac{2k^2}{\pi^3}\left[\left(\frac{\pi^2}{1} - \frac{4}{1}\right)\sin\frac{\pi x}{k} - \frac{\pi^2}{2}\sin\frac{2\pi x}{k} + \left(\frac{\pi^2}{3} - \frac{4}{3^3}\right)\sin\frac{3\pi x}{k} - \frac{\pi^2}{4}\sin\frac{4\pi x}{k} + \left(\frac{\pi^2}{5} - \frac{4}{5^3}\right)\sin\frac{5\pi x}{k} + \cdots\right] \qquad [0 < x < k]$$

$$x^2 = \frac{k^2}{3} - \frac{4k^2}{\pi^2}\left[\cos\frac{\pi x}{k} - \frac{1}{2^2}\cos\frac{2\pi x}{k} + \frac{1}{3^2}\cos\frac{3\pi x}{k} - \frac{1}{4^2}\cos\frac{4\pi x}{k} + \cdots\right] \qquad [-k < x < k]$$

$$1 - \frac{1}{3} + \frac{1}{5} - \frac{1}{7} + \cdots = \frac{\pi}{4}$$

$$1 + \frac{1}{2^2} + \frac{1}{3^2} + \frac{1}{4^2} + \cdots = \frac{\pi^2}{6}$$

$$1 - \frac{1}{2^2} + \frac{1}{3^2} - \frac{1}{4^2} + \cdots = \frac{\pi^2}{12}$$

$$1 + \frac{1}{3^2} + \frac{1}{5^2} + \frac{1}{7^2} + \cdots = \frac{\pi^2}{8}$$

$$\frac{1}{2^2} + \frac{1}{4^2} + \frac{1}{6^2} + \frac{1}{8^2} + \cdots = \frac{\pi^2}{24}$$

*SUMS OF RECIPROCAL POWERS

n	$\zeta(n) = \sum_{k=1}^{\infty} k^{-n}$				$\sum_{k=1}^{\infty} (-1)^{k-1} k^{-n}$			
1		∞			0.69314	71805	59945	30942
2	1.64493	40668	48226	43637	0.82246	70334	24113	21824
3	1.20205	69031	59594	28540	0.90154	26773	69695	71405
4	1.08232	32337	11138	19152	0.94703	28294	97245	91758
5	1.03692	77551	43369	92633	0.97211	97704	46909	30594
6	1.01734	30619	84449	13971	0.98555	10912	97435	10410
7	1.00834	92773	81922	82684	0.99259	38199	22830	28267
8	1.00407	73561	97944	33938	0.99623	30018	52647	89923
9	1.00200	83928	26082	21442	0.99809	42975	41605	33077
10	1.00099	45751	27818	08534	0.99903	95075	98271	56564
11	1.00049	41886	04119	46456	0.99951	71434	98060	75414
12	1.00024	60865	53308	04830	0.99975	76851	43858	19085
13	1.00012	27133	47578	48915	0.99987	85427	63265	11549
14	1.00006	12481	35058	70483	0.99993	91703	45979	71817
15	1.00003	05882	36307	02049	0.99996	95512	13099	23808
16	1.00001	52822	59408	65187	0.99998	47642	14906	10644
17	1.00000	76371	97637	89976	0.99999	23782	92041	01198
18	1.00000	38172	93264	99984	0.99999	61878	69610	11348
19	1.00000	19082	12716	55394	0.99999	80935	08171	67511
20	1.00000	09539	62033	87280	0.99999	90466	11581	52212
21	1.00000	04769	32986	78781	0.99999	95232	58215	54282
22	1.00000	02384	50502	72773	0.99999	97616	13230	82255
23	1.00000	01192	19925	96531	0.99999	98808	01318	43950
24	1.00000	00596	08189	05126	0.99999	99403	98892	39463
25	1.00000	00298	03503	51465	0.99999	99701	98856	96283
26	1.00000	00149	01554	82837	0.99999	99850	99231	99657
27	1.00000	00074	50711	78984	0.99999	99925	49550	48496
28	1.00000	00037	25334	02479	0.99999	99962	74753	40011
29	1.00000	00018	62659	72351	0.99999	99981	37369	41811
30	1.00000	00009	31327	43242	0.99999	99990	68682	28145
31	1.00000	00004	65662	90650	0.99999	99995	34340	33145
32	1.00000	00002	32831	18337	0.99999	99997	67169	89595
33	1.00000	00001	16415	50173	0.99999	99998	83584	85805
34	1.00000	00000	58207	72088	0.99999	99999	41792	39905
35	1.00000	00000	29103	85044	0.99999	99999	70896	18953
36	1.00000	00000	14551	92189	0.99999	99999	85448	09143
37	1.00000	00000	07275	95984	0.99999	99999	92724	04461
38	1.00000	00000	03637	97955	0.99999	99999	96362	02193
39	1.00000	00000	01818	98965	0.99999	99999	98181	01084
40	1.00000	00000	00909	49478	0.99999	99999	99090	50538
41	1.00000	00000	00454	74738	0.99999	99999	99545	25268
42	1.00000	00000	00227	37368	0.99999	99999	99772	62633

For $n > 42$, $\sum_{k=1}^{\infty} k^{-(n+1)} = \frac{1}{2}\left[1 + \sum_{k=1}^{\infty} k^{-n}\right]$, $\sum_{k=1}^{\infty} (-1)^{k-1} k^{-(n+1)} = \frac{1}{2}\left[1 + \sum_{k=1}^{\infty} (-1)^{k-1} k^{-n}\right]$

*Note: By definition, Riemann's Zeta Function is

$$\zeta(p) = \text{Zeta}(p) = 1 + \frac{1}{2^p} + \frac{1}{3^p} + \frac{1}{4^p} + \cdots = \zeta(p) = \text{Zeta}(p)$$

*SUMS OF RECIPROCAL POWERS

n	$\sum_{k=0}^{\infty} (2k+1)^{-n}$				$\sum_{k=0}^{\infty} (-1)^k (2k+1)^{-n}$			
1	∞				0.78539	81633	97448	310
2	1.23370	05501	36169	82735	0.91596	55941	77219	015
3	1.05179	97902	64644	99972	0.96894	61462	59369	380
4	1.01467	80316	04192	05455	0.98894	45517	41105	336
5	1.00452	37627	95139	61613	0.99615	78280	77088	064
6	1.00144	70766	40942	12191	0.99868	52222	18438	135
7	1.00047	15486	52376	55476	0.99955	45078	90539	909
8	1.00015	51790	25296	11930	0.99984	99902	46829	657
9	1.00005	13451	83843	77259	0.99994	96841	87220	090
10	1.00001	70413	63044	82549	0.99998	31640	26196	877
11	1.00000	56660	51090	10935	0.99999	43749	73823	699
12	1.00000	18858	48583	11958	0.99999	81223	50587	882
13	1.00000	06280	55421	80232	0.99999	93735	83771	841
14	1.00000	02092	40519	21150	0.99999	97910	87248	735
15	1.00000	00697	24703	12929	0.99999	99303	40842	624
16	1.00000	00232	37157	37916	0.99999	99767	75950	903
17	1.00000	00077	44839	45587	0.99999	99922	57782	104
18	1.00000	00025	81437	55666	0.99999	99974	19086	745
19	1.00000	00008	60444	11452	0.99999	99991	39660	745
20	1.00000	00002	86807	69746	0.99999	99997	13213	274
21	1.00000	00000	95601	16531	0.99999	99999	04403	029
22	1.00000	00000	31866	77514	0.99999	99999	68134	064
23	1.00000	00000	10622	20241	0.99999	99999	89377	965
24	1.00000	00000	03540	72294	0.99999	99999	96459	311
25	1.00000	00000	01180	23874	0.99999	99999	98819	768
26	1.00000	00000	00393	41247	0.99999	99999	99606	589
27	1.00000	00000	00131	13740	0.99999	99999	99868	863
28	1.00000	00000	00043	71245	0.99999	99999	99956	288
29	1.00000	00000	00014	57081	0.99999	99999	99985	429
30	1.00000	00000	00004	85694	0.99999	99999	99995	143
31	1.00000	00000	00001	61898	0.99999	99999	99998	381
32	1.00000	00000	00000	53966	0.99999	99999	99999	460
33	1.00000	00000	00000	17989	0.99999	99999	99999	820
34	1.00000	00000	00000	05996	0.99999	99999	99999	940
35	1.00000	00000	00000	01999	0.99999	99999	99999	980
36	1.00000	00000	00000	00666	0.99999	99999	99999	993
37	1.00000	00000	00000	00222	0.99999	99999	99999	998
38	1.00000	00000	00000	00074	0.99999	99999	99999	999
39	1.00000	00000	00000	00025				
40	1.00000	00000	00000	00008				
41	1.00000	00000	00000	00003				
42	1.00000	00000	00000	00001				

*This table is related to Riemann's Zeta Function, where

$$1 + \frac{1}{2^p} + \frac{1}{3^p} + \frac{1}{4^p} + \cdots = \zeta(p) = \text{Zeta}(p)$$

*FOURIER EXPANSIONS FOR BASIC PERIODIC FUNCTIONS

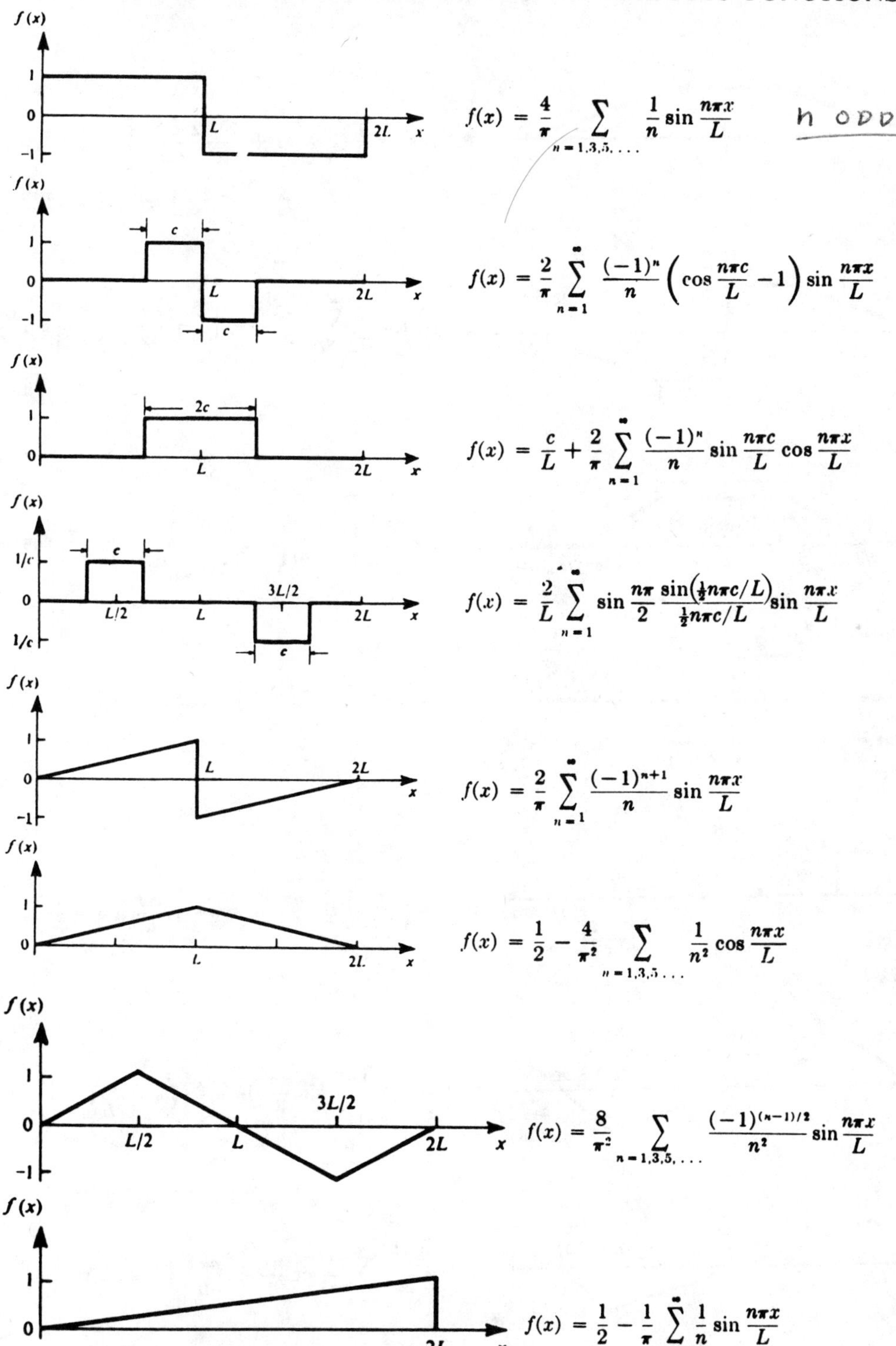

$$f(x) = \frac{4}{\pi} \sum_{n=1,3,5,\ldots} \frac{1}{n} \sin \frac{n\pi x}{L}$$

$$f(x) = \frac{2}{\pi} \sum_{n=1}^{\infty} \frac{(-1)^n}{n} \left(\cos \frac{n\pi c}{L} - 1 \right) \sin \frac{n\pi x}{L}$$

$$f(x) = \frac{c}{L} + \frac{2}{\pi} \sum_{n=1}^{\infty} \frac{(-1)^n}{n} \sin \frac{n\pi c}{L} \cos \frac{n\pi x}{L}$$

$$f(x) = \frac{2}{L} \sum_{n=1}^{\infty} \sin \frac{n\pi}{2} \frac{\sin(\frac{1}{2} n\pi c/L)}{\frac{1}{2} n\pi c/L} \sin \frac{n\pi x}{L}$$

$$f(x) = \frac{2}{\pi} \sum_{n=1}^{\infty} \frac{(-1)^{n+1}}{n} \sin \frac{n\pi x}{L}$$

$$f(x) = \frac{1}{2} - \frac{4}{\pi^2} \sum_{n=1,3,5\ldots} \frac{1}{n^2} \cos \frac{n\pi x}{L}$$

$$f(x) = \frac{8}{\pi^2} \sum_{n=1,3,5,\ldots} \frac{(-1)^{(n-1)/2}}{n^2} \sin \frac{n\pi x}{L}$$

$$f(x) = \frac{1}{2} - \frac{1}{\pi} \sum_{n=1}^{\infty} \frac{1}{n} \sin \frac{n\pi x}{L}$$

*Extracted from graphs and formulas, pages 372, 373, *Differential Equations in Engineering Problems,* Salvadori and Schwarz, published by Prentice-Hall, Inc., 1954.

FOURIER EXPANSIONS FOR BASIC PERIODIC FUNCTIONS (Continued)

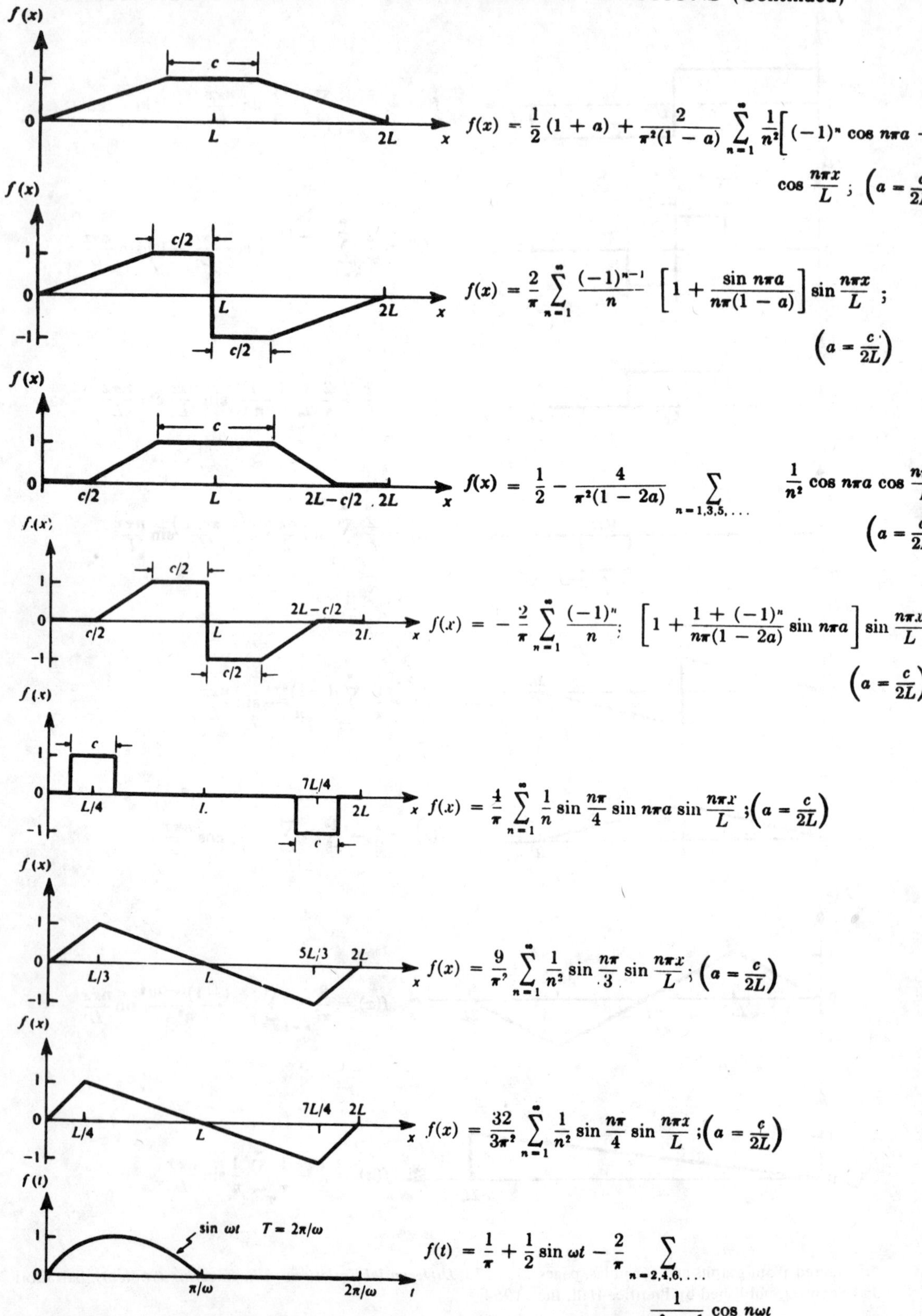

$$f(x) = \frac{1}{2}(1 + a) + \frac{2}{\pi^2(1 - a)} \sum_{n=1}^{\infty} \frac{1}{n^2}\left[(-1)^n \cos n\pi a - \cos\frac{n\pi x}{L}\right]; \quad \left(a = \frac{c}{2L}\right)$$

$$f(x) = \frac{2}{\pi} \sum_{n=1}^{\infty} \frac{(-1)^{n-1}}{n}\left[1 + \frac{\sin n\pi a}{n\pi(1 - a)}\right] \sin\frac{n\pi x}{L}; \quad \left(a = \frac{c}{2L}\right)$$

$$f(x) = \frac{1}{2} - \frac{4}{\pi^2(1 - 2a)} \sum_{n=1,3,5,\ldots} \frac{1}{n^2} \cos n\pi a \cos\frac{n\pi x}{L} \quad \left(a = \frac{c}{2L}\right)$$

$$f(x) = -\frac{2}{\pi} \sum_{n=1}^{\infty} \frac{(-1)^n}{n}\left[1 + \frac{1 + (-1)^n}{n\pi(1 - 2a)} \sin n\pi a\right] \sin\frac{n\pi x}{L} \quad \left(a = \frac{c}{2L}\right)$$

$$f(x) = \frac{4}{\pi} \sum_{n=1}^{\infty} \frac{1}{n} \sin\frac{n\pi}{4} \sin n\pi a \sin\frac{n\pi x}{L}; \left(a = \frac{c}{2L}\right)$$

$$f(x) = \frac{9}{\pi^2} \sum_{n=1}^{\infty} \frac{1}{n^2} \sin\frac{n\pi}{3} \sin\frac{n\pi x}{L}; \left(a = \frac{c}{2L}\right)$$

$$f(x) = \frac{32}{3\pi^2} \sum_{n=1}^{\infty} \frac{1}{n^2} \sin\frac{n\pi}{4} \sin\frac{n\pi x}{L}; \left(a = \frac{c}{2L}\right)$$

$$f(t) = \frac{1}{\pi} + \frac{1}{2}\sin\omega t - \frac{2}{\pi} \sum_{n=2,4,6,\ldots} \frac{1}{n^2 - 1} \cos n\omega t$$

*Extracted from graphs and formulas, pages 372, 373, *Differential Equations in Engineering Problems*, Salvadori and Schwarz, published by Prentice-Hall, Inc., 1954.

BERNOULLI AND EULER NUMBERS—POLYNOMIALS

There are numerous sets of defined numbers and polynomials, among which the more important ones are those classified under the names of Bernoulli and Euler.

The *Bernoulli Polynomials* are generated by the defining condition

$$\frac{te^{tx}}{e^t - 1} = B_0(x) + B_1(x)t + B_2(x)\frac{t^2}{2!} + B_3(x)\frac{t^3}{3!} + \cdots$$

$$B_0(x) = 1, \quad B_1(x) = x - \tfrac{1}{2}, \quad B_2(x) = x^2 - x + \tfrac{1}{6}$$

$$B_3(x) = x^3 - \frac{3}{2}x^2 + \frac{x}{2}, \quad B_4(x) = x^4 - 2x^3 + x^2 - \tfrac{1}{30}, \ldots$$

The following useful relations should be noted

$$B_n'(x) = nB_{n-1}(x), \qquad \int_a^x B_n(x)\,dx = \frac{1}{n+1}\left[B_{n+1}(x) - B_{n+1}(a)\right]$$

The Bernoulli numbers $B_0, B_1, B_2, \ldots,$ can be obtained by putting $x = 0$ in the respective polynomials. A simpler method is to use the generating function

$$\frac{x}{e^x - 1} = \sum_{n=0}^{\infty} \frac{B_n x^n}{n!}$$

and read the coefficients from the expansion. Here,

$$B_0 = 1, \quad B_1 = -\tfrac{1}{2}, \quad B_2 = \tfrac{1}{6}, \quad B_4 = -\tfrac{1}{30},$$

$$B_6 = \tfrac{1}{42}, \quad B_8 = -\tfrac{1}{30}, \ldots B_{2n+1} = 0 \qquad (n \geq 1)$$

An important application for the Bernoulli coefficients is in its use involved for the Euler-Maclaurin Sum Formula:

$$\sum_{x=1}^{n-1} f(x) = \int_1^n f(x)\,dx + \left[\sum_{i=1}^{\infty} \frac{B_i}{i!} f^{(i-1)}(x)\right]_{x=1}^{x=n}$$

$$\doteq \left[\int f(x)\,dx - \frac{1}{2}f(x) + \frac{1}{12}f'(x) - \frac{1}{720}f'''(x) + \frac{1}{30{,}240}f^{(V)}(x) - \frac{1}{1{,}209{,}600}f^{(VII)}(x)\right]_{x=1}^{x=n}$$

$$\sum_{x=1}^{n-1} \sqrt{x} = \sqrt{n}\left\{\frac{2}{3}n - \frac{1}{2} + \frac{1}{24n} - \frac{1}{1{,}920n^3} + \frac{1}{9{,}216n^5} - \frac{11}{163{,}840n^7}\right\} - 0.207{,}886{,}224{,}977{,}355$$

correct to 12 places for $n \geq 10$.

$$\log_e(x!) = \log_e \Gamma(x+1)$$

$$= \left(x + \frac{1}{2}\right)\log_e x - x + \frac{1}{12x} - \frac{1}{360x^3} + \frac{1}{1{,}260x^5} - \frac{1}{1{,}680x^7} + 0.918{,}938{,}533{,}205$$

accurate to 12 places for $x \geq 10$.

$$f(x)\,(\text{digamma function}) = \frac{d \log \Gamma(x)}{dx}$$

$$= 1 + \frac{1}{2} + \frac{1}{3} + \cdots + \frac{1}{x-1} - \gamma$$

(Euler Constant) for x integer.

By Euler-Maclaurin

$$f(x) = \log_e x - \frac{1}{2x} - \frac{1}{12x^2} + \frac{1}{120x^4} - \frac{1.}{252x^6} + \frac{1}{240x^8} - \frac{5}{660x^{10}} + \frac{691}{32{,}760x^{12}}$$

correct to 12 places for $x \geq 10$.

The Euler numbers together with their respective polynomials can be generated from

$$\frac{2e^{tx}}{e^t + 1} = \sum_{i=0.}^{\infty} E_i(x)\,t^i$$

and the relation

$$x^n = \frac{1}{2}\,[E_n(x+1) + E_n(x)]$$

The Euler polynomials are

$$E_0(x) = 1, \quad E_1(x) = x - \frac{1}{2}, \quad E_2(x) = x^2 - x$$

$$E_3(x) = x^3 - \frac{3}{2}x^2 + \frac{1}{4}, \quad E_4(x) = x^4 - 2x^3 + x$$

$$E_5(x) = x^5 - \frac{5}{2}x^4 + \frac{5}{2}x^2 - \frac{1}{2}$$

Tables which follow record the first fifteen polynomials of $B_k(x)$ and $E_k(x)$. The first sixty Bernoulli and Euler numbers are given in a separate table. The value of $x^n/n!$ is also important and this is given in a succeeding table for values of x from 1 to 9 and n from 1 to 50.

COEFFICIENTS b_k OF THE BERNOULLI POLYNOMIALS $B_n(x) = \sum_{k=0}^{n} b_k x^k$

$n\backslash k$	0	1	2	3	4	5	6	7	8	9	10	11	12	13	14	15
0	1															
1	$-\frac{1}{2}$	1														
2	$\frac{1}{6}$	-1	1													
3	0	$\frac{1}{2}$	$-\frac{3}{2}$	1												
4	$-\frac{1}{30}$	0	1	-2	1											
5	0	$-\frac{1}{6}$	0	$\frac{5}{3}$	$-\frac{5}{2}$	1										
6	$\frac{1}{42}$	0	$-\frac{1}{2}$	0	$\frac{5}{2}$	-3	1									
7	0	$\frac{1}{6}$	0	$-\frac{7}{6}$	0	$\frac{7}{2}$	$-\frac{7}{2}$	1								
8	$-\frac{1}{30}$	0	$\frac{2}{3}$	0	$-\frac{7}{3}$	0	$\frac{14}{3}$	-4	1							
9	0	$-\frac{3}{10}$	0	2	0	$-\frac{21}{5}$	0	6	$-\frac{9}{2}$	1						
10	$\frac{5}{66}$	0	$-\frac{3}{2}$	0	5	0	-7	0	$\frac{15}{2}$	-5	1					
11	0	$\frac{5}{6}$	0	$-\frac{11}{2}$	0	11	0	-11	0	$\frac{55}{6}$	$-\frac{11}{2}$	1				
12	$-\frac{691}{2730}$	0	5	0	$-\frac{33}{2}$	0	22	0	$-\frac{33}{2}$	0	11	-6	1			
13	0	$-\frac{691}{210}$	0	$\frac{65}{3}$	0	$-\frac{429}{10}$	0	$\frac{286}{7}$	0	$-\frac{143}{6}$	0	13	$-\frac{13}{2}$	1		
14	$\frac{7}{6}$	0	$-\frac{691}{30}$	0	$\frac{455}{6}$	0	$-\frac{1001}{10}$	0	$\frac{143}{2}$	0	$-\frac{1001}{30}$	0	$\frac{91}{6}$	-7	1	
15	0	$\frac{35}{2}$	0	$-\frac{691}{6}$	0	$\frac{455}{2}$	0	$-\frac{429}{2}$	0	$\frac{715}{6}$	0	$-\frac{91}{2}$	0	$\frac{35}{2}$	$-\frac{15}{2}$	1

COEFFICIENTS e_k OF THE EULER POLYNOMIALS $E_n(x) = \sum_{k=0}^{n} e_k x^k$

$n\backslash k$	0	1	2	3	4	5	6	7	8	9	10	11	12	13	14	15
0	1															
1	$-\frac{1}{2}$	1														
2	0	-1	1													
3	$\frac{1}{4}$	0	$-\frac{3}{2}$	1												
4	0	1	0	-2	1											
5	$-\frac{1}{2}$	0	$\frac{5}{2}$	0	$-\frac{5}{2}$	1										
6	0	-3	0	5	0	-3	1									
7	$\frac{17}{8}$	0	$-\frac{21}{2}$	0	$\frac{35}{4}$	0	$-\frac{7}{2}$	1								
8	0	17	0	-28	0	14	0	-4	1							
9	$-\frac{31}{2}$	0	$\frac{153}{2}$	0	-63	0	21	0	$-\frac{9}{2}$	1						
10	0	-155	0	255	0	-126	0	30	0	-5	1					
11	$\frac{691}{4}$	0	$-\frac{1705}{2}$	0	$\frac{2805}{4}$	0	-231	0	$\frac{165}{4}$	0	$-\frac{11}{2}$	1				
12	0	2073	0	-3410	0	1683	0	-396	0	55	0	-6	1			
13	$-\frac{5461}{2}$	0	$\frac{26949}{2}$	0	$-\frac{22165}{2}$	0	$\frac{7293}{2}$	0	$-\frac{1287}{2}$	0	$\frac{143}{2}$	0	$-\frac{13}{2}$	1		
14	0	-38227	0	62881	0	-31031	0	7293	0	-1001	0	91	0	-7	1	
15	$\frac{929569}{16}$	0	$-\frac{573405}{2}$	0	$\frac{943215}{4}$	0	$-\frac{155155}{2}$	0	$\frac{109395}{8}$	0	$-\frac{3003}{2}$	0	$\frac{455}{4}$	0	$-\frac{15}{2}$	1

BERNOULLI NUMBERS

$$B_n = N/D$$

n	N	D	B_n
0	1	1	(0) 1.0000 00000
1	−1	2	(−1) − 5.0000 00000*
2	1	6	(−1) 1.6666 66667
4	−1	30	(−2) − 3.3333 33333
6	1	42	(−2) 2.3809 52381
8	−1	30	(−2) − 3.3333 33333
10	5	66	(−2) 7.5757 57576
12	−691	2730	(−1) − 2.5311 35531
14	7	6	(0) 1.1666 66667
16	−3617	510	(0) − 7.0921 56863
18	43867	798	(1) 5.4971 17794
20	−1 74611	330	(2) − 5.2912 42424
22	8 54513	138	(3) 6.1921 23188
24	−2363 64091	2730	(4) − 8.6580 25311
26	85 53103	6	(6) 1.4255 17167
28	−2 37494 61029	870	(7) − 2.7298 23107
30	861 58412 76005	14322	(8) 6.0158 08739
32	−770 93210 41217	510	(10) − 1.5116 31577
34	257 76878 58367	6	(11) 4.2961 46431
36	−26315 27155 30534 77373	19 19190	(13) − 1.3711 65521
38	2 92999 39138 41559	6	(14) 4.8833 23190
40	−2 61082 71849 64491 22051	13530	(16) − 1.9296 57934
42	15 20097 64391 80708 02691	1806	(17) 8.4169 30476
44	−278 33269 57930 10242 35023	690	(19) − 4.0338 07185
46	5964 51111 59391 21632 77961	282	(21) 2.1150 74864
48	−560 94033 68997 81768 62491 27547	46410	(23) − 1.2086 62652
50	49 50572 05241 07964 82124 77525	66	(24) 7.5008 66746
52	−80116 57181 35489 95734 79249 91853	1590	(26) − 5.0387 78101
54	29 14996 36348 84862 42141 81238 12691	798	(28) 3.6528 77648
56	−2479 39292 93132 26753 68541 57396 63229	870	(30) − 2.8498 76930
58	84483 61334 88800 41862 04677 59940 36021	354	(32) 2.3865 42750
60	−121 52331 40483 75557 20403 04994 07982 02460 41491	567 86730	(34) − 2.1399 94926

*The floating decimal point notation is used here. For example for $n = 1$, $B_1 = -\frac{1}{2} = -.500000 = (-5.00000)(10^{-1}) = (-1) - 5.0000,0000$.

EULER NUMBERS

n	E_n
0	1
2	−1
4	5
6	−61
8	1385
10	−50521
12	27 02765
14	−1993 60981
16	1 93915 12145
18	−240 48796 75441
20	37037 11882 37525
22	−69 34887 43931 37901
24	15514 53416 35570 86905
26	−40 87072 50929 31238 92361
28	12522 59641 40362 98654 68285
30	−44 15438 93249 02310 45536 82821
32	17751 93915 79539 28943 66647 89665
34	−80 72329 92358 87898 06216 82474 53281
36	41222 06033 95177 02122 34707 96712 59045
38	−234 89580 52704 31082 52017 82857 61989 47741
40	1 48511 50718 11498 00178 77156 78140 58266 84425
42	−1036 46227 33519 61211 93979 57304 74518 59763 10201
44	7 94757 94225 97592 70360 80405 10088 07061 95192 73805
46	−6667 53751 66855 44977 43502 84747 73748 19752 41076 84661
48	60 96278 64556 85421 58691 68574 28768 43153 97653 90444 35185
50	−60532 85248 18862 18963 14383 78511 16490 88103 49822 51468 15121
52	650 61624 86684 60884 77158 70634 08082 29834 83644 23676 53855 76565
54	−7 54665 99390 08739 09806 14325 65889 73674 42122 40024 71169 98586 45581
56	9420 32189 64202 41204 20228 62376 90583 22720 93888 52599 64600 93949 05945
58	−126 22019 25180 62187 19903 40923 72874 89255 48234 10611 91825 59406 99649 20041
60	181089 11496 57923 04965 45807 74165 21586 88733 48734 92363 14106 00809 54542 31325

BERNOULLI AND EULER POLYNOMIALS

$x^n/n!$

$n\backslash x$	2	3	4	5
1	(0) 2.0000 00000	(0) 3.0000 00000	(0) 4.0000 00000	(0) 5.0000 00000
2	(0) 2.0000 00000	(0) 4.5000 00000	(0) 8.0000 00000	(1) 1.2500 00000
3	(0) 1.3333 33333	(0) 4.5000 00000	(1) 1.0666 66667	(1) 2.0833 33333
4	(− 1) 6.6666 66667	(0) 3.3750 00000	(1) 1.0666 66667	(1) 2.6041 66667
5	(− 1) 2.6666 66667	(0) 2.0250 00000	(0) 8.5333 33333	(1) 2.6041 66667
6	(− 2) 8.8888 88889	(0) 1.0125 00000	(0) 5.6888 88889	(1) 2.1701 38889
7	(− 2) 2.5396 82540	(− 1) 4.3392 85714	(0) 3.2507 93651	(1) 1.5500 99206
8	(− 3) 6.3492 06349	(− 1) 1.6272 32143	(0) 1.6253 96825	(0) 9.6881 20040
9	(− 3) 1.4109 34744	(− 2) 5.4241 07143	(− 1) 7.2239 85891	(0) 5.3822 88911
10	(− 4) 2.8218 69489	(− 2) 1.6272 32144	(− 1) 2.8895 94356	(0) 2.6911 44455
11	(− 5) 5.1306 71797	(− 3) 4.4379 05844	(− 1) 1.0507 61584	(0) 1.2232 47480
12	(− 6) 8.5511 19662	(− 3) 1.1094 76461	(− 2) 3.5025 38614	(− 1) 5.0968 64499
13	(− 6) 1.3155 56871	(− 4) 2.5603 30295	(− 2) 1.0777 04189	(− 1) 1.9603 32500
14	(− 7) 1.8793 66959	(− 5) 5.4864 22060	(− 3) 3.0791 54825	(− 2) 7.0011 87499
15	(− 8) 2.5058 22612	(− 5) 1.0972 84412	(− 4) 8.2110 79534	(− 2) 2.3337 29166
16	(− 9) 3.1322 78264	(− 6) 2.0574 08272	(− 4) 2.0527 69883	(− 3) 7.2929 03644
17	(−10) 3.6850 33252	(− 7) 3.6307 20481	(− 5) 4.8300 46785	(− 3) 2.1449 71660
18	(−11) 4.0944 81391	(− 8) 6.0512 00801	(− 5) 1.0733 43730	(− 4) 5.9582 54611
19	(−12) 4.3099 80412	(− 9) 9.5545 27582	(− 6) 2.2596 71011	(− 4) 1.5679 61740
20	(−13) 4.3099 80413	(− 9) 1.4331 79137	(− 7) 4.5193 42021	(− 5) 3.9199 04350
21	(−14) 4.1047 43250	(−10) 2.0473 98768	(− 8) 8.6082 70516	(− 6) 9.3331 05595
22	(−15) 3.7315 84772	(−11) 2.7919 07410	(− 8) 1.5651 40093	(− 6) 2.1211 60362
23	(−16) 3.2448 56324	(−12) 3.6416 18361	(− 9) 2.7219 82772	(− 7) 4.6112 18179
24	(−17) 2.7040 46937	(−13) 4.5520 22952	(−10) 4.5366 37953	(− 8) 9.6067 04540
25	(−18) 2.1632 37550	(−14) 5.4624 27543	(−11) 7.2586 20726	(− 8) 1.9213 40908
26	(−19) 1.6640 28884	(−15) 6.3028 01010	(−11) 1.1167 10881	(− 9) 3.6948 86362
27	(−20) 1.2326 13988	(−16) 7.0031 12233	(−12) 1.6543 86490	(−10) 6.8423 82151
28	(−22) 8.8043 85630	(−17) 7.5033 34535	(−13) 2.3634 09271	(−10) 1.2218 53956
29	(−23) 6.0719 90089	(−18) 7.7620 70209	(−14) 3.2598 74857	(−11) 2.1066 44751
30	(−24) 4.0479 93393	(−19) 7.7620 70209	(−15) 4.3464 99810	(−12) 3.5110 74585
31	(−25) 2.6116 08641	(−20) 7.5116 80847	(−16) 5.6083 86851	(−13) 5.6630 23524
32	(−26) 1.6322 55401	(−21) 7.0422 00795	(−17) 7.0104 83564	(−14) 8.8484 74257
33	(−28) 9.8924 56972	(−22) 6.4020 00722	(−18) 8.4975 55834	(−14) 1.3406 77918
34	(−29) 5.8190 92337	(−23) 5.6488 24167	(−19) 9.9971 24511	(−15) 1.9715 85173
35	(−30) 3.3251 95620	(−24) 4.8418 49286	(−19) 1.1425 28515	(−16) 2.8165 50246
36	(−31) 1.8473 30900	(−25) 4.0348 74405	(−20) 1.2694 76128	(−17) 3.9118 75343
37	(−33) 9.9855 72436	(−26) 3.2715 19788	(−21) 1.3724 06625	(−18) 5.2863 18032
38	(−34) 5.2555 64439	(−27) 2.5827 78779	(−22) 1.4446 38552	(−19) 6.9556 81619
39	(−35) 2.6951 61251	(−28) 1.9867 52908	(−23) 1.4816 80567	(−20) 8.9175 40539
40	(−36) 1.3475 80626	(−29) 1.4900 64681	(−24) 1.4816 80567	(−20) 1.1146 92567
41	(−38) 6.5735 64028	(−30) 1.0902 91230	(−25) 1.4455 42017	(−21) 1.3593 81180
42	(−39) 3.1302 68584	(−32) 7.7877 94498	(−26) 1.3767 06682	(−22) 1.6183 10928
43	(−40) 1.4559 38876	(−33) 5.4333 44999	(−27) 1.2806 57379	(−23) 1.8817 56893
44	(−42) 6.6179 03983	(−34) 3.7045 53408	(−28) 1.1642 33981	(−24) 2.1383 60106
45	(−43) 2.9412 90659	(−35) 2.4697 02271	(−29) 1.0348 74650	(−25) 2.3759 55673
46	(−44) 1.2788 22026	(−36) 1.6106 75395	(−31) 8.9989 09998	(−26) 2.5825 60514
47	(−46) 5.4417 95855	(−37) 1.0280 90677	(−32) 7.6586 46807	(−27) 2.7474 04803
48	(−47) 2.2674 14940	(−39) 6.4255 66736	(−33) 6.3822 05674	(−28) 2.8618 80003
49	(−49) 9.2547 54855	(−40) 3.9340 20450	(−34) 5.2099 63815	(−29) 2.9202 85717
50	(−50) 3.7019 10942	(−41) 2.3604 12270	(−35) 4.1679 71052	(−30) 2.9202 85717

BERNOULLI AND EULER POLYNOMIALS, RIEMANN ZETA FUNCTION

$x^n/n!$

$n \backslash x$	6	7	8	9
1	(0) 6.0000 00000	(0) 7.0000 00000	(0) 8.0000 00000	(0) 9.0000 00000
*2	(1) 1.8000 00000	(1) 2.4500 00000	(1) 3.2000 00000	(1) 4.0500 00000
3	(1) 3.6000 00000	(1) 5.7166 66667	(1) 8.5333 33333	(2) 1.2150 00000
4	(1) 5.4000 00000	(2) 1.0004 16667	(2) 1.7066 66667	(2) 2.7337 50000
5	(1) 6.4800 00000	(2) 1.4005 83333	(2) 2.7306 66667	(2) 4.9207 50000
6	(1) 6.4800 00000	(2) 1.6340 13889	(2) 3.6408 88889	(2) 7.3811 25000
7	(1) 5.5542 85714	(2) 1.6340 13889	(2) 4.1610 15873	(2) 9.4900 17857
8	(1) 4.1657 14286	(2) 1.4297 62153	(2) 4.1610 15873	(3) 1.0676 27009
9	(1) 2.7771 42857	(2) 1.1120 37230	(2) 3.6986 80776	(3) 1.0676 27009
10	(1) 1.6662 85714	(1) 7.7842 60610	(2) 2.9589 44621	(2) 9.6086 43080
11	(0) 9.0888 31169	(1) 4.9536 20388	(2) 2.1519 59724	(2) 7.8616 17066
12	(0) 4.5444 15584	(1) 2.8896 11893	(2) 1.4346 39816	(2) 5.8962 12799
13	(0) 2.0974 22577	(1) 1.5559 44865	(1) 8.8285 52715	(2) 4.0819 93476
14	(− 1) 8.9889 53903	(0) 7.7797 24327	(1) 5.0448 87266	(2) 2.6241 38663
15	(− 1) 3.5955 81561	(0) 3.6305 38019	(1) 2.6906 06542	(2) 1.5744 83198
16	(− 1) 1.3483 43085	(0) 1.5883 60383	(1) 1.3453 03271	(1) 8.8564 67988
17	(− 2) 4.7588 57949	(− 1) 6.5403 07461	(0) 6.3308 38921	(1) 4.6887 18347
18	(− 2) 1.5862 85983	(− 1) 2.5434 52902	(0) 2.8137 06187	(1) 2.3443 59173
19	(− 3) 5.0093 24157	(− 2) 9.3706 15954	(0) 1.1847 18395	(1) 1.1104 85924
20	(− 3) 1.5027 97247	(− 2) 3.2797 15584	(− 1) 4.7388 73579	(0) 4.9971 86660
21	(− 4) 4.2937 06421	(− 2) 1.0932 38528	(− 1) 1.8052 85173	(0) 2.1416 51426
22	(− 4) 1.1710 10841	(− 3) 3.4784 86224	(− 2) 6.5646 73356	(− 1) 8.7613 01286
23	(− 5) 3.0548 10892	(− 3) 1.0586 69721	(− 2) 2.2833 64645	(− 1) 3.4283 35286
24	(− 6) 7.6370 27230	(− 4) 3.0877 86685	(− 3) 7.6112 15485	(− 1) 1.2856 25732
25	(− 6) 1.8328 86535	(− 5) 8.6458 02719	(− 3) 2.4355 88956	(− 2) 4.6282 52637
26	(− 7) 4.2297 38158	(− 5) 2.3277 16117	(− 4) 7.4941 19863	(− 2) 1.6020 87451
27	(− 8) 9.3994 18129	(− 6) 6.0348 19562	(− 4) 2.2204 79959	(− 3) 5.3402 91503
28	(− 8) 2.0141 61028	(− 6) 1.5087 04890	(− 5) 6.3442 28454	(− 3) 1.7165 22269
29	(− 9) 4.1672 29712	(− 7) 3.6417 01460	(− 5) 1.7501 31987	(− 4) 5.3271 38075
30	(−10) 8.3344 59424	(− 8) 8.4973 03406	(− 6) 4.6670 18634	(− 4) 1.5981 41423
31	(−10) 1.6131 21179	(− 8) 1.9187 45930	(− 6) 1.2043 91905	(− 5) 4.6397 65421
32	(−11) 3.0246 02211	(− 9) 4.1972 56723	(− 7) 3.0109 79764	(− 5) 1.3049 34025
33	(−12) 5.4992 76746	(−10) 8.9032 71836	(− 8) 7.2993 44881	(− 6) 3.5589 10976
34	(−13) 9.7046 06022	(−10) 1.8330 26555	(− 8) 1.7174 92913	(− 7) 9.4206 46701
35	(−13) 1.6636 46746	(−11) 3.6660 53108	(− 9) 3.9256 98086	(− 7) 2.4224 52008
36	(−14) 2.7727 44578	(−12) 7.1284 36600	(−10) 8.7237 73527	(− 8) 6.0561 30022
37	(−15) 4.4963 42559	(−12) 1.3486 23141	(−10) 1.8862 21303	(− 8) 1.4731 12708
38	(−16) 7.0994 88250	(−13) 2.4843 05785	(−11) 3.9709 92217	(− 9) 3.4889 51151
39	(−16) 1.0922 28962	(−14) 4.4590 10384	(−12) 8.1456 25061	(−10) 8.0514 25733
40	(−17) 1.6383 43443	(−15) 7.8032 68172	(−12) 1.6291 25012	(−10) 1.8115 70790
41	(−18) 2.3975 75770	(−15) 1.3322 65298	(−13) 3.1787 80512	(−11) 3.9766 18807
42	(−19) 3.4251 08241	(−16) 2.2204 42162	(−14) 6.0548 20021	(−12) 8.5213 26014
43	(−20) 4.7792 20803	(−17) 3.6146 73288	(−14) 1.1264 78144	(−12) 1.7835 33352
44	(−21) 6.5171 19276	(−18) 5.7506 16594	(−15) 2.0481 42079	(−13) 3.6481 36401
45	(−22) 8.6894 92369	(−19) 8.9454 03592	(−16) 3.6411 41473	(−14) 7.2962 72804
46	(−22) 1.1334 12048	(−19) 1.3612 57068	(−17) 6.3324 19955	(−14) 1.4275 31635
47	(−23) 1.4469 08998	(−20) 2.0274 04144	(−17) 1.0778 58716	(−15) 2.7335 71217
48	(−24) 1.8086 36247	(−21) 2.9566 31045	(−18) 1.7964 31193	(−16) 5.1254 46033
49	(−25) 2.2146 56629	(−22) 4.2237 58634	(−19) 2.9329 48887	(−17) 9.4140 84548
50	(−26) 2.6575 87955	(−23) 5.9132 62088	(−20) 4.6927 18219	(−17) 1.6945 35219

*The floating decimal point notation is used here. For example (1) 1.8000,000 = (1.8000,000) 10.

STIRLING NUMBERS

Stirling numbers

The Stirling numbers are used for reducing factorials such as $x^{(n)}$ to polynomials in x and vice versa.

Stirling numbers of the first kind

The factorial polynomial $x^{(n)}$ is defined and represented by

$$x^{(n)} = x(x-1)(x-2)\cdots(x-n+1)$$

where $x^{(0)}$ is 1 by definition.

If n is a non-negative integer, then

$$x^{(n)} = s_{n1}x + s_{n2}x^2 + \cdots + s_{nn}x^n.$$

Here the numbers $s_{n1}, s_{n2}, s_{n3}, \ldots,$ are Stirling numbers of the first kind. A table listing these numbers with an example using same follows.

STIRLING NUMBERS OF THE FIRST KIND

n	s_{n1}	s_{n2}	s_{n3}	s_{n4}	s_{n5}	s_{n6}	s_{n7}	s_{n8}
1	1	0	0	0	0	0	0	0
2	−1	1	0	0	0	0	0	0
3	2	−3	1	0	0	0	0	0
4	−6	11	−6	1	0	0	0	0
5	24	−50	35	−10	1	0	0	0
6	−120	274	−225	85	−15	1	0	0
7	720	−1 764	1 624	−735	175	−21	1	0
8	−5 040	13 068	−13 132	6 769	−1 960	322	−28	1

The table may be continued by using the recurrence formula

$$s_{ni} = s_{n-1,i-1} - (n-1)s_{n-1,i}, \quad i = 1, 2, \ldots, n$$

where $s_{n0} = 0$ for all n.

Example

Express $3x^{(3)} + 2x^{(1)}$ using Stirling's numbers of the first kind

$$\begin{aligned} 3x^{(3)} &= 3(2x - 3x^2 + x^3) \\ &= 6x - 9x^2 + 3x^3 \\ 2x^{(1)} &= 2x \\ \therefore 3x^{(3)} + 2x^{(1)} &= 8x - 9x^2 + 3x^3 \end{aligned}$$

This may be verified by carrying out the indicated operations as follows

$$\begin{aligned} 3x^{(3)} + 2x^{(1)} &= 3(x)(x-1)(x-2) + 2x \\ &= 3x^3 - 9x^2 + 6x + 2x \\ &= 8x - 9x^2 + 3x^3 \end{aligned}$$

Stirling numbers of the second kind

For every non-negative integer n the function defined by x^n can be expressed as a linear combination of factorial powers of x not higher than the n'th. In other words

$$x^n = t_{n1}x^{(1)} + t_{n2}x^{(2)} + \cdots + t_{nn}x^{(n)}.$$

The numbers $t_{n1}, t_{n2}, \ldots, t_{nn}$ are called Stirling's numbers of the second kind.

A table listing these numbers with an example using same follows.

STIRLING NUMBERS OF THE SECOND KIND

n	t_{n1}	t_{n2}	t_{n3}	t_{n4}	t_{n5}	t_{n6}	t_{n7}	t_{n8}
1	1	0	0	0	0	0	0	0
2	1	1	0	0	0	0	0	0
3	1	3	1	0	0	0	0	0
4	1	7	6	1	0	0	0	0
5	1	15	25	10	1	0	0	0
6	1	31	90	65	15	1	0	0
7	1	63	301	350	140	21	1	0
8	1	127	966	1 701	1 050	266	28	1

This table may be continued by using the recurrence formula

$$t_{ni} = it_{n-1,i} + t_{n-1,i-1}, \quad i = 1, 2, \ldots, n$$

where $t_{n0} = 0$ for all n.

Example

Express $2x^3 - 3x^2 + x + 2$ by use of factorial powers

$$2x^3 = 2[x^{(1)} + 3x^{(2)} + x^{(3)}] = 2x^{(1)} + 6x^{(2)} + 2x^{(3)}$$

$$-3x^2 = -3[x^{(1)} + x^{(2)}] = -3x^{(1)} - 3x^{(2)}$$

$$x = x^{(1)} = x^{(1)}$$

$$+2 = +2x^{(0)} = 2x^{(0)}$$

$$\therefore 2x^3 - 3x^2 + x + 2 = 2 + 3x^{(2)} + 2x^{(3)}$$

Stirling's approx. to $n!$

$$n! \approx \sqrt{2\pi n}\, e^{-n} n^n \left(1 + \frac{1}{12n}\right)$$

YIELDS $1! \approx 0.9990$

$$\ln n! \approx n \ln n - n + \frac{1}{2}\ln n + \overbrace{\frac{1}{2}\ln 2\pi}^{0.92\ldots} + \ln\left(1 + \frac{1}{12n}\right)$$

$$\approx n \ln n - n \quad \text{WITHIN } 5\% \text{ FOR } n \geq 23$$

" 1% " $n \geq 91$

" 10^{-3} " 753

" 10^{-4} " 6820

" 10^{-6} " $616{,}000$

Better approx. is $n! \approx \sqrt{2\pi n}\, e^{-n} n^n e^{1/12n} = \sqrt{2\pi n}\, \exp\{-n + \frac{1}{12n} + n\ln n\}$

WITHIN 10^{-6} FOR $n \geq 15$

$$\ln n! \approx n \ln n - n + \frac{1}{2}\ln n + \frac{1}{2}\ln 2\pi + \frac{1}{12n}$$

DIFFERENTIAL EQUATIONS

SPECIAL FORMULAS

Certain types of differential equations occur sufficiently often to justify the use of formulas for the corresponding particular solutions. The following set of tables I to XIV covers all first, second and nth order ordinary linear differential equations with constant coefficients for which the right members are of the form $P(x)e^{rx}\sin sx$ or $P(x)e^{rx}\cos sx$, where r and s are constants and $P(x)$ is a polynomial of degree n.

When the right member of a reducible linear partial differential equation with constant coefficients is not zero, particular solutions for certain types of right members are contained in tables XV to XXI. In these tables both F and P are used to denote polynomials, and it is assumed that no denominator is zero. In any formula the roles of x and y may be reversed throughout, changing a formula in which x dominates to one in which y dominates. Tables XIX, XX, XXI are applicable whether the equations are reducible or not. The symbol $\binom{m}{n}$ stands for $\dfrac{m!}{(m-n)!n!}$ and is the $n+1$ st coefficient in the expansion of $(a+b)^m$. Also $0! = 1$ by definition.

The tables as herewith given are those contained in the text *Differential Equations* by Ginn and Company (1955) and are published with their kind permission and that of the author, Professor Frederick H. Steen.

Solution of Linear Differential Equations with Constant Coefficients

Any linear differential equation with constant coefficients may be written in the form

$$p(D)y = R(x),$$

where D is the differential operation

$$Dy = \frac{dy}{dx},$$

$p(D)$ is a polynomial in D,
y is the dependent variable,
x is the independent variable,
$R(x)$ is an arbitrary function of x.

A power of D represents repeated differentiation, that is

$$D^n y = \frac{d^n y}{dx^n}.$$

For such an equation, the general solution may be written in the form

$$y = y_c + y_p,$$

where y_p is any particular solution, and y_c is called the *complementary function*. This complementary function is defined as the general solution of the *homogeneous equation*, which is the original differential equation with the right side replaced by zero, i.e.

$$p(D)y = 0$$

The complementary function y_c may be determined as follows:

1. Factor the polynomial $p(D)$ into real and complex linear factors, just as if D were a variable instead of an operator.

2. For each non-repeated linear factor of the form $(D - a)$, where a is real, write down a term of the form

$$ce^{ax},$$

where c is an arbitrary constant.

3. For each repeated real linear factor of the form $(D - a)^n$, write down n terms of the form

$$c_1e^{ax} + c_2xe^{ax} + c_3x^2e^{ax} + \cdots + c_nx^{n-1}e^{ax},$$

where the c_i's are arbitrary constants.

4. For each non-repeated conjugate complex pair of factors of the form $(D - a + ib)(D - a - ib)$, write down 2 terms of the form

$$c_1e^{ax}\cos bx + c_2e^{ax}\sin bx$$

5. For each repeated conjugate complex pair of factors of the form $(D - a + ib)^n(D - a - ib)^n$, write down $2n$ terms of the form

$$c_1e^{ax}\cos bx + c_2e^{ax}\sin bx + c_3xe^{ax}\cos bx + c_4xe^{ax}\sin bx + \cdots$$
$$+ c_{2n-1}x^{n-1}e^{ax}\cos bx + c_{2n}x^{n-1}e^{ax}\sin bx$$

6. The sum of all the terms thus written down is the complementary function y_c.

To find the particular solution y_p, use the following tables, as shown in the examples. For cases not shown in the tables, there are various methods of finding y_p. The most general method is called *variation of parameters*. The following example illustrates the method:

Find y_p for $(D^2 - 4)y = e^x$.

This example can be solved most easily by use of equation 63 in the tables following. However it is given here as an example of the method of variation of parameters.

The complementary function is

$$y_c = c_1e^{2x} + c_2e^{-2x}$$

To find y_p, replace the constants in the complementary function with unknown functions,

$$y_p = ue^{2x} + ve^{-2x}.$$

We now prepare to substitute this assumed solution into the original equation. We begin by taking all the necessary derivatives:

$$y_p = ue^{2x} + ve^{-2x}$$
$$y_p' = 2ue^{2x} - 2ve^{-2x} + u'e^{2x} + v'e^{-2x}$$

For each derivative of y_p except the highest, we set the sum of all the terms containing u' and v' to 0. Thus the above equation becomes

$$u'e^{2x} + v'e^{-2x} = 0 \quad \text{and} \quad y_p' = 2ue^{2x} - 2ve^{-2x}$$

Continuing to differentiate, we have

$$y_p'' = 4ue^{2x} + 4ve^{-2x} + 2u'e^{2x} - 2v'e^{-2x}$$

When we substitute into the original equation, all the terms not containing u' or v' cancel out. This is a consequence of the method by which y_p was set up.

Thus all that is necessary is to write down the terms containing u' or v' in the highest order derivative of y_p, multiply by the constant coefficient of the highest power of D in $p(D)$, and set it equal to $R(x)$. Together with the previous terms in u' and v' which were set equal to 0, this gives us as many linear equations in the first derivatives of the unknown functions as there are unknown functions. The first derivatives may then be solved for by algebra, and the unknown functions found by integration. In the present example, this becomes

$$u'e^{2x} + v'e^{-2x} = 0$$
$$2u'e^{2x} - 2v'e^{-2x} = e^x.$$

We eliminate v' and u' separately, getting

$$4u'e^{2x} = e^x$$
$$4v'e^{-2x} = -e^x.$$

Thus

$$u' = \tfrac{1}{4}e^{-x}$$
$$v' = -\tfrac{1}{4}e^{3x}.$$

Therefore, by integrating

$$u = -\tfrac{1}{4}e^{-x}$$
$$v = -\tfrac{1}{12}e^{3x}.$$

A constant of integration is not needed, since we need only one particular solution. Thus

$$y_p = ue^{2x} + ve^{-2x} = -\tfrac{1}{4}e^{-x}e^{2x} - \tfrac{1}{12}e^{3x}e^{-2x}$$
$$= -\tfrac{1}{4}e^x - \tfrac{1}{12}e^x = -\tfrac{1}{3}e^x,$$

and the general solution is

$$y = y_c + y_p = c_1e^{2x} + c_2e^{-2x} - \tfrac{1}{3}e^x.$$

The following examples illustrate the use of the tables.

Example 1. Solve $(D^2 - 4)y = \sin 3x$.

Substitution of $q = -4$, $s = 3$ in formula 24 gives

$$y_p = \frac{\sin 3x}{-9 - 4},$$

wherefore the general solution is

$$y = c_1e^{2x} + c_2e^{-2x} - \frac{\sin 3x}{13}.$$

Example 2. Obtain a particular solution of $(D^2 - 4D + 5)y = x^2e^{3x}\sin x$.

Applying formula 40 with $a = 2$, $b = 1$, $r = 3$, $s = 1$, $P(x) = x^2$, $s + b = 2$, $s - b = 0$, $a - r = -1$, $(a - r)^2 + (s + b)^2 = 5$, $(a - r)^2 + (s - b)^2 = 1$, we have

$$y_p = \frac{e^{3x}\sin x}{2}\left[\left(\frac{2}{5} - \frac{0}{1}\right)x^2 + \left(\frac{2(-1)2}{25} - \frac{2(-1)0}{1}\right)2x\right.$$
$$\left. + \left(\frac{3\cdot 1\cdot 2 - 2^3}{125} - \frac{3\cdot 1\cdot 0 - 0}{1}\right)2\right]$$
$$- \frac{e^{3x}\cos x}{2}\left[\left(\frac{-1}{5} - \frac{-1}{1}\right)x^2 + \left(\frac{1-4}{25} - \frac{1-0}{1}\right)2x\right.$$
$$\left. + \left(\frac{-1 - 3(-1)4}{125} - \frac{-1 - 3(-1)0}{1}\right)2\right]$$
$$= \left(\tfrac{1}{5}x^2 - \tfrac{4}{25}x - \tfrac{2}{125}\right)e^{3x}\sin x + \left(-\tfrac{2}{5}x^2 + \tfrac{28}{25}x - \tfrac{136}{125}\right)e^{3x}\cos x.$$

The special formulas effect a very considerable saving of time in problems of this type.

Example 3. Obtain a particular solution of $(D^2 - 4D + 5)y = x^2e^{2x} \cos x$. (Compare with Example 2.)

Formula 40 is not applicable here since for this equation $r = a$, $s = b$, wherefore the denominator $(a - r)^2 + (s - b)^2 = 0$. We turn instead to formula 44. Substituting $a = 2, b = 1, P(x) = x^2$ and replacing sin by cos, cos by $-\sin$, we obtain

$$y_p = \frac{e^{2x} \cos x}{4}(x^2 - \tfrac{2}{4}) + \frac{e^{2x} \sin x}{2} \int (x^2 - \tfrac{1}{2})\, dx$$

$$= \left(\frac{x^2}{4} - \frac{1}{8}\right) e^{2x} \cos x + \left(\frac{x^3}{6} - \frac{x}{4}\right) e^{2x} \sin x,$$

which is the required solution.

Example 4. Find z_p for $(D_x - 3D_y)z = \ln(y + 3x)$.

Referring to Table XV we note that formula 69 (not 68) is applicable. This gives

$$z_p = x \ln(y + 3x).$$

It is easily seen that $-\dfrac{y}{3}\ln(y + 3x)$ would serve equally well.

Example 5. Solve $(D_x + 2D_y - 4)z = y \cos(y - 2x)$.

Since R in formula 76 contains a polynomial in x, not y, we rewrite the given equation in the form

$$(D_y + \tfrac{1}{2}D_x - 2)z = \tfrac{1}{2}y \cos(y - 2x).$$

Then

$$z_c = e^{2y}F(x - \tfrac{1}{2}y) = e^{2y}f(2x - y),$$

and by the formula

$$z_p = -\tfrac{1}{2}\cos(y - 2x)\cdot\left(\frac{y}{2} + \frac{\frac{1}{2}}{2}\right)$$

$$= -\tfrac{1}{8}(2y + 1)\cos(y - 2x).$$

Example 6. Find z_p for $(D_x + 4D_y)^3 z = (2x - y)^2$.

Using formula 79, we obtain

$$z_p = \frac{\int\int\int u^2 du^3}{[2 + 4(-1)]^3} = \frac{u^5}{5\cdot 4\cdot 3\cdot(-8)} = -\frac{(2x - y)^5}{480}.$$

Example 7. Find z_p for $(D_x^3 + 5D_x^2D_y - 7D_x + 4)z = e^{2x+3y}$.

By formula 87

$$z_p = \frac{e^{2x+3y}}{2^3 + 5\cdot 2^2\cdot 3 - 7\cdot 2 + 4} = \frac{e^{2x+3y}}{58}.$$

Example 8. Find z_p for

$$(D_x^4 + 6D_x^3D_y + D_xD_y + D_y^2 + 9)z = \sin(3x + 4y).$$

Since every term in the left member is of *even* degree in the two operators D_x and D_y, formula 90 is applicable.

It gives

$$z_p = \frac{\sin(3x + 4y)}{(-9)^2 + 6(-9)(-12) + (-12) + (-16) + 9}$$

$$= \frac{\sin(3x + 4y)}{710}.$$

DIFFERENTIAL EQUATIONS

TABLE I: $(D - a)y = R$

R	y_p
1. e^{rx}	$\dfrac{e^{rx}}{r-a}$
2. $\sin sx$*	$-\dfrac{a \sin sx + s \cos sx}{a^2+s^2} = -\dfrac{1}{\sqrt{a^2+s^2}} \sin\left(sx + \tan^{-1}\dfrac{s}{a}\right)$
3. $P(x)$	$-\dfrac{1}{a}\left[P(x) + \dfrac{P'(x)}{a} + \dfrac{P''(x)}{a^2} + \cdots + \dfrac{P^{(n)}(x)}{a^n}\right]$
4. $e^{rx} \sin sx$*	Replace a by $a - r$ in formula 2 and multiply by e^{rx}.
5. $P(x)e^{rx}$	Replace a by $a - r$ in formula 3 and multiply by e^{rx}.
6. $P(x) \sin sx$*	$-\sin sx \left[\dfrac{a}{a^2+s^2}P(x) + \dfrac{a^2-s^2}{(a^2+s^2)^2}P'(x) + \dfrac{a^3-3as^2}{(a^2+s^2)^3}P''(x) + \cdots + \dfrac{a^k - \binom{k}{2}a^{k-2}s^2 + \binom{k}{4}a^{k-4}s^4 - \cdots}{(a^2+s^2)^k}P^{(k-1)}(x) + \cdots\right]$ $-\cos sx \left[\dfrac{s}{a^2+s^2}P(x) + \dfrac{2as}{(a^2+s^2)^2}P'(x) + \dfrac{3a^2s-s^3}{(a^2+s^2)^3}P''(x) + \cdots + \dfrac{\binom{k}{1}a^{k-1}s - \binom{k}{3}a^{k-3}s^3 + \cdots}{(a^2+s^2)^k}P^{(k-1)}(x) + \cdots\right]$
7. $P(x)e^{rx} \sin sx$*	Replace a by $a - r$ in formula 6 and multiply by e^{rx}.
8. e^{ax}	xe^{ax}
9. $e^{ax} \sin sx$*	$-\dfrac{e^{ax} \cos sx}{s}$
10. $P(x)e^{ax}$	$e^{ax} \int P(x)\, dx$
11. $P(x)e^{ax} \sin sx$*	$\dfrac{e^{ax} \sin sx}{s}\left[\dfrac{P'(x)}{s^3} - \dfrac{P'''(x)}{s^3} + \dfrac{P^{v}(x)}{s^5} - \cdots\right] - \dfrac{e^{ax} \cos sx}{s}\left[P(x) - \dfrac{P''(x)}{s^2} + \dfrac{P^{iv}(x)}{s^4} - \cdots\right]$

* For $\cos sx$ in R replace "sin" by "cos" and "cos" by "− sin" in y_p.

$$D^n = \frac{d^n}{dx^n} \qquad \binom{m}{n} = \frac{m!}{(m-n)!n!} \qquad 0! = 1$$

DIFFERENTIAL EQUATIONS (Continued)

TABLE II: $(D - a)^2 y = R$

R	y_p
12. e^{rx}	$\dfrac{e^{rx}}{(r-a)^2}$
13. $\sin sx$*	$\dfrac{1}{(a^2+s^2)^2}[(a^2-s^2)\sin sx + 2as\cos sx] = \dfrac{1}{a^2+s^2}\sin\left(sx + \tan^{-1}\dfrac{2as}{a^2-s^2}\right)$
14. $P(x)$	$\dfrac{1}{a^2}\left[P(x) + \dfrac{2P'(x)}{a} + \dfrac{3P''(x)}{a^2} + \cdots + \dfrac{(n+1)P^{(n)}(x)}{a^n}\right]$
15. $e^{rx}\sin sx$*	Replace a by $a - r$ in formula 13 and multiply by e^{rx}.
16. $P(x)e^{rx}$	Replace a by $a - r$ in formula 14 and multiply by e^{rx}.
17. $P(x)\sin sx$*	$\sin sx\left[\dfrac{a^2-s^2}{(a^2+s^2)^2}P(x) + 2\dfrac{a^3-3as^2}{(a^2+s^2)^3}P'(x) + 3\dfrac{a^4-6a^2s^2+s^4}{(a^2+s^2)^4}P''(x) + \cdots + (k-1)\dfrac{a^k - \binom{k}{2}a^{k-2}s^2 + \binom{k}{4}a^{k-4}s^4 - \cdots}{(a^2+s^2)^k}P^{(k-2)}(x) + \cdots\right]$ $+ \cos sx\left[\dfrac{2as}{(a^2+s^2)^2}P(x) + 2\dfrac{3a^2s-s^3}{(a^2+s^2)^3}P'(x) + 3\dfrac{4a^3s-4as^3}{(a^2+s^2)^4}P''(x) + \cdots + (k-1)\dfrac{\binom{k}{1}a^{k-1}s - \binom{k}{3}a^{k-3}s^3 + \cdots}{(a^2+s^2)^k}P^{(k-2)}(x) + \cdots\right]$
18. $P(x)e^{rx}\sin sx$*	Replace a by $a - r$ in formula 17 and multiply by e^{rx}.
19. e^{ax}	$\frac{1}{2}x^2e^{ax}$
20. $e^{ax}\sin sx$*	$-\dfrac{e^{ax}\sin sx}{s^2}$
21. $P(x)e^{ax}$	$e^{ax}\displaystyle\iint P(x)dx\,dx$
22. $P(x)e^{ax}\sin sx$*	$-\dfrac{e^{ax}\sin sx}{s^2}\left[P(x) - \dfrac{3P''(x)}{s^2} + \dfrac{5P^{iv}(x)}{s^4} - \dfrac{7P^{vi}(x)}{s^6} + \cdots\right] - \dfrac{e^{ax}\cos sx}{s^2}\left[\dfrac{2P'(x)}{s} - \dfrac{4P'''(x)}{s^3} + \dfrac{6P^{v}(x)}{s^5} - \cdots\right]$

* For $\cos sx$ in R replace "sin" by "cos" and "cos" by "$-$ sin" in y_p.

DIFFERENTIAL EQUATIONS (Continued)

TABLE III: $(D^2 + q)y = R$

R	y_p
23. e^{rx}	$\dfrac{e^{rx}}{r^2+q}$
24. $\sin sx$*	$\dfrac{\sin sx}{-s^2+q}$
25. $P(x)$	$\dfrac{1}{q}\left[P(x) - \dfrac{P''(x)}{q} + \dfrac{P^{iv}(x)}{q^2} - \cdots + (-1)^k \dfrac{P^{(2k)}(x)}{q^k} \cdots\right]$
26. $e^{rx}\sin sx$*	$\dfrac{(r^2 - s^2 + q)e^{rx}\sin sx - 2\,rse^{rx}\cos sx}{(r^2-s^2+q)^2 + (2rs)^2} = \dfrac{e^{rx}}{\sqrt{(r^2-s^2+q)^2+(2rs)^2}}\sin\left[sx - \tan^{-1}\dfrac{2rs}{r^2-s^2+q}\right]$
27. $P(x)e^{rx}$	$\dfrac{e^{rx}}{r^2+q}\left[P(x) - \dfrac{2r}{r^2+q}P'(x) + \dfrac{3r^2-q}{(r^2+q)^2}P''(x) - \dfrac{4r^3-4qr}{(r^2+q)^3}P'''(x) + \cdots + (-1)^{k-1}\dfrac{\binom{k}{1}r^{k-1} - \binom{k}{3}r^{k-3}q + \binom{k}{5}r^{k-5}q^2 - \cdots}{(r^2+q)^{k-1}}P^{(k-1)}(x) + \cdots\right]$
28. $P(x)\sin sx$*	$\dfrac{\sin sx}{(-s^2+q)}\left[P(x) - \dfrac{3s^2+q}{(-s^2+q)^2}P''(x) + \dfrac{5s^4+10s^2q+q^2}{(-s^2+q)^4}P^{iv}(x) + \cdots + (-1)^k\dfrac{\binom{2k+1}{1}s^{2k} + \binom{2k+1}{3}s^{2k-2}q + \binom{2k+1}{5}s^{2k-4}q^2 + \cdots}{(-s^2+q)^{2k}}P^{(2k)}(x) + \cdots\right]$ $- \dfrac{s\cos sx}{(-s^2+q)}\left[\dfrac{2P'(x)}{(-s^2+q)} - \dfrac{4s^2+4q}{(-s^2+q)^3}P'''(x) + \cdots + (-1)^{k+1}\dfrac{\binom{2k}{1}s^{2k-2} + \binom{2k}{3}s^{2k-4}q + \cdots}{(-s^2+q)^{2k-1}}P^{(2k-1)}(x) + \cdots\right]$

TABLE IV: $(D^2 + b^2)y = R$

R	y_p
29. $\sin bx$*	$-\dfrac{x\cos bx}{2b}$
30. $P(x)\sin bx$*	$\dfrac{\sin bx}{(2b)^2}\left[P(x) - \dfrac{P''(x)}{(2b)^2} + \dfrac{P^{iv}(x)}{(2b)^4} - \cdots\right] - \dfrac{\cos bx}{2b}\int\left[P(x) - \dfrac{P''(x)}{(2b)^2} + \cdots\right]dx$

* For $\cos sx$ in R replace "sin" by "cos" and "cos" by "− sin" in y_p.

DIFFERENTIAL EQUATIONS (Continued)

TABLE V: $(D^2 + pD + q)y = R$

R	y_p
31. e^{rx}	$\dfrac{e^{rx}}{r^2 + pr + q}$
32. $\sin sx$*	$\dfrac{(q - s^2)\sin sx - ps\cos sx}{(q - s^2)^2 + (ps)^2} = \dfrac{1}{\sqrt{(q - s^2)^2 + (ps)^2}}\sin\left(sx - \tan^{-1}\dfrac{ps}{q - s^2}\right)$
33. $P(x)$	$\dfrac{1}{q}\left[P(x) - \dfrac{p}{q}P'(x) + \dfrac{p^2 - q}{q^2}P''(x) - \dfrac{p^3 - 2pq}{q^3}P'''(x) + \cdots + (-1)^n \dfrac{p^n - \binom{n-1}{1}p^{n-2}q + \binom{n-2}{2}p^{n-4}q^2 - \cdots}{q^n}P^{(n)}(x)\right]$
34. $e^{rx}\sin sx$*	Replace p by $p + 2r$, q by $q + pr + r^2$ in formula 32 and multiply by e^{rx}.
35. $P(x)e^{rx}$	Replace p by $p + 2r$, q by $q + pr + r^2$ in formula 33 and multiply by e^{rx}.

TABLE VI: $(D - b)(D - a)y = R$

R	y_p
36. $P(x)\sin sx$*	$\dfrac{\sin sx}{b - a}\left[\left(\dfrac{a}{a^2 + s^2} - \dfrac{b}{b^2 + s^2}\right)P(x) + \left(\dfrac{a^2 - s^2}{(a^2 + s^2)^2} - \dfrac{b^2 - s^2}{(b^2 + s^2)^2}\right)P'(x) + \left(\dfrac{a^3 - 3as^2}{(a^2 + s^2)^3} - \dfrac{b^3 - 3bs^2}{(b^2 + s^2)^3}\right)P''(x) + \cdots\right]$ $+ \dfrac{\cos sx}{b - a}\left[\left(\dfrac{s}{a^2 + s^2} - \dfrac{s}{b^2 + s^2}\right)P(x) + \left(\dfrac{2as}{(a^2 + s^2)^2} - \dfrac{2bs}{(b^2 + s^2)^2}\right)P'(x) + \left(\dfrac{3a^2s - s^3}{(a^2 + s^2)^3} - \dfrac{3b^2s - s^3}{(b^2 + s^2)^3}\right)P''(x) + \cdots\right]$†
37. $P(x)e^{rx}\sin sx$*	Replace a by $a - r$, b by $b - r$ in formula 36 and multiply by e^{rx}.
38. $P(x)e^{ax}$	$\dfrac{e^{ax}}{a - b}\left[\int P(x)dx + \dfrac{P(x)}{(b - a)} + \dfrac{P'(x)}{(b - a)^2} + \dfrac{P''(x)}{(b - a)^3} + \cdots + \dfrac{P^{(n)}(x)}{(b - a)^{n+1}}\right]$

* For $\cos sx$ in R replace "sin" by "cos" and "cos" by "$-$sin" in y_p.

† For additional terms, compare with formula 6.

DIFFERENTIAL EQUATIONS (Continued)

TABLE VII: $(D^2 - 2aD + a^2 + b^2)y = R$

R	y_p
39. $P(x)\sin sx$*	$\frac{\sin sx}{2b}\left[\left(\frac{s+b}{a^2+(s+b)^2} - \frac{s-b}{a^2+(s-b)^2}\right)P(x) + \left(\frac{2a(s+b)}{[a^2+(s+b)^2]^2} - \frac{2a(s-b)}{[a^2+(s-b)^2]^2}\right)P'(x) + \left(\frac{3a^2(s+b)-(s+b)^3}{[a^2+(s+b)^2]^3} - \frac{3a^2(s-b)-(s-b)^3}{[a^2+(s-b)^2]^3}\right)P''(x) + \cdots\right]$ $- \frac{\cos sx}{2b}\left[\left(\frac{a}{a^2+(s+b)^2} - \frac{a}{a^2+(s-b)^2}\right)P(x) + \left(\frac{a^2-(s+b)^2}{[a^2+(s+b)^2]^2} - \frac{a^2-(s-b)^2}{[a^2+(s-b)^2]^2}\right)P'(x) + \left(\frac{a^3-3a(s+b)^2}{[a^2+(s+b)^2]^3} - \frac{a^3-3a(s-b)^2}{[a^2+(s-b)^2]^3}\right)P''(x) + \cdots\right]^{\dagger}$
40. $P(x)e^{rx}\sin sx$*	Replace a by $a - r$ in formula 39 and multiply by e^{rx}.
41. $P(x)e^{ax}$	$\frac{e^{ax}}{b^2}\left[P(x) - \frac{P''(x)}{b^2} + \frac{P^{iv}(x)}{b^4} - \cdots\right]$
42. $e^{ax}\sin sx$*	$\frac{e^{ax}\sin sx}{-s^2+b^2}$
43. $e^{ax}\sin bx$*	$-\frac{xe^{ax}\cos bx}{2b}$
44. $P(x)e^{ax}\sin bx$*	$\frac{e^{ax}\sin bx}{(2b)^2}\left[P(x) - \frac{P''(x)}{(2b)^2} + \frac{P^{iv}(x)}{(2b)^4} - \cdots\right] - \frac{e^{ax}\cos bx}{2b}\int\left[P(x) - \frac{P''(x)}{(2b)^2} + \frac{P^{iv}(x)}{(2b)^4} - \cdots\right]dx$

* For $\cos sx$ in R replace "sin" by "cos" and "cos" by "$-$ sin" in y_p.

† For additional terms, compare with formula 6.

DIFFERENTIAL EQUATIONS (Continued)

TABLE VIII: $f(D)y = [D^n + a_{n-1}D^{n-1} + \cdots + a_1D + a_0]y = R$

R	y_p
45. e^{rx}	$\dfrac{e^{rx}}{f(r)}$
46. $\sin sx$*	$\dfrac{[a_0 - a_2s^2 + a_4s^4 - \cdots]\sin sx - [a_1s - a_3s^3 + a_5s^5 + \cdots]\cos sx}{[a_0 - a_2s^2 + a_4s^4 - \cdots]^2 + [a_1s - a_3s^3 + a_5s^5 - \cdots]^2}$

TABLE IX: $f(D^2)y = R$

R	y_p
47. $\sin sx$*	$\dfrac{\sin sx}{f(-s^2)} = \dfrac{\sin sx}{a_0 - a_2s^2 + \cdots \pm s^{2n}}$

TABLE X: $(D - a)^n y = R$

R	y_p
48. e^{rx}	$\dfrac{e^{rx}}{(r-a)^n}$
49. $\sin sx$*	$\dfrac{(-1)^n}{(a^2+s^2)^n}\left\{\left[a^n - \binom{n}{2}a^{n-2}s^2 + \binom{n}{4}a^{n-4}s^4 - \cdots\right]\sin sx + \left[\binom{n}{1}a^{n-1}s - \binom{n}{3}a^{n-3}s^3 + \cdots\right]\cos sx\right\}$
50. $P(x)$	$\dfrac{(-1)^n}{a^n}\left[P(x) + \binom{n}{1}\dfrac{P'(x)}{a} + \binom{n+1}{2}\dfrac{P''(x)}{a^2} + \binom{n+2}{3}\dfrac{P'''(x)}{a^3} + \cdots\right]$
51. $e^{rx}\sin sx$*	Replace a by $a - r$ in formula 49 and multiply by e^{rx}.
52. $e^{rx}P(x)$	Replace a by $a - r$ in formula 50 and multiply by e^{rx}.

* For $\cos sx$ in R replace "sin" by "cos" and "cos" by "$-$ sin" in y_p.

DIFFERENTIAL EQUATIONS (Continued)

53. $P(x)\sin sx$*

$$(-1)^n \sin sx\left[A_nP(x) + \binom{n}{1}A_{n+1}P'(x) + \binom{n+1}{2}A_{n+2}P''(x) + \binom{n+2}{3}A_{n+3}P'''(x) + \cdots\right]$$
$$+ (-1)^n \cos sx\left[B_nP(x) + \binom{n}{1}B_{n+1}P'(x) + \binom{n+1}{2}B_{n+2}P''(x) + \binom{n+2}{3}B_{n+3}P'''(x) + \cdots\right]$$
$$A_1 = \frac{a}{a^2+s^2},\ A_2 = \frac{a^2-s^2}{(a^2+s^2)^2},\ \cdots,\ A_k = \frac{a^k - \binom{k}{2}a^{k-2}s^2 + \binom{k}{4}a^{k-4}s^4 - \cdots}{(a^2+s^2)^k}.$$
$$B_1 = \frac{s}{a^2+s^2},\ B_2 = \frac{2as}{(a^2+s^2)^2},\ \cdots,\ B_k = \frac{\binom{k}{1}a^{k-1}s - \binom{k}{3}a^{k-3}s^3 + \cdots}{(a^2+s^2)^k}.$$

54. $P(x)e^{rx}\sin sx$* Replace a by $a - r$ in formula 53 and multiply by e^{rx}.

55. $e^{ax}P(x)$ $\quad e^{ax}\iint\cdots\int P(x)dx^n$

56. $P(x)e^{ax}\sin sx$*

$$\frac{(-1)^{\frac{n-1}{2}}e^{ax}\sin sx}{s^n}\left[\binom{n}{n-1}\frac{P'(x)}{s} - \binom{n+2}{n-1}\frac{P'''(x)}{s^3} + \binom{n+4}{n-1}\frac{P^{\text{v}}(x)}{s^5} - \cdots\right]$$
$$+ \frac{(-1)^{\frac{n+1}{2}}e^{ax}\cos sx}{s^n}\left[\binom{n-1}{n-1}P(x) - \binom{n+1}{n-1}\frac{P''(x)}{s^2} + \binom{n+3}{n-1}\frac{P^{\text{iv}}(x)}{s^4} - \cdots\right] \quad (n \text{ odd})$$

$$\frac{(-1)^{\frac{n}{2}}e^{ax}\sin sx}{s^n}\left[\binom{n-1}{n-1}P(x) - \binom{n+1}{n-1}\frac{P''(x)}{s^2} + \binom{n+3}{n-1}\frac{P^{\text{iv}}(x)}{s^4} - \cdots\right]$$
$$+ \frac{(-1)^{\frac{n}{2}}e^{ax}\cos sx}{s^n}\left[\binom{n}{n-1}\frac{P'(x)}{s} - \binom{n+2}{n-1}\frac{P'''(x)}{s^3} + \binom{n+4}{n-1}\frac{P^{\text{v}}(x)}{s^5} - \cdots\right] \quad (n \text{ even})$$

TABLE XI: $(D - a)^n f(D)y = R$

57. e^{ax} $\quad \dfrac{x^n}{n!}\cdot\dfrac{e^{ax}}{f(a)}$

* For $\cos sx$ in R replace "sin" by "cos" and "cos" by "$-$ sin" in y_p.

DIFFERENTIAL EQUATIONS (Continued)

TABLE XII: $(D^2 + q)^n y = R$

R	y_p
58. e^{rx}	$e^{rx}/(r^2 + q)^n$
59. $\sin sx$*	$\sin sx/(q - s^2)^n$
60. $P(x)$	$\frac{1}{q^n}\left[P(x) - \binom{n}{1}\frac{P''(x)}{q} + \binom{n+1}{2}\frac{P^{\text{iv}}(x)}{q^2} - \binom{n+2}{3}\frac{P^{\text{vi}}(x)}{q^3} + \cdots\right]$
61. $e^{rx}\sin sx$*	$\frac{e^{rx}}{(A^2 + B^2)^n}\{[A^n - \binom{r}{2}A^{n-2}B^2 + \binom{n}{4}A^{n-4}B^4 - \cdots]\sin sx - [\binom{n}{1}A^{n-1}B - \binom{n}{3}A^{n-3}B^3 + \cdots]\cos sx\}$ $A = r^2 - s^2 + q, \quad B = 2rs$

TABLE XIII: $(D^2 + b^2)^n y = R$

R	y_p
62. $\sin bx$*	$(-1)^{\frac{n+1}{2}}\frac{x^n \cos bx}{n!(2b)^n}$ (n odd), $\quad (-1)^{\frac{n}{2}}\frac{x^n \sin bx}{n!(2b)^n}$ (n even)

TABLE XIV: $(D^n - q)y = R$

R	y_p
63. e^{rx}	$e^{rx}/(r^n - q)$
64. $P(x)$	$-\frac{1}{q}\left[P(x)\frac{P^{(n)}(x)}{q} + \frac{P^{(2n)}(x)}{q^2} + \cdots\right]$
65. $\sin sx$*	$-\frac{q \sin sx + (-1)^{\frac{n-1}{2}} s^n \cos sx}{q^2 + s^{2n}}$ (n odd), $\quad \frac{\sin sx}{(-s^2)^{n/2} - q}$ (n even)
66. $e^{rx}\sin sx$*	$\frac{Ae^{rx}\sin sr - Be^{rx}\cos sx}{A^2 + B^2} = \frac{e^{rx}}{\sqrt{A^2 + B^2}}\sin\left(sx - \tan^{-1}\frac{B}{A}\right)$ $A = [r^n - \binom{n}{2}r^{n-2}s^2 + \binom{n}{4}r^{n-4}s^4 - \cdots] - q, \quad B = [\binom{n}{1}r^{n-1}s - \binom{n}{3}r^{n-3}s^3 + \cdots]$

* For $\cos sx$ in R replace "sin" by "cos" and "cos" by "$-$ sin" in y_p.

DIFFERENTIAL EQUATIONS (Continued)

TABLE XV: $(D_x + mD_y)z = R$

R	z_p
67. e^{ax+by}	$\dfrac{e^{ax+by}}{a+mb}$
68. $f(ax+by)$	$\dfrac{\int f(u)du}{a+mb}$, $u = ax + by$
69. $f(y-mx)$	$xf(y-mx)$
70. $\phi(x, y)f(y-mx)$	$f(y-mx)\int \phi(x, a+mx)dx$ $(a = y - mx$ after integration)

TABLE XVI: $(D_x + mD_y - k)z = R$

R	z_p
71. e^{ax+by}	$\dfrac{e^{ax+by}}{a+mb-k}$
72. $\sin(ax+by)$*	$-\dfrac{(a+bm)\cos(ax+by) + k\sin(ax+by)}{(a+bm)^2+k^2}$
73. $e^{\alpha x+\beta y}\sin(ax+by)$*	Replace k in 72 by $k - \alpha - m\beta$ and multiply by $e^{\alpha x+\beta y}$
74. $e^{xk}f(ax+by)$	$\dfrac{e^{kx}\int f(u)du}{a+mb}$, $u = ax + by$
75. $f(y-mx)$	$-\dfrac{f(y-mx)}{k}$
76. $P(x)f(y-mx)$	$-\dfrac{1}{k}f(y-mx)\left[P(x) + \dfrac{P'(x)}{k} + \dfrac{P''(x)}{k^2} + \cdots + \dfrac{P^{(n)}(x)}{k^n}\right]$
77. $e^{kx}f(y-mx)$	$xe^{kx}f(y-mx)$

* For $\cos(ax+by)$ replace "sin" by "cos," and "cos" by "– sin" in z_p.

$$D_x = \frac{\partial}{\partial x}; \quad D_y = \frac{\partial}{\partial y}; \quad D_x^k D_y^r = \frac{\partial^{k+r}}{\partial_x^k \partial_y^r}$$

DIFFERENTIAL EQUATIONS (Continued)

TABLE XVII: $(D_x + mD_y)^n z = R$

R	z_p
78. e^{ax+by}	$\dfrac{e^{ax+by}}{(a+mb)^n}$
79. $f(ax+by)$	$\dfrac{\int\int\cdots\int f(u)du^n}{(a+mb)^n}$, $u = ax+by$
80. $f(y-mx)$	$\dfrac{x^n}{n!}f(y-mx)$
81. $\phi(x, y)f(y+mx)$	$f(y-mx)\int\int\cdots\int\phi(x, a+mx)dx^n$ $(a = y - mx$ after integration)

TABLE XVIII: $(D_x + mD_y - k)^n z = R$

R	z_p
82. e^{ax+by}	$\dfrac{e^{ax+by}}{(a+mb-k)^n}$
83. $f(y-mx)$	$\dfrac{(-1)^n f(y-mx)}{k^n}$
84. $P(x)f(y-mx)$	$\dfrac{(-1)^n}{k^n}f(y-mx)\left[P(x) + \binom{n}{1}\dfrac{P'(x)}{k} + \binom{n+1}{2}\dfrac{P''(x)}{k^2} + \binom{n+2}{3}\dfrac{P'''(x)}{k^3} + \cdots\right]$
85. $e^{kx}f(ax+by)$	$\dfrac{e^{kx}\int\int\cdots\int f(u)du^n}{(a+mb)^n}$, $u = ax+by$
86. $e^{kx}f(y-mx)$	$\dfrac{x^n}{n!}e^{kx}f(y-mx)$

DIFFERENTIAL EQUATIONS (Continued)

TABLE XIX: $[D_x^n + a_1 D_x^{n-1} D_y + a_2 D_x^{n-2} D_y^2 + \cdots + a^n D_y^n]z = R$

87. e^{ax+by}	$\dfrac{e^{ax+by}}{a + a_1 a^{n-1} b + a_2 a^{n-2} b^2 + \cdots + a_n b^n}$
88. $f(ax+by)$	$\dfrac{\int\int \cdots \int f(u) du^n}{a^n + a_1 a^{n-1} b + a_2 a^{n-2} b^2 + \cdots + a^n b^n}, \ (u = ax + by)$

TABLE XX: $F(D_x, D_y)z = R$

89. e^{ax+by}	$\dfrac{e^{ax+by}}{F(a, b)}$

TABLE XXI: $F(D_x^2, D_x D_y, D_y^2)z = R$

90. $\sin(ax + by)$*	$\dfrac{\sin(ax + by)}{F(-a^2, -ab, -b^2)}$

* For $\cos(ax + by)$ replace "sin" by "cos", and "cos" by "−sin" in z_p.

THE NUMERICAL SOLUTION OF DIFFERENTIAL EQUATIONS

Introduction

The aim is to produce a numerical approximation to the solution of the differential equation, correct to a specified number of decimals.

The principal factors affecting accuracy are

a) the form of the equation,
b) the interval of tabulation,
c) the formula chosen to effect the integration.

The main types of equation considered here are classified as follows

TABLE I

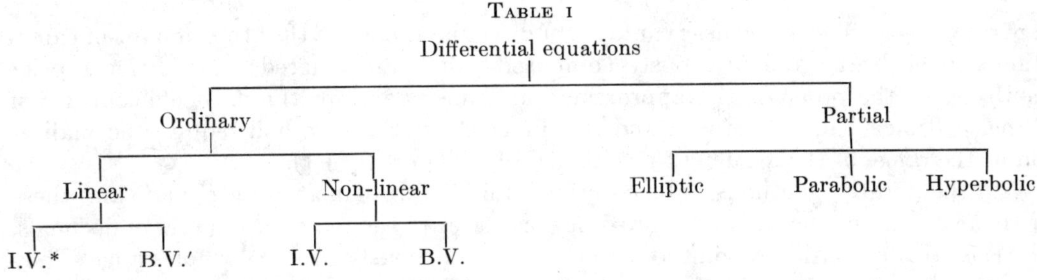

I.V.* = Initial value problems, sometimes referred to as Marching problems.
B.V.' = Boundary value problems, sometimes referred to as Jury problems.

Ordinary differential equations

With initial value problems, all the conditions required to determine uniquely the constants arising from the integration procedure are given at one point. With boundary value problems these conditions are given at two or more points.

For ordinary boundary value problems the solution can usually be based upon initial value techniques as follows. One of the boundary points is chosen as the principal point and the variables which are given at the other boundary point(s) are estimated at this point. This is repeated for a number of different estimates and a combination of the resulting solutions is chosen so that the numerical values satisfy both the differential equation and also all the original boundary conditions.

The principal methods of solving ordinary initial value differential equations to be considered are classified as follows. Purely analytical techniques will not be dealt with.

In the direct methods the value at any point is obtained from the value at the preceeding point without any iteration. In the case of the Taylor's series approach it is a simple matter to include a running check on previously computed values. With the Runge-Kutta and Chebyshev techniques checking is more difficult. For linear differential equations, or any for which repeated differentiation is practical, the Taylors series method can be used with a

THE NUMERICAL SOLUTION OF DIFFERENTIAL EQUATIONS (Continued)

TABLE II

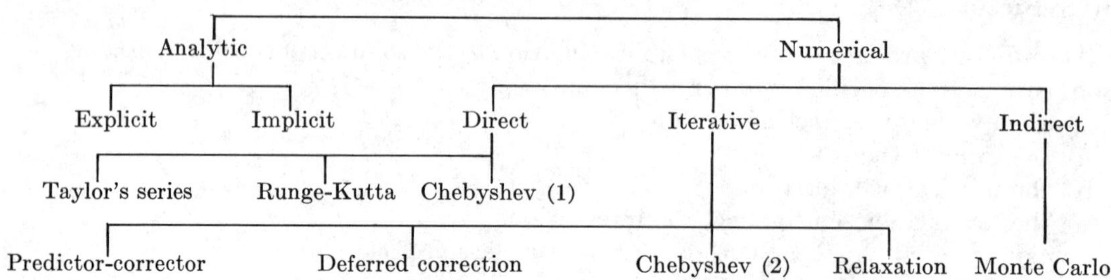

large step length. The basic disadvantages of direct methods are that the amount of calculation can be heavy and previously computed values are ignored. In addition a prior knowledge of the order of the approximation is necessary for the most efficient use of the basic Runga-Kutta approach and the direct Chebyshev method requires normalization of the range of the argument to either $(-1, +1)$ or $(0, +1)$.

With the iterative methods to be described, initial approximations are made and these are then refined until successive approximations agree to the required number of decimals. This is carried out either a point at a time or simultaneously over the whole range. They have the advantage of being self correcting in the sense that small computational errors are eliminated in subsequent iterations.

In this group of methods the predictor-corrector methods are particularly suitable for desk machine operation, especially with the more complicated non-linear differential equation. Their accuracy is controlled by the corrector formula which has associated with it an indication of the maximum interval to ensure convergence. These formulae are expressed either in terms of ordinates or backward and or central differences.

The deferred corrector methods are particularly suitable for desk machine work and are highly recommended. Relaxation methods are seldom used for initial value problems but they are naturally applicable to boundary value problems and in fact an initial value problem has to be converted to a boundary value one if relaxation is to be used.

Some predictor-corrector methods can be modified to be non-iterative but all will require special starting techniques to get under way.

General observations applicable to all methods

There will always be an optimum interval "h" of the argument (independent variable) and the method chosen should be able to take into account the need for changing this as the solution proceeds.

Build-up error due to rounding-off of differences can cause trouble. Two or three guarding figures should be held and there should always be running and terminal checks incorporated in any solution.

The stability of the method i.e. the convergence to the wanted solution, depends upon the differential equation as well as the method chosen. Starting with the last computed value and integrating in the reverse direction back to the original origin will check this.

The truncation error due to using a finite approximation to an infinite series will reduce the number of significant decimals obtained. However, this can be corrected at a later stage.

THE NUMERICAL SOLUTION OF DIFFERENTIAL EQUATIONS (Continued)

Any order differential equation can be reduced to a number of simultaneous differential equations of the first order and so the solution of first order differential equations is considered in detail. In some cases there will however be a loss of convenience if such a reduction is employed.

e.g. Given $y'' = xy - \sin x$ with $y(0) = 1, \quad y'(0) = 0.$

This reduces to

$$\begin{aligned} y' &= w & &\text{with } y(0) = 1 \\ z' &= v & &z(0) = 0 \\ v' &= -z & &v(0) = 1 \\ w' &= xy - z & &w(0) = 0 \end{aligned}$$

DIRECT METHODS

Taylor's series

Applicable to all linear equations and those non-linear equations whose higher derivatives can be easily found.

e.g. $$y' = f(x,y) \qquad \text{i.e.} \quad y^{(1)} = f(x,y^{(0)})$$

then $$y^{(n+1)} = \frac{\partial y^{(n)}}{\partial x} + y^{(1)} \frac{\partial y^{(n)}}{\partial y}$$

(since $y^{(n)}$ can be reduced to a function of x and y only.)

Finally $$x_1 = x_0 + h$$

and $$y_1 = y_0 + \frac{hy_0^{(1)}}{1!} + \frac{h^2y_0^{(2)}}{2!} + \frac{h^3y_0^{(3)}}{3!} + \frac{h^4y_0^{(4)}}{4!} + \cdots$$

In practice the computational procedure is simplified by introducing the reduced derivatives defined by

$$\tau^{(n)} \equiv \frac{h^n y^{(n)}}{n!} \qquad (\tau^{(0)} \equiv y)$$

from which it follows that $$\frac{d\tau^{(n)}}{dx} = \frac{(n+1)\tau^{(n+1)}}{h}$$

For linear equations a recurrence relationship is obtained by means of Liebnitz's theorem to generate any particular $\tau^{(n)}$ from the preceeding τ's. Thus, given the differential equation and the initial conditions, the sequence $\tau^{(0)}, \tau^{(1)}\ \tau^{(2)}, \tau^{(3)}, \ldots, \tau^{(n)}$ is computed until the terms cease to have any significance with regard to the number of places of decimals required.

If the Taylor's series is re-arranged then not only can the new function value be computed but a running check on the previously computed value be obtained with a minimum of effort. This can also be applied to as many of the derivatives as are desired.

$$\begin{aligned} y_1 &= (\tau_0^{(0)} + \tau_0^{(2)} + \tau_0^{(4)} + \tau_0^{(6)} \cdots) + (\tau_0^{(1)} + \tau_0^{(3)} + \tau_0^{(5)} + \cdots) \\ \tau_1^{(1)} &= (2\tau_0^{(2)} + 4\tau_0^{(4)} + 6\tau_0^{(6)} + \cdots) + (\tau_0^{(1)} + 3\tau_0^{(3)} + 5\tau_0^{(5)} + \cdots) \end{aligned}$$

The checking formulae corresponding to the above pair are

$$\begin{aligned} y_{-1} &= (\tau_0^{(0)} + \tau_0^{(2)} + \tau_0^{(4)} + \tau_0^{(6)} \cdots) - (\tau_0^{(1)} + \tau_0^{(3)} + \tau_0^{(5)} \cdots) \\ \tau_{-1}^{(1)} &= (2\tau_0^{(2)} + 4\tau_0^{(4)} + 6\tau_0^{(6)} + \cdots) - (\tau_0^{(1)} + 3\tau_0^{(3)} + 5\tau_0^{(5)} + \cdots) \end{aligned}$$

THE NUMERICAL SOLUTION OF DIFFERENTIAL EQUATIONS (Continued)

It will be seen that the sums inside the brackets which are used to evaluate the y_1 and the $\tau_1^{(1)}$ are also used to evaluate the y_{-1} and the $\tau_{-1}^{(1)}$. A shift of origin from (x_0,y_0) to (x_1,y_1) is then made, in accordance with the usual procedure for advancing the solution of a differential equation, and the process repeated. If too many terms are necessary, before the τ's assume negligible proportions, then the interval of tabulation "h" should be reduced. Conversely, if only a few terms have significance, then "h" should be increased.

Computational layout for one such stage

x_0			y_0	
	$\tau_0^{(0)}$	$\tau_0^{(1)}$		
	$\tau_0^{(2)}$	$\tau_0^{(3)}$		
	$\tau_0^{(4)}$	$\tau_0^{(5)}$		
	. . .	. . . .		
	$\tau_0^{(2n)}$	$\tau_0^{(2n+1)}$		
x_1	$\sum_{r=0}^{n} \tau_0^{(2n)}$	$\sum_{r=0}^{n} \tau_0^{(2n+1)}$	y_1	y_{-1}
	$\sum_{r=0}^{n} 2r\tau_0^{(2r)}$	$\sum_{r=0}^{n} (2r+1)\tau_0^{(2r+1)}$	$\tau_1^{(1)}$	$\tau_{-1}^{(1)}$

Method of Runge-Kutta

An nth order Runge-Kutta process has an error of $0(h^{n+1})$. If $y' = f(x,y)$ then $y_1 = y_0 + K$, where K is the weighted mean of "n" evaluations of $h \cdot y'$ taken at specified points. There are an infinity of such formulae and the weights, and the constants necessary to determine the specified points, are tabulated for some simple cases only.

At any stage of the numerical solution, only the function value at the beginning of the stage need be known. This means that no special starting technique is necessary, but against this must be set the fact that the function must be evaluated "n" times per step. Further, previously computed values are not used to check the computation. Error analysis is not easy and the usual check is simply one of repeating the calculations at a different interval (one simple formula involving an error check is given at the end). Alternatively a finite difference check can be used.

If the functions are not smooth or if a frequent change in step length, particularly when such a change is expected to be a decrease, is envisaged then this process is of particular importance.

Second order systems Error $= 0(h^3)$

Given $y' = f(x,y)$ and y_0
Then $x_1 = x_0 + h$ and $y_1 = y_0 + K$
where $K = \omega_1 k_1 + \omega_2 k_2 \qquad (\omega_1 + \omega_2 = 1)$
and $k_r = h \cdot f(x_0 + a_r h,\ y_0 + b_r k_1)$

THE NUMERICAL SOLUTION OF DIFFERENTIAL EQUATIONS (Continued)

Six typical systems are listed

(i)

r	a_r	b_r	ω_r
1	0	0	-1
2	$\frac{1}{4}$	$\frac{1}{4}$	2

(ii)

r	a_r	b_r	ω_r
1	0	0	0
2	$\frac{1}{2}$	$\frac{1}{2}$	1

(iii)

r	a_r	b_r	ω_r
1	0	0	$\frac{2}{3}$
2	$\frac{3}{4}$	$\frac{3}{4}$	$\frac{1}{3}$

(iv)

r	a_r	b_r	ω_r
1	0	0	$\frac{1}{2}$
2	1	1	$\frac{1}{2}$

(v)

r	a_r	b_r	ω_r
1	0	0	$\frac{1}{4}$
2	$\frac{2}{3}$	$\frac{2}{3}$	$\frac{3}{4}$

(vi)

r	a_r	b_r	ω_r
1	0	0	$-\frac{1}{2}$
2	$\frac{1}{3}$	$\frac{1}{3}$	$1\frac{1}{2}$

Third order systems Error $= 0(h^4)$

Given $y' = f(x,y)$ and y_0
Then $x_1 = x_0 + h$ and $y_1 = y_0 + K$
where $K = \omega_1 k_1 + \omega_2 k_2 + \omega_3 k_3$ $(\omega_1 + \omega_2 + \omega_3 = 1)$
and $k_r = h \cdot f(x_0 + a_r h,\ y_0 + b_r k_1 + c_r k_2)$
Four typical systems are listed

(i)

r	a_r	b_r	c_r	ω_r
1	0	0	0	$\frac{2}{3}$
2	$\frac{1}{4}$	$\frac{1}{4}$	0	$-\frac{4}{3}$
3	$\frac{1}{2}$	$\frac{1}{10}$	$\frac{2}{5}$	$\frac{5}{3}$

(ii)

r	a_r	b_r	c_r	ω_r
1	0	0	0	$\frac{1}{4}$
2	$\frac{1}{3}$	$\frac{1}{3}$	0	0
3	$\frac{2}{3}$	0	$\frac{2}{3}$	$\frac{3}{4}$

(iii)

r	a_r	b_r	c_r	ω_r
1	0	0	0	$\frac{1}{6}$
2	$\frac{1}{2}$	$\frac{1}{2}$	0	$\frac{2}{3}$
3	1	-1	2	$\frac{1}{6}$

(iv)

r	a_r	b_r	c_r	ω_r
1	0	0	0	$\frac{1}{4}$
2	$\frac{2}{3}$	$\frac{2}{3}$	0	$\frac{1}{2}$
3	$\frac{2}{3}$	$-\frac{1}{3}$	1	$\frac{1}{4}$

Fourth order systems Error $= (h^5)$

Given $y' = f(x,y)$ and y_0
Then $x_1 = x_0 + h$ and $y_1 = y_0 + K$
where $K = \omega_1 k_1 + \omega_2 k_2 + \omega_3 k_3 + \omega_4 k_4$ $(\omega_1 + \omega_2 + \omega_3 + \omega_4 = 1)$
and $k_r = h \cdot f(x_0 + a_r,\ y_0 + b_r k_1 + c_r k_2 + d_r k_3)$

THE NUMERICAL SOLUTION OF DIFFERENTIAL EQUATIONS (Continued)

Four typical systems are listed. The third is of particular importance both from the point of view of simplicity of numerical constants and of and the sequential relationships between the various k's.

(i)

r	a_r	b_r	c_r	d_r	ω_r
1	0	0	0	0	$\frac{1}{8}$
2	$\frac{1}{3}$	$\frac{1}{3}$	0	0	$\frac{3}{8}$
3	$\frac{2}{3}$	$-\frac{1}{3}$	1	0	$\frac{3}{8}$
4	1	1	-1	1	$\frac{1}{8}$

(ii)

r	a_r	b_r	c_r	d_r	ω_r
1	0	0	0	0	$\frac{1}{6}$
2	1	1	0	0	0
3	$\frac{1}{2}$	$\frac{3}{8}$	$\frac{1}{8}$	0	$\frac{2}{3}$
4	1	$-\frac{1}{2}$	$-\frac{1}{2}$	2	$\frac{1}{6}$

(iii)

r	a_r	b_r	c_r	d_r	ω_r
1	0	0	0	0	$\frac{1}{6}$
2	$\frac{1}{2}$	$\frac{1}{2}$	0	0	$\frac{1}{3}$
3	$\frac{1}{2}$	0	$\frac{1}{2}$	0	$\frac{1}{3}$
4	1	0	0	1	$\frac{1}{6}$

(iv)

r	a_r	b_r	c_r	d_r	ω_r
1	0	0	0	0	0
2	$\frac{1}{2}$	$\frac{1}{2}$	0	0	$\frac{2}{3}$
3	0	$-\frac{1}{2}$	$\frac{1}{2}$	0	$\frac{1}{6}$
4	1	$-\frac{1}{2}$	$1\frac{1}{2}$	1	$\frac{1}{6}$

Fifth order systems Error $= 0(h^6)$

The number of terms required to form the weighted mean "K" increases, for orders greater than the fourth and, in addition, the coefficients are much less easy to handle.

The following system requires one additional term.

Given $y' = f(x,y)$ and y_0

Then $x_1 = x_0 + h$, and $y_1 = y_0 + K$

where $K = \omega_1 k_1 + \omega_2 k_2 + \omega_3 k_3 + \omega_4 k_4 + \omega_5 k_5 + \omega_6 k_6$

$$(\omega_1 + \omega_2 + \omega_3 + \omega_4 + \omega_5 + \omega_6 = 1)$$

and $k_r = h \cdot f(x_0 + a_r h,\ y_0 + b_r k_1 + c_r k_2 + d_r k_3 + e_r k_4 + f_r k_5)$

r	a_r	b_r	c_r	d_r	e_r	f_r	ω_r
1	0	0	0	0	0	0	23/192
2	$\frac{1}{3}$	$\frac{1}{3}$	0	0	0	0	0
3	$\frac{2}{5}$	$\frac{4}{25}$	$\frac{16}{25}$	0	0	0	125/192
4	1	$\frac{1}{4}$	-3	$\frac{15}{4}$	0	0	0
5	$\frac{2}{3}$	$\frac{2}{27}$	$\frac{10}{9}$	$\frac{50}{81}$	$\frac{8}{81}$	0	$-27/64$
6	$\frac{4}{5}$	$\frac{2}{25}$	$\frac{12}{25}$	$\frac{2}{15}$	$\frac{8}{75}$	0	125/192

THE NUMERICAL SOLUTION OF DIFFERENTIAL EQUATIONS (Continued)

An example of computational layout

Consider formula (iii) of the fourth order systems. On substituting the values from the table the resulting formulae are as follows

$$y' = f(x,y)$$
$$x_1 = x_0 + h$$
$$k_1 = h \cdot f(x_0, y_0)$$
$$k_2 = h \cdot f(x_0 + \tfrac{1}{2}h,\ y_0 + \tfrac{1}{2}k_1)$$
$$k_3 = h \cdot f(x_0 + \tfrac{1}{2}h,\ y_0 + \tfrac{1}{2}k_2)$$
$$k_4 = h \cdot f(x_0 + h,\ y_0 + k_3)$$
$$y_1 = y_0 + \tfrac{1}{6}(k_1 + 2k_2 + 2k_3 + k_4)$$

and the layout of one state of the calculation is

x	y	$k_r = hy'$	K
$\underline{x_0}$	$\underline{y_0}$	k_1	
$x_0 + \frac{1}{2}h$	$y_0 + \frac{1}{2}k_1$	k_2	
$x_0 + \frac{1}{2}h$	$y_0 + \frac{1}{2}k_2$	k_3	
$x_0 + h$	$y_0 + k_3$	k_4	
			$(k_1 + 2k_2 + 2k_3 + k_4)$
$x_1 = x_0 + h$	$y_1 = y_0 + K$		

The starting values for this stage are underlined.

Extension of the Runge-Kutta approach to simultaneous differential equations

Consider again table (iii) of the fourth order systems. It can be extended to cover any number of simultaneous D.E's. The extension to two is shown in detail and the general case can be easily deduced from it.

Given $y' = f(x,y,z)$

and $z' = g(x,y,z)$

$x_1 = x_0 + h$

we obtain $y_1 = y_0 + K$

and $z_1 = z_0 + M$

where $K = \omega_1 k_1 + \omega_2 k_2 + \omega_3 k_3 + \omega_4 k_4$

$M = \omega_1 m_1 + \omega_2 m_2 + \omega_3 m_3 + \omega_4 m_4$

and $k_1 = h \cdot f(x_0,\ y_0,\ z_0)$

$k_2 = h \cdot f(x_0 + \frac{1}{2}h,\ y_0 + \frac{1}{2}k_1,\ z_0 + \frac{1}{2}m_1)$

$k_3 = h \cdot f(x_0 + \frac{1}{2}h,\ y_0 + \frac{1}{2}k_2,\ z_0 + \frac{1}{2}m_2)$

$k_4 = h \cdot f(x_0 + h,\ y_0 + k_3,\ z_0 + m_3)$

THE NUMERICAL SOLUTION OF DIFFERENTIAL EQUATIONS (Continued)

$$\text{and}\quad m_1 = h \cdot g(x_0, y_0, z_0)$$
$$m_2 = h \cdot g(x_0 + \tfrac{1}{2}h,\ y_0 + \tfrac{1}{2}k_1,\ z_0 + \tfrac{1}{2}m_1)$$
$$m_3 = h \cdot g(x_0 + \tfrac{1}{2}h,\ y_0 + \tfrac{1}{2}k_2,\ z_0 + \tfrac{1}{2}m_2)$$
$$m_4 = h \cdot g(x_0 + h,\ y_0 + k_3,\ z_0 + m_3)$$

and the computational layout will be similar to the previous one

x	y	z	$k_r = hf$ $K = \Sigma\omega_r k_r$	$m_r = hg$ $M = \Sigma\omega_r m_r$
x_0	y_0	z_0	k_1	m_1
$x_0 + \frac{1}{2}h$	$y_0 + \frac{1}{2}k_1$	$z_0 + \frac{1}{2}m_1$	k_2	m_2
$x_0 + \frac{1}{2}h$	$y_0 + \frac{1}{2}k_2$	$z_0 + \frac{1}{2}m_2$	k_3	m_3
$x_0 + h$	$y_0 + k_3$	$z_0 + m_3$	k_4	m_4
			"K"	"M"
$x_1 = x_0 + h$	$y_1 = y_0 + K$	$z_1 = z_0 + M$		

The use of Runge-Kutta techniques for higher order differential equations

A schematic for the general second order differential equation is shown.

Given $y'' = f(x, y, y')$ and y_0, y_0' the reduced derivatives can be used i.e.

$$\tau^{(n)} \equiv \frac{h^n}{n!} \cdot y^{(n)}$$

$$\text{Then}\quad y'' = f\left(x, y, \frac{\tau^{(1)}}{h}\right)$$

$$\text{and}\quad x_1 = x_0 + h$$
$$y_1 = y_0 + \tau_0^{(1)} + K$$
$$\tau_1^{(1)} = \tau_0^{(1)} + K'$$

$$\text{where}\quad K = \sum_{r=1}^{4} \omega_r k_r \qquad \left(1 = \sum_{r=1}^{4} \omega_r\right)$$

$$\text{and}\quad K' = \sum_{r=1}^{4} \omega_r' k_r \qquad \left(1 = \sum_{r=1}^{4} \omega_r'\right)$$

$$k_r = \frac{h^2}{2} \cdot f(x_0 + a_r h,\ y_0 + b_r\tau_0^{(1)} + c_r k_1 + d_r k_2 + e_r k_3,\ \{\tau_0^{(1)} + g_r k_{r-1}\}/h)$$

THE NUMERICAL SOLUTION OF DIFFERENTIAL EQUATIONS (Continued)

r	a_r	b_r	c_r	d_r	e_r	g_r	ω_r	ω_r'
1	0	0	0	0	0	0	0	$\frac{1}{3}$
2	$\frac{1}{2}$	$\frac{1}{2}$	$\frac{1}{4}$	0	0	1	$\frac{1}{3}$	$\frac{2}{3}$
3	$\frac{1}{2}$	$\frac{1}{2}$	$\frac{1}{4}$	0	0	1	$\frac{1}{3}$	$\frac{3}{2}$
4	1	1	0	0	1	2	$\frac{1}{3}$	$\frac{1}{3}$

The computational layout being as follows

x	y	τ	k, K, K'
x_0	y_0	$\tau_0^{(1)}$	k_1
$x_0 + \frac{1}{2}h$	$y_0 + \frac{1}{2}\tau_0^{(1)} + \frac{1}{4}k_1$	$\tau_0^{(1)} + k_1$	k_2
$x_0 + \frac{1}{2}h$	$y_0 + \frac{1}{2}\tau_0^{(1)} + \frac{1}{4}k_1$	$\tau_0^{(1)} + k_2$	k_3
$x_0 + h$	$y_0 + \tau_0^{(1)} + k_3$	$\tau_0^{(1)} + 2k_3$	k_4
			K
			K'

$$x_1 = x_0 + h \qquad y_1 = y_0 + \tau_0^{(1)} + K \tau_1^{(1)} = \tau_0^{(1)} + K'$$

where
$$K = \omega_1 k_1 + \omega_2 k_2 + \omega_3 k_3 + \omega_4 k_4$$
$$K' = \omega_1' k_1 + \omega_2' k_2 + \omega_3' k_3 + \omega_4' k_4$$

Chebyshev Polynomials (1)

The Chebyshev polynomial of degree r in x where $-1 \leqq x \leqq 1$ is denoted by $T_r \equiv T_r(x)$.

$$T_r(x) \equiv \cos\,(r \cdot \cos^{-1} x)$$
$$\text{and } 2x \cdot T_r = T_{r+1} + T_{r-1}$$

The shifted Chebyshev polynomials of degree r in x where $0 \leqq x \leqq 1$ are denoted by $T_r^* \equiv T_r^*(x)$.

$$T_r^*(x) \equiv \cos\,(r \cdot \cos^{-1}(2x - 1))$$
$$\text{and } \quad 2(2x - 1) \cdot T_r^* = T_{r+1}^* + T_{r-1}^*$$

N.B. $T_r^*(x^2) = T_{2r}(x)$

Both types of polynomials can be used for linear differential equations over a finite interval. The equation is assumed to have polynomial coefficients—if not then suitable polynomial approximations should be inserted. The finite range must first be normalized to either $(-1,1)$ or $(0,1)$.

a) Chebyshev polynomials

$$T_0 = 1$$
$$T_r(1) = 1$$
$$T_r(-1) = (-1)^r$$
$$T_{2r}(0) = (-1)^r$$
$$T_{2r+1}(0) = 0$$

THE NUMERICAL SOLUTION OF DIFFERENTIAL EQUATIONS (Continued)

The solution of the equation $y^{(1)} = f(x,y)$ is assumed to have the form

$$y = \tfrac{1}{2}a_0 + a_1T_1 + a_2T_2 + a_3T_3 + a_4T_4 + \cdots$$

where $a_r = 0$ for all $r > N$

i.e.
$$y = \tfrac{1}{2}a_0 + a_1T_1 + a_2T_2 + \cdots + a_NT_N.$$

similarly
$$y^{(1)} = \tfrac{1}{2}a_0^{(1)} + \sum_{r=1}^{N} a_r^{(1)}T_r \text{ etc.}$$

Writing $C_r(y^{(s)}) \equiv$ the coefficient of T_r in the expansion of $y^{(s)}$ (the s th derivative of y with respect to x)

$$C_r(y^{(s)}) = a_r^{(s)} \tag{1}$$

Further
$$C_r(xy) = \tfrac{1}{2}(a_{r+1} + a_{r-1})$$

and
$$C_r(x^2y) = \tfrac{1}{4}(a_{r+2} + 2a_r + a_{r-2})$$

in general
$$C_r(x^py^{(s)}) = 2^{-p}\sum_{i=0}^{p}\binom{p}{i}a^{(s)}_{|r-p+2i|} \qquad \begin{pmatrix} r = 0,1,2 \ldots \\ s = 1,2, \ldots \end{pmatrix} \tag{2}$$

In addition the Chebyshev recurrence relationship itself yields

$$a_{r-1}^{(s)} = 2r \cdot a_r^{(s-1)} + a_{r+1}^{(s)} \tag{3}$$

Method of solution

(i) Obtain a relationship between the a's and the $a^{(1)}$'s from $C_r\{y^{(1)} - f(x,y)\} = 0$, e.g. If $(x^2 + 1)y^{(1)} - 4xy = 0$.

Then
$$a_{r-2}^{(1)} + 6a_r^{(1)} + a_{r+2}^{(1)} - 8a_{r-1} - 8a_{r+1} = 0 \tag{4}$$

since
$$C_r\{x^2y^{(1)}\} + C_r\{y^{(1)}\} - 4C_r\{xy\} = 0$$

(ii) From (3) with $s = 1$.

$$a_{r-1}^{(1)} = 2ra_r + a_{r+1}^{(1)} \tag{5}$$

(iii) Substitute (5) into (4) in such a manner that the suffix of one of the a's is *less* than all the other suffices. The above example yields

$$2(r - 5)a_{r-1} = 8a_{r+1} - 7a_r^{(1)} - a_{r+2}^{(1)} \tag{4a}$$

(iv) Equations (4a) and (5) are then used to generate the a's and $a^{(1)}$'s by recurrence *backwards* from some arbitrarily chosen values of a_N and $a_N^{(1)}$.

e.g. Set $a_N = 1$, $a_N^{(1)} = 1$ and $a_r = a_r^{(1)} = 0$, all $r > N$

The values found will however be far too large, in general, so a scale factor has to be determined.

(v) The initial conditions will be given either at the origin or at the extremities of the range.

THE NUMERICAL SOLUTION OF DIFFERENTIAL EQUATIONS (Continued)

Noting that

$$y(0) = \tfrac{1}{2}a_0 - a_2 + a_4 - a_6 + a_8 \cdots$$

$$y(1) = \tfrac{1}{2}a_0 + a_1 + a_2 + a_3 + a_4 \cdots$$

$$y(-1) = \tfrac{1}{2}a_0 - a_1 + a_2 - a_3 + a_4 \cdots$$

If $y(0)$ is given then the scale factor is

$$K = y(0)/(\tfrac{1}{2}a_0 - a_1 + a_2 - a_3 + a_4 \cdots)$$

where the values inside the brackets are the values just computed.

(vi) The values for $a_r^{(1)}$ just computed, when multiplied by K, should satisfy the given equation at the origin

i.e. $\quad y^{(1)} = f(x,y) \quad$ at $x = 0, \quad y = y(0),$

or $\quad y^{(1)}(0) = K(\tfrac{1}{2}a_0^{(1)} - a_2^{(1)} + a_4^{(1)} - a_6^{(1)} + \cdots)$

This in general will not be the case and so the computation is repeated with different starting values for the a_N and the $a_N^{(1)}$. A linear combination of the two sets is then chosen so that both the a's *and* the $a^{(1)}$'s satisfy the given conditions.

The method can be extended to equations of higher degree if the necessary $a_r^{(s)}$ are tabulated. This will, of course, necessitate a linear combination of $s + 1$ sets of results.

(b) Shifted Chebyshev polynomials

Fundamental values

$$T_0^* = 1$$

$$T_r^*(1) = 1$$

$$T_{2r}^*(\tfrac{1}{2}) = (-1)^r$$

$$T_{2r+1}^*(\tfrac{1}{2}) = 0$$

$$T_{2r}^*(0) = 1$$

$$T_{2r+1}^*(0) = -1$$

And correspondingly

$$y = \tfrac{1}{2}a_0 + a_1T_1^* + a_2T_2^* + \cdots + a_NT_N^*$$

$$C_r^*(y^{(s)}) = a_r^{(s)}$$

$$C_r^*(x^py^{(s)}) = 2^{-2p}\sum_{i=0}^{2p}\binom{2p}{i}a_{|r-p+i|}^{(s)} \qquad \begin{pmatrix} r = 0,1,2 \cdots \\ s = 1,2, \cdots \end{pmatrix}$$

also $\quad a_{r-1}^{(s)} = 4r \cdot a_r^{(s-1)} + a_{r+1}^{(s)}$

In addition the initial values yield

$$y(0) = \tfrac{1}{2}a_0 - a_1 + a_2 - a_3 + \cdots$$

$$y(\tfrac{1}{2}) = \tfrac{1}{2}a_0 - a_2 + a_4 - a_6 + \cdots$$

$$y(1) = \tfrac{1}{2}a_0 + a_1 + a_2 + a_3 + \cdots$$

The computational procedure being identical with that of the preceding section.

THE NUMERICAL SOLUTION OF DIFFERENTIAL EQUATIONS (Continued)

ITERATIVE METHODS

With the iterative methods of solution of differential equations, two fundamental approaches exist

a) Point iteration.
b) Block iteration.

In the former, the solution at each point is refined as far as is necessary before the next point is considered. The latter, on the other hand, involves successive sweeps over the whole range of integration until no further improvement is detected.

With either group of methods the number of iterations should be small, i.e. if more than three or four applications of the iterative formula in question are necessary then the interval of tabulation should be reduced—in practice this is usually taken to mean halved.

The restriction on the number of iterations does not apply to the Chebyshev method since here the coefficients of an approximating polynomial are being determined and thus the iterations are continued until the required number of significant places is obtained. The solution of the differential equation is then obtained by direct substitution into the polynomial.

When only one starting value is given, as is usual at the beginning of the tabulation of a first order equation, then special starting techniques are necessary for the point iteration approach as a knowledge of the function at several consecutive points is necessary before it can be evaluated at the next point. Either one of the direct methods previously tabulated can be invoked or an iterative starting formula used.

With the iterative predictor-corrector methods, the starting formulae are of the same type.

Doubling the interval of tabulation does not give any trouble with the point iteration schemes whereas halving the interval requires the use of an interpolation routine.

The block iteration schemes require a fresh start to be made if the interval is to be changed.

Iterative Starting Formulae

Notation used:

$$x_r \equiv x_0 + rh$$
$$y_r \equiv y(x_r)$$
$$f_r \equiv f(x_r, y_r) \equiv hy'_r$$

$E_r \equiv$ The truncation error in the expression for y_r.

The superscript (ν) denotes the νth iterate.

Since the iterative starting formulae require initial values to be inserted the first iterative values are obtained from

$$y_r^{(1)} = y_0 + rf_0 \qquad r = 1(1)n - 1$$

This enables "n" starting values to be determined, including the given value y_0, for use with the following groups of formulae.

It will be seen that the values of y_0 and f_0 are unaltered as they will be the given values; f_0 being simply a function of h, x_0 and y_0 determined by the differential equation in question. Further, as each of the other values are refined, they are used in the succeeding equations.

E.g., anticipating the table associated with $n = 3$ in group 1, we obtain, on writing out the formulae in full

ITERATIVE STARTING FORMULAE (Continued)

$$y_1^{(\nu+1)} = \left(y_0 + \frac{5}{12}f_0\right) + \frac{(8f_1^{(\nu)} - f_2^{(\nu)})}{12} + \frac{\delta^3}{24}$$

$$y_2^{(\nu+1)} = \left(y_0 + \frac{1}{3}f_0\right) + \frac{4}{3}f_1^{(\nu+1)} + \frac{1}{3}f_2^{(\nu)} - \frac{\delta^4}{90}$$

The worst error term in the above is $\delta^3/24$ and thus for the contribution from this source to be less than 1/2 unit in the last significant figure $\delta^3/24 < 1/2$ i.e. $\delta^3 < 12$. If this is not so then the interval h must be reduced.

The coefficients A_{rs} (and A^*_{rs}) can be expressed as rational fractions in the form A_{rs} = Numerator of $A_{rs}/D_r \equiv$ Num. A_{rs}/D_r. The tabulation will therefore be of

Num. A_{rs}	D_r	E_r
$s = 0(1)n-1$ $r = 1(1)n-1$	$r = 1(1)n-1$	$r = 1(1)n-1$

where E_r indicates the principal error term associated with f_r.

Group 1

$$y_1^{(\nu+1)} = y_0 + A_{10}f_0 + \sum_{s=1}^{n-1} A_{1s}f_s^{(\nu)} + E_1$$

$$y_r^{(\nu+1)} = y_0 + A_{r0}f_0 + \sum_{s=1}^{r-1} A_{rs}f_s^{(\nu+1)} + \sum_{s=r}^{n-1} A_{rs}f_s^{(\nu)} + E_r, \quad r = 2(1)n-1$$

$n = 2$

r \ s	0	1	D_r	E_r
1	1	1	2	$-\frac{1}{12}\delta^2$

Requires $\delta^2 < 6$

$n = 3$

r \ s	0	1	2	D_r	E_r
1	5	8	−1	12	$\frac{1}{24}\delta^3$
2	1	4	1	3	$-\frac{1}{90}\delta^4$

Requires $\delta^3 < 12$

$n = 4$

r \ s	0	1	2	3	D_r	E_r
1	9	19	−5	1	24	$-\frac{19}{720}\delta_4$
2	1	4	1	0	3	$-\frac{1}{90}\delta^4$
3	3	9	9	3	8	$-\frac{3}{80}\delta^4$

Requires $\delta^4 < 13$

ITERATIVE STARTING FORMULAE (Continued)

$n = 5$

r \ s	0	1	2	3	4	D_r	E_r
	Num. A_{rs}						
1	251	646	−264	106	−19	720	$\frac{3}{160}\delta^5$
2	29	124	24	4	−1	90	$\frac{1}{90}\delta^5$
3	27	102	72	42	−3	80	$\frac{3}{160}\delta^5$
4	14	64	24	64	14	45	$-\frac{1}{105}\delta^6$

Requires $\delta^5 < 27$

$n = 6$

r \ s	0	1	2	3	4	5	D_r	E_r
	Num. A_{rs}							
1	493	1337	−618	302	−83	9	1440	$-\frac{14}{695}\delta^6$
2	28	129	14	14	−6	1	90	$-\frac{1}{105}\delta^6$
3	51	219	114	114	−21	3	160	$-\frac{3}{224}\delta^6$
4	28	128	48	128	28	0	90	$-\frac{1}{105}\delta^6$
5	95	375	250	250	375	95	288	$-\frac{1}{42}\delta^6$

Requires $\delta^6 < 21$

Group 2

The formulae of this group involve the function values as well as the derivatives and consequently tend to have smaller truncation errors.

$$y_r^{(\nu+1)} = \sum_{s=0}^{r-1} A^*_{rs} y_s^{(\nu+1)} + \sum_{s=0}^{r-1} A_{rs} f_s^{(\nu+1)} + \sum_{s=r}^{n-1} A_{rs} f_s^{(\nu)} + E_r, \qquad r = 1(1)n-1$$

$n = 4$

r \ s	0	1	2	0	1	2	3	D_r	E_r
	Num. A^*_{rs}			Num. A_{rs}					
1	24	0	0	9	19	−5	1	24	$-\frac{19}{720}\delta^4$
2	33	24	0	10	57	24	−1	57	$\frac{1}{342}\delta^5$
3	11	27	−27	3	27	27	3	11	$-\frac{1}{513}\delta^6$

Requires $\delta^4 < 18$

ITERATIVE STARTING FORMULAE (Continued)

$n = 5$

r \ s	Num. A^*_{rs} 0	1	2	3	Num. A_{rs} 0	1	2	3	4	D_r	E_r
1	720	0	0	0	251	646	−264	106	−19	720	$\frac{3}{160}\delta^5$
2	990	1440	0	0	281	2056	1176	−104	11	2430	$-\frac{1}{729}\delta^6$
3	550	2160	−1350	0	141	1656	2376	456	−9	1360	$\frac{1}{2539}\delta^7$
4	25	160	0	−160	6	96	216	96	6	25	$-\frac{1}{2625}\delta^8$

Requires $\delta^5 < 27$

$n = 6$

r \ s	Num. A^*_{rs} 0	1	2	3	4
1	1440	0	0	0	0
2	24390	53280	0	0	0
3	9550	51570	−17550	0	0
4	3425	29560	14040	−35960	0
5	137	1625	2000	−2000	−1625

r \ s	Num. A_{rs} 0	1	2	3	4	5	D_r	E_r
1	475	1427	−798	482	−173	27	1440	$-\frac{1}{70}\delta^6$
2	6589	58528	41608	−5752	1223	−136	77670	$\frac{31}{290}\delta^7$
3	2337	33687	62352	16512	−693	45	43570	$-\frac{1}{9072}\delta^8$
4	786	15540	45720	29280	3210	−36	11065	$\frac{1}{16800}\delta^9$
5	30	750	3000	3000	750	30	137	$-\frac{1}{12659}\delta^{10}$

Requires $\delta^6 < 35$

Group 3

In this group neither y_r nor f_r appears on the right hand side i.e. $A^*_{ss} \equiv A_{ss} \equiv 0$.

$$y_r^{(\nu+1)} = \sum_{s=0}^{r-1} A^*_{rs} y_s^{(\nu+1)} + \sum_{s=r+1}^{n-1} A^*_{rs} y^{(\nu)} + \sum_{s=0}^{r-1} A_{rs} f^{(v+1)} + \sum_{s=r+1}^{n-1} A_{rs} f^{(v)} + E_r$$

ITERATIVE STARTING FORMULAE (Continued)

$n = 3$

$r \backslash s$	Num. A^*_{rs} 0	1	2	Num. A_{rs} 0	1	2	D_r	E_r
1	2	0	2	1	0	−1	4	$\frac{1}{24}\delta^3$
2	5	−4	0	2	4	0	1	$\frac{1}{12}\delta^3$

Requires $\delta^3 < 6$

$n = 4$

$r \backslash s$	Num. A^*_{rs} 0	1	2	3	Num. A_{rs} 0	1	2	3	D_r	E_r
1	8	0	0	19	3	0	−27	−6	27	$\frac{1}{180}\delta^5$
2	19	0	0	8	6	27	0	−3	27	$\frac{1}{180}\delta^5$
3	10	9	−18	0	3	18	9	0	1	$\frac{1}{20}\delta^5$

Requires $\delta^5 < 10$

$n = 5$

$r \backslash s$	Num. A^*_{rs} 0	1	2	3	4	Num. A_{rs} 0	1	2	3	4	D_r	E_r
1	19	0	−216	224	69	6	0	−216	−192	−18	96	$\frac{1}{1120}\delta^7$
2	11	16	0	16	11	3	24	0	−24	−3	54	$\frac{1}{2520}\delta^7$
3	69	224	−216	0	19	18	192	216	0	−6	96	$\frac{1}{805}\delta^7$
4	47	192	−108	−128	0	12	144	216	48	0	3	$\frac{1}{70}\delta^7$

Requires $\delta^7 < 35$

Group 4

In this group the starting values are symmetrically placed with respect to the origin. These formulae usually have a better truncation error and their coefficients exhibit a certain degree of skew symmetry.

Assuming that y_r, and hence f_r, are found before y_{-r} and f_{-r} the formulae used are, for $N \equiv (n - 1)/2$ where n denotes the odd number of points involved, as follows

$$y_r^{(\nu+1)} = y_0 + \sum_{s=-N}^{-r} A_{rs} f_s^{(\nu)} + \sum_{s=-r+1}^{r-1} A_{rs} f_s^{(\nu+1)} + \sum_{s=r}^{N} A_{rs} f_s^{(\nu)} + E_r$$

which is followed immediately by

$$y_{-r}^{(\nu+1)} = y_0 + \sum_{s=-N}^{-r} A_{rs} f_s^{(\nu)} + \sum_{s=-r+1}^{r} A_{rs} f_s^{(\nu+1)} + \sum_{s=r+1}^{N} A_{rs} f_s^{(r)} + E_r$$

ITERATIVE STARTING FORMULAE (Continued)

The formulae are tabulated in the order in which they are to be used i.e. $r = 1, -1, 2, -2, 3, -3$.

$n = 3$

	Num. A_{rs}			D_r	E_r
$r \backslash s$	-1	0	1		
1	-1	8	5	12	$-\frac{1}{24}\delta^3$
-1	-5	-8	1	12	$-\frac{1}{24}\delta^3$

Requires $\delta^3 < 12$

$n = 5$

	Num. A_{rs}					D_r	E_r
$r \backslash s$	-2	-1	0	1	2		
1	11	-74	456	346	-19	720	$\frac{1}{144}\delta^5$
-1	19	-346	-456	74	-11	720	$\frac{1}{144}\delta^5$
2	-1	4	24	124	29	90	$-\frac{1}{90}\delta^5$
-2	-29	-124	-24	-4	1	90	$-\frac{1}{90}\delta^5$

Requires $\delta^5 < 45$

$n = 7$

	Num. A_{rs}							D_r	E_r
$r \backslash s$	-3	-2	-1	0	1	2	3		
1	-191	1608	-6771	37504	30819	-2760	271	60480	$\frac{1}{633}\delta^7$
-1	-271	2760	-30819	-37504	6771	-1608	191	60480	$\frac{1}{633}\delta^7$
2	5	-30	33	1328	4863	1398	-37	3780	$\frac{1}{8172}\delta^7$
-2	37	-1398	-4863	-1328	-33	30	-5	3780	$\frac{1}{8172}\delta^7$
3	-29	216	-729	2176	1161	3240	685	2240	$\frac{9}{896}\delta^7$
-3	-685	-3240	-1161	-2176	729	-216	29	2240	$\frac{9}{896}\delta^7$

Requires $\delta^7 < 50$

Group 5

As group 4, with the addition of function values. The same observations apply. y_0 and f_0 do not alter from cycle to cycle. Defining N the same as for group 4, the formulae used are:

ITERATIVE STARTING FORMULAE (Continued)

$$y_r^{(\nu+1)} = \sum_{s=-N}^{-r} A_{rs}^* y_s^{(\nu)} + \sum_{s=-r+1}^{r-1} A_{rs}^* y_s^{(\nu+1)} + \sum_{s=r}^{N} A_{rs}^* y_s^{(\nu)}$$

$$+ \sum_{s=-N}^{-r} A_{rs} f_s^{(\nu)} + \sum_{s=-r+1}^{r-1} A_{rs} f_s^{(\nu+1)} + \sum_{s=r}^{N} A_{rs} f_s^{(\nu)} + E_r$$

$n = 5$

Num. A_{rs}^*

$r \backslash s$	−2	−1	0	1	2
1	−85	0	540	0	−135
−1	−135	0	540	0	−85.
2	0	1360	1350	−2160	0
−2	0	−2160	1350	1360	0

$r \backslash s$	−2	−1	0	1	2	D_r	E_r
	Num. A_{rs}						
1	−24	−144	216	336	36	320	$-\frac{1}{1643}\delta^7$
−1	−36	−336	−216	144	24	320	$-\frac{1}{1643}\delta^7$
2	−9	456	2376	1656	141	550	$-\frac{1}{1067}\delta^7$
−2	−141	−1656	−2376	−456	9	550	$-\frac{1}{1067}\delta^7$

Requires $\delta^7 < 533$

The coefficients become unwieldy for higher values of n.

Continuing the solution

Two formulae are used to continue the solution once sufficient starting values have been obtained. The first, the "predictor," is used to obtain an estimate of the next value. It is used once only and can be appreciably less accurate than the second, the "corrector," which determines the precision.

The corrector is used iteratively until sufficient significant figures are obtained.

Predictors and correctors can involve either backward differences, central differences or function values.

Predictor-corrector formulae involving backward differences

A typical predictor formula is

$$y_1 = y_0 + \left(1 + \frac{1}{2}\nabla + \frac{5}{12}\nabla^2 + \frac{3}{8}\nabla^3 + \frac{251}{720}\nabla^4 + \frac{95}{288}\nabla^5 + \frac{19087}{60480}\nabla^6 + \cdots\right) f_0$$

this is followed by

$$f_1 = h \cdot y_1',$$

where y' is defined as a function of x and y.

PREDICTOR-CORRECTOR METHODS

This predicted value of f_1 is used to generate all the backward differences ∇_1^r for $r = 1(1)6$.

These are then used with a corrector formula such as

$$y_1 = y_0 + \left(1 - \frac{1}{2}\nabla - \frac{1}{12}\nabla^2 - \frac{1}{24}\nabla^3 - \frac{19}{720}\nabla^4 - \frac{3}{160}\nabla^5 - \frac{863}{60480}\nabla^6 - \cdots\right)f_1$$

to recompute y_1 and thus f_1. Note that all the backward differences ∇_1^r must then be recomputed before the corrector is applied once more.

The numerical coefficients tend to be rather high and round-off errors can become troublesome if differences beyond ∇^6 are significant.

An alternative method of *prediction* when one particular column of differences is nearly constant (say for example $\nabla^6 \sim k$) is to use the recurrence formula

$$\nabla_1^{r-1} = \nabla_1^r + \nabla_0^{r-1}$$

to generate the required differences. The overall effect is as if

$$f_1 = (1 + \nabla + \nabla^2 + \nabla^3 + \nabla^4 + \nabla^5]f_0 + k$$

had been used as a predictor.

Even when ∇^6 is not constant it can usually be estimated with sufficient accuracy from the preceding entries in the column, i.e., it is extrapolated from a knowledge of the previous differences, and is then used in the predictor

$$f_1 = (1 + \nabla + \nabla^2 + \nabla^3 + \nabla^4 + \nabla^5)f_0 + \nabla^6 f_1$$

The corrector quoted in the example is the first of a group of such formulae having the general form

$$y_1 = y_r + \sum_{s=0} A_{rs}\nabla_s f_1 \qquad r < 0.$$

Truncating the infinite series at $s = 6$, we have

A_{rs}

$r \backslash s$	0	1	2	3	4	5	6
0	1	$\frac{-1}{2}$	$\frac{-1}{12}$	$\frac{-1}{24}$	$\frac{-19}{720}$	$\frac{-3}{160}$	$\frac{-863}{60480}$
−1	2	−2	$\frac{1}{3}$	0	$\frac{-1}{90}$	$\frac{-1}{90}$	$\frac{-37}{3780}$
−2	3	$\frac{-9}{2}$	$\frac{9}{4}$	$\frac{-3}{8}$	$\frac{-3}{80}$	$\frac{-3}{160}$	$\frac{-29}{2240}$
−3	4	−8	$\frac{20}{3}$	$\frac{-8}{3}$	$\frac{14}{45}$	0	$\frac{-8}{945}$
−4	5	$\frac{-25}{2}$	$\frac{175}{12}$	$\frac{-225}{24}$	$\frac{425}{144}$	$\frac{-95}{288}$	$\frac{-275}{12096}$
−5	6	−18	27	−24	$\frac{123}{10}$	$\frac{-33}{10}$	$\frac{41}{140}$

PREDICTOR-CORRECTOR METHODS (Continued)

Predictor-corrector formulae involving central differences

In these formulae, the predictors and the correctors have the same form. In fact the corrector has the same coefficients as the predictor on the previous line of the table with the addition of one more term, together with an estimate of the error.

The formulae involved are

Predictor $$y_1 = y_{1-2r} + \sum_{s=0}^{r-1} A_{rs}\delta^{2s}f_{1-r}$$

Corrector $$y_1 = y_{3-2r} + \sum_{s=0}^{r-1} B_{rs}\delta^{2s}f_{2-r} + \text{Error } (\equiv B_{rr}\delta^{2r})$$

A_{rs}

r \ s	0	1	2	3	4
2	4	$\frac{8}{3}$			
3	6	9	$\frac{33}{10}$		
4	8	$\frac{64}{3}$	$\frac{868}{45}$	$\frac{736}{189}$	
5	10	$\frac{125}{3}$	$\frac{875}{18}$	$\frac{17225}{756}$	$\frac{20225}{4536}$

B_{rs}

r \ s	0	1	2	3	4	5
2	2	$\frac{1}{3}$	$\frac{-1}{90}$			
3	4	$\frac{8}{3}$	$\frac{14}{45}$	$\frac{-8}{945}$		
4	6	9	$\frac{33}{10}$	$\frac{41}{140}$	$\frac{-9}{1400}$	
5	8	$\frac{64}{3}$	$\frac{868}{45}$	$\frac{736}{189}$	$\frac{3956}{14175}$	$\frac{-2368}{467775}$

e.g. Suppose that enough starting values have been obtained to indicate that the sixth differences are of the order of 50 (say).

The corrector corresponding to $r = 3$ has an error term of magnitude $8\delta^6/945$ which is less than $\frac{1}{2}$ when. $\delta^6 = 50$.

Therefore label the last computed value y_0 and use the predictor

$$y_1 = y_{-5} + 6f_{-2} + 9\delta^2_{-2} + \frac{33}{10}\,\delta^4_{-2} \quad (r = 3 \text{ in the above table}).$$

From this value of y_1, generate f_1 $(= hy_1')$ and carry the differences back to the sixth differences column.

PREDICTOR-CORRECTOR METHODS (Continued)

The corrector to use is then

$$y_1 = y_{-3} + 4f_{-1} + \frac{8}{3}\delta^2_{-1} + \frac{14}{45}\delta^4_{-1} \qquad (r = 3 \text{ above}).$$

Note that the error term should strictly be expressed in terms of δ^6_{-1} and so an adequate check cannot be carried out until the next term has been computed.

Note also that, in common with all predictor-corrector methods, any form of predictor can be used, as the refinement depends only upon the corrector.

Predictor-corrector formulae involving function values

The general form of the simplest type of predictor is

$$y_1 = y_r + \sum_{s=r}^{0} A_{rs}f_s + E_r \qquad < 0$$

r \ s	0	−1	−2	−3	−4	−5	E_r
0	1						$\frac{1}{2}\delta$
−1	2	0					$\frac{1}{3}\delta^2$
−2	$\frac{9}{4}$	0	$\frac{3}{4}$				$\frac{3}{8}\delta^3$
−3	$\frac{8}{3}$	$-\frac{4}{3}$	$\frac{8}{3}$	0			$\frac{14}{45}\delta^4$
−4	$\frac{55}{24}$	$\frac{5}{24}$	$\frac{5}{24}$	$\frac{55}{24}$	0		$\frac{95}{144}\delta^5$
−5	$\frac{33}{10}$	$-\frac{42}{10}$	$\frac{78}{10}$	$-\frac{42}{10}$	$\frac{33}{10}$	0	$\frac{41}{140}\delta^6$

(Columns 0 to −5 give A_{rs}.)

An alternative group of predictor formulae have the form

$$y_1 = y_0 + \sum_{s=r}^{0} A_{rs}f_s + E_r \qquad (r < 0)$$

r \ s	0	−1	−2	−3	−4	E_r
−1	$\frac{3}{2}$	$-\frac{1}{2}$				$\frac{5}{12}\delta^2$
−2	$\frac{23}{12}$	$-\frac{4}{3}$	$\frac{5}{12}$			$\frac{3}{8}\delta^3$
−3	$\frac{55}{24}$	$-\frac{59}{24}$	$\frac{37}{24}$	$-\frac{3}{8}$		$\frac{25}{72}\delta^4$
−4	$\frac{1901}{720}$	$-\frac{1387}{360}$	$\frac{327}{90}$	$-\frac{637}{360}$	$\frac{251}{720}$	$\frac{95}{288}\delta^5$

(Columns 0 to −4 give A_{rs}.)

The correctors fall roughly into main groups. The first group is similar to the predictors tabulated above in that the emphasis is upon a weighted mean of the derivatives rather than the function values.

PREDICTOR-CORRECTOR METHODS (Continued)

Group 1

$$y_1 = y_r + \sum_{s=r}^{0} A_{rs} f_s + E_r \qquad (r < 0)$$

$r \backslash s$	A_{rs}: 1	0	-1	-2	-3	-4	-5	E_r
0	$\frac{1}{2}$	$\frac{1}{2}$						$\frac{-1}{12}\delta^2$
-1	$\frac{1}{3}$	$\frac{4}{3}$	$\frac{1}{3}$					$\frac{-1}{90}\delta^4$
-2	$\frac{3}{8}$	$\frac{9}{8}$	$\frac{9}{8}$	$\frac{3}{8}$				$\frac{-3}{80}\delta^4$
-3	$\frac{14}{45}$	$\frac{64}{45}$	$\frac{24}{45}$	$\frac{64}{45}$	$\frac{14}{45}$			$\frac{-1}{105}\delta^6$
-4	$\frac{95}{288}$	$\frac{375}{288}$	$\frac{125}{144}$	$\frac{125}{144}$	$\frac{375}{288}$	$\frac{95}{288}$		$\frac{-1}{44}\delta^6$
-5	$\frac{41}{140}$	$\frac{54}{35}$	$\frac{27}{140}$	$\frac{68}{35}$	$\frac{27}{140}$	$\frac{54}{35}$	$\frac{41}{140}$	$\frac{-1}{140}\delta^8$
-6	$\frac{3}{10}$	$\frac{3}{2}$	$\frac{3}{10}$	$\frac{9}{5}$	$\frac{3}{10}$	$\frac{3}{2}$	$\frac{3}{10}$	$\frac{-1}{140}\delta^6$

The second group involves the computation of a weighted mean of the function value as well as the derivatives.

Group 2

$$y_1 = \sum_{s=-4}^{0} A^*_{rs} y_s + \sum_{s=-4}^{1} A_{rs} f_s + E_r$$

$r \backslash s$	A^*_{rs}: -4	-3	-2	-1	0
1	0	0	0	-1	2
2	0	0	1	$\frac{27}{11}$	$-\frac{27}{11}$
3	0	1	$\frac{32}{5}$	0	$-\frac{32}{5}$
4	1	$\frac{1625}{137}$	$\frac{2000}{137}$	$-\frac{2000}{137}$	$-\frac{1625}{137}$
5	0	0	0	$\frac{1}{5}$	$\frac{4}{5}$
6	0	0	0	$-\frac{19}{13}$	$\frac{32}{13}$
7	0	0	$\frac{32}{5}$	$-\frac{27}{5}$	0
8	0	0	$\frac{136}{55}$	$\frac{27}{11}$	$-\frac{216}{55}$

PREDICTOR-CORRECTOR METHODS (Continued)

r \ s	-4	-3	-2	-1	0	1	E_r
	A_{rs}						
1	0	0	0	$\frac{1}{2}$	0	$\frac{-1}{2}$	$\frac{-1}{12}\delta^3$
2	0	0	$\frac{3}{11}$	$\frac{27}{11}$	$\frac{27}{11}$	$\frac{3}{11}$	$\frac{-1}{513}\delta^6$
3	0	$\frac{6}{25}$	$\frac{96}{25}$	$\frac{216}{25}$	$\frac{96}{25}$	$\frac{6}{25}$	$\frac{-1}{2625}\delta^8$
4	$\frac{30}{137}$	$\frac{750}{137}$	$\frac{3000}{137}$	$\frac{3000}{137}$	$\frac{750}{137}$	$\frac{30}{137}$	$\frac{-1}{12659}\delta^{10}$
5	0	0	0	0	$\frac{4}{5}$	$\frac{2}{5}$	$\frac{-1}{30}\delta^3$
6	0	0	$\frac{4}{39}$	-1	0	$\frac{17}{39}$	$\frac{-19}{390}\delta^4$
7	0	0	$\frac{12}{5}$	$\frac{27}{5}$	0	$\frac{3}{5}$	$\frac{-1}{30}\delta^4$
8	0	$\frac{-9}{550}$	$\frac{228}{275}$	$\frac{1188}{275}$	$\frac{828}{275}$	$\frac{141}{550}$	$\frac{-1}{1067}\delta^7$

Iterative methods

Should it become necessary to halve the interval, because the error term is becoming significant or the number of iterations required is becoming excessive, then interpolation becomes necessary.

This interpolation, actually subtabulation to halves, is based upon previously computed values and the formulae used can take several forms. One such form is as follows

$$y_k = \frac{1}{D_r} \sum_{s=-n}^{0} A_{rs} y_s + E_r, \qquad (r = 1(1)n)$$

where $k \equiv (1 - 2r)/2$, and E_r denotes the principal error term.

$n = 2$

r \ s	0	-1	-2	D_r	E_r
	A_{rs}				
1	3	6	-1	8	$\frac{-1}{16}\nabla_0^3$
2	-1	6	3	8	$\frac{1}{16}\nabla_0^3$

$n = 3$

r \ s	0	-1	-2	-3	D_r	E_r
	A_{rs}					
1	5	15	-5	1	16	$\frac{-5}{128}\nabla_0^4$
2	-1	9	9	-1	16	$\frac{3}{128}\nabla_0^4$
3	1	-5	15	5	16	$\frac{-5}{128}\nabla_0^4$

DEFERRED-CORRECTOR METHODS

$n = 4$	A_{rs}						
$r \backslash s$	0	−1	−2	−3	−4	D_r	E_r
1	35	140	−70	28	−5	128	$\frac{-7}{256}\nabla_0^5$
2	−5	60	90	−20	3	128	$\frac{3}{256}\nabla_0^5$
3	3	−20	90	60	−5	128	$\frac{-3}{256}\nabla_0^5$
4	−5	28	−70	140	35	128	$\frac{7}{256}\nabla_0^5$

For a given "n" there will be $2n + 1$ values available at the new interval to continue the solution.

Deferred-correction methods

One of the most powerful of the iterative methods is that of "deferred correction." This is a block iteration method, a complete set of function values being obtained from a recurrence relationship, their differences formed and corrections applied from these differences.

Central difference corrections are the most useful. Let the first order differential equation be

$$y' + g(x) \cdot y = k(x)$$

The first recurrence relation, assuming an initial y only given, is

$$y_1\left(1 + \frac{1}{2}hg_1\right) - y_0\left(1 - \frac{1}{2}hg_0\right) = \frac{1}{2}h(k_0 + k_1)$$

This is used to generate y_r for both positive and negative values of r if required. When the required range of values of y_r has been covered, with sufficient additional terms to generate all the differences required, a difference table is formed and the correction terms $C(y_{\frac{1}{2}})$ tabulated from

$$C(y_{\frac{1}{2}}) \equiv \frac{1}{12}\delta_{\frac{1}{2}}^3 - \frac{1}{120}\delta_{\frac{1}{2}}^5 + \frac{1}{840}\delta_{\frac{1}{2}}^7 - \frac{1}{5040}\delta_{\frac{1}{2}}^9 \cdots$$

The correction terms $C(y_{\frac{1}{2}})$ are corrections to the recurrence relation. The corrections to the function values y_r are η_r i.e. True value = Approximate value + η

where
$$\eta_1\left(1 + \frac{1}{2}hg_1\right) - \eta_0\left(1 - \frac{1}{2}hg_0\right) + C(y_{\frac{1}{2}}) = 0$$

In this formula the initial η_0 is zero and the recurrence relation can be used in either direction.

New $C(y_{\frac{1}{2}})$ are then computed and the correction process repeated until there is no change in the values of $C(y_{\frac{1}{2}})$. At this point the computed values and the differences derived there-from should be checked for consistency with the differential equation. The method of deferred correction is particularly useful for linear differential equations as its convergence is extremely rapid, even for very large intervals. If more starting values are known formulae which have even faster convergence can be employed. Two such formulae are as follows: Two starting values given

DEFERRED-CORRECTOR METHODS (Continued)

Predictor $\quad y_1\left(1+\frac{1}{3}hg_1\right)+\frac{4}{3}hg_0y_0-y_{-1}\left(1-\frac{1}{3}hg_{-1}\right)=\frac{1}{3}h(k_1+4k_0+k_{-1})$

$$C(y_0)\equiv\frac{1}{180}(\delta^5_{-\frac{1}{2}}+\delta^5_{\frac{1}{2}})-\frac{1}{630}(\delta^7_{-\frac{1}{2}}+\delta^7_{\frac{1}{2}})+\frac{1}{2510}(\delta^9_{-\frac{1}{2}}+\delta^9_{\frac{1}{2}})\cdots$$

Corrector $\quad \eta_1\left(1+\frac{1}{3}hg_1\right)+\frac{4}{3}hg_0\eta_0-\eta_{-1}\left(1-\frac{1}{3}hg_{-1}\right)+C(y_0)=0.$

Three starting values given

Predictor $\quad y_1+y_0(9+6hg_0)-y_{-1}(9-6hg_{-1})-y_{-2}=6h(k_0+k_{-1})$

$$C(y_{-\frac{1}{2}})\equiv-\frac{1}{10}\delta^5_{-\frac{1}{2}}+\frac{1}{70}\delta^7_{-\frac{1}{2}}-\frac{1}{420}\delta^9_{-\frac{1}{2}}+\cdots$$

Corrector $\quad \eta_1+\eta_0(9+6hg_0)-\eta_{-1}(9-6hg_{-1})-\eta_{-2}+C(y_{-\frac{1}{2}})=0$

The use of the deferred correction technique for higher order differential equations.

Second order differential equations.

Two starting values given e.g. $y''+m\cdot y'+g\cdot y=k$ where y_0 and y_{-1} are known and $y\equiv y(x)$; $m\equiv m(x)$; $g\equiv g(x)$; $k\equiv k(x)$.

Predictor $\quad y_1\left(1+\frac{1}{2}hm_0\right)-y_0(2-h^2g_0)+y_{-1}\left(1-\frac{1}{2}hm_0\right)=h^2k_0$

$$C(y_0)\equiv\left[-\frac{1}{12}\delta^4_0+\frac{1}{90}\delta^6_0-\frac{1}{560}\delta^8_0+\frac{1}{3150}\delta^{10}_0\right]$$
$$+hg_0\left[-\frac{1}{12}(\delta^3_{\frac{1}{2}}+\delta^3_{-\frac{1}{2}})+\frac{1}{60}(\delta^5_{\frac{1}{2}}+\delta^5_{-\frac{1}{2}})-\frac{1}{280}(\delta^7_{\frac{1}{2}}+\delta^7_{-\frac{1}{2}})+\frac{1}{1260}(\delta^9_{\frac{1}{2}}+\delta^9_{-\frac{1}{2}})\right].$$

Corrector $\quad \eta_1\left(1+\frac{1}{2}hm_0\right)-\eta_0(2-h^2g_0)+\eta_{-1}\left(1-\frac{1}{2}hm_0\right)+C(y_0)=0.$

Three starting values require the known functions $m(x)$, $g(x)$ $g(x)$ and $k(x)$ to be tabulated at half way points and the advantage in speed of convergence is relatively slight.

Predictor $\quad y_1\left(1-\frac{1}{12}hm_{-\frac{1}{2}}-\frac{1}{8}h^2g_{-\frac{1}{2}}\right)-y_0\left(1-\frac{9}{4}hm_{-\frac{1}{2}}-\frac{9}{8}h^2g_{-\frac{1}{2}}\right)$

$$-y_{-1}\left(1+\frac{9}{4}hm_{-\frac{1}{2}}-\frac{9}{8}h^2g_{-\frac{1}{2}}\right)+y_{-2}\left(1+\frac{1}{12}hm_{-\frac{1}{2}}-\frac{1}{8}h^2g_{-\frac{1}{2}}\right)=2h^2k_{-\frac{1}{2}}$$

$$C(y_{-\frac{1}{2}})\equiv\left[\frac{-5}{24}(\delta^4_0+\delta^4_{-1})+\frac{259}{5760}(\delta^6_0+\delta^6_{-1})-\frac{3229}{322560}(\delta^8_0+\delta^8_{-1})\right.$$
$$\left.+\frac{117469}{51609600}(\delta^{10}_0+\delta^{10}_{-1})\right]+hm_{-\frac{1}{2}}\left[\frac{3}{320}\delta^5_{-\frac{1}{2}}-\frac{5}{3584}\delta^7_{-\frac{1}{2}}+\frac{35}{147456}\delta^9_{-\frac{1}{2}}\right]$$
$$+h^2g_{-\frac{1}{2}}\left[\frac{3}{128}(\delta^4_0+\delta^4_{-1})-\frac{5}{1024}(\delta^6_0+\delta^6_{-1})+\frac{35}{32768}(\delta^8_0+\delta^8_{-1})\right]$$

Corrector $\quad \eta_1\left(1-\frac{1}{12}hm_{-\frac{1}{2}}-\frac{1}{8}h^2g_{-\frac{1}{2}}\right)-\eta_0\left(1-\frac{9}{4}hm_{-\frac{1}{2}}-\frac{9}{8}h^2g_{-\frac{1}{2}}\right)$

$$-\eta_{-1}\left(1+\frac{9}{4}hm_{-\frac{1}{2}}-\frac{9}{8}h^2g_{-\frac{1}{2}}\right)+\eta_{-2}\left(1+\frac{1}{12}hm_{-\frac{1}{2}}-\frac{1}{8}h^2g_{-\frac{1}{2}}\right)+C(y_{-\frac{1}{2}})=0$$

CHEBYSHEV METHODS

In general, for second order differential equations, recurence relations involving an even number of known values are simpler than those involving an odd number of such values. The formulae for four starting values are given below and they are seen to be much simpler than those for three values. Further the correction terms should have more rapid convergence properties.

Predictor
$$y_1(1 + hm_{-1}) - y_0(16 + 8hm_{-1}) + y_{-1}(30 - 12h^2g_{-1}) + y_{-2}(16 - 8hm_{-1}) + y_{-3}(1 - hm_{-1}) = -12h^2k_{-1}$$

$$C(y_{-1}) \equiv \left[-\frac{4}{30}\delta^6_{-1} + \frac{3}{140}\delta^8_{-1} - \frac{2}{525}\delta^{10}_1 + \frac{1}{1386}\delta^{12}_{-1}\right] + hm_{-1}\left[-\frac{1}{5}(\delta^5_{-\frac{1}{2}} + \delta^5_{-1\frac{1}{2}}) + \frac{3}{70}(\delta^7_{-\frac{1}{2}} + \delta^7_{-1\frac{1}{2}}) - \frac{1}{105}(\delta^9_{-\frac{1}{2}} + \delta^9_{-1\frac{1}{2}}) + \frac{1}{462}(\delta^{11}_{-\frac{1}{2}} + \delta^{11}_{-1\frac{1}{2}})\right].$$

Corrector
$$\eta_1(1 + hm_{-1}) - \eta_0(16 + 8hm_{-1}) + \eta_{-1}(30 - 12h^2g_{-1}) + \eta_{-2}(16 - 8hm_{-1}) + \eta_{-3}(1 - hm_{-1}) + C(y_{-1}) = 0$$

Iterative Chebyshev

The recurrence relationships used are identical with those set up in the direct Chebyshev method. The first value of $a_0^{(s)}$ is obtained from the first term of the series for the initial value of $y^{(s)}$, all the other terms being initially set at zero. e.g. Consider the case when $a_r^{(s)}$ are required for $s = 0,1,2$ and 3. (The initial values of y, y', y'' being known at $x = -1$.) The computational sequence is as follows

Set all $a_r^{(3)} = 0$
Obtain $a_0^{(2)}$ from $y''(-1) = \frac{1}{2}a_0^{(2)}$
Use the Chebyshev recurrence formula to obtain $a_1^{(1)}$ from this value of $a_0^{(2)}$
Obtain $a_0^{(1)}$ from $y'(-1) = \frac{1}{2}a_0^{(1)} - a_1^{(1)}$
Use the Chebyshev formula again to obtain both $a_1^{(0)}$ and $a_2^{(0)}$
Obtain $a_0^{(0)}$ from $y(-1) = \frac{1}{2}a_1^{(0)} - a_1^{(0)} + a_2^{(0)}$
Now utilise the recurrence relation given by the differential equation to generate $a_0^{(3)}$.

The cycle is now repeated each series being increased by one term if desired. The determining factor being the number of significant decimals required in the result. The method is in general slower than the direct Chebyshev process but it has the merit that no predetermined knowledge of the number of terms required is necessary—further computational errors die away as the method is truly iterative, whereas in the direct approach they can nullify all the working.

Direct Chebyshev method

This method is suitable for a differential equation of any order. The range of the arguments must however be normalized to $-1 \leqq x \leqq 1$ (or $0 \leqq x \leqq 1$ if the "shifted" Chebyshev polynomials are to be used).

A polynomial approximation to the analytical solution is sought, the degree of the polynomial N being decided beforehand.

CHEBYSHEV METHODS (Continued)

Assume
$$y(x) = \tfrac{1}{2}a_0 + \sum_{r=1}^{N} a_r \cdot T_r(x)$$
$$y'(x) = \tfrac{1}{2}a_0' + \sum_{r=1}^{N} a_r' \cdot T_r(x)$$
$$y''(x) = \tfrac{1}{2}a_0'' + \sum_{r=1}^{N} a_r'' \cdot T_r(x) \qquad \text{etc.}$$

where the a's are constants.

Writing $y^{(0)}$, $y^{(1)}$, $y^{(2)}$ for $y(x)$, $y'(x)$, $y''(x)$ etc., $a_r^{(0)}$, $a_r^{(1)}$, $a_r^{(2)}$ for a_r, a_r', a_r'', etc., and T_r for $T_r(x)$.

The recurrence relationship for the Chebyshev polynomials can be written as

$$2xT_r = T_{|r-1|} + T_{r+1} \qquad (r = 0,1,2, \ldots)$$

further
$$2\int T_r \cdot dx = \frac{T_{r+1}}{(r+1)} - \frac{T_{r-1}}{(r-1)} \qquad (r > 1)$$
$$= \tfrac{1}{2}T_2 \qquad (r = 1)$$
$$= 2T_1 \qquad (r = 0)$$

Thus $\quad 2r \cdot a_r^{(s)} = a_{r-1}^{(s+1)} - a_{r+1}^{(s+1)} \qquad (r > 1)$

If, in addition, the notation $C_r(x^p y^{(s)})$ is introduced to denote the coefficient of T_r in the expansion of $x^p y^{(s)}$ *and twice this coefficient when* $r = 0$ then

$$C_r(y^{(0)}) = a_r$$
$$C_r(y^{(s)}) = a_r^{(s)}$$
$$C_r(x^p y^{(s)}) = \frac{1}{2^p} \sum_{k=0}^{p} \binom{p}{k} \cdot a_{|r-p+2k|}^{(s)}$$

Method of use of Chebyshev polynomials

Suppose the differential equation is

$$f(x) \cdot y'' + g(x) \cdot y' + h(x) \cdot y = 0$$

i.e.
$$fy^{(2)} + gy^{(1)} + hy^{(0)} = 0$$

where f, g and h are polynomials in x (or can be approximated to by such polynomials).

Then

$$C_r(fy^{(2)}) + C_r(gy^{(1)}) + C_r(hy^{(0)}) = 0$$

will yield a recurrence relationship based upon the differential equation. The Chebyshev polynomials themselves yield the second recurrence relationship, namely

$$2r \cdot a_r^{(s)} = a_{r-1}^{(s+1)} - a_{r+1}^{(s+1)}$$

To start the procedure, assume that

$$a_r^{(s)} = 0 \qquad r > N$$

and assign arbitrary integer values to $a_N^{(1)}$ and $a_N^{(2)}$. The recurrence relationships are used to generate $a_r^{(s)}$ for $r = N-1$, $N-2$, etc.

CHEBYSHEV METHODS (Continued)

The formulae for the initial conditions are

$$y^{(s)}(-1) = \tfrac{1}{2}a_0^{(s)} - a_1^{(s)} + a_2^{(s)} - a_3^{(s)} + \cdots$$
$$y^{(s)}(0) = \tfrac{1}{2}a_0^{(s)} - a_2^{(s)} + a_4^{(s)} - a_6^{(s)} + \cdots$$
$$y^{(s)}(1) = \tfrac{1}{2}a_0^{(s)} + a_1^{(s)} + a_2^{(s)} + a_3^{(s)} + \cdots$$

It will be found that the series obtained do not satisfy the given initial conditions. Therefore a second set of coefficients must be obtained by repeating the process for different arbitrary starting values and a linear combination of the two sets will yield the required solution.

$$T_0 = 1 \qquad\qquad T_{r+1} - 2xT_r + T_{r-1} = 0$$
$$T_1 = x$$
$$T_2 = 2x^2 - 1$$
$$T_3 = 4x^3 - 3x$$
$$T_4 = 8x^4 - 8x^2 + 1$$
$$T_5 = 16x^5 - 20x^3 + 5x$$
$$T_6 = 32x^6 - 48x^4 + 18x^2 - 1$$

If the "shifted" Chebyshev polynomials T_r^* are to be used then

$$y(x) = \tfrac{1}{2}a_0 + \sum_{r=1}^{N} a_r T_r^*$$
$$4r \cdot a_r^{(s)} = a_{|r-1|}^{(s+1)} - a_{r+1}^{(s+1)}$$
$$C_r(x^p y^{(s)}) = \frac{1}{2^p} \sum_{k=0}^{2p} \binom{2p}{k} a_{|r-p+k|}^{(s)}$$

with

$$y^{(s)}(0) = \tfrac{1}{2}a_0^{(s)} - a_1^{(s)} + a_2^{(s)} - a_3^{(s)} \cdots$$
$$y^{(s)}(\tfrac{1}{2}) = \tfrac{1}{2}a_0^{(s)} - a_2^{(s)} + a_4^{(s)} - a_6^{(s)} \cdots$$
$$y^{(s)}(1) = \tfrac{1}{2}a_0^{(s)} + a_1^{(s)} + a_2^{(s)} + a_3^{(s)} \cdots$$
$$T_r^*(x) = T_r(2x - 1) = \cos\{r \cdot \arccos(2x - 1)\}$$

$$T_0^* = 1 \qquad\qquad (4x - 2)T_r^* = T_{r-1}^* + T_{r+1}^*$$
$$T_1^* = 2x - 1$$
$$T_2^* = 8x^2 - 8x + 1$$
$$T_3^* = 32x^3 - 48x^2 + 18x - 1$$
$$T_4^* = 128x^4 - 256x^3 + 160x^2 - 32x + 1$$

The approach is otherwise the same.

INDIRECT METHODS

MONTE CARLO

Applicable to first order equations i.e. quadrature type, for which an upper bound can be set for the function. e.g. $y' = f(x)$; $a \leq x \leq b$; $0 \leq f(x) < e$.

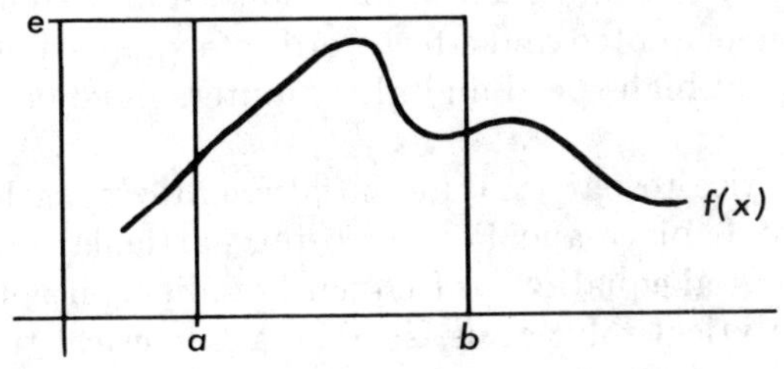

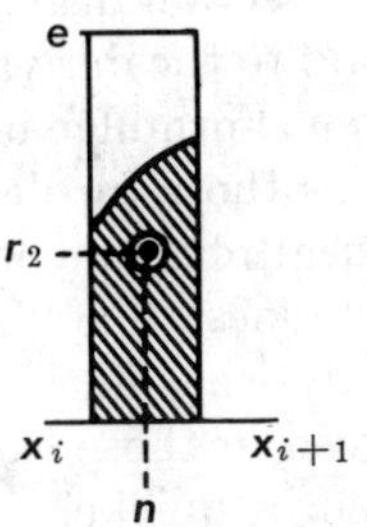

The range $[a,b]$ is subdivided into n intervals, not necessarily equal, such that $x_0 = a$ and $x_n = b$. For the sub-interval $[x_i,x_{i+1}]$ the procedure is as follows:

Set N_i equal to zero and after each trial increase N_i by unity so that it will always indicate the number of completed trials.

Set T_i equal to zero and after each trial increase T_i by unity *only if the trial was successful* otherwise allow it to retain its previous value.

A trial consists of selecting two random numbers r_1 and r_2. The first of these is selected from a uniform (rectangular) distribution in the range $[x_i,x_{i+1}]$ and the second from a uniform distribution in the range $[0,e]$. If $r_2 \leq f(r_1)$ then the trial is deemed to be successful—otherwise it is a failure.

As $N_i \to \infty$,
$$\frac{T_i}{N_i} e(x_{i+1} - x_i) \to \int_{x_i}^{x_{i+1}} f(x) \cdot dx$$

and hence
$$y_{i+1} = y_i + \frac{T_i e(x_{i+1} - x_i)}{N_i} \quad \text{as } N_i \to \infty.$$

If the function has a negative lower bound such that $-d \leq f(x) \leq e$ then a simple axis transformation is necessary i.e. replace $f(x)$ by $d + f(x)$ and e by $(e + d)$ in the above, giving

$$y_{i+1} = y_i + \frac{T_i(e + d)(x_{i+1} - x_i)}{N_i} \quad \text{as } N_i \to \infty.$$

In general, the range of values for which a uniform distribution is required will not be standard and a transformation will be necessary. Such a transformation is obtained as follows:

if r^* is a random number from a uniform distribution in the range $[a,b]$ then r a random number from a uniform distribution in the range $[s,t]$ can be obtained from

$$r = \left(\frac{bs - at}{b - a}\right) - r^* \left(\frac{s - t}{b - a}\right)$$

In particular, if the given range is $[0, 1]$ then

$$r = s + r(t - s)$$

while if the given range is $[-1, -1]$ then

$$r = \tfrac{1}{2}(s + t) + \tfrac{1}{2}r(t - s).$$

In all *Monte Carlo* techniques the number of trials must be large—usually several thousand are required.

PARTIAL DIFFERENTIAL EQUATIONS

Introduction

In the section dealing with finite differences and their applications, some attention has been paid to the derivation of the formulae quoted and a few words of explanation of their use given. Formulae of interest to users of high speed digital computers have been given, as well as those for desk computation.

Physical phenomena frequently give rise to partial differential equations, rather than ordinary ones. The "field" equations of Laplace and Poisson being particular examples. The analytical solution of partial differential equations is frequently tedious, if not impossible, even in those cases where an analytical solution is feasible, a numerical treatment may well be quicker.

Partial differential equations in practice are often reduced to simultaneous equations in two dimensions, one of which is frequently time-like. It is this type of equation which will be considered here.

The classification of partial differential equations is, by analogy with the conic sections, into elliptic, parabolic and hyperbolic equations.

The physical criterion for this grouping is one of boundary conditions. It is also convenient to define the *characteristics* of the equation under consideration. If these are real and distinct then the classification is that of hyperbolic, if real and equal then it is parabolic, while if they are complex it is elliptic.

Consider partial differential equations in two dimensions.

a) Second order equation.
Notation used:

$$p = \frac{\partial u}{\partial x}; \quad q = \frac{\partial u}{\partial y}; \quad r = \frac{\partial^2 u}{\partial x^2}; \quad s = \frac{\partial^2 u}{\partial x \partial y}; \quad t = \frac{\partial^2 u}{\partial y^2}$$

A, B, C and D are functions of u, x, y, p and q.

The physical problem is determined by a second order partial differential equation, the general form of which is

$$A \cdot r + B \cdot s + C \cdot t = D$$

The geometry of the space in which the equation holds yields two more equations:

$$dp = d\left(\frac{\partial u}{\partial x}\right) = r \cdot dx + s \cdot dy; \qquad \text{and similarly, } dq = s \cdot dx + t \cdot dy$$

In matrix form these can be expressed as

$$\begin{bmatrix} A & B & C \\ dx & dy & 0 \\ 0 & dx & dy \end{bmatrix} \cdot \begin{bmatrix} r \\ s \\ t \end{bmatrix} = \begin{bmatrix} D \\ dp \\ dq \end{bmatrix}$$

PARTIAL DIFFERENTIAL EQUATIONS (Continued)

At any point at which the determinant of the 3×3 matrix is zero

i.e.
$$\begin{vmatrix} A & B & C \\ dx & dy & 0 \\ 0 & dx & dy \end{vmatrix} = 0,$$

a necessary condition for the equations to have a solution is that the rank of the following matrix is two.

$$\begin{bmatrix} A & B & C & D \\ dx & dy & 0 & dp \\ 0 & dx & dy & dq \end{bmatrix}$$

This condition can be stated in an alternative fashion as follows. The equations have a solution if any of the following three determinental equations are satisfied

$$\begin{vmatrix} A & B & D \\ dx & dy & dp \\ 0 & dx & dq \end{vmatrix} = 0$$

$$\begin{vmatrix} A & D & C \\ dx & dp & 0 \\ 0 & dq & dy \end{vmatrix} = 0$$

$$\begin{vmatrix} D & B & C \\ dp & dy & 0 \\ dq & dx & dy \end{vmatrix} = 0.$$

The equation
$$\begin{vmatrix} A & B & C \\ dx & dy & 0 \\ 0 & dx & dy \end{vmatrix} = 0$$

can be written in the form $$A\left(\frac{dy}{dx}\right)^2 - B\left(\frac{dy}{dx}\right) + C = 0$$

and thus for the discriminant Δ defined by

$$\Delta \equiv B^2 - 4AC$$

PARTIAL DIFFERENTIAL EQUATIONS (Continued)

there exist three possibilities

(i) $\Delta > 0$, in which case two real distinct values can be obtained for $\frac{dy}{dx}$, corresponding to real characteristics, and the equation is said to be hyperbolic.

(ii) $\Delta = 0$, when the equation is said to be parabolic.

(iii) $\Delta < 0$, when the equation is said to be elliptic.

Simple examples of these equations are

Hyperbolic $$\frac{\partial^2 u}{\partial x^2} - \frac{\partial^2 u}{\partial y^2} = 0$$

Parabolic $$\frac{\partial^2 u}{\partial x^2} - \frac{\partial u}{\partial y} = 0$$

Elliptic $$\frac{\partial^2 u}{\partial x^2} + \frac{\partial^2 u}{\partial y^2} = 0$$

In the above simple forms A, B, C and D are constants while in fact they can depend upon the solution, and its first derivatives, as well as position in the region. Thus the classification of the equation *can* depend upon the solution itself.

b) Two simultaneous first order equations.

The techniques are similar to those employed in section (*a*) above.

A, B, C, D and E are functions of $u, v, x, y, \frac{\partial u}{\partial x}, \frac{\partial u}{\partial y}, \frac{\partial v}{\partial x}, \frac{\partial v}{\partial y}$ and the general form of the equations are

$$A_1 \frac{\partial u}{\partial x} + B_1 \frac{\partial u}{\partial y} + C_1 \frac{\partial v}{\partial x} + D_1 \frac{\partial v}{\partial y} = E_1$$

$$A_2 \frac{\partial u}{\partial x} + B_2 \frac{\partial u}{\partial y} + C_2 \frac{\partial v}{\partial x} + D_2 \frac{\partial v}{\partial y} = E_2.$$

The geometry of the space yields, as before

$$du = \frac{\partial u}{\partial x} \cdot dx + \frac{\partial u}{\partial y} \cdot dy$$

$$dv = \frac{\partial v}{\partial x} \cdot dx + \frac{\partial v}{\partial y} \cdot dy$$

The discriminant Δ now being given by

$$\Delta \equiv (A_1D_2 - A_2D_1 + B_1C_2 - B_2C_1) - 4(A_1C_2 - A_2C_1)(B_1D_2 - B_2D_1).$$

Δ decides the class into which the two equations fall in the same manner as in the previous case.

i.e. Hyperbolic if $\Delta > 0$
Parabolic if $\Delta = 0$
Elliptic if $\Delta < 0$.

An elegant method for the solution of hyperbolic equations, based on the determination of the equations of the characteristics, exists but will not be discussed further here.

PARTIAL DIFFERENTIAL EQUATIONS (Continued)

Elliptic equations are best treated numerically by iteration. This enables both the interval to be adjusted arbitrarily and also the powerful convergence techniques of numerical analysis to be invoked.

Parabolic and hyperbolic equations on the other hand can be covered by step-by-step methods. These latter fall into two main groups. On the one hand, there are those methods which yield a direct solution, and, on the other, those which yield the solution implicitly.

In general, however, elliptic equations arise when the boundary conditions are specified on a contour enclosing a region, which need not be simply connected, i.e. elliptic equations are of the boundary value type.

The hyperbolic equations are of the initial value type, and require the solution over a domain having an open boundary.

For parabolic equations the boundary is also open, but in the main only in one direction, the boundary conditions themselves being mixed.

Elliptic Partial Differential Equations

A square mesh of side "h" is superimposed upon the region to enable the points to be labelled with respect to an arbitrary origin.

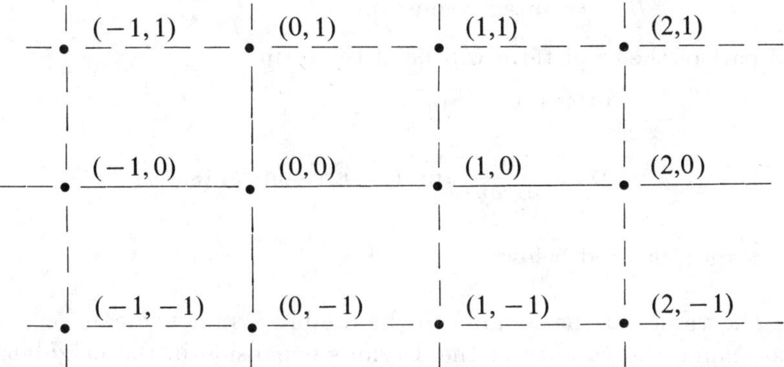

This presupposes that the boundary is formed from the sides of such a mesh. If such is not the case, the procedure must be modified at the boundary. This will be discussed at the end of this section.

The emphasis here will be on the equations of Poisson and Laplace, although the methods involved can easily be extended to any elliptic partial differential equation. In essence, the Taylor's series expansion for a number of mesh points is used to express the partial derivatives occurring in the equation at the origin, in terms of the function values at neighboring points, and higher partial derivatives at the origin. These function values are used to replace the differential equation by a finite difference equation, which is cast in the form of an iteration formula, the higher derivatives consituting the truncation error term.

Consider the equation

$$\nabla^2 u(x,y) = f(x,y)$$

where

$$\nabla^2 \equiv \frac{\partial^2}{\partial x^2} + \frac{\partial^2}{\partial y^2}$$

ELLIPTIC DIFFERENTIAL EQUATIONS

The differential equation is to be replaced by the iteration formula having the form

$$A_{00}u_{00} = \sum_{s=-n}^{n} \sum_{r=-n}^{n}{}^{*} A_{rs}u_{rs} - Bh^2 f_{00} + \text{error term}$$

(the asterisk attached to the double summation indicating that the term involving u_{00} is to be omitted from the sum). In practice n will be restricted to 1 or 2.

Notation used:

$$x_r = x_0 + rh$$

$$y_r = y_0 + sh$$

$$(r,s) = (x_r, y_s)$$

$$f_{00} = f(x_0, y_0)$$

$$u_{rs} = u(x_r, y_s)$$

A_{rs} = Integer weight associated with u_{rs}

B = Integer constant.

The principal part of the error term will be of the form

$$h^4(\theta\nabla^4 + \phi\mathfrak{D}^4)u_{00}$$

where $\mathfrak{D}^4 \equiv \dfrac{\partial^4}{\partial x^2 \partial y^2}$ and θ, ϕ are constants.

A_{rs}, B, θ and ϕ are tabulated below.

Restricting the values of the weights to the doubly symmetric case, i.e. $A_{rs} = A_{sr} = A_{|r|\,|s|}$, and assuming the validity of the Taylor's expansion in the neighborhood of u_{00}, integer values can be obtained for A_{rs} and B, and rational fractions for θ and ϕ. However, the assumptions of the existence of the Taylor's expansion implies that $u(x,y)$ can be adequately represented by a polynomial in x and y over the region in question. Thus if a square region, centered on an arbitrary origin, is considered, the terms involving odd powers of either variable will contribute nothing to the sum providing the coordinates involved are specially selected.

The significant part of the expansion is therefore

$$u_{rs} = \left\{1 + \frac{h^2}{2!}\left(r^2\frac{\partial^2}{\partial x^2} + s^2\frac{\partial^2}{\partial y^2}\right) + \frac{h^4}{4!}\left(r^4\frac{\partial^4}{\partial x^4} + 6r^2s^2\frac{\partial^4}{\partial x^2\partial y^2} + s^4\frac{\partial^4}{\partial y^4}\right)\cdots\right\} u_{00}$$

The coordinate groupings that are used are as follows:

(i) (0,1), (1,0), (0,−1) and (−1,0)
(ii) (1,1), (1,−1), (−1,1) and (−1,−1)
(iii) (0,2), (2,0), (0,−2) and (−2,0)
(iv) (2,2), (2,−2), (−2,2) and (−2,−2)
(v) (2,1), (1,2), (−1,2), (−2,1), (−2,−1), (−1,−2), (1,−2) and (2,−1).

ELLIPTIC DIFFERENTIAL EQUATIONS (Continued)

Group 1. The general form is

A_{rs}

r \ s	−1	0	1
1	1	α	1
0	α	$4(1+\alpha)$	α
−1	1	α	1

$B = (\alpha + 2)$
$\theta = -(\alpha + 2)/12$
$\phi = (\alpha - 4)/6$

in particular

A_{rs}

r \ s	−1	0	1
1	1	0	1
0	0	4	0
−1	1	0	1

$B = 2$
$\theta = -1/6$
$\phi = -2/3$

A_{rs}

r \ s	−1	0	1
1	1	2	1
0	2	12	2
−1	1	2	1

$B = 4$
$\theta = -1/3$
$\phi = -1/3$

A_{rs}

r \ s	−1	0	1
1	1	3	1
0	3	16	3
−1	1	3	1

$B = 5$
$\theta = -5/12$
$\phi = -1/6$

A_{rs}

r \ s	−1	0	1
1	1	4	1
0	4	20	4
−1	1	4	1

$B = 6$
$\theta = -1/2$
$\phi = 0$

A_{rs}

r \ s	−1	0	1
1	1	6	1
0	6	28	6
−1	1	6	1

$B = 8$
$\theta = -2/3$
$\phi = 1/3$

ELLIPTIC DIFFERENTIAL EQUATIONS (Continued)

Group 2. The general form is

A_{rs}

$r \backslash s$	-1	0	1
1	α	1	α
0	1	$4(1+\alpha)$	1
-1	α	1	α

$B = (1 + 2\alpha)$

$\theta = -(1 + 2\alpha)/12$

$\phi = (1 - 4\alpha)/6$

In particular

A_{rs}

$r \backslash s$	-1	0	1
1	0	1	0
0	1	4	1
-1	0	1	0

$B = 1$

$\theta = -1/2$

$\phi = 1/6$

A_{rs}

$r \backslash s$	-1	0	1
1	1	1	1
0	1	8	1
-1	1	1	1

$B = 3$

$\theta = -1/4$

$\phi = -1/2$

A_{rs}

$r \backslash s$	-1	0	1
1	2	1	2
0	1	12	1
-1	2	1	2

$B = 5$

$\theta = -5/12$

$\phi = -7/6$

A_{rs}

$r \backslash s$	-1	0	1
1	4	1	4
0	1	20	1
-1	4	1	4

$B = 9$

$\theta = -3/4$

$\phi = -5/3$

A_{rs}

$r \backslash s$	-1	0	1
1	8	1	8
0	1	36	1
-1	8	1	8

$B = 17$

$\theta = -17/12$

$\phi = -31/6$

ELLIPTIC DIFFERENTIAL EQUATIONS (Continued)

Group 3. The general form is

A_{rs}

r \ s	1	0	−1
1	β	α	β
0	α	$4(\alpha+\beta)$	α
−1	β	α	β

$B = \alpha + 2\beta$

$\theta = -(\alpha + 2\beta)/12$

$\phi = (\alpha - 4\beta)/6$

In particular

A_{rs}

r \ s	−1	0	1
1	2	3	2
0	3	20	3
−1	2	3	2

$B = 7$

$\theta = -7/2$

$\phi = -5/6$

A_{rs}

r \ s	−1	0	1
1	2	5	2
0	5	28	5
−1	2	5	2

$B = 9$

$\theta = -3/4$

$\phi = -1/2$

A_{rs}

r \ s	−1	0	1
1	3	4	3
0	4	28	4
−1	3	4	3

$B = 10$

$\theta = -5/6$

$\phi = -4/3$

A_{rs}

r \ s	−1	0	1
1	3	5	3
0	5	32	5
−1	3	5	3

$B = 11$

$\theta = -11/12$

$\phi = -7/3$

A_{rs}

r \ s	−1	0	1
1	4	5	4
0	5	36	5
−1	4	5	4

$B = 13$

$\theta = -13/12$

$\phi = -11/6$

ELLIPTIC DIFFERENTIAL EQUATIONS (Continued)

While negative weights can be used in all groups care must be taken that the subtraction of comparable terms does not reduce the number of significant decimals left in the result.

The above formulae are of the nine-point type. To increase the accuracy, twenty-five-point formulae can be used. For these the error term is taken to be

$$\{h^4(\theta\nabla^4 + \phi\mathfrak{D}^4) + h^6(\eta\nabla^6 + \gamma\mathfrak{D}^4\nabla^2)\}u_{00}$$

The twenty-five-point formulae which form the basic set are

A_{rs}

r \ s	−2	−1	0	1	2	
2	0	0	0	0	0	$B = 1$
1	0	0	1	0	0	$\theta = -1/12$
0	0	1	4	1	0	$\phi = 1/6$
−1	0	0	1	0	0	$\eta = -1/360$
−2	0	0	0	0	0	$\gamma = 1/120$

A_{rs}

r \ s	−2	−1	0	1	2	
2	0	0	0	0	0	$B = 2$
1	0	1	0	1	0	$\theta = -1/6$
0	0	0	4	0	0	$\phi = -2/3$
−1	0	1	0	1	0	$\eta = -1/180$
−2	0	0	0	0	0	$\gamma = -1/15$

A_{rs}

r \ s	−2	−1	0	1	2	
2	0	1	0	1	0	$B = 10$
1	1	0	0	0	1	$\theta = -17/6$
0	0	0	8	0	0	$\phi = -7/3$
−1	1	0	0	0	1	$\eta = -13/36$
−2	0	1	0	1	0	$\gamma = -7/12$

ELLIPTIC DIFFERENTIAL EQUATIONS (Continued)

A_{rs}

r \ s	−2	−1	0	1	2
2	0	0	1	0	0
1	0	0	0	0	0
0	1	0	4	0	1
−1	0	0	0	0	0
−2	0	0	1	0	0

$B = 4$
$\theta = -4/3$
$\phi = 8/3$
$\eta = -8/45$
$\gamma = 8/15$

A_{rs}

r \ s	−2	−1	0	1	2
2	1	0	0	0	1
1	0	0	0	0	0
0	0	0	4	0	0
−1	0	0	0	0	0
−2	1	0	0	0	1

$B = 8$
$\theta = -8/3$
$\phi = -32/3$
$\eta = -16/45$
$\gamma = -64/15$

Any linear combination of the above five sets can be used.

Group 4 The general form is

A_{rs}

r \ s	−2	−1	0	1	2
2	μ	κ	ρ	κ	μ
1	κ	β	α	β	κ
0	ρ	α	$4(\alpha+\beta+\rho+\mu+2\kappa)$	α	ρ
−1	κ	β	α	β	κ
−2	μ	κ	ρ	κ	μ

$B = \alpha + 2\beta + 4\rho + 8\mu + 10\kappa$
$\theta = -(\alpha + 2\beta + 16\rho + 32\mu + 34\kappa)/12$
$\phi = (\alpha - 4\beta + 16\rho - 64\mu - 14\kappa)/6$
$\eta = -(\alpha + 2\beta + 64\rho + 128\mu + 130\kappa)/360$
$\gamma = (\alpha - 8\beta + 64\rho - 512\mu - 70\kappa)/120$

In particular θ and ϕ can be reduced to zero simultaneously if either *a*) $\rho = -1$, $\alpha = 16$, $\mu = 0 = \beta = \kappa$.

A_{rs}

r \ s	−2	−1	0	1	2
2	0	0	−1	0	0
1	0	0	16	0	0
0	−1	16	60	16	−1
−1	0	0	16	0	0
−2	0	0	−1	0	0

$B = 12$
$\theta = 0$
$\phi = 0$
$\eta = 2/15$
$\gamma = -2/5$

ELLIPTIC DIFFERENTIAL EQUATIONS (Continued)

or $b)$ $\mu = -1$, $\beta = 16$, $\alpha = 0 = \rho = \kappa$.

A_{rs}

r \ s	−2	−1	0	1	2	
2	−1	0	0	0	−1	$B = 24$
1	0	16	0	16	0	$\theta = 0$
0	0	0	60	0	0	$\phi = 0$
−1	0	16	0	16	0	$\eta = 4/15$
−2	−1	0	0	0	−1	$\gamma = 16/5$

or $c)$ $\rho = -1 = \mu$, $\alpha = 16 = \beta$.

A_{rs}

r \ s	−2	−1	0	1	2	
2	−1	0	−1	0	−1	$B = 36$
1	0	16	16	16	0	$\theta = 0$
0	−1	16	120	16	−1	$\phi = 0$
−1	0	16	16	16	0	$\eta = 3/5$
−2	−1	0	−1	0	−1	$\gamma = 14/5$

When working with digital computers it is frequently advantageous to have, as weights, powers of two. The following sets of weights yield this, and in addition the overall divisor A_{00} is also a power of two.

A_{rs}

r \ s	−2	−1	0	1	2	
2	0	1	2	1	0	$B = 14$
1	1	0	0	0	1	$\theta = -11/2$
0	2	0	16	0	2	$\phi = 3$
−1	1	0	0	0	1	$\eta = -43/60$
−2	1	1	2	1	0	$\gamma = 29/60$

A_{rs}

r \ s	−2	−1	0	1	2	
2	0	1	2	1	0	$B = 22$
1	1	4	0	4	1	$\theta = -37/6$
0	2	0	32	0	2	$\phi = 1/3$
−1	1	4	0	4	1	$\eta = -133/180$
−2	0	1	2	1	0	$\gamma = 13/60$

ELLIPTIC DIFFERENTIAL EQUATIONS (Continued)

A_{rs}

$r \backslash s$	−2	−1	0	1	2
2	0	1	2	1	0
1	1	0	4	0	1
0	2	4	32	4	2
−1	1	0	4	0	1
−2	0	1	2	1	0

$B = 18$
$\theta = -35/6$
$\phi = 11/3$
$\eta = -131/180$
$\gamma = 31/60$

A_{rs}

$r \backslash s$	−2	−1	0	1	2
2	0	1	2	1	0
1	1	4	8	4	1
0	2	8	64	8	2
−1	1	4	8	4	1
−2	0	1	2	1	0

$B = 34$
$\theta = -41/3$
$\phi = 5/3$
$\eta = -137/180$
$\gamma = 17/60$

A_{rs}

$r \backslash s$	−2	−1	0	1	2
2	0	0	1	0	0
1	0	1	2	1	0
0	1	2	16	2	1
−1	0	1	2	1	0
−2	1	0	1	0	0

$B = 8$
$\theta = -5/3$
$\phi = 7/3$
$\eta = -17/90$
$\gamma = 29/60$

A_{rs}

$r \backslash s$	−2	−1	0	1	2
2	0	1	0	1	0
1	1	1	1	1	1
0	0	1	16	1	0
−1	1	1	1.	1	1
−2	0	1	0	1	0

$B = 13$
$\theta = -37/12$
$\phi = -17/6$
$\eta = -133/360$
$\gamma = -77/120$

ELLIPTIC DIFFERENTIAL EQUATIONS (Continued)

For desk machine working the value of A_{00} should be chosen so as to simplify the process of division. Some typical sets of weights in which A_{00} is 100 are

A_{rs}

$r \backslash s$	-2	-1	0	1	2	
2	1	7	10	7	1	$B = 118$
1	7	0	0	0	7	$\theta = -215/6$
0	10	0	100	0	10	$\phi = -1/6$
-1	7	0	0	0	7	$\eta = -839/180$
-2	1	7	10	7	1	$\gamma = -181/60$

A_{rs}

$r \backslash s$	-2	-1	0	1	2	
2	0	1	5	1	0	$B = 56$
1	1	8	10	8	1	$\theta = -35/3$
0	5	10	100	10	5	$\phi = 22/3$
-1	1	8	10	8	1	$\eta = -119/90$
-2	0	1	5	1	0	$\gamma = 49/30$

A_{rs}

$r \backslash s$	-2	-1	0	1	2	
2	0	5	0	5	0	$B = 70$
1	5	5	10	5	5	$\theta = -95/6$
0	0	10	100	10	0	$\phi = -40/3$
-1	5	5	10	5	5	$\eta = -67/36$
-2	0	5	0	5	0	$\gamma = -19/6$

Starting the solution

Any rough estimate can be used to start the solution. In the absence of such an estimate all the interior points can be set at zero. In the case of Laplace's equation, straight lines can be drawn through any point to cut the boundary. The estimated value of the function at that point is then obtained by simple ratio and proportion.

Continuing the solution.

A systematic sweep through the region is essential, although the actual direction of travel chosen is immaterial for all practical purposes. It is important however that once the value at a specific point has been recomputed the new value should immediately replace the old.

Refining the mesh.

Once all the values at the mesh points have converted to the degree of accuracy permitted by the interval "h," the mesh can be subdivided in the following fashion and convergence to more accurate values obtained.

ELLIPTIC DIFFERENTIAL EQUATIONS (Continued)

Fig. 1 shows the mesh of side h, near a boundary where it is required to produce a mesh of side $\frac{1}{2}h$

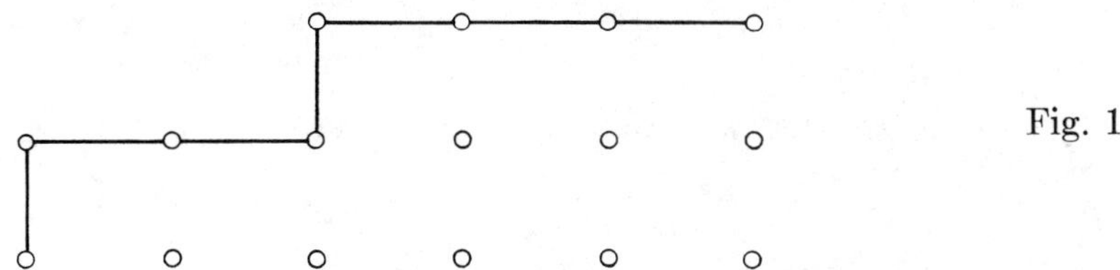

Fig. 1

The points marked with a + in Fig. 2 can be inserted by means of the first formula of Group 1.

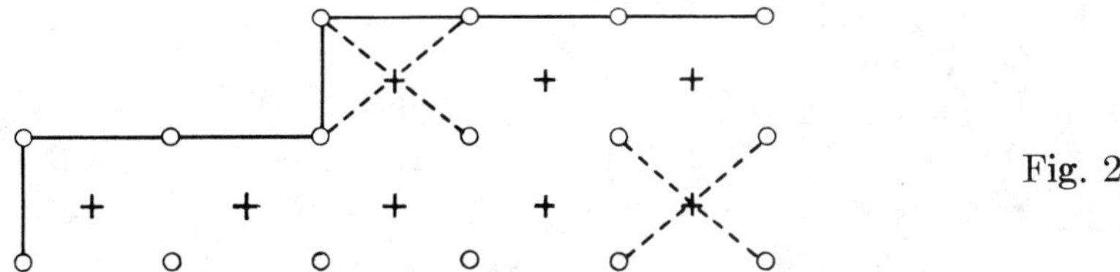

Fig. 2

The points marked with a • in Fig. 3 can then be inserted by means of the first formula of Group 2.

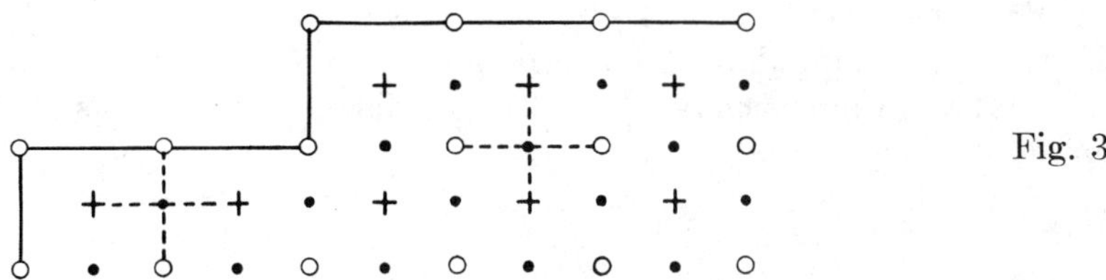

Fig. 3

Finally the boundary conditions enable the values at the new mesh points on the boundary to be inserted. These are indicated by ● in Fig. 4.

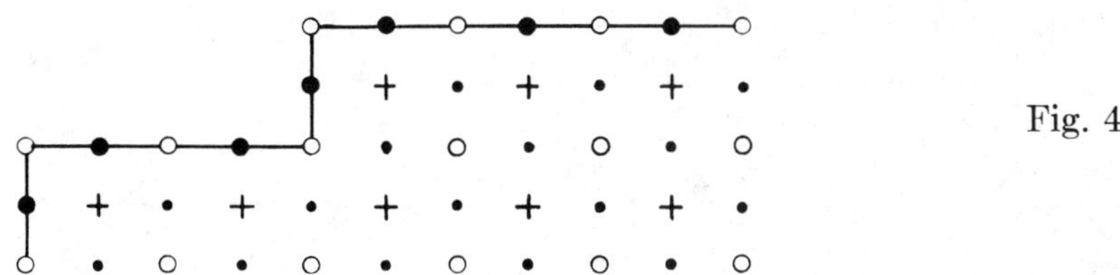

Fig. 4

Improving the numerical solution

(i) Aitken's δ^2 process.

If $u^{(1)}$, $u^{(2)}$ and $u^{(3)}$ are three successive approximations for a given h to u, then a better approximation is given by

$$u = u^{(3)} - \frac{[u^{(3} - u^{(2)}]^2}{u^{(3)} - 2u^{(2)} + u^{(1)}}$$

(ii) Richardson's deferred approach to the limit.

ELLIPTIC DIFFERENTIAL EQUATIONS (Continued)

If $u_{(1)}$ is the value obtained for a mesh size h and $u_{(2)}$ is the value when the mesh size is $\frac{1}{2}h$ then a better value for u is given by

$$u = u_{(2)} + \frac{1}{3}[u_{(2)} - u_{(1)}]$$

If, in addition, $u_{(3)}$ is the value obtained for a mesh size $\frac{1}{4}h$ then the improved value is given by

$$u = u_{(3)} + \frac{4}{9}[u_{(3)} - u_{(2)}] - \frac{1}{45}[u_{(3)} - u_{(1)}]$$

It is not necessary to refine the mesh over the entire region. In most cases only specially selected areas will need refining.

(iii) Difference correction.

The technique here is the same as is described in the section on ordinary differential equations. The derivatives in the error term are replaced by their finite difference equivalents, in both the x and y directions, and these constitute the difference correction.

Non-rectangular boundaries.

In some cases a transformation of coordinates will produce a boundary which will fit a rectangular mesh at the expense of increased complexity of the partial differential equation.

In others a mesh which is not rectangular can be employed, the usual alternatives being triangular and hexagonal.

Otherwise the rectangular mesh can be preserved for the mean body of the calculation and the iterative formula modified on the boundary. Thus two basic forms will be needed at a curved boundary.

Case 1.

$$A_0 u_0 = -Bf_0 + \sum_{r=1}^{4} A_r u_r + 0(h^3)$$

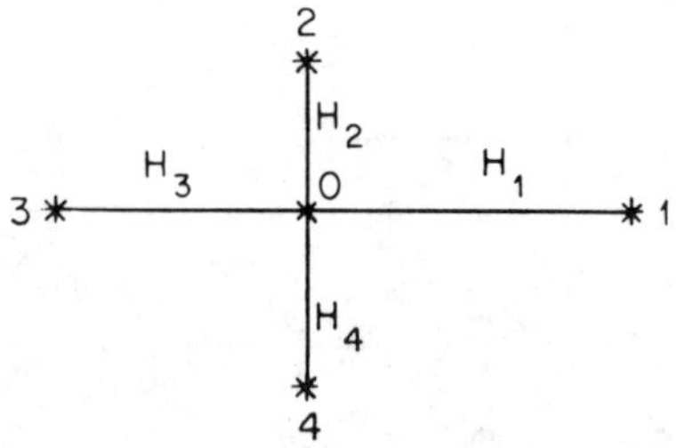

2
H_2
3 H_3 0 H_1 1
H_4
4

$$h \geqq H_i \qquad (i = 1,2,3,4)$$

$$h_i = H_i/h$$

$$A_0 = (h_1 + h_3)(h_2 + h_4)(h_1h_3 + h_2h_4)$$

$$B = \tfrac{1}{2}h_1h_2h_3h_4(h_1 + h_3)(h_2 + h_4)$$

$$A_1 = h_2h_3h_4(h_2 + h_4)$$

$$A_2 = h_3h_4h_1(h_3 + h_1)$$

$$A_3 = h_4h_1h_2(h_4 + h_2)$$

$$A_4 = h_1h_2h_3(h_1 + h_3)$$

ELLIPTIC DIFFERENTIAL EQUATIONS (Continued)

Case 2

The formulae are the same as in Case 1; it is only the orientation of the mesh which is changed

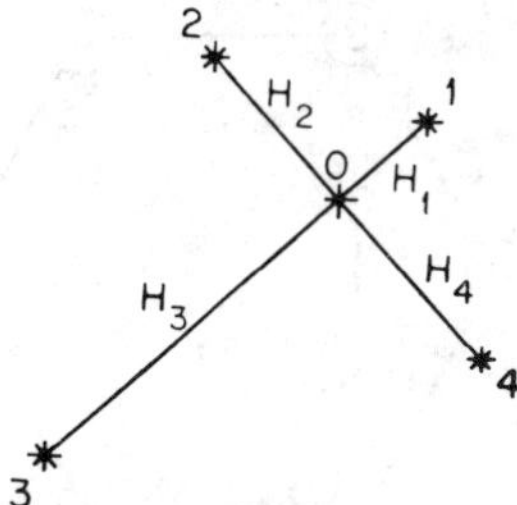

Triangular mesh formulae for Poisson's equation

The points are numbered as shown

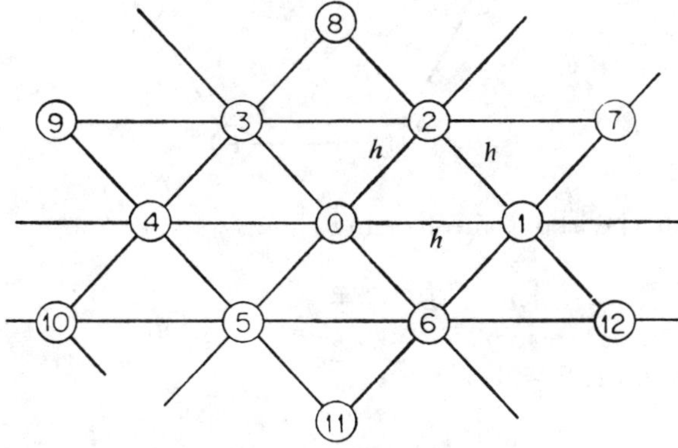

The iteration formula for the above mesh is

$$A_0 = \Sigma A_r u_r - Bh^2 f_0 + \epsilon$$

a) Summation extends over $r = 1(1)6$

$$A_0 = 6, \qquad A_r = 1, \qquad B = 3/2$$

$$\epsilon = -\frac{32}{32} h^4 \nabla_0^4$$

b) Summation extends over $r = 7(1)12$

$$A_0 = 6, \qquad A_r = 1, \qquad B = 9/2$$

$$\epsilon = -\frac{27}{32} h^4 \nabla_0^4$$

c) Summation extends over $r = 1(1)12$

$$A_0 = 48, \qquad A_r = 9 \qquad (r = 1(1)6)$$

$$B = 8 \qquad A_r = -1 \qquad (r = 7(1)12)$$

$$\epsilon = -\frac{h^6}{3840}\left(11 \frac{\partial^6}{\partial x^6} + 15 \frac{\partial^6}{\partial x^4 \partial y^2} + 45 \frac{\partial^6}{\partial x^2 \partial y^4} + 9 \frac{\partial^6}{\partial y^6}\right) u_0$$

ELLIPTIC DIFFERENTIAL EQUATIONS (Continued)

Hexagonal mesh formulae for Poisson's equation

The points are numbered as shown

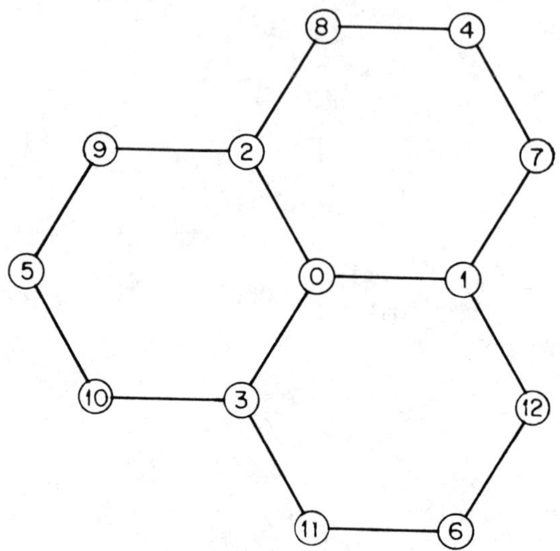

The iteration formula for the above mesh being

$$A_0 = \Sigma A_r u_r - Bh^2 f_0 + \epsilon$$

a) Summation extends over $r = 1(1)3$

$$A_0 = 3, \qquad A_r = 1, \qquad B = 3/4$$

$$\epsilon = -\frac{h^3}{8}\left(\frac{\partial^3}{\partial x^3} - 3\frac{\partial^3}{\partial x \partial y^2}\right) u_0$$

b) Summation extends over $r = 4(1)6$

$$A_0 = 3, \qquad A_r = 1, \qquad B = 3$$

$$\epsilon = -h^3\left(\frac{\partial^3}{\partial x^3} - 3\frac{\partial^3}{\partial x \partial y^2}\right) u_0$$

c) Summation extends over $r = 1(1)6$

$$A_0 = 21, \qquad A_r = 8 \qquad (r = 1(1)3)$$

$$B = 3, \qquad A_r = -1 \qquad (r = 4(1)6)$$

$$\epsilon = \frac{h^4}{8}\left(3\frac{\partial^4}{\partial x^4} + \frac{\partial^4}{\partial x^2 \partial y^2} + 3\frac{\partial^4}{\partial y^4}\right) u_0$$

PARABOLIC DIFFERENTIAL EQUATIONS

Parabolic and Hyperbolic Partial Differential Equations

While this section will be restricted to partial differential equations with two independent variables, the techniques involved can easily be extended to those with three or more.

For simplicity, a rectangular mesh will be superimposed upon the region of integration. The pivotal points will have co-ordinates (x_s,t_r) where, for the intervals "h" and "k,"

$$x_s = x_0 + sh$$

$$t_r = t_0 + rk$$

$$f_{s,r} = f(x_s,t_r).$$

If the derivatives, with respect to either x or t, are replaced by their finite difference approximations, the result will be a set of simultaneous ordinary differential equations.

If, however, all the derivatives are replaced by their relevant approximations, a difference equation corresponding to the original differential equations is obtained. It should be noted, of course, that the approximation chosen will not be unique. Some typical examples are as follows,

$$\left(\frac{\partial f}{\partial t}\right)_{s,r} = (f_{s,r+1} - f_{s,r})/k$$

or

$$= (f_{s,r} - f_{s,r-1})/k$$

or

$$= (f_{s,r+1} - f_{s,r-1})/2k$$

$$\left(\frac{\partial^2 f}{\partial x^2}\right)_{s,r} = (f_{s+1,r} - 2f_{s,r} + f_{s-1,r})/h^2$$

or

$$= (2f_{s,r} - 5f_{s-1,r} + 4f_{s-2,r} - f_{s-3,r})/h^2$$

or

$$= (11f_{s+1,r} - 20f_{s,r} + 6f_{s-1,r} + 4f_{s-2,r} - f_{s-3,r})/12h^2$$

and so forth.

Parabolic Partial Differential Equations

The examples in this section will be concerned in the main, with the simple but nevertheless typical one-dimensional heat equation given by:

$$\frac{\partial f}{\partial t} = \frac{\partial^2 f}{\partial x^2}$$

with the initial conditions $f = I(x)$ $[x_0 \leqslant x \leqslant x_n,\ t = 0]$

and the boundary conditions $f = B(t)$ $[t > 0,\quad x = x_0]$

$f = B^*(t)$ $[t > 0,\quad x = x_n]$

PARABOLIC DIFFERENTIAL EQUATIONS (Continued)

Without loss of generality, x_0 can be replaced by 0 so that

$$x_s = sh, \quad t_r = rk,$$

and
$$I(x_s) = f_{s,0}, \quad B(t_r) = f_{0,r}, \quad B^*(t_r) = f_{n,r}.$$

Two decisions have now to be made.

(*a*) Is the relationship between the values of "*f*" at the pivotal points to be an implicit one or an explicit one? That is to say, if the solution is being determined in the direction of t increasing, is the value at $f_{s,r}$ going to involve other values for which $t = rk$, or is it only going to involve those points for which $t < rk$?

(*b*) How many points are going to be involved at any one time?

A small number of points will mean less computational labor, but generally at the expense of the truncation error.

The explicit formulae are more direct than the implicit ones, but may require the mesh size to be unacceptably small in order to ensure the stability of the solution. In some cases the explicit form is unstable for all meshes. On the other hand, the implicit formulae are stable but they require the solution of large numbers of simultaneous equations.

Some typical explicit formulae

i.
$$f_{s,r+1} = f_{s,r} + (k/h^2)(f_{s-1,r} - 2f_{s,r} + f_{s+1,r})$$

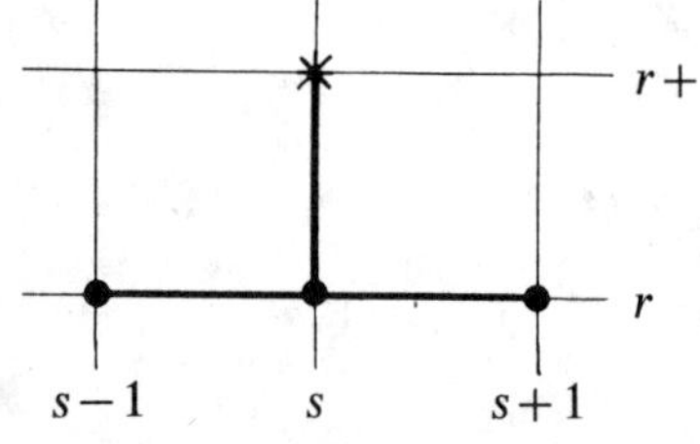

Stable if $0 < k/h^2 \leqslant 1/2$

Truncation error $= 0(k) + 0(h^2)$

In particular when $k/h^2 = 1/2$ the point pattern is simplified by the omission of the pivot $f_{s,r}$.

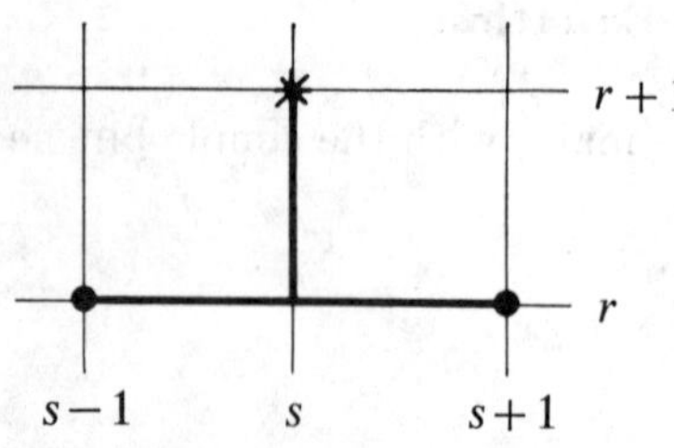

$$f_{s,r+1} = \tfrac{1}{2}(f_{s-1,r} + f_{s+1,r})$$

Milne has shown that an improved truncation error results when $k/h^2 = 1/6$

$$f_{s,r+1} = (1/6)(f_{s-1,r} + 4f_{s,r} + f_{s+1,r})$$

Truncation error $= 0(k^2) + 0(h^4)$

PARABOLIC DIFFERENTIAL EQUATIONS (Continued)

ii.
$$f_{s,r+1} = \frac{(h^2 - 2k)f_{s,r-1} + 2k(f_{s-1,r} + f_{s+1,r})}{(h^2 + 2k)}$$

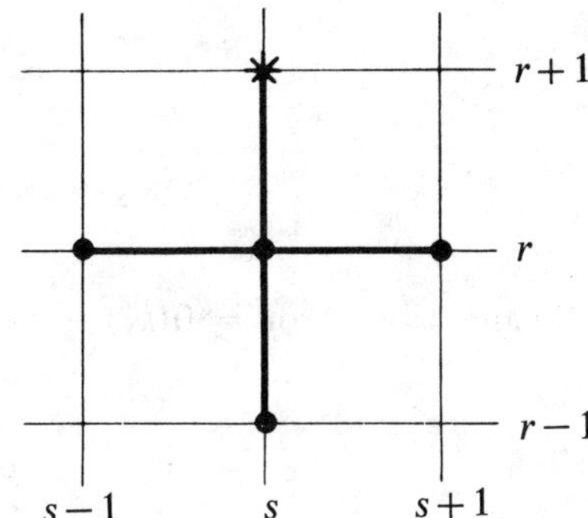

Always stable, but $f_{s,r-1}$ is required.

Truncation error $= 0(k) + 0(h^2) + 0(k/h)$

iii. $f_{s,r+1} = f_{s,r} - (1/12)(k/h^2)(f_{s-2,r} - 16f_{s-1,r} + 30f_{s,r} - 16f_{s+1,r} + f_{s+2,r})$

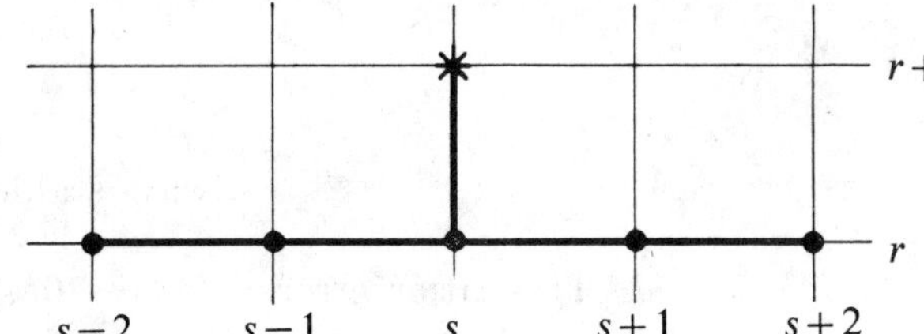

Stable if $0 < (k/h^2) \leq 3/8$

Truncation error $= 0(k) + 0(h^4)$

If the above mesh is used for the main body of the computation, then unsymmetric formulae are required at the ends of the range of s. The formula for the left hand side is shown below. (The formula for the right hand side is the mirror image.)

$$f_{s,r+1} = f_{s,r} + (1/12)(k/h^2)(11f_{s-1,r} - 20f_{s,r} + 6f_{s+1,r} + 4f_{s+2,r} - f_{s+3,r})$$

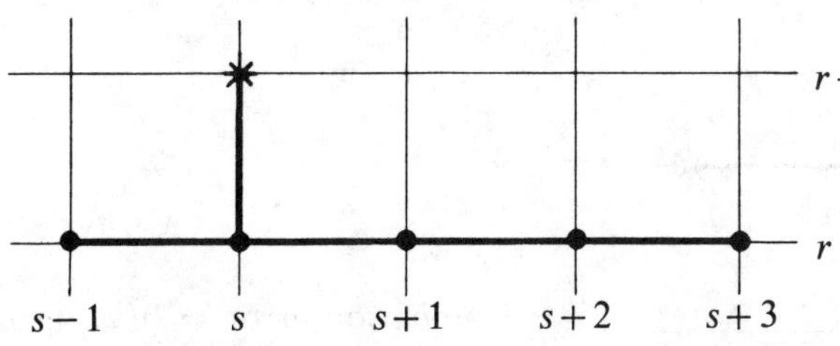

Stable if $0 < (k/h^2) \leq 2/7$

Truncation error $= 0(h^3) + 0(k)$

The truncation error of the unsymmetrical case will of course determine the overall truncation error if all three formulae are to be used in the one problem. The same observation will hold for the stability requirements.

One way of reducing the truncation error is to use additional, previously computed values; i.e., instead of using only those for which $t = rk$, increase the number of pivotal values involved by incorporating those for which $t = (r - 1)k$, $(r - 2)k$ etc. This method was used in example ii. However, it is generally unwise as such formulae tend toward high instability.

Some typical implicit formulae

These have the merit of stability but require the solution of a number of simultaneous equations. However, when matrix methods are used, often only one inversion is required. This simplifies the problem considerably.

PARABOLIC DIFFERENTIAL EQUATIONS (Continued)

i. $(k/h^2)(f_{s+1,r+1} + f_{s-1,r+1} + f_{s+1,r} + f_{s-1,r})$

$$- 2[1 + (k/h^2)]f_{s,r+1} + 2[1 - (k/h^2)]f_{s,r} = 0$$

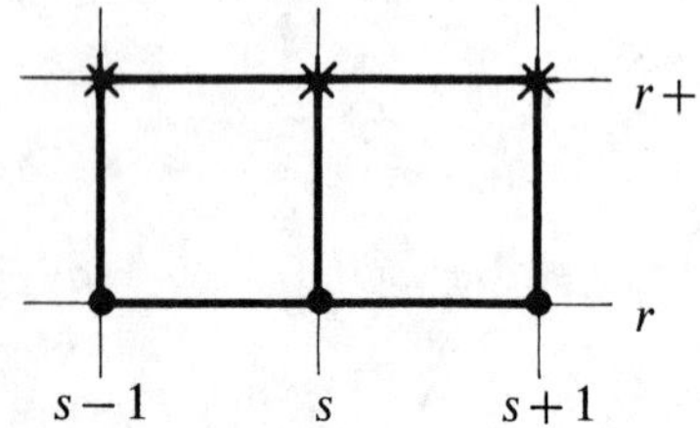

Always stable

Truncation error $= 0(k^2) + 0(h^2)$

ii. $$f_{s,r+1} - f_{s,r} = (k/h^2)(f_{s-1,r+1} - 2f_{s,r+1} + f_{s+1,r+1})$$

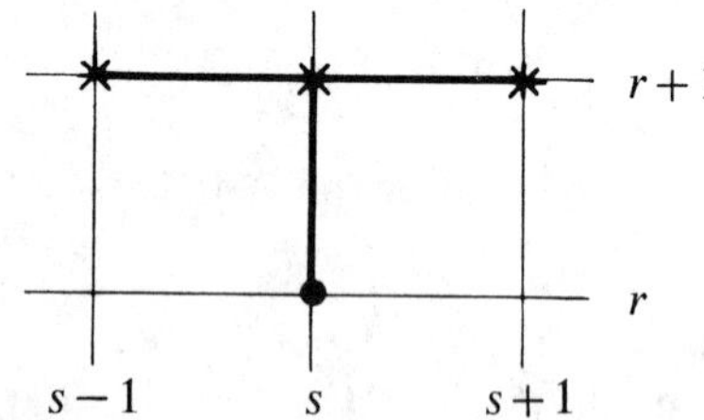

Always stable

Truncation error $= 0(k) + 0(h^2))$

iii. $12(f_{s,r+1} - f_{s,r}) + (k/h^2)(f_{s-2,r+1} - 16f_{s-1,r+1} + 30f_{s,r+1}$

$$- 16f_{s+1,r+1} + f_{s+2,r+1}) = 0$$

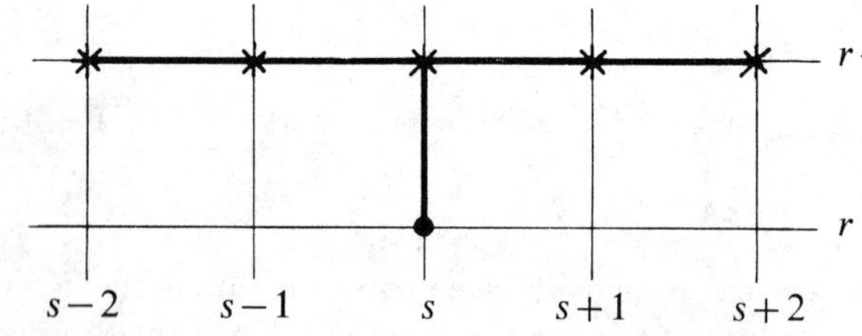

Always stable

Truncation error $= 0(k) + 0(h^4)$

As in the explicit case, unsymmetrical forms of the above formula are required at the extremities of the range. Only one will be quoted here (the other is merely a mirror image):

$$12(f_{s,r+1} - f_{s,r}) = (k/h^2)(11f_{s-1,r+1} - 20f_{s,r+1} + 6f_{s+1,r+1} + 4f_{s+2,r+1} - f_{s+3,r+1})$$

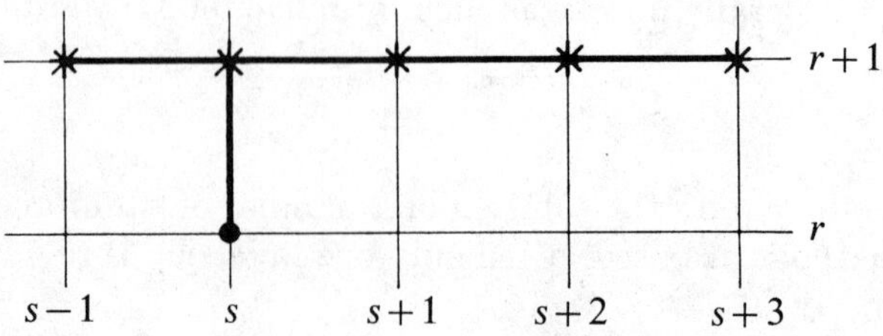

Always stable

Truncation error $= 0(k) + 0(h^3)$

HYPERBOLIC DIFFERENTIAL EQUATIONS (Continued)

iv. $$[18(k/h^2) - 1](f_{s,r+1} - f_{s,r}) - [6(k/h^2) - 1](f_{s,r} - f_{s,r-1}) = 12(k^2/h^4)(f_{s+1,r+1} - 2f_{s,r+1} + f_{s-1,r+1})$$

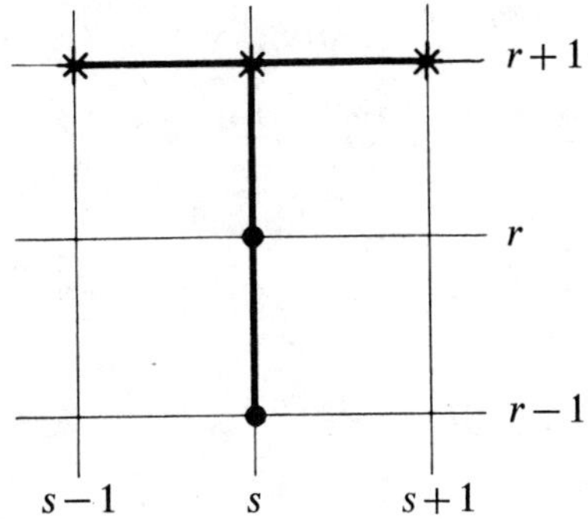

Always stable

Truncation error $= 0(k^2) + 0(h^4)$

v. An implicit method in its own right which can be used as a check on (iv) above is given by

$$(3f_{s+1,r+1} - 4f_{s+1,r} + f_{s+1,r-1}) + (30f_{s,r+1} - 40f_{s,r} + 10f_{s,r-1}) + (3f_{s-1,r+1} - 4f_{s,r+1} + f_{s-1,r-1}) = 24(k/h^2)(f_{s+1,r+1} - 2f_{s,r+1} + f_{s-1,r+1})$$

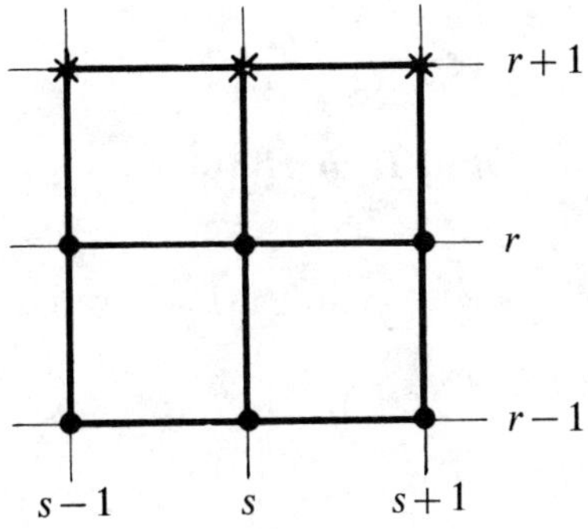

Always stable

Truncation error $= 0(k^2) + 0(h^4)$

Hyperbolic Partial Differential Equations

The methods used in the parabolic case carry over into the hyperbolic case; therefore, only one example of each will be quoted.

Consider

$$(\partial^2 f/\partial x^2) = (\partial^2 f/\partial t^2)$$

with appropriate boundary conditions.

An explicit formula:

$$f_{s,r+1} = 2f_{s,r} - f_{s,r-1} + (k^2/h)(f_{s-1,r} - 2f_{s,r} + f_{s+1,r})$$

PARABOLIC AND HYPERBOLIC DIFFERENTIAL EQUATIONS (Continued)

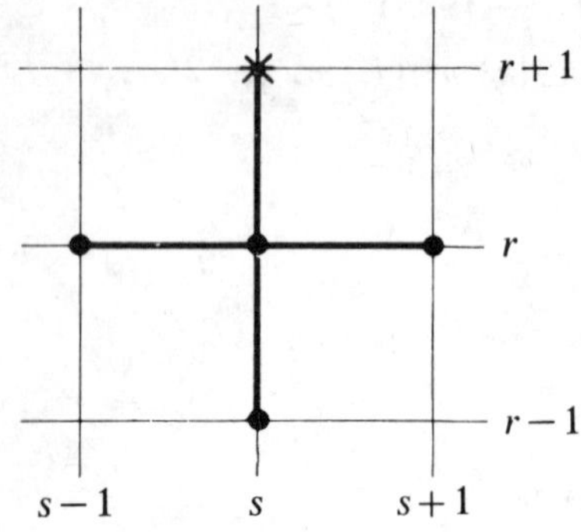

Stable if $0 < (k/h)^2 \leq 1$

Truncation error $= 0(h^4) + 0(k^4)$

An implicit formula:

$$(f_{s,r+1} - 2f_{s,r} + f_{s,r-1}) = (k/h)^2(f_{s-1,r+1} - 2f_{s,r+1} + f_{s+1,r+1})$$

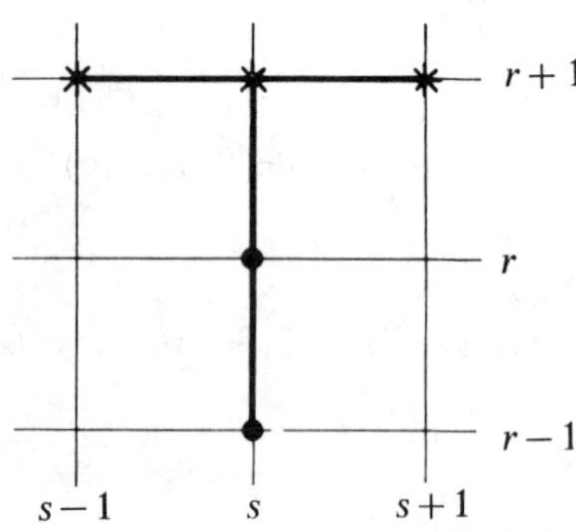

Always stable

Truncation error $= 0(h^4) + 0(k^4)$

Mixed Boundary Conditions for Parabolic and Hyperbolic Differential Equations

When the boundary conditions involve a derivative some slight modifications are necessary to take this condition into account.

Example 1 $\quad \dfrac{\partial f}{\partial t} = \dfrac{\partial^2 f}{\partial x^2};$

$$f = I(x), \qquad [t = 0, \quad 0 \leq x \leq nh]$$

$$f = B(t), \qquad [t > 0, \quad x = 0]$$

$$f + A(\partial f/\partial x) = B^*(t), \qquad [t > 0, \quad x = nh]$$

The derivative can be replaced by a finite difference approximation as follows

$$f_{n,r} + \frac{A(f_{n,r} - f_{n-1,r})}{k} = B^*(t_r)$$

which can be rewritten as $f_{n,r} = \dfrac{(k/A)B^*(t_r) + f_{n-1,r}}{(1 + k/A)}$

This enables the end points to be computed. A more accurate approximation to the derivative can be used of course, e.g.,

$$f_{n,r} + \frac{A(6f_{n,r} - 10f_{n-1,r} + 5f_{n-2,r} - f_{n-3,r})}{6k} = B^*(t_r)$$

PARABOLIC AND HYPERBOLIC DIFFERENTIAL EQUATIONS (Continued)

Example 2. $\dfrac{\partial f}{\partial t} = \dfrac{\partial^2 f}{\partial x^2};\qquad f + A\dfrac{\partial f}{\partial t} = I(x),\qquad [t = 0,\ 0 \le x \le nh]$

$$f = B(t),\qquad [t > 0,\quad x = 0]$$

$$f = B^*(t),\qquad [t > 0,\quad x = nh]$$

In this example the unconditionally stable form (ii) of the explicit set can be used. This is rewritten to give

$$f_{s,-1} = \frac{(h^2 + 2k)}{(h^2 - 2k)}\left(f_{s,1} - \frac{2k}{(h^2 + 2k)}(f_{s-1,0} + f_{s+1,0})\right)$$

The initial conditions yield

$$f_{s,0} + (A/2k)(f_{s\,1} - f_{s-1}) = I(x_s)$$

or

$$f_{s,-1} = (2k/A)[f_{s,0} - I(x_s)] + f_{s,1}$$

Eliminating $f_{s,-1}$ from these two equations will yield an expression for $f_{s,1}[1 \le s \le (n - 1)]$ and thereafter explicit formula (ii) can be used without modification.

MONTE CARLO TECHNIQUES FOR PARTIAL DIFFERENTIAL EQUATIONS

(i) Elliptic Equations.

The two dimensional case will be discussed here—the extension to more than two dimensions follows by increasing the number of the directions of the random walks by two for each added dimension.

The finite difference approximations

$$\frac{\partial^2 u}{\partial x^2} = \frac{u_{i+1,j} - 2u_{i,j} + u_{i-1,j}}{h_1^2}$$

and

$$\frac{\partial^2 u}{\partial y^2} = \frac{u_{i,j+1} - 2u_{i,j} + u_{i,j-1}}{h_2^2}$$

when substituted into $\nabla^2 u = 0$, yield

$$u_{i,j} = \frac{h_2^2 u_{i+1,j}}{2(h_1^2 + h_2^2)} + \frac{h_2^2 u_{i-1,j}}{2(h_1^2 + h_2^2)} + \frac{h_1^2 u_{i,j+1}}{2(h_1^2 + h_2^2)} + \frac{h_1^2 u_{i,j-1}}{2(h_1^2 + h_2^2)}$$

and, defining

$$p_1 = p_2 = \frac{h_2^2}{2(h_1^2 + h_2^2)}$$

$$p_3 = p_4 = \frac{h_1^2}{2(h_1^2 + h_2^2)}$$

It can be seen that $\sum_{i=1}^{4} p_i = 1$, which is the condition for carrying out a random walk along the four directions x increasing, x decreasing, y increasing and y decreasing. Thus to determine $u_{i,j}$ the value of u at the point with coordinates (x_i, y_j) the procedure is as follows; Set both T and N to zero initially.

For each walk completed add 1 to N.

A walk is completed when the boundary is reached and the value of u *at that point on the boundary* has been added to T. As the number of completed walks increases, $N \to \infty$ and $\frac{T}{N} \to u_{i,j}$.

For the equation $\nabla^2 u = f$ use the relationship

$$\frac{h_1^2 h_2^2}{2(h_1^2 + h_2^2)} f_{i,j} + u_{i,j} = p_1 u_{i+1,j} + p_2 u_{i-1,j} + p_3 u_{i,j+1} + p_4 u_{i,j-1}$$

As before the expression on the left hand side is the limiting value of the ratio $\frac{T}{N}$ as $N \to \infty$.

(ii) Parabolic equations.

Formula (ii) on page 685 yields, as a finite difference representation of

$$\frac{\partial f_{s,r}}{\partial t} = \frac{\partial^2 f_{s,r}}{\partial x^2}$$

$$f_{s,r+1} = \left(\frac{h^2 - 2k}{h^2 + 2k}\right) f_{s,r-1} + \left(\frac{2k}{h^2 + 2k}\right) f_{s-1,r} + \left(\frac{2k}{h^2 + 2k}\right) f_{s+1,r}$$

This can be written as

$$f_{s,r+1} = p_1 f_{s-1,r} + p_2 f_{s+1,r} + p_3 f_{s,r-1}$$

where as before $\sum_{r=1}^{3} p_r = 1$

MONTE CARLO TECHNIQUES (Continued)

for $$p_1 = p_2 = \left(\frac{2k}{h^2 + 2k}\right); \qquad p_3 = \left(\frac{h^2 - 2k}{h^2 + 2k}\right).$$

The three directions are x increasing, x decreasing and t decreasing. In other words progress is restricted to be either towards the boundaries (along the x directions) or back towards the initial conditions along the direction of t decreasing. As before, N will be the total number of completed walks and T will be the sum of the boundary values or initial values reached.

THE LAPLACE TRANSFORM

DR. R. E. GASKELL

If $F(t)$ is a piecewise continuous real-valued function of the real variable $t(0 \leqq t < \infty)$, and if $F(t)$ is of exponential order, that is if $|F(t)| < Me^{at}(t > T;\ M,\ a,\ T$ positive constants) then the *Laplace transform* of $F(t)$,

$$L\{F(t)\} = f(s) = \int_0^\infty e^{-st}F(t)\,dt, \tag{1}$$

exists in the half-plane of the complex variable s for which the real part of s is greater than some fixed value s_0, i.e., $R(s) \geq s_0$. Furthermore

$$F(t) = \frac{1}{2\pi i}\int_{a-i\infty}^{a+i\infty} e^{st}f(s)\,ds, \tag{2}$$

where $a > s_0$. The important property

$$\begin{aligned} L\{F^{(r)}(t)\} &= \int_0^\infty e^{-st}\left(\frac{d^rF}{dt^r}\right)dt \\ &= s^r f(s) - \sum_{n=0}^{r-1} s^{r-1-n}F^{(n)}(+0) \end{aligned} \tag{3}$$

makes the Laplace transform very useful for solving linear differential equations with constant coefficients, and many boundary value problems.

	$F(t)$	$f(s)$
1	$F(t)$	$\int_0^\infty e^{-st}F(t)\,dt$
2	$AF(t) + BG(t)$	$Af(s) + Bg(s)$
3	$F'(t)$	$sf(s) - F(+0)$
4	$F^{(n)}(t)$	$s^n f(s) - s^{n-1}F(+0) - s^{n-2}F'(+0) - \cdots - F^{(n-1)}(+0)$
5	$\int_0^t F(\tau)\,d\tau$	$\frac{1}{s} f(s)$
6	$\int_0^t \int_0^\tau F(\lambda)\,d\lambda\,d\tau$	$\frac{1}{s^2} f(s)$
7	$\int_0^t F_1(t - \tau)F_2(\tau)\,d\tau = F_1 * F_2$	$f_1(s)\,f_2(s)$
8	$tF(t)$	$-f'(s)$
9	$t^n F(t)$	$(-1)^n f^{(n)}(s)$
10	$\frac{1}{t} F(t)$	$\int_s^\infty f(x)\,dx$
11	$e^{at}F(t)$	$f(s - a)$
12	$F(t - b)$, where $F(t) = 0$ when $t < 0$	$e^{-bs} f(s)$
13	$\frac{1}{c} F\left(\frac{t}{c}\right)$	$f(cs)$
14	$\frac{1}{c} e^{(bt)/c} F\left(\frac{t}{c}\right)$	$f(cs - b)$
15	$F(t + a) = F(t)$	$\dfrac{\int_0^a e^{-st}F(t)\,dt}{1 - e^{-as}}$
16	$F(t + a) = -F(t)$	$\dfrac{\int_0^a e^{-st}F(t)\,dt}{1 + e^{-as}}$
17	$F_1(t)$, the half-wave rectification of $F(t)$ in No. 16	$\dfrac{f(s)}{1 - e^{-as}}$
18	$F_2(t)$, the full-wave rectification of $F(t)$ in No. 16	$f(s) \coth \frac{as}{2}$
19	$\sum_1^m \frac{p(a_n)}{q'(a_n)} e^{a_n t}$	$\frac{p(s)}{q(s)}, q(s) = (s - a_1)(s - a_2)\cdots(s - a_m)$
20	$e^{at} \sum_{n=1}^{r} \frac{\phi^{(r-n)}(a)}{(r - n)!} \frac{t^{n-1}}{(n - 1)!} + \cdots$	$\frac{p(s)}{q(s)} = \frac{\phi(s)}{(s - a)^r}$

[1]These tables of Laplace Operations, Laplace Transforms, and Finite Fourier sine and cosine transforms were taken from "Modern Operational Mathematics in Engineering", by permission of the author, R. V. Churchill, and the publisher, McGraw-Hill Book Company, Inc.

LAPLACE TRANSFORMS

	$f(s)$	$F(t)$
1	$\frac{1}{s}$	$\mu(t)$, unit step function
2	$\frac{1}{s^2}$	t
3	$\frac{1}{s^n}\ (n = 1, 2, \ldots)$	$\frac{t^{n-1}}{(n-1)!}$
4	$\frac{1}{\sqrt{s}}$	$\frac{1}{\sqrt{\pi t}}$
5	$s^{-3/2}$	$2\sqrt{\frac{t}{\pi}}$
6	$s^{-[n+(1/2)]}\ (n = 1, 2, \ldots)$	$\frac{2^n t^{n-(1/2)}}{1 \cdot 3 \cdot 5 \cdots (2n-1)\sqrt{\pi}}$
7	$\frac{\Gamma(k)}{s^k}\ (k \geq 0)$	t^{k-1}
8	$\frac{1}{s-a}$	e^{at}
9	$\frac{1}{(s-a)^2}$	te^{at}
10	$\frac{1}{(s-a)^n}\ (n = 1, 2, \ldots)$	$\frac{1}{(n-1)!} t^{n-1} e^{at}$
11	$\frac{\Gamma(k)}{(s-a)^k}\ (k \geq 0)$	$t^{k-1} e^{at}$
12*	$\frac{1}{(s-a)(s-b)}$	$\frac{1}{a-b}(e^{at} - e^{bt})$
13*	$\frac{s}{(s-a)(s-b)}$	$\frac{1}{a-b}(ae^{at} - be^{bt})$
14*	$\frac{1}{(s-a)(s-b)(s-c)}$	$-\frac{(b-c)e^{at} + (c-a)e^{bt} + (a-b)e^{ct}}{(a-b)(b-c)(c-a)}$
15	$\frac{1}{s^2 + a^2}$	$\frac{1}{a}\sin at$
16	$\frac{s}{s^2 + a^2}$	$\cos at$
17	$\frac{1}{s^2 - a^2}$	$\frac{1}{a}\sinh at$
18	$\frac{s}{s^2 - a^2}$	$\cosh at$

*Here a, b, and (in 14) c represent distinct constants.

LAPLACE TRANSFORMS (Continued)

	$f(s)$	$F(t)$
19	$\dfrac{1}{s(s^2+a^2)}$	$\dfrac{1}{a^2}(1-\cos at)$
20	$\dfrac{1}{s^2(s^2+a^2)}$	$\dfrac{1}{a^3}(at-\sin at)$
21	$\dfrac{1}{(s^2+a^2)^2}$	$\dfrac{1}{2a^3}(\sin at - at\cos at)$
22	$\dfrac{s}{(s^2+a^2)^2}$	$\dfrac{t}{2a}\sin at$
23	$\dfrac{s^2}{(s^2+a^2)^2}$	$\dfrac{1}{2a}(\sin at + at\cos at)$
24	$\dfrac{s^2-a^2}{(s^2+a^2)^2}$	$t\cos at$
25	$\dfrac{s}{(s^2+a^2)(s^2+b^2)}\ (a^2\neq b^2)$	$\dfrac{\cos at - \cos bt}{b^2-a^2}$
26	$\dfrac{1}{(s-a)^2+b^2}$	$\dfrac{1}{b}e^{at}\sin bt$
27	$\dfrac{s-a}{(s-a)^2+b^2}$	$e^{at}\cos bt$
27.1	$\dfrac{1}{[(s+a)^2+b^2]^n}$	$\dfrac{-e^{-at}}{4^{n-1}b^{2n}}\sum_{r=1}^{n}\binom{2n-r-1}{n-1}(-2t)^{r-1}\dfrac{d^r}{dt^r}[\cos(bt)]$
27.2	$\dfrac{s}{[(s+a)^2+b^2]^n}$	$\dfrac{e^{-at}}{4^{n-1}b^{2n}}\left\{\sum_{r=1}^{n}\binom{2n-r-1}{n-1}(-2t)^{r-1}\dfrac{d^r}{dt^r}[a\cos(bt)+b\sin(bt)] -2b\sum_{r=1}^{n-1}r\binom{2n-r-2}{n-1}(-2t)^{r-1}\dfrac{d^r}{dt^r}[\sin bt]\right\}$
28	$\dfrac{3a^2}{s^3+a^3}$	$e^{-at}-e^{(at)/2}\left(\cos\dfrac{at\sqrt{3}}{2}-\sqrt{3}\sin\dfrac{at\sqrt{3}}{2}\right)$
29	$\dfrac{4a^3}{s^4+4a^4}$	$\sin at\cosh at - \cos at\sinh at$
30	$\dfrac{s}{s^4+4a^4}$	$\dfrac{1}{2a^2}\sin at\sinh at$

LAPLACE TRANSFORMS (Continued)

	$f(s)$	$F(t)$
31	$\dfrac{1}{s^4 - a^4}$	$\dfrac{1}{2a^3}(\sinh at - \sin at)$
32	$\dfrac{s}{s^4 - a^4}$	$\dfrac{1}{2a^2}(\cosh at - \cos at)$
33	$\dfrac{8a^3 s^2}{(s^2 + a^2)^3}$	$(1 + a^2 t^2)\sin at - \cos at$
34*	$\dfrac{1}{s}\left(\dfrac{s-1}{s}\right)^n$	$L_n(t) = \dfrac{e^t}{n!}\dfrac{d^n}{dt^n}(t^n e^{-t})$
35	$\dfrac{s}{(s-a)^{3/2}}$	$\dfrac{1}{\sqrt{\pi t}} e^{at}(1 + 2at)$
36	$\sqrt{s-a} - \sqrt{s-b}$	$\dfrac{1}{2\sqrt{\pi t^3}}(e^{bt} - e^{at})$
37	$\dfrac{1}{\sqrt{s} + a}$	$\dfrac{1}{\sqrt{\pi t}} - a e^{a^2 t}\operatorname{erfc}(a\sqrt{t})$
38	$\dfrac{\sqrt{s}}{s - a^2}$	$\dfrac{1}{\sqrt{\pi t}} + a e^{a^2 t}\operatorname{erf}(a\sqrt{t})$
39	$\dfrac{\sqrt{s}}{s + a^2}$	$\dfrac{1}{\sqrt{\pi t}} - \dfrac{2a}{\sqrt{\pi}} e^{-a^2 t}\displaystyle\int_0^{a\sqrt{t}} e^{\lambda^2}\,d\lambda$
40	$\dfrac{1}{\sqrt{s}(s - a^2)}$	$\dfrac{1}{a} e^{a^2 t}\operatorname{erf}(a\sqrt{t})$
41	$\dfrac{1}{\sqrt{s}(s + a^2)}$	$\dfrac{2}{a\sqrt{\pi}} e^{-a^2 t}\displaystyle\int_0^{a\sqrt{t}} e^{\lambda^2}\,d\lambda$
42	$\dfrac{b^2 - a^2}{(s - a^2)(b + \sqrt{s})}$	$e^{a^2 t}[b - a\operatorname{erf}(a\sqrt{t})] - b e^{b^2 t}\operatorname{erfc}(b\sqrt{t})$
43	$\dfrac{1}{\sqrt{s}(\sqrt{s} + a)}$	$e^{a^2 t}\operatorname{erfc}(a\sqrt{t})$
44	$\dfrac{1}{(s + a)\sqrt{s + b}}$	$\dfrac{1}{\sqrt{b - a}} e^{-at}\operatorname{erf}(\sqrt{b - a}\sqrt{t})$
45	$\dfrac{b^2 - a^2}{\sqrt{s}(s - a^2)(\sqrt{s} + b)}$	$e^{a^2 t}\left[\dfrac{b}{a}\operatorname{erf}(a\sqrt{t}) - 1\right] + e^{b^2 t}\operatorname{erfc}(b\sqrt{t})$
46†	$\dfrac{(1 - s)^n}{s^{n+(1/2)}}$	$\dfrac{n!}{(2n)!\sqrt{\pi t}} H_{2n}(\sqrt{t})$
47	$\dfrac{(1 - s)^n}{s^{n+(3/2)}}$	$-\dfrac{n!}{\sqrt{\pi}(2n + 1)!} H_{2n+1}(\sqrt{t})$

*$L_n(t)$ is the Laguerre polynomial of degree n.

†$H_n(x)$ is the Hermite polynomial, $H_n(x) = e^{x^2}\dfrac{d^n}{dx^n}(e^{-x^2})$.

LAPLACE TRANSFORMS (Continued)

	$f(s)$	$F(t)$
48†	$\dfrac{\sqrt{s+2a}}{\sqrt{s}} - 1$	$ae^{-at}[I_1(at) + I_0(at)]$
49	$\dfrac{1}{\sqrt{s+a}\sqrt{s+b}}$	$e^{-(1/2)(a+b)t} I_0\left(\dfrac{a-b}{2}t\right)$
50	$\dfrac{\Gamma(k)}{(s+a)^k(s+b)^k}\ (k \geq 0)$	$\sqrt{\pi}\left(\dfrac{t}{a-b}\right)^{k-(1/2)} e^{-(1/2)(a+b)t} I_{k-(1/2)}\left(\dfrac{a-b}{2}t\right)$
51	$\dfrac{1}{(s+a)^{1/2}(s+b)^{3/2}}$	$te^{-(1/2)(a+b)t}\left[I_0\left(\dfrac{a-b}{2}t\right) + I_1\left(\dfrac{a-b}{2}t\right)\right]$
52	$\dfrac{\sqrt{s+2a}-\sqrt{s}}{\sqrt{s+2a}+\sqrt{s}}$	$\dfrac{1}{t}e^{-at}I_1(at)$
53	$\dfrac{(a-b)^k}{(\sqrt{s+a}+\sqrt{s+b})^{2k}}\ (k > 0)$	$\dfrac{k}{t}e^{-(1/2)(a+b)t}I_k\left(\dfrac{a-b}{2}t\right)$
54	$\dfrac{(\sqrt{s+a}+\sqrt{s})^{-2\nu}}{\sqrt{s}\sqrt{s+a}}\ (\nu > -1)$	$\dfrac{1}{a^\nu}e^{-(1/2)(at)}I_\nu\left(\dfrac{1}{2}at\right)$
55	$\dfrac{1}{\sqrt{s^2+a^2}}$	$J_0(at)$
56	$\dfrac{(\sqrt{s^2+a^2}-s)^\nu}{\sqrt{s^2+a^2}}\ (\nu > -1)$	$a^\nu J_\nu(at)$
57	$\dfrac{1}{(s^2+a^2)^k}\ (k > 0)$	$\dfrac{\sqrt{\pi}}{\Gamma(k)}\left(\dfrac{t}{2a}\right)^{k-(1/2)} J_{k-(1/2)}(at)$
58	$(\sqrt{s^2+a^2}-s)^k (k > 0)$	$\dfrac{ka^k}{t}J_k(at)$
59	$\dfrac{(s-\sqrt{s^2-a^2})^\nu}{\sqrt{s^2-a^2}}\ (\nu > -1)$	$a^\nu I_\nu(at)$
60	$\dfrac{1}{(s^2-a^2)^k}\ (k > 0)$	$\dfrac{\sqrt{\pi}}{\Gamma(k)}\left(\dfrac{t}{2a}\right)^{k-(1/2)} J_{k-(1/2)}(at)$
61	$\dfrac{e^{-ks}}{s}$	$S_k(t) = \begin{cases} 0 \text{ when } 0 < t < k \\ 1 \text{ when } t > k \end{cases}$
62	$\dfrac{e^{-ks}}{s^2}$	$\begin{cases} 0 \text{ when } 0 < t < k \\ t - k \text{ when } t > k \end{cases}$
63	$\dfrac{e^{-ks}}{s^\mu}\ (\mu > 0)$	$\begin{cases} 0 & \text{when } 0 < t < k \\ \dfrac{(t-k)^{\mu-1}}{\Gamma(\mu)} & \text{when } t > k \end{cases}$
64	$\dfrac{1-e^{-ks}}{s}$	$\begin{cases} 1 \text{ when } 0 < t < k \\ 0 \text{ when } t > k \end{cases}$

†$I_n(x) = i^{-n}J_n(ix)$, where J_n is Bessel's function of the first kind.

LAPLACE TRANSFORMS (Continued)

	$f(s)$	$F(t)$
65	$\dfrac{1}{s(1-e^{-ks})} = \dfrac{1+\coth\frac{1}{2}ks}{2s}$	$S(k,t) = n$ when $(n-1)k < t < nk \ (n = 1, 2, \ldots)$
66	$\dfrac{1}{s(e^{ks}-a)}$	$\begin{cases} 0 \text{ when } 0 < t < k \\ 1 + a + a^2 + \cdots + a^{n-1} \\ \quad \text{when } nk < t < (n+1)k \ (n = 1, 2, \ldots) \end{cases}$
67	$\dfrac{1}{s}\tanh ks$	$M(2k,t) = (-1)^{n-1}$ when $2k(n-1) < t < 2kn$ $(n = 1, 2, \ldots)$
68	$\dfrac{1}{s(1+e^{-ks})}$	$\dfrac{1}{2}M(k,t) + \dfrac{1}{2} = \dfrac{1-(-1)^n}{2}$ when $(n-1)k < t < nk$
69*	$\dfrac{1}{s^2}\tanh ks$	$H(2k,t)$
70	$\dfrac{1}{s\sinh ks}$	$2S(2k, t+k) - 2 = 2(n-1)$ when $(2n-3)k < t < (2n-1)k \quad (t > 0)$
71	$\dfrac{1}{s\cosh ks}$	$M(2k, t+3k) + 1 = 1 + (-1)^n$ when $(2n-3)k < t < (2n-1)k \quad (t > 0)$
72	$\dfrac{1}{s}\coth ks$	$2S(2k,t) - 1 = 2n - 1$ when $2k(n-1) < t < 2kn$
73	$\dfrac{k}{s^2+k^2}\coth\dfrac{\pi s}{2k}$	$\|\sin kt\|$
74	$\dfrac{1}{(s^2+1)(1-e^{-\pi s})}$	$\begin{cases} \sin t \text{ when } (2n-2)\pi < t < (2n-1)\pi \\ 0 \quad \text{when } (2n-1)\pi < t < 2n\pi \end{cases}$
75	$\dfrac{1}{s}e^{-k/s}$	$J_0(2\sqrt{kt})$
76	$\dfrac{1}{\sqrt{s}}e^{-k/s}$	$\dfrac{1}{\sqrt{\pi t}}\cos 2\sqrt{kt}$
77	$\dfrac{1}{\sqrt{s}}e^{k/s}$	$\dfrac{1}{\sqrt{\pi t}}\cosh 2\sqrt{kt}$
78	$\dfrac{1}{s^{3/2}}e^{-k/s}$	$\dfrac{1}{\sqrt{\pi k}}\sin 2\sqrt{kt}$
79	$\dfrac{1}{s^{3/2}}e^{k/s}$	$\dfrac{1}{\sqrt{\pi k}}\sinh 2\sqrt{kt}$
80	$\dfrac{1}{s^{\mu}}e^{-k/s} \ (\mu > 0)$	$\left(\dfrac{t}{k}\right)^{(\mu-1)/2} J_{\mu-1}(2\sqrt{kt})$

$^*H(2k,t) = k + (r-k)(-1)^n$ where $t = 2kn + r$; $0 \le r < 2k$; $n = 0, 1, 2, \ldots$.

LAPLACE TRANSFORMS (Continued)

	$f(s)$	$F(t)$
81	$\frac{1}{s^{\mu}} e^{k/s}(\mu > 0)$	$\left(\frac{t}{k}\right)^{(\mu-1)/2} I_{\mu-1}(2\sqrt{kt})$
82	$e^{-k\sqrt{s}}(k > 0)$	$\frac{k}{2\sqrt{\pi t^3}} \exp\left(-\frac{k^2}{4t}\right)$
83	$\frac{1}{s} e^{-k\sqrt{s}}(k \geqq 0)$	$\operatorname{erfc}\left(\frac{k}{2\sqrt{t}}\right)$
84	$\frac{1}{\sqrt{s}} e^{-k\sqrt{s}}(k \geqq 0)$	$\frac{1}{\sqrt{\pi t}} \exp\left(-\frac{k^2}{4t}\right)$
85	$s^{-3/2} e^{-k\sqrt{s}}(k \geqq 0)$	$2\sqrt{\frac{t}{\pi}} \exp\left(-\frac{k^2}{4t}\right) - k \operatorname{erfc}\left(\frac{k}{2\sqrt{t}}\right)$
86	$\frac{ae^{-k\sqrt{s}}}{s(a+\sqrt{s})}(k \geqq 0)$	$-e^{ak}e^{a^2t} \operatorname{erfc}\left(a\sqrt{t} + \frac{k}{2\sqrt{t}}\right) + \operatorname{erfc}\left(\frac{k}{2\sqrt{t}}\right)$
87	$\frac{e^{-k\sqrt{s}}}{\sqrt{s}(a+\sqrt{s})}(k \geqq 0)$	$e^{ak}e^{a^2t} \operatorname{erfc}\left(a\sqrt{t} + \frac{k}{2\sqrt{t}}\right)$
88	$\frac{e^{-k\sqrt{s(s+a)}}}{\sqrt{s(s+a)}}$	$\begin{cases} 0 & \text{when } 0 < t < k \\ e^{-(1/2)(at)} I_0(\frac{1}{2}a\sqrt{t^2-k^2}) & \text{when } t > k \end{cases}$
89	$\frac{e^{-k\sqrt{s^2+a^2}}}{\sqrt{s^2+a^2}}$	$\begin{cases} 0 & \text{when } 0 < t < k \\ J_0(a\sqrt{t^2-k^2}) & \text{when } t > k \end{cases}$
90	$\frac{e^{-k\sqrt{s^2-a^2}}}{\sqrt{s^2-a^2}}$	$\begin{cases} 0 & \text{when } 0 < t < k \\ I_0(a\sqrt{t^2-k^2}) & \text{when } t > k \end{cases}$
91	$\frac{e^{-k(\sqrt{s^2+a^2}-s)}}{\sqrt{s^2+a^2}}(k \geqq 0)$	$J_0(a\sqrt{t^2+2kt})$
92	$e^{-ks} - e^{-k\sqrt{s^2+a^2}}$	$\begin{cases} 0 & \text{when } 0 < t < k \\ \frac{ak}{\sqrt{t^2-k^2}} J_1(a\sqrt{t^2-k^2}) & \text{when } t > k \end{cases}$
93	$e^{-k\sqrt{s^2+a^2}} - e^{-ks}$	$\begin{cases} 0 & \text{when } 0 < t < k \\ \frac{ak}{\sqrt{t^2-k^2}} I_1(a\sqrt{t^2-k^2}) & \text{when } t > k \end{cases}$
94	$\frac{a^{\nu}e^{-k\sqrt{s^2-a^2}}}{\sqrt{s^2+a^2}(\sqrt{s^2+a^2}+s)^{\nu}}$ $(\nu > -1)$	$\begin{cases} 0 & \text{when } 0 < t < k \\ \left(\frac{t-k}{t+k}\right)^{(1/2)\nu} J_{\nu}(a\sqrt{t^2-k^2}) & \text{when } t > k \end{cases}$
95	$\frac{1}{s}\log s$	$\Gamma'(1) - \log t\ [\Gamma'(1) = -0.5772]$
96	$\frac{1}{s^k}\log s\ (k > 0)$	$t^{k-1}\left\{\frac{\Gamma'(x)}{[\Gamma(k)]^2} - \frac{\log t}{\Gamma(k)}\right\}$
97	$\frac{\log s}{s-a}(a > 0)$	$e^{at}[\log a - \operatorname{Ei}(-at)]$

LAPLACE TRANSFORMS (Continued)

	$f(s)$	$F(t)$
98	$\frac{\log s}{s^2+1}$	$\cos t\,\mathrm{Si}(t) - \sin t\,\mathrm{Ci}(t)$
99	$\frac{s \log s}{s^2+1}$	$-\sin t\,\mathrm{Si}(t) - \cos t\,\mathrm{Ci}(t)$
100	$\frac{1}{s}\log(1+ks)(k>0)$	$-\mathrm{Ei}\left(-\frac{t}{k}\right)$
101	$\log\frac{s-a}{s-b}$	$\frac{1}{t}(e^{bt}-e^{at})$
102	$\frac{1}{s}\log(1+k^2s^2)$	$-2\mathrm{Ci}\left(\frac{t}{k}\right)$
103	$\frac{1}{s}\log(s^2+a^2)\quad(a>0)$	$2\log a - 2\mathrm{Ci}(at)$
104	$\frac{1}{s^2}\log(s^2+a^2)\quad(a>0)$	$\frac{2}{a}[at\log a + \sin at - at\,\mathrm{Ci}(at)]$
105	$\log\frac{s^2+a^2}{s^2}$	$\frac{2}{t}(1-\cos at)$
106	$\log\frac{s^2-a^2}{s^2}$	$\frac{2}{t}(1-\cosh at)$
107	$\arctan\frac{k}{s}$	$\frac{1}{t}\sin kt$
108	$\frac{1}{s}\arctan\frac{k}{s}$	$\mathrm{Si}(kt)$
109	$e^{k^2s^2}\operatorname{erfc}(ks)\quad(k>0)$	$\frac{1}{k\sqrt{\pi}}\exp\left(-\frac{t^2}{4k^2}\right)$
110	$\frac{1}{s}e^{k^2s^2}\operatorname{erfc}(ks)\quad(k>0)$	$\operatorname{erf}\left(\frac{t}{2k}\right)$
111	$e^{ks}\operatorname{erfc}(\sqrt{ks})\quad(k>0)$	$\frac{\sqrt{k}}{\pi\sqrt{t}(t+k)}$
112	$\frac{1}{\sqrt{s}}\operatorname{erfc}(\sqrt{ks})$	$\begin{cases}0 & \text{when } 0<t<k\\(\pi t)^{-1/2} & \text{when } t>k\end{cases}$
113	$\frac{1}{\sqrt{s}}e^{ks}\operatorname{erfc}(\sqrt{ks})(k>0)$	$\frac{1}{\sqrt{\pi(t+k)}}$
114	$\operatorname{erf}\left(\frac{k}{\sqrt{s}}\right)$	$\frac{1}{\pi t}\sin(2k\sqrt{t})$
115	$\frac{1}{\sqrt{s}}e^{k^2/s}\operatorname{erfc}\left(\frac{k}{\sqrt{s}}\right)$	$\frac{1}{\sqrt{\pi t}}e^{-2k\sqrt{t}}$

LAPLACE TRANSFORMS (Continued)

	$f(s)$	$F(t)$
115.1	$-e^{as}\,\mathrm{Ei}(-as)$	$\frac{1}{t+a}$; $(a > 0)$
115.2	$\frac{1}{a} + se^{as}\,\mathrm{Ei}(-as)$	$\frac{1}{(t+a)^2}$; $(a > 0)$
115.3	$\left[\frac{\pi}{2} - \mathrm{Si}(s)\right]\cos s + \mathrm{Ci}(s)\sin s$	$\frac{1}{t^2+1}$
116*	$K_0(ks)$	$\begin{cases} 0 & \text{when } 0 < t < k \\ (t^2 - k^2)^{-1/2} & \text{when } t > k \end{cases}$
117	$K_0(k\sqrt{s})$	$\frac{1}{2t}\exp\left(-\frac{k^2}{4t}\right)$
118	$\frac{1}{s}e^{ks}K_1(ks)$	$\frac{1}{k}\sqrt{t(t+2k)}$
119	$\frac{1}{\sqrt{s}}K_1(k\sqrt{s})$	$\frac{1}{k}\exp\left(-\frac{k^2}{4t}\right)$
120	$\frac{1}{\sqrt{s}}e^{k/s}K_0\left(\frac{k}{s}\right)$	$\frac{2}{\sqrt{\pi t}}K_0(2\sqrt{2kt})$
121	$\pi e^{-ks}I_0(ks)$	$\begin{cases} [t(2k - t)]^{-1/2} & \text{when } 0 < t < 2k \\ 0 & \text{when } t > 2k \end{cases}$
122†	$e^{-ks}I_1(ks)$	$\begin{cases} \frac{k - t}{\pi k\sqrt{t(2k - t)}} & \text{when } 0 < t < 2k \\ 0 & \text{when } t > 2k \end{cases}$

*$K_n(x)$ is Bessel's function of the second kind for the imaginary argument.

†Several additional transforms, especially those involving other Bessel functions, can be found in the tables by G. A. Campbell and R. M. Foster, "Fourier Integrals for Practical Applications", or "Vol. 1, Bateman Manuscript Project, Transform Tables, McGraw-Hill, 1955", or N. W. McLachlan and P. Humbert, "Formulaire pour le calcul symbolique". In the tables by Campbell and Foster, only those entries containing the condition $0 < g$ or $k < g$, where g is our t, are Laplace transforms.

THE Z TRANSFORM

B. Girling

When $F(t)$, a continuous function of time, is sampled at regular intervals of period T the usual Laplace transform techniques are modified. The diagramatic form of a simple sampler together with its associated input-output waveforms is shown below

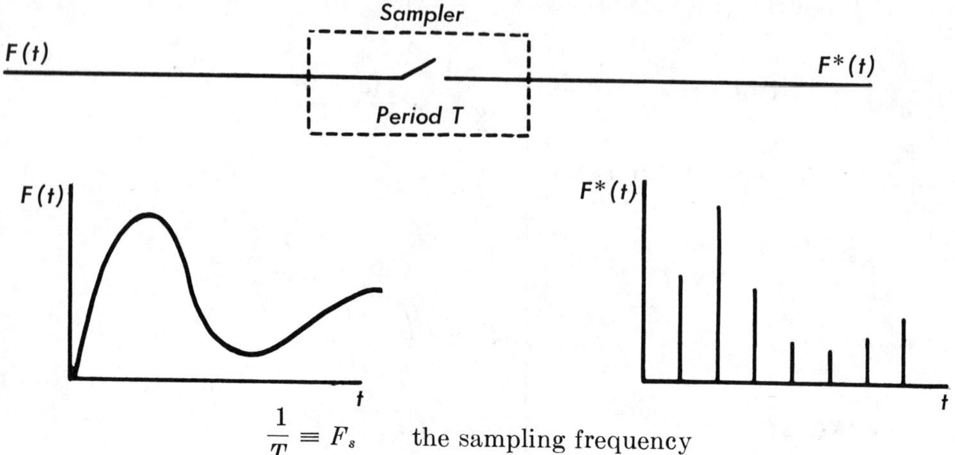

$$\frac{1}{T} \equiv F_s \qquad \text{the sampling frequency}$$

Defining the set of impulse functions $\delta_T(t)$ by

$$\delta_T(t) \equiv \sum_{n=0}^{\infty} \delta(t - nT)$$

the input-output relationship of the sampler becomes

$$\begin{aligned} F^*(t) &= F(t) \cdot \delta_T(t) \\ &= \sum_{n=0}^{\infty} F(nT) \cdot \delta(t - nT). \end{aligned}$$

While for a given $F(t)$ and T the $F^*(t)$ is unique, the converse is not true.

The Laplace transform can be used to define $F^*(s)$ as follows

$$\begin{aligned} L\{F^*(t)\} &\equiv f^*(s) \\ &= \sum_{n=0}^{\infty} F(nT) \cdot e^{-nTs}. \end{aligned}$$

The variable 'z' is introduced by means of the transformation

$$z = e^{Ts}$$

and since any function of s can now be replaced by a corresponding function of z we have

$$f(z) = \sum_{n=0}^{\infty} F(nT) \cdot z^{-n}$$

where

$$f^*(s) \equiv f(z)$$

and

$$s = \frac{1}{T} \ln Z$$

The Z operator can now be defined in terms of the Laplace operator by the relationship

$$Z\{F(t)\} \equiv L\{F^*(t)\}$$

THE Z TRANSFORM (Continued)

An alternative definition (quoted without proof) is

$$Z\{F(t)\} = \sum \text{residues of} \left[\left(\frac{1}{1 - e^{Tx}z^{-1}}\right) \cdot f(z)\right]$$

The inverse z transform

$$\begin{aligned} Z^{-1}\{f(z)\} &\equiv F^*(t) \\ &= \frac{1}{2\pi j} \oint f(z) \cdot z^{n-1}\, dz \end{aligned}$$

where the contour of integration encloses all the singularities of the integrand.

In the following table Greek letters denote constants.

$F(t)$	$f(z) = Z\{F(t)\}$
$\alpha F(t)$	$\alpha f(z)$
$F(t) + G(t)$	$f(z) + g(z)$
$F(t + T)$	$zf(z) - zF(0)$
$F(t + 2T)$	$z^2 f(z) - z^2 F(0) - zF(T)$
$F(t + mT)$	$z^m f(z) - \sum_{r=0}^{m-1} z^{m-r} F(rT)$
	$= z^m f(z)$ when $F(rT) = 0, 0 \leq r \leq m - 1$
$F(t - mT)$	$z^{-m} f(z)$
$e^{\alpha t} F(t)$	$f(e^{-\alpha T} z)$
$e^{-\alpha t} F(t)$	$f(e^{\alpha T} z)$
$t \cdot F(t)$	$-Tz \frac{d}{dz} f(z)$
$t^{-1} F(t)$	$-\frac{1}{T} \int_0^z \frac{f(z)}{z}\, dz$
$\sum_{m=0}^{T/t} F(mT)$	$\left(\frac{z}{z - 1}\right) f(z)$

The following limits are also valid

$$\lim_{t \to 0} F(t) = \lim_{z \to \infty} f(z)$$

$$\lim_{t \to \infty} F^*(t) = \lim_{z \to 1} \left[\left(\frac{z - 1}{z}\right) f(z)\right]$$

In the table which follows, the Heavyside unit step function is defined by

$$H(t - nT) \equiv \begin{cases} 1; t \geq nT \\ 0; t < nT. \end{cases}$$

THE Z TRANSFORM (Continued)

$F(t)$	$f(z)$
$\delta(t)$	1
$\delta(t - mT)$	$\frac{1}{z^m}$
$H(t)$	$\frac{z}{z-1}$
$H(t - T)$	$\frac{1}{z-1}$
$H(t - mT)$	$\frac{z}{z^m \cdot (z-1)}$
$H(t) - H(t - T)$	1
$H(t) - H(t - 2T)$	$1 + \frac{1}{z}$
$H(t - mT) - H(t - \overline{m+1}T)$	$\frac{1}{z} m$
$\frac{T}{t} H(t - T)$	$\ln\left(\frac{z}{z-1}\right)$
t	$\frac{Tz}{(z-1)^2}$
t^2	$\frac{T^2 z(z+1)}{(z-1)^3}$
t^3	$\frac{T^3 z(z^2 + 4z + 1)}{(z-1)^4}$
t^n	$(-1)^n \lim_{\chi \to 0} \frac{\partial^n}{\partial \chi^n}\left(\frac{z}{z - e^{-\chi T}}\right)$
$1 - a^{\omega t}$	$\frac{z(1 - a^{\omega T})}{(z-1)(z - a^{\omega T})}$
$a^{\omega t}$	$\frac{z}{(z - a^{\omega T})}$
$ta^{\omega t}$	$\frac{Tza^{\omega T}}{(z - a^{\omega T})^2}$
$t^2 a^{\omega t}$	$\frac{T^2 a^{\omega T} z(z + a^{\omega T})}{(z - a^{\omega T})^3}$
$\sin \omega t$	$\frac{z \sin \omega T}{z^2 - 2z \cos \omega T + 1}$
$\cos \omega t$	$\frac{z(z - \cos \omega T)}{z^2 - 2z \cos \omega T + 1}$
$\sinh \omega t$	$\frac{z \sinh \omega T}{z^2 - 2z \cosh \omega T + 1}$
$\cosh \omega t$	$\frac{z(z - \cosh \omega T)}{z^2 - 2z \cosh \omega T + 1}$
$e^{-\alpha t} \sin \omega t$	$\frac{ze^{-\alpha T} \sin \omega T}{z^2 - 2ze^{-T} \cos \omega T + e^{-2\alpha T}}$
$e^{-\alpha t} \cos \omega t$	$\frac{z(z - e^{-\alpha T} \cos \omega T)}{z^2 - 2ze^{-T} \cos \omega T + e^{-2\alpha T}}$
$e^{-\alpha t} \sinh \omega t$	$\frac{ze^{-\alpha T} \sinh \omega T}{z^2 - 2ze^{-\alpha T} \cosh \omega T + e^{-2\alpha T}}$
$e^{-\alpha t} \cosh \omega t$	$\frac{z(z - e^{-\alpha T} \cosh \omega T)}{z^2 - 2ze^{-\alpha T} \cosh \omega T + e^{-2\alpha T}}$

THE Z TRANSFORM

$F(t)$	$f(z)$
$-\frac{1}{a}[\delta(t) - a^{t/T}]$	$\frac{1}{z - a}$
$\frac{1}{(a-b)}\left[a^{\left(\frac{t}{T}-1\right)} - b^{\left(\frac{T}{t}-1\right)}\right]$	$\frac{1}{(z-a)(z-b)}$
$\frac{1}{(a-b)}\left[a^{\frac{t}{T}} - b^{\frac{t}{T}}\right]$	$\frac{z}{(z-a)(z-b)}$
$\frac{1}{(a-b)}\left[(a-c)a^{\left(\frac{t}{T}-1\right)} - (b-c)b^{\left(\frac{t}{T}-1\right)}\right]$	$\frac{z-c}{(z-a)(z-b)}$
$\frac{1}{(a-b)}\left[a^{\left(\frac{t}{T}+1\right)} - b^{\left(\frac{T}{t}+1\right)}\right]$	$\frac{z^2}{(z-a)(z-b)}$
$\left(\frac{T}{t}\right)!$	$e^{1/z}$
$\left(\frac{T}{2T}\right)!$	$\cosh(z^{-\frac{1}{2}})$

Methods of evaluating inverse *z* transforms.

(1) Cauchy's residue theorem.

For $t = nT$,

$$G(nT) = \sum_{\text{all } z_k} [\text{residues of } g(z)z^{n-1} \text{ at } z_k]$$

where the z_k define all the poles of $g(z)z^{n-1}$.

(2) Partial fractions.

Expand $g(z)/z$ into partial fractions. The product of z with each of the partial fractions will then be recognizable from the standard forms in the table of z transforms. Note however that the continuous functions obtained are only valid at the sampling instants.

(3) Power series expansion by long division using detached coefficients.

$g(z)$ is expanded into a power series in z^{-1} and the coefficient of the term in z^{-n} is the value of $g(nT)$ i.e. the value of $G(t)$ at the nth sampling instant.

The z transform as a means of determining approximately the inverse Laplace transform.

Since

$$z \equiv e^{Ts}$$

$$s^{-1} = \frac{T}{2}\left[\frac{1}{v} - \frac{v}{3} - \frac{4v^3}{45} - \frac{44v^5}{945} - \cdots\right]$$

where

$$v \equiv \frac{1 - z^{-1}}{1 + z^{-1}},$$

the series being very rapid in its convergence. Given $g(s)$, to find its inverse Laplace transform the following operations are carried out:-

(i) Divide the numerator and denominator of $g(s)$ by the highest power of s yielding as an alternative form for $g(s)$ the quotient of two polynomials in s^{-1}.

(ii) Chose as a numerical value of T, that which makes $2\pi/T$ much larger than the imaginary part of the poles of $G(s)$.

(iii) Substitute into the alternative form for $g(s)$ obtained in (i) above the expansion for s^{-n} determined from the following short table of approximations.

THE Z TRANSFORM (Continued)

Do not at this stage insert the numerical value for T as tabulations with different intervals may be required.
(iv) Divide by T.
(v) Insert the chosen value for T and divide the numerator by the denominator.
(vi) The coefficient of z^{-n} is the required value of the function at $t = nT$.

s^{-n}	z transform (approximate)
s^{-1}	$\frac{T}{2}\left[\frac{1+z^{-1}}{1-z^{-1}}\right]$
s^{-2}	$\frac{T^2}{12}\left[\frac{1+10z^{-1}+z^{-2}}{(1-z^{-1})^2}\right]$
s^{-3}	$\frac{T^3}{3}\left[\frac{z^{-1}+z^{-2}}{(1-z^{-1})^3}\right]$
s^{-4}	$\frac{T^4}{144}\left[\frac{1+20z^{-1}+102z^{-2}+20z^{-3}+z^{-4}}{(1-z^{-1})^4}\right]$
s^{-5}	$\frac{T^5}{24}\left[\frac{z^{-1}+11z^{-2}+11z^{-3}+z^{-4}}{(1-z^{-1})^5}\right]$
s^{-6}	$\frac{T^6}{4}\left[\frac{z^{-2}+2z^{-3}+z^{-4}}{(1-z^{-1})^6}\right]$
s^{-7}	$\frac{T^7}{8}\left[\frac{z^{-2}+3z^{-3}+3z^{-4}+z^{-5}}{(1-z^{-1})^7}\right]$

THE FINITE FOURIER TRANSFORM

DR. R. E. GASKELL

For a piecewise continuous function $F(x)$ over a finite interval $0 \leqq x \leqq \pi$, the *finite Fourier cosine transform* of $F(x)$ is

$$f_c(n) = \int_0^\pi F(x) \cos nx \, dx \quad (n = 0, 1, 2, \ldots). \tag{4}$$

If x ranges over the interval $0 \leqq x \leqq L$, the substitution $x' = \pi x/L$ allows the use of this definition, also. The inverse transform is written

$$\bar{F}(x) = \frac{1}{\pi} f_c(0) - \frac{2}{\pi} \sum_{n=1}^{\infty} f_c(n) \cos nx \quad (0 < x < \pi) \tag{5}$$

where $\bar{F}(x) = \dfrac{[F(x + 0) + F(x - 0)]}{2}$. We observe that $\bar{F}(x) = F(x)$ at points of continuity. The formula

$$\begin{aligned} f_c^{(2)}(n) &= \int_0^\pi F''(x) \cos nx \, dx \\ &= -n^2 f_c(n) - F'(0) + (-1)^n F'(\pi) \end{aligned} \tag{6}$$

makes the finite Fourier cosine transform useful in certain boundary value problems.

Analogously, the *finite Fourier sine transform* of $F(x)$ is

$$f_s(n) = \int_0^\pi F(x) \sin nx \, dx \quad (n = 1, 2, 3, \ldots) \tag{7}$$

and

$$\bar{F}(x) = \frac{2}{\pi} \sum_{n=1}^{\infty} f_s(n) \sin nx \quad (0 < x < \pi) \tag{8}$$

Corresponding to (6) we have

$$\begin{aligned} f_s^{(2)}(n) &= \int_0^\pi F''(x) \sin nx \, dx \\ &= -n^2 f_s(n) - nF(0) - n(-1)^n F(\pi). \end{aligned} \tag{9}$$

FOURIER TRANSFORMS

If $F(x)$ is defined for $x \geqq 0$ and is piecewise continuous over any finite interval, and if

$$\int_0^\infty F(x) \, \mathrm{dx}$$

is absolutely convergent, then

$$f_c(\alpha) = \sqrt{\frac{2}{\pi}} \int_0^\infty F(x) \cos(\alpha x) \, dx \tag{10}$$

is the *Fourier cosine transform* of $F(x)$. Furthermore,

$$\bar{F}(x) = \sqrt{\frac{2}{\pi}} \int_0^\infty f_c(\alpha) \cos(\alpha x) \, d\alpha. \tag{11}$$

If $\lim_{x\to\infty} \dfrac{d^n F}{dx^n} = 0$, an important property of the Fourier cosine transform

$$f_c^{(2r)}(\alpha) = \sqrt{\frac{2}{\pi}} \int_0^\infty \left(\frac{d^{2r}F}{dx^{2r}}\right) \cos(\alpha x)\, dx$$

$$= -\sqrt{\frac{2}{\pi}} \sum_{n=0}^{r-1} (-1)^n a_{2r-2n-1} \alpha^{2n} + (-1)^r \alpha^{2r} f_c(\alpha) \tag{12}$$

where $\lim_{x\to 0} \dfrac{d^r F}{dx^r} = a_r$, makes it useful in the solution of many problems.

Under the same conditions,

$$f_s(\alpha) = \sqrt{\frac{2}{\pi}} \int_0^\infty F(x) \sin(\alpha x)\, dx \tag{13}$$

defines the *Fourier sine transform* of $F(x)$, and

$$\bar{F}(x) = \sqrt{\frac{2}{\pi}} \int_0^\infty f_s(\alpha) \sin(\alpha x)\, d\alpha. \tag{14}$$

Corresponding to (12) we have

$$f_s^{(2r)}(\alpha) = \sqrt{\frac{2}{\pi}} \int_0^\infty \frac{d^{2r}F}{dx^{2r}} \sin(\alpha x)\, dx$$

$$= -\sqrt{\frac{2}{\pi}} \sum_{n=1}^{r} (-1)^n \alpha^{2n-1} a_{2r-2n} + (-1)^{r-1} \alpha^{2r} f_s(\alpha). \tag{15}$$

Similarly, if $F(x)$ is defined for $-\infty < x < \infty$, and if $\int_{-\infty}^{\infty} F(x)\, dx$ is absolutely convergent, then

$$f(\alpha) = \frac{1}{\sqrt{2\pi}} \int_{-\infty}^{\infty} F(x) e^{i\alpha x}\, dx \tag{16}$$

is the *Fourier transform* of $F(x)$, and

$$\bar{F}(x) = \frac{1}{\sqrt{2\pi}} \int_{-\infty}^{\infty} f(\alpha) e^{-i\alpha x}\, d\alpha. \tag{17}$$

Also, if

$$\lim_{|x|\to\infty} \left|\frac{d^n F}{dx^n}\right| = 0 \quad (n = 1, 2, \ldots, r-1),$$

then

$$f^{(r)}(\alpha) = \frac{1}{\sqrt{2\pi}} \int_{-\infty}^{\infty} F^{(r)}(x) e^{i\alpha x}\, dx = (-i\alpha)^r f(\alpha). \tag{18}$$

FOURIER TRANSFORMS
FINITE SINE TRANSFORMS

	$f_s(n)$	$F(x)$
1	$f_s(n) = \int_0^\pi F(x) \sin nx \, dx \; (n = 1, 2, \cdots)$	$F(x)$
2	$(-1)^{n+1} f_s(n)$	$F(\pi - x)$
3	$\frac{1}{n}$	$\frac{\pi - x}{\pi}$
4	$\frac{(-1)^{n+1}}{n}$	$\frac{x}{\pi}$
5	$\frac{1 - (-1)^n}{n}$	1
6	$\frac{2}{n^2} \sin \frac{n\pi}{2}$	$\begin{cases} x & \text{when } 0 < x < \pi/2 \\ \pi - x & \text{when } \pi/2 < x < \pi \end{cases}$
7	$\frac{(-1)^{n+1}}{n^3}$	$\frac{x(\pi^2 - x^2)}{6\pi}$
8	$\frac{1 - (-1)^n}{n^3}$	$\frac{x(\pi - x)}{2}$
9	$\frac{\pi^2(-1)^{n-1}}{n} - \frac{2[1 - (-1)^n]}{n^3}$	x^2
10	$\pi(-1)^n \left(\frac{6}{n^3} - \frac{\pi^2}{n}\right)$	x^3
11	$\frac{n}{n^2 + c^2} [1 - (-1)^n e^{c\pi}]$	e^{cx}
12	$\frac{n}{n^2 + c^2}$	$\frac{\sinh c(\pi - x)}{\sinh c\pi}$
13	$\frac{n}{n^2 - k^2} \; (k \neq 0, 1, 2, \cdots)$	$\frac{\sin k(\pi - x)}{\sin k\pi}$
14	$\begin{cases} \frac{\pi}{2} & \text{when } n = m \\ 0 & \text{when } n \neq m \end{cases} \quad (m = 1, 2, \cdots)$	$\sin mx$
15	$\frac{n}{n^2 - k^2} [1 - (-1)^n \cos k\pi] \quad (k \neq 1, 2, \cdots)$	$\cos kx$
16	$\begin{cases} \frac{n}{n^2 - m^2} [1 - (-1)^{n+m}] & \text{when } n \neq m = 1, 2, \cdots \\ 0 & \text{when } n = m \end{cases}$	$\cos mx$
17	$\frac{n}{(n^2 - k^2)^2} \; (k \neq 0, 1, 2, \cdots)$	$\frac{\pi \sin kx}{2k \sin^2 k\pi} - \frac{x \cos k(\pi - x)}{2k \sin k\pi}$
18	$\frac{b^n}{n} \; (\lvert b \rvert \leqq 1)$	$\frac{2}{\pi} \arctan \frac{b \sin x}{1 - b \cos x}$
19	$\frac{1 - (-1)^n}{n} b^n \quad (\lvert b \rvert \leq 1)$	$\frac{2}{\pi} \arctan \frac{2b \sin x}{1 - b^2}$

FINITE COSINE TRANSFORMS

	$f_c(n)$	$F(x)$
1	$f_c(n) = \int_0^\pi F(x)\cos nx\,dx \quad (n = 0, 1, 2, \cdots)$	$F(x)$
2	$(-1)^n f_c(n)$	$F(\pi - x)$
3	0 when $n = 1, 2, \cdots$; $f_c(0) = \pi$	1
4	$\frac{2}{n}\sin\frac{n\pi}{2}$; $f_c(0) = 0$	$\begin{cases} 1 & \text{when } 0 < x < \pi/2 \\ -1 & \text{when } \pi/2 < x < \pi \end{cases}$
5	$-\frac{1-(-1)^n}{n^2}$; $f_c(0) = \frac{\pi^2}{2}$	x
6	$\frac{(-1)^n}{n^2}$; $f_c(0) = \frac{\pi^2}{6}$	$\frac{x^2}{2\pi}$
7	$\frac{1}{n^2}$; $f_c(0) = 0$	$\frac{(\pi - x)^2}{2\pi} - \frac{\pi}{6}$
8	$3\pi^2\frac{(-1)^n}{n^2} - 6\frac{1-(-1)^n}{n^4}$; $f_c(0) = \frac{\pi^4}{4}$	x^3
9	$\frac{(-1)^n e^c\pi - 1}{n^2 + c^2}$	$\frac{1}{c}e^{cx}$
10	$\frac{1}{n^2 + c^2}$	$\frac{\cosh c(\pi - x)}{c\sinh c\pi}$
11	$\frac{k}{n^2 - k^2}[(-1)^n\cos\pi k - 1]$ $(k \neq 0, 1, 2, \cdots)$	$\sin kx$
12	$\frac{(-1)^{n+m} - 1}{n^2 - m^2}$; $f_c(m) = 0 \quad (m = 1, 2, \cdots)$	$\frac{1}{m}\sin mx$
13	$\frac{1}{n^2 - k^2} \quad (k \neq 0, 1, 2, \cdots)$	$-\frac{\cos k(\pi - x)}{k\sin k\pi}$
14	0 when $n = 1, 2, \cdots$; $f_c(m) = \frac{\pi}{2} \quad (m = 1, 2, \cdots)$	$\cos mx$

FOURIER SINE TRANSFORMS[1]

	$F(x)$	$f_s(\alpha)$
1	$\begin{cases} 1 & (0 < x < a) \\ 0 & (x > a) \end{cases}$	$\sqrt{\frac{2}{\pi}}\left[\frac{1 - \cos\alpha}{\alpha}\right]$
2	$x^{p-1}\ (0 < p < 1)$	$\sqrt{\frac{2}{\pi}}\frac{\Gamma(p)}{\alpha^p}\sin\frac{p\pi}{2}$
3	$\begin{cases} \sin x & (0 < x < a) \\ 0 & (x > a) \end{cases}$	$\frac{1}{\sqrt{2\pi}}\left[\frac{\sin[a(1-\alpha)]}{1-\alpha} - \frac{\sin[a(1+\alpha)]}{1+\alpha}\right]$
4	e^{-x}	$\sqrt{\frac{2}{\pi}}\left[\frac{\alpha}{1+\alpha^2}\right]$
5	$xe^{-x^2/2}$	$\alpha e^{-\alpha^2/2}$
6	$\cos\frac{x^2}{2}$	$\sqrt{2}\left[\sin\frac{\alpha^2}{2}C\left(\frac{\alpha^2}{2}\right) - \cos\frac{\alpha^2}{2}S\left(\frac{\alpha^2}{2}\right)\right]^*$
7	$\sin\frac{x^2}{2}$	$\sqrt{2}\left[\cos\frac{\alpha^2}{2}C\left(\frac{\alpha^2}{2}\right) + \sin\frac{\alpha^2}{2}S\left(\frac{\alpha^2}{2}\right)\right]^*$

*$C(y)$ and $S(y)$ are the Fresnel integrals

$$C(y) = \frac{1}{\sqrt{2\pi}}\int_0^y \frac{1}{\sqrt{t}}\cos t\,dt,$$

$$S(y) = \frac{1}{\sqrt{2\pi}}\int_0^y \frac{1}{\sqrt{t}}\sin t\,dt.$$

[1] More extensive tables of the Fourier sine and cosine transforms can be found in Fritz Oberhettinger, "Tabellen zur-Fourier Transformation," Springer (1957).

FOURIER COSINE TRANSFORMS

	$F(x)$	$f_c(\alpha)$
1	$\begin{cases} 1 & (0 < x < a) \\ 0 & (x > a) \end{cases}$	$\sqrt{\frac{2}{\pi}}\frac{\sin a\alpha}{\alpha}$
2	$x^{p-1}\quad (0 < p < 1)$	$\sqrt{\frac{2}{\pi}}\frac{\Gamma(p)}{\alpha^p}\cos\frac{p\pi}{2}$
3	$\begin{cases} \cos x & (0 < x < a) \\ 0 & (x > a) \end{cases}$	$\frac{1}{\sqrt{2\pi}}\left[\frac{\sin[a(1-\alpha)]}{1-\alpha} + \frac{\sin[a(1+\alpha)]}{1+\alpha}\right]$
4	e^{-x}	$\sqrt{\frac{2}{\pi}}\left(\frac{1}{1+\alpha^2}\right)$
5	$e^{-x^2/2}$	$e^{-\alpha^2/2}$
6	$\cos\frac{x^2}{2}$	$\cos\left(\frac{\alpha^2}{2} - \frac{\pi}{4}\right)$
7	$\sin\frac{x^2}{2}$	$\cos\left(\frac{\alpha^2}{2} + \frac{\pi}{4}\right)$

FOURIER TRANSFORMS[1]

	$F(x)$	$f(\alpha)$
1	$\dfrac{\sin ax}{x}$	$\begin{cases} \sqrt{\dfrac{\pi}{2}} & \lvert\alpha\rvert < a \\ 0 & \lvert\alpha\rvert > a \end{cases}$
2	$\begin{cases} e^{iwx} & (p < x < q) \\ 0 & (x < p,\ x > q) \end{cases}$	$\dfrac{i}{\sqrt{2\pi}} \dfrac{e^{ip(w+\alpha)} - e^{iq(w+\alpha)}}{(w+\alpha)}$
3	$\begin{cases} e^{-cx+iwx} & (x > 0) \\ 0 & (x < 0) \end{cases} \quad (c > 0)$	$\dfrac{i}{\sqrt{2\pi}(w + \alpha + ic)}$
4	$e^{-px^2} \quad R(p) > 0$	$\dfrac{1}{\sqrt{2p}} e^{-\alpha^2/4p}$
5	$\cos px^2$	$\dfrac{1}{\sqrt{2p}} \cos\left[\dfrac{\alpha^2}{4p} - \dfrac{\pi}{4}\right]$
6	$\sin px^2$	$\dfrac{1}{\sqrt{2p}} \cos\left[\dfrac{\alpha^2}{4p} + \dfrac{\pi}{4}\right]$
7	$\lvert x\rvert^{-p} \quad (0 < p < 1)$	$\sqrt{\dfrac{2}{\pi}} \dfrac{\Gamma(1-p)\sin\dfrac{p\pi}{2}}{\lvert\alpha\rvert^{(1-p)}}$
8	$\dfrac{e^{-a\lvert x\rvert}}{\sqrt{\lvert x\rvert}}$	$\dfrac{\sqrt{\sqrt{(a^2+\alpha^2)} + a}}{\sqrt{a^2+\alpha^2}}$
9	$\dfrac{\cosh ax}{\cosh \pi x} \quad (-\pi < a < \pi)$	$\sqrt{\dfrac{2}{\pi}} \dfrac{\cos\dfrac{a}{2}\cosh\dfrac{\alpha}{2}}{\cosh\alpha + \cos a}$
10	$\dfrac{\sinh ax}{\sinh \pi x} \quad (-\pi < a < \pi)$	$\dfrac{1}{\sqrt{2\pi}} \dfrac{\sin a}{\cosh\alpha + \cos a}$
11	$\begin{cases} \dfrac{1}{\sqrt{a^2 - x^2}} & (\lvert x\rvert < a) \\ 0 & (\lvert x\rvert > a) \end{cases}$	$\sqrt{\dfrac{\pi}{2}} J_0(a\alpha)$
12	$\dfrac{\sin[b\sqrt{a^2+x^2}]}{\sqrt{a^2+x^2}}$	$\begin{cases} 0 & (\lvert\alpha\rvert > b) \\ \sqrt{\dfrac{\pi}{2}} J_0(a\sqrt{b^2-\alpha^2}) & (\lvert\alpha\rvert < b) \end{cases}$
13	$\begin{cases} P_n(x) & (\lvert x\rvert < 1) \\ 0 & (\lvert x\rvert > 1) \end{cases}$	$\dfrac{i^n}{\sqrt{\alpha}} J_{n+\frac{1}{2}}(\alpha)$
14	$\begin{cases} \dfrac{\cos[b\sqrt{a^2-x^2}]}{\sqrt{a^2-x^2}} & (\lvert x\rvert < a) \\ 0 & (\lvert x\rvert > a) \end{cases}$	$\sqrt{\dfrac{\pi}{2}} J_0(a\sqrt{a^2+b^2})$
15	$\begin{cases} \dfrac{\cosh[b\sqrt{a^2-x^2}]}{\sqrt{a^2-x^2}} & (\lvert x\rvert < a) \\ 0 & (\lvert x\rvert > a) \end{cases}$	$\sqrt{\dfrac{\pi}{2}} J_0(a\sqrt{\alpha^2-b^2})$

[1]More extensive tables of Fourier transforms can be found in W. Magnus and F. Oberhettinger, "Formulas and Theorems of the Special Functions of Mathematical Physics," pp. 116–120. Chelsea (1949).

The following functions appear among the entries of the tables on transforms.

Function	Definition	Name
$Ei(x)$	$\int_{-\infty}^{x} \frac{e^v}{v}\,dv$; or sometimes defined as $-Ei(-x) = \int_{x}^{\infty} \frac{e^{-v}}{v}\,dv$	See index for Sine, Cosine, and Exponential Integral tables
$Si(x)$	$\int_{0}^{x} \frac{\sin v}{v}\,dv$	See index for Sine, Cosine, and Exponential Integral tables
$Ci(x)$	$\int_{\infty}^{x} \frac{\cos v}{v}\,dv$; or sometimes defined as negative of this integral	See index for Sine, Cosine, and Exponential Integral tables
$erf(x)$	$\frac{2}{\sqrt{\pi}} \int_{0}^{x} e^{-v^2}\,dv$	Error function
$erfc(x)$	$1 - erf(x) = \frac{2}{\sqrt{\pi}} \int_{x}^{\infty} e^{-v^2}\,dv$	Complementary function to error function
$L_n(x)$	$\frac{e^x}{n!} \frac{d^n}{dx^n} (x^n e^{-x}),\quad n = 0, 1, \cdots$	Laguerre polynomial of degree n

ORTHOGONAL POLYNOMIALS

I

Name: Legendre *Symbol*: $P_n(x)$ *Interval*: $[-1, 1]$

Differential Equation: $(1 - x^2)y'' - 2xy' + n(n + 1)y = 0$

$$y = P_n(x)$$

Explicit Expression: $P_n(x) = \dfrac{1}{2^n} \displaystyle\sum_{m=0}^{[n/2]} (-1)^m \binom{n}{m}\binom{2n - 2m}{n} x^{n-2m}$

Recurrence Relation: $(n + 1)P_{n+1}(x) = (2n + 1)xP_n(x) - nP_{n-1}(x)$

Weight: 1 *Standardization*: $P_n(1) = 1$

Norm: $\displaystyle\int_{-1}^{+1} [P_n(x)]^2\,dx = \frac{2}{2n + 1}$

Rodrigues' Formula: $P_n(x) = \dfrac{(-1)^n}{2^n n!} \dfrac{d^n}{dx^n} \{(1 - x^2)^n\}$

Generating Function: $R^{-1} = \displaystyle\sum_{n=0}^{\infty} P_n(x) z^n;\ -1 < x < 1,\ |z| < 1,$

$$R = \sqrt{1 - 2xz + z^2}.$$

Inequality: $|P_n(x)| \le 1,\ -1 \le x \le 1.$

II

Name: Tschebysheff, First Kind *Symbol*: $T_n(x)$ *Interval*: $[-1, 1]$

Differential Equation: $(1 - x^2)y'' - xy' + n^2y = 0$

$$y = T_n(x)$$

Explicit Expression: $\dfrac{n}{2} \displaystyle\sum_{m=0}^{[n/2]} (-1)^m \frac{(n - m - 1)!}{m!(n - 2m)!} (2x)^{n-2m} = \cos(n \arccos x) = T_n(x)$

Recurrence Relation: $T_{n+1}(x) = 2xT_n(x) - T_{n-1}(x)$

Weight: $(1 - x^2)^{-1/2}$ *Standardization*: $T_n(1) = 1$

Norm: $\displaystyle\int_{-1}^{+1} (1 - x^2)^{-1/2}[T_n(x)]^2\,dx = \begin{cases} \pi/2, & n \ne 0 \\ \pi, & n = 0 \end{cases}$

Rodrigues' Formula: $\dfrac{(-1)^n(1 - x^2)^{1/2}\sqrt{\pi}}{2^{n+1}\Gamma(n + \frac{1}{2})} \dfrac{d^n}{dx^n} \{(1 - x^2)^{n-(1/2)}\} = T_n(x)$

Generating Function: $\dfrac{1 - xz}{1 - 2xz + z^2} = \displaystyle\sum_{n=0}^{\infty} T_n(x) z^n,\ -1 < x < 1,\ |z| < 1.$

Inequality: $|T_n(x)| \le 1,\ -1 \le x \le 1.$

III

Name: Tschebysheff, Second Kind *Symbol*: $U_n(x)$ *Interval*: $[-1, 1]$

Differential Equation: $(1 - x^2)y'' - 3xy' + n(n + 2)y = 0$

$$y = U_n(x)$$

Explicit Expression: $$U_n(x) = \sum_{m=0}^{[n/2]} (-1)^m \frac{(m - n)!}{m!(n - 2m)!}(2x)^{n-2m}$$

$$U_n(\cos\theta) = \frac{\sin[(n + 1)\theta]}{\sin\theta}$$

Recurrence Relation: $U_{n+1}(x) = 2xU_n(x) - U_{n-1}(x)$

Weight: $(1 - x^2)^{1/2}$ *Standardization*: $U_n(1) = n + 1$

Norm: $$\int_{-1}^{+1} (1 - x^2)^{1/2}[U_n(x)]^2\,dx = \frac{\pi}{2}$$

Rodrigues' Formula: $$U_n(x) = \frac{(-1)^n(n + 1)\sqrt{\pi}}{(1 - x^2)^{1/2}2^{n+1}\Gamma(n + \frac{3}{2})}\frac{d^n}{dx^n}\{(1 - x^2)^{n+(1/2)}\}$$

Generating Function: $$\frac{1}{1 - 2xz + z^2} = \sum_{n=0}^{\infty} U_n(x)z^n, \quad -1 < x < 1, \quad |z| < 1.$$

Inequality: $|U_n(x)| \leq n + 1, \; -1 \leq x \leq 1.$

IV

Name: Jacobi *Symbol*: $P_n^{(\alpha,\beta)}(x)$ *Interval*: $[-1, 1]$

Differential Equation:

$$(1 - x^2)y'' + [\beta - \alpha - (\alpha + \beta + 2)x]y' + n(n + \alpha + \beta + 1)y = 0$$

$$y = P_n^{(\alpha,\beta)}(x)$$

Explicit Expression: $$P_n^{(\alpha,\beta)}(x) = \frac{1}{2^n}\sum_{m=0}^{n}\binom{n + \alpha}{m}\binom{n + \beta}{n - m}(x - 1)^{n-m}(x + 1)^m$$

Recurrence Relation: $$\begin{aligned} 2(n + 1)(n + \alpha + \beta + 1)(2n + \alpha + \beta)P_{n+1}^{(\alpha,\beta)}(x) \\ = (2n + \alpha + \beta + 1)[(\alpha^2 - \beta^2) + (2n + \alpha + \beta + 2) \\ \times (2n + \alpha + \beta)x]P_n^{(\alpha,\beta)}(x) \\ - 2(n + \alpha)(n + \beta)(2n + \alpha + \beta + 2)P_{n-1}^{(\alpha,\beta)}(x) \end{aligned}$$

Weight: $(1 - x)^\alpha(1 + x)^\beta$; $\alpha, \beta > 1$ *Standardization*: $P_n^{(\alpha,\beta)}(x) = \binom{n + \alpha}{n}$

Norm: $$\int_{-1}^{+1} (1 - x)^\alpha(1 + x)^\beta[P_n^{(\alpha,\beta)}(x)]^2\,dx = \frac{2^{\alpha+\beta+1}\Gamma(n + \alpha + 1)\Gamma(n + \beta + 1)}{(2n + \alpha + \beta + 1)n!\Gamma(n + \alpha + \beta + 1)}$$

Rodrigues' Formula: $$P_n^{(\alpha,\beta)}(x) = \frac{(-1)^n}{2^n n!(1 - x)^\alpha(1 + x)^\beta}\frac{d^n}{dx^n}\{(1 - x)^{n+\alpha}(1 + x)^{n+\beta}\}$$

IV (Continued)

Generating Function: $R^{-1}(1 - z + R)^{-\alpha}(1 + z + R)^{-\beta} = \sum_{n=0}^{\infty} 2^{-\alpha-\beta} P_n^{(\alpha,\beta)}(x) z^n,$

$$R = \sqrt{1 - 2xz + z^2}, \ |z| < 1$$

Inequality:
$$\max_{-1 \le x \le 1} |P_n^{(\alpha,\beta)}(x)| = \begin{cases} \binom{n+q}{n} \sim n^q \text{ if } q = \max(\alpha,\beta) \ge -\frac{1}{2} \\ |P_n^{(\alpha,\beta)}(x')| \sim n^{-1/2} \text{ if } q < -\frac{1}{2} \\ x' \text{ is one of the two maximum points nearest} \\ \dfrac{\beta - \alpha}{\alpha + \beta + 1} \end{cases}$$

V

Name: Generalized Laguerre *Symbol*: $L_n^{(\alpha)}(x)$ *Interval*: $[0, \infty]$

Differential Equation: $xy'' + (\alpha + 1 - x)y' + ny = 0$

$$y = L_n^{(\alpha)}(x)$$

Explicit Expression: $L_n^{(\alpha)}(x) = \sum_{m=0}^{n} (-1)^m \binom{n+\alpha}{n-m} \frac{1}{m!} x^m$

Recurrence Relation: $(n + 1)L_{n+1}^{(\alpha)}(x) = [(2n + \alpha + 1) - x]L_n^{(\alpha)}(x) - (n + \alpha)L_{n-1}^{(\alpha)}(x)$

Weight: $x^\alpha e^{-x}, \alpha > -1$ *Standardization*: $L_n^{(\alpha)}(x) = \frac{(-1)^n}{n!} x^n + \cdots$

Norm: $\int_0^\infty x^\alpha e^{-x} [L_n^{(\alpha)}(x)]^2 dx = \frac{\Gamma(n + \alpha + 1)}{n!}$

Rodrigues' Formula: $L_n^{(\alpha)}(x) = \frac{1}{n! x^\alpha e^{-x}} \frac{d^n}{dx^n} \{x^{n+\alpha} e^{-x}\}$

Generating Function: $(1 - z)^{-\alpha-1} \exp\left(\frac{xz}{z-1}\right) = \sum_{n=0}^{\infty} L_n^{(\alpha)}(x) z^n$

Inequality: $|L_n^{(\alpha)}(x)| \le \frac{\Gamma(n + \alpha + 1)}{n!\Gamma(\alpha + 1)} e^{x/2}; \quad x \ge 0, \ \alpha > 0$

$$|L_n^{(\alpha)}(x)| \le \left[2 - \frac{\Gamma(\alpha + n + 1)}{n!\Gamma(\alpha + 1)}\right] e^{x/2}; \quad x \ge 0, \ -1 < \alpha < 0$$

$$Si(x) = \int_0^x \frac{\sin v}{v} dv; \qquad Ci(x) = \int_\infty^x \frac{\cos v}{v} dv;$$

$$Ei(x) = \int_{-\infty}^x \frac{e^v}{v} dv; \qquad E_1(x) = -Ei(-x) = \int_x^\infty \frac{e^{-v}}{v} dv$$

Interpolation can be by iterative linear interpolation or by means of Everett's Formula using throwback to second differences. With this in view the modified second differences have been tabulated where relevant.

Everett's Formula.

$$f_p = (1 - p)f_0 + pf_1 + E_2\delta_{m0}^2 + F_2\delta_{m1}^2$$

SINE, COSINE, AND EXPONENTIAL INTEGRALS

x	$Si(x)$	δ_m^2	$Ci(x)$	δ_m^2	$Ei(x)$	δ_m^2	$E_1(x) = -Ei(-x)$	δ_m^2
.00	.00000		$-\infty$		$-\infty$		$+\infty$	
.01	.01000	0	-4.02798		-4.01790		4.03790	
.02	.02000	0	-3.33482		-3.31476		3.35476	
.03	.03000	0	-2.92957		-2.89912		2.95912	
.04	.04000	-2	-2.64206		-2.60126		2.68126	
.05	.04999	2	-2.41914		-2.36788		2.46790	
.06	.05999	-2	-2.23709		-2.17528		2.29531	
.07	.06998	0	-2.08327		-2.01080		2.15084	
.08	.07997	0	-1.95011		-1.86688		2.02694	
.09	.08996	-2	-1.83275		-1.73866		1.91874	
.10	.09994	2	-1.72787	-997	-1.62281	-989	1.82292	990
.11	.10993	-3	-1.63308	-829	-1.51696	-816	1.73711	816
.12	.11990	2	-1.54665	-695	-1.41935	-687	1.65954	688
.13	.12988	-1	-1.46723	-594	-1.32866	-582	1.58890	585
.14	.13985	-1	-1.39379	-514	-1.24384	-505	1.52415	503
.15	.14981	0	-1.32552	-449	-1.16409	-436	1.46446	441
.16	.15977	0	-1.26176	-394	-1.08873	-384	1.40919	384
.17	.16973	-1	-1.20196	-349	-1.01723	-341	1.35778	342
.18	.17968	-1	-1.14567	-314	-.94915	-302	1.30980	303
.19	.18962	0	-1.09253	-281	-.88410	-270	1.26486	272
.20	.19956	-1	-1.04221	-254	-.82176	-244	1.22265	245
.21	.20949	-1	-.99444	-231	-.76187	-222	1.18290	223
.22	.21941	1	-.94899	-212	-.70420	-199	1.14538	201
.23	.22933	-3	-.90566	-194	-.64853	-184	1.10988	186
.24	.23923	0	-.86427	-177	-.59470	-166	1.07624	167
.25	.24913	0	-.82466	-166	-.54254	-155	1.04428	158
.26	.25903	-3	-.78671	-153	-.49193	-142	1.01389	142
.27	.26891	-1	-.75029	-142	-.44274	-131	.98493	134
.28	.27878	1	-.71529	-132	-.39486	-122	.95731	122
.29	.28865	-3	-.68161	-124	-.34820	-113	.93092	115
.30	.29850	1	-.64917	-117	-.30267	-105	.90568	107
.31	.30835	-1	-.61790	-107	-.25819	-96	.88151	100
.32	.31819	-2	-.58771	-103	-.21468	-94	.85834	92
.33	.32801	-1	-.55855	-97	-.17210	-83	.83610	90
.34	.33782	1	-.53036	-91	-.13036	-81	.81475	82
.35	.34763	-3	-.50308	-86	-.08943	-76	.79422	77
.36	.35742	-1	-.47666	-84	-.04926	-69	.77446	74
.37	.36720	-3	-.45107	-76	-.00979	-67	.75544	69
.38	.37696	1	-.42625	-76	.02901	-63	.73711	66
.39	.38672	-3	-.40218	-70	.06718	-57	.71944	60
.40	.39646	-1	-.37881	-67	.10477	-58	.70238	59
.41	.40619	-1	-.35611	-65	.14179	-53	.68591	56
.42	.41591	-2	-.33406	-60	.17828	-50	.67000	51
.43	.42561	-2	-.31262	-61	.21427	-47	.65461	51
.44	.43529	1	-.29178	-54	.24979	-45	.63973	48
.45	.44497	-2	-.27149	-56	.28486	-43	.62533	46
.46	.45463	-2	-.25175	-52	.31950	-40	.61139	43
.47	.46427	-1	-.23253	-48	.35374	-39	.59788	41
.48	.47390	-2	-.21380	-50	.38759	-36	.58478	42
.49	.48351	-1	.19556	-46	.42108	-35	.57209	36
.50	.49311	-2	-.17778	-45	.45422	-33	.55977	37

SINE, COSINE, AND EXPONENTIAL INTEGRALS (Continued)

x	$Si(x)$	δ_m^2	$Ci(x)$	δ_m^2	$Ei(x)$	δ_m^2	$E_1(x) = -Ei(-x)$	δ_m^2
.50	.49311	−2	−.17778	−45	.45422	−33	.55977	37
.51	.50269	−2	−.16045	−43	.48703	−31	.54782	35
.52	.51225	−1	−.14355	−42	.51953	−30	.53622	33
.53	.52180	−2	−.12707	−40	.55173	−28	.52495	32
.54	.53133	−2	−.11099	−39	.58365	−29	.51400	31
.55	.54084	−2	−.09530	−38	.61529	−24	.50336	30
.56	.55033	−1	−.07999	−36	.64668	−25	.49302	28
.57	.55981	−2	−.06504	−35	.67782	−23	.48296	27
.58	.56927	−2	−.05044	−35	.70873	−23	.47317	27
.59	.57871	−2	−.03619	−33	.73941	−21	.46365	25
.60	.58813	−2	−.02227	−34	.76988	−20	.45438	24
.61	.59753	−2	−.00868	−29	.80015	−19	.44535	24
.62	.60691	−2	.00461	−33	.83023	−19	.43656	23
.63	.61627	−2	.01758	−28	.86012	−17	.42800	20
.64	.62561	0	.03026	−29	.88984	−17	.41965	23
.65	.63494	−3	.04265	−28	.91939	−16	.41152	20
.66	.64424	−3	.05476	−28	.94878	−15	.40359	19
.67	.65351	0	.06659	−25	.97802	−14	.39585	21
.68	.66277	−2	.07816	−28	1.00712	−14	.38831	18
.69	.67201	−3	.08946	−25	1.03608	−13	.38095	18
.70	.68122	−2	.10051	−24	1.06491	−13	.37377	17
.71	.69041	−2	.11132	−25	1.09361	−11	.36676	17
.72	.69958	−2	.12188	−24	1.12220	−11	.35992	16
.73	.70873	−3	.13220	−21	1.15068	−10	.35324	15
.74	.71785	−2	.14230	−25	1.17906	−12	.34671	16
.75	.72695	−2	.15216	−20	1.20733	−9	.34034	15
.76	.73603	−3	.16181	−22	1.23551	−9	.33412	12
.77	.74508	−2	.17124	−21	1.26360	−8	.32803	16
.78	.75411	−2	.18046	−21	1.29161	−8	.32209	13
.79	.76312	−3	.18947	−20	1.31954	−7	.31628	13
.80	.77210	−3	.19828	−20	1.34740	−9	.31060	12
.81	.78105	−2	.20689	−20	1.37518	−6	.30504	13
.82	.78998	−3	.21530	−17	1.40290	−6	.29961	12
.83	.79888	−2	.22353	−19	1.43056	−6	.29430	11
.84	.80776	−3	.23157	−19	1.45816	−5	.28910	12
.85	.81661	−2	.23942	−16	1.48571	−3	.28402	11
.86	.82544	−3	.24710	−18	1.51322	−7	.27905	10
.87	.83424	−3	.25460	−18	1.54067	−2	.27418	10
.88	.84301	−3	.26192	−15	1.56809	−4	.26941	12
.89	.85175	−2	.26908	−17	1.59547	−4	.26475	8
.90	.86047	−3	.27607	−17	1.62281	−1	.26018	10
.91	.86916	−3	.28289	−14	1.65013	−5	.25571	10
.92	.87782	−1	.28956	−18	1.67741	0	.25134	7
.93	.88646	−5	.29606	−13	1.70468	−4	.24705	9
.94	.89506	−1	.30242	−18	1.73192	0	.24285	9
.95	.90364	−3	.30861	−13	1.75915	−2	.23874	8
.96	.91219	−5	.31466	−15	1.78636	−1	.23471	8
.97	.92070	−1	.32056	−14	1.81356	−1	.23076	8
.98	.92919	−3	.32632	−15	1.84075	−1	.22689	8
.99	.93765	−3	.33193	−14	1.86793	2	.22310	7
1.00	.94608	−3	.33740	−13	1.89512	−2	.21938	8

SINE, COSINE, AND EXPONENTIAL INTEGRALS (Continued)

x	$Si(x)$	δ_m^2	$Ci(x)$	δ_m^2	$Ei(x)$	$E_1(x) = -Ei(-x)$
1.0	.94608	−301	.33740	−1376	1.89512	.21938
1.1	1.02869	−324	.38487	−1181	2.16738	.18599
1.2	1.10805	−345	.42046	−1025	2.44209	.15841
1.3	1.18396	−365	.44574	−897	2.72140	.13545
1.4	1.25623	−382	.46201	−789	3.00721	.11622
1.5	1.32468	−396	.47036	−695	3.30129	.10002
1.6	1.38918	−409	.47173	−612	3.60532	.08631
1.7	1.44959	−419	.46697	−538	3.92096	.07465
1.8	1.50582	−427	.45681	−470	4.24987	.06471
1.9	1.55778	−433	.44194	−408	4.59371	.05620
2.0	1.60541	−436	.42298	−350	4.95423	.04890
2.1	1.64870	−436	.40051	−296	5.33324	.04261
2.2	1.68762	−435	.37507	−246	5.73261	.03719
2.3	1.72221	−431	.34718	−198	6.15438	.03250
2.4	1.75249	−425	.31729	−153	6.60067	.02844
2.5	1.77852	−416	.28587	−111	7.07377	.02491
2.6	1.80039	−406	.25334	−71	7.57611	.02185
2.7	1.81821	−394	.22008	−34	8.11035	.01918
2.8	1.83210	−379	.18649	1	8.67930	.01686
2.9	1.84220	−363	.15290	33	9.28602	.01482
3.0	1.84865	−346	.11963	63	9.93383	.01305
3.1	1.85166	−327	.08699	91	10.6263	.01149
3.2	1.85140	−306	.05526	116	11.3673	.01013
3.3	1.84808	−285	.02468	139	12.1610	.00894
3.4	1.84191	−262	−.00452	159	13.0121	.00789
3.5	1.83313	−239	−.03213	177	13.9254	.00697
3.6	1.82195	−215	−.05797	192	14.9063	.00616
3.7	1.80862	−191	−.08190	205	15.9606	.00545
3.8	1.79339	−166	−.10378	216	17.0948	.00482
3.9	1.77650	−141	−.12350	224	18.3157	.00427
4.0	1.75820	−116	−.14098	230	19.6309	.00378
4.1	1.73874	−91	−.15617	234	21.0485	.00335
4.2	1.71837	−67	−.16901	236	22.5774	.00297
4.3	1.69732	−44	−.17951	235	24.2274	.00263
4.4	1.67583	−21	−.18766	232	26.0090	.00234
4.5	1.65414	2	−.19349	228	27.9337	.00207
4.6	1.63246	23	−.19705	221	30.0141	.00184
4.7	1.61101	43	−.19839	213	32.2639	.00164
4.8	1.58998	62	−.19760	204	34.6979	.00145
4.9	1.56956	79	−.19478	193	37.3325	.00129
5.0	1.54993	95	−.19003	181	40.1853	.00115
5.1	1.53125	110	−.18348	167	43.2757	.00102
5.2	1.51367	123	−.17525	153	46.6249	.00091
5.3	1.49732	134	−.16551	137	50.2557	.00081
5.4	1.48230	144	−.15439	121	54.1935	.00072
5.5	1.46872	152	−.14205	105	58.4655	.00064
5.6	1.45667	159	−.12867	88	63.1018	.00057
5.7	1.44620	164	−.11441	71	68.1350	.00051
5.8	1.43736	167	−.09944	54	73.6008	.00045
5.9	1.43018	168	−.08393	37	79.5382	.00040

SINE, COSINE, AND EXPONENTIAL INTEGRALS (Continued)

x	$Si(x)$	δ_m^2	$Ci(x)$	δ_m^2	$Ei(x)$	$E_1(x) = -Ei(-x)$
6.0	1.42469	168	−.06806	20	85.9898	.00036
6.1	1.42087	166	−.05198	3	93.0020	.00032
6.2	1.41871	163	−.03587	−13	100.626	.00029
6.3	1.41817	158	−.01989	−28	108.916	.00026
6.4	1.41922	152	−.00418	−43	117.935	.00023
6.5	1.42179	145	.01110	−56	127.747	.00020
6.6	1.42582	137	.02582	−69	138.426	.00018
6.7	1.43121	128	.03986	−81	150.050	.00016
6.8	1.43787	117	.05308	−92	162.707	.00014
6.9	1.44570	106	.06539	−101	176.491	.00013
7.0	1.45460	94	.07670	−109	191.505	.00012
7.1	1.46443	82	.08691	−116	207.863	.00010
7.2	1.47509	69	.09596	−122	225.688	.00009
7.3	1.48644	56	.10379	−127	245.116	.00008
7.4	1.49834	43	.11036	−130	266.296	.00007
7.5	1.51068	30	.11563	−131	289.388	.00007
7.6	1.52331	16	.11960	−132	314.572	.00006
7.7	1.53611	3	.12225	−131	342.040	.00005
7.8	1.54894	−10	.12359	−129	372.006	.00005
7.9	1.56167	−22	.12364	−126	404.701	.00004
8.0	1.57419	−34	.12.	−122	440.380	.00004
8.1	1.58637	−45	.12002	−116	479.322	.00003
8.2	1.59810	−55	.11644	−110	521.831	.00003
8.3	1.60928	−65	.11177	−103	568.242	.00003
8.4	1.61981	−74	.10607	−94	618.919	.00002
8.5	1.62960	−82	.09943	−86	674.264	.00002
8.6	1.63857	−89	.09194	−76	734.714	.00002
8.7	1.64665	−95	.08368	−66	800.749	.00002
8.8	1.65379	−100	.07476	−56	872.895	.00002
8.9	1.65993	−104	.06528	−45	951.728	.00001
9.0	1.66504	−106	.05535	−35	1037.88	.00001
9.1	1.66908	−108	.04507	−24	1132.04	.00001
9.2	1.67205	−109	.03455	−13	1234.96	.00001
9.3	1.67393	−108	.02391	−2	1347.48	.00001
9.4	1.67473	−107	.01325	9	1470.51	.00001
9.5	1.67446	−104	.00268	19	1605.03	.00001
9.6	1.67316	−101	−.00771	29	1752.14	.00001
9.7	1.67084	−96	−.01780	38	1913.05	.00001
9.8	1.66757	−91	−.02752	47	2089.05	.00001
9.9	1.66338	−85	−.03676	55	2281.58	.00000
10.0	1.65835	−78	−.04546	63	2492.23	.00000

SINE, COSINE, AND EXPONENTIAL INTEGRALS (Continued)

x	$Si(x)$	δ_m^2	$Ci(x)$	δ_m^2
10.0	1.65835	−2001	−.04546	1616
10.5	1.62294	−944	−.07828	2255
11.0	1.57831	229	−.08956	2323
11.5	1.53571	1252	−.07857	1849
12.0	1.50497	1898	−.04978	987
12.5	1.49234	2051	−.01141	−35
13.0	1.49936	1720	.02676	−968
13.5	1.52291	1007	.05576	−1610
14.0	1.55621	116	.06940	−1828
14.5	1.59072	−745	.06554	−1604
15.0	1.61819	−1371	.04628	−1019
15.5	1.63258	−1638	.01719	−231
16.0	1.63130	−1500	−.01420	559
16.5	1.61563	−1020	−.04031	1173
17.0	1.59014	−326	−.05524	1471
17.5	1.56146	407	−.05610	1406
18.0	1.53661	999	−.04348	1014
18.5	1.52128	1327	−.02111	400
19.0	1.51863	1320	.00515	−276
19.5	1.52863	1000	.02883	−849
20.0	1.54824	462	.04442	−1195
20.5	1.57232	−163	.04859	−1240
21.0	1.59489	−717	.04089	−985
21.5	1.61063	−1077	.02373	−511
22.0	1.61608	−1161	.00164	67
22.5	1.61041	−967	−.01986	599
23.0	1.59546	−550	−.03566	969
23.5	1.57521	−19	−.04221	1090
24.0	1.55474	493	−.03833	946
24.5	1.53896	870	−.02538	579
25.0	1.53148	1020	−.00685	96
25.5	1.53376	921	.01262	−399
26.0	1.54487	605	.02830	−776
26.5	1.56180	161	.03657	−954
27.0	1.58029	−314	.03572	−894
27.5	1.59581	−700	.02630	−626
28.0	1.60475	−781	.01087	−214
28.5	1.60581	−1062	−.00664	232
29.0	1.59731	−530	−.02195	610
29.5	1.58331	−275	−.03144	826
30.0	1.56676	160	−.03303	837
30.5	1.55172	534	−.02660	649
31.0	1.54177	765	−.01395	307
31.5	1.53914	808	.00166	−94
32.0	1.54424	653	.01639	−464
32.5	1.55560	345	.02670	−705
33.0	1.57028	−33	.03026	−777
33.5	1.58466	−397	.02639	−653
34.0	1.59526	−651	.01626	−377
34.5	1.59964	−742	.00250	−22

SINE, COSINE, AND EXPONENTIAL INTEGRALS (Continued)

x	$Si(x)$	δ_m^2	$Ci(x)$	δ_m^2
35.0	1.59692	−650	−.01148	335
35.5	1.58797	−408	−.02228	592
36.0	1.57511	−70	−.02741	709
36.5	1.56156	273	−.02576	646
37.0	1.55061	541	−.01792	429
37.5	1.54483	673	−.00596	117
38.0	1.54549	638	.00713	−216
38.5	1.55226	449	.01816	−490
39.0	1.56334	157	.02451	−638
39.5	1.57594	−164	.02476	−625
40.0	1.58698	−437	.01902	−467
40.5	1.59384	−603	.00881	−195
41.0	1.59494	−611	−.00328	114
41.5	1.59018	−479	−.01429	391
42.0	1.58083	−230	−.02157	563
42.5	1.56927	68	−.02346	599
43.0	1.55835	343	−.01962	486
43.5	1.55070	527	−.01112	262
44.0	1.54809	581	−.00011	−26
44.5	1.55104	494	.01067	−295
45.0	1.55872	286	.01863	−491
45.5	1.56915	18	.02190	−562
46.0	1.57976	−254	.01979	−496
46.5	1.58796	−451	.01293	−313
47.0	1.59184	−544	.00307	−52
47.5	1.59052	−496	−.00731	207
48.0	1.58445	−331	−.01571	416
48.5	1.57520	−93	−.02013	521
49.0	1.56506	168	−.01957	494
49.5	1.55652	381	−.01428	351
50.0	1.55162	495	−.00563	125
50.5	1.55146	491	.00422	−128
51.0	1.55600	365	.01285	−345
51.5	1.56404	154	.01819	−471
52.0	1.57356	−88	.01902	−484
52.5	1.58224	−310	.01522	−379
53.0	1.58797	−446	.00779	−182
53.5	1.58943	−477	−.00139	54
54.0	1.58633	−385	−.01006	272
54.5	1.57953	−210	−.01613	422
55.0	1.57072	21	−.01817	463
55.5	1.56210	236	−.01577	396
56.0	1.55574	393	−.00958	232
56.5	1.55314	454	−.00117	12
57.0	1.55488	401	.00737	−202
57.5	1.56046	251	.01398	−369
58.0	1.56845	44	.01707	−437
58.5	1.57687	−168	.01597	−403
59.0	1.58368	−340	.01101	−272
59.5	1.58724	−423	.00345	−72

SINE, COSINE, AND EXPONENTIAL INTEGRALS (Continued)

x	$Si(x)$	δ_m^2	$Ci(x)$	δ_m^2
60.0	1.58675	−404	−.00481	136
60.5	1.58239	−287	−.01177	311
61.0	1.57528	−99	−.01576	408
61.5	1.56721	103	−.01585	402
62.0	1.56013	284	−.01209	301
62.5	1.55576	389	−.00545	127
63.0	1.55511	401	.00241	−75
63.5	1.55829	311	.00956	−256
64.0	1.56445	151	.01427	−368
64.5	1.57206	−44	.01545	−395
65.0	1.57925	−227	.01285	−321
65.5	1.58427	−350	.00717	−173
66.0	1.58594	−388	−.00017	16
66.5	1.58390	−329	−.00736	199
67.0	1.57871	−193	−.01265	331
67.5	1.57167	−11	−.01478	378
68.0	1.56452	171	−.01329	335
68.5	1.55900	309	−.00859	211
69.0	1.55643	369	−.00187	37
69.5	1.55739	338	.00521	−144
70.0	1.56159	229	.01092	−285
70.5	1.56798	61	.01390	−359
71.0	1.57496	−116	.01345	−340
71.5	1.58083	−263	.00974	−242
72.0	1.58418	−346	.00371	−85
72.5	1.58422	−340	−.00314	90
73.0	1.58100	−253	−.00913	240
73.5	1.57535	−109	−.01282	332
74.0	1.56866	64	−.01334	339
74.5	1.56258	218	−.01061	267
75.0	1.55858	317	−.00533	127
75.5	1.55761	336	.00117	−39
76.0	1.55986	275	.00730	−193
76.5	1.56474	147	.01158	−302
77.0	1.57103	−15	.01298	−331
77.5	1.57718	−169	.01121	−283
78.0	1.58171	−287	.00673	−165
78.5	1.58351	−323	.00067	−9
79.0	1.58221	−290	−.00548	149
79.5	1.57814	−182	−.01021	265
80.0	1.57233	−31	−.01240	317
80.5	1.56622	126	−.01155	294
81.0	1.56131	246	−.00789	195
81.5	1.55876	309	−.00236	53
82.0	1.55917	296	.00368	−102
82.5	1.56241	209	.00875	−229
83.0	1.56765	73	.01163	−300
83.5	1.57359	−77	.01164	−296
84.0	1.57879	−212	.00882	−220
84.5	1.58197	−287	.00389	−93

SINE, COSINE, AND EXPONENTIAL INTEGRALS (Continued)

x	$Si(x)$	δ_m^2	$Ci(x)$	δ_m^2
85.0	1.58240	−296	−.00193	55
85.5	1.58000	−232	−.00722	192
86.0	1.57538	−110	−.01068	274
86.5	1.56970	36	−.01151	295
87.0	1.56436	169	−.00952	239
87.5	1.56064	264	−.00524	129
88.0	1.55944	291	.00027	−14
88.5	1.56102	246	.00565	−149
89.0	1.56496	144	.00960	−250
89.5	1.57028	5	.01116	−285
90.0	1.57566	−128	.00999	−254
90.5	1.57981	−236	.00639	−157
91.0	1.58171	−280	.00128	−26
91.5	1.58093	−255	−.00408	108
92.0	1.57770	−174	−.00840	220
92.5	1.57281	−44	−.01062	273
93.0	1.56749	90	−.01023	259
93.5	1.56303	203	−.00736	185
94.0	1.56051	264	−.00272	63
94.5	1.56052	263	.00253	−70
95.0	1.56304	192	.00711	−186
95.5	1.56741	83	.00991	−256
96.0	1.57257	−52	.01026	−261
96.5	1.57724	−170	.00811	−203
97.0	1.58029	−244	.00401	−98
97.5	1.58100	−262	−.00103	33
98.0	1.57921	−210	−.00576	150
98.5	1.57540	−114	−.00905	234
99.0	1.57050	13	−.01010	260
99.5	1.56572	135	−.00867	218
100.0	1.56223	222	−.00515	127

GAMMA FUNCTION AND ITS NATURAL LOGARITHM

n	$\Gamma(n)$	$\log_e \Gamma(n)$		n	$\Gamma(n)$	$\log_e \Gamma(n)$	
1.000	1.00000 00	0.00000 00	0.000	1.250	0.90640 25	−0.09827 18	0.250
1.005	0.99713 85	−0.00286 56	0.005	1.255	0.90538 58	−0.09939 42	0.255
1.010	0.99432 59	−0.00569 03	0.010	1.260	0.90439 71	−0.10048 67	0.260
1.015	0.99156 13	−0.00847 45	0.015	1.265	0.90343 63	−0.10154 97	0.265
1.020	0.98884 42	−0.01121 85	0.020	1.270	0.90250 31	−0.10258 32	0.270
1.025	0.98617 40	−0.01392 25	0.025	1.275	0.90159 72	−0.10358 74	0.275
1.030	0.98355 00	−0.01658 69	0.030	1.280	0.90071 85	−0.10456 25	0.280
1.035	0.98097 16	−0.01921 18	0.035	1.285	0.89986 67	−0.10550 87	0.285
1.040	0.97843 82	−0.02179 77	0.040	1.290	0.89904 16	−0.10642 60	0.290
1.045	0.97594 93	−0.02434 46	0.045	1.295	0.89824 30	−0.10731 47	0.295
1.050	0.97350 43	−0.02685 31	0.050	1.300	0.89747 07	−0.10817 48	0.300
1.055	0.97110 26	−0.02932 32	0.055	1.305	0.89672 45	−0.10900 66	0.305
1.060	0.96874 36	−0.03175 53	0.060	1.310	0.89600 42	−0.10981 02	0.310
1.065	0.96642 70	−0.03414 95	0.065	1.315	0.89530 96	−0.11058 57	0.315
1.070	0.96415 20	−0.03650 63	0.070	1.320	0.89464 05	−0.11133 34	0.320
1.075	0.96191 83	−0.03882 57	0.075	1.325	0.89399 67	−0.11205 32	0.325
1.080	0.95972 53	−0.04110 82	0.080	1.330	0.89337 81	−0.11274 54	0.330
1.085	0.95757 25	−0.04335 38	0.085	1.335	0.89278 44	−0.11341 02	0.335
1.090	0.95545 95	−0.04556 29	0.090	1.340	0.89221 55	−0.11404 76	0.340
1.095	0.95338 57	−0.04773 57	0.095	1.345	0.89167 12	−0.11465 78	0.345
1.100	0.95135 08	−0.04987 24	0.100	1.350	0.89115 14	−0.11524 09	0.350
1.105	0.94935 42	−0.05197 33	0.105	1.355	0.89065 59	−0.11579 71	0.355
1.110	0.94739 55	−0.05403 86	0.110	1.360	0.89018 45	−0.11632 65	0.360
1.115	0.94547 43	−0.05606 86	0.115	1.365	0.88973 71	−0.11682 92	0.365
1.120	0.94359 02	−0.05806 33	0.120	1.370	0.88931 35	−0.11730 55	0.370
1.125	0.94174 27	−0.06002 32	0.125	1.375	0.88891 36	−0.11775 53	0.375
1.130	0.93993 14	−0.06194 83	0.130	1.380	0.88853 71	−0.11817 88	0.380
1.135	0.93815 60	−0.06383 90	0.135	1.385	0.88818 41	−0.11857 62	0.385
1.140	0.93641 61	−0.06569 54	0.140	1.390	0.88785 43	−0.11894 76	0.390
1.145	0.93471 12	−0.06751 77	0.145	1.395	0.88754 76	−0.11929 32	0.395
1.150	0.93304 09	−0.06930 62	0.150	1.400	0.88726 38	−0.11961 29	0.400
1.155	0.93140 50	−0.07106 11	0.155	1.405	0.88700 29	−0.11990 70	0.405
1.160	0.92980 31	−0.07278 25	0.160	1.410	0.88676 47	−0.12017 57	0.410
1.165	0.92823 47	−0.07447 07	0.165	1.415	0.88654 90	−0.12041 89	0.415
1.170	0.92669 96	−0.07612 58	0.170	1.420	0.88635 58	−0.12063 68	0.420
1.175	0.92519 74	−0.07774 81	0.175	1.425	0.88618 49	−0.12082 97	0.425
1.180	0.92372 78	−0.07933 78	0.180	1.430	0.88603 62	−0.12099 74	0.430
1.185	0.92229 05	−0.08089 51	0.185	1.435	0.88590 97	−0.12114 03	0.435
1.190	0.92088 50	−0.08242 01	0.190	1.440	0.88580 51	−0.12125 84	0.440
1.195	0.91951 12	−0.08391 30	0.195	1.445	0.88572 23	−0.12135 18	0.445
1.200	0.91816 87	−0.08537 41	0.200	1.450	0.88566 14	−0.12142 06	0.450
1.205	0.91685 73	−0.08680 35	0.205	1.455	0.88562 21	−0.12146 50	0.455
1.210	0.91557 65	−0.08820 14	0.210	1.460	0.88560 43	−0.12148 50	0.460
1.215	0.91432 62	−0.08956 79	0.215	1.465	0.88560 80	−0.12148 08	0.465
1.220	0.91310 59	−0.09090 34	0.220	1.470	0.88563 31	−0.12145 25	0.470
1.225	0.91191 56	−0.09220 78	0.225	1.475	0.88567 95	−0.12140 02	0.475
1.230	0.91075 49	−0.09348 15	0.230	1.480	0.88574 70	−0.12132 40	0.480
1.235	0.90962 34	−0.09472 46	0.235	1.485	0.88583 56	−0.12122 40	0.485
1.240	0.90852 11	−0.09593 72	0.240	1.490	0.88594 51	−0.12110 03	0.490
1.245	0.90744 75	−0.09711 96	0.245	1.495	0.88607 56	−0.12095 30	0.495
	$y!$	$\log_e y!$	y		$y!$	$\log_e y!$	y

GAMMA FUNCTION AND ITS NATURAL LOGARITHM (Continued)

n	$\Gamma(n)$	$\log_e \Gamma(n)$		n	$\Gamma(n)$	$\log_e \Gamma(n)$	
1.500	0.88622 69	−0.12078 22	0.500	1.750	0.91906 25	−0.08440 11	0.750
1.505	0.88639 90	−0.12058 81	0.505	1.755	0.92020 92	−0.08315 42	0.755
1.510	0.88659 17	−0.12037 07	0.510	1.760	0.92137 49	−0.08188 83	0.760
1.515	0.88680 50	−0.12013 02	0.515	1.765	0.92255 95	−0.08060 34	0.765
1.520	0.88703 88	−0.11986 66	0.520	1.770	0.92376 31	−0.07929 96	0.770
1.525	0.88729 30	−0.11958 00	0.525	1.775	0.92498 57	−0.07797 70	0.775
1.530	0.88756 76	−0.11927 06	0.530	1.780	0.92622 73	−0.07663 56	0.780
1.535	0.88786 25	−0.11893 84	0.535	1.785	0.92748 79	−0.07527 55	0.785
1.540	0.88817 77	−0.11858 35	0.540	1.790	0.92876 75	−0.07389 69	0.790
1.545	0.88851 30	−0.11820 61	0.545	1.795	0.93006 61	−0.07249 96	0.795
1.550	0.88886 83	−0.11780 61	0.550	1.800	0.93138 38	−0.07108 39	0.800
1.555	0.88924 38	−0.11738 39	0.555	1.805	0.93272 05	−0.06964 97	0.805
1.560	0.88963 92	−0.11693 93	0.560	1.810	0.93407 63	−0.06819 72	0.810
1.565	0.89005 45	−0.11647 25	0.565	1.815	0.93545 11	−0.06672 64	0.815
1.570	0.89048 97	−0.11598 37	0.570	1.820	0.93684 51	−0.06523 73	0.820
1.575	0.89094 48	−0.11547 28	0.575	1.825	0.93825 82	−0.06373 01	0.825
1.580	0.89141 96	−0.11494 01	0.580	1.830	0.93969 04	−0.06220 48	0.830
1.585	0.89191 41	−0.11438 55	0.585	1.835	0.94114 18	−0.06066 15	0.835
1.590	0.89242 82	−0.11380 92	0.590	1.840	0.94261 24	−0.05910 01	0.840
1.595	0.89296 20	−0.11321 13	0.595	1.845	0.94410 22	−0.05752 09	0.845
1.600	0.89351 53	−0.11259 18	0.600	1.850	0.94561 12	−0.05592 38	0.850
1.605	0.89408 82	−0.11195 08	0.605	1.855	0.94713 95	−0.05430 89	0.855
1.610	0.89468 06	−0.11128 85	0.610	1.860	0.94868 70	−0.05267 63	0.860
1.615	0.89529 24	−0.11060 49	0.615	1.865	0.95025 39	−0.05102 60	0.865
1.620	0.89592 37	−0.10990 01	0.620	1.870	0.95184 02	−0.04935 81	0.870
1.625	0.89657 43	−0.10917 41	0.625	1.875	0.95344 58	−0.04767 27	0.875
1.630	0.89724 42	−0.10842 72	0.630	1.880	0.95507 09	−0.04596 97	0.880
1.635	0.89793 35	−0.10765 93	0.635	1.885	0.95671 53	−0.04424 94	0.885
1.640	0.89864 20	−0.10687 05	0.640	1.890	0.95837 93	−0.04251 16	0.890
1.645	0.89936 98	−0.10606 10	0.645	1.895	0.96006 28	−0.04075 66	0.895
1.650	0.90011 68	−0.10523 07	0.650	1.900	0.96176 58	−0.03898 43	0.900
1.655	0.90088 30	−0.10437 99	0.655	1.905	0.96348 85	−0.03719 48	0.905
1.660	0.90166 84	−0.10350 85	0.660	1.910	0.96523 07	−0.03538 81	0.910
1.665	0.90247 29	−0.10261 66	0.665	1.915	0.96699 27	−0.03356 44	0.915
1.670	0.90329 65	−0.10170 44	0.670	1.920	0.96877 43	−0.03172 36	0.920
1.675	0.90413 92	−0.10077 19	0.675	1.925	0.97057 57	−0.02986 59	0.925
1.680	0.90500 10	−0.09981 92	0.680	1.930	0.97239 69	−0.02799 12	0.930
1.685	0.90588 19	−0.09884 63	0.685	1.935	0.97423 80	−0.02609 97	0.935
1.690	0.90678 18	−0.09785 34	0.690	1.940	0.97609 89	−0.02419 14	0.940
1.695	0.90770 08	−0.09684 05	0.695	1.945	0.97797 98	−0.02226 63	0.945
1.700	0.90863 87	−0.09580 77	0.700	1.950	0.97988 07	−0.02032 45	0.950
1.705	0.90959 57	−0.09475 51	0.705	1.955	0.98180 16	−0.01836 61	0.955
1.710	0.91057 17	−0.09368 27	0.710	1.960	0.98374 25	−0.01639 11	0.960
1.715	0.91156 66	−0.09259 06	0.715	1.965	0.98570 37	−0.01439 95	0.965
1.720	0.91258 06	−0.09147 89	0.720	1.970	0.98768 50	−0.01239 15	0.970
1.725	0.91361 35	−0.09034 77	0.725	1.975	0.98968 65	−0.01036 70	0.975
1.730	0.91466 54	−0.08919 70	0.730	1.980	0.99170 84	−0.00832 62	0.980
1.735	0.91573 62	−0.08802 69	0.735	1.985	0.99375 06	−0.00626 90	0.985
1.740	0.91682 60	−0.08683 75	0.740	1.990	0.99581 33	−0.00419 55	0.990
1.745	0.91793 48	−0.08562 89	0.745	1.995	0.99789 64	−0.00210 59	0.995
				2.000	1.00000 00	0.00000 00	1.000
	$y!$	$\log_e y!$	y		$y!$	$\log_e y!$	y

GAMMA FUNCTION FOR COMPLEX ARGUMENTS

$x = 1.0$

y	$\mathcal{R} \ln \Gamma(z) = \log_e \lvert\Gamma(z)\rvert$	$\mathcal{I} \ln \Gamma(z)$	y	$\mathcal{R} \ln \Gamma(z) = \log_e \lvert\Gamma(z)\rvert$	$\mathcal{I} \ln \Gamma(z)$
0.0	0.00000	0.00000	5.0	− 6.13032	3.81590
0.1	−0.00820	−0.05732	5.1	− 6.27750	3.97816
0.2	−0.03248	−0.11230	5.2	− 6.42487	4.14238
0.3	−0.07195	−0.16282	5.3	− 6.57243	4.30850
0.4	−0.12529	−0.20716	5.4	− 6.72016	4.47650
0.5	−0.19095	−0.24406	5.5	− 6.86807	4.64634
0.6	−0.26729	−0.27274	5.6	− 7.01614	4.81799
0.7	−0.35277	−0.29283	5.7	− 7.16437	4.99142
0.8	−0.44598	−0.30423	5.8	− 7.31275	5.16659
0.9	−0.54571	−0.30707	5.9	− 7.46128	5.34348
1.0	−0.65092	−0.30164	6.0	− 7.60996	5.52205
1.1	−0.76078	−0.28827	6.1	− 7.75877	5.70229
1.2	−0.87459	−0.26733	6.2	− 7.90772	5.88415
1.3	−0.99177	−0.23922	6.3	− 8.05680	6.06762
1.4	−1.11186	−0.20430	6.4	− 8.20601	6.25267
1.5	−1.23448	−0.16294	6.5	− 8.35534	6.43928
1.6	−1.35931	−0.11547	6.6	− 8.50478	6.62742
1.7	−1.48609	−0.06220	6.7	− 8.65434	6.81707
1.8	−1.61460	−0.00342	6.8	− 8.80402	7.00821
1.9	−1.74464	+0.06061	6.9	− 8.95380	7.20081
2.0	−1.87608	0.12965	7.0	− 9.10368	7.39486
2.1	−2.00876	0.20346	7.1	− 9.25367	7.59033
2.2	−2.14258	0.28185	7.2	− 9.40375	7.78720
2.3	−2.27744	0.36461	7.3	− 9.55394	7.98546
2.4	−2.41324	0.45159	7.4	− 9.70421	8.18508
2.5	−2.54991	0.54260	7.5	− 9.85458	8.38605
2.6	−2.68738	0.63751	7.6	−10.00504	8.58835
2.7	−2.82259	0.73617	7.7	−10.15558	8.79197
2.8	−2.96448	0.83844	7.8	−10.30621	8.99687
2.9	−3.10402	0.94421	7.9	−10.45692	9.20306
3.0	−3.24414	1.05335	8.0	−10.60771	9.41051
3.1	−3.38483	1.16577	8.1	−10.75858	9.61920
3.2	−3.52603	1.28135	8.2	−10.90952	9.82913
3.3	−3.66773	1.40001	8.3	−11.06054	10.04027
3.4	−3.80988	1.52165	8.4	−11.21163	10.25262
3.5	−3.95247	1.64619	8.5	−11.36280	10.46615
3.6	−4.09546	1.77355	8.6	−11.51403	10.68086
3.7	−4.23884	1.90365	8.7	−11.66533	10.89673
3.8	−4.38259	2.03642	8.8	−11.81669	11.11374
3.9	−4.52668	2.17179	8.9	−11.96812	11.33189
4.0	−4.67110	2.30970	9.0	−12.11962	11.55116
4.1	−4.81583	2.45008	9.1	−12.27117	11.77153
4.2	−4.96086	2.59287	9.2	−12.42279	11.99301
4.3	−5.10618	2.73803	9.3	−12.57446	12.21557
4.4	−5.25176	2.88549	9.4	−12.72619	12.43920
4.5	−5.39761	3.03520	9.5	−12.87798	12.66390
4.6	−5.54370	3.18711	9.6	−13.02982	12.88964
4.7	−5.69002	3.34118	9.7	−13.18172	13.11643
4.8	−5.83658	3.49737	9.8	−13.33367	13.34424
4.9	−5.98335	3.65562	9.9	−13.48568	13.57307
5.0	−6.13032	3.81590	10.0	−13.63773	13.80291

Linear interpolation will yield about three figures; eight-point interpolation will yield about eight figures.

$\mathcal{R} \ln \Gamma(z) = \log_e \lvert\Gamma(z)\rvert$ $\qquad$ $\mathcal{I} \ln \Gamma(z) = \arg \Gamma(z)$

GAMMA FUNCTION FOR COMPLEX ARGUMENTS

$x = 1.1$

y	$\Re \ln \Gamma(z) = \log_e \lvert \Gamma(z) \rvert$	$\Im \ln \Gamma(z)$	y	$\Re \ln \Gamma(z) = \text{l}_e \lvert \Gamma(z) \rvert$	$\Im \ln \Gamma(z)$
0.0	−0.04987	0.00000	5.0	− 5.96894	3.96199
0.1	−0.05702	−0.04207	5.1	− 6.11415	4.12447
0.2	−0.07824	−0.08231	5.2	− 6.25960	4.28889
0.3	−0.11291	−0.11905	5.3	− 6.40527	4.45521
0.4	−0.16008	−0.15087	5.4	− 6.55114	4.62340
0.5	−0.21859	−0.17666	5.5	− 6.69723	4.79343
0.6	−0.28719	−0.19566	5.6	− 6.84351	4.96526
0.7	−0.36464	−0.20740	5.7	− 6.98998	5.13886
0.8	−0.44979	−0.21167	5.8	− 7.13664	5.31419
0.9	−0.54158	−0.20844	5.9	− 7.28347	5.49124
1.0	−0.63909	−0.19782	6.0	− 7.43048	5.66997
1.1	−0.74154	−0.18001	6.1	− 7.57765	5.85035
1.2	−0.84826	−0.15525	6.2	− 7.72498	6.03236
1.3	−0.95869	−0.12384	6.3	− 7.87247	6.25198
1.4	−1.07235	−0.08605	6.4	− 8.02011	6.40116
1.5	−1.18886	−0.04217	6.5	− 8.16790	6.58790
1.6	−1.30787	+0.00752	6.6	− 8.31582	6.77617
1.7	−1.42911	0.06276	6.7	− 8.46389	6.96595
1.8	−1.55234	0.12330	6.8	− 8.61208	7.15720
1.9	−1.67734	0.18890	6.9	− 8.76041	7.34992
2.0	−1.80396	0.25936	7.0	− 8.90886	7.54408
2.1	−1.93203	0.33446	7.1	− 9.05744	7.73966
2.2	−2.06143	0.41402	7.2	− 9.20613	7.93664
2.3	−2.19204	0.49787	7.3	− 9.35494	8.13501
2.4	−2.32376	0.58583	7.4	− 9.50387	8.33473
2.5	−2.45650	0.67775	7.5	− 9.65290	8.53580
2.6	−2.59018	0.77350	7.6	− 9.80203	8.73820
2.7	−2.72474	0.87293	7.7	− 9.95128	8.94191
2.8	−2.86010	0.97592	7.8	−10.10062	9.14690
2.9	−2.99623	1.08236	7.9	−10.25006	9.35318
3.0	−3.13305	1.19214	8.0	−10.39960	9.56071
3.1	−3.27054	1.30514	8.1	−10.54923	9.76950
3.2	−3.40864	1.42128	8.2	−10.69895	9.97950
3.3	−3.54732	1.54045	8.3	−10.84876	10.19073
3.4	−3.68655	1.66258	8.4	−10.99866	10.40315
3.5	−3.82629	1.78758	8.5	−11.14864	10.61676
3.6	−3.96651	1.91537	8.6	−11.29870	10.83154
3.7	−4.10720	2.04589	8.7	−11.44885	11.04749
3.8	−4.24832	2.17905	8.8	−11.59908	11.26457
3.9	−4.38985	2.31479	8.9	−11.74938	11.48279
4.0	−4.53178	2.45304	9.0	−11.89976	11.70213
4.1	−4.67408	2.59376	9.1	−12.05021	11.92257
4.2	−4.81673	2.73687	9.2	−12.20074	12.14411
4.3	−4.95972	2.88233	9.3	−12.35133	12.36673
4.4	−5.10303	3.03008	9.4	−12.50200	12.59043
4.5	−5.24665	3.18007	9.5	−12.65273	12.81519
4.6	−5.39057	3.33225	9.6	−12.80353	13.04099
4.7	−5.53477	3.48657	9.7	−12.95439	13.26784
4.8	−5.67924	3.64300	9.8	−13.10532	13.49571
4.9	−5.82396	3.80148	9.9	−13.25631	13.72460
5.0	−5.96894	3.96199	10.0	−13.40736	13.95449

Linear interpolation will yield about three figures; eight-point interpolation will yield about eight figures.

$\Re \ln \Gamma(z) = \log_e \lvert \Gamma(z) \rvert$ $\qquad$ $\Im \ln \Gamma(z) = \arg \Gamma(z)$

GAMMA FUNCTION FOR COMPLEX ARGUMENTS

$x = 1.2$

y	$\mathcal{R} \ln \Gamma(z) = \log_e \lvert \Gamma(z) \rvert$	$\mathcal{I} \ln \Gamma(z)$	y	$\mathcal{R} \ln \Gamma(z) = \log_e \lvert \Gamma(z) \rvert$	$\mathcal{I} \ln \Gamma(z)$
0.0	−0.08537	0.00000	5.0	− 5.80732	4.10610
0.1	−0.09170	−0.02866	5.1	− 5.95058	4.26883
0.2	−0.11051	−0.05586	5.2	− 6.09410	4.43349
0.3	−0.14135	−0.08026	5.3	− 6.23789	4.60005
0.4	−0.18352	−0.10066	5.4	− 6.38192	4.76847
0.5	−0.23614	−0.11611	5.5	− 6.52619	4.93871
0.6	−0.29825	−0.12588	5.6	− 6.67069	5.11075
0.7	−0.36885	−0.12949	5.7	− 6.81541	5.28455
0.8	−0.44698	−0.12664	5.8	− 6.96035	5.46009
0.9	−0.53174	−0.11721	5.9	− 7.10549	5.63732
1.0	−0.62233	−0.10119	6.0	− 7.25083	5.81623
1.1	−0.71804	−0.07869	6.1	− 7.39636	5.99679
1.2	−0.81823	−0.04984	6.2	− 7.54209	6.17898
1.3	−0.92238	−0.01484	6.3	− 7.68799	6.36275
1.4	−1.03001	+0.02611	6.4	− 7.83407	6.54810
1.5	−1.14074	0.07278	6.5	− 7.98031	6.73500
1.6	−1.25421	0.12496	6.6	− 8.12673	6.92342
1.7	−1.37015	0.18241	6.7	− 8.27330	7.11334
1.8	−1.48830	0.24494	6.8	− 8.42002	7.30474
1.9	−1.60844	0.31234	6.9	− 8.56690	7.49759
2.0	−1.73039	0.38443	7.0	− 8.71393	7.69189
2.1	−1.85398	0.46101	7.1	− 8.86109	7.88760
2.2	−1.97907	0.54192	7.2	− 9.00840	8.08471
2.3	−2.10553	0.62700	7.3	− 9.15584	8.28319
2.4	−2.23326	0.71610	7.4	− 9.30341	8.48304
2.5	−2.36215	0.80908	7.5	− 9.45110	8.68422
2.6	−2.49211	0.90579	7.6	− 9.59892	8.88673
2.7	−2.62308	1.00613	7.7	− 9.74687	9.09055
2.8	−2.75497	1.10996	7.8	− 9.89493	9.29566
2.9	−2.88773	1.21718	7.9	−10.04310	9.50204
3.0	−3.02130	1.32769	8.0	−10.19139	9.70968
3.1	−3.15563	1.44138	8.1	−10.33978	9.91856
3.2	−3.29066	1.55816	8.2	−10.48828	10.12866
3.3	−3.42637	1.67794	8.3	−10.63689	10.33998
3.4	−3.56270	1.80064	8.4	−10.78559	10.55250
3.5	−3.69962	1.92618	8.5	−10.93440	10.76620
3.6	−3.83711	2.05448	8.6	−11.08330	10.98107
3.7	−3.97513	2.18547	8.7	−11.23229	11.19710
3.8	−4.11364	2.31909	8.8	−11.38138	11.41427
3.9	−4.25264	2.45526	8.9	−11.53056	11.63257
4.0	−4.39209	2.59393	9.0	−11.67983	11.85199
4.1	−4.53197	2.73504	9.1	−11.82918	12.07251
4.2	−4.67226	2.87853	9.2	−11.97861	12.29413
4.3	−4.81294	3.02434	9.3	−12.12813	12.51683
4.4	−4.95399	3.17243	9.4	−12.27773	12.74060
4.5	−5.09541	3.32274	9.5	−12.42741	12.96543
4.6	−5.23716	3.47523	9.6	−12.57717	13.19130
4.7	−5.37924	3.62986	9.7	−12.72700	13.41822
4.8	−5.52164	3.78657	9.8	−12.87691	13.64616
4.9	−5.66433	3.94533	9.9	−13.02689	13.87511
5.0	−5.80732	4.10610	10.0	−13.17694	14.10508

Linear interpolation will yield about three figures; eight-point interpolation will yield about eight figures.

$\mathcal{R} \ln \Gamma(z) = \log_e \lvert \Gamma(z) \rvert$ $\qquad$ $\mathcal{I} \ln \Gamma(z) = \arg \Gamma(z)$

GAMMA FUNCTION FOR COMPLEX ARGUMENTS

$x = 1.3$

y	$\Re \ln \Gamma(z) = \log_e \lvert \Gamma(z) \rvert$	$\mathscr{I} \ln \Gamma(z)$	y	$\Re \ln \Gamma(z) = \log_e \lvert \Gamma(z) \rvert$	$\mathscr{I} \ln \Gamma(z)$
0.0	−0.10817	−0.00000	5.0	− 5.64541	4.24824
0.1	−0.11384	−0.01672	5.1	− 5.78673	4.41126
0.2	−0.13070	−0.03226	5.2	− 5.92835	4.57621
0.3	−0.15843	−0.04550	5.3	− 6.07027	4.74303
0.4	−0.19649	−0.05545	5.4	− 6.21246	4.91171
0.5	−0.24421	−0.06127	5.5	− 6.35492	5.08220
0.6	−0.30082	−0.06230	5.6	− 6.49765	5.25448
0.7	−0.36553	−0.05805	5.7	− 6.64063	5.42852
0.8	−0.43755	−0.04821	5.8	− 6.78385	5.60428
0.9	−0.15610	−0.03257	5.9	− 6.92730	5.78173
1.0	−0.60048	−0.01108	6.0	− 7.07099	5.96085
1.1	−0.69007	+0.01628	6.1	− 7.21489	6.14161
1.2	−0.78427	0.04942	6.2	− 7.35901	6.32399
1.3	−0.88259	0.08822	6.3	− 7.50333	6.50796
1.4	−0.98459	0.13255	6.4	− 7.64785	6.69349
1.5	−1.08987	0.18224	6.5	− 7.79257	6.88057
1.6	−1.19810	0.23711	6.6	− 7.93747	7.06916
1.7	−1.30899	0.29700	6.7	− 8.08255	7.25925
1.8	−1.42227	0.36172	6.8	− 8.22781	7.45081
1.9	−1.53773	0.43111	6.9	− 8.37325	7.64383
2.0	−1.65518	0.50500	7.0	− 8.51885	7.83828
2.1	−1.77443	0.58323	7.1	− 8.66461	8.03414
2.2	−1.89533	0.66565	7.2	− 8.81053	8.23139
2.3	−2.01776	0.75212	7.3	− 8.95660	8.43002
2.4	−2.14159	0.84250	7.4	− 9.10282	8.63000
2.5	−2.26672	0.93666	7.5	− 9.24919	8.83132
2.6	−2.39305	1.03448	7.6	− 9.39569	9.03396
2.7	−2.52049	1.13583	7.7	− 9.54234	9.23791
2.8	−2.64898	1.24062	7.8	− 9.68912	9.44314
2.9	−2.77843	1.34873	7.9	− 9.83603	9.64964
3.0	−2.90879	1.46006	8.0	− 9.98307	9.85740
3.1	−3.04001	1.57453	8.1	−10.13023	10.06639
3.2	−3.17202	1.69204	8.2	−10.27751	10.27661
3.3	−3.30478	1.81251	8.3	−10.42491	10.48804
3.4	−3.43826	1.93586	8.4	−10.57243	10.70067
3.5	−3.57240	2.06201	8.5	−10.72006	10.91447
3.6	−3.70717	2.19090	8.6	−10.86780	11.12944
3.7	−3.84255	2.32244	8.7	−11.01564	11.34557
3.8	−3.97849	2.45658	8.8	−11.16360	11.56284
3.9	−4.11497	2.59324	8.9	−11.31165	11.78123
4.0	−4.25196	2.73239	9.0	−11.45981	12.00075
4.1	−4.38945	2.87394	9.1	−11.60806	12.22136
4.2	−4.52739	3.01786	9.2	−11.75641	12.44307
4.3	−4.66578	3.16408	9.3	−11.90485	12.66585
4.4	−4.80460	3.31256	9.4	−12.05339	12.88971
4.5	−4.94382	3.46324	9.5	−12.20202	13.11462
4.6	−5.08342	3.61609	9.6	−12.35073	13.34058
4.7	−5.22340	3.77106	9.7	−12.49953	13.56757
4.8	−5.36374	3.92810	9.8	−12.64842	13.79559
4.9	−5.50441	4.08717	9.9	−12.79739	14.02463
5.0	−5.64541	4.24824	10.0	−12.94644	14.25466

Linear interpolation will yield about three figures: eight-point interpolation will yield about eight figures.

$\Re \ln \Gamma(z) = \log_e \lvert \Gamma(z) \rvert$ $\mathscr{I} \ln \Gamma(z) = \arg \Gamma(z)$

GAMMA FUNCTION FOR COMPLEX ARGUMENTS

$x = 1.4$

y	$\Re \ln \Gamma(z) = \log_e \lvert \Gamma(z) \rvert$	$\Im \ln \Gamma(z)$	y	$\Re \ln \Gamma(z) = \log_e \lvert \Gamma(z) \rvert$	$\Im \ln \Gamma(z)$
0.0	−0.11961	0.00000	5.0	−5.48320	4.38843
0.1	−0.12473	−0.00597	5.1	−5.62259	4.55178
0.2	−0.14000	−0.01097	5.2	−5.76231	4.71704
0.3	−0.16516	−0.01406	5.3	−5.90236	4.88417
0.4	−0.19979	−0.01439	5.4	−6.04273	5.05314
0.5	−0.24337	−0.01125	5.5	−6.18340	5.22391
0.6	−0.29530	−0.00402	5.6	−6.32436	5.39646
0.7	−0.35492	+0.00776	5.7	−6.46560	5.57076
0.8	−0.42158	0.02442	5.8	−6.60712	5.74677
0.9	−0.49463	0.04618	5.9	−6.74889	5.92447
1.0	−0.57345	0.07318	6.0	−6.89093	6.10383
1.1	−0.65748	0.10546	6.1	−7.03321	6.28482
1.2	−0.74620	0.14300	6.2	−7.17572	6.46742
1.3	−0.83914	0.18576	6.3	−7.31847	6.65160
1.4	−0.93588	0.23363	6.4	−7.46144	6.83734
1.5	−1.03606	0.28650	6.5	−7.60463	7.02462
1.6	−1.13934	0.34426	6.6	−7.74803	7.21341
1.7	−1.24543	0.40674	6.7	−7.89163	7.40369
1.8	−1.35407	0.47383	6.8	−8.03543	7.59543
1.9	−1.46505	0.54537	6.9	−8.17943	7.78863
2.0	−1.57816	0.62123	7.0	−8.32360	7.98325
2.1	−1.69322	0.70126	7.1	−8.46797	8.17928
2.2	−1.81008	0.78534	7.2	−8.61250	8.37670
2.3	−1.92859	0.87334	7.3	−8.75721	8.57549
2.4	−2.04863	0.96513	7.4	−8.90209	8.77563
2.5	−2.17009	1.06059	7.5	−9.04713	8.97710
2.6	−2.29287	1.15962	7.6	−9.19233	9.17989
2.7	−2.41687	1.26211	7.7	−9.33768	9.38398
2.8	−2.54201	1.36795	7.8	−9.48318	9.58935
2.9	−2.66822	1.47704	7.9	−9.62883	9.79599
3.0	−2.79544	1.58930	8.0	−9.77462	10.00388
3.1	−2.92359	1.70464	8.1	−9.92056	10.21300
3.2	−3.05262	1.82297	8.2	−10.06662	10.42335
3.3	−3.18249	1.94421	8.3	−10.21282	10.63490
3.4	−3.31315	2.06828	8.4	−10.35915	10.84765
3.5	−3.44454	2.19512	8.5	−10.50561	11.06157
3.6	−3.57664	2.32465	8.6	−10.65219	11.27666
3.7	−3.70940	2.45681	8.7	−10.79889	11.49290
3.8	−3.84280	2.59153	8.8	−10.94571	11.71028
3.9	−3.97679	2.72876	8.9	−11.09264	11.92878
4.0	−4.11135	2.86842	9.0	−11.23969	12.14840
4.1	−4.24646	3.01048	9.1	−11.38685	12.36912
4.2	−4.38209	3.15488	9.2	−11.53411	12.59093
4.3	−4.51821	3.30156	9.3	−11.68148	12.81381
4.4	−4.65480	3.45047	9.4	−11.82896	13.03776
4.5	−4.79184	3.60158	9.5	−11.97653	13.26277
4.6	−4.92932	3.75483	9.6	−12.12421	13.48882
4.7	−5.06721	3.91018	9.7	−12.27198	13.71591
4.8	−5.20549	4.06759	9.8	−12.41985	13.94401
4.9	−5.34416	4.22702	9.9	−12.56781	14.17313
5.0	−5.48320	4.38843	10.0	−12.71586	14.40326

Linear interpolation will yield about three figures; eight-point interpolation will yield about eight figures.

$\Re \ln \Gamma(z) = \log_e \lvert \Gamma(z) \rvert$ $\qquad$ $\Im \ln \Gamma(z) = \arg \Gamma(z)$

GAMMA FUNCTION FOR COMPLEX ARGUMENTS

$x = 1.5$

y	$\Re \ln \Gamma(z) = \log_e \lvert \Gamma(z) \rvert$	$\Im \ln \Gamma(z)$	y	$\Re \ln \Gamma(z) = \log_e \lvert \Gamma(z) \rvert$	$\Im \ln \Gamma(z)$
0.0	−0.12078	0.00000	5.0	−5.32063	4.52667
0.1	−0.12545	0.00379	5.1	−5.45810	4.69038
0.2	−0.13939	0.00839	5.2	−5.59594	4.85599
0.3	−0.16238	0.01461	5.3	−5.73414	5.02346
0.4	−0.19412	0.02315	5.4	−5.87269	5.19275
0.5	−0.23419	0.03467	5.5	−6.01158	5.36384
0.6	−0.28208	0.04969	5.6	−6.15078	5.53669
0.7	−0.33728	0.06867	5.7	−6.29030	5.71128
0.8	−0.39924	0.09192	5.8	−6.43012	5.88757
0.9	−0.46739	0.11969	5.9	−6.57023	6.06555
1.0	−0.54122	0.15214	6.0	−6.71062	6.24517
1.1	−0.62022	0.18936	6.1	−6.85128	6.42641
1.2	−0.70392	0.23137	6.2	−6.99221	6.60926
1.3	−0.79189	0.27817	6.3	−7.13339	6.79368
1.4	−0.88376	0.32970	6.4	−7.27482	6.97966
1.5	−0.97915	0.38589	6.5	−7.41649	7.16716
1.6	−1.07776	0.44666	6.6	−7.55839	7.35616
1.7	−1.17932	0.51190	6.7	−7.70051	7.54666
1.8	−1.28355	0.58149	6.8	−7.84286	7.73861
1.9	−1.39024	0.65532	6.9	−7.98542	7.93200
2.0	−1.49920	0.73328	7.0	−8.12818	8.12682
2.1	−1.61023	0.81525	7.1	−8.27115	8.32304
2.2	−1.72317	0.90111	7.2	−8.41431	8.52064
2.3	−1.83790	0.99076	7.3	−8.55766	8.71961
2.4	−1.95426	1.08407	7.4	−8.70120	8.91992
2.5	−2.07215	1.18096	7.5	−8.84491	9.12156
2.6	−2.19146	1.28131	7.6	−8.98881	9.32452
2.7	−2.31210	1.38503	7.7	−9.13287	9.52877
2.8	−2.43398	1.49202	7.8	−9.27710	9.73429
2.9	−2.55701	1.60219	7.9	−9.42149	9.94108
3.0	−2.68114	1.71547	8.0	−9.56604	10.14912
3.1	−2.80629	1.83176	8.1	−9.71075	10.35839
3.2	−2.93240	1.95098	8.2	−9.85560	10.56888
3.3	−3.05942	2.07307	8.3	−10.00060	10.78057
3.4	−3.18730	2.19794	8.4	−10.14575	10.99345
3.5	−3.31598	2.32553	8.5	−10.29104	11.20751
3.6	−3.44544	2.45578	8.6	−10.43646	11.42273
3.7	−3.57563	2.58862	8.7	−10.58202	11.63909
3.8	−3.70650	2.72398	8.8	−10.72771	11.85659
3.9	−3.83804	2.86182	8.9	−10.87352	12.07522
4.0	−3.97020	3.00207	9.0	−11.01946	12.29495
4.1	−4.10296	3.14469	9.1	−11.16553	12.51579
4.2	−4.23629	3.28962	9.2	−11.31171	12.73771
4.3	−4.37016	3.43680	9.3	−11.45801	12.96070
4.4	−4.50455	3.58620	9.4	−11.60442	13.18476
4.5	−4.63943	3.73777	9.5	−11.75095	13.40987
4.6	−4.77480	3.89147	9.6	−11.89759	13.63603
4.7	−4.91061	4.04724	9.7	−12.04433	13.86321
4.8	−5.04687	4.20506	9.8	−12.19118	14.09142
4.9	−5.18355	4.36488	9.9	−12.33814	14.32064
5.0	−5.32063	4.52667	10.0	−12.48519	14.55086

Linear interpolation will yield about three figures; eight-point interpolation will yield about eight figures.

$\Re \ln \Gamma(z) = \log_e \lvert \Gamma(z) \rvert$ $\Im \ln \Gamma(z) = \arg \Gamma(z)$

GAMMA FUNCTION FOR COMPLEX ARGUMENTS
$x = 1.6$

y	$\Re \ln \Gamma(z) = \log_e \lvert \Gamma(z) \rvert$	$\Im \ln \Gamma(z)$	y	$\Re \ln \Gamma(z) = \log_e \lvert \Gamma(z) \rvert$	$\Im \ln \Gamma(z)$
0.0	−0.11259	0.00000	5.0	−5.15767	4.66299
0.1	−0.11688	0.01272	5.1	−5.29324	4.82710
0.2	−0.12969	0.02614	5.2	−5.42921	4.99309
0.3	−0.15085	0.04093	5.3	−5.56558	5.16093
0.4	−0.18012	0.05771	5.4	−5.70233	5.33058
0.5	−0.21716	0.07706	5.5	−5.83944	5.50201
0.6	−0.26156	0.09944	5.6	−5.97690	5.67519
0.7	−0.31289	0.12528	5.7	−6.11470	5.85010
0.8	−0.37069	0.15489	5.8	−6.25283	6.02670
0.9	−0.43449	0.18851	5.9	−6.39128	6.20497
1.0	−0.50382	0.22633	6.0	−6.53004	6.38488
1.1	−0.57826	0.26845	6.1	−6.66910	6.56641
1.2	−0.65737	0.31495	6.2	−6.80844	6.74953
1.3	−0.74077	0.36584	6.3	−6.94806	6.93422
1.4	−0.82810	0.42111	6.4	−7.08795	7.12044
1.5	−0.91903	0.48071	6.5	−7.22811	7.30819
1.6	−1.01326	0.54460	6.6	−7.36852	7.49744
1.7	−1.11052	0.61269	6.7	−7.50917	7.68816
1.8	−1.21057	0.68490	6.8	−7.65007	7.88034
1.9	−1.31318	0.76114	6.9	−7.79120	8.07396
2.0	−1.41816	0.84132	7.0	−7.93256	8.26899
2.1	−1.52532	0.92534	7.1	−8.07413	8.46541
2.2	−1.63450	1.01310	7.2	−8.21592	8.66322
2.3	−1.74556	1.10450	7.3	−8.35792	8.86238
2.4	−1.85836	1.19946	7.4	−8.50012	9.06288
2.5	−1.97279	1.29786	7.5	−8.64252	9.26471
2.6	−2.08874	1.39963	7.6	−8.78511	9.46785
2.7	−2.20610	1.50467	7.7	−8.92789	9.67227
2.8	−2.32478	1.61291	7.8	−9.07085	9.87797
2.9	−2.44472	1.72425	7.9	−9.21399	10.08493
3.0	−2.56582	1.83862	8.0	−9.35730	10.29314
3.1	−2.68802	1.95594	8.1	−9.50079	10.50257
3.2	−2.81127	2.07613	8.2	−9.64443	10.71321
3.3	−2.93549	2.19914	8.3	−9.78824	10.92505
3.4	−3.06063	2.32488	8.4	−9.93221	11.13808
3.5	−3.18666	2.45329	8.5	−10.07633	11.35228
3.6	−3.31351	2.58432	8.6	−10.22060	11.56764
3.7	−3.44116	2.71790	8.7	−10.36501	11.78415
3.8	−3.56955	2.85396	8.8	−10.50958	12.00178
3.9	−3.69866	2.99247	8.9	−10.65428	12.22054
4.0	−3.82845	3.13336	9.0	−10.79911	12.44041
4.1	−3.95889	3.27659	9.1	−10.94409	12.66137
4.2	−4.08994	3.42209	9.2	−11.08919	12.88341
4.3	−4.22159	3.56983	9.3	−11.23442	13.10653
4.4	−4.35380	3.71977	9.4	−11.37978	13.33071
4.5	−4.48655	3.87184	9.5	−11.52526	13.55594
4.6	−4.61982	4.02602	9.6	−11.67086	13.78220
4.7	−4.75359	4.18227	9.7	−11.81658	14.00950
4.8	−4.88783	4.34053	9.8	−11.96242	14.23782
4.9	−5.02253	4.50079	9.9	−12.10836	14.46714
5.0	−5.15767	4.66299	10.0	−12.25442	14.69747

Linear interpolation will yield about three figures; eight-point interpolation will yield about eight figures.

$\Re \ln \Gamma(z) = \log_e \lvert \Gamma(z) \rvert$ $\qquad$ $\Im \ln \Gamma(z) = \arg \Gamma(z)$

GAMMA FUNCTION FOR COMPLEX ARGUMENTS

$x = 1.7$

y	$\Re \ln \Gamma(z) = \log_e \lvert \Gamma(z) \rvert$	$\Im \ln \Gamma(z)$	y	$\Re \ln \Gamma(z) = \log_e \lvert \Gamma(z) \rvert$	$\Im \ln \Gamma(z)$
0.0	−0.09581	0.00000	5.0	−4.99429	4.79739
0.1	−0.09977	0.02096	5.1	−5.12797	4.96193
0.2	−0.11161	0.04251	5.2	−5.26209	5.12834
0.3	−0.13121	0.06524	5.3	−5.39664	5.29658
0.4	−0.15835	0.08971	5.4	−5.53159	5.46662
0.5	−0.19275	0.11639	5.5	−5.66694	5.63842
0.6	−0.23410	0.14573	5.6	−5.80267	5.81197
0.7	−0.28203	0.17811	5.7	−5.93877	5.98722
0.8	−0.33614	0.21382	5.8	−6.07523	6.16416
0.9	−0.39604	0.25313	5.9	−6.21203	6.34276
1.0	−0.46133	0.29620	6.0	−6.34916	6.52299
1.1	−0.53162	0.34317	6.1	−6.48662	6.70482
1.2	−0.60653	0.39413	6.2	−6.62439	6.88823
1.3	−0.68572	0.44913	6.3	−6.76246	7.07320
1.4	−0.76885	0.50817	6.4	−6.90083	7.25971
1.5	−0.85561	0.57125	6.5	−7.03948	7.44773
1.6	−0.94574	0.63833	6.6	−7.17840	7.63724
1.7	−1.03895	0.70936	6.7	−7.31760	7.82821
1.8	−1.13503	0.78428	6.8	−7.45705	8.02064
1.9	−1.23376	0.86303	6.9	−7.59676	8.21449
2.0	−1.33493	0.94553	7.0	−7.73671	8.40976
2.1	−1.43838	1.03169	7.1	−7.87690	8.60641
2.2	−1.54395	1.12145	7.2	−8.01733	8.80443
2.3	−1.65148	1.21470	7.3	−8.15798	9.00381
2.4	−1.76084	1.31138	7.4	−8.29885	9.20452
2.5	−1.87192	1.41140	7.5	−8.43994	9.40656
2.6	−1.98459	1.51468	7.6	−8.58123	9.60989
2.7	−2.09877	1.62113	7.7	−8.72273	9.81451
2.8	−2.21435	1.73069	7.8	−8.86443	10.02040
2.9	−2.33125	1.84328	7.9	−9.00632	10.22754
3.0	−2.44940	1.95882	8.0	−9.14840	10.43592
3.1	−2.56872	2.07724	8.1	−9.29066	10.64553
3.2	−2.68915	2.19848	8.2	−9.43310	10.85634
3.3	−2.81063	2.32247	8.3	−9.57572	11.06835
3.4	−2.93310	2.44914	8.4	−9.71851	11.28154
3.5	−3.05650	2.57844	8.5	−9.86147	11.49590
3.6	−3.18080	2.71031	8.6	−10.00459	11.71142
3.7	−3.30594	2.84468	8.7	−10.14787	11.92807
3.8	−3.43189	2.98151	8.8	−10.29130	12.14586
3.9	−3.55861	3.12074	8.9	−10.43489	12.36476
4.0	−3.68605	3.26232	9.0	−10.57863	12.58476
4.1	−3.81420	3.40620	9.1	−10.72251	12.80586
4.2	−3.94301	3.55233	9.2	−10.86654	13.02804
4.3	−4.07245	3.70068	9.3	−11.01071	13.25129
4.4	−4.20251	3.85118	9.4	−11.15501	13.47560
4.5	−4.33315	4.00381	9.5	−11.29945	13.70096
4.6	−4.46434	4.15852	9.6	−11.44402	13.92735
4.7	−4.59608	4.31527	9.7	−11.58871	14.15477
4.8	−4.72833	4.47402	9.8	−11.73354	14.38321
4.9	−4.86107	4.63474	9.9	−11.87848	14.61265
5.0	−4.99429	4.79739	10.0	−12.02355	14.84309

Linear interpolation will yield about three figures; eight-point interpolation will yield about eight figures.

$\Re \ln \Gamma(z) = \log_e \lvert \Gamma(z) \rvert$ $\Im \ln \Gamma(z) = \arg \Gamma(z)$

GAMMA FUNCTION FOR COMPLEX ARGUMENTS

$x = \mathbf{1.8}$

y	$\Re \ln \Gamma(z) = \log_e \lvert \Gamma(z) \rvert$	$\Im \ln \Gamma(z)$	y	$\Re \ln \Gamma(z) = \log_e \lvert \Gamma(z) \rvert$	$\Im \ln \Gamma(z)$
0.0	−0.07108	0.00000	5.0	− 4.83046	4.92990
0.1	−0.07477	0.02859	5.1	− 4.96227	5.09491
0.2	−0.08578	0.05769	5.2	− 5.09455	5.26177
0.3	−0.10401	0.08783	5.3	− 5.22729	5.43044
0.4	−0.12929	0.11946	5.4	− 5.36046	5.60089
0.5	−0.16140	0.15305	5.5	− 5.49407	5.77310
0.6	−0.20007	0.18897	5.6	− 5.62808	5.94703
0.7	−0.24498	0.22758	5.7	− 5.76249	6.12266
0.8	−0.29581	0.26917	5.8	− 5.89728	6.29997
0.9	−0.35222	0.31396	5.9	− 6.03244	6.47891
1.0	−0.41385	0.36216	6.0	− 6.16796	6.65948
1.1	−0.48036	0.41390	6.1	− 6.30383	6.84164
1.2	−0.55143	0.46928	6.2	− 6.44004	7.02538
1.3	−0.62673	0.52837	6.3	− 6.57657	7.21066
1.4	−0.70597	0.59120	6.4	− 6.71341	7.39746
1.5	−0.78885	0.65778	6.5	− 6.85057	7.58577
1.6	−0.87511	0.72810	6.6	− 6.98802	7.77556
1.7	−0.96452	0.80213	6.7	− 7.12576	7.96682
1.8	−1.05685	0.87984	6.8	− 7.26378	8.15951
1.9	−1.15188	0.96117	6.9	− 7.40207	8.35362
2.0	−1.24944	1.04606	7.0	− 7.54063	8.54914
2.1	−1.34934	1.13446	7.1	− 7.67944	8.74603
2.2	−1.45143	1.22629	7.2	− 7.81851	8.94430
2.3	−1.55557	1.32148	7.3	− 7.95782	9.14391
2.4	−1.66161	1.41997	7.4	− 8.09737	9.34485
2.5	−1.76944	1.52168	7.5	− 8.23715	9.54710
2.6	−1.87895	1.62655	7.6	− 8.37715	9.75064
2.7	−1.99003	1.73449	7.7	− 8.51737	9.95547
2.8	−1.10259	1.84545	7.8	− 8.65781	10.16156
2.9	−2.21654	1.95935	7.9	− 8.79846	10.36891
3.0	−2.33181	2.07613	8.0	− 8.93931	10.57748
3.1	−2.44832	2.19573	8.1	− 9.08036	10.78728
3.2	−2.56599	2.31808	8.2	− 9.22160	10.99827
3.3	−2.68478	2.44311	8.3	− 9.36303	11.21046
3.4	−2.80462	2.57078	8.4	− 9.50465	11.42383
3.5	−2.92546	2.70103	8.5	− 9.64645	11.63836
3.6	−3.04724	2.83379	8.6	− 9.78842	11.85405
3.7	−3.16992	2.96901	8.7	− 9.93057	12.07087
3.8	−3.29346	3.10665	8.8	−10.07288	12.28881
3.9	−3.41782	3.24666	8.9	−10.21536	12.50787
4.0	−3.54296	3.38897	9.0	−10.35800	12.72803
4.1	−3.66884	3.53356	9.1	−10.50080	12.94928
4.2	−3.79543	3.68037	9.2	−10.64375	13.17161
4.3	−3.92271	3.82935	9.3	−10.78686	13.39500
4.4	−4.05063	3.98048	9.4	−10.93011	13.61945
4.5	−4.17918	4.13370	9.5	−11.07350	13.84494
4.6	−4.30833	4.28897	9.6	−11.21704	14.07147
4.7	−4.43806	4.44627	9.7	−11.36072	14.29902
4.8	−4.56833	4.60554	9.8	−11.50453	14.52759
4.9	−4.66914	4.76676	9.9	−11.64848	14.75716
5.0	−4.83046	4.92990	10.0	−11.79255	14.98772

Linear interpolation will yield about three figures; eight-point interpolation will yield about eight figures.

$\Re \ln \Gamma(z) = \log_e \lvert \Gamma(z) \rvert$ $\qquad$ $\Im \ln \Gamma(z) = \arg \Gamma(z)$

GAMMA FUNCTION FOR COMPLETE ARGUMENTS

$x = 1.9$

y	$\mathcal{R}\ln\Gamma(z) = \log_e\lvert\Gamma(z)\rvert$	$\mathcal{I}\ln\Gamma(z)$	y	$\mathcal{R}\ln\Gamma(z) = \log_e\lvert\Gamma(z)\rvert$	$\mathcal{I}\ln\Gamma(z)$
0.0	−0.03898	0.00000	5.0	− 4.66613	5.06053
0.1	−0.04242	0.03569	5.1	− 4.79608	5.22604
0.2	−0.05270	0.07184	5.2	− 4.92655	5.39337
0.3	−0.06975	0.10890	5.3	− 5.05749	5.56251
0.4	−0.09340	0.14727	5.4	− 5.18891	5.73341
0.5	−0.12349	0.18736	5.5	− 5.32078	5.90605
0.6	−0.15978	0.22952	5.6	− 5.45309	6.08040
0.7	−0.20201	0.27408	5.7	− 5.58582	6.25643
0.8	−0.24990	0.32129	5.8	− 5.71896	6.43413
0.9	−0.30316	0.37139	5.9	− 5.85250	6.61345
1.0	−0.36148	0.42457	6.0	− 5.98642	6.79438
1.1	−0.42456	0.48098	6.1	− 6.12071	6.97690
1.2	−0.49210	0.54071	6.2	− 6.25536	7.16097
1.3	−0.56382	0.60386	6.3	− 6.39036	7.34659
1.4	−0.63944	0.67047	6.4	− 6.52570	7.53372
1.5	−0.71870	0.74056	6.5	− 6.66136	7.72234
1.6	−0.80136	0.81416	6.6	− 6.79735	7.92143
1.7	−0.88718	0.89124	6.7	− 6.93364	8.10398
1.8	−0.97596	0.97178	6.8	− 7.07023	8.29696
1.9	−1.06749	1.05574	6.9	− 7.20712	8.49135
2.0	−1.16160	1.14310	7.0	− 7.34429	8.68713
2.1	−1.25812	1.23379	7.1	− 7.48173	8.88429
2.2	−1.35688	1.32777	7.2	− 7.61945	9.08281
2.3	−1.45775	1.42497	7.3	− 7.75742	9.28267
2.4	−1.56060	1.52534	7.4	− 7.89565	9.48385
2.5	−1.66529	1.62881	7.5	− 8.03413	9.68634
2.6	−1.77174	1.73533	7.6	− 8.17285	9.89012
2.7	−1.87982	1.84483	7.7	− 8.31180	10.09517
2.8	−1.98944	1.95726	7.8	− 8.45099	10.30148
2.9	−2.10052	2.07255	7.9	− 8.59039	10.50904
3.0	−2.21298	2.19064	8.0	− 8.73002	10.71782
3.1	−2.32674	2.31147	8.1	− 8.86986	10.92782
3.2	−2.44173	2.43498	8.2	− 9.00991	11.13902
3.3	−2.55788	2.56113	8.3	− 9.15016	11.35140
3.4	−2.67515	2.68985	8.4	− 9.29060	11.56496
3.5	−2.79346	2.82109	8.5	− 9.43125	11.77967
3.6	−2.91278	2.95480	8.6	− 9.57208	11.99554
3.7	−3.03304	3.09093	8.7	− 9.71309	12.21253
3.8	−3.15422	3.22944	8.8	− 9.85429	12.43065
3.9	−3.27626	3.37026	8.9	− 9.99567	12.64988
4.0	−3.39912	3.51336	9.0	−10.13721	12.87021
4.1	−3.52277	3.65870	9.1	−10.27893	13.09162
4.2	−3.64718	3.80622	9.2	−10.42081	13.31410
4.3	−3.77231	3.95589	9.3	−10.56286	13.53765
4.4	−3.89814	4.10768	9.4	−10.70506	13.76225
4.5	−4.02462	4.26153	9.5	−10.84742	13.98789
4.6	−4.15175	4.41741	9.6	−10.98993	14.21457
4.7	−4.27948	4.57528	9.7	−11.13259	14.44226
4.8	−4.40781	4.73512	9.8	−11.27539	14.67097
4.9	−4.53670	4.89688	9.9	−11.41834	14.90067
5.0	−4.66613	5.06053	10.0	−11.56143	15.13137

Linear interpolation will yield about three figures; eight-point interpolation will yield about eight figures.

$\mathcal{R}\ln\Gamma(z) = \log_e\lvert\Gamma(z)\rvert$ $\mathcal{I}\ln\Gamma(z) = \arg\Gamma(z)$

GAMMA FUNCTION FOR COMPLEX ARGUMENTS

$x = 2.0$

y	$\mathcal{R} \ln \Gamma(z) = \log_e \lvert \Gamma(z) \rvert$	$\mathcal{I} \ln \Gamma(z)$	y	$\mathcal{R} \ln \Gamma(z) = \log_e \lvert \Gamma(z) \rvert$	$\mathcal{I} \ln \Gamma(z)$
0.0	−0.00000	0.00000	5.0	− 4.50128	5.18930
0.1	−0.00322	0.04235	5.1	− 4.62940	5.35534
0.2	−0.01287	0.08509	5.2	− 4.75806	5.52319
0.3	−0.02886	0.12864	5.3	− 4.88723	5.69281
0.4	−0.05108	0.17335	5.4	− 5.01690	5.86419
0.5	−0.07937	0.21959	5.5	− 5.14706	6.03729
0.6	−0.11355	0.26768	5.6	− 5.27768	6.21208
0.7	−0.15338	0.31790	5.7	− 5.40874	6.38855
0.8	−0.19863	0.37052	5.8	− 5.54025	6.56665
0.9	−0.24904	0.42574	5.9	− 5.67217	6.74638
1.0	−0.30435	0.48376	6.0	− 5.80450	6.92770
1.1	−0.36429	0.54471	6.1	− 5.93723	7.11059
1.2	−0.42859	0.60873	6.2	− 6.07033	7.29503
1.3	−0.49700	0.67588	6.3	− 6.20381	7.48100
1.4	−0.56927	0.74625	6.4	− 6.33765	7.66847
1.5	−0.64516	0.81985	6.5	− 6.47184	7.85743
1.6	−0.72443	0.89673	6.6	− 6.60636	8.04785
1.7	−0.80689	0.97687	6.7	− 6.74122	8.23971
1.8	−0.89231	1.06028	6.8	− 6.87639	8.43299
1.9	−0.98053	1.14693	6.9	− 7.01188	8.62768
2.0	−1.07136	1.23680	7.0	− 7.14767	8.82376
2.1	−1.16464	1.32984	7.1	− 7.28375	9.02120
2.2	−1.26022	1.42601	7.2	− 7.42012	9.21999
2.3	−1.35796	1.52528	7.3	− 7.55677	9.42012
2.4	−1.45773	1.62759	7.4	− 7.69369	9.62156
2.5	−1.55941	1.73289	7.5	− 7.83087	9.82430
2.6	−1.66288	1.84113	7.6	− 7.96831	10.02832
2.7	−1.76806	1.95226	7.7	− 8.10600	10.23362
2.8	−1.87484	2.06621	7.8	− 8.24394	10.44016
2.9	−1.98313	2.18294	7.9	− 8.38211	10.64794
3.0	−2.09285	2.30240	8.0	− 8.52052	10.85695
3.1	−2.20393	2.42452	8.1	− 8.65915	11.06716
3.2	−2.31629	2.54926	8.2	− 8.79801	11.27857
3.3	−2.42988	2.67657	8.3	− 8.93708	11.49117
3.4	−2.54462	2.80640	8.4	− 9.07637	11.70493
3.5	−2.66047	2.93869	8.5	− 9.21586	11.91984
3.6	−2.77736	3.07340	8.6	− 9.35555	12.13590
3.7	−2.89526	3.21048	8.7	− 9.49544	12.35308
3.8	−3.01411	3.34989	8.8	− 9.63553	12.57138
3.9	−3.13386	3.49159	8.9	− 9.77580	12.79079
4.0	−3.25449	3.63552	9.0	− 9.91626	13.01130
4.1	−3.37595	3.78164	9.1	−10.05689	13.23288
4.2	−3.49821	3.92993	9.2	−10.19771	13.45553
4.3	−3.62123	4.08033	9.3	−10.33870	13.67925
4.4	−3.74498	4.23281	9.4	−10.47986	13.90401
4.5	−3.86943	4.38732	9.5	−10.62118	14.12981
4.6	−3.99455	4.54385	9.6	−10.76267	14.35664
4.7	−4.12032	4.70234	9.7	−10.90431	14.58449
4.8	−4.24672	4.86277	9.8	−11.04611	14.81335
4.9	−4.37371	5.02510	9.9	−11.18807	15.04320
5.0	−4.50128	5.18930	10.0	−11.33017	15.27404

Linear interpolation will yield about three figures; eight-point interpolation will yield about eight figures.

$\mathcal{R} \ln \Gamma(z) = \log_e \lvert \Gamma(z) \rvert$ $\qquad$ $\mathcal{I} \ln \Gamma(z) = \arg \Gamma(z)$

1. Bessel's differential equations for a real variable x is

$$x^2 \frac{d^2y}{dx^2} + x\frac{dy}{dx} + (x^2 - n^2)y = 0$$

2. When n is not an integer, two independent solutions of the equation are $J_n(x)$ and $J_{-n}(x)$, where

$$J_n(x) = \sum_{k=0}^{\infty} \frac{(-1)^k}{k!\,\Gamma(n+k+1)}\left(\frac{x}{2}\right)^{n+2k}$$

3. If n is an integer $J_{-n}(x) = (-1)^n J_n(x)$, where

$$J_n(x) = \frac{x^n}{2^n n!}\left\{1 - \frac{x^2}{2^2 \cdot 1!(n+1)} + \frac{x^4}{2^4 \cdot 2!(n+1)(n+2)} - \frac{x^6}{2^6 \cdot 3!(n+1)(n+2)(n+3)} + \cdots\right\}$$

4. For $n = 0$ and $n = 1$, this formula becomes

$$J_0(x) = 1 - \frac{x^2}{2^2(1!)^2} + \frac{x^4}{2^4(2!)^2} - \frac{x^6}{2^6(3!)^2} + \frac{x^8}{2^8(4!)^2} - \cdots$$

$$J_1(x) = \frac{x}{2} - \frac{x^3}{2^3 \cdot 1!2!} + \frac{x^5}{2^5 \cdot 2!3!} - \frac{x^7}{2^7 \cdot 3!4!} + \frac{x^9}{2^9 \cdot 4!5!} - \cdots$$

5. When x is large and positive, the following asymptotic series may be used

$$J_0(x) = \left(\frac{2}{\pi x}\right)^{\frac{1}{2}}\left\{P_0(x)\cos\left(x - \frac{\pi}{4}\right) - Q_0(x)\sin\left(x - \frac{\pi}{4}\right)\right\}$$

$$J_1(x) = \left(\frac{2}{\pi x}\right)^{\frac{1}{2}}\left\{P_1(x)\cos\left(x - \frac{3\pi}{4}\right) - Q_1(x)\sin\left(x - \frac{3\pi}{4}\right)\right\},$$

where

$$P_0(x) \sim 1 - \frac{1^2 \cdot 3^2}{2!(8x)^2} + \frac{1^2 \cdot 3^2 \cdot 5^2 \cdot 7^2}{4!(8x)^4} - \frac{1^2 \cdot 3^2 \cdot 5^2 \cdot 7^2 \cdot 9^2 \cdot 11^2}{6!(8x)^6} + \cdots$$

$$Q_0(x) \sim -\frac{1^2}{1!8x} + \frac{1^2 \cdot 3^2 \cdot 5^2}{3!(8x)^3} - \frac{1^2 \cdot 3^2 \cdot 5^2 \cdot 7^2 \cdot 9^2}{5!(8x)^5} + - \cdots$$

$$P_1(x) \sim 1 + \frac{1^2 \cdot 3 \cdot 5}{2!(8x)^2} - \frac{1^2 \cdot 3^2 \cdot 5^2 \cdot 7 \cdot 9}{4!(8x)^4} + \frac{1^2 \cdot 3^2 \cdot 5^2 \cdot 7^2 \cdot 9^2 \cdot 11 \cdot 13}{6!(8x)^6} - + \cdots$$

$$Q_1(x) \sim \frac{1 \cdot 3}{1!8x} - \frac{1^2 \cdot 3^2 \cdot 5 \cdot 7}{3!(8x)^3} + \frac{1^2 \cdot 3^2 \cdot 5^2 \cdot 7^2 \cdot 9 \cdot 11}{5!(8x)^6} - \cdots$$

[In $P_1(x)$ the signs alternate from + to − after the first term]

6. If $x > 25$, it is convenient to use the formulas

$$J_0(x) = A_0(x)\sin x + B_0(x)\cos x$$
$$J_1(x) = B_1(x)\sin x - A_1(x)\cos x,$$

where

$$A_0(x) = \frac{P_0(x) - Q_0(x)}{(\pi x)^{\frac{1}{2}}} \quad \text{and} \quad A_1(x) = \frac{P_1(x) - Q_1(x)}{(\pi x)^{\frac{1}{2}}}$$

$$B_0(x) = \frac{P_0(x) + Q_0(x)}{(\pi x)^{\frac{1}{2}}} \quad \text{and} \quad B_1(x) = \frac{P_1(x) + Q_1(x)}{(\pi x)^{\frac{1}{2}}}$$

7. The zeros of $J_0(x)$, and $J_1(x)$

If $j_{0,s}$ and $j_{1,s}$ are the s'th zeros of $J_0(x)$ and $J_1(x)$ respectively, and if $a = 4s - 1$, $b = 4s + 1$

$$j_{0,s} \sim \frac{1}{4}\pi a\left\{1 + \frac{2}{\pi^2 a^2} - \frac{62}{3\pi^4 a^4} + \frac{15{,}116}{15\pi^6 a^6} - \frac{12{,}554{,}474}{105\pi^8 a^8} + \frac{8{,}368{,}654{,}292}{315\pi^{10} a^{10}} - + \cdots\right\}$$

$$j_{1,s} \sim \frac{1}{4}\pi b\left\{1 - \frac{6}{\pi^2 b^2} + \frac{6}{\pi^4 b^4} - \frac{4716}{5\pi^6 b^6} + \frac{3{,}902{,}418}{35\pi^8 b^8} - \frac{8{,}952{,}167{,}324}{35\pi^{10} b^{10}} + \cdots\right\}$$

$$J_1(j_{0,s}) \sim \frac{(-1)^{s+1}2^{\frac{3}{2}}}{\pi a^{\frac{1}{2}}}\left\{1 - \frac{56}{3\pi^4 a^4} + \frac{9664}{5\pi^6 a^6} - \frac{7{,}381{,}280}{21\pi^8 a^8} + \cdots\right\}$$

$$J_0(j_{1,s}) \sim \frac{(-1)^{s}2^{\frac{3}{2}}}{\pi b^{\frac{1}{2}}}\left\{1 + \frac{24}{\pi^4 b^4} - \frac{19{,}584}{10\pi^6 b^6} + \frac{2{,}466{,}720}{7\pi^8 b^8} - \cdots\right\}$$

8. Table of zeros for $J_0(x)$ and $J_1(x)$

$$J_0(\alpha_n) = 0 \qquad J_1(\beta_n) = 0$$

Roots α_n	$J_1(\alpha_n)$	Roots β_n	$J_0(\beta_n)$
2.4048	0.5191	0.0000	1.0000
5.5201	−0.3403	3.8317	−0.4028
8.6537	0.2715	7.0156	0.3001
11.7915	−0.2325	10.1735	−0.2497
14.9309	0.2065	13.3237	0.2184
18.0711	−0.1877	16.4706	−0.1965
21.2116	0.1733	19.6159	0.1801

9. Recurrence formulas

$$J_{n-1}(x) + J_{n+1}(x) = \frac{2n}{x}J_n(x) \qquad nJ_n(x) + xJ_n'(x) = xJ_{n-1}(x)$$

$$J_{n-1}(x) - J_{n+1}(x) = 2J_n'(x) \qquad nJ_n(x) - xJ_n'(x) = xJ_{n+1}(x)$$

10. If J_n is written for $J_n(x)$ and $J_n^{(k)}$ is written for $\frac{d^k}{dx^k}\{J_n(x)\}$, then the following derivative relationships are important

$$J_0^{(r)} = -J_1^{(r-1)}$$

$$J_0^{(2)} = -J_0 + \frac{1}{x} J_1 = \frac{1}{2}(J_2 - J_0)$$

$$J_0^{(3)} = \frac{1}{x} J_0 + \left(1 - \frac{2}{x^2}\right) J_1 = \frac{1}{4}(-J_3 + 3J_1)$$

$$J_0^{(4)} = \left(1 - \frac{3}{x^2}\right) J_0 - \left(\frac{2}{x} - \frac{6}{x^3}\right) J_1 = \frac{1}{8}(J_4 - 4J_2 + 3J_0), \text{ etc.}$$

11. Half order Bessel functions

$$J_{\frac{1}{2}}(x) = \sqrt{\frac{2}{\pi x}} \sin x$$

$$J_{-\frac{1}{2}}(x) = \sqrt{\frac{2}{\pi x}} \cos x$$

$$J_{n+\frac{3}{2}}(x) = -x^{n+\frac{1}{2}} \frac{d}{dx} \{x^{-(n+\frac{1}{2})} J_{n+\frac{1}{2}}(x)\}$$

$$J_{n-\frac{1}{2}}(x) = x^{-(n+\frac{1}{2})} \frac{d}{dx} \{x^{n+\frac{1}{2}} J_{n+\frac{1}{2}}(x)\}$$

n	$\left(\frac{\pi x}{2}\right)^{\frac{1}{2}} J_{n+\frac{1}{2}}(x)$	$\left(\frac{\pi x}{2}\right)^{\frac{1}{2}} J_{-(n+\frac{1}{2})}(x)$
0	$\sin x$	$\cos x$
1	$\frac{\sin x}{x} - \cos x$	$-\frac{\cos x}{x} - \sin x$
2	$\left(\frac{3}{x^2} - 1\right)\sin x - \frac{3}{x}\cos x$	$\left(\frac{3}{x^2} - 1\right)\cos x + \frac{3}{x}\sin x$
3	$\left(\frac{15}{x^3} - \frac{6}{x}\right)\sin x - \left(\frac{15}{x^2} - 1\right)\cos x$	$-\left(\frac{15}{x^3} - \frac{6}{x}\right)\cos x - \left(\frac{15}{x^2} - 1\right)\sin x$
	etc.	

12. Additional solutions to Bessel's equation are

$Y_n(x)$ (also called Weber's function, and sometimes denoted by $N_n(x)$)
$H_n^{(1)}(x)$ and $H_n^{(2)}(x)$ (also called Hankel functions)

These solutions are defined as follows

$$Y_n(x) = \begin{cases} \dfrac{J_n(x)\cos(n\pi) - J_{-n}(x)}{\sin(n\pi)} & n \text{ not an integer} \\ \lim\limits_{\nu \to n} \dfrac{J_\nu(x)\cos(\nu\pi) - J_{-\nu}(x)}{\sin(\nu\pi)} & n \text{ an integer} \end{cases}$$

$$H_n^{(1)}(x) = J_n(x) + iY_n(x)$$
$$H_n^{(2)}(x) = J_n(x) - iY_n(x)$$

The additional properties of these functions may all be derived from the above relations and the known properties of $J_n(x)$.

13. Complete solutions to Bessel's equation may be written as

$$c_1 J_n(x) + c_2 J_{-n}(x) \qquad \text{if } n \text{ is not an integer,}$$

or

$$\left.\begin{array}{l} c_1 J_n(x) + c_2 Y_n(x) \\ \\ c_1 H_n^{(1)}(x) + c_2 H_n^{(2)}(x) \end{array}\right\} \text{for any value of } n$$

or

14. The modified (or hyperbolic) Bessel's differential equation is

$$x^2 \frac{d^2 y}{dx^2} + x \frac{dy}{dx} - (x^2 + n^2) y = 0$$

15. When n is not an integer, two independent solutions of the equation are $I_n(x)$ and $I_{-n}(x)$, where

$$I_n(x) = \sum_{k=0}^{\infty} \frac{1}{k!\Gamma(n + k + 1)} \left(\frac{x}{2}\right)^{n+2k}$$

16. If n is an integer,

$$I_n(x) = I_{-n}(x) = \frac{x^n}{2^n n!} \left\{1 + \frac{x^2}{2^2 \cdot 1!(n + 1)} + \frac{x^4}{2^4 \cdot 2!(n + 1)(n + 2)} + \frac{x^6}{2^6 \cdot 3!(n + 1)(n + 2)(n + 3)} + \cdots\right\}$$

17. For $n = 0$ and $n = 1$, this formula becomes

$$I_0(x) = 1 + \frac{x^2}{2^2(1!)^2} + \frac{x^4}{2^4(2!)^2} + \frac{x^6}{2^6(3!)^2} + \frac{x^8}{2^8(4!)^2} + \cdots$$

$$I_1(x) = \frac{x}{2} + \frac{x^3}{2^3 \cdot 1!2!} + \frac{x^5}{2^5 \cdot 2!3!} + \frac{x^7}{2^7 \cdot 3!4!} + \frac{x^9}{2^9 \cdot 4!5!} + \cdots$$

18. Another solution to the modified Bessel's equation is

$$K_n(x) = \begin{cases} \dfrac{1}{2}\pi \dfrac{I_{-n}(x) - I_n(x)}{\sin(n\pi)} & n \text{ not an integer} \\ \\ \lim\limits_{\nu \to n} \dfrac{1}{2}\pi \dfrac{I_{-\nu}(x) - I_\nu(x)}{\sin(\nu\pi)} & n \text{ an integer} \end{cases}$$

This function is linearly independent of $I_n(x)$ for all values of n. Thus the complete solution to the modified Bessel's equation may be written as

$$c_1 I_n(x) + c_2 I_{-n}(x) \qquad n \text{ not an integer}$$

or

$$c_1 I_n(x) + c_2 K_n(x) \qquad \text{any } n$$

19. The following relations hold among the various Bessel functions:

$$I_n(z) = i^{-m} J_m(iz)$$

$$Y_n(iz) = (i)^{n+1} I_n(z) - \frac{2}{\pi} i^{-n} K_n(z)$$

Most of the properties of the modified Bessel function may be deduced from the known properties of $J_n(x)$ by use of these relations and those previously given.

20. Recurrence formulas

$$I_{n-1}(x) - I_{n+1}(x) = \frac{2n}{x} I_n(x) \qquad I_{n-1}(x) + I_{n+1}(x) = 2I_n'(x)$$

$$I_{n-1}(x) - \frac{n}{x} I_n(x) = I_n'(x) \qquad I_n'(x) = I_{n+1}(x) + \frac{n}{x} I_n(z)$$

BESSEL FUNCTIONS FOR SPHERICAL COORDINATES

$$j_n(x) = \sqrt{\frac{\pi}{2x}}\, J_{(n+\frac{1}{2})}(x),\quad y_n(x) = \sqrt{\frac{\pi}{2x}}\, Y_{(n+\frac{1}{2})}(x) = (-1)^{n+1}\sqrt{\frac{\pi}{2x}}\, J_{-(n+\frac{1}{2})}(x)$$

x	$j_0(x)$	$y_0(x)$	$j_1(x)$	$y_1(x)$	$j_2(x)$	$y_2(x)$
0.0	1.0000	$-\infty$	0.0000	$-\infty$	0.0000	$-\infty$
0.1	0.9983	−9.9500	0.0333	−100.50	0.0007	−3005.0
0.2	0.9933	−4.9003	0.0664	−25.495	0.0027	−377.52
0.4	0.9735	−2.3027	0.1312	−6.7302	0.0105	−48.174
0.6	0.9411	−1.3756	0.1929	−3.2337	0.0234	−14.793
0.8	0.8967	−0.8709	0.2500	−1.9853	0.0408	−6.5740
1.0	0.8415	−0.5403	0.3012	−1.3818	0.0620	−3.6050
1.2	0.7767	−0.3020	0.3453	−1.0283	0.0865	−2.2689
1.4	0.7039	−0.1214	0.3814	−0.7906	0.1133	−1.5728
1.6	0.6247	+0.0182	0.4087	−0.6133	0.1416	−1.1682
1.8	0.5410	0.1262	0.4268	−0.4709	0.1703	−0.9111
2.0	0.4546	0.2081	0.4354	−0.3506	0.1984	−0.7340
2.2	0.3675	0.2675	0.4345	−0.2459	0.2251	−0.6028
2.4	0.2814	0.3072	0.4245	−0.1534	0.2492	−0.4990
2.6	0.1983	0.3296	0.4058	−0.0715	0.2700	−0.4121
2.8	0.1196	0.3365	0.3792	+0.0005	0.2867	−0.3359
3.0	+0.0470	0.3300	0.3457	0.0630	0.2986	−0.2670
3.2	−0.0182	0.3120	0.3063	0.1157	0.3054	−0.2035
3.4	−0.0752	0.2844	0.2622	0.1588	0.3066	−0.1442
3.6	−0.1229	0.2491	0.2150	0.1921	0.3021	−0.0890
3.8	−0.1610	0.2082	0.1658	0.2158	0.2919	−0.0378
4.0	−0.1892	0.1634	0.1161	0.2301	0.2763	+0.0091
4.2	−0.2075	0.1167	0.0673	0.2353	0.2556	0.0514
4.4	−0.2163	0.0698	+0.0207	0.2321	0.2304	0.0884
4.6	−0.2160	+0.0244	−0.0226	0.2213	0.2013	0.1200
4.8	−0.2075	−0.0182	−0.0615	0.2037	0.1691	0.1456
5.0	−0.1918	−0.0567	−0.0951	0.1804	0.1347	0.1650
5.2	−0.1699	−0.0901	−0.1228	0.1526	0.0991	0.1781
5.4	−0.1431	−0.1175	−0.1440	0.1213	0.0631	0.1850
5.6	−0.1127	−0.1385	−0.1586	0.0880	+0.0277	0.1856
5.8	−0.0801	−0.1527	−0.1665	0.0538	−0.0060	0.1805
6.0	−0.0466	−0.1600	−0.1678	+0.0199	−0.0373	0.1700
6.2	−0.0134	−0.1607	−0.1629	−0.0125	−0.0654	0.1547
6.4	+0.0182	−0.1552	−0.1523	−0.0425	−0.0896	0.1353
6.6	0.0472	−0.1440	−0.1368	−0.0690	−0.1094	0.1126
6.8	0.0727	−0.1278	−0.1172	−0.0915	−0.1243	0.0875
7.0	0.0939	−0.1077	−0.0943	−0.1092	−0.1343	0.0609
7.2	0.1102	−0.0845	−0.0692	−0.1220	−0.1391	0.0337
7.4	0.1215	−0.0593	−0.0429	−0.1295	−0.1388	+0.0068
7.6	0.1274	−0.0331	−0.0163	−0.1317	−0.1338	−0.0189
7.8	0.1280	−0.0069	+0.0095	−0.1289	−0.1244	−0.0427
8.0	0.1237	+0.0182	0.0336	−0.1214	−0.1111	−0.0637

BESSEL FUNCTION J_0

T	.00	.01	.02	.03	.04
0.0	1.0000000	.9999750	.9999000	.9997750	.9996000
.1	.9975016	.9969773	.9964032	.9957795	.9951060
.2	.9900250	.9890054	.9879366	.9868187	.9856518
.3	.9776262	.9761189	.9745634	.9729597	.9713081
.4	.9603982	.9584145	.9563838	.9543065	.9521825
.5	.9384698	.9360245	.9335339	.9309983	.9284179
.6	.9120049	.9091162	.9061843	.9032094	.9001918
.7	.8812009	.8778904	.8745391	.8711471	.8677147
.8	.8462874	.8425797	.8388338	.8350500	.8312284
.9	.8075238	.8034465	.7993339	.7951863	.7910039
1.0	.7651977	.7607810	.7563321	.7518513	.7473390
1.1	.7196220	.7148985	.7101461	.7053653	.7005564
1.2	.6711327	.6661371	.6611163	.6560706	.6510004
1.3	.6200860	.6148549	.6096023	.6043287	.5990343
1.4	.5668551	.5614267	.5559807	.5505176	.5450376
1.5	.5118277	.5062414	.5006415	.4950285	.4894026
1.6	.4554022	.4496983	.4439850	.4382625	.4325313
1.7	.3979849	.3922044	.3864185	.3806276	.3748321
1.8	.3399864	.3341705	.3283532	.3225351	.3167166
1.9	.2818186	.2760083	.2702008	.2643965	.2585959
2.0	.2238908	.2181268	.2123697	.2066198	.2008776
2.1	.1666070	.1609293	.1552625	.1496068	.1439626
2.2	.1103623	.1048098	.0992720	.0937491	.0882416
2.3	.0555398	.0501501	.0447787	.0394259	.0340921
2.4	.0025077	−.0026834	−.0078527	−.0129999	−.0181247
2.5	−.0483838	−.0533423	−.0582758	−.0631839	−.0680664
2.6	−.0968050	−.1014992	−.1061654	−.1108031	−.1154123
2.7	−.1424494	−.1468500	−.1512198	−.1555585	−.1598658
2.8	−.1850360	−.1891165	−.1931636	−.1971771	−.2011568
2.9	−.2243115	−.2280481	−.2317491	−.2354142	−.2390434
3.0	−.2600520	−.2634239	−.2667583	−.2700551	−.2733140
3.1	−.2920643	−.2950541	−.2980048	−.3009162	−.3037884
3.2	−.3201882	−.3227815	−.3253345	−.3278471	−.3303193
3.3	−.3442963	−.3464823	−.3486272	−.3507308	−.3527931
3.4	−.3642956	−.3660670	−.3677967	−.3694845	−.3711306
3.5	−.3801277	−.3814805	−.3827914	−.3840603	−.3852873
3.6	−.3917690	−.3927027	−.3935947	−.3944449	−.3952533
3.7	−.3992302	−.3997479	−.4002242	−.4006593	−.4010532
3.8	−.4025564	−.4026643	−.4027318	−.4027588	−.4027456
3.9	−.4018260	−.4015339	−.4012023	−.4008316	−.4004218
4.0	−.3971498	−.3964704	−.3957530	−.3949979	−.3942053
4.1	−.3886697	−.3876188	−.3865318	−.3854088	−.3842500
4.2	−.3765571	−.3751534	−.3737157	−.3722440	−.3707386
4.3	−.3610111	−.3592761	−.3575093	−.3557108	−.3538810
4.4	−.3422568	−.3402143	−.3381424	−.3360414	−.3339116
4.5	−.3205425	−.3182185	−.3158678	−.3134908	−.3110877
4.6	−.2961378	−.2935603	−.2909591	−.2883344	−.2856866
4.7	−.2693308	−.2665295	−.2637076	−.2608653	−.2580029
4.8	−.2404253	−.2374315	−.2344201	−.2313916	−.2283462
4.9	−.2097383	−.2065842	−.2034158	−.2002335	−.1970377

BESSEL FUNCTION J_0

T	.05	.06	.07	.08	.09
0.0	.9993751	.9991002	.9987754	.9984006	.9979760
.1	.9943829	.9936102	.9927880	.9919164	.9909953
.2	.9844359	.9831713	.9818579	.9804958	.9790853
.3	.9696087	.9678615	.9660667	.9642245	.9623350
.4	.9500121	.9477955	.9455328	.9432242	.9408698
.5	.9257928	.9231233	.9204096	.9176518	.9148501
.6	.8971316	.8940292	.8908846	.8876982	.8844702
.7	.8642423	.8607300	.8571780	.8535868	.8499565
.8	.8273695	.8234734	.8195405	.8155711	.8115654
.9	.7867871	.7825361	.7782514	.7739332	.7695819
1.0	.7427956	.7382212	.7336163	.7289813	.7243164
1.1	.6957198	.6908557	.6859646	.6810469	.6761028
1.2	.6459061	.6407880	.6356466	.6304822	.6252952
1.3	.5937196	.5883850	.5830309	.5776576	.5722655
1.4	.5395413	.5340289	.5285010	.5229579	.5174000
1.5	.4837644	.4781143	.4724526	.4667797	.4610961
1.6	.4267919	.4210446	.4152898	.4095280	.4037595
1.7	.3690325	.3632292	.3574225	.3516128	.3458007
1.8	.3108980	.3050797	.2992623	.2934460	.2876313
1.9	.2527992	.2470071	.2412197	.2354376	.2296612
2.0	.1951434	.1894177	.1837008	.1779931	.1722950
2.1	.1383305	.1327106	.1271035	.1215095	.1159290
2.2	.0827499	.0772742	.0718150	.0663726	.0609474
2.3	.0287776	.0234828	.0182081	.0129538	.0077202
2.4	−.0232267	−.0283057	−.0333611	−.0383929	−.0434005
2.5	−.0729229	−.0777531	−.0825567	−.0873334	−.0920829
2.6	−.1199924	−.1245434	−.1290648	−.1335565	−.1380181
2.7	−.1641414	−.1683852	−.1725967	−.1767759	−.1809224
2.8	−.2051024	−.2090137	−.2128905	−.2167325	−.2205396
2.9	−.2426364	−.2461931	−.2497131	−.2531964	−.2566427
3.0	−.2765350	−.2797178	−.2828623	−.2859683	−.2890357
3.1	−.3066211	−.3094142	−.3121675	−.3148811	−.3175547
3.2	−.3327508	−.3351416	−.3374917	−.3398009	−.3420691
3.3	−.3548140	−.3567934	−.3587314	−.3606277	−.3624825
3.4	−.3727349	−.3742972	−.3758177	−.3772963	−.3787330
3.5	−.3864724	−.3876155	−.3887167	−.3897760	−.3907934
3.6	−.3960201	−.3967452	−.3974287	−.3980707	−.3986712
3.7	−.4014061	−.4017178	−.4019887	−.4022187	−.4024079
3.8	−.4026921	−.4025986	−.4024651	−.4022918	−.4020787
3.9	−.3999730	−.3994854	−.3989591	−.3983943	−.3977912
4.0	−.3933752	−.3925079	−.3916035	−.3906622	−.3896842
4.1	−.3830556	−.3818259	−.3805609	−.3792610	−.3779263
4.2	−.3691998	−.3676276	−.3660225	−.3643845	−.3627140
4.3	−.3520200	−.3501281	−.3482056	−.3462527	−.3442697
4.4	−.3317533	−.3295666	−.3273519	−.3251095	−.3228396
4.5	−.3086589	−.3062045	−.3037249	−.3012204	−.2986913
4.6	−.2830159	−.2803228	−.2776073	−.2748700	−.2721110
4.7	−.2551208	−.2522193	−.2492987	−.2463592	−.2434014
4.8	−.2252843	−.2222062	−.2191122	−.2160027	−.2128779
4.9	−.1938286	−.1906067	−.1873722	−.1841255	−.1808669

BESSEL FUNCTION J_0

T	.00	.01	.02	.03	.04
5.0	−.1775968	−.1743154	−.1710232	−.1677205	−.1644075
5.1	−.1443347	−.1409599	−.1375776	−.1341882	−.1307919
5.2	−.1102904	−.1068561	−.1034176	−.0999753	−.0965297
5.3	−.0758031	−.0723430	−.0688822	−.0654211	−.0619598
5.4	−.0412101	−.0377578	−.0343082	−.0308615	−.0274180
5.5	−.0068439	−.0034323	−.0000266	.0033730	.0067661
5.6	.0269709	.0303098	.0336398	.0369605	.0402716
5.7	.0599200	.0631556	.0663792	.0695906	.0727894
5.8	.0917026	.0948055	.0978937	.1009668	.1040245
5.9	.1220334	.1249761	.1279015	.1308091	.1336987
6.0	.1506453	.1534022	.1561393	.1588562	.1615527
6.1	.1772914	.1798391	.1823646	.1848678	.1873484
6.2	.2017472	.2040644	.2063574	.2086262	.2108705
6.3	.2238120	.2258800	.2279222	.2299383	.2319283
6.4	.2433106	.2451134	.2468888	.2486369	.2503573
6.5	.2600946	.2616188	.2631145	.2645817	.2660201
6.6	.2740434	.2752785	.2764843	.2776606	.2788074
6.7	.2850647	.2860032	.2869117	.2877901	.2886385
6.8	.2930956	.2937327	.2943394	.2949159	.2954620
6.9	.2981020	.2984359	.2987395	.2990127	.2992557
7.0	.3000793	.3001111	.3001128	.3000846	.3000264
7.1	.2990514	.2987851	.2984893	.2981641	.2978096
7.2	.2950707	.2945131	.2939268	.2933119	.2926686
7.3	.2882169	.2873774	.2865103	.2856158	.2846939
7.4	.2785962	.2774868	.2763512	.2751894	.2740018
7.5	.2663397	.2649748	.2635853	.2621712	.2607329
7.6	.2516018	.2499982	.2483717	.2467225	.2450508
7.7	.2345591	.2327355	.2308910	.2290257	.2271400
7.8	.2154078	.2133848	.2113430	.2092828	.2072042
7.9	.1943618	.1921618	.1899452	.1877126	.1854639
8.0	.1716508	.1692974	.1669299	.1645488	.1621542
8.1	.1475175	.1450356	.1425423	.1400378	.1375223
8.2	.1222153	.1196308	.1170375	.1144357	.1118256
8.3	.0960061	.0933456	.0906789	.0880063	.0853282
8.4	.0691573	.0664476	.0637345	.0610183	.0582992
8.5	.0419393	.0392076	.0364752	.0337424	.0310094
8.6	.0146230	.0118963	.0091717	.0064492	.0037293
8.7	−.0125227	−.0152177	−.0179081	−.0205935	−.0232739
8.8	−.0392338	−.0418710	−.0445011	−.0471237	−.0497387
8.9	−.0652532	−.0678075	−.0703522	−.0728869	−.0754116
9.0	−.0903336	−.0927808	−.0952160	−.0976390	−.1000496
9.1	−.1142392	−.1165565	−.1188596	−.1211483	−.1234224
9.2	−.1367484	−.1389144	−.1410642	−.1431976	−.1453143
9.3	−.1576552	−.1596503	−.1616274	−.1635862	−.1655265
9.4	−.1767716	−.1785781	−.1803648	−.1821316	−.1838783
9.5	−.1939287	−.1955308	−.1971117	−.1986712	−.2002092
9.6	−.2089787	−.2103628	−.2117244	−.2130634	−.2143797
9.7	−.2217955	−.2229502	−.2240814	−.2251890	−.2262730
9.8	−.2322760	−.2331923	−.2340844	−.2349521	−.2357955
9.9	−.2403411	−.2410124	−.2416590	−.2422808	−.2428777

BESSEL FUNCTION J_0

T	.05	.06	.07	.08	.09
5.0	−.1610847	−.1577524	−.1544109	−.1510606	−.1477018
5.1	−.1273892	−.1239803	−.1205657	−.1171456	−.1137204
5.2	−.0930810	−.0896295	−.0861757	−.0827198	−.0792621
5.3	−.0584989	−.0550386	−.0515792	−.0481211	−.0446646
5.4	−.0239781	−.0205422	−.0171104	−.0136833	−.0102610
5.5	.0101524	.0135315	.0169033	.0202673	.0236233
5.6	.0435728	.0468638	.0501444	.0534141	.0566727
5.7	.0759753	.0791482	.0823076	.0854533	.0885851
5.8	.1070666	.1100928	.1131028	.1160964	.1190734
5.9	.1365701	.1394230	.1422573	.1450725	.1478686
6.0	.1642286	.1668837	.1695178	.1721306	.1747218
6.1	.1898062	.1922411	.1946529	.1970413	.1994061
6.2	.2130901	.2152848	.2174546	.2195991	.2217183
6.3	.2338920	.2358292	.2377398	.2396237	.2414807
6.4	.2520501	.2537151	.2553522	.2569612	.2585420
6.5	.2674298	.2688106	.2701625	.2714853	.2727789
6.6	.2799246	.2810122	.2820700	.2830981	.2840964
6.7	.2894568	.2902449	.2910029	.2917307	.2924283
6.8	.2959779	.2964633	.2969185	.2973434	.2977379
6.9	.2994685	.2996510	.2998033	.2999254	.3000174
7.0	.2999383	.2998204	.2996727	.2994953	.2992881
7.1	.2974258	.2970128	.2965707	.2960996	.2955996
7.2	.2919969	.2912970	.2905689	.2898128	.2890288
7.3	.2837448	.2827687	.2817656	.2807358	.2796793
7.4	.2727883	.2715492	.2702846	.2689947	.2676797
7.5	.2592704	.2577839	.2562736	.2547397	.2531824
7.6	.2433568	.2416407	.2399026	.2381429	.2363617
7.7	.2252341	.2233081	.2213622	.2193967	.2174119
7.8	.2051076	.2029932	.2008612	.1987118	.1965453
7.9	.1831996	.1809198	.1786247	.1763147	.1739900
8.0	.1597463	.1573255	.1548919	.1524459	.1499876
8.1	.1349963	.1324598	.1299132	.1273568	.1247907
8.2	.1092075	.1065816	.1039483	.1013077	.0986602
8.3	.0826448	.0799563	.0772630	.0745652	.0718632
8.4	.0555775	.0528534	.0501273	.0473994	.0446699
8.5	.0282765	.0255440	.0228121	.0200812	.0173513
8.6	.0010122	−.0017019	−.0044128	−.0071200	−.0098234
8.7	−.0259489	−.0286182	−.0312816	−.0339388	−.0365896
8.8	−.0523457	−.0549445	−.0575350	−.0601167	−.0626896
8.9	−.0779258	−.0804295	−.0829224	−.0854042	−.0878747
9.0	−.1024475	−.1048325	−.1072044	−.1095629	−.1119080
9.1	−.1256816	−.1279258	−.1301548	−.1323684	−.1345663
9.2	−.1474143	−.1494972	−.1515629	−.1536113	−.1556421
9.3	−.1674482	−.1693511	−.1712351	−.1730999	−.1749455
9.4	−.1856048	−.1873109	−.1889965	−.1906615	−.1923056
9.5	−.2017255	−.2032202	−.2046929	−.2061437	−.2075723
9.6	−.2156732	−.2169439	−.2181915	−.2194161	−.2206174
9.7	−.2273333	−.2283698	−.2293823	−.2303710	−.2313356
9.8	−.2366145	−.2374090	−.2381789	−.2389243	−.2396451
9.9	−.2434497	−.2439968	−.2445190	−.2450163	−.2454885

BESSEL FUNCTION J_0

T	.00	.01	.02	.03	.04
10.0	−.2459358	−.2463580	−.2467551	−.2471272	−.2474743
10.1	−.2490297	−.2492011	−.2493474	−.2494687	−.2495649
10.2	−.2496171	−.2495385	−.2494350	−.2493067	−.2491536
10.3	−.2477168	−.2473914	−.2470416	−.2466674	−.2462690
10.4	−.2433718	−.2428051	−.2422148	−.2416008	−.2409633
10.5	−.2366482	−.2358483	−.2350255	−.2341800	−.2333120
10.6	−.2276350	−.2266119	−.2255670	−.2245006	−.2234127
10.7	−.2164427	−.2152085	−.2139539	−.2126791	−.2113843
10.8	−.2032020	−.2017708	−.2003208	−.1988522	−.1973650
10.9	−.1880622	−.1864501	−.1848208	−.1831745	−.1815115
11.0	−.1711903	−.1694147	−.1676238	−.1658178	−.1639968
11.1	−.1527683	−.1508483	−.1489149	−.1469684	−.1450089
11.2	−.1329919	−.1309477	−.1288922	−.1268256	−.1247483
11.3	−.1120685	−.1099213	−.1077650	−.1055999	−.1034261
11.4	−.0902145	−.0879865	−.0857517	−.0835102	−.0812623
11.5	−.0676539	−.0653678	−.0630771	−.0607821	−.0584830
11.6	−.0446157	−.0422945	−.0399711	−.0376456	−.0353184
11.7	−.0213313	−.0189982	−.0166653	−.0143327	−.0120006
11.8	.0019672	.0042889	.0066082	.0089249	.0112388
11.9	.0250494	.0273370	.0296200	.0318980	.0341710
12.0	.0476893	.0499204	.0521447	.0543619	.0565718
12.1	.0696668	.0718199	.0739640	.0760989	.0782245
12.2	.0907701	.0928245	.0948680	.0969003	.0989212
12.3	.1107980	.1127342	.1146576	.1165679	.1184651
12.4	.1295610	.1313609	.1331462	.1349167	.1366724
12.5	.1468841	.1485309	.1501615	.1517758	.1533737
12.6	.1626073	.1640860	.1655471	.1669905	.1684160
12.7	.1765879	.1778851	.1791636	.1804231	.1816637
12.8	.1887014	.1898058	.1908904	.1919550	.1929997
12.9	.1988424	.1997446	.2006261	.2014869	.2023269
13.0	.2069261	.2076187	.2082899	.2089399	.2095684
13.1	.2128882	.2133659	.2138219	.2142562	.2146687
13.2	.2166859	.2169457	.2171835	.2173995	.2175935
13.3	.2182981	.2183389	.2183579	.2183551	.2183304
13.4	.2177252	.2175484	.2173499	.2171299	.2168884
13.5	.2149892	.2145981	.2141858	.2137525	.2132981
13.6	.2101332	.2095332	.2089128	.2082720	.2076107
13.7	.2032208	.2024195	.2015986	.2007581	.1998982
13.8	.1943356	.1933424	.1923305	.1913002	.1902515
13.9	.1835799	.1824059	.1812145	.1800059	.1787801
14.0	.1710735	.1697317	.1683739	.1670002	.1656108
14.1	.1569529	.1554577	.1539481	.1524242	.1508861
14.2	.1413694	.1397368	.1380914	.1364334	.1347629
14.3	.1244877	.1227348	.1209709	.1191961	.1174107
14.4	.1064841	.1046292	.1027650	.1008919	.0990100
14.5	.0875449	.0856069	.0836617	.0817095	.0797504
14.6	.0678641	.0658629	.0638565	.0618450	.0598288
14.7	.0476418	.0455977	.0435503	.0415000	.0394470
14.8	.0270823	.0250157	.0229481	.0208795	.0188102
14.9	.0063915	.0043232	.0022558	.0001896	−.0018753

BESSELL FUNCTION J_0

T	.05	.06	.07	.08	.09
10.0	−.2477962	−.2480931	−.2483649	−.2486116	−.2488332
10.1	−.2496361	−.2496822	−.2497034	−.2496996	−.2496708
10.2	−.2489758	−.2487732	−.2485460	−.2482942	−.2480177
10.3	−.2458463	−.2453994	−.2449285	−.2444335	−.2439146
10.4	−.2403022	−.2396178	−.2389101	−.2381792	−.2374252
10.5	−.2324214	−.2315085	−.2305732	−.2296158	−.2286364
10.6	−.2223036	−.2211732	−.2200218	−.2188495	−.2176565
10.7	−.2100695	−.2087349	−.2073807	−.2060071	−.2046141
10.8	−.1958594	−.1943357	−.1927939	−.1912343	−.1896570
10.9	−.1798318	−.1781356	−.1764232	−.1746947	−.1729504
11.0	−.1621611	−.1603109	−.1584463	−.1565675	−.1546748
11.1	−.1430367	−.1410520	−.1390549	−.1370458	−.1350247
11.2	−.1226602	−.1205618	−.1184532	−.1163346	−.1142063
11.3	−.1012439	−.0990535	−.0968552	−.0946491	−.0924354
11.4	−.0790083	−.0767484	−.0744828	−.0722117	−.0699353
11.5	−.0561801	−.0538735	−.0515636	−.0492505	−.0469344
11.6	−.0329897	−.0306597	−.0283286	−.0259967	−.0236642
11.7	−.0096694	−.0073391	−.0050101	−.0026825	−.0003567
11.8	.0135496	.0158571	.0181610	.0204612	.0227574
11.9	.0364386	.0387007	.0409570	.0432074	.0454516
12.0	.0587743	.0609690	.0631559	.0653346	.0675049
12.1	.0803406	.0824468	.0845431	.0866292	.0887050
12.2	.1009306	.1029283	.1049140	.1068877	.1088491
12.3	.1203489	.1222191	.1240756	.1259182	.1277467
12.4	.1384129	.1401382	.1418482	.1435426	.1452212
12.5	.1549550	.1565195	.1580671	.1595977	.1611111
12.6	.1698236	.1712131	.1725845	.1739374	.1752719
12.7	.1828851	.1840872	.1852700	.1864334	.1875772
12.8	.1940242	.1950286	.1960127	.1969764	.1979197
12.9	.2031460	.2039441	.2047213	.2054773	.2062123
13.0	.2101755	.2107612	.2113253	.2118679	.2123888
13.1	.2150594	.2154284	.2157755	.2161009	.2164043
13.2	.2177657	.2179159	.2180443	.2181508	.2182354
13.3	.2182839	.2182156	.2181256	.2180138	.2178803
13.4	.2166254	.2163409	.2160349	.2157076	.2153590
13.5	.2128227	.2123263	.2118092	.2112712	.2107125
13.6	.2069293	.2062276	.2055059	.2047641	.2040024
13.7	.1990189	.1981203	.1972026	.1962659	.1953102
13.8	.1891845	.1880993	.1869962	.1858751	.1847363
13.9	.1775373	.1762777	.1750014	.1737085	.1723991
14.0	.1642058	.1627855	.1613498	.1598991	.1584334
14.1	.1493340	.1477681	.1461885	.1445954	.1429890
14.2	.1330800	.1313851	.1296782	.1279596	.1262293
14.3	.1156147	.1138085	.1119921	.1101658	.1083297
14.4	.0971195	.0952206	.0933135	.0913984	.0894754
14.5	.0777848	.0758127	.0738344	.0718500	.0698599
14.6	.0578079	.0557827	.0537533	.0517198	.0496826
14.7	.0373914	.0353334	.0332733	.0312113	.0291476
14.8	.0167404	.0146704	.0126003	.0105303	.0084606
14.9	−.0039386	−.0060002	−.0080597	−.0101171	−.0121721

BESSEL FUNCTION J_0

T	.00	.01	.02	.03	.04
15.0	−.0142245	−.0162741	−.0183207	−.0203641	−.0224042
15.1	−.0345619	−.0365725	−.0385782	−.0405787	−.0425738
15.2	−.0544208	−.0563729	−.0583180	−.0602560	−.0621868
15.3	−.0736075	−.0754820	−.0773477	−.0792045	−.0810521
15.4	−.0919362	−.0937150	−.0954833	−.0972409	−.0989876
15.5	−.1092307	−.1108968	−.1125507	−.1141923	−.1158215
15.6	−.1253260	−.1268636	−.1283875	−.1298977	−.1313938
15.7	−.1400702	−.1414649	−.1428446	−.1442091	−.1455583
15.8	−.1533257	−.1545646	−.1557872	−.1569934	−.1581832
15.9	−.1649705	−.1660422	−.1670966	−.1681336	−.1691532
16.0	−.1748991	−.1757940	−.1766708	−.1775294	−.1783697
16.1	−.1830237	−.1837341	−.1844257	−.1850984	−.1857523
16.2	−.1892749	−.1897949	−.1902956	−.1907770	−.1912390
16.3	−.1936024	−.1939279	−.1942339	−.1945203	−.1947870
16.4	−.1959748	−.1961039	−.1962134	−.1963031	−.1963732
16.5	−.1963807	−.1963133	−.1962262	−.1961196	−.1959935
16.6	−.1948279	−.1945657	−.1942843	−.1939836	−.1936637
16.7	−.1913435	−.1908905	−.1904186	−.1899280	−.1894187
16.8	−.1859739	−.1853355	−.1846791	−.1840045	−.1833120
16.9	−.1787834	−.1779672	−.1771337	−.1762830	−.1754152
17.0	−.1698543	−.1688694	−.1678682	−.1668508	−.1658174
17.1	−.1592853	−.1581425	−.1569846	−.1558116	−.1546238
17.2	−.1471911	−.1459027	−.1446004	−.1432844	−.1419548
17.3	−.1337006	−.1322802	−.1308473	−.1294021	−.1279449
17.4	−.1189559	−.1174182	−.1158697	−.1143105	−.1127408
17.5	−.1031104	−.1014715	−.0998235	−.0981664	−.0965004
17.6	−.0863279	−.0846047	−.0828740	−.0811360	−.0793908
17.7	−.0687804	−.0669904	−.0651947	−.0633935	−.0615870
17.8	−.0506464	−.0488078	−.0469654	−.0451192	−.0432696
17.9	−.0321095	−.0302408	−.0283701	−.0264976	−.0246235
18.0	−.0133558	−.0114757	−.0095956	−.0077155	−.0058357
18.1	.0054270	.0072997	.0091706	.0110396	.0129064
18.2	.0240523	.0258990	.0277422	.0295815	.0314169
18.3	.0423358	.0441384	.0459356	.0477272	.0495131
18.4	.0600979	.0618387	.0635723	.0652987	.0670175
18.5	.0771648	.0788268	.0804800	.0821243	.0837595
18.6	.0933708	.0949380	.0964948	.0980411	.0995768
18.7	.1085595	.1100167	.1114621	.1128956	.1143171
18.8	.1225853	.1239187	.1252390	.1265461	.1278398
18.9	.1353152	.1365122	.1376949	.1388632	.1400170
19.0	.1466294	.1476788	.1487129	.1497316	.1507347
19.1	.1564231	.1573152	.1581912	.1590508	.1598942
19.2	.1646067	.1653335	.1660435	.1667364	.1674124
19.3	.1711073	.1716625	.1722003	.1727206	.1732233
19.4	.1758692	.1762481	.1766092	.1769525	.1772779
19.5	.1788538	.1790536	.1792354	.1793991	.1795449
19.6	.1800407	.1800603	.1800618	.1800454	.1800109
19.7	.1794272	.1792673	.1790895	.1788939	.1786806
19.8	.1770286	.1766917	.1763373	.1759654	.1755761
19.9	.1728777	.1723681	.1718414	.1712978	.1707374

BESSEL FUNCTION J_0

T	.05	.06	.07	.08	.09
15.0	−.0244406	−.0264732	−.0285019	−.0305263	−.0325464
15.1	−.0445634	−.0465472	−.0485250	−.0504967	−.0524620
15.2	−.0641100	−.0660256	−.0679333	−.0698330	−.0717245
15.3	−.0828904	−.0847192	−.0865384	−.0883477	−.0901471
15.4	−.1007233	−.1024478	−.1041610	−.1058626	−.1075525
15.5	−.1174380	−.1190418	−.1206326	−.1222103	−.1237748
15.6	−.1328759	−.1343438	−.1357973	−.1372363	−.1386606
15.7	−.1468921	−.1482104	−.1495130	−.1507998	−.1520708
15.8	−.1593565	−.1605130	−.1616528	−.1627757	−.1638816
15.9	−.1701552	−.1711396	−.1721063	−.1730551	−.1739861
16.0	−.1791917	−.1799952	−.1807802	−.1815467	−.1822945
16.1	−.1863871	−.1870029	−.1875996	−.1881773	−.1887357
16.2	−.1916815	−.1921047	−.1925084	−.1928926	−.1932572
16.3	−.1950342	−.1952616	−.1954694	−.1956576	−.1958260
16.4	−.1964235	−.1964543	−.1964653	−.1964567	−.1964285
16.5	−.1958479	−.1956827	−.1954981	−.1952941	−.1950707
16.6	−.1933247	−.1929665	−.1925892	−.1921930	−.1917777
16.7	−.1888908	−.1883442	−.1877792	−.1871958	−.1865940
16.8	−.1826015	−.1818732	−.1811272	−.1803635	−.1795822
16.9	−.1745303	−.1736286	−.1727100	−.1717747	−.1708227
17.0	−.1647680	−.1637027	−.1626217	−.1615250	−.1604129
17.1	−.1534212	−.1522039	−.1509721	−.1497260	−.1484656
17.2	−.1406118	−.1392556	−.1378862	−.1365038	−.1351086
17.3	−.1264757	−.1249947	−.1235021	−.1219979	−.1204825
17.4	−.1111607	−.1095704	−.1079700	−.1063598	−.1047399
17.5	−.0948257	−.0931425	−.0914510	−.0897512	−.0880435
17.6	−.0776387	−.0758799	−.0741145	−.0723426	−.0705645
17.7	−.0597754	−.0579587	−.0561374	−.0543114	−.0524810
17.8	−.0414167	−.0395607	−.0377018	−.0358402	−.0339760
17.9	−.0227480	−.0208713	−.0189935	−.0171149	−.0152356
18.0	−.0039563	−.0020776	−.0001997	.0016771	.0035528
18.1	.0147709	.0166330	.0184923	.0203487	.0222021
18.2	.0332481	.0350750	.0368974	.0387152	.0405280
18.3	.0512931	.0530669	.0548345	.0565956	.0583501
18.4	.0687288	.0704323	.0721278	.0738151	.0754942
18.5	.0853855	.0870020	.0886089	.0902061	.0917935
18.6	.1011017	.1026157	.1041186	.1056103	.1070906
18.7	.1157264	.1171234	.1185079	.1198798	.1212390
18.8	.1291201	.1303868	.1316397	.1328789	.1341041
18.9	.1411562	.1422807	.1433904	.1444851	.1455648
19.0	.1517223	.1526942	.1536503	.1545905	.1555148
19.1	.1607211	.1615315	.1623253	.1631025	.1638630
19.2	.1680713	.1687130	.1693375	.1699448	.1705347
19.3	.1737084	.1741760	.1746258	.1750580	.1754725
19.4	.1775854	.1778749	.1781466	.1784003	.1786361
19.5	.1796726	.1797823	.1798739	.1799475	.1800031
19.6	.1799585	.1798881	.1797998	.1796935	.1795693
19.7	.1784494	.1782006	.1779340	.1776498	.1773480
19.8	.1751695	.1747456	.1743044	.1738459	.1733704
19.9	.1701602	.1695663	.1689557	.1683285	.1676848

BESSEL FUNCTION J_0

T	.00	.01	.02	.03	.04
20.0	.1670247	.1663482	.1656554	.1649464	.1642212
20.1	.1595361	.1587003	.1578491	.1569825	.1561007
20.2	.1504946	.1495086	.1485082	.1474935	.1464645
20.3	.1399977	.1388722	.1377333	.1365812	.1354160
20.4	.1281571	.1269038	.1256384	.1243611	.1230720
20.5	.1150970	.1137290	.1123504	.1109612	.1095615
20.6	.1009532	.0994848	.0980071	.0965203	.0950246
20.7	.0858717	.0843178	.0827563	.0811872	.0796108
20.8	.0700069	.0683835	.0667540	.0651187	.0634776
20.9	.0535204	.0518439	.0501631	.0484780	.0467889
21.0	.0365791	.0348665	.0331512	.0314334	.0297133
21.1	.0193536	.0176221	.0158896	.0141564	.0124226
21.2	.0020167	.0002836	−.0014488	−.0031802	−.0049105
21.3	−.0152587	−.0169762	−.0186913	−.0204036	−.0221131
21.4	−.0323011	−.0339860	−.0356667	−.0373431	−.0390149
21.5	−.0489420	−.0505777	−.0522076	−.0538314	−.0554492
21.6	−.0650179	−.0665883	−.0681513	−.0697068	−.0712545
21.7	−.0803714	−.0818612	−.0833421	−.0848140	−.0862767
21.8	−.0948530	−.0962478	−.0976323	−.0990063	−.1003699
21.9	−.1083229	−.1096091	−.1108838	−.1121469	−.1133981
22.0	−.1206515	−.1218169	−.1229697	−.1241096	−.1252366
22.1	−.1317214	−.1327550	−.1337749	−.1347809	−.1357731
22.2	−.1414282	−.1423203	−.1431977	−.1440605	−.1449085
22.3	−.1496814	−.1504237	−.1511507	−.1518622	−.1525582
22.4	−.1564055	−.1569913	−.1575612	−.1581151	−.1586529
22.5	−.1615403	−.1619646	−.1623724	−.1627639	−.1631389
22.6	−.1650419	−.1653010	−.1655435	−.1657694	−.1659785
22.7	−.1668827	−.1669748	−.1670502	−.1671089	−.1671509
22.8	−.1670515	−.1669765	−.1668849	−.1667766	−.1666517
22.9	−.1655542	−.1653135	−.1650564	−.1647829	−.1644931
23.0	−.1624128	−.1620096	−.1615903	−.1611551	−.1607040
23.1	−.1576658	−.1571047	−.1565282	−.1559363	−.1553290
23.2	−.1513673	−.1506547	−.1499273	−.1491852	−.1484286
23.3	−.1435868	−.1427303	−.1418600	−.1409758	−.1400779
23.4	−.1344080	−.1334168	−.1324128	−.1313959	−.1303663
23.5	−.1239282	−.1228128	−.1216856	−.1205467	−.1193963
23.6	−.1122573	−.1110293	−.1097907	−.1085417	−.1072823
23.7	−.0995165	−.0981886	−.0968514	−.0955051	−.0941498
23.8	−.0858371	−.0844229	−.0830008	−.0815710	−.0801337
23.9	−.0713592	−.0698732	−.0683807	−.0668820	−.0653773
24.0	−.0562303	−.0546874	−.0531398	−.0515874	−.0500306
24.1	−.0406036	−.0390195	−.0374323	−.0358420	−.0342487
24.2	−.0246367	−.0230274	−.0214166	−.0198042	−.0181905
24.3	−.0084900	−.0068717	−.0052533	−.0036351	−.0020171
24.4	.0076750	.0092863	.0108961	.0125040	.0141101
24.5	.0236974	.0252857	.0268708	.0284525	.0300307
24.6	.0394183	.0409677	.0425125	.0440524	.0455872
24.7	.0546823	.0561777	.0576669	.0591498	.0606261
24.8	.0693393	.0707661	.0721851	.0735964	.0749998
24.9	.0832460	.0845901	.0859252	.0872512	.0885679

BESSEL FUNCTION J_0

T	.05	.06	.07	.08	.09
20.0	.1634800	.1627228	.1619498	.1611609	.1603563
20.1	.1552037	.1542916	.1533645	.1524226	.1514659
20.2	.1454214	.1443643	.1432932	.1422083	.1411098
20.3	.1342379	.1330469	.1318432	.1306269	.1293982
20.4	.1217712	.1204589	.1191352	.1178002	.1164541
20.5	.1081516	.1067316	.1053016	.1038618	.1024123
20.6	.0935202	.0920071	.0904855	.0889557	.0874177
20.7	.0780272	.0764366	.0748391	.0732348	.0716240
20.8	.0618310	.0601790	.0585217	.0568595	.0551923
20.9	.0450960	.0433993	.0416991	.0399956	.0382888
21.0	.0279911	.0262669	.0245408	.0228132	.0210841
21.1	.0106883	.0089538	.0072192	.0054847	.0037505
21.2	−.0066394	−.0083669	−.0100928	−.0118168	−.0135388
21.3	−.0238196	−.0255230	−.0272229	−.0289194	−.0306122
21.4	−.0406821	−.0423444	−.0440018	−.0456539	−.0473007
21.5	−.0570606	−.0586656	−.0602640	−.0618556	−.0634403
21.6	−.0727945	−.0743264	−.0758502	−.0773658	−.0788729
21.7	−.0877302	−.0891742	−.0906086	−.0920333	−.0934482
21.8	−.1017228	−.1030649	−.1043961	−.1057162	−.1070252
21.9	−.1146375	−.1158648	−.1170800	−.1182829	−.1194734
22.0	−.1263506	−.1274514	−.1285390	−.1296133	−.1306741
22.1	−.1367512	−.1377151	−.1386649	−.1396004	−.1405215
22.2	−.1457416	−.1465597	−.1473628	−.1481509	−.1489238
22.3	−.1532387	−.1539035	−.1545527	−.1551861	−.1558037
22.4	−.1591747	−.1596803	−.1601697	−.1606428	−.1610997
22.5	−.1634974	−.1638394	−.1641649	−.1644738	−.1647662
22.6	−.1661710	−.1663467	−.1665058	−.1666481	−.1667738
22.7	−.1671761	−.1671846	−.1671764	−.1671515	−.1671098
22.8	−.1665101	−.1663520	−.1661773	−.1659861	−.1657784
22.9	−.1641870	−.1638645	−.1635258	−.1631710	−.1627999
23.0	−.1602370	−.1597541	−.1592555	−.1587412	−.1582113
23.1	−.1547065	−.1540688	−.1534159	−.1527480	−.1520651
23.2	−.1476574	−.1468718	−.1460719	−.1452576	−.1444293
23.3	−.1391664	−.1382414	−.1373029	−.1363511	−.1353861
23.4	−.1293241	−.1282695	−.1272024	−.1261232	−.1250317
23.5	−.1182343	−.1170611	−.1158767	−.1146811	−.1134747
23.6	−.1060127	−.1047331	−.1034435	−.1021441	−.1008351
23.7	−.0927856	−.0914128	−.0900313	−.0886415	−.0872434
23.8	−.0786889	−.0772369	−.0757778	−.0743117	−.0728388
23.9	−.0638667	−.0623503	−.0608283	−.0593009	−.0577681
24.0	−.0484694	−.0469040	−.0453345	−.0437612	−.0421842
24.1	−.0326527	−.0310540	−.0294530	−.0278496	−.0262441
24.2	−.0165757	−.0149599	−.0133433	−.0117260	−.0101082
24.3	−.0003997	.0012172	.0028332	.0044483	.0060623
24.4	.0157141	.0173158	.0189152	.0205120	.0221062
24.5	.0316054	.0331762	.0347430	.0363057	.0378642
24.6	.0471169	.0486412	.0501601	.0516733	.0531807
24.7	.0620958	.0635587	.0650146	.0664635	.0679051
24.8	.0763951	.0777822	.0791610	.0805313	.0818930
24.9	.0898752	.0911731	.0924613	.0937397	.0950082

BESSEL FUNCTION J_0

T	.00	.01	.02	.03	.04
25.0	.0962668	.0975152	.0987534	.0999812	.1011985
25.1	.1082757	.1094164	.1105457	.1116634	.1127696
25.2	.1191571	.1201791	.1211887	.1221858	.1231703
25.3	.1288072	.1297009	.1305812	.1314482	.1323016
25.4	.1371348	.1378917	.1386345	.1393632	.1400777
25.5	.1440622	.1446753	.1452738	.1458574	.1464263
25.6	.1495258	.1499896	.1504383	.1508718	.1512900
25.7	.1534770	.1537875	.1540825	.1543621	.1546260
25.8	.1558824	.1560371	.1561761	.1562995	.1564071
25.9	.1567239	.1567219	.1567041	.1566707	.1566216
26.0	.1559993	.1558411	.1556673	.1554781	.1552734
26.1	.1537218	.1534096	.1530822	.1527396	.1523819
26.2	.1499200	.1494575	.1489803	.1484883	.1479817
26.3	.1446375	.1440300	.1434082	.1427724	.1421225
26.4	.1379327	.1371866	.1364271	.1356543	.1348682
26.5	.1298776	.1290010	.1281119	.1272102	.1262962
26.6	.1205577	.1195597	.1185503	.1175293	.1164970
26.7	.1100704	.1089615	.1078422	.1067125	.1055726
26.8	.0985245	.0973162	.0960986	.0948719	.0936361
26.9	.0860391	.0847437	.0834404	.0821291	.0808102
27.0	.0727419	.0713727	.0699969	.0686145	.0672259
27.1	.0587684	.0573393	.0559049	.0544655	.0530212
27.2	.0442603	.0427857	.0413073	.0398254	.0383400
27.3	.0293640	.0278587	.0263513	.0248417	.0233302
27.4	.0142293	.0127085	.0111870	.0096650	.0081426
27.5	−.0009922	−.0025133	−.0040336	−.0055529	−.0070711
27.6	−.0161486	−.0176548	−.0191587	−.0206602	−.0221590
27.7	−.0310890	−.0325655	−.0340381	−.0355069	−.0369715
27.8	−.0456656	−.0470977	−.0485245	−.0499460	−.0513619
27.9	−.0597344	−.0611079	−.0624749	−.0638351	−.0651885
28.0	−.0731570	−.0744586	−.0757523	−.0770380	−.0783155
28.1	−.0858021	−.0870191	−.0882269	−.0894254	−.0906146
28.2	−.0975466	−.0986670	−.0997772	−.1008770	−.1019663
28.3	−.1082765	−.1092896	−.1102915	−.1112819	−.1122609
28.4	−.1178886	−.1187847	−.1196686	−.1205403	−.1213995
28.5	−.1262911	−.1270617	−.1278193	−.1285638	−.1292952
28.6	−.1334046	−.1340423	−.1346665	−.1352769	−.1358737
28.7	−.1391625	−.1396616	−.1401466	−.1406174	−.1410739
28.8	−.1435124	−.1438684	−.1442098	−.1445367	−.1448490
28.9	−.1464158	−.1466256	−.1468206	−.1470008	−.1471663
29.0	−.1478488	−.1479107	−.1479578	−.1479901	−.1480077
29.1	−.1478020	−.1477161	−.1476154	−.1475000	−.1473698
29.2	−.1462813	−.1460487	−.1458017	−.1455402	−.1452642
29.3	−.1433065	−.1429303	−.1425399	−.1421354	−.1417168
29.4	−.1389125	−.1383968	−.1378675	−.1373245	−.1367680
29.5	−.1331479	−.1324983	−.1318357	−.1311601	−.1304717
29.6	−.1260746	−.1252980	−.1245092	−.1237083	−.1228952
29.7	−.1177675	−.1168722	−.1159655	−.1150475	−.1141184
29.8	−.1083137	−.1073089	−.1062938	−.1052683	−.1042327
29.9	−.0978112	−.0967073	−.0955941	−.0944718	−.0933404

BESSEL FUNCTION J_0

T	.05	.06	.07	.08	.09
25.0	.1024052	.1036012	.1047863	.1059606	.1071237
25.1	.1138641	.1149467	.1160175	.1170762	.1181228
25.2	.1241421	.1251011	.1260472	.1269803	.1279003
25.3	.1331415	.1339678	.1347803	.1355790	.1363639
25.4	.1407778	.1414636	.1421350	.1427920	.1434344
25.5	.1469803	.1475194	.1480436	.1485527	.1490468
25.6	.1516929	.1520805	.1524527	.1528096	.1531510
25.7	.1548744	.1551072	.1553244	.1555261	.1557120
25.8	.1564991	.1565755	.1566361	.1566811	.1567103
25.9	.1565569	.1564766	.1563807	.1562691	.1561420
26.0	.1550532	.1548177	.1545667	.1543004	.1540187
26.1	.1520090	.1516211	.1512182	.1508004	.1503676
26.2	.1474605	.1469248	.1463745	.1458099	.1452308
26.3	.1414587	.1407810	.1400895	.1393842	.1386652
26.4	.1340689	.1332565	.1324311	.1315928	.1307416
26.5	.1253699	.1244315	.1234809	.1225183	.1215439
26.6	.1154534	.1143986	.1133328	.1122561	.1111686
26.7	.1044225	.1032625	.1020925	.1009128	.0997234
26.8	.0923915	.0911380	.0898759	.0886053	.0873263
26.9	.0794837	.0781497	.0768084	.0754599	.0741044
27.0	.0658310	.0644300	.0630232	.0616105	.0601922
27.1	.0515721	.0501184	.0486602	.0471977	.0457310
27.2	.0368513	.0353596	.0338648	.0323671	.0308668
27.3	.0218170	.0203021	.0187857	.0172680	.0157491
27.4	.0066199	.0050970	.0035743	.0020517	.0005295
27.5	−.0085880	−.0101035	−.0116175	−.0131298	−.0146402
27.6	−.0236551	−.0251483	−.0266384	−.0281254	−.0296090
27.7	−.0384319	−.0398880	−.0413395	−.0427864	−.0442285
27.8	−.0527722	−.0541768	−.0555754	−.0569680	−.0583543
27.9	−.0665348	−.0678741	−.0692060	−.0705306	−.0718476
28.0	−.0795847	−.0808455	−.0820977	−.0833414	−.0845762
28.1	−.0917943	−.0929644	−.0941248	−.0952754	−.0964160
28.2	−.1030451	−.1041131	−.1051704	−.1062168	−.1072522
28.3	−.1132283	−.1141841	−.1151281	−.1160602	−.1169804
28.4	−.1222464	−.1230807	−.1239024	−.1247114	−.1255077
28.5	−.1300134	−.1307184	−.1314101	−.1320884	−.1327532
28.6	−.1364566	−.1370257	−.1375809	−.1381221	−.1386494
28.7	−.1415162	−.1419442	−.1423579	−.1427572	−.1431420
28.8	−.1451468	−.1454299	−.1456984	−.1459522	−.1461914
28.9	−.1473170	−.1474530	−.1475741	−.1476805	−.1477720
29.0	−.1480104	−.1479983	−.1479714	−.1479297	−.1478733
29.1	−.1472250	−.1470655	−.1468913	−.1467026	−.1464992
29.2	−.1449738	−.1446690	−.1443498	−.1440163	−.1436686
29.3	−.1412842	−.1408377	−.1403771	−.1399028	−.1394145
29.4	−.1361980	−.1356145	−.1350177	−.1344077	−.1337843
29.5	−.1297704	−.1290564	−.1283297	−.1275905	−.1268387
29.6	−.1220701	−.1212331	−.1203842	−.1195236	−.1186514
29.7	−.1131781	−.1122268	−.1112647	−.1102917	−.1093080
29.8	−.1031870	−.1021313	−.1010658	−.0999905	−.0989056
29.9	−.0922000	−.0910508	−.0898929	−.0887264	−.0875514

BESSEL FUNCTION J_0

T	.00	.01	.02	.03	.04
30.0	−.0863680	−.0851764	−.0839766	−.0827689	−.0815533
30.1	−.0741014	−.0728342	−.0715601	−.0702793	−.0689919
30.2	−.0611363	−.0598064	−.0584709	−.0571301	−.0557839
30.3	−.0476042	−.0462250	−.0448417	−.0434543	−.0420631
30.4	−.0336418	−.0322273	−.0308101	−.0293902	−.0279678
30.5	−.0193897	−.0179541	−.0165171	−.0150790	−.0136398
30.6	−.0049909	−.0035485	−.0021062	−.0006641	.0007775
30.7	.0094108	.0108456	.0122788	.0137103	.0151400
30.8	.0236718	.0250846	.0264945	.0279013	.0293049
30.9	.0376503	.0390273	.0403999	.0417680	.0431315
31.0	.0512082	.0525356	.0538574	.0551733	.0564834
31.1	.0642115	.0654763	.0667343	.0679851	.0692287
31.2	.0765325	.0777224	.0789042	.0800776	.0812426
31.3	.0880507	.0891540	.0902480	.0913326	.0924077
31.4	.0986537	.0996597	.1006553	.1016406	.1026154

BESSEL FUNCTION J_0

T	.05	.06	.07	.08	.09
30.0	−.0803300	−.0790990	−.0778605	−.0766147	−.0753616
30.1	−.0676981	−.0663979	−.0650914	−.0637790	−.0624605
30.2	−.0544327	−.0530764	−.0517153	−.0503495	−.0489791
30.3	−.0406681	−.0392695	−.0378674	−.0364620	−.0350535
30.4	−.0265431	−.0251162	−.0236873	−.0222565	−.0208239
30.5	−.0121997	−.0107588	−.0093174	−.0078755	−.0064332
30.6	.0022186	.0036590	.0050986	.0065371	.0079746
30.7	.0165677	.0179934	.0194167	.0208376	.0222560
30.8	.0307051	.0321017	.0334947	.0348839	.0362691
30.9	.0444903	.0458442	.0471930	.0485367	.0498752
31.0	.0577873	.0590850	.0603764	.0616614	.0629398
31.1	.0704650	.0716939	.0729152	.0741288	.0753347
31.2	.0823992	.0835472	.0846864	.0858168	.0869383
31.3	.0934732	.0945292	.0955752	.0966114	.0976377
31.4	.1035796	.1045332	.1054759	.1064079	.1073289

BESSEL FUNCTION J_1

T	.00	.01	.02	.03	.04
0.0	0.0000000	.0049999	.0099995	.0149983	.0199960
.1	.0499375	.0549169	.0598921	.0648628	.0698286
.2	.0995008	.1044223	.1093358	.1142412	.1191381
.3	.1483188	.1531455	.1579607	.1627641	.1675553
.4	.1960266	.2007225	.2054034	.2100689	.2147188
.5	.2422685	.2467987	.2513105	.2558035	.2602774
.6	.2867010	.2910319	.2953412	.2996284	.3038932
.7	.3289957	.3330955	.3371705	.3412203	.3452448
.8	.3688420	.3726806	.3764916	.3802745	.3840292
.9	.4059495	.4094991	.4130184	.4165071	.4199649
1.0	.4400506	.4432858	.4464882	.4496577	.4527939
1.1	.4709024	.4738003	.4766634	.4794913	.4822840
1.2	.4982891	.5008297	.5033336	.5058006	.5082305
1.3	.5220232	.5241895	.5263174	.5284070	.5304580
1.4	.5419477	.5437255	.5454638	.5471625	.5488215
1.5	.5579365	.5593150	.5606532	.5619508	.5632079
1.6	.5698959	.5708676	.5717984	.5726881	.5735368
1.7	.5777652	.5783259	.5788453	.5793235	.5797604
1.8	.5815170	.5816656	.5817731	.5818396	.5818649
1.9	.5811571	.5808962	.5805946	.5802523	.5798695
2.0	.5767248	.5760601	.5753554	.5746109	.5738267
2.1	.5682921	.5672326	.5661342	.5649970	.5638212
2.2	.5559630	.5545208	.5530410	.5515239	.5499696
2.3	.5398725	.5380627	.5362170	.5343358	.5324190
2.4	.5201853	.5180259	.5158327	.5136058	.5113456
2.5	.4970941	.4946060	.4920863	.4895351	.4869528
2.6	.4708183	.4680247	.4652020	.4623503	.4594700
2.7	.4416014	.4385280	.4354281	.4323020	.4291500
2.8	.4097092	.4063837	.4030346	.3996622	.3962667
2.9	.3754275	.3718794	.3683108	.3647218	.3611130
3.0	.3390590	.3353194	.3315626	.3277886	.3239979
3.1	.3009211	.2970226	.2931100	.2891837	.2852440
3.2	.2613432	.2573192	.2532845	.2492394	.2451844
3.3	.2206635	.2165481	.2124255	.2082960	.2041599
3.4	.1792259	.1750538	.1708779	.1666987	.1625163
3.5	.1373775	.1331835	.1289892	.1247949	.1206010
3.6	.0954655	.0912841	.0871059	.0829311	.0787602
3.7	.0538340	.0496993	.0455712	.0414500	.0373359
3.8	.0128210	.0087665	.0047218	.0006872	−.0033369
3.9	−.0272440	−.0311861	−.0351151	−.0390308	−.0429330
4.0	−.0660433	−.0698418	−.0736243	−.0773905	−.0811401
4.1	−.1032733	−.1068987	−.1105054	−.1140928	−.1176609
4.2	−.1386469	−.1420717	−.1454750	−.1488565	−.1522160
4.3	−.1718966	−.1750950	−.1782695	−.1814198	−.1845457
4.4	−.2027755	−.2057242	−.2086467	−.2115429	−.2144125
4.5	−.2310604	−.2337384	−.2363882	−.2390097	−.2416027
4.6	−.2565528	−.2589416	−.2613006	−.2636296	−.2659284
4.7	−.2790807	−.2811647	−.2832174	−.2852387	−.2872286
4.8	−.2984999	−.3002661	−.3019999	−.3037013	−.3053702
4.9	−.3146947	−.3161332	−.3175386	−.3189107	−.3202495

BESSEL FUNCTION J_1

T	.05	.06	.07	.08	.09
0.0	.0249922	.0299865	.0349786	.0399680	.0449545
.1	.0747893	.0797443	.0846933	.0896360	.0945720
.2	.1240260	.1289046	.1337735	.1386325	.1434810
.3	.1723340	.1770997	.1818522	.1865911	.1913160
.4	.2193525	.2239699	.2285705	.2331540	.2377201
.5	.2647318	.2691665	.2735811	.2779752	.2823486
.6	.3081355	.3123547	.3165506	.3207230	.3248715
.7	.3492436	.3532164	.3571629	.3610829	.3649760
.8	.3877554	.3914529	.3951213	.3987603	.4023699
.9	.4233917	.4267871	.4301509	.4334829	.4367829
1.0	.4558968	.4589660	.4620014	.4650027	.4679698
1.1	.4850413	.4877629	.4904486	.4930984	.4957119
1.2	.5106233	.5129786	.5152965	.5175766	.5198189
1.3	.5324703	.5344439	.5363785	.5382741	.5401305
1.4	.5504407	.5520200	.5535593	.5550586	.5565177
1.5	.5644245	.5656003	.5667354	.5678298	.5688833
1.6	.5743443	.5751108	.5758362	.5765204	.5771634
1.7	.5801562	.5805107	.5808241	.5810962	.5813272
1.8	.5818493	.5817926	.5816951	.5815566	.5813772
1.9	.5794463	.5789825	.5784784	.5779341	.5773495
2.0	.5730028	.5721393	.5712364	.5702942	.5693127
2.1	.5626069	.5613543	.5600635	.5587345	.5573677
2.2	.5483784	.5467502	.5450854	.5433841	.5416464
2.3	.5304671	.5284801	.5264582	.5244016	.5223106
2.4	.5090521	.5067256	.5043663	.5019745	.4995503
2.5	.4843396	.4816957	.4790214	.4763168	.4735824
2.6	.4565613	.4536245	.4506598	.4476676	.4446480
2.7	.4259723	.4227693	.4195412	.4162882	.4130109
2.8	.3928485	.3894079	.3859452	.3824607	.3789547
2.9	.3574845	.3538368	.3501700	.3464846	.3427808
3.0	.3201909	.3163677	.3125289	.3086746	.3048052
3.1	.2812912	.2773257	.2733478	.2693579	.2653563
3.2	.2411197	.2370457	.2329627	.2288711	.2247712
3.3	.2000177	.1958696	.1917161	.1875574	.1833938
3.4	.1583313	.1541439	.1499545	.1457634	.1415709
3.5	.1164079	.1122159	.1080253	.1038365	.0996498
3.6	.0745934	.0704312	.0662737	.0621215	.0579748
3.7	.0332293	.0291307	.0250402	.0209582	.0168850
3.8	−.0073502	−.0113524	−.0153432	−.0193223	−.0232894
3.9	−.0468212	−.0506953	−.0545548	−.0583995	−.0622291
4.0	−.0848728	−.0885884	−.0922865	−.0959669	−.0996292
4.1	−.1212093	−.1247378	−.1282461	−.1317339	−.1352009
4.2	−.1555532	−.1588679	−.1621598	−.1654287	−.1686744
4.3	−.1876469	−.1907233	−.1937745	−.1968005	−.1998009
4.4	−.2172552	−.2200710	−.2228596	−.2256209	−.2283545
4.5	−.2441671	−.2467026	−.2492091	−.2516864	−.2541344
4.6	−.2681970	−.2704352	−.2726428	−.2748196	−.2769657
4.7	−.2891868	−.2911133	−.2930080	−.2948707	−.2967014
4.8	−.3070064	−.3086098	−.3101805	−.3117182	−.3132230
4.9	−.3215549	−.3228269	−.3240653	−.3252702	−.3264415

BESSEL FUNCTION J_1

T	.00	.01	.02	.03	.04
5.0	−.3275791	−.3286831	−.3297533	−.3307898	−.3317925
5.1	−.3370972	−.3378626	−.3385940	−.3392915	−.3399550
5.2	−.3432230	−.3436489	−.3440409	−.3443991	−.3447234
5.3	−.3459608	−.3460493	−.3461043	−.3461259	−.3461140
5.4	−.3453448	−.3451009	−.3448242	−.3445147	−.3441725
5.5	−.3414382	−.3408699	−.3402696	−.3396376	−.3389739
5.6	−.3343328	−.3334507	−.3325379	−.3315946	−.3306208
5.7	−.3241477	−.3229651	−.3217534	−.3205126	−.3192429
5.8	−.3110277	−.3095607	−.3080661	−.3065442	−.3049952
5.9	−.2951424	−.2934090	−.2916501	−.2898658	−.2880563
6.0	−.2766839	−.2747045	−.2727017	−.2706758	−.2686269
6.1	−.2558648	−.2536618	−.2514378	−.2491929	−.2469275
6.2	−.2329166	−.2305141	−.2280930	−.2256536	−.2231961
6.3	−.2080869	−.2055105	−.2029180	−.2003100	−.1976865
6.4	−.1816375	−.1789139	−.1761771	−.1734274	−.1706650
6.5	−.1538413	−.1509984	−.1481451	−.1452818	−.1424086
6.6	−.1249802	−.1220465	−.1191054	−.1161571	−.1132019
6.7	−.0953421	−.0923468	−.0893469	−.0863427	−.0833346
6.8	−.0652187	−.0621909	−.0591615	−.0561307	−.0530989
6.9	−.0349021	−.0318710	−.0288412	−.0258130	−.0227866
7.0	−.0046828	−.0016773	.0013241	.0043211	.0073134
7.1	.0251533	.0281050	.0310498	.0339875	.0369177
7.2	.0543274	.0571980	.0600589	.0629100	.0657511
7.3	.0825704	.0853335	.0880844	.0908230	.0935488
7.4	.1096251	.1122557	.1148718	.1174730	.1200593
7.5	.1352484	.1377232	.1401811	.1426220	.1450456
7.6	.1592138	.1615109	.1637892	.1660484	.1682883
7.7	.1813127	.1834125	.1854916	.1875497	.1895868
7.8	.2013569	.2032417	.2051041	.2069439	.2087611
7.9	.2191794	.2208337	.2224642	.2240708	.2256533
8.0	.2346363	.2360471	.2374329	.2387936	.2401291
8.1	.2476078	.2487643	.2498950	.2509996	.2520782
8.2	.2579986	.2588928	.2597605	.2606016	.2614159
8.3	.2657393	.2663657	.2669651	.2675375	.2680829
8.4	.2707863	.2711419	.2714704	.2717718	.2720460
8.5	.2731220	.2732065	.2732640	.2732946	.2732981
8.6	.2727548	.2725706	.2723596	.2721221	.2718580
8.7	.2697190	.2692707	.2687964	.2682961	.2677699
8.8	.2640737	.2633687	.2626384	.2618831	.2611028
8.9	.2559024	.2549502	.2539740	.2529738	.2519497
9.0	.2453118	.2441244	.2429143	.2416816	.2404263
9.1	.2324307	.2310222	.2295925	.2281416	.2266698
9.2	.2174087	.2157950	.2141618	.2125092	.2108375
9.3	.2004139	.1986130	.1967943	.1949582	.1931047
9.4	.1816322	.1796634	.1776789	.1756789	.1736637
9.5	.1612644	.1591486	.1570192	.1548765	.1527208
9.6	.1395248	.1372840	.1350319	.1327688	.1304950
9.7	.1166386	.1142959	.1119443	.1095840	.1072154
9.8	.0928401	.0904193	.0879920	.0855585	.0831189
9.9	.0683698	.0658953	.0634167	.0609343	.0584484

BESSEL FUNCTION J_1

T	.05	.06	.07	.08	.09
5.0	−.3327613	−.3336963	−.3345974	−.3354646	−.3362979
5.1	−.3405846	−.3411802	−.3417418	−.3422695	−.3427632
5.2	−.3450140	−.3452707	−.3454938	−.3456831	−.3458388
5.3	−.3460688	−.3459903	−.3458785	−.3457337	−.3455557
5.4	−.3437977	−.3433905	−.3429508	−.3424788	−.3419746
5.5	−.3382786	−.3375518	−.3367938	−.3360045	−.3351841
5.6	−.3296168	−.3285826	−.3275185	−.3264245	−.3253009
5.7	−.3179445	−.3166176	−.3152623	−.3138787	−.3124672
5.8	−.3034193	−.3018166	−.3001874	−.2985318	−.2968501
5.9	−.2862220	−.2843629	−.2824793	−.2805715	−.2786396
6.0	−.2665553	−.2644612	−.2623449	−.2602066	−.2580464
6.1	−.2446417	−.2423358	−.2400101	−.2376649	−.2353003
6.2	−.2207209	−.2182281	−.2157181	−.2131910	−.2106472
6.3	−.1950479	−.1923944	−.1897264	−.1870440	−.1843476
6.4	−.1678903	−.1651035	−.1623049	−.1594949	−.1566736
6.5	−.1395260	−.1366341	−.1337333	−.1308238	−.1279060
6.6	−.1102401	−.1072720	−.1042978	−.1013179	−.0983326
6.7	−.0803228	−.0773076	−.0742893	−.0712681	−.0682445
6.8	−.0500663	−.0470332	−.0440000	−.0409669	−.0379341
6.9	−.0197623	−.0167404	−.0137213	−.0107051	−.0076922
7.0	.0103007	.0132828	.0162594	.0192302	.0221949
7.1	.0398402	.0427547	.0456609	.0485586	.0514476
7.2	.0685817	.0714017	.0742109	.0770089	.0797955
7.3	.0962619	.0989617	.1016482	.1043211	.1069802
7.4	.1226303	.1251857	.1277255	.1302494	.1327571
7.5	.1474518	.1498404	.1522110	.1545636	.1568979
7.6	.1705088	.1727096	.1748906	.1770516	.1791923
7.7	.1916026	.1935970	.1955697	.1975208	.1994499
7.8	.2105554	.2123267	.2140749	.2157999	.2175014
7.9	.2272116	.2287457	.2302553	.2317403	.2332007
8.0	.2414393	.2427241	.2439835	.2452173	.2464254
8.1	.2531307	.2541570	.2551569	.2561306	.2570778
8.2	.2622036	.2629644	.2636985	.2644056	.2650859
8.3	.2686012	.2690924	.2695566	.2699936	.2704035
8.4	.2722931	.2725131	.2727059	.2728717	.2730104
8.5	.2732747	.2732244	.2731472	.2730432	.2729124
8.6	.2715674	.2712504	.2709069	.2705372	.2701412
8.7	.2672179	.2666402	.2660368	.2654079	.2647535
8.8	.2602976	.2594677	.2586131	.2577339	.2568303
8.9	.2509019	.2498306	.2487357	.2476176	.2464762
9.0	.2391487	.2378489	.2365270	.2351833	.2338178
9.1	.2251772	.2236640	.2221304	.2205765	.2190026
9.2	.2091467	.2074370	.2057087	.2039620	.2021970
9.3	.1912342	.1893468	.1874427	.1855221	.1835852
9.4	.1716335	.1695884	.1675288	.1654548	.1633666
9.5	.1505523	.1483711	.1461775	.1439718	.1417542
9.6	.1282106	.1259159	.1236111	.1212965	.1189722
9.7	.1048385	.1024537	.1000612	.0976613	.0952542
9.8	.0806737	.0782229	.0757669	.0733059	.0708401
9.9	.0559592	.0534670	.0509720	.0484745	.0459746

BESSEL FUNCTION J_1

T	.00	.01	.02	.03	.04
10.0	.0434727	.0409691	.0384638	.0359573	.0334497
10.1	.0183955	.0158874	.0133801	.0108741	.0083694
10.2	−.0066157	−.0091038	−.0115886	−.0140698	−.0165471
10.3	−.0313178	−.0337618	−.0362001	−.0386324	−.0410586
10.4	−.0554728	−.0578492	−.0602176	−.0625779	−.0649296
10.5	−.0788500	−.0811364	−.0834125	−.0856782	−.0879333
10.6	−.1012287	−.1034034	−.1055659	−.1077159	−.1098532
10.7	−.1223994	−.1244424	−.1264711	−.1284855	−.1304852
10.8	−.1421666	−.1440590	−.1459354	−.1477956	−.1496394
10.9	−.1603497	−.1620744	−.1637815	−.1654708	−.1671422
11.0	−.1767853	−.1783270	−.1798496	−.1813530	−.1828371
11.1	−.1913283	−.1926735	−.1939984	−.1953028	−.1965868
11.2	−.2038531	−.2049904	−.2061063	−.2072008	−.2082738
11.3	−.2142550	−.2151751	−.2160729	−.2169486	−.2178019
11.4	−.2224506	−.2231462	−.2238192	−.2244693	−.2250966
11.5	−.2283786	−.2288450	−.2292883	−.2297085	−.2301055
11.6	−.2320005	−.2322350	−.2324463	−.2326344	−.2327992
11.7	−.2333002	−.2333026	−.2332817	−.2332378	−.2331707
11.8	−.2322847	−.2320568	−.2318060	−.2315324	−.2312361
11.9	−.2289832	−.2285292	−.2280528	−.2275541	−.2270334
12.0	−.2234471	−.2227732	−.2220777	−.2213608	−.2206225
12.1	−.2157490	−.2148637	−.2139578	−.2130314	−.2120846
12.2	−.2059820	−.2048957	−.2037900	−.2026649	−.2015206
12.3	−.1942588	−.1929838	−.1916907	−.1903795	−.1890506
12.4	−.1807102	−.1792605	−.1777942	−.1763114	−.1748122
12.5	−.1654838	−.1638750	−.1622513	−.1606127	−.1589594
12.6	−.1487423	−.1469916	−.1452276	−.1434505	−.1416606
12.7	−.1306622	−.1287877	−.1269019	−.1250050	−.1230971
12.8	−.1114316	−.1094528	−.1074646	−.1054674	−.1034612
12.9	−.0912483	−.0891854	−.0871153	−.0850381	−.0829541
13.0	−.0703181	−.0681921	−.0660609	−.0639249	−.0617841
13.1	−.0488525	−.0466847	−.0445140	−.0423405	−.0401645
13.2	−.0270667	−.0248789	−.0226902	−.0205009	−.0183113
13.3	−.0051775	−.0029912	−.0008063	.0013771	.0035587
13.4	.0165990	.0187622	.0209219	.0230780	.0252301
13.5	.0380493	.0401683	.0422817	.0443894	.0464911
13.6	.0589646	.0610188	.0630655	.0651044	.0671353
13.7	.0791428	.0811125	.0830728	.0850233	.0869640
13.8	.0983905	.1002570	.1021121	.1039558	.1057877
13.9	.1165249	.1182704	.1200029	.1217222	.1234282
14.0	.1333752	.1349834	.1365770	.1381560	.1397201
14.1	.1487844	.1502404	.1516805	.1531045	.1545122
14.2	.1626107	.1639013	.1651747	.1664308	.1676695
14.3	.1747291	.1758426	.1769380	.1780149	.1790734
14.4	.1850317	.1859585	.1868661	.1877546	.1886237
14.5	.1934295	.1941616	.1948740	.1955664	.1962389
14.6	.1998527	.2003843	.2008956	.2013866	.2018572
14.7	.2042513	.2045785	.2048851	.2051711	.2054365
14.8	.2065956	.2067165	.2068167	.2068963	.2069553
14.9	.2068762	.2067910	.2066853	.2065590	.2064124

BESSEL FUNCTION ~~J_0~~ J_1

T	.05	.06	.07	.08	.09
10.0	.0309412	.0284322	.0259229	.0234135	.0209043
10.1	.0058663	.0033652	.0008662	−.0016305	−.0041246
10.2	−.0190205	−.0214895	−.0239540	−.0264137	−.0288684
10.3	−.0434783	−.0458914	−.0482976	−.0506967	−.0530885
10.4	−.0672727	−.0696068	−.0719318	−.0742475	−.0765537
10.5	−.0901775	−.0924107	−.0946326	−.0968431	−.0990418
10.6	−.1119776	−.1140889	−.1161870	−.1182715	−.1203424
10.7	−.1324701	−.1344401	−.1363949	−.1383343	−.1402583
10.8	−.1514668	−.1532774	−.1550711	−.1568479	−.1586075
10.9	−.1687954	−.1704305	−.1720471	−.1736452	−.1752247
11.0	−.1843017	−.1857467	−.1871720	−.1885774	−.1899629
11.1	−.1978500	−.1990926	−.2003142	−.2015150	−.2026946
11.2	−.2093252	−.2103549	−.2113628	−.2123488	−.2133129
11.3	−.2186329	−.2194415	−.2202277	−.2209912	−.2217322
11.4	−.2257010	−.2262825	−.2268410	−.2273766	−.2278891
11.5	−.2304793	−.2308300	−.2311575	−.2314617	−.2317427
11.6	−.2329407	−.2330591	−.2331542	−.2332261	−.2332747
11.7	−.2330806	−.2329674	−.2328312	−.2326720	−.2324898
11.8	−.2309170	−.2305754	−.2302111	−.2298243	−.2294150
11.9	−.2264905	−.2259255	−.2253387	−.2247299	−.2240994
12.0	−.2198629	−.2190821	−.2182803	−.2174574	−.2166136
12.1	−.2111175	−.2101303	−.2091230	−.2080958	−.2070488
12.2	−.2003572	−.1991749	−.1979738	−.1967540	−.1955156
12.3	−.1877039	−.1863397	−.1849580	−.1835591	−.1821432
12.4	−.1732969	−.1717656	−.1702185	−.1686557	−.1670774
12.5	−.1572917	−.1556097	−.1539136	−.1522036	−.1504797
12.6	−.1398581	−.1380431	−.1362158	−.1343765	−.1325252
12.7	−.1211786	−.1192494	−.1173100	−.1153604	−.1134008
12.8	−.1014463	−.0994229	−.0973912	−.0953513	−.0933036
12.9	−.0808634	−.0787663	−.0766630	−.0745538	−.0724387
13.0	−.0596388	−.0574892	−.0553356	−.0531781	−.0510170
13.1	−.0379861	−.0358056	−.0336232	−.0314391	−.0292535
13.2	−.0161215	−.0139317	−.0117422	−.0095532	−.0073649
13.3	.0057383	.0079157	.0100907	.0122630	.0144326
13.4	.0273781	.0295218	.0316610	.0337954	.0359249
13.5	.0485866	.0506758	.0527583	.0548341	.0569029
13.6	.0691581	.0711725	.0731784	.0751755	.0771637
13.7	.0888946	.0908150	.0927250	.0946243	.0965129
13.8	.1076079	.1094160	.1112119	.1129955	.1147665
13.9	.1251207	.1267995	.1284645	.1301156	.1317525
14.0	.1412691	.1428030	.1443216	.1458248	.1473124
14.1	.1559036	.1572785	.1586368	.1599783	.1613030
14.2	.1688906	.1700940	.1712797	.1724475	.1735973
14.3	.1801133	.1811346	.1821372	.1831209	.1840858
14.4	.1894735	.1903038	.1911147	.1919059	.1926775
14.5	.1968915	.1975240	.1981364	.1987287	.1993008
14.6	.2023074	.2027371	.2031464	.2035352	.2039035
14.7	.2056813	.2059054	.2061089	.2062918	.2064540
14.8	.2069936	.2070113	.2070084	.2069849	.2069408
14.9	.2062452	.2060577	.2058498	.2056215	.2053729

BESSEL FUNCTION J_1

T	.00	.01	.02	.03	.04
15.0	.2051040	.2048149	.2045057	.2041762	.2038267
15.1	.2013102	.2008214	.2003130	.1997849	.1992373
15.2	.1955454	.1948631	.1941618	.1934416	.1927027
15.3	.1878794	.1870115	.1861255	.1852216	.1842998
15.4	1784003	.1773565	.1762958	.1752183	.1741239
15.5	.1672132	.1660051	.1647812	.1635418	.1622868
15.6	.1544396	.1530801	.1517062	.1503181	.1489160
15.7	.1402157	.1387193	.1372099	.1356878	.1341533
15.8	.1246913	.1230735	.1214444	.1198043	.1181532
15.9	.1080279	.1063054	.1045735	.1028322	.1010818
16.0	.0903972	.0885878	.0867707	.0849461	.0831142
16.1	.0719794	.0701015	.0682178	.0663284	.0644336
16.2	.0529615	.0510340	.0491027	.0471677	.0452291
16.3	.0335351	.0315774	.0296179	.0276565	.0256936
16.4	.0138947	.0119264	.0099581	.0079901	.0060224
16.5	−.0057642	−.0077236	−.0096811	−.0116364	−.0135893
16.6	−.0252471	−.0271783	−.0291056	−.0310289	−.0329479
16.7	−.0443624	−.0462465	−.0481248	−.0499972	−.0518635
16.8	−.0629232	−.0647418	−.0665528	−.0683562	−.0701516
16.9	−.0807493	−.0824847	−.0842110	−.0859278	−.0876351
17.0	−.0976685	−.0993042	−.1009291	−.1025429	−.1041456
17.1	−.1135188	−.1150392	−.1165472	−.1180427	−.1195255
17.2	−.1281497	−.1295403	−.1309172	−.1322802	−.1336293
17.3	−.1414233	−.1426712	−.1439041	−.1451219	−.1463246
17.4	−.1532162	−.1543097	−.1553872	−.1564487	−.1574939
17.5	−.1634200	−.1643493	−.1652617	−.1661571	−.1670354
17.6	−.1719427	−.1726995	−.1734387	−.1741601	−.1748638
17.7	−.1787096	−.1792874	−.1798469	−.1803883	−.1809113
17.8	−.1836635	−.1840575	−.1844329	−.1847898	−.1851280
17.9	−.1867654	−.1869728	−.1871614	−.1873313	−.1874824
18.0	−.1879949	−.1880146	−.1880156	−.1879979	−.1879614
18.1	−.1873502	−.1871831	−.1869975	−.1867933	−.1865706
18.2	−.1848479	−.1844967	−.1841273	−.1837398	−.1833341
18.3	.1805231	−.1799922	−.1794437	−.1788776	−.1782940
18.4	−.1744283	−.1737241	−.1730029	−.1722648	−.1715100
18.5	−.1666336	−.1657639	−.1648780	−.1639762	−.1630586
18.6	−.1572254	−.1561996	−.1551588	−.1541031	−.1530325
18.7	−.1463053	−.1451345	−.1439499	−.1427515	−.1415396
18.8	−.1339897	−.1326863	−.1313703	−.1300419	−.1287013
18.9	−.1204080	−.1189855	−.1175520	−.1161074	−.1146521
19.0	−.1057014	−.1041747	−.1026383	−.1010925	−.0995375
19.1	−.0900216	−.0884062	−.0867828	−.0851517	−.0835128
19.2	−.0735290	−.0718414	−.0701476	−.0684476	−.0667417
19.3	−.0563913	−.0546486	−.0529015	−.0511499	−.0493942
19.4	−.0387816	−.0370015	−.0352186	−.0334331	−.0316452
19.5	−.0208771	−.0190773	−.0172765	−.0154749	−.0136727
19.6	−.0028566	−.0010551	.0007456	.0025454	.0043439
19.7	.0151006	.0168860	.0186688	.0204488	.0222259
19.8	.0328168	.0345684	.0363156	.0380584	.0397965
19.9	.0501174	.0518181	.0535127	.0552011	.0568832

BESSEL FUNCTION J_1

T	.05	.06	.07	.08	.09
15.0	.2034571	.2030675	.2026580	.2022286	.2017793
15.1	.1986703	.1980838	.1974780	.1968530	.1962088
15.2	.1919450	.1911688	.1903740	.1895608	.1887292
15.3	.1833603	.1824032	.1814285	.1804363	.1794269
15.4	.1730130	.1718855	.1707417	.1695816	.1684054
15.5	.1610165	.1597310	.1584304	.1571149	.1557846
15.6	.1474999	.1460700	.1446265	.1431695	.1416992
15.7	.1326063	.1310471	.1294758	.1278927	.1262978
15.8	.1164915	.1148192	.1131365	.1114436	.1097407
15.9	.0993224	.0975542	.0957774	.0939922	.0921987
16.0	.0812751	.0794291	.0775764	.0757170	.0738513
16.1	.0625336	.0606285	.0587186	.0568040	.0548849
16.2	.0432872	.0413423	.0393944	.0374437	.0354906
16.3	.0237294	.0217640	.0197976	.0178305	.0158628
16.4	.0040553	.0020891	.0001238	−.0018403	−.0038031
16.5	−.0155397	−.0174874	−.0194322	−.0213739	−.0233122
16.6	−.0348625	−.0367725	−.0386776	−.0405778	−.0424728
16.7	−.0537236	−.0555771	−.0574241	−.0592642	−.0610973
16.8	−.0719390	−.0737182	−.0754890	−.0772512	−.0790047
16.9	−.0893326	−.0910202	−.0926978	−.0943651	−.0960221
17.0	−.1057370	−.1073169	−.1088852	−.1104417	−.1119863
17.1	−.1209956	−.1224528	−.1238970	−.1253279	−.1267455
17.2	−.1349642	−.1362850	−.1375914	−.1388833	−.1401607
17.3	−.1475120	−.1486840	−.1498406	−.1509815	−.1521067
17.4	−.1585228	−.1595354	−.1605315	−.1615110	−.1624738
17.5	−.1678966	−.1687405	−.1695672	−.1703765	−.1711684
17.6	−.1755497	−.1762176	−.1768677	−.1774997	−.1781137
17.7	−.1814160	−.1819024	−.1823704	−.1828199	−.1832509
17.8	−.1854476	−.1857486	−.1860308	−.1862944	−.1865392
17.9	−.1876147	−.1877283	−.1878231	−.1878991	−.1879564
18.0	−.1879062	−.1878323	−.1877397	−.1876285	−.1874987
18.1	−.1863295	−.1860699	−.1857919	−.1854956	−.1851809
18.2	−.1829104	−.1824688	−.1820091	−.1815316	−.1810362
18.3	−.1776930	−.1770745	−.1764388	−.1757858	−.1751156
18.4	−.1707385	−.1699503	−.1691457	−.1683247	−.1674873
18.5	−.1621252	−.1611761	−.1602115	−.1592314	−.1582360
18.6	−.1519474	−.1508476	−.1497334	−.1486049	−.1474621
18.7	−.1403142	−.1390754	−.1378235	−.1365584	−.1352805
18.8	−.1273486	−.1259838	−.1246073	−.1232190	−.1218192
18.9	−.1131861	−.1117096	−.1102227	−.1087256	−.1072185
19.0	−.0979733	−.0964002	−.0948183	−.0932278	−.0916289
19.1	−.0818666	−.0802130	−.0785523	−.0768846	−.0752101
19.2	−.0650301	−.0633128	−.0615901	−.0598622	−.0581292
19.3	−.0476344	−.0458709	−.0441036	−.0423329	−.0405588
19.4	−.0298551	−.0280629	−.0262688	−.0244730	−.0226757
19.5	−.0118701	−.0100672	−.0082642	−.0064613	−.0046587
19.6	.0061411	.0079368	.0097307	.0115228	.0133128
19.7	.0239998	.0257705	.0275377	.0293012	.0310610
19.8	.0415298	.0432580	.0449810	.0466988	.0484109
19.9	.0585587	.0602276	.0618897	.0635447	.0651926

BESSEL FUNCTION J_1

T	.00	.01	.02	.03	.04
20.0	.0668331	.0684662	.0700916	.0717092	.0733189
20.1	.0828010	.0843506	.0858911	.0874222	.0889438
20.2	.0978664	.0993176	.1007583	.1021881	.1036071
20.3	.1118844	.1132233	.1145503	.1158652	.1171679
20.4	.1247210	.1259349	.1271356	.1283231	.1294972
20.5	.1362547	.1373321	.1383953	.1394442	.1404787
20.6	.1463774	.1473083	.1482241	.1491247	.1500099
20.7	.1549955	.1557714	.1565314	.1572754	.1580034
20.8	.1620307	.1626447	.1632421	.1638229	.1643871
20.9	.1674209	.1678675	.1682972	.1687099	.1691055
21.0	.1711203	.1713960	.1716544	.1718957	.1721196
21.1	.1731003	.1732031	.1732886	.1733568	.1734076
21.2	.1733493	.1732790	.1731915	.1730868	.1729649
21.3	.1718730	.1716312	.1713724	.1710966	.1708039
21.4	.1686941	.1682839	.1678572	.1674139	.1669541
21.5	.1638521	.1632784	.1626887	.1620831	.1614616
21.6	.1574027	.1566720	.1559260	.1551648	.1543884
21.7	.1494174	.1485376	.1476434	.1467348	.1458121
21.8	.1399825	.1389630	.1379301	.1368840	.1358246
21.9	.1291982	.1280498	.1268892	.1257164	.1245316
22.0	.1171778	.1159125	.1146362	.1133490	.1120511
22.1	.1040461	.1026770	.1012982	.0999099	.0985123
22.2	.0899387	.0884797	.0870126	.0855375	.0840545
22.3	.0749998	.0734659	.0719255	.0703785	.0688253
22.4	.0593815	.0577884	.0561902	.0545871	.0529793
22.5	.0432420	.0416056	.0399658	.0383228	.0366766
22.6	.0267439	.0250807	.0234157	.0217492	.0200812
22.7	.0100524	.0083791	.0067056	.0050322	.0033591
22.8	−.0066657	−.0083325	−.0099978	−.0116614	−.0133231
22.9	−.0232443	−.0248881	−.0265288	−.0281660	−.0297998
23.0	−.0395193	−.0411239	−.0427237	−.0443186	−.0459083
23.1	−.0553305	−.0568801	−.0584233	−.0599601	−.0614902
23.2	−.0705228	−.0720022	−.0734738	−.0749374	−.0763929
23.3	−.0849479	−.0863427	−.0877284	−.0891046	−.0904714
23.4	−.0984658	−.0997626	−.1010489	−.1023245	−.1035894
23.5	−.1109461	−.1121324	−.1133070	−.1144697	−.1156206
23.6	−.1222693	−.1233337	−.1243854	−.1254242	−.1264500
23.7	−.1323277	−.1332602	−.1341790	−.1350840	−.1359752
23.8	−.1410266	−.1418185	−.1425959	−.1433588	−.1441070
23.9	−.1482855	−.1489295	−.1495583	−.1501720	−.1507705
24.0	−.1540381	−.1545284	−.1550031	−.1554621	−.1559054
24.1	−.1582335	−.1585659	−.1588823	−.1591828	−.1594672
24.2	−.1608365	−.1610084	−.1611641	−.1613036	−.1614270
24.3	−.1618278	−.1618381	−.1618321	−.1618101	−.1617718
24.4	−.1612042	−.1610534	−.1608865	−.1607037	−.1605049
24.5	−.1589784	−.1586687	−.1583432	−.1580021	−.1576454
24.6	−.1551791	−.1547142	−.1542340	−.1537387	−.1532282
24.7	−.1498504	−.1492356	−.1486061	−.1479620	−.1473034
24.8	−.1430514	−.1422934	−.1415214	−.1407357	−.1399362
24.9	− 1348557	−.1339625	−.1330564	−.1321373	−.1312054

BESSEL FUNCTION J_1

T	.05	.06	.07	.08	.09
20.0	.0749204	.0765137	.0780985	.0796748	.0812423
20.1	.0904558	.0919581	.0934504	.0949326	.0964047
20.2	.1050150	.1064117	.1077972	.1091712	.1105336
20.3	.1184583	.1197362	.1210015	.1222542	.1234941
20.4	.1306578	.1318048	.1329381	.1340576	.1351631
20.5	.1414986	.1425040	.1434946	.1444704	.1454314
20.6	.1508798	.1517342	.1525730	.1533962	.1542037
20.7	.1587152	.1594109	.1600904	.1607535	.1614003
20.8	.1649347	.1654655	.1659796	.1664769	.1669573
20.9	.1694841	.1698456	.1701900	.1705173	.1708274
21.0	.1723263	.1725157	.1726878	.1728426	.1729801
21.1	.1734412	.1734574	.1734563	.1734379	.1734022
21.2	.1728258	.1726695	.1724960	.1723054	.1720977
21.3	.1704943	.1701678	.1698245	.1694644	.1690876
21.4	.1664779	.1659852	.1654763	.1649511	.1644097
21.5	.1608242	.1601711	.1595023	.1588180	.1581181
21.6	.1535970	.1527906	.1519694	.1511334	.1502827
21.7	.1448752	.1439243	.1429595	.1419808	.1409885
21.8	.1347522	.1336669	.1325687	.1314578	.1303342
21.9	.1233350	.1221266	.1209065	.1196749	.1184320
22.0	.1107426	.1094237	.1080944	.1067550	.1054055
22.1	.0971055	.0956897	.0942649	.0928314	.0913893
22.2	.0825638	.0810655	.0795598	.0780469	.0765268
22.3	.0672658	.0657003	.0641290	.0625520	.0609695
22.4	.0513669	.0497501	.0481290	.0465039	.0448748
22.5	.0350275	.0333756	.0317212	.0300643	.0284051
22.6	.0184119	.0167415	.0150702	.0133982	.0117255
22.7	.0016863	.0000141	−.0016574	−.0033279	−.0049974
22.8	−.0149827	−.0166401	−.0182951	−.0199476	−.0215974
22.9	−.0314299	−.0330561	−.0346783	−.0362964	−.0379101
23.0	−.0474927	−.0490718	−.0506452	−.0522129	−.0537747
23.1	−.0630134	−.0645298	−.0660391	−.0675411	−.0690357
23.2	−.0778402	−.0792791	−.0807094	−.0821311	−.0835440
23.3	−.0918286	−.0931761	−.0945136	−.0958412	−.0971586
23.4	−.1048434	−.1060864	−.1073183	−.1085390	−.1097483
23.5	−.1167594	−.1178861	−.1190005	−.1201026	−.1211922
23.6	−.1274628	−.1284624	−.1294488	−.1304219	−.1313815
23.7	−.1368525	−.1377157	−.1385648	−.1393997	−.1402203
23.8	−.1448406	−.1455593	−.1462633	−.1469523	−.1476264
23.9	−.1513536	−.1519214	−.1524738	−.1530107	−.1535322
24.0	−.1563330	−.1567448	−.1571408	−.1575209	−.1578852
24.1	−.1597356	−.1599880	−.1602243	−.1604445	−.1606486
24.2	−.1615342	−.1616253	−.1617002	−.1617589	−.1618015
24.3	−.1617175	−.1616470	−.1615604	−.1614577	−.1613390
24.4	−.1602902	−.1600595	−.1598130	−.1595506	−.1592724
24.5	−.1572730	−.1568852	−.1564818	−.1560629	−.1556287
24.6	−.1527026	−.1521619	−.1516064	−.1510359	−.1504505
24.7	−.1466304	−.1459431	−.1452414	−.1445255	−.1437955
24.8	−.1391230	−.1382963	−.1374562	−.1366026	−.1357357
24.9	−.1302607	−.1293035	−.1283337	−.1273515	−.1263570

BESSEL FUNCTION J_1

T	.00	.01	.02	.03	.04
25.0	−.1253502	−.1243314	−.1233006	−.1222578	−.1212033
25.1	−.1146348	−.1135009	−.1123561	−.1112005	−.1100344
25.2	−.1028206	−.1015834	−.1003365	−.0990801	−.0978143
25.3	−.0900295	−.0887017	−.0873655	−.0860211	−.0846686
25.4	−.0763926	−.0749876	−.0735757	−.0721571	−.0707318
25.5	−.0620485	−.0605808	−.0591076	−.0576290	−.0561453
25.6	−.0471429	−.0456272	−.0441075	−.0425840	−.0410569
25.7	−.0318259	−.0302775	−.0287266	−.0271735	−.0256183
25.8	−.0162515	−.0146859	−.0131194	−.0115523	−.0099846
25.9	−.0005755	.0009916	.0025581	.0041237	.0056883
26.0	.0150457	.0165989	.0181497	.0196982	.0212441
26.1	.0304572	.0319809	.0335009	.0350169	.0365288
26.2	.0455065	.0469857	.0484597	.0499283	.0513914
26.3	.0600453	.0614655	.0628791	.0642858	.0656857
26.4	.0739309	.0752783	.0766176	.0779488	.0792716
26.5	.0870278	.0882891	.0895412	.0907838	.0920169
26.6	.0992087	.1003718	.1015244	.1026665	.1037979
26.7	.1103559	.1114095	.1124517	.1134822	.1145010
26.8	.1203624	.1212965	.1222182	.1231273	.1240238
26.9	.1291329	.1299386	.1307311	.1315103	.1322760
27.0	.1365847	.1372546	.1379105	.1385524	.1391803
27.1	.1426487	.1431765	.1436899	.1441887	.1446729
27.2	.1472696	.1476507	.1480168	.1483681	.1487043
27.3	.1504068	.1506378	.1508537	.1510544	.1512400
27.4	.1520345	.1521137	.1521776	.1522264	.1522600
27.5	1521419	.1520691	.1519811	.1518779	.1517597
27.6	.1507335	.1505100	.1502714	.1500180	.1497496
27.7	.1478289	.1474573	.1470712	.1466705	.1462553
27.8	.1434622	.1429469	.1424175	.1418741	.1413166
27.9	.1376822	.1370288	.1363619	.1356817	.1349881
28.0	.1305515	.1297669	.1289697	.1281599	.1273375
28.1	.1221457	.1212383	.1203191	.1193882	.1184457
28.2	.1125531	.1115323	.1105007	.1094584	.1084056
28.3	.1018732	.1007495	.0996162	.0984734	.0973211
28.4	.0902161	.0890012	.0877778	.0865461	.0853062
28.5	.0777014	.0764076	.0751066	.0737985	.0724836
28.6	.0644564	.0630968	.0617315	.0603604	.0589838
28.7	.0506155	.0492040	.0477880	.0463678	.0449434
28.8	.0363185	.0348692	.0334170	.0319619	.0305041
28.9	.0217093	.0202368	.0187628	.0172874	.0158109
29.0	.0069342	.0054533	.0039723	.0024914	.0010108
29.1	−.0078591	−.0093337	−.0108069	−.0122786	−.0137485
29.2	−.0225233	−.0239770	−.0254278	−.0268756	−.0283202
29.3	−.0369130	−.0383314	−.0397454	−.0411550	−.0425600
29.4	−.0508859	−.0522549	−.0536183	−.0549758	−.0563273
29.5	−.0643044	−.0656106	−.0669098	−.0682019	−.0694868
29.6	−.0770368	−.0782674	−.0794898	−.0807039	−.0819095
29.7	−.0889586	−.0901017	−.0912354	−.0923596	−.0934742
29.8	−.0999541	−.1009985	−.1020325	−.1030559	−.1040687
29.9	−.1099168	−.1108525	−.1117768	−.1126896	−.1135909

BESSEL FUNCTION J_1

T	.05	.06	.07	.08	.09
25.0	−.1201371	−.1190593	−.1179701	−.1168695	−.1157577
25.1	−.1088577	−.1076706	−.1064732	−.1052656	−.1040481
25.2	−.0965392	−.0952550	−.0939618	−.0926597	−.0913489
25.3	−.0833083	−.0819401	−.0805643	−.0791811	−.0777904
25.4	−.0693000	−.0678618	−.0664174	−.0649670	−.0635107
25.5	−.0546566	−.0531630	−.0516647	−.0501618	−.0486544
25.6	−.0395262	−.0379922	−.0364551	−.0349148	−.0333717
25.7	−.0240612	−.0225022	−.0209416	−.0193795	−.0178161
25.8	−.0084164	−.0068481	−.0052797	−.0037114	−.0021433
25.9	.0072517	.0088138	.0103744	.0119333	.0134905
26.0	.0227873	.0243276	.0258649	.0273991	.0289299
26.1	.0380365	.0395398	.0410387	.0425328	.0440221
26.2	.0528487	.0543003	.0557458	.0571853	.0586185
26.3	.0670784	.0684639	.0698420	.0712126	.0725757
26.4	.0805861	.0818920	.0831893	.0844778	.0857573
26.5	.0932404	.0944540	.0956578	.0968516	.0980353
26.6	.1049185	.1060282	.1071269	.1082144	.1092908
26.7	.1155080	.1165030	.1174861	.1184571	.1194159
26.8	.1249076	.1257786	.1266367	.1274818	.1283139
26.9	.1330283	.1337670	.1344920	.1352034	.1359010
27.0	.1397940	.1403935	.1409788	.1415498	.1421065
27.1	.1451425	.1455974	.1460376	.1464631	.1468738
27.2	.1490257	.1493320	.1496233	.1498995	.1501607
27.3	.1514104	.1515656	.1517056	.1518304	.1519401
27.4	.1522783	.1522814	.1522693	.1522420	.1521996
27.5	.1516263	.1514779	.1513143	.1511358	.1509421
27.6	.1494665	.1491684	.1488556	.1485281	.1481858
27.7	.1458256	.1453815	.1449231	.1444504	.1439634
27.8	.1407453	.1401601	.1395612	.1389485	.1383222
27.9	.1342813	.1335613	.1328283	.1320822	.1313233
28.0	.1265027	.1256556	.1247963	.1239248	.1230412
28.1	.1174918	.1165264	.1155498	.1145619	.1135630
28.2	.1073424	.1062688	.1051849	.1040910	.1029870
28.3	.0961595	.0949888	.0938089	.0926201	.0914225
28.4	.0840582	.0828022	.0815384	.0802669	.0789879
28.5	.0711619	.0698335	.0684987	.0671574	.0658099
28.6	.0576018	.0562145	.0548221	.0534247	.0520224
28.7	.0435151	.0420829	.0406470	.0392075	.0377646
28.8	.0290438	.0275811	.0261161	.0246491	.0231801
28.9	.0143332	.0128547	.0113754	.0098954	.0084150
29.0	−.0004694	−.0019490	−.0034280	−.0049060	−.0063831
29.1	−.0152165	−.0166825	−.0181463	−.0196078	−.0210669
29.2	−.0297615	−.0311993	−.0326335	−.0340639	−.0354905
29.3	−.0439603	−.0453557	−.0467461	−.0481314	−.0485113
29.4	−.0576728	−.0590121	−.0603450	−.0616715	−.0629913
29.5	−.0707643	−.0720343	−.0732966	−.0745513	−.0757980
29.6	−.0831065	−.0842948	−.0854742	−.0866448	−.0878063
29.7	−.0945791	−.0956741	−.0967593	−.0978344	−.0988994
29.8	−.1050708	−.1060620	−.1070423	−.1080116	−.1089698
29.9	−.1144805	−.1153584	−.1162245	−.1170787	−.1179209

BESSEL FUNCTION J_1

T	.00	.01	.02	.03	.04
30.0	−.1187511	−.1195691	−.1203749	−.1211684	−.1219495
30.1	−.1263727	−.1270653	−.1277450	−.1284116	−.1290653
30.2	−.1327098	−.1332705	−.1338177	−.1343514	−.1348714
30.3	−.1377037	−.1381273	−.1385370	−.1389328	−.1393145
30.4	−.1413090	−.1415918	−.1418604	−.1421147	−.1423547
30.5	−.1434943	−.1436339	−.1437592	−.1438701	−.1439666
30.6	−.1442426	−.1442382	−.1442194	−.1441861	−.1441385
30.7	−.1435512	−.1434032	−.1432409	−.1430644	−.1428736
30.8	−.1414315	−.1411419	−.1408382	−.1405206	−.1401891
30.9	−.1379094	−.1374814	−.1370399	−.1365848	−.1361162
31.0	−.1330243	−.1324628	−.1318882	−.1313006	−.1307001
31.1	−.1268294	−.1261403	−.1254388	−.1247250	−.1239990
31.2	−.1193904	−.1185810	−.1177600	−.1169275	−.1160836
31.3	−.1107855	−.1098642	−.1089323	−.1079897	−.1070368
31.4	−.1011040	−.1000804	−.0990471	−.0980043	−.0969520

BESSEL FUNCTION J_1

T	.05	.06	.07	.08	.09
30.0	−.1227182	−.1234744	−.1242180	−.1249490	−.1256672
30.1	−.1297058	−.1303332	−.1309473	−.1315482	−.1321357
30.2	−.1353778	−.1358705	−.1363495	−.1368148	−.1372661
30.3	−.1396822	−.1400358	−.1403753	−.1407007	−.1410119
30.4	−.1425805	−.1427920	−.1429891	−.1431719	−.1433403
30.5	−.1440486	−.1441162	−.1441695	−.1442083	−.1442327
30.6	−.1440765	−.1440001	−.1439094	−.1438043	−.1436849
30.7	−.1426686	−.1424495	−.1422161	−.1419687	−.1417071
30.8	−.1398436	−.1394843	−.1391112	−.1387243	−.1383237
30.9	−.1356342	−.1351388	−.1346300	−.1341080	−.1335727
31.0	−.1300868	−.1294606	−.1288217	−.1281702	−.1275060
31.1	−.1232608	−.1225106	−.1217483	−.1209742	−.1201882
31.2	−.1152284	−.1143619	−.1134843	−.1125956	−.1116960
31.3	−.1060733	−.1050996	−.1041158	−.1031218	−.1021178
31.4	−.0958903	−.0948194	−.0937394	−.0926504	−.0915524

HYPERBOLIC BESSEL FUNCTIONS

$$I_m(x) = i^{-m} J_m(ix)$$

x	$I_0(x)$	$I_1(x)$	$I_2(x)$
0.0	1.0000	0.0000	0.0000
0.1	1.0025	0.0501	0.0012
0.2	1.0100	0.1005	0.0050
0.4	1.0404	0.2040	0.0203
0.6	1.0920	0.3137	0.0464
0.8	1.1665	0.4329	0.0844
1.0	1.2661	0.5652	0.1357
1.2	1.3937	0.7147	0.2026
1.4	1.5534	0.8861	0.2875
1.6	1.7500	1.0848	0.3940
1.8	1.9896	1.3172	0.5260
2.0	2.2796	1.5906	0.6889
2.2	2.6291	1.9141	0.8891
2.4	3.0493	2.2981	1.1342
2.6	3.5533	2.7554	1.4337
2.8	4.1573	3.3011	1.7994
3.0	4.8808	3.9534	2.2452
3.2	5.7472	4.7343	2.7883
3.4	6.7848	5.6701	3.4495
3.6	8.0277	6.7927	4.2540
3.8	9.5169	8.1404	5.2325
4.0	11.302	9.7595	6.4222
4.2	13.442	11.706	7.8684
4.4	16.010	14.046	9.6258
4.6	19.093	16.863	11.761
4.8	22.794	20.253	14.355
5.0	27.240	24.336	17.506
5.2	32.584	29.254	21.332
5.4	39.009	35.182	25.978
5.6	46.738	42.328	31.620
5.8	56.038	50.946	38.470
6.0	67.234	61.342	46.787
6.2	80.718	73.886	56.884
6.4	96.962	89.026	69.141
6.6	116.54	107.30	84.021
6.8	140.14	129.38	102.08
7.0	168.59	156.04	124.01
7.2	202.92	188.25	150.63
7.4	244.34	227.17	182.94
7.6	294.33	274.22	222.17
7.8	354.68	331.10	269.79
8.0	427.56	399.87	327.60

Taken from Vibration and Sound with the permission of Philip Morse, author, and McGraw-Hill Book Company, Inc., publisher.

SURFACE ZONAL HARMONICS

$$P_n(x) = \frac{1}{2^n n!}\frac{d^n}{dx^n}(x^2-1)^n$$

x	$P_1(x)$	$P_2(x)$	$P_3(x)$	$P_4(x)$	$P_5(x)$
0	0	−.5	0	.375	0
.01	.01	−.49985	−.0149975	.37463	.018741
.02	.02	−.4994	−.02998	.37350	.037430
.03	.03	−.49865	−.0449325	.37163	.056014
.04	.04	−.4976	−.05984	.36901	.074441
.05	.05	−.49625	−.0746875	.36565	.092659
.06	.06	−.4946	−.08946	.36156	.11062
.07	.07	−.49265	−.10414	.35673	.12826
.08	.08	−.4904	−.11872	.35118	.14555
.09	.09	−.48785	−.13318	.34491	.16242
.10	.10	−.485	−.1475	.33794	.17883
.11	.11	−.48185	−.16167	.33027	.19473
.12	.12	−.4784	−.17568	.32191	.21008
.13	.13	−.47465	−.18951	.31287	.22482
.14	.14	−.4706	−.20314	.30318	.23891
.15	.15	−.46625	−.21656	.29284	.25232
.16	.16	−.4616	−.22976	.28187	.26499
.17	.17	−.45665	−.24272	.27028	.27688
.18	.18	−.4514	−.25542	.25809	.28796
.19	.19	−.44585	−.26785	.24533	.29818
.20	.20	−.44	−.28	.232	.30752
.21	.21	−.43385	−.29185	.21813	.31593
.22	.22	−.4274	−.30338	.20375	.32339
.23	.23	−.42065	−.31458	.18887	.32986
.24	.24	−.4136	−.32544	.17352	.33531
.25	.25	−.40625	−.33594	.15771	.33972
.26	.26	−.3986	−.34606	.14149	.34307
.27	.27	−.39065	−.35579	.12488	.34532
.28	.28	−.3824	−.36512	.10789	.34647
.29	.29	−.37385	−.37403	.09057	.34650
.30	.30	−.365	−.3825	.07294	.34539
.31	.31	−.35585	−.39052	.05503	.34312
.32	.32	−.3464	−.39808	.03688	.33970
.33	.33	−.33665	−.40516	.01851	.33512
.34	.34	−.3266	−.41174	−.00004	.32937
.35	.35	−.31625	−.41781	−.01872	.32245
.36	.36	−.3056	−.42336	−.03752	.31438
.37	.37	−.29465	−.42837	−.05638	.30514
.38	.38	−.2834	−.43282	−.07528	.29477
.39	.39	−.27185	−.43670	−.09416	.28326
.40	.40	−.26	−.44	−.113	.27064
.41	.41	−.24785	−.44270	−.13175	.25693
.42	.42	−.2354	−.44478	−.15036	.24215
.43	.43	−.22265	−.44623	−.16880	.22633
.44	.44	−.2096	−.44704	−.18702	.20951
.45	.45	−.19625	−.44719	−.20497	.19172
.46	.46	−.1826	−.44666	−.22261	.17301
.47	.47	−.16865	−.44544	−.23989	.15341
.48	.48	−.1544	−.44352	−.25676	.13298
.49	.49	−.13985	−.44088	−.27316	.11177

SURFACE ZONAL HARMONICS (Continued)

x	$P_6(x)$	$P_7(x)$	$P_8(x)$	$P_9(x)$	$P_{10}(x)$
0	−.3125	−0	.2734	0	−.2461
.01	−.31184	−.021855	.2725	.0246	−.2447
.02	−.30988	−.043593	.2695	.0489	−.2407
.03	−.30661	−.065094	.2646	.0729	−.2340
.04	−.30205	−.086244	.2578	.0961	−.2247
.05	−.29622	−.10693	.2492	.1186	−.2130
.06	−.28913	−.12703	.2387	.1400	−.1989
.07	−.28081	−.14644	.2265	.1601	−.1826
.08	−.27130	−.16506	.2126	.1789	−.1642
.09	−.26063	−.18278	.1972	.1960	−.1440
.10	−.24883	−.19949	.1803	.2114	−.1221
.11	−.23595	−.21511	.1621	.2249	−.0989
.12	−.22204	−.22955	.1426	.2364	−.0745
.13	−.20715	−.24271	.1221	.2457	−.0492
.14	−.19133	−.25453	.1006	.2529	−.0233
.15	−.17465	−.26492	.0783	.2577	.0020
.16	−.15716	−.27383	.0554	.2601	.0293
.17	−.13894	−.28119	.0319	.2602	.0553
18	−.12005	−.28695	.0082	.2579	.0808
.19	−.10057	−.29107	−.0157	.2531	.1055
.20	−.080576	−.29352	−.0396	.2460	.1291
.21	−.060144	−.29426	−.0632	.2365	.1513
.22	−.039357	−.29327	−.0865	.2247	.1718
.23	−.018300	−.29055	−.1093	.2108	.1905
.24	.002941	−.28610	−.1313	.1948	.2070
.25	.024277	−.27992	−.1525	.1768	.2212
.26	.045618	−.27203	−.1725	.1571	.2329
.27	.066872	−.26246	−.1914	.1357	.2419
.28	.087947	−.25124	−.2089	.1129	.2480
.29	.10875	−.23843	−.2248	.0888	.2513
.30	.12918	−.22407	−.2391	.0637	.2515
.31	.14915	−.20824	−.2515	.0378	.2487
.32	.16856	−.19100	−.2621	.0114	.2428
.33	.18732	−.17244	−.2706	−.0154	.2339
.34	.20534	−.15266	−.2770	−.0422	.2220
.35	.22251	−.13176	−.2812	−.0688	.2073
.36	.23875	−.10984	−.2831	−.0948	.1899
.37	.25397	−.087036	−.2826	−.1201	.1699
.38	.26808	−.063467	−.2798	−.1444	.1476
.39	.28100	−.039271	−.2746	−.1674	.1231
.40	.29264	−.014590	−.2670	−.1888	.0968
.41	.30291	.010424	−.2570	−.2083	.0690
.42	.31176	.035614	−.2447	−.2258	.0401
.43	.31909	.060820	−.2302	−.2410	.0103
.44	32486	.085873	−.2134	−.2537	−.0200
.45	.32898	.11060	−.1945	−.2637	−.0503
.46	.33141	.13483	−.1737	−.2708	−.0803
.47	.33209	.15838	−.1510	−.2749	−.1095
.48	.33098	.18107	−.1267	−.2758	−.1375
.49	.32804	.20272	−.1008	−.2735	−.1639

SURFACE ZONAL HARMONICS (Continued)

x	$P_1(x)$	$P_2(x)$	$P_3(x)$	$P_4(x)$	$P_5(x)$
0.50	0.50	−0.125	−0.4375	−0.28906	0.08984
.51	.51	−.10985	−.43337	−.30440	.06726
.52	.52	−.0944	−.42848	−.31912	.04409
.53	53	−.07865	−.42281	−.33317	.02041
.54	.54	−.0626	−.41634	−.34649	−.00372
55	.55	−.04625	−.40906	−.35904	−.02819
.56	.56	−.0296	−.40096	−.37074	−.05294
.57	.57	−.01265	−.39202	−.38155	−.07786
.58	.58	.0046	−.38222	−.39140	−.10285
.59	.59	.02215	−.37155	−.40024	−.12781
.60	.60	.04	−.36	−.408	−.15264
.61	.61	.05815	−.34755	−.41462	−.17721
.62	.62	.0766	−.33418	−.42004	−.20142
.63	.63	.09535	−.31988	−.42418	−.22512
.64	.64	.1144	−.30464	.42700	−.24819
.65	.65	.13375	−.28844	−.42841	−.27049
.66	.66	.1534	−.27126	−.42836	−.29188
.67	.67	.17335	−.25309	−.42676	−.31220
.68	.68	.1936	−.23392	−.42356	−.33131
.69	.69	.21415	−.21373	−.41869	−.34903
.70	.70	.235	−.1925	−.41206	−.36520
.71	.71	.25615	−.17022	−.40361	−.37964
.72	.72	.2776	−.14688	−.39327	−.39217
.73	.73	.29935	−.12246	−.38095	−.40260
.74	.74	.3214	−.09694	−.36659	−.41074
.75	.75	.34375	−.07031	−.35010	−.41638
.76	.76	.3664	−.04256	−.33140	−.41931
.77	.77	.38935	−.01367	−.31043	−.41932
.78	.78	.4126	.01638	−.28709	−.41618
.79	.79	.43615	.04760	−.26131	−.40966
.80	.80	.46	.08	−.233	−.39952
.81	.81	.48415	.11360	−.20208	−.38552
.82	.82	.5086	.14842	−.16847	−.36739
.83	.83	.53335	.18447	−.13207	−.34489
.84	.84	.5584	.22176	−.09281	−.31774
.85	.85	.58375	.26031	−.05060	−.28566
.86	.86	.6094	.30014	−.00534	−.24838
.87	.87	.63535	.34126	.04305	−.20559
.88	.88	.6616	.38368	.09467	−.15699
.89	.89	.68815	.42742	.14960	−.10228
.90	.90	.715	.4725	.20794	−.04114
.91	.91	.74215	.51893	.26978	0.02676
.92	.92	.7696	.56672	.33522	.10175
.93	.93	.79735	.61589	.40435	.18417
.94	.94	.8254	.66646	.47728	.27438
.95	.95	.85373	.71844	.55409	.37274
.96	.96	.8824	.77184	.63489	.47962
.97	.97	.91135	.82668	.71978	.59539
.98	.98	.9406	.88298	.80886	.72045
.99	.99	.97015	.94075	.90223	.85518
100	1	1	1	1	1

SURFACE ZONAL HARMONICS (Continued)

x	$P_6(x)$	$P_7(x)$	$P_8(x)$	$P_9(x)$	$P_{10}(x)$
.50	0.32324	0.22314	−0.0736	−0.2679	−0.1882
.51	.31655	.24217	−.0454	−.2590	−.2101
.52	.30796	.25961	−.0163	−.2468	−.2291
.53	.29747	.27530	.0133	−.2314	−.2450
.54	.28506	.28906	.0432	−.2128	−.2573
.55	.27077	.30074	.0732	−.1913	−.2658
.56	.25460	.31016	.1029	−.1669	−.2701
.57	.23660	.31719	.1320	−.1399	−.2702
.58	.21681	.32169	.1601	−.1105	−.2659
.59	.19528	.32353	.1870	−.0791	−.2570
.60	.17210	.32260	.2133	−.0450	−.2433
.61	.14733	.31880	.2357	−.0118	−.2258
.63	.12109	.31207	.2568	.0234	−.2036
.63	.093475	.30232	.2753	.0589	−.1773
.64	.064623	.28954	.2909	.0943	−.1471
.65	.034675	.27371	.3032	.1290	−.1136
.66	.003790	.25483	.3120	.1625	−.0771
.67	−.027853	.23295	.3170	.1941	−.0382
.68	−.060059	.20813	.3179	.2233	.0024
.69	−.092615	.18049	.3145	.2495	.0440
.70	−.12529	.15016	.3067	.2721	.0858
.71	−.15782	.11731	.2943	.2904	.1269
.72	−.18994	.082166	.2771	.3039	.1663
.73	−.22136	.044990	.2553	.3120	.2030
.74	−.25175	.006087	.2287	.3143	.2360
.75	−.28078	−.034184	.1976	.3103	.2644
.76	−.30807	−.075411	.1621	.2997	.2869
.77	−.33325	−.11713	.1225	.2823	.3027
.78	−.35589	−.15881	.0791	.2578	.3108
.79	−.37557	−.19987	.0326	.2263	.3103
.80	−.39180	−.23965	−.0167	.1879	.3005
.81	−.40409	−.27743	−.0678	.1429	.2810
.82	−.41193	−.31240	−.1199	.0920	.2513
.83	−.41475	−.34368	−.1720	.0359	.1942
.84	−.41198	−.37033	−.2228	−.0243	.1617
.85	−.40300	−.39130	−.2710	−.0873	.1029
.86	−.38716	−.40545	−.3150	−.1513	.0362
.87	−.36379	−.41156	−.3530	−.2143	−.0366
.88	−.33217	−.40829	−.3830	−.2738	−.1130
.89	−.29156	−.39423	−.4028	−.3267	−.1899
.90	−.24116	−.36782	−.4097	−.3695	−.2631
.91	−.18018	−.32743	−.4010	−.3983	−.3277
.92	−.10774	−.27129	−.3737	−.4083	−.3773
.93	−.02295	−.19749	−.3243	−.3941	−.4046
.94	.07512	−.10404	−.2491	−.3498	−.4006
.95	.18745	.01123	−.1440	−.2684	−.3549
.96	.31506	.15060	−.0046	−.1422	−.2552
.97	.45899	.31650	.1740	.0375	−.0875
.98	.62035	.51151	.3971	.2804	.1647
.99	.80029	.73838	.6704	.5973	.5201
1.00	1	1	1	1	1

SURFACE ZONAL HARMONICS (Continued)

θ deg.	$P_1(\cos\theta)$	$P_2(\cos\theta)$	$P_3(\cos\theta)$	$P_4(\cos\theta)$	$P_5(\cos\theta)$
0	1	1	1	1	1
1	0.99985	0.99954	0.99909	0.99848	0.99772
2	.99939	.99817	.99635	.99392	.99088
3	.99863	.99589	.99179	.98634	.97954
4	.99756	.99270	.98543	.97577	.96377
5	.99619	.98861	.97728	.96227	.94368
6	.99452	.98361	.96736	.94589	.91939
7	.99255	.97772	.95569	.92670	.89108
8	.99027	.97095	.94232	.90480	.85893
9	.98769	.96329	.92726	.88026	.82315
10	.98481	.95477	.91057	.85321	.78399
11	.98163	.94539	.89228	.82376	.74170
12	.97815	.93516	.87244	.79204	.69656
13	.97437	.92410	.85111	.75819	.64888
14	.97030	.91221	.82833	.72235	.59895
15	.96593	.89952	.80416	.68470	.54713
16	.96126	.88604	.77868	.64537	.49373
17	.95630	.87178	.75194	.60456	.43911
18	.95106	.85676	.72401	.56244	.38363
19	.94552	.84101	.69497	.51918	.32763
20	.93969	.82453	.66488	.47498	.27149
21	.93358	.80736	.63384	.43002	.21556
22	.92718	.78950	.60190	.38450	.16019
23	.92050	.77099	.56917	.33862	.10573
24	.91355	.75185	.53572	.29256	.05252
25	.90631	.73209	.50163	.24653	.00088
26	.89879	.71175	.46699	.20072	−.04887
27	.89101	.69084	.43190	.15531	−.09642
28	.88295	.66939	.39644	.11051	−.14151
29	.87462	.64744	.36069	.06649	−.18388
30	.86603	.62500	.32476	.02344	−.22327
31	.85717	.60210	.28873	−.01847	−.25949
32	.84805	.57878	.25269	−.05907	−.29233
33	.83867	.55505	.21673	−.09820	−.32163
34	.82904	.53095	.18094	−.13570	−.34726
35	.81915	.50652	.14542	−.17142	−.36910
36	.80902	.48176	.11025	−.20524	−.38707
37	.79864	.45673	.07551	−.23701	−.40113
38	.78801	.43144	.04129	−.26664	−.41124
39	.77715	.40593	.00769	−.29400	−.41741
40	.76604	.38024	−.02523	−.31900	−.41968
41	.75471	.35438	−.05738	−.34157	−.41811
42	.74314	.32840	−.08869	−.36163	−.41279
43	.73135	.30232	−.11907	−.37913	−.40385
44	.71934	.27617	−.14845	−.39401	−.39141
45	.70711	.25000	−.17678	−.40625	−.37565
46	.69466	.22383	−.20397	−.41582	−.35677
47	.68200	.19768	−.22997	−.42273	−.33496
48	.66913	.17160	−.25471	−.42696	−.31048
49	.65606	.14562	−.27815	−.42856	−.28357

For $P_n(\cos\theta)$, use the definition of $P_n(x)$ given on the first page of this table, and put $x = \cos\theta$.

SURFACE ZONAL HARMONICS (Continued)

θ°	$P_6(\cos\theta)$	$P_7(\cos\theta)$	$P_8(\cos\theta)$	$P_9(\cos\theta)$	$P_{10}(\cos\theta)$
0	1	1	1	1	1
1	0.99680	0.99574	0.9945	0.9932	0.9916
2	.98725	.98301	.9782	.9728	.9668
3	.97142	.96198	.9513	.9393	.9260
4	.94947	.93291	.9142	.8933	.8704
5	.92160	.89616	.8675	.8358	.8012
6	.88808	.85220	.8121	.7680	.7203
7	.84922	.80158	.7487	.6911	.6296
8	.80538	.74493	.6784	.6069	.5312
9	.75698	.68296	.6024	.5168	.4277
10	.70447	.61644	.5218	.4228	.3214
11	.64833	.54619	.4380	.3266	.2150
12	.58909	.47307	.3522	.2302	.1108
13	.52729	.39798	.2657	.1353	.0113
14	.46350	.32183	.1799	.0437	−.0813
15	.39831	.24554	.0962	−.0428	−.1651
16	.33229	.17001	.0157	−.1227	−.2381
17	.26606	.09614	−.0604	−.1946	−.2992
18	.20020	.02477	−.1310	−.2573	−.3471
19	.13529	−.04327	−.1951	−.3100	−.3813
20	.07190	−.10723	−.2518	−.3517	−.4013
21	.01059	−.16640	−.3005	−.3821	−.4073
22	−.04813	−.22017	−.3407	−.4009	−.3997
23	−.10376	−.26800	−.3718	−.4082	−.3793
24	−.15585	−.30942	−.3936	−.4042	−.3473
25	−.20398	−.34408	−.4062	−.3896	−.3052
26	−.24779	−.37172	−.4096	−.3650	−.2547
27	−.28694	−.39216	−.4041	−.3515	−.1975
28	−.32117	−.40534	−.3900	−.2902	−.1358
29	−.35025	−.41130	−.3680	−.2424	−.0716
30	−.37402	−.41018	−.3388	−.1896	−.0070
31	−.39238	−.40221	−.3031	−.1332	.0559
32	−.40527	−.38771	−.2619	−.0749	.1150
33	−.41269	−.36710	−.2162	−.0161	.1689
34	−.41471	−.34086	−.1670	.0415	.2157
35	−.41145	−.30956	−.1154	.0966	.2542
36	−.40307	−.27382	−.0627	.1476	.2833
37	−.38980	−.23432	−.0098	.1935	.3024
38	−.37191	−.19178	.0421	.2331	.3111
39	−.34972	−.14695	.0919	.2655	.3093
40	−.32357	−.10060	.1386	.2900	.2973
41	−.29387	−.05351	.1814	.3062	.2758
42	−.26104	−.00645	.2194	.3137	.2455
43	−.22554	.03982	.2519	.3127	.2077
44	−.18784	.08455	.2784	.3031	.1637
45	−.14844	.12706	.2983	.2855	.1151
46	−.10783	.16668	.3115	.2605	.0635
47	−.06654	.20283	.3176	.2288	.0107
48	−.02508	.23497	.3167	.1915	−.0416
49	.01606	.26263	.3090	.1495	−.0918

SURFACE ZONAL HARMONICS (Continued)

$\theta°$	$P_1(\cos\theta)$	$P_2(\cos\theta)$	$P_3(\cos\theta)$	$P_4(\cos\theta)$	$P_5(\cos\theta)$
50	0.64279	0.11976	−0.30022	−0.42753	−0.25449
51	.62932	.09407	−.32088	−.42394	−.22353
52	.61566	.06856	−.34009	−.41784	−.19097
53	.60182	.04327	−.35781	−.40929	−.15712
54	.58779	.01824	−.37399	−.39837	−.12229
55	.57358	−.00652	−.38861	−.38519	−.08679
56	.55919	−.03095	−.40164	−.36983	−.05093
57	.54464	−.05505	−.41307	−.35241	−.01503
58	.52992	−.07878	−.42286	−.33306	.02060
59	.51504	−.10210	−.43100	−.31189	.05566
60	.50000	−.12500	−.43750	−.28906	.08984
61	.48481	−.14744	−.44234	−.26471	.12287
62	.46947	−.16939	−.44552	−.23899	.15446
63	.45399	−.19084	−.44706	−.21205	.18436
64	.43837	−.21175	−.44695	−.18407	.21232
65	.42262	−.23209	−.44522	−.15521	.23811
66	.40674	−.25185	−.44188	−.12564	.26152
67	.39073	−.27099	−.43696	−.09554	.28238
68	.37461	−.28950	−.43049	−.06508	.30051
69	.35837	−.30736	−.42249	−.03444	.31577
70	.34202	−.32453	−.41301	−.00380	.32807
71	.32557	−.34101	−.40208	.02667	.33730
72	.30902	−.35676	−.38975	.05680	.34340
73	.29237	−.37178	−.37608	.08641	.34634
74	.27564	−.38604	−.36110	.11534	.34611
75	.25882	−.39952	−.34488	.14343	.34273
76	.24192	−.41221	−.32749	.17051	.33624
77	.22495	−.42410	−.30897	.19644	.32672
78	.20791	−.43516	−.28940	.22107	.31425
79	.19081	−.44539	−.26885	.24427	.29897
80	.17365	−.45477	−.24738	.26590	.28102
81	.15643	−.46329	−.22508	.28585	.26056
82	.13917	−.47095	−.20202	.30401	.23777
83	.12187	−.47772	−.17828	.32027	.21288
84	.10453	−.48361	−.15394	.33455	.18610
85	.08716	−.48861	−.12908	.34677	.15766
86	.06976	−.49270	−.10379	.35686	.12784
87	.05234	−.49589	−.07815	.36476	09688
88	.03490	−.49817	−.05224	.37044	.06507
89	.01745	−.49954	−.02617	.37386	.03268
90	0	−.50000	0	.37500	0

SURFACE ZONAL HARMONICS (Continued)

$\theta°$	$P_6(\cos\theta)$	$P_7(\cos\theta)$	$P_8(\cos\theta)$	$P_9(\cos\theta)$	$P_{10}(\cos\theta)$
50	0.05638	0.28543	0.2947	0.1041	−0.1381
51	.09539	.30308	.2742	.0565	−.1792
52	.13265	.31535	.2480	.0080	−.2137
53	.16772	.32213	.2167	−.0400	−.2407
54	.20020	.32336	.1812	−.0863	−.2594
55	.22972	.31910	.1422	−.1296	−.2692
56	.25597	.30949	.1005	−.1689	−.2700
57	.27866	.29475	.0572	−.2032	−.2617
58	.29756	.27518	.0131	−.2315	−.2449
59	.31246	.25117	−.0309	−.2533	−.2201
60	.32324	.22314	−.0736	−.2679	−.1882
61	.32980	.19162	−.1144	−.2751	−.1504
62	.33210	.15715	−.1523	−.2747	−.1080
63	.33016	.12034	−.1865	−.2669	−.0624
64	.32403	.08181	−.2163	−.2518	−.0151
65	.31383	.04222	−.2411	−.2300	.0323
66	.29971	.00223	−.2605	−.2022	.0783
67	.28189	−.03748	−.2741	−.1690	.1213
68	.26062	−.07627	−.2816	−.1315	.1599
69	.23617	−.11348	−.2829	−.0906	.1929
70	.20888	−.14853	−.2780	−.0476	.2193
71	.17910	−.18082	−.2671	−.0035	.2382
72	.14721	−.20986	−.2504	.0404	.2491
73	.11363	−.23516	−.2283	.0829	.2516
74	.07878	−.25634	−.2014	.1230	.2457
75	.04310	−.27305	−.1702	.1595	.2316
76	.00704	−.28504	−.1355	.1915	.2099
77	−.02896	−.29214	−.0979	.2181	.1813
78	−.06444	−.29424	−.0583	.2387	.1468
79	−.09897	−.29133	−.0176	.2526	.1074
80	−.13212	−.28348	.0233	.2596	.0647
81	−.16348	−.27083	.0636	.2595	.0199
82	−.19267	−.25360	.1024	.2523	−.0254
83	−.21933	−.23211	.1389	.2383	−.0698
84	−.24313	.20671	.1722	.2177	−.1118
85	−.26378	−.17784	.2017	.1913	−.1499
86	−.28103	−.14598	.2268	.1597	−.1830
87	−.29467	−.11168	.2469	.1237	−.2099
88	−.30454	−.07551	.2615	.0844	−.2298
89	−.31050	−.03807	.2704	.0428	−.2420
90	−.31250	0	.2734	0	−.2461

For larger tables, see "Report of the British Association for the Advancement of Science," 1879, pp. 54–57, Philosophical Magazine, Dec., 1891, "Tafeln der Besselschen, Theta-, Kugel- und anderer Funktionen" by K. Hayashi and "Fünfstellige Funktionentafeln" by K. Hayashi.

SURFACE ZONAL HARMONICS—FIRST DERIVATIVES

$$\frac{d}{d\theta} P_n(\cos \theta)$$

$\theta°$	$\frac{d}{d\theta} P_1$	$\frac{d}{d\theta} P_2$	$\frac{d}{d\theta} P_3$	$\frac{d}{d\theta} P_4$	$\frac{d}{d\theta} P_5$	$\frac{d}{d\theta} P_6$	$\frac{d}{d\theta} P_7$
0	0	0	0	0	0	0	0
1	−0.0175	−0.0523	−0.1047	−0.1744	−0.2615	−0.3659	−0.488
2	−.0349	−.1046	−.2091	−.3480	−.5213	−.7284	−.969
3	−.0523	−.1568	−.3129	−.5201	−.7775	−1.0841	−1.438
4	−.0698	−.2088	−.4160	−.6901	−1.0286	−1.4295	−1.890
5	−.0872	−.2605	−.5180	−.8570	−1.2728	−1.7614	−2.317
6	−.1045	−.3119	−.6186	−1.0197	−1.5085	−2.0768	−2.715
7	−.1219	−.3629	−.7176	−1.1781	−1.7341	−2.3727	−3.080
8	−.1392	−.4135	−.8148	−1.3315	−1.9481	−2.6465	−3.405
9	−.1564	−.4635	−.9099	−1.4789	−2.1492	−2.8954	−3.689
10	−.1736	−.5130	−1.0026	−1.6199	−2.3360	−3.1174	−3.926
11	−.1908	−.5619	−1.0928	−1.7537	−2.5074	−3.3104	−4.116
12	−.2079	−.6101	−1.1801	−1.8799	−2.6621	−3.4729	−4.254
13	−.2250	−.6576	−1.2643	−1.9978	−2.7993	−3.6034	−4.341
14	−.2419	−.7042	−1.3453	−2.1069	−2.9181	−3.7008	−4.376
15	−.2588	−.7500	−1.4229	−2.2069	−3.0178	−3.7646	−4.358
16	−.2756	−.7949	−1.4968	−2.2973	−3.0978	−3.7943	−4.288
17	−.2924	−.8388	−1.5668	−2.3777	−3.1576	−3.7899	−4.169
18	−.3090	−.8817	−1.6328	−2.4478	−3.1970	−3.7518	−4.001
19	−.3256	−.9235	−1.6946	−2.5073	−3.2158	−3.6806	−3.788
20	−.3420	−.9642	−1.7520	−2.5560	−3.2141	−3.5774	−3.534
21	−.3584	−1.0037	−1.8050	−2.5937	−3.1920	−3.4435	−3.241
22	−.3746	−1.0420	−1.8534	−2.6203	−3.1497	−3.2804	−2.915
23	−.3907	−1.0790	−1.8970	−2.6357	−3.0877	−3.0917	−2.561
24	−.4067	−1.1147	−1.9357	−2.6400	−3.0067	−2.8749	−2.183
25	−.4226	−1.1491	−1.9696	−2.6330	−2.9073	−2.6371	−1.787
26	−.4384	−1.1820	−1.9984	−2.6150	−2.7903	−2.3794	−1.378
27	−.4540	−1.2135	−2.0222	−2.5861	−2.6568	−2.1046	−.963
28	−.4695	−1.2436	−2.0408	−2.5464	−2.5077	−1.8156	−.548
29	−.4848	−1.2721	−2.0542	−2.4961	−2.3443	−1.5155	−.136
30	−.5000	−1.2990	−2.0625	−2.4357	−2.1680	−1.2077	.263
31	−.5150	−1.3244	−2.0654	−2.3654	−1.9799	−.8953	.647
32	−.5299	−1.3482	−2.0635	−2.2855	−1.7817	−.5815	1.010
33	−.5446	−1.3703	−2.0562	−2.1966	−1.5748	−.2697	1.347
34	−.5592	−1.3908	−2.0437	−2.0991	−1.3608	.0370	1.654
35	−.5736	−1.4095	−2.0262	−1.9934	−1.1413	.3354	1.927
36	−.5878	−1.4266	−2.0036	−1.8802	−.9179	.6225	2.162
37	−.6018	−1.4419	−1.9761	−1.7600	−.6924	.8955	2.357
38	−.6157	−1.4554	−1.9438	−1.6334	−.4664	1.1516	2.510
39	−.6293	−1.4672	−1.9066	−1.5011	−.2415	1.3885	2.620
40	−.6428	−1.4772	−1.8648	−1.3637	−.0194	1.6038	2.684
41	−.6561	−1.4854	−1.8185	−1.2219	.1983	1.7955	2.705
43	−.6691	−1.4918	−1.7678	−1.0764	.4100	1.9620	2.681
43	−.6820	−1.4963	−1.7129	−.9279	.6142	2.1017	2.614
44	−.6947	−1.4991	−1.6539	−.7772	.8095	2.2136	2.506
45	−.7071	−1.5000	−1.5910	−.6250	.9943	2.2969	2.359
46	−.7193	−1.4991	−1.5244	−.4720	1.1677	2.3510	2.176
47	−.7314	−1.4963	−1.4542	−.3190	1.3282	2.3757	1.961
48	−.7431	−1.4918	−1.3808	−.1668	1.4749	2.3715	1.717
49	−.7547	−1.4854	−1.3042	−.0160	1.6067	2.3382	1.449

SURFACE ZONAL HARMONICS—FIRST DERIVATIVES (Continued)

$$\frac{d}{d\theta} P_n(\cos\theta)$$

$\theta°$	$\frac{d}{d\theta}P_1$	$\frac{d}{d\theta}P_2$	$\frac{d}{d\theta}P_3$	$\frac{d}{d\theta}P_4$	$\frac{d}{d\theta}P_5$	$\frac{d}{d\theta}P_6$	$\frac{d}{d\theta}P_7$
50	−0.7660	−1.4772	−1.2248	0.1327	1.7229	2.2772	1.161
51	−.7771	−1.4672	−1.1427	.2784	1.8225	2.1892	.859
52	−.7880	−1.4554	−1.0581	.4205	1.9053	2.0760	.546
53	−.7986	−1.4419	−.9714	.5584	1.9704	1.9387	.230
54	−.8090	−1.4266	−.8828	.6914	2.0173	1.7797	−.088
55	−.8192	−1.4095	−.7925	.8188	2.0474	1.6009	−.399
56	−.8290	−1.3908	−.7007	.9401	2.0587	1.4046	−.700
57	−.8387	−1.3703	−.6078	1.0546	2.0521	1.1936	−.986
58	−.8480	−1.3482	−.5140	1.1620	2.0281	.9699	−1.252
59	−.8572	−1.3244	−.4196	1.2617	1.9866	.7369	−1.495
60	−.8660	−1.2990	−.3248	1.3532	1.9283	.4973	−1.711
61	−.8746	−1.2721	−.2299	1.4361	1.8538	.2540	−1.896
62	−.8829	−1.2436	−.1351	1.5101	1.7640	.0099	−2.048
63	−.8910	−1.2135	−.0409	1.5748	1.6596	−.2321	−2.165
64	−.8988	−1.1820	.0527	1.6306	1.5420	−.4691	−2.244
65	−.9063	−1.1491	.1454	1.6755	1.4114	−.6983	−2.286
66	−.9135	−1.1147	.2368	1.7111	1.2698	−.9170	−2.290
67	−.9205	−1.0790	.3268	1.7366	1.1187	−1.1226	−2.255
68	−.9272	−1.0420	.4149	1.7520	.9580	−1.3129	−2.183
69	−.9336	−1.0037	.5011	1.7574	.7906	−1.4855	−2.076
70	−.9397	−.9642	.5851	1.7525	.6173	−1.6386	−1.984
71	−.9455	−.9235	.6666	1.7376	.4397	−1.7704	−1.762
72	−.9511	−.8817	.7455	1.7131	.2594	−1.8794	−1.561
73	−.9563	−.8388	.8214	1.6787	.0776	−1.9646	−1.334
74	−.9613	−.7949	.8941	1.6349	−.1037	−2.0248	−1.088
75	−.9659	−.7500	.9636	1.5819	−.2833	−2.0596	−.824
76	−.9703	−.7042	1.0295	1.5200	−.4595	−2.0687	−.548
77	−.9744	−.6576	1.0918	1.4498	−.6309	−2.0520	−.264
78	−.9781	−.6101	1.1501	1.3714	−.7961	−2.0098	.023
79	−.9816	−.5619	1.2044	1.2854	−.9537	−1.9428	.309
80	−.9848	−.5130	1.2545	1.1923	−1.1023	−1.8519	.589
81	−.9877	−.4635	1.3003	1.0296	−1.2407	−1.7382	.858
82	−.9903	−.4135	1.3416	.9870	−1.3679	−1.6032	1.112
83	−.9925	−.3629	1.3783	.8758	−1.4831	−1.4487	1.348
84	−.9945	−.3119	1.4103	.7598	−1.5841	−1.2760	1.559
85	−.9962	−.2605	1.4376	.6396	−1.6715	−1.0881	1.745
86	−.9976	−.2088	1.4599	.5160	−1.7440	−.8868	1.901
87	−.9986	−.1568	1.4774	.3895	−1.8009	−.6747	2.024
88	−.9994	−.1046	1.4899	.2608	−1.8420	−.4544	2.115
89	−.9998	−.0523	1.4975	.1308	−1.8667	−.2286	2.169
90	−1.0000	0	1.5000	0	−1.8750	0	2.187

Note that

$$\frac{d}{dx} P_n(x) = -\frac{1}{\sin\theta}\frac{d}{d\theta} P_n(\cos\theta)$$

where

$$x = \cos\theta$$

LEGENDRE FUNCTIONS

In the following, m and n are positive integers.

1. The differential equation $(1 - z^2)\dfrac{d^2w}{dz^2} - 2z\dfrac{dw}{dz} + n(n+1)z = 0$ is known as Legendre's differential equation. The Legendre polynomials $P_0(z)$, $P_1(z)$, $P_2(z)$, . . . are solutions of this equation, and are referred to as Surface Zonal Harmonics, for which tables are herewith given.

2. The solution of Legendre's equation can be stated as

$$w = AP_n(z) + BQ_n(z) \qquad |z| < 1$$

or

$$w = AP_n(z) + B\mathfrak{Q}_n(z) \qquad |z| > 1,$$

where A and B are arbitrary constants, $P_n(z)$ the Legendre polynomials, $Q_n(z)$ and $\mathfrak{Q}_n(z)$ Legendre functions of the second kind.

3. The differential equation $(1 - z^2)\dfrac{d^2u}{dz^2} - 2z\dfrac{du}{dz} + \left[n(n+1) - \dfrac{m^2}{1 - z^2}\right]u = 0$ which is obtained from the Legendre equation in 1. after it has been differentiated m times and u replaced by $(1 - z^2)^{m/2}\left(\dfrac{d^mw}{dz^m}\right)$ is referred to as the "associated Legendre differential equation."

4. The solution for the "associated Legendre differential equation" is given by

$$u = AP_n^m(z) + BQ_n^m(z),$$

where A and B are arbitrary constants and $P_n^m(z)$, $Q_n^m(z)$ are called associated Legendre functions.

5. $$P_n(z) = \sum_{1=0}^{m} (-1)^r \frac{(2n - 2r)!}{2^n(r!)(n - r_r)!(n - 2r)!} z^{n-2r} \qquad m = \tfrac{1}{2}n,\ n \text{ even}$$

$$m = \tfrac{1}{2}(n - 1),\ n \text{ odd}$$

$$Q_n(z) = \tfrac{1}{2}P_n(z)\log_e\frac{1+z}{1-z} - \sum_{r=1}^{n}\frac{1}{r}P_{r-1}(z)P_{n-r}(z)$$

$$= P_n(z)Q_0(z) - \sum_{r=1}^{n}\frac{1}{r}P_{r-1}(z)P_{n-r}(z)$$

$$Q_0(z) = \tfrac{1}{2}\log_e\frac{1+z}{1-z} = \tanh^{-1} z$$

$$\mathfrak{Q}_n(z) = \tfrac{1}{2}P_n(z)\log_e\frac{z+1}{z-1} - \sum_{r=1}^{n}\frac{1}{r}P_{r-1}(z)\,P_{n-r}(z)$$

$$= P_n(z)\mathfrak{Q}_0(z) - \sum_{r=1}^{n}\frac{1}{r}P_{r-1}(z)P_{n-r}(z)$$

$$\mathfrak{Q}_0(z) = \tfrac{1}{2}\log_e\frac{z+1}{z-1} = \text{ctnh}^{-1} z$$

LEGENDRE FUNCTIONS (Continued)

6.

n	$P_n(z)$	$Q_n(z)$
0	1	$\frac{\log_e}{2}\frac{(1+z)}{(1-z)}$
1	z	$zQ_0(z) - 1$
2	$\frac{1}{2}(3z^2 - 1)$	$\frac{1}{2}(3z^2 - 1)Q_0(z) - \frac{3}{2}z$
3	$\frac{1}{2}(5z^3 - 3z)$	$\frac{1}{2}(5z^3 - 3z)Q_0(z) - \frac{5}{2}z^2 + \frac{2}{3}$
4	$\frac{1}{8}(35z^4 - 30z^2 + 3)$	$\frac{1}{8}(35z^4 - 30z^2 + 3)Q_0(z) - \frac{35}{8}z^3 + \frac{55}{24}z$
5	$\frac{1}{8}(63z^5 - 70z^3 + 15z)$	$\frac{1}{8}(63z^5 - 70z^3 + 15z)Q_0(z) - \frac{63}{8}z^4 + \frac{49}{8}z^2 - \frac{8}{15}$

7. $$\frac{1}{(1 - 2xZ + Z^2)^{1/2}} = \sum P_n(x)Z^n$$

If $x = \cos\theta = \frac{1}{2}(e^{i\theta} + e^{-i\theta})$, $i = \sqrt{-1}$, then

$$[1 - Z(e^{i\theta} + e^{-i\theta}) + Z^2]^{-1/2} = \sum_{n=0}^{\infty} P_n(\cos\theta)Z^n$$

$$P_0(\cos\theta) = 1$$

$$P_1(\cos\theta) = \cos\theta$$

$$P_2(\cos\theta) = \frac{1}{4}(3\cos 2\theta + 1)$$

$$P_3(\cos\theta) = \frac{1}{8}(5\cos 3\theta + 3\cos\theta)$$

$$P_4(\cos\theta) = \frac{1}{64}(35\cos 4\theta + 20\cos 2\theta + 9)$$

$$P_n(\cos\theta) = \frac{(2n)!}{2^{2n}(n!)^2}\left[\cos n\theta + 1/1\,\frac{n}{2n-1}\cos(n-2)\theta + \frac{1\cdot 3}{1\cdot 2}\,\frac{n(n-1)}{(2n-1)(2n-3)}\cos(n-4)\theta + \frac{1\cdot 3\cdot 5}{1\cdot 2\cdot 3}\,\frac{n(n-1)(n-2)}{(2n-1)(2n-3)(2n-5)}\cos(n-6)\theta + \cdots\right]$$

LEGENDRE FUNCTIONS (Continued)

8. $P_{2n+1}(0) = 0$

$$P_{2n}(0) = (-1)^n \frac{1 \cdot 3 \cdot 5 \cdot \cdot \cdot (2n-1)}{2 \cdot 4 \cdot 6 \cdot 8 \cdot \cdot \cdot (2n)}$$

$$P_n(1) = 1$$

$$P_n(-1) = (-1)^n$$

$$P_n(z) = \frac{1}{2^n n!} \frac{d^n}{dz^n} (z^2 - 1)^n \text{ (Rodrigues' formula)}$$

$$P_n^m(z) = (1 - z^2)^{m/2} \frac{d^m[P_n(z)]}{dz^m}$$

$$Q_n^m(z) = (1 - z^2)^{m/2} \frac{d^m[Q_n(z)]}{dz^m}$$

9. Important recurrence and orthogonality relations (note that w_n stands for either P_n or Q_n in the following)

$$(n+1)[w_{n+1}(z)] - (2n+1)z[w_n(z)] + n[w_{n-1}(z)] = 0$$

$$z\frac{d[w_n(z)]}{dz} - z\frac{d[w_{n-1}(z)]}{dz} = n[w_n(z)]$$

$$\frac{d[w_{n+1}(z)]}{dz} - z\frac{d[w_n(z)]}{dz} = (n+1)[w_n(z)]$$

$$\frac{d[w_{n+1}(z)]}{dz} - \frac{d}{dz}[w_{n-1}(z)] = (2n+1)[w_n(z)]$$

$$(z^2 - 1)\frac{d}{dz}[w_n(z)] = nz[w_n(z)] - n[w_{n-1}(z)]$$

$$\int_{-1}^{1} [P_m(z)][P_n(z)]dz \begin{cases} = 0 & m \neq n \\ = \dfrac{2}{2n+1} & m = n \end{cases}$$

$$\int_{-1}^{1} [P_n^m(z)][P_r^m(z)]dz \begin{cases} = 0 & r \neq n \\ = \dfrac{2}{2n+1}\dfrac{(n+m)!}{(n-m)!} & r = n > m. \end{cases}$$

ELLIPTIC INTEGRALS OF THE FIRST, SECOND AND THIRD KIND

An elliptic integral has the form $\int R(x, \sqrt{f(x)}\, dx$, where R represents a rational function and $f(x) = a + bx + cx^2 + dx^3 + ex^4$, an algebraic function of the third or fourth degree.

1. Elliptic integrals of the *first kind* are represented by

$$F(k, \phi) = \int_0^{\phi} \frac{d\Phi}{\sqrt{1 - k^2 \sin^2 \Phi}}$$
$$= \int_0^{x} \frac{d\xi}{\sqrt{(1 - \xi^2)(1 - k^2\xi^2)}}, \quad x = \sin \phi,\ k^2 < 1.$$

2. Elliptic integrals of the second kind are represented by

$$E(k, \phi) = \int_0^{\phi} \sqrt{1 - k^2 \sin^2 \Phi}\, d\Phi$$
$$= \int_0^{x} \frac{\sqrt{1 - k^2\xi^2}}{\sqrt{1 - \xi^2}}\, d\xi, \quad x = \sin \phi,\ k^2 < 1.$$

3. Elliptic integrals of the third kind are represented as

$$\pi(k, n, \phi) = \int_0^{\phi} \frac{d\Phi}{(1 + n \sin^2 \Phi)\sqrt{1 - k^2 \sin^2 \Phi}},$$
$$k^2 < 1,\ n \text{ an integer.}$$

Elliptic integrals of the third kind are also presented as

$$\pi_1(k, n, x) = \int_0^{x} \frac{d\xi}{(1 + n\xi^2)\sqrt{(1 - \xi^2)(1 - k^2\xi^2)}},$$
$$x = \sin \phi,\ k^2 < 1,\ n \text{ an integer.}$$

4. The complete integrals are

$$K = F\left(k, \frac{\pi}{2}\right) = \frac{\pi}{2}\left[1 + \left(\frac{1}{2}\right)^2 k^2 + \left(\frac{3}{2 \cdot 4}\right)^2 k^4 + \left(\frac{3 \cdot 5}{2 \cdot 4 \cdot 6}\right)^2 k^6 + \cdots\right]$$

$$E = E\left(k, \frac{\pi}{2}\right) = \frac{\pi}{2}\left[1 - \left(\frac{1}{2^2}\right) k^2 - \left(\frac{3^2}{2^2 \cdot 4^2}\right)\frac{k^4}{3} - \left(\frac{3^2 \cdot 5^2}{2^2 \cdot 4^2 \cdot 6^2}\right)\frac{k^6}{5} - \left(\frac{3^2 \cdot 5^2 \cdot 7^2}{2^2 \cdot 4^2 \cdot 6^2 \cdot 8^2}\right)\frac{k^8}{7} - \cdots\right].$$

$$K' = F\left(\sqrt{1 - k^2}, \frac{\pi}{2}\right),\quad E' = E\left(\sqrt{1 - k^2}, \frac{\pi}{2}\right)$$

5. The following relation holds between K, K', E, E', namely

$$KE' + EK' - KK' = \frac{\pi}{2} \qquad \text{Legendre's relation}$$

E: see 2 above. $E' = \int_0^{\pi/2} (1 - k'^2 \sin^2 \phi)^{\frac{1}{2}} d\phi$. $k' = \sqrt{(1 - k^2)}$

$$K = \int_0^{\pi/2} (1 - k^2 \sin^2 \phi)^{-\frac{1}{2}} d\phi \qquad K' = \int_0^{\pi/2} (1 - k'^2 \sin^2 \phi)^{-\frac{1}{2}} d\phi$$

ELLIPTIC INTEGRALS OF THE FIRST, SECOND AND THIRD KIND (Continued)

6. To evaluate elliptic integrals for values outside the range contained in the following tables, these relations are useful

$$F(k, \pi) = 2K;\ E(k, \pi) = 2E$$

$$F(k, \phi + m\pi) = mF(k, \pi) + F(k, \phi) = 2mK + F(k, \phi) \qquad m = 0, 1, 2, 3, \ldots$$

$$E(k, \phi + m\pi) = mE(k, \pi) + E(k, \phi) = 2mE + E(k, \phi) \qquad m = 0, 1, 2, 3, \ldots$$

7. If $u = F(k, \phi) = \int_0^\phi \frac{d\Phi}{\sqrt{1 - k^2 \sin^2 \Phi}}$ $(k^2 < 1)$,

$= $ elliptic integral of the first kind.

$$u = \int_0^x \frac{dx}{\sqrt{(1 - \xi^2)(1 - k^2\xi^2)}}, \text{ where } x = \sin \phi.$$

ϕ is called the amplitude of u or am u.
k is called the modulus.

$$k' = \sqrt{1 - k^2} = \text{the complementary modulus.}$$

$\sin \phi = \text{sn } u = x \qquad \tan \phi = \text{tn } u = \dfrac{x}{\sqrt{1 - x^2}}$.

$\cos \phi = \text{cn } u = \sqrt{1 - x^2}$. $\qquad \Delta\phi = \text{dn } u = \sqrt{1 - k^2x^2}$.

am $0 = 0$. $\qquad$ sn $0 = 0$.

cn $0 = 1$. $\qquad$ dn $0 = 1$.

am $(-u) = -$am u. $\qquad$ sn $(-u) = -$sn u.

cn $(-u) = $ cn u. $\qquad$ dn $(-u) = $ dn u.

tn $(-u) = -$tn u.

$\text{sn}^2\, u + \text{cn}^2\, u = 1$.

$\text{dn}^2\, u + k^2\, \text{sn}^2\, u = 1$.

$\text{dn}^2\, u - k^2\, \text{cn}^2\, u = 1 - k^2 = k'^2$.

$$\text{sn } u = u - (1 + k^2)\frac{u^3}{3!} + (1 + 14k^2 + k^4)\frac{u^5}{5!} - (1 + 135k^2 + 135k^4 + k^6)\frac{u^7}{7!} + \cdots$$

Periods: $4k$ and $2ik'$

$$\text{cn } u = 1 - \frac{u^2}{2!} + (1 + 4k^2)\frac{u^4}{4!} - (1 + 44k^2 + 16k^4)\frac{u^6}{6!} + \cdots$$

Periods: $4k$ and $2k + 2ik'$

$$\text{dn } u = 1 - k^2\frac{u^2}{2!} + k^2(4 + k^2)\frac{u^4}{4!} - k^2(16 + 44k^2 + k^4)\frac{u^6}{6!} + \cdots$$

Periods: $2k$ and $4ik'$

*ELLIPTIC INTEGRALS OF THE FIRST KIND: $F(k, \phi)$

$$F(k, \phi) = \int_0^\phi \frac{d\Phi}{\sqrt{1 - k^2 \sin^2 \Phi}}, \qquad \theta = \sin^{-1} k$$

ϕ \ θ	5°	10°	15°	20°	25°	30°	35°	40°	45°
1°	0.0175	0.0175	0.0175	0.0175	0.0175	0.0175	0.0175	0.0175	0.0175
2°	0.0349	0.0349	0.0349	0.0349	0.0349	0.0349	0.0349	0.0349	0.0349
3°	0.0524	0.0524	0.0524	0.0524	0.0524	0.0524	0.0524	0.0524	0.0524
4°	0.0698	0.0698	0.0698	0.0698	0.0698	0.0698	0.0698	0.0698	0.0698
5°	0.0873	0.0873	0.0873	0.0873	0.0873	0.0873	0.0873	0.0873	0.0873
6°	0.1047	0.1047	0.1047	0.1047	0.1048	0.1048	0.1048	0.1048	0.1048
7°	0.1222	0.1222	0.1222	0.1222	0.1222	0.1222	0.1223	0.1223	0.1223
8°	0.1396	0.1396	0.1397	0.1397	0.1397	0.1397	0.1398	0.1398	0.1399
9°	0.1571	0.1571	0.1571	0.1572	0.1572	0.1572	0.1573	0.1573	0.1574
10°	0.1745	0.1746	0.1746	0.1746	0.1747	0.1748	0.1748	0.1749	0.1750
11°	0.1920	0.1920	0.1921	0.1921	0.1922	0.1923	0.1924	0.1925	0.1926
12°	0.2095	0.2095	0.2095	0.2096	0.2097	0.2098	0.2099	0.2101	0.2102
13°	0.2269	0.2270	0.2270	0.2271	0.2272	0.2274	0.2275	0.2277	0.2279
14°	0.2444	0.2444	0.2445	0.2446	0.2448	0.2450	0.2451	0.2453	0.2456
15°	0.2618	0.2619	0.2620	0.2621	0.2623	0.2625	0.2628	0.2630	0.2633
16°	0.2793	0.2794	0.2795	0.2797	0.2799	0.2802	0.2804	0.2808	0.2811
17°	0.2967	0.2968	0.2970	0.2972	0.2975	0.2978	0.2981	0.2985	0.2989
18°	0.3142	0.3143	0.3145	0.3148	0.3151	0.3154	0.3159	0.3163	0.3167
19°	0.3317	0.3318	0.3320	0.3323	0.3327	0.3331	0.3336	0.3341	0.3347
20°	0.3491	0.3493	0.3495	0.3499	0.3503	0.3508	0.3514	0.3520	0.3526
21°	0.3666	0.3668	0.3671	0.3675	0.3680	0.3685	0.3692	0.3699	0.3706
22°	0.3840	0.3842	0.3846	0.3851	0.3856	0.3863	0.3871	0.3879	0.3887
23°	0.4015	0.4017	0.4021	0.4027	0.4033	0.4041	0.4049	0.4059	0.4068
24°	0.4190	0.4192	0.4197	0.4203	0.4210	0.4219	0.4229	0.4239	0.4250
25°	0.4364	0.4367	0.4372	0.4379	0.4387	0.4397	0.4408	0.4420	0.4433
26°	0.4539	0.4542	0.4548	0.4556	0.4565	0.4576	0.4588	0.4602	0.4616
27°	0.4714	0.4717	0.4724	0.4732	0.4743	0.4755	0.4769	0.4784	0.4800
28°	0.4888	0.4893	0.4899	0.4909	0.4921	0.4934	0.4950	0.4967	0.4985
29°	0.5063	0.5068	0.5075	0.5086	0.5099	0.5114	0.5132	0.5150	0.5170
30°	0.5238	0.5243	0.5251	0.5263	0.5277	0.5294	0.5313	0.5334	0.5356
31°	0.5412	0.5418	0.5427	0.5440	0.5456	0.5475	0.5496	0.5519	0.5543
32°	0.5587	0.5593	0.5603	0.5617	0.5635	0.5656	0.5679	0.5704	0.5731
33°	0.5762	0.5769	0.5780	0.5795	0.5814	0.5837	0.5862	0.5890	0.5920
34°	0.5937	0.5944	0.5956	0.5973	0.5994	0.6018	0.6046	0.6077	0.6109
35°	0.6111	0.6119	0.6133	0.6151	0.6173	0.6200	0.6231	0.6264	0.6300
36°	0.6286	0.6295	0.6309	0.6329	0.6353	0.6383	0.6416	0.6452	0.6491
37°	0.6461	0.6470	0.6486	0.6507	0.6534	0.6565	0.6602	0.6641	0.6684
38°	0.6636	0.6646	0.6662	0.6685	0.6714	0.6749	0.6788	0.6831	0.6877
39°	0.6810	0.6821	0.6839	0.6864	0.6895	0.6932	0.6975	0.7021	0.7071
40°	0.6985	0.6997	0.7016	0.7043	0.7076	0.7116	0.7162	0.7213	0.7267
41°	0.7160	0.7173	0.7193	0.7222	0.7258	0.7301	0.7350	0.7405	0.7463
42°	0.7335	0.7348	0.7370	0.7401	0.7440	0.7486	0.7539	0.7598	0.7661
43°	0.7510	0.7524	0.7548	0.7580	0.7622	0.7671	0.7728	0.7791	0.7859
44°	0.7685	0.7700	0.7725	0.7760	0.7804	0.7857	0.7918	0.7986	0.8059
45°	0.7859	0.7876	0.7903	0.7940	0.7987	0.8044	0.8109	0.8181	0.8260

*** For useful information about these tables see preceding page.**

*ELLIPTIC INTEGRALS OF THE FIRST KIND: $F(k, \phi)$ (Continued)

$$F(k, \phi) = \int_0^\phi \frac{d\Phi}{\sqrt{1 - k^2 \sin^2 \Phi}}, \qquad \theta = \sin^{-1} k$$

ϕ \ θ	50°	55°	60°	65°	70°	75°	80°	85°	90°
1°	0.0175	0.0175	0.0175	0.0175	0.0175	0.0175	0.0175	0.0175	0.0175
2°	0.0349	0.0349	0.0349	0.0349	0.0349	0.0349	0.0349	0.0349	0.0349
3°	0.0524	0.0524	0.0524	0.0524	0.0524	0.0524	0.0524	0.0524	0.0524
4°	0.0698	0.0699	0.0699	0.0699	0.0699	0.0699	0.0699	0.0699	0.0699
5°	0.0873	0.0873	0.0873	0.0874	0.0874	0.0874	0.0874	0.0874	0.0874
6°	0.1048	0.1048	0.1049	0.1049	0.1049	0.1049	0.1049	0.1049	0.1049
7°	0.1224	0.1224	0.1224	0.1224	0.1224	0.1225	0.1225	0.1225	0.1225
8°	0.1399	0.1399	0.1400	0.1400	0.1400	0.1401	0.1401	0.1401	0.1401
9°	0.1575	0.1575	0.1576	0.1576	0.1577	0.1577	0.1577	0.1577	0.1577
10°	0.1751	0.1751	0.1752	0.1753	0.1753	0.1754	0.1754	0.1754	0.1754
11°	0.1927	0.1928	0.1929	0.1930	0.1930	0.1931	0.1931	0.1932	0.1932
12°	0.2103	0.2105	0.2106	0.2107	0.2108	0.2109	0.2109	0.2110	0.2110
13°	0.2280	0.2282	0.2284	0.2285	0.2286	0.2287	0.2288	0.2288	0.2289
14°	0.2458	0.2460	0.2462	0.2464	0.2465	0.2466	0.2467	0.2468	0.2468
15°	0.2636	0.2638	0.2641	0.2643	0.2645	0.2646	0.2647	0.2648	0.2648
16°	0.2814	0.2817	0.2820	0.2823	0.2825	0.2827	0.2828	0.2829	0.2830
17°	0.2993	0.2997	0.3000	0.3003	0.3006	0.3008	0.3010	0.3011	0.3012
18°	0.3172	0.3177	0.3181	0.3185	0.3188	0.3191	0.3193	0.3194	0.3195
19°	0.3352	0.3357	0.3362	0.3367	0.3371	0.3374	0.3377	0.3378	0.3379
20°	0.3533	0.3539	0.3545	0.3550	0.3555	0.3559	0.3561	0.3563	0.3564
21°	0.3714	0.3721	0.3728	0.3734	0.3740	0.3744	0.3747	0.3749	0.3750
22°	0.3896	0.3904	0.3912	0.3919	0.3926	0.3931	0.3935	0.3937	0.3938
23°	0.4078	0.4088	0.4097	0.4105	0.4113	0.4119	0.4123	0.4126	0.4127
24°	0.4261	0.4272	0.4283	0.4292	0.4301	0.4308	0.4313	0.4316	0.4317
25°	0.4446	0.4458	0.4470	0.4481	0.4490	0.4498	0.4504	0.4508	0.4509
26°	0.4630	0.4645	0.4658	0.4670	0.4681	0.4690	0.4697	0.4701	0.4702
27°	0.4816	0.4832	0.4847	0.4861	0.4873	0.4884	0.4891	0.4896	0.4897
28°	0.5003	0.5021	0.5038	0.5053	0.5067	0.5079	0.5087	0.5092	0.5094
29°	0.5190	0.5210	0.5229	0.5247	0.5262	0.5275	0.5285	0.5291	0.5293
30°	0.5379	0.5401	0.5422	0.5442	0.5459	0.5474	0.5484	0.5491	0.5493
31°	0.5568	0.5593	0.5617	0.5639	0.5658	0.5674	0.5686	0.5693	0.5696
32°	0.5759	0.5786	0.5812	0.5837	0.5858	0.5876	0.5889	0.5898	0.5900
33°	0.5950	0.5980	0.6010	0.6037	0.6060	0.6080	0.6095	0.6104	0.6107
34°	0.6143	0.6176	0.6208	0.6238	0.6265	0.6287	0.6303	0.6313	0.6317
35°	0.6336	0.6373	0.6408	0.6441	0.6471	0.6495	0.6513	0.6525	0.6528
36°	0.6531	0.6571	0.6610	0.6647	0.6679	0.6706	0.6726	0.6739	0.6743
37°	0.6727	0.6771	0.6814	0.6854	0.6890	0.6919	0.6941	0.6955	0.6960
38°	0.6925	0.6973	0.7019	0.7063	0.7102	0.7135	0.7159	0.7175	0.7180
39°	0.7123	0.7176	0.7227	0.7275	0.7318	0.7353	0.7380	0.7397	0.7403
40°	0.7323	0.7380	0.7436	0.7488	0.7535	0.7575	0.7604	0.7623	0.7629
41°	0.7524	0.7586	0.7647	0.7704	0.7756	0.7799	0.7831	0.7852	0.7859
42°	0.7727	0.7794	0.7860	0.7922	0.7979	0.8026	0.8062	0.8084	0.8092
43°	0.7931	0.8004	0.8075	0.8143	0.8204	0.8256	0.8295	0.8320	0.8328
44°	0.8136	0.8215	0.8293	0.8367	0.8433	0.8490	0.8533	0.8560	0.8569
45°	0.8343	0.8428	0.8512	0.8592	0.8665	0.8727	0.8774	0.8804	0.8814

*ELLIPTIC INTEGRALS OF THE FIRST KIND: $F(k, \phi)$ (Continued)

$$F(k, \phi) = \int_0^{\phi} \frac{d\Phi}{\sqrt{1 - k^2 \sin^2 \Phi}}, \qquad \theta = \sin^{-1} k$$

ϕ \ θ	5°	10°	15°	20°	25°	30°	35°	40°	45°
46°	0.8034	0.8052	0.8080	0.8120	0.8170	0.8230	0.8300	0.8378	0.8462
47°	0.8209	0.8227	0.8258	0.8300	0.8353	0.8418	0.8492	0.8575	0.8666
48°	0.8384	0.8403	0.8436	0.8480	0.8537	0.8606	0.8685	0.8773	0.8870
49°	0.8559	0.8579	0.8614	0.8661	0.8721	0.8794	0.8878	0.8972	0.9076
50°	0.8734	0.8756	0.8792	0.8842	0.8905	0.8982	0.9072	0.9173	0.9283
51°	0.8909	0.8932	0.8970	0.9023	0.9090	0.9172	0.9267	0.9374	0.9491
52°	0.9084	0.9108	0.9148	0.9204	0.9275	0.9361	0.9462	0.9575	0.9701
53°	0.9259	0.9284	0.9326	0.9385	0.9460	0.9551	0.9658	0.9778	0.9912
54°	0.9434	0.9460	0.9505	0.9567	0.9646	0.9742	0.9855	0.9982	1.0124
55°	0.9609	0.9637	0.9683	0.9748	0.9832	0.9933	1.0052	1.0187	1.0337
56°	0.9784	0.9813	0.9862	0.9930	1.0018	1.0125	1.0250	1.0393	1.0552
57°	0.9959	0.9989	1.0041	1.0112	1.0204	1.0317	1.0449	1.0600	1.0768
58°	1.0134	1.0166	1.0219	1.0295	1.0391	1.0509	1.0648	1.0807	1.0985
59°	1.0309	1.0342	1.0398	1.0477	1.0578	1.0702	1.0848	1.1016	1.1204
60°	1.0484	1.0519	1.0577	1.0660	1.0766	1.0896	1.1049	1.1226	1.1424
61°	1.0659	1.0695	1.0757	1.0843	1.0953	1.1089	1.1250	1.1436	1.1646
62°	1.0834	1.0872	1.0936	1.1026	1.1141	1.1284	1.1452	1.1648	1.1868
63°	1.1009	1.1049	1.1115	1.1209	1.1330	1.1478	1.1655	1.1860	1.2093
64°	1.1184	1.1225	1.1295	1.1392	1.1518	1.1674	1.1859	1.2073	1.2318
65°	1.1359	1.1402	1.1474	1.1575	1.1707	1.1869	1.2063	1.2288	1.2545
66°	1.1534	1.1579	1.1654	1.1759	1.1896	1.2065	1.2267	1.2503	1.2773
67°	1.1709	1.1756	1.1833	1.1943	1.2085	1.2262	1.2472	1.2719	1.3002
68°	1.1884	1.1932	1.2013	1.2127	1.2275	1.2458	1.2678	1.2936	1.3232
69°	1.2059	1.2109	1.2193	1.2311	1.2465	1.2655	1.2885	1.3154	1.3464
70°	1.2234	1.2286	1.2373	1.2495	1.2655	1.2853	1.3092	1.3372	1.3697
71°	1.2410	1.2463	1.2553	1.2680	1.2845	1.3051	1.3299	1.3592	1.3931
72°	1.2585	1.2640	1.2733	1.2864	1.3036	1.3249	1.3507	1.3812	1.4167
73°	1.2760	1.2817	1.2913	1.3049	1.3226	1.3448	1.3715	1.4033	1.4403
74°	1.2935	1.2994	1.3093	1.3234	1.3417	1.3647	1.3924	1.4254	1.4640
75°	1.3110	1.3171	1.3273	1.3418	1.3608	1.3846	1.4134	1.4477	1.4879
76°	1.3285	1.3348	1.3454	1.3603	1.3800	1.4045	1.4344	1.4700	1.5118
77°	1.3460	1.3525	1.3634	1.3788	1.3991	1.4245	1.4554	1.4923	1.5359
78°	1.3636	1.3702	1.3814	1.3974	1.4183	1.4445	1.4765	1.5147	1.5600
79°	1.3811	1.3879	1.3995	1.4159	1.4374	1.4645	1.4976	1.5372	1.5842
80°	1.3986	1.4056	1.4175	1.4344	1.4566	1.4846	1.5187	1.5597	1.6085
81°	1.4161	1.4234	1.4356	1.4530	1.4758	1.5046	1.5399	1.5823	1.6328
82°	1.4336	1.4411	1.4536	1.4715	1.4950	1.5247	1.5611	1.6049	1.6572
83°	1.4512	1.4588	1.4717	1.4901	1.5143	1.5448	1.5823	1.6276	1.6817
84°	1.4687	1.4765	1.4897	1.5086	1.5335	1.5649	1.6035	1.6502	1.7062
85°	1.4862	1.4942	1.5078	1.5272	1.5527	1.5850	1.6248	1.6730	1.7308
86°	1.5037	1.5120	1.5259	1.5457	1.5720	1.6052	1.6461	1.6957	1.7554
87°	1.5212	1.5297	1.5439	1.5643	1.5912	1.6253	1.6673	1.7184	1.7801
88°	1.5388	1.5474	1.5620	1.5829	1.6105	1.6454	1.6886	1.7412	1.8047
89°	1.5563	1.5651	1.5801	1.6015	1.6297	1.6656	1.7099	1.7640	1.8294
90°	1.5738	1.5828	1.5981	1.6200	1.6490	1.6858	1.7312	1.7868	1.8541

*ELLIPTIC INTEGRALS OF THE FIRST KIND: $F(k, \phi)$ (Continued)

$$F(k, \phi) = \int_0^{\phi} \frac{d\Phi}{\sqrt{1 - k^2 \sin^2 \Phi}}, \qquad \theta = \sin^{-1} k$$

ϕ \ θ	50°	55°	60°	65°	70°	75°	80°	85°	90°
46°	0.8552	0.8643	0.8734	0.8821	0.8900	0.8968	0.9019	0.9052	0.9063
47°	0.8761	0.8860	0.8958	0.9053	0.9139	0.9212	0.9269	0.9304	0.9316
48°	0.8973	0.9079	0.9185	0.9287	0.9381	0.9461	0.9523	0.9561	0.9575
49°	0.9186	0.9300	0.9415	0.9525	0.9627	0.9714	0.9781	0.9824	0.9838
50°	0.9401	0.9523	0.9647	0.9766	0.9876	0.9971	1.0044	1.0091	1.0107
51°	0.9617	0.9748	0.9881	1.0010	1.0130	1.0233	1.0313	1.0364	1.0381
52°	0.9835	0.9976	1.0118	1.0258	1.0387	1.0499	1.0587	1.0642	1.0662
53°	1.0055	1.0205	1.0359	1.0509	1.0649	1.0771	1.0866	1.0927	1.0948
54°	1.0277	1.0437	1.0602	1.0764	1.0915	1.1048	1.1152	1.1219	1.1242
55°	1.0500	1.0672	1.0848	1.1022	1.1186	1.1331	1.1444	1.1517	1.1542
56°	1.0725	1.0908	1.1097	1.1285	1.1462	1.1619	1.1743	1.1823	1.1851
57°	1.0952	1.1147	1.1349	1.1551	1.1743	1.1914	1.2049	1.2136	1.2167
58°	1.1180	1.1389	1.1605	1.1822	1.2030	1.2215	1.2362	1.2458	1.2492
59°	1.1411	1.1632	1.1864	1.2097	1.2321	1.2522	1.2684	1.2789	1.2826
60°	1.1643	1.1879	1.2125	1.2376	1.2619	1.2837	1.3014	1.3129	1.3170
61°	1.1877	1.2128	1.2392	1.2660	1.2922	1.3159	1.3352	1.3480	1.3524
62°	1.2113	1.2379	1.2661	1.2949	1.3231	1.3490	1.3701	1.3841	1.3890
63°	1.2351	1.2633	1.2933	1.3242	1.3547	1.3828	1.4059	1.4214	1.4268
64°	1.2591	1.2890	1.3209	1.3541	1.3870	1.4175	1.4429	1.4599	1.4659
65°	1.2833	1.3149	1.3489	1.3844	1.4199	1.4532	1.4810	1.4998	1.5065
66°	1.3076	1.3411	1.3773	1.4153	1.4536	1.4898	1.5203	1.5411	1.5485
67°	1.3321	1.3675	1.4060	1.4467	1.4880	1.5274	1.5610	1.5840	1.5923
68°	1.3568	1.3942	1.4351	1.4786	1.5232	1.5661	1.6030	1.6287	1.6379
69°	1.3817	1.4212	1.4646	1.5111	1.5591	1.6059	1.6466	1.6752	1.6856
70°	1.4068	1.4484	1.4944	1.5441	1.5959	1.6468	1.6918	1.7237	1.7354
71°	1.4320	1.4759	1.5246	1.5777	1.6335	1.6891	1.7388	1.7745	1.7877
72°	1.4574	1.5036	1.5552	1.6118	1.6720	1.7326	1.7876	1.8277	1.8427
73°	1.4830	1.5315	1.5862	1.6465	1.7113	1.7774	1.8384	1.8837	1.9008
74°	1.5087	1.5597	1.6175	1.6818	1.7516	1.8237	1.8915	1.9427	1.9623
75°	1.5345	1.5882	1.6492	1.7176	1.7927	1.8715	1.9468	2.0050	2.0276
76°	1.5606	1.6168	1.6812	1.7540	1.8347	1.9207	2.0047	2.0711	2.0973
77°	1.5867	1.6457	1.7136	1.7909	1.8777	1.9716	2.0653	2.1414	2.1721
78°	1.6130	1.6748	1.7462	1.8284	1.9215	2.0240	2.1288	2.2164	2.2528
79°	1.6394	1.7040	1.7792	1.8664	1.9663	2.0781	2.1954	2.2969	2.3404
80°	1.6660	1.7335	1.8125	1.9048	2.0119	2.1339	2.2653	2.3836	2.4362
81°	1.6926	1.7631	1.8461	1.9438	2.0584	2.1913	2.3387	2.4775	2.5421
82°	1.7193	1.7929	1.8799	1.9831	2.1057	2.2504	2.4157	2.5795	2.6603
83°	1.7462	1.8228	1.9140	2.0229	2.1537	2.3110	2.4965	2.6911	2.7942
84°	1.7731	1.8528	1.9482	2.0630	2.2024	2.3731	2.5811	2.8136	2.9487
85°	1.8001	1.8830	1.9826	2.1035	2.2518	2.4366	2.6694	2.9487	3.1313
86°	1.8271	1.9132	2.0172	2.1442	2.3017	2.5013	2.7612	3.0978	3.3547
87°	1.8542	1.9435	2,0519	2.1852	2.3520	2.5670	2.8561	3.2620	3.6425
88°	1.8813	1.9739	2.0867	2.2263	2.4026	2.6336	2.9537	3.4412	4.0481
89°	1.9084	2.0043	2.1216	2.2675	2.4535	2.7007	3.0530	3.6328	4.7413
90°	1.9356	2.0347	2.1565	2.3088	2.5046	2.7681	3.1534	3.8317	——

ELLIPTIC INTEGRALS OF THE SECOND KIND: $E(k, \phi)$

$$E(k, \phi) = \int_0^{\phi} \sqrt{(1 - k^2 \sin^2 \Phi)}\, d\Phi, \quad \theta = \sin^{-1} k$$

ϕ \ θ	5°	10°	15°	20°	25°	30°	35°	40°	45°
1°	0.0175	0.0175	0.0175	0.0175	0.0175	0.0175	0.0175	0.0175	0.0175
2°	0.0349	0.0349	0.0349	0.0349	0.0349	0.0349	0.0349	0.0349	0.0349
3°	0.0524	0.0524	0.0524	0.0524	0.0524	0.0524	0.0524	0.0523	0.0523
4°	0.0698	0.0698	0.0698	0.0698	0.0698	0.0698	0.0698	0.0698	0.0698
5°	0.0873	0.0873	0.0873	0.0873	0.0872	0.0872	0.0872	0.0872	0.0872
6°	0.1047	0.1047	0.1047	0.1047	0.1047	0.1047	0.1047	0.1046	0.1046
7°	0.1222	0.1222	0.1222	0.1221	0.1221	0.1221	0.1221	0.1220	0.1220
8°	0.1396	0.1396	0.1396	0.1396	0.1395	0.1395	0.1395	0.1394	0.1394
9°	0.1571	0.1571	0.1570	0.1570	0.1570	0.1569	0.1569	0.1568	0.1568
10°	0.1745	0.1745	0.1745	0.1744	0.1744	0.1743	0.1742	0.1742	0.1741
11°	0.1920	0.1920	0.1919	0.1918	0.1918	0.1917	0.1916	0.1915	0.1914
12°	0.2094	0.2094	0.2093	0.2093	0.2092	0.2091	0.2089	0.2088	0.2087
13°	0.2269	0.2268	0.2268	0.2267	0.2265	0.2264	0.2263	0.2261	0.2259
14°	0.2443	0.2443	0.2442	0.2441	0.2439	0.2437	0.2436	0.2433	0.2431
15°	0.2618	0.2617	0.2616	0.2615	0.2613	0.2611	0.2608	0.2606	0.2603
16°	0.2792	0.2791	0.2790	0.2788	0.2786	0.2784	0.2781	0.2778	0.2775
17°	0.2967	0.2966	0.2964	0.2962	0.2959	0.2956	0.2953	0.2949	0.2946
18°	0.3141	0.3140	0.3138	0.3136	0.3133	0.3129	0.3125	0.3121	0.3116
19°	0.3316	0.3314	0.3312	0.3309	0.3305	0.3301	0.3296	0.3291	0.3286
20°	0.3490	0.3489	0.3486	0.3483	0.3478	0.3473	0.3468	0.3462	0.3456
21°	0.3665	0.3663	0.3660	0.3656	0.3651	0.3645	0.3639	0.3632	0.3625
22°	0.3839	0.3837	0.3834	0.3829	0.3823	0.3817	0.3809	0.3802	0.3793
23°	0.4013	0.4011	0.4007	0.4002	0.3996	0.3988	0.3980	0.3971	0.3961
24°	0.4188	0.4185	0.4181	0.4175	0.4168	0.4159	0.4150	0.4139	0.4129
25°	0.4362	0.4359	0.4354	0.4348	0.4339	0.4330	0.4319	0.4308	0.4296
26°	0.4537	0.4533	0.4528	0.4520	0.4511	0.4500	0.4488	0.4475	0.4462
27°	0.4711	0.4707	0.4701	0.4693	0.4682	0.4670	0.4657	0.4643	0.4628
28°	0.4886	0.4881	0.4874	0.4865	0.4854	0.4840	0.4825	0.4809	0.4793
29°	0.5060	0.5055	0.5048	0.5037	0.5025	0.5010	0.4993	0.4975	0.4957
30°	0.5234	0.5229	0.5221	0.5209	0.5195	0.5179	0.5161	0.5141	0.5120
31°	0.5409	0.5403	0.5394	0.5381	0.5366	0.5348	0.5327	0.5306	0.5283
32°	0.5583	0.5577	0.5567	0.5553	0.5536	0.5516	0.5494	0.5470	0.5446
33°	0.5757	0.5751	0.5740	0.5725	0.5706	0.5684	0.5660	0.5634	0.5607
34°	0.5932	0.5924	0.5912	0.5896	0.5876	0.5852	0.5826	0.5797	0.5768
35°	0.6106	0.6098	0.6085	0.6067	0.6045	0.6019	0.5991	0.5960	0.5928
36°	0.6280	0.6272	0.6258	0.6238	0.6214	0.6186	0.6155	0.6122	0.6087
37°	0.6455	0.6445	0.6430	0.6409	0.6383	0.6353	0.6319	0.6283	0.6245
38°	0.6629	0.6619	0.6602	0.6580	0.6552	0.6519	0.6483	0.6444	0.6403
39°	0.6803	0.6792	0.6775	0.6750	0.6720	0.6685	0.6646	0.6604	0.6559
40°	0.6977	0.6966	0.6947	0.6921	0.6888	0.6851	0.6808	0.6763	0.6715
41°	0.7152	0.7139	0.7119	0.7091	0.7056	0.7016	0.6970	0.6921	0.6870
42°	0.7326	0.7313	0.7291	0.7261	0.7224	0.7180	0.7132	0.7079	0.7024
43°	0.7500	0.7486	0.7463	0.7431	0.7391	0.7345	0.7293	0.7237	0.7178
44°	0.7674	0.7659	0.7634	0.7600	0.7558	0.7508	0.7453	0.7393	0.7330
45°	0.7849	0.7832	0.7806	0.7770	0.7725	0.7672	0.7613	0.7549	0.7482

ELLIPTIC INTEGRALS OF THE SECOND KIND: $E(k, \phi)$ (Continued)

$$E(k, \phi) = \int_0^{\phi} \sqrt{(1 - k^2 \sin^2 \Phi)}\, d\Phi, \quad \theta = \sin^{-1} k$$

ϕ \ θ	50°	55°	60°	65°	70°	75°	80°	85°	90°
1°	0.0175	0.0175	0.0175	0.0175	0.0175	0.0175	0.0175	0.0175	0.0175
2°	0.0349	0.0349	0.0349	0.0349	0.0349	0.0349	0.0349	0.0349	0.0349
3°	0.0523	0.0523	0.0523	0.0523	0.0523	0.0523	0.0523	0.0523	0.0523
4°	0.0698	0.0698	0.0698	0.0698	0.0698	0.0698	0.0698	0.0698	0.0698
5°	0.0872	0.0872	0.0872	0.0872	0.0872	0.0872	0.0872	0.0872	0.0872
6°	0.1046	0.1046	0.1046	0.1046	0.1046	0.1045	0.1045	0.1045	0.1045
7°	0.1220	0.1220	0.1219	0.1219	0.1219	0.1219	0.1219	0.1219	0.1219
8°	0.1394	0.1393	0.1393	0.1393	0.1392	0.1392	0.1392	0.1392	0.1392
9°	0.1567	0.1566	0.1566	0.1566	0.1565	0.1565	0.1565	0.1564	0.1564
10°	0.1740	0.1739	0.1739	0.1738	0.1738	0.1737	0.1737	0.1737	0.1736
11°	0.1913	0.1912	0.1911	0.1910	0.1909	0.1909	0.1908	0.1908	0.1908
12°	0.2085	0.2084	0.2083	0.2082	0.2081	0.2080	0.2080	0.2079	0.2079
13°	0.2258	0.2256	0.2254	0.2253	0.2252	0.2251	0.2250	0.2250	0.2250
14°	0.2429	0.2427	0.2425	0.2424	0.2422	0.2421	0.2420	0.2419	0.2419
15°	0.2601	0.2598	0.2596	0.2594	0.2592	0.2590	0.2589	0.2588	0.2588
16°	0.2771	0.2768	0.2765	0.2763	0.2761	0.2759	0.2757	0.2757	0.2756
17°	0.2942	0.2938	0.2935	0.2932	0.2929	0.2927	0.2925	0.2924	0.2924
18°	0.3112	0.3107	0.3103	0.3099	0.3096	0.3094	0.3092	0.3091	0.3090
19°	0.3281	0.3276	0.3271	0.3267	0.3263	0.3260	0.3258	0.3256	0.3256
20°	0.3450	0.3444	0.3438	0.3433	0.3429	0.3425	0.3422	0.3421	0.3420
21°	0.3618	0.3611	0.3604	0.3598	0.3593	0.3589	0.3586	0.3584	0.3584
22°	0.3785	0.3777	0.3770	0.3763	0.3757	0.3752	0.3749	0.3747	0.3746
23°	0.3952	0.3943	0.3935	0.3927	0.3920	0.3915	0.3911	0.3908	0.3907
24°	0.4118	0.4108	0.4098	0.4090	0.4082	0.4076	0.4071	0.4068	0.4067
25°	0.4284	0.4272	0.4261	0.4251	0.4243	0.4236	0.4230	0.4227	0.4226
26°	0.4449	0.4436	0.4423	0.4412	0.4402	0.4394	0.4389	0.4385	0.4384
27°	0.4613	0.4598	0.4584	0.4572	0.4561	0.4552	0.4545	0.4541	0.4540
28°	0.4776	0.4760	0.4744	0.4730	0.4718	0.4708	0.4701	0.4696	0.4695
29°	0.4938	0.4920	0.4903	0.4887	0.4874	0.4863	0.4855	0.4850	0.4848
30°	0.5100	0.5080	0.5061	0.5044	0.5029	0.5016	0.5007	0.5002	0.5000
31°	0.5261	0.5239	0.5218	0.5199	0.5182	0.5169	0.5159	0.5152	0.5150
32°	0.5421	0.5396	0.5373	0.5352	0.5334	0.5319	0.5308	0.5301	0.5299
33°	0.5580	0.5553	0.5528	0.5505	0.5485	0.5468	0.5456	0.5449	0.5446
34°	0.5738	0.5709	0.5681	0.5656	0.5634	0.5616	0.5603	0.5595	0.5592
35°	0.5895	0.5863	0.5833	0.5806	0.5782	0.5762	0.5748	0.5739	0.5736
36°	0.6051	0.6017	0.5984	0.5954	0.5928	0.5907	0.5891	0.5881	0.5878
37°	0.6207	0.6169	0.6134	0.6101	0.6073	0.6050	0.6032	0.6022	0.6018
38°	0.6361	0.6321	0.6282	0.6247	0.6216	0.6191	0.6172	0.6160	0.6157
39°	0.6515	0.6471	0.6429	0.6391	0.6357	0.6330	0.6310	0.6297	0.6293
40°	0.6667	0.6620	0.6575	0.6533	0.6497	0.6468	0.6446	0.6432	0.6428
41°	0.6818	0.6767	0.6719	0.6674	0.6636	0.6604	0.6580	0.6566	0.6561
42°	0.6969	0.6914	0.6862	0.6814	0.6772	0.6738	0.6712	0.6697	0.6691
43°	0.7118	0.7059	0.7003	0.6952	0.6907	0.6870	0.6843	0.6826	0.6820
44°	0.7266	0.7204	0.7144	0.7088	0.7040	0.7000	0.6971	0.6953	0.6947
45°	0.7414	0.7346	0.7282	0.7223	0.7171	0.7129	0.7097	0.7078	0.7071

ELLIPTIC INTEGRALS OF THE SECOND KIND: $E(k, \phi)$ (Continued)

$$E(k, \phi) = \int_0^{\phi} \sqrt{(1 - k^2 \sin^2 \Phi)}\, d\Phi, \quad \theta = \sin^{-1} k$$

ϕ \ θ	5°	10°	15°	20°	25°	30°	35°	40°	45°
46°	0.8023	0.8006	0.7977	0.7939	0.7891	0.7835	0.7772	0.7704	0.7633
47°	0.8197	0.8179	0.8149	0.8108	0.8057	0.7998	0.7931	0.7858	0.7782
48°	0.8371	0.8352	0.8320	0.8277	0.8223	0.8160	0.8089	0.8012	0.7931
49°	0.8545	0.8525	0.8491	0.8446	0.8389	0.8322	0.8247	0.8165	0.8079
50°	0.8719	0.8698	0.8663	0.8614	0.8554	0.8483	0.8404	0.8317	0.8227
51°	0.8894	0.8871	0.8834	0.8783	0.8719	0.8644	0.8560	0.8469	0.8373
52°	0.9068	0.9044	0.9004	0.8951	0.8884	0.8805	0.8716	0.8620	0.8518
53°	0.9242	0.9217	0.9175	0.9119	0.9048	0.8965	0.8872	0.8770	0.8663
54°	0.9416	0.9389	0.9345	0.9287	0.9212	0.9125	0.9026	0.8919	0.8806
55°	0.9590	0.9562	0.9517	0.9454	0.9376	0.9284	0.9181	0.9068	0.8949
56°	0.9764	0.9735	0.9687	0.9622	0.9540	0.9443	0.9335	0.9216	0.9091
57°	0.9938	0.9908	0.9858	0.9789	0.9703	0.9602	0.9488	0.9363	0.9232
58°	1.0112	1.0080	1.0028	0.9956	0.9866	0.9760	0.9641	0.9510	0.9372
59°	1.0286	1.0253	1.0198	1.0123	1.0029	1.9918	0.9793	0.9656	0.9511
60°	1.0460	1.0426	1.0368	1.0290	1.0191	1.0076	0.9945	0.9801	0.9650
61°	1.0634	1.0598	1.0538	1.0456	1.0354	1.0233	1.0096	0.9946	0.9787
62°	1.0808	1.0771	1.0708	1.0623	1.0516	1.0389	1.0246	1 0090	0.9924
63°	1.0982	1.0943	1.0878	1.0789	1.0678	1.0546	1.0397	1.0233	1.0060
64°	1.1156	1.1115	1.1048	1.0955	1.0839	1.0702	1.0547	1.0376	1.0195
65°	1.1330	1.1288	1.1218	1.1121	1.1001	1.0858	1.0696	1.0518	1.0329
66°	1.1504	1.1460	1.1387	1.1287	1.1162	1.1013	1.0845	1.0660	1.0463
67°	1.1678	1.1632	1.1557	1.1453	1.1323	1.1168	1.0993	1.0801	1.0596
68°	1.1852	1.1805	1.1726	1.1618	1.1483	1.1323	1.1141	1.0941	1.0728
69°	1.2026	1.1977	1.1896	1.1784	1.1644	1.1478	1.1289	1.1081	1.0859
70°	1.2200	1.2149	1.2065	1.1949	1.1804	1.1632	1.1436	1.1221	1.0990
71°	1.2374	1.2321	1.2234	1.2114	1.1964	1.1786	1.1583	1.1359	1.1120
72°	1.2548	1.2493	1.2403	1.2280	1.2124	1.1939	1.1729	1.1498	1.1250
73°	1.2722	1.2666	1.2573	1.2445	1.2284	1.2093	1.1875	1.1636	1.1379
74°	1.2896	1.2838	1.2742	1.2609	1.2443	1.2246	1.2021	1.1773	1.1507
75°	1.3070	1.3010	1.2911	1.2774	1.2603	1.2399	1.2167	1.1910	1.1635
76°	1.3244	1.3182	1.3080	1.2939	1.2762	1.2552	1.2312	1.2047	1.1762
77°	1.3418	1.3354	1.3249	1.3104	1.2921	1.2704	1.2457	1.2183	1.1889
78°	1.3592	1.3526	1.3417	1.3268	1.3080	1.2856	1.2601	1.2319	1.2015
79°	1.3765	1.3698	1.3586	1.3432	1.3239	1.3009	1.2746	1.2454	1.2141
80°	1.3939	1.3870	1.3755	1.3597	1.3398	1.3161	1.2890	1.2590	1.2266
81°	1.4113	1.4042	1.3924	1.3761	1.3556	1.3312	1.3034	1.2725	1.2391
82°	1.4287	1.4214	1.4093	1.3925	1.3715	1.3464	1.3177	1.2859	1.2516
83°	1.4461	1.4386	1.4261	1.4090	1.3873	1.3616	1.3321	1.2994	1.2640
84°	1.4635	1.4558	1.4430	1.4254	1.4032	1.3767	1.3464	1.3128	1.2765
85°	1.4809	1.4729	1.4598	1.4418	1.4190	1.3919	1.3608	1.3262	1.2889
86°	1.4983	1.4901	1.4767	1.4582	1.4348	1.4070	1.3751	1.3396	1.3012
87°	1.5156	1.5073	1.4936	1.4746	1.4507	1.4221	1.3894	1.3530	1.3136
88°	1.5330	1.5245	1.5104	1.4910	1.4665	1.4372	1.4037	1.3664	1.3260
89°	1.5504	1.5417	1.5273	1.5074	1.4823	1.4523	1.4180	1.3798	1.3383
90°	1.5678	1.5589	1.5442	1.5238	1.4981	1.4675	1.4323	1.3931	1.3506

ELLIPTIC INTEGRALS OF THE SECOND KIND: $E(k, \phi)$ (Continued)

$$E(k, \phi) = \int_0^{\phi} \sqrt{(1 - k^2 \sin^2 \Phi)}\, d\Phi, \quad \theta = \sin^{-1} k$$

ϕ \ θ	50°	55°	60°	65°	70°	75°	80°	85°	90°
46°	0.7560	0.7488	0.7419	0.7356	0.7301	0.7255	0.7221	0.7200	0.7193
47°	0.7705	0.7628	0.7555	0.7488	0.7429	0.7380	0.7344	0.7321	0.7314
48°	0.7849	0.7768	0.7690	0.7618	0.7555	0.7502	0.7464	0.7440	0.7431
49°	0.7992	0.7905	0.7822	0.7746	0.7679	0.7623	0.7581	0.7556	0.7547
50°	0.8134	0.8042	0.7954	0.7872	0.7801	0.7741	0.7697	0.7670	0.7660
51°	0.8275	0.8177	0.8084	0.7997	0.7921	0.7858	0.7811	0.7781	0.7771
52°	0.8414	0.8311	0.8212	0.8120	0.8039	0.7972	0.7922	0.7891	0.7880
53°	0.8553	0.8444	0.8339	0.8241	0.8155	0.8084	0.8031	0.7998	0.7986
54°	0.8690	0.8575	0.8464	0.8361	0.8270	0.8194	0.8137	0.8102	0.8090
55°	0.8827	0.8705	0.8588	0.8479	0.8382	0.8302	0.8242	0.8204	0.8192
56°	0.8962	0.8834	0.8710	0.8595	0.8493	0.8408	0.8344	0.8304	0.8290
57°	0.9096	0.8961	0.8831	0.8709	0.8601	0.8511	0.8443	0.8401	0.8387
58°	0.9230	0.9088	0.8950	0.8822	0.8707	0.8612	0.8540	0.8496	0.8480
59°	0.9362	0.9213	0.9068	0.8932	0.8812	0.8711	0.8635	0.8588	0.8572
60°	0.9493	0.9336	0.9184	0.9042	0.8914	0.8808	0.8728	0.8677	0.8660
61°	0.9623	0.9459	0.9299	0.9149	0.9015	0.8903	0.8817	0.8764	0.8746
62°	0.9752	0.9580	0.9412	0.9254	0.9113	0.8995	0.8905	0.8849	0.8829
63°	0.9880	0.9700	0.9524	0.9358	0.9210	0.9085	0.8990	0.8930	0.8910
64°	1.0007	0.9818	0.9634	0.9460	0.9304	0.9173	0.9072	0.9009	0.8988
65°	1.0133	0.9936	0.9743	0.9561	0.9397	0.9258	0.9152	0.9086	0.9063
66°	1.0258	1.0052	0.9850	0.9659	0.9487	0.9341	0.9230	0.9159	0.9135
67°	1.0383	1.0167	0.9956	0.9756	0.9576	0.9422	0.9305	0.9230	0.9205
68°	1.0506	1.0281	1.0061	0.9852	0.9662	0.9501	0.9377	0.9299	0.9272
69°	1.0628	1.0394	1.0164	0.9946	0.9747	0.9578	0.9447	0.9364	0.9336
70°	1.0750	1.0506	1.0266	1.0038	0.9830	0.9652	0.9514	0.9427	0.9397
71°	1.0871	1.0617	1.0367	1.0129	0.9911	0.9724	0.9579	0.9487	0.9455
72°	1,0991	1.0727	1.0467	1.0218	0.9990	0.9794	0.9642	0.9544	0.9511
73°	1.1110	1.0836	1.0565	1.0306	1.0067	0.9862	0.9702	0.9599	0.9563
74°	1.1228	1.0944	1.0662	1.0392	1.0143	0.9928	0.9759	0.9650	0.9613
75°	1.1346	1.1051	1.0759	1.0477	1.0217	0.9992	0.9814	0.9699	0.9659
76°	1.1463	1.1158	1.0854	1.0561	1.0290	1.0053	0.9867	0.9745	0.9703
77°	1.1580	1.1263	1.0948	1.0643	1.0361	1.0113	0.9917	0.9789	0.9744
78°	1.1695	1.1368	1.1041	1.0724	1.0430	1.0171	0.9965	0.9829	0.9781
79°	1.1811	1.1472	1.1133	1.0805	1.0498	1.0228	1.0011	0.9867	0.9816
80°	1.1926	1.1576	1.1225	1.0884	1.0565	1.0282	1.0054	0.9902	0.9848
81°	1.2040	1.1678	1.1316	1.0962	1.0630	1.0335	1.0096	0.9935	0.9877
82°	1.2154	1.1781	1.1406	1.1040	1.0695	1.0387	1.0135	0.9965	0.9903
83°	1.2267	1.1883	1.1495	1.1116	1.0758	1.0437	1.0173	0.9992	0.9925
84°	1.2381	1.1984	1.1584	1.1192	1.0821	1.0486	1.0209	1.0017	0.9945
85°	1.2493	1.2085	1.1673	1.1267	1.0882	1.0534	1.0244	1.0039	0.9962
86°	1.2606	1.2186	1.1761	1.1342	1.0944	1.0581	1.0277	1.0060	0.9976
87°	1.2719	1.2286	1.1848	1.1417	1.1004	1.0628	1.0309	1.0078	0.9986
88°	1.2831	1.2386	1.1936	1.1491	1.1064	1.0673	1.0340	1.0095	0.9994
89°	1.2943	1.2487	1.2023	1.1565	1.1124	1.0719	1.0371	1.0111	0.9998
90°	1.3055	1.2587	1.2111	1.1638	1.1184	1.0764	1.0401	1.0127	1.0000

COMPLETE ELLIPTIC INTEGRALS

$$K = \int_0^{\pi/2} \frac{d\Phi}{\sqrt{1 - k^2 \sin^2 \Phi}} = F\left(k, \frac{\pi}{2}\right)$$

$\sin^{-1} k$	K	$\log K$	$\sin^{-1} k$	K	$\log K$
0°	1.5708	0.196120	**40°**	1.7868	0.252068
1	1.5709	0.196153	41	1.7992	0.255085
2	1.5713	0.196252	42	1.8122	0.258197
3	1.5719	0.196418	43	1.8256	0.261406
4	1.5727	0.196649	44	1.8396	0.264716
5	1.5738	0.196947	**45**	1.8541	0.268127
6	1.5751	0.197312	46	1.8691	0.271644
7	1.5767	0.197743	47	1.8848	0.275267
8	1.5785	0.198241	48	1.9011	0.279001
9	1.5805	0.198806	49	1.9180	0.282848
10	1.5828	0.199438	**50**	1.9356	0.286811
11	1.5854	0.200137	51	1.9539	0.290895
12	1.5882	0.200904	52	1.9729	0.295101
13	1.5913	0.201740	53	1.9927	0.299435
14	1.5946	0.202643	54	2.0133	0.303901
15	1.5981	0.203615	**55**	2.0347	0.308504
16	1.6020	0.204657	56	2.0571	0.313247
17	1.6061	0.205768	57	2.0804	0.318138
18	1.6105	0.206948	58	2.1047	0.323182
19	1.6151	0.208200	59	2.1300	0.328384
20	1.6200	0.209522	**60**	2.1565	0.333753
21	1.6252	0.210916	61	2.1842	0.339295
22	1.6307	0.212382	62	2.2132	0.345020
23	1.6365	0.213921	63	2.2435	0.350936
24	1.6426	0.215533	64	2.2754	0.357053
25	1.6490	0.217219	**65**	2.3088	0.363384
26	1.6557	0.218981	66	2.3439	0.369940
27	1.6627	0.220818	67	2.3809	0.376736
28	1.6701	0.222732	68	2.4198	0.383787
29	1.6777	0.224723	69	2.4610	0.391112
30	1.6858	0.226793	**70**	2.5046	0.398730
31	1.6941	0.228943	71	2.5507	0.406665
32	1.7028	0.231173	72	2.5998	0.414943
33	1.7119	0.233485	73	2.6521	0.423596
34	1.7214	0.235880	74	2.7081	0.432660
35	1.7312	0.238359	**75**	2.7681	0.442176
36	1.7415	0.240923	76	2.8327	0.452196
37	1.7522	0.243575	77	2.9026	0.462782
38	1.7633	0.246315	78	2.9786	0.474008
39	1.7748	0.249146	79	3.0617	0.485967
40	1.7868	0.252068	**80**	3.1534	0.498777

COMPLETE ELLIPTIC INTEGRALS (Continued)

$$K = \int_0^{\pi/2} \frac{d\Phi}{\sqrt{1 - k^2 \sin^2 \Phi}} = F\left(k, \frac{\pi}{2}\right)$$

$\sin^{-1} k$	K	$\log K$	$\sin^{-1} k$	K	$\log K$
80°	3.1534	0.498777	**85°**	3.8317	0.583396
81	3.2553	0.512591	86	4.0528	0.607751
82	3.3699	0.527613	87	4.3387	0.637355
83	3.5004	0.544120	88	4.7427	0.676027
84	3.6519	0.562514	89	5.4349	0.735192
85	3.8317	0.583396	**90**	∞	∞

Values of K for $\sin^{-1} k$ = 85° to 89° by 0.1° and 89° to 90° by minutes

$\sin^{-1} k$	K	$\log K$	$\sin^{-1} k$	K	$\log K$
85.0°	3.832	0.58343	**89° 0′**	5.435	0.73520
85.1	3.852	0.58569	89 2	5.469	0.73791
85.2	3.872	0.58794	89 4	5.504	0.74068
85.3	3.893	0.59028	89 6	5.540	0.74351
85.4	3.914	0.59262	89 8	5.578	0.74648
85.5	3.936	0.59506	**89 10**	5.617	0.74950
85.6	3.958	0.59748	89 12	5.658	0.75266
85.7	3.981	0.59999	89 14	5.700	0.75587
85.8	4.004	0.60249	89 16	5.745	0.75929
85.9	4.028	0.60509	89 18	5.791	0.76275
86.0	4.053	0.60778	**89 20**	5.840	0.76641
86.1	4.078	0.61045	89 22	5.891	0.77019
86.2	4.104	0.61321	89 24	5.946	0.77422
86.3	4.130	0.61595	89 26	6.003	0.77837
86.4	4.157	0.61878	89 28	6.063	0.78269
86.5	4.185	0.62170	**89 30**	6.128	0.78732
86.6	4.214	0.62469	89 32	6.197	0.79218
86.7	4.244	0.62778	89 34	6.271	0.79734
86.8	4.274	0.63083	89 36	6.351	0.80284
86.9	4.306	0.63407	89 38	6.438	0.80875
87.0	4.339	0.63739	**89 40**	6.533	0.81511
87.1	4.372	0.64068	89 41	6.584	0.81849
87.2	4.407	0.64414	89 42	6.639	0.82210
87.3	4.444	0.64777	89 43	6.696	0.82582
87.4	4.481	0.65137	89 44	6.756	0.82969
87.5	4.520	0.65514	**89 45**	6.821	0.83385
87.6	4.561	0.65916	89 46	6.890	0.83822
87.7	4.603	0.66304	89 47	6.964	0.84286
87.8	4.648	0.66727	89 48	7.044	0.84782
87.9	4.694	0.67154	89 49	7.131	0.85315
88.0	4.743	0.67605	**89 50**	7.226	0.85890
88.1	4.794	0.68070	89 51	7.332	0.86522
88.2	4.848	0.68556	89 52	7.449	0.87210
88.3	4.905	0.69064	89 53	7.583	0.87984
88.4	4.965	0.69592	89 54	7.737	0.88857
88.5	5.030	0.70157	**89 55**	7.919	0.89867
88.6	5.099	0.70749	89 56	8.143	0.91078
88.7	5.173	0.71374	89 57	8.430	0.92583
88.8	5.253	0.72041	89 58	8.836	0.94626
88.9	5.340	0.72754	89 59	9.529	0.97905
89.0	5.435	0.73520	**90 0**	∞	∞

COMPLETE ELLIPTIC INTEGRALS (Continued)

$$E = \int_0^{\pi/2} \sqrt{1 - k^2 \sin^2 \Phi} \cdot d\Phi = E\left(k, \frac{\pi}{2}\right)$$

$\sin^{-1} k$	E	log E	$\sin^{-1} k$	E	log E
0°	1.5708	0.196120	**45°**	1.3506	0.130541
1	1.5707	0.196087	46	1.3418	0.127690
2	1.5703	0.195988	47	1.3329	0.124788
3	1.5697	0.195822	48	1.3238	0.121836
4	1.5689	0.195591	49	1.3147	0.118836
5	1.5678	0.195293	**50**	1.3055	0.115790
6	1.5665	0.194930	51	1.2963	0.112698
7	1.5649	0.194500	52	1.2870	0.109563
8	1.5632	0.194004	53	1.2776	0.106386
9	1.5611	0.193442	54	1.2681	0.103169
10	1.5589	0.192815	**55**	1.2587	0.099915
11	1.5564	0.192121	56	1.2492	0.096626
12	1.5537	0.191362	57	1.2397	0.093303
13	1.5507	0.190537	58	1.2301	0.089950
14	1.5476	0.189646	59	1.2206	0.086569
15	1.5442	0.188690	**60**	1.2111	0.083164
16	1.5405	0.187668	61	1.2015	0.079738
17	1.5367	0.186581	62	1.1920	0.076293
18	1.5326	0.185428	63	1.1826	0.072834
19	1.5283	0.184210	64	1.1732	0.069364
20	1.5238	0.182928	**65**	1.1638	0.065889
21	1.5191	0.181580	66	1.1545	0.062412
22	1.5141	0.180168	67	1.1453	0.058937
23	1.5090	0.178691	68	1.1362	0.055472
24	1.5037	0.177150	69	1.1272	0.052020
25	1.4981	0.175545	**70**	1.1184	0.048589
26	1.4924	0.173876	71	1.1096	0.045183
27	1.4864	0.172144	72	1.1011	0.041812
28	1.4803	0.170348	73	1.0927	0.038481
29	1.4740	0.168489	74	1.0844	0.035200
30	1.4675	0.166567	**75**	1.0764	0.031976
31	1.4608	0.164583	76	1.0686	0.028819
32	1.4539	0.162537	77	1.0611	0.025740
33	1.4469	0.160429	78	1.0538	0.022749
34	1.4397	0.158261	79	1.0468	0.019858
35	1.4323	0.156031	**80**	1.0401	0.017081
36	1.4248	0.153742	81	1.0338	0.014432
37	1.4171	0.151393	82	1.0278	0.011927
38	1.4092	0.148985	83	1.0223	0.009584
39	1.4013	0.146519	84	1.0172	0.007422
40	1.3931	0.143995	**85**	1.0127	0.005465
41	1.3849	0.141414	86	1.0086	0.003740
42	1.3765	0.138778	87	1.0053	0.002278
43	1.3680	0.136086	88	1.0026	0.001121
44	1.3594	0.133340	89	1.0008	0.000326
45	1.3506	0.130541	**90**	1.0000	0.000000

VECTOR ANALYSIS

Definitions

Any quantity which is completely determined by its magnitude is called a *scalar*. Examples of such are mass, density, temperature, etc. Any quantity which is completely determined by its magnitude and direction is called a *vector*. Examples of such are velocity, acceleration, force, etc. A vector quantity is represented by a directed line segment, the length of which represents the magnitude of the vector. A vector quantity is usually represented by a boldfaced letter such as $\mathbf{V}$. Two vectors $\mathbf{V}_1$ and $\mathbf{V}_2$ are equal to one another if they have equal magnitudes and are acting in the same directions. A negative vector, written as $-\mathbf{V}$, is one which acts in the opposite direction to $\mathbf{V}$, but is of equal magnitude to it. If we represent the magnitude of $\mathbf{V}$ by v, we write $\mathbf{V} = v$. A vector parallel to $\mathbf{V}$, but equal to the reciprocal of its magnitude is written as $\mathbf{V}^{-1}$ or as $\frac{1}{\mathbf{V}}$.

The *unit vector* $\frac{\mathbf{V}}{v}$ $(v \neq 0)$ is that vector which has the same direction as $\mathbf{V}$, but has a magnitude of unity (sometimes represented as $\mathbf{V}_0$ or $\hat{\mathbf{v}}$).

Vector Algebra

The vector sum of $\mathbf{V}_1$ and $\mathbf{V}_2$ is represented by $\mathbf{V}_1 + \mathbf{V}_2$. The vector sum of $\mathbf{V}_1$ and $-\mathbf{V}_2$, or the difference of the vector $\mathbf{V}_2$ from $\mathbf{V}_1$ is represented by $\mathbf{V}_1 - \mathbf{V}_2$.

If r is a scalar, then $r\mathbf{V} = \mathbf{V}r$, and represents a vector r times the magnitude of $\mathbf{V}$, in the same direction as $\mathbf{V}$ if r is positive, and in the opposite direction if r is negative. If r and s are scalars, $\mathbf{V}_1$, $\mathbf{V}_2$, $\mathbf{V}_3$, vectors, then the following rules of scalars and vectors hold:

$$\mathbf{V}_1 + \mathbf{V}_2 = \mathbf{V}_2 + \mathbf{V}_1$$

$$(r + s)\mathbf{V}_1 = r\mathbf{V}_1 + s\mathbf{V}_1; \qquad r(\mathbf{V}_1 + \mathbf{V}_2) = r\mathbf{V}_1 + r\mathbf{V}_2$$

$$\mathbf{V}_1 + (\mathbf{V}_2 + \mathbf{V}_3) = (\mathbf{V}_1 + \mathbf{V}_2) + \mathbf{V}_3 = \mathbf{V}_1 + \mathbf{V}_2 + \mathbf{V}_3$$

Vectors in Space

A plane is described by two distinct vectors $\mathbf{V}_1$ and $\mathbf{V}_2$. Should these vectors not intersect each other, then one is displaced parallel to itself until they do (fig. 1.) Any other vector $\mathbf{V}$ lying in this plane is given by

$$\mathbf{V} = r\mathbf{V}_1 + s\mathbf{V}_2$$

VECTOR ANALYSIS (Continued)

A *position vector* specifies the position in space of a point relative to a fixed origin. If therefore $\mathbf{V}_1$ and $\mathbf{V}_2$ are the position vectors of the points A and B, relative to the origin O, then any point P on the line AB has a position vector $\mathbf{V}$ given by

$$\mathbf{V} = r\mathbf{V}_1 + (1 - r)\mathbf{V}_2$$

The scalar "r" can be taken as the parametric representation of P since $r = 0$ implies $P = B$ and $r = 1$ implies $P = A$. (fig. 2). If P divides the line AB in the ratio $r:s$ then

$$\mathbf{V} = \left(\frac{r}{r+s}\right)\mathbf{V}_1 + \left(\frac{s}{r+s}\right)\mathbf{V}_2$$

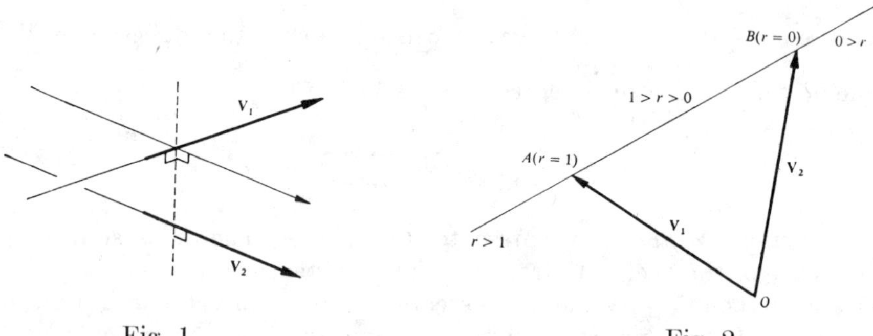

Fig. 1. Fig. 2.

The vectors $\mathbf{V}_1, \mathbf{V}_2, \mathbf{V}_3, \ldots, \mathbf{V}_n$ are said to be *linearly dependent* if there exist scalars $r_1, r_2, r_3, \ldots, r_n$, not all zero, such that

$$r_1\mathbf{V}_1 + r_2\mathbf{V}_2 + \cdots + r_n\mathbf{V}_n = 0$$

A vector $\mathbf{V}$ is linearly dependent upon the set of vectors $\mathbf{V}_1, \mathbf{V}_2, \mathbf{V}_3, \ldots, \mathbf{V}_n$ if

$$\mathbf{V} = r_1\mathbf{V}_1 + r_2\mathbf{V}_2 + r_3\mathbf{V}_3 + \cdots + r_n\mathbf{V}_n$$

Three vectors are linearly dependent if and only if they are co-planar.

All points in space can be uniquely determined by linear dependence upon three *base vectors* i.e. three vectors any one of which is linearly independent of the other two. The simplest set of base vectors are the unit vectors along the coordinate Ox, Oy and Oz axes. These are usually designated by $\mathbf{i}$, $\mathbf{j}$ and $\mathbf{k}$ respectively.

If $\mathbf{V}$ is a vector in space, and a, b and c are the respective magnitudes of the projections of the vector along the axes then

$$\mathbf{V} = a\mathbf{i} + b\mathbf{j} + c\mathbf{k}$$

and

$$v = \sqrt{a^2 + b^2 + c^2}$$

and the direction cosines of $\mathbf{V}$ are

$$\cos\alpha = a/v, \quad \cos\beta = b/v, \quad \cos\gamma = c/v.$$

The law of addition yields

$$\mathbf{V}_1 + \mathbf{V}_2 = (a_1 + a_2)\mathbf{i} + (b_1 + b_2)\mathbf{j} + (c_1 + c_2)\mathbf{k}$$

The Scalar, Dot, or Inner Product of Two Vectors $\mathbf{V}_1$ and $\mathbf{V}_2$

This product is represented as $\mathbf{V}_1 \cdot \mathbf{V}_2$ and is defined to be equal to $v_1 v_2 \cos\theta$, where θ is the angle from $\mathbf{V}_1$ to $\mathbf{V}_2$, i.e.,

$$\mathbf{V}_1 \cdot \mathbf{V}_2 = v_1 v_2 \cos\theta$$

The following rules apply for this product:

$$\mathbf{V}_1 \cdot \mathbf{V}_2 = a_1 a_2 + b_1 b_2 + c_1 c_2 = \mathbf{V}_2 \cdot \mathbf{V}_1$$

It should be noted that this verifies that scalar multiplication is commutative.

$$\begin{aligned}(\mathbf{V}_1 + \mathbf{V}_2) \cdot \mathbf{V}_3 &= \mathbf{V}_1 \cdot \mathbf{V}_3 + \mathbf{V}_2 \cdot \mathbf{V}_3 \\ \mathbf{V}_1 \cdot (\mathbf{V}_2 + \mathbf{V}_3) &= \mathbf{V}_1 \cdot \mathbf{V}_2 + \mathbf{V}_1 \cdot \mathbf{V}_3\end{aligned}$$

If $\mathbf{V}_1$ is perpendicular to $\mathbf{V}_2$ then $\mathbf{V}_1 \cdot \mathbf{V}_2 = 0$, and if $\mathbf{V}_1$ is parallel to $\mathbf{V}_2$ then $\mathbf{V}_1 \cdot \mathbf{V}_2 = v_1 v_2 = rw_1^2$

In particular

$$\mathbf{i} \cdot \mathbf{i} = \mathbf{j} \cdot \mathbf{j} = \mathbf{k} \cdot \mathbf{k} = 1,$$

and

$$\mathbf{i} \cdot \mathbf{j} = \mathbf{j} \cdot \mathbf{k} = \mathbf{k} \cdot \mathbf{i} = 0$$

The Vector or Cross Product of Vectors $\mathbf{V}_1$ and $\mathbf{V}_2$

This product is represented as $\mathbf{V}_1 \times \mathbf{V}_2$ and is defined to be equal to $v_1 v_2 (\sin\theta)\mathbf{1}$, where θ is the angle from $\mathbf{V}_1$ to $\mathbf{V}_2$ and $\mathbf{1}$ is a unit vector perpendicular to the plane of $\mathbf{V}_1$ and $\mathbf{V}_2$ and so directed that a right-handed screw driven in the direction of $\mathbf{1}$ would carry $\mathbf{V}_1$ into $\mathbf{V}_2$, i.e.,

$$\mathbf{V}_1 \times \mathbf{V}_2 = v_1 v_2 (\sin\theta)\mathbf{1}$$

and

$$\tan\theta = \frac{|\mathbf{V}_1 \times \mathbf{V}_2|}{\mathbf{V}_1 \cdot \mathbf{V}_2}$$

The following rules apply for vector products:

$$\begin{aligned}\mathbf{V}_1 \times \mathbf{V}_2 &= -\mathbf{V}_2 \times \mathbf{V}_1 \\ \mathbf{V}_1 \times (\mathbf{V}_2 + \mathbf{V}_3) &= \mathbf{V}_1 \times \mathbf{V}_2 + \mathbf{V}_1 \times \mathbf{V}_3 \\ (\mathbf{V}_1 + \mathbf{V}_2) \times \mathbf{V}_3 &= \mathbf{V}_1 \times \mathbf{V}_3 + \mathbf{V}_2 \times \mathbf{V}_3 \\ \mathbf{V}_1 \times (\mathbf{V}_2 \times \mathbf{V}_3) &= \mathbf{V}_2(\mathbf{V}_3 \cdot \mathbf{V}_1 - \mathbf{V}_3(\mathbf{V}_1 \cdot \mathbf{V}_2) \\ \mathbf{i} \times \mathbf{i} = \mathbf{j} \times \mathbf{j} = \mathbf{k} \times \mathbf{k} &= 0.1 \text{ (zero vector)} \\ &= \mathbf{0}\end{aligned}$$

$$\mathbf{i} \times \mathbf{j} = \mathbf{k}, \qquad \mathbf{j} \times \mathbf{k} = \mathbf{i}, \qquad \mathbf{k} \times \mathbf{i} = \mathbf{j}$$

If $\mathbf{V}_1 = a_1\mathbf{i} + b_1\mathbf{j} + c_1\mathbf{k}$, $\quad \mathbf{V}_2 = a_2\mathbf{i} + b_2\mathbf{j} + c_2\mathbf{k}$, $\quad \mathbf{V}_3 = a_3\mathbf{i} + b_3\mathbf{j} + c_3\mathbf{k}$,

then

$$\mathbf{V}_1 \times \mathbf{V}_2 = \begin{vmatrix} \mathbf{i} & \mathbf{j} & \mathbf{k} \\ a_1 & b_1 & c_1 \\ a_2 & b_2 & c_2 \end{vmatrix} = (b_1 c_2 - b_2 c_1)\mathbf{i} + (c_1 a_2 - c_2 a_1)\mathbf{j} + (a_1 b_2 - a_2 b_1)\mathbf{k}$$

It should be noted that, since $\mathbf{V}_1 \times \mathbf{V}_2 = -\mathbf{V}_2 \times \mathbf{V}_1$, the vector product is not commutative.

Scalar Triple Product

There is only one possible interpretation of the expression $\mathbf{V}_1 \cdot \mathbf{V}_2 \times \mathbf{V}_3$ and that is $\mathbf{V}_1 \cdot (\mathbf{V}_2 \times \mathbf{V}_3)$ which is obviously a scalar.

Further $\mathbf{V}_1 \cdot (\mathbf{V}_2 \times \mathbf{V}_3) = (\mathbf{V}_1 \times \mathbf{V}_2) \cdot \mathbf{V}_3 = \mathbf{V}_2 \cdot (\mathbf{V}_3 \times \mathbf{V}_1)$

$$= \begin{vmatrix} a_1 & b_1 & c_1 \\ a_2 & b_2 & c_2 \\ a_3 & b_3 & c_3 \end{vmatrix}$$

$$= v_1 v_2 v_3 \cos \phi \sin \theta,$$

Where θ is the angle between $\mathbf{V}_2$ and $\mathbf{V}_3$ and ϕ is the angle between $\mathbf{V}_1$ and the normal to the plane of $\mathbf{V}_2$ and $\mathbf{V}_3$.

This product is called the *scalar triple product* and is written as $[\mathbf{V}_1\mathbf{V}_2\mathbf{V}_3]$.

The determinant indicates that it can be considered as the volume of the parallelepiped whose three determining edges are $\mathbf{V}_1$, $\mathbf{V}_2$ and $\mathbf{V}_3$.

It also follows that cyclic permutation of the subscripts does not change the value of the scalar triple product so that

$$[\mathbf{V}_1\mathbf{V}_2\mathbf{V}_3] = [\mathbf{V}_2\mathbf{V}_3\mathbf{V}_1] = [\mathbf{V}_3\mathbf{V}_1\mathbf{V}_2]$$

$$\text{but} \quad [\mathbf{V}_1\mathbf{V}_2\mathbf{V}_3] = -[\mathbf{V}_2\mathbf{V}_1\mathbf{V}_3] \quad \text{etc.} \qquad \text{and} \quad [\mathbf{V}_1\mathbf{V}_1\mathbf{V}_2] \equiv 0 \text{ etc.}$$

Given three non-coplanar reference vectors $\mathbf{V}_1$, $\mathbf{V}_2$ and $\mathbf{V}_3$, the *reciprocal system is* given by $\mathbf{V}_1^*$, $\mathbf{V}_2^*$ and $\mathbf{V}_3^*$, where

$$1 = v_1 v_1^* = v_2 v_2^* = v_3 v_3^*$$

$$0 = v_1 v_2^* = v_1 v_3^* = v_2 v_1^* \quad \text{etc.}$$

$$\mathbf{V}_1^* = \frac{\mathbf{V}_2 \times \mathbf{V}_3}{[\mathbf{V}_1\mathbf{V}_2\mathbf{V}_3]}, \qquad \mathbf{V}_2^* = \frac{\mathbf{V}_3 \times \mathbf{V}_1}{[\mathbf{V}_1\mathbf{V}_2\mathbf{V}_3]}, \qquad \mathbf{V}_3^* = \frac{\mathbf{V}_1 \times \mathbf{V}_2}{[\mathbf{V}_1\mathbf{V}_2\mathbf{V}_3]}$$

The system $\mathbf{i}, \mathbf{j}, \mathbf{k}$ is its own reciprocal.

Vector Triple Product

The product $\mathbf{V}_1 \times (\mathbf{V}_2 \times \mathbf{V}_3)$ defines the *vector triple product*. Obviously, in this case, the brackets are vital to the definition.

$$\mathbf{V}_1 \times (\mathbf{V}_2 \times \mathbf{V}_3) = (\mathbf{V}_1 \cdot \mathbf{V}_3)\mathbf{V}_2 - (\mathbf{V}_1 \cdot \mathbf{V}_2)\mathbf{V}_3$$

$$= \begin{vmatrix} \mathbf{i} & \mathbf{j} & \mathbf{k} \\ a_1 & b_1 & c_1 \\ \begin{vmatrix} b_2 & c_2 \\ b_3 & c_3 \end{vmatrix} & \begin{vmatrix} c_2 & a_2 \\ c_3 & a_3 \end{vmatrix} & \begin{vmatrix} a_2 & b_2 \\ a_3 & b_3 \end{vmatrix} \end{vmatrix}$$

i.e. it is a vector, perpendicular to $\mathbf{V}_1$, lying in the plane of $\mathbf{V}_2$, $\mathbf{V}_3$.

Similarly

$$(\mathbf{V}_1 \times \mathbf{V}_2) \times \mathbf{V}_3 = \begin{vmatrix} \mathbf{i} & \mathbf{j} & \mathbf{k} \\ \begin{vmatrix} b_1 & c_1 \\ b_2 & c_2 \end{vmatrix} & \begin{vmatrix} c_1 & a_1 \\ c_2 & a_2 \end{vmatrix} & \begin{vmatrix} a_1 & b_1 \\ a_2 & b_2 \end{vmatrix} \\ a_3 & b_3 & c_3 \end{vmatrix}$$

$$\mathbf{V}_1 \times (\mathbf{V}_2 \times \mathbf{V}_3) + \mathbf{V}_2 \times (\mathbf{V}_3 \times \mathbf{V}_1) + \mathbf{V}_3 \times (\mathbf{V}_1 + \mathbf{V}_2) \equiv 0$$

If $\mathbf{V}_1 \times (\mathbf{V}_2 \times \mathbf{V}_3) = (\mathbf{V}_1 \times V_2) \times \mathbf{V}_3$ then $\mathbf{V}_1$, $\mathbf{V}_2$, $\mathbf{V}_3$ form an *orthogonal set.* Thus **i**, **j**, **k** form an orthogonal set.

Geometry of the Plane, Straight Line and Sphere

The position vectors of the fixed points A, B, C, D relative to O are $\mathbf{V}_1$, $\mathbf{V}_2$, $\mathbf{V}_3$, $\mathbf{V}_4$ and the position vector of the variable point P is $\mathbf{V}$.

The vector form of the equation of the straight line through A parallel to $\mathbf{V}_2$ is

$$\mathbf{V} = \mathbf{V}_1 + r\mathbf{V}_2$$

or

$$(\mathbf{V} - \mathbf{V}_1) = r\mathbf{V}_2$$

or

$$(\mathbf{V} - \mathbf{V}_1) \times \mathbf{V}_2 = 0$$

while that of the plane through A perpendicular to $\mathbf{V}_2$ is

$$(\mathbf{V} - \mathbf{V}_1) \cdot \mathbf{V}_2 = 0$$

The equation of the line AB is

$$\mathbf{V} = r\mathbf{V}_1 + (1 - r)\mathbf{V}_2$$

and those of the bisectors of the angles between $\mathbf{V}_1$ and $\mathbf{V}_2$ are

$$\mathbf{V} = r\left(\frac{\mathbf{V}_1}{v} \pm \frac{\mathbf{V}_2}{v_2}\right)$$

or

$$\mathbf{V} = r(\hat{\mathbf{v}}_1 \pm \hat{\mathbf{v}}_2)$$

The perpendicular from C to the line through A parallel to $\mathbf{V}_2$ has as its equation

$$\mathbf{V} = \mathbf{V}_1 - \mathbf{V}_3 - \hat{\mathbf{v}}_2 \cdot (\mathbf{V}_1 - \mathbf{V}_3)\hat{\mathbf{v}}_2.$$

The condition for the intersection of the two lines,

$$\mathbf{V} = \mathbf{V}_1 + r\mathbf{V}_3$$

and

$$\mathbf{V} = \mathbf{V}_2 + s\mathbf{V}_4$$

is

$$[(\mathbf{V}_1 - \mathbf{V}_2)\mathbf{V}_3\mathbf{V}_4] = 0.$$

The common perpendicular to the above two lines is the line of intersection of the two planes

$$[(\mathbf{V} - \mathbf{V}_1)\mathbf{V}_3(\mathbf{V}_3 \times \mathbf{V}_4)] = 0$$

and

$$[(\mathbf{V} - \mathbf{V}_2)\mathbf{V}_4(\mathbf{V}_3 \times \mathbf{V}_4)] = 0$$

and the length of this perpendicular is

$$\frac{[(\mathbf{V}_1 - \mathbf{V}_2)\mathbf{V}_3\mathbf{V}_4]}{|\mathbf{V}_3 \times \mathbf{V}_4|}.$$

The equation of the line perpendicular to the plane ABC is

$$\mathbf{V} = \mathbf{V}_1 \times \mathbf{V}_2 + \mathbf{V}_2 \times \mathbf{V}_3 + \mathbf{V}_3 \times \mathbf{V}_1$$

and the distance of the plane from the origin is

$$\frac{[\mathbf{V}_1\mathbf{V}_2\mathbf{V}_3]}{|(\mathbf{V}_2 - \mathbf{V}_1) \times (\mathbf{V}_3 - \mathbf{V}_1)|}.$$

In general the vector equation

$$\mathbf{V} \cdot \mathbf{V}_2 = r$$

defines the plane which is perpendicular to $\mathbf{V}_2$, and the perpendicular distance from A to this plane is

$$\frac{r - \mathbf{V}_1 \cdot \mathbf{V}_2}{v_2}.$$

The distance from A, measured along a line parallel to $\mathbf{V}_3$, is

$$\frac{r - \mathbf{V}_1 \cdot \mathbf{V}_2}{\mathbf{V}_2 \cdot \hat{\mathbf{v}}_3} \quad \text{or} \quad \frac{r - \mathbf{V}_1 \cdot \mathbf{V}_2}{v_2 \cos\theta}$$

where θ is the angle beween $\mathbf{V}_2$ and $\mathbf{V}_3$.
(If this plane contains the point C then $r = \mathbf{V}_3 \cdot \mathbf{V}_2$ and if it passes through the origin then $r = 0$.)

Given two planes

$$\mathbf{V} \cdot \mathbf{V}_1 = r$$

$$\mathbf{V} \cdot \mathbf{V}_2 = s$$

then any plane through the line of intersection of these two planes is given by

$$\mathbf{V} \cdot (\mathbf{V}_1 + \lambda \mathbf{V}_2) = r + \lambda s$$

where λ is a scalar parameter. In particular $\lambda = \pm v_1/v_2$ yields the equation of the two planes bisecting the angle between the given planes.

The plane through A parallel to the plane of $\mathbf{V}_2$, $\mathbf{V}_3$ is

$$\mathbf{V} = \mathbf{V}_1 + r\mathbf{V}_2 + s\mathbf{V}_3$$

$$\text{or} \quad (\mathbf{V} - \mathbf{V}_1) \cdot \mathbf{V}_2 \times \mathbf{V}_3 = 0$$

$$\text{or} \quad [\mathbf{V}\mathbf{V}_2\mathbf{V}_3] - [\mathbf{V}_1\mathbf{V}_2\mathbf{V}_3] = 0$$

so that the expansion in rectangular Cartesian coordinates yields

$$\begin{vmatrix} (x - a_1) & (y - b_1) & (z - c_1) \\ a_2 & b_2 & c_2 \\ a_3 & b_3 & c_3 \end{vmatrix} = 0 \qquad (\mathbf{V} \equiv x\mathbf{i} + y\mathbf{j} + z\mathbf{k})$$

which is obviously the usual linear equation in x, y and z.

The plane through AB parallel to $\mathbf{V}_3$ is given by

$$[(\mathbf{V} - \mathbf{V}_1)(\mathbf{V}_1 - \mathbf{V}_2)\mathbf{V}_3] = 0$$

$$\text{or} \quad [\mathbf{V}\mathbf{V}_2\mathbf{V}_3] - [\mathbf{V}\mathbf{V}_1\mathbf{V}_3] - [\mathbf{V}_1\mathbf{V}_2\mathbf{V}_3] = 0$$

The plane through the three points A, B and C is

$$\mathbf{V} = \mathbf{V}_1 + s(\mathbf{V}_2 - \mathbf{V}_1) + t(\mathbf{V}_3 - \mathbf{V}_1)$$

$$\text{or} \quad \mathbf{V} = r\mathbf{V}_1 + s\mathbf{V}_2 + t\mathbf{V}_3 \qquad (r + s + t \equiv 1)$$

$$\text{or} \quad [(\mathbf{V} - \mathbf{V}_1)(\mathbf{V}_1 - \mathbf{V}_2)(\mathbf{V}_2 - \mathbf{V}_3)] = 0$$

$$\text{or} \quad [\mathbf{V}\mathbf{V}_1\mathbf{V}_2] + [\mathbf{V}\mathbf{V}_2\mathbf{V}_3] + [\mathbf{V}\mathbf{V}_3\mathbf{V}_1] - [\mathbf{V}_1\mathbf{V}_2\mathbf{V}_3] = 0$$

For four points A, B, C, D to be coplanar, then

$$r\mathbf{V}_1 + s\mathbf{V}_2 + t\mathbf{V}_3 + u\mathbf{V}_4 \equiv 0 \equiv r + s + t + u$$

The following formulae relate to a sphere when the vectors are taken to lie in three dimensional space and to a circle when the space is two dimensional. For a circle in three dimensions take the intersection of the sphere with a plane.

The equation of a sphere with center O and radius OA is

$$\mathbf{V} \cdot \mathbf{V} = v_1^2 \qquad (\text{not } \mathbf{V} = \mathbf{V}_1)$$

or

$$(\mathbf{V} - \mathbf{V}_1) \cdot (\mathbf{V} + \mathbf{V}_1) = 0$$

while that of a sphere with center B radius v_1 is

$$(\mathbf{V} - \mathbf{V}_2) \cdot (\mathbf{V} - \mathbf{V}_2) = v_1^2$$

or

$$\mathbf{V} \cdot (\mathbf{V} - 2\mathbf{V}_2) = v_1^2 - v_2^2$$

If the above sphere passes through the origin then

$$\mathbf{V} \cdot (\mathbf{V} - 2\mathbf{V}_2) = 0$$

(note that in two dimensional polar coordinates this is simply)

$$r = 2a \cdot \cos\theta$$

while in three dimensional Cartesian coordinates it is

$$x^2 + y^2 + z^2 - 2(a_2x + b_2y + c_2x) = 0.$$

The equation of a sphere having the points A and B as the extremities of a diameter is

$$(\mathbf{V} - \mathbf{V}_1) \cdot (\mathbf{V} - \mathbf{V}_2) = 0.$$

The square of the length of the tangent from C to the sphere with center B and radius v_1 is given by

$$(\mathbf{V}_3 - \mathbf{V}_2) \cdot (\mathbf{V}_3 - \mathbf{V}_2) = v_1^2$$

The condition that the plane $\mathbf{V} \cdot \mathbf{V}_3 = s$ is tangential to the sphere $(\mathbf{V} - \mathbf{V}_2) \cdot (\mathbf{V} - \mathbf{V}_2) = v_1^2$ is

$$(s - \mathbf{V}_3 \cdot \mathbf{V}_2) \cdot (s - \mathbf{V}_3 \cdot \mathbf{V}_2) = v_1^2 v_3^2.$$

The equation of the tangent plane at D, on the surface of sphere $(\mathbf{V} - \mathbf{V}_2) \cdot (\mathbf{V} - \mathbf{V}_2) = v_1^2$, is

$$(\mathbf{V} - \mathbf{V}_4) \cdot (\mathbf{V}_4 - \mathbf{V}_2) = 0$$

or

$$\mathbf{V} \cdot \mathbf{V}_4 - \mathbf{V}_2 \cdot (\mathbf{V} + \mathbf{V}_4) = v_1^2 - v_2^2$$

The condition that the two circles $(\mathbf{V} - \mathbf{V}_2) \cdot (\mathbf{V} - \mathbf{V}_2) = v_1^2$ and $(\mathbf{V} - \mathbf{V}_4) \cdot (\mathbf{V} - \mathbf{V}_4) = v_3^2$ intersect orthogonally is clearly

$$(\mathbf{V}_2 - \mathbf{V}_4) \cdot (\mathbf{V}_2 - \mathbf{V}_4) = v_1^2 + v_3^2$$

The polar plane of D with respect to the circle

$$(\mathbf{V} - \mathbf{V}_2) \cdot (\mathbf{V} - \mathbf{V}_2) = v_1^2 \quad \text{is}$$

$$\mathbf{V} \cdot \mathbf{V}_4 - \mathbf{V}_2 \cdot (\mathbf{V} + \mathbf{V}_4) = v_1^2 - v_2^2$$

Any sphere through the intersection of the two spheres $(\mathbf{V} - \mathbf{V}_2) \cdot (\mathbf{V} - \mathbf{V}_2) = v_1^2$ and $(\mathbf{V} - \mathbf{V}_4) \cdot (\mathbf{V} - \mathbf{V}_4) = v_3^2$ is given by

$$(\mathbf{V} - \mathbf{V}_2) \cdot (\mathbf{V} - \mathbf{V}_2) + \lambda(\mathbf{V} - \mathbf{V}_4) \cdot (\mathbf{V} - \mathbf{V}_4) = v_1^2 + \lambda v_3^2$$

while the radical plane of two such spheres is

$$\mathbf{V} \cdot (\mathbf{V}_2 - \mathbf{V}_4) = -\tfrac{1}{2}(v_1^2 - v_2^2 - v_3^2 + v_4^2)$$

Differentiation of Vectors

If $\mathbf{V}_1 = a_1\mathbf{i} + b_1\mathbf{j} + c_1\mathbf{k}$, and $\mathbf{V}_2 = a_2\mathbf{i} + b_2\mathbf{j} + c_2\mathbf{k}$, and if $\mathbf{V}_1$ and $\mathbf{V}_2$ are functions of the scalar t, then

$$\frac{d}{dt}(\mathbf{V}_1 + \mathbf{V}_2 + \cdots) = \frac{d\mathbf{V}_1}{dt} + \frac{d\mathbf{V}_2}{dt} + \cdots,$$

$$\text{where} \quad \frac{d\mathbf{V}_1}{dt} = \frac{da_1}{dt}\mathbf{i} + \frac{db_1}{dt}\mathbf{j} + \frac{dc_1}{dt}\mathbf{k}, \text{ etc.}$$

$$\frac{d}{dt}(\mathbf{V}_1 \cdot \mathbf{V}_2) = \frac{d\mathbf{V}_1}{dt} \cdot \mathbf{V}_2 + \mathbf{V}_1 \cdot \frac{d\mathbf{V}_2}{dt}$$

$$\frac{d}{dt}(\mathbf{V}_1 \times \mathbf{V}_2) = \frac{d\mathbf{V}_1}{dt} \times \mathbf{V}_2 + \mathbf{V}_1 \times \frac{d\mathbf{V}_2}{dt}$$

$$\mathbf{V} \cdot \frac{d\mathbf{V}}{dt} = v \cdot \frac{dv}{dt}$$

In particular, if $\mathbf{V}$ is a vector of constant length then the right hand side of the last equation is identically zero showing that $\mathbf{V}$ is perpendicular to its derivative.

The derivatives of the triple products are

$$\frac{d}{dt}[\mathbf{V}_1\mathbf{V}_2\mathbf{V}_3] = \left[\left(\frac{d\mathbf{V}_1}{dt}\right)\mathbf{V}_2\mathbf{V}_3\right] + \left[\mathbf{V}_1\left(\frac{d\mathbf{V}_2}{dt}\right)\mathbf{V}_3\right] + \left[\mathbf{V}_1\mathbf{V}_2\left(\frac{d\mathbf{V}_3}{dt}\right)\right]$$

$$\text{and} \quad \frac{d}{dt}\left\{\mathbf{V}_1 \times (\mathbf{V}_2 \times \mathbf{V}_3)\right\} = \left(\frac{d\mathbf{V}_1}{dt}\right) \times (\mathbf{V}_2 \times \mathbf{V}_3) + \mathbf{V}_1 \times \left(\left(\frac{d\mathbf{V}_2}{dt}\right) \times \mathbf{V}_3\right) + \mathbf{V}_1 \times \left(\mathbf{V}_2 \times \left(\frac{d\mathbf{V}_3}{dt}\right)\right).$$

Geometry of Curves in Space

s = the *length of arc*, measured from some fixed point on the curve (fig. 3).
$\mathbf{V}_1$ = the position vector of the point A on the curve
$\mathbf{V}_1 + \delta\mathbf{V}_1$ = the position vector of the point P in the neighborhood of A
$\hat{\mathbf{t}}$ = the *unit tangent* to the curve at the point A, measured in the direction of s increasing.

The *normal plane* is that plane which is perpendicular to the unit tangent. The principal normal is defined as the intersection of the normal plane with the plane defined by $\mathbf{V}_1$ and $\mathbf{V}_1 + \delta\mathbf{V}_1$ in the limit as $\delta\mathbf{V}_1 - 0$.

$\hat{\mathbf{n}}$ = the *unit normal* (principal) at the point A. The plane defined by $\hat{\mathbf{t}}$ and $\hat{\mathbf{n}}$ is called the *osculating plane* (alternatively plane of curvature or local plane).

ρ = the radius of curvature at A
$\delta\theta$ = the angle subtended at the origin by $\delta\mathbf{V}_1$.

$$\kappa = \frac{d\theta}{ds} = \frac{1}{\rho}$$

$\hat{\mathbf{b}}$ = the *unit binormal* i.e. the unit vector which is parallel to $\hat{\mathbf{t}} \times \hat{\mathbf{n}}$ at the point A:
λ = the *torsion* of the curve at A

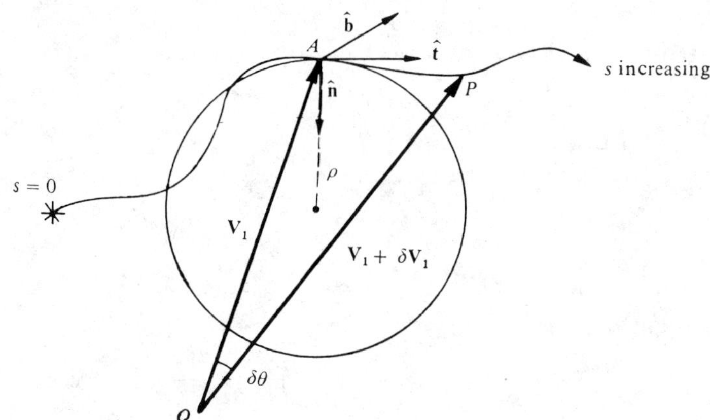

Figure 3.

Frenet's Formulae:

$$\frac{d\hat{\mathbf{t}}}{ds} = \kappa\hat{\mathbf{n}}$$

$$\frac{d\hat{\mathbf{n}}}{ds} = -\kappa\hat{\mathbf{t}} + \lambda\hat{\mathbf{b}}$$

$$\frac{d\hat{\mathbf{b}}}{ds} = -\lambda\hat{\mathbf{n}}$$

The following formulae are also applicable:

Unit tangent $\quad \hat{\mathbf{t}} = \dfrac{d\mathbf{V}_1}{ds}$

Equation of the tangent $\quad (\mathbf{V} - \mathbf{V}_1) \times \hat{\mathbf{t}} = 0$

or $\quad \mathbf{V} = \mathbf{V}_1 + q\hat{\mathbf{t}}$

Unit normal $\quad \hat{\mathbf{n}} = \dfrac{1 d^2\mathbf{V}_1}{\kappa ds^2}$

Equation of the normal plane $\quad (\mathbf{V} - \mathbf{V}_1) \cdot \hat{\mathbf{t}} = 0$

Equation of the normal $\quad (\mathbf{V} - \mathbf{V}_1) \times \hat{\mathbf{n}} = 0$

or $\quad \mathbf{V} = \mathbf{V}_1 + r\hat{\mathbf{n}}$

Unit binormal $\quad \hat{\mathbf{b}} = \hat{\mathbf{t}} \times \hat{\mathbf{n}}$

Equation of the binormal $\quad (\mathbf{V} - \mathbf{V}_1) \times \hat{\mathbf{b}} = 0$

or $\quad \mathbf{V} = \mathbf{V}_1 + u\hat{\mathbf{b}}$

or $\quad \mathbf{V} = \mathbf{V}_1 + w\dfrac{d\mathbf{V}_1}{ds} \times \dfrac{d^2\mathbf{V}_1}{ds^2}$

Equation of the osculating plane:

$$[(\mathbf{V} - \mathbf{V}_1)\hat{\mathbf{t}}\hat{\mathbf{n}}] = 0$$

or

$$\left[(\mathbf{V} - \mathbf{V}_1)\left(\frac{d\mathbf{V}_1}{ds}\right)\left(\frac{d^2\mathbf{V}_1}{ds^2}\right)\right] = 0$$

A *geodetic line* on a surface is a curve, the osculating plane of which is everywhere normal to the surface.

The differential equation of the geodetic is

$$[\hat{\mathbf{n}} d\mathbf{V}_1\, d^2\mathbf{V}_1] = 0$$

Differential Operators—Rectangular Coordinates

$$dS = \frac{\partial S}{\partial x} \cdot dx + \frac{\partial S}{\partial y} \cdot dy + \frac{\partial S}{\partial z} \cdot dz$$

By definition

$$\nabla \equiv \text{del} \equiv \mathbf{i}\frac{\partial}{\partial x} + \mathbf{j}\frac{\partial}{\partial y} + \mathbf{k}\frac{\partial}{\partial z}$$

$$\nabla^2 \equiv \text{Laplacian} \equiv \frac{\partial^2}{\partial x^2} + \frac{\partial^2}{\partial y^2} + \frac{\partial^2}{\partial z^2}$$

If S is a scalar function, then

$$\nabla S \equiv \text{grad } S \equiv \frac{\partial S}{dx}\mathbf{i} + \frac{\partial S}{dy}\mathbf{j} + \frac{\partial S}{dz}\mathbf{k}$$

Grad S defines both the direction and magnitude of the maximum rate of increase of S at any point. Hence the name *gradient* and also its vectorial nature. ∇S is independent of the choice of rectangular coordinates.

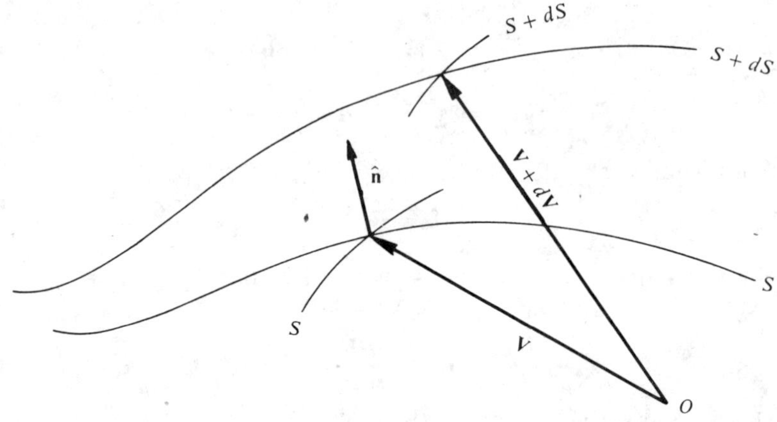

Figure 4

$$\nabla S = \frac{\partial S}{\partial n}\,\hat{\mathbf{n}}$$

where $\hat{\mathbf{n}}$ is the unit normal to the surface S = constant, in the direction of S increasing. The total derivative of S at a point having the position vector $\mathbf{V}$ is given by (fig. 4)

$$dS = \frac{\partial S}{\partial n}\,\hat{\mathbf{n}} \cdot d\mathbf{V}$$

$$= d\mathbf{V} \cdot \nabla S$$

and the directional derivative of S in the direction of $\mathbf{U}$ is

$$\mathbf{U} \cdot \nabla S = \mathbf{U} \cdot (\nabla S) = (\mathbf{U} \cdot \nabla)S$$

Similarly the directional derivative of the vector $\mathbf{V}$ in the direction of $\mathbf{U}$ is

$$(\mathbf{U} \cdot \nabla)\mathbf{V}$$

The *distributive* law holds for finding a gradient. Thus if S and T are scalar functions

$$\nabla(S + T) = \nabla S + \nabla T$$

The *associative* law becomes the rule for differentiating a product:

$$\nabla(ST) = S\nabla T + T\nabla S$$

If **V** is a vector function with the magnitudes of the components parallel to the three coordinate axes V_x, V_y, V_z, then

$$\nabla \cdot \mathbf{V} \equiv \text{div } \mathbf{V} \equiv \frac{\partial V_x}{\partial x} + \frac{\partial V_y}{\partial y} + \frac{\partial V_z}{\partial z}$$

The divergence obeys the distributive law. Thus, if **V** and **U** are vectors functions, then

$$\nabla \cdot (\mathbf{V} + \mathbf{U}) = \nabla \cdot \mathbf{V} + \nabla \cdot \mathbf{U}$$
$$\nabla \cdot (S\mathbf{V}) = (\nabla S) \cdot \mathbf{V} + S(\nabla \cdot \mathbf{V})$$
$$\nabla \cdot (\mathbf{U} \times \mathbf{V}) = \mathbf{V} \cdot (\nabla \times \mathbf{U}) - \mathbf{U} \cdot (\nabla \times \mathbf{V})$$

As with the gradient of a scalar, the divergence of a vector is invariant under a transformation from one set of rectangular coordinates to another.

$$\begin{aligned}\nabla \times \mathbf{V} &\equiv \text{curl } \mathbf{V} \text{ (sometimes } \nabla \wedge \mathbf{V} \text{ or rot } \mathbf{V}) \\ &\equiv \left(\frac{\partial \mathbf{V}_z}{\partial y} - \frac{\partial \mathbf{V}_y}{\partial z}\right)\mathbf{i} + \left(\frac{\partial \mathbf{V}_x}{\partial z} - \frac{\partial \mathbf{V}_z}{\partial x}\right)\mathbf{j} + \left(\frac{\partial \mathbf{V}_y}{\partial x} - \frac{\partial \mathbf{V}_x}{\partial y}\right)\mathbf{k} \\ &= \begin{vmatrix} \mathbf{i} & \mathbf{j} & \mathbf{k} \\ \dfrac{\partial}{\partial x} & \dfrac{\partial}{\partial y} & \dfrac{\partial}{\partial z} \\ V_x & V_y & V_z \end{vmatrix}\end{aligned}$$

The *curl* (or *rotation*) of a vector is a vector which is invariant under a transformation from one set of rectangular coordinates to another.

$$\nabla \times (\mathbf{U} + \mathbf{V}) = \nabla \times \mathbf{U} + \nabla \times \mathbf{V}$$
$$\nabla \times (S\mathbf{V}) = (\nabla S) \times \mathbf{V} + S(\nabla \times \mathbf{V})$$
$$\nabla \times (\mathbf{U} \times \mathbf{V}) = (\mathbf{V} \cdot \nabla)\mathbf{U} - (\mathbf{U} \cdot \nabla)\mathbf{V} + \mathbf{U}(\nabla \cdot \mathbf{V}) - \mathbf{V}(\nabla \cdot \mathbf{U})$$

$$\begin{aligned}\text{grad } (\mathbf{U} \cdot \mathbf{V}) &= \nabla(\mathbf{U} \cdot \mathbf{V}) \\ &= (\mathbf{V} \cdot \nabla)\mathbf{U} + (\mathbf{U} \cdot \nabla)\mathbf{V} + \mathbf{V} \times (\nabla \times \mathbf{U}) + \mathbf{U} \times (\nabla \times \mathbf{V})\end{aligned}$$

If

$$\mathbf{V} = V_x\mathbf{i} + V_y\mathbf{j} + V_z\mathbf{k}$$
$$\nabla \cdot \mathbf{V} = \nabla V_x \cdot \mathbf{i} + \nabla V_y \cdot \mathbf{j} + \nabla V_z \cdot \mathbf{k}$$

and

$$\nabla \times \mathbf{V} = \nabla V_x \times \mathbf{i} + \nabla V_y \times \mathbf{j} + \nabla V_z \times \mathbf{k}$$

The operator ∇ can be used more than once. The number of possibilities where ∇ is used twice are

$$\nabla \cdot (\nabla \theta) \equiv \text{div grad } \theta$$
$$\nabla \times (\nabla \theta) \equiv \text{curl grad } \theta$$
$$\nabla(\nabla \cdot \mathbf{V}) \equiv \text{grad div } \mathbf{V}$$
$$\nabla \cdot (\nabla \times \mathbf{V}) \equiv \text{div curl } \mathbf{V}$$
$$\nabla \times (\nabla \times \mathbf{V}) \equiv \text{curl curl } \mathbf{V}$$

Thus: div grad $S \equiv \nabla \cdot (\nabla S) \equiv$ Laplacian $S \equiv \nabla^2 S$

$$\equiv \frac{\partial^2 S}{\partial x^2} + \frac{\partial^2 S}{\partial y^2} + \frac{\partial^2 S}{\partial z^2}$$

curl grad $S \equiv 0$; curl curl $\mathbf{V} \equiv$ grad div $\mathbf{V} - \nabla^2 \mathbf{V}$;

div curl $\mathbf{V} \equiv 0$

Taylor's expansion in three dimensions can be written

$$f(\mathbf{V} + \boldsymbol{\varepsilon}) = e^{\boldsymbol{\varepsilon} \cdot \nabla} f(\mathbf{V})$$

where $\mathbf{V} = x\mathbf{i} + y\mathbf{j} + z\mathbf{k}$

and $\boldsymbol{\varepsilon} = h\mathbf{i} + l\mathbf{j} + m\mathbf{k}$

(note the analogy with $f_p = e^{phD} f_0$ in finite difference methods).

Orthogonal Curvilinear Coordinates

If at a point P there exist three uniform point functions u, v and w so that the surfaces $u =$ const., $v =$ const., and $w =$ const., intersect in three distinct curves through P then the surfaces are called the *coordinate surfaces* through P. The three lines of intersection are referred to as the *coordinate lines* and their tangents a, b, and c as the *coordinate axes.* When the coordinate axes form an orthogonal set the system is said to define *orthogonal curvilinear coordinates* at P.

Consider an infinitesimal volume enclosed by the surfaces u, v, w, $u + du$, $v + dv$, and $w + dw$ (fig. 5).

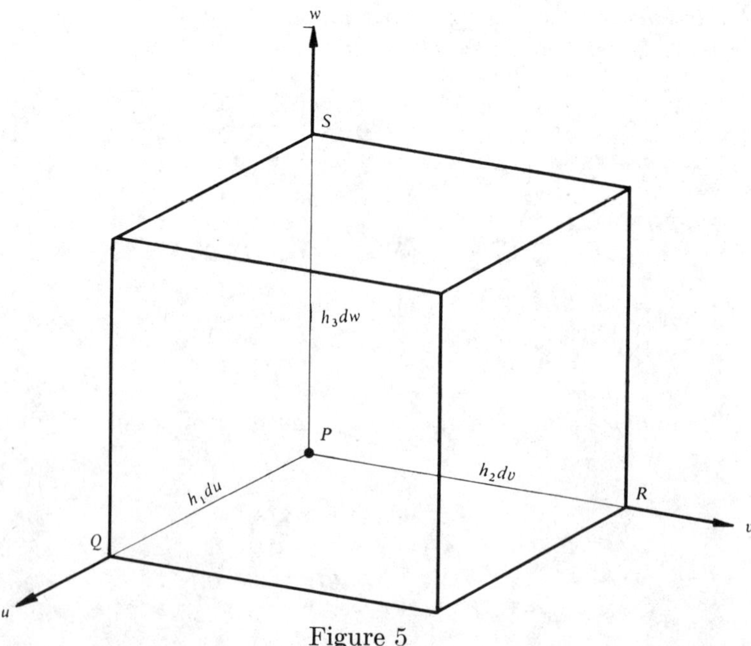

Figure 5

The surface $PRS \equiv u =$ const., and the face of the curvilinear figure immediately opposite this is $u + du =$ const. etc.

In terms of these surface constants

$$\begin{aligned} P &= P(u,v,w) \\ Q &= Q(u + du,v,w) \\ R &= R(u,v + dv,w) \\ S &= S(u,v,w + dw) \end{aligned} \qquad \text{and} \qquad \begin{aligned} PQ &\doteq h_1 du \\ PR &= h_2 dv \\ PS &= h_3 dw \end{aligned}$$

where h_1, h_2, and h_3 are functions of u, v, and w.

In rectangular Cartesians **i**, **j**, **k**

$$h_1 = 1, \qquad h_2 = 1, \qquad h_3 = 1.$$

$$\frac{\hat{\mathbf{a}}}{h_1}\frac{\partial}{\partial u} = \mathbf{i}\frac{\partial}{\partial x}, \qquad \frac{\hat{\mathbf{b}}}{h_2}\frac{\partial}{\partial v} = \mathbf{j}\frac{\partial}{\partial y}, \qquad \frac{\hat{\mathbf{c}}}{h_3}\frac{\partial}{\partial w} = \mathbf{k}\frac{\partial}{\partial z}.$$

In cylindrical coordinates $\hat{\mathbf{r}}$, $\hat{\boldsymbol{\phi}}$, $\hat{\mathbf{k}}$

$$h_1 = 1, \qquad h_2 = r \qquad h_3 = 1.$$

$$\frac{\hat{\mathbf{a}}}{h_1}\frac{\partial}{\partial u} = \hat{\mathbf{r}}\frac{\partial}{\partial r}, \qquad \frac{\hat{\mathbf{b}}}{h_2}\frac{\partial}{\partial v} = \frac{\hat{\boldsymbol{\phi}}}{r}\frac{\partial}{\partial \phi} \qquad \frac{\hat{\mathbf{c}}}{h_3}\frac{\partial}{\partial w} = \hat{\mathbf{k}}\frac{\partial}{\partial z}$$

In spherical coordinates $\hat{\mathbf{r}}$, $\hat{\boldsymbol{\theta}}$, $\hat{\boldsymbol{\phi}}$

$$h_1 = 1, \qquad h_2 = r, \qquad h_3 = r \sin\theta$$

$$\frac{\hat{\mathbf{a}}}{h_1}\frac{\partial}{\partial u} = \hat{\mathbf{r}}\frac{\partial}{\partial r}, \qquad \frac{\mathbf{b}}{h_2}\frac{\partial}{\partial v} = \frac{\hat{\boldsymbol{\phi}}}{r}\frac{\partial}{\partial \theta}, \qquad \frac{\hat{\mathbf{c}}}{h_3}\frac{\partial}{\partial w} = \frac{\hat{\boldsymbol{\phi}}}{r\sin\theta}\frac{\partial}{\partial \phi}$$

The general expressions for grad, div and curl together with those for ∇^2 and the directional derivative are, in orthogonal curvilinear coordinates, given by

$$\nabla S = \frac{\hat{\mathbf{a}}}{h_1}\frac{\partial S}{\partial u} + \frac{\hat{\mathbf{b}}}{h_2}\frac{\partial S}{\partial v} + \frac{\hat{\mathbf{c}}}{h_3}\frac{\partial S}{\partial w}$$

$$(\mathbf{V}\cdot\nabla)S = \frac{V_1}{h_1}\frac{\partial S}{\partial u} + \frac{V_2}{h_2}\frac{\partial S}{\partial v} + \frac{V_3}{h_3}\frac{\partial S}{\partial w}$$

$$\nabla\cdot\mathbf{V} = \frac{1}{h_1h_2h_3}\left\{\frac{\partial}{\partial u}(h_2h_3V_1) + \frac{\partial}{\partial v}(h_3h_1V_2) + \frac{\partial}{\partial w}(h_1h_2V_3)\right.$$

$$\nabla\times\mathbf{V} = \frac{\hat{\mathbf{a}}}{h_2h_3}\left\{\frac{\partial}{\partial v}(h_3V_3) - \frac{\partial}{\partial w}(h_2V_2)\right\} + \frac{\hat{\mathbf{b}}}{h_3h_1}\left\{\frac{\partial}{\partial w}(h_1V_1) - \frac{\partial}{\partial u}(h_3V_3)\right\} + \frac{\hat{\mathbf{c}}}{h_1h_2}\left\{\frac{\partial}{\partial u}(h_2V_2) - \frac{\partial}{\partial v}(h_1V_1)\right\}$$

$$\nabla^2 S = \frac{1}{h_1h_2h_3}\left\{\frac{\partial}{\partial u}\left(\frac{h_2h_3}{h_1}\frac{\partial S}{\partial u}\right) + \frac{\partial}{\partial v}\left(\frac{h_3h_1}{h_2}\frac{\partial S}{\partial v}\right) + \frac{\partial}{\partial w}\left(\frac{h_1h_2}{h_3}\frac{\partial S}{\partial w}\right)\right\}$$

FORMULAS OF VECTOR ANALYSIS

	Rectangular coordinates	Cylindrical coordinates	Spherical coordinates
Conversion to rectangular coordinates		$x = r\cos\varphi \quad y = r\sin\varphi \quad z = z$	$x = r\cos\varphi\sin\theta \quad y = r\sin\varphi\sin\theta$ $z = r\cos\theta$
Gradient	$\nabla\phi = \frac{\partial\phi}{\partial x}\mathbf{i} + \frac{\partial\phi}{\partial y}\mathbf{j} + \frac{\partial\phi}{\partial z}\mathbf{k}$	$\nabla\phi = \frac{\partial\phi}{\partial r}\mathbf{r} + \frac{1}{r}\frac{\partial\phi}{\partial\varphi}\boldsymbol{\phi} + \frac{\partial\phi}{\partial z}\mathbf{k}$	$\nabla\phi = \frac{\partial\phi}{\partial r}\mathbf{r} + \frac{1}{r}\frac{\partial\phi}{\partial\theta}\boldsymbol{\theta} + \frac{1}{r\sin\theta}\frac{\partial\phi}{\partial\varphi}\boldsymbol{\phi}$
Divergence	$\nabla\cdot\mathbf{A} = \frac{\partial A_x}{\partial x} + \frac{\partial A_y}{\partial y} + \frac{\partial A_z}{\partial z}$	$\nabla\cdot\mathbf{A} = \frac{1}{r}\frac{\partial(rA_r)}{\partial r} + \frac{1}{r}\frac{\partial A\varphi}{\partial\varphi} + \frac{\partial A_z}{\partial z}$	$\nabla\cdot\mathbf{A} = \frac{1}{r^2}\frac{\partial(r^2A_r)}{\partial r} + \frac{1}{r\sin\theta}\frac{\partial(A_\theta\sin\theta)}{\partial\theta} + \frac{1}{r\sin\theta}\frac{\partial A_\varphi}{\partial\varphi}$
Curl	$\nabla\times\mathbf{A} = \begin{vmatrix} \mathbf{i} & \mathbf{j} & \mathbf{k} \\ \frac{\partial}{\partial x} & \frac{\partial}{\partial y} & \frac{\partial}{\partial z} \\ A_x & A_y & A_z \end{vmatrix}$	$\nabla\times\mathbf{A} = \begin{vmatrix} \frac{1}{r}\mathbf{r} & \boldsymbol{\phi} & \frac{1}{r}\mathbf{k} \\ \frac{\partial}{\partial r} & \frac{\partial}{\partial\varphi} & \frac{\partial}{\partial z} \\ A_r & rA_\varphi & A_z \end{vmatrix}$	$\nabla\times\mathbf{A} = \begin{vmatrix} \frac{\mathbf{r}}{r^2\sin\theta} & \frac{\boldsymbol{\theta}}{r\sin\theta} & \frac{\boldsymbol{\phi}}{r} \\ \frac{\partial}{\partial r} & \frac{\partial}{\partial\theta} & \frac{\partial}{\partial\varphi} \\ A_r & rA_\theta & rA_\varphi\sin\theta \end{vmatrix}$
Laplacian	$\nabla^2\phi = \frac{\partial^2\phi}{\partial x^2} + \frac{\partial^2\phi}{\partial y^2} + \frac{\partial^2\phi}{\partial z^2}$	$\nabla^2\phi = \frac{1}{r}\frac{\partial}{\partial r}\left(r\frac{\partial\phi}{\partial r}\right) + \frac{1}{r^2}\frac{\partial^2\phi}{\partial\varphi^2} + \frac{\partial^2\phi}{\partial z^2}$	$\nabla^2\phi = \frac{1}{r^2}\frac{\partial}{\partial r}\left(r^2\frac{\partial\phi}{\partial r}\right) + \frac{1}{r^2\sin\theta}\frac{\partial}{\partial\theta}\left(\sin\theta\frac{\partial\phi}{\partial\theta}\right) + \frac{1}{r^2\sin^2\theta}\frac{\partial^2\phi}{\partial\varphi^2}$

Transformation of Integrals

s = the distance along some curve "C" in space and is measured from some fixed point.
S = a surface area
V = a volume contained by a specified surface
$\hat{\mathbf{t}}$ = the unit tangent to C at the point P
$\hat{\mathbf{n}}$ = the unit outward pointing normal
F = some vector function
ds = the vector element of curve ($= \hat{\mathbf{t}}\, ds$)
dS = the vector element of surface ($= \hat{\mathbf{n}}\, dS$)

Then
$$\int_{(c)} \mathbf{F} \cdot \hat{\mathbf{t}}\, ds = \int_{(c)} \mathbf{F} \cdot \mathbf{ds}$$

and when
$$\mathbf{F} = \nabla\phi$$
$$\int_{(c)} (\nabla\phi) \cdot \hat{\mathbf{t}}\, ds = \int_{(c)} d\phi$$

Gauss' Theorem (Green's Theorem)

When S defines a closed region having a volume V

$$\iiint_{(v)} (\nabla \cdot \mathbf{F})\, dV = \iint_{(s)} (\mathbf{F} \cdot \hat{\mathbf{n}})\, dS = \iint_{(s)} \mathbf{F} \cdot d\mathbf{S}$$

also
$$\iiint_{(v)} (\nabla\phi)\, dV = \iint_{(s)} \phi\hat{\mathbf{n}}\, dS$$

and
$$\iiint_{(v)} (\nabla \times \mathbf{F})\, dV = \iint_{(s)} (\hat{\mathbf{n}} \times \mathbf{F})\, dS$$

Stokes' Theorem

When C is closed and bounds the open surface S.

$$\iint_{(s)} \hat{\mathbf{n}} \cdot (\nabla \times \mathbf{F})\, dS = \int_{(c)} \mathbf{F} \cdot \mathbf{ds}$$

also
$$\iint_{(s)} (\hat{\mathbf{n}} \times \nabla\phi)\, dS = \int_{(c)} \phi\, \mathbf{ds}$$

Green's Theorem

$$\iint_{(s)} (\nabla\phi \cdot \nabla\theta)\, dS = \iint_{(s)} \phi\hat{\mathbf{n}} \cdot (\nabla\theta)\, dS = \iiint_{(v)} \phi(\nabla^2\theta)\, dV$$
$$= \iint_{(s)} \theta \cdot \hat{\mathbf{n}}(\nabla\phi)\, dS = \iiint_{(v)} \theta(\nabla^2\phi)\, dV$$

ASTRODYNAMICS: BASIC ORBITAL EQUATIONS

SAMUEL HERRICK

University of California, Los Angeles

Abstracted from *Astrodynamics* by Samuel Herrick, published by Van Nostrand Reinhold 1970. For more complete discussion refer to *Astrodynamics*.

1. **Astrodynamics** includes those portions of *celestial mechanics* that bear directly upon the determination of orbits and the integration of ephemerides of natural and artificial objects in interplanetary and sublunar space. For the most part, but not exclusively, it is founded upon the law of universal gravitation,

$$\mathbf{f}_{ij} = -\mathbf{f}_{ji} = k^2 m_i m_j \mathbf{r}_{ij}/r_{ij}^3 \tag{1}$$

where m_i and m_j are two masses, $\mathbf{r}_{ij}$ is the vector directed from m_i to m_j, r_{ij} is its magnitude, $\mathbf{f}_{ij}$ is the vector attraction of m_j for m_i, $\mathbf{f}_{ji}$ of m_i for m_j, and k^2 is the constant of gravitation.

The laboratory value of k^2, $G = 6.673 \times 10^{-23}$ km^3/gm sec^2, is totally inadequate to the determination of a value of k^2 that can be used in astrodynamics, and it is determined instead from planetary and satellite observations, and for the units of mass that are required by astrodynamics, as follows:

	Heliocentric Orbits		Geocentric Orbits		
Unit of length	astronomical unit (au)	km	geo-unit (gu)	km	
Unit of mass	Sun's mass ($m_\odot$)	$m_\odot$	Earth's mass ($m_\oplus$)	$m_\oplus$	(2)
Unit of time	ephemeris day (day)	sec	ephem. sec (sec)	sec	
k	0.017 202 09895	3.643×10^5	0.001 239 444	631.35	
k^2	$295.912\,2083 \times 10^{-6}$	13.271×10^{10}	$1.536\,221 \times 10^{-6}$	398603	

The au and gu are approximately the mean distance of the earth from the sun and the earth's equatorial radius, respectively, but are defined in such ways as to make the corresponding values of k exact. Each of these units is essential to a "canonical" set, as discussed in §2.

2. **The equations of motion,** for a system of n mutually attracting bodies, are derived from equations (1) and the second law of motion. Assuming that the masses are constant, we may write them in their *inertial form* as follows:

$$\left.\begin{aligned} m_1 \frac{d^2\mathbf{r}_1}{dt^2} &= \mathbf{f}_{12} + \sum_{j=3}^{n} \mathbf{f}_{1j} = k^2 m_1 m_2 \frac{\mathbf{r}_{12}}{r_{12}^3} + k^2 m_1 \sum_{j=3}^{n} m_j \frac{\mathbf{r}_{1j}}{r_{1j}^3} \\ m_2 \frac{d^2\mathbf{r}_2}{dt^2} &= \mathbf{f}_{21} + \sum_{j=3}^{n} \mathbf{f}_{2j} = -k^2 m_1 m_2 \frac{\mathbf{r}_{12}}{r_{12}^3} + k^2 m_2 \sum_{j=3}^{n} m_j \frac{\mathbf{r}_{2j}}{r_{2j}^3} \\ &\cdots\cdots\cdots\cdots\cdots\cdots \end{aligned}\right\} \tag{3}$$

From these equations we may obtain the well known ten integrals of the n-body problem, which tell us (a) that the center of mass moves uniformly in a straight line, and so may be visualized as the inertial origin, (b) that the total angular momentum of the system is constant, and (c) that the total energy of the system is constant.

The *relative motion form* of the equations of motion is derived from equations (3) and

$$\mathbf{r} = \mathbf{r}_{12} = \mathbf{r}_2 - \mathbf{r}_1 \tag{4}$$

especially for an m_1 that has, like the Sun, a dominant effect on the motion of an m_2:

$$\frac{d^2\mathbf{r}}{dt^2} = -k^2(m_1 + m_2)\frac{\mathbf{r}}{r^3} + k^2 \sum_{j=3}^{n} m_j \left(\frac{\mathbf{r}_{2j}}{r_{21}^3} - \frac{\mathbf{r}_{1j}}{r_{1j}^3}\right) \tag{5}$$

It is thus simply that $m_1 + m_2$ replaces the $m_1 m_2$ of equations (3); clearly there is no occasion for the surprise or puzzlement this replacement sometimes arouses.

The first term of equation (5) is often called the *two-body term*. The remaining terms are the *perturbations* (for relative motion and the n-body problem). The perturbation terms can also include the effects of thrust, drag, an equatorial bulge, the departures of relativity mechanics from Newtonian mechanics, etc. Even when we restrict the disturbing forces to those of the n-body problem, the perturbations can differ from those shown in equation (5), if a "two-body term" is constructed for an origin other than m_1.

The first term of equation (5) may be simplified, both theoretically and computationally, by the absorption of the gravitation-mass constant into a "canonical time,"

$$\tau = k_*(t - t_0) \tag{6}$$

and by the use of such definitions as

$$\dot{\mathbf{r}} = \frac{d\mathbf{r}}{dt} = k_*\dot{\mathbf{r}}_\tau = k_*\frac{d\mathbf{r}}{d\tau}, \qquad \ddot{\mathbf{r}} = \frac{d^2\mathbf{r}}{dt^2} = k_*^2\ddot{\mathbf{r}}_\tau = k_*^2\frac{d^2\mathbf{r}}{d\tau^2}, \text{ etc.} \tag{7}$$

where

$$k_* = k\sqrt{m_1 + m_2} \qquad \mu = k^2(m_1 + m_2) = k_*^2\mu_\tau \qquad \mu_\tau = 1 \tag{8}$$

For other forms of the equations of motion the constant μ has slightly different values.

It is best to think of τ as having the dimensions of time; the dimensions of $\sqrt{\mu}$ are "left behind" in $\mu_\tau = 1$ when the derivatives in equations hereafter are expressed with canonical units of time. When m_2 is negligible as compared with m_1, as it usually is, the canonical units of time are:

$$\left.\begin{array}{lrl} \text{au} - \text{m}_\odot - \text{day system:} & 1 \text{ kaday} & = 58.132\,44087 \text{ days} \\ \text{gu} - \text{m}_\oplus - \text{sec system:} & 1 \text{ kesec} & = 806.813\,3776 \text{ sec} \\ & & = 13.446\,8896 \text{ min} \end{array}\right\} \tag{9}$$

The practical advantage of canonical units of time and distance is that they tend to keep the magnitudes of position, velocity, and acceleration vectors all in the neighborhood of unity, thus reducing, for example, the likelihood of machine errors in floating-decimal-point operations.

3. The two-body problem includes, of course, only the first term of the equations of relative motion (5), which we may now write, with the aid of equations (6), (7), (8) as follows:

$$\ddot{\mathbf{r}} = -\mu\mathbf{r}/r^3 \tag{10}$$

We seek integrals of the two-body problem first by means of

$$\mathbf{r} \times \ddot{\mathbf{r}} = 0 \tag{11}$$

of which the x component is

$$y\ddot{z} - z\ddot{y} = 0 \tag{12}$$

so that

$$\mathbf{r} \times \dot{\mathbf{r}} = \mathbf{h} = h\mathbf{W} \tag{13}$$

and

$$y\dot{z} - z\dot{y} = h_x = hW_x \tag{14}$$

where **W** is a constant unit vector normal to the orbit plane, thus established as a fixed plane, and where, with v the *true anomaly* or angle in the orbit plane measured from the perifocus to **r**,

$$h = r^2\dot{v} = \sqrt{\mu p} \tag{15}$$

expresses *Kepler's second law*, or law of areas: that the areal velocity (or angular momentum) is constant.

The orientation of the orbit plane, embodied in **W**, is often expressed by the *inclination*, i, and the *longitude of the (ascending) node*, ☊.

The relationship of h to p, the *parameter* or semi latus rectum, may be established along with Kepler's first law in several different ways. When one is interested in the use as well as in the existence of integrals of the two-body problem, it is well to proceed by seeking additional functions that satisfy the differential equation (10) and so lead to additional simple integrations such as are found in equations (12) and (14). One such function is found in

$$\sqrt{\mu}\, D = r\dot{r} = \dot{\mathbf{r}} \cdot \dot{\mathbf{r}} \qquad \sqrt{\mu}\, \dot{D} = \dot{s}^2 - \mu/r \qquad \ddot{D} = -\mu D/r^3 \tag{16}$$

where

$$r^2 = \mathbf{r} \cdot \mathbf{r} \qquad \dot{s}^2 = \dot{\mathbf{r}} \cdot \dot{\mathbf{r}} = \dot{r}^2 + r^2\dot{v}^2 \tag{17}$$

so that

$$\mathbf{r}\ddot{D} - D\ddot{\mathbf{r}} = 0 \tag{18}$$

$$\mathbf{r}\dot{D} - D\dot{\mathbf{r}} = \mathbf{a} = e\sqrt{\mu}\,\mathbf{P} \tag{19}$$

where **a** is a constant vector, whose relationship to e, the eccentricity, and P, a unit vector directed to the perifocus, is established by

$$\sqrt{\mu}\,(\mathbf{r} \cdot \mathbf{a}) = \sqrt{\mu}\,(r^2\dot{D} - Dr\dot{r}) = r^2(\dot{s}^2 - \dot{r}^2) - \mu r = r^4\dot{v}^2 - \mu r = h^2 - \mu r \tag{20}$$

or

$$h^2/\mu = r - (\mathbf{r} \cdot \mathbf{a})/\sqrt{\mu} = r[1 - (|\mathbf{a}|/\sqrt{\mu}) \cos (\mathbf{a},\mathbf{r})] \tag{21}$$

which is the polar equation of the conic,

$$p = r(1 - e \cos v) \tag{22}$$

so that we have established *Kepler's first law*, that the orbit is a conic with the Sun (or the Earth) at one focus. We have also established the facts that **a** must be directed to perifocus and that

$$|\mathbf{a}| = e\sqrt{\mu} \qquad h^2 = \mu p \tag{23}$$

which are the evaluations needed for equations (15) and (19).

The magnitude of **a** must not be confused with the constant a, which for the ellipse (cf. fig 1) we call the *mean distance* or *semi major axis*, and which is related to p, the *parameter*, and q, the *perifocal distance*, for all of the conics, and r_A, the *apofocal distance*, for the ellipse, by

$$q = a(1 - e) \qquad r_A = a(1 + e) \tag{24}$$

$$p = a(1 - e^2) = q(1 + e) = r_A(1 - e) \tag{25}$$

The orthogonal unit vectors **W** and **P** (cf. equations 13 and 19) are joined by the unit vector **Q**, which is orthogonal to each of them, and is defined and obtained by

$$\mathbf{Q} = \mathbf{W} \times \mathbf{P} \tag{26}$$

or by one of the relationships in

$$\mathbf{b} = e\sqrt{\mu p}\,\mathbf{Q} = \mathbf{r}\dot{H} - H\dot{\mathbf{r}} \tag{27}$$

where the latter relationship may be obtained by integration, since

$$H = r - p \qquad \dot{H} = \dot{r} \qquad \ddot{H} = -\mu H/r^3 \tag{28}$$

The orientation vectors **P** and **Q**, like **W**, are functions of i and Ω, but also of ω, the *argument of perifocus*, which is an angle measured from the node and defines the orientation of the perifocus, and of the major axis or line of apsides, in the orbit plane. They are used efficiently in the calculation of x, y, z, referred to general axes, from

$$x_\omega = r \cos v \qquad y_\omega = r \sin v \tag{29}$$

referred to orbital axes (cf. fig. 1 and §§5, 6), by means of

$$\mathbf{r} = x_\omega \mathbf{P} + y_\omega \mathbf{Q} \tag{30}$$

4. The vis-viva or energy integral may be obtained as follows:

$$2\dot{\mathbf{r}} \cdot \ddot{\mathbf{r}} = -2\mu \mathbf{r} \cdot \dot{\mathbf{r}}/r^3 = -2\mu \dot{r}/r^2 \tag{31}$$

$$\dot{s}^2 = \dot{\mathbf{r}} \cdot \dot{\mathbf{r}} = \mu(2/r + \alpha) \qquad \alpha = -1/a \tag{32}$$

where the constant α is evaluated from simple conditions known, for example, at perifocus.

Equations (15), (16), (17), (25), and (32) may now be recognized as sources of information about the nature of the conic determined by given position and velocity, **r** and $\dot{\mathbf{r}}$:

	Ellipse	Parabola	Hyperbola
General	$\frac{1}{a} > 0,\ 0 \le e \le 1$	$\frac{1}{a} = 0,\ e = 1$	$\frac{1}{a} < 0,\ e \ge 1$
Circular	$D = 0,\ e = 0$	—	—
Rectilinear	$\frac{1}{a} > 0,\ p = 0,\ e = 1$	$\frac{1}{a} = 0,\ p = 0,\ e = 1$	$\frac{1}{a} < 0,\ p = 0,\ e = 1$

(33)

The vis-viva integral is used also to evaluate useful factors for converting canonical units to conventional units. When m_2 is negligible as compared with m_1, they are:

Heliocentric	Geocentric		
$V = V_s \dot{s}_\tau$ $\mathbf{V} = V_s \dot{\mathbf{r}}_\tau$	$V = V_e \dot{s}_\tau$ $\mathbf{V} = V_e \dot{\mathbf{r}}_\tau$		(34)
$V_s = 29.785$ 18.508 57,897 97,720	$V_e = 7.905$ 4.912 15,367 25,936	km/sec mi/sec knots ft/sec	(35)

The constants V_s and V_e may be recognized as being approximately the velocities, respectively, of the Earth in its orbit, and of a circular satellite of the Earth hypothetically at one equatorial radius or geo-unit. The velocity of escape from the Earth, in the latter circumstances, is

$$V_{pe} = V_e \sqrt{2} \tag{36}$$

Two special values of $\dot{s}$ are of interest: $\dot{s}_q$ corresponding to $r = q$ and $\dot{s}_A$ corresponding to $r = r_A$,

$$\left.\begin{aligned} \dot{s}_q/\sqrt{\mu} &= \dot{s}_{\tau q}/\sqrt{\mu_\tau} = \sqrt{\frac{2}{q} - \frac{1}{a}} = \sqrt{\frac{1+e}{q}} = \sqrt{\frac{1}{a}\left(\frac{1+e}{1-e}\right)} \\ \dot{s}_A/\sqrt{\mu} &= \dot{s}_{\tau A}/\sqrt{\mu_\tau} = \sqrt{\frac{2}{r_A} - \frac{1}{a}} = \sqrt{\frac{1-e}{r_A}} = \sqrt{\frac{1}{a}\left(\frac{1-e}{1+e}\right)} \end{aligned}\right\} \tag{37}$$

5. The ellipse has special formulae in terms of E, the *eccentric anomaly*, an extremely useful angle that is defined by fig. 1:

$$\left.\begin{aligned} r &= a(1 - e\cos E) \\ r\cos v &= x_\omega = a(\cos E - e) \\ r\sin v &= y_\omega = \sqrt{ap}\,\sin E \end{aligned}\right\} \tag{38}$$

From equations (6), (15), and (38) we derive

$$\frac{r^2\dot{v}}{\sqrt{p}} = \sqrt{\mu} = a^{\frac{1}{2}}(1 - e\cos E)\dot{E} = r\sqrt{a}\,\dot{E} \tag{39}$$

and *Kepler's equation*,

$$M = M_0 + n(t - t_0) = n(t - T) = E - e\sin E \tag{40}$$

where M is the *mean anomaly*, M_0 is the *mean anomaly at the epoch*, t_0, T is the *time of perifocal passage*, and n is the *mean motion* or mean angular velocity:

$$n = \dot{M} = \sqrt{\mu}/a^{\frac{3}{2}} \tag{41}$$

We find from equations (38), (40), and (41) that the three anomalies, v, E, M are together at 0, π, 2π, . . . , and that the corresponding values of $t - T$ are 0, $\frac{1}{2}P$, P, . . . , where P is the *period* expressed in the conventional unit of time:

$$P = 2\pi/n = 2\pi a^{\frac{3}{2}}/\sqrt{\mu} \tag{42}$$

Equation (42) is an expression of *Kepler's third law*, that the squares of the periods are proportional to the cubes of the mean distances.

On next page is a table of comparative formulae and definitions.

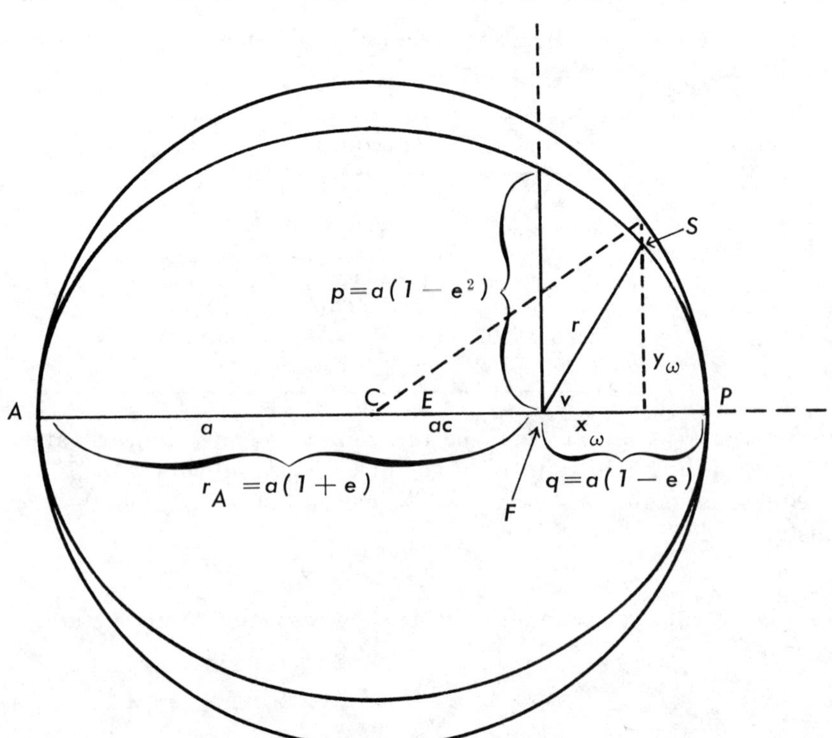

6. A table of comparative formulae and definitions:

General	Ellipse	Parabola	Hyperbola
	$0 \leq e \leq 1$	$e = 1$	$e \geq 1$
	$1/a > 0$	$1/a = 0$	$1/a < 0$
$p = q(1 + e)$	$p \leq 2q$	$p = 2q$	$p \geq 2q$
$M = M_0 + n(t - t_0)$	$\begin{cases} n = a^{-\frac{3}{2}}\sqrt{\mu} \\ M = E - e\sin E \end{cases}$	$n = \sqrt{\mu}$	$n = (-a)^{-\frac{3}{2}}\sqrt{\mu}$
$= n(t - T)$		$M = qD + \frac{1}{6}D^3$	$M = e\sinh F - F$
$r = p/(1 + e\cos v)$	$r = a(1 - e\cos E)$	$r = q + \frac{1}{2}D^2$	$r = -a(e\cosh F - 1)$
$\dot{r}\sqrt{p/\mu} = e\sin v$	$D = \dfrac{r\dot{r}}{\sqrt{\mu}} = \sqrt{a}\, e\sin E$	$D = \dfrac{r\dot{r}}{\sqrt{\mu}}$	$D = \dfrac{r\dot{r}}{\sqrt{\mu}}$
		$= \sqrt{2q}\tan\dfrac{v}{2}$	$= \sqrt{-a}\, e\sinh F$
$r\dot{v}\sqrt{p/\mu} = 1 + e\cos v$	$r^2\dot{v} = \sqrt{\mu p}$	$r^2\dot{v} = \sqrt{2\mu q}$	$r^2\dot{v} = \sqrt{\mu p}$
	$= \sqrt{\mu a(1 - e^2)}$		$= \sqrt{-\mu a(e^2 - 1)}$
$x_\omega = r\cos v$	$x_\omega = a(\cos E - e)$	$x_\omega = q - \frac{1}{2}D^2$	$x_\omega = -a(\cosh F - e)$
$y_\omega = r\sin v$	$y_\omega = \sqrt{ap}\sin E$	$y_\omega = \sqrt{2q}\, D$	$y_\omega = \sqrt{-ap}\sinh F$
$\dot{x}_\omega\sqrt{p/\mu} = -\sin v$	$r\dot{x}_\omega = -\sqrt{\mu a}\sin E$	$r\dot{x}_\omega = -\sqrt{\mu}\, D$	$r\dot{x}_\omega = -\sqrt{-\mu a}\sinh F$
$\dot{y}_\omega\sqrt{p/\mu} = \cos v + e$	$r\dot{y}_\omega = \sqrt{\mu p}\cos E$	$r\dot{y}_\omega = \sqrt{2\mu q}$	$r\dot{y}_\omega = \sqrt{\mu p}\cosh F$

Circular Ellipse*	Rectilinear		
	Ellipse	Parabola	Hyperbola
$e = 0$	$e = 1$	$e = 1$	$e = 1$
$r = p = q = a$	$p = 2q = 0$	$p = 2q = 0$	$p = 2q = 0$
$M = E$	$M = E - \sin E$	$M = \frac{1}{6}D^3$	$M = \sinh F - F$
$\dot{s} = r\dot{v} = \sqrt{\mu/a}$	$r = a(1 - \cos E)$	$r = \frac{1}{2}D^2$	$r = -a(\cosh F - 1)$
$D = r\dot{r}/\sqrt{\mu} = 0$	$D = r\dot{r}/\sqrt{\mu} = \sqrt{a}\sin E$	$D = r\dot{r}/\sqrt{\mu}$	$D = r\dot{r}/\sqrt{\mu} = \sqrt{-a}\sinh F$
$x_\omega = a\cos E$	$x_\omega = -r$	$x_\omega = -r$	$x_\omega = -r$
$y_\omega = a\sin E$	$y_\omega = 0$	$y_\omega = 0$	$y_\omega = 0$
$\dot{x}_\omega = -\dot{s}\sin E$	$\dot{x}_\omega = -\dot{r}$	$\dot{x}_\omega = -\dot{r}$	$\dot{x}_\omega = -\dot{r}$
$\dot{y}_\omega = \dot{s}\cos E$	$\dot{y}_\omega = 0$	$\dot{y}_\omega = 0$	$\dot{y}_\omega = 0$

* M, E, x_ω, y_ω, $\dot{x}_\omega$, $\dot{y}_\omega$ are formally related as shown, but all are indeterminate!

Barker's equation, the parabolic equivalent of Kepler's equation in the foregoing table, is often written

$$\frac{\sqrt{\mu}\,(t - T)}{q^{\frac{3}{2}}\sqrt{2}} = \tan\tfrac{1}{2}v + \tfrac{1}{3}\tan^3\tfrac{1}{2}v \tag{43}$$

but, so written, it is indeterminate for the rectilinear orbit ($q = 0$, $\tan\frac{1}{2}v = \infty$).

Unified variables, parameters, and formulae are designed to care for ellipses ($-1/a = \alpha < 0$), parabolas ($\alpha = 0$), and hyperbolas ($\alpha > 0$), including rectilinear ones, but not circles or nearly-circular ellipses. They are illustrated by the following table, which adds elliptic definitions for illumination but not for calculation:

Unified Formulae	*Elliptic Definitions*
$\sqrt{\mu}\,(t - T) = \tilde{M} = q\tilde{X} + e\tilde{U}$	$\tilde{X} = \sqrt{a}\,E$
$r = q + e\tilde{C}$	
$\tilde{U} = \dfrac{\tilde{X}^3}{3!} + \alpha\dfrac{\tilde{X}^5}{5!} + \alpha^2\dfrac{\tilde{X}^7}{7!} + \cdots$	$\tilde{U} = a^{3/2}(E - \sin E)$
$\tilde{C} = \dfrac{\tilde{X}^2}{2!} + \alpha\dfrac{\tilde{X}^4}{4!} + \alpha^2\dfrac{\tilde{X}^6}{6!} + \cdots$	$\tilde{C} = a(1 - \cos E)$
$x_\omega = q - \tilde{C}$	
$\dfrac{y_\omega}{\sqrt{p}} = -\dfrac{r\dot{x}_\omega}{\sqrt{\mu}} = \tilde{S} = \tilde{X} + \alpha\tilde{U}$	$\tilde{S} = \sqrt{a}\sin E$
$\dfrac{r\dot{y}_\omega}{\sqrt{\mu p}} = \tilde{K} = 1 + \alpha\tilde{C}$	$\tilde{K} = \cos E$

The definitions in terms of parabolic and hyperbolic variables may be inferred.

Universal variables, parameters, and formulae are designed to care for the foregoing and also circles and nearly-circular ellipses. They are illustrated by

Universal Formulae	*Elliptic Definitions*
$D_0 = r_0\dot{r}_0$	$D_0 = \sqrt{a}\,e\sin E_0$
$c_0 = 1 + \alpha r_0$	$c_0 = e\cos E_0$
$\sqrt{\mu}\,(t - t_0) = \hat{M} = r_0\hat{X} + D_0\hat{C} + c_0\hat{U}$	$\hat{E} = E - E_0$
$r = r_0 + D_0\hat{S} + c_0\hat{C}$	$\hat{X} = \sqrt{a}\,\hat{E}$
$\hat{U} = \dfrac{\hat{X}^3}{3!} + \alpha\dfrac{\hat{X}^5}{5!} + \alpha^2\dfrac{\hat{X}^7}{7!} + \cdots$	$\hat{U} = a^{3/2}(\hat{E} - \sin\hat{E})$
$\hat{C} = \dfrac{\hat{X}^2}{2!} + \alpha\dfrac{\hat{X}^4}{4!} + \alpha^2\dfrac{\hat{X}^6}{6!} + \cdots$	$\hat{C} = a(1 - \cos\hat{E})$
$\hat{S} = \hat{X} + \alpha\hat{U}$	$\hat{S} = \sqrt{a}\sin\hat{E}$
$f = 1 - \hat{C}/r_0 \qquad \sqrt{\mu}\,g = \hat{M} - \hat{U}$	
$\dot{g} = 1 - \hat{C}/r \qquad \dot{f} = -\sqrt{\mu}\,\hat{S}/rr_0$	
$\mathbf{r} = f\mathbf{r}_0 + g\dot{\mathbf{r}}_0 \qquad \dot{\mathbf{r}} = \dot{f}\mathbf{r}_0 + \dot{g}\dot{\mathbf{r}}_0$	

It is evident that universal variables reduce to unified variables when $t_0 = T$, and $M_0 = 0$, $E_0 = 0$, $r_0 = q$, $D_0 = 0$, $c_0 = e$.

ASTRODYNAMICAL TERMINOLOGY, NOTATION, AND USAGE

Samuel Herrick

University of California, Los Angeles

Abstracted from *Astrodynamics* by Samuel Herrick, published by Van Nostrand Reinhold 1970. For more complete discussion of the philosophy of selection of these symbols refer to *Astrodynamics*.

INTRODUCTION

Standardization of notation and terminology, or more properly, attention to established conventions and usage, contributes to understanding in all fields of science.

In the compilation of a standard listing of terms and symbols it is often implied, unwisely, that conformity with it is to be required thereafter of all "respectable" writing in the field. The philosophy adopted in the present listing, on the contrary, is that it should be based upon usage, that it should present alternatives sanctioned either by usage or by logical development where conflicts have come into being, and that it should be a reference list rather than a required list.

Typical of the conflicting usages in astrodynamics, encountered in celestial mechanics even before it had fully established contact with such fields as propulsion, aerodynamics, optimization, are those between v, true anomaly, and v, velocity, and between a, mean distance, and a, acceleration. In accordance with the foregoing expression of philosophy, we report all these usages but indicate that $\dot{\mathbf{r}}$, $\dot{s}$, or V should be preferred for velocity, and that $\ddot{\mathbf{r}}$, g, or A should be preferred for acceleration, because of the dominant use of v and a in the other senses.

The central core of astrodynamics is celestial mechanics, and an important function of the present listing is the providing of an entrée to the literature of celestial mechanics, much of which is still untapped in astrodynamical research. In consequence the present listing leans heavily upon astronomical usage, and upon the 1938 listing of the International Astronomical Union, *Transactions*, *6*, 345–348, but with departures from the latter in such matters as the following:

(1) The preferred designations for the geocentric distance (or "range") and for an increment or correction or residual are ρ and Δ rather than Δ and δ. The I.A.U. ignored long-established astronomical usage in such standard and familiar quantities as $\rho \cos \delta \, \Delta\alpha$ and $\rho\Delta\delta$, converting them into the whimsical $\Delta \cos \delta \, \delta\alpha$ and $\Delta\delta\delta$.

(2) The "direction cosines" associated with α and δ are designated as the components of a unit vector, $\mathbf{L}$, ($L_x = \cos \delta \cos \alpha$, $L_y = \cos \delta \sin \alpha$, $L_z = \sin \delta$), rather than by the l, m, n of the I.A.U. listing, or by the older sets: λ, μ, ν; α, β, γ. The use of vectors in celestial mechanics has become almost universal since 1938.

(3) The unit vector normal to the orbit plane is designated by $\mathbf{W}(W_x, W_y, W_z)$. The I.A.U. designations R_x, R_y, R_z have come into conflict with its designations for the "solar coordinates," X, Y, Z, because the vector notation for both sets would be $\mathbf{R}$, and because the two vectors sometimes appear in the same equation. Accordingly we prefer $\mathbf{R}(X,Y,Z)$ for the solar coordinates (or more generally the coordinates of the dynamical center referred to the observer).

The listing of terms and symbols is organized as follows:

(1) There is one alphabetical listing for terms and alphabetical symbols.

(2) Capital letters precede the corresponding lower case letters.

(3) Vector notation follows non-vector notation.

(4) Greek letters are alphabetized with Roman letters according to their usual trans-

literations (or the spellings of their names), thus changing the Greek order for γ, ζ, η, θ, φ, χ, ψ, ω.

(5) Vector triads are listed as groups under the first letter of triads, with cross references appearing under the other two letters.

(6) Terms are alphabetized according to spelling, and cross-referenced with notations, and vice versa.

(7) Non-alphabetical symbols are listed separately at the end. This listing makes no effort to be complete, or to include standard mathematical symbols.

(8) Terminology and definitions are primarily those that are peculiar to celestial mechanics or for which differences in usage exist, and do not include general terms such as "conic," "focus," "directrix," "least squares," "standard deviation," "celestial sphere," "ecliptic," etc., with which all users of the list may be assumed to be familiar.

(9) Alternative notations and usages are usually indicated with the author's preference or a cross reference introduced by "prefer" or by "see," "also," or "q.v." ("which see").

A

A, a	Acceleration; prefer g or A (see introduction and "g")
A, Az	Azimuth (q.v.); also Z
$A = A\left[\frac{km}{au}\right]$, $A\ [km]$	Length conversion factor, astronomical units to kilometers; au expressed in km.
A	Cross-sectional area
A, B	Constants in dynamical equation of first approximation in orbit determination: $\rho = A - B/r^3$
A, B, C	Principal moments of inertia
A_1, A_3	Ratios of sectors or time intervals, used in expressions for ratios of triangles (see "c_1, c_3"); also c_1^0, c_3^0: $$A_1 = \tau_{23}/\tau_{13};\ A_3 = \tau_{12}/\tau_{13}$$
$A.\ E.$	*American Ephemeris and Nautical Almanac* or (British) *Astronomical Ephemeris*
$\mathbf{A}(A_x, A_y, A_z)$	See "**L, A, D**"
$\mathbf{A}_h(A_{xh}, A_{yh}, A_{zh})$	See "**L, A$_h$, D$_h$**"
A, B	$\mathbf{A} = a\mathbf{P}$; $\mathbf{B} = b\mathbf{Q}$. Used in $\mathbf{r} = \mathbf{A}(\cos E - e) + \mathbf{B}\sin E$ Not recommended in place of $a\mathbf{P}$, $b\mathbf{Q}$ because of increasing emphasis on $\dot{\mathbf{r}}$.
a	Mean distance or semi major axis of ellipse, semi transverse axis of hyperbola.
a	Altitude (*q.v.*); prefer h
au, *a.u.*, or A.U.	Astronomical unit, approximately equal to semi major axis of Earth's orbit, but accurately defined by Gaussian $k = k_s$.
a_e	Earth's equatorial radius
a, b, h	$\mathbf{a} = e\sqrt{\mu}\,\mathbf{P}$, $\mathbf{b} = e\sqrt{\mu p}\,\mathbf{Q}$, $\mathbf{h} = \sqrt{\mu p}\,\mathbf{W}$.
α (alpha)	Right ascension
α	Azimuth (*q.v.*); prefer A
α	Time required for light to travel unit distance; also τ_A: $$\alpha_s = 0.005\ 7755\ \text{day/au}$$ $$\alpha_e = 0.021\ 275\ \text{sec/gu}$$
α	Flattening; prefer f (q.v.)

α — $= -1/a$; constant of integration in vis-viva integral.

α, β — ϵ, δ (q.v.) in Lambert-Euler equation (q.v.); astronomers have usually used ϵ, δ; mathematicians starting with Whittaker have tended to use α, β

α, β, γ — See "**L, A, D**"

adjoint method — Method for obtaining the partial differential coefficients used in differential correction, rendezvous, etc. Adjoint to "linearized Encke" or "variational orbit" method.

algorithm — Algorism, misspelled because of mistakenly assumed association with the root of *arithmetic;* correctly applied to arithmetic using concept of zero; incorrectly substituted for *collection of formulae* by some programmers.

altitude (h) — Angular distance above the horizon. Also called "angular altitude," "elevation," or "angular elevation" by those who have grown up in the traditions of aeronautics rather than astronomy or navigation.

altitude (H) — Linear height above sea level; prefer "height."

angular velocity (ω) — ω should be used preferably only for earth's rotation with respect to an inertially fixed direction. Differs slightly from the rate of change of sidereal time, because the latter is referred to a moving equinox.

angular velocity ($\dot{v}, \dot{u}, \dot{l}, \dot{\theta}$) — $\dot{v} = \dot{u} = \dot{l} = \dot{\theta} = \sqrt{\mu p}/r^2$ in two-body motion, and also in perturbation theory, provided the perturbations of the origins of v, u, l are absorbed in $v^{`} = \dfrac{dv}{dt} - \dot{v}$, etc.; usually $\theta^{`} = 0, \dfrac{d\theta}{dt} = \dot{\theta}$

anomaly — See "eccentric anomaly," "mean anomaly," "true anomaly"

apapsis, apoapsis, apex, apocenter — Apofocus (q.v.)

apofocal distance (r_A) — The distance from the focus to the apofocus: $a(1 + e) = r_A$

apofocus — Point in orbit farthest from focus of attraction; the farther apsis (q.v.). Also: apogee in geocentric orbit, aphelion in heliocentric orbit, etc., apocenter, apex; avoid the incorrectly formed and misleading term "apoapsis."

apsis — Point on orbit where radius vector is maximum or minimum (apofocus or perifocus). Line of apsides: line joining apofocus and perifocus. (Say ap′-si-deez!)

argument of latitude (u) — Angle measured in orbit plane from node to object; $u = v + \omega$

argument of perifocus (ω) — Angle measured in orbit plane from node to perifocus.

associated Legendre function — See P_n^m

azimuth (A) — Angle measured in horizon plane eastward from north point: also Z, α is unacceptable; also, in infrequent astronomical usage, measured from the south point.

B

B	See "A, B"
B	Argument in Barker's Tables; see "Barker's equation"; $$B = \frac{\sqrt{\mu}}{k_s}\frac{t-T}{q^{\frac{3}{2}}} = \frac{\sqrt{2}}{k_s}(\tan \tfrac{1}{2}v + \tfrac{1}{3}\tan^3 \tfrac{1}{2}v)$$
B_i'	Coefficients used in ratios of triangles (see "c_1, c_3") $$B_i' = \tau_{13}^2 A_i(1 - A_i^2)/6$$
B_1, B_2, B_3	Coefficients used in Gibbs's expressions for the ratios of triangles (see "c_1, c_3"): $$B_1 = (\tau_{12}^2 + \tau_{12}\tau_{23} - \tau_{23}^2)/12$$ $$B_2 = (\tau_{12}^2 + 3\tau_{12}\tau_{23} + \tau_{23}^2)/12$$ $$B_3 = (-\tau_{12}^2 + \tau_{12}\tau_{23} + \tau_{23}^2)/12$$
$\mathbf{B}$	See "$\mathbf{A}$, $\mathbf{B}$"
b	$= a\sqrt{1-e^2}$; semi minor axis of ellipse.
b	Latitude referred to the orbital plane.
b	Heliocentric β (q.v.)
$\mathbf{b}(b_x, b_y, b_z)$	$= e\sqrt{\mu p}\,\mathbf{Q}$; see $\mathbf{a}$, $\mathbf{b}$, $\mathbf{h}$. Note: $$\lvert\mathbf{b}\rvert = e\sqrt{\mu p} \neq b = a\sqrt{1-e^2}$$
β (beta)	Latitude, celestial
Barker's equation	Equation relating time, t, and true anomaly, v, for parabolic orbits (see also B): $$\frac{\sqrt{\mu}\,(t-T)}{q^{\frac{3}{2}}\sqrt{2}} = \tan \tfrac{1}{2}v + \tfrac{1}{3}\tan^3 \tfrac{1}{2}v$$ or (see D) $$\sqrt{\mu}\,(t-T) = qD + \tfrac{1}{6}D^3,\ D = \sqrt{2q}\tan \tfrac{1}{2}v$$
barycenter	Center of mass of two or more bodies.
burnout	See "initial, injection, insertion."

C

C	$= a_e/\sqrt{1 - e^2 \sin^2 \varphi}$, quantity tabulated in the almanacs for finding geocentric coordinates; see also S, x_c, y_c
C	$= rr_0[1 - \cos(v - v_0)]$; also $C_{ij} = r_i r_j[1 - \cos(v_j - v_i)]$; see also S, K, κ, σ
C	$= \cos E - e$; prefer next definition.
C	$= 1 - \cos E = (r - q)/ae$
$\tilde{C}$	$= a(1 - \cos E) = -a(\cosh F - 1)$, etc., unified variable.
$\hat{C}$	$= a[1 - \cos(E - E_0)] = -a[\cosh(F - F_0) - 1]$, etc., universal variable.
C_D	Shape factor for drag calculations
C_n, $C_{n,m}$ or C_l, $C_{l,m}$	Coefficients of zonal and sectoral harmonics in the gravitational potential of a spheroid, defined under "potential of spheroid."

$c = c\left[\frac{\text{km}}{\text{sec}}\right] = c\left[\frac{\text{km}}{\text{ls}}\right]$ — Velocity of light $\left[\frac{\text{km}}{\text{sec}}\right]$; distance conversion factor $\left[\frac{\text{kilometers}}{\text{light second}}\right]$

c — $= 1 + \alpha r = 1 - \frac{r}{a} = e \cos E = e \cosh F$

c — Distance from center of ellipse to focus; $c = ae$.

c — Length of chord; see s

c_1, c_3 — Ratios of the triangles: also n_1, n_3, or n_1/n_2, n_3/n_2. May be series or closed expressions (see also "d_1, d_2, d_3"), including:

$$c_1 = \frac{\mathbf{r}_2 \times \mathbf{r}_3}{\mathbf{r}_1 \times \mathbf{r}_3} = \frac{r_2 r_3 \sin(v_3 - v_2)}{r_1 r_3 \sin(v_3 - v_1)}$$

$$c_3 = \frac{\mathbf{r}_1 \times \mathbf{r}_2}{\mathbf{r}_1 \times \mathbf{r}_3} = \frac{r_1 r_2 \sin(v_2 - v_1)}{r_1 r_3 \sin(v_3 - v_1)}$$

$$c_i = A_i + B_i'/r_2^3 \qquad i = 1, 3$$

$$c_i = A_i \frac{1 + B_i/r_i^3}{1 - B_2/r_2^3} \qquad i = 1, 3$$

The last equations are "Gibbs's expressions."

c_1^0, c_3^0 — Ratios of the time intervals; see "A_1, A_3"

$\mathbf{c}$ — $= \sqrt{\mu p}\,\mathbf{W}$, angular momentum vector; prefer $\mathbf{h}$ (see "**a, b, h**")

χ (chi) — This symbol is often used in astrodynamics, but has no dominant use.

canonical time variable — See "τ"

canonical variables — Variables satisfying Hamilton's canonical equation (q.v., also "Delaunay variables").

climb angle (γ) — Flight-path angle

center of attraction — Focus of attraction (q.v.)

compression — Flattening (q.v.)

Cowell's method — Process in special perturbations in which the total acceleration of the orbiting body is numerically integrated to produce the position and velocity. The term "Cowell's method" has been applied to the "second-sum" or "Gauss-Jackson" integration formula, but Cowell did not actually use that formula. It is better to adopt the definition above, permitting the use of any integration formula with "Cowell's method," including the Runge-Kutta formulae. The term "Cowell's method" is usually, though not necessarily, limited to integrations in rectangular coordinates.

D

D — "Parabolic anomaly" analogous to the eccentric and hyperbolic anomalies (see "Barker's equation"):

$$r\dot{r}/\sqrt{\mu} = D = \sqrt{2q}\tan\tfrac{1}{2}v,\ r = q + \tfrac{1}{2}D^2$$

$D, \dot{D}, \ddot{D}$ — $D = r\dot{r}/\sqrt{\mu} = \sqrt{a}\, e \sin E = \sqrt{-a}\, e \sinh F$;

$$\frac{\dot{D}}{\sqrt{\mu}} = \frac{1}{r} - \frac{1}{a};\ \ddot{D} = -\mu\frac{D}{r^3}$$

$\mathbf{D}(D_x, D_y, D_z)$ — See "**L**, **A**, **D**"

$\mathbf{D}_h$ — See "**L**, $\mathbf{A}_h$, $\mathbf{D}_h$"

d_1, d_2, d_3 — Coefficients for $\dot{\mathbf{r}}_2 = -d_1\mathbf{r}_1 + d_2\mathbf{r}_2 + d_3\mathbf{r}_3$. They may be series or closed expressions (see "c_1, c_3").

Δ(delta) — Denotes an increment, residual, correction, or error: e.g., $\Delta\rho$, $\Delta\alpha$, $\Delta\delta$, Δa, Δe, Δk_s, $\Delta\mu$. Also δ; prefer δ for "virtual" increment and for numerical-analysis central-difference formulae (see "$\delta, \delta^2 \ldots$"); prefer Δ when used with α and δ (see introduction).

Δ — Topocentric distance of object; prefer $\rho = |\boldsymbol{\varrho}|$; see $\boldsymbol{\varrho}$ and the introduction.

$\Delta, \Delta^2 \ldots$ — Differences in numerical tables when subscript runs down a forewards diagonal

ΔL — $= \sqrt{(\Delta L_x)^2 + (\Delta L_y)^2 + (\Delta L_z)^2} = \sqrt{(\Delta\alpha \cos\delta)^2 = (\Delta\delta)^2}$ = arc residual

ΔT or Δt — = E.T. − U.T.; time correction to eliminate irregularities of U.T., given in almanacs.

$\Delta X, \Delta Y, \Delta Z$ — Corrections to solar coordinates to eliminate parallax, i.e., to translate from center of earth to point of observation.

$$\Delta X = \Delta_{xy}\cos\theta;\ \Delta Y = \Delta_{xy}\sin\theta;\ \text{and}$$
$$\Delta Z = -r_c \sin\varphi' = -x_c$$

Δ_{xy} — $\Delta_{xy} = -r_c\cos\varphi' = -y_c$; tabulated in almanacs with ΔZ, for active observatories.

δ (delta) — Declination

δ — $= h - \vartheta$, auxiliary quantity used in the Lambert-Euler equation (see: ϵ, ϑ, h); also β (see "α, β")

δ — Denotes a "virtual" increment: e.g., a virtual displacement, $\delta\mathbf{r}$. See also "Δ," "virtual increment."

$\delta, \delta^2 \ldots$ — Differences in numerical tables when subscript runs horizontally

$\nabla, \nabla^2 \ldots$ (del or nabla) — Differences in numerical tables when subscript runs up a backwards diagonal

declination (δ) — Topocentric or geocentric coordinate measured north (+) or south (−) from celestial equator along an hour circle. See "right ascension."

Delaunay variables:

$$l = M \qquad L = \sqrt{\mu a}$$
$$g = \omega \qquad G = \sqrt{\mu p}$$
$$h = \Omega \qquad H = \sqrt{\mu p}\cos i$$

differential correction — A method of correcting the elements of an orbit by means of residuals (observation minus computation) sometimes called the Newton-Raphson process.

disturbing function — Perturbing function (q.v.)

dot-variation / dot-derivative — = {two-body variation / two-body derivative}

E

E — Eccentric anomaly (q.v.)

$E = E\left[\dfrac{gm}{m_\oplus}\right]$ or $E[gm]$ — Mass conversion factor, earth masses to grams; mass of the Earth in grams.

E or E_{II}	Elliptic integral of the second kind
$\hat{E}$	$= 2\hat{g} = E - E_0$
E.T., E.D.	Ephemeris Time, Ephemeris Date (q.v.)
$\mathbf{E}(E_x,E_y,E_z)$	See "**Z, E, N**"
e	Eccentricity (q.v.)
e	Earth's meridional eccentricity; $e^2 = 2f - f^2$ where f is the flattening. See f
e	Flattening; prefer f (q.v.)
e	Subscript designating the reference orbit in Encke's method; see o
ϵ (epsilon)	Obliquity of ecliptic
ϵ	Error; mean square error
ϵ	$= h + \hat{g}$, auxiliary quantity used in the Lambert-Euler development (see δ, $\hat{g}$, h); also α (see "α, β").
ϵ	Flattening; prefer f (q.v.)
η (eta)	See "$\boldsymbol{\rho}(\xi, \eta, \zeta)$"
eccentric anomaly (E)	Angle at center of ellipse between major axis and radius of auxiliary circle through a point which has same abscissa as a given point on ellipse. In connection with an ellipsoid of revolution the eccentric anomaly is also known as the "parametric," "geometric," "reduced," or "eccentric" latitude. Also u; prefer E because of notation $u = v + \omega$
eccentricity (e)	Constant ratio between the distance from a point on a conic to the focus and the distance from the point to the directrix.
ellipticity	Flattening (q.v.)
elements of orbit	Selected set of six independent constants of integration of the equations of motion.
elevation (h)	Angular distance above the horizon; prefer "altitude" (q.v.)
Encke's method	Process in special perturbations by which the perturbed departures from the reference orbit, $\boldsymbol{\rho}(\xi,\eta,\zeta)$ and $\dot{\boldsymbol{\rho}}(\dot{\xi},\dot{\eta},\dot{\zeta})$, are integrated from $\ddot{\boldsymbol{\rho}}(\ddot{\xi},\ddot{\eta},\ddot{\zeta})$ and used to correct the reference position and velocity into the actual position and velocity. May be used to obtain partial differential coefficients, in either linearized or non-linear form.
Encke terms	The differences of the two-body terms in the equations for $\ddot{\rho}(\ddot{\xi},\ddot{\eta},\ddot{\zeta})$ in Encke's method of special perturbations: e.g., $$\mu\left(\frac{x_e}{r_e^3} - \frac{x}{r^3}\right) = \frac{\mu}{r_e^3}(fqx - \xi)$$ Also called "indirect terms," but then confused with the terms of the equations of motion that represent the accelerations of the origin.
ephemeris	A table of calculated positions of an object, with equidistant dates as arguments. Plural: ephemerides. (Say ef-eh-mer′-i-deez!)

ephemeris second	Unit of time associated with Ephemeris Time; $$1 \text{ ephemeris second} = \frac{1 \text{ tropical year (1900.0)}}{31{,}556{,}925.9747}$$
Ephemeris Time (E.T.)	= U.T. + ΔT, uniform time used as independent variable in the ephemerides of the planets and other objects in space. ΔT is listed in almanacs.
Ephemeris Date (E.D.)	= Julian Date + ΔT − 2400 000.^{d}5
epoch (t_0)	Arbitrary instant of time for which the constants of an orbit are specified.
Eulerian angles	Orientation angles (q.v.) used for 2000 years before Euler.

F

F	Hyperbolic anomaly
F	$= \frac{1}{2}(v + v_0)$ Also: $F_{ij} = \frac{1}{2}(v_j = v_i)$
F	The negative of the Hamiltonian function, also called the force function; in astronomy usually $$F = \Phi - T = \frac{\mu}{2a} + R$$
F or E_I	Elliptic integral of the first kind
f, g	Series or closed expressions used in the formulae: $$\mathbf{r} = f\mathbf{r}_0 + g\dot{\mathbf{r}}_0$$ $$\mathbf{r}_j = f_j\mathbf{r}_0 + g_j\dot{\mathbf{r}}_0 = f_{ij}\mathbf{r}_i + g_{ij}\dot{\mathbf{r}}_i$$
f	Flattening; also ϵ, e, α. Prefer f, with ϵ as second choice, because of confusion with closely associated eccentricity: $$f = 1 - \frac{b}{a} \qquad e = \sqrt{1 - \left(\frac{b}{a}\right)^2}$$
f or $\hat{f}$	$= \frac{1}{2}(v - v_0)$. Prefer $\hat{f}$ because of close association with primary f, g above. Also: $\hat{f}_{ij} = \frac{1}{2}(v_j - v_i)$
f	Function (of q) used in Encke's method: $$1 + 2q = r^2/r_e^2 \qquad fq = 1 - (1 + 2q)^{-\frac{3}{2}}$$
f	Frequency
f	True anomaly. Prefer v because of confusion with f's above.
flattening	See f; also called: oblateness, ellipticity, compression.
flight-path angle (γ)	Angle between velocity vector and plane perpendicular to radius vector; also "climb angle."
focus of attraction	Principal or dynamical focus, occupied by principal attracting mass: prefer to "center of attraction."
focus, empty	Secondary focus of conic, unoccupied.

G

G	Constant of gravitation in cgs or mks system. In astrodynamics the kilometer is preferred to the meter or the centimeter; accordingly the units of G are km^3/gm sec^2. Prefer k^2 for other systems.
G	$= \frac{1}{2}(E + E_0)$. Also: $G_{ij} = \frac{1}{2}(E_j + E_i)$
G	Velocity; see "V, v"

G	$= \sqrt{\mu p}$; prefer $\sqrt{\mu p}$ or h, except when using Delaunay variables (q.v.)
G.C.T., G.M.T.	Greenwich Civil Time, Greenwich Mean Time (q.v.)
GHA	Greenwich hour angle; also t_G; see "hour angle."
g	See "f, g"
g, g_e	Terrestrial acceleration of gravity; g_e = equatorial value.
g	Acceleration in general; prefer to a; also A.
g or $\mathfrak{g}$	$= \frac{1}{2}(E - E_0)$. Prefer $\mathfrak{g}$ to avoid confusion with primary f, g above. Also: $\mathfrak{g}_{ij} = \frac{1}{2}(E_j - E_0)$.
g	Argument of perifocus; prefer ω except when using "Delaunay variables" (q.v.)
gu	Geo-unit, or geocentric unit, nearly equal to the Earth's equatorial radius, a_e. Analogous to the astronomical unit of the heliocentric system.
γ (gamma)	Flight-path angle (q.v.)
γ	$=$ sin i or sin $\frac{1}{2}i$ in lunar and planetary theory
γ	Normal acceleration of gravity
Gauss-Jackson formula	Formula for numerically integrating second-order differential equations; also: second-sum formula, Σ^2 formula; see "Cowell's method."
geoid	The mean sea level surface of the earth. See "spheroid."
Gibb's expressions	See "c_1, c_3"
grave-variation grave-derivative	$=$ perturbed variation $=$ perturbed derivative
Greenwich Civil Time (G.C.T.)	$=$ Universal Time (U.T.)
Greenwich Mean Time (G.M.T.)	$=$ U.T. $+ 12^h$ before 1925.0 $=$ U.T. after 1925.0

H

H or h	Height above sea level, or height above the geoid; see "altitude"; prefer H.
H	$= \sqrt{\mu p} \cos i$; see "Delaunay variables."
H	Hamiltonian function
H or HA or h	Hour angle (q.v.)
h	Altitude (q.v.) or "elevation"
h	$= \sqrt{\mu p}$; angular momentum (see "**a**, **b**, **h**"); also G, as one of the Delaunay variables (q.v.)
h	Longitude of ascending node; prefer Ω except when using Delaunay variables (q.v.)
h	Interval of argument in integration tables or in any interpolation table. Also: w.
h	Auxiliary quantity used in the Lambert-Euler formulae (see also G, $\mathfrak{g}$, ϵ, δ): $$\cos h = e \cos G$$
$\mathbf{h}(h_x, h_y, h_z)$	$= \sqrt{\mu p}\,\mathbf{W}$; angular momentum vector.
Hamiltonian function (H)	$H = \sum_s p_s \dot{q}_s - L$; also $-H = F$ (q.v.)
Hamilton's canonical equations	$\dot{q}_s = \frac{\partial H}{\partial p_s} \qquad \dot{p}_s = -\frac{\partial H}{\partial q_s}$

harmonics	Mathematical terms that indicate differences in the surface shape between a reference spheroid (e.g., for the Earth) and a sphere. "Zonal" harmonics indicate latitudinal differences; "sectoral" harmonics indicate longitudinal differences; "tesseral" harmonics include zonal and sectoral harmonics.
height (H)	Linear height above sea level; prefer to "altitude" (q.v.)
Hohmann orbit	Minimum energy heliocentric orbit tangent to the earth's orbit at one apsis and to that of the target planet at the other.
hour angle (t)	Angle measured in equator plane westward from local meridian, usually 0^h to 24^h; prefer (in order): t, HA, LHA, H; see "GHA"; the use of h for hour angle rather than angular altitude is undisciplined!
hour circles	Great circles that pass through the celestial pole, sometimes incorrectly referred to as "meridians."
hyperbolic anomaly (F)	Angle in hyperbola corresponding to E in the ellipse and D in the parabola: $F = iE$ where $i = \sqrt{-1}$, and $M = e \sinh F - F$. Also H.

I

I, **J**, **K**	$\mathbf{I}(I_x, I_y, I_z)$ is the unit vector along the x-axis in the reference plane (directed to ♈ in equatorial and ecliptic references). **J** is directed along the y-axis, 90° east from **I** in the reference plane. **K** forms a right-handed orthogonal set with **I**, **J**. Also: **i**, **j**, **k**; prefer **I**, **J**, **K** in association with **P**, **Q**, **W** or **U**, **V**, **W**, etc.
i	$= \sqrt{-1}$
i, I, ι (iota) / inclination	i is standard; I and ι are convenient when it has a perturbed variation; angle measured from the reference plane (ecliptic or equatorial) to the orbit plane. Limits of i: 0° and 180°; or 0° and 90° if "direct" and "retrograde" are specified.
inertial axes	Axes which are fixed in direction, but which may have a moving origin (e.g., Sun or Earth) rather than an inertial one (e.g., the center of mass in the n-body problem).
indirect terms	The terms of the equations of motion (e.g., the relative-motion form) that represent the accelerations of the origin (e.g., the Sun); see "Encke terms"
initial / injection / insertion	These adjectives are sometimes used synonymously, along with burnout, launch, takeoff; writers should be careful to specify distinctions they wish to make.

J

J_n	Coefficients of zonal harmonics in the gravitational potential of a spheroid of revolution, defined under "potential of spheroid"
$J_{n,m}$	See "potential of spheroid"
J.D., J.E.D.	Julian Date, Julian Ephemeris Date (q.v.)
$\mathbf{J}(J_x, J_y, J_z)$	See "**I**, **J**, **K**"
jovicentric	Referred to the center of Jupiter; also: Jupiter-centered.
Julian Date (J.D.)	Universal Time (q.v.) expressed by a continuing day count.
Julian Ephemeris Date (J.E.D.)	Ephemeris Time (q.v.) similarly expressed: $$\text{J.E.D.} = \text{J.D.} + \Delta T$$

K

K — $= rr_0 \cos (v - v_0)$. See also: C, S, κ, σ.

$\tilde{K}$ — $= \cos E = \cosh F$, etc., unified variable.

$\hat{K}$ — $= \cos (E - E_0) = \cosh (F - F_0)$, etc., universal variable.

$\mathbf{K}(K_x, K_y, K_z)$ — Unit vector perpendicular to reference plane; see "**I**, **J**, **K**"

k, k^2 — Gravitational constant, its value depending on units of time, mass and distance involved. See: k_s, k_e.

k_e — Geocentric gravitational constant:

$$k_e = 0.001\,239\,444[\mathrm{gu}^{\frac{3}{2}}/\mathrm{m}_{\oplus}^{\frac{1}{2}}\ \mathrm{sec}]$$

k_s or k — Gaussian or heliocentric gravitational constant:

$$k_s = 0.017\,202\,09895[\mathrm{au}^{\frac{3}{2}}/\mathrm{m}_{\odot}^{\frac{1}{2}}\ \mathrm{day}]$$

κ (kappa) — Auxiliary quantity used in development of closed expressions for f and g, in Lambert-Euler equation, etc.:

$$\kappa = \pm \sqrt{2(rr_0 + xx_0 + yy_0 + zz_0)} = 2\sqrt{rr_0} \cos \tfrac{1}{2}(v - v_0)$$

See also: σ, C, K, S.

kaday — The unit of $\tau = k_*(t - t_0)$ for heliocentric orbits; when

k_*^2 and $\mu = k_s^2$, it is k_s^{-1} days $= 58.132\,44087$ days

kesec — The unit of $\tau = k_*(t - t_0)$ for geocentric orbits; it has also been called the "kemin" because k_e was first based upon the minute rather than the second; when k_*^2 and $\mu = k_e^2$, it is

$$k_e^{-1}\ \text{sec.} = 13.446\,8896\ \text{min.}$$

Kepler's equation — $M = E - e \sin E$, and variants thereof.

Kepler's first law — $r = p/(1 + e \cos v) = a(1 - e \cos E)$

Kepler's second law — $r^2 \dot{v} = \sqrt{\mu p}$

Kepler's third law — $\mu P^2 = (2\pi)^2 a^3$ or $n^2 a^3 = \mu$

L

L, L_r — Mean longitude: $L = L_0 + n(t - t_0) = M + \omega + \Omega$; for retrograde orbits: $L_r = M + \omega - \Omega$. See "$l$, l_r" and "$\tilde{\omega}$, $\tilde{\omega}_r$"

L — Lagrangian function (q.v.)

L — $= \sqrt{\mu a}$; see "Delaunay variables."

L — Longitude, terrestrial (q.v.); prefer λ.

LHA — Local hour angle; also t; see "hour angle"

L, A, D — $\mathbf{L}(L_x, L_y, L_z)$ is the unit vector directed from the observational reference point to the object. **A** is a unit vector perpendicular to **L**, parallel to the equatorial plane, and directed with increasing right ascension, α. **D** is a unit vector such that **L**, **A**, **D** make up a right-handed orthogonal set, and so is directed with increasing declination, δ; so also $d\mathbf{L} = \mathbf{A} \cos \delta\, d\alpha + \mathbf{D} d\delta$:

$$\begin{aligned} L_x &= \cos \delta \cos \alpha & A_x &= -\sin \alpha & D_x &= -\sin \delta \cos \alpha \\ L_y &= \cos \delta \sin \alpha & A_y &= \cos \alpha & D_y &= -\sin \delta \sin \alpha \\ L_z &= \sin \delta & A_z &= 0 & D_z &= \cos \delta \end{aligned}$$

Alt. to L_x, L_y, L_z: l, m, n; λ, μ, ν; α, β, γ; see introduction.

L, $\mathbf{A}_h$, $\mathbf{D}_h$	$\mathbf{L}(L_{xh}, L_{yh}, L_{zh})$ is the unit vector directed from the observational reference point to the object. $\mathbf{A}_h$ is a unit vector perpendicular to **L**, parallel to the horizon plane, and directed with clockwise increasing azimuth, A. $\mathbf{D}_h$ is a unit vector such that **L**, $\mathbf{A}_h$, $\mathbf{D}_h$ make up a left-handed orthogonal set, and so is directed with increasing altitude, h; so also $d\mathbf{L} = \mathbf{A}_h \cos h\, dA + \mathbf{D}_h dh$: $$\begin{array}{lll} L_{xh} = \cos h \cos A & A_{xh} = -\sin A & D_{xh} = -\sin h \cos A \\ L_{yh} = \cos h \sin A & A_{yh} = +\cos A & D_{yh} = -\sin h \sin A \\ L_{zh} = \sin h & A_{zh} = 0 & D_{zh} = +\cos h \end{array}$$ See "**Z**, **E**, **N**"
l, l_r	True longitude: $l = v + \omega + \Omega$; for retrograde motion: $l_r = v + \omega - \Omega$. See "$L$, L_r" and "$\tilde{\omega}$, $\tilde{\omega}_r$"
l	Mean anomaly. Prefer M, except when using Delaunay variables (q.v.)
l, m, n	Direction cosines. For preferred alternate see "**L**, **A**, **D**"
Λ (lambda)	$= E + \omega + \Omega$, to accompany $l = v + \omega + \Omega$, $L = M + \omega + \Omega$, $\tilde{\omega} = \omega + \Omega$
λ (lambda)	Longitude, celestial (q.v.)
λ	Longitude, terrestrial (q.v.); also L
λ, μ, ν	Direction cosines. For preferred alternate see "**L**, **A**, **D**"
Lagrange's equations	$\dfrac{d}{dt}\dfrac{\partial T}{\partial \dot{q}_s} - \dfrac{\partial T}{\partial q_s} = Q_s$, where T denotes the kinetic energy of the body. If the forces involved have a potential function, then the equations have the form: $$\frac{d}{dt}\frac{\partial L}{\partial \dot{q}_s} - \frac{\partial L}{\partial q_s} = 0$$ where L is the Lagrangian function.
Lagrangian function (L)	Kinetic energy minus potential energy, usually. In astronomy $L = T + \Phi$; see "potential function (Φ)"
Lambert-Euler equation	In elliptic motion often $n(t - t_0) = (\epsilon - \sin \epsilon) - (\delta - \sin \delta)$; ϵ, δ also α, β (q.v.)
Lambert's equation	In orbit determination the dynamical equation $\rho = A - B/r^3$
latitude, celestial (β)	Angle measured north or south from the ecliptic along a secondary great circle perpendicular to it; see "longitude, celestial."
latitude, terrestrial	Angle measured in a meridian plane from the Earth's equator to, e.g., the point of observation; see "φ, φ_g, φ_a" and φ'
latitude, argument of (u)	$= v + \omega$; see $U = M + \omega$, $\Upsilon = E + \omega$
launch	See "initial, injection, insertion"
Legendre polynomials	See P_n.
light time	Interval during which light travels from the observed object to the observer. See α.
line of nodes	Line formed by the intersection of the reference plane and the orbit plane.
longitude, celestial (λ)	Angle measured from vernal equinox along the ecliptic to the secondary circle passing through the object, geocentric or topocentric. See "latitude, celestial."

longitude, terrestrial (λ) — Angle measured from Greenwich meridian along the equator, east or west, to the meridian through, e.g., the point of observation. For geodesy and perturbation theory λ should be taken as positive eastward, negative westward. Also: L.

longitude, mean (L) — $= M + \omega + \Omega$; see "L, L_r" and $\Lambda = E + \omega + \Omega$

longitude, true (l) — $= v + \omega + \Omega$; see "l, l_r"

longitude of perifocus ($\tilde{\omega}$) — $= \omega + \Omega$; see "$\tilde{\omega}$, $\tilde{\omega}_r$"

M

M, M_0 — Mean anomaly (q.v.), mean anomaly at the epoch

m_1 or M — Mass at focus of attraction; prefer m_1 because of mean anomaly

m_2 or m — Mass of orbiting body

m_i, m_j — Masses of particles, planets, etc.

$m_{\odot}$, $m_{\oplus}$, $m_{☾}$, etc. — Masses of Sun, Earth, Moon, etc. See astronomical symbols at end.

m_0 — Initial mass of a vehicle, before takeoff

m, m_0 — See "magnitude"

μ (mu) — Mass function. In the two-body problem, $\mu = k^2(m_1 + m_2)$. More generally, $\mu = k^2(1 + m)$, where m may include some perturbation effects as well as $m_1 + m_2 - 1$.

μ — Ratio of mass of Moon to mass of Earth

μ, $1 - \mu$ — Masses of principal bodies in restricted 3-body problem; μ is usually the smaller of the two masses.

μ — Index of refraction

μ — Mean angular motion; prefer n (q.v.)

μ_a — Ratio of the mass of the atmosphere to the mass of Earth.

magnitude (m,m_0) — Measures of the brightness of an object, often calculated from

$$m = g + 5 \log \rho + 5 \log r + C\beta^0$$

where g is a constant determined from observed values, ρ and r are geocentric and heliocentric distances, β^0 is the phase angle (in degrees), and C is a constant; for m_0: $r = a$, $\rho = a - 1$, $\beta^0 = 0$ (a hypothetical "mean" opposition).

mean anomaly (M) — Angle measured from perifocus through which a body would have travelled if it were moving uniformly at its mean angular velocity, n:

$$M = M_0 + n(t - t_0) = n(t - T)$$

where M_0 is the value of M at epoch t_0 and T is the time of perifocal passage. Also l; prefer M except when using Delaunay variables (q.v.)

mean distance (a) — Semi major axis (q.v.)

mean motion, mean angular motion, mean angular velocity — See "n"

meridian, celestial — The hour circle that passes through the zenith.

N

n — Mean angular motion of a body in orbit; see "mean anomaly"; also sometimes μ, but prefer n because of the notation:

$$n = \sqrt{\mu}\, a^{-\frac{3}{2}}$$

n_1, n_2, n_3	Ratios of the triangles (n_1,n_3), or quantities proportional to the triangles (n_1,n_2,n_3); prefer c_1, c_3 (q.v.)
ν (nu)	True anomaly; prefer v; the rather curious use of ν may come from misreading of classical French works printed from type-fonts in which there are two italic v's. One is used at the beginnings of words; the other, used internally and in equations, closely resembles the Greek ν (e.g., Tisserand, v. 1, pp. 3, 25, 26).
n-body problem	Orbit problem involving only the mutual gravitational attractions of n bodies.
node	Intersection of orbit with a reference plane.
node (☊)	Angle measured in reference plane from the vernal equinox to the ascending node of the orbit; "node" is an abridgement of "longitude of the ascending node"; $☊_\epsilon$ may be used for the angle in the ecliptic plane, and $☊_\alpha$ for that in the equatorial plane. Note that the node symbol (☊) is not the same as the capital omega (Ω), though the latter may be used when the type-font lacks ☊.
north celestial pole (P_n)	Point where the Earth's axis of rotation extended from the northern hemisphere intersects the celestial sphere.
north point (N)	Origin for azimuth in horizon system of coordinates, defined by an intersection of horizon with the celestial meridian, a great circle through the observer's zenith and the north celestial pole.
nutation	The periodic terms of the precession-nutation complex that describes the motion of the Earth's axis of rotation.

O

o	Subscript designating a quantity or value associated with the epoch, t_0.
o	Subscript sometimes used to designate a quantity or value associated with the Encke or other reference orbit; prefer subscript e or R to avoid confusion with the preceding use of subscript o.
Ω (omega)	See "node (☊)"
Ω	Perturbing function; prefer R
ω (omega)	Argument of perifocus
ω	Angular velocity
$\tilde{\omega}$, $\tilde{\omega}_r$	Longitude of perifocus: $\tilde{\omega} = \omega + ☊$; for retrograde orbits $\tilde{\omega}_r = \omega - ☊$; also π, π_r (see L, L_r). If you want to be intelligible, call $\tilde{\omega}$ "omega-tilde," not "πi."
oblateness	See f and "flattening"; some would prefer to use "oblateness" for J_2, and restrict it thereto.
obliquity of the ecliptic (ϵ)	Inclination of the ecliptic plane to the equatorial plane. Its true value for a given date (varying slightly from $23\frac{1}{2}°$) may be found with the solar coordinates in the national ephemerides. Its mean value for a given date may be found in *Planetary Coordinates*.

orbit — A path in space; more specifically, a set of constants defining it; the set can include physical constants as well as the six elements or constants of integration; also "state vector" or "state."

orbit — Period, or revolution; this recent introduction is expressive in popular discussions or news reporting, but should be avoided, in favor of the standard terms, in precise scientific writing.

orientation angles — The elements i, Ω, ω; see "Eulerian angles"

osculating orbit — An approximation to the actual path of an object derived from the actual position and velocity at a given moment; usually a two-body orbit.

osculating variation / osculating derivative — = two-body variation / two-body derivative (q.v.), but are more general terms that can be used whether or not the reference solution is a two-body orbit.

P

P — Period of revolution

P_n — Legendre polynomial, used in the development of the potential function in series form:

$$P_0(x) = 1 \qquad P_1(x) = x$$
$$P_2(x) = \tfrac{1}{2}(3x^2 - 1) \qquad P_3(x) = \tfrac{1}{2}(5x^3 - 3x)$$
$$P_4(x) = \tfrac{1}{8}(35x^4 - 30x^2 + 3)$$
$$P_n(x) = \frac{2n-1}{n} xP_{n-1}(x) - \frac{n-1}{n} P_{n-2}(x)$$

P_n^m or P_{nm} — Associated Legendre functions defined by

$$P_n^m(x) = (1 - x^2)^{\frac{1}{2}m} \frac{d^m}{dx^m} P_n(x)$$

P, **Q**, **W** — $\mathbf{P}(P_x,P_y,P_z)$ is the unit vector directed toward the perifocus. **Q** is directed 90° from **P** in the direction of motion of the object in the orbital plane. **W** is perpendicular to the orbital plane, thus forming a right-handed orthogonal set with **P** and **Q**.

p, p_i — See "parameter"

p_s — Component of generalized momentum: $p_s = \dfrac{\partial L}{\partial \dot{q}_s}$

p_α — Annual precession in right ascension:

$$p_\alpha = m^s + n^s \sin\alpha \tan\delta$$

p_δ — Annual precession in declination:

$$p_\delta = n'' \cos\alpha$$

Φ (phi) — Potential function (q.v.)

φ, φ_g, φ_a (phi) — φ is used for geodetic latitude (φ_g), which is the angle between the normal to a reference spheroid and the plane of the equator. φ is also used for astronomical latitude (φ_a), which is the angle between the direction of gravity and the plane of the equator, and which differs from the geodetic latitude by "station error" or "local anomaly." φ_g is asso-

	ciated with the reference spheroid; φ_a, with the geoid. Both φ_g and φ_a are sometimes called, indiscriminately, "geographic latitude."
φ'	Geocentric latitude: angle between a radius from the center of reference spheroid and the plane of its equator.
φ	Angle of eccentricity: $\sin \varphi = e$, $\cos \varphi = \sqrt{1 - e^2}$ Convenient in logarithmic calculation, but now exists only as a confusing holdover. Used as argument in some tables.
$\Pi_\odot$ (pi)	Solar parallax, i.e., Sun's mean geocentric equatorial horizontal parallax; also $\pi_\odot$
$\Pi_{\text{☾}}$	Moon's varying geocentric equatorial horizontal parallax; also $\pi_{\text{☾}}$
π, π_r (pi)	Longitude of perifocus: $\pi = \omega + \text{☊}$; for retrograde motion, $\pi_r = \omega - \text{☊}$. Prefer $\tilde{\omega}$, $\tilde{\omega}_r$ (q.v.)
ψ (psi)	Elongation; angle at the observational reference point between the directions of the dynamical center and the observed object, i.e., between the vectors **R** and **L**
parallax	Geocentric equatorial horizontal parallax is the angle at an object subtended by the earth's equatorial radius and is a measure of distance of objects in the solar system. Heliocentric parallax is the angle at an object subtended by the radius of the earth's orbit and is a measure of distance of objects outside the solar system: Π, π
parameter	Not generally used in its broad mathematical senses, but specialized in astrodynamics to: (1) The semi latus rectum of a conic; usually designated by p (2) A selected two-body element or variable (e.g., a or M), or a constant of integration in an analytic solution of some problem, used as a variable in the method of variation of parameters; sometimes designated by p_i
periapsis, pericenter	Perifocus (q.v.)
perifocal distance (q)	$= a(1 - e)$, distance between focus and perifocus
perifocus	Point in orbit nearest focus of attraction; the nearer apsis; perigee in geocentric orbit, perihelion in heliocentric orbit, etc.; also "pericenter;" "periapsis" should be avoided. See "apofocus"
perifocus, argument of (ω)	Angle measured in orbit plane from node to perifocus.
perifocus, longitude of ($\tilde{\omega}$)	$\tilde{\omega} = \omega + \text{☊}$; for retrograde motion, $\tilde{\omega}_r = \omega - \text{☊}$; also π, π_r.
perturbation (1)	Deviation from a reference orbit in force, acceleration, velocity, position, etc., because of effects not included therein. "General perturbations" involve series expansion before integration. "Special perturbations" are integrated numerically. "Secular perturbations" vary directly with time.
perturbation (2)	The engineer's term for an arbitrary variation of initial conditions, now sanctioned by time; prefer "variant" whenever there is a possibility of confusion between the two usages.

perturbed differentiation — In the method of variation of parameters, a technique (alternative to that of Hamiltonian mechanics) in which the equations for the variations are obtained by differentiations in which the "two-body derivatives" are omitted. See "perturbed variation," "two-body variation"

perturbed variation / perturbed derivative / perturbative — In the method of variation of parameters, that part of the total derivative or variation that is added by the perturbations:

$\dfrac{df}{dt} = \dot{f} + f^{\cdot}$, where $f^{\cdot}$ is the perturbed part and $\dot{f}$ is the two-body part of the total derivative. Also called the "grave-derivative" (or "grave-variation"). See "perturbed differentiation," "two-body variation"

perturbing function (R) — That portion of the potential function (Φ) which takes into account perturbations caused by conservative forces, e.g., $\Phi = \dfrac{\mu}{r} + R$, where $\dfrac{\mu}{r}$ denotes the two-body portion of Φ. Also: Ω; disturbing function. See "potential function (Φ)."

planetary aberration — Discrepancy between calculation and observation that results when "light time" is neglected.

precession — The secular term of the motion of the Earth's axis of rotation with respect to inertial axes.

potential function (Φ) or "potential" — Function proportional to the potential energy, usually negatively proportional in astronomical and astrodynamical practice; also U or V; e.g., two-body potential: $\Phi = \mu/r$.

potential of spheroid — Potential function associated with, e.g.:

(a) spheroid of revolution:

$$\Phi = \frac{\mu}{r}\left[1 - \sum_{n=1}^{\infty} J_n \left(\frac{a_e}{r}\right)^n P_n\,(\sin\delta)\right]$$

(b) general spheroid:

$$\Phi = \frac{\mu}{r}\left\{1 + \sum_{n=1}^{\infty}\sum_{m=0}^{n}\left(\frac{a_e}{r}\right)^n P_n^m\,(\sin\delta)[C_{n,m}\cos m\lambda + S_{n,m}\sin m\lambda]\right\}$$

where a_e is the equatorial radius, δ is the geocentric (or planetocentric) declination or latitude, and the attracted object has negligible mass. Also: $-J_n = C_{n,0} = C_n$; $C_{n,m}^2 + S_{n,m}^2 = J_{n,m}^2$; $\delta = \varphi = \beta$; $\Phi = U = V$; $\mu = k^2m$; $n,m = l,m$. See P_n, P_n^m and *IAU Trans.*, **11B** (1961), 173-174. Form (b) is often replaced by one of several forms in which the coefficients are fully normalized.

Q

Q_s — Generalized force: $Q_s = \ddot{\mathbf{r}} \cdot \dfrac{\partial \mathbf{r}}{\partial q_s}$

$\mathbf{Q}(Q_x, Q_y, Q_z)$ — See "**P, Q, W**"

$\tilde{\mathbf{Q}}$ — $= \sqrt{p}\,\mathbf{Q}$, a vector that is determinate for rectilinear orbits, for which $\mathbf{Q}$ is not.

q, q_2 — Perifocal distance: $q = a(1 - e)$. Apofocal distance: $q_2 = a(1 + e) = r_A$; prefer r_A.

q — Auxiliary quantity used in Encke's method in special perturbations:

$$q = \xi(x_e + \tfrac{1}{2}\xi) + \eta(y_e + \tfrac{1}{2}\eta) + \zeta(z_e + \tfrac{1}{2}\zeta)/r^2e$$

q_s — Generalized coordinate, or angle or position variable in a canonical set (see "Hamilton's canonical equations")

R

R — Magnitude of $\mathbf{R}$; distance from point of observation to focus of attraction.

R — Geocentric distance; prefer r_c (q.v.)

R — Perturbing function (or "disturbing function")

$R = R\left[\frac{km}{gu}\right] = R[km]$ — Length conversion factor, geo-units to kilometers; gu expressed in km.

$\mathbf{R}(X,Y,Z)$ — Position vector directed from the point of observation to the focus of attraction.

$\mathbf{R}(R_x,R_y,R_z)$ — See "$\mathbf{W}(W_x,W_y,W_z)$," which should be preferred because of conflict with $\mathbf{R}(X,Y,Z)$; see introduction also.

$\tilde{\mathbf{R}}(\tilde{X},\tilde{Y},\tilde{Z})$ — Vector from the focus of attraction to the point of observation, used in place of $-\mathbf{R}$ when the conventional $\mathbf{R}$ orientation seems inappropriate.

r,r_i,r_j — Distance of object from focus of attraction; see "radius vector."

r_2 — Distance of the object from the secondary focus of a conic.

r_A — $= a(1 + e)$; apofocal distance

r_c — $= \sqrt{x_c^2 + y_c^2}$; distance from center of Earth to a point at or near the surface; also ρ, R; prefer r_c; see "x_c, y_c"

r_{ij} — Distance from m_i to m_j.

$\mathbf{r}(x,y,z)$ — Radius vector, directed from the focus of attraction to the object whose motion is under study.

ρ (rho) — Topocentric distance or "range" of any object: $\rho = |\boldsymbol{\rho}|$; also Δ (see introduction).

$\boldsymbol{\rho}(\xi,\eta,\zeta)$ — Topocentric vector or similar vector directed from the point of observation to the object whose motion is under study, e.g., $\xi = x + X$

$\boldsymbol{\rho}(\xi,\eta,\zeta)$ — Difference in position between the actual orbit and a reference orbit in Encke's method of special perturbations, e.g., $\xi = x - x_e$

radius vector (r) — Established mathematical term for $r = p/(1 + e\cos v)$; not a vector in the modern usage of the term; plurals: radius vectors, radii vectores.

range (ρ) — See ρ, $\boldsymbol{\rho}$

ratios of the triangles — In the orbit methods of Gauss, Olbers, et al., the ratios of the triangles formed by the radii and chords are calculated from the corresponding time intervals, with greater or less approximation; see "c_1, c_3"

ratios of sectors or time intervals — See "A_1, A_3"

ratio of sector to triangle ($\bar{y}$) — $\bar{y} = \sqrt{\mu p}\,(t-t_0)/S$

rectification — The process of shifting to a new (usually osculating) reference orbit in Encke's method when the Encke terms associated with the old reference orbit become too large.

rectilinear orbit — Orbit for which $e = 1$, $p = 0$, $q = 0$.

reference orbit — An orbit, often a fixed two-body ellipse, to which perturbative effects are added to obtain the perturbed orbit; see "Encke's method," $\boldsymbol{\rho}$

representation — Calculation of position of a body for a given date (usually the date of an observation) from a given set of elements. Usually compared with an observation to determine validity of the elements, i.e., whether a differential correction is required.

residuals — Differences between given data (O) that are observed, measured, or otherwise determined, and the corresponding "represented" data (C) that are calculated from conjectured values of the parameters. See "differential correction" and "representation."

right ascension (α) — Topocentric or geocentric coordinate measured along celestial equator eastward, usually 0^h to 24^h, from vernal equinox to hour circle through object. As a subscript α may be used to indicate equatorial coordinates. See "declination (δ)"

Runge-Kutta formulae — Formulae for the numerical integration of differential equations.

S

S — $= C(1 - e^2) = a_e(1 - e^2)/\sqrt{1 - e^2 \sin^2 \varphi}$, quantity tabulated in the almanacs for finding geocentric coordinates; see also C, x_c, y_c

S — $= rr_0 \sin (v - v_0)$, twice the area of the triangle; also $S_{ij} = r_i r_j \sin (v_j - v_i)$; see "$c_1$, c_3," "ratios of the triangles," "ratio of sector to triangle"; see also C, K, κ, σ

S — $= \sqrt{1 - e^2} \sin E$; prefer next definition.

S — $= \sin E = D/e\sqrt{a}$

$\tilde{S}$ — $= \sqrt{a} \sin E = \sqrt{-a} \sinh F$, unified variable.

$\hat{S}$ — $= \sqrt{a} \sin (E - E_0) = \sqrt{-a} \sinh (F - F_0)$, universal variable.

$S = S\left[\frac{gm}{m_\odot}\right]$ or $S[gm]$ — Mass conversion factor, solar masses to grams; mass of Sun in grams.

S — Determining function of a canonical transformation

$S_{n,m}$ or $S_{l,m}$ — Coefficient of a sectoral harmonic; see "potential of spheroid"

S, T, W — $\mathbf{S}(S_x,S_y,S_z)$ is the tangential unit vector: $\mathbf{S} = \dot{\mathbf{r}}/\dot{s}$. The normal unit vector $\mathbf{T}$ is also the orbital plane, 90° from $\mathbf{S}$ in the same sense as the direction of motion. $\mathbf{W}$ forms a right-handed orthogonal set with $\mathbf{S}$, $\mathbf{T}$.

$\dot{s}$	Magnitude of the velocity vector, $\dot{\mathbf{r}}$: $$\dot{s}^2 = \dot{x}^2 + \dot{y}^2 + \dot{z}^2 = \dot{\mathbf{r}} \cdot \dot{\mathbf{r}}$$ See s and "vis-viva integral"; also see "V, v"
s	Length of arc between two positions on an orbit.
s	Length of the chord between two positions $\mathbf{r}_0$ and $\mathbf{r}$ on an orbit; also c: $$s^2 = (x - x_0)^2 + (y - y_0)^2 + (z - z_0)^2$$
Σ, Σ^2 (sigma)	Sums in numerical integration tables and formulae; also $$\Sigma^2 \ddot{x}_i = \Delta^{-2}\ddot{x}_{i+1} = \nabla^{-2}\ddot{x}_{i-1}, \text{ etc.}$$
Σ^2 formula	See "second-sum formula."
σ (sigma)	$= 2\sqrt{rr_0} \sin \frac{1}{2}(v - v_0)$; also $\sigma_{ij} = 2\sqrt{r_i r_j} \sin \frac{1}{2}(v_j - v_i)$; used in development of closed expressions for f and g, in Lambert-Euler equation, etc., see also κ, C, K, S
σ	Density
σ	Standard deviation
second-difference formula	Formula for numerical integration of second-order differential equations; also "δ^2 formula," "Adams-Moulton formula"
second-sum formula	Formula for numerical integration of second-order differential equations; also "Gauss-Jackson formula," "Σ^2 formula"
sectoral harmonics	See "harmonics"
selenocentric	Having the center of the moon as focus of attraction or origin of coordinates; prefer "moon-centered"
semi latus rectum (p)	$= a(1 - e^2)$, one half of the chord passing through a focus and perpendicular to the major axis; also "semi-parameter" or "parameter" (q.v.)
semi major axis (a)	One-half the distance from the perifocus to the apofocus of an ellipse; also "mean distance"
sidereal time (θ)	The hour angle of the vernal equinox; the right ascension of the celestial meridian.
solar parallax ($\Pi_\odot$)	The angle that (in radians) is equal to the ratio of equatorial radius to the astronomical unit.
spheroid	A symmetrical mathematical approximation to the geoid: e.g., an ellipsoid of revolution.

T

T	Time of perifocal passage
T	Period; prefer P because of preceding well-established usage
T	Time measured in Julian centuries from some epoch such as 1900.0
T	Kinetic energy, or K. E. per unit mass in many astrodynamic applications
$\mathbf{T}(T_x, T_y, T_z)$	See "**S**, **T**, **W**"
t	Time; true time (i.e., the time the light left the object). Distinguished from "observed time" by $t = t_{\text{obs}} - \alpha\rho$.
t	Hour angle or local hour angle; also H or HA or LHA

t_0 — The epoch (q.v.)

t_{obs} — Time of observation; see t (true time)

τ (tau) — Canonical or normalized time variable: $\tau = k_*(t - t_0)$, or $\tau_j = k_*(t_j - t_0)$, where k_* may have the numerical value of $\sqrt{\mu}$, but not its dimension or units. Alternative notations are:

(a) $\tau_{ij} = k_*(t_j - t_i)$ (the "unambiguous convention")

(b) $\begin{cases} \tau_1 = k_*(t_1 - t_2) \\ \tau_3 = k_*(t_3 - t_2) \end{cases}$ (the "Laplacian convention")

(c) $\begin{cases} \tau_1 = k_*(t_3 - t_2) \\ \tau_2 = k_*(t_3 - t_1) \\ \tau_3 = k_*(t_2 - t_1) \end{cases}$ (the "Gaussian convention")

The unambiguous convention is to be preferred when there might be confusion between the other two.

τ_A — Time required for light to travel one astronomical unit; also α_s, α (q.v.)

θ (theta) — Sidereal time

θ_g — Greenwich sidereal time

θ — Angle measured in the orbital plane to the radius vector from an origin that is not subject to longitudinal perturbed variation

θ — Flight-path angle (q.v.); prefer γ

takeoff — See "initial, injection, insertion"

tesseral harmonics — See "harmonics"

topocentric — Referred to the point of observation

true anomaly (v) — Angle measured at the focus from the perifocal vector **P** to the radius vector **r**; also f, but prefer v (see "f")

triaxial ellipsoid — Ellipsoidal figure having three orthogonal axes of symmetry

two-body variation / two-body derivative — In the method of variation of parameters, that part of the total derivative or variation that is attributed to the object at the focus of attraction:

$\dfrac{df}{dt} = \dot{f} + f^{\backprime}$, where $\dot{f}$ is the two-body part and $f^{\backprime}$ is the perturbed part of the total derivative. Also called the "dot-variation" or "osculating variation" (or " . . . derivative"). When the parameters p_i are all constants of the two body problem (or osculating solution), we may recognize

$$\dot{f} = \frac{\partial f}{\partial t} \qquad f^{\backprime} = \sum_{i=1}^{6} \frac{\partial f}{\partial p_i}\frac{dp_i}{dt} \qquad \frac{dp_i}{dt} = p_i^{\backprime}$$

When the parameters include a two-body variable such as the mean anomaly M,

$$\dot{f} = \frac{\partial f}{\partial M}\dot{M} \qquad f^{\backprime} = \sum_{i=1}^{5} \frac{\partial f}{\partial p_i} p_i^{\backprime} + \frac{\partial f}{\partial M} M^{\backprime} \qquad \dot{M} = n$$

U

U — $= M + \omega = U_0 + n(t - t_0)$, mean argument of latitude

U — $= E - \sin E$

$\tilde{U}$ — $= a^{\frac{3}{2}}(E - \sin E) = (-a)^{\frac{3}{2}}(\sinh F - F)$, etc., unified variable

$\hat{U}$ — $= a^{\frac{3}{2}}(E - E_0) - \sin(E - E_0) = (-a)^{\frac{3}{2}} \sinh(F - F_0) - (F - F_0)$, etc., universal variable.

U — Potential function (q.v.)

U.T. — Universal time (q.v.)

U, V, W — $\mathbf{U}(U_x, U_y, U_z)$ is the unit vector directed along the radius vector from the dynamical center to the object. **V** is directed 90° from **U** in the direction of motion of the object in the orbital plane. **W** is perpendicular to the orbital plane, thus forming a right-handed orthogonal set with **U**, **V**; see "**P, Q, W**"

u — $= v + \omega$, (true) argument of latitude

u — Eccentric anomaly; prefer E because of preceding well-established usage.

Υ (upsilon) — $= E + \omega$, to accompany $u = v + \omega$, $U = M + \omega$

Υ — See "vernal equinox (♈)"

υ (upsilon) — Generally not used because of similarity to u

Universal Time (U.T.) — = G.C.T. = G.M.T. after 1925.0
= G.M.T. − 12^h before 1925.0
= E.T. − ΔT

V

V, v — $= \dot{s} = ds/dt$, velocity per unit of t; also "G" (for Geschwindigkeit); prefer $\dot{s}$, V (see introduction)

V_c — $= \sqrt{\mu/r}$, circular velocity

V_p — $= \sqrt{2\mu/r}$, parabolic velocity, or velocity of escape

V_∞ — $= \sqrt{-\mu/a}$, velocity at infinity, imaginary for elliptic orbits

V — Potential function (q.v.)

$\mathbf{V}(V_x, V_y, V_z)$ — See "**U, V, W**"

$\hat{\mathbf{V}}$ — $= \sqrt{p}\,\mathbf{V}$, a vector that is determinate for rectilinear orbits, for which **V** is not.

v — True anomaly; also f (q.v.)

$\dot{v}$ — Angular velocity (q.v.)

variant calculations — Method of calculating the partial differential coefficients that are used in differential correction formulae, rendezvous correction, etc.; see "perturbation (2)"

variation — Used in astrodynamics as the rate of change in the value of a parameter with respect to time. It is noted that this definition is unlike the general usage of the word, in which "variation" indicates a modification or deviation. See "perturbed variation" and "two-body variation"

variation of parameters — Perturbation technique in which the constants of the 2-body orbit, or of the analytical solution of an approximation to an actual problem, are allowed to vary.

vernal equinox (♈) — That intersection of the ecliptic and celestial equator where the Sun crosses from south to north (approximately March 21). Note that the vernal equinox symbol ♈ is not the capital upsilon (Υ), which should not be used for it unless the type-font lacks ♈; distinction between ♈ and Υ is more stringent than that between ☊ and Ω; see "node."

vertical circle — Great circle passing through the zenith.

virtual increment — An increment, as a "virtual displacement," $\delta\mathbf{r}$, that is not necessarily nor usually the same as a time-displacement along a dynamical path, e.g., as in the establishment of Lagrange's equations. Of the same character as a differential correction.

vis-viva integral — $\dot{s}^2 = \mu\left(\frac{2}{r} - \frac{1}{a}\right)$; also "energy integral"

W

$\mathbf{W}(W_x, W_y, W_z)$ — Unit vector perpendicular to orbit plane; see "**P**, **Q**, **W**"; "**U**, **V**, **W**"; "$\mathbf{R}(R_x,R_y,R_z)$"; introduction.

w — Interval of argument in integration tables or in any interpolation table. Also: h

X, Y, Z

X, Y, Z — See "$\mathbf{R}(X,Y,Z)$"

X — $= E$ or F or D

$\tilde{X}$ — $= \sqrt{a}\,E$ or $\sqrt{-a}\,F$ or D, unified variable

$\hat{X}$ — $= \sqrt{a}\,(E - E_0)$ or $\sqrt{-a}\,(F - F_0)$ or $D - D_0$, universal variable

Z — Azimuth (q.v.)

Z, **E**, **N** — Right-handed orthogonal set of unit vectors directed to zenith, east point, north point, useful in rotation from equatorial to horizon system: e.g.,

$$L_{xh} = \mathbf{L} \cdot \mathbf{N} \qquad L_{yh} = \mathbf{L} \cdot \mathbf{E} \qquad L_{zh} = \mathbf{L} \cdot \mathbf{Z}$$

where

$$\mathbf{Z} = \begin{Bmatrix} \cos\varphi \cos\theta \\ \cos\varphi \sin\theta \\ \sin\varphi \end{Bmatrix} \quad \mathbf{E} = \begin{Bmatrix} -\sin\theta \\ +\cos\theta \\ 0 \end{Bmatrix} \quad \mathbf{N} = \begin{Bmatrix} -\sin\varphi\cos\theta \\ -\sin\varphi\sin\theta \\ +\cos\varphi \end{Bmatrix}$$

See: "**L**, **A**, **D**"; "**L**, $\mathbf{A}_h$, $\mathbf{D}_h$"

x, y, z — See "$\mathbf{r}(x,y,z)$"

x_c, y_c — Coordinates referred to the geometrical center and major axis of an ellipse; geocentric coordinates in the plane of a meridian:

$$\left.\begin{aligned} x_c &= r_c \cos\varphi' = C\cos\varphi \\ y_c &= r_c \sin\varphi' = S\sin\varphi \end{aligned}\right\} \text{ See } C, S, \varphi, \varphi'$$

x_ω, y_ω — Coordinates referred to the focus and transverse axis of a conic: $x_\omega = r\cos v$, $y_\omega = r\sin v$

$\bar{y}$ — $= \sqrt{\mu p}\,(t - t_0)/S$, ratio of sector to triangle

z — $= 90 - h$, zenith distance

ξ, η, ζ (xi, eta, zeta) — See "$\boldsymbol{\rho}(\xi,\eta,\zeta)$"

zenith — Intersection (upper) of direction of gravity and the celestial sphere

zonal harmonics — See "harmonics"

Astronomical symbols in common use in astrodynamics:

☉	Sun	♈	Vernal equinox (q.v.)
☿	Mercury	☊	Node (q.v.); also Ω
♀	Venus	☄	Comet
⊕	Earth	☾	Moon
♂	Mars		
♃ or ♃	Jupiter		
♄	Saturn		
⛢ or ♅	Uranus		
♆	Neptune		
♇	Pluto		

Special uses of mathematical symbols:

˙, ¨	Derivatives with respect to τ; restricted to "two-body variations" (q.v.) in method of variation of parameters.
ˋ, ˋˋ	"Perturbed variations" (q.v.) in method of variation of parameters.
∇	Del or nabla; listed under D

Acknowledgements for helpful suggestions and comments are due especially to J. D. Anderson, R. M. L. Baker, Jr., R. A. Broucke, C. J. Cohen, C. G. Jacchia, W. M. Kaula, D. G. King-Hele, J. D. Mulholland, C. Oesterwinter, D. A. Pierce, E. Rabe; M. S. W. Keesey was very helpful in an earlier stage of the compilation.

TABLES OF SOLID ANGLES

COMPILED BY DR. A. VICTOR MASKET

Abridged from U.S. Atomic Energy Commission Report TID-14975 (July, 1962), *Tables of Solid Angles*, by A. V. H. Masket and William C. Rodgers; composed at the Computation Center, University of North Carolina, Chapel Hill, North Carolina; abridgment prepared at General Atomic Division, General Dynamics Corporation, San Diego 12, California; reproduced with permission of the U.S. Atomic Energy Commission.

I. SOLID ANGLE SUBTENDED BY THE CIRCULAR DISC

Let the point in space at which the solid angle is subtended be 0, an origin of coordinate systems. The normal to the plane of the disc from 0 defines the z-axis and the normal to the symmetry axis of the disc defines the x-axis. *All lengths are considered normalized relative to the radius of the disc as a unit of length.* The solid angle is represented by the symbol $\Omega(1;\rho,z)$, where ρ is the distance (along the x-axis) from 0 to the symmetry axis and z is the distance (along the z-axis) from 0 to the plane of the disc. From the general expression for solid angles given by $\Omega = \iint \sin\theta \, d\theta \, d\phi$ in spherical polar coordinates, three contour integrals have been derived, which correspond to the cases with ρ less than, equal to, or greater than the radius of the disc as a unit of length.

Case 1. $\rho < 1$; the z-axis pierces the interior of the disc.

The contour integral employed for digital computation is

$$\Omega(1;\rho,z) = 2 \cdot \int_0^{\pi} f_1^2[1 + f_1^2 + (1 + f_1^2)^{1/2}]^{-1} \, d\phi \qquad (\rho < 1), \tag{1}$$

where $f_1 = S_1/z$, $S_1 = \rho\cos\phi + (1 - \rho^2\sin^2\phi)^{1/2}$. When $\rho = 0$, $\Omega(1;0,z) = 2\pi[1 - z/(1+z^2)^{1/2}]$, exactly.

Case 2. $\rho = 1$; the z-axis touches the periphery of the disc.

$$\Omega(1;1,z) = 2 \cdot \int_0^{\pi/2} f_2^2[1 + f_2^2 + (1 + f_2^2)^{1/2}]^{-1} \, d\phi, \tag{2}$$

with $f_2 = (2\cos\phi)/z$. In elliptic integrals, the formula is given exactly by $\Omega(1;1,z) = \pi - kzK(k)$, where $k^2 = 4/(z^2 + 4)$ and $K(k) \equiv F(\pi/2,k)$ is the complete elliptic integral of the first kind.

Case 3. $\rho > 1$; the z-axis is external to the disc.

The computational form, where $S_2 = \rho\cos\phi - (1 - \rho^2\sin^2\phi)^{1/2}$, is

$$\underset{(\rho>1)}{\Omega(1;\rho,z)} = 2z \int_0^{\arcsin(1/\rho)} \frac{(S_1^2 - S_2^2)\, d\phi}{[(z^2 + S_1^2)(z^2 + S_2^2)]^{1/2}[(z^2 + S_1^2)^{1/2} + (z^2 + S_2^2)^{1/2}]}. \tag{3}$$

The solid angle subtended by the circular disc is expressed most generally in elliptic integrals. In conformity with the definitions adopted for ρ and z,

$$(4) \qquad \Omega(1;\rho,z) = 2\pi - 2[k'K(k) \sin \alpha + K(k) \cdot E(k',\alpha) - F(k',\alpha)\{K(k) - E(k)\}],$$

where $k^2 = 4\rho/[(1 + \rho)^2 + z^2]$, $(k')^2 = 1 - k^2$, $\alpha = (\pi - \phi)/2$, and $K(k)$ and $E(k)$ are the complete elliptic integrals, and $F(k',\alpha)$ and $E(k',\alpha)$ are the corresponding (incomplete) elliptic integrals of the first and second kind. The formula given by (4) is valid for all values of ρ and z; it is not convenient for computation. When $z > 100$, it is practical to use $\Omega \sim (\pi/d^2) \cdot \cos \gamma$ where d is the normalized distance from 0 to the center of the disc and γ is the projection angle. Accurate approximations are discussed in the references and literature.

II. SOLID ANGLE SUBTENDED BY THE RIGHT CIRCULAR CYLINDER

The tabulated values consist of the solid angles subtended only by the lateral surface at points in the plane of a circular flat (base) external to a cylinder of given radius and height. It will be shown that these values alone or taken in conjunction with the values of solid angles subtended by the circular disc are sufficient to cover the general case of any point in space external to the right circular cylinder. *All lengths are normalized with respect to the radius of the cylinder as a unit of length.* The variables of the table are h, the normalized height of the cylinder, and ρ, the normalized distance from the point (at which the solid angle is subtended) to the axis of the cylinder. With these definitions, the solid angle is represented by $\Omega(1,h;\rho)$.

Let the point in the plane of a flat be 0, the origin of coordinate systems. The z-axis is the normal to the plane and the x-axis is the line in the plane from 0 to the axis of the cylinder. The solid angle subtended at 0 by the lateral surface of the cylinder is given by the contour integral.

$$(5) \qquad \Omega(1,h;\rho) = 2h \int_0^{\arcsin(1/\rho)} [h^2 + S_2^2]^{-1/2}\, d\phi \qquad [\rho \geqq 1],$$

where ϕ is the azimuthal angle of the associated spherical polar system.

If a point is not located in a plane determined by a flat of the cylinder, then it is either interior or exterior to the space between the planes of the flats. In the first case, a plane through the point normal to the cylindrical axis determines two smaller cylinders and two solid angles in accordance with (5) which are added. For the second case, where a flat is viewed from 0 in addition to the lateral surface, let the normalized height of the cylinder be h and the normalized distance to the plane of the viewed flat be H. Then the total solid angle is

$$(6) \qquad \Omega = \Omega(1,h + H;\rho) - \Omega(1,H;\rho) + \Omega(1;\rho,H),$$

where ρ is the normalized distance from 0 to the axis of the cylinder. Approximations are given in the references.

III. USE OF TABLES

A. Reading the Entry Numbers (steradians)

Each entry is given to six significant figures followed by an exponent for the power of 10 as a multiplier. For example,

$$3.73950\text{–}3 \equiv 3.73950 \times 10^{-3} \equiv 0.00373950;$$
$$5.65798\text{–}0 \equiv 5.65798 \times 10^{0} \equiv 5.65798.$$

B. Solid Angle Subtended by a Disc

To find the solid angle subtended by an end-window counter 3 cm in radius at a point 15 cm above its plane and 18 cm from its axis, we have

$$\Omega(1;\rho,z) = \Omega(1;18/3,15/3) = \Omega(1;6,5) = 0.0331611.$$

C. Solid Angle Subtended by the Lateral Surface of a Right Circular Cylinder

To find the solid angle subtended by a Geiger counter 3 cm in radius and 30 cm long at a point in the plane of a flat at 25.5 cm from the axis, we have

$$\Omega(1,h;\rho) = \Omega(1,30/3;25.5/3) = \Omega(1,10;8.5) = 0.186909.$$

D. Fraction of Radiation Entering a Circular Aperture from a Distributed Plane Source

Consider the problem of estimating the fraction of radiation entering a circular end-window counter from a thin disc source whose plane is parallel to the window. Let b be the normalized radius of the source relative to the counter, let s be the normalized distance between the (parallel) axes of the source and window, and let z be the normalized distance between the planes of the window and source.

Case 1. $s = 0$: coaxial.

Divide the source into N concentric annuli of width $\Delta\rho_i$, $i = 1,2,3, \ldots, N$. Each point in a given annulus has subtended approximately the same solid angle $\Omega(1;\rho_i,z)$ by the window. The area of each annulus is $2\pi\rho_i\Delta\rho_i$. Hence the fraction is given by the summation

$$G_1 = \frac{1}{2\pi b^2}\sum_{i=1}^{N}\Omega(1;\rho_i,z)\rho_i\Delta\rho_i. \tag{6}$$

Case 2. $s \geqq b$: non-coaxial and no vertical overlap.

Divide the source into N concentric strips using the point common to the axis of the window and the plane of the source as a center. The area of each strip is $2\alpha_i\rho_i\Delta\rho_i$, where $\alpha_i = \arccos\,[(\rho_i^2 + S^2 - b^2)/2S\rho_i]$. The fraction is given by

$$G_2 = \frac{1}{2\pi^2 b^2}\sum_{i=1}^{N}\Omega(1;\rho_i,z)\alpha_i\rho_i\Delta\rho_i. \tag{7}$$

Case 3. $s < b$: non-coaxial with vertical overlap. Apply Cases 1 and 2.

In the above calculations, self-absorption, scattering, non-uniformity of source strength, and non-isotropic emission have been omitted. Corrections for these effects cannot be treated here.

I. SOLID ANGLE SUBTENDED BY THE CIRCULAR DISC OF UNIT RADIUS

ρ \ z	0.2	0.4	0.6	0.8	1.0	1.2	1.4	1.6	1.8	2.0
0.0	5.05095 0	3.94967 0	3.05051 0	2.35811 0	1.84030 0	1.45631 0	1.17035 0	9.55055–1	7.90693–1	6.63334–1
0.1	5.04234 0	3.93659 0	3.03737 0	2.34716 0	1.83198 0	1.45024 0	1.16598 0	9.51907–1	7.88406–1	6.61650–1
0.2	5.01561 0	3.89649 0	2.99753 0	2.31425 0	1.80709 0	1.43213 0	1.15296 0	9.42532–1	7.81591–1	6.56633–1
0.3	4.96793 0	3.82665 0	2.92981 0	2.25924 0	1.76590 0	1.40231 0	1.13157 0	9.27131–1	7.70390–1	6.48380–1
0.4	4.89368 0	3.72227 0	2.83236 0	2.18209 0	1.70895 0	1.36136 0	1.10227 0	9.06036–1	7.55034–1	6.37049–1
0.5	4.78279 0	3.57610 0	2.70304 0	2.08304 0	1.63710 0	1.31011 0	1.06569 0	8.79705–1	7.35838–1	6.22856–1
0.6	4.61712 0	3.37830 0	2.54009 0	1.96307 0	1.55174 0	1.24970 0	1.02267 0	8.48708–1	7.13192–1	6.06068–1
0.7	4.36323 0	3.11751 0	2.34335 0	1.82425 0	1.45479 0	1.18157 0	9.74194–1	8.13717–1	6.87550–1	5.86992–1
0.8	3.95926 0	2.78525 0	2.11602 0	1.67016 0	1.34883 0	1.10741 0	9.21391–1	7.75480–1	6.59413–1	5.65969–1
0.9	3.31094 0	2.38629 0	1.86638 0	1.50600 0	1.23698 0	1.02917 0	8.65483–1	7.34795–1	6.29313–1	5.43360–1
1.00	2.40515 0	1.95112 0	1.60798 0	1.33824 0	1.12269 0	9.48851–1	8.07727–1	6.92475–1	5.97794–1	5.19535–1
1.25	7.83966–1	1.02735 0	1.02474 0	9.45919–1	8.48210–1	7.51604–1	6.63084–1	5.84636–1	5.16227–1	4.57034–1
1.50	3.30050–1	5.40110–1	6.29125–1	6.43292–1	6.19102–1	5.77621–1	5.30021–1	4.82015–1	4.36391–1	3.94380–1
1.75	1.74764–1	3.11680–1	3.97052–1	4.37844–1	4.47467–1	4.37793–1	4.17265–1	3.91356–1	3.63433–1	3.35486–1
2.00	1.05541–1	1.95924–1	2.62557–1	3.04563–1	3.25802–1	3.31582–1	3.26809–1	3.15351–1	3.00008–1	2.82710–1
2.25	6.93038–2	1.31484–1	1.81639–1	2.17930–1	2.41045–1	2.53071–1	2.56518–1	2.53750–1	2.46739–1	2.37015–1
2.50	4.82364–2	9.27336–2	1.30625–1	1.60381–1	1.81769–1	1.95484–1	2.02713–1	2.04791–1	2.02982–1	1.98380–1
2.75	3.50529–2	6.79748–2	9.70218–2	1.21083–1	1.39726–1	1.53087–1	1.61683–1	1.66228–1	1.67492–1	1.66206–1
3.00	2.63375–2	5.13818–2	7.40283–2	9.34934–2	1.09364–1	1.21567–1	1.30297–1	1.35917–1	1.38876–1	1.39646–1
3.25	2.03247–2	3.98250–2	5.77746–2	7.36211–2	8.70214–2	9.78406–2	1.06119–1	1.12032–1	1.15831–1	1.17811–1
3.50	1.60321–2	3.15170–2	4.59620–2	5.89727–2	7.02786–2	7.97380–2	8.73258–2	9.31121–2	9.72349–2	9.98737–2
3.75	1.28802–2	2.53850–2	3.71704–2	4.79510–2	5.75168–2	6.57396–2	7.25687–2	7.80213–2	8.21674–2	8.51134–2
4.00	1.05108–2	2.07566–2	3.04914–2	3.95055–2	4.76364–2	5.47745–2	6.08632–2	6.58932–2	6.98951–2	7.29297–2
4.25	8.69349–3	1.71951–2	2.53256–2	3.29283–2	3.98770–2	4.60809–2	5.14851–2	5.60685–2	5.98397–2	6.28312–2
4.50	7.27511–3	1.44086–2	2.12670–2	2.77316–2	3.37040–2	3.91096–2	4.38989–2	4.80465–2	5.15489–2	5.44214–2
4.75	6.15128–3	1.21961–2	1.80334–2	2.35721–2	2.87348–2	3.34605–2	3.77061–2	4.14456–2	4.46693–2	4.73822–2
5.00	5.24891–3	1.04165–2	1.54251–2	2.02039–2	2.46916–2	2.88383–2	3.26071–2	3.59732–2	3.89245–2	4.14597–2
5.25	4.51573–3	8.96845–3	1.32977–2	1.74477–2	2.13695–2	2.50223–2	2.83747–2	3.14042–2	3.40977–2	3.64505–2
5.50	3.91368–3	7.77790–3	1.15450–2	1.51706–2	1.86154–2	2.18459–2	2.48352–2	2.75636–2	3.00179–2	3.21920–2
5.75	3.41457–3	6.78986–3	1.00879–2	1.32731–2	1.63135–2	1.91814–2	2.18542–2	2.43144–2	2.65498–2	2.85533–2
6	2.99721–3	5.96290–3	8.86648–3	1.16792–2	1.43749–2	1.69305–2	1.93270–2	2.15491–2	2.35856–2	2.54291–2
7	1.87242–3	3.73056–3	5.56046–3	7.34874–3	9.08291–3	1.07515–2	1.23445–2	1.38533–2	1.52706–2	1.65909–2
8	1.24791–3	2.48860–3	3.71496–3	4.92012–3	6.09755–3	7.24114–3	8.34531–3	9.40506–3	1.04159–2	1.13744–2
9	8.73357–4	1.74274–3	2.60424–3	3.45403–3	4.28845–3	5.10400–3	5.89743–3	6.66576–3	7.40627–3	8.11658–3
10	6.35074–4	1.26782–3	1.89593–3	2.51716–3	3.12931–3	3.73028–3	4.31810–3	4.89090–3	5.44696–3	5.98473–3
11	4.76253–4	9.51068–4	1.42302–3	1.89070–3	2.35275–3	2.80785–3	3.25473–3	3.69220–3	4.11913–3	4.53449–3
12	3.66318–4	7.31708–4	1.09525–3	1.45604–3	1.81318–3	2.16581–3	2.51311–3	2.85426–3	3.18853–3	3.51520–3
13	2.87802–4	5.74985–4	8.60932–4	1.14503–3	1.42670–3	1.70533–3	1.98038–3	2.25129–3	2.51755–3	2.77865–3
14	2.30230–4	4.60033–4	6.88984–4	9.16665–4	1.14266–3	1.36655–3	1.58797–3	1.80653–3	2.02184–3	2.23358–3
15	1.87054–4	3.73806–4	5.59956–4	7.45206–4	9.29263–4	1.11183–3	1.29264–3	1.47142–3	1.64789–3	1.82180–3
16	1.54039–4	3.07860–4	4.61245–4	6.13980–4	7.65850–4	9.16646–4	1.06616–3	1.21420–3	1.36056–3	1.50506–3
18	1.08092–4	2.16063–4	3.23793–4	4.31162–4	5.38052–4	6.44345–4	7.49926–4	8.54683–4	9.58505–4	1.06128–3
20	7.87496–5	1.57428–4	2.35964–4	3.14288–4	3.92328–4	4.70016–4	5.47293–4	6.24061–4	7.00285–4	7.75889–4
22	5.91384–5	1.18233–4	1.77239–4	2.36113–4	2.94811–4	3.53290–4	4.11508–4	4.69421–4	5.26989–4	5.84171–4
24	4.55356–5	9.10426–5	1.36493–4	1.81857–4	2.27108–4	2.72217–4	3.17157–4	3.61899–4	4.06417–4	4.50685–4
26	3.58052–5	7.15912–5	1.07339–4	1.43030–4	1.78644–4	2.14162–4	2.49568–4	2.84841–4	3.19964–4	3.54918–4
28	2.86614–5	5.73096–5	8.59314–5	1.14514–4	1.43043–4	1.71507–4	1.99892–4	2.28186–4	2.56376–4	2.84450–4
30	2.32987–5	4.65880–5	6.98587–5	9.31014–5	1.16307–4	1.39465–4	1.62568–4	1.85607–4	2.08571–4	2.31453–4
32	1.91948–5	3.83828–5	5.75573–5	7.67115–5	9.58388–5	1.14930–4	1.33984–4	1.52990–4	1.71943–4	1.90836–4
34	1.60009–5	3.19968–5	4.79828–5	6.39537–5	7.99048–5	9.58308–5	1.11725–4	1.27587–4	1.43409–4	1.59187–4

I. SOLID ANGLE SUBTENDED BY THE CIRCULAR DISC OF UNIT RADIUS (Continued)

ρ \ z	2.2	2.4	2.6	2.8	3.0	3.2	3.4	3.6	3.8	4.0
0.0	5.63184–1	4.83322–1	4.18802–1	3.66046–1	3.22432–1	2.86011–1	2.55314–1	2.29224–1	2.06877–1	1.87600–1
0.1	5.61928–1	4.82371–1	4.18072–1	3.65479–1	3.21986–1	2.85655–1	2.55028–1	2.28991–1	2.06686–1	1.87442–1
0.2	5.58183–1	4.79535–1	4.15895–1	3.63785–1	3.20651–1	2.84592–1	2.54172–1	2.28295–1	2.06114–1	1.86969–1
0.3	5.52015–1	4.74860–1	4.12302–1	3.60987–1	3.18446–1	2.82833–1	2.52755–1	2.27142–1	2.05167–1	1.86185–1
0.4	5.43533–1	4.68421–1	4.07345–1	3.57123–1	3.15396–1	2.80399–1	2.50791–1	2.25542–1	2.03853–1	1.85096–1
0.5	5.32887–1	4.60320–1	4.01098–1	3.52243–1	3.11538–1	2.77315–1	2.48300–1	2.23511–1	2.02182–1	1.83710–1
0.6	5.20258–1	4.50686–1	3.93649–1	3.46411–1	3.06918–1	2.73615–1	2.45307–1	2.21067–1	2.00169–1	1.82038–1
0.7	5.05858–1	4.39664–1	3.85102–1	3.39700–1	3.01590–1	2.69339–1	2.41841–1	2.18232–1	1.97830–1	1.80093–1
0.8	4.89922–1	4.27418–1	3.75571–1	3.32194–1	2.95612–1	2.64529–1	2.37934–1	2.15029–1	1.95183–1	1.77888–1
0.9	4.72698–1	4.14123–1	3.65182–1	3.23982–1	2.89051–1	2.59235–1	2.33622–1	2.11487–1	1.92250–1	1.75441–1
1.00	4.54444–1	3.99960–1	3.54063–1	3.15158–1	2.81976–1	2.53507–1	2.28944–1	2.07635–1	1.89052–1	1.72768–1
1.25	4.05975–1	3.61950–1	3.23942–1	2.91052–1	2.62506–1	2.37644–1	2.15914–1	1.96849–1	1.80060–1	1.65219–1
1.50	3.56380–1	3.22360–1	2.92080–1	2.65206–1	2.41380–1	2.20250–1	2.01493–1	1.84813–1	1.69952–1	1.56679–1
1.75	3.08635–1	2.83459–1	2.60212–1	2.38952–1	2.19628–1	2.02126–1	1.86307–1	1.72020–1	1.59117–1	1.47457–1
2.00	2.64732–1	2.46887–1	2.29671–1	2.13367–1	1.98115–1	1.83967–1	1.70915–1	1.58920–1	1.47920–1	1.37847–1
2.25	2.25711–1	2.13635–1	2.01347–1	1.89221–1	1.77498–1	1.66324–1	1.55778–1	1.45895–1	1.36679–1	1.28115–1
2.50	1.91871–1	1.84146–1	1.75727–1	1.66995–1	1.58222–1	1.49599–1	1.41249–1	1.33253–1	1.25657–1	1.18484–1
2.75	1.63015–1	1.58461–1	1.52983–1	1.46922–1	1.40544–1	1.34045–1	1.27566–1	1.21210–1	1.15049–1	1.09128–1
3.00	1.38676–1	1.36366–1	1.33064–1	1.29055–1	1.24571–1	1.19798–1	1.14877–1	1.09917–1	1.04999–1	1.00181–1
3.25	1.18276–1	1.17512–1	1.15782–1	1.13313–1	1.10300–1	1.06904–1	1.03256–1	9.94607–2	9.55989–2	9.17341–2
3.50	1.01228–1	1.01499–1	1.00879–1	9.95452–2	9.76532–2	9.53378–2	9.27128–2	8.98729–2	8.68954–2	8.38425–2
3.75	8.69879–2	8.79296–2	8.80774–2	8.75641–2	8.65120–2	8.50306–2	8.32157–2	8.11498–2	7.89024–2	7.65315–2
4.00	7.50785–2	7.64353–2	7.70987–2	7.71673–2	7.67350–2	7.58888–2	7.47074–2	7.32606–2	7.16088–2	6.98043–2
4.25	6.50933–2	6.66885–2	6.76860–2	6.81579–2	6.81754–2	6.78065–2	6.71146–2	6.61575–2	6.49867–2	6.36476–2
4.50	5.66938–2	5.84071–2	5.96093–2	6.03525–2	6.06901–2	6.06747–2	6.03566–2	5.97829–2	5.89966–2	5.80367–2
4.75	4.96011–2	5.13519–2	5.26677–2	5.35858–2	5.41457–2	5.43878–2	5.43515–2	5.40748–2	5.35933–2	5.29395–2
5.00	4.35870–2	4.53227–2	4.66888–2	4.77118–2	4.84209–2	4.88467–2	4.90199–2	4.89708–2	4.87284–2	4.83198–2
5.25	3.84656–2	4.01524–2	4.15256–2	4.26035–2	4.34075–2	4.39608–2	4.42869–2	4.44101–2	4.43535–2	4.41397–2
5.50	3.40854–2	3.57030–2	3.70541–2	3.81516–2	3.90107–2	3.96487–2	4.00840–2	4.03355–2	4.04218–2	4.03613–2
5.75	3.03226–2	3.18596–2	3.31699–2	3.42622–2	3.51474–2	3.58383–2	3.63490–2	3.66942–2	3.68889–2	3.69480–2
6	2.70765–2	2.85276–2	2.97854–2	3.08556–2	3.17460–2	3.24660–2	3.30263–2	3.34383–2	3.37138–2	3.38650–2
7	1.78102–2	1.89259–2	1.99368–2	2.08432–2	2.16464–2	2.23489–2	2.29537–2	2.34649–2	2.38870–2	2.42250–2
8	1.22774–2	1.31225–2	1.39081–2	1.46332–2	1.52974–2	1.59009–2	1.64443–2	1.69288–2	1.73559–2	1.77274–2
9	8.79463–3	9.43866–3	1.00473–2	1.06194–2	1.11544–2	1.16516–2	1.21111–2	1.25329–2	1.29173–2	1.32651–2
10	6.50280–3	6.99995–3	7.47512–3	7.92743–3	8.35619–3	8.76087–3	9.14110–3	9.49667–3	9.82754–3	1.01338–2
11	4.93731–3	5.32675–3	5.70202–3	6.06247–3	6.40752–3	6.73671–3	7.04966–3	7.34608–3	7.62581–3	7.88872–3
12	3.83361–3	4.14314–3	4.44326–3	4.73345–3	5.01328–3	5.28237–3	5.54039–3	5.78708–3	6.02223–3	6.24568–3
13	3.03414–3	3.28358–3	3.52656–3	3.76273–3	3.99173–3	4.21328–3	4.42710–3	4.63298–3	4.83072–3	5.02016–3
14	2.44139–3	2.64497–3	2.84401–3	3.03824–3	3.22741–3	3.41129–3	3.58966–3	3.76235–3	3.92918–3	4.09003–2
15	1.99292–3	2.16099–3	2.32581–3	2.48717–3	2.64488–3	2.79874–3	2.94861–3	3.09433–3	3.23577–3	3.37280–3
16	1.64751–3	1.78775–3	1.92560–3	2.06091–3	2.19353–3	2.32333–3	2.45016–3	2.57392–3	2.69448–3	2.81177–3
18	1.16291–3	1.26329–3	1.36232–3	1.45992–3	1.55599–3	1.65044–3	1.74320–3	1.83419–3	1.92333–3	2.01056–3
20	8.50809–4	9.24986–4	9.98357–4	1.07086–3	1.14245–3	1.21307–3	1.28266–3	1.35118–3	1.41857–3	1.48481–3
22	6.40926–4	6.97214–4	7.52999–4	8.08242–4	8.62906–4	9.16958–4	9.70363–4	1.02309–3	1.07510–3	1.12637–3
24	4.94674–4	5.38361–4	5.81718–4	6.24722–4	6.67349–4	7.09574–4	7.51375–4	7.92729–4	8.33615–4	8.74012–4
26	3.89686–4	4.24249–4	4.58591–4	4.92695–4	5.26543–4	5.60120–4	5.93410–4	6.26397–4	6.59066–4	6.91403–4
28	3.12394–4	3.40196–4	3.67846–4	3.95329–4	4.22636–4	4.49754–4	4.76672–4	5.03379–4	5.29864–4	5.56117–4
30	2.54243–4	2.76933–4	2.99514–4	3.21977–4	3.44313–4	3.66516–4	3.88576–4	4.10485–4	4.32236–4	4.53822–4
32	2.09663–4	2.28416–4	2.47090–4	2.65679–4	2.84176–4	3.02574–4	3.20870–4	3.39055–4	3.57126–4	3.75075–4
34	1.74916–4	1.90591–4	2.06207–4	2.21760–4	2.37245–4	2.52656–4	2.67991–4	2.83245–4	2.98413–4	3.13491–4

I. SOLID ANGLE SUBTENDED BY THE CIRCULAR DISC OF UNIT RADIUS (Continued)

ρ \ z	4.2	4.4	4.6	4.8	5.0	5.2	5.4	5.6	5.8	6.0
0.0	1.70864–1	1.56245–1	1.43405–1	1.32070–1	1.22015–1	1.13057–1	1.05042–1	9.78444–2	9.13568–2	8.54895–2
0.1	1.70732–1	1.56134–1	1.43311–1	1.31990–1	1.21947–1	1.12998–1	1.04991–1	9.78001–2	9.13181–2	8.54556–2
0.2	1.70337–1	1.55802–1	1.43031–1	1.31715–1	1.21742–1	1.12822–1	1.04839–1	9.76675–2	9.12022–2	8.53539–2
0.3	1.69683–1	1.55252–1	1.42565–1	1.31355–1	1.21403–1	1.12529–1	1.04585–1	9.74470–2	9.10096–2	8.51848–2
0.4	1.68773–1	1.54487–1	1.41917–1	1.30803–1	1.20929–1	1.12121–1	1.04232–1	9.71398–2	9.07410–2	8.49490–2
0.5	1.67615–1	1.53512–1	1.41091–1	1.30098–1	1.20325–1	1.11601–1	1.03781–1	9.67470–2	9.03975–2	8.46475–2
0.6	1.66216–1	1.52333–1	1.40092–1	1.29245–1	1.19594–1	1.10970–1	1.03234–1	9.62704–2	8.99806–2	8.42812–2
0.7	1.64586–1	1.50959–1	1.38925–1	1.28249–1	1.18738–1	1.10231–1	1.02593–1	9.57120–2	8.94919–2	8.38517–2
0.8	1.62737–1	1.49397–1	1.37598–1	1.27115–1	1.17763–1	1.09388–1	1.01862–1	9.50742–2	8.89333–2	8.33605–2
0.9	1.60681–1	1.47658–1	1.36118–1	1.25848–1	1.16673–1	1.08446–1	1.01043–1	9.43597–2	8.83070–2	8.28094–2
1.00	1.58431–1	1.45752–1	1.34494–1	1.24456–1	1.15474–1	1.07408–1	1.00140–1	9.35711–2	8.76153–2	8.22003–2
1.25	1.52054–1	1.40333–1	1.29863–1	1.20478–1	1.12039–1	1.04428–1	9.75433–2	9.12984–2	8.56186–2	8.04395–2
1.50	1.44798–1	1.34134–1	1.24540–1	1.15885–1	1.08058–1	1.00962–1	9.45137–2	8.86386–2	8.32765–2	7.83691–2
1.75	1.36909–1	1.27354–1	1.18687–1	1.10810–1	1.03639–1	9.70999–2	9.11250–2	8.56557–2	8.06400–2	7.60318–2
2.00	1.28628–1	1.20185–1	1.12459–1	1.05380–1	9.88928–2	9.29323–2	8.74532–2	8.24103–2	7.77626–2	7.34729–2
2.25	1.20174–1	1.12819–1	1.06019–1	9.97372–2	9.39249–2	8.85487–2	8.35739–2	7.89673–2	7.46983–2	7.07383–2
2.50	1.11738–1	1.05414–1	9.95043–2	9.39840–2	8.88347–2	8.40338–2	7.95591–2	7.53883–2	7.14998–2	6.78732–2
2.75	1.03474–1	9.81042–2	9.30224–2	8.82267–2	8.37106–2	7.94638–2	7.54748–2	7.17304–2	6.82170–2	6.49209–2
3.00	9.55037–2	9.09961–2	8.66758–2	8.25520–2	7.86288–2	7.49060–2	7.13801–2	6.80455–2	6.48950–2	6.19210–2
3.25	8.79142–2	8.41742–2	8.05402–2	7.70290–2	7.36525–2	7.04170–2	6.73257–2	6.43789–2	5.15745–2	5.89096–2
3.50	8.07635–2	7.76969–2	7.46722–2	7.17117–2	6.88315–2	6.60431–2	6.33540–2	6.07690–2	5.82901–2	5.59182–2
3.75	7.40847–2	7.16008–2	6.91105–2	6.66385–2	6.42035–2	6.18200–2	5.94987–2	5.72474–2	5.50709–2	5.29731–2
4.00	6.78909–2	6.59056–2	6.38787–2	6.18352–2	5.97951–2	5.77745–2	5.57858–2	5.38388–2	5.19407–2	5.00967–2
4.25	6.21797–2	6.06169–2	5.89879–2	5.73167–2	5.56236–2	5.39250–2	5.22343–2	5.05622–2	5.89176–2	4.73067–2
4.50	5.69376–2	5.57298–2	5.44393–2	5.30890–2	5.16981–2	5.02828–2	4.88567–2	4.74312–2	4.60152–2	4.46165–2
4.75	5.21432–2	5.12311–2	5.02266–2	4.91506–2	4.80209–2	4.68531–2	4.56604–2	4.44540–2	4.32434–2	4.20364–2
5.00	4.77702–2	4.71028–2	4.63381–2	4.54949–2	4.45896–2	4.36364–2	4.26481–2	4.16355–2	4.06079–2	3.95730–2
5.25	4.37897–2	4.33234–2	4.27586–2	4.21119–2	4.13980–2	4.06300–2	3.98196–2	3.89770–2	3.81114–2	3.72301–2
5.50	4.01716–2	3.98694–2	3.94702–2	3.89884–2	3.84369–2	3.78276–2	3.71713–2	3.64774–2	3.57543–2	3.50094–2
5.75	3.68860–2	3.67169–2	3.64540–2	3.61096–2	3.56953–2	3.52215–2	3.46979–2	3.41331–2	3.35351–2	3.29105–2
6	3.39038–2	3.38419–2	3.36905–2	3.34602–2	3.31611–2	3.28023–2	3.23925–2	3.19395–2	3.14505–2	3.09318–2
7	2.44839–2	2.46694–2	2.47868–2	2.48416–2	2.48393–2	2.47850–2	2.46838–2	2.45406–2	2.43598–2	2.41460–2
8	1.80456–2	1.83128–2	1.85314–2	1.87043–2	1.88341–2	1.89237–2	1.89757–2	1.89930–2	1.89782–2	1.89341–2
9	1.35769–2	1.38536–2	1.40966–2	1.43070–2	1.44862–2	1.46355–2	1.47565–2	1.48505–2	1.49192–2	1.49641–2
10	1.04157–2	1.06733–2	1.09075–2	1.11186–2	1.13073–2	1.14742–2	1.16201–2	1.17457–2	1.18520–2	1.19398–2
11	8.13482–3	8.36415–3	8.57687–3	8.77317–3	8.95332–3	9.11764–3	9.26648–3	9.40025–3	9.51941–3	9.62443–3
12	6.45734–3	6.65714–3	6.84510–3	7.02127–3	7.18572–3	7.33860–3	7.48005–3	7.61027–3	7.72949–3	7.83797–3
13	5.20118–3	5.37369–3	5.53764–3	5.69301–3	5.83979–3	5.97804–3	6.10778–3	6.22911–3	6.34215–3	6.44701–3
14	4.24474–3	4.39326–3	4.53550–3	4.67141–3	4.80096–3	4.92412–3	5.04092–3	5.15138–3	5.25553–3	5.35343–3
15	3.50533–3	3.63325–3	3.75650–3	3.87503–3	3.98878–3	4.09772–3	4.20184–3	4.30112–3	4.39557–3	4.48521–3
16	2.92568–3	3.03613–3	3.14306–3	3.24641–3	3.34614–3	3.44219–3	3.53455–3	3.62319–3	3.70811–3	3.78929–3
18	2.09582–3	2.17906–3	2.26021–3	2.33924–3	2.41611–3	2.49077–3	2.56319–3	2.63335–3	2.70123–2	2.76681–3
20	1.54983–3	1.61361–3	1.67610–3	1.73728–3	1.79711–3	1.85556–3	1.91260–3	1.96822–3	2.02239–3	2.07510–3
22	1.17687–3	1.22658–3	1.27546–3	1.32349–3	1.37065–3	1.41692–3	1.46227–3	1.50670–3	1.55016–3	1.59267–3
24	9.13900–4	9.53259–4	9.92072–4	1.03032–3	1.06799–3	1.10506–3	1.14152–3	1.17735–3	1.21253–3	1.24707–4
26	7.23392–4	7.55021–4	7.86277–4	8.17146–4	8.47617–4	8.77677–4	9.07316–4	9.36522–4	9.65289–4	9.93601–3
28	5.82128–4	6.07885–4	6.33381–4	6.58606–4	6.83550–4	7.08206–4	7.32564–4	7.56616–4	7.80358–4	8.03776–4
30	4.75233–4	4.96463–4	5.17506–4	5.38354–4	5.59000–4	5.79438–4	5.99662–4	6.19665–4	6.39445–4	6.58989–4
32	3.92897–4	4.10587–4	4.28140–4	4.45550–4	4.62812–4	4.79921–4	4.96873–4	5.13663–4	5.30285–4	5.46736–4
34	3.28475–4	3.43360–4	3.58143–4	3.72820–4	3.87386–4	4.01839–4	4.16173–4	4.30386–4	4.44474–4	4.58434–4

I. SOLID ANGLE SUBTENDED BY THE CIRCULAR DISC OF UNIT RADIUS (Continued)

ρ \ z	6.2	6.4	6.6	6.8	7.0	7.2	7.4	7.6	7.8	8.0
0.0	8.01664-2	7.53226-2	7.09026-2	6.68585-2	6.31492-2	5.97388-2	5.65962-2	5.36942-2	5.10090-2	4.85195-2
0.1	8.01365-2	7.52962-2	7.08791-2	6.68376-2	6.31305-2	5.97221-2	5.65812-2	5.36807-2	5.09968-2	4.85085-2
0.2	8.00469-2	7.52170-2	7.08088-2	6.67750-2	6.30746-2	5.96720-2	5.65362-2	5.36402-2	5.09602-2	4.84753-2
0.3	7.98980-2	7.50852-2	7.06919-2	6.66709-2	6.29816-2	5.95887-2	5.64613-2	5.35727-2	5.08992-2	4.84201-2
0.4	7.96902-2	7.49015-2	7.05288-2	6.65256-2	6.28518-2	5.94723-2	5.63568-2	5.34785-2	5.08141-2	4.83430-2
0.5	7.94244-2	7.46662-2	7.03199-2	6.63396-2	6.26855-2	5.93233-2	5.62228-2	5.33577-2	5.07050-2	4.82442-2
0.6	7.91014-2	7.43804-2	7.00660-2	6.61133-2	6.24833-2	5.91419-2	5.60598-2	5.32107-2	5.05721-2	4.81238-2
0.7	7.87224-2	7.40448-2	6.97679-2	6.58475-2	6.22456-2	5.89288-2	5.58681-2	5.30379-2	5.04159-2	4.79822-2
0.8	7.82888-2	7.36607-2	6.94264-2	6.55430-2	6.19732-2	5.86844-2	5.56482-2	5.28396-2	5.02365-2	4.78197-2
0.9	7.78021-2	7.32292-2	6.90427-2	6.52007-2	6.16668-2	5.84095-2	5.54008-2	5.26163-2	5.00346-2	4.76365-2
1.00	7.72637-2	7.27518-2	6.86178-2	6.48214-2	6.13272-2	5.81045-2	5.61262-2	5.23685-2	4.98103-2	4.74331-2
1.25	7.57054-2	7.13679-2	6.73849-2	6.37197-2	6.03399-2	5.72172-2	5.43267-2	5.16462-2	4.91563-2	4.68396-2
1.50	7.38690-2	6.97338-2	6.59264-2	6.24141-2	5.91679-2	5.61624-2	5.33749-2	5.07853-2	4.83758-2	4.61304-2
1.75	7.17904-2	6.78799-2	6.42680-2	6.09265-2	5.78301-2	5.49562-2	5.22848-2	4.97979-2	4.74793-2	4.53147-2
2.00	6.95082-2	6.58387-2	6.24376-2	5.92809-2	5.63469-2	5.36163-2	5.10716-2	4.86969-2	4.64783-2	4.44027-2
2.25	6.70614-2	6.36438-2	6.04640-2	5.75020-2	5.47399-2	5.21614-2	4.97515-2	1.74968-2	4.53851-2	4.34049-2
2.50	6.44889-2	6.13289-2	5.83764-2	5.56152-2	5.30310-2	5.06105-2	4.83413-2	4.72121-2	4.42125-2	4.23328-2
2.75	6.18284-2	5.89266-2	5.62029-2	5.36450-2	5.12417-2	4.89825-2	4.68574-2	4.48572-2	4.29733-2	4.11975-2
3.00	5.91147-2	5.64674-2	5.39704-2	5.16150-2	4.93929-2	4.72958-2	4.53161-2	4.34465-2	4.16802-2	4.00103-2
3.25	5.63797-2	5.39798-2	5.17045-2	4.95480-2	4.75045-2	4.55682-2	4.37333-2	4.19942-2	4.03458-2	3.87827-2
3.50	5.36519-2	5.14893-2	4.94279-2	4.74643-2	4.55950-2	4.38161-2	4.21236-2	4.05136-2	3.89820-2	3.75251-2
3.75	5.09552-2	4.90178-2	4.71605-2	4.53820-2	4.36807-2	4.20543-2	4.05005-2	3.90166-2	3.75998-2	3.62475-2
4.00	4.83105-2	4.65845-2	4.49200-2	4.33174-2	4.17765-2	4.02965-2	3.88764-2	3.75146-2	3.62095-2	3.49592-2
4.25	4.57347-2	4.42054-2	4.27213-2	4.12843-2	3.98953-2	3.85546-2	3.72622-2	3.60177-2	3.48203-2	3.36689-2
4.50	4.32409-2	4.18931-2	4.05767-2	3.92943-2	3.80479-2	3.68387-2	3.56676-2	3.45349-2	3.34406-2	3.23842-2
4.75	4.08395-2	3.96580-2	3.84961-2	3.73571-2	3.62437-2	3.51578-2	3.41008-2	3.30739-2	3.20774-2	3.11117-2
5.00	3.85377-2	3.75075-2	3.64871-2	3.54803-2	3.44900-2	3.35189-2	3.25688-2	3.16412-2	3.07372-2	2.98574-2
5.25	3.63400-2	3.54468-2	3.45552-2	3.36693-2	3.27926-2	3.19278-2	3.10771-2	3.02425-2	2.94253-2	2.86266-2
5.50	3.42491-2	3.34791-2	3.27042-2	3.19286-2	3.11558-2	3.03889-2	2.96303-2	2.88821-2	2.81460-2	2.74234-2
5.75	3.22657-2	3.16061-2	3.09363-2	3.02607-2	2.95827-2	2.89055-2	2.82317-2	2.75635-2	2.69029-2	2.62513-2
6	3.03892-2	2.98280-2	2.92527-2	2.86673-2	2.80753-2	2.74800-2	2.68839-2	2.62895-2	2.56987-2	2.51131-2
7	2.39030-2	2.36346-2	2.33443-2	2.30352-2	2.27104-2	2.23724-2	2.20237-2	2.16664-2	2.13026-2	2.09340-2
8	1.88631-2	1.87676-2	1.86502-2	1.85128-2	1.83577-2	1.81868-2	1.80019-2	1.78048-2	1.75970-2	1.73801-2
9	1.49867-2	1.49886-2	1.49712-2	1.49359-2	1.48841-2	1.48173-2	1.47366-2	1.46433-2	1.45386-2	1.44236-2
10	1.20100-2	1.20634-2	1.21010-2	1.21237-2	1.21324-2	1.21279-2	1.21111-2	1.20829-2	1.20440-2	1.19953-2
11	9.71581-3	9.79407-3	9.85976-3	9.91341-3	9.95559-3	9.98686-3	1.00077-2	1.00188-2	1.00206-2	1.00137-2
12	7.93598-3	8.02381-3	8.10179-3	8.17023-3	8.22949-3	8.27989-3	8.32181-3	8.35559-3	8.38159-3	8.40019-3
13	6.54384-3	6.63280-3	6.71408-3	6.78786-3	6.85434-3	6.91375-3	6.96629-3	7.01220-3	7.05171-3	7.08506-3
14	5.44515-5	5.53078-3	5.61041-3	5.68416-3	5.75213-3	5.81447-3	5.87130-3	5.92276-3	5.96901-3	6.01019-3
15	4.57008-3	4.65020-3	4.72563-3	4.79643-3	4.86267-3	4.92441-3	4.98173-3	5.03473-3	5.08351-3	5.12814-3
16	3.86675-3	3.94050-3	4.01056-3	4.07696-3	4.13974-3	4.19892-3	4.25457-3	4.30673-3	4.35546-3	4.40082-3
18	2.83008-3	2.89102-3	2.94964-3	3.00594-3	3.05991-3	3.11158-3	3.16094-3	3.20801-3	3.25282-3	3.29539-3
20	2.12632-3	2.17605-3	2.22427-3	2.27098-3	2.31616-3	2.35983-3	2.40197-3	2.44259-3	2.48169-3	2.51927-3
22	1.63421-3	1.67476-3	1.71430-3	1.75283-3	1.79034-3	1.82682-3	1.86227-3	1.89669-3	1.93007-3	1.96240-3
24	1.28096-3	1.31416-3	1.34668-3	1.37851-3	1.40964-3	1.44006-3	1.46977-3	1.49876-3	1.52702-3	1.55456-3
26	1.02145-3	1.04883-3	1.07574-3	1.10215-3	1.12808-3	1.15351-3	1.17844-3	1.20285-3	1.22676-3	1.25014-3
28	8.26865-4	8.49621-4	8.72036-4	8.94105-4	9.15822-4	9.37180-4	9.58177-4	9.78806-4	9.99064-4	1.01895-3
30	6.78295-4	6.97359-4	7.16175-4	7.34739-4	7.53044-4	7.71089-4	7.88868-4	8.06378-4	8.23615-4	8.40575-4
32	5.63012-4	5.79109-4	5.95022-4	6.10747-4	6.26282-4	6.41622-4	6.56765-4	6.71707-4	6.68444-4	7.00975-4
34	4.72262-4	4.85955-4	4.99510-4	5.12924-4	5.26194-4	5.39318-4	5.52292-4	5.65114-4	5.77781-4	5.90291-4

I. SOLID ANGLE SUBTENDED BY THE CIRCULAR DISC OF UNIT RADIUS (Continued)

ρ \ z	8.2	8.4	8.6	8.8	9.0	9.2	9.4	9.6	9.8	10,0
0.0	4.62073-2	4.40560-2	4.20510-2	4.01794-2	3.84296-2	3.67914-2	3.52555-2	3.38135-2	3.24580-2	3.11823-2
0.1	4.61973-2	4.40468-2	4.20426-2	4.01718-2	3.84227-2	3.67850-2	3.52496-2	3.38081-2	3.24531-2	3.11777-2
0.2	4.61672-2	4.40195-2	4.20177-2	4.01490-2	3.84018-2	3.67659-2	3.52320-2	3.37919-2	3.24381-2	3.11639-2
0.3	4.61171-2	4.39739-2	4.19761-2	4.01110-2	3.83670-2	3.67340-2	3.52029-2	3.37650-2	3.24133-2	3.11409-2
0.4	4.60471-2	4.39102-2	4.19180-2	4.00579-2	3.83185-2	3.66895-2	3.51618-2	3.37273-2	3.23786-2	3.11089-2
0.5	4.59573-2	4.38285-2	4.18436-2	3.99899-2	3.82562-2	3.66323-2	3.51093-2	3.36790-2	3.23340-2	3.10677-2
0.6	4.58480-2	4.37291-2	4.17529-2	3.99070-2	3.81802-2	3.65627-2	3.50453-2	3.36201-2	3.22797-2	3.10175-2
0.7	4.57194-2	4.36120-2	4.16461-2	3.98094-2	3.80908-2	3.64806-2	3.49699-2	3.35506-2	3.22156-2	3.09584-2
0.8	4.55718-2	4.34775-2	4.15234-2	3.96972-2	3.79881-2	3.63864-2	3.48832-2	3.34708-2	3.21420-2	3.08904-2
0.9	4.54053-2	4.33260-2	4.13851-2	3.95707-2	3.78722-2	3.62800-2	3.47855-2	3.33808-2	3.20589-2	3.08136-2
1.00	4.52204-2	4.31575-2	4.12313-2	3.94301-2	3.77433-2	3.61617-2	3.46766-2	3.32805-2	3.19664-2	3.07281-2
1.25	4.46805-2	4.26654-2	4.07818-2	3.90187-2	3.73662-2	3.58153-2	3.43579-2	3.29868-2	3.16953-2	3.04775-2
1.50	4.40348-2	4.20762-2	4.02432-2	3.85254-2	3.69135-2	3.53992-2	3.39748-2	3.26335-2	3.13690-2	3.01757-2
1.75	4.32912-2	4.13969-2	3.96215-2	3.79554-2	3.63901-2	3.49176-2	3.35311-2	3.22240-2	3.09905-2	2.98253-2
2.00	4.24585-2	4.06353-2	3.89236-2	3.73148-2	3.58011-2	3.43752-2	3.30308-2	3.17618-2	3.05629-2	2.94290-2
2.25	4.15462-2	3.97997-2	3.81569-2	3.66101-2	3.51524-2	3.37771-2	3.24785-2	3.12512-2	3.00901-2	2.89906-2
2.50	4.05642-2	3.88988-2	3.73290-2	3.58482-2	3.44501-2	3.31288-2	3.18792-2	3.06964-2	2.95759-2	2.85133-2
2.75	3.95225-2	3.79414-2	3.64478-2	3.50359-2	3.37003-2	3.24357-2	3.12377-2	3.01018-2	2.90241-2	2.80006-2
3.00	3.84310-2	3.69365-2	3.55213-2	3.41805-2	3.29093-2	3.17036-2	3.05590-2	2.94720-2	2.84388-2	2.74563-2
3.25	3.73000-2	3.58931-2	3.45575-2	3.32891-2	3.20839-2	3.09383-2	2.98486-2	2.88118-2	2.78246-2	2.68842-2
3.50	3.61389-2	3.48198-2	3.35642-2	3.23687-2	3.12302-2	3.01455-2	2.91116-2	2.81258-2	2.71856-2	2.62881-2
3.75	3.49567-2	3.37247-2	3.25487-2	3.14261-2	3.03542-2	2.93306-2	2.83528-2	2.74186-2	2.65257-2	2.56719-2
4.00	3.37619-2	3.26155-2	3.15181-2	3.04675-2	2.94618-2	2.84990-2	2.75773-2	2.66946-2	2.58492-2	2.50392-2
4.25	3.25624-2	3.14995-2	3.04789-2	2.94991-2	2.85585-2	2.76558-2	2.67894-2	2.59579-2	2.51598-2	2.43935-2
4.50	3.13653-2	3.03833-2	2.94373-2	2.85264-2	2.76496-2	2.68057-2	2.59938-2	2.52127-2	2.44613-2	2.37383-2
4.75	3.01768-2	2.92726-2	2.83986-2	2.75544-2	2.67394-2	2.59530-2	2.51843-2	2.44626-2	2.37571-2	2.30767-2
5.00	2.90025-2	2.81725-2	2.73676-2	2.65877-2	2.58325-2	2.51016-2	2.43946-2	2.37110-2	2.30503-2	2.24118-2
5.25	2.78473-2	2.70880-2	2.63490-2	2.56306-2	2.49327-2	2.42553-2	2.35983-2	2.29613-2	2.23442-2	2.17463-2
5.50	2.67154-2	2.60229-2	2.53464-2	2.46865-2	2.40434-2	2.34174-2	2.28083-2	2.22163-2	2.16412-2	2.10828-2
5.75	2.56102-2	2.49805-2	2.43631-2	2.37587-2	2.31677-2	2.25906-2	2.20275-2	2.14786-2	2.09440-2	2.04236-2
6	2.45344-2	2.39636-2	2.34018-2	2.28497-2	2.23081-2	2.17775-2	2.12582-2	2.07505-2	2.02547-2	1.97707-2
7	2.05622-2	2.01886-2	1.98145-2	1.94409-2	1.90688-2	1.86992-2	1.83326-2	1.79697-2	1.76112-2	1.72572-2
8	1.71553-2	1.69240-2	1.66873-2	1.64463-2	1.62019-2	1.59549-2	1.57062-2	1.54565-2	1.52064-2	1.49563-2
9	1.42993-2	1.41667-2	1.40268-2	1.38804-2	1.37283-2	1.35713-2	1.34101-2	1.32454-2	1.30776-2	1.29076-2
10	1.19375-2	1.18713-2	1.17974-2	1.17166-2	1.16294-2	1.15364-2	1.14382-2	1.13353-2	1.12282-2	1.11176-2
11	9.99873-3	9.97597-3	9.94604-3	9.90942-3	9.86657-3	9.81795-3	9.76400-3	9.70514-3	9.64175-3	9.57422-3
12	8.41172-3	8.41656-3	8.41505-3	8.40753-3	8.39434-3	8.37582-3	9.35228-3	8.32404-3	8.29141-3	8.25465-2
13	7.11248-3	7.13422-3	7.15051-3	7.16160-3	7.16773-3	7.16912-3	7.16603-3	7.15866-3	7.14725-3	7.13201-3
14	6.04647-3	6.07800-3	6.10496-3	6.12749-3	6.14578-3	6.15998-3	6.17025-3	6.17678-3	6.17971-3	6.17921-3
15	5.16875-3	5.20542-3	5.23829-3	5.26744-3	5.29301-3	5.31510-3	5.33382-3	5.34931-3	5.36167-3	5.37102-3
16	4.44288-3	4.48170-3	4.51736-3	4.54993-3	4.57949-3	4.60612-3	4.62990-3	4.65091-3	4.66923-3	4.68495-3
18	3.33573-3	3.37388-3	3.40987-3	3.44372-3	3.47548-3	3.50518-3	3.53285-3	3.55854-3	3.58229-3	3.60413-3
20	2.55534-3	2.58992-3	2.62301-3	2.65462-3	2.68476-3	2.71347-3	2.74074-3	2.76659-3	2.79106-3	2.81414-3
22	1.99370-3	2.02396-3	2.05318-3	2.08137-3	2.10853-3	2.13466-3	2.15978-3	2.18389-3	2.20699-3	2.22910-3
24	1.57137-3	1.60744-3	1.63278-3	1.65739-3	1.68127-3	1.70441-3	1.72682-3	1.74850-3	1.76945-3	1.78968-3
26	1.27300-3	1.29534-3	1.31716-3	1.33844-3	1.35919-3	1.37941-3	1.39910-3	1.41825-3	1.43688-3	1.45496-3
28	1.03845-3	1.05757-3	1.07631-3	1.09466-3	1.11262-3	1.13019-3	1.14736-3	1.16415-3	1.18053-3	1.19651-3
30	8.57256-4	8.73654-4	8.89767-4	9.05592-4	9.21127-4	9.36371-4	9.51321-4	9.65975-4	9.80333-4	9.94393-4
32	7.15297-4	7.29406-4	7.43301-4	7.56979-4	7.70439-4	7.83678-4	7.96695-4	8.09488-4	8.22055-4	8.34397-4
34	6.02643-4	6.14833-4	6.26860-4	6.38721-4	6.50415-4	6.61941-4	6.73296-4	6.84479-4	6.95489-4	7.06326-4

I. SOLID ANGLE SUBTENDED BY THE CIRCULAR DISC OF UNIT RADIUS (Continued)

ρ \ z	11	12	13	14	15	16	17	18	19	20
0.0	2.58037-2	2.17036-2	1.85072-2	1.59675-2	1.39163-2	1.22360-2	1.08424-2	9.67389-3	8.68443-3	7.83929-3
0.1	2.58006-2	2.17014-2	1.85056-2	1.59662-2	1.39153-2	1.22353-2	1.08419-2	9.67344-3	8.68408-3	7.83899-3
0.2	2.57911-2	2.16947-2	1.85007-2	1.59626-2	1.39126-2	1.22332-2	1.08402-2	9.67211-3	8.68300-3	7.83812-3
0.3	2.57754-2	2.16836-2	1.84926-2	1.59566-2	1.39080-2	1.22296-2	1.08374-2	9.66988-3	8.68121-3	7.83665-3
0.4	2.57534-2	2.16680-2	1.84812-2	1.59481-2	1.39016-2	1.22246-2	1.08335-2	9.66677-3	8.67869-3	7.83461-3
0.5	2.57251-2	2.16479-2	1.84667-2	1.59372-2	1.38933-2	1.22182-2	1.08285-2	9.66277-3	8.67547-3	7.83198-3
0.6	2.56907-2	2.16235-2	1.84489-2	1.59240-2	1.38832-2	1.22104-2	1.08223-2	9.65788-3	8.67153-3	7.82877-3
0.7	2.56500-2	2.15947-2	1.84279-2	1.59083-2	1.38713-2	1.22012-2	1.08151-2	9.65211-3	8.66688-3	7.82497-3
0.8	2.56033-2	2.15615-2	1.84037-2	1.58903-2	1.38576-2	1.21906-2	1.08067-2	9.64546-3	8.66151-3	7.82060-3
0.9	2.55505-2	2.15240-2	1.83763-2	1.58699-2	1.38421-2	1.21786-2	1.07973-2	9.63793-3	8.65544-3	7.81565-3
1.00	2.54916-2	2.14822-2	1.83458-2	1.58471-2	1.38247-2	1.21651-2	1.07867-2	9.62951-3	8.64865-3	7.81011-3
1.25	2.53188-2	2.13593-2	1.82561-2	1.57801-2	1.37737-2	1.21256-2	1.07556-2	9.60471-3	8.62864-3	7.79379-3
1.50	2.51101-2	2.12106-2	1.81474-2	1.56988-2	1.37117-2	1.20775-2	1.07178-2	9.57453-3	8.60427-3	7.77390-3
1.75	2.48672-2	2.10371-2	1.80202-2	1.56036-2	1.36390-2	1.20211-2	1.06734-2	9.53907-3	8.57562-3	7.75051-3
2.00	2.45914-2	2.08395-2	1.78751-2	1.54946-2	1.35556-2	1.19563-2	1.06221-2	9.49837-3	8.54272-3	7.72368-3
2.25	2.42852-2	2.06195-2	1.77132-2	1.53730-2	1.34625-2	1.18838-2	1.05649-2	9.45265-3	8.50573-3	7.69344-3
2.50	2.39506-2	2.03784-2	1.75354-2	1.52391-2	1.33599-2	1.18038-2	1.05017-2	9.40198-3	8.46471-3	7.65986-3
2.75	2.35893-2	2.01171-2	1.73420-2	1.50931-2	1.32478-2	1.17163-2	1.04324-2	9.34649-3	8.41974-3	7.62304-3
3.00	2.32037-2	1.98372-2	1.71343-2	1.49359-2	1.31266-2	1.16216-2	1.03575-2	9.28636-3	8.37096-3	7.58305-3
3.25	2.27964-2	1.95402-2	1.69131-2	1.47681-2	1.29971-2	1.15202-2	1.02770-2	9.22171-3	8.31846-3	7.53998-3
3.50	2.23696-2	1.92275-2	1.66794-2	1.45902-2	1.28596-2	1.14124-2	1.01912-2	9.15272-3	8.26236-3	7.49393-3
3.75	2.19257-2	1.89009-2	1.64344-2	1.44032-2	1.27146-2	1.12984-2	1.01004-2	9.07957-3	8.20280-3	7.44498-3
4.00	2.14672-2	1.85618-2	1.61790-2	1.42076-2	1.25627-2	1.11786-2	1.00048-2	9.00242-3	8.13992-3	7.39323-3
4.25	2.09963-2	1.82118-2	1.59143-2	1.40043-2	1.24041-2	1.10533-2	9.90500-3	8.92150-3	8.07387-3	7.33880-3
4.50	2.05153-2	1.78524-2	1.56414-2	1.37938-2	1.22396-2	1.09230-2	9.80063-3	8.83700-3	8.00477-3	7.28179-3
4.75	2.00262-2	1.74851-2	1.53611-2	1.35770-2	1.20696-2	1.07881-2	9.69227-3	8.74910-3	7.93279-3	7.22232-3
5.00	1.95314-2	1.71113-2	1.50747-2	1.33545-2	1.18945-2	1.06487-2	9.58018-3	8.65801-3	7.85808-3	7.16052-3
5.25	1.90326-2	1.67323-2	1.47829-2	1.31270-2	1.17150-2	1.05055-2	9.46466-3	8.56394-3	7.78079-3	7.09649-3
5.50	1.85318-2	1.63496-2	1.44869-2	1.28953-2	1.15315-2	1.03587-3	9.34596-3	8.46710-3	7.70107-3	7.03033-3
5.75	1.80306-2	1.59645-2	1.41875-2	1.26599-2	1.13445-2	1.02086-2	9.22437-3	8.36768-3	7.61910-3	6.96220-3
6	1.75307-2	1.55780-2	1.38856-2	1.24216-2	1.11546-2	1.00557-2	9.10016-3	8.26590-3	7.53501-3	6.89221-3
7	1.55713-2	1.40401-2	1.26691-2	1.14513-2	1.03741-2	9.42281-3	8.58240-3	7.83921-3	7.18077-3	6.59605-3
8	1.37257-2	1.25564-2	1.14713-2	1.04791-2	9.58060-3	8.77072-3	8.04325-3	7.39062-3	6.80522-3	6.27977-3
9	1.20370-2	1.11663-2	1.03261-2	9.53327-3	8.79628-3	8.11783-3	7.49719-3	6.93168-3	6.41759-3	5.95072-3
10	1.05231-2	9.89186-3	9.25505-3	8.63315-3	8.03868-3	7.47864-3	6.95626-3	6.47231-3	6.02601-3	5.61563-3
11	9.18660-3	8.74249-3	8.27073-3	7.79187-3	7.31996-3	6.86414-3	6.43004-3	6.02070-3	5.63744-3	5.28034-3
12	8.01888-3	7.71835-3	7.37790-3	7.01642-3	6.64783-3	6.28199-3	5.92569-3	5.58331-3	5.25753-3	4.94971-3
13	7.00581-3	6.81362-3	6.57595-3	6.30927-3	6.02638-3	5.73695-3	5.44809-3	5.16484-3	4.89063-3	4.62770-3
14	6.13076-3	6.01915-3	5.86083-3	5.66965-3	5.45688-3	5.23144-3	5.00025-3	4.76848-3	4.53994-3	4.31732-3
15	5.37672-3	5.32420-3	5.22639-3	5.09465-3	4.93859-3	4.76613-3	4.58366-3	4.39620-3	4.20760-3	4.02076-3
16	4.72752-3	4.71769-3	4.66549-3	4.58004-3	4.46938-3	4.34033-3	4.19862-3	4.04890-3	3.89490-3	3.73950-3
18	3.68640-3	3.72795-3	3.73463-3	3.71212-3	3.66576-3	3.60038-3	3.52028-3	3.42913-3	3.33008-3	3.22572-3
20	2.90978-3	2.97464-3	3.01202-3	3.02537-3	3.01809-3	2.99345-3	2.95448-3	2.90392-3	2.84421-3	2.77746-3
22	2.32514-3	2.39814-3	2.44991-3	2.48250-3	2.49803-3	2.49863-3	2.48638-3	2.46323-3	2.43101-3	2.39133-3
24	1.88015-3	1.95338-3	2.01033-3	2.05219-3	2.08026-3	2.09590-3	2.10051-3	2.09544-3	2.08201-3	2.06141-3
26	1.53749-3	1.60705-3	1.66415-3	1.70943-3	1.74369-3	1.76778-3	1.78263-3	1.78915-3	1.78827-3	1.78088-3
28	1.27054-3	1.33474-3	1.38935-3	1.43473-3	1.47132-3	1.49967-3	1.52037-3	1.53404-3	1.54134-3	1.54288-3
30	1.06019-3	1.11851-3	1.16939-3	1.21303-3	1.24968-3	1.27967-3	1.30337-3	1.32121-3	1.33362-3	1.34106-3
32	8.92664-4	9.45152-4	9.91852-4	1.03284-3	1.06825-3	1.09827-3	1.12315-3	1.14316-3	1.15859-3	1.16976-3
34	7.57846-4	8.04868-4	8.47346-4	8.85295-4	9.18776-4	9.47896-4	9.72800-4	9.93667-4	1.01070-3	1.02411-3

I. SOLID ANGLE SUBTENDED BY THE CIRCULAR DISC OF UNIT RADIUS (Continued)

ρ \ z	22	26	30	40	50	60	70	80	90	100
0.0	6.48085–3	4.64218–3	3.48775–3	1.96256–3	1.25626–3	8.72483–4	6.41043–4	4.90816–4	3.87815–4	3.14136–4
0.1	6.48065–3	4.64207–3	3.48769–3	1.96256–3	1.25625–3	8.72479–4	6.41041–4	4.90815–4	3.87814–4	3.14135–4
0.2	6.48005–3	4.64177–3	3.48752–3	1.96250–3	1.25623–3	8.72468–4	6.41035–4	4.90812–4	3.87812–4	3.14134–4
0.3	6.47905–3	4.64125–3	3.48723–3	1.96241–3	1.25619–3	8.72450–4	6.41026–4	4.90806–4	3.87809–4	3.14132–4
0.4	6.47765–3	4.64053–3	3.48683–3	1.96228–3	1.25614–3	8.72425–4	6.41012–4	4.90798–4	3.87804–4	3.14128–4
0.5	6.47586–3	4.63961–3	3.48630–3	1.96212–3	1.25607–3	8.72393–4	6.40995–4	4.90788–4	3.87797–4	3.14124–4
0.6	6.47366–3	4.63848–3	3.48567–3	1.96192–3	1.25599–3	8.72353–4	6.40973–4	4.90775–4	3.87790–4	3.14119–4
0.7	6.47107–3	4.63715–3	3.48492–3	1.96168–3	1.25589–3	8.72306–4	6.40948–4	4.90761–4	3.87780–4	3.14113–4
0.8	6.46808–3	4.63562–3	3.48405–3	1.96140–3	1.25578–3	8.72252–4	6.40919–4	4.90744–4	3.87770–4	3.14106–4
0.9	6.46469–3	4.63388–3	3.48306–3	1.96109–3	1.25565–3	8.72190–4	6.40886–4	4.90724–4	3.87758–4	3.14098–4
1.00	6.46089–3	4.63192–3	3.48196–3	1.96074–3	1.25551–3	8.72120–4	6.40847–4	4.90701–4	3.87743–4	3.14089–4
1.25	6.44972–3	4.62618–3	3.47871–3	1.95971–3	1.25509–3	8.71817–4	6.40738–4	4.90638–4	3.87704–4	3.14063–4
1.50	6.43610–3	4.61916–3	3.47475–3	1.95845–3	1.25457–3	8.71668–4	6.40603–4	4.90559–4	3.87654–4	3.14030–4
1.75	6.42006–3	4.61090–3	3.47007–3	1.95696–3	1.25396–3	8.71373–4	6.40444–4	4.90466–4	3.87596–4	3.13992–4
2.00	6.40164–3	4.60142–3	3.46470–3	1.95525–3	1.25327–3	8.71034–4	6.40261–4	4.90358–4	3.87529–4	3.13948–4
2.25	6.38086–3	4.59069–3	3 45861–3	1.95331–3	1.25247–3	8.70649–4	6.40053–4	4.90236–4	3.87453–4	3.13898–4
2.50	6.35778–3	4.57871–3	3.45181–3	1.95115–3	1.25154–3	8.70218–4	6.39822–4	4.90100–4	3.87368–4	3.13842–4
2.75	6.33242–3	4.56556–3	3.44434–3	1.94876–3	1.25056–3	8.69743–4	6.39565–4	4.89950–4	3.87274–4	3.13780–4
3.00	6.30482–3	4.55123–3	3.43619–3	1.94616–3	1.24949–3	8.69225–4	6.39282–4	4.89784–4	3.87171–4	3.13713–4
3.25	6.27507–3	4.53574–3	3.42736–3	1.94333–3	1.24833–3	8.68661–4	6.38977–4	4.89605–4	3.87059–4	3.13639–4
3.50	6.24321–3	4.51911–3	3.41788–3	1.94028–3	1.24706–3	8.68052–4	6.38649–4	4.89411-4	3.86938–4	3.13560–4
3.75	6.20927–3	4.50136–3	3.40773–3	1.93702–3	1.24572–3	8.67399–4	6.38296–4	4.89204–4	3.86808–4	3.13475–4
4.00	6.17334–3	4.48252–3	3.39695–3	1.93353–3	1.24428–3	8.66703–4	6.37919–4	4.88983–4	3.86670–4	3.13384–4
4.25	6.13546–3	4.46260–3	3.38553–3	1.92984–3	1.24275–3	8.65962–4	6.37517–4	4.88747–4	3.86523–4	3.13287–4
4.50	6.09570–3	4.44163–3	3.37349–3	1.92593–3	1.24114–3	8.65178–4	6.37093–4	4.88497–4	3.86366–4	3.13184–4
4.75	6.05414–3	4.41964–3	3.36083–3	1.92182–3	1.23943–3	8.64350–4	6.36644–4	4.88233–4	3.86201–4	3.13075–4
5.00	6.01086–3	4.39668–3	3.34759–3	1.91751–3	1.23764–3	8.63479–4	6.36171–4	4.87956–4	3.86027–4	3.12961–4
5.25	5.96589–3	4.37273–3	3.33375–3	1.91299–3	1.23576–3	8.62564–4	6.35675–4	4.87664–4	3.85844–4	3.12841–4
5.50	5.91932–3	4.34786–3	3.31933–3	1.90826–3	1.23379–3	8.61607–4	6.35155–4	4.87359–4	3.85654–4	3.12716–4
5.75	5.87124–3	4.32207–3	3.30436–3	1.90333–3	1.23174–3	8.60607–4	6.34612–4	4.87039–4	3.85454–4	3.12585–4
6	5.82171–3	4.29540–3	3.28885–3	1.89822–3	1.22960–3	8.59565–4	6.34047–4	4.86706–4	3.85245–4	3.12448–4
7	5.61063-3	4.18064–3	3.22162–3	1.87584–3	1.22021–3	5.54978–4	6.31551–4	4.85235–4	3.84324–4	3.11842–4
8	5.38247–3	4.05444–3	3.14684–3	1.85055–3	1.20953–3	8.49735–4	6.28692–4	4.83548–4	3.83266–4	3.11145–4
9	5.14198–3	3.91890–3	3.06552–3	1.82258–3	1.19764–3	8.43859-4	6.25477–4	4.81648–4	3.82072–4	3.10359–4
10	4.89367–3	3.77612–3	2.97879–3	1.79214–3	1.18454–3	8.37371–4	6.21917–4	4.79538–4	3.80745–4	3.09484–4
11	4.64164–3	3.62813–3	2.88741–3	1.75948–3	1.17034–3	8.30295–4	6.18020–4	4.77224–4	3.79287-4	3.08521–4
12	4.38948–1	3.47681–3	2.79265–3	1.72482–3	1.15511–3	8.22659–4	6.13800–4	4.74711–4	3.77702–4	3.07473–4
13	4.14026–3	3.32389–3	2.69539–3	1.68841–3	1.13893–3	8.14491–4	6.09266–4	4.72004–4	3.75990–4	3.06340–4
14	3.89651–3	3.17092–3	2.59653–3	1.65051–3	1.12185–3	8.05819–4	6.04431–4	4.69109–4	3.74157–4	3.05123–4
15	3.66024–3	3.01926–3	2.49689–3	1.61136–3	1.10399–3	7.96676–4	5.99310–4	4.66033–4	3.72204–4	3.03826–4
16	3.43295–3	2.87006–3	2.39723–3	1.57119–3	1.08543–3	7.87092–4	5.93915–4	4.62782–4	3.70134–4	3.02449–4
18	3.00925–3	2.58263–3	2.20044–3	1.48869–3	1.04648–3	7.66730–4	5.82360–4	4.55791–4	3.65662–4	2.99466–4
20	2.62993–3	2.31411–3	2.01048–3	1.40470–3	1.00563–3	7.44990–4	5.69882–4	4.48163–4	3.60770–4	2.96190–4
22	2.29533–3	2.06752–3	1.83037–3	1.32068–3	9.63516–4	7.22132–4	5.56601–4	4.39987–4	3.55489–4	2.92639–4
24	2.00320–3	1.84398–3	1.66202–3	1.23786–3	9.20632–4	6.98409–4	5.42637–4	4.31314–4	3.49851–4	2.88830–4
26	1.74991–3	1.64331–3	1.50644–3	1.15722–3	8.77511–4	6.74062–4	5.28109–4	4.22205–4	3.43889–4	2.84782–4
28	1.53121–3	1.46444–3	1.36392–3	1.07951–3	8.34596–4	6.49315–4	5.13133–4	4.12722–4	3.37638–4	2.80515–4
30	1.34281–3	1.30580–3	1.23426–3	1.00528–3	7.92265–4	6.24378–4	4.97820–4	4.02926–4	3.31133–4	2.76049–4
32	1.18065–3	1.16560–3	1.11692–3	9.34894–4	7.50829–4	5.99437–4	4.82275–4	3.92875–4	3.24407–4	2.71406–4
34	1.04102–3	1.04193–3	1.01112–3	8.68556–4	7.10546–4	5.74662–4	4.66596–4	3.82627–4	3.17493–4	2.66604–4

II. SOLID ANGLE SUBTENDED BY THE LATERAL SURFACE OF A RIGHT CIRCULAR CYLINDER OF UNIT RADIUS

ρ \ h	0.2	0.4	0.6	0.8	1.0	1.2	1.4	1.6	1.8	2.0
1.2	1.20776 0	1.63842 0	1.79396 0	1.86300 0	1.89875 0	1.91940 0	1.93233 0	1.94092 0	1.94691 0	1.95124 0
1.4	6.13287–1	1.01195 0	1.23118 0	1.35275 0	1.42410 0	1.46860 0	1.49786 0	1.51798 0	1.53234 0	1.54292 0
1.6	3.73447–1	6.70631–1	8.76125–1	1.01135 0	1.10050 0	1.16064 0	1.20240 0	1.23226 0	1.25419 0	1.27070 0
1.8	2.53539–1	4.73988–1	6.46727–1	7.74499–1	8.67005–1	9.33995–1	9.83056–1	1.01957 0	1.04721 0	1.06851 0
2.0	1.84468–1	3.52378–1	4.93869–1	6.07035–1	6.94977–1	7.62546–1	8.14449–1	8.54569–1	8.85882–1	9.10596–1
2.2	1.40759–1	2.72331–1	3.88319–1	4.86135–1	5.66248–1	6.30794–1	6.82443–1	7.23759–1	7.56935–1	7.83743–1
2.4	1.11208–1	2.16910–1	3.12895–1	3.96886–1	4.68412–1	5.28240–1	5.77774–1	6.18602–1	6.52243–1	6.80031–1
2.6	9.02264–1	1.76951–1	2.57326–1	3.29544–1	3.92871–1	4.47430–1	4.93885–1	5.33170–1	5.66289–1	5.94203–1
2.8	7.47567–2	1.47179–1	2.15285–1	2.77684–1	3.33637–1	3.82981–1	4.25973–1	4.63128–1	4.95086–1	5.22511–1
3.0	6.30036–2	1.24390–1	1.82744–1	2.37000–1	2.86501–1	3.30975–1	3.70462–1	4.05220–1	4.35639–1	4.62166–1
3.2	5.38531–2	1.06550–1	1.57055–1	2.04550–1	2.48476–1	2.88537–1	3.24663–1	3.56960–1	3.85650–1	4.11026–1
3.4	4.65828–2	9.23163–2	1.36428–1	1.78281–1	2.17413–1	2.53539–1	2.86539–1	3.16430–1	3.43328–1	3.67416–1
3.6	4.07061–2	8.07745–2	1.19617–1	1.56734–1	1.91744–1	2.24390–1	2.54535–1	2.82143–1	3.07265–1	3.30009–1
3.8	3.58856–2	7.12827–2	1.05736–1	1.38850–1	1.70310–1	1.99891–1	2.27452–1	2.52935–1	2.76347–1	2.97746–1
4.0	3.18806–2	6.33806–2	9.41421–2	1.23849–1	1.52241–1	1.79123–1	2.04362–1	2.27888–1	2.49682–1	2.69771–1
4.2	2.85157–2	5.67305–2	8.43592–2	1.11146–1	1.36877–1	1.61381–1	1.84538–1	2.06275–1	2.26558–1	2.45391–1
4.4	2.56606–2	5.10800–2	7.60285–2	1.00297–1	1.23708–1	1.46114–1	1.67407–1	1.87514–1	2.06397–1	2.24044–1
4.6	2.32166–2	4.62377–2	6.88760–2	9.09597–2	1.12339–1	1.32888–1	1.52512–1	1.71140–1	1.88731–1	2.05265–1
4.8	2.11080–2	4.20558–2	6.26892–2	8.28657–2	1.02459–1	1.21362–1	1.39488–1	1.56774–1	1.73177–1	1.88673–1
5.0	1.92758–2	3.84191–2	5.73018–2	7.58047–2	9.38206–2	1.11257–1	1.28039–1	1.44108–1	1.59422–1	1.73954–1
5.2	1.76735–2	3.52364–2	5.25815–2	6.96082–2	8.62255–2	1.02354–1	1.17926–1	1.32890–1	1.47204–1	1.50844–1
5.4	1.62640–2	3.24349–2	4.84223–2	6.41411–2	7.95131–2	9.44694–2	1.08951–1	1.22910–1	1.36310–1	1.49123–1
5.6	1.50174–2	2.99559–2	4.47388–2	5.92933–2	7.35522–2	8.74557–2	1.00952–1	1.13997–1	1.26558–1	1.38608–1
5.8	1.39094–2	2.77516–2	4.14609–2	5.49747–2	6.82353–2	8.11901–2	9.37934–2	1.06006–1	1.17797–1	1.29142–1
6.0	1.29202–2	2.57827–2	3.85310–2	5.11113–2	6.34731–2	7.55706–2	8.73633–2	9.88163–2	1.09901–1	1.20594–1
6.5	1.08655–2	2.16906–2	3.24357–2	4.30628–2	5.35358–2	6.38213–2	7.38892–2	8.37127–2	9.32689–2	1.02539–1
7.5	7.99611–3	1.59708–2	2.39028–2	3.17717–2	3.95575–2	4.72417–2	5.48067–2	6.22366–2	6.95169–2	7.66348–2
8.5	6.13162–3	1.22508–2	1.83452–2	2.44028–2	3.04119–2	3.63613–2	4.22402–2	4.80388–2	5.37478–2	5.93586–2
9.5	4.85156–3	9.69542–3	1.45239–2	1.93296–2	2.41050–2	2.88431–2	3.35371–2	3.81804–2	4.27670–2	4.72910–2
10.5	3.93468–3	7.86433–3	1.17840–2	1.56887–2	1.95736–2	2.34340–2	2.72654–2	3.10634–2	3.48237–2	3.85425–2
11.5	3.25539–2	6.50737–3	9.75254–3	1.29875–2	1.62091–2	1.94139–2	2.25988–2	2.57608–2	2.88969–2	3.20044–2
12.5	2.73811–3	5.47382–3	8.20474–3	1.09285–2	1.36428–2	1.63452–2	1.90336–2	2.17058–2	2.43596–2	2.69931–2
13.5	2.33509–3	4.66845–3	6.99833–3	9.32303–3	1.16408–2	1.39501–2	1.62491–2	1.85364–2	2.08103–2	2.30692–2
14.5	2.01498–3	4.02868–3	6.03980–3	8.04707–3	1.00492–2	1.20450–2	1.40332–2	1.60127–2	1.79821–2	1.99405–2
15.5	1.75649–3	3.51201–3	5.26558–3	7.01623–3	8.76301–3	1.05050–2	1.22412–2	1.39707–2	1.56926–2	1.74061–2
16.5	1.54476–3	3.08876–3	4.63127–3	6.17152–3	7.70879–3	9.24233–3	1.07714–2	1.22954–2	1.38134–2	1.53249–2
17.5	1.36914–3	2.73769–3	4.10507–3	5.47068–3	6.83395–3	8.19430–3	9.55116–3	1.09040–2	1.22522–2	1.35952–2
18.5	1.22187–3	2.44326–3	3.66373–3	4.88279–3	6.09999–3	7.31487–3	8.52697–3	9.73585–3	1.09411–2	1.21422–2
19.5	1.09715–3	2.19392–3	3.28993–3	4.38483–3	5.47822–3	6.56974–3	7.65903–3	8.74572–3	9.82945–3	1.09099–2
20.5	9.90599–4	1.98089–3	2.97057–3	3.95934–3	4.94688–3	5.93290–3	6.91710–3	7.89919–3	8.87886–3	9.85584–3
21.5	8.98856–4	1.79746–3	2.69556–3	3.59291–3	4.48925–3	5.38435–3	6.27794–3	7.16980–3	8.05967–3	8.94732–3
22.5	8.19295–4	1.63838–3	2.45705–3	3.28510–3	4.09231–3	4.90848–3	5.72341–3	6.53690–3	7.34874–3	8.15873–3
23.5	7.49852–4	1.49953–2	2.24886–3	2.99766–3	3.74577–3	4.49301–3	5.23921–3	5.98420–3	6.72781–3	7.46988–3
24.5	6.88880–4	1.37761–3	2.06605–3	2.75405–3	3.44146–3	4.12814–3	4.81394–3	5.49872–3	6.18234–4	6.86465–3
25.5	6.35054–4	1.26998–3	1.90466–3	2.53897–3	3.17278–3	3.80596–3	4.43841–3	5.06998–3	5.70056–3	6.33004–3
26	6.10478–4	1.22084–3	1.83097–3	2.44076–3	3.05009–3	3.65884–3	4.26690–3	4.87417–3	5.48052–3	6.08584–3
28	5.25174–4	1.05026–3	1.57518–3	2.09984–3	2.62417–3	3.14807–3	3.67146–3	4.19425–3	4.71638–3	5.23774–3
30	4.56578–4	9.13093–4	1.36948–3	1.82567–3	2.28161–3	2.73722–3	3.19246–3	3.64724–3	4.10152–3	4.55522–3
32	4.00597–4	8.01144–4	1.20195–3	1.60189–3	2.00200–3	2.40185–3	2.80142–3	3.20063–3	3.59946–3	3.99785–3
34	3.54315–4	7.08591–4	1.06279–3	1.41687–3	1.77080–3	2.12454–3	2.47805–3	2.83129–3	3.18422–3	3.53680–3

II. SOLID ANGLE SUBTENDED BY THE LATERAL SURFACE OF A RIGHT CIRCULAR CYLINDER OF UNIT RADIUS (Continued)

ρ \ h	2.2	2.4	2.6	2.8	3.0	3.2	3.4	3.6	3.8	4.0
1.2	1.95448 0	1.95695 0	1.95889 0	1.96043 0	1.96168 0	1.96271 0	1.96356 0	1.96427 0	1.96488 0	1.96540 0
1.4	1.55092 0	1.55710 0	1.56198 0	1.56589 0	1.56907 0	1.57169 0	1.57387 0	1.57571 0	1.57727 0	1.57861 0
1.6	1.28338 0	1.29332 0	1.30123 0	1.30763 0	1.31287 0	1.31721 0	1.32084 0	1.32392 0	1.32654 0	1.32879 0
1.8	1.08519 0	1.09845 0	1.10913 0	1.11786 0	1.12506 0	1.13106 0	1.13612 0	1.14042 0	1.14410 0	1.14727 0
2.0	9.30329–1	9.46267–1	9.59281–1	9.70018–1	9.78962–1	9.86479–1	9.92850–1	9.98291–1	1.00297 0	1.00702 0
2.2	8.05568–1	8.23482–1	8.38309–1	8.50681–1	8.61087–1	8.69905–1	8.77430–1	8.83896–1	8.89486–1	8.94347–1
2.4	7.03083–1	7.22308–1	7.38436–1	7.52051–1	7.63616–1	7.73500–1	7.81997–1	7.89345–1	7.95734–1	8.01317–1
2.6	6.17769–1	6.37724–1	6.54688–1	6.69173–1	6.81600–1	6.92314–1	7.01595–1	7.09674–1	7.16739–1	7.22945–1
2.8	5.46042–1	5.66254–1	5.83655–1	5.98680–1	6.11697–1	6.23016–1	6.32898–1	6.41557–1	6.49175–1	6.55902–1
3.0	4.85262–1	5.05365–1	5.22878–1	5.38162–1	5.51530–1	5.63253–1	5.73564–1	5.82661–1	5.90712–1	5.97861–1
3.2	4.33412–1	4.53135–1	4.70508–1	4.85822–1	4.99338–1	5.11289–1	5.21877–1	5.31282–1	5.39655–1	5.47130–1
3.4	3.88917–1	4.08069–1	4.25113–1	4.40277–1	4.53777–1	4.65806–1	4.76541–1	4.86136–1	4.94731–1	5.02444–1
3.6	3.50524–1	3.68981–1	3.85559–1	4.00437–1	4.13789–1	4.25775–1	4.36544–1	4.46231–1	4.54957–1	4.62830–1
3.8	3.17228–1	3.34912–1	3.50932–1	3.65426–1	3.78529–1	3.90375–1	4.01087–1	4.10781–1	4.19561–1	4.27523–1
4.0	2.88211–1	3.05085–1	3.20489–1	3.34528–1	3.47309–1	3.58939–1	3.69520–1	3.79150–1	3.87919–1	3.95909–1
4.2	2.62807–1	2.78859–1	2.93616–1	3.07155–1	3.19561–1	3.30918–1	3.41311–1	3.50820–1	3.59523–1	3.67491–1
4.4	2.40469–1	2.55706–1	2.69803–1	2.82817–1	2.94812–1	3.05856–1	3.16016–1	3.25359–1	3.33951–1	3.41854–1
4.6	2.20744–1	2.35188–1	2.48627–1	2.61104–1	2.72666–1	2.83368–1	2.93263–1	3.02406–1	3.10853–1	3.18654–1
4.8	2.03257–1	2.16936–1	2.29730–1	2.41669–1	2.52789–1	2.63130–1	2.72737–1	2.81655–1	2.89928–1	2.97600–1
5.0	1.87694–1	2.00643–1	2.12811–1	2.24219–1	2.34894–1	2.44866–1	2.54171–1	2.62844–1	2.70922–1	2.78444–1
5.2	1.73793–1	1.86049–1	1.97616–1	2.08507–1	2.18740–1	2.28340–1	2.37333–1	2.45749–1	2.53618–1	2.60971–1
5.4	1.61334–1	1.72935–1	1.83927–1	1.94317–1	2.04118–1	2.13348–1	2.22026–1	2.30178–1	2.37826–1	2.44998–1
5.6	1.50130–1	1.61115–1	1.71560–1	1.81469–1	1.90849–1	1.99714–1	2.08079–1	2.15962–1	2.23384–1	2.30366–1
5.8	1.40023–1	1.50429–1	1.60356–1	1.69804–1	1.78778–1	1.87286–1	1.95341–1	2.02956–1	2.10148–1	2.16934–1
6.0	1.30878–1	1.40742–1	1.50179–1	1.59188–1	1.67771–1	1.75933–1	1.83683–1	1.91033–1	1.97994–1	2.04581–1
6.5	1.11507–1	1.20161–1	1.28494–1	1.36500–1	1.44178–1	1.51528–1	1.58554–1	1.65260–1	1.71654–1	1.77743–1
7.5	8.35794–2	9.03412–2	9.69127–2	1.03288–1	1.09462–1	1.15433–1	1.21199–1	1.26759–1	1.32115–1	1.37269–1
8.5	6.48637–2	7.02563–2	7.55304–2	8.06810–2	8.57040–2	9.05959–2	9.53542–2	9.99771–2	1.04463–1	1.08813–1
9.5	5.17473–2	5.61309–2	6.04376–2	6.46633–2	6.88047–2	7.28588–2	7.68232–2	8.06957–2	8.44749–2	8.81595–2
10.5	4.22160–2	4.58407–2	4.94134–2	5.29312–2	5.63913–2	5.97915–2	6.31296–2	6.64037–2	6.96124–2	7.27542–2
11.5	3.50805–2	3.81227–2	4.11287–2	4.40962–2	4.70231–2	4.99077–2	5.27481–2	5.55429–2	5.82907–2	6.09902–2
12.5	2.96042–2	3.21910–2	3.47518–2	3.72850–2	3.97889–2	4.22621–2	4.47031–2	4.71109–2	4.94841–2	5.18218–2
13.5	2.53119–2	2.75367–2	2.97425–2	3.19279–2	3.40916–2	3.62326–2	3.83497–2	4.04420–2	4.25085–2	4.45484–2
14.5	2.18866–2	2.38194–2	2.57379–2	2.76410–2	2.95278–2	3.13975–2	3.32491–2	3.50818–2	3.68949–2	3.86876–2
15.5	1.91103–2	2.08043–2	2.24873–2	2.41586–2	2.58174–2	2.74630–2	2.90947–2	3.07118–2	3.23138–2	3.38999–2
16.5	1.68292–2	1.83256–2	1.98135–2	2.12923–2	2.27613–2	2.42200–2	2.56678–2	2.71043–2	2.85288–2	2.99409–2
17.5	1.49326–2	1.62637–2	1.75882–2	1.89054–2	2.02149–2	2.15163–2	2.28091–2	2.40928–2	2.53670–2	2.66314–2
18.5	1.33387–2	1.45304–2	1.57166–2	1.68971–2	1.80714–2	1.92392–2	2.04001–2	2.15537–2	2.26997–2	2.38377–2
19.5	1.19866–2	1.30594–2	1.41278–2	1.51916–2	1.62504–2	1.73038–2	1.83517–2	1.93937–2	2.04294–2	2.14588–2
20.5	1.08298–2	1.18006–2	1.27677–2	1.37311–2	1.46904–2	1.56454–2	1.65957–2	1.75412–2	1.84817–2	1.94168–2
21.5	9.83250–3	1.07150–2	1.15946–2	1.24710–2	1.33441–2	1.42135–2	1.50792–2	1.59409–2	1.67984–2	1.76515–2
22.5	8.96668–3	9.77238–3	1.05757–2	1.13763–2	1.21742–2	1.29691–2	1.37608–2	1.45491–2	1.53340–2	1.61152–2
23.5	8.21023–3	8.94871–3	9.68514–3	1.04194–2	1.11513–2	1.18807–2	1.26074–2	1.33313–2	1.40522–2	1.47701–2
24.5	7.54551–3	8.22480–3	8.90236–3	9.57806–3	1.02518–2	1.09234–2	1.15927–2	1.22597–2	1.29241–2	1.35860–2
25.5	6.95829–3	7.58519–3	8.21063–3	8.83449–3	9.45665–3	1.00770–2	1.06955–2	1.13119–2	1.19262–2	1.25382–2
26	6.69003–3	7.29298–3	7.89457–3	8.49470–3	9.09327–3	9.69016–3	1.02853–2	1.08785–2	1.14698–2	1.20590–2
28	5.75827–3	6.27787–3	6.79647–3	7.31399–3	7.83055–3	8.34547–3	8.85927–3	9.37168–3	9.88262–3	1.03920–2
30	5.00829–3	5.46066–3	5.91228–3	6.36307–3	6.81299–3	7.26196–3	7.70994–3	8.15687–3	8.60268–3	9.04732–3
32	4.39574–3	4.79310–3	5.18988–3	5.58603–3	5.98150–3	6.37624–3	6.77022–3	7.16338–3	7.55568–3	7.94708–3
34	3.88901–3	4.24079–3	4.59212–3	4.94296–3	5.29326–3	5.64300–3	5.99213–3	6.34063–3	6.68845–3	7.03556–3

II. SOLID ANGLE SUBTENDED BY THE LATERAL SURFACE OF A RIGHT CIRCULAR CYLINDER OF UNIT RADIUS (Continued)

ρ \ h	4.2	4.4	4.6	4.8	5.0	5.2	5.4	5.6	5.8	6.0
1.2	1.96584 0	1.96623 0	1.96657 0	1.96687 0	1.96713 0	1.96736 0	1.96757 0	1.96775 0	1.96792 0	1.96807 0
1.4	1.57977 0	1.58077 0	1.58165 0	1.58242 0	1.58310 0	1.58371 0	1.58425 0	1.58473 0	1.58517 0	1.58556 0
1.6	1.33073 0	1.33243 0	1.33392 0	1.33522 0	1.33638 0	1.33741 0	1.33833 0	1.33916 0	1.33990 0	1.34057 0
1.8	1.15002 0	1.15242 0	1.15453 0	1.15639 0	1.15805 0	1.15952 0	1.16083 0	1.16201 0	1.16308 0	1.16404 0
2.0	1.01055 0	1.01364 0	1.01636 0	1.01877 0	1.02091 0	1.02282 0	1.02453 0	1.02607 0	1.02746 0	1.02872 0
2.2	8.98599-1	9.02335-1	9.05635-1	9.08563-1	9.11171-1	9.13504-1	9.15599-1	9.17486-1	9.19192-1	9.20739-1
2.4	8.06220-1	8.10546-1	8.14380-1	8.17791-1	8.20839-1	8.23571-1	8.26030-1	8.28250-1	8.30259-1	8.32085-1
2.6	7.28420-1	7.33271-1	7.37585-1	7.41437-1	7.44887-1	7.47989-1	7.50787-1	7.53318-1	7.55615-1	7.57704-1
2.8	6.61866-1	6.67172-1	6.71909-1	6.76152-1	6.79965-1	6.83403-1	6.86512-1	6.89330-1	6.91893-1	6.94229-1
3.0	6.04229-1	6.09919-1	6.15019-1	6.19604-1	6.23737-1	6.27475-1	6.30863-1	6.33942-1	6.36749-1	6.39312-1
3.2	5.53821-1	5.59826-1	5.65230-1	5.70106-1	5.74517-1	5.78516-1	5.82152-1	5.85465-1	5.88491-1	5.91261-1
3.4	5.09382-1	5.15636-1	5.21287-1	5.26405-1	5.31049-1	5.35274-1	5.39126-1	5.42644-1	5.45866-1	5.48821-1
3.6	4.69945-1	4.76387-1	4.82231-1	4.87543-1	4.92381-1	4.96795-1	5.00831-1	5.04527-1	5.07919-1	5.11038-1
3.8	4.34752-1	4.41326-1	4.47314.1	4.52777-1	4.57768-1	4.62337-1	4.66526-1	4.70373-1	4.73913-1	4.77174-1
4.0	4.03198-1	4.09855-1	4.15941-1	4.21513-1	4.26622-1	4.31313-1	4.35627-1	4.39599-1	4.43263-1	4.46646-1
4.2	3.74791-1	3.81485-1	3.87630-1	3.93276-1	3.98469-1	3.03252-1	4.07663-1	4.11736-1	4.15501-1	4.18987-1
4.4	3.49125-1	3.55819-1	3.61986-1	3.67672-1	3.72920-1	3.77768-1	3.82251-1	3.86402-1	3.90249-1	3.93818-1
4.6	3.25862-1	3.32523-1	3.38682-1	3.44379-1	3.49654-1	3.54542-1	3.59074-1	3.63282-1	3.67191-1	3.70826-1
4.8	3.04716-1	3.11317-1	3.17440-1	3.23124-1	3.28403-1	3.33308-1	3.37869-1	3.42114-1	3.46067-1	3.49752-1
5.0	2.85445-1	2.91962-1	2.98028-1	3.03676-1	3.08938-1	3.13840-1	3.18412-1	3.22677-1	3.26658-1	3.30378-1
5.2	2.67839-1	2.74254-1	2.80244-1	2.85839-1	2.91065-1	2.95948-1	3.00513-1	3.04782-1	3.08778-1	3.12519-1
5.4	2.51720-1	2.58017-1	2.63916-1	2.69441-1	2.74616-1	2.79465-1	2.84010-1	2.88270-1	2.92266-1	2.96015-1
5.6	2.36930-1	2.43098-1	2.48893-1	2.54336-1	2.59448-1	2.64250-1	2.68761-1	2.73000-1	2.76984-1	2.80731-1
5.8	2.23333-1	2.29364-1	2.35044-1	2.40395-1	2.45432-1	2.50176-1	2.54643-1	2.58850-1	2.62814-1	2.66548-1
6.0	2.10810-1	2.16696-1	2.22256-1	2.27505-1	2.32460-1	2.37136-1	2.41550-1	2.45716-1	2.49649-1	2.53362-1
6.5	1.83538-1	1.89047-1	1.94282-1	1.99254-1	2.03974-1	2.08454-1	2.12704-1	2.16736-1	2.20561-1	2.24190-1
7.5	1.42222-1	1.46980-1	1.51545-1	1.55923-3	1.60119-1	1.64139-1	1.67987-1	1.71671-1	1.75195-1	1.78567-1
8.5	1.13026-1	1.17102-1	1.21044-1	1.24853-1	1.28530-1	1.32079-1	1.35501-1	1.38801-1	1.41980-1	1.45042-1
9.5	9.17487-2	9.52421-2	9.86398-2	1.01942-1	1.05149-1	1.08261-1	1.11281-1	1.14208-1	1.17045-1	1.19793-1
10.5	7.58282-2	7.88336-2	8.17698-2	8.46364-2	8.74333-2	9.01605-2	9.28183-2	9.54072-2	9.79275-2	1.00380-1
11.5	6.36405-2	6.62407-2	6.87901-2	7.12881-2	7.37342-2	7.61283-2	7.84701-2	8.07597-2	8.29971-2	8.51825-2
12.5	5.41231-2	5.63872-2	5.86133-2	6.08008-2	6.29492-2	6.50582-2	6.71274-2	6.91567-2	7.11457-2	7.30946-2
13.5	4.65608-2	4.85450-2	5.05005-2	5.24265-2	5.43227-2	5.61886-2	5.80238-2	5.98281-2	6.16012-2	6.33428-2
14.5	4.04594-2	4.22095-2	4.39374-2	4.56427-2	4.73248-2	4.89833-2	5.06179-2	5.22283-2	5.38141-2	5.53752-2
15.5	3.54698-2	3.70228-2	3.85585-2	4.00764-2	4.15762-2	4.30574-2	4.45197-2	4.59628-2	4.73865-2	4.87904-2
16.5	3.13402-2	3.27262-2	3.40984-2	3.54566-2	3.68004-2	3.81294-2	3.94434-2	4.07419-2	4.20249-2	4.32920-2
17.5	2.78855-2	2.91290-2	3.03616-2	3.15828-2	3.27925-2	3.39903-2	3.51759-2	3.63492-2	3.75098-2	3.86575-2
18.5	2.49675-2	2.60887-2	2.72011-2	2.83043-2	2.93981-2	3.04822-2	3.15565-2	3.26206-2	3.36745-2	3.47178-2
19.5	2.24813-2	2.34969-2	2.45053-2	2.55062-2	2.64994-2	2.74846-2	2.84618-2	2.94306-2	3.03909-2	3.13425-2
20.5	2.03464-2	1.12703-2	2.21881-2	2.30998-2	2.40052-2	2.49040-2	2.57961-2	2.66812-2	2.75593-2	2.84302-2
21.5	1.84999-2	1.93436-2	2.01824-2	2.10160-2	2.18443-2	2.26672-2	2.34844-2	2.42959-2	2.51015-2	2.59010-2
22.5	1.68925-2	1.76658-2	1.84350-2	1.91999-2	1.99603-2	2.07162-2	2.14673-2	2.22136-2	2.29549-2	2.36911-2
23.5	1.54847-2	1.61960-2	1.69038-2	1.76079-2	1.83083-2	1.90048-2	1.96973-2	2.03856-2	2.10698-2	2.17497-2
24.5	1.42451-2	1.49013-2	1.55546-2	1.62048-2	1.68518-2	1.74954-2	1.81357-2	1.87725-2	1.94057-2	2.00352-2
25.5	1.31479-2	1.37552-2	1.43599-2	1.49620-2	1.55613-2	1.61579-2	1.67515-2	1.73421-2	1.79296-2	1.85140-2
26	1.26461-2	1.32308-2	1.38133-2	1.43932-2	1.49707-2	1.55455-2	1.61176-2	1.66870-2	1.72535-2	1.78170-2
28	1.08998-2	1.14059-2	1.19103-2	1.24128-2	1.29134-2	1.34121-2	1.39087-2	1.44033-2	1.48957-2	1.53859-2
30	9.49073-3	9.93286-3	1.03737-2	1.08131-2	1.12510-2	1.16875-2	1.21224-2	1.25558-2	1.29874-2	1.34174-2
32	8.33753-3	8.72698-3	9.11541-3	9.50276-3	9.88899-3	1.02741-2	1.06579-2	1.10406-2	1.14219-2	1.18020-2
34	7.38193-3	7.72752-3	8.07230-3	8.41624-3	8.75929-3	9.10143-3	9.44263-3	9.78286-3	1.01221-2	1.04603-2

II. SOLID ANGLE SUBTENDED BY THE LATERAL SURFACE OF A RIGHT CIRCULAR CYLINDER OF UNIT RADIUS (Continued)

ρ \ h	6.2	6.4	6.6	6.8	7.0	7.2	7.4	7.6	7.8	8.0
1.2	1.96821 0	1.96833 0	1.96844 0	1.96855 0	1.96864 0	1.96873 0	1.96881 0	1.96888 0	1.96895 0	1.96901 0
1.4	1.58592 0	1.58624 0	1.58654 0	1.58681 0	1.58705 0	1.58728 0	1.58749 0	1.58768 0	1.58786 0	1.58802 0
1.6	1.34118 0	1.34173 0	1.34224 0	1.34270 0	1.34312 0	1.34351 0	1.34386 0	1.34419 0	1.34450 0	1.34478 0
1.8	1.16492 0	1.16571 0	1.16644 0	1.16710 0	1.16771 0	1.16827 0	1.16878 0	1.16926 0	1.16970 0	1.17011 0
2.0	1.02986 0	1.03090 0	1.03185 0	1.03272 0	1.03352 0	1.03425 0	1.03493 0	1.03555 0	1.03613 0	1.03667 0
2.2	9.22145–1	9.23428–1	9.24601–1	9.25676–1	9.26663–1	9.27573–1	9.28412–1	9.29188–1	9.29907–1	9.30574–1
2.4	8.33747–1	8.35265–1	8.36655–1	8.37930–1	8.39103–1	8.40184–1	8.41182–1	8.42106–1	8.42963–1	8.43758–1
2.6	7.59610–1	7.61353–1	7.62951–1	7.64419–1	7.65771–1	7.67018–1	7.68171–1	7.69239–1	7.70230–1	7.71151–1
2.8	6.96363–1	6.98318–1	7.00113–1	7.01765–1	7.03287–1	7.04694–1	7.05995–1	7.07202–1	7.08323–1	7.09365–1
3.0	6.41658–1	6.43811–1	6.45791–1	6.47615–1	6.49299–1	6.50856–1	6.52299–1	6.53638–1	6.54883–1	6.56043–1
3.2	5.93801–1	5.96137–1	5.98287–1	6.00272–1	6.02107–1	6.03806–1	6.05382–1	6.06847–1	6.08210–1	6.09480–1
3.4	5.51537–1	5.54038–1	5.56345–1	5.58478–1	5.60453–1	5.62284–1	5.63985–1	5.65567–1	5.67041–1	5.68417–1
3.6	5.13911–1	5.16561–1	5.19011–1	5.21279–1	5.23382–1	5.25334–1	5.27151–1	5.28843–1	5.30421–1	5.31895–1
4.8	4.80185–1	4.82968–1	4.85545–1	4.87934–1	4.90154–1	4.92218–1	4.94141–1	4.95934–1	4.97608–1	4.99174–1
4.0	4.49776–1	4.52676–1	4.55365–1	4.57863–1	4.60188–1	4.62352–1	4.64372–1	4.66258–1	4.68021–1	4.69672–1
4.2	4.22219–1	4.25218–1	4.28006–1	4.30600–1	4.33018–1	4.35273–1	4.37379–1	4.39349–1	4.41193–1	4.42922–1
4.4	3.97134–1	4.00219–1	4.03091–1	4.05768–1	4.08267–1	4.10602–1	4.12786–1	4.14832–1	4.16749–1	4.18549–1
4.6	3.74211–1	3.77366–1	3.80309–1	3.83057–1	3.85627–1	3.88032–1	3.90285–1	3.92397–1	3.94381–1	3.96245–1
4.8	3.53191–1	3.56402–1	3.59403–1	3.62211–1	3.64841–1	3.67306–1	3.69619–1	3.71791–1	3.73832–1	3.75753–1
5.0	3.33856–1	3.37111–1	3.40159–1	3.43015–1	3.45695–1	3.48211–1	3.50575–1	3.52798–1	3.54891–1	3.56863–1
5.2	3.16024–1	3.19310–1	3.22393–1	3.25288–1	3.28008–1	3.30566–1	3.32974–1	3.35241–1	3.37378–1	3.39393–1
5.4	2.99535–1	3.02842–1	3.05951–1	3.08875–1	3.11626–1	3.14218–1	3.16661–1	3.18965–1	3.21139–1	3.23193–1
5.6	2.84256–1	2.87574–1	2.90698–1	2.93642–1	2.96417–1	2.99034–1	3.01505–1	3.03839–1	3.06045–1	3.08130–1
5.8	2.70068–1	2.73387–1	2.76519–1	2.79474–1	2.82265–1	2.84902–1	2.87394–1	2.89751–1	2.91982–1	2.94094–1
6.0	2.56869–1	2.60182–1	2.63313–1	2.66272–1	2.69072–1	2.71720–1	2.74228–1	2.76602–1	2.78853–1	2.80987–1
6.5	2.27632–1	2.30899–1	2.33999–1	2.36941–1	2.39735–1	2.42388–1	2.44909–1	2.47304–1	2.49581–1	2.51747–1
7.5	1.81791–1	1.84875–1	1.87823–1	1.90642–1	1.93338–1	1.95916–1	1.98381–1	2.00738–1	2.02993–1	2.05150–1
8.5	1.47990–1	1.50828–1	1.53560–1	1.56188–1	1.58716–1	1.61148–1	1.63487–1	1.65737–1	1.67901–1	1.69982–1
9.5	1.22454–1	1.25029–1	1.27521–1	1.29931–1	1.32262–1	1.34516–1	1.36695–1	1.38800–1	1.40835–1	1.42802–1
10.5	1.02766–1	1.05085–1	1.07339–1	1.09530–1	1.11658–2	1.13724–1	1.15729–1	1.17676–1	1.19565–1	1.21398–1
11.5	8.73162–2	8.93986–2	9.14302–2	9.34115–2	9.53431–2	9.72257–2	9.90600–2	1.00847–1	1.02587–1	1.04281–1
12.5	7.50033–2	7.68719–2	7.87005–2	8.04895–2	8.22391–2	8.39495–2	8.56212–2	8.72546–2	8.88502–2	9.04085–2
13.5	6.50530–2	6.67316–2	6.83786–2	6.99941–2	7.15781–2	7.31308–2	7.46524–2	7.61429–2	7.76028–2	7.90322–2
14.5	5.69114–2	5.84225–2	5.99084–2	6.13691–2	6.28045–2	6.42147–2	6.55997–2	6.69595–2	6.82944–2	6.96043–2
15.5	5.01745–2	5.15384–2	5.28822–2	5.42056–2	5.55086–2	5.67911–2	5.80530–2	5.92945–2	6.05155–2	6.17160–2
16.5	4.45431–2	4.57780–2	4.69964–2	4.81984–2	4.93837–2	5.05523–2	5.17040–2	5.28390–2	5.39570–2	5.50582–2
17.5	3.97922–2	4.09137–2	4.20218–2	4.31164–2	4.41973–2	4.52645–2	4.63178–2	4.73572–2	4.83826–2	4.93940–2
18.5	3.57504–2	3.67721–2	3.77829–2	3.87824–2	3.97707–2	4.07477–2	4.17131–2	4.26669–2	4.36091–2	4.45396–2
19.5	3.22853–2	3.32191–2	3.41438–2	3.50592–2	3.59652–2	3.68618–2	3.77487–2	3.86260–2	3.94935–2	4.03511–2
20.5	2.92938–2	3.01498–2	3.09982–2	3.18389–2	3.26717–2	3.34965–2	3.43133–2	3.51220–2	3.59224–2	3.67145–2
21.5	2.63943–2	2.74814–2	2.82620–2	2.90362–2	2.98037–2	3.05645–2	3.13185–2	3.20655–2	3.28057–2	3.35387–2
22.5	2.44222–2	2.51478–2	2.58681–2	2.65829–2	2.72920–2	2.79954–2	2.86930–2	2.93848–2	3.00707–2	3.07505–2
23.5	2.24251–2	2.30960–2	2.37623–2	2.44239–2	2.50807–2	2.57326–2	2.63796–2	2.70216–2	2.76585–2	2.82902–2
24.5	2.06609–2	2.12828–2	2.19007–2	2.25146–2	2.31243–2	2.37299–2	2.43313–2	2.49283–2	2.55210–2	2.61092–2
25.5	1.90951–2	1.96729–2	2.02473–2	2.08183–2	2.13857–2	2.19494–2	2.25096–2	2.30660–2	2.36186–2	2.41674–2
26	1.83775–2	1.89350–2	1.94893–2	2.00403–2	2.05881–2	2.11325–5	2.16735–2	2.22111–2	2.27451–2	2.32755–2
28	1.58738–2	1.63594–2	1.68426–2	1.73234–2	1.78017–2	1.82775–2	1.87506–2	1.92212–2	1.96890–2	2.01541–2
30	1.38457–2	1.42722–2	1.46968–2	1.51196–2	1.55405–2	1.59594–2	1.63763–2	1.67912–2	1.72040–2	1.76147–2
32	1.21807–2	1.25580–2	1.29338–2	1.33083–2	1.36812–2	1.40526–2	1.44224–2	1.47907–2	1.51573–2	1.55222–2
34	1.07974–2	1.11334–2	1.14683–2	1.18020–2	1.21346–2	1.24659–2	1.27960–2	1.31248–2	1.34524–2	1.37786 2

II. SOLID ANGLE SUBTENDED BY THE LATERAL SURFACE OF A RIGHT CIRCULAR CYLINDER OF UNIT RADIUS (Continued)

ρ \ h	8.2	8.4	8.6	8.8	9.0	9.2	9.4	9.6	9.8	10.0
1.2	1.96907 0	1.96912 0	1.96917 0	1.96922 0	1.96926 0	1.96931 0	1.96934 0	1.96938 0	1.96941 0	1.96945 0
1.4	1.58817 0	1.58832 0	1.58845 0	1.58857 0	1.58869 0	1.58880 0	1.58890 0	1.58899 0	1.58908 0	1.58917 0
1.6	1.34504 0	1.34529 0	1.34552 0	1.34573 0	1.34593 0	1.34611 0	1.34629 0	1.34645 0	1.34660 0	1.34675 0
1.8	1.17049 0	1.17084 0	1.17117 0	1.17148 0	1.17176 0	1.17203 0	1.17228 0	1.17252 0	1.17274 0	1.17295 0
2.0	1.03717 0	1.03764 0	1.03807 0	1.03847 0	1.03885 0	1.03921 0	1.03954 0	1.03985 0	1.04014 0	1.04042 0
2.2	9.31194-1	9.31772-1	9.32311-1	9.32814-1	9.33286-1	9.33727-1	9.34141-1	9.34530-1	9.34896-1	9.35241-1
2.4	8.44499-1	8.45188-1	8.45832-1	8.46434-1	8.46997-1	8.47525-1	8.48021-1	8.48487-1	8.48925-1	8.49338-1
2.6	7.72008-1	7.72808-1	7.73555-1	7.74254-1	7.74908-2	7.75522-1	7.76098-1	7.76640-1	7.77150-1	7.77631-1
2.8	7.10337-1	7.11244-1	7.12091-1	7.12884-1	7.13628-1	7.14325-1	7.14981-1	7.15597-1	7.16178-1	7.16725-1
3.0	6.57124-1	6.58134-1	6.59078-1	6.59963-1	6.60793-1	6.61572-1	6.62304-1	6.62993-1	6.63643-1	6.64256-1
3.2	6.10666-1	6.11775-1	6.12813-1	6.13786-1	6.14698-1	6.15556-1	6.16363-1	6.17123-1	6.17840-1	6.18516-1
3.4	5.69702-1	5.70905-1	5.72032-1	5.73089-1	5.74081-1	5.75015-1	5.75893-1	5.76721-1	5.77502-1	5.78240-1
3.6	5.33274-1	5.34565-1	5.35776-1	5.36913-1	5.37982-1	5.38987-1	5.39935-1	5.40828-1	5.41671-1	5.42468-1
3.8	5.00641-1	5.02015-1	5.03306-1	5.04518-1	5.05659-1	5.06733-1	5.07745-1	5.08701-1	5.09603-1	5.10457-1
4.0	4.71219-1	4.72672-1	4.74036-1	4.75319-1	4.76528-1	4.77666-1	4.78741-1	4.79755-1	4.80715-1	4.81622-1
4.2	4.44545-1	4.46069-1	4.47502-1	4.78852-1	4.50123-1	4.51323-1	4.52456-1	4.53526-1	4.54539-1	4.55497-1
4.4	4.20240-1	4.21830-1	4.23327-1	4.24738-1	4.26069-1	4.27325-1	4.28512-1	4.29635-1	4.30698-1	4.31706-1
4.6	3.97998-1	3.99648-1	4.01204-1	4.02671-1	4.04056-1	4.05365-1	4.06603-1	4.07775-1	4.08886-1	4.09938-1
4.8	3.77562-1	3.79268-1	3.80876-1	3.82395-1	3.83831-1	3.85188-1	3.86474-1	3.87691-1	3.88846-1	3.89941-1
5.0	3.58722-1	3.60476-1	3.62132-1	3.63698-1	3.65180-1	3.66582-1	3.67911-1	3.69170-1	3.70366-1	3.71501-1
5.2	3.41296-1	3.43094-1	3.44793-1	3.46402-1	3.47924-1	3.49367-1	3.50736-1	3.52034-1	3.53267-1	3.54439-1
5.4	3.25134-1	3.26970-1	3.28707-1	3.30353-1	3.31913-1	3.33393-1	3.34798-1	3.36131-1	3.37399-1	3.38605-1
5.6	3.10104-1	3.11973-1	3.13744-1	3.15424-1	3.17017-1	3.18530-1	3.19967-1	3.21333-1	3.22632-1	3.23869-1
5.8	2.96096-1	2.97993-1	2.99793-1	3.01501-1	3.03124-1	3.04665-1	3.06132-1	3.07527-1	3.08855-1	3.10120-1
6.0	2.83010-1	2.84931-1	2.86756-1	2.88489-1	2.90137-1	2.91705-1	2.93197-1	2.94617-1	2.95971-1	2.97262-1
6.5	2.53808-1	2.55769-1	2.57637-1	2.59416-1	2.61112-1	2.62729-1	2.64272-1	2.65744-1	2.67150-1	2.68493-1
7.5	2.07215-1	2.09190-1	2.11081-1	2.12892-1	2.14626-1	2.16288-1	2.17880-1	2.19406-1	2.20869-1	2.22272-1
8.5	1.71984-1	1.73909-1	1.75761-1	1.77543-1	1.79257-1	1.80907-1	1.82494-1	1.84022-1	1.85492-1	1.86909-1
9.5	1.44702-1	1.46537-1	1.48310-1	1.50024-1	1.51679-1	1.53278-1	1.54823-1	1.56316-1	1.57758-1	1.59152-1
10.5	1.23176-1	1.24901-1	1.26574-1	1.28196-1	1.29769-1	1.31294-1	1.32773-1	1.34206-1	1.35596-1	1.36944-1
11.5	1.05930-1	1.07536-1	1.09098-1	1.10618-1	1.12096-1	1.13535-1	1.14934-1	1.16294-1	1.17618-1	1.18904-1
12.5	9.19300-2	9.34152-2	9.48648-2	9.62794-2	9.76595-2	9.90058-2	1.00319-1	1.01600-1	1.02849-1	1.04066-1
13.5	8.04315-2	8.18010-2	8.31411-2	8.44520-2	8.57343-2	8.69884-2	8.82145-2	8.94134-2	9.05852-2	9.17306-2
14.5	7.08894-2	7.21500-2	7.33863-2	7.45984-2	7.57866-2	7.69511-2	7.80923-2	7.92105-2	8.03058-2	8.13787-2
15.5	6.28961-2	6.40559-2	6.51954-2	6.63149-2	6.74144-2	6.84942-2	6.95543-2	7.05949-2	7.16163-2	7.26186-2
16.5	5.61424-2	5.72098-2	5.82604-2	5.92942-2	6.03113-2	6.13117-2	6.22956-2	6.32631-2	6.42142-2	6.51492-2
17.5	5.03913-2	5.13746-2	5.23437-2	5.32989-2	5.42399-2	5.51670-2	5.60801-2	5.69792-2	5.78646-2	5.87361-2
18.5	4.54583-2	4.63653-2	4.72604-2	4.81436-2	4.90150-2	4.98746-2	5.07223-2	5.15582-2	5.23823-2	5.31947-2
19.5	4.11989-2	4.20367-2	4.28646-2	4.36824-2	4.44902-2	4.52879-2	4.60756-2	4.68531-2	4.76206-2	4.83781-2
20.5	3.74983-2	3.82736-2	3.90404-2	3.97988-2	4.05486-2	4.12898-2	4.20224-2	4.27464-2	4.34618-2	4.41685-2
21.5	3.42647-2	3.49835-2	3.56951-2	3.63994-2	3.70964-2	3.77860-2	3.84683-2	3.91432-2	3.98106-2	4.04706-2
22.5	3.14242-2	3.20919-2	3.27533-2	3.34085-2	3.40574-2	3.47000-2	3.53363-2	3.59662-2	3.65897-2	3.72067-2
23.5	2.89167-2	2.95380-2	3.01539-2	3.07645-2	3.13696-2	3.19693-2	3.25635-2	3.31522-2	3.37353-2	3.43129-2
24.5	2.66929-2	2.72721-2	2.78467-2	2.84166-2	2.89819-2	2.95424-2	3.00982-2	3.06491-2	3.11952-2	3.17365-2
25.5	2.47122-2	2.52532-2	2.57901-2	2.63231-2	2.68519-2	2.73766-2	2.78972-2	2.84136-2	2.89258-2	2.94337-2
26	2.38023-2	2.43254-2	2.48449-2	2.53605-2	2.58723-2	2.63803-2	2.68844-2	2.73846-2	2.78809-2	2.83732-2
28	2.06165-2	2.10761-2	2.15328-2	2.19866-2	2.24375-2	2.28854-2	2.33304-2	2.37723-2	2.42112-2	2.46470-2
30	1.80232-2	1.84296-2	1.88338-2	1.92357-2	1.96353-2	2.00327-2	2.04277-2	2.08203-2	2.12106-2	2.15984-2
32	1.58855-2	1.62471-2	1.66069-2	1.69650-2	1.73212-2	1.76757-2	1.80283-2	1.83790-2	1.87278-2	1.90748-2
34	1.41035-2	1.44271-2	1.47492-2	1.50699-2	1.53892-2	1.57071-2	1.60235-2	1.63383-2	1.66517-2	1.69635-2

II. SOLID ANGLE SUBTENDED BY THE LATERAL SURFACE OF A RIGHT CIRCULAR CYLINDER OF UNIT RADIUS (Continued)

ρ \ h	11	12	13	14	15	16	17	18	19	20
1.2	1.96958 0	1.96968 0	1.96976 0	1.96983 0	1.96988 0	1.96992 0	1.96995 0	1.96998 0	1.97001 0	1.97003 0
1.4	1.58952 0	1.58979 0	1.59000 0	1.59016 0	1.59030 0	1.59041 0	1.59050 0	1.59058 0	1.59064 0	1.59069 0
1.6	1.34735 0	1.34782 0	1.34818 0	1.34846 0	1.34870 0	1.34889 0	1.34904 0	1.34917 0	1.34929 0	1.34938 0
1.8	1.17383 0	1.17450 0	1.17503 0	1.17544 0	1.17578 0	1.17606 0	1.17628 0	1.17648 0	1.17664 0	1.17678 0
2.0	1.04159 0	1.04248 0	1.04317 0	1.04372 0	1.04417 0	1.04453 0	1.04484 0	1.04509 0	1.04531 0	1.04549 0
2.2	9.36695-1	9.37807-1	9.38675-1	9.39365-1	9.39923-1	9.40381-1	9.40761-1	9.41079-1	9.41349-1	9.41580-1
2.4	8.51083-1	8.52418-1	8.53462-1	8.54292-1	8.54964-1	8.55516-1	8.55974-1	8.56358-1	8.56683-1	8.56962-1
2.6	7.79665-1	7.81223-1	7.82442-1	7.83414-1	7.84200-1	7.84846-1	7.85382-1	7.85833-1	7.86214-1	7.86541-1
2.8	7.19044-1	7.20824-1	7.22218-1	7.23330-1	7.24231-1	7.24971-1	7.25587-1	7.26103-1	7.26542-1	7.26916-1
3.0	6.66855-1	6.68852-1	6.70420-1	6.71671-1	6.72687-1	6.73521-1	6.74215-1	6.74798-1	6.75293-1	6.75716-1
3.2	6.21388-1	6.23600-1	6.25338-1	6.26728-1	6.27856-1	6.28784-1	6.29556-1	6.30205-1	6.30756-1	6.31228-1
3.4	5.81377-1	5.83798-1	5.85704-1	5.87230-1	5.88470-1	5.89490-1	5.90340-1	5.91055-1	5.91663-1	5.92183-1
3.6	5.45862-1	5.48487-1	5.50556-1	5.52216-1	5.53565-1	5.54678-1	5.55605-1	5.56385-1	5.57048-1	5.57616-1
3.8	5.14097-1	5.16920-1	5.19150-1	5.20940-1	5.22398-1	5.23600-1	5.24603-1	5.25448-1	5.26166-1	5.26782-1
4.0	4.85500-1	4.88513-1	4.90898 1	4.92816-1	4.94380-1	4.95671-1	4.96749-1	4.97658 2	4.98431-1	4.99094-1
4.2	4.59601-1	4.62799-1	4.65334-1	4.67376-1	4.69044-1	4.70422-1	4.71574-1	4.72545-1	4.73373-1	4.74082-1
4.4	4.36025-1	4.39399-1	4.42080-1	4.44243-1	4.46011-1	4.47475-1	4.48699-1	4.49732-1	4.50613-1	4.51369-1
4.6	4.14461-1	4.18004-1	4.20825-1	4.23106-1	4.24973-1	4.26520-1	4.27815-1	4.28909-1	4.29842-1	4.30643-1
4.8	3.94656-1	3.98360-1	4.01316-1	4.03710-1	4.05672-1	4.07300-1	4.08665-1	4.09819-1	4.10803-1	4.11649-1
5.0	3.76396-1	3.80253-1	3.83339-1	3.85841-1	3.87897-1	3.89604-1	3.91036-1	3.92249-1	3.93284-1	3.94174-1
5.2	3.59504-1	3.63505-1	3.66714-1	3.69322-1	3.71467-1	3.73252-1	3.74750-1	3.76020-1	3.77104-1	3.78038-1
5.4	3.43827-1	3.47965-1	3.51291-1	3.54000-1	3.56232-1	3.58091-1	3.59654-1	3.60979-1	3.62113-1	3.63089-1
5.6	3.29236-1	3.33503-1	3.36941-1	3.39746-1	3.42062-1	3.43993-1	3.45618-1	3.46998-1	3.48179-1	3.49198-1
5.8	3.15621-1	3.20008-1	3.23552-1	3.26450-1	3.28845-1	3.30846-1	3.32532-1	3.33966-1	3.35193-1	3.36252-1
6.0	3.02886-1	3.07384-1	3.11028-1	3.14014-1	3.16487-1	3.18555-1	3.20300-1	3.21785-1	3.23058-1	3.24157-1
6.5	2.74376-1	2.79120-1	2.82988-1	2.86175-1	2.88826-1	2.91051-1	2.92935-1	2.94543-1	2.95924-1	2.97120-1
7.5	2.28488-1	2.33581-1	2.37791-1	2.41298-1	2.44243-1	2.46736-1	2.48862-1	2.50686-1	2.52261-1	2.53630-1
8.5	1.93246-1	1.98524-1	2.02945-1	2.06671-1	2.09832-1	2.12530-1	2.14847-1	2.16848-1	2.18586-1	2.20103-1
9.5	1.65452-1	1.70779-1	1.75302-1	1.79157-1	1.82461-1	1.85305-1	1.87766-1	1.89906-1	1.91775-1	1.93415-1
10.5	1.43092-1	1.48367-1	1.52901-1	1.56811-1	1.60193-1	1.63131-1	1.65693-1	1.67935-1	1.69905-1	1.71643-1
11.5	1.24825-1	1.29971-1	1.34449-1	1.38352-1	1.41761-1	1.44746-1	1.47369-1	1.49681-1	1.51724-1	1.53537-1
12.5	1.09712-1	1.14680-1	1.19051-1	1.22899-1	1.26291-1	1.29286-1	1.31937-1	1.34289-1	1.36381-1	1.38246-1
13.5	9.70772-2	1.01834-1	1.06063-1	1.09821-1	1.13162-1	1.16136-1	1.18787-1	1.21154-1	1.23272-1	1.25171-1
14.5	8.64177-2	9.09463-2	9.50095-2	9.86522-2	1.01917-1	1.04846-1	1.07474-1	1.09835-1	1.11960-1	1.13875-1
15.5	7.73521-1	8.16447-2	8.55292-2	8.90399-2	9.22107-2	9.50741-2	9.76608-2	9.99992-2	1.02115-1	1.04032-1
16.5	6.95864-2	7.36430-2	7.73426-2	8.07111-2	8.37747-2	8.65597-2	8.90911-2	9.13926-2	9.34864-2	9.53927-2
17.5	6.28905-2	6.67161-2	7.02299-2	7.34510-2	7.63996-2	7.90965-2	8.15620-2	8.38159-2	8.58768-2	8.77622-2
18.5	5.70820-2	6.06853-2	6.40160-2	6.70883-2	6.99176-2	7.25201-2	7.49123-2	7.71105-2	7.91301-2	8.09861-2
19.5	5.20154-2	5.54067-2	5.85598-2	6.14847-2	6.41931-2	6.66977-2	6.90116-2	7.11481-2	7.31200-2	7.49401-2
20.5	4.75729-2	5.07639-2	5.37463-2	5.65273-2	5.91155-2	6.15207-2	6.37533-2	6.58240-2	6.77436-2	6.95226-2
21.5	4.36589-2	4.66616-2	4.94814-2	5.21232-2	5.45934-2	5.68993-2	5.90492-2	6.10517-2	6.29156-2	6.46497-2
22.5	4.01949-2	4.30213-2	4.56871-2	4.81955-2	5.05509-2	5.27589-2	5.48259-2	5.67589-2	5.85649-2	6.02515-2
23.5	3.71163-2	3.97781-2	4.22987-2	4.46797-2	4.69243-2	4.90367-2	5.10217-2	5.28849-2	5.46320-2	5.62691-2
24.5	3.43692-2	3.68778-2	3.92617-2	4.15219-2	4.36603-2	4.56799-2	4.75844-2	4.93782-2	5.10660-2	5.26528-2
25.5	3.19088-2	3.42748-2	3.65306-2	3.86764-2	4.07133-2	4.26435-2	4.44696-2	4.61952-2	4.78239-2	4.93598-2
26	3.07741-2	3.30726-2	3.52674-2	3.73584-2	3.93464-2	4.12332-2	4.30211-2	4.47131-2	4.63125-2	4.78230-2
28	2.67791-2	2.88310-2	3.08012-2	3.26887-2	3.44934-2	3.62159-2	3.78574-2	3.94197-2	4.09046-2	4.23147-2
30	2.35008-2	2.53397-2	2.71135-2	2.88210-2	3.04614-2	3.20348-2	3.35416-2	3.49826-2	3.63590-2	3.76724-2
32	2.07801-2	2.24347-2	2.40370-2	2.55855-2	2.70794-2	2.85184-2	2.99023-2	3.12314-2	3.25064-2	3.37281-2
34	1.84991-2	1.99938-2	2.14460-2	2.28543-2	2.42179-2	2.55360-2	2.68084-2	2.80350-2	2.92161-2	3.03521-2

II. SOLID ANGLE SUBTENDED BY THE LATERAL SURFACE OF A RIGHT CIRCULAR CYLINDER OF UNIT RADIUS (Continued)

ρ \ h	22	26	30	40	50	60	70	80	90	100
1.2	1.97006 0	1.97011 0	1.97014 0	1.97017 0	1.97019 0	1.97020 0	1.97021 0	1.97021 0	1.97021 0	1.97021 0
1.4	1.59078 0	1.59090 0	1.59098 0	1.59108 0	1.59112 0	1.59115 0	1.59116 0	1.59117 0	1.59118 0	1.59119 0
1.6	1.34953 0	1.34974 0	1.34987 0	1.35004 0	1.35012 0	1.35016 0	1.35019 0	1.35021 0	1.35022 0	1.35023 0
1.8	1.17700 0	1.17730 0	1.17749 0	1.17774 0	1.17786 0	1.17792 0	1.17796 0	1.17798 0	1.17800 0	1.17801 0
2.0	1.04579 0	1.04619 0	1.04644 0	1.04677 0	1.04692 0	1.04701 0	1.04706 0	1.04709 0	1.04711 0	1.04713 0
2.2	9.41951–1	9.42453–1	9.42769–1	9.43186–1	9.43380–1	9.43485–1	9.43548–1	9.43589–1	9.43617–1	9.43638–1
2.4	8.57409–1	8.58016–1	8.58397–1	8.58901–1	8.59135–1	8.59262–1	8.59339–1	8.59388–1	8.59422–1	8.59447–1
2.6	7.87066–1	7.87778–1	7.88226-1	7.88818–1	7.89093–1	7.89242–1	7.89333–1	7.89391–1	7.89431–1	7.89460–1
2.8	7.27520–1	7.28338–1	7.28853–1	7.29535–1	7.29851–1	7.30023 1	7.30127–1	7.30194–1	7.30240–1	7.30273–1
3.0	6.76398–1	6.77323–1	6.77906–1	6.78678–1	6.79036–1	6.79231–1	6.79348–1	6.79424–1	6.79477–1	6.79514–1
3.2	6.31988–1	6.33021–1	6.33672–1	6.34534–1	6.34934–1	6.35152–1	6.35283–1	6.35368–1	6.35427–1	6.35469–1
3.4	5.93021–1	5.94161–1	5.94880–1	5.95832–1	5.96275–1	5.96516–1	5.96661–1	5.96756–1	5.96820–1	5.96867–1
3.6	5.58533–1	5.59779–1	5.60566–1	5.61609–1	5.62095–1	5.62359–1	5.62518–1	5.62622–1	5.62693–1	5.62743–1
3.8	5.27776–1	5.29129–1	5.29984–1	5.31118–1	5.31646–1	5.31933–1	5.32107–1	5.32219–1	5.32297–1	5.32352–1
4.0	5.00164–1	5.01624–1	5.02546–1	5.03771–1	5.04342–1	5.04652–1	5.04840–1	5.04962–1	5.05045–1	5.05105–1
4.2	4.75229–1	4.76794–1	4.77783–1	4.79100–1	4.79713–1	4.80047–1	4.80248–1	4.80380–1	4.80469–1	4.80534–1
4.4	4.52591–1	4.54260–1	4.55317–1	4.56724–1	4.57380–1	4.57737–1	4.57953–1	4.58093–1	4.58189–1	4.58258–1
4.6	4.31940–1	4.33713–1	4.34837–1	4.36334–1	4.37032–1	4.37413–1	4.37643–1	4.37793–1	4.37895–1	4.37969–1
4.8	4.13020–1	4.14896–1	4.16086–1	4.17673–1	4.18414–1	4.18818–1	4.19063–1	4.19221–1	4.19330–1	4.19408–1
5.0	3.95617–1	3.97595–1	3.98851–1	4.00528–1	4.01312–1	4.01739–1	4.01997–1	4.02166–1	4.02281–1	4.02363–1
5.2	3.79553–1	3.81631–1	3.82953–1	3.84719–1	3.85545–1	3.85996–1	3.86268–1	3.86446–1	3.86567–1	3.86654–1
5.4	3.64674–1	3.66852–1	3.68238–1	3.70093–1	3.70962–1	3.71436–2	3.71723–1	3.71909–1	3.72037–1	3.72129–1
5.6	3.50852–1	3.53128–1	3.54579–1	3.56522–1	3.57433–1	3.57930–1	3.58231–1	3.58427–1	3.58561–1	3.58658–1
5.8	3.37975–1	3.40348–1	3.41862–1	3.43893–1	3.44846–1	3.45367–1	3.45682–1	3.45887–1	3.46028–1	3.46129–1
6.0	3.25947–1	3.28415–1	3.29993–1	3.32111–1	3.33106–1	3.33650–1	3.33979–1	3.34193–1	3.34340–1	3.34446–1
6.5	2.99070–1	3.01771–1	3.03503–1	3.05837–1	3.06936 1	3.07538–1	3.07903–1	3.08140–1	3.08303–1	3.08420–1
7.5	2.55877–1	2.59013–1	2.61041–1	2.63793–1	2.65097–1	2.65813–1	2.66248–1	2.66531–1	2.66726–1	2.66866–1
8.5	2.22609–1	2.26138–1	2.28440–1	2.31591–1	2.33094–1	2.33923–1	2.34427–1	2.34756–1	2.34982–1	2.35145–1
9.5	1.96141–1	2.00019–1	2.02572–1	2.06099–1	2.07795–1	2.08734–1	2.09307–1	2.09681–1	2.09939–1	2.10123–1
10.5	1.74552–1	1.78731–1	1.81512–1	1.85391–1	1.87273–1	1.88320–1	1.88959-1	1.89378–1	1.89667–1	1.89874–1
11.5	1.56591–1	1.61027–1	1.64010–1	1.68216–1	1.70276–1	1.71427–1	1.72133–1	1.72595–1	1.72915–1	1.73144–1
12.5	1.41411–1	1.46059–1	1.49220–1	1.53727–1	1.55956–1	1.57209–1	1.57978–1	1.58484–1	1.58834–1	1.59085–1
13.5	1.28415–1	1.33232–1	1.36547–1	1.41328–1	1.43718–1	1.45068–1	1.45901–1	1.46449–1	1.46828–1	1.47102–1
14.5	1.17169–1	1.22117–1	1.25561–1	1.30589–1	1.33130–1	1.34575–1	1.35469–1	1.36059–1	1.36467–1	1.36762–1
15.5	1.07352–1	1.12393–1	1.15943–1	1.21192–1	1.23876–1	1.25410–1	1.26364–1	1.26994–1	1.27432–1	1.27748–1
16.5	9.87155–2	1.03817–1	1.07453–1	1.12896–1	1.15712–1	1.17333–1	1.18344–1	1.19015–1	1.19481–1	1.19818–1
17.5	9.10694–2	9.62022–2	9.99031–2	1.05515–1	1.08454–1	1.10158–1	1.11224–1	1.11933–1	1.12427–1	1.12785–1
18.5	8.42621–2	8.93999–2	9.31475–2	9.89041–2	1.01956–1	1.03738–1	1.04858–1	1.05605–1	1.06126-1	1.06504–1
19.5	7.81718-2	8.32920–2	8.70693–2	9.29470–2	9.61031–2	9.79583–2	9.91302–2	9.99138–2	1.00462–1	1.00859–1
20.5	7.26993–2	7.77824–2	8.15742–2	8.75508–2	9.08012–3	9.27260–2	9.39474–2	9.47666–2	9.53407–2	9.57578–2
21.5	6.77631–2	7.27924–2	7.65849–2	8.26398–2	8.59752–2	8.79652–2	8.92341–2	9.00877–2	9.06873–2	9.11235–2
22.5	6.32951–2	6.82567–2	7.20379–2	7.81515–2	8.15629–2	8.36139–2	8.49281–2	8.58149–2	8.64392–2	8.68943–2
23.5	5.92381–2	6.41207–2	6.78797–2	7.40341–2	7.75126–2	7.96204–2	8.09777–2	8.18966–2	8.25451–2	8.30184–2
24.5	5.55437–2	6.03378–2	6.40655–2	7.02440–2	7.37812–2	7.59416–2	7.73398–2	7.82896–2	7.89615–2	7.94528–2
25.5	5.21704–2	5.68689–2	6.05571–2	6.67446–2	7.03324–2	7.25412–2	7.39780–2	7.49576–2	7.56521–1	7.61610–2
26	5.05929–2	5.52413–2	5.89072–2	6.50939–2	6.87040–2	7.09355–2	7.23909–2	7.33849–2	7.40905–2	7.46079–2
28	4.49208–2	4.93585–2	5.29206–2	5.90727–2	6.27538–2	6.50659–2	6.65901–2	6.76386–2	6.83867–2	6.89374–2
30	4.01167–2	4.43342–2	4.77750–2	5.38500–2	5.75745–2	5.99521–2	6.15364–2	6.26346-2	6.34223–2	6.40044–2
32	3.60162–2	4.00114–2	4.33199–2	4.92838–2	5.30278–2	5.54564–2	5.70926–2	5.82354–2	5.90598–2	5.96715–2
34	2.24916–2	3.62675–2	3.94376–2	4.52648–2	4.90070–2	5.14733–2	5.31534–2	5.43361–2	5.51942–2	5.58335–2

FINITE DIFFERENCES

Uniform interval h.

If a function $f(x)$ is tabulated at a uniform interval h, that is, for arguments given by $x_n = x_0 + nh$, where n is an integer, then the function $f(x)$ may be denoted by f_n.

This can be generalized so that for all values of p, and in particular for $0 \leqq p \leqq 1$,

$$f(x_0 + ph) = f(x_p) = f_p \; ,$$

where the argument designated x_0 can be chosen quite arbitrarily.

The notation $x = a(h)b(H)C$ means that the arguments run from a to b inclusive at the uniform interval h and thereafter continue to C at the uniform interval H.

The statement $f(x)$ $4D$ means that the function $f(x)$ is to be tabulated and the answer rounded off to four places of decimals.

The statement $f(x)$ $5S$ means that the function $f(x)$ is to be tabulated with the answer being rounded off to five significant figures.

The following table lists and defines the standard operators used in numerical analysis.

Symbol	Function	Definition
E	Displacement	$Ef_p = f_{p+1}$
Δ	Forward difference	$\Delta f_p = f_{p+1} - f_p$
∇	Backward difference	$\nabla f_p = f_p - f_{p-1}$
$\wedge\!\!\!\!\mid$	Divided difference	See page 821
δ	Central difference	$\delta f_p = f_{p+\frac{1}{2}} - f_{p-\frac{1}{2}}$
μ	Average	$\mu f_p = \frac{1}{2}(f_{p+\frac{1}{2}} + f_{p-\frac{1}{2}})$
Δ^{-1}	Backward sum	$\Delta^{-1} f_p = \Delta^{-1} f_{p-1} + f_{p-1}$
∇^{-1}	Forward sum	$\nabla^{-1} f_p = \nabla^{-1} f_{p-1} + f_p$
δ^{-1}	Central sum	$\delta^{-1} f_p = \delta^{-1} f_{p-1} + f_{p-\frac{1}{2}}$
D	Differentiation	$Df_p = \frac{d}{dx} f(x) = \frac{1}{h} \cdot \frac{d}{dp} f_p$
$I(=D^{-1})$	Integration	$If_p = \int^{x_p} f(x)dx = h \int^p f_p dp$
$J(=\Delta D^{-1})$	Definite integration	$Jf_p = h \int_p^{p+1} f_p dp$

I, Δ^{-1}, ∇^{-1} and δ^{-1} all imply the existence of an arbitrary constant which is determined by the initial conditions of the problem.

Where no confusion can arise the f can be omitted as, for example in writing Δ_p for Δf_p.

Higher differences are formed by successive operations, e.g.,

$$\begin{aligned} \Delta^2 f_p &= \Delta_p^2 \\ &= \Delta \cdot \Delta_p \\ &= \Delta(f_{p+1} - f_p) \\ &= \Delta_{p+1} - \Delta_p \\ &= f_{p+2} - f_{p+1} - f_{p+1} + f_p \; . \end{aligned}$$

Thus $\Delta_p^2 = f_{p+2} - 2f_{p+1} + f_p$.

Note that $f_p \equiv \Delta_p^0 \equiv \nabla_p^0 \equiv \delta_p^0$.

The disposition of the differences and sums relative to the function values is as shown (the arguments are omitted in these cases in the interests of clarity). In manuscript working double spacing is recommended.

Forward difference scheme

Δ_{-1}^{-2}		f_{-2}		Δ_{-3}^{2}	
	Δ_{-1}^{-1}		Δ_{-2}		Δ_{-3}^{3}
Δ_{0}^{-2}		f_{-1}		Δ_{-2}^{2}	
	Δ_{0}^{-1}		Δ_{-1}		Δ_{-2}^{3}
Δ_{1}^{-2}		f_{0}		Δ_{-1}^{2}	
	Δ_{1}^{-1}		Δ_{0}		Δ_{-1}^{3}
Δ_{2}^{-2}		f_{1}		Δ_{0}^{2}	
	Δ_{2}^{-1}		Δ_{1}		Δ_{0}^{3}
Δ_{3}^{-2}		f_{2}		Δ_{1}^{2}	

Backward difference scheme

∇_{-3}^{-2}		f_{-2}		∇_{-1}^{2}	
	∇_{-2}^{-1}		∇_{-1}		∇_{0}^{3}
∇_{-2}^{-2}		f_{-1}		∇_{0}^{2}	
	∇_{-1}^{-1}		∇_{0}		∇_{1}^{3}
∇_{-1}^{-2}		f_{0}		∇_{1}^{2}	
	∇_{0}^{-1}		∇_{1}		∇_{2}^{3}
∇_{0}^{-2}		f_{1}		∇_{2}^{2}	
	∇_{1}^{-1}		∇_{2}		∇_{3}^{3}
∇_{1}^{-2}		f_{2}		∇_{3}^{2}	

Central difference scheme

δ_{-2}^{-2}		f_{-2}		δ_{-2}^{2}		δ_{-2}^{4}
	$\delta_{-1\frac{1}{2}}^{-1}$		$\delta_{-1\frac{1}{2}}$		$\delta_{-1\frac{1}{2}}^{3}$	
δ_{-1}^{-2}		f_{-1}		δ_{-1}^{2}		δ_{-1}^{4}
	$\delta_{-\frac{1}{2}}^{-1}$		$\delta_{-\frac{1}{2}}$		$\delta_{-\frac{1}{2}}^{3}$	
δ_{0}^{-2}		f_{0}		δ_{0}^{2}		δ_{0}^{4}
	$\delta_{\frac{1}{2}}^{-1}$		$\delta_{\frac{1}{2}}$		$\delta_{\frac{1}{2}}^{3}$	
δ_{1}^{-2}		f_{1}		δ_{1}^{2}		δ_{1}^{4}
	$\delta_{1\frac{1}{2}}^{-1}$		$\delta_{1\frac{1}{2}}$		$\delta_{1\frac{1}{2}}^{3}$	
δ_{2}^{-2}		f_{2}		δ_{2}^{2}		δ_{2}^{4}

In the forward difference scheme the subscripts are seen to move forward into the difference table and no fractional subscripts occur. In the backward difference scheme the subscripts lie on diagonals slanting backwards into the table while in the central difference scheme the subscripts maintain their position and the odd order subscripts are fractional.

All three however are merely alternative ways of labeling the same numerical quantities as any difference is the result of subtracting the number diagonally above it in the preceding column from that diagonally below it in the preceding column or, alternatively, it is the sum of the number diagonally above it in the subsequent column with that immediately above it in its own column.

In general $\Delta^n_{p-\frac{1}{2}n} \equiv \delta^n_p \equiv \nabla^n_{p+\frac{1}{2}n}$.

If a polynomial of degree r is tabulated exactly i.e., without any round-off errors, then the rth differences are constant.

Tabulate $f(x)3D$ for $x = -.4\ (.2)\ .8$ where $f(x) = x^3 - .3x^2 + .07x - .123$.

x	f	δ	δ^2	δ^3	δ^4
$-.4$	$-.263$				
		106			
$-.2$	$-.157$		-72		
		34		48	
$+0$	$-.123$		-24		0
		10		48	
.2	$-.113$		$+24$		0
		34		48	
.4	$-.079$		72		0
		106		48	
.6	$+.027$		120		
		226			
.8	.253				

Had the function been rounded off to two places of decimals then the third differences would not have been constant. The differences are conventionally expressed in units of the least significant decimal as shown. This obviates the need to write down non-significant zeros. In any well tabulated function the differences decrease in magnitude.

The effect of an error term e in any function value is superimposed upon the difference table in a manner shown in the following table:

f	δ	δ^2	δ^3	δ^4	δ^5
0		0		0	
	0		0		$+1e$
0		0		$+1e$	
	0		$+1e$		$-5e$
0		$+1e$		$-4e$	
	$+1e$		$-3e$		$+10e$
$+e$		$-2e$		$+6e$	
	$-1e$		$+3e$		$-10e$
0		$+1e$		$-4e$	
	0		$-1e$		$+5e$
0		0		$+1e$	
	0		0		$-1e$
0		0		0	

In particular since round-off can contribute an error of half a unit (of the least significant decimal) in the function values, reference to the table of binomial coefficients will show that the maximum effect of this upon the twelfth differences (say) is $924(\frac{1}{2}) = 462$. This could give trouble if the problem does not require this to be multiplied by a small constant and is in fact one of the reasons why numerical differentiation is unreliable.

The following table enables the simpler operators to be expressed in terms of the others:

	E	Δ	δ,μ	∇
E	—	$1+\Delta$	$1+\mu\delta+\frac{1}{2}\delta^2$	$(1-\nabla)^{-1}$
Δ	$E-1$	—	$\mu\delta+\frac{1}{2}\delta^2$	$\nabla(1-\nabla)^{-1}$
δ	$E^{\frac{1}{2}}-E^{-\frac{1}{2}}$	$\Delta(1+\Delta)^{-\frac{1}{2}}$	$2(\mu^2-1)^{\frac{1}{2}}$	$\nabla(1-\nabla)^{-\frac{1}{2}}$
∇	$-E^{-1}$	$\Delta(1+\Delta)^{-1}$	$\mu\delta-\frac{1}{2}\delta^2$	—
μ	$\frac{1}{2}(E^{\frac{1}{2}}+E^{-\frac{1}{2}})$	$\frac{1}{2}(2+\Delta)(1+\Delta)^{-\frac{1}{2}}$	$(1+\frac{1}{4}\delta^2)^{\frac{1}{2}}$	$\frac{1}{2}(2-\nabla)(1-\nabla)^{-\frac{1}{2}}$

In addition to the above there are other identities by means of which the above table can be extended, viz.,

$$E = e^{hD} = \Delta\nabla^{-1}$$

$$\mu = E^{-\frac{1}{2}} + \tfrac{1}{2}\delta = E^{\frac{1}{2}} - \tfrac{1}{2}\delta = \cosh\left(\tfrac{1}{2}hD\right)$$

$$\delta = E^{-\frac{1}{2}}\Delta = E^{\frac{1}{2}}\nabla = (\Delta\nabla)^{\frac{1}{2}} = 2\sinh\left(\tfrac{1}{2}hD\right).$$

Note the emergence of Taylor's series from

$$\begin{aligned} f_p &= E^p f_0 \\ &= e^{phD} f_0 \\ &= f_0 + phDf_0 + \frac{1}{2!}p^2h^2D^2f_0 + \cdots. \end{aligned}$$

FUNCTION BUILD-UP FROM DIFFERENCES

Backward differences will be used here since the notation is simpler in this case.

If a function is tabulated as far as f_0 say, then the difference table can be built up. Assuming the differences up to and including ∇_0'' have been formed, if ∇_1'' is known it follows from the definition of backward differences that f_1 can be found from

$$f_1 = f_0 + \nabla_0 + \nabla_0^2 + \nabla_0^3 + \nabla_0^4 + \cdots + \nabla_0^{n-1} + \nabla_1^n .$$

For example, in the cubic tabulated earlier if we take $x_0 = .8$ then $f_0 = .253$ and f_1 can be built-up using the above scheme as follows:

$$\begin{aligned} f_1 &= .253 + (226 + 120 + 48)(10^{-3}) \\ &= .647 \end{aligned}$$

INTERPOLATION

Finite difference interpolation entails taking a given set of points and fitting a function to them. This function is usualy a polynomial, although it is rarely found explicitly as such. Central differences formulae are in general to be preferred to those involving either forward or backward differences, although the latter may be essential at the extremities of a table where the central differences may not extend far enough.

The most useful of the central differences formulae are those of *Bessel* and *Everett*, the latter particularly so since it only utilizes even order differences. Nevertheless all central difference formulae stopping at the same difference are equivalent.

Notation used:

$$\Delta_{-m_r}^r \equiv \delta_{(r/2)-m_r}^r \equiv \nabla_{r-m_r}^r$$

$$p = \frac{x - x_0}{h} = 1 - q \quad (\text{usually } 0 \leqq p \leqq 1) .$$

Any number of finite difference interpolation formulae can be obtained from the following schematic (shown for central differences only, but easily extensible by the above identities):

$$\begin{aligned} y_p &= \sum_{r=0} \binom{p + m_{r-1}}{r} \cdot \delta_{(r/2)-m_r}^r \\ &= f_{-m_0} + \binom{p + m_0}{1} \delta_{\frac{1}{2}-m_1} + \binom{p + m_1}{2} \delta_{1-m_2}^2 + \binom{p + m_2}{3} \delta_{1\frac{1}{2}-m_3}^3 \cdots , \end{aligned}$$

where the m_r are integers so chosen that the binomial coefficient term $\binom{p + m_{r-1}}{r}$ contains all the linear factors occurring in the preceding term $\binom{p + m_{r-2}}{r - 1}$. Obviously $m_{r+1} = m_r$ or $1 + m_r$.

Further formulae are obtained by taking linear combinations of the formulae obtained from the above schematic, or by using the standard identity transformations upon chosen terms of one of the formulae.

Another way of obtaining interpolation formulae is to apply the standard identity transformations to the operator E in the operational form of the interpolation formula, viz., $f_p = E^p f_0$.

Newton's forward formula

$$f_p = f_0 + p\Delta_0 + \frac{1}{2!} p(p - 1)\Delta_0^2 + \frac{1}{3!} p(p - 1)(p - 2)\Delta_0^3 \cdots \quad 0 \leqq p \leqq 1$$

Newton's backward formula

$$f_p = f_0 + p\nabla_0 + \frac{1}{2!}p(p+1)\nabla_0^2 + \frac{1}{3!}p(p+1)(p+2)\nabla_0^3 \cdots \qquad 0 \leqq p \leqq 1$$

Gauss' forward formula

$$f_p = f_0 + p\delta_{\frac{1}{2}} + G_2\delta_0^2 + G_3\delta_{\frac{1}{2}}^3 + G_4\delta_0^4 + G_5\delta_{\frac{1}{2}}^5 \cdots \qquad 0 \leqq p \leqq 1$$

Gauss' backward formula

$$f_p = f_0 + p\delta_{-\frac{1}{2}} + G_2^*\delta_0^2 + G_3\delta_{-\frac{1}{2}}^3 + G_4^*\delta_0^4 + G_5\delta_{-\frac{1}{2}}^5 \cdots \qquad 0 \leqq p \leqq 1$$

In the above $G_{2n} = \binom{p+n-1}{2n}$

$$G_{2n}^* = \binom{p+n}{2n}$$

$$G_{2n+1} = \binom{p+n}{2n+1}$$

Stirling's formula

$$f_p = f_0 + \tfrac{1}{2}p(\delta_{\frac{1}{2}} + \delta_{-\frac{1}{2}}) + \tfrac{1}{2}p^2\delta_0^2 + S_3(\delta_{\frac{1}{2}}^3 + \delta_{-\frac{1}{2}}^3) + S_4\delta_0^4 + \cdots \qquad -\tfrac{1}{2} \leqq p \leqq \tfrac{1}{2}$$

Steffenson's formula

$$f_p = f_0 + \tfrac{1}{2}p(p+1)\delta_{\frac{1}{2}} - \tfrac{1}{2}(p-1)p\delta_{-\frac{1}{2}} + (S_3 + S_4)\delta_{\frac{1}{2}}^3 + (S_3 - S_4)\delta_{-\frac{1}{2}}^3 \cdots \quad -\tfrac{1}{2} \leqq p \leqq \tfrac{1}{2}.$$

In the above $S_{2n+1} = \frac{1}{2}\binom{p+n}{2n+1}$

$$S_{2n+2} = \frac{p}{2n+2}\binom{p+n}{2n+1}$$

$$S_{2n+1} + S_{2n+2} = \binom{p+n+1}{2n+2}$$

$$S_{2n+1} - S_{2n+2} = -\binom{p+n}{2n+2}$$

Bessel's formula

$$f_p = f_0 + p\delta_{\frac{1}{2}} + B_2(\delta_0^2 + \delta_1^2) + B_3\delta_{\frac{1}{2}}^3 + B_4(\delta_0^4 + \delta_1^4) + B_5\delta_{\frac{1}{2}}^5 + \cdots \qquad 0 \leqq p \leqq 1$$

Everett's formula

$$f_p = (1-p)f_0 + pf_1 + E_2\delta_0^2 + F_2\delta_1^2 + E_4\delta_0^4 + F_4\delta_1^4 + E_6\delta_0^6 + F_6\delta_1^6 + \cdots \qquad 0 \leqq p \leqq 1$$

The coefficients in the above two formulae are related to each other and to the coefficients in the Gaussian formulae by the identities

$$\begin{aligned}
B_{2n} &\equiv \tfrac{1}{2}G_{2n} \equiv \tfrac{1}{2}(E_{2n} + F_{2n}) \\
B_{2n+1} &\equiv G_{2n+1} - \tfrac{1}{2}G_{2n} \equiv \tfrac{1}{2}(F_{2n} - E_{2n}) \\
E_{2n} &\equiv G_{2n} - G_{2n+1} \equiv B_{2n} - B_{2n+1} \\
F_{2n} &\equiv G_{2n+1} \equiv B_{2n} + B_{2n+1}
\end{aligned}$$

Also for $q \equiv 1 - p$ the following symmetrical relationships hold:

$$\begin{aligned}
B_{2n}(p) &\equiv B_{2n}(q) \\
B_{2n+1}(p) &\equiv -B_{2n+1}(q) \\
E_{2n}(p) &\equiv F_{2n}(q) \\
F_{2n}(p) &\equiv E_{2n}(q)
\end{aligned}$$

as can be seen from the tables of these coefficients.

p	B_2	B_3	B_4	B_5	B_6	B_7	$(1-p)$
0.00	−0.000000	0.0000000	0.000000	−0.000000	−0.00000	0.00000	1.00
0.01	2475	8085	415	81	8	1	0.99
0.02	4900	15680	825	158	17	2	0.98
0.03	7275	22795	1230	231	25	3	0.97
0.04	9600	29440	1631	300	33	4	0.96
0.05	−0.011875	0.0035625	0.002026	−0.000365	−0.00041	0.00005	0.95
0.06	14100	41360	2416	425	49	6	0.94
0.07	16275	46655	2801	482	57	7	0.93
0.08	18400	51520	3180	534	64	8	0.92
0.09	20475	55965	3552	583	72	8	0.91
0.10	−0.022500	0.0060000	0.003919	−0.000627	−0.00080	0.00009	0.90
0.11	24475	63635	4279	667	87	10	0.89
0.12	26400	66880	4632	704	94	10	0.88
0.13	28275	69745	4979	737	101	11	0.87
0.14	30100	72240	5319	766	109	11	0.86
0.15	−0.031875	0.0074375	0.005651	−0.000791	−0.00115	0.00012	0.85
0.16	33600	76160	5976	813	122	12	0.84
0.17	35275	77605	6294	831	129	12	0.83
0.18	36900	78720	6604	845	135	12	0.82
0.19	38475	79515	6906	856	142	13	0.81
0.20	−0.040000	0.0080000	0.007200	−0.000864	−0.00148	0.00013	0.80
0.21	41475	80185	7486	868	154	13	0.79
0.22	42900	80080	7763	870	160	13	0.78
0.23	44275	79695	8033	868	165	13	0.77
0.24	45600	79040	8293	862	171	13	0.76
0.25	−0.046875	0.0078125	0.008545	−0.000554	−0.00176	0.00013	0.75
0.26	48100	76960	8788	844	181	12	0.74
0.27	49275	75555	9022	830	186	12	0.73
0.28	50400	73920	9247	814	191	12	0.72
0.29	51475	72065	9462	795	196	12	0.71
0.30	−0.052500	0.0070000	0.009669	−0.000773	−0.00200	0.00011	0.70
0.31	53475	67735	9866	750	204	11	0.69
0.32	54400	65280	10053	724	208	11	0.68
0.33	55275	62645	10231	696	212	10	0.67
0.34	56100	59840	10399	666	216	10	0.66
0.35	−0.056875	0.0056875	0.010557	−0.000633	−0.00219	0.00009	0.65
0.36	57600	53760	10706	600	222	9	0.64
0.37	58275	50505	10844	564	225	8	0.63
0.38	58900	47120	10973	527	228	8	0.62
0.39	59475	43615	11092	488	231	7	0.61
0.40	−0.060000	0.0040000	0.011200	−0.000448	−0.00233	0.00007	0.60
0.41	60475	36285	11298	407	235	6	0.59
0.42	60900	32480	11386	364	237	5	0.58
0.43	61275	28595	11464	321	239	5	0.57
0.44	61600	24640	11532	277	240	4	0.56
0.45	−0.061875	0.0020625	0.011589	−0.000232	−0.00241	0.00003	0.55
0.46	62100	16560	11635	186	242	3	0.54
0.47	62275	12455	11672	140	243	2	0.53
0.48	62400	8320	11698	94	244	1	0.52
0.49	62475	4165	11714	47	244	1	0.51
0.50	−0.062500	0.0000000	0.011719	−0.000000	−0.00244	0.00000	0.50
$(1-p)$	B_2	$-B_3$	B_4	$-B_5$	B_6	$-B_7$	p

Bessel's formula (unmodified)

$$(a)\ f_p = f_0 + p\delta_{\frac{1}{2}} + B_2(\delta_0^2 + \delta_1^2) + B_3\delta_{\frac{1}{2}}^3 + B_4(\delta_0^4 + \delta_1^4) + B_5\delta_{\frac{1}{2}}^5 + B_6(\delta_0^6 + \delta_1^6) + B_7\delta_{\frac{1}{2}}^7 + \cdots .$$

The end terms can be neglected,i.e.,they contribute less than half a unit in the last place of decimals, if the differences are less than the values shown in the following list:

Neglect	$B_7\delta_{\frac{1}{2}}^7$ if	$\delta^7 < 3500$
	$B_6(\delta_0 + \delta_1)$	$\delta^6 < 100$
	$B_5\delta_{\frac{1}{2}}^5$	$\delta^5 < 500$
	$B_4(\delta_0^4 + \delta_1^4)$	$\delta^4 < 20$
	$B_3\delta_{\frac{1}{2}}^3$	$\delta^3 < 60$
	$B_2(\delta_0^2 + \delta_1^2)$	$\delta^2 < 4$

A powerful simplification is that of throwback. If the higher differences are less than a specified value they can be combined with the lower order ones as follows:

If $\delta^6 < 10{,}000$ use $\delta_m^4 \equiv \delta^4 - 0.20697\delta^6$

and if $\delta^6 < 1{,}000$* use $\delta_m^4 \equiv \delta^4 - 0.2\delta^6$
in the modified form of Bessel's formula

$$(b)\ f_p = f_0 + p\delta_{\frac{1}{2}} + B_2(\delta_0^2 + \delta_1^2) + B_3\delta_{\frac{1}{2}}^3 + B_4(\delta_{m0}^4 + \delta_{m1}^4) + B_5\delta_{\frac{1}{2}}^5 .$$

If $\delta^4 < 1{,}000$ and $\delta^6 < 1{,}000$* use $\delta_m^2 \equiv \delta^2 - 0.18393\delta^4 + 0.03808\delta^6$
and if $\delta^4 < 1{,}000$* use $\delta_m^2 \equiv \delta^2 - 0.18393\delta^4$
in the modified form of Bessel's formula

$$(c)\ f_p = f_0 + p\delta_{\frac{1}{2}} + B_2(\delta_{m0}^2 + \delta_{m1}^2) + B_3\delta_{\frac{1}{2}}^3 .$$

* Higher even order differences are negligible. Throwback involving odd order differences is of little practical use.

p	E_2	F_2	E_4	F_4	E_6	F_6	$(1-p)$
0.00	−0.000000	−0.0000000	0.000000	0.000000	−0.00000	−0.00000	1.00
0.01	32835	16665	496	33	9	7	0.99
0.02	64680	33320	983	67	19	14	0.98
0.03	95545	49955	1461	100	28	21	0.97
0.04	125440	66560	1931	133	37	29	0.96
0.05	−0.0154375	−0.0083125	0.002391	0.001661	−0.00046	−0.00036	0.95
0.06	182360	99640	2842	199	55	43	0.94
0.07	209405	116095	3283	232	64	50	0.93
0.08	235520	132480	3714	264	72	57	0.92
0.09	260715	148785	4135	297	80	64	0.91
0.10	−0.0285000	−0.0165000	0.004546	0.003292	−0.00089	−0.00070	0.90
0.11	308385	181115	4946	361	97	77	0.89
0.12	330880	197120	5336	393	105	84	0.88
0.13	352495	213005	5716	424	112	91	0.87
0.14	373240	228760	6085	455	120	97	0.86
0.15	−0.0393125	−0.0244375	0.006442	0.004860	−0.00127	−0.00104	0.85
0.16	412160	259840	6789	516	134	110	0.84
0.17	430355	275145	7125	546	141	117	0.83
0.18	447720	290280	7449	576	148	123	0.82
0.19	464265	305235	7762	605	154	129	0.81
0.20	−0.0480000	−0.0320000	0.008064	0.006336	−0.00161	−0.00135	0.80
0.21	494935	334565	8354	662	167	141	0.79
0.22	509080	348920	8633	689	172	147	0.78
0.23	522445	363055	8900	716	178	153	0.77
0.24	535040	376960	9156	743	184	158	0.76
0.25	−0.0546875	−0.0390625	0.009399	0.007690	−0.00189	−0.00164	0.75
0.26	557960	404040	9632	794	194	169	0.74
0.27	568305	417195	9852	819	199	174	0.73
0.28	577920	430080	10060	843	203	179	0.72
0.29	586815	442685	10257	867	207	184	0.71
0.30	−0.0595000	−0.0455000	0.010442	0.008895	−0.00212	−0.00189	0.70
0.31	602485	467015	10615	912	215	193	0.69
0.32	609280	478720	10777	933	219	198	0.68
0.33	615395	490105	10927	954	222	202	0.67
0.34	620840	501160	11065	973	226	206	0.66
0.35	−0.0625625	−0.0511875	0.011191	0.009924	−0.00229	−0.00210	0.65
0.36	629760	522240	11305	1011	231	213	0.64
0.37	633255	532245	11408	1028	234	217	0.63
0.38	636120	541880	11500	1045	236	220	0.62
0.39	638365	551135	11580	1060	238	223	0.61
0.40	−0.0640000	−0.0560000	0.011648	0.010752	−.00240	−0.00226	0.60
0.41	641035	568465	11705	1089	241	229	0.59
0.42	641480	576520	11751	1102	242	232	0.58
0.43	641345	584155	11785	1114	243	234	0.57
0.44	640640	591360	11808	1125	244	236	0.56
0.45	−0.0639375	−0.0598125	0.011820	0.011357	−0.00245	−0.00238	0.55
0.46	637560	604440	11822	1145	245	240	0.54
0.47	635205	610295	11812	1153	245	241	0.53
0.48	632320	615680	11792	1160	245	242	0.52
0.49	628915	620585	11760	1167	245	243	0.51
0.50	−0.0625000	−0.0625000	0.011719	0.011719	−0.00244	−0.00244	0.50
$(1-p)$	F_2	E_2	F_4	E_4	F_6	E_6	p

Everett's formula (unmodified)

(a) $f_p = (1-p)f_0 + pf_1 + E_2\delta_0^2 + F_2\delta_1^2 + E_4\delta_0^4 + F_4\delta_1^4 + E_6\delta_0^6 + F_6\delta_1^6 + \cdots$

As with Bessel's formula, end terms can be neglected as follows:

Neglect	$E_6\delta_0^6 + F_6\delta_1^6$	if	$\delta^6 < 100$
	$E_4\delta_0^4 + F_4\delta_1^4$		$\delta^4 < 20$
	$E_2\delta_0^2 + F_2\delta_1^2$		$\delta^2 < 4$

The corresponding throwback formulae are

(b) $f_p = (1-p)f_0 + pf_1 + E_2\delta_0^2 + F_2\delta_1^2 + E_4\delta_{m0}^4 + F_4\delta_{m1}^4,$

where $\delta_m^4 \equiv \delta^4 - 0.20697\delta^6$ providing $\delta^6 < 10{,}000$*

or $\delta_m^4 \equiv \delta^4 - 0.2\delta^6$ providing $\delta^6 < 1{,}000$*;

(c) $f_p = (1-p)f_0 + pf_1 + E_2\delta_{m0}^4 + F_2\delta_{m1}^2$

where $\delta_m^2 \equiv \delta^2 - 0.18393\delta^4 + 0.03808\delta^6$ if $\delta^4, \delta^6 < 1{,}000$*.

Generalized Throwback

Using 0.184 instead of the more precise value of 0.18393 we obtain a very simple but nevertheless accurate interpolation formulae

$$f \equiv (1-p)f_0 + pf_1 + E_2\delta_{m0}^2 + F_2\delta_{m1}^2 + P_4\theta_0^4 + Q_4\theta_1^4,$$

where $\delta_m^2 \equiv \delta^2 - 0.184\delta^4 + 0.03882\delta^6 - 0.0083\delta^8 + 0.0019\delta^{10}$

and $100\theta^4 \equiv \delta^4 - 0.2783\delta^6 + 0.0685\delta^8 - 0.0168\delta^{10}$

and $P_4 \equiv 100(E_4 + 0.184E_2); \quad Q_4 \equiv 100(F_4 + 0.184F_2).$

Symmetric formulae for interpolation to halves (sub-tabulation to halves)

When $p = \frac{1}{2}$ is substituted into Bessel's interpolation formula the odd coefficients B_{2n+1} all become zero and the resulting formula is

$$f_{\frac{1}{2}} = \left(1 - \frac{1}{8}\delta^2 + \frac{3}{128}\delta^4 - \frac{5}{1024}\delta^6 + \frac{35}{32768}\delta^8 - \cdots\right)\mu f_{\frac{1}{2}}\,.$$

If this is truncated after one, two, three terms etc. the following formulae and error terms are obtained:

Formula	Error
$f_{\frac{1}{2}} = \frac{1}{2}(f_0 + f_1)$	$\frac{-1}{8}\mu\delta_{\frac{1}{2}}^2$
$f_{\frac{1}{2}} = (-f_{-1} + 9f_0 + 9f_1 - f_2)/16$	$\frac{+3}{128}\mu\delta_{\frac{1}{2}}^4$
$f_{\frac{1}{2}} = (3f_{-2} - 25f_{-1} + 150f_0 + 150f_1 - 25f_2 + 3f_3)/256$	$\frac{-5}{1024}\mu\delta_{\frac{1}{2}}^6$
$f_{\frac{1}{2}} = (-5f_{-3} + 49f_{-2} - 245f_{-1} + 1225f_0 + 1225f_1 - 245f_2 + 49f_3 - 5f_4)/2048$	$\frac{+35}{32768}\mu\delta_{\frac{1}{2}}^8$

---

* Higher order differences are negligible.

Similarly, unsymmetric formulae for subtabulation to halves can be obtained from Newton's formula by substituting $p = \frac{1}{2}$ and truncating, e.g.,

Error

$$f_{\frac{1}{2}} = \left(1 + \frac{1}{2}\Delta - \frac{1}{8}\Delta^2 + \frac{1}{16}\Delta^3 - \frac{5}{128}\Delta^4 + \frac{7}{256}\Delta^5 \cdots\right) f_0$$

$$f_{\frac{1}{2}} = (3f_0 + 6f_1 - f_2)/8 \qquad \frac{+1}{16}\Delta_0^3$$

$$f_{\frac{1}{2}} = (5f_0 + 15f_1 - 5f_2 + f_3)/16 \qquad \frac{-5}{128}\Delta_0^4$$

$$f_{\frac{1}{2}} = (35f_0 + 140f_1 - 70f_2 + 28f_3 - 5f_4)/128 \qquad \frac{+7}{256}\Delta_0^5$$

The formula obtained by considering the first two terms only of the series is identical with the first one derived from Bessel's formula.

Interpolation techniques which do not require the function to be tabulated for equal interval of the argument

a) Lagrangian Polynomials

The interpolated value is obtained from a set of points which bridge the required point. The nearer the argument of the required value is to the center of the range of arguments of the function values used the better.

For an odd number of points $$f(x) = \sum_{r=-n}^{n} L_r f_r \,.$$

For an even number of points $$f(x) = \sum_{r=-n}^{n+1} L_r f_r \,.$$

The interpolating polynomial is of degree $2n$ in the first case and $2n + 1$ in the second. Further, it must pass through the given points $f_{-n}, f_{1-n}, f_{2-n}, \cdots, f_n$ or f_{1+n} as the case may be.

The coefficient L_r associated with the function value f_r is given by

$$L_r = \frac{(x - x_{-n})(x - x_{1-n}) \cdots (x - x_{r-1})(x - x_{r+1}) \cdots (x - x_{n-1})(x - x_n)}{(x_r - x_{-n})(x_r - x_{1-n}) \cdots (x_r - x_{r-1})(x_r - x_{r+1}) \cdots (x_r - x_{n-1})(x_r - x_n)}$$

for the first case with an additional factor $(x - x_{1+n})/(x_r - x_{1+n})$ in the second.

The error involved is equal to the $(2n + 2)$th. derivative, at some point in the range, multiplied by

$$(x - x_{-n})(x - x_{1-n}) \cdots (x - x_{1-n})/(2n + 2)!$$

in the case of the even number of points, and the $(2n + 1)$th. derivative, again at some point in the range, multiplied by

$$(x - x_{-n})(x - x_{1-n}) \cdots (x - x_n)/(2n + 1)!$$

in the other case.

In both cases the multipliers are functions which oscillate with an amplitude which increases substantially as the argument of the required point departs from the center of the range. When the arguments are spaced at equal intervals h and the point x is nearer to x_0 than to the other points, then $L_r(p)$ can be tabulated for $p = (x - x_0)/h$.

If the degree of the approximating polynomial is known the method is very powerful, otherwise it is best avoided.

b) Divided differences

The layout of a divided difference table is similar to that of an ordinary finite difference table.

$$\begin{array}{cccccc}
x_{-1} & f_{-1} & & \Delta\!\!\!\!\Delta^2_{-1} & & \Delta\!\!\!\!\Delta^4_{-1} \\
 & & \Delta\!\!\!\!\Delta_{-\frac{1}{2}} & & \Delta\!\!\!\!\Delta_{-\frac{1}{2}} & \\
x_0 & f_0 & & \Delta\!\!\!\!\Delta^2_0 & & \Delta\!\!\!\!\Delta^4_0 \\
 & & \Delta\!\!\!\!\Delta_{\frac{1}{2}} & & \Delta\!\!\!\!\Delta^3_{\frac{1}{2}} & \\
x_1 & f_1 & & \Delta\!\!\!\!\Delta^2_1 & & \Delta\!\!\!\!\Delta^4_1
\end{array}$$

where the Δ's are defined as follows:

$$\Delta_r^0 \equiv f_r, \qquad \Delta_{r+\frac{1}{2}} \equiv (f_{r+1} - f_r)/(x_{r+1} - x_r),$$

and in general $$\Delta_r^{2n} \equiv (\Delta_{r+\frac{1}{2}}^{2n-1} - \Delta_{r-\frac{1}{2}}^{2n-1})/(x_{r+n} - x_{r-n})$$

and $$\Delta_{r+\frac{1}{2}}^{2n+1} \equiv (\Delta_{r+1}^{2n} - \Delta_r^{2n})/(x_{r+1+n} - x_{r-n}).$$

Divided differences can with advantage be replaced by adjusted divided differences.

c) *Adjusted divided differences*

These differences are a modified form of the above and, when the interval of tabulation becomes constant, they reduce to ordinary central differences. Formulae analogous to all the existing formulae involving forward, central and backward differences can be obtained by means of Sheppard's rules which are stated below.

x_{-1}	p_{-1}	f_{-1}		δ_{-1}^2		δ_{-1}^4
			$\delta_{-\frac{1}{2}}$		$\delta_{-\frac{1}{2}}^3$	
x_0	p_0	f_0		δ_0^2		δ_0^4
			$\delta_{\frac{1}{2}}$		$\delta_{\frac{1}{2}}^3$	
x_1	p_1	f_1		δ_1^2		δ_1^4
			$\delta_{1\frac{1}{2}}$		$\delta_{1\frac{1}{2}}^3$	
x_2	p_2	f_2		δ_2^2		δ_2^4

where h is the total range of arguments divided by the number of intervals in that range, the result being rounded to a suitable figure and $x_k \cong p_k \cdot h$ or $(x/h)_k = p_k + w_k$ where the w's are used to shift the origin, if necessary, and to round-off the p's.

Define $$\delta_r \equiv \frac{1}{(p_{r+\frac{1}{2}} - p_{r-\frac{1}{2}})} (f_{r+\frac{1}{2}} - f_{r-\frac{1}{2}})$$

and $$\delta_r^2 \equiv \frac{2}{(p_{r+1} - p_{r-1})} (\delta_{r+\frac{1}{2}} - \delta_{r-\frac{1}{2}}).$$

In general $$\delta_r^{2n+1} \equiv \frac{2n+1}{(p_{r+\frac{1}{2}+n} - p_{r-\frac{1}{2}-n})} (\delta_{r+\frac{1}{2}}^{2n} - \delta_{r-\frac{1}{2}}^{2n}),$$

$$\delta_r^{2n} \equiv \frac{2n}{(p_{r+n} - p_{r-n})} (\delta_{r+\frac{1}{2}}^{2n-1} - \delta_{r+\frac{1}{2}}^{2n-1}).$$

The formula will be in the form

$$f(x) = f_0 + \sum_{r=1}^{N} \left\{ \prod (p - p_j \cdot) \frac{\delta_s^r}{r!} \right\},$$

where the product part of each term of the formula consists of r factors and the p_j's and s are chosen as follows:

a) When $r = 1$, $j = 0$ and $s = +\frac{1}{2}$, or $-\frac{1}{2}$, thereafter:—

b) r is increased by unity and s is either increased or decreased by $\frac{1}{2}$. Thus the p's, involved in the new divided difference of the series, are all those which are necessary for the computation of the divided difference used in the *preceding term* plus one more added either to the beginning or to the end of the set.

c) The product part of the term consists of factors $(p - p_j)$ such that all the p's necessary for the evaluation of the *previous* divided difference are involved. For example, Gauss' forward difference interpolation formula in terms of adjusted divided differences is obtained by following the path shown below

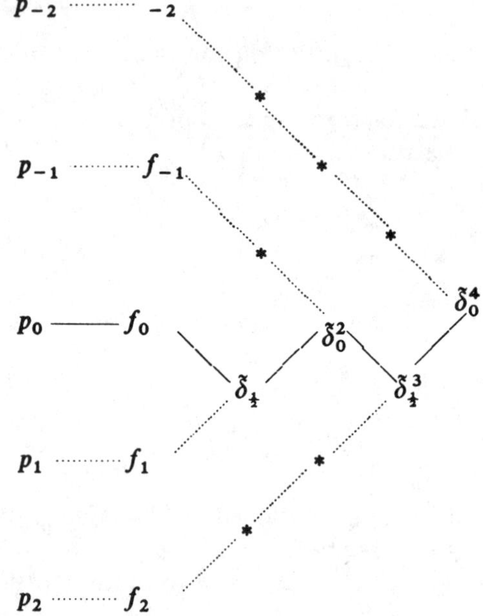

The solid line indicates the path chosen through the difference table.

$$f(x) = f_0 + (p - p_0)\tilde{\delta}_{\frac{1}{2}} + \frac{1}{2!}\cdot(p - p_0)(p - p_1)\tilde{\delta}_0^2 + \frac{1}{3!}\cdot(p - p_0)(p - p_1)(p - p_{-1})\tilde{\delta}_{\frac{1}{2}}^3$$
$$+ \frac{1}{4!}\cdot(p - p_0)(p - p_1)(p - p_{-1})(p - p_2)\tilde{\delta}_0^4 \cdots .$$

Similarly Gauss' backward difference interpolation formula is

$$f(x) = f_0 + (p - p_0)\tilde{\delta}_{-\frac{1}{2}} + \frac{1}{2!}\cdot(p - p_0)(p - p_{-1})\tilde{\delta}_0^2 + \frac{1}{3!}\cdot(p - p_0)(p - p_{-1})(p - p_1)\tilde{\delta}_{-\frac{1}{2}}^3$$
$$+ \frac{1}{4!}\cdot(p - p_0)(p - p_{-1})(p - p_1)(p - p_{-2})\tilde{\delta}_0^4 \cdots ,$$

and Stirling's formula is

$$f(x) = f_0 + \frac{(p - p_0)}{2}[\tilde{\delta}_{\frac{1}{2}} + \tilde{\delta}_{-\frac{1}{2}}] + \frac{1}{2!}\cdot\frac{(p - p_0)}{2}[(p - p_{-1}) + (p - p_1)]\tilde{\delta}_0^2$$
$$+ \frac{1}{3!}\cdot\frac{(p - p_1)(p - p_0)(p - p_{-1})}{2}[\tilde{\delta}_{\frac{1}{2}}^3 + \tilde{\delta}_{-\frac{1}{2}}^3]$$
$$+ \frac{1}{4!}\cdot\frac{(p - p_1)(p - p_0)(p - p_{-1})}{2}[(p - p_{-2}) + (p - p_2)]\tilde{\delta}_0^4 \cdots$$

Replacing the odd order differences, in Gauss' forward formula, by combinations of the previous even order differences the formula analogous to Everett's is obtained, i.e.,

$$f(x) = \left[1 - \frac{(p-p_0)}{(p_1-p_0)}\right] f_0 + \frac{(p-p_0)}{(p_1-p_0)} f_1 - \frac{1}{2!}\frac{(p-p_0)(p-p_1)(p-p_2)}{(p_2-p_{-1})}\delta_0^2$$
$$+ \frac{1}{2!}\frac{(p-p_0)(p-p_1)(p-p_{-1})}{(p_2-p_{-1})}\delta_1^2 - \cdots .$$

d) Iterative linear interpolation

Neville's modification of Aiken's method of iterative linear interpolation is one of the most powerful methods of interpolation when the arguments are unevenly spaced as no prior knowledge of the order of the approximating polynomial is necessary nor is a difference table required.

The values obtained are successive approximations to the required result and the process terminates when there is no appreciable change. These values are of course useless if a new interpolation is required when the procedure must be started afresh.

Defining

$$f_{r,s} \equiv \frac{(x_s - x)f_r - (x_r - x)f_s}{(x_s - x_r)}$$

$$f_{r,s,t} \equiv \frac{(x_t - x)f_{r,s} - (x_r - x)f_{s,t}}{(x_t - x_r)}$$

$$f_{r,s,t,u} \equiv \frac{(x_u - x)f_{r,s,t} - (x_r - x)f_{s,t,u}}{(x_u - x_r)},$$

the computation is laid out as follows:

$$\begin{array}{cccccc}
x_{-1} & (x_{-1} - x) & f_{-1} & & & \\
 & & & f_{-1,0} & & \\
x_0 & (x_0 - x) & f_0 & & f_{-1\,0,1} & \\
 & & & f_{0,1} & & f_{-1\,0\,1,2} \\
x_1 & (x_1 - x) & f_1 & & f_{0,1,2} & \\
 & & & f_{1,2} & & \\
x_2 & (x_2 - x) & f_2 & & &
\end{array}$$

As the iterates tend to their limit the common leading figures can be omitted.

e) Gauss' trigonometric interpolation formula

This is of greatest value when the function is periodic,i.e.,a Fourier series expansion is possible.

$$f(x) = \sum_{r=0}^{n} C_r f_r ,$$

where $C_r = N_r(x)/N_r(x_r)$ and

$$N_r(x) = \left[\sin\frac{(x-x_0)}{2}\right]\left[\sin\frac{(x-x_1)}{2}\right]\cdots\left[\sin\frac{(x-x_{r-1})}{2}\right]$$
$$\left[\sin\frac{(x-x_{r+1})}{2}\right]\cdots\left[\sin\frac{(x-x_n)}{2}\right].$$

This is similar to the Lagrangian formula.

f) Reciprocal differences

These are used when the quotient of two polynomials will give a better representation of the interpolating function than a simple polynomial expression.

A convenient layout is as shown below:

$$\begin{array}{llllll}
x_{-1} & f_{-1} & & & & \\
 & & \rho_{-\frac{1}{2}} & & & \\
x_0 & f_0 & & \rho_0^2 & & \\
 & & \rho_{\frac{1}{2}} & & \rho_{\frac{1}{2}}^3 & \\
x_1 & f_1 & & \rho_1^2 & & \rho_1^4 \\
 & & \rho_{1\frac{1}{2}} & & \rho_{1\frac{1}{2}}^3 & \\
x_2 & f_2 & & \rho_2^2 & & \\
 & & \rho_{2\frac{1}{2}} & & & \\
x_3 & f_3 & & & &
\end{array}$$

where
$$\rho_{r+\frac{1}{2}} \equiv \frac{x_{r+1} - x_r}{f_{r+1} - f_r}$$

and
$$\rho_r^2 \equiv \frac{x_{r+1} - x_{r-1}}{f_{r+\frac{1}{2}} - f_{r-\frac{1}{2}}} + f_r$$

In general
$$\rho_{r+\frac{1}{2}}^{2n+1} \equiv \frac{x_{r+n+1} - x_{r-n}}{\rho_{r+1}^{2n} - \rho_r^{2n}} + \rho_{r+\frac{1}{2}}^{2n-1}$$

$$\rho_r^{2n} \equiv \frac{x_{r+n} - x_{r-n}}{\rho_{r+\frac{1}{2}}^{2n-1} - \rho_{r-\frac{1}{2}}^{2n-1}} + \rho_r^{2n-2} .$$

The interpolation formula is expressed in the form of a continued fraction expansion.

The expansion corresponding to Newton's forward difference interpolation formula, in the sense of the differences involved, is

$$f(x) = f_0 + \cfrac{(x - x_0)}{\rho_{\frac{1}{2}} + \cfrac{(x_2 - x_1)}{\rho_1 - f_0 + \cfrac{(x - x_2)}{\rho_{1\frac{1}{2}}^3 - \rho_{\frac{1}{2}} + \cfrac{(x_4 - x_3)}{\rho_2^4 - \rho_1^2 + \cfrac{(x - x_4)}{\text{etc.}}}}}}$$

while that corresponding to Gauss' forward formula is

$$f(x) = f_0 + \cfrac{(x - x_0)}{\rho_{\frac{1}{2}} + \cfrac{(x_2 - x_1)}{\rho_0^2 - f_0 + \cfrac{(x_3 - x_{-1})}{\rho_{\frac{1}{2}}^3 - \rho_{\frac{1}{2}} + \cfrac{(x_4 - x_2)}{\rho_0^4 - \rho_0^2 + \cfrac{(x - x_{-2})}{\text{etc.}}}}}}$$

Inverse interpolation

Any method of interpolation which does not require the arguments to be evenly spaced will be satisfactory, by simply interchanging the roles of the arguments and the function values.

Alternative (i) Sub-tabulate the function until linear interpolation is adequate and simple ratio and proportion will yield the result.

Alternative (ii) Find an approximate value for p from

$$p \cong (f_p - f_0)/(f_1 - f_0)$$

and then iterate using *any* of the standard interpolation formulae;
e.g., Bessel

$$p = [f_p - f_0 - B_2(\delta_0^2 + \delta_1^2) - B_3\delta_{\frac{1}{2}}^3 - \cdots \text{ etc.}]/\delta_{\frac{1}{2}}$$

where the Bessel's coefficients on the right hand side are evaluated for the approximate p.

Everrett

$$p = [f_p - f_0 - E_2\delta_0^2 - F_2\delta_1^2 - E_4\delta^4 - F_4\delta_0^4 \cdots]/\delta_{\frac{1}{2}}$$

Newton forward

$$p = [f_p - f_0 - \frac{1}{2!}p(p-1)\Delta_0^2 - \frac{1}{3!}p(p-1)(p-2)\Delta_0^3 \cdots]/\Delta_0$$

Newton backward

$$p = [f_p - f_0 - \frac{1}{2!}p(p+1)\nabla_0^2 - \frac{1}{3!}p(p+1)(p+2)\nabla_0^3 \cdots]/\nabla_0 .$$

Note that the divisors are in fact identical.

DIFFERENTIATION FORMULAS

The notation used is defined as follows:

$$h^n \frac{d^n}{dx^n} f(x_0 + ph) \equiv h^n \frac{d^n}{dx^n} f_p \equiv \frac{d^n}{dp^n} f_p$$

Derivatives at tabular points.

$$hf_0^{(1)} = \left(\Delta - \frac{\Delta^2}{2} + \frac{\Delta^3}{3} - \frac{\Delta^4}{4} + \frac{\Delta^5}{5} - \frac{\Delta^6}{6} \cdots\right) f_0$$

$$= \left(\mu\delta - \tfrac{1}{6}\mu\delta^3 + \tfrac{1}{30}\mu\delta^5 - \tfrac{1}{140}\mu\delta^7 \cdots\right) f_0$$

$$= \left(\nabla + \frac{\nabla^2}{2} + \frac{\nabla^3}{3} + \frac{\nabla^4}{4} + \frac{\nabla^5}{5} + \frac{\nabla^6}{6} \cdots\right) f_0$$

$$h^2 f_0^{(2)} = \left(\Delta^2 - \Delta^3 + \tfrac{11}{12}\Delta^4 - \tfrac{5}{6}\Delta^5 + \tfrac{137}{180}\Delta^6 - \tfrac{7}{10}\Delta^7 \cdots\right) f_0$$

$$= \left(\delta^2 - \tfrac{1}{12}\delta^4 + \tfrac{1}{90}\delta^6 - \tfrac{1}{560}\delta^8 \cdots\right) f_0$$

$$= \left(\nabla^2 + \nabla^3 + \tfrac{11}{12}\nabla^4 + \tfrac{5}{6}\nabla^5 + \tfrac{137}{180}\nabla^6 + \tfrac{7}{10}\nabla^7 \cdots\right) f_0$$

$$h^3 f_0^{(3)} = \left(\Delta^3 - \tfrac{3}{2}\Delta^4 + \tfrac{21}{12}\Delta^5 - \tfrac{45}{24}\Delta^6 + \tfrac{348}{180}\Delta^7 - \tfrac{79}{40}\Delta^8 \cdots\right) f_0$$

$$= \left(\mu\delta^3 - \tfrac{1}{4}\mu\delta^5 + \tfrac{7}{120}\mu\delta^7 - \tfrac{41}{3024}\mu\delta^9 \cdots\right) f_0$$

$$= \left(\nabla^3 + \tfrac{3}{2}\nabla^4 + \tfrac{21}{12}\nabla^5 + \tfrac{45}{24}\nabla^6 + \tfrac{348}{180}\nabla^7 + \tfrac{79}{40}\nabla^8 \cdots\right) f_0$$

Higher derivatives can be found by direct multiplication of the above formulae e.g.

$$h^5 f_0^{(5)} = h^3 \frac{d^3}{dp^3}\left(h^2 \frac{d^2}{dp^2} f_0\right)$$

$$\therefore h^5 f_0^{(5)} = \left(\mu\delta^3 - \tfrac{1}{4}\mu\delta^5 + \tfrac{7}{120}\mu\delta^7 \cdots\right)\left(\delta^2 - \tfrac{1}{12}\delta^4 + \tfrac{1}{90}\delta^6 - \tfrac{1}{560}\delta^8 \cdots\right) f_0$$

$$= \left(\mu\delta^5 - \tfrac{1}{3}\mu\delta^7 + \tfrac{13}{144}\mu\delta^9 \cdots\right) f_0$$

The identity $\mu^2 \equiv 1 + \frac{1}{4}\delta^2$ is used to ensure that no powers of 'μ,' other than the first, appear in any formula.

$$h^4 f_0^{(4)} = \left(\delta^4 - \tfrac{1}{6}\delta^6 + \tfrac{7}{240}\delta^8 - \tfrac{41}{7560}\delta^{10} \cdots\right) f_0$$

$$h^6 f_0^{(6)} = \left(\delta^6 - \tfrac{1}{4}\delta^8 + \tfrac{13}{240}\delta^{10} \cdots\right) f_0$$

$$h^7 f_0^{(7)} = \left(\mu\delta^7 - \tfrac{5}{12}\mu\delta^9 + \tfrac{31}{240}\mu\delta^{11} \cdots\right) f_0$$

The preceding formulae can be truncated, as in the examples, and the differences expressed in terms of the function values. The resulting formulae are of the type

$$h^n f_0^{(n)} = \sum_{r=i}^{k} a_r \cdot f_r + \epsilon$$

where

$$\sum_{r=i}^{k} a_r = 0$$

and the principal part of the error 'ϵ' is the first neglected difference. It is still necessary for the differences to be converging.

In the following formulae only the order of the differences constituting the error term are indicated.

$$
\begin{aligned}
hf_0^{(1)} &\simeq (f_1 - f_0) - \tfrac{1}{2}\Delta^2\\
&\simeq \tfrac{1}{2}(-f_2 + 4f_1 - 3f_0) + \tfrac{1}{3}\Delta^3\\
&\simeq \tfrac{1}{6}(2f_3 - 9f_2 + 18f_1 - 11f_0) - \tfrac{1}{4}\Delta^4\\
&\simeq \tfrac{1}{2}(f_1 - f_{-1}) - \tfrac{1}{6}\mu\delta^3\\
&\simeq \tfrac{1}{12}(-f_2 + 8f_1 - 8f_{-1} + f_2) + \tfrac{1}{30}\mu\delta^5\\
&\simeq \tfrac{1}{60}(f_3 - 9f_2 + 45f_1 - 45f_{-1} + 9f_{-2} - f_{-3}) - \tfrac{1}{140}\mu\delta^7\\
&\simeq (f_0 - f_{-1}) + \tfrac{1}{2}\nabla^2\\
&\simeq \tfrac{1}{2}(3f_0 - 4f_{-1} + f_{-2}) + \tfrac{1}{3}\nabla^3\\
&\simeq \tfrac{1}{6}(11f_0 - 18f_{-1} + 9f_{-2} - 2f_{-3}) + \tfrac{1}{4}\nabla^4\\
h^2f_0^{(2)} &\simeq (f_2 - 2f_1 + f_0) - \Delta^3\\
&\simeq (-f_3 + 4f_2 - 5f_1 + 2f_0) + \tfrac{11}{12}\Delta^4\\
&\simeq (11f_4 - 56f_3 + 114f_2 - 104f_1 + 35f_0) - \tfrac{5}{6}\Delta^5\\
&\simeq (f_1 - 2f_0 + f_{-1}) - \tfrac{1}{12}\delta^4\\
&\simeq \tfrac{1}{12}(-f_2 + 16f_1 - 30f_0 + 16f_{-1} - f_{-2}) + \tfrac{1}{90}\delta^6\\
&\simeq \tfrac{1}{180}(2f_3 - 27f_2 + 270f_1 - 490f_0 + 270f_{-1} - 27f_{-2} + 2f_{-3}) - \tfrac{1}{560}\delta^8\\
&\simeq (f_0 - 2f_{-1} + f_{-2}) + \nabla^3\\
&\simeq (2f_0 - 5f_{-1} + 4f_{-2} - f_{-3}) + \tfrac{11}{12}\nabla^4\\
&\simeq (35f_0 - 104f_{-1} + 114f_{-2} - 56f_{-3} + 11f_{-4}) + \tfrac{5}{6}\nabla^5\\
h^3f_0^{(3)} &\simeq (f_3 - 3f_2 + 3f_1 - f_0) - \tfrac{3}{2}\Delta^4\\
&\simeq \tfrac{1}{2}(-3f_4 + 14f_3 - 24f_2 + 18f_1 - 5f_0) + \tfrac{21}{12}\Delta^5\\
&\simeq \tfrac{1}{2}(f_2 - 2f_1 + 2f_{-1} - f_{-2}) - \tfrac{1}{4}\mu\delta^5\\
&\simeq \tfrac{1}{8}(-f_3 + 8f_2 - 13f_1 + 13f_{-1} - 8f_{-2} + f_{-3}) + \tfrac{7}{120}\mu\delta^7\\
&\simeq (f_0 - 3f_{-1} + 3f_{-2} - f_{-3}) + \tfrac{3}{2}\nabla^4\\
&\simeq \tfrac{1}{2}(5f_0 - 18f_{-1} + 24f_{-2} - 14f_{-3} + 3f_{-4}) + \tfrac{21}{12}\nabla^5
\end{aligned}
$$

For derivatives at half-way points there exist analagous formulae. The operators act upon f_0 and f_1, in the case of the forward and backward differences, and upon $f_{1/2}$ in the case of the central differences δ_0^{2n+1} being meaningless.

$$
\begin{aligned}
hf_{1/2}^{(1)} &= (\Delta - \tfrac{1}{24}\Delta^3 + \tfrac{1}{24}\Delta^4 - \tfrac{71}{1920}\Delta^5 \cdots) f_0\\
&= (\delta - \tfrac{1}{24}\delta^3 + \tfrac{3}{640}\delta^5 - \tfrac{5}{7168}\delta^7 \cdots) f_{1/2}\\
&= (\nabla - \nabla^2 + \tfrac{23}{24}\nabla^3 + \tfrac{11}{12}\nabla^4 + \tfrac{563}{640}\nabla^5 \cdots) f_0\\
&= (\nabla - \tfrac{1}{24}\nabla^3 - \tfrac{1}{24}\nabla^4 - \tfrac{71}{1920}\nabla^5 \cdots) f_1\\
h^2f_{1/2}^{(2)} &= (\Delta^2 - \tfrac{1}{12}\Delta^4 + \tfrac{1}{12}\Delta^5 - \tfrac{13}{180}\Delta^6 \cdots) f_0\\
&= (\mu\delta^2 - \tfrac{5}{24}\mu\delta^4 + \tfrac{259}{5760}\mu\delta^6 - \tfrac{3229}{322560}\mu\delta^8 \cdots) f_{1/2}\\
&= (\nabla^2 - \tfrac{1}{12}\nabla^4 - \tfrac{1}{12}\nabla^5 - \tfrac{13}{180}\nabla^6 \cdots) f_1\\
h^3f_{1/2}^{(3)} &= (\Delta^3 - \tfrac{1}{8}\Delta^5 + \tfrac{1}{8}\Delta^6 - \tfrac{203}{1920}\Delta^7 \cdots) f_0\\
&= (\delta^3 - \tfrac{1}{8}\delta^5 + \tfrac{37}{1920}\delta^7 - \tfrac{2229}{967680}\delta^9 \cdots) f_{1/2}\\
&= (\nabla^3 - \tfrac{1}{8}\nabla^5 - \tfrac{1}{8}\nabla^6 - \tfrac{203}{1920}\nabla^7 \cdots) f_1
\end{aligned}
$$

As for tabular values, these formulae can be used to yield higher derivatives by multiplication of the operator series. Derivative formulae involving function values can also be obtained by truncation of the above series.

$$
\begin{aligned}
hf^{(1)}_{1/2} &\simeq \tfrac{1}{24}(-f_3 + 3f_2 + 21f_1 - 23f_0) - \tfrac{1}{24}\Delta^4 \\
&\simeq \tfrac{1}{24}(f_4 - 5f_3 + 9f_2 + 13f_1 - 22f_0) - \tfrac{71}{1920}\Delta^5 \\
&\simeq (f_1 - f_0) - \tfrac{1}{24}\delta^3 \\
&\simeq \tfrac{1}{24}(-f_2 + 27f_1 - 27f_0 + f_{-1}) + \tfrac{3}{640}\delta^5 \\
&\simeq \tfrac{1}{1920}(3f_3 - 95f_2 + 2190f_1 - 2190f_0 + 95f_{-1} - 3f_{-2}) - \tfrac{5}{7168}\delta^7 \\
&\simeq (f_0 - f_{-1}) + \nabla^2 \\
&\simeq \tfrac{1}{24}(23f_1 - 21f_0 - 3f_{-1} + f_{-2}) - \tfrac{1}{24}\nabla^4 \\
&\simeq \tfrac{1}{24}(22f_1 - 17f_0 - 9f_{-1} + 5f_{-2} - f_{-3}) - \tfrac{71}{1920}\nabla^5 \\
h^2 f^{(2)}_{1/2} &\simeq (f_2 - 2f_1 + f_0) - \tfrac{1}{12}\Delta^4 \\
&\simeq \tfrac{1}{12}(-f_4 + 4f_3 + 6f_2 - 20f_1 + 11f_0) + \tfrac{1}{12}\Delta^5 \\
&\simeq \tfrac{1}{2}(f_2 - f_1 - f_0 + f_{-1}) - \tfrac{5}{24}\mu\delta^4 \\
&\simeq \tfrac{1}{48}(5f_3 + 33f_2 - 38f_1 - 38f_0 + 33f_{-1} + 5f_{-2}) + \tfrac{259}{5760}\mu\delta^6 \\
&\simeq (f_1 - 2f_0 + f_{-1}) - \tfrac{1}{12}\nabla^4 \\
&\simeq \tfrac{1}{12}(11f_1 - 20f_0 + 6f_{-1} + 4f_{-2} - f_{-3}) - \tfrac{1}{12}\nabla^5 \\
h^3 f^{(3)}_{1/2} &\simeq (f_3 - 3f_2 + 3f_1 - f_0) - \tfrac{1}{8}\Delta^5 \\
&\simeq (f_2 - 3f_1 + 3f_0 - f_{-1}) - \tfrac{1}{8}\delta^5 \\
&\simeq \tfrac{1}{8}(-f_3 + 13f_2 - 34f_1 + 34f_0 - 13f_{-1} + f_{-2}) + \tfrac{37}{1920}\delta^7 \\
&\simeq (f_1 - 3f_0 + 3f_{-1} - f_{-2}) - \tfrac{1}{8}\nabla^5
\end{aligned}
$$

To calculate the derivative at an intermediate point 'p' the values of the derivative at tabular and half-way points can be evaluated and then interpolated, or the following formulae which involve the derivatives of Bessel's coefficients can be used. B_r' denotes the first derivative and B_r'' the second etc. Note that the formula for the fourth derivative involves $B_4'' - \frac{1}{24}$ which is the same as B_6^{IV}

$$
\begin{aligned}
hf_p' &= \delta_{1/2} + \tfrac{1}{2}(p - \tfrac{1}{2})(\delta_0^2 + \delta_1^2) + B_3'\delta_{1/2}^3 + B_4'(\delta_0^4 + \delta_1^4) + B_5'\delta_{1/2}^5 \\
&\quad + B_6'(\delta_0^6 + \delta_1^6) + B_7'\delta_{1/2}^7 + \cdots \\
h^2 f_p'' &= \delta_0^2 + p\delta_{1/2}^3 + B_4''(\delta_0^4 + \delta_1^4) + B_5''\delta_{1/2}^5 + B_6''(\delta_0^6 + \delta_1^6) + B_7''\delta_{1/2}^7 + \cdots \\
h^3 f_p''' &= \delta_{1/2}^3 + \tfrac{1}{2}(p - \tfrac{1}{2})(\delta_0^4 + \delta_1^4) + B_5'''\delta_{1/2}^5 + B_6'''(\delta_0^6 + \delta_1^6) + B_7'''\delta_{1/2}^7 + \cdots \\
h^4 f_p &= \delta_0^4 + p\delta_{1/2}^5 + (B_4'' - 0.417)(\delta_0^6 + \delta_1^6) + \cdots
\end{aligned}
$$

To find the value of p at which a derivative has a given value a technique of successive approximation using these tables can be employed as an alternative to inverse interpolation in a table of the appropriate derivative.

Starting values for this procedure are given by

$$p \simeq \tfrac{1}{2} + 2(h^n f_p^{(n)} - \delta_{1/2}^n)/(\delta_0^{n+1} + \delta_1^{n+1}) \quad \text{when } n \text{ is odd}$$

and

$$p \simeq (h^n f_p^{(n)} - \delta_0^n)/\delta_{1/2}^{n+1} \quad \text{when } n \text{ is even}$$

DIFFERENTIATION TABLES FOR NON-TABULAR POINTS

p	B'_3	B'_4	B'_5	B'_6	B'_7	$(1-p)$
.00	+.083333	+.041667	−.00833	−.0083	+.0012	1.00
.01	78383	41238	792	83	11	.99
.02	73533	40784	750	82	11	.98
.03	68783	40306	709	81	10	.97
.04	64133	39805	667	81	10	.96
.05	+.059583	+.039281	−.00626	−.0080	+.0009	.95
.06	55133	38735	585	79	9	.94
.07	50783	38166	544	78	8	.93
.08	46533	37576	504	77	7	.92
.09	42383	36965	464	76	7	.91
.10	+.038333	+.036333	−.00425	−.0075	+.0006	.90
.11	34383	35682	385	74	6	.89
.12	30533	35011	347	72	5	.88
.13	26783	34321	309	71	5	.87
.14	23133	33612	271	70	4	.86
.15	+.019583	+.032885	−.00234	−.0068	+.0004	.85
.16	16133	32141	198	67	3	.84
.17	12783	31380	162	66	3	.83
.18	9533	30603	128	64	2	.82
.19	6383	29809	93	63	1	.81
.20	+.003333	+.029000	−.00060	−.0061	+.0001	.80
.21	+ 383	28176	− 27	59	+ 1	.79
.22	− 2467	27337	+ 4	58	− 0	.78
.23	5217	26485	35	56	0	.77
.24	7867	25619	65	54	1	.76
.25	−.010417	+.024740	+.00094	−.0052	−.0001	.75
.26	12867	23848	123	51	2	.74
.27	15217	22944	150	49	2	.73
.28	17467	22029	176	47	3	.72
.29	19617	21103	201	45	3	.71
.30	−.021667	+.020167	+.00225	−.0043	−.0003	.70
.31	23617	19220	249	41	4	.69
.32	25467	18264	271	39	4	.68
.33	27217	17299	292	37	4	.67
.34	28867	16325	311	35	5	.66
.35	−.030417	+.015344	+.00330	−.0033	−.0005	.65
.36	31867	14355	348	31	5	.64
.37	33217	13359	364	29	5	.63
.38	34467	12356	380	27	6	.62
.39	35617	11347	394	24	6	.61
.40	−.036667	+.010333	+.00407	−.0022	−.0006	.60
.41	37617	9314	418	20	6	.59
.42	38467	8291	429	18	6	.58
.43	39217	7263	438	16	7	.57
.44	39867	6232	446	13	7	.56
.45	−.040417	+.005198	+.00453	−.0011	−.0007	.55
.46	40867	4161	459	9	7	.54
.47	41217	3123	463	7	7	.53
.48	41467	2083	466	4	7	.52
.49	41617	1042	468	2	7	.51
.50	−.041667	+.000000	+.00469	−.0000	−.0007	.50
$(1-p)$	B'_3	$-B'_4$	B'_5	$-B'_6$	B'_7	p

DIFFERENTIATION TABLES FOR NON-TABULAR POINTS

p	B_4''	B_5''	B_6''	B_7''	$(1-p)$
.00	−.041667	+.04167	+.0056	−.0056	1.00
.01	44142	4164	62	56	.99
.02	46567	4157	68	56	.98
.03	48942	4145	74	56	.97
.04	51267	4128	80	57	.96
.05	−.053542	+.04106	+.0086	−.0057	.95
.06	55767	4080	91	57	.94
.07	57942	4050	97	56	.93
.08	60067	4015	103	56	.92
.09	62142	3976	108	56	.91
.10	−.064167	+.03933	+.0113	−.0056	.90
.11	66142	3886	119	55	.89
.12	68067	3835	124	55	.88
.13	69942	3781	129	54	.87
.14	71767	3722	134	54	.86
.15	−.073542	+.03660	+.0139	−.0053	.85
.16	75267	3595	143	52	.84
.17	76942	3526	148	51	.83
.18	78567	3454	152	51	.82
.19	80142	3378	157	50	.81
.20	−.081667	+.03300	+.0161	−.0049	.80
.21	83142	3219	165	48	.79
.22	84567	3134	169	46	.78
.23	85942	3047	173	45	.77
.24	87267	2957	176	44	.76
.25	−.088542	+.02865	+.0180	−.0043	.75
.26	89767	2770	184	42	.74
.27	90942	2672	187	40	.73
.28	92067	2573	190	39	.72
.29	93142	2471	193	37	.71
.30	−.094167	+.02367	+.0196	−.0036	.70
.31	95142	2261	199	34	.69
.32	96067	2153	201	33	.68
.33	96942	2043	204	31	.67
.34	97767	1932	206	29	.66
.35	−.098542	+.01819	+.0209	−.0028	.65
.36	99267	1704	211	26	.64
.37	99942	1588	213	24	.63
.38	100567	1471	214	23	.62
.39	101142	1353	216	21	.61
.40	−.101667	+.01233	+.0218	−.0019	.60
.41	102142	1113	219	17	.59
.42	102567	991	220	15	.58
.43	102942	869	221	13	.57
.44	103267	746	222	11	.56
.45	−.103542	+.00623	+.0223	−.0010	.55
.46	103767	499	224	8	.54
.47	103942	375	224	6	.53
.48	104067	250	225	4	.52
.49	104142	125	225	2	.51
.50	−.104167	+.00000	+.0225	−.0000	.50
$(1-p)$	B_4''	$-B_5''$	B_6''	$-B_7''$	p

DIFFERENTIATION TABLES FOR NON-TABULAR POINTS

p	B_5'''	B_6'''	B_7'''	$(1-p)$
.00	−.00000	+.0625	−.0042	1.00
.01	495	617	33	.99
.02	980	608	25	.98
.03	1455	599	17	.97
.04	1920	590	9	.96
.05	−.02375	+.0580	−.0001	.95
.06	2820	571	+ 7	.94
.07	3255	561	14	.93
.08	3680	551	22	.92
.09	4095	540	29	.91
.10	−.04500	+.0530	+.0037	.90
.11	4895	519	44	.89
.12	5280	508	51	.88
.13	5655	497	58	.87
.14	6020	486	65	.86
.15	−.06375	+.0475	+.0071	.85
.16	6720	463	78	.84
.17	7055	451	84	.83
.18	7380	439	90	.82
.19	7695	427	96	.81
.20	−.08000	+.0415	+.0102	.80
.21	8295	403	108	.79
.22	8580	390	114	.78
.23	8855	377	119	.77
.24	9120	365	124	.76
.25	−.09375	+.0352	+.0129	.75
.26	9620	338	134	.74
.27	9855	325	139	.73
.28	10080	312	143	.72
.29	10295	299	148	.71
.30	−.10500	+.0285	+.0152	.70
.31	10695	271	156	.69
.32	10880	258	159	.68
.33	11055	244	163	.67
.34	11220	230	166	.66
.35	−.11375	+.0216	.0169	.65
.36	11520	202	172	.64
.37	11655	188	175	.63
.38	11780	174	178	.62
.39	11895	159	180	.61
.40	−.12000	+.0145	+.0182	.60
.41	12095	131	184	.59
.42	12180	116	186	.58
.43	12255	102	188	.57
.44	12320	87	189	.56
.45	−.12375	+.0073	+.0190	.55
.46	12420	58	191	.54
.47	12455	44	192	.53
.48	12480	29	192	.52
.49	12495	15	193	.51
.50	−.12500	+.0000	+.0193	.50
$(1-p)$	B_5'''	B_6'''	B_7'''	p

Caution

Numerical differentiation is error magnifying and many guarding figures are usually necessary. This is due to two factors namely:

a) the interval h enters each formula in the form $1/h^n$

b) each formula starts with the nth order difference.

Both of these combine to make numerical differentiation unreliable. e.g. Suppose the formula $f^{(2)} = (\Delta^2 - \Delta^3 + \frac{11}{12}\Delta^4)/h^2$ is used with $h = 0.1$ and an error of 5×10^{-4} can be tolerated in the derivative. For an error of e in each of the tabulated values

$$\begin{aligned} 5 \times 10^{-4} &= \tfrac{1}{0.01}(2e + 3e + \tfrac{11}{12}6e) \\ &= 100(\tfrac{21}{2}e) \\ &= 1050e \end{aligned}$$

and therefore $e \simeq 5 \times 10^{-7}$ i.e. the function must be tabulated to *six* places of decimals before the second derivative can be correct to *three*.

If, on the other hand, h had been unity in magnitude then

$$5 \times 10^{-4} = 10.5e$$

i.e. $e \simeq 5 \times 10^{-5}$ and the function would need only one guarding figure to produce the required degree of accuracy in the second derivative.

Obviously this effect is magnified for higher derivatives. In any case, the largest value of h, which permits the function to be well represented by a converging table of differences, should be used.

INTEGRATION FORMULAS

Integration is a smoothing procedure and as such errors in tabulated values tend to be cancelled out.

Notation used:

$$\int_{x_0}^{x_n} f(x)\cdot dx = h\int_0^n f_p \cdot dp$$

where

$$f_p \equiv f(x_0 + ph) \equiv f(x)$$

$$\begin{aligned} \int_0^n f_p \cdot dp &= \tfrac{1}{2}f_0 + f_1 + f_2 + \cdots + f_{n-1} + \tfrac{1}{2}f_n + \tfrac{1}{12}(\Delta_0 - \nabla_n) - \tfrac{1}{24}(\Delta_0^2 + \nabla_n^2) \\ &\quad + \tfrac{19}{720}(\Delta_0^3 - \nabla_n^3) - \tfrac{3}{160}(\Delta_0^4 + \nabla_n^4) \cdots \\ &= \tfrac{1}{2}f_0 + f_1 + f_2 + \cdots + f_{n-1} + \tfrac{1}{2}f_n + \tfrac{1}{12}(\mu\delta_0 - \mu\delta_n) \\ &\quad - \tfrac{11}{720}(\mu\delta_0^3 - \mu\delta_n^3) + \tfrac{191}{60480}(\mu\delta_0^5 - \mu\delta_n^5) - \cdots \end{aligned}$$

In particular

$$\begin{aligned} \int_0^1 f_p \cdot dp &= (1 + \tfrac{1}{2}\Delta - \tfrac{1}{12}\Delta^2 + \tfrac{1}{24}\Delta^3 - \tfrac{19}{720}\Delta^4 + \tfrac{3}{160}\Delta^5 - \tfrac{863}{60480}\Delta^6 + \cdots) f_0 \\ &= (1 + \tfrac{1}{2}\nabla + \tfrac{5}{12}\nabla^2 + \tfrac{3}{8}\nabla^3 + \tfrac{251}{720}\nabla^4 + \tfrac{95}{288}\nabla^5 + \tfrac{19087}{60480}\nabla^6 \cdots) f_0 \\ &= (1 - \tfrac{1}{2}\nabla - \tfrac{1}{12}\nabla^2 - \tfrac{1}{24}\nabla^3 - \tfrac{19}{720}\nabla^4 - \tfrac{3}{160}\nabla^5 - \tfrac{863}{60480}\nabla^6 \cdots) f_1 \\ &= (\mu - \tfrac{1}{12}\mu\delta^2 + \tfrac{11}{720}\mu\delta^4 - \tfrac{191}{60480}\mu\delta^6 + \cdots) f_{1/2} \end{aligned}$$

$$\begin{aligned} \int_0^{n+(1/2)} f_p \cdot dp &= \tfrac{1}{2}f_0 + f_1 + f_2 \cdots + f_n + (\tfrac{1}{12}\mu\delta - \tfrac{11}{720}\mu\delta^3 + \tfrac{191}{60480}\mu\delta^5 \cdots) f_0 \\ &\quad + (\tfrac{1}{24}\delta - \tfrac{17}{5760}\delta^3 + \tfrac{367}{967680}\delta^5 \cdots) f_{n+(1/2)} \end{aligned}$$

In particular

$$\int_0^{1/2} f_p \cdot dp = \tfrac{1}{2} f_0 + \left(\tfrac{1}{8}\delta - \tfrac{1}{24}\mu\delta^2 + \tfrac{1}{384}\delta^3 + \tfrac{11}{1440}\mu\delta^4 - \tfrac{13}{46080}\delta^5 \cdots\right) f_{1/2}$$

Symmetric range integrals

$$\int_{-1/2}^{1/2} f_p \cdot dp = \left(1 + \tfrac{1}{24}\delta^2 - \tfrac{17}{5760}\delta^4 + \tfrac{367}{967680}\delta^6 - \tfrac{27859}{464486400}\delta^8 \cdots\right) f_0$$

$$\int_{-1}^{1} f_p \cdot dp = 2\left(1 + \tfrac{1}{6}\delta^2 - \tfrac{1}{180}\delta^4 + \tfrac{1}{1512}\delta^6 - \tfrac{23}{226800}\delta^8 + \cdots\right) f_0$$

$$\int_{-2}^{2} f_p \cdot dp = 4\left(1 + \tfrac{2}{3}\delta^2 + \tfrac{7}{90}\delta^4 - \tfrac{2}{945}\delta^6 + \tfrac{13}{56700}\delta^8 - \cdots\right) f_0$$

$$\int_{-3}^{3} f_p \cdot dp = 6\left(1 + \tfrac{3}{2}\delta^2 + \tfrac{11}{20}\delta^4 + \tfrac{41}{840}\delta^6 - \tfrac{3}{2800}\delta^8 + \cdots\right) f_0$$

$$\int_{-4}^{4} f_p \cdot dp = 8\left(1 + \tfrac{8}{3}\delta^2 + \tfrac{217}{90}\delta^4 + \tfrac{92}{189}\delta^6 + \tfrac{989}{28350}\delta^8 + \cdots\right) f_0$$

The differences in the above formulae can be replaced by their function values, after truncation of the series, if extensive differencing is to be avoided.

When this is done, the formulae take on the forms

$$\sum_m A_m \cdot f_m$$

Some typical formulae of this type, used in the iterative starting of the solution of ordinary differential equations, can be found on pages 627. These are essentially quadrature formulae as are the predictor corrector formulae of pages 634.

When the limits involved in the integrals are not tabular or half-way points, formulae analagous to Bessel's interpolation formula can be used, the corresponding coefficients being designated by B^* etc. Single integration.

$$\int_0^{p} f_p \cdot dp = \tfrac{1}{2}p(f_0 + f_1)' + B_1^*\delta_{1/2} + B_2^*(\delta_0^2 + \delta_1^2) + B_3^*\delta_{1/2}^3 + B_4^*(\delta_0^4 + \delta_1^4) + B_5^*\delta_{1/2}^5 + B_6^*(\delta_0^6 + \delta_1^6)$$

$$\iint_0^{p} f_p \cdot d^2p = B_0^{**}(f_0 + f_1) + B_1^{**}\delta_{1/2} + B_2^{**}(\delta_0 + \delta_1) + B_3^{**}\delta_{1/2} + B_4^{**}(\delta_0 + \delta_1) + B_5^{**}\delta_{1/2}$$

INTEGRATION TABLES FOR NON-TABULAR POINTS

p	B_1^*	B_2^*	B_3^*	B_4^*	B_5^*	B_6^*
.00	−.000000	−.000000	+.000000	+.00000	−.00000	−.0000
.01	4950	12	4	0	0	0
.02	9800	49	16	1	0	0
.03	14550	110	35	2	0	0
.04	19200	195	61	3	1	0
.05	−.023750	−.000302	+.000094	+.00005	−.00001	−.0000
.06	28200	432	133	7	1	0
.07	32550	584	177	10	2	0
.08	36800	757	226	13	2	0
.09	40950	952	279	16	3	0
.10	−.045000	−.001167	+.000338	+.00020	−.00003	−.0000
.11	48950	1402	399	24	4	0
.12	52800	1656	465	29	5	1
.13	56550	1929	533	33	6	1
.14	60200	2221	604	39	6	1
.15	−.063750	−.002531	+.000677	+.00044	−.00007	−.0001
.16	67200	2859	753	50	8	1
.17	70550	3203	830	56	9	1
.18	73800	3564	908	62	10	1
.19	76950	3941	987	69	10	1
.20	−.080000	−.004333	+.001067	+.00076	−.00011	−.0002
.21	82950	4741	1147	84	12	2
.22	85800	5163	1227	91	13	2
.23	88550	5599	1307	99	14	2
.24	91200	6048	1386	107	15	2
.25	−.093750	−.006510	+.001465	+.00116	−.00016	−.0002
.26	96200	6985	1542	124	16	3
.27	98550	7472	1619	133	17	3
.28	100800	7971	1693	142	18	3
.29	102950	8480	1766	152	19	3
.30	−.105000	−.009000	+.001838	+.00161	−.00020	−.0003
.31	106950	9530	1906	171	20	4
.32	108800	10069	1973	181	21	4
.33	110550	10618	2037	191	22	4
.34	112200	11175	2098	202	23	4
.35	−.113750	−.011740	+.002157	+.00212	−.00023	−.0004
.36	115200	12312	2212	223	24	5
.37	116550	12891	2264	233	24	5
.38	117800	13477	2313	244	25	5
.39	118950	14069	2358	255	25	5
.40	−.120000	−.014667	+.002400	+.00266	−.00026	−.0005
.41	120950	15269	2438	278	26	6
.42	121800	15876	2473	289	27	6
.43	122550	16487	2503	301	27	6
.44	123200	17101	2530	312	27	6
.45	−.123750	−.017719	+.002552	+.00324	−.00028	−.0007
.46	124200	18339	2571	335	28	7
.47	124550	18961	2585	347	28	7
.48	124800	19584	2596	359	28	7
.49	124950	20208	2602	370	28	8
.50	−.125000	−.020833	+.002604	+.00382	−.00028	−.0008

INTEGRATION TABLES FOR NON-TABULAR POINTS (Continued)

p	B_1^*	B_2^*	B_3^*	B_4^*	B_5^*	B_6^*
.50	−.125000	−.020833	+.002604	+.00382	−.00028	−.0008
.51	124950	21458	2602	394	28	8
.52	124800	22083	2596	405	28	8
.53	124550	22706	2585	417	28	9
.54	124200	23328	2571	429	28	9
.55	−.123750	−.023948	+.002552	+.00440	−.00028	−.0009
.56	123200	24565	2530	452	27	9
.57	122550	25180	2503	463	27	10
.58	121800	25791	2473	475	27	10
.59	120950	26398	2438	486	26	10
.60	−.120000	−.027000	+.002400	+.00497	−.00026	−.0010
.61	118950	27597	2358	509	25	11
.62	117800	28189	2313	520	25	11
.63	116550	28775	2264	530	24	11
.64	115200	29355	2212	541	24	11
.65	−.113750	−.029927	+.002157	+.00552	−.00023	−.0011
.66	112200	30492	2098	562	23	12
.67	110550	31049	2037	573	22	12
.68	108800	31597	1973	583	21	12
.69	106950	32137	1906	593	20	12
.70	−.105000	−.032667	+.001838	+.00603	−.00020	−.0012
.71	102950	33187	1766	612	19	13
.72	100800	33696	1693	621	18	13
.73	98550	34194	1619	631	17	13
.74	96200	34681	1542	640	16	13
.75	−.093750	−.035156	+.001465	+.00648	−.00016	−.0013
.76	91200	35619	1386	657	15	14
.77	88550	36068	1307	665	14	14
.78	85800	36504	1227	673	13	14
.79	82950	36926	1147	680	12	14
.80	−.080000	−.037333	+.001067	+.00688	−.00011	−.0014
.81	76950	37726	987	695	10	14
.82	73800	38103	908	701	10	15
.83	70550	38464	830	708	9	15
.84	67200	38808	753	714	8	15
.85	−.063750	−.039135	+.000677	+.00720	−.00007	−.0015
.86	60200	39445	604	725	6	15
.87	56550	39737	533	730	6	15
.88	52800	40011	465	735	5	15
.89	48950	40265	399	740	4	15
.90	−.045000	−.040500	+.000337	+.00744	−.00003	−.0015
.91	40950	40715	279	748	3	15
.92	36800	40909	226	751	2	16
.93	32550	41083	177	754	2	16
.94	28200	41235	133	757	1	16
.95	−.023750	−.041365	+.000094	+.00759	−.00001	−.0016
.96	19200	41472	61	761	1	16
.97	14550	41556	35	762	0	16
.98	9800	41617	16	763	0	16
.99	4950	41654	4	764	0	16
1.00	−.000000	−.041667	+.000000	+.00764	−.00000	−.0016

INTEGRATION TABLES FOR NON-TABULAR POINTS

p	B_0^{**}	B_1^{**}	B_2^{**}	B_3^{**}	B_4^{**}	B_5^{**}
.00	+.000000	−.000000	−.000000	+.00000	+.00000	−.0000
.01	25	25	0	0	0	0
.02	100	99	0	0	0	0
.03	225	220	1	0	0	0
.04	400	389	3	0	0	0
.05	+.000625	−.000604	−.000005	+.00000	+.00000	−.0000
.06	900	864	9	0	0	0
.07	1225	1168	14	0	0	0
.08	1600	1515	20	1	0	0
.09	2025	1904	29	1	0	0
.10	+.002500	−.002333	−.000040	+.00001	+.00001	−.0000
.11	3025	2803	52	2	1	0
.12	3600	3312	68	2	1	0
.13	4225	3859	86	2	1	0
.14	4900	4443	106	3	2	0
.15	+.005625	−.005063	−.000130	+.00004	+.00002	−.0000
.16	6400	5717	157	4	3	0
.17	7225	6406	187	5	3	0
.18	8100	7128	221	6	4	0
.19	9025	7882	259	7	4	0
.20	+.010000	−.008667	−.000300	+.00008	+.00005	−.0000
.21	11025	9481	345	9	6	0
.22	12100	10325	395	10	7	0
.23	13225	11197	449	12	8	0
.24	14400	12096	507	13	9	0
.25	+.015625	−.013021	−.000570	+.00014	+.00010	−.0000
.26	16900	13971	637	16	11	0
.27	18225	14945	709	17	12	0
.28	19600	15941	787	19	14	0
.29	21025	16960	869	21	15	0
.30	+.022500	−.018000	−.000956	+.00023	+.00017	−.0000
.31	24025	19060	1049	25	19	0
.32	25600	20139	1147	26	20	0
.33	27225	21235	1250	28	22	0
.34	28900	22349	1359	31	24	0
.35	+.030625	−.023479	−.001474	+.00033	+.00026	−.0000
.36	32400	24624	1594	35	28	0
.37	34225	25783	1720	37	31	0
.38	36100	26955	1852	39	33	0
.39	38025	28138	1990	42	36	0
.40	+.040000	−.029333	−.002133	+.00044	+.00038	−.0000
.41	42025	30538	2283	47	41	0
.42	44100	31752	2439	49	44	1
.43	46225	32974	2601	51	47	1
.44	48400	34203	2768	54	50	1
.45	+.050625	−.035438	−.002943	+.00057	+.00053	−.0001
.46	52900	36677	3123	59	56	1
.47	55225	37921	3309	62	60	1
.48	57600	39168	3502	64	63	1
.49	60025	40417	3701	67	67	1
.50	+.062500	−.041667	−.003906	+.00069	+.00071	−.0001

INTEGRATION TABLES FOR NON-TABULAR POINTS (Continued)

p	B_0^{**}	B_1^{**}	B_2^{**}	B_3^{**}	B_4^{**}	B_5^{**}
.50	+.062500	−.041667	−.003906	+.00069	+.00071	−.0001
.51	65025	42916	4118	72	74	1
.52	67600	44165	4335	75	78	1
.53	70225	45412	4559	77	83	1
.54	72900	46656	4790	80	87	1
.55	+.075625	−.047896	−.005026	+.00082	+.00091	−.0001
.56	78400	49131	5268	85	96	1
.57	81225	50359	5517	87	100	1
.58	84100	51581	5772	90	105	1
.59	87025	52795	6033	92	110	1
.60	+.090000	−.054000	−.006300	+.00095	+.00115	−.0001
.61	93025	55195	6573	97	120	1
.62	96100	56379	6852	100	125	1
.63	99225	57550	7137	102	130	1
.64	102400	58709	7427	104	135	1
.65	+.105625	−.059854	−.007724	+.00106	+.00141	−.0001
.66	108900	60984	8026	108	146	1
.67	112225	62098	8334	110	152	1
.68	115600	63195	8647	112	158	1
.69	119025	64274	8966	114	164	1
.70	+.122500	−.065333	−.009290	+.00116	+.00170	−.0001
.71	126025	66373	9619	118	176	1
.72	129600	67392	9953	120	182	1
.73	133225	68389	10293	121	188	1
.74	136900	69363	10637	123	195	1
.75	+.140625	−.070313	−.010986	+.00125	+.00201	−.0001
.76	144400	71237	11340	126	207	1
.77	148225	72136	11699	127	214	1
.78	152100	73008	12062	129	221	1
.79	156025	73852	12429	130	228	1
.80	+.160000	−.074667	−.012800	+.00131	+.00234	−.0001
.81	164025	75451	13175	132	241	1
.82	168100	76205	13554	133	248	1
.83	172225	76927	13937	134	255	1
.84	176400	77616	14324	134	262	1
.85	+.180625	−.078271	−.014713	+.00135	+.00270	−.0001
.86	184900	78891	15106	136	277	1
.87	189225	79474	15502	136	284	1
.88	193600	80021	15901	137	291	1
.89	198025	80530	16302	137	299	1
.90	+.202500	−.081000	−.016706	+.00138	+.00306	−.0001
.91	207025	81430	17112	138	314	1
.92	211600	81819	17520	138	321	1
.93	216225	82165	17930	138	329	1
.94	220900	82469	18342	139	336	1
.95	+.225625	−.082729	−.018755	+.00139	+.00344	−.0001
.96	230400	82944	19169	139	351	1
.97	235225	83113	19584	139	359	1
.98	240100	83235	20000	139	367	1
.99	245025	83308	20417	139	374	1
1.00	+.250000	−.083333	−.020833	+.00139	+.00382	−.0001

GAUSS-TYPE WEIGHTS ABSCISSAE

Quadrature formulae using unevenly spaced ordinates. Some high accuracy formulae involve the use of abscissae which are unevenly spaced throughout the interval. The interval itself is usually subject to a transformation and various 'weighting functions' are involved.

To change the interval (a, b) in x to that of (θ, ϕ) in t the transformation

$$t = \frac{(a\theta - b\phi)}{(a - b)} + \frac{(\theta - \phi)x}{(a - b)}$$

is used.

Some of the more popular weighting functions and their associated intervals are given below. The general formulae are of two main types derived from

$$\int_a^b W(x) \cdot F(x) \cdot dx = \int_\theta^\phi \omega(t) \cdot f(t) \cdot dt$$

namely

$$\int_\theta^\phi \omega(t) \cdot f(t) \cdot dt = \sum_t \{H_t \cdot f(x_t)\}$$

and

$$\int_\theta^\phi \omega(t) \cdot f(t) \cdot dt = H \sum_t \{f(x_t) \pm f(-x_t)\}$$

$\omega(t)$	(θ, ϕ)	x_t are the zeros of:
1	$(-1, 1)$	$P_n(x)$
$t^{1/2}$	$(0, 1)$	$x^{-1/2} \cdot P_{2n+1}(\sqrt{x})$
$t^{-1/2}$	$(0, 1)$	$P_n(\sqrt{x})$
$(1 - t^2)^{1/2}$	$(-1, 1)$	$S_n(x)$
$(1 - t^2)^{-1/2}$	$(-1, 1)$	$T_n(x)$
e^{-t}	$(0, \infty)$	$L_n(x)$
e^{-t^2}	$(-\infty, \infty)$	$H_n(x)$

Some useful tables of abscissae x_t and their corresponding H_t are as follows.

Gaussian Quadrature.

$$\int_{-1}^{1} f(x) \cdot dx = \sum_{1}^{n} \{H_t \cdot f(x_t)\}$$

The x_t occur in pairs symmetrically placed with respect to the origin.

$\pm x_t$		H_t
	$n = 2$	
0.5773503		1.0000000
	$n = 3$	
0.7745967		0.5555556
0.0000000		0.8888889
	$n = 4$	
0.8611363		0.3478548
0.3399810		0.6521452
	$n = 5$	
0.9061798		0.2369269
0.5384693		0.4786287
0.0000000		0.5688889
	$n = 6$	
0.9324695		0.1713245
0.6612094		0.3607616
0.2386192		0.4679139
	$n = 7$	
0.9491079		0.1294850
0.7415312		0.2797054
0.4058452		0.3818301
0.0000000		0.4179592
	$n = 8$	
0.9602899		0.1012285
0.7966665		0.2223810
0.5255324		0.3137066
0.1834346		0.3626838
	$n = 9$	
0.9681602		0.0812744
0.8360311		0.1806482
0.6133714		0.2606107
0.3242534		0.3123471
0.0000000		0.3302394
	$n = 10$	
0.9739065		0.0666713
0.8650634		0.1494513
0.6794096		0.2190864
0.4333954		0.2692602
0.1488743		0.2955242

$\pm x_i$	H_i
$n = 11$	
0.9782287	0.0556686
0.8870626	0.1255804
0.7301520	0.1862902
0.5190961	0.2331938
0.2695432	0.2628045
0.0000000	0.2729251
$n = 12$	
0.9815606	0.0471753
0.9041173	0.1069393
0.7699027	0.1600783
0.5873180	0.2031674
0.3678315	0.2334925
0.1253334	0.2491470
$n = 13$	
0.9841831	0.0404840
0.9175984	0.0921215
0.8015781	0.1388735
0.6423493	0.1781460
0.4484928	0.2078160
0.2304583	0.2262832
0.0000000	0.2325516
$n = 14$	
0.9862838	0.0351195
0.9284349	0.0801581
0.8272013	0.1215186
0.6872929	0.1572032
0.5152486	0.1855384
0.3191124	0.2051985
0.1080549	0.2152639
$n = 15$	
0.9879925	0.0307532
0.9372734	0.0703660
0.8482066	0.1071592
0.7244177	0.1395707
0.5709722	0.1662692
0.3941513	0.1861610
0.2011941	0.1984315
0.0000000	0.2025782
$n = 16$	
0.9894009	0.0271525
0.9445750	0.0622535
0.8656312	0.0951585
0.7554044	0.1246290
0.6178762	0.1495960
0.4580168	0.1691565
0.2816036	0.1826034
0.0950125	0.1894506

Laguerre Quadrature.

$$\int_0^{\infty} e^{-x} \cdot f(x) \cdot dx = \sum_1^n \{H_t \cdot f(x_t)\}$$

x_t	H_t
$n = 2$	
0.5857864	0.8535534
3.4142136	0.1464466
$n = 3$	
0.4157746	0.7110930
2.2942804	0.2785177
6.2899451	0.0103893
$n = 4$	
0.3225477	0.6031541
1.7457611	0.3574187
4.5366203	0.0388879
9.3950709	0.0005393
$n = 5$	
0.2635603	0.5217556
1.4134031	0.3986668
3.5964258	0.0759424
7.0858100	0.0036118
12.6408008	0.0000234
$n = 6$	
0.2228466	0.4589647
1.1889321	0.4170008
2.9927363	0.1133734
5.7751436	0.0103992
9.8374674	0.0002610
15.9828740	0.0000009

Hermitian Quadrature.

$$e^{-x^2} \cdot f(x) \cdot dx = H_t \cdot f(x_t)$$

The abscissae are symmetrically placed with respect to the origin.

$\pm x_t$		H_t
	$n = 2$	
0.7071068		0.8862269
	$n = 3$	
0.0000000		1.1816359
1.2247449		0.2954090
	$n = 4$	
0.5246476		0.8049141
1.6506801		0.0813128
	$n = 5$	
0.0000000		0.9453087
0.9585725		0.3936193
2.0201829		0.0199532
	$n = 6$	
0.4360774		0.7246296
1.3358491		0.1570673
2.3506050		0.0045300
	$n = 7$	
0.0000000		0.8102646
0.8162879		0.4256073
1.6735516		0.0545156
2.6519614		0.0009718
	$n = 8$	
0.3811870		0.6611470
1.1571937		0.2078023
1.9816568		0.0170780
2.9306374		0.0001996
	$n = 9$	
0.0000000		0.7202352
0.7235510		0.4326516
1.4685533		0.0884745
2.2665806		0.0049436
3.1909932		0.0000396
	$n = 10$	
0.3429013		0.6108626
1.0366108		0.2401386
1.7566836		0.0338744
2.5327317		0.0013436
3.4361591		0.0000076

$\pm x_t$	H_t
$n = 11$	
0.0000000	0.6547593
0.6568096	0.4293598
1.3265571	0.1172279
2.0259480	0.0119114
2.7832901	0.0003468
3.6684708	0.0000014
$n = 12$	
0.3142404	0.5701352
0.9477884	0.2604923
1.5976826	0.0516080
2.2795071	0.0039054
3.0206370	0.0000857
3.8897249	0.0000003

Radau Quadrature.

$$\int_{-1}^{1} f(x)\cdot dx = H_1\cdot f(-1) + H_n\cdot f(1) + \sum_{2}^{n-1} \{H_t\cdot f(x_t)\}$$

$\pm x_t$		H_t
	$n = 2$	
1.0000000		1.0000000
	$n = 3$	
1.0000000		0.3333333
0.0000000		1.3333333
	$n = 4$	
1.0000000		0.1666667
0.4472136		0.8333333
	$n = 5$	
1.0000000		0.1000000
0.6546537		0.5444444
0.0000000		0.7111111
	$n = 6$	
1.0000000		0.0666667
0.7650553		0.3784750
0.2852315		0.5548584
	$n = 7$	
1.0000000		0.0476190
0.8302239		0.2768260
0.4688488		0.4317454
0.0000000		0.4876190
	$n = 8$	
1.0000000		0.0357143
0.8717402		0.2107042
0.5917002		0.3411227
0.2092992		0.4124588
	$n = 9$	
1.0000000		0.0277778
0.8997580		0.1654954
0.6771863		0.2745387
0.3631175		0.3464285
0.0000000		0.3715193
	$n = 10$	
1.0000000		0.0222222
0.9195339		0.1333060
0.7387739		0.2248893
0.4779249		0.2920427
0.1652790		0.3275398

$\pm x_i$	H_i
	$n = 11$
1.0000000	0.0181818
0.9340014	0.1096123
0.7844835	0.1871699
0.5652353	0.2480481
0.2957581	0.2868791
0.0000000	0.3002176

Chebyshev-Radau Quadrature.

$$\int_{-1}^{1} x \cdot f(x) \cdot dx = H \cdot \sum_{1}^{n} \{f(x_t) - f(-x_t)\}$$

x_t		H
	$n = 1$	
0.7745967		0.4303315
	$n = 2$	
0.5002990		0.2393715
0.8922365		
	$n = 3$	
0.4429861		
0.7121545		0.1599145
0.9293066		
	$n = 4$	
0.3549416		
0.6433097		0.1223363
0.7783202		
0.9481574		

Chebyshev Quadrature.

$$\int_{-1}^{1} f(x) \cdot dx = \frac{2}{n} \cdot \sum_{1}^{n} f(x_t)$$

The abscissae are skew symmetric with respect to the origin.

x_t	x_t
$n = 2$	$n = 6$
−0.5773503	−0.8662468
+0.5773503	−0.4225187
$n = 3$	−0.2666354
−0.7071068	+0.2666354
0.0000000	+0.4225187
+0.7071068	+0.8662468
$n = 4$	$n = 7$
−0.7946545	−0.8838617
−0.1875925	−0.5296568
+0.1875925	−0.3239118
+0.7946545	0.0000000
$n = 5$	+0.3239118
−0.8324975	+0.5296568
−0.3745414	+0.8838617
0.0000000	
+0.3745414	
+0.8324975	

Probability and Statistics

DESCRIPTIVE STATISTICS

a) Ungrouped Data

The formulas of this section designated as *a)* apply to a random sample of size n, denoted by x_i, $i = 1, 2, \ldots, n$.

b) Grouped Data

The formulas of this section designated as *b)* apply to data grouped into a frequency distribution having class marks x_i, $i = 1, 2, \ldots, k$, and corresponding class frequencies f_i, $i = 1, 2, \ldots, k$. The total number of observations given by

$$n = \sum_{i=1}^{k} f_i$$

In the formulas that follow, c denotes the width of the class interval, x_o denotes one of the class marks taken to be the computing origin, and $u_i = \dfrac{x_i - x_o}{c}$. Then coded class marks are obtained by replacing the original class marks with the integers $\ldots, -3, -2, -1, 0, 1, 2, 3, \ldots$ where 0 corresponds to class mark x_o in the original scale.

Mean (Arithmetic Mean)

a) $$\bar{x} = \frac{1}{n}\sum_{i=1}^{n} x_i = \frac{x_1 + x_2 + \cdots + x_n}{n}$$

b.1) $$\bar{x} = \frac{1}{n}\sum_{i=1}^{k} f_i x_i = \frac{f_1x_1 + f_2x_2 + \cdots + f_kx_k}{n}$$

If data is coded

b.2) $$\bar{x} = x_o + c\,\frac{\sum_{i=1}^{k} f_i u_i}{n}$$

Weighted Mean (Weighted Arithmetic Mean)

If with each value x_i is associated a weighting factor $w_i \geq 0$, then $\sum_{i=1}^{n} w_i$ is the total weight, and

a) $$\bar{x} = \frac{\sum_{i=1}^{n} w_i x_i}{\sum_{i=1}^{n} w_i} = \frac{w_1x_1 + w_2x_2 + \cdots + w_nx_n}{w_1 + w_2 + \cdots + w_n}$$

Geometric Mean

a) $$\text{G.M.} = \sqrt[n]{x_1 \cdot x_2 \cdots x_n}$$

In logarithmic form

$$\log(\text{G.M.}) = \frac{1}{n}\sum_{i=1}^{n} \log x_i = \frac{\log x_1 + \log x_2 + \cdots + \log x_n}{n}$$

b) $\text{G.M.} = \sqrt[n]{x_1^{f_1} \cdot x_2^{f_2} \cdots x_k^{f_k}}$

In logarithmic form

$$\log(\text{G.M.}) = \frac{1}{n}\sum_{i=1}^{k} f_i \log x_i = \frac{f_1 \log x_1 + f_2 \log x_2 + \cdots + f_k \log x_k}{n}$$

Harmonic Mean

a) $$\text{H.M.} = \frac{n}{\sum_{i=1}^{n} \frac{1}{x_i}} = \frac{n}{\frac{1}{x_1} + \frac{1}{x_2} + \cdots + \frac{1}{x_n}}$$

b) $$\text{H.M.} = \frac{n}{\sum_{i=1}^{k} \frac{f_i}{x_i}} = \frac{n}{\frac{f_1}{x_1} + \frac{f_2}{x_2} + \cdots + \frac{f_k}{x_k}}$$

Relation Between Arithmetic, Geometric, and Harmonic Mean

H.M. $\leq$ G.M. $\leq \bar{x}$, (Equality sign holds only if all sample values are identical.)

Mode

a) A mode M_o of a sample of size n is a value which occurs with greatest frequency, i.e., it is the most common value. A mode may not exist, and even if it does exist it may not be unique.

b) $$M_o = L + C\frac{\Delta_1}{\Delta_1 + \Delta_2},$$

where L is the lower class boundary of the modal class (class containing the mode),
Δ_1 is the excess of modal frequency over frequency of next lower class,
Δ_2 is the excess of modal frequency over frequency of next higher class.

Median

a) If the sample is arranged in ascending order of magnitude, then the median M_d is given by the $\frac{n+1}{2}$ *nd* value. When n is odd, the median is the middle value of the set of ordered data; when n is even, the median is usually taken as the mean of the two middle values of the set of ordered data.

b) $$M_d = L + c\,\frac{\frac{n}{2} - F_c}{f_m},$$

where L is lower class boundary of median class (class containing the median),
F_c is the sum of the frequencies of all classes lower than the median class,
f_m is the frequency of the median class.

Empirical Relation Between Mean, Median, and Mode

Mean-Mode = 3 (Mean-Median)

Quartiles

a) If the data is arranged in ascending order of magnitude, the *j*th quartile Q_j, $j = 1, 2$, or 3, is given by the $\frac{j(n+1)}{4}$ th value. It may be necessary to interpolate between successive values.

b) The *j*th quartile Q_j, $j = 1, 2$, or 3, is obtained from formula *b*) for the median by counting $\frac{jn}{4}$ cases starting at the bottom of the distribution.

Deciles

a) If the sample is arranged in ascending order of magnitude, the *j*th decile D_j, $j = 1, 2, \ldots$, or 9, is given by the $\frac{j(n+1)}{10}$ th value. It may be necessary to interpolate between successive values.

b) The *j*th decile D_j, $j = 1, 2, \ldots$, or 9, is obtained from formula *b*) for the median by counting $\frac{jn}{10}$ cases starting at the bottom of the distribution.

Percentiles

a) If the sample is arranged in ascending order of magnitude, the *j*th percentile P_j, $j = 1, 2, \ldots$, or 99 is given by the $\frac{j(n+1)}{100}$ th value. It may be necessary to interpolate between successive values.

b) The *j*th percentile P_j, $j = 1, 2, \ldots$, or 99, is obtained from formulas *b*) for the median by counting $\frac{jn}{100}$ cases starting at the bottom of the distribution.

Mean Deviation

a) $$\text{M.D.} = \frac{1}{n}\sum_{i=1}^{n} |x_i - \bar{x}|$$

or

$$\text{M.D.} = \frac{1}{n}\sum_{i=1}^{n} |x_i - M_d|$$

where $\bar{x}$ is the mean and M_d is the median of the sample.

b) $$\text{M.D.} = \frac{1}{n}\sum_{i=1}^{k} f_i|x_i - \bar{x}|$$

or

$$\text{M.D.} = \frac{1}{n}\sum_{i=1}^{k} f_i|x_i - M_d|$$

where $\bar{x}$ is the mean and M_d the median of the sample.

Standard Deviation

a) $s = \sqrt{\dfrac{\sum_{i=1}^{n}(x_i - \bar{x})^2}{n-1}}$, where $\bar{x}$ is the mean of the sample.

For computational purposes,

$$s = \sqrt{\frac{\sum_{i=1}^{n} x_i^2 - n\bar{x}^2}{n-1}}$$

$$s = \sqrt{\frac{n\sum_{i=1}^{n} x_i^2 - \left(\sum_{i=1}^{n} x_i\right)^2}{n(n-1)}}$$

b) $s = \sqrt{\dfrac{\sum_{i=1}^{k} f_i(x_i - \bar{x})^2}{n-1}}$, where $\bar{x}$ is the mean of the sample.

For computational purposes ·

$$s = \sqrt{\frac{\sum_{i=1}^{k} f_i x_i^2 - n\bar{x}^2}{n-1}}$$

$$s = \sqrt{\frac{n\sum_{i=1}^{k} f_i x_i^2 - \left(\sum_{i=1}^{k} f_i x_i\right)^2}{n(n-1)}}$$

If data is coded,

$$s = c\sqrt{\frac{n\sum_{i=1}^{k} f_i u_i^2 - \left(\sum_{i=1}^{k} f_i u_i\right)^2}{n(n-1)}}$$

Variance

The variance is the square of the standard deviation.

Range

The range of a set of values is the difference between the largest and smallest values in the set.

Root Mean Square

a) $\text{R.M.S.} = \left[\dfrac{1}{n}\sum_{i=1}^{n} x_i^2\right]$

b) $\text{R.M.S.} = \left[\dfrac{1}{n}\sum_{i=1}^{k} f_i x_i^2\right]$

Interquartile Range

$$Q_3 - Q_1,$$

where Q_1 and Q_3 are the first and third quartiles.

Quartile Deviation (Semi-Interquartile Range)

$$\frac{Q_3 - Q_1}{2},$$

where Q_1 and Q_3 are the first and third quartiles.

Coefficient of Variation

$$V = \frac{100s}{\bar{x}},$$

where $\bar{x}$ is the mean and s the standard deviation of the sample.

Coefficient of Quartile Variation

$$V = 100\,\frac{Q_3 - Q_1}{Q_3 + Q_1},$$

where Q_1 and Q_3 are the first and third quartiles.

Standardized Variable (Standard Scores)

$$z = \frac{x_i - \bar{x}}{s},$$

where $\bar{x}$ is the mean and s the standard deviation of the sample.

Moments

a) The rth moment about the origin is given by

$$m'_r = \frac{1}{n}\sum_{i=1}^{n} x_i^{\,r}$$

The rth moment about the mean $\bar{x}$ is given by

$$m_r = \frac{1}{n}\sum_{i=1}^{n} (x_i - \bar{x})^r$$

If $\sum_{i=1}^{n} (x_i - \bar{x})^r$ is expanded by use of the binomial theorem, moments about the mean may be expressed in terms of moments about the origin.

b) The rth moment about the origin is given by

$$m'_r = \frac{1}{n}\sum_{i=1}^{k} f_i x_i^{\,r}$$

The rth moment about the mean $\bar{x}$ is given by

$$m_r = \frac{1}{n}\sum_{i=1}^{k} f_i (x_i - \bar{x})^r$$

If $\sum_{i=1}^{k} f_i(x_i - \bar{x})^r$ is expanded by use of the binomial theorem, moments about the mean may be expressed in terms of moments about the origin.

If data is coded

$$m'_r = c^r \frac{\sum_{i=1}^{k} f_i u_i^r}{n}$$

Coefficient of Skewness

$$\alpha_3 = \frac{m_3}{(m_2)^{3/2}},$$

where m_2 and m_3 are the second and third moments about the mean of the sample.

Coefficient of Momental Skewness

$$\frac{\alpha_3}{2} = \frac{m_3}{2(m_2)^{3/2}},$$

where m_2 and m_3 are the second and third moments about the mean of the sample.

Pearson's First Coefficient of Skewness

$$S_{k_1} = \frac{3(\bar{x} - M_o)}{s},$$

where $\bar{x}$ is the mean, M_o the mode, and s the standard deviation of the sample.

Pearson's Second Coefficient of Skewness

$$S_{k_2} = \frac{3(\bar{x} - M_d)}{s},$$

where $\bar{x}$ is the mean, M_d the median, and s the standard deviation of the sample.

Quartile Coefficient of Skewness

$$S_{k_Q} = \frac{Q_3 - 2Q_2 + Q_1}{Q_3 - Q_1},$$

where Q_1, Q_2, and Q_3 are the first, second, and third quartiles.

Coefficient of Kurtosis

$$\alpha_4 = \frac{m_4}{(m_2)^2},$$

where m_2 and m_4 are the second and fourth moments about the mean of the sample.

Coefficient of Excess (Kurtosis)

$$\alpha_4 - 3 = \frac{m_4}{(m_2)^2} - 3,$$

where m_2 and m_4 are the second and fourth moments about the mean of the sample.

Sheppards Corrections for Grouping

Let all class intervals be of equal length c. If the distribution of x has a high order of contact with the x-axis at both tails, (i.e., if the distribution of x has tails which are very nearly tangent to the x-axis), one may improve the grouped data approximation to the variance by adding Sheppard's correction $-\frac{c^2}{12}$. Thus

$$\text{corrected variance} = \text{grouped data variance} - \frac{c^2}{12}$$

Analogous corrections for grouped data sample moments

$$m'_r = \frac{1}{n}\sum_{i=1}^{k} f_i x_i^r \qquad \text{and} \qquad m_r = \frac{1}{n}\sum_{i=1}^{k} f_i (x_i - \bar{x})^r$$

yield improved estimates m'_{r_c} and m_{r_c} given by

$$m'_{1_c} = m'_1 \qquad m_{1_c} = m_1$$

$$m'_{2_c} = m'_2 - \frac{c^2}{12} \qquad m_{2_c} = m_2 - \frac{c^2}{12}$$

$$m'_{3_c} = m'_3 - \frac{c^2}{4} m'_1 \qquad m_{3_c} = m_3$$

$$m'_{4_c} = m'_4 - \frac{c^2}{2} m'_1 + \frac{7c^4}{240} \qquad m_{4_c} = m_4 - \frac{c^2}{2} m_2 + \frac{7c^4}{240}$$

Curve Fitting, Regression, and Correlation

The following formulas apply to a set of n ordered pairs $\{(x_i,y_i)\}$, $i = 1, 2, \ldots, n$. The assumptions of normal regression analysis are that the x's are fixed variables, and the y's are independent random variables having normal distributions with common variance σ^2. The assumptions of normal correlation analysis are that $\{(x_i,y_i)\}$ constitute a random sample from a bivariate normal population.

Curve Fitting

1. Polynomial Function

$$y = b_0 + b_1 x + b_2 x^2 + \cdots + b_m x^m$$

For a polynomial function fit by the method of least squares, the values of $b_o, b_1, \ldots, b_m$ are obtained by solving the system of $m + 1$ normal equations

$$\begin{aligned} nb_o + b_1\Sigma x_i + b_2\Sigma x_i^2 + \cdots + b_m\Sigma x_i^m &= \Sigma y_i \\ b_o\Sigma x_i + b_i\Sigma x_i^2 + b_2\Sigma x_i^3 + \cdots + b_m\Sigma x_i^{m+1} &= \Sigma x_i y_i \\ \cdots\cdots\cdots\cdots\cdots\cdots\cdots\cdots\cdots\cdots \\ b_o\Sigma x_i^m + b_1\Sigma x_i^{m+1} + b_2\Sigma x_i^{m+2} + \cdots + b_m\Sigma x_i^{2m} &= \Sigma x_i^m y_i \end{aligned}$$

2. Straight Line

$$y = b_0 + b_1 x$$

For a straight line fit by the method of least squares, the values b_o and b_1 are obtained by solving the normal equations

$$\begin{aligned} nb_0 + b_1\Sigma x_i &= \Sigma y_i \\ b_0\Sigma x_i + b_1\Sigma x_i^2 &= \Sigma x_i y_i \end{aligned}$$

The solutions of these normal equations are

$$b_1 = \frac{n\Sigma x_i y_i - (\Sigma x_i)(\Sigma y_i)}{n\Sigma x_i^2 - (\Sigma x_i)^2}$$

$$b_0 = \frac{\Sigma y_i}{n} - b_1\frac{\Sigma x_i}{n} = \bar{y} - b_1\bar{x}$$

3. Exponential Curve

$$y = ab^x$$

or

$$\log y = \log a + (\log b)x$$

For an exponential curve fit by the method of least squares, the values $\log a$ and $\log b$ are obtained by fitting a straight line to the set of ordered pairs $\{(x_i, \log y_i)\}$.

4. Power Function

$$y = ax^b$$

or

$$\log y = \log a + b \log x$$

For a power function fit by the method of least squares, the values $\log a$ and b are obtained by fitting a straight line to the set or ordered pairs $\{(\log x_i, \log y_i)\}$.

Regression and Correlation

1. Simple Linear Regression.

For a regression of y on x

$$E(y/x) = b_0 + b_1 x,$$

where $E(y/x)$ is the mean of the distribution of y for a given x.

Standard Error of Estimate

$$s_e = \sqrt{\frac{\Sigma[y_i - (b_0 + b_1 x_i)]^2}{n - 2}},$$

where b_0 and b_1 are given by

$$b_1 = \frac{n\Sigma x_i y_i - (\Sigma x_i)(\Sigma y_i)}{n\Sigma x_i^2 - (\Sigma x_i)^2}$$

$$b_0 = \frac{\Sigma y_i}{n} - b_1 \frac{\Sigma x_i}{n} = \bar{y} - b_1 \bar{x}$$

2. Correlation.

An estimate of the population correlation coefficient ρ is given by

$$r = \frac{\Sigma(x_i - \bar{x})(y_i - \bar{y})}{\sqrt{[\Sigma(x_i - \bar{x})^2][\Sigma(y_i - \bar{y})^2]}}$$

or by the computing formula

$$r = \frac{n\Sigma x_i y_i - (\Sigma x_i)(\Sigma y_i)}{\sqrt{[n\Sigma x_i^2 - (\Sigma x_i)^2][n\Sigma y_i^2 - (\Sigma y_i)]^2}}$$

For grouped data

$$r = \frac{n\Sigma f x_i y_i - (\Sigma f_x x_i)(\Sigma f_y y_i)}{\sqrt{[n\Sigma f_x x_i^2 - (\Sigma f_x x_i)^2][n\Sigma f_y y_i^2 - (\Sigma f_y y_i)^2]}}$$

where f_x and f_y denote the frequencies corresponding to the class marks x and y, and f denotes the frequency of the corresponding cell of the correlation table.

If the data is coded

$$r = \frac{n\Sigma fuv - (\Sigma f_u u)(\Sigma f_v v)}{\sqrt{[n\Sigma f_u u^2 - (\Sigma f_u u)^2][n\Sigma f_v v^2 - (\Sigma f_v v)^2]}}$$

where the u's and v's are coded class marks. The frequencies f_u and f_v are defined analogous to f_x and f_y.

PROBABILITY

Definitions

A sample space S associated with an experiment is a set S of elements such that any outcome of the experiment corresponds to one and only one element of the set. An event E is a subset of a sample space S. An element in a sample space is called a sample point or a simple event (Unit subset of S).

Definition of Probability

If an experiment can occur in n mutually exclusive and equally likely ways, and if exactly m of these ways correspond to an event E, then the probability of E is given by

$$P(E) = \frac{m}{n}.$$

If E is a subset of S, and if to each unit subset of S, a non-negative number, called its probability, is assigned, and if E is the union of two or more different simple events, then the probability of E, denoted by $P(E)$, is the sum of the probabilities of those simple events whose union is E.

Marginal and Conditional Probability

Suppose a sample space S is partioned into rs disjoint subsets where the general subset is denoted by $E_i \cap F_j$. Then the marginal probability of E_i is defined as

$$P(E_i) = \sum_{j=1}^{s} P(E_i \cap F_j)$$

and the marginal probability of F_j is defined as

$$P(F_j) = \sum_{i=1}^{r} P(E_i \cap F_j)$$

The conditional probability of E_i, given that F_j has occurred, is defined as

$$P(E_i/F_j) = \frac{P(E_i \cap F_j)}{P(F_j)}, \qquad P(F_j) \neq 0$$

and that of F_j, given that E_i has occurred, is defined as

$$P(F_j/E_i) = \frac{P(E_i \cap F_j)}{P(E_i)}, \qquad P(E_i) \neq 0 .$$

Probability Theorems

1. If ϕ is the null set, $P(\phi) = 0$.
2. If S is the sample space, $P(S) = 1$.
3. If E and F are two events

$$P(E \cup F) = P(E) + P(F) - P(E \cap F).$$

4. If E and F are mutually exclusive events,

$$P(E \cup F) = P(E) + P(F).$$

5. If E and E' are complementary events,

$$P(E) = 1 - P(E').$$

6. The conditional probability of an event E, given an event F, is denoted by $P(E/F)$ and is defined as

$$P(E/F) = \frac{P(E \cap F)}{P(F)},$$

where $P(F) \neq 0$.

7. Two events E and F are said to be independent if and only if

$$P(E \cap F) = P(E) \cdot P(F).$$

E is said to be statistically independent of F if $P(E/F) = P(E)$ and $P(F/E) = P(F)$.

8. The events $E_1, E_2, \ldots, E_n$ are called mutually independent for all combinations if and only if every combination of these events taken any number at a time is independent.

9. *Bayes Theorem.*

If $E_1, E_2, \ldots, E_n$ are n mutually exclusive events whose union is the sample space S, and E is any arbitrary event of S such that $P(E) \neq 0$, then

$$P(E_k/E) = \frac{P(E_k) \cdot P(E/E_k)}{\sum_{j=1}^{n} [P(E_j) \cdot P(E/E_j)]}$$

EXAMPLE: Two unbiased six sided dice, one red and one green, are tossed and the number of dots appearing on their upper faces is observed.

The sample space S consists of 36 elements, i.e.

$$S = \{(1, 1), (1, 2), \ldots, (6, 5), (6, 6)\}.$$

1. What is the probability of throwing a seven, denoted by the event A?

$$A = \{(1, 6)\} \cup \{(2, 5)\} \cup \{(3, 4)\} \cup \{(4, 3)\} \cup \{(5, 2)\} \cup \{(6, 1)\}$$
$$P(A) = \tfrac{1}{36} + \tfrac{1}{36} + \tfrac{1}{36} + \tfrac{1}{36} + \tfrac{1}{36} + \tfrac{1}{36} = \tfrac{1}{6}$$

2. What is the probability of throwing a seven or a ten? Denote by A the event "throwing a 7" and by B the event "throwing a ten".

$$P(A) = \tfrac{1}{6} \qquad P(A \cup B) = P(A) + P(B)$$
$$P(B) = \tfrac{1}{12} \qquad = \tfrac{1}{6} + \tfrac{1}{12}$$
$$= \tfrac{1}{4}$$

3. What is the probability that the red die shows a number less than or equal to three and the green die shows a number greater than or equal to five? Denote by C the event "red die shows number ≤ 3" and by D the event "green die shows number ≥ 5".

$$C \cap D = \{(1, 5), (2, 5), (3, 5), (1, 6), (2, 6), (3, 6)\}$$
$$P(C \cap D) = \tfrac{6}{36} = \tfrac{1}{6}$$

Note:

$$P(C) = \tfrac{18}{36} = \tfrac{1}{2}$$
$$P(D) = \tfrac{12}{36} = \tfrac{1}{3}$$
$$P(C) \cdot P(D) = \tfrac{1}{2} \cdot \tfrac{1}{3} = \tfrac{1}{6}$$

Thus $P(C \cap D) = P(C) \cdot P(D)$ and events C and D are independent.

4. What is the probability that the green die shows a one, given that the sum of the numbers on the two dice is less than four? Denote by E the event "green die shows 1" and by F the event "sum of numbers on dice <4".

$$\begin{aligned} E &= \{(1, 1), (1, 2), (1, 3), (1, 4), (1, 5), (1, 6)\} \\ F &= \{(1, 1), (1, 2), (2, 1)\} \\ E \cap F &= \{(1, 1), (1, 2)\} \\ P(E/F) &= \frac{P(E \cap F)}{P(F)} = \frac{\frac{2}{36}}{\frac{3}{36}} = \frac{2}{3} \end{aligned}$$

5. What is the probability that the sum of the numbers on the two dice is not seven?

$$\begin{aligned} P(A) &= \tfrac{1}{6} \\ P(A') &= 1 - \tfrac{1}{6} = \tfrac{5}{6} \end{aligned}$$

Random Variable

A function whose domain is a sample space S and whose range is some set of real numbers is called a random variable, denoted by **X**. The function **X** transforms sample points of S into points on the x-axis. **X** will be called a discrete random variable if it is a random variable that assumes only a finite or denumerable number of values on the x-axis. **X** will be called a continuous random variable if it assumes a continuum of values on the x-axis.

Probability Function (Discrete Case)

The random variable **X** will be called a discrete random variable if there exists a function f such that $f(x_i) \geq 0$ and $\sum_i f(x_i) = 1$ for $i = 1, 2, 3, \ldots$ and such that for any event E,

$$P(E) = P[\mathbf{X} \text{ is in } E] = \sum_E f(x)$$

where $\sum_E$ means sum $f(x)$ over those values x_i that are in E and where $f(x) = P[\mathbf{X} = x]$. The probability that the value of **X** is some real number x, is given by $f(x) = P[\mathbf{X} = x]$, where f is called the probability function of the random variable **X**.

Cumulative Distribution Function (Discrete Case)

The probability that the value of a random variable **X** is less than or equal to some real number x is defined as

$$\begin{aligned} F(x) &= P(\mathbf{X} \leq x) \\ &= \Sigma f(x_i), \qquad -\infty < x < \infty, \end{aligned}$$

where the summation extends over those values of i such that $x_i \leq x$.

Probability Density (Continuous Case)

The random variable **X** will be called a continuous random variable if there exists a function f such that $f(x) \geq 0$ and $\int_{-\infty}^{\infty} f(x)\,dx = 1$ for all x in interval $-\infty < x < \infty$ and such that for any event E

$$P(E) = P(\mathbf{X} \text{ is in } E) = \int_E f(x)\,dx.$$

$f(x)$ is called the probability density of the random variable **X**. The probability that **X** assumes any given value of x is equal to zero and the probability that it assumes a value on the interval from a to b, including or excluding either end point, is equal to

$$\int_a^b f(x)\,dx.$$

Cumulative Distribution Function (Continuous Case)

The probability that the value of a random variable **X** is less than or equal to some real number x is defined as

$$\begin{aligned} F(x) &= P(\mathbf{X} \le x), \qquad -\infty < x < \infty \\ &= \int_{-\infty}^{x} f(x)\,dx. \end{aligned}$$

From the cumulative distribution, the density, if it exists, can be found from

$$f(x) = \frac{dF(x)}{dx}.$$

From the cumulative distribution

$$\begin{aligned} P(a \le \mathbf{X} \le b) &= P(\mathbf{X} \le b) - P(\mathbf{X} \le a) \\ &= F(b) - F(a) \end{aligned}$$

Mathematical Expectation

A. Expected Value

Let **X** be a random variable with density $f(x)$. Then the expected value of **X**, $E(\mathbf{X})$, is defined to be

$$E(\mathbf{X}) = \sum_x x f(x)$$

if **X** is discrete and

$$E(\mathbf{X}) = \int_{-\infty}^{\infty} x f(x)\,dx$$

if **X** is continuous. The expected value of a function g of a random variable **X** is defined as

$$E[g(\mathbf{X})] = \sum_x g(x) \cdot f(x)$$

if **X** is discrete and

$$E[g(\mathbf{X})] = \int_{-\infty}^{\infty} g(x) \cdot f(x)\,dx$$

if **X** is continuous.

Theorems

1. $E[a\mathbf{X} + b\mathbf{Y}] = aE(\mathbf{X}) + bE(\mathbf{Y})$
2. $E[\mathbf{X} \cdot \mathbf{Y}] = E(\mathbf{X}) \cdot E(\mathbf{Y})$ if **X** and **Y** are statistically independent.

B. Moments

a. Moments About the Origin. The moments about the origin of a probability distribution are the expected values of the random variable which has the given distribution. The rth moment of **X**, usually denoted by μ'_r, is defined as

$$\mu'_r = E[\mathbf{X}^r] = \sum_x x^r f(x)$$

if **X** is discrete and

$$\mu'_r = E[\mathbf{X}^r] = \int_{-\infty}^{\infty} x^r f(x)\, dx$$

if **X** is continuous.

The first moment, μ'_1, is called the mean of the random variable **X** and is usually denoted by μ.

b. Moments About the Mean. The rth moment about the mean, usually denoted by μ_r, is defined as

$$\mu_r = E[(\mathbf{X} - \mu)^r] = \sum_x (x - \mu)^r f(x)$$

if **X** is discrete and

$$\mu_r = E[(\mathbf{X} - \mu)^r] = \int_{-\infty}^{\infty} (x - \mu)^r f(x)\, dx$$

if **X** is continuous.

The second moment about the mean, μ_2, is given by

$$\mu_2 = E[(\mathbf{X} - \mu)^2] = \mu'_2 - \mu^2$$

and is called the variance of the random variable **X**, and is denoted by σ^2. The square root of the variance, σ, is called the standard deviation.

Theorems

1. $\sigma^2_{c\mathbf{X}} = c^2\sigma^2_{\mathbf{X}}$
2. $\sigma^2_{c+\mathbf{X}} = \sigma^2_{\mathbf{X}}$
3. $\sigma^2_{a\mathbf{X}+b} = a^2\sigma^2_{\mathbf{X}}$

c. Factorial Moments. The rth factorial moment of a probability distribution is defined as

$$\mu'_{(r)} = E[\mathbf{X}^{[r]}] = \sum_x x^{[r]} f(x)$$

if **X** is discrete and

$$\mu'_{(r)} = E[\mathbf{X}^{[r]}] = \int_{-\infty}^{\infty} x^{[r]} f(x)\, dx$$

if **X** is continuous, where the symbol $x^{[r]}$ denotes the factorial expression

$$x^{[r]} = x(x-1)(x-2)\cdots(x-r+1),\ r = 1, 2, 3, \ldots .$$

C. Generating Functions

a. Moment Generating Functions. The moment generating function (m.g.f.) of the random variable **X** is defined as

$$m_x(t) = E(e^{t\mathbf{X}}) = \sum_x e^{tx} f(x)$$

if **X** is discrete and

$$m_x(t) = E(e^{t\mathbf{X}}) = \int_{-\infty}^{\infty} e^{tx} f(x)\, dx$$

if **X** is continuous.

$E(e^{t\mathbf{X}})$ is the expected value of $e^{t\mathbf{X}}$. If $m_x(t)$ and its derivatives exist, $|t| < h^2$, the rth moment about the origin is

$$\mu'_r = m_x^{(r)}(0), \qquad r = 0, 1, 2, \ldots$$

where $m_x^{(r)}(0)$ is the rth derivative of $m_x(t)$ with respect to t, evaluated at $t = 0$. For

$$m_x(t) = E(e^{t\mathbf{X}})$$
$$= E\left[1 + \mathbf{X}t + \frac{(\mathbf{X}t)^2}{2!} + \cdots\right]$$
$$= 1 + \mu'_1 t + \mu'_2 \frac{t^2}{2} + \cdots$$

Thus, the moments μ'_r appear as coefficients of $\frac{t^r}{r!}$, and $m_x(t)$ may be regarded as generating the moments ν_r. The moments μ_r may be generated by the generating function

$$M_x(t) = E[e^{t(\mathbf{X}-\mu)}] = e^{-\mu t}E(e^{t\mathbf{X}}) = e^{-\mu t}m_x(t) .$$

b. Factorial Moment Generating Function. The factorial moment generating function is defined as

$$E(t^{\mathbf{X}}) = \sum_x t^x f(x) \quad \text{(probability generation function)}$$

if **X** is discrete and

$$E(t^{\mathbf{X}}) = \int_{-\infty}^{\infty} t^x f(x)\, dx$$

if **X** is continuous.

The rth factorial moment is obtained from the factorial moment generating function by differentiating it r times with respect to t and then evaluating the result when $t = 1$.

Theorems

1. If c is a constant, the m.g.f. of $c + \mathbf{X}$ is $e^{ct}m_x(t)$.
2. If c is a constant, the m.g.f. of $c\mathbf{X}$ is $m_x(ct)$.
3. If $\mathbf{Y} = \sum_{i=1}^{n} \mathbf{X}_i$, and $m_x(t)$ is the m.g.f. of $\mathbf{X}_i$, where $\mathbf{X}_1, \ldots, \mathbf{X}_n$ is a random sample from $f(x)$, then the m.g.f. of **Y** is $[m_x(t)]^n$.

D. Cumulant Generating Function

Let $m_x(t)$ be a m.g.f. If $\ln m_x(t)$ can be expanded in the form

$$c(t) = \ln m_x(t) = \kappa_1 t + \kappa_2 \frac{t^2}{2!} + \kappa_3 \frac{t^3}{3!} + \cdots + \kappa_r \frac{t^r}{r!} + \cdots,$$

then $c(t)$ is called the cumulant generating function (semi-invariant generating function) and κ_r are called the cumulants (semi-invariants) of a distribution.

$$\kappa_r = c^{(r)}(0)$$

where $c^{(r)}(0)$ is the rth derivative of $c(t)$ with respect to t evaluated at $t = 0$.

E. Characteristic Functions

The characteristic function of a distribution is defined as

$$\phi(t) = E(e^{it\mathbf{X}}) = \sum_x e^{itx} \cdot f(x)$$

if **X** is discrete and

$$\phi(t) = E(e^{it\mathbf{X}}) = \int_{-\infty}^{\infty} e^{itx} \cdot f(x)\, dx$$

if **X** is continuous.

Here t is a real number, $i^2 = -1$, and $e^{it\mathbf{X}} = \cos(t\mathbf{X}) + i \sin(t\mathbf{X})$. The characteristic function also generates moments, if they exist for

$$i^r \mu'_r = \phi^{(r)}(0)$$

where $\phi^{(r)}(0)$ is the rth derivative of $\phi(t)$ with respect to t evaluated at $t = 0$.

Multivariate Distributions

A. Discrete Case

The k-dimensional random variable $(\mathbf{X}_1, \mathbf{X}_2, \ldots, \mathbf{X}_k)$ is a k-dimensional discrete random variable if it assumes values only at a finite or denumerable number of points $(x_1, x_2, \ldots, x_k)$. Define

$$P[\mathbf{X}_1 = x_1, \mathbf{X}_2 = x_2, \ldots, \mathbf{X}_k = x_k] = f(x_1, x_2, \ldots, x_k)$$

for every value that the random variable can assume. $f(x_1, x_2, \ldots, x_k)$ is called the joint density of the k-dimensional random variable. If E is any subset of the set of values that the random variable can assume, then

$$P(E) = P[(\mathbf{X}_1, \mathbf{X}_2, \ldots, \mathbf{X}_k) \text{ is in } E] = \sum_E f(x_1, x_2, \ldots, x_k)$$

where the sum is over all those points in E. The cumulative distribution is defined as

$$F(x_1, x_2, \ldots, x_k) = \sum_{x_1} \sum_{x_2} \cdots \sum_{x_k} f(x_1, x_2, \ldots, x_k).$$

B. Continuous Case

The k random variables $\mathbf{X}_1, \mathbf{X}_2, \ldots, \mathbf{X}_k$ are said to be jointly distributed if there exists a function f such that $f(x_1, x_2, \ldots, x_k) \geq 0$ for all $-\infty < x_i < \infty$, $i = 1, 2, \ldots, k$ and such that for any event E

$$\begin{aligned} P(E) &= P[(\mathbf{X}_1, \mathbf{X}_2, \ldots, \mathbf{X}_k) \text{ is in } E] \\ &= \int_E f(x_1, x_2, \ldots, x_k)\, dx_1\, dx_2 \cdots dx_k. \end{aligned}$$

$f(x_1, x_2, \ldots, x_k)$ is called the joint density of the random variables $\mathbf{X}_1, \mathbf{X}_2, \ldots, \mathbf{X}_k$.

The cumulative distribution is defined as

$$F(x_1, x_2, \ldots, x_k) = \int_{-\infty}^{x_1} \int_{-\infty}^{x_2} \cdots \int_{-\infty}^{x_k} f(x_1, x_2, \ldots, x_k)\, dx_k \cdots dx_2\, dx_1 .$$

Given the cumulative distribution, the density may be found by

$$f(x_1, x_2, \ldots, x_k) = \frac{\partial}{\partial x_1} \cdot \frac{\partial}{\partial x_2} \cdots \frac{\partial}{\partial x_k} F(x_1, x_2, \ldots, x_k) .$$

Moments

The rth moment of $\mathbf{X}_i$, say, is defined as

$$E(\mathbf{X}_i^r) = \sum_{x_1} \sum_{x_2} \cdots \sum_{x_k} x_i^r f(x_1, x_2, \ldots, x_k)$$

if the $\mathbf{X}_i$ are discrete and

$$E(\mathbf{X}_i^r) = \int_{-\infty}^{\infty} \int_{-\infty}^{\infty} \cdots \int_{-\infty}^{\infty} x_i^r f(x_1, x_2, \ldots, x_k)\, dx_k \cdots dx_2\, dx_1$$

if the $\mathbf{X}_i$ are continuous.

Joint moments about the origin are defined as

$$E(\mathbf{X}_1^{r_1}\mathbf{X}_2^{r_2} \cdots \mathbf{X}_k^{r_k})$$

where $r_1 + r_2 + \cdots + r_k$ is the order of the moment.

Joint moments about the mean are defined as

$$E[(\mathbf{X}_1 - \mu_1)^{r_1}(\mathbf{X}_2 - \mu_2)^{r_2} \cdots (\mathbf{X}_k - \mu_k)^{r_k}].$$

Marginal and Conditional Distributions

If the random variables $\mathbf{X}_1, \mathbf{X}_2, \ldots, \mathbf{X}_k$ have the joint density function $f(x_1, x_2, \ldots, x_k)$, then the marginal distribution of the subset of the random variables, say, $\mathbf{X}_1, \mathbf{X}_2, \ldots, \mathbf{X}_p$ $(p < k)$, is given by

$$g(x_1, x_2, \ldots, x_p) = \sum_{x_{p+1}} \sum_{x_{p+2}} \cdots \sum_{x_k} f(x_1, x_2, \ldots, x_k)$$

if the X's are discrete, and

$$g(x_1, x_2, \ldots, x_p) = \int_{-\infty}^{\infty} \int_{-\infty}^{\infty} \cdots \int_{-\infty}^{\infty} f(x_1, x_2, \ldots, x_k)\, dx_{p+1} \cdots dx_{k-1}\, dx_k$$

if the X's are continuous.

The conditional distribution of a certain subset of the random variables is the joint distribution of this subset under the condition that the remaining variables are given certain values. The conditional distribution of $\mathbf{X}_1, \mathbf{X}_2, \ldots, \mathbf{X}_p$ given $\mathbf{X}_{p+1}, \mathbf{X}_{p+2}, \ldots, \mathbf{X}_k$ is

$$h(x_1, x_2, \ldots, x_p | x_{p+1}, x_{p+2}, \ldots, x_k) = \frac{f(x_1, x_2, \ldots, x_k)}{g(x_{p+1}, x_{p+2}, \ldots, x_k)}$$

if $g(x_{p+1}, x_{p+2}, \ldots, x_k) \neq 0$.

The variance σ_{ii} of $\mathbf{X}_i$ and the covariance σ_{ij} of $\mathbf{X}_i$ and $\mathbf{X}_j$ are given by

$$\sigma_{ii} = \sigma_i^2 = E[(\mathbf{X}_i - \mu_i)^2]$$

and

$$\sigma_{ij} = \rho_{ij}\sigma_i\sigma_j = E[(\mathbf{X}_i - \mu_i)(\mathbf{X}_j - \mu_j)]$$

where ρ_{ij} is the correlation coefficient and σ_i and σ_j are the standard deviations of $\mathbf{X}_i$ and $\mathbf{X}_j$.

A joint m.g.f. is defined as

$$m(t_1, t_2, \ldots, t_k) = E[e^{t_1\mathbf{X}_1 + t_2\mathbf{X}_2 + \cdots + t_k\mathbf{X}_k}]$$

if it exists for all values of t_i such that $|t_i| < h^2$.

The rth moment of $\mathbf{X}_i$ may be obtained by differentiating the m.g.f. r times with respect to t_i and then evaluating the result when all t's are set equal to zero. Similarly, a joint moment would be found by differentiatins the m.g.f. r_1 times with respect to t_1, $\ldots$, r_k times with respect to t_k, and then evaluating the result when all t's are set equal to zero.

Probability Distributions

A. Discrete Case

1. *Discrete Uniform Distribution.* If the random variable $\mathbf{X}$ has a probability function given by

$$P(\mathbf{X} = x) = f(x) = \frac{1}{n}, \qquad x = x_1, x_2, \ldots, x_n,$$

then the variable $\mathbf{X}$ is said to possess a discrete uniform distribution.

Properties

When $x_i = i$ for $i = 1, 2, \ldots,$ and n

$$\text{Mean} = \mu = \frac{n+1}{2}$$

$$\text{Variance} = \sigma^2 = \frac{n^2-1}{12}$$

$$\text{Standard Deviation} = \sigma = \sqrt{\frac{n^2-1}{12}}$$

$$\text{Moment Generating Function} = m_x(t) = \frac{e^t(1-e^{nt})}{n(1-e^t)}$$

2. *Binomial Distribution.* If the random variable $\mathbf{X}$ has a probability function given by

$$P(\mathbf{X} = x) = f(x) = \binom{n}{x} \theta^x (1-\theta)^{n-x}, \qquad x = 0, 1, 2, \ldots, n$$

where

$$\binom{n}{x} = \frac{n!}{x!(n-x)!},$$

then the variable $\mathbf{X}$ is said to possess a binomial distribution. $f(x)$ is the general term of the expansion of $[\theta + (1-\theta)]^n$.

Properties

$$\text{Mean} = \mu = n\theta$$

$$\text{Variance} = \sigma^2 = n\theta(1-\theta)$$

$$\text{Standard Deviation} = \sigma = \sqrt{n\theta(1-\theta)}$$

$$\text{Moment Generating Function} = m_{\mathbf{x}}(t) = [\theta e^t + (1-\theta)]^n$$

3. *Geometric Distribution.* If the random variable $\mathbf{X}$ has a probability function given by

$$P(\mathbf{X} = x) = f(x) = \theta(1-\theta)^{x-1}, \qquad x = 1, 2, 3, \ldots,$$

then the variable $\mathbf{X}$ is said to possess a geometric distribution.

Properties

$$\text{Mean} = \mu = \frac{1}{\theta}$$

$$\text{Variance} = \sigma^2 = \frac{1-\theta}{\theta^2}$$

$$\text{Standard Deviation} = \sigma = \sqrt{\frac{1-\theta}{\theta^2}}$$

$$\text{Moment Generating Function} = m_x(t) = \frac{\theta e^t}{1 - e^t(1-\theta)}$$

4. *Multinomial Distribution.* If a set of random variables $\mathbf{X}_1, \mathbf{X}_2, \ldots, \mathbf{X}_n$ has a probability function given by

$$P(\mathbf{X}_1 = x_1, \mathbf{X}_2 = x_2, \ldots, \mathbf{X}_n = x_n) = f(x_1, x_2, \ldots, x_n) = \frac{N!}{\prod_{i=1}^{n} x_i!} \prod_{i=1}^{n} \theta_i^{x_i}$$

where x_i are positive integers and each $\theta_i > 0$ for $i = 1, 2, \ldots, n$ and

$$\sum_{i=1}^{n} \theta_i = 1, \qquad \sum_{i=1}^{n} x_i = N,$$

then the joint distribution of $\mathbf{X}_1, \mathbf{X}_2, \ldots, \mathbf{X}_n$ is called the multinomial distribution. $f(x_1, x_2, \ldots, x_n)$ is the general term of the expansion of $(\theta_1 + \theta_2 + \cdots + \theta_n)^N$.

Properties

$$\text{Mean of } \mathbf{X}_i = \mu_i = N\theta_i$$
$$\text{Variance of } \mathbf{X}_i = \sigma_i^2 = N\theta_i(1 - \theta_i)$$
$$\text{Covariance of } \mathbf{X}_i \text{ and } \mathbf{X}_j = \sigma_{ij}^2 = -N\theta_i\theta_j$$
$$\text{Joint Moment Generating Function} = (\theta_1 e^{t_1} + \cdots + \theta_n e^{t_n})^N$$

5. *Poisson Distribution.* If the random variable $\mathbf{X}$ has a probability function given by

$$P(\mathbf{X} = x) = f(x) = \frac{e^{-\lambda}\lambda^x}{x!}, \qquad \lambda > 0, x = 0, 1, \ldots,$$

then the variable $\mathbf{X}$ is said to possess a Poisson distribution.

Properties

$$\text{Mean} = \mu = \lambda$$
$$\text{Variance} = \sigma^2 = \lambda$$
$$\text{Standard Deviation} = \sigma = \sqrt{\lambda}$$
$$\text{Moment Generating Function} = m_x(t) = e^{\lambda(e^t - 1)}$$

6. *Hypergeometric Distribution.* If the random variable $\mathbf{X}$ has a probability function given by

$$P(\mathbf{X} = x) = f(x) = \frac{\binom{k}{x}\binom{N-k}{n-x}}{\binom{N}{n}}, \qquad x = 0, 1, 2, \ldots, [n, k],$$

where $[n, k]$ means the smaller of the two numbers n, k, then the variable $\mathbf{X}$ is said to possess a hypergeometric distribution.

Properties

$$\text{Mean} = \mu = \frac{kn}{N}$$

$$\text{Variance} = \sigma^2 = \frac{k(N-k)n(N-n)}{N^2(N-1)}$$

$$\text{Standard Deviation} = \sigma = \sqrt{\frac{k(N-k)n(N-n)}{N^2(N-1)}}$$

7. *Negative Binomial Distribution.* If the random variable $\mathbf{X}$ has a probability function given by

$$\text{P}(\mathbf{X} = x) = f(x) = \binom{x + r - 1}{r - 1}\theta^r(1 - \theta)^x, \qquad x = 0, 1, 2, \ldots;$$

then the variable $\mathbf{X}$ is said to possess a negative binomial distribution, known also as the Pascal or Pólya distribution.

Properties

$$\text{Mean} = \mu = \frac{r}{\theta}$$

$$\text{Variance} = \sigma^2 = \frac{r}{\theta}\left(\frac{1}{\theta} - 1\right) = \frac{r(1-\theta)}{\theta^2}$$

$$\text{Standard Deviation} = \sigma = \sqrt{\frac{r}{\theta}\left(\frac{1}{\theta} - 1\right)} = \sqrt{\frac{r(1-\theta)}{\theta^2}}$$

$$\text{Moment Generating Function} = m_x(t) = e^{tr}\theta^r[1 - (1-\theta)e^t]^{-r}$$

B. Continuous Case

1. *Uniform Distribution.* A random variable **X** is said to be distributed as the uniform distribution if the density function is given by

$$f(x) = \frac{1}{\beta - \alpha}, \qquad \alpha < x < \beta,$$

where α and β are parameters with $\alpha < \beta$.

Properties

$$\text{Mean} = \mu = \frac{\alpha + \beta}{2}$$

$$\text{Variance} = \sigma^2 = \frac{(\beta - \alpha)^2}{12}$$

$$\text{Standard Deviation} = \sigma = \sqrt{\frac{(\beta - \alpha)^2}{12}}$$

$$\text{Moment Generating Function} = m_x(t) = \frac{2}{(\beta - \alpha)t} \sin\left[\frac{(\beta - \alpha)t}{2}\right] e^{\frac{\alpha+\beta}{2}t}.$$

2. *Normal Distribution.* A random variable **X** is said to be normally distributed if its density function is given by

$$f(x) = \frac{1}{\sqrt{2\pi}\,\sigma} e^{-(x-\mu)^2/2\sigma^2}, \qquad -\infty < x < \infty$$

where μ and σ are parameters, called the mean and the standard deviation or the random variable **X**, respectively.

Properties

$$\text{Mean} = \mu$$

$$\text{Variance} = \sigma^2$$

$$\text{Standard Deviation} = \sigma$$

$$\text{Moment Generating Function} = m_x(t) = e^{t\mu + \frac{\sigma^2 t^2}{2}}$$

Cumulative Distribution

$$F(x) = \int_{-\infty}^{x} \frac{1}{\sqrt{2\pi}\,\sigma} e^{-(x-\mu)^2/2\sigma^2}\, dx$$

Set $y = \dfrac{x - \mu}{\sigma}$ to obtain the cumulative standard normal.

3. *Gamma Distribution.* A random variable **X** is said to be distributed as the gamma distribution if the density function is given by

$$f(x) = \frac{1}{\Gamma(\alpha + 1)\beta^{\alpha+1}} x^\alpha e^{-x/\beta}, \qquad 0 < x < \infty$$

where α and β are parameters with $\alpha > -1$ and $\beta > 0$.

Properties

$$\text{Mean} = \mu = \beta(\alpha + 1)$$

$$\text{Variance} = \sigma^2 = \beta^2(\alpha + 1)$$

$$\text{Standard Deviation} = \sigma = \beta\sqrt{\alpha + 1}$$

$$\text{Moment Generating Function} = m_x(t) = (1 - \beta t)^{-(\alpha+1)}, \qquad t < \frac{1}{\beta}.$$

4. *Exponential Distribution.* A random variable $\mathbf{X}$ is said to be distributed as the exponential distribution if the density function is given by

$$f(x) = \frac{1}{\theta} e^{-x/\theta}, \qquad x > 0$$

where θ is a parameter and $\theta > 0$.

Properties

$$\begin{aligned} \text{Mean} &= \mu = \theta \\ \text{Variance} &= \sigma^2 = \theta^2 \\ \text{Standard Deviation} &= \sigma = \sqrt{\theta^2} \\ \text{Moment Generating Function} &= m_x(t) = (1 - \theta t)^{-1} \end{aligned}$$

5. *Beta Distribution.* A random variable $\mathbf{X}$ is said to be distributed as the beta distribution if the density function is given by

$$f(x) = \frac{\Gamma(\alpha + \beta + 2)}{\Gamma(\alpha + 1)\Gamma(\beta + 1)} x^{\alpha}(1 - x)^{\beta}, \qquad 0 < x < 1$$

where α and β are parameters with $\alpha > -1$ and $\beta > -1$.

Properties

$$\text{Mean} = \mu = \frac{\alpha + 1}{\alpha + \beta + 2}$$

$$\text{Variance} = \sigma^2 = \frac{(\alpha + 1)(\beta + 1)}{(\alpha + \beta + 2)^2(\alpha + \beta + 3)}$$

$$r\text{th moment about the origin} = \nu_r = \frac{\Gamma(\alpha + \beta + 2)\Gamma(\alpha + r + 1)}{\Gamma(\alpha + \beta + r + 2)\Gamma(\alpha + 1)}.$$

Sampling Distributions

Population—A finite or infinite set of elements of a random variable $\mathbf{X}$.

Random Sample—If the random variables $\mathbf{X}_1, \mathbf{X}_2, \ldots, \mathbf{X}_n$ have a joint density,

$$g(x_1, x_2, \ldots, x_n) = f(x_1)f(x_2) \cdots f(x_n)$$

where the density of each $\mathbf{X}_i$ is $f(x)$, then $\mathbf{X}_1, \mathbf{X}_2, \ldots, \mathbf{X}_n$ is said to be a random sample of size n from the population with density $f(x)$.

Sampling Distributions

A random sample is selected from a population in which the form of the probability function is known, and from the joint density of the random variables a distribution, called the sampling distribution, of a function of the random variables is derived.

1. *Chi-Square Distribution.* If $\mathbf{Y}_1, \mathbf{Y}_2, \ldots, \mathbf{Y}_n$ are normally and independently distributed with mean 0 and variance 1, then

$$\chi^2 = \sum_{i=1}^{n} \mathbf{Y}_i^2$$

is distributed as Chi-Square (χ^2) with n degrees of freedom. The density function is given by

$$f(\chi^2) = \frac{(\chi^2)^{\frac{1}{2}(n-2)}}{2^{\frac{n}{2}}\Gamma\left(\frac{n}{2}\right)} e^{-\chi^2/2}, \qquad 0 < \chi^2 < \infty.$$

Properties

$$\text{Mean} = \mu = n$$
$$\text{Variance} = \sigma^2 = 2n$$

Reproductive Property of χ^2 - Distribution

If $\chi_1^2, \chi_2^2, \ldots, \chi_k^2$ are independently distributed according to χ^2 - distributions with $n_1, n_2, \ldots, n_k$ degrees of freedom, respectively, then $\sum_{j=1}^{k} \chi_j^2$ is distributed according to a χ^2 - distribution with $n = \sum_{j=1}^{k} n_j$ degrees of freedom.

2. *Snedecor's F-Distribution.* If a random variable **X** is distributed as χ^2 with m degrees of freedom (χ_m^2) and a random variable **Y** is distributed as χ^2 with n degrees of freedom (χ_n^2) and if **X** and **Y** are independent, then $F = \dfrac{\mathbf{X}/m}{\mathbf{Y}/n}$ is distributed as Snedecor's F with m and n degrees of freedom, denoted by $F(m, n)$. The density function of the F-distribution is given by

$$f(F) = \frac{\Gamma\left(\frac{m+n}{2}\right)\left(\frac{m}{n}\right)^{m/2} F^{(m-2)/2}}{\Gamma\left(\frac{m}{2}\right)\Gamma\left(\frac{n}{2}\right)\left(1 + \frac{m}{n}F\right)^{(m+n)/2}}, \qquad 0 < F < \infty.$$

Properties

$$\text{Mean} = \mu = \frac{n}{n-2}, \qquad n > 2$$
$$\text{Variance} = \sigma^2 = \frac{2n^2(m+n-2)}{m(n-2)^2(n-4)}, \qquad n > 4 .$$

The transformation $w = \dfrac{mF/n}{1 + \dfrac{mF}{n}}$ transforms the F-density into a Beta density.

3. *Student's t-Distribution.* If a random variable **X** is normally distributed with mean 0 and variance σ^2, and if $\mathbf{Y}^2/\sigma^2$ is distributed as χ^2 with n degrees of freedom and if **X** and **Y** are independent, then

$$t = \frac{\mathbf{X}\sqrt{n}}{\mathbf{Y}}$$

is distributed as Student's t with n degrees of freedom. The density function is given by

$$f(t) = \frac{\Gamma\left(\frac{n+1}{2}\right)}{\sqrt{n\pi}\,\Gamma\left(\frac{n}{2}\right)\left(1 + \frac{t^2}{n}\right)^{\frac{1}{2}(n+1)}}, \qquad -\infty < t < \infty .$$

Properties

$$\text{Mean} = \mu = 0$$
$$\text{Variance} = \sigma^2 = \frac{n}{n-2}, \qquad n > 2 .$$

THE NORMAL PROBABILITY FUNCTION AND RELATED FUNCTIONS

This table gives values of:

a) $f(x)$ = the probability density of a standardized random variable

$$= \frac{1}{\sqrt{2\pi}} e^{-\frac{1}{2}x^2}$$

For negative values of x, one uses the fact that $f(-x) = f(x)$.

b) $F(x)$ = the cumulative distribution function of a standardized normal random variable

$$= \int_{-\infty}^{x} \frac{1}{\sqrt{2\pi}} e^{-\frac{1}{2}t^2}\, dt$$

For negative values of x, one uses the relationship $F(-x) = 1 - F(x)$. Values of x corresponding to a few special values of $F(x)$ are given in a separate table following the main table.

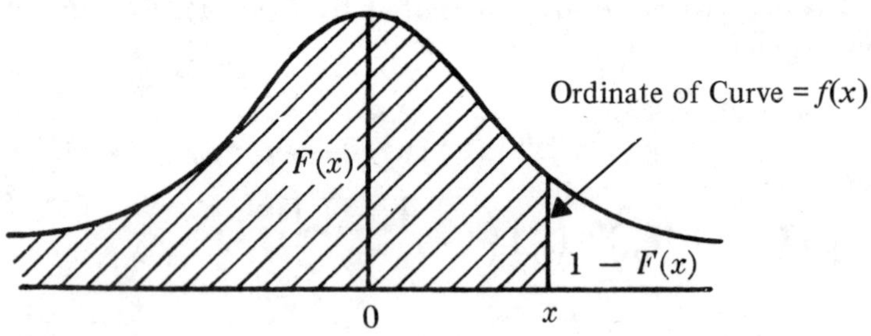

c) $f'(x)$ = the first derivative of $f(x)$ with respect to x

$$= -\frac{x}{\sqrt{2\pi}} e^{-\frac{1}{2}x^2} = -xf(x)$$

d) $f''(x)$ = the second derivative of $f(x)$ with respect to x

$$= \frac{(x^2 - 1)}{\sqrt{2\pi}} e^{-\frac{1}{2}x^2} = (x^2 - 1)f(x)$$

e) $f'''(x)$ = the third derivative of $f(x)$ with respect to x

$$= \frac{3x - x^3}{\sqrt{2\pi}} e^{-\frac{1}{2}x^2} = (3x - x^3)f(x)$$

f) $f^{iv}(x)$ = the fourth derivative of $f(x)$ with respect to x

$$= \frac{x^4 - 6x^2 + 3}{\sqrt{2\pi}} e^{-\frac{1}{2}x^2} = (x^4 - 6x^2 + 3)f(x)$$

THE NORMAL PROBABILITY FUNCTION AND RELATED FUNCTIONS (Continued)

It should be noted that other probability integrals can be evaluated by the use of these tables. For example,

$$\int_0^x f(t)dt = \tfrac{1}{2}\,\mathrm{erf}\left(\frac{x}{\sqrt{2}}\right),$$

where $\mathrm{erf}\left(\frac{x}{\sqrt{2}}\right)$ represents the error function associated with the normal curve.

To evaluate erf (2.3) one proceeds as follows: Since $\frac{x}{\sqrt{2}} = 2.3$, one finds $x = (2.3)(\sqrt{2}) = 3.25$. In the entry opposite $x = 3.25$, the value 0.9994 is given. Subtracting 0.5000 from the tabular value, one finds the value 0.4994. Thus erf $(2.3) = 2(0.4994) = 0.9988$.

NORMAL PROBABILITY FUNCTION AND RELATED FUNCTIONS (Continued)

x	$F(x)$	$1 - F(x)$	$f(x)$	$f'(x)$	$f''(x)$	$f'''(x)$	$f^{iv}(x)$
.00	.5000	.5000	.3989	−.0000	−.3989	.0000	1.1968
.01	.5040	.4960	.3989	−.0040	−.3989	.0120	1.1965
.02	.5080	.4920	.3989	−.0080	−.3987	.0239	1.1956
.03	.5120	.4880	.3988	−.0120	−.3984	.0359	1.1941
.04	.5160	.4840	.3986	−.0159	−.3980	.0478	1.1920
.05	.5199	.4801	.3984	−.0199	−.3975	.0597	1.1894
.06	.5239	.4761	.3982	−.0239	−.3968	.0716	1.1861
.07	.5279	.4721	.3980	−.0279	−.3960	.0834	1.1822
.08	.5319	.4681	.3977	−.0318	−.3951	.0952	1.1778
.09	.5359	.4641	.3973	−.0358	−.3941	.1070	1.1727
.10	.5398	.4602	.3970	−.0397	−.3930	.1187	1.1671
.11	.5438	.4562	.3965	−.0436	−.3917	.1303	1.1609
.12	.5478	.4522	.3961	−.0475	−.3904	.1419	1.1541
.13	.5517	.4483	.3956	−.0514	−.3889	.1534	1.1468
.14	.5557	.4443	.3951	−.0553	−.3873	.1648	1.1389
.15	.5596	.4404	.3945	−.0592	−.3856	.1762	1.1304
.16	.5636	.4364	.3939	−.0630	−.3838	.1874	1.1214
.17	.5675	.4325	.3932	−.0668	−.3819	.1986	1.1118
.18	.5714	.4286	.3925	−.0707	−.3798	.2097	1.1017
.19	.5753	.4247	.3918	−.0744	−.3777	.2206	1.0911
.20	.5793	.4207	.3910	−.0782	−.3754	.2315	1.0799
.21	.5832	.4168	.3902	−.0820	−.3730	.2422	1.0682
.22	.5871	.4129	.3894	−.0857	−.3706	.2529	1.0560
.23	.5910	.4090	.3885	−.0894	−.3680	.2634	1.0434
.24	.5948	.4052	.3876	−.0930	−.3653	.2737	1.0302
.25	.5987	.4013	.3867	−.0967	−.3625	.2840	1.0165
.26	.6026	.3974	.3857	−.1003	−.3596	.2941	1.0024
.27	.6064	.3936	.3847	−.1039	−.3566	.3040	0.9878
.28	.6103	.3897	.3836	−.1074	−.3535	.3138	0.9727
.29	.6141	.3859	.3825	−.1109	−.3504	.3235	0.9572
.30	.6179	.3821	.3814	−.1144	−.3471	.3330	0.9413
.31	.6217	.3783	.3802	−.1179	−.3437	.3423	0.9250
.32	.6255	.3745	.3790	−.1213	−.3402	.3515	0.9082
.33	.6293	.3707	.3778	−.1247	−.3367	.3605	0.8910
.34	.6331	.3669	.3765	−.1280	−.3330	.3693	0.8735
.35	.6368	.3632	.3752	−.1313	−.3293	.3779	0.8556
.36	.6406	.3594	.3739	−.1346	−.3255	.3864	0.8373
.37	.6443	.3557	.3726	−.1378	−.3216	.3947	0.8186
.38	.6480	.3520	.3712	−.1410	−.3176	.4028	0.7996
.39	.6517	.3483	.3697	−.1442	−.3135	.4107	0.7803
.40	.6554	.3446	.3683	−.1473	−.3094	.4184	0.7607
.41	.6591	.3409	.3668	−.1504	−.3051	.4259	0.7408
.42	.6628	.3372	.3653	−.1534	−.3008	.4332	0.7206
.43	.6664	.3336	.3637	−.1564	−.2965	.4403	0.7001
.44	.6700	.3300	.3621	−.1593	−.2920	.4472	0.6793
.45	.6736	.3264	.3605	−.1622	−.2875	.4539	0.6583
.46	.6772	.3228	.3589	−.1651	−.2830	.4603	0.6371
.47	.6808	.3192	.3572	−.1679	−.2783	.4666	0.6156
.48	.6844	.3156	.3555	−.1707	−.2736	.4727	0.5940
.49	.6879	.3121	.3538	−.1734	−.2689	.4785	0.5721
.50	.6915	.3085	.3521	−.1760	−.2641	.4841	0.5501

NORMAL PROBABILITY FUNCTION AND RELATED FUNCTIONS (Continued)

x	$F(x)$	$1 - F(x)$	$f(x)$	$f'(x)$	$f''(x)$	$f'''(x)$	$f^{\text{iv}}(x)$
.50	.6915	.3085	.3521	−.1760	−.2641	.4841	.5501
.51	.6950	.3050	.3503	−.1787	−.2592	.4895	.5279
.52	.6985	.3015	.3485	−.1812	−.2543	.4947	.5056
.53	.7019	.2981	.3467	−.1837	−.2493	.4996	.4831
.54	.7054	.2946	.3448	−.1862	−.2443	.5043	.4605
.55	.7088	.2912	.3429	−.1886	−.2392	.5088	.4378
.56	.7123	.2877	.3410	−.1920	−.2341	.5131	.4150
.57	.7157	.2843	.3391	−.1933	−.2289	.5171	.3921
.58	.7190	.2810	.3372	−.1956	−.2238	.5209	.3691
.59	.7224	.2776	.3352	−.1978	−.2185	.5245	.3461
.60	.7257	.2743	.3332	−.1999	−.2133	.5278	.3231
.61	.7291	.2709	.3312	−.2020	−.2080	.5309	.3000
.62	.7324	.2676	.3292	−.2041	−.2027	.5338	.2770
.63	.7357	.2643	.3271	−.2061	−.1973	.5365	.2539
.64	.7389	.2611	.3251	−.2080	−.1919	.5389	.2309
.65	.7422	.2578	.3230	−.2099	−.1865	.5411	.2078
.66	.7454	.2546	.3209	−.2118	−.1811	.5431	.1849
.67	.7486	.2514	.3187	−.2136	−.1757	.5448	.1620
.68	.7517	.2483	.3166	−.2153	−.1702	.5463	.1391
.69	.7549	.2451	.3144	−.2170	−.1647	.5476	.1164
.70	.7580	.2420	.3123	−.2186	−.1593	.5486	.0937
.71	.7611	.2389	.3101	−.2201	−.1538	.5495	.0712
.72	.7642	.2358	.3079	−.2217	−.1483	.5501	.0487
.73	.7673	.2327	.3056	−.2231	−.1428	.5504	.0265
.74	.7704	.2296	.3034	−.2245	−.1373	.5506	.0043
.75	.7734	.2266	.3011	−.2259	−.1318	.5505	−.0176
.76	.7764	.2236	.2989	−.2271	−.1262	.5502	−.0394
.77	.7794	.2206	.2966	−.2284	−.1207	.5497	−.0611
.78	.7823	.2177	.2943	−.2296	−.1153	.5490	−.0825
.79	.7852	.2148	.2920	−.2307	−.1098	.5481	−.1037
.80	.7881	.2119	.2897	−.2318	−.1043	.5469	−.1247
.81	.7910	.2090	.2874	−.2328	−.0988	.5456	−.1455
.82	.7939	.2061	.2850	−.2337	−.0934	.5440	−.1660
.83	.7967	.2033	.2827	−.2346	−.0880	.5423	−.1862
.84	.7995	.2005	.2803	−.2355	−.0825	.5403	−.2063
.85	.8023	.1977	.2780	−.2363	−.0771	.5381	−.2260
.86	.8051	.1949	.2756	−.2370	−.0718	.5358	−.2455
.87	.8079	.1921	.2732	−.2377	−.0664	.5332	−.2646
.88	.8106	.1894	.2709	−.2384	−.0611	.5305	−.2835
.89	.8133	.1867	.2685	−.2389	−.0558	.5276	−.3021
.90	.8159	.1841	.2661	−.2395	−.0506	.5245	−.3203
.91	.8186	.1814	.2637	−.2400	−.0453	.5212	−.3383
.92	.8212	.1788	.2613	−.2404	−.0401	.5177	−.3559
.93	.8238	.1762	.2589	−.2408	−.0350	.5140	−.3731
.94	.8264	.1736	.2565	−.2411	−.0299	.5102	−.3901
.95	.8289	.1711	.2541	−.2414	−.0248	.5062	−.4066
.96	.8315	.1685	.2516	−.2416	−.0197	.5021	−.4228
.97	.8340	.1660	.2492	−.2417	−.0147	.4978	−.4387
.98	.8365	.1635	.2468	−.2419	−.0098	.4933	−.4541
.99	.8389	.1611	.2444	−.2420	−.0049	.4887	−.4692
1.00	.8413	.1587	.2420	−.2420	.0000	.4839	−.4839

NORMAL PROBABILITY FUNCTION AND RELATED FUNCTIONS (Continued)

x	$F(x)$	$1 - F(x)$	$f(x)$	$f'(x)$	$f''(x)$	$f'''(x)$	$f^{iv}(x)$
1.00	.8413	.1587	.2420	−.2420	.0000	.4839	−.4839
1.01	.8438	.1562	.2396	−.2420	.0048	.4790	−.4983
1.02	.8461	.1539	.2371	−.2419	.0096	.4740	−.5122
1.03	.8485	.1515	.2347	−.2418	.0143	.4688	−.5257
1.04	.8508	.1492	.2323	−.2416	.0190	.4635	−.5389
1.05	.8531	.1469	.2299	−.2414	.0236	.4580	−.5516
1.06	.8554	.1446	.2275	−.2411	.0281	.4524	−.5639
1.07	.8577	.1423	.2251	−.2408	.0326	.4467	−.5758
1.08	.8599	.1401	.2227	−.2405	.0371	.4409	−.5873
1.09	.8621	.1379	.2203	−.2401	.0414	.4350	−.5984
1.10	.8643	.1357	.2179	−.2396	.0458	.4290	−.6091
1.11	.8665	.1335	.2155	−.2392	.0500	.4228	−.6193
1.12	.8686	.1314	.2131	−.2386	.0542	.4166	−.6292
1.13	.8708	.1292	.2107	−.2381	.0583	.4102	−.6386
1.14	.8729	.1271	.2083	−.2375	.0624	.4038	−.6476
1.15	.8749	.1251	.2059	−.2368	.0664	.3973	−.6561
1.16	.8770	.1230	.2036	−.2361	.0704	.3907	−.6643
1.17	.8790	.1210	.2012	−.2354	.0742	.3840	−.6720
1.18	.8810	.1190	.1989	−.2347	.0780	.3772	−.6792
1.19	.8830	.1170	.1965	−.2339	.0818	.3704	−.6861
1.20	.8849	.1151	.1942	−.2330	.0854	.3635	−.6926
1.21	.8869	.1131	.1919	−.2322	.0890	.3566	−.6986
1.22	.8888	.1112	.1895	−.2312	.0926	.3496	−.7042
1.23	.8907	.1093	.1872	−.2303	.0960	.3425	−.7094
1.24	.8925	.1075	.1849	−.2293	.0994	.3354	−.7141
1.25	.8944	.1056	.1826	−.2283	.1027	.3282	−.7185
1.26	.8962	.1038	.1804	−.2273	.1060	.3210	−.7224
1.27	.8980	.1020	.1781	−.2262	.1092	.3138	−.7259
1.28	.8997	.1003	.1758	−.2251	.1123	.3065	−.7291
1.29	.9015	.0985	.1736	−.2240	.1153	.2992	−.7318
1.30	.9032	.0968	.1714	−.2228	.1182	.2918	−.7341
1.31	.9049	.0951	.1691	−.2216	.1211	.2845	−.7361
1.32	.9066	.0934	.1669	−.2204	.1239	.2771	−.7376
1.33	.9082	.0918	.1647	−.2191	.1267	.2697	−.7388
1.34	.9099	.0901	.1626	−.2178	.1293	.2624	−.7395
1.35	.9115	.0885	.1604	−.2165	.1319	.2550	−.7399
1.36	.9131	.0869	.1582	−.2152	.1344	.2476	−.7400
1.37	.9147	.0853	.1561	−.2138	.1369	.2402	−.7396
1.38	.9162	.0838	.1539	−.2125	.1392	.2328	−.7389
1.39	.9177	.0823	.1518	−.2110	.1415	.2254	−.7378
1.40	.9192	.0808	.1497	−.2096	.1437	.2180	−.7364
1.41	.9207	.0793	.1476	−.2082	.1459	.2107	−.7347
1.42	.9222	.0778	.1456	−.2067	.1480	.2033	−.7326
1.43	.9236	.0764	.1435	−.2052	.1500	.1960	−.7301
1.44	.9251	.0749	.1415	−.2037	.1519	.1887	−.7274
1.45	.9265	.0735	.1394	−.2022	.1537	.1815	−.7243
1.46	.9279	.0721	.1374	−.2006	.1555	.1742	−.7209
1.47	.9292	.0708	.1354	−.1991	.1572	.1670	−.7172
1.48	.9306	.0694	.1334	−.1975	.1588	.1599	−.7132
1.49	.9319	.0681	.1315	−.1959	.1604	.1528	−.7089
1.50	.9332	.0668	.1295	−.1943	.1619	.1457	−.7043

NORMAL PROBABILITY FUNCTION AND RELATED FUNCTIONS (Continued)

x	$F(x)$	$1 - F(x)$	$f(x)$	$f'(x)$	$f''(x)$	$f'''(x)$	$f^{\text{iv}}(x)$
1.50	.9332	.0668	.1295	−.1943	.1619	.1457	−.7043
1.51	.9345	.0655	.1276	−.1927	.1633	.1387	−.6994
1.52	.9357	.0643	.1257	−.1910	.1647	.1317	−.6942
1.53	.9370	.0630	.1238	−.1894	.1660	.1248	−.6888
1.54	.9382	.0618	.1219	−.1877	.1672	.1180	−.6831
1.55	.9394	.0606	.1200	−.1860	.1683	.1111	−.6772
1.56	.9406	.0594	.1182	−.1843	.1694	.1044	−.6710
1.57	.9418	.0582	.1163	−.1826	.1704	.0977	−.6646
1.58	.9429	.0571	.1145	−.1809	.1714	.0911	−.6580
1.59	.9441	.0559	.1127	−.1792	.1722	.0846	−.6511
1.60	.9452	.0548	.1109	−.1775	.1730	.0781	−.6441
1.61	.9463	.0537	.1092	−.1757	.1738	.0717	−.6368
1.62	.9474	.0526	.1074	−.1740	.1745	.0654	−.6293
1.63	.9484	.0516	.1057	−.1723	.1751	.0591	−.6216
1.64	.9495	.0505	.1040	−.1705	.1757	.0529	−.6138
1.65	.9505	.0495	.1023	−.1687	.1762	.0468	−.6057
1.66	.9515	.0485	.1006	−.1670	.1766	.0408	−.5975
1.67	.9525	.0475	.0989	−.1652	.1770	.0349	−.5891
1.68	.9535	.0465	.0973	−.1634	.1773	.0290	−.5806
1.69	.9545	.0455	.0957	−.1617	.1776	.0233	−.5720
1.70	.9554	.0446	.0940	−.1599	.1778	.0176	−.5632
1.71	.9564	.0436	.0925	−.1581	.1779	.0120	−.5542
1.72	.9573	.0427	.0909	−.1563	.1780	.0065	−.5452
1.73	.9582	.0418	.0893	−.1546	.1780	.0011	−.5360
1.74	.9591	.0409	.0878	−.1528	.1780	−.0042	−.5267
1.75	.9599	.0401	.0863	−.1510	.1780	−.0094	−.5173
1.76	.9608	.0392	.0848	−.1492	.1778	−.0146	−.5079
1.77	.9616	.0384	.0833	−.1474	.1777	−.0196	−.4983
1.78	.9625	.0375	.0818	−.1457	.1774	−.0245	−.4887
1.79	.9633	.0367	.0804	−.1439	.1772	−.0294	−.4789
1.80	.9641	.0359	.0790	−.1421	.1769	−.0341	−.4692
1.81	.9649	.0351	.0775	−.1403	.1765	−.0388	−.4593
1.82	.9656	.0344	.0761	−.1386	.1761	−.0433	−.4494
1.83	.9664	.0336	.0748	−.1368	.1756	−.0477	−.4395
1.84	.9671	.0329	.0734	−.1351	.1751	−.0521	−.4295
1.85	.9678	.0322	.0721	−.1333	.1746	−.0563	−.4195
1.86	.9686	.0314	.0707	−.1316	.1740	−.0605	−.4095
1.87	.9693	.0307	.0694	−.1298	.1734	−.0645	−.3995
1.88	.9699	.0301	.0681	−.1281	.1727	−.0685	−.3894
1.89	.9706	.0294	.0669	−.1264	.1720	−.0723	−.3793
1.90	.9713	.0287	.0656	−.1247	.1713	−.0761	−.3693
1.91	.9719	.0281	.0644	−.1230	.1705	−.0797	−.3592
1.92	.9726	.0274	.0632	−.1213	.1697	−.0832	−.3492
1.93	.9732	.0268	.0620	−.1196	.1688	−.0867	−.3392
1.94	.9738	.0262	.0608	−.1179	.1679	−.0900	−.3292
1.95	.9744	.0256	.0596	−.1162	.1670	−.0933	−.3192
1.96	.9750	.0250	.0584	−.1145	.1661	−.0964	−.3093
1.97	.9756	.0244	.0573	−.1129	.1651	−.0994	−.2994
1.98	.9761	.0239	.0562	−.1112	.1641	−.1024	−.2895
1.99	.9767	.0233	.0551	−.1096	.1630	−.1052	−.2797
2.00	.9772	.0228	.0540	−.1080	.1620	−.1080	−.2700

NORMAL PROBABILITY FUNCTION AND RELATED FUNCTIONS (Continued)

x	$F(x)$	$1 - F(x)$	$f(x)$	$f'(x)$	$f''(x)$	$f'''(x)$	$f^{iv}(x)$
2.00	.9772	.0228	.0540	−.1080	.1620	−.1080	−.2700
2.01	.9778	.0222	.0529	−.1064	.1609	−.1106	−.2603
2.02	.9783	.0217	.0519	−.1048	.1598	−.1132	−.2506
2.03	.9788	.0212	.0508	−.1032	.1586	−.1157	−.2411
2.04	.9793	.0207	.0498	−.1016	.1575	−.1180	−.2316
2.05	.9798	.0202	.0488	−.1000	.1563	−.1203	−.2222
2.06	.9803	.0197	.0478	−.0985	.1550	−.1225	−.2129
2.07	.9808	.0192	.0468	−.0969	.1538	−.1245	−.2036
2.08	.9812	.0188	.0459	−.0954	.1526	−.1265	−.1945
2.09	.9817	.0183	.0449	−.0939	.1513	−.1284	−.1854
2.10	.9821	.0179	.0440	−.0924	.1500	−.1302	−.1765
2.11	.9826	.0174	.0431	−.0909	.1487	−.1320	−.1676
2.12	.9830	.0170	.0422	−.0894	.1474	−.1336	−.1588
2.13	.9834	.0166	.0413	−.0879	.1460	−.1351	−.1502
2.14	.9838	.0162	.0404	−.0865	.1446	−.1366	−.1416
2.15	.9842	.0158	.0396	−.0850	.1433	−.1380	−.1332
2.16	.9846	.0154	.0387	−.0836	.1419	−.1393	−.1249
2.17	.9850	.0150	.0379	−.0822	.1405	−.1405	−.1167
2.18	.9854	.0146	.0371	−.0808	.1391	−.1416	−.1086
2.19	.9857	.0143	.0363	−.0794	.1377	−.1426	−.1006
2.20	.9861	.0139	.0355	−.0780	.1362	−.1436	−.0927
2.21	.9864	.0136	.0347	−.0767	.1348	−.1445	−.0850
2.22	.9868	.0132	.0339	−.0754	.1333	−.1453	−.0774
2.23	.9871	.0129	.0332	−.0740	.1319	−.1460	−.0700
2.24	.9875	.0125	.0325	−.0727	.1304	−.1467	−.0626
2.25	.9878	.0122	.0317	−.0714	.1289	−.1473	−.0554
2.26	.9881	.0119	.0310	−.0701	.1275	−.1478	−.0484
2.27	.9884	.0116	.0303	−.0689	.1260	−.1483	−.0414
2.28	.9887	.0113	.0297	−.0676	.1245	−.1486	−.0346
2.29	.9890	.0110	.0290	−.0664	.1230	−.1490	−.0279
2.30	.9893	.0107	.0283	−.0652	.1215	−.1492	−.0214
2.31	.9896	.0104	.0277	−.0639	.1200	−.1494	−.0150
2.32	.9898	.0102	.0270	−.0628	.1185	−.1495	−.0088
2.33	.9901	.0099	.0264	−.0616	.1170	−.1496	−.0027
2.34	.9904	.0096	.0258	−.0604	.1155	−.1496	.0033
2.35	.9906	.0094	.0252	−.0593	.1141	−.1495	.0092
2.36	.9909	.0091	.0246	−.0581	.1126	−.1494	.0149
2.37	.9911	.0089	.0241	−.0570	.1111	−.1492	.0204
2.38	.9913	.0087	.0235	−.0559	.1096	−.1490	.0258
2.39	.9916	.0084	.0229	−.0548	.1081	−.1487	.0311
2.40	.9918	.0082	.0224	−.0538	.1066	−.1483	.0362
2.41	.9920	.0080	.0219	−.0527	.1051	−.1480	.0412
2.42	.9922	.0078	.0213	−.0516	.1036	−.1475	.0461
2.43	.9925	.0075	.0208	−.0506	.1022	−.1470	.0508
2.44	.9927	.0073	.0203	−.0496	.1007	−.1465	.0554
2.45	.9929	.0071	.0198	−.0486	.0992	−.1459	.0598
2.46	.9931	.0069	.0194	−.0476	.0978	−.1453	.0641
2.47	.9932	.0068	.0189	−.0467	.0963	−.1446	.0683
2.48	.9934	.0066	.0184	−.0457	.0949	−.1439	.0723
2.49	.9936	.0064	.0180	−.0448	.0935	−.1432	.0762
2.50	.9938	.0062	.0175	−.0438	.0920	−.1424	.0800

NORMAL PROBABILITY FUNCTION AND RELATED FUNCTIONS (Continued)

x	F(x)	1 − F(x)	f(x)	f′(x)	f″(x)	f‴(x)	$f^{iv}(x)$
2.50	.9938	.0062	.0175	−.0438	.0920	−.1424	.0800
2.51	.9940	.0060	.0171	−.0429	.0906	−.1416	.0836
2.52	.9941	.0059	.0167	−.0420	.0892	−.1408	.0871
2.53	.9943	.0057	.0163	−.0411	.0878	−.1399	.0905
2.54	.9945	.0055	.0158	−.0403	.0864	−.1389	.0937
2.55	.9946	.0054	.0155	−.0394	.0850	−.1380	.0968
2.56	.9948	.0052	.0151	−.0386	.0836	−.1370	.0998
2.57	.9949	.0051	.0147	−.0377	.0823	−.1360	.1027
2.58	.9951	.0049	.0143	−.0369	.0809	−.1350	.1054
2.59	.9952	.0048	.0139	−.0361	.0796	−.1339	.1080
2.60	.9953	.0047	.0136	−.0353	.0782	−.1328	.1105
2.61	.9955	.0045	.0132	−.0345	.0769	−.1317	.1129
2.62	.9956	.0044	.0129	−.0338	.0756	−.1305	.1152
2.63	.9957	.0043	.0126	−.0330	.0743	−.1294	.1173
2.64	.9959	.0041	.0122	−.0323	.0730	−.1282	.1194
2.65	.9960	.0040	.0119	−.0316	.0717	−.1270	.1213
2.66	.9961	.0039	.0116	−.0309	.0705	−.1258	.1231
2.67	.9962	.0038	.0113	−.0302	.0692	−.1245	.1248
2.68	.9963	.0037	.0110	−.0295	.0680	−.1233	.1264
2.69	.9964	.0036	.0107	−.0288	.0668	−.1220	.1279
2.70	.9965	.0035	.0104	−.0281	.0656	−.1207	.1293
2.71	.9966	.0034	.0101	−.0275	.0644	−.1194	.1306
2.72	.9967	.0033	.0099	−.0269	.0632	−.1181	.1317
2.73	.9968	.0032	.0096	−.0262	.0620	−.1168	.1328
2.74	.9969	.0031	.0093	−.0256	.0608	−.1154	.1338
2.75	.9970	.0030	.0091	−.0250	.0597	−.1141	.1347
2.76	.9971	.0029	.0088	−.0244	.0585	−.1127	.1356
2.77	.9972	.0028	.0086	−.0238	.0574	−.1114	.1363
2.78	.9973	.0027	.0084	−.0233	.0563	−.1100	.1369
2.79	.9974	.0026	.0081	−.0227	.0552	−.1087	.1375
2.80	.9974	.0026	.0079	−.0222	.0541	−.1073	.1379
2.81	.9975	.0025	.0077	−.0216	.0531	−.1059	.1383
2.82	.9976	.0024	.0075	−.0211	.0520	−.1045	.1386
2.83	.9977	.0023	.0073	−.0206	.0510	−.1031	.1389
2.84	.9977	.0023	.0071	−.0201	.0500	−.1017	.1390
2.85	.9978	.0022	.0069	−.0196	.0490	−.1003	.1391
2.86	.9979	.0021	.0067	−.0191	.0480	−.0990	.1391
2.87	.9979	.0021	.0065	−.0186	.0470	−.0976	.1391
2.88	.9980	.0020	.0063	−.0182	.0460	−.0962	.1389
2.89	.9981	.0019	.0061	−.0177	.0451	−.0948	.1388
2.90	.9981	.0019	.0060	−.0173	.0441	−.0934	.1385
2.91	.9982	.0018	.0058	−.0168	.0432	−.0920	.1382
2.92	.9982	.0018	.0056	−.0164	.0423	−.0906	.1378
2.93	.9983	.0017	.0055	−.0160	.0414	−.0893	.1374
2.94	.9984	.0016	.0053	−.0156	.0405	−.0879	.1369
2.95	.9984	.0016	.0051	−.0152	.0396	−.0865	.1364
2.96	.9985	.0015	.0050	−.0148	.0388	−.0852	.1358
2.97	.9985	.0015	.0048	−.0144	.0379	−.0838	.1352
2.98	.9986	.0014	.0047	−.0140	.0371	−.0825	.1345
2.99	.9986	.0014	.0046	−.0137	.0363	−.0811	.1337
3.00	.9987	.0013	.0044	−.0133	.0355	−.0798	.1330

NORMAL PROBABILITY FUNCTION AND RELATED FUNCTIONS (Continued)

x	$F(x)$	$1 - F(x)$	$f(x)$	$f'(x)$	$f''(x)$	$f'''(x)$	$f^{iv}(x)$
3.00	.9987	.0013	.0044	−.0133	.0355	−.0798	.1330
3.01	.9987	.0013	.0043	−.0130	.0347	−.0785	.1321
3.02	.9987	.0013	.0042	−.0126	.0339	−.0771	.1313
3.03	.9988	.0012	.0040	−.0123	.0331	−.0758	.1304
3.04	.9988	.0012	.0039	−.0119	.0324	−.0745	.1294
3.05	.9989	.0011	.0038	−.0116	.0316	−.0732	.1285
3.06	.9989	.0011	.0037	−.0113	.0309	−.0720	.1275
3.07	.9989	.0011	.0036	−.0110	.0302	−.0707	.1264
3.08	.9990	.0010	.0035	−.0107	.0295	−.0694	.1254
3.09	.9990	.0010	.0034	−.0104	.0288	−.0682	.1243
3.10	.9990	.0010	.0033	−.0101	.0281	−.0669	.1231
3.11	.9991	.0009	.0032	−.0099	.0275	−.0657	.1220
3.12	.9991	.0009	.0031	−.0096	.0268	−.0645	.1208
3.13	.9991	.0009	.0030	−.0093	.0262	−.0633	.1196
3.14	.9992	.0008	.0029	−.0091	.0256	−.0621	.1184
3.15	.9992	.0008	.0028	−.0088	.0249	−.0609	.1171
3.16	.9992	.0008	.0027	−.0086	.0243	−.0598	.1159
3.17	.9992	.0008	.0026	−.0083	.0237	−.0586	.1146
3.18	.9993	.0007	.0025	−.0081	.0232	−.0575	.1133
3.19	.9993	.0007	.0025	−.0079	.0226	−.0564	.1120
3.20	.9993	.0007	.0024	−.0076	.0220	−.0552	.1107
3.21	.9993	.0007	.0023	−.0074	.0215	−.0541	.1093
3.22	.9994	.0006	.0022	−.0072	.0210	−.0531	.1080
3.23	.9994	.0006	.0022	−.0070	.0204	−.0520	.1066
3.24	.9994	.0006	.0021	−.0068	.0199	−.0509	.1053
3.25	.9994	.0006	.0020	−.0066	.0194	−.0499	.1039
3.26	.9994	.0006	.0020	−.0064	.0189	−.0488	.1025
3.27	.9995	.0005	.0019	−.0062	.0184	−.0478	.1011
3.28	.9995	.0005	.0018	−.0060	.0180	−.0468	.0997
3.29	.9995	.0005	.0018	−.0059	.0175	−.0458	.0983
3.30	.9995	.0005	.0017	−.0057	.0170	−.0449	.0969
3.31	.9995	.0005	.0017	−.0055	.0166	−.0439	.0955
3.32	.9995	.0005	.0016	−.0054	.0162	−.0429	.0941
3.33	.9996	.0004	.0016	−.0052	.0157	−.0420	.0927
3.34	.9996	.0004	.0015	−.0050	.0153	−.0411	.0913
3.35	.9996	.0004	.0015	−.0049	.0149	−.0402	.0899
3.36	.9996	.0004	.0014	−.0047	.0145	−.0393	.0885
3.37	.9996	.0004	.0014	−.0046	.0141	−.0384	.0871
3.38	.9996	.0004	.0013	−.0045	.0138	−.0376	.0857
3.39	.9997	.0003	.0013	−.0043	.0134	−.0367	.0843
3.40	.9997	.0003	.0012	−.0042	.0130	−.0359	.0829
3.41	.9997	.0003	.0012	−.0041	.0127	−.0350	.0815
3.42	.9997	.0003	.0012	−.0039	.0123	−.0342	.0801
3.43	.9997	.0003	.0011	−.0038	.0120	−.0334	.0788
3.44	.9997	.0003	.0011	−.0037	.0116	−.0327	.0774
3.45	.9997	.0003	.0010	−.0036	.0113	−.0319	.0761
3.46	.9997	.0003	.0010	−.0035	.0110	−.0311	.0747
3.47	.9997	.0003	.0010	−.0034	.0107	−.0304	.0734
3.48	.9997	.0003	.0009	−.0033	.0104	−.0297	.0721
3.49	.9998	.0002	.0009	−.0032	.0101	−.0290	.0707
3.50	.9998	.0002	.0009	−.0031	.0098	−.0283	.0694

NORMAL PROBABILITY FUNCTION AND RELATED FUNCTIONS (Continued)

x	$F(x)$	$1 - F(x)$	$f(x)$	$f'(x)$	$f''(x)$	$f'''(x)$	$f^{iv}(x)$
3.50	.9998	.0002	.0009	−.0031	.0098	−.0283	.0694
3.51	.9998	.0002	.0008	−.0030	.0095	−.0276	.0681
3.52	.9998	.0002	.0008	−.0029	.0093	−.0269	.0669
3.53	.9998	.0002	.0008	−.0028	.0090	−.0262	.0656
3.54	.9998	.0002	.0008	−.0027	.0087	−.0256	.0643
3.55	.9998	.0002	.0007	−.0026	.0085	−.0249	.0631
3.56	.9998	.0002	.0007	−.0025	.0082	−.0243	.0618
3.57	.9998	.0002	.0007	−.0024	.0080	−.0237	.0606
3.58	.9998	.0002	.0007	−.0024	.0078	−.0231	.0594
3.59	.9998	.0002	.0006	−.0023	.0075	−.0225	.0582
3.60	.9998	.0002	.0006	−.0022	.0073	−.0219	.0570
3.61	.9998	.0002	.0006	−.0021	.0071	−.0214	.0559
3.62	.9999	.0001	.0006	−.0021	.0069	−.0208	.0547
3.63	.9999	.0001	.0005	−.0020	.0067	−.0203	.0536
3.64	.9999	.0001	.0005	−.0019	.0065	−.0198	.0524
3.65	.9999	.0001	.0005	−.0019	.0063	−.0192	.0513
3.66	.9999	.0001	.0005	−.0018	.0061	−.0187	.0502
3.67	.9999	.0001	.0005	−.0017	.0059	−.0182	.0492
3.68	.9999	.0001	.0005	−.0017	.0057	−.0177	.0481
3.69	.9999	.0001	.0004	−.0016	.0056	−.0173	.0470
3.70	.9999	.0001	.0004	−.0016	.0054	−.0168	.0460
3.71	.9999	.0001	.0004	−.0015	.0052	−.0164	.0450
3.72	.9999	.0001	.0004	−.0015	.0051	−.0159	.0440
3.73	.9999	.0001	.0004	−.0014	.0049	−.0155	.0430
3.74	.9999	.0001	.0004	−.0014	.0048	−.0150	.0420
3.75	.9999	.0001	.0004	−.0013	.0046	−.0146	.0410
3.76	.9999	.0001	.0003	−.0013	.0045	−.0142	.0401
3.77	.9999	.0001	.0003	−.0012	.0043	−.0138	.0392
3.78	.9999	.0001	.0003	−.0012	.0042	−.0134	.0382
3.79	.9999	.0001	.0003	−.0012	.0041	−.0131	.0373
3.80	.9999	.0001	.0003	−.0011	.0039	−.0127	.0365
3.81	.9999	.0001	.0003	−.0011	.0038	−.0123	.0356
3.82	.9999	.0001	.0003	−.0010	.0037	−.0120	.0347
3.83	.9999	.0001	.0003	−.0010	.0036	−.0116	.0339
3.84	.9999	.0001	.0003	−.0010	.0034	−.0113	.0331
3.85	.9999	.0001	.0002	−.0009	.0033	−.0110	.0323
3.86	.9999	.0001	.0002	−.0009	.0032	−.0107	.0315
3.87	.9999	.0001	.0002	−.0009	.0031	−.0104	.0307
3.88	.9999	.0001	.0002	−.0008	.0030	−.0100	.0299
3.89	1.0000	.0000	.0002	−.0008	.0029	−.0098	.0292
3.90	1.0000	.0000	.0002	−.0008	.0028	−.0095	.0284
3.91	1.0000	.0000	.0002	−.0008	.0027	−.0092	.0277
3.92	1.0000	.0000	.0002	−.0007	.0026	−.0089	.0270
3.93	1.0000	.0000	.0002	−.0007	.0026	−.0086	.0263
3.94	1.0000	.0000	.0002	−.0007	.0025	−.0084	.0256
3.95	1.0000	.0000	.0002	−.0006	.0024	−.0081	.0250
3.96	1.0000	.0000	.0002	−.0006	.0023	−.0079	.0243
3.97	1.0000	.0000	.0002	−.0006	.0022	−.0076	.0237
3.98	1.0000	.0000	.0001	−.0006	.0022	−.0074	.0230
3.99	1.0000	.0000	.0001	−.0006	.0021	−.0072	.0224
4.00	1.0000	.0000	.0001	−.0005	.0020	−.0070	.0218

x	1.282	1.645	1.960	2.326	2.576	3.090
$F(x)$	.90	.95	.975	.99	.995	.999
$2[1 - F(x)]$	.20	.10	.05	.02	.01	.002

FACTORS FOR COMPUTING PROBABLE ERRORS

The probable error of a series of n measures $a_1, a_2, a_3 \cdots a_n$, the mean of which is m, is given by the expression,

$$e = \frac{0.6745}{\sqrt{n-1}} \sqrt{(m-a_1)^2 + (m-a_2)^2 + \cdots + (m-a_n)^2}\ .$$

The probable error of the mean is,

$$E = \frac{0.6745}{\sqrt{n(n-1)}} \sqrt{(m-a_1)^2 + (m-a_2)^2 + \cdots + (m-a_n)^2}$$

The following approximate equations are convenient forms for computation,

$$e = 0.8453 \frac{\Sigma d}{\sqrt{n(n-1)}}$$

$$E = 0.8453 \frac{\Sigma d}{n\sqrt{n-1}}\ .$$

The symbol Σd represents the arithmetical sum of the deviations.

For convenience in computing the probable error the value of several of the factors involved is given for values of n from 2 to 100.

FACTORS FOR COMPUTING PROBABLE ERRORS (Continued)

n	$\frac{1}{\sqrt{n}}$	$\frac{1}{\sqrt{n(n-1)}}$	$\frac{.6745}{\sqrt{n-1}}$	$\frac{.6745}{\sqrt{n(n-1)}}$	$\frac{.8453}{n\sqrt{n-1}}$	$\frac{.8453}{\sqrt{n(n-1)}}$
2	.707107	.707107	.6745	.4769	.4227	.5978
3	.577350	.408248	.4769	.2754	.1993	.3451
4	.500000	.288675	.3894	.1947	.1220	.2440
5	.447214	.223607	.3372	.1508	.0845	.1890
6	.408248	.182574	.3016	.1231	.0630	.1543
7	.377964	.154303	.2754	.1041	.0493	.1304
8	.353553	.133631	.2549	.0901	.0399	.1130
9	.333333	.117851	.2385	.0795	.0332	.0996
10	.316228	.105409	.2248	.0711	.0282	.0891
11	.301511	.095346	.2133	.0643	.0243	.0806
12	.288675	.087039	.2034	.0587	.0212	.0736
13	.277350	.080064	.1947	.0540	.0188	.0677
14	.267261	.074125	.1871	.0500	.0167	.0627
15	.258199	.069007	.1803	.0465	.0151	.0583
16	.250000	.064550	.1742	.0435	.0136	.0546
17	.242536	.060634	.1686	.0409	.0124	.0513
18	.235702	.057166	.1636	.0386	.0114	.0483
19	.229416	.054074	.1590	.0365	.0105	.0457
20	.223607	.051299	.1547	.0346	.0097	.0434
21	.218218	.048795	.1508	.0329	.0090	.0412
22	.213201	.046524	.1472	.0314	.0084	.0393
23	.208514	.044455	.1438	.0300	.0078	.0376
24	.204124	.042563	.1406	.0287	.0073	.0360
25	.200000	.040825	.1377	.0275	.0069	.0345
26	.196116	.039223	.1349	.0265	.0065	.0332
27	.192450	.037743	.1323	.0255	.0061	.0319
28	.188982	.036370	.1298	.0245	.0058	.0307
29	.185695	.035093	.1275	.0237	.0055	.0297
30	.182574	.033903	.1252	.0229	.0052	.0287
31	.179605	.032791	.1231	.0221	.0050	.0277
32	.176777	.031750	.1211	.0214	.0047	.0268
33	.174078	.030773	.1192	.0208	.0045	.0260
34	.171499	.029854	.1174	.0201	.0043	.0252
35	.169031	.028989	.1157	.0196	.0041	.0245
36	.166667	.028172	.1140	.0190	.0040	.0238
37	.164399	.027400	.1124	.0185	.0038	.0232
38	.162221	.026669	.1109	.0180	.0037	.0225
39	.160128	.025976	.1094	.0175	.0035	.0220
40	.158114	.025318	.1080	.0171	.0034	.0214
41	.156174	.024693	.1066	.0167	.0033	.0209
42	.154303	.024098	.1053	.0163	.0031	.0204
43	.152499	.023531	.1041	.0159	.0030	.0199
44	.150756	.022990	.1029	.0155	.0029	.0194
45	.149071	.022473	.1017	.0152	.0028	.0190
46	.147442	.021979	.1005	.0148	.0027	.0186
47	.145865	.021507	.0994	.0145	.0027	.0182
48	.144338	.021054	.0984	.0142	.0026	.0178
49	.142857	.020620	.0974	.0139	.0025	.0174
50	.141421	.020203	.0964	.0136	.0024	.0171

FACTORS FOR COMPUTING PROBABLE ERRORS (Continued)

n	$\frac{1}{\sqrt{n}}$	$\frac{1}{\sqrt{n\ (n\text{-}1)}}$	$\frac{.6745}{\sqrt{n\text{-}1}}$	$\frac{.6745}{\sqrt{n(n\text{-}1)}}$	$\frac{.8453}{n\sqrt{n\text{-}1}}$	$\frac{.8453}{\sqrt{n(n\text{-}1)}}$
50	.141421	.020203	.0964	.0136	.0024	.0171
51	.140028	.019803	.0954	.0134	.0023	.0167
52	.138675	.019418	.0945	.0131	.0023	.0164
53	.137361	.019048	.0935	.0129	.0022	.0161
54	.136083	.018692	.0927	.0126	.0022	.0158
55	.134840	.018349	.0918	.0124	.0021	.0155
56	.133631	.018019	.0910	.0122	.0020	.0152
57	.132453	.017700	.0901	.0119	.0020	.0150
58	.131306	.017392	.0893	.0117	.0019	.0147
59	.130189	.017095	.0886	.0115	.0019	.0145
60	.129099	.016807	.0878	.0113	.0018	.0142
61	.128037	.016529	.0871	.0112	.0018	.0140
62	.127000	.016261	.0864	.0110	.0018	.0138
63	.125988	.016001	.0857	.0108	.0017	.0135
64	.125000	.015749	.0850	.0106	.0017	.0133
65	.124035	.015504	.0843	.0105	.0016	.0131
66	.123091	.015268	.0837	.0103	.0016	.0129
67	.122169	.015038	.0830	.0101	.0016	.0127
68	.121268	.014815	.0824	.0100	.0015	.0125
69	.120386	.014599	.0818	.0099	.0015	.0123
70	.119523	.014389	.0812	.0097	.0015	.0122
71	.118678	.014185	.0806	.0096	.0014	.0120
72	.117851	.013986	.0801	.0094	.0014	.0118
73	.117041	.013793	.0795	.0093	.0014	.0117
74	.116248	.013606	.0789	.0092	.0013	.0115
75	.115470	.013423	.0784	.0091	.0013	.0113
76	.114708	.013245	.0779	.0089	.0013	.0112
77	.113961	.013072	.0773	.0088	.0013	.0111
78	.113228	.012904	.0769	.0087	.0012	.0109
79	.112509	.012739	.0764	.0086	.0012	.0108
80	.111803	.012579	.0759	.0085	.0012	.0106
81	.111111	.012423	.0754	.0084	.0012	.0105
82	.110432	.012270	.0749	.0083	.0012	.0104
83	.109764	.012121	.0745	.0082	.0011	.0103
84	.109109	.011976	.0740	.0081	.0011	.0101
85	.108465	.011835	.0736	.0080	.0011	.0100
86	.107833	.011696	.0732	.0079	.0011	.0099
87	.107211	.011561	.0727	.0078	.0011	.0098
88	.106600	.011429	.0723	.0077	.0010	.0097
89	.106000	.011300	.0719	.0076	.0010	.0096
90	.105409	.011173	.0715	.0075	.0010	.0094
91	.104828	.011050	.0711	.0075	.0010	.0093
92	.104257	.010929	.0707	.0074	.0010	.0092
93	.103695	.010811	.0703	.0073	.0010	.0091
94	.103142	.010695	.0699	.0072	.0009	.0090
95	.102598	.010582	.0696	.0071	.0009	.0089
96	.102062	.010471	.0692	.0071	.0009	.0089
97	.101535	.010363	.0688	.0070	.0009	.0088
98	.101015	.010257	.0685	.0069	.0009	.0087
99	.100504	.010152	.0681	.0039	.0009	.0086
100	.100000	.010050	.0678	.0068	.0008	.0085

PROBABILITY OF OCCURRENCE OF DEVIATIONS

The significance of deviations is indicated by this table. The probability of occurrence of deviations as great as or greater than any specific value is given for various ratios of deviation to probable error and also with respect to the standard deviation. The probability of occurrence is stated in per cent or chances in 100. The odds against occurrence are also stated. The probable error is 0.6745 × the standard deviation.

Ratio, dev. to P.E.	Probable occurrence %	Odds against, to 1	Ratio dev. to std. dev.	Probable occurrence %	Odds against, to 1
1.0	50.00	1.00	0.67449	50.00	1.00
1.1	45.81	1.18	0.7	48.39	1.07
1.2	41.83	1.39	0.8	42.37	1.36
1.3	38.06	1.63	0.9	36.81	1.72
1.4	34.50	1.90	1.0	31.73	2.15
1.5	31.17	2.21	1.1	27.13	2.69
1.6	28.05	2.57	1.2	23.01	3.35
1.7	25.15	2.98	1.3	19.36	4.17
1.8	22.47	3.45	1.4	16.15	5.19
1.9	20.00	4.00	1.5	13.36	6.48
2.0	17.73	4.64	1.6	10.96	8.12
2.1	15.67	5.38	1.7	8.91	10.22
2.2	13.78	6.25	1.8	7.19	12.92
2.3	12.08	7.28	1.9	5.74	16.41
2.4	10.55	8.48	2.0	4.55	20.98
2.5	9.18	9.90	2.1	3.57	26.99
2.6	7.95	11.58	2.2	2.78	34.96
2.7	6.86	13.58	2.3	2.14	45.62
2.8	5.89	15.96	2.4	1.64	59.99
2.9	5.05	18.82	2.5	1.24	79.52
3.0	4.30	22.24	2.6	.932	106.3
3.1	3.65	26.37	2.7	.693	143.2
3.2	3.09	31.36	2.8	.511	194.7
3.3	2.60	37.42	2.9	.373	267.0
3.4	2.18	44.80	3.0	.270	369.4
3.5	1.82	53.82	3.1	.194	515.7
3.6	1.52	64.89	3.2	.137	726.7
3.7	1.26	78.53	3.3	.0967	1033.
3.8	1.04	95.38	3.4	.0674	1483.
3.9	.853	116.3	3.5	.0465	2149.
4.0	.698	142.3	3.6	.0318	3142.
4.1	.569	174.9	3.7	.0216	4637.
4.2	.461	215.8	3.8	.0145	6915.
4.3	.373	267.2	3.9	.00962	10394.
4.4	.300	332.4	4.0	.00634	15772.
4.5	.240	415.0	5.0	5.73×10^{-5}	1.744×10^{6}
4.6	.192	520.4	6.0	2.0×10^{-7}	5.0×10^{8}
4.7	.152	655.3	7.0	2.6×10^{-10}	3.9×10^{11}
4.8	.121	828.3			
4.9	.0950	1052.			
5.0	.0745	1341.			
6.0	.0052	19300.			
7.0	.00023	4.27×10^{5}			
8.0	6.8×10^{-6}	1.47×10^{7}			
9.0	1.3×10^{-7}	7.30×10^{8}			
10.0	1.5×10^{-9}	6.5×10^{10}			

Valid for samples of size 30 or greater.

INDIVIDUAL TERMS, BINOMIAL DISTRIBUTION

The $(x + 1)^{st}$ term in the expansion of the binomial $[\theta + (1 - \theta)]^n$ is given by

$$f(x;n,\theta) = \binom{n}{x} \theta^x (1 - \theta)^{n-x}, \qquad x = 0, 1, 2, \ldots, n.$$

This is the probability of exactly x successes in n independent binomial trials with probability of success on a single trial equal to θ. This table contains the individual terms of $f(x;n,\theta)$ for specified choices of x, n, and θ.

For $\theta > 0.5$, the value of $\binom{n}{x} \theta^x (1 - \theta)^{n-x}$ is found by using the table entry for $\binom{n}{n-x} (1 - \theta)^{n-x} \theta^x$.

INDIVIDUAL TERMS, BINOMIAL DISTRIBUTION (Continued)

n	x	.05	.10	.15	.20	.25	.30	.35	.40	.45	.50
						θ					
1	0	.9500	.9000	.8500	.8000	.7500	.7000	.6500	.6000	.5500	.5000
	1	.0500	.1000	.1500	.2000	.2500	.3000	.3500	.4000	.4500	.5000
2	0	.9025	.8100	.7225	.6400	.5625	.4900	.4225	.3600	.3025	.2500
	1	.0950	.1800	.2550	.3200	.3750	.4200	.4550	.4800	.4950	.5000
	2	.0025	.0100	.0225	.0400	.0625	.0900	.1225	.1600	.2025	.2500
3	0	.8574	.7290	.6141	.5120	.4219	.3430	.2746	.2160	.1664	.1250
	1	.1354	.2430	.3251	.3840	.4219	.4410	.4436	.4320	.4084	.3750
	2	.0071	.0270	.0574	.0960	.1406	.1890	.2389	.2880	.3341	.3750
	3	.0001	.0010	.0034	.0080	.0156	.0270	.0429	.0640	.0911	.1250
4	0	.8145	.6561	.5220	.4096	.3164	.2401	.1785	.1296	.0915	.0625
	1	.1715	2916	.3685	.4096	.4219	.4116	.3845	.3456	.2995	.2500
	2	.0135	.0486	.0975	.1536	.2109	.2646	.3105	.3456	.3675	.3750
	3	.0005	.0036	.0115	.0256	.0469	.0756	.1115	.1536	.2005	.2500
	4	.0000	.0001	.0005	.0016	.0039	.0081	.0150	.0256	.0410	.0625
5	0	.7738	.5905	.4437	.3277	.2373	.1681	.1160	.0778	.0503	.0312
	1	.2036	.3280	.3915	.4096	.3955	.3602	.3124	.2592	.2059	.1562
	2	.0214	.0729	.1382	.2048	.2637	.3087	.3364	.3456	.3369	.3125
	3	.0011	.0081	.0244	.0512	.0879	.1323	.1811	.2304	.2757	.3125
	4	.0000	.0004	.0022	.0064	.0146	.0284	.0488	.0768	.1128	.1562
	5	.0000	.0000	.0001	.0003	.0010	.0024	.0053	.0102	.0185	.0312
6	0	.7351	.5314	.3771	.2621	.1780	.1176	.0754	.0467	.0277	.0156
	1	.2321	.3543	.3993	.3932	.3560	.3025	.2437	.1866	.1359	.0938
	2	.0305	.0984	.1762	.2458	.2966	.3241	.3280	.3110	.2780	.2344
	3	.0021	.0146	.0415	.0819	.1318	.1852	.2355	.2765	.3032	.3125
	4	.0001	.0012	.0055	.0154	.0330	.0595	.0951	.1382	.1861	.2344
	5	.0000	.0001	.0004	.0015	.0044	.0102	.0205	.0369	.0609	.0938
	6	.0000	.0000	.0000	.0001	.0002	.0007	.0018	.0041	.0083	.0156
7	0	.6983	.4783	.3206	.2097	.1335	.0824	.0490	.0280	.0152	.0078
	1	.2573	.3720	.3960	.3670	.3115	.2471	.1848	.1306	.0872	.0547
	2	.0406	.1240	.2097	.2753	.3115	.3177	.2985	.2613	.2140	.1641
	3	.0036	.0230	.0617	.1147	.1730	.2269	.2679	.2903	.2918	.2734
	4	.0002	.0026	.0109	.0287	.0577	.0972	.1442	.1935	.2388	.2734
	5	.0000	.0002	.0012	.0043	.0115	.0250	.0466	.0774	.1172	.1641
	6	.0000	.0000	.0001	.0004	.0013	.0036	.0084	.0172	.0320	.0547
	7	.0000	.0000	.0000	.0000	.0001	.0002	.0006	.0016	.0037	.0078
8	0	.6634	.4305	.2725	.1678	.1001	.0576	.0319	.0168	.0084	.0039
	1	.2793	.3826	.3847	.3355	.2670	.1977	.1373	.0896	.0548	.0312
	2	.0515	.1488	.2376	.2936	.3115	.2965	.2587	.2090	.1569	.1094
	3	.0054	.0331	.0839	.1468	.2076	.2541	.2786	.2787	.2568	.2188
	4	.0004	.0046	.0185	.0459	.0865	.1361	.1875	.2322	.2627	.2734
	5	.0000	.0004	.0026	.0092	.0231	.0467	.0808	.1239	.1719	.2188
	6	.0000	.0000	.0002	.0011	.0038	.0100	.0217	.0413	.0703	.1094
	7	.0000	.0000	.0000	.0001	.0004	.0012	.0033	.0079	.0164	.0312
	8	.0000	.0000	.0000	.0000	.0000	.0001	.0002	.0007	.0017	.0039

Linear interpolations with respect to p will in general be accurate at most to two decimal places.

INDIVIDUAL TERMS, BINOMIAL DISTRIBUTION (Continued)

n	x	θ .05	.10	.15	.20	.25	.30	.35	.40	.45	.50
9	0	.6302	.3874	.2316	.1342	.0751	.0404	.0207	.0101	.0046	.0020
	1	.2985	.3874	.3679	.3020	.2253	.1556	.1004	.0605	.0339	.0176
	2	.0629	.1722	.2597	.3020	.3003	.2668	.2162	.1612	.1110	.0703
	3	.0077	.0446	.1069	.1762	.2336	.2668	.2716	.2508	.2119	.1641
	4	.0006	.0074	.0283	.0661	.1168	.1715	.2194	.2508	.2600	.2461
	5	.0000	.0008	.0050	.0165	.0389	.0735	.1181	.1672	.2128	.2461
	6	.0000	.0001	.0006	.0028	.0087	.0210	.0424	.0743	.1160	.1641
	7	.0000	.0000	.0000	.0003	.0012	.0039	.0098	.0212	.0407	.0703
	8	.0000	.0000	.0000	.0000	.0001	.0004	.0013	.0035	.0083	.0176
	9	.0000	.0000	.0000	.0000	.0000	.0000	.0001	.0003	.0008	.0020
10	0	.5987	.3487	.1969	.1074	.0563	.0282	.0135	.0060	.0025	.0010
	1	.3151	.3874	.3474	.2684	.1877	.1211	.0725	.0403	.0207	.0098
	2	.0746	.1937	.2759	.3020	.2816	.2335	.1757	.1209	.0763	.0439
	3	.0105	.0574	.1298	.2013	.2503	.2668	.2522	.2150	.1665	.1172
	4	.0010	.0112	.0401	.0881	.1460	.2001	.2377	.2508	.2384	.2051
	5	.0001	.0015	.0085	.0264	.0584	.1029	.1536	.2007	.2340	.2461
	6	.0000	.0001	.0012	.0055	.0162	.0368	.0689	.1115	.1596	.2051
	7	.0000	.0000	.0001	.0008	.0031	.0090	.0212	.0425	.0746	.1172
	8	.0000	.0000	.0000	.0001	.0004	.0014	.0043	.0106	.0229	.0439
	9	.0000	.0000	.0000	.0000	.0000	.0001	.0005	.0016	.0042	.0098
	10	.0000	.0000	.0000	.0000	.0000	.0000	.0000	.0001	.0003	.0010
11	0	.5688	.3138	.1673	.0859	.0422	.0198	.0088	.0036	.0014	.0004
	1	.3293	.3835	.3248	.2362	.1549	.0932	.0518	.0266	.0125	.0055
	2	.0867	.2131	.2866	.2953	.2581	.1998	.1395	.0887	.0513	.0269
	3	.0137	.0710	.1517	.2215	.2581	.2568	.2254	.1774	.1259	.0806
	4	.0014	.0158	.0536	.1107	.1721	.2201	.2428	.2365	.2060	.1611
	5	.0001	.0025	.0132	.0388	.0803	.1321	.1830	.2207	.2360	.2256
	6	.0000	.0003	.0023	.0097	.0268	.0566	.0985	.1471	.1931	.2256
	7	.0000	.0000	.0003	.0017	.0064	.0173	.0379	.0701	.1128	.1611
	8	.0000	.0000	.0000	.0002	.0011	.0037	.0102	.0234	.0462	.0806
	9	.0000	.0000	.0000	.0000	.0001	.0005	.0018	.0052	.0126	.0269
	10	.0000	.0000	.0000	.0000	.0000	.0000	.0002	.0007	.0021	.0054
	11	.0000	.0000	.0000	.0000	.0000	.0000	.0000	.0000	.0002	.0005
12	0	.5404	.2824	.1422	.0687	.0317	.0138	.0057	.0022	.0008	.0002
	1	.3413	.3766	.3012	.2062	.1267	.0712	.0368	.0174	.0075	.0029
	2	.0988	.2301	.2924	.2835	.2323	.1678	.1088	.0639	.0339	.0161
	3	.0173	.0852	.1720	.2362	.2581	.2397	.1954	.1419	.0923	.0537
	4	.0021	.0213	.0683	.1329	.1936	.2311	.2367	.2128	.1700	.1208
	5	.0002	.0038	.0193	.0532	.1032	.1585	.2039	.2270	.2225	.1934
	6	.0000	.0005	.0040	.0155	.0401	.0792	.1281	.1766	.2124	.2256
	7	.0000	.0000	.0006	.0033	.0115	.0291	.0591	.1009	.1489	.1934
	8	.0000	.0000	.0001	.0005	.0024	.0078	.0199	.0420	.0762	.1208
	9	.0000	.0000	.0000	.0001	.0004	.0015	.0048	.0125	.0277	.0537
	10	.0000	.0000	.0000	.0000	.0000	.0002	.0008	.0025	.0068	.0161
	11	.0000	.0000	.0000	.0000	.0000	.0000	.0001	.0003	.0010	.0029
	12	.0000	.0000	.0000	.0000	.0000	.0000	.0000	.0000	.0001	.0002

INDIVIDUAL TERMS, BINOMIAL DISTRIBUTION (Continued)

n	x	θ = .05	.10	.15	.20	.25	.30	.35	.40	.45	.50
13	0	.5133	.2542	.1209	.0550	.0238	.0097	.0037	.0013	.0004	.0001
	1	.3512	.3672	.2774	.1787	.1029	.0540	.0259	.0113	.0045	.0016
	2	.1109	.2448	.2937	.2680	.2059	.1388	.0836	.0453	.0220	.0095
	3	.0214	.0997	.1900	.2457	.2517	.2181	.1651	.1107	.0660	.0349
	4	.0028	.0277	.0838	.1535	.2097	.2337	.2222	.1845	.1350	.0873
	5	.0003	.0055	.0266	.0691	.1258	.1803	.2154	.2214	.1989	.1571
	6	.0000	.0008	.0063	.0230	.0559	.1030	.1546	.1968	.2169	.2095
	7	.0000	.0001	.0011	.0058	.0186	.0442	.0833	.1312	.1775	.2095
	8	.0000	.0000	.0001	.0011	.0047	.0142	.0336	.0656	.1089	.1571
	9	.0000	.0000	.0000	.0001	.0009	.0034	.0101	.0243	.0495	.0873
	10	.0000	.0000	.0000	.0000	.0001	.0006	.0022	.0065	.0162	.0349
	11	.0000	.0000	.0000	.0000	.0000	.0001	.0003	.0012	.0036	.0095
	12	.0000	.0000	.0000	.0000	.0000	.0000	.0000	.0001	.0005	.0016
	13	.0000	.0000	.0000	.0000	.0000	.0000	.0000	.0000	.0000	.0001
14	0	.4877	.2288	.1028	.0440	.0178	.0068	.0024	.0008	.0002	.0001
	1	.3593	.3559	.2539	.1539	.0832	.0407	.0181	.0073	.0027	.0009
	2	.1229	.2570	.2912	.2501	.1802	.1134	.0634	.0317	.0141	.0056
	3	.0259	.1142	.2056	.2501	.2402	.1943	.1366	.0845	.0462	.0222
	4	.0037	.0349	.0998	.1720	.2202	.2290	.2022	.1549	.1040	.0611
	5	.0004	.0078	.0352	.0860	.1468	.1963	.2178	.2066	.1701	.1222
	6	.0000	.0013	.0093	.0322	.0734	.1262	.1759	.2066	.2088	.1833
	7	.0000	.0002	.0019	.0092	.0280	.0618	.1082	.1574	.1952	.2095
	8	.0000	.0000	.0003	.0020	.0082	.0232	.0510	.0918	.1398	.1833
	9	.0000	.0000	.0000	.0003	.0018	.0066	.0183	.0408	.0762	.1222
	10	.0000	.0000	.0000	.0000	.0003	.0014	.0049	.0136	.0312	.0611
	11	.0000	.0000	.0000	.0000	.0000	.0002	.0010	.0033	.0093	.0222
	12	.0000	.0000	.0000	.0000	.0000	.0000	.0001	.0005	.0019	.0056
	13	.0000	.0000	.0000	.0000	.0000	.0000	.0000	.0001	.0002	.0009
	14	.0000	.0000	.0000	.0000	.0000	.0000	.0000	.0000	.0000	.0001
15	0	.4633	.2059	.0874	.0352	.0134	.0047	.0016	.0005	.0001	.0000
	1	.3658	.3432	.2312	.1319	.0668	.0305	.0126	.0047	.0016	.0005
	2	.1348	.2669	.2856	.2309	.1559	.0916	.0476	.0219	.0090	.0032
	3	.0307	.1285	.2184	.2501	.2252	.1700	.1110	.0634	.0318	.0139
	4	.0049	.0428	.1156	.1876	.2252	.2186	.1792	.1268	.0780	.0417
	5	.0006	.0105	.0449	.1032	.1651	.2061	.2123	.1859	.1404	.0916
	6	.0000	.0019	.0132	.0430	.0917	.1472	.1906	.2066	.1914	.1527
	7	.0000	.0003	.0030	.0138	.0393	.0811	.1319	.1771	.2013	.1964
	8	.0000	.0000	.0005	.0035	.0131	.0348	.0710	.1181	.1647	.1964
	9	.0000	.0000	.0001	.0007	.0034	.0116	.0298	.0612	.1048	.1527
	10	.0000	.0000	.0000	.0001	.0007	.0030	.0096	.0245	.0515	.0916
	11	.0000	.0000	.0000	.0000	.0001	.0006	.0024	.0074	.0191	.0417
	12	.0000	.0000	.0000	.0000	.0000	.0001	.0004	.0016	.0052	.0139
	13	.0000	.0000	.0000	.0000	.0000	.0000	.0001	.0003	.0010	.0032
	14	.0000	.0000	.0000	.0000	.0000	.0000	.0000	.0000	.0001	.0005
	15	.0000	.0000	.0000	.0000	.0000	.0000	.0000	.0000	.0000	.0000

INDIVIDUAL TERMS, BINOMIAL DISTRIBUTION (Continued)

						θ					
n	x	.05	.10	.15	.20	.25	.30	.35	.40	.45	.50
16	0	.4401	.1853	.0743	.0281	.0100	.0033	.0010	.0003	.0001	.0000
	1	.3706	.3294	.2097	.1126	.0535	.0228	.0087	.0030	.0009	.0002
	2	.1463	.2745	.2775	.2111	.1336	.0732	.0353	.0150	.0056	.0018
	3	.0359	.1423	.2285	.2463	.2079	.1465	.0888	.0468	.0215	.0085
	4	.0061	.0514	.1311	.2001	.2252	.2040	.1553	.1014	.0572	.0278
	5	.0008	.0137	.0555	.1201	.1802	.2099	.2008	.1623	.1123	.0667
	6	.0001	.0028	.0180	.0550	.1101	.1649	.1982	.1983	.1684	.1222
	7	.0000	.0004	.0045	.0197	.0524	.1010	.1524	.1889	.1969	.1746
	8	.0000	.0001	.0009	.0055	.0197	.0487	.0923	.1417	.1812	.1964
	9	.0000	.0000	.0001	.0012	.0058	.0185	.0442	.0840	.1318	.1746
	10	.0000	.0000	.0000	.0002	.0014	.0056	.0167	.0392	.0755	.1222
	11	.0000	.0000	.0000	.0000	.0002	.0013	.0049	.0142	.0337	.0667
	12	.0000	.0000	.0000	.0000	.0000	.0002	.0011	.0040	.0115	.0278
	13	.0000	.0000	.0000	.0000	.0000	.0000	.0002	.0008	.0029	.0085
	14	.0000	.0000	.0000	.0000	.0000	.0000	.0000	.0001	.0005	.0018
	15	.0000	.0000	.0000	.0000	.0000	.0000	.0000	.0000	.0001	.0002
	16	.0000	.0000	.0000	.0000	.0000	.0000	.0000	.0000	.0000	.0000
17	0	.4181	.1668	.0631	.0225	.0075	.0023	.0007	.0002	.0000	.0000
	1	.3741	.3150	.1893	.0957	.0426	.0169	.0060	.0019	.0005	.0001
	2	.1575	.2800	.2673	.1914	.1136	.0581	.0260	.0102	.0035	.0010
	3	.0415	.1556	.2359	.2393	.1893	.1245	.0701	.0341	.0144	.0052
	4	.9076	.0605	.1457	.2093	.2209	.1868	.1320	.0796	.0411	.0182
	5	.0010	.0175	.0668	.1361	.1914	.2081	.1849	.1379	.0875	.0472
	6	.0001	.0039	.0236	.0680	.1276	.1784	.1991	.1839	.1432	.0944
	7	.0000	.0007	.0065	.0267	.0668	.1201	.1685	.1927	.1841	.1484
	8	.0000	.0001	.0014	.0084	.0279	.0644	.1134	.1606	.1883	.1855
	9	.0000	.0000	.0003	.0021	.0093	.0276	.0611	.1070	.1540	.1855
	10	.0000	.0000	.0000	.0004	.0025	.0095	.0263	.0571	.1008	.1484
	11	.0000	.0000	.0000	.0001	.0005	.0026	.0090	.0242	.0525	.0944
	12	.0000	.0000	.0000	.0000	.0001	.0006	.0024	.0081	.0215	.0472
	13	.0000	.0000	.0000	.0000	.0000	.0001	.0005	.0021	.0068	.0182
	14	.0000	.0000	.0000	.0000	.0000	.0000	.0001	.0004	.0016	.0052
	15	.0000	.0000	.0000	.0000	.0000	.0000	.0000	.0001	.0003	.0010
	16	.0000	.0000	.0000	.0000	.0000	.0000	.0000	.0000	.0000	.0001
	17	.0000	.0000	.0000	.0000	.0000	.0000	.0000	.0000	.0000	.0000
18	0	.3972	.1501	.0536	.0180	.0056	.0016	.0004	.0001	.0000	.0000
	1	.3763	.3002	.1704	.0811	.0338	.0126	.0042	.0012	.0003	.0001
	2	.1683	.2835	.2556	.1723	.0958	.0458	.0190	.0069	.0022	.0006
	3	.0473	.1680	.2406	.2297	.1704	.1046	.0547	.0246	.0095	.0031
	4	.0093	.0700	.1592	.2153	.2130	.1681	.1104	.0614	.0291	.0117
	5	.0014	.0218	.0787	.1507	.1988	.2017	.1664	.1146	.0666	.0327
	6	.0002	.0052	.0301	.0816	.1436	.1873	.1941	.1655	.1181	.0708
	7	.0000	.0010	.0091	.0350	.0820	.1376	.1792	.1892	.1657	.1214
	8	.0000	.0002	.0022	.0120	.0376	0811	.1327	.1734	.1864	.1669
	9	.0000	.0000	.0004	.0033	.0139	.0380	.0794	.1284	.1694	.1855
	10	.0000	.0000	.0001	.0008	.0042	.0149	.0385	.0771	.1248	.1669
	11	.0000	.0000	.0000	.0001	.0010	.0046	.0151	.0374	.0742	.1214

INDIVIDUAL TERMS, BINOMIAL DISTRIBUTION (Continued)

n	x	θ .05	.10	.15	.20	.25	.30	.35	.40	.45	.50
18	12	.0000	.0000	.0000	.0000	.0002	.0012	.0047	.0145	.0354	.0708
	13	.0000	.0000	.0000	.0000	.0000	.0002	.0012	.0045	.0134	.0327
	14	.0000	.0000	.0000	.0000	.0000	.0000	.0002	.0011	.0039	.0117
	15	.0000	.0000	.0000	.0000	.0000	.0000	.0000	.0002	.0009	.0031
	16	.0000	.0000	.0000	.0000	.0000	.0000	.0000	.0000	.0001	.0006
	17	.0000	.0000	.0000	.0000	.0000	.0000	.0000	.0000	.0000	.0001
	18	.0000	.0000	.0000	.0000	.0000	.0000	.0000	.0000	.0000	.0000
19	0	.3774	.1351	.0456	.0144	.0042	.0011	.0003	.0001	.0000	.0000
	1	.3774	.2852	.1529	.0685	.0268	.0093	.0029	.0008	.0002	.0000
	2	.1787	.2852	.2428	.1540	.0803	.0358	.0138	.0046	.0013	.0003
	3	.0533	.1796	.2428	.2182	.1517	.0869	.0422	.0175	.0062	.0018
	4	.0112	.0798	.1714	.2182	.2023	.1491	.0909	.0467	.0203	.0074
	5	.0018	.0266	.0907	.1636	.2023	.1916	.1468	.0933	.0497	.0222
	6	.0002	.0069	.0374	.0955	.1574	.1916	.1844	.1451	.0949	.0518
	7	.0000	.0014	.0122	.0443	.0974	.1525	.1844	.1797	.1443	.0961
	8	.0000	.0002	.0032	.0166	.0487	.0981	.1489	.1797	.1771	.1442
	9	.0000	.0000	.0007	.0051	.0198	.0514	.0980	.1464	.1771	.1762
	10	.0000	.0000	.0001	.0013	.0066	.0220	.0528	.0976	.1449	.1762
	11	.0000	.0000	.0000	.0003	.0018	.0077	.0233	.0532	.0970	.1442
	12	.0000	.0000	.0000	.0000	.0004	.0022	.0083	.0237	.0529	.0961
	13	.0000	.0000	.0000	.0000	.0001	.0005	.0024	.0085	.0233	.0518
	14	.0000	.0000	.0000	.0000	.0000	.0001	.0006	.0024	.0082	.0222
	15	.0000	.0000	.0000	.0000	.0000	.0000	.0001	.0005	.0022	.0074
	16	.0000	.0000	.0000	.0000	.0000	.0000	.0000	.0001	.0005	.0018
	17	.0000	.0000	.0000	.0000	.0000	.0000	.0000	.0000	.0001	.0003
	18	.0000	.0000	.0000	.0000	.0000	.0000	.0000	.0000	.0000	.0000
	19	.0000	.0000	.0000	.0000	.0000	.0000	.0000	.0000	.0000	.0000
20	0	.3585	.1216	.0388	.0115	.0032	.0008	.0002	.0000	.0000	.0000
	1	.3774	.2702	.1368	.0576	.0211	.0068	.0020	.0005	.0001	.0000
	2	.1887	.2852	.2293	.1369	.0669	.0278	.0100	.0031	.0008	.0002
	3	.0596	.1901	.2428	.2054	.1339	.0716	.0323	.0123	.0040	.0011
	4	.0133	.0898	.1821	.2182	.1897	.1304	.0738	.0350	.0139	.0046
	5	.0022	.0319	.1028	.1746	.2023	.1789	.1272	.0746	.0365	.0148
	6	.0003	.0089	.0454	.1091	.1686	.1916	.1712	.1244	.0746	.0370
	7	.0000	.0020	.0160	.0545	.1124	.1643	.1844	.1659	.1221	.0739
	8	.0000	.0004	.0046	.0222	.0609	.1144	.1614	.1797	.1623	.1201
	9	.0000	.0001	.0011	.0074	.0271	.0654	.1158	.1597	.1771	.1602
	10	.0000	.0000	.0002	.0020	.0099	.0308	.0686	.1171	.1593	.1762
	11	.0000	.0000	.0000	.0005	.0030	.0120	.0336	.0710	.1185	.1602
	12	.0000	.0000	.0000	.0001	.0008	.0039	.0136	.0355	.0727	.1201
	13	.0000	.0000	.0000	.0000	.0002	.0010	.0045	.0146	.0366	.0739
	14	.0000	.0000	.0000	.0000	.0000	.0002	.0012	.0049	.0150	.0370
	15	.0000	.0000	.0000	.0000	.0000	.0000	.0003	.0013	.0049	.0148
	16	.0000	.0000	.0000	.0000	.0000	.0000	.0000	.0003	.0013	.0046
	17	.0000	.0000	.0000	.0000	.0000	.0000	.0000	.0000	.0002	.0011
	18	.0000	.0000	.0000	.0000	.0000	.0000	.0000	.0000	.0000	.0002
	19	.0000	.0000	.0000	.0000	.0000	.0000	.0000	.0000	.0000	.0000
	20	.0000	.0000	.0000	.0000	.0000	.0000	.0000	.0000	.0000	.0000

CUMULATIVE TERMS, BINOMIAL DISTRIBUTION

For the binomial probability function $f(x;n,\theta)$ the probability of observing x' or more successes is given by

$$\sum_{x=x'}^{n} \binom{n}{x} \theta^x(1-\theta)^{n-x},$$

This table contains the values of $\sum_{x=x'}^{n} \binom{n}{x} \theta^x(1-\theta)^{n-x}$ for specified values of n, x', and θ. If $\theta > 0.5$, the values for $\sum_{x=x'}^{n} \binom{n}{x} \theta^x(1-\theta)^{n-x}$ are obtained using the corresponding results obtained from

$$1 - \sum_{x=n-x'+1}^{n} \binom{n}{x} (1-\theta)^x\theta^{n-x}$$

The cumulative binomial distribution is related to the incomplete beta function as follows:

$$\sum_{x=x'}^{n} \binom{n}{x} \theta^x(1-\theta)^{n-x} = I_\theta(x', n-x'+1),$$

$$\begin{aligned}\sum_{x=0}^{x'-1} \binom{n}{x} \theta^x(1-\theta)^{n-x} &= 1 - I_\theta(x', n-x'+1)\\ &= 1 - \int_0^\theta u^{x'-1}(1-u)^{n-x'}\,du \Big/ \int_0^1 u^{x'-1}(1-u)^{n-x'}\,du\,.\end{aligned}$$

The cumulative binomial distribution is related to the cumulative negative binomial distribution as follows:

$$1 - \sum_{x'=0}^{r-1} \binom{x+r}{x'} \theta^{x'}(1-\theta)^{x+r-x'} = \sum_{x'=0}^{x} \binom{x'+r-1}{r-1} \theta^r(1-\theta)^{x'}$$

or

$$\sum_{x'=r}^{x+r} \binom{x+r}{x'} \theta^{x'}(1-\theta)^{x+r-x'} = \sum_{x'=0}^{x} \binom{x'+r-1}{r-1} \theta^r(1-\theta)^{x'}.$$

CUMULATIVE TERMS, BINOMIAL DISTRIBUTION (Continued)

						θ					
n	x'	.05	.10	.15	.20	.25	.30	.35	.40	.45	.50
2	1	.0975	.1900	.2775	.3600	.4375	.5100	.5775	.6400	.6975	.7500
	2	.0025	.0100	.0225	.0400	.0625	.0900	.1225	.1600	.2025	.2500
3	1	.1426	.2710	.3859	.4880	.5781	.6570	.7254	.7840	.8336	.8750
	2	.0072	.0280	.0608	.1040	.1562	.2160	.2818	.3520	.4252	.5000
	3	.0001	.0010	.0034	.0080	.0156	.0270	.0429	.0640	.0911	.1250
4	1	.1855	.3439	.4780	.5904	.6836	.7599	.8215	.8704	.9085	.9375
	2	.0140	.0523	.1095	.1808	.2617	.3483	.4370	.5248	.6090	.6875
	3	.0005	.0037	.0120	.0272	.0508	.0837	.1265	.1792	.2415	.3125
	4	.0000	.0001	.0005	.0016	.0039	.0081	.0150	.0256	.0410	.0625
5	1	.2262	.4095	.5563	.6723	.7627	.8319	.8840	.9222	.9497	.9688
	2	.0226	.0815	.1648	.2627	.3672	.4718	.5716	.6630	.7438	.8125
	3	.0012	.0086	.0266	.0579	.1035	.1631	.2352	.3174	.4069	.5000
	4	.0000	.0005	.0022	.0067	.0156	.0308	.0540	.0870	.1312	.1875
	5	.0000	.0000	.0001	.0003	.0010	.0024	.0053	.0102	.0185	.0312
6	1	.2649	.4686	.6229	.7379	.8220	.8824	.9246	.9533	.9723	.9844
	2	.0328	.1143	.2235	.3447	.4661	.5798	.6809	.7667	.8364	.8906
	3	.0022	.0158	.0473	.0989	.1694	.2557	.3529	.4557	.5585	.6562
	4	.0001	.0013	.0059	.0170	.0376	.0705	.1174	.1792	.2553	.3438
	5	.0000	.0001	.0004	.0016	.0046	.0109	.0223	.0410	.0692	.1094
	6	.0000	.0000	.0000	.0001	.0002	.0007	.0018	.0041	.0083	.0156
7	1	.3017	.5217	.6794	.7903	.8665	.9176	.9510	.9720	.9848	.9922
	2	.0444	.1497	.2834	.4233	.5551	.6706	.7662	.8414	.8976	.9375
	3	.0038	.0257	.0738	.1480	.2436	.3529	.4677	.5801	.6836	.7734
	4	.0002	.0027	.0121	.0333	.0706	.1260	.1998	.2898	.3917	.5000
	5	.0000	.0002	.0012	.0047	.0129	.0288	.0556	.0963	.1529	.2266
	6	.0000	.0000	.0001	.0004	.0013	.0038	.0090	.0188	.0357	.0625
	7	.0000	.0000	.0000	.0000	.0001	.0002	.0006	.0016	.0037	.0078
8	1	.3366	.5695	.7275	.8322	.8999	.9424	.9681	.9832	.9916	.9961
	2	.0572	.1869	.3428	.4967	.6329	.7447	.8309	.8936	.9368	.9648
	3	.0058	.0381	.1052	.2031	.3215	.4482	.5722	.6846	.7799	.8555
	4	.0004	.0050	.0214	.0563	.1138	.1941	.2936	.4059	.5230	.6367
	5	.0000	.0004	.0029	.0104	.0273	.0580	.1061	.1737	.2604	.3633
	6	.0000	.0000	.0002	.0012	.0042	.0113	.0253	.0498	.0885	.1445
	7	.0000	.0000	.0000	.0001	.0004	.0013	.0036	.0085	.0181	.0352
	8	.0000	.0000	.0000	.0000	.0000	.0001	.0002	.0007	.0017	.0039
9	1	.3698	.6126	.7684	.8658	.9249	.9596	.9793	.9899	.9954	.9980
	2	.0712	.2252	.4005	.5638	.6997	.8040	.8789	.9295	.9615	.9805
	3	.0084	.0530	.1409	.2618	.3993	.5372	.6627	.7682	.8505	.9102
	4	.0006	.0083	.0339	.0856	.1657	.2703	.3911	.5174	.6386	.7461
	5	.0000	.0009	.0056	.0196	.0489	.0988	.1717	.2666	.3786	.5000
	6	.0000	.0001	.0006	.0031	.0100	.0253	.0536	.0994	.1658	.2539
	7	.0000	.0000	.0000	.0003	.0013	.0043	.0112	.0250	.0498	.0898
	8	.0000	.0000	.0000	.0000	.0001	.0004	.0014	.0038	.0091	.0195
	9	.0000	.0000	.0000	.0000	.0000	.0000	.0001	.0003	.0008	.0020

Linear interpolation will be accurate at most to two decimal places.

CUMULATIVE TERMS, BINOMIAL DISTRIBUTION (Continued)

n	x'	θ .05	.10	.15	.20	.25	.30	.35	.40	.45	.50
10	1	.4013	.6513	.8031	.8926	.9437	.9718	.9865	.9940	.9975	.9990
	2	.0861	.2639	.4557	.6242	.7560	.8507	.9140	.9536	.9767	.9893
	3	.0115	.0702	.1798	.3222	.4744	.6172	.7384	.8327	.9004	.9453
	4	.0010	.0128	.0500	.1209	.2241	.3504	.4862	.6177	.7340	.8281
	5	.0001	.0016	.0099	.0328	.0781	.1503	.2485	.3669	.4956	.6230
	6	.0000	.0001	.0014	.0064	.0197	.0473	.0949	.1662	.2616	.3770
	7	.0000	.0000	.0001	.0009	.0035	.0106	.0260	.0548	.1020	.1719
	8	.0000	.0000	.0000	.0001	.0004	.0016	.0048	.0123	.0274	.0547
	9	.0000	.0000	.0000	.0000	.0000	.0001	.0005	.0017	.0045	.0107
	10	.0000	.0000	.0000	.0000	.0000	.0000	.0000	.0001	.0003	.0010
11	1	.4312	.6862	.8327	.9141	.9578	.9802	.9912	.9964	.9986	.9995
	2	.1019	.3026	.5078	.6779	.8029	.8870	.9394	.9698	.9861	.9941
	3	.0152	.0896	.2212	.3826	.5448	.6873	.7999	.8811	.9348	.9673
	4	.0016	.0185	.0694	.1611	.2867	.4304	.5744	.7037	.8089	.8867
	5	.0001	.0028	.0159	.0504	.1146	.2103	.3317	.4672	.6029	.7256
	6	.0000	.0003	.0027	.0117	.0343	.0782	.1487	.2465	.3669	.5000
	7	.0000	.0000	.0003	.0020	.0076	.0216	.0501	.0994	.1738	.2744
	8	.0000	.0000	.0000	.0002	.0012	.0043	.0122	.0293	.0610	.1133
	9	.0000	.0000	.0000	.0000	.0001	.0006	.0020	.0059	.0148	.0327
	10	.0000	.0000	.0000	.0000	.0000	.0000	.0002	.0007	.0022	.0059
	11	.0000	.0000	.0000	.0000	.0000	.0000	.0000	.0000	.0002	.0005
12	1	.4596	.7176	.8578	.9313	.9683	.9862	.9943	.9978	.9992	.9998
	2	.1184	.3410	.5565	.7251	.8416	.9150	.9576	.9804	.9917	.9968
	3	.0196	.1109	.2642	.4417	.6093	.7472	.8487	.9166	.9579	.9807
	4	.0022	.0256	.0922	.2054	.3512	.5075	.6533	.7747	.8655	.9270
	5	.0002	.0043	.0239	.0726	.1576	.2763	.4167	.5618	.6956	.8062
	6	.0000	.0005	.0046	.0194	.0544	.1178	.2127	.3348	.4731	.6128
	7	.0000	.0001	.0007	.0039	.0143	.0386	.0846	.1582	.2607	.3872
	8	.0000	.0000	.0001	.0006	.0028	.0095	.0255	.0573	.1117	.1938
	9	.0000	.0000	.0000	.0001	.0004	.0017	.0056	.0153	.0356	.0730
	10	.0000	.0000	.0000	.0000	.0000	.0002	.0008	.0028	.0079	.0193
	11	.0000	.0000	.0000	.0000	.0000	.0000	.0001	.0003	.0011	.0032
	12	.0000	.0000	.0000	.0000	.0000	.0000	.0000	.0000	.0001	.0002
13	1	.4867	.7458	.8791	.9450	.9762	.9903	.9963	.9987	.9996	.9999
	2	.1354	.3787	.6017	.7664	.8733	.9363	.9704	9874	.9951	.9983
	3	.0245	.1339	.3080	.4983	.6674	.7975	.8868	.9421	.9731	.9888
	4	.0031	.0342	.1180	.2527	.4157	.5794	.7217	.8314	.9071	.9539
	5	.0003	.0065	.0342	.0991	.2060	.3457	.4995	.6470	.7721	.8666
	6	.0000	.0009	.0075	.0300	.0802	.1654	.2841	.4256	.5732	.7095
	7	.0000	.0001	.0013	.0070	.0243	.0624	.1295	.2288	.3563	.5000
	8	.0000	.0000	.0002	.0012	.0056	.0182	.0462	.0977	.1788	.2905
	9	.0000	.0000	.0000	.0002	.0010	.0040	.0126	.0321	.0698	.1334
	10	.0000	.0000	.0000	.0000	.0001	.0007	.0025	.0078	.0203	.0461
	11	.0000	.0000	.0000	.0000	.0000	.0001	.0003	.0013	.0041	.0112
	12	.0000	.0000	.0000	.0000	.0000	.0000	.0000	.0001	.0005	.0017
	13	.0000	.0000	.0000	.0000	.0000	.0000	.0000	.0000	.0000	.0001

CUMULATIVE TERMS, BINOMIAL DISTRIBUTION (Continued)

		θ									
n	x'	.05	.10	.15	.20	.25	.30	.35	.40	.45	.50
14	1	.5123	.7712	.8972	.9560	.9822	.9932	.9976	.9992	.9998	.9999
	2	.1530	.4154	.6433	.8021	.8990	.9525	.9795	.9919	.9971	.9991
	3	.0301	.1584	.3521	.5519	.7189	.8392	.9161	.9602	.9830	.9935
	4	.0042	.0441	.1465	.3018	.4787	.6448	.7795	.8757	.9368	.9713
	5	.0004	.0092	.0467	.1298	.2585	.4158	.5773	.7207	.8328	.9102
	6	.0000	.0015	.0115	.0439	.1117	.2195	.3595	.5141	.6627	.7880
	7	.0000	.0002	.0022	.0116	.0383	.0933	.1836	.3075	.4539	.6047
	8	.0000	.0000	.0003	.0024	.0103	.0315	.0753	.1501	.2586	.3953
	9	.0000	.0000	.0000	.0004	.0022	.0083	.0243	.0583	.1189	.2120
	10	.0000	.0000	.0000	.0000	.0003	.0017	.0060	.0175	.0426	.0898
	11	.0000	.0000	.0000	.0000	.0000	.0002	.0011	.0039	.0114	.0287
	12	.0000	.0000	.0000	.0000	.0000	.0000	.0001	.0006	.0022	.0065
	13	.0000	.0000	.0000	.0000	.0000	.0000	.0000	.0001	.0003	.0009
	14	.0000	.0000	.0000	.0000	.0000	.0000	.0000	.0000	.0000	.0001
15	1	.5367	.7941	.9126	.9648	.9866	.9953	.9984	.9995	.9999	1.0000
	2	.1710	.4510	.6814	.8329	.9198	.9647	.9858	.9948	.9983	.9995
	3	.0362	.1841	.3958	.6020	.7639	.8732	.9383	.9729	.9893	.9963
	4	.0055	.0556	.1773	.3518	.5387	.7031	.8273	.9095	.9576	.9824
	5	.0006	.0127	.0617	.1642	.3135	.4845	.6481	.7827	.8796	.9408
	6	.0001	.0022	.0168	.0611	.1484	.2784	.4357	.5968	.7392	.8491
	7	.0000	.0003	.0036	.0181	.0566	.1311	.2452	.3902	.5478	.6964
	8	.0000	.0000	.0006	.0042	.0173	.0500	.1132	.2131	.3465	.5000
	9	.0000	.0000	.0001	.0008	.0042	.0152	.0422	.0950	.1818	.3036
	10	.0000	.0000	.0000	.0001	.0008	.0037	.0124	.0338	.0769	.1509
	11	.0000	.0000	.0000	.0000	.0001	.0007	.0028	.0093	.0255	.0592
	12	.0000	.0000	.0000	.0000	.0000	.0001	.0005	.0019	.0063	.0176
	13	.0000	.0000	.0000	.0000	.0000	.0000	.0001	.0003	.0011	.0037
	14	.0000	.0000	.0000	.0000	.0000	.0000	.0000	.0000	.0001	.0005
	15	.0000	.0000	.0000	.0000	.0000	.0000	.0000	.0000	.0000	.0000
16	1	.5599	.8147	.9257	.9719	.9900	.9967	.9990	.9997	.9999	1.0000
	2	.1892	.4853	.7161	.8593	.9365	.9739	.9902	.9967	.9990	.9997
	3	.0429	.2108	.4386	.6482	.8029	.9006	.9549	.9817	.9934	.9979
	4	.0070	.0684	.2101	.4019	.5950	.7541	.8661	.9349	.9719	.9894
	5	.0009	.0170	.0791	.2018	.3698	.5501	.7108	.8334	.9147	.9616
	6	.0001	.0033	.0235	.0817	.1897	.3402	.5100	.6712	.8024	.8949
	7	.0000	.0005	.0056	.0267	.0796	.1753	.3119	.4728	.6340	.7228
	8	.0000	.0001	.0011	.0070	.0271	.0744	.1594	.2839	.4371	.5982
	9	.0000	.0000	.0002	.0015	.0075	.0257	.0671	.1423	.2559	.4018
	10	.0000	.0000	.0000	.0002	.0016	.0071	.0229	.0583	.1241	.2272
	11	.0000	.0000	.0000	.0000	.0003	.0016	.0062	.0191	.0486	.1051
	12	.0000	.0000	.0000	.0000	.0000	.0003	.0013	.0049	.0149	.0384
	13	.0000	.0000	.0000	.0000	.0000	.0000	.0002	.0009	.0035	.0106
	14	.0000	.0000	.0000	.0000	.0000	.0000	.0000	.0001	.0006	.0021
	15	.0000	.0000	.0000	.0000	.0000	.0000	.0000	.0000	.0001	.0003
	16	.0000	.0000	.0000	.0000	.0000	.0000	.0000	.0000	.0000	.0000

CUMULATIVE TERMS, BINOMIAL DISTRIBUTION (Continued)

n	x′	θ .05	.10	.15	.20	.25	.30	.35	.40	.45	.50
17	1	.5819	.8332	.9369	.9775	.9925	.9977	.9993	.9998	1.0000	1.0000
	2	.2078	.5182	.7475	.8818	.9499	.9807	.9933	.9979	.9994	.9999
	3	.0503	.2382	.4802	.6904	.8363	.9226	.9673	.9877	.9959	.9988
	4	.0088	.0826	.2444	.4511	.6470	.7981	.8972	.9536	.9816	.9936
	5	.0012	.0221	.0987	.2418	.4261	.6113	.7652	.8740	.9404	.9755
	6	.0001	.0047	.0319	.1057	.2347	.4032	.5803	.7361	.8529	.9283
	7	.0000	.0008	.0083	.0377	.1071	.2248	.3812	.5522	.7098	.8338
	8	.0000	.0001	.0017	.0109	.0402	.1046	.2128	.3595	.5257	.6855
	9	.0000	.0000	.0003	.0026	.0124	.0403	.0994	.1989	.3374	.5000
	10	.0000	.0000	.0000	.0005	.0031	.0127	.0383	.0919	.1834	.3145
	11	.0000	.0000	.0000	.0001	.0006	.0032	.0120	.0348	.0826	.1662
	12	.0000	.0000	.0000	.0000	.0001	.0007	.0030	.0106	.0301	.0717
	13	.0000	.0000	.0000	.0000	.0000	.0001	.0006	.0025	.0086	.0245
	14	.0000	.0000	.0000	.0000	.0000	.0000	.0001	.0005	.0019	.0064
	15	.0000	.0000	.0000	.0000	.0000	.0000	.0000	.0001	.0003	.0012
	16	.0000	.0000	.0000	.0000	.0000	.0000	.0000	.0000	.0000	.0001
	17	.0000	.0000	.0000	.0000	.0000	.0000	.0000	.0000	.0000	.0000
18	1	.6028	.8499	.9464	.9820	.9944	.9984	.9996	.9999	1.0000	1.0000
	2	.2265	.5497	.7759	.9009	.9605	.9858	.9954	.9987	.9997	.9999
	3	.0581	.2662	.5203	.7287	.8647	.9400	.9764	.9918	.9975	.9993
	4	.0109	.0982	.2798	.4990	.6943	.8354	.9217	.9672	.9880	.9962
	5	.0015	.0282	.1206	.2836	.4813	.6673	.8114	.9058	.9589	.9846
	6	.0002	.0064	.0419	.1329	.2825	.4656	.6450	.7912	.8923	.9519
	7	.0000	.0012	.0118	.0513	.1390	.2783	.4509	.6257	.7742	.8811
	8	.0000	.0002	.0027	.0163	.0569	.1407	.2717	.4366	.6085	.7597
	9	.0000	.0000	.0005	.0043	.0193	.0596	.1391	.2632	.4222	.5927
	10	.0000	.0000	.0001	.0009	.0054	.0210	.0597	.1347	.2527	.4073
	11	.0000	.0000	.0000	.0002	.0012	.0061	.0212	.0576	.1280	.2403
	12	.0000	.0000	.0000	.0000	.0002	.0014	.0062	.0203	.0537	.1189
	13	.0000	.0000	.0000	.0000	.0000	.0003	.0014	.0058	.0183	.0481
	14	.0000	.0000	.0000	.0000	.0000	.0000	.0003	.0013	.0049	.0154
	15	.0000	.0000	.0000	.0000	.0000	.0000	.0000	.0002	.0010	.0038
	16	.0000	.0000	.0000	.0000	.0000	.0000	.0000	.0000	.0001	.0007
	17	.0000	.0000	.0000	.0000	.0000	.0000	.0000	.0000	.0000	.0001
	18	.0000	.0000	.0000	.0000	.0000	.0000	.0000	.0000	.0000	.0000
19	1	.6226	.8649	.9544	.9856	.9958	.9989	.9997	.9999	1.0000	1.0000
	2	.2453	.5797	.8015	.9171	.9690	.9896	.9969	.9992	.9998	1.0000
	3	.0665	.2946	.5587	.7631	.8887	.9538	.9830	.9945	.9985	.9996
	4	.0132	.1150	.3159	.5449	.7369	.8668	.9409	.9770	.9923	.9978
	5	.0020	.0352	.1444	.3267	.5346	.7178	.8500	.9304	.9720	.9904
	6	.0002	.0086	.0537	.1631	.3322	.5261	.7032	.8371	.9223	.9682
	7	.0000	.0017	.0163	.0676	.1749	.3345	.5188	.6919	.8273	.9165
	8	.0000	.0003	.0041	.0233	.0775	.1820	.3344	.5122	.6831	.8204
	9	.0000	.0000	.0008	.0067	.0287	.0839	.1855	.3325	.5060	.6762
	10	.0000	.0000	.0001	.0016	.0089	.0326	.0875	.1861	.3290	.5000

CUMULATIVE TERMS, BINOMIAL DISTRIBUTION (Continued)

n	x'	.05	.10	.15	.20	θ .25	.30	.35	.40	.45	.50
19	11	.0000	.0000	.0000	.0003	.0023	.0105	.0347	.0885	.1841	.3238
	12	.0000	.0000	.0000	.0000	.0005	.0028	.0114	.0352	.0871	.1796
	13	.0000	.0000	.0000	.0000	.0001	.0006	.0031	.0116	.0342	.0835
	14	.0000	.0000	.0000	.0000	.0000	.0001	.0007	.0031	.0109	.0318
	15	.0000	.0000	.0000	.0000	.0000	.0000	.0001	.0006	.0028	.0096
	16	.0000	.0000	.0000	.0000	.0000	.0000	.0000	.0001	.0005	.0022
	17	.0000	.0000	.0000	.0000	.0000	.0000	.0000	.0000	.0001	.0004
	18	.0000	.0000	.0000	.0000	.0000	.0000	.0000	.0000	.0000	.0000
	19	.0000	.0000	.0000	.0000	.0000	.0000	.0000	.0000	.0000	.0000
20	1	.6415	.8784	.9612	.9885	.9968	.9992	.9998	1.0000	1.0000	1.0000
	2	.2642	.6083	.8244	.9308	.9757	.9924	.9979	.9995	.9999	1.0000
	3	.0755	.3231	.5951	.7939	.9087	.9645	.9879	.9964	.9991	.9998
	4	.0159	.1330	.3523	.5886	.7748	.8929	.9556	.9840	.9951	.9987
	5	.0026	.0432	.1702	.3704	.5852	.7625	.8818	.9490	.9811	.9941
	6	.0003	.0113	.0673	.1958	.3828	.5836	.7546	.8744	.9447	.9793
	7	.0000	.0024	.0219	.0867	.2142	.3920	.5834	.7500	.8701	.9423
	8	.0000	.0004	.0059	.0321	.1018	.2277	.3990	.5841	.7480	.8684
	9	.0000	.0001	.0013	.0100	.0409	.1133	.2376	.4044	.5857	.7483
	10	.0000	.0000	.0002	.0026	.0139	.0480	.1218	.2447	.4086	.5881
	11	.0000	.0000	.0000	.0006	.0039	.0171	.0532	.1275	.2493	.4119
	12	.0000	.0000	.0000	.0001	.0009	.0051	.0196	.0565	.1308	.2517
	13	.0000	.0000	.0000	.0000	.0002	.0013	.0060	.0210	.0580	.1316
	14	.0000	.0000	.0000	.0000	.0000	.0003	.0015	.0065	.0214	.0577
	15	.0000	.0000	.0000	.0000	.0000	.0000	.0003	.0016	.0064	.0207
	16	.0000	.0000	.0000	.0000	.0000	.0000	.0000	.0003	.0015	.0059
	17	.0000	.0000	.0000	.0000	.0000	.0000	.0000	.0000	.0003	.0013
	18	.0000	.0000	.0000	.0000	.0000	.0000	.0000	.0000	.0000	.0002
	19	.0000	.0000	.0000	.0000	.0000	.0000	.0000	.0000	.0000	.0000
	20	.0000	.0000	.0000	.0000	.0000	.0000	.0000	.0000	.0000	.0000
21	1	.6594	.8906	.9671	.9908	.9976	.9994	.9999	1.0000	1.0000	1.0000
	2	.2830	.6353	.8450	.9424	.9810	.9944	.9996	.9997	.9999	1.0000
	3	.0849	.3516	.6295	.8213	.9255	.9729	.9914	.9976	.9994	.9999
	4	.0189	.1520	.3887	.6296	.8083	.9144	.9669	.9890	.9969	.9993
	5	.0032	.0522	.1975	.4140	.6326	.8016	.9076	.9630	.9874	.9967
	6	.0004	.0144	.0827	.2307	.4334	.6373	.7991	.9043	.9611	.9867
	7	.0000	.0033	.0287	.1085	.2564	.4495	.6433	.7998	.9036	.9608
	8	.0000	.0006	.0083	.0431	.1299	.2770	.4635	.6505	.8029	.9054
	9	.0000	.0001	.0020	.0144	.0561	.1477	.2941	.4763	.6587	.8083
	10	.0000	.0000	.0004	.0041	.0206	.0676	.1632	.3086	.4883	.6682
	11	.0000	.0000	.0001	.0010	.0064	.0264	.0772	.1744	.3210	.5000
	12	.0000	.0000	.0000	.0002	.0017	.0087	.0313	.0849	.1841	.3318
	13	.0000	.0000	.0000	.0000	.0004	.0024	.0108	.0352	.0908	.1917
	14	.0000	.0000	.0000	.0000	.0001	.0006	.0031	.0123	.0379	.0946
	15	.0000	.0000	.0000	.0000	.0000	.0001	.0007	.0036	.0132	.0392
	16	.0000	.0000	.0000	.0000	.0000	.0000	.0001	.0008	.0037	.0133
	17	.0000	.0000	.0000	.0000	.0000	.0000	.0000	.0002	.0008	.0036
	18	.0000	.0000	.0000	.0000	.0000	.0000	.0000	.0000	.0001	.0007
	19	.0000	.0000	.0000	.0000	.0000	.0000	.0000	.0000	.0000	.0001

CUMULATIVE TERMS, BINOMIAL DISTRIBUTION (Continued)

n	x'	θ .05	.10	.15	.20	.25	.30	.35	.40	.45	.50
21	20	.0000	.0000	.0000	.0000	.0000	.0000	.0000	.0000	.0000	.0000
	21	.0000	.0000	.0000	.0000	.0000	.0000	.0000	.0000	.0000	.0000
22	1	.6765	.9015	.9720	.9926	.9982	.9966	.9999	1.0000	1.0000	1.0000
	2	.3018	.6608	.8633	.9520	.9851	.9959	.9990	.9998	1.0000	1.0000
	3	.0948	.3800	.6618	.8455	.9394	.9793	.9399	.9984	.9997	.9999
	4	.0222	.1719	.4248	.6680	.8376	.9319	.9755	.9924	.9980	.9996
	5	.0040	.0621	.2262	.4571	.6765	.8355	.9284	.9734	.9917	.9978
	6	.0006	.0182	.0999	.2674	.4832	.6866	.8371	.9278	.9729	.9915
	7	.0001	.0044	.0368	.1330	.3006	.5058	.6978	.8416	.9295	.9738
	8	.0000	.0009	.0114	.0561	.1615	.3287	.5264	.7102	.8482	.9331
	9	.0000	.0001	.0030	.0201	.0746	.1865	.3534	.5460	.7236	.8569
	10	.0000	.0000	.0007	.0061	.0295	.0916	.2084	.3756	.5650	.7383
	11	.0000	.0000	.0001	.0016	.0100	.0387	.1070	.2281	.3963	.5841
	12	.0000	.0000	.0000	.0003	.0029	.0140	.0474	.1207	.2457	.4159
	13	.0000	.0000	.0000	.0001	.0007	.0043	.0180	.0551	.1328	.2617
	14	.0000	.0000	.0000	.0000	.0001	.0011	.0058	.0215	.0617	.1431
	15	.0000	.0000	.0000	.0000	.0000	.0002	.0015	.0070	.0243	.0669
	16	.0000	.0000	.0000	.0000	.0000	.0000	.0003	.0019	.0080	.0262
	17	.0000	.0000	.0000	.0000	.0000	.0000	.0001	.0004	.0021	.0085
	18	.0000	.0000	.0000	.0000	.0000	.0000	.0000	.0001	.0005	.0022
	19	.0000	.0000	.0000	.0000	.0000	.0000	.0000	.0000	.0001	.0004
	20	.0000	.0000	.0000	.0000	.0000	.0000	.0000	.0000	.0000	.0001
	21	.0000	.0000	.0000	.0000	.0000	.0000	.0000	.0000	.0000	.0000
	22	.0000	.0000	.0000	.0000	.0000	.0000	.0000	.0000	.0000	.0000
23	1	.6926	.9114	.9762	.9941	.9987	.9997	1.0000	1.0000	1.0000	1.0000
	2	.3206	.6849	.8796	.9602	.9884	.9970	.9993	.9999	1.0000	1.0000
	3	.1052	.4080	.6920	.8668	.9508	.9843	.9957	.9990	1.0000	1.0000
	4	.0258	.1927	.4604	.7035	.8630	.9462	.9819	.9948	.9988	.9998
	5	.0049	.0731	.2560	.4993	.7168	.8644	.9449	.9810	.9945	.9987
	6	.0008	.0226	.1189	.3053	.5315	.7312	.8691	.9460	.9814	.9947
	7	.0001	.0058	.0463	.1598	.3463	.5601	.7466	.8760	.9490	.9827
	8	.0000	.0012	.0152	.0715	.1963	.3819	.5864	.7627	.8848	.9534
	9	.0000	.0002	.0042	.0273	.0963	.2291	.4140	.6116	.7797	.8950
	10	.0000	.0000	.0010	.0089	.0408	.1201	.2592	.4438	.6364	.7976
	11	.0000	.0000	.0002	.0025	.0149	.0546	.1425	.2871	.4722	.6612
	12	.0000	.0000	.0000	.0006	.0046	.0214	.0682	.1636	.3135	.5000
	13	.0000	.0000	.0000	.0001	.0012	.0072	.0283	.0813	.1836	.3388
	14	.0000	.0000	.0000	.0000	.0003	.0021	.0100	.0349	.0937	.2024
	15	.0000	.0000	.0000	.0000	.0001	.0005	.0030	.0128	.0411	.1050
	16	.0000	.0000	.0000	.0000	.0000	.0001	.0008	.0040	.0153	.0466
	17	.0000	.0000	.0000	.0000	.0000	.0000	.0002	.0010	.0048	.0173
	18	.0000	.0000	.0000	.0000	.0000	.0000	.0000	.0002	.0012	.0053
	19	.0000	.0000	.0000	.0000	.0000	.0000	.0000	.0000	.0002	.0013
	20	.0000	.0000	.0000	.0000	.0000	.0000	.0000	.0000	.0000	.0002
	21	.0000	.0000	.0000	.0000	.0000	.0000	.0000	.0000	.0000	.0000
	22	.0000	.0000	.0000	.0000	.0000	.0000	.0000	.0000	.0000	.0000
	23	.0000	.0000	.0000	.0000	.0000	.0000	.0000	.0000	.0000	.0000

CUMULATIVE TERMS, BINOMIAL DISTRIBUTION (Continued)

n	x'	θ .05	.10	.15	.20	.25	.30	.35	.40	.45	.50
24	1	.7080	.9202	.9798	.9953	.9990	.9998	1.0000	1.0000	1.0000	1.0000
	2	.3391	.7075	.8941	.9669	.9910	.9978	.9995	.9999	1.0000	1.0000
	3	.1159	.4357	.7202	.8855	.9602	.9881	.9970	.9993	.9999	1.0000
	4	.0298	.2143	.4951	.7361	.8850	.9576	.9867	.9965	.9992	.9999
	5	.0060	.0851	.2866	.5401	.7534	.8889	.9578	.9866	.9964	.9992
	6	.0010	.0277	.1394	.3441	.5778	.7712	.8956	.9600	.9873	.9967
	7	.0001	.0075	.0572	.1889	.3926	.6114	.7894	.9040	.9636	.9887
	8	.0000	.0017	.0199	.0892	.2338	.4353	.6425	.8081	.9137	.9680
	9	.0000	.0003	.0059	.0362	.1213	.2750	.4743	.6721	.8270	.9242
	10	.0000	.0001	.0015	.0126	.0547	.1528	.3134	.5109	.7009	.8463
	11	.0000	.0000	.0003	.0038	.0213	.0742	.1833	.3498	.5461	.7294
	12	.0000	.0000	.0001	.0010	.0072	.0314	.0942	.2130	.3849	.5806
	13	.0000	.0000	.0000	.0002	.0021	.0115	.0423	.1143	.2420	.4194
	14	.0000	.0000	.0000	.0000	.0005	.0036	.0164	.0535	.1341	.2706
	15	.0000	.0000	.0000	.0000	.0001	.0010	.0055	.0217	.0648	.1537
	16	.0000	.0000	.0000	.0000	.0000	.0002	.0016	.0075	.0269	.0758
	17	.0000	.0000	.0000	.0000	.0000	.0000	.0004	.0022	.0095	.0320
	18	.0000	.0000	.0000	.0000	.0000	.0000	.0001	.0005	.0028	.0113
	19	.0000	.0000	.0000	.0000	.0000	.0000	.0000	.0001	.0007	.0033
	20	.0000	.0000	.0000	.0000	.0000	.0000	.0000	.0000	.0001	.0008
	21	.0000	.0000	.0000	.0000	.0000	.0000	.0000	.0000	.0000	.0001
	22	.0000	.0000	.0000	.0000	.0000	.0000	.0000	.0000	.0000	.0000
	23	.0000	.0000	.0000	.0000	.0000	.0000	.0000	.0000	.0000	.0000
	24	.0000	.0000	.0000	.0000	.0000	.0000	.0000	.0000	.0000	.0000
25	1	.7226	.9282	.9828	.9962	.9992	.9999	1.0000	1.0000	1.0000	1.0000
	2	.3576	.7288	.9069	.9726	.9930	.9984	.9997	.9999	1.0000	1.0000
	3	.1271	.4629	.7463	.9018	.9679	.9910	.9979	.9996	.9999	1.0000
	4	.0341	.2364	.5289	.7660	.9038	.9668	.9903	.9976	.9995	.9999
	5	.0072	.0980	.3179	.5793	.7863	.9095	.9680	.9905	.9977	.9995
	6	.0012	.0334	.1615	.3833	.6217	.8065	.9174	.9706	.9914	.9980
	7	.0002	.0095	.0695	.2200	.4389	.6593	.8266	.9264	.9742	.9927
	8	.0000	.0023	.0255	.1091	.2735	.4882	.6939	.8464	.9361	.9784
	9	.0000	.0005	.0080	.0468	.1494	.3231	.5332	.7265	.8660	.9461
	10	.0000	.0001	.0021	.0173	.0713	.1894	.3697	.5754	.7576	.8852
	11	.0000	.0000	.0005	.0056	.0297	.0978	.2288	.4142	.6157	.7878
	12	.0000	.0000	.0001	.0015	.0107	.0442	.1254	.2677	.4574	.6550
	13	.0000	.0000	.0000	.0004	.0034	.0175	.0604	.1538	.3063	.5000
	14	.0000	.0000	.0000	.0001	.0009	.0060	.0255	.0778	.1827	.3450
	15	.0000	.0000	.0000	.0000	.0002	.0018	.0093	.0344	.0960	.2122
	16	.0000	.0000	.0000	.0000	.0000	.0005	.0029	.0132	.0440	.1148
	17	.0000	.0000	.0000	.0000	.0000	.0001	.0008	.0043	.0174	.0539
	18	.0000	.0000	.0000	.0000	.0000	.0000	.0002	.0012	.0058	.0216
	19	.0000	.0000	.0000	.0000	.0000	.0000	.0000	.0003	.0016	.0073
	20	.0000	.0000	.0000	.0000	.0000	.0000	.0000	.0001	.0004	.0020
	21	.0000	.0000	.0000	.0000	.0000	.0000	.0000	.0000	.0001	.0005
	22	.0000	.0000	.0000	.0000	.0000	.0000	.0000	.0000	.0000	.0001
	23	.0000	.0000	.0000	.0000	.0000	.0000	.0000	.0000	.0000	.0000
	24	.0000	.0000	.0000	.0000	.0000	.0000	.0000	.0000	.0000	.0000

CUMULATIVE TERMS, BINOMIAL DISTRIBUTION (Continued)

		θ									
n	x'	.05	.10	.15	.20	.25	.30	.35	.40	.45	.50
25	25	.0000	.0000	.0000	.0000	.0000	.0000	.0000	.0000	.0000	.0000
30	1	.7854	.9576	.9924	.9988	.9998	1.0000	1.0000	1.0000	1.0000	1.0000
	2	.4465	.8163	.9520	.9895	.9980	.9997	1.0000	1.0000	1.0000	1.0000
	3	.1878	.5886	.8486	.9558	.9894	.9979	.9997	1.0000	1.0000	1.0000
	4	.0608	.3526	.6783	.8773	.9626	.9907	.9981	.9997	1.0000	1.0000
	5	.0156	.1755	.4755	.7448	.9021	.9698	.9925	.9985	.9998	1.0000
	6	.0033	.0732	.2894	.5725	.7974	.9234	.9767	.9943	.9989	.9998
	7	.0006	.0258	.1526	.3930	.6519	.8405	.9414	.9828	.9960	.9993
	8	.0001	.0078	.0698	.2392	.4857	.7186	.8762	.9565	.9879	.9974
	9	.0000	.0020	.0278	.1287	.3264	.5685	.7753	.9060	.9688	.9919
	10	.0000	.0005	.0097	.0611	.1966	.4112	.6425	.8237	.9306	.9786
	11	.0000	.0001	.0029	.0256	.1057	.2696	.4922	.7085	.8650	.9506
	12	.0000	.0000	.0008	.0095	.0507	.1593	.3452	.5689	.7673	.8998
	13	.0000	.0000	.0002	.0031	.0216	.0845	.2198	.4215	.6408	.8192
	14	.0000	.0000	.0000	.0009	.0082	.0401	.1263	.2855	.4975	.7077
	15	.0000	.0000	.0000	.0002	.0027	.0169	.0652	.1754	.3552	.5722
	16	.0000	.0000	.0000	.0001	.0008	.0064	.0301	.0971	.2309	.4278
	17	.0000	.0000	.0000	.0000	.0002	.0021	.0124	.0481	.1356	.2923
	18	.0000	.0000	.0000	.0000	.0001	.0006	.0045	.0212	.0714	.1808
	19	.0000	.0000	.0000	.0000	.0000	.0002	.0014	.0083	.0334	.1002
	20	.0000	.0000	.0000	.0000	.0000	.0000	.0004	.0029	.0138	.0494
	21	.0000	.0000	.0000	.0000	.0000	.0000	.0001	.0009	.0050	.0214
	22	.0000	.0000	.0000	.0000	.0000	.0000	.0000	.0002	.0016	.0081
	23	.0000	.0000	.0000	.0000	.0000	.0000	.0000	.0000	.0004	.0026
	24	.0000	.0000	.0000	.0000	.0000	.0000	.0000	.0000	.0001	.0007
	25	.0000	.0000	.0000	.0000	.0000	.0000	.0000	.0000	.0000	.0002
	26	.0000	.0000	.0000	.0000	.0000	.0000	.0000	.0000	.0000	.0000
	27	.0000	.0000	.0000	.0000	.0000	.0000	.0000	.0000	.0000	.0000
	28	.0000	.0000	.0000	.0000	.0000	.0000	.0000	.0000	.0000	.0000
	29	.0000	.0000	.0000	.0000	.0000	.0000	.0000	.0000	.0000	.0000
	30	.0000	.0000	.0000	.0000	.0000	.0000	.0000	.0000	.0000	.0000
35	1	.8339	.9750	.9966	.9996	1.0000	1.0000	1.0000	1.0000	1.0000	1.0000
	2	.5280	.8776	.9757	.9960	.9995	.9999	1.0000	1.0000	1.0000	1.0000
	3	.2542	.6937	.9130	.9810	.9967	.9995	.9999	1.0000	1.0000	1.0000
	4	.0958	.4690	.7912	.9395	.9864	.9976	.9997	1.0000	1.0000	1.0000
	5	.0290	.2693	.6193	.8565	.9590	.9909	.9984	.9998	1.0000	1.0000
	6	.0073	.1316	.4311	.7279	.9024	.9731	.9942	.9990	.9999	1.0000
	7	.0015	.0552	.2652	.5672	.8080	.9350	.9830	.9966	.9995	.9999
	8	.0003	.0200	.1438	.4007	.6777	.8674	.9581	.9898	.9981	.9997
	9	.0000	.0063	.0689	.2550	.5257	.7659	.9110	.9740	.9943	.9991
	10	.0000	.0017	.0292	.1457	.3737	.6354	.8349	.9425	.9848	.9970
	11	.0000	.0004	.0110	.0747	.2419	.4900	.7284	.8877	.9646	.9917
	12	.0000	.0001	.0037	.0344	.1421	.3484	.5981	.8048	.9271	.9795
	13	.0000	.0000	.0011	.0142	.0756	.2271	.4577	.6943	.8656	.9552
	14	.0000	.0000	.0003	.0053	.0363	.1350	.3240	.5639	.7767	.9123
	15	.0000	.0000	.0001	.0018	.0158	.0731	.2109	.4272	.6624	.8447
	16	.0000	.0000	.0000	.0005	.0062	.0359	.1256	.2997	.5315	.7502

CUMULATIVE TERMS, BINOMIAL DISTRIBUTION (Continued)

n	x'	θ .05	.10	.15	.20	.25	.30	.35	.40	.45	.50
35	17	.0000	.0000	.0000	.0001	.0022	.0160	.0682	.1935	.3976	.6321
	18	.0000	.0000	.0000	.0000	.0007	.0064	.0336	.1143	.2751	.5000
	19	.0000	.0000	.0000	.0000	.0002	.0023	.0150	.0615	.1749	.3679
	20	.0000	.0000	.0000	.0000	.0001	.0008	.0061	.0300	.1016	.2498
	21	.0000	.0000	.0000	.0000	.0000	.0002	.0022	.0133	.0536	.1553
	22	.0000	.0000	.0000	.0000	.0000	.0001	.0007	.0053	.0255	.0877
	23	.0000	.0000	.0000	.0000	.0000	.0000	.0002	.0019	.0109	.0448
	24	.0000	.0000	.0000	.0000	.0000	.0000	.0001	.0006	.0042	.0205
	25	.0000	.0000	.0000	.0000	.0000	.0000	.0000	.0002	.0014	.0083
	26	.0000	.0000	.0000	.0000	.0000	.0000	.0000	.0000	.0004	.0030
	27	.0000	.0000	.0000	.0000	.0000	.0000	.0000	.0000	.0001	.0009
	28	.0000	.0000	.0000	.0000	.0000	.0000	.0000	.0000	.0000	.0003
	29	.0000	.0000	.0000	.0000	.0000	.0000	.0000	.0000	.0000	.0001
	30	.0000	.0000	.0000	.0000	.0000	.0000	.0000	.0000	.0000	.0000
	31	.0000	.0000	.0000	.0000	.0000	.0000	.0000	.0000	.0000	.0000
	32	.0000	.0000	.0000	.0000	.0000	.0000	.0000	.0000	.0000	.0000
	33	.0000	.0000	.0000	.0000	.0000	.0000	.0000	.0000	.0000	.0000
	34	.0000	.0000	.0000	.0000	.0000	.0000	.0000	.0000	.0000	.0000
	35	.0000	.0000	.0000	.0000	.0000	.0000	.0000	.0000	.0000	.0000
40	1	.8715	.9852	.9985	.9999	1.0000	1.0000	1.0000	1.0000	1.0000	1.0000
	2	.6009	.9195	.9879	.9985	.9999	1.0000	1.0000	1.0000	1.0000	1.0000
	3	.3233	.7772	.9514	.9921	.9990	.9999	1.0000	1.0000	1.0000	1.0000
	4	.1381	.5769	.8698	.9715	.9953	.9994	.9999	1.0000	1.0000	1.0000
	5	.0480	.3710	.7367	.9241	.9840	.9974	.9997	1.0000	1.0000	1.0000
	6	.0139	.2063	.5675	.8387	.9567	.9914	.9987	.9999	1.0000	1.0000
	7	.0034	.0995	.3933	.7141	.9038	.9762	.9956	.9994	.9999	1.0000
	8	.0007	.0419	.2441	.5629	.8180	.9447	.9876	.9979	.9998	1.0000
	9	.0001	.0155	.1354	.4069	.7002	.8890	.9697	.9939	.9991	.9999
	10	.0000	.0051	.0672	.2682	.5605	.8041	.9356	.9844	.9973	.9997
	11	.0000	.0015	.0299	.1608	.4161	.6913	.8785	.9648	.9926	.9989
	12	.0000	.0004	.0120	.0875	.2849	.5594	.7947	.9291	.9821	.9968
	13	.0000	.0001	.0043	.0432	.1791	.4228	.6857	.8715	.9614	.9917
	14	.0000	.0000	.0014	.0194	.1032	.2968	.5592	.7888	.9249	.9808
	15	.0000	.0000	.0004	.0079	.0544	.1926	.4279	.6826	.8674	.9597
	16	.0000	.0000	.0001	.0029	.0262	.1151	.3054	.5598	.7858	.9231
	17	.0000	.0000	.0000	.0010	.0116	.0633	.2022	.4319	.6815	.8659
	18	.0000	.0000	.0000	.0003	.0047	.0320	.1239	.3115	.5609	.7852
	19	.0000	.0000	.0000	.0001	.0017	.0148	.0699	.2089	.4349	.6821
	20	.0000	.0000	.0000	.0000	.0006	.0063	.0363	.1298	.3156	.5627
	21	.0000	.0000	.0000	.0000	.0002	.0024	.0173	.0744	.2130	.4373
	22	.0000	.0000	.0000	.0000	.0000	.0009	.0075	.0392	.1331	.3179
	23	.0000	.0000	.0000	.0000	.0000	.0003	.0030	.0189	.0767	.2148
	24	.0000	.0000	.0000	.0000	.0000	.0001	.0011	.0083	.0405	.1341
	25	.0000	.0000	.0000	.0000	.0000	.0000	.0004	.0034	.0196	.0769
	26	.0000	.0000	.0000	.0000	.0000	.0000	.0001	.0012	.0086	.0403
	27	.0000	.0000	.0000	.0000	.0000	.0000	.0000	.0004	.0034	.0192
	28	.0000	.0000	.0000	.0000	.0000	.0000	.0000	.0001	.0012	.0083
	29	.0000	.0000	.0000	.0000	.0000	.0000	.0000	.0000	.0004	.0032

CUMULATIVE TERMS, BINOMIAL DISTRIBUTION (Continued)

n	x′	.05	.10	.15	.20	θ .25	.30	.35	.40	.45	.50
40	30	.0000	.0000	.0000	.0000	.0000	.0000	.0000	.0000	.0001	.0011
	31	.0000	.0000	.0000	.0000	.0000	.0000	.0000	.0000	.0000	.0003
	32	.0000	.0000	.0000	.0000	.0000	.0000	.0000	.0000	.0000	.0001
	33	.0000	.0000	.0000	.0000	.0000	.0000	.0000	.0000	.0000	.0000
	34	.0000	.0000	.0000	.0000	.0000	.0000	.0000	.0000	.0000	.0000
	35	.0000	.0000	.0000	.0000	.0000	.0000	.0000	.0000	.0000	.0000
	36	.0000	.0000	.0000	.0000	.0000	.0000	.0000	.0000	.0000	.0000
	37	.0000	.0000	.0000	.0000	.0000	.0000	.0000	.0000	.0000	.0000
	38	.0000	.0000	.0000	.0000	.0000	.0000	.0000	.0000	.0000	.0000
	39	.0000	.0000	.0000	.0000	.0000	.0000	.0000	.0000	.0000	.0000
	40	.0000	.0000	.0000	.0000	.0000	.0000	.0000	.0000	.0000	.0000
45	1	.9006	.9913	.9993	1.0000	1.0000	1.0000	1.0000	1.0000	1.0000	1.0000
	2	.6650	.9476	.9940	.9995	1.0000	1.0000	1.0000	1.0000	1.0000	1.0000
	3	.3923	.8410	.9735	.9968	.9997	1.0000	1.0000	1.0000	1.0000	1.0000
	4	.1866	.6711	.9215	.9871	.9984	.9999	1.0000	1.0000	1.0000	1.0000
	5	.0729	.4729	.8252	.9618	.9941	.9993	.9999	1.0000	1.0000	1.0000
	6	.0239	.2923	.6858	.9098	.9821	.9974	.9997	1.0000	1.0000	1.0000
	7	.0066	.1585	.5218	.8232	.9554	.9920	.9990	.9999	1.0000	1.0000
	8	.0016	.0757	.3606	.7025	.9059	.9791	.9967	.9996	1.0000	1.0000
	9	.0003	.0320	.2255	.5593	.8275	.9529	.9909	.9988	.9999	1.0000
	10	.0001	.0120	.1274	.4120	.7200	.9066	.9780	.9964	.9996	1.0000
	11	.0000	.0040	.0651	.2795	.5911	.8353	.9531	.9906	.9987	.9999
	12	.0000	.0012	.0302	.1741	.4543	.7380	.9104	.9784	.9964	.9996
	13	.0000	.0003	.0127	.0995	.3252	.6198	.8453	.9554	.9910	.9988
	14	.0000	.0001	.0048	.0521	.2159	.4912	.7563	.9164	.9799	.9967
	15	.0000	.0000	.0017	.0250	.1327	.3653	.6467	.8570	.9591	.9920
	16	.0000	.0000	.0005	.0110	.0753	.2538	.5248	.7751	.9238	.9822
	17	.0000	.0000	.0002	.0044	.0395	.1642	.4017	.6728	.8698	.9638
	18	.0000	.0000	.0000	.0017	.0191	.0986	.2887	.5564	.7944	.9324
	19	.0000	.0000	.0000	.0006	.0085	.0549	.1940	.4357	.6985	.8837
	20	.0000	.0000	.0000	.0002	.0035	.0283	.1215	.3214	.5869	.8144
	21	.0000	.0000	.0000	.0001	.0013	.0135	.0708	.2223	.4682	.7243
	22	.0000	.0000	.0000	.0000	.0005	.0060	.0382	.1436	.3526	.6170
	23	.0000	.0000	.0000	.0000	.0001	.0024	.0191	.0865	.2494	.5000
	24	.0000	.0000	.0000	.0000	.0000	.0009	.0089	.0483	.1650	.3830
	25	.0000	.0000	.0000	.0000	.0000	.0003	.0038	.0250	.1017	.2757
	26	.0000	.0000	.0000	.0000	.0000	.0001	.0015	.0120	.0582	.1856
	27	.0000	.0000	.0000	.0000	.0000	.0000	.0005	.0053	.0308	.1163
	28	.0000	.0000	.0000	.0000	.0000	.0000	.0002	.0021	.0150	.0676
	29	.0000	.0000	.0000	.0000	.0000	.0000	.0001	.0008	.0068	.0362
	30	.0000	.0000	.0000	.0000	.0000	.0000	.0000	.0003	.0028	.0178
	31	.0000	.0000	.0000	.0000	.0000	.0000	.0000	.0001	.0010	.0080
	32	.0000	.0000	.0000	.0000	.0000	.0000	.0000	.0000	.0004	.0033
	33	.0000	.0000	.0000	.0000	.0000	.0000	.0000	.0000	.0001	.0012
	34	.0000	.0000	.0000	.0000	.0000	.0000	.0000	.0000	.0000	.0004
	35	.0000	.0000	.0000	.0000	.0000	.0000	.0000	.0000	.0000	.0001
	36	.0000	.0000	.0000	.0000	.0000	.0000	.0000	.0000	.0000	.0000

CUMULATIVE TERMS, BINOMIAL DISTRIBUTION (Continued)

n	x'	.05	.10	.15	.20	θ .25	.30	.35	.40	.45	.50
45	37	.0000	.0000	.0000	.0000	.0000	.0000	.0000	.0000	.0000	.0000
	38	.0000	.0000	.0000	.0000	.0000	.0000	.0000	.0000	.0000	.0000
	39	.0000	.0000	.0000	.0000	.0000	.0000	.0000	.0000	.0000	.0000
	40	.0000	.0000	.0000	.0000	.0000	.0000	.0000	.0000	.0000	.0000
	41	.0000	.0000	.0000	.0000	.0000	.0000	.0000	.0000	.0000	.0000
	42	.0000	.0000	.0000	.0000	.0000	.0000	.0000	.0000	.0000	.0000
	43	.0000	.0000	.0000	.0000	.0000	.0000	.0000	.0000	.0000	.0000
	44	.0000	.0000	.0000	.0000	.0000	.0000	.0000	.0000	.0000	.0000
	45	.0000	.0000	.0000	.0000	.0000	.0000	.0000	.0000	.0000	.0000

INDIVIDUAL TERMS, POISSON DISTRIBUTION

The Poisson probability function is given by

$$f(x;\lambda) = \frac{\lambda^x e^{-\lambda}}{x!}, \qquad \lambda > 0,\ x = 0, 1, 2, \ldots .$$

This table contains the individual terms of $f(x;\lambda)$ for specified values of x and λ.

INDIVIDUAL TERMS, POISSON DISTRIBUTION (Continued)

x	λ 0.1	0.2	0.3	0.4	0.5	0.6	0.7	0.8	0.9	1.0
0	.9048	.8187	.7408	.6703	.6065	.5488	.4966	.4493	.4066	.3679
1	.0905	.1637	.2222	.2681	.3033	.3293	.3476	.3595	.3659	.3679
2	.0045	.0164	.0333	.0536	.0758	.0988	.1217	.1438	.1647	.1839
3	.0002	.0011	.0033	.0072	.0126	.0198	.0284	.0383	.0494	.0613
4	.0000	.0001	.0003	.0007	.0016	.0030	.0050	.0077	.0111	.0153
5	.0000	.0000	.0000	.0001	.0002	.0004	.0007	.0012	.0020	.0031
6	.0000	.0000	.0000	.0000	.0000	.0000	.0001	.0002	.0003	.0005
7	.0000	.0000	.0000	.0000	.0000	.0000	.0000	.0000	.0000	.0001

x	λ 1.1	1.2	1.3	1.4	1.5	1.6	1.7	1.8	1.9	2.0
0	.3329	.3012	.2725	.2466	.2231	.2019	.1827	.1653	.1496	.1353
1	.3662	.3614	.3543	.3452	.3347	.3230	.3106	.2975	.2842	.2707
2	.2014	.2169	.2303	.2417	.2510	.2584	.2640	.2678	.2700	.2707
3	.0738	.0867	.0998	.1128	.1255	.1378	.1496	.1607	.1710	.1804
4	.0203	.0260	.0324	.0395	.0471	.0551	.0636	.0723	.0812	.0902
5	.0045	.0062	.0084	.0111	.0141	.0176	.0216	.0260	.0309	.0361
6	.0008	.0012	.0018	.0026	.0035	.0047	.0061	.0078	.0098	.0120
7	.0001	.0002	.0003	.0005	.0008	.0011	.0015	.0020	.0027	.0034
8	.0000	.0000	.0001	.0001	.0001	.0002	.0003	.0005	.0006	.0009
9	.0000	.0000	.0000	.0000	.0000	.0000	.0001	.0001	.0001	.0002

x	λ 2.1	2.2	2.3	2.4	2.5	2.6	2.7	2.8	2.9	3.0
0	.1225	.1108	.1003	.0907	.0821	.0743	.0672	.0608	.0550	.0498
1	.2572	.2438	.2306	.2177	.2052	.1931	.1815	.1703	.1596	.1494
2	.2700	.2681	.2652	.2613	.2565	.2510	.2450	.2384	.2314	.2240
3	.1890	.1966	.2033	.2090	.2138	.2176	.2205	.2225	.2237	.2240
4	.0992	.1082	.1169	.1254	.1336	.1414	.1488	.1557	.1622	.1680
5	.0417	.0476	.0538	.0602	.0668	.0735	.0804	.0872	.0940	.1008
6	.0146	.0174	.0206	.0241	.0278	.0319	.0362	.0407	.0455	.0504
7	.0044	.0055	.0068	.0083	.0099	.0118	.0139	.0163	.0188	.0216
8	.0011	.0015	.0019	.0025	.0031	.0038	.0047	.0057	.0068	.0081
9	.0003	.0004	.0005	.0007	.0009	.0011	.0014	.0018	.0022	.0027
10	.0001	.0001	.0001	.0002	.0002	.0003	.0004	.0005	.0006	.0008
11	.0000	.0000	.0000	.0000	.0000	.0001	.0001	.0001	.0002	.0002
12	.0000	.0000	.0000	.0000	.0000	.0000	.0000	.0000	.0000	.0001

x	λ 3.1	3.2	3.3	3.4	3.5	3.6	3.7	3.8	3.9	4.0
0	.0450	.0408	.0369	.0334	.0302	.0273	.0247	.0224	.0202	.0183
1	.1397	.1304	.1217	.1135	.1057	.0984	.0915	.0850	.0789	.0733
2	.2165	.2087	.2008	.1929	.1850	.1771	.1692	.1615	.1539	.1465
3	.2237	.2226	.2209	.2186	.2158	.2125	.2087	.2046	.2001	.1954
4	.1734	.1781	.1823	.1858	.1888	.1912	.1931	.1944	.1951	.1954
5	.1075	.1140	.1203	.1264	.1322	.1377	.1429	.1477	.1522	.1563
6	.0555	.0608	.0662	.0716	.0771	.0826	.0881	.0936	.0989	.1042
7	.0246	.0278	.0312	.0348	.0385	.0425	.0466	.0508	.0551	.0595
8	.0095	.0111	.0129	.0148	.0169	.0191	.0215	.0241	.0269	.0298
9	.0033	.0040	.0047	.0056	.0066	.0076	.0089	.0102	.0116	.0132

INDIVIDUAL TERMS, POISSON DISTRIBUTION (Continued)

x	λ = 3.1	3.2	3.3	3.4	3.5	3.6	3.7	3.8	3.9	4.0
10	.0010	.0013	.0016	.0019	.0023	.0028	.0033	.0039	.0045	.0053
11	.0003	.0004	.0005	.0006	.0007	.0009	.0011	.0013	.0016	.0019
12	.0001	.0001	.0001	.0002	.0002	.0003	.0003	.0004	.0005	.0006
13	.0000	.0000	.0000	.0000	.0001	.0001	.0001	.0001	.0002	.0002
14	.0000	.0000	.0000	.0000	.0000	.0000	.0000	.0000	.0000	.0001

x	λ = 4.1	4.2	4.3	4.4	4.5	4.6	4.7	4.8	4.9	5.0
0	.0166	.0150	.0136	.0123	.0111	.0101	.0091	.0082	.0074	.0067
1	.0679	.0630	.0583	.0540	.0500	.0462	.0427	.0395	.0365	.0337
2	.1393	.1323	.1254	.1188	.1125	.1063	.1005	.0948	.0894	.0842
3	.1904	.1852	.1798	.1743	.1687	.1631	.1574	.1517	.1460	.1404
4	.1951	.1944	.1933	.1917	.1898	.1875	.1849	.1820	.1789	.1755
5	.1600	.1633	.1662	.1687	.1708	.1725	.1738	.1747	.1753	.1755
6	.1093	.1143	.1191	.1237	.1281	.1323	.1362	.1398	.1432	.1462
7	.0640	.0686	.0732	.0778	.0824	.0869	.0914	.0959	.1002	.1044
8	.0328	.0360	.0393	.0428	.0463	.0500	.0537	.0575	.0614	.0653
9	.0150	.0168	.0188	.0209	.0232	.0255	.0280	.0307	.0334	.0363
10	.0061	.0071	.0081	.0092	.0104	.0118	.0132	.0147	.0164	.0181
11	.0023	.0027	.0032	.0037	.0043	.0049	.0056	.0064	.0073	.0082
12	.0008	.0009	.0011	.0014	.0016	.0019	.0022	.0026	.0030	.0034
13	.0002	.0003	.0004	.0005	.0006	.0007	.0008	.0009	.0011	.0013
14	.0001	.0001	.0001	.0001	.0002	.0002	.0003	.0003	.0004	.0005
15	.0000	.0000	.0000	.0000	.0001	.0001	.0001	.0001	.0001	.0002

x	λ = 5.1	5.2	5.3	5.4	5.5	5.6	5.7	5.8	5.9	6.0
0	.0061	.0055	.0050	.0045	.0041	.0037	.0033	.0030	.0027	.0025
1	.0311	.0287	.0265	.0244	.0225	.0207	.0191	.0176	.0162	.0149
2	.0793	.0746	.0701	.0659	.0618	.0580	.0544	.0509	.0477	.0446
3	.1348	.1293	.1239	.1185	.1133	.1082	.1033	.0985	.0938	.0892
4	.1719	.1681	.1641	.1600	.1558	.1515	.1472	.1428	.1383	.1339
5	.1753	.1748	.1740	.1728	.1714	.1697	.1678	.1656	.1632	.1606
6	.1490	.1515	.1537	.1555	.1571	.1584	.1594	.1601	.1605	.1606
7	.1086	.1125	.1163	.1200	.1234	.1267	.1298	.1326	.1353	.1377
8	.0692	.0731	.0771	.0810	.0849	.0887	.0925	.0962	.0998	.1033
9	.0392	.0423	.0454	.0486	.0519	.0552	.0586	.0620	.0654	.0688
10	.0200	.0220	.0241	.0262	.0285	.0309	.0334	.0359	.0386	.0413
11	.0093	.0104	.0116	.0129	.0143	.0157	.0173	.0190	.0207	.0225
12	.0039	.0045	.0051	.0058	.0065	.0073	.0082	.0092	.0102	.0113
13	.0015	.0018	.0021	.0024	.0028	.0032	.0036	.0041	.0046	.0052
14	.0006	.0007	.0008	.0009	.0011	.0013	.0015	.0017	.0019	.0022
15	.0002	.0002	.0003	.0003	.0004	.0005	.0006	.0007	.0008	.0009
16	.0001	.0001	.0001	.0001	.0001	.0002	.0002	.0002	.0003	.0003
17	.0000	.0000	.0000	.0000	.0000	.0000	.0001	.0001	.0001	.0001

INDIVIDUAL TERMS, POISSON DISTRIBUTION (Continued)

	λ									
x	6.1	6.2	6.3	6.4	6.5	6.6	6.7	6.8	6.9	7.0
0	.0022	.0020	.0018	.0017	.0015	.0014	.0012	.0011	.0010	.0009
1	.0137	.0126	.0116	.0106	.0098	.0090	.0082	.0076	.0070	.0064
2	.0417	.0390	.0364	.0340	.0318	.0296	.0276	.0258	.0240	.0223
3	.0848	.0806	.0765	.0726	.0688	.0652	.0617	.0584	.0552	.0521
4	.1294	.1249	.1205	.1162	.1118	.1076	.1034	.0992	.0952	.0912
5	.1579	.1549	.1519	.1487	.1454	.1420	.1385	.1349	.1314	.1277
6	.1605	.1601	.1595	.1586	.1575	.1562	.1546	.1529	.1511	.1490
7	.1399	.1418	.1435	.1450	.1462	.1472	.1480	.1486	.1489	.1490
8	.1066	.1099	.1130	.1160	.1188	.1215	.1240	.1263	.1284	.1304
9	.0723	.0757	.0791	.0825	.0858	.0891	.0923	.0954	.0985	.1014
10	.0441	.0469	.0498	.0528	.0558	.0588	.0618	.0649	.0679	.0710
11	.0245	.0265	.0285	.0307	.0330	.0353	.0377	.0401	.0426	.0452
12	.0124	.0137	.0150	.0164	.0179	.0194	.0210	.0227	.0245	.0264
13	.0058	.0065	.0073	.0081	.0089	.0098	.0108	.0119	.0130	.0142
14	.0025	.0029	.0033	.0037	.0041	.0046	.0052	.0058	.0064	.0071
15	.0010	.0012	.0014	.0016	.0018	.0020	.0023	.0026	.0029	.0033
16	.0004	.0005	.0005	.0006	.0007	.0008	.0010	.0011	.0013	.0014
17	.0001	.0002	.0002	.0002	.0003	.0003	.0004	.0004	.0005	.0006
18	.0000	.0001	.0001	.0001	.0001	.0001	.0001	.0002	.0002	.0002
19	.0000	.0000	.0000	.0000	.0000	.0000	.0000	.0001	.0001	.0001

	λ									
x	7.1	7.2	7.3	7.4	7.5	7.6	7.7	7.8	7.9	8.0
0	.0008	.0007	.0007	.0006	.0006	.0005	.0005	.0004	.0004	.0003
1	.0059	.0054	.0049	.0045	.0041	.0038	.0035	.0032	.0029	.0027
2	.0208	.0194	.0180	.0167	.0156	.0145	.0134	.0125	.0116	.0107
3	.0492	.0464	.0438	.0413	.0389	.0366	.0345	.0324	.0305	.0286
4	.0874	.0836	.0799	.0764	.0729	.0696	.0663	.0632	.0602	.0573
5	.1241	.1204	.1167	.1130	.1094	.1057	.1021	.0986	.0951	.0916
6	.1468	.1445	.1420	.1394	.1367	.1339	.1311	.1282	.1252	.1221
7	.1489	.1486	.1481	.1474	.1465	.1454	.1442	.1428	.1413	.1396
8	.1321	.1337	.1351	.1363	.1373	.1382	.1388	.1392	.1395	.1396
9	.1042	.1070	.1096	.1121	.1144	.1167	.1187	.1207	.1224	.1241
10	.0740	.0770	.0800	.0829	.0858	.0887	.0914	.0941	.0967	.0993
11	.0478	.0504	.0531	.0558	.0585	.0613	.0640	.0667	.0695	.0722
12	.0283	.0303	.0323	.0344	.0366	.0388	.0411	.0434	.0457	.0481
13	.0154	.0168	.0181	.0196	.0211	.0227	.0243	.0260	.0278	.0296
14	.0078	.0086	.0095	.0104	.0113	.0123	.0134	.0145	.0157	.0169
15	.0037	.0041	.0046	.0051	.0057	.0062	.0069	.0075	.0083	.0090
16	.0016	.0019	.0021	.0024	.0026	.0030	.0033	.0037	.0041	.0045
17	.0007	.0008	.0009	.0010	.0012	.0013	.0015	.0017	.0019	.0021
18	.0003	.0003	.0004	.0004	.0005	.0006	.0006	.0007	.0008	.0009
19	.0001	.0001	.0001	.0002	.0002	.0002	.0003	.0003	.0003	.0004
20	.0000	.0000	.0001	.0001	.0001	.0001	.0001	.0001	.0001	.0002
21	.0000	.0000	.0000	.0000	.0000	.0000	.0000	.0000	.0001	.0001

INDIVIDUAL TERMS, POISSON DISTRIBUTION (Continued)

x	λ 8.1	8.2	8.3	8.4	8.5	8.6	8.7	8.8	8.9	9.0
0	.0003	.0003	.0002	.0002	.0002	.0002	.0002	.0002	.0001	.0001
1	.0025	.0023	.0021	.0019	.0017	.0016	.0014	.0013	.0012	.0011
2	.0100	.0092	.0086	.0079	.0074	.0068	.0063	.0058	.0054	.0050
3	.0269	.0252	.0237	.0222	.0208	.0195	.0183	.0171	.0160	.0150
4	.0544	.0517	.0491	.0466	.0443	.0420	.0398	.0377	.0357	.0337
5	.0882	.0849	.0816	.0784	.0752	.0722	.0692	.0663	.0635	.0607
6	.1191	.1160	.1128	.1097	.1066	.1034	.1003	.0972	.0941	.0911
7	.1378	.1358	.1338	.1317	.1294	.1271	.1247	.1222	.1197	.1171
8	.1395	.1392	.1388	.1382	.1375	.1366	.1356	.1344	.1332	.1318
9	.1256	.1269	.1280	.1290	.1299	.1306	.1311	.1315	.1317	.1318
10	.1017	.1040	.1063	.1084	.1104	.1123	.1140	.1157	.1172	.1186
11	.0749	.0776	.0802	.0828	.0853	.0878	.0902	.0925	.0948	.0970
12	.0505	.0530	.0555	.0579	.0604	.0629	.0654	.0679	.0703	.0728
13	.0315	.0334	.0354	.0374	.0395	.0416	.0438	.0459	.0481	.0504
14	.0182	.0196	.0210	.0225	.0240	.0256	.0272	.0289	.0306	.0324
15	.0098	.0107	.0116	.0126	.0136	.0147	.0158	.0169	.0182	.0194
16	.0050	.0055	.0060	.0066	.0072	.0079	.0086	.0093	.0101	.0109
17	.0024	.0026	.0029	.0033	.0036	.0040	.0044	.0048	.0053	.0058
18	.0011	.0012	.0014	.0015	.0017	.0019	.0021	.0024	.0026	.0029
19	.0005	.0005	.0006	.0007	.0008	.0009	.0010	.0011	.0012	.0014
20	.0002	.0002	.0002	.0003	.0003	.0004	.0004	.0005	.0005	.0006
21	.0001	.0001	.0001	.0001	.0001	.0002	.0002	.0002	.0002	.0003
22	.0000	.0000	.0000	.0000	.0001	.0001	.0001	.0001	.0001	.0001

x	λ 9.1	9.2	9.3	9.4	9.5	9.6	9.7	9.8	9.9	10
0	.0001	.0001	.0001	.0001	.0001	.0001	.0001	.0001	.0001	.0000
1	.0010	.0009	.0009	.0008	.0007	.0007	.0006	.0005	.0005	.0005
2	.0046	.0043	.0040	.0037	.0034	.0031	.0029	.0027	.0025	.0023
3	.0140	.0131	.0123	.0115	.0107	.0100	.0093	.0087	.0081	.0076
4	.0319	.0302	.0285	.0269	.0254	.0240	.0226	.0213	.0201	.0189
5	.0581	.0555	.0530	.0506	.0483	.0460	.0439	.0418	.0398	.0378
6	.0881	.0851	.0822	.0793	.0764	.0736	.0709	.0682	.0656	.0631
7	.1145	.1118	.1091	.1064	.1037	.1010	.0982	.0955	.0928	.0901
8	.1302	.1286	.1269	.1251	.1232	.1212	.1191	.1170	.1148	.1126
9	.1317	.1315	.1311	.1306	.1300	.1293	.1284	.1274	.1263	.1251
10	.1198	.1210	.1219	.1228	.1235	.1241	.1245	.1249	.1250	.1251
11	.0991	.1012	.1031	.1049	.1067	.1083	.1098	.1112	.1125	.1137
12	.0752	.0776	.0799	.0822	.0844	.0866	.0888	.0908	.0928	.0948
13	.0526	.0549	.0572	.0594	.0617	.0640	.0662	.0685	.0707	.0729
14	.0342	.0361	.0380	.0399	.0419	.0439	.0459	.0479	.0500	.0521
15	.0208	.0221	.0235	.0250	.0265	.0281	.0297	.0313	.0330	.0347
16	.0118	.0127	.0137	.0147	.0157	.0168	.0180	.0192	.0204	.0217
17	.0063	.0069	.0075	.0081	.0088	.0095	.0103	.0111	.0119	.0128
18	.0032	.0035	.0039	.0042	.0046	.0051	.0055	.0060	.0065	.0071
19	.0015	.0017	.0019	.0021	.0023	.0026	.0028	.0031	.0034	.0037

INDIVIDUAL TERMS, POISSON DISTRIBUTION (Continued)

x	λ 9.1	9.2	9.3	9.4	9.5	9.6	9.7	9.8	9.9	10
20	.0007	.0008	.0009	.0010	.0011	.0012	.0014	.0015	.0017	.0019
21	.0003	.0003	.0004	.0004	.0005	.0006	.0006	.0007	.0008	.0009
22	.0001	.0001	.0002	.0002	.0002	.0002	.0003	.0003	.0004	.0004
23	.0000	.0001	.0001	.0001	.0001	.0001	.0001	.0001	.0002	.0002
24	.0000	.0000	.0000	.0000	.0000	.0000	.0000	.0001	.0001	.0001

x	λ 11	12	13	14	15	16	17	18	19	20
0	.0000	.0000	.0000	.0000	.0000	.0000	.0000	.0000	.0000	.0000
1	.0002	.0001	.0000	.0000	.0000	.0000	.0000	.0000	.0000	.0000
2	.0010	.0004	.0002	.0001	.0000	.0000	.0000	.0000	.0000	.0000
3	.0037	.0018	.0008	.0004	.0002	.0001	.0000	.0000	.0000	.0000
4	.0102	.0053	.0027	.0013	.0006	.0003	.0001	.0001	.0000	.0000
5	.0224	.0127	.0070	.0037	.0019	.0010	.0005	.0002	.0001	.0001
6	.0411	.0255	.0152	.0087	.0048	.0026	.0014	.0007	.0004	.0002
7	.0646	.0437	.0281	.0174	.0104	.0060	.0034	.0018	.0010	.0005
8	.0888	.0655	.0457	.0304	.0194	.0120	.0072	.0042	.0024	.0013
9	.1085	.0874	.0661	.0473	.0324	.0213	.0135	.0083	.0050	.0029
10	.1194	.1048	.0859	.0663	.0486	.0341	.0230	.0150	.0095	.0058
11	.1194	.1144	.1015	.0844	.0663	.0496	.0355	.0245	.0164	.0106
12	.1094	.1144	.1099	.0984	.0829	.0661	.0504	.0368	.0259	.0176
13	.0926	.1056	.1099	.1060	.0956	.0814	.0658	.0509	.0378	.0271
14	.0728	.0905	.1021	.1060	.1024	.0930	.0800	.0655	.0514	.0387
15	.0534	.0724	.0885	.0989	.1024	.0992	.0906	.0786	.0650	.0516
16	.0367	.0543	.0719	.0866	.0960	.0992	.0963	.0884	.0772	.0646
17	.0237	.0383	.0550	.0713	.0847	.0934	.0963	.0936	.0863	.0760
18	.0145	.0256	.0397	.0554	.0706	.0830	.0909	.0936	.0911	.0844
19	.0084	.0161	.0272	.0409	.0557	.0699	.0814	.0887	.0911	.0888
20	.0046	.0097	.0177	.0286	.0418	.0559	.0692	.0798	.0866	.0888
21	.0024	.0055	.0109	.0191	.0299	.0426	.0560	.0684	.0783	.0846
22	.0012	.0030	.0065	.0121	.0204	.0310	.0433	.0560	.0676	.0769
23	.0006	.0016	.0037	.0074	.0133	.0216	.0320	.0438	.0559	.0669
24	.0003	.0008	.0020	.0043	.0083	.0144	.0226	.0328	.0442	.0557
25	.0001	.0004	.0010	.0024	.0050	.0092	.0154	.0237	.0336	.0446
26	.0000	.0002	.0005	.0013	.0029	.0057	.0101	.0164	.0246	.0343
27	.0000	.0001	.0002	.0007	.0016	.0034	.0063	.0109	.0173	.0254
28	.0000	.0000	.0001	.0003	.0009	.0019	.0038	.0070	.0117	.0181
29	.0000	.0000	.0001	.0002	.0004	.0011	.0023	.0044	.0077	.0125
30	.0000	.0000	.0000	.0001	.0002	.0006	.0013	.0026	.0049	.0083
31	.0000	.0000	.0000	.0000	.0001	.0003	.0007	.0015	.0030	.0054
32	.0000	.0000	.0000	.0000	.0001	.0001	.0004	.0009	.0018	.0034
33	.0000	.0000	.0000	.0000	.0000	.0001	.0002	.0005	.0010	.0020
34	.0000	.0000	.0000	.0000	.0000	.0000	.0001	.0002	.0006	.0012
35	.0000	.0000	.0000	.0000	.0000	.0000	.0000	.0001	.0003	.0007
36	.0000	.0000	.0000	.0000	.0000	.0000	.0000	.0001	.0002	.0004
37	.0000	.0000	.0000	.0000	.0000	.0000	.0000	.0000	.0001	.0002
38	.0000	.0000	.0000	.0000	.0000	.0000	.0000	.0000	.0000	.0001
39	.0000	.0000	.0000	.0000	.0000	.0000	.0000	.0000	.0000	.0001

CUMULATIVE TERMS, POISSON DISTRIBUTION

This table contains the values of

$$\sum_{x=x'}^{\infty} \frac{e^{-\lambda}\lambda^x}{x!}$$

for specified values of x' and λ. The cumulative Poisson distribution and the cumulative chi-square (χ^2) distribution are related as follows:

$$\sum_{x=0}^{x'-1} \frac{e^{-\lambda}\lambda^x}{x!} = 1 - F(\chi^2)$$

$$= \frac{1}{2^{\frac{n}{2}}\Gamma\left(\frac{n}{2}\right)} \int_{\chi^2}^{\infty} x^{\frac{n}{2}-1} e^{-\frac{x}{2}}\, dx$$

where $\lambda = \frac{1}{2}\chi^2$ and $x' = \frac{1}{2}n$.

CUMULATIVE TERMS, POISSON DISTRIBUTION (Continued)

x'	λ 0.1	0.2	0.3	0.4	0.5	0.6	0.7	0.8	0.9	1.0
0	1.0000	1.0000	1.0000	1.0000	1.0000	1.0000	1.0000	1.0000	1.0000	1.0000
1	.0952	.1813	.2592	.3297	.3935	.4512	.5034	.5507	.5934	.6321
2	.0047	.0175	.0369	.0616	.0902	.1219	.1558	.1912	.2275	.2642
3	.0002	.0011	.0036	.0079	.0144	.0231	.0341	.0474	.0629	.0803
4	.0000	.0001	.0003	.0008	.0018	.0034	.0058	.0091	.0135	.0190
5	.0000	.0000	.0000	.0001	.0002	.0004	.0008	.0014	.0023	.0037
6	.0000	.0000	.0000	.0000	.0000	.0000	.0001	.0002	.0003	.0006
7	.0000	.0000	.0000	.0000	.0000	.0000	.0000	.0000	.0000	.0001

x'	λ 1.1	1.2	1.3	1.4	1.5	1.6	1.7	1.8	1.9	2.0
0	1.0000	1.0000	1.0000	1.0000	1.0000	1.0000	1.0000	1.0000	1.0000	1.0000
1	.6671	.6988	.7275	.7534	.7769	.7981	.8173	.8347	.8504	.8647
2	.3010	.3374	.3732	.4082	.4422	.4751	.5068	.5372	.5663	.5940
3	.0996	.1205	.1429	.1665	.1912	.2166	.2428	.2694	.2963	.3233
4	.0257	.0338	.0431	.0537	.0656	.0788	.0932	.1087	.1253	.1429
5	.0054	.0077	.0107	.0143	.0186	.0237	.0296	.0364	.0441	.0527
6	.0010	.0015	.0022	.0032	.0045	.0060	.0080	.0104	.0132	.0166
7	.0001	.0003	.0004	.0006	.0009	.0013	.0019	.0026	.0034	.0045
8	.0000	.0000	.0001	.0001	.0002	.0003	.0004	.0006	.0008	.0011
9	.0000	.0000	.0000	.0000	.0000	.0000	.0001	.0001	.0002	.0002

x'	λ 2.1	2.2	2.3	2.4	2.5	2.6	2.7	2.8	2.9	3.0
0	1.0000	1.0000	1.0000	1.0000	1.0000	1.0000	1.0000	1.0000	1.0000	1.0000
1	.8775	.8892	.8997	.9093	.9179	.9257	.9328	.9392	.9450	.9502
2	.6204	.6454	.6691	.6916	.7127	.7326	.7513	.7689	.7854	.8009
3	.3504	.3773	.4040	.4303	.4562	.4816	.5064	.5305	.5540	.5768
4	.1614	.1806	.2007	.2213	.2424	.2640	.2859	.3081	.3304	.3528
5	.0621	.0725	.0838	.0959	.1088	.1226	.1371	.1523	.1682	.1847
6	.0204	.0249	.0300	.0357	.0420	.0490	.0567	.0651	.0742	.0839
7	.0059	.0075	.0094	.0116	.0142	.0172	.0206	.0244	.0287	.0335
8	.0015	.0020	.0026	.0033	.0042	.0053	.0066	.0081	.0099	.0119
9	.0003	.0005	.0006	.0009	.0011	.0015	.0019	.0024	.0031	.0038
10	.0001	.0001	.0001	.0002	.0003	.0004	.0005	.0007	.0009	.0011
11	.0000	.0000	.0000	.0000	.0001	.0001	.0001	.0002	.0002	.0003
12	.0000	.0000	.0000	.0000	.0000	.0000	.0000	.0000	.0001	.0001

x'	λ 3.1	3.2	3.3	3.4	3.5	3.6	3.7	3.8	3.9	4.0
0	1.0000	1.0000	1.0000	1.0000	1.0000	1.0000	1.0000	1.0000	1.0000	1.0000
1	.9550	.9592	.9631	.9666	.9698	.9727	.9753	.9776	.9798	.9817
2	.8153	.8288	.8414	.8532	.8641	.8743	.8838	.8926	.9008	.9084
3	.5988	.6201	.6406	.6603	.6792	.6973	.7146	.7311	.7469	.7619
4	.3752	.3975	.4197	.4416	.4634	.4848	.5058	.5265	.5468	.5665

CUMULATIVE TERMS, POISSON DISTRIBUTION (Continued)

	λ									
x'	3.1	3.2	3.3	3.4	3.5	3.6	3.7	3.8	3.9	5.0
5	.2018	.2194	.2374	.2558	.2746	.2936	.3128	.3322	.3516	.3712
6	.0943	.1054	.1171	.1295	.1424	.1559	.1699	.1844	.1994	.2149
7	.0388	.0446	.0510	.0579	.0653	.0733	.0818	.0909	.1005	.1107
8	.0142	.0168	.0198	.0231	.0267	.0308	.0352	.0401	.0454	.0511
9	.0047	.0057	.0069	.0083	.0099	.0117	.0137	.0160	.0185	.0214
10	.0014	.0018	.0022	.0027	.0033	.0040	.0048	.0058	.0069	.0081
11	.0004	.0005	.0006	.0008	.0010	.0013	.0016	.0019	.0023	.0028
12	.0001	.0001	.0002	.0002	.0003	.0004	.0005	.0006	.0007	.0009
13	.0000	.0000	.0000	.0001	.0001	.0001	.0001	.0002	.0002	.0003
14	.0000	.0000	.0000	.0000	.0000	.0000	.0000	.0000	.0001	.0001

	λ									
x'	4.1	4.2	4.3	4.4	4.5	4.6	4.7	4.8	4.9	5.0
0	1.0000	1.0000	1.0000	1.0000	1.0000	1.0000	1.0000	1.0000	1.0000	1.0000
1	.9834	.9850	.9864	.9877	.9889	.9899	.9909	.9918	.9926	.9933
2	.9155	.9220	.9281	.9337	.9389	.9437	.9482	.9523	.9561	.9596
3	.7762	.7898	.8026	.8149	.8264	.8374	.8477	.8575	.8667	.8753
4	.5858	.6046	.6228	.6406	.6577	.6743	.6903	.7058	.7207	.7350
5	.3907	.4102	.4296	.4488	.4679	.4868	.5054	.5237	.5418	.5595
6	.2307	.2469	.2633	.2801	.2971	.3142	.3316	.3490	.3665	.3840
7	.1214	.1325	.1442	.1564	.1689	.1820	.1954	.2092	.2233	.2378
8	.0573	.0639	.0710	.0786	.0866	.0951	.1040	.1133	.1231	.1334
9	.0245	.0279	.0317	.0358	.0403	.0451	.0503	.0558	.0618	.0681
10	.0095	.0111	.0129	.0149	.0171	.0195	.0222	.0251	.0283	.0318
11	.0034	.0041	.0048	.0057	.0067	.0078	.0090	.0104	.0120	.0137
12	.0011	.0014	.0017	.0020	.0024	.0029	.0034	.0040	.0047	.0055
13	.0003	.0004	.0005	.0007	.0008	.0010	.0012	.0014	.0017	.0020
14	.0001	.0001	.0002	.0002	.0003	.0003	.0004	.0005	.0006	.0007
15	.0000	.0000	.0000	.0001	.0001	.0001	.0001	.0001	.0002	.0002
16	.0000	.0000	.0000	.0000	.0000	.0000	.0000	.0000	.0001	.0001

	λ									
x'	5.1	5.2	5.3	5.4	5.5	5.6	5.7	5.8	5.9	6.0
0	1.0000	1.0000	1.0000	1.0000	1.0000	1.0000	1.0000	1.0000	1.0000	1.0000
1	.9939	.9945	.9950	.9955	.9959	.9963	.9967	.9970	.9973	.9975
2	.9628	.9658	.9686	.9711	.9734	.9756	.9776	.9794	.9811	.9826
3	.8835	.8912	.8984	.9052	.9116	.9176	.9232	.9285	.9334	.9380
4	.7487	.7619	.7746	.7867	.7983	.8094	.8200	.8300	.8396	.8488
5	.5769	.5939	.6105	.6267	.6425	.6579	.6728	.6873	.7013	.7149
6	.4016	.4191	.4365	.4539	.4711	.4881	.5050	.5217	.5381	.5543
7	.2526	.2676	.2829	.2983	.3140	.3297	.3456	.3616	.3776	.3937
8	.1440	.1551	.1665	.1783	.1905	.2030	.2159	.2290	.2424	.2560
9	.0748	.0819	.0894	.0974	.1056	.1143	.1234	.1328	.1426	.1528

CUMULATIVE TERMS, POISSON DISTRIBUTION (Continued)

x	λ 5.1	5.2	5.3	5.4	5.5	5.6	5.7	5.8	5.9	6.0
10	.0356	.0397	.0441	.0488	.0538	.0591	.0648	.0708	.0772	.0839
11	.0156	.0177	.0200	.0225	.0253	.0282	.0314	.0349	.0386	.0426
12	.0063	.0073	.0084	.0096	.0110	.0125	.0141	.0160	.0179	.0201
13	.0024	.0028	.0033	.0038	.0045	.0051	.0059	.0068	.0078	.0088
14	.0008	.0010	.0012	.0014	.0017	.0020	.0023	.0027	.0031	.0036
15	.0003	.0003	.0004	.0005	.0006	.0007	.0009	.0010	.0012	.0014
16	.0001	.0001	.0001	.0002	.0002	.0002	.0003	.0004	.0004	.0005
17	.0000	.0000	.0000	.0001	.0001	.0001	.0001	.0001	.0001	.0002
18	.0000	.0000	.0000	.0000	.0000	.0000	.0000	.0000	.0000	.0001

x'	λ 6.1	6.2	6.3	6.4	6.5	6.6	6.7	6.8	6.9	7.0
0	1.0000	1.0000	1.0000	1.0000	1.0000	1.0000	1.0000	1.0000	1.0000	1.0000
1	.9978	.9980	.9982	.9983	.9985	.9986	.9988	.9989	.9990	.9991
2	.9841	.9854	.9866	.9877	.9887	.9897	.9905	.9913	.9920	.9927
3	.9423	.9464	.9502	.9537	.9570	.9600	.9629	.9656	.9680	.9704
4	.8575	.8658	.8736	.8811	.8882	.8948	.9012	.9072	.9129	.9182
5	.7281	.7408	.7531	.7649	.7763	.7873	.7978	.8080	.8177	.8270
6	.5702	.5859	.6012	.6163	.6310	.6453	.6594	.6730	.6863	.6993
7	.4098	.4258	.4418	.4577	.4735	.4892	.5047	.5201	.5353	.5503
8	.2699	.2840	.2983	.3127	.3272	.3419	.3567	.3715	.3864	.4013
9	.1633	.1741	.1852	.1967	.2084	.2204	.2327	.2452	.2580	.2709
10	.0910	.0984	.1061	.1142	.1226	.1314	.1404	.1498	.1505	.1695
11	.0469	.0514	.0563	.0614	.0668	.0726	.0786	.0849	.0916	.0985
12	.0224	.0250	.0277	.0307	.0339	.0373	.0409	.0448	.0490	.0534
13	.0100	.0113	.0127	.0143	.0160	.0179	.0199	.0221	.0245	.0270
14	.0042	.0048	.0055	.0063	.0071	.0080	.0091	.0102	.0115	.0128
15	.0016	.0019	.0022	.0026	.0030	.0034	.0039	.0044	.0050	.0057
16	.0006	.0007	.0008	.0010	.0012	.0014	.0016	.0018	.0021	.0024
17	.0002	.0003	.0003	.0004	.0004	.0005	.0006	.0007	.0008	.0010
18	.0001	.0001	.0001	.0001	.0002	.0002	.0002	.0003	.0003	.0004
19	.0000	.0000	.0000	.0000	.0001	.0001	.0001	.0001	.0001	.0001

x'	λ 7.1	7.2	7.3	7.4	7.5	7.6	7.7	7.8	7.9	8.0
0	1.0000	1.0000	1.0000	1.0000	1.0000	1.0000	1.0000	1.0000	1.0000	1.0000
1	.9992	.9993	.9993	.9994	.9994	.9995	.9995	.9996	.9996	.9997
2	.9933	.9939	.9944	.9949	.9953	.9957	.9961	.9964	.9967	.9970
3	.9725	.9745	.9764	.0781	.9797	.9812	.9826	.9839	.9851	.9862
4	.9233	.9281	.9326	.9368	.9409	.9446	.9482	.9515	.9547	.9576
5	.8359	.8445	.8527	.8605	.8679	.8751	.8819	.8883	.8945	.9004
6	.7119	.7241	.7360	.7474	.7586	.7693	.7797	.7897	.7994	.8088
7	.5651	.5796	.5940	.6080	.6218	.6354	.6486	.6616	.6743	.6866
8	.4162	.4311	.4459	.4607	.4754	.4900	.5044	.5188	.5330	.5470
9	.2840	.2973	.3108	.3243	.3380	.3518	.3657	.3796	.3935	.4075
10	.1798	.1904	.2012	.2123	.2236	.2351	.2469	.2589	.2710	.2834
11	.1058	.1133	.1212	.1293	.1378	.1465	.1555	.1648	.1743	.1841
12	.0580	.0629	.0681	.0735	.0792	.0852	.0915	.0980	.1048	.1119
13	.0297	.0327	.0358	.0391	.0427	.0464	.0504	.0546	.0591	.0638
14	.0143	.0159	.0176	.0195	.0216	.0238	.0261	.0286	.0313	.0342

CUMULATIVE TERMS, POISSON DISTRIBUTION (Continued)

x'	λ = 7.1	7.2	7.3	7.4	7.5	7.6	7.7	7.8	7.9	8.0
15	.0065	.0073	.0082	.0092	.0103	.0114	.0127	.0141	.0156	.0173
16	.0028	.0031	.0036	.0041	.0046	.0052	.0059	.0066	.0074	.0082
17	.0011	.0013	.0015	.0017	.0020	.0022	.0026	.0029	.0033	.0037
18	.0004	.0005	.0006	.0007	.0008	.0009	.0011	.0012	.0014	.0016
19	.0002	.0002	.0002	.0003	.0003	.0004	.0004	.0005	.0006	.0006
20	.0001	.0001	.0001	.0001	.0001	.0001	.0002	.0002	.0002	.0003
21	.0000	.0000	.0000	.0000	.0000	.0000	.0001	.0001	.0001	.0001

x'	λ = 8.1	8.2	8.3	8.4	8.5	8.6	8.7	8.8	8.9	9.0
0	1.0000	1.0000	1.0000	1.0000	1.0000	1.0000	1.0000	1.0000	1.0000	1.0000
1	.9997	.9997	.9998	.9998	.9998	.9998	.9998	.9998	.9999	.9999
2	.9972	.9975	.9977	.9979	.9981	.9982	.9984	.9985	.9987	.9988
3	.9873	.9882	.9891	.9900	.9907	.9914	.9921	.9927	.9932	.9938
4	.9604	.9630	.9654	.9677	.9699	.9719	.9738	.9756	.9772	.9788
5	.9060	.9113	.9163	.9211	.9256	.9299	.9340	.9379	.9416	.9450
6	.8178	.8264	.8347	.8427	.8504	.8578	.8648	.8716	.8781	.8843
7	.6987	.7104	.7219	.7330	.7438	.7543	.7645	.7744	.7840	.7932
8	.5609	.5746	.5881	.6013	.6144	.6272	.6398	.6522	.6643	.6761
9	.4214	.4353	.4493	.4631	.4769	.4906	.5042	.5177	.5311	.5443
10	.2959	.3085	.3212	.3341	.3470	.3600	.3731	.3863	.3994	.4126
11	.1942	.2045	.2150	.2257	.2366	.2478	.2591	.2706	.2822	.2940
12	.1193	.1269	.1348	.1429	.1513	.1600	.1689	.1780	.1874	.1970
13	.0687	.0739	.0793	.0850	.0909	.0971	.1035	.1102	.1171	.1242
14	.0372	.0405	.0439	.0476	.0514	.0555	.0597	.0642	.0689	.0739
15	.0190	.0209	.0229	.0251	.0274	.0299	.0325	.0353	.0383	.0415
16	.0092	.0102	.0113	.0125	.0138	.0152	.0168	.0184	.0202	.0220
17	.0042	.0047	.0053	.0059	.0066	.0074	.0082	.0091	.0101	.0111
18	.0018	.0021	.0023	.0027	.0030	.0034	.0038	.0043	.0048	.0053
19	.0008	.0009	.0010	.0011	.0013	.0015	.0017	.0019	.0022	.0024
20	.0003	.0003	.0004	.0005	.0005	.0006	.0007	.0008	.0009	.0011
21	.0001	.0001	.0002	.0002	.0002	.0002	.0003	.0003	.0004	.0004
22	.0000	.0000	.0001	.0001	.0001	.0001	.0001	.0001	.0002	.0002
23	.0000	.0000	.0000	.0000	.0000	.0000	0000	.0000	.0001	.0001

x'	λ = 9.1	9.2	9.3	9.4	9.5	9.6	9.7	9.8	9.9	10
0	1.0000	1.0000	1.0000	1.0000	1.0000	1.0000	1.0000	1.0000	1.0000	1.0000
1	.9999	.9999	.9999	.9999	.9999	.9999	.9999	.9999	1.0000	1.0000
2	.9989	.9990	.9991	.9991	.9992	.9993	.9993	.9994	.9995	.9995
3	.9942	.9947	.9951	.9955	.9958	.9962	.9965	.9967	.9970	.9972
4	.9802	.9816	.9828	.9840	.9851	.9862	.9871	.9880	.9889	.9897
5	.9483	.9514	.9544	.9571	.9597	.9622	.9645	.9667	.9688	.9707
6	.8902	.8959	.9014	.9065	.9115	.9162	.9207	.9250	.9290	.9329
7	.8022	.8108	.8192	.8273	.8351	.8426	.8498	.8567	.8634	.8699
8	.6877	.6990	.7101	.7208	.7313	.7416	.7515	.7612	.7706	.7798
9	.5574	.5704	.5832	.5958	.6082	.6204	.6324	.6442	.6558	.6672

CUMULATIVE TERMS, POISSON DISTRIBUTION (Continued)

x'	λ 9.1	9.2	9.3	9.4	9.5	9.6	9.7	9.8	9.9	10
10	.4258	.4389	.4521	.4651	.4782	.4911	.5040	.5168	.5295	.5421
11	.3059	.3180	.3301	.3424	.3547	.3671	.3795	.3920	.4045	.4170
12	.2068	.2168	.2270	.2374	.2480	.2588	.2697	.2807	.2919	.3032
13	.1316	.1393	.1471	.1552	.1636	.1721	.1809	.1899	.1991	.2084
14	.0790	.0844	.0900	.0958	.1019	.1081	.1147	.1214	.1284	.1355
15	.0448	.0483	.0520	.0559	.0600	.0643	.0688	.0735	.0784	.0835
16	.0240	.0262	.0285	.0309	.0335	.0362	.0391	.0421	.0454	.0487
17	.0122	.0135	.0148	.0162	.0177	.0194	.0211	.0230	.0249	.0270
18	.0059	.0066	.0073	.0081	.0089	.0098	.0108	.0119	.0130	.0143
19	.0027	.0031	.0034	.0038	.0043	.0048	.0053	.0059	.0065	.0072
20	.0012	.0014	.0015	.0017	.0020	.0022	.0025	.0028	.0031	.0035
21	.0005	.0006	.0007	.0008	.0009	.0010	.0011	.0013	.0014	.0016
22	.0002	.0002	.0003	.0003	.0004	.0004	.0005	.0005	.0006	.0007
23	.0001	.0001	.0001	.0001	.0001	.0002	.0002	.0002	.0003	.0003
24	.0000	.0000	.0000	.0000	.0001	.0001	.0001	.0001	.0001	.0001

x'	λ 11	12	13	14	15	16	17	18	19	20
0	1.0000	1.0000	1.0000	1.0000	1.0000	1.0000	1.0000	1.0000	1.0000	1.0000
1	1.0000	1.0000	1.0000	1.0000	1.0000	1.0000	1.0000	1.0000	1.0000	1.0000
2	.9998	.9999	1.0000	1.0000	1.0000	1.0000	1.0000	1.0000	1.0000	1.0000
3	.9988	.9995	.9998	.9999	1.0000	1.0000	1.0000	1.0000	1.0000	1.0000
4	.9951	.9977	.9990	.9995	.9998	.9999	1.0000	1.0000	1.0000	1.0000
5	.9849	.9924	.9963	.9982	.9991	.9996	.9998	.9999	1.0000	1.0000
6	.9625	.9797	.9893	.9945	.9972	.9986	.9993	.9997	.9998	.9999
7	.9214	.9542	.9741	.9858	.9924	.9960	.9979	.9990	.9995	.9997
8	.8568	.9105	.9460	.9684	.9820	.9900	.9946	.9971	.9985	.9992
9	.7680	.8450	.9002	.9379	.9626	.9780	.9874	.9929	.9961	.9979
10	.6595	.7576	.8342	.8906	.9301	.9567	.9739	.9846	.9911	.9950
11	.5401	.6528	.7483	.8243	.8815	.9226	.9509	.9696	.9817	.9892
12	.4207	.5384	.6468	.7400	.8152	.8730	.9153	.9451	.9653	.9786
13	.3113	.4240	.5369	.6415	.7324	.8069	.8650	.9083	.9394	.9610
14	.2187	.3185	.4270	.5356	.6368	.7255	.7991	.8574	.9016	.9339
15	.1460	.2280	.3249	.4296	.5343	.6325	.7192	.7919	.8503	.8951
16	.0926	.1556	.2364	.3306	.4319	.5333	.6285	.7133	.7852	.8435
17	.0559	.1013	.1645	.2441	.3359	.4340	.5323	.6250	.7080	.7789
18	.0322	.0630	.1095	.1728	.2511	.3407	.4360	.5314	.6216	.7030
19	.0177	.0374	.0698	.1174	.1805	.2577	.3450	.4378	.5305	.6186
20	.0093	.0213	.0427	.0765	.1248	.1878	.2637	.3491	.4394	.5297
21	.0047	.0116	.0250	.0479	.0830	.1318	.1945	.2693	.3528	.4409
22	.0023	.0061	.0141	.0288	.0531	.0892	.1385	.2009	.2745	.3563
23	.0010	.0030	.0076	.0167	.0327	.0582	.0953	.1449	.2069	.2794
24	.0005	.0015	.0040	.0093	.0195	.0367	.0633	.1011	.1510	.2125
25	.0002	.0007	.0020	.0050	.0112	.0223	.0406	.0683	.1067	.1568
26	.0001	.0003	.0010	.0026	.0062	.0131	.0252	.0446	.0731	.1122
27	.0000	.0001	.0005	.0013	.0033	.0075	.0152	.0282	.0486	.0779
28	.0000	.0001	.0002	.0006	.0017	.0041	.0088	.0173	.0313	.0525
29	.0000	.0000	.0001	.0003	.0009	.0022	.0050	.0103	.0195	.0343

CUMULATIVE TERMS, POISSON DISTRIBUTION (Continued)

x'	λ 11	12	13	14	15	16	17	18	19	20
30	.0000	.0000	.0000	.0001	.0004	.0011	.0027	.0059	.0118	.0218
31	.0000	.0000	.0000	.0001	.0002	.0006	.0014	.0033	.0070	.0135
32	.0000	.0000	.0000	.0000	.0001	.0003	.0007	.0018	.0040	.0081
33	.0000	.0000	.0000	.0000	.0000	.0001	.0004	.0010	.0022	.0047
34	.0000	.0000	.0000	.0000	.0000	.0001	.0002	.0005	.0012	.0027
35	.0000	.0000	.0000	.0000	.0000	.0000	.0001	.0002	.0006	.0015
36	.0000	.0000	.0000	.0000	.0000	.0000	.0000	.0001	.0003	.0008
37	.0000	.0000	.0000	.0000	.0000	.0000	.0000	.0001	.0002	.0004
38	.0000	.0000	.0000	.0000	.0000	.0000	.0000	.0000	.0001	.0002
39	.0000	.0000	.0000	.0000	.0000	.0000	.0000	.0000	.0000	.0001
40	.0000	.0000	.0000	.0000	.0000	.0000	.0000	.0000	.0000	.0001

PERCENTAGE POINTS, STUDENT'S t-DISTRIBUTION

This table gives values of t such that

$$F(t) = \int_{-\infty}^{t} \frac{\Gamma\left(\frac{n+1}{2}\right)}{\sqrt{n\pi}\,\Gamma\left(\frac{n}{2}\right)} \left(1 + \frac{x^2}{n}\right)^{-\frac{n+1}{2}} dx$$

for n, the number of degrees of freedom, equal to 1, 2, . . . , 30, 40, 60, 120, ∞; and for $F(t)$ = 0.60, 0.75, 0.90, 0.95, 0.975, 0.99, 0.995, and 0.9995. The t-distribution is symmetrical, so that $F(-t) = 1 - F(t)$

n \ F	.60	.75	.90	.95	.975	.99	.995	.9995
1	.325	1.000	3.078	6.314	12.706	31.821	63.657	636.619
2	.289	.816	1.886	2.920	4.303	6.965	9.925	31.598
3	.277	.765	1.638	2.353	3.182	4.541	5.841	12.924
4	.271	.741	1.533	2.132	2.776	3.747	4.604	8.610
5	.267	.727	1.476	2.015	2.571	3.365	4.032	6.869
6	.265	.718	1.440	1.943	2.447	3.143	3.707	5.959
7	.263	.711	1.415	1.895	2.365	2.998	3.499	5.408
8	.262	.706	1.397	1.860	2.306	2.896	3.355	5.041
9	.261	.703	1.383	1.833	2.262	2.821	3.250	4.781
10	.260	.700	1.372	1.812	2.228	2.764	3.169	4.587
11	.260	.697	1.363	1.796	2.201	2.718	3.106	4.437
12	.259	.695	1.356	1.782	2.179	2.681	3.055	4.318
13	.259	.694	1.350	1.771	2.160	2.650	3.012	4.221
14	.258	.692	1.345	1.761	2.145	2.624	2.977	4.140
15	.258	.691	1.341	1.753	2.131	2.602	2.947	4.073
16	.258	.690	1.337	1.746	2.120	2.583	2.921	4.015
17	.257	.689	1.333	1.740	2.110	2.567	2.898	3.965
18	.257	.688	1.330	1.734	2.101	2.552	2.878	3.922
19	.257	.688	1.328	1.729	2.093	2.539	2.861	3.883
20	.257	.687	1.325	1.725	2.086	2.528	2.845	3.850
21	.257	.686	1.323	1.721	2.080	2.518	2.831	3.819
22	.256	.686	1.321	1.717	2.074	2.508	2.819	3.792
23	.256	.685	1.319	1.714	2.069	2.500	2.807	3.767
24	.256	.685	1.318	1.711	2.064	2.492	2.797	3.745
25	.256	.684	1.316	1.708	2.060	2.485	2.787	3.725
26	.256	.684	1.315	1.706	2.056	2.479	2.779	3.707
27	.256	.684	1.314	1.703	2.052	2.473	2.771	3.690
28	.256	.683	1.313	1.701	2.048	2.467	2.763	3.674
29	.256	.683	1.311	1.699	2.045	2.462	2.756	3.659
30	.256	.683	1.310	1.697	2.042	2.457	2.750	3.646
40	.255	.681	1.303	1.684	2.021	2.423	2.704	3.551
60	.254	.679	1.296	1.671	2.000	2.390	2.660	3.460
120	.254	.677	1.289	1.658	1.980	2.358	2.617	3.373
∞	.253	.674	1.282	1.645	1.960	2.326	2.576	3.291

* This table is abridged from the "Statistical Tables" of R. A. Fisher and Frank Yates published by Oliver & Boyd, Ltd., Edinburgh and London, 1938. It is here published with the kind permission of the authors and their publishers.

PERCENTAGE POINTS, CHI-SQUARE DISTRIBUTION

This table gives values of χ^2 such that

$$F(\chi^2) = \int_0^{\chi^2} \frac{1}{2^{\frac{n}{2}}\Gamma\left(\frac{n}{2}\right)} x^{\frac{n-2}{2}} e^{-\frac{x}{2}} dx$$

for n, the number of degrees of freedom, equal to 1, 2, . . . , 30. For $n > 30$, a normal approximation is quite accurate. The expression $\sqrt{2\chi^2} - \sqrt{2n-1}$ is approximately normally distributed as the standard normal distribution. Thus χ^2_α, the α-point of the distribution, may be computed by the formula

$$\chi^2_\alpha = \tfrac{1}{2}[x_\alpha + \sqrt{2n-1}]^2,$$

where x_α is the α-point of the cumulative normal distribution. For even values of n, $F(\chi^2)$ can be written as

$$1 - F(\chi^2) = \sum_{x=0}^{x'-1} \frac{e^{-\lambda}\lambda^x}{x!}$$

with $\lambda = \frac{1}{2}\chi^2$ and $x' = \frac{1}{2}n$. Thus the cumulative Chi-Square distribution is related to the cumulative Poisson distribution.

Another approximate formula for large n

$$\chi^2_\alpha = n\left(1 - \frac{2}{9n} + z_\alpha\sqrt{\frac{2}{9n}}\right)^3$$

n = degrees of freedom
z_α = the normal deviate, (the value of x for which $F(x)$ = the desired percentile).

x	1.282	1.645	1.960	2.326	2.576	3.090
$F(x)$	.90	.95	.975	.99	.995	.999

$\chi^2_{.99} = 60[1 - 0.00370 + 2.326(0.06086)]^3 = 88.4$ is the 99th percentile for 60 degrees of freedom.

PERCENTAGE POINTS, CHI-SQUARE DISTRIBUTION (Continued)

$$F(\chi^2) = \int_0^{\chi^2} \frac{1}{2^{\frac{n}{2}}\Gamma\left(\frac{n}{2}\right)} x^{\frac{n-2}{2}} e^{-\frac{x}{2}}\, dx$$

n \ F	.005	.010	.025	.050	.100	.250	.500	.750	.900	.950	.975	.990	.995
1	.0000393	.000157	.000982	.00393	.0158	.102	.455	1.32	2.71	3.84	5.02	6.63	7.88
2	.0100	.0201	.0506	.103	.211	.575	1.39	2.77	4.61	5.99	7.38	9.21	10.6
3	.0717	.115	.216	.352	.584	1.21	2.37	4.11	6.25	7.81	9.35	11.3	12.8
4	.207	.297	.484	.711	1.06	1.92	3.36	5.39	7.78	9.49	11.1	13.3	14.9
5	.412	.554	.831	1.15	1.61	2.67	4.35	6.63	9.24	11.1	12.8	15.1	16.7
6	.676	.872	1.24	1.64	2.20	3.45	5.35	7.84	10.6	12.6	14.4	16.8	18.5
7	.989	1.24	1.69	2.17	2.83	4.25	6.35	9.04	12.0	14.1	16.0	18.5	20.3
8	1.34	1.65	2.18	2.73	3.49	5.07	7.34	10.2	13.4	15.5	17.5	20.1	22.0
9	1.73	2.09	2.70	3.33	4.17	5.90	8.34	11.4	14.7	16.9	19.0	21.7	23.6
10	2.16	2.56	3.25	3.94	4.87	6.74	9.34	12.5	16.0	18.3	20.5	23.2	25.2
11	2.60	3.05	3.82	4.57	5.58	7.58	10.3	13.7	17.3	19.7	21.9	24.7	26.8
12	3.07	3.57	4.40	5.23	6.30	8.44	11.3	14.8	18.5	21.0	23.3	26.2	28.3
13	3.57	4.11	5.01	5.89	7.04	9.30	12.3	16.0	19.8	22.4	24.7	27.7	29.8
14	4.07	4.66	5.63	6.57	7.79	10.2	13.3	17.1	21.1	23.7	26.1	29.1	31.3
15	4.60	5.23	6.26	7.26	8.55	11.0	.4.3	18.2	22.3	25.0	27.5	30.6	32.8
16	5.14	5.81	6.91	7.96	9.31	11.9	15.3	19.4	23.5	26.3	28.8	32.0	34.3
17	5.70	6.41	7.56	8.67	10.1	12.8	16.3	20.5	24.8	27.6	30.2	33.4	35.7
18	6.26	7.01	8.23	9.39	10.9	13.7	17.3	21.6	26.0	28.9	31.5	34.8	37.2
19	6.84	7.63	8.91	10.1	11.7	14.6	18.3	22.7	27.2	30.1	32.9	36.2	38.6
20	7.43	8.26	9.59	10.9	12.4	15.5	19.3	23.8	28.4	31.4	34.2	37.6	40.0
21	8.03	8.90	10.3	11.6	13.2	16.3	20.3	24.9	29.6	32.7	35.5	38.9	41.4
22	8.64	9.54	11.0	12.3	14.0	17.2	21.3	26.0	30.8	33.9	36.8	40.3	42.8
23	9.26	10.2	11.7	13.1	14.8	18.1	22.3	27.1	32.0	35.2	38.1	41.6	44.2
24	9.89	10.9	12.4	13.8	15.7	19.0	23.3	28.2	33.2	36.4	39.4	43.0	45.6
25	10.5	11.5	13.1	14.6	16.5	19.9	24.3	29.3	34.4	37.7	40.6	44.3	46.9
26	11.2	12.2	13.8	15.4	17.3	20.8	25.3	30.4	35.6	38.9	41.9	45.6	48.3
27	11.8	12.9	14.6	16.2	18.1	21.7	26.3	31.5	36.7	40.1	43.2	47.0	49.6
28	12.5	13.6	15.3	16.9	18.9	22.7	27.3	32.6	37.9	41.3	44.5	48.3	51.0
29	13.1	14.3	16.0	17.7	19.8	23.6	28.3	33.7	39.1	42.6	45.7	49.6	52.3
30	13.8	15.0	16.8	18.5	20.6	24.5	29.3	34.8	40.3	43.8	47.0	50.9	53.7

PERCENTAGE POINTS, F-DISTRIBUTION

This table gives values of F such that

$$F(F) = \int_0^F \frac{\Gamma\left(\frac{m+n}{2}\right)}{\Gamma\left(\frac{m}{2}\right)\Gamma\left(\frac{n}{2}\right)} m^{\frac{m}{2}} n^{\frac{n}{2}} x^{\frac{m-2}{2}} (n + mx)^{-\frac{m+n}{2}} dx$$

for selected values of m, the number of degrees of freedom of the numerator of F; and for selected values of n, the number of degrees of freedom of the denominator of F. The table also provides values corresponding to $F(F) = .10, .05, .025, .01, .005, .001$ since $F_{1-\alpha}$ for m and n degrees of freedom is the reciprocal of F_α for n and m degrees of freedom. Thus

$$F_{.05}(4, 7) = \frac{1}{F_{.95}(7, 4)} = \frac{1}{6.09} = .164 .$$

PERCENTAGE POINTS, *F*-DISTRIBUTION (Continued)

$$F(F) = \int_0^F \frac{\Gamma\left(\frac{m+n}{2}\right)}{\Gamma\left(\frac{m}{2}\right)\Gamma\left(\frac{n}{2}\right)} m^{\frac{m}{2}} n^{\frac{n}{2}} x^{\frac{m}{2}-1} (n+mx)^{-\frac{m+n}{2}}\, dx = .90$$

n \ m	1	2	3	4	5	6	7	8	9	10	12	15	20	24	30	40	60	120	∞
1	39.86	49.50	53.59	55.83	57.24	58.20	58.91	59.44	59.86	60.19	60.71	61.22	61.74	62.00	62.26	62.53	62.79	63.06	63.33
2	8.53	9.00	9.16	9.24	9.29	9.33	9.35	9.37	9.38	9.39	9.41	9.42	9.44	9.45	9.46	9.47	9.47	9.48	9.49
3	5.54	5.46	5.39	5.34	5.31	5.28	5.27	5.25	5.24	5.23	5.22	5.20	5.18	5.18	5.17	5.16	5.15	5.14	5.13
4	4.54	4.32	4.19	4.11	4.05	4.01	3.98	3.95	3.94	3.92	3.90	3.87	3.84	3.83	3.82	3.80	3.79	3.78	3.76
5	4.06	3.78	3.62	3.52	3.45	3.40	3.37	3.34	3.32	3.30	3.27	3.24	3.21	3.19	3.17	3.16	3.14	3.12	3.10
6	3.78	3.46	3.29	3.18	3.11	3.05	3.01	2.98	2.96	2.94	2.90	2.87	2.84	2.82	2.80	2.78	2.76	2.74	2.72
7	3.59	3.26	3.07	2.96	2.88	2.83	2.78	2.75	2.72	2.70	2.67	2.63	2.59	2.58	2.56	2.54	2.51	2.49	2.47
8	3.46	3.11	2.92	2.81	2.73	2.67	2.62	2.59	2.56	2.54	2.50	2.46	2.42	2.40	2.38	2.36	2.34	2.32	2.29
9	3.36	3.01	2.81	2.69	2.61	2.55	2.51	2.47	2.44	2.42	2.38	2.34	2.30	2.28	2.25	2.23	2.21	2.18	2.16
10	3.29	2.92	2.73	2.61	2.52	2.46	2.41	2.38	2.35	2.32	2.28	2.24	2.20	2.18	2.16	2.13	2.11	2.08	2.06
11	3.23	2.86	2.66	2.54	2.45	2.39	2.34	2.30	2.27	2.25	2.21	2.17	2.12	2.10	2.08	2.05	2.03	2.00	1.97
12	3.18	2.81	2.61	2.48	2.39	2.33	2.28	2.24	2.21	2.19	2.15	2.10	2.06	2.04	2.01	1.99	1.96	1.93	1.90
13	3.14	2.76	2.56	2.43	2.35	2.28	2.23	2.20	2.16	2.14	2.10	2.05	2.01	1.98	1.96	1.93	1.90	1.88	1.85
14	3.10	2.73	2.52	2.39	2.31	2.24	2.19	2.15	2.12	2.10	2.05	2.01	1.96	1.94	1.91	1.89	1.86	1.83	1.80
15	3.07	2.70	2.49	2.36	2.27	2.21	2.16	2.12	2.09	2.06	2.02	1.97	1.92	1.90	1.87	1.85	1.82	1.79	1.76
16	3.05	2.67	2.46	2.33	2.24	2.18	2.13	2.09	2.06	2.03	1.99	1.94	1.89	1.87	1.84	1.81	1.78	1.75	1.72
17	3.03	2.64	2.44	2.31	2.22	2.15	2.10	2.06	2.03	2.00	1.96	1.91	1.86	1.84	1.81	1.78	1.75	1.72	1.69
18	3.01	2.62	2.42	2.29	2.20	2.13	2.08	2.04	2.00	1.98	1.93	1.89	1.84	1.81	1.78	1.75	1.72	1.69	1.66
19	2.99	2.61	2.40	2.27	2.18	2.11	2.06	2.02	1.98	1.96	1.91	1.86	1.81	1.79	1.76	1.73	1.70	1.67	1.63
20	2.97	2.59	2.38	2.25	2.16	2.09	2.04	2.00	1.96	1.94	1.89	1.84	1.79	1.77	1.74	1.71	1.68	1.64	1.61
21	2.96	2.57	2.36	2.23	2.14	2.08	2.02	1.98	1.95	1.92	1.87	1.83	1.78	1.75	1.72	1.69	1.66	1.62	1.59
22	2.95	2.56	2.35	2.22	2.13	2.06	2.01	1.97	1.93	1.90	1.86	1.81	1.76	1.73	1.70	1.67	1.64	1.60	1.57
23	2.94	2.55	2.34	2.21	2.11	2.05	1.99	1.95	1.92	1.89	1.84	1.80	1.74	1.72	1.69	1.66	1.62	1.59	1.55
24	2.93	2.54	2.33	2.19	2.10	2.04	1.98	1.94	1.91	1.88	1.83	1.78	1.73	1.70	1.67	1.64	1.61	1.57	1.53
25	2.92	2.53	2.32	2.18	2.09	2.02	1.97	1.93	1.89	1.87	1.82	1.77	1.72	1.69	1.66	1.63	1.59	1.56	1.52
26	2.91	2.52	2.31	2.17	2.08	2.01	1.96	1.92	1.88	1.86	1.81	1.76	1.71	1.68	1.65	1.61	1.58	1.54	1.50
27	2.90	2.51	2.30	2.17	2.07	2.00	1.95	1.91	1.87	1.85	1.80	1.75	1.70	1.67	1.64	1.60	1.57	1.53	1.49
28	2.89	2.50	2.29	2.16	2.06	2.00	1.94	1.90	1.87	1.84	1.79	1.74	1.69	1.66	1.63	1.59	1.56	1.52	1.48
29	2.89	2.50	2.28	2.15	2.06	1.99	1.93	1.89	1.86	1.83	1.78	1.73	1.68	1.65	1.62	1.58	1.55	1.51	1.47
30	2.88	2.49	2.28	2.14	2.05	1.98	1.93	1.88	1.85	1.82	1.77	1.72	1.67	1.64	1.61	1.57	1.54	1.50	1.46
40	2.84	2.44	2.23	2.09	2.00	1.93	1.87	1.83	1.79	1.76	1.71	1.66	1.61	1.57	1.54	1.51	1.47	1.42	1.38
60	2.79	2.39	2.18	2.04	1.95	1.87	1.82	1.77	1.74	1.71	1.66	1.60	1.54	1.51	1.48	1.44	1.40	1.35	1.29
120	2.75	2.35	2.13	1.99	1.90	1.82	1.77	1.72	1.68	1.65	1.60	1.55	1.48	1.45	1.41	1.37	1.32	1.26	1.19
∞	2.71	2.30	2.08	1.94	1.85	1.77	1.72	1.67	1.63	1.60	1.55	1.49	1.42	1.38	1.34	1.30	1.24	1.17	1.00

$F = \frac{s_1^2}{s_2^2} = \frac{S_1}{m} \Big/ \frac{S_2}{n}$, where $s_1^2 = S_1/m$ and $s_2^2 = S_2/n$ are independent mean squares estimating a common variance σ^2 and based on m and n degrees of freedom, respectively.

PERCENTAGE POINTS, *F*-DISTRIBUTION (Continued)

$$F(F) = \int_0^F \frac{\Gamma\left(\frac{m+n}{2}\right)}{\Gamma\left(\frac{m}{2}\right)\Gamma\left(\frac{n}{2}\right)} m^{\frac{m}{2}} n^{\frac{n}{2}} x^{\frac{m}{2}-1} (n+mx)^{-\frac{m+n}{2}} dx = .95$$

n \ m	1	2	3	4	5	6	7	8	9	10	12	15	20	24	30	40	60	120	∞
1	161.4	199.5	215.7	224.6	230.2	234.0	236.8	238.9	240.5	241.9	243.9	245.9	248.0	249.1	250.1	251.1	252.2	253.3	254.3
2	18.51	19.00	19.16	19.25	19.30	19.33	19.35	19.37	19.38	19.40	19.41	19.43	19.45	19.45	19.46	19.47	19.48	19.49	19.50
3	10.13	9.55	9.28	9.12	9.01	8.94	8.89	8.85	8.81	8.79	8.74	8.70	8.66	8.64	8.62	8.59	8.57	8.55	8.53
4	7.71	6.94	6.59	6.39	6.26	6.16	6.09	6.04	6.00	5.96	5.91	5.86	5.80	5.77	5.75	5.72	5.69	5.66	5.63
5	6.61	5.79	5.41	5.19	5.05	4.95	4.88	4.82	4.77	4.74	4.68	4.62	4.56	4.53	4.50	4.46	4.43	4.40	4.36
6	5.99	5.14	4.76	4.53	4.39	4.28	4.21	4.15	4.10	4.06	4.00	3.94	3.87	3.84	3.81	3.77	3.74	3.70	3.67
7	5.59	4.74	4.35	4.12	3.97	3.87	3.79	3.73	3.68	3.64	3.57	3.51	3.44	3.41	3.38	3.34	3.30	3.27	3.23
8	5.32	4.46	4.07	3.84	3.69	3.58	3.50	3.44	3.39	3.35	3.28	3.22	3.15	3.12	3.08	3.04	3.01	2.97	2.93
9	5.12	4.26	3.86	3.63	3.48	3.37	3.29	3.23	3.18	3.14	3.07	3.01	2.94	2.90	2.86	2.83	2.79	2.75	2.71
10	4.96	4.10	3.71	3.48	3.33	3.22	3.14	3.07	3.02	2.98	2.91	2.85	2.77	2.74	2.70	2.66	2.62	2.58	2.54
11	4.84	3.98	3.59	3.36	3.20	3.09	3.01	2.95	2.90	2.85	2.79	2.72	2.65	2.61	2.57	2.53	2.49	2.45	2.40
12	4.75	3.89	3.49	3.26	3.11	3.00	2.91	2.85	2.80	2.75	2.69	2.62	2.54	2.51	2.47	2.43	2.38	2.34	2.30
13	4.67	3.81	3.41	3.18	3.03	2.92	2.83	2.77	2.71	2.67	2.60	2.53	2.46	2.42	2.38	2.34	2.30	2.25	2.21
14	4.60	3.74	3.34	3.11	2.96	2.85	2.76	2.70	2.65	2.60	2.53	2.46	2.39	2.35	2.31	2.27	2.22	2.18	2.13
15	4.54	3.68	3.29	3.06	2.90	2.79	2.71	2.64	2.59	2.54	2.48	2.40	2.33	2.29	2.25	2.20	2.16	2.11	2.07
16	4.49	3.63	3.24	3.01	2.85	2.74	2.66	2.59	2.54	2.49	2.42	2.35	2.28	2.24	2.19	2.15	2.11	2.06	2.01
17	4.45	3.59	3.20	2.96	2.81	2.70	2.61	2.55	2.49	2.45	2.38	2.31	2.23	2.19	2.15	2.10	2.06	2.01	1.96
18	4.41	3.55	3.16	2.93	2.77	2.66	2.58	2.51	2.46	2.41	2.34	2.27	2.19	2.15	2.11	2.06	2.02	1.97	1.92
19	4.38	3.52	3.13	2.90	2.74	2.63	2.54	2.48	2.42	2.38	2.31	2.23	2.16	2.11	2.07	2.03	1.98	1.93	1.88
20	4.35	3.49	3.10	2.87	2.71	2.60	2.51	2.45	2.39	2.35	2.28	2.20	2.12	2.08	2.04	1.99	1.95	1.90	1.84
21	4.32	3.47	3.07	2.84	2.68	2.57	2.49	2.42	2.37	2.32	2.25	2.18	2.10	2.05	2.01	1.96	1.92	1.87	1.81
22	4.30	3.44	3.05	2.82	2.66	2.55	2.46	2.40	2.34	2.30	2.23	2.15	2.07	2.03	1.98	1.94	1.89	1.84	1.78
23	4.28	3.42	3.03	2.80	2.64	2.53	2.44	2.37	2.32	2.27	2.20	2.13	2.05	2.01	1.96	1.91	1.86	1.81	1.76
24	4.26	3.40	3.01	2.78	2.62	2.51	2.42	2.36	2.30	2.25	2.18	2.11	2.03	1.98	1.94	1.89	1.84	1.79	1.73
25	4.24	3.39	2.99	2.76	2.60	2.49	2.40	2.34	2.28	2.24	2.16	2.09	2.01	1.96	1.92	1.87	1.82	1.77	1.71
26	4.23	3.37	2.98	2.74	2.59	2.47	2.39	2.32	2.27	2.22	2.15	2.07	1.99	1.95	1.90	1.85	1.80	1.75	1.69
27	4.21	3.35	2.96	2.73	2.57	2.46	2.37	2.31	2.25	2.20	2.13	2.06	1.97	1.93	1.88	1.84	1.79	1.73	1.67
28	4.20	3.34	2.95	2.71	2.56	2.45	2.36	2.29	2.24	2.19	2.12	2.04	1.96	1.91	1.87	1.82	1.77	1.71	1.65
29	4.18	3.33	2.93	2.70	2.55	2.43	2.35	2.28	2.22	2.18	2.10	2.03	1.94	1.90	1.85	1.81	1.75	1.70	1.64
30	4.17	3.32	2.92	2.69	2.53	2.42	2.33	2.27	2.21	2.16	2.09	2.01	1.93	1.89	1.84	1.79	1.74	1.68	1.62
40	4.08	3.23	2.84	2.61	2.45	2.34	2.25	2.18	2.12	2.08	2.00	1.92	1.84	1.79	1.74	1.69	1.64	1.58	1.51
60	4.00	3.15	2.76	2.53	2.37	2.25	2.17	2.10	2.04	1.99	1.92	1.84	1.75	1.70	1.65	1.59	1.53	1.47	1.39
120	3.92	3.07	2.68	2.45	2.29	2.17	2.09	2.02	1.96	1.91	1.83	1.75	1.66	1.61	1.55	1.50	1.43	1.35	1.25
∞	3.84	3.00	2.60	2.37	2.21	2.10	2.01	1.94	1.88	1.83	1.75	1.67	1.57	1.52	1.46	1.39	1.32	1.22	1.00

$F = \frac{s_1^2}{s_2^2} = \frac{S_1}{m} \Big/ \frac{S_2}{n}$, where $s_1^2 = S_1/m$ and $s_2^2 = S_2/n$ are independent mean squares estimating a common variance σ^2 and based on m and n degrees of freedom, respectively.

PERCENTAGE POINTS, *F*-DISTRIBUTION (Continued)

$$F(F) = \int_0^F \frac{\Gamma\left(\frac{m+n}{2}\right)}{\Gamma\left(\frac{m}{2}\right)\Gamma\left(\frac{n}{2}\right)} m^{\frac{m}{2}} n^{\frac{n}{2}} x^{\frac{m}{2}-1} (n+mx)^{-\frac{m+n}{2}}\, dx = .975$$

n \ m	1	2	3	4	5	6	7	8	9	10	12	15	20	24	30	40	60	120	∞
1	647.8	799.5	864.2	899.6	921.8	937.1	948.2	956.7	963.3	968.6	976.7	984.9	993.1	997.2	1001	1006	1010	1014	1018
2	38.51	39.00	39.17	39.25	39.30	39.33	39.36	39.37	39.39	39.40	39.41	39.43	39.45	39.46	39.46	39.47	39.48	39.49	39.50
3	17.44	16.04	15.44	15.10	14.88	14.73	14.62	14.54	14.47	14.42	14.34	14.25	14.17	14.12	14.08	14.04	13.99	13.95	13.90
4	12.22	10.65	9.98	9.60	9.36	9.20	9.07	8.98	8.90	8.84	8.75	8.66	8.56	8.51	8.46	8.41	8.36	8.31	8.26
5	10.01	8.43	7.76	7.39	7.15	6.98	6.85	6.76	6.68	6.62	6.52	6.43	6.33	6.28	6.23	6.18	6.12	6.07	6.02
6	8.81	7.26	6.60	6.23	5.99	5.82	5.70	5.60	5.52	5.46	5.37	5.27	5.17	5.12	5.07	5.01	4.96	4.90	4.85
7	8.07	6.54	5.89	5.52	5.29	5.12	4.99	4.90	4.82	4.76	4.67	4.57	4.47	4.42	4.36	4.31	4.25	4.20	4.14
8	7.57	6.06	5.42	5.05	4.82	4.65	4.53	4.43	4.36	4.30	4.20	4.10	4.00	3.95	3.89	3.84	3.78	3.73	3.67
9	7.21	5.71	5.08	4.72	4.48	4.32	4.20	4.10	4.03	3.96	3.87	3.77	3.67	3.61	3.56	3.51	3.45	3.39	3.33
10	6.94	5.46	4.83	4.47	4.24	4.07	3.95	3.85	3.78	3.72	3.62	3.52	3.42	3.37	3.31	3.26	3.20	3.14	3.08
11	6.72	5.26	4.63	4.28	4.04	3.88	3.76	3.66	3.59	3.53	3.43	3.33	3.23	3.17	3.12	3.06	3.00	2.94	2.88
12	6.55	5.10	4.47	4.12	3.89	3.73	3.61	3.51	3.44	3.37	3.28	3.18	3.07	3.02	2.96	2.91	2.85	2.79	2.72
13	6.41	4.97	4.35	4.00	3.77	3.60	3.48	3.39	3.31	3.25	3.15	3.05	2.95	2.89	2.84	2.78	2.72	2.66	2.60
14	6.30	4.86	4.24	3.89	3.66	3.50	3.38	3.29	3.21	3.15	3.05	2.95	2.84	2.79	2.73	2.67	2.61	2.55	2.49
15	6.20	4.77	4.15	3.80	3.58	3.41	3.29	3.20	3.12	3.06	2.96	2.86	2.76	2.70	2.64	2.59	2.52	2.46	2.40
16	6.12	4.69	4.08	3.73	3.50	3.34	3.22	3.12	3.05	2.99	2.89	2.79	2.68	2.63	2.57	2.51	2.45	2.38	2.32
17	6.04	4.62	4.01	3.66	3.44	3.28	3.16	3.06	2.98	2.92	2.82	2.72	2.62	2.56	2.50	2.44	2.38	2.32	2.25
18	5.98	4.56	3.95	3.61	3.38	3.22	3.10	3.01	2.93	2.87	2.77	2.67	2.56	2.50	2.44	2.38	2.32	2.26	2.19
19	5.92	4.51	3.90	3.56	3.33	3.17	3.05	2.96	2.88	2.82	2.72	2.62	2.51	2.45	2.39	2.33	2.27	2.20	2.13
20	5.87	4.46	3.86	3.51	3.29	3.13	3.01	2.91	2.84	2.77	2.68	2.57	2.46	2.41	2.35	2.29	2.22	2.16	2.09
21	5.83	4.42	3.82	3.48	3.25	3.09	2.97	2.87	2.80	2.73	2.64	2.53	2.42	2.37	2.31	2.25	2.18	2.11	2.04
22	5.79	4.38	3.78	3.44	3.22	3.05	2.93	2.84	2.76	2.70	2.60	2.50	2.39	2.33	2.27	2.21	2.14	2.08	2.00
23	5.75	4.35	3.75	3.41	3.18	3.02	2.90	2.81	2.73	2.67	2.57	2.47	2.36	2.30	2.24	2.18	2.11	2.04	1.97
24	5.72	4.32	3.72	3.38	3.15	2.99	2.87	2.78	2.70	2.64	2.54	2.44	2.33	2.27	2.21	2.15	2.08	2.01	1.94
25	5.69	4.29	3.69	3.35	3.13	2.97	2.85	2.75	2.68	2.61	2.51	2.41	2.30	2.24	2.18	2.12	2.05	1.98	1.91
26	5.66	4.27	3.67	3.33	3.10	2.94	2.82	2.73	2.65	2.59	2.49	2.39	2.28	2.22	2.16	2.09	2.03	1.95	1.88
27	5.63	4.24	3.65	3.31	3.08	2.92	2.80	2.71	2.63	2.57	2.47	2.36	2.25	2.19	2.13	2.07	2.00	1.93	1.85
28	5.61	4.22	3.63	3.29	3.06	2.90	2.78	2.69	2.61	2.55	2.45	2.34	2.23	2.17	2.11	2.05	1.98	1.91	1.83
29	5.59	4.20	3.61	3.27	3.04	2.88	2.76	2.67	2.59	2.53	2.43	2.32	2.21	2.15	2.09	2.03	1.96	1.89	1.81
30	5.57	4.18	3.59	3.25	3.03	2.87	2.75	2.65	2.57	2.51	2.41	2.31	2.20	2.14	2.07	2.01	1.94	1.87	1.79
40	5.42	4.05	3.46	3.13	2.90	2.74	2.62	2.53	2.45	2.39	2.29	2.18	2.07	2.01	1.94	1.88	1.80	1.72	1.64
60	5.29	3.93	3.34	3.01	2.79	2.63	2.51	2.41	2.33	2.27	2.17	2.06	1.94	1.88	1.82	1.74	1.67	1.58	1.48
120	5.15	3.80	3.23	2.89	2.67	2.52	2.39	2.30	2.22	2.16	2.05	1.94	1.82	1.76	1.69	1.61	1.53	1.43	1.31
∞	5.02	3.69	3.12	2.79	2.57	2.41	2.29	2.19	2.11	2.05	1.94	1.83	1.71	1.64	1.57	1.48	1.39	1.27	1.00

$F = \frac{s_1^2}{s_2^2} = \frac{S_1}{m} \Big/ \frac{S_2}{n}$, where $s_1^2 = S_1/m$ and $s_2^2 = S_2/n$ are independent mean squares estimating a common variance σ^2 and based on m and n degrees of freedom, respectively.

PERCENTAGE POINTS, *F*-DISTRIBUTION (Continued)

$$F(F) = \int_0^F \frac{\Gamma\left(\frac{m+n}{2}\right)}{\Gamma\left(\frac{m}{2}\right)\Gamma\left(\frac{n}{2}\right)} m^{\frac{m}{2}} n^{\frac{n}{2}} x^{\frac{m}{2}-1} (n+mx)^{-\frac{m+n}{2}} dx = .99$$

n \ m	1	2	3	4	5	6	7	8	9	10	12	15	20	24	30	40	60	120	∞
1	4052	4999.5	5403	5625	5764	5859	5928	5982	6022	6056	6106	6157	6209	6235	6261	6287	6313	6339	6366
2	98.50	99.00	99.17	99.25	99.30	99.33	99.36	99.37	99.39	99.40	99.42	99.43	99.45	99.46	99.47	99.47	99.48	99.49	99.50
3	34.12	30.82	29.46	28.71	28.24	27.91	27.67	27.49	27.35	27.23	27.05	26.87	26.69	26.60	26.50	26.41	26.32	26.22	26.13
4	21.20	18.00	16.69	15.98	15.52	15.21	14.98	14.80	14.66	14.55	14.37	14.20	14.02	13.93	13.84	13.75	13.65	13.56	13.46
5	16.26	13.27	12.06	11.39	10.97	10.67	10.46	10.29	10.16	10.05	9.89	9.72	9.55	9.47	9.38	9.29	9.20	9.11	9.02
6	13.75	10.92	9.78	9.15	8.75	8.47	8.26	8.10	7.98	7.87	7.72	7.56	7.40	7.31	7.23	7.14	7.06	6.97	6.88
7	12.25	9.55	8.45	7.85	7.46	7.19	6.99	6.84	6.72	6.62	6.47	6.31	6.16	6.07	5.99	5.91	5.82	5.74	5.65
8	11.26	8.65	7.59	7.01	6.63	6.37	6.18	6.03	5.91	5.81	5.67	5.52	5.36	5.28	5.20	5.12	5.03	4.95	4.86
9	10.56	8.02	6.99	6.42	6.06	5.80	5.61	5.47	5.35	5.26	5.11	4.96	4.81	4.73	4.65	4.57	4.48	4.40	4.31
10	10.04	7.56	6.55	5.99	5.64	5.39	5.20	5.06	4.94	4.85	4.71	4.56	4.41	4.33	4.25	4.17	4.08	4.00	3.91
11	9.65	7.21	6.22	5.67	5.32	5.07	4.89	4.74	4.63	4.54	4.40	4.25	4.10	4.02	3.94	3.86	3.78	3.69	3.60
12	9.33	6.93	5.95	5.41	5.06	4.82	4.64	4.50	4.39	4.30	4.16	4.01	3.86	3.78	3.70	3.62	3.54	3.45	3.36
13	9.07	6.70	5.74	5.21	4.86	4.62	4.44	4.30	4.19	4.10	3.96	3.82	3.66	3.59	3.51	3.43	3.34	3.25	3.17
14	8.86	6.51	5.56	5.04	4.69	4.46	4.28	4.14	4.03	3.94	3.80	3.66	3.51	3.43	3.35	3.27	3.18	3.09	3.00
15	8.68	6.36	5.42	4.89	4.56	4.32	4.14	4.00	3.89	3.80	3.67	3.52	3.37	3.29	3.21	3.13	3.05	2.96	2.87
16	8.53	6.23	5.29	4.77	4.44	4.20	4.03	3.89	3.78	3.69	3.55	3.41	3.26	3.18	3.10	3.02	2.93	2.84	2.75
17	8.40	6.11	5.18	4.67	4.34	4.10	3.93	3.79	3.68	3.59	3.46	3.31	3.16	3.08	3.00	2.92	2.83	2.75	2.65
18	8.29	6.01	5.09	4.58	4.25	4.01	3.84	3.71	3.60	3.51	3.37	3.23	3.08	3.00	2.92	2.84	2.75	2.66	2.57
19	8.18	5.93	5.01	4.50	4.17	3.94	3.77	3.63	3.52	3.43	3.30	3.15	3.00	2.92	2.84	2.76	2.67	2 58	2.49
20	8.10	5.85	4.94	4.43	4.10	3.87	3.70	3.56	3.46	3.37	3.23	3.09	2.94	2.86	2.78	2.69	2.61	2.52	2.42
21	8.02	5.78	4.87	4.37	4.04	3.81	3.64	3.51	3.40	3.31	3.17	3.03	2.88	2.80	2.72	2.64	2.55	2.46	2.36
22	7.95	5.72	4.82	4.31	3.99	3.76	3.59	3.45	3.35	3.26	3.12	2.98	2.83	2.75	2.67	2.58	2.50	2.40	2.31
23	7.88	5.66	4.76	4.26	3.94	3.71	3.54	3.41	3.30	3.21	3.07	2.93	2.78	2.70	2.62	2.54	2.45	2.35	2.26
24	7.82	5.61	4.72	4.22	3.90	3.67	3.50	3.36	3.26	3.17	3.03	2.89	2.74	2.66	2.58	2.49	2.40	2.31	2.21
25	7.77	5.57	4.68	4.18	3.85	3.63	3.46	3.32	3.22	3.13	2.99	2.85	2.70	2.62	2.54	2.45	2.36	2.27	2.17
26	7.72	5.53	4.64	4.14	3.82	3.59	3.42	3.29	3.18	3.09	2.96	2.81	2.66	2.58	2.50	2.42	2.33	2.23	2.13
27	7.68	5.49	4.60	4.11	3.78	3.56	3.39	3.26	3.15	3.06	2.93	2.78	2.63	2.55	2.47	2.38	2.29	2.20	2.10
28	7.64	5.45	4.57	4.07	3.75	3.53	3.36	3.23	3.12	3.03	2.90	2.75	2.60	2.52	2.44	2.35	2.26	2.17	2 06
29	7.60	5.42	4.54	4.04	3.73	3.50	3.33	3.20	3.09	3.00	2.87	2.73	2.57	2.49	2.41	2.33	2.23	2 14	2.03
30	7.56	5.39	4.51	4.02	3.70	3.47	3.30	3.17	3.07	2.98	2.84	2.70	2.55	2.47	2.39	2.30	2.21	2.11	2.01
40	7.31	5.18	4.31	3.83	3.51	3.29	3.12	2.99	2.89	2.80	2.66	2.52	2.37	2.29	2.20	2.11	2.02	1.92	1.80
60	7.08	4.98	4.13	3.65	3.34	3.12	2.95	2.82	2.72	2.63	2.50	2.35	2.20	2.12	2.03	1.94	1.84	1.73	1.60
120	6.85	4.79	3.95	3.48	3.17	2.96	2.79	2.66	2.56	2.47	2.34	2.19	2.03	1.95	1.86	1.76	1.66	1.53	1.38
∞	6.63	4.61	3.78	3.32	3.02	2.80	2.64	2.51	2.41	2.32	2.18	2.04	1.88	1.79	1.70	1.59	1.47	1.32	1.00

$F = \frac{s_1^2}{s_2^2} = \frac{S_1}{m} \Big/ \frac{S_2}{n}$, where $s_1^2 = S_1/m$ and $s_2^2 = S_2/n$ are independent mean squares estimating a common variance σ^2 and based on m and n degrees of freedom, respectively.

PERCENTAGE POINTS, F-DISTRIBUTION (Continued)

$$F(F) = \int_0^F \frac{\Gamma\left(\frac{m+n}{2}\right)}{\Gamma\left(\frac{m}{2}\right)\Gamma\left(\frac{n}{2}\right)} m^{\frac{m}{2}} n^{\frac{n}{2}} x^{\frac{m}{2}-1} (n+mx)^{-\frac{m+n}{2}} dx = .995$$

n \ m	1	2	3	4	5	6	7	8	9	10	12	15	20	24	30	40	60	120	∞
1	16211	20000	21615	22500	23056	23437	23715	23925	24091	24224	24426	24630	24836	24940	25044	25148	25253	25359	25465
2	198.5	199.0	199.2	199.2	199.3	199.3	199.4	199.4	199.4	199.4	199.4	199.4	199.4	199.5	199.5	199.5	199.5	199.5	199.5
3	55.55	49.80	47.47	46.19	45.39	44.84	44.43	44.13	43.88	43.69	43.39	43.08	42.78	42.62	42.47	42.31	42.15	41.99	41.83
4	31.33	26.28	24.26	23.15	22.46	21.97	21.62	21.35	21.14	20.97	20.70	20.44	20.17	20.03	19.89	19.75	19.61	19.47	19.32
5	22.78	18.31	16.53	15.56	14.94	14.51	14.20	13.96	13.77	13.62	13.38	13.15	12.90	12.78	12.66	12.53	12.40	12.27	12.14
6	18.63	14.54	12.92	12.03	11.46	11.07	10.79	10.57	10.39	10.25	10.03	9.81	9.59	9.47	9.36	9.24	9.12	9.00	8.88
7	16.24	12.40	10.88	10.05	9.52	9.16	8.89	8.68	8.51	8.38	8.18	7.97	7.75	7.65	7.53	7.42	7.31	7.19	7.08
8	14.69	11.04	9.60	8.81	8.30	7.95	7.69	7.50	7.34	7.21	7.01	6.81	6.61	6.50	6.40	6.29	6.18	6.06	5.95
9	13.61	10.11	8.72	7.96	7.47	7.13	6.88	6.69	6.54	6.42	6.23	6.03	5.83	5.73	5.62	5.52	5.41	5.30	5.19
10	12.83	9.43	8.08	7.34	6.87	6.54	6.30	6.12	5.97	5.85	5.66	5.47	5.27	5.17	5.07	4.97	4.86	4.75	4.64
11	12.23	8.91	7.60	6.88	6.42	6.10	5.86	5.68	5.54	5.42	5.24	5.05	4.86	4.76	4.65	4.55	4.44	4.34	4.23
12	11.75	8.51	7.23	6.52	6.07	5.76	5.52	5.35	5.20	5.09	4.91	4.72	4.53	4.43	4.33	4.23	4.12	4.01	3.90
13	11.37	8.19	6.93	6.23	5.79	5.48	5.25	5.08	4.94	4.82	4.64	4.46	4.27	4.17	4.07	3.97	3.87	3.76	3.65
14	11.06	7.92	6.68	6.00	5.56	5.26	5.03	4.86	4.72	4.60	4.43	4.25	4.06	3.96	3.86	3.76	3.66	3.55	3.44
15	10.80	7.70	6.48	5.80	5.37	5.07	4.85	4.67	4.54	4.42	4.25	4.07	3.88	3.79	3.69	3.58	3.48	3.37	3.26
16	10.58	7.51	6.30	5.64	5.21	4.91	4.69	4.52	4.38	4.27	4.10	3.92	3.73	3.64	3.54	3.44	3.33	3.22	3.11
17	10.38	7.35	6.16	5.50	5.07	4.78	4.56	4.39	4.25	4.14	3.97	3.79	3.61	3.51	3.41	3.31	3.21	3.10	2.98
18	10.22	7.21	6.03	5.37	4.96	4.66	4.44	4.28	4.14	4.03	3.86	3.68	3.50	3.40	3.30	3.20	3.10	2.99	2.87
19	10.07	7.09	5.92	5.27	4.85	4.56	4.34	4.18	4.04	3.93	3.76	3.59	3.40	3.31	3.21	3.11	3.00	2.89	2.78
20	9.94	6.99	5.82	5.17	4.76	4.47	4.26	4.09	3.96	3.85	3.68	3.50	3.32	3.22	3.12	3.02	2.92	2.81	2.69
21	9.83	6.89	5.73	5.09	4.68	4.39	4.18	4.01	3.88	3.77	3.60	3.43	3.24	3.15	3.05	2.95	2.84	2.73	2.61
22	9.73	6.81	5.65	5.02	4.61	4.32	4.11	3.94	3.81	3.70	3.54	3.36	3.18	3.08	2.98	2.88	2.77	2.66	2.55
23	9.63	6.73	5.58	4.95	4.54	4.26	4.05	3.88	3.75	3.64	3.47	3.30	3.12	3.02	2.92	2.82	2.71	2.60	2.48
24	9.55	6.66	5.52	4.89	4.49	4.20	3.99	3.83	3.69	3.59	3.42	3.25	3.06	2.97	2.87	2.77	2.66	2.55	2.43
25	9.48	6.60	5.46	4.84	4.43	4.15	3.94	3.78	3.64	3.54	3.37	3.20	3.01	2.92	2.82	2.72	2.61	2.50	2.38
26	9.41	6.54	5.41	4.79	4.38	4.10	3.89	3.73	3.60	3.49	3.33	3.15	2.97	2.87	2.77	2.67	2.56	2.45	2.33
27	9.34	6.49	5.36	4.74	4.34	4.06	3.85	3.69	3.56	3.45	3.28	3.11	2.93	2.83	2.73	2.63	2.52	2.41	2.25
28	9.28	6.44	5.32	4.70	4.30	4.02	3.81	3.65	3.52	3.41	3.25	3.07	2.89	2.79	2.69	2.59	2.48	2.37	2.29
29	9.23	6.40	5.28	4.66	4.26	3.98	3.77	3.61	3.48	3.38	3.21	3.04	2.86	2.76	2.66	2.56	2.45	2.33	2.24
30	9.18	6.35	5.24	4.62	4.23	3.95	3.74	3.58	3.45	3.34	3.18	3.01	2.82	2.73	2.63	2.52	2.42	2.30	2.18
40	8.83	6.07	4.98	4.37	3.99	3.71	3.51	3.35	3.22	3.12	2.95	2.78	2.60	2.50	2.40	2.30	2.18	2.06	1.93
60	8.49	5.79	4.73	4.14	3.76	3.49	3.29	3.13	3.01	2.90	2.74	2.57	2.39	2.29	2.19	2.08	1.96	1.83	1.69
120	8.18	5.54	4.50	3.92	3.55	3.28	3.09	2.93	2.81	2.71	2.54	2.37	2.19	2.09	1.98	1.87	1.75	1.61	1.43
∞	7.88	5.30	4.28	3.72	3.35	3.09	2.90	2.74	2.62	2.52	2.36	2.19	2.00	1.90	1.79	1.67	1.53	1.36	1.00

$F = \frac{s_1^2}{s_2^2} = \frac{S_1}{m} \Big/ \frac{S_2}{n}$, where $s_1^2 = S_1/m$ and $s_2^2 = S_2/n$ are independent mean squares estimating a common variance σ^2 and based on m and n degrees of freedom, respectively.

PERCENTAGE POINTS, *F*-DISTRIBUTION (Continued)

$$F(F) = \int_0^F \frac{\Gamma\left(\frac{m+n}{2}\right)}{\Gamma\left(\frac{m}{2}\right)\Gamma\left(\frac{n}{2}\right)} m^{\frac{m}{2}} n^{\frac{n}{2}} x^{\frac{m}{2}-1} (n+mx)^{-\frac{m+n}{2}} dx = .999$$

n \ m	1	2	3	4	5	6	7	8	9	10	12	15	20	24	30	40	60	120	∞
1	4053*	5000*	5404*	5625*	5764*	5859*	5929*	5981*	6023*	6056*	6107*	6158*	6209*	6235*	6261*	6287*	6313*	6340*	6366*
2	998.5	999.0	999.2	999.2	999.3	999.3	999.4	999.4	999.4	999.4	999.4	999.4	999.4	999.5	999.5	999.5	999.5	999.5	999.5
3	167.0	148.5	141.1	137.1	134.6	132.8	131.6	130.6	129.9	129.2	128.3	127.4	126.4	125.9	125.4	125.0	124.5	124.0	123.5
4	74.14	61.25	56.18	53.44	51.71	50.53	49.66	49.00	48.47	48.05	47.41	46.76	46.10	45.77	45.43	45.09	44.75	44.40	44.05
5	47.18	37.12	33.20	31.09	29.75	28.84	28.16	27.64	27.24	26.92	26.42	25.91	25.39	25.14	24.87	24.60	24.33	24.06	23.79
6	35.51	27.00	23.70	21.92	20.81	20.03	19.46	19.03	18.69	18.41	17.99	17.56	17.12	16.89	16.67	16.44	16.21	15.99	15.75
7	29.25	21.69	18.77	17.19	16.21	15.52	15.02	14.63	14.33	14.08	13.71	13.32	12.93	12.73	12.53	12.33	12.12	11.91	11.70
8	25.42	18.49	15.83	14.39	13.49	12.86	12.40	12.04	11.77	11.54	11.19	10.84	10.48	10.30	10.11	9.92	9.73	9.53	9.33
9	22.86	16.39	13.90	12.56	11.71	11.13	10.70	10.37	10.11	9.89	9.57	9.24	8.90	8.72	8.55	8.37	8.19	8.00	7.81
10	21.04	14.91	12.55	11.28	10.48	9.92	9.52	9.20	8.96	8.75	8.45	8.13	7.80	7.64	7.47	7.30	7.12	6.94	6.76
11	19.69	13.81	11.56	10.35	9.58	9.05	8.66	8.35	8.12	7.92	7.63	7.32	7.01	6.85	6.68	6.52	6.35	6.17	6.00
12	18.64	12.97	10.80	9.63	8.89	8.38	8.00	7.71	7.48	7.29	7.00	6.71	6.40	6.25	6.09	5.93	5.76	5.59	5.42
13	17.81	12.31	10.21	9.07	8.35	7.86	7.49	7.21	6.98	6.80	6.52	6.23	5.93	5.78	5.63	5.47	5.30	5.14	4.97
14	17.14	11.78	9.73	8.62	7.92	7.43	7.08	6.80	6.58	6.40	6.13	5.85	5.56	5.41	5.25	5.10	4.94	4.77	4.60
15	16.59	11.34	9.34	8.25	7.57	7.09	6.74	6.47	6.26	6.08	5.81	5.54	5.25	5.10	4.95	4.80	4.64	4.47	4.31
16	16.12	10.97	9.00	7.94	7.27	6.81	6.46	6.19	5.98	5.81	5.55	5.27	4.99	4.85	4.70	4.54	4.39	4.23	4.06
17	15.72	10.66	8.73	7.68	7.02	6.56	6.22	5.96	5.75	5.58	5.32	5.05	4.78	4.63	4.48	4.33	4.18	4.02	3.85
18	15.38	10.39	8.49	7.46	6.81	6.35	6.02	5.76	5.56	5.39	5.13	4.87	4.59	4.45	4.30	4.15	4.00	3.84	3.67
19	15.08	10.16	8.28	7.26	6.62	6.18	5.85	5.59	5.39	5.22	4.97	4.70	4.43	4.29	4.14	3.99	3.84	3.68	3.51
20	14.82	9.95	8.10	7.10	6.46	6.02	5.69	5.44	5.24	5.08	4.82	4.56	4.29	4.15	4.00	3.86	3.70	3.54	3.38
21	14.59	9.77	7.94	6.95	6.32	5.88	5.56	5.31	5.11	4.95	4.70	4.44	4.17	4.03	3.88	3.74	3.58	3.42	3.26
22	14.38	9.61	7.80	6.81	6.19	5.76	5.44	5.19	4.99	4.83	4.58	4.33	4.06	3.92	3.78	3.63	3.48	3.32	3.15
23	14.19	9.47	7.67	6.69	6.08	5.65	5.33	5.09	4.89	4.73	4.48	4.23	3.96	3.82	3.68	3.53	3.38	3.22	3.05
24	14.03	9.34	7.55	6.59	5.98	5.55	5.23	4.99	4.80	4.64	4.39	4.14	3.87	3.74	3.59	3.45	3.29	3.14	2.97
25	13.88	9.22	7.45	6.49	5.88	5.46	5.15	4.91	4.71	4.56	4.31	4.06	3.79	3.66	3.52	3.37	3.22	3.06	2.89
26	13.74	9.12	7.36	6.41	5.80	5.38	5.07	4.83	4.64	4.48	4.24	3.99	3.72	3.59	3.44	3.30	3.15	2.99	2.82
27	13.61	9.02	7.27	6.33	5.73	5.31	5.00	4.76	4.57	4.41	4.17	3.92	3.66	3.52	3.38	3.23	3.08	2.92	2.75
28	13.50	8.93	7.19	6.25	5.66	5.24	4.93	4.69	4.50	4.35	4.11	3.86	3.60	3.46	3.32	3.18	3.02	2.86	2.69
29	13.39	8.85	7.12	6.19	5.59	5.18	4.87	4.64	4.45	4.29	4.05	3.80	3.54	3.41	3.27	3.12	2.97	2.81	2.64
30	13.29	8.77	7.05	6.12	5.53	5.12	4.82	4.58	4.39	4.24	4.00	3.75	3.49	3.36	3.22	3.07	2.92	2.76	2.59
40	12.61	8.25	6.60	5.70	5.13	4.73	4.44	4.21	4.02	3.87	3.64	3.40	3.15	3.01	2.87	2.73	2.57	2.41	2.23
60	11.97	7.76	6.17	5.31	4.76	4.37	4.09	3.87	3.69	3.54	3.31	3.08	2.83	2.69	2.55	2.41	2.25	2.08	1.89
120	11.38	7.32	5.79	4.95	4.42	4.04	3.77	3.55	3.38	3.24	3.02	2.78	2.53	2.40	2.26	2.11	1.95	1.76	1.54
∞	10.83	6.91	5.42	4.62	4.10	3.74	3.47	3.27	3.10	2.96	2.74	2.51	2.27	2.13	1.99	1.84	1.66	1.45	1.00

* Multiply these entries by 100.

RANDOM UNITS

Use of Table. If one wishes to select a random sample of N items from a universe of M items, the following procedure may be applied. ($M > N$.)

1. Decide upon some arbitrary scheme of selecting entries from the table. For example, one may decide to use the entries in the first line, second column; second line, third column; third line, fourth column; etc.

2. Assign numbers to each of the items in the universe from 1 to M. Thus, if $M = 500$, the items would be numbered from 001 to 500, and therefore, each designated item is associated with a three digit number.

3. Decide upon some arbitrary scheme of selecting positional digits from each entry chosen according to Step 1. Thus, if $M = 500$, one may decide to use the first, third, and fourth digit of each entry selected, and as a consequence a three digit number is created for each entry choice.

4. If the number formed is $\leqq M$, the correspondingly designated item in the universe is chosen for the random sample of N items. If a number formed is $>M$ or is a repeated number of one already chosen, it is passed over and the next desirable number is taken. This process is continued until the random sample of N items is selected.

A TABLE OF 14,000 RANDOM UNITS

Line/Col.	(1)	(2)	(3)	(4)	(5)	(6)	(7)	(8)	(9)	(10)	(11)	(12)	(13)	(14)
1	10480	15011	01536	02011	81647	91646	69179	14194	62590	36207	20969	99570	91291	90700
2	22368	46573	25595	85393	30995	89198	27982	53402	93965	34095	52666	19174	39615	99505
3	24130	48360	22527	97265	76393	64809	15179	24830	49340	32081	30680	19655	63348	58629
4	42167	93093	06243	61680	07856	16376	39440	53537	71341	57004	00849	74917	97758	16379
5	37570	39975	81837	16656	06121	91782	60468	81305	49684	60672	14110	06927	01263	54613
6	77921	06907	11008	42751	27756	53498	18602	70659	90655	15053	21916	81825	44394	42880
7	99562	72905	56420	69994	98872	31016	71194	18738	44013	48840	63213	21069	10634	12952
8	96301	91977	05463	07972	18876	20922	94595	56869	69014	60045	18425	84903	42508	32307
9	89579	14342	63661	10281	17453	18103	57740	84378	25331	12566	58678	44947	05585	56941
10	85475	36857	43342	53988	53060	59533	38867	62300	08158	17983	16439	11458	18593	64952
11	28918	69578	88231	33276	70997	79936	56865	05859	90106	31595	01547	85590	91610	78188
12	63553	40961	48235	03427	49626	69445	18663	72695	52180	20847	12234	90511	33703	90322
13	09429	93969	52636	92737	88974	33488	36320	17617	30015	08272	84115	27156	30613	74952
14	10365	61129	87529	85689	48237	52267	67689	93394	01511	26358	85104	20285	29975	89868
15	07119	97336	71048	08178	77233	13916	47564	81056	97735	85977	29372	74461	28551	90707
16	51085	12765	51821	51259	77452	16308	60756	92144	49442	53900	70960	63990	75601	40719
17	02368	21382	52404	60268	89368	19885	55322	44819	01188	65255	64835	44919	05944	55157
18	01011	54092	33362	94904	31273	04146	18594	29852	71585	85030	51132	01915	92747	64951
19	52162	53916	46369	58586	23216	14513	83149	98736	23495	64350	94738	17752	35156	35749
20	07056	97628	33787	09998	42698	06691	76988	13602	51851	46104	88916	19509	25625	58104
21	48663	91245	85828	14346	09172	30168	90229	04734	59193	22178	30421	61666	99904	32812
22	54164	58492	22421	74103	47070	25306	76468	26384	58151	06646	21524	15227	96909	44592
23	32639	32363	05597	24200	13363	38005	94342	28728	35806	06912	17012	64161	18296	22851
24	29334	27001	87637	87308	58731	00256	45834	15398	46557	41135	10367	07684	36188	18510
25	02488	33062	28834	07351	19731	92420	60952	61280	50001	67658	32586	86679	50720	94953
26	81525	72295	04839	96423	24878	82651	66566	14778	76797	14780	13300	87074	79666	95725
27	29676	20591	68086	26432	46901	20849	89768	81536	86645	12659	92259	57102	80428	25280
28	00742	57392	39064	66432	84673	40027	32832	61362	98947	96067	64760	64584	96096	98253
29	05366	04213	25669	26422	44407	44048	37937	63904	45766	66134	75470	66520	34693	90449
30	91921	26418	64117	94305	26766	25940	39972	22209	71500	64568	91402	42416	07844	69618
31	00582	04711	87917	77341	42206	35126	74087	99547	81817	42607	43808	76655	62028	76630
32	00725	69884	62797	56170	86324	88072	76222	36086	84637	93161	76038	65855	77919	88006
33	69011	65797	95876	55293	18988	27354	26575	08625	40801	59920	29841	80150	12777	48501
34	25976	57948	29888	88604	67917	48708	18912	82271	65424	69774	33611	54262	85963	03547
35	09763	83473	73577	12908	30883	18317	28290	35797	05998	41688	34952	37888	38917	88050
36	91567	42595	27958	30134	04024	86385	29880	99730	55536	84855	29080	09250	79656	73211
37	17955	56349	90999	49127	20044	59931	06115	20542	18059	02008	73708	83517	36103	42791
38	46503	18584	18845	49618	02304	51038	20655	58727	28168	15475	56942	53389	20562	87338
39	92157	89634	94824	78171	84610	82834	09922	25417	44137	48413	25555	21246	35509	20468
40	14577	62765	35605	81263	39667	47358	56873	56307	61607	49518	89656	20103	77490	18062
41	98427	07523	33362	64270	01638	92477	66969	98420	04880	45585	46565	04102	46880	45709
42	34914	63976	88720	82765	34476	17032	87589	40836	32427	70002	70663	88863	77775	69348
43	70060	28277	39475	46473	23219	53416	94970	25832	69975	94884	19661	72828	00102	66794
44	53976	54914	06990	67245	68350	82948	11398	42878	80287	88267	47363	46634	06541	97809
45	76072	29515	40980	07391	58745	25774	22987	80059	39911	96189	41151	14222	60697	59583
46	90725	52210	83974	29992	65831	38857	50490	83765	55657	14361	31720	57375	56228	41546
47	64364	67412	33339	31926	14883	24413	59744	92351	97473	89286	35931	04110	23726	51900
48	08962	00358	31662	25388	61642	34072	81249	35648	56891	69352	48373	45578	78547	81788
49	95012	68379	93526	70765	10593	04542	76463	54328	02349	17247	28865	14777	62730	92277
50	15664	10493	20492	38391	91132	21999	59516	81652	27195	48223	46751	22923	32261	85653

A TABLE OF 14,000 RANDOM UNITS (Continued)

Line/Col.	(1)	(2)	(3)	(4)	(5)	(6)	(7)	(8)	(9)	(10)	(11)	(12)	(13)	(14)
51	16408	81899	04153	53381	79401	21438	83035	92350	36693	31238	59649	91754	72772	02338
52	18629	81953	05520	91962	04739	13092	97662	24822	94730	06496	35090	04822	86772	98289
53	73115	35101	47498	87637	99016	71060	88824	71013	18735	20286	23153	72924	35165	43040
54	57491	16703	23167	49323	45021	33132	12544	41035	80780	45393	44812	12515	98931	91202
55	30405	83946	23792	14422	15059	45799	22716	19792	09983	74353	68668	30429	70735	25499
56	16631	35006	85900	98275	32388	52390	16815	69298	82732	38480	73817	32523	41961	44437
57	96773	20206	42559	78985	05300	22164	24369	54224	35083	19687	11052	91491	60383	19746
58	38935	64202	14349	82674	66523	44133	00697	35552	35970	19124	63318	29686	03387	59846
59	31624	76384	17403	53363	44167	64486	64758	75366	76554	31601	12614	33072	60332	92325
60	78919	19474	23632	27889	47914	02584	37680	20801	72152	39339	34806	08930	85001	87820
61	03931	33309	57047	74211	63445	17361	62825	39908	05607	91284	68833	25570	38818	46920
62	74426	33278	43972	10119	89917	15665	52872	73823	73144	88662	88970	74492	51805	99378
63	09066	00903	20795	95452	92648	45454	09552	88815	16553	51125	79375	97596	16296	66092
64	42238	12426	87025	14267	20979	04508	64535	31355	86064	29472	47689	05974	52468	16834
65	16153	08002	26504	41744	81959	65642	74240	56302	00033	67107	77510	70625	28725	34191
66	21457	40742	29820	96783	29400	21840	15035	34537	33310	06116	95240	15957	16572	06004
67	21581	57802	02050	89728	17937	37621	47075	42080	97403	48626	68995	43805	33386	21597
68	55612	78095	83197	33732	05810	24813	86902	60397	16489	03264	88525	42786	05269	92532
69	44657	66999	99324	51281	84463	60563	79312	93454	68876	25471	93911	25650	12682	73572
70	91340	84979	46949	81973	37949	61023	43997	15263	80644	43942	89203	71795	99533	50501
71	91227	21199	31935	27022	84067	05462	35216	14486	29891	68607	41867	14951	91696	85065
72	50001	38140	66321	19924	72163	09538	12151	06878	91903	18749	34405	56087	82790	70925
73	65390	05224	72958	28609	81406	39147	25549	48542	42627	45233	57202	94617	23772	07896
74	27504	96131	83944	41575	10573	08619	64482	73923	36152	05184	94142	25299	84387	34925
75	37169	94851	39117	89632	00959	16487	65536	49071	39782	17095	02330	74301	00275	48280
76	11508	70225	51111	38351	19444	66499	71945	05422	13442	78675	84081	66938	93654	59894
77	37449	30362	06694	54690	04052	53115	62757	95348	78662	11163	81651	50245	34971	52924
78	46515	70331	85922	38329	57015	15765	97161	17869	45349	61796	66345	81073	49106	79860
79	30986	81223	42416	58353	21532	30502	32305	86482	05174	07901	54339	58861	74818	46942
80	63798	64995	46583	09765	44160	78128	83991	42865	92520	83531	80377	35909	81250	54238
81	82486	84846	99254	67632	43218	50076	21361	64816	51202	88124	41870	52689	51275	83556
82	21885	32906	92431	09060	64297	51674	64126	62570	26123	05155	59194	52799	28225	85762
83	60336	98782	07408	53458	13564	59089	26445	29789	85205	41001	12535	12133	14645	23541
84	43937	46891	24010	25560	86355	33941	25786	54990	71899	15475	95434	98227	21824	19585
85	97656	63175	89303	16275	07100	92063	21942	18611	47348	20203	18534	03862	78095	50136
86	03299	01221	05418	38982	55758	92237	26759	86367	21216	98442	08303	56613	91511	75928
87	79626	06486	03574	17668	07785	76020	79924	25651	83325	88428	85076	72811	22717	50585
88	85636	68335	47539	03129	65651	11977	02510	26113	99447	68645	34327	15152	55230	93448
89	18039	14367	61337	06177	12143	46609	32989	74014	64708	00533	35398	58408	13261	47908
90	08362	15656	60627	36478	65648	16764	53412	09013	07832	41574	17639	82163	60859	75567
91	79556	29068	04142	16268	15387	12856	66227	38358	22478	73373	88732	09443	82558	05250
92	92608	82674	27072	32534	17075	27698	98204	63863	11951	34648	88022	56148	34925	57031
93	23982	25835	40055	67006	12293	02753	14827	22235	35071	99704	37543	11601	35503	85171
94	09915	96306	05908	97901	28395	14186	00821	80703	70426	75647	76310	88717	37890	40129
95	50937	33300	26695	62247	69927	76123	50842	43834	86654	70959	79725	93872	28117	19233
96	42488	78077	69882	61657	34136	79180	97526	43092	04098	73571	80799	76536	71255	64239
97	46764	86273	63003	93017	31204	36692	40202	35275	57306	55543	53203	18098	47625	88684
98	03237	45430	55417	63282	90816	17349	88298	90183	36600	78406	06216	95787	42579	90730
99	86591	81482	52667	61583	14972	90053	89534	76036	49199	43716	97548	04379	46370	28672
100	38534	01715	94964	87288	65680	43772	39560	12918	86537	62738	19636	51132	25739	56947

A TABLE OF 14,000 RANDOM UNITS (Continued)

Line/Col.	(1)	(2)	(3)	(4)	(5)	(6)	(7)	(8)	(9)	(10)	(11)	(12)	(13)	(14)
101	13284	16834	74151	92027	24670	36665	00770	22878	02179	51602	07270	76517	97275	45960
102	21224	00370	30420	03883	96648	89428	41583	17564	27395	63904	41548	49197	82277	24120
103	99052	47887	81085	64933	66279	80432	65793	83287	34142	13241	30590	97760	35848	91983
104	00199	50993	98603	38452	87890	94624	69721	57484	67501	77638	44331	11257	71131	11059
105	60578	06483	28733	37867	07936	98710	98539	27186	31237	80612	44488	97819	70401	95419
106	91240	18312	17441	01929	18163	69201	31211	54288	39296	37318	65724	90401	79017	62077
107	97458	14229	12063	59611	32249	90466	33216	19358	02591	54263	88449	01912	07436	50813
108	35249	38646	34475	72417	60514	69257	12489	51924	86871	92446	36607	11458	30440	52639
109	38980	46600	11759	11900	46743	27860	77940	39298	97838	95145	32378	68038	89351	37005
110	10750	52745	38749	87365	58959	53731	89295	59062	39404	13198	59960	70408	29812	83126
111	36247	27850	73958	20673	37800	63835	71051	84724	52492	22342	78071	17456	96104	18327
112	70994	66986	99744	72438	01174	42159	11392	20724	54322	36923	70009	23233	65438	59685
113	99638	94702	11463	18148	81386	80431	90628	52506	02016	85151	88598	47821	00265	82525
114	72055	15774	43857	99805	10419	76939	25993	03544	21560	83471	43989	90770	22965	44247
115	24038	65541	85788	55835	38835	59399	13790	35112	01324	39520	76210	22467	83275	32286
116	74976	14631	35908	28221	39470	91548	12854	30166	09073	75887	36782	00268	97121	57676
117	35553	71628	70189	26436	63407	91178	90348	55359	80392	41012	36270	77786	89578	21059
118	35676	12797	51434	82976	42010	26344	92920	92155	58807	54644	58581	95331	78629	73344
119	74815	67523	72985	23183	02446	63594	98924	20633	58842	85961	07648	70164	34994	67662
120	45246	88048	65173	50989	91060	89894	36063	32819	68559	99221	49475	50558	34698	71800
121	76509	47069	86378	41797	11910	49672	88575	97966	32466	10083	54728	81972	58975	30761
122	19689	90332	04315	21358	97248	11188	39062	63312	52496	07349	79178	33692	57352	72862
123	42751	35318	97513	61537	54955	08159	00337	80778	27507	95478	21252	12746	37554	97775
124	11946	22681	45045	13964	57517	59419	58045	44067	58716	58840	45557	96345	33271	53464
125	96518	48688	20996	11090	48396	57177	83867	86464	14342	21545	46717	72364	86954	55580
126	35726	58643	76869	84622	39098	36083	72505	92265	23107	60278	05822	46760	44294	07672
127	39737	42750	48968	70536	84864	64952	38404	94317	65402	13589	01055	79044	19308	83623
128	97025	66492	56177	04049	80312	48028	26408	43591	75528	65341	49044	95495	81256	53214
129	62814	08075	09788	56350	76787	51591	54509	49295	85830	59860	30883	89660	96142	18354
130	25578	22950	15227	83291	41737	79599	96191	71845	86899	70694	24290	01551	80092	82118
131	68763	69576	88991	49662	46704	63362	56625	00481	73323	91427	15264	06969	57048	54149
132	17900	00813	64361	60725	88974	61005	99709	30666	26451	11528	44323	34778	60342	60388
133	71944	60227	63551	71109	05624	43836	58254	26160	32116	63403	35404	57146	10909	07346
134	54684	93691	85132	64399	29182	44324	14491	55226	78793	34107	30374	48429	51376	09559
135	25946	27623	11258	65204	52832	50880	22273	05554	99521	73791	85744	29276	70326	60251
136	01353	39318	44961	44972	91766	90262	56073	06606	51826	18893	83448	31915	97764	75091
137	99083	88191	27662	99113	57174	35571	99884	13951	71057	53961	61448	74909	07322	80960
138	52021	45406	37945	75234	24327	86978	22644	87779	23753	99926	63898	54886	18051	96314
139	78755	47744	43776	83098	03225	14281	83637	55984	13300	52212	58781	14905	46502	04472
140	25282	69106	59180	16257	22810	43609	12224	25643	89884	31149	85423	32581	34374	70873
141	11959	94202	02743	86847	79725	51811	12998	76844	05320	54236	53891	70226	38632	84776
142	11644	13792	98190	01424	30078	28197	55583	05197	47714	68440	22016	79204	06862	94451
143	06307	97912	68110	59812	95448	43244	31262	88880	13040	16458	43813	89416	42482	33939
144	76285	75714	89585	99296	52640	46518	55486	90754	88932	19937	57119	23251	55619	23679
145	55322	07589	39600	60866	63007	20007	66819	84164	61131	81429	60676	42807	78286	29015
146	78017	90928	90220	92503	83375	26986	74399	30885	88567	29169	72816	53357	15428	86932
147	44768	43342	20696	26331	43140	69744	82928	24988	94237	46138	77426	39039	55596	12655
148	25100	19336	14605	86603	51680	97678	24261	02464	86563	74812	60069	71674	15478	47642
149	83612	46623	62876	85197	07824	91392	58317	37726	84628	42221	10268	20692	15699	29167
150	41347	81666	82961	60413	71020	83658	02415	33322	66036	98712	46795	16308	28413	05417

A TABLE OF 14,000 RANDOM UNITS (Continued)

Line/Col.	(1)	(2)	(3)	(4)	(5)	(6)	(7)	(8)	(9)	(10)	(11)	(12)	(13)	(14)
151	38128	51178	75096	13609	16110	73533	42564	59870	29399	67834	91055	89917	51096	89011
152	60950	00455	73254	96067	50717	13878	03216	78274	65863	37011	91283	33914	91303	49326
153	90524	17320	29832	96118	75792	25326	22940	24904	80523	38928	91374	55597	97567	38914
154	49897	18278	67160	39408	97056	43517	84426	59650	20247	19293	02019	14790	02852	05819
155	18494	99209	81060	19488	65596	59787	47939	91225	98768	43688	00438	05548	09443	82897
156	65373	72984	30171	37741	70203	94094	87261	30056	58124	70133	18936	02138	59372	09075
157	40653	12843	04213	70925	95360	55774	76439	61768	52817	81151	52188	31940	54273	49032
158	51638	22238	56344	44587	83231	50317	74541	07719	25472	41602	77318	15145	57515	07633
159	69742	99303	62578	83575	30337	07488	51941	84316	42067	49692	28616	29101	03013	73449
160	58012	74072	67488	74580	47992	69482	58624	17106	47538	13452	22620	24260	40155	74716
161	18348	19855	42887	08279	43206	47077	42637	45606	00011	20662	14642	49984	94509	56380
162	59614	09193	58064	29086	44385	45740	70752	05663	49081	26960	57454	99264	24142	74648
163	75688	28630	39210	52897	62748	72658	98059	67202	72789	01869	13496	14663	87645	89713
164	13941	77802	69101	70061	35460	34576	15412	81304	58757	35498	94830	75521	00603	97701
165	96656	86420	96475	86458	54463	96419	55417	41375	76886	19008	66877	35934	59801	00497
166	03363	82042	15942	14549	38324	87094	19069	67590	11087	68570	22591	65232	85915	91499
167	70366	08390	69155	25496	13240	57407	91407	49160	07379	34444	94567	66035	38918	65708
168	47870	36605	12927	16043	53257	93796	52721	73120	48025	76074	95605	67422	41646	14557
169	79504	77606	22761	30518	28373	73898	30550	76684	77366	32276	04690	61667	64798	66276
170	46967	74841	50923	15339	37755	98995	40162	89561	69199	42257	11647	47603	48779	97907
171	14558	50769	35444	59030	87516	48193	02945	00922	48189	04724	21263	20892	92955	90251
172	12440	25057	01132	38611	28135	68089	10954	10097	54243	06460	50856	65435	79377	53890
173	32293	29938	68653	10497	98919	46587	77701	99119	93165	67788	17638	23097	21468	36992
174	10640	21875	72462	77981	56550	55999	87310	69643	45124	00349	25748	00844	96831	30651
175	47615	23169	39571	56972	20628	21788	51736	33133	72696	32605	41569	76148	91544	21121
176	16948	11128	71624	72754	49084	96303	27830	45817	67867	18062	87453	17226	72904	71474
177	21258	61092	66634	70335	92448	17354	83432	49608	66520	06442	59664	20420	39201	69549
178	15072	48853	15178	30730	47481	48490	41436	25015	49932	20474	53821	51015	79841	32405
179	99154	57412	09858	65671	70655	71479	63520	31357	56968	06729	34465	70685	04184	25250
180	08759	61089	23706	32994	35426	36666	63988	98844	37533	08269	27021	45886	22835	78451
181	67323	57839	61114	62192	47547	58023	64630	34886	98777	75442	95592	06141	45096	73117
182	09255	13986	84834	20764	72206	89393	34548	93438	88730	61805	78955	18952	46436	58740
183	36304	74712	00374	10107	85061	69228	81969	92216	03568	39630	81869	52824	50937	27954
184	15884	67429	86612	47367	10242	44880	12060	44309	46629	55105	66793	93173	00480	13311
185	18745	32031	35303	08134	33925	03044	59929	95418	04917	57596	24878	61733	92834	64454
186	72934	40086	88292	65728	38300	42323	64068	98373	48971	09049	59943	36538	05976	82118
187	17626	02944	20910	57662	80181	38579	24580	90529	52303	50436	29401	57824	86039	81062
188	27117	61399	50967	41399	81636	16663	15634	79717	94696	59240	25543	97989	63306	90946
189	93995	18678	90012	63645	85701	85269	62263	68331	00389	72571	15210	20769	44686	96176
190	67392	89421	09623	80725	62620	84162	87368	29560	00519	84545	08004	24526	41252	14521
191	04910	12261	37566	80016	21245	69377	50420	85658	55263	68667	78770	04533	14513	18099
192	81453	20283	79929	59839	23875	13245	46808	74124	74703	35769	95588	21014	37078	39170
193	19480	75790	48539	23703	15537	48885	02861	86587	74539	65227	90799	58789	96257	02708
194	21456	13162	74608	81011	55512	07481	93551	72189	76261	91206	89941	15132	37738	59284
195	89406	20912	46189	76376	25538	87212	20748	12831	57166	35026	16817	79121	18929	40628
196	09866	07414	55977	16419	01101	69343	13305	94302	80703	57910	36933	57771	42546	03003
197	86541	24681	23421	13521	28000	94917	07423	57523	97234	63951	42876	46829	09781	58160
198	10414	96941	06205	72222	57167	83902	07460	69507	10600	08858	07685	44472	64220	27040
199	49942	06683	41479	58982	56288	42853	92196	20632	62045	78812	35895	51851	83534	10689
200	23995	68882	42291	23374	24299	27024	67460	94783	40937	16961	26053	78749	46704	21983

CRITICAL VALUES FOR THE SIGN TEST

The observations in a random sample of size n from X and those of the same size from Y are paired according to the order of observation: (X_i, Y_i), $i = 1, 2, \ldots, n$. The differences $d_i = X_i - Y_i$ are calculated for each of the n pairs. The null hypothesis is that the difference d_i has a distribution with median zero, i.e., the true proportion of positive (negative) signs is equal to $p = \frac{1}{2}$. Thus the test is whether X and Y have the same median. The probability of x positive (negative) signs is given by the binomial probability function

$$f(x) = f(x;n, p = \tfrac{1}{2}) = \binom{n}{x}\left(\frac{1}{2}\right)^n .$$

This table gives the critical value k such that

$$P(x \leq k) = \sum_{x=0}^{k} \binom{n}{x}\left(\frac{1}{2}\right)^n < \frac{\alpha}{2} .$$

For a one-tailed test with significance level α enter the table in the column headed by 2α.

CRITICAL VALUES FOR THE SIGN TEST (Continued)

(Two-tail percentage points for the binomial for $p = .5$)

n	1%	5%	10%	25%	n	1%	5%	10%	25%
1					46	13	15	16	18
2					47	14	16	17	19
3				0	48	14	16	17	19
4				0	49	15	17	18	19
5			0	0	50	15	17	18	20
6		0	0	1	51	15	18	19	20
7		0	0	1	52	16	18	19	21
8	0	0	1	1	53	16	18	20	21
9	0	1	1	2	54	17	19	20	22
10	0	1	1	2	55	17	19	20	22
11	0	1	2	3	56	17	20	21	23
12	1	2	2	3	57	18	20	21	23
13	1	2	3	3	58	18	21	22	24
14	1	2	3	4	59	19	21	22	24
15	2	3	3	4	60	19	21	23	25
16	2	3	4	5	61	20	22	23	25
17	2	4	4	5	62	20	22	24	25
18	3	4	5	6	63	20	23	24	26
19	3	4	5	6	64	21	23	24	26
20	3	5	5	6	65	21	24	25	27
21	4	5	6	7	66	22	24	25	27
22	4	5	6	7	67	22	25	26	28
23	4	6	7	8	68	22	25	26	28
24	5	6	7	8	69	23	25	27	29
25	5	7	7	9	70	23	26	27	29
26	6	7	8	9	71	24	26	28	30
27	6	7	8	10	72	24	27	28	30
28	6	8	9	10	73	25	27	28	31
29	7	8	9	10	74	25	28	29	31
30	7	9	10	11	75	25	28	29	32
31	7	9	10	11	76	26	28	30	32
32	8	9	10	12	77	26	29	30	32
33	8	10	11	12	78	27	29	31	33
34	9	10	11	13	79	27	30	31	33
35	9	11	12	13	80	28	30	32	34
36	9	11	12	14	81	28	31	32	34
37	10	12	13	14	82	28	31	33	35
38	10	12	13	14	83	29	32	33	35
39	11	12	13	15	84	29	32	33	36
40	11	13	14	15	85	30	32	34	36
41	11	13	14	16	86	30	33	34	37
42	12	14	15	16	87	31	33	35	37
43	12	14	15	17	88	31	34	35	38
44	13	15	16	17	89	31	34	36	38
45	13	15	16	18	90	32	35	36	39

For values of n larger than 90, approximate values of r may be found by taking the nearest integer less than $(n - 1)/2 - k\sqrt{n + 1}$, where k is 1.2879, 0.9800, 0.8224, 0.5752 for the 1, 5, 10, 25% values, respectively.

FACTORS FOR COMPUTING CONTROL LIMITS

A. Control Charts for Measurement

If the process mean and standard deviation, μ and σ, are known, and it is assumed that the underlying distribution is normal, it is possible to assert with probability $1 - \alpha$ that the mean of a random sample of size n will fall between $\bar{x} - z_{\alpha/2}\frac{\sigma}{\sqrt{n}}$ and $\bar{x} + z_{\alpha/2}\frac{\sigma}{\sqrt{n}}$. These two limits on $\bar{x}$ provide upper and lower control limits. In actual practice, μ and σ are usually unknown and it is necessary to estimate their values from a large sample taken while the process is "in control." The central line of an $\bar{x}$-chart is given by μ and the lower and upper three-sigma control limits are given by $\mu - A\sigma$ and $\mu + A\sigma$, respectively, where $A = \frac{3}{\sqrt{n}}$ and n is the sample size. Where the population parameters are unknown, it is necessary to estimate these parameters on the basis of preliminary samples. If k samples are used, each of size n, denote the mean of the ith sample by $\bar{x}_i$ and the grand mean of the k sample means by $\bar{\bar{x}}$, i.e.

$$\bar{\bar{x}} = \frac{1}{k}\sum_{i=1}^{k} \bar{x}_i \ .$$

Denote the range of the ith sample by R_i and by $\bar{R}$ the mean of the k sample ranges, i.e.

$$\bar{R} = \frac{1}{k}\sum_{i=1}^{k} R_i \ .$$

Since $\bar{\bar{x}}$ is an unbiased estimate of the population mean μ, the central line for the $\bar{x}$-chart is given by $\bar{\bar{x}}$. The statistic R does not provide an unbiased estimate of σ, but $A_2\bar{R}$ is an unbiased estimate of $\frac{3\sigma}{\sqrt{n}}$. The constant multiplier A_2 depends on the assumption of normality. Thus, the central line and the lower and upper three sigma limits, LCL and UCL, for an $\bar{x}$-chart (with μ and σ estimated from past date) are given by

$$\begin{aligned} \text{central line} &= \bar{\bar{x}} \\ \text{LCL} &= \bar{\bar{x}} - A_2\bar{R} \\ \text{UCL} &= \bar{\bar{x}} + A_2\bar{R} \ . \end{aligned}$$

The central line and control limits of an R chart are based on the distribution of the range of samples of size n from a normal population. The mean and standard deviation of the sampling distribution of R are given by $d_2\sigma$ and $d_3\sigma$, respectively, when σ is known. Here d_2 and d_3 are constants which depend on the size of the sample. The set of control chart values for an R chart (with σ known) is given by

$$\begin{aligned} \text{central line} &= d_2\sigma \\ \text{LCL} &= D_1\sigma \\ \text{UCL} &= D_2\sigma, \end{aligned}$$

where $D_1 = d_2 - 3d_3$ and $D_2 = d_2 + 3d_3$.

If σ is unknown, the control chart values for an R chart are given by

$$\begin{aligned}\text{central line} &= \bar{R}\\ \text{LCL} &= D_3\bar{R}\\ \text{UCL} &= D_4\bar{R},\end{aligned}$$

where $D_3 = \dfrac{D_1}{d_2}$ and $D_4 = \dfrac{D_2}{d_2}$.

The central line and control limits of an s-chart are based on estimates obtained from the samples. A pooled estimate of the population variance is obtained from the k samples, i.e.

$$s_p^2 = \frac{\sum_i (n_i - 1)s_i^2}{\sum_i (n_i - 1)}, \qquad i = 1, 2, \ldots, k .$$

If the sample sizes are all equal, the pooled estimate is

$$s_p^2 = \frac{1}{k}\sum_i s_i^2 .$$

The control chart values for an s-chart are given by

$$\begin{aligned}\text{central line} &= C_2' s_p\\ \text{LCL} &= B_2' s_p\\ \text{UCL} &= B_4' s_p .\end{aligned}$$

If one uses the biased estimator of the variance s_p', as is often done in quality control work, the control chart values are given by

$$\begin{aligned}\text{central line} &= c_2 s_p'\\ \text{LCL} &= B_2 s_p'\\ \text{UCL} &= B_4 s_p' .\end{aligned}$$

B. Control Charts for Attributes

Control limits for a fraction-defective chart are based on the sampling theory for proportions, using the normal curve approximation to the binomial. If k samples are taken, the estimator of p is given by

$$\bar{p} = \frac{\sum_i x_i}{\sum_i n_i}, \qquad i = 1, 2, \ldots, k$$

where x_i is the number of defectives in the ith sample of size n_i. The central line and control limits of a fraction defective chart based on analysis of past data are given by

$$\begin{aligned}\text{central line} &= \bar{p}\\ \text{LCL} &= \bar{p} - 3\sqrt{\frac{\bar{p}(1-\bar{p})}{n_i}}\\ \text{UCL} &= \bar{p} + 3\sqrt{\frac{\bar{p}(1-\bar{p})}{n_i}} .\end{aligned}$$

When the sample sizes are approximately equal, n_i is replaced by $\bar{n} = \frac{1}{k}\sum_i n_i$.

Equivalent to the p chart for the fraction defective is the control chart for the number of defective. Here, if p is estimated by $\bar{p}$, the control chart values for a number-of-defectives chart are given by

$$\begin{aligned} \text{central line} &= \bar{n}\bar{p} \\ \text{LCL} &= \bar{n}\bar{p} - 3\sqrt{\bar{n}\bar{p}(1-\bar{p})} \\ \text{UCL} &= \bar{n}\bar{p} + 3\sqrt{\bar{n}\bar{p}(1-\bar{p})}\ . \end{aligned}$$

In many cases it is necessary to control the number of defects per unit C, where C is taken to be a value of a random variable having a Poisson distribution. If k is the number of units available for estimating λ, the parameter of the Poisson distribution, and if C_i is the number of defects in the ith unit, then λ is estimated by

$$\bar{C} = \frac{1}{k}\sum_{i=1}^{k} C_i\ ,$$

and the control-chart values for the C-chart are

$$\begin{aligned} \text{central line} &= \bar{C} \\ \text{LCL} &= \bar{C} - 3\sqrt{\bar{C}} \\ \text{UCL} &= \bar{C} + 3\sqrt{\bar{C}} \end{aligned}$$

This table presents values of the factors for computing control limits for various sample sizes n.

Number of observations in sample, n	$\bar{X}$ chart		R chart			s chart			$\hat{\sigma}$ chart (biased)		
	Factors for control limits		Factor for central line	Factors for control limits		Factor for central line	Factors for control limits		Factor for central line	Factors for control limits	
	A	A_2	d_2	D_3	D_4	c_2'	B_2'	B_4'	c_2	B_2	B_4
2	2.121	1.880	1.128	0	3.267	0.798	0	2.298	0.5642	0	3.267
3	1.732	1.023	1.693	0	2.575	0.886	0	2.111	0.7236	0	2.568
4	1.500	0.729	2.059	0	2.282	0.921	0	1.982	0.7979	0	2.266
5	1.342	0.577	2.326	0	2.115	0.940	0	1.889	0.8407	0	2.089
6	1.225	0.483	2.534	0	2.004	0.951	0.085	1.817	0.8686	0.030	1.970
7	1.134	0.419	2.704	0.076	1.924	0.960	0.158	1.762	0.8882	0.118	1.882
8	1.061	0.373	2.847	0.136	1.864	0.965	0.215	1.715	0.9027	0.185	1.815
9	1.000	0.337	2.970	0.184	1.816	0.969	0.262	1.676	0.9139	0.239	1.761
10	0.949	0.308	3.078	0.223	1.777	0.973	0.302	1.644	0.9227	0.284	1.716
11	0.905	0.285	3.173	0.256	1.744	0.976	0.336	1.616	0.9300	0.321	1.679
12	0.866	0.266	3.258	0.284	1.716	0.977	0.365	1.589	0.9359	0.354	1.646
13	0.832	0.249	3.336	0.308	1.692	0.980	0.392	1.568	0.9410	0.382	1.618
14	0.802	0.235	3.407	0.329	1.671	0.981	0.414	1.548	0.9453	0.406	1.594
15	0.775	0.223	3.472	0.348	1.652	0.982	0.434	1.530	0.9490	0.428	1.572
16	0.750	0.212	3.532	0.364	1.636	0.984	0.454	1.514	0.9523	0.448	1.552
17	0.728	0.203	3.588	0.379	1.621	0.984	0.469	1.499	0.9551	0.466	1.534
18	0.707	0.194	3.640	0.392	1.608	0.986	0.486	1.486	0.9576	0.482	1.518
19	0.688	0.187	3.689	0.404	1.596	0.986	0.500	1.472	0.9599	0.497	1.503
20	0.671	0.180	3.735	0.414	1.586	0.987	0.513	1.461	0.9619	0.510	1.490
21	0.655	0.173	3.778	0.425	1.575	0.988	0.525	1.451	0.9638	0.523	1.477
22	0.640	0.167	3.819	0.434	1.566	0.988	0.536	1.440	0.9655	0.534	1.466
23	0.626	0.162	3.858	0.443	1.557	0.989	0.546	1.432	0.9670	0.545	1.455
24	0.612	0.157	3.895	0.452	1.548	0.989	0.556	1.422	0.9684	0.555	1.445
25	0.600	0.153	3.931	0.459	1.541	0.990	0.566	1.414	0.9696	0.565	1.435

NUMBER OF PERMUTATIONS $P(n,m)$

This table contains the number of permutations of n distinct things taken m at a time, given by

$$P(n,m) = \frac{n!}{(n-m)!} = n(n-1) \cdot\cdot\cdot (n-m+1)$$

n \ m	0	1	2	3	4	5	6	7	8	9	10
0	1										
1	1	1									
2	1	2	2								
3	1	3	6	6							
4	1	4	12	24	24						
5	1	5	20	60	120	120					
6	1	6	30	120	360	720	720				
7	1	7	42	210	840	2520	5040	5040			
8	1	8	56	336	1680	6720	20160	40320	40320		
9	1	9	72	504	3024	15120	60480	1 81440	3 62880	3 62880	
10	1	10	90	720	5040	30240	1 51200	6 04800	18 14400	36 28800	36 28800
11	1	11	110	990	7920	55440	3 32640	16 63200	66 52800	199 58400	399 16800
12	1	12	132	1320	11880	95040	6 65280	39 91680	199 58400	798 33600	2395 00800
13	1	13	156	1716	17160	1 54440	12 35520	86 48640	518 91840	2594 59200	10378 36800
14	1	14	182	2184	24024	2 40240	21 62160	172 97280	1210 80960	7264 85760	36324 28800
15	1	15	210	2730	32760	3 60360	36 03600	324 32400	2594 59200	18162 14400	1 08972 86400

n \ m	11	12	13	14	15
8					
9					
10					
11	399 16800				
12	4790 01600	4790 01600			
13	31135 10400	62270 20800	62270 20800		
14	1 45297 15200	4 35891 45600	8 71782 91200	8 71782 91200	
15	5 44864 32000	21 79457 28000	65 38371 84000	130 76743 68000	130 76743 68000

NUMBER OF COMBINATIONS

$$\binom{n}{m} = C(n,m)$$

NUMBER OF COMBINATIONS $C(n,m)$

This table contains the number of combinations of n distinct things taken m at a time, given by

$$\binom{n}{m} = C(n,m) = \frac{n!}{m!(n-m)!} = \frac{P(n,r)}{m!}\ .$$

For values missing from the above table, use the relation $\binom{n}{m} = \binom{n}{n-m}$, e.g. $\binom{20}{12} = \binom{20}{8} = 125970$. $\binom{n}{m}$ is also referred to as a binomial coefficient. A recursion relation for the binomial coefficients is

$$\binom{n+1}{m+1} = \binom{n}{m} + \binom{n}{m+1}$$

NUMBER OF COMBINATIONS

$$\binom{n}{m} = C(n, m)$$

n \ m	0	1	2	3	4	5	6	7	8
1	1	1							
2	1	2	1						
3	1	3	3	1					
4	1	4	6	4	1				
5	1	5	10	10	5	1			
6	1	6	15	20	15	6	1		
7	1	7	21	35	35	21	7	1	
8	1	8	28	56	70	56	28	8	1
9	1	9	36	84	126	126	84	36	9
10	1	10	45	120	210	252	210	120	45
11	1	11	55	165	330	462	462	330	165
12	1	12	66	220	495	792	924	792	495
13	1	13	78	286	715	1287	1716	1716	1287
14	1	14	91	364	1001	2002	3003	3432	3003
15	1	15	105	455	1365	3003	5005	6435	6435
16	1	16	120	560	1820	4368	8008	11440	12870
17	1	17	136	680	2380	6188	12376	19448	24310
18	1	18	153	816	3060	8568	18564	31824	43758
19	1	19	171	969	3876	11628	27132	50388	75582
20	1	20	190	1140	4845	15504	38760	77520	1 25970
21	1	21	210	1330	5985	20349	54264	1 16280	2 03490
22	1	22	231	1540	7315	26334	74613	1 70544	3 19770
23	1	23	253	1771	8855	33649	1 00947	2 45157	4 90314
24	1	24	276	2024	10626	42504	1 34596	3 46104	7 35471
25	1	25	300	2300	12650	53130	1 77100	4 80700	10 81575
26	1	26	325	2600	14950	65780	2 30230	6 57800	15 62275
27	1	27	351	2925	17550	80730	2 96010	8 88030	22 20075
28	1	28	378	3276	20475	98280	3 76740	11 84040	31 08105
29	1	29	406	3654	23751	1 18755	4 75020	15 60780	42 92145
30	1	30	435	4060	27405	1 42506	5 93775	20 35800	58 52925
31	1	31	465	4495	31465	1 69911	7 36281	26 29575	78 88725
32	1	32	496	4960	35960	2 01376	9 06192	33 65856	105 18300
33	1	33	528	5456	40920	2 37336	11 07568	42 72048	138 84156
34	1	34	561	5984	46376	2 78256	13 44904	53 79616	181 56204
35	1	35	595	6545	52360	3 24632	16 23160	67 24520	235 35820
36	1	36	630	7140	58905	3 76992	19 47792	83 47680	302 60340
37	1	37	666	7770	66045	4 35897	23 24784	102 95472	386 08020
38	1	38	703	8436	73815	5 01942	27 60681	126 20256	489 03492
39	1	39	741	9139	82251	5 75757	32 62623	153 80937	615 23748
40	1	40	780	9880	91390	6 58008	38 38380	186 43560	769 04685
41	1	41	820	10660	101270	7 49398	44 96388	224 81940	955 48245
42	1	42	861	11480	111930	8 50668	52 45786	269 78328	1180 30185
43	1	43	903	12341	123410	9 62598	60 96454	322 24114	1450 08513
44	1	44	946	13244	135751	10 86008	70 59052	383 20568	1772 32627
45	1	45	990	14190	148995	12 21759	81 45060	453 79620	2155 53195
46	1	46	1035	15180	163185	13 70754	93 66819	535 24680	2609 32815
47	1	47	1081	16215	178365	15 33939	107 37573	628 91499	3144 57495
48	1	48	1128	17296	194580	17 12304	122 71512	736 29072	3773 48994
49	1	49	1176	18424	211876	19 06884	139 83816	859 00584	4509 78066
50	1	50	1225	19600	230300	21 18760	158 90700	998 84400	5368 78650

NUMBER OF COMBINATIONS

$$\binom{n}{m} = C(n, m)$$

n \ m	9	10	11	12	13
9	1				
10	10	1			
11	55	11	1		
12	220	66	12	1	
13	715	286	78	13	1
14	2002	1001	364	91	14
15	5005	3003	1365	455	105
16	11440	8008	4368	1820	560
17	24310	19448	12376	6188	2380
18	48620	43758	31824	18564	8568
19	92378	92378	75582	50388	27132
20	1 67960	1 84756	1 67960	1 25970	77520
21	2 93930	3 52716	3 52716	2 93930	2 03490
22	4 97420	6 46646	7 05432	6 46646	4 97420
23	8 17190	11 44066	13 52078	13 52078	11 44066
24	13 07504	19 61256	24 96144	27 04156	24 96144
25	20 42975	32 68760	44 57400	52 00300	52 00300
26	31 24550	53 11735	77 26160	96 57700	104 00600
27	46 86825	84 36285	130 37895	173 83860	200 58300
28	69 06900	131 23110	214 74180	304 21755	374 42160
29	100 15005	200 30010	345 97290	518 95935	678 63915
30	143 07150	300 45015	546 27300	864 93225	1197 59850
31	201 60075	443 52165	846 72315	1411 20525	2062 53075
32	280 48800	645 12240	1290 24480	2257 92840	3474 73600
33	385 67100	925 61040	1935 36720	3548 17320	5731 66440
34	524 51256	1311 28140	2860 97760	5483 54040	9279 83760
35	706 07460	1835 79396	4172 25900	8344 51800	14763 37800
36	941 43280	2541 86856	6008 05296	12516 77700	23107 89600
37	1244 03620	3483 30136	8549 92152	18524 82996	35624 67300
38	1630 11640	4727 33756	12033 22288	27074 75148	54149 50296
39	2119 15132	6357 45396	16760 56044	39107 97436	81224 25444
40	2734 38880	8476 60528	23118 01440	55868 53480	1 20332 22880
41	3503 43565	11210 99408	31594 61968	78986 54920	1 76200 76360
42	4458 91810	14714 42973	42805 61376	1 10581 16888	2 55187 31280
43	5639 21995	19173 34783	57520 04349	1 53386 78264	3 65768 48168
44	7098 30508	24812 56778	76693 39132	2 10906 82613	5 19155 26432
45	8861 63135	31901 87286	1 01505 95910	2 87600 21745	7 30062 09045
46	11017 16330	40763 50421	1 33407 83196	3 89106 17655	10 17662 30790
47	13626 49145	51780 66751	1 74171 33617	5 22514 00851	14 06768 48445
48	16771 06640	65407 15896	2 25952 00368	6 96685 34468	19 29282 49296
49	20544 55634	82178 22536	2 91359 16264	9 22637 34836	26 25967 83764
50	25054 33700	1 02722 78170	3 73537 38800	12 13996 51100	35 48605 18600

NUMBER OF COMBINATIONS

$$\binom{n}{m} = C(n,m)$$

n \ m	14	15	16	17	18	19
14	1					
15	15	1				
16	120	16	1			
17	680	136	17	1		
18	3060	816	153	18	1	
19	11628	3876	969	171	19	1
20	38760	15504	4845	1140	190	20
21	1 16280	54264	20349	5985	1330	210
22	3 19770	1 70544	74613	26334	7315	1540
23	8 17190	4 90314	2 45157	1 00947	33649	8855
24	19 61256	13 07504	7 35471	3 46104	1 34596	42504
25	44 57400	32 68760	20 42975	10 81575	4 80700	1 77100
26	96 57700	77 26160	53 11735	31 24550	15 62275	6 57800
27	200 58300	173 83860	130 37895	84 36285	46 86825	22 20075
28	401 16600	374 42160	304 21755	214 74180	131 23110	69 06900
29	775 58760	775 58760	678 63915	518 95935	345 97290	200 30010
30	1454 22675	1551 17520	1454 22675	1197 59850	864 93225	546 27300
31	2651 82525	3005 40195	3005 40195	2651 82525	2062 53075	1411 20525
32	4714 35600	5657 22720	6010 80390	5657 22720	4714 35600	3473 73600
33	8188 09200	10371 58320	11668 03110	11668 03110	10371 58320	8188 09200
34	13919 75640	18559 67520	22039 61430	23336 06220	22039 61430	18559 67520
35	23199 59400	32479 43160	40599 28950	45375 67650	45375 67650	40599 28950
36	37962 97200	55679 02560	73078 72110	85974 96600	90751 35300	85974 96600
37	61070 86800	93641 99760	1 28757 74670	1 59053 68710	1 76726 31900	1 76726 31900
38	96695 54100	1 54712 86560	2 22399 74430	2 87811 43380	3 35780 00610	3 53452 63800
39	1 50845 04396	2 51408 40660	3 77112 60990	5 10211 17810	6 23591 43990	6 89232 64410
40	2 32069 29840	4 02253 45056	6 28521 01650	8 87323 78800	11 33802 61800	13 12824 08400
41	3 52401 52720	6 34322 74896	10 30774 46706	15 15844 80450	20 21126 40600	24 46626 70200
42	5 28602 29080	9 86724 27616	16 65097 21602	25 46619 27156	35 36971 21050	44 67753 10800
43	7 83789 60360	15 15326 56696	26 51821 49218	42 11716 48758	60 83590 48206	80 04724 31850
44	11 49558 08528	22 99116 17056	41 67148 05914	68 63537 97976	102 95306 96964	140 88314 80056
45	16 68713 34960	34 48674 25584	64 66264 22970	110 30686 03890	171 58844 94940	243 83621 77020
46	23 98775 44005	51 17387 60544	99 14938 48554	174 96950 26860	281 89530 98830	415 42466 71960
47	34 16437 74795	75 16163 04549	150 32326 09098	274 11888 75414	456 86481 25690	697 31997 70790
48	48 23206 23240	109 32600 79344	225 48489 13647	424 44214 84512	730 98370 01104	1154 18478 96480
49	67 52488 72536	157 55807 02584	334 81089 92991	649 92703 98159	1155 42584 85616	1885 16848 97584
50	93 78456 56300	225 08295 75120	492 36896 95575	984 73793 91150	1805 35288 83775	3040 59433 83200

NUMBER OF COMBINATIONS

$$\binom{n}{m} = C(n,m)$$

n \ m	20	21	22	23	24	25
20	1					
21	21	1				
22	231	22	1			
23	1771	253	23	1		
24	10626	2024	276	24	1	
25	53130	12650	2300	300	25	1
26	2 30230	65780	14950	2600	325	26
27	8 88030	2 96010	80730	17550	2925	351
28	31 08105	11 84040	3 76740	98280	20475	3276
29	100 15005	42 92145	15 60780	4 75020	1 18755	23751
30	300 45015	143 07150	58 52925	20 35800	5 93775	1 42506
31	846 72315	443 52165	201 60075	78 88725	26 29575	7 36281
32	2257 92840	1290 24480	645 12240	280 48800	105 18300	33 65856
33	5731 66440	3548 17320	1935 36720	925 61040	385 67100	138 84156
34	13919 75640	9279 83760	5483 54040	2860 97760	1311 28140	524 51256
35	32479 43160	23199 59400	14763 37800	8344 51800	4172 25900	1835 79396
36	73078 72110	55679 02560	37962 97200	23107 89600	12516 77700	6008 05296
37	1 59053 68710	1 28757 74670	93641 99760	61070 86800	35624 67300	18524 82996
38	3 35780 00610	2 87811 43380	2 22399 74430	1 54712 86560	96695 54100	54149 50296
39	6 89232 64410	6 23591 43990	5 10211 17810	3 77112 60990	2 51408 40660	1 50845 04396
40	13 78465 28820	13 12824 08400	11 33802 61800	8 87323 78800	6 28521 01650	4 02253 45056
41	26 91289 37220	26 91289 37220	24 46626 70200	20 21126 40600	15 15844 80450	10 30774 46706
42	51 37916 07420	53 82578 74440	51 37916 07420	44 67753 10800	35 36971 21050	25 46619 27156
43	96 05669 18220	105 20494 81860	105 20494 81860	96 05669 18220	80 04724 31850	60 83590 48206
44	176 10393 50070	201 26164 00080	210 40989 63720	201 26164 00080	176 10393 50070	140 88314 80056
45	316 98708 30126	377 36557 50150	411 67153 63800	411 67153 63800	377 36557 50150	316 98708 30126
46	560 82330 07146	694 35265 80276	789 03711 13950	823 34307 27600	789 03711 13950	694 35265 80276
47	976 24796 79106	1255 17595 87422	1483 38976 94226	1612 38018 41550	1612 38018 41550	1483 38976 94226
48	1673 56794 49896	2231 42392 66528	2738 56572 81648	3095 76995 35776	3224 76036 83100	3095 76995 35776
49	2827 75273 46376	3904 99187 16424	4969 98965 48176	5834 33568 17424	6320 53032 18876	6320 53032 18876
50	4712 92122 43960	6732 74460 62800	8874 98152 64600	10804 32533 65600	12154 86600 36300	12641 06064 37752

LOGARITHMS OF THE BINOMIAL COEFFICIENTS

This table contains the logarithms of the binomial coefficients $\binom{n}{m}$; also referred to as the number of combinations of n distinct things taken m at a time. For values missing from the table, use the relation

$$\binom{n}{m} = \frac{n!}{m!(n-m)!} = \binom{n}{n-m}$$

LOGARITHMS OF THE BINOMIAL COEFFICIENTS

m	$n = 1$	$n = 2$	$n = 3$	$n = 4$	$n = 5$	$n = 6$	$n = 7$	$n = 8$	$n = 9$	$n = 10$
1	.0000	.3010	.4771	.6021	.6990	.7782	.8451	.9031	.9542	1.0000
2			.4771	.7782	1.0000	1.1761	1.3222	1.4472	1.5563	1.6532
3					1.0000	1.3010	1.5441	1.7482	1.9243	2.0792
4							1.5441	1.8451	2.1004	2.3222
5									2.1004	2.4014

m	$n = 11$	$n = 12$	$n = 13$	$n = 14$	$n = 15$	$n = 16$	$n = 17$	$n = 18$	$n = 19$	$n = 20$
1	1.0414	1.0792	1.1139	1.1461	1.1761	1.2041	1.2304	1.2553	1.2788	1.3010
2	1.7404	1.8195	1.8921	1.9590	2.0212	2.0792	2.1335	2.1847	2.2330	2.2788
3	2.2175	2.3424	2.4564	2.5611	2.6580	2.7482	2.8325	2.9117	2.9863	3.0569
4	2.5185	2.6946	2.8543	3.0004	3.1351	3.2601	3.3766	3.4857	3.5884	3.6853
5	2.6646	2.8987	3.1096	3.3015	3.4776	3.6403	3.7916	3.9329	4.0655	4.1904
6	2.6646	2.9657	3.2345	3.4776	3.6994	3.9035	4.0926	4.2687	4.4335	4.5884
7			3.2345	3.5355	3.8085	4.0584	4.2889	4.5028	4.7023	4.8894
8					3.8085	4.1096	4.3858	4.6411	4.8784	5.1003
9							4.3858	4.6868	4.9656	5.2252
10									4.9656	5.2666

m	$n = 21$	$n = 22$	$n = 23$	$n = 24$	$n = 25$	$n = 26$	$n = 27$	$n = 28$	$n = 29$	$n = 30$
1	1.3222	1.3424	1.3617	1.3802	1.3979	1.4150	1.4314	1.4472	1.4624	1.4771
2	2.3222	2.3636	2.4031	2.4409	2.4771	2.5119	2.5453	2.5775	2.6085	2.6385
3	3.1239	3.1875	3.2482	3.3062	3.3617	3.4150	3.4661	3.5153	3.5628	3.6085
4	3.7771	3.8642	3.9472	4.0264	4.1021	4.1746	4.2443	4.3112	4.3757	4.4378
5	4.3085	4.4205	4.5270	4.6284	4.7253	4.8181	4.9070	4.9925	5.0747	5.1538
6	4.7345	4.8728	5.0041	5.1290	5.2482	5.3622	5.4713	5.5760	5.6767	5.7736
7	5.0655	5.2318	5.3894	5.5392	5.6819	5.8181	5.9484	6.0734	6.1933	6.3087
8	5.3085	5.5048	5.6905	5.8666	6.0341	6.1938	6.3464	6.4925	6.6327	6.7674
9	5.4682	5.6967	5.9123	6.1164	6.3103	6.4948	6.6709	6.8393	7.0007	7.1556
10	5.5474	5.8107	6.0585	6.2925	6.5144	6.7252	6.9262	7.1180	7.3017	7.4778
11	5.5474	5.8485	6.1310	6.3973	6.6491	6.8880	7.1152	7.3319	7.5390	7.7374
12			6.1310	6.4320	6.7160	6.9849	7.2401	7.4832	7.7151	7.9370
13					6.7160	7.0171	7.3023	7.5734	7.8316	8.0783
14							7.3023	7.6033	7.8896	8.1626
15									7.8896	8.1907

LOGARITHMS OF THE BINOMIAL COEFFICIENTS

m	n = 31	n = 32	n = 33	n = 34	n = 35	n = 36	n = 37	n = 38	n = 39	n = 40
1	1.4914	1.5052	1.5185	1.5315	1.5441	1.5563	1.5682	1.5798	1.5911	1.6021
2	2.6675	2.6955	2.7226	2.7490	2.7745	2.7993	2.8235	2.8470	2.8698	2.8921
3	3.6527	3.6955	3.7369	3.7770	3.8159	3.8537	3.8904	3.9261	3.9609	3.9948
4	4.4978	4.5558	4.6119	4.6663	4.7190	4.7702	4.8198	4.8681	4.9151	4.9609
5	5.2302	5.3040	5.3754	5.4444	5.5114	5.5763	5.6394	5.7007	5.7602	5.8182
6	5.8670	5.9572	6.0444	6.1287	6.2104	6.2895	6.3664	6.4410	6.5136	6.5841
7	6.4199	6.5271	6.6306	6.7308	6.8277	6.9216	7.0126	7.1011	7.1870	7.2705
8	6.8970	7.0219	7.1425	7.2590	7.3717	7.4809	7.5867	7.6893	7.7890	7.8860
9	7.3045	7.4479	7.5862	7.7198	7.8489	7.9738	8.0948	8.2122	8.3262	8.4369
10	7.6469	7.8096	7.9664	8.1177	8.2638	8.4052	8.5420	8.6746	8.8033	8.9282
11	7.9277	8.1107	8.2868	8.4565	8.6204	8.7787	8.9320	9.0804	9.2243	9.3640
12	8.1496	8.3537	8.5500	8.7391	8.9214	9.0975	9.2678	9.4326	9.5923	9.7472
13	8.3144	8.5408	8.7583	8.9675	9.1692	9.3638	9.5518	9.7336	9.9097	10.0804
14	8.4235	8.6734	8.9132	9.1436	9.3655	9.5794	9.7858	9.9854	10.1785	10.3656
15	8.4779	8.7526	9.0158	9.2686	9.5116	9.7457	9.9715	10.1895	10.4004	10.6045
16	8.4779	8.7789	9.0670	9.3432	9.6085	9.8638	10.1098	10.3471	10.5765	10.7983
17			9.0670	9.3680	9.6568	9.9344	10.2015	10.4591	10.7078	10.9481
18					9.6568	9.9579	10.2473	10.5261	10.7949	11.0545
19							10.2473	10.5483	10.8384	11.1182
20									10.8384	11.1394

m	n = 41	n = 42	n = 43	n = 44	n = 45	n = 46	n = 47	n = 48	n = 49	n = 50
1	1.6128	1.6232	1.6335	1.6435	1.6532	1.6628	1.6721	1.6812	1.6902	1.6990
2	2.9138	2.9350	2.9557	2.9759	2.9956	3.0149	3.0338	3.0523	3.0704	3.0881
3	4.0278	4.0599	4.0914	4.1220	4.1520	4.1813	4.2099	4.2379	4.2654	4.2923
4	5.0055	5.0489	5.0914	5.1327	5.1732	5.2127	5.2513	5.2891	5.3261	5.3623
5	5.8747	5.9298	5.9834	6.0358	6.0870	6.1370	6.1858	6.2336	6.2803	6.3261
6	6.6529	6.7198	6.7851	6.8487	6.9109	6.9716	7.0309	7.0889	7.1456	7.2011
7	7.3518	7.4310	7.5082	7.5834	7.6569	7.7286	7.7986	7.8670	7.9340	7.9995
8	7.9802	8.0720	8.1614	8.2485	8.3336	8.4165	8.4976	8.5767	8.6542	8.7299
9	8.5445	8.6492	8.7512	8.8506	8.9475	9.0421	9.1344	9.2246	9.3127	9.3989
10	9.0496	9.1677	9.2827	9.3947	9.5038	9.6103	9.7142	9.8156	9.9148	10.0117
11	9.4996	9.6315	9.7598	9.8848	10.0065	10.1252	10.2410	10.3540	10.4644	10.5723
12	9.8976	10.0437	10.1858	10.3241	10.4588	10.5901	10.7181	10.8430	10.9650	11.0842
13	10.2460	10.4069	10.5632	10.7153	10.8634	11.0076	11.1482	11.2854	11.4193	11.5501
14	10.5470	10.7231	10.8942	11.0605	11.2224	11.3800	11.5336	11.6833	11.8295	11.9721
15	10.8023	10.9942	11.1805	11.3616	11.5377	11.7090	11.8760	12.0387	12.1974	12.3523
16	11.0132	11.2214	11.4235	11.6198	11.8107	11.9963	12.1770	12.3531	12.5248	12.6923
17	11.1807	11.4060	11.6245	11.8365	12.0426	12.2430	12.4379	12.6278	12.8129	12.9933
18	11.3056	11.5486	11.7842	12.0126	12.2345	12.4501	12.6598	12.8639	13.0627	13.2566
19	11.3886	11.6501	11.9033	12.1489	12.3871	12.6185	12.8434	13.0623	13.2754	13.4830
20	11.4300	11.7108	11.9825	12.2458	12.5010	12.7488	12.9896	13.2236	13.4514	13.6733
21	11.4300	11.7310	12.0220	12.3038	12.5768	12.8416	13.0987	13.3486	13.5916	13.8282
22			12.0220	12.3231	12.6146	12.8971	13.1713	13.4375	13.6964	13.9482
23					12.6146	12.9156	13.2075	13.4908	13.7660	14.0336
24							13.2075	13.5085	13.8008	14.0848
25									13.8008	14.1018

LOGARITHMS OF THE BINOMIAL COEFFICIENTS

m	$n = 51$	$n = 52$	$n = 53$	$n = 54$	$n = 55$	$n = 56$	$n = 57$	$n = 58$	$n = 59$	$n = 60$
1	1.7076	1.7160	1.7243	1.7324	1.7404	1.7482	1.7559	1.7634	1.7709	1.7782
2	3.1055	3.1225	3.1392	3.1556	3.1717	3.1875	3.2030	3.2183	3.2333	3.2480
3	4.3186	4.3444	4.3697	4.3945	4.4189	4.4428	4.4663	4.4893	4.5120	4.5343
4	5.3978	5.4325	5.4666	5.5000	5.5328	5.5650	5.5966	5.6276	5.6581	5.6881
5	6.3709	6.4148	6.4578	6.5000	6.5414	6.5820	6.6219	6.6611	6.6995	6.7373
6	7.2555	7.3087	7.3609	7.4121	7.4622	7.5115	7.5598	7.6072	7.6538	7.6995
7	8.0636	8.1264	8.1879	8.2482	8.3073	8.3653	8.4222	8.4781	8.5329	8.5868
8	8.8040	8.8765	8.9476	9.0172	9.0855	9.1524	9.2181	9.2826	9.3459	9.4080
9	9.4832	9.5657	9.6466	9.7257	9.8033	9.8794	9.9541	10.0273	10.0992	10.1698
10	10.1065	10.1992	10.2900	10.3790	10.4661	10.5515	10.6353	10.7175	10.7982	10.8773
11	10.6778	10.7811	10.8821	10.9810	11.0779	11.1729	11.2660	11.3573	11.4470	11.5349
12	11.2007	11.3147	11.4262	11.5353	11.6422	11.7469	11.8496	11.9503	12.0490	12.1459
13	11.6778	11.8028	11.9250	12.0446	12.1617	12.2764	12.3889	12.4991	12.6072	12.7132
14	12.1115	12.2477	12.3809	12.5113	12.6388	12.7638	12.8862	13.0062	13.1238	13.2392
15	12.5036	12.6514	12.7959	12.9372	13.0755	13.2109	13.3436	13.4735	13.6009	13.7259
16	12.8558	13.0155	13.1716	13.3242	13.4735	13.6196	13.7627	13.9029	14.0403	14.1750
17	13.1694	13.3413	13.5093	13.6735	13.8341	13.9912	14.1450	14.2957	14.4433	14.5880
18	13.4456	13.6301	13.8104	13.9864	14.1586	14.3270	14.4918	14.6532	14.8113	14.9662
19	13.6854	13.8829	14.0757	14.2640	14.4481	14.6280	14.8041	14.9765	15.1453	15.3107
20	13.8895	14.1004	14.3061	14.5070	14.7033	14.8952	15.0829	15.2665	15.4463	15.6224
21	14.0586	14.2833	14.5024	14.7163	14.9252	15.1293	15.3289	15.5241	15.7152	15.9022
22	14.1933	14.4322	14.6651	14.8924	15.1142	15.3309	15.5427	15.7499	15.9525	16.1509
23	14.2940	14.5476	14.7948	15.0358	15.2710	15.5007	15.7251	15.9444	16.1590	16.3689
24	14.3610	14.6298	14.8917	15.1470	15.3960	15.6390	15.8764	16.1083	16.3351	16.5569
25	14.3944	14.6790	14.9561	15.2261	15.4894	15.7462	15.9969	16.2418	16.4812	16.7153
26	14.3944	14.6954	14.9883	15.2736	15.5515	15.8226	16.0871	16.3454	16.5977	16.8444
27			14.9883	15.2894	15.5826	15.8683	16.1471	16.4192	16.6849	16.9445
28					15.5826	15.8836	16.1771	16.4634	16.7429	17.0159
29							16.1771	16.4781	16.7718	17.0586
30									16.7718	17.0729

m	$n = 61$	$n = 62$	$n = 63$	$n = 64$	$n = 65$	$n = 66$	$n = 67$	$n = 68$	$n = 69$	$n = 70$
1	1.7853	1.7924	1.7993	1.8062	1.8129	1.8195	1.8261	1.8325	1.8388	1.8451
2	3.2625	3.2767	3.2907	3.3045	3.3181	3.3314	3.3446	3.3576	3.3703	3.3829
3	4.5562	4.5777	4.5989	4.6198	4.6403	4.6605	4.6804	4.7000	4.7193	4.7383
4	5.7175	5.7465	5.7750	5.8030	5.8306	5.8578	5.8845	5.9108	5.9368	5.9623
5	6.7745	6.8110	6.8469	6.8822	6.9170	6.9512	6.9849	7.0180	7.0507	7.0829
6	7.7445	7.7887	7.8322	7.8749	7.9170	7.9584	7.9991	8.0392	8.0787	8.1177
7	8.6398	8.6918	8.7429	8.7932	8 8427	8.8914	8.9393	8.9865	9.0330	9.0787
8	9.4691	9.5291	9.5880	9.6460	9.7031	9.7592	9.8144	9.8688	9.9223	9.9750
9	10.2391	10.3072	10.3742	10.4400	10.5047	10.5684	10.6310	10.6927	10.7534	10.8131
10	10.9551	11.0315	11.1066	11.1803	11.2529	11.3242	11.3944	11.4635	11.5315	11.5985
11	11.6213	11.7061	11.7894	11.8713	11.9519	12.0310	12.1089	12.1856	12.2610	12.3352
12	12.2411	12.3345	12.4263	12.5164	12.6051	12.6922	12.7779	12.8623	12.9452	13.0269
13	12.8173	12.9195	13.0199	13.1185	13.2154	13.3107	13.4044	13.4965	13.5872	13.6764
14	13.3524	13.4636	13.5727	13.6799	13.7853	13.8888	13.9906	14.0907	14.1892	14.2861
15	13.8484	13.9687	14.0868	14.2028	14.3168	14.4287	14.5388	14.6470	14.7535	14.8582
16	14.3071	14.4367	14.5640	14.6889	14.8116	14.9322	15.0507	15.1672	15.2818	15.3945
17	14.7298	14.8690	15.0056	15.1397	15.2714	15.4007	15.5278	15.6527	15.7756	15.8964
18	15.1180	15.2670	15.4131	15.5565	15.6973	15.8356	15.9715	16.1050	16.2363	16.3654
19	15.4727	15.6317	15.7875	15.9405	16.0907	16.2381	16.3829	16.5253	16.6651	16.8027
20	15.7950	15.9641	16.1300	16.2927	16.4524	16.6092	16.7632	16.9144	17.0631	17.2092

LOGARITHMS OF THE BINOMIAL COEFFICIENTS

m	n = 61	n = 62	n = 63	n = 64	n = 65	n = 66	n = 67	n = 68	n = 69	n = 70
21	16.0855	16.2651	16.4412	16.6139	16.7834	16.9497	17.1130	17.2734	17.4311	17.5860
22	16.3452	16.5355	16.7220	16.9050	17.0844	17.2605	17.4334	17.6031	17.7699	17.9337
23	16.5745	16.7758	16.9731	17.1665	17.3562	17.5422	17.7249	17.9042	18.0802	18.2532
24	16.7741	16.9867	17.1949	17.3991	17.5992	17.7955	17.9881	18.1772	18.3628	18.5451
25	16.9443	17.1685	17.3881	17.6032	17.8140	18.0208	18.2236	18.4227	18.6181	18.8099
26	17.0857	17.3217	17.5529	17.7793	18.0011	18.2186	18.4319	18.6412	18.8465	19.0482
27	17.1984	17.4467	17.6897	17.9277	18.1608	18.3893	18.6133	18.8330	19.0486	19.2603
28	17.2827	17.5436	17.7989	18.0487	18.2935	18.5332	18.7682	18.9987	19.2247	19.4466
29	17.3388	17.6127	17.8805	18.1426	18.3993	18.6506	18.8969	19.1383	19.3751	19.6074
30	17.3668	17.6541	17.9349	18.2096	18.4784	18.7417	18.9996	19.2523	19.5001	19.7431
31	17.3668	17.6679	17.9620	18.2497	18.5311	18.8066	19.0764	19.3407	19.5998	19.8538
32			17.9620	18.2631	18.5575	18.8455	19.1275	19.4038	19.6744	19.9397
33					18.5575	18.8585	19.1531	19.4415	19.7241	20.0010
34							19.1531	19.4541	19.7489	20.0377
35									19.7489	20.0499

m	n = 71	n = 72	n = 73	n = 74	n = 75	n = 76	n = 77	n = 78	n = 79	n = 80
1	1.8513	1.8573	1.8633	1.8692	1.8751	1.8808	1.8865	1.8921	1.8976	1.9031
2	3.3953	3.4076	3.4196	3.4315	3.4433	3.4548	3.4663	3.4776	3.4887	3.4997
3	4.7571	4.7755	4.7938	4.8117	4.8295	4.8470	4.8642	4.8812	4.8981	4.9147
4	5.9875	6.0123	6.0368	6.0609	6.0847	6.1082	6.1314	6.1542	6.1768	6.1991
5	7.1146	7.1459	7.1767	7.2071	7.2370	7.2666	7.2957	7.3245	7.3529	7.3809
6	8.1560	8.1938	8.2310	8.2678	8.3040	8.3397	8.3749	8.4097	8.4440	8.4778
7	9.1238	9.1682	9.2120	9.2552	9.2977	9.3397	9.3811	9.4219	9.4622	9.5020
8	10.0269	10.0781	10.1285	10.1782	10.2271	10.2754	10.3231	10.3701	10.4165	10.4622
9	10.8720	10.9300	10.9871	11.0435	11.0990	11.1537	11.2077	11.2609	11.3135	11.3653
10	11.6644	11.7293	11.7933	11.8564	11.9185	11.9798	12.0402	12.0998	12.1586	12.2166
11	12.4083	12.4803	12.5513	12.6212	12.6900	12.7579	12.8249	12.8909	12.9560	13.0203
12	13.1073	13.1865	13.2645	13.3413	13.4170	13.4917	13.5652	13.6378	13.7094	13.7799
13	13.7642	13.8507	13.9359	14.0198	14.1024	14.1839	14.2642	14.3434	14.4215	14.4985
14	14.3815	14.4754	14.5679	14.6590	14.7487	14.8371	14.9243	15.0102	15.0949	15.1784
15	14.9613	15.0628	15.1626	15.2610	15.3579	15.4534	15.5475	15.6403	15.7317	15.8219
16	15.5054	15.6145	15.7220	15.8278	15.9320	16.0346	16.1358	16.2355	16.3338	16.4307
17	16.0153	16.1322	16.2474	16.3607	16.4724	16.5823	16.6907	16.7974	16.9027	17.0064
18	16.4924	16.6173	16.7403	16.8613	16.9805	17.0979	17.2136	17.3275	17.4398	17.5505
19	16.9379	17.0710	17.2019	17.3308	17.4576	17.5826	17.7057	17.8269	17.9464	18.0641
20	17.3529	17.4942	17.6333	17.7701	17.9048	18.0374	18.1680	18.2967	18.4235	18.5484
21	17.7382	17.8880	18.0353	18.1803	18.3230	18.4634	18.6017	18.7379	18.8721	19.0044
22	18.0948	18.2532	18.4089	18.5621	18.7129	18.8613	19.0075	19.1514	19.2931	19.4328
23	18.4233	18.5904	18.7547	18.9164	19.0755	19.2320	19.3861	19.5378	19.6873	19.8345
24	18.7243	18.9004	19.0735	19.2438	19.4113	19.5761	19.7383	19.8980	20.0553	20.2102
25	18.9984	19.1837	19.3568	19.5448	19.7209	19.8941	20.0646	20.2324	20.3977	20.5604
26	19.2462	19.4408	19.6320	19.8200	20.0049	20.1867	20.3657	20.5417	20.7151	20.8858
27	19.4681	19.6722	19.8728	20.0699	20.2637	20.4543	20.6419	20.8264	21.0080	21.1868
28	19.6644	19.8783	20.0884	20.2948	20.4978	20.6974	20.8937	21.0868	21.2769	21.4639
29	19.8354	20.0593	20.2792	20.4952	20.7075	20.9162	21.1215	21.3234	21.5220	21.7175
30	19.9816	20.2157	20.4455	20.6713	20.8931	21.1112	21.3256	21.5364	21.7439	21.9480
31	20.1030	20.3475	20.5876	20.8234	21.0550	21.2826	21.5063	21.7263	21.9427	22.1556
32	20.1999	20.4552	20.7057	20.9517	21.1933	21.4307	21.6639	21.8933	22.1188	22.3406
33	20.2725	20.5387	20.8000	21.0564	21.3082	21.5556	21.7986	22.0375	22.2724	22.5034
34	20.3208	20.5983	20.8706	21.1377	21.4000	21.6576	21.9106	22.1592	22.4037	22.6440
35	20.3449	20.6340	20.9176	21.1957	21.4687	21.7368	22.0000	22.2586	22.5128	22.7627

LOGARITHMS OF THE BINOMIAL COEFFICIENTS

m	n = 71	n = 72	n = 73	n = 74	n = 75	n = 76	n = 77	n = 78	n = 79	n = 80
36	20.3449	20.6459	20.9410	21.2305	21.5145	21.7932	22.0670	22.3358	22.6000	22.8596
37			20.9410	21.2421	21.5374	21.8271	22.1115	22.3908	22.6652	22.9348
38					21.5374	21.8384	22.1338	22.4238	22.7087	22.9885
39							22.1338	22.4348	22.7304	23.0207
40									22.7304	23.0314

m	n = 81	n = 82	n = 83	n = 84	n = 85	n = 86	n = 87	n = 88	n = 89	n = 90
1	1.9085	1.9138	1.9191	1.9243	1.9294	1.9345	1.9395	1.9445	1.9494	1.9542
2	3.5105	3.5213	3.5319	3.5423	3.5527	3.5629	3.5730	3.5830	3.5928	3.6026
3	4.9311	4.9472	4.9632	4.9790	4.9946	5.0100	5.0253	5.0403	5.0552	5.0700
4	6.2211	6.2428	6.2643	6.2854	6.3064	6.3271	6.3475	6.3677	6.3877	6.4074
5	7.4086	7.4359	7.4629	7.4896	7.5159	7.5419	7.5676	7.5930	7.6181	7.6430
6	8.5113	8.5443	8.5769	8.6090	8.6408	8.6722	8.7033	8.7339	8.7643	8.7942
7	9.5412	9.5800	9.6182	9.6560	9.6934	9.7302	9.7667	9.8027	9.8382	9.8734
8	10.5074	10.5520	10.5960	10.6394	10.6824	10.7248	10.7667	10.8081	10.8490	10.8894
9	11.4165	11.4669	11.5168	11.5660	11.6146	11.6626	11.7100	11.7569	11.8032	11.8490
10	12.2738	12.3303	12.3860	12.4411	12.4954	12.5491	12.6021	12.6545	12.7063	12.7574
11	13.0837	13.1462	13.2080	13.2689	13.3291	13.3885	13.4472	13.5052	13.5625	13.6191
12	13.8496	13.9183	13.9861	14.0531	14.1191	14.1844	14.2489	14.3125	14.3754	14.4376
13	14.5745	14.6494	14.7234	14.7964	14.8685	14.9397	15.0100	15.0794	15.1480	15.2157
14	15.2609	15.3422	15.4224	15.5016	15.5797	15.6569	15.7331	15.8083	15.8827	15.9561
15	15.9108	15.9986	16.0851	16.1706	16.2549	16.3381	16.4203	16.5015	16.5816	16.6608
16	16.5263	16.6205	16.7135	16.8053	16.8959	16.9853	17.0735	17.1607	17.2468	17.3318
17	17.1087	17.2096	17.3092	17.4074	17.5043	17.5999	17.6943	17.7876	17.8796	17.9705
18	17.6596	17.7673	17.8734	17.9782	18.0815	18.1835	18.2842	18.3836	18.4817	18.5786
19	18.1802	18.2947	18.4076	18.5190	18.6288	18.7373	18.8443	18.9499	19.0542	19.1572
20	18.6716	18.7930	18.9127	19.0308	19.1473	19.2623	19.3757	19.4877	19.5983	19.7074
21	19.1347	19.2632	19.3899	19.5148	19.6380	19.7596	19.8796	19.9980	20.1149	20.2303
22	19.5704	19.7061	19.8398	19.9717	20.1018	20.2301	20.3567	20.4817	20.6050	20.7267
23	19.9795	20.1225	20.2634	20.4024	20.5394	20.6746	20.8079	20.9395	21.0693	21.1975
24	20.3628	20.5131	20.6614	20.8075	20.9516	21.0937	21.2339	21.3722	21.5087	21.6434
25	20.7207	20.8786	21.0343	21.1877	21.3390	21.4882	21.6353	21.7804	21.9236	22.0650
26	21.0539	21.2195	21.3827	21.5436	21.7022	21.8585	22.0127	22.1648	22.3148	22.4629
27	21.3629	21.5364	21.7073	21.8757	22.0416	22.2053	22.3667	22.5258	22.6828	22.8377
28	21.6481	21.8296	22.0083	22.1844	22.3579	22.5290	22.6977	22.8640	23.0280	23.1899
29	21.9100	22.0996	22.2862	22.4702	22.6514	22.8300	23.0061	23.1797	23.3510	23.5199
30	22.1489	22.3467	22.5415	22.7334	22.9225	23.1088	23.2924	23.4735	23.6520	23.8281
31	22.3651	22.5714	22.7744	22.9744	23.1715	23.3656	23.5569	23.7455	23.9315	24.1149
32	22.5589	22.7738	22.9853	23.1936	23.3987	23.6008	23.8000	23.9963	24.1898	24.3806
33	22.7306	22.9542	23.1743	23.3911	23.6045	23.8147	24.0218	24.2259	24.4271	24.6255
34	22.8804	23.1130	23.3418	23.5671	23.7890	24.0075	24.2227	24.4348	24.6438	24.8499
35	23.0084	23.2501	23.4880	23.7220	23.9525	24.1794	24.4029	24.6231	24.8401	25.0540
36	23.1149	23.3659	23.6129	23.8559	24.0952	24.3307	24.5626	24.7911	25.0162	25.2381
37	23.1999	23.4605	23.7168	23.9690	24.2172	24.4615	24.7020	24.9389	25.1723	25.4023
38	23.2635	23.5339	23.7998	24.0613	24.3186	24.5719	24.8212	25.0667	25.3085	25.5468
39	23.3060	23.5863	23.8619	24.1330	24.3996	24.6620	24.9203	25.1746	25.4250	25.6717
40	23.3271	23.6177	23.9033	24.1841	24.4603	24.7321	24.9995	25.2627	25.5219	25.7772
41	23.3271	23.6828	23.9240	24.2148	24.5008	24.7821	25.0588	25.3312	25.5994	25.8634
42			23.9240	24.2250	24.5210	24.8120	25.0983	25.3801	25.6573	25.9303
43					24.5210	24.8220	25.1181	25.4093	25.6960	25.9781
44							25.1181	25.4191	25.7153	26.0068
45									25.7153	26.0163

LOGARITHMS OF THE BINOMIAL COEFFICIENTS

m	$n = 91$	$n = 92$	$n = 93$	$n = 94$	$n = 95$	$n = 96$	$n = 97$	$n = 98$	$n = 99$	$n = 100$
1	1.9590	1.9638	1.9685	1.9731	1.9777	1.9823	1.9868	1.9912	1.9956	2.0000
2	3.6123	3.6218	3.6312	3.6406	3.6498	4.6590	3.6680	3.6770	3.6858	3.6946
3	5.0845	5.0989	5.1132	5.1272	5.1412	5.1550	5.1686	5.1821	5.1955	5.2087
4	6.4269	6.4463	6.4653	6.4842	6.5029	6.5214	6.5397	6.5578	6.5757	6.5934
5	7.6675	7.6918	7.7158	7.7395	7.7630	7.7862	7.8092	7.8319	7.8544	7.8767
6	8.8238	8.8531	8.8821	8.9107	8.9391	8.9671	8.9948	9.0223	9.0494	9.0763
7	9.9082	9.9425	9.9765	10.0101	10.0434	10.0762	10.1088	10.1410	10.1728	10.2043
8	10.9294	10.9689	11.0079	11.0466	11.0848	11.1225	11.1599	11.1969	11.2335	11.2697
9	11.8942	11.9389	11.9831	12.0268	12.0700	12.1128	12.1551	12.1969	12.2383	12.2793
10	12.8080	12.8580	12.9074	12.9562	13.0045	13.0523	13.0996	13.1463	13.1925	13.2383
11	13.6751	13.7304	13.7851	13.8391	13.8926	13.9454	13.9977	14.0494	14.1005	14.1512
12	14.4990	14.5597	14.6197	14.6790	14.7377	14.7956	14.8530	14.9097	14.9658	15.0214
13	15.2827	15.3488	15.4142	15.4789	15.5428	15.6060	15.6685	15.7303	15.7914	15.8519
14	16.0287	16.1003	16.1712	16.2412	16.3105	16.3789	16.4466	16.5136	16.5798	16.6453
15	16.7391	16.8163	16.8927	16.9682	17.0429	17.1167	17.1896	17.2618	17.3331	17.4037
16	17.4157	17.4987	17.5807	17.6617	17.7418	17.8210	17.8993	17.9767	18.0533	18.1290
17	18.0604	18.1491	18.2368	18.3234	18.4090	18.4937	18.5773	18.6601	18.7419	18.8228
18	18.6743	18.7689	18.8623	18.9546	19.0458	19.1360	19.2252	19.3133	19.4005	19.4866
19	19.2589	19.3594	19.4586	19.5567	19.6536	19.7494	19.8440	19.9376	20.0302	20.1217
20	19.8152	19.9216	20.0268	20.1307	20.2334	20.3348	20.4351	20.5342	20.6322	20.7292
21	20.3442	20.4568	20.5679	20.6777	20.7862	20.8934	20.9994	21.1041	21.2076	21.3100
22	20.8469	20.9656	21.0828	21.1986	21.3130	21.4261	21.5378	21.6482	21.7573	21.8652
23	21.3240	21.4490	21.5723	21.6942	21.8146	21.9336	22.0511	22.1673	22.2821	22.3956
24	21.7763	21.9076	22.0372	22.1653	22.2917	22.4167	22.5401	22.6621	22.7827	22.9019
25	22.2045	22.3422	22.4781	22.6124	22.7450	22.8761	23.0055	23.1334	23.2598	23.3847
26	22.6090	22.7533	22.8957	23.0363	23.1752	23.3123	23.4479	23.5818	23.7141	23.8448
27	22.9906	23.1415	23.2904	23.4374	23.5827	23.7261	23.8678	24.0077	24.1460	24.2827
28	23.3496	23.5072	23.6628	23.8164	23.9680	24.1178	24.2657	24.4118	24.5562	24.6989
29	23.6865	23.8510	24.0133	24.1735	24.3317	24.4879	24.6421	24.7945	24.9451	25.0938
30	24.0018	24.1732	24.3423	24.5093	24.6741	24.8368	24.9975	25.1563	25.3130	25.4679
31	24.2958	24.4742	24.6503	24.8241	24.9957	25.1650	25.3322	25.4974	25.6605	25.8217
32	24.5688	24.7544	24.9376	25.1183	25.2967	25.4728	25.6466	25.8183	25.9879	26.1554
33	24.9211	25.0141	25.2044	25.3922	25.5775	25.7604	25.9410	26.1194	26.2954	26.4694
34	25.0531	25.2534	25.4511	25.6460	25.8384	26.0283	26.2157	26.4008	26.5835	26.7640
35	25.2649	25.4728	25.6778	25.8801	26.0797	26.2766	26.4710	26.6629	26.8524	27.0394
36	25.4568	25.6724	25.8850	26.0947	26.3015	26.5057	26.7071	26.9059	27.1022	27.2961
37	25.6289	25.8523	26.0726	26.2899	26.5042	26.7156	26.9242	27.1301	27.3334	27.5340
38	25.7815	26.0129	26.2410	26.4660	26.6878	26.9067	27.1226	27.3357	27.5460	27.7536
39	25.9147	26.1543	26.3903	26.6231	26.8526	27.0790	27.3024	27.5228	27.7402	27.9549
40	26.0287	26.2765	26.5207	26.7614	26.9988	27.2329	27.4638	27.6916	27.9163	28.1382
41	26.1235	26.3797	26.6322	26.8810	27.1264	27.3683	27.6068	27.8422	28.0744	28.3036
42	26.1992	26.4640	26.7249	26.9820	27.2355	27.4854	27.7318	27.9748	28.2146	28.4512
43	26.2559	26.5295	26.7990	27.0646	27.3263	27.5843	27.8387	28.0895	28.3370	28.5811
44	26.2937	26.5763	26.8545	27.1287	27.3989	27.6651	27.9276	28.1865	28.4417	28.6935
45	26.3126	26.6043	26.8915	27.1745	27.4532	27.7279	27.9987	28.2656	28.5289	28.7885
46	26.3126	26.6136	26.9100	27.2019	27.3894	27.7727	28.0519	28.3272	28.5985	28.8661
47			26.9100	27.2110	27.5075	27.7996	28.0874	28.3711	28.6507	28.9264
48					27.5075	27.8086	28.1051	28.3974	28.6855	28.9694
49							28.1051	28.4062	28.7028	28.9953
50									28.7028	29.0039

VALUES OF $\frac{1}{2}$ p(1 − p)

p	0	1	2	3	4	5	6	7	8	9	10	p
	0.	0.	0.	0.	0.	0.	0.	0.	0.	0.	0.	
.00	00000	0004995	000998	0014955	001992	0024875	002982	0034755	003968	0044595	00495	.99
.01	00495	0054395	005928	0064155	006902	0073875	007872	0083555	008838	0093195	00980	.98
.02	00980	0102795	010758	0112355	011712	0121875	012662	0131355	013608	0140795	01455	.97
.03	01455	0150195	015488	0159555	016422	0168875	017352	0178155	018278	0187395	01920	.96
.04	01920	0196595	020118	0205755	021032	0214875	021942	0223955	022848	0232995	02375	.95
.05	02375	0241995	024648	0250955	025542	0259875	026432	0268755	027318	0277595	02820	.94
.06	02820	0286395	029078	0295155	029952	0303875	030822	0312555	031688	0321195	03255	.93
.07	03255	0329795	033408	0338355	034262	0346875	035112	0355355	035958	0363795	03680	.92
.08	03680	0372195	037638	0380555	038472	0388875	039302	0397155	040128	0405395	04095	.91
.09	04095	0413595	041768	0421755	042582	0429875	043392	0437955	044198	0445995	04500	.90
.10	04500	0453995	045798	0461955	046592	0469875	047382	0477755	048168	0485595	04895	.89
.11	04895	0493395	049728	0501155	050502	0508875	051272	0516555	052038	0524195	05280	.88
.12	05280	0531795	053558	0539355	054312	0546875	055062	0554355	055808	0561795	05655	.87
.13	05655	0569195	057288	0576555	058022	0583875	058752	0591155	059478	0598395	06020	.86
.14	06020	0605595	060918	0612755	061632	0619875	062342	0626955	063048	0633995	06375	.85
.15	06375	0640995	064448	0647955	065142	0654875	065832	0661755	066518	0668595	06720	.84
.16	06720	0675395	067878	0682155	068552	0688875	069222	0695555	069888	0702195	07055	.83
.17	07055	0708795	071208	0715355	071862	0721875	072512	0728355	073158	0734795	07380	.82
.18	07380	0741195	074438	0747555	075072	0753875	075702	0760155	076328	0766395	07695	.81
.19	07695	0772595	077568	0778755	078182	0784875	078792	0790955	079398	0796995	08000	.80
.20	08000	0802995	080598	0808955	081192	0814875	081782	0820755	082368	0826595	08295	.79
.21	08295	0832395	083528	0838155	084102	0843875	084672	0849555	085238	0855195	08580	.78
.22	08580	0860795	086358	0866355	086912	0871875	087462	0877355	088008	0882795	08855	.77
.23	08855	0888195	089088	0893555	089622	0898875	090152	0904155	090678	0909395	09120	.76
.24	09120	0914595	091718	0919755	092232	0924875	092742	0929955	093248	0934995	09375	.75
.25	09375	0939995	094248	0944955	094742	0949875	095232	0954755	095718	0959595	09620	.74
.26	09620	0964395	096678	0969155	097152	0973875	097622	0978555	098088	0983195	09855	.73
.27	09855	0987795	099008	0992355	099462	0996875	099912	1001355	100358	1005795	10080	.72
.28	10080	1010195	101238	1014555	101672	1018875	102102	1023155	102528	1027395	10295	.71
.29	10295	1031595	103368	1035755	103782	1039875	104192	1043955	104598	1047995	10500	.70
.30	10500	1051995	105398	1055955	105792	1059875	106182	1063755	106568	1067595	10695	.69
.31	10695	1071395	107328	1075155	107702	1078875	108072	1082555	108438	1086195	10880	.68
.32	10880	1089795	109158	1093355	109512	1096875	109862	1100355	110208	1103795	11055	.67
.33	11055	1107195	110888	1110555	111222	1113875	111552	1117155	111788	1120395	11220	.66
.34	11220	1123595	112518	1126755	112832	1129875	113142	1132955	113448	1135995	11375	.65
.35	11375	1138995	114048	1141955	114342	1144875	114632	1147755	114918	1150595	11520	.64
.36	11520	1153395	115478	1156155	115752	1158875	116022	1161555	116288	1164195	11655	.63
.37	11655	1166795	116808	1169355	117062	1171875	117312	1174355	117558	1176795	11780	.62
.38	11780	1179195	118038	1181555	118272	1183875	118502	1186155	118728	1188395	11895	.61
.39	11895	1190595	119168	1192755	119382	1194875	119592	1196955	119798	1197995	12000	.60
.40	12000	1200995	120198	1202955	120392	1204875	120582	1206755	120768	1208595	12095	.59
.41	12095	1210395	121128	1212155	121302	1213875	121472	1215555	121638	1217195	12180	.58
.42	12180	1218795	121958	1220355	122112	1221875	122262	1223355	122408	1224795	12255	.57
.43	12255	1226195	122688	1227555	122822	1228875	122952	1230155	123078	1231395	12320	.56
.44	12320	1232595	123318	1233755	123432	1234875	123542	1235955	123648	1236995	12375	.55
.45	12375	1237995	123848	1238955	123942	1239875	124032	1240755	124118	1241595	12420	.54
.46	12420	1242395	124278	1243155	124352	1243875	124422	1244555	124488	1245195	12455	.53
.47	12455	1245795	124608	1246355	124662	1246875	124712	1247355	124758	1247795	12480	.52
.48	12480	1248195	124838	1248555	124872	1248875	124902	1249155	124928	1249395	12495	.51
.49	12495	1249595	124968	1249755	124982	1249875	124992	1249955	124998	1249995	12500	.50
p	10	9	8	7	6	5	4	3	2	1	0	p

THE NUMBER OF EACH DAY OF THE YEAR

Day of Mo.	Jan.	Feb.	Mar.	Apr.	May	Jun.	Jul.	Aug.	Sep.	Oct.	Nov.	Dec.	Day of Mo.
1	1	32	60	91	121	152	182	213	244	274	305	335	1
2	2	33	61	92	122	153	183	214	245	275	306	336	2
3	3	34	62	93	123	154	184	215	246	276	307	337	3
4	4	35	63	94	124	155	185	216	247	277	308	338	4
5	5	36	64	95	125	156	186	217	248	278	309	339	5
6	6	37	65	96	126	157	187	218	249	279	310	340	6
7	7	38	66	97	127	158	188	219	250	280	311	341	7
8	8	39	67	98	128	159	189	220	251	281	312	342	8
9	9	40	68	99	129	160	190	221	252	282	313	343	9
10	10	41	69	100	130	161	191	222	253	283	314	344	10
11	11	42	70	101	131	162	192	223	254	284	315	345	11
12	12	43	71	102	132	163	193	224	255	285	316	346	12
13	13	44	72	103	133	164	194	225	256	286	317	347	13
14	14	45	73	104	134	165	195	226	257	287	318	348	14
15	15	46	74	105	135	166	196	227	258	288	319	349	15
16	16	47	75	106	136	167	197	228	259	289	320	350	16
17	17	48	76	107	137	168	198	229	260	290	321	351	17
18	18	49	77	108	138	169	199	230	261	291	322	352	18
19	19	50	78	109	139	170	200	231	262	292	323	353	19
20	20	51	79	110	140	171	201	232	263	293	324	354	20
21	21	52	80	111	141	172	202	233	264	294	325	355	21
22	22	53	81	112	142	173	203	234	265	295	326	356	22
23	23	54	82	113	143	174	204	235	266	296	327	357	23
24	24	55	83	114	144	175	205	236	267	297	328	358	24
25	25	56	84	115	145	176	206	237	268	298	329	359	25
26	26	57	85	116	146	177	207	238	269	299	330	360	26
27	27	58	86	117	147	178	208	239	270	300	331	361	27
28	28	59	87	118	148	179	209	240	271	301	332	362	28
29	29	*	88	119	149	180	210	241	272	302	333	363	29
30	30		89	120	150	181	211	242	273	303	334	364	30
31	31		90		151		212	243		304		365	31

* In leap years, after February 28, add 1 to the tabulated number.

TABLE FOR CONVERSION OF CALENDAR DATE TO JULIAN DATE

DR. E. H. CHERRINGTON, JR.

CHRONOLOGY I

Julian Date = 2,430,000 + Tabular Entry + Day of Month.

Year	Jan.	Feb.	Mar.	Apr.	May	June	July	Aug.	Sept.	Oct.	Nov.	Dec.
1950	3282	3313	3341	3372	3402	3433	3463	3494	3525	3555	3586	3616
1951	3647	3678	3706	3737	3767	3798	3828	3859	3890	3920	3951	3981
1952	4012	4043	4072	4103	4133	4164	4194	4225	4256	4286	4317	4347
1953	4378	4409	4437	4468	4498	4529	4559	4590	4621	4651	4682	4712
1954	4743	4774	4802	4833	4863	4894	4924	4955	4986	5016	5047	5077
1955	5108	5139	5167	5198	5228	5259	5289	5320	5351	5381	5412	5442
1956	5473	5504	5533	5564	5594	5625	5655	5686	5717	5747	5778	5808
1957	5839	5870	5898	5929	5959	5990	6020	6051	6082	6112	6143	6173
1958	6204	6235	6263	6294	6324	6355	6385	6416	6447	6477	6508	6538
1959	6569	6600	6628	6659	6689	6720	6750	6781	6812	6842	6873	6903
1960	6934	6965	6994	7025	7055	7086	7116	7147	7178	7208	7239	7269
1961	7300	7331	7359	7390	7420	7451	7481	7512	7543	7573	7604	7634
1962	7665	7696	7724	7755	7785	7816	7846	7877	7908	7938	7969	7999
1963	8030	8061	8089	8120	8150	8181	8211	8242	8273	8303	8334	8364
1964	8395	8426	8455	8486	8516	8547	8577	8608	8639	8669	8700	8730
1965	8761	8792	8820	8851	8881	8912	8942	8973	9004	9034	9065	9095
1966	9126	9157	9185	9216	9246	9277	9307	9338	9369	9399	9430	9460
1967	9491	9522	9550	9581	9611	9642	9672	9703	9734	9764	9795	9825
1968	9856	9887	9916	9947	9977	*0008	*0038	*0069	*0100	*0130	*0161	*0191
*1969	0222	0253	0281	0312	0342	0373	0403	0434	0465	0495	0526	0556
*1970	0587	0618	0646	0677	0707	0738	0768	0799	0830	0860	0891	0921
*1971	0952	0983	1011	1042	1072	1103	1133	1164	1195	1225	1256	1286
*1972	1317	1348	1377	1408	1438	1469	1499	1530	1561	1591	1622	1652
*1973	1683	1714	1742	1773	1803	1834	1864	1895	1926	1956	1987	2017
*1974	2048	2079	2107	2138	2168	2199	2229	2260	2291	2321	2352	2382
*1975	2413	2444	2472	2503	2533	2564	2594	2625	2656	2686	2717	2747
*1976	2778	2809	2838	2869	2899	2930	2960	2991	3022	3052	3083	3113
*1977	3144	3175	3203	3234	3264	3295	3325	3356	3387	3417	3448	3478
*1978	3509	3540	3568	3599	3629	3660	3690	3721	3752	3782	3813	3843
*1979	3874	3905	3933	3964	3994	4025	4055	4086	4117	4147	4178	4208

*From June, 1968, through October 1995, add 10,000 to computed date.
(The Julian Day begins at 7:00 a.m., EST.)

TABLE FOR CONVERSION OF CALENDAR DATE TO JULIAN DATE

DR. E. H. CHERRINGTON, JR.

CHRONOLOGY I

Julian Date = 2,430,000 + Tabular Entry + Day of Month.

Year	Jan.	Feb.	Mar.	Apr.	May	June	July	Aug.	Sept.	Oct.	Nov.	Dec.
*1980	4239	4270	4299	4330	4360	4391	4421	4452	4483	4513	4544	4574
*1981	4605	4636	4664	4695	4725	4756	4786	4817	4848	4878	4909	4939
*1982	4970	5001	5029	5060	5090	5121	5151	5182	5213	5243	5274	5304
*1983	5335	5366	5394	5425	5455	5486	5516	5547	5578	5608	5639	5669
*1984	5700	5731	5760	5791	5821	5852	5882	5913	5944	5974	6005	6035
*1985	6066	6097	6125	6156	6186	6217	6247	6278	6309	6339	6370	6400
*1986	6431	6462	6490	6521	6551	6582	6612	6643	6674	6704	6735	6765
*1987	6796	6827	6855	6886	6916	6947	6977	7008	7039	7069	7100	7130
*1988	7161	7192	7221	7252	7282	7313	7343	7374	7405	7435	7466	7496
*1989	7527	7558	7586	7617	7647	7678	7708	7739	7770	7800	7831	7861
*1990	7892	7923	7951	7982	8012	8043	8073	8104	8135	8165	8196	8226
*1991	8257	8288	8316	8347	8377	8408	8438	8469	8500	8530	8561	8591
*1992	8622	8653	8682	8713	8743	8774	8804	8835	8866	8896	8927	8957
*1993	8988	9019	9047	9078	9108	9139	9169	9200	9231	9261	9292	9322
*1994	9353	9384	9412	9443	9473	9504	9534	9565	9596	9626	9657	9687
*1995	9718	9749	9777	9808	9838	9869	9899	9930	9961	9991	**0022	**0052
**1996	0083	0114	0143	0174	0204	0235	0265	0296	0327	0357	0388	0418
**1997	0449	0480	0508	0539	0569	0600	0630	0661	0692	0722	0753	0783
**1998	0814	0845	0873	0904	0934	0965	0995	1026	1057	1087	1118	1148
**1999	1179	1210	1238	1269	1299	1330	1360	1391	1422	1452	1483	1513
**2000	1544	1575	1604	1635	1665	1696	1726	1757	1788	1818	1849	1879
**2001	1910	1941	1969	2000	2030	2061	2091	2122	2153	2183	2214	2244
**2002	2275	2306	2334	2365	2395	2426	2456	2487	2518	2548	2579	2609
**2003	2640	2671	2699	2730	2760	2791	2821	2852	2883	2913	2944	2974
**2004	3005	3036	3065	3096	3126	3157	3187	3218	3249	3279	3310	3340
**2005	3371	3402	3430	3461	3491	3522	3552	3583	3614	3644	3675	3705
**2006	3736	3767	3795	3826	3856	3887	3917	3948	3979	4009	4040	4070
**2007	4101	4132	4160	4191	4221	4252	4282	4313	4344	4374	4405	4435
**2008	4466	4497	4526	4557	4587	4618	4648	4679	4710	4740	4771	4801
**2009	4832	4863	4891	4922	4952	4983	5013	5044	5075	5105	5136	5166

*From June, 1968, through October, 1995, add 10,000 to computed date.
**From November, 1995, through December, 2009, add 20,000 to computed date.
(The Julian Day begins at 7:00 a.m., EST.)

TABLE FOR CONVERSION OF HOURS, MINUTES, AND SECONDS INTO DECIMALS OF A DAY.

DR. E. H. CHERRINGTON, JR.

CHRONOLOGY II

For converting Hours, Minutes, and Seconds into Decimals of a Day.

Hours.	Decimal.	Min.	Decimal.	Min.	Decimal.	Sec.	Decimal.	Sec.	Decimal.
1	0.0416+	1	.000694+	31	.021527+	1	.0000116	31	.0003588
2	.0833+	2	.001388+	32	.022222+	2	.0000231	32	.0003704
3	.1250+	3	.002083+	33	.022916+	3	.0000347	33	.0003819
4	.1666+	4	.002777+	34	.023611+	4	.0000463	34	.0003935
5	.2083+	5	.003472+	35	.024305+	5	.0000579	35	.0004051
6	.2500+	6	.004166+	36	.025000+	6	.0000694	36	.0004167
7	0.2916+	7	.004861+	37	.025694+	7	.0000810	37	.0004282
8	.3333+	8	.005555+	38	.026388+	8	.0000925	38	.0004398
9	.3750+	9	.006250+	39	.027083+	9	.0001042	39	.0004514
10	.4166+	10	.006944+	40	.027777+	10	.0001157	40	.0004630
11	.4583+	11	.007638+	41	.028472+	11	.0001273	41	.0004745
12	.5000+	12	.008333+	42	.029166+	12	.0001389	42	.0004861
13	0.5416+	13	.009027+	43	.029861+	13	.0001505	43	.0004977
14	.5833+	14	.009722+	44	.030555+	14	.0001620	44	.0005093
15	.6250+	15	.010416+	45	.031250+	15	.0001736	45	.0005208
16	.6666+	16	.011111+	46	.031944+	16	.0001852	46	.0005324
17	.7083+	17	.011805+	47	.032638+	17	.0001968	47	.0005440
18	.7500+	18	.021500+	48	.033333+	18	.0002083	48	.0005556
19	0.7916+	19	.013194+	49	.034027+	19	.0002199	49	.0005671
20	.8333+	20	.013888+	50	.034722+	20	.0002315	50	.0005787
21	.8750+	21	.014583+	51	.035416+	21	.0002431	51	.0005903
22	.9166+	22	.015277+	52	.036111+	22	.0002546	52	.0006019
23	0.9583+	23	.015972+	53	.036805+	23	.0002662	53	.0006134
24	1.0000+	24	.016666+	54	.037500+	24	.0002778	54	.0006250
		25	.017361+	55	.038194+	25	.0002894	55	.0006366
		26	.018055+	56	.038888+	26	.0003009	56	.0006481
		27	.018750+	57	.039583+	27	.0003125	57	.0006597
		28	.019444+	58	.040277+	28	.0003241	58	.0006713
		29	.020138+	59	.040972+	29	.0003356	59	.0006829
		30	.020833+	60	.041666+	30	.0003472	60	.0006944

The sign +, appended to numbers in this table, signifies that the last figure repeats to infinity.

INTEREST TABLES

Simple Interest

If P is the principal placed at interest at a rate i (expressed as a decimal), for a period of n years

The **amount,**

$$A = P(1 + ni)$$

Present value,

$$P = \frac{A}{1 + ni}$$

Compound Interest

At interest compounded annually the **amount,—**

$$A = P(1 + i)^n$$

At interest compounded q times per year,—

$$A = P\left(1 + \frac{i}{q}\right)^{nq}$$

At interest compounded annually the **present value,—**

$$P = \frac{A}{(1 + i)^n} = A(1 + i)^{-n} = Av^n. \quad v = \frac{1}{1 + i}$$

At interest compounded q times per year,—

$$P = A\left(1 + \frac{i}{q}\right)^{-nq}$$

* The **amount of an annuity of 1 per annum,—**

$$s_{\overline{n}|i} = \frac{(1 + i)^n - 1}{i}$$

* The **present value of an annuity,—**

$$a_{\overline{n}|i} = \frac{1 - (1 + i)^{-n}}{i}$$

The **annuity whose present value is 1,—**

$$\frac{1}{a_{\overline{n}|i}} = \frac{1}{s_{\overline{n}|i}} + i = \frac{i}{(1 - v^n)}$$

Compound amount of 1 for fractional periods,—$(1+i)^{1/p}$

Nominal rate convertible p times per year equivalent to effective rate i,—

$$j_p = p\left[(1+i)^{1/p} - 1\right]$$

Amount for year of p deposits of $1/p$, p times per year,—i/j_p

*This information can be constructed by use of tables $(1 + i)^n$ and $(1 + i)^{-n}$ as may be observed from the formula structures. For example $s_{\overline{50}|.12} = \frac{(1.12)^{50} - 1}{.12}$. Here $(1.12)^{50}$ is obtained from the entries $(1 + i)^n$, where $i = .12$ and $n = 50$ and is equal to 289.00218983. Accordingly $s_{\overline{50}|.12} = \frac{288.00218983}{.12} = 2400.018249$. Similarily $a_{\overline{n}|i}$ may be evaluated by use of entries $(1 + i)^{-n}$.

AMOUNT AT COMPOUND INTEREST $(1 + i)^n$*

The following table gives the amount after a term of n periods on unit original principal at rate of interest i.

Periods	Rate i					Periods	Rate i				
n	0.0025(1/4%)	0.004167(5/12%)	0.005(1/2%)	0.005833(7/12%)	0.0075(3/4%)	n	0.0025(1/4%)	0.004167(5/12%)	0.005(1/2%)	0.005833(7/12%)	0.0075(3/4%)
1	1.00250000	1.00416667	1.00500000	1.00583333	1.00750000	51	1.13580414	1.23622022	1.28964194	1.34532504	1.46385411
2	1.00500625	1.00835069	1.01002500	1.01170069	1.01505625	52	1.23864365	1.24137114	1.29609014	1.35317277	1.47483301
3	1.00751877	1.01255216	1.01507513	1.01760228	1.02266917	53	1.14149026	1.24654352	1.30257060	1.36106628	1.48589426
4	1.01003756	1.01677112	1.02015050	1.02353830	1.03033919	54	1.14434398	1.25173745	1.30908346	1.36900583	1.49703847
5	1.01256266	1.02100767	1.02525125	1.02950894	1.03806673	55	1.14720484	1.25695302	1.31562887	1.37699170	1.50826626
6	1.01509406	1.02526187	1.03037751	1.03551440	1.04585224	56	1.15007285	1.26219033	1.32220702	1.38502415	1.51957825
7	1.01763180	1.02953379	1.03552940	1.04155490	1.05369613	57	1.15294804	1.26744946	1.32881805	1.39310346	1.53097509
8	1.02017588	1.03382352	1.04070704	1.04763064	1.06159885	58	1.15583041	1.27273050	1.33546214	1.40122990	1.54245740
9	1.02272632	1.03813111	1.04591058	1.05374182	1.06956084	59	1.15871998	1.27803354	1.34213946	1.40940374	1.55402583
10	1.02528313	1.04245666	1.05114013	1.05988865	1.07758255	60	1.16161678	1.28335868	1.34885015	1.41762526	1.56568103
11	1.02784634	1.04680023	1.05639583	1.06607133	1.08566441	61	1.16452082	1.28870601	1.35559440	1.42589474	1.57742363
12	1.03041596	1.05116190	1.06167781	1.07229008	1.09380690	62	1.16743213	1.29407561	1.36237238	1.43421246	1.58925431
13	1.03299200	1.05554174	1.06698620	1.07854511	1.10201045	63	1.17035071	1.29946760	1.36918424	1.44257870	1.60117372
14	1.03557448	1.05993983	1.07232113	1.08483662	1.11027553	64	1.17327658	1.30488204	1.37603016	1.45099374	1.61318252
15	1.03816341	1.06435625	1.07768274	1.09116483	1.11860259	65	1.17620977	1.31031905	1.38291031	1.45945787	1.62528139
16	1.04075882	1.06879106	1.08307115	1.09752996	1.12699211	66	1.17915030	1.31577872	1.38982486	1.46797138	1.63747100
17	1.04336072	1.07324436	1.08848651	1.10393222	1.13544455	67	1.18209817	1.32126113	1.39677399	1.47653454	1.64975203
18	1.04596912	1.07771621	1.09392894	1.11037182	1.14396039	68	1.18505342	1.32676638	1.40375785	1.48514766	1.66212517
19	1.04858404	1.08220670	1.09939858	1.11684899	1.15254009	69	1.18801605	1.33229458	1.41077664	1.49381102	1.67459111
20	1.05120550	1.08671589	1.10489558	1.12336395	1.16118414	70	1.19098609	1.33784580	1.41783053	1.50252492	1.68715055
21	1.05383352	1.09124387	1.11042006	1.12991690	1.16989302	71	1.19396356	1.34342016	1.42491968	1.51128965	1.69980418
22	1.05646810	1.09579072	1.11597216	1.13650808	1.17866722	72	1.19694847	1.34901774	1.43204428	1.52010550	1.71255271
23	1.05910927	1.10035652	1.12155202	1.14313771	1.18750723	73	1.19994084	1.35463865	1.43920450	1.52897279	1.72539685
24	1.06175704	1.10494134	1.12715978	1.14980602	1.19641353	74	1.20294069	1.36028298	1.44640052	1.53789179	1.73833733
25	1.06441144	1.10954526	1.13279558	1.15651322	1.20538663	75	1.20594804	1.36595082	1.45363252	1.54686283	1.75137486
26	1.06707247	1.11416836	1.13845955	1.16325955	1.21442703	76	1.20896291	1.37164229	1.46090069	1.55588620	1.76451017
27	1.06974015	1.11881073	1.14415185	1.17004523	1.22353523	77	1.21198532	1.37735746	1.46820519	1.56496220	1.77774400
28	1.07241450	1.12347244	1.14987261	1.17687049	1.23271175	78	1.21501528	1.38309645	1.47554622	1.57409115	1.79107708
29	1.07509553	1.12815358	1.15562197	1.18373557	1.24195709	79	1.21805282	1.38885935	1.48292395	1.58327334	1.80451915
30	1.07778327	1.13285422	1.16140008	1.19064069	1.25127176	80	1.22109795	1.39464627	1.49033857	1.59250910	1.81804398
31	1.08047773	1.13757444	1.16720708	1.19758610	1.26065630	81	1.22415070	1.40045729	1.49779026	1.60179874	1.83167931
32	1.08317892	1.14231434	1.17304312	1.20457202	1.27011122	82	1.22721108	1.40629253	1.50527921	1.61114257	1.84541691
33	1.08588687	1.14707398	1.17890833	1.21159869	1.27963706	83	1.23027910	1.41215209	1.51280561	1.62054090	1.85925753
34	1.08860159	1.15185346	1.18480288	1.21866634	1.28923434	84	1.23335480	1.41803605	1.52036964	1.62999405	1.87320196
35	1.09132309	1.15665284	1.19072689	1.22577523	1.29890359	85	1.23643819	1.42394454	1.52797148	1.63950235	1.88725098
36	1.09405140	1.16147223	1.19668052	1.23292559	1.30864537	86	1.23952928	1.42987764	1.53561134	1.64906612	1.90140536
37	1.09678653	1.16631170	1.20266393	1.24011765	1.31846021	87	1.24262811	1.43583546	1.54328940	1.65868567	1.91566590
38	1.09952850	1.17117133	1.20867725	1.24735167	1.32834866	88	1.24573468	1.44181811	1.55100585	1.66836134	1.93003339
39	1.10227732	1.17605121	1.21472063	1.25462789	1.33831128	89	1.24884901	1.44782568	1.55876087	1.67809344	1.94450865
40	1.10503301	1.18095142	1.22079424	1.26194655	1.34834861	90	1.25197114	1.45385829	1.56655468	1.68788232	1.95909246
41	1.10779559	1.18587206	1.22689821	1.26930791	1.35846123	91	1.25510106	1.45991603	1.57438745	1.69772830	1.97378565
42	1.11056508	1.19081319	1.23303270	1.27671220	1.36864969	92	1.25823882	1.46599902	1.58225939	1.70763172	1.98858905
43	1.11334149	1.19577491	1.23919786	1.28415969	1.37891456	93	1.26138441	1.47210735	1.59017069	1.71759290	2.00350346
44	1.11612485	1.20075731	1.24539385	1.29165062	1.38925642	94	1.26453787	1.47824113	1.59812154	1.72761219	2.01852974
45	1.11891516	1.20576046	1.25162082	1.29918525	1.39967584	95	1.26769922	1.48440047	1.60611215	1.73768993	2.03366871
46	1.12171245	1.21078446	1.25787892	1.30676383	1.41017341	96	1.27086847	1.49058547	1.61414271	1.74782646	2.04892123
47	1.12451673	1.21582940	1.26416832	1.31438662	1.42074971	97	1.27404564	1.49679624	1.62221342	1.75802211	2.06428814
48	1.12732802	1.22089536	1.27048916	1.32205388	1.43140533	98	1.27723075	1.50303289	1.63032449	1.76827724	2.07977030
49	1.13014634	1.22598242	1.27684161	1.32976586	1.44214087	99	1.28042383	1.50929553	1.63847611	1.77859219	2.09536858
50	1.13297171	1.23109068	1.28322581	1.33752283	1.45295693	100	1.28362489	1.51558426	1.64666849	1.78896731	2.11108384

*The [illegible] table may be constructed from this table by use of the formula $(1 + i)^n - 1$ See page 1004.

AMOUNT AT COMPOUND INTEREST $(1 + i)^n$ (Continued)

Periods n	Rate i 0.01(1%)	0.001125(1 1/8%)	0.0125(1 1/4%)	0.015(1 1/2%)	0.0175(1 3/4%)
1	1.01000000	1.01125000	1.01250000	1.01500000	1.01750000
2	1.02010000	1.02262656	1.02515625	1.03022500	1.03530625
3	1.03030100	1.03413111	1.03797070	1.04567838	1.05342411
4	1.04060401	1.04576509	1.05094534	1.06136355	1.07185903
5	1.05101005	1.05752994	1.06408215	1.07728400	1.09061656
6	1.06152015	1.06942716	1.07738318	1.09344326	1.10970235
7	1.07213535	1.08145821	1.09085047	1.10984491	1.12912215
8	1.08285671	1.09362462	1.10448610	1.12649259	1.14888178
9	1.09368527	1.10592789	1.11829218	1.14338998	1.16898721
10	1.10462213	1.11836958	1.13227083	1.16054083	1.18944449
11	1.11566835	1.13095124	1.14642422	1.17794894	1.21025977
12	1.12682503	1.14367444	1.16075452	1.19561817	1.23143931
13	1.13809328	1.15654078	1.17526395	1.21355244	1.25298950
14	1.14947421	1.16955186	1.18995475	1.23175573	1.27491682
15	1.16096896	1.18270932	1.20482918	1.25023207	1.29722786
16	1.17257864	1.19601480	1.21988955	1.26898555	1.31992935
17	1.18430443	1.20946997	1.23513817	1.28802033	1.34302811
18	1.19614748	1.22307650	1.25057739	1.30734064	1.36653111
19	1.20810895	1.23683611	1.26620961	1.32695075	1.39044540
20	1.22019004	1.25075052	1.28203723	1.34685501	1.41477820
21	1.23239194	1.26482146	1.29806270	1.36705783	1.43953681
22	1.24471586	1.27905071	1.31428848	1.38756370	1.46472871
23	1.25716302	1.29344003	1.33071709	1.40837715	1.49036146
24	1.26973465	1.30799123	1.34735105	1.42950281	1.51644279
25	1.28243200	1.32270613	1.36419294	1.45094535	1.54298054
26	1.29525631	1.33758657	1.38124535	1.47270953	1.56998269
27	1.30820888	1.35263442	1.39851092	1.49480018	1.59745739
28	1.32129097	1.46785156	1.41599230	1.51722218	1.62541290
29	1.33450388	1.38323989	1.43369221	1.53998051	1.65385762
30	1.34784892	1.39880134	1.45161336	1.56308022	1.68280013
31	1.36132740	1.41453785	1.46975853	1.58652642	1.71224913
32	1.37494068	1.43045140	1.48813051	1.61032432	1.74221349
33	1.38869009	1.44654398	1.50673214	1.63447918	1.77270223
34	1.40257699	1.46281760	1.52556629	1.65899637	1.80372452
35	1.41660276	1.47927430	1.54463587	1.68388132	1.83528970
36	1.43076878	1.49591613	1.56394382	1.70913954	1.86740727
37	1.44507647	1.51274519	1.58349312	1.73477663	1.90008689
38	1.45952724	1.52976357	1.60328678	1.76079828	1.93333841
39	1.47412251	1.54697341	1.62332787	1.78721025	1.96717184
40	1.48886373	1.56437687	1.64361946	1.81401841	2.00159734
41	1.50375237	1.58197611	1.66416471	1.84122868	2.03662530
42	1.51878989	1.59977334	1.68496677	1.86884712	2.07226624
43	1.53397779	1.61777079	1.70602885	1.89687982	2.10853090
44	1.54931757	1.63597071	1.72735421	1.92533302	2.14543019
45	1.56481075	1.65437538	1.74894614	1.95421301	2.18297522
46	1.58045885	1.67298710	1.77080797	1.98352621	2.22117728
47	1.59626344	1.69180821	1.79294306	2.01327910	2.26004789
48	1.61222608	1.71084105	1.81535485	2.04347829	2.29959872
49	1.62834834	1.73008801	1.83804679	2.07413046	2.33984170
50	1.64463182	1.74955150	1.86102237	2.10524242	2.38078893

Periods n	Rate i 0.01(1%)	0.001125(1 1/8%)	0.0125(1 1/4%)	0.015(1 1/2%)	0.0175(1 3/4%)
51	1.66107814	1.76923395	1.88428515	2.13682106	2.42245274
52	1.67768892	1.78913784	1.90783872	2.16887337	2.46484566
53	1.69446581	1.80926564	1.93168670	2.20140647	2.50798046
54	1.71141047	1.82961988	1.95583279	2.23442757	2.55187012
55	1.72852457	1.85020310	1.98028070	2.26794398	2.59652785
56	1.74580982	1.87101788	2.00503420	2.30196314	2.64196708
57	1.76326792	1.89206684	2.03009713	2.33649259	2.68820151
58	1.78090060	1.91335259	2.05547335	2.37153998	2.73524503
59	1.79870960	1.93487780	2.08116676	2.40711308	2.78311182
60	1.81669670	1.95664518	2.10718135	2.44321978	2.83181628
61	1.83486367	1.97865744	2.13352111	2.47986807	2.88137306
62	1.85321230	2.00091733	2.16019013	2.51706609	2.93179709
63	1.87174443	2.02342765	2.18719250	2.55482208	2.98310354
64	1.89046187	2.04619121	2.21453241	2.59314442	3.03530785
65	1.90936649	2.06921087	2.24221407	2.63204158	3.08842574
66	1.92846015	2.09248949	2.27024174	2.67152221	3.14247319
67	1.94774475	2.11602999	2.29861976	2.71159504	3.19746647
68	1.96722220	2.13983533	2.32735251	2.75226896	3.25342213
69	1.98689442	2.16390848	2.35644442	2.79355300	3.31035702
70	2.00676337	2.18825245	2.38589997	2.83545629	3.36828827
71	2.02683100	2.21287029	2.41572372	2.87798814	3.42723331
72	2.04709931	2.23776508	2.44592027	2.92115696	3.48720990
73	2.06757031	2.26293994	2.47649427	2.96497533	3.54823607
74	2.08824601	2.28839801	2.50745045	3.00944996	3.61033020
75	2.10912847	2.31414249	2.53879358	3.05459171	3.67351098
76	2.13021975	2.34017659	2.57052850	3.10041059	3.73779742
77	2.15152195	2.36650358	2.60266011	3.14691674	3.80320888
78	2.17303717	2.39312675	2.63519336	3.19412050	3.86976503
79	2.19476754	2.42004942	2.66813327	3.24203230	3.93748592
80	2.21671522	2.44727498	2.70148494	3.29066279	4.00639192
81	2.23888237	2.47480682	2.73525350	3.34002273	4.07650378
82	2.26127119	2.50264840	2.76944417	3.39012307	4.14784260
83	2.28388390	2.53080319	2.80406222	3.44097492	4.22042984
84	2.30672274	2.55927473	2.83911300	3.49258954	4.29428737
85	2.32978997	2.58806657	2.87460191	3.54497838	4.36943740
86	2.35308787	2.61718232	2.91053444	3.59815306	4.44590255
87	2.37661875	2.64662562	2.94691612	3.65212535	4.52370584
88	2.40038494	2.67640016	2.98375257	3.70690723	4.60287070
89	2.42438879	2.70650966	3.02104948	3.76251084	4.68342093
90	2.44863267	2.73695789	3.05881260	3.81894851	4.76538080
91	2.47311900	2.76774867	3.09704775	3.87623273	4.84877496
92	2.49785019	2.79888584	3.13576085	3.93437622	4.93362853
93	2.52282869	2.83037331	3.17495786	3.99339187	5.01996703
94	2.54805698	2.86221501	3.21464483	4.05329275	5.10781645
95	2.57353755	2.89441492	3.25482789	4.11409214	5.19720324
96	2.59927293	2.92697709	3.29551324	4.17580352	5.28815429
97	2.62526565	2.95990559	3.33670716	4.23844057	5.38069699
98	2.65151831	2.99320452	3.37841600	4.30201718	5.47485919
99	2.67803349	3.02687807	3.42064620	4.36654744	5.57066923
100	2.70481383	3.06093045	3.46340427	4.43204565	5.66815594

AMOUNT AT COMPOUND INTEREST $(1 + i)^n$ (Continued)

Periods n	Rate i 0.02(2%)	0.0225(2 1/4%)	0.025(2 1/2%)	0.0275(2 3/4%)	Periods n	Rate i 0.02(2%)	0.0225(2 1/4%)	0.025(2 1/2%)	0.0275(2 3/4%)
1	1.02000000	1.02250000	1.02500000	1.02750000	51	2.74541979	3.11049244	3.52303644	3.98908562
2	1.04040000	1.04550625	1.05062500	1.05575625	52	2.80032819	3.18047852	3.61111235	4.09878547
3	1.06120800	1.06903014	1.07689063	1.08478955	53	2.85633475	3.25203929	3.70139016	4.21150208
4	1.08243216	1.09308332	1.10381289	1.11462126	54	2.91346144	3.32521017	3.79392491	4.32721838
5	1.10408080	1.11767769	1.13140821	1.14527334	55	2.97173067	3.40002740	3.88877303	4.44631964
6	1.12616242	1.14282544	1.15969342	1.17676836	56	3.03116529	3.47652802	3.98599236	4.56859343
7	1.14868567	1.16853901	1.18868575	1.20912949	57	3.09178859	3.55474990	4.08564217	4.69422975
8	1.17165938	1.19483114	1.21840290	1.24238055	58	3.15362436	3.63473177	4.18778322	4.82332107
9	1.19509257	1.22171484	1.24886297	1.27654602	59	3.21669685	3.71651324	4.29247780	4.95596239
10	1.21899442	1.24920343	1.28008454	1.31165103	60	3.28103079	3.80013479	4.39978975	5.09225136
11	1.24337431	1.27731050	1.31208666	1.34772144	61	3.34665140	3.88563782	4.50978449	5.23228827
12	1.26824179	1.30604999	1.34488882	1.38478378	62	3.41358443	3.97306467	4.62252910	5.37617620
13	1.29360663	1.33543611	1.37851104	1.42286533	63	3.48185612	4.06245862	4.73809233	5.52402105
14	1.31947876	1.36548343	1.41297382	1.46199413	64	3.55149324	4.15386394	4.85654464	5.67593162
15	1.34586834	1.39620680	1.44829817	1.50219896	65	3.62252311	4.24732588	4.97795826	5.83201974
16	1.37278571	1.42762146	1.48450562	1.54350944	66	3.69497357	4.34289071	5.10240721	5.99240029
17	1.40024142	1.45974294	1.52161826	1.58595595	67	3.76887304	4.44060576	5.22996739	6.15719130
18	1.42824625	1.49258716	1.55965872	1.62956973	68	3.84425050	4.54051939	5.36071658	6.32651406
19	1.45681117	1.52617037	1.59865019	1.67438290	69	3.92113551	4.64268107	5.49473449	6.50049319
20	1.48594740	1.56050920	1.63861644	1.72042843	70	3.99955822	4.74714140	5.63210286	6.67925676
21	1.51566634	1.59562066	1.67958185	1.76774021	71	4.07954939	4.85395208	5.77290543	6.86293632
22	1.54597967	1.63152212	1.72157140	1.81635307	72	4.16114038	4.96316600	5.91722806	7.05166706
23	1.57689926	1.66823137	1.76461068	1.86630278	73	4.24436318	5.07483723	6.06515876	7.24558791
24	1.60843725	1.70576658	1.80872595	1.91762610	74	4.32925045	5.18902107	6.21678773	7.44484158
25	1.64060599	1.74414632	1.85394410	1.97036082	75	4.41583546	5.30577405	6.37220743	7.64957472
26	1.67341811	1.78338962	1.90029270	2.02454575	76	4.50415216	5.42515396	6.53151261	7.85993802
27	1.70688648	1.82351588	1.94780002	2.08022075	77	4.59423521	5.54721993	6.69480043	8.07608632
28	1.74102421	1.86454499	1.99649602	2.13742682	78	4.68611991	5.67203237	6.86217044	8.29817869
29	1.77584469	1.90649725	2.04640739	2.19620606	79	4.77984231	5.79965310	7.03372470	8.52637861
30	1.81136158	1.94939344	2.09756758	2.25660173	80	4.87543916	5.93014530	7.20956782	8.76085402
31	1.84758882	1.99325479	2.15000677	2.31865828	81	4.97294794	6.06357357	7.38980701	9.00177751
32	1.88454059	2.03810303	2.20375694	2.38242138	82	5.07240690	6.20000397	7.57455219	9.24932639
33	1.92223140	2.08396034	2.25885086	2.44793797	83	5.17385504	6.33950406	7.76391599	9.50368286
34	1.96067603	2.13084945	2.31532213	2.51525626	84	5.27733214	6.48214290	7.95801389	9.76503414
35	1.99988955	2.17879356	2.37320519	2.58442581	85	5.38287878	6.62799112	8.15695424	10.03357258
36	2.03988734	2.22781642	2.43253532	2.65549752	86	5.49053636	6.77712092	8.36088834	10.30949583
37	2.08068509	2.27794229	2.49334870	2.72852370	87	5.60034708	6.92960614	8.56991055	10.59300696
38	2.12229879	2.32919599	2.55568242	2.80355810	88	5.71235402	7.08552228	8.78415832	10.88421465
39	2.16474477	2.38160290	2.61957448	2.88065595	89	5.82660110	7.24494653	9.00376228	11.18363331
40	2.20803966	2.43518897	2.68506384	2.95987399	90	5.94313313	7.40795782	9.22885633	11.49118322
41	2.25220046	2.48998072	2.75219043	3.04127052	91	6.06199579	7.57463688	9.45957774	11.80719076
42	2.29724447	2.54600528	2.82099520	3.12490546	92	6.18323570	7.74506621	9.69606718	12.13188851
43	2.34318936	2.60329040	2.89152008	3.21084036	93	6.30690042	7.91933020	9.93846886	12.46551544
44	2.39005214	2.66186444	2.96380808	3.29913847	94	6.43303843	8.09751512	10.18693058	12.80831711
45	2.43785421	2.72175639	3.03790328	3.38986478	95	6.56169920	8.27970921	10.44160385	13.16054584
46	2.48661129	2.78299590	3.11385086	3.48308606	96	6.69293318	8.46600267	10.70264395	13.52246085
47	2.53634352	2.84561331	3.19169713	3.57887093	97	6.82679184	8.65648773	10.97021004	13.89432852
48	2.58707039	2.90963961	3.27148956	3.67728988	98	6.96332768	8.85125871	11.24446530	14.27642255
49	2.63881179	2.97510650	3.35327680	3.77841535	99	7.10359423	9.05041203	11.52557693	14.66902417
50	2.69158803	3.04204640	3.43710872	3.88232177	100	7.24464612	9.25404630	11.81371635	15.07242234

AMOUNT AT COMPOUND INTEREST $(1 + i)^n$ (Continued)

Periods	Rate i				Periods	Rate i			
n	0.03(3%)	0.0325(3 1/4%)	0.035(3 1/2%)	0.0375(3 3/4%)	n	0.03(3%)	0.0325(3 1/4%)	0.035(3 1/2%)	0.0375(3 3/4%)
1	1.03000000	1.03250000	1.03500000	1.03750000	51	4.51542320	5.10967263	5.78039930	6.53722412
2	1.06090000	1.06605625	1.07122500	1.07640625	52	4.65088590	5.27573700	5.98271327	6.78237003
3	1.09272700	1.10070308	1.10871788	1.11677148	53	4.79041247	5.44719845	6.19210824	7.03670890
4	1.12550881	1.13647593	1.14752300	1.15865042	54	4.93412485	5.62423240	6.40883202	7.30058549
5	1.15927407	1.17341140	1.18768631	1.20209981	55	5.08214859	5.80701995	6.63314114	7.57435744
6	1.19405230	1.21154727	1.22925533	1.24717855	56	5.23461305	5.99574810	6.86530108	7.85839585
7	1.22987387	1.25092255	1.27227926	1.29394774	57	5.39165144	6.19060991	7.10558662	8.15308569
8	1.26677008	1.29157754	1.31680904	1.34247078	58	5.55340098	6.39180473	7.35428215	8.45882640
9	1.30477318	1.33355381	1.36289735	1.39281344	59	5.72000301	6.59953839	7.61168203	8.77603239
10	1.34391638	1.37689430	1.41059876	1.44504394	60	5.89160310	6.81402339	7.87809090	9.10513361
11	1.38423387	1.42164337	1.45996972	1.49923309	61	6.06835120	7.03547915	8.15382408	9.44657612
12	1.42576089	1.46784678	1.51106866	1.55545433	62	6.25040173	7.26413222	8.43920793	9.80082272
13	1.46853371	1.51555180	1.56395606	1.61378387	63	6.43791379	7.50021651	8.73458020	10.16835358
14	1.51258972	1.56480723	1.61869452	1.67430076	64	6.63105120	7.74397355	9.04029051	10.54966684
15	1.55796742	1.61566347	1.67534883	1.73708704	65	6.82998273	7.99565269	9.35670068	10.94527934
16	1.60470644	1.66817253	1.73398604	1.80222781	66	7.03488222	8.25551140	9.68418520	11.35572732
17	1.65284763	1.72238814	1.79467555	1.86981135	67	7.24592868	8.52381552	10.02313168	11.78156709
18	1.70243306	1.77836575	1.85748920	1.93992927	68	7.46330654	8.80083953	10.37394129	12.22337586
19	1.75350605	1.83616264	1.92250132	2.01267662	69	7.68720574	9.08686681	10.73702924	12.68175245
20	1.80611123	1.89583792	1.98978886	2.08815200	70	7.91782191	9.38218999	11.11282526	13.15731817
21	1.86029457	1.95745266	2.05943147	2.16645770	71	8.15535657	9.68711116	11.50177414	13.65071760
22	1.91610341	2.02106987	2.13151158	2.24769986	72	8.40001727	10.00194227	11.90433624	14.16261951
23	1.97358651	2.08675464	2.20611448	2.33198860	73	8.65201778	10.32700540	12.32098801	14.69371774
24	2.03279411	2.15457416	2.28332849	2.41943818	74	8.91157832	10.66263307	12.75222259	15.24473126
25	2.09377793	2.22459782	2.36324498	2.51016711	75	9.17892567	11.00916865	13.19855038	15.81640961
26	2.15659127	2.29689725	2.44595856	2.60429838	76	9.45429344	11.36696663	13.66049964	16.40952497
27	2.22128901	2.37154641	2.53156711	2.70195956	77	9.73792224	11.73639304	14.13861713	17.02488216
28	2.28792768	2.44862167	2.62017196	2.80328305	78	10.03005991	12.11782582	14.63346873	17.66331524
29	2.35656551	2.52820188	2.71187798	2.90840616	79	10.33096171	12.51165516	15.14564013	18.32568956
30	2.42726247	2.61036844	2.80679370	3.01747139	80	10.64089056	12.91828395	15.67573754	19.01290292
31	2.50008035	2.69520541	2.90503148	3.13062657	81	10.96011727	13.33812818	16.22438835	19.72588678
32	2.57508276	2.78279959	2.00670759	3.24802507	82	11.28892079	13.77161734	16.79224195	20.46560754
33	2.65233524	2.87324058	3.11194235	3.36982601	83	11.62758842	14.21919491	17.37997041	21.23306782
34	2.73190530	2.96662089	3.22086033	3.49619448	84	11.97641607	14.68131874	17.98826938	22.02930786
35	2.81386245	3.06303607	3.33359045	3.62730178	85	12.33570855	15.15846160	18.61785881	22.85540691
36	2.89827833	2.16258475	3.45026611	3.76332559	86	12.70577981	15.65111160	19.26948387	23.71248466
37	2.98522668	3.26536875	3.57102543	3.90445030	87	13.08695320	16.15977273	19.94391580	24.60170284
38	3.07478348	3.37149323	3.69601132	4.05086719	88	13.47956180	16.68496534	20.64195285	25.52426670
39	3.16702698	3.48106676	3.82537171	4.20277471	89	13.88394865	17.22722672	21.36442120	26.48142670
40	3.26203779	3.59420143	3.95925972	4.36037876	90	14.30046711	17.78711158	22.11217595	27.47448020
41	3.35989893	3.71101298	4.09783381	4.52389296	91	14.72948112	18.36519271	22.88610210	28.50477321
42	3.46069589	3.83162090	4.24125799	4.69353895	92	15.17136556	18.96206147	23.68711568	29.57370220
43	3.56451677	3.95614858	4.38970202	4.86954666	93	15.62650652	19.57832847	24.51616473	30.68271603
44	3.67145227	4.08472341	4.54334160	5.05215466	94	16.09530172	20.21462415	25.37423049	31.83331789
45	3.78159584	4.21747692	4.70235855	5.24161046	95	16.57816077	20.87159943	26.26232856	33.02706731
46	3.89504372	4.35454492	4.86694110	5.43817085	96	17.07550559	21.54992641	27.18151006	34.26558233
47	4.01189503	4.49606763	5.03728404	5.64210226	97	17.58777076	22.25029902	28.13286291	35.55054167
48	4.13225188	4.64218983	5.21358898	5.85368109	98	18.11540388	22.97343374	29.11751311	36.88368698
49	4.25621944	4.79306100	5.39606459	6.07319413	99	18.65886600	23.72007034	30.13662607	38.26682524
50	4.38390602	4.94883548	5.58492686	6.30093891	100	19.21863198	24.49097262	31.19140798	39.70183119

AMOUNT AT COMPOUND INTEREST $(1 + i)^n$ (Continued)

Periods	Rate i				Periods	Rate i			
n	0.04(4%)	0.0425(4 1/4%)	0.045(4 1/2%)	0.0475(4 3/4%)	n	0.04(4%)	0.0425(4 1/4%)	0.045(4 1/2%)	0.0475(4 3/4%)
1	1.04000000	1.04250000	1.04500000	1.04750000	51	7.39095068	8.35370714	9.43910490	10.66241577
2	1.08160000	1.08680625	1.09202500	1.09725625	52	7.68658871	8.70873969	9.86386463	11.16888052
3	1.12486400	1.13299552	1.14116613	1.14937592	53	7.99405226	9.07886113	10.30773853	11.69940235
4	1.16985856	1.18114783	1.19251860	1.20397128	54	8.31381435	9.46471273	10.77158677	12.25512396
5	1.21665290	1.23134661	1.24618194	1.26115991	55	8.64636692	9.86696302	11.25630817	12.83724235
6	1.26531902	1.28367884	1.30226012	1.32106501	56	8.99222160	10.28630895	11.76284204	13.44701136
7	1.31593178	1.33823519	1.36086183	1.38381560	57	9.35191046	10.72347708	12.29216993	14.08574440
8	1.36856905	1.39511018	1.42210061	1.44954684	58	9.72598688	11.17922485	12.84531758	14.75481726
9	1.42331181	1.45440237	1.48609514	1.51840031	59	10.11502635	11.65434191	13.42335687	15.45567108
10	1.48024428	1.51621447	1.55296942	1.59052433	60	10.51962741	12.14965144	14.02740793	16.18981545
11	1.53945406	1.58065358	1.62285305	1.66607423	61	10.94041250	12.66601163	14.65864129	16.95883169
12	1.60103222	1.64783136	1.69588143	1.74521276	62	11.37802900	13.20431712	15.31828014	17.76437619
13	1.66507351	1.71786419	1.77219610	1.82811037	63	11.83315016	13.76550060	16.00760275	18.60818406
14	1.73167645	1.79087342	1.85194492	1.91494561	64	12.30647617	14.35053438	16.72794487	19.49207281
15	1.80094351	1.86698554	1.93528244	2.00590552	65	12.79873522	14.96043209	17.48070239	20.41794626
16	1.87298125	1.94633243	2.02237015	2.10118604	66	13.31068463	15.59625045	18.26733400	21.38779871
17	1.94790050	2.02905156	2.11337681	2.20099237	67	13.84311201	16.25909109	19.08936403	22.40371915
18	2.02581652	2.11528625	2.20847877	2.30553951	68	14.39683649	16.95010247	19.94838541	23.46789581
19	2.10684918	2.20518591	2.30786031	2.41505264	69	14.97270995	17.67048182	20.84606276	24.58262086
20	2.19112314	2.29890631	2.41171402	2.52976764	70	15.57161835	18.42147730	21.78413558	25.75029535
21	2.27876807	2.39660983	2.52024116	2.64993160	71	16.19448308	19.20439008	22.76442168	26.97343438
22	2.36991879	2.49846575	2.63365201	2.77580335	72	16.84226241	20.02057666	23.78882066	28.25467251
23	2.46471554	2.60465054	2.75216635	2.90765401	73	17.51595290	20.87145117	24.85931759	29.59676946
24	2.56330416	2.71534819	2.87601383	2.04576758	74	18.21659102	21.75848784	25.97798688	31.00261601
25	2.66583633	2.83075049	3.00543446	3.19044154	75	18.94525466	22.68322358	27.14699629	32.47524027
26	2.77246978	2.95105739	3.14067901	3.34198751	76	19.70306485	23.64726058	28.36861112	34.01781418
27	2.88336858	3.07647732	3.28200956	3.50073192	77	20.49118744	24.65226915	29.64519862	35.63366035
28	2.99870332	3.20722761	3.42969999	3.66701668	78	21.31083494	25.69999059	30.97923256	37.32625922
29	3.11865145	3.34353478	3.58403649	3.84119998	79	22.16326834	26.79224019	32.37329802	39.09925653
30	3.24339751	3.48563501	3.74531813	4.02365698	80	23.04979907	27.93091040	33.83009643	40.95647122
31	3.37313341	3.63377450	3.91385745	4.21478068	81	23.97179103	29.11797409	35.35245077	42.90190360
32	3.50805875	3.78820992	4.08998104	4.41498276	82	24.93066267	30.35548799	36.94331106	44.93974402
33	3.64838110	3.94920884	4.27403018	4.62469445	83	25.92788918	31.64559623	38.60576006	47.07438186
34	3.79431634	4.11705021	4.46636154	4.84436743	84	26.96500475	32.99053407	40.34301926	49.31041500
35	3.94608899	4.29202485	4.66734781	5.07447488	85	28.04360494	34.39263177	42.15845513	51.65265971
36	4.10393255	4.47443590	4.87737846	5.31551244	86	29.16534914	35.85431862	44.05558561	54.10616105
37	4.26808986	4.66459943	5.09686049	5.56799928	87	30.33196310	37.37812716	46.03808696	56.67620370
38	4.43881345	4.86284491	5.32621921	5.83247925	88	31.54524163	38.96669757	48.10980087	59.36832338
39	4.61636599	5.06951581	5.56589908	6.10952201	89	32.80705129	40.62278221	50.27474191	62.18831874
40	4.80102063	5.28497024	5.81636454	6.39972431	90	34.11933334	42.34925046	52.53710530	65.14226388
41	4.99306145	5.50958147	6.07810094	6.70371121	91	35.48410668	44.14909360	54.90127503	68.23652141
42	5.19278391	5.74373868	6.35161548	7.02213750	92	36.90347094	46.02543008	57.37183241	71.47775618
43	5.40049527	5.98784758	6.63743818	7.35568903	93	38.37960978	47.98151086	59.95356487	74.87294960
44	5.61651508	6.24233110	6.93612290	7.70508426	94	39.91479417	50.02072507	62.65147529	78.42941470
45	5.84117568	6.50763017	7.24824843	8.07107576	95	41.51138594	52.14660588	65.47079168	82.15481190
46	6.07482271	6.78420445	7.57441961	8.45445186	96	43.17184138	54.36283664	68.41697730	86.05716547
47	6.31781562	7.07253314	7.91526849	8.85603832	97	44.89871503	56.67325719	71.49574128	90.14488083
48	6.57052824	7.37311580	8.27145557	9.27670014	98	46.69466363	59.08187062	74.71304964	94.42676267
49	6.83334937	7.68647322	8.64367107	9.71734340	99	48.56245018	61.59285012	78.07513687	98.91203389
50	7.10668335	8.01314834	9.03263627	10.17891721	100	50.50494818	64.21054625	81.58851803	103.61035550

AMOUNT AT COMPOUND INTEREST $(1 + i)^n$ (Continued)

Periods	Rate *i*				Periods	Rate *i*			
n	0.05(5%)	0.0525(5 1/4%)	0.055(5 1/2%)	0.0575(5 3/4%)	*n*	0.05(5%)	0.0525(5 1/4%)	0.055(5 1/2%)	0.0575(5 3/4%)
1	1.05000000	1.05250000	1.05500000	1.05750000	51	12.04076978	13.59337600	15.34176907	17.31008412
2	1.10250000	1.10775625	1.11302500	1.11830625	52	12.64280826	14.30702824	16.18556637	18.30541396
3	1.15762500	1.16591345	1.17424138	1.18260886	53	13.27494868	15.05814722	17.07577252	19.35797526
4	1.21550625	1.22712391	1.23882465	1.25060887	54	13.93869611	15.84869995	18.01494001	20.47105884
5	1.27628156	1.29154791	1.30696001	1.32251888	55	14.63563092	16.68075670	19.00576171	21.64814472
6	1.34009564	1.35935418	1.37884281	1.39856371	56	15.36741246	17.55649643	20.05107860	22.89291304
7	1.40710042	1.43072027	1.45467916	1.47898113	57	16.13578309	18.47821249	21.15388793	24.20925554
8	1.47745544	1.50583309	1.53468651	1.56402254	58	16.94257224	19.44831864	22.31735176	25.60128774
9	1.55132822	1.58488933	1.61909427	1.65395384	59	17.78970085	20.46935537	23.54480611	27.07336178
10	1.62889463	1.66809602	1.70814446	1.74905618	60	18.67918589	21.54399653	24.83977045	28.63008008
11	1.71033936	1.75567106	1.80209240	1.84962692	61	19.61314519	22.67505635	26.20595782	30.27630969
12	1.79585633	1.84784379	1.90120749	1.95598046	62	20.59380245	23.86549681	27.64728550	32.01719750
13	1.88564914	1.94485559	2.00577390	2.06844934	63	21.62349257	25.11843539	29.16788620	33.85818635
14	1.97993160	2.04696050	2.11609146	2.18738518	64	22.70466720	26.43715325	30.77211994	35.80503207
15	2.07892818	2.15442593	2.23247649	2.31315982	65	23.83990056	27.82510379	32.46458654	37.86382141
16	2.18287459	2.26753329	2.35526270	2.44616651	66	25.03189559	29.28592174	34.25013880	40.04099114
17	2.29201832	2.38657879	2.48480215	2.58682109	67	26.28349037	30.82343263	36.13389643	42.34334813
18	2.40661923	2.51187418	2.62146627	2.73556330	68	27.59766488	32.44166285	38.12126074	44.77809065
19	2.52695020	2.64374757	2.76564691	2.89285819	69	28.97754813	34.14485015	40.21793008	47.35283086
20	2.65329771	2.78254432	2.91775749	3.05919754	70	30.42642554	35.93745478	42.42991623	50.07561864
21	2.78596259	2.92862789	3.07823415	3.23510140	71	31.94774681	37.82417115	44.76356163	52.95496671
22	2.92526072	2.08238086	3.24753703	3.42111973	72	33.54513415	39.80994014	47.22555751	55.99987729
23	3.07152376	3.24420585	3.42615157	3.61783411	73	35.22239086	41.89996200	49.82296318	59.21987024
24	3.22509994	3.41452666	3.61458990	3.82585957	74	36.98351040	44.09971000	52.56322615	62.62501278
25	3.38635494	3.59378931	3.81339235	4.04584650	75	38.83268592	46.41494478	55.45420359	66.22595101
26	3.55567269	3.78246325	4.02312893	4.27848267	76	40.77432022	48.85172938	58.50418479	70.03394320
27	3.73345632	3.98104257	4.24440102	4.52449542	77	42.81303623	51.41644517	61.72191495	74.06089493
28	3.92012914	4.19004731	4.47784307	4.78465391	78	44.95368804	54.11580854	65.11662027	78.31939639
29	4.11613560	4.41002479	4.72412444	5.05977151	79	47.20137244	56.95688849	68.69803439	82.82276168
30	4.32194238	4.64155109	4.98395129	5.35070837	80	49.56144107	59.94712514	72.47642628	87.58507048
31	4.53803949	4.88523252	5.25806861	5.65837410	81	52.03951312	63.09434921	76.46262973	92.62121203
32	4.76494147	5.14170723	5.54726238	5.98373061	82	54.64148878	66.40680254	80.66807436	97.94693172
33	5.00318854	5.41164686	5.85236181	6.32779512	83	57.37356322	69.89315967	85.10481845	103.57888029
34	5.25334797	5.69575832	6.17424171	6.69164334	84	60.24224138	73.56255055	89.78558347	109.53466591
35	5.51601537	5.99478563	6.51382501	7.07641284	85	63.25435344	77.42458446	94.72379056	115.83290920
36	5.79181614	6.30951188	6.87208538	7.48330657	86	66.41707112	81.48937514	99.93359904	122.49330148
37	6.08140694	6.64076125	7.25005008	7.91359670	87	69.73792467	85.76756734	105.42994698	129.53666632
38	6.38547729	6.98940122	7.64880283	8.36862851	88	73.22482091	90.27036462	111.22859407	136.98502463
39	6.70475115	7.35634478	8.06948699	8.84982465	89	76.88606195	95.00955877	117.34616674	144.86166354
40	7.03998871	7.74255288	8.51330877	9.35868957	90	80.73036505	99.99756060	123.80020591	153.19120920
41	7.39198815	8.14903691	8.98154076	9.89681422	91	84.76688330	105.24743253	130.60921724	161.99970373
42	7.76158756	8.57686135	9.47552550	10.46588104	92	89.00522747	110.77292274	137.79272419	171.31468669
43	8.14966693	9.02714657	9.99667940	11.06766920	93	93.45548884	116.58850118	145.37132402	181.16528118
44	8.55715028	9.50107176	10.54649677	11.70406018	94	98.12826328	122.70939750	153.36674684	191.58228484
45	8.98500779	9.99987803	11.12655409	12.37704364	95	103.03467645	129.15164086	161.80191791	202.59826622
46	9.43425818	10.52487163	11.73851456	13.08872365	96	108.18641027	135.93210201	170.70102340	214.24766653
47	9.90597109	11.07742739	12.38413287	13.84132526	97	113.59573078	143.06853737	180.08957969	226.56690736
48	10.40126965	11.65899232	13.06526017	14.63720146	98	119.27551732	150.57963558	198.99450657	239.59450453
49	10.92133313	12.27108942	13.78384948	15.47884054	99	125.23929319	158.48506644	200.44420443	253.37118854
50	11.46739979	12.91532162	14.54196120	16.36887387	100	131.50125785	166.80553243	211.46863567	267.94003188

AMOUNT AT COMPOUND INTEREST $(1+i)^n$ (Continued)

Periods	Rate i				Periods	Rate i			
n	0.06(6%)	0.0625(6 1/4%)	0.065(6 1/2%)	0.0675(6 3/4%)	n	0.06(6%)	0.0625(6 1/4%)	0.065(6 1/2%)	0.0675(6 3/4%)
1	1.06000000	1.06250000	1.06500000	1.06750000	51	19.52536353	22.01789898	24.82161279	27.97448297
2	1.12360000	1.12890625	1.13422500	1.13955625	52	20.69688534	23.39401767	26.43501762	29.86276057
3	1.19101600	1.19946289	1.20794963	1.21647630	53	21.93869846	24.85614377	28.15329377	31.87849691
4	1.26247696	1.27442932	1.28646635	1.29858845	54	23.25502037	26.40965276	29.98325786	34.03029545
5	1.33822558	1.35408115	1.37008666	1.38624317	55	24.65032159	28.06025606	31.93216963	36.32734039
6	1.41851911	1.43871123	1.45914230	1.47981458	56	26.12934089	29.81402206	34.00776065	38.77943587
7	1.50363026	1.52863068	1.55398655	1.57970207	57	27.69710134	31.67739844	36.21826509	41.39704779
8	1.59384807	1.62417009	1.65499567	1.68633195	58	29.35892742	33.65723584	38.57245233	44.19134852
9	1.68947896	1.72568073	1.76257039	1.80015936	59	31.12046307	35.76081308	41.07966173	47.17426454
10	1.79084770	1.83353577	1.87713747	1.92167012	60	32.98769085	37.99586390	43.74983974	50.35852740
11	1.89829856	1.94813176	1.99915140	2.05138285	61	34.96695230	40.37060539	46.59357932	53.75772800
12	2.01219647	2.06988999	2.12909624	2.18985119	62	37.06496944	42.89376823	49.62216198	57.38637464
13	2.13292826	2.19925812	2.26748750	2.33766615	63	39.28886761	45.57462874	52.84760251	61.25995492
14	2.26090396	2.33671175	2.41487418	2.49545861	64	41.64619967	48.42304304	56.28269667	65.39500188
15	2.39655819	2.48275623	2.57184101	2.66390207	65	44.14497165	51.44948323	59.94107195	69.80916451
16	2.54035168	2.63792850	2.73901067	2.84371546	66	46.79366994	54.66507593	63.83724163	74.52128311
17	2.69277279	2.80279903	2.91704637	3.03566625	67	49.60129014	58.08164318	67.98666234	79.55146972
18	2.85433915	2.97797397	3.10665438	3.24057373	68	52.57736755	61.71174587	72.40579539	84.92119393
19	3.02559950	3.16409734	3.30858691	3.45931245	69	55.73200960	65.56872999	77.11217209	90.65337452
20	3.20713547	3.36185342	3.52364506	3.69281604	70	59.07593018	69.66677562	82.12446327	96.77247730
21	3.39956360	3.57196926	3.75268199	3.94208113	71	62.62048599	74.02094909	87.46255339	103.30461952
22	3.60353742	3.79521734	3.99660632	4.20817160	72	66.37771515	78.64725841	93.14671936	110.27768134
23	3.81974966	4.03241843	4.25638573	4.49222319	73	70.36037806	83.56271206	99.20221461	117.72142483
24	4.04893464	4.28444458	4.53305081	4.79544825	74	74.58200074	88.78538156	105.65035856	125.66762100
25	4.29187072	4.55222236	4.82769911	5.11914101	75	79.05692079	94.33446791	112.51763187	134.15018542
26	4.54938296	4.83673626	5.14149955	5.46468303	76	83.80033603	100.23037216	119.83127794	143.20532293
27	4.82234594	5.13903228	5.47569702	5.83354913	77	88.82835620	106.49477042	127.62031101	152.87168223
28	5.11168670	5.46022180	5.83161733	6.22731370	78	94.15805757	113.15069357	135.91563122	163.19052078
29	5.41838790	5.80148566	6.21067245	6.64765737	79	99.80754102	120.22261191	144.75014725	174.20588094
30	5.74349117	6.16407851	6.61436616	7.90637424	80	105.79599348	127.73652516	154.15890683	185.96477790
31	6.08810064	6.54933342	7.04429996	7.57537950	81	112.14375309	135.72005798	164.17923577	198.51740041
32	6.45338668	6.95866676	7.50217946	8.08671762	82	118.87237828	144.20256161	174.85088609	211.91732494
33	6.84058988	7.39358343	7.98982113	8.63257106	83	126.00472097	153.21522171	186.21619369	226.22174437
34	7.25102528	7.85568239	8.50915950	9.21526961	84	133.56500423	162.79117306	198.32024628	241.49171211
35	7.68608679	8.34666254	9.06225487	9.83730031	85	141.57890449	172.96562138	211.21106229	257.79240268
36	8.14725200	8.86832895	9.65130143	10.50131808	86	150.07363875	183.77597272	224.93978134	275.19338986
37	8.63608712	9.42259951	10.27863603	11.21015705	87	159.07805708	195.26197101	239.56086712	293.76894368
38	9.15425235	10.01151198	10.94674737	11.96684265	88	168.62274050	207.46584420	255.13232349	313.59834738
39	9.70350749	10.63723148	11.65828595	12.77460453	89	178.74010493	220.43245946	271.71592451	334.76623582
40	10.28571794	11.30205845	12.41607453	13.63689033	90	189.46451123	234.20948818	289.37745961	357.36295674
41	10.90286101	12.00843710	13.22311938	14.55738043	91	200.83238190	248.84758119	308.18699448	381.48495632
42	11.55703267	12.75896442	14.08262214	15.54000361	92	212.88232482	264.40055501	328.21914912	407.23519087
43	12.25045463	13.55639970	14.99799258	16.58895385	93	225.65526431	280.92558970	349.55339382	434.72356626
44	12.98548191	14.40367468	15.97286209	17.70870824	94	239.19458017	298.48343906	372.27436441	464.06740698
45	13.76461083	15.30390434	17.01109813	18.90404604	95	253.54625498	317.13865400	396.47219810	495.39195695
46	14.59048748	16.26039837	18.11681951	20.18006915	96	268.75903028	336.95981987	422.24289098	528.83091405
47	15.46591673	17.27667326	19.29441278	21.54222382	97	284.88457209	358.01980861	449.68867889	564.52700074
48	16.39387173	18.35646534	20.54854961	22.99632392	98	301.97764642	380.39604665	478.91844302	602.63257329
49	17.37750403	19.50374443	21.88420533	24.54857579	99	320.09630520	404.17079957	510.04814181	643.31027199
50	18.42015427	20.72272845	23.30667868	26.20560466	100	339.30208351	429.43147454	543.20127103	686.73371535

AMOUNT AT COMPOUND INTEREST $(1 + i)^n$ (Continued)

Periods	Rate *i*				Periods	Rate *i*			
n	0.07(7%)	0.0725(7 1/4%)	0.075(7 1/2%)	0.0775(7 3/4%)	*n*	0.07(7%)	0.0725(7 1/4%)	0.075(7 1/2%)	0.0775(7 3/4%)
1	1.07000000	1.07250000	1.07500000	1.07750000	51	31.51901682	35.50277856	39.97897698	45.00711970
2	1.14490000	1.15025625	1.15562500	1.16100625	52	33.72534799	38.07673001	42.97740026	48.49517148
3	1.22504300	1.23364983	1.24229687	1.25098423	53	36.08612235	40.83729294	46.20070528	52.25354726
4	1.31079601	1.32308944	1.33546914	1.34793551	54	38.61215092	43.79799667	49.66575817	56.30319718
5	1.40255173	1.41901343	1.43562933	1.45240051	55	41.31500148	46.97335143	53.39069004	60.66669496
6	1.50073035	1.52189190	1.54330153	1.56496155	56	44.20705159	50.37891941	57.39499179	65.36836382
7	1.60578148	1.63222906	1.65904914	1.68624608	57	47.30154520	54.03139107	61.69961617	70.43441201
8	1.71818618	1.75056567	1.78347783	1.81693015	58	50.61265336	57.94866692	66.32708739	75.89307895
9	1.83845921	1.87748168	1.91723866	1.95774223	59	54.15553910	62.14994527	71.30161894	81.77479256
10	1.96715136	2.01359910	2.06103156	2.10946726	60	57.94642683	66.65581630	76.64924036	88.11233899
11	2.10485195	2.15958504	2.21560893	2.27295097	61	62.00267671	71.48836299	82.39793339	94.94104526
12	2.25219159	2.31615495	2.38177960	2.44910467	62	66.34286408	76.67126930	88.57777839	102.29897627
13	2.40984500	2.48407618	2.56041307	2.63891028	63	70.98686457	82.22993633	95.22111177	110.22714693
14	2.57853415	2.66417171	2.75244405	2.84342583	64	75.95594509	88.19160671	102.36269515	118.76975081
15	2.75903154	2.85732416	2.95887735	3.06379133	65	81.27286124	94.58549820	110.03989729	127.97440650
16	2.95216375	3.06448016	3.18079315	3.30123516	66	86.96196153	101.44294682	118.29288959	137.89242301
17	3.15881521	3.28665497	3.41935264	3.55708088	67	93.04929884	108.79756046	127.16485631	148.57908579
18	3.37993228	3.52493745	3.67580409	3.83275465	68	99.56274976	116.68538360	136.70222053	160.09396494
19	3.61652754	3.78049542	3.95148940	4.12979313	69	106.53214224	125.14507391	146.95488707	172.50124722
20	3.86968446	4.05458134	4.24785110	4.44985210	70	113.98939220	134.21809176	157.97650360	185.87009388
21	4.14056237	4.34853849	4.56643993	4.79471564	71	121.96864965	143.94890342	169.82474137	200.27502616
22	4.43040174	4.66380753	4.90892293	5.16630610	72	130.50645513	154.38519891	182.56159697	215.79634068
23	4.74052986	5.00193357	5.27709215	5.56669482	73	139.64190699	165.57812584	196.25371675	232.52055709
24	5.07236695	5.36457375	5.67287406	5.99811367	74	149.41684047	177.58253996	210.97274550	250.54090026
25	5.42743264	5.75350535	6.09833961	6.46296748	75	159.87601931	190.45727411	226.79570141	269.95782003
26	5.80735292	6.17063449	6.55571508	6.96384746	76	171.06734066	204.26542648	243.80537902	290.87955108
27	6.21386763	6.61800549	7.04739371	7.50354564	77	183.04205451	219.07466990	262.09078245	313.42271629
28	6.64883836	7.09781089	7.57594824	8.08507043	78	195.85499832	234.95758347	281.74759113	337.71297680
29	7.11425705	7.61240218	8.14414436	8.71166339	79	209.56484820	251.99200827	302.87866046	363.88573251
30	7.61225504	8.16430134	8.75495519	9.38681730	80	224.23438758	270.26142887	325.59456000	392.08687677
31	8.14511290	8.75621318	9.41157683	10.11429564	81	239.93079471	289.85538246	350.01415200	422.47360972
32	8.71527080	9.39103864	10.11744509	10.89815355	82	256.72595034	310.86989769	376.26521340	455.21531448
33	9.32533975	10.07188894	10.87625347	11.74276045	83	274.69676686	333.40796527	404.48510440	490.49450135
34	9.97811354	10.80210089	11.69197248	12.65282439	84	293.92554054	357.58004275	434.82148723	528.50782521
35	10.67658148	11.58525320	12.56887042	13.63341828	85	314.50032838	383.50459585	467.43309878	569.46718166
36	11.42394219	12.42518406	13.51153570	14.69000819	86	336.51535137	411.30867905	502.49058119	613.60088824
37	12.22361814	13.32600990	14.52490088	15.82848383	87	360.07142596	441.12855828	540.17737477	661.15495708
38	13.07927141	14.29214562	15.61426844	17.05519132	88	385.27642578	473.11037876	580.69067788	712.39446625
39	13.99482041	15.32832618	16.78533858	18.37696865	89	412.24577558	507.41088122	624.24247872	767.60503738
40	14.97445784	16.43962983	18.04423897	19.80118372	90	441.10297988	544.19817011	671.06066463	827.09442778
41	16.02266989	17.63150299	19.39755689	21.33577546	91	471.98018847	583.65253744	721.39021447	891.19424593
42	17.14425678	18.90978696	20.85237366	22.98929806	92	505.01880166	625.96734640	775.49448056	960.26179999
43	18.34435475	20.28074651	22.41630168	24.77096866	93	540.37011778	671.34997902	833.65656660	1034.68208949
44	19.62845959	21.75110063	24.09752431	26.69071873	94	578.19602602	720.02285250	896.18080910	1114.86995143
45	21.00245176	23.32805543	25.90483863	28.75924943	95	618.66974784	772.22450930	963.39436978	1201.27237266
46	22.47262338	25.01933945	27.84770153	30.98809126	96	661.97663019	828.21078623	1035.64894751	1294.37098155
47	24.04570702	26.83324156	29.93627915	33.38966833	97	708.31499430	888.25606823	1113.32261858	1394.68473262
48	25.72890651	28.77865157	32.18150008	35.97736763	98	757.89704390	952.65463317	1196.82181497	1502.77279939
49	27.52992997	30.86510381	34.59511259	38.76561362	99	810.94983698	1021.72209408	1286.58345109	1619.23769135
50	29.45702506	33.10282284	37.18974603	41.76994868	100	867.71632557	1095.79694590	1383.07720993	1744.72861243

AMOUNT AT COMPOUND INTEREST $(1 + i)^n$ (Continued)

Periods n	Rate i 0.08(8%)	0.0825(8 1/4%)	0.085(8 1/2%)	0.0875(8 3/4%)	Periods n	Rate i 0.08(8%)	0.0825(8 1/4%)	0.085(8 1/2%)	0.0875(8 3/4%)
1	1.08000000	1.08250000	1.08500000	1.08750000	51	50.65374151	56.99321500	64.10865233	72.09290850
2	1.16640000	1.17180625	1.17722500	1.18265625	52	54.70604084	61.69515523	69.55788778	78.40103799
3	1.25971200	1.26848027	1.27728912	1.28613867	53	59.08252410	66.78500554	75.47030824	85.26112882
4	1.36048896	1.37312989	1.38585870	1.39867581	54	63.80912603	72.29476850	81.88528444	92.72147759
5	1.46932808	1.48641310	1.50365669	1.52105994	55	68.91385611	78.25908690	88.84553362	100.83460688
6	1.58687432	1.60904218	1.63146751	1.65415268	56	74.42696460	84.71546157	96.39740398	109.65763498
7	1.71382427	1.74178816	1.77014225	1.79889104	57	80.38112177	91.70448715	104.59118332	119.25267804
8	1.85093021	1.88548569	1.92060434	1.95629401	58	86.81161151	99.27010734	113.48143390	129.68728737
9	1.99900463	2.04103826	2.08385571	2.12746974	59	93.75654043	107.45989119	123.12735578	141.03492502
10	2.15892500	2.20942391	2.26098344	2.31362334	60	101.25706367	116.32533222	133.59318102	153.37548096
11	2.33163900	2.39170139	2.45316703	2.51606538	61	109.35762876	125.92217213	144.94860141	166.79583554
12	2.51817012	2.58901675	2.66168623	2.73622110	62	118.10623906	136.31075133	157.26923253	181.39047115
13	2.71962373	2.80261063	2.88792956	2.97564045	63	127.55473819	147.55638831	170.63711729	197.26213737
14	2.93719362	3.03382601	3.13340357	3.23600898	64	137.75911724	159.72979035	185.14127226	214.52257439
15	3.17216911	3.28411666	3.39974288	3.51915977	65	148.77984662	172.90749805	200.87828041	233.29329965
16	3.42594264	3.55505628	3.68872102	3.82708625	66	160.68223435	187.17236664	217.95293424	253.70646337
17	3.70001805	3.84834842	4.00226231	4.16195630	67	173.53681310	202.61408689	236.47893365	275.90577892
18	3.99601950	4.16583717	4.34245461	4.52612747	68	187.41975815	219.32974905	256.57964301	300.04753457
19	4.31570106	4.50951873	4.71156325	4.92216363	69	202.41333880	237.42445335	278.38891267	326.30169385
20	4.66095714	4.88155403	5.11204612	5.35285295	70	218.60640590	257.01197075	302.05197024	354.85309206
21	5.03383372	5.28428224	5.54657005	5.82122758	71	236.09491837	278.21545834	327.72638771	385.90273762
22	5.43654041	5.72023552	6.01802850	6.33058499	72	254.98251184	301.16823365	355.58313067	419.66922716
23	5.87146365	6.19215495	6.52956092	6.88451118	73	275.38111279	326.01461293	385.80769678	456.39028454
24	6.34118074	6.70300774	7.08457360	7.48690591	74	297.41160181	352.91081850	418.60135100	496.32443443
25	6.84847520	7.25600587	7.68676236	8.14201017	75	321.20452996	382.02596102	454.18246584	539.75282245
26	7.39635321	7.85462636	8.34013716	8.85443606	76	346.90089236	413.64310281	492.78797543	586.98119441
27	7.98806147	8.50263303	9.04904881	9.62919922	77	374.65296374	447.66040879	534.67495335	638.34204892
28	8.62710639	9.20410026	9.81821796	10.47175415	78	404.62520084	484.59239251	580.12232438	694.19697820
29	9.31727490	9.96343853	10.65276649	11.38803264	79	436.99521691	524.57126490	629.43272195	754.93921379
30	10.06265689	10.78542221	11.55825164	12.38448549	80	471.95483426	567.84839425	682.93450332	820.99639500
31	10.86766944	11.67521954	12.54070303	13.46812797	81	509.71122101	614.69588677	740.98393610	892.83357956
32	11.73708300	12.63842515	13.60666279	14.64658917	82	550.48811869	665.40829743	803.96757067	970.95651777
33	12.67604964	13.68109523	14.76322913	15.92816573	83	594.52716818	720.30448197	872.30481418	1055.91521308
34	13.69013361	14.80978558	16.01810360	17.32188023	84	642.08934164	779.72960173	946.45072338	1148.30779422
35	14.78534429	16.03159290	17.37964241	18.83754475	85	693.45648897	844.05729388	1026.89903487	1248.78472622
36	15.96817184	17.35419931	18.85691201	20.48582991	86	748.93300808	913.69202062	1114.18545283	1358.05338976
37	17.24562558	18.78592075	20.45974953	22.27834003	87	808.84764873	989.07161232	1208.89121633	1476.88306137
38	18.62527563	20.33575921	22.19882824	24.22769478	88	873.55546063	1070.67002034	1311.64696971	1606.11032924
39	20.11529768	22.01345935	24.08572865	26.34761807	89	943.43989748	1159.00029702	1423.13696214	1746.64498304
40	21.72452150	23.82956975	26.13301558	28.65303466	90	1018.91508928	1254.61782152	1544.10360392	1899.47641906
41	23.46248322	25.79550925	28.35432190	31.16017519	91	1100.42829642	1358.12379180	1675.35241025	2065.68060573
42	25.33948187	27.92363876	30.76443927	33.88669052	92	1188.46256013	1470.16900462	1817.75736512	2246.42765873
43	27.36664042	30.22733896	33.37941660	36.85177594	93	1283.53956494	1591.45794750	1972.26674116	2442.99007887
44	29.55597166	32.72109442	36.21666702	40.07630633	94	1386.22273014	1722.75322817	2139.90941416	2656.75171077
45	31.92044939	35.42058471	39.29508371	43.58298314	95	1497.12054855	1864.88036950	2321.80171436	2889.21748546
46	34.47408534	38.34278295	42.63516583	47.39649416	96	1616.89019244	2018.73299998	2519.15486008	3142.02401544
47	37.23201217	41.50606255	46.25915492	51.54368740	97	1746.24140783	2185.27847248	2733.28302319	3416.95111679
48	40.21057314	44.93031271	50.19118309	56.05376005	98	1885.94072046	2365.56394646	2965.61208016	3715.93433951
49	43.42741899	48.63706351	54.45743365	60.95846405	99	2036.81597809	2560.72297204	3217.68910698	4041.07859422
50	46.90161251	52.64962124	59.08631551	66.29232966	100	2199.76125634	2771.98261723	3491.19268107	4394.67297121

AMOUNT AT COMPOUND INTEREST $(1 + i)^n$ (Continued)

Periods	Rate i				Periods	Rate i			
n	0.09(9%)	0.0925(9 1/4%)	0.095(9 1/2%)	0.0975(9 3/4%)	n	0.09(9%)	0.0925(9 1/4%)	0.095(9 1/2%)	0.0975(9 3/4%)
1	1.09000000	1.09250000	1.09500000	1.09750000	51	81.04969688	91.09482959	102.35759614	114.98229550
2	1.18810000	1.19355625	1.19902500	1.20450625	52	88.34416960	99.52110133	112.08156778	126.19306931
3	1.29502900	1.30396020	1.31293237	1.32194561	53	96.29514487	108.72680320	122.72931672	138.49689357
4	1.41158161	1.42457652	1.43766095	1.45083531	54	104.96170790	118.78403250	134.38860181	152.00034069
5	1.53862395	1.55634985	1.57423874	1.59229175	55	114.40826162	129.77155551	147.15551898	166.82037391
6	1.67710011	1.70031221	1.72379142	1.74754019	56	124.70500516	141.77542439	161.13529328	183.08536036
7	1.82803912	1.85759109	1.88755161	1.91792536	57	135.92845563	154.88965115	176.44314614	200.93618300
8	1.99256264	2.02941827	2.06686901	2.10492309	58	148.16201663	169.21694388	193.20524502	220.52746084
9	2.17189328	2.21713946	2.26322156	2.31015309	59	161.49659813	184.86951119	211.55974330	242.02888827
10	2.36736367	2.42222486	2.47822761	2.53539301	60	176.03129196	201.96994097	231.65791892	265.62670488
11	2.58042641	2.64628066	2.71365924	2.78259383	61	191.87410824	220.65216051	253.66542121	291.52530861
12	2.81266478	2.89106162	2.97145686	3.05389673	62	209.14277798	241.06248536	277.76363623	319.94902620
13	3.06580461	3.15848482	3.25374527	3.35165166	63	227.96562800	263.36076525	304.15118167	351.14405625
14	3.34172703	3.45064466	3.56285107	3.67843770	64	248.48253452	287.72163604	333.04554393	385.38060173
15	3.64248246	3.76982929	3.90132192	4.03708537	65	270.84596262	314.33588737	364.68487060	422.95521040
16	3.97030588	4.11853850	4.27194750	4.43070120	66	295.22209926	343.41195696	399.32993331	464.19334342
17	4.32763341	4.49950331	4.67778251	4.86269456	67	321.79208819	375.17756297	437.26627697	509.45219440
18	4.71712042	4.91570737	5.12217185	5.33680728	68	350.75337613	409.88148755	478.80657329	559.12378335
19	5.14166125	5.37041030	5.60877818	5.85714599	69	382.32117998	447.79552515	524.29319775	613.63835223
20	5.60441077	5.86717325	6.14161210	6.42821773	70	416.73008618	489.21661122	574.10105153	673.46809157
21	6.10880774	6.40988678	6.72506525	7.05496896	71	454.23579393	534.46914776	628.64065143	739.13123050
22	6.65860043	7.00280131	7.36394645	7.74282843	72	495.11701539	583.90754393	688.36151331	811.19652548
23	7.25787447	7.65056043	8.06352137	8.49775420	73	539.67754677	637.91899174	753.75585708	890.28818671
24	7.91108317	8.35823727	8.82955590	9.32628524	74	588.24852598	696.92649848	825.36266350	977.09128491
25	8.62308066	9.13137421	9.66836371	10.23559805	75	641.19089332	761.39219959	903.77211653	1072.35768519
26	9.39915792	9.97602633	10.58685826	11.23356886	76	698.89807372	831.82097805	989.63046761	1176.91255950
27	10.24508213	10.89880877	11.59260979	12.32884182	77	761.79890036	908.76441852	1083.64536203	1291.66153405
28	11.16713952	11.90694858	12.69390772	13.53090390	78	830.36080139	992.82512724	1186.59167142	1417.59853362
29	12.17218208	13.00834132	13.89982896	14.85016703	79	905.09327351	1084.66145150	1299.31788021	1555.81439065
30	13.26767847	14.21161289	15.22031271	16.29805832	80	986.55166813	1184.99263577	1422.75307883	1707.50629374
31	14.46176953	15.52618708	16.66624241	17.88711900	81	1075.34131826	1294.60445458	1557.91462131	1873.98815738
32	15.76332879	16.96235939	18.24953544	19.63111310	82	1172.12203690	1414.35536663	1705.91651034	2056.70200272
33	17.18202838	18.53137763	19.98324131	21.54514663	83	1277.61302022	1545.18323804	1867.97857882	2257.23044799
34	18.72841093	20.24553006	21.88164924	23.64579843	84	1392.59819204	1688.11268756	2045.43654381	2477.31041666
35	20.41396792	22.11824159	23.96040591	25.95126378	85	1517.93202933	1844.26311116	2239.75301547	2718.84818229
36	22.25122503	24.16417894	26.23664448	28.48151199	86	1654.54591197	2014.85744894	2452.52955194	2983.93588006
37	24.25383528	26.39936549	28.72912570	31.25845941	87	1803.45504404	2201.23176296	2685.51985938	3274.86962837
38	26.43668046	28.84130680	31.45839264	34.30615921	88	1965.76599801	2404.84570104	2940.64424602	3594.16941714
39	28.81598170	31.50912768	34.44693994	37.65100973	89	2142.68493783	2627.29392838	3220.00544939	3944.60093531
40	31.40942005	34.42372199	37.71939924	41.32198318	90	2335.52658223	2870.31861676	3525.90596708	4329.19952650
41	34.23626786	37.60791628	41.30274216	45.35087654	91	2545.72397464	3135.82308881	3860.86703395	4751.29648033
42	37.31753197	41.08664853	45.22650267	49.77258700	92	2774.83913235	3425.88672453	4227.64940218	5214.54788716
43	40.67610984	44.88716352	49.52302042	54.62541423	93	3024.57465426	3742.78124654	4629.27609538	5722.96630616
44	44.33695973	49.03922615	54.22770736	59.95139212	94	3296.78637315	4088.98851185	5069.05732445	6280.95552101
45	48.32728610	53.57535456	59.37933956	65.79665285	95	3593.49714673	4467.21994920	5550.61777027	6893.34868431
46	52.67674185	58.53107486	65.02037682	72.21182650	96	3916.91188994	4880.43779450	6077.92645844	7565.45018103
47	57.41764862	63.94519929	71.19731262	79.25247959	97	4269.43396003	5331.87829049	6655.32947199	8303.08157368
48	62.58523700	69.86013022	77.96105732	86.97959635	98	4653.68301643	5825.07703236	7287.58577183	9112.63202712
49	68.21790833	76.32219227	85.36735777	95.46010699	99	5072.51448791	6363.89665785	7979.90642016	10001.11364976
50	74.35752008	83.28199505	93.47725675	104.76746742	100	5529.04079183	6952.55709870	8737.99753007	10976.22223061

AMOUNT AT COMPOUND INTEREST $(1+i)^n$ (Continued)

Periods n	Rate i 0.10(10%)	0.1025(10 1/4%)	0.105(10 1/2%)	0.1075(10 3/4%)	Periods n	Rate i 0.10(10%)	0.1025(10 1/4%)	0.105(10 1/2%)	0.1075(10 3/4%)
1	1.10000000	1.10250000	1.10500000	1.10750000	51	129.12993817	144.98013678	162.73320545	182.61249057
2	1.21000000	1.21550625	1.22102500	1.22655625	52	142.04293198	159.84060080	179.82019202	202.24333330
3	1.33100000	1.34009564	1.34923262	1.35841105	53	156.24722518	176.22426238	198.70131219	223.98449163
4	1.46410000	1.47745544	1.49090205	1.50444023	54	171.87194770	194.28724927	219.56494997	248.06282448
5	1.61051000	1.62889463	1.64744677	1.66616756	55	189.05914247	214.20169232	242.61926971	274.72957811
6	1.77156100	1.79585633	1.82042868	1.84528057	56	207.96505672	236.15736578	268.09429303	304.26300776
7	1.94871710	1.97993160	2.01157369	2.04364823	57	228.76156239	260.36349578	296.24419380	336.97128109
8	2.14358881	2.18287459	2.22278892	2.26334042	58	251.63771863	287.05075409	327.34983415	373.19569381
9	2.35794769	2.40661923	2.45618176	2.50664951	59	276.80149049	316.47345639	361.72156674	413.31423090
10	2.59374246	2.65329771	2.71408085	2.77611434	60	304.48163954	348.91198567	399.70233124	457.74551072
11	2.85311671	2.92526072	2.99905934	3.07454663	61	334.92980350	384.67546420	441.67107602	506.95315312
12	3.13842838	3.22509994	3.31396057	3.40506039	62	368.42278385	424.10469928	488.04653901	561.45061708
13	3.45227121	3.55567269	3.66192643	3.77110438	63	405.26506223	467.57543095	539.29142560	621.80655842
14	3.79749834	3.92012914	4.04642870	4.17649810	64	445.79156845	515.50191263	595.91702529	688.65076345
15	4.17724817	4.32194238	4.47130371	4.62547165	65	490.37072530	568.34085867	658.48831295	762.68072052
16	4.59497299	4.76494147	4.94079060	5.12270985	66	539.40779783	626.59579669	727.62958581	844.66889797
17	5.05447028	5.25334797	5.45957362	5.67340116	67	593.34857761	690.82186585	804.03069231	935.47080451
18	5.55991731	5.79181614	6.03282885	6.28329179	68	652.68343537	761.63110709	888.45391501	1036.03391599
19	6.11590904	6.38547729	6.66627588	6.95874565	69	717.95177891	839.69829557	981.74157608	1147.40756196
20	6.72749995	7.03998871	7.36623484	7.70681081	70	789.74695680	925.76737087	1084.82444157	1270.75387487
21	7.40024994	7.76158756	8.13968950	8.53529297	71	868.72165248	1020.65852638	1198.73100794	1407.35991642
22	8.14027494	8.55715028	8.99435690	9.45283697	72	955.59381773	1125.27602534	1324.59776377	1558.65110743
23	8.95430243	9.43425818	9.93876437	10.46901694	73	1051.15319950	1240.61681793	1463.68052897	1726.20610148
24	9.84973268	10.40126965	10.98233463	11.59443626	74	1156.26851945	1367.78004177	1617.36698451	1911.77325739
25	10.83470594	11.46739979	12.13547977	12.84083816	75	1271.89537140	1507.97749605	1787.19051788	2117.28888256
26	11.91817654	12.64280826	13.40970514	14.22122826	76	1399.08490853	1662.54518940	1974.84552226	2344.89743744
27	13.10999419	13.93869611	14.81772418	15.75001030	77	1538.99339939	1832.95607131	2182.20430210	2596.97391196
28	14.42099361	15.36741246	16.37358522	17.44313641	78	1692.89273933	2020.83406862	2411.33575382	2876.14860750
29	15.86309297	16.94257224	18.09281167	19.31827357	79	1862.18201326	2227.96956065	2664.52600797	3185.33458280
30	17.44940227	18.67918589	19.99255690	21.39498798	80	2048.40021459	2456.33644062	2944.30123880	3527.75805046
31	19.19434250	20.59380245	22.09177537	23.69494919	81	2253.24023604	2708.11092579	3253.45286888	3906.99204088
32	21.11377675	22.70466720	24.41141178	26.24215623	82	2478.56425965	2985.69229568	3595.06542011	4326.99368527
33	23.22515442	25.03189559	26.97461002	29.06318802	83	2726.42068561	3291.72575599	3972.54728922	4792.14550644
34	25.54766986	27.59766488	29.80694407	32.18748073	84	2999.06275417	3629.12764597	4389.66475459	5307.30114838
35	28.10243685	30.42642554	32.93667320	35.64763491	85	3298.96902959	4001.11322969	4850.57955382	5877.83602184
36	30.91268053	33.54513415	36.39502389	39.47975567	86	3628.86593255	4411.22733573	5359.89040697	6509.70339418
37	34.00394859	36.98351040	40.21650140	43.72382940	87	3991.75252581	4863.37813764	5922.67889971	7209.49650906
38	37.40434344	40.77432022	44.43923404	48.42414106	88	4390.92777839	5361.87439675	6544.56018417	7984.51738378
39	41.14477779	44.95368804	49.10535362	53.62973622	89	4830.02055623	5911.46652242	7231.73900351	8842.85300254
40	45.25925557	49.56144107	54.26141575	59.39493287	90	5313.02261185	6517.39184097	7991.07159888	9793.45970031
41	49.78518112	54.64148878	59.95886440	65.77988815	91	5844.32487303	7185.42450466	8830.13411676	10846.25661809
42	54.76369924	60.24224138	66.25454516	72.85122613	92	6428.75736034	7921.93051639	9757.29819902	12012.22920454
43	60.24006916	66.41707112	73.21127240	80.68273294	93	7071.63309637	8733.92839432	10781.81450992	13303.54384403
44	66.26407608	73.22482091	80.89845601	89.35612673	94	7778.79640601	9629.15605474	11913.90503346	14733.67480726
45	72.89048369	80.73036505	89.39279389	98.96191035	95	8556.67604661	10616.14455035	13164.86506198	16317.54484904
46	80.17953205	89.00522747	98.77903724	109.60031571	96	9412.34365127	11704.29936676	14547.17589348	18071.68092031
47	88.19748526	98.12826328	109.15083616	121.38234965	97	10353.57801640	12903.99005186	16074.62936230	20014.38661925
48	97.01723378	108.18641027	120.61167395	134.43095224	98	11388.93581803	14226.64903217	17762.46544534	22165.93318081
49	106.71895716	119.27551732	133.27589972	148.88227961	99	12527.82939984	15684.88055797	19627.52431710	24548.77099775
50	117.39085288	131.50125785	147.26986919	164.88712466	100	13780.61233982	17292.58081516	21688.41437040	27187.76388001

AMOUNT AT COMPOUND INTEREST $(1 + i)^n$ (Continued)

Periods	Rate i				Periods	Rate i			
n	0.11(11%)	0.1125(11 1/4%)	0.115(11 1/2%)	0.1175(11 3/4%)	n	0.11(11%)	0.1125(11 1/4%)	0.115(11 1/2%)	0.1175(11 3/4%)
1	1.11000000	1.11250000	1.11500000	1.11750000	51	204.86695768	229.77406223	257.64293424	288.81791071
2	1.23210000	1.23765625	1.24322500	1.24880625	52	227.40232303	255.62364423	287.27187167	322.75401522
3	1.36763100	1.37689258	1.38619587	1.39554098	53	252.41657856	284.38130420	320.30813692	360.67761200
4	1.51807041	1.53179299	1.54560840	1.55951705	54	280.18240220	316.37420093	357.14357266	403.05723142
5	1.68505816	1.70411970	1.72335337	1.74276030	55	311.00246644	351.96629853	398.21508352	450.41645611
6	1.87041455	1.89583317	1.92153900	1.94753464	56	345.21273775	391.56250712	444.00981812	503.34038970
7	2.07616015	2.10911440	2.14251599	2.17636996	57	383.18613890	435.61328917	495.07094721	562.48288549
8	2.30453777	2.34638977	2.38890533	2.43209343	58	425.33661418	484.61978420	552.00410614	628.57462453
9	2.55803692	2.61035862	2.66362944	2.71786441	59	472.12364174	539.13950992	615.48457834	702.43214292
10	2.83942099	2.90402397	2.96994683	3.03721348	60	524.05724234	599.79270479	686.26530485	784.96791971
11	3.15175729	3.23072667	3.31149071	3.39408606	61	581.70353899	667.26938407	765.18581491	877.20165028
12	3.49845060	3.59418341	3.69231214	3.79289117	62	645.69092828	742.33718978	853.18218362	980.27284418
13	3.88328016	3.99852905	4.11692804	4.23855588	63	716.71693039	825.85012363	951.29813474	1095.45490337
14	4.31044098	4.44836357	4.59037476	4.73658620	64	795.55579274	918.75826254	1060.69742023	1224.17085452
15	4.78458949	4.94880447	5.11826786	5.29313508	65	883.06692994	1022.11856708	1182.67762356	1368.01092993
16	5.31089433	5.50554497	5.70686867	5.91507845	66	980.20429223	1137.10690587	1318.68555027	1528.75221419
17	5.89509271	6.12491878	6.36315856	6.61010017	67	1088.02676438	1265.03143278	1470.33438855	1708.38059936
18	6.54355291	6.81397214	7.09492180	7.38678694	68	1207.70970846	1407.34746897	1639.42284324	1909.11531979
19	7.26334373	7.58054401	7.91083780	8.25473440	69	1340.55777639	1565.67405923	1827.95647021	2133.43636986
20	8.06231154	8.43335521	8.82058415	9.22466569	70	1488.01913179	1741.81239089	2038.17146428	2384.11514332
21	8.94916581	9.38210767	9.83495133	10.30856391	71	1651.70123629	1937.76628487	2272.56118267	2664.24867266
22	9.93357404	10.43759478	10.96597073	11.51982017	72	1833.38837228	2155.76499192	2533.90571868	2977.29789170
23	11.02626719	11.61182420	12.22705737	12.87339904	73	2035.06109323	2398.28855351	2825.30487633	3327.13039397
24	12.23915658	12.91815442	13.63316896	14.38602343	74	2258.91781349	2668.09601578	3150.21493711	3718.06821526
25	13.58546380	14.37144679	15.20098340	16.07638118	75	2507.39877297	2968.25681755	3512.48965487	4154.94123056
26	15.07986482	15.98823456	16.94909649	17.96535597	76	2783.21263800	3302.18570953	3916.42596518	4643.14682515
27	16.73864995	17.78691094	18.89824258	20.07628530	77	3089.36602818	3673.68160185	4366.81495118	5188.71657710
28	18.57990145	19.78793843	21.07154048	22.43524882	78	3429.19629128	4086.97078206	4868.99867057	5798.39077491
29	20.62369061	22.01408150	23.49476763	25.07139056	79	3806.40788332	4546.75499504	5428.93351768	6479.70169096
30	22.89229657	24.49066567	26.19666591	28.01727895	80	4225.11275048	5058.26493198	6053.26087222	7241.06663965
31	25.41044919	27.24586555	29.20928249	31.30930923	81	4689.87515303	5627.31973683	6749.38587252	8091.89196981
32	28.20559861	30.31102543	32.56834998	34.98815306	82	5205.76141987	6260.39320722	7525.56524786	9042.68927626
33	31.30821445	33.72101579	36.31371022	39.09926104	83	5778.39517605	6964.68744304	8391.00525136	10105.20526622
34	34.75211804	37.51463007	40.48978690	43.69342422	84	6414.01864542	7748.21478038	9355.97085527	11292.56688500
35	38.57485103	41.73502595	45.14611239	48.82740156	85	7119.56069641	8619.88894317	10431.90750363	12619.44349399
36	42.81808464	46.43021637	50.33791532	54.56462125	86	7902.71237302	9589.62644928	11631.57686654	14102.22810454
37	47.52807395	51.65361571	56.12677558	60.97596424	87	8772.01073405	10668.45942482	12969.20820620	15759.23990682
38	52.75616209	57.46464748	62.58135477	68.14064004	88	9736.93191480	11868.66111011	14460.66714991	17610.95059587
39	58.55933991	63.92942032	69.77821057	76.14716525	89	10807.99442542	13203.88548500	16123.64387215	19680.23729088
40	65.00086731	71.12148010	77.80270479	85.09445716	90	11996.87381222	14689.32260206	17977.86291745	21992.66517256
41	72.15096271	79.12264661	86.75001584	95.09305588	91	13316.52993157	16341.87139480	20045.31715295	24576.80333034
42	80.08756861	88.02394436	96.72626766	106.26648994	92	14781.34822404	18180.33192671	22350.52862554	27464.57772166
43	88.89720115	97.92663810	107.84978844	118.75280251	93	16407.29652868	20225.61926847	24920.83941748	30691.66560395
44	98.67589328	108.94338488	120.25251411	132.70625681	94	18212.09914684	22501.00143617	27786.73595049	34297.93631241
45	109.53024154	121.19951568	134.08155323	148.29924198	95	20215.43005299	25032.36409774	30982.21058480	38327.94382912
46	121.57856811	134.83446120	149.50093186	165.72440292	96	22439.12735882	27848.50505873	34545.16480205	42831.47722904
47	134.95221060	150.00333808	166.69353902	185.19702026	97	24907.43136829	30981.46187784	38517.85875428	47864.17580346
48	149.79695377	166.87871362	185.86329601	206.95767014	98	27647.24881880	34466.87633910	42947.41251102	53488.21646036
49	166.27461868	185.65256890	207.23757505	231.27519638	99	30688.44618887	38344.39992725	47886.36494979	59773.08189446
50	184.56482674	206.53848290	231.06989618	258.45003195	100	34064.17526964	42658.14491906	53393.29691902	66796.41901705

AMOUNT AT COMPOUND INTEREST $(1 + i)^t$ (Continued)

Periods n	Rate i 0.12(12%)	0.1225(12 1/4%)	0.125(12 1/2%)	0.1275(12 3/4%)
1	1.12000000	1.12250000	1.12500000	1.12750000
2	1.25440000	1.26000625	1.26562500	1.27125625
3	1.40492800	1.41435702	1.42382812	1.43334142
4	1.57351936	1.58761575	1.60180664	1.61609245
5	1.76234168	1.78209868	1.80203247	1.82214424
6	1.97382269	2.00040577	2.02728653	2.05446763
7	2.21068141	2.24545547	2.28069735	2.31641225
8	2.47596318	2.52052377	2.56578451	2.61175482
9	2.77307876	2.82928793	2.88650758	2.94475356
10	3.10584821	3.17587570	3.24732103	3.32020963
11	3.47854999	3.56492048	3.65323615	3.74353636
12	3.89597599	4.00162324	4.10989067	4.22083725
13	4.36349311	4.49182208	4.62362701	4.75899400
14	4.88711229	5.04207029	5.20158038	5.36576573
15	5.47356576	5.65972390	5.85177793	6.04990086
16	6.13039365	6.35304007	6.58325017	6.82126323
17	6.86604089	7.13128748	7.40615644	7.69097429
18	7.68996580	8.00487020	8.33192600	8.67157351
19	8.61276169	8.98546680	9.37341675	9.77719913
20	9.64629309	10.08618648	10.54509384	11.02379202
21	10.80384826	11.32174433	11.86323057	12.42932550
22	12.10031006	12.70865801	13.34613439	14.01406450
23	13.55234726	14.26546861	15.01440119	15.80085773
24	15.17862893	16.01298852	16.89120134	17.81546709
25	17.00006441	17.97457961	19.00260151	20.08693914
26	19.04007214	20.17646561	21.37792670	22.64802388
27	21.32488079	22.64808265	24.05016754	25.53564693
28	23.88386649	25.42247277	27.05643848	28.79144191
29	26.74993047	28.53672569	30.43849329	32.46235075
30	29.95992212	32.03247459	34.24330495	36.60130047
31	33.55511278	35.95645272	38.52371807	41.26796628
32	37.58172631	40.36111818	43.33918283	46.52963199
33	42.09153347	45.30535516	48.75658068	52.46216006
34	47.14251748	50.85526117	54.85115327	59.15108547
35	52.79961958	57.08503066	61.70754742	66.69284887
36	59.13557393	64.07794691	69.42099085	75.19618710
37	66.23184280	71.92749541	78.09861471	84.78370095
38	74.17966394	80.73861360	87.86094155	95.59362283
39	83.08122361	90.62909377	98.84355924	107.78180974
40	93.05097044	101.73115775	111.19900415	121.52399048
41	104.21708689	114.19322458	125.09887966	137.01829926
42	116.72313732	128.18189459	140.73623962	154.48813242
43	130.72991380	143.88417667	158.32826958	174.18536930
44	146.41750346	161.50998832	178.11930327	196.39400389
45	163.98760387	181.29496188	200.38421618	221.43423939
46	183.66611634	203.50359472	225.43224320	249.66710491
47	205.70605030	228.43278507	253.61127360	281.49966078
48	230.39077633	256.41580124	285.31268280	317.39086753
49	258.03766949	287.82673689	320.97676816	357.85820315
50	289.00218983	323.08551216	361.09886417	403.48512405

Periods n	Rate i 0.12(12%)	0.1225(12 1/4%)	0.125(12 1/2%)	0.1275(12 3/4%)
51	323.68245261	362.66348740	406.23622220	454.92947736
52	362.52434692	407.08976461	457.01574997	512.93298573
53	406.02726855	456.95826077	514.14271872	578.33194141
54	454.75054078	512.93564771	578.41055856	652.06926394
55	509.32060567	575.77026456	650.71187838	735.20809509
56	570.43907835	646.30212197	732.05086317	828.94712721
57	638.89176776	725.47413191	823.55722107	934.63788593
58	715.55877989	814.34471307	926.50187370	1053.80421639
59	801.42583347	914.10194042	1042.31460792	1188.16425397
60	897.59693349	1026.07942812	1172.60393391	1339.65519636
61	1005.30856551	1151.77415807	1319.17942565	1510.46123389
62	1125.94559337	1292.86649243	1484.07685385	1703.04504121
63	1261.05906458	1451.24263775	1669.58646058	1920.18328397
64	1412.38615232	1629.01986088	1878.28476816	2165.00665267
65	1581.87249060	1828.57479383	2113.07036417	2441.04500089
66	1771.69718948	2052.57520608	2377.20415970	2752.27823850
67	1984.30085221	2304.01566882	2674.35467966	3103.19371391
68	2222.41695448	2586.25758825	3008.64901462	3498.85091244
69	2489.10698902	2903.07414281	3384.73014144	3944.95440377
70	2787.79982770	3258.70072531	3807.82140912	4447.93609025
71	3122.33580702	3657.89156416	4283.79908526	5015.04794176
72	3497.01610386	4105.98328077	4819.27397092	5654.46655433
73	3916.65803633	4608.96623266	5421.68321729	6375.41104001
74	4386.65700069	5173.56459616	6099.39361945	7188.27594761
75	4913.05584077	5807.32625919	6861.81782188	8104.78113094
76	5502.62254166	6518.72372594	7719.54504961	9138.14072513
77	6162.93724666	7317.26738237	8684.48818082	10303.25366758
78	6902.48971626	8213.63263671	9770.04920342	11616.91851020
79	7730.78848221	9219.80263471	10991.30535384	13098.07562025
80	8658.48310008	10349.22845746	12365.21852308	14768.08026183
81	9697.50107209	11617.00894350	13910.87083846	16651.01049522
82	10861.20120074	13040.09253908	15649.72969327	18774.01433336
83	12164.54534483	14637.50387511	17605.94590493	21167.70116086
84	13624.29078621	16430.59809981	19806.68914304	23866.58305887
85	15259.20568055	18443.34636704	22282.52528592	26909.57239887
86	17090.31036222	20702.65629700	25067.84094666	30340.54287973
87	19141.14760569	23238.73169339	28201.32106499	34208.96209690
88	21438.08531837	26085.47632583	31726.48619812	38570.60476425
89	24010.65555657	29280.94717574	35692.29697288	43488.35687169
90	26891.93422336	32867.86320477	40153.83409449	49033.12237283
91	30118.96633016	36894.17644735	45173.06335631	55284.84547537
92	33733.24228978	41413.71306215	50819.69627584	62333.66327348
93	37781.23136456	46486.89291226	57172.15831032	70281.20534085
94	42314.97912830	52181.53729402	64318.67809911	79242.05902181
95	47392.77662370	58573.77561253	72358.51286150	89345.42154709
96	53079.90981854	65749.06312507	81403.32696919	100736.96279434
97	59449.49899677	73803.32335789	91578.74284034	113580.92555062
98	66583.43887638	82844.23046923	103026.08569538	128062.49355832
99	74573.45154155	92992.64870171	115904.34640731	144390.46148701
100	83522.26572653	104384.24816767	130392.38970822	162800.24532660

AMOUNT AT COMPOUND INTEREST $(1 + i)^n$ (Continued)

Periods n	Rate i 0.13(13%)	0.1325(13 1/4%)	0.135(13 1/2%)	0.1375(13 3/4%)	Periods n	Rate i 0.13(13%)	0.1325(13 1/4%)	0.135(13 1/2%)	0.1375(13 3/4%)
1	1.13000000	1.13250000	1.13500000	1.13750000	51	509.33159543	570.09698594	637.95337300	713.70981916
2	1.27690000	1.28255625	1.28822500	1.29390625	52	575.54470283	645.63483658	724.07707835	811.84491930
3	1.44289700	1.45249495	1.46213537	1.47181836	53	650.36551420	731.18145242	821.82748393	923.47359570
4	1.63047361	1.64495053	1.65952365	1.67419338	54	734.91303105	828.06299487	932.77419426	1050.45121511
5	1.84243518	1.86290648	1.88355934	1.90439497	55	830.45172509	937.78134169	1058.69871049	1194.88825719
6	2.08195175	2.10974159	2.13783985	2.16624928	56	938.41044935	1062.03736946	1201.62303641	1359.18539255
7	2.35260548	2.38928235	2.42644824	2.46410856	57	1060.40380776	1202.75732092	1363.84214632	1546.07338403
8	2.65844419	2.70586226	2.75401875	2.80292349	58	1198.25630277	1362.12266594	1547.96083607	1758.65847433
9	3.00404194	3.06438901	3.12581128	3.18832547	59	1354.02962213	1542.60391918	1756.93554894	2000.47401455
10	3.39456739	3.47042055	3.54779580	3.62672022	60	1530.05347301	1746.99893847	1994.12184805	2275.53919155
11	3.83586115	3.93025128	4.02674823	4.12539425	61	1728.96042450	1978.47629782	2263.32829754	2588.42583039
12	4.33452310	4.45100957	4.57035924	4.69263596	62	1953.72527969	2240.62440728	2568.87761770	2944.33438207
13	4.89801110	5.04076834	5.18735774	5.33787340	63	2207.70956604	2537.50714124	2915.67609609	3349.18035960
14	5.53475255	5.70867015	5.88765104	6.07183099	64	2494.71180963	2873.72683746	3309.29236907	3809.69265905
15	6.25427038	6.46506894	6.68248393	6.90670775	65	2819.02434488	3254.49564342	3756.04683889	4333.52539966
16	7.06732553	7.32169057	7.58461926	7.85638007	66	3185.49750972	3685.71631617	4263.11316214	4929.38514212
17	7.98607785	8.29181457	8.60854286	8.93663233	67	3599.61218598	4174.07372806	4838.63343903	5607.17559916
18	9.02426797	9.39048001	9.77069614	10.16541928	68	4067.56177016	4727.13849703	5491.84895330	6378.16224404
19	10.19742280	10.63471861	11.08974012	11.56316443	69	4596.34480028	5353.48434789	6233.24856200	7255.15955260
20	11.52308776	12.04381882	12.58685504	13.15309953	70	5193.86962431	6062.82102398	7074.73711786	8252.74399108
21	13.02108917	13.63962482	14.28608047	14.96165072	71	5869.07267547	6866.14480966	8029.82662878	9387.49628986
22	14.71383077	15.44687510	16.21470134	17.01887769	72	6632.05212329	7775.90899694	9113.85322366	10678.27702971
23	16.62662877	17.49358606	18.40368602	19.35897338	73	7494.21889931	8806.21693904	10344.22340886	12146.54012130
24	18.78809051	19.81148621	20.88818363	22.02083222	74	8468.46735622	9973.04068346	11740.69356905	13816.68938798
25	21.23054227	22.43650813	23.70808842	25.04869665	75	9569.36811253	11294.46857402	13325.68720087	15716.48417882
26	23.99051277	25.40934546	26.90868035	28.49289243	76	10813.38596716	12790.98566008	15124.65497299	17877.50075341
27	27.10927943	28.77608373	30.54135220	32.41066514	77	12219.12614289	14485.79126004	17166.48339435	20335.65710701
28	30.63348575	32.58891482	34.66443475	36.86713160	78	13807.61254147	16405.15860199	19483.95865258	23131.80995922
29	34.61583890	36.90694604	39.34413344	41.93636220	79	15602.60217186	18578.84211675	22114.29307068	26312.43382861
30	39.11589796	41.79711639	44.65559145	47.70261200	80	17630.94045420	21040.53869722	25099.72263522	29930.39348005
31	44.20096469	47.33523431	50.68409630	54.26172115	81	19922.96271325	23828.41007461	28488.18519098	34045.82258355
32	49.94709010	53.60715286	57.52644930	61.72270781	82	22512.94786597	26985.67440949	32334.09019176	38727.12318879
33	56.44021181	60.71010061	65.29251996	70.20958013	83	25439.63108855	30561.27626875	36699.19236765	44052.10262725
34	63.77743935	68.75418894	74.10701015	79.86339740	84	28746.78313006	34610.64537436	41653.58333728	50109.26673850
35	72.06850647	77.86411898	84.11145652	90.84461454	85	32483.86493697	39196.55588646	47276.81708781	56999.29091504
36	81.43741231	88.18111474	95.46650315	103.33574904	86	36706.76737878	44390.09954142	53659.18739467	64836.69341586
37	92.02427591	99.86511244	108.35448108	117.54441453	87	41478.64713802	50271.78773066	60903.17769295	73751.73876054
38	103.98743178	113.09723984	122.98233602	133.70677153	88	46870.87126596	56932.79960497	69125.10668149	83892.60284011
39	117.50579791	128.08262412	139.58495138	152.09145262	89	52964.08453053	64476.39555263	78456.99608349	95427.83573063
40	132.78155163	145.05357182	158.42891982	173.00402735	90	59849.41551950	73019.51796335	89048.69055477	108549.16314359
41	150.04315335	164.27317008	179.81682400	196.79208111	91	67629.83953704	82694.60409349	101070.26377966	123474.67307583
42	169.54876328	186.03936512	204.09209523	223.85099227	92	76421.71867685	93651.63913588	114714.74938991	140452.44062376
43	191.59010251	210.68958100	231.64452809	254.63050370	93	86356.54210484	106060.48132139	130201.24055755	159764.65120953
44	216.49681583	238.60595048	262.91653938	289.64219796	94	97582.89257847	120113.49509647	147778.40803282	181732.29075084
45	244.64140189	270.22123892	298.41027220	329.46800018	95	110268.66861367	136028.53319675	167728.49311725	206720.48072908
46	276.44478414	306.02555308	338.69565895	374.76985021	96	124603.59553345	154052.31384532	190371.83968808	235144.54682932
47	312.38260608	346.57393886	384.41957291	426.30070461	97	140802.06295280	174464.24542983	216072.03804597	267476.92201836
48	352.99234487	392.49498576	436.31621525	484.91705149	98	159106.33113666	197580.75794928	245241.76318218	304254.99879588
49	398.88134970	444.50057137	495.21890431	551.59314607	99	179790.15418443	223760.20837756	278349.40121177	346090.06113031
50	450.73592516	503.39689708	562.07345639	627.43720366	100	203162.87422841	253408.43598758	315926.57037536	393677.44453573

AMOUNT AT COMPOUND INTEREST $(1 + i)^n$ (Continued)

Periods n	Rate i 0.14(14%)	0.1425(14 1/4%)	0.145(14 1/2%)	0.1475(14 3/4%)
1	1.14000000	1.14250000	1.14500000	1.14750000
2	1.29960000	1.30530625	1.31102500	1.31675625
3	1.48154400	1.49131239	1.50112362	1.51097780
4	1.68896016	1.70382441	1.71878655	1.73384702
5	1.92541458	1.94661938	1.96801060	1.98958946
6	2.19497262	2.22401265	2.25337214	2.28305390
7	2.50226879	2.54093445	2.58011110	2.61980435
8	2.85258642	2.90301761	2.95422721	3.00622550
9	3.25194852	3.31669762	3.38259015	3.44964376
10	3.70722131	3.78932703	3.87306572	3.95846621
11	4.22623230	4.32930613	4.43466025	4.54233998
12	4.81790482	4.94623225	5.07768599	5.21233512
13	5.49241149	5.65107035	5.81395046	5.98115455
14	6.26134910	6.45634787	6.65697328	6.86337485
15	7.13793798	7.37637744	7.62223440	7.87572264
16	8.13724930	8.42751123	8.72745839	9.03739173
17	9.27646420	9.62843158	9.99293985	10.37040701
18	10.57516918	11.00048308	11.44191613	11.90004204
19	12.05569287	12.56805192	13.10099397	13.65529824
20	13.74348987	14.35899932	15.00063810	15.66945474
21	15.66757845	16.40515672	17.17573062	17.98069931
22	17.86103944	18.74289155	19.66621156	20.63285246
23	20.36158496	21.41375360	22.51781224	23.67619819
24	23.21220685	24.46521349	25.78289502	27.16843743
25	26.46191581	27.95150641	29.52141479	31.17578195
26	30.16658403	31.93459607	33.80201994	35.77420979
27	34.38990579	36.48527601	38.70331283	41.05090573
28	39.20449260	41.68442784	44.31529319	47.10591432
29	44.69312156	47.62445881	50.74101070	54.05403669
30	50.95015858	54.41094419	58.09845725	62.02700710
31	58.08318078	62.16450374	66.52273355	71.17599065
32	66.21482609	71.02294552	76.16852992	81.67444927
33	75.48490175	81.14371526	87.21296676	93.72143053
34	86.05278799	92.70669468	99.85884694	107.54534154
35	98.10017831	105.91739867	114.33837974	123.40827941
36	111.83420328	121.01062798	130.91744481	141.61100063
37	127.49099173	138.25464247	149.90047430	162.49862322
38	145.33973058	157.95592902	171.63604308	186.46717015
39	165.68729286	180.46464891	196.52326932	213.97107774
40	188.88351386	206.18086138	225.01914337	245.53181171
41	215.32720580	235.56163412	257.64691916	281.74775394
42	245.47301461	269.12916699	295.00572244	323.30554764
43	279.83923665	307.48007328	337.78155220	370.99311592
44	319.01672979	351.29598372	386.75987726	425.71460052
45	363.67907196	401.35566140	442.84005947	488.50750409
46	414.59414203	458.54884315	507.05186809	560.56236095
47	472.63732191	523.89205330	580.57438896	643.24530919
48	538.80654698	598.54667090	664.75767536	738.12399229
49	614.23946356	683.83957150	761.14753829	846.99728115
50	700.23298846	781.28671044	871.51393134	971.92938012

Periods n	Rate i 0.14(14%)	0.1425(14 1/4%)	0.145(14 1/2%)	0.1475(14 3/4%)
51	798.26560684	892.62006668	997.88345139	1115.28896369
52	910.02279180	1019.81842618	1142.57655184	1279.79408584
53	1037.42598265	1165.14255191	1308.25015186	1468.56371350
54	1182.66562023	1331.17536556	1497.94642388	1685.17686124
55	1348.23880706	1520.86785515	1715.14865534	1933.74044827
56	1536.99224004	1737.59152451	1963.84521036	2218.96716439
57	1752.17115365	1985.19831675	2248.60276586	2546.26482114
58	1997.47511516	2268.08907689	2574.65016692	2921.83888226
59	2277.12163128	2591.29177035	2947.97444112	3352.81011739
60	2595.91865966	2960.55084762	3375.43073508	3847.34960971
61	2959.34727202	3382.42934341	3864.86819167	4414.83367714
62	3373.65589010	3864.42552484	4425.27407946	5066.02164452
63	3845.96771471	4415.10616213	5066.93882098	5813.25983708
64	4384.40319477	5044.25879024	5801.64495002	6670.71566305
65	4998.21964204	5763.06566784	6642.88346778	7654.64622335
66	5697.97039193	6584.30252551	7606.10157060	8783.70654130
67	6495.68624680	7522.56563540	8708.98629834	10079.30325614
68	7405.08232135	8594.53123844	9971.78931160	11566.00048642
69	8441.79384634	9819.25193992	11417.69876178	13271.98555817
70	9623.64498483	11218.49534136	13073.26508224	15229.60342800
71	10970.95528270	12817.13092750	14968.88851916	17475.96993363
72	12506.88902228	14643.57208467	17139.37735444	20053.67549884
73	14257.85348540	16730.28110674	19624.58707084	23011.59263492
74	16253.95297336	19114.34616445	22470.15219611	26405.80254857
75	18529.50638963	21838.14049288	25728.32426454	30300.65842448
76	21123.63728417	24950.07551312	29458.93128290	34770.00554209
77	24080.94650396	28505.46127374	33730.47631892	39898.58135955
78	27452.27901451	32567.48950524	38621.39538517	45783.62211008
79	31295.59807655	37208.35675974	44221.49771602	52536.70637132
80	35676.98180726	42510.54759800	50633.61488484	60285.87056109
81	40671.75926028	48568.30063072	57975.48904314	69178.03646885
82	46365.80555672	55489.28347060	66381.93495440	79381.79684801
83	52857.01833466	63396.50636516	76007.31552279	91090.61188309
84	60257.00090151	72430.50852219	87028.37627359	104526.47713584
85	68692.98102772	82751.85598660	99647.49083326	119944.13251338
86	78309.99837160	94543.99546469	114096.37700408	137635.89205910
87	89273.39814363	108016.51481841	130640.35166967	157937.18613782
88	101771.67388373	123408.86818004	149583.20266178	181232.92109315
89	116019.70822746	140994.63189569	171272.76704773	207964.77695439
90	132262.46737930	161086.36694083	196107.31826966	238639.58155516
91	150779.21281240	184041.17422990	224542.87941876	273838.91983454
92	171888.30260614	210267.04155766	257101.59693448	314230.16051014
93	195952.66497100	240230.09497962	294381.32848997	360579.10918538
94	223386.03806694	274462.88351422	337066.62112102	413764.52779023
95	254660.08339631	313573.84441499	385941.28118357	474794.79563929
96	290312.49507180	358258.11724413	441902.76695519	544827.02799608
97	330956.24438185	409309.89895142	505978.66816369	625189.01462550
98	377290.11859531	467636.55955200	579345.57504742	717404.39428276
99	430110.73519865	534274.76928815	663350.68342930	823221.54243947
100	490326.23812646	610408.92391172	759536.53252655	944646.71994929

AMOUNT AT COMPOUND INTEREST $(1 + i)^n$ (Continued)

Periods	Rate i			
n	0.15(15%)	0.1525(15 1/4%)	0.155(15 1/2%)	0.1575(15 3/4%)
1	1.15000000	1.15250000	1.15500000	1.15750000
2	1.32250000	1.32825625	1.33402500	1.33980625
3	1.52087500	1.53081533	1.54079887	1.55082573
4	1.74900625	1.76426467	1.77962270	1.79508079
5	2.01135719	2.03331503	2.05546422	2.07780601
6	2.31306077	2.34339557	2.37406117	2.40506046
7	2.66001988	2.70076339	2.74204066	2.78385748
8	3.05902286	3.11262981	3.16705696	3.22231503
9	3.51787629	3.58730586	3.65795078	3.72982965
10	4.04555774	4.13437000	4.22493316	4.31727782
11	4.65239140	4.76486142	4.87979780	4.99724908
12	5.35025011	5.49150279	5.63616645	5.78431581
13	6.15278762	6.32895697	6.50977225	6.69534555
14	7.07570576	7.29412291	7.51878695	7.74986247
15	8.13706163	8.40647665	8.68419893	8.97046581
16	9.35762087	9.68846434	10.03024977	10.38331418
17	10.76126400	11.16595515	11.58493848	12.01868616
18	12.37545361	12.86876331	13.38060394	13.91162923
19	14.23177165	14.83124971	15.45459756	16.10271083
20	16.36653739	17.09301529	17.85006018	18.63888779
21	18.82151800	19.69970013	20.61681950	21.57451262
22	21.64474570	22.70390440	23.81242653	24.97249835
23	24.89145756	26.16624982	27.50335264	28.90566685
24	28.62517619	30.15660291	31.76637230	33.45830937
25	32.91895262	34.75548486	36.69016000	38.72799310
26	37.85679551	40.05569630	42.37713481	44.82765201
27	43.53531484	46.16418999	48.94559070	51.88800720
28	50.06561207	53.20422896	56.53215726	60.06036834
29	57.57545388	61.31787387	65.29464163	69.51987635
30	66.21177196	70.66884964	75.41531109	80.46925688
31	76.14353775	81.44584921	87.10468431	93.14316484
32	87.56506841	93.86634121	100.60591037	107.81321330
33	100.69982867	108.18095825	116.19982648	124.79379439
34	115.80480298	124.67855438	134.21079959	144.44881701
35	133.17552342	143.69203393	155.01347352	167.19950569
36	153.15185194	165.60506910	179.04056192	193.53342783
37	176.12462973	190.85984214	206.79184901	224.01494272
38	202.54332419	219.96596806	238.84458561	259.29729620
39	232.92482281	253.51077819	275.86549638	300.13662035
40	267.86354623	292.17117187	318.62464832	347.40813805
41	308.04307817	336.72727558	368.01146881	402.12491980
42	354.24953990	388.07818510	425.05324647	465.45959466
43	407.38697088	447.26010833	490.93649968	538.76948082
44	468.49501651	515.46727485	567.03165713	623.62567405
45	538.76926899	594.07603427	654.92156398	721.84671772
46	619.58465934	684.67262949	756.43440640	835.53757576
47	712.52235824	789.08520549	873.68173939	967.13474394
48	819.40071197	909.42069933	1009.10240900	1119.45846611
49	942.31081877	1048.10735598	1165.51328239	1295.77317452
50	1083.65744158	1207.94372776	1346.16784116	1499.85744951
51	1246.20605782	1392.15514625	1554.82385654	1736.08499780
52	1433.13696649	1604.45880605	1795.82155431	2009.51838496
53	1648.10751147	1849.13877397	2074.17389522	2326.01753059
54	1895.32363819	2131.13243701	2395.67084898	2692.36529166
55	2179.62218392	2456.13013365	2766.99983058	3116.41282509
56	2506.56551151	2830.68997903	3195.88480431	3607.24784505
57	2882.55033823	3262.37020083	3691.24694898	4175.38938064
58	3314.93288897	3759.88165646	4263.39022608	4833.01320809
59	3812.17282231	4333.26360907	4924.21571112	5594.21278837
60	4383.99874566	4994.08630945	5687.46914634	6475.30130253
61	5041.59855751	5755.68447164	6569.02686402	7495.16125768
62	5797.83834113	6633.42635357	7587.22602795	8675.64915577
63	6667.51409230	7645.02387249	8763.24606228	10042.06389780
64	7667.64120615	8810.89001304	10121.54920193	11623.68896171
65	8817.78738707	10154.55074003	11690.38932823	13454.41997318
66	10140.45549513	11703.11972789	13502.39967411	15573.49111895
67	11661.52381940	13487.84548639	15595.27162360	18026.31597019
68	13410.75239231	15544.74192306	18012.53872525	20865.46073549
69	15422.36525115	17915.31506633	20804.48222767	24151.77080133
70	17735.72003883	20647.40061395	24029.17697296	27955.67470254
71	20396.07804465	23796.12920757	27753.69940376	32358.69346819
72	23455.48975135	27425.03891173	32055.52281135	37455.18768943
73	26973.81321405	31607.35734577	37024.12884711	43354.37975051
74	31019.88519616	36427.47934100	42762.86881841	50182.69456122
75	35672.86797558	41982.66994050	49391.11348526	50182.69456122
76	41023.79817192	48385.02710643	57046.73607548	58086.46895461
77	47177.36789771	55763.74374016	65888.98016718	77824.61414582
78	54253.97308237	64267.71466053	76101.77209309	90081.99087379
79	62392.06904472	74068.54114626	87897.54676752	104269.90443641
80	71750.87940143	85363.99367107	101521.66651648	120692.41438514
81	82513.51131165	98382.00270590	117257.52482654	139701.46965080
82	94890.53800839	113385.25811855	135432.44117465	161704.45112080
83	109124.11870965	130676.50998163	156424.46955672	187172.90217233
84	125492.73651610	150604.67775383	180670.26233801	216652.63426447
85	144316.64699351	173571.89111129	208674.15300041	250775.42416112
86	165964.14404254	200041.60450576	241018.64671547	290272.55346650
87	190858.76564892	230547.94919289	278376.53695637	335990.48063747
88	219487.58049626	265706.51144481	321524.90018460	388908.98133788
89	252410.71757070	306226.75444014	371361.25971322	450162.14589859
90	290272.32520630	352926.33449226	428922.25496876	521062.68387762
91	333813.17398725	406747.60050233	495405.20448892	603130.05658835
92	383885.15008533	468776.60957894	572193.01118470	698123.04050101
93	441467.92259814	540265.04253973	660882.92791833	808077.41937992
94	507688.11098786	622655.46152703	763319.78174568	935349.61293226
95	583841.32763603	717610.41940991	881634.34791625	1082667.17696909
96	671417.52678144	827046.00836992	1018287.67184327	1253187.25734172
97	772130.15579865	953170.52464633	1176122.26097898	1450564.25037304
98	887949.67916845	1098529.02965489	1358421.21143072	1679028.11980679
99	1021142.13104372	1266054.70667727	1568976.49920248	1943475.04867636
100	1174313.45070028	1459128.04944555	1812167.85657887	2249572.36884288

AMOUNT AT COMPOUND INTEREST $(1 + i)^n$ (Continued)

Periods	Rate *i*			
n	0.16(16%)	0.1625(16 1/4%)	0.165(16 1/2%)	0.1675(16 3/4%)
1	1.16000000	1.16250000	1.16500000	1.16750000
2	1.34560000	1.35140625	1.35722500	1.36305625
3	1.56089600	1.57100977	1.58116712	1.59136817
4	1.81063936	1.82629885	1.84205970	1.85792234
5	2.10034166	2.12307242	2.14599955	2.16912433
6	2.43639632	2.46807168	2.50008948	2.53245266
7	2.82621973	2.86913333	2.91260424	2.95663848
8	3.27841489	3.33536750	3.39318394	3.45187542
9	3.80296127	3.87736472	3.95305929	4.03006456
10	4.41143508	4.50743648	4.60531407	4.70510037
11	5.11726469	5.23989491	5.36519090	5.49320468
12	5.93602704	6.09137784	6.25044739	6.41331647
13	6.88579137	7.08122673	7.28177121	7.48754698
14	7.98751799	8.23192608	8.48326346	8.74171109
15	9.26552087	9.56961407	9.88300194	10.20594770
16	10.74800420	11.12467635	11.51369726	11.91544394
17	12.46768488	12.93243626	13.41345730	13.91128080
18	14.46251446	15.03395715	15.62667776	16.24142034
19	16.77651677	17.47697519	18.20507959	18.96185824
20	19.46075945	20.31698366	21.20891772	22.13796950
21	22.57448097	23.61849350	24.70838914	25.84607939
22	26.18639792	27.45649869	28.78527335	30.17529769
23	30.37622159	31.91817973	33.53484345	35.22966005
24	35.23641704	37.10488394	39.06809262	41.13062811
25	40.87424377	43.13442758	45.51432791	48.02000832
26	47.41412277	50.14377206	53.02419201	56.06335971
27	55.00038241	58.29213502	61.77318369	65.45397246
28	63.80044360	67.76460696	71.96575900	76.41751285
29	74.00851458	78.77635559	83.84010924	89.21744625
30	85.84987691	91.57751338	97.67372726	104.16136850
31	99.58585721	106.45885930	113.78989226	121.60839773
32	115.51959437	123.75842394	132.56522448	141.97780434
33	134.00272947	143.86916783	154.43848652	165.75908657
34	155.44316618	167.24790760	179.92083680	193.52373357
35	180.31407277	194.42569258	209.60777487	225.93895895
36.	209.16432441	226.01986763	244.19305773	263.78373457
37	242.63061632	262.74809612	284.48491225	307.96751011
38	281.45151493	305.44466173	331.42492277	359.55206805
39	326.48375732	355.07941927	386.11003503	419.77703945
40	378.72115849	412.77982490	449.81819081	490.08969356
41	439.31654385	479.85654644	524.03819230	572.17971723
42	509.60719087	557.83323524	610.50449402	668.01981987
43	591.14434141	648.48113596	711.23773554	779.91313970
44	685.72743603	753.85932056	828.59196190	910.54859060
45	795.44382580	876.36146015	965.30963562	1063.06547952
46	922.71483793	1018.77019742	1124.58572549	1241.12894734
47	1070.34921199	1184.32035451	1310.14237020	1449.01804602
48	1241.60508591	1376.77241211	1526.31586128	1691.72856873
49	1440.26189966	1600.49792908	1778.15797839	1975.09310399
50	1670.70380360	1860.57884256	2071.55404483	2305.92119891

Periods	Rate *i*			
n	0.16(16%)	0.1625(16 1/4%)	0.165(16 1/2%)	0.1675(16 3/4%)
51	1938.01641218	2162.92290447	2413.36046222	2692.16299973
52	2248.09903813	2514.39787645	2811.56493849	3143.10030218
53	2607.79488423	2922.98753137	3275.47315334	3669.56960279
54	3025.04206571	3397.97300522	3815.92622364	4284.22251126
55	3509.04879622	3950.14361857	4445.55405054	5001.82978190
56	4070.49660361	4592.04195659	5179.07046888	5839.63627037
57	4721.77606019	5338.24877453	6033.61709625	6817.77534565
58	5477.26022982	6205.71420039	7029.16391713	7959.75271605
59	6353.62186659	7214.14275796	8188.97596345	9293.01129599
60	7370.20136525	8386.44095613	9540.15699742	10849.59068807
61	8549.43358369	9749.23761150	11114.28290200	12666.89712832
62	9917.34295708	11333.48872336	12948.13958083	14788.60239731
63	11504.11783021	13175.18064091	15084.58261166	17265.69329886
64	13344.77668305	15316.14749506	17573.53874259	20157.69692642
65	15479.94095233	17805.02146300	20473.17263511	23534.11116160
66	17956.73150471	20698.33745074	23851.24611991	27476.07478116
67	20829.80854546	24061.81728649	27786.70172969	32078.31730701
68	24162.57791273	27971.86259554	32371.50751509	37451.43545593
69	28028.59037877	32517.29026732	37712.80625508	43724.55089480
70	32513.16483938	37801.34993576	43935.41928717	51048.41316968
71	37715.27121368	43944.06930032	51184.76346956	59599.02237560
72	43749.71460786	51084.98056162	59630.24944203	69581.85862351
73	50749.66894512	59386.28990288	69469.24059997	81236.81994295
74	58869.61597634	69036.56201210	80931.66529896	94843.98728340
75	68288.75453256	80255.00333907	94285.39007329	110730.35515337
76	79214.95525777	93296.44138167	109842.47943538	129277.68964156
77	91889.34809901	108457.11310619	127966.48854222	150931.70265652
78	106591.64379485	126081.39398594	149080.95915169	176212.76285148
79	123646.30680203	146569.62050866	173679.31741172	205728.40062911
80	143429.71589035	170387.18384132	202336.40478465	240187.90773448
81	166378.47043281	198075.10121553	235721.91157412	280419.38228001
82	192999.02570205	230262.30516305	274616.02698385	327389.62881191
83	223878.86981438	267679.92975205	319927.67143618	382227.39163790
84	259699.48898468	311177.91833676	372715.73722315	446250.47973725
85	301251.40722223	361744.33006648	434213.83386497	520997.43509324
86	349451.63237779	420527.78370229	505859.11645269	710148.81013781
87	405363.89355824	488863.54855391	589325.87066739	608264.50547136
88	470222.11652756	568303.87519392	686564.63932751	829098.73583589
89	545457.65517196	660653.25491293	799847.80481655	967972.77408841
90	632730.87999948	768009.40883628	931822.69261128	1130108.21374821
91	733967.82079940	892810.93777217	1085573.43689214	1319401.33955104
92	851402.67212730	1037892.71516015	1264693.05397934	1540401.06392584
93	987627.09966767	1206550.28137367	1473367.40788593	1798418.24213342
94	1145647.43561449	1402614.70209690	1716473.03018711	2099653.29769076
95	1328951.02531281	1630539.59118764	1999691.08016798	2451345.22505396
96	1541583.18936286	1895502.27475563	2329640.10839569	2861945.55025050
97	1788236.49966092	2203521.39440342	2714030.72628098	3341321.42991746
98	2074354.33960667	2561593.62099398	3161845.79611735	3900992.76942863
99	2406251.03394373	2977852.58440550	3683550.35247671	4554409.05830793
100	2791251.19937473	3461753.62937139	4291336.16063536	5317272.57557451

AMOUNT AT COMPOUND INTEREST $(1 + i)^n$ (Continued)

Periods n	Rate i 0.17(17%)	0.1725(17 1/4%)	0.175(17 1/2%)	0.1775(17 3/4%)	Periods n	Rate i 0.17(17%)	0.1725(17 1/4%)	0.175(17 1/2%)	0.1775(17 3/4%)
1	1.17000000	1.17250000	1.17500000	1.17750000	51	3002.47188274	3347.76851605	3731.91030408	4159.17032530
2	1.36890000	1.37475625	1.38062500	1.38650625	52	3512.89210280	3925.25858506	4384.99460730	4897.42305804
3	1.60161300	1.61190170	1.62223437	1.63261111	53	4110.08376028	4602.36569099	5152.36866357	5766.71565084
4	1.87388721	1.88995475	1.90612539	1.92239958	54	4808.79799952	5396.27377268	6054.03317970	6790.30767886
5	2.19244804	2.21597194	2.23969733	2.26362551	55	5626.29365944	6327.13099847	7113.48898614	7995.58729186
6	2.56516420	2.59822710	2.63164437	2.66541903	56	6582.76358155	7418.56109571	8358.34955872	9414.80403617
7	3.00124212	3.04642128	3.09218213	3.13853091	57	7701.83339041	8698.26288472	9821.06073150	11085.93175259
8	3.51145328	3.57192895	3.63331400	3.69562015	58	9011.14506678	10198.71323233	11539.74635951	13053.68463867
9	4.10840033	4.18808669	4.26914396	4.35159273	59	10543.03972814	11957.99126491	13559.20197242	15370.71366204
10	4.80682839	4.91053164	5.01624415	5.12400044	60	12335.35648192	14020.74475810	15932.06231759	18099.01533705
11	5.62398922	5.75759835	5.89408687	6.03351051	61	14432.36708385	16439.32322888	18720.17322317	21311.59055938
12	6.58006738	6.75078407	6.92555208	7.10445863	62	16885.86948810	19275.10648586	21996.20353723	25094.39788366
13	7.69867884	7.91529432	8.13752369	8.36550004	63	19756.46730108	22600.06235467	25845.53915624	29548.65350802
14	9.00745424	9.28068259	9.56159034	9.85037629	64	23115.06674226	26498.57311085	30368.50850859	34793.53950569
15	10.53872146	10.88160033	11.23486864	11.59881808	65	27044.62808844	31069.57697247	35682.99749759	40969.39276795
16	12.33030411	12.75867639	13.20097066	13.65760829	66	31642.21486348	36429.07900022	41927.52205967	48241.45998426
17	14.42645581	14.95954807	15.51114052	16.08183377	67	37021.39139027	42713.09512776	49264.83842011	56804.31913146
18	16.87895329	17.54007011	18.22559011	18.93635926	68	43315.02792662	50081.10403730	57886.18514363	66887.08577730
19	19.74837535	20.56573220	21.41506838	22.29756303	69	50678.58267414	58720.09448373	68016.26754376	78759.54350277
20	23.10559916	24.11332101	25.16270535	26.25538047	70	59293.94172875	68849.31078218	79919.11436392	92739.36247451
21	27.03355102	28.27286888	29.56617879	30.91561050	71	69373.91182263	80725.81689210	93904.95937761	109200.59931374
22	31.62925470	33.14993877	34.74026008	36.40324911	72	81167.47683248	94651.02030599	110338.32726869	128583.70569193
23	37.00622799	38.86830320	40.81980559	42.86482583	73	94965.94789400	110978.32130877	129647.53454071	151407.31345224
24	43.29728675	45.57308551	47.96327157	50.47333241	74	111110.15903598	130122.08173454	152335.85308534	178282.11159002
25	50.65782550	53.43444275	56.35684409	59.43234892	75	129998.88607210	152568.14083375	178994.62737527	209927.18639724
26	59.26965584	62.65188413	66.21929181	69.98159085	76	152098.69670436	178886.14512757	210318.68716594	247189.26198275
27	69.34549733	73.45933414	77.80766787	82.40332323	77	177955.47514410	209744.00516207	247124.45741998	291065.35598469
28	81.13423187	86.13106928	91.42400975	94.02991310	78	208207.90591859	245924.84605253	290371.23746848	342729.45667198
29	94.92705129	100.98867873	107.42321146	114.25272268	79	243603.24992475	288346.88199659	341186.20402547	403563.93523125
30	111.06465001	118.40922581	126.22227346	134.53258095	80	285015.80241196	338087.71914100	400893.78972992	475196.53373480
31	129.94564051	138.83481727	148.31117132	158.41211407	81	333468.48882200	396406.67819283	471050.20293266	559543.91847273
32	152.03639940	162.78382325	174.26562630	186.53026432	82	390158.13192173	464786.83018109	553483.98844587	658862.96400163
33	177.88258730	190.86403276	204.76211090	219.63938623	83	456485.01434843	544962.55838732	650343.68642390	775811.14011192
34	208.12262714	223.78807841	240.59548031	258.62537729	84	534087.46678766	638968.59970914	764153.83154809	913537.61748179
35	243.50347375	262.39152193	282.69968936	304.53138176	85	624882.33614156	749190.68315896	897880.75206900	1075666.99458481
36	284.89906429	307.65405947	332.17213500	358.58570202	86	731112.33328563	878426.07600389	1055009.88368108	1266597.88612361
37	333.33190522	360.72438472	390.30225862	422.23466413	87	855401.42994419	1029954.57411456	1239636.61332526	1491419.01091055
38	389.99832910	422.94934109	458.60515388	497.18131701	88	1000819.67303470	1207621.73814932	1456573.02065718	1756145.88534717
39	456.29804505	495.90810243	538.86105581	585.43100078	89	1170959.01745060	1415936.48798007	1711473.29927219	2067861.77999630
40	533.86871271	581.45225010	633.16174058	689.34500342	90	1370022.05041720	1660185.53215664	2010981.12664482	2434907.24594564
41	624.62639387	681.75276324	743.96504518	811.70374152	91	1602925.79898812	1946567.53645366	2362902.82380767	2867103.28210099
42	730.81288083	799.35511490	874.15892808	955.78115564	92	1875423.18481610	2282350.43649191	2776410.81797401	3376014.11467391
43	855.05107057	937.24387221	1027.13674050	1125.43231077	93	2194245.12623484	2676055.88678676	3262282.71111946	3975256.62002853
44	1000.40975257	1098.91844017	1206.88567009	1325.19654593	94	2567266.79769476	3137675.52725748	3833182.18556536	4680864.67008360
45	1170.47941051	1288.48187110	1418.09066235	1560.41893283	95	3003702.15330287	3678924.55570940	4503989.06803930	5511718.14902343
46	1369.46091029	1510.74499387	1666.25652826	1837.39329341	96	3514331.51936436	4313539.04156927	5292187.15494618	6490048.12047509
47	1602.26926504	1771.34850531	1957.85142071	2163.53060299	97	4111767.87765630	5057624.52623996	6218319.90706176	7642031.66185942
48	1874.65504010	2076.90612247	2300.47541933	2547.55728502	98	4810768.41685787	5930064.75701636	7306525.89079756	8998492.28183946
49	2193.34639691	2435.17242860	2703.05861771	2999.74870311	99	5628599.04772371	6953000.92760168	8585167.92168713	10595724.66186596
50	2566.21528439	2855.23967253	3176.09387581	3532.20409792	100	6585460.88583674	8152393.58761296	10087572.30798238	12476465.78934716

AMOUNT AT COMPOUND INTEREST $(1 + i)^n$ (Continued)

Periods n	Rate i 0.18(18%)	0.1825(18 1/4%)	0.185(18 1/2%)	0.1875(18 3/4%)
1	1.18000000	1.18250000	1.18500000	1.18750000
2	1.39240000	1.39830625	1.40422500	1.41015625
3	1.64303200	1.65349714	1.66400662	1.67456055
4	1.93877776	1.95526037	1.97184785	1.98854065
5	2.28775776	2.31209539	2.33663970	2.36139202
6	2.69955415	2.73405279	2.76891805	2.80415303
7	3.18547390	3.23301743	3.28116789	3.32993172
8	3.75885920	3.82304311	3.88818395	3.95429391
9	4.43545386	4.52074848	4.60749798	4.69572402
10	5.23383555	5.34578507	5.45988510	5.57617228
11	6.17592595	6.32139085	6.46996385	6.62170458
12	7.28759263	7.47504468	7.66690716	7.86327419
13	8.59935930	8.83924033	9.08528498	9.33763810
14	10.14724397	10.45240170	10.76606270	11.08844524
15	11.97374789	12.35996501	12.75778430	13.16752873
16	14.12902251	14.61565862	15.11797440	15.63644036
17	16.67224656	17.28301632	17.91479966	18.56827293
18	19.67325094	20.43716679	21.22903760	22.04982410
19	23.21443611	24.16694973	25.15640955	26.18416612
20	27.39303460	28.57741806	29.81034532	31.09369727
21	32.32378083	33.79279686	35.32525921	36.92376551
22	38.14206138	39.95998228	41.86043216	43.84697154
23	45.00763243	47.25267905	49.60461211	52.06827871
24	53.10900627	55.87629298	58.78146535	61.83108096
25	62.66862740	66.07371645	69.65603644	73.42440864
26	73.94898033	78.13216970	82.54240318	87.19148526
27	87.25979679	92.39129067	97.81274777	103.53988875
28	102.96656021	109.25270122	115.90810611	122.95361789
29	121.50054105	129.19131919	137.35110574	146.00742125
30	143.37063844	152.76873494	162.76106030	173.38381273
31	169.17735336	180.64902906	192.87185646	205.89327762
32	199.62927696	213.61747687	228.55314990	244.49826717
33	235.56254681	252.60266640	270.83548263	290.34169227
34	277.96380524	298.70265302	320.94004692	344.78075956
35	327.99729018	353.21588719	380.31395560	409.42715198
36	387.03680242	417.67778660	450.67203738	486.19474298
37	456.70342685	493.90398266	534.04636430	577.35625729
38	538.91004369	584.04145949	632.84494170	685.61055553
39	635.91385155	690.62902585	749.92125591	814.16253469
40	750.37834483	816.66882307	888.65668825	966.81800995
41	885.44644690	965.71088328	1053.05817558	1148.09638681
42	1044.82680734	1141.95311948	1247.87393806	1363.36445934
43	1232.89563266	1350.35956378	1478.73061660	1618.99529547
44	1454.81684654	1596.80018417	1752.29578067	1922.55691337
45	1716.68387891	1888.21621778	2076.47050010	2283.03633462
46	2025.68697712	2232.81567753	2460.61754262	2711.10564736
47	2390.31063300	2640.30453867	2915.83178800	3219.43795624
48	2820.56654694	3122.16011698	3455.26066878	3823.08257304
49	3328.26852539	3691.95433833	4094.48389250	4539.91055548
50	3927.35685996	4365.73600508	4851.96341262	5391.14378464

Periods n	Rate i 0.18(18%)	0.1825(18 1/4%)	0.185(18 1/2%)	0.1875(18 3/4%)
51	4634.28109475	5162.48282600	5749.57664395	6401.98324426
52	5468.45169180	6104.63594175	6813.24832308	7602.35510256
53	6452.77299633	7218.73200112	8073.69926285	9027.79668429
54	7614.27213567	8536.15059132	9567.33362648	10720.50856259
55	8984.84112009	10093.99807424	11337.29034738	12730.60391807
56	10602.11252170	11936.15272279	13434.68906164	15117.59215271
57	12510.49277561	14114.50059469	15920.10653805	17952.14068135
58	14762.38147522	16690.39695323	18865.32624759	21318.16705910
59	17419.61014076	19736.39439719	22355.41160339	25315.32338268
60	20555.13996610	23338.28637468	26491.16275002	30061.94651693
61	24255.06516000	27597.52363806	31392.02785877	35698.56148886
62	28620.97688880	32634.07170200	37199.55301264	42392.04176802
63	33772.75272878	38589.78978762	44081.47031998	50340.54959953
64	39851.84821996	45632.42642386	52236.54232918	59779.40264944
65	47025.18089955	53960.34424621	61900.30266007	70988.04064621
66	55489.71346147	63808.10707114	73351.85865219	84298.29826737
67	65477.86188454	75453.08661163	86921.95250284	100104.22919250
68	77263.87702375	89223.27491825	103002.51371587	118873.77216609
69	91171.37488803	105506.52259083	122057.97875330	141162.60444724
70	107582.22236787	124761.46296365	144638.70482267	167630.59278109
71	126947.02239409	147530.42995452	171396.86521486	199061.32892755
72	149797.48642503	174454.73342122	203105.28527961	236385.32810146
73	176761.03398153	206292.72227059	240679.76305634	280707.57712049
74	208578.02009821	243941.14408497	285205.51922176	333340.24783058
75	246122.06371589	288460.40288048	337968.54027778	395841.54429881
76	290424.03518474	341104.42640617	400492.72022917	470061.83385484
77	342700.36151800	403355.98422530	474583.87347157	558198.42770263
78	404386.42659124	476968.45134641	562381.89006381	662860.63289687
79	477175.98337766	564015.19371713	666422.53972561	787147.00156503
80	563067.66038564	666947.96657051	789710.70957485	934737.06435847
81	664419.83925505	788665.97046963	935807.19084620	1110000.26392569
82	784015.41032096	932597.51008033	1108931.52115274	1318125.31341175
83	925138.18417874	1102796.55566999	1314083.85256600	1565273.80967645
84	1091663.05733091	1304056.92707977	1557189.36529071	1858762.64899079
85	1288162.40765047	1542047.31627182	1845269.39786949	2207280.64567656
86	1520031.64102756	1823470.95149143	2186644.23647535	2621145.76674092
87	1793637.33641252	2156254.40013862	2591173.42022329	3112610.59800484
88	2116492.05696677	2549770.82816391	3070540.50296459	3696225.08513074
89	2497460.62722079	3015104.00430383	3638590.49601304	4389267.28859276
90	2947003.54012054	3565360.48508928	4311729.73777545	5212254.90520390
91	3477464.17734223	4216038.77361807	5109399.73926391	6189552.69992963
92	4103407.72926383	4985465.84980337	6054638.69102773	7350093.83116643
93	4842021.12053132	5895313.36739248	7174746.84886786	8728236.42451013
94	5713584.92222696	6971208.05694161	8502075.01590842	10364780.75410578
95	6742030.20822781	8243453.52733345	10074958.89385147	12308177.14550061
96	7955595.64570881	9747883.79607180	11938826.28921398	14615960.36028197
97	9387602.86193640	11526872.58885490	14147509.15271857	17356452.92783483
98	11077371.37708495	13630526.83632091	16764798.34597150	20610787.85180386
99	13071298.22496024	16118097.98394948	19866286.03997623	24475310.57401708
100	15424131.90545308	19059650.86602026	23541548.95737182	29064431.30664527

AMOUNT AT COMPOUND INTEREST $(1 + i)^n$ (Continued)

Periods	Rate *i*					Periods	Rate *i*				
n	0.19(19%)	0.1925(19 1/4%)	0.195(19 1/2%)	0.1975(19 3/4%)	0.20(20%)	*n*	0.19(19%)	0.1925(19 1/4%)	0.195(19 1/2%)	0.1975(19 3/4%)	0.20(20%)
1	1.19000000	1.19250000	1.19500000	1.19750000	1.20000000	51	7126.80754362	7931.90991805	8825.98468450	9818.64654475	10920.52578000
2	1.41610000	1.42205625	1.42802500	1.43400625	1.44000000	52	8480.90097691	9458.80257727	10547.05169798	11757.82923734	13104.63093600
3	1.68515900	1.69580208	1.70648987	1.71722248	1.72800000	53	10092.27216252	11279.62207339	12603.72677908	14080.00051172	15725.55712320
4	2.00533921	2.02224398	2.03925540	2.05637393	2.07360000	54	12009.80387340	13450.94932252	15061.45350100	16860.80061278	18870.66854784
5	2.38635366	2.41152594	2.43691020	2.46250778	2.48832000	55	14291.66660934	16040.25706711	17998.43693370	20190.80873380	22644.80225741
6	2.83976086	2.87574469	2.91210769	2.94885306	2.98598400	56	17007.08326512	19128.00655253	21508.13213577	24178.49345873	27173.76270890
7	3.37931542	3.42932554	3.47996869	3.53125154	3.58318080	57	20238.42908549	22810.14781389	25702.21790225	28953.74591683	32608.51525067
8	4.02138535	4.08947071	4.15856259	4.22867372	4.29981696	58	24083.73061173	27201.10126806	30714.15039318	34672.11073540	39130.21830081
9	4.78544856	4.87669382	4.96948229	5.06383678	5.15978035	59	28659.63942796	32437.31326216	36703.40971986	41519.85260564	46956.26196097
10	5.69468379	5.81545738	5.93853134	6.06394454	6.19173642	60	34104.97091927	38681.49606513	43860.57461523	49720.02349526	56347.51435317
11	6.77667371	6.93493292	7.09654495	7.26157359	7.43008371	61	40584.91539394	46127.68405767	52413.38666520	59539.72813557	67617.01722380
12	8.06424172	8.26990751	8.48037122	8.69573437	8.91610045	62	48296.04931878	55007.26323877	62633.99706491	71298.82444235	81140.42066856
13	9.59644764	9.86186471	10.13404361	10.41314191	10.69932054	63	57472.29868935	65596.16141223	74847.62649257	85380.34226971	97368.50480227
14	11.41977269	11.76027366	12.11018211	12.46973744	12.83918465	64	68392.03544033	78223.42248408	89442.91365862	102242.95986798	116842.20576272
15	13.58952950	14.02412634	14.47166762	14.93251059	15.40702157	65	81386.52217399	93281.43131227	106884.28182205	122435.94444190	140210.64691527
16	16.17154011	16.72377067	17.29364281	17.88168143	18.48842589	66	96849.96138705	111238.10683988	127726.71677735	146617.04346918	168252.77629832
17	19.24413273	19.94309652	20.66590315	21.41331351	22.18611107	67	115251.45405059	132651.44240656	152633.42654893	175573.90955434	201903.33155799
18	22.90051795	23.78214260	24.69575427	25.64244293	26.62333328	68	137149.23032020	158186.84506982	182396.94472597	210249.75669132	242283.99786959
19	27.25161636	28.36020505	29.51142635	30.70682540	31.94799994	69	163207.58408104	188637.81274576	217964.34894754	251774.08363786	290740.79744350
20	32.42942347	33.81954452	35.26615449	36.77142342	38.33759992	70	194217.02505644	224950.59169932	260467.39699230	301499.46515634	348888.95693220
21	38.59101393	40.32980684	42.14305461	44.03377955	46.00511991	71	231118.25981716	268253.58060144	311258.53940580	361045.60952472	418666.74831864
22	45.92330658	48.09329466	50.36095026	52.73045101	55.20614389	72	275030.72918242	319892.39486721	371953.95458994	432352.11740585	502400.09798237
23	54.64873482	57.35125388	60.18133557	63.14471508	66.24737267	73	327286.56772708	381471.68087915	444484.97573497	517741.66059350	602880.11757885
24	65.03199444	68.39137025	71.91669600	75.61579631	79.49684720	74	389471.01559523	454904.97944839	531159.54600329	619995.63856072	723456.14109462
25	77.38807338	81.55670902	85.94045172	90.54991608	95.39621664	75	463470.50855832	542474.18799220	634735.65747393	742444.77717646	868147.36931354
26	92.09180733	97.25637551	102.69883981	108.43352451	114.47545997	76	551529.90518440	646900.46918070	758509.11068135	889077.62066881	1041776.84317625
27	109.58925072	115.97822780	122.72511357	129.84914560	137.37055197	77	656320.58716944	771428.80949799	906418.38726421	1064670.45075090	1250132.21181149
28	130.41120836	138.30403665	146.65651072	155.49435185	164.84466236	78	781021.49873163	919928.85532635	1083169.97278074	1274942.86477420	1500158.65417379
29	155.18933794	164.92756370	175.25453031	186.20448635	197.81359483	79	929415.58349064	1097015.15997667	1294388.11747298	1526744.08056711	1800190.38500855
30	184.67531215	196.67611971	209.42916372	222.97987240	237.37631380	80	1106004.54435386	1308190.57827218	1546793.80038021	1828276.03647911	2160228.46201026
31	219.76362146	234.53627276	250.26785064	267.01839720	284.85157656	81	1316145.40778110	1560017.26458958	1848418.59145435	2189360.55368374	2592274.15441231
32	261.51870954	279.68450526	299.07008152	319.75453064	341.82189187	82	1566213.03525950	1860320.58802307	2208860.21678795	2621759.26303627	3110728.98529477
33	311.20726435	333.52377253	357.38874741	382.90605045	410.18627025	83	1863793.51195881	2218432.30121751	2639587.95906160	3139556.71748594	3732874.78235373
34	370.33664458	397.72709874	427.07955316	458.52999541	492.22352430	84	2217914.27923098	2645480.51920188	3154307.61107861	3759619.16918941	4479449.73882447
35	440.70060705	474.28956524	510.36006602	549.08966950	590.66822915	85	2639317.99228487	3154735.51914825	3769397.59523893	4502143.95510432	5375339.68658937
36	524.43372239	565.59030655	609.88027890	657.53487923	708.80187499	86	3140788.41081899	3762022.10658428	4504430.12631052	5391317.38623742	6450407.62390724
37	624.07612965	674.46644056	728.80693328	787.39801788	850.56224998	87	3737538.20887460	4486211.36210176	5382794.00094107	6456102.57001930	7740789.14868868
38	742.65059428	804.30123037	870.92428527	942.90912641	1020.67469998	88	4447670.46856077	5349807.04930634	6432438.83112458	7731182.82759812	9288586.97842642
39	883.75420719	959.12921722	1040.75452090	1129.13367888	1224.80963997	89	5292727.85758731	6379644.90629782	7686764.40319387	9258091.43604874	11146304.37411169
40	1051.66750656	1143.76159154	1243.70165248	1352.13758046	1469.77156797	90	6298346.15052890	7607726.55076014	9185683.46181668	11086564.49466836	13375565.24893404
41	1251.48433281	1363.93569791	1486.22347471	1619.18475260	1763.72588156	91	7495031.91912939	9072213.91178147	10976891.73687093	13276160.98236537	16050678.29872083
42	1489.26635604	1626.49331975	1776.03705228	1938.97374123	2116.47105788	92	8919087.98376397	10818615.08979940	13117385.62556076	15898202.77638252	19260813.95846499
43	1772.22696369	1939.59328381	2122.36427748	2321.92105513	2539.76526945	93	10613714.70067912	12901198.49458578	15675275.82254510	19038097.82471807	23112976.75015799
44	2108.95008679	2312.96499094	2536.22531159	2780.50046351	3047.71832334	94	12630320.49380815	15384679.20479354	18731954.60794139	22798122.14509988	27735572.10018957
45	2509.65060328	2758.21075169	3030.78924734	3329.64930506	3657.26198801	95	15030081.38763170	18346229.95171629	22384685.75648996	27300751.26875710	33282686.52022748
46	2986.48421790	3289.16632140	3621.79315058	3987.25504281	4388.71438561	96	17885796.85128171	21877879.21742167	26749699.47900548	32692649.64433662	39939223.82427297
47	3553.91621930	3922.33083826	4328.04281494	4774.73791376	5266.45726273	97	21284098.25302524	26089370.96677534	31965890.87741155	39149447.94909309	47927068.58912755
48	4229.16030097	4677.37952463	5172.01116385	5717.74865173	6319.74871528	98	25328076.92110002	31111574.87787958	38199239.59850679	46881463.91903896	57512482.30695304
49	5032.70075815	5577.77508312	6180.55334080	6847.00401044	7583.69845834	99	30140411.53610902	37100553.04187139	45648091.32021561	56140553.04304915	69014978.76834364
50	5988.91390220	6651.49678662	7385.76124226	8199.28730251	9100.43815000	100	35867089.72796973	44242409.50243164	54549469.12765764	67228312.26905134	82817974.52201234

PRESENT VALUE $1/(1 + i)^n$*

The following table gives the value of unit amount due in n years at rate of interest i, compounded annually, $1/(1 + i)^n = v^n$

Periods n	Rate i: 0.0025(1/4%)	0.004167(5/12%)	0.005(1/2%)	0.005833(7/12%)	0.0075(3/4%)
1	0.99750623	0.99585062	0.99502488	0.99420050	0.99255583
2	0.99501869	0.99171846	0.99007450	0.98843463	0.98516708
3	0.99253734	0.98760345	0.98514876	0.98270220	0.97783333
4	0.99006219	0.98350551	0.98024752	0.97700301	0.97055417
5	0.98759321	0.97942457	0.97537067	0.97133688	0.96332920
6	0.98513038	0.97536057	0.97051808	0.96570361	0.95615802
7	0.98267370	0.97131343	0.96568963	0.96010301	0.94904022
8	0.98022314	0.96728308	0.96088520	0.95453489	0.94197540
9	0.97777869	0.96326946	0.95610468	0.94899906	0.93496318
10	0.97534034	0.95927249	0.95134794	0.94349534	0.92800315
11	0.97290807	0.95529211	0.94661487	0.93802354	0.92109494
12	0.97048187	0.95132824	0.94190534	0.93258347	0.91423815
13	0.96806171	0.94738082	0.93721924	0.92717495	0.90743241
14	0.96564759	0.94344978	0.93255646	0.92179779	0.90067733
15	0.96323949	0.93953505	0.92791688	0.91645182	0.89397254
16	0.96083740	0.93563657	0.92330037	0.91113686	0.88731766
17	0.95844130	0.93175426	0.91870684	0.90585272	0.88071231
18	0.95605117	0.92788806	0.91413616	0.90059922	0.87415614
19	0.95366700	0.92403790	0.90958822	0.89537619	0.86764878
20	0.95128878	0.92020372	0.90506290	0.89018346	0.86118985
21	0.94891649	0.91638544	0.90056010	0.88502084	0.85477901
22	0.94655011	0.91258301	0.89607971	0.87988815	0.84841589
23	0.94418964	0.90879636	0.89162160	0.87478524	0.84210014
24	0.94183505	0.90502542	0.88718567	0.86971192	0.83583140
25	0.93948634	0.90127013	0.88277181	0.86466802	0.82960933
26	0.93714348	0.89753042	0.87837991	0.85965338	0.82343358
27	0.93480646	0.89380623	0.87400986	0.85466782	0.81730380
28	0.93247527	0.89009749	0.86966155	0.84971117	0.81121966
29	0.93014990	0.88640414	0.86533488	0.84478327	0.80518080
30	0.92783032	0.88272611	0.86102973	0.83988394	0.79918690
31	0.92551653	0.87906335	0.85674600	0.83501303	0.79323762
32	0.92320851	0.87541578	0.85248358	0.83017037	0.78733262
33	0.92090624	0.87178335	0.84824237	0.82535580	0.78147158
34	0.91860972	0.86816599	0.84402226	0.82056914	0.77565418
35	0.91631892	0.86456365	0.83982314	0.81581025	0.76988008
36	0.91403384	0.86097624	0.83564492	0.81107896	0.76414896
37	0.91175445	0.85740373	0.83148748	0.80637510	0.75846051
38	0.90948075	0.85384604	0.82735073	0.80169853	0.75281440
39	0.90721272	0.85030311	0.82323455	0.79704907	0.74721032
40	0.90495034	0.84677488	0.81913886	0.79242659	0.74164796
41	0.90269361	0.84326129	0.81506354	0.78783091	0.73612701
42	0.90044250	0.83976228	0.81100850	0.78326188	0.73064716
43	0.89819701	0.83627779	0.80697363	0.77871935	0.72520809
44	0.89595712	0.83280776	0.80295884	0.77420316	0.71980952
45	0.89372281	0.82935212	0.79896402	0.76971317	0.71445114
46	0.89149407	0.82591083	0.79498907	0.76524922	0.70913264
47	0.88927090	0.82248381	0.79103390	0.76081115	0.70385374
48	0.88705326	0.81907102	0.78709841	0.75639883	0.69861414
49	0.88484116	0.81567238	0.78318250	0.75201209	0.69341353
50	0.88263457	0.81228785	0.77928607	0.74765079	0.68825165

Periods n	Rate i: 0.0025(1/4%)	0.004167(5/12%)	0.005(1/2%)	0.005833(7/12%)	0.0075(3/4%)
51	0.88043349	0.80891736	0.77540902	0.74331479	0.68312819
52	0.87823790	0.80556086	0.77155127	0.73900393	0.67804286
53	0.87604778	0.80221828	0.76771270	0.73471808	0.67299540
54	0.87386312	0.79888957	0.76389324	0.73045708	0.66798551
55	0.87168391	0.79557468	0.76009277	0.72622079	0.66301291
56	0.86951013	0.79227354	0.75631122	0.72200907	0.65807733
57	0.86734178	0.78898610	0.75254847	0.71782178	0.65317849
58	0.86517883	0.78571230	0.74880445	0.71365877	0.64831612
59	0.86302128	0.78245208	0.74507906	0.70951990	0.64348995
60	0.86086911	0.77920539	0.74137220	0.70540504	0.63869970
61	0.85872230	0.77597217	0.73768378	0.70131404	0.63394511
62	0.85658085	0.77275237	0.73401371	0.69724677	0.62922592
63	0.85444474	0.76954593	0.73036190	0.69320308	0.62454185
64	0.85231395	0.76635279	0.72672826	0.68918285	0.61989266
65	0.85018848	0.76317291	0.72311269	0.68518593	0.61527807
66	0.84806831	0.76000621	0.71951512	0.68121219	0.61069784
67	0.84595343	0.75685266	0.71593544	0.67726150	0.60615170
68	0.84384382	0.75371219	0.71237357	0.67333372	0.60163940
69	0.84173947	0.75058476	0.70882943	0.66942872	0.59716070
70	0.83964037	0.74747030	0.70530291	0.66554637	0.59271533
71	0.83754650	0.74436876	0.70179394	0.66168653	0.58830306
72	0.83545786	0.74128009	0.69830243	0.65784908	0.58392363
73	0.83337442	0.73820424	0.69482829	0.65403388	0.57957681
74	0.83129618	0.73514115	0.69137143	0.65024081	0.57526234
75	0.82922312	0.73209078	0.68793177	0.64646973	0.57097999
76	0.82715523	0.72905306	0.68450923	0.64272053	0.56672952
77	0.82509250	0.72602794	0.68110371	0.63899306	0.56251069
78	0.82303491	0.72301537	0.67771513	0.63528723	0.55832326
79	0.82098246	0.72001531	0.67434342	0.63160288	0.55416701
80	0.81893512	0.71702770	0.67098847	0.62793989	0.55004170
81	0.81689289	0.71405248	0.66765022	0.62429816	0.54594710
82	0.81485575	0.71108960	0.66432858	0.62067754	0.54188297
83	0.81282369	0.70813902	0.66102346	0.61707792	0.53784911
84	0.81079670	0.70520069	0.65773479	0.61349917	0.53384527
85	0.80877476	0.70227454	0.65446248	0.60994118	0.52987123
86	0.80675787	0.69936054	0.65120644	0.60640382	0.52592678
87	0.80474600	0.69645863	0.64796661	0.60288698	0.52201169
88	0.80273915	0.69356876	0.64474290	0.59939054	0.51812575
89	0.80073731	0.69069088	0.64153522	0.59591437	0.51426873
90	0.79874046	0.68782495	0.63834350	0.59245836	0.51044043
91	0.79674859	0.68497090	0.63516766	0.58902240	0.50664063
92	0.79476168	0.68212870	0.63200763	0.58560636	0.50286911
93	0.79277973	0.67929829	0.62886331	0.58221014	0.49912567
94	0.79080273	0.67647962	0.62573464	0.57883361	0.79541009
95	0.78883065	0.67367265	0.62262153	0.57547666	0.49172217
96	0.78686349	0.67087733	0.61952391	0.57213918	0.48806171
97	0.78490124	0.66809361	0.61644170	0.56882106	0.48442850
98	0.78294388	0.66532143	0.61337483	0.56552218	0.48082233
99	0.78099140	0.66256076	0.61032321	0.56224243	0.47724301
100	0.77904379	0.65981155	0.60728678	0.55898171	0.47369033

*$a_{\overline{n}|i}$ may be constructed from these tables by use of the formula $\frac{1-(1+i)^{-n}}{i}$. See page 1004.

PRESENT VALUE $1/(1+i)^n$ (Continued)

Periods n	Rate i 0.01(1%)	0.001125(1 1/8%)	0.0125(1 1/4%)	0.015(1 1/2%)	0.0175(1 3/4%)
1	0.99009901	0.98887515	0.98765432	0.98522167	0.98280098
2	0.98029605	0.97787407	0.97546106	0.97066175	0.96589777
3	0.97059015	0.96699537	0.96341833	0.95631699	0.94928528
4	0.96098034	0.95623770	0.95152428	0.94218423	0.93295851
5	0.95146569	0.94559970	0.93977706	0.92826033	0.91691254
6	0.94204524	0.93508005	0.92817488	0.91454219	0.90114254
7	0.93271805	0.92467743	0.91671593	0.90102679	0.88564378
8	0.92348322	0.91439054	0.90539845	0.88771112	0.87041157
9	0.91433982	0.90421808	0.89422069	0.87459224	0.85544135
10	0.90528695	0.89415880	0.88318093	0.86166723	0.84072860
11	0.89632372	0.88421142	0.87227746	0.84893323	0.82626889
12	0.88744923	0.87437470	0.86150860	0.83638742	0.81205788
13	0.87866260	0.86464742	0.85087269	0.82402702	0.79809128
14	0.86996297	0.85502835	0.84036809	0.81184928	0.78436490
15	0.86134947	0.84551629	0.82999318	0.79985150	0.77087459
16	0.85282126	0.83611005	0.81974635	0.78803104	0.75761631
17	0.84437749	0.82680846	0.80962602	0.77638526	0.74458605
18	0.83601731	0.81761034	0.79963064	0.76491159	0.73177990
19	0.82773992	0.80851455	0.78975866	0.75360747	0.71919401
20	0.81954447	0.79951995	0.78000855	0.74247042	0.70682458
21	0.81143017	0.79062542	0.77037881	0.73149795	0.69466789
22	0.80339621	0.78182983	0.76086796	0.72068763	0.68272028
23	0.79544179	0.77313210	0.75147453	0.71003708	0.67097817
24	0.78756613	0.76453112	0.74219707	0.69954392	0.65943800
25	0.77976844	0.75602583	0.73303414	0.68920583	0.64809632
26	0.77204796	0.74761516	0.72398434	0.67092052	0.63694970
27	0.76440392	0.73929806	0.71504626	0.66898574	0.62599479
28	0.75683557	0.73107348	0.70621853	0.65909925	0.61522829
29	0.74934215	0.72294040	0.69749978	0.64935887	0.60464697
30	0.74192292	0.71489780	0.68888867	0.63976243	0.59424764
31	0.73457715	0.70694467	0.68038387	0.63030781	0.58402716
32	0.72730411	0.69908002	0.67198407	0.62099292	0.57398247
33	0.72010307	0.69130287	0.66368797	0.61181568	0.56411053
34	0.71297334	0.68361223	0.65549429	0.60277407	0.55440839
35	0.70591420	0.67600715	0.64740177	0.59386608	0.54487311
36	0.69892495	0.66848667	0.63940916	0.58508974	0.53550183
37	0.69200490	0.66104986	0.63151522	0.57644309	0.52629172
38	0.68515337	0.65369578	0.62371873	0.56792423	0.51724002
39	0.67836967	0.64642352	0.61601850	0.55953126	0.50834400
40	0.67165314	0.63923216	0.60841334	0.55126232	0.49960098
41	0.66500311	0.63212080	0.60090206	0.54311559	0.49100834
42	0.65841892	0.62508855	0.59348352	0.53508925	0.48256348
43	0.65189992	0.61813454	0.58615656	0.52718153	0.47426386
44	0.64544546	0.61125789	0.57892006	0.51939067	0.46610699
45	0.63905492	0.60445774	0.57177290	0.51171494	0.45809040
46	0.63272764	0.59773324	0.56471397	0.50415265	0.45021170
47	0.62646301	0.59108355	0.55774219	0.49670212	0.44246850
48	0.62026041	0.58450784	0.55085649	0.48936170	0.43485848
49	0.61411921	0.57800528	0.54405579	0.48212975	0.42737934
50	0.60803882	0.57157506	0.53733905	0.47500468	0.42002883

Periods n	Rate i 0.01(1%)	0.001125(1 1/8%)	0.0125(1 1/4%)	0.0175(1 3/4%)	
51	0.60201864	0.56521637	0.53070524	0.46798491	0.41280475
52	0.59605806	0.55892843	0.52415332	0.46106887	0.40570492
53	0.59015649	0.55271044	0.51768229	0.45425505	0.39872719
54	0.58431336	0.54656162	0.51129115	0.44754192	0.39186947
55	0.57852808	0.54048120	0.50497892	0.44092800	0.38512970
56	0.57280008	0.53446843	0.49874461	0.43441182	0.37850585
57	0.56712879	0.52852256	0.49258727	0.42799194	0.37199592
58	0.56151365	0.52264282	0.48650594	0.42166694	0.36559796
59	0.55595411	0.51682850	0.48049970	0.41543541	0.35931003
60	0.55044962	0.51107887	0.47456760	0.40929597	0.35313025
61	0.54499962	0.50539319	0.46870874	0.40324726	0.34705676
62	0.53960358	0.49977077	0.46292222	0.39728794	0.34108772
63	0.53426097	0.49421090	0.45720713	0.39141669	0.33522135
64	0.52897126	0.48871288	0.45156259	0.38563221	0.32945587
65	0.52373392	0.48327602	0.44598775	0.37993321	0.32378956
66	0.51854844	0.47789965	0.44048173	0.37431843	0.31822069
67	0.51341429	0.47258309	0.43504368	0.36878663	0.31274761
68	0.50833099	0.46732568	0.42967277	0.36333658	0.30736866
69	0.50329801	0.46212675	0.42436817	0.35796708	0.30208222
70	0.49831486	0.45698566	0.41912905	0.35267692	0.29688670
71	0.49338105	0.45190177	0.41395462	0.34746495	0.29178054
72	0.48849609	0.44687443	0.40884407	0.34233000	0.28676221
73	0.48365949	0.44190302	0.40379661	0.33727093	0.28183018
74	0.47887078	0.43698692	0.39881147	0.33228663	0.27698298
75	0.47412949	0.43212551	0.39388787	0.32737599	0.27221914
76	0.46943514	0.42731818	0.38902506	0.32253793	0.26753724
77	0.46478726	0.42256433	0.38422228	0.31777136	0.26293586
78	0.46018541	0.41786337	0.37947879	0.31307523	0.25841362
79	0.45562912	0.41321470	0.37479387	0.30844850	0.25396916
80	0.45111794	0.40861775	0.37016679	0.30389015	0.24960114
81	0.44665142	0.40407194	0.36559683	0.29939916	0.24530825
82	0.44222913	0.39957670	0.36108329	0.29497454	0.24108919
83	0.43785063	0.39513148	0.35662547	0.29061531	0.23694269
84	0.43351547	0.39073570	0.35222268	0.28632050	0.23286751
85	0.42922324	0.38638882	0.34787426	0.28208917	0.22886242
86	0.42497350	0.38209031	0.34357951	0.27792036	0.22492621
87	0.42076585	0.37783961	0.33933779	0.27381316	0.22105770
88	0.41659985	0.37363621	0.33514843	0.26976666	0.21725572
89	0.41247510	0.36947956	0.33101080	0.26577996	0.21351914
90	0.40839119	0.36536916	0.32692425	0.26185218	0.20984682
91	0.40434771	0.36130448	0.32288814	0.25798245	0.20623766
92	0.40034427	0.35728503	0.31890187	0.25416990	0.20269057
93	0.39638046	0.35331029	0.31496481	0.25041369	0.19920450
94	0.39245590	0.34937976	0.31107636	0.24671300	0.19577837
95	0.38857020	0.34549297	0.30723591	0.24306699	0.19241118
96	0.38472297	0.34164941	0.30344287	0.23947487	0.18910190
97	0.38091383	0.33784861	0.29969666	0.23593583	0.18584953
98	0.37714241	0.33409010	0.29599670	0.23244909	0.18265310
99	0.37340832	0.33037340	0.29234242	0.22901389	0.17951165
100	0.36971121	0.32669805	0.28873326	0.22562944	0.17642422

PRESENT VALUE $1/(1+i)^n$ (Continued)

Periods n	0.02(2%)	0.0225(2 1/4%)	0.025(2 1/2%)	0.0275(2 3/4%)	Periods n	0.02(2%)	0.0225(2 1/4%)	0.025(2 1/2%)	0.0275(2 3/4%)
1	0.98039216	0.97799511	0.97560976	0.97323601	51	0.36424302	0.32149250	0.28384606	0.25068402
2	0.96116878	0.95647444	0.95181440	0.94718833	52	0.35710100	0.31441810	0.27692298	0.24397471
3	0.94232233	0.93542732	0.92859941	0.92183779	53	0.35009902	0.30749936	0.27016876	0.23744497
4	0.92384543	0.91484335	0.90595064	0.89716573	54	0.34323433	0.30073287	0.26357928	0.23109000
5	0.90573081	0.89471232	0.88385429	0.87315400	55	0.33650425	0.29411528	0.25715052	0.22490511
6	0.88797138	0.87502427	0.86229687	0.84978491	56	0.32990613	0.28764330	0.25087855	0.21888575
7	0.87056018	0.85576946	0.84126524	0.82704128	57	0.32343738	0.28131374	0.24475956	0.21302749
8	0.85349037	0.83693835	0.82074657	0.80490635	58	0.31709547	0.27512347	0.23878982	0.20732603
9	0.83675527	0.81852161	0.80072836	0.78336385	59	0.31087791	0.26906940	0.23296568	0.20177716
10	0.82034830	0.80051013	0.78119840	0.76239791	60	0.30478227	0.26314856	0.22728359	0.19637679
11	0.80426304	0.78289499	0.76214478	0.74199310	61	0.29880614	0.25735801	0.22174009	0.19112097
12	0.78849318	0.76566748	0.74355589	0.72213440	62	0.29294720	0.25169487	0.21633179	0.18600581
13	0.77303253	0.74881905	0.72542038	0.70280720	63	0.28720314	0.24615635	0.21105541	0.18102755
14	0.75787502	0.73234137	0.70772720	0.68399728	64	0.28157170	0.24073971	0.20590771	0.17618253
15	0.74301473	0.71622628	0.69046556	0.66569078	65	0.27605069	0.23544226	0.20088557	0.17146718
16	0.72844581	0.70046580	0.67362493	0.64787424	66	0.27063793	0.23026138	0.19598593	0.16687804
17	0.71416256	0.68505212	0.65719506	0.63053454	67	0.26533130	0.22519450	0.19120578	0.16241172
18	0.70015937	0.66997763	0.64115691	0.61365892	68	0.26012873	0.22023912	0.18654223	0.15806493
19	0.68643076	0.65523484	0.62552772	0.59723496	69	0.25502817	0.21539278	0.18199241	0.15383448
20	0.67297133	0.64081647	0.61027094	0.58125057	70	0.25002761	0.21065309	0.17755358	0.14971726
21	0.65977582	0.62671538	0.59538629	0.56569398	71	0.24512511	0.20601769	0.17322300	0.14571023
22	0.64683904	0.61292457	0.58086467	0.55055375	72	0.24031874	0.20148429	0.16899805	0.14181044
23	0.63415592	0.59943724	0.56669724	0.53581874	73	0.23560661	0.19705065	0.16487615	0.13801503
24	0.62172149	0.58624668	0.55287535	0.52147809	74	0.23098687	0.19271458	0.16085478	0.13432119
25	0.60953087	0.57334639	0.53939059	0.50752126	75	0.22645771	0.18847391	0.15693149	0.13072622
26	0.59757928	0.56072997	0.52623472	0.49393796	76	0.22201737	0.18432657	0.15310389	0.12722747
27	0.58586204	0.54839117	0.51339973	0.48071821	77	0.21766408	0.18027048	0.14936965	0.12382235
28	0.57437455	0.53632388	0.50087778	0.46785227	78	0.21339616	0.17630365	0.14572649	0.12050837
29	0.56311231	0.52452213	0.48866125	0.45533068	79	0.20921192	0.17242411	0.14217218	0.11728309
30	0.55207089	0.51298008	0.47674269	0.44314421	80	0.20510973	0.16862993	0.13870457	0.11414412
31	0.54124597	0.50169201	0.46511481	0.43128391	81	0.20108797	0.16491925	0.13532153	0.11108917
32	0.53063330	0.49065233	0.45377055	0.41974103	82	0.19714507	0.16129022	0.13202101	0.10811598
33	0.52022873	0.47985558	0.44270298	0.40850708	83	0.19327948	0.15774105	0.12880098	0.10522237
34	0.51002817	0.46929641	0.43190534	0.39757380	84	0.18948968	0.15426997	0.12565949	0.10240620
35	0.50002761	0.45896960	0.42137107	0.38693314	85	0.18577420	0.15087528	0.12259463	0.09966540
36	0.49022315	0.44887002	0.41109372	0.37657727	86	0.18213157	0.14755528	0.11960452	0.09699795
37	0.48061093	0.43899268	0.40106705	0.36649856	87	0.17856036	0.14430835	0.11668733	0.09440190
38	0.47118719	0.42933270	0.39128492	0.35668959	88	0.17505918	0.14113286	0.11384130	0.09187533
39	0.46194822	0.41988528	0.38174139	0.34714316	89	0.17162665	0.13802724	0.11106468	0.08941638
40	0.45289042	0.41064575	0.37243062	0.33785222	90	0.16826142	0.13498997	0.10835579	0.08702324
41	0.44401021	0.40160954	0.36334695	0.32880995	91	0.16496217	0.13201953	0.10571296	0.08469415
42	0.43530413	0.39277216	0.35448483	0.32000968	92	0.16172762	0.12911445	0.10313460	0.08242740
43	0.42676875	0.38412925	0.34583886	0.31144495	93	0.15855649	0.12627331	0.10061912	0.08022131
44	0.41840074	0.37567653	0.33740376	0.30310944	94	0.15544754	0.12349468	0.09816500	0.07807427
45	0.41019680	0.36740981	0.32917440	0.29499702	95	0.15239955	0.12077719	0.09577073	0.07598469
46	0.40215373	0.35932500	0.32114576	0.28710172	96	0.14941132	0.11811950	0.09343486	0.07395104
47	0.39426836	0.35141809	0.31331294	0.27941773	97	0.14648169	0.11552029	0.09115596	0.07197181
48	0.38653761	0.34368518	0.30567116	0.27193940	98	0.14360950	0.11297828	0.08893264	0.07004556
49	0.37895844	0.33612242	0.29821576	0.26466122	99	0.14079363	0.11049221	0.08676355	0.06817086
50	0.37152788	0.32872608	0.29094221	0.25757783	100	0.13803297	0.10806084	0.08464737	0.06634634

PRESENT VALUE $1/(1 + i)^n$ (Continued)

Periods	Rate *i*				Periods	Rate *i*			
n	0.03(3%)	0.0325(3 1/4%)	0.035(3 1/2%)	0.0375(3 3/4%)	*n*	0.03(3%)	0.0325(3 1/4%)	0.035(3 1/2%)	0.0375(3 3/4%)
1	0.97087379	0.96852300	0.96618357	0.96385542	51	0.22146318	0.19570725	0.17299843	0.15297013
2	0.94259591	0.93803681	0.93351070	0.92901727	52	0.21501280	0.18954698	0.16714824	0.14744109
3	0.91514166	0.90851022	0.90194271	0.89543834	53	0.20875029	0.18358061	0.16149589	0.14211189
4	0.88848705	0.87991305	0.87144223	0.86307310	54	0.20267019	0.17780204	0.15603467	0.13697532
5	0.86260878	0.85221603	0.84197317	0.83187768	55	0.19676717	0.17220537	0.15075814	0.13202440
6	0.83748426	0.82539083	0.81350064	0.80180981	56	0.19103609	0.16678486	0.14566004	0.12725243
7	0.81309151	0.79941000	0.78599096	0.77282874	57	0.18547193	0.16153497	0.14073433	0.12265295
8	0.78940923	0.77424698	0.75941156	0.74489517	58	0.18006984	0.15645034	0.13597520	0.11821971
9	0.76641673	0.74987601	0.73373097	0.71797125	59	0.17482508	0.15152575	0.13137701	0.11394671
10	0.74409391	0.72627216	0.70891881	0.69202048	60	0.16973309	0.14675617	0.12693431	0.10982815
11	0.72242128	0.70341129	0.68494571	0.66700769	61	0.16478941	0.14213673	0.12264184	0.10585846
12	0.70137988	0.68127002	0.66178330	0.64289898	62	0.15998972	0.13766269	0.11849453	0.10203225
13	0.68095134	0.65982568	0.63940415	0.61966167	63	0.15532982	0.13332948	0.11448747	0.09834434
14	0.66111781	0.63905635	0.61778179	0.59726426	64	0.15080565	0.12913267	0.11061591	0.09478972
15	0.64186195	0.61894078	0.59689062	0.57567639	65	0.14641325	0.12506796	0.10687528	0.09136359
16	0.62316694	0.59945838	0.57670591	0.55486881	66	0.14214879	0.12113120	0.10326114	0.08806129
17	0.60501645	0.58058923	0.55720378	0.53481331	67	0.13800853	0.11731835	0.09976922	0.08487835
18	0.58739461	0.56231402	0.53836114	0.51548271	68	0.13398887	0.11362552	0.09639538	0.08181046
19	0.57028603	0.54461407	0.52015569	0.49685080	69	0.13008628	0.11004893	0.09313563	0.07885346
20	0.55367575	0.52747125	0.50256588	0.47889234	70	0.12629736	0.10658492	0.08998612	0.07600333
21	0.53754928	0.51086804	0.48557090	0.46158298	71	0.12261880	0.10322995	0.08694311	0.07325622
22	0.52189250	0.49478745	0.46915063	0.44489926	72	0.11904737	0.09998058	0.08400300	0.07060841
23	0.50669175	0.47921302	0.45328563	0.42881856	73	0.11557998	0.09683349	0.08116232	0.06805630
24	0.49193374	0.46412884	0.43795713	0.41331910	74	0.11221357	0.09378546	0.07841770	0.06559643
25	0.47760557	0.44951945	0.42314699	0.39837985	75	0.10894521	0.09083338	0.07576590	0.06322547
26	0.46369473	0.43536993	0.40883767	0.38398058	76	0.10577205	0.08797422	0.07320376	0.06094022
27	0.45018906	0.42166579	0.39501224	0.37010176	77	0.10269131	0.08520505	0.07072827	0.05873756
28	0.43707675	0.40839302	0.38165434	0.35672459	78	0.09970030	0.08252305	0.06833650	0.05661451
29	0.42434636	0.39553803	0.36874815	0.34389093	79	0.09679641	0.07992548	0.06602560	0.05456821
30	0.41198676	0.38308768	0.35627841	0.33140331	80	0.09397710	0.07740966	0.06379285	0.05259586
31	0.39998715	0.37102923	0.34423035	0.31942487	81	0.09123990	0.07497304	0.06163461	0.05069481
32	0.38833703	0.35935035	0.33258971	0.30787940	82	0.08858243	0.07261311	0.05955131	0.04886246
33	0.37702625	0.34803908	0.32134271	0.29675123	83	0.08600236	0.07032747	0.05753750	0.04709635
34	0.36604490	0.33708385	0.31047605	0.28602528	84	0.08349743	0.06811377	0.05559178	0.04539407
35	0.35538340	0.32647346	0.29997686	0.27568702	85	0.08106547	0.06596976	0.05371187	0.04375332
36	0.35503243	0.31619706	0.28983272	0.26572242	86	0.07870434	0.06389323	0.05189553	0.04217188
37	0.33498294	0.30624413	0.28003161	0.25611800	87	0.07641198	0.06188206	0.05014060	0.04064759
38	0.32522615	0.29660448	0.27056194	0.24686072	88	0.07418639	0.05993420	0.04844503	0.03917840
39	0.31575355	0.28726826	0.26141250	0.23793805	89	0.07202562	0.05804765	0.04680679	0.03776232
40	0.30655684	0.27822592	0.25257247	0.22933788	90	0.06992779	0.05622048	0.04522395	0.03639741
41	0.29762800	0.26946820	0.24403137	0.22104855	91	0.06789105	0.05445083	0.04369464	0.03508184
42	0.28895922	0.26098615	0.23577910	0.21305885	92	0.06591364	0.05273688	0.04221704	0.03381383
43	0.28054294	0.25277109	0.22780590	0.20535793	93	0.06399383	0.05107688	0.04079841	0.03259164
44	0.27237178	0.24481462	0.22010231	0.19793535	94	0.06212993	0.04946914	0.03941006	0.03141363
45	0.26443862	0.23710859	0.21265924	0.19078106	95	0.06032032	0.04791200	0.03807735	0.03027820
46	0.25673653	0.22964512	0.20546787	0.18388536	96	0.05856342	0.04640387	0.03678971	0.02918380
47	0.24925876	0.22241658	0.19851968	0.17723890	97	0.05685769	0.04494322	0.03554562	0.02812897
48	0.24199880	0.21541558	0.19180645	0.17083268	98	0.05520164	0.04352854	0.03434359	0.02711226
49	0.23495029	0.20863494	0.18532024	0.16465800	99	0.05359383	0.04215839	0.03318221	0.02613230
[illegible]	[illegible]	[illegible]	0.17905337	0.15870651	100	0.05203284	0.04083137	0.03206011	0.02518776

PRESENT VALUE $1/(1+i)^n$ (Continued)

Periods n	0.04(4%)	0.0425(4 1/4%)	0.045(4 1/2%)	0.0475(4 3/4%)	Periods n	0.04(4%)	0.0425(4 1/4%)	0.045(4 1/2%)	0.0475(4 3/4%)
1	0.96153846	0.95923261	0.95693780	0.95465894	51	0.13530059	0.11970733	0.10594225	0.09378738
2	0.92455621	0.92012721	0.91572995	0.91136414	52	0.13009672	0.11482718	0.10138014	0.08953449
3	0.88899636	0.88261603	0.87629660	0.87003737	53	0.12509300	0.11014597	0.09701449	0.08547445
4	0.85480419	0.84663408	0.83856134	0.83058460	54	0.12028173	0.10565561	0.09283683	0.08159852
5	0.82192711	0.81211902	0.80245105	0.79292086	55	0.11565551	0.10134831	0.08883970	0.07789835
6	0.79031453	0.77901105	0.76789574	0.75696502	56	0.11120722	0.09721660	0.08501347	0.07436597
7	0.75991781	0.74725281	0.73482846	0.72263964	57	0.10693002	0.09325333	0.08135260	0.07099376
8	0.73069021	0.71678926	0.70318513	0.68987077	58	0.10281733	0.08945164	0.07784938	0.06777448
9	0.70258674	0.68756764	0.67290443	0.65858785	59	0.09886282	0.08580493	0.07449701	0.06470117
10	0.67556417	0.65953730	0.64392768	0.62872349	60	0.09506040	0.08230689	0.07128901	0.06176723
11	0.64958093	0.63264969	0.61619874	0.60021335	61	0.09140423	0.07895145	0.06821915	0.05896633
12	0.62459705	0.60685822	0.58966386	0.57299604	62	0.08788868	0.07573281	0.06528148	0.05629244
13	0.60057409	0.58211819	0.56427164	0.54701293	63	0.08450835	0.07264538	0.06247032	0.05373980
14	0.57747508	0.55838676	0.53997286	0.52220804	64	0.08125803	0.06968382	0.05978021	0.05130291
15	0.55526450	0.53562279	0.51672044	0.49852797	65	0.07813272	0.06684299	0.05720594	0.04897652
16	0.53390818	0.51378685	0.49446932	0.47592169	66	0.07512762	0.06411793	0.05474253	0.04675563
17	0.51337325	0.49284110	0.47317639	0.45434051	67	0.07223809	0.06150405	0.05238519	0.04463545
18	0.49362812	0.47274926	0.45280037	0.43373796	68	0.06945970	0.05899669	0.05012937	0.04261140
19	0.47464242	0.45347650	0.43330179	0.41406965	69	0.06678818	0.05659155	0.04797069	0.04067915
20	0.45638695	0.43498945	0.41464286	0.39529322	70	0.06421940	0.05428446	0.04590497	0.03883451
21	0.43883360	0.41725607	0.39678743	0.37736823	71	0.06174942	0.05207143	0.04392820	0.03707351
22	0.41295539	0.40024563	0.37970089	0.36025607	72	0.05937445	0.04994861	0.04203655	0.03539238
23	0.40572633	0.38392866	0.36335013	0.34391987	73	0.05709081	0.04791234	0.04022637	0.03378747
24	0.39012147	0.36827689	0.34770347	0.32832446	74	0.05489501	0.04595908	0.03849413	0.03225534
25	0.37511680	0.35326321	0.33273060	0.31343624	75	0.05278367	0.04408544	0.03683649	0.03079269
26	0.36068923	0.33886159	0.31840248	0.29922314	76	0.05075353	0.04228820	0.03525023	0.02939636
27	0.34681657	0.32504709	0.30469137	0.28565455	77	0.04880147	0.04056422	0.03373228	0.02806335
28	0.33347747	0.31179577	0.29157069	0.27270124	78	0.04692449	0.03891052	0.03227969	0.02679079
29	0.32065141	0.29908467	0.27901502	0.26033531	79	0.04511970	0.03732424	0.03088965	0.02557593
30	0.30831867	0.28689177	0.26700002	0.24853013	80	0.04338433	0.03580263	0.02955948	0.02441617
31	0.29646026	0.27519594	0.25550241	0.23726027	81	0.04171570	0.03434305	0.02828658	0.02330899
32	0.28505794	0.26397692	0.24449991	0.22650145	82	0.04011125	0.03294297	0.02706850	0.02225202
33	0.27409417	0.25321527	0.23397121	0.21623050	83	0.03856851	0.03159997	0.02590287	0.02124298
34	0.26355209	0.24289235	0.22389589	0.20642530	84	0.03708510	0.03031173	0.02478744	0.02027969
35	0.25341547	0.23299026	0.21425444	0.19706473	85	0.03565875	0.02907600	0.02372003	0.01936009
36	0.24366872	0.22349186	0.20502817	0.18812862	86	0.03428726	0.02789064	0.02269860	0.01848218
37	0.23429685	0.21438068	0.19619921	0.17959772	87	0.03296852	0.02675361	0.02172115	0.01764409
38	0.22528543	0.20564094	0.18775044	0.17134367	88	0.03170050	0.02566294	0.02078579	0.01684400
39	0.21662061	0.19725750	0.17966549	0.16367893	89	0.03048125	0.02461673	0.01989070	0.01608019
40	0.20828904	0.18921582	0.17192870	0.17145367	90	0.02930890	0.02361317	0.01903417	0.01535102
41	0.20027793	0.18150199	0.16452507	0.14917110	91	0.02818163	0.02265052	0.01821451	0.01465491
42	0.19257493	0.17410263	0.15744026	0.14240678	92	0.02709772	0.02172712	0.01743016	0.01399037
43	0.18516820	0.16700492	0.15066054	0.13594919	93	0.02605550	0.02084136	0.01667958	0.01335596
44	0.17804635	0.16019657	0.14417276	0.12978443	94	0.02505337	0.01999171	0.01596132	0.01275032
45	0.17119841	0.15366577	0.13796437	0.12389922	95	0.02408978	0.01917670	0.01527399	0.01217214
46	0.16461386	0.14740122	0.13202332	0.11828088	96	0.02316325	0.01839492	0.01461626	0.01162018
47	0.15828256	0.14139206	0.12633810	0.11291731	97	0.02227235	0.01764501	0.01398685	0.01109325
48	0.15219476	0.13562787	0.12089771	0.10779695	98	0.02141572	0.01692567	0.01338454	0.01059022
49	0.14634112	0.13009868	0.11569158	0.10290878	99	0.02059204	0.01623565	0.01280817	0.01010999
50	0.14071262	0.12479489	0.11070965	0.09824228	100	0.01980004	0.01557377	0.01225663	0.00965154

PRESENT VALUE $1/(1+i)^n$ (Continued)

Periods n	Rate i 0.05(5%)	0.0525(5 1/4%)	0.055(5 1/2%)	0.0575(5 3/4%)
1	0.95238095	0.95011876	0.94786730	0.94562648
2	0.90702948	0.90272567	0.89845242	0.89420944
3	0.86383760	0.85769660	0.85161366	0.84558812
4	0.82270247	0.81491363	0.80721674	0.79961051
5	0.78352617	0.77426473	0.76513434	0.75613287
6	0.74621540	0.73564345	0.72524583	0.71501927
7	0.71068133	0.69894865	0.68743681	0.67614115
8	0.67683936	0.66408423	0.65159887	0.63937697
9	0.64460892	0.63095888	0.61762926	0.60461180
10	0.61391325	0.59948588	0.58543058	0.57173692
11	0.58467929	0.56958278	0.55491050	0.54064957
12	0.55683742	0.54117129	0.52598152	0.51125255
13	0.53032135	0.51417699	0.49856068	0.48345395
14	0.50506795	0.48852921	0.47256937	0.45716685
15	0.48101710	0.46416077	0.44793305	0.43230908
16	0.45811152	0.44100786	0.42458109	0.40880291
17	0.43629669	0.41900984	0.40244653	0.38657486
18	0.41552065	0.39810911	0.38146590	0.36555542
19	0.39573396	0.37825094	0.36157906	0.34567889
20	0.37688948	0.35938331	0.34272896	0.32688311
21	0.35894236	0.34145683	0.32486158	0.30910932
22	0.34184987	0.32442454	0.30792567	0.29230196
23	0.32557131	0.30824185	0.29187267	0.27640847
24	0.31006791	0.29286636	0.27665656	0.26137917
25	0.29530277	0.27825783	0.26223370	0.24716706
26	0.28124073	0.26437798	0.24856275	0.23372772
27	0.26784832	0.25119048	0.23560450	0.22101912
28	0.25509364	0.23866079	0.22332181	0.20900153
29	0.24294632	0.22675609	0.21167944	0.19763738
30	0.23137745	0.21544522	0.20064402	0.18689114
31	0.22035947	0.20469855	0.19018390	0.17672921
32	0.20986617	0.19448793	0.18026910	0.16711982
33	0.19987254	0.18478663	0.17087119	0.15803293
34	0.19035480	0.17556925	0.16196321	0.14944012
35	0.18129029	0.16681164	0.15341963	0.14131454
36	0.17265741	0.15849087	0.14551624	0.13363077
37	0.16443563	0.15058515	0.13793008	0.12636479
38	0.15660536	0.14307377	0.13073941	0.11949389
39	0.14914797	0.13593708	0.12392362	0.11299659
40	0.14204568	0.12915637	0.11746314	0.10685257
41	0.13528160	0.12271389	0.11133947	0.10104262
42	0.12883962	0.11659277	0.10553504	0.09554857
43	0.12270440	0.11077698	0.10003322	0.09035326
44	0.11686133	0.10525128	0.09481822	0.08544044
45	0.11129651	0.10000122	0.08987509	0.08079474
46	0.10599668	0.09501304	0.08518965	0.07640164
47	0.10094921	0.09027367	0.08074849	0.07224742
48	0.09614211	0.08577071	0.07653885	0.06831907
49	0.09156391	0.08149236	0.07254867	0.06460432

Periods n	Rate i 0.05(5%)	0.0525(5 1/4%)	0.055(5 1/2%)	0.0575(5 3/4%)
51	0.08305117	0.07356524	0.06518153	0.05776979
52	0.07909635	0.06989572	0.06178344	0.05462865
53	0.07532986	0.06640923	0.05856250	0.05165830
54	0.07174272	0.06309666	0.05550948	0.04884945
55	0.06832640	0.05994932	0.05261562	0.04619333
56	0.06507276	0.05695897	0.04987263	0.04368164
57	0.06197406	0.05411779	0.04727263	0.04130652
58	0.05902291	0.05141833	0.04480818	0.03906054
59	0.05621230	0.04885352	0.04247221	0.03693668
60	0.05353552	0.04641664	0.04025802	0.03492830
61	0.05098621	0.04410132	0.03815926	0.03302912
62	0.04855830	0.04190150	0.03616992	0.03123321
63	0.04624600	0.03981140	0.03428428	0.02953495
64	0.04404381	0.03782556	0.03249695	0.02792904
65	0.04194648	0.03593877	0.03080279	0.02641044
66	0.03994903	0.03414610	0.02919696	0.02497441
67	0.03804670	0.03244285	0.02767485	0.02361646
68	0.03623495	0.03082456	0.02623208	0.02233235
69	0.03450948	0.02928699	0.02486453	0.02111806
70	0.03286617	0.02782612	0.02356828	0.01996980
71	0.03130111	0.02643812	0.02233960	0.01888397
72	0.02981058	0.02511935	0.02117498	0.01785718
73	0.02839103	0.02386637	0.02007107	0.01688622
74	0.02703908	0.02267589	0.01902471	0.01596806
75	0.02575150	0.02154478	0.01803290	0.01509982
76	0.02452524	0.02047010	0.01709279	0.01427879
77	0.02335737	0.01944903	0.01620170	0.01350240
78	0.02224512	0.01847889	0.01535706	0.01276823
79	0.02118582	0.01755714	0.01455646	0.01207398
80	0.02017698	0.01668137	0.01379759	0.01141747
81	0.01921617	0.01584928	0.01307828	0.01079666
82	0.01830111	0.01505870	0.01239648	0.01020961
83	0.01742963	0.01430755	0.01175022	0.00965448
84	0.01659965	0.01359387	0.01113765	0.00912953
85	0.01580919	0.01291579	0.01055701	0.00863313
86	0.01505637	0.01227154	0.01000664	0.00816371
87	0.01433940	0.01165942	0.00948497	0.00771982
88	0.01365657	0.01107783	0.00899049	0.00730007
89	0.01300626	0.01052526	0.00852180	0.00690314
90	0.01238691	0.01000024	0.00807753	0.00652779
91	0.01179706	0.00950142	0.00765643	0.00617285
92	0.01123530	0.00902748	0.00725728	0.00583721
93	0.01070028	0.00857718	0.00687894	0.00551982
94	0.01019074	0.00814934	0.00652032	0.00521969
95	0.00970547	0.00774284	0.00618040	0.00493588
96	0.00924331	0.00735661	0.00585820	0.00466750
97	0.00880315	0.00698966	0.00555279	0.00441371
98	0.00838395	0.00664100	0.00526331	0.00417372
99	0.00798471	0.00630974	0.00498892	0.00394678
	[illegible]	[illegible]	0.00472883	0.00373218

PRESENT VALUE $1/(1+i)^n$ (Continued)

Periods n	Rate i 0.06(6%)	0.0625(6 1/4%)	0.065(6 1/2%)	0.0675(6 3/4%)	Periods n	Rate i 0.06(6%)	0.0625(6 1/4%)	0.065(6 1/2%)	0.0675(6 3/4%)
1	0.94339623	0.94117647	0.93896714	0.93676815	51	0.05121544	0.04541759	0.04028747	0.03574686
2	0.88999644	0.88581315	0.88165928	0.87753457	52	0.04831645	0.04274597	0.03782861	0.03348652
3	0.83961928	0.83370649	0.82784909	0.82204643	53	0.04558156	0.04023150	0.03551982	0.03136911
4	0.79209366	0.78466493	0.77732309	0.77006692	54	0.04300147	0.03786494	0.03335195	0.02938558
5	0.74725817	0.73850817	0.72988084	0.72137416	55	0.04056742	0.03563759	0.03131638	0.02752748
6	0.70496054	0.69506652	0.68533412	0.67576034	56	0.03827115	0.03354126	0.02940505	0.02578686
7	0.66505711	0.65418025	0.64350621	0.63303076	57	0.03610486	0.03156825	0.02761038	0.02415631
8	0.62741237	0.61569906	0.60423119	0.59300305	58	0.03406119	0.02971129	0.02592524	0.02262886
9	0.59189846	0.57948147	0.56735323	0.55550637	59	0.03213320	0.02796357	0.02434295	0.02119800
10	0.55839478	0.54539432	0.53272604	0.52038068	60	0.03031434	0.02631865	0.02285723	0.01985761
11	0.52678753	0.51331230	0.50021224	0.48747605	61	0.02859843	0.02477050	0.02146218	0.01860198
12	0.49696936	0.48311746	0.46968285	0.45665203	62	0.02697965	0.02331341	0.02015229	0.01742574
13	0.46883902	0.45469879	0.44101676	0.42777708	63	0.02545250	0.02194203	0.01892233	0.01632388
14	0.44230096	0.42795180	0.41410025	0.40072794	64	0.02401179	0.02065133	0.01776745	0.01529169
15	0.41726506	0.40277817	0.38882652	0.37538917	65	0.02265264	0.01943654	0.01668305	0.01432477
16	0.39364628	0.37908533	0.36509533	0.35165262	66	0.02137041	0.01829322	0.01566484	0.01341899
17	0.37136442	0.35678619	0.34281251	0.32941698	67	0.02016077	0.01721714	0.01470877	0.01257048
18	0.35034379	0.33579877	0.32188969	0.30858733	68	0.01901959	0.01620437	0.01381105	0.01177562
19	0.33051301	0.31604590	0.30224384	0.28907478	69	0.01794301	0.01525117	0.01296812	0.01103103
20	0.31180473	0.29745497	0.28379703	0.27079605	70	0.01692737	0.01435404	0.01217664	0.01033352
21	0.29415540	0.27995762	0.26647608	0.25367312	71	0.01596921	0.01350969	0.01143346	0.00968011
22	0.27750510	0.26348952	0.25021228	0.23763289	72	0.01506530	0.01271500	0.01073565	0.00906802
23	0.26179726	0.24799014	0.23494111	0.22260693	73	0.01421254	0.01196706	0.01008042	0.00849463
24	0.24697855	0.23340248	0.22060198	0.20853108	74	0.01340806	0.01126312	0.00946518	0.00795750
25	0.23299863	0.21967292	0.20713801	0.19534527	75	0.01264911	0.01060058	0.00888750	0.00745433
26	0.21981003	0.20675099	0.19449579	0.18299323	76	0.01193313	0.00997702	0.00834507	0.00698298
27	0.20736795	0.19458917	0.18262515	0.17142223	77	0.01125767	0.00939013	0.00783574	0.00654143
28	0.19563014	0.18314274	0.17147902	0.16058289	78	0.01062044	0.00883777	0.00735751	0.00612781
29	0.18455674	0.17236964	0.16101316	0.15042893	79	0.01001928	0.00831790	0.00690846	0.00574033
30	0.17411013	0.16223025	0.15118607	0.14091703	80	0.00945215	0.00782861	0.00648681	0.00537736
31	0.16425484	0.15268729	0.14195875	0.13200659	81	0.00891713	0.00736811	0.00609090	0.00503734
32	0.15495740	0.14370569	0.13329460	0.12365957	82	0.00841238	0.00693469	0.00571916	0.00471882
33	0.14618622	0.13525241	0.12515925	0.11584034	83	0.00793621	0.00652677	0.00537010	0.00442044
34	0.13791153	0.12729639	0.11752042	0.10851554	84	0.00748699	0.00614284	0.00504235	0.00414093
35	0.13010522	0.11980837	0.11034781	0.10165391	85	0.00706320	0.00578150	0.00473460	0.00387909
36	0.12274077	0.11276081	0.10361297	0.09522614	86	0.00666340	0.00544141	0.00444563	0.00363381
37	0.11579318	0.10612783	0.09728917	0.08920482	87	0.00628622	0.00512132	0.00417430	0.00340404
38	0.10923885	0.09988501	0.09135134	0.08356423	88	0.00593040	0.00482007	0.00391953	0.00318879
39	0.10305552	0.09400942	0.08577590	0.07828031	89	0.00559472	0.00453654	0.00368031	0.00298716
40	0.09722219	0.08847946	0.08054075	0.07333050	90	0.00527803	0.00426968	0.00345569	0.00279828
41	0.09171905	0.08327478	0.07562512	0.06869368	91	0.00497928	0.00401852	0.00324478	0.00262134
42	0.08652740	0.07837627	0.07100950	0.06435005	92	0.00469743	0.00378214	0.00304674	0.00245558
43	0.08162962	0.07376590	0.06667559	0.06028108	93	0.00443154	0.00355966	0.00286079	0.00230031
44	0.07700908	0.06942673	0.06260619	0.05646939	94	0.00418070	0.00335027	0.00268619	0.00215486
45	0.07265007	0.06534280	0.05878515	0.05289873	95	0.00394405	0.00315319	0.00252224	0.00201860
46	0.06853781	0.06149911	0.05519733	0.04955384	96	0.00372081	0.00296771	0.00236831	0.00189096
47	0.06465831	0.05788151	0.05182848	0.04642046	97	0.00351019	0.00279314	0.00222376	0.00177139
48	0.06099840	0.05447672	0.04866524	0.04348521	98	0.00331150	0.00262884	0.00208804	0.00165939
49	0.05754566	0.05127221	0.04569506	0.04073556	99	0.00312406	0.00247420	0.00196060	0.00155446
50	0.05428836	0.04825619	0.04290616	0.03815978	100	0.00294723	0.00232866	0.00184094	0.00145617

PRESENT VALUE $1/(1+i)^n$ (Continued)

Periods n	Rate i 0.07(7%)	0.0725(7 1/4%)	0.075(7 1/2%)	0.0775(7 3/4%)	Periods n	Rate i 0.07(7%)	0.0725(7 1/4%)	0.075(7 1/2%)	0.0775(7 3/4%)
1	0.93457944	0.93240093	0.93023256	0.92807425	51	0.03172688	0.02816681	0.02501315	0.02221871
2	0.87343873	0.86937150	0.86533261	0.86132181	52	0.02965129	0.02626276	0.02326804	0.02062061
3	0.81629788	0.81060280	0.80496057	0.79937059	53	0.02771148	0.02448742	0.02164469	0.01913746
4	0.76289521	0.75580680	0.74880053	0.74187525	54	0.02589858	0.02283209	0.02013460	0.01776098
5	0.71298618	0.70471497	0.69655863	0.68851532	55	0.02420428	0.02128867	0.01872986	0.01648351
6	0.66634222	0.65707689	0.64796152	0.63899333	56	0.02262083	0.01984957	0.01742312	0.01529792
7	0.62274974	0.61265911	0.60275490	0.59303326	57	0.02114096	0.01850776	0.01620756	0.01419761
8	0.58200910	0.57124392	0.56070223	0.55037889	58	0.01975791	0.01725665	0.01507680	0.01317643
9	0.54393374	0.53262837	0.52158347	0.51079247	59	0.01846533	0.01609012	0.01402493	0.01222871
10	0.50834929	0.49662319	0.48519393	0.47405334	60	0.01725732	0.01500244	0.01304644	0.01134915
11	0.47509280	0.46305192	0.45134319	0.43995670	61	0.01612834	0.01398829	0.01213623	0.01053285
12	0.44401196	0.43175004	0.41985413	0.40831248	62	0.01507321	0.01304270	0.01128951	0.00977527
13	0.41496445	0.40256414	0.39056198	0.37894430	63	0.01408711	0.01216102	0.01050187	0.00907218
14	0.38781724	0.37535118	0.36331347	0.35168844	64	0.01316553	0.01133895	0.00976918	0.00841965
15	0.36244602	0.34997779	0.33796602	0.32639299	65	0.01230423	0.01057245	0.00908761	0.00781406
16	0.33873460	0.32631962	0.31438699	0.30291692	66	0.01149928	0.00985776	0.00845359	0.00725203
17	0.31657439	0.30426072	0.29245302	0.28112940	67	0.01074699	0.00919138	0.00786381	0.00673042
18	0.29586392	0.28369298	0.27204932	0.26090895	68	0.01004392	0.00857005	0.00731517	0.00624633
19	0.27650833	0.26451560	0.25306913	0.24214288	69	0.00938684	0.00799073	0.00680481	0.00579706
20	0.25841900	0.24663459	0.23541315	0.22472657	70	0.00877275	0.00745056	0.00633006	0.00538010
21	0.24151309	0.22996232	0.21898897	0.20856294	71	0.00819883	0.00694691	0.00588842	0.00499313
22	0.22571317	0.21441708	0.20371067	0.19356190	72	0.00766246	0.00647730	0.00547760	0.00463400
23	0.21094688	0.19992269	0.18949830	0.17963981	73	0.00716117	0.00603945	0.00509544	0.00430070
24	0.19714662	0.18640810	0.17627749	0.16671908	74	0.00669269	0.00563118	0.00473995	0.00399136
25	0.18424918	0.17380709	0.16397906	0.15472769	75	0.00625485	0.00525052	0.00440925	0.00370428
26	0.17219549	0.16205789	0.15253866	0.14359878	76	0.00584565	0.00489559	0.00410163	0.00343785
27	0.16093037	0.15110293	0.14189643	0.13327033	77	0.00546323	0.00456465	0.00381547	0.00319058
28	0.15040221	0.14088851	0.13199668	0.12368476	78	0.00510582	0.00425609	0.00354928	0.00296109
29	0.14056282	0.13136458	0.12278761	0.11478864	79	0.00477179	0.00396838	0.00330165	0.00274812
30	0.13136712	0.12248446	0.11422103	0.10653238	80	0.00445962	0.00370012	0.00307130	0.00255046
31	0.12277301	0.11420462	0.10625212	0.09886996	81	0.00416787	0.00345000	0.00285703	0.00236701
32	0.11474113	0.10648449	0.09883918	0.09175866	82	0.00389520	0.00321678	0.00265770	0.00219676
33	0.10723470	0.09928624	0.09194343	0.08515885	83	0.00364038	0.00299933	0.00247228	0.00203876
34	0.10021934	0.09257458	0.08552877	0.07903374	84	0.00340222	0.00279658	0.00229979	0.00189212
35	0.09366294	0.08631663	0.07956164	0.07334918	85	0.00317965	0.00260753	0.00213934	0.00175603
36	0.08753546	0.08048171	0.07401083	0.06807348	86	0.00297163	0.00243126	0.00199009	0.00162972
37	0.08180884	0.07504122	0.06884729	0.06317724	87	0.00277723	0.00226691	0.00185124	0.00151250
38	0.07645686	0.06996850	0.06404399	0.05863317	88	0.00259554	0.00211367	0.00172209	0.00140372
39	0.07145501	0.06523870	0.05957580	0.05441594	89	0.00242574	0.00197079	0.00160194	0.00130275
40	0.06678038	0.06082862	0.05541935	0.05050203	90	0.00226704	0.00183757	0.00149018	0.00120905
41	0.06241157	0.05671666	0.05155288	0.04686963	91	0.00211873	0.00171335	0.00138621	0.00112209
42	0.05832857	0.05288267	0.04795617	0.04349850	92	0.00198012	0.00159753	0.00128950	0.00104138
43	0.05451268	0.04930785	0.04461039	0.04036984	93	0.00185058	0.00148954	0.00119953	0.00096648
44	0.05094643	0.04597468	0.04149804	0.03746621	94	0.00172952	0.00138884	0.00111585	0.00089697
45	0.04761349	0.04286684	0.03860283	0.03477142	95	0.00161637	0.00129496	0.00103800	0.00083245
46	0.04449859	0.03996908	0.03590961	0.03227046	96	0.00151063	0.00120742	0.00096558	0.00077258
47	0.04158747	0.03726721	0.03340428	0.02994938	97	0.00141180	0.00112580	0.00089821	0.00071701
48	0.03886679	0.03474798	0.03107375	0.02779525	98	0.00131944	0.00104970	0.00083555	0.00066544
49	0.03632410	0.03239905	0.02890582	0.02579606	99	0.00123312	0.00097874	0.00077725	0.00061757
50	0.03394776	0.03020890	0.02688913	0.02394066	100	0.00115245	0.00091258	0.00072303	0.00057316

PRESENT VALUE $1/(1+i)^n$ (Continued)

Periods n	Rate i 0.08(8%)	0.0825(8 1/4%)	0.085(8 1/2%)	0.0875(8 3/4%)
1	0.92592593	0.92378753	0.92165899	0.91954023
2	0.85733882	0.85338340	0.84945529	0.84555423
3	0.79383224	0.78834494	0.78290810	0.77752114
4	0.73502985	0.72826322	0.72157428	0.71496196
5	0.68058320	0.67276048	0.66504542	0.65743629
6	0.63016963	0.62148775	0.61294509	0.60453912
7	0.58349040	0.57412263	0.56492635	0.55589804
8	0.54026888	0.53036732	0.52066945	0.51117061
9	0.50024897	0.48994672	0.47987968	0.47004194
10	0.46319349	0.45260667	0.44228542	0.43222247
11	0.42888286	0.41811240	0.40763633	0.39744595
12	0.39711376	0.38624702	0.37570168	0.36546754
13	0.36769792	0.35681018	0.34626883	0.33606211
14	0.34046104	0.32961679	0.31914178	0.30902263
15	0.31524170	0.30449588	0.29413989	0.28415874
16	0.29189047	0.28128950	0.27109667	0.26129539
17	0.27026895	0.25985173	0.24985869	0.24027162
18	0.25024903	0.24004779	0.23028450	0.22093942
19	0.23171206	0.22175315	0.21224378	0.20316269
20	0.21454821	0.20485280	0.19561639	0.18681627
21	0.19865575	0.18924046	0.18029160	0.17178507
22	0.18394051	0.17481798	0.16616738	0.15796328
23	0.17031528	0.16149467	0.15314965	0.14525360
24	0.15769934	0.14918676	0.14115176	0.13356652
25	0.14601790	0.13781687	0.13009378	0.12281979
26	0.13520176	0.12731350	0.11990210	0.11293774
27	0.12518682	0.11761063	0.11050885	0.10385080
28	0.11591372	0.10864723	0.10185148	0.09549498
29	0.10732752	0.10036696	0.09387233	0.08781148
30	0.09937733	0.09271774	0.08651828	0.08074619
31	0.09201605	0.08565149	0.07974035	0.07424937
32	0.08520005	0.07912378	0.07349341	0.06827528
33	0.07888893	0.07309356	0.06773586	0.06278187
34	0.07304531	0.06752292	0.06242936	0.05773045
35	0.06763454	0.06237683	0.05753858	0.05308547
36	0.06262458	0.05762294	0.05303095	0.04881423
37	0.05798572	0.05323135	0.04887645	0.04488665
38	0.05369048	0.04917446	0.04504742	0.04127508
39	0.04971341	0.04542675	0.04151836	0.03795410
40	0.04603093	0.04196467	0.03826577	0.03490032
41	0.04262123	0.03876644	0.03526799	0.03209225
42	0.03946411	0.03581195	0.03250506	0.02951011
43	0.03654084	0.03308263	0.02995858	0.02713573
44	0.03383411	0.03056132	0.02761160	0.02495240
45	0.03132788	0.02823217	0.02544848	0.02294474
46	0.02900730	0.02608053	0.02345482	0.02109861
47	0.02685861	0.02409287	0.02161734	0.01940102
48	0.02486908	0.02225669	0.01992382	0.01784002
49	0.02302693	0.02056045	0.01836297	0.01640461
50	0.02132123	0.01899349	0.01692439	0.01508470

Periods n	Rate i 0.08(8%)	0.0825(8 1/4%)	0.085(8 1/2%)	0.0875(8 3/4%)
51	0.01974188	0.01754595	0.01559852	0.01387099
52	0.01827952	0.01620873	0.01437651	0.01275493
53	0.01692548	0.01497342	0.01325024	0.01172867
54	0.01567174	0.01383226	0.01221221	0.01078499
55	0.01451087	0.01277807	0.01125549	0.00991723
56	0.01343599	0.01180422	0.01037372	0.00911929
57	0.01244073	0.01090459	0.00956104	0.00838556
58	0.01151920	0.01007353	0.00881201	0.00771086
59	0.01066592	0.00930580	0.00812167	0.00709044
60	0.00987585	0.00859658	0.00748541	0.00651995
61	0.00914431	0.00794141	0.00689900	0.00599535
62	0.00846695	0.00733618	0.00635852	0.00551297
63	0.00783977	0.00677707	0.00586039	0.00506940
64	0.00725905	0.00626057	0.00540128	0.00466151
65	0.00672134	0.00578344	0.00497814	0.00428645
66	0.00622346	0.00534267	0.00458815	0.00394156
67	0.00576247	0.00493549	0.00422871	0.00362443
68	0.00533562	0.00455935	0.00389743	0.00333281
69	0.00494039	0.00421187	0.00359210	0.00306465
70	0.00457443	0.00389087	0.00331069	0.00281807
71	0.00423558	0.00359434	0.00305133	0.00259133
72	0.00392184	0.00332040	0.00281228	0.00238283
73	0.00363133	0.00306735	0.00259196	0.00219111
74	0.00336234	0.00283358	0.00238891	0.00201481
75	0.00311328	0.00261762	0.00220176	0.00185270
76	0.00288267	0.00241813	0.00202927	0.00170363
77	0.00266914	0.00223384	0.00187030	0.00156656
78	0.00247142	0.00206359	0.00172377	0.00144051
79	0.00228835	0.00190632	0.00158873	0.00132461
80	0.00211885	0.00176103	0.00146427	0.00121803
81	0.00196190	0.00162682	0.00134956	0.00112003
82	0.00181657	0.00150284	0.00124383	0.00102991
83	0.00168201	0.00138830	0.00114639	0.00094705
84	0.00155742	0.00128250	0.00105658	0.00087085
85	0.00144205	0.00118475	0.00097381	0.00080078
86	0.00133523	0.00109446	0.00089752	0.00073635
87	0.00123633	0.00101105	0.00082720	0.00067710
88	0.00114475	0.00093399	0.00076240	0.00062262
89	0.00105995	0.00086281	0.00070267	0.00057253
90	0.00098144	0.00079706	0.00064762	0.00052646
91	0.00090874	0.00073631	0.00059689	0.00048410
92	0.00084142	0.00068019	0.00055013	0.00044515
93	0.00077910	0.00062835	0.00050703	0.00040933
94	0.00072138	0.00058047	0.00046731	0.00037640
95	0.00066795	0.00053623	0.00043070	0.00034611
96	0.00061847	0.00049536	0.00039696	0.00031827
97	0.00057266	0.00045761	0.00036586	0.00029266
98	0.00053024	0.00042273	0.00033720	0.00026911
99	0.00049096	0.00039051	0.00031078	0.00024746
100	0.00045459	0.00036075	0.00028644	0.00022755

PRESENT VALUE $1/(1+i)^n$ (Continued)

Periods n	Rate i 0.09(9%)	0.0925(9 1/4%)	0.095(9 1/2%)	0.0975(9 3/4%)
1	0.91743119	0.91533181	0.91324201	0.91116173
2	0.84167999	0.83783232	0.83401097	0.83021570
3	0.77218348	0.76689457	0.76165385	0.75646077
4	0.70842521	0.70196299	0.69557429	0.68925811
5	0.64993139	0.64252906	0.63522767	0.62802561
6	0.59626733	0.58812728	0.58011659	0.57223290
7	0.54703424	0.53833161	0.52978684	0.52139672
8	0.50186628	0.49275204	0.48382360	0.47507674
9	0.46042778	0.45103162	0.44184803	0.43287175
10	0.42241081	0.41284359	0.40351419	0.39441617
11	0.38753285	0.37788887	0.36850611	0.35937692
12	0.35553473	0.34589370	0.33653626	0.32745050
13	0.32617865	0.31660751	0.30733813	0.29836036
14	0.29924647	0.28980092	0.28067410	0.27185454
15	0.27453804	0.26526400	0.25632337	0.24770346
16	0.25186976	0.24280458	0.23408527	0.22569791
17	0.23107318	0.22224675	0.21377651	0.20564730
18	0.21199374	0.20342952	0.19522969	0.18737795
19	0.19448967	0.18620551	0.17829195	0.17073162
20	0.17843089	0.17043983	0.16282370	0.15556411
21	0.16369806	0.15600900	0.14869744	0.14174407
22	0.15018171	0.14280000	0.13579675	0.12915177
23	0.13778139	0.13070938	0.12401530	0.11767815
24	0.12640494	0.11964245	0.11325598	0.10722383
25	0.11596784	0.10951254	0.10343012	0.09769825
26	0.10639251	0.10024031	0.09445673	0.08901891
27	0.09760781	0.09175315	0.08626185	0.08111062
28	0.08954845	0.08398457	0.07877795	0.07390489
29	0.08215454	0.07687375	0.07194333	0.06733931
30	0.07537114	0.07036499	0.06570167	0.06135700
31	0.06914783	0.06440731	0.06000153	0.05590615
32	0.06343838	0.05895406	0.05479592	0.05093955
33	0.05820035	0.05396253	0.05004193	0.04641417
34	0.05339481	0.04939362	0.04570039	0.04229081
35	0.04898607	0.04521155	0.04173552	0.03853377
36	0.04494135	0.04138357	0.03811463	0.03511050
37	0.04123059	0.03787970	0.03480788	0.03199134
38	0.03782623	0.03467249	0.03178802	0.02914928
39	0.03470296	0.03173684	0.02903015	0.02655971
40	0.03183758	0.02904973	0.02651156	0.02420019
41	0.02920879	0.02659015	0.02421147	0.02205029
42	0.02679706	0.02433881	0.02211093	0.02009138
43	0.02458446	0.02227808	0.02019263	0.01830650
44	0.02255455	0.02039184	0.01844076	0.01668018
45	0.02069224	0.01866530	0.01684087	0.01519834
46	0.01898371	0.01708494	0.01537979	0.01384815
47	0.01741625	0.01563839	0.01404547	0.01261790
48	0.01597821	0.01431432	0.01282692	0.01149695
49	0.01465891	0.01310235	0.01171408	0.01047558
50	0.01344854	0.01199300	0.01069779	0.00954495

Periods n	Rate i 0.09(9%)	0.0925(9 1/4%)	0.095(9 1/2%)	0.0975(9 3/4%)
51	0.01233811	0.01097757	0.00976967	0.00869699
52	0.01131937	0.01004812	0.00892207	0.00792437
53	0.01038474	0.00919736	0.00814801	0.00722038
54	0.00952728	0.00841864	0.00744111	0.00657893
55	0.00874063	0.00770585	0.00679553	0.00599447
56	0.00801892	0.00705341	0.00620597	0.00546193
57	0.00735681	0.00645621	0.00566755	0.00497670
58	0.00674937	0.00590957	0.00517584	0.00453458
59	0.00619208	0.00540922	0.00472680	0.00413174
60	0.00568081	0.00495123	0.00431671	0.00376468
61	0.00521175	0.00453202	0.00394220	0.00343023
62	0.00478142	0.00414830	0.00360018	0.00312550
63	0.00438663	0.00379707	0.00328784	0.00284783
64	0.00402443	0.00347558	0.00300259	0.00259484
65	0.00369214	0.00318131	0.00274209	0.00236432
66	0.00338728	0.00291195	0.00250419	0.00215427
67	0.00310760	0.00266540	0.00228694	0.00196289
68	0.00285101	0.00243973	0.00208853	0.00178851
69	0.00261560	0.00223316	0.00190733	0.00162962
70	0.00239963	0.00204408	0.00174185	0.00148485
71	0.00220150	0.00187102	0.00159073	0.00135294
72	0.00201972	0.00171260	0.00145273	0.00123275
73	0.00185296	0.00156760	0.00132669	0.00112323
74	0.00169996	0.00143487	0.00121159	0.00102345
75	0.00155960	0.00131338	0.00110647	0.00093252
76	0.00143082	0.00120218	0.00101048	0.00084968
77	0.00131268	0.00110040	0.00092281	0.00077420
78	0.00120430	0.00100723	0.00084275	0.00070542
79	0.00110486	0.00092195	0.00076963	0.00064275
80	0.00101363	0.00084389	0.00070286	0.00058565
81	0.00092994	0.00077244	0.00064188	0.00053362
82	0.00085315	0.00070704	0.00058620	0.00048622
83	0.00078271	0.00064717	0.00053534	0.00044302
84	0.00071808	0.00059238	0.00048889	0.00040366
85	0.00065879	0.00054222	0.00044648	0.00036780
86	0.00060440	0.00049631	0.00040774	0.00033513
87	0.00055449	0.00045429	0.00037237	0.00030536
88	0.00050871	0.00041583	0.00034006	0.00027823
89	0.00046670	0.00038062	0.00031056	0.00025351
90	0.00042817	0.00034839	0.00028362	0.00023099
91	0.00039282	0.00031890	0.00025901	0.00021047
92	0.00036038	0.00029190	0.00023654	0.00019177
93	0.00033063	0.00026718	0.00021602	0.00017473
94	0.00030333	0.00024456	0.00019728	0.00015921
95	0.00027828	0.00022385	0.00018016	0.00014507
96	0.00025530	0.00020490	0.00016453	0.00013218
97	0.00023422	0.00018755	0.00015026	0.00012044
98	0.00021488	0.00017167	0.00013722	0.00010974
99	0.00019714	0.00015714	0.00012531	0.00009999
100	0.00018086	0.00014383	0.00011444	0.00009111

PRESENT VALUE $1/(1+i)^n$ (Continued)

Periods n	Rate i 0.10(10%)	0.1025(10 1/4%)	0.105(10 1/2%)	0.1075(10 3/4%)	Periods n	Rate i 0.10(10%)	0.1025(10 1/4%)	0.105(10 1/2%)	0.1075(10 3/4%)
1	0.90909091	0.90702948	0.90497738	0.90293454	51	0.00774414	0.00689750	0.00614503	0.00547608
2	0.82644628	0.82270247	0.81898405	0.81529078	52	0.00704013	0.00625623	0.00556111	0.00494454
3	0.75131480	0.74621540	0.74116204	0.73615420	53	0.00640011	0.00567459	0.00503268	0.00446459
4	0.68301346	0.67683936	0.67073487	0.66469905	54	0.00581829	0.00514702	0.00455446	0.00403124
5	0.62092132	0.61391325	0.60699989	0.60017973	55	0.00528935	0.00466850	0.00412168	0.00363994
6	0.56447393	0.55683742	0.54932116	0.54192301	56	0.00480850	0.00423446	0.00373003	0.00328663
7	0.51315812	0.50506795	0.49712323	0.48932100	57	0.00437136	0.00384078	0.00337559	0.00296761
8	0.46650738	0.45811152	0.44988527	0.44182483	58	0.00397397	0.00348370	0.00305484	0.00267956
9	0.42409762	0.41552065	0.40713599	0.39893890	59	0.00361270	0.00315982	0.00276456	0.00241947
10	0.38554329	0.37688948	0.36844886	0.36021571	60	0.00328427	0.00286605	0.00250186	0.00218462
11	0.35049390	0.34184987	0.33343788	0.32525121	61	0.00298570	0.00259959	0.00226413	0.00197257
12	0.31863082	0.31006791	0.30175374	0.29368055	62	0.00271427	0.00235791	0.00204898	0.00178110
13	0.28966438	0.28124073	0.27308031	0.26517431	63	0.00246752	0.00213869	0.00185428	0.00160822
14	0.26333125	0.25509364	0.24713150	0.23943504	64	0.00224320	0.00193986	0.00167809	0.00145211
15	0.23939205	0.23137745	0.22364842	0.21619417	65	0.00203927	0.00175951	0.00151863	0.00131116
16	0.21762914	0.20986617	0.20239676	0.19520918	66	0.00185388	0.00159593	0.00137433	0.00118390
17	0.19784467	0.19035480	0.18316449	0.17626111	67	0.00168535	0.00144755	0.00124373	0.00106898
18	0.17985879	0.17265741	0.16575972	0.15915225	68	0.00153214	0.00131297	0.00112555	0.00096522
19	0.16350799	0.15660536	0.15000879	0.14370406	69	0.00139285	0.00119090	0.00101860	0.00087153
20	0.14864363	0.14204568	0.13575456	0.12975536	70	0.00126623	0.00108018	0.00092181	0.00078693
21	0.13513057	0.12883962	0.12285481	0.11716059	71	0.00115112	0.00097976	0.00083422	0.00071055
22	0.12284597	0.11686133	0.11118082	0.10578835	72	0.00104647	0.00088867	0.00075495	0.00064158
23	0.11167816	0.10599668	0.10061613	0.09551995	73	0.00095134	0.00080605	0.00068321	0.00057931
24	0.10152560	0.09614211	0.09105532	0.08624826	74	0.00086485	0.00073111	0.00061829	0.00052307
25	0.09229600	0.08720373	0.08240301	0.07787654	75	0.00078623	0.00066314	0.00055954	0.00047230
26	0.08390545	0.07909635	0.07457286	0.07031741	76	0.00071475	0.00060149	0.00050637	0.00042646
27	0.07627768	0.07174272	0.06748675	0.06349202	77	0.00064978	0.00054557	0.00045825	0.00038506
28	0.06934335	0.06507276	0.06107398	0.05732914	78	0.00059070	0.00049485	0.00041471	0.00034769
29	0.06303941	0.05902291	0.05527057	0.05176446	79	0.00053700	0.00044884	0.00037530	0.00031394
30	0.05730855	0.05353552	0.05001861	0.04673992	80	0.00048819	0.00040711	0.00033964	0.00028347
31	0.05209868	0.04855830	0.04526571	0.04220309	81	0.00044381	0.00036926	0.00030737	0.00025595
32	0.04736244	0.04404381	0.04096445	0.03810662	82	0.00040346	0.00033493	0.00027816	0.00023111
33	0.04305676	0.03994903	0.03707190	0.03440779	83	0.00036678	0.00030379	0.00025173	0.00020867
34	0.03914251	0.03623495	0.03354923	0.03106798	84	0.00033344	0.00027555	0.00022781	0.00018842
35	0.03558410	0.03286617	0.03036129	0.02805235	85	0.00030313	0.00024993	0.00020616	0.00017013
36	0.03234918	0.20981058	0.02747628	0.02532944	86	0.00027557	0.00022669	0.00018657	0.00015362
37	0.02940835	0.02703908	0.02486542	0.02287082	87	0.00025052	0.00020562	0.00016884	0.00013871
38	0.02673486	0.02452524	0.02250264	0.02065086	88	0.00022774	0.00018650	0.00015280	0.00012524
39	0.02430442	0.02224512	0.02036438	0.01864637	89	0.00020704	0.00016916	0.00013828	0.00011309
40	0.02209493	0.02017698	0.01842930	0.01683645	90	0.00018822	0.00015344	0.00012514	0.00010211
41	0.02008630	0.01830111	0.01667810	0.01520221	91	0.00017111	0.00013917	0.00011325	0.00009220
42	0.01826027	0.01659965	0.01509330	0.01372660	92	0.00015555	0.00012623	0.00010249	0.00008325
43	0.01660025	0.01505637	0.01365910	0.01239423	93	0.00014141	0.00011450	0.00009275	0.00007517
44	0.01509113	0.01365657	0.01236118	0.01119117	94	0.00012855	0.00010385	0.00008394	0.00006787
45	0.01371921	0.01238691	0.01118658	0.01010490	95	0.00011687	0.00009420	0.00007596	0.00006128
46	0.01247201	0.01123530	0.01012361	0.00912406	96	0.00010624	0.00008544	0.00006874	0.00005534
47	0.01133819	0.01019074	0.00916163	0.00823843	97	0.00009658	0.00007750	0.00006221	0.00004996
48	0.01030745	0.00924331	0.00829107	0.00743876	98	0.00008780	0.00007029	0.00005630	0.00004511
49	0.00937041	0.00838395	0.00750323	0.00671672	99	0.00007982	0.00006376	0.00005095	0.00004074
50	0.00851855	0.00760449	0.00679026	0.00606475	100	0.00007257	0.00005783	0.00004611	0.00003678

PRESENT VALUE $1/(1+i)^n$ (Continued)

Periods n	Rate i 0.11(11%)	0.1125(11 1/4%)	0.115(11 1/2%)	0.1175(11 3/4%)	Periods n	Rate i 0.11(11%)	0.1125(11 1/4%)	0.115(11 1/2%)	0.1175(11 3/4%)
1	0.90090090	0.89887640	0.89686099	0.89485459	51	0.00488122	0.00435210	0.00388134	0.00346239
2	0.81162243	0.80797879	0.80435963	0.80076473	52	0.00439749	0.00391200	0.00348102	0.00309833
3	0.73119138	0.72627307	0.72139877	0.71656799	53	0.00396170	0.00351641	0.00312199	0.00277256
4	0.65873097	0.65282973	0.64699441	0.64122415	54	0.00356910	0.00316081	0.00279999	0.00248104
5	0.59345133	0.58681324	0.58026405	0.57380237	55	0.00321541	0.00284118	0.00251121	0.00222017
6	0.53464084	0.52747257	0.52041619	0.51346969	56	0.00289676	0.00255387	0.00225220	0.00198673
7	0.48165841	0.47413265	0.46674097	0.45948070	57	0.00260970	0.00229561	0.00201991	0.00177783
8	0.43392650	0.42618665	0.41860177	0.41116841	58	0.00235108	0.00206347	0.00181158	0.00159090
9	0.39092477	0.38308912	0.37542760	0.36793594	59	0.00211809	0.00185481	0.00162474	0.00142363
10	0.35218448	0.34434977	0.33670636	0.32924916	60	0.00190819	0.00166724	0.00145716	0.00127394
11	0.31728331	0.30952789	0.30197880	0.29463013	61	0.00171909	0.00149865	0.00130687	0.00113999
12	0.28584082	0.27822731	0.27083301	0.26365112	62	0.00154873	0.00134710	0.00117208	0.00102012
13	0.25751426	0.25009197	0.24289956	0.23592941	63	0.00139525	0.00121087	0.00105120	0.00091286
14	0.23199482	0.22480177	0.21784714	0.21112252	64	0.00125698	0.00108843	0.00094278	0.00081688
15	0.20900435	0.20206901	0.19537860	0.18892395	65	0.00113242	0.00097836	0.00084554	0.00073099
16	0.18829220	0.18163506	0.17522744	0.16905947	66	0.00102020	0.00087942	0.00075833	0.00065413
17	0.16963262	0.16326747	0.15715466	0.15128364	67	0.00091910	0.00079049	0.00068012	0.00058535
18	0.15282218	0.14675728	0.14094588	0.13537686	68	0.00082801	0.00071056	0.00060997	0.00052380
19	0.13767764	0.13191665	0.12640886	0.12114260	69	0.00074596	0.00063870	0.00054706	0.00046873
20	0.12403391	0.11857677	0.11337118	0.10840501	70	0.00067203	0.00057411	0.00049064	0.00041944
21	0.11174226	0.10658586	0.10167818	0.09700672	71	0.00060544	0.00051606	0.00044003	0.00037534
22	0.10066870	0.09580751	0.09119120	0.08680691	72	0.00054544	0.00046387	0.00039465	0.00033588
23	0.09069252	0.08611911	0.08178583	0.07767956	73	0.00049139	0.00041696	0.00035394	0.00030056
24	0.08170498	0.07741044	0.07335052	0.06951191	74	0.00044269	0.00037480	0.00031744	0.00026896
25	0.07360809	0.06958242	0.06578522	0.06220305	75	0.00039882	0.00033690	0.00028470	0.00024068
26	0.06631359	0.06254599	0.05900020	0.05566269	76	0.00035930	0.00030283	0.00025533	0.00021537
27	0.05974197	0.05622112	0.05291497	0.04981001	77	0.00032369	0.00027221	0.00022900	0.00019273
28	0.05382160	0.05053584	0.04745738	0.04457272	78	0.00029161	0.00024468	0.00020538	0.00017246
29	0.04848793	0.04542547	0.04256267	0.03988610	79	0.00026271	0.00021994	0.00018420	0.00015433
30	0.04368282	0.04083188	0.03817280	0.03569226	80	0.00023668	0.00019770	0.00016520	0.00013810
31	0.03935389	0.03670282	0.03423569	0.03193938	81	0.00021323	0.00017770	0.00014816	0.00012358
32	0.03545395	0.03299130	0.03070466	0.02858110	82	0.00019209	0.00015973	0.00013288	0.00011059
33	0.03194050	0.02965510	0.02753781	0.02557593	83	0.00017306	0.00014358	0.00011918	0.00009896
34	0.02877522	0.02665627	0.02469759	0.02288674	84	0.00015591	0.00012906	0.00010688	0.00008855
35	0.02592363	0.02396069	0.02215030	0.02048030	85	0.00014046	0.00011601	0.00009586	0.00007924
36	0.02335462	0.02153770	0.01986574	0.01832689	86	0.00012654	0.00010428	0.00008597	0.00007091
37	0.02104020	0.01935973	0.01781681	0.01639990	87	0.00011400	0.00009373	0.00007711	0.00006345
38	0.01895513	0.01740200	0.01597920	0.01467553	88	0.00010270	0.00008426	0.00006915	0.00005678
39	0.01707670	0.01564225	0.01433112	0.01313247	89	0.00009252	0.00007574	0.00006202	0.00005081
40	0.01538441	0.01406045	0.01285302	0.01175165	90	0.00008336	0.00006808	0.00005562	0.00004547
41	0.01385983	0.01263861	0.01152738	0.01051601	91	0.00007509	0.00006119	0.00004989	0.00004069
42	0.01248633	0.01136055	0.01033845	0.00941030	92	0.00006765	0.00005500	0.00004474	0.00003641
43	0.01124895	0.01021173	0.00927216	0.00842085	93	0.00006095	0.00004944	0.00004013	0.00003258
44	0.01013419	0.00917908	0.00831583	0.00753544	94	0.00005491	0.00004444	0.00003599	0.00002916
45	0.00912990	0.00825086	0.00745815	0.00674312	95	0.00004947	0.00003995	0.00003228	0.00002609
46	0.00822513	0.00741650	0.00668892	0.00603411	96	0.00004457	0.00003591	0.00002895	0.00002335
47	0.00741003	0.00666652	0.00599903	0.00539965	97	0.00004015	0.00003228	0.00002596	0.00002089
48	0.00667570	0.00599238	0.00538030	0.00483191	98	0.00003617	0.00002901	0.00002328	0.00001870
49	0.00601415	0.00538641	0.00482538	0.00432385	99	0.00003259	0.00002608	0.00002088	0.00001673
50	0.00541815	0.00484171	0.00432769	0.00386922	100	0.00002936	0.00002344	0.00001873	0.00001497

PRESENT VALUE $1/(1+i)^n$ (Continued)

Periods n	Rate i 0.12(12%)	0.1225(12 1/4%)	0.125(12 1/2%)	0.1275(12 3/4%)
1	0.89285714	0.89086860	0.88888889	0.88691796
2	0.79719388	0.79364686	0.79012346	0.78662347
3	0.71178025	0.70703506	0.70233196	0.69767048
4	0.63551808	0.62987533	0.62429508	0.61877648
5	0.56742686	0.56113615	0.55492896	0.54880397
6	0.50663112	0.49989858	0.49327018	0.48674410
7	0.45234922	0.44534395	0.43846239	0.43170208
8	0.40388323	0.39674294	0.38974434	0.38288433
9	0.36061002	0.35344582	0.34643942	0.33958699
10	0.32197324	0.31487378	0.30794615	0.30118580
11	0.28747610	0.28051117	0.27372991	0.26712710
12	0.25667509	0.24989859	0.24331547	0.23691982
13	0.22917419	0.22262681	0.21628042	0.21012844
14	0.20461981	0.19833123	0.19224926	0.18636669
15	0.18269626	0.17668706	0.17088823	0.16529196
16	0.16312166	0.15740496	0.15190065	0.14660041
17	0.14564434	0.14022713	0.13502280	0.13002254
18	0.13003959	0.12492395	0.12002027	0.11531932
19	0.11610678	0.11129082	0.10668468	0.10227878
20	0.10366677	0.09914550	0.09483083	0.09071289
21	0.09255961	0.08832561	0.08429407	0.08045489
22	0.08264251	0.07868651	0.07492806	0.07135689
23	0.07378796	0.07009934	0.06660272	0.06328770
24	0.06588210	0.06244930	0.05920242	0.05613100
25	0.05882331	0.05563412	0.05262437	0.04978359
26	0.05252081	0.04956269	0.04677722	0.04415396
27	0.04689358	0.04415385	0.04157975	0.03916094
28	0.04186927	0.03933528	0.03695978	0.03473254
29	0.03738327	0.03504256	0.03285314	0.03080492
30	0.03337792	0.03121832	0.02920279	0.02732143
31	0.02980172	0.02781142	0.02595803	0.02423187
32	0.02660868	0.02477632	0.02307381	0.02149168
33	0.02375775	0.02207245	0.02051005	0.01906136
34	0.02121227	0.01966365	0.01823116	0.01690586
35	0.01893953	0.01751773	0.01620547	0.01499411
36	0.01691029	0.01560599	0.01440486	0.01329855
37	0.01509848	0.01390289	0.01280432	0.01179472
38	0.01348078	0.01238565	0.01138162	0.01046095
39	0.01203641	0.01103398	0.01011700	0.00927800
40	0.01074680	0.00982983	0.00899289	0.00822883
41	0.00959536	0.00875709	0.00799368	0.00729830
42	0.00856728	0.00780141	0.00710549	0.00647299
43	0.00764936	0.00695003	0.00631599	0.00574101
44	0.00682978	0.00619157	0.00561421	0.00509181
45	0.00609802	0.00551587	0.00499041	0.00451601
46	0.00544466	0.00491392	0.00443592	0.00400533
47	0.00486131	0.00437766	0.00394304	0.00355240
48	0.00434045	0.00389992	0.00350493	0.00315069
49	0.00387540	0.00347431	0.00311549	0.00279440
50	0.00346018	0.00309516	0.00276932	0.00247841

Periods n	Rate i 0.12(12%)	0.1225(12 1/4%)	0.125(12 1/2%)	0.1275(12 3/4%)
51	0.00308945	0.00275738	0.00246162	0.00219814
52	0.00275844	0.00245646	0.00218811	0.00194957
53	0.00246289	0.00218838	0.00194499	0.00172911
54	0.00219901	0.00194956	0.00172888	0.00153358
55	0.00196340	0.00173680	0.00153678	0.00136016
56	0.00175304	0.00154726	0.00136603	0.00120635
57	0.00156521	0.00137841	0.00121424	0.00106993
58	0.00139751	0.00122798	0.00107933	0.00094894
59	0.00124778	0.00109397	0.00095940	0.00084163
60	0.00111409	0.00097458	0.00085280	0.00074646
61	0.00099472	0.00086823	0.00075805	0.00066205
62	0.00088814	0.00077348	0.00067382	0.00058718
63	0.00079298	0.00068906	0.00059895	0.00052078
64	0.00070802	0.00061387	0.00053240	0.00046189
65	0.00063216	0.00054687	0.00047325	0.00040966
66	0.00056443	0.00048719	0.00042066	0.00036334
67	0.00050396	0.00043402	0.00037392	0.00032225
68	0.00044996	0.00038666	0.00033238	0.00028581
69	0.00040175	0.00034446	0.00029544	0.00025349
70	0.00035871	0.00030687	0.00026262	0.00022482
71	0.00032027	0.00027338	0.00023344	0.00019940
72	0.00028596	0.00024355	0.00020750	0.00017685
73	0.00025532	0.00021697	0.00018444	0.00015685
74	0.00022796	0.00019329	0.00016395	0.00013912
75	0.00020354	0.00017220	0.00014573	0.00012338
76	0.00018173	0.00015340	0.00012954	0.00010943
77	0.00016226	0.00013666	0.00011515	0.00009706
78	0.00014488	0.00012175	0.00010235	0.00008608
79	0.00012935	0.00010846	0.00009098	0.00007635
80	0.00011549	0.00009663	0.00008087	0.00006771
81	0.00010312	0.00008608	0.00007189	0.00006006
82	0.00009207	0.00007669	0.00006390	0.00005327
83	0.00008221	0.00006832	0.00005680	0.00004724
84	0.00007340	0.00006086	0.00005049	0.00004190
85	0.00006553	0.00005422	0.00004488	0.00003716
86	0.00005851	0.00004830	0.00003989	0.00003296
87	0.00005224	0.00004303	0.00003546	0.00002923
88	0.00004665	0.00003834	0.00003152	0.00002593
89	0.00004165	0.00003415	0.00002802	0.00002299
90	0.00003719	0.00003042	0.00002490	0.00002039
91	0.00003320	0.00002710	0.00002214	0.00001809
92	0.00002964	0.00002415	0.00001968	0.00001604
93	0.00002647	0.00002151	0.00001749	0.00001423
94	0.00002363	0.00001916	0.00001555	0.00001262
95	0.00002110	0.00001707	0.00001382	0.00001119
96	0.00001884	0.00001521	0.00001228	0.00000993
97	0.00001682	0.00001355	0.00001092	0.00000880
98	0.00001502	0.00001207	0.00000971	0.00000781
99	0.00001341	0.00001075	0.00000863	0.00000693
100	0.00001197	0.00000958	0.00000767	0.00000614

PRESENT VALUE $1/(1 + i)^n$ (Continued)

Periods	Rate i				Periods	Rate i			
n	0.13(13%)	0.1325(13 1/4%)	0.135(13 1/2%)	0.1375(13 3/4%)	n	0.13(13%)	0.1325(13 1/4%)	0.135(13 1/2%)	0.1375(13 3/4%)
1	0.88495575	0.88300221	0.88105727	0.87912088	51	0.00196336	0.00175409	0.00156751	0.00140113
2	0.78314668	0.77969290	0.77626191	0.77285352	52	0.00173748	0.00154886	0.00138107	0.00123176
3	0.69305016	0.68847055	0.68393120	0.67943167	53	0.00153760	0.00136765	0.00121680	0.00108287
4	0.61331873	0.60792102	0.60258255	0.59730256	54	0.00136071	0.00120764	0.00107207	0.00095197
5	0.54275994	0.53679560	0.53090974	0.52510115	55	0.00120416	0.00106635	0.00094456	0.00083690
6	0.48031853	0.47399170	0.46776188	0.46162739	56	0.00106563	0.00094159	0.00083221	0.00073573
7	0.42506064	0.41853572	0.41212501	0.40582628	57	0.00094304	0.00083142	0.00073322	0.00064680
8	0.37615986	0.36956796	0.36310573	0.35677035	58	0.00083455	0.00073415	0.00064601	0.00056862
9	0.33288483	0.32632933	0.31991695	0.31364427	59	0.00073854	0.00064825	0.00056917	0.00049988
10	0.29458835	0.28814952	0.28186515	0.27573122	60	0.00065357	0.00057241	0.00050147	0.00043946
11	0.26069765	0.25443666	0.24833934	0.24240107	61	0.00057838	0.00050544	0.00044183	0.00038634
12	0.23070589	0.22466813	0.21880118	0.21309985	62	0.00051184	0.00044630	0.00038928	0.00033964
13	0.20416450	0.19838246	0.19277637	0.18734052	63	0.00045296	0.00039409	0.00034297	0.00029858
14	0.18067655	0.17517215	0.16984702	0.16469497	64	0.00040085	0.00034798	0.00030218	0.00026249
15	0.15989075	0.15467739	0.14964495	0.14478678	65	0.00035473	0.00030727	0.00026624	0.00023076
16	0.14149624	0.13658048	0.13184577	0.12728508	66	0.00031392	0.00027132	0.00023457	0.00020287
17	0.12521791	0.12060086	0.11616368	0.11189898	67	0.00027781	0.00023957	0.00020667	0.00017834
18	0.11081231	0.10649083	0.10234685	0.09837273	68	0.00024585	0.00021154	0.00018209	0.00015678
19	0.09806399	0.09403164	0.09017344	0.08648152	69	0.00021756	0.00018679	0.00016043	0.00013783
20	0.08678229	0.08303014	0.07944796	0.07602771	70	0.00019253	0.00016494	0.00014135	0.00012117
21	0.07679849	0.07331580	0.06999821	0.06683754	71	0.00017038	0.00014564	0.00012454	0.00010652
22	0.06796327	0.06473801	0.06167243	0.05875828	72	0.00015078	0.00012860	0.00010972	0.00009365
23	0.06014448	0.05716381	0.05433694	0.05165563	73	0.00013344	0.00011356	0.00009667	0.00008233
24	0.05322521	0.05047577	0.04787396	0.04541154	74	0.00011809	0.00010027	0.00008517	0.00007238
25	0.04710195	0.04457022	0.04217970	0.03992224	75	0.00010450	0.00008854	0.00007504	0.00006363
26	0.04168314	0.03935560	0.03716273	0.03509647	76	0.00009248	0.00007818	0.00006612	0.00005594
27	0.03688774	0.03475108	0.03274249	0.03085404	77	0.00008184	0.00006903	0.00005825	0.00004917
28	0.03264402	0.03068528	0.02884801	0.02712443	78	0.00007242	0.00006096	0.00005132	0.00004323
29	0.02888851	0.02709517	0.02541675	0.02384565	79	0.00006409	0.00005382	0.00004522	0.00003800
30	0.02556505	0.02392510	0.02239361	0.02096321	80	0.00005672	0.00004753	0.00003984	0.00003341
31	0.02262394	0.02112591	0.01973005	0.01842920	81	0.00005019	0.00004197	0.00003510	0.00002937
32	0.02002119	0.01865423	0.01738331	0.01620149	82	0.00004442	0.00003706	0.00003093	0.00002582
33	0.01771786	0.01647172	0.01531569	0.01424307	83	0.00003931	0.00003272	0.00002725	0.00002270
34	0.01567953	0.01454457	0.01349400	0.01252138	84	0.00003479	0.00002889	0.00002401	0.00001996
35	0.01387569	0.01284289	0.01188899	0.01100781	85	0.00003078	0.00002551	0.00002115	0.00001754
36	0.01227937	0.01134030	0.01047488	0.00967719	86	0.00002724	0.00002253	0.00001864	0.00001542
37	0.01086670	0.01001351	0.00922897	0.00850742	87	0.00002411	0.00001989	0.00001642	0.00001356
38	0.00961655	0.00884195	0.00813125	0.00747905	88	0.00002134	0.00001756	0.00001447	0.00001192
39	0.00851022	0.00780746	0.00716410	0.00657499	89	0.00001888	0.00001551	0.00001275	0.00001048
40	0.00753117	0.00689400	0.00631198	0.00578021	90	0.00001671	0.00001369	0.00001123	0.00000921
41	0.00666475	0.00608742	0.00556121	0.00508151	91	0.00001479	0.00001209	0.00000989	0.00000810
42	0.00589801	0.00537521	0.00489975	0.00446726	92	0.00001309	0.00001068	0.00000872	0.00000712
43	0.00521948	0.00474632	0.00431696	0.00392726	93	0.00001158	0.00000943	0.00000768	0.00000626
44	0.00461901	0.00419101	0.00380349	0.00345254	94	0.00001025	0.00000833	0.00000677	0.00000550
45	0.00408762	0.00370067	0.00335109	0.00303520	95	0.00000907	0.00000735	0.00000596	0.00000484
46	0.00361736	0.00326770	0.00295250	0.00266830	96	0.00000803	0.00000649	0.00000525	0.00000425
47	0.00320120	0.00288539	0.00260132	0.00234576	97	0.00000710	0.00000573	0.00000463	0.00000374
48	0.00283292	0.00254780	0.00229192	0.00206221	98	0.00000629	0.00000506	0.00000408	0.00000329
49	0.00250701	0.00224972	0.00201931	0.00181293	99	0.00000556	0.00000447	0.00000359	0.00000289
50	0.00221859	0.00198650	0.00177913	0.00159378	100	0.00000492	0.00000395	0.00000317	0.00000254

PRESENT VALUE $1/(1+i)^n$ (Continued)

Periods	Rate i				Periods	Rate i			
n	0.14(14%)	0.1425(14 1/4%)	0.145(14 1/2%)	0.1475(14 3/4%)	n	0.14(14%)	0.1425(14 1/4%)	0.145(14 1/2%)	0.1475(14 3/4%)
1	0.87719298	0.87527352	0.87336245	0.87145969	51	0.00125272	0.00112030	0.00100212	0.00089663
2	0.76946753	0.76610374	0.76276196	0.75944200	52	0.00109887	0.00098057	0.00087521	0.00078138
3	0.67497152	0.67055032	0.66616765	0.66182309	53	0.00096392	0.00085826	0.00076438	0.00068094
4	0.59208028	0.58691494	0.58180581	0.57675215	54	0.00084555	0.00075122	0.00066758	0.00059341
5	0.51936866	0.51371111	0.50812734	0.50261625	55	0.00074171	0.00065752	0.00058304	0.00051713
6	0.45558655	0.44963773	0.44377934	0.43800981	56	0.00065062	0.00057551	0.00050921	0.00045066
7	0.39963732	0.39355600	0.38758021	0.38170789	57	0.00057072	0.00050373	0.00044472	0.00039273
8	0.35055905	0.34446915	0.33849800	0.33264304	58	0.00050063	0.00044090	0.00038840	0.00034225
9	0.30750794	0.30150472	0.29563144	0.28988501	59	0.00043915	0.00038591	0.00033922	0.00029826
10	0.26974381	0.26389910	0.25819340	0.25262310	60	0.00038522	0.00033777	0.00029626	0.00025992
11	0.23661738	0.23098390	0.22549642	0.22015085	61	0.00033791	0.00029565	0.00025874	0.00022651
12	0.20755910	0.20217409	0.19694010	0.19185259	62	0.00029641	0.00025877	0.00022597	0.00019739
13	0.18206939	0.17695763	0.17200009	0.16719180	63	0.00026001	0.00022650	0.00019736	0.00017202
14	0.15970999	0.15488633	0.15021842	0.14570092	64	0.00022808	0.00019825	0.00017236	0.00014991
15	0.14009648	0.13556790	0.13119513	0.12697248	65	0.00020007	0.00017352	0.00015054	0.00013064
16	0.12289165	0.11865899	0.11458090	0.11065139	66	0.00017550	0.00015188	0.00013147	0.00011385
17	0.10779969	0.10385908	0.10007065	0.09642823	67	0.00015395	0.00013293	0.00011482	0.00009921
18	0.09456113	0.09090510	0.08739795	0.08403332	68	0.00013504	0.00011635	0.00010028	0.00008646
19	0.08294836	0.07956683	0.07633009	0.07323165	69	0.00011846	0.00010184	0.00008758	0.00007535
20	0.07276172	0.06964274	0.06666383	0.06381843	70	0.00010391	0.00008914	0.00007649	0.00006566
21	0.06382607	0.06095644	0.05822169	0.05561519	71	0.00009115	0.00007802	0.00006681	0.00005722
22	0.05598778	0.05335356	0.05084863	0.04846640	72	0.00007996	0.00006829	0.00005835	0.00004987
23	0.04911209	0.04669896	0.04440929	0.04223651	73	0.00007014	0.00005977	0.00005096	0.00004346
24	0.04308078	0.04087436	0.03878540	0.03680742	74	0.00006152	0.00005232	0.00004450	0.00003787
25	0.03779016	0.03577625	0.03387372	0.03207618	75	0.00005397	0.00004579	0.00003887	0.00003300
26	0.03314926	0.03131400	0.02958403	0.02795310	76	0.00004734	0.00004008	0.00003395	0.00002876
27	0.02907830	0.02740832	0.02583758	0.02436000	77	0.00004153	0.00003508	0.00002965	0.00002506
28	0.02550728	0.02398977	0.02256557	0.02122876	78	0.00003643	0.00003071	0.00002589	0.00002184
29	0.02237481	0.02099761	0.01970792	0.01850001	79	0.00003195	0.00002688	0.00002261	0.00001903
30	0.01962702	0.01837866	0.01721216	0.01612201	80	0.00002803	0.00002352	0.00001975	0.00001659
31	0.01721669	0.01608635	0.01503246	0.01404968	81	0.00002459	0.00002059	0.00001725	0.00001446
32	0.01510236	0.01407996	0.01312878	0.01224373	82	0.00001892	0.00001577	0.00001316	0.00001098
33	0.01324768	0.01232381	0.01146618	0.01066992	83	0.00002157	0.00001802	0.00001506	0.00001260
34	0.01162077	0.01078671	0.01001414	0.00929840	84	0.00001660	0.00001381	0.00001149	0.00000957
35	0.01019366	0.00944132	0.00874597	0.00810318	85	0.00001456	0.00001208	0.00001004	0.00000727
36	0.00894181	0.00826374	0.00763840	0.00706160	86	0.00001277	0.00001058	0.00000876	0.00000834
37	0.00784369	0.00723303	0.00667109	0.00615390	87	0.00001120	0.00000926	0.00000765	0.00000633
38	0.00688043	0.00633088	0.00582628	0.00536287	88	0.00000983	0.00000810	0.00000669	0.00000552
39	0.00603547	0.00554125	0.00508846	0.00467353	89	0.00000862	0.00000709	0.00000584	0.00000481
40	0.00529427	0.00485011	0.00444407	0.00407279	90	0.00000756	0.00000621	0.00000510	0.00000419
41	0.00464410	0.00424517	0.00388128	0.00354927	91	0.00000663	0.00000543	0.00000445	0.00000365
42	0.00407377	0.00371569	0.00338976	0.00309305	92	0.00000582	0.00000476	0.00000389	0.00000318
43	0.00357348	0.00325224	0.00296049	0.00269547	93	0.00000510	0.00000416	0.00000340	0.00000277
44	0.00313463	0.00284660	0.00258558	0.00234899	94	0.00000448	0.00000364	0.00000297	0.00000242
45	0.00274968	0.00249156	0.00225815	0.00204705	95	0.00000393	0.00000319	0.00000259	0.00000211
46	0.00241200	0.00218079	0.00197218	0.00178392	96	0.00000344	0.00000279	0.00000226	0.00000184
47	0.00211579	0.00190879	0.00172243	0.00155462	97	0.00000302	0.00000244	0.00000198	0.00000160
48	0.00185595	0.00167071	0.00150431	0.00135479	98	0.00000265	0.00000214	0.00000173	0.00000139
49	0.00162803	0.00146233	0.00131381	0.00118064	99	0.00000232	0.00000187	0.00000151	0.00000121
50	0.00142810	0.00127994	0.00114743	0.00102888	100	0.00000204	0.00000164	0.00000132	0.00000106

PRESENT VALUE $1/(1 + i)^n$ (Continued)

Periods *n*	Rate *i* 0.15(15%)	0.1525(15 1/4%)	0.155(15 1/2%)	0.1575(15 3/4%)	Periods *n*	Rate *i* 0.15(15%)	0.1525(15 1/4%)	0.155(15 1/2%)	0.1575(15 3/4%)
1	0.86956522	0.86767896	0.86580087	0.86393089	51	0.00080244	0.00071831	0.00064316	0.00057601
2	0.75614367	0.75286678	0.74961114	0.74637657	52	0.00069777	0.00062326	0.00055685	0.00049763
3	0.65751623	0.65324666	0.64901397	0.64481778	53	0.00060676	0.00054079	0.00048212	0.00042992
4	0.57175325	0.56680838	0.56191686	0.55707799	54	0.00052761	0.00046923	0.00041742	0.00037142
5	0.49717674	0.49180771	0.48650810	0.48127688	55	0.00045880	0.00040714	0.00036140	0.00032088
6	0.43232760	0.42673120	0.42121914	0.41578996	56	0.00039895	0.00035327	0.00031290	0.00027722
7	0.37593704	0.37026568	0.36469189	0.35921379	57	0.00034692	0.00030653	0.00027091	0.00023950
8	0.32690177	0.32127174	0.31575056	0.31033589	58	0.00030167	0.00026597	0.00023456	0.00020691
9	0.28426241	0.27876073	0.27337711	0.26810876	59	0.00026232	0.00023077	0.00020308	0.00017876
10	0.24718471	0.24187482	0.23669014	0.23162744	60	0.00022810	0.00020024	0.00017583	0.00015443
11	0.21494322	0.20986969	0.20492652	0.20011010	61	0.00019835	0.00017374	0.00015223	0.00013342
12	0.18690715	0.18209952	0.17742556	0.17288129	62	0.00017248	0.00015075	0.00013180	0.00011527
13	0.16252796	0.15800392	0.15361521	0.14935749	63	0.00014998	0.00013080	0.00011411	0.00009958
14	0.14132866	0.13709668	0.13300018	0.12903455	64	0.00013042	0.00011350	0.00009880	0.00008603
15	0.12289449	0.11895590	0.11515167	0.11147693	65	0.00011341	0.00009848	0.00008554	0.00007433
16	0.10686477	0.10321553	0.09969841	0.09630836	66	0.00009861	0.00008545	0.00007406	0.00006421
17	0.09292589	0.08955795	0.08631897	0.08320377	67	0.00008575	0.00007414	0.00006412	0.00005547
18	0.08080512	0.07770754	0.07473504	0.07188231	68	0.00007457	0.00006433	0.00005552	0.00004793
19	0.07026532	0.06742520	0.06470566	0.06210134	69	0.00006484	0.00005582	0.00004807	0.00004140
20	0.06110028	0.05850343	0.05602222	0.05365127	70	0.00005638	0.00004843	0.00004162	0.00003577
21	0.05313068	0.05076219	0.04850409	0.04635099	71	0.00004903	0.00004202	0.00003603	0.00003090
22	0.04620059	0.04404529	0.04199488	0.04004405	72	0.00004263	0.00003646	0.00003120	0.00002670
23	0.04017443	0.03821717	0.03635920	0.03459529	73	0.00003707	0.00003164	0.00002701	0.00002307
24	0.03493428	0.03316023	0.03147983	0.02988794	74	0.00003224	0.00002745	0.00002338	0.00001993
25	0.03037764	0.02877244	0.02725526	0.02582112	75	0.00002803	0.00002382	0.00002025	0.00001722
26	0.02641534	0.02496524	0.02359763	0.02230766	76	0.00002438	0.00002067	0.00001753	0.00001487
27	0.02296986	0.02166181	0.02043085	0.01927228	77	0.00002120	0.00001793	0.00001518	0.00001285
28	0.01997379	0.01879550	0.01768905	0.01664991	78	0.00001843	0.00001556	0.00001314	0.00001110
29	0.01736851	0.01630846	0.01531519	0.01438438	79	0.00001603	0.00001350	0.00001138	0.00000959
30	0.01510305	0.01415051	0.01325991	0.01242711	80	0.00001394	0.00001171	0.00000985	0.00000829
31	0.01313309	0.01227810	0.01148044	0.01073616	81	0.00001212	0.00001016	0.00000853	0.00000716
32	0.01142008	0.01065345	0.00993977	0.00927530	82	0.00001054	0.00000882	0.00000738	0.00000618
33	0.00993050	0.00924377	0.00860586	0.00801322	83	0.00000916	0.00000765	0.00000639	0.00000534
34	0.00863522	0.00802063	0.00745097	0.00692287	84	0.00000797	0.00000664	0.00000553	0.00000462
35	0.00750889	0.00695933	0.00645105	0.00598088	85	0.00000693	0.00000576	0.00000479	0.00000399
36	0.00652947	0.00603846	0.00558533	0.00516707	86	0.00000603	0.00000500	0.00000415	0.00000345
37	0.00567780	0.00523945	0.00483578	0.00446399	87	0.00000524	0.00000434	0.00000359	0.00000298
38	0.00493722	0.00454616	0.00418682	0.00385658	88	0.00000456	0.00000376	0.00000311	0.00000257
39	0.00429323	0.00394461	0.00362495	0.00333182	89	0.00000396	0.00000327	0.00000269	0.00000222
40	0.00373324	0.00342265	0.00313849	0.00287846	90	0.00000345	0.00000283	0.00000233	0.00000192
41	0.00324630	0.00296976	0.00271731	0.00248679	91	0.00000300	0.00000246	0.00000202	0.00000166
42	0.00282287	0.00257680	0.00235265	0.00214841	92	0.00000260	0.00000213	0.00000175	0.00000143
43	0.00245467	0.00223584	0.00203692	0.00185608	93	0.00000227	0.00000185	0.00000151	0.00000124
44	0.00213449	0.00193999	0.00176357	0.00160353	94	0.00000197	0.00000161	0.00000131	0.00000107
45	0.00185608	0.00168329	0.00152690	0.00138534	95	0.00000171	0.00000139	0.00000113	0.00000092
46	0.00161398	0.00146055	0.00132199	0.00080244	96	0.00000149	0.00000121	0.00000098	0.00000080
47	0.00140346	0.00126729	0.00114458	0.00103398	97	0.00000130	0.00000105	0.00000085	0.00000069
48	0.00122040	0.00109960	0.00099098	0.00089329	98	0.00000113	0.00000091	0.00000074	0.00000060
49	0.00106122	0.00095410	0.00085799	0.00077174	99	0.00000098	0.00000079	0.00000064	0.00000051
50	0.00092280	0.00082785	0.00074285	0.00066673	100	0.00000085	0.00000069	0.00000055	0.00000044

PRESENT VALUE $1/(1 + i)^n$ (Continued)

Periods	Rate i				Periods	Rate i			
n	0.16(16%)	0.1625(16 1/4%)	0.165(16 1/2%)	0.1675(16 3/4%)	n	0.16(16%)	0.1625(16 1/4%)	0.165(16 1/2%)	0.1675(16 3/4%)
1	0.86206897	0.86021505	0.85836910	0.85653105	51	0.00051599	0.00046234	0.00041436	0.00037145
2	0.74316290	0.73996994	0.73679751	0.73364544	52	0.00044482	0.00039771	0.00035567	0.00031816
3	0.64065767	0.63653328	0.63244421	0.62839020	53	0.00038347	0.00034212	0.00030503	0.00027251
4	0.55229110	0.54755551	0.54287057	0.53823563	54	0.00033057	0.00029429	0.00026206	0.00023341
5	0.47611302	0.47101549	0.46598332	0.46101553	55	0.00028498	0.00025316	0.00022494	0.00019993
6	0.41044225	0.40517462	0.39998568	0.39487411	56	0.00024567	0.00021777	0.00019308	0.00017124
7	0.35382953	0.34853731	0.34333535	0.33822194	57	0.00021178	0.00018733	0.00016574	0.00014668
8	0.30502546	0.29981704	0.29470846	0.28969759	58	0.00018257	0.00016114	0.00014226	0.00012563
9	0.26295298	0.25790713	0.25296863	0.24713498	59	0.00015739	0.00013862	0.00012212	0.00010761
10	0.22668360	0.22185559	0.21714046	0.21253532	60	0.00013568	0.00011924	0.00010482	0.00009217
11	0.19541690	0.19084352	0.18638666	0.18204310	61	0.00011697	0.00010257	0.00008997	0.00007895
12	0.16846284	0.16416647	0.15998855	0.15592557	62	0.00010083	0.00008823	0.00007723	0.00006762
13	0.14522659	0.14121847	0.13732923	0.13355509	63	0.00008693	0.00007590	0.00006629	0.00005792
14	0.12519534	0.12147825	0.11787916	0.11438408	64	0.00007494	0.00006529	0.00005690	0.00004961
15	0.10792701	0.10449742	0.10118383	0.09798208	65	0.00006460	0.00005616	0.00004884	0.00004249
16	0.09304053	0.08989026	0.08685307	0.08392470	66	0.00005569	0.00004831	0.00004193	0.00003640
17	0.08020735	0.07732495	0.07455199	0.07188411	67	0.00004801	0.00004156	0.00003599	0.00003117
18	0.06914427	0.06651609	0.06399313	0.06157097	68	0.00004139	0.00003575	0.00003089	0.00002670
19	0.05960713	0.05721814	0.05492972	0.05273745	69	0.00003568	0.00003075	0.00002652	0.00002287
20	0.05138546	0.04921990	0.04714998	0.04517126	70	0.00003076	0.00002645	0.00002276	0.00001959
21	0.04429781	0.04233970	0.04047208	0.03869059	71	0.00002651	0.00002276	0.00001954	0.00001678
22	0.03818776	0.03642125	0.03473999	0.03313969	72	0.00002286	0.00001958	0.00001677	0.00001437
23	0.03292049	0.03133011	0.02981973	0.02838517	73	0.00001970	0.00001684	0.00001439	0.00001231
24	0.02837973	0.02695063	0.02559634	0.02431278	74	0.00001699	0.00001449	0.00001236	0.00001054
25	0.02446528	0.02318334	0.02197110	0.02082465	75	0.00001464	0.00001246	0.00001061	0.00000903
26	0.02109076	0.01994266	0.01885932	0.01783696	76	0.00001262	0.00001072	0.00000910	0.00000774
27	0.01818169	0.01715497	0.01618825	0.01527791	77	0.00001088	0.00000922	0.00000781	0.00000663
28	0.01567387	0.01475697	0.01389550	0.01308601	78	0.00000938	0.00000793	0.00000671	0.00000567
29	0.01351196	0.01269416	0.01192747	0.01120857	79	0.00000809	0.00000682	0.00000576	0.00000486
30	0.01164824	0.01091971	0.01023817	0.00960049	80	0.00000697	0.00000587	0.00000494	0.00000416
31	0.01004159	0.00939330	0.00878813	0.00822312	81	0.00000601	0.00000505	0.00000424	0.00000357
32	0.00865654	0.00808026	0.00754346	0.00704335	82	0.00000518	0.00000434	0.00000364	0.00000305
33	0.00746253	0.00695076	0.00647507	0.00603285	83	0.00000447	0.00000374	0.00000313	0.00000262
34	0.00643322	0.00597915	0.00555800	0.00516732	84	0.00000385	0.00000321	0.00000268	0.00000224
35	0.00554588	0.00514335	0.00477082	0.00442597	85	0.00000332	0.00000276	0.00000230	0.00000192
36	0.00478093	0.00442439	0.00409512	0.00379098	86	0.00000286	0.00000238	0.00000198	0.00000164
37	0.00412149	0.00380593	0.00351512	0.00324710	87	0.00000247	0.00000205	0.00000170	0.00000141
38	0.00355301	0.00327392	0.00201727	0.00278124	88	0.00000213	0.00000176	0.00000146	0.00000121
39	0.00306294	0.00281627	0.00258994	0.00238222	89	0.00000183	0.00000151	0.00000125	0.00000103
40	0.00264047	0.00242260	0.00222312	0.00204044	90	0.00000158	0.00000130	0.00000107	0.00000088
41	0.00227626	0.00208396	0.00190826	0.00174770	91	0.00000136	0.00000112	0.00000092	0.00000076
42	0.00196230	0.00179265	0.00163799	0.00149696	92	0.00000117	0.00000096	0.00000079	0.00000065
43	0.00169163	0.00154206	0.00140600	0.00128219	93	0.00000101	0.00000083	0.00000068	0.00000056
44	0.00145831	0.00132651	0.00120687	0.00109824	94	0.00000087	0.00000071	0.00000058	0.00000048
45	0.00125716	0.00114108	0.00103594	0.00094068	95	0.00000075	0.00000061	0.00000050	0.00000041
46	0.00109376	0.00098158	0.00088922	0.00090572	96	0.00000065	0.00000053	0.00000043	0.00000035
47	0.00093427	0.00084437	0.00076328	0.00069012	97	0.00000056	0.00000045	0.00000037	0.00000030
48	0.00080541	0.00072643	0.00065517	0.00059111	98	0.00000048	0.00000039	0.00000032	0.00000026
49	0.00069432	0.00062481	0.00056238	0.00050631	99	0.00000042	0.00000034	0.00000027	0.00000022
50	0.00059855	0.00053747	0.00048273	0.00043367	100	0.00000036	0.00000029	0.00000023	0.00000019

PRESENT VALUE $1/(1 + i)^n$ (Continued)

Periods	Rate *i*				Periods	Rate *i*			
n	0.17(17%)	0.1725(17 1/4%)	0.175(17 1/2%)	0.1775(17 3/4%)	*n*	0.17(17%)	0.1725(17 1/4%)	0.175(17 1/2%)	0.1775(17 3/4%)
1	0.85470085	0.85287846	0.85106383	0.72123728	51	0.00033306	0.00029871	0.00026796	0.00024043
2	0.73051355	0.72740168	0.72430964	0.72123728	52	0.00028467	0.00025476	0.00022805	0.00020419
3	0.62437056	0.62038522	0.61643374	0.61251574	53	0.00024330	0.00021728	0.00019409	0.00017341
4	0.53365005	0.52911320	0.52462446	0.52018322	54	0.00020795	0.00018531	0.00016518	0.00014727
5	0.45611115	0.45126925	0.44648890	0.44176919	55	0.00017774	0.00015805	0.00014058	0.00012507
6	0.38983859	0.38487783	0.37999055	0.37517553	56	0.00015191	0.00013480	0.00011964	0.00010622
7	0.33319538	0.32825401	0.32339622	0.31862041	57	0.00012984	0.00011497	0.00010182	0.00009020
8	0.28478237	0.27996078	0.27523082	0.27059058	58	0.00011097	0.00009805	0.00008666	0.00007661
9	0.24340374	0.23877252	0.23423900	0.22980092	59	0.00009485	0.00008363	0.00007375	0.00006506
10	0.20803738	0.20364394	0.19935234	0.19156001	60	0.00008107	0.00007132	0.00006277	0.00005525
11	0.17780973	0.17368353	0.16966156	0.16574099	61	0.00006929	0.00006083	0.00005342	0.00004692
12	0.15197413	0.14813094	0.14439282	0.10485668	62	0.00005922	0.00005188	0.00004546	0.00003985
13	0.12989242	0.12633769	0.12288751	0.11953858	63	0.00005062	0.00004425	0.00003869	0.00003384
14	0.11101916	0.10775070	0.10458511	0.10151896	64	0.00004326	0.00003774	0.00003293	0.00002874
15	0.09488817	0.09189825	0.08900861	0.08621568	65	0.00003698	0.00003219	0.00002802	0.00002441
16	0.08110100	0.07837804	0.07575201	0.07321926	66	0.00003160	0.00002745	0.00002385	0.00002073
17	0.06931709	0.06684694	0.06446979	0.06219196	67	0.00002701	0.00002341	0.00002030	0.00001760
18	0.05924538	0.05701231	0.05486791	0.05280846	68	0.00002309	0.00001997	0.00001728	0.00001495
19	0.05063708	0.04862458	0.04669609	0.04484795	69	0.00001973	0.00001703	0.00001470	0.00001270
20	0.04327955	0.04147085	0.03974135	0.03808743	70	0.00001687	0.00001452	0.00001251	0.00001078
21	0.03699107	0.03536960	0.03382243	0.03234601	71	0.00001441	0.00001239	0.00001065	0.00000916
22	0.03161630	0.03016597	0.02878505	0.02747008	72	0.00001232	0.00001057	0.00000906	0.00000778
23	0.02702248	0.02572790	0.02449791	0.02332915	73	0.00001053	0.00000901	0.00000771	0.00000660
24	0.02309614	0.02194278	0.02084929	0.01981244	74	0.00000900	0.00000769	0.00000656	0.00000561
25	0.01974029	0.01871452	0.01774407	0.01682585	75	0.00000769	0.00000655	0.00000559	0.00000476
26	0.01687204	0.01596121	0.01510134	0.01428947	76	0.00000657	0.00000559	0.00000475	0.00000405
27	0.01442055	0.01361297	0.01285220	0.01213543	77	0.00000562	0.00000477	0.00000405	0.00000344
28	0.01232525	0.01161021	0.01093805	0.01030610	78	0.00000480	0.00000407	0.00000344	0.00000292
29	0.01053440	0.00990210	0.00930898	0.00875253	79	0.00000411	0.00000347	0.00000293	0.00000248
30	0.00900376	0.00844529	0.00792253	0.00743314	80	0.00000351	0.00000296	0.00000249	0.00000210
31	0.00769553	0.00720280	0.00674258	0.00631265	81	0.00000300	0.00000252	0.00000212	0.00000179
32	0.00657737	0.00614312	0.00573837	0.00536106	82	0.00000256	0.00000215	0.00000181	0.00000152
33	0.00562169	0.00523933	0.00488372	0.00455292	83	0.00000219	0.00000183	0.00000154	0.00000129
34	0.00480486	0.00446851	0.00415635	0.00386660	84	0.00000187	0.00000157	0.00000131	0.00000109
35	0.00410672	0.00381110	0.00353732	0.00328373	85	0.00000160	0.00000133	0.00000111	0.00000093
36	0.00351002	0.00325040	0.00301049	0.00279973	86	0.00000137	0.00000114	0.00000095	0.00000079
37	0.00300001	0.00277220	0.00256212	0.00236835	87	0.00000117	0.00000097	0.00000081	0.00000067
38	0.00256411	0.00236435	0.00218052	0.00201134	88	0.00000100	0.00000083	0.00000069	0.00000057
39	0.00219155	0.00201650	0.00185577	0.00170814	89	0.00000085	0.00000071	0.00000058	0.00000048
40	0.00187312	0.00171983	0.00157938	0.00145065	90	0.00000073	0.00000060	0.00000050	0.00000041
41	0.00160096	0.00146681	0.00134415	0.00123198	91	0.00000062	0.00000051	0.00000042	0.00000035
42	0.00136834	0.00125101	0.00114396	0.00104626	92	0.00000053	0.00000044	0.00000036	0.00000030
43	0.00116952	0.00106696	0.00097358	0.00088855	93	0.00000046	0.00000037	0.00000031	0.00000025
44	0.00099959	0.00090999	0.00082858	0.00075461	94	0.00000039	0.00000032	0.00000026	0.00000021
45	0.00085435	0.00077611	0.00070517	0.00064085	95	0.00000033	0.00000027	0.00000022	0.00000018
46	0.00073021	0.00066193	0.00060015	0.00054425	96	0.00000028	0.00000023	0.00000019	0.00000015
47	0.00062411	0.00056454	0.00051076	0.00046221	97	0.00000024	0.00000020	0.00000016	0.00000013
48	0.00053343	0.00048149	0.00043469	0.00039253	98	0.00000021	0.00000017	0.00000014	0.00000011
49	0.00045592	0.00041065	0.00036995	0.00033336	99	0.00000018	0.00000014	0.00000012	0.00000009
50	0.00038968	0.00035023	0.00031485	0.00028311	100	0.00000015	0.00000012	0.00000010	0.00000008

PRESENT VALUE $1/(1+i)^n$ (Continued)

Periods	Rate i				Periods	Rate i			
n	0.18(18%)	0.1825(18 1/4%)	0.185(18 1/2%)	0.1875(18 3/4%)	n	0.18(18%)	0.1825(18 1/4%)	0.185(18 1/2%)	0.1875(18 3/4%)
1	0.84745763	0.84566596	0.84388186	0.84210526	51	0.00021578	0.00019371	0.00017393	0.00015620
2	0.71818443	0.71515092	0.71213659	0.70914127	52	0.00018287	0.00016381	0.00014677	0.00013154
3	0.60863087	0.60477879	0.60095915	0.59717160	53	0.00015497	0.00013853	0.00012386	0.00011077
4	0.51578888	0.51144084	0.50713852	0.50288135	54	0.00013133	0.00011715	0.00010452	0.00009328
5	0.43710922	0.43250811	0.42796500	0.42347903	55	0.00011130	0.00009907	0.00008820	0.00007855
6	0.37043154	0.36575738	0.36115189	0.35661392	56	0.00009432	0.00008378	0.00007443	0.00006615
7	0.31392503	0.30930857	0.30476953	0.30030646	57	0.00007993	0.00007085	0.00006281	0.00005570
8	0.26603816	0.26157173	0.25718948	0.25288965	58	0.00006774	0.00005991	0.00005301	0.00004691
9	0.22545607	0.22120231	0.21703753	0.21295970	59	0.00005741	0.00005067	0.00004473	0.00003950
10	0.19106447	0.18706326	0.18315404	0.17933449	60	0.00004865	0.00004285	0.00003775	0.00003326
11	0.16191904	0.15819303	0.15456037	0.15101852	61	0.00004123	0.00003624	0.00003186	0.00002801
12	0.13721953	0.13377846	0.13043069	0.12717349	62	0.00003494	0.00003064	0.00002688	0.00002359
13	0.11628773	0.11313189	0.11006809	0.10709346	63	0.00002961	0.00002591	0.00002269	0.00001986
14	0.09854893	0.09567179	0.09288447	0.09018397	64	0.00002509	0.00002191	0.00001914	0.00001673
15	0.08351604	0.08090638	0.07838352	0.07594439	65	0.00002127	0.00001853	0.00001616	0.00001409
16	0.07077630	0.06841977	0.06614643	0.06395317	66	0.00001802	0.00001567	0.00001363	0.00001186
17	0.05997992	0.05786027	0.05581977	0.05385530	67	0.00001527	0.00001325	0.00001150	0.00000999
18	0.05083044	0.04893046	0.04710529	0.04535184	68	0.00001294	0.00001121	0.00000971	0.00000841
19	0.04307664	0.04137883	0.03975130	0.03819102	69	0.00001097	0.00000948	0.00000819	0.00000708
20	0.03650563	0.03499266	0.03354540	0.03216086	70	0.00000930	0.00000802	0.00000691	0.00000597
21	0.03093698	0.02959211	0.02830836	0.02708283	71	0.00000788	0.00000678	0.00000583	0.00000502
22	0.02621778	0.02502504	0.02388891	0.02280659	72	0.00000668	0.00000573	0.00000492	0.00000423
23	0.02221845	0.02116282	0.02015942	0.01920555	73	0.00000566	0.00000485	0.00000415	0.00000356
24	0.01882920	0.01789668	0.01701217	0.01617310	74	0.00000479	0.00000410	0.00000351	0.00000300
25	0.01595695	0.01513461	0.01435626	0.01361945	75	0.00000406	0.00000347	0.00000296	0.00000253
26	0.01352284	0.01279883	0.01211499	0.01146901	76	0.00000344	0.00000293	0.00000250	0.00000213
27	0.01146003	0.01082353	0.01022362	0.00965811	77	0.00000292	0.00000248	0.00000211	0.00000179
28	0.00971189	0.00915309	0.00862752	0.00813315	78	0.00000247	0.00000210	0.00000178	0.00000151
29	0.00823042	0.00774046	0.00728061	0.00684897	79	0.00000210	0.00000177	0.00000150	0.00000127
30	0.00697493	0.00654584	0.00614398	0.00576755	80	0.00000178	0.00000150	0.00000127	0.00000107
31	0.00591096	0.00553560	0.00518479	0.00485689	81	0.00000151	0.00000127	0.00000107	0.00000090
32	0.00500929	0.00468126	0.00437535	0.00409001	82	0.00000128	0.00000107	0.00000090	0.00000076
33	0.00424516	0.00395879	0.00369228	0.00344422	83	0.00000108	0.00000091	0.00000076	0.00000064
34	0.00359759	0.00334781	0.00311585	0.00290039	84	0.00000092	0.00000077	0.00000064	0.00000054
35	0.00304881	0.00283113	0.00262941	0.00244244	85	0.00000078	0.00000065	0.00000054	0.00000045
36	0.00258373	0.00239419	0.00221891	0.00205679	86	0.00000066	0.00000055	0.00000046	0.00000038
37	0.00218960	0.00202469	0.00187250	0.00173203	87	0.00000056	0.00000046	0.00000039	0.00000032
38	0.00185560	0.00171221	0.00158017	0.00145855	88	0.00000047	0.00000039	0.00000033	0.00000027
39	0.00157254	0.00144796	0.00133347	0.00122826	89	0.00000040	0.00000033	0.00000027	0.00000023
40	0.00133266	0.00122449	0.00112529	0.00103432	90	0.00000034	0.00000028	0.00000023	0.00000019
41	0.00112937	0.00103551	0.00094962	0.00087101	91	0.00000029	0.00000024	0.00000020	0.00000016
42	0.00095710	0.00087569	0.00080136	0.00073348	92	0.00000024	0.00000020	0.00000017	0.00000014
43	0.00081110	0.00074054	0.00067626	0.00061767	93	0.00000021	0.00000017	0.00000014	0.00000011
44	0.00068737	0.00062625	0.00057068	0.00052014	94	0.00000018	0.00000014	0.00000012	0.00000010
45	0.00058252	0.00052960	0.00048159	0.00043801	95	0.00000015	0.00000012	0.00000010	0.00000008
46	0.00049366	0.00044787	0.00040640	0.00036885	96	0.00000013	0.00000010	0.00000008	0.00000007
47	0.00041836	0.00037874	0.00034296	0.00031061	97	0.00000011	0.00000009	0.00000007	0.00000006
48	0.00035454	0.00032029	0.00028941	0.00026157	98	0.00000009	0.00000007	0.00000006	0.00000005
49	0.00030046	0.00027086	0.00024423	0.00022027	99	0.00000008	0.00000006	0.00000005	0.00000004
50	0.00025462	0.00022906	0.00020610	0.00018549	100	0.00000006	0.00000005	0.00000004	0.00000003

PRESENT VALUE $1/(1 + i)^n$ (Continued)

Periods	Rate i				
n	0.19(19%)	0.1925(19 1/4%)	0.195(19 1/2%)	0.1975(19 3/4%)	0.20(20%)
1	0.84033613	0.83857442	0.83682008	0.83507307	0.83333333
2	0.70616482	0.70320706	0.70026785	0.69734703	0.69444444
3	0.59341581	0.58969146	0.58599820	0.58233572	0.57870370
4	0.49866875	0.49450017	0.49037507	0.48629288	0.48225309
5	0.41904937	0.41467520	0.41035570	0.40609009	0.40187757
6	0.35214233	0.34773602	0.34339389	0.33911490	0.33489798
7	0.29591792	0.29160253	0.28735891	0.28318572	0.27908165
8	0.24867052	0.24453042	0.24046770	0.23648077	0.23256804
9	0.20896683	0.20505696	0.20122820	0.19747872	0.19380670
10	0.17560238	0.17195552	0.16839180	0.16490916	0.16150558
11	0.14756502	0.14419750	0.14091364	0.13771120	0.13458799
12	0.12400422	0.12092034	0.11791937	0.11499891	0.11215665
13	0.10420523	0.10140070	0.09867729	0.09603250	0.09346388
14	0.08756742	0.08503203	0.08257514	0.08019415	0.07788657
15	0.07358606	0.07130569	0.06910054	0.06696798	0.06490547
16	0.06183703	0.05979513	0.05782472	0.05592315	0.05408789
17	0.05196389	0.05014266	0.04838888	0.04669992	0.04507324
18	0.04366713	0.04204836	0.04049279	0.03899784	0.03756104
19	0.03669507	0.03526068	0.03388518	0.03256605	0.03130086
20	0.03083619	0.02956870	0.02835580	0.02719503	0.02608405
21	0.02591277	0.02479556	0.02372870	0.02270984	0.02173671
22	0.02177544	0.02079292	0.01985665	0.01896437	0.01811393
23	0.01829869	0.01743641	0.01661645	0.01583664	0.01509494
24	0.01537705	0.01462173	0.01390498	0.01322475	0.01257912
25	0.01292189	0.01226141	0.01163596	0.01104363	0.01048260
26	0.01085873	0.01028210	0.00973721	0.00922224	0.00873550
27	0.00912498	0.00862231	0.00814829	0.00770124	0.00727958
28	0.00766805	0.00723045	0.00681865	0.00643110	0.00606632
29	0.00644374	0.00606327	0.00570599	0.00537044	0.00505526
30	0.00541491	0.00508450	0.00477488	0.00448471	0.00421272
31	0.00455034	0.00426373	0.00399572	0.00374506	0.00351060
32	0.00382382	0.00357546	0.00334370	0.00312740	0.00292550
33	0.00321329	0.00299829	0.00279807	0.00261161	0.00243792
34	0.00270025	0.00251429	0.00234148	0.00218088	0.00203160
35	0.00226911	0.00210842	0.00195940	0.00182120	0.00169300
36	0.00190682	0.00176806	0.00163967	0.00152083	0.00141083
37	0.00160237	0.00148265	0.00137211	0.00127001	0.00117569
38	0.00134653	0.00124332	0.00114821	0.00106055	0.00097974
39	0.00113154	0.00104261	0.00096084	0.00088563	0.00081645
40	0.00095087	0.00087431	0.00080405	0.00073957	0.00068038
41	0.00079905	0.00073317	0.00067285	0.00061759	0.00056698
42	0.00067147	0.00061482	0.00056305	0.00051574	0.00047248
43	0.00056426	0.00051557	0.00047117	0.00043068	0.00039374
44	0.00047417	0.00043235	0.00039429	0.00035965	0.00032811
45	0.00039846	0.00036255	0.00032995	0.00030033	0.00027343
46	0.00033484	0.00030403	0.00027611	0.00025080	0.00022786
47	0.00028138	0.00025495	0.00023105	0.00020944	0.00018988
48	0.00023645	0.00021379	0.00019335	0.00017489	0.00015823
49	0.00019870	0.00017928	0.00016180	0.00014605	0.00013186
50	0.00016698	0.00015034	0.00013540	0.00012196	0.00010988

Periods	Rate i				
n	0.19(19%)	0.1925(19 1/4%)	0.195(19 1/2%)	0.1975(19 3/4%)	0.20(20%)
51	0.00014032	0.00012607	0.00011330	0.00010185	0.00009157
52	0.00011791	0.00010572	0.00009481	0.00008505	0.00007631
53	0.00009909	0.00008866	0.00007934	0.00007102	0.00006359
54	0.00008327	0.00007434	0.00006639	0.00005931	0.00005299
55	0.00006997	0.00006234	0.00005556	0.00004953	0.00004416
56	0.00005880	0.00005228	0.00004649	0.00004136	0.00003680
57	0.00004941	0.00004384	0.00003891	0.00003454	0.00003067
58	0.00004152	0.00003676	0.00003256	0.00002884	0.00002556
59	0.00003489	0.00003083	0.00002725	0.00002408	0.00002130
60	0.00002932	0.00002585	0.00002280	0.00002011	0.00001775
61	0.00002464	0.00002168	0.00001908	0.00001680	0.00001479
62	0.00002071	0.00001818	0.00001597	0.00001403	0.00001232
63	0.00001740	0.00001524	0.00001336	0.00001171	0.00001027
64	0.00001462	0.00001278	0.00001118	0.00000978	0.00000856
65	0.00001229	0.00001072	0.00000936	0.00000817	0.00000713
66	0.00001033	0.00000899	0.00000783	0.00000682	0.00000594
67	0.00000868	0.00000754	0.00000655	0.00000570	0.00000495
68	0.00000729	0.00000632	0.00000548	0.00000476	0.00000413
69	0.00000613	0.00000530	0.00000459	0.00000397	0.00000344
70	0.00000515	0.00000445	0.00000384	0.00000332	0.00000287
71	0.00000433	0.00000373	0.00000321	0.00000277	0.00000239
72	0.00000364	0.00000313	0.00000269	0.00000231	0.00000199
73	0.00000306	0.00000262	0.00000225	0.00000193	0.00000166
74	0.00000257	0.00000220	0.00000188	0.00000161	0.00000138
75	0.00000216	0.00000184	0.00000158	0.00000135	0.00000115
76	0.00000181	0.00000155	0.00000132	0.00000112	0.00000096
77	0.00000152	0.00000130	0.00000110	0.00000094	0.00000080
78	0.00000128	0.00000109	0.00000092	0.00000078	0.00000067
79	0.00000108	0.00000091	0.00000077	0.00000065	0.00000056
80	0.00000090	0.00000076	0.00000065	0.00000055	0.00000046
81	0.00000076	0.00000064	0.00000054	0.00000046	0.00000039
82	0.00000064	0.00000054	0.00000045	0.00000038	0.00000032
83	0.00000054	0.00000045	0.00000038	0.00000032	0.00000027
84	0.00000045	0.00000038	0.00000032	0.00000027	0.00000022
85	0.00000038	0.00000032	0.00000027	0.00000022	0.00000019
86	0.00000032	0.00000027	0.00000022	0.00000019	0.00000016
87	0.00000027	0.00000022	0.00000019	0.00000015	0.00000013
88	0.00000022	0.00000019	0.00000016	0.00000013	0.00000011
89	0.00000019	0.00000016	0.00000013	0.00000011	0.00000009
90	0.00000016	0.00000013	0.00000011	0.00000009	0.00000007
91	0.00000013	0.00000011	0.00000009	0.00000008	0.00000006
92	0.00000011	0.00000009	0.00000008	0.00000006	0.00000005
93	0.00000009	0.00000008	0.00000006	0.00000005	0.00000004
94	0.00000008	0.00000006	0.00000005	0.00000004	0.00000004
95	0.00000007	0.00000005	0.00000004	0.00000004	0.00000003
96	0.00000006	0.00000005	0.00000004	0.00000003	0.00000003
97	0.00000005	0.00000004	0.00000003	0.00000003	0.00000002
98	0.00000004	0.00000003	0.00000003	0.00000002	0.00000002
99	0.00000003	0.00000003	0.00000002	0.00000002	0.00000001
100	0.00000003	0.00000002	0.00000002	0.00000001	0.00000001

AMOUNT OF ANNUITY $[(1+i)^n - 1]/i$ *

The following table gives the amount of an annuity of unit value per period after a term of n periods at rate of interest of i per period; usually indicated as $s_{\overline{n}|i}$.

Periods	Rate i				
n	.0025 ($\frac{1}{4}$%)	.004167 ($\frac{5}{12}$%)	.005 ($\frac{1}{2}$%)	.005833 ($\frac{7}{12}$%)	.0075 ($\frac{3}{4}$%)
1	1.0000 0000	1.0000 0000	1.0000 0000	1.0000 0000	1.0000 0000
2	2.0025 0000	2.0041 6667	2.0050 0000	2.0058 3333	2.0075 0000
3	3.0075 0625	3.0125 1736	3.0150 2500	3.0175 3403	3.0225 5625
4	4.0150 2502	4.0250 6952	4.0301 0013	4.0351 3631	4.0452 2542
5	5.0250 6258	5.0418 4064	5.0502 5063	5.0586 7460	5.0755 6461
6	6.0376 2523	6.0628 4831	6.0755 0188	6.0881 8354	6.1136 3135
7	7.0527 1930	7.0881 1018	7.1058 7939	7.1236 9794	7.1594 8358
8	8.0703 5110	8.1176 4397	8.1414 0879	8.1652 5285	8.2131 7971
9	9.0905 2697	9.1514 6749	9.1821 1583	9.2128 8349	9.2747 7856
10	10.1132 5329	10.1895 9860	10.2280 2641	10.2666 2531	10.3443 3940
11	11.1385 3642	11.2320 5526	11.2791 6654	11.3265 1396	11.4219 2194
12	12.1663 8277	12.2788 5549	12.3355 6237	12.3925 8529	12.5075 8636
13	13.1967 9872	13.3300 1739	13.3972 4018	13.4648 7537	13.6013 9325
14	14.2297 9072	14.3855 5913	14.4642 2639	14.5434 2048	14.7034 0370
15	15.2653 6520	15.4454 9896	15.5365 4752	15.6282 5710	15.8136 7923
16	16.3035 2861	16.5098 5520	16.6142 3026	16.7194 2193	16.9322 8183
17	17.3442 8743	17.5786 4627	17.6973 0141	17.8169 5189	18.0592 7394
18	18.3876 4815	18.6518 9063	18.7857 8791	18.9208 8411	19.1947 1849
19	19.4336 1727	19.7296 0684	19.8797 1685	20.0312 5593	20.3386 7888
20	20.4822 0131	20.8118 1353	20.9791 1544	21.1481 0493	21.4912 1897
21	21.5334 0682	21.8985 2942	22.0840 1101	22.2714 6887	22.6524 0312
22	22.5872 4033	22.9897 7330	23.1944 3107	23.4013 8577	23.8222 9614
23	23.6437 0843	24.0855 6402	24.3104 0323	24.5378 9386	25.0009 6336
24	24.7028 1770	25.1859 2053	25.4319 5524	25.6810 3157	26.1884 7059
25	25.7645 7475	26.2908 6187	26.5591 1502	26.8308 3759	27.3848 8411
26	26.8289 8619	27.4004 0713	27.6919 1059	27.9873 5081	28.5902 7075
27	27.8960 5865	28.5145 7549	28.8303 7015	29.1506 1036	29.8046 9778
28	28.9657 9880	29.6333 8622	29.9745 2200	30.3206 5558	31.0282 3301
29	30.0382 1330	30.7568 5866	31.1243 9461	31.4975 2607	32.2609 4476
30	31.1133 0883	31.8850 1224	32.2800 1658	32.6812 6164	33.5029 0184
31	32.1910 9210	33.0178 6646	33.4414 1666	33.8719 0233	34.7541 7361
32	33.2715 6983	34.1554 4090	34.6086 2375	35.0694 8843	36.0148 2991
33	34.3547 4876	35.2977 5524	35.7816 6686	36.2740 6045	37.2849 4113
34	35.4406 3563	36.4448 2922	36.9605 7520	37.4856 5913	38.5645 7819
35	36.5292 3722	37.5966 8268	38.1453 7807	38.7043 2548	39.8538 1253
36	37.6205 6031	38.7533 3552	39.3361 0497	39.9301 0071	41.1527 1612
37	38.7146 1171	39.9148 0775	40.5327 8549	41.1630 2630	42.4613 6149
38	39.8113 9824	41.0811 1945	41.7354 4942	42.4031 4395	43.7798 2170
39	40.9109 2674	42.2522 9078	42.9441 2666	43.6504 9562	45.1081 7037
40	42.0132 0405	43.4283 4199	44.1588 4730	44.9051 2352	46.4464 8164
41	43.1182 3706	44.6092 9342	45.3796 4153	46.1670 7007	47.7948 3026
42	44.2260 3265	45.7951 6547	46.6065 3974	47.4363 7798	49.1532 9148
43	45.3365 9774	46.9859 7866	47.8395 7244	48.7130 9018	50.5219 4117
44	46.4499 3923	48.1817 5357	49.0787 7030	49.9972 4988	51.9008 5573
45	47.5660 6408	49.3825 1088	50.3241 6415	51.2889 0050	53.2901 1215
46	48.6849 7924	50.5882 7134	51.5757 8498	52.5880 8575	54.6897 8799
47	49.8066 9169	51.7990 5581	52.8336 6390	53.8948 4959	56.0999 6140
48	50.9312 0842	53.0148 8521	54.0978 3222	55.2092 3621	57.5207 1111
49	52.0585 3644	54.2357 8056	55.3683 2138	56.5312 9009	58.9521 1644
50	53.1886 8278	55.4617 6298	56.6451 6299	57.8610 5595	60.3942 5732

*For $s_{\overline{n}|i}$ where rates i = 0% to 20% are lacking in the 1/4% scaling use tables $(1+i)^n$ and construct information by use of formula. See page 1004.

AMOUNT OF ANNUITY $[(1+i)^n - 1]/i$ (Continued)

Periods	Rate i				
n	.0025 ($\frac{1}{4}$%)	.004167 ($\frac{5}{12}$%)	.005 ($\frac{1}{2}$%)	.005833 ($\frac{7}{12}$%)	.0075 ($\frac{3}{4}$%)
50	53.1886 828	55.4617 630	56.6451 630	57.8610 559	60.3942 573
51	54.3216 545	56.6928 537	57.9283 888	59.1985 788	61.8472 142
52	55.4574 586	57.9290 739	59.2180 307	60.5439 038	63.3110 684
53	56.5961 023	59.1704 450	60.5141 209	61.8970 766	64.7859 014
54	57.7375 925	60.4169 885	61.8166 915	63.2581 429	66.2717 956
55	58.8819 365	61.6687 260	63.1257 750	64.6271 487	67.7688 341
56	60.0291 413	62.9256 790	64.4414 038	66.0041 404	69.2771 003
57	61.1792 142	64.1878 694	65.7636 109	67.3891 646	70.7966 786
58	62.3321 622	65.4553 188	67.0924 289	68.7822 680	72.3276 537
59	63.4879 926	66.7280 493	68.4278 911	70.1834 979	73.8701 111
60	64.6467 126	68.0060 828	69.7700 305	71.5929 016	75.4241 369
61	65.8083 294	69.2894 415	71.1188 807	73.0105 269	76.9898 180
62	66.9728 502	70.5781 475	72.4744 751	74.4364 216	78.5672 416
63	68.1402 824	71.8722 231	73.8368 474	75.8706 341	80.1564 959
64	69.3106 331	73.1716 907	75.2060 317	77.3132 128	81.7576 696
65	70.4839 096	74.4765 728	76.5820 618	78.7642 065	83.3708 521
66	71.6601 194	75.7868 918	77.9649 721	80.2236 644	84.9961 335
67	72.8392 697	77.1026 706	79.3547 970	81.6916 358	86.6336 045
68	74.0213 679	78.4239 317	80.7515 710	83.1681 703	88.2833 566
69	75.2064 213	79.7506 981	82.1553 288	84.6533 180	89.9454 817
70	76.3944 374	81.0829 926	83.5661 055	86.1471 290	91.6200 729
71	77.5854 235	82.4208 384	84.9839 360	87.6496 539	93.3072 234
72	78.7793 870	83.7642 586	86.4088 557	89.1609 436	95.0070 276
73	79.9763 355	85.1132 763	87.8409 000	90.6810 491	96.7195 803
74	81.1762 763	86.4679 150	89.2801 045	92.2100 219	98.4449 771
75	82.3792 170	87.8281 980	90.7265 050	93.7479 137	100.1833 145
76	83.5851 651	89.1941 488	92.1801 375	95.2947 765	101.9346 893
77	84.7941 280	90.5657 911	93.6410 382	96.8506 627	103.6991 995
78	86.0061 133	91.9431 485	95.1092 434	98.4156 249	105.4769 435
79	87.2211 286	93.3262 450	96.5847 896	99.9897 160	107.2680 206
80	88.4391 814	94.7151 044	98.0677 136	101.5729 894	109.0725 307
81	89.6602 793	96.1097 506	99.5580 521	103.1654 985	110.8905 747
82	90.8844 300	97.5102 079	101.0558 424	104.7672 972	112.7222 540
83	92.1116 411	98.9165 004	102.5611 216	106.3784 398	114.5676 709
84	93.3419 202	100.3286 525	104.0739 272	107.9989 807	116.4269 284
85	94.5752 750	101.7466 886	105.5942 969	109.6289 748	118.3001 304
86	95.8117 132	103.1706 331	107.1222 683	111.2684 771	120.1873 814
87	97.0512 425	104.6005 108	108.6578 797	112.9175 432	122.0887 867
88	98.2938 706	106.0363 462	110.2011 691	114.5762 289	124.0044 526
89	99.5396 053	107.4781 643	111.7521 749	116.2445 902	125.9344 860
90	100.7884 543	108.9259 900	113.3109 358	117.9226 837	127.8789 947
91	102.0404 254	110.3798 483	114.8774 905	119.6105 660	129.8380 871
92	103.2955 265	111.8397 643	116.4518 779	121.3082 943	131.8118 728
93	104.5537 653	113.3057 634	118.0341 373	123.0159 260	133.8004 618
94	105.8151 497	114.7778 707	119.6243 080	124.7335 189	135.8039 653
95	107.0796 876	116.2561 118	121.2224 295	126.4611 311	137.8224 951
96	108.3473 868	117.7405 123	122.8285 417	128.1988 210	139.8561 638
97	109.6182 553	119.2310 978	124.4426 844	129.9466 475	141.9050 850
98	110.8923 009	120.7278 940	126.0648 978	131.7046 696	143.9693 731
99	112.1695 317	122.2309 269	127.6952 223	133.4729 468	146.0491 434
100	113.4499 555	123.7402 224	129.3336 984	135.2515 390	148.1445 120

AMOUNT OF ANNUITY $[(1 + i)^n - 1]/i$ (Continued)

Periods	Rate i				
n	.01 (1%)	.01125 (1⅛%)	.0125 (1¼%)	.015 (1½%)	.0175 (1¾%)
1	1.0000 0000	1.0000 0000	1.0000 0000	1.0000 0000	1.0000 0000
2	2.0100 0000	2.0112 5000	2.0125 0000	2.0150 0000	2.0175 0000
3	3.0301 0000	3.0338 7656	3.0376 5625	3.0452 2500	3.0528 0625
4	4.0604 0100	4.0680 0767	4.0756 2695	4.0909 0338	4.1062 3036
5	5.1010 0501	5.1137 7276	5.1265 7229	5.1522 6693	5.1780 8939
6	6.1520 1506	6.1713 0270	6.1906 5444	6.2295 5093	6.2687 0596
7	7.2135 3521	7.2407 2986	7.2680 3762	7.3229 9419	7.3784 0831
8	8.2856 7056	8.3221 8807	8.3588 8809	8.4328 3911	8.5075 3045
9	9.3685 2727	9.4158 1269	9.4633 7420	9.5593 3169	9.6564 1224
10	10.4622 1254	10.5217 4058	10.5816 6637	10.7027 2167	10.8253 9945
11	11.5668 3467	11.6401 1016	11.7139 3720	11.8632 6249	12.0148 4394
12	12.6825 0301	12.7710 6140	12.8603 6142	13.0412 1143	13.2251 0371
13	13.8093 2804	13.9147 3584	14.0211 1594	14.2368 2960	14.4565 4303
14	14.9474 2132	15.0712 7662	15.1963 7988	15.4503 8205	15.7095 3253
15	16.0968 9554	16.2408 2848	16.3863 3463	16.6821 3778	16.9844 4935
16	17.2578 6449	17.4235 3780	17.5911 6382	17.9323 6984	18.2816 7721
17	18.4304 4314	18.6195 5260	18.8110 5336	19.2013 5539	19.6016 0656
18	19.6147 4757	19.8290 2257	20.0461 9153	20.4893 7572	20.9446 3468
19	20.8108 9504	21.0520 9907	21.2967 6893	21.7967 1636	22.3111 6578
20	22.0190 0399	22.2889 3519	22.5629 7854	23.1236 6710	23.7016 1119
21	23.2391 9403	23.5396 8571	23.8450 1577	24.4705 2211	25.1163 8938
22	24.4715 8598	24.8045 0717	25.1430 7847	25.8375 7994	26.5559 2620
23	25.7163 0183	26.0835 5788	26.4573 6695	27.2251 4364	28.0206 5490
24	26.9734 6485	27.3769 9790	27.7880 8403	28.6335 2080	29.5110 1637
25	28.2431 9950	28.6849 8913	29.1354 3508	30.0630 2361	31.0274 5915
26	29.5256 3150	30.0076 9526	30.4996 2802	31.5139 6896	32.5704 3969
27	30.8208 8781	31.3452 8183	31.8808 7337	32.9866 7850	34.1404 2238
28	32.1290 9669	32.6979 1625	33.2793 8429	34.4814 7867	35.7378 7977
29	33.4503 8766	34.0657 6781	34.6953 7659	35.9987 0085	37.3632 9267
30	34.7848 9153	35.4490 0769	36.1290 6880	37.5386 8137	39.0171 5029
31	36.1327 4045	36.8478 0903	37.5806 8216	39.1017 6159	40.6999 5042
32	37.4940 6785	38.2623 4688	39.0504 4069	40.6882 8801	42.4121 9955
33	38.8690 0853	39.6927 9829	40.5385 7120	42.2986 1233	44.1544 1305
34	40.2576 9862	41.1393 4227	42.0453 0334	43.9330 9152	45.9271 1527
35	41.6602 7560	42.6021 5987	43.5708 6963	45.5920 8789	47.7308 3979
36	43.0768 7836	44.0814 3417	45.1155 0550	47.2759 6921	49.5661 2949
37	44.5076 4714	45.5773 5030	46.6794 4932	48.9851 0874	51.4335 3675
38	45.9527 2361	47.0900 9549	48.2629 4243	50.7198 8538	53.3336 2365
39	47.4122 5085	48.6198 5906	49.8662 2921	52.4806 8366	55.2669 6206
40	48.8863 7336	50.1668 3248	51.4895 5708	54.2678 9391	57.2341 3390
41	50.3752 3709	51.7312 0934	53.1331 7654	56.0819 1232	59.2357 3124
42	51.8789 8946	53.3131 8545	54.7973 4125	57.9231 4100	61.2723 5654
43	53.3977 7936	54.9129 5879	56.4823 0801	59.7919 8812	63.3446 2278
44	54.9317 5715	56.5307 2957	58.1883 3686	61.6888 6794	65.4531 5367
45	56.4810 7472	58.1667 0028	59.9156 9108	63.6142 0096	67.5985 8386
46	58.0458 8547	59.8210 7566	61.6646 3721	65.5684 1398	69.7815 5908
47	59.6263 4432	61.4940 6276	63.4354 4518	67.5519 4018	72.0027 3637
48	61.2226 0777	63.1858 7097	65.2283 8824	69.5652 1929	74.2627 8425
49	62.8348 3385	64.8967 1201	67.0437 4310	71.6086 9758	76.5623 8298
50	64.4631 8218	66.6268 0002	68.8817 8989	73.6828 2804	78.9022 2468

AMOUNT OF ANNUITY $[(1 + i)^n - 1]/i$ (Continued)

Periods	Rate i				
n	.01 (1%)	.01125 (1⅛%)	.0125 (1¼%)	.015 (1½%)	.0175 (1¾%)
50	64.4631 822	66.6268 000	68.8817 899	73.6828 280	78.9022 247
51	66.1078 140	68.3763 515	70.7428 123	75.7880 705	81.2830 136
52	67.7688 921	70.1455 855	72.6270 974	77.9248 915	83.7054 663
53	69.4465 811	71.9347 233	74.5349 361	80.0937 649	86.1703 120
54	71.1410 469	73.7439 890	76.4666 228	82.2951 714	88.6782 925
55	72.8524 573	75.5736 088	78.4224 556	84.5295 989	91.2301 626
56	74.5809 819	77.4238 119	80.4027 363	86.7975 429	93.8266 904
57	76.3267 917	79.2948 298	82.4077 705	89.0995 061	96.4686 575
58	78.0900 597	81.1868 966	84.4378 676	91.4359 987	99.1568 590
59	79.8709 603	83.1002 492	86.4933 410	93.8075 386	101.8921 041
60	81.6696 699	85.0351 270	88.5745 078	96.2146 517	104.6752 159
61	83.4863 666	86.9917 722	90.6816 891	98.6578 715	107.5070 322
62	85.3212 302	88.9704 297	92.8152 102	101.1377 396	110.3884 052
63	87.1744 425	90.9713 470	94.9754 003	103.6548 057	113.3202 023
64	89.0461 869	92.9947 746	97.1625 928	106.2096 277	116.3033 058
65	90.9366 488	95.0409 659	99.3771 253	108.8027 722	119.3386 137
66	92.8460 153	97.1101 767	101.6193 393	111.4348 137	122.4270 394
67	94.7744 755	99.2026 662	103.8895 811	114.1063 359	125.5695 126
68	96.7222 202	101.3186 962	106.1882 008	116.8179 310	128.7669 791
69	98.6894 424	103.4585 315	108.5155 533	119.5701 999	132.0204 012
70	100.6763 368	105.6224 400	110.8719 978	122.3637 529	135.3307 583
71	102.6831 002	107.8106 925	113.2578 977	125.1992 092	138.6990 465
72	104.7099 312	110.0235 628	115.6736 215	128.0771 974	142.1262 798
73	106.7570 305	112.2613 278	118.1195 417	130.9983 553	145.6134 897
74	108.8246 008	114.5242 678	120.5960 360	133.9633 307	149.1617 258
75	110.9128 468	116.8126 658	123.1034 864	136.9727 806	152.7720 560
76	113.0219 753	119.1268 083	125.6422 800	140.0273 723	156.4455 670
77	115.1521 951	121.4669 849	128.2128 085	143.1277 829	160.1833 644
78	117.3037 170	123.8334 885	130.8154 686	146.2746 997	163.9865 733
79	119.4767 542	126.2266 152	133.4506 620	149.4688 202	167.8563 383
80	121.6715 217	128.6466 646	136.1187 953	152.7108 525	171.7938 242
81	123.8882 369	131.0939 396	138.8202 802	156.0015 153	175.8002 162
82	126.1271 193	133.5687 464	141.5555 337	159.3415 380	179.8767 200
83	128.3883 905	136.0713 948	144.3249 779	162.7316 611	184.0245 625
84	130.6722 744	138.6021 980	147.1290 401	166.1726 360	188.2449 924
85	132.9789 971	141.1614 727	149.9681 531	169.6652 255	192.5392 798
86	135.3087 871	143.7495 393	152.8427 550	173.2102 039	196.9087 172
87	137.6618 750	146.3667 216	155.7532 895	176.8083 569	201.3546 197
88	140.0384 937	149.0133 472	158.7002 056	180.4604 823	205.8783 256
89	142.4388 787	151.6897 474	161.6839 581	184.1673 895	210.4811 962
90	144.8632 675	154.3962 571	164.7050 076	187.9299 004	215.1646 172
91	147.3119 001	157.1332 149	167.7638 202	191.7488 489	219.9299 980
92	149.7850 191	159.9009 636	170.8608 680	195.6250 816	224.7787 729
93	152.2828 693	162.6998 495	173.9966 288	199.5594 578	229.7124 015
94	154.8056 980	165.5302 228	177.1715 867	203.5528 497	234.7323 685
95	157.3537 550	168.3924 378	180.3862 315	207.6061 425	239.8401 850
96	159.9272 926	171.2868 527	183.6410 594	211.7202 346	245.0373 882
97	162.5265 655	174.2138 298	186.9365 726	215.8960 381	250.3255 425
98	165.1518 311	177.1737 354	190.2732 798	220.1344 787	255.7062 395
99	167.8033 494	180.1669 399	193.6516 958	224.4364 959	261.1810 987
100	170.4813 829	183.1938 180	197.0723 420	228.8030 433	266.7517 679

AMOUNT OF ANNUITY $[(1+i)^n - 1]/i$ (Continued)

Periods	Rate i				
n	.02 (2%)	.0225 (2¼%)	.025 (2½%)	.0275 (2¾%)	.03 (3%)
1	1.0000 0000	1.0000 0000	1.0000 0000	1.0000 0000	1.0000 0000
2	2.0200 0000	2.0225 0000	2.0250 0000	2.0275 0000	2.0300 0000
3	3.0604 0000	3.0680 0625	3.0756 2500	3.0832 5625	3.0909 0000
4	4.1216 0800	4.1370 3639	4.1525 1563	4.1680 4580	4.1836 2700
5	5.2040 4016	5.2301 1971	5.2563 2852	5.2826 6706	5.3091 3581
6	6.3081 2096	6.3477 9740	6.3877 3673	6.4279 4040	6.4684 0988
7	7.4342 8338	7.4906 2284	7.5474 3015	7.6047 0876	7.6624 6218
8	8.5829 6905	8.6591 6186	8.7361 1590	8.8138 3825	8.8923 3605
9	9.7546 2843	9.8539 9300	9.9545 1880	10.0562 1880	10.1591 0613
10	10.9497 2100	11.0757 0784	11.2033 8177	11.3327 6482	11.4638 7931
11	12.1687 1542	12.3249 1127	12.4834 6631	12.6444 1585	12.8077 9569
12	13.4120 8973	13.6022 2177	13.7955 5297	13.9921 3729	14.1920 2956
13	14.6803 3152	14.9082 7176	15.1404 4179	15.3769 2107	15.6177 9045
14	15.9739 3815	16.2437 0788	16.5189 5284	16.7997 8639	17.0863 2416
15	17.2934 1692	17.6091 9130	17.9319 2666	18.2617 8052	18.5989 1389
16	18.6392 8525	19.0053 9811	19.3802 2483	19.7639 7948	20.1568 8130
17	20.0120 7096	20.4330 1957	20.8647 3045	21.3074 8892	21.7615 8774
18	21.4123 1238	21.8927 6251	22.3863 4871	22.8934 4487	23.4144 3537
19	22.8405 5863	23.3853 4966	23.9460 0743	24.5230 1460	25.1168 6844
20	24.2973 6980	24.9115 2003	25.5446 5761	26.1973 9750	26.8703 7449
21	25.7833 1719	26.4720 2923	27.1832 7405	27.9178 2593	28.6764 8572
22	27.2989 8354	28.0676 4989	28.8628 5590	29.6855 6615	30.5367 8030
23	28.8449 6321	29.6991 7201	30.5844 2730	31.5019 1921	32.4528 8370
24	30.4218 6247	31.3674 0338	32.3490 3798	33.3682 2199	34.4264 7022
25	32.0302 9972	33.0731 6996	34.1577 6393	35.2858 4810	36.4592 6432
26	33.6709 0572	34.8173 1628	36.0117 0803	37.2562 0892	38.5530 4225
27	35.3443 2383	36.6007 0590	37.9120 0073	39.2807 5467	40.7096 3352
28	37.0512 1031	38.4242 2178	39.8598 0075	41.3609 7542	42.9309 2252
29	38.7922 3451	40.2887 6677	41.8562 9577	43.4984 0224	45.2188 5020
30	40.5680 7921	42.1952 6402	43.9027 0316	45.6946 0831	47.5754 1571
31	42.3794 4079	44.1446 5746	46.0002 7074	47.9512 1003	50.0026 7818
32	44.2270 2961	46.1379 1226	48.1502 7751	50.2698 6831	52.5027 5852
33	46.1115 7020	48.1760 1528	50.3540 3445	52.6522 8969	55.0778 4128
34	48.0338 0160	50.2599 7563	52.6128 8531	55.1002 2765	57.7301 7652
35	49.9944 7763	52.3908 2508	54.9282 0744	57.6154 8391	60.4620 8181
36	51.9943 6719	54.5696 1864	57.3014 1263	60.1999 0972	63.2759 4427
37	54.0342 5453	56.7974 3506	59.7339 4794	62.8554 0724	66.1742 2259
38	56.1149 3962	59.0753 7735	62.2272 9664	65.5839 3094	69.1594 4927
39	58.2372 3841	61.4045 7334	64.7829 7906	68.3874 8904	72.2342 3275
40	60.4019 8318	63.7861 7624	67.4025 5354	71.2681 4499	75.4012 5973
41	62.6100 2284	66.2213 6521	70.0876 1737	74.2280 1898	78.6632 9753
42	64.8622 2330	68.7113 4592	72.8398 0781	77.2692 8950	82.0231 9645
43	67.1594 6777	71.2573 5121	75.6608 0300	80.3941 9496	85.4838 9234
44	69.5026 5712	73.8606 4161	78.5523 2308	83.6050 3532	89.0484 0911
45	71.8927 1027	76.5225 0605	81.5161 3116	86.9041 7379	92.7198 6139
46	74.3305 6447	79.2442 6243	84.5540 3443	90.2940 3857	96.5014 5723
47	76.8171 7576	82.0272 5834	87.6678 8530	93.7771 2463	100.3965 0095
48	79.3535 1928	84.8728 7165	90.8595 8243	97.3559 9556	104.4083 9598
49	81.9405 8966	87.7825 1126	94.1310 7199	101.0332 8544	108.5406 4785
50	84.5794 0145	90.7576 1776	97.4843 4879	104.8117 0079	112.7968 6729

AMOUNT OF ANNUITY $[(1+i)^n - 1]/i$ (Continued)

Periods	Rate i				
n	.02 (2%)	.0225 (2¼%)	.025 (2½%)	.0275 (2¾%)	.03 (3%)
50	84.5794 015	90.7576 178	97.4843 488	104.8117 008	112.7968 673
51	87.2709 895	93.7996 642	100.9214 575	108.6940 226	117.1807 733
52	90.0164 093	96.9101 566	104.4444 939	112.6831 082	121.6961 965
53	92.8167 375	100.0906 351	108.0556 063	116.7818 937	126.3470 824
54	95.6730 722	103.3426 744	111.7569 965	120.9933 957	131.1374 949
55	98.5865 337	106.6678 846	115.5509 214	125.3207 141	136.0716 197
56	101.5582 643	110.0679 120	119.4396 944	129.7670 337	141.1537 683
57	104.5894 296	113.5444 400	123.4256 868	134.3356 272	146.3883 814
58	107.6812 182	117.0991 899	127.5113 289	139.0298 569	151.7800 328
59	110.8348 426	120.7339 217	131.6991 121	143.8531 780	157.3334 338
60	114.0515 394	124.4504 349	135.9915 900	148.8091 404	163.0534 368
61	117.3325 702	128.2505 697	140.3913 797	153.9013 917	168.9450 399
62	120.6792 216	132.1362 075	144.9011 642	159.1336 800	175.0133 911
63	124.0928 060	136.1092 722	149.5236 933	164.5098 562	181.2637 928
64	127.5746 622	140.1717 308	154.2617 856	170.0338 773	187.7017 066
65	131.1261 554	144.3255 948	159.1183 303	175.7098 089	194.3327 578
66	134.7486 785	148.5729 207	164.0962 885	181.5418 286	201.1627 406
67	138.4436 521	152.9158 114	169.1986 957	187.5342 289	208.1976 228
68	142.2125 251	157.3564 171	174.4286 631	193.6914 202	215.4435 515
69	146.0567 756	161.8969 365	179.7893 797	200.0179 343	222.9068 580
70	149.9779 111	166.5396 176	185.2841 142	206.5184 275	230.5940 637
71	153.9774 694	171.2867 590	190.9162 171	213.1976 842	238.5118 856
72	158.0570 188	176.1407 111	196.6891 225	220.0606 205	246.6672 422
73	162.2181 591	181.1038 771	202.6063 506	227.1122 876	255.0672 595
74	166.4625 223	186.1787 143	208.6715 093	234.3578 755	263.7192 773
75	170.7917 728	191.3677 354	214.8882 970	241.8027 171	272.6308 556
76	175.2076 082	196.6735 094	221.2605 045	249.4522 918	281.8097 813
77	179.7117 604	202.0986 634	227.7920 171	257.3122 298	291.2640 747
78	184.3059 956	207.6458 833	234.4868 175	265.3883 162	301.0019 969
79	188.9921 155	213.3179 157	241.3489 880	273.6864 948	311.0320 568
80	193.7719 578	219.1175 688	248.3827 126	282.2128 735	321.3630 185
81	198.6473 970	225.0477 141	255.5922 805	290.9737 275	332.0039 091
82	203.6203 449	231.1112 876	262.9820 875	299.9755 050	342.9640 264
83	208.6927 518	237.3112 916	270.5566 397	309.2248 314	354.2529 472
84	213.8666 068	243.6507 957	278.3205 557	318.7285 142	365.8805 356
85	219.1439 390	250.1329 386	286.2785 695	328.4935 484	377.8569 517
86	224.5268 178	256.7609 297	294.4355 338	338.5271 209	390.1926 602
87	230.0173 541	263.5380 506	302.7964 221	348.8366 168	402.8984 400
88	235.6177 012	270.4676 567	311.3663 327	359.4296 237	415.9853 932
89	241.3300 552	277.5531 790	320.1504 910	370.3139 384	429.4649 550
90	247.1566 563	284.7981 255	329.1542 533	381.4975 717	443.3489 037
91	253.0997 894	292.2060 834	338.3831 096	392.9887 549	457.6493 708
92	259.1617 852	299.7807 202	347.8426 873	404.7959 457	472.3788 519
93	265.3450 209	307.5257 865	357.5387 545	416.9278 342	487.5502 174
94	271.6519 214	315.4451 166	367.4772 234	429.3933 496	503.1767 240
95	278.0849 598	323.5426 318	377.6641 540	442.2016 667	519.2720 257
96	284.6466 590	331.8223 410	388.1057 578	455.3622 126	535.8501 865
97	291.3395 922	340.2883 437	398.8084 018	468.8846 734	552.9256 920
98	298.1663 840	348.9448 314	409.7786 118	482.7790 019	570.5134 628
99	305.1297 117	357.7960 901	421.0230 771	497.0554 245	588.6288 667
100	312.2323 059	366.8465 021	432.5486 540	511.7244 487	607.2877 327

AMOUNT OF ANNUITY $[(1+i)^n - 1]/i$ (Continued)

Periods	Rate i				
n	.035 (3½ %)	.04 (4 %)	.045 (4½ %)	.05 (5 %)	.055 (5½ %)
1	1.0000 000	1.0000 000	1.0000 000	1.0000 000	1.0000 000
2	2.0350 000	2.0400 000	2.0450 000	2.0500 000	2.0550 000
3	3.1062 250	3.1216 000	3.1370 250	3.1525 000	3.1680 250
4	4.2149 429	4.2464 640	4.2781 911	4.3101 250	4.3422 664
5	5.3624 659	5.4163 226	5.4707 097	5.5256 313	5.5810 910
6	6.5501 522	6.6329 755	6.7168 917	6.8019 128	6.8880 510
7	7.7794 075	7.8982 945	8.0191 518	8.1420 085	8.2668 938
8	9.0516 868	9.2142 263	9.3800 136	9.5491 089	9.7215 730
9	10.3684 958	10.5827 953	10.8021 142	11.0265 643	11.2562 595
10	11.7313 932	12.0061 071	12.2882 094	12.5778 925	12.8753 538
11	13.1419 919	13.4863 514	13.8411 788	14.2067 872	14.5834 982
12	14.6019 616	15.0258 055	15.4640 318	15.9171 265	16.3855 907
13	16.1130 303	16.6268 377	17.1599 133	17.7129 828	18.2867 981
14	17.6769 864	18.2919 112	18.9321 094	19.5986 320	20.2925 720
15	19.2956 809	20.0235 876	20.7840 543	21.5785 636	22.4086 635
16	20.9710 297	21.8245 311	22.7193 367	23.6574 918	24.6411 400
17	22.7050 157	23.6975 124	24.7417 069	25.8403 664	26.9964 027
18	24.4996 913	25.6454 129	26.8550 837	28.1323 847	29.4812 048
19	26.3571 805	27.6712 294	29.0635 625	30.5390 039	32.1026 711
20	28.2796 818	29.7780 786	31.3714 228	33.0659 541	34.8683 180
21	30.2694 707	31.9692 017	33.7831 368	35.7192 518	37.7860 755
22	32.3289 022	34.2479 698	36.3033 780	38.5052 144	40.8643 097
23	34.4604 137	36.6178 886	38.9370 300	41.4304 751	44.1118 467
24	36.6665 282	39.0826 041	41.6891 963	44.5019 989	47.5379 983
25	38.9498 567	41.6459 083	44.5652 101	47.7270 988	51.1525 882
26	41.3131 017	44.3117 446	47.5706 446	51.1134 538	54.9659 805
27	43.7590 602	47.0842 144	50.7113 236	54.6691 264	58.9891 094
28	46.2906 273	49.9675 830	53.9933 332	58.4025 828	63.2335 105
29	48.9107 993	52.9662 863	57.4230 332	62.3227 119	67.7113 535
30	51.6226 773	56.0849 378	61.0070 697	66.4388 475	72.4354 780
31	54.4294 710	59.3283 353	64.7523 878	70.7607 899	77.4194 293
32	57.3345 025	62.7014 687	68.6662 452	75.2988 294	82.6774 979
33	60.3412 101	66.2095 274	72.7562 263	80.0637 708	88.2247 603
34	63.4531 524	69.8579 085	77.0302 565	85.0669 594	94.0771 221
35	66.6740 127	73.6522 249	81.4966 180	90.3203 074	100.2513 638
36	70.0076 032	77.5983 138	86.1639 658	95.8363 227	106.7651 888
37	73.4578 693	81.7022 464	91.0413 443	101.6281 389	113.6372 742
38	77.0288 947	85.9703 363	96.1382 048	107.7095 458	120.8873 242
39	80.7249 060	90.4091 497	101.4644 240	114.0950 231	128.5361 271
40	84.5502 777	95.0255 157	107.0303 231	120.7997 742	136.6056 141
41	88.5095 375	99.8265 363	112.8466 876	127.8397 630	145.1189 228
42	92.6073 713	104.8195 978	118.9247 885	135.2317 511	154.1004 636
43	96.8486 293	110.0123 817	125.2764 040	142.9933 387	163.5759 891
44	101.2383 313	115.4128 770	131.9138 422	151.1430 056	173.5726 685
45	105.7816 729	121.0293 920	138.8499 651	159.7001 559	184.1191 653
46	110.4840 314	126.8705 677	146.0982 135	168.6851 637	195.2457 194
47	115.3509 725	132.9453 904	153.6726 331	178.1194 218	206.9842 339
48	120.3882 566	139.2632 060	161.5879 016	188.0253 929	219.3683 668
49	125.6018 456	145.8337 343	169.8593 572	198.4266 626	232.4336 270
50	130.9979 102	152.6670 837	178.5030 283	209.3479 957	246.2174 764

AMOUNT OF ANNUITY $[(1 + i)^n - 1]/i$ (Continued)

Periods	Rate i				
n	.06 (6%)	.065 (6½%)	.07 (7%)	.075 (7½%)	.08 (8%)
1	1.0000 000	1.0000 000	1.0000 000	1.0000 000	1.0000 000
2	2.0600 000	2.0650 000	2.0700 000	2.0750 000	2.0800 000
3	3.1836 000	3.1992 250	3.2149 000	3.2306 250	3.2464 000
4	4.3746 160	4.4071 746	4.4399 430	4.4729 219	4.5061 120
5	5.6370 930	5.6936 410	5.7507 390	5.8083 910	5.8666 010
6	6.9753 185	7.0637 276	7.1532 907	7.2440 203	7.3359 290
7	8.3938 376	8.5228 699	8.6540 211	8.7873 219	8.9228 034
8	9.8974 679	10.0768 565	10.2598 026	10.4463 710	10.6366 276
9	11.4913 160	11.7318 522	11.9779 887	12.2298 488	12.4875 578
10	13.1807 949	13.4944 225	13.8164 480	14.1470 875	14.4865 625
11	14.9716 426	15.3715 600	15.7835 993	16.2081 191	16.6454 875
12	16.8699 412	17.3707 114	17.8884 513	18.4237 280	18.9771 265
13	18.8821 377	19.4998 076	20.1406 429	20.8055 076	21.4952 966
14	21.0150 659	21.7672 951	22.5504 879	23.3659 207	24.2149 203
15	23.2759 699	24.1821 693	25.1290 220	26.1183 647	27.1521 139
16	25.6725 281	26.7540 103	27.8880 536	29.0772 421	30.3242 830
17	28.2128 798	29.4930 210	30.8402 173	32.2580 352	33.7502 257
18	30.9056 525	32.4100 674	33.9990 325	35.6773 879	37.4502 437
19	33.7599 917	35.5167 218	37.3789 648	39.3531 919	41.4462 632
20	36.7855 912	38.8253 087	40.9954 923	43.3046 813	45.7619 643
21	39.9927 267	42.3489 537	44.8651 768	47.5525 324	50.4229 214
22	43.3922 903	46.1016 357	49.0057 392	52.1189 724	55.4567 552
23	46.9958 277	50.0982 420	53.4361 409	57.0278 953	60.8932 956
24	50.8155 774	54.3546 278	58.1766 708	62.3049 874	66.7647 592
25	54.8645 120	58.8876 786	63.2490 377	67.9778 615	73.1059 400
26	59.1563 827	63.7153 777	68.6764 704	74.0762 011	79.9544 151
27	63.7057 657	68.8568 772	74.4838 233	80.6319 162	87.3507 684
28	68.5281 116	74.3325 743	80.6976 909	87.6793 099	95.3388 298
29	73.6397 983	80.1641 916	87.3465 293	95.2552 582	103.9659 362
30	79.0581 862	86.3748 640	94.4607 863	103.3994 025	113.2832 111
31	84.8016 774	92.9892 302	102.0730 414	112.1543 577	123.3458 680
32	90.8897 780	100.0335 302	110.2181 543	121.5659 345	134.2135 374
33	97.3431 647	107.5357 096	118.9334 251	131.6833 796	145.9506 204
34	104.1837 546	115.5255 308	128.2587 648	142.5596 331	158.6266 701
35	111.4347 799	124.0346 903	138.2368 784	154.2516 056	172.3168 037
36	119.1208 667	133.0969 451	148.9134 598	166.8204 760	187.1021 480
37	127.2681 187	142.7482 466	160.3374 020	180.3320 117	203.0703 198
38	135.9042 058	153.0268 826	172.5610 202	194.8569 126	220.3159 454
39	145.0584 581	163.9736 300	185.6402 916	210.4711 810	238.9412 210
40	154.7619 656	175.6319 159	199.6351 120	227.2565 196	259.0565 187
41	165.0476 836	188.0479 904	214.6095 698	245.3007 586	280.7810 402
42	175.9505 446	201.2711 098	230.6322 397	264.6983 155	304.2435 234
43	187.5075 772	215.3537 320	247.7764 965	285.5506 891	329.5830 053
44	199.7580 319	230.3517 245	266.1208 513	307.9669 908	356.9496 457
45	212.7435 138	246.3245 866	285.7493 108	332.0645 151	386.5056 174
46	226.5081 246	263.3356 848	306.7517 626	357.9693 537	418.4260 668
47	241.0986 121	281.4525 043	329.2243 860	385.8170 553	452.9001 521
48	256.5645 288	300.7469 170	353.2700 930	415.7533 344	490.1321 643
49	272.9584 006	321.2954 666	378.9989 995	447.9348 345	530.3427 374
50	290.3359 046	343.1796 720	406.5289 295	482.5299 471	573.7701 564

PRESENT VALUE OF ANNUITY $[1 - (1 + i)^{-n}]/i$*

The following table gives the present value of an annuity of unit value per period for a term of n periods at rate of interest i per period; usually indicated as $a_{\overline{n}|i}$.

Periods	Rate i				
n	.0025 (1/4 %)	.004167 (5/12 %)	.005 (1/2 %)	.005833 (7/12 %)	.0075 (3/4 %)
1	0.9975 0623	0.9958 5062	0.9950 2488	0.9942 0050	0.9925 5583
2	1.9925 2492	1.9875 6908	1.9850 9938	1.9826 3513	1.9777 2291
3	2.9850 6227	2.9751 7253	2.9702 4814	2.9653 3732	2.9555 5624
4	3.9751 2446	3.9586 7804	3.9504 9566	3.9423 4034	3.9261 1041
5	4.9627 1766	4.9381 0261	4.9258 6633	4.9136 7722	4.8894 3961
6	5.9478 4804	5.9134 6318	5.8963 8441	5.8793 8083	5.8455 9763
7	6.9305 2174	6.8847 7661	6.8620 7404	6.8394 8384	6.7946 3785
8	7.9107 4487	7.8520 5970	7.8229 5924	7.7940 1874	7.7366 1325
9	8.8885 2357	8.8153 2916	8.7790 6392	8.7430 1780	8.6715 7642
10	9.8638 6391	9.7746 0165	9.7304 1186	9.6865 1314	9.5995 7958
11	10.8367 7198	10.7298 9376	10.6770 2673	10.6245 3667	10.5206 7452
12	11.8072 5384	11.6812 2200	11.6189 3207	11.5571 2014	11.4349 1267
13	12.7753 1555	12.6286 0283	12.5561 5131	12.4842 9509	12.3423 4508
14	13.7409 6314	13.5720 5261	13.4887 0777	13.4060 9288	13.2430 2242
15	14.7042 0264	14.5115 8766	14.4166 2465	14.3225 4470	14.1369 9495
16	15.6650 4004	15.4472 2422	15.3399 2502	15.2336 8156	15.0243 1261
17	16.6234 8133	16.3789 7848	16.2586 3186	16.1395 3427	15.9050 2492
18	17.5795 3250	17.3068 6654	17.1727 6802	17.0401 3350	16.7791 8107
19	18.5331 9950	18.2309 0443	18.0823 5624	17.9355 0969	17.6468 2984
20	19.4844 8828	19.1511 0815	18.9874 1915	18.8256 9315	18.5080 1969
21	20.4334 0477	20.0674 9359	19.8879 7925	19.7107 1398	19.3627 9870
22	21.3799 5488	20.9800 7661	20.7840 5896	20.5906 0213	20.2112 1459
23	22.3241 4452	21.8888 7297	21.6756 8055	21.4653 8738	21.0533 1473
24	23.2659 7957	22.7938 9839	22.5628 6622	22.3350 9930	21.8891 4614
25	24.2054 6591	23.6951 6853	23.4456 3803	23.1997 6732	22.7187 5547
26	25.1426 0939	24.5926 9895	24.3240 1794	24.0594 2070	23.5421 8905
27	26.0774 1585	25.4865 0517	25.1980 2780	24.9140 8852	24.3594 9286
28	27.0098 9112	26.3766 0266	26.0676 8936	25.7637 9968	25.1707 1251
29	27.9400 4102	27.2630 0680	26.9330 2424	26.6085 8295	25.9758 9331
30	28.8678 7134	28.1457 3291	27.7940 5397	27.4484 6689	26.7750 8021
31	29.7933 8787	29.0247 9626	28.6507 9997	28.2834 7993	27.5683 1783
32	30.7165 9638	29.9002 1205	29.5032 8355	29.1136 5030	28.3556 5045
33	31.6375 0262	30.7719 9540	30.3515 2592	29.9390 0610	29.1371 2203
34	32.5561 1234	31.6401 6139	31.1955 4818	30.7595 7524	29.9127 7621
35	33.4724 3126	32.5047 2504	32.0353 7132	31.5753 8549	30.6826 5629
36	34.3864 6510	33.3657 0128	32.8710 1624	32.3864 6445	31.4468 0525
37	35.2982 1955	34.2231 0501	33.7025 0372	33.1928 3955	32.2052 6576
38	36.2077 0030	35.0769 5105	34.5298 5445	33.9945 3808	32.9580 8016
39	37.1149 1302	35.9272 5416	35.3530 8900	34.7915 8716	33.7052 9048
40	38.0198 6336	36.7740 2904	36.1722 2786	35.5840 1374	34.4469 3844
41	38.9225 5697	37.6172 9033	36.9872 9141	36.3718 4465	35.1830 6545
42	39.8229 9947	38.4570 5261	37.7982 9991	37.1551 0653	35.9137 1260
43	40.7211 9648	39.2933 3040	38.6052 7354	37.9338 2588	36.6389 2070
44	41.6171 5359	40.1261 3816	39.4082 3238	38.7080 2904	37.3587 3022
45	42.5108 7640	40.9554 9028	40.2071 9640	39.4777 4221	38.0731 8136
46	43.4023 7048	41.7814 0111	41.0021 8547	40.2429 9143	38.7823 1401
47	44.2916 4137	42.6038 8492	41.7932 1937	41.0038 0258	39.4861 6775
48	45.1786 9464	43.4229 5594	42.5803 1778	41.7602 0141	40.1847 8189
49	46.0635 3580	44.2386 2832	43.3635 0028	42.5122 1349	40.8781 9542
50	46.9461 7037	45.0509 1617	44.1427 8635	43.2598 6428	41.5664 4707

*For $a_{\overline{n}|i}$ where rates i = 0% to 20% are lacking in 1/4% scaling use tables $(1 + i)^{-n}$ and construct information by use of formula. See page 1004.

PRESENT VALUE OF ANNUITY $[1 - (1 + i)^{-n}]/i$ (Continued)

Periods	Rate i				
n	.0025 ($\frac{1}{4}$ %)	.004167 ($\frac{5}{12}$ %)	.005 ($\frac{1}{2}$ %)	.005833 ($\frac{7}{12}$ %)	.0075 ($\frac{3}{4}$ %)
50	46.9461 7037	45.0509 1617	44.1427 8635	43.2598 6428	41.5664 4707
51	47.8266 0386	45.8598 3353	44.9181 9537	44.0031 7907	42.2495 7525
52	48.7048 4176	46.6653 9439	45.6897 4664	44.7421 8301	42.9276 1812
53	49.5808 8953	47.4676 1267	46.4574 5934	45.4769 0108	43.6006 1351
54	50.4547 5265	48.2665 0224	47.2213 5258	46.2073 5816	44.2685 9902
55	51.3264 3656	49.0620 7692	47.9814 4535	46.9335 7895	44.9316 1193
56	52.1959 4669	49.8543 5046	48.7377 5657	47.6555 8802	45.5896 8926
57	53.0632 8847	50.6433 3656	49.4903 0505	48.3734 0980	46.2428 6776
58	53.9284 6730	51.4290 4885	50.2391 0950	49.0870 6856	46.8911 8388
59	54.7914 8858	52.2115 0093	50.9841 8856	49.7965 8846	47.5346 7382
60	55.6523 5769	52.9907 0632	51.7255 6075	50.5019 9350	48.1733 7352
61	56.5110 7999	53.7666 7850	52.4632 4453	51.2033 0754	48.8073 1863
62	57.3676 6083	54.5394 3087	53.1972 5824	51.9005 5431	49.4365 4455
63	58.2221 0557	55.3089 7680	53.9276 2014	52.5937 5739	50.0610 8640
64	59.0744 1952	56.0753 2959	54.6543 4840	53.2829 4024	50.6809 7906
65	59.9246 0800	56.8385 0250	55.3774 6109	53.9681 2617	51.2962 5713
66	60.7726 7631	57.5985 0871	56.0969 7621	54.6493 3836	51.9069 5497
67	61.6186 2974	58.3553 6137	56.8129 1165	55.3265 9986	52.5131 0667
68	62.4624 7355	59.1090 7357	57.5252 8522	55.9999 3358	53.1147 4607
69	63.3042 1302	59.8596 5832	58.2341 1465	56.6693 6230	53.7119 0677
70	64.1438 5339	60.6071 2862	58.9394 1756	57.3349 0867	54.3046 2210
71	64.9813 9989	61.3514 9738	59.6412 1151	57.9965 9520	54.8929 2516
72	65.8168 5774	62.0927 7748	60.3395 1394	58.6544 4427	55.4768 4880
73	66.6502 3216	62.8309 8172	61.0343 4222	59.3084 7815	56.0564 2561
74	67.4815 2834	63.5661 2287	61.7257 1366	59.9587 1896	56.6316 8795
75	68.3107 5146	64.2982 1365	62.4136 4543	60.6051 8869	57.2026 6794
76	69.1379 0670	65.0272 6670	63.0981 5466	61.2479 0922	57.7693 9746
77	69.9629 9920	65.7532 9464	63.7792 5836	61.8869 0229	58.3319 0815
78	70.7860 3411	66.4763 1002	64.4569 7350	62.5221 8952	58.8902 3141
79	71.6070 1657	67.1963 2533	65.1313 1691	63.1537 9239	59.4443 9842
80	72.4259 5169	67.9133 5303	65.8023 0539	63.7817 3229	59.9944 4012
81	73.2428 4458	68.6274 0550	66.4699 5561	64.4060 3044	60.5403 8722
82	74.0577 0033	69.3384 9511	67.1342 8419	65.0267 0798	61.0822 7019
83	74.8705 2402	70.0466 3413	67.7953 0765	65.6437 8590	61.6201 1930
84	75.6813 2072	70.7518 3482	68.4530 4244	66.2572 8507	62.1539 6456
85	76.4900 9548	71.4541 0936	69.1075 0491	66.8672 2625	62.6838 3579
86	77.2968 5335	72.1534 6991	69.7587 1135	67.4736 3007	63.2097 6257
87	78.1015 9935	72.8499 2854	70.4066 7796	68.0765 1706	63.7317 7427
88	78.9043 3850	73.5434 9730	71.0514 2086	68.6759 0759	64.2499 0002
89	79.7050 7581	74.2341 8818	71.6929 5608	69.2718 2197	64.7641 6875
90	80.5038 1627	74.9220 1313	72.3312 9958	69.8642 8033	65.2746 0918
91	81.3005 6486	75.6069 8403	72.9664 6725	70.4533 0273	65.7812 4981
92	82.0953 2654	76.2891 1272	73.5984 7487	71.0389 0910	66.2841 1892
93	82.8881 0628	76.9684 1101	74.2273 3818	71.6211 1923	66.7832 4458
94	83.6789 0901	77.6448 9063	74.8530 7282	72.1999 5284	67.2786 5467
95	84.4677 3966	78.3185 6329	75.4756 9434	72.7754 2950	67.7703 7685
96	85.2546 0315	78.9894 4062	76.0952 1825	73.3475 6869	68.2584 3856
97	86.0395 0439	79.6575 3422	76.7116 5995	73.9163 8975	68.7428 6705
98	86.8224 4827	80.3228 5566	77.3250 3478	74.4819 1193	69.2236 8938
99	87.6034 3967	80.9854 1642	77.9353 5799	75.0441 5436	69.7009 3239
100	88.3824 8346	81.6452 2797	78.5426 4477	75.6031 3607	70.1746 2272

PRESENT VALUE OF ANNUITY $[1 - (1 + i)^{-n}]/i$ (Continued)

Periods	Rate i				
n	.01 (1%)	.01125 ($1\frac{1}{8}$%)	.0125 ($1\frac{1}{4}$%)	.015 ($1\frac{1}{2}$%)	.0175 ($1\frac{3}{4}$%)
1	0.9900 9901	0.9888 7515	0.9876 5432	0.9852 2167	0.9828 0098
2	1.9703 9506	1.9667 4923	1.9631 1538	1.9558 8342	1.9486 9875
3	2.9409 8521	2.9337 4460	2.9265 3371	2.9122 0042	2.8979 8403
4	3.9019 6555	3.8899 8230	3.8780 5798	3.8543 8465	3.8309 4254
5	4.8534 3124	4.8355 8200	4.8178 3504	4.7826 4497	4.7478 5508
6	5.7954 7647	5.7706 6205	5.7460 0992	5.6971 8717	5.6489 9762
7	6.7281 9453	6.6953 3948	6.6627 2585	6.5982 1396	6.5346 4139
8	7.6516 7775	7.6097 3002	7.5681 2429	7.4859 2508	7.4050 5297
9	8.5660 1758	8.5139 4810	8.4623 4498	8.3605 1732	8.2604 9432
10	9.4713 0453	9.4081 0690	9.3455 2591	9.2221 8455	9.1012 2291
11	10.3676 2825	10.2923 1832	10.2178 0337	10.0711 1779	9.9274 9181
12	11.2550 7747	11.1666 9302	11.0793 1197	10.9075 0521	10.7395 4969
13	12.1337 4007	12.0313 4044	11.9301 8466	11.7315 3222	11.5376 4097
14	13.0037 0304	12.8863 6880	12.7705 5275	12.5433 8150	12.3220 0587
15	13.8650 5252	13.7318 8509	13.6005 4592	13.3432 3301	13.0928 8046
16	14.7178 7378	14.5679 9514	14.4202 9227	14.1312 6405	13.8504 9677
17	15.5622 5127	15.3948 0360	15.2299 1829	14.9076 4931	14.5950 8282
18	16.3982 6858	16.2124 1395	16.0295 4893	15.6725 6089	15.3268 6272
19	17.2260 0850	17.0209 2850	16.8193 0759	16.4261 6837	16.0460 5673
20	18.0455 5297	17.8204 4845	17.5993 1613	17.1686 3879	16.7528 8130
21	18.8569 8313	18.6110 7387	18.3696 9495	17.9001 3673	17.4475 4919
22	19.6603 7934	19.3929 0371	19.1305 6291	18.6208 2437	18.1302 6948
23	20.4558 2113	20.1660 3580	19.8820 3744	19.3308 6145	18.8012 4764
24	21.2433 8726	20.9305 6693	20.6242 3451	20.0304 0537	19.4606 8565
25	22.0231 5570	21.6865 9276	21.3572 6865	20.7196 1120	20.1087 8196
26	22.7952 0366	22.4342 0792	22.0812 5299	21.3986 3172	20.7457 3166
27	23.5596 0759	23.1735 0598	22.7962 9925	22.0676 1746	21.3717 2644
28	24.3164 4316	23.9045 7946	23.5025 1778	22.7267 1671	21.9869 5474
29	25.0657 8530	24.6275 1986	24.2000 1756	23.3760 7558	22.5916 0171
30	25.8077 0822	25.3424 1766	24.8889 0623	24.0158 3801	23.1858 4934
31	26.5422 8537	26.0493 6233	25.5692 9010	24.6461 4582	23.7698 7650
32	27.2695 8947	26.7484 4236	26.2412 7418	25.2671 3874	24.3438 5897
33	27.9896 9255	27.4397 4522	26.9049 6215	25.8789 5442	24.9079 6951
34	28.7026 6589	28.1233 5745	27.5604 5644	26.4817 2849	25.4623 7789
35	29.4085 8009	28.7993 6460	28.2078 5822	27.0755 9458	26.0072 5100
36	30.1075 0504	29.4678 5127	28.8472 6737	27.6606 8431	26.5427 5283
37	30.7995 0994	30.1289 0114	29.4787 8259	28.2371 2740	27.0690 4455
38	31.4846 6330	30.7825 9692	30.1025 0133	28.8050 5163	27.5862 8457
39	32.1630 3298	31.4290 2044	30.7185 1983	29.3645 8288	28.0946 2857
40	32.8346 8611	32.0682 5260	31.3269 3316	29.9158 4520	28.5942 2955
41	33.4996 8922	32.7003 7340	31.9278 3522	30.4589 6079	29.0852 3789
42	34.1581 0814	33.3254 6195	32.5213 1874	30.9940 5004	29.5678 0136
43	34.8100 0806	33.9435 9649	33.1074 7530	31.5212 3157	30.0420 6522
44	35.4554 5352	34.5548 5438	33.6863 9536	32.0406 2223	30.5081 7221
45	36.0945 0844	35.1593 1212	34.2581 6825	32.5523 3718	30.9662 6261
46	36.7272 3608	35.7570 4536	34.8228 8222	33.0564 8983	31.4164 7431
47	37.3536 9909	36.3481 2891	35.3806 2442	33.5531 9195	31.8589 4281
48	37.9739 5949	36.9326 3674	35.9314 8091	34.0425 5365	32.2938 0129
49	38.5880 7871	37.5106 4202	36.4755 3670	34.5246 8339	32.7211 8063
50	39.1961 1753	38.0822 1708	37.0128 7575	34.9996 8807	33.1412 0946

PRESENT VALUE OF ANNUITY $[1 - (1 + i)^{-n}]/i$ (Continued)

Periods	Rate i				
n	.01 (1%)	.01125 (1$\frac{1}{8}$%)	.0125 (1$\frac{1}{4}$%)	.015 (1$\frac{1}{2}$%)	.0175 (1$\frac{3}{4}$%)
50	39.1961 1753	38.0822 1708	37.0128 7575	34.9996 8807	33.1412 0946
51	39.7981 3617	38.6474 3345	37.5435 8099	35.4676 7298	33.5540 1421
52	40.3941 9423	39.2063 6188	38.0677 3431	35.9287 4185	33.9597 1913
53	40.9843 5072	39.7590 7232	38.5854 1660	36.3829 9690	34.3584 4632
54	41.5686 6408	40.3056 3394	39.0967 0776	36.8305 3882	34.7503 1579
55	42.1471 9216	40.8461 1514	39.6016 8667	37.2714 6681	35.1354 4550
56	42.7199 9224	41.3805 8358	40.1004 3128	37.7058 7863	35.5139 5135
57	43.2871 2102	41.9091 0613	40.5930 1855	38.1338 7058	35.8859 4727
58	43.8486 3468	42.4317 4896	41.0795 2449	38.5555 3751	36.2515 4523
59	44.4045 8879	42.9485 7746	41.5600 2419	38.9709 7292	36.6108 5526
60	44.9550 3841	43.4596 5633	42.0345 9179	39.3802 6889	36.9639 8552
61	45.5000 3803	43.9650 4952	42.5033 0054	39.7835 1614	37.3110 4228
62	46.0396 4161	44.4648 2029	42.9662 2275	40.1808 0408	37.6521 3000
63	46.5739 0258	44.9590 3119	43.4234 2988	40.5722 2077	37.9873 5135
64	47.1028 7385	45.4477 4407	43.8749 9247	40.9578 5298	38.3168 0723
65	47.6266 0777	45.9310 2009	44.3209 8022	41.3377 8618	38.6405 9678
66	48.1451 5621	46.4089 1975	44.7614 6195	41.7121 0461	38.9588 1748
67	48.6585 7050	46.8815 0284	45.1965 0563	42.0808 9125	39.2715 6509
68	49.1669 0149	47.3488 2852	45.6261 7840	42.4442 2783	39.5789 3375
69	49.6701 9949	47.8109 5527	46.0505 4656	42.8021 9490	39.8810 1597
70	50.1685 1435	48.2679 4094	46.4696 7562	43.1548 7183	40.1779 0267
71	50.6618 9539	48.7198 4270	46.8836 3024	43.5023 3678	40.4696 8321
72	51.1503 9148	49.1667 1714	47.2924 7431	43.8446 6677	40.7564 4542
73	51.6340 5097	49.6086 2016	47.6962 7093	44.1819 3771	41.0382 7560
74	52.1129 2175	50.0456 0708	48.0950 8240	44.5142 2434	41.3152 5857
75	52.5870 5124	50.4777 3259	48.4889 7027	44.8416 0034	41.5874 7771
76	53.0564 8638	50.9050 5077	48.8779 9533	45.1641 3826	41.8550 1495
77	53.5212 7364	51.3276 1510	49.2622 1761	45.4819 0962	42.1179 5081
78	53.9814 5905	51.7454 7847	49.6416 9640	45.7949 8485	42.3763 6443
79	54.4370 8817	52.1586 9317	50.0164 9027	46.1034 3335	42.6303 3359
80	54.8882 0611	52.5673 1092	50.3866 5706	46.4073 2349	42.8799 3474
81	55.3348 5753	52.9713 8286	50.7522 5389	46.7067 2265	43.1252 4298
82	55.7770 8666	53.3709 5957	51.1133 3717	47.0016 9720	43.3663 3217
83	56.2149 3729	53.7660 9104	51.4699 6264	47.2923 1251	43.6032 7486
84	56.6484 5276	54.1568 2674	51.8221 8532	47.5786 3301	43.8361 4237
85	57.0776 7600	54.5432 1557	52.1700 5958	47.8607 2218	44.0650 0479
86	57.5026 4951	54.9253 0588	52.5136 3909	48.1386 4254	44.2899 3099
87	57.9234 1535	55.3031 4549	52.8529 7688	48.4124 5571	44.5109 8869
88	58.3400 1520	55.6767 8169	53.1881 2531	48.6822 2237	44.7282 4441
89	58.7524 9030	56.0462 6126	53.5191 3611	48.9480 0234	44.9417 6355
90	59.1608 8148	56.4116 3041	53.8460 6035	49.2098 5452	45.1516 1037
91	59.5652 2919	56.7729 3490	54.1689 4850	49.4678 3696	45.3578 4803
92	59.9655 7346	57.1302 1992	54.4878 5037	49.7220 0686	45.5605 3860
93	60.3619 5392	57.4835 3021	54.8028 1518	49.9724 2055	45.7597 4310
94	60.7544 0982	57.8329 0997	55.1138 9154	50.2191 3355	45.9555 2147
95	61.1429 8002	58.1784 0294	55.4211 2744	50.4622 0054	46.1479 3265
96	61.5277 0299	58.5200 5235	55.7245 7031	50.7016 7541	46.3370 3455
97	61.9086 1682	58.8579 0096	56.0242 6698	50.9376 1124	46.5228 8408
98	62.2857 5923	59.1919 9106	56.3202 6368	51.1700 6034	46.7055 3718
99	62.6591 6755	59.5223 6446	56.6126 0610	51.3990 7422	46.8850 4882
100	63.0288 7877	59.8490 6251	56.9013 3936	51.6247 0367	47.0614 7304

PRESENT VALUE OF ANNUITY $[1 - (1 + i)^{-n}]/i$ (Continued)

Periods	Rate i				
n	.02 (2%)	.0225 (2¼%)	.025 (2½%)	.0275 (2¾%)	.03 (3%)
1	0.9803 9216	0.9779 9511	0.9756 0976	0.9732 3601	0.9708 7379
2	1.9415 6094	1.9344 6955	1.9274 2415	1.9204 2434	1.9134 6970
3	2.8838 8327	2.8698 9687	2.8560 2356	2.8422 6213	2.8286 1135
4	3.8077 2870	3.7847 4021	3.7619 7421	3.7394 2787	3.7170 9840
5	4.7134 5951	4.6794 5253	4.6458 2850	4.6125 8186	4.5797 0719
6	5.6014 3089	5.5544 7680	5.5081 2536	5.4623 6678	5.4171 9144
7	6.4719 9107	6.4102 4626	6.3493 9060	6.2894 0806	6.2302 8296
8	7.3254 8144	7.2471 8461	7.1701 3717	7.0943 1441	7.0196 9219
9	8.1622 3671	8.0657 0622	7.9708 6553	7.8776 7826	7.7861 0892
10	8.9825 8501	8.8662 1635	8.7520 6393	8.6400 7616	8.5302 0284
11	9.7868 4805	9.6491 1134	9.5142 0871	9.3820 6926	9.2526 2411
12	10.5753 4122	10.4147 7882	10.2577 6460	10.1042 0366	9.9540 0399
13	11.3483 7375	11.1635 9787	10.9831 8497	10.8070 1086	10.6349 5533
14	12.1062 4877	11.8959 3924	11.6909 1217	11.4910 0814	11.2960 7314
15	12.8492 6350	12.6121 6551	12.3813 7773	12.1566 9892	11.9379 3509
16	13.5777 0931	13.3126 3131	13.0550 0266	12.8045 7315	12.5611 0203
17	14.2918 7188	13.9976 8343	13.7121 9772	13.4351 0769	13.1661 1847
18	14.9920 3125	14.6676 6106	14.3533 6363	14.0487 6661	13.7535 1308
19	15.6784 6201	15.3228 9590	14.9788 9134	14.6460 0157	14.3237 9911
20	16.3514 3334	15.9637 1237	15.5891 6229	15.2272 5213	14.8774 7486
21	17.0112 0916	16.5904 2775	16.1845 4857	15.7929 4612	15.4150 2414
22	17.6580 4820	17.2033 5232	16.7654 1324	16.3434 9987	15.9369 1664
23	18.2922 0412	17.8027 8955	17.3321 1048	16.8793 1861	16.4436 0839
24	18.9139 2560	18.3890 3624	17.8849 8583	17.4007 9670	16.9355 4212
25	19.5234 5647	18.9623 8263	18.4243 7642	17.9083 1795	17.4131 4769
26	20.1210 3576	19.5231 1260	18.9506 1114	18.4022 5592	17.8768 4242
27	20.7068 9780	20.0715 0376	19.4640 1087	18.8829 7413	18.3270 3147
28	21.2812 7236	20.6078 2764	19.9648 8866	19.3508 2640	18.7641 0823
29	21.8443 8466	21.1323 4977	20.4535 4991	19.8061 5708	19.1884 5459
30	22.3964 5555	21.6453 2985	20 9302 9259	20.2493 0130	19.6004 4135
31	22.9377 0152	22.1470 2186	21.3954 0741	20.6805 8520	20.0004 2849
32	23.4683 3482	22.6376 7419	21.8491 7796	21.1003 2623	20.3887 6553
33	23.9885 6355	23.1175 2977	22.2918 8094	21.5088 3332	20.7657 9178
34	24.4985 9172	23.5868 2618	22.7237 8628	21.9064 0712	21.1318 3668
35	24.9986 1933	24.0457 9577	23.1451 5734	22.2933 4026	21.4872 2007
36	25.4888 4248	24.4946 6579	23.5562 5107	22.6699 1753	21.8322 5250
37	25.9694 5341	24.9336 5848	23.9573 1812	23.0364 1609	22.1672 3544
38	26.4406 4060	25.3629 9118	24.3486 0304	23.3931 0568	22.4924 6159
39	26.9025 8883	25.7828 7646	24.7303 4443	23.7402 4884	22.8082 1513
40	27.3554 7924	26.1935 2221	25.1027 7505	24.0781 0106	23.1147 7197
41	27.7994 8945	26.5951 3174	25.4661 2200	24.4069 1101	23.4123 9997
42	28.2347 9358	26.9879 0390	25.8206 0683	24.7269 2069	23.7013 5920
43	28.6615 6233	27.3720 3316	26.1664 4569	25.0383 6563	23.9819 0213
44	29.0799 6307	27.7477 0969	26.5038 4945	25.3414 7507	24.2542 7392
45	29.4901 5987	28.1151 1950	26.8330 2386	25.6364 7209	24.5187 1254
46	29.8923 1360	28.4744 4450	27.1541 6962	25.9235 7381	24.7754 4907
47	30.2865 8196	28.8258 6259	27.4674 8255	26.2029 9154	25.0247 0783
48	30.6731 1957	29.1695 4777	27.7731 5371	26.4749 3094	25.2667 0664
49	31.0520 7801	29.5056 7019	28.0713 6947	26.7395 9215	25.5016 5693
50	31.4236 0589	29.8343 9627	28.3623 1168	26.9971 6998	25.7297 6401

PRESENT VALUE OF ANNUITY $[1 - (1 + i)^{-n}]/i$ (Continued)

Periods	Rate i				
n	.02 (2%)	.0225 (2¼%)	.025 (2½%)	.0275 (2¾%)	.03 (3%)
50	31.4236 0589	29.8343 9627	28.3623 1168	26.9971 6998	25.7297 6401
51	31.7878 4892	30.1558 8877	28.6461 5774	27.2478 5400	25.9512 2719
52	32.1449 4992	30.4703 0687	28.9230 8072	27.4918 2871	26.1662 3999
53	32.4950 4894	30.7778 0623	29.1932 4948	27.7292 7368	26.3749 9028
54	32.8382 8327	31.0785 3910	29.4568 2876	27.9603 6368	26.5776 6047
55	33.1747 8752	31.3726 5438	29.7139 7928	28.1852 6879	26.7744 2764
56	33.5046 9365	31.6602 9768	29.9648 5784	28.4041 5454	26.9654 6373
57	33.8281 3103	31.9416 1142	30.2096 1740	28.6171 8203	27.1509 3566
58	34.1452 2650	32.2167 3489	30.4484 0722	28.8245 0806	27.3310 0549
59	34.4561 0441	32.4858 0429	30.6813 7290	29.0262 8522	27.5058 3058
60	34.7608 8668	32.7489 5285	30.9086 5649	29.2226 6201	27.6755 6367
61	35.0596 9282	33.0063 1086	31.1303 9657	29.4137 8293	27.8403 5307
62	35.3526 4002	33.2580 0573	31.3467 2836	29.5997 8879	28.0003 4279
63	35.6398 4316	33.5041 6208	31.5577 8377	29.7808 1634	28.1556 7261
64	35.9214 1486	33.7449 0179	31.7636 9148	29.9569 9887	28.3064 7826
65	36.1974 6555	33.9803 4405	31.9645 7705	30.1284 6605	28.4528 9152
66	36 4681 0348	34.2106 0543	32.1605 6298	30.2953 4409	28.5950 4031
67	36.7334 3478	34.4357 9993	32.3517 6876	30.4577 5581	28.7330 4884
68	36.9935 6351	34.6560 3905	32.5383 1099	30.6158 2074	28.8670 3771
69	37.2485 9168	34.8714 3183	32.7203 0340	30.7696 5522	28.9971 2399
70	37.4986 1929	35.0820 8492	32.8978 5698	30.9193 7247	29.1234 2135
71	37.7437 4441	35.2881 0261	33.0710 7998	31.0650 8270	29.2460 4015
72	37.9840 6314	35.4895 8691	33.2400 7803	31.2068 9314	29.3650 8752
73	38.2196 6975	35.6866 3756	33.4049 5417	31.3449 0816	29.4806 6750
74	38.4506 5662	35.8793 5214	33.5658 0895	31.4792 2936	29.5928 8107
75	38.6771 1433	36.0678 2605	33.7227 4044	31.6099 5558	29.7018 2628
76	38.8991 3170	36.2521 5262	33.8758 4433	31.7371 8304	29.8075 9833
77	39.1167 9578	36.4324 2310	34.0252 1398	31.8610 0540	29.9102 8964
78	39.3301 9194	36.6087 2675	34.1709 4047	31.9815 1377	30.0099 8994
79	39.5394 0386	36.7811 5085	34.3131 1265	32.0987 9685	30.1067 8635
80	39.7445 1359	36.9497 8079	34.4518 1722	32.2129 4098	30.2007 6345
81	39.9456 0156	37.1147 0004	34.5871 3875	32 3240 3015	30.2920 0335
82	40.1427 4663	37.2759 9026	34.7191 5976	32.4321 4613	30.3805 8577
83	40.3360 2611	37.4337 3130	34.8479 6074	32 5373 6850	30.4665 8813
84	40.5255 1579	37.5880 0127	34.9736 2023	32.6397 7469	30.5500 8556
85	40.7112 8999	37.7388 7655	35.0962 1486	32.7394 4009	30.6311 5103
86	40.8934 2156	37.8864 3183	35.2158 1938	32.8364 3804	30.7098 5537
87	41.0719 8192	38.0307 4018	35.3325 0671	32.9308 3994	30.7862 6735
88	41.2470 4110	38.1718 7304	35.4463 4801	33.0227 1527	30.8604 5374
89	41.4186 6774	38.3099 0028	35.5574 1269	33.1121 3165	30.9324 7936
90	41.5869 2916	38.4448 9025	35.6657 6848	33.1991 5489	31.0024 0714
91	41.7518 9133	38.5769 0978	35.7714 8144	33.2838 4905	31.0702 9820
92	41.9136 1895	38.7060 2423	35.8746 1604	33.3662 7644	31.1362 1184
93	42.0721 7545	38.8322 9754	35.9752 3516	33.4464 9776	31.2002 0567
94	42.2276 2299	38.9557 9221	36.0734 0016	33.5245 7202	31.2623 3560
95	42.3800 2254	39.0765 6940	36.1691 7089	33.6005 5671	31.3226 5592
96	42.5294 3386	39.1946 8890	36.2626 0574	33.6745 0775	31.3812 1934
97	42.6759 1555	39.3102 0920	36.3537 6170	33.7464 7956	31.4380 7703
98	42.8195 2505	39.4231 8748	36.4426 9434	33.8165 2512	31.4932 7867
99	42.9603 1867	39.5336 7968	36.5294 5790	33.8846 9598	31.5468 7250
100	43.0983 5164	39.6417 4052	36.6141 0526	33.9510 4232	31.5989 0534

PRESENT VALUE OF ANNUITY $[1 - (1 + i)^{-n}]/i$ (Continued)

Periods	Rate i				
n	.035 (3½%)	.04 (4%)	.045 (4½%)	.05 (5%)	.055 (5½%)
1	0.9661 8357	0.9615 3846	0.9569 3780	0.9523 8095	0.9478 6730
2	1.8996 9428	1.8860 9467	1.8726 6775	1.8594 1043	1.8463 1971
3	2.8016 3698	2.7750 9103	2.7489 6435	2.7232 4803	2.6979 3338
4	3.6730 7921	3.6298 9522	3.5875 2570	3.5459 5050	3.5051 5012
5	4.5150 5238	4.4518 2233	4.3899 7674	4.3294 7667	4.2702 8448
6	5.3285 5302	5.2421 3686	5.1578 7248	5.0756 9207	4.9955 3031
7	6.1145 4398	6.0020 5467	5.8927 0094	5.7863 7340	5.6829 6712
8	6.8739 5554	6.7327 4487	6.5958 8607	6.4632 1276	6.3345 6599
9	7.6076 8651	7.4353 3161	7.2687 9050	7.1078 2168	6.9521 9525
10	8.3166 0532	8.1108 9578	7.9127 1818	7.7217 3493	7.5376 2583
11	9.0015 5104	8.7604 7671	8.5289 1692	8.3064 1422	8.0925 3633
12	9.6633 3433	9.3850 7376	9.1185 8078	8.8632 5164	8.6185 1785
13	10.3027 3849	9.9856 4785	9.6828 5242	9.3935 7299	9.1170 7853
14	10.9205 2028	10.5631 2293	10.2228 2528	9.8986 4094	9.5896 4790
15	11.5174 1090	11.1183 8743	10.7395 4573	10.3796 5804	10.0375 8094
16	12.0941 1681	11.6522 9561	11.2340 1505	10.8377 6956	10.4621 6203
17	12.6513 2059	12.1656 6885	11.7071 9143	11.2740 6625	10.8646 0856
18	13.1896 8173	12.6592 9697	12.1599 9180	11.6895 8690	11.2460 7447
19	13.7098 3742	13.1339 3940	12.5932 9359	12.0853 2086	11.6076 5352
20	14.2124 0330	13.5903 2634	13.0079 3645	12.4622 1034	11.9503 8248
21	14.6979 7420	14.0291 5995	13.4047 2388	12.8211 5271	12.2752 4406
22	15.1671 2484	14.4511 1533	13.7844 2476	13.1630 0258	12.5831 6973
23	15.6204 1047	14.8568 4167	14.1477 7489	13.4885 7388	12.8750 4239
24	16.0583 6760	15.2469 6314	14.4954 7837	13.7986 4179	13.1516 9895
25	16.4815 1459	15.6220 7994	14.8282 0896	14.0939 4457	13.4139 3266
26	16.8903 5226	15.9827 6918	15.1466 1145	14.3751 8530	13.6624 9541
27	17.2853 6451	16.3295 8575	15.4513 0282	14.6430 3362	13.8980 9991
28	17.6670 1885	16.6630 6322	15.7428 7351	14.8981 2726	14.1214 2172
29	18.0357 6700	16.9837 1463	16.0218 8853	15.1410 7358	14.3331 0116
30	18.3920 4541	17.2920 3330	16.2888 8854	15.3724 5103	14.5337 4517
31	18.7362 7576	17.5884 9356	16.5443 9095	15.5928 1050	14.7239 2907
32	19.0688 6547	17.8735 5150	16.7888 9086	15.8026 7667	14.9041 9817
33	19.3902 0818	18.1476 4567	17.0228 6207	16.0025 4921	15.0750 6936
34	19.7006 8423	18.4111 9776	17.2467 5796	16.1929 0401	15.2370 3257
35	20.0006 6110	18.6646 1323	17.4610 1240	16.3741 9429	15.3905 5220
36	20.2904 9381	18.9082 8195	17.6660 4058	16.5468 5171	15.5360 6843
37	20.5705 2542	19.1425 7880	17.8622 3979	16.7112 8734	15.6739 9851
38	20.8410 8736	19.3678 6423	18.0499 9023	16.8678 9271	15.8047 3793
39	21.1024 9987	19.5844 8484	18.2296 5572	17.0170 4067	15.9286 6154
40	21.3550 7234	19.7927 7388	18.4015 8442	17.1590 8635	16.0461 2469
41	21.5991 0371	19.9930 5181	18.5661 0949	17.2943 6796	16.1574 6416
42	21.8348 8281	20.1856 2674	18.7235 4975	17.4232 0758	16.2629 9920
43	22.0626 8870	20.3707 9494	18.8742 1029	17.5459 1198	16.3630 3242
44	22.2827 9102	20.5488 4129	19.0183 8305	17.6627 7331	16.4578 5063
45	22.4954 5026	20.7200 3970	19.1563 4742	17.7740 6982	16.5477 2572
46	22.7009 1813	20.8846 5356	19.2883 7074	17.8800 6650	16.6329 1537
47	22.8994 3780	21.0429 3612	19.4147 0884	17.9810 1571	16.7136 6386
48	23.0912 4425	21.1951 3088	19.5356 0654	18.0771 5782	16.7902 0271
49	23.2765 6450	21.3414 7200	19.6512 9813	18.1687 2173	16.8627 5139
50	23.4556 1787	21.4821 8462	19.7620 0778	18.2559 2546	16.9315 1790

PRESENT VALUE OF ANNUITY $[1 - (1 + i)^{-n}]/i$ (Continued)

Periods	Rate i				
n	.06 (6%)	.065 (6½%)	.07 (7%)	.075 (7½%)	.08 (8%)
1	0.9433 9623	0.9389 6714	0.9345 7944	0.9302 3256	0.9259 2593
2	1.8333 9267	1.8206 2642	1.8080 1817	1.7955 6517	1.7832 6475
3	2.6730 1195	2.6484 7551	2.6243 1604	2.6005 2574	2.5770 9699
4	3.4651 0561	3.4257 9860	3.3872 1126	3.3493 2627	3.3121 2684
5	4.2123 6379	4.1556 7944	4.1001 9744	4.0458 8490	3.9927 1004
6	4.9173 2433	4.8410 1356	4.7665 3966	4.6938 4642	4.6228 7966
7	5.5823 8144	5.4845 1977	5.3892 8940	5.2966 0132	5.2063 7006
8	6.2097 9381	6.0887 5096	5.9712 9851	5.8573 0355	5.7466 3894
9	6.8016 9227	6.6561 0419	6.5152 3225	6.3788 8703	6.2468 8791
10	7.3600 8705	7.1888 3022	7.0235 8154	6.8640 8096	6.7100 8140
11	7.8868 7458	7.6890 4246	7.4986 7434	7.3154 2415	7.1389 6426
12	8.3838 4394	8.1587 2532	7.9426 8630	7.7352 7827	7.5360 7802
13	8.8526 8296	8.5997 4208	8.3576 5074	8.1258 4026	7.9037 7594
14	9.2949 8393	9.0138 4233	8.7454 6799	8.4891 5373	8.2442 3698
15	9.7122 4899	9.4026 6885	9.1079 1401	8.8271 1975	8.5594 7869
16	10.1058 9527	9.7677 6418	9.4466 4860	9.1415 0674	8.8513 6916
17	10.4772 5969	10.1105 7670	9.7632 2299	9.4339 5976	9.1216 3811
18	10.8276 0348	10.4324 6638	10.0590 8691	9.7060 0908	9.3718 8714
19	11.1581 1649	10.7347 1022	10.3355 9524	9.9590 7821	9.6035 9920
20	11.4699 2122	11.0185 0725	10.5940 1425	10.1944 9136	9.8181 4741
21	11.7640 7662	11.2849 8333	10.8355 2733	10.4134 8033	10.0168 0316
22	12.0415 8172	11.5351 9562	11.0612 4050	10.6171 9101	10.2007 4366
23	12.3033 7898	11.7701 3673	11.2721 8738	10.8066 8931	10.3710 5895
24	12.5503 5753	11.9907 3871	11.4693 3400	10.9829 6680	10.5287 5828
25	12.7833 5616	12.1978 7673	11.6535 8318	11.1469 4586	10.6747 7619
26	13.0031 6619	12.3923 7251	11.8257 7867	11.2994 8452	10.8099 7795
27	13.2105 3414	12.5749 9766	11.9867 0904	11.4413 8095	10.9351 6477
28	13.4061 6428	12.7464 7668	12.1371 1125	11.5733 7763	11.0510 7849
29	13.5907 2102	12.9074 8984	12.2776 7407	11.6961 6524	11.1584 0601
30	13.7648 3115	13.0586 7591	12.4090 4118	11.8103 8627	11.2577 8334
31	13.9290 8599	13.2006 3465	12.5318 1419	11.9166 3839	11.3497 9939
32	14.0840 4339	13.3339 2925	12.6465 5532	12.0154 7757	11.4349 9944
33	14.2302 2961	13.4590 8850	12.7537 9002	12.1074 2099	11.5138 8837
34	14.3681 4114	13.5766 0892	12.8540 0936	12.1929 4976	11.5869 3367
35	14.4982 4636	13.6869 5673	12.9476 7230	12.2725 1141	11.6545 6822
36	14.6209 8713	13.7905 6970	13.0352 0776	12.3465 2224	11.7171 9279
37	14.7367 8031	13.8878 5887	13.1170 1660	12.4153 6952	11.7751 7851
38	14.8460 1916	13.9792 1021	13.1934 7345	12.4794 1351	11.8288 6899
39	14.9490 7468	14.0649 8611	13.2649 2846	12.5389 8931	11.8785 8240
40	15.0462 9687	14.1455 2687	13.3317 0884	12.5944 0866	11.9246 1333
41	15.1380 1592	14.2211 5199	13.3941 2041	12.6459 6155	11.9672 3457
42	15.2245 4332	14.2921 6149	13.4524 4898	12.6939 1772	12.0066 9867
43	15.3061 7294	14.3588 3708	13.5069 6167	12.7385 2811	12.0432 3951
44	15.3831 8202	14.4214 4327	13.5579 0810	12.7800 2615	12.0770 7362
45	15.4558 3209	14.4802 2842	13.6055 2159	12.8186 2898	12.1084 0150
46	15.5243 6990	14.5354 2575	13.6500 2018	12.8545 3858	12.1374 0880
47	15.5890 2821	14.5872 5422	13.6916 0764	12.8879 4287	12.1642 6741
48	15.6500 2661	14.6359 1946	13.7304 7443	12.9190 1662	12.1891 3649
49	15.7075 7227	14.6816 1451	13.7667 9853	12.9479 2244	12.2121 6341
50	15.7618 6064	14.7245 2067	13.8007 4629	12.9748 1157	12.2334 8464

ANNUITY WHOSE PRESENT VALUE IS 1

$$\frac{1}{a_{\overline{n}|i}} = \frac{i}{1-(1+i)^{-n}} = \frac{i}{1-v^n} = \frac{1}{s_{\overline{n}|i}} + i$$

Periods n	Rate i: 0.0025(1/4%)	0.004167(5/12%)	0.005(1/2%)	0.005833(7/12%)	0.0075(3/4%)
1	1.00250000	1.00416667	1.00500000	1.00583333	1.00750000
2	0.50187578	0.50312717	0.50375312	0.50437924	0.50563200
3	0.33500139	0.33611496	0.33667221	0.33722976	0.33834579
4	0.25156445	0.25260958	0.25313279	0.25365644	0.25470501
5	0.20150250	0.20250693	0.20300997	0.20351357	0.20452242
6	0.16812803	0.16910564	0.16959546	0.17008594	0.17106891
7	0.14428928	0.14524800	0.14572854	0.14620986	0.14717488
8	0.12641035	0.12735512	0.12782886	0.12830352	0.12925552
9	0.11250462	0.11343876	0.11390736	0.11437698	0.11531929
10	0.10138015	0.10230596	0.10277057	0.10323632	0.10417123
11	0.09227840	0.09319757	0.09365903	0.09412175	0.09505094
12	0.08469370	0.08560748	0.08606643	0.08652675	0.08745148
13	0.07827595	0.07918532	0.07964224	0.08010064	0.08102188
14	0.07277510	0.07368082	0.07413609	0.07459295	0.07551146
15	0.06800777	0.06891045	0.06936436	0.06982000	0.07073639
16	0.06383642	0.06473655	0.06518937	0.06564401	0.06655879
17	0.06015587	0.06105387	0.06150579	0.06195966	0.06287321
18	0.05688433	0.05778053	0.05823173	0.05868499	0.05959766
19	0.05395722	0.05485191	0.05530253	0.05575532	0.05666740
20	0.05132288	0.05221630	0.05266645	0.05311889	0.05403063
21	0.04893947	0.04983183	0.05028163	0.05073383	0.05164543
22	0.04677278	0.04766427	0.04811380	0.04856585	0.04947748
23	0.04479455	0.04568531	0.04613465	0.04658663	0.04749846
24	0.04298121	0.04387139	0.04432061	0.04477258	0.04568474
25	0.04131298	0.04220270	0.04265186	0.04310388	0.04401650
26	0.03977312	0.04066247	0.04111163	0.04156376	0.04247693
27	0.03834736	0.03923645	0.03968565	0.04013793	0.04105176
28	0.03702347	0.03791239	0.03836167	0.03881415	0.03972871
29	0.03579093	0.03667974	0.03712914	0.03758186	0.03849723
30	0.03464059	0.03552936	0.03597892	0.03643191	0.03734816
31	0.03356440	0.03445330	0.03490304	0.03535633	0.03627352
32	0.03255569	0.03344458	0.03389453	0.03434815	0.03526634
33	0.03160806	0.03249708	0.03294727	0.03340124	0.03432048
34	0.03071620	0.03160540	0.03205586	0.03251020	0.03343053
35	0.02987533	0.03076476	0.03121550	0.03167024	0.03259170
36	0.02908121	0.02997090	0.03042194	0.03087710	0.03179973
37	0.02833004	0.02922003	0.02967139	0.03012698	0.03105082
38	0.02761843	0.02850875	0.02896045	0.02941649	0.03034157
39	0.02694335	0.02783402	0.02828607	0.02874258	0.02966893
40	0.02630204	0.02719310	0.02764552	0.02810251	0.02903016
41	0.02569204	0.02658352	0.02703631	0.02749379	0.02842276
42	0.02511112	0.02600303	0.02645622	0.02691420	0.02784452
43	0.02455724	0.02544961	0.02590320	0.02636170	0.02729338
44	0.02402855	0.02492141	0.02537541	0.02583443	0.02676751
45	0.02352339	0.02441675	0.02487117	0.02533073	0.02626521
46	0.02304022	0.02393409	0.02438894	0.02484905	0.02578495
47	0.02257762	0.02347204	0.02392733	0.02438798	0.02532532
48	0.02213433	0.02302929	0.02348503	0.02394624	0.02488504
49	0.02170915	0.02260468	0.02306087	0.02352265	0.02446292
50	0.02130099	0.02219711	0.02265376	0.02311612	0.02405787

Periods n	Rate i: 0.0025(1/4%)	0.004167(5/12%)	0.005(1/2%)	0.005833(7/12%)	0.0075(3/4%)
51	0.02090886	0.02180557	0.02226269	0.02272563	0.02366888
52	0.02053184	0.02142916	0.02188675	0.02235027	0.02329503
53	0.02016906	0.02106700	0.02152507	0.02198919	0.02293546
54	0.01981974	0.02071830	0.02117686	0.02164157	0.02258938
55	0.01948314	0.02038234	0.02081439	0.02130671	0.02225605
56	0.01915858	0.02005843	0.02051797	0.02098390	0.02193478
57	0.01884542	0.01974593	0.02020598	0.02067251	0.02162496
58	0.01854308	0.01944426	0.01990481	0.02037196	0.02132597
59	0.01825101	0.01915287	0.01961392	0.02008170	0.02103727
60	0.01796869	0.01887123	0.01933280	0.01980120	0.02075836
61	0.01769564	0.01859888	0.01906096	0.01952999	0.02048873
62	0.01743142	0.01833536	0.01879796	0.01926762	0.02022795
63	0.01717561	0.01808025	0.01854337	0.01901366	0.01997560
64	0.01692780	0.01783315	0.01829681	0.01876773	0.01973127
65	0.01668764	0.01759371	0.01805789	0.01852946	0.01949460
66	0.01645476	0.01736156	0.01782627	0.01829848	0.01926524
67	0.01622886	0.01713639	0.01760163	0.01807449	0.01904286
68	0.01600961	0.01691788	0.01738366	0.01785716	0.01882716
69	0.01579674	0.01670574	0.01717206	0.01764622	0.01861785
70	0.01558996	0.01649971	0.01696657	0.01744138	0.01841464
71	0.01538902	0.01629952	0.01676693	0.01724239	0.01821728
72	0.01519368	0.01610493	0.01657289	0.01704901	0.01802554
73	0.01500370	0.01591572	0.01638422	0.01686100	0.01783917
74	0.01481887	0.01573165	0.01620070	0.01667814	0.01765796
75	0.01463898	0.01555253	0.01602214	0.01650024	0.01748170
76	0.01446385	0.01537816	0.01584832	0.01632709	0.01731020
77	0.01429327	0.01520836	0.01567908	0.01615851	0.01714328
78	0.01412708	0.01504295	0.01551423	0.01599432	0.01698074
79	0.01396511	0.01488177	0.01535360	0.01583436	0.01682244
80	0.01380721	0.01472464	0.01519704	0.01567847	0.01666821
81	0.01365321	0.01457144	0.01504439	0.01552650	0.01651790
82	0.01350298	0.01442200	0.01489552	0.01537830	0.01637136
83	0.01335639	0.01427620	0.01475028	0.01523373	0.01622847
84	0.01321330	0.01413391	0.01460855	0.01509268	0.01608908
85	0.01307359	0.01399500	0.01447021	0.01495501	0.01595308
86	0.01293714	0.01385935	0.01433513	0.01482060	0.01582034
87	0.01280384	0.01372685	0.01420320	0.01468935	0.01569076
88	0.01267357	0.01359740	0.01407431	0.01456115	0.01556423
89	0.01254625	0.01347088	0.01394837	0.01443588	0.01544064
90	0.01242177	0.01334721	0.01382527	0.01431347	0.01531989
91	0.01230004	0.01322629	0.01370493	0.01409380	0.01520190
92	0.01218096	0.01310803	0.01358724	0.01407679	0.01508657
93	0.01206446	0.01299234	0.01347213	0.01396236	0.01497382
94	0.01195044	0.01287915	0.01335950	0.01385042	0.01486356
95	0.01183884	0.01276836	0.01324930	0.01374090	0.01475571
96	0.01172957	0.01265992	0.01314143	0.01363372	0.01465020
97	0.01162257	0.01255374	0.01303583	0.01352880	0.01454696
98	0.01151776	0.01244976	0.01293242	0.01342608	0.01444592
99	0.01141508	0.01234790	0.01283115	0.01332549	0.01434701
100	0.01131446	0.01224811	0.01273194	0.01322696	0.01425017

ANNUITY WHOSE PRESENT VALUE IS 1

$$a^{-1}_{\overline{n}|i} = \frac{i}{1 - v^n} = s^{-1}_{\overline{n}|i} + i \text{ (Continued)}$$

Periods	Rate i					Periods	Rate i				
n	0.01(1%)	0.001125(1 1/8%)	0.0125(1 1/4%)	0.015(1 1/2%)	0.0175(1 3/4%)	n	0.01(1%)	0.001125(1 1/8%)	0.0125(1 1/4%)	0.015(1 1/2%)	0.0175(1 3/4%)
1	1.01000000	1.01125000	1.01250000	0.01500000	1.01750000	51	0.02512680	0.02587494	0.02663571	0.02819469	0.02980269
2	0.50751244	0.50845323	0.50939441	0.51127792	0.51316295	52	0.02475603	0.02550606	0.02626897	0.02783287	0.02944665
3	0.34002211	0.34086130	0.34170117	0.34338296	0.34506746	53	0.02439956	0.02515149	0.02591653	0.02748537	0.02910492
4	0.25628109	0.25707058	0.25786102	0.25944479	0.26103237	54	0.02405658	0.02481043	0.02557760	0.02715138	0.02877672
5	0.20603980	0.20680034	0.20756211	0.20908932	0.21062142	55	0.02372637	0.02448213	0.02525145	0.02683018	0.02846129
6	0.17254837	0.17329034	0.17403381	0.17552521	0.17702256	56	0.02340824	0.02416592	0.02493739	0.02652106	0.02815795
7	0.14862828	0.14935762	0.15008872	0.15155616	0.15303059	57	0.02310156	0.02386116	0.02463478	0.02622341	0.02786606
8	0.13069029	0.13141071	0.13213314	0.13358402	0.13504292	58	0.02280573	0.02356726	0.02434303	0.02593661	0.02758503
9	0.11674036	0.11745432	0.11817055	0.11960982	0.12105813	59	0.02252020	0.02328366	0.02406158	0.02566012	0.02731430
10	0.10558208	0.10629131	0.10700307	0.10843418	0.10987534	60	0.02224445	0.02300985	0.02378993	0.02539343	0.02705336
11	0.09645408	0.09715984	0.09786839	0.09929384	0.10073038	61	0.02197800	0.02274534	0.02352758	0.02513604	0.02680172
12	0.08884879	0.08955203	0.09025831	0.09167999	0.09311377	62	0.02172041	0.02248969	0.02327410	0.02488751	0.02655892
13	0.08241482	0.08311626	0.08382100	0.08524036	0.08667283	63	0.02147125	0.02224247	0.02302904	0.02464741	0.02632455
14	0.07690117	0.07760138	0.07830515	0.07972332	0.08115562	64	0.02123013	0.02200329	0.02279203	0.02441534	0.02609821
15	0.07212378	0.07282321	0.07352646	0.07494436	0.07637739	65	0.02099667	0.02177178	0.02256268	0.02419094	0.02587952
16	0.06794460	0.06864363	0.06934672	0.07076508	0.07219958	66	0.02077052	0.02154758	0.02234065	0.02397386	0.02566813
17	0.06425806	0.06495698	0.06566023	0.06707966	0.06851623	67	0.02055136	0.02133037	0.02212560	0.02376376	0.02546372
18	0.06098205	0.06168113	0.06238479	0.06380578	0.06524492	68	0.02033889	0.02111985	0.02191724	0.02356033	0.02526597
19	0.05805175	0.05875120	0.05945548	0.06087847	0.06232061	69	0.02013280	0.02091571	0.02171527	0.02336329	0.02507459
20	0.05541531	0.05611531	0.05682039	0.05824574	0.05969122	70	0.01993282	0.02071769	0.02151941	0.02317235	0.02488930
21	0.05303075	0.05373145	0.05443749	0.05586550	0.05731464	71	0.01973870	0.02052552	0.02132941	0.02298727	0.02470985
22	0.05086372	0.05156525	0.05227238	0.05370332	0.05515638	72	0.01955019	0.02033896	0.02114501	0.02280779	0.02453600
23	0.04888584	0.04958833	0.05029666	0.05173075	0.05318796	73	0.01936706	0.02015779	0.02096600	0.02263368	0.02436750
24	0.04707347	0.04777701	0.04848665	0.04992410	0.05138565	74	0.01918910	0.01998177	0.02079215	0.02246473	0.02420413
25	0.04540675	0.04611144	0.04682247	0.04826345	0.04972952	75	0.01901609	0.01981072	0.02062325	0.02230072	0.02404570
26	0.04386888	0.04457479	0.04528729	0.04673196	0.04820269	76	0.01884784	0.01964442	0.02045910	0.02214146	0.02389200
27	0.04244553	0.04315273	0.04386677	0.04531527	0.04679079	77	0.01868416	0.01948269	0.02029953	0.02198676	0.02374285
28	0.04112444	0.04183299	0.04254863	0.04400108	0.04548151	78	0.01852488	0.01932536	0.02014436	0.02183645	0.02359806
29	0.03989502	0.04060498	0.04132228	0.04277878	0.04426424	79	0.01836983	0.01917226	0.01999341	0.02169036	0.02345748
30	0.03874811	0.03945953	0.04017854	0.04163919	0.04312975	80	0.01821885	0.01902323	0.01984652	0.02154832	0.02332093
31	0.03767573	0.03838866	0.03910942	0.04057430	0.04207005	81	0.01807179	0.01887812	0.01970356	0.02141019	0.02318828
32	0.03667089	0.03738535	0.03810791	0.03957710	0.04107812	82	0.01792851	0.01873678	0.01956437	0.02127583	0.02305936
33	0.03572744	0.03644349	0.03716786	0.03864144	0.04014779	83	0.01778887	0.01859908	0.01942881	0.02114509	0.02293406
34	0.03483997	0.03555763	0.03628387	0.03776189	0.03927363	84	0.01765273	0.01846489	0.01929675	0.02101784	0.02281223
35	0.03400368	0.03472299	0.03545111	0.03693363	0.03845082	85	0.01751998	0.01833409	0.01916808	0.02089396	0.02269375
36	0.03321431	0.03393529	0.03466533	0.03615240	0.03767507	86	0.01739050	0.01820654	0.01904267	0.02077333	0.02257850
37	0.03246805	0.03319072	0.03392270	0.03541437	0.03694257	87	0.01726418	0.01808215	0.01892041	0.02065584	0.02246636
38	0.03176150	0.03248589	0.03321983	0.03471613	0.03624990	88	0.01714089	0.01796081	0.01880119	0.02054138	0.02235724
39	0.03109160	0.03181773	0.03255365	0.03405463	0.03559399	89	0.01702056	0.01784240	0.01868491	0.02042984	0.02225102
40	0.03045560	0.03118349	0.03192141	0.03342710	0.03497209	90	0.01690306	0.01772684	0.01857146	0.02032113	0.02214760
41	0.02985102	0.03058069	0.03132063	0.03283106	0.03438170	91	0.01678832	0.01761403	0.01846076	0.02021516	0.02204690
42	0.02927563	0.03000709	0.03074906	0.03226426	0.03382057	92	0.01667624	0.01750387	0.01835272	0.02011182	0.02194882
43	0.02872737	0.02946064	0.03020466	0.03172465	0.03328666	93	0.01656673	0.01739629	0.01824724	0.02001104	0.02185327
44	0.02820441	0.02893949	0.02968557	0.03121038	0.03277810	94	0.01645971	0.01729119	0.01814425	0.01991273	0.02176017
45	0.02770505	0.02844197	0.02919012	0.03071976	0.03229321	95	0.01635511	0.01718851	0.01804366	0.01981681	0.02166944
46	0.02722775	0.02796652	0.02871675	0.03025125	0.03183043	96	0.01625284	0.01708816	0.01794541	0.01972321	0.02158101
47	0.02677111	0.02751173	0.02826406	0.02980342	0.03138836	97	0.01615284	0.01699007	0.01784941	0.01963186	0.02149480
48	0.02633384	0.02707632	0.02783075	0.02937500	0.03096569	98	0.01605503	0.01689418	0.01775560	0.01954268	0.02141074
49	0.02591474	0.02665910	0.02741563	0.02896478	0.03056124	99	0.01595936	0.01680041	0.01766391	0.01945560	0.02132876
50	0.02551273	0.02625898	0.02701763	0.02857168	0.03017391	100	0.01586574	0.01670870	0.01757428	0.01937057	0.02124880

ANNUITY WHOSE PRESENT VALUE IS 1

$$a^{-1}_{\overline{n}|i} = \frac{i}{1 - v^n} = s^{-1}_{\overline{n}|i} + i \text{ (Continued)}$$

Periods n	Rate i 0.02(2%)	0.0225(2 1/4%)	0.025(2 1/2%)	0.0275(2 3/4%)	Periods n	Rate i 0.02(2%)	0.0225(2 1/4%)	0.025(2 1/2%)	0.0275(2 3/4%)
1	1.02000000	1.02250000	1.02500000	1.02750000	51	0.03145856	0.03316102	0.03490870	0.03670014
2	0.51504950	0.51693758	0.51882716	0.52071825	52	0.03110909	0.03281884	0.03457446	0.03637444
3	0.34675467	0.34844458	0.35013717	0.35183243	53	0.03077392	0.03249094	0.03425449	0.03606297
4	0.26262375	0.26421893	0.26581788	0.26742059	54	0.03045226	0.03217654	0.03394799	0.03576491
5	0.21215839	0.21370021	0.21524686	0.21679832	55	0.03014337	0.03187489	0.03365419	0.03547953
6	0.17852581	0.18003496	0.18154997	0.18307083	56	0.02984656	0.03158530	0.03337243	0.03520612
7	0.15451196	0.15600025	0.15749543	0.15899747	57	0.02956120	0.03130712	0.03310204	0.03494404
8	0.13650980	0.13798462	0.13946735	0.14095795	58	0.02928667	0.03103977	0.03284244	0.03469270
9	0.12251544	0.12398170	0.12545689	0.12694095	59	0.02902243	0.03078268	0.03259307	0.03445153
10	0.11132653	0.11278768	0.11425876	0.11573972	60	0.02876797	0.03053533	0.03235340	0.03422002
11	0.10217794	0.10363649	0.10510596	0.10658629	61	0.02852278	0.03029724	0.03212294	0.03399767
12	0.09455960	0.09601740	0.09748713	0.09896871	62	0.02828643	0.03006795	0.03190126	0.03378402
13	0.08811835	0.08957686	0.09104827	0.09253252	63	0.02805848	0.02984704	0.03168790	0.03357866
14	0.08260197	0.08406230	0.08553652	0.08702457	64	0.02783855	0.02963411	0.03148249	0.03338118
15	0.07782547	0.07928852	0.08076646	0.08225917	65	0.02762624	0.02942878	0.03128463	0.03319120
16	0.07365013	0.07511663	0.07659899	0.07809710	66	0.02742122	0.02923070	0.03109398	0.03300837
17	0.06996984	0.07144039	0.07292777	0.07443186	67	0.02722316	0.02903955	0.03091021	0.03283236
18	0.06670210	0.06817720	0.06967008	0.07118063	68	0.02703173	0.02885500	0.03073300	0.03266285
19	0.06378177	0.06526182	0.06676062	0.06827802	69	0.02684665	0.02867677	0.03056206	0.03249955
20	0.06115672	0.06264207	0.06414713	0.06567173	70	0.02666765	0.02850458	0.03039712	0.03234218
21	0.05878477	0.06027572	0.06178733	0.06331941	71	0.02649446	0.02833816	0.03023790	0.03219048
22	0.05663140	0.05812821	0.05964661	0.06118640	72	0.02632683	0.02817728	0.03008417	0.03204420
23	0.05466810	0.05617097	0.05769638	0.05924410	73	0.02616454	0.02802169	0.02993568	0.03190311
24	0.05287110	0.05438023	0.05591282	0.05746863	74	0.02600736	0.02787118	0.02979222	0.03176698
25	0.05122044	0.05273599	0.05427592	0.05583997	75	0.02585508	0.02772554	0.02965358	0.03163560
26	0.04969923	0.05122134	0.05276875	0.05434116	76	0.02570751	0.02758457	0.02951956	0.03150878
27	0.04829309	0.04982188	0.05137687	0.05295776	77	0.02556447	0.02744808	0.02938997	0.03138633
28	0.04698967	0.04852525	0.05008793	0.05167738	78	0.02542576	0.02731589	0.02926463	0.03126806
29	0.04577836	0.04732081	0.04889127	0.05048935	79	0.02529123	0.02718784	0.02914338	0.03115382
30	0.04464992	0.04619934	0.04777764	0.04938442	80	0.02516071	0.02706376	0.02902605	0.03104342
31	0.04359635	0.04515280	0.04673900	0.04835453	81	0.02503405	0.02694350	0.02891248	0.03093674
32	0.04261061	0.04417415	0.04576831	0.04739263	82	0.02491110	0.02682692	0.02880254	0.03083361
33	0.04168653	0.04325722	0.04485938	0.04649253	83	0.02479173	0.02671387	0.02869608	0.03073389
34	0.04081867	0.04239655	0.04400675	0.04564875	84	0.02467581	0.02660423	0.02859298	0.03063747
35	0.04000221	0.04158731	0.04320558	0.04485645	85	0.02456321	0.02649787	0.02849310	0.03054420
36	0.03923285	0.04082522	0.04245158	0.04411132	86	0.02445381	0.02639467	0.02839633	0.03045397
37	0.03850678	0.04010643	0.04174090	0.04340953	87	0.02434750	0.02629452	0.02830255	0.03036667
38	0.03782057	0.03942753	0.04107012	0.04274764	88	0.02424416	0.02619730	0.02821165	0.03028219
39	0.03717114	0.03878543	0.04043615	0.04212256	89	0.02414370	0.02610291	0.02812353	0.03020041
40	0.03655575	0.03817738	0.03983623	0.04153151	90	0.02404602	0.02601126	0.02803809	0.03012125
41	0.03597188	0.03760087	0.03926786	0.04097200	91	0.02395101	0.02592224	0.02795523	0.03004460
42	0.03541729	0.03705364	0.03872876	0.04044175	92	0.02385859	0.02583577	0.02787486	0.02997038
43	0.03488993	0.03653364	0.03821688	0.03993871	93	0.02376868	0.02575176	0.02779690	0.02989850
44	0.03438794	0.03603901	0.03773037	0.03946100	94	0.02368118	0.02567012	0.02772126	0.02982887
45	0.03390962	0.03556805	0.03726751	0.03900693	95	0.02359602	0.02559078	0.02764786	0.02976141
46	0.03345342	0.03511921	0.03682676	0.03857493	96	0.02351313	0.02551366	0.02757662	0.02969605
47	0.03301792	0.03469107	0.03640669	0.03816358	97	0.02343242	0.02543868	0.02750747	0.02963272
48	0.03260184	0.03428233	0.03600599	0.03777158	98	0.02335383	0.02536578	0.02744034	0.02957134
49	0.03220396	0.03389179	0.03562348	0.03739773	99	0.02327729	0.02529489	0.02737517	0.02951185
50	0.03182321	0.03351836	0.03525806	0.03704092	100	0.02320274	0.02522594	0.02731188	0.02945418

ANNUITY WHOSE PRESENT VALUE IS 1

$$a^{-1}_{\overline{n}|i} = \frac{i}{1 - v^n} = s^{-1}_{\overline{n}|i} + i \text{ (Continued)}$$

Periods n	Rate i: 0.03(3%)	0.0325(3 1/4%)	0.035(3 1/2%)	0.0375(3 3/4%)	Periods n	Rate i: 0.03(3%)	0.0325(3 1/4%)	0.035(3 1/2%)	0.0375(3 3/4%)
1	1.03000000	1.03250000	1.03500000	1.03750000	51	0.03853382	0.04040817	0.04232156	0.04427235
2	0.52261084	0.52450492	0.52640049	0.52829755	52	0.03821718	0.04010103	0.04202429	0.04398523
3	0.35353036	0.35523095	0.35693418	0.35864005	53	0.03791471	0.03980797	0.04174100	0.04371199
4	0.26902705	0.27063723	0.27225114	0.27386875	54	0.03762558	0.03952819	0.04147090	0.04345183
5	0.21835457	0.21991560	0.22148137	0.22305189	55	0.03734907	0.03926095	0.04121323	0.04320398
6	0.18459750	0.18612997	0.18766821	0.18921219	56	0.03708447	0.03900553	0.04096730	0.04296775
7	0.16050635	0.16202204	0.16354449	0.16507370	57	0.03683114	0.03876131	0.04073245	0.04274249
8	0.14245639	0.14396263	0.14547665	0.14699839	58	0.03658848	0.03852767	0.04050810	0.04252760
9	0.12843386	0.12993555	0.13144601	0.13296517	59	0.03635593	0.03830405	0.04029366	0.04232251
10	0.11723051	0.11873107	0.12024137	0.12176134	60	0.03613296	0.03808993	0.04008862	0.04212670
11	0.10807745	0.10957936	0.11109197	0.11261521	61	0.03591908	0.03788483	0.03989249	0.04193967
12	0.10046209	0.10196719	0.10348395	0.10501230	62	0.03571385	0.03768827	0.03970480	0.04176097
13	0.09402954	0.09553925	0.09706157	0.09859642	63	0.03551682	0.03749983	0.03952513	0.04159016
14	0.08852634	0.09004176	0.09157073	0.09311317	64	0.03532760	0.03731912	0.03935308	0.04142684
15	0.08376658	0.08528858	0.08682507	0.08837595	65	0.03514581	0.03714574	0.03918826	0.04127063
16	0.07961085	0.08114013	0.08268483	0.08424483	66	0.03497110	0.03697935	0.03903031	0.04112118
17	0.07595253	0.07748966	0.07904313	0.08061280	67	0.03480313	0.03681962	0.03887892	0.04097816
18	0.07270870	0.07425415	0.07581684	0.07739662	68	0.03464159	0.03666622	0.03873375	0.04084124
19	0.06981388	0.07136804	0.07294033	0.07453058	69	0.03448618	0.03651886	0.03859453	0.04071013
20	0.06721571	0.06877888	0.07036108	0.07196210	70	0.03433663	0.03637727	0.03846095	0.04058456
21	0.06487178	0.06644424	0.06803659	0.06964862	71	0.03419266	0.03624117	0.03833277	0.04046426
22	0.06274739	0.06432936	0.06593207	0.06755531	72	0.03405404	0.03611033	0.03820973	0.04034898
23	0.06081390	0.06240555	0.06401880	0.06565339	73	0.03392053	0.03598451	0.03809160	0.04023848
24	0.05904742	0.06064891	0.06227283	0.06391890	74	0.03379191	0.03586347	0.03797816	0.04013255
25	0.05742787	0.05903933	0.06067404	0.06233169	75	0.03366796	0.03574702	0.03786919	0.04003098
26	0.05593829	0.05755981	0.05920540	0.06087470	76	0.03354849	0.03563496	0.03776450	0.03993356
27	0.05456421	0.05619588	0.05785241	0.05953343	77	0.03343331	0.03552709	0.03766390	0.03984011
28	0.05329323	0.05493512	0.05660265	0.05829540	78	0.03332224	0.03542323	0.03756721	0.03975045
29	0.05211467	0.05376682	0.05544538	0.05714991	79	0.03321510	0.03532323	0.03747426	0.03966442
30	0.05101926	0.05268172	0.05437133	0.05608762	80	0.03311175	0.03522690	0.03738489	0.03958184
31	0.04999893	0.05167172	0.05337240	0.05510046	81	0.03301201	0.03513411	0.03729894	0.03950258
32	0.04904662	0.05072976	0.05244150	0.05418131	82	0.03291576	0.03504471	0.03721628	0.03942647
33	0.04815612	0.04984961	0.05157242	0.05332395	83	0.03282284	0.03495855	0.03713676	0.03935340
34	0.04732196	0.04902581	0.05075966	0.05252287	84	0.03273313	0.03487550	0.03706025	0.03928323
35	0.04653929	0.04825348	0.04999835	0.05177320	85	0.03264650	0.03479545	0.03698662	0.03921582
36	0.04580379	0.04752831	0.04928416	0.05107060	86	0.03256284	0.03471826	0.03691576	0.03915107
37	0.04511162	0.04684645	0.04861325	0.05041122	87	0.03248202	0.03464383	0.03684756	0.03908887
38	0.04445934	0.04620445	0.04798214	0.04979159	88	0.03240393	0.03457205	0.03678190	0.03902910
39	0.04384385	0.04559920	0.04738775	0.04920860	89	0.03232848	0.03450281	0.03671868	0.03897166
40	0.04326238	0.04502794	0.04682728	0.04865946	90	0.03225556	0.03443601	0.03665781	0.03891646
41	0.04271241	0.04448814	0.04629822	0.04814164	91	0.03218508	0.03437156	0.03659919	0.03886340
42	0.04219167	0.04397753	0.04579828	0.04765286	92	0.03211694	0.03430937	0.03654273	0.03881240
43	0.04169811	0.04349403	0.04532539	0.04719106	93	0.03205107	0.03424935	0.03648834	0.03876336
44	0.04122985	0.04303579	0.04487768	0.04675434	94	0.03198737	0.03419142	0.03643594	0.03871622
45	0.04078518	0.04260108	0.04445343	0.04634098	95	0.03192577	0.03413550	0.03638546	0.03867088
46	0.04036254	0.04218835	0.04405108	0.04594943	96	0.03186619	0.03408151	0.03633682	0.03862729
47	0.03996051	0.04179616	0.04366919	0.04557824	97	0.03180856	0.03402939	0.03628995	0.03858537
48	0.03957777	0.04142320	0.04330646	0.04522609	98	0.03175281	0.03397906	0.03624478	0.03854504
49	0.03921314	0.04106828	0.04296167	0.04489179	99	0.03169886	0.03393045	0.03620124	0.03850626
50	0.03886549	0.04073027	0.04263371	0.04457422	100	0.03164667	0.03388351	0.03615927	0.03846895

ANNUITY WHOSE PRESENT VALUE IS 1

$$a_{\overline{n}|i}^{-1} = \frac{i}{1 - v^n} = s_{\overline{n}|i}^{-1} + i \text{ (Continued)}$$

Periods n	Rate i 0.04(4%)	0.0425(4 1/4%)	0.045(4 1/2%)	0.0475(4 3/4%)	Periods n	Rate i 0.04(4%)	0.0425(4 1/4%)	0.045(4 1/2%)	0.0475(4 3/4%)
1	1.04000000	1.04250000	1.04500000	1.04750000	51	0.04625885	0.04827940	0.05033232	0.05241595
2	0.53019608	0.53209608	0.53399756	0.53590049	52	0.04598212	0.04801322	0.05007679	0.05217111
3	0.36034854	0.36205965	0.36377336	0.36548967	53	0.04571915	0.04776064	0.04983469	0.05193950
4	0.27549005	0.27711502	0.27874365	0.28037592	54	0.04546910	0.04752084	0.04960519	0.05172030
5	0.22462711	0.22620704	0.22779164	0.22938090	55	0.04523124	0.04729307	0.04938754	0.05151276
6	0.19076190	0.19231731	0.19387839	0.19544512	56	0.04500487	0.04707663	0.04918105	0.05131618
7	0.16660961	0.16815221	0.16970147	0.17125735	57	0.04478932	0.04687086	0.04898506	0.05112990
8	0.14852783	0.15006493	0.15160965	0.15316196	58	0.04458401	0.04667517	0.04879897	0.05095334
9	0.13449299	0.13602944	0.13757447	0.13912803	59	0.04438836	0.04648898	0.04862221	0.05078591
10	0.12329094	0.12483012	0.12637882	0.12793699	60	0.04420185	0.04631178	0.04845426	0.05062710
11	0.11414904	0.11569338	0.11724818	0.11881337	61	0.04402398	0.04614306	0.04829462	0.05047641
12	0.10655217	0.10810349	0.10966619	0.11124019	62	0.04385430	0.04598237	0.04814284	0.05033339
13	0.10014373	0.10170340	0.10327535	0.10485950	63	0.04369237	0.04582929	0.04799848	0.05019761
14	0.09466897	0.09623806	0.09782032	0.09941565	64	0.04353780	0.04568339	0.04786115	0.05006867
15	0.08994110	0.09152043	0.09311381	0.09472113	65	0.04339019	0.04554432	0.04773047	0.04994619
16	0.08582000	0.08741022	0.08901537	0.09063531	66	0.04324921	0.04541171	0.04760608	0.04982982
17	0.08219852	0.08380017	0.08541758	0.08705063	67	0.04311451	0.04528522	0.04748765	0.04971924
18	0.07899333	0.08060681	0.08223690	0.08388343	68	0.04298578	0.04516456	0.04737487	0.04961413
19	0.07613862	0.07776427	0.07940734	0.08106766	69	0.04286272	0.04504942	0.04726745	0.04951420
20	0.07358175	0.07521983	0.07687614	0.07855047	70	0.04274506	0.04493952	0.04716511	0.04941917
21	0.07128011	0.07293083	0.07460057	0.07628907	71	0.04263253	0.04483460	0.04706759	0.04932879
22	0.06919881	0.07086234	0.07254565	0.07424846	72	0.04252489	0.04473442	0.04697465	0.04924282
23	0.06730906	0.06898552	0.07068249	0.07239969	73	0.04242190	0.04463875	0.04688606	0.04916103
24	0.06558683	0.06727631	0.06898703	0.07071867	74	0.04232334	0.04454736	0.04680159	0.04908320
25	0.06401196	0.06571452	0.06743903	0.06918513	75	0.04222900	0.04446004	0.04672104	0.04900912
26	0.06256738	0.06428306	0.06602137	0.06778192	76	0.04213869	0.04437661	0.04664422	0.04893862
27	0.06123854	0.06296736	0.06471946	0.06649444	77	0.04205221	0.04429687	0.04657094	0.04887150
28	0.06001298	0.06175492	0.06352081	0.06531016	78	0.04196939	0.04422065	0.04650104	0.04880759
29	0.05887993	0.06063500	0.06241461	0.06421829	79	0.04189007	0.04414778	0.04643434	0.04874674
30	0.05783010	0.05959825	0.06139154	0.06320945	80	0.04181408	0.04407811	0.04637069	0.04868879
31	0.05685535	0.05863654	0.06044345	0.06227550	81	0.04174127	0.04401149	0.04630995	0.04863360
32	0.05594859	0.05774275	0.05956320	0.06140929	82	0.04167150	0.04394777	0.04625197	0.04858103
33	0.05510357	0.05691064	0.05874453	0.06060455	83	0.04160463	0.04388682	0.04619663	0.04853094
34	0.05431477	0.05613469	0.05798191	0.05985574	84	0.04154054	0.04382852	0.04614379	0.04848322
35	0.05357732	0.05540999	0.05727045	0.05915794	85	0.04147909	0.04377274	0.04609334	0.04843776
36	0.05288688	0.05473220	0.05660578	0.05850680	86	0.04142018	0.04371936	0.04604516	0.04839443
37	0.05223957	0.05409745	0.05598402	0.05789843	87	0.04136370	0.04366828	0.04599915	0.04835315
38	0.05163192	0.05350225	0.05540169	0.05732932	88	0.04130953	0.04361940	0.04595522	0.04831380
39	0.05106083	0.05294350	0.05485567	0.05679637	89	0.04125758	0.04357262	0.04591325	0.04827629
40	0.05052349	0.05241839	0.05434315	0.05629675	90	0.04120775	0.04352783	0.04587316	0.04824054
41	0.05001738	0.05192438	0.05386158	0.05582791	91	0.04115995	0.04348496	0.04583486	0.04820646
42	0.04954020	0.05145918	0.05340868	0.05538756	92	0.04111410	0.04344391	0.04579827	0.04817397
43	0.04908989	0.05102071	0.05298235	0.05497362	93	0.04107010	0.04340461	0.04576331	0.04814300
44	0.04866454	0.05060708	0.05258071	0.05458418	94	0.04102789	0.04336698	0.04572991	0.04811346
45	0.04826246	0.05021657	0.05220202	0.05421751	95	0.04098738	0.04333094	0.04569799	0.04808530
46	0.04788205	0.04984760	0.05184471	0.05387203	96	0.04094850	0.04329643	0.04566749	0.04805845
47	0.04752189	0.04949873	0.05150734	0.05354630	97	0.04091119	0.04326338	0.04563834	0.04803284
48	0.04718065	0.04916864	0.05118858	0.05323900	98	0.04087538	0.04323173	0.04561048	0.04800842
49	0.04685712	0.04885612	0.05088722	0.05294891	99	0.04084100	0.04320140	0.04558385	0.04798513
50	0.04655020	0.04856005	0.05060215	0.05267490	100	0.04080800	0.04317236	0.04555839	0.04796292

ANNUITY WHOSE PRESENT VALUE IS 1

$$a_{\overline{n}|i}^{-1} = \frac{i}{1 - v^n} = s_{\overline{n}|i}^{-1} + i \text{ (Continued)}$$

Periods n	0.05(5%)	0.0525(5 1/4%)	0.055(5 1/2%)	0.0575(5 3/4%)	Periods n	0.05(5%)	0.0525(5 1/4%)	0.055(5 1/2%)	0.0575(5 3/4%)
1	1.05000000	1.05250000	1.05500000	1.05750000	51	0.05452867	0.05666886	0.05883495	0.06102543
2	0.53780488	0.53971072	0.54161800	0.54352673	52	0.05429450	0.05644528	0.05862186	0.06082266
3	0.36720856	0.36893004	0.37065407	0.37238067	53	0.05407334	0.05623449	0.05842130	0.06063215
4	0.28201183	0.28365136	0.28529449	0.28694120	54	0.05386438	0.05603566	0.05823245	0.06045310
5	0.23097480	0.23257332	0.23417644	0.23578414	55	0.05366686	0.05584805	0.05805458	0.06028475
6	0.19701747	0.19859542	0.20017895	0.20176803	56	0.05348010	0.05567096	0.05788698	0.06012642
7	0.17281982	0.17438885	0.17596442	0.17754648	57	0.05330343	0.05550374	0.05772900	0.05997746
8	0.15472181	0.15628918	0.15786401	0.15944628	58	0.05313626	0.05534579	0.05758006	0.05983728
9	0.14069008	0.14226057	0.14383946	0.14542670	59	0.05297802	0.05519655	0.05743959	0.05970532
10	0.12950457	0.13108152	0.13266777	0.13426327	60	0.05282818	0.05505549	0.05730707	0.05958107
11	0.12038889	0.12197467	0.12357065	0.12517676	61	0.05268627	0.05492214	0.05718202	0.05946405
12	0.11282541	0.11442178	0.11602923	0.11764767	62	0.05255183	0.05479604	0.05706400	0.05935381
13	0.10645577	0.10806405	0.10968426	0.11131631	63	0.05242442	0.05467676	0.05695258	0.05924994
14	0.10102397	0.10264516	0.10427912	0.10592574	64	0.05230365	0.05456391	0.05684737	0.05915206
15	0.09634229	0.09797715	0.09962560	0.10128751	65	0.05218915	0.05445712	0.05674800	0.05905979
16	0.09226991	0.09391903	0.09558254	0.09726029	66	0.05208057	0.05435605	0.05665413	0.05897281
17	0.08869914	0.09036298	0.09204197	0.09373597	67	0.05197758	0.05426036	0.05656544	0.05889079
18	0.08554622	0.08722511	0.08891992	0.09063045	68	0.05187986	0.05416976	0.05648163	0.05881344
19	0.08274501	0.08443921	0.08615006	0.08787734	69	0.05178715	0.05408396	0.05640242	0.05874049
20	0.08024259	0.08195228	0.08367933	0.08542350	70	0.05169915	0.05400269	0.05632754	0.05867166
21	0.07799611	0.07972143	0.08146478	0.08322590	71	0.05161563	0.05392569	0.05625675	0.05860673
22	0.07597051	0.07771153	0.07947123	0.08124934	72	0.05153633	0.05385275	0.05618982	0.05854546
23	0.07413682	0.07589358	0.07766965	0.07946472	73	0.05146103	0.05378362	0.05612652	0.05848764
24	0.07247090	0.07424339	0.07603580	0.07784779	74	0.05138953	0.05371811	0.05606665	0.05843306
25	0.07095246	0.07274066	0.07454935	0.07637817	75	0.05132161	0.05365601	0.05601002	0.05838155
26	0.06956432	0.07136817	0.07319307	0.07503860	76	0.05125709	0.05359714	0.05595645	0.05833292
27	0.06829186	0.07011129	0.07195228	0.07381439	77	0.05119580	0.05354133	0.05590577	0.05828701
28	0.06712253	0.06895744	0.07081440	0.07269293	78	0.05113756	0.05348841	0.05585781	0.05824367
29	0.06604551	0.06789578	0.06976857	0.07166336	79	0.05108222	0.05343822	0.05581243	0.05820274
30	0.06505144	0.06691693	0.06880539	0.07071624	80	0.05102962	0.05339063	0.05576948	0.05816409
31	0.06413212	0.06601270	0.06791665	0.06984336	81	0.05097963	0.05334549	0.05572884	0.05812758
32	0.06328042	0.06517593	0.06709519	0.06903754	82	0.05093211	0.05330267	0.05569036	0.05809311
33	0.06249004	0.06440032	0.06633469	0.06829246	83	0.05088694	0.05326205	0.05565395	0.05806054
34	0.06175545	0.06368030	0.06562958	0.06760253	84	0.05084399	0.05322351	0.05561947	0.05802978
35	0.06107171	0.06301096	0.06497493	0.06696282	85	0.05080316	0.05318695	0.05558683	0.05800073
36	0.06043446	0.06238791	0.06436635	0.06636893	86	0.05076433	0.05315226	0.05555593	0.05797328
37	0.05983979	0.06180725	0.06379993	0.06581694	87	0.05072740	0.05311934	0.05552667	0.05794734
38	0.05928423	0.06126548	0.06327217	0.06530335	88	0.05069228	0.05308810	0.05549896	0.05792284
39	0.05876462	0.06075946	0.06277991	0.06482500	89	0.05065888	0.05305845	0.05547273	0.05789969
40	0.05827816	0.06028637	0.06232034	0.06437907	90	0.05062711	0.05303032	0.05544788	0.05787781
41	0.05782229	0.05984365	0.06189090	0.06396299	91	0.05059689	0.05300361	0.05542435	0.05785714
42	0.05739471	0.05942899	0.06148927	0.06357445	92	0.05056815	0.05297826	0.05540207	0.05783761
43	0.05699333	0.05904031	0.06111337	0.06321135	93	0.05054080	0.05295420	0.05538096	0.05781915
44	0.05661625	0.05867569	0.06076128	0.06287179	94	0.05051478	0.05293136	0.05536097	0.05780171
45	0.05626173	0.05833341	0.06043127	0.06255404	95	0.05049003	0.05290967	0.05534204	0.05778522
46	0.05592820	0.05801189	0.06012175	0.06225650	96	0.05046648	0.05288908	0.05532410	0.05776964
47	0.05561421	0.05770966	0.05983129	0.06197773	97	0.05044407	0.05286954	0.05530711	0.05775491
48	0.05531843	0.05742542	0.05955854	0.06171641	98	0.05042274	0.05285098	0.05529101	0.05774099
49	0.05503965	0.05715793	0.05930230	0.06147131	99	0.05040245	0.05283336	0.05527577	0.05772784
50	0.05477674	0.05690609	0.05906145	0.06124133	100	0.05038314	0.05281664	0.05526132	0.05771540

ANNUITY WHOSE PRESENT VALUE IS 1

$$a_{\overline{n}|i}^{-1} = \frac{i}{1 - v^n} = s_{\overline{n}|i}^{-1} + i \text{ (Continued)}$$

Periods	Rate i			
n	0.06(6%)	0.0625(6 1/4%)	0.065(6 1/2%)	0.0675(6 3/4%)
1	1.06000000	1.06250000	1.06500000	1.06750000
2	0.54543689	0.54734848	0.54926150	0.55117594
3	0.37410981	0.37584149	0.37757570	0.37931243
4	0.28859149	0.29024534	0.29190274	0.29356367
5	0.23739640	0.23901321	0.24063454	0.24226037
6	0.20336263	0.20496273	0.20656831	0.20817934
7	0.17913502	0.18072999	0.18233137	0.18393912
8	0.16103594	0.16263296	0.16423730	0.16584891
9	0.14702224	0.14862603	0.15023803	0.15185820
10	0.13586796	0.13748179	0.13910469	0.14073662
11	0.12679294	0.12841911	0.13005521	0.13170116
12	0.11927703	0.12091722	0.12256817	0.12422978
13	0.11296011	0.11461555	0.11628256	0.11796102
14	0.10758491	0.10925653	0.11094048	0.11263666
15	0.10296276	0.10465123	0.10635278	0.10806729
16	0.09895214	0.10065795	0.10237757	0.10411086
17	0.09544480	0.09716831	0.09890633	0.10065868
18	0.09235654	0.09409799	0.09585461	0.09762621
19	0.08962086	0.09138040	0.09315575	0.09494670
20	0.08718456	0.08896227	0.09075640	0.09256670
21	0.08500455	0.08680045	0.08861333	0.09044294
22	0.08304557	0.08485962	0.08669120	0.08854002
23	0.08127848	0.08311061	0.08496078	0.08682866
24	0.07967900	0.08152909	0.08339770	0.08528446
25	0.07822672	0.08009462	0.08198148	0.08388691
26	0.07690435	0.07878989	0.08069480	0.08261865
27	0.07569717	0.07760015	0.07952288	0.08146489
28	0.07459255	0.07651276	0.07845305	0.08041294
29	0.07357961	0.07551680	0.07747440	0.07945186
30	0.07264891	0.07460284	0.07657744	0.07857215
31	0.07179222	0.07376261	0.07575393	0.07776557
32	0.07100234	0.07298892	0.07499665	0.07702486
33	0.07027293	0.07227543	0.07429924	0.07634368
34	0.06959843	0.07161653	0.07365610	0.07571641
35	0.06897386	0.07100726	0.07306226	0.07513808
36	0.06839483	0.07044324	0.07251332	0.07460428
37	0.06785743	0.06992051	0.07200534	0.07411106
38	0.06735812	0.06943557	0.07153480	0.07365492
39	0.06689377	0.06898526	0.07109854	0.07323268
40	0.06646154	0.06856675	0.07069373	0.07284150
41	0.06605886	0.06817746	0.07031779	0.07247884
42	0.06568342	0.06781509	0.06996842	0.07214236
43	0.06533312	0.06747754	0.06964352	0.07182999
44	0.06500606	0.06716290	0.06934119	0.07153981
45	0.06470050	0.06686944	0.06905968	0.07127010
46	0.06441485	0.06659557	0.06879743	0.07101928
47	0.06414768	0.06633985	0.06855300	0.07078591
48	0.06389765	0.06610096	0.06832505	0.07056869
49	0.06366356	0.06587769	0.06811240	0.07036642
50	0.06344429	0.06566893	0.06791393	0.07017798

Periods	Rate i			
n	0.06(6%)	0.0625(6 1/4%)	0.065(6 1/2%)	0.0675(6 3/4%)
51	0.06323880	0.06547366	0.06772861	0.07000236
52	0.06304617	0.06529092	0.06755553	0.06983865
53	0.06286551	0.06511987	0.06739382	0.06968599
54	0.06269602	0.06495970	0.06724267	0.06954358
55	0.06253696	0.06480966	0.06710137	0.06941070
56	0.06238765	0.06466908	0.06696923	0.06928669
57	0.06224744	0.06453733	0.06684563	0.06917091
58	0.06211574	0.06441382	0.06672999	0.06906281
59	0.06199200	0.06429800	0.06662177	0.06896185
60	0.06187572	0.06418938	0.06652047	0.06886754
61	0.06176642	0.06408748	0.06642564	0.06877943
62	0.06166366	0.06399187	0.06633684	0.06869710
63	0.06156704	0.06390214	0.06625367	0.06862015
64	0.06147615	0.06381792	0.06617577	0.06854822
65	0.06139066	0.06373886	0.06610280	0.06848097
66	0.06131022	0.06366463	0.06603442	0.06841810
67	0.06123454	0.06359492	0.06597034	0.06835931
68	0.06116330	0.06352945	0.06591029	0.06830433
69	0.06109625	0.06346796	0.06585400	0.06825290
70	0.06103313	0.06341019	0.06580124	0.06820480
71	0.06097370	0.06335592	0.06575177	0.06815979
72	0.06091774	0.06330492	0.06570539	0.06811769
73	0.06086505	0.06325700	0.06566190	0.06807830
74	0.06081542	0.06321196	0.06562112	0.06804144
75	0.06076867	0.06316963	0.06558287	0.06800695
76	0.06072463	0.06312985	0.06554699	0.06797467
77	0.06068315	0.06309245	0.06551335	0.06794445
78	0.06064407	0.06305729	0.06548178	0.06791618
79	0.06060724	0.06302423	0.06545217	0.06788971
80	0.06057254	0.06299315	0.06542440	0.06786493
81	0.06053984	0.06296392	0.06539833	0.06784174
82	0.06050903	0.06293644	0.06537388	0.06782003
83	0.06047998	0.06291060	0.06535094	0.06779970
84	0.06045261	0.06288630	0.06532941	0.06778067
85	0.06042681	0.06286344	0.06530921	0.06776286
86	0.06040249	0.06284195	0.06529026	0.06774618
87	0.06037956	0.06282173	0.06527247	0.06773056
88	0.06035795	0.06280271	0.06525577	0.06771593
89	0.06033757	0.06278483	0.06524010	0.06770224
90	0.06031836	0.06276800	0.06522540	0.06768941
91	0.06030025	0.06275217	0.06521160	0.06767741
92	0.06028318	0.06273728	0.06519864	0.06766616
93	0.06026708	0.06272327	0.06518649	0.06765563
94	0.06025189	0.06271010	0.06517507	0.06764577
95	0.06023758	0.06269770	0.06516436	0.06763653
96	0.06022408	0.06268603	0.06515431	0.06762788
97	0.06021135	0.06267506	0.06514487	0.06761978
98	0.06019935	0.06266474	0.06513601	0.06761219
99	0.06018803	0.06265502	0.06512769	0.06760509
100	0.06017736	0.06264588	0.06511988	0.06759843

ANNUITY WHOSE PRESENT VALUE IS 1

$$a_{\overline{n}|i}^{-1} = \frac{i}{1 - v^n} = s_{\overline{n}|i}^{-1} + i \text{ (Continued)}$$

Periods	Rate i				Periods	Rate i			
n	0.07(7%)	0.0725(7 1/4%)	0.075(7 1/2%)	0.0775(7 3/4%)	n	0.07(7%)	0.0725(7 1/4%)	0.075(7 1/2%)	0.0775(7 3/4%)
1	1.07000000	1.07250000	1.07500000	1.07750000	51	0.07229365	0.07460128	0.07692411	0.07926108
2	0.55309179	0.55500905	0.55692771	0.55884777	52	0.07213901	0.07445540	0.07678668	0.07913174
3	0.38105167	0.38279340	0.38453763	0.38628434	53	0.07199509	0.07431990	0.07665927	0.07901209
4	0.29522812	0.29689607	0.29856751	0.30024243	54	0.07186110	0.07419400	0.07654112	0.07890137
5	0.24389069	0.24552548	0.24716472	0.24880838	55	0.07173633	0.07407700	0.07643155	0.07879888
6	0.20979580	0.21141766	0.21304489	0.21467748	56	0.07162011	0.07396824	0.07632991	0.07870401
7	0.18555322	0.18717363	0.18880032	0.19043325	57	0.07151183	0.07386711	0.07623559	0.07861616
8	0.16746776	0.16909381	0.17072702	0.17236735	58	0.07141093	0.07377308	0.07614807	0.07853481
9	0.15348647	0.15512281	0.15676716	0.15841948	59	0.07131689	0.07368561	0.07606683	0.07845946
10	0.14237750	0.14402729	0.14568593	0.14735335	60	0.07122923	0.07360424	0.07599142	0.07838966
11	0.13335690	0.13502237	0.13669747	0.13838216	61	0.07114749	0.07352854	0.07592140	0.07832499
12	0.12590199	0.12758470	0.12927783	0.13098130	62	0.07107127	0.07345809	0.07585638	0.07826506
13	0.11965085	0.12135194	0.12306420	0.12478752	63	0.07100019	0.07339253	0.07579600	0.07820953
14	0.11434494	0.11606522	0.11779737	0.11954129	64	0.07093388	0.07333150	0.07573992	0.07815806
15	0.10979462	0.11153465	0.11328724	0.11505225	65	0.07087203	0.07327469	0.07568782	0.07811036
16	0.10585765	0.10761780	0.10939116	0.11117757	66	0.07081431	0.07322180	0.07563942	0.07806614
17	0.10242519	0.10420570	0.10600003	0.10780800	67	0.07076046	0.07317256	0.07559446	0.07802514
18	0.09941260	0.10121358	0.10302896	0.10485853	68	0.07071021	0.07312670	0.07555268	0.07798713
19	0.09675301	0.09857449	0.10041090	0.10226202	69	0.07066331	0.07308399	0.07551386	0.07795189
20	0.09439293	0.09623484	0.09809219	0.09996473	70	0.07061953	0.07304422	0.07547778	0.07791921
21	0.09228900	0.09415124	0.09602937	0.09792314	71	0.07057866	0.07300717	0.07544425	0.07788891
22	0.09040577	0.09228816	0.09418687	0.09610161	72	0.07054051	0.07297267	0.07541308	0.07786081
23	0.08871393	0.09061624	0.09253528	0.09447070	73	0.07050490	0.07294052	0.07538412	0.07783474
24	0.08718902	0.08911101	0.09105008	0.09300585	74	0.07047164	0.07291057	0.07535719	0.07781057
25	0.08581052	0.08775190	0.08971067	0.09168643	75	0.07044060	0.07288267	0.07533216	0.07778815
26	0.08456103	0.08652149	0.08849961	0.09049497	76	0.07041160	0.07285668	0.07530889	0.07776735
27	0.08342573	0.08540494	0.08740204	0.08941658	77	0.07038453	0.07283245	0.07528726	0.07774806
28	0.08239193	0.08438951	0.08640520	0.08843849	78	0.07035924	0.07280989	0.07526714	0.07773017
29	0.08144865	0.08346425	0.08549811	0.08754971	79	0.07033563	0.07278885	0.07524844	0.07771357
30	0.08058640	0.08261962	0.08467124	0.08674069	80	0.07031357	0.07276926	0.07523106	0.07769817
31	0.07979691	0.08184734	0.08391628	0.08600313	81	0.07029297	0.07275099	0.07521489	0.07768388
32	0.07907292	0.08114017	0.08322599	0.08532974	82	0.07027373	0.07273397	0.07519986	0.07767062
33	0.07840807	0.08049172	0.08259397	0.08471416	83	0.07025576	0.07271811	0.07518588	0.07765833
34	0.07779674	0.07989637	0.08201461	0.08415075	84	0.07023897	0.07270332	0.07517288	0.07764692
35	0.07723396	0.07934915	0.08148291	0.08363452	85	0.07022329	0.07268954	0.07516079	0.07763633
36	0.07671531	0.07884563	0.08099447	0.08316106	86	0.07020863	0.07267670	0.07514955	0.07762651
37	0.07623685	0.07838187	0.08054533	0.08272643	87	0.07019495	0.07266472	0.07513910	0.07761740
38	0.07579505	0.07795435	0.08013197	0.08232710	88	0.07018216	0.07265357	0.07512938	0.07760894
39	0.07538676	0.07755991	0.07975124	0.08195993	89	0.07017021	0.07264316	0.07512034	0.07760110
40	0.07500914	0.07719571	0.07940031	0.08162208	90	0.07015905	0.07263347	0.07511193	0.07759381
41	0.07465962	0.07685920	0.07907663	0.08131102	91	0.07014863	0.07262443	0.07510411	0.07758706
42	0.07433591	0.07654807	0.07877789	0.08102444	92	0.07013888	0.07261601	0.07509684	0.07758079
43	0.07403590	0.07626023	0.07850201	0.08076028	93	0.07012978	0.07260815	0.07509007	0.07757497
44	0.07375769	0.07599379	0.07824710	0.08051665	94	0.07012128	0.07260083	0.07508378	0.07756958
45	0.07349957	0.07574704	0.07801146	0.08029186	95	0.07011333	0.07259401	0.07507793	0.07756457
46	0.07325996	0.07551840	0.07779354	0.08008436	96	0.07010590	0.07258764	0.07507249	0.07755992
47	0.07303744	0.07530646	0.07759190	0.07989274	97	0.07009897	0.07258171	0.07506743	0.07755561
48	0.07283070	0.07510992	0.07740527	0.07971572	98	0.07009248	0.07257618	0.07506272	0.07755161
49	0.07263853	0.07492758	0.07723247	0.07955213	99	0.07008643	0.07257103	0.07505834	0.07754789
50	0.07245985	0.07475837	0.07707241	0.07940091	100	0.07008076	0.07256622	0.07505427	0.07754444

ANNUITY WHOSE PRESENT VALUE IS 1

$$a_{\overline{n}|i}^{-1} = \frac{i}{1 - v^n} = s_{\overline{n}|i}^{-1} + i \text{ (Continued)}$$

Periods	Rate i				Periods	Rate i			
n	0.08(8%)	0.0825(8 1/4%)	0.085(8 1/2%)	0.0875(8 3/4%)	n	0.08(8%)	0.0825(8 1/4%)	0.085(8 1/2%)	0.0875(8 3/4%)
1	1.08000000	1.08250000	1.08500000	1.08750000	51	0.08161116	0.08397339	0.08634688	0.08873078
2	0.56076923	0.56269208	0.56461631	0.56654192	52	0.08148959	0.08385925	0.08623983	0.08863048
3	0.38803351	0.38978515	0.39153925	0.39329579	53	0.08137735	0.08375409	0.08614139	0.08853844
4	0.30192080	0.30360263	0.30528789	0.30697657	54	0.08127370	0.08365717	0.08605087	0.08845398
5	0.25045645	0.25210892	0.25376575	0.25542694	55	0.08117796	0.08356784	0.08596761	0.08837645
6	0.21631539	0.21795860	0.21960708	0.22126082	56	0.08108952	0.08348548	0.08589101	0.08830528
7	0.19207240	0.19371774	0.19536922	0.19702683	57	0.08100780	0.08340955	0.08582053	0.08823994
8	0.17401476	0.17566921	0.17733065	0.17899906	58	0.08093227	0.08333952	0.08575568	0.08817994
9	0.16007971	0.16174781	0.16342372	0.16510740	59	0.08086247	0.08327494	0.08569599	0.08812484
10	0.14902949	0.15071430	0.15240771	0.15410966	60	0.08079795	0.08321537	0.08564106	0.08807424
11	0.14007634	0.14177996	0.14349293	0.14521519	61	0.08073830	0.08316041	0.08559049	0.08802776
12	0.13269502	0.13441890	0.13615286	0.13789681	62	0.08068314	0.08310971	0.08554393	0.08798506
13	0.12652181	0.12826696	0.13002287	0.13178944	63	0.08063214	0.08306292	0.08550107	0.08794583
14	0.12129685	0.12306394	0.12484244	0.12663222	64	0.08058497	0.08301975	0.08546160	0.08790979
15	0.11682954	0.11861900	0.12042046	0.12223380	65	0.08054135	0.08297991	0.08542526	0.08787668
16	0.11297687	0.11478892	0.11661354	0.11845059	66	0.08050099	0.08294314	0.08539179	0.08784625
17	0.10962943	0.11146415	0.11331198	0.11517274	67	0.08046367	0.08290920	0.08536097	0.08781829
18	0.10670210	0.10855946	0.11043041	0.11231476	68	0.08042914	0.08287787	0.08533258	0.08779260
19	0.10412763	0.10600750	0.10790140	0.10980912	69	0.08039719	0.08284895	0.08530643	0.08776898
20	0.10185221	0.10375437	0.10567097	0.10760176	70	0.08036764	0.08282225	0.08528234	0.08774728
21	0.09983225	0.10175643	0.10369541	0.10564890	71	0.08034029	0.08279760	0.08526016	0.08772733
22	0.09803207	0.09997794	0.10193892	0.10391471	72	0.08031498	0.08277485	0.08523972	0.08770900
23	0.09642217	0.09838936	0.10037193	0.10236954	73	0.08029157	0.08275383	0.08522089	0.08769214
24	0.09497796	0.09696605	0.09896975	0.10098871	74	0.08026989	0.08273443	0.08520354	0.08767665
25	0.09367878	0.09568733	0.09771168	0.09975145	75	0.08024984	0.08271652	0.08518756	0.08766241
26	0.09250713	0.09453567	0.09658017	0.09864020	76	0.08023128	0.08269998	0.08517284	0.08764932
27	0.09144810	0.09349614	0.09556025	0.09763999	77	0.08021410	0.08268470	0.08515927	0.08763729
28	0.09048891	0.09255595	0.09463914	0.09673799	78	0.08019820	0.08267060	0.08514677	0.08762623
29	0.08961854	0.09170406	0.09380577	0.09592315	79	0.08018349	0.08265757	0.08513526	0.08761606
30	0.08882743	0.09093091	0.09305058	0.09518590	80	0.08016987	0.08264554	0.08512465	0.08760671
31	0.08810728	0.09022818	0.09236524	0.09451789	81	0.08015726	0.08263443	0.08511487	0.08759811
32	0.08745081	0.08958859	0.09174247	0.09391186	82	0.08014559	0.08262417	0.08510586	0.08759021
33	0.08685163	0.08900575	0.09117588	0.09336140	83	0.08013479	0.08261469	0.08509755	0.08758295
34	0.08630411	0.08847402	0.09065984	0.09286090	84	0.08012479	0.08260594	0.08508990	0.08757627
35	0.08580326	0.08798844	0.09018937	0.09240538	85	0.08011553	0.08259786	0.08508285	0.08757012
36	0.08534467	0.08754458	0.08976006	0.09199044	86	0.08010696	0.08259039	0.08507636	0.08756448
37	0.08492440	0.08713850	0.08936799	0.09161216	87	0.08009903	0.08258350	0.08507037	0.08755929
38	0.08453894	0.08676671	0.08900966	0.09126705	88	0.08009168	0.08257713	0.08506485	0.08755451
39	0.08418513	0.08642606	0.08868193	0.09095200	89	0.08008489	0.08257124	0.08505977	0.08755012
40	0.08386016	0.08611373	0.08838201	0.09066421	90	0.08007859	0.08256581	0.08505508	0.08754609
41	0.08356149	0.08582722	0.08810737	0.09040118	91	0.08007277	0.08256079	0.08505077	0.08754238
42	0.08328684	0.08556422	0.08785576	0.09016065	92	0.08006737	0.08255615	0.08504679	0.08753897
43	0.08303414	0.08532270	0.08762512	0.08994060	93	0.08006238	0.08255187	0.08504312	0.08753583
44	0.08280152	0.08510079	0.08741363	0.08973921	94	0.08005775	0.08254792	0.08503974	0.08753295
45	0.08258728	0.08489682	0.08721961	0.08955481	95	0.08005347	0.08254426	0.08503663	0.08753030
46	0.08238991	0.08470926	0.08704154	0.08938592	96	0.08004951	0.08254089	0.08503375	0.08752786
47	0.08220799	0.08453673	0.08687807	0.08923118	97	0.08004584	0.08253777	0.08503111	0.08752562
48	0.08204027	0.08437797	0.08672795	0.08908936	98	0.08004244	0.08253489	0.08502867	0.08752355
49	0.08188557	0.08423184	0.08659005	0.08895934	99	0.08003930	0.08253223	0.08502642	0.08752166
50	0.08174286	0.08409730	0.08646334	0.08884013	100	0.08003638	0.08252977	0.08502435	0.08751992

ANNUITY WHOSE PRESENT VALUE IS 1

$$a_{\overline{n}|i}^{-1} = \frac{i}{1 - v^n} = s_{\overline{n}|i}^{-1} + i \text{ (Continued)}$$

Periods n	0.09(9%)	0.0925(9 1/4%)	0.095(9 1/2%)	0.0975(9 3/4%)
1	1.09000000	1.09250000	1.09500000	1.09750000
2	0.56846890	0.57039725	0.57232697	0.57425805
3	0.39505476	0.39681615	0.39857997	0.40034619
4	0.30866866	0.31036414	0.31206300	0.31376523
5	0.25709246	0.25876229	0.26043642	0.26211482
6	0.22291978	0.22458395	0.22625328	0.22792777
7	0.19869052	0.20036026	0.20203603	0.20371779
8	0.18067438	0.18235658	0.18404561	0.18574144
9	0.16679880	0.16849786	0.17020454	0.17191878
10	0.15582009	0.15753894	0.15926615	0.16100166
11	0.14694666	0.14868726	0.15043693	0.15219558
12	0.13965066	0.14141432	0.14318771	0.14497074
13	0.13356656	0.13535414	0.13715206	0.13896022
14	0.12843317	0.13024517	0.13206809	0.13390182
15	0.12405888	0.12589556	0.12774370	0.12960315
16	0.12029991	0.12216133	0.12403470	0.12591985
17	0.11704625	0.11893232	0.12083078	0.12274145
18	0.11421229	0.11612281	0.11804610	0.11998198
19	0.11173041	0.11366506	0.11561284	0.11757352
20	0.10954648	0.11150487	0.11347670	0.11546170
21	0.10761663	0.10959832	0.11159370	0.11360248
22	0.10590499	0.10790947	0.10992784	0.11195981
23	0.10438188	0.10640860	0.10844938	0.11050389
24	0.10302256	0.10507095	0.10713351	0.10920990
25	0.10180625	0.10387569	0.10595939	0.10805698
26	0.10071536	0.10280523	0.10490940	0.10702747
27	0.09973491	0.10184456	0.10396852	0.10610635
28	0.09885205	0.10098083	0.10312389	0.10528076
29	0.09805572	0.10020298	0.10236444	0.10453963
30	0.09733635	0.09950142	0.10168058	0.10387336
31	0.09668560	0.09886781	0.10106399	0.10327363
32	0.09609619	0.09829488	0.10050739	0.10273318
33	0.09556173	0.09777625	0.10000441	0.10224565
34	0.09507660	0.09730631	0.09954945	0.10180543
35	0.09463584	0.09688010	0.09913756	0.10140762
36	0.09423505	0.09649323	0.09876437	0.10104784
37	0.09387033	0.09614182	0.09842600	0.10072224
38	0.09353820	0.09582240	0.09811901	0.10042739
39	0.09323555	0.09553188	0.09784032	0.10016023
40	0.09295961	0.09526750	0.09758719	0.09991804
41	0.09270789	0.09502678	0.09735716	0.09969838
42	0.09247814	0.09480750	0.09714803	0.09949907
43	0.09226837	0.09460768	0.09695783	0.09931817
44	0.09207675	0.09442551	0.09678478	0.09915390
45	0.09190165	0.09425938	0.09662729	0.09900471
46	0.09174160	0.09410783	0.09648390	0.09886915
47	0.09159525	0.09396953	0.09635333	0.09874597
48	0.09146139	0.09384330	0.09623439	0.09863399
49	0.09133893	0.09372806	0.09612603	0.09853218
50	0.09122687	0.09362282	0.09602728	0.09843960

Periods n	0.09(9%)	0.0925(9 1/4%)	0.095(9 1/2%)	0.0975(9 3/4%)
51	0.09112430	0.09352670	0.09593728	0.09835540
52	0.09103041	0.09343889	0.09585523	0.09827880
53	0.09094443	0.09335865	0.09578042	0.09820911
54	0.09086570	0.09328534	0.09571220	0.09814569
55	0.09079359	0.09321833	0.09564999	0.09808799
56	0.09072754	0.09315707	0.09559325	0.09803546
57	0.09066702	0.09310108	0.09554149	0.09798766
58	0.09061157	0.09304989	0.09549426	0.09794414
59	0.09056076	0.09300307	0.09545118	0.09790452
60	0.09051419	0.09296027	0.09541187	0.09786844
61	0.09047151	0.09292112	0.09537599	0.09783560
62	0.09043240	0.09288532	0.09534325	0.09780569
63	0.09039654	0.09285257	0.09531337	0.09777846
64	0.09036366	0.09282261	0.09528611	0.09775365
65	0.09033352	0.09279521	0.09526122	0.09773107
66	0.09030589	0.09277014	0.09523850	0.09771050
67	0.09028056	0.09274721	0.09521776	0.09769176
68	0.09025732	0.09272623	0.09519883	0.09767469
69	0.09023602	0.09270703	0.09518154	0.09765915
70	0.09021649	0.09268947	0.09516576	0.09764499
71	0.09019857	0.09267339	0.09515136	0.09763209
72	0.09018214	0.09265869	0.09513821	0.09762034
73	0.09016708	0.09264523	0.09512620	0.09760964
74	0.09015326	0.09263292	0.09511524	0.09759989
75	0.09014058	0.09262165	0.09510523	0.09759101
76	0.09012896	0.09261134	0.09509609	0.09758291
77	0.09011830	0.09260190	0.09508775	0.09757554
78	0.09010852	0.09259326	0.09508013	0.09756883
79	0.09009955	0.09258536	0.09507317	0.09756271
80	0.09009132	0.09257813	0.09506682	0.09755713
81	0.09008377	0.09257151	0.09506102	0.09755206
82	0.09007685	0.09256545	0.09505572	0.09754743
83	0.09007050	0.09255990	0.09505088	0.09754321
84	0.09006467	0.09255483	0.09504647	0.09753937
85	0.09005933	0.09255018	0.09504243	0.09753587
86	0.09005443	0.09254593	0.09503875	0.09753269
87	0.09004993	0.09254204	0.09503539	0.09752978
88	0.09004581	0.09253848	0.09503232	0.09752713
89	0.09004202	0.09253522	0.09502951	0.09752472
90	0.09003855	0.09253224	0.09502695	0.09752253
91	0.09003537	0.09252951	0.09502461	0.09752053
92	0.09003245	0.09252701	0.09502248	0.09751870
93	0.09002977	0.09252472	0.09502053	0.09751704
94	0.09002731	0.09252263	0.09501874	0.09751553
95	0.09002505	0.09252071	0.09501712	0.09751415
96	0.09002298	0.09251896	0.09501563	0.09751289
97	0.09002019	0.09251735	0.09501428	0.09751174
98	0.09001934	0.09251588	0.09501304	0.09751070
99	0.09001775	0.09251454	0.09501191	0.09750975
100	0.09001628	0.09251331	0.09501087	0.09750888

ANNUITY WHOSE PRESENT VALUE IS 1

$$a^{-1}_{\overline{n}|i} = \frac{i}{1 - v^n} = s^{-1}_{\overline{n}|i} + i \text{ (Continued)}$$

Periods	Rate i				Periods	Rate i			
n	0.10(10%)	0.1025(10 1/4%)	0.105(10 1/2%)	0.1075(10 3/4%)	n	0.10(10%)	0.1025(10 1/4%)	0.105(10 1/2%)	0.1075(10 3/4%)
1	1.10000000	1.10250000	1.10500000	1.10750000	51	0.10078046	0.10321190	0.10564922	0.10809192
2	0.57619048	0.57812426	0.58005938	0.58199585	52	0.10070900	0.10314530	0.10558718	0.10803418
3	0.40211480	0.40388581	0.40565920	0.40743495	53	0.10064413	0.10308496	0.10553110	0.10798210
4	0.31547080	0.31717972	0.31889196	0.32060751	54	0.10058523	0.10303030	0.10548041	0.10793511
5	0.26379748	0.26548438	0.26717550	0.26887081	55	0.10053175	0.10298077	0.10543457	0.10789272
6	0.22960738	0.23129209	0.23298187	0.23467671	56	0.10048317	0.10293588	0.10539312	0.10785448
7	0.20540550	0.20709914	0.20879867	0.21050405	57	0.10043906	0.10289520	0.10535564	0.10781997
8	0.18744402	0.18915331	0.19086928	0.19259187	58	0.10039898	0.10285833	0.10532174	0.10778883
9	0.17364054	0.17536976	0.17710638	0.17885037	59	0.10036258	0.10282491	0.10529108	0.10776072
10	0.16274539	0.16449730	0.16625732	0.16802538	60	0.10032951	0.10279461	0.10526335	0.10773536
11	0.15396314	0.15573954	0.15752470	0.15931855	61	0.10029946	0.10276715	0.10523827	0.10771247
12	0.14676332	0.14856535	0.15037675	0.15219742	62	0.10027217	0.10274226	0.10521559	0.10769181
13	0.14077852	0.14260686	0.14444512	0.14629320	63	0.10024736	0.10271969	0.10519506	0.10767316
14	0.13574622	0.13760119	0.13946659	0.14134230	64	0.10022482	0.10269922	0.10517650	0.10765633
15	0.13147378	0.13335544	0.13524800	0.13715131	65	0.10020434	0.10268067	0.10515970	0.10764114
16	0.12781662	0.12972486	0.13164440	0.13357508	66	0.10018573	0.10266384	0.10514450	0.10762742
17	0.12466413	0.12659866	0.12854485	0.13050252	67	0.10016882	0.10264859	0.10513075	0.10761504
18	0.12193022	0.12389064	0.12586302	0.12784716	68	0.10015345	0.10263476	0.10511832	0.10760386
19	0.11954687	0.12153267	0.12353069	0.12554071	69	0.10013948	0.10262221	0.10510706	0.10759377
20	0.11745962	0.11947023	0.12149327	0.12352848	70	0.10012678	0.10261084	0.10509688	0.10758466
21	0.11562439	0.11765916	0.11970652	0.12176620	71	0.10011524	0.10260052	0.10508767	0.10757644
22	0.11400506	0.11606331	0.11813426	0.12021762	72	0.10010476	0.10259117	0.10507933	0.10756901
23	0.11257181	0.11465282	0.11674659	0.11885282	73	0.10009522	0.10258269	0.10507179	0.10756231
24	0.11129978	0.11340278	0.11551858	0.11764684	74	0.10008656	0.10257499	0.10506496	0.10755626
25	0.11016807	0.11229231	0.11442932	0.11657875	75	0.10007868	0.10256802	0.10505878	0.10755080
26	0.10915904	0.11130372	0.11346112	0.11563086	76	0.10007153	0.10256169	0.10505320	0.10754586
27	0.10825764	0.11042197	0.11259894	0.11478813	77	0.10006502	0.10255595	0.10504814	0.10754141
28	0.10745101	0.10963420	0.11182990	0.11403768	78	0.10005911	0.10255075	0.10504356	0.10753739
29	0.10672807	0.10892933	0.11114293	0.11336846	79	0.10005373	0.10254603	0.10503942	0.10753376
30	0.10607925	0.10829778	0.11052848	0.11277090	80	0.10004884	0.10254175	0.10503567	0.10753048
31	0.10549621	0.10773125	0.10997824	0.11223674	81	0.10004440	0.10253786	0.10503228	0.10752752
32	0.10497172	0.10722249	0.10948499	0.11175875	82	0.10004036	0.10253434	0.10502921	0.10752485
33	0.10449941	0.10676517	0.10904241	0.11133064	83	0.10003669	0.10253115	0.10502644	0.10752244
34	0.10407371	0.10635372	0.10864495	0.11094690	84	0.10003335	0.10252825	0.10502393	0.10752026
35	0.10368971	0.10598326	0.10828776	0.11060266	85	0.10003032	0.10252562	0.10502165	0.10751829
36	0.10334306	0.10564947	0.10796652	0.11029368	86	0.10002756	0.10252324	0.10501959	0.10751652
37	0.10302994	0.10534853	0.10767744	0.11001616	87	0.10002506	0.10252108	0.10501773	0.10751491
38	0.10274692	0.10507704	0.10741717	0.10976678	88	0.10002278	0.10251912	0.10501605	0.10751347
39	0.10249098	0.10483200	0.10718271	0.10954257	89	0.10002071	0.10251734	0.10501452	0.10751216
40	0.10225941	0.10461073	0.10697141	0.10934091	90	0.10001883	0.10251573	0.10501314	0.10751098
41	0.10204980	0.10441083	0.10678090	0.10915947	91	0.10001711	0.10251427	0.10501189	0.10760991
42	0.10185999	0.10423018	0.10660908	0.10899615	92	0.10001556	0.10251294	0.10501076	0.10750895
43	0.10168805	0.10406687	0.10645407	0.10884910	93	0.10001414	0.10251174	0.10500974	0.10750808
44	0.10153224	0.10391918	0.10631417	0.10871667	94	0.10001286	0.10251065	0.10500881	0.10750730
45	0.10139100	0.10378558	0.10618788	0.10859737	95	0.10001169	0.10250966	0.10500798	0.01750659
46	0.10126295	0.10366470	0.10607385	0.10848987	96	0.10001063	0.10250876	0.10500722	0.10750595
47	0.10114682	0.10355531	0.10597087	0.10839299	97	0.10000966	0.10250794	0.10500653	0.10750537
48	0.10104148	0.10345628	0.10587784	0.10830566	98	0.10000878	0.10250721	0.10500591	0.10750485
49	0.10094590	0.10336662	0.10579380	0.10822693	99	0.10000798	0.10250654	0.10500535	0.10750438
50	0.10085917	0.10328543	0.10571785	0.10815594	100	0.10000726	0.10250593	0.10500484	0.10750395

ANNUITY WHOSE PRESENT VALUE IS 1

$$a^{-1}_{\overline{n}|i} = \frac{i}{1 - v^n} = s^{-1}_{\overline{n}|i} + i \text{ (Continued)}$$

Periods	Rate *i*			
n	0.11(11%)	0.1125(11 1/4%)	0.115(11 1/2%)	0.1175(11 3/4%)
1	1.11000000	1.11250000	1.11500000	1.11750000
2	0.58393365	0.58587278	0.58781324	0.58975502
3	0.40921307	0.41099354	0.41277636	0.41456150
4	0.32232635	0.32404848	0.32577388	0.32750254
5	0.27057031	0.27227397	0.27398177	0.27569370
6	0.23637656	0.23808142	0.23979125	0.24150602
7	0.21221527	0.21393228	0.21565505	0.21738354
8	0.19432105	0.19605678	0.19779902	0.19954772
9	0.18060166	0.18236022	0.18412597	0.18589888
10	0.16980143	0.17158539	0.17337721	0.17517682
11	0.16112101	0.16293200	0.16475144	0.16657927
12	0.15402729	0.15586625	0.15771422	0.15957110
13	0.14815099	0.15001840	0.15189530	0.15378160
14	0.14322820	0.14512417	0.14703008	0.14894582
15	0.13906524	0.14098964	0.14292436	0.14486928
16	0.13551675	0.13746923	0.13943238	0.14140603
17	0.13247148	0.13445157	0.13644259	0.13844437
18	0.12984287	0.13184994	0.13386817	0.13589736
19	0.12756250	0.12959585	0.13164053	0.13369632
20	0.12557564	0.12763448	0.12970478	0.13178629
21	0.12383793	0.12592145	0.12801648	0.13012278
22	0.12231310	0.12442041	0.12653927	0.12866939
23	0.12097118	0.12310138	0.12524311	0.12739607
24	0.11978721	0.12193938	0.12410302	0.12627781
25	0.11874024	0.12091345	0.12309803	0.12529365
26	0.11781258	0.12000589	0.12221044	0.12442588
27	0.11698916	0.11920165	0.12142521	0.12365948
28	0.11625715	0.11848788	0.12072951	0.12298163
29	0.11560547	0.11785355	0.12011230	0.12238131
30	0.11502460	0.11728914	0.11956410	0.12184907
31	0.11450627	0.11678639	0.11907667	0.12137670
32	0.11404329	0.11633815	0.11864289	0.12095709
33	0.11362938	0.11593816	0.11825653	0.12058405
34	0.11325905	0.11558096	0.11791215	0.12025218
35	0.11292749	0.11526175	0.11760499	0.11995675
36	0.11263044	0.11497633	0.11733086	0.11969361
37	0.11236416	0.11472097	0.11708610	0.11945912
38	0.11212535	0.11449240	0.11686745	0.11925006
39	0.11191107	0.11428872	0.11667204	0.11906360
40	0.11171873	0.11428772	0.11649734	0.11889724
41	0.11154601	0.11410436	0.11634111	0.11874876
42	0.11139086	0.11379275	0.11620134	0.11861621
43	0.11125146	0.11366067	0.11607628	0.11849785
44	0.11112617	0.11354221	0.11596434	0.11839214
45	0.11101354	0.11343594	0.11586413	0.11829770
46	0.11091227	0.11334059	0.11577441	0.11821331
47	0.11082119	0.11325502	0.11569405	0.11813790
48	0.11073926	0.11317821	0.11562208	0.11807051
49	0.11066556	0.11310925	0.11555761	0.11801026

Periods	Rate *i*			
n	0.11(11%)	0.1125(11 1/4%)	0.115(11 1/2%)	0.1175(11 3/4%)
51	0.11053957	0.11299175	0.11544809	0.11790824
52	0.11048586	0.11294183	0.11540172	0.11786519
53	0.11043752	0.11289699	0.11536015	0.11782668
54	0.11039401	0.11285672	0.11532290	0.11779225
55	0.11035484	0.11282054	0.11528952	0.11776145
56	0.11031957	0.11278805	0.11525959	0.11773391
57	0.11028782	0.11275885	0.11523276	0.11770927
58	0.11025923	0.11273262	0.11520871	0.11768723
59	0.11023348	0.11270905	0.11518715	0.11766751
60	0.11021030	0.11268788	0.11516782	0.11764988
61	0.11018943	0.11266885	0.11515049	0.11763410
62	0.11017062	0.11265175	0.11513495	0.11761999
63	0.11015369	0.11263639	0.11512101	0.11760736
64	0.11013844	0.11262258	0.11510852	0.11759606
65	0.11012471	0.11261017	0.11509732	0.11758595
66	0.11011234	0.11259902	0.11508727	0.11757691
67	0.11010119	0.11258900	0.11507827	0.11756882
68	0.11009116	0.11257999	0.11507019	0.11756158
69	0.11008212	0.11257190	0.11506295	0.11755510
70	0.11007397	0.11256462	0.11505645	0.11754931
71	0.11006664	0.11255809	0.11505063	0.11754412
72	0.11006003	0.11255221	0.11504540	0.11753948
73	0.11005408	0.11254693	0.11504072	0.11753533
74	0.11004872	0.11254218	0.11503652	0.11753161
75	0.11004389	0.11253791	0.11503275	0.11752829
76	0.11003954	0.11253408	0.11502937	0.11752531
77	0.11003562	0.11253063	0.11502634	0.11752265
78	0.11003209	0.11252753	0.11502362	0.11752027
79	0.11002891	0.11252475	0.11502119	0.11751814
80	0.11002604	0.11252225	0.11501900	0.11751623
81	0.11002346	0.11252000	0.11501704	0.11751452
82	0.11002113	0.11251797	0.11501528	0.11751300
83	0.11001904	0.11251616	0.11501371	0.11751163
84	0.11001715	0.11251452	0.11501229	0.11751041
85	0.11001545	0.11251305	0.11501102	0.11750931
86	0.11001392	0.11251173	0.11500989	0.11750833
87	0.11001254	0.11251055	0.11500887	0.11750746
88	0.11001130	0.11250948	0.11500795	0.11750667
89	0.11001018	0.11250852	0.11500713	0.11750597
90	0.11000917	0.11250766	0.11500640	0.11750534
91	0.11000826	0.11250688	0.11500574	0.11750478
92	0.11000744	0.11250619	0.11500515	0.11750428
93	0.11000670	0.11250556	0.11500461	0.11750383
94	0.11000604	0.11250500	0.11500414	0.11750343
95	0.11000544	0.11250449	0.11500371	0.11750307
96	0.11000490	0.11250404	0.11500333	0.11750274
97	0.11000442	0.11250363	0.11500299	0.11750245
98	0.11000398	0.11250326	0.11500268	0.11750220
99	0.11000358	0.11250293	0.11500240	0.11750197

ANNUITY WHOSE PRESENT VALUE IS 1

$$a_{\overline{n}|i}^{-1} = \frac{i}{1 - v^n} = s_{\overline{n}|i}^{-1} + i \text{ (Continued)}$$

Periods	Rate i				Periods	Rate i			
n	0.12(12%)	0.1225(12 1/4%)	0.125(12 1/2%)	0.1275(12 3/4%)	n	0.12(12%)	0.1225(12 1/4%)	0.125(12 1/2%)	0.1275(12 3/4%)
1	1.12000000	1.12250000	1.12500000	1.12750000	51	0.12037188	0.12283871	0.12530846	0.12778088
2	0.59169811	0.59364252	0.59558824	0.59753525	52	0.12033193	0.12280166	0.12527411	0.12774906
3	0.41634898	0.41813877	0.41993088	0.42172528	53	0.12029628	0.12276866	0.12524360	0.12772084
4	0.32923444	0.33096957	0.33270791	0.33444946	54	0.12026446	0.12273929	0.12521648	0.12769583
5	0.27740973	0.27912985	0.28085404	0.28258228	55	0.12023607	0.12271313	0.12519239	0.12767366
6	0.24322572	0.24495031	0.24667978	0.24841410	56	0.12021073	0.12268983	0.12517099	0.12765400
7	0.21911774	0.22085759	0.22260308	0.22435416	57	0.12018812	0.12266909	0.12515197	0.12763656
8	0.20130284	0.20306434	0.20483219	0.20660632	58	0.12016794	0.12265061	0.12513506	0.12762111
9	0.18767889	0.18946595	0.19126000	0.19306101	59	0.12014992	0.12263416	0.12512004	0.12760740
10	0.17698416	0.17879917	0.18062178	0.18245193	60	0.12013384	0.12261950	0.12510669	0.12759524
11	0.16841540	0.17025977	0.17211228	0.17397287	61	0.12011949	0.12260645	0.12509483	0.12758447
12	0.16143681	0.16331125	0.16519434	0.16708598	62	0.12010667	0.12259482	0.12508428	0.12757491
13	0.15567720	0.15758197	0.15949582	0.16141865	63	0.12009523	0.12258447	0.12507491	0.12756643
14	0.15087125	0.15280625	0.15475071	0.15670450	64	0.12008502	0.12257524	0.12506659	0.12755892
15	0.14682424	0.14878911	0.15076375	0.15274802	65	0.12007591	0.12256703	0.12505918	0.12755225
16	0.14339002	0.14538419	0.14738839	0.14940246	66	0.12006777	0.12255971	0.12505260	0.12754634
17	0.14045673	0.14247949	0.14451248	0.14655552	67	0.12006051	0.12255319	0.12504676	0.12754110
18	0.13793731	0.13998783	0.14204873	0.14411980	68	0.12005402	0.12254738	0.12504156	0.12753645
19	0.13576300	0.13784037	0.13992820	0.14202627	69	0.12004823	0.12254221	0.12503694	0.12753233
20	0.13387878	0.13598200	0.13809573	0.14021974	70	0.12004306	0.12253760	0.12503284	0.12752867
21	0.13224009	0.13436815	0.13650671	0.13865551	71	0.12003845	0.12253350	0.12502919	0.12752543
22	0.13081051	0.13296234	0.13512463	0.13729709	72	0.12003432	0.12252984	0.12502594	0.12752255
23	0.12955996	0.13173450	0.13391940	0.13611437	73	0.12003065	0.12252658	0.12502306	0.12752000
24	0.12846344	0.13065960	0.13286599	0.13508230	74	0.12002736	0.12252368	0.12502050	0.12751774
25	0.12749997	0.12971667	0.13194344	0.13417996	75	0.12002443	0.12252110	0.12501822	0.12751573
26	0.12665186	0.12888804	0.13113409	0.13338968	76	0.12002181	0.12251879	0.12501619	0.12751395
27	0.12590409	0.12815870	0.13042295	0.13269652	77	0.12001947	0.12251674	0.12501440	0.12751238
28	0.12524387	0.12751587	0.12979728	0.13208774	78	0.12001739	0.12251492	0.12501280	0.12751098
29	0.12466021	0.12694860	0.12924614	0.13155246	79	0.12001552	0.12251329	0.12501137	0.12750973
30	0.12414366	0.12644748	0.12876016	0.13108133	80	0.12001386	0.12251184	0.12501011	0.12750863
31	0.12368606	0.12600436	0.12833123	0.13066629	81	0.12001238	0.12251055	0.12500899	0.12750766
32	0.12328033	0.12561221	0.12795235	0.13030037	82	0.12001105	0.12250939	0.12500799	0.12750679
33	0.12292031	0.12526490	0.12761744	0.12997755	83	0.12000987	0.12250837	0.12500710	0.12750602
34	0.12260064	0.12495711	0.12732121	0.12969256	84	0.12000881	0.12250746	0.12500631	0.12750534
35	0.12231662	0.12468418	0.12705905	0.12944085	85	0.12000786	0.12250664	0.12500561	0.12750474
36	0.12206414	0.12444204	0.12682692	0.12921842	86	0.12000702	0.12250592	0.12500499	0.12750420
37	0.12183959	0.12422712	0.12662130	0.12902178	87	0.12000627	0.12250527	0.12500443	0.12750373
38	0.12163980	0.12403627	0.12643908	0.12884787	88	0.12000560	0.12250470	0.12500394	0.12750331
39	0.12146197	0.12386674	0.12627755	0.12869402	89	0.12000500	0.12250418	0.12500350	0.12750293
40	0.12130363	0.12371611	0.12613431	0.12855788	90	0.12000446	0.12250373	0.12500311	0.12750260
41	0.12116260	0.12358222	0.12600726	0.12843737	91	0.12000398	0.12250332	0.12500277	0.12750231
42	0.12103696	0.12346319	0.12589454	0.12833068	92	0.12000356	0.12250296	0.12500246	0.12750205
43	0.12092500	0.12335734	0.12579452	0.12823621	93	0.12000318	0.12250264	0.12500219	0.12750181
44	0.12082521	0.12326319	0.12570574	0.12815253	94	0.12000284	0.12250235	0.12500194	0.12750161
45	0.12073625	0.12317944	0.12562693	0.12807840	95	0.12000253	0.12250209	0.12500173	0.12750143
46	0.12065694	0.12310493	0.12555696	0.12801273	96	0.12000226	0.12250186	0.12500154	0.12750127
47	0.12058621	0.12303862	0.12549483	0.12795455	97	0.12000202	0.12250166	0.12500136	0.12750112
48	0.12052312	0.12297961	0.12543966	0.12790298	98	0.12000180	0.12250148	0.12500121	0.12750100
49	0.12046686	0.12292709	0.12539065	0.12785728	99	0.12000161	0.12250132	0.12500108	0.12750088
50	0.12041666	0.12288033	0.12534713	0.12781678	100	0.12000144	0.12250117	0.12500096	0.12750078

ANNUITY WHOSE PRESENT VALUE IS 1

$$a^{-1}_{\overline{n}|i} = \frac{i}{1 - v^n} = s^{-1}_{\overline{n}|i} + i \text{ (Continued)}$$

Periods	Rate i				Periods	Rate i			
n	0.12(13%)	0.1325(13 1/4%)	0.135(13 1/2%)	0.1375(13 3/4%)	n	0.13(13%)	0.1325(13 1/4%)	0.135(13 1/2%)	0.1375(13 3/4%)
1	1.13000000	1.13250000	1.13500000	1.13750000	51	0.13025574	0.13273282	0.13521195	0.13769293
2	0.59948357	0.60143318	0.60338407	0.60533626	52	0.13022627	0.13270554	0.13518670	0.13766958
3	0.42352197	0.42532095	0.42712219	0.42892571	53	0.13020020	0.13268146	0.13516447	0.13764906
4	0.33619420	0.33794211	0.33969319	0.34144742	54	0.13017713	0.13266021	0.13514488	0.13763102
5	0.28431454	0.28605082	0.28779110	0.28953534	55	0.13015673	0.13264144	0.13512764	0.13761517
6	0.25015323	0.25189717	0.25364587	0.25539932	56	0.13013868	0.13262488	0.13511244	0.13760124
7	0.22611080	0.22787298	0.22964066	0.23141380	57	0.13012271	0.13261026	0.13509906	0.13758899
8	0.20838672	0.21017333	0.21196611	0.21376502	58	0.13010858	0.13259735	0.13508727	0.13757823
9	0.19486890	0.19668364	0.19850517	0.20033343	59	0.13009608	0.13258595	0.13507688	0.13756877
10	0.18428956	0.18613459	0.18798698	0.18984665	60	0.13008502	0.13257589	0.13506773	0.13756045
11	0.17584145	0.17771797	0.17960232	0.18149445	61	0.13007523	0.13256700	0.13505967	0.13755314
12	0.16898608	0.17089456	0.17281132	0.17473627	62	0.13006657	0.13255916	9.13505257	0.13754672
13	0.16335034	0.16529079	0.16723990	0.16919756	63	0.13005891	0.13255224	0.13504632	0.13754107
14	0.15866750	0.16063958	0.16262063	0.16461052	64	0.13005213	0.13254612	0.13504081	0.13753610
15	0.15474178	0.15674489	0.15875722	0.16077862	65	0.13004613	0.13254073	0.13503595	0.13753174
16	0.15142624	0.15345958	0.15550232	0.15755431	66	0.13004082	0.13253596	0.13503167	0.13752790
17	0.14860844	0.15067106	0.15274321	0.15482473	67	0.13003613	0.13253175	0.13502791	0.13752453
18	0.14620085	0.14829171	0.15039216	0.15250204	68	0.13003197	0.13252804	0.13502459	0.13752156
19	0.14413439	0.14625235	0.14837993	0.15051693	69	0.13002829	0.13252475	0.13502166	0.13751895
20	0.14235379	0.14449766	0.14665113	0.14881399	70	0.13002503	0.13252186	0.13501908	0.13751666
21	0.14081433	0.14298291	0.14516101	0.14734841	71	0.13002215	0.13251930	0.13501681	0.13751465
22	0.13947948	0.14167153	0.14387300	0.14608362	72	0.13001960	0.13251704	0.13501481	0.13751288
23	0.13831913	0.14053343	0.14275698	0.14498953	73	0.13001735	0.13251505	0.13501305	0.13751132
24	0.13730826	0.13954357	0.14178795	0.14404113	74	0.13001535	0.13251329	0.13501150	0.13750995
25	0.13642593	0.13868104	0.14094502	0.14321757	75	0.13001359	0.13251173	0.13501013	0.13750875
26	0.13565451	0.13792825	0.14021061	0.14250129	76	0.13001202	0.13251036	0.13500893	0.13750769
27	0.13497907	0.13727029	0.13956987	0.14187749	77	0.13001064	0.13250915	0.13500786	0.13750676
28	0.13438693	0.13669451	0.13901017	0.14133359	78	0.13000942	0.13250808	0.13500693	0.13750594
29	0.13386722	0.13619009	0.13852075	0.14085887	79	0.13000833	0.13250713	0.13500610	0.13750523
30	0.13341065	0.13574778	0.13809239	0.14044416	80	0.13000737	0.13250630	0.13500538	0.13750459
31	0.13300919	0.13535959	0.13771717	0.14008159	81	0.13000653	0.13250556	0.13500474	0.13750404
32	0.13265593	0.13501867	0.13738826	0.13976439	82	0.13000577	0.13250491	0.13500418	0.13750355
33	0.13234487	0.13471906	0.13709978	0.13948672	83	0.13000511	0.13250434	0.13500368	0.13750312
34	0.13207081	0.13445560	0.13684661	0.13924352	84	0.13000452	0.13250383	0.13500324	0.13750274
35	0.13182922	0.13422382	0.13662432	0.13903042	85	0.13000400	0.13250338	0.13500286	0.13750241
36	0.13161616	0.13401982	0.13642908	0.13884362	86	0.13000354	0.13250298	0.13500252	0.13750212
37	0.13142819	0.13384021	0.13625752	0.13867981	87	0.13000313	0.13250264	0.13500222	0.13750186
38	0.13126229	0.13368201	0.13610672	0.13853612	88	0.13000277	0.13250233	0.13500195	0.13750164
39	0.13111582	0.13354263	0.13597413	0.13841004	89	0.13000245	0.13250206	0.13500172	0.13750144
40	0.13098648	0.13341980	0.13585753	0.13829940	90	0.13000217	0.13250181	0.13500152	0.13750127
41	0.13087223	0.13331152	0.13575496	0.13820228	91	0.13000192	0.13250160	0.13500134	0.13750111
42	0.13077129	0.13321606	0.13566472	0.13811700	92	0.13000170	0.13250141	0.13500118	0.13750098
43	0.13068209	0.13313189	0.13558532	0.13804213	93	0.13000151	0.13250125	0.13500104	0.13750086
44	0.13060326	0.13305765	0.13551543	0.13797637	94	0.13000133	0.13250110	0.13500091	0.13750076
45	0.13053357	0.13299216	0.13545392	0.13791861	95	0.13000118	0.13250097	0.13500080	0.13750067
46	0.13047196	0.13293439	0.13539977	0.13786787	96	0.13000104	0.13250086	0.13500071	0.13750058
47	0.13041749	0.13288342	0.13535209	0.13782330	97	0.13000092	0.13250076	0.13500062	0.13750051
48	0.13036933	0.13283845	0.13531012	0.13778414	98	0.13000082	0.13250067	0.13500055	0.13750045
49	0.13032673	0.13279876	0.13527316	0.13774973	99	0.13000072	0.13250059	0.13500049	0.13750040
50	0.13028906	0.13276374	0.13524061	0.13771950	100	0.13000064	0.13250052	0.13500043	0.13750035

ANNUITY WHOSE PRESENT VALUE IS 1

$$a_{\overline{n}|i}^{-1} = \frac{i}{1 - v^n} = s_{\overline{n}|i}^{-1} + i \text{ (Continued)}$$

Periods	Rate i				Periods	Rate i			
n	0.14(14%)	0.1425(14 1/4%)	0.145(14 1/2%)	0.1475(14 3/4%)	n	0.14(14%)	0.1425(14 1/4%)	0.145(14 1/2%)	0.1475(14 3/4%)
1	1.14000000	1.14250000	1.14500000	1.14750000	51	0.14017560	0.14265982	0.14514545	0.14763237
2	0.60728972	0.60924446	0.61120047	0.61315774	52	0.14015401	0.14263987	0.14512702	0.14761534
3	0.43073148	0.43253950	0.43434976	0.43616225	53	0.14013508	0.14262241	0.14511092	0.14760051
4	0.34320478	0.34496527	0.34672887	0.34849557	54	0.14011848	0.14260713	0.14509686	0.14758758
5	0.29128355	0.29303569	0.29479175	0.29655171	55	0.14010392	0.14259376	0.14508459	0.14757632
6	0.25715750	0.25892037	0.26068791	0.26246010	56	0.14009115	0.14258206	0.14507387	0.14756650
7	0.23319238	0.23497635	0.23676570	0.23856038	57	0.14007995	0.14257182	0.14506451	0.14755795
8	0.21557002	0.21738107	0.21919813	0.22102115	58	0.14007012	0.14256286	0.14505634	0.14755050
9	0.20216838	0.20400997	0.20585814	0.20771284	59	0.14006151	0.14255501	0.14504920	0.14754401
10	0.19171354	0.19358759	0.19546874	0.19735692	60	0.14005395	0.14254815	0.14504297	0.14753835
11	0.18339427	0.18530171	0.18721669	0.18913914	61	0.14004732	0.14254214	0.14503753	0.14753342
12	0.17666933	0.17861039	0.18055938	0.18251621	62	0.14004151	0.14253688	0.14503277	0.14752912
13	0.17116366	0.17313811	0.17512079	0.17711161	63	0.14003641	0.14253228	0.14502862	0.14752538
14	0.16660914	0.16861637	0.17063208	0.17265616	64	0.14003194	0.14252826	0.14502500	0.14752211
15	0.16280896	0.16484811	0.16689593	0.16895229	65	0.14002802	0.14252473	0.14502183	0.14751927
16	0.15961540	0.16168543	0.16376426	0.16585172	66	0.14002457	0.14252165	0.14501907	0.14751679
17	0.15691544	0.15901517	0.16112376	0.16324105	67	0.14002156	0.14251895	0.14501665	0.14751464
18	0.15462115	0.15674931	0.15888634	0.16103206	68	0.14001891	0.14251658	0.14501454	0.14751275
19	0.15266316	0.15481841	0.15698249	0.15915520	69	0.14001659	0.14251451	0.14501270	0.14751111
20	0.15098600	0.15316697	0.15535667	0.15755491	70	0.14001455	0.14251270	0.14501109	0.14750969
21	0.14954486	0.15175015	0.15396405	0.15618633	71	0.14001276	0.14251112	0.14500969	0.14750844
22	0.14830317	0.15053139	0.15276805	0.15501292	72	0.14001119	0.14250973	0.14500846	0.14750736
23	0.14723081	0.14948059	0.15173860	0.15400462	73	0.14000982	0.14250852	0.14500739	0.14750641
24	0.14630284	0.14857282	0.15085081	0.15313656	74	0.14000861	0.14250746	0.14500645	0.14750559
25	0.14549841	0.14778727	0.15008390	0.15238803	75	0.14000756	0.14250653	0.14500564	0.14750487
26	0.14480001	0.14710649	0.14942046	0.15174165	76	0.14000663	0.14250571	0.14500492	0.14750424
27	0.14419288	0.14651575	0.14884582	0.15118281	77	0.14000581	0.14250500	0.14500430	0.14750370
28	0.14366449	0.14600257	0.14834755	0.15069916	78	0.14000510	0.14250438	0.14500375	0.14750322
29	0.14320417	0.14555634	0.14791510	0.15028018	79	0.14000447	0.14250383	0.14500328	0.14750281
30	0.14280279	0.14516799	0.14753947	0.14991696	80	0.14000392	0.14250335	0.14500286	0.14750245
31	0.14245256	0.14482978	0.14721297	0.14960186	81	0.14000344	0.14250293	0.14500250	0.14750213
32	0.14214675	0.14453505	0.14692900	0.14932834	82	0.14000302	0.14250257	0.14500218	0.14750186
33	0.14187958	0.14427806	0.14668188	0.14909079	83	0.14000265	0.14250225	0.14500191	0.14750162
34	0.14164604	0.14405387	0.14646674	0.14888439	84	0.14000232	0.14250197	0.14500167	0.14750141
35	0.14144181	0.14385821	0.14627935	0.14870498	85	0.14000204	0.14250172	0.14500146	0.14750123
36	0.14126315	0.14368739	0.14611609	0.14854899	86	0.14000179	0.14250151	0.14500127	0.14750107
37	0.14110680	0.14353822	0.14597380	0.14841332	87	0.14000157	0.14250132	0.14500111	0.14750093
38	0.14096993	0.14340790	0.14584976	0.14829529	88	0.14000138	0.14250115	0.14500097	0.14750081
39	0.14085010	0.14329403	0.14574160	0.14819258	89	0.14000121	0.14250101	0.14500085	0.14750071
40	0.14074514	0.14319451	0.14564727	0.14810319	90	0.14000106	0.14250088	0.14500074	0.14750062
41	0.14065321	0.14310752	0.14556498	0.14802538	91	0.14000093	0.14250077	0.14500065	0.14750054
42	0.14057266	0.14303146	0.14549319	0.14795764	92	0.14000081	0.14250068	0.14500056	0.14750047
43	0.14050208	0.14296496	0.14543055	0.14789866	93	0.14000071	0.14250059	0.14500049	0.14750041
44	0.14044023	0.14290680	0.14537588	0.14784729	94	0.14000063	0.14250052	0.14500043	0.14750036
45	0.14038602	0.14285593	0.14532817	0.14780256	95	0.14000055	0.14250045	0.14500038	0.14750031
46	0.14033850	0.14281144	0.14528653	0.14776360	96	0.14000048	0.14250040	0.14500033	0.14750027
47	0.14029684	0.14277252	0.14525018	0.14772966	97	0.14000042	0.14250035	0.14500029	0.14750024
48	0.14026032	0.14273848	0.14521845	0.14770010	98	0.14000037	0.14250030	0.14500025	0.14750021
49	0.14022830	0.14270869	0.14519075	0.14767435	99	0.14000033	0.14250027	0.14500022	0.14750018
50	0.14020022	0.14268263	0.14516657	0.14765192	100	0.14000029	0.14250023	0.14500019	0.14750016

ANNUITY WHOSE PRESENT VALUE IS 1

$$a_{\overline{n}|i}^{-1} = \frac{i}{1 - v^n} = s_{\overline{n}|i}^{-1} + i \text{ (Continued)}$$

Periods	Rate i			
n	0.15(15%)	0.1525(15 1/4%)	0.155(15 1/2%)	0.1575(15 3/4%)
1	1.15000000	1.15250000	1.15500000	1.15750000
2	0.61511628	0.61707607	0.61903712	0.62099942
3	0.43797696	0.43979389	0.44161302	0.44343435
4	0.35026535	0.35203821	0.35381412	0.35559308
5	0.29831555	0.30008326	0.30185481	0.30363019
6	0.26423691	0.26601831	0.26780429	0.26959482
7	0.24036036	0.24216562	0.24397611	0.24579181
8	0.22285009	0.22468491	0.22652558	0.22837204
9	0.20957402	0.21144162	0.21331560	0.21519591
10	0.19925206	0.20115412	0.20306301	0.20497869
11	0.19106898	0.19300614	0.19495054	0.19690210
12	0.18448078	0.18645300	0.18843279	0.19042007
13	0.17911046	0.18111723	0.18313183	0.18515416
14	0.17468849	0.17672895	0.17877743	0.18083381
15	0.17101705	0.17309009	0.17517126	0.17726045
16	0.16794769	0.17005201	0.17216453	0.17428511
17	0.16536686	0.16750105	0.16964345	0.17179390
18	0.16318629	0.16534885	0.16751958	0.16969831
19	0.16133635	0.16352576	0.16572323	0.16792859
20	0.15976147	0.16197616	0.16419878	0.16642913
21	0.15841679	0.16065521	0.16290138	0.16515510
22	0.15726577	0.15952639	0.16179454	0.16407003
23	0.15627839	0.15855970	0.16084832	0.16314401
24	0.15542983	0.15773038	0.16003797	0.16235238
25	0.15469940	0.15701778	0.15934293	0.16167462
26	0.15406981	0.15640468	0.15874603	0.16109362
27	0.15352648	0.15587657	0.15823283	0.16059503
28	0.15305713	0.15542122	0.15779118	0.16016676
29	0.15265133	0.15502827	0.15741078	0.15979860
30	0.15230020	0.15468893	0.15708290	0.15948190
31	0.15199618	0.15439569	0.15680013	0.15920930
32	0.15173280	0.15414215	0.15655613	0.15897454
33	0.15150452	0.15392283	0.15634549	0.15877228
34	0.15130675	0.15373304	0.15616357	0.15859795
35	0.15113485	0.15356874	0.15600641	0.15844766
36	0.15098586	0.15342646	0.15587059	0.15831804
37	0.15085653	0.15330322	0.15575319	0.15820623
38	0.15074426	0.15319646	0.15565169	0.15810976
39	0.15064676	0.15310393	0.15556391	0.15802652
40	0.15056209	0.15302375	0.15548800	0.15795467
41	0.15048853	0.15295424	0.15542233	0.15789265
42	0.15042463	0.15289398	0.15536552	0.15783910
43	0.15036911	0.15284173	0.15531637	0.15779288
44	0.15032086	0.15279642	0.15527384	0.15775296
45	0.15027893	0.15275713	0.15523703	0.15771849
46	0.15024249	0.15272306	0.15520518	0.15768873
47	0.15021082	0.15269351	0.15517761	0.15766302
48	0.15018328	0.15266787	0.15515375	0.15764082
49	0.15015935	0.15264564	0.15513310	0.15762164
[illegible]	[illegible]	[illegible]	[illegible]	[illegible]

Periods	Rate i			
n	0.15(15%)	0.1525(15 1/4%)	0.155(15 1/2%)	0.1575(15 3/4%)
51	0.15012046	0.15260962	0.15509975	0.15759077
52	0.15010474	0.15259511	0.15508636	0.15757842
53	0.15009107	0.15258252	0.15507476	0.15756774
54	0.15007918	0.15257159	0.15506473	0.15755852
55	0.15006885	0.15256211	0.15505604	0.15755056
56	0.15005987	0.15255389	0.15504852	0.15754367
57	0.15005206	0.15254676	0.15504200	0.15753773
58	0.15004526	0.15254057	0.15503636	0.15753260
59	0.15003936	0.15253520	0.15503148	0.15752816
60	0.15003422	0.15253054	0.15502726	0.15752433
61	0.15002976	0.15252650	0.15502360	0.15752102
62	0.15002588	0.15252299	0.15502043	0.15751816
63	0.15002250	0.15251995	0.15501779	0.15751569
64	0.15001957	0.15251731	0.15501532	0.15751355
65	0.15001701	0.15251502	0.15501326	0.15751171
66	0.15001479	0.15251303	0.15501148	0.15751011
67	0.15001286	0.15251131	0.15500994	0.15750874
68	0.15001119	0.15250981	0.15500861	0.15750755
69	0.15000973	0.15250851	0.15500745	0.15750652
70	0.15000846	0.15250739	0.15500645	0.15750563
71	0.15000735	0.15250641	0.15500559	0.15750487
72	0.15000640	0.15250556	0.15500484	0.15750421
73	0.15000556	0.15250482	0.15500419	0.15750363
74	0.15000484	0.15250419	0.15500362	0.15750314
75	0.15000420	0.15250363	0.15500314	0.15750271
76	0.15000366	0.15250315	0.15500272	0.15750234
77	0.15000318	0.15250273	0.15500235	0.15750202
78	0.15000276	0.15250237	0.15500204	0.15750175
79	0.15000240	0.15250206	0.15500176	0.15750151
80	0.15000209	0.15250179	0.15500153	0.15750130
81	0.15000182	0.15250155	0.15500132	0.15750113
82	0.15000158	0.15250134	0.15500114	0.15750097
83	0.15000137	0.15250117	0.15500099	0.15750084
84	0.15000120	0.15250101	0.15500086	0.15750073
85	0.15000104	0.15250088	0.15500074	0.15750063
86	0.15000090	0.15250076	0.15500074	0.15750054
87	0.15000079	0.15250066	0.15500056	0.15750047
88	0.15000068	0.15250057	0.15500048	0.15750040
89	0.15000059	0.15250050	0.15500042	0.15750035
90	0.15000052	0.15250043	0.15500036	0.15750030
91	0.15000045	0.15250037	0.15500031	0.15750026
92	0.15000039	0.15250033	0.15500027	0.15750023
93	0.15000034	0.15250028	0.15500023	0.15750019
94	0.15000030	0.15250024	0.15500020	0.15750017
95	0.15000026	0.15250021	0.15500018	0.15750015
96	0.15000022	0.15250018	0.15500015	0.15750013
97	0.15000019	0.15250016	0.15500013	0.15750011
98	0.15000017	0.15250014	0.15500011	0.15750009
99	0.15000015	0.15250012	0.15500010	0.15750008
100	0.15000013	0.15250010	0.15500009	0.15750007

ANNUITY WHOSE PRESENT VALUE IS 1

$$a_{\overline{n}|i}^{-1} = \frac{i}{1 - v^n} = s_{\overline{n}|i}^{-1} + i \text{ (Continued)}$$

Periods n	Rate i: 0.16(16%)	0.1625(16 1/4%)	0.165(16 1/2%)	0.1675(16 3/4%)	Periods n	Rate i: 0.16(16%)	0.1625(16 1/4%)	0.165(16 1/2%)	0.1675(16 3/4%)
1	1.16000000	1.16250000	1.16500000	1.16750000	51	0.16008260	0.16257516	0.16506840	0.16756224
2	0.62296296	0.62492775	0.62689376	0.62886101	52	0.16007120	0.16256465	0.16505871	0.16755331
3	0.44525787	0.44708357	0.44891145	0.45074149	53	0.16006138	0.16255561	0.16505039	0.16754566
4	0.35737507	0.35916008	0.36094810	0.36273912	54	0.16005291	0.16254784	0.16504325	0.16753911
5	0.30540938	0.30719236	0.30897911	0.31076962	55	0.16004561	0.16254115	0.16503712	0.16753349
6	0.27138987	0.27318942	0.27499344	0.27680191	56	0.16003932	0.16253540	0.16503187	0.16752869
7	0.24761268	0.24943869	0.25126981	0.25310600	57	0.16003389	0.16253045	0.16502735	0.16752457
8	0.23022426	0.23208220	0.23394581	0.23581505	58	0.16002922	0.16252619	0.16502348	0.16752105
9	0.21708249	0.21897529	0.22087426	0.22277935	59	0.16002519	0.16252253	0.16502015	0.16751803
10	0.20690108	0.20883013	0.21076578	0.21270795	60	0.16002171	0.16251938	0.16501730	0.16751544
11	0.19886075	0.20082642	0.20279903	0.20477852	61	0.16001872	0.16251667	0.16501485	0.16751322
12	0.19241473	0.19441670	0.19642589	0.19844221	62	0.16001613	0.16251434	0.16501274	0.16751133
13	0.18718411	0.18922158	0.19126648	0.19331870	63	0.16001391	0.16251233	0.16501094	0.16750970
14	0.18289797	0.18496981	0.18704920	0.18913604	64	0.16001199	0.16251061	0.16500939	0.16750831
15	0.17935752	0.18146235	0.18357480	0.18569476	65	0.16001034	0.16250913	0.16500806	0.16750712
16	0.17641362	0.17854990	0.18069381	0.18284523	66	0.16000891	0.16250785	0.16500692	0.16750610
17	0.17395225	0.17611834	0.17829203	0.18047315	67	0.16000768	0.16250675	0.16500594	0.16750522
18	0.17188485	0.17407906	0.17628076	0.17848979	68	0.16000662	0.16250581	0.16500510	0.16750447
19	0.17014166	0.17236225	0.17459019	0.17682532	69	0.16000571	0.16250500	0.16500438	0.16750383
20	0.16866703	0.17091229	0.17316471	0.17542413	70	0.16000492	0.16250430	0.16500376	0.16750328
21	0.16741617	0.16968439	0.17195956	0.17424151	71	0.16000424	0.16250370	0.16500322	0.16750281
22	0.16635264	0.16864216	0.17093840	0.17324116	72	0.16000366	0.16250318	0.16500277	0.16750241
23	0.16544658	0.16775581	0.17007149	0.17239342	73	0.16000315	0.16250274	0.16500238	0.16750206
24	0.16467339	0.16700078	0.16933434	0.17167387	74	0.16000272	0.16250235	0.16500204	0.16750177
25	0.16401262	0.16635670	0.16870667	0.17106231	75	0.16000234	0.16250202	0.16500175	0.16750151
26	0.16344723	0.16580662	0.16817160	0.17054195	76	0.16000202	0.16250174	0.16500150	0.16750130
27	0.16296294	0.16533634	0.16771501	0.17009875	77	0.16000174	0.16250150	0.16500129	0.16750111
28	0.16254775	0.16493392	0.16732506	0.16972097	78	0.16000150	0.16250129	0.16500111	0.16750095
29	0.16219153	0.16458932	0.16699179	0.16939872	79	0.16000129	0.16250111	0.16500095	0.16750081
30	0.16188568	0.16429404	0.16670677	0.16912367	80	0.16000112	0.16250095	0.16500082	0.16750070
31	0.16162295	0.16404089	0.16646290	0.16888879	81	0.16000096	0.16250082	0.16500070	0.16750060
32	0.16139714	0.16382374	0.16625413	0.16868813	82	0.16000083	0.16250071	0.16500060	0.16750051
33	0.16120298	0.16363740	0.16607535	0.16851664	83	0.16000071	0.16250061	0.16500052	0.16750044
34	0.16103598	0.16347746	0.16592220	0.16837002	84	0.16000062	0.16250052	0.16500044	0.16750038
35	0.16089229	0.16334012	0.16579096	0.16824465	85	0.16000053	0.16250045	0.16500038	0.16750032
36	0.16076862	0.16322216	0.16567847	0.16813741	86	0.16000046	0.16250039	0.16500033	0.16750028
37	0.16066217	0.16312083	0.16558204	0.16804566	87	0.16000039	0.16250033	0.16500028	0.16750024
38	0.16057051	0.16303376	0.16549936	0.16796716	88	0.16000034	0.16250029	0.16500024	0.16750020
39	0.16049158	0.16295894	0.16542845	0.16789997	89	0.16000029	0.16250025	0.16500021	0.16750017
40	0.16042359	0.16289463	0.16536763	0.16584247	90	0.16000025	0.16250021	0.16500018	0.16750015
41	0.16036503	0.16283935	0.16531546	0.16779325	91	0.16000022	0.16250018	0.16500015	0.16750013
42	0.16031458	0.16279183	0.16527071	0.16775112	92	0.16000019	0.16250016	0.16500013	0.16750011
43	0.16027112	0.16275097	0.16523232	0.16771504	93	0.16000016	0.16250013	0.16500011	0.16750009
44	0.16023367	0.16271584	0.16519937	0.16768416	94	0.16000014	0.16250012	0.16500010	0.16750008
45	0.16020140	0.16268564	0.16517111	0.16765771	95	0.16000012	0.16250010	0.16500008	0.16750007
46	0.16017359	0.16265966	0.16514685	0.16763507	96	0.16000010	0.16250009	0.16500007	0.16750006
47	0.16014962	0.16263733	0.16512604	0.16761568	97	0.16000009	0.16250007	0.16500006	0.16750005
48	0.16012897	0.16261812	0.16510817	0.16759907	98	0.16000008	0.16250006	0.16500005	0.16750004
49	0.16011117	0.16260159	0.16509284	0.16758485	99	0.16000007	0.16250005	0.16500004	0.16750004
50	0.16009583	0.16258739	0.16507969	0.16757267	100	0.16000006	0.16250005	0.16500004	0.16750003

ANNUITY WHOSE PRESENT VALUE IS 1

$$a_{\overline{n}|i}^{-1} = \frac{i}{1 - v^n} = s_{\overline{n}|i}^{-1} + i \text{ (Continued)}$$

Periods	Rate i			
n	0.17(17%)	0.1725(17 1/4%)	0.175(17 1/2%)	0.1775(17 3/4%)
1	0.17000000	1.17250000	1.17500000	1.17750000
2	0.63082949	0.63279919	0.63477011	0.63674225
3	0.45257368	0.45440802	0.45624451	0.45808312
4	0.36453311	0.36633008	0.36813000	0.36993287
5	0.31256386	0.31436183	0.31616349	0.31796883
6	0.27861480	0.28043210	0.28225376	0.28407978
7	0.25494724	0.25679349	0.25864473	0.26050090
8	0.23768989	0.23957028	0.24145618	0.24334756
9	0.22469051	0.22660769	0.22853083	0.23045990
10	0.21465660	0.21661165	0.21857305	0.22054073
11	0.20676479	0.20875779	0.21075744	0.21276366
12	0.20046558	0.20249591	0.20453311	0.20657711
13	0.19537814	0.19744471	0.19951831	0.20159884
14	0.19123022	0.19333162	0.19544013	0.19755564
15	0.18782209	0.18995669	0.19209841	0.19424715
16	0.18500401	0.18717002	0.18934312	0.19152319
17	0.18266157	0.18485713	0.18705970	0.18926913
18	0.18070600	0.18292922	0.18515930	0.18739610
19	0.17906745	0.18131643	0.18357210	0.18583429
20	0.17769036	0.17996323	0.18224257	0.18452821
21	0.17653004	0.17882497	0.18112613	0.18343334
22	0.17555025	0.17786548	0.18018668	0.18251366
23	0.17472141	0.17705526	0.17939480	0.18173984
24	0.17401917	0.17637005	0.17872632	0.18108779
25	0.17342343	0.17578982	0.17816131	0.18053770
26	0.17291747	0.17529797	0.17768326	0.18007315
27	0.17248736	0.17488065	0.17727842	0.17968050
28	0.17212144	0.17452629	0.17693533	0.17934838
29	0.17180992	0.17422520	0.17664438	0.17906729
30	0.17154455	0.17396922	0.17639751	0.17882926
31	0.17131839	0.17375150	0.17618796	0.17862761
32	0.17112556	0.17356624	0.17601001	0.17845672
33	0.17096109	0.17340854	0.17585884	0.17831184
34	0.17082077	0.17327428	0.17573040	0.17818898
35	0.17070102	0.17315993	0.17562123	0.17808478
36	0.17059880	0.17306252	0.17552843	0.17799638
37	0.17051154	0.17297953	0.17544952	0.17792138
38	0.17043702	0.17290882	0.17538243	0.17785773
39	0.17037338	0.17284855	0.17532536	0.17780371
40	0.17031903	0.17279718	0.17527683	0.17775786
41	0.17027260	0.17275340	0.17523554	0.17771895
42	0.17023294	0.17271607	0.17520042	0.17768591
43	0.17019905	0.17268425	0.17517054	0.17765786
44	0.17017010	0.17265712	0.17514512	0.17763404
45	0.17014536	0.17263398	0.17512349	0.17761382
46	0.17012423	0.17261426	0.17510509	0.17759666
47	0.17010617	0.17259744	0.17508943	0.17758208
48	0.17009073	0.17258310	0.17507610	0.17756970
49	0.17007754	0.17257087	0.17506477	0.17755919

Periods	Rate i			
n	0.17(17%)	0.1725(17 1/4%)	0.175(17 1/2%)	0.1775(17 3/4%)
51	0.17005664	0.17255154	0.17504691	0.17754269
52	0.17004841	0.17254396	0.17503992	0.17753625
53	0.17004137	0.17253749	0.17503397	0.17753079
54	0.17003536	0.17253197	0.17502891	0.17752614
55	0.17003022	0.17252727	0.17502460	0.17752220
56	0.17002583	0.17252326	0.17502094	0.17751886
57	0.17002208	0.17251983	0.17501782	0.17751601
58	0.17001887	0.17251692	0.17501517	0.17751360
59	0.17001613	0.17251443	0.17501291	0.17751155
60	0.17001378	0.17251230	0.17501098	0.17750981
61	0.17001178	0.17251049	0.17500935	0.17750833
62	0.17001007	0.17250895	0.17500796	0.17750707
63	0.17000861	0.17250763	0.17500677	0.17750601
64	0.17000735	0.17250651	0.17500576	0.17750510
65	0.17000629	0.17250555	0.17500490	0.17750433
66	0.17000537	0.17250474	0.17500417	0.17750368
67	0.17000459	0.17250404	0.17500355	0.17750312
68	0.17000392	0.17250344	0.17500302	0.17750265
69	0.17000335	0.17250294	0.17500257	0.17750255
70	0.17000287	0.17250251	0.17500219	0.17750191
71	0.17000245	0.17250214	0.17500186	0.17750163
72	0.17000209	0.17250182	0.17500159	0.17750138
73	0.17000179	0.17250155	0.17500135	0.17750117
74	0.17000153	0.17250133	0.17500115	0.17750100
75	0.17000131	0.17250113	0.17500098	0.17750085
76	0.17000112	0.17250096	0.17500083	0.17750072
77	0.17000096	0.17250082	0.17500071	0.17750061
78	0.17000082	0.17250070	0.17500060	0.17750052
79	0.17000070	0.17250060	0.17500051	0.17750044
80	0.17000060	0.17250051	0.17500044	0.17750037
81	0.17000051	0.17250044	0.17500037	0.17750032
82	0.17000044	0.17250037	0.17500032	0.17750027
83	0.17000037	0.17250032	0.17500027	0.17750023
84	0.17000032	0.17250027	0.17500023	0.17750019
85	0.17000027	0.17250023	0.17500019	0.17750017
86	0.17000023	0.17250020	0.17500017	0.17750014
87	0.17000020	0.17250017	0.17500014	0.17750012
88	0.17000017	0.17250014	0.17500012	0.17750010
89	0.17000015	0.17250012	0.17500010	0.17750009
90	0.17000012	0.17250010	0.17500009	0.17750007
91	0.17000011	0.17250009	0.17500007	0.17750006
92	0.17000009	0.17250008	0.17500006	0.17750005
93	0.17000008	0.17250006	0.17500005	0.17750004
94	0.17000007	0.17250005	0.17500005	0.17750004
95	0.17000006	0.17250005	0.17500004	0.17750003
96	0.17000005	0.17250004	0.17500003	0.17750003
97	0.17000004	0.17250003	0.17500003	0.17750002
98	0.17000004	0.17250003	0.17500002	0.17750002
99	0.17000003	0.17250002	0.17500002	0.17750002

ANNUITY WHOSE PRESENT VALUE IS 1

$$a_{\overline{n}|i}^{-1} = \frac{i}{1 - v^n} = s_{\overline{n}|i}^{-1} + i \text{ (Continued)}$$

Periods	Rate i				Periods	Rate i			
n	0.18(18%)	0.1825(18 1/4%)	0.185(18 1/2%)	0.1875(18 3/4%)	n	0.18(18%)	0.1825(18 1/4%)	0.185(18 1/2%)	0.1875(18 3/4%)
1	1.18000000	1.18250000	1.18500000	1.18750000	51	0.18003885	0.18253536	0.18503218	0.18752929
2	0.63871560	0.64069015	0.64266590	0.64464286	52	0.18003292	0.18252990	0.18502716	0.18752467
3	0.45992386	0.46176672	0.46361168	0.46545874	53	0.18002790	0.18252528	0.18502292	0.18752077
4	0.37173867	0.37354739	0.37535902	0.37717354	54	0.18002364	0.18252138	0.18501934	0.18751749
5	0.31977784	0.32159050	0.32340678	0.32522668	55	0.18002004	0.18251808	0.18501632	0.18751473
6	0.28591013	0.28774478	0.28958370	0.29142688	56	0.18001698	0.18251529	0.18501377	0.18751240
7	0.26236200	0.26422798	0.26609881	0.26797446	57	0.18001439	0.18251293	0.18501162	0.18751045
8	0.24524436	0.24714655	0.24905409	0.25096694	58	0.18001219	0.18251094	0.18500981	0.18750880
9	0.23239482	0.23433557	0.23628208	0.23823431	59	0.18001033	0.18250925	0.18500828	0.18750741
10	0.22251464	0.22449471	0.22648089	0.22847311	60	0.18000876	0.18250782	0.18500698	0.18750624
11	0.21477639	0.21679555	0.21882106	0.22085287	61	0.18000742	0.18250661	0.18500589	0.18750525
12	0.20862781	0.21068513	0.21274900	0.21481932	62	0.18000629	0.18250559	0.18500497	0.18750442
13	0.20368621	0.20578032	0.20788107	0.20998838	63	0.18000533	0.18250473	0.18500420	0.18750372
14	0.19967806	0.20180726	0.20394315	0.20608562	64	0.18000452	0.18250400	0.18500354	0.18750314
15	0.19640278	0.19856519	0.20073426	0.20290987	65	0.18000383	0.18250338	0.18500299	0.18750264
16	0.19371008	0.19590369	0.19810386	0.20031049	66	0.18000324	0.18250286	0.18500252	0.18750222
17	0.19148527	0.19370800	0.19593717	0.19817265	67	0.18000275	0.18250242	0.18500213	0.18750187
18	0.18963946	0.19188923	0.19414527	0.19640744	68	0.18000233	0.18250205	0.18500180	0.18750158
19	0.18810284	0.19037760	0.19265842	0.19494515	69	0.18000197	0.18250173	0.18500152	0.18750133
20	0.18681998	0.18911773	0.19142130	0.19373054	70	0.18000167	0.18250146	0.18500128	0.18750112
21	0.18574643	0.18806525	0.19038962	0.19271939	71	0.18000142	0.18250124	0.18500108	0.18750094
22	0.18484626	0.18718429	0.18952761	0.19187604	72	0.18000120	0.18250105	0.18500091	0.18750079
23	0.18409020	0.18644572	0.18880622	0.19117156	73	0.18000102	0.18250088	0.18500077	0.18750067
24	0.18345430	0.18582566	0.18820172	0.19058231	74	0.18000086	0.18250075	0.18500065	0.18750056
25	0.18291883	0.18530451	0.18769459	0.19008891	75	0.18000073	0.18250063	0.18500055	0.18750047
26	0.18246748	0.18486607	0.18726876	0.18967539	76	0.18000062	0.18250054	0.18500046	0.18750040
27	0.18208672	0.18449691	0.18691091	0.18932856	77	0.18000053	0.18250045	0.18500039	0.18750034
28	0.18176528	0.18418587	0.18660998	0.18903747	78	0.18000045	0.18250038	0.18500033	0.18750028
29	0.18149377	0.18392365	0.18635679	0.18879304	79	0.18000038	0.18250032	0.18500028	0.18750024
30	0.18126431	0.18370249	0.18614366	0.18858769	80	0.18000032	0.18250027	0.18500023	0.18750020
31	0.18107030	0.18351587	0.18596419	0.18841511	81	0.18000027	0.18250023	0.18500020	0.18750017
32	0.18090621	0.18335835	0.18581300	0.18827003	82	0.18000023	0.18250020	0.18500017	0.18750014
33	0.18076739	0.18322535	0.18568560	0.18814802	83	0.18000019	0.18250017	0.18500014	0.18750012
34	0.18064990	0.18311303	0.18557823	0.18804541	84	0.18000016	0.18250014	0.18500012	0.18750010
35	0.18055046	0.18301815	0.18548772	0.18795908	85	0.18000014	0.18250012	0.18500010	0.18750008
36	0.18046628	0.18293799	0.18541141	0.18788644	86	0.18000012	0.18250010	0.18500008	0.18750007
37	0.18039499	0.18287025	0.18534706	0.18782532	87	0.18000010	0.18250008	0.18500007	0.18750006
38	0.18033463	0.18281301	0.18529279	0.18777388	88	0.18000009	0.18250007	0.18500006	0.18750005
39	0.18028350	0.18276464	0.18524702	0.18773058	89	0.18000007	0.18250006	0.18500005	0.18750004
40	0.18024020	0.18272374	0.18520841	0.18769414	90	0.18000006	0.18250005	0.18500004	0.18750004
41	0.18020352	0.18268918	0.18517585	0.18766346	91	0.18000005	0.18250004	0.18500004	0.18750003
42	0.18017244	0.18265995	0.18514837	0.18763763	92	0.18000004	0.18250004	0.18500003	0.18750003
43	0.18014612	0.18263525	0.18512519	0.18761588	93	0.18000004	0.18250003	0.18500003	0.18750002
44	0.18012381	0.18261436	0.18510564	0.18759758	94	0.18000003	0.18250003	0.18500002	0.18750002
45	0.18010491	0.18259670	0.18508914	0.18758216	95	0.18000003	0.18250002	0.18500002	0.18750002
46	0.18008890	0.18258177	0.18507521	0.18756919	96	0.18000002	0.18250002	0.18500002	0.18750001
47	0.18007534	0.18256915	0.18506347	0.18755826	97	0.18000002	0.18250002	0.18500001	0.18750001
48	0.18006384	0.18225847	0.18505356	0.18754906	98	0.18000002	0.18250001	0.18500001	0.18750001
49	0.18005410	0.18254945	0.18504519	0.18754131	99	0.18000001	0.18250001	0.18500001	0.18750001
50	0.18004584	0.18254181	0.18503814	0.18753479	100	0.18000001	0.18250001	0.18500001	0.18750001

ANNUITY WHOSE PRESENT VALUE IS 1

$$a_{\overline{n}|i}^{-1} = \frac{i}{1 - v^n} = s_{\overline{n}|i}^{-1} + i \text{ (Continued)}$$

Periods	Rate i				
n	0.19(19%)	0.1925(19 1/4%)	0.195(19 1/2%)	0.1975(19 3/4%)	0.20(20%)
1	1.19000000	1.19250000	1.19500000	1.19750000	1.20000000
2	0.64662100	0.64860034	0.65058087	0.65256257	0.65454545
3	0.46730789	0.46915913	0.47101245	0.47286783	0.47472527
4	0.37899094	0.38081121	0.38263434	0.38446031	0.38628912
5	0.32705017	0.32887723	0.33070785	0.33254202	0.33437970
6	0.29327429	0.29512591	0.29698170	0.29884166	0.30070575
7	0.26985490	0.27174010	0.27363002	0.27552464	0.27742393
8	0.25288506	0.25480841	9.25673694	0.25867063	0.26060942
9	0.24019220	0.24215571	0.24412479	0.24609939	0.24807946
10	0.23047131	0.23247543	0.23448542	0.23650122	0.23852276
11	0.22289090	0.22493508	0.22698533	0.22904159	0.23110379
12	0.21689602	0.21897902	0.22106822	0.22316357	0.22526496
13	0.21210215	0.21422229	0.21634870	0.21848130	0.22062000
14	0.20823456	0.21038988	0.21255147	0.21471923	0.21689306
15	0.20509191	0.20728026	0.20947482	0.21167548	0.21388212
16	0.20252345	0.20474261	0.20696786	0.20919907	0.21143614
17	0.20041431	0.20266201	0.20491564	0.20717506	0.20944015
18	0.19867559	0.20094960	0.20322932	0.20551463	0.20780539
19	0.19723765	0.19953577	0.20183936	0.20414830	0.20646245
20	0.19604529	0.19836541	0.20069075	0.20302117	0.20535653
21	0.19505440	0.19739451	0.19973956	0.20208942	0.20444394
22	0.19422943	0.19658763	0.19895049	0.20131787	0.20368962
23	0.19354156	0.19591607	0.19829496	0.20067807	0.20306526
24	0.19296727	0.19535645	0.19774971	0.20014689	0.20254787
25	0.19248730	0.19488962	0.19729573	0.19970547	0.20211873
26	0.19208581	0.19449987	0.19691743	0.19933835	0.20176250
27	0.19174971	0.19417423	0.19660197	0.19903280	0.20146659
28	0.19146819	0.19390200	0.19633877	0.19877836	0.20122067
29	0.19123225	0.19367430	0.19611905	0.19856639	0.20101619
30	0.19103443	0.19348377	0.19593557	0.19838972	0.20084611
31	0.19086852	0.19332428	0.19578229	0.19824243	0.20070459
32	0.19072931	0.19319075	0.19565421	0.19811960	0.20058682
33	0.19061249	0.19307891	0.19554716	0.19801714	0.20048877
34	0.19051444	0.19298522	0.19545766	0.19793167	0.20040715
35	0.19043211	0.19290673	0.19538283	0.19786034	0.20033917
36	0.19036299	0.19284096	0.19532026	0.19780082	0.20028256
37	0.19030494	0.19278583	0.19526793	0.19775115	0.20023542
38	0.19025619	0.19273964	0.19522416	0.19770968	0.20019614
39	0.19021524	0.19270091	0.19518754	0.19767507	0.20016342
40	0.19018084	0.19266845	0.19515692	0.19764617	0.20013617
41	0.19015194	0.19264124	0.19513129	0.19762205	0.20011346
42	0.19012767	0.19261843	0.19510986	0.19760191	0.20009454
43	0.19010727	0.19259930	0.19509192	0.19758510	0.20007878
44	0.19009013	0.19258326	0.19507692	0.19757106	0.20006564
45	0.19007574	0.19256982	0.19506436	0.19755933	0.20005470
46	0.19006364	0.19255854	0.19505386	0.19754955	0.20004558
47	0.19005348	0.19254909	0.19504507	0.19754137	0.20003798
48	0.19004494	0.19254116	0.19503771	0.19753455	0.20003165
49	0.19003776	0.19253452	0.19503156	0.19752885	0.20002638
50	0.19003173	0.19252895	0.19502641	0.19752409	0.20002198

Periods	Rate i				
n	0.19(19%)	0.1925(19 1/4%)	0.195(19 1/2%)	0.1975(19 3/4%)	0.20(20%)
51	0.19002666	0.19252427	0.19502210	0.19752012	0.20001832
52	0.19002241	0.19252035	0.19501849	0.19751680	0.20001526
53	0.19001883	0.19251707	0.19501547	0.19751403	0.20001272
54	0.19001582	0.19251431	0.19501295	0.19751171	0.20001060
55	0.19001330	0.19251200	0.19501083	0.19750978	0.20000883
56	0.19001117	0.19251006	0.19500907	0.19750817	0.20000736
57	0.19000939	0.19250844	0.19500759	0.19750682	0.20000613
58	0.19000789	0.19250708	0.19500635	0.19750570	0.20000511
59	0.19000663	0.19250593	0.19500531	0.19750476	0.20000426
60	0.19000557	0.19250498	0.19500445	0.19750397	0.20000355
61	0.19000468	0.19250417	0.19500372	0.19750332	0.20000296
62	0.19000393	0.19250350	0.19500311	0.19750277	0.20000246
63	0.19000331	0.19250293	0.19500261	0.19750231	0.20000205
64	0.19000278	0.19250246	0.19500218	0.19750193	0.20000171
65	0.19000233	0.19250206	0.19500182	0.19750161	0.20000143
66	0.19000196	0.19250173	0.19500153	0.19750135	0.20000119
67	0.19000165	0.19250145	0.19500128	0.19750112	0.20000099
68	0.19000139	0.19250122	0.19500107	0.19750094	0.20000083
69	0.19000116	0.19250102	0.19500089	0.19750078	0.20000069
70	0.19000098	0.19250086	0.19500075	0.19750066	0.20000057
71	0.19000082	0.19250072	0.19500063	0.19750055	0.20000048
72	0.19000069	0.19250060	0.19500052	0.19750046	0.20000040
73	0.19000058	0.19250050	0.19500044	0.19750038	0.20000033
74	0.19000049	0.19250042	0.19500037	0.19750032	0.20000028
75	0.19000041	0.19250035	0.19500031	0.19750027	0.20000023
76	0.19000034	0.19250030	0.19500026	0.19750022	0.20000019
77	0.19000029	0.19250025	0.19500022	0.19750019	0.20000016
78	0.19000024	0.19250021	0.19500018	0.19750015	0.20000013
79	0.19000020	0.19250018	0.19500015	0.19750013	0.20000011
80	0.19000017	0.19250015	0.19500013	0.19750011	0.20000009
81	0.19000014	0.19250012	0.19500011	0.19750009	0.20000008
82	0.19000012	0.19250010	0.19500009	0.19750008	0.20000006
83	0.19000010	0.19250009	0.19500007	0.19750006	0.20000005
84	0.19000009	0.19250007	0.19500006	0.19750005	0.20000004
85	0.19000007	0.19250006	0.19500005	0.19750004	0.20000004
86	0.19000006	0.19250005	0.19500004	0.19750004	0.20000003
87	0.19000005	0.19250004	0.19500004	0.19750003	0.20000003
88	0.19000004	0.19250004	0.19500003	0.19750003	0.20000002
89	0.19000004	0.19250003	0.19500003	0.19750002	0.20000002
90	0.19000003	0.19250003	0.19500002	0.19750002	0.20000001
91	0.19000003	0.19250002	0.19500002	0.19750001	0.20000001
92	0.19000002	0.19250002	0.19500001	0.19750001	0.20000001
93	0.19000002	0.19250001	0.19500001	0.19750001	0.20000001
94	0.19000002	0.19250001	0.19500001	0.19750001	0.20000001
95	0.19000001	0.19250001	0.19500001	0.19750001	0.20000001
96	0.19000001	0.19250001	0.19500001	0.19750001	0.20000001
97	0.19000001	0.19250001	0.19500001	0.19750001	0.20000000
98	0.19000001	0.19250001	0.19500001	0.19750000	0.20000000
99	0.19000001	0.19250001	0.19500000	0.19750000	0.20000000
100	0.19000001	0.19250000	0.19500000	0.19750000	0.20000000

COMPOUND AMOUNT OF 1 FOR FRACTIONAL PERIODS, $(1 + i)^{1/p}$

p	i = ¼%	5/12%	1/2%	7/12%	3/4%
2	1.0012492	1.0020812	1.0024969	1.0029124	1.0037430
3	1.0008326	1.0013870	1.0016639	1.0019407	1.0024938
4	1.0006244	1.0010400	1.0012477	1.0014552	1.0018697
6	1.0004162	1.0006932	1.0008316	1.0009699	1.0012461
12	1.0002089	1.0003466	1.0004157	1.0004848	1.0006229
13	1.0001921	1.0003199	1.0003837	1.0004475	1.0005749
26	1.0000960	1.0001599	1.0001919	1.0002237	1.0002874
52	1.0000480	1.0000800	1.0000959	1.0001119	1.0001437
365	1.0000068	1.0000114	1.0000137	1.0000159	1.0000205

p	1%	1 1/8%	1 1/4%	1 1/2%	1 3/4%
2	1.0049876	1.0056093	1.0062306	1.0074721	1.0087121
3	1.0033223	1.0037360	1.0041494	1.0049752	1.0057996
4	1.0024907	1.0028008	1.0031105	1.0037291	1.0043466
6	1.0016598	1.0018663	1.0020726	1.0024845	1.0028956
12	1.0008295	1.0009327	1.0010357	1.0012415	1.0014468
13	1.0007657	1.0008609	1.0009560	1.0011459	1.0013354
26	1.0003828	1.0004304	1.0004779	1.0005728	1.0006675
52	1.0001914	1.0002152	1.0002389	1.0002864	1.0003337
365	1.0000273	1.0000307	1.0000340	1.0000408	1.0000475

p	2%	2 1/4%	2 1/2%	2 3/4%	3%
2	1.0099505	1.0111874	1.0124228	1.0136568	1.01488916
3	1.0066227	1.0074444	1.0082648	1.0090839	1.00990163
4	1.0049629	1.0055782	1.0061922	1.0068052	1.00741707
6	1.0033059	1.0037153	1.0041239	1.0045317	1.00493862
12	1.0016516	1.0018559	1.0020598	1.0022633	1.00246627
13	1.0015244	1.0017130	1.0019012	1.0020890	1.00227634
26	1.0007619	1.0008562	1.0009502	1.0010440	1.00113752
52	1.0003809	1.0004280	1.0004750	1.0005218	1.00056860
365	1.0000543	1.0000610	1.0000676	1.0000743	1.00008099

p	3 1/4%	3 1/2%	3 3/4%	4%	4 1/4%
2	1.01612007	1.01734950	1.01857744	1.01980390	1.02102889
3	1.01071805	1.01153314	1.01234693	1.01315940	1.01397058
4	1.00802781	1.00863745	1.00924598	1.00985341	1.01045974
6	1.00534474	1.00575004	1.00615452	1.00655820	1.00696106
12	1.00266881	1.00287090	1.00307254	1.00327374	1.00347450
13	1.00246326	1.00264977	1.00283586	1.00302153	1.00320680
26	1.00123087	1.00132401	1.00141692	1.00150963	1.00160212
52	1.00061525	1.00066178	1.00070821	1.00075453	1.00080074
365	1.00008763	1.00009425	1.00010087	1.00010746	1.00011404

p	4 1/2%	4 3/4%	5%	5 1/4%	5 1/2%
2	1.02225242	1.02347447	1.02469508	1.02591423	1.02713193
3	1.01478046	1.01558905	1.01639636	1.01720238	1.01800713
4	1.01106499	1.01166915	1.01227223	1.01287424	1.01347517
6	1.00736312	1.00776438	1.00816485	1.00856452	1.00896339
12	1.00367481	1.00387468	1.00407412	1.00427313	1.00447170
13	1.00339165	1.00357610	1.00376014	1.00394378	1.00412701
26	1.00169439	1.00178645	1.00187831	1.00196995	1.00206138
52	1.00084684	1.00089283	1.00093871	1.00098449	1.00103016
365	1.00012060	1.00012715	1.00013368	1.00014020	1.00014670

p	5 3/4%	6%	6 1/4%	6 1/2%	6 3/4%
2	1.02834819	1.02956301	1.03077641	1.03198837	1.03319892
3	1.01881061	1.01961282	1.02041378	1.02121347	1.02201192
4	1.01407504	1.01467385	1.01527159	1.01586828	1.01646393
6	1.00936149	1.00975879	1.01015532	1.01055107	1.01094605
12	1.00466984	1.00486755	1.00506483	1.00526169	1.00545813
13	1.00430985	1.00449228	1.00467432	1.00485597	1.00503722
26	1.00215261	1.00224363	1.00233444	1.00242504	1.00251545
52	1.00107572	1.00112118	1.00116654	1.00121179	1.00125693
365	1.00015318	1.00015965	1.00016611	1.00017255	1.00017897

p	7%	7 1/4%	7 1/2%	7 3/4%	8%
2	1.03440804	1.03561576	1.03682207	1.03802697	1.03923048
3	1.02280912	1.02360508	1.02439981	1.02519330	1.02598557
4	1.01705853	1.01765208	1.01824460	1.01883609	1.01942655
6	1.01134026	1.01173370	1.01212638	1.01251830	1.01290946
12	1.00565415	1.00584974	1.00604492	1.00623968	1.00643403
13	1.00521808	1.00539855	1.00557863	1.00575833	1.00593764
26	1.00260564	1.00269564	1.00278544	1.00287503	1.00296443
52	1.00130197	1.00134691	1.00139175	1.00143648	1.00148112
365	1.00018538	1.00019178	1.00019816	1.00020452	1.00021087

p	8 1/4%	8 1/2%	8 3/4%	9%	9 1/4%
2	1.04043260	1.04163333	1.04283268	1.04403065	1.04522725
3	1.02677661	1.02756644	1.02835506	1.02914247	1.02992867
4	1.02001598	1.02060440	1.02119179	1.02177818	1.02236356
6	1.01329986	1.01368952	1.01407843	1.01446659	1.01485402
12	1.00662797	1.00682149	1.00701461	1.00720732	1.00739963
13	1.00611657	1.00629512	1.00647328	1.00665107	1.00682849
26	1.00305362	1.00314262	1.00323142	1.00332003	1.00340844
52	1.00152565	1.00157008	1.00161441	1.00165864	1.00170277
365	1.00021721	1.00022353	1.00022984	1.00023613	1.00024241

COMPOUND AMOUNT OF 1 FOR FRACTIONAL PERIODS $(1 + i)^{1/p}$ (Continued)

p	9 1/2%	9 3/4%	10%	10 1/4%	10 1/2%
2	1.04642248	1.04761634	1.04880885	1.05000000	1.05118980
3	1.03071368	1.03149749	1.03228012	1.03306155	1.03384181
4	1.02294793	1.02353131	1.02411369	1.02469508	1.02527548
6	1.01524070	1.01562665	1.01601187	1.01639636	1.01678012
12	1.00759153	1.00778304	1.00797414	1.00816485	1.00835516
13	1.00700553	1.00718220	1.00735849	1.00753442	1.00770998
26	1.00349665	1.00358467	1.00367250	1.00376014	1.00384759
52	1.00174680	1.00179073	1.00183457	1.00187831	1.00192195
365	1.00024867	1.00025492	1.00026116	1.00026738	1.00027359

p	10 3/4%	11%	11 1/4%	11 1/2%	11 3/4%
2	1.05237826	1.05356538	1.05475116	1.05593560	1.05711873
3	1.03462089	1.03539881	1.03617555	1.03695113	1.03772555
4	1.02585489	1.02643333	1.02701079	1.02758727	1.02816279
6	1.01716316	1.01754548	1.01792708	1.01830797	1.01868815
12	1.00854507	1.00873459	1.00892373	1.00911247	1.00930082
13	1.00788517	1.00806000	1.00823447	1.00840857	1.00858231
26	1.00393485	1.00402191	1.00410879	1.00419548	1.00428199
52	1.00196549	1.00200894	1.00205229	1.00209555	1.00213871
365	1.00027978	1.00028596	1.00029212	1.00029828	1.00030441

p	12%	12 1/4%	12 1/2%	12 3/4%	13%
2	1.05830052	1.05948101	1.06066017	1.06183803	1.06301458
3	1.03849882	1.03927094	1.04004191	1.04081174	1.04158044
4	1.02873734	1.02931094	1.02988357	1.03045525	1.03102598
6	1.01906762	1.01944639	1.01982445	1.02020181	1.02057848
12	1.00948879	1.00967638	1.00986358	1.01005040	1.01023684
13	1.00875570	1.00892873	1.00910140	1.00927372	1.00944569
26	1.00436831	1.00445444	1.00454039	1.00462616	1.00471174
52	1.00218177	1.00222475	1.00226763	1.00231041	1.00235310
365	1.00031054	1.00031665	1.00032275	1.00032883	1.00033490

p	13 1/4%	13 1/2%	13 3/4%	14%	14 1/4%
2	1.06418983	1.06536379	1.06653645	1.06770783	1.06887792
3	1.04234800	1.04311443	1.04387974	1.04464393	1.04540700
4	1.03159577	1.03216461	1.03273252	1.03329948	1.03386552
6	1.02095445	1.02132974	1.02170433	1.02207824	1.02245146
12	1.01042291	1.01060860	1.01079391	1.01097885	1.01116342
13	1.00961730	1.00978857	1.00995949	1.01013006	1.01030029
26	1.00479715	1.00488237	1.00496741	1.00505227	1.00513695
52	1.00239570	1.00243821	1.00248063	1.00252295	1.00256519
365	1.00034096	1.00034700	1.00035303	1.00035905	1.00036505

p	14 1/2%	14 3/4%	15%	15 1/4%	15 1/2%
2	1.07004673	1.07121426	1.07238053	1.07354553	1.07470926
3	1.04616896	1.04692981	1.04768955	1.04844820	1.04920575
4	1.03443063	1.03499481	1.03555808	1.03612042	1.03668185
6	1.02282401	1.02319588	1.02356707	1.02393760	1.02430745
12	1.01134762	1.01153145	1.01171492	1.01189802	1.01208075
13	1.01047017	1.01063972	1.01080892	1.01097778	1.01114630
26	1.00522146	1.00530578	1.00538993	1.00547391	1.00555771
52	1.00260733	1.00264938	1.00269134	1.00273322	1.00277500
365	1.00037104	1.00037702	1.00038298	1.00038893	1.00039487

p	15 3/4%	16%	16 1/4%	16 1/2%	16 3/4%
2	1.07587174	1.07703296	1.07819293	1.07935166	1.08050914
3	1.04996221	1.05071757	1.05147186	1.05222506	1.05297719
4	1.03724237	1.03780199	1.03836069	1.03891850	1.03947542
6	1.02467663	1.02504516	1.02541302	1.02578022	1.02614677
12	1.01226313	1.01244514	1.01262679	1.01280809	1.01298903
13	1.01131449	1.01148235	1.01164986	1.01181705	1.01198391
26	1.00564133	1.00572479	1.00580807	1.00589117	1.00597411
52	1.00281670	1.00285831	1.00289983	1.00294126	1.00298261
365	1.00040080	1.00040671	1.00041261	1.00041850	1.00042438

COMPOUND AMOUNT OF 1 FOR FRACTIONAL PERIODS $(1+i)^{1/p}$ (Continued)

p	17%	17 1/4%	17 1/2%	17 3/4%	18%
2	1.08166538	1.08282039	1.08397417	1.08512672	1.08627805
3	1.05372824	1.05447823	1.05522715	1.05597501	1.05672181
4	1.04003143	1.04058656	1.04114080	1.04169416	1.04224664
6	1.02651266	1.02687790	1.02724250	1.02760644	1.02796975
12	1.01316961	1.01334984	1.01352972	1.01370925	1.01388843
13	1.01215044	1.01231663	1.01248251	1.01264805	1.01281328
26	1.00605687	1.00613947	1.00622190	1.00630416	1.00638625
52	1.00302387	1.00306504	1.00310612	1.00314713	1.00318804
365	1.00043024	1.00043609	1.00044193	1.00044775	1.00045357

p	18 1/4%	18 1/2%	18 3/4%	19%	19 1/4%
2	1.08742816	1.08857705	1.08972474	1.09087121	1.09201648
3	1.05746755	1.05821225	1.05895590	1.05969850	1.06044007
4	1.04279823	1.04334896	1.04389881	1.04444780	1.04499593
6	1.02833241	1.02869444	1.02905583	1.02941658	1.02977671
12	1.01406726	1.01424575	1.01442389	1.01460169	1.01477914
13	1.01297817	1.01314275	1.01330701	1.01347095	1.01363457
26	1.00646817	1.00654993	1.00663152	1.00671294	1.00679421
52	1.00322887	1.00326962	1.00331028	1.00335086	1.00339135
365	1.00045937	1.00046516	1.00047093	1.00047670	1.00048245

p	19 1/2%	19 3/4%	20%
2	1.09316056	1.09430343	1.09544512
3	1.06118060	1.06192010	1.06265857
4	1.04554319	1.04608959	1.04663514
6	1.03013620	1.03049507	1.03085332
12	1.01495626	1.01513303	1.01530947
13	1.01379788	1.01396087	1.01412354
26	1.00687530	1.00695624	1.00703701
52	1.00343176	1.00347209	1.00351234
365	1.00048819	1.00049392	1.00049964

NOMINAL RATES CONVERTIBLE p TIMES PER YEAR EQUIVALENT TO EFFECTIVE RATE i GIVEN IN HEADING,

$$j_p = p[(1+i)^{\frac{1}{p}} - 1]$$

p	1/4	5/12	1/2	7/12	3/4
1/4	.0025 094	.0041 928	.0050 376	.0058 846	.0075 848
1/2	.0025 032	.0041 754	.0050 125	.0058 504	.0075 281
2	.0024 984	.0041 623	.0049 938	.0058 249	.0074 860
4	.0024 977	.0041 602	.0049 907	.0058 206	.0074 790
6	.0024 974	.0041 595	.0049 896	.0058 192	.0074 767
12	.0024 971	.0041 587	.0049 886	.0058 178	.0074 743
13	.0024 971	.0041 587	.0049 885	.0058 177	.0074 742
52	.0024 969	.0041 582	.0049 878	.0058 167	.0074 725
365	.0024 969	.0041 580	.0049 876	.0058 164	.0074 721
∞	.0024 969	.0041 580	.0049 875	.0058 164	.0074 720

p	1	1 1/8	1 1/4	1 1/2	1 3/4
1/4	.0101 510	.0114 413	.0127 363	.0153 409	.0179 648
1/2	.0100 500	.0113 133	.0125 781	.0151 125	.0176 531
2	.0099 751	.0112 185	.0124 612	.0149 442	.0174 241
4	.0099 627	.0112 029	.0124 418	.0149 164	.0173 863
6	.0099 586	.0111 976	.0124 354	.0149 071	.0173 737
12	.0099 545	.0111 924	.0124 290	.0148 979	.0173 612
13	.0099 541	.0111 920	.0124 285	.0148 971	.0173 602
52	.0099 513	.0111 884	.0124 240	.0148 907	.0173 515
365	.0099 505	.0111 874	.0124 227	.0148 889	.0173 490
∞	.0099 503	.0111 872	.0124 225	.0148 886	.0173 486

p	2	2 1/4	2 1/2	2 3/4	3
1/4	.0206 080	.0232 708	.0259 532	.0286 553	.0313 772
1/2	.0202 000	.0227 531	.0253 125	.0278 781	.0304 500
2	.0199 010	.0223 748	.0248 457	.0273 135	.0297 783
4	.0198 517	.0223 126	.0247 690	.0272 209	.0296 683
6	.0198 353	.0222 919	.0247 435	.0271 901	.0296 317
12	.0198 190	.0222 713	.0247 180	.0271 594	.0295 952
13	.0198 177	.0222 697	.0247 161	.0271 570	.0295 924
52	.0198 064	.0222 554	.0246 985	.0271 358	.0295 672
365	.0198 032	.0222 513	.0246 934	.0271 297	.0295 600
∞	.0198 026	.0222 506	.0246 926	.0271 287	.0295 588

p	3 1/2	4	4 1/2	5	5 1/2
1/4	.0368 808	.0424 647	.0481 297	.0538 766	.0597 062
1/2	.0356 125	.0408 000	.0460 125	.0512 500	.0565 125
2	.0346 990	.0396 078	.0445 048	.0493 902	.0542 639
4	.0345 498	.0394 136	.0442 600	.0490 889	.0539 007
6	.0345 002	.0393 492	.0441 787	.0489 891	.0537 804
12	.0344 508	.0392 849	.0440 977	.0488 895	.0536 604
13	.0344 470	.0392 799	.0440 915	.0488 818	.0536 512
52	.0344 128	.0392 355	.0440 355	.0488 131	.0535 683
365	.0344 031	.0392 228	.0440 195	.0487 934	.0535 447
∞	.0344 014	.0392 207	.0440 169	.0487 902	.0535 408

p	6	6 1/2	7	7 1/2	8
1/4	.0656 193	.0716 166	.0776 990	.0838 673	.0901 223
1/2	.0618 000	.0671 125	.0724 500	.0778 125	.0832 000
2	.0591 260	.0639 767	.0688 161	.0736 441	.0784 610
4	.0586 954	.0634 731	.0682 341	.0729 784	.0777 062
6	.0585 528	.0633 064	.0680 416	.0727 583	.0774 567
12	.0584 106	.0631 403	.0678 497	.0725 390	.0772 084
13	.0583 997	.0631 276	.0678 350	.0725 222	.0771 893
52	.0583 016	.0630 129	.0677 027	.0723 710	.0770 180
365	.0582 736	.0629 802	.0676 649	.0723 278	.0769 692
∞	.0582 689	.0629 748	.0676 586	.0723 207	.0769 610

AMOUNT FOR YEAR OF p DEPOSITS OF $1/p, p$ TIMES PER YEAR, i/j_p

p	1/4	5/12	1/2	7/12	3/4
1/4	0.9962 5	0.9937 717	0.9925 312	0.9912 924	0.9888 201
1/2	0.9974 8	0.9979 210	0.9975 062	0.9970 918	0.9962 640
2	1.0006 246	1.0010 406	1.0012 484	1.0014 562	1.0018 715
4	1.0009 370	1.0015 611	1.0018 730	1.0021 848	1.0028 081
6	1.0010 412	1.0017 347	1.0020 813	1.0024 278	1.0031 205
12	1.0011 453	1.0019 083	1.0022 896	1.0026 708	1.0034 329
13	1.0011 533	1.0019 216	1.0023 056	1.0026 895	1.0034 569
52	1.0012 254	1.0020 418	1.0024 498	1.0028 577	1.0036 732
365	1.0012 461	1.0020 762	1.0024 911	1.0029 058	1.0037 351
∞	1.0012 495	1.0020 819	1.0024 979	1.0029 138	1.0037 453

p	1	1 1/8	1 1/4	1 1/2	1 3/4
1/4	0.9851 244	0.9832 823	0.9814 441	0.9777 791	0.9741 295
1/2	0.9950 249	0.9944 065	0.9937 888	0.9925 558	0.9913 259
2	1.0024 938	1.0028 046	1.0031 153	1.0037 360	1.0043 618
4	1.0037 422	1.0042 089	1.0046 754	1.0056 076	1.0065 388
6	1.0041 586	1.0046 773	1.0051 958	1.0062 319	1.0072 671
12	1.0045 751	1.0051 458	1.0057 163	1.0068 565	1.0079 957
13	1.0046 071	1.0051 819	1.0057 564	1.0069 046	1.0080 518
52	1.0048 956	1.0055 063	1.0061 169	1.0073 372	1.0085 564
365	1.0049 780	1.0055 991	1.0062 199	1.0074 608	1.0087 007
∞	1.0049 917	1.0056 145	1.0062 371	1.0074 814	1.0087 247

p	2	2 1/4	2 1/2	2 3/4	3
1/4	0.9704 950	0.9668 757	0.9632 715	0.9596 824	0.9561 082
1/2	0.9900 990	0.9888 752	0.9876 543	0.9864 365	0.9852 217
2	1.0049 752	1.0055 937	1.0062 114	1.0068 284	1.0074 446
4	1.0074 686	1.0083 984	1.0093 268	1.0102 542	1.0111 807
6	1.0083 013	1.0093 344	1.0103 667	1.0113 979	1.0124 282
12	1.0091 339	1.0102 711	1.0114 072	1.0125 424	1.0136 766
13	1.0091 980	1.0103 431	1.0114 873	1.0126 305	1.0137 727
52	1.0097 747	1.0109 919	1.0122 082	1.0134 234	1.0146 376
365	1.0099 396	1.0111 775	1.0124 143	1.0136 502	1.0148 850
∞	1.0099 670	1.0112 083	1.0124 486	1.0136 878	1.0149 261

p	3 1/2	4	4 1/2	5	5 1/2
1/4	0.9490 046	0.9419 6	0.9349 7	0.9280 5	0.9211 8
1/2	0.9828 010	0.9803 922	0.9779 951	0.9756 098	0.9732 360
2	1.0086 748	1.0099 020	1.0111 262	1.0123 475	1.0135 660
4	1.0130 309	1.0148 774	1.0167 203	1.0185 594	1.0203 950
6	1.0144 858	1.0165 396	1.0185 895	1.0206 357	1.0226 781
12	1.0159 420	1.0182 035	1.0204 611	1.0227 148	1.0249 647
13	1.0160 541	1.0183 316	1.0206 051	1.0228 748	1.0251 407
52	1.0170 632	1.0194 847	1.0219 623	1.0243 160	1.0267 259
365	1.0173 517	1.0198 145	1.0222 733	1.0247 282	1.0271 793
∞	1.0173 997	1.0198 693	1.0223 349	1.0247 967	1.0272 546

p	6	6 1/2	7	7 1/2	8
1/4	0.9143 7	0.9076 1	0.9009 1	0.8942 7	0.8876 8
1/2	0.9708 738	0.9685 230	0.9661 836	0.9638 554	0.9615 385
2	1.0147 815	1.0159 942	1.0172 040	1.0184 110	1.0196 152
4	1.0222 269	1.0240 552	1.0258 800	1.0277 013	1.0295 190
6	1.0247 168	1.0267 517	1.0287 830	1.0308 106	1.0328 346
12	1.0272 107	1.0294 529	1.0316 914	1.0339 262	1.0361 572
13	1.0274 027	1.0296 609	1.0319 154	1.0341 661	1.0364 131
52	1.0291 319	1.0315 340	1.0339 324	1.0363 270	1.0387 179
365	1.0296 265	1.0320 699	1.0345 095	1.0369 453	1.0393 774
∞	1.0297 087	1.0321 589	1.0346 053	1.0370 480	1.0394 870

COMMISSIONERS 1941 STANDARD ORDINARY MORTALITY TABLE*

x	l_x	d_x	p_x	$\overset{\circ}{e}_x$	x	l_x	d_x	p_x	$\overset{\circ}{e}_x$
0	1 023 102	23 102	.977 42	62.33	**50**	810 900	9 990	.987 68	21.37
1	1 000 000	5 770	.994 23	62.76	51	800 910	10 628	.986 73	20.64
2	994 230	4 116	.995 86	62.12	52	790 282	11 301	.985 70	19.91
3	990 114	3 347	.996 62	61.37	53	778 981	12 020	.984 57	19.19
4	986 767	2 950	.997 01	60.58	54	766 961	12 770	.983 35	18.48
5	983 817	2 715	.997 24	59.76	**55**	754 191	13 560	.982 02	17.78
6	981 102	2 561	.997 39	58.92	56	740 631	14 390	.980 57	17.10
7	978 541	2 417	.997 53	58.08	57	726 241	15 251	.979 00	16.43
8	976 124	2 255	.997 69	57.22	58	710 990	16 147	.977 29	15.77
9	973 869	2 065	.997 88	56.35	59	694 843	17 072	.975 43	15.13
10	971 804	1 914	.998 03	55.47	**60**	677 771	18 022	.973 41	14.50
11	969 890	1 852	.998 09	54.58	61	659 749	18 988	.971 22	13.88
12	968 038	1 859	.998 08	53.68	62	640 761	19 979	.968 82	13.27
13	966 179	1 913	.998 02	52.78	63	620 782	20 958	.966 24	12.69
14	964 266	1 996	.997 93	51.89	64	599 824	21 942	.963 42	12.11
15	962 270	2 069	.997 85	50.99	**65**	577 882	22 907	.960 36	11.55
16	960 201	2 103	.997 81	50.10	66	554 975	23 842	.957 04	11.01
17	958 098	2 156	.997 75	49.21	67	531 133	24 730	.953 44	10.48
18	955 942	2 199	.997 70	48.32	68	506 403	25 553	.949 54	9.97
19	953 743	2 260	.997 63	47.43	69	480 850	26 302	.945 30	9.47
20	951 483	2 312	.997 57	46.54	**70**	454 548	26 955	.940 70	8.99
21	949 171	2 382	.997 49	45.66	71	427 593	27 481	.935 73	8.52
22	946 789	2 452	.997 41	44.77	72	400 112	27 872	.930 34	8.08
23	944 337	2 531	.997 32	43.88	73	372 240	28 104	.924 50	7.64
24	941 806	2 609	.997 23	43.00	74	344 136	28 154	.918 19	7.23
25	939 197	2 705	.997 12	42.12	**75**	315 982	28 009	.911 36	6.82
26	936 492	2 800	.997 01	41.24	76	287 973	27 651	.903 98	6.44
27	933 692	2 904	.996 89	40.36	77	260 322	27 071	.896 01	6.07
28	930 788	3 025	.996 75	39.49	78	233 251	26 262	.887 41	5.72
29	927 763	3 154	.996 60	38.61	79	206 989	25 224	.878 14	5.38
30	924 609	3 292	.996 44	37.74	**80**	181 765	23 966	.868 15	5.06
31	921 317	3 437	.996 27	36.88	81	157 799	22 502	.857 40	4.75
32	917 880	3 598	.996 08	36.01	82	135 297	20 857	.845 84	4.46
33	914 282	3 767	.995 88	35.15	83	114 440	19 062	.833 43	4.18
34	910 515	3 961	.995 65	34.29	84	95 378	17 157	.820 12	3.91
35	906 554	4 161	.995 41	33.44	**85**	78 221	15 185	.805 87	3.66
36	902 393	4 386	.995 14	32.59	86	63 036	13 198	.790 63	3.42
37	898 007	4 625	.994 85	31.75	87	49 838	11 245	.774 37	3.19
38	893 382	4 878	.994 54	30.91	88	38 593	9 378	.757 00	2.98
39	888 504	5 162	.994 19	30.08	89	29 215	7 638	.738 56	2.77
40	883 342	5 459	.993 82	29.25	**90**	21 577	6 063	.719 01	2.58
41	877 883	5 785	.993 41	28.43	91	15 514	4 681	.698 27	2.39
42	872 098	6 131	.992 97	27 62	92	10 833	3 506	.676 36	2.21
43	865 967	6 503	.992 49	26.81	93	7 327	2 540	.653 34	2.03
44	859 464	6 910	.991 96	26.01	94	4 787	1 776	.629 00	1.84
45	852 554	7 340	.991 39	25.21	**95**	3 011	1 193	.603 79	1.63
46	845 214	7 801	.990 77	24.43	96	1 818	813	.552 81	1.37
47	837 413	8 299	.990 09	23.65	97	1 005	551	.451 74	1.08
48	829 114	8 822	.989 36	22.88	98	454	329	.275 33	.78
49	820 292	9 392	.988 55	22.12	99	125	125	.000 00	.50

* Reproduced by permission of the Actuarial Society of America.

COMMUTATION COLUMNS 2 1/2%

Commissioners 1941 Standard Ordinary Mortality Table*

x	D_x	N_x	C_x	M_x	$1 + a_x$	A_x
0	1 023 102.00	31 374 230	22 538.536 6	257 876.88	30.665 8	0.252 054
1	975 609.76	30 351 128	5 491.969 1	235 338.35	31.109 9	.241 222
2	946 322.43	29 375 518	3 822.115 2	229 846.38	31.041 8	.242 884
3	919 419.28	28 429 196	3 032.216 8	226 024.26	30.920 8	.245 834
4	893 962.20	27 509 776	2 607.370 2	222 992.05	30.772 9	.249 442
5	869 550.88	26 615 814	2 341.136 0	220 384.68	30.608 7	.253 447
6	846 001.18	25 746 263	2 154.480 3	218 043.54	30.432 9	.257 734
7	823 212.53	24 900 262	1 983.744 5	215 889.06	30.247 7	.262 252
8	801 150.42	24 077 050	1 805.642 5	213 905.32	30.053 1	.266 998
9	779 804.53	23 275 899	1 613.174 7	212 099.67	29.848 4	.271 991
10	759 171.73	22 496 095	1 458.745 1	210 486.50	29.632 4	.277 258
11	739 196.60	21 736 923	1 377.065 5	209 027.75	29.406 1	.282 777
12	719 790.36	20 997 726	1 348.556 5	207 650 69	29.172 0	.288 488
13	700 885.94	20 277 936	1 353.882 1	206 302.13	28.931 9	.294 345
14	682 437.28	19 577 050	1 378.169 3	204 948.25	28.687 0	.300 318
15	664 414.29	18 894 613	1 393.730 0	203 570.08	28.438 0	.306 390
16	646 815.33	18 230 198	1 382.081 2	202 176.35	28.184 5	.312 572
17	629 657.27	17 583 383	1 382.353 7	200 794.27	27.925 3	.318 895
18	612 917.42	16 953 726	1 375.535 5	199 411.91	27.660 7	.325 349
19	596 592.68	16 340 808	1 379.212 3	198 036.38	27.390 2	.331 946
20	580 662.42	15 744 216	1 376.533 1	196 657.17	27.114 2	.338 677
21	565 123.40	15 163 553	1 388.619 6	195 280.63	26.832 3	.345 554
22	549 956.28	14 598 430	1 389.541 6	193 897 01	25.544 7	.352 568
23	535 153.17	14 048 474	1 399.327 5	192 507.47	26.251 3	.359 724
24	520 701.32	13 513 320	1 407.270 0	191 108.14	25.952 2	.367 021
25	506 594.02	12 992 619	1 423.464 9	189 700.88	25.647 0	.374 463
26	492 814.61	12 486 025	1 437.519 2	188 277.41	25.336 2	.382 045
27	479 357.22	11 993 210	1 454.549 1	186 839.89	25.019 4	.389 772
28	466 211.03	11 513 853	1 478.200 3	185 385.34	24.696 7	.397 643
29	453 361.83	11 047 642	1 503.646 4	183 907.14	24.368 3	.405 652
30	440 800.58	10 594 280	1 531.158 0	182 403.50	24.034 2	.413 800
31	428 518.18	10 153 480	1 559.609 4	180 872.34	23.694 4	.422 088
32	416 506.91	9 724 962	1 592.845 3	179 312.73	23 348 9	.430 516
33	404 755.37	9 308 455	1 626.987 4	177 719.88	22.997 7	.439 080
34	393 256.29	8 903 699	1 669.050 8	176 092.90	22.641 0	.447 781
35	381 995.63	8 510 443	1 710.561 0	174 423.84	22.278 9	.456 612
36	370 968.10	8 128 447	1 759.080 1	172 713.28	21.911 4	.465 574
37	360 161.02	7 757 479	1 809.692 8	170 954.20	21.538 9	.474 660
38	349 566.90	7 397 318	1 862.134 5	169 144.51	21.161 4	.483 869
39	339 178.75	7 047 751	1 922.486 9	167 282.38	20.778 9	.493 198
40	328 983.61	6 708 573	1 983.511 0	165 359.89	20.391 8	.502 639
41	318 976.11	6 379 589	2 050.694 7	163 376.38	20.000 2	.512 190
42	309 145.51	6 060 613	2 120.338 1	161 325.68	19.604 4	.521 844
43	299 485.04	5 751 467	2 194.136 7	159 205.35	19.204 5	.531 597
44	289 986.39	5 451 982	2 274.595 1	157 011.21	18.800 8	.541 443
45	280 638.95	5 161 996	2 357.209 9	154 736.61	18.393 7	.551 373
46	271 436.89	4 881 357	2 444.154 2	152 379.40	17.983 4	.561 381
47	262 372.33	4 609 920	2 536.765 0	149 935.25	17.570 1	.571 460
48	253 436.24	4 347 548	2 630.859 4	147 398.48	17.154 4	.581 600
49	244 624.00	4 094 112	2 732.529 2	144 767.62	16.736 3	.591 796
50	235 925.04	3 849 488	2 835.622 1	142 035.10	16.316 6	.602 035

COMMUTATION COLUMNS 2 1/2% (Continued)
Commissioners 1941 Standard Ordinary Mortality Table*

x	D_x	N_x	C_x	M_x	$1 + a_x$	A_x
51	227 335.15	3 613 563	2 943.137 4	139 199.47	15.895 3	.612 310
52	218 847.25	3 386 227	3 053.177 2	136 256.34	15.473 0	.622 609
53	210 456.33	3 167 380	3 168.222 9	133 203.16	15.050 1	.632 925
54	202 155.03	2 956 924	3 283.812 1	130 034.94	14.627 0	.643 244
55	193 940.61	2 754 769	3 401.913 1	126 751.12	14.204 2	.653 556
56	185 808.43	2 560 828	3 522.090 1	123 349.21	13.782 1	.663 852
57	177 754.43	2 375 020	3 641.783 5	119 827.12	13 361 2	.674 116
58	169 777.17	2 197 265	3 761.696 8	116 185.34	12.942 1	.684 340
59	161 874.57	2 027 488	3 880.185 4	112 423.64	12.525 1	.694 511
60	154 046.23	1 865 614	3 996.199 9	108 543.46	12.110 7	.704 616
61	146 292.80	1 711 567	4 107.708 0	104 547.26	11.699 6	.714 644
62	138 616.97	1 565 275	4 216.676 0	100 439.55	11.292 1	.724 583
63	131 019.40	1 426 658	4 315.413 8	96 222.87	10.888 9	.734 417
64	123 508.39	1 295 638	4 407.831 2	91 907.46	10.490 3	.744 139
65	116 088.15	1 172 130	4 489.449 7	87 499.63	10.096 9	.753 734
66	108 767.29	1 056 042	4 558.728 2	83 010.18	9.709 2	.763 191
67	101 555.70	947 274.4	4 613.189 3	78 451.45	9.327 6	.772 497
68	94 465.545	845 718.7	4 650.452 1	73 838.26	8.952 7	.781 642
69	87 511.050	751 253.1	4 670.014 3	69 187.81	8.584 7	.790 618
70	80 706.625	663 742.1	4 669.226 0	64 517.79	8.224 1	.799 411
71	74 068.942	583 035.4	4 644.235 4	59 848.57	7.871 5	.808 012
72	67 618.148	508 966.5	4 595.428 1	55 204.33	7.527 1	.816 413
73	61 373.498	441 348.3	4 520.662 7	50 608.90	7.191 2	.824 605
74	55 355.921	379 974.8	4 418.249 2	46 088.24	6.864 2	.832 580
75	49 587.526	324 618.9	4 288.286 9	41 669.99	6.546 4	.840 332
76	44 089.787	275 031.4	4 130.220 2	37 381.70	6.238 0	.847 854
77	38 884.206	230 941.6	3 944.961 8	33 251.48	5.939 2	.855 141
78	33 990.850	192 057.4	3 733.725 8	29 306.52	5.650 3	.862 189
79	29 428.077	158 066.6	3 498.684 1	25 572.80	5.371 3	.868 993
80	25 211.636	128 638.5	3 243.115 8	22 074.11	5.102 3	.875 553
81	21 353.602	103 426.8	2 970.736 8	18 831.00	4.843 5	.881 865
82	17 862.047	82 073.24	2 686.402 0	15 860.26	4.594 8	.887 931
83	14 739.984	64 211.19	2 395.321 2	13 173.86	4.356 3	.893 750
84	11 985.151	49 471.21	2 103.356 1	10 778.54	4.127 7	.899 324
85	9 589.474 6	37 486.06	1 816.194 6	8 675.180	3.909 1	.904 656
86	7 539.390 5	27 896.58	1 540.039 4	6 858.986	3.700 1	.909 753
87	5 815.463 2	20 357.19	1 280.145 4	5 318.946	3.500 5	.914 621
88	4 393.477 3	14 541.73	1 041.564 6	4 038.801	3.309 8	.919 272
89	3 244.754 6	10 148.25	827.621 52	2 997.236	3.127 6	.923 717
90	2 337.992 9	6 903.496	640.937 68	2 169.615	2.952 7	.927 982
91	1 640.030 9	4 565.503	482.773 06	1 528.677	2.783 8	.932 103
92	1 117.257 1	2 925.472	352.770 63	1 045.904	2.618 4	.936 136
93	737.236 29	1 808.215	249.339 10	693.133 5	2.452 7	.940 178
94	469.915 86	1 070.979	170.088 82	443.794 4	2.279 1	.944 413
95	288.365 67	601.062 8	111.467 79	273.705 6	2.084 4	.949 162
96	169.864 58	312.697 2	74.109 795	162.237 8	1.840 9	.955 101
97	91.611 740	142.832 6	49.001 885	88.128 0	1.559 1	.961 973
98	40.375 419	51.220 9	28.545 208	39.126 1	1.268 6	.969 058
99	10.845 444	10.845 4	10.580 921	10.580 9	1.000 0	.975 610

* Reproduced by permission of the Actuarial Society of America.

COMMISSIONERS 1958 STANDARD ORDINARY MORTALITY TABLE*

Age, x	Number Living l_x	Number Dying d_x	Deaths per 1,000
0	10,000,000	70,800	7.08
1	9,929,200	17,475	1.76
2	9,911,725	15,066	1.52
3	9,896,659	14,449	1.46
4	9,882,210	13,835	1.40
5	9,868,375	13,322	1.35
6	9,855,053	12,812	1.30
7	9,842,241	12,401	1.26
8	9,829,840	12,091	1.23
9	9,817,749	11,879	1.21
10	9,805,870	11,865	1.21
11	9,794,005	12,047	1.23
12	9,781,958	12,325	1.26
13	9,769,633	12,896	1.32
14	9,756,737	13,562	1.39
15	9,743,175	14,225	1.46
16	9,728,950	14,983	1.54
17	9,713,967	15,737	1.62
18	9,698,230	16,390	1.69
19	9,681,840	16,846	1.74
20	9,664,994	17,300	1.79
21	9,647,694	17,655	1.83
22	9,630,039	17,912	1.86
23	9,612,127	18,167	1.89
24	9,593,960	18,324	1.91
25	9,575,636	18,481	1.93
26	9,557,155	18,732	1.96
27	9,538,423	18,981	1.99
28	9,519,442	19,324	2.03
29	9,500,118	19,760	2.08
30	9,480,358	20,193	2.13
31	9,460,165	20,718	2.19
32	9,439,447	21,239	2.25
33	9,418,208	21,850	2.32
34	9,396,358	22,551	2.40
35	9,373,807	23,528	2.51
36	9,350,279	24,685	2.64
37	9,325,594	26,112	2.80
38	9,299,482	27,991	3.01
39	9,271,491	30,132	3.25
40	9,241,359	32,622	3.53
41	9,208,737	35,362	3.84
42	9,173,375	38,253	4.17
43	9,135,122	41,382	4.53
44	9,093,740	44,741	4.92
45	9,048,999	48,412	5.35
46	9,000,587	52,473	5.83
47	8,948,114	56,910	6.36
48	8,891,204	61,794	6.95
49	8,829,410	67,104	7.60
50	8,762,306	72,902	8.32
51	8,689,404	79,160	9.11
52	8,610,244	85,758	9.96
53	8,524,486	92,832	10.89
54	8,431,654	100,337	11.90
55	8,331,317	108,307	13.00
56	8,223,010	116,849	14.21
57	8,106,161	125,970	15.54
58	7,980,191	135,663	17.00
59	7,844,528	145,830	18.59
60	7,698,698	156,592	20.34
61	7,542,106	167,736	22.24
62	7,374,370	179,271	24.31
63	7,195,099	191,174	26.57
64	7,003,925	203,394	29.04
65	6,800,531	215,917	31.75
66	6,584,614	228,749	34.74
67	6,355,865	241,777	38.04
68	6,114,088	254,835	41.68
69	5,859,253	267,241	45.61
70	5,592,012	278,426	49.79
71	5,313,586	287,731	54.15
72	5,025,855	294,766	58.65
73	4,731,089	299,289	63.26
74	4,431,800	301,894	68.12
75	4,129,906	303,011	73.37
76	3,826,895	303,014	79.18
77	3,523,881	301,997	85.70
78	3,221,884	299,829	93.06
79	2,922,055	295,683	101.19
80	2,626,372	238,848	109.98
81	2,337,524	278,983	119.35
82	2,058,541	265,902	129.17
83	1,792,639	249,858	139.38
84	1,542,781	231,433	150.01
85	1,311,348	211,311	161.14
86	1,100,037	190,108	172.82
87	909,929	168,455	185.13
88	741,474	146,997	198.25
89	594,477	126,303	212.46
90	468,174	106,809	228.14
91	361,365	88,813	245.77
92	272,552	72,480	265.93
93	200,072	57,881	289.30
94	142,191	45,026	316.66
95	97,165	34,128	351.24
96	63,037	25,250	400.56
97	37,787	18,456	488.42
98	19,331	12,916	668.15
99	6,415	6,415	1,000.00

* Reproduced by permission of the Actuarial Society of America.

COMMUTATION COLUMNS 2 1/2%*

Commissioners 1958 Standard Ordinary Mortality Table

Age, x	D_x	N_x	C_x	M_x
0	10,000,000.0000	324,850,104.9680	69,073.1710	2,076,826.7172
1	9,687,024.4290	314,850,104.9680	16,632.9566	2,007,753.5462
2	9,434,122.5838	305,163,080.5390	13,990.2787	1,991,120.5896
3	9,190,031.7084	295,728,957.9552	13,090.0808	1,977,130.3109
4	8,952,794.4741	286,538,926.2468	12,228.1241	1,964,040.2301
5	8,722,205.5791	277,586,131.7727	11,487.5189	1,951,812.1060
6	8,497,981.3556	268,863,926.1936	10,778.2903	1,940,324.5871
7	8,279,935.2370	260,365,944.8380	10,178.0782	1,929,546.2968
8	8,067,807.4636	252,086,009.6010	9,681.6066	1,919,368.2186
9	7,861,350.0557	244,018,202.1374	9,279.8558	1,909,686.6120
10	7,660,329.9546	236,156,852.0817	9,042.8478	1,900,406.7562
11	7,464,449.7860	228,496,522.1271	8,957.6178	1,891,363.9084
12	7,273,432.4866	221,032,072.3411	8,940.8062	1,882,406.2906
13	7,087,090.8833	213,758,639.8545	9,126.8500	1,873,465.4844
14	6,905,108.1581	206,671,548.9712	9,364.0939	1,864,338.6344
15	6,727,326.7826	199,766,440.8131	9,582.3146	1,854,974.5405
16	6,553,663.2627	193,039,114.0305	9,846.7536	1,845,392.2259
17	6,383,971.1254	186,485,450.7678	10,090.0279	1,835,545.4723
18	6,218,174.4633	180,101,479.6424	10,252.3993	1,825,455.4444
19	6,056,259.3006	173,883,305.1791	10,280.6243	1,815,203.0451
20	5,898,264.9735	167,827,045.8785	10,300.1828	1,804,922.4208
21	5,744,104.7377	161,928,780.9050	10,255.1657	1,794,622.2380
22	5,593,749.4258	156,184,676.1673	10,150.6810	1,784,367.0723
23	5,447,165.8414	150,590,926.7415	10,044.0865	1,774,216.3913
24	5,304,263.9929	145,143,760.9001	9,883.7932	1,764,172.3048
25	5,165,007.9517	139,839,496.9072	9,725.3439	1,754,288.5116
26	5,029,306.7854	134,674,488.9555	9,617.0037	1,744,563.1677
27	4,897,023.7928	129,645,182.1701	9,507.1611	1,734,946.1640
28	4,768,076.9758	124,748,158.3773	9,442.8900	1,725,439.0029
29	4,642,339.5370	119,980,081.4015	9,420.4356	1,715,996.1129
30	4,519,691.3751	115,337,741.8645	9,392.0634	1,706,575.6773
31	4,400,062.8465	110,818,050.4894	9,401.2183	1,697,183.6139
32	4,283,343.0569	106,417,987.6429	9,402.5686	1,687,782.3956
33	4,169,468.7479	102,134,644.5860	9,437.1317	1,678,379.8270
34	4,058,337.1968	97,965,175.8381	9,502.3390	1,668,942.6953
35	3,949,851.0856	93,906,838.6413	9,672.2130	1,659,440.3563
36	3,843,840.9771	89,956,987.5557	9,900.3401	1,649,768.1433
37	3,740,188.4751	86,113,146.5786	10,217.2318	1,639,867.8032
38	3,638,747.0704	82,372,958.1035	10,685.3232	1,629,650.5714
39	3,539,311.8617	78,734,211.0331	11,222.0794	1,618,965.2482
40	3,441,765.0620	75,194,899.1714	11,853.1042	1,607,743.1688
41	3,345,966.5023	71,753,134.1094	12,535.2926	1,595,890.0646
42	3,251,822.2774	68,407,167.6071	13,229.3739	1,583,354.7720
43	3,159,280.1784	65,155,345.3297	13,962.4424	1,570,125.3981
44	3,068,262.0685	61,996,065.1513	14,727.5918	1,556,162.9557
45	2,978,698.8164	58,927,803.0828	15,547.3085	1,541,435.3639
46	2,890,500.3526	55,949,104.2664	16,440.4699	1,525,888.0554
47	2,803,559.9048	53,058,603.9138	17,395.7457	1,509,447.5855
48	2,717,784.6405	50,255,044.0090	18,427.9447	1,492,051.8398
49	2,633,069.2135	47,537,259.3685	19,523.3861	1,473,623.8951

COMMUTATION COLUMNS 2 1/2%* (Continued)
Commissioners 1958 Standard Ordinary Mortality Table

Age, x	D_x	N_x	C_x	M_x
50	2,549,324.6723	44,904,190.1550	20,692.9455	1,454,100.5090
51	2,466,453.0891	42,354,865.4827	21,921.2231	1,433,407.5635
52	2,384,374.4270	39,888,412.3936	23,169.1325	1,411,486.3404
53	2,303,049.8123	37,504,037.9666	24,468.5917	1,388,317.2079
54	2,222,409.2905	35,200,988.1543	25,801.7117	1,363,848.6162
55	2,142,402.4988	32,978,578.8638	27,171.9031	1,338,046.9045
56	2,062,976.8254	30,836,176.3650	28,599.9098	1,310,875.0014
57	1,984,060.3996	28,773,199.5396	30,080.3536	1,282,275.0916
58	1,905,588.3725	26,789,139.1400	31,604.8230	1,252,194.7380
59	1,827,505.7998	24,883,550.7675	33,144.7659	1,220,589.9150
60	1,749,787.7198	23,056,044.9677	34,722.7242	1,187,445.1491
61	1,672,387.2632	21,306,257.2479	36,286.6291	1,152,722.4249
62	1,595,310.6622	19,633,869.9847	37,836.1144	1,116,435.7958
63	1,518,564.5694	18,038,559.3225	39,364.2006	1,078,599.6814
64	1,442,162.1578	16,519,994.7531	40,858.9196	1,039,235.4808
65	1,366,128.5462	15,077,832.5953	42,316.6940	998,376.5612
66	1,290,491.6985	13,711,704.0491	43,738.1310	956,059.8672
67	1,215,278.1249	12,421,212.3506	45,101.6207	912,321.7362
68	1,140,535.6099	11,205,934.2257	46,378.0358	867,220.1155
69	1,066,339.5743	10,065,398.6158	47,449.5963	820,842.0797
70	992,881.7500	8,999,059.0415	48,229.7870	773,392.4834
71	920,435.3077	8,006,177.2915	48,625.9779	725,162.6964
72	849,359.6946	7,085,741.9838	48,599.8832	676,536.7185
73	780,043.7396	6,236,382.2892	48,142.0663	627,936.8353
74	712,876.2140	5,456,338.5496	47,376.6752	579,794.7690
75	648,112.3021	4,743,462.3356	46,392.1628	532,418.0938
76	585,912.5111	4,095,350.0335	45,261.0951	486,025.9310
77	526,360.8716	3,509,437.5224	44,008.9628	440,764.8359
78	469,513.8465	2,983,076.6508	42,627.3426	396,755.8731
79	415,434.9294	2,513,562.8043	41,012.5834	354,128.5305
80	364,289.7989	2,098,127.8749	39,087.3533	313,115.9471
81	316,317.3241	1,733,838.0760	36,831.6174	274,028.5938
82	271,770.6619	1,417,520.7519	34,248.4382	237,196.9764
83	230,893.6600	1,145,750.0900	31,397.0289	202,948.5382
84	193,865.0736	914,856.4300	28,372.4430	171,551.5093
85	160,764.2229	720,991.3564	25,273.7507	143,179.0663
86	131,569.3974	560,227.1335	22,183.1949	117,905.3156
87	106,177.1855	428,657.7361	19,177.1362	95,722.1207
88	84,410.3641	322,480.5506	16,326.1748	76,544.9845
89	66,025.3978	238,070.1865	13,685.6613	60,218.8097
90	50,729.3636	172,044.7887	11,291.0955	46,533.1484
91	38,200.9638	121,315.4251	9,159.6932	35,242.0529
92	28,109.5415	83,114.4613	7,292.8738	26,082.3597
93	20,131.0686	55,004.9198	5,681.8884	18,789.4859
94	13,958.1795	34,873.8512	4,312.1729	13,107.5975
95	9,305.5630	20,915.6717	3,188.7449	8,795.4246
96	5,889.8533	11,610.1087	2,301.6880	5,606.6797
97	3,444.5103	5,720.2554	1,641.3408	3,304.9917
98	1,719.1569	2,275.7451	1,120.6380	1,663.6509
99	556.5882	556.5882	543.0129	543.0129

* Reproduced by permission of the Actuarial Society of America.

CONVERSION FACTORS

(From various U. S. Government and IUPAC publications and from calculations based on values given in these publications)

To convert from	To	Multiply by
Abamperes	Amperes	10
"	E.M. cgs. units of current	1
"	E.S. cgs. units	2.997930×10^{10}
"	Faradays (chem.)/sec	1.036377×10^{-4}
"	Faradays (phys.)/sec	1.036086×10^{-4}
"	Statamperes	2.997930×10^{10}
Abamperes/cm	E.M. cgs. units of surface charge density	1
"	E.S. cgs. units	2.997930×10^{10}
Abamperes/sq. cm	Amperes/circ. mil	5.0670748×10^{-5}
"	Amperes/sq. cm	10
"	Amperes/sq. inch	64.516
Abampere-turns	Ampere-turns	10
Abampere-turns/cm	Ampere-turns/cm	10
Abcoulombs	Ampere-hours	0.0027777
"	Coulombs	10
"	Electronic charges	6.24196×10^{19}
"	E.M. cgs. units of charge	1
"	E.S. cgs. units	2.997930×10^{10}
"	Faradays (chem.)	1.036377×10^{-4}
"	Faradays (phys.)	1.036086×10^{-4}
"	Statcoulombs	2.997930×10^{10}
Abfarads	E.M. cgs. units of capacitance	1
"	E.S. cgs. units	8.987584×10^{20}
"	Farads	1×10^{9}
"	Microfarads	1×10^{15}
"	Statfarads	8.987584×10^{20}
Abhenries	E.M. cgs. units of induction	1
"	E.S. cgs. units	1.112646×10^{-21}
"	Henries	1×10^{-9}
Abmhos	E.M. cgs. units of conductance	1
"	E.S. cgs. units	8.987584×10^{20}
"	Megamhos	1000
"	Mhos	1×10^{9}
"	Statmhos	8.987584×10^{20}
Abohms	E.M. cgs. units of resistance	1
"	Megohms	1×10^{-15}
"	Microhms	0.001
"	Ohms	1×10^{-9}
"	Statohms	1.112646×10^{-21}
Abohm-cm	Circ. mil-ohms/ft	0.0060153049
"	E.M. cgs. units of resistivity	1
"	Microhm-inches	0.00039370079
"	Ohm-cm	1×10^{-9}
Abvolts	Microvolts	0.01
"	Millivolts	1×10^{-5}
"	Volts	1×10^{-8}
"	Volts (Int.)	9.99670×10^{-9}
Abvolts/cm	E.M. cgs. units of electric field intensity	1
"	E.S. cgs. units	3.335635×10^{-11}
"	Volts/cm	1×10^{-8}
"	Volts/inch	2.54×10^{-8}
"	Volts/meter	1×10^{-6}
Acres	Sq. cm	40468564
"	Sq. ft	43560
"	Sq. ft. (U.S. Survey)	43559.826
"	Sq. inches	6272640
"	Sq. kilometers	0.0040468564
"	Sq. links (Gunter's)	1×10^{5}
"	Sq. meters	4046.8564
"	Sq. miles (statute)	0.0015625
"	Sq. perches	160
"	Sq. rods	160
"	Sq. yards	4840
Acre-feet	Cu. feet	43560
"	Cu. meters	1233.4818
"	Cu. yards	1613.333
Acre-inches	Cu. feet	3630
"	Cu. meters	102.79033
Acre-inches	Gallons (U.S.)	27154.286
Amperes	Abamperes	0.1
"	Amperes (Int.)	1.000165
"	Cgs. units of current	1
"	Mks. units of current	1
"	Coulombs/sec	1
"	Coulombs (Int.)/sec	1.000165
"	Faradays (chem.)/sec	1.036377×10^{-5}
"	Faradays (phys.)/sec	1.036086×10^{-5}
"	Statamperes	2.997930×10^{9}
Amperes (Int.)	Amperes	0.999835
"	Coulombs/sec	0.999835
"	Coulombs (Int.)/sec	1
"	Faradays (chem.)/sec	1.03623×10^{-5}
"	Faradays (phys.)/sec	1.03592×10^{-5}
Amperes/meter	Cgs. units of surface current density	0.01
"	E.M. cgs. units	0.001
"	E.S. cgs. units	2.997930×10^{7}
"	Mks. units	1
Amperes/sq. meter	Cgs. units of volume current density	0.0001
"	E.M. cgs. units	1×10^{-5}
"	E.S. cgs. units	299793.0
"	Mks. units	1
Amperes/sq. mil	Abamperes/sq. cm	15500.031
"	Amperes/sq. cm	1.5500031×10^{5}
Ampere-hours	Abcoulombs	360
"	Coulombs	3600
"	Faradays (chem.)	0.373096
"	Faradays (phys.)	0.372991
Ampere-turns	Cgs. units of magnetomotive force	1.2566371
"	E.M. cgs. units	1.2566371
"	E.S. cgs. units	3.767310×10^{10}
"	Gilberts	1.2566371
Ampere-turns/weber	Cgs. units of reluctance	1.256637×10^{-8}
"	E.M. cgs. units	1.256637×10^{-8}
"	E.S. cgs. units	1.129413×10^{13}
"	Gilberts/maxwell	1.256637×10^{-8}
Ångström units	Centimeters	1×10^{-8}
"	Inches	3.9370079×10
"	Microns	0.0001
"	Millimicrons	0.1
"	Wave length of orange-red line of krypton 86	0.000165076373
"	Wave length of red line of cadmium	0.000155316413
Ares	Acres	0.024710538
"	Sq. dekameters	1
"	Sq. feet	1076.3910
"	Sq. ft. (U.S. Survey)	1076.3867
"	Sq. meters	100
"	Sq. miles	3.8610216×10
Atmospheres	Bars	1.01325
"	Cm. of Hg (0°C.)	76
"	Cm. of H_2O (4°C.)	1033.26
"	Dynes/sq. cm	1.01325×10^{6}
"	Ft. of H_2O (39.2°F.)	33.8995
"	Grams/sq. cm	1033.23
"	In. of Hg (32°F.)	29.9213
"	Kg./sq. cm	1.03323
"	Mm. of Hg (0°C.)	760
"	Pounds/sq. inch	14.6960
"	Tons (short)/sq. ft	1.05811
"	Torrs	760
Atomic mass units (chem.)*	Electron volts	9.31395×10^{8}
"	Grams	1.66024×10^{-2}
Atomic mass units (phys.)*	Electron volts	9.31141×10^{8}
"	Grams	1.65979×10^{-2}

CONVERSION FACTORS (Continued)

To convert from	To	Multiply by
Bags (Brit.)	Bushels (Brit.)	3
Barns	Sq. cm	1×10^{-24}
Barrels (Brit.)	Bags (Brit.)	1.5
"	Barrels (U.S., dry)	1.415404
"	Barrels (U.S., liq.)	1.372513
"	Bushels (Brit.)	4.5
"	Bushels (U.S.)	4.644253
"	Cu. feet	5.779568
"	Cu. meters	0.1636591
"	Gallons (Brit.)	36
"	Liters	163.6546
Barrels (petroleum, U.S.)	Cu. feet	5.614583
"	Gallons (U.S.)	42
"	Liters	158.98284
Barrels (U.S., dry)	Barrels (U.S. liq.)	0.969696
"	Bushels (U.S.)	3.2812195
"	Cu. feet	4.083333
"	Cu. inches	7056
"	Cu. meters	0.11562712
"	Quarts (U.S., dry)	105
Barrels (U.S., liq.)	Barrels (U.S., dry)	1.03125
"	Barrels (wine)	1
"	Cu. feet	4.2109375
"	Cu. inches	7276.5
"	Cu. meters	0.11924047
"	Gallons (Brit.)	26.22925
"	Gallons (U.S., liq.)	31.5
"	Liters	119.23713
Bars	Atmospheres	0.986923
"	Baryes	1×10^{6}
"	Cm. of Hg (0°C.)	75.0062
"	Dynes/sq. cm	1×10^{6}
"	Ft. of H_2O (60°F.)	33.4883
"	Grams/sq. cm	1019.716
"	In. of Hg (32°F.)	29.5300
"	Kg./sq. cm	1.019716
"	Millibars	1000
"	Pounds/sq. inch	14.5038
Baryes	Atmospheres	9.86923×10^{-7}
"	Bars	1×10^{-6}
"	Dynes/sq. cm	1
"	Grams/sq. cm	0.001019716
"	Millibars	0.001
Bels	Decibels	10
Board feet	Cu. cm	2359.7372
"	Cu. feet	0.083333
"	Cu. inches	144
Bolts of cloth	Linear feet	120
"	Meters	36.576
Bougie decimales	Candles (Int.)	1.00
B.t.u	B.t.u. (IST.)**	0.999346
"	B.t.u. (mean)	0.998563
"	B.t.u. (39°F.)	0.994982
"	B.t.u. (60°F.)	0.999689
"	Cal. *gm*	251.99576
"	Cal., *gm.* (IST.)	251.831
"	Cal., *gm.* (mean)	251.634
"	Cal., *gm.* (20°C.)	252.122
"	Cu. cm.-atm	10405.6
"	Ergs	1.05435×10^{10}
"	Foot-poundals	25020.1
"	Foot-pounds	777.649
"	Gram-cm	1.07514×10^{7}
"	Hp.-hours	0.000392752
"	Hp.-years	4.48347×10^{-8}
"	Joules	1054.35
"	Joules (Int.)	1054.18
"	Kg.-meters	107.514
"	Kw.-hours	0.000292875
B.t.u	Kw.-hours (Int.)	0.000292827
"	Liter-atm	10.4053
"	Tons of refrig. (U.S. std.)	3.46995×10^{-8}
"	Watt-seconds	1054.35
"	Watt-seconds (Int.)	1054.18
B.t.u. (IST.)	B.t.u	1.00065
B.t.u. (mean)	B.t.u	1.00144
"	B.t.u. (IST.)	1.00078
"	B.t.u. (39°F.)	0.996415
"	B.t.u. (60°F.)	1.00113
"	Hp.-hours	0.000393317
"	Joules	1055.87
"	Kg.-meters	107.669
"	Kw.-hours	0.000293297
"	Kw.-hours (Int.)	0.000293248
"	Liter-atm	10.4203
"	Watt-hours	0.293297
"	Watt-hours (Int.)	0.293248
B.t.u. (39°F.)	B.t.u	1.00504
"	B.t.u. (IST.)	1.00439
"	B.t.u. (mean)	1.00360
"	B.t.u. (60°F)	1.00473
"	Joules	1059.67
B.t.u. (60°F.)	B.t.u	1.00031
"	B.t.u. (IST.)	0.999657
"	B.t.u. (mean)	0.998873
"	B.t.u. (39°F.)	0.995291
B.t.u./hr	Cal., *kg.*/hr	0.251996
"	Ergs/sec	2.928751×10^{6}
"	Foot-pounds/hr	777.649
"	Horsepower	0.000392752
"	Horsepower (boiler)	2.98563×10^{-5}
"	Horsepower (electric)	0.000392594
"	Horsepower (metric)	0.000398199
"	Kilowatts	0.000292875
"	Lb. ice melted/hr	0.0069714
"	Tons of refrig. (U.S. comm.)	8.32789×10^{-5}
"	Watts	0.292875
B.t.u./min	Cal., *kg.*/min	0.251996
"	Ergs/sec	1.75725×10^{8}
"	Foot-pounds/min	777.649
"	Horsepower	0.0235651
"	Horsepower (boiler)	0.00179138
"	Horsepower (electric)	0.0235556
"	Horsepower (metric)	0.0238920
"	Joules/sec	17.5725
"	Kg.-meters/min	107.514
"	Kilowatts	0.0175725
"	Lb. ice melted/hr	0.41828
"	Tons of refrig. (U.S. comm.)	0.00499673
"	Watts	17.5725
B.t.u. (mean)/min	B.t.u. (mean)/hr	60
"	Cal., *kg.* (mean)/hr	15.1197
"	Cal., *kg.* (mean)/min	0.251996
"	Ergs/sec	1.75978×10^{8}
"	Foot-pounds/min	778.768
"	Horsepower	0.0235990
"	Horsepower (boiler)	0.00179396
"	Horsepower (electric)	0.0235895
"	Horsepower (metric)	0.0239264
"	Joules/sec	17.5978
"	Kg.-meters/min	107.669
"	Kilowatts	0.0175978
"	Lb. ice-melted/hr	0.41888
B.t.u./lb	Cal., *gm.*/gram	0.555555
"	Cu. cm.-atm./gram	22.9405
"	Cu. ft.-atm./lb	0.367471
"	Cu. ft.-(lb./sq. in.)/lb	5.40034
"	Foot-pounds/lb	777.649
"	Hp.-hr./lb	0.000392752

* Barrel (Brit., liq.) = Barrel (Brit., dry)

** International Steam Table.

CONVERSION FACTORS (Continued)

To convert from	To	Multiply by
B.t.u./lb	Joules/gram	2.32444
B.t.u. (mean)/lb	Cal., *gm.* (mean)/gram	0.555555
"	Cu. cm.-atm./gram	22.9735
"	Foot-pounds/lb	778.768
"	Hp.-hr./lb	0.000393317
"	Joules/gram	2.32779
B.t.u./sec	B.t.u./hr	3600
"	B.t.u./min	60
"	Cal., *kg.*/hr	907.185
"	Cal., *kg.*/min	15.1197
"	Cheval-vapeur	1.43352
"	Ergs/sec	1.05435×10^{10}
"	Foot-pounds/sec	777.649
"	Horsepower	1.41391
"	Horsepower (boiler)	0.107483
"	Horsepower (electric)	1.41334
"	Horsepower (metric)	1.43352
"	Kg.-meters/sec	107.514
"	Kilowatts	1.05435
"	Kilowatts (Int.)	1.05418
"	Watts	1054.35
"	Watts (Int.)	1054.18
B.t.u. (mean)/sec	Ergs/sec	1.05587×10^{10}
"	Foot-pounds/sec	778.768
"	Horsepower	1.41594
"	Horsepower (boiler)	0.107637
"	Horsepower (electric)	1.41537
"	Horsepower (metric)	1.43558
"	Watts	1055.87
B.t.u./sq. ft	Cal., *gm.*/sq. cm	0.271246
B.t.u./sq.ft. × min.)	Hp./sq. ft	0.0235651
"	Kw./sq. ft	0.0175725
"	Watts/sq. in	0.122031
Buckets (Brit.)	Cu. cm	18184.35
"	Gallons (Brit.)	4
Bushels (Brit.)	Bags (Brit.)	0.333333
"	Bushels (U.S.)	1.032056
"	Cu. cm	36368.70
"	Cu. feet	1.284348
"	Cu. inches	2219.354
"	Dekaliters	3.636768
"	Gallons (Brit.)	8
"	Hectoliters	0.3636768
"	Liters	36.36768
Bushels (U.S.)*	Barrels (U.S.), dry	0.3047647
"	Bushels (Brit.)	0.9689395
"	Cu. cm	35239.07
"	Cu. feet	1.244456
"	Cu. inches	2150.42
"	Cu. meters	0.03523907
"	Cu. yards	0.04609096
"	Gallons (U.S., dry)	8
"	Gallons (U.S., liq.)	9.309177
"	Liters	35.23808
"	Ounces (U.S., fluid)	1191.575
"	Pecks (U.S.)	4
"	Pints (U.S., dry)	64
"	Quarts (U.S., dry)	32
"	Quarts (U.S., liq.)	37.23671
Butts (Brit.)	Bushels (U.S.)	13.53503
"	Cu. feet	16.84375
"	Cu. meters	0.4769619
"	Gallons (U.S.)	126
Cable lengths	Fathoms	120
"	Feet	720
"	Meters	219.456

To convert from	To	Multiply by
Calories, *gm.***	B.t.u	0.0039683207
"	B.t.u. (IST.)	0.00396573
"	B.t.u. (mean)	0.00396262
"	B.t.u. (39°F.)	0.00394841
"	B.t.u. (60°F.)	0.00396709
"	Cal., *gm.* (IST.)	0.999346
"	Cal., *gm.* (mean)	0.998563
"	Cal., *gm.* (15°C.)	0.999570
"	Cal., *gm.* (20°C.)	1.00050
"	Cal., *kg*	0.001
"	Cal., *kg.* (IST.)	0.000999346
"	Cal., *kg.* (mean)	0.000998563
"	Cal., *kg.* (15°C.)	0.000999570
"	Cal., *kg.* (20°C.)	0.00100050
"	Cu. cm.-atm	41.2929
"	Cu. ft.-atm	0.00145824
"	Ergs	4.184×10^{7}
"	Foot-poundals	99.2878
"	Foot-pounds	3.08596
"	Gram-cm	42664.9
"	Hp.-hours	$1.55857 \times 10^{-}$
"	Joules	4.184
"	Joules (Int.)	4.18331
"	Kg.-meters	0.426649
"	Kw.-hours	1.162222 × 10
"	Liter-atm	0.0412917
"	Watt-hours	0.001162222
"	Watt-hours (Int.)	0.00116203
"	Watt-seconds	4.184
Calories, *gm.* (mean)	B.t.u	0.00397403
"	Cal., *gm*	1.00144
"	Cal., *gm.* (IST.)	1.00078
"	Cal., *gm.* (20°C.)	1.00194
"	Cal., *kg.* (mean)	0.001
"	Cu. cm.-atm	41.3523
"	Cu. ft.-atm	0.00146034
"	Ergs	4.19002×10^{7}
"	Foot-poundals	99.4308
"	Foot-pounds	3.09040
"	Hp.-hours	$1.56081 \times 10^{-}$
"	Joules	4.19002
"	Joules (Int.)	4.18933
"	Kg.-meters	0.427263
"	Kw.-hours	$1.16390 \times 10^{-}$
"	Liter-atm	0.0413511
"	Watt-seconds	4.19002
Calories, *gm.* (15°C.)	B.t.u	0.00397003
"	Cal., *gm*	1.00043
"	Cal., *gm.* (IST.)	0.999776
"	Cal., *gm.* (mean)	0.998992
"	Cal., *gm.* (20°C.)	1.00093
"	Joules	4.18580
"	Joules (Int.)	4.18511
Calories, *gm.* (20°C.)	B.t.u	0.00396633
"	Cal., *gm*	0.999498
"	Cal., *gm.* (IST.)	0.998845
"	Cal., *gm.* (mean)	0.998061
"	Cal., *gm.* (15°C.)	0.999068
"	Joules	4.18190
"	Joules (Int.)	4.18121
Calories, *kg*	B.t.u	3.9683207
"	B.t.u. (IST.)	3.96573
"	B.t.u. (mean)	3.96262
"	B.t.u. (60°F.)	3.96709
"	Cal., *gm*	1000
"	Cal., *kg.* (mean)	0.998563
"	Cal., *kg.* (15°C.)	0.999570
"	Cal., *kg.* (20°C.)	1.00050
"	Cu. cm.-atm	41292.86

* Stricken or struck bushel. A heaped bushel for apples of 2747.715 cu. inches was established by the U.S. Court of Customs Appeals on Feb. 15, A heaped bushel equal to $1\frac{1}{4}$ stricken bushels is also known.

** This is the calorie as defined by the U.S. National Bureau of Standards and is equal to 4.18400 joules.

CONVERSION FACTORS (Continued)

To convert from	To	Multiply by
ories, *kg*.	Ergs	4.184×10^{10}
"	Foot-poundals	99287.8
"	Foot-pounds	3085.96
"	Gram-cm	4.26649×10^{7}
"	Hp.-hours	0.00155857
"	Joules	4184
"	Kw.-hours	0.001162222
"	Liter-atm	41.2917
"	Watt-hours	1.162222
ories, *kg*. (mean)	B.t.u.	3.97403
"	B.t.u. (IST.)	3.97144
"	B.t.u. (mean)	3.9683207
"	B.t.u. (60°F.)	3.97280
"	Cal., *gm*	1001.44
"	Cal., *gm*. (IST.)	1000.78
"	Cal., *gm*. (mean)	1000
"	Cal., *gm*. (15°C.)	1000.10
"	Cal., *gm*. (20°C.)	1001.94
"	Ergs	4.19002×10^{10}
"	Foot-poundals	99430.8
"	Foot-pounds	3090.40
"	Gram-cm	4.27263×10^{7}
"	Hp.-hours	0.00156081
"	Joules	4190.02
"	Kg.-meters	427.263
"	Kw.-hours (Int.)	0.00116370
"	Liter-atm	41.3511
"	Watt-hours	1.16390
., *gm*./°C	B.t.u./°F	0.00220462
"	Joules/°F	2.324444
"	Joules (Int.)/°F	2.32406
., *gm*./gram	B.t.u./lb	1.8
"	Foot-pounds/lb	1399.77
"	Joules/gram	4.184
"	Watt-hours/gram	0.001162222
., *gm*./(gram × °C)	B.t.u./(lb. × °C.)	1.8
"	B.t.u./(lb. × °F.)	1
"	Cal., *kg*./(kg. × °C.)	1
"	Joules/(gram × °C.)	4.184
"	Joules/(lb. × °F.)	1054.35
., *gm*./hr	B.t.u./hr	0.0039683207
"	Ergs/sec	11622.222
"	Watts	0.001162222
., *gm*. (mean)/hr	B.t.u. (mean)/hr	0.0039683207
"	Ergs/sec	11639.0
"	Watts	0.00116390
., *kg*./hr	Watts	1.162222
., *gm*./min	B.t.u./min	0.0039683207
"	Ergs/sec	697333.3
"	Watts	0.069733
., *gm*. (mean)/min	B.t.u. (mean)/min	0.0039683207
"	Ergs/sec	698337
"	Joules/sec	0.0698337
"	Watts	0.0698337
., *kg*./min	Kg. ice melted/min	0.012548
"	Lb. ice melted/min	0.027665
"	Watts	69.7333
., *gm*./sec	B.t.u./sec	0.0039683207
"	Ergs/sec	4.184×10^{7}
"	Foot-pounds/sec	3.08596
"	Horsepower	0.00561084
"	Watts	4.184
., *gm*. (mean)/sec	Ergs/sec	4.19002×10^{7}
"	Watts	4.19002
., *gm*./(sec. × q. cm.)	B.t.u./(hr. × sq. ft.)	13272.1
"	Cal., *gm*./(hr. × sq. cm.)	3600
"	Watts/sq. cm	4.184
., *gm*./(sec. × q. cm. × °C.)	B.t.u./(hr. × sq. ft. × °F.)	7373.38
., *gm*./sq. cm	B.t.u./sq. ft	3.68669

To convert from	To	Multiply by
Cal., *gm*.-cm./(hr. × sq. cm. × °C.)	B.t.u.-ft./(hr. × sq. ft. × °F.)	0.0671969
"	B.t.u.-inch/(hr. × sq. ft. × °F.)	0.806363
Cal., *gm*.-cm./sq. cm	B.t.u.-inch/sq. ft	1.4514530
Cal., *gm*.-sec	Planck's constant	6.31531×10^{33}
Cal., *gm*.-sec./Avog. No. (chem.)*	Planck's constant	1.04849×10^{10}
Cal., *gm*.-sec./Avog. No. (phys.)*	Planck's constant	1.04821×10^{10}
Candles (English)	Candles (Int.)	1.04
"	Hefner units	1.16
Candles (German)	Candles (English)	1.01
"	Candles (Int.)	1.05
"	Hefner units	1.17
Candles (Int.)	Candles (English)	0.96
"	Candles (German)	0.95
"	Candles (pentane)	1.00
"	Hefner units	1.11
"	Lumens (Int.)/steradian	1
Candles (pentane)	Candles (Int.)	1.00
Candles/sq. cm	Candles/sq. inch	6.4516
"	Candles/sq. meter	10000
"	Foot-lamberts	2918.6351
"	Lamberts	3.1415927
Candles/sq. ft	Candles/sq. inch	0.0069444
"	Candles/sq. meter	10.763910
"	Foot-lamberts	3.1415927
"	Lamberts	0.0033815822
Candles/sq. inch	Candles/sq. cm	0.15500031
"	Candles/sq. foot	144
"	Foot-lamberts	452.38934
"	Lamberts	0.48694784
Candle power (spher.)	Lumens	12.566370
Carats (parts of gold per 24 of mixture)	Milligrams/gram	41.6666
Carats (1877)	Grains	3.168
"	Milligrams	205.3
Carats (metric)	Grains	3.08647
"	Grams	0.2
"	Milligrams	200
Carcel units	Candles (Int.)	9.61
Centals	Kilograms	45.359237
"	Pounds	100
Centares	Ares	0.01
"	Sq. feet	10.763910
"	Sq. inches	1550.0031
"	Sq. meters	1
"	Sq. yards	1.1959900
Centigrams	Grains	0.15432358
"	Grams	0.01
Centiliters	Cu. cm	10.00028
"	Cu. inches	0.6102545
"	Liters	0.01
"	Ounces (U.S., fluid)	0.3381497
Centimeters	Ångström units	1×10^{8}
"	Feet	0.032808399
"	Feet (U.S. Survey)	0.032808333
"	Hands	0.098425197
"	Inches	0.39370079
"	Links (Gunter's)	0.049709695
"	Links (Ramden's)	0.032808399
"	Meters	0.01
"	Microns	10000
"	Miles (naut., Int.)	5.3995680×10^{-6}
"	Miles (statute)	6.2137119×10^{-6}
"	Millimeters	10
"	Millimicrons	1×10^{7}
"	Mils	393.70079
"	Picas (printer's)	2.3710630
"	Points (printer's)	28.452756

For factors for C = 12 scale or those which can be derived from same table on Values for General Physical Constants.

CONVERSION FACTORS (Continued)

To convert from	To	Multiply by
Centimeters	Rods	0.0019883878
"	Wave length of orange-red line of krypton 86	16507.6373
"	Wave length of red line of cadmium	15531.6413
"	Yards	0.010936133
Cm. of Hg (0°C.)	Atmospheres	0.013157895
"	Bars	0.0133322
"	Dynes/sq. cm	13332.2
"	Ft. of H_2O (4°C.)	0.446050
"	Ft. of H_2O (60°F.)	0.446474
"	In. of Hg (0°C.)	0.39370079
"	Kg./sq. meter	135.951
"	Pounds/sq. ft	27.8450
"	Pounds/sq. inch	0.193368
"	Torrs	10
Cm. of H_2O (4°C.)	Atmospheres	0.000967814
"	Dynes/sq. cm	980.638
"	Pounds/sq. inch	0.0142229
Centimeters/sec	Feet/min	1.9685039
"	Feet/sec	0.032808399
"	Kilometers/hr	0.036
"	Kilometers/min	0.0006
"	Knots (Int.)	0.019438445
"	Meters/min	0.6
"	Miles/hr	0.022369363
"	Miles/min	0.00037282272
Cm./(sec. × sec.)	Kilometers/(hr. × sec.)	0.036
"	Miles/(hr. × sec.)	0.022369363
Centimeters/year	Inches/year	0.39370079
Centipoises	Grams/(cm. × sec.)	0.01
"	Poises	0.01
"	Pound/(ft. × hr.)	2.4190883
"	Pounds/(ft. × sec.)	0.00067196898
Centistokes	Stokes	0.01
Chains (Gunter's)	Centimeters	2011.68
"	Chains (Ramden's)	0.66
"	Feet	66
"	Feet (U.S. Survey)	65.999868
"	Furlongs	0.1
"	Inches	792
"	Links (Gunter's)	100
"	Links (Ramden's)	66
"	Meters	20.1168
"	Miles (statute)	0.0125
"	Rods	4
"	Yards	22
Chains (Ramden's)	Centimeters	3048
"	Chains (Gunter's)	1.515151
"	Feet	100
"	Feet (U.S. Survey)	99.999800
Cheval-vapeur	Horsepower (metric)	1
Cheval-vapeur-heures	Joules	2647795
Circles	Degrees	360
"	Grades	400
"	Minutes	21600
"	Radians	6.2831853
"	Signs	12
Circular inches	Circular mm	645.16
"	Sq. cm	5.0670748
"	Sq. inches	0.78539816
Circular mm	Sq. cm	0.0078539816
"	Sq. inches	0.0012173696
"	Sq. mm	0.78539816
Circular mils	Circular inches	1×10^{-6}
"	Sq. cm	5.0670748×10^{-6}
"	Sq. inches	7.8539816×10^{-7}
"	Sq. mm	0.00050670748
"	Sq. mils	0.78539816
Circumferences	Degrees	360
"	Grades	400

To convert from	To	Multiply by
Circumferences	Minutes	21600
"	Radians	6.2831853
"	Seconds	1296000
Cords	Cord-feet	8
"	Cu. feet	128
"	Cu. meters	3.6245734
Cord-feet	Cords	0.125
"	Cu. feet	16
Coulombs	Abcoulombs	0.1
"	Ampere-hours	0.0002777
"	Ampere-seconds	1
"	Coulombs (Int.)	1.000165
"	Electronic charge	6.24196×10^{18}
"	E.M. cgs. units of electric charge	0.1
"	E.S. cgs. units of electric charge	2.997930×10^{9}
"	Faradays (chem.)	1.036377×10^{-5}
"	Faradays (phys.)	1.036086×10^{-5}
"	Mks. units of electric charge	1
"	Statcoulombs	2.997930×10^{9}
Coulombs/cu. meter	E.M. cgs. units of volume charge density	1×10^{-7}
"	E.S. cgs. units	2997.930
Coulombs/sq. cm	Abcoulombs/sq. cm	0.1
"	Cgs. units of polarization, *and* surface charge density	1
Cubic centimeters	Board feet	0.00042377600
"	Bushels (Brit.)	2.749617×10^{-5}
"	Bushels (U.S.)	2.837759×10^{-5}
"	Cu. feet	3.5314667×10^{-5}
"	Cu. inches	0.061023744
"	Cu. meters	1×10^{-6}
"	Cu. yards	1.3079506×10^{-6}
"	Drachms (Brit., fluid)	0.28156080
"	Drams (U.S., fluid)	0.27051218
"	Gallons (Brit.)	0.0002199694
"	Gallons (U.S., dry)	0.00022702075
"	Gallons (U.S., liq.)	0.00026417205
"	Gills (Brit.)	0.007039020
"	Gills (U.S.)	0.0084535058
"	Liters	0.000999972
"	Ounces (Brit., fluid)	0.03519510
"	Ounces (U.S., fluid)	0.033814023
"	Pints (U.S., dry)	0.0018161660
"	Pints (U.S., liq.)	0.0021133764
"	Quarts (Brit.)	0.0008798775
"	Quarts (U.S., dry)	0.00090808298
"	Quarts (U.S., liq.)	0.0010566882
Cu. cm./gram	Cu. ft./lb	0.016018463
Cu. cm./sec	Cu. ft./min	0.0021188800
"	Gal. (U.S.)/min	0.015850323
"	Gal. (U.S.)/sec	0.00026417205
Cu. cm.-atm	B.t.u	9.61019×10^{-5}
"	B.t.u. (mean)	9.59637×10^{-5}
"	Cal., *gm*	0.0242173
"	Cal., *gm.* (mean)	0.0241824
"	Cu. ft.-atm	3.5314667×10^{-5}
"	Joules	0.101325
"	Watt-hours	2.81458×10^{-5}
Cu. cm.-atm./gram	B.t.u./lb	0.0435911
"	Cal., *gm.*/gram	0.0242173
"	Cu. ft.-(lb./sq. in.)/lb	0.235406
"	Ft.-lb./lb	33.8985
"	Joules/gram	0.101325
"	Kg.-meters/gram	0.0103323
"	Kw.-hr./gram	2.81458×10^{-8}
Cubic decimeters	Cu. cm	1000
"	Cu. feet	0.035316667
"	Cu. inches	61.023744
"	Cu. meters	0.001

CONVERSION FACTORS (Continued)

To convert from	To	Multiply by
Cubic decimeters	Cu. yards	0.0013079506
"	Liters	0.999972
Cubic dekameters	Cu. decimeters	1×10^6
"	Cu. feet	35314.667
"	Cu. inches	6.1023744×10^7
"	Cu. meters	1000
"	Liters	999972
Cubic feet	Acre-feet	2.2956841×10^{-5}
"	Board feet	12
"	Bushels (Brit.)	0.7786049
"	Bushels (U.S.)	0.80356395
"	Cords (wood)	0.0078125
"	Cord-feet	0.0625
"	Cu. centimeters	28316.847
"	Cu. meters	0.028316847
"	Gallons (U.S., dry)	6.4285116
"	Gallons (U.S., liq.)	7.4805195
"	Liters	28.31605
"	Ounces (Brit., fluid)	996.6143
"	Ounces (U.S., fluid)	957.50649
"	Pints (U.S., liq.)	59.844156
"	Quarts (U.S., dry)	25.714047
"	Quarts (U.S., liq.)	29.922078
Cu. ft. of H_2O (39.2°F.)	Pounds of H_2O	62.4262
Cu. ft. ot H_2O (60°F.)	Pounds of H_2O	62.3663
Cu. ft./hr	Acre-feet/hr	2.2956841×10^{-5}
"	Cu. cm./sec	7.8657907
"	Cu. ft./day	24
"	Gal. (U.S.)/hr	7.4805195
"	Liters/hr	28.31605
Cu. ft./min	Acre-feet/hr	0.0013774105
"	Acre-feet/min	2.2956841×10^{-5}
"	Cu. cm./sec	471.94744
"	Cu. ft./hr	60
"	Gal. (U.S.)/min	7.4805195
"	Liters/sec	0.4719342
Cu. ft./lb	Cu. cm./gram	62.427961
"	Millimeters/gram	62.42621
Cu. ft./sec	Acre-inches/hr	0.99173553
"	Cu. cm./sec	28316.847
"	Cu. yards/min	2.222222
"	Gal. (U.S.)/min	448.83117
"	Liters/min	1698.963
"	Liters/sec	28.31605
Cu. ft. of H_2O (60°F.)/sec	Lb. of H_2O/min	3741.98
Cu. ft.-atm	B.t.u.	2.72130
"	Cal., *gm*	685.756
"	Cu. cm.-atm	28316.847
"	Cu. ft.-(lb/sq. in.)	14.6960
"	Foot-pounds	2116.22
"	Hp.-hours	0.00106880
"	Joules	2869.20
"	Kg.-meters	292.577
"	Kw.-hours	0.000797001
Cubic inches	Barrels (Brit.)	0.0001001292
"	Barrels (U.S., dry)	0.00014172336
"	Board feet	0.0069444
"	Bushels (Brit.)	0.0004505815
"	Bushels (U.S.)	0.00046502544
"	Cu. cm	16.387064
"	Cu. feet	0.00057870370
"	Cu. meters	1.6387064×10^{-5}
"	Cu. yards	2.1433470×10^{-5}
"	Drams (U.S., fluid)	4.4329004
"	Gallons (Brit.)	0.003604652
"	Gallons (U.S., dry)	0.0037202035
"	Gallons (U.S., liq.)	0.0043290043
"	Liters	0.01638661
"	Milliliters	16.38661
"	Ounces (Brit., fluid)	0.5767444
Cubic inches	Ounces (U.S., fluid)	0.55411255
"	Pecks (U.S.)	0.0018601017
"	Pints (U.S., dry)	0.029761628
"	Pints (U.S., liq.)	0.034632035
"	Quarts (U.S., dry)	0.014880814
"	Quarts (U.S., liq.)	0.017316017
Cu. in. of H_2O (4°C.)	Pounds of H_2O	0.0361263
Cu. in. of H_2O (60°F.)	Pounds of H_2O	0.0360916
Cubic meters	Acre-feet	0.00081071319
"	Barrels (Brit.)	6.110261
"	Barrels (U.S., dry)	8.648490
"	Barrels (U.S., liq.)	8.3864145
"	Bushels (Brit.)	27.49617
"	Bushels (U.S.)	28.377593
"	Cu. cm	1×10^6
"	Cu. feet	35.314667
"	Cu. inches	61023.74
"	Cu. yards	1.3079506
"	Gallons (Brit.)	219.9694
"	Gallons (U.S., liq.)	264.17205
"	Hogshead	4.1932072
"	Liters	999.972
"	Pints (U.S., liq.)	2113.3764
"	Quarts (U.S., liq.)	1056.6882
"	Steres	1
Cu. meters/min	Gal. (Brit.)/min	219.9694
"	Gal. (U.S.)/min	264.1721
"	Liters/min	999.972
Cu. millimeters	Cu. cm	0.001
"	Cu. inches	6.1023744×10^{-5}
"	Cu. meters	1×10^{-9}
"	Minims (Brit.)	0.01689365
"	Minims (U.S.)	0.016230731
Cu. yards	Bushels (Brit.)	21.02233
"	Bushels (U.S.)	21.696227
"	Cu. cm	764554.86
"	Cu. feet	27
"	Cu. inches	46.656
"	Cu. meters	0.76455486
"	Gallons (Brit.)	168.1787
"	Gallons (U.S., dry)	173.56981
"	Gallons (U.S., liq.)	201.97403
"	Liters	764.5335
"	Quarts (Brit.)	672.7146
"	Quarts (U.S., dry)	694.27926
"	Quarts (U.S., liq.)	807.89610
Cu. yd./min	Cu. ft./sec	0.45
"	Gal. (U.S.)/sec	3.3662338
"	Liters/sec	12.74222
Cubits	Centimeters	45.72
"	Feet	1.5
"	Inches	18
Daltons (chem.)	Grams	1.66024×10^{-24}
Daltons (phys.)	Grams	1.65979×10^{-24}
Days (mean solar)	Days (sidereal)	1.00273791
"	Hours (mean solar)	24
"	Hours (sidereal)	24.065710
"	Years (calendar)	0.0027397260
"	Years (sidereal)	0.0027378031
"	Years (tropical)	0.0027379093
Days (sidereal)	Days (mean solar)	0.99726957
"	Hours (mean solar)	23.934470
"	Hours (sidereal)	24
"	Minutes (mean solar)	1436.0682
"	Minute (sidereal)	1440
"	Second (sidereal)	86400
"	Years (calendar)	0.0027322454
"	Years (sidereal)	0.0027303277
"	Years (tropical)	0.0027304336
Decibels	Bels	0.1
Decimeters	Centimeters	10

CONVERSION FACTORS (Continued)

To convert from	To	Multiply by
Decimeters	Feet	0.32808399
"	Feet (U.S. Survey)	0.328083333
"	Inches	3.9370079
"	Meters	0.1
Decisteres	Cu. meters	0.1
Degrees	Circles	0.0027777
"	Minutes	60
"	Quadrants	0.0111111
"	Radians	0.017453293
"	Seconds	3600
Degrees/cm	Radians/cm	0.017453293
Degrees/foot	Radians/cm	0.00057261458
Degrees/inch	Radian/cm	0.0068713750
Degrees/min	Degrees/sec	0.0166666
"	Radians/sec	0.00029088821
"	Revolutions/sec	4.629629×10^{-5}
Degrees/sec	Radians/sec	0.017453293
"	Revolutions/min	0.166666
"	Revolutions/sec	0.0027777
Dekaliters	Pecks (U.S.)	1.135136
"	Pints (U.S., dry)	18.16217
Dekameters	Centimeters	1000
"	Feet	32.808399
"	Feet (U.S. Survey)	32.808333
"	Inches	393.70079
"	Kilometers	0.01
"	Meters	10
"	Yards	10.93613
Demals	Gram-equiv./cu. decimeter	1
Drachms (Brit., fluid)	Cu. cm	3.551631
"	Cu. inches	0.2167338
"	Drams (U.S., fluid)	0.9607594
"	Milliliters	3.551531
Drams (apoth. *or* troy)	Drams (avdp.)	2.1942857
"	Grains	60
"	Grams	3887.9346
"	Ounces (apoth. *or* troy)	0.125
"	Ounces (avdp.)	0.13714286
"	Scruples (apoth.)	3
Drams (avdp.)	Drams (apoth. *or* troy)	0.455729166
"	Grains	27.34375
"	Grams	1.7718452
"	Ounces (apoth. *or* troy)	0.056966146
"	Ounces (avdp.)	0.0625
"	Pennyweights	1.1393229
"	Pounds (apoth. *or* troy)	0.0047471788
"	Pounds (avdp.)	0.00390625
"	Scruples (apoth.)	1.3671875
Drams (U.S., fluid)	Cu. cm	3.6967162
"	Cu. inches	0.22558594
"	Drachms (Brit., fluid)	1.040843
"	Gills (U.S.)	0.03125
"	Milliliters	3.696588
"	Minims (U.S.)	60
"	Ounces (U.S., fluid)	0.125
"	Pints (U.S., liq.)	0.0078125
Dynes	Grains	0.01573663
"	Grams	0.001019716
"	Newtons	0.00001
"	Poundals	7.2330138×10^{-5}
"	Pounds	2.248089×10^{-6}
Dynes/cm	Ergs/sq. cm	1
"	Ergs/sq. mm	0.01
"	Grams/cm	0.001019716
"	Poundals/inch	0.00018371855
Dynes/cu. cm	Grams/cu. cm	0.001019716
"	Poundals/cu. inch	0.0011852786
Dynes/sq. cm	Atmospheres	9.86923×10^{-7}
"	Bars	1×10^{-6}
"	Baryes	1
"	Cm. of Hg (0°C.)	7.50062×10^{-5}
Dynes/sq. cm	Cm. of H_2O (4°C.)	0.001019745
"	Grams/sq. cm	0.001019716
"	In. of Hg (32°F.)	2.95300×10^{-5}
"	In. of H_2O (4°C.)	0.000401474
"	Kg./sq. meter	0.01019716
"	Poundals/sq. in	0.00046664510
"	Pounds/sq. in	1.450377×10^{-5}
Dyne-centimeters	Ergs	1
"	Foot-poundals	2.3730360×10^{-6}
"	Foot-pounds	7.37562×10^{-8}
"	Gram-cm	0.001019716
"	Inch-pounds	8.85075×10^{-7}
"	Kg.-meters	1.019716×10^{-8}
"	Newton-meters	1×10^{-7}
Electron volts	Ergs	1.60209×10^{-12}
"	Grams	1.78253×10^{-33}
Electronic charges	Abcoulombs	1.60209×10^{-20}
"	Coulombs	1.60209×10^{-19}
"	Statcoulombs	4.80296×10^{-10}
Electronic charges/kg	Statcoulombs/dyne	4.89766×10^{-16}
E.S. cgs. units of induction flux	E.M. cgs. units	2.997930×10^{10}
E.S. cgs. units of magnetic charge	E.M. cgs. units	2.997930×10^{10}
E.S. cgs. units of magnetic field intensity	E.M. cgs. units	3.335635×10^{-11}
Ells	Centimeters	114.3
"	Inches	45
Ergs	B.t.u	9.48451×10^{-11}
"	Cal., *gm*	2.39006×10^{-8}
"	Cal., *kg*	2.39006×10^{-11}
"	Cal., *kg*. (20°C.)	2.39126×10^{-11}
"	Cu. cm.-atm	9.86923×10^{-7}
"	Cu. ft.-atm	3.48529×10^{-11}
"	Cu. ft.-(lb./sq. in.)	5.12196×10^{-10}
"	Dyne-cm	1
"	Electron volts	6.24196×10^{11}
"	Foot-poundals	$2.3730360 \times 10^{-}$[illegible]
"	Foot-pounds	7.37562×10^{-8}
"	Gram-cm	0.001019716
"	Joules	1×10^{-7}
"	Joules (Int.)	9.99835×10^{-8}
"	Kw.-hours	2.777777×10^{-14}
"	Kg.-meters	1.019716×10^{-8}
"	Liter-atm	9.86895×10^{-10}
"	Watt-sec	1×10^{-7}
Ergs/(gram-mol. × °C.)	Foot-pounds/(lb.-mol. × °F.)	1.85863×10^{-5}
Ergs/sec	B.t.u./min	5.69071×10^{-9}
"	Cal., *gm*./min	1.43403×10^{-6}
"	Dyne-cm./sec	1
"	Foot-pounds/min	4.42537×10^{-6}
"	Gram-cm./sec	0.001019716
"	Horsepower	1.34102×10^{-10}
"	Joules/sec	1×10^{-7}
"	Kilowatts	1×10^{-10}
"	Watts	1×10^{-7}
Ergs/sq. cm	Dynes/cm	1
"	Ergs/sq. mm	0.01
Ergs/sq. mm	Dynes/cm	100
"	Ergs/sq. cm	100
Erg-sec	Planck's constant	1.50932×10^{26}
Farads	Abfarads	1×10^{-9}
"	E.M. cgs. units	1×10^{-9}
"	E.S. cgs. units	8.987584×10^{11}
"	Farads (Int.)	1.000495
"	Microfarads	1×10^{6}
"	Statfarads	8.98758×10^{11}
Farads (Int.)	Farads	0.999505
Fathoms	Centimeters	182.88

CONVERSION FACTORS (Continued)

To convert from	To	Multiply by
Fathoms	Feet	6
"	Inches	72
"	Meters	1.8288
"	Miles (naut., Int.)	0.00098747300
"	Miles (statute)	0.001136363
"	Yards	2
Feet	Centimeters	30.48
"	Chains (Gunter's)	0.01515151
"	Fathoms	0.166666
"	Feet (U.S. Survey)	0.99999800
"	Furlongs	0.00151515
"	Inches	12
"	Meters	0.3048
"	Microns	304800
"	Miles (naut., Int.)	0.00016457883
"	Miles (statute)	0.000189393
"	Rods	0.060606
"	Ropes (Brit.)	0.05
"	Yards	0.333333
Feet (U.S. Survey)	Centimeters	30.480061
"	Chains (Gunter's)	0.015151545
"	Chains (Ramden's)	0.010000020
"	Feet	1.0000020
"	Inches	12.000024
"	Links (Gunter's)	1.5151545
"	Links (Ramden's)	1.0000020
"	Meters	0.30480061
"	Miles (statute)	0.00018939432
"	Rods	0.060606182
"	Yards	0.33333400
Feet of air (1 atm., 60°F.)	Atmospheres	3.6083×10^{-5}
"	Ft. of Hg (32°F.)	0.00089970
"	Ft. of H_2O (60°F.)	0.0012244
"	In. of Hg (32°F.)	0.0010796
"	Pounds/sq. inch	0.00053027
Feet of Hg (32°F.)	Cm. of Hg (0°C.)	30.48
"	Ft. of H_2O (60°F.)	13.6085
"	In. of H_2O (60°F.)	163.302
"	Ounces/sq. inch	94.3016
"	Pounds/sq. inch	5.89385
Feet of H_2O (4°C.)	Atmospheres	0.0294990
"	Cm. of Hg (0°C.)	2.24192
"	Dynes/sq. cm	29889.8
"	Grams/sq. cm	30.4791
"	In. of Hg (32°F.)	0.882646
"	Kg./sq. meter	304.791
"	Pounds/sq. inch	0.433515
Feet/hour	Cm./hr	30.48
"	Cm./min	0.508
"	Cm./sec	0.0084666
"	Feet/min	0.0166666
"	Inches/hr	12
"	Kilometers/hr	0.0003048
"	Kilometers/min	5.08×10^{-6}
"	Knots (Int.)	0.0001645788
"	Miles/hr	0.000189393
"	Miles/min	3.156565×10^{-6}
"	Miles/sec	5.2609428×10^{-8}
Feet/minute	Cm./sec	0.508
"	Feet/sec	0.0166666
"	Kilometers/hr	0.018288
"	Meters/min	0.3048
"	Meters/sec	0.00508
"	Miles/hr	0.01136363
Feet/second	Cm./sec	30.48
"	Kilometers/hr	1.09728
"	Kilometers/min	0.018288
"	Meters/min	18.288
"	Miles/hr	0.68181818
"	Miles/min	0.01136363
Feet/(sec. × sec.)	Kilometers/(hr. × sec.)	1.09728
Feet/(sec. × sec.)	Meters/(sec. × sec.)	0.3048
"	Miles/(hr. × sec.)	0.68181818
Firkins (Brit.)	Bushels (Brit.)	1.125
"	Cu. cm	40914.79
"	Cu. feet	1.444892
"	Firkins (U.S.)	1.200949
"	Gallons (Brit.)	9
"	Liters	40.91364
"	Pints (Brit.)	72
Firkins (U.S.)	Barrels (U.S., dry)	0.29464286
"	Barrels (U.S., liq.)	0.28571429
"	Bushels (U.S.)	0.96678788
"	Cu. feet	1.203125
"	Firkins (Brit.)	0.8326747
"	Liters	34.06775
"	Pints (U.S., liq.)	72
Foot-candles	Lumens/sq. ft	1
"	Lumens/sq. meter	10.763910
"	Lux	10.763910
"	Milliphots	1.0763910
Foot-lamberts	Candles/sq. cm	0.00034262591
"	Candles/sq. ft	0.31830989
"	Millilamberts	1.0763910
"	Lamberts	0.0010763910
"	Lumens/sq. ft	1
Foot-poundals	B.t.u	3.99678×10^{-5}
"	B.t.u. (IST.)	3.99417×10^{-5}
"	B.t.u. (mean)	3.99104×10^{-5}
"	Cal., *gm*	0.0100717
"	Cal., *gm.* (IST.)	0.0100651
"	Cal., *gm.* (mean)	0.0100573
"	Cu. cm.-atm	0.415890
"	Cu. ft.-atm	1.46870×10^{-5}
"	Dyne-cm	4.2140110×10^{5}
"	Ergs	4.2140110×10^{5}
"	Foot-pounds	0.0310810
"	Hp.-hours	1.56974×10^{-8}
"	Joules	0.042140110
"	Joules (Int.)	0.0421332
"	Kg.-meters	0.00429710
"	Kw.-hours	1.17056×10^{-8}
"	Liter-atm	0.000415879
Foot-pounds	B.t.u	0.00128593
"	B.t.u. (IST.)	0.00128509
"	B.t.u. (mean)	0.00128408
"	Cal., *gm*	0.324048
"	Cal., *gm.* (IST.)	0.323836
"	Cal., *gm.* (mean)	0.323582
"	Cal., *gm.* (20°C.)	0.324211
"	Cal., *kg*	0.000324048
"	Cal., *kg.* (IST.)	0.000323836
"	Cal., *kg.* (mean)	0.000323582
"	Cu. ft.-atm	0.000472541
"	Dyne-cm	1.35582×10^{7}
"	Ergs	1.35582×10^{7}
"	Foot-poundals	32.1740
"	Gram-cm	13825.5
"	Hp.-hours	5.05050×10^{-7}
"	Joules	1.35582
"	Kg.-meters	0.138255
"	Kw.-hours	3.76616×10^{-7}
"	Kw.-hours (Int.)	3.76554×10^{-7}
"	Liter-atm	0.0133805
"	Newton-meters	1.3558180
"	Lb. H_2O evap. from and at 212°F	1.3245×10^{-6}
"	Watt-hours	0.000376616
Foot-pounds/hr	B.t.u./min	2.14321×10^{-5}
"	B.t.u. (mean)/min	2.14013×10^{-5}
"	Cal., gm./min	0.00540080
"	Cal., *gm.* (mean)/min	0.00539304
"	Ergs/min	2.25970×10^{5}

CONVERSION FACTORS.(Continued)

To convert from	To	Multiply by
Foot-pounds/hr	Foot-pounds/min	0.0166666
"	Horsepower	5.050505×10^{-7}
"	Horsepower (metric)	5.12055×10^{-7}
"	Kilowatts	3.76616×10^{-7}
"	Watts	0.000376616
"	Watts (Int.)	0.000376554
Foot-pounds/min	B.t.u./sec	2.14321×10^{-5}
"	B.t.u. (mean)/sec	2.14013×10^{-5}
"	Cal., *gm.*/sec	0.00540080
"	Cal., *gm.* (mean)/sec	0.00539304
"	Ergs/sec	2.25970×10^{5}
"	Foot-pounds/sec	0.0166666
"	Horsepower	3.030303×10^{-5}
"	Horsepower (metric)	3.07233×10^{-5}
"	Joules/sec	0.0225970
"	Joules (Int.)/sec	0.0225932
"	Kilowatts	2.25970×10^{-5}
"	Watts	0.0225970
Foot-pounds/lb	B.t.u./lb	0.00128593
"	B.t.u. (IST.)/lb	0.00128509
"	B.t.u. (mean)/lb	0.00128408
"	Cal., *gm.*/gm	0.000714404
"	Cal., *gm.* (IST.)/gram	0.000713937
"	Cal., *gm.* (mean)/gram	0.000713377
"	Hp.-hr./lb	5.05050×10^{-7}
"	Joules/gram	0.00298907
"	Kg.-meters/gram	0.000304800
"	Kw.-hr./gram	8.30296×10^{-10}
Foot-pounds/sec	B.t.u./min	0.0771556
"	B.t.u. (mean)/min	0.0770447
"	B.t.u./sec	0.00128593
"	B.t.u. (mean)/sec	0.00128408
"	Cal., *gm.*/sec	0.324048
"	Cal., *gm.* (mean)/sec	0.323582
"	Ergs/sec	1.35582×10^{7}
"	Gram-cm./sec	13825.5
"	Horsepower	0.00181818
"	Joules/sec	1.35582
"	Kilowatts	0.00135582
"	Watts	1.35582
"	Watts (Int.)	1.35559
Furlongs	Centimeters	20116.8
"	Chains (Gunter's)	10
"	Chains (Ramden's)	6.6
"	Feet	660
"	Inches	7920
"	Meters	201.168
"	Miles (naut., Int.)	0.10862203
"	Miles (statute)	0.125
"	Rods	40
"	Yards	220
Gallons (Brit.)	Barrels (Brit.)	0.027777
"	Bushels (Brit.)	0.125
"	Cu. centimeters	4546.087
"	Cu. feet	0.1605436
"	Cu. inches	277.4193
"	Drachms (Brit. fluid)	1280
"	Firkins (Brit.)	0.111111
"	Gallons (U.S., liq.)	1.200949
"	Gills (Brit.)	32
"	Liters	4.545960
"	Minims (Brit.)	76800
"	Ounces (Brit., fluid)	160
"	Ounces (U.S., fluid)	153.7215
"	Pecks (Brit.)	0.5
"	Lb. of H_2O (62°F.)	10
Gallons (U.S., dry)	Barrels (U.S., dry)	0.038095592
"	Barrels (U.S., liq.)	0.036941181
"	Bushels (U.S.)	0.125
"	Cu. centimeters	4404.8828
"	Cu. feet	0.15555700
Gallons (U.S., dry)	Cu. inches	268.8025
"	Gallons (U.S., liq.)	1.16364719
"	Liters	4.404760
Gallons (U.S., liq.)	Acre-feet	3.0688833×10^{-6}
"	Barrels (U.S., liq.)	0.031746032
"	Barrels (petroleum, U.S.)	0.023809524
"	Bushels (U.S.)	0.10742088
"	Cu. centimeters	3785.4118
"	Cu. feet	0.133680555
"	Cu. inches	231
"	Cu. meters	0.0037854118
"	Cu. yards	0.0049511317
"	Gallons (Brit.)	0.8326747
"	Gallons (U.S., dry)	0.85936701
"	Gallons (wine)	1
"	Gills (U.S.)	32
"	Liters	3.785306
"	Minims (U.S.)	61440
"	Ounces (U.S., fluid)	128
"	Pints (U.S., liq.)	8
"	Quarts (U.S., liq.)	4
Gallons (U.S.) of H_2O (4°C.)	Lb. of H_2O	8.34517
Gallons (U.S.) of H_2O (60°F.)	Lb. of H_2O	8.33717
Gallons (U.S.)/day	Cu. ft./hr	0.0055700231
Gallons (Brit.)/hr	Cu. meters/min	7.576812×10^{-5}
Gallons (U.S.)/hr	Acre-feet/hr	3.0688833×10^{-6}
"	Cu. ft./hr	0.1336805
"	Cu. meters/min	6.3090197×10^{-5}
"	Cu. yd./min	8.2518861×10^{-5}
"	Liters/hr	3.785306
Gal. (Brit.)/sec	Cu. cm./sec	4546.087
Gal. (U.S.)/sec	Cu. cm./sec	3785.4118
"	Cu. ft./min	8.020833
"	Cu. yd./min	0.29706790
"	Liters/min	227.1183
Gammas	Grams	1×10^{-6}
"	Micrograms	1
Gausses	E.M. cgs. units of magnetic flux density	1
"	E.S. cgs. units	3.335635×10^{-11}
"	Gausses (Int.)	0.999670
"	Maxwells/sq. cm	1
"	Lines/sq. cm	1
"	Lines/sq. inch	6.4516
Gausses (Int.)	Gausses	1.000330
Gausses/oersted	E.M. cgs. units of permeability	1
"	E.S. cgs. units	1.112646×10^{-21}
Geepounds	Slugs	1
"	Kilograms	14.5939
Gigameters	Meters	1×10^{9}
Gilberts	Abampere-turns	0.079577472
"	Ampere-turns	0.79577472
"	E.M. cgs. units of mmf., *or* magnetic potential	1
"	E.S. cgs. units	2.997930×10^{10}
"	Gilberts (Int.)	1.000165
Gilberts (Int.)	Gilberts	0.999835
Gilberts/cm	Ampere-turns/cm	0.79577472
"	Ampere-turns/in	2.0212678
"	Oersteds	1
Gilberts/maxwell	Ampere-turns/weber	7.957747×10^{7}
"	E.M. cgs. units of reluctance	1
"	E.S. cgs. units	8.987584×10^{20}
Gills (Brit.)	Cu. cm	142.0652
"	Gallons (Brit.)	0.03125
"	Gills (U.S.)	1.200949
"	Liters	0.1420613
"	Ounces (Brit., fluid)	5

CONVERSION FACTORS (Continued)

To convert from	To	Multiply by
Gills (Brit.)	Ounces (U.S., fluid)	4.803764
"	Pints (Brit.)	0.25
Gills (U.S.)	Cu. cm	118.29412
"	Cu. inches	7.21875
"	Drams (U.S., fluid)	32
"	Gallons (U.S., liq.)	0.03125
"	Gills (Brit.)	0.8326747
"	Liters	0.1182908
"	Minims (U.S.)	1920
"	Ounces (U.S., fluid)	4
"	Pints (U.S., liq.)	0.25
"	Quarts (U.S., liq.)	0.125
Grades	Circles	0.0025
"	Circumferences	0.0025
"	Degrees	0.9
"	Minutes	54
"	Radians	0.015707963
"	Revolutions	0.0025
"	Seconds	3240
Grains	Carats (metric)	0.32399455
"	Drams (apoth. *or* troy)	0.016666
"	Drams (avdp.)	0.036571429
"	Dynes	63.5460
"	Grams	0.06479891
"	Milligrams	64.79891
"	Ounces (apoth. *or* troy)	0.0020833
"	Ounces (avdp.)	0.0022857143
"	Pennyweights	0.041666
"	Pounds (apoth. *or* troy)	0.000173611
"	Pounds (avdp.)	0.00014285714
"	Scruples (apoth.)	0.05
"	Tons (metric)	6.479891×10^{-8}
Grains/cu. ft	Grams/cu. meter	2.2883519
Grains/gal. (U.S.)	Parts/million*	17.11854
"	Pounds/million gal	142.8571
Grams	Carats (metric)	5
"	Decigrams	10
"	Dekagrams	0.1
"	Drams (apoth. *or* troy)	0.25720597
"	Drams (avdp.)	0.56438339
"	Dynes	980.665
"	Grains	15.432358
"	Kilograms	0.001
"	Micrograms	1×10^{6}
"	Myriagrams	0.0001
"	Ounces (apoth. *or* troy)	0.032150737
"	Ounces (avdp.)	0.035273962
"	Pennyweights	0.64301493
"	Poundals	0.0709316
"	Pounds (apoth. *or* troy)	0.0026792289
"	Pounds (avdp.)	0.0022046226
"	Scruples (apoth.)	0.77161792
"	Tons (metric)	1×10^{-6}
Grams/cm	Dynes/cm	980.665
"	Grams/inch	2.54
"	Kg./km	100
"	Kg./meter	0.1
"	Poundals/inch	0.180166
"	Pounds/ft	0.067196898
"	Pounds/inch	0.0055997415
"	Tons (metric)/km	0.1
Grams/(cm. × sec.)	Poises	1
"	Lb./(ft. × sec.)	0.06719690
Grams/cu. cm	Dynes/cu. cm	980.665
"	Grains/milliliter	15.43279
"	Grams/milliliter	1.000028
"	Poundals/cu. inch	1.16236
"	Pounds/circ. mil-ft	3.4049170×10^{-7}
"	Pounds/cu. ft	62.427961
"	Pounds/cu. inch	0.036127292
"	Pounds/gal. (Brit.)	10.02241
Grams/cu. cm	Pounds/gal. (U.S., dry)	9.7111064
"	Pounds/gal. (U.S., liq.)	8.3454044
Grams/cu. meter	Grains/cu. ft	0.43699572
Grams/liter	Parts/million*	1000
"	Lb./ cu. ft	0.06242621
"	Lb./gal. (U.S.)	8.345171
Grams/milliliter	Grams/cu. cm	0.999972
"	Pounds/cu. ft	62.42621
"	Pounds/gallon (U.S.)	8.345171
Grams/sq. cm	Atmospheres	0.000967841
"	Bars	0.000980665
"	Cm. of Hg. (0°C.)	0.0735559
"	Dynes/sq. cm	980.665
"	In. of Hg (32°F.)	0.0289590
"	Kg./sq. meter	10
"	Mm. of Hg (0°C.)	0.735559
"	Poundals/sq. inch	0.457623
"	Pounds/sq. inch	0.014223343
Grams/ton (long)	Milligrams/kg	0.98420653
Grams/ton (short)	Milligrams/kg	1.1023113
Grams-cm	B.t.u.	9.30113×10^{-8}
"	B.t.u. (IST.)	9.29505×10^{-8}
"	B.t.u. (mean)	9.28776×10^{-8}
"	Cal., *gm*	2.34385×10^{-5}
"	Cal., *gm.* (IST.)	2.34231×10^{-5}
"	Cal., *gm.* (mean)	2.34048×10^{-5}
"	Cal., *gm.* (15°C.)	2.34284×10^{-5}
"	Cal., *gm*, (20°C.)	2.34502×10^{-5}
"	Cal., *kg*	2.34385×10^{-8}
"	Cal., *kg.* (IST.)	2.34231×10^{-8}
"	Cal., *kg.* (mean)	2.34048×10^{-8}
"	Dyne-cm	980.665
"	Ergs	980.665
"	Foot-poundals	0.00232715
"	Foot-pounds	7.2330138×10^{-5}
"	Hp.-hours	3.65303×10^{-11}
"	Joules	9.80665×10^{-5}
"	Kw.-hours	2.72407×10^{-11}
"	Kw.-hours (Int.)	2.72362×10^{-11}
"	Newton-meters	9.80665×10^{-5}
"	Watt-hours	2.72407×10^{-8}
Gram-cm./sec	B.t.u./sec	9.30113×10^{-8}
"	Cal., *gm.*/sec	2.34385×10^{-5}
"	Ergs-sec	980.665
"	Foot-pounds/sec	7.2330138×10^{-5}
"	Horsepower	1.31509×10^{-7}
"	Joules/sec	9.80665×10^{-5}
"	Kilowatts	9.80665×10^{-8}
"	Kilowatts (Int.)	9.80503×10^{-8}
"	Watts	9.80665×10^{-5}
Gram/sq. cm	Pounds/sq. inch	0.000341717
Gram wt.-sec,/sq. cm	Poises	980.665
Gravitational constants	Cm./(sec. × sec.)	980.621
"	Ft./(sec. × sec.)	32.1725
Hands	Centimeters	10.16
"	Inches	4
Hectares	Acres	2.4710538
"	Ares	100
"	Sq. cm	1×10^{8}
"	Sq. feet	107639.10
"	Sq. meters	10000
"	Sq. miles	0.0038610216
"	Sq. rods	395.36861
Hectograms	Grams	100
"	Poundals	7.09316
"	Pounds (apoth *or* troy)	0.26792289
"	Pounds (avdp.)	0.22046226
Hectoliters	Bushels (Brit.)	2.749694
"	Bushels (U.S.)	2.837839

* Based on density of 1 gram/ml.

CONVERSION FACTORS (Continued)

To convert from	To	Multiply by
Hectoliters	Cu. cm	1.00028×10^{5}
"	Cu. feet	3.531566
"	Gallons (U.S., liq.)	26.41794
"	Liters	100
"	Ounces (U.S.) fluid	3381.497
"	Pecks (U.S.)	11.35136
Hectometers	Centimeters	10000
"	Decimeters	1000
"	Dekameters	10
"	Feet	328.08399
"	Meters	100
"	Rods	19.883878
"	Yards	109.3613
Hectowatts	Watts	100
Hefner units	Candles (English)	0.86
"	Candles (German)	0.85
"	Candles (Int.)	0.90
"	10-cp. pentane candles	0.090
Henries	Abhenries	1×10^{9}
"	E.M. cgs. units	1×10^{9}
"	E.S. cgs. units	1.112646×10^{-12}
"	Henries (Int.)	0.999505
"	Millihenries	1000
"	Mks. (r *or* nr) units	1
"	Stathenries	1.112646×10^{-12}
Henries (Int.)	Henries	1.000495
Henries/meter	Cgs. units of permeability	795774.72
"	E.M. cgs. units	795774.72
"	E.S. cgs. units	8.854156×10^{-16}
"	Gausses/oersted	795774.72
"	Mks. (nr) units	0.079577472
"	Mks. (r) units	1
Hogsheads	Butts (Brit.)	0.5
"	Cu. feet	8.421875
"	Cu. inches	14553
"	Cu. meters	0.23848094
"	Gallons (Brit.)	52.458505
"	Gallons (U.S.)	63
"	Gallons (wine)	63
"	Liters	238.47427
Horsepower*	B.t.u. (mean)/hr	2542.48
"	B.t.u./min	42.4356
"	B.t.u. (mean)/sec	0.706243
"	Cal., *gm.*/hr	6.41616×10^{5}
"	Cal., *gm.* (IST.)/hr	6.41196×10^{5}
"	Cal., *gm.* (mean)/hr	6.40693×10^{5}
"	Cal., *gm.*/min	10693.6
"	Cal., *gm.* (IST.)/min	10686.6
"	Cal., *gm.* (mean)/min	10678.2
"	Ergs/sec	7.45700×10^{9}
"	Foot-pounds/hr	1980000
"	Foot-pounds/min	33000
"	Foot-pounds/sec	550
"	Horsepower (boiler)	0.0760181
"	Horsepower (electric)	0.999598
"	Horsepower (metric)	1.01387
"	Joules/sec	745.700
"	Kilowatts	0.745700
"	Kilowatts (Int.)	0.745577
"	Tons of refrig. (U.S., comm.)	0.21204
"	Watts	745.700
Horsepower (boiler)	B.t.u. (mean)/hr	33445.7
"	Cal., *gm.*/min	140671.6
"	Cal., *gm.* (mean)/min	140469.4
"	Cal., *gm.* (15°C.)/min	140611.1
"	Cal., *gm.*, (20°C.)/min	140742.2
"	Ergs/sec	9.80950×10^{10}
"	Foot-pounds/min	434107
"	Horsepower	13.1548
"	Horsepower (electric)	13.1495
Horsepower (boiler)	Horsepower (metric)	13.3372
"	Horsepower (water)	13.1487
"	Joules/sec	9809.50
"	Kilowatts	9.80950
"	Lb. H_2O evap. per hr. from and at 212°F	34.5
Horsepower (electric)	B.t.u./hr	2547.16
"	B.t.u. (IST.)/hr	2545.50
"	B.t.u. (mean)/hr	2543.50
"	Cal., *gm.*/sec	178.298
"	Cal., *kg.*/hr	641.874
"	Ergs/sec	7.46×10^{9}
"	Foot-pounds/min	33013.3
"	Foot-pounds/sec	550.221
"	Horsepower	1.00040
"	Horsepower (boiler)	0.0760487
"	Horsepower (metric)	1.0142777
"	Horsepower (water)	0.999942
"	Joules/sec	746
"	Kilowatts	0.746
"	Watts	746
Horsepower (metric)	B.t.u./hr	2511.31
"	B.t.u. (IST.)/hr	2509.66
"	B.t.u. (mean)/hr	2507.70
"	Cal., *gm.*/hr	6.32838×10^{5}
"	Cal., *gm.* (IST.)/hr	6.32425×10^{5}
"	Cal., *gm.* (mean)/hr	6.31929×10^{5}
"	Ergs/sec	7.35499×10^{9}
"	Foot-pounds/min	32548.6
"	Foot-pounds/sec	542.476
"	Horsepower	0.986320
"	Horsepower (boiler)	0.0749782
"	Horsepower (electric)	0.985923
"	Horsepower (water)	0.985866
"	Kg.-meters/sec	75
"	Kilowatts	0.735499
"	Watts	735.499
Horsepower (water)	Foot-pounds/min	33015.2
"	Horsepower	1.00046
"	Horsepower (boiler)	0.0760531
"	Horsepower (electric)	1.00006
"	Horsepower (metric)	1.01434
"	Kilowatts	0.746043
Horsepower-hours	B.t.u.	2546.14
"	B.t.u. (IST.)	2544.47
"	B.t.u. (mean)	2542.48
"	Cal., *gm*	641616
"	Cal., *gm.* (IST.)	641196
"	Cal., *gm.* (mean)	640693
"	Foot-pounds	1.98×10^{6}
"	Joules	2.68452×10^{6}
"	Kg.-meters	273745
"	Kw.-hours	0.745700
"	Watt-hours	745.700
Hp.-hr./lb	B.t.u./lb	2546.14
"	Cal., *gm.*/gram	1414.52
"	Cu. ft.-(lb./sq. in.)/lb	13750
"	Foot-pounds/lb	1980000
"	Joules/gram	5918.35
Hours (mean solar)	Days (mean solar)	0.0416666
"	Days (sidereal)	0.041780746
"	Hours (sidereal)	1.00273791
"	Minutes (mean solar)	60
"	Minutes (sidereal)	60.164275
"	Seconds (mean solar)	3600
"	Seconds (sidereal)	3609.8565
"	Weeks (mean calendar)	0.0059523809
Hours (sidereal)	Days (mean solar)	0.41552899
"	Days (sidereal)	0.0416666
"	Hours (mean solar)	0.99726957
"	Minutes (mean solar)	59.836174

* Mechanical horsepower, equal to 550 ft.-lb./sec.

CONVERSION FACTORS (Continued)

To convert from	To	Multiply by
rs (sidereal)	Minutes (sidereal)	60
ndredweights (long)	Kilograms	50.802345
"	Pounds	112
"	Quarters (Brit., long)	4
"	Quarters (U.S., long)	0.2
"	Tons (long)	0.05
ndredweights short)	Kilograms	45.359237
"	Pounds (advp.)	100
"	Quarters (Brit., short)	4
"	Quarters (U.S., short)	0.2
"	Tons (long)	0.044642857
"	Tons (metric)	0.045359237
"	Tons (short)	0.05
hes	Ångström units	2.54×10^8
	Centimeters	2.54
	Chains (Gunter's)	0.00126262
	Cubits	0.055555
	Fathoms	0.013888
	Feet	0.083333
	Feet (U.S. Survey)	0.083333167
	Links (Gunter's)	0.126262
	Links (Ramden's)	0.083333
	Meters	0.0254
	Mils	1000
	Picas (printer's)	6.0225
	Points (printer's)	72.27000
	Wave length of orange-red line of krypton 86	41929.399
	Wave length of the red line of cadmium	39450.369
	Yards	0.027777
es of Hg (32°F.)	Atmospheres	0.0334211
"	Bars	0.0338639
"	Dynes/sq. cm	33863.9
"	Ft. of air (1 atm., 60°F.)	926.24
"	Ft. of H_2O (39.2°F.)	1.132957
"	Grams/sq. cm	34.5316
"	Kg./sq. meter	345.316
"	Mm. of Hg (60°C.)	25.4
"	Ounces/sq. inch	7.85847
es of Hg (32°F.)	Pounds/sq. ft	70.7262
es of Hg (60°F.)	Atmospheres	0.0333269
"	Dynes/sq. cm	39768.5
"	Grams/sq. cm	34.4343
"	Mm. of Hg (60°F.)	25.4
"	Ounces/sq. inch	7.83633
"	Pounds/sq. ft	70.5269
es of H_2O(4°C.)	Atmospheres	0.0024582
	Dynes/sq. cm	2490.82
"	In. of Hg (32°F.)	0.0735539
"	Kg./sq. meter	25.3993
"	Ounces/sq. ft	83.2350
"	Ounces/sq. inch	0.578020
"	Pouncs/sq. ft	5.20218
"	Pounds/sq. inch	0.03612628
es/hr	Cm./hr	2.54
"	Feet/hr	0.0833333
	Miles/hr	1.578282×10^{-5}
es/min	Cm./hr	152.4
"	Feet/hr	5
"	Miles/hr	0.000946969
es (abs.)	B.t.u	0.000948451
"	B.t.u. (IST.)	0.000947831
"	B.t.u. (mean)	0.000947088
"	Cal., *gm*	0.239006
"	Cal., *gm.* (IST.)	0.238849
"	Cal., *gm.* (mean)	0.238662
"	Cal., *gm.* (15°C.)	0.238903
"	Cal., *gm.* (20°C.)	0.239126

To convert from	To	Multiply by
Joules (abs)	Cal., *kg.* (mean)	0.000238662
" "	Cu. ft.-atm	0.000348529
" "	Ergs	1×10^7
" "	Foot-poundals	23.730360
" "	Foot-pounds	0.737562
" "	Gram-cm	10197.16
" "	Hp.-hours	3.72506×10^{-7}
" "	Joules (Int.)	0.999835
" "	Kg.-meters	0.1019716
" "	Kw.-hours	2.7777×10^{-7}
" "	Liter-atm	0.00986895
" "	Volt-coulombs (Int.)	0.999835
" "	Watt-hours (abs.)	0.0002777777
" "	Watt-hours (Int.)	0.000277732
" "	Watt-sec	1
" "	Watt-sec. (Int.)	0.999835
Joules (Int.)	B.t.u	0.000948608
"	B.t.u. (IST.)	0.000947988
"	B.t.u. (mean)	0.000947244
"	Cal. *gm*	0.239045
"	Cal., *gm.* (IST.)	0.238888
"	Cal., *gm.* (mean)	0.238702
"	C.h.u	0.000527004
"	C.h.u. (IST.)	0.000526660
"	C.h.u. (mean)	0.000526247
"	Cu. cm.-atm	9.87086
"	Cu. ft.-atm	0.000348586
"	Dyne-cm	1.000165×10^7
"	Ergs	1.000165×10^7
"	Foot-poundals	23.73428
"	Foot-pounds	0.737684
"	Gram-cm	10198.8
"	Joules (abs.)	1.000165
"	Kw.-hours	2.77824×10^{-7}
"	Liter-atm	0.00987058
"	Volt-coulombs	1.000165
"	Volt-coulombs (Int.)	1
"	Watt-sec	1.000165
"	Watt-sec. (Int.)	1
Joules/(abcoulomb × °F.)	Joules/(coulomb × °C.)	0.18
Joules/amp.-hr	Joules/abcoulomb	0.002777
"	Joules/statcoulomb	9.265653×10^{-14}
Joules/coulomb	Joules/abcoulomb	10
"	Volts	1
Joules/(coulomb × °F.)	Joules/(coulomb × °C.)	1.8
Joules/°C	B.t.u./°F	0.000526917
"	Cal., *gm.*/°C	0.239006
"	Cal., *gm.* (mean)/°C	0.238662
Joules/electronic charge	Joules/abcoulomb	6.24196×10^{19}
Joules/(electronic charge × °C.)	Joules/(coulomb × °C.)	6.24196×10^{18}
Joules/(gram × °C.)	B.t.u./(lb. × °F.)	0.239006
"	Cal., *gm.*/(gram × °C.)	0.239006
Joules (Int.)/(gram °C.)	B.t.u./(lb. × °F.)	0.239045
"	Cal., *gm.* (mean)/(gram × °C.)	0.238702
Joules/sec. (abs.)	B.t.u./min	0.0569071
"	Cal., *gm.*/min	14.3403
"	Cal., *kg.*/min	0.0143403
"	Cal., *kg.* (mean)/min	0.0143197
"	Dyne-cm./sec	1×10^7
"	Ergs/sec	1×10^7
"	Foot-pounds/sec	0.737562
"	Gram-cm./sec	10197.16
"	Horsepower	0.00134102
"	Watts	1
"	Watts (Int.)	0.999835
Joules (Int.)/sec	B.t.u./min	0.0569165

CONVERSION FACTORS (Continued)

To convert from	To	Multiply by
Joules (Int.)/sec	B.t.u. (mean)/min	0.0568347
"	Cal., *gm.*/min	14.3427
"	Cal., *kg.*/min	0.0143427
"	Dyne-cm./sec	1.000165×10^{7}
"	Ergs/sec	1.000165×10^{7}
"	Foot-pounds/min	44.2610
"	Foot-pounds/sec	0.737684
"	Gram-cm./sec	10198.8
"	Horsepower	0.00134124
"	Watts	1.000165
"	Watts (Int.)	1
Kilderkins (Brit.)	Cu. cm	81829.57
"	Cu. feet	2.889784
"	Cu. inches	4993.55
"	Cu. meters	0.08182957
"	Gallons (Brit.)	18
Kilograms	Drams (apoth. *or* troy)	257.20597
"	Drams (avdp.)	564.38339
"	Dynes	980665
"	Grains	15432.358
"	Hundredweights (long)	0.019684131
"	Hundredweights (short)	0.022046226
"	Ounces (apoth. *or* troy)	32.150737
"	Ounces (avdp.)	35.273962
"	Pennyweights	643.01493
"	Poundals	70.931635
"	Pounds (apoth. *or* troy)	2.6792289
"	Pounds (avdp.)	2.2046226
"	Quarters (Brit., long)	0.078736522
"	Quarters (U.S. long)	0.0039368261
"	Scruples (apoth.)	771.61792
"	Slugs	0.06852177
"	Tons (long)	0.00098420653
"	Tons (metric)	0.001
"	Tons (short)	0.0011023113
Kilograms/cu. meter	Grams/cu. cm	0.001
"	Lb./cu. ft	0.062427961
"	Lb./cu. inch	3.6127292×10^{-5}
Kg. of ice melted/hr	Tons of refrig. (U.S., comm.)	0.026336
Kilograms/sq. cm	Atmospheres	0.967841
"	Bars	0.980665
"	Cm. of Hg (0°C.)	73.5559
"	Dynes/sq. cm	980665
"	Ft. of H_2O (39.2°F.)	32.8093
"	In. of Hg (32°F.)	28.9590
"	Pounds/sq. inch	14.223343
Kilograms/sq. meter	Atmospheres	9.67841×10^{-5}
"	Bars	9.80665×10^{-5}
"	Dynes/sq. cm	98.0665
"	Ft. of H_2O (39.2°F.)	0.00328093
"	Grams/sq. cm	0.1
"	In. of Hg. (32°F.)	0.00289590
"	Mm. of Hg (0°C.)	0.0735559
"	Pounds/sq. ft	0.20481614
"	Pounds/sq. in	0.0014223343
Kilograms/sq. mm	Pounds/sq. ft	204816.14
"	Pounds/sq. in	1422.3343
"	Tons (short)/sq. in	0.71116716
Kilogram sq. cm	Pounds sq. ft	0.0023730360
"	Pounds sq. in	0.34171719
Kilogram-meters	B.t.u. (mean)	0.00928776
"	Cal., *gm.* (mean)	2.34048
"	Cal., *kg.* (mean)	0.00234048
"	Cu. ft.-atm	0.00341790
"	Dynes-cm	9.80665×10^{7}
"	Ergs	9.80665×10^{7}
"	Foot-poundals	232.715
"	Foot-pounds	7.23301
"	Gram-cm	100000

To convert from	To	Multiply by
Kilogram-meters	Hp.-hours	3.65304×10^{-6}
"	Joules	9.80665
"	Joules (Int.)	9.80503
"	Kw.-hours	2.72407×10^{-6}
"	Liter-atm	0.0967814
"	Newton-meters	9.80665
"	Watt-hours	0.00272407
"	Watt-hours (Int.)	0.00272362
Kilogram-meters/sec	Watts	9.80665
Kilolines	Maxwells	1000
"	Webers	1×10^{-5}
Kiloliters	Cu. centimeters	1.000028×10^{6}
"	Cu. feet	35.31566
"	Cu. inches	61025.45
"	Cu. meters	1.000028
"	Cu. yards	1.307987
"	Gallons (Brit.)	219.9755
"	Gallons (U.S., dry)	227.0271
"	Gallons (U.S., liq.)	264.1794
"	Liters	1000
Kilometers	Astronomical units	6.68878×10^{-9}
"	Centimeters	100000
"	Feet	3280.8399
"	Feet (U.S. Survey)	3280.833
"	Light years	1.05702×10^{-13}
"	Meters	1000
"	Miles (naut., Int.)	0.53995680
"	Miles (statute)	0.62137119
"	Myriameters	0.1
"	Rods	198.83878
"	Yards	1093.6133
Kilometers/hr	Cm./sec	27.7777
"	Feet/hr	3280.8399
"	Feet/min	54.680665
"	Knots (Int.)	0.53995680
"	Meters/sec	0.277777
"	Miles (statute)/hr	0.62137119
Kilometers/(hr. × sec.)	Cm./(sec. × sec.)	27.7777
"	Ft./(sec. × sec.)	0.91134442
"	Meters/(sec. × sec.)	0.277777
Kilometers/min	Cm./sec	1666.666
"	Feet/min	3280.8399
"	Kilometers/hr	60
"	Knots (Int.)	32.397408
"	Miles/hr	37.282272
"	Miles/min	0.62137119
Kilovolts/cm	Abvolts/cm	1×10^{11}
"	Microvolts/meter	1×10^{11}
"	Millivolts/meter	1×10^{8}
"	Statvolts/cm	3.335635
"	Volts/inch	2540
Kilowatts	B.t.u./hr	3414.43
"	B.t.u. (IST.)/hr	3412.19
"	B.t.u. (mean)/hr	3409.52
"	B.t.u. (mean)/min	56.8253
"	B.t.u. (mean)/sec	0.947088
"	Cal., *gm.* (mean)/hr	859184
"	Cal., *gm.* (mean)/min	14319.7
"	Cal., *gm.* (mean)/sec	238.662
"	Cal., *kg.* (mean)/hr	859.184
"	Cal., *kg.* (mean)/min	14.3197
"	Cal., *kg.* (mean)/sec	0.238662
"	Cu. ft.-atm./hr	1254.70
"	Ergs/sec	1×10^{10}
"	Foot-poundals/min	1.42382×10^{6}
"	Foot-pounds/hr	2.65522×10^{6}
"	Foot-pounds/min	44253.7
"	Foot-pounds/sec	737.562
"	Gram-cm./sec	1.019716×10^{7}
"	Horsepower	1.34102

CONVERSION FACTORS (Continued)

To convert from	To	Multiply by
Kilowatts	Horsepower (boiler)	0.101942
"	Horsepower (electric)	1.34048
"	Horsepower (metric)	1.35962
"	Joules/hr	3.6×10^6
"	Joules (IST.)/hr	3.59941×10^6
"	Joules/sec	1000
"	Kg.-meters/hr	3.67098×10^5
"	Kilowatts (Int.)	0.999835
"	Watts (Int.)	999.835
Kilowatts (Int.)	B.t.u./hr	3414.99
"	B.t.u. (IST.)/hr	3412.76
"	B.t.u. (mean)/hr	3410.08
"	B.t.u. (mean)/min	56.8347
"	B.t.u. (mean)/sec	0.947244
"	Cal., *gm.* (mean)/hr	859326
"	Cal., *gm.* (mean)/min	14322.1
"	Cal., *kg.*/hr	860.563
"	Cal., *kg.* (IST.)/hr	860
"	Cal., *kg.* (mean)/hr	859.326
"	Cu. cm.-atm./hr	3.55351×10^7
"	Cu. ft.-atm./hr	1254.91
"	Ergs/sec	1.000165×10^{10}
"	Foot-poundals/min	1.42406×10^6
"	Foot-pounds/min	44261.0
"	Foot-pounds/sec	737.684
"	Gram-cm./sec	1.01988×10^7
"	Horsepower	1.34124
"	Horsepower (boiler)	0.101959
"	Horsepower (electric)	1.34070
"	Horsepower (metric)	1.35985
"	Joules/hr	3.60059×10^6
"	Joules (Int.)/hr	3.6×10^6
"	Kg.-meters/hr	367158
"	Kilowatts	1.000165
Kilowatt-hours	B.t.u. (mean)	3409.52
"	Cal., *gm.* (mean)	859184
"	Foot-pounds	2.65522×10^6
"	Hp.-hours	1.34102
"	Joules	3.6×10^6
"	Kg.-meters	367098
"	Lb. H_2O evap. from and at 212°F	3.5168
"	Watt-hours	1000
"	Watt-hours (Int.)	999.835
Kilowatt-hours (Int.)	B.t.u. (mean)	3410.08
"	Cal., *gm.* (IST.)	860000
"	Cal., *gm.* (mean)	859326
"	Cu. cm.-atm	3.55351×10^7
"	Cu. ft.-atm	1254.91
"	Foot-pounds	2.65566×10^6
"	Hp.-hours	1.34124
"	Joules	3.60059×10^6
"	Joules (Int.)	3.6×10^6
"	Kg.-meters	367158
Kw.-hr./gram	B.t.u./(lb	1.54876×10^6
"	B.t.u. (IST.)/lb	1.54774×10^6
"	B.t.u. (mean)/lb	1.54653×10^6
"	Cal., *gm.*/gram	860421
"	Cal., *gm.* (mean)/gram	859184
"	Cu. cm.-atm./gram	3.55292×10^7
"	Cu. ft.-atm./lb	569124
"	Hp.-hr./lb	608.277
"	Joules/gram	3.6×10^6
Knots (Int.)	Cm./sec	51.4444
"	Feet/hr	6076.1155
"	Feet/min	101.26859
"	Feet/sec	1.6878099
"	Kilometers/hr	1.852
"	Meters/min	30.8666
"	Meters/sec	0.514444
"	Miles (naut., Int.)/hr	1
"	Miles (statute)/hr	1.1507794

To convert from	To	Multiply by
Lamberts	Candles/sq. cm	0.31830989
"	Candles/sq. ft	295.71956
"	Candles/sq. inch	2.0536081
"	Foot-lamberts	929.0304
"	Lumens/sq. cm	1
Lasts (Brit.)	Liters	2909.414
Leagues (naut., Brit.)	Feet	18240
"	Kilometers	5.559552
"	Leagues (naut., Int.)	1.0006393
"	Leagues (statute)	1.151515
"	Miles (statute)	3.454545
Leagues (naut., Int.)	Fathoms	3038.0577
"	Feet	18228.346
"	Kilometers	5.556
"	Leagues (statute)	1.1507794
"	Miles (statute)	3.4523383
Leagues (statute)	Fathoms	2640
"	Feet	15840
"	Kilometers	4.828032
"	Leagues (naut., Int.)	0.86897625
"	Miles (naut., Int.)	2.6069287
"	Miles (statute)	3
Light years	Astronomical units	63279.5
"	Kilometers	9.46055×10^{12}
"	Miles (statute)	5.87851×10^{12}
Lines	Maxwells	1
Lines (Brit.)	Centimeters	0.211666
"	Inches	0.083333
Lines/sq. cm	Gausses	1
Lines/sq. inch	Gausses	0.15500031
"	Webers/sq. inch	1×10^{-8}
Links (Gunter's)	Chains (Gunter's)	0.01
"	Feet	0.66
"	Feet (U.S. Survey)	0.65999868
"	Inches	7.92
"	Meters	0.201168
"	Miles (statute)	0.000125
"	Rods	0.04
Links (Ramden's)	Centimeters	30.48
"	Chains (Ramdens)	0.01
"	Feet	1
"	Inches	12
Liters	Bushels (Brit.)	0.02749694
"	Bushels (U.S.)	0.02837839
"	Cu. centimeters	1000
"	Cu. feet	0.03531566
"	Cu. inches	61.02545
"	Cu. meters	0.001000028
"	Cu. yards	0.001307987
"	Drams (U.S., fluid)	270.5198
"	Gallons (Brit.)	0.2199755
"	Gallons (U.S., dry)	0.2270271
"	Gallons (U.S., liq.)	0.2641794
"	Gills (Brit.)	7.039217
"	Gills (U.S.)	8.453742
"	Hogsheads	0.004193325
"	Minims (U.S.)	16231.19
"	Ounces (Brit., fluid)	35.19609
"	Ounces (U.S., fluid)	33.81497
"	Pecks (Brit.)	0.1099878
"	Pecks (U.S.)	0.1135136
"	Pints (Brit.)	1.759804
"	Pints (U.S., dry)	1.816217
"	Pints (U.S., liq.)	2.113436
"	Quarts. (Brit.)	0.8799021
"	Quarts (U.S., dry)	0.9081084
"	Quarts (U.S., liq.)	1.056718
Liters/min	Cu. ft./min	0.03531566
"	Cu. ft./sec	0.0005885943
"	Gal. (U.S., liq.)/min	0.2641794
Liters/sec	Cu. ft./min	2.118939
"	Cu. ft./sec	0.03531566

CONVERSION FACTORS (Continued)

To convert from	To	Multiply by
Liters/sec	Cu. yards/min	0.07847923
"	Gal. (U.S., liq.)/min	15.85077
"	Gal. (U.S., liq.)/sec	0.2641794
Liter-atm	B.t.u.	0.0961045
"	B.t.u. (IST.)	0.0960417
"	B.t.u. (mean)	0.0959664
"	Cal., *gm*	24.2179
"	Cal., *gm*. (IST.)	24.2021
"	Cal., *gm*. (mean)	24.1831
"	Cu. ft.-atm	0.0353157
"	Foot-poundals	2404.55
"	Foot-pounds	74.7356
"	Hp.-hours	3.77452×10^{-5}
"	Joules	101.328
"	Joules (Int.)	101.311
"	Kg.-meters	10.3326
"	Kw.-hours	2.81466×10^{-5}
Liter-atm. (lat. 45°)	Joules	101.323
Lumens	Candle power (spher.)	0.079577472
Lumens (at 5550 Å)	Watts	0.0014705882
Lumens/sq. cm	Lamberts	1
"	Phots	1
Lumens/(sq. cm. × steradian)	Lamberts	3.1415927
Lumens/sq. ft	Foot-candles	1
"	Foot-lamberts	1
"	Lumens/sq. meter	10.763910
Lumens/(sq. ft. × steradian)	Millilamberts	3.3815822
Lumens/sq. meter	Foot-candles	0.09290304
"	Lumens/sq. ft	0.09290304
"	Phots	0.0001
Lux	Foot-candles	0.09290304
"	Lumens/sq. meter	1
"	Phots	0.0001
Maxwells	E.M. cgs. units of induction flux	1
"	E.S. cgs. units	3.335635×10^{-11}
"	Gauss-sq. cm	1
"	Lines	1
"	Maxwells (Int.)	0.999670
"	Volt-seconds	1×10^{-8}
"	Webers	1×10^{-8}
Maxwells (Int.)	Maxwells	1.000330
Maxwells/sq. cm	Maxwells/sq. in	6.4516
"	Maxwells (Int.)/sq. cm	0.999670
Maxwells (Int.)/sq. cm	Maxwells/sq. cm	1.000330
Maxwells/sq. inch	Maxwells/sq. cm	0.15500031
Megalines	Maxwells	1×10^{6}
Megmhos/cm	Abmhos/cm	0.001
"	Megmhos/inch cube	2.54
"	(Microhm-cm.)$^{-1}$	1
Megmhos/inch	Megmhos/cm	0.39370079
"	(Microhm-inches)$^{-1}$	1
Megohms	Microhms	1×10^{12}
"	Ohms	1×10^{6}
"	Statohms	1.112646×10^{-6}
Megohms^{-1}	Micromhos	1
Meters	Ångström units	1×10^{10}
"	Centimeters	100
"	Chains (Gunter's)	0.049709695
"	Chains (Ramden's)	0.032808399
"	Fathoms	0.54680665
"	Feet	3.2808399
"	Feet (U.S. Survey)	3.280833
"	Furlongs	0.0049709695
"	Inches	39.370079
"	Kilometers	0.001
"	Links (Gunter's)	4.9709695

To convert from	To	Multiply by
Meters	Links (Ramden's)	3.2808399
"	Megameters	1×10^{-6}
"	Miles (naut., Brit.)	0.00053961182
"	Miles (naut., Int.)	0.00053995680
"	Miles (statute)	0.00062137119
"	Millimeters	1000
"	Millimicrons	1×10^{9}
"	Mils	39370.079
"	Rods	0.19883878
"	Yards	1.0936133
Meters of Hg (0°C.)	Atmospheres	1.3157895
"	Ft. of H_2O (60°F.)	44.6474
"	In. of Hg (32°F.)	39.370079
"	Kg./sq. cm	1.35951
"	Pounds/sq. inch	19.3368
Meters/hr	Feet/hr	3.2808399
"	Feet/min	0.054680665
"	Knots (Int.)	0.00053995680
"	Miles (statute)/hr	0.00062137119
Meters/min	Cm./sec	1.666666
"	Feet/min	3.2808399
"	Feet/sec	0.054680665
"	Kilometers/hr	0.06
"	Knots (Int.)	0.032397408
"	Miles (statute)/hr	0.037282272
Meters/sec	Feet/min	196.85039
"	Feet/sec	3.2808399
"	Kilometers/hr	3.6
"	Kilometers/min	0.06
"	Miles (statute)/hr	2.2369363
Meters/(sec. × sec.)	Kilometers/(hr. × sec.)	3.6
"	Miles/(hr. × sec.)	2.2369363
Meter-candles	Lumens/sq. meter	1
Mhos	Abmhos	1×10^{-9}
"	Cgs. units of conductance	1
"	E.M. cgs. units	1×10^{-9}
"	E.S. cgs. units	8.987584×10^{11}
"	Mhos (Int.)	1.000495
"	Mks. (r *or* nr) units	1
"	Ohms^{-1}	1
"	Siemen's units	1
"	Statmhos	8.987584×10^{11}
Mhos (Int.)	Abmhos	9.99505×10^{-10}
"	Mhos	0.999505
Mhos/meter	Abmhos/cm	1×10^{-11}
"	Mhos (Int.)/meter	1.000495
Mho-ft./circ. mil	Mhos/cm	6.0153049×10^{6}
Microfarads	Abfarads	1×10^{-15}
"	Farads	1×10^{-6}
"	Statfarads	8.987584×10^{5}
Micrograms	Grams	1×10^{-6}
"	Milligrams	0.001
Microhenries	Henries	1×10^{-6}
"	Stathenries	1.112646×10^{-18}
Microhms	Abohms	1000
"	Megohms	1×10^{-12}
"	Ohms	1×10^{-6}
"	Statohms	1.112646×10^{-18}
Microhm-cm	Abohm-cm	1000
"	Circ. mil-ohms/ft	6.0153049
"	Microhm-inches	0.39370079
"	Ohm-cm	1×10^{-6}
Microhm-inches	Circ. mil-ohms/ft	15.278875
"	Michrom-cm	2.54
Micromicrofarads	Farads	1×10^{-12}
Micromicrons	Ångström units	0.01
"	Centimeters	1×10^{-10}
"	Inches	3.9370079×10^{-1}
"	Meters	1×10^{-12}
"	Microns	1×10^{-6}
Microns	Ångström units	10000

CONVERSION FACTORS (Continued)

To convert from	To	Multiply by
Microns	Centimeters	0.0001
"	Feet	3.2808399×10^{-6}
"	Inches	3.9370079×10^{-5}
"	Meters	1×10^{-6}
"	Millimeters	0.001
"	Millimicrons	1000
Miles (naut., Brit.)	Cable lengths (Brit.)	8.4444
"	Fathoms	1013.333
"	Feet	6080
"	Meters	1853.184
"	Miles (Adm., Brit.)	1
"	Miles (naut., Int.)	1.0006393
"	Miles (statute)	1.151515
Miles (naut., Int.)	Cable lengths	8.4390493
"	Fathoms	1012.6859
"	Feet	6076.1155
"	Feet (U.S. Survey)	6076.1033
"	Kilometers	1.852
"	Leagues (naut., Int.)	0.333333
"	Meters	1852
"	Miles (geographical)	1
"	Miles (naut. Brit.)	0.99936110
"	Miles (statute)	1.1507794
Miles (statute)	Centimeters	160934.4
"	Chains (Gunter's)	80
"	Chains (Ramden's)	52.8
"	Feet	5280
"	Feet (U.S. Survey)	5279.9894
"	Furlongs	8
"	Inches	63360
"	Kilometers	1.609344
"	Light years	1.70111×10^{-13}
"	Links (Gunter's)	8000
"	Meters	1609.344
"	Miles (naut., Brit.)	0.86842105
"	Miles (naut., Int.)	0.86897624
"	Myriameters	0.1609344
"	Rods	320
"	Yards	1760
Miles/hr	Cm./sec	44.704
"	Feet/hr	5280
"	Feet/min	88
"	Feet/sec	1.466666
"	Kilometers/hr	1.609344
"	Knots (Int.)	0.86897624
"	Meters/min	26.8224
"	Miles/min	0.0166666
Miles/(hr. × min.)	Cm./(sec. × sec.)	0.7450666
Miles/(hr. × sec.)	Cm./(sec. × sec.)	44.704
"	Ft./(sec. × sec.)	1.466666
"	Kilometers/(hr. × sec.)	1.609344
"	Meters/(sec. × sec.)	0.44704
Miles/min	Cm./sec	2682.24
"	Feet/hr	316800
"	Feet/sec	88
"	Kilometers/min	1.609344
"	Knots (Int.)	52.138574
"	Meters/min	1609.344
"	Miles/hr	60
Millibars	Atmospheres	0.000986923
"	Bars	0.001
"	Baryes	1000
"	Dynes/sq. cm	1000
"	Grams/sq. cm	1.019716
"	In. of Hg (32°F.)	0.0295300
"	Pounds/sq. ft	2.088543
"	Pounds/sq. inch	0.0145038
Milligrams	Carats (1877)	0.004871
"	Carats (metric)	0.005
"	Drams (apoth. *or* troy)	0.00025720597
"	Drams (advp.)	0.00056438339
Milligrams	Grains	0.015432358
"	Grams	0.001
"	Ounces (apoth. *or* troy)	3.2150737×10^{-5}
"	Ounces (avdp.)	3.5273962×10^{-5}
"	Pennyweights	0.00064301493
"	Pounds (apoth. *or* troy)	2.6792289×10^{-6}
"	Pounds (avdp.)	2.2046226×10^{-6}
"	Scruples (apoth.)	0.00077161792
Milligrams/assay ton	Milligrams/kg	34.285714
"	Ounces (troy)/ton (avdp.)	1
Milligrams/gm	Dynes/cm	0.980665
"	Pounds/inch	5.5997415×10^{-6}
Milligrams/gram	Carats (parts gold per 24 of mixture)	0.024
"	Grams/ton (short)	907.18474
"	Milligrams/assay ton	29.166666
"	Ounces (avdp.)/ton (long)	35.84
"	Ounces (avdp.)/ton (short)	32
"	Ounces (troy)/ton (long)	32.6666
"	Ounces (troy)/ton (short)	29.1666
Milligrams/inch	Dynes/cm	0.386089
"	Dynes/inch	0.980665
"	Grams/cm	0.00039370079
"	Grams/inch	0.0001
Milligrams/kg	Pounds (avdp.)/ton (short)	0.002
Milligrams/liter	Grains/gal. (U.S.)	0.05841620
"	Grams/liter	0.001
"	Parts/million*	1
"	Lb./cu. ft	6.242621×10^{-5}
Milligrams/mm	Dynes/cm	9.80665
Millihenries	Abhenries	1×10^{6}
"	Henries	0.001
"	Stathenries	1.112646×10^{-15}
Millilamberts	Candles/sq. cm	0.00031830989
"	Candles/sq. inch	0.0020536081
"	Foot-lamberts	0.9290304
"	Lamberts	0.001
"	Lumens/sq. cm	0.001
"	Lumens/sq. ft	0.9290304
Milliliters	Cu. cm	1.000028
"	Cu. inches	0.06102545
"	Drams (U.S., fluid)	0.2705198
"	Gills (U.S.)	0.008453742
"	Liters	0.001
"	Minims (U.S.)	16.23119
"	Ounces (Brit., fluid)	0.03519609
"	Ounces (U.S., fluid)	0.03381497
"	Pints (Brit.)	0.001759804
"	Pints (U.S., liq.)	0.002113436
Millimeters	Ångström units	1×10^{7}
"	Centimeters	0.1
"	Decimeters	0.01
"	Dekameters	0.0001
"	Feet	0.0032808399
"	Inches	0.039370079
"	Meters	0.001
"	Microns	1000
"	Mils	39.370079
"	Wave length of orange-red line of krypton 86	1650.76373
"	Wave length of red line of cadmium	1553.16413
Millimeters of Hg (0°C.)	Atmospheres	0.0013157895
"	Bars	0.00133322
"	Dynes/sq. cm	1333.224
"	Grams/sq. cm	1.35951
"	Kg./sq. meter	13,5951
"	Pounds/sq. ft	2.78450
"	Pounds/sq. inch	0.0193368
"	Torrs	1

* Density of 1 gram per milliliter of solvent.

CONVERSION FACTORS (Continued)

To convert from	To	Multiply by
Millimicrons	Ångström units	10
''	Centimeters	1×10^{-7}
''	Inches	3.9370079×10^{-8}
''	Microns	0.001
''	Millimeters	1×10^{-6}
Milliphots	Foot-candles	0.9290304
''	Lumens/sq. ft	0.9290304
''	Lumens/sq. meter	10
''	Lux	10
''	Phots	0.001
Millivolts	Statvolts	3.335635×10^{-6}
''	Volts	0.001
Minims (Brit.)	Cu. cm	0.05919385
''	Cu. inches	0.003612230
''	Milliliters	0.05919219
''	Ounces (Brit., fluid)	0.0020833333
''	Scruples (Brit., fluid)	0.05
Minims (U.S.)	Cu. cm	0.061611520
''	Cu. inches	0.0037597656
''	Drams (U.S., fluid)	0.0166666
''	Gallons (U.S., liq.)	1.6276042×10^{-5}
''	Gills (U.S.)	0.0005208333
''	Liters	6.160979×10^{-5}
''	Milliliters	0.06160979
''	Ounces (U.S., fluid)	0.002083333
''	Pints (U.S., liq.)	0.0001302083
Minutes (angular)	Degrees	0.0166666
''	Quadrants	0.000185185
''	Radians	0.00029088821
''	Seconds (angular)	60
Minutes (mean solar)	Days (mean solar)	0.0006944444
''	Days (sidereal)	0.00069634577
''	Hours (mean solar)	0.0166666
''	Hours (sidereal)	0.016712298
''	Minutes (sidereal)	1.00273791
Minutes (sidereal)	Days (mean solar)	0.00069254831
''	Minutes (mean solar)	0.99726957
''	Months (mean calendar)	2.2768712×10^{-5}
''	Seconds (sidereal)	60
Minutes/cm	Radians/cm	0.00029088821
Months (lunar)	Days (mean solar)	29.530588
''	Hours (mean solar)	708.73411
''	Minutes (mean solar)	42524.047
''	Second (mean solar)	2.5514428×10^{6}
''	Weeks (mean calendar)	4.2186554
Months (mean calendar)	Days (mean solar)	30.416666
''	Hours (mean solar)	730
''	Months (lunar)	1.0300055
''	Weeks (mean calendar)	4.3452381
''	Years (calendar)	0.08333333
''	Years (sidereal)	0.083274845
''	Years (tropical)	0.083278075
Myriagrams	Grams	10000
''	Kilograms	10
''	Pounds (avdp.)	22.046226
Newtons	Dynes	100000
''	Pounds	0.22480894
Newton-meters	Dyne-cm	1×10^{7}
''	Gram-cm	10197.162
''	Kg.-meters	0.10197162
''	Pound-feet	0.73756215
Noggins (Brit.)	Cu. cm	142.0652
''	Gallons (Brit.)	0.03125
''	Gills (Brit.)	1
Oersteds	Ampere-turns/inch	2.0212678
''	Ampere-turns/meter	79.577472
''	E.M. cgs. units of magnetic field intensity	1
''	E.S. cgs. units	2.997930×10^{10}
Oersteds	Gilberts/cm	1
''	Oersteds (Int.)	1.000165
Oersteds (Int.)	Oersteds	0.999835
Ohms	Abohms	1×10^{9}
''	Cgs. units of resistance	1
''	Megohms	1×10^{-6}
''	Microhms	1×10^{6}
''	Ohms (Int.)	0.999505
''	Statohms	1.112646×10^{-12}
Ohms (Int.)	Ohms	1.000495
Ohms (mil, foot)	Circ. mil-ohms/ft	1
''	Ohm-cm	1.6624261×10^{-7}
Ohm-cm	Circ. mil-ohms/ft	6.0153049×10^{6}
''	Microhm-cm	1×10^{6}
''	Ohm-inches	0.39370079
Ohm-inches	Ohm-cm	2.54
Ohm-meters	Abohm-cm	1×10^{11}
''	E.M. cgs. units	1×10^{11}
''	E.S. cgs. units	1.112646×10^{-10}
''	Mks. units	1
''	Statohm-cm	1.112646×10^{-10}
Ounces (apoth. *or* troy)	Dekagrams	1.7554286
''	Drams (apoth. *or* troy)	8
''	Drams (avdp.)	17.554286
''	Grains	480
''	Grams	31.103486
''	Milligrams	31103.486
''	Ounces (avdp.)	1.0971429
''	Pennyweights	20
''	Pounds (apoth. *or* troy)	0.0833333
''	Pounds (avdp.)	0.068571429
''	Scruples (apoth.)	24
''	Tons (short)	3.4285714×10^{-5}
Ounces (avdp.)	Drams (apoth. *or* troy)	7.291666
''	Drams (avdp.)	16
''	Grains	437.5
''	Grams	28.349523
''	Hundredweights (long)	0.00055803571
''	Hundredweights (short)	0.000625
''	Ounces (apoth. *or* troy)	0.9114583
''	Pennyweights	18.229166
''	Pounds (apoth. *or* troy)	0.075954861
''	Pounds (avdp.)	0.0625
''	Scruples (apoth.)	21.875
''	Tons (long)	2.7901786×10^{-5}
''	Tons (metric)	2.8349527×10^{-5}
''	Tons (short)	3.125×10^{-5}
Ounces (Brit., fluid)	Cu. cm	28.41305
''	Cu. inches	1.733870
''	Drachms (Brit., fluid)	8
''	Drams (U.S., fluid)	7.686075
''	Gallons (Brit.)	0.00625
''	Milliliters	28.41225
''	Minims (Brit.)	480
''	Ounces (U.S., fluid)	0.9607594
Ounces (U.S., fluid)	Cu. cm	29.573730
''	Cu. inches	1.8046875
''	Cu. meters	2.9573730×10^{-5}
''	Drams (U.S., fluid)	8
''	Gallons (U.S., dry)	0.0067138047
''	Gallons (U.S., liq.)	0.0078125
''	Gills (U.S.)	0.25
''	Liters	0.029572702
''	Minims (U.S.)	480
''	Ounces (Brit., fluid)	1.040843
''	Pints (U.S., liq.)	0.0625
''	Quarts (U.S., liq.)	0.03125
Ounces/sq. inch	Dynes/sq. cm	4309.22
''	Grams/sq. cm	4.3941849
''	In. of H_2O (39.2°F.)	1.73004
''	In. of H_2O (60°F.)	1.73166

CONVERSION FACTORS (Continued)

To convert from	To	Multiply by
Ounces/sq. inch	Pounds/sq. ft	9
"	Pounds/sq. inch	0.0625
Ounces (avdp.)/ton (long)	Milligrams/kg	27.901786
Ounces (avdp.)/ton (short)	Milligrams/kg	31.25
Paces	Centimeters	76.2
"	Chains (Gunter's)	0.0378788
"	Chains (Ramden's)	0.025
"	Feet	2.5
"	Hands	7.5
"	Inches	30
"	Ropes (Brit.)	0.125
Palms	Centimeters	7.62
"	Chains (Ramden's)	0.0025
"	Cubits	0.1666666
"	Feet	0.25
"	Hands	0.75
"	Inches	3
Parsecs	Kilometers	3.08374×10^{13}
"	Miles (statute	1.91615×10^{13}
Parts/million*	Grains/gal. (Brit.)	0.07015488
"	Grains/gal. (U.S.)	0.05841620
"	Grams/liter	0.001
"	Milligrams/liter	1
Pecks (Brit.)	Bushels (Brit.)	0.25
"	Coombs (Brit.)	0.0625
"	Cu. cm	9092.175
"	Cu. inches	554.8385
"	Gallons (Brit.)	2
"	Gills (Brit.)	64
"	Hogsheads	0.03812537
"	Kilderkins (Brit.)	0.111111
"	Liters	9.091920
"	Pints (Brit.)	16
"	Quarterns (Brit., dry)	4
"	Quarters (Brit., dry)	0.03125
"	Quarts (Brit.)	8
"	Quarts (U.S., dry)	8.256449
Pecks (U.S.)	Barrels (U.S., dry)	0.076191185
"	Bushels (U.S.)	0.25
"	Cu. cm	8809.7675
"	Cu. feet	0.311114005
"	Cu. inches	537.605
"	Gallons (U.S., dry)	2
"	Gallons (U.S., liq.)	2.3272944
"	Liters	8.809521
"	Pints (U.S., dry)	16
"	Quarts (U.S., dry)	8
Pennyweights	Drams (apoth. *or* troy)	0.4
"	Drams (avdp.)	0.87771429
"	Grains	24
"	Grams	1.55517384
"	Ounces (apoth. *or* troy)	0.05
"	Ounces (avdp.)	0.054857143
"	Pounds (apoth. *or* troy)	0.0041666
"	Pounds (avdp.)	0.0034285714
Perches (masonry)	Cu. feet	24.75
Phots	Foot-candles	929.0304
"	Lumens/sq. cm	1
"	Lumens/sq. meter	10000
"	Lux	10000
Picas (printer's)	Centimeters	0.42175176
"	Inches	0.166044
Pints (Brit.)	Cu. cm	568.26092
"	Gallons (Brit.)	0.125
"	Gills (Brit.)	4
"	Gills (U.S.)	4.803797
"	Liters	0.5682450
Pints (Brit.)	Minims (Brit.)	9600
"	Ounces (Brit., fluid)	20
"	Pints (U.S., dry)	1.032056
"	Pints (U.S., liq.)	1.200949
"	Quarts (Brit.)	0.5
"	Scruples (Brit., fluid)	480
Pints (U.S., dry)	Bushels (U.S.)	0.015625
"	Cu. cm	550.61047
"	Cu. inches	33.6003125
"	Gallons, (U.S., dry)	0.125
"	Gallons (U.S., liq.)	0.14545590
"	Liters	0.5505951
"	Pecks (U.S.)	0.0625
"	Quarts (U.S., dry)	0.5
Pints (U.S., liq.)	Cu. cm	473.17647
"	Cu. feet	0.016710069
"	Cu. inches	28.875
"	Cu. yards	0.00061889146
"	Drams (U.S., fluid)	128
"	Gallons (U.S., liq.)	0.125
"	Gills (U.S.)	4
"	Liters	0.4731632
"	Milliliters	473.1632
"	Minims (U.S.)	7680
"	Ounces (U.S., fluid)	16
"	Pints (Brit.)	0.8326747
"	Quarts (U.S., liq.)	0.5
Planck's constant	Erg-seconds	6.6255×10^{-27}
"	Joule-seconds	6.6255×10^{-34}
"	Joule-sec./Avog. No. (chem.)	3.9905×10^{-10}
Points (printer's)	Centimeters	0.03514598
"	Inches	0.013837
"	Picas	0.0833333
Poises	Cgs. units of absolute viscosity	1
"	Grams/(cm. × sec.)	1
Poise-cu. cm./gram	Sq. cm./sec	1
Poise-cu. ft./lb	Sq. cm./sec	62.427960
Poise-cu. in./gram	Sq. cm./sec	16.387064
Poles/sq. cm	E.M. cgs. units of magnetization	1
Pottles (Brit.)	Gallons (Brit.)	0.5
"	Liters	2.272980
Poundals	Dynes	13825.50
"	Grams	14.09808
"	Pounds (avdp.)	0.0310810
Pounds (apoth. *or* troy)	Drams (apoth. *or* troy)	96
"	Drams (avdp.)	210.65143
"	Grains	5760
"	Grams	373.24172
"	Kilograms	0.37324172
"	Ounces (apoth. *or* troy)	12
"	Ounces (avdp.)	13.165714
"	Pennyweights	240
"	Pounds (avdp.)	0.8228571
"	Scruples (apoth.)	288
"	Tons (long)	0.00036734694
"	Tons (metric)	0.00037324172
"	Tons (short)	0.00041142857
Pounds (avdp.)	Drams (apoth. *or* troy)	116.6666
"	Drams (avdp.)	256
"	Grains	7000
"	Grams	453.59237
"	Hundredweights (long)	0.00892857
"	Hundredweights (short)	0.01
"	Kilograms	0.45359237
"	Ounces (apoth. *or* troy)	14.583333
"	Ounces (avdp.)	16

* Based on density of 1 gram/ml. for the solvent.

CONVERSION FACTORS (Continued)

To convert from	To	Multiply by
Pounds (avdp.)	Pennyweights	291.6666
"	Poundals	32.1740
"	Pounds (apoth. *or* troy)	1.215277
"	Scruples (apoth.)	350
"	Slugs	0.0310810
"	Tons (long)	0.00044642857
"	Tons (metric)	0.00045359237
"	Tons (short)	0.0005
Pounds of H_2O evap. from and at 212°F	B.t.u.	970.9
"	B.t.u. (IST.)	970.2
"	B.t.u. (mean)	969.4
"	Joules	1.0237×10^6
"	Joules (Int.)	1.0234×10^6
Pounds/cu. ft	Grams/cu. cm	0.016018463
"	Kg./cu. meter	16.018463
Pounds/cu. inch	Grams/cu. cm	27.679905
"	Grams/liter	27.68068
"	Kg./cu. meter	27679.905
Pounds/gal. (Brit.)	Pounds/cu. ft	6.228839
Pounds/gal. (U.S., liq.)	Grams/cu. cm	0.11982643
"	Pounds/cu. ft	7.4805195
Pounds/inch	Grams/cm	178.57967
"	Grams/ft	5443.1084
"	Grams/inch	453.59237
"	Ounces/cm	6.2992
"	Ounces/inch	16
"	Pounds/meter	39.370079
Pounds/minute	Kilograms/hr	27.2155422
"	Kilograms/min	0.45359237
Pounds of H_2O (39.2°F.)/min	Cu. ft./min	0.01601891
"	Gal. (U.S.)/min	0.1198298
"	Liters/min	0.45359237
Pounds/sq. ft	Atmospheres	0.000472541
"	Bars	0.000478803
"	Cm. of Hg (0°C.)	0.0359131
"	Dynes/sq. cm	478.803
"	Ft. of air (1 atm., 60°F.)	13.096
"	Grams/sq. cm	0.48824276
"	In. of Hg (32°F.)	0.0141390
"	In. of H_2O (39.2°F.)	0.192227
"	Kg./sq. meter	4.8824276
"	Mm. of Hg (0°C.)	0.359131
Pounds/sq. inch	Atmospheres	0.0680460
"	Bars	0.0689476
"	Cm. of Hg (0°C.)	5.17149
"	Cm. of H_2O (4°C.)	70.3089
"	Dynes/sq. cm	68947.6
"	Grams/sq. cm	70.306958
"	In. of Hg (32°F.)	2.03602
"	In. of H_2O (39.2°F.)	27.6807
"	Kg./sq. cm	0.070306958
"	Mm. of Hg (0°C.)	51.7149
Pound wt.-sec./sq. ft.	Poises	478.803
Pound wt.-sec./sq. in.	Poises	68947.6
Puncheons (Brit.)	Cu. meters	0.31797510
"	Gallons (Brit.)	69.94467
"	Gallons (U.S.)	84
Quadrants	Minutes	5400
"	Radians	1.5707963
Quarterns (Brit., dry)	Buckets (Brit.)	0.125
"	Bushels (Brit.)	0.0625
"	Cu. cm	2273.044
"	Gallons (Brit.)	0.5
"	Liters	2.272980
"	Pecks (Brit.)	0.25
Quarterns (Brit., liq.)	Cu. cm	142.0652
Quarterns (Brit., liq.)	Gallons (Brit.)	0.03125
"	Liters	0.1420613
Quarters (U.S., long)	Kilograms	254.0117272
"	Pounds (avdp.)	560
Quarters (U.S., short)	Kilograms	226.796185
"	Pounds	500
Quarts (Brit.)	Cu. cm	1136.522
"	Cu. inches	69.35482
"	Gallons (Brit.)	0.25
"	Gallons (U.S., liq.)	0.3002373
"	Liters	1.136490
"	Quarts (U.S., dry)	1.032056
"	Quarts (U.S., liq.)	1.200949
Quarts (U.S., dry)	Bushels (U.S.)	0.03125
"	Cu. cm	1101.2209
"	Cu. feet	0.038889251
"	Cu. inches	67.200625
"	Gallons (U.S., dry)	0.25
"	Gallons (U.S., liq.)	0.29091180
"	Liters	1.1011901
"	Pecks (U.S.)	0.125
"	Pints (U.S., dry)	2
Quarts (U.S., liq.)	Cu. cm	946.35295
"	Cu. feet	0.033420136
"	Cu. inches	57.75
"	Drams (U.S., fluid)	256
"	Gallons (U.S., dry)	0.21484175
"	Gallons (U.S., liq.)	0.25
"	Gills (U.S.)	8
"	Liters	0.9463264
"	Ounces (U.S., fluid)	32
"	Pints (U.S., liq.)	2
"	Quarts (Brit.)	0.8326747
"	Quarts (U.S., dry)	0.8593670
Quintals (metric)	Grams	100000
"	Hundredweights (long)	1.9684131
"	Kilograms	100
"	Pounds (avdp.)	220.46226
Radians	Circumferences	0.15915494
"	Degrees	57.295779
"	Minutes	3437.7468
"	Quadrants	0.63661977
"	Revolutions	0.15915494
"	Seconds	206264.81
Radians/cm	Degrees/cm	57.295779
"	Degrees/ft	1746.3754
"	Degrees/inch	145.53128
"	Minutes/cm	3437.7468
Radians/sec	Degrees/sec	57.295779
"	Revolutions/min	9.5492966
"	Revolutions/sec	0.15915494
Radians/(sec. × sec.)	Revolutions/(min. × min.)	572.95779
"	Revolutions/(min. × sec.)	9.5492966
"	Revolutions/(sec. × sec.)	0.15915494
Register tons	Cu. feet	100
"	Cu. meters	2.8316847
Revolutions	Degrees	360
"	Grades	400
"	Quadrants	4
"	Radians	6.2831853
Reyns	Centipoises	6.89476×10^6
Rhes	$Poises^{-1}$	1
Rods	Centimeters	502.92
"	Chains (Gunter's)	0.25
"	Chains (Ramden's)	0.165
"	Feet	16.5
"	Feet (U.S. Survey)	16.499967
"	Furlongs	0.025
"	Inches	198
"	Links (Gunter's)	25

CONVERSION FACTORS (Continued)

To convert from	To	Multiply by
Rods	Links (Ramden's)	16.5
"	Meters	5.0292
"	Miles (statute)	0.003125
"	Perches	1
"	Yards	5.5
Rods (Brit., volume)	Cu. feet	1000
"	Cu. meters	28.316847
Roods (Brit.)	Acres	0.25
"	Ares	10.117141
"	Sq. perches	40
"	Sq. yards	1210
Ropes (Brit.)	Feet	20
"	Meters	6.096
"	Yards	6.6666666
Scruples (apoth.)	Drams (apoth. *or* troy)	0.333333
"	Drams (avdp.)	0.73142857
"	Grains	20
"	Grams	1.2959782
"	Ounces (apoth. *or* troy)	0.041666
"	Ounces (avdp.)	0.045714286
"	Pennyweights	0.833333
"	Pounds (apoth. *or* troy)	0.003472222
"	Pounds (avdp.)	0.0028571429
Scruples (Brit., fluid)	Minims (Brit.)	20
Seams (Brit.)	Bushels (Brit.)	8
"	Cu. feet	10.27479
"	Liters	290.9414
Seconds (angular)	Degrees	0.000277777
"	Minutes	0.0166666
"	Radians	4.8481368×10^{-6}
Seconds (mean solar)	Days (mean solar)	1.1574074×10^{-5}
"	Days (sidereal)	1.1605763×10^{-5}
"	Hours (mean solar)	0.0002777777
"	Hours (sidereal)	0.00027853831
"	Minutes (mean solar)	0.0166666
"	Minutes (sidereal)	0.016712298
"	Seconds (sidereal)	1.00273791
Seconds (sidereal)	Days (mean solar)	1.1542472×10^{-5}
"	Days (sidereal)	1.1574074×10^{-5}
"	Hours (mean solar)	0.00027701932
"	Hours (sidereal)	0.000277777
"	Minutes (mean solar)	0.016621159
"	Minutes (sidereal)	0.0166666
"	Seconds (mean solar)	0.99726957
Siemen's units	*Same as* Mhos	
Skeins	Feet	360
"	Meters	109.728
Slugs	Geepounds	1
"	Kilograms	14.5939
"	Pounds (avdp.)	32.1740
Slugs/cu. ft.	Grams/cu. cm	0.515379
Space (entire)	Hemispheres	2
"	Steradians	12.566371
Spans	Centimeters	22.86
"	Fathoms	0.125
"	Feet	0.75
"	Inches	9
"	Quarters (Brit. linear)	1
Spherical right angles	Hemispheres	0.25
"	Spheres	0.125
"	Steradians	1.5707963
Sq. centimeters	Ares	1×10^{-6}
"	Circ. mm	127.32395
"	Circ. mils	197352.52
"	Sq. chains (Gunter's)	2.4710538×10^{-7}
"	Sq. chains (Ramden's)	1.0763910×10^{-7}
"	Sq. decimeters	0.01
"	Sq. feet	0.0010763910
"	Sq. ft. (U.S. Survey)	0.0010763867
"	Sq. inches	0.15500031
"	Sq. meters	0.0001
Sq. centimeters	Sq. mm	100
"	Sq. mils	155000.31
"	Sq. rods	3.9536861×10^{-6}
"	Sq. yards	0.00011959900
Sq. chains (Gunter's)	Acres	0.1
"	Sq. feet	4356
"	Sq. ft. (U.S. Survey)	4355.9826
"	Sq. inches	627264
"	Sq. links (Gunter's)	10000
"	Sq. meters	404.68564
"	Sq. miles	0.00015625
"	Sq. rods	16
"	Sq. yards	484
Sq. chains (Ramden's)	Acres	0.22956841
"	Sq. feet	10000
"	Sq. ft. (U.S. Survey)	9999.9600
"	Sq. inches	1.44×10^{6}
"	Sq. links (Ramden's)	10000
"	Sq. meters	929.0304
"	Sq. miles	0.00035870064
"	Sq. rods	36.730946
"	Sq. yards	1111.111
Sq. decimeters	Sq. cm	100
"	Sq. inches	15.500031
Square degrees	Steradians	0.00030461742
Sq. dekameters	Acres	0.024710538
"	Ares	1
"	Sq. meters	100
"	Sq. yards	119.59900
Sq. feet	Acres	2.295684×10^{-5}
"	Ares	0.0009290304
"	Sq. cm	929.0304
"	Sq. chains (Gunter's)	0.00022956841
"	Sq. ft. (U.S. Survey)	0.99999600
"	Sq. inches	144
"	Sq. links (Gunter's)	2.2956841
"	Sq. meters	0.09290304
"	Sq. miles	3.5870064×10^{-8}
"	Sq. rods	0.0036730946
"	Sq. yards	0.111111
Sq. feet (U.S. Survey)	Acres	$2.29569330 \times 10^{-5}$
"	Sq. centimeters	929.03412
"	Sq. chains (Ramden's)	0.00010000040
"	Sq. feet	1.0000040
Sq. hectometers	Sq. meters	10000
Sq. inches	Circ. mils	1273239.5
"	Sq. cm	6.4516
"	Sq. chains (Gunter's)	1.5942251×10^{-6}
"	Sq. decimeters	0.064516
"	Sq. feet	0.0069444
"	Sq. ft. (U.S. Survey)	0.0069444167
"	Sq. links (Gunter's)	0.01594225
"	Sq. meters	0.00064516
"	Sq. miles	$2.4909767 \times 10^{-10}$
"	Sq. mm	645.16
"	Sq. mils	1×10^{6}
Sq. inches/sec	Sq. cm./hr.	23225.76
"	Sq. cm./sec	6.4516
"	Sq. ft./min	0.416666
Sq. kilometers	Acres	247.10538
"	Sq. feet	1.0763910×10^{7}
"	Sq. ft. (U.S. Survey)	1.0763867×10^{7}
"	Sq. inches	1.5500031×10^{9}
"	Sq. meters	1×10^{6}
"	Sq. miles	0.38610216
"	Sq. yards	1.1959900×10^{6}
Sq. links (Gunter's)	Acres	1×10^{-5}
"	Sq. cm	404.68564
"	Sq. chains (Gunter's)	0.0001
"	Sq. feet	0.4356
"	Sq. ft. (U.S. Survey)	0.43559826
"	Sq. Inches	62.7264

CONVERSION FACTORS (Continued)

To convert from	To	Multiply by
Sq. links (Ramden's)	Acres	2.2956841×10^{-5}
"	Sq. feet	1
Sq. meters	Acres	0.00024710538
"	Ares	0.01
"	Hectares	0.0001
"	Sq. cm	10000
"	Sq. feet	10.763910
"	Sq. inches	1550.0031
"	Sq. kilometers	1×10^{-6}
"	Sq. links (Gunter's)	24.710538
"	Sq. links (Ramden's)	10.763910
"	Sq. miles	3.8610216×10^{-7}
"	Sq. mm	1×10^{6}
"	Sq. rods	0.039536861
"	Sq. yards	1.1959900
Sq. miles	Acres	640
"	Hectares	258.99881
"	Sq. chains (Gunter's)	6400
"	Sq. feet	2.7878288×10^{7}
"	Sq. ft. (U.S. Survey)	2.78288×10^{7}
"	Sq. kilometers	2.5899881
"	Sq. meters	2589988.1
"	Sq. rods	102400
"	Sq. yards	3.0976×10^{6}
Sq. millimeters	Circ. mm	1.2732395
"	Circ. mils	1973.5252
"	Sq. cm	0.01
"	Sq. inches	0.0015500031
"	Sq. meters	1×10^{-6}
Sq. mils	Circ. mils	1.2732395
"	Sq. cm	6.4516×10^{-6}
"	Sq. inches	1×10^{-6}
"	Sq. mm	0.00064516
Sq. rods	Acres	0.00625
"	Ares	0.2529285264
"	Hectares	0.002529285264
"	Sq. cm	252928.5264
"	Sq. feet	272.25
"	Sq. ft. (U.S. Survey)	272.24891
"	Sq. inches	39204
"	Sq. links (Gunter's)	625
"	Sq. links (Ramden's)	272.25
"	Sq. meters	25.29285264
"	Sq. miles	9.765625×10^{-6}
"	Sq. yards	30.25
Sq. yards	Acres	0.00020661157
"	Ares	0.0083612736
"	Hectares	8.3612736×10^{-5}
"	Sq. cm	8361.2736
"	Sq. chains (Gunter's)	0.0020661157
"	Sq. chains (Ramden's)	0.0009
"	Sq. feet	9
"	Sq. ft. (U.S. Survey)	8.9999640
"	Sq. inches	1296
"	Sq. links (Gunter's)	20.661157
"	Sq. links (Ramden's)	9
"	Sq. meters	0.83612736
"	Sq. miles	$3.228305785 \times 10^{-7}$
"	Sq. perches (Brit.)	0.033057851
"	Sq. rods	0.033057851
Statamperes	Abamperes	3.335635×10^{-11}
"	Amperes	3.335635×10^{-10}
"	E.M. cgs. units of current	3.335635×10^{-11}
"	E.S. cgs. units	1
Statcoulombs	Ampere-hours	9.265653×10^{-14}
"	Coulombs	3.335635×10^{-10}
"	Electronic charges	2.082093×10^{9}
"	E.M. cgs. units of electric charge	3.335635×10^{-11}
Statfarads	E.M. cgs. units of capacitance	1.112646×10^{-21}
"	E.S. cgs. units	1
Statfarads	Farads	1.112646×10^{-12}
"	Microfarads	1.112646×10^{-6}
Stathenries	Abhenries	8.987584×10^{20}
"	E.M. cgs. units of inductance	8.987584×10^{20}
"	E.S. cgs. units	1
"	Henries	8.987584×10^{11}
"	Millihenries	8.987584×10^{14}
Statohms	Abohms	8.987584×10^{20}
"	E.S. cgs. units	1
"	Ohms	8.987584×10^{11}
Statvolts	Abvolts	2.997930×10^{10}
"	Volts	299.7930
Statvolts/cm	Volts/cm	299.7930
"	Volts/inch	761.4742
Statvolts/inch	Volts/cm	118.0287
Steradians	Hemispheres	0.15915494
"	Solid angles	0.079577472
"	Spheres	0.079577472
"	Spher. right angles	0.63661977
"	Square degrees	3282.8063
Steres	Cubic meters	1
"	Decisteres	10
"	Dekasteres	0.1
"	Liters	999.972
Stilbs	Candles/sq. cm	1
"	Candles/sq. inch	6.4516
"	Lamberts	3.1415927
Stokes	Cgs. units of kinematic viscosity	1
"	Sq. cm./sec	1
"	Sq. inches/sec	0.15500031
"	Poise cu. cm./gram	1
Stones (Brit., legal)	Centals (Brit.)	0.14
Tons (long)	Dynes	9.96402×10^{8}
"	Hundredweights (long)	20
"	Hundredweights (short)	22.4
"	Kilograms	1016.0469
"	Ounces (avdp.)	35840
"	Pounds (apoth. *or* troy)	2722.22
"	Pounds (avdp.)	2240
"	Tons (metric)	1.0160469
"	Tons (short)	1.12
Tons (metric)	Dynes	9.80665×10^{8}
"	Grams	1×10^{6}
"	Hundredweights (short)	22.046226
"	Kilograms	1000
"	Ounces (avdp.)	35273.962
"	Pounds (apoth. *or* troy)	2679.2289
"	Pounds (avdp.)	2204.6226
"	Tons (long)	0.98420653
"	Tons (short)	1.1023113
Tons (short)	Dynes	8.89644×10^{8}
"	Hundredweights (short)	20
"	Kilograms	907.18474
"	Ounces (avdp.)	32000
"	Pounds (apoth. *or* troy)	2430.555
"	Pounds (avdp.)	2000
"	Tons (long)	0.89285714
"	Tons (metric)	0.90718474
Tons of refrig. (U.S., comm.)	B.t.u. (IST.)/hr	12000
"	B.t.u. (IST.)/min	200
"	Cal., *kg.* (IST.)/hr	3023.949
"	Horsepower	4.71611
"	Kg. of ice melted/hr	37.971
"	Lb. of ice melted/hr	83.711
Tons of refrig. (U.S., std.)	B.t.u. (IST.)	288000
"	B.t.u. (mean)	287774
"	Cal., *kg.* (IST.)	72574.8

CONVERSION FACTORS (Continued)

To convert from	To	Multiply by
Tons of refrig. (U.S., std.)	Cal., *kg*. (mean)	72517.9
"	Lb. of ice melted	2009.1
Tons (long)/sq. ft	Atmospheres	1.05849
"	Dynes/sq. cm	1.07252 × 10^6
"	Grams/sq. cm	1093.6638
"	Pounds/sq. ft	2240
Tons (short)/sq. ft	Atmospheres	0.945082
"	Dynes/sq. cm	957.605
"	Grams/sq. cm	976.486
"	Pounds/sq. inch	13.8888
Tons (long)/sq. in	Atmospheres	152.423
"	Dynes/sq. cm	1.54443 × 10^8
"	Grams/sq. cm	157487.59
Tons (short)/sq. in	Dynes/sq. cm	1.37895 × 10^8
"	Kg./sq. mm	1406.139
"	Pounds/sq. inch	2000
Torrs (*or* Tors)	Millimeters of Hg (0°C.)	1
Townships (U.S.)	Acres	23040
"	Sections	36
"	Sq. miles	36
Tuns	Gallons (U.S.)	252
"	Hogsheads	4
Volts	Abvolts	1 × 10^8
"	Mks. (r *or* nr) units	1
"	Statvolts	0.003335635
"	Volts (Int.)	0.999670
Volts (Int.)	Volts	1.000330
Volts/°C	Joules/(coulomb × °C.)	1
Volt-coulombs	Joules (Int.)	0.999835
Volt-coulombs (Int.)	Joules	1.000165
Volt-electronic charge-seconds	Planck's constant*	2.41814 × 10^{14}
Volt-faraday (chem.)-seconds	Planck's constant*	1.45650 × 10^{38}
Volt-faraday (phys.)-seconds	Planck's constant	1.45690 × 10^{38}
Volt-seconds	Maxwells	1 × 10^8
Watts	B.t.u./hr	3.41443
"	B.t.u. (mean)/hr	3.40952
"	B.t.u. (mean)/min	0.0568253
"	B.t.u./sec	0.000948451
"	B.t.u. (mean)/sec	0.000947088
"	Cal., *gm*./hr	860.421
"	Cal., *gm*. (mean)/hr	859.184
"	Cal., *gm*. (20°C.)/hr	860.853
"	Cal., *gm*./min	14.3403
"	Cal., *gm*. (IST.)/min	14.3310
"	Cal., *gm*. (mean)/min	14.3197
"	Cal., *kh*./min	0.0143403
"	Cal., *kg*. (IST.)/min	0.0143310
"	Cal., *kg*. (mean)/min	0.0143197
"	Ergs/sec	1 × 10^7
"	Foot-pounds/min	44.2537
"	Horsepower	0.00134102
"	Horsepower (boiler)	0.000101942
"	Horsepower (elec.)	0.00134048
"	Horsepower (metric)	0.00135962
"	Joules/sec	1
"	Kilowatts	0.001
"	Liter-atm./hr	35.5282
Watts (Int.)	B.t.u./hr	3.41499
"	B.t.u. (mean)/hr	3.41008
"	B.t.u./min	0.569165
"	B.t.u. (mean)/min	0.0568347
"	Cal., *gm*./hr	860.563
"	Cal., *gm*. (mean)/hr	859.326
"	Cal., *kg*./min	0.0143427
"	Cal., *kg*. (IST.)/min	0.0143333
"	Cal., *kg*. (mean)/min	0.0143221
Watts (Int.)	Ergs/sec	1.000165 × 10^7
"	Joules (Int.)/sec	1
"	Watts	1.000165
Watts/sq. cm	B.t.u./(hr. × sq. ft.)	3172.10
"	Cal., *gm*./(hr. × sq. cm.)	860.421
"	Ft.-lb./(min. × sq. ft.)	41113.1
Watts/sq. in	B.t.u./(hr. × sq. ft.)	491.677
"	Cal., *gm*./(hr. × sq. cm.)	133.365
"	Ft.-lb./(min. × sq. ft.)	6372.54
Watt-hours	B.t.u	3.41443
"	B.t.u. (mean)	3.40952
"	Cal., *gm*	860.421
Watt-hours	Cal., *kg*. (mean)	0.859184
"	Cal., *gm*. (mean)	859.184
"	Foot-pounds	2655.22
"	Hp.-hours	0.00134102
"	Joules	3600
"	Joules (Int.)	3599.41
"	Kg.-meters	367.098
"	Kw.-hours	0.001
"	Watt-hours (Int.)	0.999835
Watt-sec	Foot-pounds	0.737562
"	Gram-cm	10197.16
"	Joules	1
"	Liter-atm	0.00986895
"	Volt-coulombs	1
Wave length of orange-red line of krypton 86	Ångström units	6057.80211
"	Millimeters	0.000605780211
Wave length of red line of cadmium	Ångström units	6438.4696
"	Millimeters	0.00064384696
Webers	Cgs. units of induction flux	1 × 10^8
"	E.M. cgs. units of induction flux	1 × 10^8
"	Lines	1 × 10^8
"	Maxwells	1 × 10^8
"	Mks. units of induction flux	1
"	Mks. units of magnetic charge	0.079577472
"	Mks. units of magnetic charge	1
"	Volt-seconds	1
Webers/sq. cm	Gausses	1 × 10^8
"	Lines	1 × 10^8
"	Lines/sq. inch	6.4516 × 10^8
Webers/sq. in	Gausses	1.5500031 × 10^7
Weeks (mean calendar)	Days (mean solar)	7
"	Days (sidereal)	7.0191654
"	Hours (mean solar)	168
"	Hours (sidereal)	168.45997
"	Minutes (mean solar)	10080
"	Minutes (sidereal)	10107.598
"	Months (lunar)	0.23704235
"	Months (mean calendar)	0.23013699
"	Years (calendar)	0.019178082
"	Years (sidereal)	0.019164622
"	Years (tropical)	0.019165365
Weys (Brit., mass.)	Pounds (avdp.)	252
Yards	Centimeters	91.44
"	Chains (Gunter's)	0.4545454
"	Chains (Ramden's)	0.03
"	Cubits	2
"	Fathoms	0.5
"	Feet	3
"	Feet (U.S. Survey)	2.9999940
"	Furlongs	0.00454545
"	Inches	36
"	Meters	0.9144
"	Poles (Brit.)	0.181818

* For factors for C = 12 scale or those which can be derived from same see table on Values for General Physical Constants.

CONVERSION FACTORS (Continued)

To convert from	To	Multiply by	To convert from	To	Multiply by
Yards	Quarters (Brit., linear)	4	Years (sidereal)	Days (sidereal)	366.25640
''	Rods	0.181818	''	Years (calendar)	1.0007024
''	Spans	4	''	Years (tropical)	1.0000388
Years (calendar)	Days (mean solar)	365	Years (tropical)	Days (mean solar)	365.24219
''	Hours (mean solar)	8760	''	Days (sidereal)	366.24219
''	Minutes (mean solar)	525600	''	Hours (mean solar)	8765.8126
''	Months (lunar)	12.360065	''	Hours (sidereal)	8789.8126
''	Months (mean calendar)	12	''	Months (mean calendar)	12.007963
''	Seconds (mean solar)	3.1536×10^7	''	Seconds (mean solar)	3.1556926×10^7
''	Weeks (mean calendar)	52.142857	''	Seconds (sidereal)	3.1643326×10^7
''	Years (sidereal)	0.99929814	''	Weeks (mean calendar)	52.177456
''	Years (tropical)	0.99933690	''	Years (calendar)	1.0006635
Years (leap)	Days (mean solar)	366	''	Years (sidereal)	0.99996121
Years (sidereal)	Days (mean solar)	365.25636			

GREEK ALPHABET

Greek letter	Greek name	English equivalent	Greek letter	Greek name	English equivalent
Α α	Alpha	a	Ν ν	Nu	n
Β β	Beta	b	Ξ ξ	Xi	x
Γ γ	Gamma	g	Ο ο	Omicron	ŏ
Δ δ	Delta	d	Π π	Pi	p
Ε ϵ	Epsilon	ĕ	Ρ ρ	Rho	r
Ζ ζ	Zeta	z	Σ σ ς	Sigma	s
Η η	Eta	ē	Τ τ	Tau	t
Θ θ ϑ	Theta	th	Υ υ	Upsilon	u
Ι ι	Iota	i	Φ ϕ φ	Phi	ph
Κ κ	Kappa	k	Χ χ	Chi	ch
Λ λ	Lambda	l	Ψ ψ	Psi	ps
Μ μ	Mu	m	Ω ω	Omega	ō

MATHEMATICAL SYMBOLS AND ABBREVIATIONS

Symbols and Abbreviations of Commercial Arithmetic

#	Number (if writ-	Apr.	April	Dr.	Debit, debtor,
	ten before a	a/s	Account sales		doctor
	numeral);	Aug.	August	E	East
	pounds	av.	Average	ea.	Each
	(weight), lb. (if	avoir.	Avoirdupois	e.g.	(exempli gratia)
	written after a	bal.	Balance		for example
	numeral.)	bbl. *or*	Barrel	etc.	And so forth
@	At, as "@ 5¢ per	brl.		ex.	Example, exer-
	C," for "at 5	bk.	Bank, book		cise, express
	cents per hun-	bl.	Bale	exch.	Exchange
	dred."	B/L	Bill of lading	exp.	Expense
%	Per cent; per	bu.	Bushel	F	Fahrenheit
	hundred.	bx.	Box	Feb.	February
¢	Cents (placed	C	(centum)	f.o.b.	Free on board
	after figures)		hundred	Fri.	Friday
$	Dollars, (pre-	cd.	Cord	frt.	Freight
	fixed before	cg.	Centigram	ft. *or* f.	Foot
	figures).	ch.	Chain	gal.	Gallon
£	Pounds (British	chg.	Charge	gi.	Gill
	currency)	c.i.f.	Carriage and in-	gr.	Grain
✓	Check mark		surance free.	gro.	Gross
&	And, as in	ck.	Check	gr. gro.	Great gross
	"Smith, Jones	cm.	Centimeter	guar.	Guarantee
	& Co."	cml.	Commercial	hf.	Half
c/o	Care of	Co.	Company,	hhd.	Hogshead
A	Acre		county	hr.	Hour
a/c	Account	c.o.d.	Cash on delivery	i.e.	(id est) that is
acct.	Account	coll.	Collection	in.	Inch, inches
ad val	(ad valorem),	com.	Commission	ins.	Insurance
	according to	cr.	Credit, creditor,	*inst.*	(instant) the
	value		crate		present month
A.M.	(ante meridiem)	cs.	Case	int.	Interest
or	in the morning,	c. *or*	Cent	inv.	Invoice
a.m.	between mid-	ct.		inv'y	Inventory
	night and the	cu.	Cubic	Jan.	January
	following noon.	cwt.	Hundredweight	kg.	Keg, kilogram
	12:00 A.M. is	*d*	Pence (British	km.	Kilometer
	noon, better		currency)	lb. lbs.	Pound, pounds
	12:00 M, 12:01	da.	Day	lp	List price
	A.M. is one	Dec.	December	ltd.	Limited
	minute after	dept.	Department	L.S.	(locus sigillis)
	midnight.	dft.	Draft		place for the
amt.	Amount	disc.	Discount		seal
ans.	Answer	dm.	Decimeter	M	(mille) thou-
ap	Apothecaries'	do.	Ditto		sand; meridiem
	weight or	doz.	Dozen		as in 12:00 M
	measure	dr.	Dram	m.	Mill, meter

MATHEMATICAL SYMBOLS AND ABBREVIATIONS (Continued)

Symbols and Abbreviations of Commercial Arithmetic

Mar.	March	P.M. *or* p.m.	(post meridiem) in the afternoon, between noon and the following midnight. 12:00 P.M. is midnight, 12:01 P.M. is one minute after noon.	Sat.	Saturday
mdse.	Merchandise	pp.	Pages	sec.	Second
mi.	Mile	pr.	Pair	sec'y	Secretary
min.	Minute	*prox.*	(proximo) in the following month	Sept.	September
mm.	Millimeter	pt.	Pint, point	set.	Settlement
mo.	Month	pwt. (*or* dwt.)	Pennyweight	sig.	Signed, signature
Mon.	Monday	qr.	Quire	sq.	Square
mortg.	Mortgage	qt.	Quart	stk.	Stock
N, NE, NW, etc.	North, Northeast, Northwest, etc.	rd.	Rod, road	Sun.	Sunday
no. *or* numb.	Number	rec'd	Received	T.	Ton
Nov.	November	rec't	Receipt	temp.	Temperature
Oct.	October	rm.	Ream	Thu.	Thursday
O.K.	Correct	S, SE, SW, etc.	South, Southeast, Southwest, etc.	treas.	Treasurer, treasury
oz.	Ounce			Tues.	Tuesday
p.	Page			*ult.*	(ultimo) in the last month
par.	Paragraph			*via*	By way of
pay't	Payment			viz.	(videlicet) namely
pc.	Piece			vol.	Volume
pd.	Paid			W	West
per	By, by the, as in "per C," "per M," "per doz."			Wed.	Wednesday
pfd.	Preferred			wk.	Week
pk.	Peck, pecks			wt.	Weight
pkg.	Package			yd.	Yard
				yr.	Year

Symbols Belonging to Plane Geometry

In blackboard presentation combined with oral discussion a variety of informal abbreviations is acceptable. But in written reports it is best to use only standard abbreviations, preferably those which are found on a typewriter. Avoid characters not sufficiently distinctive as to remain legible and unambiguous when written hurriedly. When used as nouns, the plural may be indicated by adding "s"; when used as relations the copula "is" and preposition "to" may be deleted.

A°	*Angle* associated with the point or vector A. Use "rt°" for "right angle," and "st°" for "straight angle." Avoid ∠ which resembles symbol for "less than." Use "arc" rather than "⌢" to avoid confusion with bar for "line segment."
∠	Angle
∡	Angles

MATHEMATICAL SYMBOLS AND ABBREVIATIONS (Continued)

⊙	Circle. ⊙ A(B) designates the circle with center A passing through B. ⊙ (AB) has AB as diameter, and ⊙ (ABC) circumscribes △ABC. ⊙A, r has center A, radius r.
Ⓢ	Circles.
A:B(S)	Collinear. The line AB passes through the point S. Use $\overline{\text{AB}}$(S) for "S lies between A and B."
≅	Congruent. (Same shape and same size).
A:B(S,r)	Point of division. "S divides AB in the ratio AS:BS = *r*." Use (S, −) for internal division, and (S, +) for external division.
=	Equal, equivalent. The quantity measured may be indicated in parenthesis at the right of the statement, as "(area)," "(vol.)," "(angle)." Avoid pictogram $\stackrel{\frown}{-}$.
$\overline{\overline{\wedge}}$	Perspective. See ABC:DEF(S) below.
$\overline{\wedge}$	Projective. See ABC::DEF below.
a/b	Fraction (with numerator a, denominator b). Same meaning as quotient "a ÷ b" (with dividend a, divisor b) and ratio a:b (with antecedent a, consequent b).
$\underset{\cdot}{\sim}$	Homothetic (Similar and perspective). "ABC $\underset{\cdot}{\sim}$ DEF" indicates that corresponding lines, such as BC and EF, are parallel. Use "$\underset{\cdot}{\sim}$ (S, r)" for "the homothecy with center S and ratio r."
AB	Line, segment, vector. When further distinctions are desired, use $\overline{\text{AB}}$ (or *AB*) for the length of the line segment (distance between A and B), use $\overrightarrow{\text{AB}}$ for the vector (displacement from A toward B), and use A:B for the infinite line (determined by the two points).
∥	Parallel.
▱	Parallelogram (vertices are named in counterclockwise order).
⊥	Perpendicular.
ABC::DEF	Projective (related by a sequence of perspectivities). Same meaning as pictogram "$\overline{\wedge}$."
ABC:DEF(S)	Perspective. The lines A:D and B:E and C:F are concurrent at the point S (center of perspectivity). Same meaning as pictogram "$\overline{\overline{\wedge}}$."
QED	End of proof.
QEF	End of construction.
℞	Cross ratio, anharmonic ratio ℞(AC, BD) = AB · CD/AD · CB. ℞ = −1 means "harmonic" and may be used for collinear points (range), concurrent lines (pencil), or cyclic quadrangle. ℞(CA, BD) = 1/℞(AC, BD) = ℞(AC, DB). Also ℞(AB, CD) = 1 − ℞(AC, BD).
∼	Similar. △ABC ∼ △DEF means that A corresponds to D, B to E, and C to F, respectively.
□	Square. Use "rect" for "rectangle," "quad" for

MATHEMATICAL SYMBOLS AND ABBREVIATIONS (Continued)

	"quadrilateral" (or quadrangle), and "trap" for "trapezoid" rather than pictograms.
ABC $\tilde{\sim}$ DEF(r = 1)	Displaced by translation. (The triangles are congruent and similarly placed).
$\triangle$	Triangle. Use "isos" for "isosceles triangle" (naming vertex opposite base first).
$\dot{\text{B}}$	Vertex. In using triliteral symbol ABC, A is on the initial side, B is the vertex, C is on the terminal side. The vertex may be marked with a caret "ABC" or by superscript position "A^{B}C."

General Mathematical Symbols and Abbreviations

"Bold-face" or gothic type—To indicate vectors corresponding italic type indicates magnitude of the vector. In manuscript and at the blackboard, bold-faced type is variously indicated by wavy underscoring, or enclosure in a circle, or even by wavy overscoring. Some persons use German type.

Half-spaces—In writing numbers with many recorded digits, half-spaces (rather than commas or other marks) may well be used to separate convenient groups of digits. Thus $\pi = 3.14159\ 26536-$.

Superscripts—To indicate: **1.** powers, as in x^2, $(a - x)^n$, etc. In modern practice $a^0 = 1$ always by definition (except for $a = 0$). Also in ∞', ∞^2, etc., indicating number of degrees of freedom. Wherever the context restricts the value of a to non-negative (real) values and n to positive integers, $a^{1/n}$ means the non-negative (real) nth root of a. For complex numbers, x^y is defined as $e^{y\,(\log x)}$, where the principal value of $\log x$ is to be taken. In tables 0.0^5314 may be used to indicate 0.00000314. Note special use of $\sin^n x$ for $(\sin x)^n$ except for $n = -1$, also for $\cos^n x$, etc. **2.** symbolic powers, or order of iteration, as in T^n or in D^n $(= d^n/dx^n)$, or in inverse functions as in $\sin^{-1}$, $\cos^{-1}$, $\sinh^{-1}$, etc. **3.** order of differentiation, as in y' ("y prime"), y'' ("y second," or "y double prime"), $\cdots$, $y^{(N)}$, $\cdot\cdot$ **4.** feet and inches, as in 3′4″. **5.** degrees, minutes, seconds, as in 34°5′17″. Do not omit ° for common angles. Write 0°, 30°, 45°, 60°, etc., not 0, 30, 45, 60, etc. Do not use superscript, ^{r}, for radians. Write $180° = \pi$ rad, but write $\cos(\pi/3)$ for $\cos 60°$. **6.** days, hours, minutes, seconds, as in $10^d3^h27^m5.3^s$. **7.** degrees of temperature as in 104°. Where C (for Centigrade) or F (for Fahrenheit) is given, recent usage approves the omission of the °, thus 100C = 212F, and −40C = −40F. **8.** For use with integral sign $\int$, and with vertical bar |, see these symbols. **9.** Vertex, as in A^BC.

Dot-accents—To indicate derivatives with respect to time, (Newton's notation), as in $\dot{x}$ for x-component of velocity, and $\ddot{x}$ for x-component of acceleration.

Subscripts—To indicate: **1.** position in a sequence, set, or matrix, as in a_1, a_2, a_3, $\cdots a_n$, $\cdots$, or $a_0x^n + a_1x^{n-1} + \cdots + a_rx^{n-r} + \cdots a_n$ or in $\begin{pmatrix} a_{11}a_{12}a_{13} \\ a_{21}a_{22}a_{23} \end{pmatrix}$. **2.** general distinguishing mark. Two subscripts may be written adjacently without commas as a_{11} and to be read "a sub one one," not "a sub eleven." A subscript is sometimes enclosed in parentheses as in $F_{(0)1}$ where such distinction seems demanded. For special uses see associated symbols.

MATHEMATICAL SYMBOLS AND ABBREVIATIONS (Continued)

Juxtaposition—To indicate: **1.** the algebraic product, as in $2bxy$. **2.** the logical product as in AB where A and B are given classes, and in symbolic logic. Also written with centrally placed dot as $A \cdot B$. **3.** the group product, as in ST (the result of performing first S, then T, or in aH, the co-set consisting for given a of all operations ah, where h is in H. **4.** general operational or functional combination as in dy/dx, sin x, log x, max y, lim x_n, etc. **5.** sequence of points or other elements determining a geometric figure, as line AB, parallelogram $ABCD$, angle ABC, etc. **6.** sum of products (in tensor notation) when index appears as subscript for one factor and superscript for another. Thus a_ix^i means Σa_ix^i, in tensor notation.

( )—Parentheses ("round brackets") to indicate: **1.** aggregation, as in $(a + b) \cdot (a - b) = a^2 - b^2$. **2.** argument of function, as in $f(x)$, $g(x,y)$, etc. **3.** sequence or set, as in $a = (a_i)$, $x = (x_{ij})$, (x,y,z), etc. **4.** matrix, as in $\begin{pmatrix} a_{11}a_{12}a_{13} \\ a_{21}a_{22}a_{23} \end{pmatrix}$ also written as $\begin{Vmatrix} a_{11}a_{12}a_{13} \\ a_{21}a_{22}a_{23} \end{Vmatrix}$. **5.** permutation (or substitution) in group theory as in $\begin{pmatrix} a_1a_2a_3 \\ b_1b_2b_3 \end{pmatrix}$ where a_i is replaced by $b_i (i = 1,2,3)$. **6.** binomial coefficient, as in $\binom{n}{r} = n!/[r!(n - r)!]$. This is also designated by $C_{n,r}$ or ${}_nC_r$. For n,r, positive integers, $\binom{-n}{r} = (-1)^r\binom{n + r - 1}{r}$, $\binom{n}{-r} = 0$, by definition. **7.** cycle or cylic permutation (in group theory) as in (a_1, a_2, a_3) for $\begin{pmatrix} a_1a_2a_3 \\ a_2a_3a_1 \end{pmatrix}$. **8.** greatest common divisor, as in $(30,42) = 6$, $(7,5) = 1$. **9.** inner product as in $(ab) = \Sigma_i a_ib_i$. **10.** segment or open interval, as in (a,b), for system of values of x, where $a < x < b$.

Superscript $^{( )}$—To indicate: **1.** general index as distinguished from exponent. **2.** index of order of derivative as in $y, y', \cdots, y^{(n)}, \cdots$. **3.** "factorial," as in $x^{(r)} = x(x - 1) \cdots (x - r + 1)$. By definition $x^{(-r)} = 1/[(x + 1) \cdot (x + 2) \cdots (x + r)]$.

[ ]—Brackets ("square brackets"), to indicate: **1.** aggregation. **2.** argument of function as with ( ). **3.** greatest integer in, as $[2] = 2$, $[-7/3] = -3$. **4.** inner product (for coefficients in normal equations in the method of least squares), as in $[aa]$, $[XY]$, etc. **5.** outer product of vectors. Other notations are $V\ \boldsymbol{ab}$ and $\boldsymbol{a} \times \boldsymbol{b}$. **6.** divided difference (in formal interpolation). $[x_i] = y_i$, $[x_i, x_{i+1}] = (y_{i+1} - y_i)/(x_{i+1} - x_i)$, $\cdots [x_i, x_{i+1}, \cdots, x_{i+r}] = ([x_{i+1}, \cdots, x_{i+r}] - [x_i, \cdots, x_{i+r-1}])/(x_{i+r} - x_i)$. **7.** range of points (in projective geometry) as in $[P]$. **8.** base (basis) of Abelian group, as in $[a,b, \cdots, k]$. **9.** module or ideal, as $[2] = [0, \pm 2, \pm 4, \cdots, \pm 2n, \cdots]$. **10.** Christoffel symbol, as in $\begin{bmatrix} mn \\ p \end{bmatrix} = \frac{1}{2}\left(\frac{\partial g_{pm}}{\partial x^n} + \frac{\partial g_{pn}}{\partial x^m} = \frac{\partial g_{mn}}{\partial x^p}\right)$. **11.** closed interval, as in $[a,b]$ for system of values of x where $a \leqq x \leqq b$.

Subscript note. The use of adjacent subscripts, rather than of indices placed directly below is recommended on account of its availability for running text, and its economy of space and of expense in type setting. Thus use

$\sum_i$ rather than $\sum\limits_i$, $\int_a^b$ rather than $\int\limits_a^b$, etc.

MATHEMATICAL SYMBOLS AND ABBREVIATIONS (Continued)

$\{\ \}$—Braces ("curly brackets") to indicate: **1.**:aggregation, as in $\{(x-a)(x-b)\}^2$. **2.** class of (in theory of aggregates), where the general element only is mentioned, as in $\{a_i\} = [a_1, a_2, a_3]$, $(i = 1,2,3)$. **3.** Christoffel symbol, as in $\left\{ {m\ n \atop p} \right\} = g^{rp}\left[{m\ n \atop p} \right]$. **4.** the members in a uniquely determined set, as in $\{x_1, x_2, \cdots x_n\}$.

$<\ >$—Angle brackets to indicate: **1.** aggregation. **2.** closed interval as with $[\ ]$. **3.** Angle between as in $A \cdot B = AB \cos <A, B>$. Use sparingly because of ambiguity with signs of inequality.

$|\ |$—Vertical bars, to indicate: **1.** absolute value (modulus of complex number). as $|a + ib|^2 = a^2 + b^2$. **2.** magnitude of (for vectors) as $a = |a|$. **3.** determinant, as in $\begin{vmatrix} a & b \\ c & d \end{vmatrix} = ad - bc$. The use of the notation $|a_{ij}|$ for the determinant of the matrix (a_{ij}), is common but is ambiguous. The notation det (a_{ij}) may be used for this determinant.

$\|\ \|$—Double bars, to indicate: **1.** matrix as in $\begin{Vmatrix} a_{11}a_{12}a_{13} \\ a_{21}a_{22}a_{23} \end{Vmatrix}$ or in $\|a_{ij}\|$. **2.** generalized length (for metrical spaces), as in $\|f\|^2 = \int f^2(x)dx$. **3.** To indicate the length of a vector, as in $\|X\|$.

(], [).—For intervals, as $(a,b]$, for system of values of x for which $a < x \leqq b$, and $[a,b)$, for system of values of x for which $a \leqq x < b$. Similarly $(a,b >$, and $< a,b)$ are sometimes used for these respectively.

$\langle\ \rangle$—Brackets ("bent brackets"), to indicate an ordered set of objects, as in $\langle x, y, z\rangle$.

$\vdash$—Is deducible from.

$\models$—Is valid, or is a tautology (in logic), as in $\models A$ (A is valid).

$\cup$—(is) the union, or sum or join (of), as in $A \cup B$.

$\cap$—(is) the intersection, or product (of), as in $A \cap B$.

$\supset$—**1.** contains (or containing) as proper sub-class. **2.** implies (or implying).

$\supseteq$—contains (or containing) as sub-class. (Some writers use, $\supset$, for this.)

$\subset$—(is) contained as proper sub-class within.

$\subseteq$—(is) contained as sub-class within. (Some writers use, $\subset$, for this.)

$\equiv$—**1.** (is) identical with. $\equiv_x$ indicates (is) identical with for all values of x for which both members are defined. **2.** (is) congruent to (with respect to indicated modulus) as in $a \equiv b \pmod m$. **3.** (is) equivalent to (in formal logic).

$=$—(is) equal (to).

$\wedge$—and (in logic).

$\vee$—or (in logic), as in $p \vee q$ (p or q), and/or

$\veebar$—exclusive or, as in $p \veebar q$ (p or q, but not both).

$<$—(is) less than.

$>$—(is greater than. Use "less than" preferably as in "$a < x < b$" rather than "$b > x > a$" since former corresponds to order of real numbers on x-axis.

$\leqq$ *or* $\leq$—(is) less than or equal to. Sometimes read (in the case of real numbers) as "(is) not greater than."

$\geqq$ *or* $\geq$—(is) greater than or equal to. Sometimes read (in the case of real numbers) as "(is) not less than."

MATHEMATICAL SYMBOLS AND ABBREVIATIONS (Continued)

$\not\equiv$—**1.** (is) not identically equal (to). (Not "identically unequal to"), is unequal to for at least one value. **2.** (is) not congruent (to).

$\neq$—(is) not equal (to).

$\lessgtr$—**1.** (is) not equal (to), (for real quantities). **2.** (Sometimes when explained by context) less than or greater than respectively.

$\sim$—**1.** (is) formally, asymptotically, or approximately equal to. (The context should make the meaning specific.) Do not use $\doteq$ for "approximately equal to." **2.** (is) similar (to). **3.** not (in some works on formal logic). **4.** used to indicate the equivalence of two matrices.

Use "iff" for "if and only if." Alternately "$^{(n)}$ as" (read "as" may be used for "necessary and sufficient."

$\rightarrow$—**1.** approaches (as a limit), as in $\lim\limits_{x \to a} f(x) = b$, $f(x) \to b$, as $x \to a$, etc. (Do not use $\doteq$). Not usually employed with long expressions. **2.** leads to, validates, implies (in logic). **3.** corresponds to. **4.** "if . . . , then . . . " relation (in logic).

$\leftrightarrow$—**1.** mutually implies (in logic). **2.** in one-to-one correspondence with, corresponds reciprocally to. **3.** if and only if, also iff.

$\downarrow$—decreases monotonically to a limit.

$\uparrow$—increases monotonically to a limit.

Superscript $\rightarrow$—directed line as $\overrightarrow{AB}$.

|—Vertical bar, to indicate: **1.** value at, as in $f(x)|_a = f(a)$, or $f(x)|_{x=a} = f(a)$. **2.** value between as in $f(x)|_a^b = f(b) - f(a)$. **3.** is a divisor of, divides (in number theory) as $3|6$, $3\underline{/6}$ for "3 divides 6" or use solidus 6/3 to indicate that this quotient is an integer, or $(x - a)|(x^2 - a^2)$. **4.** inner product (with parentheses) as $(a|b) = \Sigma_i a_i b_i$.

\ or /—Stroke, mark of cancellation as in $\cancel{3}x = \overset{2}{\cancel{6}}$, $x + \cancel{3} = \overset{4}{\cancel{7}}$.

/—Solidus, or oblique rule, to indicate: **1.** actual or symbolic division, as in $3/7$, $(x - a)/(x - b)$, d/dx, dy/dx, d^2y/dx^2. Do not write ambiguously $a - b/c - d$, but $(a - b)/(c - d)$ or $a - (b/c) - d$, as may be intended. Do not write a/bc but $(a/b)c$ or $a/(bc)$ as intended. Write a proportion as $a/b = c/d$ not $a:b::c:d$. Where A, B, C, D designate displayed expressions, write the proportion as $\frac{A}{B} = \frac{C}{D}\cdot$ In commercial typing in place of $8\frac{5}{12}$, it is usual to write 8-5/12. The solidus form a/b, adapted to running text should be used where conveniently possible, rather than the displayed form $\frac{a}{b}\cdot$ **2.** quotient or factor group (in group theory) as G/H, (where H is a normal subgroup of G). **3.** per, as in ft/sec. **4.** discount symbol, as in Cash 6, 4/5, 2/30, n/90 indicating 6% discount for immediate payment, 4% discount if paid within 5 days, 2% discount if paid within 30 days, no discount thereafter, but face amount of bill is due (net) not later than the 90th day. **5.** shilling, (in British currency) as 3/6d, or 10/ –.

Superscript $\overline{}$—Vinculum. This may be regarded as obsolescent for general use as a mark of aggregation due to its unsuitability for monotype setting. In conjunction with the radical sign it is widely used, but may often be avoided. There is little logical or historical basis for using $\sqrt{2}$ rather than

MATHEMATICAL SYMBOLS AND ABBREVIATIONS (Continued)

$\sqrt{2}$. For a longer expression, one may write $\sqrt{(x^2 + a^2)}$ rather than $\sqrt{x^2 + a^2}$. Instead of $\sqrt{x - a}\,(x - b)$, one might write $(x - b)\sqrt{(x - a)}$. In geometry, the vinculum may be used for line segments as in $\overline{AB}$.

Superscript $\overline{}$—Bar. To indicate: **1.** complex conjugate of, as $\bar{z}$. This is somewhat inconvenient for "upper extended" letters and capitals as, $\bar{b}$, $\bar{h}$, $\bar{X}$, etc. Also indicated by *conj*, as $conj\,(x + iy) = x - iy$ or by use of a "star" as in z^*. **2.** arithmetic mean value of, as in $\bar{x}, = \sum_i x_i/n$. **3.** closure of (in topology), as in $\bar{E}$ (the closure of E). **4.** "least upper," as in $\overline{\lim}$, and $\bar{B}$ for least upper limit and least upper bound, respectively. See "sup." **5.** repeating decimal, as in $1.\overline{14} = 1.141414 \cdots$.

Subscript $\underline{}$—To indicate: **1.** italics (in manuscript). **2.** "greatest lower," as in $\underline{\lim}$, and $\underline{B}$, for greatest lower limit, and greatest lower bound respectively. See "inf."

—— —horizontal rule, sign of division, as in $\dfrac{x - a}{x + a}$. Ordinarily the solidus form, adapted to running text, is preferred, as in $(x - a)/(x + a)$. When numerator and denominator are both complicated, the displayed form using horizontal rule may be avoided by writing "A/B, where $A = \cdots$, and $B = \cdots$."

$-$ (centrally placed)—Minus sign. To indicate: **1.** subtraction as in $7 - 2 = 5$, $a^2 - b^2 = (a - b)(a + b)$. **2.** overestimate, as in 3.5−. **3.** approach through negative values as in $-\infty$ and -0. **4.** region where variable indicated by context is negative (in graphs). **5.** logical difference (in theory of classes). **6.** in $(-)^n$, the sign expressed by $(-1)^n$.

. (on line)—**1.** decimal point. In the decimal representation of a number between 0 and 1, the cipher, 0, should (except in tables) appear before the decimal point. Thus 0.314 not .314. Notation by powers of 10, ("scientific notation") is recommended, especially when recording approximate values; thus to four significant figures, 3.140×10^9 and 3.140×10^{-6}. **2.** (sometimes used in quoting bond prices) as in 95.17 for $95^{17}/_{32}$. **3.** (sometimes used in recording mental age) as in 12.3 for 12 yr. 3 mo. **4.** (in symbolic logic) as mark of punctuation separating terms, also as, "and."

:—Colon. To indicate: **1.** hours, in recording time, as in 4:10 p.m. **2.** ratio (an obsolescent form) as in $a{:}b$. The form a/b is preferred. **3.** (in symbolic logic) as mark of punctuation separating groups of terms, as in $p \cdot p \supset q : \supset : q$. **4.** Indicates infinite line $A{:}B$, collinearity $A{:}B(S)$, and perspectivity. **5.** such that (in logic). $\cdot$ (centrally placed)—**1.** mark of algebraic multiplication, particularly where mere juxtaposition would be ambiguous, as in $\overline{AB} \cdot \overline{CD}$. **2.** (for vectors), the mark of inner or dot multiplication as in $\mathbf{a} \cdot \mathbf{b} = ab \cos \angle \mathbf{ab}$. Other notations are $(\mathbf{ab})$ and $S\,\mathbf{ab}$.

. (placed above line)—to indicate repeating decimal, used in pairs, as in $1.\dot{7}3\dot{5} = 1.735735735 \cdots$.

$\cdots$ (preferably centrally placed)—"three dots" meaning "and so forth," or "and so forth up to." Particularly when the reader knows what has been omitted, as in the sequence of natural integers, etc. as in $1, 2 \cdots, n, \cdots$; or $a_0, a_1, \cdots, a_n, \cdots$, or $1, \cdots, m$.

MATHEMATICAL SYMBOLS AND ABBREVIATIONS (Continued)

$\therefore$—hence, therefore. Inverted $\because$ since.

:., ::, ::., etc.—(in symbolic logic), marks of punctuation stronger than · and :

,,—ditto.

$\circ$ (centrally placed)—indicates (in classes or sets) the composite of functions, as in $g \circ f$.

+ plus sign. To indicate: **1.** addition, as in $2 + 3$, $a + b$, 10^{a+1}. **2.** underestimate, as in 3.5+. **3.** continued fraction as in $a_0 + \frac{1}{a_1+} \frac{1}{a_2+} \cdots$ for $a_0 + \cfrac{1}{a_1 + \cfrac{1}{a_2+}}$. **4.** approach through positive values as in $+\infty$, and in $+0$ **5.** region where variable indicated by the context is positive, (in graphs). **6.** logical addition (in theory of classes). **7.** " . . or . . or both," (in formal logic). Note: In writing series indicate sign before and after dots of omission, as $a_0 + a_1 + \cdots + a_n$, or $1 - \frac{1}{2} + \frac{1}{3} + \cdots + (-1)^{n-1}\frac{1}{n}$. **8.** in abstract group theory a group or co-set may be expressed as the sum of its elements.

$\oplus$—denoting a logical operation of summation between elements of a set, such as the sum of vectors.

$\pm$—**1.** "plus or minus." The repeated appearance of $\pm$ as in $\pm a \pm b \pm c$ is ambiguous. In many cases the sign $\pm$ before a term which appears repeatedly is intended to indicate the systematic use of the positive determination or of the negative determination throughout. Thus one may write $(a \pm b)^3 = a^3 \pm 3a^2b + 3ab^2 \pm b^3$. Where the context restricts the value of a to non-negative (real) values, $\sqrt{a}$ means the non-negative (real) square root of a. Hence when both signs are desired, write $\pm$ before the radical. For roots of a quadratic equation $ax^2 + bx + c = 0$, use $\pm$ as in $\frac{-b \pm \sqrt{b^2 - 4ac}}{2a}$. **2.** (in theory of observation), "with a probable error of." As in 17.2 ± 0.5 cm.

$\mp$—"minus or plus respectively." Used in context where $\pm$ has appeared previously, as in $a^3 \pm b^3 = (a \pm b)(a^2 \mp ab + b^2)$. Here upper signs are to be taken throughout, or else lower signs. The notation $\pm a \mp b \pm c$ is ambiguous, meaning perhaps one of the four values $\pm(a - b) \pm c$, or one of the two values $\pm(a - b + c)$.

$\times$—**1.** times, (sign of algebraic multiplication). Used chiefly in arithmetic, as in $2 \times 2 = 4$, 7.3×10^4. **2.** (for vectors) the sign of outer or cross multiplication. **3.** (for classes) the Cartesian product. Thus $A \times B$ is the class of all ordered pairs (a,b) where a is an element of A, and b of B.

$\otimes$—denoting a logical operation of multiplication between elements of a set, such as the product of vectors.

$\div$—sign of division. Used chiefly in arithmetic. Should be replaced by solidus, /, where convenient.

$\sqrt{\ }\ \sqrt[n]{\ }$—square root of, nth root of. (See discussion under "superscript," and under "vinculum"). By custom, for a positive, $\sqrt{(-a)}$ means usually $i\sqrt{a}$, rather than $-i\sqrt{a}$, but the latter unambiguous forms are preferred.

!—"factorial," as in $3! = 1 \cdot 2 \cdot 3 = 6$. $0! = 1$, (by definition). The ele-

MATHEMATICAL SYMBOLS AND ABBREVIATIONS (Continued)

mentary arithmetic definition of factorial n, may be replaced in favor of the definition as a special case of the Gamma function $\Gamma(x)$. For $-1 < n$, $n! = \Gamma(n+1), = \int_0^\infty x^n e^{-x} dx$. For n a large natural number, Stirling's asymptotic formula (extended) yields $n! \sim \sqrt{2n\pi}\,(n/e)^n \left(1 + \frac{1}{12n} + \frac{1}{288n^2} - \frac{139}{51840n^3} - \cdots\right)$. Note: Do not use the obsolescent form $\underline{|n}$ for $n!$

$\int, \int_a^b, \int_a^x, \int\int, \int_c$—Integral signs. (use preferably bold-face type)

$$\int_a^b \int_c^d f(x,y)dxdy \text{ denotes } \int_a^b \left(\int_c^d f(x,y)dx\right)dy, = \int_a^b dy \int_c^d dxf(x,y).$$

$\oint$—curvilinear integral over closed path free from singularities. (Use preferably bold-face type.)

§—section, or article.

⁋ *or* ¶—paragraph.

$\propto$—varies as. Instead of $y \propto x$ one may write $y = kx$, k being the constant factor of proportionality.

∇—nabla—To indicate: **1.** linear vector operator $\left(\frac{\partial}{\partial x}, \frac{\partial}{\partial y}, \frac{\partial}{\partial z}\right)$ as used also in divergence, gradient, and curl (or rotation). **2.** backward difference (in interpolation theory) $\nabla a_n = a_n - a_{n-1}$.

∇^2—Laplace operator. Sometimes written Δ.

⊡—Four dimensional differential operator of D'Alembert and Poincare [= point carré].

$^\circ$, $'$, $''$, $'''$, $\cdots$, (N) $\cdots$, (superscript)—superscript numbers. See "superscripts."

∞—infinity. Use $+\infty$ or $-\infty$ respectively, where direction of approach along real numbers is to be indicated. Otherwise use ∞ rather than $\pm\infty$. Note: "$n \to \infty$," may be read "as n increases without bound." Lim $f(x) = \infty$ means "Lim $1/f(x) = 0$," so that whenever one variable "becomes infinite" the meaning is that its reciprocal "becomes zero."

$\in$—(is) a member (of), as in $x \in A$. Several objects can be denoted as members in a set A by $x_1, x_2, \cdots x_n \in A$.

$\notin$—(is) not a member (of).

$\mathcal{P}$—denotes a power set, as in $\mathcal{P}(A)$ which denotes the set of all subsets of a set A.

$\aleph$—Aleph (initial Hebrew letter) transfinite cardinal number, in particular that of all real numbers. $\aleph_0$ (aleph null) first transfinite cardinal. Use of Hebrew letters for transfinite numerals is recommended: "aleph" (omit the null) for the cardinal of the set of integers; $\beth$ "beth" for the cardinal of the continuum (number of points on a line); and $\gimel$ "gimel" for the cardinal of functions of a real variable.

$\mathfrak{F}$—the Farey series. $\mathfrak{F}_n$ denotes the Farey series of order n.

α—Alpha. To indicate: **1.** (in analytic geometry of 3 dimensions), direction angle with X-axis. **2.** angular acceleration. **3.** (in statistics) $\alpha_0 = 1$, $\alpha_1 = 0$, $\alpha_2 = 1$, $\alpha_3 = \mu_3/\sigma^3$, $\alpha_4 = \mu_4/\sigma^4 = \mu_4/\mu_2{}^2$. **4.** (in mathematical astronomy), right ascension (also indicated by R.A.) **5.** angle of triangle

MATHEMATICAL SYMBOLS AND ABBREVIATIONS (Continued)

at A, opposite side a. **6.** root of algebraic equation as in $a(x - \alpha)(x - \beta)(x - \gamma) = 0$.

B—(Greek Beta)—$B(m,n) = \Gamma(m)\Gamma(n)/\Gamma(m+n)$, (Eulerian Beta-function).

β—Beta. To indicate: **1.** (in analytic geometry of 3 dimensions), direction angle with Y-axis. **2.** (in statistics), $\beta_1 = \alpha_3{}^2 = \mu_3{}^2/\mu_2{}^3$, $\beta_2 = \alpha_4 = \mu_4/\mu_2{}^2$. **3.** angle of triangle at B, opposite side b. **4.** root of algebraic equation. See α.

Γ—$\Gamma(x)$ Gamma-function. See "!" Among numerous definitions equivalent for positive real values of x, are the two following: (i) $\Gamma(x) = \lim_{n\to\infty} \dfrac{1 \cdot 2 \cdots n}{x(x+1)\cdots(x+n-1)} n^{x-1}$, $x > 0$. (ii) $\Gamma(x) = \int_0^\infty e^{-t}t^{x-1}dt$, $R(x) > 0$.

γ—Gamma. To indicate: **1.** (in analytic geometry of 3 dimensions) direction angle with Z-axis. **2.** Euler or Mascheroni constant. (Also indicated by C.) $\gamma = \lim_{n\to\infty}\left(\frac{1}{1} + \frac{1}{2} + \cdots + \frac{1}{n} - \log n\right) = 0.57721\ 56649\ 01532\ 86060\ 65120 \cdots$ **3.** angle of triangle at C, opposite side c. **4.** radius of geodesic curvature. **5.** universal constant of gravitation $= 6.670 \times 10^{-8}$ cm^3/(gm. sec^2). **6.** root of algebraic equation, see α.

Δ—Delta. To indicate: **1.** triangle (in plane geometry). (For right triangle, write rtΔ, not $\measuredangle$). Also area of triangle. **2.** increment, as in Δx, Δf, $f(x + \Delta x)$, etc. **3.** forward difference (interpolation theory). $\Delta a_n = a_{n+1} - a_n$. **4.** Laplacian operator $\frac{\partial^2}{\partial x^2} + \frac{\partial^2}{\partial y^2}$, or $\frac{\partial^2}{\partial x^2} + \frac{\partial^2}{\partial y^2} + \frac{\partial^2}{\partial z^2}$, also designated by ∇^2. **5.** selected square root of the discriminant of a given polynomial, as in $\Delta^2 = b^2 - 4ac$ for the polynomial $ax^2 + bx + c$. **6.** triangular number, (of form $(n^2 - n)/2$). **7.** Legendre's radical, $\Delta(\varphi)^2 = 1 - k^2 \sin^2 \varphi$.

δ—Delta. To indicate: **1.** positive constant dependent upon ϵ that may be chosen initially as near to zero as desired. (In theory of limits, of continuity, etc.) **2.** variation of. **3.** (in interpolational theory) central difference $\delta y_{c+i+\frac{1}{2}} = y_{c+i+1} - y_{c+i}$. **4.** (in mathematical astronomy) apparent declination. **5.** number of double points, or nodes (Plücker number). **6.** (in statistics) deviation. **7.** force of interest, $e^\delta = 1 + i$, (in mathematics of finance). **8.** Kronecker Delta, $\delta_{ij} = 0$ for $i \neq j$, $= 1$ for $i = j$. In tensor notation, also $\delta_i{}^j$. One has $\delta_{ij} = \binom{o}{i - j}$ or $C_{0,i-i}$. **9.** unit elongation (in strength of materials).

∂—curly d. To indicate: **1.** partial differentiation as in $\partial f(x,y)/\partial x$. **2.** The Jacobian operator, as in $\partial(u,v,w)/\partial(x,y,z)$. This is variously represented, sometimes as $J\left(\frac{u,v,w}{x,y,z}\right)$ etc. **3.** a specified square root of the discriminant D. (also sometimes as Δ.)

ϵ—Epsilon. To indicate: **1.** positive constant, that may be chosen initially independently as near to zero as desired. (In theory of limits, of continuity etc.) **2.** primitive root of unity. **3.** (is) member of, (relation of element to containing class). **4.** eccentricity of conic section, usually better designated by e. **5.** (in mathematical astronomy) obliquity of ecliptic. **6.** an ϵ-number is a transfinite ordinal or a certain limiting type. Note: Do not use ϵ or $\mathcal{E}$ for Napierian base except in engineering. See e.

MATHEMATICAL SYMBOLS AND ABBREVIATIONS (Continued)

$\ni$—(Inverted Epsilon). Such that.

$\notin$—(is not a member of.)

ζ—Zeta. To indicate: **1.** Riemann Zeta-function $\zeta(s) = \Sigma_{n=1}^{\infty} n^{-s}$ (for $s > 1$). **2.** (in statistics) a test of linearity. $\zeta = \eta^2 - r^2$.

η—Eta. To indicate: **1.** general variable or unknown constant, analogous to y as in case of moving system of coordinates, etc. Used in ordered set (ξ,η,ζ). **2.** a confocal coordinate. See ξ. **3.** (in statistics), correlation ratio. **4.** order-type of the aggregate of all rational numbers.

Θ—Theta. To indicate: **1.** Theta-function. See ϑ. **2.** absolute temperature (where t is used for time).

θ—Theta. To indicate: **1.** general angular displacement (in trigonometry and analytic geometry). **2.** (in plane polar coordinates (r cis θ)) angle from initial ray to radius vector, $x = r \cos \theta$, $y = r \sin \theta$. **3.** (in cylindrical coordinates ($r\ \theta\ z$)) angle from initial radial half-plane to radial half-plane containing radius vector. **4.** (in spherical coordinates (r,θ,φ)) co-latitude (measured from zenith) and (in astronomy) zenith distance. This notation is traditional in mathematical physics. In texts on analytic geometry, usage varies, θ being employed frequently for the longitude. See φ. **5.** Theta-function. See ϑ. **6.** (in formal theory of operations) a displacement operator. **7.** ordinary temperature (when t is used for time).

ϑ—Theta-function. Definitions and notations vary widely. For elliptic Theta-functions the notation given here is that followed by Whittaker and Watson. Here $\boldsymbol{\vartheta_i}$ designates the value of $\vartheta_i(o,q)$, ϑ_i', the value of $d\vartheta_i(z,q)/dz$. $(i = 1,2,3,4)$. These are defined by $\vartheta_1(z,q) = 2q^{\frac{1}{4}} \sin Z - 2q^{\frac{9}{4}} \sin 3z + 2q^{\frac{25}{4}} \sin 5z - \cdots$ $\vartheta_2(z,q) = 2q^{\frac{1}{4}} \sin z + 2q^{\frac{9}{4}} \sin 3z + 2q^{\frac{25}{4}} \sin 5z + \cdots$ $\vartheta_3(z,q) = 1 + 2q \cos 2z + 2q^4 \cos 4z + 2q^9 \cos 6z + \cdots$ $\vartheta_4(z,q) = 1 - 2q \cos 2z + 2q^4 \cos 4z - 2q^9 \cos 6z + \cdots$.

ι—Iota, number of inflexions (a Plücker number).

ι—inverted Iota, (in formal logic) the unique element fulfilling description stated.

κ—Kappa, number of cusps. (A Plücker number).

Λ—Lambda, sometimes used for null-class.

λ—Lambda. To indicate: **1.** general linear parameter (e.g., in a pencil), as in $F + \lambda G$. **2.** running index, as in $x_\lambda (\lambda = 1, 2, \cdots)$. **3.** longitude (in mathematical astronomy). **4.** characteristic value (as in λ_i) in theory of linear differential equations of second order, linear integral equations, etc. **5.** order-type of the aggregate of all real numbers. **6.** In solid analytic geometry (λ, μ, ν) are direction cosines.

μ—Mu. To indicate: **1.** running index, usually used with λ, as in $x_\lambda y_\mu$. **2.** general linear parameter, when used with λ, as in $\lambda F + \mu G$. **3.** (in statistics) moment, about the arithmetic mean, as in $\mu_k = \Sigma_i f_i(x_i - \bar{x})^k$, $\mu_2 = \sigma^2 =$ variance. **4.** μ_x, force of mortality $= -d(\log_e l_x)/dx$. **5.** (in number theory) inversion function of Moebius and Mertens. **6.** Various physical constants such as permeability, reduced mass. **7.** Integrating factor which converts integrable expression into exact differential form.

ν—Nu. To indicate: **1.** running index, usually with λ and μ. **2.** (in statistics) moment about arbitrary origin A (in "short method") as in $\nu_k = \Sigma_i f_i(x_i - A)^k$. **3.** Frequency.

ξ—Xi. To indicate: **1.** general variable, or unknown constant analogous to

MATHEMATICAL SYMBOLS AND ABBREVIATIONS (Continued)

x, as in moving systems of coordinates, etc. **2.** a confocal coordinate as in (i) confocal ellipses and hyperbolas, $\frac{x^2}{a^2-\lambda}+\frac{y^2}{b^2-\lambda}=1,\ -\infty<\xi<b^2<\eta<a^2$. (ii) confocal ellipsoids and hyperboloids of revolution, $\frac{x^2}{a^2-\lambda}+\frac{y^2+z^2}{b^2-\lambda}=1,\ -\infty<\xi<b^2<\eta<a^2$. (iii) confocal parabolas, $y^2+2\lambda.(x-\lambda)=0,\ -\infty<\xi<0<\eta<+\infty$. (iv) confocal paraboloids of revolution, $y^2+z^2+2\lambda(x-\lambda)=0,\ -\infty<\xi<0<\eta<+\infty$. (v) confocal ellipsoids and hyperboloids, $\frac{x^2}{a^2-\lambda}+\frac{y^2}{b^2-\lambda}+\frac{z^2}{c^2-\lambda}=1,\ -\infty<\xi<c^2<\eta<b^2<\zeta<a^2$. (vi) confocal paraboloids, $\frac{x^2}{a^2-\lambda}+\frac{y^2}{b^2-\lambda}=2z-\lambda$, $-\infty<\xi<b^2<\eta<a^2<\zeta<+\infty$. (ξ,η,ζ), Xi, eta, zeta, are used as alternates for rectangular coordinates (x,y,z).

Π, $\Pi_i\Pi_{i=m}^n$, $\Pi_{(R)}$ or $\underset{(R)}{\Pi}$—**1.** product of terms with index i, or j, etc. ranging from m to n, or over R. Do not use $\underset{i}{\Pi}$, $\overset{n}{\underset{i=m}{\Pi}}$, etc. (Bold-faced type preferred.) **2.** (in some formal logical treatments) "for every."

Π_{ij}, Π_{ijk}, $\cdots$ —$\Pi_i\Pi_j$, $\Pi_i\Pi_j\Pi_k$, etc.

π—Pi. **1.** the ratio of the length of circumference of a circle, to the diameter. $\pi = 3.14159\ 26535\ 89793\ 23846\cdots$ **2.** general notation for plane, projectivity, projective, period, etc.

ρ—Rho. **1.** radius of geodesic curvature. **2.** proportionality factor, as in $\rho X_i = \Sigma_j a_{ij} x_j$.

$\sum$, $\sum_i$, $\sum_{i=m}^n$, $\sum_{(R)}$ or $\underset{(R)}{\sum}$—Sigma. **1.** summation, sum of terms of index i, or j, etc., ranging from m to n, or over range R. Do not use $\sum\limits_i$, $\sum\limits_{i=m}^n$, etc. (bold-faced type preferred). **2.** (on some formal logical treatments) "for at least one." **3.** (in number theory) $\Sigma_{d/n}$ summation extended over all divisors of n. **4.** (in mathematical astronomy) Σ-pt is the intersection of the meridian with the equator.

Σ_{ij}, Σ_{ijk} $\cdots$ —$\Sigma_i\Sigma_j$, $\Sigma_i\Sigma_j\Sigma_k$, etc.

σ—Sigma. To indicate: **1.** radius of torsion. **2.** (in statistics) standard deviation $\sigma^2 N = \Sigma_i(x_i - \bar{x})^2 f_i$. **3.** (in number theory), $\sigma_k(n)$ = sum of kth powers of divisors of n. **4.** any one of several analogous Sigma-functions. The simplest elliptic Sigma-function $\sigma(x)$ is related to the Weierstrassian $\wp$ function by $\wp u = -d^2 \log \sigma u/du^2$. One has $\sigma u = u\left\{1 - \frac{g_2}{2}\frac{u^4}{5!} - 6g_3\frac{u^6}{7!} - \frac{9}{4}g_2^2\frac{u^8}{9!} - 18g_2g_3\frac{u^{10}}{11!} - \cdots\right\}$ **5.** proportionality factor, usually used with ρ.

τ—Tau. To indicate: **1.** number of bitangents (a Plücker number). **2.** time (when t is used for temperature). **3.** torsion (of curve in space).

Υ—Upsilon. (In mathematical astronomy) vernal equinox.

ϕ—Phi. **1.** Used to indicate the Golden Ratio or Section, $\phi = 1.61803\ 39887\ 49894\cdots$. **2.** Used to denote Euler's function, where $\phi(m)$ is the number

MATHEMATICAL SYMBOLS AND ABBREVIATIONS (Continued)

of positive integers not greater than and prime to m. **3.** The null (or empty) set containing no elements.

φ—Phi. To indicate: **1.** general functional symbol, especially for polynomials. **2.** (in spherical coordinates, (r,θ,φ), longitude from x to y in right-handed system. In some astronomical work the z-axis points to the zenith, θ is the zenith-distance, and φ is the latitude. In some works on analytic geometry the roles of θ and φ are interchanged, although the system given is traditional mathematical physics. $x = r \sin \theta \cos \varphi$, $y = r \sin \theta \sin \varphi$, $z = r \cos \theta$. **3.** (in geocentric coordinates, (r,φ,λ), φ = latitude (not co-latitude as with spherical coordinates). **4.** (in plane polar coordinates (r,φ).) Used chiefly as specialized case of spherical coordinates $(r,\pi/2,\varphi)$. See θ. **5.** inclination of plane curve, $\tan \varphi = dy/dx = m$. **6.** (in number theory.) Euler's function or indicatrix. $\varphi(n)$ is the number of positive integers not exceeding n and prime to n. **7.** $\varphi_i(x)$, characteristic function, see λ_i. **8.** argument in Legendre's elliptic integrals. $E(k,\varphi) = \int_0^\varphi \Delta(\varphi)d\varphi$, $F(k,\varphi) = \int_0^\varphi d\varphi/\Delta(\varphi)$. **9.** the normal probability function of Laplace and Gauss in the form $\varphi(t) = \frac{1}{\sqrt{2\pi}} e^{\frac{-t^2}{2}}$. **10.** $\varphi_n(x)$, sometimes used for Bernoulli polynomial. See $B_n(x)$.

χ—Chi. (In statistical theory) χ^2 is a measure of goodness of fit, devised by Karl Pearson. $\chi^2 = \Sigma(f_i - Np_i)^2/(Np_i)$ for N items, with p_i the probability of appearances and f_i the frequency for items in an ith class.

ψ—Psi. To indicate: **1.** general functional symbol (usually with φ). **2.** angle from radius vector to tangent of plane curve. **3.** (with geocentric coordinates in mathematical astronomy), co-latitude. See φ.

Ω—Omega. To indicate: **1.** a certain annihilator in the theory of binary concomitants. **2.** (with subscript) transfinite ordinals of certain minimal type. **3.** (in geometry of the triangle) Ω,Ω', the Brocard points.

ω—Omega. To indicate: **1.** angular velocity. **2.** first transfinite ordinal, order-type of the aggregate of all natural numbers. **3.** imaginary cube root of unity, related to i, by $\omega = (-1 + i\sqrt{3})/2$. **4.** ω_1, ω_2, ω_3, half-periods of Weierstrassian $\wp$-function. **5.** (in geometry of the triangle) Brocard angle. $\cot \omega = \cot A + \cot B + \cot C$.

A—**1.** A vertex, and the associated angle of ΔABC. See α. **2.** A_{ij}, algebraic complement of a_{ij} in determinant, D. $A_{ij} = dD/da_{ij}$. **3.** (in astronomy) azimuth. **4.** (in astronomy) astronomical unit, mean geocentric distance to the sun. **5.** acres. **6.** area.

Å—ångstrom.

$\forall_x$—For every x.

AM—arithmetic mean. See also superscript $^{-}$.

Ans.—answer.

AP—arithmetic progression.

Ax—axiom.

a—**1.** (in elementary algebra) initial term in arithmetic or geometric progression. **2.** (with subscript) coefficient in Fourier series, $(a_0/2) + \Sigma_{n=1}^{\infty}(a_n \cos nx + b_n \sin nx)$. **3.** (with two indices) element in matrix or determinant, as in $\begin{pmatrix} a_{11} & a_{12} & a_{13} \\ a_{21} & a_{22} & a_{23} \end{pmatrix}$ or $\begin{vmatrix} a_{11} & a_{12} \\ a_{21} & a_{22} \end{vmatrix}$. **4.** (in elementary geometry) apothegm. **5.** (in

MATHEMATICAL SYMBOLS AND ABBREVIATIONS (Continued)

geometry or triangle) first side-line, also length of first side of the triangle. **6.** (in elementary analytic geometry) x-intercept. **7.** (in elementary analytic geometry) semi-major axis of ellipse or semi-transverse axis of hyperbola, etc. Thus latus rectum of central conic is $2p = 2b^2/a$, eccentricity $e = c/a$. For central conics and quadrics the a is sometimes associated with x, as in $(x^2/a^2) + (y^2/b^2) = 1$, even when a may be less than b, or as in $-(x^2/a^2) + (y^2/b^2) = 1$, where a is the semi-conjugate axis.

abs—absolute value of.

acc—acceleration.

am—amplitude function. $\varphi = \text{am } u$, where $u = F = \int_0^\varphi d\varphi/\Delta(\varphi)$.

amp—amplitude of vibration.

antilog—antilogarithm.

approx—approximate(ly).

arc (in "arc sin" etc.)—inverse. Also written $\sin^{-1}$ etc. Do not use "arc" for inverse of hyperbolic functions. Write $\sinh^{-1}$, etc.

arg—argument. Angle in polar coordinates r cis θ, and in complex number. $re^{i\theta}$ for r and θ real. Better than "arc" and more legible than superscript -1 to indicate inverse hyperbolic functions, though "$\sinh^{-1} x$" is often used.

av—average.

B—**1.** (With subscripts) Bernoulli numbers and polynomials. To indicate what usage among many is being followed in any given case, authors would do well to list the values of the first few Bernoulli numbers as for example, $B_1 = \frac{1}{2}$, $B_2 = \frac{1}{6}$, $B_3 = 0$, $B_4 = -\frac{1}{30}$, etc. The Bernoulli polynomials, are defined as $B_n(x) = \Sigma_{r=0}^{n}\binom{n}{r}B_rx^{n-r}$, and satisfy $B_n(x+1) - B_n(x) = nx^{n-1}$ with the choice of notation for Bernoulli numbers given above. (Also designated by $\varphi_n(x)$.) See page 613 for a second definition of Bernoulli numbers. **2.** bound (general symbol). $\bar{B}$, $\underline{B}$ designate least upper, and greatest lower bound respectively. Preferred notations are "sup" and "inf" respectively. **3.** a vertex and the associated angle of ΔABC. See β. **4.** (in elementary solid geometry), area of base of a solid.

b—**1.** (in elementary geometry) length of base of plane figure. b,b', parallel bases of trapezoid. **2.** (in elementary analytic geometry) y-intercept. **3.** (in elementary analytic geometry) semi-axis. See a. **4.** (in geometry of the triangle) second side-line, also length of second side of the triangle.

bei(z)—Thomson-Bessel function, $\text{bei}(z) = \frac{(\frac{1}{2}z)^2}{(2!)^2} - \frac{(\frac{1}{2}z)_6}{(6!)^2} + \frac{(\frac{1}{2}z)^{10}}{(10!)^2} - \cdots$

$\text{ber}(z) \pm i\,\text{bei}(z) = J_0(zi\sqrt{\pm i}) = I_0(z\sqrt{\pm i})$.

ber(z)—Thomson-Bessel function. (See bei(z).) $\text{ber}(z) = 1 - \frac{(\frac{1}{2}z)^4}{(4!)^2} + \frac{(\frac{1}{2}z)^8}{(8!)^2} - \cdots$.

(B)-space—A Banach space.

B_m^s—An s-dimensional Betti group modulo m (m prime).

B_0^s—An s-dimensional Betti group relative to the group of integers.

C—**1.** arbitrary constant of integration. **2.** (in elementary geometry) circumference of circle; also, circle. **3.** general symbol for curve. **4.** (with subscripts) combination, as in $C_{n,r}$ or ${}_nC_r$, the number of combinations of n things taken r at a time (without repetitions). The form ${}_nC_r$ or even ${}^nC_r$ is

MATHEMATICAL SYMBOLS AND ABBREVIATIONS (Continued)

widely used but the notation $C_{n,r}$ or $C(n,r)$ or $\binom{n}{r}$ is to be preferred. **5.** Roman numeral for "hundred." **6.** Euler or Mascheroni constant. Also designated by γ. (See γ.) **7.** (chiefly as subscript) contour of integration. **8.** Centigrade, degree Centigrade, as −52C.

C_{I}—(in topology) class of spaces satisfying first axiom of countability, or property of satisfying first axiom of countability.

C_{II}—Same as above for second axiom of countability.

$\underline{C}$—(underlined) the set of complex numbers.

$\underset{\sim}{C}$—(wavy underline) indicates congruence of two n-square matrices.

Ci—cosine integral function, $Ci(x) = \int_{\infty}^{x} (\cos u/u)du$.

c—**1.** (in geometry of triangle) third side-line, also length of third side of triangle. **2.** (in elementary analytic geometry), z-intercept. **3.** (in elementary analytic geometry) semi-axis. (See a.)

cis—$\operatorname{cis} \theta = \cos \theta + i \sin \theta = e^{i\theta} = exp(i\theta)$. Used to distinguish polar coordinates $r \operatorname{cis} \theta$ from rectangular coordinates. Using Gaussian coordinates, $r \operatorname{cis} \theta = x + iy$. In de Moivres' formula $\operatorname{cis}^n \theta = \operatorname{cis} n\theta$.

cn—cosine amplitude function. (Jacobian elliptic function.)

colog—cologarithm, = log of reciprocal = negative of log (especially with base 10).

conj—(complex) conjugate (of). See superscript —.

cos—cosine = sine of the complementary angle.

$\cos^{-1}$—inverse cosine (of). Also written arc cos. $\theta = \cos^{-1} x$ means $\cos \theta = x$. Arc cos x (with capital A) indicates principal value, $0 < \theta < \pi$.

cosh—hyperbolic cosine (of). Do not write Cos nor $\mathfrak{Cof}$ (in German letters).

$\cosh^{-1}$—inverse hyperbolic cosine of. Also written as arg cosh.

covers—coversed sine or coversine.

csc—cosecant (of). Reciprocal of sine.

$\csc^{-1}$—inverse cosecant (of). Also written arc csc.

ctn also cot—cotangent (of). Reciprocal of tangent.

ctn^{-1}—inverse cotangent (of). Also written as $\cot^{-1}$, arc cot, or arc ctn.

ctnh—hyperbolic cotangent (of). Also written as coth.

ctnh^{-1}—inverse hyperbolic cotangent of. Also written as arg coth or arg ctnh.

cu—cubic. In arithmetic "cubic feet" is written "cu ft" but "ft^3" is better.

cum—cumulative.

D—**1.** differential operator, as in $Dy = y'$, $d_x f(x,y) = \partial f/\partial x$. **2.** Roman numeral for "five hundred." **3.** general symbol for denominator, or for determinant. **4.** discriminant of binary form, or of polynomial. **5.** (in statistical theory), (with subscripts 0, 1, $\cdots$, 10), decile marks. **6.** $D(a_1, a_1, \cdots, a_n)$ sometimes designates the Vandermonde determinant.

$$\begin{vmatrix} 1 a_1 & \cdots & a_1^{n-1} \\ \cdots & \cdots & \cdots \\ 1 a_n & \cdots & a_n^{n-1} \end{vmatrix}$$

Def—definition.

Dem—demonstration, proof.

d—**1.** (in elementary algebra), common difference in arithmetic progression. **2.** differential operator, as in d^2y/dx^2. **3.** (in elementary geometry) diameter. **4.** (as superscript) days, as in $2^d3^h17^m$. **5.** (British currency) pence.

MATHEMATICAL SYMBOLS AND ABBREVIATIONS (Continued)

deg—degree, degrees. In navigation use 3 digits measured clockwise from north, so that 090 is due east. In trigonometry use cis θ to indicate direction measured counterclockwise from the positive x-axis, so that cis 90 is the direction of the positive y-axis.

det—determinant of, as in det (a_{ii}).

div—divergence of, also indicated by $\nabla\cdot$ (with nabla and dot).

dn—dn(z), a Jacobian elliptic function.

$\exists$—(there) exists.

$\exists|$—there exist uniquely.

E—**1.** E,F,G fundamental differential quantities of first order for surfaces. **2.** $E(k,\varphi)$, Legendre's normal elliptic integral, of the second kind, $E = E(k,\varphi)$ $\int_0^\varphi \Delta(\varphi)d\varphi$. **3.** east. **4.** (in Euler's polyhedral formula) number of edges of polyhedron. **5.** displacement operator, $E(f(x)) = f(x+1)$. **6.** (in mathematical astronomy) equation of time.

Ei—exponential integral function $Ei(x) = \int_{-\infty}^{x} \frac{e^v dv}{v}$.

Eq—equation.

Ex—exercise.

e—**1.** base of natural (or Napierian) logarithms. In place of e^A one may write exp A. $e = 2.71828\ 18284\ 59045\ 23536 \cdots \log_{10} e = M = 0.43429\ 44819\ 03251\ 82765 \cdots \log_e 10 = 1/M = 2.30258\ 50929\ 94045\ 68402$. In some engineering work, where e is used otherwise, the base of natural logarithms is designated by ϵ. **2.** eccentricity of a conic. **3.** $e_1 = \wp(\omega_1)$, $e_2 = \wp(\omega_2)$, $e_3 = \wp(\omega_3)$, for Weierstrassian elliptic functions. **4.** (in mathematical astronomy) eccentricity of earth's orbit. **5.** (in elementary solid geometry) length of lateral edge (of right pyramid, prism, etc.).

eq—equivalent to.

erf—error function, $\text{erf}(x) = \frac{2}{\sqrt{\pi}} \int_0^x e^{-v^2} dv$.

erfc—complementary function to error function, $\text{erfc}(x) = 1 - \text{erf}(x) = \frac{2}{\sqrt{\pi}} \int_x^\infty e^{-v^2} dv$.

exp—exponential function of, as in $\exp(a^2 + b^2)$ for $e^{a^2+b^2}$. One could also write this "e^u where $u = a^2 + b^2$."

exsec—Exsecant

F—Force.

F—**1.** general symbol for function or functional. **2.** the second fundamental differential quantity of first order for surfaces. See E. **3.** (in Euler's polyhedral formula) number of faces of polyhedron. **4.** $F(k,\varphi)$, Legendre's normal elliptic integral of the first kind, $F = F(k,\varphi) \int_0^\varphi d\varphi/\Delta(\varphi)$. See Δ and E. φ is here the amplitude of F, $\varphi = \text{am } F$. **5.** $F(a, b; c: x)$, hypergeometric function. **6.** Fahrenheit, degree Fahrenheit, as in 70F. **7.** falsity (in logic).

FS—Fourier series.

Fig.—figure.

Fr—frontier set of.

f—**1.** general symbol for function or functional. **2.** f_i, frequency of X_i in

MATHEMATICAL SYMBOLS AND ABBREVIATIONS (Continued)

univariate table. **3.** frequency of vibration. **4.** feet as in f/s, feet per second. Preferable ft/sec.

ft—feet. See also f and $'$.

G—**1.** general symbol for group. **2.** the third fundamental differential quantity of first order for surfaces. **3.** $G(x_1, \cdots, x_n; \xi_1, \cdots, \xi_n)$ Green's function for two points in n-space. **4.** (constant) linear group of points on an algebraic curve. See g. **5.** gravitational constant.

G.C.T.—Greenwich civil time.

GCD—greatest common divisor.

GCS—greatest common subgroup.

GF—Galois field, as in GF(p^n).

GM—geometric mean.

GP—geometric progression.

g—**1.** general function symbol, used with f. **2.** (terrestrial) gravitational attraction. **3.** (variable) linear group of points on an algebraic curve. See G. **4.** g_i, frequency of y_i in bivariate table. **5.** general coefficient in tensor, as in $g_{ij}^k x^i y^i z_k$.

gd—Gudermannian. $e^u = \tan\left(\frac{\pi}{4} + \frac{1}{2}\text{gd u}\right)$. If $\theta = \text{gd u}$ then

$$\sin\theta = \tanh u, \quad \cos\theta = \operatorname{sech} u$$
$$\tan\theta = \sinh u, \quad \cot\theta = \operatorname{csch} u$$
$$\sec\theta = \cosh u, \quad \csc\theta = \coth u$$

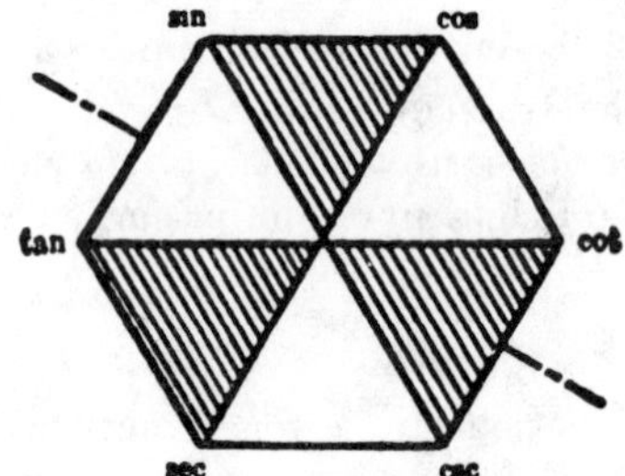

and corresponds to flipping the trigonometric hexagon on an axis perpendicular to cos-sec diagonal. Every vertex of the hexagon is the product of its adjacent vertices, (exhibiting 6 products, and 12 quotients, such as tan/sec = sin) opposite vertices are reciprocals, and in each shaded sector the Pythagorean identity holds. These identities also hold for hyperbolic functions except that Pythagorean sum is at right (so that $\tanh^2 + \operatorname{sech}^2 = 1$).

grad—gradient of. Also written ∇.

H—**1.** general symbol for subgroup, as in G/H, particularly for normal (or self-conjugate) subgroup. **2.** Hessian, as $H(f) = \det\frac{\partial^2 f}{\partial x_i \partial x_j}$. **3.** $H_n(x)$, nth Hermite polynomial, $H_n(x) = e^{x^2}D^n e^{-x^2}$. **4.** $H(P_1, P_3; P_2, P_4)$ is the proposition that P_1 and P_3 separate harmonically P_2 and P_4. **5.** orthocenter of triangle where altitudes meet. **6.** $H(q_1, \cdots, q_n, t, p_1, \cdots, p_n)$ Hamiltonian function (in mathematical physics). **7.** mean curvature of surface, $H = EG - F^2$. **8.** (in mathematical astronomy) hour-angle. Also designated by t. **9.** Hilbert space, also $\mathfrak{H}$.

$\mathfrak{H}$—Hilbert space.

HCF—highest common factor.

MATHEMATICAL SYMBOLS AND ABBREVIATIONS (Continued)

HM—harmonic mean.

HP—(in elementary algebra) harmonic progression.

Hyp.—hypothesis.

h—**1.** (in interpolation theory), distance, between uniformly spaced ordinates. **2.** class interval, in x. **3.** increment of x. Also Δx. **4.** (as superscript) hours as in $8^{h}15^{m}$. **5.** (terminal) hyperbolic, as in sinh, $\tanh^{-1}$, etc. **6.** altitude ("height").

hav—haversine (of). $\text{hav } x = (\text{vers } x)/2 = (1 - \cos x)/2$.

I—**1.** general symbol for interval, and for definite integral. **2.** Roman numeral for "one." Write IV not IIII, IX not VIIII. **3.** $I(\ )$, imaginary part of, also designated by $\mathfrak{I}(\ )$, or $\text{Im}(\ )$. **4.** $I_n(z)$, the Bessel function "of imaginary argument," $I_n(z) = \sum_{m=0}^{\infty} \frac{(\frac{1}{2})z^{n+2m}}{m!\Gamma(n+m+1)}$. **5.** (in geometry of triangle) in center where angle bisectors meet. **6.** sometimes used for the identity as an operation in a group.

$\mathfrak{I}$—(Black letter) imaginary part of. See I.

Im—imaginary part of. See I.

i—**1.** running index, as in a_i, $i = 1, \cdots, n$. **2.** one of the two imaginary square roots of -1. (In electrical engineering, when it is used for the current, j is used for this imaginary unit.)

$\boldsymbol{i}$—$\boldsymbol{i}, \boldsymbol{j}, \boldsymbol{k}$, unit vectors in a right-handed rectangular system. $\boldsymbol{i} \times \boldsymbol{j} = \boldsymbol{k}$, $\boldsymbol{j} \times \boldsymbol{k} = \boldsymbol{i}$, $\boldsymbol{k} \times \boldsymbol{i} = \boldsymbol{j}$.

inf.—("infimum") greatest lower bound.

J—**1.** Jacobian as in $J\left(\frac{u,v,w}{x,y,z,}\right)$, also designated by $\partial(u,v,w)/\partial(x,y,z)$. **2.** $J_n(z)$, Bessel coefficient of order n. **3.** Jacobian curve. **4.** Jacobian group of points in a linear series of groups of points on an algebraic curve.

j—**1.** running index, with i. **2.** (sometimes in electrical engineering) $\sqrt{(-1)}$. See i.

$\boldsymbol{j}$—a unit vector. See $\boldsymbol{i}$.

K—**1.** specific curvature of surface. **2.** kernel of integral equation, as in $u(x) = f(x) + \lambda \int_a^b \bar{K}(x,t)u(t)dt$. K_n, the nth iterated kernel is then defined recursively by $K_1 \equiv K$, $K_n(x,y) = \int_a^b K_{n-1}(x,t)K(t,y)dt$. Also designated by $K^{(n)}$. **3.** K, iK', periods for Jacobian elliptic functions. **4.** Symmedian (Lemoine) center of triangle. **5.** $K_n(z)$, the Bessel function ("second solution") $\lim_{\nu \to n} \frac{(-1)^n}{2}\left[\frac{I_{-\nu}(z) - I_\nu(z)}{\nu - n}\right]$

k—**1.** proportionality factor for variation. Do not use $y \propto x$. **2.** running index, used with h. **3.** k,k', modulus and complementary modulus respectively of Jacobian elliptic functions. **4.** class interval, in y. **5.** increment in y. Also Δy.

$\boldsymbol{k}$—a unit vector. See $\boldsymbol{i}$.

kei, ker—Bessel functions defined as real for real z and satisfying, $\text{ker}(z) + i\,\text{kei}(z) = K_0(z\sqrt{\pm i})$.

L—**1.** general symbol for linear function. **2.** general symbol for linear system, or linear space. **3.** limit, as in $\underset{n\to\infty}{L}\ e^{-n} = 0$. See "lim." **4.** Roman nu-

MATHEMATICAL SYMBOLS AND ABBREVIATIONS (Continued)

meral for "fifty." **5.** length. Also designated by l. **6.** first fundamental differential quantity of second order for surfaces. **7.** Lexis ratio, $\sigma/\sqrt{pq/s}$, for sets of s objects each. **8.** $L_n(x)$, nth Laguerre polynomial, $L_n(x)e^{-x}n! = D^n(e^{-x}x^n)$.

LCD—lowest common denominator.

LCM—lowest (or least) common multiple.

l—**1.** running index. **2.** (in elementary algebra) last term (of arithmetic or geometric progression). **3.** $l = \cos \alpha$, directional cosine, (with x-axis). **4.** length. Also designated by L.

lat.—latitude.

li—logarithmic integral or integral logarithm function, $li(x) = \int_0^x du/\log u$.

lim—limit ( ), $\overline{\lim}$, least upper limit, $\underline{\lim}$, greatest lower limit.

ln—(Sometimes) natural logarithm of.

log—logarithm (of). In theoretical work, the natural base, e, is understood; in numerical computation with tables, the base 10 is understood unless otherwise specified. Some writers use "ln" for natural logarithm of. Where ambiguity is otherwise likely, indicate the base, thus $\log_b x$, $\log_{10} x$, $\log_b x$.

long—longitude.

M—**1.** Roman numeral, thousand. **2.** arithmetic mean. Also designated by superscript bar, as $\bar{X}$, or by AM. **3.** centroid (in geometry of triangle) where medians meet. **4.** second fundamental differential quantity of second order for surfaces.

M.D.—mean deviation.

Md—median.

Mm—mid-mean, arithmetic mean of data in range Q_1 to Q_3.

Mo—mode.

m—**1.** general symbol for natural number, or integer, usually used with n. **2.** slope of line, dy/dx. **3.** $m = \cos \beta$, directional cosine (with y-axis). **4.** m_a, m_b, m_c, median lines, lengths of medians of triangle. **5.** (as subscript) meridian measurement. **6.** class of an algebraic plane curve (a Plücker number). **7.** (superscript) minutes, as in $10^d13^h5^m$. **8.** meters.

max—maximum (of).

meas—measure (of).

mi—miles.

min—**1.** minimum (of). **2.** minutes. **3.** mi. n. = nautical miles (at equator each minute of longitude corresponds to one nautical mile).

mod—**1.** modulus (of), as in $\text{mod}(re^{i\theta}) = r$. **2.** modulo as in $7 \equiv -3 (\text{mod } 5)$.

N—**1.** third fundamental differential quantity of second order for surfaces. **2.** north. **3.** total frequency in a statistical distribution.

n—**1.** general symbol for natural number or integer. **2.** $n = \cos \gamma$, directional cosine (with z-axis). **3.** order or degree of plane algebraic curve (a Plücker number). **4.** (in elementary algebra) number of terms in finite progression. **5.** (sometimes) total frequency. See N. **6.** outward normal to surface, as in $\cos(\theta, n)$.

O—**1.** origin of coordinates. **2.** circumcenter of triangle where the perpendicular bisectors of the sides meet. **3.** of comparable order with, as $\Sigma_{n=0}^{N} n = O(N)^2$.

o—of inferior order to, as $\log n = o(n)$.

MATHEMATICAL SYMBOLS AND ABBREVIATIONS (Continued)

ord—order.

P—**1.** general symbol for polynomial, in particular the interpolational polynomial. **2.** product moment. Also expressed by p. **3.** general symbol for point. The common notation indicating the coordinate system used as $P(x,y,z)$, $P(r,\theta,\varphi)$, etc., is not recommended. **4.** general probability distribution function, in particular any one of Pearson's standard types, or Poisson's forms. **5.** $P_n(x)$, Legendre polynomial, $2^n n! P_n(x) = D^n(x^2 - 1)^n$. **6.** total force due to pressure. **7.** function, as in $Pdx + Qdy$, and in $y' + P(x)y = Q(x)$. **8.** $P_{n,r}$ or ${}_nP_r$, or $P(n,r)$, number of permutations of n distinct things taken r at a time (without repetitions), $n!/(n - r)!$. **9.** general potential due to finite number of particles. In Newtonian case $P = \Sigma_i m_i/r_i$. See U and W. **10.** horizontal parallax. **11.** north celestial pole.

PE—probable error. See $\pm$.

$P(E)$—Probability of E.

$P(E \wedge F)$—Probability of both E and F.

$P(E\backslash F)$—Probability of E, given F.

$P(E \veebar F)$—Probability of E or F, but not both.

$P(E \vee F)$—Probability of E or F or both.

Post.—postulate.

Prob.—**1.** problem. **2.** probability.

Prop.—proposition.

PS—power series.

Pt.—point.

$\wp$—Weierstrassian elliptic function.

p—**1.** general symbol for prime number. **2.** semi-latus rectum. **3.** probability ratio. **4.** genus (or deficiency) of a plane algebraic curve (a Plücker number). **5.** perpendicular distance from origin to given line or plane. **6.** p_n, numerator of nth convergent of continued fraction $a_0 + \frac{1}{a_1+}\frac{1}{a_2+} \cdots$ $(p_{-1} = 0,\ p_0 = 1)$, $p_1 = a_0$, $p_{n+1} = a_n p_n + p_{n-1}$. **7.** p_{12}, p_{13}, p_{14}, p_{23}, p_{24}, p_{34}, line coordinates (point system). **8.** sometimes "per," as in "*rpm*," revolutions per minute. This use is not recommended. **9.** perimeter. Use s for semiperimeter. **10.** $p(n)$, total number of partitions of n. **11.** p_i, impulse component. **12.** the genus of an orientable surface.

p—page. pp—pages.

pos.—positive.

Q—**1.** general symbol for quadratic form or quadratic manifold. **2.** Q_1, Q_3 first and third quartile marks. (Q_2 is the median, Md.) **3.** function, as in $Pdx + Qdy$, and in $y' + P(x)y = Q(x)$. **4.** $Q(P_1, P_2, P_3; P_4, P_5, P_6)$, quadrangular set of six points.

$\underline{Q}$—(underlined) the set of rational numbers.

$\underline{Q}^+$—(underlined) the set of positive rational numbers.

QD—quartile deviation.

Q.E.D.—("Quod erat demonstrandum") which was to be proved.

Q.E.F.—("Quod erat faciendum") which was to be constructed.

q—**1.** complementary probability, $q = 1 - p$. **2.** q_n, denominator of nth convergent of continued fraction, $a_0 + \frac{1}{a_1+}\frac{1}{a_2+} \cdots$ $(q_0 = 0)$, $q_1 = 1$, $q_{n+1} = a_n q_n + q_{n-1}$. **3.** q_{12}, q_{13}, q_{14}, q_{23}, q_{24}, q_{34}, line coordinates (plane sys-

MATHEMATICAL SYMBOLS AND ABBREVIATIONS (Continued)

tem). **4.** q_1, q_2, quartile distances from the median. **5.** q_i, force component. **6.** the number of cross-caps on a non-orientable surface.

R—**1.** general symbol for remainder. $R_n(x)$, remainder after n terms in power-series in x. **2.** radius, in particular circumradius of triangle. **3.** real part of. Also indicated by $\mathfrak{R}$ (Black letter), and Re. **4.** general symbol for range of variable, as in $\int_{(R)}$.

R—(underlined) the set of real numbers.

R^+—(underlined) the set of positive real numbers.

$\mathfrak{R}$ —(Black letter) real value of. See R.

R. A.—right ascension. Also designated by α.

R_m^s—An s-dimensional Betti number modulo m (m prime).

R_0^s—An s-dimensional Betti number relative to the group of integers.

Re—real value of. See R.

RMS—root-mean-square. $\sqrt{(\Sigma_{i=1}^n x^2/n)}$.

r—**1.** general running index, as in rth term. **2.** radius (see R), in particular in radius of triangle. **3.** coefficient of linear correlation, correlation coefficient. **4.** (in elementary algebra) common ratio between successive terms in a geometric progression. **5.** distance, in polar and in spherical coordinates; projected distances in cylindrical coordinates. **6.** (sometimes) revolutions, as in *r.p.m.* (revolutions per minute). **7.** (sometimes) (superscript) radians, as in $2\pi^{(r)}$. (This usage is not recommended.) **8.** the number of boundary curves on a surface.

rad—**1.** radius. **2.** radians. Where no units are indicated angles are measured in radians, sin 30 means "sine of 30 radians," not sin 30°. Do not write sin a^r or sin $a^{(r)}$ for sin a (where a is measured in radians), since superscript r is sometimes interpreted as "revolutions."

rot—rotation or curl of vector. Also designated by $\nabla\times$ (nabla and cross).

S—**1.** general symbol for space, as S_n, space of n dimension. See L. **2.** general symbol for sum, as in S_n, sum of first n terms of given sequence or series. See s. **3.** south. **4.** standard error of estimate. **5.** radius of spherical curvature of curve. **6.** a set (any collection of distinguishable objects).

∂S—Boundary of a set S.

Si—1st sine-integral function $\text{Si}(x) = \int_0^x (\sin u/u)du$. See si.

S.T.—sidereal time.

s.—**1.** general running index. Used with r. **2.** general symbol for sum, as in s_n sum of first n terms of sequence or series. See S. **3.** (in elementary algebra), sum of arithmetic or of geometric progression. **4.** $s_k(n)$, sum of kth powers of first n natural numbers. **5.** s_k, sum of the kth powers of the roots of an algebraic equation. **6.** arc length. **7.** slant height. **8.** (usually as superscript) seconds. **9.** semi-interquartile range. **10.** semi-perimeter of triangle, $s = (a + b + c)/2$. **11.** (sometimes) number of individuals in sample. **12.** root-mean-square deviation about arbitrarily assumed origin, (in "short method").

sec—secant (of). Reciprocal of cosine.

$\sec^{-1}$—inverse secant (of). Also designated by arc sec.

sech—hyperbolic secant (of). Do not use Sec or $\mathfrak{Sec}$ (in German letters).

sech^{-1}—inverse hyperbolic secant of. Also written as arg sech.

MATHEMATICAL SYMBOLS AND ABBREVIATIONS (Continued)

si—2nd sine-integral function, $\beta_i(x) = \int_\infty^x (\sin u/u)du$. See Si.

sgn—(signum) sign (of), more generally for z complex, $\operatorname{sgn} z = z/|z|$.

Use $\sigma_1 = \begin{pmatrix} + & + \\ - & - \end{pmatrix}$ for the signum of the sine (in the corresponding quadrant).

Use $\sigma_2 = \begin{pmatrix} - & + \\ - & + \end{pmatrix}$ for the signum of the cosine.

and $\sigma_3 = \begin{pmatrix} - & + \\ + & - \end{pmatrix}$ for the signum of the tangent.

Thus $d(\text{arc sin } x)/dx = \sigma_2/\sqrt{1 - x^2}$;
$d(\text{arc cos } x)/dx = -\sigma_1/\sqrt{1 - x^2}$
and $\sigma_1/\sigma_2 = \sigma_1\sigma_2 = \sigma_3$.

sin—sine (of). Do not use Sin. In general triangle $\sin A = a/d$ where d is circumdiameter. Identity $d = a/\sin A = b/\sin B = c/\sin C$ is the "law of sines."

$\sin^{-1}$—inverse sine (of). Also designated arc sin.

sinh—hyperbolic sine (of).

$\sinh^{-1}$—inverse hyperbolic sine of—also written as arg sinh.

sk—skewness of frequency distribution.

sn—(Jacobian elliptic function), sine amplitude.

sq—square.

sup—(supremum) least upper bound. Sometimes designated by L.U.B., or l.u.b.

T—**1.** total time, (as in time of flight of projectile). **2.** clock time. **3.** general symbol for transformation, T^n, nth iterate of T. **4.** tons. **5.** (in logic), true or truth. **6.** absolute temperature.

Th—theorem.

t—**1.** general variable or parameter. **2.** time. (In navigation) use 4 digits 0230 for 2:30 a.m., and 1430 for 2:30 p.m. See τ. See "dot accent." **3.** ordinary temperature. See T, τ, θ.

tan—tangent (of).

$\tan^{-1}$—inverse tangent (of). Also designated by arc tan.

tanh—hyperbolic tangent (of). Do not use Tan or $\mathfrak{Tan}$ (in German letters).

$\tanh^{-1}$—inverse hyperbolic tangent (of). Do not use arc tanh. Also written as arg tanh.

T_0-space—A topological space such that for distinct x and y there is either a neighborhood of x not containing y or a neighborhood of y not containing x.

T_1-space—A topological space such that for distinct x and y there is a neighborhood of x not containing y.

T_2-space—A Hausdorff topological space.

T_3-space—A T_2-space which is regular.

T_4-space—A T_2-space which is normal.

U-general (line surface or volume) potential. In Newtonian case, $U = \int_{(L)} \frac{\mu dL}{r}$, or $\int_{(S)} \frac{\mu dS}{r}$, or $\int_{(V)} \frac{\mu dV}{r}$. See P and W.

u—**1.** general variable, especially dependent variable or real part thereof. See w. **2.** u_n, Lucas' function $u_n = (\alpha^n - \beta^n)/(\alpha - \beta)$, α, β, roots of given quadratic.

V—**1.** Roman numeral, five. **2.** (in Euler's polyhedral formula) number of

MATHEMATICAL SYMBOLS AND ABBREVIATIONS (Continued)

vertices. **3.** general harmonic function. **4.** volume. **5.** coefficient of variation.

$\mathbf{v}$—(linear) velocity.

v—**1.** general variable, especial dependent variable, used with u, or coefficient of pure imaginary part thereof. See w. **2.** v_n, Lucas' function $\alpha^n + \beta^n$, α, β, roots of given quadratic. See u_n. **3.** speed.

vers—versed sine or versine.

W—**1.** west. **2.** general potential of double layer. In Newtonian case, $W = \int_{(S)} \nu \frac{\partial}{\partial n}\left(\frac{1}{r}\right) dS, = -\int_{(S)} \frac{\nu}{r^2} \cos(r,n)\, dS$. See P and U. **3.** Wronskian,

$$W = \begin{vmatrix} y_1 \; y_1' & \cdots & y_1^{(n-1)} \\ \cdots & \cdots & \cdots \\ y_n \; y_n' & \cdots & y_n^{(n-1)} \end{vmatrix}.$$

4. total weight. Also designated by Wt. **5.** work (or energy). Also designated by Wk.

Wt.—weight. See W.

w—**1.** general symbol for variable, particularly dependent variable. Used with u and v as in $J\left(\frac{u,v,w}{x,y,z}\right)$, etc. **2.** dependent complex variable, $w = u + iv$, or $w(z) = u(x,y) + iv(x,y)$ where $z = x + iy$, x,y,u,v, real.

X—**1.** Roman numeral, ten; as in XCIII, etc. **2.** X_i, original numerical data for x-variates, (previous to change of origin or scale). **3.** general symbol for variable point, particularly on x-axis. **4.** function, as in $Xdx + Ydy + Zdz$ or $\frac{dx}{X} = \frac{dy}{Y} = \frac{dz}{Z}$.

x—**1.** general symbol for independent variable or unknown. **2.** first rectangular coordinate. In space a right-handed coordinate system such as indicated by

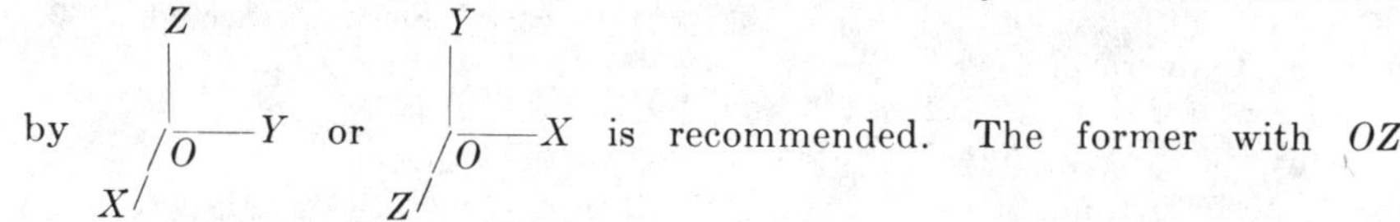

is recommended. The former with OZ directed to the zenith is more common. **3.** real part of independent complex variable $z = x + iy$.

Y—**1.** Y_i, original numerical data for y-variates (previous to change of origin or scale). **2.** function as in $Xdx + Ydy + Zdz$, or $\frac{dx}{X} = \frac{dy}{Y} = \frac{dz}{Z}$.

y—**1.** general symbol for dependent (real) variable. **2.** general symbol for second independent variable, r unknown. **3.** second rectangular coordinate. See x. **4.** pure imaginary coefficient, in independent complex variable, $z = x + iy$.

Z—zenith.

$\underline{Z}$—(underlined) the set of integers.

$\underline{Z}^+$—(underlined) the set of positive integers.

Z.T.—zone time.

z—**1.** independent complex variable, $z = x + iy$. **2.** general symbol for third independent variable or unknown. **3.** third rectangular coordinate. See x. **4.** axial coordinate in cylindrical coordinates (r, θ, z). **5.** zenith distance, $(90^0 - h)$. **6.** Fisher's z statistic.

MATHEMATICAL SYMBOLS AND ABBREVIATIONS (Continued)

Selected Symbols Used in Financial and Actuarial Theory

α—net premium in first policy year.

β—net premium in each policy year after the first.

δ—force of interest (and of discount) $e^\delta = 1 + i$.

ω—terminal recorded age in mortality table, "limiting" age.

A_x—commutation symbol, present value of net single premium for whole life policy on (x), $= M_x/D_x$.

$a_{\overline{n}|i}$—present value of annuity for unit periodic payment, $= (1 - v^n)/i$.

$a_{\overline{n}|i}^{(p)}$—present value of an annuity of 1 per annum at interest rate i, payable p times a year (in installments of $1/p$ each).

a_x—commutation symbol, present value of (ordinary) whole life annuity on (x) of unit annual payment.

$a_x^{(m)}$—present value of whole life annuity on (x), of unit annual payment, payable m times a year.

$n|a_x$—present value of whole life annuity, on (x), deferred n years for unit annual payment.

$a_{x\overline{n}|}$—present value of temporary life annuity, on (x), for n years, for unit annual payment.

$a_{\infty i}$—present value of perpetuity at rate of interest, i.

$\mathrm{a}_{\infty \mathrm{i}}$—present value of perpetuity due at rate of interest, i.

a_{x}—commutation symbol present value of whole life annuity due, on (x), of unit annual payment.

$\mathrm{a}_{\overline{\mathrm{n}}|\mathrm{i}}$—present value of an annuity due, for unit periodic payments at interest rate i.

B—book value.

B_k—book value, after k years.

C—**1.** original cost. **2.** capitalized cost.

C_x—commutation symbol, $= v^{x+1}d_x$.

c_x—natural premium at age x, $= C_x/D_x$.

D—total simple discount.

D_b—total bank discount.

D_c—cash discount, where for example, "terms 4/10, 2/30, n/90" designates 4% discount for payment within 10 days, 2% thereafter but within 30 days, payable net in 90 days.

D_t—trade discount.

D_x—commutation symbol, $= v^x l_x$.

d—rate of simple discount, $= 1 - v = (1 + i)^{-1} = i/(1 + i)$.

d_x—number dying between ages x and $x + 1$, according to mortality table (for 100,000 alive at age 10).

E—present value of expectation.

${}_nE_x$—present value of an n-year pure endowment to (x), $= v^n l_{x+n}/l_x$.

e_x—curtate expectation of life for (x), $= (l_x + \cdots + l_\omega)/l_x$.

$\overset{o}{e}_x$—complete expectation of life, for (x), $= e_x + \frac{1}{2}$.

F—**1.** net-cost-rate factor. **2.** face value of bond.

I—total interest (ordinary simple).

I'—total interest (exact, simple).

i—rate of interest (effective).

MATHEMATICAL SYMBOLS AND ABBREVIATIONS (Continued)

$j_{(p)}$—nominal rate of interest convertible p times a year corresponding to effective annual rate i, $= p[(1 + i)^{1/p} - 1]$.

k_x—valuation symbol, $= C_x/D_{x+1}$.

${}_nk_x$—valuation symbol, $= (M_x - M_{x+n})/D_{x+n}$, $[{}_1k_x = k_x]$, accumulated cost of insurance.

L—**1.** list price (for trade discount). **2.** salvage value.

l_x—number living at age x according to mortality table.

$\mathbb{N}_x$—("open bar N"), commutation symbol, $= D_x + D_{x+1} + \cdots + D\omega$.

n—number of conversion periods (for compound interest).

P—principal.

P_b—bank proceeds.

P_x—net annual premium for ordinary whole life policy, on (x), for unit annual payment $= A_x/(1 + a_x)$.

${}_rP_x$—net annual premium for r-payment life policy, $= M_x/(\mathbb{N}_x - \mathbb{N}_{x+r})$.

p—**1.** probability of success. **2.** interest period/payment interval.

p_x—probability of living for another year for a person of age x (according to mortality table).

${}_np_x$—probability that a person aged x will live n years, $= l_{x+n}/l_x$.

p_{xy}—the probability that (x) and (y) will survive jointly for one year.

${}_np_{xy}$—the probability that (x) and (y) will survive jointly for n years.

q—probability of failure, $= 1 - p$.

q_x—probability of dying within a year for a person of age x (according to mortality table).

$|_nq_x$—probability that (x) will die within n years.

${}_n|q_x$—the probability that (x) will die between ages $x + n$ and $x + n + 1$.

R—**1.** periodic payment or "rent." **2.** replacement cost. **3.** repair charge (annual).

S—amount.

S—**1.** scrap value. **2.** amount of sinking fund. **3.** subscription price. **4.** amount (at simple interest).

s—amount (at compound interest) for unit principal $= (1 + i)^n$.

$s_{\overline{n}|i}$—amount of annuity for unit periodic payment, $= [(1 + i)^n - 1]/i$.

$s_{\infty i}$—amount of perpetuity at interest rate i.

s_∞—amount of perpetuity due.

$\mathrm{s}_{\overline{n}|i}$—amount of an annuity due, for unit periodic payments.

$s^{(p)}_{\overline{n}|i}$—amount at end of n years of annuity of 1 per annum at interest rate i, payable p times a year (in installments of $1/p$ each).

u_x—valuation symbol, $= D_x/D_{x+1}$.

${}_nu_x$—valuation symbol, $(\mathbb{N}_x - \mathbb{N}_{x+n})/D_{x+n}$, $[{}_1u_x = u_x]$, accumulated value of individual survivors payments.

V—purchase price (of bond).

${}_tV_x$—terminal reserve of tth policy year on policy on (x) of unit annual payment.

v—present value (at compound interest) for unit principal, $= (1 + i)^{-1}$.

W—wearing value, = original cost minus scrap value.

x—age of insured (to nearest year).

(x)—a person aged x (for use with mortality table).

$x]$—where $[x] + t$ indicates a life, now aged $x + t$, who was accepted for insurance t years ago at age x.

Index

A

B

continued from p1143

D

Go to p1144

Cont'd from p. 1141

C

Go to p1142

Cont'd from p. 1142

E

F

G

H

I

J

L

M

N

O

P

Q

R

S

Stirling's approx. to n! 620

T

U

Index